D1207248

GRANT & HACKH'S

CHEMICAL DICTIONARY

GRANT & HACKH'S
CHEMICAL DICTIONARY

[*American, International, European and British Usage*]

Containing the Words Generally Used in Chemistry, and Many of the Terms Used in the Related Sciences of Physics, Medicine, Engineering, Biology, Pharmacy, Astrophysics, Agriculture, Mineralogy, etc.

Based on Recent Scientific Literature

FIFTH EDITION
Completely Revised and Edited by

ROGER GRANT

M.A., D. de l'U., Ph.D., C. Chem., M.R.S.C. *Consultant*

CLAIRE GRANT

M.B., B.S., M.R.C.P.E. *Medical Practitioner*

McGRAW-HILL BOOK COMPANY

New York St. Louis San Francisco Auckland Bogotá Hamburg Johannesburg London Madrid Mexico Milan Montreal New Delhi Panama Paris São Paulo Singapore Sydney Tokyo Toronto

Library of Congress Cataloging-in-Publication Data

Hackh, Ingo W. D. (Ingo Waldemar Dagobert), 1890–1938.
Grant & Hackh's chemical dictionary.

Rev. ed. of: Chemical dictionary. 4th ed. 1969.
1. Chemistry—Dictionaries. I. Grant, Roger L.
II. Grant, Claire. III. Title. IV. Title: Grant &
Hackh's chemical dictionary. V. Title: Chemical
dictionary.
QD5.H3 1987 540′.3 86-7496
ISBN 0-07-024067-1

1234567890 DOCDOC 8943210987

ISBN 0-07-024067-1

The previous edition of this book was *Hackh's Chemical Dictionary*, 4th ed., published by McGraw-Hill in 1969. It was prepared by Dr. Julius Grant from a *Chemical Dictionary* compiled by Ingo W. D. Hackh. The current, or 5th, edition of this book was prepared by Dr. Roger L. Grant, whose father prepared the 4th edition.

The editors for this book were Betty J. Sun and Susan Thomas, the designer was Naomi Auerbach, and the production supervisor was Teresa F. Leaden. It was set in Palatino by University Graphics, Inc.

Printed and bound by R. R. Donnelley & Sons Company.

CONTENTS

Index of Figures and Tables	vii
Preface	ix
Acknowledgments	x
Explanatory Notes	xi
Abbreviations Commonly Used in This Book	xii
Disclaimer	xiii
Chemical Dictionary	1

INDEX OF FIGURES AND TABLES

Figure No.	Title	Page No.
1	Alchemical and Chemical Symbols	20
2	Atomic Planes	57
3	Atomic Structure and Dependent Properties	58
4	Forms of Bacteria	64
5	The Benzene Ring	74
6	The Carbon Cycle	113
7	Ceramic Ware	123
8	Color Diagram	146
9	Types of Conformation	150
10	Crystal	160
11	Crystal Coordinates	161
12	Crystal Structure	161
13	Crystal Systems	161
14	Hazard Warning Symbols	276
15	The Iodine Cycle	308
16	Classification of Monosaccharides	377
17	The Orthorhombic Crystal System	413
18	Classification of Polymer Molecules	462
19	Enantiomers	525
20	Starches	551
21	Specific Volume of Water	627

Table No.	Title	Page No.
1	Common Abbreviations and Symbols	2
2	Percentage Abundance of Chemical Elements	4
3	Common Acids, Radicals, and Salts	11
4	Organic Acid Series	13
5	U.S. Air Quality Standards	18
6	Alizarin Dyes (Anthraquinones)	23
7	Chemical Classification of Alkaloids	23
8	Prinicpal Amino Acids	32
9	Magnitudes of Analysis	38
10	Common Forms of Asbestos	54
11	Standard Atomic Weights	59
12	Bead Test Colors	70
13	Benzene Series	74
14	Bioelement Composition of Organisms	84
15	Biology-Related Sciences	85
16	Oxidation States of Chromium	135
17	Coinage Metals	144
18	Colloidal Systems	145
19	Wavelength Range of Colors	146
20	Important Constants	152
21	Critical Temperatures and Presures of Some Common Gases	158
22	Cytochromes	170
23	Range of Sound Levels	173
24	Types, Uses, and Properties of Elastomers	202
25	Electrical Units	204
26	Standard Electrode Potentials	204
27	Electron Configuration and Oxidation States of the Elements	207
28	Historical Table of Elements	209
29	Energy Conversion Factors	213
30	Classification of Enzymes	214
31	Dimensions of Some Common Fibers	234
32	Classification of Filters	235
33	Flavanols	238
34	Flavones	238
35	Formula Signs	244
36	Properties of Liquid Fuels	248

37	Fungicides	250
38	Geologic Eras	259
39	Typical Glass Compositions	261
40	The Greek Alphabet	270
41	Units of Water Hardness	275
42	Herbicides	281
43	Hydrocarbon Series	291
44	Properties of Common Indicators	302
45	Insecticides	305
46	Oxidation States of Iron	310
47	Isoelectric Points of Various Proteins	313
48	Classification and Examples of Isomerism	314
49	Natural Isotopes	315
50	Various Magnitudes of Length	335
51	Oxidation States of Manganese	352
52	Various Magnitudes of Mass	355
53	Contents Analysis of Milk from Various Mammals	371
54	Composition and Hardness of Some Nickel Alloys	391
55	Isomers of Nitrotoluidine	398
56	Basics of IUPAC Nomenclature	399
57	Number Systems	404
58	Minimum Perceptible Concentration of Various Odors	406
59	Subatomic Particles	424
60	Periodic Table of the Elements	433
61	Petroleum Fractions	436
62	Classification of Plastics	456
63	Fatal Percentages of Poison Vapors	459
64	Water-Soluble Polysaccharides	463
65	Specific Impulses of Various Propellants	474
66	Classification of Proteins	477
67	Quantum Relation of Radiation and Energy	486
68	Quinine-Related Compounds	488
69	Electromagnetic Radiation	492
70	Names for Radicals and Ions	493
71	Radon Decay Chain	496
72	Refractive Index n_D of Various Compounds at 23°C	501
73	Formulas and Properties of Some Refractory Materials	501
74	Mechanisms of Ion-Exchange Resins	503
75	Resistivity of Various Elements	504
76	Effects on Various Compounds of Roasting at Selected Temperatures	509
77	Oxidation States of Ruthenium	513
78	Base SI Units	527
79	Derived SI Units with Special Names	527
80	Multiples of SI Units	528
81	Traditional Units with SI Equivalents	529
82	Units Allowed in Conjunction with SI System	530
83	Silicic Acids	530
84	Composition of Soils	540
85	USP Solubility Classification	541
86	Effect of Heat on Sulfur	562
87	Sweetness Factors of Some Compounds Relative to Sucrose = 1.0	565
88	Internationally Approved Symbols for Physical Quantities	566
89	Percent of Tannin in Various Tannin Materials	574
90	Oxidation States of Tellurium	577
91	International Reference Temperatures	578
92	Temperature Magnitudes, °C	578
93	Color Scale of Temperatures	578
94	Thermal Conductivity λ of Various Materials at 25°C	585
95	Magnitudes of Time	592
96	Toxic Concentrations of Various Substances	597
97	Analyses of Turpentines	607
98	Vacuum Achievable by Various Methods	614
99	Oxidation States of Vanadium	615
100	Magnitudes of Velocity	618
101	Vitamins	622
102	Physical Constants of Water and of Heavy Water	627
103	Source and Composition of Various Waxes	628

PREFACE

The eighteen years that have passed since the previous edition of this dictionary are notable for the breadth and degree of change in chemistry and her sister sciences. Fields of science that were in their relative infancy at the time of the previous edition are now on the steep part of their growth curve. Examples are biotechnology, ecology, organometallics, immunology, space science, cytotoxic drugs, lasers, small-scale technologies, robotics, and alternative forms of energy. Significant advances have also been made in the more mature scientific fields, such as plastics, elastomers, pesticides, antibiotics, electronics, nuclear power, and waste recycling. Computers merit particular mention. Their pervading influence stretches from permitting development of the hitherto impossible to controlling conventional chemical processes with a new degree of precision and subtlety. The economic and social background is also very different, with much greater emphasis on the role of energy and of men and women—both as workers and as people.

These developments alone would have justified a major revision of the previous edition. However, there has also been a parallel revolution in the nomenclature of chemistry, in particular, and of some related sciences. Greater travel and easier communications had augmented the international character of chemistry to the point where progress was being retarded for lack of an internationally accepted nomenclature. Further, the increasing power of computers allowed a means of storing and handling the burgeoning quantities of chemical literature.

To this end, the International Union of Pure and Applied Chemistry (IUPAC) has, over the time span since the last edition, provided a comprehensive and flexible system in both organic and inorganic chemistry, and it continues to develop this system. On the practical side, the American Chemical Society has developed a nomenclature similar to that of IUPAC but more suitable for the *Chemical Abstracts* computerized information retrieval system. These two systems, together with IUPAC's recommended usage of terminology and symbols, are regarded as correct international usage in this edition.

Certain other branches of science lend themselves well to formalization, and there has been parallel and cooperative development in these areas by international organizations. Particularly notable, and thus correspondingly well represented in this new edition, are the SI system of the General Conference on Weights and Measures, the physical quantities and symbols of the International Organization for Standardization (ISO) and of the International Union of Pure and Applied Physics (IUPAP), enzyme names of the International Union of Biochemistry (IUB), and the pesticides nomenclature of the ISO.

These and numerous other more minor influences add to the complexity of the lexicographers' task. Our solution has been to augment considerably the internationally recognized areas and to use them as the basic framework for the dictionary. On this structure, and commanding equal importance, are the other forms of usage, which range from the non-SI (eminently the retention in the United States of the old "English" system of units) to the obsolete.

In order to accommodate all these requirements and yet not exceed the previous edition's length of roughly 55,000 entries, judicious pruning has been necessary. This has been achieved by a combination of consolidating related information under a single heading, streamlining the format, and reducing repetition. A few areas, particularly the former use in medicine of inorganic and traditional plant-derived chemicals, have been largely culled. In all, about 50 percent of the previous entries have either been modified to some extent or else deleted.

Another change with this edition is that of editor. For more than 50 years my father, Julius Grant, has worked with evident affection on the dictionary—first with Ingo Hackh and then by himself on the subsequent three editions. Our taking over the reins has been considerably facilitated by his help. In fact, between the dictionary's ranking among my earliest recollections and my having contributed to the previous edition, it is almost a member of the family. My wife's contribution on the medical side has been a most welcome complement to my efforts. New editors bring new emphases. Besides the increased international character mentioned above and the strengthening of the medical side, we have also amplified the scientific aspects that are mainly of interest to those in industry.

We would heartily like to thank numerous friends, in both the U.K. and the U.S., for sharing their specific professional expertise with us in a variety of fields. Acknowledgment is also due our computer. Itself a member of the "family" that was chiefly responsible for most of the new entries and changes, it alone enabled us to do justice to the editing they required.

A dictionary is written for its readers—whose interests the editors can only guess at. We will therefore be pleased to receive from readers any suggestions or comments of a conceptual or factual nature.

Acknowledgments

Material has been included at numerous points in the text from the following publications by permission of the publishers: *Nomenclature of Organic Chemistry,* IUPAC, Pergamon Press, Oxford, 1979. *Nomenclature of Inorganic Chemistry,* IUPAC, Butterworth, London, 1970. *Quantities, Units and Symbols,* The Royal Society, London, 1981. *Enzyme Nomenclature,* International Union of Biochemistry, Academic Press, New York, 1978.

The use of portions of the text of USP XX-NF XV is by permission of the USP Convention. The Convention is not responsible for any inaccuracy of quotation or for false or misleading implication that may arise from separation of excerpts from the original context or by obsolescence resulting from publication of a Supplement.

Material has been reproduced from the British Pharmacopoeia (13th edition) by permission of Her Majesty's Stationery Office.

London, England ROGER GRANT CLAIRE GRANT

EXPLANATORY NOTES

Spelling. American usage is given precedence, but British forms are also listed: e.g., "sulfur" and "sulphur," respectively.

Capitals are always used in the definitions for the initial letter of the first word of a sentence.

Chemical nomenclature. The two systems used internationally, namely that recommended by the International Union of Pure and Applied Chemistry (IUPAC) and that used in *Chemical Abstracts* (CA), are considered correct usage. IUPAC names are indicated by *, and CA (where different from or a less common form of IUPAC) by †. Where usage is allowed but discouraged by IUPAC (e.g., "-ic" and "-ous," as in ferric and ferrous), no * is used. See *nomenclature* for sources and other information.

Asterisks (*) are used to indicate internationally acceptable terminology or symbols. For chemical nomenclature the recommending body is IUPAC. For other fields, the main recommending body is generally as follows: SI system—CGPM; physical quantities and symbols—ISO and IUPAP; enzymes—IUB; pesticides—ISO.

Drugs nomenclature. Entries are generally to be found under the name used by the U.S. Pharmacopoeia (USP) or National Formulary (NF). Names used in the European Pharmacopoeia (EP) and British Pharmacopoeia (BP) are cross-referenced to the USP name. However, where a compound's significance as a chemical outweighs that as a drug, the entry will be found under the chemical name. Although British and other non-American trademarks are listed under the USP or NF name, the products may not meet the USP specification.

Ring systems are represented by one of the three following methods and are numbered as shown.

Line formula

```
┌─────────────────────────────────────┐
CH : CH · CH : CH · CH : CH
1    2    3    4    5    6
```

Square formula

```
1    2    3
CH : CH · CH
|         ||
CH : CH · CH
6    5    4
```

Geometric formula

```
          ¹CH
        /    \\
  ₆HC          CH₂
    ||          |
  ₅HC          CH₃
        \    //
          ₄CH
```

Synonyms are given under each definition in approximate order of importance. Some of them are also listed separately in the dictionary, with a cross-reference to the entry under which the actual definition is to be found. Thus, "**brassil** Pyrite." means either (1) for "brassil" see "pyrite" or (2) same as "pyrite."

Commercial names are referred to the scientific synonym, unless the compound is of special commercial importance; e.g., "soda ash" is defined under that heading.

Italics are used only as follows: (1) according to custom (thus "*Acacia*" the plant, but "acacia" the gum; or for the letters *o*-, *m*-, *p*- used as abbreviations for *ortho, meta,* and *para,* respectively); (2) for cross-reference (e.g., "**acaricide** A pesticide used to kill mites. Cf. *insecticide*"); (3) for the titles of publications, etc.; (4) for certain symbols, where custom demands (see Table 88 under "symbols").

Some words may be found only as subentries of a main entry. For example, "Agent Orange" and "balancing agent" are both found only as subentries of **agent**.

For subentries such as "balancing agent" that do not begin with the same word as the main entry or that for other reasons cannot be found through a simple alphabetical search, direct cross-references take the form of "See *balancing agent* under *agent*."

When cross-references of comparison (cf.) or additional definition (q.v.) are made, only the portion of the word that occurs in the main entry is given in italics, as in the following examples: (1) "cf. balancing *agent*"; (2) "the buffer is a balancing *agent,* q.v." For the same reasons, some synonyms in the definitions may also be given partly in italics, for example, the entry for 3-phenylchromone is "**chromone 3-phenyl ~** Iso*flavone,*" indicating that "isoflavone" can be found under the main heading **flavone**.

Compound words should generally be sought under the heading of the first of the words. Thus "soda process" is to be found under "soda", and not under "process". Cross-references are sometimes given under the second word, and in some cases where importance or general custom demand, the actual definition will be found under the latter, and a cross-reference under the former.

Where the word defined is continually repeated in its main definition or subdefinition, it is represented by its initial letter followed by a period, so long as the meaning is not thereby obscured. This abbreviation can also represent the plural, where the sense so requires.

Subentries are shown in boldface type. If the main entry word is to be repeated in front of the subentry word, it is shown by its initial letter as follows and spaced appropriately:

dibenzyl . . . d.amine	Read as *dibenzylamine*
acetoacetic . . . a. acid	Read as *acetoacetic acid*

If the main entry is to be repeated after the subentry word, it is shown by the symbol ~ and spaced appropriately as follows:

agent . . . balancing ~	Read as *balancing agent*
acetone . . . acetonyl ~	Read as *acetonylacetone*
acridine . . . diamino ~ chloride	Read as *diaminoacridine chloride*

Sub-subentries, shown in smaller boldface type, refer to the immediately preceding bold subentry.

acetamido . . . a.naphthol . . . 1,2- ~	Read as *1,2-acetamidonaphthol*

Trade names and registered trademarks have a capital initial letter. The listing of a word as a free chemical term does not necessarily mean that it has not been adopted as a trade name or trademark. However, there is no intention to use a term in a generic sense if it is in fact a trade name or trademark. Where the legal status of a trade name or trademark is important, the reader should check the current status. British and other non-American trademarks for drugs are listed under the USP or NF name, but the products may not meet the USP or NF specification.

Where an entry is both a trademark or trade name and a generic name, the entry generally has a capitalized initial and is followed by the trademark or trade name definition first; then, after "(Not cap.)," by the generic definition.

Letters in parentheses in or at the end of a word defined indicate an alternative method of spelling. Thus, "flavin(e)."

Water of crystallization is shown with a center dot: e.g., $Na_2SO_4 \cdot 10H_2O$.

The usual order of presenting or describing a compound is: Name. Formula = molecular weight. Synonyms. Occurrence, preparation, or type of substance. Color; state (e.g., liquid) at room temperature; density (of more important compounds); melting point or boiling point (in °C); solubility (in water unless otherwise specified; mention of solvents other than water implies low solubility in water); chemical, industrial, and medicinal uses.

Temperatures are all in °C unless otherwise stated.

Symbols are defined in the appropriate places (see especially the initial entries under each letter of the alphabet) and under the entries *constant, nomenclature, SI,* and *symbol.*

Abbreviations Commonly Used in This Book

Å	angstrom unit.
α	specific rotation (e.g., $[\alpha]_D^{20}$ is the value of α for the D line at 20°C).
abbrev.	abbreviation.
Ac	acetyl, CH_3CO- (e.g., AcOH is acetic acid).
Ar	aromatic.
at.	atomic (e.g., at. wt., at. no.)
b.	boiling point in °C (e.g., $b_{600mm}60$ means boils at 60°C under a pressure of 600 mm of mercury).
BP	British Pharmacopoeia.
Bu	butyl, C_4H_9- (e.g., BuOH is butyl alcohol).
Bz	benzoyl, PhCO− (e.g., BzOH is benzoic acid).
ca.	approximately (circa).
Cf.	compare.
CGPM	General Conference on Weights and Measures (Conférence Générale des Poids et Mesures).
C.I.	Color Index.
cryst.	crystalline or crystallizing.
d.	density at room temperature. $d_{0°}$ for density at 0°C. $d_{25/4}$ for density at 25°C relative to water at 4°C (=1.000). $d_{air=1}$ for density compared with air.
decomp.	decompose, -osing, -oses, or -osed (e.g., decomp.20 means decomposes at 20°C).
EP	European Pharmacopoeia (correctly: Eur.P.).
Et	ethyl, C_2H_5- (e.g., EtOH is ethyl alcohol).
insol.	insoluble in water.
ISO	International Organization for Standardization.
IUB	International Union of Biochemistry.
IUPAC	International Union of Pure and Applied Chemistry.
IUPAP	International Union of Pure and Applied Physics.
M	a monovalent metal.
M	molar, molarity.
m.	melting point in °C (used similarly to b., boiling point).
Me	methyl, CH_3- (e.g., MeOH is methyl alcohol).
N	normal, normality.

n	refractive index (e.g., $[n]_D^{20}$ is the value for the D line at 20°C).
NF	National Formulary.
no.	number (e.g., at. no.).
Ph	phenyl, C_6H_5- (e.g., PhOH is phenol).
pl.	plural.
Pr	propyl, C_3H_7- (PrOH is propyl alcohol).
q.v.	which see (in cross-references).
R	a monovalent radical. For R^2, R_2, R'', etc., see *organic radical* under *radical.*
sapon.	saponification (e.g., sapon. val.)
sol.	soluble in water.
unsap.	unsaponifiable (e.g., unsap. matter).
USP	United States Pharmacopoeia.
val.	value (e.g., iodine val.).
vol.	volume.
wt.	weight (e.g., at. wt.).
X	a halogen.

A

A Symbol for ampere.
A Symbol for (1) area; (2) Helmholtz function; (3) nucleon (mass) number.
A_r Symbol for relative atomic mass *(atomic weight).*
°**A** Symbol for degree absolute. See *temperature.*
Å Symbol for *angstrom,* q.v.
a Symbol for (1) atto (10^{-18}); (2) year (annum).
a-, an- Prefix indicating without or not; as, *an*hydrous.
a Symbol for thermal diffusivity.
α See *alpha.*
A acid 1,7-Dihydroxynaphthalene-3,6-disulfonic acid.
abaca Manila hemp. The inner fiber of *Musa textilis,* a banana species of the Philippine Islands.
abati drying oven A constant-temperature oven with xylene bath (100–200°C).
abaxial Not in the line of axis.
Abbe, Ernst (1840–1905) German physicist. **A. condenser** An arrangement of lenses to increase the illumination of an object under the microscope. **A. refractometer** A device for rapid, direct determination of refractive index. **A. theory** The limit of microscopic visibility is determined by $b = \lambda/2n \sin \mu$, where b is the breadth of the object, $n \sin \mu$ the aperture, and λ the wavelength of the light.
abbreviations See Table 1, on p. 2, Table 25 on p. 204, Table 88 on p. 566, and *SI units.*
Abderhalden, Emil (1877–1950) Swiss-born German chemist, noted for physiological and pathological research.
Abegg, Richard (1869–1910) German chemist noted for his theory of valence.
Abel, Sir Frederick Augustus (1827–1902) English chemist noted for research on explosives and petroleum. **A. fuse** See *fuse.* **A. heat test** A test for the stability of nitroglycerin; the sample is heated with a potassium iodide-starch paper which must not be colored by the products evolved within a certain time. **A. reagent** A 10% solution of chromium iodide used for etching steel in microanalysis. **A. tester** An apparatus for the determination of the flash point of oils.
abelmoschus See *musk seed.*
abeo- (Latin *abeo:* "I go away") Prefix indicating a naturally occurring compound that arises, by bond migration, from the structure following the prefix.
aberration Deviation from the normal. **astronomical ~** The apparent angular displacement of light from a star. **chromatic ~** The unequal refraction of different wavelengths of light. **spherical ~** The deviation of light passing through a lens, or reflected from a mirror, caused by the inequality in the degree of convergence. See aplanatic *focus.*
abhesive Nonadhesive.
abienic acid Abieninic acid.
abieninic acid $C_{13}H_{20}O_2$ = 208.3. Abienic acid. An acid resin from *Abies pectinata,* European silver fir.
Abies A genus of evergreen trees, the Coniferae (Pinaceae) or firs, which yield turpentine, resin, and pitch.
abietadienoic acid* **7,13-~** Abietic acid. **8,13-~** Palustric acid.
abietene $C_{19}H_{30}$ = 258.4 Diterebentyl. A liquid hydrocarbon, b.340–345, from the resin of *Pinus sabiniana.* Cf. *colophene.* **a. sulfonic acid** Black paste produced by the slow action of sulfuric acid on a. at 10°C. A textile industry wetting agent.
abietic acid $C_{19}H_{29}COOH$ = 302.5. 7,13-Abietadien-18-oic acid*, sylvic acid. An acid from the resin of pine species (colophony). Yellow leaflets, m.172, insoluble in water, used in varnishes and driers. Cf. *pimaric acid, Steele acid.* **dehydro ~** $C_{19}H_{27}COOH$ = 300.4 An oxidized resinic acid produced by treating a. with sulfur. **dihydro ~** $C_{19}H_{31}COOH$ = 304.5. A reduced resinic acid produced by heating a. with hydrogen under pressure.
abietic anhydride $C_{44}H_{26}O_4$ = 618.7. The anhydride of abietic acid; the chief constituent of colophony.
abietin (1) $C_{53}H_{76}O_8$ = 841.2. A crystalline resin from the pitch of European silver fir. (2) Abietene. (3) Coniferin.
abietinic acid $C_{19}H_{28}O_2$ = 288.4. A crystalline acid from rosin; 3 isomeric forms.
abietinolic acid $C_{16}H_{24}O_2$ = 248.4. An acid resin from the pitch of European silver fir.
abietite $C_4H_8O_6$ = 152.1. A tetrose sugar from the needles of European silver fir.
abiogenesis Autogenesis. Spontaneous generation of life from lifeless matter. Cf. *biogenesis.*
abiosis Absence of life.
ablation The sacrificial pyrolytic vaporization of solid surfaces as a means of dissipating heat generated; as, on space craft shields on reentering the earth's atmosphere.
Abney clinometer A pocket altimeter for measuring heights or slopes.
abnormal Irregular; different from normal.
ABO test The simplest (antigen) blood grouping test. See *blood groups.*
abradant Abrasive.
abrasion Mechanical wearing away.
abrasive A grinding or polishing material.
abraum salts (German *abräumen:* "to put away") The potash salts of Stassfurt; as, kainite.
abrazite Gismondite.
abriachanite Amorphous blue *asbestos.*
abric acid $C_{21}H_{24}NO_3$ = 338.4. An acid from the abrus seeds or prayer beads of *Abrus precatorius.*
abrin Jequiritin. A poisonous mixture of albumose and paraglobulin from abrus seeds; an ophthalmic irritant.
abrotanum Southern wood. The leaves are used as an aromatic.
abrotine $C_{21}H_{22}NO_2$ = 320.4. An alkaloid from southern wood, *Artemisia abrotanum* (Compositae).
abrus **a. root** Indian licorice. The roots of *A. precatorius* (Leguminosae) of India, tropical Africa, and Brazil; a licorice substitute. **a. seeds** Jequirity, prayer or jumble beads, love peas, crab's eye. Formerly used in ophthalmology, and in India as a weight (rati).
abs. Abbreviation for absolute (temperature).
abscess A collection of pus in the tissues of the body.
abscissa The coordinate of a point measured along the horizontal (x) axis. See *coordinates.*
ABS copolymers Generic term for thermoplastic resins made

TABLE 1. COMMON ABBREVIATIONS AND SYMBOLS

Where multiple abbreviations are given, they are alternatives if separated by a comma. If they are separated by a semicolon, the first is preferred. The same abbreviation is used for the singular and plural form of nouns. See also the table of symbols on p. 566.

absolute	abs.
adenosine 5′-triphosphate	ATP
alternating current	a.c., ac
ampere	A; amp
angstrom	Å
anhydrous	anhyd.
approximate, -ly	approx., *ca.*
aqueous	aq.
Assignee (patent titles only)	Assee.
atmospher(e), (es), (ic)	atm
atomic number	at. no.
atomic weight	at. wt.
average	av., avg.; ave.
barrel	bbl
Baumé	Bé
biochemical oxygen demand	B.O.D.
boiling point	b.p., bp
British thermal unit	Btu; BThU
butyl	Bu
calculated	calc.
Calorie (large)	kcal
calorie (small)	cal
candela	cd
centimeter	cm
centimeter-gram-second system	cgs
centipoise	cP
centistoke	cSt
chemical oxygen demand	C.O.D.
coefficient	coeff.
concentrat(ed), (ion)	conc.
constant	const.
corrected	corr.
coulomb	C
crystal,-lize(d)	cryst. (-d.)
crystalli(ne),(zation)	crystn.
cubic centimeter	cm^3, mL, ml; cc
cubic meter	m^3; cu m
curie	Ci
cycles per second	Hz; cps
decibel	dB
decompos(e),(ed),(ition)	decomp., (-d.), (-n.)
degree Celsius (centigrade)	°C
density	ρ; d.
deoxyribonucleic acid	DNA
deriv(ative), (ed), (ation)	deriv., (-d.), (-n.)
diameter	dia.; diam.
dilut(e),(ed),(ion)	dil., (-td.), (-n.)
direct current	d.c., dc
dyne	dyn
electromagnetic unit	e.m.u, emu
electromotive force	e.m.f., emf
electron spin resonance	e.s.r., esr
electronvolt	eV
electrostatic unit	e.s.u, esu
equivalent	equiv.
equivalent to	≡
ethyl	Et
ethylenediaminetetraacetic acid	EDTA
farad	F
feet, foot	ft
feet per minute	ft/min, fpm
footcandle	fc
foot-pound-second system	fps
freezing point	f.p., fp
gallon	gal
gas chromatography	g.c.
gas-liquid chromatography	g.l.c.
gram	g
hectare	ha
hemoglobin	Hb
hertz	Hz
horsepower	hp; HP
hour	h; hr
hydrogen ion concentration	$[H^+]$
inch	in
indefinite	indef.
infrared	i.r., ir; IR
inorganic	inorg.
insoluble	insol.
iso-	iso-, *i-*
joule	J
kelvin	K
kilogram	kg
kilovolt	kV
kilowatt	kW
kilowatthour	kWh
liter, litre	L, l
maximum, maxima	max.
megacycle	MHz; Mc
melting point	m.p., mp
meter, metre	m
meter-kilogram-second system	mks
methyl	Me
microgram	μg
micrometer (micron)	μm
micron (see *micrometer*)	—
microsecond	μs
milliampere	mA
milligram	mg
milliliter	mL, ml, cm^3; cc
millimeter	mm
millimicron (see *nanometer*)	—
millisecond	ms
millivolt	mV
minute	min
molar solution	*M*
mole	mol
molecular weight	mol. wt.
namely	viz.
nanometer (millimicron)	nm
newton	N
normal solution	*N*
nuclear magnetic resonance	n.m.r., nmr
number	no.
organic	org.
ounce	oz
part(s)	pt.(s.)
parts per million	$/10^6$; ppm
pascal	Pa
percent, per cent (age)	%
phenyl	Ph
potential difference	p.d.
pounds per square inch	lb/in^2, psi
pounds per square inch absolute	psia
pounds per square inch gage	psig
precipitat(e), (ed), (ion)	ptt., (-d), (-n)
pressure	*p*; *P*
propyl	Pr

TABLE 1. COMMON ABBREVIATIONS AND SYMBOLS (*Continued*)

qualitative(ly)	qual.
quantitative(ly)	quant.
recrystalliz(e), (ation)	recryst., (-n.)
refractive index	*n*
relative analgesia	R.A.
relative humidity	r.h.; R.H.
respiratory quotient	r.q.
revolutions per minute	r/min; rpm
ribonucleic acid	RNA
roentgen unit	R
root mean square	r.m.s., rms
saponification value	sap. val.
second (time only)	s
secondary	*sec-*; *s-*
siemens	S
soluble	sol.
specific	sp.
specific gravity	sp. gr.
square centimeter	cm^2; sq cm
standard (normal) temperature and pressure	s.t.p.; STP
stoke	St
temperature	temp., t, T
tertiary	*tert-*
thin-layer chromatography	t.l.c.
ton (qualify, if necessary)	t
ultraviolet	u.v.; UV
vacuum	vac.
vapor density	v.d.
viscosity, dynamic	η
viscosity, kinematic	ν
volt	V
volume	vol., V, v
watt	W
wavelength	λ
weight	wt.
weight/weight	w/w

from acrylonitrile, butadiene, and styrene; light in weight and resistant to chemicals and tensile stress.

absinthe (1) Absinthium. (2) A strong alcoholic beverage containing anise and wormwood oil. **a. oil** Wormwood oil.

absinthic acid An acid of wormseed and related to succinic acid.

absinthiin $C_{15}H_{20}O_4$ = 264.3. A crystalline, poisonous principle, m.68, from absinthium.

absinthin, absynthin $C_{40}H_{56}O_8 \cdot H_2O$ = 682.9. A glucoside from the dried leaves and flowering tops of *Artemisia absinthium*, wormwood. Brown powder, m.122, insoluble in water.

absinthium Vermouth, wormseed, matterwood. The bitter dried leaves and flowering tops of *Artemisia absinthium* (Compositae). **oil of ~** Wormwood oil.

absinthol Thujenol.

absolute (1) An actual condition, independent, unrestrained, and nonrelative. (2) Pure, refined. **a. alcohol** Ethanol(100%). **a. boiling point** Critical temperature. The temperature above which the liquid phase cannot exist. **a. density** The density or specific gravity reduced to standard conditions; e.g., with gases, 760 mm pressure and 0°C. **a. humidity** The water-vapor content of the atmosphere in g/m^3. **a. temperature** See *temperature*. **a. units** Base units. **a. zero** The temperature at which molecular movement ceases. Absolute zero, 0 K (or 0°A) = −273.15°C = −459.7°F = −219°R.

absorbance* Optical density. $\log I_0/I$, where I_0 is the intensity of the incident ray, and I the intensity of the transmitted ray. Cf. *extinction coefficient*.

absorbate The absorbed phase in *chromatography*, q.v.

absorbent (1) Able to take up gases or liquids. (2) An agent which imbibes or attracts moisture or gases. **light-~** Pertaining to any solid, liquid, or gaseous substance which converts certain radiation falling on it into other forms of energy (thermal, electronic, etc.).

a. cotton A purified fat-free *cotton*.

absorbing power See *absorption coefficient*.

absorptiometer (1) An apparatus (bunsen) to measure the absorption of a gas by a liquid. (2) A device for regulating the thickness of a liquid in spectrophotometry.

absorption The apparent disappearance of one or more substances or forces by being taken into another substance or transformed into another form of energy. Cf. *adsorption*. (1) *Chemical*. Holding by cohesion or capillary action in the pores of a solid. (2) *Physical*. Retention of the energy of radiation, such as heat or light, by a solid, liquid, or gas. The energy increases that of individual atoms and molecules, with the effect (e.g., temperature increase or ionization) depending on the wavelength of the radiation and the nature of the material. (3) *Physiological*. The transformation of nonliving into living matter, i.e., food to protoplasm. **heat of ~** The number of joules (or cal) evolved when a gas is absorbed by a liquid.

a. apparatus An absorptiometer or other device for the absorption of gases or vapors. **a. band** A group of light wavelengths which are absorbed by molecules. Cf. *a. spectrum*. **a. cell** A small glass or quartz cup with parallel walls filled with a solution for the production of absorption spectra; or a solid smeared on a plate, as in reflectance spectroscopy. **a. coefficient** (1) *Chemical*. The amount of a gas at s.t.p., which will saturate a quantity of a liquid solvent. Ostwald a.c.: ratio of the concentration of a gas in the liquid to its concentration in the gas phase. (2) *Radiation*. Linear a.c.* The constant a in the equation $I = I_0 \cdot e^{-acd}$, where I_0 and I are the intensities of the incident and transmitted light, respectively; and c is the concentration, d the thickness of the solution. (3) *X-rays*. The constant a in the formula $I = I_0 \cdot e^{-ad}$. **a. of food** See *metabolism*. **a. of gases** See *absorption (1)*. **a. law** See *Beer's law*, *Dalton's law* (3), *Henry's law*. **a. lines** See *a. sprectrum*. **a. paper** A fat-free filter paper used for the Adams determination of fat in milk. **a. ray** A light ray that, when impinging on matter, can be transformed into thermal or kinetic energy, or electromagnetic radiation of a different wavelength (fluorescence, phosphorescence), induce photochemical reactions, decompose molecules (photodecomposition), or eventually cause cancer (X-rays, γ-rays). **a. spectrum** The image produced by rays when passing in succession through an absorbing medium and a spectroscope. In general, *molecules* absorb groups of wavelengths, giving *bands*; *atoms* absorb single wavelengths, giving *lines*. Cf. *spectrum*. **a. tube** Glass apparatus used in the laboratory for the absorption of gases by liquids or solids. **a. value** Iodine number.

absorptive power Capacity to absorb, expressed numerically.

absorptivity Extinction coefficient.
abstergent Detergent.
abstract A summary of essential features.
absynthin Absinthin.
abundance of elements The estimated relative proportions of chemical elements in the earth. See Table 2 below.
Ac (1) Symbol for actinium. (2) Abbreviation for (1) acetyl; (2) acetate; (3) acyl radical.
ac, a.c. Abbreviation for alternating current.
ac- Prefix indicating substitution in an alicyclic nucleus. Cf. *ar-*.
acacatechin $C_{15}H_{14}O_6$ = 290.3. 4-Catechol-3,5,7-trihydroxychroman. A constituent of catechu.
acacia Gum arabic. The dried, gummy exudation of *Acacia senegal* (Leguminosae). Cf. *babool, catechu.* Oval beads used extensively in pharmacy and in adhesives (N.F., B.P.).
acacin $C_{15}H_{15}O_6$ = 291.3. A catechin present in *Anacardium occidentale,* Lin. See *kauri.*
acacipetalin $C_{11}H_{19}NO_6$ = 261.3. A cyanogenetic glucoside, m.176, from *Acacia* species.
acadialite A variety of chabazite.
acajou The fragrant heartwood of *Cedrela brasiliensis* (mahogany). **a. balsam** An extract of the seeds of *Anacardium occidentale* (mahogany nuts); a blistering agent. **a. nut** Semecarpus.
acalypha The Indian herb *A. indica;* a substitute for ipecac.
Acanthaceae A family of tropical plants comprising 175 genera and 1,400 species. Cf. *adhatotic acid, ibogaine.*
acanthite Ag_2S. A mineral, black needles, d.7.2.
acari A large order of arachnids, commonly called mites.
acaricide A pesticide used to kill mites.
acaroid resin Earth shellac, yellow resin, grass tree gum, botany bay gum, aceroides gum, yacca gum. The exudation of the stems of *Xanthorrhoea hastilis* (Liliaceae) of Australia.
acceleration (1) *Chemical.* An increase in the speed of a chemical reaction. Cf. *catalysis.* (2) *Physical.* The increase in velocity per second, in m/s^2. Cf. *force.* **a. of free fall*** *g.* Force of gravity. The tendency of objects to move toward the center of the earth. g = 9.80665 m/s^2 = 32.174 ft/s^2. Cf. *Helmert's equation.*
accelerator A substance which increases the speed of a chemical reaction. Cf. *catalyst, retarder.* **impregnation ~** Introfier. **linear ~** Linac. A particle a. with stages along its (straight) tube to accelerate particles progressively. Stanford Linear Accelerator Center (SLAC) can accelerate electrons up to an energy of 30 GeV. Cf. *synchrotron.* **particle ~** A long pipe structure, maintained under high vacuum, that electromagnetically accelerates subatomic particles. They can be made to collide, with the resulting particle fragments studied. **rubber ~** A compound which hastens and improves the curing of rubber, e.g., thiocarbanilide. Cf. *vulcanization.* **super ~** A quick-acting vulcanizing agent, e.g., thiuram. **ultra ~** Superaccelerator.
acceptable explosives Unstable chemicals and mixtures which may be transported on railways and ships, subject to certain restrictions.
acceptor See *induced reaction.*
accessory Supplementary in function. **a. food factors** Vitamins.
accident frequency rate The number of (time-costing) accidents per 100,000 person working hours.
accretion The increase in size of a body by external addition.
accroides gum Acaroid resin.
accumulator A device for storing electricity; usually based on a reversible chemical reaction. **steam ~** A vessel in which steam is stored under pressure.
a. metal An alloy: Pb 90, Sb 75, Sn 9.25%.
accuracy The systematic bias which is the difference between the "true" value and that obtained from a very large number of tests. Cf. *precision.*
ace- (1) Prefix indicating relationship to acetylene (or the ethylene radical). (2) The group —C—C— attached to a bicyclic system; as, the atoms 1 and 2 in acenaphthene. Cf. *aci-*.
acedicon An alkaloidal isomer of acetylated codeine, m.152.
acenaphthene* $C_{10}H_6(CH_2)_2$ = 154.2. Ethylenenaphthalene.

TABLE 2. PERCENTAGE ABUNDANCE OF CHEMICAL ELEMENTS

Order of abundance	Lithosphere		Hydrosphere		Atmosphere		Meteorites: Iron		Meteorites: Stone	
1	O	47.33	O	85.79	N	75.53	Fe	72.06	O	35.43
2	Si	27.74	H	10.67	O	23.02	O	10.10	Fe	23.32
3	Al	7.85	Cl	2.07	Ar	1.40	Ni	6.50	Si	18.03
4	Fe	4.50	Na	1.14	H	0.02	Si	5.20	Mg	13.60
5	Ca	3.47	Mg	0.14	C	0.01	Mg	3.80	S	1.80
6	Na	2.46	Ca	0.05	Kr	0.01	S	0.49	Ca	1.72
7	K	2.46	S	0.05	Xe	0.005	Ca	0.46	Na	1.64
8	Mg	2.24	K	0.04	Remainder	0.005	Co	0.44	Al	1.53
9	Ti	0.46	N	0.02		—	Al	0.39	Ni	1.52
10	H	0.22	Br	0.01		—	Na	0.17	Cr	0.32
11	C	0.19	C	0.01		—	P	0.14	Mn	0.23
12	P	0.12	I	0.006		—	Cr	0.09	K	0.17
13	S	0.12	Fe	0.002		—	C	0.04	C	0.15
14	Mn	0.08	Remainder	0.002		—	K	0.04	Co	0.12
15	Ba	0.08		—		—	Mn	0.03	Ti	0.11
16	F	0.07		—		—	Ti	0.01	P	0.11
17	Cl	0.06		—		—	Cu	0.01	H	0.09
18	Cu	0.03		—		—	Remainder	0.03	Cl	0.09
19	N	0.02		—		—		—	Cu	0.01
20	Sr	0.02		—		—		—	Remainder	0.01
	Remainder	0.48		—		—		—		—

A hydrocarbon from coal tar distillates. Colorless needles, m.95, insoluble in water; used in organic synthesis. **bi ~** Biacene. **1,2-dioxo ~** Acenaphthenequinone.
a. dione Acenaphthenequinone.

acenaphthenequinone $C_{12}H_6O_2$ = 182.2. Dioxoacenaphthene. Colorless crystals, m.261, soluble in alcohol.

acenaphthenone $C_{12}H_8O$ = 168.2. Colorless crystals, m.121, soluble in alcohol.

acenaphthenyl* The radical $C_{12}H_9$—, from acenaphthene.

acenaphthylene* $C_{12}H_8$ = 152.2. An unsaturated hydrocarbon from acenaphthene. Colorless crystals, m.92, insoluble in water. **1,2-dihydro ~** Acenaphthene.

-acene Suffix indicating 5 or more fused benzene rings in a linear configuration.

Acer A genus of broad-leaved deciduous trees commonly known as the maples. **A. saccharum** Sugar maple.

acerdese Manganite.

acerdol Calcium permanganate.

aceric acid Impure malic acid from the sap of the maple *(Acer rubrum).*

aceritol See *acertannin.*

aceroides gum Misnomer for *acaroid resin.*

acertannin $C_{20}H_{20}O_{13}(+2 \text{ or } 4H_2O)$ = 468.4. A *pyrogallol tannin,* q.v. Hydrolysis by tannase produces gallic acid and aceritol.

acesulfame potassium 6-Methyl-1,2,3-oxathiozane-4-one-2,2-dioxide. A sweetener, 200 times as sweet as sucrose.

acet (1) Indicating the group MeC≡; as in the acetyl radical, MeCO—. (2) The acetyl* radical.

acetal $MeCH(OEt)_2$ = 118.2. 1,1-Diethoxyethane*, diethylacetal. Colorless liquid, d.0.831, b.103, slightly soluble in water; solvent and intermediate in chemical synthesis. Cf. *acetals.* **amino ~** See *aminoacetal.* **dichloro ~** See *dichloroacetal.* **trichloro ~** See *trichloroacetal.*
a. diethyl Acetal.

acetaldehyde* $CH_3\cdot CHO$ = 44.1. Ethanal*, aldehyde, ethyl aldehyde, acetic aldehyde. Colorless aromatic liquid, b.20.8, soluble in water, alcohol, or ether. Used as a solvent and reducing agent (silvering mirrors), and in the manufacture of organic compounds. **amino ~*** $H_2N\cdot CH_2\cdot CHO$ = 59.1. Glycine aldehyde. Readily polymerizes. Stable in conc. acid solutions. **benzoyl ~** See *Benzoylacetaldehyde*.* **benzylidene ~** Cinnamaldehyde*. **hydroxy ~*** Glycolaldehyde*. **met(a) ~** Metaldehyde. **oxo ~** Pyruvaldehyde*. ***para-* ~** Paraldehyde. **pentyl ~** Hexanal*. **phenyl ~*** α-Tolualdehyde. **tribromo ~** Bromal. **trichloro ~** Chloral. **trimethyl ~** Pivaldehyde*.
a. ammonia $CH_3\cdot CHOH\cdot NH_2$ = 61.1. 1-Aminoethanol*. Addition compound of aldehyde and ammonia. Solid, m.97; soluble in water. **a. cyanohydrin** $CH_3\cdot CHOH\cdot CN$ = 71.1. Liquid, b.183 (decomp.), soluble in water. **a. semicarbazone** $Me\cdot CH{:}N\cdot NHCONH_2$ = 101.1. Solid, m.162.

acetaldol Aldol.

acetaldoxime Aldoxime.

acetaldoxine $Me\cdot CHNO$ = 58.1. Colorless crystals or liquid, d.0.9645, m.13, b.114, soluble in alcohol.

acetalphenaphthylamine Acetnaphthalide.

acetals* Compounds containing the group $=C(OR)_2$. **hemi ~** Compounds containing the group =C(OH)OR, as glucose. **ketone ~** Acetals*, ketals.

acetamide* $Me\cdot CO\cdot NH_2$ = 59.1. Ethanamide. Colorless crystals, m.82, soluble in water; used in organic synthesis. **acetyl ~** Diacetamide*. **benzyl ~** *α-* **~** Hydrocinnamamide. ***N-* ~** Benzylacetamide. **benzylidene ~** Cinnamamide. **bromo ~** Acetbromamide. **cyan ~** $CN\cdot CH_2\cdot CO\cdot NH_2$ = 84.1. Colorless crystals, m.118. **di ~** See *diacetamide.* **dichloro ~*** $CHCl_2\cdot CO\cdot NH_2$ = 128.0. Colorless crystals, m.98, soluble in water. **hydroxy ~*** Glycolamide. **phenyl ~† *N-*** Acetanilide. *α-* Toluamide.
a. chloride $Me\cdot CCl_2\cdot NH_2$ = 114.0, b.90. **a. nitrate** $MeCO\cdot NH_3\cdot ONO_2$ = 122.1. Colorless crystals, formed by the action of nitric acid on a.

acetamidine* $C_2H_6N_2$ = 58.1. Ethanamidine, m.166.

acetamido* Indicating the MeCO·NH— radical. **a. ethylsalicylic acid** Benzacetin. **a.naphthol** $C_{12}H_{11}O_2$ = 201.2. **1,2- ~** White leaflets, m.235. **1,4- ~** White needles, m.187. **a.phenetol** Phenacetin. **a.phenol** $C_6H_4(OH)\cdot NH\cdot CO\cdot Me$ = 151.2. **1,2- ~** White leaflets, m.201. **1,3- ~** Colorless needles, m.149. **1,4- ~** Acetaminophen.

acetamino The acetamido* radical.

acetaminophen $C_6H_4(OH)\cdot NH\cdot CO\cdot Me$ = 151.2. 4-Hydroxyacetanilide, paracetamol, Liquiprin, Panadol, Tylenol. White, bitter crystals, m.171, soluble in water; an analgesic and antipyretic for moderate pain (USP, EP, BP).

acetanilide $PhNH\cdot CO\cdot Me$ = 135.2. *N*-phenylacetamide†, Antifebrin. Colorless leaflets, 114, soluble in water; formerly used as antipyretic, antirheumatic; a preservative. **aceto ~** $MeCOCH_2\cdot CONHPh$ = 177.2. Colorless crystals, m.85. **acetyl ~** Diacetanilide. **amino ~*** $NH_2C_6H_4\cdot NH\cdot CO\cdot Me$ = 150.2. **para ~** Colorless crystals, m.160. **bromo ~*** Acetbromoanilide. **di ~** $PhN(MeCO)_2$ = 177.2. Colorless leaflets, m.37. **ar-ethoxy ~** Acetophenetide. **p-ethoxy-** Phenacetin. **ar-methoxy ~** Acetaniside. **ar-methyl ~** Acetotoluide. ***α-*phenyl ~** α-Toluanilide.

acetaniside $C_6H_4(OMe)NH\cdot COMe$ = 165.2. Methoxyacetanilide. ***ortho-* ~** Colorless crystals, m.80, soluble in water.

acetarsol $C_6H_3\cdot OH\cdot (NHCOMe)AsO\cdot (OH)_2$ = 275.1. Stovarsol, acetarsone, 3-acetamido-4-hydroxyphenyl-arsonic acid. White powder, slightly soluble in water; an antiprotozoan; formerly used to treat syphilis. Cf. *carbarsone.*

acetarsone Acetarsol.

acetate Ac. A salt of acetic acid containing the CH_3COO— radical. A. are readily decomp. by strong acids or heat.

acetazolamide $C_4H_6O_3N_4S_2$ = 222.2. *N*-(5-sulfamoyl-1,3,4-thia-diazol-2-yl) acetamide. White crystals, m.258, slightly soluble in water. An inhibitor of carbonic anhydrase; a diuretic, used to treat glaucoma (USP, BP).

acetbromamide $CH_2Br\cdot CONH_2$ = 138.0. Bromoacetamide. Colorless leaflets, m.108, soluble in ether. ***N-* ~** $CH_3CONHBr$.

acetbromoanilide $BrC_6H_4NHCOCH_3$ = 214.1. ***para-* ~** Colorless needles, m.165.

acetene Ethylene*.

acetenyl Ethynyl*.

Acetest Trademark of tablet for testing for acetone in urine; contains nitroprusside.

acet extract See *extract.*

acethydrazide A compound containing the $NH_2\cdot NH\cdot CO\cdot CH_2$— radical. Cf. *hydrazide.*

acetic Describing compounds containing acetyl, CH_3CO—. **a. aldehyde** Acetaldehyde*. **a. anhydride*.** $(MeCO)_2O$ = 102.1. Ethanoic anhydride, acetyl oxide, acetic acid anhydride. Colorless liquid, b.137, soluble in alcohol; a reagent. **a. ester, a. ether** Ethyl acetate*. **a. peracid** Peracetic acid*. **a. peroxide** $(CH_3CO)_2O_2$. An explosive derivative of a. anhydride.

acetic acid* $CH_3\cdot COOH$ = 60.1. Ethanoic acid, ethylic acid, vinegar acid, acetone carboxylic acid (USP, NF, BP). (1) 99.5% (glacial). Clear, colorless liquid or crystalline mass miscible with water or alcohol, d.1.054, m.16, b.118. A reagent, solvent, precipitant, and general neutralizing and acidifying agent. (2) 36% (common). A clear, colorless liquid, d.1.045, miscible with water or alcohol. (3) 5.1–6.3% (dilute). Official test solution of the NF. A. is produced by the oxidation of alcohol (vinegar), and in the pyroligneous acid resulting from the destructive distillation of wood. **aceto ~** See *acetoacetic acid.* **acetyl ~** Acetoacetic acid*. **activated ~** Acetyl *coenzyme* A. **allyl ~** 4-Pentenoic acid*. **amino ~** Glycine*. **aminobutyl ~** Norleucine. ***m*-aminophenyl ~** $NH_2 \cdot C_6H_4 \cdot CH_2 \cdot COOH$. Colorless crystals, m.149. **benzamido ~** Hippuric acid*. **benzoyl ~** PhC:OCH_2COOH = 164.2. 3-Oxo-3-phenylpropanoic acid. Colorless needles, m.103. **benzyl ~** Hydrocinnamic acid. **benzylidene ~** Cinnamic acid*. **bromo ~*** $C_2H_3O_2Br$ = 139.0. Bromoethanoic acid. Colorless hexagons, m.50, soluble in water. **butyl ~** Hexanoic acid*. ***tert*- ~** $MeCMe_2 \cdot CH_2COOH$ = 116.2. Pseudocaproic acid. Liquid, m.-11, b.190. **carbamido ~** Hydantoic acid*. **chloro-*** $C_2H_3O_2Cl$ = 94.5. Chloroethanoic acid. Rhombic crystals, m.51. **di ~** (1) Acetoacetic acid. (2) See *diacetic acid.* **diacetyl ~** Diacetic acid. **dibromo ~*** $C_2H_2O_2Br_2$ = 217.8. Dibromoethanoic acid. Colorless crystals, m.48. **dicarbamido ~** Allantoic acid. **dichloro ~*** $C_2H_2O_2Cl_2$ = 128.9. Dichloroethanoic acid. Colorless liquid, b.190. **diethyl ~** $Et_2CH \cdot COOH$ = 116.2. Carboxypentane. Colorless liquid, b.190. **dihydroxy ~** Glyoxylic acid*. **diiodo ~*** $C_2H_2O_2I_2$ = 311.8. Diiodoethanoic acid. Yellow crystals, m.110. **dimethyl ~** 2-Methyl*propionic acid**. **dimethylene ~** $EtCMe_2 \cdot COOH$ = 116.2. Colorless liquid, b.187; an isomer of hexanoic acid. **diphenyl ~*** $Ph_2CHCOOH$ = 212.2. Colorless needles, m.148. **diphenylene ~** Fluorenecarboxylic acid. **ethoxy ~** Ethylglycolic acid. **ethyl ~** Butanoic acid*. **formyl ~** Malonaldehydic acid*. **furfuryl ~** Furonic acid. **hydrazi- ~** $(NH)_2CH \cdot COOH$ = 88.1. **hydroxy ~** Glycolic acid. **imino ~** $NH(CH_2COOH)_2$ = 133.1. Rhombic crystals, m.227. **iodo ~*** $C_2H_3O_2I$ = 185.9. Iodoethanoic acid. Yellow crystals, m.82. **isobutyl ~** Dimethylbutanoic acid*. **isopentyl ~** $C_7H_{14}O_2$ = 130.2. Colorless liquid, d.0.910, b.209. **isopropyl ~** Iso*valeric acid.* **isopropylidene ~** Senecioic acid. **mercapto ~** $HS \cdot CH_2COOH$ = 92.1. 2-Mercaptoethanoic acid, thioglycolic acid. Colorless liquid, m.−16. **methoxy ~*** $MeOCH_2COOH$ = 90.1. Methylglycolic acid. Colorless liquid, d.1.1768. **methyl ~** Propionic acid*. **methylamino ~** Sarcosine*. **nitro ~** See *nitroacid.* **oxo ~** Pyruvic acid*. **oxybis ~** Diglycolic acid. **per ~*** See *peracetic acid.* **phenyl ~*** See *phenylacetic acid.* **pyro ~** Pyroligneous acid. **sulfo ~** $HSO_3 \cdot CH_2 \cdot COOH$ = 140.1. Sulfoethanoic acid. Colorless crystals, m.86. **thiobis ~** Thiodiglycolic acid. **thio ~** MeCOSH = 76.12. Ethanethiolic acid, thiolacetic acid. Colorless liquid, b.93. **triethyl ~** CEt_3COOH = 144.2. Isocaprylic acid. Solid, m.40, b.202. **trimethyl ~** Pivalic acid*. **triphenyl ~** $Ph_3C \cdot COOH$ = 288.3. Colorless crystals. m.264. **ureido ~** Hydantoic acid*. **vinyl ~** 3-Butenoic acid*.

a. a. amide Acetamide*. **a. a. amine** Acetamide*. **a. a. aldehyde** Acetaldehyde*. **a. a. anhydride** See *acetic.* **a. a. ester** Ethyl acetate*. **a. a. ether** Ethyl acetate*. **a. a. peracid** Peracetic acid.

acetidin Ethyl acetate*.

acetifier An apparatus to hasten acetification, the production of vinegar from fermented liquids by atmospheric oxidation.

acetimeter Hydrometer (obsolete).

acetimetry The determination of acetic acid in vinegar by titration with alkali.

acetimidoyl* The radical $CH_3C(:NH)$-.

acetin Esters obtained by reaction of glycerol and acetic acid; e.g., **mono ~** $C_3H_5(OH)_2(O \cdot COMe)$ = 134.1. Acetin. Colorless liquid, d.1.221, b.240. Used in the manufacture of dynamite and as a solvent for dyes. **di ~** See *diacetin.* **tri ~** See *triacetin.*

acetnaphthalide $MeCONHC_{10}H_7$ = 185.2. *N*-Acetyl-1-naphthylamine. Colorless crystals, b.159. 2-Acetyl-2-naphthylamine. Colorless leaflets, m.132.

aceto Acetyl* is preferred form, except in acceptable trivial names.

acetoacetate* A salt containing the $MeCOCH_2COO$—radical, from acetoacetic acid.

acetoacetic a. acid* $Me \cdot CO \cdot CH_2 \cdot COOH$ = 102.1. 3-Oxobutanoic acid†, acetylacetic acid, diacetic acid. Colorless liquid, decomp. 100, miscible with water. Derived from acetyl coenzyme A in the body and increases during diabetes. Cf. *acetone.* See *ketone bodies.* **a. ester** (1) Acetoacetic ether. (2) Compounds derived from acetoacetic acid by replacing the acid hydrogen by an organic radical. Two isomeric forms: ketonic form, $R-CH_2 \cdot CO \cdot CH_2COO-R$; enolic form, $R-CH_2 \cdot C(OH):CH \cdot COO-R$. **a. ester condensation** The formation of a. esters by metallic sodium from the corresponding alkyl ester. **a. ester decomposition** The decomposition of a. esters. Strong acids or weak alkalies split the ester to a ketone alcohol and carbon dioxide; strong bases decompose it to acids (acetic acid, etc.). **a. ester synthesis** The formation of organic compounds by hydrolyzing their a. esters. **a. ether** $MeCOCH_2COOEt$ = 130.11. Ethylacetoacetate*, ethyl-3-oxobutanoate. Colorless liquid, b.181; used in organic synthesis.

acetoacetyl-* 1,3-Dioxobutyl†. The radical $Me \cdot CO \cdot CH_2 \cdot CO-$.

acetobenzoic anhydride Benzoyl acetate*.

acetobrom- Acetbrom-.

acetocaustin Trichloroacetic acid*.

acetochloral Chloral.

acetocinnamene Benzylidene acetone.

acetoglyceral $C_5H_{10}O_3$ = 118.1. Glycerol ethylidene ether. Colorless liquid, d.1.081, b.184.

acetoin* $MeCO \cdot CHOH \cdot Me$ = 88.1. 3-Hydroxy-2-butanone, methylacetylcarbinol. A condensation product of 2 molecules of acetic acid. Colorless liquid, b.141.

acetol 1-Hydroxy-2-*propanone**.

acetoluide Acetotoluide.

acetolysis The breaking up of an organic molecule by acetic anhydride or acetic acid.

aceton acid Acetonic acid.

acetonamine Organic compounds containing the NH_2- and $CO-$ radicals. **di ~** $NH_2 \cdot CMe_2 \cdot CH_2 \cdot CO \cdot Me$.

acetonaphthone Methyl naphthyl *ketone**. **1-hydroxy-2- ~** $MeCOC_{10}H_6OH$ = 186.2. 2-Acetyl-1-naphthol. Yellow needles, m.100.

acetone* Me_2CO = 58.1. Dimethyl ketone*, 2-propanone, methylacetal, dimethylketal. Colorless ethereal liquid, d.0.792, b.57, miscible with water; a constituent of wood spirit. Produced by fermentation and in the body during diabetes by decarboxylation of acetoacetic acid. A solvent, precipitant for albumin, reagent (USP, BP). **acetonyl ~** See *acetonylacetone.* **amido ~** Aminoacetone. **benzylidene ~** PhCH:$CH \cdot COMe$ = 146.2. 4-Phenyl-3-buten-2-one, methyl styryl ketone*, benzalacetone. Colorless crystals, m.41; used in organic synthesis. **bromo ~*** $CH_2Br \cdot CO \cdot Me$ = 137.0. Monobromacetone. Colorless liquid, b.140. **chloro ~*** $CH_2Cl \cdot CO \cdot Me$ = 92.5. Monochloracetone. Colorless liquid, b.119. **diamino ~*** $CO(CH_2NH_2)_2$ = 88.1. **dibenzylidene ~** Styryl ketone*. **dichloro ~*** *alpha*-$CHCl_2 \cdot CO \cdot CH_3$ = 127.0. Colorless liquid, b.190. *beta*-$CH_2Cl \cdot CO \cdot CH_2Cl$.

Colorless liquid, b.120. **diethoxy ~*** $CO(CH_2OEt)_2$. **dihydroxy ~** $CO(CH_2OH)_2$. A constituent of suntan lotions. **diisonitroso ~** $CO(CH:NOH)_2$ = 116.1. Colorless crystals, m.144. **dimethoxy ~*** $CO(CH_2OMe)_2$. **hydroxy ~** Acetol. **meta- ~** Diethyl ketone*. **monobromo ~** See *bromoacetone.* **monochloro ~** See *chloroacetone.* **oxalyldi ~** $Me \cdot CO \cdot CH_2 \cdot CO \cdot CO \cdot CH_2 \cdot CO \cdot Me$ = 170.2. Colorless crystals, m.120, insoluble in water. **thiocyanato ~** $MeCO \cdot CH_2 \cdot SCN$ = 115.2. Colorless liquid, d.1.180, soluble in water.

a. acid Acetonic acid. **a. alcohol** 1-Hydroxy-2-*propanone**. **a. bodies** See *ketone bodies.* **a. bromoform** $CBr_3 \cdot COHMe_2$ = 310.8. Brometone. Colorless crystals; a sedative. **a. carboxylic acid** Acetic acid*. **a. chloride** $MeCCl_2Me$= 113.0. 2,2,-Dichloropropane*, b.69. **a. chloroform** Chlorbutanol. **a. collodium** Collodion. **a. cyanhydrin** $Me_2C \cdot OH \cdot CN$ = 85.1, b.120. **a. diacetic acid** Hydrochelidonic acid. **a. dicarboxylic acid** $CO \cdot (CH_2COOH)_2$ = 146.1. β-Ketoglutaric acid, 3-oxopentanedioic acid*. Colorless crystals, m.130. **a. oil** The oily residue from the distillation of crude a. Used as a solvent, as a denaturating agent, and in the purification of anthracene. **a. phenylhydrazone** $Me_2C:N \cdot NHPh$ = 148.2. Crystals, m.16. **a. semicarbazone** $Me_2C:N \cdot NH \cdot CONH_2$ = 115.1, m.190. **a. sodium bisulfite** $NaSO_3 \cdot CMe_2OH$. Colorless crystals, soluble in water. **a. sugar** Isopropylidine sugar. A condensation compound of sugar and a. containing a $-O \cdot CMe_2 \cdot O-$ group.

acetonic acid $Me_2C(OH)COOH$ = 104.1. Acetone acid, butyllactic acid, 2-hydroxyisobutyric acid. Colorless crystals, m.79, soluble in water.

acetonitrile* $CH_3 \cdot CN$ = 41.1. Methyl cyanide, ethane nitrile*, cyanomethane. From coal tar and the residue of molasses. Colorless liquid, d.0.783, b.81, soluble in water; used in organic synthesis and perfumery. **allyl ~** See *allylacetonitrile.* **benzoyl ~** $PhCOCH_2CN$ = 145.2. Cyanoacetophenone, β-ketohydrocinnamonitrile. White leaflets, m.80. **phenyl ~** Benzyl cyanide*. **triethoxy ~** $(EtO)_3C \cdot CN$. Cyanoformic ester, oxalnitrilic ethyl ester. Colorless liquid, b.160. **trinitro ~** $CN \cdot C(NO_2)_3$ = 176.1. Trinitroethanenitrile*. (1) Colorless crystals, m.41. (2) White leaflets, m.42. **vinyl ~** Allyl cyanide*.

acetonitrolic acid $Me(NO_2)C:NOH$ = 104.1. Ethylnitrolic acid. Yellow rhombs, m.88 (decomp.), soluble in water.

acetonyl* 2-Oxopropyl†. The radical $CH_3 \cdot CO \cdot CH_2-$ **a.acetone** $MeCO(CH_2)_2 \cdot CO \cdot Me$ = 114.1. 2,5-Hexanedione*, γ-diketohexane. Colorless liquid of pleasant odor, b.188; used as a solvent. **a.amine** 1-Amino-2-propanone*. **a.urea** $C_5H_8O_2N_2$ = 128.1, b.175, soluble in water.

acetonylidene* 2-Oxopropylidene†. The radical MeCOCH=.

acetophenetide $EtO \cdot C_6H_4 \cdot NH \cdot COMe$ = 179.2. ***ortho-* ~** *N*-Acetyl-*o*-phenetidine. Colorless leaflets, m.79.

acetophenetidin Phenacetin.

acetophenine $C_{24}H_{19}N$ = 321.4. Colorless crystals, m.135, slightly soluble in alcohol.

acetophenone* $Me \cdot CO \cdot Ph$ = 120.2 Methyl phenyl ketone*, 1-phenylethanone†. Transparent crystals, m.21; an orange perfume. **allyl ~** $Ph \cdot CO \cdot CH_2 \cdot CH_2 \cdot CH:CH_2$ = 160.2. Colorless liquid, b.236. **amino ~** See *aminoacetophenone.* **benzyl ~** Phenyl*propiophenone.* **benzylidene ~** Chalcone*. **chloro ~** See *tear gases.* **cyano ~** Benzoyl acetonitrile. **ar-hydroxy ~** See *hydroxy.* **ω-hydroxy ~** $PhCOCH_2 \cdot OH$ = 136.2. Phenacyl alcohol. Colorless crystals, m.86. **3′,4′-methylenedioxy ~** Acetopiperone. **nitroso ~** Benzoylformoxime. **phenyl ~** A. acetone.

a. acetone $PhCO \cdot CH_2 \cdot CH_2 \cdot COMe$ = 176.2. Phenylacetylacetone. **a. carboxylic acid** Acetylbenzoic acid. **a. oxime** $Me \cdot C(NOH) \cdot Ph$ = 135.2. Colorless crystals, m.59.

acetophenones The homologs of acetophenone:

$PhCOCH_2-R$	Acetylated benzenes
$Ph-R-COMe$	Phenylated fatty ketones
$R-C_6H_4 \cdot COMe$	Nucleus-substituted phenones

acetopiperone $MeOC \cdot C_6H_3(O \cdot CH_2O)$ = 164.2. 3′,4′-Methylenedioxyacetophenone. Colorless crystals, m.87.

acetosalicylic acid Acetylsalicylic acid.

acetothienone Methyl thienyl *ketone**.

acetotoluide $Me \cdot C_6H_4 \cdot NHCOMe$ = 149.2. Acetotoluidine. ***ortho-* ~** Colorless crystals, m.107; an antipyretic. ***meta-* ~** Colorless crystals, m.65. ***para-*** Colorless needles, m.151. ***N*-methyl ~** $MeCON(Me) \cdot C_6H_4Me$ = 163.2. *N*-Acetyl-*N*-methyl-*o*-toluidine. White crystals, m.56.

acetoveratrone $C_{10}H_{12}O_3$ = 180.2. 3,4-Dimethoxy-1-methylketobenzene. Colorless crystals, m.48; used in perfumery.

acetoxime $Me_2C:NOH$ = 73.1. Propanoxime, 2-propanone oxime*. Colorless prisms, m.58.

acetoximes A group of ketoximes derived from acetoximes; two isomeric forms.

```
  N·OH        HO·N
   ||           ||
R′—C —R     R′— C—R
```

acetoxy* Acetyloxy†. The radical CH_3COO-.

acetoxylide $MeCONHC_6H_3Me_2$ = 163.2. 2,4-Dimethylacetanilide. White needles, m.129.

acetphenetid Acetophenetide.

acettoluide Acetotoluide.

acetum Latin for vinegar.

aceturic acid Acetylglycine.

acetyl* Ac. Ethanoyl, acet. The acyl radical CH_3CO-. **a. hydrate** Acetic acid. **a. number** A measure of the amount of oxyacids and alcohols in a vegetable or animal fat. The mg of potassium hydroxide to neutralize the acetic acid obtained by saponification of 1g of a fat which has been acetylated with acetic anhydride. Cf. *Cook formula.*

acetylacetonate $M[CH(MeCO)_2]_n$. A chelate of acetylacetone formed with many metals, e.g., aluminum, beryllium, and the rare earth metals. Used for extraction in organic solvents.

acetylacetone $MeCO \cdot CH_2 \cdot COMe$ = 100.1. 2,4-Pentanedione*. Colorless liquid, b.137; used in organic synthesis.

acetylamine Acetamide*.

acetylamino* The radical $CH_3 \cdot CO \cdot NH-$, derived from acetamide. **a. acetic acid** Acetyl glycine. **a. benzoic acid** $MeCONH \cdot C_6H_4 \cdot COOH$ = 179.2. ***ortho-* ~** Colorless needles, m.185. ***meta-* ~** Colorless crystals, m.243 (decomp.) ***para-* ~** Colorless needles, m.250. **a. phenetole** Acetophenetide.

acetyl anhydride Benzoyl acetate*.

acetylate, acetylation Introducing an acetyl radical into an organic molecule; e.g., by boiling glacial acetic acid.

acetylbenzene Acetophenone*.

acetylbenzoate Benzoyl acetate*.

acetylbenzoic acid $MeCO \cdot C_6H_4COOH$ = 164.2. Acetophenone carboxylic acid. ***ortho-* ~** Crystals, m.114. ***para-* ~** Crystals, m.205 (sublimes).

acetylbiuret $MeCO \cdot NH \cdot CO \cdot NH \cdot CO \cdot NH_2$ = 145.11. Colorless needles, m.193.

acetyl bromide $Me \cdot COBr$ = 122.9. Ethanoyl bromide. Colorless fuming liquid, b.77, decomp. in water. **bromo ~** $CH_2Br \cdot COBr$ = 201.8. Bromoethanoyl bromide. Colorless liquid, b.148, decomp. in water.

acetylcannabinol $MeCO \cdot O \cdot C_{21}H_{29}O$. Obtained by heating cannabinol with acetic acid anhydride.

acetylcarbamide Acetylurea.

acetylcarbazole $C_{12}H_8N \cdot COCH_3$ = 209.2. Crystals, m.70.
acetylcarbinol 1-Hydroxy-2-*propanone**.
acetyl chloride* Me·COCl = 78.5. Ethanoyl chloride. Colorless fuming liquid, b.55, decomp. in water; a reagent. **chloro ~** $CH_2Cl \cdot COCl$ = 112.9. Chloroethanoyl chloride. Colorless liquid, b.107, decomp. in water. **dichloro ~** $CHCl_2 \cdot COCl$ = 147.4. Dichloroethanoyl chloride. Colorless liquid, b.105 decomp. in water. **trichloro ~** CCl_3COCl = 181.8. Colorless liquid, b.118, decomp. in water. **trimethyl ~** Pivaloyl chloride.
acetylcholine $C_7H_{16}O_2NCl$ = 181.7. (2-Acetoxyethyl)trimethylammonium chloride, Vagusstoff. White, hygroscopic crystals, m.150; a miotic in eye surgery and important neuromuscular transmitter.
acetyl-CoA See *coenzyme*. **a. synthetase*** Enzyme catalyzing the formation of acetyl-CoA plus AMP and pyrophosphate from acetate, CoA, and ATP.
acetyl cyanide Me·COCN = 69.1. Pyruvonitrile. A colorless liquid, b.93.
acetylcysteine $C_5H_9O_3NS$ = 163.2. *N*-Acetyl-L-cysteine, Airbron. White crystals, m.108, soluble in water. A mucolytic for chest infections and antidote for acetaminophen overdose (USP, BP).
acetylene (1)*HC≡CH = 26.0. Ethine, ethyne. The simplest compound containing a triple bond. Colorless gas with garlic odor, d.0.91, b.−83, slightly soluble in water, soluble in alcohol. Used for oxyacetylene welding and cutting of metals and in organic synthesis. Produced by the action of water on calcium carbide. (2) A member of the alkyne* series. (3) The ethanediylidene* radical. **butyl ~** Hexyne*. **carboxy ~** Propiolic acid*. **di ~** See *diacetylene**. **dimethyl ~** 2-Butyne*. **diphenyl ~** Tolane. **ethyl ~** Butyne*. **methyl ~** Propyne*. **methyl phenyl ~** PhC≡CMe. Phenylallylene. Colorless liquid, b.185. **pentyl ~** Heptyne*. **phenyl ~** PhC≡CH. Ethynylbenzene. Colorless liquid, b.139. **propyl ~** Pentyne*.
a. acids $C_nH_{2n-3}COOH$. Acids which contain the −C:C− group; e.g., propiolic acid*, HC:C·COOH.
a.alcohols Organic compounds which contain both an OH− and a −C:C− group; e.g., 2-propyn-1-ol*, $HC:C \cdot CH_2OH$.
a.carboxylic acid Propiolic acid*. **a. dicarboxylic acid** HOOC·C:C·COOH = 114.0. Butynedioic acid*. Colorless crystals, m.178, insoluble in water. Cf. *glutinic acid*. **a. dichloride** Dichloroethylene*. **a. dinitrile** NC·C:C·CN = 76.1. Carbon subnitride. **a. series** Alkynes*. **a. stones** Acetylith. **a. tetrabromide** $CHBr_2 \cdot CHBr_2$ = 345.7. Tetrabromoethane*, Muthmann's liquid. Yellow oil, b_{36mm}136, insoluble in water. Used in separation of mineral mixtures and in microscopy. **a. tetrachloride** Tetrachloroethane*.
a.urea $C_2H_2(CON_2H_2)_2$ = 142.1. A liquid, decomp. 300.
acetyl fluoride* MeCOF = 62.03. Ethanoyl fluoride. Colorless liquid, d.1.0369, b.10; used in organic synthesis.
acetylformic acid Pyruvic acid*.
acetylglycine $MeCO \cdot NH \cdot CH_2 \cdot COOH$ = 117.1. Acetylaminoacetic acid, aceturic acid. Colorless crystals, m.206.
acetylide* A derivative of acetylene with the H atoms replaced by metals; e.g., cuprous acetylide, Cu_2C_2. **a. ion*** The anion C_2^{2-}.
acetylidene The hypothetical isomer of acetylene, $C{=}CH_2$, in which one atom is divalent.
acetylimino* The radical $CH_3.CO \cdot N{=}$.
acetyl iodide* MeCOI = 169.9. Ethanoyl iodide. Brown, fuming liquid, d.1.98, b.107; used for acetylation.
acetylith Acetylene stones. Sugar-coated granules of calcium carbide used to generate acetylene.
acetylization Acetylation.
acetylize Acetylate.
acetyl peroxide $(MeCO)_2O_2$ = 118.1. Ethanoyl peroxide. Colorless leaflets, m.30, slightly soluble in water.
acetylphenol $PhO \cdot COCH_3$ = 136.2. Colorless liquid, b.193.
acetylphenylenediamine Amino*acetanilide*.
acetylpropyl alcohol $MeCO \cdot CH_2CH_2 \cdot CH_2OH$ = 102.1. Colorless liquid, d.1.0159, b.208, miscible with water; a solvent.
acetylrosaniline $C_{20}H_{18}N_3COMe$ = 343.4. Red crystals, m.218, insoluble in water.
acetylsalicylate A compound containing the MeCOO·C_6H_4COO− radical, from acetylsalicylic acid.
acetylsalicylic acid See *aspirin*.
acetylthymol $C_{12}H_{16}O_2$ = 192.3. Thymylacetate. Yellow liquid, b.244; an antiseptic.
acetyltropeine $MeCO \cdot C_8H_{14}ON$ = 183.3. A syrup, b.236, soluble in water.
acetylurea $MeCO \cdot NH \cdot CONH_2$ = 102.1. Acetylcarbamide. Colorless crystals, m.218, slightly soluble in water.
Achard, Franz Karl (1753–1821) A German pioneer in the manufacture of beet sugar.
achilleic acid 1,2,3-Propenetricarboxylic acid obtained from *Achillea millefolium*.
achilleine $C_{20}H_{38}O_{15}N_2$ = 546.5. An alkaloid from *Achillea millefolium*, yarrow, or milfoil. Brownish-red bitter masses, soluble in water.
achilletin $C_{11}H_{17}O_4N$ = 227.3. A split product produced from achilleine by sulfuric acid.
achiral* Describing a molecule that, in a given configuration, is identical with its mirror image. Cf. *chiral*. See *stereoisomerism*.
achmatite, achmite An epidote (Ural Mountains).
achondrite A meteorite containing chondrules, of variable composition, but usually of low metal content. See *chondrite*.
achrodextrins Achroodextrins.
achroite A colorless tourmaline from Elbe.
achromatic Transmitting white light without breaking it up into colored rays. Cf. *apochromatic*. **a. lens** A combination of lenses of different refractive indices to correct chromatic aberration by bringing two spectral color rays to a common focus. A combination of a convex lens of crown glass with a concave lens of flint glass. **a. objectives** A system of lenses which brings light of different wavelengths to a common focus; i.e., avoids chromatic aberration. **a. substage** A device for microscopes to correct chromatic aberration. **a. system** A combination of prisms or lenses with a common focus for different colors; e.g., an achromatic lens.
achromic Without color. **a. method** The evaluation of amylase preparations from the a. period. **a. period** The time for a 1% starch solution to reach the a. point; i.e., become transformed into achroodextrin. **a. point** The stage in the hydrolysis of starch by enzymes when iodine no longer produces a blue color.
achromobacteria Small Gram-negative bacteria, usually motile. They produce no pigment or gas on gelatin, but a brown color on potato.
achroodextrins $C_{36}H_{62}O_{31}$ = 990.9. Intermediate split products formed after erythrodextrin by the hydrolysis of starch or dextrin; termed alpha-, beta, etc. They are not colored blue by iodine, and are converted into glucose by boiling with dilute acid.
aci- Prefix indicating acid character, as in *aci*-phenylnitromethane, PHCH:NO(OH). Cf. *ace-*.
acicular Slender, needlelike, or hairlike in shape, as crystals or bacteria. **a. oxolepidene** Dibenzoyl *stilbene*.
acid (1) A chemical compound which yields hydrogen ions

when dissolved in water, whose hydrogen can be replaced by metals or basic radicals, or which reacts with bases to form salts and water (neutralization). Antonym: base. (2) An extension of the term includes substances dissolved in media other than water. Cf. *base, ammonia system.* (3) A substance which gives up protons *(proton donor)* or reacts as $A \rightleftharpoons B + H^+$, where A is acid and B is base, e.g., $HAc \rightleftharpoons Ac^- + H^+$. **A-, α-, β-, etc.** See *A* and *sulfonic* acids. **acrylic ~** **alcohol ~** **aldehyde ~** An organic a. containing aldehyde and carboxyl radicals. **aliphatic ~** An a. derived from acyclic chain compounds, e.g., fatty acids. **amic ~*** Amido a. An organic a. containing the $-CONH_2$ group and the carboxyl radical. **amino ~** An organic a. containing the $-NH_2$ and $-COOH$ groups. **aromatic ~** An organic compound containing a closed chain (e.g., benzene ring) and the carboxyl group. **battery ~** Electrolyte a. Sulfuric a. of d.1.150–1.835. **bromo ~** An organic a. containing bromine. **Brönsted ~** A proton donor, as H_3O^+. **carboxylic ~** See *organic acid.* **carbylic ~** See *carbylic acid.* **chloro ~** An organic a. containing chlorine. **dibasic ~** A. containing 2 H atoms which are replaceable by metals or basic radicals, e.g., oxalic acid. **effective ~** See *acidity.* **electrolyte ~** Battery a. **fatty ~** See *fatty acids.* **halogen ~** An inorganic a. containing a halogen, e.g., HCl, $HClO_3$. **haloid ~** An inorganic a. containing a halogen element and no oxygen, e.g., HCl. **hexabasic ~** An a. containing 6 H atoms replaceable by metals or basic radicals. **hydrogen ~** Hydracid. **hydroxy ~** An organic a. containing one hydroxy group and one or more carboxyl groups, e.g., *phenol acids.* **inorganic ~** A compound of hydrogen with a nonmetal, other than carbon, or an acid radical containing no carbon. **labile ~** A grouping, as $-SO_3H$, the acidic properties of which depend on the other groups present. **Lewis ~** A compound that accepts an electron pair. It shares the electron pair donated by a Lewis base. **magic ~** See *magic acid.* **meta ~** (1) See orth*acid (1).* (2) An organic, aromatic a. with the radical in the *meta* position. **mineral ~** Inorganic a. **mixed ~** A mixture of nitric and sulfuric acids used for nitration. **monoamino ~** An organic a. containing one amino group and carboxyl group(s). **monobasic ~** An a. having 1 H atom replaceable by metals or basic radicals. **monohydroxy ~** An organic a. containing 1 OH group and carboxyl group(s). **nitro ~** See *nitroacid,, nitroic acids,* etc. **normal ~** (1) inorganic ortho a. (2) A straight-chain fatty a. (3) A solution containing one gram-equivalent a. per liter, e.g., 1 *N* HCl. **organic ~** A carbon compound containing carboxyl group(s) $-COOH$; indicated by the prefix *carboxy-,* or the suffixes *carboxylic, carbonyl,* or simply *-oic* or *-oyl.* For examples, see *acids* (tables). **ortho ~** (1) A. containing a multiple of H_2O higher than the meta form of that a.; e.g., H_3PO_4, orthophosphoric acid, and HPO_3, metaphosphoric a. (2) An organic a. containing a benzene ring with radicals in the *ortho* position. **oxy ~** (1) An inorganic a. containing oxygen, thereby differing from the haloid or hydrogen acids. Those with the suffix *-ous* have less oxygen than those with the suffix *-ic.* (2) An organic a. containing an oxygen group (e.g., an ether) in addition to the carboxyl radical. **oxygen ~** Oxy a. (inorganic). **oxyhalogen ~** An a. containing oxygen and a halogen; as, $HClO_3$. **para ~** An organic a. with the benzene ring radicals in the *para* position. **peroxy ~*** A. containing the group $-C(:O)OOH$. **phenol ~** See *phenol acids.* **plant ~** A. derived from vegetable sources. **poly ~** See *polyacid.* **polybasic ~** An a. containing more than 1 H atom replaceable by metals or basic radicals. **pseudo ~** (1) A tautomeric form of an organic a. (2) An a. containing hydroxyl radicals. **pyro ~** An a. produced by the action of heat; usually an intermediate compound between acids and acid anhydrides. **racemic ~** An organic a. consisting of a molecular mixture of optically active dextro and levo compounds; e.g., tartaric acid. **resin ~** See *resin acid.* **rubber ~** See *rubber acid.* **soldering ~** Hydrochloric a. of suitable strength used for soldering. **Spekker ~** A mixture of equal volumes of sulfuric a. (d.1.84) and phosphoric a. (d.1.75) used to dissolve iron alloys for analysis, particularly with the Spekker photoelectric absorptiometer. **Steele ~** See *Steele acid.* **sulfinic ~** An acidic organic compound in which an H atom is replaced by the $-SO_2H$ group. Named from the hydrocarbon; e.g., ethanesulfinic acid*, $EtSO_2H$. **sulfonic ~** An acidic organic compound in which an H atom is replaced by the SO_3H group. Named from the hydrocarbon; e.g., ethanesulfonic acid*, $EtSO_3H$. **thio ~** See *thio acid.*

a. of air Carbon dioxide. **a. albuminate** A metaprotein soluble in weak alkali. **a. of amber** Succinic a. **a. anhydride** An a. from which one or more molecules of water have been removed; e.g., SO_3 is the anhydride of H_2SO_4, MeCO·O·OCMe is the anhydride of MeCOOH. **a. of ants** Formic a*. **a. of apples** Malic a*. **a. of benzoin** Benzoic a*. **a. bordeaux** Bordeaux B. **a. brown** Naphthylamine brown. **a. capacity** The number of OH^- ions which a molecule of a base yields in an aqueous solution; thus NaOH, one, $Ca(OH)_2$, two. **a. chloride** A compound containing the $-COCl$ radical; e.g., acetyl chloride, MeCOCl. **a. dye** A dye which acts as a base and is used in an acid bath. **a. esters** The derivatives of polyvalent organic acids of which some of the acid H atoms are replaced by a radical (R); hence they contain the COOH and COOR groups. **a. fuchsin** A. magenta II, a. rosein. The di- and trisulfonic acids of rosaniline and pararosaniline; a stain and indicator. **a. function** Hydrogen ion. **a. group** The carboxyl radical $-COOH$, present in all organic acids. **a. halide*** An organic compound containing the $-COX$ radical, in which X is a halogen (Cl, Br, I, or F); as, acid chlorides, R-oyl chloride*, $R-COCl$. **a. hydrazide** An organic compound containing the $-CONHNH_2$ radical; as, acethydrazide, $MeCONHNH_2$. **a. hydrogen** The hydrogen of the $-COOH$ group in organic compounds replaceable by metals, alkyls, aryls, or basic radicals. **a. infraprotein** An acid metaprotein. **a. ion** An anion with 1 or more replaceable H atoms. See Table 3 on p. 11. **a. of lemon** Citric a. **a. of milk** Lactic a. **a. mordant dye** An a. dye requiring a mordant to fix it on fibers. **a. nitriles** Nitriles. **a. number** A measure of free fatty acids in animal and vegetable fats. The mg of KOH necessary to neutralize the free fatty acids in 1g of fat. **a. peroxide** An organic compound containing the $-CO·O-O·CO-$ radical; e.g., acetyl peroxide $(C_2H_3O)_2O_2$. Cf. *a. anhydride.* **a.-proof material** Any substance resisting the corroding effect of a.; as, glassware. **a. pump** A pressure pump used to draw a. or ammonia from carboys. **a. radical** (1) A radical derived from a mono- or polybasic a. by subtracting one or more H atoms. See Table 3 on p. 11. (2) The acyl radical, sometimes known as the acid radical, has the suffix *carbonyl* or *-oyl;* as, benzoyl, $PhCO-$. **a. rain** Rain containing quantities of sulfuric a. and nitric a., originating from industrial, domestic, and automobile air pollution, that are harmful to forests and lake fish life. **a. rock** A rock containing more than 60% Si. **a. salts** Compounds derived from a. and bases in which only a part of the hydrogen of the a. is replaced by a basic radical, e.g., an acid sulfate $NaHSO_4$. They are correctly designated by hydrogen, dihydrogen, etc.; traditionally by the prefix bi-, as, bisulfate. **a. solution** An aqueous solution containing more hydrogen ions than hydroxyl ions. See *hydrogen ion concentration.* **a. of sugar** Oxalic acid. **a. sulfate** Hydrogensulfate*, bisulfate, disulfate. An inorganic compound containing the $-HSO_4$ radical derived from sulfuric a. **a. value** (1) Acidity expressed in terms of normality. (2) A. number.

acidation (1) Acidylation. Conversion into an acid. (2) Acidification; making a solution acidic.

acidify To add an acid to a solution until the pH value falls below 7.0.

acidimetry The titration of an acid with a standard alkali solution. See *quantitative analysis.*

acidity (1) Sourness. See *taste.* (2) An excess of hydrogen ions in aqueous solution; measured by (*a*) the intensity or *degree* of acidity, expressed as *pH* value, q.v.; (*b*) the *amount* of acidity, expressed as *normality,* q.v. Antonym: alkalinity. (3) The power of a base to unite with one or more equivalents of an acid. Antonym: basicity. **amount of ~** The normality or percentage of an acid as determined by titration (effective acid). **degree of ~** The strength of an acid expressed by its hydrogen ion concentration. Cf. *pH.*

acidium* Indicating a cation formed by adding protons to the acid of the anion; as, $H_2NO_3^+$, the nitrate acidium ion.

acidosis A metabolic state in which the acidity of the body fluids (e.g., blood) is above the normal level.

acids See Tables 3 and 4 on pp. 11–13.

acidulate Acidify.

acidulation Acidation (2).

acidum Latin for acid. **a. aceticum** Acetic acid. **a. benzoicum** Benzoic acid; etc.

acidylation Acylation. The process of introducing an acid radical into an organic compound, e.g., acetylation (acetyl radical).

acieral An aluminum alloy containing Cu 3-6, Fe 0.1–1.4, Mn 0–1.5, Mg 0.5–0.9, Si 0–0.4%. Cf. *aerometal.*

aci*-nitro compound Isonitro c. A colored isomer of a nitro compound containing the OH(O:)N=group.

acivinyl alcohols Unsaturated *ketols.*

Ackermann automatic reckoner A device to determine the dry substance of milk from its specific gravity and fat content.

Acker process The manufacture of sodium hydroxide by electrolysis of molten salt using molten lead as cathode.

acme burner A bunsen burner with regulators for gas and air, constructed so that the flame cannot strike back.

acmite $NaFeSi_2O_6$. Aegirite. A rock-forming monoclinic pyroxene, d.3.53, hardness 6–6.5, mol. vol. 65.5; occurs as a brownish, greenish, or black *silica* mineral, q.v., in Norway, and in boiler scales.

acocantherin A crystalline glucoside from *Acocanthera abyssinica.* The active principle of the shashi arrow poison of eastern Africa, related to ouabain.

acolytine Lyaconine. An alkaloid of *Aconitum.*

aconic acid $C_5H_4O_4$ = 128.1. Formylsuccinic acid lactone, 4,5-dihydro-5-oxo-3-furancarboxylic acid†. Colorless, triclinic crystals, m.164, sparingly soluble in water.

aconine $C_{25}H_{41}O_9N$ = 499.6. An amorphous alkaloid from the root of aconite. **acetylbenzoyl ~** Aconitine. **pseudo ~** Pseudaconitine.

aconitase A. hydratase*. An enzyme which catalyzes the conversion of a citrate into a *cis*-aconitate.

aconite Aconitum, monkshood, wolf's bane, blue rocket, friar's cowl, *Aconitum napellus (Ranunculaceae).* **a. alkaloids** Alkaloids from *Aconitum* species, e.g.:

Aconine	$C_{25}H_{41}O_9N$
Indaconine	$C_{27}H_{47}O_9N$
Pyraconitine	$C_{32}H_{41}O_9N$
Aconitine	$C_{34}H_{47}O_{11}N$
Japaconitine	$C_{34}H_{49}O_{11}N$
Indaconitine	$C_{34}H_{47}O_{10}N$
Pseudaconitine	$C_{36}H_{51}O_{12}N$

a. leaves The dried leaves of *A. napellus,* used similarly to a. *root.*

aconitic acid 1,2,3-Propenetricarboxylic acid*.

aconitine $C_{34}H_{47}O_{11}N$ = 645.7. Acetylbenzoylaconine. An extremely poisonous alkaloid from the root of *Aconitum napellus.* Colorless prisms, or amorphous powder, m.195, slightly soluble in water; a circulatory sedative. Cf. *aconite alkaloids.* **pseudo ~** Pseudaconitine.

a. arsenate Colorless crystals, soluble in water. **a. phosphate** $C_{34}H_{47}O_{11}N \cdot H_3PO_4$ = 743.7. White crystals, soluble in water. **a. salicylate** $C_{34}H_{47}O_{11}N \cdot C_7H_6O_3$ = 783.9. White crystals, soluble in water. **a. sulfate** $(C_{34}H_{47}O_{11}N)_2 \cdot H_2SO_4$ = 1389.6. Colorless (-)-rotatory crystals, soluble in water.

Aconitum A genus of poisonous plants of the Ranunculaceae family. See *aconite.*

acoretin A neutral resin obtained by the oxidation of the aqueous extract of sweet flagroot, *Acorus calamus.*

acorin $C_{36}H_{60}O_6$ = 588.9. A glucoside from calamus; the rhizome of *Acorus calamus,* sweet flag (Araceae); used in perfumery.

acorn The fruit of the oak *Quercus robur;* an astringent. **a. flour** Racahout. **a. sugar** Quercitol.

acoustics The study of sound and its effects.

acovenoside A Venenatin. A cardiac glycoside from the bark and wood of *Acokanthera venenata,* G. Don. Crystalline plates, m.222.

ACP Calcium hydrogenphosphate for use in foods, e.g., baking powders.

acqua Italian for "water."

acquired immunity The resistance of an organism resulting from an attack by an infectious disease. Also artificially produced by treatment with a vaccine or serum.

acraldehyde Acrylaldehyde*.

acre A surface measure: 1 acre = 0.4047 hectare = 4 rods = 160 poles = 1/640 sq mile.

Acree-Rosenheim reaction A reaction used to test for protein. A test solution plus dilute formaldehyde is layered on concentrated sulfuric acid; a purple ring indicates proteins.

acrid Pungent, bitter, burning, or irritating; as some burning plastics.

acridic acid $C_9H_5N(COOH)_2$ = 217.2. Acridinic acid, 2,3-quinolinedicarboxylic acid. Colorless crystals, decomp. 130; an oxidation product of acridine.

acridine* $(C_6H_4)_2N \cdot CH$ = 179.2. A tricyclic, heterocyclic hydrocarbon obtained from coal tar. Colorless leaflets, m.109, soluble in water. Used in the synthesis of dyes and drugs. Cf. *chrysaniline.* **diamino ~ chloride** Acriflavine. **diaminodimethyl ~** A yellow dyestuff. **diamino ~ sulfate** Proflavine.

a. dye *para* ~ Derivatives of a. (relative to the methine carbon) as, *acriflavine,* characterized by fluorescent solutions.

acridinic acid Acridic acid.

acridinyl* The radical $C_{13}H_8N-$, from acridine.

acridone* $(C_6H_4)_2NH \cdot CO$ = 195.2. Colorless crystals, m.354, insoluble in water.

acriflavine $C_{14}H_{14}N_3Cl$ = 259.7. 4,8-Diamino-1-methylacridine chloride, trypaflavine,

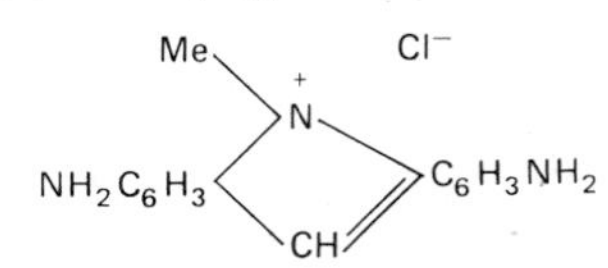

Brown crystals, soluble in water (fluorescent solution); an antiseptic and disinfectant. **a. hydrochloride** A more soluble

TABLE 3. COMMON ACIDS, RADICALS, AND SALTS

Names in parentheses are former names. Those separated by a comma are alternative names. See also *amino acids* (Table 8) and *radicals* (Table 70).

Acid	Radical or Ion	Salt
Acetic	CH_3COO-	Acetate
Adipic	$-OOC(CH_2)_4COO-$	Adipate
Anisic	$CH_3O\cdot C_6H_4\cdot COO-$	Anisate
Arsenic	$AsO_4\equiv$	Orthoarsenate
	AsO_3-	Metaarsenate
Arsenious	$AsO_3\equiv$	Orthoarsenite
	AsO_2-	Metaarsenite
	As_2O_5≣	Pyroarsenite
Benzoic	$PhCOO-$	Benzoate
Boric	BO_3	Orthoborate
	$B_4O_7=$	Tetraborate
	BO_2-	Metaborate
	B_2O_5≣	Diborate
Bromic	BrO_3-	Bromate
Bromous	BrO_2-	Bromite
Butanoic, butyric	$Me(CH_2)_2COO-$	Butanoate, butyrate
Capric	See *decanoic acid*	
Caproic	See *hexanoic acid*	
Caprylic	See *octanoic acid*	
Carbonic	$CO_3=$	Carbonate
	HCO_3-	Hydrogencarbonate (bicarbonate)
Chloric	ClO_3-	Chlorate
Chlorous	ClO_2-	Chlorite
Chromic	$CrO_4=$	Chromate
	$Cr_2O_7=$	Dichromate
Chromous	CrO_2-	Chromite
Cinnamic	$PhCH{:}CHCOO-$	Cinnamate
Citric	$C_3H_4(OH)(COO-)_3$	Citrate
Cyanic	$NCO-$	Cyanate
Cyanoacetic	$NC\cdot CH_2\cdot COO-$	Cyanoacetate
Decanoic (capric)	$Me(CH_2)_8COO-$	Decanoate (caprate)
Diphosphonic	$H_2P_2O_5=$	Diphosphonate
Disulfuric	$S_2O_7=$	Disulfate
Disulfurous	$S_2O_5=$	Disulfite
Dithionic	$S_2O_6=$	Dithionate
Dithionous	$S_2O_4=$	Dithionite
Dodecanoic (lauric)	$C_{11}H_{23}COO-$	Dodecanoate (laurate)
Ferricyanic	See *hexacyanoferric(III) acid*	
Ferrocyanic	See *hexacyanoferric(II) acid*	
Fluorosilicic	See *hexafluorosilicic acid*	
Formic	$HCOO-$	Formate
Fulminic	$CNO-$	Fulminate
Fumaric	$-OOCCH{:}CHCOO-$	Fumarate
Gallic	$C_6H_2(OH)_3COO-$	Gallate
Glutaric	$-OOC(CH_2)_3COO-$	Glutarate
Glycolic	$HOCH_2COO-$	Glycolate
Hexacyanoferric(II) (ferrocyanic)	$Fe(CN)_6$≣	Hexacyanoferrate(II) (ferrocyanide)
Hexacyanoferric(III) (ferricyanic)	$Fe(CN)_6\equiv$	Hexacyanoferrate(III) (ferricyanide)
Hexadecanoic, palmitic	$C_{15}H_{31}COO-$	Hexadecanoate, palmitate
Hexafluorosilicic	$SiF_6=$	Hexafluorosilicate
Hexanoic (caproic)	$Me(CH_2)_4COO-$	Hexanoate (caproate)
Hydrobromic	$Br-$	Bromide
Hydrochloric	$Cl-$	Chloride
Hydrocyanic	$CN-$	Cyanide
Hydrofluoric	$F-$	Fluoride
Hydroiodic	$I-$	Iodide
Hydroxybenzoic	HOC_6H_4COO-	Hydroxybenzoate
Hypochlorous	$ClO-$	Hypochlorite
Iodic	IO_3-	Iodate
Iodous	$IO_2=$	Iodite
Isocyanic	$OCN-$	Isocyanate
Isothiocyanic	$SCN-$	Isothiocyanate

(*continued*)

TABLE 3. COMMON ACIDS, RADICALS, AND SALTS (*Continued*)

Acid	Radical or Ion	Salt
Lactic	MeCH(OH)COO−	Lactate
Lauric	See *dodecanoic acid*	
Levulinic	$MeCO(CH_2)_2COO-$	Levulinate
Maleic	−OOCCH:CHCOO−	Maleate
Malic	$-OOCCH_2CHOHCOO-$	Malate
Malonic	$-OOC \cdot CH_2 \cdot COO-$	Malonate
Manganic	$MnO_4{=}$	Manganate
Molybdic	$MoO_4{=}$	Molybdate
Nitric	NO_3-	Nitrate
Nitrous	NO_2-	Nitrite
Octadecanoic, stearic	$C_{17}H_{35}COO-$	Octadecanoate, stearate
Octanoic (caprylic)	$Me(CH_2)_6COO-$	Octanoate (caprylate)
Oleic	$C_{17}H_{33}COO-$	Oleate
Oxalic	$-OOC \cdot COO-$	Oxalate
	$HOOC \cdot COO-$	Hydrogenoxalate (bioxalate)
Palmitic, hexadecanoic	$C_{15}H_{31}COO-$	Palmitate, hexadecanoate
Pentanoic, valeric	C_4H_9COO-	Pentanoate, valerate
Perchloric	ClO_4-	Perchlorate
Periodic	IO_4-	Periodate
Pertechnetic	TcO_4-	Pertechnate
Phosphinic	H_2PO_2-	Phosphinate, phosphinic acid
Phosphonic, phosphorous	$HPO_3{=}$	Phosphonate, phosphonic acid
	$PO_3{\equiv}$	Phosphite
Phosphoric	$PO_4{\equiv}$	Orthophosphate
	PO_3-	Metaphosphate
	H_2PO_4-	Dihydrogenphosphate
	HPO_4-	Hydrogenphosphate
	$P_2O_7{\equiv}$	Diphosphate, pyrophosphate
Phthalic	$C_6H_4(COO-)_2$	Phthalate
Picric	See *trinitrophenol* (in text)	
Propanoic, propionic	C_2H_5COO-	Propanoate, propionate
Pyrogallic	See *trihydroxybenzene* (in text)	
Pyruvic	$MeCO \cdot COO-$	Pyruvate
Rhenic	$ReO_4{=}$	Rhenate
Salicylic	HOC_6H_4COO-	Salicylate
Selenic	$SeO_4{=}$	Selenate
Selenious	$SeO_3{=}$	Selenite
Silicic	$SiO_4{\equiv}$	Orthosilicate
	$SiO_3{=}$	Metasilicate
Stearic, octadecanoic	$C_{17}H_{35}COO-$	Stearate, octadecanoate
Succinic	$-OOC \cdot CH_2CH_2 \cdot COO-$	Succinate
Sulfanilic	$NH_2 \cdot C_6H_4 \cdot SO_3-$	Sulfanilate
Sulfuric	$SO_4{=}$	Sulfate
	HSO_4-	Hydrogensulfate (bisulfate)
Sulfurous	$SO_3{=}$	Sulfite
	HSO_3-	Hydrogensulfite (bisulfite)
Tartaric	$(CHOH)_2 \cdot (COO-)_2$	Tartrate
	$(CHOH)_2COOH \cdot COO-$	Hydrogentartrate (bitartrate)
Telluric	$TeO_4{=}$	Tellurate(VI)
Tellurous	$TeO_3{=}$	Tellurite
Thioacetic	MeCOS−	Thioacetate
Thiocyanic	NCS−	Thiocyanate
Thiosulfurous	$S_2O_2{=}$	Thiosulfite
Titanic	$TiO_4{\equiv}$	Orthotitanate
Tungstic	$WO_4{=}$	Tungstate
Uranic	$U_2O_7{=}$	Uranate
Valeric, pentanoic	C_4H_9COO-	Valerate, pentanoate
Vanillic	$MeO \cdot C_6H_3(OH)COO-$	Vanillate

(*continued*)

TABLE 4. ORGANIC ACID SERIES

$C_nH_{2n+1}COOH$	Fatty acids (saturated)
$C_nH_{2n-1}COOH$	Acrylic acids, alicyclic acids
$C_nH_{2n-3}COOH$	Acetylene acids
$Ar(COOH)_n$	Aromatic acids, arene acids
$C_nH_{2n}(COOH)_2$	Oxalic acids
$C_nH_{2n-2}(COOH)_2$	Fumaric acids
$C_nH_{2n-4}(COOH)_2$	Acetylenedicarboxylic acids
$(HO)_nR\cdot COOH$	Alcohol acids, phenol acids
$R(CO)_nCOOH$	Ketone acids
$(NH_2)_nR\cdot COOH$	Amino acids
$R\cdot COSH$, $R\cdot CSOH$	Thiocarbonic acids

a. containing 1 mole HCl. **a. neutral** A less soluble a. without HCl.

acrifoline $C_{16}H_{23}O_2N$ = 261.4. A minor alkaloid from *L. annotinum.*

Acrilan Trademark for a synthetic fiber made from acrylonitrile.

acrimonious Bitter, caustic, sharp; as, a taste.

acrinyl *p*-Hydroxybenzyl. The $HO\cdot C_6H_4\cdot CH_2-$ radical. **a. isothiocyanate** $C_7H_7O\cdot NCS$ = 165.2. Occurs in white mustard oil and is a split product of sinalbin. **a. thiocyanate** $C_7H_7O\cdot SCN$ = 165.2. Oily, colorless liquid in white mustard oil; insoluble in water.

Acrocomia The coyol palm of S. America.

acrodextrin Achroodextrin.

acrol The allylidene* radical.

acrolactic acid Glucic acid.

acroleic acid Acrylic acid*.

acrolein Acrylaldehyde*. **furfur ~** See *furfuracrolein.* **met ~** See *metacrolein.*

acromelin $C_{17}H_{16}O_9$ = 364.3. A lactone obtained from *Physcia acromela.*

acrometer A hydrometer for determining the specific gravity of oils.

acronize A preparation of chlortetracycline used to preserve poultry and fish.

Acronol Trademark for basic dyestuffs.

acropeptides Complex amino acids produced by the action of hot glycerol on gelatin.

acrose **β- ~** $C_6H_{12}O_6$ = 180.2. A synthetic carbohydrate obtained by the action of dilute sodium hydroxide on glycerose.

acrosite Pyrargyrite.

acryl The radical C_3H_3O- from acrylaldehyde.

acrylaldehyde* $CH_2{:}CH\cdot CHO$ = 56.1. Acrylic aldehyde, acrolein, 2-propenal*, allyl aldehyde. A decomposition product of glycerol and glycerides. A yellow, irritant pungent liquid, d.0.84, b.52.4, soluble in water; used in organic synthesis. Flammable. Cf. *allylidene, acryl.* **dimethyl ~** 2-Methyl-2-butenal*. **furfur ~** Furfuracrolein. **met ~** Metacrolein. **methyl ~** Crotonaldehyde*. **2-naphthyl ~** $C_{13}H_{10}O$ = 182.2. Yellow needles, m.48. **phenyl ~** Cinnamaldehyde.

acrylate A salt of acrylic acid containing the $C_3H_3O_2-$ radical. **cyano ~** A group of strong adhesives; synthetic polymers based on alkyl 2-cyanoacrylates.

acrylic acid $CH_2{:}CH\cdot COOH$ = 72.1. Acroleic acid, 2-propenoic acid*, propene acid, ethylenecarboxylic acid. An oxidation product of acrylaldehyde. A pungent, colorless liquid, m.10, b.140, soluble in water. **dimethyl ~** See *butenoic acid.* **3-ethyl ~** 2-Pentenoic acid*. **imidoazole ~** Urocanic acid. **methyl ~** **2-** $CH_2{:}CMe\cdot COOH$ = 86.1. Methacrylic acid. An isomer of butenoic acid, m.16. **3-** See *crotonic* and iso*crotonic acid.* **phenyl ~** **2-** Atropic acid*. **3-** Cinnamic acid. **propyl ~** Hexenoic acid*.

acrylic acids Olefin acids. A series of unsaturated aliphatic acids of general formula $C_nH_{2n-1}COOH$ e.g., acrylic acid, $C_2H_3\cdot COOH$; butenoic acid, $C_3H_5\cdot COOH$.

acrylic aldehyde Acrylaldehyde*.

acrylonitrile* Vinylcyanide*. **poly ~** See *polyacrilonitrile.*

acrylophenone $CH_2{:}CH\cdot COPh$ = 132.2. Phenylvinyl ketone*. An isomer of cinnamaldehyde. **β-phenyl ~** Chalcone*.

acryloyl* 1-Oxo-2-propenyl†, arylyl. The radical $CH_2{:}CH\cdot CO-$, from acrylic acid.

Actaea A genus of ranunculaceous herbs.

acterol An irradiated ergosterol, q.v.

actiduins A group of antibiotics from the *Streptomycetes.* They are yellow to red in color and fluoresce yellow, especially in organic solvents. They contain C, H, N, and S, and sometimes Cl; m. exceeds 350 (decomp.). Sparingly soluble in water.

actin A fibrous protein constituent of muscle material. See *myosin.*

actiniasterol A sterol, m.145, from sea anemone, *Anemonia sulcata.*

actinic Pertaining to radiation; especially light that produces a chemical change. **a. rays** Rays in the violet and ultraviolet regions which produce chemical changes. Cf. *irradiation, ultraviolet.*

actinides Actinoids*.

actinism The study of chemical changes produced by radiation. Cf. *photochemistry, excitation, activation.*

actinium* Ac. Atomic weight of isotope with longest half-life (^{227}Ac) = 227.03. A radioactive element, at. no. 89, discovered (1899) by DeBierne in pitchblende and identical with the anemium of Giesel (1902). Metallic, m.1050, b.3250. It is trivalent and the first element in the actinoids. See *radioactive elements.* **a. D** The isotope lead-207. **a. emanation** Radon-219. **a. K** Francium*. **a. series** See *radioactive elements.*

actinodaphnine $C_{18}H_{17}O_4N$ = 311.3. An alkaloid from *Actinodaphne* species (Lauraceae), resembling laurotetanine. Cf. *bulbocapnine.*

actinogram, actinograph X-ray photograph.

actinoids* Actinides. The 15 elements of atomic number 89–103. The theoretical analogs of the lanthanoids. **trans ~** The elements, at. no. 104–120, following the a. Their isotopes have very short lives. The creation of elements up to 109 has been claimed. See *rutherfordium, hahnium.* **super ~** The elements following the transactinoids.

actinolite $Ca(MgFe)(SiO_4)_3$. A green, rock-forming amphibole resembling tremolite; fibrous variety is asbestos.

actinometry The measurement of light intensity.
actinomycetin An antibiotic substance from cultures of *Streptomyces albus.*
actinomycin D Dactinomycin.
actinon Early name for radon-219.
actinouranium Early name for uranium-235. **a. series** See *radioactive elements.*
actinozoa The phylum Coelenterata, or jellyfish, which have starlike structures.
action The physical concept of activity. **chemical ~** A reaction in which the atoms of a molecule or molecules are rearranged. **electronic ~** The change of an electron from one to another energy level. Cf. *excitation.* **physical ~** A transformation of matter which does not affect molecular structure.
activated Rendered active, reactive, or excited. **a. atom** See excited *atom.* **a. carbon** Charcoal produced by the destructive distillation of vegetable matter, e.g., nutshells, with or without the addition of chemicals. Used in powdered form to decolorize sugar solutions, oils, etc., or in granular form as an adsorbent in gas masks and for the recovery of solvent vapors; used in treatment of poisoning, particularly by drugs, when either it is given by mouth or blood is perfused through it (USP, EP, BP). Cf. *Norit, revivification.* **a. molecule** A molecule with one or more excited atoms. Cf. *irradiation, excitation.* **a. sludge** The oxidized and flocculent sediment of sewage which contains bacteria. **a. s. process** Sewage is agitated in contact with air, thereby causing oxidation and flocculation by bacterial action; it is left to settle in separation tanks and yields an essentially harmless effluent.
activation (1) A method by which a metallic catalyst is rendered active or is regenerated, e.g., heating platinum sponge. Cf. *revivification.* (2) The transformation of an inactive enzyme into an active enzyme by the creation of a transient substrate-enzyme complex. Cf. *kinase.* (3) Excitation. (4) Irradiation. (5) A. of carbon, e.g., by heating with steam, or sulfuric acid. **energy of ~** E_a. The energy required to initiate a reaction or process; sometimes greater than that required to sustain it. Derived from the Arrhenius equation, $E_a = RT^2(\delta \ln k/\delta T)_p$. It is thus related to the dependence of the rate constant on temperature at constant pressure.
activator (1) A catalyst. (2) A substance used in flotation to produce a coating having metallic properties, as, sodium sulfide for lead carbonate ores. (3) In electronics, describing a component, such as a transistor, that produces gain. Cf. *passive.*
activatory See *phase.*
active (1) Dynamic or working, as opposed to static or inert, as in metabolism. (2) Having optical properties, as an asymmetric carbon atom. Cf. *optical activity.* **surface-~** See *surfactant.*
a. center That part of an enzyme molecule which forms an activated complex with the substrate. **a. deposit** The formation of a radioactive layer on a substance exposed to radioelements. **a. immunity** The stimulation of an organism to produce antibodies against infection by microorganisms. **a. immunization** The processes by which the protective agencies of an organism are made resistant to bacterial invasion. **a. mass** Amount-of-substance *concentration.* **a. oxygen test** A test for rancidity in fats, by the liberation of iodine from potassium iodide in acetic acid. **a. principle** The substance responsible for the physiological action of a drug; e.g., an alkaloid.
activity (1) The rate in watts at which work is performed. (2) The ratio of the escaping tendency *(fugacity)* of two phases at the same temperature. A correction applied to the concentration of a strong electrolyte to satisfy *Ostwald's dilution law,* q.v. (3) A measure of interionic forces. (4)* The decay of a radionuclide. See *becquerel.* **amylolytic ~** Digestive power of amylase. **excited ~** Active deposit. **ionic ~** Thermodynamic concentration. In a dilute solution which obeys the gas laws, the i. a. equals the concentration; in other solutions, the value which ensures that the gas laws hold. **optical ~** The capacity of a substance to rotate the plane of polarized light. **peptic ~** Digestive power of pepsin. **radio ~** See *radioactivity.* **tryptic ~** Digestive power of trypsin.
a. of activated carbon The percentage of carbon disulfide vapor absorbed by carbon (generally 50%).
actomyosin A combination of *actin* and *myosin,* q.v., which comprises the tractile muscle system.
actor A compound which takes part in both primary and secondary reactions. See *induced reaction.*
acute Quick, short, or sharp. Cf. *chronic.* **a. poisoning** See *poisoning.*
acyclic* Describing organic compounds which contain no ring system, as, the alkanes. Synonym: Aliphatic (chains). Antonym: cyclic, aromatic (rings).
acyl* An organic radical derived from an organic acid by the removal of the hydroxyl group; e.g., R·C(O)— is the a. radical of R·CO·OH. See *acetyl, benzenesulfonyl, benzoyl,* etc. **a. derivative** An organic compound containing an a. radical, e.g., amides, $R\cdot CO\cdot NH_2$.
acylals* Generic term for compounds of the type $R'CH(OCOR'')_2$.
acylamines *N*-substituted primary and secondary amides. More specifically, monoacylamines and diacylamines, respectively.
acylation Acidylation.
acyloins* α-Hydroxy ketones of the type R·CO·CHOH·R. Formed by condensation of aldehydes, as, Ph·CO·CHOH·Ph, benzoin.
aczol An ammoniacal solution of zinc and copper phenolates; a wood preservative.
adamant A hard mineral, as, diamond.
adamantane $C_{10}H_{16}$ = 136.2. Diamantane. ***sym*-~** Tricyclodecane. White crystals, m.207 (subl.). Its derivatives are used to make plastics heat- and chemical-resistant.
adamantine Diamond. **a. boron** See *boron.* **a. spar** A dark gray, smoky variety of corundum from India; green in transmitted light.
adamellose An igneous andesite-diorite rock containing hornblende, feldspar, quartz, chlorite, agnetite, apatite, and rutile (Pigeon Point, Minn.).
Adam galactometer A graduated buret with two glass bulbs, used in milk analysis.
adamine Adamite.
adamite Zn_2HAsO_5. Adamine. A native arsenate; yellow orthorhombic crystals (Chile, Greece).
Adamkiewicz reaction Protein solutions give a violet ring when layered on glacial acetic acid and concentrated sulfuric acid.
adamsite (1) A greenish-black mica. (2) Diphenylamine chlorarsine. Adansonia. *Adansonia digitata* (Bombacaceae), the baobab tree of Africa, yields edible boui or monkey bread. The bark is an emollient; the dried leaves, lalo, are an antipyretic.
adansonine An alkaloid from the bark and leaves of *Adansonia digitata.* Colorless white crystals; a febrifuge.

adaptation The advantageous adjustment of an organism to a change in its surrounding.
adapter A tapered glass tube used to connect a retort or condenser with the receiving vessel.
adatom An atom adsorbed on a surface so that it will migrate over the surface like a two-dimensional gas. Cf. *adion.*
addiction Describing dependence, either physiological or psychological (or both), on a drug or chemical substance. Tolerance develops; abrupt withdrawal causes unpleasant symptoms. **a.-producing drugs** Drugs subjected to international control by the World Health Organization because of their a.-producing powers. See *barbiturates, cocaine, heroin.*
addition A chemical reaction which involves no change of valency; usually the union of two binary molecules to form a more complex compound, as, $HCl + NH_3 = NH_4Cl$. **a. compound** Adduct. An inorganic compound formed by addition, e.g., NH_4Cl.
additive Added to. **a. nomenclature*** See *nomenclature.* **a. property** A property of a molecule which is the sum of the individual properties of the atoms or linkages composing it; thus, the molecular refraction of a molecule is the sum of the atomic refractions of its atoms.
adduct Addition group or compound.
adduction Oxidation.
adelgesin $C_{23}H_{28}O_{15}$ = 544.5. Light brown needles, m.205. A glucosidal constituent of the bark of "pineapple" gall, produced by *Adelges abietes.*
adelite $MgCaHAsO_5$. A native arsenate of the wagnerite group; monoclinic gray crystals.
adelomorphic cell Any living cell of uncertain or indefinite shape; e.g., an ameba.
adenantherine A crystalline alkaloid resembling physostigmine, obtained from *Adenanthera pavonia* (Leguminosae).
adenase Adenine deaminase*. An enzyme in animal tissues which hydrolyzes adenine to hypoxanthine.
adenine* $C_5H_5N_5$ = 135.1. 6-Aminopurine*. A purine base derived from nucleic acid and found in the pancreas and spleen. White needles, m.360, slightly soluble in water. **a. deaminase*** See *adenase.* **a. nucleotide** Adenosine triphosphate.
adenocarpine $C_{19}H_{24}ON_2$ = 296.4. An alkaloid from the *Adenocarpus* species, m.(hydrochloride) approx. 140. Cf. *santiaguine.*
adenoid Resembling or pertaining to (1) glands; (2) lymphoid tissue.
adenos A fine quality of cotton from Asia Minor.
adenosine $C_{10}H_{13}N_5O_4$ = 267.2. Adenine riboside. A nucleoside consisting of adenine and (+)-ribose. Colorless needles, m.229, soluble in water. **a. 5′-diphosphate** ADP*. A metabolic breakdown product of a. 5′-triphosphate, with liberation of energy. **a. 5′-phosphate, a. phosphoric acid** Adenylic acid. **a. 5′-triphosphate** ATP*, a. nucleotide. An important body phosphorylating agent; provides the energy for muscle contraction. See *myosin.*
adenylic acid $C_{10}H_{13}N_5O_4 \cdot HPO_3$ = 347.2. Adenosine 5′-phosphate, AMP*. A nucleotide from red blood corpuscles and yeast, consisting of adenine, ribose, and phosphoric acid. White needles, m.197. Cf. *guanylic acid.*
adenylyl* Adenyl. The $C_{10}H_{12}O_7N_4P \cdot NH-$ radical, from adenylic acid.
adeps Any animal grease or fat, e.g., lard. **a. lanae** Lanolin. **a. mineralis** Petrolatum.
adfluxion Affluxion.
adglutinate Agglutinate.
adhatotic acid An organic acid from the leaves of *Adhatoda vasica* (Acanthaceae). Cf. *vasicinone.*
adhere To be attached to or to stick to another substance. Cf. *cohere.*
adhesion The attraction or force which holds unlike molecules together. Cf. *cohesion, wetting, adsorption.*
adhesive Any substance that sticks or binds materials together. Can be classified as (1) carbohydrate-based (e.g., starch, gums, cellulose acetate), (2) protein-based (bones, casein), (3) other natural sources (sodium silicate, rubber latex), (4) synthetic (epoxy resins). **a. meter** An apparatus to determine the relative viscosity of road materials.
adiabatic A change occurring without a loss or gain of heat. Cf. *Reech's theorem.* **a. calorimeter** An instrument for the study of reactions in which there is a minimum loss of heat. **a. elasticity** The modulus of elasticity of a gas undergoing a change in volume without transfer of heat; e.g., in explosions. **a. expansion** The expansion of a gas without the production of a cooling effect. **a. process** An experiment carried out so that no heat can leave or enter the system.
adiactinic A substance which does not transmit photochemically active radiation.
adicity Valency.
adinole A mineral consisting of metallic silver and albite.
adion An adsorbed ion. Cf. *adatom.*
adipaldehyde* $CHO(CH_2)_4CHO$ = 114.1. Hexanedial*, adipic dialdehyde. Oily liquid, b.94.
adipamide* $NH_2CO(CH_2)_4CONH_2$ = 144.2. Adipic diamide, hexanediamide*. An isolog of succinamide, m.220.
adipic acid* $COOH(CH_2)_4COOH$ = 146.1. Adipinic acid, hexanedioic acid*. Colorless needles, m. 149, sparingly soluble in water. Used principally in nylon manufacture and for plasticizers. **tetrahydroxy ~** See *saccharic acid.* **a. dialdehyde** Adipaldehyde*. **a. diamide** Adipamide*. **a. ketone** Cyclopentanone*.
adipinic acid Adipic acid*.
adipoin $C_6H_{10}O_2$ = 114.1. 2-Hydroxycyclohexanone*. White powder, m.91, slightly soluble in water.
adipoyl* 1,6-Dioxo-1,6-hexanediyl†. The radical $-OC(CH_2)_4CO-$, from adipic acid. **a. chloride*** $C_6H_8O_2Cl_2$ = 183.0. Hexanedioyl dichloride. Colorless liquid, b_{10mm}113.
adjacent (position) Adjoining, near. The consecutive arrangement of radicals in an organic ring compound, as, the 1,2,3 position in the benzene ring.
adjective Supplemental or accessory. **a. dyes** Dyes requiring a mordant.
adjoining Adjacent.
adjuvant An auxilliary drug which assists the action of another drug.
Adler benzidine reaction A sensitive test for blood in urine. Cf. *benzidine test.*
adlumine $C_{21}H_{21}NO_6$ = 383.4. An alkaloid from *Adlumia fungosa* (Papaveraceae). Cf. *bicuculline.*
adnephrine Epinephrine.
adnic Admiralty nickel. An alloy: Cu 70, Ni 29, Sn1%.
adobe A soil of arid regions, formed by the disintegration of rocks, consisting essentially of CaO, SiO_2, CO_2, and Fe_2O_3. Brown, claylike masses used in Egypt and New Mexico for building and pottery.
adonidin $C_{24}H_{42}O_9$ = 474.6. A glucoside from *Adonis vernalis.* A hygroscopic, yellow, very bitter powder; a heart stimulant. Cf. *picroadonidine, digoxin.*
adonin $C_{24}H_{40}O_9$ = 472.6. A glucoside from *Adonis vernalis.*
adonis False hellebore, pheasant's eye. The dried overground

leaves and stems of *A. vernalis* (Ranunculaceae). The fluid extract is a heart stimulant.
adonite $C_5H_{12}O_5$ = 152.1. Adonitol. A pentahydric alcohol, from *Adonis vernalis*. Colorless prisms, m.102, soluble in water.
adonitol Adonite.
ADP* Symbol for adenosine 5′-diphosphate.
adrenal gland An endocrine gland above each kidney, consisting of 2 histologically and functionally separate parts: the cortex, which secretes corticoid hormones and small quantities of sex hormones, and the medulla, which secretes epinephrine and norepinephrine.
Adrenalin Trademark for adrenaline, epinephrine.
adrenaline BP name for epinephrine.
Adrenine Trade name for epinephrine.
adrenolutine A fluorescent oxidation product of epinephrine.
Adriamycin Trademark for doxorubicin. **a. hydrochloride** Doxorubicin.
adsorbate That which is adsorbed.
adsorbent A substance that adsorbs, as, charcoal. Cf. *absorbent.* **crystallogenetic ~** A crystalline substance which, when dehydrated without losing its shape, becomes porous and, thus, absorbent, e.g., chabazite.
adsorption The ability of a substance *(adsorbent)* to hold or concentrate gases, liquids, or dissolved substances *(adsorbate)* upon its surface; cf. *adhesion, sorption, absorption, desorption, persorption, wetting.* **anomal ~** A. which does not follow the a. isotherm, as with certain colloidal dyes. **apolar ~** Nonpolar a. **co ~** A. in which two substances are held on a surface, which adsorbs neither alone. **differential ~** See *differential adsorption.* **heat of ~** The joules (or cal) liberated during a. **negative ~** A. in which the surface concentration of the adsorbate is lowered by preferential a. of the liquid. **nonpolar ~** Apolar. A. in which nonelectrolytes or equivalent ions of electrolytes are held on a surface. **oriented ~** A. in which the molecules are grouped on the surface in a definite direction. **polar ~** A. in which definite anions or cations are held in nonequivalent amounts. **positive ~** A. in which the surface concentration of the adsorbate is relatively high. **preferential ~** Pronounced a. of one substance compared with another of similar physical properties. **specific ~** (1) Preferential a. (2) The quantity of adsorbate on 1 cm^2 of surface.
a. analysis Separation of mixtures by the different adsorbability of the components. Cf. *chromatographic analysis.* **a. catalysis** A chemical reaction in which the adsorbent acts as a catalyst. **a. coefficient** The quantity x of the *a. isotherm.* **a. colorimetry** The a. of colored substances from solution on a white adsorbent, and comparison of the separated and dried colored powders with standards, similarly prepared, in visible or ultraviolet light (a. fluorimetry). **a. displacement** The replacement of one adsorbate on a surface by another. **a. equilibrium** The distribution of molecules on a surface and in the surrounding medium. Cf. *a. isotherm.* **a. exponent** The quantity $1/n$ of the a. isotherm. **a. fluorimetry** See *a. colorimetry.* **a. isotherm** The approximate empirical relationship existing between the concentration x, held upon the surface, and the amount c, which is not adsorbed: $x/m = \alpha \cdot c^{1/n}$, where m is the amount of adsorbent, and α and $1/n$ are experimental constants. **a. potential** The work obtainable when an adsorbate is brought into the a. space.
adsorptive Adsorbate.
adstringent Astringent.
adubiri A fish poison from *Paulowilhelmia speciosa,* used in Ghana.
adularia A variety of orthoclase.
adularin A potassium silicate of uncertain composition.
adulterant A substance of cheaper or inferior quality, sometimes harmful, added to an article, compound, or food.
adulteration (1) The fraudulent addition of a foreign substance, especially harmful preservatives, to food products. (2) The removal of an essential constituent of a substance, as cream from milk.
advection Heating or cooling effects due to horizontal currents in air or water. Cf. *convection.*
adventitious Not typical or normal; accidental.
æ- See (1) *ae-*; (2) also *e-*.
aegerite Wurtzite.
aegirite Acmite.
aenigmatite Black, triclinic, amphibole metasilicate of sodium and ferrous iron, containing some titanium instead of silicon, d.3.8.
aeolotropic Anisotropic.
aeonite Wurtzite.
aerated water A water artificially impregnated with oxygen or carbon dioxide, as, soda water.
aerator A machine to oxygenate effluent and thus reduce its B.O.D.
aerobe An organism that requires an atmosphere of oxygen for respiration. Cf. *anaerobe.*
aerobic bacteria Certain bacteria which require gaseous oxygen for growth. Cf. *anaerobic.*
aerobioscope A device to determine the number of bacteria in air.
aerobiosis Life sustained in an atmosphere containing oxygen.
aerodynamics Pneumatics. The study of the motion of gases.
Aerofloat Trademark for flotation agents of the dithiophosphoric acid type.
aeroklinoscope An air cell used to float algae on water.
aerolite A meteoric stone of silicates. Cf. *siderolite.*
aerometal An alloy: Al, with Cu 0.2–4, Fe 0.3–1.3, Mn 0–1.2, Mg 0–3, Zn 0–3, Si 0.5–1.0%. Cf. *acieral.*
aerometer An instrument to determine the density of gases.
aeron An aluminum alloy with Cu 1.5–2.0, Si 1.0, Mn 0.75%.
aeronomy The study of the upper atmosphere and especially its radiation and electromagnetic properties.
aerophone An apparatus to amplify sound waves.
aeroplankton Organisms (pollen, bacteria, etc.) carried by air.
aeroscope A glass apparatus to obtain bacteria from air.
aerosiderite An iron meteorite. Cf. *siderite.*
aerosite Pyrargyrite.
Aerosol (1) Trade name for a wetting agent of the sulfonated dicarboxylic acid ester type; e.g., **A.-OT** is sodium bis(2-ethylhexyl) succinic sulfonate. (2) (not cap.) A *colloidal system,* q.v., with gas as the surrounding medium; as, smokes and fogs. **a. pressure packaging** A method of dispensing cosmetics, medicaments, and similar products in the form of a gas-liquid aerosol spray. A homogeneous mixture of the product and a liquified gas (e.g., dichlorofluoroethane) in the pack under pressure is released by opening a valve.
aerosphere Atmosphere.
aerostatics The study of gases in mechanical equilibrium. Cf. *hydrostatics.*
aerotherapeutics Therapy by varying the pressure or composition of the atmosphere in which the patient lives.
aerotonometer An instrument to determine the pressure of gases in blood.

aerotropic Attracted by air; as, of an organism.
aerugo (1) Cupric subacetate. (2) The oxide or rust of a metal.
aeschynite Resinous, black, orthorhombic native niobates of calcium, iron, cerium, lanthanum, titanium, and thorium.
aescinic acid $C_{24}H_{40}O_2$ = 360.3. Capsuloesic acid. A monobasic acid from the seeds of *Aesculus hippocastanum*, horse chestnut.
aescorcin Escorcin.
aesculetinic acid Esculetinic acid.
aethan Ethane (German).
aether (1) Official Latin and German for "ether." (2) The hypothetical universal medium through which electromagnetic waves (e.g., light and heat) are propagated.
aetherische oel German for essential oil.
aethyl Ethyl (German).
aethyleni Official Latin for "ethylene."
aethylis Official Latin for "ethyl."
aetioporphyrin See *etioporphyrin III.*
Afcodur Trademark for a polyvinyl chloride plastic, m. approx. 120. It is 5.5 times lighter than steel, and resistant to corrosion and flame.
affination A centrifugal filtration process in which most of the molasses contaminating raw sugars is removed.
affinity The selective tendency of elements to combine with one element rather than another when physicochemical conditions are equal or slightly in favor of the "rejected" element. See *affinity curve, reaction isotherm.* **electron ~** See *electron.*
a. bond See *bond.* **a. constant** The ratio F_a/F_b, where F_a is the intrinsic tendency of substance *a* to decompose, and F_b the tendency of substance *b* to combine. See *mass action.* **a. curve** A graph obtained when the heat of formation is plotted against the types of combining atoms. The peak shown by C:C is considered to be due to affinity and thus explains the many carbon compounds.
affluxion Adfluxion. Flowing or coming together.
aflatoxin A toxin associated with moldy peanuts, produced by *Aspergillus flavus.*
African **A. kino** Kino. **A. pepper** Capsicum. **A. saffron** Carthamus.
after Behind; following in point of time. **a. contraction, a. expansion** The percentage change in length of a brick or refractory after 2 h at 1410°C.
aftitalite $KNaSO_4$. A mineral from Italy.
Ag Symbol for silver (argentum).
agalmatolite Pyrophyllite.
agamy Agamogenesis. Asexual reproduction. Cf. gamogenesis (sexual reproduction).
agaphite A Persian turquoise.
agar (-agar) Agal-agal, Bengal isinglass, Japanese gelatin, Japanese isinglass. The dried mucilaginous substance from marine algae or seaweeds such as *Gelidium corneum, Gracilaria lichenoides, Gigartina speciosa* (Rhodophyceae). Bundles of thin, transparent membranes insoluble in cold water, and slowly swelling and soluble in hot water. Used as a solid bacterial culture medium, as a medicine for constipation (NF), as a glue, and as a chemical to make silk transparent. Basically, agars for bacterial culture are formulated as **plain ~**, but with variations for the nutrient (peptone). **blood ~** A. plus sterile blood. **chocolate ~** A. and heated blood, which turns a brown color. **litmus milk ~** A. plus milk colored with litmus to detect acidity. **MacConkey's ~** See *MacConkey's medium.* **plain ~** Contains 15 g agar, 10 g peptone, 5 g sodium chloride and 1000 ml bouillon stock solution, neutralized with sodium hydroxide, and filtered while hot.
agaric *Agaricus albus,* touchwood, spunk, tinder. The dried fruit body of the fungus *Polyporus officinalis* (Polyponaceae), which grows on the *Larix* species. **fly ~** See *Amanita.* **surgeon's ~** Amadou. **white ~** Agaric.
a. acid $C_{16}H_{33}\cdot CH(OH)\cdot C(OH)COOH\cdot CH_2COOH\cdot 1\frac{1}{2}H_2O$ = 411.6. Laricic acid, agaricic acid, agaricinum. Colorless microcrystals, m.140, slightly soluble in water; the active principle of agaric, used as an astringent. **a. mineral** Rock milk. A soft, white deposit of microscopic crystals of calcite.
Agaricaceae Mushrooms or toadstools (4,600 species); some are poisonous *(Amanita),* others edible *(Agaricus, Cantharellus).*
agaricic acid Agaric acid.
agaricin An alcoholic extract of white *Agaric* or *Polyporus officinalis.* Brown powder, soluble in water, containing impure agaric acid; used as an antiperspirant.
agaricinum Agaric acid.
agaricol $C_{10}H_{16}O$ = 152.1. A monohydric alcohol in white agaric; colorless powder, m.223.
agarose A neutral galactose polymer, responsible for the gel strength of agar.
agarythrine Yellow, bitter alkaloid in the fungus *Agaricus rubra,* oxidized to the red-coloring matter of fungi.
agate SiO_2. A cryptocrystalline quartz or chalcedony, composed of colored layers or clouds. Used as a semiprecious stone, for ornaments, for pebbles in ball mills, and for the knife edges of chemical balances. **blood ~** Hemachate. **Iceland ~** An obsidian. **oriental ~** A translucent gem variety. **white ~** Chalcedony.
Agathis A conifer of Australia and Malaysia; source of kauri resin.
Agave American aloe. A genus of Central American plants (Amaryllidaceae). The fibers are used for threads and ropes; the leaves and juice as a diuretic and in adhesives. The fermented juice (pulque) is popular in Mexico; from it is distilled a spirit (mescal). Cf. *henequen, sisal.*
agedoite Asparagine.
ageing Aging.
Agene process The bleaching of flour with nitrogen trichloride. It can produce a crystalline toxic peptide, and is prohibited in some countries.
agent A substance or force that effects a change. Cf. *reagent.* **balancing ~** See *buffer, poised.* **catalytic ~** See *catalyst.* **emulsifying ~** See *emulsifier.* **freezing ~** See *freezing mixture.* **frothing ~** See *frothing agent.* **oxidizing ~** See *oxidizing agent.* **reducing ~** See *reducing agent.* **refining ~** Substances, e.g., sulfuric acid or caustic soda, used to purify organic compounds. **refrigerating ~** The gases ammonia, sulfur dioxide, and methyl chloride, used in refrigerators. **vulcanizing ~** See *vulcanization.* **warning ~** An odorous substance used to indicate danger in mines, or to detect leakage in pipes.
A. Orange See *trichlorophenoxyacetic acid.*
AgeRite Trademark for aldol-1-naphthylamine; an antiaging agent for rubber. Cf. *ohmoil.*
Agfa Trade name for dyes, fine chemicals, etc. (Aktien Gesellschaft für Anilin Farben). **A. silk** A viscose rayon.
agglomeration A cluster or accumulation of particles or substances.
agglutinant A substance which causes agglomeration.
agglutinate (1) To cause agglutination. (2) The product of agglutination.
agglutination The clumping together of cells or bacteria by the interaction of bacteria and the corresponding immune serum; used in serodiagnosis.
agglutinins A group of substances formed in the blood as a

result of bacterial infection or inoculation which cause the clumping together of the bacteria. Cf. *hemagglutinins.*
agglutinogen A substance in bacteria which stimulates the production of agglutinins in animals.
aggregate See *concrete.*
aggregation The gathering of units into a whole; the clustering together of particles. **state of ~** The physical form of matter: solid, liquid, gaseous, or colloidal.
aggressivity A quality soil characterized by a corrosive action on metals.
aging, ageing (1) Any irreversible change in living matter which occurs as a function of time. (2) Natural or artificial maturing or ripening; as of cheese. **a. test** A test involving exaggerated conditions of a.; used to determine rapidly how a material will behave over a period of time.
agitator A device to keep liquids in motion.
aglucone The nonsugar part of a glucoside.
aglycone That component of a glycoside, e.g., plant pigment, which is not a sugar.
agmatine $NH_2C(NH) \cdot NH(CH_2)_4NH_2 = 130.2$. 4-Aminobutylguanidine†. An amine isolated from herring spawn and ergot.
agnetite A mineral constituent of adamellose.
agnin Lanolin.
Agricola, Georg Bauer (1490–1555) German physician and alchemist.
agrimonia oil The volatile oil of agrimony.
agrimony A rosaceous plant, *Agrimonia eupatoria;* formerly a tonic and astringent.
Agrinite Trade name for a *tankage,* q.v., containing 8¼% nitrogen.
Agripol Norepol. Trademark for a rubber substitute made by condensing a polyhydric alcohol with polymerized vegetable oil acids.
agrochemistry The study of soil nutrition chemistry, and fertility, as distinct from its formation and classification.
agrostemma Corncockle. The poisonous plant *A. githago* (Caryophyllaceae). **a. saponin** $C_{35}H_{54}O_{10}$ *(Brandl)* or $C_{29}H_{44}O_4$ *(Wedekind & Schicke).* A hemolytic substance, m.286. It causes paralysis.
agucarina Saccharin.
ague tree Sassafras.
aguilarite Ag_4SSe. A black mineral sulfoselenide.
Agulhon's reagent A 0.1 *N* solution of chromic acid in dilute nitric acid, used to titrate primary alcohols.
Aich's metal An alloy containing Cu 56, Zn 42, Fe 1%, which is very malleable at red heat.
AIDS Acquired immune deficiency syndrome. See *immuno-.* Due to virus termed Human T-cell Lymphotropic Virus type III (also, Lymphadenopathy Associated Virus).
aikinite $CuPbBiS_3$. Aikenite. A black, orthorhombic mineral.
ailanthic acid A bitter, nitrogenous acid from the bark of *Ailanthus excelsa,* "tree of heaven" (Simarubaceae).
air The mixture of gases that forms the earth's *atmosphere,* q.v. **alkaline ~** Ammonia. **azotic ~** Nitrogen. **dephlogisticated ~** Oxygen (obsolete). **enriched ~** A. to which oxygen has been added. **expired ~** Warm a. coming from the lungs, having less oxygen and more carbon dioxide and moisture than normal a. **fixed ~** Carbon dioxide (obsolete). **inspired ~** A. which is taken into the lungs. **liquid ~** A clear liquid obtained by alternate compression and expansion of refrigerated a., b. −181. Used commercially for the manufacture of oxygen, in the production of low temperatures, and as an external cure for poison ivy or poison oak inflammations. **mephitic ~** Carbon dioxide (obsolete). **phlogisticated ~** A. from which oxygen has been removed by a burning substance (obsolete). **reserve ~** A. that can still be exhaled after an ordinary expiration. **residual ~** (1) A. that remains in the lungs after the most complete expiration possible. (2) A. that remains in an evacuated container, e.g., an incandescent lamp bulb. **tidal ~** A. taken in and given out at each respiration. **vital ~** Oxygen.
a. bath A drying oven utilizing a current of heated air for maintaining a desired temperature. **a. compressor** A pump which compresses atmospheric a. **a.-conditioning** The control of humidity and temperature of air by filtering and washing with water of definite temperature, either for human comfort (45–55% humidity) or for industries (viscose, paper). **a.-dry** To expose to air without heating. **a. gas** Producer gas. **a. liquefying apparatus** An arrangement of high-pressure pump, valves, and expansion orifices for the liquefaction of air. **a. meter** Anemometer. **a. pollution** Caused by industrial, domestic, and automobile fuels. Legislation is usually based either on the best practicable means of control or on quality objectives. See Table 5 for U.S. health and welfare (i.e., corrosion, visibility, plant life) standards. **a. pump** A mechanical device to compress or withdraw air, e.g., blowers, vacuum pumps. **a.proof** Hermetically sealed, or gastight. **a. salt** An early term for the efflorescence of bricks. **a. sampler** A device to take samples of dust and bacteria from air. **a. showers** Showers, mainly of electrons and protons, with some more penetrating particles, which occur in the air.
Airbron Trademark for acetylcysteine.

TABLE 5. U.S. AIR QUALITY STANDARDS
Measured at 25°C and 1013 mbar

Pollutant	Period	Standard ($\mu g/m^3$)		Method of measurement
		Health	Welfare	
Particulate	24-h avg.	260	150	Large volume sampling
	Annual geom. mean	75	60	
Sulfur dioxide	3-h avg.	—	1300	*p*-Rosaniline colorimetry
	24-h avg.	365	260	
	Annual arith. mean	80	60	
Ozone	1-h avg.	120	80	Photometry/gas phase titration
Carbon monoxide	8-h avg.	10		Nondispersive i.r. spectrometry
Hydrocarbons	3-h avg.	160		Hydrogen flame ionization
Nitrogen dioxide	Annual arith. mean	100		Chemiluminescence
Lead	3-month avg.	1.5		Atomic absorption analysis

airosol Aerosol.
ajacine $C_{15}H_{21}O_4N$ = 279.3. An alkaloid, m.143, from larkspur, *Delphinium ajacis.* **a. hydrochloride** $C_{15}H_{21}O_4N \cdot HCl$ = 315.8. Colorless crystals, m.93, soluble in water.
ajakol, ajacol Thanatol.
ajava The ripe fruit of *Ammi* or *Carum copticum,* (Umbelliferae) of India; a carminative.
Ajax metal A bearing alloy: Ni 25–50, Fe 30–70, Cu 5–20%.
ajowan Ajava. **a. oil** The essential oil from the seeds of *Carum copticum,* ajava. It resembles caraway seed oil, d.0.900–0.930, $[\alpha]_D$ +1.3; contains thymol and cymene.
akazgine An alkaloid from the bark and leaves of African *Strychnos akazga* (Loganiaceae). Colorless crystals resembling strychnine in their effects on the body.
akcethin Thioacetin.
akee oil A nondrying, yellow fat from *Blighia sapida* (Sapindaceae) from Jamaica.
akermanite $2CaO \cdot MgO \cdot 2SiO_2$. A melilite.
Akineton Trademark for biperiden.
akuammine $C_{22}H_{26}O_4N_2$ = 382.5. An alkaloid from the seeds of *Picralima kleineana* (Apocynaceae) of Gabon.
Akulon Trade name for a polyamide synthetic fiber.
akund See *Calotropis floss.*
akundarol $C_{38}H_{61}O \cdot OH$ = 550.9. A sterol from akanda, *Calotropis gigantea* (Asclepiadaceae). Cf. *calotropin.*
Al Symbol for aluminum.
-al* Suffix indicating an aldehyde structure.
Ala* Symbol for alanine.
alabamine Ab. Ekaiodine. The original name for astatine, element 85, discovered in monazites and by the magneto-optic method by Allison et al. (1931); a homotope of iodine.
alabandite MnS. Manganblende. A black, isometric, native sulfide.
alabaster $CaSO_4 \cdot 2H_2O$. A fine-grained, colorless, compact gypsum. See *selenite.*
alabastrine Naphthalene.
alacreatine $C_4H_7O_2N_3$ = 129.1. Lactoylguanidine, guanidinopropanoic acid. An isomer of creatine, formed by the union of alanine and cyanamide.
alamandite Almandite.
alangine An alkaloid from the bark of *Alangium lamarckii* (Cornaceae) of India; a febrifuge and emetic.
alanine* $C_3H_7O_2N$ = 89.1. Ala*, lactamine, 2-aminopropanoic acid*. White crystals, **(±)-~** m.295. **(*R*)-~** m.297. **(*S*)-~** m.295. (subl, 200), soluble in water. Synthesized from acetaldehyde, ammonia, and hydrochloric acid; a constituent of many proteins. See poly*peptide.* Stereoisomers:

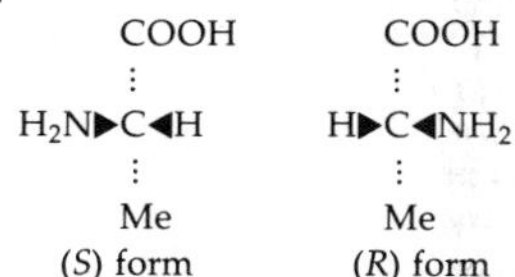

(*S*) form (*R*) form

β-~* 3-Aminopropanoic acid*. **benzoyl ~** $PhCONH \cdot CHMe \cdot COOH$ = 193.2. α-Benzamidopropionic acid. Colorless crystals, m.163. **dithiobis ~** Cystine. **hydrozy ~** Serine. **hydroxyphenyl ~** Tyrosine. **imidazolyl ~** Histidine. **indolyl ~** Tryptophan*. **mercapto ~** Cysteine. **phenyl ~** See *phenylalanine.* **salicyl ~** *o*-Tyrosine.
alant **a. acid anhydride.** See *helenin.* **a. camphor.** Helenin. **a. starch.** Inulin.
alantic acid $C_{15}H_{22}O_3$ = 250.3. Inulic acid. A monobasic acid from the roots of *Inula elecampane.* **a. a. anhydride** $C_{15}H_{20}O_2$ = 232.3. Colorless crystals or incrustation from the roots of *Inula elecampane.*

alantin Inulin.
alantolactone See *helenin.*
alanyl* 2-Amino-1-oxopropyl†. The radical $MeCHNH_2 \cdot CO-$, from alanine. **β-~** 3-Amino-1-oxypropyl†. The radical $NH_2(CH_2)_2CO-$.
a. alanine $C_6H_{12}N_2O_3$ = 160.2. A dipeptide. White needles, m. 276. slightly soluble in water.
alaskite An igneous rhyolite-granite rock containing quartz and feldspar (aplite).
alaskose A rhyolite containing phenocrysts of quartz and feldspar.
Alastra Trademark for a viscose fiber.
alaterite Mineral caoutchouc.
albamine Alabamine.
albane $C_{10}H_{16}O$ = 152.2. A crystalline resin obtained by boiling gutta-percha in alcohol.
albanose A leucitic rock containing melilite, diopside, and magnetite.
albarium A white lime obtained by burning marble; used for stucco.
albata Nickel silver. White alloy of Cu, Ni, and Zn.
albedo The degree of whiteness of a reflecting surface, expressed as the ratio of the reflected light to the incident light: e.g.,

Black velvet	0.004
Moist earth	0.08
Blue paper	0.26
Pinewood	0.40
White paper	0.60
Snow	0.78

Cf. *brightness.*
Albene (1) Trademark for a cellulose fiber. (2) (not cap.) An insoluble white precipitate obtained by boiling melam in water.
alberene A fine grade of soapstone used for electrical insulation.
albertite Albert coal. A black, brittle, asphaltumlike hydrocarbon with conchoidal fracture, d.1.1; a fuel.
Albertol Trademark for a phenolformaldehyde plastic modified with colophony. Cf. *Bakelite.*
Albertus Magnus (1203–1280) Albrecht Graf von Bollstädt. German alchemist noted for his writings.
albite $NaAlSi_3O_8$. Pericline. A white, triclinic soda-feldspar common in granite and other rocks, d.2.605, hardness 6–6.5, mol. vol. 101.1. Cf. *anemosite.*
albocarbon Naphthalene*.
alboferrin Phosphoalbuminate of iron. Brown powder; a former ferruginous tonic.
albopannin $C_{21}H_{24}O_7$ = 388.4. White needles, m.147, from the rhizome of *Aspidium,* male fern.
Albucid Trademark for sulfacetamide sodium.
albumate Albuminate.
albumen White of egg. The liquid white of fresh eggs; chiefly albumin.
albumin (1) $C_{72}H_{112}N_{18}O_{22}S$. A protein in the white of egg, blood, lymph, chyle, and many other animal and vegetable tissues and fluids, soluble in water, coagulates on heating, and hydrolyzes to amino acids; white or yellow scales (USP). (2) A simple *protein,* q.v., insoluble in pure water or dilute salt solution and coagulable by heat. Cf. *globulin.* **blood ~** Serum a. A. prepared from blood and used in the manufacture of photographic papers, in textile printing, in the leather industry, and as a clarifying agent. **egg ~** Ovalbumin. A. prepared from eggs, used in foodstuffs and as a clarifying agent. **milk ~** Whey albumin. A. prepared

from milk by coagulating the casein, used in the manufacture of adhesives, varnishes, and ivory substitutes. **nucleo ~** See *nucleoalbumin.* **osso ~** A protein from ossein. **ov ~** Egg a. **phospho ~** An albuminous substance which contains phosphorus. **serum ~** Blood a. **whey ~** Milk a.

albuminate A compound of albumin with another substance (often basic), as, metals, creosote, quinine. **acid ~** A metaprotein soluble in weak alkali. **alkali ~** A compound of albumin and alkali metals.

albuminoid (1) Resembling albumin. (2) Scleroproteins. Simple proteins, insoluble in neutral solvents; e.g., gelatin.

albuminometer A graduated glass tube for the quantitative determination of albumin in urine.

albuminose A decomposition product of fibrin.

albuminuria The presence of albumin in urine, a condition which may be either physiological or pathological.

albumoscope A U-shaped glass tube for the detection of albumin in urine. One arm is filled with nitric acid and the other with urine; at the point of contact a white ring forms.

albumose A cleavage product of albumin formed during its hydrolysis. Cf. *artose.*

alburnitas A disease of trees which hinders the transformation of sapwood into heartwood.

alburnum The sapwood of a tree.

alcahest Alkahest.

alcamines Aminoalcohols. Cf. *Alkamine.*

alcapton Homogentisic acid.

alchemical Pertaining to alchemy. **a. symbols.** See *alchemistic symbols.*

alchemistic **a. period** About 300–1550 A.D. **a. symbols** The characters or ideographs used by the alchemists. They indicate the relationship of astrological and alchemical speculations. See Fig. 1.

alchemy The empirical stage of chemical knowledge, characterized by speculative theories. The chief aim of alchemy was the transmutation of base metals into gold and the search for the alkahest, or philosopher's stone, which was supposed to confer eternal youth.

Alcian Trademark for dyestuff onium salts, made by reacting a tertiary amine with the chloromethyl derivatives of insoluble phthalocyanine colors. Used in textile printing.

Alclad Trademark for a strong, light aluminum alloy coated with pure aluminum to resist corrosion.

Alcobon Trademark for flucytosine.

Alcocheck See *breath alcohol.*

alcohol (1) Ethanol*, ethyl a.* (2) See *alcohols.* **absolute ~** See *ethanol,* 100%. **acetone ~** See 1-hydroxy-2-*propanone.* **allyl ~** See *allylalcohol.* **amyl ~** See *Pentyl alcohol.* **benzyl ~** See *benzyl alcohol.* **breath ~** See *breath alcohol.* **butyl ~, butyric ~** See *butyl alcohol**. **capryl ~** Octanol*. **cetyl ~** Hexadecyl alcohol*. **cinnamyl ~** See *cinnamyl alcohol.* **dehydrated ~** See *ethanol, (1).* **denatured ~** See *ethanol.* **ethyl ~*** Ethanol*. **ethylene ~** See *glycol.* **ethylic ~** See *ethanol.* **glycyl ~** See *glycerol.* **grain ~** See *ethanol.* **hexadecyl ~*** See *hexadecyl alcohol.* **isobutyl ~** See *butyl alcohol.* **isopentylic ~** See *pentyl alcohol.* **isopropyl ~*** See *isopropyl alcohol.* **methyl ~*** See *methanol.* **octoic ~, octylic ~** Octanol*. **palmityl ~** Hexadecyl alcohol*. **pentyl ~*** See *pentyl alcohol.* **phenylallyl ~** See *cinnamyl alcohol.* **rubbing ~** Denatured a. containing 70% v/v alcohol. **solid ~** A gelatinous mixture of ethanol with solid fatty acids or with soaps; a fuel. **sulfur ~** See *carbon disulfide.* **wood ~** See *methanol.*

a. acid See *alcohol acids.* **a. aldehyde** HO·R·CHO; as, aldol. **a. amide** $HO \cdot R \cdot CONH_2$. Hydroxyamide; as, glycolamide. **a. amine** $HO \cdot R \cdot NH_2$. Hydroxyamine. **a. ether** R·O·R·OH. Hydroxy ether; as, diethyline, Cellosolve.

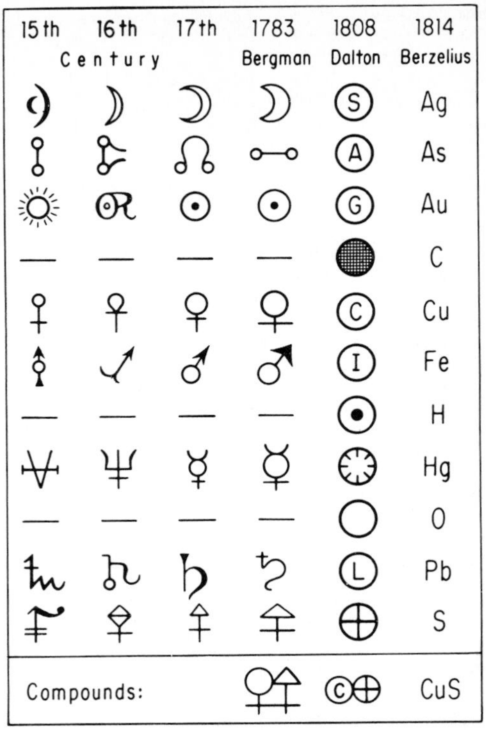

15th	16th	17th	1783	1808	1814
Century			Bergman	Dalton	Berzelius
				S	Ag
				A	As
				G	Au
—	—	—	—		C
				C	Cu
				I	Fe
—	—	—	—		H
					Hg
—	—	—	—		O
				L	Pb
					S
Compounds:					CuS

Fig. 1. Alchemical and chemical symbols.

a. fuel A blend of ethyl, methyl, or butyl alcohol with benzene, petrol, acetone, and/or ether. **a. ketone** R·CO·R·OH. Hydroxyketone; as, ketol. **a. phenols** ar-Hydroxyphenols. Compounds containing hydroxy groups attached to both a ring and a side chain, as *o*-hydroxygenzyl a., salicyl a., $C_6H_4OH \cdot CH_2OH$.

alcohol acids Aliphatic, monohydroxy, monobasic acids; as, carbonic acid, hydroxyformic acid, HO·COOH. Aliphatic, dihydroxy, monobasic acids; as glyoxylic acid, dihydroxyacetic acid, $(HO)_2CH \cdot COOH$. Aliphatic, polyhydroxy, monobasic acids; as, erythric acid, $CH_2OH(CHOH)_2COOH$. Aliphatic, monohydroxy, dibasic acids; as, tartronic acid, HOOC·CHOH·COOH. Aliphatic, dihydroxy, dibasic acids; as, tartaric acid, HOOC·$(CHOH)_2COOH$. Aliphatic, polyhydroxy, dibasic acids; as, trihydroxyglutaric acid, $HOOC(CHOH)_3COOH$. Aliphatic, monohydroxy, tribasic acids; as, citric acid, $HO(COOH)C(CH_2 \cdot COOH)_2$. Aromatic, monohydroxy, monobasic acids; as, mandelic acid, PhCHOHCOOH.

alcoholate* A compound derived from an alcohol by replacing the hydroxyl —H by a base; as, sodium ethanolate, EtONa.

alcoholic (1) Describing a preparation containing an alcohol (usually ethanol), e.g., a. extract. (2) Describing a reaction which forms an alcohol, e.g., a fermentation. (3) Dissolved in alcohol; a. potash.

alcoholometer An apparatus for estimating the alcohol content of a liquid, e.g., a hydrometer.

alcohols* R·OH. Alkyloxides. Alkyl compounds containing a hydroxyl group. Classified according to (1) relation of the carbon atom: primary a., $R \cdot CH_2OH$; secondary a., R_2CHOH; tertiary a., R_3COH; (2) the number of OH groups; as,

		Prefix		*Suffix*
$R \cdot OH$	mono-	-hydroxy		ol
$R \cdot (OH)_2$	di-			diol
$R(OH)_3$	tri-	-hydric	a. or	triol
$R(OH)_4$	tetra-	-basic		tetrol
$R(OH)_5$	penta-			pentol
$R(OH)_n$	poly-			—

See *aromatic alcohols, phenols,* and *alcohol acids.* **aldehyde ~** Compounds containing the —CHO and —OH groups. **aromatic ~** Cyclic compounds containing the —OH group in a side chain; cf. *phenols.* **primary ~** Compounds containing the group $-CH_2OH$. **secondary ~** Compounds containing the =CHOH group. **tertiary ~** Compounds containing the ≡COH group.

a. of crystallization The a. contained in a crystalline salt in a molecule; e.g., $KOH \cdot 2C_2H_6O$.

alcoholysis The cleavage of a C—C bond by the addition of an alcohol: $R \cdot CH_2 \cdot R' + R''OH \rightarrow R''OCH_2R + R'H$. Cf. *hydrolysis.*

Alcolmeter See *breath alcohol.*

alcopol Trade name for a surfactant of the dioctylsulfosuccinate type.

alcosol A sol in alcohol.

alcumite A corrosion-resistant alloy: Cu 88–90, Al 7.5, Fe 28.35, Ni 1%.

alcyl Alicyclic. An aliphatic-cyclic radical; a saturated aromatic radical.

Aldactone Trademark for spironolactone.

aldalcoketose A carbohydrate containing the aldehyde (—CHO), alcohol (—OH), and carbonyl (=CO) radicals.

aldebaranium Thulium*.

aldehydase Aldehyde oxidase*, which forms acids from aldehydes.

aldehyde (1) Acetaldehyde*. (2) See *aldehydes.* **acetic ~** Acetaldehyde*. **anisic ~** Anisaldehyde*. **cinnamic ~** Cinnamaldehyde*. **cuminic ~** Cumic a. **heptylic ~** Heptanal*. **met ~** See *metaldehyde.* **oenanthic ~** Heptanal*. **par ~** See *paraldehyde.* **propionic ~** Propionaldehyde*. **pyromucic ~** Furaldehyde*. **salicylic ~** Salicylaldehyde.

a. ammonia (1) A compound formed by the combination of an a. and ammonia. Crystalline, decomp. on warming with dilute acid; used for the purification of aldehydes. (2) $MeCH(OH)NH_2$ = 61.1. Colorless crystals, m.97, soluble in water. **a. condensation** See *aldol condensation.* **a. group** The —CHO radical, in which the H is not replaceable by a positive radical, but can be replaced by negative atoms or groups. Cf. *aldehydes.* **a. ketone** $R \cdot CO \cdot R \cdot CHO$. Ketoaldehyde. **a. oxidase*** See *aldehydase.*

aldehydene Acetylene*.

aldehydes* Organic compounds containing the —CHO radical, oxidized to acids and reduced to alcohols. A. are indicated by the prefix *oxo-** (for O of CO) or *formyl-** (for CHO), or by the suffix *-al**, *-dial**, *-trial**, *-(carb)aldehyde**, etc. **di ~** Compounds containing 2 a. groups. **olefin ~** Compounds containing a double bond and the a. group. **paraffin ~** Compounds containing the a. group attached to a saturated aliphatic chain. **thio ~** Compounds containing the —CHS group.

-aldehydic Suffix indicating that one COOH group in a dicarboxylic acid, with a trivial name, has been changed into a CHO group; as malonaldehydic acid, $OHC \cdot CH_2COOH$. **a. hydrogen** The H atom of the aldehyde group; not readily replaced by metals.

aldehydine C_5H_3NMeEt = 121.2. 2-Ethyl-5-methylpyridine. Colorless liquid, $d_{23} \cdot 0.9918$, b.173, insoluble in water.

aldicarb* See *insecticides,* Table 45.

aldime R—CH(:NH). An acid imine.

aldobionic acids Oxidized trisaccharides; as, gluco-β-glucuronic acid, from the hydrolysis of flaxseed mucilage.

aldohexose* A hexose containing the aldehyde group; e.g., glucose. Cf. *ketohexose.*

aldoketenes See *ketenes.*

aldol $Me \cdot CHOH \cdot CH_2 \cdot CHO$ = 88.1. (1) β-Hydroxybutyric aldehyde, 3-hydroxybutanal*. A condensation product of acetaldehyde. Colorless liquid, d.1.109, soluble in hot water. Its solution leaves a polymer, paraldol, on evaporation. Cf. *paraldehyde, metaldehyde.* (2) One of a class of condensation products formed from an aldehyde. **a. condensation** The polymerization of an aldehyde in presence of dilute acid or alkali, e.g., aldol formation. The aldol polymer is stabler than the meta and para polymers. Three types: (1) true aldol condensation: $R_2CO + H \cdot CH_2COR \rightarrow R_2C(OH) \cdot CH_2 \cdot COR \rightarrow R_2C{:}CH \cdot COR$. (2) Cannizzaro reaction: $2R \cdot CHO \rightarrow R \cdot COOH + R \cdot CH_2OH$. (3) Claisen condensation: $2R \cdot COOR' \rightarrow RC(OH){:}CHCOOR' + R'OH$.

aldolase* See *enzymes,* Table 30.

aldonic acids Acids produced by gentle oxidation of the corresponding aldose; as, gluconic acid from glucose.

aldopentose* A pentose containing the aldehyde group; as, arabinose.

aldose* A carbohydrate containing the aldehyde group. Cf. *ketose, sugar.*

aldoxime C_2H_5ON = 59.1. Acetaldoxime. Colorless liquid, b.115, soluble in water; used in organic synthesis. Isomeric forms:

```
 N·OH          HO·N
  ||             ||
 Me C H        Me C H
 trans          cis
```

aldoximes* Organic compounds containing the —C(H):NOH group. Stereoisomers:

```
 HO·N          N·OH
  ||            ||
 R·C·H         R·C·H
  cis          trans
```

Form *C*-, *N*- and *O*-substituted compounds.

aldrey A noncorroding aluminum alloy, used for transmission lines: Mg 0.4, Si 0.6, Fe 0.3%.

aldrin* See *insecticide,* Table 45 on p. 305.

alembic (1) Ancient name for a retort. (2) Figuratively, anything that purifies.

aletris False unicorn, starwort, blazing star, colic root, star grass, bitter grass, devil's bit. The dried rhizomes of *Aletris farinosa* (Haemodoraceae) of the United States.

Aleurites (1)The Chinese wood or tung oil plant. (2) A genus of trees (Euphorbiaceae) of the warmer zones of Asia which yield oil; as, *A. cordata,* tung oil.

aleuritic acid $C_{16}H_{32}O_5$ = 304.4. 9,10,16-Trihydroxypalmitic acid, m. 102, from the shellac of *Aleurites montana.*

aleurometer A cylinder for testing the baking capacity of flour from the expansion of its gluten.

aleuronate A vegetable protein food. A tasteless, yellow powder. **a. powder** Baked flour mixed with cooked starch for injection into the pleural cavity of animals (rabbits) to stimulate the production of leucocytes. Cf. *leucocyte.*

aleurone Protein grains in the endosperm of ripe seeds.

Alexander tester An apparatus to determine the gel strength of glue.
alexandrite $Al_2O_3 \cdot BeO$. A green variety of chrysoberyl with red fluorescence.
alexandrolite A clay containing chromium.
alexin Former term for complement.
alfa Halfa. Local name for *esparto* grass, q.v., from Algeria and Tunisia.
alfalfa Lucerne. The plant *Medicago sativa* (Leguminosae); a fodder.
alfalfone $C_{21}H_{42}O$ = 310.3. A ketone from alfalfa.
alferric Containing alumina and ferric oxide. **a. minerals** Igneous rocks containing aluminum and iron; intermediate between the salic and femic groups.
alga (Pl. algae). Unicellular or polycellular plants (Thallophytae), a class of cryptograms, which live in fresh or salt water and are distinguished from fungi by the presence of chlorophyll and response to photosynthesis; as, seaweeds, kelps, agar-agar. Cf. *lichen.* Classification by color: Cyanophyceae (blue-green or fission a.), Phaeophyceae (brown), Chlorophyceae (green), Rhodophyceae (red), Bacillariophyceae (yellow a., diatoms). **calcareous ~** A. containing calcium, e.g., *Lithothamnium,* whose skeletons consist of calcite 95, magnesite 5%. **siliceous ~** Slimy ooze (Si 70%) formed on the seabed by diatoms, radiolarians, and algae skeletons. Cf. *oceanic sediments.*
algarites The naturally occurring protobitumins formed during the acid hydrolysis of algae. They consist of algarose with metallic salts, sulfur, and nitrogen derivatives.
algaroba Algarobilla. The sweet pods of *Prosopsis dulcis,* of S. America; used for tanning and for preparation of gum.
algarose A protobitumen formed by acid hydrolysis of carbohydrates.
algin Alginic acid, Norgine. A protein of marine algae, a by-product in the preparation of iodine from kelps, principally *Laminaria digitata.* Used as a fabric dressing, as a thickener for jellies, and in mucilages. **a. fiber** $(C_6H_2O_6)_n$. Seaweed rayon. A polymer of D-mannuronic acid, mol. wt. 48,000–185,000.
Alginate (1) Trade name for an algin synthetic fiber. (2) (Not cap.) Any of the salts of algin, e.g., sodium alginate; used as protective colloids, for thickening solutions, for forming films on drying, for textile dressing, for making impressions in dentistry, and for use in synthetic fibers.
alginic acid Algin.
algodonite Cu_6As. A steel-gray native arsenide.
algorithm In computer applications, a prescribed set of well-defined instructions or procedural steps for the solution of a specific (e.g., process control) problem.
algulose Almost pure cellulose obtained in the extraction of iodine from kelp.
alibate To cover with a protective layer of aluminum.
alicyclic Aliphatic-cyclic. The group of cyclic organic compounds derived from the corresponding aliphatic compounds by ring formation and having a saturated ring, as, the cycloparaffins. **a. acids** A group of acids whose molecules contain a saturated ring, as, cyclopropanecarboxylic acid, $CH_2 \cdot CH_2 \cdot CH \cdot COOH$. **a. compound** A compound containing a carbon ring whose properties are aliphatic rather than aromatic.
alignment chart Nomograph.
aliphatic Acyclic. Pertaining to an open-chain carbon compound; as, alkanes. Cf. *aromatic, alicyclic.* **a. acids** Fatty acids. The organic acids derived from the a. hydrocarbons. See *acid.* **a. amino group** The NH_2 radical, when attached to a chain. **a. amino group apparatus** Van Slyke apparatus. **a. compounds** See *organic compounds.* **a. hydrocarbon** A compound of carbon and hydrogen having an open chain, as in the alkanes. See *hydrocarbon, nomenclature.*
alipite A nickel mineral containing silicates, magnesium, and free silica.
aliquot A part which when multiplied by an integer makes a whole. Thus, in chemistry, a sample representing a definite proportion of the whole to be tested. **a. sampling** Taking a representative sample; as by quartering.
alismin A principle extracted from *Alisma plantago,* the water plantain.
alite $8CaO \cdot SiO_2Al_2O_3$(?). The primary crystalline constituent of Portland cement clinker, also found native. Cf. *belite.*
alival $CH_2I \cdot CHOH \cdot CH_2OH$ = 202.0. Iodopropylene glycol. Colorless powder; an iodoform substitute.
alizaramide $C_6H_4(CO)_2C_6H_2(OH)NH_2$ = 239.2. Aminohydroxyanthraquinone. Derived from alizarin by heating with ammonia water. Brown needles, sublime 150, insoluble in water.
alizarate (1) A derivative of alizarin in which the H of both OH groups is replaced by a metal. Cf. *lake.* (2) Phthalate*.
alizaric acid Phthalic acid*.
alizarin $C_{14}H_8O_4$ = 240.2. Anthraquinonic acid, 1,2-dihydroxyanthraquinone*, 1,2-dihydroxy-9,10-

$$C_6H_4\langle{}^{CO}_{CO}\rangle C_6H_2\langle{}^{OH}_{OH}$$

anthracenedione†, alizarine (in dye names). The red color of *Rubia tinctorium,* dyers' madder. Synthetically prepared from anthracene and the hydrolysis of rubianic acid. Orange crystals, m.290, insoluble in cold water; used to dye fabrics and in the manufacture of dyestuffs; also as a reagent and indicator (alkalies, red; acids, yellow, changing at pH 9.5). **a. black** Naphthazarine. A black a. dye, made by the reduction of 1,8-dinitronaphthalene. **a. blue** $C_{17}H_9O_4N$ = 291.3. Anthracene blue. Brown-violet needles, m.270, soluble in alcohol. Blue coloring matter, dye and pH indicator (acids, green; alkalies, blue, changing at pH 12). **a. blue amide** $C_{17}H_{10}O_3N_2$ = 290.3. Aminohydroxyanthraquinone quinoline. A derivative of a. blue in which one —OH is replaced by a NH_2 group; a dye. **a. brown** Anthragallol. **a.carboxylic acid** $C_{14}H_7O_4 \cdot COOH$ = 284.2. 2-Carboxyl a. Red triclinic crystals, m.305, soluble in water; used to manufacture dyestuffs. **a. carmine** Sodium a. sulfonate. **a. cyanine** Mordant and acid dyes based on 1,2,4,5,8-pentahydroxyanthraquinone. **a. dyes** Dyes derived from anthraquinone (see Table 6); used as sodium salts of their sulfonic acids. **a. fluorine blue** AFB. A reagent for the determination of fluorides and lanthanum. **a. red** Sodium sulfalizarate, a. sodium monosulfonate. Brown powder, soluble in water; an indicator for titrating acids (yellow) and bases (red); changes at pH 5.5. **a. sulfonate** Any salt of a.sulfonic acid. **a.sulfonic acid** $C_{14}H_7O_4 \cdot SO_3H$ = 320.1. 1,2-Dihydroxy-7-sulfonic anthraquinone. Orange crystals, soluble in water; used to manufacture dyes. **a. yellow** Sodium *p*-nitraniline salicylate. An indicator changing at pH 11.1 from yellow (acid) to purple (basic).
alizarinic acid Phthalic acid*.
Alk Abbreviation for (1) alkaloid, (2) alkyl.
alkadienes* Acyclic hydrocarbons with 2 double bonds.
alkadiynes* Acyclic hydrocarbons with 2 triple bonds.
alkahest See *alchemy.*
alkalamides Organic compounds derived from ammonia by replacing H atoms by acid or basic radicals.
alkalescent Slightly alkaline.
alkali [Pl. alkalies (U.S.), alkalis (U.K.)]. Usually a hydroxide

TABLE 6. ALIZARIN DYES (ANTHRAQUINONES)

Alizarin	1,2-Dihydroxyanthraquinone
Quinizarin	1,4-Dihydroxyanthraquinone
Anthrarufin	1,5-Dihydroxyanthraquinone
Chrysazin	1,8-Dihydroxyanthraquinone
Anthraflavin	2,6-Dihydroxyanthraquinone
Anthragallol	1,2,3-Trihydroxyanthraquinone
Purpurin	1,2,4-Trihydroxyanthraquinone
Quinalizarin	1,2,5,8-Tetrahydroxyanthraquinone
Alizarin cyanine	1,2,4,5,8-Pentahydroxyanthraquinone

of lithium, sodium, potassium, rubidium or cesium; but also the carbonates of these metals and ammonia, and the amines. **mineral ~** An inorganic base. **vegetable ~** An organic base. **volatile ~** Ammonia*.

a. blue The sodium salt of a triphenylrosanilinesulfonic acid, used as indicator, changing at pH 12.5 from blue (acid) to red (basic), and in dyeing wool. **a. cell** See *photoelectric cell.* **a. earth metals** See *alkaline.* **a. earths** The oxides of Ca and Ba (lime, baryta). **a. metals*** The elements of the first group of the periodic system: Li, Na, K, Rb, Cs, Fr. They are strongly electropositive, have a low specific gravity and melting point, are silver-white and ductile, and react vigorously with water, liberating hydrogen and forming hydroxides. **a. metaprotein** A protein cleavage product, soluble in alkalies. **a. reaction** Basic reaction. See *alkaline reaction.* **a. waste** Waste calcium sulfide from the LeBlanc process.

alkalide Compound containing an alkali metal as both cation and anion; as, Na^{+}cryptate-222·Na^{-}. Cf. *electride.*

alkalimeter Calcimeter.

alkaline Producing hydroxyl ions in an aqueous solution; as, any base. **a.-earth metals*** The elements Ca, Sr, Ba, Ra of the second group in the periodic table. **a. reaction** The color changes caused by alkalies, or the hydroxyl ion; as, red litmus turning blue, colorless phenolphthalein turning red. See *indicator.* **a. solution** A solution containing more hydroxyl ions than hydrogen ions. See *hydrogen ion concentration.* **a. tide** The reduced acidity of tissue fluids and urine which follows eating.

alkalinity Having an excess of hydroxyl ions in aqueous solution. Cf. *base, acidity, hydrogen ion.*

alkalize To make alkaline or basic. Antonym: acidify.

alkaloid An organic nitrogenous base (see Table 7). Many a. of medical importance occur in the animal and vegetable kingdoms, and some have been synthesized. **acylindole ~** A. from apocyanaceous plants (genus *Tabernaemontana*), derived from tetracyclic 2-acylindoles; as, vobasine. **animal ~** A base derived from animals; as, xanthine. **cadaveric ~** A cleavage product from putrefaction of animal tissues; as, betaine. See *ptomaine.* **putrefactive ~** Cadaveric a. **vegetable ~** An a. from vegetable tissues.

alkaloidal Pertaining to alkaloids.

alkalometry (1) The medication of alkaloids for therapeutic purposes. (2) The determination of alkaloids.

alkalosis A metabolic state in which the alkalinity of the body fluids (e.g., blood) is above the normal level.

Alkamine Trade name for detergents containing a fatty acid amide sulfonate. Cf. *alcamines.*

alkanes* Methane series, paraffins. Generic name for the group of aliphatic hydrocarbons C_nH_{2n+2}; denoted by the suffix -ane. Tetracontane, $C_{40}H_{82}$, has theoretically over 62×10^{12} possible isomeric forms.

alkanet Alkanna root. The root of *Alkanna tinctoria* (Boraginaceae); a red coloring for oils or fats.

alkanization The combination of isobutane and butenes from petroleum, in presence of sulfuric acid, to form isooctane and trimethylbutene.

alkannin $C_{16}H_{16}O_5$ = 288.3. Shikonin. Red powder,

TABLE 7. CHEMICAL CLASSIFICATION OF ALKALOIDS

Derivatives of	Example	Occurring in
1. Pyridine	Piperine	Pepper
	Conine, trigonelline	Hemlock
	Arecoline, arecaidine	Arecanut
	Nicotine	Tobacco
2. Pyrrolidine	Atropine, hyoscyamine, sparteine	Solanaceae
	Cocaine	Coca leaves
3. Quinoline	Quinine, cinchonine, quinidine, cinchonidine	Cinchona
	Strychnine, brucine	Nux vomica
4. Isoquinoline	Papaverine, morphine, codeine, thebaine	Morphium
	Narcotine, hydrastine	Hydrastis
	Laudanosine	Opium
5. Purine	Caffeine, theobromine, xanthine	Coffee, tea, beets
6. Glyoxaline	Pilocarpine, ergotoxine, ergometrine	Jaborandus, ergot
7. Amines	Asparagine, leucine, betaine, choline	
8. Glucosides	Solanine	
9. Others	Aconitine, yohimbine	

See also *aconite; Areca; opium;* and *Solanaceae, spartium* and *staphisagria alkaloids.*

softening at 100, insoluble in water. A color for fats or oils, and an indicator. **a. extract** Extract alkanet. A crude a. **a. paper** Boettger's paper. A filter paper stained with a. and used as a test for alkalies (green), carbonates (blue), and acids (red).
Alkanol Trademark for a group of wetting agents.
alkanolamine Alkylolamide.
alkapton Homogentisic acid.
alkaptonuria See inborn error of *metabolism.*
alkargen Cacodylic acid.
alkasal A mixture of aluminum salicylate and potassium acetate.
Alkathene Trademark for a polyethylene plastic.
alkazid process The removal of sulfur dioxide from coal gas by absorption in cold α-aminopropionic acid solution, and subsequent liberation by heat as a source of sulfur. Cf. *Thylox process.*
alkenes* See *olefin(e).*
alkenyl An unsaturated, univalent aliphatic radical.
alkines Alkynes*.
alkones Terpenes*. See *hydrocarbon series.*
alkoxides Organic compounds in which the H of the hydroxyl group is replaced by a metal, as in alcoholates.
alkoxy* An alkyl radical attached to the remainder of the molecule by oxygen; as, methoxy.
alkoxylate An acyclic polyether with the repeating group $-[CH(R)CH_2O]-$, as in polyethylene glycol.
alkyd resins A group of adhesive resins made from unsaturated acids (phthalic anhydride) and glycerol; protective and decorative coatings and bonding materials.
alkyl $C_nH_{2n+1}-$. Alphyl, aphyl. A monovalent radical derived from an aliphatic hydrocarbon by removal of 1 H; as, methyl-. Cf. *alkyle, alkylene.* **a. halide** R·Cl. **a. nitrite** $R \cdot NO_2$. **a. oxide** See *ethers.*
alkylamine An amine-containing alkyl group attached to aminonitrogen.
alkylation Substitution of an aliphatic hydrocarbon radical for a H atom in a cyclic compound; as, the introduction of a side chain into an aromatic compound. **acid ~** or **hot ~** Introducing the alkyl into the amino or hydroxy group by heating under pressure with alcohol in presence of mineral acid. **basic ~** or **cold ~** Treatment with alkyl sulfate in presence of sodium hydroxide.
alkylbenzylideneoxime Compound derived from benzylideneoxime. Forms *C*-, *N*- and *O*-substituted compounds. Cf. *aldoxime.*
alkyle (1) Alkene*. (2) Alkylide. A compound of a metal (Al, Pb, Hg, Zn, etc.) and an alkyl radical; as, tetraethylplumbane.
alkylene $C_nH_{2n-1}-$. An alkene radical, as, ethylene. **a. oxides** (1) Alcohol ethers. The aliphatic compounds that contain both the primary alcohol and the ether groups; as, diethyline. (2) Epihydrins. Alicyclic compounds which are the ethers of glycols; as, ethylene oxide.
alkyl halides $C_nH_{2n+1}X$. Alkylogen. A combination of an alkyl and halogen grouping.
alkylide See *alkyle* (2).
alkylidene $C_nH_{2n}=$. A divalent organic radical derived from an aliphatic hydrocarbon; as, ethylidene, in which 2 H atoms are taken from the same C atom.
alkylogens Alkyl halides.
alkylolamide (-ine) Alkanolamine. A compound of an amide and an alkyl group; in particular, ethanol amines and dodecanoic or coconut fatty acids. The di ~ are stabilizers in liquid detergents.
alkyls See *alkyl.*
alkynes* Acyclic hydrocarbons with one triple bond of general formula C_nH_{2n-2}; as, acetylene. Cf. *alkadiynes.*

allanic acid $C_4H_5O_5N_5$ = 203.1. A monobasic crystalline acid obtained by the oxidation of allantoin with nitric acid.
allanite Bagrationite. Orthite. A dark monoclinic orthite of the zoisite group which resembles epidote, in which a rare-earth metal partly replaces the aluminum and iron.
allantal Sheet aluminum coated with an Al-Cu alloy.
allantoic acid $(NH_2CONH)_2CHCOOH$ = 176.1. Diureidoacetic acid. A crystalline monobasic acid obtained by boiling allantoin in alkaline solution.
allantoin $C_4H_6O_3N_4$ = 158.1. Glyoxylic diureide. Glyoxuric acid diureide. Colorless crystals, m.227 (decomp.), soluble in hot water. Produced by the oxidation of uric acid and widely distributed in nature (urine of animals, seeds and roots of plants). It hastens epithelial formation.
allantoxanic acid $C_4H_3O_4N_3$ = 157.1. Colorless crystals obtained by the action of potassium hexacyanoferrate(III) on allantoin.
allanturic acid NH·CO·NH·(CO)·CHOH = 116.1. (ring closed from first NH to CHOH) Glyoxalylurea. On reduction and hydrolysis, it yields hydantoin, hydantoic acid, and, finally, glycine and ammonium carbonate.
allelomorph (1) One of a mixture of isomers in a solution which will first separate or crystallize out. (2) One or more forms of a gene occupying the same position on a chromosome.
allelomorphism Desmotropism.
allelotrope A member of a system of 2 isomeric or desmotropic substances in equilibrium.
allelotropism The existence of 2 isomeric or desmotropic substances in an unstable equilibrium, so that either rearrangement of the atoms may occur.
allemontite $SbAs_3$. A rhombohedric, gray or reddish native arsenide.
allene* Propadiene*. **ethyl ~** 1,2-Pentadiene*. **methyl ~** 1,2-Butadiene*.
Allen's test A modification of Fehling's test for sugar in urine.
allergen Anaphylactin. A toxic substance which causes hypersensitivity or anaphylaxis; as, bee sting.
allergia Allergy.
allergic protein An extract prepared from a vegetable or animal substance, used for diagnosis or desensitization; e.g., pollen extract in hay fever.
allergy Allergia. Hypersensitivity to an antigen. **bacterial ~** A. to a particular bacterium. **bronchial ~** A. of bronchial mucosa to inhaled allergens; e.g., house dust mites causing asthma. **contact ~** A. to substances applied to the skin; e.g., some metals in watch straps or clothes fasteners. **food ~** A. to some substances in food.
allicin $C_6H_{10}OS_2$ = 162.3. A water-soluble oil; the principal antibacterial constituent of onion oil.
alligation If two substances when mixed retain their specific values (*A* and *B*) of a property, the value for the mixture can be calculated from the formula $(aA + bB)/(a + b)$, where a and b are the proportions. Cf. *additive property.*
Allihn condenser A condenser with a condensing surface of glass bulbs.
Allium A genus of plants (Liliaceae) which contain a pungent, volatile oil (allyl sulfide). Their bulbs are used as a food, condiment, and aid to digestion; as, *A. cepa* (common onion), *A. porrum* (leek), *A. sativum* (garlic).
allo- Prefix (Greek "other") applied to the more stable form of 2 isomers.
allochroite A green to black manganese *garnet,* q.v.
allochromy A *phosphorescence* or *radiation* effect, q.v., in which the wavelength of the emitted light differs from that of the incident light.

allocinnamic acid PhCH:CH·COOH = 148.2. White crystals; 3 allotropic forms, m.42, 58, and 68.
alloclasite A steel-gray sulfide of cobalt, bismuth, and arsenic with some iron.
allocyanine Neocyanine.
alloisomerism Stereoisomerism.
allomaleic acid Fumaric acid*.
allomerism Similarity of crystalline form with a difference of chemical composition.
allomerization The dehydrogenation of chlorophyll in alcoholic solution by the action of atmospheric oxygen.
allomone A compound or mixture that adaptively favors the emitting species; as, a floral scent attracting pollinating insects. Cf. *pheromone.*
allomorphism Similarity of chemical composition but difference in crystalline form, especially of minerals; as, $CaCO_3$ in calcite and aragonite.
allomucic acid $HOOC(CHOH)_4COOH$ = 210.1. An optically inactive acid derived from D- and L-allose, m.166–171, soluble in hot water.
allopalladium See *palladium.*
allophanamide Biuret*.
allophane $Al_2SiO_5 \cdot 4H_2O$. An amorphous, native hydrated calcium silicate from Grafenthäl, Germany.
allophanic **a. acid*** $NH_2 \cdot CONH \cdot COOH$ = 104.1 (Aminocarbonyl)carbamic acid†, ureacarbonic acid, carbamoylcarbamic acid. A hypothetical acid known as its esters. **a. amide** Biuret*.
allopurinol $C_5H_4ON_4$ = 136.1. 1*H*-pyrazolo[3,4-*d*]pyrimidin-4-ol, isopurinol, Zyloprim, Zyloric. White powder, slightly soluble in water. Used to prevent gout by increasing excretion of uric acid (USP, BP).
allose $C_6H_{12}O_6$ = 180.2. A synthetic hexose.
allotelluric acid See *telluric acid.*
allotoxin A substance formed by living tissue, which tends to counteract the toxic effect of bacteria.
allotrope One of two or more isomeric forms of an element, as, red and white phosphorus.
allotropic Occurring in 2 or more isomeric forms; e.g., carbon as graphite, diamond, or charcoal.
a. modification Allotrope.
allotropism Allotropy.
allotropy Allotropism. A change in the properties of an element without a change of *state,* q.v.; isomerism of a chemical element which occurs in different amorphous and crystalline forms, as with sulfur. Cf. *monotropic, pseudomonotropy.* **dynamic ~** Tautomerism.
alloxan $CO(NH \cdot CO)_2 \cdot CO$ = 142.1. Mesoxalylurea, pyrimidinetetrone, "erythric acid" (Brugnatelli). Two crystalline forms: 1 mole H_2O, small, colorless crystals; 4 moles H_2O, large, colorless, efflorescent prisms; decomp. 170, soluble in water. It appears in the intestinal mucus during diarrhea. A. causes reversible diabetes in experimental animals. **oxime ~** Violuric acid.
alloxanic acid $C_4H_4O_5N_2$ = 160.1. 4-Hydroxy-2,5-dioxo-4-imidazolidinecarboxylic acid*. Coloress crystals obtained from alloxan by treatment with alkalies; decomp. by heat.
alloxanthin Alloxantin.
alloxantin $C_8H_6O_8N_4$ = 286.2. A purine base. Colorless prisms or rhombs, soluble in water. **tetramethyl ~** Amalinic acid.
alloxazine $C_{10}H_6O_2N_4$ = 214.2. A heterocyclic compound in some plant pigments. **iso ~** Flavin.
alloxuric bodies Purine bases.
alloy A mixture of two or more metallic elements (or nonmetals, as C, Te, P) which has a metallic appearance and which is either:

1. A molecular mixture, microscopically homogeneous; as,
   (*a*) A solid solution of A in B or B in A, in which case the properties are intermediate between those of A and B.
   (*b*) A metallic compound, A_xB_y, with properties differing from A and B.
   (*c*) A combination of (*a*) and (*b*); as, A in A_xB_y.
2. A colloidal mixture, microscopically heterogeneous; as, two or more phases consisting of crystals:
   (*d*) of metallic elements, [A] + [B].
   (*e*) of metallic compounds, $[A_xB_y] + [AB_z]$.
   (*f*) of solid solutions, [A in B] + [B in A].
   (*g*) of combinations of *d, e,* and *f;* as, [A] + $[A_xB_y]$ + [A in B], etc.

Generally the m. of an alloy is lower than that of its highest-melting constituent. A. are usually described in terms of their constituents, in decreasing order of amounts present, e.g., Heusler's *alloy,* q.v., as a copper-manganese-aluminum a. Cf. *intermetallic compound.* **antifriction ~** An a. sufficiently plastic to fit itself to the shape of a shaft; usually Sn or Pb alloyed with Cu or Sb. **ceramic ~** See *ceramic alloys.* **coinage ~** A. used for coins. Cf. *gold* alloys, *silver* alloys. **eutectic ~** An a. which has the lowest constant melting point, i.e., whose constituents are present in such proportions that it solidifies completely at the *eutectic temperature,* q.v. **fusible ~** Low-fusing a. **heavy ~** An a. of W 90, Ni 7.5, Cu 2.5%. It has a high density (16.5–17.0 g/cm^3) and tensile strength equal to that of steel. **Heusler's ~** A nonferrous, magnetic a. which contains nonmagnetic metals; as, Cu 60, Mn 20, Al 14 pts. **light ~** An a. containing Al as distinct from Mg. Cf. *ultralight a.* **low ~** An a. containing a large proportion of one constituent. **low-fusing ~** Fusible a. An a. melting below the m. of tin. The lowest-melting a. (15°C) consists of Ga 88, Sn 12%. Others are Newton's, Bi 8, Pb 5, Sn 3% (m.94); Lipowitz's, Bi 5, Pb 2.7, Sn 1.3, Cd 1.0% (m. 65). Cf. *D'Arcet metal, Wood's* alloy, *rose* metal. **master ~** An a. containing a known amount of a particular metal, used instead of the pure metal for making alloys. **mercury ~** Amalgam. **pyrophoric ~** A Fe-Ce a. that sparks when rubbed with a file. **ultralight ~** An a. containing magnesium, as distinct from aluminum. Cf. *light alloy.*
alloying Producing an alloy.
allspice Pimenta.
allulose Psicose.
alluvium The deposit carried down by a river.
allyl* 2-Propenyl†. The radical $-CH_2 \cdot CH{:}CH_2$. Misnomer for 1-propenyl, $-CH{:}CH \cdot Me$. **a. acetate** $MeCOOC_3H_5$ = 100.1. Colorless liquid, b.103, slightly soluble in water. **a.acetic acid** $C_3H_5 \cdot CH_2 \cdot COOH$ = 100.1. Colorless liquid, b.187, slightly soluble in water. **a.acetone** $C_3H_5 \cdot CH_2 \cdot CO \cdot CH_3$ = 98.1. Colorless liquid, b.129, insoluble in water. **a.acetonitrile** $C_3H_5CH_2CN$ = 81.1. Colorless liquid, b.140, insoluble in water. **a. alcohol*** $CH_2{:}CH \cdot CH_2OH$ = 58.1. Propenol, propenyl alcohol. Colorless liquid, b.96, insoluble in water. Used in organic synthesis, in chemical warfare, as an antiseptic. **a. aldehyde** Acrylaldehyde*. **a.amine** $CH_2{:}CH \cdot CH_2NH_2$ = 57.1. 2-Propenylamine*. Yellow oil, b.53, miscible with water; prepared from mustard oil. **methyl ~** $C_3H_5 \cdot NH \cdot CH_3$ = 71.1. Colorless liquid, b.65, miscible with water. **phenyl ~** Allylaniline. **a.aniline** $C_3H_5 \cdot NH \cdot C_6H_5$ = 133.2. *N*-Phenylallylamine. Colorless liquid, b.208, slightly soluble in water. **a. benzoate** $C_6H_5 \cdot COOC_3H_5$ = 162.2. Colorless liquid, b.228. **a. bromide*** $CH_2{:}CH \cdot CH_2Br$ = 121.0. Bromopentene. Colorless liquid, b.70, insoluble in water. **a. butyrate** $CH_3(CH_2)_2COOC_3H_5$ = 128.2. Colorless liquid, b.142, miscible with alcohol. **a. chloride*** $CH_2{:}CH \cdot CH_2Cl$ = 76.5. Chloroallylene, 3-chloropropene*. Colorless liquid, b.46; insoluble in water. **a. cinnamate*** $C_6H_5 \cdot$

$CH:CH \cdot COOC_3H_5$ = 188.2. Colorless crystals, b.285, insoluble in water. **a. cyanamide** Sinamine. **a. cyanide*** $CH_2:CH \cdot CH_2CN$ = 67.1. 3-Butanenitrile. A colorless liquid, b.119; soluble in alcohol. **a. disulfide** $C_3H_5 \cdot S \cdot S \cdot C_3H_5$ = 146.3. Colorless liquid, b.138, in garlic. **a. ether** $(CH_2:CH \cdot CH_2)_2O$ = 98.1. Diallyloxide. Colorless liquid, b.94, slightly soluble in water. **methyl ~** $CH_2:CH \cdot CH_2 \cdot O \cdot CH_3$ = 72.1. Colorless liquid, b.46, slightly soluble in water. **a. ethyl ether*** $C_3H_5 \cdot O \cdot C_2H_5$ = 86.1. A colorless liquid, b.64. **a. fluoride*** $CH_2:CH \cdot CH_2F$ = 60.1. 3-Fluoropropene*. Colorless gas, b.−10, slightly soluble in water. **a. formate*** $HCOOC_3H_5$ = 86.1. Colorless liquid, b.82, insoluble in water, soluble in alcohol. **a. iodide*** $CH_2:CH \cdot CH_2I$ = 168.0. 3-Iodopropene. Yellow liquid, b.101, insoluble in water. **a. isocyanide*** $CH_2:CH \cdot CH_2NC$ = 67.1. Colorless liquid, b.100, slightly soluble in water. **a. isopentyl ether** $C_3H_5 \cdot O \cdot C_5H_{11}$ = 128.2. Allylisoamyl oxide. Colorless liquid, b.120, slightly soluble in water. **a. isothiocyanate*** A. mustard oil. **a.malonic acid** $C_3H_5 \cdot CH:(COOH)_2$ = 144.1. Colorless crystals, m.103, (decomp.), soluble in water. **a.methanol** $CH_2:CH \cdot (CH_2)_2OH$ = 72.1. 3-Butenol*. Colorless liquid. **methyl ~** $CH_2CH \cdot CH_2 \cdot CH(OH)Me$ = 86.1. Colorless liquid, b.115, soluble in water. **a. mustard oil** $CH_2:CH \cdot CH_2NCS$ = 99.1. A. isothiocyanate. Colorless liquid, d.1.017, b.151, slightly soluble in water. It occurs in mustard seeds *(Sinapis nigra)* as *potassium myronate,* q.v., combined in the form of glucoside, and in horseradish; a vesicant; a poison gas used in World War II. **a. oxalate** $(COOC_3H_5)_2$ = 170.2. Colorless liquid, b.217, insoluble in water. **a.phenyl cinchonine ester** Antiquinol. **a. phenyl ether*** $C_6H_5 \cdot O \cdot C_3H_5$ = 134.2. Colorless liquid, b.192, insoluble in water. **a. phenylurea** $Ph \cdot NH \cdot CO \cdot NH \cdot C_3H_5$ = 176.2. Colorless crystals, m.115. **a.pyridine** $C_5H_4N \cdot C_3H_5$ = 119.2. Colorless liquid, b.190. **a.pyrocatechol methylene ester** Safrole. **a. sucrose** A substance formed by the action of a. chloride with sucrose at 85°C. Used in heat-, water-, and grease-resistant coatings. **a. sulfide** $(C_3H_5)_2S$ = 114.2. Bis(2-propenyl) sulfide*, diallyl sulfide, Yellow liquid, b.138, slightly soluble in water; a constituent of garlic, formerly used medicinally. **a. thiocyanate*** $CH_2:CH \cdot CH_2SCN$ = 99.2. Colorless liquid, b.161, insoluble in water. **a. thioether** A. sulfide. **a.thiourea** $C_3H_5 \cdot NH \cdot CS \cdot NH_2$ = 116.2. Colorless, monoclinic crystals of garlic odor, m.74, soluble in water. Used in treatment of scar tissue, in photography, and as a reagent. **a. tribromide** $(CH_2Br)_2 \cdot CHBr$ = 280.8. Tribromoallylene, 1,2,3-tribromopropane*. Colorless crystals or oil, m.18, b.219, soluble in alcohol; formerly an antispasmodic and sedative. **a. trisulfide** $(C_3H_5)_2S_3$ = 178.3. Diallyl trisulfide. Colorless liquid, b.140, insoluble in water.

allylene Propyne*. **a. dichloride** 1,2-Dichloro*propene**. **a. oxide** 1,2-Epoxy*propene**. Colorless liquid, b.62, slightly soluble in water.

allylidene* The $CH_2:CH \cdot CH=$ radical, from acrylaldehyde.

allylin $CH_2:CH \cdot CH_2 \cdot O \cdot CH_2 \cdot CHOH \cdot CH_2OH$ = 132.1. Allyloxyglycerol. Heavy liquid, b.230, formed on heating glycerol with oxalic acid.

almadina Euphorbia gum.

almandine Almandite.

almandite (1) $Fe_3Al_2Si_3O_{12}$. Almandine. An isometric, red-brown to black garnet, d.3.9–4.2, mol. wt. 499.1, mol. vol. 118.0. (2) A violet variety of *spinel,* q.v.

Almen test The detection of carbohydrates by the reduction of alkaline bismuth ions.

almond The dried ripe seeds of *Prunus amygdalus, communis* (Rosaceae) of S. Europe and California. **bitter ~** The seeds of variety *amara,* used in the manufacture of a. oil and amygdalin and for flavoring. **bitter ~ water** A solution containing hydrocyanic acid, ammononium cyanide, and mandelonitrile. **sweet ~** The seeds of variety *dulcis,* used in the manufacture of oil.

a. cake Crude a. meal. **a. camphor** Benzoin. **a. furnace** A remelting *furnace* of the reverberatory type, q.v. **a. meal** The residue remaining after the epxression of oil from almonds; used in cosmetics, cooking, confectionery, and perfumes, **a. oil artificial ~** Benzaldehyde*. **artificial bitter ~** Nitrobenzene*. **bitter ~** An oil obtained from bitter a. by maceration with water and distillation. Colorless liquid, d.1.045–1.060, b.180, soluble in water; a flavoring. **sweet ~** An oil obtained by expressing sweet a. Yellow liquid, d.0.915–0.920, soluble in alcohol; a perfume and a lubricant for delicate mechanisms.

alneon An alloy of Al with Zn (10–20%) and small amounts of Cu and Ni; used for castings.

Alnus Alnus bark. The astringent bark of the American alder, *A. serrulata* (Betulaceae); contains tannin.

alochrysine $C_{15}H_8O_5$ = 268.2. An oxidation product of barbaloin, obtained with chromic acid.

aloe(s) (1) A genus of plants (Lilaceae); the juices are the aloes of commerce, and the fibers are used for cords and nets. (2) An extremely bitter, black, shining resin from the juice of the leaves of several a. species; a purgative. **American ~** Agave. **Barbados ~** The resin from *A. vulgaris* of Jamaica and Barbados (EP, BP). **Cape ~** The resin from *A. ferox* and other species of S. Africa; used in the manufacture of a brown dye (USP). **hepatic ~** The resin from *A. chinensis* and other species from the Dutch Indies. **Socotrine ~** The resin from *A. perryi* from Socotra.

a. emodin $C_{15}H_{10}O_5$ = 270.2. A crystalline constituent of aloes, cascara sagrada, frangula, and senna leaves. Yellow crystals, m.224; a purgative.

aloeresic acid $C_{30}H_{32}O_{14}$ = 616.6. A yellow-brown, microcrystalline powder from Cape aloes.

aloetic acid $C_{15}H_6O_{13}N_4$ = 450.2. Orange powder obtained by treating aloin with nitric acid.

aloewood Eaglewood, lignum, aloe. A fragrant, resinous heartwood from *Aquilaria agallocha* (Asia). An incense and perfume.

aloin $C_{21}H_{22}O_9$ = 418.4. A neutral, bitter, purgative principle of aloes. Yellow prisms, m.147, soluble in water, named according to its origin: barbaloin, nataloin, etc.

alopecia The disease of hair loss.

Aloxite Trademark for artificial aluminum oxide products made by fusing materials high in alumina (as bauxite). White to red crystals; used as abrasives, filters, and refractories. Cf. *Alundum.*

alpaca The long, silky, lustrous wool from the alpaca llama *(Auchenia paco).*

alpax A light Al eutectic alloy with Si 13%; used for molding, as it has a low shrinkage.

alpha α. First letter of the Greek alphabet. α (not italic) is symbol for α-particle. α (italic) is symbol for (1) angle of optical rotation, (2) cubic and linear expansivity, (3) degree of dissociation, (4) fine structure *constant.* **a. acid** 2,8-Naphthylaminesulfonic acid. **a. cellulose** See *cellulose.* **a. derivatives** *Correctly used:* (1) In conjunctive nomenclature, to indicate a substituent on the carbon atom next to the principal group of a side chain of a cyclic constituent; as, α-chloro-3-quinolineacetic acid, $C_9H_6N \cdot CHCl \cdot COOH$. (2) In sugars, to indicate a group located below the plane of the ring, as α-D-glucopyranose, rather than above it (β). *Formerly used* (1) as in correct usage just shown, but in (1) above not limited to cyclic constituent side chains; as, α-hydroxypropionic acid, Me·

CHOH·COOH. Still commonly used for α-amino acids. (2) A substitution product of a polycyclic compound, in which the substituents are attached to the C atom closest to the C atom shared by both rings; e.g., in naphthalene, the 1, 4, 5, or 8 position. (3) Substitution products of heterocyclic compounds in which the substituent is attached to the C atom closest to the heterocyclic atom; e.g., in pyridine, the 2 or 6 position. (4) The first of a series of derivatives to be discovered; no structural significance. *Note.* For compounds beginning with *alpha-* under the main heading. **a. flock** Finely divided, purified wood cellulose; used as a filler for rubber. **a. particles** Positively charged helium nuclei emitted from radioactive substances at 32,000 km/s. Their mass equals that of a helium atom; when their charge is neutralized by the capture of 2 negative electrons, they form helium atoms. Cf. *alphatopic change.* **a. position** See *alpha derivatives.* **a. rays** A. particles.

alphafetoprotein A protein present in increased amounts in the amniotic fluid in early pregnancy when the fetus has a spina bifida or similar abnormality; or, normally, in multiple pregnancy. See *amniocentesis.* **alphametrin** An insecticide. See Table 45 on p. 305.

alphatopic Pertaining to 2 radioactive elements or isotopes which differ by an α particle and an atomic weight of 4.00. **a. change** The spontaneous disintegration of certain radioactive elements in which an α particle is emitted and the element thereby moves 2 places lower in the periodic table. See *radioactive* elements.

alphazurine An indicator changing at pH 6.0 from purple (acid) to green (alkaline); used to improve the end point of methyl red.

alphyl Alkylphenyl. A radical having both aromatic and aliphatic structures; as, benzyl. Cf. *aralkyl.*

alpinin $C_{17}H_{12}O_6$ = 312.3. A crystalline constituent of galangal, the rhizome of *Alpinia officinarum* (Zingiberaceae).

alquifou Potters' lead. Black lead ore. A mineral zinc sulfide; produces a green pottery glaze.

Alsace gum Dextrin.

alstonite $BaCO_3 \cdot CaCO_3$. Neotype. Barytocalcite.

altaite PbTe. Occurs native in tin ores as white isometric crystals (Central Asia).

Alter, David (1807–1881). An American physician, discoverer of spectrum analysis.

alternaric acid $C_{21}H_{30}O_8$ = 410.5. An antifungistatic substance from *Alternaria solani,* Ell.

alternating current a.c., ac. An electric current, the direction of flow of which periodically changes rapidly. Cf. *alternator, commutator, rectifier.* **pure ~** An a.c. which is constant in frequency and output and gives a true sine-wave curve.

alternation (1) A series of changes following periodically or in turn. **a. of generations** (2) The successive change from asexual to sexual reproduction and back again, observed in low forms of life, as mosses. **a. law** The arc spectra of the elements have alternately odd and even multiplicities when passing from the first to the higher groups of the periodic table. **a. rule** Pauli's principle.

alternator Synchronous generator. A generator of a.c. current.

althea, althaea Marshmallow, hollyhock, The dried roots, flowers, and leaves of *Altheae officinalis* (Malvaceae); formerly an emollient and demulcent.

altheine Asparagine.

althionic acid Ethyl hydrogensulfate*.

altimeter An instrument to measure heights.

altiscope Periscope.

altitude The vertical elevation above any given point. Cf. *coordinates;* in particular above sea level. **a. gage** A barometer for determining a.

altrose $C_6H_{12}O_6$ = 180.2. A hexose, isomeric with glucose.

aludel A pear-shaped vessel open at each end used to connect other vessels. **a. furnace** A furnace used to reduce mercurial ores.

alum (1) Generic for double salts of the general formula $M'_2SO_4 \cdot M'''_2(SO_4)_3 \cdot 24H_2O$, or $M'M'''(SO_4)_2 \cdot 12H_2O$. M′ is monovalent and may be Na, K, Rb, Cs, NH_4, Tl, Ag, hydroxylamine, or the radical of an organic quaternary base, (e.g., NMe_4). M‴ is trivalent and may be Fe, Cr, Al, Mn, In, Tl, Ga, V, Co, Ti, Rh, etc. SeO_4 or TeO_4 may replace SO_4. (2) Double salts of aluminum sulfate and the sulfate of a monovalent metal (M′); hence: $M'_2SO_4 \cdot Al_2(SO_4)_3 \cdot 24H_2O$. (3) The original name of ammonium alum, $(NH_4)_2SO_4 \cdot Al_2(SO_4)_3 \cdot 24H_2O$. Cf. pseudo*alum.* (4) Erroneously, aluminum sulfate; as, papermaker's a. (5) Generally, aluminum potassium sulfate. **ammonium ~** Aluminum ammonium sulfate. **ammonium chrome ~** Ammonium chromium sulfate. **ammonium iron ~** Ammonium iron sulfate. **burnt ~** Aluminum potassium sulfate, dehydrated at 200°. **cesium ~** Aluminum cesium sulfate. **chrome ~** Chromium sodium sulfate. **common ~** Aluminum potassium sulfate. **copper ~** Cupric aluminate. **iron ~** A native ferric potassium sulfate. **manganese ~** Ammonium manganese sulfate. **neutral ~** Alunite. **official ~** Either ammonium or potassium alum (BP). **pearl ~** Specially prepared a. for paper manufacture. **pickel ~** A. prepared for canneries. **porous ~** Aluminum sodium sulfate. **potassium ~** Aluminum potassium sulfate (USP, EP, BP) **potassiumchrome ~** Chromium potassium sulfate. **potassium manganese ~** Manganese potassium sulfate. **pseudo ~** A. containing a divalent in place of the univalent metallic sulfate; as, $MnSO_4 \cdot Al_2(SO_4)_3 \cdot 24H_2O$; not isomorphous with the alums. **roman ~** A native aluminum and iron sulfate from Tolfar, Italy. **rubidium ~** Aluminum rubidium sulfate. **sodium ~** Aluminum sodium sulfate. **thallium ~** Aluminum thallium sulfate. **true ~** A. containing aluminum.

a. flour Aluminum potassium sulfate. **a. hematoxylin solution** A stain (1 pt. hematoxylin, 100 pts. saturated aqueous ammonium alum, 0.5 pt. thymol, 300 pts. water). **a. meal** Aluminum potassium sulfate. **a. root** The root of *Heuchera americana,* which contains gallic acid and tannin; an astringent. **a. shale** A clay containing iron pyrites and aluminum silicate; source of ammonium alum. **a. stone** Alunite.

Alumel Trademark for an alloy: Ni 94, Mn 2.5, Al 2.0, Si 1.0, Fe 0.5%; used in thermocouples.

alumian $Al_2O(SO_4)_2$. A white, native, basic aluminum sulfate.

alumina Aluminum oxide, Al_2O_3. α-Corundum. **ferric ~** A colloidal solution of $Al(OH)_3$ and $FeCl_3$; a coagulant in water purification. **lime ~** Essonite. **natural ~** Corundum. Cf. *ruby, sapphire.*

a. cream The hydroxides of aluminum. A clarifying agent, used in optical experiments. **a. mordant** A fixing agent used in dyeing; usually aluminum compounds. **a. white** Soluble a., transparent a., a. hydrate; a white pigment in paints and inks.

aluminate Indicating aluminum as the central atom(s) in an anion, as (1) Compounds derived from $Al(OH)_3$. **ortho ~** M_3AlO_3, as, Na_3AlO_3, sodium aluminate. **meta ~** $MAlO_2$. Salts existing only in solution; as, AlO_2^- ion. (2) A combination of Al_2O_3 with a metallic oxide; as, $MgAl_2O_4$, spinel.

aluminic acid H_3AlO_3. A hypothetical tautomer of aluminum hydroxide, $Al(OH)_3 \rightleftharpoons H_3AlO_3$.
aluminiferous A rock yielding or containing Al.
aluminite $Al_2(SO_4)(OH)_4 \cdot 7H_2O$. Native, soft, white, monoclinic hydrous aluminum sulfate. Cf. *websterite.*
aluminium Aluminum* (U.K. usage).
aluminoferric A mixture of aluminum sulfate and a ferrous salt, used to coagulate sewage.
aluminum* Al = 26.98154. Aluminium. Metal and element of at. no. 13. One of the most abundant metals, isolated in an impure form by Oersted (1825), and in the pure form by Wöhler (1827) from sodium and aluminum trichloride. Silver-white, light, ductile metal, d_{20} 2.70, m.660, b.2460, soluble in acids or alkalies; readily oxidized and covered with a fine protective film of aluminum oxide. Used extensively for cooking utensils, a. foil, airplanes, boats, automobiles. A. forms many important alloys, e.g., *magnalium, electron, Duralumin.* A. has a valency of 3 and forms one series of compounds. Aluminates are derived from aluminum hydroxide (aluminum valency 3):

Aluminum ion . . . . Al^{3+}
Orthoaluminate . . . . AlO_3^{3-}
Metaaluminate . . . . AlO_2^-
Alums . . . . $M'Al(SO_4)_2 \cdot 12H_2O$

primary ~ Virgin a. A. obtained direct from ore. **secondary ~** A. from scrap metal. **triethyl ~*** Et_3Al = 114.1. A. ethyl, a. triethide. Colorless liquid, b.190; decomp. explosively in water to ethane and a. hydroxide. **trimethyl ~*** Me_3Al = 72.1. A. methyl, a. trimethide. Colorless liquid, b.130; decomp. in water to methane and a. hydroxide. **virgin ~** Primary a.

a. acetate $(CH_3COO)_3Al$ = 204.1. Amorphous white powder, decomp. by heat, soluble in water. A 5% aqueous solution is used as a gargle, astringent, or antiseptic (USP). **a. acetate, basic** $Al(C_2H_3O_2)_2OH$ = 162.1. A. subacetate. White crystals; used as a mordant and disinfectant, and for embalming. **a. ammonium sulfate*** $Al_2(NH_4)_2(SO_4)_4 \cdot 24H_2O$ = 906.6. Ammonium alum. Colorless, regular crystals, m.94, soluble in water. Used for water purification and in baking powder, foam fire extinguishers, and electroplating; an astringent. **a. alkyls** Compounds of aluminum and alkyl radicals, e.g., trimethylaluminums. **a. alloys** Some m. (in °C) are:

With	90% Al 10%	80% Al 20%	70% Al 30%	60% Al 40%	50% Al 50%
Ag	625	615	600	590	580
Au	675	740	800	855	915
Cu	630	600	560	540	580
Fe	860	1015	1110	1145	1145
Sb	750	840	925	945	950
Sn	645	635	625	620	605

a. arsenate* $AlAsO_4$ = 165.9. White powder, soluble in acids. **a. benzoate*** $Al(C_7H_5O_2)_3$ = 390.3. White crystals, soluble in water. **a. bifluoride** $(Al_2F_6)_3(HF)_4 \cdot 10H_2O$. A. acid fluoride. White crystals, soluble in water. **a. borate** $2Al_2O_3 \cdot B_2O_3 \cdot 3H_2O$ = 327.6. White granules, soluble in water; used in glass and porcelain. **a. bromate*** $Al(BrO_3)_3 \cdot 9H_2O$ = 572.8. Colorless, hygroscopic crystals, m.62, decomp. 100, soluble in water. **a. bromide*** $AlBr_3 \cdot 6H_2O$ = 374.8. Colorless or yellow hygroscopic crystals, m.93, soluble in water; used in organic synthesis. **anhydrous ~** $AlBr_3$ = 266.7. Yellow fuming scales; used in organic synthesis. **a. bronze** An alloy: Cu 90, Al 10%, m.1050. **a. butoxide** $Al(OC_4H_9)_3$ = 246.3. White powder, m.102, decomp. in water. **a. carbide*** Al_4C_3 = 144.0. Yellow hexagons, decomp. in water; used to generate methane. **a. carbonate*** $Al_2(CO_3)_3$ = 234.0. White lumps, insoluble in water; a mild antiseptic. **a. caseinate** Yellow powder, insoluble in water. **a. cesium rubidium sulfate*** $Al_2CsRb(SO_4)_4 \cdot 24H_2O$ = 1089. Cesium rubidium alum. Colorless crystals, soluble in water. **a. cesium sulfate*** $AlCs(SO_4)_2 \cdot 12H_2O$ = 568.2. Cesium alum. Colorless crystals, m.117, slightly soluble in water. **a. chloride*** $AlCl_3 \cdot 6H_2O$ = 241.4. Colorless crystals, soluble in water. **anhydrous ~** $AlCl_3$ = 133.3. Yellow crystals, m.180, soluble in water. Used in Friedel-Crafts reaction; as an antiperspirant (USP); industrially as a catalytic agent and in refining petroleum; and as a reagent for naphthalene. **a. dichromate*** $Al_2(Cr_2O_7)_3$ = 701.9. Red crystals, soluble in water. **a. ethoxide** $Al(OC_2H_5)_3$ = 162.2. Triethoxyaluminum. White powder, m.134, soluble in hot water; a reagent for water. **a. ethyl** Triethylaluminum*. **a. fluoride*** $AlF_3 \cdot 3\frac{1}{2}H_2O$ = 143.0. Colorless crystals, slightly soluble in water, losing 2 H_2O at 120. **anhydrous ~** AlF_3 = 84.0. White powder, soluble in water; used in the glass industry. **a. fluorosilicate** A. hexafluorosilicate*. **a. gluconate** See *gluconate.* **A. hexafluorosilicate(IV)*** $Al_2(SiF_6)_3$ = 480.2. A. fluo(ro)silicate, a. silicofluoride. White powder, insoluble in water; used in the glass and enamel industries. **a. hydrate, a. hydroxide*** $Al(OH)_3$ = 78.0. White powder, insoluble in water; occurs native as gibbsite, hydrargylite, and zirlite. With bases it forms salts; see *aluminic* acid. Used as an antacid and demulcent for indigestion and peptic ulcer (USP, BP); an astringent, dusting powder, mordant, filter aid, and neutralizing agent. **basic ~** Includes AlO(OH), e.g., *boehmite;* $Al_2O(OH)_4$, e.g., *bauxite;* also mixtures with the normal hydroxide. **a. hydroxide gel** An aqueous suspension of a. h. and alumina equivalent to 3.6–4.4% $Al(OH)_3$, with preservative and flavoring; an antacid. **a. iodide*** $AlI_3 \cdot 6H_2O$ = 515.8. White crystals, soluble in water. **anhydrous ~** AlI_3 = 407.7. Brown crystals, m.180, soluble in water; used in organic synthesis. **a. lactate*** $Al(C_3H_5O_3)_3$ = 294.2. Yellow powder, soluble in water. **a. minerals** A. is the most abundant metal and is present in all rocks (except the limestones and sandstones), chiefly in silicates, e.g., feldspar, clays, micas, sillimanite,

Bauxite . . . . $Al_2O_3 \cdot 2H_2O$
Bayerite . . . . $Al_2O_3 \cdot H_2O$
Boehmite, diaspore . . . . $AlO \cdot OH$
Corundum . . . . Al_2O_3
Cryolite . . . . $AlF_3 \cdot 3NaF$
Gibbsite . . . . $Al(OH)_3$
Hydragillite . . . . $Al_2O_3 \cdot 3H_2O$

andalusite, cyanite. The principal ores are corundum, bauxite, and cryolite. **a. nitrate*** $Al(NO_3)_3 \cdot 9H_2O$ = 375.1. Colorless, rhombs, m.73, decomp. 150, soluble in water. **anhydrous ~** $Al(NO_3)_3$ = 213.0. White powder, soluble in water; used in the textile and leather industries. **a. nitride*** AlN = 41.0. Yellow crystals, m.2150, decomp. by water. **a. oleate*** $Al(C_{18}H_{33}O_2)_3$ = 871.4. White powder, soluble in alcohol. Used in the manufacture of waterproofing materials, and as a paint drier. **a. oxalate*** $Al_2(COO)_6$ = 318.0. White powder, insoluble in water. **a. oxide** Al_2O_3 = 101.0. Alumina, corundum. White powder, or colorless hexagons, d.3.75, m.2020, insoluble in water. Crystalline forms: α- or corundum (the common form); β-, a mixed oxide, $M_2O.Al_2O_3$, with alkali metals; γ-, by heating over 1000°C. An abrasive, refractory, and filtering material. Cf. *Alundum, borolon.* **a. palmitate***

$Al(C_{16}H_{31}O_2)_3$ = 793.2. White granules, insoluble in water; a lubricant and waterproofing material. **a. phenolate*** $Al(OC_6H_5)_3$ = 306.3. A. phenoxide. Colorless powder, m.265, decomp. in water. **a. phosphate** $AlPO_4$ = 122.1. Colorless hexagons or white powder, insoluble in water. Occurs in native variscite, sphaerite, and turquoise. Used in pharmacy, as a cement, and in ceramics. **a. phosphinate*** $Al(H_2PO_2)_3$ = 221.9. A. hypophosphite. Colorless crystals, soluble in water. **a. potassium chloride*** $AlCl_3$. KCl = 207.9. White crystals, soluble in water. **a. potassium sulfate*** $Al_2K_2(SO_4)_4 \cdot 24H_2O$ = 948.7. Potassium alum, potash, alum, common alum, kalinite. White powder or transparent cubes, m.84, soluble in water; used as a mordant, in glue and cements. **a. potassium tartrate** $KAl(C_4H_4O_6)_3$ = 510.3. White powder, soluble in water. **a. powder** A medicinal protective paste (BP). **a. propoxide** $Al(OC_3H_7)_3$ = 204.2. White powder, m.106, b.248, decomp. in water. **a. rectifier** An apparatus to convert an alternating current into a pulsating direct current when the former is connected to 2 aluminum plates in sodium bicarbonate solution. **a. resinate** $Al(C_{44}H_{63}O_5)_3$ = 2043. Brown soft mass, insoluble in water; used for waterproofing and as a drier in varnish. **a. rubidium sulfate*** $AlRb(SO_4)_2 \cdot 12H_2O$ = 520.7. Rubidium alum. Colorless crystals, m.99, slightly soluble in water. **a. salicylate*** $Al(C_6H_4OH \cdot COO)_3$ = 438.3. Salamin. Pink powder, insoluble in water, soluble in alkalies. **a. silicate*** $Al_2(SiO_3)_3$ = 282.2. White masses, insoluble in water; used in the glass industry as a refractory lining. Natural forms are cyanite, andalusite, sillimanite, mullite. Cf. *kaolin.* **a. sodium chloride*** $AlCl_3 \cdot NaCl$ = 191.7. Colorless crystals, m.183, soluble in water. **a. sodium fluoride*** $AlF_3 \cdot 3NaF$ = 209.9. Colorless crystals, sparingly soluble in water. **a. sodium sulfate*** $AlNa(SO_4)_2 \cdot 12H_2O$ = 458.3. Sodium alum. Colorless rhombs, m.61, soluble in water; used for water purification, **a. stearate*** $Al(C_{18}H_{35}O_2)_3$ = 877.4. Obtained by saponifying animal fat and treatment of the soap with alum. Gray mass, soluble in warm water. Used in lubricating and waterproofing compounds. **a. sulfate*** $Al_2(SO_4)_3 \cdot 18H_2O$ = 666.4. Colorless, monoclinic crystals, decomp. by heat, soluble in water; used as alum. **anhydrous ~** $Al_2(SO_4)_3$ = 342.1. White crystals, decomp. 77, soluble in water. Used as a reagent for dyes; in wine; as an antiseptic, caustic, or astringent (USP, EP, BP); and in the leather, paper, and dye industries. **a. sulfide*** Al_2S_3 = 150.1. Yellow crystals, m.1100, decomp. by water or acids. **a. tartrate*** $Al_2(C_4H_4O_6)_3$ = 498.2. White powder, insoluble in water, soluble in ammonia or acids. **a. thallium sulfate*** $AlTl(SO_4)_2 \cdot 12H_2O$ = 639.6. Thallium alum. Colorless crystals, slightly soluble in water. **a. thiocyanate*** $Al(SCN)_3$ = 201.2. A. sulfocyanide, a. rhodanate. Yellow powder, slightly soluble in water; used in the textile industry. **a. zinc sulfate*** $Al_2Zn(SO_4)_4$ = 503.6. Zinc alum. Colorless crystals, powder, or sticks; soluble in water; a caustic.

alumite Alunite.

alums See *alum.*

Alundum Trademark for a pure, crystalline, granular aluminum oxide, d.3.9–4.0, m.2050. Used as reagent for the determination of carbon in steel; as an abrasive, a basic refractory, filtering material; and for chemical apparatus. **A. cement** A carrier for catalysts; prepared by heating Al_2O_3 90, SiO_2 5, CaO 5 pts. **A. crucible** (1) A highly refractive crucible for the electric furnace, used for melting platinum. (2) A porous filtering crucible. **A. filter cone** An A. filter cone which fits into a funnel. **A. muffle** An A. muffle for metal assay.

alunite $Al_2(SO_4)_3(Na, K) \cdot 6(OH)$. Alumite. Neutral alum. Alum stone. Native, white rhombic hexagons; hardness 3.5–4.0; sparingly soluble in water.

alunogen $Al_2(SO_4)_3 \cdot 18H_2O$. Native, silky-white, monoclinic crystals.

Alurate Trademark for a brand of allylisopropylbarbituric acid; an oral sedative-hypnotic.

alveograph An instrument for evaluating the qualities of baking dough. A disk of nonyeasted dough is clamped across an air inlet, and the pressure required to burst it is measured.

alveolus Extremity of respiratory tract, located in the lung and consisting of single-cell lined sacs (about 3×10^8 cells/lung), where O_2/CO_2 exchange occurs with the blood.

alvite (ZnHfTh)SiO_4. A mineral.

Am (1) Abbreviation for NH_4 group (ammonium). (2) Symbol for americium.

amadou Surgeons' agaric, punk, tinder. The fungus *Boletus igniarius,* found on old tree trunks; used as tinder, and as a styptic.

amalgam (1) An alloy of mercury, generally solid or semiliquid; e.g., sodium a. (2) An alloy of silver and mercury. **dental ~** See *dental amalgam.* **native ~** An alloy of mercury with silver or gold; e.g., $AuHg_3$, occurring as minerals, as *arquerite.*

amalgamation The formation of an alloy of a metal with mercury. **a. process** A method of extracting noble metals from ores by alloying them with mercury.

amalgamator An apparatus to extract gold from ores.

amalic acid Amalinic acid.

amalinic acid $C_{12}H_{17}O_8N_4$ = 345.3. Amalic acid, tetramethylalloxantin. Colorless crystals, m.245, soluble in hot water.

amandin A globulin from fruit kernels, e.g., peach seeds.

Amanita A genus of fungi (mushrooms), many of which are poisonous; as, *A. phalloides,* death head fungus, death cap, containing *phallin,* q.v.; *A. muscaria,* fly agaric, fly amanita, containing muscarine; *A. verna,* the "destroying angel." **a. toxin** A protein from *A. phalloides,* death head fungus; a cause of mushroom poisoning.

amanitine $Me_3N(OH)CHOH \cdot Me$ = 121.2. (1) An alkaloid of *Amanita* species, as, *A. pantherina* and *A. muscaria;* identical with neurine. (2) $Me_3N(OH)CH_2CH_2OH$ = 121.1. Isocholine. An alkaloid from mushrooms *(Agaricus muscaria),* oxidized to muscarine. Cf. *choline.*

amaranth $NaSO_3 \cdot C_{10}H_6 \cdot N{:}N \cdot C_{10}H_4(SO_3Na)_2OH$ = 604.5. C.I. Food Red 9. A red dye used as a food color.

Amaranthaceae A family of plants comprising 40 genera and 500 species of weeds, pot plants, and fodder plants.

amargosa bark The root bark of the goatbush, *Castela nicholsoni* (Simarubaceae); a source of castelamarin.

amarine $C_{21}H_{18}N_2 \cdot \frac{1}{2}H_2O$ = 303.4. Triphenyldihydroimidazole. Bitter, colorless prisms, anhydrous at 130, insoluble in water, formed by the action of ammonia on benzoic acid; occurs in oil of bitter almonds.

amaroid Bitter principles, in plants other than alkaloids, glucosides, or tannins; as, quassin, chamomillin.

amaron $C_{28}H_{20}N_2$ = 384.5. Tetraphenylpyrazine, benzoin imide. Colorless needles, m.245, insoluble in water.

Amaryllidaceae A family of plants comprising about 75 genera and 700 species. Many are ornamental tropical and subtropical plants; as narcissus, agave. Cf. *lycorine.*

amatol An explosive (80 pts. ammonium nitrate, 20 pts. trinitrotoluene) used in coal mines.

amazonite Amazon stone, microcline. A bright green aluminum potassium silicate; precious stone.

amber $C_{10}H_{16}O$ (?). Electrum, succinum. A fossilized, bituminous resin, d.1.1; brittle, transparent brown masses. Used as a precious stone and in experiments on static electricity. It occurs as succinite and gedanite. **artificial ~** Rosin. **Baltic ~** Succinite. **Canadian ~** Chernawinite. **synthetic ~** A formaldehyde-phenol or -urea resin.
a. acid Succinic acid. **a. oil** A brown essential oil of empyreumatic and balsamic odor distilled from a., d.0.915–0.975, miscible with alcohol. **a. seed** Musk seed.

ambergris The opaque, gray wax formed in the intestines of the sperm whale, d.0.7–0.92, m.60, b.100, soluble in alcohol or ether; used in perfumery. Cf. *ambrein, spermaceti.*

Amberg swimming cup A perforated porcelain tumbler with cork stopper, used for washing microscope specimens by floating them in a solvent.

amberite (1) A smokeless powder composed of guncotton, paraffin, and barium nitrate. (2) Compressed amber scrap used for electrical insulation.

Amberlite Trademark for certain ion-exchange resins.

amblygonite $LiAlFPO_4$. Native, green, triclinic crystals from California.

amblystegite Enstatite.

amboyna wood A wood, *Pterospernum indicum,* used in southeast Asia for inlay work.

ambrein $C_{23}H_{40}O$ = 332.6. A cholesterin from ambergris.

ambrette Musk seed.

ambrettolic acid $Me(CH_2)_7CH{:}COH \cdot (CH_2)_5 \cdot COOH$ = 270.4. 7-Hydroxy-7-hexadecenoic acid*.

ambrion An insulating material; asbestos impregnated with pitch.

ambrite A fossilized resin (New Zealand); greasy, yellow masses, sometimes used for semiprecious stones.

Ambrosia A genus of composite-flowered herbs, whose pollen causes hay fever, as *A. artemisifolia,* or ragweed.

ameba Amoeba. (1) A unicellular animal without definite shape, resembling a mass of moving jelly, with all the features of life: growth, reproduction, metabolism, locomotion. (2) (cap. A). A genus of protozoa of the family Amebidae. Some a. cause disease; as, amebic dysentery.

amebicide An agent that destroys amebas.

ameboid Assuming various shapes, like a moving ameba.

amelairoside Piceoside. A glucoside from *Amelancher vulgaris* (Rosaceae).

American **A. aloe** Agave. **A. ginseng** Ginseng. **A. hellebore** Veratrum. **A. ipecac** Gillenia. **A. melting point** An arbitrary value for the m. of paraffin wax which is 3°F. higher than the standard m. as determined by the ASTM method. **A. saffron** Carthamus. **A. veratrum** Veratrum. **A. wormseed oil** Chenopodium oil.

American **A. Chemical Society** Organized in 1876. Publishers of the *Journal of the Am. Chem. Soc., Chemical Abstracts, Industrial and Engineering Chemistry,* and several specialized journals. Secretary, 1155 16th St., N.W., Washington, D.C. **A. National Standards Institute** ANSI. Formerly United States of America Standards Institute (USASI), and before that, American Standards Association (ASA). **A. Society for Testing and Materials** See *ASTM.*

americium* Am. Atomic weight of isotope with longest half-life (^{243}Am) = 243.06, m.1000, b.2600. Forms Am^{3+}, Am^{4+}, AmO_2^{+}, and AmO_2^{2+} compounds. Cf. *curium.*

amesite A ferromagnesium aluminum silicate.

Ames moisture tester A paraffin-jacketed container for the determination of moisture in butter.

amethocaine hydrochloride BP name for tetracaine hydrochloride.

amethyst A purple, quartz precious stone from India, Brazil. **oriental ~** Al_2O_3. A purple, native alumina. **true ~** Oriental a.

amianthus Earth flax, mountain flax. A fine, silky asbestos.

Amicar Trademark for aminocaproic acid.

amicron A particle of diameter less than 5 nm. See *micron.*

amidase* An enzyme which splits ammonia from a carboxylic acid amide. Cf. *urease.*

amidation The process of forming an amide, by (1) heating of the corresponding ammonium salt, (2) reaction of ammonia on an acid chloride, (3) reaction of ammonia with an ester. Cf. *amination.*

amide* (1) An organic compound containing the $-CO \cdot NH_2$ radical; as, formamide, $H \cdot CO \cdot NH_2$; benzamide, $Ph \cdot CO \cdot NH_2$. A. are derived from acids by replacement of $-OH$ by $-NH_2$; as, $-COOH \rightarrow -CONH_2$; or from ammonia by the replacement of H by an acyl group: $NH_3 \rightarrow NH_2 \cdot OCR$. (2) Ammonobases. Compounds in which H of NH_3 is replaced by a metal; as, sodium amide, $NaNH_2$. **alkyl ~** $R \cdot CO \cdot NHR$. *N*-compounds obtained by treating an acid chloride or acid anhydride with an amine and sodium hyodroxide. Cf. *anilides.* **hydroxy ~** A compound containing both the $-OH$ and $-CONH_2$ groups. **oxo ~** A compound containing the radical $-CO \cdot CONH_2$. **thio ~** A compound containing the $-CSNH_2$ group.
a. chloride A chlorinated amide, $R \cdot CCl_2 \cdot NR_2$ derived from an alkyl a., which changes readily to the imine chloride, $R \cdot C{:}NH \cdot Cl$. **a. group** The carbamoyl group, $-CO \cdot NH_2$, which confers weakly basic properties. **a. oxime*** Amidoxime, oxamidine. A compound containing the a. group, $-C({:}NOH)NH_2$; e.g., formamide oxime, $HC(NOH)NH_2$.

amidin A transparent solution of starch in water.

amidines* Carboxamidines*. Compounds containing the radical $-C({:}NH)NH_2$; as, formamidine, $H \cdot C({:}NH)NH_2$. Derived from the amides by replacement of O by the imino residue, $>NH$ or $>NR$. Cf. *phasotropy.* **di ~** Compounds containing 2 amidine groups; e.g., oxalamidine, $NH_2(HN{:})C \cdot C({:}NH)NH_2$.

amidino* Aminoiminomethyl†, guanyl. The radical $H_2N(HN{:})C-$. **a.urea** $NH_2CO \cdot NH \cdot C({:}NH)NH_2$ = 102.1.

amido* (1) Infix indicating groups of the type $R \cdot CO \cdot NH-$, $R' \cdot CO \cdot NR^2-$, or (sulfon-a.) $R \cdot SO_2 \cdot NH-$. (2) Amino-*.
a.oxime Amide oxime*.

amidogen Amido.

amidol $(NH_2)_2C_6H_3OH \cdot HCl$ = 160.6. 2,4-Diaminophenol hydrochloride. Colorless crystals, slightly soluble in water; a photographic developer.

amidone Methadone hydrochloride.

amidoxalyl Oxamoyl*.

amidoxime Amide oxime*.

amidrazone* A compound of formula $RC(NH_2){:}N \cdot NH_2$ or $RC({:}NH) \cdot NH \cdot NH_2$; as, benzamide hydrazone, $PhC(NH_2){:}N \cdot NH_2$. See *hydrazidine.*

amigen A protein hydrolysis product used in amino acid therapy.

Amilan Trade name for a polyamide synthetic fiber.

Amilar Trade name for a polyester synthetic fiber.

amiloride hydrochloride $C_6H_8ON_7Cl \cdot HCl$ = 266.1. Pale yellow powder, soluble in water. A diuretic (BP).

amination The formation of an amine by (1) reduction of a nitro compound, (2) reduction of a cyanide, (3) oxidation of an amide, (4) treatment of isocyanate with alkali. Cf. *amidation.*

amine (1) See *amines.* (2) Suffix indicating an $-NH_2$ group.
a. oxidase* Monoamine oxidase.

amines A group of compounds derived from ammonia by substituting organic radicals for the hydrogens; as, H_2NR, primary*; RHNR, secondary*; R_2NR, tertiary*; R_4NOH, quaternary (ammonium). Cf. *amino, amide, arsine, phosphine.* **di ~** A. containing 2 NH_2 groups. **diacyl ~*** Symmetrical compounds of formula $HN(CO \cdot R)_2$. **filming ~** A. which form monomolecular films on hot surfaces, thereby promoting drop-type condensation; e.g., the use of *n*-octadecylamine inside drying cylinders. **hydroxynitroso ~** Compounds containing the radical $-N \cdot NO(OH)$. **metallic ~** An *amide* (2), q.v. **neutralizing ~** A. used to neutralize boiler feed water, the excess being steam-volatile; as, cyclohexamine. **pressor ~** A protein derivative having vasoconstrictory effects; as, tyramine. **primary ~*** Compounds in which 1 H is replaced by a radical; e.g., NH_2CH_3. **quarternary ~** Tetraalkyl ammonium bases. Compounds derived from ammonium hydroxide containing 4 radicals; e.g., $N(CH_3)_4OH$. **secondary ~*** Compounds in which 2 H are replaced by a radical; e.g., $NH(CH_3)_2$. **tertiary ~*** Compounds in which 3 H are replaced by radicals; e.g., $N(CH_3)_3$. **thionyl ~** A compound containing the $-N{:}SO$ radical. **tri ~*** A compound containing 3 NH_2 groups. **triacyl ~*** Symmetrical compounds of formula $N(CO \cdot R)_3$.

aminic acid Formic acid*.

amino* The $-NH_2$ group indicated by the prefix *amino** or suffix *amine**; as, aminomethane or methylamine. For a list of principal amino acids, see Table 8, next page. **a.acetal** $NH_2 \cdot CH_2 \cdot CH(OEt)_2$ = 133.2. Colorless needles m.163, soluble in water. **a.acetanilide** $NH_2 \cdot C_6H_4 \cdot NHCOMe$ = 150.2. Acetylphenylenediamine. Colorless needles. ***ortho-* ~** m.165. ***meta-* ~** m.90. ***para-* ~** m.160, slightly soluble in water. **a.acetic acid** Glycine*. **a.acetone** NH_2CH_2COMe = 73.1. Colorless needles, m.188 (decomp.), soluble in water. **a.acetophenone** $NH_2 \cdot C_6H_4 \cdot COMe$ = 135.2. *p*-Aminophenyl methyl ketone. Yellow powder, m.105, soluble in water. ***ortho-* ~** b.251. **a.acetyl†** See glycyl. **a. acids** See *amino acids.* **a. alcohols** Alcamines. **a.anthraquinone** See *anthraquinone.* **a.azobenzene** $PhN{:}NC_6H_4NH_2$ = 197.2. **4- ~** Aniline yellow. Yellow needles, m.125, slightly soluble in water. An intermediate in the preparation of dyes and medicinals; an indicator changing at pH 2.5 from orange (acid) to yellow (alkaline). **a.azobenzene chlorhydrate** $C_{12}H_{11}N_3 \cdot HCl$ = 233.7. Blue needles, slightly soluble in water; used in dye manufacture. **a.azonaphthalene*** $C_{10}H_7N{:}NC_{10}H_6NH_2$ = 297.4. **2-amino-1,1′-azonaphthalene** Red needles, m.184, slightly soluble in alcohol. **4-amino-1,1′-azonaphthalene** Red needles, m.159. **a.azotoluene*** $MeC_6H_4N{:}NC_6H_3(NH_2)Me$ = 225.3. **4-amino-2,2′-azotoluene.** Red crystals, m.100, insoluble in water. **a.azotoluene hydrochloride** $MeC_6H_4N{:}NC_6H_3(NH_2)Me \cdot HCl$. Colorless crystals, soluble in water. Its 4 isomeric forms are used in organic synthesis. **a.barbituric acid** Uramil. **a.benzaldehyde** See *benzaldehyde.* **a.benzamide** See *benzamide.* **a.benzene** Aniline*. **a.benzenesulfonic acid *ortho-* ~** Orthanilic acid. ***meta-* ~** Metanilic acid. ***para-N-* ~** Sulfanilic acid. **a.benzoic acids** See *benzoic acid.* **2-a.benzoyl†** See *anthraniloyl.* **a.benzoylformic acid** Isatic acid. **a.caproic acid** A. hexanoic acid*. **a.carbonyl†** See *carbamoyl.* **(a.carbonyl)amino†** See *ureido.* **2-(a.carbonyl)hydrazino†** See *semicarbazido.* **a.dracilic acid** See *p*-amino*benzoic acid.* **a.ethanol** Ethanolamine. **a.ethionic acid** Taurine. **(2-a.ethyl)sulfonyl†** See *tauryl.* **a. F acid** 2-Naphthylamine-7-sulfonic acid. **a.formic acid** See *carbamic acid.* **a. G acid** 2-Naphthylamine-6,8-disulfonic acid. **a.glutaric acid*** Glutamic acid*. **a.hexanoic acid*** **6-** $C_6H_{13}O_2N$ = 131.2. (ε-)A.caproic acid (USP, BP), EACA, Amicar, Epsikapron. White crystals, soluble in water. Inhibits fibrinolysis; used to treat and prevent some hemorrhages. **a.iminomethyl†** See *amidino.* **(a.iminomethyl)amino†** See *guanidino.* **a.ketone** A compound containing both the a. and carbonyl groups; as, NH_2CH_2COMe, aminoacetone. **a.mandelic acid lactam** Dioxindole. **a.naphtholdisulfonic acid** H acid. **2-a.-1-oxopropyl†** See *alanyl.* **a.phenol** $NH_2 \cdot C_6H_4OH$ = 109.1 **1,2- ~** Aminophenol base. Colorless rhombs, m.179, slightly soluble in water. **1,3- ~** Colorless crystals, m.123, slightly soluble in water. **1,4- ~** Rodinal. Colorless leaflets, m.184 (decomp.), slightly soluble in water. A developer in photography and reducing agent in the dye industry. **a.phosphoric acid*** Phosphamic acid. **a.phthalic hydrazide** Luminol. **a.propionic acid** Alanine*. **a.purine** Adenine*. **2-a.pyridine*** $C_5H_6N_2$ = 94.1. Yellow lumps, m.55 (sublimes), soluble in water; an intermediate in organic syntheses. **a.quinoline*** $C_9H_6N \cdot NH_2$ = 144.2. 7 isomers. **a. R acid** 2-Naphthylamine-3,6-disulfonic acid. **a.salicylate sodium** $C_7H_6O_3NNa \cdot 2H_2O$ = 211.1. Sodium aminosalicylate. Cream crystals, with sweet saline taste, soluble in water (slow decomp.); a tuberculostatic antibacterial (USP, BP). **a.succinamic acid** Asparagine*. **a.sulfuric acid** Sulfamic acid*. **a.thiophene** Thiophenine. **a.toluene** Toluidine*. **a.urea** Semicarbazide*. **a.xylene** Xylidine.

amino acids* $NH_2-R-COOH$. R is usually an aliphatic radical. A. a. have both basic and acidic properties and form the units of peptides and proteins. For IUPAC symbols, see Table 8 on p. 32. **alpha ~** or **primary ~** $NH_2 \cdot CHR \cdot COOH$. The commoner amino acids. **beta ~** or **secondary ~** $NH_2 \cdot CHR \cdot CH_2 \cdot COOH$. **essential ~** 8 a. a. that are essential for protein synthesis in the body, normal health, and growth of children (see Table 8 on p. 32 for meaning of symbols): Ile, Leu, Lys, Met, Phe, Thr, Trp, Val. They are not synthesized by the body, so must be obtained from food. **gamma ~** or **tertiary ~** $NH_2 \cdot CHR \cdot CH_2 \cdot CH_2 \cdot COOH$. **delta ~** or **quaternary ~** $NH_2CHR \cdot CH_2CH_2 \cdot CH_2 \cdot COOH$. **glucogenic ~** A. a. (as, aspartic, glutamic, cystine) that are metabolized to other than acetyl-CoA (e.g., to α-oxoglutarate or pyruvate). **ketogenic ~** A. a. (as, leucine, tyrosine, phenylalanine) that are metabolized to acetyl-CoA.

a. residue See *peptide.*

aminoacrine hydrochloride $C_{13}H_{10}N_2 \cdot HCl \cdot H_2O$ = 248.7. 5-Aminoacridine hydrochloride. Yellow, bitter crystals, slightly soluble in water.

aminoglycosides Bacteriocidal antibiotics derived from *Streptomyces* and *Microspora.* Contain 2-deoxystreptamine and act by interfering with protein synthesis of ribosomes in bacteria. Are active against Gram-negative bacilli. Examples include streptomycin, neomycin, kanamycin, gentamicin.

aminohippuric acid $C_9H_{10}N_2O_3$ = 194.2. White crystals, discolor in light, m.198, soluble in water; used to assess renal function (USP).

aminoid A protein hydrolysis product used in amino acid therapy.

aminoketones Aromatic compounds containing the $-NHC(O)R$ radical; e.g., formanilide, $C_6H_5 \cdot NHCHO$. Cf. *peptide.*

aminooxy* The radical $H_2N \cdot O-$.

aminophenols Aromatic compounds containing both the $-OH$ and $-NH_2$ groups attached to the benzene ring; e.g., aminophenol, $HO \cdot C_6H_4 \cdot NH_2$.

aminophylline $C_{16}H_{24}N_{10}O_4 \cdot 2HO$ = 420.4. Theophylline ethylenediamine. Phyllocontin, Somophyllin. Yellow, bitter powder, with ammoniacal odor, soluble in water (decomp.).

TABLE 8. PRINCIPAL AMINO ACIDS

Abbreviation	Name	Formula
	1. *Monoaminomonocarboxylic acids:* $NH_2\cdot R\cdot COOH$	
	Carbamic acid	$NH_2\cdot COOH$
Gly	Glycine, aminoethanoic acid	$NH_2\cdot CH_2\cdot COOH$
Ala	Alanine, 2-aminopropanoic acid	$NH_2\cdot CHMe\cdot COOH$
Val	Valine, 2-amino-3-dimethylbutanoic acid	$NH_2\cdot CH\cdot COOH(CHMe_2)$
Leu	Leucine, 2-amino-4-methylpentanoic acid	$NH_2\cdot CH\cdot COOH(CH_2\cdot CHMe_2)$
Ile	Isovaline, 2-amino-2-methylbutanoic acid	$NH_2\cdot CMe\cdot COOH(CH_2Me)$
Phe	Phenylalanine, 2-amino-3-phenylpropanoic acid	$NH_2\cdot CH\cdot COOH(CH_2Ph)$
Tyr	Tyrosine, 2-amino-3-*p*-hydroxyphenylpropanoic acid	$NH_2\cdot CH\cdot COOH(CH_2\cdot PhOH)$
Ser	Serine, 2-amino-3-hydroxypropanoic acid	$NH_2\cdot CH\cdot COOH(CH_2OH)$
Cys	Cysteine, 2-amino-3-mercaptopropanoic acid	$NH_2\cdot CH\cdot COOH(CH_2SH)$
Met	Methionine, 2-amino-4-methylthiobutanoic acid	$NH_2\cdot CH\cdot COOH[(CH_2)_2SMe)]$
Ile	Isoleucine, 2-amino-3-methylpentanoic acid	$NH_2\cdot CH\cdot COOH(CHMe\cdot Et)$
Thr	Threonine, 2-amino-3-hydroxybutanoic acid	$NH_2\cdot CH\cdot COOH(CHOH\cdot Me)$
	2. *Monoaminodicarboxylic acids:* $NH_2\cdot R(COOH)_2$	
Asp	Aspartic acid, 2-aminobutanedioic acid	$NH_2\cdot CH\cdot COOH(CH_2\cdot COOH)$
Glu	Glutamic acid, 2-aminopentanedioic acid	$NH_2\cdot CH\cdot COOH[(CH_2)_2COOH]$
	3. *Diaminomonocarboxylic acids:* $(NH_2)_2R\cdot COOH$	
Arg	Arginine, 2-amino-5-guanidinovaleric acid	$NH_2\cdot CH\cdot COOH[(CH_2)_3\cdot NH\cdot CNH(NH_2)]$
Lys	Lysine, 2,6-diaminohexanoic acid	$NH_2\cdot CH\cdot COOH[(CH_2)_4NH_2]$
	Ornithine, 2,5-diaminopentanoic acid	$NH_2\cdot CH\cdot COOH[(CH_2)_3NH_2]$
Asn	Asparagine, 2,4-diamino-4-oxobutanoic acid	$NH_2\cdot CH\cdot COOH(CH_2\cdot CONH_2)$
	Citrulline, 2-amino-5-ureidovaleric acid	$NH_2\cdot CH\cdot COOH[(CH_2)_3NH\cdot CONH_2]$
Gln	Glutamine, 2,5-diamino-5-oxopentanoic acid	$NH_2\cdot CH\cdot COOH[(CH_2)_2CONH_2]$
	4. *Heterocyclic amino acids*	
His	Histidine, 2-amino-3-imidazolylpropanoic acid	$NH_2\cdot CH\cdot COOH(CH_2\cdot C_3H_3N_2)$
Trp	Tryptophan, 2-amino-3,3′-indolylpropanoic acid	$NH_2\cdot CH\cdot COOH[CH_2\cdot (C_2H_2N):C_6H_4]$
Pro	Proline, 2-pyrrolidinecarboxylic acid	$NH\cdot CH\cdot COOH(CH_2\cdot CH_2\cdot CH_2)$ (ring: NH joined to terminal CH_2)

Loses carbon dioxide in air to form theophylline; relaxes muscles in bronchi, stimulates heart and respiration; a diuretic, used in heart failure and asthma (USP, EP, BP).

Aminosol Trademark for a mixture of free amino acids (70–80%) with glucose and mineral salts; used for intravenous feeding.

aminostratigraphy Technique used to date samples 10^5–10^6 years old; based on measuring the extent of racemization, as in the gradual epimerization of L-isoleucine to D-alloisoleucine

aminyl* Suffix indicating a free radical formed by loss of a H atom from a base whose name ends in amine; as, difluoroaminyl.

amithiozone Thiacetazone.

amitosis Cell division with no apparent change in the structure of the cell nucleus. Cf. *karyokinesis.*

amitriptyline hydrochloride $C_{20}H_{23}N\cdot HCl$ = 313.9. White, bitter powder, soluble in water; an antidepressant of the *tricyclic* group, q.v. (USP, EP, BP).

ammelide $C_3H_4O_2N_4$ = 128.1. Cyanin monamide, aminocyanuric acid, cyanuramide, 6-amino-1,3,5-triazine-2,4-dione. Colorless powder, decomp. if heated, insoluble in H_2O.

ammeline $N:(CNH_2:N)_2:C\cdot OH$ = 127.1. Diaminocyanuric acid, cyanurodiamide, cyanin diamide, 4,6-diamino-1,3,5-triazin-2-ol. Colorless, deliquescent needles, decomp. by heat, insoluble in water.

ammeter Ampere meter. An instrument for measuring electric current.

ammine* Affix indicating the NH_3 group as a neutral ligand in a coordination compound; as in *ammines.* Thus diammine (2 groups), etc.

ammines* Ammoniates, ammonates, ammino compounds, metal ammines, ammono. Complex inorganic metal-ammonia compounds, e.g., $[CoCl(NH_3)_5]Cl_2$. Cf. *aqua ions, ammonia system, crystal ammonia.*

ammiolite A native mercury antimonate.

ammite A sandstone or oolite.

ammonate See *crystal ammonia.*

ammonia* (1) NH_3 = 17.03. Colorless gas, $d_{air=1}$ 0.719, m.−77.7, b.−33.4, very soluble in water; forming ammonium hydroxide. It has a strong characteristic odor and is used as a refrigerant, as a fertilizer applied directly to irrigation water, and as a reagent in the chemical industries. (2) Incorrect term for a solution of a. in water (NH_4OH); see *ammonium hydroxide.* **anhydrous ~** Liquid a. Colorless liquid, $d_{-32°}$ 0.6382. Used in organic synthesis, as a solvent, and in refrigeration. **cracked ~** Catalytically decomposed a. vapor (N and H); used to provide an inert atmosphere. **crystal ~** See *crystal ammonia.* **hydroxy ~** Hydroxylamine. **liquid ~** Anhydrous a. **substituted ~** See *amine, amide, ammines, amino, ammonium.*

a. absorption apparatus A device for absorbing a.; as Folin's tubes. **a. amalgam** An amalgam of mercury and ammonia. **a. liquor** Gas liquor; obtained by scrubbing coke plant gases with water and used in the *Solvay process,* q.v., and in the manufacture of a. and ammonium salts. **a. nitrogen** The nitrogen in an organic compound especially proteins, which is in the form of a NH_2 or >NH radical (amino- or imino-nitrogen, respectively). **a. soda process.** A. process (Brunner and Mond, 1874) for the manufacture of sodium carbonate from salt, ammonia, limestone, and coke. Cf. *Solvay process.* **a. system** A system of acids, bases, and salts with liquid a. as solvent instead of water. Thus

$$\underset{\text{Ammonio base}}{M\cdot NH_2} + \underset{\text{Ammonio acid}}{HX} \underset{\text{ammonolysis}}{\overset{\text{neutralization}}{\rightleftharpoons}} \underset{\text{Ammonio salt}}{MX + H\cdot NH_2}$$

a. tube Faraday tube. **a. solution, a. water.** Ammonium hydroxide.

ammoniac Ammoniacum. **sal ~** Ammonium chloride*. **a. oil** Dark, yellow essential oil from ammoniacum, d.0.891, b.250–290, slightly dextrorotatory.

ammoniacal Pertaining to ammonia; as, its odor.

ammoniacum Gum ammoniac. A resinous gum from *Dorema ammoniacum,* (Umbelliferae) of Iran and Northern India. Formerly an expectorant for bronchitis; used externally in plasters and as a porcelain cement.

ammoniate (1) Ammine. (2) An organic nitrogen fertilizer.

ammoniated **a. iron** Ammonium iron(III) chloride*. **a. mercury** See *mercury.* **a. superphosphate** A superphosphate treated with (1) ammonia or (2) dissolved bone and nitrogenous compounds; a fertilizer.

ammonification The enrichment of the soil with ammonia or ammonium compounds; e.g., the production of NH_3 from proteins and decaying organic matter by soil bacteria.

ammonifying (1) Producing or (2) adding ammonia.

ammonio* Ammono. Prefix indicating the ammonium ion NH_4^+, including forms where 1 or more H atoms are substituted.

ammonite A fossil shell.

ammonium* Am. The a. cation NH_4^+, which has basic properties. Cf. *ammonio, tetraethylammonium.* **a. abietadienoate*** $NH_4C_{20}H_{29}O_2$ = 319.5. A. silv(in)ate. Yellow microcrystals, slightly soluble in water. **a. acetate*** CH_3COONH_4 = 77.1. Colorless hygroscopic crystals, m.89, very soluble in water. Used as an antidote in formaldehyde poisoning, as a reagent, and in separating Pb, Ca, Ba, and Sr sulfates. **a. aldehyde** See *aldehyde* ammonia. **a. alum** Aluminum a. sulfate*. **a. aminosulfonate** $NH_4SO_3NH_2$ = 114.1. Deliquescent crystals, soluble in water. **a. anacardate** Brown syrup, used to color hair. **a. antimoniate*** $NH_4SbO_3\cdot 2H_2O$ = 223.8. Colorless crystals, decomp. by heat, soluble in water. **a. arsenate*** $(NH_4)_3AsO_4\cdot 3H_2O$ = 247.1. Colorless crystals, soluble in water. **acid ~** $(NH_4)_2HAsO_4$ = 176.0. White crystals, soluble in water. **a. arsenite*** NH_4AsO_2 = 125.0. Colorless prisms, soluble in water. **a. aurichloride** A. tetrachloroaurate(III)*. **a. auricyanide** A. tetracyanoaurate(III)*. **a. aurocyanide** A. tricyanoaurate(II)*. **a. benzoate*** $NH_4\cdot C_7H_5O_2$ = 139.2. White crystals, decomp. 198, soluble in water. **a. bicarbonate** A. hydrogencarbonate*. **a. bichromate** A. dichromate*. **a. bifluoride** A. hydrogendifluoride*. **a. bioxalate** A. hydrogenoxalate*. **a. biphosphate** A. dihydrogen*phosphate**. **a. bisulfate** A. hydrogensulfate*. **a. bisulfite** A. hydrogensulfite*. **a. bitartrate** A. hydrogentartrate*. **a. borate** A. peroxoborate*. **a. bromide*** NH_4Br = 97.9. Colorless crystals, sublime on heating, soluble in water; used as a reagent and in the manufacture of photographic plates. **a. bromoplatinate** A. hexabromoplatinate(IV)*. **a. carbamate*** $NH_2\cdot COONH_4$ = 78.1. A. carbaminate. White crystals, soluble in water. Cf. *a. carbamate carbonate.* **a. carbamate carbonate** $NH_4HCO_3\cdot NH_4NH_2CO_2$ = 157.1. Hartshorn salt. The a. carbonate of commerce. Colorless crystals, m.85 (sublimes), soluble in water; contains 30–32% ammonia. Used as a reagent in separating the alkaline earth metals from magnesium; in the manufacture of cocoa, baking powder, rubber goods; and in the dye industry. **a. carbazotate** A. picrate. **a. carbonate*** $(NH_4)_2CO_3\cdot H_2O$ = 114.1. Colorless plates, decomp. 85, soluble in water; a reagent. **a. carnallite** The mineral $MgCl_2\cdot NH_4Cl\cdot 6H_2O$. **a. caseinate** Eucasin. White powder, soluble in water, either *mono-* or *di-* (0.011 or 0.022% NH_3, respectively). **a. cesium chloride*** $CsCl\cdot 3NH_4Cl$ = 328.8. White crystals, soluble in water. **a. chlorate*** NH_4ClO_3 = 101.5. Colorless, explosive, monoclinic crystals, soluble in water. **a. chloraurate** A. tetrachloroaurate(III)*. **a. chloride*** NH_4Cl = 53.5. Sal ammoniac, salmiac, ammoniak. Colorless, regular or tetragonal crystals, or white granules, dissociates 350, sublimes without melting, soluble in water. Used as a reagent in qualitative analysis; as an expectorant, stimulant, diuretic; used to render urine acid (USP, EP, BP); also for filling dry batteries; for soldering flux; in textile printing. **a. chloroiridate** A. hexachloroiridate(IV)*. **a. chloropalladate** A. hexachloropalladate(IV)*. **a. chloropalladite** A. tetrachloropalladate(II)*. **a. chloroplatinate** A. hexachloroplatinate(IV)*. **a. chloroplatinite** A. tetrachloroplatinate(II)*. **a. chlorostannate** A. hexachlorostannate(IV)*. **a. chromate*** $(NH_4)_2CrO_4$ = 152.1. Yellow needles, decomp. 185, soluble in water; a reagent and mordant. **a. chromium sulfate*** $Cr_2(SO_4)_3(NH_4)_2SO_4\cdot 24H_2O$ = 956.6. Green octagons, soluble in water. **a. citrate*** $(NH_4)_3C_6H_5O_7$ = 243.2. White, deliquescent powder, soluble in water; a reagent. **a. cobalt sulfate*** $CoSO_4\cdot(NH_4)_2SO_4\cdot 6H_2O$ = 395.2. Cobaltous ammonium sulfate. Red crystals, soluble in water. **a. copper(II) chloride*** $CuCl_2\cdot 2NH_4Cl\cdot 2H_2O$ = 277.5. Ammoniocupric chloride. Blue crystals, soluble in water; a reagent for carbon in steel. **a. crystals** A. carbamate carbonate. **a. cuprate** Deep blue ammoniacal solution of cupric hydroxide, used for waterproofing fabrics. **a. cyanate*** NH_4CNO = 60.1. Colorless crystals, soluble in cold water, decomp. in hot water. **a. cyanide*** NH_4CN = 44.1. Colorless, regular crystals, decomp. 37, soluble in water. **a. dichromate*** $(NH_4)_2Cr_2O_7$ = 252.1. A. bichromate. Orange crystals, decomp. by heat, soluble in water. Used as a reagent and in the manufacture of inks, glass, leather, and fireworks. **a. dithiocarbamate*** $NH_2\cdot CS\cdot SNH_4$ = 110.2. Yellow prisms, soluble in water. **a. dithiocarbonate** $NH_4S\cdot CO\cdot SH$ = 111.2. Yellow crystals, a substitute for hydrogen sulfide in qualitative analysis. **a. dithionate*** $(NH_4)_2S_2O_6$ = 196.2. Colorless, monoclinic crystals, soluble in water. **a. embeliate** $NH_4C_9H_{13}O_3$ = 187.2. Violet-gray powder, soluble in water; a teniacide. **a. ethyl sulfate** $(NH_4)(C_2H_5)SO_4$ = 143.2. Ethyl a. sulfate*. Colorless crystals, m.99, soluble in water. **a. ferrichloride** A. iron(III) chloride*. **a. ferricitrate** A. iron(III) citrate*. **a. ferricyanide** A. hexacyanoferrate(III)*. **a. ferrocyanide** A. hexacyanoferrate(II)*. **a. ferrum** See *a. iron* salts. **a. fluoride*** NH_4F = 37.0. Malt salt. Colorless hexagons, sublime if heated, soluble in water. Used as a reagent and for etching glass. **a. fluosilicate** A. hexafluorosilicate(IV)*. **a. formate*** $H\cdot COONH_4$ = 63.1. Colorless crystals, m.116, decomp. 180, soluble in water. **a. gallate*** $NH_4C_7H_5O_5\cdot H_2O$ = 205.2. Yellow crystals, soluble in water. **a. heptamolybdate** A. molybdate basic. **a.**

hexabromoplatinate(IV)* $(NH_4)_2[PtBr_6]$ = 710.6. A. platinic bromide. Red regular crystals, decomp. by heat, slightly soluble in water. **a. hexachloroiridate(IV)*** $(NH_4)_2[IrCl_6]$ = 441.0. A. iridi(c)chloride. Red powder, slightly soluble in water. **a. hexachloropalladate(IV)*** $(NH_4)_2[PdCl_6]$ = 355.2. A palladi(c)chloride. Brown crystals, decomp. by heat, sparingly soluble in water. **a. hexachloroplatinate(IV)*** $(NH_4)_2[PtCl_6]$ = 443.9. Yellow regular crystals, decomp. by heat, slightly soluble in water. **a. hexachlorostannate(IV)*** $(NH_4)_2[SnCl_6]$ = 367.5. White crystals, soluble in water. **a. hexacyanoferrate(II)*** $(NH_4)_4[Fe(CN)_6]\cdot 3H_2O$ = 338.2. A. ferrocyanide. Green-yellow crystals, soluble in water; a reagent. **a. hexacyanoferrate(III)*** $2(NH_4)_3[Fe(CN)_6]\cdot H_2O$ = 550.2. A. ferricyanide. Red crystals, soluble in water; a reagent. **a. hexafluorosilicate(IV)*** $(NH_4)_2[SiF_6]$ = 178.2. Colorless rhombs, soluble in water. **a. hippurate*** $NH_4H(C_9H_8O_3N)_2\cdot H_2O$ = 393.4. Colorless crystals, soluble in water. **a. hydrogencarbonate*** NH_4HCO_3 = 79.1. A. bicarbonate. Colorless rhombs, decomp. 40–60, soluble in water. **a. hydrogendifluoride*** NH_4HF_2 = 57.0. A. bifluoride, a. acid fluoride, matt salt. Colorless hexagons, soluble in water. Used in the analysis of silicates, in the glass and porcelain industry for etching, and as a preservative. **a. hydrogenoxalate*** HOOC·$COONH_4$ = 107.1. A bi(n)oxalate, a. acid oxalate. Colorless crystals, soluble in water; an ink eraser and reagent. **a. hydrogenphosphate*** See *a. phosphate.* **a. hydrogensulfate*** NH_4HSO_4 = 115.1. A. bisulfate, a. acid sulfate. Colorless crystals, soluble in water; a reagent. **a. hydrogensulfide*** NH_4SH = 51.1. Colorless crystals, soluble in water; a reagent for metals. **a. h. solution** A solution prepared by passing hydrogen sulfide through a. hydroxide. Yellow alkaline liquid; a reagent. **a. hydrogensulfite*** NH_4HSO_3 = 99.1. A bisulfite. Colorless crystals, soluble in water; a preservative. **a. hydrogentartrate*** $NH_4HC_4H_4O_6$ = 167.1. A. bitartrate, a. acid tartrate. Colorless crystals, soluble in water. Used in detection of calcium and in baking powders. **a. hydrosulfide** A. hydrogensulfide*. **a. hydroxide*** NH_4OH = 35.0. Ammonia water, ammonia solution, spirits of hartshorn. An aqueous solution of ammonia. (1) 35%. Concentrated ammonium hydroxide. Clear, colorless liquid, d.0.880, of strong, characteristic odor and alkaline reaction; a reagent, solvent, precipitant, and neutralizing agent; used in pharmacy as 31.5–33.5% w/w strength (NF). (2) 20%, d.0.925. (3) 10% Dilute a. h., d.0.960. A common reagent. **a. ichthyolsulfonate** Ichthyol. **a. iodate*** NH_4IO_3 = 192.9. Colorless rhombs, decomp. 125, soluble in water. **a. iodide*** NH_4I = 144.9. Colorless regular crystals or powder, sublimes when heated, soluble in water. A reagent for acetone; used in photography. **a. iridichloride** A. hexachloroiridate(IV)*. **a. iron(II) bromide*** $(NH_4)_2FeBr_4$ = 411.5. Ferroammonium bromide. Brown powder, soluble in water. **a. iron(II) sulfate*** $(NH_4)_2Fe(SO_4)_2\cdot 6H_2O$ = 392.1. Ferroammonium sulfate, Mohr's salt. Green crystals, soluble in water; used in photography and as a volumetric standard. **a. iron(III) chloride*** $(NH_4)FeCl_4$ = 215.7. Ammoniated iron, ferriammonium chloride, ammonio chloride of iron, ammonium ferratum chloratum, flores martialis, aes martis. Orange crystals of metallic taste, soluble in water. **a. iron(III) chromate*** $(NH_4)Fe(CrO_4)_2$ = 305.9. Ferriammonium chromate. Brown crystals, soluble in water. **a. iron(III) citrate*** $Fe(NH_4)_3(C_6H_5O_7)_2$ = 488.2. Ferriammonium citrate. Brown scales, soluble in water; used to treat anemia. **green ~** Green iron, ammonium citrate. Green scales, turning brown in light, soluble in water. **a. iron(III) oxalate*** $(NH_4)_3Fe(C_2O_4)_3\cdot 3H_2O$ = 428.1. Ferriammonium oxalate. Green crystals, soluble in water. Used in photography and blueprint paper. **a. iron(III) sulfate*** $(NH_4)Fe(SO_4)_2\cdot 12H_2O$ = 482.2. Ferriammonium sulfate, iron and ammonium alum. Violet octahedra, soluble in water. Used as a styptic, astringent, reagent, and indicator, and in the dye industry. **a. iron(III) tartrate*** $NH_3Fe(C_4H_4O_6)_2$ = 369.0. Ferriammonium tartrate, iron and ammonium tartrate. Brown crystals, soluble in water; a styptic. **a. lactate*** $NH_4C_3H_5O_3$ = 107.1. Colorless syrup, miscible with water. **a. linoleate** $C_{17}H_{31}COONH_4$ = 297.5. Soft mass; an emulsifier for fats, and a waterproofing, glazing, and polishing agent. **a. magnesium arsenate*** $MgNH_4AsO_4\cdot 6H_2O$ = 289.4. Colorless tetragons, decomp. by heat, slightly soluble in water. **a. magnesium chloride*** $NH_4Cl\cdot MgCl_2\cdot 6H_2O$ = 256.8. White crystals, soluble in water. **a. magnesium phosphate*** $NH_4MgPO_4\cdot 6H_2O$ = 245.4. Colorless tetragons, slightly soluble in water. **a. magnesium sulfate*** $(NH_4)_2SO_4\cdot MgSO_4\cdot 6H_2O$ = 372.7. Colorless crystals, soluble in water. **a. malate*** $NH_4C_4H_5O_5$ = 151.1. Rhombic crystals, m.161, soluble in water. **a. manganese(II) phosphate*** $NH_4MnPO_4\cdot H_2O$ = 186.0. Manganous ammonium phosphate. Colorless crystals, soluble in water. **a. manganese(II) sulfate*** $(NH_4)_2Mn(SO_4)_2$ = 283.1. Manganous alum. Pink crystals, soluble in water; a reagent. **a. mellitate** $C_6(COONH_4)_6\cdot 9H_2O$ = 605. Colorless crystals, soluble in water. An emulsifier for fats and waxes and a waterproofing, glazing, and polishing agent. **a. metaphosphate*** See *a. phosphate.* **a. metavanadate*** A. vanadate. **a. molybdate*** $(NH_4)_2MoO_4$ = 196.0. Colorless, monoclinic crystals, decomp. by water. Used as a reagent and in the manufacture of pigments. **basic ~** $(NH_4)_6Mo_7O_{24}\cdot 4H_2O$ = 1236. A. heptamolybdate. Green-yellow crystals, decomp. by heat, soluble in a. chloride solution. Used as a reagent and in the preparation of pigments. **a. molybdate solution** (6%) Yellow liquid, with a slight odor of nitric acid; a reagent for orthophosphates and arsenates. **a. mucate** $(NH_4)_2C_6H_8O_8$ = 244.2. Colorless crystals, soluble in water. **a. nickel(II) chloride*** $NH_4NiCl_3\cdot 6H_2O$ = 291.2. Nickelous ammonium chloride. Green rhombs, soluble in water; used in nickel plating. **a. nickel(II) sulfate*** $(NH_4)_2Ni(SO_4)_2\cdot 6H_2O$ = 395.0. Nickelous ammonium sulfate. Green crystals, soluble in water; used in nickel plating and as a reagent. **a. nitrate*** NH_4NO_3 = 80.0. German saltpeter, Norway saltpeter. Colorless tetragons, m.170, decomp. 200, soluble in water. Used as an oxidizing agent, as a flux for metals, in the preparation of dinitrogen oxide, in freezing mixtures, and in explosives. **a. nitrite*** NH_4NO_2 = 64.0. Colorless crystals, decomp. by heat into water and nitrogen, soluble in water. **a. oleate*** $NH_4\cdot C_{18}H_{33}O_2$ = 299.5. Ammonia soap. Colorless gel, soluble in alcohol; a cleanser. **a. oxalate*** $(COONH_4)_2\cdot H_2O$ = 142.1. Colorless prisms, soluble in water. A reagent used in the separation of Ca; as a precipitant for Ba, Zn, Pb; and in determination of quinine. **a. oxalurate** $NH_4C_3H_3N_2O_4$ = 149.1. Yellow crystals, soluble in hot water. **a. oxamate*** $NH_2\cdot CO\cdot COONH_4$ = 106.1. Colorless crystals, soluble in water. **a. pallidichloride** A. hexachloropalladate(IV)*. **a. palmitate*** $NH_4C_{16}H_{31}O_2$ = 273.5 Soft masses, soluble in water; a cleanser. **a. perchlorate*** NH_4ClO_4 = 117.5. White prisms, decomp. by heat, soluble in water; used in explosives. **a. perchromate** $(NH_4)_3CrO_8$ = 234.1. Yellow rhombs, decomp. 50, soluble in water. **a. permanganate*** NH_4MnO_4 = 137.0. Purple rhombs, decomp. by heat, soluble in water; used in pyrotechnics. **a. peroxoborate*** NH_4BO_3 = 76.8. Colorless crystals, soluble in water. **a. peroxodisulfate*** $(NH_4)_2S_2O_8$ = 228.2. Colorless, monoclinic crystals, decomp.

by heat, soluble in water. A preservative, a reagent for albumen and indican, an oxidizing agent in analysis; used in photography as a reducer, and in electroplating. **a. phenolsulfonate*** $C_6H_4(OH)SO_3 \cdot NH_4$ = 191.0. White crystals, soluble in water. **a. phosphate** **a. dihydrogenphosphate*** $NH_4H_2PO_4$ = 115.0. A. biphosphate, monoammonium phosphate, diacid a. phosphate, Ammo-Phos. Colorless tetragons, soluble in water. Used as a reagent and fertilizer and with sodium hydrogencarbonate in baking powder. **diammonium hydrogenphosphate*** $(NH_4)_2HPO_4$ = 132.2. Diammonium orthophosphate, ammonium diphosphate. Colorless, monoclinic crystals, soluble in water. A precipitant for Mg, Zn, Ni, and U; a fertilizer and fireproofing material. **a. metaphosphate*** NH_4PO_3 = 97.0. Colorless crystals, soluble in water. **triammonium phosphate*** $(NH_4)_3PO_4$ = 149.1. Tribasic a. phosphate. Colorless crystals, soluble in water; a culture medium. **a. pyrophosphate*** $(NH_4)_4P_2O_7$ = 246.1. White crystals, soluble in water. **a. phosphite** **a. dihydrogen phosphite** $NH_4H_2PO_3$ = 99.0. Colorless crystals, m.123, decomp. 150, soluble in water. **a. hydrogen phosphite** $(NH_4)_2HPO_3$ = 116.1. Diammonium phosphite. Colorless crystals, soluble in water; a reducing agent. **a. phosphomolybdate** $(NH_4)_3[PMo_{14}O_{40}]3H_2O$ = 2122. Yellow crystals, slightly soluble in water; a reagent for alkaloids. **a. phosphotungstate, a. phosphowolframate** A. tungstophosphate*. **a. phthalate** $C_6H_4(COONH_4)_2$ = 200.2. Colorless crystals, soluble in water. **a. picramate** $C_6H_2NH_2(NO_2)_2 \cdot ONH_4$ = 216.2. Brown crystals, soluble in water, used in bacteriology. **a. picrate** $NH_4C_6H_2O(NO_2)_3$ = 246.1. A. carbazotate, ammonium picronitrate. Yellow, explosive crystals, decomp. by heat, soluble in water or alcohol. **a. picrocarminate solution** Red-staining solution used to differentiate the nucleus and cytoplasm of cells. **a. picronitrate** A. picrate. **a. platinic bromide** A. hexabromoplatinate(IV)*. **a. platinic chloride** A. hexachloroplatinate(IV)*. **a. selenate*** $(NH_4)_2SeO_4$ = 179.0. White crystals, soluble in water. **a. selenate (acid)** NH_4HSeO_4 = 162.0. Colorless crystals, soluble in water. **a. selenite*** $(NH_4)_2SeO_3$ = 163.0. White crystals, soluble in water; used as a reagent for alkaloids and in red glass. **a. sesquicarbonate** $2NH_4CO_3(NH_4)_2CO_3 \cdot H_2O$ = 270.2. White crystals, decomp. by heat. **a. silver nitrate solution** A solution of 1 g silver nitrate in 20 ml water, ammonia being added until the precipitate just dissolves; produces a silver mirror on heating with certain reducing agents. **a. sodium hydrogenphosphate*** $NH_4NaHPO_4 \cdot 4H_2O$ = 209.1. Microcosmic salt, triphosphate, acid ammonium phosphate. Colorless crystals, decomp. by heat, soluble in water; an analytical reagent; forms a molten bead that is colored characteristically by impurities. **a. sodium sulfate*** NH_4NaSO_4 = 137.1. Colorless crystals, soluble in water. **a. stannic chloride** A. hexachlorostannate(IV)*. **a. stearate*** $C_{17}H_{35} \cdot COONH_4$ = 301.6. White, soaplike mass, soluble in hot alcohol. **a. succinate*** $C_2H_4(COONH_4)_2$ = 152.1. Colorless crystals, soluble in water. **a. sulfate*** $(NH_4)_2SO_4$ = 132.1. Colorless rhombs, m.140, decomp. 280, soluble in water. Used as a precipitant for proteins, in separation of Ca and Sr, and in the manufacture of fertilizers. **acid ~** A. hydrogensulfate. **a. sulfide*** $(NH_4)_2S$ = 68.1. White crystals, decomp. by heat, soluble in water; a reagent. **a. sulfide solution** An aqueous solution of a. hydrogensulfide. **a. sulfite*** $(NH_4)_2SO_3 \cdot H_2O$ = 134.1. White, monoclinic crystals, decomp. by heat, soluble in water. **acid ~** A. hydrogensulfite. **a. tartrate*** $C_2H_4O_2(COONH_4)_2$ = 184.1. Colorless, monoclinic crystals. Two forms: refractive indices α = 1.55 and β = 1.581, soluble in water; an expectorant. **acid ~** $NH_4HC_4H_4O_6$ = 167.1. A. bitartrate, a. hydrogentartrate*. Monoclinic prisms: refractive indices 1.519, 1.561, and 1.591; soluble in water. **a. tellurate*** $(NH_4)_2TeO_4$ = 227.7. White powder, soluble in acids; a reagent for glucosides and alkaloids. **a. tetrachloroaurate(III)*** $NH_4[AuCl_4]$ = 356.8. A. aurichloride. Yellow plates, decomp. 100, soluble in water. **a. tetrachloropalladate(II)*** $(NH_4)_2[PdCl_4]$ = 284.3. A chloropalladite, a. palladous chloride. Red crystals, soluble in water. **a. tetrachloroplatinate(II)*** $(NH_4)_2[PtCl_4]$ = 373.0. Yellow, regular crystals, slightly soluble in water. **a. thioacetate*** CH_3COSNH_4 = 93.1. Yellow, crystalline masses, soluble in water; a reagent. **a. thioacetate solution** A 30% aqueous solution; a reagent in organic analysis and substitute for hydrogen sulfide. **a. thiocyanate*** NH_4CNS = 90.1. A. rhodanate. Colorless, deliquescent, monoclinic crystals, m.159; at 170 it changes to thiourea; soluble in water. A reagent; also used in the dye and textile industry. **a. thiostannate** $(NH_4)_2SnS_3$ = 251. **a. thiosulfate*** $(NH_4)_2S_2O_3$ = 148.2. A. hyposulfite. Colorless rhombs, decomp. 150, soluble in water. **a. tricyanoaurate(II)*** $NH_4[Au(CN)_3]$ = 293.1. A. aurocyanide. Colorless plates, decomp. 150, soluble in water. **a. trithiocarbonate*** $(NH_4)_2CS_3$ = 144.3. Yellow crystals, soluble in water. **a. tungstate*** $(NH_4)_2WO_4$ = 283.9. Colorless crystals, soluble in water. **metatungstate** $(NH_4)_6H_2W_{12}O_{40} \cdot nH_2O$. Colorless octahedra, soluble in water **paratungstate** $(NH_4)_6W_7O_{24} \cdot 6H_2O$ = 188.7. Colorless rhombs, slightly soluble in water. Used to prepare a. tungstophosphate. **a. tungstophosphate*** $(NH_4)_3PO_4 \cdot 12WO_2 \cdot 3H_2O$ = 279.3. A. phosphotungstate, a. phosphowolframate. White powder, soluble in water; a reagent. **a. uranate** $(NH_4)_2U_2O_7$ = 624.1. Yellow-red powder used for coloring porcelain. **a. uranylcarbonate*** $UO_2CO_3 \cdot 2(NH_4)_2CO_3$ = 522.2. Yellow crystals, soluble in water, used in ceramics and glass. **a. uranylfluoride*** $UO_2F_2 \cdot 3NH_4F$ = 419.1. Uranium ammonium fluoride. Green, fluorescent crystals, soluble in water; used in X-ray screens. **a. urate** $(NH_4)C_5H_3N_4O_3$ = 185.1. Colorless crystals, soluble in water. **a. valerate*** $C_4H_9 \cdot COONH_4$ = 119.2. Colorless, hygroscopic crystals, soluble in water; a hypnotic and sedative. **a. vanadate** NH_4VO_3 = 117.0. A. metavanadate*. White powder, soluble in water; used in the textile industry and in the manufacture of vanadium catalysts, photographic dyes, indelible ink, and ceramics. **a. wolframate*** A. tungstate*.

ammono Ammonio*.

ammonocarbonous acid Hydrocyanic acid.

ammonolysis (1) A reaction which corresponds with hydrolytic dissociation: $NH_3 \rightarrow H^+ + NH_2^-$. (2) The cleavage of a bond by the addition of ammonia; $R-R' + NH_3 \rightarrow RNH_2 + HR'$. (3) Treatment (of oils) with hot ammonia gas under pressure.

Ammo-Phos Trademark for ammonium phosphate fertilizer containing ammonia 13 and available phosphorus pentoxide 48%.

amniocentesis The removal of some amniotic fluid from the membranous sac surrounding a fetus; the sample can be tested for fetal chromosomal and metabolic abnormality.

amniotin Estrone.

amobarbital $C_{11}H_{18}N_2O_3$ = 226.3. 5-Ethyl-5-isopentyl barbituric acid, amylobarbital, Amytal. White, bitter crystals, m.157, slightly soluble in water; a hypnotic and sedative (USP). **a. sodium** $C_{11}H_{17}N_2O_3Na$ = 248.3. White, hygroscopic, bitter granules; used as a. (USP, EP, BP).

amodiaquine (hydrochloride) $C_{20}H_{22}ON_3Cl \cdot 2HCl \cdot 2H_2O$ = 464.8. Yellow, bitter crystals, m.158, soluble in water, an antimalarial (USP, BP).
amoeba Ameba.
amoeboid Resembling an ameba. **a. movement** The rolling movement of the protoplasmic jelly of a unicellular organism.
amoil Pentyl phthalate used for high-vacuum pumps.
amorphism The noncrystalline condition of a solid substance; due to an irregular molecular assembly.
amorphous (1) Unorganized. (2) Describing a solid substance which does not crystallize and is without definite geometrical shape. (3) In bacteriology, any bacteria without visible differentiation in structure.
amosite A long-fibered variety of Transvaal asbestos.
amoxy The pentyloxy* radical.
AMP* Symbol for adenosine 5′-monophosphate. See *adenylic acid.*
ampelite A graphite schist containing SiO_2 53.8, Al_2O_3 23.5%, and sulfur (Pyrenees). Used as refractory up to 1850°C.
Ampére, Andrè Marie (1775–1836) French physicist; developed molecular theory.
ampere* A*. An SI base unit. The constant current that produces a force of 2×10^{-7} N/m length between 2 parallel conductors placed 1 m apart in a vacuum. 1 amp = 1 coulomb/s = 1 volt through 1 ohm = 10^{-1} emu (cgs) = $2{\cdot}998 \times 10^9$ esu. **micro ~** One-millionth of an a. **milli ~** One-thousandth of an a.
a.meter Ammeter. **a.-volt** See *volt-ampere.*
amperometer Ammeter.
amperometry Chemical analysis by methods which involve measurements of electric currents. Cf. *conductometric analysis, polarograph, potentiometer.*
amphetamine $C_6H_5 \cdot CH_2 \cdot CH(NH_2)Me$ = 135.2. (±)-α-Methylphenethylamine, benzedrine, speed. Colorless liquid. Its vapors shrink nasal mucosa. **a. sulfate** $(C_9H_{13}N)_2 \cdot H_2SO_4$ = 368.5. Benzedrine sulfate. White, bitter powder; used for its stimulatory effects, particularly on nervous system; use is restricted owing to addictive nature.
amphi- Prefix (Greek) meaning "on both sides" or "both." **amphi position** Formerly, the 2,6-hydrogen atoms in 2 fused hexatomic rings; as, naphthalene.
amphiboles $M_x(SiO_3)_y$. Rock-forming minerals with occluded water; M is Ca, Mg, Fe, or the alkali metals, e.g., crocidolite. Elongated, fibrous, black or dark green crystals. Cf. *silica minerals.*
amphibolites Metamorphic rocks derived from argillaceous limestones and related to the glaucophane schists.
amphichroic, amphichromatic Describing mixed indicators whose color changes neutralize each other; e.g., litmus and congo red.
Amphicol Trademark for chloramphenicol.
amphigene Leucite.
amphiphatic Describing a detergent or wetting agent which greatly reduces the surface tension of water. Part of the molecule is hydrophilic, and the other consists of straight or branched long hydrocarbon chains; e.g., hexadecyltrimethyl bromide, $[C_{16}H_{33}Me_3N]^+Br^-$.
amphiphile A substance containing both polar, water-soluble and hydrophobic, water-insoluble groups; e.g., $C_{12}H_{25} \cdot OH$, where $C_{12}H_{25}$ is hydrophobic.
amphiprotic Able to lose or gain a proton. Cf. *acid, base.*
ampholyte Amphoteric electrolyte. A substance which in solution yields H^+ or OH^-, i.e., donates or accepts a proton, according to whether it is in an acid or basic solution. Cf. *isoelectric point, zwitterion.*
ampholytoid Amphoteric colloid. A particle in suspension capable of adsorbing H^+ or OH^-, depending on the pH value.
amphoteric Describing substances having both acid and basic properties; as, the amino acids, $NH_2 \cdot R \cdot COOH$. **a. hydroxides** The hydroxides of some metals which may dissociate to H^+ or OH^- ions; as, $2Al(OH)_3 \rightleftharpoons H^+ + AlO_2 + H_2O$ (acid) or $Al^{3+} + 3OH^-$(base). **a. sulfides** The sulfides of the metallic nonmetals, which may react as weak acids or weak bases; e.g., $H_3(AsS_3) \rightleftharpoons As(SH)_3$.
amphotericin B $C_{47}H_{73}O_{17}N$ = 924.1. Fungizone. A patented polyene antifungal antibiotic, produced by *Streptomyces nodosus.* Orange powder, insoluble in water (USP, BP).
amphotropine $[(CH_2)_6N_4]_2 \cdot C_8H_{14}(COOH)_2$ = 480. Hexamethylenamine camphorate. Colorless crystals.
ampicillin $C_{16}H_{19}O_4N_3S$ = 349.4. (6*R*)-6-(α-phenyl-D-glycylamino)penicillanic acid, Penbritin. White, bitter crystals, soluble in water; a broad-spectrum antibiotic, used for bacterial infections (USP, BP). Also used as the sodium salt and trihydrate (BP).
amplifier (1) A magnifier. (2) A transistor (or vacuum tube), by which weak electric currents or voltages are strengthened; as in radio reception.
amplitude The maximum displacement of an oscillation, vibration, or wave.
ampoule, ampul A small, sealed glass vial, e.g., sterilized solution for injection.
amrad gum A gummy exudation from elephant apple. *Feronia elephantum* (Rutaceae), of India.
amurea The bitter, watery residue from crushing olives; a pesticide.
amyctic (Greek: "to scratch") An irritating stimulant, especially for skin.
amygdala The seeds of *Prunus amygdalus,* containing amygdalin. **a. amara** See *bitter* under *almond.* **a. dulcis** See *sweet* under *almond.*
amygdalic acid (1) Mandelic acid. (2) $C_{20}H_{28}O_{13}$ = 476.4. Gentiobioside. A glucoside from almonds.
amygdalin $C_{20}H_{27}O_{11}N \cdot 3H_2O$ = 511.5. D-Mandelonitrile glucoside, amygdaloside. A glucoside from bitter almonds and wild cherry bark. Colorless crystals, m.210, soluble in water. Hydrolyzes to 2 moles glucose, hydrocyanic acid, and benzaldehyde; an expectorant and source of oil or bitter almonds. See *laetrile.*
amygdalinic acid Mandelic acid.
amygdaloid Shaped like an almond.
amygdophenin $EtOC_6H_4NH \cdot CO \cdot CHOH \cdot C_6H_5$ = 271.3. Phenetidine amygdalate. Gray crystals; an analgesic.
amyl The pentyl* radical, $Me(CH_2)_4-$. **iso ~** 3-Methylbutyl†. The isopentyl* radical, $Me_2CH(CH_2)_2-$. ***tert*- ~** The *tert*-pentyl* radical, $Et \cdot C(Me)_2-$.
a.acetate Iso*pentyl acetate**. **a.acetic ether** Iso*pentyl acetate**. **a. alcohol** See *pentyl alcohol.* **a. hydride** Pentane*. **a. nitrite** (1) Pentyl nitrite*. (2) A mixture of nitrous acid, and 2- and 3-methyl butyl ester. Yellow volatile liquid, b.96. A vasodilator, used as inhalant and in angina, and as an antidote for cyanide poisoning (USP).
amylan A levorotatory gum in malt and barley; does not reduce Fehling's solution.
amylase* A group of enzymes that hydrolyze 1,4-α-D-glucosidic linkages. ***α*- ~ *** Diastase, ptyalase, ptyalin. Endohydrolyzes molecules containing 3 or more glucose units; e.g., starch to glucose, maltose and oligosaccharides. ***β*- ~ *** Diastase. Removes maltose units from nonreducing ends of chains; as, starch to maltose and dextrins. **gluco ~**

Exo-1,4-α-D-glucosidase*. Successively hydrolyzes nonreducing ends with release of β-D-glucose.

amylate A compound of starch.

amylene $MeCH_2CH_2CH:CH_2$ = 70.11. **1-~** 1-Pentene*. **iso~** or **2-~** 2-Pentene*.

amylic alcohol Pentyl alcohol*.

amylidene The pentylidene* radical.

amyline The cellulose membrane of starch granules.

amylo- Pertaining to starch. (Greek amylum.)

amylobacter A bacillus acting on starch and causing butanoic acid fermentation.

amylobarbital Amobarbital.

amylobarbitone EP, BP name for amobarbital.

amylocellulose A starch-cellulose complex which encloses the starch granulose of plants.

amyloclastic Amylolytic.

amylodextrin Soluble *starch.*

amyloform An antiseptic mixture of starch and formaldehyde.

amylogen A soluble starch.

amylograph An instrument to measure the baking quality of starch in terms of its viscosity at various temperatures.

amyloid (1) $(C_6H_{10}O_5)_n$ = $(162.11)_n$. An explosive, m.42, slightly soluble in water. (2) Colloidal cellulose, Guignet cellulose. Parchment paper formed by the action of sulfuric acid on cellulose. (3) An intermediate in the lignification of woody tissues.

amyloin Maltodextrin.

amylolysis The conversion of starch into sugar by boiling dilute acids (hydrolysis) or by enzymes.

amylolytic Amyloclastic. Capable of transforming starch into sugar. **a. activity** The digestive power of amylase. **a. enzyme** Amylase*. **a. fermentation** See *fermentation.*

amylopectin The gel constituent of starch paste. It comprises 75% of starch substance and has a branched or laminated structure. Its shorter α-D-(1,4)-glucose chains are linked laterally by α-(1,6)-glycosidic bonds. Cf. *amylose.*

amylose The sol constituent of starch paste. It comprises about 25% of starch substance and consists of unbranched chains of D-glucose residues, members of which are linked internally by α-(1→4)-glycosidic bonds. Cf. *amylopectin.*

amylum (1) Latin name for starch. (2) Corn *starch.*

amyrin **α-~** Ilicic alcohol.

Amyris A genus of tropical trees and shrubs producing fragrant resins and gums, e.g., *A. elemifera* (elemi) of Mexico.

Amytal Trademark for amobarbital.

an- Prefix indicating without or not.

-an Suffix indicating (1) a sugar body, glucoside, or gum. (2) The systematic name of a saturated natural product containing heteroatoms. Cf. *-en, -ane.*

An Symbol for actinon.

ana- Prefix (Greek) indicating again, along, over, through, without, or against. **ana position** The positions of the 2 H atoms attached to the first and fifth C atom of two condensed hexatomic rings; as, naphthalene.

anabasine $C_{10}H_{14}N_2$ = 162.2. 1,2-(3-Pyridyl)piperidine. An isomer of nicotine from *Anabasis aphylla* (Chenopodiaceae) of Central Africa. Colorless liquid, d.1.048, b.280, soluble in water. **(±)-~** Neonicotine.

anabolic (steroid) A substance that increases or causes anabolism, in particular the growth and development of muscles; as, androgens.

anabolism Synthetic metabolism. The metabolic construction of larger and more complex molecules from smaller, simpler ones; as, proteins from amino acids.

Anacardiaceae The cashew family, 60 genera and 500 species of trees or shrubs with gummy, milky, or resinous juices, often poisonous; e.g., *Rhus toxicodendron,* poison ivy, poison oak; *Anacardium occidentale,* cashew nuts. See also *Rhus, quebracho, mangiferin.*

anacardic acid $C_{22}H_{32}O_3$ = 344.5. An acid from the seeds of *Anacardium occidentale.* Brown crystals, m.26, soluble in alcohol; an anthelmintic.

anacardin $C_{15}H_{14}O_6$ = 290.3. A catechin present in *Anacardium occidentale* Lin. Cf. *kauri.*

anacardium Cashew nut, caje nut. The dried, edible fruit of *A. occidentale,* a shrub of tropical America. Cf. *acajou balsam, Cardol.*

anaclastic Having refracting powers.

anaerobe A bacterium which can grow and divide in an oxygen-free atmosphere and derives the oxygen it requires from other compounds. **facultative ~** A bacterium which prefers an atmosphere containing oxygen, but is not dependent on oxygen gas for its metabolism. **obligatory ~** A bacterium which cannot live in an atmosphere containing oxygen.

anaerobic Able to grow in an oxygen-free atmosphere. **a. culture medium** Culture medium for bacteria that are obligatory anaerobes; as, *Clostridium tetani.* Oxygen is removed by (1) adding a reducing agent, e.g., sodium thioglycollate; (2) extracting oxygen (physically or chemically); (3) growing bacteria in the depth of the medium (stab culture).

anaesthesia Anesthesia.

anaesthesin(e) Benzocaine.

anagyrine $C_{15}H_{18}ON_2$ = 242.3. An alkaloid from the seeds of *Anagyris foetida* (Leguminosae). A brittle, resinlike mass, b.245. It resembles cytisine and sparteine in structure.

analeptic (1) A drug that restores health. (2) More specifically, a drug that stimulates the central nervous system, and especially the respiratory and vasomotor centers.

analgen(e) (1) $C_7H_5O \cdot NHC_6H_2(OEt):C_3H_3N$ = 292.3. 5-Benzamido-8-ethoxyquinoline, benzanalgen, quinalgene, labordin. Colorless crystals, m.208, insoluble in water. (2) $C_{13}H_{14}O_2N_2$ = 230.3. 5-Acetylamino-8-ethoxyquinoline. Colorless crystals, m.155.

analgesia The relief of pain, as by drugs. **relative ~** R.A. Term, particularly in dentistry, for the use of a mixture of nitrous oxide and oxygen as an analgesic.

analgesic A drug that relieves pain without causing loss of consciousness, either by direct action on nerve centers (brain), or by diminishing the conductivity of the sensory nerve fibers.

analgesine Antipyrine.

analgic Painless.

analog computer See *computer.*

analogs, analogues Analogous series. Compounds with similar electronic structures but different atoms; as *isostere* and *isolog.*

analogy A similarity or a likeness in properties. Cf. *homology, isolog.*

Analoids Patented tablets containing exact quantities of reagents, used in chemical analysis. Cf. *Fixanal.*

analyser Analyzer.

analysis (1) Assay. The determination, detection, or examination of a substance. (2) Breaking down or splitting into simpler constituents. (3) The reverse of synthesis. **activation ~** The use of a nuclear reactor to produce neutrons for bombarding small particles of material (usually biological) to be analyzed, followed by measurement of the induced radioactivity to provide an estimate of the concentrations of elements present; as, arsenic in hair.

autocorrelation ~ See *autocorrelation analysis.* **bio ~** (1) The detection of substances with the aid of microorganisms; e.g., by observing the selective action of yeasts on sugars or by ascertaining the minimum amount of the sample under test that will inhibit growth of an organism or produce specific disease symptoms in an experimental animal. (2) The determination of the strength of substances (e.g., hormones) from their effects on animals. **biochemical ~** The chemical examination of biological material. **blowpipe ~** The detection of metallic elements and acid radicals by means of the blowpipe. **chromatographic ~** See *chromatographic analysis.* **clinical ~** The examination of body fluids and tissues for the diagnosis of diseases. **colorimetric ~** The quantitative a. of materials by means of the color intensity at wavelength(s) of light characteristic of a substance. **complexometric ~** See *complexometric titration.* **conductometric ~** See *conductometric analysis.* **differential thermal ~** A technique in which the temperature difference between the sample being tested and a reference substance is measured as a function of temperature, while both are subjected to a controlled temperature change program. **diffusion ~** See *diffusion analysis.* **dry ~** A. without the use of solutions. **electro ~** See *electrodeposition analysis.* **elementary ~** The determination of the constituents and molecular combination, as, of an organic compound by combustion; e.g., C as CO_2, H as H_2O. **gas ~** A. of gas mixtures by measuring the volumes before and after treatment with selective absorbing agents. **gravimetric ~** The determination of the composition of a substance by weighing its constituents directly or indirectly. **headspace ~** A. of gases or vapor in equilibrium with a sample. It avoids interference by other constituents of the sample and is a simple sampling method. **iodimetric ~** Titration of oxidizing substances with standard sodium thiosulfate and acid potassium iodide solutions. **mechanical ~** See *mechanical analysis.* **mechanochemical ~** Qualitative microanalysis based on the color produced when solid reactants are rubbed together in a mortar. **micro ~** (1) See Table 9. (2) Microscopic identification of substances (e.g., starch) and characteristic reaction products (precipitates, crystals). (3) Modifications on the small scale of the processes of quantitative analysis. **multiregression ~** See *regression analysis.* **nephelometric ~** Measurements of turbidity to determine amount of precipitate. **organic ~** Elementary analysis. **proximate ~** The determination of the chemical nature of a sample with regard to molecular combination. **qualitative ~** The detection of the kind or nature of an element or compound in a substance. **quantitative ~** The determination of the amount or quantity of an element or compound in a substance. **rational ~** See *rational analysis.* **regression ~** See *regression analysis.* **ring-oven ~** Spot analysis on a filter paper in which substances to be determined are concentrated in well-defined rings by a ring-shaped oven. **screen ~** See *screen analysis.* **spectroscopic ~** See *spectroscopic analysis.* **spectrum ~** The detection of elements and binary compounds by their characteristic radiation as observed through the spectroscope. **spot ~** See *spot analysis.* **technical ~** Practical or empirical methods used in industry for evaluating materials. **thermodynamic ~** The measurement of a component of a gas mixture by passing the mixture through a calibrated orifice for a known time, into an evacuated vessel, and measuring the increase in pressure in the latter. **ultimate ~** Elementary analysis. **volumetric ~** The determination of elements and compounds in a substance by titration with standard solutions. **wet ~** An a. made with solutions. See also *bead test, biological assay, calorimetry, chromatographic a., colorimetry, electrolysis, flame tests, reaction, spectroscopy, thermal, titration.*

TABLE 9. MAGNITUDES OF ANALYSIS
Expressed as Size of Sample Used

Magnitude name	Size (g)
Macroanalysis*	more than 0.1
Mesoanalysis*, semimicroanalysis	0.1 to 0.01
Microanalysis*	10^{-2} to 10^{-3}
Submicroanalysis*	10^{-3} to 10^{-4}
Ultramicroanalysis*	less than 10^{-4}

analyte A substance that is being analyzed.

analytical Pertaining to analysis. **a. balance** See *analytical* under *balance.* **a. chemistry** See *analytical* under *chemistry.* **a. metabolism** Catabolism. **a. reactions** The characteristic reactions of elements or ions, used for their identification or determination; e.g., precipitate formation, color changes. **a. weights** The standardized weights of an analytical balance.

analyzer (1) A device which indicates a certain condition, change, or phenomenon. (2) The nicol prism of a polariscope nearest to the eyepiece. (3) The first tower of a coffee still. **curve ~** A device for determining the slope of a graph. **micropolar ~** An optical attachment for a microscope for the determination of polarization in crystals. **polarization ~** The nicol prism in a polariscope nearest to the eye. Cf. *polarizer.*

anamirtin (1) $C_{10}H_{26}O_{10}$. A glucoside from the fruits of *Anamirta paniculata.* (2) $C_{19}H_{24}O_{10}$. A glyceride from fishberries, *Cocculus indicus.* Cf. *cocculin.*

anamorphism A change caused by the action of pressure, water, or heat on rock. Cf. *catamorphism.*

anamorphoscope A mirror which corrects distorted images.

anamorphosis The distortion of objects by mirrors.

anaphase A stage of cell division following the division of the nucleus.

anaphe A wild silk, resembling tussah.

anaphoresis The antonym of *cataphoresis,* q.v.

anaphrodisiac A drug which is said to diminish sexual desire.

anaphylactic shock The sometimes fatal condition of low blood pressure and constriction of bronchi occurring in severe anaphylactic states.

anaphylactin Allergen.

anaphylatoxin A poison produced in anaphylaxis caused by the injection of proteins.

anaphylaxis (1) Extreme hypersensitivity, of a person or animal, to second and subsequent antigen or protein injections; as, to penicillin, wasp or bee stings, or vaccine prepared from horse serum. (2) Anaphylactic shock.

Anaplas Trademark for a thermoplastic material made from recovered waste plastics.

anarcotine A nonnarcotic alkaloid in Indian opium.

anatase TiO_2. Octahedrite. A native titanium oxide.

Anaxagoras (500–428 B.C.) The first Greek scientist. He distinguished physical from psychical phenomena and assumed indivisible parts of matter.

anchietine An alkaloid from the root of *Anchietea salutaris* (Violaceae).

anchoic acid Azelaic acid.

anchored compound An organic compound attached to certain cells by a characteristic radical. See *chemotherapy.*

anchoring group The salt-forming radical of a dyestuff.

anchovy A small edible fish of the herring family. **a. pear** The fruit of a Jamaican tree used in the manufacture of pickles.
anchusic acid An acid from the root of *Alkanna tinctoria* (Boraginaceae).
anchusin $C_{35}H_{40}O_8$ = 588.7. A red coloring matter from the root of *Alkanna tinctoria.* Cf. *alkanet, alkannin.*
Ancobon, Anocotil Trademarks for flucytosine.
andalusite Al_2SiO_5. A native aluminum silicate, gray or pink rhombic prisms, used for gems; decomp. at 1410 to mullite. Cf. *silica.*
andesine $NaCaAl_3Si_5O_{16}$. A native, triclinic feldspar. Cf. *andose.*
andesite A rock-forming mineral consisting of andesine with hornblende and mica.
andorite $AgPbSb_3S_6$. A native, orthorhombic sulfostibide.
andose A volcanic rock of andesite with diorite.
andradite An *iron garnet,* q.v.
Andrews, Thomas (1813–1885) Irish worker on the critical temperature and pressure of gases.
androgen A hormone or substance that causes development or maintenance of male sex organs and secondary sex characteristics. A. also controls growth of spermatozoa.
androkin Androsterone.
andromedotoxin A poisonous principle of mountain laurel, *Kalmia latifolia* (Ericaceae).
andrometoxin A poisonous principle from *Andromeda, Azalea,* and *Rhododendron* species. Colorless crystals, m.228, soluble in alcohol.
Andropogon A genus of grasses, some of which yield essential oils used in perfumery; e.g., *A.* (or *Cymbopogon*) *citratus,* lemongrass oil.
androstane* $C_{19}H_{32}$ = 260.5. A solid hydrocarbon, parent substance of the steroids. **des-*A*-~** $C_{15}H_{26}$ = 206.4. A. less the *A* (i.e., 1,2,3,4-carbon) ring. See *steroid.*
a. ring Cholane ring. The saturated 4-ring nucleus of the steroid structure, in many important biochemical compounds. Always assumed to be (8β, 9α, 10β, 13β, 14α) unless otherwise stated. R is H for androstane; other radicals give pregnane (R = Et), steroids, etc. The 5α-form:

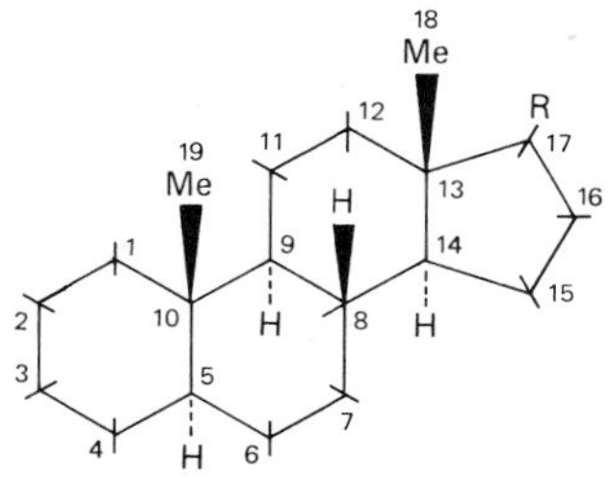

androsterone $C_{19}H_{30}O_2$ = 290.4. Androstane-3-(α)-ol-17-one, androtin, androkin. Colorless crystals, m.178, soluble in benzene. A male sex hormone occurring naturally in the testes, male urine, pregnant women, women with masculinizing tumors; also synthetically prepared. Cf. *estrone.* **dehydro ~** $C_{19}H_{28}O_2$ = 288.4. Colorless crystals, m.148, soluble in water. A male hormone occurring in male urine and women with adrenal tumors; also synthetically prepared.
androtin Androsterone.
-ane Suffix indicating (1)* the saturated hydrocarbon series (cyclo)alkanes, (2)* certain hydrides and related compounds, as silane, (3)* The semisystematic name of a saturated natural product whose parent structure is a hydrocarbon or carbocyclic.
anechoic Acoustically neutral; without echoes.
anemium Actinium.
anemometer An apparatus to measure wind velocity or pressure. Cf. *pitot tube.*
anemone camphor A principle from pulsatilla *(Anemone pulsatilla)* which splits into anemonin and anemonic acid.
anemonic acid $C_{10}H_{10}O_5$ = 210.2. Yellow crystals from anemone camphor.
anemoninic acid $C_{10}H_{12}O_6$ = 228.2. Colorless crystals derived from anemonic acid, m.189, soluble in water.
anemosite A mixed feldspar; sodium anorthite with albite.
aneroid See *aneroid* under *barometer.* **a. battery** Dry cell.
anesthesia Loss of feeling. **local ~** Loss of feeling at a definite part of the body.
anesthesin Benzocaine.
anesthesiophore radical The benzoyl group, PhCO—, which produces anesthetic effects.
anesthetic (1) Without feeling. (2) A substance used to produce anesthesia. (3) Pertaining to anesthesia. **general ~** Abolition of all sensation, including pain and touch, in which unconsciousness of different degrees is produced. A. is often inhaled. **local ~** Anesthesia produced in a region or part of the body by injection into that part or around the nerve supplying it. **surface ~** Topical a. Local a. applied directly to mucous membrane or open wound; used in ophthalmology, and in the nose and throat.
anesthetizing valve A device for mixing an anesthetic with air for respiration in animal experiments.
anethole* $MeO\cdot C_6H_4\cdot CH{:}CHMe$ = 148.2. Anethol, 1-methoxy-4-(1-propenyl)benzene†, *p*-allyl phenyl methyl ether, anise camphor, *p*-propenylanisole. A constituent of anise and fennel oils. Colorless leaflets, or aromatic liquid with fragrant odor, m.22, slightly soluble in water. Used as a reagent for lignin, flavoring agent, carminative, and in microscopy. **a. dibromide** Needles, m.67. **a. glycol** $C_{10}H_{14}O_3$ = 182.2. (1*RS*,2*RS*)-form, m.63; (1*RS*,2*SR*)-form, m.115. **a. picrate** Orange needles, m.70 (decomp.).
anethoquinine Quinine anisate.
anethum Garden dill, dill seeds. The fruits of *Paucedanum (Anethum) graveolens* (Umbelliferae); a carminative and condiment.
aneurin(e) hydrochloride Thiamine hydrochloride.
angelica The dried herb of *A.* or *Archongelica officinalis* (Umbelliferae). **a. lactone** $C_5H_6O_2$ = 98.1. **alpha-~** Colorless crystals, m.18, or colorless liquid, b.167. **beta-~** Colorless liquid, b.208; a flavoring. **a. root** The rhizomes and roots of *Angelica* species; the fluid extract is a diuretic, diaphoretic, and stimulant. **a. root oil** Colorless, essential oil, b.60–70. Its chief constituents are phellandrene and valeric acid; a flavoring. **Japanese ~** Colorless crystals, m.62; a flavoring. **a. seed** The ripe, dried fruits of *Angelica* species; a carminative. **a. seed oil** An essential oil from angelica seeds. Its chief constituents are phellandrene and valeric acid; a flavoring. **a. tree** The shrub *Xanthoxylum americanum* (Rutaceae) of the United States. The crushed bark smells like angelica.
angelic acid (*Z*)-2-Methyl-2-*butenoic acid**. **hydro ~** 2-Methyl*butanoic acid**. **iso ~** (*E*)-2-Methyl-2-*butenoic acid**.
angelicic acid (*Z*)-2-Methyl-2-*butenoic acid**.
angelicin $C_{11}H_6O_3$ = 186.2. A constituent of the roots of *Angelica archangelica.* Colorless crystals, m.138.
angico gum Brazilian gum, Para gum. A mucilaginous

secretion of *Piptadenia rigida* (Leguminosae) of Brazil. Cf. *cebil gum.*
angiology The science of blood and lymph vessels.
angioneurosin Nitroglycerin.
angiosperm A flowering plant whose seeds are enclosed in a fruit; e.g., the peapod. Cf. *gymnosperm.*
angle The inclination between two converging lines where they meet. **acute ~** An a. less than 90°. **adjacent ~** An a. which has one line common with another a. **complementary ~** The complement of an a. is 90° less the a. **critical ~** See *critical angle.* **meter ~** See *meter angle.* **oblique ~** An a. that is not a right a. **obtuse ~** An a. greater than 90°. **right ~** An a. of 90°. **supplementary ~** The supplement of an a. is 180° less the a.
a. thermometer An L-shaped thermometer.
anglesite $PbSO_4$. Occurs native in colored orthorhombic crystals. Cf. *sardinianite.*
angora A long, silky, curly goat's wool.
angostura A S. American tree, *Galipea cusparia, Cusparia febrifuga,* (Rutaceae). **a. alkaloids** The alkaloids of a.; as, angosturine. **a. bark** Cusparia bark, carony bark. The bark of a.; a bitter. **a. oil** An essential oil from the bark of a. Yellow, aromatic liquid, d.0.930–0.960, soluble in alcohol. It contains galipene, galipol, cadinene, and pinene; used in flavorings.
angosturine $C_{10}H_{40}O_{14}N$ = 398.4. An alkaloid from angostura bark. Colorless crystals, m.85; a bitter.
Ångström, A. J. (1814–1874) Swedish optical physicist.
angstrom Å. A unit of wavelength; 1 Å = 10^{-7} mm = 10^{-10} m = 1 am. (atom meter.) **international ~** I.A. The wavelength of the red line of cadmium = 6438.4696 I.A. in air at 15°C. Thence the I.A. = 10^{-10} m. Cf. *kX.*
angular Having sharp angles. **a. acceleration** A. acceleration (α) = $(\omega_t - \omega_0)/t$, where ω_t is the a. speed after time *t*, and ω_0 the initial a. speed; unit is rad/s^2. **a. aperture** The largest angle subtended by a wave surface transmitted by an objective. **a. momentum** Spin. The product of the a. speed and moment of inertia of a body expressed in g/cm·s. **a. motion** The motion of a line, fixed at one end in one plane, relative to a straight line through the center of rotation. **a. speed** A. velocity. The ratio $\omega = \theta/t$, where θ is the angle traversed in time *t*; unit is rad/s.
angustione $C_{11}H_{16}O_3$ = 196.2. A cyclohexane triketone isolated from the oil of *Backhousia angustifolia* (Myrtaceae). Colorless liquid, $b_{15mm}129$.
anhalamine $C_9H_7(OCH_3)_2OH \cdot NH$ = 209.2. An alkaloid from *mescal buttons,* q.v.
anhaline Hordenine.
anhalonidine $C_{12}H_{17}O_3N$ = 223.3. 1,2,3,4-Tetrahydro-8-hydroxy-6,7-methoxy-1-methylisoquinoline. An alkaloid from mescal buttons. Colorless octahedra, m.154, soluble in water.
anhalonine $C_{12}H_{15}O_3N$ = 221.3. 1,2,3,4-Tetrahydro-6-methoxy-1-methyl-7,8-methylenedioxyquinoline. An extremely poisonous alkaloid from *Anhalonium* species (Cactaceae). Colorless needles, m.254, soluble in water.
Anhalonium A Mexican cactus species, *A. lewinii* (peyotl), containing narcotic alkaloids. Cf. *mescal buttons.* **a. alkaloids** See *anhalamine, anhalonine, lophophorine, mescaline, pellotine.*
anhidrotic Antihidrotic. A drug which reduces perspiration.
anhydride* A compound (usually an acid) from which water has been removed; as, $H_2XO_3 - H_2O = XO_2$. **acid ~** The oxides of nonmetals, which form acids with water. **basic ~** The oxides of metals which yield bases with water. **inner ~** A ring compound formed by the abstraction of water; as, lactones. Cf. *acetic-, chromic-,* etc.
Anhydrite (1) Trade name for a desiccant containing chiefly anhydrous calcium sulfate. (2) (not cap.) $CaSO_4$. A native anhydrous calcium sulfate; gray, orthorhombic masses. **insoluble ~** β-Calcium sulfate. **soluble ~** γ-Calcium sulfate. Dehydrated bassanite, solubility 0.5% (20°C). Cf. *muriacite, tripestone, vulpinite.*
anhydro-* Prefix to compounds denoting that one or more water molecules have been removed from the prefixed compound. Cf. *anhydrous.*
anhydroecgonine $C_9H_{13}O_2N$ = 167.2. Ecgonidine. Colorless crystals, m.235 (decomp.), soluble in water.
anhydroformaldehyde aniline PhN:CH_2 = 105.1. A solid, m.120, insoluble in water.
anhydroglycochloral Chloralose.
Anhydrone Trade name for anhydrous magnesium perchlorate prepared by igniting the trihydrate; a powerful desiccant.
anhydrosynthesis The theoretical coupling of a group with another compound, with the subsequent elimination of water. Cf. *derivative.*
anhydrotimboine Timbonine.
anhydrous Describing a compound that has lost all its water. Cf. *anhydro-.*
anibine $C_{11}H_9O_3N$ = 213.3. 4-Methoxy-6(3-pyridyl)-2*H*-pyran-2-one. An alkaloid from the wood of the S. American rosewood, *Aniba duckei.*
anilides (1)*Phenyl-substituted amides, thus containing the C_6H_5NH- radical, from aniline; e.g., benzanilide, Ph·NH·CO·Ph. (2) Sometimes applied to compounds containing the $NH_2C_6H_4-$ group (anilinate); as, arsenic acid anilide (arsanilic acid). **acet ~** Acetanilide*. **form ~** $C_6H_5NH \cdot CHO$. Colorless crystals, m.46.
anilinate $NH_2C_6H_4M$. A compound of aniline and a metal. Cf. *anilides.*
aniline* $C_6H_5NH_2$ = 93.1. Phenylamine, benzenamine†, aminobenzene, aniline oil, benzidam. A pale brown liquid, darkening with age, m.−6, b.184, slightly soluble in water. Used as a reagent for aldehydes, chloroform, fusel oil, phenols, etc.; in bacteriology, for preparing staining solutions; and in the dye and rubber industries for organic synthesis and in the manufacture of resins and varnishes. Cf. *anilides, s anilinate, nitrobenzene reduction.* **acetyl ~** Acetanilide*. **allyl ~** See *allylaniline.* **amino ~** Phenylenediamine. **aminodimethyl ~** Dimethylphenylenediamine. **benzal ~, benzilidene ~** See *benzylidene.* **benzoyl ~** Benzanilide. **benzyl ~** See *benzylaniline.* **bi-** Benzidine*. **bromo ~*** $NH_2C_6H_4Br$ = 172.0. Aminobromobenzene. ***ortho-* ~** Colorless crystals, m.31, soluble in alcohol. ***meta-* ~** Colorless crystals, m.18, soluble in alcohol. ***para-* ~** Colorless rhombs, m.66 (decomp.), insoluble in water. **chloro ~*** $NH_2C_6H_4Cl$ = 127.6. Aminochlorobenzene. ***ortho-* ~** Colorless liquid, b.207, soluble in water. ***meta-* ~** Colorless liquid, b.230. ***para-* ~** Rhombs, m.70, soluble in hot water. **cyano ~** cyanoanilide. **diacetyl ~** Diacetanilide. **dibenzyl ~** See *benzyl aniline.* **dichloro ~*** $NH_2C_6H_3Cl_2$ = 162.0. Aminodichlorobenzene. **2,4- ~** Colorless needles, m.63, soluble in alcohol. **3,4- ~** Colorless needles, m.72, soluble in alcohol. **3,5- ~** Colorless needles, m.51, soluble in alcohol. **diethyl ~** $C_6H_5N(C_2H_5)_2$ = 149.2. Yellow liquid, m.38, sparingly soluble in water. ***N*-diethylnitro ~*** $NO_2C_6H_4NEt_2$ = 194.2. ***ortho-* ~** Soluble oil, b.290. ***para-* ~** Colorless needles, m.77, soluble in hot alcohol. ***N*-diethylnitroso ~*** $C_6H_4(NO)NEt_2$ = 178.2. ***para-* ~** Colorless needles, m.84, slightly soluble in water. **dimethyl ~**

$C_6H_5N(CH_3)_2$ = 121.2. Yellow liquid, m.2.5, slightly soluble in water. **ar-~** Xylidine*. ***N*-~** See *dimethylaniline* **dimethylamino ~** Dimethylphenylenediamine. ***N*-dimethylnitro ~** $NO_2C_6H_4NMe_2$ = 168.2. ***ortho*-~** A liquid, $b_{32mm}151$. ***meta*-~** Red prisms, m.60, insoluble in water. ***para*-~** Yellow needles, m.164, insoluble in water. ***N*-dimethylnitroso ~** $C_6H_4(NO)NMe_2$ = 150.2. ***para*-~** Green scales, m.88, slightly soluble in water. **d. hydrochloride** $C_6H_4(NO)NMe_2 \cdot HCl$ = 186.7. Yellow needles, soluble in water; used in dye manufacture. **dinitro ~*** $NH_2 \cdot C_6H_3(NO_2)_2$ = 185.1. **2,3-~** Orange needles, m.127, soluble in EtOH. **2,4-~** Yellow needles, m.188. **2,6-~** Yellow needles, m.141. **3,5-~** Yellow needles, m.163. **diphenyl ~** Triphenylamine*. **ethoxy ~** Phenetidine*. **ethyl ~** $C_6H_5NH(C_2H_5)$ = 121.2. Colorless liquid, b.205, sparingly soluble in water. **formyl ~** Formanilide. **hexahydro ~** Cyclohexylamine. **hydroxy ~** **ar-~** Aminophenol. ***N*-~** Phenylhydroxylamine*. **iodo ~** $NH_2C_6H_4I$ = 219.0. Aminoiodobenzene. ***ortho*-~** Colorless needles, m.60, sparingly soluble in water. ***meta*-~** Colorless leaflets, m.26, insoluble in water. ***para*-~** Colorless needles, m.63, insoluble in water. ***N*-isopentyl ~** $PhNHC_5H_{11}$ = 163.3. A liquid, b.260. **isopropyl ~** Cumidine. **methenyltri ~** Leucaniline. **methoxy ~** Anisidine*. **methyl ~** **ar ~** Toluidine*. ***N*-~** $C_6H_5NHCH_3$ = 107.2. Yellow liquid, m.196, slightly soluble in water. **nitro ~** $NH_2 \cdot C_6H_4 \cdot NO_2$ = 138.1. Aminonitrobenzene. ***N*-~** Phenylnitramine, nitranilide. Yellow crystals, m.46 (explode 98), soluble in water. **1,2-~** Yellow rhombs, m.72. **1,3-~** Yellow needles, m.112. **1,4-~** Yellow needles, m.147. **nitroso ~** $NH_2C_6H_4NO$ = 122.0. ***para*-~** Steel-blue needles, m.173, soluble in alcohol. ***p*-nitrosodiethyl ~** $(C_2H_5)_2N \cdot C_6H_4NO$ = 178.2. Colorless needles, m.84, slightly soluble in water. ***p*-nitrosodimethyl ~** $(CH_3)_2N \cdot C_6H_4NO$ = 150.2. Green scales, m.88, slightly soluble in water. **pentachloro ~*** $NH_2C_6Cl_5$ = 265.4. Colorless needles, m.232, soluble in alcohol. **phenyl ~** ***ar*-~** Amino*biphenyl*. ***N*-~** Diphenylamine*. **propionyl ~** See *propionylaniline.* **thio ~** See *thioaniline.* **thionyl ~** Sulfinylaniline. **tribromo ~*** See *tribromoaniline.* **trimethyl ~*** **2,4,5-~** Pseudocumidene. **2,4,6-~** Mesidine. **trinitro ~*** Picramide*.

a. acetate $C_6H_5NH_2 \cdot CH_3COOH$ = 153.2. Colorless liquid, soluble in water; a reagent for furaldehyde. **a. azo-2-naphthol** Sudan yellow. **a. black** Nigrosine. **a. blue** A mixture of the salts of triphenylrosanilinesulfonic acids. Blue powder, soluble in water; a dye for cotton and silk. **a. brown** Bismark brown. **a. chloride** A. hydrochloride. **a. colors** A. dyes. **a. dyes** (1) Artificial or synthetic coloring matters. (2) Dyes derived from benzene or aniline. **a. fluoride** A. hydrofluoride. **a. hydrobromide** $C_6H_5NH_2 \cdot HBr$ = 174.0. Gray crystals, soluble in water. **a. hydrochloride** $C_6H_5NH_2 \cdot HCl$ = 129.6. A. salt. Colorless crystals, m.198, soluble in water. Used as reagent for chlorates and for preparation of dyestuffs. **a. hydrofluoride** $C_6H_5NH_2 \cdot HF$ = 113.1. Colorless crystals, soluble in water. **a. hydroiodide** $C_6H_5NH_2 \cdot HI$ = 221.0. Yellow crystals, soluble in water. **a. hydrosilicofluoride** $(C_6H_5NH_2)_2 \cdot H_2SiF_6$ = 330.3. Colorless crystalline powder, soluble in water; a reagent. **a. nitrate** $C_6H_5NH_2 \cdot HNO_3$ = 156.1. Colorless crystals, decomp. 190, soluble in water. **a. oil** Crude a. **a. orange** Victoria orange. **a. oxalate** $(C_6H_5NH_2)_2C_2H_2O_4$ = 276.3. Colorless crystals, soluble in water. **a. point** The critical solution temperature of a mixture of a. and a water-insoluble liquid; used to determine petroleum in mixtures by comparing their a. p. with those of known mixtures. **a. printing** Early name for a flexographic printing process, with transparent inks. **a. purple** Mauvein. The first a. dye (W. H. Perkin, 1856). **a. red** Fuchsin. **a. salt** A. hydrochloride. **a. sulfate** $(C_6H_5NH_2)_2 \cdot H_2SO_4$ = 284.3. White crystals, soluble in water; a stimulant. **a.sulfonic acid.** ***ortho*-~** Orthanilic acid. ***meta*-~** Metanilic acid. ***para*-~** Sulfanilic acid. **a. tribromide** Tribromoaniline. **a. yellow** $C_6H_5N:NC_6H_4 \cdot NH_2$ = 197.2. C.I. Solvent Yellow 1. A yellow dye.

anilinium* The cation $Ph \cdot NH_3^+$.

anilino* The radical C_6H_5NH-.

anilism Poisoning produced by aniline vapors.

aniluvitonic acid $C_{11}H_9O_2N$ = 187.2. Methylcinchoninic acid. Colorless crystals, m.24.

animal A living organism, capable of locomotion and requiring organic food. **a. alkaloids** Basic organic compounds formed by the decomposition of animal matter. See *leucomaines, ptomaine.* **a. charcoal** See *charcoal.* **a. dyes** Coloring matters of a. origin; as, cochineal. **a. fibers** Textile fibers of a. origin; as, silk. **a. oil** (1) A fat or oil of a. origin. (2) Bone oil. **a. poisons** Toxic substances of a. origin; as, the toxins from snakes. **a. products** Substances of a. species; as, secretions (e.g., lanolin from sheep) and tissues (e.g., sponge from *Spongia* species).

animalcule A protozoon or microscopically small animal in ponds or stagnant water. Cf. *ameba, paramecium.*

anime Animi resin.

animikite Ag_9Sb, occurring native in the Lake Superior region.

animi resin A fossil copal from E. Africa, used for varnishes and lacquers.

aninsulin An antigenic, nonhypoglycemic substance made by heating insulin with formaldehyde.

anion* The negatively charged atom or radical liberated at the anode during electrolysis, e.g., Cl^-. Cf. *cation, ion.*

aniono* General term for anionic ligands.

anionotropy A case of *ionotropy,* q.v., in which a OH^- or X^- group breaks off from a molecule and leaves a positive ion in a state of dynamic equilibrium. Cf. *prototropy.*

anisacetone $MeO \cdot C_6H_4CH_2COMe$ = 164.2. *p*-Methoxyphenylacetone. A constituent of anise oil.

anisal Anisylidene. Methoxybenzylidene. The radical $MeOC_6H_4CH=$, from anisaldehyde.

anisalcohol $MeO \cdot C_6H_4 \cdot CH_2OH$ = 138.2. Methoxybenzyl alcohol. Colorless needles, m.45, insoluble in water.

anisaldehyde* $MeO \cdot C_6H_4 \cdot CHO$ = 136.2. Anisyl aldehyde, *p*-methoxybenzaldehyde. A colorless liquid, b.245, sparingly soluble in water; used in perfumery. **3-hydroxy ~** Iso*vanillin.*

anisaldoxime $MeO \cdot C_6H_4 \cdot CH:NOH$ = 151.2. Anisaldehyde oxime*. Colorless crystals. **(*E*)-~** or **α-~** m.64. **(Z)-~** or **β-~** m.133.

anisate* $MeO \cdot C_6H_4 \cdot COOM$ A salt of anisic acid.

anise Aniseed, a. fruit. The dried ripe fruit of *Pimpinella anisum* (Umbelliferae). A condiment, expectorant, and carminative (EP, BP). **star ~** The seeds of *Illicium verum* (Magnoliaceae), the source of anise oil.

a. bark oil An essential oil from the bark of a Malagasy Rep. tree. Yellow, spicy oil, soluble in alcohol, chief constituent is estragole. **a. camphor** Anethole*. **a. fruit** Anise. **a. oil** The essential oil from the seeds of *Pimpinella anisum* (Umbelliferae). Yellow liquid, b.210, soluble in alcohol; a flavoring and carminative. It contains anethole and estragole (NF, BP). **star ~** Illicium oil. An essential oil from the seeds of *Illicium verum* (Magnolaceae). Colorless liquid, m.14–18, soluble in alcohol; a flavoring containing anethole and estragole. **a. seed oil** A. oil **a. water** Aqua anisi. A saturated solution of a. oil in water.

anisic **a. acid*** $MeO \cdot C_6H_4 \cdot COOH$ = 152.2. *p*-Methoxybenzoic acid*, draconic acid. Colorless, monoclinic crystals, m.184, slightly soluble in water. **hydroxymethyl ~** Everninic acid. **3-methoxy ~** Veratric acid. **a. alcohol** Anisalcohol. **a. aldehyde** Anisaldehyde*.

anisidine* $NH_2C_6H_4OMe$ = 123.2. Methoxyaniline, methyloxyaniline. ***ortho-* ~** Brown oil, m.6, b.224, slightly soluble in water. ***meta-* ~** Volatile liquid, b.251. ***para-* ~** Colorless needles, m.58, soluble in alcohol. **bi ~**, **di ~** Bianisidine.

a. value A measure of the oxidation history of a refined oil (as, off-flavor) in terms of a.-reactive unsaturated aldehydes.

anisil $MeOC_6H_4CO \cdot COC_6H_4OMe$ = 270.3. Bianisaldehyde*. Colorless crystals, m.133.

anisole* $C_6H_5 \cdot OMe$ = 108.1. Methoxybenzene*, methyl phenyl ether. Colorless liquid, b.155, insoluble in water; used in perfumery. **acetamido ~** Acetaniside. ***p*-allyl ~** Estragole. **amino ~** Anisidine. **azoxy ~** See *azoxyanisole.* **bromo ~** BrC_6H_4OMe = 187.0. Bromomethoxybenzene*, bromophenyl methyl ether. **1,2- ~** An oil, b.222. **1,4- ~** Colorless crystals, m.11. **dinitro ~** $(NO_2)_2C_6H_3OMe$ = 198.1. Dinitromethoxybenzene. Yellow leaflets, m.89. **hydroxy ~** Guaiacol*. **methoxy ~** Veratrole. **nitro ~** See *nitroanisole.* **1-propenyl ~** Anethole. **trinitro ~** See *trinitroanisole.* **vinyl ~** $CH_2{:}CH \cdot C_6H_4 \cdot OMe$ = 134.2. Methoxystyrene. **1,2- ~** Colorless liquid, b.195. **1,3- ~** An oil, b.90. **1,4- ~** Colorless liquid, b.204, insoluble in water.

anisomeric Not isomeric.

anisonitrile $MeO \cdot C_6H_4CN$ = 133.1. Methoxybenzonitrile. **1,4- ~** Colorless crystals, m.60.

anisotonic Not isotonic.

anisotropic (1) Having different physical properties in different directions, as crystals. (2) Doubly refractive; not optically homogeneous. Antonym: isotropic. **a. liquid** Liquid *crystal.*

anisoyl* Methoxybenzoyl†. The radical $MeO \cdot C_6H_4 \cdot CO-$, from anisic acid. **bi ~**, **di ~** Anisil.

a. chloride* $MeO \cdot C_6H_4 \cdot COCl$ = 170.6. ***para-* ~** Anisyl chloride. Colorless needles, m.22 insoluble in water.

anisum Anise.

anisyl The methoxyphenyl* radical. **a. alcohol** Anisalcohol. **a. amine** Anisidine*. **a. chloride** Anisoyl chloride*.

anisylidene Anisal.

anitin Anytin.

anitol Anytol.

ankerite $(CaMgFeMn)CO_3$. A native carbonate.

annabergite $Ni_3As_2O_2 \cdot 8H_2O$. Nickel bloom, nickel ocher. Occurs as green, monoclinic crystals in Nevada.

annaline A native calcium sulfate.

annatto Annotto, annotta, arnotta, orleana, roucou. The orange coloring matter from the pulp of the fruit seeds of the evergreen *Bixa orellana* (Bixaceae). Used to color cheese and milk and to dye textiles. Cf. *bixin.*

anneal To temper by heating.

annealing The tempering of glass or metals by heating, then cooling, to render them less brittle. See *tempering.* **a. color** The tints of steel during a. **a. cup** A fine clay crucible for silica fusions.

annerodite A native uranium and yttrium niobate; e.g., black orthorhombic crystals.

annotto Annatto.

annular (1) Ring-shaped. (2) A ring-shaped space.

annulene Generic name for large, ring-conjugated, aromatic compounds.

anode Posode. The positive pole or electrode of a battery, vacuum tube, or electrolyzing circuit.

anodic Pertaining to the anode. **a. oxidation** The protection of metals, as, aluminum, against corrosion by producing a thin surface film of oxide on the metal, which is made the anode in a chromic acid bath.

anodyne A drug which relieves pain and causes sedation or relief from anxiety; as, morphia.

anol $MeCH{:}CH \cdot C_6H_4 \cdot OH$ = 134.2. *p*-Propenylphenol. Crystals, m.98; a constituent of essential oils.

anolyte The liquid in the immediate neighborhood of the anode during electrolysis. Cf. *catholyte.*

anomalous Contradictory. Cf. *anomalous* under *liquid, anomalous* under *viscosity.*

anomaly An abnormal or irregular type or form.

anomite Biotite, containing lithium.

anonaceine An alkaloid from *Hylopeia aethiopica* (Anonaceae).

anorectic drug A drug that reduces appetite; used to treat obesity.

anorganic Inorganic.

anorthic Triclinic.

anorthite $CaAl_2Si_2O_3$. A triclinic feldspar. **sodium ~** A. containing albite. Cf. *silica, anemosite.*

anorthoclase A triclinic sodium-potassium feldspar.

anorthosite An igneous gabbro feldspar.

anoxemia Lack of blood oxygen, as in mountain sickness and diseases of the respiratory system.

anoxyscope An apparatus to demonstrate the necessity of oxygen for plant growth.

anserine $C_{10}H_{16}O_3N_4$ = 240.3. *N*-*β*-alanyl-3-methylhistidine. A dipeptide in the muscles of birds, reptiles, and fishes, m.238; a homolog of carnosine.

ANSI See *American National Standards Institute.*

ant **a. amber, a. butter, a. incense** A dirty-white resin collected by ants in their nests, chiefly in European conifer woods.

Antabuse Trademark for disulfiram.

antacid A substance that neutralizes acids or relieves acidity, e.g., aluminum hydroxide.

antagonism Counteraction or opposition. **biological ~** Inhibition of the toxic effect of certain substances by other substances, e.g., certain ions.

antagonist Physiological *antidote.*

antalkaline A substance that neutralizes alkalies or relieves alkalinity, e.g., acetic acid.

antarcticite $CaCl_2 \cdot 6H_2O$. Crystals found in an antarctic lake, believed to prevent its freezing.

anthelmintic Helminthic. A drug that expels intestinal worms. Cf. *vermicide.*

anthemane $C_{18}H_{38}$ = 254.5. Octadecane; a paraffin in chamomile flowers.

anthemidine An antispasmodic alkaloid from mayweed *(Anthemis cotula).*

anthemis Roman chamomile, ground apple. The flower heads of *Anthemis nobilis* (Compositae); an antispasmodic.

anthemol $C_{10}H_{16}O$ = 152.2. Chamomile camphor, a constituent of anthemis.

anther The pollen-bearing part of a flower.

antheraxanthin $C_{40}H_{56}O_3$ = 584.9. A carotenoid, m.211, from the anthers of *Lilium tigrinum* (Liliaceae).

anthesterol Taraxasterol.

anthion Potassium peroxosulfate.
Anthisan Trademark for pyrilamine maleate.
antho- Prefix (Greek "flower") indicating a relationship to flowers.
anthocyanase Enzyme used to reduce the pigment coloration in berries; as in the manufacture of jelly.
anthocyanidins Derivatives of 3,5,7-trihydroxyflavylium chloride occurring in the red, blue, and purple colorings of flowers. Cf. *flavanols.*
anthocyanins Glucosides comprising the soluble colors of blue, red, and violet flowers; as, delphinin. Cf. *flavanols.*
anthocyans A group of red, blue, or violet plant colors.
anthophyllite $MgFe(SiO_3)$. A brown, orthorhombic amphibole.
anthoxanthins Glucosides comprising the colors of yellow flowers; as, quercetin.
anthra-* Prefix indicating the anthracene ring in a polycyclic compound in addition to the prefixed base component.
anthracene* $C_{14}H_{10}$ = 178.2. Anthracin, *p*-naphthalene, anthracene oil. A hydrocarbon from coal tar distillation.

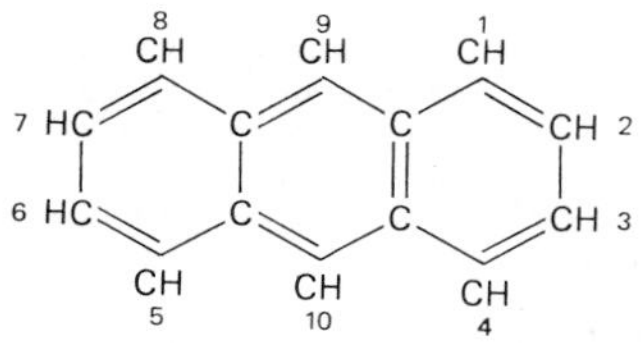

Colorless, blue-fluorescent needles, m.216, insoluble in water; used in the manufacture of alizarin dyes. Commercial a. contains carbazole and phenanthrene. **amino ~** Anthramine. **diamino ~** Anthradiamine. **dibromo ~*** $C_6H_4 \cdot C_2Br_2C_6H_4$ = 336.0. **9,10- ~** Yellow crystals, m.221, insoluble in water. **dichloro ~*** $C_6H_4 \cdot C_2Cl_2 \cdot C_6H_4$ = 247.1. **9,10- ~** Yellow needles, m.209, insoluble in water. **dihydro ~*** $C_6H_4(CH_2)_2C_6H_4$ = 180.2. Hydroanthracene, diphenylene dimenthylene. Colorless, triclinic crystals, m.109, insoluble in water. **dihydrodioxo ~** Anthraquinone. **dihydrooxo ~** Anthrone. **dihydroxy ~*** See *anthradiol.* **dimethyl ~*** $C_{14}H_8(CH_3)_2$ = 206.3. **2,3- ~** Colorless leaflets, m.246. **2,4- ~** Colorless needles, m.71. **ethyl ~*** $C_{16}H_{14}$ = 206.3. **9- ~** Colorless leaflets, m.59, insoluble in water. **hexahydro ~** $C_{14}H_{16}$ = 184.3. Colorless leaflets, insoluble in water. **hydro ~** Dihydroanthracene. **hydroxy ~** **1-,2-, or 3- ~** Anthrol. **9- ~** Anthranol. **hydroxyoxo ~** Oxanthranol. **methyl ~** $C_{14}H_9CH_3$ = 192.3. **2-** Colorless scales, m.200, soluble in alcohol. **9-** White crystals, m.82, slightly soluble in water. **nitro ~** $C_{14}H_9NO_2$ = 223.2 **1-** α-Nitrosoanthrone. Yellow needles, m.146, insoluble in water. **peri ~** Chrysazol. **tetradecahydro ~** A. perhydride. **tetrahydroxy ~** Anthratetrol. **trihydroxy ~** See *anthratriol, anthrarobin.*

a. blue Alizarin blue. **a.carboxylic acid** Anthroic acid. **a.diol** Anthradiol **a.dione** Anthraquinone*. **a.diyl†** See *anthrylene.* **a. oil** A product of coal tar distillation (above 270° C) containing a. with carbazole, phenanthrene, etc. **a. perhydride** $C_6H_8 \cdot (CH_2)_2 \cdot C_6H_8$ = 188.3. Saturated anthracene. Colorless needles, m.88, soluble in alcohol **a.sulfonic acid** Anthraquinonesulfonic acid. **a.tetrol** Anthratetrol. **a. tetrones** Anthradiquinones. A group of ketone derivatives containing 4 O atoms attached to the a. ring. **a.triol** Anthratriol. **a. violet** Gallein.
anthracenol Anthrol*.
anthracenone Anthrone*.
anthrachrysone $C_{14}H_8O_6$ = 272.2. Anthraquinone 1,3,5,7-tetrol, anthrachrysazin. Orange crystals, m.360, insoluble in water.
anthracine (1) A ptomaine produced by the anthrax bacillus. (2) Anthracene. Cf. *anthrazine.*
anthracite Stone coal. A bright, lustrous, hard, brittle mineral *coal,* q.v., containing 2-8% volatile carbon, d.1.3-1.8. **meta ~** Contains less than 2% volatiles. **semi ~** Contains 8-14% volatiles.
anthracometer A device to determine the amount of carbon dioxide in gas mixtures.
anthraconite Stinkstone. An earthy amorphous form of bituminous calcium carbonate.
Anthra-derm Trademark for anthralin.
anthradiamine $C_{14}H_8(NH_2)_2$ = 208.3. Anthracenediamine*. Yellow needles, m.160.
anthradiol $C_{14}H_{10}O_2$ = 210.2. Dihydroxyanthracene, oxanthranol, anthracenediol*. **1,2- ~** 1,2-Anthracenediol*. Green leaflets, m.160. **1,5- ~** Rufol. **1,8- ~** Chrysazol. **2,6- ~** Flavol.
anthradiquinones Anthracenetetrones.
anthraflavin $C_{14}H_8O_4$ = 240.2 **2,6- ~** Dihydroxyanthraquinone*. Yellow needles, m.330, insoluble in water. **iso ~** See *isoanthraflavic acid.*
anthragallol $C_{14}H_8O_5$ = 256.2. 1,2,3-Trihydroxanthraquinone, alizarin brown. Brown needles, m.310, soluble in alkalies (green color); a dye and dye intermediate.
anthraglucorhein A cathartic glucoside from *Rheum* species; a brown powder, soluble in alcohol.
anthraglucosagradin A glucoside from *Cascara sagrada.* Brown powder, soluble in alcohol; a cathartic.
anthraglucosennin A glucoside from *Cassia angustifolia.* Brown powder, soluble in alcohol; a cathartic.
anthrahydroquinone Anthradiol.
anthraldehyde $C_{14}H_9 \cdot CHO$ = 206.2. **9- ~** Orange crystals, m.105, insoluble in water.
anthralin $C_{14}H_{10}O_3$ = 226.2. 1,8,9-Anthracenetriol (and tautomers), Dithranol, Anthra-derm. Yellow powder, m.178, insoluble in water. An ointment for skin diseases, e.g., psoriasis (USP, BP).
anthramine $C_{14}H_{11}N$ = 193.17. Aminoanthracene, anthrylamine. **2- ~** Yellow needles, m.237, sparingly soluble in water. **5- ~** Yellow crystals, m.146, soluble in alcohol; a dye intermediate.
anthranil $C_6H_4 \cdot (CHON)$ = 119.1. Anthroxan, *o*-aminobenzoic acid lactam, anthranilic acid lactam. Colorless crystals or liquid, m.18, soluble in alkalies. **a. aldehydye** *o*-Amino*benzaldehyde.* **a. carbonic acid** Isatoic acid.
anthranilate* A salt of anthranilic acid.
anthranilic acid* $NH_2 \cdot C_6H_4 \cdot COOH$ = 137.1. 2-Aminobenzoic acid*. Yellow crystals, m.144, soluble in water; used in the manufacture of dyes and perfumes. Cf. *anthraniloyl.* ***N*-carboxy ~** Isatoic acid. **dinitro ~** See *chrysalicic* acid, *chrysanisic* acid. **oxalyl ~** Kynuric acid.
anthranilic di-*N*-propylaniline An indicator changing at pH 5.5 from red (acid) to yellow (basic).
anthranilo The radical $C_6H_4(CO \cdot N)-$, from anthranil. **a. nitrile** $NH_2 \cdot C_6H_4 \cdot CN$ = 118.1. *o*-Aminophenyl cyanide. Yellow prisms, m.50.
anthraniloyl* 2-Aminobenzoyl†. The radical *o*-$H_2N \cdot C_6H_4 \cdot CO-$, from anthranilic acid.
anthranol $C_{14}H_{10}O$ = 194.2. 9-Hydroxyanthracene*. Yellow needles, decomp. 160, soluble in ether. Cf. *anthrone.* **dioxy ~** Anthrarobin.
anthranone Anthrone*.

anthranthrenes $C_{26}H_{16}$ = 328.4. Hexacyclic hydrocarbons, consisting of 2 fused anthracene rings.
anthranylamine Anthramine.
anthraparazene $C_{30}H_{18}N_2$ = 406.5. The compound

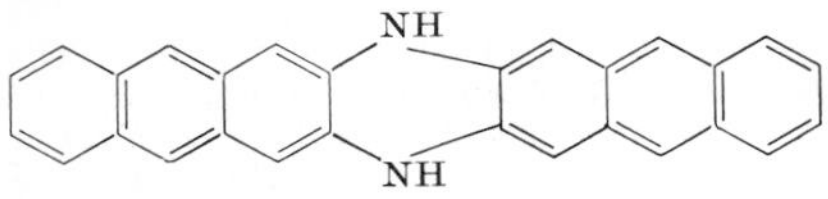

anthrapurpurin $C_{14}H_8O_5$ = 256.2 **1,2,7-** Trihydroxyanthraquinone*. Orange needles, m.370, slightly soluble in water. Used in the manufacture of dyes. **a. diacetate** Purgatol.
anthrapyridine **beta-~** Benz[*g*]isoquinoline. Yellow crystals, m.170, soluble in water.
anthraquinol Anthradiol.
anthraquinoline $C_{17}H_{11}N$ = 229.3. Naphtho[2,3-*f*]quinoline. Colorless crystals, m.170, insoluble in water.
anthraquinone* $C_6H_4\cdot(CO)_2\cdot C_6H_4$ = 208.2. Anthracenedione†, dihydrodioxoanthracene. **9,10-~** Yellow needles, insoluble in water. A constituent of *Cassia, Aloe, Mirabilis,* and *Rumex* species; manufactured by oxidation of anthracene. Used in the manufacture of alizarin and other dyes. **amino~** $C_{14}H_7O_2NH_2$ = 223.2. Anthraquinoylamine **1-~** Red, iridescent needles, m.242, insoluble in water; used in dyestuff synthesis. **2-~** Red needles, m.302, insoluble in water, soluble in alcohol; used in organic synthesis. **2-amino-1-hydroxy~** $NH_2\cdot C_{14}H_6O_2\cdot OH$ = 239.2. *β*-Alizarinamide. Brown needles. m.226 **bromo~*** $C_{14}H_7O_2Br$ = 287.1. Yellow crystals. **1-~** m.188. **2-~** m.204. **chloro~*** $C_{14}H_7O_2Cl$ = 242.7. Yellow needles. **1-~** m.162. **2-~** m.211, insoluble in water. **dihydroxy~** **1,2-~** Alizarin. **1,3-~** Xanthopurpurin. **1,4-~** Quinizarin. **1,5-~** Anthrarufin. **1,8-~** Chrysazin. **2,3-~** Hystazarin. Yellow needles, m.260. **2,6-~** Anthraflavin. **2,7-~** Isoanthraflavic acid. **1,8-dihydroxy-3-methyl~** Chrysophanic acid. **hexahydroxy~*** Rufigallic acid. **hydroxy~*** $C_{14}H_7O_2\cdot OH$ = 224.2. Yellow leaflets, m.302, soluble in water. **methyl~*** $C_{14}H_7O_2Me$ = 222.2. Yellow crystals, m.177, soluble in alcohol. **nitro~*** $C_{14}H_7O_2\cdot NO_2$ = 253.2. Yellow needles, m.228, insoluble in water. **pentahydroxy~** Alizarin cyanine. **tetrahydroxy~** **1,2,5,6-~** Rufiopin. **1,2,5,8-~** Quinalizarin. **1,3,5,7-~** Anthrachrysone. **trihydroxy~** **1,2,3-~** Anthragallol. **1,2,4-~** Purpurin. **1,2,6-~** Flavopurpurin. **1,2,7-~** Anthrapurpurin. **trihydroxymethyl~** 1,3,8,6-Emodin.
a.acridine Naphthacridinedione. **a.acridone** Naphthacridinetrione. **a. dyes** See *alizarin.* **a.sulfonic acid** Beta acid, alizarinsulfonic acid. An intermediate in the manufacture of alizarin. Cf. *silver salt.* **a.tetrol** Tetrahydroxy~ **a.triol** Trihydroxy~.
anthraquinonic acid Alizarin.
anthrarobin $C_{14}H_{10}O_3$ = 226.2. 1,2,10-Anthracenetriol*, desoxyalizarin. Yellow granules, m.208, insoluble in water. Used to treat skin diseases, and as a substitute for chrysarobin.
anthrarobinate A salt of anthrarobin.
anthrarufin $C_{14}H_8O_4$ = 240.2. 1,5-Dihydroxyanthraquinone*. Yellow leaflets, m.280, soluble in water.
anthratetrol $C_{14}H_6(OH)_4$ = 242.2. Tetrahydroxyanthracene. A group of tetrahydroxy derivatives of anthracene.
anthratriol $C_{14}H_7(OH)_3$ = 226.2. Trihydroxyanthracene. A group of anthracenetriol compounds; as, anthrarobin.
anthraxolite A metamorphic coal resembling anthracite, the end product in the metamorphosis of petroleum.
anthraxylon A layer formation in coal corresponding with the larger pieces of woody peat. Similar to, and sometimes confused with, vitrain. See *coal.*
anthrazine $(C_{14}H_8)_2N_2$ = 380.4. Anthracenediazine. m.400. Cf. *anthracine.*
anthroic acid $C_{14}H_9\cdot COOH$ = 222.2. Anthracenecarboxylic acid*. **1-~** Yellow needles, m.260 (sublime), insoluble in water. **2-~** Yellow leaflets, m.280 (sublime), insoluble in water. **5-~** Yellow needles, decomp. 206, soluble in water or alcohol. A. a. are used in organic synthesis and in the manufacture of dyes.
anthrol* $C_{14}H_9OH$ = 194.15. Hydroxyanthracene. **1-~** Colorless needles, m.153. **2-~** or **3-~** Colorless needles, decomp. 200, soluble in alcohol. **9-~** Anthranol.
anthrone* $C_6H_4\cdot CO\cdot C_6H_4\cdot CH_2$ = 194.2. 9(10*H*)-Anthracenone*, anthranone. **10-hydroxy~** Oxanthranol. **nitroso~** Nitro*anthracene.*
anthroxan Anthranil. **a.aldehyde.** $C_8H_5O_2N$ = 147.1. Colorless crystals, m.72.
anthryl* Anthracenyl†. The 5 isomeric radicals $C_{14}H_9-$, derived from anthracene. **a.amine** Anthramine.
anthrylene* Anthracenediyl†. The 11 isomeric radicals $-C_{14}H_8-$, derived from anthracene.
anti- Prefix (Greek) indicating against or opposed.
anti position Former term for *E, trans,* and *transoid* locations with respect to a double bond. See *stereoisomer.*
antiabrin An antibody formed in the blood after the injection of abrin.
antiagglutinin An antibody which prevents agglutination.
antiaggressin An antibody.
antialbumid A decomposition product of albumin formed during gastric and pancreatic digestion.
antialbuminate Antialbumate, parapeptone. An incompletely digested albumin. Cf. *syntonin.*
antianaphylactin A substance counteracting anaphylactin. Cf. *allergen.*
antiantibody A substance formed in the blood of an organism after injection of an antibody.
antiantidote A substance which opposes the action of an antidote.
antiantienzyme A substance which opposes the action of an antienzyme.
antiar The milky juice of the upas (Javanese) or ipoh (Malaysian) tree, *Antiaris toxicaria* (Malacca, Java); an arrow poison.
antiarin (1) $C_{14}H_{20}O_5$ = 268.16. A glucoside of antiar. An arrow muscle poison from southeast Asia. (2) $C_{14}H_{20}O_5\cdot 2H_2O$ = 288.3. The active principle of antiar.
antiarthritic A drug that relieves symptoms of arthritis; as, aspirin. Cf. *antipodagric.*
antiasthmatic A drug relieving asthma; e.g., salbutamol.
antibacterial A substance that checks the growth of bacteria. Cf. *antiseptic, disinfectant.*
antiberberin A black liquid prepared from rice and used to treat beriberi.
antiberiberi vitamin Thiamine.
antibilious A drug that relieves a bilious condition.
antibiote An early term for antibiotic.
antibiotics Soluble substances, produced by growing microorganisms, which kill or inhibit growth of other microorganisms. Principal classes are (see under class name) (1) aminoglycosides, (2) cephalosporins, (3) chloramphenicols,

(4) fusidic acid, (5) macrolides, (6) penicillins, (7) polymixin, (8) tetracyclines. **antitumor ~** A. isolated from several species of streptomyces, with pronounced adverse effects on cell growth by interfering with action of DNA; as, dactinomycin, bleomycin sulfate.

antiblennorrhagic A drug used to prevent or treat gonorrhea; e.g., penicillin.

antibody An immunoglobulin formed in the body fluids of animals after an injection; it counteracts the effects of the injected substance. Examples include antitoxins and precipitins.

antibonding See *σ-bond* under *bond.*

anticachectic A drug used to treat malnutrition.

anticatalase A substance opposing the action of a catalase.

anticatalyzer A substance that inhibits the action of a catalyst.

anticatarrhals A drug subduing the inflammation of mucous membranes.

anticathode Target. An electrode in a vacuum tube placed relative to the cathode so that the rays from the latter impinge on it. See *X-ray tube.*

antichlor A chemical used to remove excess chlorine after bleaching; e.g., sodium thiosulfate.

anticoagulant (1) A drug that reduces clotting of blood, e.g., warfarin. (2) Preventing coagulation.

anticoagulin Anticoagulant.

anticomplement A substance formed by using cells to combine with the *complement,* q.v.

anticytotoxin A substance opposing the action of cytotoxins.

antidiabetic A drug used to treat diabetes.

antidimmer A preparation which prevents moisture from accumulating on glass.

antidote An agent counteracting or neutralizing the action of a poison. **chemical ~** A substance which precipitates or alters a poison, as, oils, soaps, egg white. **mechanical ~** A means of removing a poison from the system; e.g., stomach pump. **physiological ~** A drug which counteracts a poison by having an opposite effect. **universal ~** A mixture of 2 pts. charcoal, 1 pt. magnesium oxide, and 1 pt. tannin in water.

antiemetic A drug which prevents or reduces vomiting; e.g., promethazine theoclate.

antienzyme A substance that inhibits enzyme action.

Antifebrin A proprietary brand of acetanilide.

antiformin Hychlorite. A strongly alkaline solution of sodium hypochlorite; a disinfectant.

antifreeze A substance added to the radiator water of automobiles to prevent freezing; e.g., ethylene glycol.

antigalactic A drug to diminish milk secretion.

antigen Immunogen. An injection or substance which causes the formation of antibodies; as, (1) *toxins:* (*a*) bacterial (diphtheria), (*b*) vegetable (ricin), (*c*) animal (snake venom); (2) *enzymes* (lipase); (3) precipitinogens, (animal proteins); (4) agglutinogens (bacteria); (5) opsogens (aggressins); (6) *lysogens* (animal cells).

antiglobulin A precipitin that coagulates globulin.

antihemolytic An agent that prevents hemolysis.

antihemophilic factor An extract of human venous plasma; white powder. Used to treat hemophilia (USP).

antihidrotic Anhidrotic.

antihistaminic A substance that counteracts the effects of an excess of histamine in the body; e.g., chlorcyclizine hydrochloride.

antihypo Potassium percarbonate, used in photography.

antiknock See *knock.*

antilab Antirennin.

antilithic A drug that prevents the formation of urinary stones; e.g., allopurinol.

antiluetic Antisyphilitic.

antilysin A substance formed in blood which destroys bacterial lysins.

antilyssic A drug used to treat rabies.

antimalarial A drug relieving or preventing malaria; e.g., quinine.

antimalum A mixture of essential oils, chiefly camphor oil, used externally for rheumatism.

antimers Enantiomers.

antimonate Antimoniate, stibnate, stibiate, stibate. (1)* Indicating antimony as the central atom(s) in an anion; as, the salts of antimonous acid [antimonate(III)* or antimonite]; as, Na_3SbO_3. (2) A salt of antimonic acids; probably mixed oxides, as; **ortho ~** antimonate, M_3SbO_4; **meta ~** $M_4Sb_2O_5$ or $MSbO_3$; **di ~, pyro ~,** $M_4Sb_2O_7$.

antimonial **a. glass** Antimonous sulfide. **a. lead** Hard lead. An alloy: Pb 85, Sb 15%; resistant to sulfuric acid. **a. nickel sulfide** Ullmannite. **a. saffron** Antimonous oxysulfide. **a. silver** Ag_3Sb. Dycrasite. Native silver antimonide.

antimoniate Antimonate.

antimonic Antimony(V)* or (5+)*. A compound of pentavalent antimony. **a. acid** **ortho ~** H_3SbO_4 = 188.8. White powder, d.6.6, decomp. by heat, slightly soluble in water. **meta ~** $HSbO_3$ = 170.8. White powder, sparingly soluble in water. **pyro ~** $H_4Sb_2O_7$ = 359.5. Colorless powder, slightly soluble in water. **a. chloride** $SbCl_5$ = 299.0. Antimony pentachloride*. Fuming yellow liquid, b_{30mm}92, decomp. in water, soluble in acids; a reagent for alkaloids. **a. fluoride** SbF_5 = 216.7. Antimony pentafluoride*. An oil, $d_{22°}$2.99, soluble in water. **a. oxide** Sb_2O_5 = 323.5. A. acid anhydride, antimony peroxide. Yellow powder, insoluble in water. **a. oxychloride** $SbOCl_3$ = 244.1. Antimony trichloride oxide*. Yellow powder, decomp. by heat, insoluble in water. **a. sulfide** Sb_2S_5 = 403.8. Antimony pentasulfide. Orange powder, $d_{0°}$4.120, insoluble in water.

antimonide* Stibide. A metal derivative of stibine, SbH_3.

antimonii Official Latin for "of antimony"; antimonic or antimonous compounds.

antimonine Antimony lactate.

antimonious Antimonous.

antimonite (1) Antimonate(III)*. (2) Antimony glance. **thio ~** See *thioantimonite.*

antimonium Antimony.

antimono* 1,2-Distibenediyl†. The divalent group —Sb:Sb—.

antimonous Antimony(III)* or (3+)*. Stibous, stibious, stibnous, antimonious. Describing compounds of trivalent antimony. **a. acid.** **ortho ~** H_3SbO_3 = 172.8. **meta ~** $HSbO_2$ = 154.8. A hypothetical acid which forms salts. **di ~, pyro ~** $H_4Sb_2O_5$ = 327.5. White solid; forms the trioxide when heated. **a. basic chloride** Antimony oxychloride. **a. bromide** $SbBr_3$ = 361.5. Yellow rhombs, m.94, hydrolyzes in water. **a. chloride** $SbCl_3$ = 228.1. Butter of antimony. White rhombs, m.73, sparingly soluble in water. Antimony butter is a fuming acid. Used as a caustic, a mordant; also for manufacturing antimony salts, and for staining iron and copper articles. **a. fluoride** SbF_3 = 178.7. Grayish octahedra, m.292, soluble in water. Cf. *DeHaën salt.* Used in ceramics and as a mordant. **a. fluoride and ammonium sulfate** De Haën salt. **a. hydride** Stibine*. **a. iodide** SbI_3 = 502.5. 3 allotropes. (1) Red, *hexagonal* crystals; (2) red, *monoclinic* crystals; (3) yellow, *rhombic* crystals, m.171, decomp. in water. **a. nickel** NiSb. A native alloy.

a. oxalate $SbO \cdot (C_2O_4)_2$ = 313.8. Antimonyl oxalate. White powder, soluble in acids; a mordant. **a. oxide** Sb_2O_3 = 291.5. Valentinite, antimony bloom, a. glass, Sb_4O_6. Colorless rhombs, m. 655, insoluble in water. A powerful reducing agent. **a. oxychloride** SbOCl = 173.2. Antimony chloride oxide*, Algaroth, mercurius vitae, basic antimony chloride. White, regular crystals, insoluble in water. Used in the manufacture of tartar emetic. **a. oxyiodide** SbOI = 264.6. Antimony iodide oxide*. Yellow crystals, insoluble in water. **a. oxysulfide** $Sb_2S_3 \cdot Sb_2O_3$ = 631.2. Antimony flowers, antimony red, antimony vermilion, antimonial saffron. A double salt of a. sulfide and oxide, containing some $SbOS_2$ (crocus metallorum). Brown powder, vulcanizer for rubber. **a. sulfate** $Sb_2(SO_4)_3$ = 531.7. White powder, soluble in acids. **a. sulfide** Sb_2S_3 = 339.7. Stibnite, antimony red, black antimony sulfide. Black powder or red crystals, m.555, insoluble in water; a pigment for the manufacture of safety matches; an *opacifier*, q.v.

antimony* Sb = 121.75. Stibium. A metal of the nitrogen family and element, at. no. 51. A rhombohedral, blue-white, brittle, lustrous substance, d.6.62, m.630, b.1750, soluble in concentrated sulfuric acid or aqua regia. Basilius Valentinus described its preparation and properties in 1450. Occurs native, as metal, as sulfide (stibnite), and as antimonides and sulfantimonides of the heavy metals. Allotropes: (1) ordinary metallic of β-antimony, Sb_1; (2) an unstable, yellow or α-antimony, Sb_4; (3) amorphous, black antimony, Sb_2, d.5.3, which heat changes to (1). A. is tri- and pentavalent and resembles arsenic and phosphorus in chemical behavior.

Compounds derived from antimony:

From trivalent antimony

Antimony(III)*, antimony(3+)*, antimonous	Sb^{3+}
Stibine*	SbR_3
Antimonate(III)*, antimonite	SbO_3^{3-}
Antimonyl	SbO^+

From pentavalent antimony

Antimony(V)*, antimony(5+)*, antimonic	Sb^{5+}
Stibonium*	R_4SbX
Metaantimonate	SbO_3^-
Orthoantimonate	SbO_4^{3-}

A. is used extensively in alloys (Britannia metal, type metal, pewter) and in the preparation of antimony compounds used in medicine and as pigments. **black ~** Antimonous sulfide. **butter of ~** Antimonous chloride. **flowers of ~** Antimonous oxysulfide. **gray ~** A. glance. **red ~** (1) Kermesite. (2) Antimonous oxysulfide. **white ~** Valentinite. **a. (III)*, a. (3+)*** See also *antimonous.* **a. (V)*, a. (5+)*** See also *antimonic.* **a. anhydride** See *antimonous* or *antimonic oxide.* **a. arsenide** Allemontite. **a. ash** See *a. oxides.* **a. black** Antimonous sulfide. **a. blende** Antimony glance. **a. bloom** Antimonous oxide. **a. butter** Antimonous chloride. **a. cesium chloride*** Cs_6SbCl_9 = 1238. Yellow crystals, soluble in water. **a. chlorides** See *antimonous(ic)* chloride and oxychloride. **a. crocus** Antimonous oxysulfide formed by the deflagration of equal parts of antimonous sulfide and saltpeter. **a. flowers** Antimonous oxysulfide. **a. fluoride** See *antimonous(ic)* fluoride. **a. glance** Antimonite, stibnite, a. blende, gray a. A native antimonous sulfide. **a. glass** Antimonous oxide. **a. lactate*** $Sb(C_3H_5O_3)_3$ = 389.0. Antimonine. Yellow crystals, soluble in water; a dye mordant. **a. minerals** A. is closely associated with arsenic and bismuth minerals; also occurs as the antimonides and sulfantimonides of the heavy metals, and as stibnite, kermesite, valentinite, and cervantite. Cf. *plagionite, antimonous* nickel. **a. mordants** See *antimonous* fluoride, *a. lactate* and *a. potassium tartrate.* **a. needles** Antimonous sulfide. **a. ocher** Stibiconite. **a. oxides** See *antimonous oxide, antimony tetraoxide, antimonic oxide.* **a. pentachloride*** Antimonic chloride. **a. pentafluoride*** Antimonic fluoride. **a. pentamethyl** Pentamethyl*stibine**. **a. pentasulfide*** Antimonic sulfide. **a. pentoxide, a. peroxide** Antimonic oxide. **a. persulfide** Antimonic sulfide. **a. potassium oxalate*** $SbK_3(C_2O_4)_3 \cdot 6H_2O$ = 611.2. White powder, soluble in water; a dye mordant. **a. potassium tartrate** $Sb_2K_2C_8H_4O_{12} \cdot 3H_2O$ = 667.8. Tartar emetic, antimonyl potassium tartrate, tartar stibiatus. Colorless octahedra, soluble in water. Formerly an emetic; used for schistosomiasis(USP). A dye mordant for textiles and leather and an analytical reagent. **a. red** Antimonous oxysulfide. **a. regulus** Metallic a. **a. salt** DeHaën salt. **a. sodium sulfate** $SbF_3 \cdot Na_2SO_4$. A. trifluoride sodium sulfate. **a. sodium tartrate** $(SbO)NaC_4H_4O_6$ = 308.8. Plimmer's salt. Colorless, transparent crystals, soluble in water. Used to treat schistosomiasis. **a. sulfate** Antimonous sulfate. **a. sulfide** See *antimonous(ic)* sulfide. **a. sulfuret** Antimonous sulfide. **a. tetraoxide*** Sb_2O_4 = 307.5. Cervanite, cervantite. Colorless powder, insoluble in water. **a. tribromide*** Antimonous bromide. **a. trichloride*** Antimonous chloride. **a. triethyl** Triethyl*stibine**. **a. trifluoride*** Antimonous fluoride. **a. trifluoride sodium sulfate** $SbF_3 \cdot Na_2SO_4$ = 320.8. White crystals, soluble in water. **a. trimethyl** Trimethyl*stibine**. **a. trioxide*** Antimonous oxide. **a. triphenyl** Triphenyl*stibine**. **a. trisulfate*** Antimonous sulfate. **a. trisulfide** Antimonous sulfide. **a. vermilion** Antimonous oxysulfide. **a. white** Antimonous oxide. **a. yellow** Basic *lead antimonate.*

antimonyl The basic radical Sb(O)—. **a. aniline tartrate** Yellow crystals, a trypanosomicide. **a. oxalate** Antimonous oxalate **a. potassium tartrate** Antimony potassium tartrate. **a. sulfate** $(SbO)_2SO_4$ = 371.6. White powder. **basic ~** $(SbO)_2SO_4 \cdot Sb_2(OH)_4$ = 683.1. White powder.

antimycin An antibiotic.

antineuralgic A drug used to treat neuralgia; e.g., aspirin.

antinonnin $C_6H_2(NO_2)_2CH_3ONa$ = 220.1. Sodium 2-methyl-4,6-dinitrophenol, victoria yellow. Yellow, m.87.

antioxidant Age resistor, antioxygen. A substance to retard deterioration by oxidation, e.g., of rubber (hydroquinone) or of food.

antioxygen Antioxidant.

antiparasite Antiparasitic.

antiparasitic A drug that inhibits the growth of vegetable and animal parasites.

antipepsin A substance that opposes the action of pepsin.

antiphlogistic A drug that inhibits the progress of inflammation. **a. theory** A theory (Lavoisier) contradicting the older *phlogiston theory,* q.v.

antipodagric A drug used to treat gout; e.g., colchicum.

antiprismo* Indicating 8 atoms bound into a regular antiprism.

antopyonin Sodium tetraborate*.

antipyrene Antipyrine.

antipyretic A drug relieving fever, as, aspirin.

antipyrine $C_{11}H_{12}ON_2$ = 188.2. Antipyrene, analgesine, phenazone, 2,3-dimethyl-1-phenyl-3-pyrazol-5-one, methozan, parodyn, phenylon, oxydimethyl quinizine, dimethyloxyquinizine, pyrazin, metozine. White powder, m.111, soluble in water. Formerly an antipyretic and analgesic (USP). A reagent for nitrous and nitric acid. **amino ~**

$C_{11}H_{13}ON_3$ = 203.2. 4-Amino-2,3-dimethyl-1-phenyl-5-pyrazolone. Colorless crystals, m.109, soluble in water. **homo ~** Homoantipyrine. **methyl ~** $C_{12}N_{14}ON_2$ = 384.2. 1-Phenyl-2,3,4-trimethylpyrazolone. Colorless crystals, m.82. **monobromo ~** Bromopyrine. **nitroso ~** $C_{11}H_{11}O_2N_2$ = 203.2. 2,3-Dimethyl-4-nitroso-1-phenylpyrazolone.

a. benzoate Benzopyrine. **a. camphorate** A compound of a. and camphor. **a. monobromide** Bromopyrine. **a. tannate** A compound of a. and tannin. Yellow powder, insoluble in water.

antipyrotic A drug to treat burns.

antipyryl A radical derived from antipyrine.

OC·NPh·NMe·CMe:C–
3 2 1 5 4

antiquinol Allylphenyl cinchonine ester. Yellow crystals, m.30, insoluble in water; an antiseptic and analgesic.

antirachitic Effective against rickets; as, cod-liver oil. See *vitamin,* Table 101.

antirennin A constituent of blood serum opposing the enzyme rennin.

antirheumatic A compound that relieves rheumatism; e.g., aspirin.

antirheumatin A compound of sodium salicylate and methylene blue. Blue crystals, soluble in water; an antirheumatic.

antiricin A substance formed in the blood after injection of ricin.

antirobin A substance formed in the blood after injection of robin.

antiscorbutic A compound to treat scurvy by virtue of its ascorbic acid content; e.g., lime juice.

antiseptic A substance that opposes sepsis, putrefaction, or decay by inhibiting growth of or destroying microorganisms; e.g., alcohol, chlorhexidine, chlorine, phenol. Cf. *antibiotic, disinfectant, germicide.*

antispasmodic A drug relieving convulsions or spasms; e.g., probanthine.

antistat A substance that prevents the formation of static electricity, e.g., on film.

antisudorific A drug that prevents excessive perspiration. Cf. *anhidrotic.*

antisyphilitic Antiluetic. A drug to treat syphilis; e.g., penicillin; formerly, arsenic compounds.

antithrombin Substance in blood which prevents coagulation by inhibiting thrombin.

antitoxic Antidote.

antitoxin (1) An antibody formed in the blood after injection of a (usually bacterial) toxin. (2) A preparation for treatment of infection by toxin-producing bacteria. A. is obtained from blood of animals injected with bacterial toxin, or immune human plasma. **diphtheria ~** A. for treating diphtheria (USP, EP, BP). **mixed antigas gangrene ~** Preparation containing 3 types of a. for treating infection due to *Clostridii* (EP, BP). **tetanus ~** A. for treating tetanus; usually from human plasma pooled from donors immunized against tetanus (USP, EP, BP). Cf. *antigen, antibody.*

antivenin A globulin antivenom from horses immunized with snake venom.

antivenom An antitoxin against snake venom.

antiviral Describing a substance used to combat virus diseases.

antodyne See *phenoxy*propanediol.

ant oil Furaldehyde*.

antozone Monoatomic oxygen.

antu Abbreviation for 1-naphthylthiourea; a rat poison.

anxiolytic Drug used to reduce anxiety and tension.

anysin Ichthyol.

anytin Anitin. Brown-black, germicidal, sulfurated hydrocarbon derived from ichthyol.

anytol Anitol. A compound of anytin and aromatic phenols and alcohols. **cresol ~** Metasol, metacresolanytol; a germicide. **eucalyptol ~** Eucasol; a germicide and disinfectant.

AOAC Association of Official Analytical Chemists.

APA Antipernicious anemia factor; see *vitamin* B_{12}.

apa- Apo-.

apaconitine An alkaloid derived from aconitine.

apatite $3Ca_3(PO_4)_2CaF_2$. A native crystalline calcium fluorophosphate fertilizer from Norway and Canada. Cf. *phosphorite, fluorapatite.* **hydroxy ~** See *hydroxyapatite.* **octa ~** $Ca_8H_2(PO_4)_6 \cdot 5H_2O$.

apatropine Apoatropine.

APC (1) (U.S. usage.) Combination analgesic tablet of aspirin, phenacetin, and caffeine. Cf. *ASA compound.* (2) (U.K. usage.) Trademark for combination of aspirin, paracetamol, and codeine.

aperient A laxative or mild purgative.

aperiodic See *aperiodic* under *balance.*

aperture (1) An opening into or through a body. (2) The diameter of the open part of an objective or lens.

apex The highest point of a cone, curve, or orbit.

aphalerite See *wurtzite.*

aphanesite $2CuO \cdot As_2O_3 \cdot 4H_2O$. Native copper arsenate.

aphelion The position of a celestial body when farthest from the sun. Cf. *perihelion.*

aphicide A substance to kill *Aphis.* Cf. *insecticide.*

aphrite Earth foam, foam spar. A white pearly variety of calcite.

aphrizite Black *tourmaline,* q.v.

aphrodine Yohimbine.

aphrodisiac A substance reputed to arouse sexual desires. e.g., yohimbine.

aphrodite $4MgO \cdot SiO_{10}H_6$. A native basic silicate; meerschaumlike masses.

aphrosiderite $(Fe,Al)Si_4O_{20} \cdot 5H_2O$. Olive-green, soft, hexagonal scales.

aphthitalite $NaKSO_4$. Native colorless hexagons. Cf. *glaserite.*

aphthonite A gray tetrahedritic silver ore.

aphyl Alkyl.

API gravity See *gravity.*

apigenin(idin) $C_{15}H_{10}O_4$ = 254.2. 4′,5,7-Trihydroxyflavone. A plant flavone.

apiin $C_{26}H_{28}O_{14}$ = 564.5. A glucoside from the seeds of *Apium petroselinum.* Yellow crystals, m.228, soluble in hot water.

apiol(e) $C_{12}H_{14}O_4$ = 222.2. Parsley camphor, 2,5-dimethoxy-3,4-methylenedioxy-1-allylbenzene*, occurring in the seeds of *Petroselinum* (parsley) and *Apium* (celery). Colorless needles of parsley odor, m.30, insoluble in water. **iso ~** Colorless crystals, m.56. **liquid ~** The green oily oleoresin of parsley seeds, soluble in alcohol.

apiolic acid $C_{10}H_{10}O_6$ = 226.2. Colorless crystals, m.175.

apionol $C_6H_2(OH)_4$ = 142.1. Phentetrol, 1,2,3,4-tetrahydroxybenzene*. Colorless crystals, m.161. **dimethoxy ~** $C_6H_2(OH)_2(OMe)_2$ = 170.2. 3,4,-Dihydroxy-1,2-dimethoxybenzene*. Colorless crystals, m.106, soluble in alcohol. **tetramethoxy ~** $C_6H_2(OMe)_4$ = 198.2. *v*-Tetramethoxybenzene. Colorless crystals m.81.

apiose $C_5H_{10}O_5$ = 150.1. A tetrose from the glucoside of celery.
aplite (1) A pink or red cobalt-silver ore from Canada. See *alaskite.* (2) SiO_2 60, Al_2O_3 24, CaO 6, Na_2O 6, K_2O 3, Fe_2O_3 0.2%. A ceramic composed chiefly of plagioclase.
aplome A yellow-green manganese garnet.
aplotaxene $C_{17}H_{28}$ = 232.4. 1,8,11,14-Heptadecatetraene*. An aliphatic hydrocarbon from costus oil.
APNS Thorin.
apo-, apa- Prefix (Greek "from") indicating a derived compound, e.g., apomorphine derived from morphine.
apoatropine $C_{17}H_{21}O_2N$ = 271.4. Apatropine, atropamine. An alkaloid derived from atropine by the action of nitric acid. Colorless prisms. m.61, insoluble in water.
apocamphoric acid Campho acid.
apocholic acid $C_{24}H_{38}O_4$ = 390.6. Crystals, m.176. A dehydration product of cholic acid.
apochromatic A lens combination, more effective than an achromatic lens in correcting chromatic aberration by bringing color rays to a common focus.
apocodeine $C_{18}H_{19}O_2N$ = 281.4. An alkaloid derived from codeine; colorless gum, m.124, sparingly soluble in water.
apocrenic acid $C_{24}H_{12}O_{12}$ = 492.3. Brown solid derived from humus by oxidation.
Apocynacae Dogbanes. Tropical and subtropical plants with a milky juice, often poisonous; e.g., roots: *Apocynum cannabinum* (Canadian hemp), *Apocynum androsaemifolium* (dogbane); barks: *Aspidosperma quebracho blanco* (quebracho), *Alstonia scholaris* (chlorogenine); leaves: *Nerium odorum* (karabin), *Urechites suberecta* (urechitine); seeds: *Strophanthus kombe* (strophanthus), *Acocanthera venenata* (ouabain).
apocynamarin $C_{14}H_{18}O_3$ = 234.3. A glucoside from the rhizome of *Apocynum cannabinum.* Cf. *cynotoxin.*
apocynein An active glucoside from *Apocynum cannabinum,* Canadian hemp. White crystals, soluble in alcohol. Cf. *apocynin.*
apocynin A resinoid from *Apocynum cannabinum.* Brown, amorphous powder which contains apocynein.
apocynum Canadian hemp, black Indian hemp, Indian physic, dogbane. The dried rhizome and roots of *Apocynum cannabinum.*
apogee The point in its orbit at which the moon is farthest from the earth. Cf. *perigee.*
Apollinaris water A German mineral water.
apomorphine $C_{17}H_{17}O_2N$ = 267.3. An alkaloid derived from morphine. Colorless mass, becoming green on exposure, m.195 (decomp), sparringly soluble in water; an expectorant and emetic. **a. hydrochloride** $C_{17}H_{17}O_2N \cdot HCl$ = 303.8. Gray, monoclinic prisms, m.270, soluble in water; a hypnotic, emetic, and expectorant (USP, EP, BP).
apomyelin A phospholipid prepared from brain matter.
aponal $NH_2COOCMe_2Et$ = 131.2. ***tert-*~** Pentyl carbamate. Colorless crystals; a hypnotic.
apophylite $KCa_4H_{16}Si_8O_{28}$. Native white or pinkish octahedra.
apopinol Apiole.
apoquinamine $C_{19}H_{22}ON_2$ = 294.4. An alkaloid derived from quinamine.
aporetin A resin from the roots of rhubarb. Brown powder, insoluble in ether, soluble in alkalies.
aposafranone $C_{18}H_{12}ON_2$ = 272.3. Benzene indenone, 3-aminophenylphenazine. Red crystals m.242, soluble in water; a dye.
apothecaries' weights A system of weights and measures formerly used in compounding medicines. The troy pound of 5,760 grains is the standard, and is subdivided into 12 ounces, the ounce into 8 drams, the dram into 60 grains. For fluid measure the quart of 42 fluid ounces is subdivided into 2 pints, the pint into 16 fluid ounces, the fluid ounce into 8 drams, and the dram into 60 minims. Abbreviations:

♏ ....minim = 0.0616 ml
gtt ....gutta, a drop
℈ ....scruple = 20 grains = 1.296 g
ʒ ....dram = 60 grains = 3.8888 g
℥ ....ounce = 480 grains = 31.103 g
= 8 drams = 29.573 ml
lb ....libra, pound = 5,760 grains
= 373.2417 g
O ....octarius, pint = 7,680 minims
= 473.16 ml

apotoxicarol $C_{18}H_{16}O$ = 248.3. A colorless, phenolic hydrolysis product, m.247, from toxicarol.
apparatus An instrument or device used for experiments, operations, or manufacture.
appertisation (French) The preservation of foods by the use of heat alone.
apple acid Malic acid*. **a. oil** Pentyl valerate*.
Appleton, Sir Edward Victor (1893–1965) British scientist, Nobel prize winner (1947); established the existence of the Heaviside layer. **A. layer** An ionized layer in the atmosphere which reflects downward any short radio waves that have penetrated the Heaviside layer below it.
applicator A device for the local administration of a remedy; e.g., a radium applicator.
approximate Near; not exact.
aprotic Not yielding or accepting a proton. **a. substance** One which acts as neither an acid nor a base. Cf. *acid, base.*
apyonin Auramine.
aq. Abbreviation for aqueous.
aqua Water (Latin). Cf. *aquo.* Denotes (1) water; (2) an aqueous solution or infusion; (3)* H_2O as a neutral ligand in a coordination compound; (4) in formulas (+*aqua*), a variable amount of water, (5) water of crystallization. **a. anethyl** See *dill water.* **a. carui** See *caraway water.* **a. compound** Hydrate. **a.culture** Production of crops, as fish, from aquatic sources. **a. destillata** *Distilled water.* **a. fervens** Hot water. **a. fluvialis** River water. **a. fontana** Spring or well water. **a. fortis** Nitric acid. **a. ions** Complex ions which contain several molecules of water; as, $[M(H_2O)_8]^{++}$, where M is a divalent metal. Cf. *hydration, hydrogen ion, base.* **a. laurocerasi** *Cherry laurel* water. **a. marina** Sea water. **a. nivialis** Snow water. **a. pluvialis** Rainwater. **a. pura** Pure water. **a. regia** A mixture of 3 pts. HCl and 1 pt. HNO_3. A solvent for noble metals. **a. tepida** Warm water. **a. vitae** Brandy or whisky.
Aquadag Trademark for a colloidal suspension of graphite in water; a lubricant and an electrically conductive coating.
aquamarine A sea-green beryl.
aquation Introduction of water molecules into other molecules. Cf. *hydration, solvation.*
aquatone Watery. **a. solution** A solution with water as solvent.
aquifer A subterranean, water-bearing formation.
aquinite Chloropicrin.
aquo Pertaining to water. Cf. *aqua.*
A.R. Abbreviation for analytical reagent.
Ar Symbol for (1)* the element argon; (2) an aryl group.
ar- Prefix indicating substitution in the *aromatic* nucleus. Cf. *ac-.*
araban $(C_5H_8O_4)_n$ = 132.1*n*. Arabinan. An arabinose polysaccharide in the mucilage of Malvaceae.
arabate A salt of arabic acid.

arabic acid Arabitic acid. A polysaccharide, white powder precipitated from gum arabic solution by alcohol.
arabin $C_{10}H_{18}O_9$ = 282.2. Amorphous powder, soluble in water, from gums. **a. water** A solution of the water-soluble constituents of gums in hydnocarpus oil.
arabinogalactan An adhesive obtained by leaching larch wood.
arabinose $CHO(CHOH)_3 \cdot CH_2OH$ = 150.1. Pectin sugar, pectinose, gum sugar. The pentoses obtained by hydrolysis of acacia. D-~ White rhombs, m.160, soluble in water; a culture medium for certain bacteria.
arabinulose $C_5H_{10}O_5$ = 150.1. The ketopentose corresponding with arabinose.
arabite Arabitol.
arabitic acid Arabic acid.
arabitol $CH_2OH \cdot (CHOH)_3 \cdot CH_2OH$ = 152.1. Arabinitol, lyxitol, arabite. An alcohol derived from arabinose. Colorless, sweet D- and L-form crystals, m.102, soluble in water.
arabonic acid $CH_2OH \cdot (CHOH)_3 \cdot COOH$ = 166.1. Arabinonic acid. A hydroxy acid, derived from arabinose. D-~, L-~ Colorless crystals, m.116, soluble in water.
Araceae Aroideae.
arachic acid Icosanoic acid*.
arachidic acid Icosanoic acid*. **a. alcohol** Icosyl alcohol*.
arachidonic acid $C_{20}H_{32}O_2$ = 304.5. 5,8,11,14-Icosatetraenoic acid*. A liquid, unsaturated acid, b.163, in lard and mammal fat.
arachin A globulin from peanuts (24%), *Arachis hypogoea.* Contains chiefly arginine and glutamic acid with other amino acids.
arachis See *peanut.* **a. oil** Peanut oil, groundnut oil. A fixed oil from peanuts, d_{25} 0.913. See *Bellier's test.*
arachnidism Poisoning from the bite of the black widow spider, *Latrodectus mactans.*
arachnolysin A hemolytic principle of spider poison.
arack Arrack.
araeometer Hydrometer.
aragonite $CaCO_3$. Pisolite. Needle spar. A native calcium carbonate. Cf. *calcite.*
Aralac Trade name for a fibrous product made from skim milk casein. Cf. *lanital.*
Araldite Trademark for an epoxy resin used for adhesives, lamination, and surface coatings and castings.
aralia Nard, American spikenard. The dried rhizomes and roots of *A. racemosa* (Araliaceae); a diaphoretic.
Araliaceae Aromatic plants which yield *Panax quinquefolium* (ginseng), *Aralia nudicaulis* (false sarsaparilla), *Aralia racemosa* (American spikenard), *Aralia hispida* (dwarf elder).
aralin A glucoside from the fresh leaves of *Aralia* species.
aralkyl Arylated alkyl. A radical in which an alkyl H atom is substituted by an aryl group. Cf. *alphyl.*
aramid U.S. Federal Trade Commission term for a manufactured fiber, in which at least 85% of the amide linkages are attached directly to 2 aromatic rings. Intended to distinguish aromatic polyamide fibers, as Nomex and Kevlar, from aliphatic analogs, such as nylon.
Aramine Trademark for metaraminol (bi)tartrate.
aranein A homeopathic liquid from spider abdomens.
araphite A magnetic basalt from Colorado.
araroba Goa powder.
Arasan Trademark for tetramethyl thiuram disulfide; a seed disinfectant.
arbacin A protein in the sperm of the sea urchin *(Arbacia).*
arbor Official Latin for "tree." **a. Dianae** Silver tree. Arborescent silver formed on adding mercury to a silver salt solution **a. Saturni** Leadtree. Arborescent lead formed on adding zinc to a lead salt solution. **a.vitae** Thuja.
arborescent Describing a branched, treelike growth of bacteria or crystals.
arbusterol A fat from *Arbutus* species.
arbutin $C_{12}H_{16}O_7$ = 272.2. Ursin, arbutoside. A glucoside from the leaves of *Arbutus* and other Ericaceae and Pyrolaceae. Colorless crystals, m.195, soluble in water, hydrolyzed to hydroquinone and glucose; a diuretic and urinary disinfectant.
Arbutus A genue of shrubs and trees (Ericaceae): *A. menziesii* (madrone), *A. unedo* (European arbutus), *A. Arctostaphylos uvaursi* (bearberry, arbutin).
arc (1) A portion of a curved line. (2) A. lamp. **electric ~** The discharge of an electric current between electrodes. **mercury ~** An electric a. in mercury vapor. **oscillating ~** A discharge which changes its position.
a. furnace See *furnace.* **a. lamp** A luminous discharge between two electrodes. **a. resistance** Tracking. **a. spectrum** The spectrum produced by arcing between electrodes of the element under investigation. Utilized in DC arc spectroscopy. Cf. *spark spectrum, alternation law.*
arcadian nitrate Sodium nitrate made from synthetic ammonia and soda; a fertilizer. **a. sulfate** Ammonium sulfate fertilizer.
arcaine $NH_2 \cdot C(:NH)NH \cdot (CH_2)_4 \cdot NH \cdot C(:NH) \cdot NH_2$ = 172.2. 1,4-Diguanidinobutane. An animal base from the mussel Noah's Ark *(Areca Noae)*; lowers the sugar content of the blood.
arcanite Glaserite.
arcanum (1) A nostrum or secret medicine. (2) Potassium sulfate.
archaeometry The application of the physical sciences to archaeology.
archil Orchil.
Archimedes (300 B.C.) Greek mathematician of Syracuse. **A. bridge** A wooden platform across the pan of a balance, used for weighing solids immersed in a liquid. **A. principle** A body immersed in a liquid will lose weight equal to that of the liquid it displaces. Used to determine the specific gravity of dense, irregular bodies.
archon The poisonous radical of proteins.
archyl The 2-propynyl* radical.
arciform Curved, bow-shaped.
arcilla Argol.
arcing The conversion of electric energy into light by a current between two electrodes. It differs from sparking in that it depends on ionization of the vapor of the electrodes, and not on that of the gas between them.
arconium A hypothetical element, at. wt. 2.9, similar to *nebulium* and *coronium,* q.v.
Arcton Freon (U.K. usage).
arcual Arc-shaped or arched.
ardennite $M_4Al_4H_5VSi_4O_{22}$. A native vanadosilicate. Brown, orthorhombic crystals.
ardometer An *optical pyrometer,* q.v.
are (1) Latin for "area." (2) The metric unit of surface: 1 are = 100 m^2 = 119.6 yd^2.
area A region or surface A enclosed by boundaries. $A = cl^2$, where c is a constant depending on the contour of the surface (1, if square; $\pi/4$, if round) and l the surface's length (or diameter, if circular).
areametric Pertaining to area measurement. **a. analysis** Chemical analysis by forming precipitates in definite areas, which are matched with precipitates similarly produced from standards.

Areca A genus of Asiatic and Australasian palms, as *A. catechu pinang,* the areca nut, betel nut, or catechu palm. **a. alkaloid** Betel nut alkaloids; e.g., arecoline, homoarecoline ($C_9H_{15}O_2N$), arecaine, arecaidine, guvacoline ($C_7H_{11}O_2N$), coniine ($C_8H_{17}N$). **a. nut** Pinang, *betel nut,* q.v. The dried seeds of *A. catechu,* a southeast Asian palm. It is an astringent and anthelmintic, and contains alkaloids, an orange coloring matter (arecin), and chavicol. **a. red** Arecin.

arecaidine $C_7H_{11}O_2N$ = 141.2. 1,2,5,6-Tetrahydro-1-methyl-3-pyridinecarboxylic acid. An alkaloid from arecoline. Colorless scales, m.233 (decomp.), soluble in water.

arecaine $C_7H_{11}O_2N$ = 141.2. *N*-Methylguvacine. An alkaloid from the seeds of *Areca catechu* (betel nuts). Colorless crystals, m.213, soluble in water; an anthelmintic, used in veterinary medicine.

arecaline Arecoline.

arecane Arecoline.

arecin $C_{23}H_{26}ON_2$ = 346.5. Areca red. A red coloring matter from betel nuts.

arecoline $C_8H_{13}O_2N$ = 155.2. Arekane, arecaline. An alkaloid in betel nuts, the methyl ester of arecaidine. Yellow oil, d.2.02, $b_{17mm}94$, insoluble in water. **homo ~** See *homoarecoline.*

a. hydrobromide $C_8H_{13}O_2N \cdot HBr$ = 236.08. Arceine. Colorless prisms, m.167, soluble in water; an anthelmintic and miotic. **a. hydrochloride** $C_8H_{13}O_2N \cdot HCl$ = 191.7. Colorless crystals, m.158, soluble in water; an anthelmintic used in veterinary medicine.

arekane Arecoline.

arendalite A microcrystalline epidote.

arene* A hydrocarbon containing at least one *aromatic,* q.v., ring; as, benzene, anthracene.

arenolite An artificial stone.

areometer Hydrometer. **a. scales** See *hydrometer scales.*

areometry Hydrometry.

arepycnometer A pycnometer for viscious liquids.

areusin A *benzylidenecoumarone* pigment, q.v.

Arfonad Trademark for trimetaphan mesylate.

Arfvedson, Johann August (1792–1841) Afzelius. Swedish chemist and mineralogist, discoverer of lithium (1817).

arfvedsonite A black soda amphibole.

Arg* Symbol for arginene.

argal Argol.

Argand, Aimé (1755–1803) Swiss physicist, noted for A. burner. **A. burner** A gas or oil burner with a chimney and regulated air supply.

argemonine $C_{21}H_{25}O_4N$ = 355.4. *N*-Methylpavine. **D-~** m.156. An alkaloid of prickly poppy *(Argemone mexicana).*

argenol Argyrol.

argental Landsbergite.

argentamine A colorless solution of 8% silver phosphate and 15% ethylenediamine; an antiseptic, astringent, and disinfectant.

argentate* Indicating silver as the central atom(s) in an anion; as, tetrafluoro ~, $[AgF_4]^-$.

argentan Nickel silver.

argenti Official Latin for "of silver."

argentic Containing silver.

argentiferous Containing silver. **a. lead** (1) An alloy of silver and lead. (2) A lead-silver sulfide.

argentiform Silver hexamethylenetetramine.

argentimetry Argentometry.

argentine Finely divided tin sponge obtained by precipitation of a tin salt solution with zinc; used in printing.

argentite Ag_2S. Silver glance, silver glanz, argyrite, vitreous silver ore. Native vitreous silver sulfide. Black, isometric crystals. Cf. *acanthite.*

argentometer A hydrometer to determine the silver content of photographic solutions.

argentometry Volumetric analysis involving the precipitation of insoluble silver salts; as, chlorides, chromates.

argentopyrites $3FeS \cdot 3FeS_2 \cdot Ag_2S$. Native silver and iron sulfide.

argentous Containing monovalent silver. See *silver.*

argentum (1) Official latin for "silver." Cf. *argenti.* (2)* Used in naming silver compounds. **a. cornu** Horn silver, chlorargyrite, cevargyrite. A native silver chloride. **a. virum** Native mercury (Pliny). **a. vitellinum** Argyrol.

argil Argol.

argilla Kaolin.

argillaceous Containing clay.

arginase* An enzyme of the intestine, mammalian liver, and spleen which transforms L-arginine to L-ornithine and urea.

arginine* Arg*. $NH_2 \cdot CH(COOH) \cdot (CH_2)_3NH \cdot C{:}NH(NH_2)$ = 174.2. 2-Amino-5-guanidinovaleric acid. An amino acid from animal and vegetable proteins (albumin and seeds). Colorless crystals, m.238, soluble in water.

argol $KH(C_4H_4O_6)$ = 188.2. Argal, argil. Crude potassium acid tartrate deposited by grapejuice during fermentation; a raw material for the manufacture of tartaric acid, a reducing agent, and an assay flux.

argon* Ar = 39.948. A gaseous element, at. no. 18, in the atmosphere and fumaroles; discovered (1894) by Ramsay and Rayleigh. **gaseous ~** $d_{air=1}1.38$, m.−189, b.−186. **liquid ~** $d_{-186°C}1.4$. A. from fumaroles and spring water. A. is absolutely inert, possesses no valency, and belongs to the zero group of the periodic table. Obtained by fractionation of liquid oxygen, and used to fill incandescent lamps, rectifiers, and vacuum tubes.

argyria Poisoning by silver or its compounds.

argyrine An alkaloid from horse chestnuts.

argyrite Argentite.

argyrodite $4Ag_2S \cdot GeS_2$. Native, monoclinic, steel-gray germanium silver sulfide.

Argyrol Vitellin silver, argenol. Trade name of a brand of silver protein (20% Ag). Brown, glistening, hygroscopic scales, soluble in water; an antiseptic.

arheol α-Santalol.

aribine $C_{12}H_{10}N_2$ = 182.2. Harman, loturine. An alkaloid from the bark of *Sickingia* (or *Arariba*) *rubra* (Rubiaceae) and the seeds of *Peganum harmala* (Rutaceae). Colorless crystals, m.237.

aristin A constituent of various Arsitolochia species.

aristochin Aristoquin, diquinine carbonic ester. White powder, insoluble in water, sparingly soluble in alcohol.

Aristolochia The birthwort family *(Aristolochiaceae).* See *serpentaria,* **a. yellow** The coloring matter of the roots and seeds of *A.* species, e.g., *A. clematitis.*

aristoquin Aristochin.

arithmetic mean See *mean.* **a. progression** A series of numbers with equal differences *d* between consecutive terms. If *a* is the first and *z* the last term, *d* the difference, and *n* the total number of terms, then $z = nd$. The sum of *n* terms is $n(a + z)/2$.

arkansite Brookite.

arkite An igneous leucite-syenite rock.

arkusite Chiolite.

armature The rotating coil and core of a dynamo.

Armon Trade name for a viscose synthetic fiber.

armoracia The fresh root of *Cochlearia armoracia,* horseradish; a condiment.
armored thermometer A thermometer in a metal case.
Armstrong, Henry Edward (1847–1937) English organic chemist and educator. **A. acid** (1) Schäffer's acid. (2) Naphthalene-1,5-disulfonic acid. **A. metal** An alloy of Mn 4–6, C 0.10, Ni 80, Cr 17.5., Cu 2.9%; used for corrosion-resisting drawn or pressed shapes.
arnatto Annatto.
Arnaudon's green Chromic phosphate.
Arndt **A. alloy** Mg 60, Cu 40%. Used in analysis to reduce nitrates to ammonia in neutral solution. **A.-Schulz rule** *Weak* stimuli greatly accelerate life processes; the *strongest* destroy them. **A. tube** A bent glass tube with 4 bulbs used for the determination of hydrogen.
Arnel Trademark for a cellulose acetate synthetic fiber.
Arnica Aster. A genus of the composite flowered plants. **a. flowers** The dried flower heads of *A. montana,* leopard's bane, wolf's bane, mountain tobacco. A feeble rubefacient. **a. oil** An essential oil from arnica flowers. Yellow aromatic liquid, d.0.906, acid val. 75.1, sapon. val. 29.9, soluble in alcohol; used in liniments. **a. root oil** Yellow oil from arnica root, d.0.990, $[\alpha]_D$ −2.
arnicin $C_{20}H_{30}O_4$ = 334.5. A glucoside from arnica roots and flowers. Yellow, bitter powder.
arnicine $C_{35}H_{54}O_7$ = 586.8. A resinous, basic principle from arnica flowers.
arnotta Annatto.
Aroclors Trademark for (1) the polychlorine derivatives of biphenyl; used as lubricants; (2) the chlorobiphenyl resins, which are odorless and resistant to light and flame; (3) yellow liquids or brown solids; nonoxidizing and thermoplastic, and not hydrolyzed by water or alkalies.
Aroideae Arum. A family of herbs with an acrid, colorless juice, fleshy corm or rhizome, and berry fruit. The principal species: *Acorus calamus* (sweet flag, calamus), *Symplocarpus foetidus* (skunk cabbage, symplocarpus), *Arisaema triphyllum* (Indian turnip, arum).
aromadendral A mixture of cumic aldehyde, phellandral, and L-cryptal from eucalyptus.
aromatic (1) Spicy, fragrant, or agreeable in odor or taste. (2) Arene*. Having one or more unsaturated C-rings; as, benzene. **a. acids** A. compounds which contain one or more carboxyl groups; e.g., benzoic acid (PhCOOH); phthalic acid $[C_6H_4(COOH)_2]$; mellitic acid $[C_6(COOH)_6]$ **a. alcohols** A. compounds which contain a hydroxyl group in a side chain; e.g., benzyl alcohol ($PhCH_2OH$); methylbenzyl alcohol ($MeC_6H_4 \cdot CH_2OH$). **a. compound, a. hydrocarbon** See *aromatic (2).* **a. series** A series of a. compounds. **a. tincture** An alcoholic solution of cinnamon, ginger, cardamom, cloves, and galangal; a carminative. **a. vinegar** A mixture of acetic acid and essential oils, used to relieve headaches.
aromatics Spicy, fragrant, and stimulating drugs, with agreeable taste due to an essential oil.
aromatin A gentian root substitute for hops.
aroxyamines Hydroxylamines with an aromatic constituent.
aroyl The radical R·CO—; R is an aromatic group; as, benzoyl, naphthoyl.
arphoalin An albumin preparation containing arsenic and phosphorus.
arquerite $Ag_{12}Hg$. A native silver amalgam (Arqueros, Chile) in isomeric crystals.
arrack A liquor distilled from fermented rice.
arrhenal Sodium methyl arsenite.
arrhenate MeHAsO·OM. A salt of arrhenic acid.
arrhenic acid Me·H·AsO·OH = 124.0. Monomethylarsinic acid. Used in organic synthesis. *Cf. cacodylic acid.*
Arrhenius, Svante August (1859–1927) Swedish chemist and physicist, Nobel prize winner (1903); noted for his theories of ions and dissociation (1883) and of cosmics. **A. equation** For a reaction, the rate constant $k = A\exp(-E_a/RT)$, where the constant A is the preexponential factor, E_a the activation energy, R the gas constant, and T the thermodynamic temperature. **A. law** A solution of high osmotic pressure conducts an electric current. **A. theory** When an electrolyte dissolves, it splits into ions (dissociation) to an extent that increases with a decrease in the concentration, and is indicated by the deviation of the solution from the van't Hoff laws. **A. viscosity formula** $\log \eta = \theta c$, where η is the viscosity of a solution, c the volume of the suspended particles present, and θ a constant.
arrow poison A poisonous plant juice (e.g., *curare* q.v.) or venom of an animal (e.g., snake) used for poisoning arrowheads. See *acocantherin, antiar, bufagin.*
arrowroot Maranta. The starch from the rhizomes of the arrowroot plant, *Maranta arundinaceae* (Marantaceae). White powder from Bermuda, used as a nonirritating food in the treatment of fever, and commercially as an adhesive and in laundries. **Bahia ~, Para ~, Rio ~** Tapioca.
arrowwood (1) Frangula. (2) Euonymus.
arsabenzol Arsphenamine.
arsane* Arsine*.
arsanilate $M_2(C_6H_6O_3NAs)$. Atoxylate. A salt of arsanilic acid.
arsanilic acid $NH_2 \cdot C_6H_4 \cdot AsO(OH)_2$ = 217.1. Atoxylic acid, *p*-aminobenzenearsonic acid*, (4-aminophenyl)arsinic acid†, arsenic acid anilide, *m*-arsenous acid anilide. Colorless crystals, m.232, soluble in ether; used in organic synthesis. The meta- and orthoacids are also known. **acetylhydroxy ~** Stovarsol. ***N*-carbamoyl ~** Carbarsone.
Arsem furnace An electric vacuum furnace.
arsenate* Indicating arsenic as the central atom(s) in an anion; as, a salt of arsenic acid containing the $AsO_4 \equiv$ radical: Na_3AsO_4. **acid ~** Hydrogenarsenate. **basic ~** An a. containing a metal oxide or hydroxide. **diacid ~** Dihydrogenarsenate. **dihydrogen ~** A salt containing the ion $H_2AsO_4^-$. **hydrogen ~** An a. containing one hydrogen; as, the $HAsO_4^{2-}$ ion. **meta ~** An a. containing the AsO_3^- ion. **pyro ~** An a. derived from $H_4As_2O_7$. **triethyl ~** $(EtO)_3AsO$ = 226.1. Colorless liquid, d.1.326, b.238.
arsenenic acid* Meta*arsenic acid.*
arsenic* As = 74.9216. An element of the phosphorus group, at. no. 33. A rhombohedral, gray, brittle nonmetal of metallic character, d_{14} 5.727, b.610 (sublimes); insoluble in water, alcohol, or acids. It occurs widely in nature as the sulfide, arsenide, and sulfarsenides of the heavy metals, and in a number of allotropic modifications analogous to those of phosphorus: (1) Metallic arsenic, α-As, crystalline, d.5.727. The common stable form. (2) Gray or black arsenic, β-As, amorphous, d.4.64, which changes at 300 to α-As. (3) Yellow arsenic, γ-As_4, d.3.7, the nonmetallic form formed at 500 by rapid condensation of arsenic vapor. It is soluble in carbon disulfide, resembles phosphorus, is photosensitive and unstable, and changes to gray arsenic on exposure to light. (4) Brown arsenic, As_8, d.2.03, an allotrope of (3) deposited from yellow arsenic solutions; changes at 180 to gray arsenic. Arsenic may be tri- or pentavalent. Compounds: derived from trivalent, negative As (arsenides, M_3As); derived from trivalent, positive As (a.(III)*, a.(3+)*, arsenous, As^{3+}); derived from pentavalent, positive As (a.(V)*, a.(5+)*,

arsenic, As^{5+}). Orpiment (arsenic sulfide) was known to the ancients as arsenicon (the "masculine one"); used to paint the sunburnt faces of men. A. is used in medicine to treat African trypanosomiasis, and a. compounds are used as pigments, and for preparation of arsenic salts. **butter of ~** Arsenous chloride. **dimethyl ~** Cacodyl. **flowers of ~** Arsenous oxide. **red ~, ruby ~** A. tetrasulfide*. **triethyl ~** Triethyl*arsine.* **white ~** Arsenolite, arsenite. A native arsenous oxide. **yellow ~** Orpiment.

a.(III)*, a.(3+)* See also *arsenous.* **a. acid ortho ~** $H_3AsO_4 \cdot \frac{1}{2}H_2O$ = 150.9. Colorless crystals, m.35.5, soluble in water; used to manufacture arsenates. **meta ~*** $HAsO_3$ = 123.9. Arsenenic acid*. White powder, soluble in water to form orthoarsenic acid. **pyro ~** $H_4As_2O_7$ = 265.9. Diarsenous acid. Colorless powder, decomp. 208, soluble in water, forming orthoarsenic acid. **a. apparatus** See *Marsh test.* **a. bromide** Arsenous bromide. **a. butter** Arsenous chloride. **a. chloride** Arsenous chloride. **a. diiodide** AsI_2 = 328.7. White crystals, decomp. 136. **a. disulfide** A. tetrasulfide*. **a. fluoride** AsF_5 = 169.9. Colorless, poisonous gas, $d._{(H_2=1)}$ 5.964, m.−80, b.−53, soluble in water. **a. glass** (1) An indefinite term for a. compounds; e.g., a. sulfides. (2) Arsenous oxide. **a. hydride** (1)Arsine*. (2) As_4H_2, AsH_2 or $(AsH)_n$. An ill-defined solid hydride, formed from water and sodium arsenide. **a. iodide** (1) A. diiodide. (2) Arsenous iodide. **a. minerals** A. is usually associated with antimony and bismuth minerals and occurs in many sulfide minerals; e.g., realgar (As_4S_4), orpiment (As_2S_3); also in the arsenides and sulfarsenides of the heavy metals. **a. oxide** (1) As_2O_5 = 229.8. A. pentaoxide, a. acid anhydride. Colorless powder, d.4.086, decomp. by heat, soluble in water. (2) See *arsenous oxide.* **a. oxychloride** See *arsenous oxychloride.* **a. pentafluoride*** A. fluoride. **a. pentaoxide*** A. oxide. **a. pentasulfide*** A. sulfide. **a. phosphide*** Arsenous phosphide. **a. ruby** A. tetrasulfide. **a. selenide*** Arsenous selenide. **a. sulfide** (1) As_2S_5 = 310.1. A. pentasulfide*. Yellow crystals, insoluble in water; a pigment. (2) See a. tetrasulfide*. (3) Arsenous sulfide. **a. tetrasulfide*** As_4S_4 = 418.0. Realgar, red a. sulfide, red orpiment, a. ruby, red a. glass. Brown, monoclinic crystals, m.307, insoluble in water. Used in textile printing, in pyrotechnics and in tanneries. **a. thiocyanate*** $As(SCN)_3$ = 249.2. **a. tribromide*** Arsenous bromide. **a. trichloride*** Arsenous chloride. **a. triethide** Triethyl*arsine*.* **a. trifluoride*** Arsenous fluoride **a. triiodide*** Arsenous iodide. **a. trimethyl** Trimethyl*arsine*.* **a. trioxide*** Arsenous oxide. **a. trisulfide*** Arsenous sulfide.

arsenical Pertaining to arsenic. **a. nickel** Niccolite. **a. pyrites** Mispickel.

arsenicals Drugs, fungicides, and insecticides whose effects depend primarily on their arsenic content.

arsenicum Latin for "arsenic."

arsenide* A metal derivative of arsine, AsH_3; as, Na_3As. Cf. *speise.*

arsenidine $CH_2(CH_2)_2AsH \cdot CH_2 \cdot CH_2$ = 146.1. Arsepidine. A heterocyclic analog of piperidine.

arsenii Official Latin for "arsenic."

arsenious Arsenous. **a. acid*** Arsenous acid†.

arsenite (1) A salt of a hypothetical arsenous acid; orthoarsenites*, K_3AsO_3; metaarsenites, $KAsO_2$; pyroarsenites, $Ca_2As_2O_5$. (2) White *arsenic.* **triethyl ~** $(EtO)_3As$ = 210.1. Colorless liquid, b.166.

arsenium The element arsenic.

arseniuretted hydrogen Arsine*.

arsenius Arsenous.

arseno* 1,2-Diarsenediyl†. The group −As:As−, isologous with the azo group.

arsenobenzene PhAs:AsPh. An analog of azobenzene.

Arsenobenzol A brand of arsphenamine.

arsenoferratin An iron albuminate containing arsenic.

arsenofuran C_4H_5As. Arsenophen. An analog of pyrrole.

arsenolite White *arsenic.*

arsenophen Arsenofuran.

arsenophenol $HO \cdot C_6H_4 \cdot As:As \cdot C_6H_4 \cdot OH$ = 336.1. 4,4′-Dihydroxyarsenobenzene. Dark yellow powder, m.200+ (decomp.). **diamino ~** Arsphenamine.

arsenopyrite FeAsS. Mispickel. A native sulfarsenide.

arsenoso The group −OAs.

arsenous Arsenic(III)* or (3+)*, arsenious. Describing a compound of trivalent arsenic. **a. acid†** H_3AsO_3 or $As(OH)_3$. A monobasic acid, from which the *arsenites,* q.v., are derived. **a. acid anhydride** A. oxide. **a. bromide** $AsBr_3$ = 314.6. Arsenic tribromide*. Yellow crystals, m.31, soluble in water. **a. chloride** $AsCl_3$ = 181.3. Arsenic trichloride, arsenic butter. Yellow liquid, d.2.205, b.130, soluble in water. **a. fluoride** AsF_3 = 131.9. Yellow liquid, d.2.73, b.63, decomp. by water. **a. hydride** Arsine*. **a. iodide** AsI_3 = 455.6. Arsenic triiodide*, arsenic iodide. Orange crystals, m.146, (sublimes), soluble in water; an antiseptic. **a. oxide** As_4O_6 or As_2O_3 = 197.8. Arsenic trioxide*, arsenic, a. acid anhydride, white arsenic. White, octahedral or amorphous mass, sublimes 218, soluble in water. Used in the pharmacy, glass, leather, and pigment industries; also in the manufacture of arsenic salts. Antidotes: gastric lavage, emetics, dimercaprol. **a. oxychloride** AsOCl = 126.4. Arsenic chloride oxide*. Brown mass, decomp. by water or heat. **a. phosphide*.** AsP = 105.9. Brown fragments, decomp. by water or heat. **a. selenide** As_2Se_3 = 386.7. Arsenic triselenide*. Brown crystals, m.360, insoluble in water. **a. sulfide** As_2S_3 = 246.0. Arsenic trisulfide*, yellow arsenic sulfide, orpiment, auripigment, king's yellow. Orange, monoclinic crystals or amorphous powder, m.310, insoluble in water; a pigment.

arsenyl The arsoryl* radical.

arsepidine Arsenidine.

arsinato-* The group $=AsO(O^-)$, from arsinic acid.

arsine (1)* AsH_3 = 77.9. Arsenous hydride, arsane*, hydrogen arsenide. Poisonous gas with strong garlic odor, $d_{(air=1)}$2.695, soluble in water; used in organic synthesis. (2) See *arsines.* **alk ~** $Me_2As \cdot O \cdot AsMe_2$ = 219.9. Cacodylic oxide. Colorless liquid, b.120. **bi ~, di ~** Diarsane*. **dichloromethyl ~*** $MeAsCl_2$ = 160.9. Methylarsenic dichloride*. Colorless liquid, b.133. **diethyl ~** (1)* Et_2AsH = 134.1. b.186, insoluble in water. (2) Tetraethyldiarsane*. **dimethyl ~*** Me_2AsH = 106.0. Cacodylhydride. Colorless liquid, b.36, miscible with alcohol. **ditertiary ~** The compound

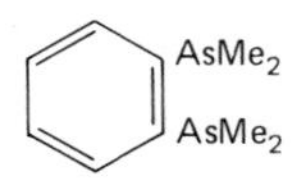

ethyl ~* $EtAsH_2$ = 106.0. Arsinoethane. Colorless liquid, b.36. **methyl ~*** $MeAsH_2$ = 92.0. Arsinomethane. A gas, b.2., soluble in alcohol. **tetraethyl ~** Tetraethyl*diarsane*.* **tetramethyl ~** Tetramethyl*diarsane*.* **triethyl ~*** Et_3As = 162.1. Arsenic triethyl. Colorless liquid, decomp. 141. **trimethyl ~*** Me_3As = 120.0. Arsenic trimethyl, arsenous methide. Colorless liquid, b.53, soluble in water. See *gosio gas.*

arsines* Arsine analogs of *phosphines,* q.v., stibines in which

the hydrogens are replaced by hydrocarbon radicals; as, R_2AsH, dialkylarsine.
arsinic acid* (1) An organic compound derived from $H_2AsO(OH)$, diaryl ~ or dialkyl ~. **aminophenyl ~** Arsanilic acid. **dimethyl ~** Cacodylic acid. **methyl ~** See *methylarsinic acid.*
arsinico-* The radical (OH)OAs=, from arsinic acid.
arsino- (1)* The H_2As- group. (2) Prefix indicating the −As:As− group. (3) The arsinico* radical.
arsinoso (1) The $(HO)_2As-$ group. (2) The HO·As= group. (3) The radical O:As−, isologous with nitroso.
arsinous acid* The acid H_2AsOH.
arsinoyl* Prefix indicating the radical $H_2As(O)-$.
arso The O_2As- group.
arsonate* A salt of arsonic acid.
arsonato-* The group $-AsO(O^-)_2$, from arsonic acid.
arsone The compound $AsO(OH)_2$.
arsonic acid* An organic compound derived from $HAsO(OH)_2$. ***p*-carbaminophenyl ~** Carbasone. **phenyl ~** Benzenearsonic acid.
arsonium* The ion AsH_4^+, an isolog of ammonium and phosphonium. **a. compounds** AsH_4X. Protonated arsine compounds. **a. hydroxide** AsH_4OH. An isolog of ammonium hydroxide and parent substance of, e.g., R_4AsOH, tetra-R-arsonium hydroxide.
arsono* The arsonic acid radical $(HO)_2OAs-$.
arsonoso (1) The HOOAs− group. (2) The HO·OAs= group.
arsonous acid* The acid $HAs(OH)_2$.
arsonoyl* Prefix indicating the HAs(O)= radical.
Arsonval, Jacques Arsene d' (1851–1940) French physicist, pioneer in high-frequency electric therapy.
arsoryl* Indicating the radical As(O)≡.
arsphenamine $As_2(C_6H_3 \cdot OH \cdot NH_2)_2 \cdot 2HCl$ = 439.0. Salvarasan, arsphenolamine hydrochloride, 3-diamino-4-dihydroxylarsenobenzene, Ehrlich 606. Yellow, hygroscopic crystals, unstable in air, soluble in water; the first effective treatment for syphilis, now superseded by penicillin. **neo ~** See *neoarsphenamine.*
arsyl The radical H_2As-, from arsine.
arsynal Sodium methyl arsenite.
artabotrine $C_{20}H_{23}O_4N$ = 341.4. Isocorydine. **D- ~** An alkaloid, m.187, from the stems and roots of *Artabotrys suaveolens* (Anonaceae).
Artane Trademark for trihexyphenidyl hydrochloride.
artarine $C_{21}H_{23}O_4N$ = 353.4. An alkaloid from artar root.
artar root A drug from the root of *Zanthoxylum senegalense* (Rutaceae), W. Africa.
Artemisia A genus of plants belonging to the aster family (Compositae); e.g., *A. absinthium,* wormwood; *A maritima,* wormseed, santonica. **a. oil** Wormwood oil.
artemisin $C_{15}H_{18}O_4$ = 262.2. Hydroxysantonin. From the seeds of *Artemisia* species. White crystals, m.200, soluble in hot water; a gastric stimulant.
arteriograph An instrument to trace and record the pulse.
arteriosclerosis Atherosclerosis (incorrect). Hardening and loss of elasticity of arterial walls. A normal aging process, accelerated in hypertension. See *atheroma.*
artery A blood vessel in which the blood passes from the heart to the organs of the body.
arthranitin $C_{58}H_{94}O_{27}$ = 1223. Cyclamin. A glucoside from arthanite *(Cyclamen europaeum).* White powder, soluble in water, m.282.
arthropods Invertebrate animals with segmented bodies and jointed limbs; as, insects, spiders, and crustaceans.
Artic Trademark for methylchloride used in refrigerators.
artificial Made by man as opposed to natural. Cf. *synthetic.*
artolinantipeptone Artose.
artose $C_{185}H_{288}O_{58}N_{50}S$ = 4172. A water-soluble albumose produced from wheat gliadin. Cf. *deuteroartose, heteroartose.*
artotype Collotype.
aruba acid A naphthenic acid extracted from Colombian gas oil by alkali.
arum (1) A genus of plants (Aroidae) whose corms yield starchy products; e.g., sage from *A. maculatum.* (2) An edible starch similar to sago from *A. maculatum* (Southern Europe).
aryl* An organic radical derived from an aromatic hydrocarbon by the removal of one atom; e.g., phenyl from benzene. Cf. *arylene, alkyl.*
arylarsonate A compound of arsenic containing aryl radicals; e.g., arsphenamine.
aryle, arylide A compound of a metal containing an aryl radical; as, PbR_4.
arylene* As aryl, but by removal of 2 atoms.
arznei German term for "drugs."
as- *asym-**.
As Symbol for arsenic.
ASA See *American National Standards Institute.* **ASA compound** U.S. trademark for combination of aspirin, phenacetin, and caffeine. Cf. *APC.*
asafetida Asafoetida. A gum resin obtained by incising the rhizomes of *Ferula asafetida* and other *Ferula* species (Umbelliferae). Soft mass of garlic odor; acrid, bitter taste; emulsifies with water; a carminative and sedative. Cf. *ferulic acid.* **milk of ~** See *milk (2) of asafetida.*
a. oil Yellow, volatile oil distilled from asafetida, d.0.975–0.990, $[\alpha]_D$ +16.
asaronic acid $(MeO)_3C_6H_2COOH$ = 212.2. 2,4,5-Trimethoxybenzoic acid*. Colorless crystals, m.144.
Asarum (1) A genus of plants of the birthwort family (Aristolochiaceae). (2) Canada snakeroot, wild ginger. The dried rhizomes and roots of *A. canadense;* a flavoring agent. **a. oil** An essential oil from asarum species. **Canadian ~** From the roots of *A. canadense.* Colorless liquid, d.0.95,$[\alpha]_D$ −3.5, soluble in alcohol. The chief constituents are asarol and methyleugenol; used in perfumery. **European ~** From the roots of *A. europaeum;* soluble in alcohol.
asaryl The radical $2,4,5\text{-}(MeO)_3C_6H_2-$.
asbestos Amianthus, earth flax, mountain cork, stone flax; fibrous actinolite. A native magnesium calcium silicate. For types, see Table 10 on p. 54. Gray masses; either compact or long, silky fibers. It is acid- and heat-resisting and may be spun or woven. **blue ~** Crocidolite. Abriachanite. **Canadian ~** Chrysotile. White a.
a. board Sheets of pressed a. fibers used for fireproofing or insulating. **a. cord** A. string to support crucibles. **a. sponge** A. impregnated with platinum salt solution and ignited; a catalyst. **a. stopper** A stopper made from a.-magnesia mixture, for high temperatures and corrosive chemicals.
asbestosis A disease that can be produced, with some forms of cancer, by inhaling a. dust. U.K. general use limits (in fiber/ml) are chrysotile, 0.5; amosite, and crocidolite, 0.2. These limits are not absolutely safe levels.
asbolane Asbolite.
asbolite Asbolane, earthy cobalt. An earthy *psilomelane* or *wad,* q.v., containing cobalt oxide.
ascaricide A drug which destroys *Ascaris* parasites (intestinal worms).
ascaridic acid Ascaridolic acid.
ascaridole $C_{10}H_{16}O_2$ = 168.2. A terpene peroxide; the active

TABLE 10. COMMON FORMS OF ASBESTOS

Type	Shape	Color	Origin	Uses
Chrysotile	Long, curly	White	Canada, Russia	95% of all asbestos products
Crocidolite	Straight, rigid	Blue	S. Africa	Little used since about 1970
Amosite	Similar to crocidolite, but thicker	Brown	S. Africa	Acid-resistant applications
Anthophyllite	As amosite	Gray	Finland	Rare
Actinolite	As amosite	Gray	Worldwide	Not processed industrially
Tremolite	As amosite	Gray		

constituent of chenopodium oil. Explosive liquid, d_{21} 0.9985, b_{8mm}97. **a. glycol** $C_{10}H_{18}O_3$ = 186.3. Colorless crystals, m.63.

ascaridolic acid $C_{10}H_{16}O_5$ = 216.2. Cineolic acid, ascaridic acid. The oxidation product of ascaridole glycol; (+)- and (−)-forms, m.138. (±)-form, m.205.

Ascaris A genus of round worms; some are intestinal parasites.

ascaryl alcohol $C_{33}H_{68}O_4$ = 528.9. A dihydric alcohol, m.84, from the fat of *Ascaris* species.

ascharite $3Mg_2B_2O_5 \cdot H_2O$. A native borate (Stassfurt).

aschistic process A process in which work is converted directly into heat.

Asclepiadaceae The milkweed family. A group of herbs, usually with a milky juice which contains caoutchouc; e.g., *Asclepias tuberosa*, asclepias root; *Gonolobus condurango*, condurango bark.

asclepias Pleurisy root. The dried roots of *A. tuberosa* (Asclepiadaceae); a diuretic and expectorant.

asclepin A resinoid from asclepias.

ascelpion $C_{20}H_{34}O_3$ = 322.5. A camphor-like principle of *A. syriaca.*

Ascomycetes A group of fungi (molds), e.g., Aspergillus, Penicillium.

ascorbic acid $C_6H_8O_6$ = 176.1. D-~ Vitamin C, cevitamic acid, antiscorbutic factor, (−)-2,4,5,6-Tetrahydroxy-3-oxohexanoic acid. Colorless crystals, m.191, soluble in water and alcohol (USP, EP, BP). A powerful reducing agent, its main function in body cells is setting the redox potential. Essential for synthesis of collagen and metabolism of tyrosine. A. a. aids the absorption of iron and is used to produce corticoids from the adrenal gland. Deficiency causes the disease scurvy. See Table 101 on p. 622. L-~ Has no vitamin effect. **iso~** Erythorbic acid.

```
CO—┐        CO—┐        CO—┐
|  |        |  |        |  |
HOC |       HOC |       HCOH |
||  O       ||  O       |   O
HOC |       HOC |       CO  |
|   |       |   |       |   |
HC——┘       HC——┘       CH——┘
|           |           |
HOCH        HCOH        HCOH
|           |           |
CH2OH       CH2OH       CH2OH
D-          L-          iso-
(Vitamin C)
```

ascorbigen Bound ascorbic acid, usually associated with an indole group.

ascus A cell in which spores are formed.

-ase Suffix indicating (1) an enzyme; e.g., amylase; (2) certain ores; e.g., orthoclase.

asebotin A glucoside from the leaves of mountain laurel *(Kalmia latifolia).*

asebotoxin A glucoside from *Andromeda japonica* (Ericaceae).

asellin $C_{25}H_{32}N_4$ = 396.5. An alkaloid from cod-liver oil.

asepsis Freedom from infection.

aseptol $C_6H_4(OH)HSO_3$ = 174.2. Phenol-2-sulfonic acid, sulfophenol, sozolic acid, *o*-sulfocarbolic acid. Red oil, d.155, soluble in water. An antiseptic; a reagent for albumin, bile pigments, nitrites, and nitrates.

aseptoline Pilocarpine.

asexual Without sex, as the lower forms of plants and animals.

ash (1) The incombustible mineral residue remaining after a substance has been incinerated. (2) See *Fraxinus.* **caustic ~** Sodium carbonate*. **causticized ~** A mixture of sodium carbonate and sodium hydroxide. **light ~** A by-product of oil combustion containing silica, alumina, iron oxide, and small amounts of alkaline oxides and carbonates. **poison ~** Ischionanthus. **pot ~** See *potash.* **prickly ~** Xanthoxylum. **soda ~** Commercial sodium carbonate.

ashing The process of burning organic matter, especially in analysis.

ashphalt Asphalt.

asiatic acid $C_{30}H_{48}O_5$ = 488.7. The aglycone of asiaticoside. Needles, m.303.

asiaticoside A glycoside from *Centella asiatica* (Umbelliferae), m.236.

asiminine An alkaloid from the seeds of papaw, *Asimina triloba* (Anonaceae); a narcotic.

askarel Generic name for a chlorinated diphenyl; a liquid dielectric.

Asn* Symbol for asparagine.

Asp* Symbol for aspartic acid.

asparacemic acid (±)-Aspartic acid*.

asparagic acid Aspartic acid*.

asparaginase* Colaspase. Enzyme, obtained from the bacteria *Escherichia coli,* that catalyzes the hydrolysis of L-asparagine to L-aspartate plus ammonia. Thus interferes with metabolism of L-asparagine in some malignant cells; used to treat leukemia.

asparagine* $NH_2 \cdot CH \cdot (COOH) \cdot (CH_2CONH_2) \cdot H_2O$ = 150.1. Asn*. 2,4-Diamino-4-oxobutanoic acid, *β*-aspargine, asparamide, agedoite, altheine, aminosuccinic acidamide, *α*-aminosuccinamic acid, aspartamic acid. An alkaloid in the sprouts of dicotyledons and in many seeds. Formed synthetically from aspartic acid. White rhombs, m.230, soluble in hot water; a constituent of many proteins and an isomer of malamide. *α*-~ or **iso~** $NH_2 \cdot CH \cdot (CH_2COOH)CONH_2$ = 140.1. 3,4-Diamino-4-oxobutanoic acid.

a. sulfate $C_4H_8O_3N_2 \cdot H_2SO_4$ = 230.2. White powder, soluble in water.

asparaginic acid Aspartic acid*.

asparaginyl* The radical $NH_2 \cdot CO \cdot CH_2 \cdot CHNH_2 \cdot CO-$, from asparagine.

asparagus The root of *Asparagus officinalis* (Liliaceae); an aperient and diuretic. Cf. *chrysoidin.*

asparamide Asparagine*.

aspargine Asparagine*.

aspartame $C_{14}H_{18}O_5N_2$ = 294.3. 3-Amino-*N*-(α-carboxyphenethyl)succinamic acid methyl ester, Nutrasweet. White crystals, $[\alpha]_D^{20}$ +15.5°. A sweetener, 180 times as sweet as sucrose.

aspartamic acid Asparagine*.

aspartase Aspartate ammonia-lyase*.

aspartate* A salt of aspartic acid. **a. aminotransferase*** A. transaminase*. An enzyme catalyzing the formation of L-glutamate plus oxaloacetate from L-aspartate and 2-oxoglutarate. **a. ammonia-lyase*** Aspartase. An enzyme which catalyzes the splitting of ammonia from L-aspartate to form the fumarate.

aspartic acid* $NH_2 \cdot CH \cdot (COOH)(CH_2 \cdot COOH)$ = 133.1. Asp*. Aminosuccinic acid, asparagi(ni)c acid, **(±)-~** Asparacemic acid. White, monoclinic prisms, m.278. **D-~** Colorless leaflets, m.251. **L-~** Rhombic leaflets, m.270, slightly soluble in water; used in organic synthesis. Important intermediate in body transamination reactions and the urea cycle, and a product of ammonia incorporation in plants.

aspartoyl* The radical $-CO \cdot CH_2 \cdot CHNH_2 \cdot CO-$, from aspartic acid.

aspartyl* The radicals from aspartic acid: **α-~** $NH_2 \cdot CH(CH_2 \cdot COOH)CO-$. **β-~** $NH_2CH(COOH)CH_2 \cdot CO-$.

aspasiolite A partly decomposed iolite.

aspergillic acid $C_{12}H_{20}O_2N_2$ = 224.3. An antibiotic produced by *Aspergillus* molds.

aspergillin Vegetable hematin. The black coloring matter of the spores of various *Aspergillus* species (molds).

Aspergillus A genus of molds; many are parasitic. Cf. *mycogalactan, tané-koji, aspergillin.*

asphalt Jews' pitch, petrolene, mineral pitch, earth pitch, Trinidad pitch, petroleum pitch. Native mixtures of hydrocarbons; amorphous, solid or semisolid, brownish-black pitch or bitumen, produced from the higher-boiling-point mineral oils by the action of oxygen; divided into *asphaltenes* and *carbenes,* q.v. Used for pavements, roofing and waterproofing materials. Cf. *parianite.* **a. base** See *petroleum.* **a. sludge** A nonconducting sludge of oxygenated bodies of high molecular weight in electric transformers. **a. stone** Natural a. A limestone naturally impregnated with bitumen. **a. testing apparatus** See *penetrometer, viscometer.* **a. thermometer** An armored thermometer graduated from 90 to 235°C.

asphaltenes That portion of asphalt or bitumen which is soluble in carbon disulfide but insoluble in paraffin oil or in ether. Cf. *carbenes, kerotenes.*

asphaltic bitumen A more accurate term for *asphalt.*

asphyxia Suffocation due to a deficiency of oxygen.

aspidinol $C_{12}H_{16}O_4$ = 224.3. An alcohol, m.161, from aspidium.

Aspidium A genus of ferns, *Filices.*

aspidosamine $C_{22}H_{28}O_2N_2$ = 352.5. An alkaloid from quebracho bark. Brown powder, m.100, soluble in alcohol; an emetic. **a. hydrochloride** $C_{22}H_{28}O_2N_2 \cdot HCl$ = 388.9. Brown powder, soluble in water; an emetic.

aspidosperma Quebracho. The bark of *Aspidosperma quebracho* (Apocynaceae) of S. America. It contains yohimbine and other alkaloids.

aspidospermine $C_{22}H_{30}O_2N_2$ = 354.5. (1) An alkaloid from the bark of *Aspidosperma quebracho.* White needles, m.206, soluble in alcohol; an antipyretic and antispasmodic. (2) Brown powder, which consists of a mixture of quebracho alkaloids. **a. citrate** $C_{22}H_{30}O_2N_2 \cdot C_6H_8O_7$ = 546.6. Orange powder, soluble in water, comprising the citrates of quebracho alkaloids. **a. hydrochloride** $C_{22}H_{30}O_2N_2 \cdot HCl$ = 391.0. Orange powder, soluble in water. **a. sulfate, amorphous** The sulfate of quebracho alkaloids. Orange-yellow powder, soluble in water.

aspirator A suction apparatus. **a. bottle** A bottle containing liquid, with an airtight stopper, a short inlet tube, and an outlet at the bottom. The flow of liquid from the outlet draws in air or gas from a source connected to the inlet tube. Cf. *siphon.*

aspirin $MeCOO \cdot C_6H_4 \cdot COOH$ = 180.2. Acetylsalicylic acid*, xaxa, acid aceticosalicylas. Colorless crystals, m.143, slightly soluble in water. A widely used analgesic; also an antirheumatic, antipyretic, antiinflammatory; reduces blood clotting tendency (USP, EP, BP). **methyl ~** Methylacetylsalicylate.

assafetida Asafetida.

assay (1) Originally, the analysis of ores and alloys. (2) Analysis in general; of pharmaceutical and official drugs in particular. **dry ~** Assaying by dry methods. **wet ~** Assaying by wet methods.

a. balance A very delicate analytical balance. **a. combination furnace** A combination of 3 furnaces for roasting sulfides, fusing in a crucible, and cupellation. **a. crucibles** Fireclay crucibles for the analysis of ores and alloys. **a. flasks** A glass vessel of shape intermediate between that of a tall beaker and an Erlenmeyer flask. **a. mill** A small crusher for pulverizing ores. **a. ton** A.T. = 29.1667 g. **a. ton system** In the analysis of gold and silver ores, the number of mg of gold or silver obtained from an assay ton equals the number of oz per short ton (2,000 lb) of ore. For a ton of 2,240 lb, A.T. = 32.6667 g. **a. ton weights** A set of weights (4 A.T. to 1/20 A.T.) used in mineralogical analysis and gold assay.

asselline A leucomaine from cod-liver oil.

assimilation Constructive metabolism, or the transformation of nonliving matter (food) into living matter (tissues).

associating Forming complexes, e.g., by polar molecules. Cf. *solvent.* **non~** Describing nonpolar molecules which do not form complexes.

association The combination, connection, or correlation of substances or functions. **molecular ~** The aggregation of similar molecules, especially in solutions. Cf. *coordinate bond* (under *bond*), *dissociation, liquid.*

astacene $C_{40}H_{48}O_4$ = 592.8. Astacin. A carotenoid produced by the oxidation of astaxanthin; a pigment from the lobster, *Astacus gammarus.*

astatic Describing forces in equilibrium. **a. couple** See *astatic galvanometer.* **a. current** An electric circuit so arranged as to be unaffected by the earth's electric field. **a. galvanometer** A galvanometer in which an astatic couple effect is produced by two equally strong magnets. **a. needle** Two magnetic needles placed above one another with reversed poles (N over S).

astatine* At. *Alabamine,* q.v., (anglo)helvetium, ekaiodine, virginium. A member of the halogen element family, at. no. 85. Atomic weight of isotope with longest half-life (^{210}At) = 209.99. Obtained from monazite sands and (Allison, etc., 1931) from bombardment of bismuth with high-energy particles and the evaporating off of the bismuth. Obtained mainly as ^{210}At; soluble in benzene or chloroform. Forms At^- compounds, and there is evidence for the AtO^-, AtO_3^-, and At^+ ions.

astaxanthin $C_{40}H_{52}O_4$ = 596.9. See *astacene.*

asteriasterol A sterol, m.70, from starfish.

asterisk* Used to indicate an excited electronic or nuclear state; as, NO*, $^{110}Ag^*$.

asterism An arrangement of *silk,* q.v., in gems, giving a starlike effect.

asterium A supposed element in the hottest stars. Cf. *nebulium.*

asterubin $Me_2NC(:NH)\cdot NH\cdot(CH_2)_2SO_3H$ = 195.2. A guanidine derivative, from the starfish *Aster rubens.*

astigmatism The inability of the eye to focus light rays from different meridians on the same point.

ASTM American Society for Testing and Materials. An organization to promote knowledge of engineering materials and to develop standards, specifications, and methods of testing for various materials. Cf. *BSI.*

Aston, Francis William (1877–1946) English physicist who developed mass spectra; Nobel prize winner (1922). **A. rule** Not more than 2 isotopes are known for any element of odd atomic number, except among the radioactive elements. **A. spectrum** A mass spectrum, from which the isotopic weights of an element are determined.

astorism The star-shaped appearance of a Laue pattern of a distorted crystal.

astrakanite, astrochanite $MgSO_4\cdot Na_2SO_4\cdot 4H_2O$. Blödite. Native magnesium sodium sulfate (Stassfurt).

astral oil Kerosine.

astringency Sour taste with contracting power, as in tannins.

astringent A drug that contracts tissues, and thereby lessens secretions. Used to treat diarrhea and some skin conditions.

astronautics The study of voyaging through space.

astronomical unit (1) The mean distance from the sun to the earth: 149,597,900 km. (2) See *parsec, light year,* and *magnitude.*

astronomy The study of the cosmos and the characteristics of celestial bodies.

astrophyllite $H_4(K,Na)_2(Fe,Mn)_5(SiTiZr)_7O_{22}$. A native silicate; bronze-yellow, orthorhombic crystals.

astrophysics The interpretation of spectral lines which indicate the movement, velocity, composition, temperature, and other characteristics of a celestial body.

asym* *as-*. Affix indicating asymmetrical.

asymmetric (1) Unsymmetric; e.g., the 1,2,4 position of benzene. (2) Pertaining to an asymmetric atom. (3) Triclinic. **a. atom.** An atom with no element of symmetry in the configuration of atoms bonded to it. **a. carbon** A C atom which has 4 different radicals or atoms tetrahedrally attached to it. Compounds containing one carbon atom are optically active; those with more than one may be, or else *meso.* **a. compound** A compound that contains one or more asymmetric atoms. Cf. *stereoisomerism.*

asymmetry Absence of any symmetry. See *chiral.*

asymptotes Rectilinear or curvilinear lines which continually approach a curve or line without touching it.

A.T. See *assay ton.*

At Symbol for the element astatine.

Atabrine Trademark for quinacrine.

atacamite, atakamite $CuOCl\cdot Cu(OH)_2$. Remolinite. A native, green hydrous copper oxychloride.

atactic See *polyvinyl chain.*

atalpo clay A Cornish colloidal china clay; a soap filler and a catalyst.

atanasin $C_{20}H_{18}O_8$ = 386.4. A flavonoid from *Brickellia squarrosa,* of Puerto Rico.

ataraxic A major tranquilizing drug, e.g., chlorpromazine hydrochloride.

-ate Suffix. (1) Formerly used for an anion with a nonmetal in its higher positive oxidation state; as, sulfate. (2)* Now used (a) in inorganic chemistry, in parallel with *-ite* and *-ide,* q.v., to refer to a negative group (anion) containing the stated central atom, as S in sulfate, irrespective of the oxidation state or ligands; the oxidation number of the central atom or ionic charge may be cited where ambiguity may arise; as, hexacyanoferrate(II), $[Fe(CN)_6]^{4-}$; (b) in organic chemistry, as the suffix of anions formed by the loss of a proton from an acid; as, benzoate. See *acid,* Table 3.

athamantin $C_{24}H_{30}O_7$ = 430.3. A bitter principle from the roots and seeds of *Athamanta oreoselinum* (Umbelliferae).

athermal A cool spring or mineral water.

atheroma Fatty substance occurring in plaques (patches) on the lining of arterial walls in the arterial disease atherosclerosis. A. causes narrowing and blockage of arteries. Chiefly consists of cholesterol, cholesterol esters, sphingomyelin, and fibrin. See *arteriosclerosis.*

atherospermine $C_{20}H_{23}O_2N$ = 309.4. An alkaloid from the bark of *Atherosperma moschatum* (Monimiaceae) of Australia; sometimes used as tea.

atisine $C_{22}H_{33}NO_2$ = 343.5. Colorless powder, m.85, slightly soluble in water.

Atkinson hemin test A method of preparing blood crystals (hemin) for microscopic examination and identification.

atm Abbreviation for atmosphere.

atmolysis The separation of gases by diffusion through porous walls; see *Graham's law.* **fractional ∼** A method for continuous diffusion of a gas, used to separate isotopes.

atmometer An instrument to determine the amount of water passing into the air by evaporation.

atmos Abbreviation for atmosphere.

atmosphere (1) The air or gases surrounding the earth. (2) The pressure exerted by the air at sea level. (3) atm, atmos: (a) Correctly, standard pressure which equals 101.325 kPa (that supporting a column of mercury 760 mm high at 0°C, of density 13.5951, and with $g = 9.80665\ m/s^2$) = 29.921 in Hg = 14.6974 psi = $1.0333\ kg/cm^2$. (b) The bar = 100.00 kPa = 750.06 mmHg = $10^{-6}\ dyn/cm^2$. **International standard ∼** An a. used for the graduation of altitude instruments: gravity, $9.8062\ m/s^2$ at all heights; pressure and temperature at zero height, 101.325 kPa and 15°C; lapse rate of temperature, 6.5°C/km to 11 km and zero above. **technical ∼** An a. equal to $1\ kg/cm^2$ = 98.0665 kPa.

atmospheric gases The percentage volume of the gases (on moisture-free air) is nitrogen, 78.08; oxygen, 20.9; argon, 0.93; carbon dioxide, 0.033; neon, 0.0018; helium, 0.00052; krypton, 0.0001; and carbon monoxide, normally less than 5 ppm.

atmotherapy Medical treatment with vapors; as, atomizers in respiratory disorders.

atom The smallest part of an element that remains unchanged during chemical reaction and is thus chemically indestructible and indivisible. It may undergo physical changes, as, excitation, disintegration, and transformation to other atoms. Cf. *radioactive elements, matter.* An a. has three fundamental numbers: (1) Mass number, giving the total number of protons plus neutrons. (2) Atomic number, giving the number of protons in the nucleus, (3) Packing fraction, which is an indication of the forces binding nuclear particles together. See also *atomic structure.* **activated ∼** Excited ∼. **asymmetric ∼** See *asymmetric atom.* **Bohr ∼** The concept of a dynamic atom derived from phenomena of radiation. Cf.

atomic structure 5. **central ~*** The central atom in a polyatomic complex, as I in ICl_4^-. See *ligand, coordination compound.* **characteristic ~*** In a polyatomic group, the atom giving character to the group, as Cl in ClO^-. **chemist's ~, cubical ~** Lewis atom. **dark ~** An a. that does not emit radiation. **dynamic ~** Bohr ~. **excited ~** Activated ~. An a. with 1 or more electrons in a higher energy level than the ground state; as occurs in sun and stars; produced by exposing vapors and gases to strong electric fields or radiation. See *excitation.* **exploding ~** An a. that rapidly disintegrates and releases much energy. Cf. *a. energy.* **fluctuating ~** Schrödinger ~. **giant ~** An entity which surrounds all matter and fills space. **ionized ~** An a. from which some of the valence electrons are removed (positive ion), or which has captured additional electrons (negative ion). **kinetic ~** Bohr ~. **labeled ~** Radioactive indicator. **Langmuir ~** An elaborated concept of Lewis a. Cf. *atomic structure 4.* **Lewis ~** The concept of a static a. based on crystal structure and chemical bonds. Cf. *atomic structure 3 and 4.* **neutral ~** An a. in which the positive nuclear charge is balanced by the negative electrons, in either the normal or excited states. Cf. *stripped a., ionized a.* **normal ~** An a. which is neither excited nor ionized; the electrons are in their lowest energy levels. **nuclear ~** Stripped ~. **physicist's ~** Bohr ~. **planetary ~** Rutherford ~. **pulsating ~** Schrödinger ~. **radiating ~** An a. in which the electrons pass from a higher to a lower energy level and thereby emit radiation. Cf. *quantum.* **recoil ~** An a. from which an α particle is rejected and which thereby recoils with a speed corresponding with its mass. Cf. *radioactivity.* **Rutherford ~** The original concept (1911) of a planetary a. which resembled a solar system. Cf. *atomic structore 1.* **Schrödinger ~** The pulsating a., consisting of an electric field of different intensities. See *atomic structure 6.* **static ~** Lewis ~. **stripped ~** An a. whose electrons have been removed by strong electric fields or extremely high temperature; supposed to exist in the interior of celestial objects and to account for their extreme densities. Cf. *spectral classification.* **tetrahedral ~** An a. in which parts of electrons are supposed to oscillate around the 4 corners of a tetrahedron. Cf. *atomic structure 4.* **tracer ~** Radioactive tracer.

a. annihilation The destruction of matter by its complete transformation into radiation; e.g., stellar energy. **a. bomb** A device for producing explosively a large amount of energy in a very short time. Two pieces of uranium (^{235}U) or plutonium (^{239}Pu) are contacted rapidly, e.g., by firing one into the other, so that together they exceed the critical mass. The mass is bombarded with thermal neutrons, which sets off a thermonuclear chain reaction. The first a. b. was exploded in New Mexico July 16, 1945. First used in warfare at Hiroshima and Nagasaki in August 1945; equivalent to 20,000 tons of TNT. **a. building** The formation of atoms from + and − electrons, a process assumed to occur in interstellar space, where pressure, temperature, and density are extremely low. **a. creation** The creation of matter by the complete transformation of energy as a link in the mass-energy cycle. **a. energy** The disintegration of atoms, either naturally, as in natural radioactivity, or artificially, as by the bombardment of atomic nuclei with protons, deuterons, α particles, neutrons, or photons, whereby new atoms are formed and energy is liberated. **a. fragment** The disintegration product of an atom that has been split by an α particle. Cf. *Wilson fog method.* **a. meter** See *angstrom.* **a. model** (1) Atomic model. (2) A model depicting certain properties of atoms, especially valency and isomerism. (3) An electrical or magnetic device which illustrates the structure and properties of the a.

atomic Pertaining to the ultimate electrical unit of an element. **a. bomb** See *atom bomb.* **a. diameter** The imaginary line connecting the extreme electron orbits through the center of the a. **a. disintegration** See *atom energy.* **a. distance** The average or equilibrium length between the centers of 2 atoms as determined from crystal structure and band-spectra data, or calculated from a. radii. **a. domain** The imaginary sphere occupied by the a. structure. **a. energy** The force which holds the atom together; the energy liberated when an atom disintegrates or is transformed. 1 g Ra ≡ 8.4 × 10^9 joules. Cf. *packing effect.* **a. excitation** See *excited atom* under *atom, excitation.* **a. evolution** Cf. *spectral classification.* **a. field** The space around an atom which cannot normally be penetrated by other atoms. See *molecular diagrams, magnetic field.* **a. fragment** See *atom fragment.* **a. frequency** Characteristic K radiation. **a. group** Radical. **a. heat** The heat required to raise the temperature of one gram atom of an element 1°C; the product of the specific heat capacity and the atomic weight of an element = 6.4. Cf. *specific heat capacity.* **a. kernel** A. nucleus. **a. mass unit** Unified atomic mass constant*. **a. meter** Angstrom unit. **a. migration** Molecular *rearrangement.* **a. model** The concept of a. structure; as, Bohr's atom. **a. nucleus** The central, positively charged part of an atom consisting of Z protons and $A - Z$ neutrons, where A is the mass number and Z the atomic number. **a. number*** Z* (German *Zahl*: number). Proton number*, ordination number. It indicates the order of the elements in the periodic system and represents the number of protons in the atomic nucleus. It is related to the frequency ν and wavelength λ of the X-ray spectrum of an element by $c/\lambda = \nu = C(Z - b)^2$, where C and b are constants, Z is the a. number and c is the speed of light. It is shown as a prefix subscript, thus, $_{16}S$. Cf. *mass number.* **equivalent a. n.** EAN. The atomic number of an element, less the number of electrons lost on ionization, plus the number of electrons gained by coordination. **a. orbital** The region for a given electronic state, over which the electronic charge is distributed. **a. oscillations** The vibrations of a. nuclei or atoms within a molecule. Cf. *activated molecule.*
a. pile See *reactor.* **a. plane** The imaginary surface which passes through a set of atoms in a space lattice indicated by the Miller indices 100, 110, 111, etc. See Fig. 2, *crystal.* **a. potential** Ionization energy. **a. properties** Those characteristics of an element that depend on *a. structure,* q.v., as opposed to molecular structure. **a. radius** (1) The distance from the a. nucleus to the valence electrons. (2) The halfway distance between like atoms. Thus the a. r. for the single-bond C is 0.77, for single-bond N is 0.70: hence the distance C—C is 1.54; C—N is 1.47. **a. refraction** See *refraction.* **a. species** (1) Atoms of the known *elements,* q.v. (2) Isotopic species. The a. structure of an isotope. Thus chlorine consists of the a. species ^{35}Cl and ^{37}Cl, or isotopes with mass numbers 35 and 37. **a. structure** The composition of atoms, based upon a speculative interpretation of chemical and physical properties of the elements. See Fig. 3 on p. 58.

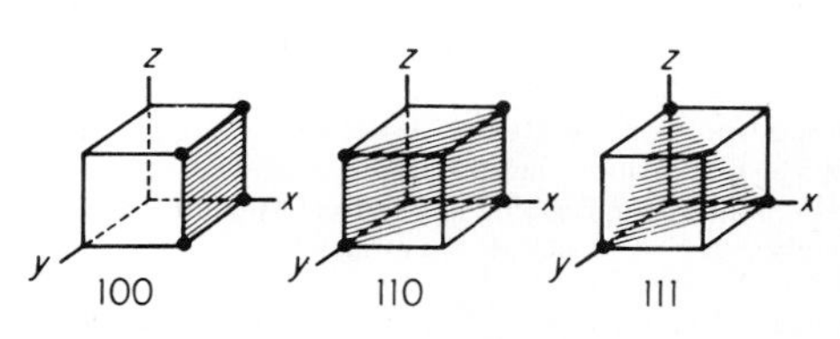

Fig. 2. Atomic planes.

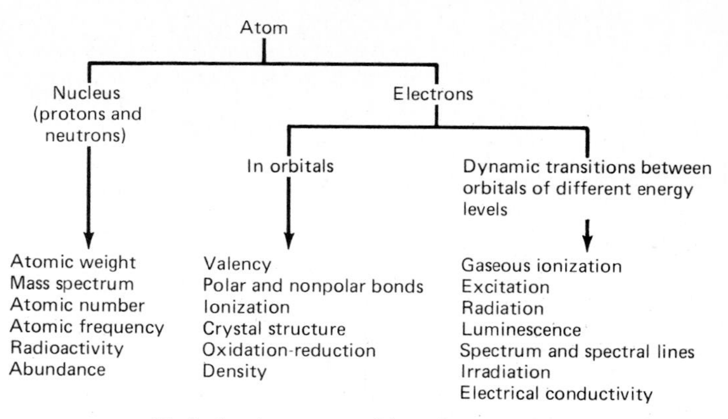

Fig. 3. Atomic structure and dependent properties.

Theories, in order of increasing development, are:

1. The atom in its ground, or normal (i.e., unexcited), state consists of an extremely small, positively charged nucleus containing the mass of the atom, surrounded by a number of electrons sufficient in number to neutralize the electric charge. The number of positive charges on the nucleus varies from 1 to 103+ and corresponds with the a. number; the number of total particles varies from 1 to 260+ and corresponds with the "isotopic weight" (Rutherford, Bohr). Cf. *periodic system.*
2. The nucleus itself, 10^{-12} cm in diameter, consists of neutrons and protons which are closely bound together (Prout, Harkins.) Cf. *packing effect, isotopes.*
3. The electrons are distributed in successive shells containing 2, 8, 18, 32, etc. (i.e., $2n^2$), electrons, respectively (Kossel, Lewis.) Cf. *orbit, shell, periodic system, Stoner quanta, Pauli's principle* (under *Pauli*).
4. The electrons oscillate in shells, usually in pairs, around centers corresponding with the corners of a cube or tetrahedron; as, in crystals (Lewis, Langmuir.) Cf. *octet, kernel, valency.*
5. The electrons exist in orbitals, which are, e.g., either spherical or dumbbell-shaped, and at different energy levels; as, in luminous gases (Bohr, Sommerfeld). Cf. *orbital, quantum, excitation.*
6. The spatial distribution of an electron in its orbital is defined by a wavelike property ("wave function"). When this is substituted in the Schrödinger wave equation, it gives the energy of the orbital. Cf. *wave mechanics.*
7. In ions and sigma bonds, the electrons are localized in the region of their particular atoms; in pi bonds, they are less localized, being shared by several atoms; in metals, they are completely delocalized ("free"), being shared by all atoms ("electron gas").

a. symbols Indices defining mass number (left upper index), atomic number (left lower), number of atoms (right lower), ionic charge (right upper). Thus ${}^{32}_{16}S_2^{2+}$ represents a doubly ionized molecule containing 2 atoms of sulfur, each of which has atomic number 16 and mass number 32. **a. theory** (1) The concept of a. structure. (2) The idea of finite particles of matter, as conceived by Democritus (400 B.C.), and established by Dalton (1808) and others. Cf. *kinetic theory.* **a. transformation** The building up of atoms from those of lower a. weight. **a. transmutation** An artificial process for changing one atom into another. Cf. *nuclear chemistry.* **a. units** The definition of standards in terms of atomic dimensions; e.g., of the meter in terms of the wavelengths of a standard krypton line in a vacuum. **a. volume** The space occupied by one gram atom of an element; the quotient of the atomic weight and the density of the solid or liquid element. **a. weight** A_r. Relative atomic mass*. The ratio of the average mass per atom of an element to 1/12 of the mass of an atom of the nuclide ^{12}C. The a. w. should be qualified by the source of the material when it is possible that its value has been changed by artificial alteration or by a difference in natural isotopic composition. Before 1960, a. weights were relative to H (= 1.0080) or O (= 16.000). For values, see Table 11.

atomicity The number of atoms in the molecule of an element.

atomization Breaking up a liquid into a fine spray or fog.

atomizer A device for atomization; a nebulizer.

atom meter Angstrom unit.

atomology The study of atomic structure.

atomsite A green slag produced on the ground surrounding the first *atom bomb,* q.v., site in New Mexico.

atopite $Ca_2Sb_2O_7$. A native, yellow, isometric antimonite.

atoxylate Arsanilate.

atoxylic acid Arsanilic acid.

ATP* Symbol for adenosine 5′-triphosphate.

atractylene $C_{15}H_{24}$ = 204.4. A sesquiterpene, d.0.927, b_{14mm}141, from the essential oil of Asian *Atractylis* (Compositae).

atractylol $C_{15}H_{26}O$ = 222.4. 3-Eudesmen-11-ol. A solid alcohol, m.75, from the essential oil of *Atractylis.*

atrazine* See *herbicide,* Table 42.

atroglyceric acid $CH_2OH·C(OH)Ph·COOH$ = 182.2. 2-Phenylglyceric acid. White crystals, m.146.

atrolactic acid $Me·C(OH)(Ph)COOH·H_2O$ = 184.2. 2-Hydroxy-2-phenylpropanoic acid*, 2-phenyllactic acid. **(*R*)-~, (*S*)-~** Colorless crystals, m.116, soluble in water.

atromentin $C_{18}H_{12}O_6$ = 324.3. 2,5-Dihydroxy-3,6-bis(4-hydroxyphenyl)quinone. A pigment from *Paxillus atromentosus.*

Atromid Trademark for clofibrate.

atronene $C_{16}H_{14}$ = 206.3. Atronol, phenyldihydronaphthalene. Colorless crystals, m.326, insoluble in water.

atronic acid $C_{17}H_{14}O_2$ = 250.3. Colorless prisms, m.164, soluble in water.

atropic acid* $Ch_2:CPh·COOH$ = 148.2. 2-Phenylpropenoic acid*, α-methylenebenzeneacetic acid†. Colorless scales, m.106, sparingly soluble in water. **iso~** Isatropic acid.

atropine $C_{17}H_{23}O_3N$ = 289.4. Coromegine, daturine,

TABLE 11. STANDARD ATOMIC WEIGHTS (1983)

International Union of Pure and Applied Chemistry. Based on the Relative Atomic Mass of ^{12}C = 12
The values apply to elements as they exist naturally on earth and to certain artificial elements.

Name	Symbol	Atomic number	Atomic weight
Actinium	Ac	89	227.03
Aluminum	Al	13	26.98154
Americium	Am	95	243.06
Antimony (Stibium)	Sb	51	121.75±3
Argon	Ar	18	39.948
Arsenic	As	33	74.9216
Astatine	At	85	209.99
Barium	Ba	56	137.33
Berkelium	Bk	97	247.07
Beryllium	Be	4	9.01218
Bismuth	Bi	83	208.9804
Boron	B	5	10.811±5
Bromine	Br	35	79.904
Cadmium	Cd	48	112.41
Calcium	Ca	20	40.078±4
Californium	Cf	98	251.08
Carbon	C	6	12.011
Cerium	Ce	58	140.12
Cesium	Cs	55	132.9054
Chlorine	Cl	17	35.453
Chromium	Cr	24	51.9961±6
Cobalt	Co	27	58.9332
Copper (Cuprum)	Cu	29	63.546±3
Curium	Cm	96	247.07
Dysprosium	Dy	66	162.50±3
Einsteinium	Es	99	252.08
Erbium	Er	68	167.26±3
Europium	Eu	63	151.96
Fermium	Fm	100	257.10
Fluorine	F	9	18.998403
Francium	Fr	87	223.02
Gadolinium	Gd	64	157.25±3
Gallium	Ga	31	69.723±4
Germanium	Ge	32	72.59±3
Gold (Aurum)	Au	79	196.9665
Hafnium	Hf	72	178.49±3
Helium	He	2	4.002602±2
Holmium	Ho	67	164.9304
Hydrogen	H	1	1.00794±7
Indium	In	49	114.82
Iodine	I	53	126.9045
Iridium	Ir	77	192.22±3
Iron (Ferrum)	Fe	26	55.847±3
Krypton	Kr	36	83.80
Lanthanum	La	57	138.9055±3
Lawrencium	Lr	103	260.11
Lead (Plumbum)	Pb	82	207.2
Lithium	Li	3	6.941±2
Lutetium	Lu	71	174.967
Magnesium	Mg	12	24.305
Manganese	Mn	25	54.9380
Mendelevium	Md	101	256.10
Mercury (Hydroargyrus)	Hg	80	200.59±3
Molybdenum	Mo	42	95.94
Neodymium	Nd	60	144.24±3
Neon	Ne	10	20.179
Neptunium	Np	93	237.05
Nickel	Ni	28	58.69
Niobium	Nb	41	92.9064
Nitrogen	N	7	14.0067
Nobelium	No	102	259.10
Osmium	Os	76	190.2
Oxygen	O	8	15.9994±3
Palladium	Pd	46	106.42
Phosphorus	P	15	30.97376
Platinum	Pt	78	195.08±3
Plutonium	Pu	94	244.06
Polonium	Po	84	208.98
Potassium (Kalium)	K	19	39.0983
Praseodymium	Pr	59	140.9077
Promethium	Pm	61	144.91
Protactinium	Pa	91	231.04
Radium	Ra	88	226.03
Radon	Rn	86	222.02
Rhenium	Re	75	186.207
Rhodium	Rh	45	102.9055
Rubidium	Rb	37	85.4678±3
Ruthenium	Ru	44	101.07±2
Samarium	Sm	62	150.36±3
Scandium	Sc	21	44.95591±1
Selenium	Se	34	78.96±3
Silicon	Si	14	28.0855±3
Silver (Argentum)	Ag	47	107.8682±3
Sodium (Natrium)	Na	11	22.98977
Strontium	Sr	38	87.62
Sulfur	S	16	32.066±6
Tantalum	Ta	73	180.9479
Technetium	Tc	43	97.907
Tellurium	Te	52	127.60±3
Terbium	Tb	65	158.9254
Thallium	Tl	81	204.383
Thorium	Th	90	232.0381
Thulium	Tm	69	168.9342
Tin (Stannum)	Sn	50	118.710±7
Titanium	Ti	22	47.88±3
Tungsten (Wolfram)	W	74	183.85±3
Uranium	U	92	238.0289
Vanadium	V	23	50.9415
Xenon	Xe	54	131.29±3
Ytterbium	Yb	70	173.04±3
Yttrium	Y	39	88.9059
Zinc	Zn	30	65.39±2
Zirconium	Zr	40	91.224±2

NOTES

a. When used with due regard to the origin and treatment of the material, and to the other footnotes, the atomic weights are reliable to ±1 in the last digit, unless otherwise stated. However, geologically exceptional specimens, which exceed the uncertainty implied for normal material, are known for the following elements: Ag, Ar, Ba, Ca, Cd, Ce, Dy, Er, Eu, Gd, H, He, In, Kr, La, Li, Lu, N, Nd, Ne, O, Os, Pb, Pd, Rb, Ru, Sm, Sr, Te, Th, U, Xe, Y, Zr.

b. The following radioactive elements lack a characteristic terrestrial isotopic composition. The atomic weight given is that of the isotope with the longest half-life, its mass number being the integer closest to the atomic weight: Ac, Am, At, Bk, Cf, Cm, Es, Fm, Fr, Lr, Md, No, Np, Pa, Pm, Po, Pu, Ra, Rn, Tc.

c. Commercially available material of the following elements may have modified isotopic composition owing to inadvertent or undisclosed isotopic separation, thus giving an atomic weight different from that in the table: B, H, Kr, Li, Ne, U, Xe.

d. The range in isotopic composition of normal terrestrial material of the following elements limits the precision of the tabulated value, but the value should be applicable to any normal material: Ar, B, C, Cu, H, He, Li, O, Pb, S, Si.

e. Thorium generally has a well-defined composition, but there are rare exceptions (notably ocean water) that contain some thorium-230.

Source: *Pure and Applied Chemistry*, June 1984

isotropine, (±)-tropine tropate, tropin tropic ester, (±)-hyoscyamine. An alkaloid from *Atropa belladonna*, deadly nightshade, or *Datura stramonium;* also synthetically prepared. Colorless needles, m.115, slightly soluble in water. An antispasmodic, mydriatic; used to reduce secretions, e.g., bronchial, before a general anesthetic. An antidote for morphine, pilocarpine, and prussic acid. The antidotes for atropine are emetics, gastric lavage, pilocarpine, morphine. **a. (hydro)iodate** $C_{17}H_{23}O_3N \cdot HIO_3$ = 465.3. Colorless crystals; soluble in water. A mydriatic (EP, BP). **a. methylnitrate** $C_{16}H_{20}O_3N(CH_3)_2NO_3$ = 366.4. A. methonitrate, Eumydrin. White crystals, soluble in water. An antispasmodic for infants with pylorospasm and whooping cough (EP, BP). **a. nitrate** $C_{17}H_{23}O_3N \cdot HNO_3$ = 352.4. Colorless crystals, soluble in water. **a. salicylate** $C_{17}H_{23}O_3N \cdot C_7H_6O_3$ = 427.5. Colorless, deliquescent powder, soluble in alcohol. **a. stearate** $C_{17}H_{23}O_3N \cdot C_{18}H_{36}O_2$ = 573.9. Colorless powder, used in ointments. **a. sulfate** $(C_{17}H_{23}O_3N)_2H_2SO_4 \cdot H_2O$ = 694.8. Colorless crystals, m.192, soluble in water; used in ophthalmology as a mydriatic; also used to quicken heart rate and reduce bronchial secretions (USP, EP, BP). **a. valerate** $C_{17}H_{23}O_3N \cdot C_5H_{10}O_2 \cdot \frac{1}{2}H_2O$ = 400.5. Colorless crystals, m.42, soluble in water.

atropoyl* 1-Oxo-2-phenyl-2-propenyl†. The radical PhC(:CH₂)CO—, from atropic acid.

atroscine $C_{17}H_{21}O_4N \cdot H_2O$ = 321.4. (±)-Hyoscine, isohyoscine, Isoscopolamine. An alkaloid from *Scopolia atropoides* (Solanaceae). Transparent crystals, soluble in water.

atroxindole C_9H_9ON = 147.2. 3-Methyloxindole. Colorless crystals, m.123.

attapulgite Palygorskite. The active element in montmorillonite clay, responsible for its catalytic polymerization of unsaturated compounds and its ion-exchange properties. A. is very absorbent and thixotropic; used in paints; used to condition fertilizers and clarify oils. Minute, rectangular, hollow needles.

attar of roses An essential oil from damascene and cabbage roses, used in perfumery.

attemperator A pipe through which water flows at a constant temperature; used for temperature control.

attenuation (1) Weakening the toxicity of a microorganism or virus. (2) The extent to which the specific gravity of a liquid is lowered by alcoholic fermentation. **a. coefficient** See *symbols,* Table 88—Group B.

atto-* a. Prefix for 10^{-18}. See *SI units.*

attraction The force that holds molecules together. **capillary ~** The force that raises or depresses a fluid in a capillary tube. **chemical ~** Affinity. **electric ~** The force that draws oppositely charged bodies together. **electron ~** The force exerted by an atomic nucleus on the electron pair; a *bond.* **gravitational ~** Gravitation. **magnetic ~** The action of a magnet on another magnet or on magnetic materials, e.g., iron particles. **mechanical ~** See *adhesion* and *cohesion.* **radiant ~** See *general relativity* under *relativity.*

attritus A botanical constituent of coal.

Atwater calorimeter A bomb calorimeter for determining the energy-heat value of foods.

at.wt. Abbreviation for atomic weight.

A.U. (1) Obsolete abbreviation for angstrom unit. (2) Abbreviation for atomic units.

Au Symbol for gold (aurum).

aubepine Anisaldehyde*.

aucubin A glucoside from greater plantain, *Plantago major.* White needles, m.180, soluble in water.

audiometer An apparatus to test hearing. Cf. *eudiometer.*

auerbachite Native, amorphous zirconium silicate.

auerlite Thorium silicophosphate (North Carolina). Cf. *monazite.*

Auer metal A mixture of mischmetal 65, Fe 35%, used as pyrophoric alloy in gas lighters and for tracer bullets.

Auer von Welsbach See *Welsbach.*

augelite A natural, basic aluminum phosphate.

Auger effect The transfer of an electron to a lower energy level without radiation emission, the excess energy being used to eject another electron of the same atom. It can occur with all elements except H and He, but is more probable for elements of atomic number lower than 35.

augite $CaMg_2Al_2Si_3O_{10}$. Malacolite. A dark aluminum pyroxene in basalt. Cf. *sahlite.*

Augmentin Trademark for combination of amoxycillin and *clavulinic acid,* q.v.

augustione Angustione.

aura (1) The current of air or breeze produced by the discharge of static electricity from a point. (2) A peculiar radiation said to emanate from living organisms. Cf. *scotography.*

auramine $Me_2N \cdot C_6H_4 \cdot C(:NH) \cdot C_6H_4 \cdot NMe_2$ = 267.4. 4,4′-Bisdimethylaminobenzophenone imide, yellow pyoktannin. Yellow scales, m.136, insoluble in water; a dye. **a. hydrochloride.** The a. of commerce. A yellow powder, soluble in water, a dye.

Aurantiaceae The orange family. See *Citrus.*

aurantiin Naringin.

aurantine An orange extract.

aurantium Orange (Latin).

aurate* Indicating gold as the central atom(s) in an anion; as, a salt of auric hydroxide containing the AuO_3^{3-} ion.

aurentia An orange aniline dye; used in light filters and biological stains.

aureolin Primulin. **a. yellow** Potassium hexanitrocobaltate(III)*; a yellow paint pigment.

Aureomycin Trademark for chlortetracycline.

auri Of gold (Latin). Cf. *aurum.*

auribromhydric acid $HAuBr_4 \cdot 5H_2O$ = 607.7. Hydrogen tetrabromoaurate(III)*. Yellow crystals, m.27, soluble in water.

auribromide Tetrabromoaurate(III)*, bromaurate. A salt of the type $M[AuBr_4]$.

auric Gold(III)* or (3+)*, aurum(III)* or (3+)*. Describing compounds of trivalent gold. **a. acid** Auric hydroxide. **a. bromide** $AuBr_3$ = 436.7. Brown powder, soluble in water. **a. chloride** $AuCl_3$ = 303.3. Red leaflets, decomp. 180, soluble in water; used in photography. **a. chloride acid** $AuCl_3 \cdot HCl \cdot 4H_2O$ = 411.8. Aurochlorohydric acid. Yellow crystals, soluble in water. **a. chloride cryst.** $AuCl_3 \cdot 2H_2O$ = 339.4. Orange leaflets, soluble in water. **a. chloride fused** $AuCl_3 \cdot HCl \cdot nH_2O$. Brown masses, soluble in water. Used in electroplating. **a. cyanide** $Au(CN)_3 \cdot 6H_2O$ = 383.1. White hygroscopic crystals, soluble in water. **a. hydroxide** $Au(OH)_3$ = 248.0. Auric acid. Brown powder, decomp. 150, insoluble in water. **a. iodide** AuI_3 = 577.7. Green powder, decomp. by heat, insoluble in water. **a. nitrate acid** $Au(NO_3)_3 \cdot HNO_3 \cdot 3H_2O$ = 500.0. Yellow triclinics, decomp. by heat, soluble in water. **a. oxide** Au_2O_3 = 441.9. Black powder, losing oxygen on heating; insoluble in water **a. potassium chloride** Potassium tetrachloroaurate(III)*. **a. sulfate** $Au_2O_3 \cdot 2SO_3 \cdot H_2O$ = 620.1. Yellow, deliquescent crystals, decomp. by heat, soluble in water. **a. sulfide** Au_2S_3 = 490.1. Brown powder, insoluble in water.

aurichalcite The mineral $2(ZnCu)Co_3 \cdot 3(ZnCu)(OH)_2$.

aurichloride Tetrachloroaurate(III)*. A salt containing the $[AuCl_4]^-$ ion.

aurichlorohydric acid Auric chloride acid.

auricyanhydric acid $HAu(CN)_4$ = 302.0. Hydrogen tetracyanoaurate(III)*. Colorless crystals, soluble in water.
auricyanide Tetracyanoaurate(III)*. A salt containing the $[Au(CN)_4]^-$ ion.
auriferous Containing gold.
auriiodide Tetraiodoaurate(III)*. A compound containing the $[AuI_4]^-$ ion.
aurin $(HOC_6H_4)_2CC_6H_4{:}O$ = 290.3. Pararosolic acid, coralline. A triphenylmethane dye. Red needles, m.220 (decomp.), insoluble in water; a textile dye and indicator; yellow (acid) to magenta (basic) at pH 7.5. Cf. *rosolic acid.*
auripigment Arsenous sulfide.
auro Aurous.
auroauric Aurum(I) aurum(III)*, gold(I) gold (III)*. A compound containing 1 atom each of mono- and trivalent gold. **a. bromide** Au_4Br_8 = 1427. A black powder, insoluble in water. **a. chloride** Au_4Cl_8 = 1072. Gold dichloride. Red crystals, insoluble in water. **a. sulfide** Au_2S_2 = 458.1. Black powder, decomp. by heat, insoluble in water.
aurobromide $MAuBr_2$. Dibromoaurate(I)*, Bromaurite.
aurochloride Dichloroaurate(III)*. A compound containing the $[AuCl_2]^-$ ion.
aurocyanide Dicyanoaurate(I)*. A compound containing the $[Au(CN)_2]^-$ ion.
aurodiamine $AuHN \cdot NH_2$ = 228.0. Fulminating gold. Green powder, explodes if struck or heated.
auroglaucin $C_{19}H_{22}O_3$ = 298.4. A pigment produced on textiles by the action of *Aspergillus glaucus,* m.152.
auromine $C_{17}H_{23}N_3$ = 269.4. Rose needles, m.96, insoluble in water.
aurone $C_{15}H_{10}O_2$ = 222.2. A benzylidene coumaranone, the coloring matter of yellow dahlia, *D. variabilis.* Orange solid, m.312 (decomp.).
aurora **a. polaris** A display of variously colored lights in the atmosphere, near the north (a. borealis) and south (a. australis) poles, with a conspicuous green line, 5577.3 Å., due to electrically charged particles being orientated by the earth's magnetic field. Intensified by sunspot acivity. **a. tube** A vacuum discharge tube of uranium glass.
aurorium A hypothetical element said to produce the characteristic lines of the aurora (obsolete). Cf. *nebulium.*
aurosulfide Thioaurate(III)*. A salt containing the $[AuS]^-$ ion.
aurothioglucose $C_6H_{11}AuO_5S$ = 392.2. Glucosylthiogold. Yellow powder, soluble in water; an antirheumatic (USP).
aurous Aurum(I)* or (1+)*. Gold (I)* or (1+)*. Describing compounds of monovalent gold. **a. bromide** AuBr = 276.9. Green powder, decomp. 115, insoluble in water. **a. bromide acid** $AuBr \cdot HBr \cdot 5H_2O$ = 447.9. Red crystals, soluble in water. **a. chloride** AuCl = 232.4. Yellow crystals, decomp. by heat or water. **a. cyanide** AuCN = 223.0. Yellow crystals, decomp. by heat, insoluble in water, soluble in potassium cyanide solutions. **a. hydroxide** AuOH = 214.0. Brown crystals, decomp. 250. **a. iodide** AuI = 323.9. Gold monoiodide*. Green powder, decomp. 120, slightly soluble in water. **a. oxide** Au_2O = 409.9. Violet powder, decomp. 250, insoluble in water. **a. potassium cyanide** Potassium dicyanoaurate(I)*. **a. sodium cyanide** Sodium dicyanoaurate(I)*. **a. sulfide** Au_2S = 426.0. Black powder, insoluble in water.
aurum* Gold* (Latin). **a. vegetable** Pipitzahoic acid.
ausonium Early name for neptunium.
austempering The tempering of steel by the transformation of the *austenite,* q.v., at a temperature within the martensitic zone.
austenite A carbon-iron γ-ferrite formed in highly carbonized steel. Cf. *martensite.*
australene Pinene.
austrium A supposed element (Przibram, 1900) shown to be gallium.
auto- Prefix (Greek) meaning "self-".
autoactivation Activation of a gland by its own secretions.
autoantibiosis Inhibition of a culture medium by the previous growth of the organism.
autocatalysis Catalysis produced by the products of a catalytic reaction.
autoclave An apparatus for heating liquids, or sterilizing at high steam pressure.
autocollimation spectroscope A comparison spectroscope.
autocorrelation analysis Mathematical technique for determining the corrrelation between variables that takes into account any time lags that occurred between input and output.
autocytolysis Autolysis.
autocytotoxin A body toxin formed by absorbed degenerated cells.
autodecomposition A decomposition autocatalysis.
autolysate, autolyzate The product of the self-liquefaction of organic cells or tissues.
autolysis The dissolution of cells by their own enzymes. Cf. *heterolysis.*
autolyze, autolyse To liquefy organic cells.
automatic Self-working. **a. buret** A buret with an overflow at the zero point. **a. pipet** A pipet with an overflow at its mark for rapid use.
automation The replacement of routine labor by instrumentation, control technniques, and power-driven equipment.
automolite Gahnite.
autooxidation (1) Oxidation by the unaided atmosphere. (2) An oxidation reaction initiated only by an inductor. See *induced reaction.*
autophytes Plants that can live on inorganic matter. Cf. *saprophyte.*
autoprotolysis Transfer of a proton from one molecule to another of the same substance.
autoradiography The use of radioactive tracer isotopes to record activity distribution on a surface by means of contact photographic film.
autotrophic Describing an organism deriving its energy by oxidation of inorganic substances.
autoxidation Autooxidation.
autunite $CaO \cdot 2UO_3P_2O_5 \cdot 8H_2O$ = 914.2. Native, yellow, orthorhombic crystals (Utah and South Dakota).
auvergnose A diorite (North Carolina).
auximone An agent that stimulates the growth of seedlings and plants; as, pantothenic acid.
auxins Phytohormones. One of the main classes of plant hormone; as, 3-indolylacetic acid. **a.a** $C_{18}H_{32}O_5$. **a.b** $C_{18}H_{30}O_4$.
auxiometer A device to measure the magnifying power of lenses.
auxochrome Formerly, functional groups of atoms that intensify the colors of a chromophore or develop a color from a chromophore. Cf. *chromophore.*
auxogluc An atom or radical which combined with glucophores yields sweet compounds.
auxograph A device to record plant growth rate.
auxotox radical The methylimine group $=N \cdot CH_3$, associated with liver degeneration.
av. (1) Avoirdupois. (2) Abbreviation for average.
ava Kava.
availability Percentage of the total available operating time that a system is able to operate; as used for manufacturing or computer equipment.

available Utilizable in a chemical reaction **a. acidity** Hydrogen ion concentration. **a. nitrogen** The soluble nitrogen of fertilizers.
avalite A claylike silicate containing chromium.
Avcoset Trademark for a viscose synthetic fiber.
Avena A genus of grasses which yield important cereals; as, *A. sativa* (common oat).
avenacein $C_{25}H_{44}O_7N_2$ = 484.6. An antibiotic pigment produced by *Fusaria.*
avenine $C_{56}H_{21}O_{18}N$ = 995.8. An alkaloid from oats.
aventurine (1) A brown venetian glass with embedded brass filings. (2) A feldspar containing iron. **a. quartz** Micaceous quartz (Urals). Cf. *imperial jade.*
Aventyl Trademark for nortriptyline hydrochloride.
avenyl A constituent of hydnocarpus oil, formerly used to treat leprosy.
average av., ave. Popular term for arithmetic *mean.*
Avertin Trademark for tribromethanol.
Avgard Trademark for an aircraft kerosene fuel additive, containing the polymer FM9, that reduces the fuel mist produced upon impact and thus the likelihood of a fire.
Avicenna [Abu Ali en Hosein Ben Abdallah] (980–1037) Iranian mathematician and alchemist.
avidin A protein-carbohydrate complex in egg white which inactivates the biotin present; it can produce dermatitis. See *biotin.*
Avisco Trademark for a viscose synthetic fiber.
avitaminosis A deficiency disease due to lack of a vitamin; as, scurvy. Cf. *hypervitaminosis.*
avocado The fruit of *Persea gratissima,* avocado or alligator pear. The seed yields a black, indelible ink. **a. oil** An oil from a. containing vitamins A, D, and E, and phytosterol and lecithin; used as an emollient. **a. sugar** A mannoketoheptose from avocado.
Avogadro, Amadeo (Conte di Quaregna) (1776–1856) Italian scientist who formulated the gas laws. **A. constant** *L,* N_A. A. number. 6.0220 × 10^{23} mol^{-1}. The number of molecules contained in 1 mole (gram molecule), determined from the spectral line structure. **A. hypothesis** A. law. **A. law** Equal volumes of all gases at the same pressure and temperature contain the same number of molecules. **A. number** A. constant. **A. theory** A. law.
avoirdupois The former English system of weights and measures. 1 lb = 16 oz = 0.4535924 kg.
Avomine Trademark for promethazine theoclate.
awaruite A native nickel-iron (67% Ni).
axestone A hard variety of jade from Cornwall.
axial Pertaining to an axis. **a. angle apparatus** Goniometer. **a. ratio** The ratio of the length to the basal side of the close-packed hexagonal structure of certain metals; usually 1.57–1.64, but 1.9 for Zn and Cd.
axin A varnish from Mexican cochineal, *Lacus axinus,* containing axinic acid.
axinic acid $C_{18}H_{28}O_2$ = 276.4. See *axin.*
axinite $H_2(Ca \cdot Fe \cdot Mn)_4(BO) \cdot Al_2(SiO_4)_5$. A plum-colored gem mineral.
axiom A self-evident proposition. Cf. *postulate.*
axis A line, imaginary or real, passing through an object, around which all parts of the object are symmetrical. Cf. *coordinates, crystal system, pinakoid.* **electric ~** The direction of a crystal that offers least resistance to an electric current. **optic ~** An imaginary line passing through the center of a lens system. **principal ~** See *principal axis.* **x-~, y-~, z-~** See *coordinates.*
axonometry Measurement of crystal axes.
Az French symbol for nitrogen (azote).

aza-* Prefix indicating the presence of N in a compound, particularly in replacement *nomenclature,* q.v., and monocyclic compounds with 3- to 10-membered rings.
azacyclo- Prefix indicating an NH— group in a saturated carbon ring.
azadarach Azadarichta, margosa bark, neem bark. The bark of the Indian lilac tree or bead tree, *Melia azadarichta* (Meliaceae) of Asia; a bitter and tonic.
azafrin $C_{27}H_{38}O_4$ = 426.6. A carotenoid pigment, m.212, from the roots of azafranilla, *Escobedia scrabifolia* (Scrophulariaceae) of S. America.
azaleine Fuchsin.
azapyrene Thebenidine.
azedarine An alkaloid from the roots of *Melia azedarach* (Asia). See *margosa oil.*
azelaic acid $HOOC(CH_2)_7COOH$ = 188.2. Nonanedioic acid*, anchoic acid, lepargylic acid. An oxidation product of oleic acid. Colorless leaflets, m.106, soluble in water. **a. a. value** A chemical constant of fats; the potassium salts of the azelaic glycerides formed.
azeotropes Any one of two or more compounds which form mixtures of constant boiling point whose distillates have the same composition as the original mixture. **negative ~** A. with minimum boiling mixtures. **positive ~** A. with maximum boiling mixtures.
azelaoyl 1,9-Dioxo-1,9-nonanediyl†. The radical $-CO(CH_2)_7CO-$, from azelaic acid.
azeotropy The property of azeotropes. Cf. *hylotropy.*
azete $(CH)_3 \cdot N$. A heterocyclic compound.
azi* Infix indicating the group —N:N—, where both N atoms are bonded to the same C atom. Cf. *azo-.*
azide* Triazo. A compound containing the $-N_3$ group. Cf. *azo-, diazo-.* **hydrogen ~** See *hydrogen azide.* **a. ion** The anion N_3^-.
azidinblue Trypan blue.
azido-* Prefix indicating the azide group, $-N_3$. **3′-Azido-3′-deoxythymidine** $C_{10}H_{13}O_4N_5$ = 267.2. AZT. An antiviral substance and DNA chain terminator, used in trials as treatment for *AIDS,* q.v.
azimid Osotriazol.
azimide Benzoylazide*.
azimido Azimino*.
azimino Azimido. The —NH·N:N— bridge. **a. benzene** Benzotriazole.
azimuth The angle between the meridian and the vertical plane through an object. Cf. *coordinates.*
azine (1) Pyridine*. (2) Compounds with the azino radical. **a. dyes** Dyes containing the grouping

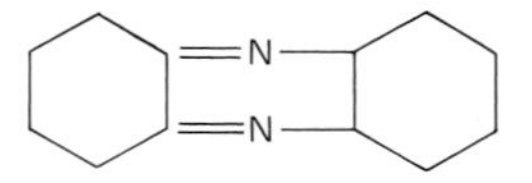

with the auxochrome *meta* to a N atom; as, indulines and safronines.
azino* The radical =N·N=.
azinphos-methyl* See *insecticides,* Table 45.
azipeten —NH·C(:O)—. Peptide amide. The basic group of poly*peptide* formation.
azlon Generic name for a synthetic protein fiber.
azo-* Infix indicating the bridge —N:N—, where each bond is to a different C atom. Cf. *azi, bisdiazo-, azote, azino, diazo-, hydrazo-.*
Azobacter Aerobic soil bacteria that oxidize atmospheric nitrogen. See *Azotobacter.*
azobenzene* PhN:NPh = 182.2. Nitrogen benzide,

diphenyldiimide. Orange leaflets, m.68, insoluble in water; used in organic synthesis. ***p*-amino ~** $C_{12}H_9N_2 \cdot NH_2$ = 197.2. **2,4-diamino ~** $C_{12}H_8N_2(NH_2)_2$ = 212.2. Chrysoidine. Pale-yellow crystals, m.118. A basic dye. **2-4-diaminoazobenzene hydrochloride** Basic orange 2, C.I. 11270, m.235. **dihydroxy ~** Azophenol. **dimethyl ~** Azotoluene. **diphenyl ~** *Coupier's blue*, q.v. **hydroxy ~** See *benzene.*

azobenzide Azobenzene*.

azobenzoic acid $(:N \cdot C_6H_4 \cdot COOH)_2$ = 270.2. ***ortho*- ~** Dark yellow needles, decomp. 237, sparingly soluble in water. ***meta*- ~** Amorphous powder, decomp. by heat, soluble in water. ***para*- ~** Red, amorphous powder, decomp. by heat, insoluble in water.

azocholoramid *N, N*-Dichloroazodicarbonamidine. A bactericide, soluble in water.

azocyanide Cyanoazo. A compound containing the $-N:N \cdot CN$ radical.

azodicarbonamide $NH_2 \cdot CON_2 \cdot CONH_2$ = 116.1. Azoformamide, decomp. 180, soluble in water.

azodiphenyl Coupier's blue.

azoisobutyronitrile $CN \cdot CMe_2 \cdot N:N \cdot CMe_2 \cdot CN$ = 164.2. Colorless crystals, m.105.

azoles Pentaatomic heterocyclic ring compounds; e.g., pyrrole.

azolitmin $C_7H_7O_4N$ = 169.1. The coloring matter of litmus indicator. Violet scales, soluble in water (acids red, alkalies blue).

azomethines* Compounds of the type R′R″C:NR.

azonaphthalene* $C_{10}H_7N:N \cdot C_{10}H_7$ = 282.3. Naphthyldiimide. **1,1′- ~** Red needles, m.190, insoluble in water. **2,2′- ~** m.204. **1,2′- ~** m.136.

azonium R_3N_2X.

azophenetole $(C_6H_4OEt)_2N_2$ = 270.3. Diethoxyazobenzene. ***ortho*- ~** m.131, insoluble in water. ***para*- ~** m. 167, insoluble in water.

azophenol $HO \cdot C_6H_4 \cdot N:N \cdot C_6H_4 \cdot OH$ = 214.2. Dihydroxyazobenzene. ***ortho*- ~** or **2,2′- ~** Yellow leaflets, m.171 (sublimes), insoluble in water. ***meta*- ~** or **3,3′- ~** Brown scales, m.205, sparingly soluble in water. ***para*- ~** or **4,4′- ~** Brown triclinics, decomp. 216, slightly soluble in water. Dyestuff intermediates.

azophenyl The phenylazo* radical.

azophenylene Phenazine*.

azorite $ZrSiO_4$. A native silicate.

azosulfonic acid A compound containing the $-N:N \cdot SO_3H$ radical.

azotate Nitrate*.

azote French for "nitrogen."

azothioprine $C_9H_7O_2N_7S$ = 277.3. 6-[(1-Methyl-4-nitroimidazol-5-yl)thio]purine, Imuran. Pale yellow powder, insoluble in water. An antineoplastic and immunosuppressant; used to prevent rejection of transplants and used in autoimmune diseases (USP, BP).

azotite Nitrate*.

Azotobacter Soil bacteria which convert atmospheric nitrogen to nitrates. Cf. *Azobacter.*

azotoluene $MeC_6H_4N:N \cdot C_6H_4Me$ = 210.3. Ditolyldiimide. **2,2′- ~** Red prisms, m.55, insoluble in water. **3,3′- ~** Orange rhombs, m.54, insoluble in water. **4,4′- ~** Yellow needles, m.114, insoluble in water.

azotometer An apparatus to determine nitrogen gasometrically.

azovan blue Evans blue.

azoxazole Furazan*.

azoxy* The $-N_2O-$ radical. Prefixed by *NNO*- or *ONN*- to indicate position of atoms.

azoxyanisole *p*-$(MeO \cdot C_6H_4N)_2O$ = 258.3. White crystals, m.117, soluble in alcohol.

azoxybenzene $Ph \cdot NON \cdot Ph$ = 198.2. Zinin. Yellow needles, m.36 (decomp.), insoluble in water.

azoxybenzoic acid $(C_6H_4COOH)_2NON$ = 286.2. **1,1′- ~** Yellow leaflets, m.248 (decomp.), sparingly soluble in water. **2,2′- ~** Yellow needles, decomp. 320, insoluble in water. **3,3′- ~** Yellow amorphous powder, decomp. 240, insoluble in water.

azoxynaphthalene $C_{10}H_7 \cdot NON \cdot C_{10}H_7$ = 298.3. **1,1′- ~** Red rhombs, m.127, insoluble in water. **2,2′- ~** m. 167, insoluble in water.

azoxytoluidine Diaminoazotoluene.

AZT 3′-Azido-3′-deoxythymidine. See *azido-*.

azulene* $C_{10}H_8$ = 128.2. A dehydrogenated azulogen. The blue coloring matter of some essential oils. Oily liquid, b.100, insoluble in water. Cf. *cerulein.*

azulin A blue dye formed by heating aniline with corallin.

azulmic acid $C_4H_5ON_5$ = 139.1. Azulmin. A brown decomposition product of cyanogen.

azulogens Sesquiterpenes in many essential oils. They give a blue color (azulin) with bromine in chloroform.

azurine (1) Theobromine sodium acetate; a diuretic. (2) A bluish-black aniline dye.

azurite $CuCO_3 \cdot Cu(OH)_2$. Blue malachite, chessylite, lazulite, lapis lazuli. A monoclinic mineral used in paints and ceramics. Cf. *malachite.*

azymic Describing (1) a reaction not caused by fermentation; (2) an enzyme that does not cause fermentation.

azymous Unfermented.

B

B Symbol for boron.
B Symbol for susceptance.
B (boldface) Symbol for magnetic flux density (magnetic induction).
b Symbol for barn.
b Symbol for (1) angular momentum, (2) breadth, (3) molality.
β See *beta.*
Ba Symbol for barium.
bababudanite $4NaFe(SiO_3)_2 \cdot 2FeSiO_3 \cdot 2MgSiO_3$. A soda amphibole (Mysore).
babassu oil Oil from the fruit kernel of the Brazilian palm tree, *Orbignya martiana* (Palmae).
babbitt (metal) A bearing alloy: Sn 65–95, Sb 8–12, Cu 1 pt. **genuine b.** The b. of Isaac Babbitt (1839): Sn 89.3, Sb 8.9, Cu 1.8%.
Babcock, Stephen Moulton (1843–1931) American chemist. **B. bottle** A graduated glass flask for the determination of fat in milk. **B. milk tester** A centrifuge used in milk fat analysis. **B. pipet** A pipet used in milk analysis.
babingtonite $(Ca,Fe,Mn)6Fe_2Si_9O_{27}$. A vitreous, greenish-black, triclinic pyroxene.
babitt Babbitt.
Babo, Clement Heinrich Lambert Freiherr von (1818–1899) German chemist. **B. absorption tube** A glass cylinder filled with glass beads, used for the absorption of gases by liquids. **B. law** The relative lowering of the vapor pressure of a solvent by a solute is the same at all temperatures.
babool, babul *Acacia arabica.* Its bark is a tannin; its gum is gum arabic.
babussu An edible oil from a Brazilian nut used for detergents.
bacca Latin for "berry."
baccarine An alkaloid from mio-mio, *Baccharis cordifolia* (Compositae) of S. America.
bacciform Berry-shaped.
bach One of 3 mass energy entities of which all subnuclear particles and atomic nuclei are composed. 1 b = 0.3233 meV. Others are zeus, 0.1877; tamaid, 25.9497 meV. Cf. *quark.*
Bachoune, Arnold (Villanovanus) (1235–1315) French alchemist who noted the dangers of puttrefaction and of copper utensils in cooking.
bacillus (1) (Cap. B.) A genus in the family *Bacillaceae* of rod-shaped gram-positive microorganisms. (2) A general term for rod-shaped organisms causing bacterial disease; as, typhoid b., tubercle b. **B. Calmette-Guérin vaccine** An attenuated strain of the bacilli used for vaccination against tuberculosis (USP, EP, BP).
bacitracin An antibiotic polypeptide produced by certain strains of *Bacillus subtilis.* White, bitter, hygroscopic powder, soluble in water (USP). **zinc ~** Hygroscopic, buff powder, slightly soluble in water; an antibiotic used for skin, eye, and ear infections.
backscatter The deflection of radiation or nuclear particles through angles greater than 90°. One application, used to measure web (as, paper or plastic) thickness, is the scattering of *β* particles from a radioisotope (as, ^{85}Kr) by the web toward a detector.
Bacon B., Francis (Baron Verulam) (1561–1626) English philosopher and exponent of the inductive method in science. **B., Roger (1214–1294)** English alchemist who studied gunpowder and combustion.
bacteria (Sing., bacterium) (1) (Cap. B.) A genus within the family *Enterobacteriaceae.* (2) Popularly, any germ or microorganism other than viruses. Correctly, unicellular, free-living microorganisms, approx. size 0.4–1.5 μm. Each b. has both cytoplasm and a nucleus contained within a rigid cell wall; the nucleus contains DNA and RNA. Cf. *virus.* **acetic acid ~** *Bacillus aceti.* A b. that oxidizes alcohol to acetic acid. **ammonifying ~** A soil b. that reduces nitrogen to ammonia. **butyric acid ~** *Bacillus butyricus.* A b. found in milk, water, dust, etc., that produces butanoic acid from fats. **chromogenic ~, chromoparous ~, chromophorous ~** B. producing colored products. **denitrifying ~** A soil b. that oxidizes ammonia to nitrogen. **lactic acid ~** A b. in air that produces lactic acid. **nitrifying ~** A soil b. that oxidizes nitrogen to nitrites and nitrates. **photo ~** A b. that causes phosphorescence, as in decaying fish.
bacterial Pertaining to bacteria. **b. action** The effects of bacteria or their metabolism; e.g., hydrolyzing, deaminizing, decarboxylating. **b. forms** The shapes of bacteria (see Fig. 4). **b. membrane** The cell walls of a bacterium. **b. poisons** See *poison.* **b. precipitation** (1) Precipitin. (2) Deposition of inorganic salts by the action of bacteria.
bactericide An agent that kills bacteria.
bactericidin A bactericidal antibody of the blood serum.
bacterin Bacterial vaccine. A sterile suspension of dead pathogenic bacteria in physiological salt solution; or a suspension of a live but attenuated strain of bacteria. Used to produce immunity by stimulating the production of antibodies.
bacteriofluorescin Fluorescent compounds produced by bacteria.
bacteriological Pertaining to bacteriology. **b. fermentation tube** A glass U-tube with a closed arm to collect gases formed

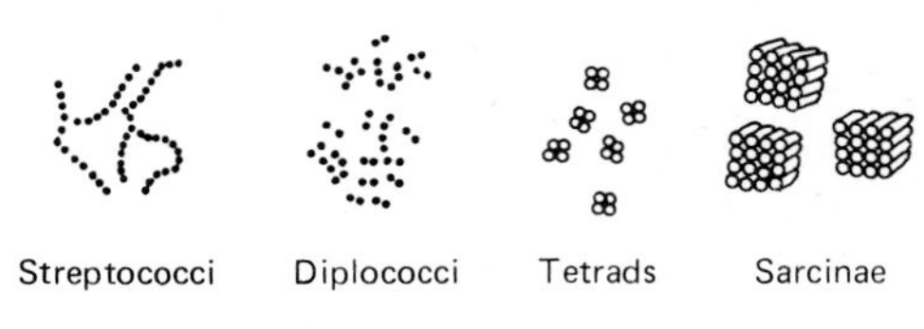

(a)

(b) (c)

Fig. 4. Forms of bacteria: (*a*) cocci (round), (*b*) bacilli (rod-shaped), (*c*) spirillae (spiral-shaped).

by bacterial action. **b. filter apparatus** A filter for bacteria which renders a solution sterile. See *Berkefeld filter.* **b. incubator** A closet at a definite temperature, in which cultures of bacteria are grown.

bacteriology The study of bacteria.

bacteriolysin A blood antibody which promotes the disintegration of bacteria.

bacteriolysis Destruction of bacteria by dissolution.

bacteriophage Phage. An ultramicroscopic, transmissible, filter-passing, lytic agent of bacteria. Known to be a virus that can infect only bacteria and is specific to each strain it infects. Used to identify and classify bacteria (phage-typing).

bacteriopurpurin $C_{41}H_{58}O_2$ = 582.9. m. 214; from purple photosynthetic bacteria.

bacteriostatic An antibiotic, q.v. *(antibiotics)*, or substance which inhibits, as distinct from killing, bacteria.

bacterium Latin (sing.) for *"bacteria."*

Bactrim Trademark for co-trimoxazole.

baddeckite A ferric iron muscovite.

baddeleyite ZrO_2. Brazilite. Zirconium oxide (Sri Lanka, Brazil).

Baden acid 2-Naphthylamine-8-sulfonic acid.

badische acid 2-Naphthylamine-8-sulfonic acid

Badouin's reagent See *Baudouin test.*

Baekeland, Leo Hendrik (1863–1944) Belgian-born American chemist; developed Bakelite and photographic processes.

bael Bengal quince, Indian bel. The fresh, unripe fruit of *Aegle marmelos* (Rutaceae); an astringent used for dysentery in India.

baeumlerite $KCl \cdot CaCl_2$. Native calcium potassium chloride (Germany).

Baeyer, J. F. W. Adolf von (1835–1917) German pioneer in the organic synthesis of arsenicals and organic compounds. **B. acid** Bayer acid. **B. strain theory** A theory that explains the relative stabilities of penta- and hexamethylene ring compounds in terms of the angles between the C atom valencies.

baffle An obstruction in the path of a gas, fluid, or forms of electromagnetic radiation.

bagasscocis, bagassosis A respiratory disease due to the dust, containing fungal spores, from dried bagasse.

bagasse Sugarcane from which the juice has been extracted; used for fuel and paper pulp.

bagrationite Allanite.

bahama white wood Canella.

Bahia arrowroot Tapioca.

baicalein $C_{15}H_{10}O_5$ = 270.2. 5,6,7-Trihydroxyflavone. A flavone from baicalin. Produced when the camphor used in the manufacture of Celluloid was replaced by nitroglycerin (Nobel).

baicalin A glucoside from the roots of *Scutellaria baicalensis* (Labiatae).

baikalite $MgO \cdot CaO \cdot 2SiO_2$. Diopside. A native silicate.

baikiain $C_6H_9O_2N$ = 127.1. L-1,2,3,6-Tetrahydro-2-pyridinecarboxylic acid. An amino acid from the heartwood of Zimbabwean teak, *Baikiaea plurijuga.* Colorless needles, decomp. 274, soluble in water.

bainite *Ferrite,* q.v., saturated with carbon; similar to martensite.

Bakelite Trademark for a synthetic resin obtained by the condensation of formaldehyde with phenols.

bakerite $8CaO \cdot 5B_2O_3 \cdot 6SiO_2$. Calcium borosilicate from the Mohave Desert.

baking b. powder A powder containing sodium hydrogencarbonate, tartaric (or other) acid, and starch filler. A substitute for yeast to produce carbon dioxide in bread making. **b. soda** Sodium hydrogencarbonate.*

BAL Common name for dimercaprol.

balaban The hydrocarbon of chicle.

balance (1) A device for weighing. (2) The harmonious adjustment of related parts; as, *nitrogen balance.* **analytical ~** A b. of sensitivity 0.1–0.01 mg. **aperiodic ~** Damped b. **assay ~** An analytical b. for metallurgy. **automatic ~** A damped b. with a single pan and built-in counterpoise, which gives an immediate scale reading without the use of weights. **chain ~** A b. in which the effects of weights are produced by altering the length of a metal chain hanging from one of the beams. **cloth testing ~** A specially constructed torsion b. for weighing textiles. **coin ~** A sensitive b. used in mints. **counter ~, counterpoise ~** A fixed weight, used as a tare or balance for a piece of apparatus or balance pan. **cream test ~** A torsion b. for milk testing. **damped ~** Aperiodic b. A b. in which the equilibrium position is rapidly attained by the damping action of a piston in a cylinder. **gas ~** A b. for the specific gravity of gases. **micro ~** A sensitive b. utilizing the torsion of a quartz fiber, sensitivity 0.001–0.000004 mg. **Mohr ~** A b. for specific gravities of solids. **moisture ~** A b. whose pan, with the sample to be weighed, hangs in an insulated oven. **projection ~** A b. in which beam movement is magnified optically. **specific gravity ~** See *Westphal balance.* **spring ~** A pan suspended by a vertical wire coil with a scale to indicate the strain. **torsion ~** A b. which measures the twisting force to turn a suspension thread or wire through a given angle. **Westphal ~** See *Westphal balance.*

b. pans Pairs of counterpoised watch glasses or metal dishes for weighing powders. **b. rider** A small platinum U-shaped wire with hook, which slides along the b. beam. **b. weights** Standardized weights of brass, aluminum, or nickel-plated metal.

balanced In equilibrium. **b. reaction** A reaction that proceeds in either direction by a variation of temperature, pressure, or concentration of the reactants. Cf. *equilibrium.*

Balard, Antoine Jerôme (1802–1876) French chemist; discovered bromine (1826) and the nature of bleaching powder.

balas ruby Spinel. Cf. *ballas.*

balata (1) The dried juice of *Minusops globosa* (Spotaceae) of W. Indies. A substitute for gutta-percha. Cf. *chicle.* (2) The hard, dense heartwood of the bully tree, *Bumelia retusa* (Sapotaceae).

baleen See *whalebone.*

balkhas(h)ite A fossil hydrocarbon wax (Lake Balkhash, Central Asia). It resembles torbanite.

ball b. clay A white, high-plasticity burning clay originally mined in balls. **b. mill** A spherical container with balls of iron or quartz for fine grinding. Cf. *pebble mill.*

ballas A mineral intermediate between carbon and diamond with a radial crystalline structure. Cf. *balas.*

Balling, Carl Joseph Napoleon German fermentation chemist. **B. degree** A specific gravity scale for liquids. (1) x° Balling = specific gravity (17.5° C) = $200/(200 \pm x)$; x = + or − according as the liquid is heavier or lighter than water, respectively. (2) Brix degree. See *hydrometer.*

ballistic galvanometer See *ballistic galvanometer* under *galvanometer.*

balloelectric Pertaining to the electric charge of atomized liquids.

ballometer An instrument to measure balloelectric charge by means of a metal plate connected to a quadrant electrometer.

ballotini Very small glass beads; a heating medium in fluid-bed drying.
balm (1) Melissa. (2) Balsam. **bee ~** Monarda.
b. of Gilead (Mecca) (1) Poplar buds. (2) The exudation of *Commiphera (Balsamodendron) opobalsamum* (Burseraceae.)
b. oil Melissa oil, lemon. Yellow essential oil, d. 0.90, from the leaves and flowering tops of *Melissa officinalis;* a flavoring.
Balmer **B. formula** The wavelength λ of the lines in the hydrogen spectrum = $3616.14\ m^2/(m^2 - 4)$, where m is an integer from 3 to 16. **B. series** A series of related lines in the hydrogen spectrum deduced from the quantum theory, and corresponding with orbital electron transitions. Cf. *energy levels, Lyman series, Paschen series, quantum number.*
balneology The study of therapy of natural waters.
balsa Tropical tree, *Ochroma pyramidale,* used for the very low density (0.1–0.25) of its wood.
balsam A plant exudate mixture of resins, essential oils, cinnamic and benzoic acid. **b. apple** The fruit of *Momordica balsamina* (Cucurbitaceae). **b. of Peru** See *Peru b.* **b. tree** Mastic.
baltimoriase Calcium, aluminum, and magnesium silicate (Baltimore County).
Baly, Edward Charles Cyril (1871–1948) British physicist. **B. spectrum tube** A graduated tube in which slides another tube. The intermediate adjustable space is filled with the solution whose absorption spectrum is to be examined.
bamboo (1) A tropical genus of treelike grasses, with woody, light cylindrical stems, growing in clumps. (2) The hollow, siliceous, coated stems of bamboo species, used as a building material and for fishing rods and paper pulp. Cf. *tabashis.* **sacred ~** *Nandina domestica* (Japan); it yields domesticine and nandinine.
b. oil Vinifera palm oil from the fruit of *Gentiliana raphia;* used to soften leather.
BAN Abbreviation for British Approved Name, as recommended by the British Pharmaceutical Commission.
banalsite $BaNa_2Al_4Si_4O_{16}$. A barium feldspar (Wales).
banana The fruit of *Musa sapientum.* Cf. *plantain, pisang wax.* **b. oil** An alcoholic solution of pentyl acetate.
bancoul nuts The seeds of *Aleurites ambina,* source of an oil resembling castor oil. Cf. *tung oil.*
Bancroft, Wilder Dwight (1867-1953) American colloid chemist.
band A compact series of spectral lines due to molecules. **b. head** The wavelength of the sharpest edge of a spectral b. **b. spectrum** See *band spectrum* under *spectrum.*
bandoline A hair gloss from tragacanth and quince seeds.
bandose A quartz-mica-hornblende diorite (Cecil County).
Bandrowski's base $(NH_2)_2C_6H_3 \cdot N \cdot C_6H_4 \cdot N \cdot C_6H_3(NH_2)_2$ = 318.4. Bisdiaminophenyl-*p*-phenylenediamine. Used in organic synthesis.
bane A pest or poison; e.g., dogbane.
Bang method The determination of glucose in urine by titration with alkaline cupric thiocyanate solution.
banisterine $C_{13}H_{12}ON_2$ = 212.3 Harmine. An alkaloid, m.257, from *Banisteria* species (Malpighiaceae) of America. Formerly used to treat Parkinson's disease.
Banting, Sir Frederick (1891–1941) Canadian biochemist, associated with the discovery of insulin.
baobab See *adansonine.*
baphiin $C_{12}H_{10}O_4$ = 218.2. Colorless crystals, decomp. by heat.
baptisia Wild indigo. The dried roots of *Baptisia tinctoria* (Leguminosae).
baptisin $C_{26}H_{32}O_{14}$ = 568.5. A glucoside, m.240, from the roots of *Baptisia tinctoria;* a purgative. Cf. *rhamnomannoside.*
baptisoid The total active principles of baptisia; a laxative.
baptitoxine (1) An alkaloid from the root of *Baptisia tinctoria.* (2) Cytisine.
bar (1) b. A unit of pressure. See *atmosphere.* Cf. *barye.* 1 bar = 10^6 dynes/cm^2 = 1.013 kg/cm^2 = 0.987 atm = 10^5 Pa. (2) Hecto*pieze.*
baras-camphor Borneol*.
barbados nuts Purging nuts. The poisonous seeds of *Jatropha curcas* (Euphorbiaceae) of W. Indies, Brazil. See *curcin.*
barbaloin $C_{21}H_{22}O_9$ = 418.4. A constituent of Barbados aloes, m.148.
barban* See *herbicides,* Table 42 on p. 281.
barbatic acid ($C_{19}H_{20}O_7$ = 360.4. A depside from *Usnea* lichens. Colorless prisms, m.186.
barberry The fruit of *Berberis vulgaris,* used in pickles. See *berberis, berberine, oxyacanthine.*
barbierite A soda orthoclase (Norway).
barbital Barbitone.
barbitone $C_8H_{12}O_3N_2$ = 184.2. Diethylbarbituric acid, 2,4,6-trioxo-5-diethylpyrimidine, Veronal, malonal, barbital. **sodium ~** $C_8H_{11}O_3N_2Na$ = 207.2. White, bitter powder, m.190, soluble in water; a hypnotic.
barbiturates Derivatives of barbituric acid; they are hypnotics, sedatives, and anticonvulsants; as, phenobarbital, butobarbital and thiopental sodium. Formerly widely used as sedatives, tranquilizers and hypnotics; now largely superseded.
barbituric acid $C_4H_4O_3N_2$ = 128.1. Malonyl urea, 2,4,6(1*H*,3*H*,5*H*)-trioxopyrimidine, pyrimidinetrione. Colorless rhombs, m.245, decomp. 260, slightly soluble in water. Cf. *hydantoin.* There are numerous derivatives; as, uramil. See *barbiturates.*
barcenite A native mercury antimonite.
Bardach reaction Acetone and iodopotassium iodide form yellow needles (instead of hexagonal iodoform crystals) in presence of proteins.
bardana Lappa.
bardane oil A semisolid oil from the seeds of burdock, *Arctium lappa.*
Barff boroglycerin A saturated solution of boric acid in glycerol; a preservative for specimens.
Barfoed, Christian Theodor (1816–1889) Danish analyst. **B. solution** A solution of 13.3 g cupric acetate in 200 ml dist. water and 5 ml 38% acetic acid. **B. test** A test for glucose (in presence of maltose), which is reduced by B. solution.
baric Barium (obsolete).
barilla The fused ash of seaweeds which consists of sodium carbonate and sulfate; formerly used to make soap, glass, etc. Cf. *kelp.* **b. de cobre** Nodules of native copper in ores.
barite $BaSO_4$. Barytes, cawk, heavy spar. A native, crystalline barium sulfate; white crystals, sometimes colored. Used in drilling muds and paints.
barium* Ba = 137.33. An element of the calcium family, at. no. 56. Baryum plutonium. A white soft metal, d.3.46, m.725, soluble in water. Widely distributed in feldspars and micas, and as sulfate and carbonate. B. is divalent and forms one series of compounds; soluble compounds are poisonous. Discovered by Scheele (1774) in barytes, and isolated by Davy (1808). **b. acetate*** $Ba(C_2H_3O_2)_2 \cdot H_2O$ = 273.4. White prisms, decomp. by heat, soluble in water. A reagent for sulfates or chromates. **b. aluminate** White powder, a water softener. **b. arsenate*** $Ba_3(AsO_4)_2$ = 689.8. Black powder, insoluble in water. **acid ~** $BaHAsO_4 \cdot H_2O$ = 295.3. B. biarsenate. Opaque crystals, anhydrous at 150. **b. benzoate*** $Ba(C_7H_5O_2)_2 \cdot 2H_2O$ = 415.6. Colorless leaflets, soluble in water. **b. bichromate** B. dichromate*. **b. bioxalate** B.

hydrogenoxalate*. **b. bioxide** B. peroxide*. **b. bisulfate** B. sulfate (acid). **b. borate** $Ba_2(BO_2)_2 \cdot 7H_2O$ = 486.4. White powder, soluble in water. **b. boride** BaB_6 = 202.2. Black, regular crystals, insoluble in water. **b. borotungstate** $2BaO \cdot B_2O_3 \cdot 9WO_3 \cdot 18H_2O$ = 2803. B. Borowolframate; colorless crystals, soluble in water. **b. bromate*** $Ba(BrO_3)_2 \cdot H_2O$ = 411.1. Monoclinic crystals, decomp. 260, slightly soluble in water. **b. bromide*** $BaBr_2$ = 297.1. Colorless crystals, m.880, soluble in water. **crystalline ~** $BaBr_2 \cdot 2H_2O$ = 333.2. White, monoclinic crystals, m.880, soluble in water. **b. butyrate*** $Ba(C_4H_7O_2)_2 \cdot 2H_2O$ = 347.6. Colorless powder, soluble in water. **b. carbide** BaC_2 = 161.4. Gray crystals, decomp. by water, yielding acetylene. **b. carbonate*** $BaCO_3$ = 197.3. Colorless rhombs, m.795, decomp. 1450, insoluble in water; occurs native as witherite. Used as a reagent and in sugar refining, case hardening, and manufacture of glass, enamel ware, and ceramics. **b. chlorate*** $Ba(ClO_3)_2 \cdot H_2O$ = 322.2. White, monoclinic crystals, m.414, soluble in water. Used as a reagent and in pyrotechnics for green fires. **b. chloride*** $BaCl_2 \cdot 2H_2O$ = 244.3. White rhombs, m.960, soluble in water; a reagent for sulfuric acid. Used in veterinary medicine as a cathartic. **b. chloroplatinate** B. hexachloroplatinate(IV)*. **b. chloroplatinite** B. tetrachloroplatinate(II)*. **b. chromate*** $BaCrO_4$ = 253.3. Yellow ultramarine, lemon yellow. Yellow scales, d.4.498, insoluble in water. Used as a pigment and in safety matches. **b. citrate*** $Ba_3(C_6H_5O_7)_2 \cdot 7H_2O$ = 916.3. Colorless powder, soluble in water. **b. cyanate*** $Ba(OCN)_2$ = 221.4. Colorless crystals, sparingly soluble in water. **b. cyanide*** $Ba(CN)_2$ = 189.4. Colorless crystals, soluble in water; used in metallurgy. **b. cyanoplatinite** B. tetracyanoplatinate(II)*. **b. dichromate** $BaCr_2O_7$ = 353.3. B. bichromate. Red prisms, soluble in water. **crystalline ~** $BaCr_2O_7 \cdot 2H_2O$ = 389.3. Orange needles. **b. dioxide** B. peroxide*. **b. diphenylamine sulfonate** An oxidation-reduction *indicator*. **b. dithionate*** $BaS_2O_6 \cdot 2H_2O$ = 333.5. B. hyposulfate. White rhombs, d.5.6, soluble in water. **b. diuranate** BaU_2O_7 = 725.4. Orange powder, soluble in acids. **b. ethylsulfate** $Ba(C_2H_5SO_4)_2 \cdot 2H_2O$ = 423.6. Colorless crystals, soluble in water; used in organic synthesis. **b. feldspar** Celsian. **b. ferrate** $BaFeO_4$ = 257.2. Purple powder, insoluble in water. **b. ferrocyanide** B. hexacyanoferrate(II)*. **b. fluoride*** BaF_2 = 175.3. White powder, m.1280, insoluble in water; used in enamels. **b. fluoride iodide*** $BaI_2 \cdot BaF_2$ = 566.5. Colorless plates, d.5.21; decomp. by water. **b. fluosilicate** B. hexafluorosilicate*. **b. formate*** $Ba(HCOO)_2$ = 227.4. B. formiate. Colorless crystals, soluble in water. **b. hexachloroplatinate(IV)*** $Ba[PtCl_6] \cdot 4H_2O$ = 617.2. Platinic b. chloride. Red, monoclinic crystals, soluble in water. **b. hexacyanoferrate(II)*** $Ba_2[Fe(CN)_6] \cdot 6H_2O$ = 594.7. B. ferrocyanide. Yellow prisms, soluble in water. **b. hexafluorosilicate(IV)*** $Ba[SiF_6]$ = 279.4. B. fluo(ro)silicate. White powder, insoluble in water. **b. hexanitride** $BaN_6 \cdot H_2O$ = 239.4. White crystals, explode when heated, soluble in water. **b. hexathiocyanatoplatinate(IV)*** $Ba[Pt(SCN)_6]$ = 680.9 Platinic b. thiocyanate. Red needles, soluble in water. **b. hydride** BaH_2 = 139.3. White crystals, m.1200, decomp. by water. **b. hydrogenoxalate*** $BaH(C_2O_4)_2 \cdot 2H_2O$ = 350.4. B. bioxalate. Colorless crystals, soluble in water. **b. hydrogenphosphate*** See *phosphate*. **b. hydrogensulfide*** $Ba(SH)_2$ = 203.5. B. hydrosulfide. Yellow powder, decomp. 250, soluble in water. **b. hydroxide*** $Ba(OH)_2 \cdot 8H_2O$ = 315.5. Caustic baryta, hydrated b. Colorless tetragons, m.78, soluble in water. Used for fusing silicates, saponifying fats, refining oils, and in manufacturing sugar. **b. hypophosphate*** $Ba_2P_2O_6$ = 432.6. White needles, soluble in water. **b. hyposulfate** B. dithionate*. **b. iodate*** $Ba(IO_3)_2 \cdot H_2O$ = 505.1. White crystals, decomp. by heat, sparingly soluble in hot water or alcohol. **b. iodide*** $BaI_2 \cdot 2H_2O$ = 427.2. Heavy rhombs, m.740, soluble in water. **b. lactate*** $Ba(C_3H_5O_3)_2$ = 315.5. Colorless crystals, soluble in water. **b. malate*** $BaC_4H_4O_5$ = 269.4. White powder, slightly soluble in water. **b. malonate*** $BaC_3H_2O_4 \cdot H_2O$ = 257.4. White powder, slightly soluble in water. **b. manganate*** $BaMnO_4$ = 256.3. Mangan green, Casseler green, Rosenstiel green. A green, nontoxic pigment. **b. meal** See *b. sulfate*. **b. mercuryiodide** See *Rohrbach's solution*. **b. metasilicate*** B. silicate. **b. methylsulfate** $Ba(CH_3SO_4)_2 \cdot 2H_2O$ = 395.5. Colorless crystals, soluble in water. **b. molybdate** $BaMoO_4$ = 297.3. White powder, insoluble in water. **b. monophosphate** See *b. phosphate*. **b. monosulfide*** B. sulfide*. **b. monoxide** B. oxide*. **b. nitrate*** $Ba(NO_3)_2$ = 261.3. White crystals, m.575 (decomp.), soluble in water. Occurs native as nitrobarite. Used as a reagent and precipitant, as a standardizing agent in soap solutions, and in pyrotechnics for green lights. **b. nitrite*** $Ba(NO_2)_2 \cdot H_2O$ = 247.4. Colorless hexagons, decomp. 115, soluble in water. **b. oleate*** $Ba(C_{18}H_{33}O_2)_2$ = 700.2. White granules, insoluble in water. **b. orthoperiodate*** $Ba_5(IO_6)_2$ = 113.2. White powder, insoluble in water. **b. oxalate*** $Ba(C_2O_4) \cdot H_2O$ = 243.4. White powder, insoluble in water. **b. oxide*** BaO = 153.3. Baryta, b. monoxide, b. protoxide. White powder or crystals, m.1923, soluble in water. Used in the glass industry, as a photographic paper coating, and in the manufacture of b. salts. **b. pentylsulfate** $Ba(C_5H_{11}SO_4)_2 \cdot 2H_2O$ = 506.7. Colorless crystals, soluble in water. **b. perchlorate*** (1) **anhydrous ~** $Ba(ClO_4)_2$. Desicchlora. (2) **tetrahydrate ~** $Ba(ClO_4)_2 \cdot 4H_2O$ = 408.3. Colorless hexagons, m.505, soluble in water; used in pyrotechnics. **b. periodate** B. orthoperiodate*. **b. permanganate*** $Ba(MnO_4)_2$ = 375.2. Violet crystals, soluble in water; used to manufacture permanganates. **b. peroxide*** BaO_2 = 169.3. B. dioxide, b. superoxide, b. binoxide. White powder, insoluble in water, decomp. acids produce hydrogen peroxide. Formerly used in the preparation of oxygen (Brin's method), as a bleach, as a reagent for iodine and indican in urine, and in the glass industry. $BaO_2 \cdot 8H_2O$ = 313.4. Colorless crystals, insoluble in water. **b. peroxodisulfate** $Ba(SO_4)_2 \cdot 4H_2O$ = 401.5. White crystals, decomp. by water. **b. phosphate** **b. orthophosphate*** $Ba_3(PO_4)_2$ = 601.9. Colorless crystals, insoluble in water. **b. hydrogenphosphate*** $BaHPO_4$ = 233.3. Acid b. phosphate. Colorless rhombs, insoluble in water. **b. monophosphate** $BaH_4(PO_4)_2$ = 331.3. Colorless, triclinic crystals, decomp. by water. **b. pyrophosphate*** $Ba_2P_2O_7$ = 448.6. White rhombs, insoluble in water. **b. phosphide** BaP_2 = 199.3. Gray masses, decomp. by water to PH_3. **b. phosphinate*** $Ba(H_2PO_2)_2 \cdot H_2O$ = 285.3. Colorless crystals, soluble in water, decomp. by heat. **b. platinic chloride** B. hexachloroplatinate(IV)*. **b. platinocyanide** B. tetracyanoplatinate(II)*. **b. platinous chloride** B. tetrachloroplatinate(II)*. **b. platinous cyanide** B. tetracyanoplatinate(II)*. **b. potassium chlorate*.** $BaK(ClO_3)_3$ = 426.8. Colorless crystals, soluble in water; used in pyrotechnics. **b. propionate** $Ba(C_3H_5O_2)_2 \cdot H_2O$ = 301.5. Colorless powder, soluble in water. **b. protoxide** B. oxide*. **b. pyrophosphate** See *b. phosphate*. **b. salicylate*** $Ba(C_7H_5O_3)_2 \cdot H_2O$ = 429.6. Colorless needles, soluble in water. **b. selenate** $BaSeO_4$ = 280.3. White, amorphous powder, insoluble in water. **b. selenite*** $BaSeO_3$ = 264.3. White powder, used in the glass industry. **b. silicate** $BaSiO_3$ = 213.4. B. metasilicate*. Colorless rhombs, m.1470, soluble in water. **b. stearate*** $Ba(C_{18}H_{35}O_2)_2$ = 704.3. White, unctuous mass; packing for pump bearings for alkalies. **b. succinate*** $BaC_4H_4O_4$ = 253.4. Colorless crystals, soluble in water. **b. sulfate*** $BaSO_4$ = 233.4. Synthetic blanc fixe.

Rhombic crystals, decomp. 1580, insoluble in water. A reagent for colloidal metals. Used in X-ray meals (USP, EP, BP) and (as "blanc fixe") in white pigments and printing inks. Cf. *lithopone,* **acid ~** $Ba(HSO_4)_2$ = 331.5. B. bisulfate. White powder. **b. sulfide b. monosulfide*** BaS = 169.4. Colorless rhombs, or yellow, phosphorescent powder, decomp. by heat. Used as a dipilatory, for the preparation of arsenic-free hydrogen sulfide gas, and in making luminous paint. Cf. *Bologna phosphorus.* **b. trisulfide*** BaS_3 = 233.5. Yellow crystals, soluble in water. **b. tetrasulfide*** BaS_4 = 265.6. Red rhombs, soluble in water. **b. sulfite*** $BaSO_3$ = 217.4. White powder, insoluble in water. **b. tartrate*** $BaC_4H_4O_6 \cdot H_2O$ = 303.4. White powder, very slightly soluble in water. **b. tetrachloroplatinate(II)*** $Ba[PtCl_4] \cdot 3H_2O$ = 528.3. B. chloroplatinite, platinous b. chloride. Orange rhombs, sparingly soluble in water. **b. tetracyanoplatinate(II)*** $Ba[Pt(CN)_4] \cdot 4H_2O$ = 508.5. B. cyanoplatinite, b. platinocyanide. Yellow crystals with bluish fluorescence, soluble in water; used in X-ray screens. **b. tetrasulfide** B. sulfide. **b. thiocyanate*** $Ba(SCN)_2 \cdot 2H_2O$ = 289.5. B. sulfocyanide. Colorless crystals, soluble in water; used in photography and dyeing. **b. thiosulfate*** $BaS_2O_3 \cdot H_2O$ = 267.5. White powder, soluble in water. **b. titanate*** $BaTiO_3$ = 233.2. White powder; a pigment. **b. triphosphate** See *b.* ortho*phosphate.* **b. trisulfide*** See *b. sulfide.* **b. trithionate** $BaS_3O_6 \cdot 2H_2O$ = 365.5. Colorless scales, soluble in water. **b. tungstate*** $BaWO_4$ = 385.2. Tungsten white. White powder, insoluble in water; a pigment. **b. value** The equivalent of the saponification value of an oil or fat (as BaO). **b. white** B. sulfate*. **b. wolframate*** B. tungstate*.

bark Any portion of a stem or root of a tree outside the cambium circle. Some are used medicinally.

Barker Index An identification system for crystal forms.

barkevite Berkevilite. A black amphibole.

Barkhausen effect The magnetization of a ferromagnetic material changes discontinuously with the magnetic field strength.

barkometer A Baumé hydrometer for tanning liquids.

barley The seeds of *Hordeum distichum.* **hulled ~** B. grain deprived of husk. **Indian ~** Sabadilla. **pearl ~** The polished and rounded b. seed, rich in starch but not protein.
b. gum Mainly pentosans, isolated from barley.

barm A suspension of yeast in a fermenting liquid.

barn b. Unit of cross section in radioactivity studies: 10^{-24} cm^2/nucleus.

barograph A self-registering barometer.

baroluminescence Luminescence induced by high pressures.

barometer An instrument to indicate atmospheric pressure. **alcohol ~** A b. using colored alcohol. **aneroid ~** A b. whose action depends on the changes in shape of an evacuated metal box, magnified by levers. **Fortin ~** A mercury b. in which the base of the scale is the point of an ivory pin adjusted so as just to touch the mercury surface of the reservoir. **glycerol ~** A b. using colored glycerol. **mercury ~** A b. using mercury.
b. tube A narrow tube closed at one end and more than 760 mm long.

barometograph A photographic record of baroscope pressure changes.

barophoresis Interaction. Diffusion at a speed dependent on extraneous forces, e.g., gravity.

baroscope A U tube with a liquid to show pressure changes.

Barosma A genus of South African evergreen plants (Rutaceae). **B. betulina** The source of *buchu.*

barosmin (1) A concentration from *Barosma betulina* or *B. crenulata;* a diuretic. (2) $C_{27}H_{30}O_{16}$ = 610.5. Rutin, rutoside, eldrin. A rhamnoglucoside from several plants, m.214 (decomp.).

barosmoid The total principles from the leaves of *Barosma betulina;* a diuretic.

barotaxis An apparent correlation between barometric pressure and the activity of an organism.

barotropism The reaction of living cells to changes in pressure.

barrandite A native iron phosphate.

barrel bbl (1) a unit of weight or volume whose magnitude varies (in U.S. gal for liquids, lb for solids):

Wine	31
Ale	36
Petroleum	42
Rosin	180
Flour	196
Butter	224
Pork, beef	200
Cement	376

(2) A cylinder-shaped, wood or sheet-metal container.

barrier A packaging material (e.g., paper, film) which acts as a b. to a specific agent (e.g., water vapor).

barringtonin $C_{18}H_{28}O_{10}$ = 404.4. A colorless, amorphous glucoside from *Barringtonia speciosa* (Myrtaceae); a cardiac poison.

Barton, Sir Derek Harold Richard (b. 1918–) British chemist, Nobel prize winner (1969); noted for work on reaction mechanisms.

baru The seeds of *Hibiscus tiliaceus* (Malvaveae), a Malaysian musk seed substitute.

barye The cgs unit of pressure; 1 barye = 1 dyne/cm^2 = 0.1000 Pa. Also a measure of noise. Cf. *bar.*

barylite $2BaO \cdot SiO_2$. Native barium silicate (Franklin, N.J.).

baryon A group of subatomic *particles,* q.v.

barysilite $Pb_3Si_2O_7$. Native lead silicate; white hexagons.

baryta Barium oxide. **calcined ~** Barium oxide. **caustic ~** Barium hydroxide. **hydrated ~** Barium hydroxide.
b. mixture A mixture (1:2) of saturated $Ba(NO_3)_2$ and $Ba(OH)_2$ used in urine analysis. **b. water** A saturated solution of barium hydroxide; an absorbent for carbon dioxide.

baryto- Prefix indicating barium in a mineral. **b. calcite** Alstonite.

baryum Barium.

basalt Igneous rock composed of particles of feldspar, augite, and iron. Cf. *diabase.*

basaluminite $2Al_2O_3 \cdot SO_3 \cdot 10H_2O$. A white mineral coating on joint faces in quarries in Northamptonshire (U.K.).

basanite Jasper used in the streak test for gold. Cf. *touchstone.*

base (1) A compound which yields hydroxyl ions in aqueous solution and which reacts with an acid to form water and a salt. (2) More generally, it includes solvents other than water. Cf. *ammoniobase* (below), *hydrocarbobase.* (3) A substance whose molecules can take up protons. (4) Radix. The number upon which a *number system,* q.v. *(number systems),* is based. **ammonio ~** A metallic amine that yields NH_2^- ions in liquid ammonia. **aqua ~** A compound that yields OH^- ions in water. **Brönsted ~** A proton acceptor, as OH^-. **hydrocarbo ~** A metallic aryl or alkyl compound which dissociates in hydrocarbons. **inorganic ~** The hydroxide of a metal. **leuco ~** See *leuco bases.* **Lewis ~** See Lewis *acid.* **nitrogenous ~** An amine characterized by the ending *-ine.* **organic ~** A carbon compound containing trivalent nitrogen. **primary ~, secondary ~, tertiary ~** See *amines.*

b. exchange The replacement of one cation adsorbed on a colloidal particle (e.g., soil) by another. **b. goods** A mixture of fertilizers, usually superphosphates and nitrogen. **b. metal** (1) A metal whose hydroxide is soluble in water. (2) A metal which oxidizes rapidly. Cf. *noble metal.* **b. unit** (1) The smallest possible repeating unit of a polymer. Cf. *mer.* (2) The 7 b. units of the SI system.

basic (1) Having properties of a base. **mono ~, di ~, tri ~, tetra ~** Having 1, 2, 3, or 4 replaceable H atoms, respectively. (2) (Caps.) A computer language, widely used because of its simplicity.

b. anhydride The oxide of a metal. **b. capacity** Basicity. The H atoms in an acid that can be replaced by a monovalent metal. **b. dyes** Salts of colorless bases and an acid. In dyeing, the free base combines with the acid constituent of an animal fiber or the acid constituent of the mordant of a vegetable fiber, thereby fixing the color. **b. hearth** See *Thomas process.* **b. lime phosphate** A superphosphate neutralized with 6% excess of calcium carbonate. **b. nitrogen** The nitrogen of a protein (basic amino acids and cystine) precipitable by phosphotungstic acid. **b. phosphate slag** B. slag. **b. principle** An alkaloid. **b. rocks** Igneous rocks consisting of free basic oxides; e.g., corundum. **b. salt** A compound of a base and acid in which not all the hydroxide of the base has been replaced by an acid radical; e.g., $Bi(OH)_2Cl$. **b. slag** Thomas slag (phosphate). A finely ground by-product of steel manufacture: P_2O_5 12–25, CaO 40–50, SiO_2 5–15% (80% of the phosphoric acid soluble in 2% citric acid).

basicity (1) Basic capacity. (2) The reciprocal of acidity: (*a*) hydroxyl ion concentration (effective alkalinity); (*b*) normality (free alkalinity). Cf. *acidity.*

basil (1) Sweet basil. The plant *Ocimum basilicum* (Labiatae); a flavoring. (2) A tanned sheepskin. **b. oil** The essential oil from the leaves of *Ocimum basilicum.* Yellow, aromatic liquid, d. 0.945–0.987, soluble in alcohol. Chief constituents eucalyptol, chavicol, linalool; a flavoring.

Basil Valentine A 15th-century monk of Erfurt (Germany) and alchemist who described antimony salts; the nitrates of bismuth, tin, mercury; the preparation of hydrochloric acid and lead acetate; and the manufacture of sulfuric acid and ammonia.

basophilic Stainable by basic dyestuffs.

bass Bast. **b. wood** Linden.

bassanite Anhydrite.

bassic acid $C_{30}H_{46}O_5$ = 486.7. A trihydroxytriterpene acid; crystals, m.320.

bassisterol $C_{27}H_{46}O$ = 386.7. An unsaponifiable alcohol in illipé *butter.*

bassora gum A mixture of colored tragacanth gums (India), partly soluble in water.

bassorin $C_6H_{10}O_5$ = 162.1. Tragacanthose. A carbohydrate from tragacanth gum. Colorless powder, producing a stiff gel and viscous paste with water.

bast Bass. The fibrous inner part of Russian *Calotropis gigantea* (Asclepiadaceae); used for ropes, mats, shoes. **b. fiber** See *bast fiber* under *fiber.*

bastite An altered pyroxene, resembling serpentine.

bastnasite $[(Ce \cdot La \cdot Di)F]CO_3$. Hamartite. A rare earth carbonate.

Bates polariscope A sugar polariscope.

bat guano Bodies and droppings of bats; a fertilizer.

bath A vessel or device for keeping objects at a desired temperature. **air ~** A b. containing air of definite temperature. **metal ~** A b. of molten metal. **oil ~, paraffin ~** A b. of molten paraffin. **sand ~** A shallow metal dish filled with sand. **steam ~** A device heated by steam. **water ~** A metallic container for keeping water at a desired temperature. Cf. *thermostat.*

bathochromic Causing or demonstrating red shift. An effect that displaces any wavelength-characterized phenomenon, such as fluorescence emission, toward longer wavelengths. Cf. *hypsochromic.*

bathocuproine $C_{26}H_{20}N_2$ = 364.4. 2,9-Dimethyl-4,7-diphenyl-1,10-phenanthroline. A reagent for copper (purple color extractable in pentyl alcohol).

batholite A fused granite having permeated sedimentary layers.

bathophenanthroline 4,7-Diphenyl-1,10-phenanthroline, Snyder's reagent. A reagent for copper (cf. *bathocuproine*) and ferrous iron (red color).

batik An ancient Oriental method of printing calico using a wax resist.

bating Puering.

batrachiolin A vitellin from frogs' eggs.

batracin A S. American arrow poison from the skin of an amphibian, *Phyllobates chocoensis.*

battery Electric cells, dynamos, or couples connected to function as a single supply. See *cell (2).* **storage ~** Accumulator.

batyl alcohol $C_{18}H_{37}OC_3H_5(OH)_2$ = 344.6. A 1,2,3-propanetriyl ether from shark liver oils. Lustrous crystals, m. 69. Cf. *selachyl alcohol.*

Baudouin test Cane sugar and boiling hydrochloric acid give a red color with sesame oil.

Baumann, Eugen (1846–1896) German chemist. **B.-Schotten reaction** Alcohols and acid chlorides in alkaline solution give an ester.

Baumé, Antoine (1728–1804) French pharmacist. **B. hydrometer** See *hydrometer.*

bausteine German for "building stones"; as, the amino acids of proteins. Cf. poly*peptide.*

bauxite $Al_2O_3 \cdot 2H_2O$. Beauxite. Native aluminum hydroxide. A gray-red claylike mineral which often contains iron; used to manufacture alum, aluminum, and firebricks. Cf. *boehmite.* **red ~** A b. from Var, France (40–45% alumina).

bay The sweet bay or laurel tree, *Laurus nobilis* (Lauraceae). **Indian ~** *L. indica.* **red ~** *Persea carolinesis.* Cf. *bayberry.*

b. leaves oil The essential oil distilled from bay leaves, *L. nobilis.* **b. oil** Myrcia oil. An essential oil from *Pimenta acris.* Yellow liquid, d. 0.980. Chief constituents: chavicol, eugenol, myrcene. **b. salt** Crude sodium chloride from the evaporation of seawater. **b. wood** Honduras (Campeche Bay) mahogany.

bayberry (1) The fruit of *Laurus nobilis,* European laurel, sweet bay. (2) The fruit of *Myrica cerifera,* wax myrtle. (3) The fruit of *Pimenta acris,* allspice or pimenta. **b. wax** Myrtle wax, laurel wax, bayberry tallow. A fat mixture, chiefly palmitin, from myrica fruits.

baycurine An alkaloid from the roots (baycuru root) of *Statice braziliensis* (Plumbaginaceae) of Brazil.

baycuru root The roots of *Statice braziliensis* (Plumbaginaceae). A powerful astringent.

Bayer acid 2-Naphthol-8-sulfonic acid.

bayerite $Al_2O_3 \cdot H_2O$. A native aluminum hydroxide.

bayldonite Native lead vanadate.

bay rum Bay-rum, spiritus myrciae. An aromatic liquid containing bay oil 8, orange oil 0.5, pimenta oil 0.5, alcohol 610, water 320 ml/liter. A refreshing ablution cosmetic.

BBC Bromobenzyl cyanide.

bbl Abbreviation for barrel(s).

B.C.G. See *Bacillus Calmette-Guérin vaccine* under *bacillus.*

B.C.R. See *Certified Reference Materials.*
bdella Hirudo.
bdellium An aromatic gum resin from *Balsamodendron africanum.* Cf. *Burseraceae.*
Be Symbol for beryllium.
Bé Abbreviation for Baumé.
beaded Describing a disjointed line of bacteria along the line of inoculation.
beading The formation of bead-shaped drops by solvent vapor.
beading oil A mixture of sweet almond oil and ammonium sulfate; used to produce artificial beading.
bead test A crystal of microcosmic salt or borax is melted to a bead in the loop of a platinum wire, dipped in the substance to be analyzed, and held in the oxidizing (O.F.) or reducing (R.F.) part of the bunsen or blowpipe flame. The color of the clear flux indicates the metals shown in Table 12.
beaker A glass, quartz, porcelain, aluminum, or copper vessel used to hold liquids. **b. flask** A wide-lipped, conical beaker or flask.
Beale's stain An aqueous carmine solution containing alcohol, ammonia, and glycerol, for staining tissue.
beans Seeds (usually of Leguminosae plants) of the bean family. **bog ~, buck ~** Menyanthes. **broad ~, common ~** Seeds of *Vicia (Faba) vulgaris.* **calabar ~** Physostigma. **castor ~** See *castor.* **French ~, kidney ~, navy ~** Seeds of *Phaseolus vulgaris.* **lima ~** Seeds of *Phaseolus lunatus.* **ordeal ~** Physostigma. **soja ~, soya ~** See *soybean.* **St. Ignatius ~** Strychnos. **vanilla ~** See *vanilla.*
bearberry Uva ursi. **b. bark** Cascara sagrada.
bearing The support of an axle or rotating shaft. **b. metal solution** A solution containing conc. HCl 400, conc. HNO_3 200 ml, KCl 40 g, water 1 liter.
bearsfoot The root of *Polymnia uvedalia* (Compositae), N. America. **English ~** The plant *Helleborus foetidus* (Ranunculaceae).
bearsweed Eriodictyon.
bearswood Cascara sagrada.
Beaudouin's regent See *Baudouin test.*
Beaufort scale Scale ranging 0–12, used to convey wind strength, with each number corresponding to a wind speed range on an anemometer. Thus 9 is a strong gale of wind speed 73–84 km/h.
Beaumé See *Baumé.*
beauxite Bauxite.
beaverite A highly hydrated lead sulfate (Utah).
bebeerine $C_{36}H_{38}O_6N_2$ = 594.7. Bauxine, bebirine, curine, cissampeline, pelosine. An alkaloid from bebeeru, the bark of Guiana *Nectandra rodiaei* (Lauraceae). White powder, m.221, soluble in alcohol.

TABLE 12. BEAD TEST COLORS

	Microcosmic salt		Borax	
Metal	O.F.	R.F.	O.F.	R.F.
Cr . .	Green	Green	Green	Green
Co . .	Blue	Blue	Blue	Blue
Cu . .	Blue	Red	Greenish-blue	Red
Fe . .	Brown	Colorless	Yellow	Green
Mn .	Violet	Colorless	Violet	Colorless
Mo .	Colorless	Green	Colorless	Brown
Ni . .	Yellow	Yellow	Brown	Gray
Ti. . .	Colorless	Violet	Colorless	Yellow
W . .	Colorless	Blue	Colorless	Brown
U . . .	Green	Green	Red	Green
V . . .	Yellow	Green	Colorless	Green

bebeeru The bark of the greenheart tree, *Nectandra rodiaei* (Lauraceae) of tropical America; used like cinchona bark. Cf. *chondroine.*
bebirine Bebeerine.
beccarite ZrO_2. An olive-green, vitreous zirconium oxide.
beche-de-mer Trepang. A coral reef slug; a Chinese delicacy.
Becher, Johann Joachim (1635–1682) German physician, adventurer, and alchemical writer; founded the phlogiston theory.
Bechhold filter An ultrafiltration apparatus to separate colloids, using disks impregnated with nitrocellulose.
bechilite Borocalcite.
Becke line The line between a solid (as glass) and a liquid of similar refractive index which appears or disappears with temperature changes of the liquid. Used to determine the refractive index of solids.
beckelite Calcium silicate containing cerium, lanthanum, neodymium, and praseodymium.
Beckmann, Ernst (1853–1923) German organic chemist. **B. apparatus** A glass apparatus to determine the molecular weight by (1) lowering the freezing point, or (2) raising the boiling point, of a solution. **B. burner** A glass tube supported over a disk in which vapors or gases are evolved; used to produce colored flames. **B. reaction, B. rearrangement** An intramolecular rearrangement, in which a ketoxime forms its isomeric amide when treated with phosphorus pentachloride. **B. thermometer** Ultrathermometer. A thermometer for the B. apparatus to read temperatures to 0.01°C over a range of about 6°C.
Becquerel, Antoine Henri (1852–1908) French physicist, Nobel prize winner (1903). **B. rays** The radiation emitted from uranium and similar substances which affects a photographic plate.
becquerel Abbrev. Bq; derived SI unit, being the activity of a radionuclide decaying at the rate of 1 spontaneous nuclear transition per second. 1 curie = 3.700×10^{10} Bq.
becquerelite $2UO_3 \cdot 3H_2O$. A hydrous, highly radioactive mineral (Congo).
beda nuts The dried ripe seeds of *Terminalia belerica* (cf. *myrobalan*) from India; a tan and black dye.
bee balm Monarda.
beech A genus of trees, *Fagus sylvatica.* **b.nut** The seed of the b.; a source of oil and b.nut cake. **b. wood creosote** The distillation product of b. wood; a preservative.
beef extract An aqueous extract of lean beef, partly desiccated. It contains the soluble fibrin and proteins; used in culture media.
beegerite $Pb_6Bi_2S_9$ or $6PbS \cdot Bi_2S_3$. Native lead sulfobismuthide.
beehive A small, circular glass shelf with a hole for collecting gases in pneumatic troughs.
beemerose A nephelite rock (New Jersey).
beer A beverage containing 3–7% alcohol; a fermented infusion or decoction of malted barley with hops. Cf. *brewing, wort.* **near ~** B. containing less than 0.5% alcohol, used during prohibition in the United States.
beerbachose A diorite from California.
Beer's law The intensity of a ray of light I, after being transmitted through a solution or gas, is related to the intensity of the incident ray I_0 by $\log_{10}(I_0/I) = \mu cd = abc$, where μ is the (molar) extinction coefficient, or a is the (molar) absorptivity of the material for that particular wavelength of light; c is the concentration of the material; and d or b is the thickness through which the light passes. Cf. *Lambert's law.*

beeswax (1) Wax from the honeycomb of bees. It contains cerolein, heptacosanoic acid, triacontyl alcohol, melissic acid, and alkanes. (2) White wax, cera alba. The bleached wax from the honeycomb of bees, d.0.966, m.63; used in pharmacy and industry.

beet *Beta vulgaris* (Chenopodiaceae), cultivated for its root (sugar beet) containing up to 20% sugar. **b. slop** The liquid product remaining after extraction of b. sugar; a fertilizer. **b. sugar** Sugar extracted from beet. See *sucrose.*

Beggiatoaceae A family of bacteria; motile cells in sheathless threads containing sulfur granules.

behenic acid $Me(CH_2)_{20}COOH$ = 340.6. Docosanoic acid*. A constituent of ben oil and the roots of *Centaurea behen.* Colorless needles, m.84, insoluble in water.

behenolic acid $Me(CH_2)_7C{:}C(CH_2)_{11}COOH$ = 336.6. 1,3-Docosynoic acid*. White needles, m.58, insoluble in water.

behenolyl The radical $C_{21}H_{39}CO-$, from behenolic acid. **b. amide** $C_{22}H_{41}ON$ = 335.6. Colorless crystals, m.90. **b. chloride** $C_{22}H_{39}OCl$ = 355.0. Colorless crystals, m.29.

beidellite $Al_2O_3 \cdot 3SiO_2 \cdot 4H_20$. An inactive base-exchange clay (Colorado).

Beilby layer The amorphous, atomic surface layer formed on a metal when it is polished, or when crystal surfaces are rubbed together.

Beilstein, Friedrich Konrad (1838–1906) Russian-born German chemist; compiler of **B.'s Handbuch,** a multivolume general reference book of *organic compounds,* q.v.

bel B. A unit of comparative power or loudness. Two amounts of power P_1 and P_2, e.g., in a telecommunications circuit, differ by $\log_{10} (P_2/P_1)$ bel. The smallest change in loudness that the ear of a young and healthy person can detect. Cf. *decibel, sound.*

Belastran Trademark for a viscose synthetic fiber.

belcherose A pyroxenite rock (Massachusetts).

beldongrite Native manganese iron oxide.

Belimat Trademark for a viscose synthetic fiber.

belite $2CaO \cdot SiO_2$. A crystalline constituent of portland cement clinker. Cf. *alite.*

Bell, Jacob (1810–1859) British pharmacist, founder of the Pharmaceutical Society of Great Britain and the *Pharmaceutical Journal.*

bell A hollow, cup-shaped vessel. **b. curve** Normal probability curve. **b. glass** B. jar. A glass b. with or without tubulations, to cover specimens and for vacuum experiments. **b. metal** A b. alloy: Cu 80, Sn 20%, d.8.7, m.890. **b. metal ore** Stannite.

belladonna Deadly nightshade, banewort, poison cherry, death's herb. *Atropa belladonna* (Solanaceae). **b. leaves** The dried leaves of *A. belladonna;* a narcotic, antispasmodic, anodyne (USP, EP, BP). **b. roots** The roots of *A. belladonna;* an antispasmodic.

belladonnine $C_{17}H_{21}O_2N$ = 271.4. An alkaloid from belladonna. Cf. *atropine.*

Bellier's test A means of detecting over 2% of arachis oil in a saponified oil, by the formation of icosanoic acid.

belonesite $MgO \cdot MoO_3$. A molybdenum ore; needle-shaped crystals.

belonites Small, needle-shaped crystals in volcanic rock.

Bemberg silk Trademark for an artificial fiber made by a cuprammonium process.

Benacol Trademark for dicyclomine hydrochloride.

Benadon Trademark for pyridoxine.

Benadryl Trademark for diphenhydramine.

Benamid Trademark for probenecid.

benazolin* See *herbicides,* Table 42 on p. 281.

Bence-Jones protein A protein in the urine of patients with myeloma.

bendrofluazide EP, BP name for bendroflumethiazide.

bendroflumethiazide $C_{15}H_{14}O_4N_3F_3S_2$ = 421.4. Bendrofluazide, Aprinox, Naturetin. White crystals, m.225, insoluble in water; a diuretic (USP, EP, BP).

bends Painful condition of divers during decompression due to previously dissolved nitrogen forming bubbles in the blood. Prevented by slow decompression and breathing helium-oxygen mixture.

bene Sesame oil.

Benedict, Stanley Rossiter (1884–1936) American biochemist. **B.'s solution** A solution of sodium and potassium tartrate, sodium carbonate, and copper sulfate; used to detect and determine reducing sugars.

beneficiation The improvement or refining of ores, e.g., by grinding, screening, concentration, etc.

bengal **B. isinglass** Agar. **b. kino** *Butea* gum. **b. lights** A mixture of sulfur, sugar, and potassium nitrate to which either barium or strontium salts are added to give a green or red light; used in pyrotechnics.

benitoite $BaTiSi_3O_9$. A mineral found only at San Benito, Calif.

benjamin See *benzoin gum.*

benodanil* See *fungicides,* Table 37 on p. 250.

ben oil The expressed oil from the seeds of *Moringa pterygosperma* and *M. oleifera* (Moringaceae). It is a bland laxative used for extracting odors. Cf. *benne oil.*

benne oil Sesame oil

benomyl* See *fungicides,* Table 37 on p. 250.

bensylidyne The benzylidyne* radical.

bentazone* See *herbicides,* Table 42 on p. 281.

benthal decomposition The overall stabilization of sludge deposits in streams due to combined aerobic and anaerobic mechanisms.

benthonics The study of the ocean floor.

bentonite Sodium montmorillonite. A clay (Pacific coast states) which swells 12-fold when wetted with water and has strong adsorbing properties. Used as a drilling mud and in pharmacy (NF, EP, BP). **b. magna** A pharmaceutical suspending agent, made by stirring b. into 20 times its weight of hot water.

benz- Prefix indicating the phenylene* radical.

benzacetin (1) $C_{11}H_{13}O_4N$ = 223.2. Acetamidoethylsalicylic acid. Colorless crystals, m.190. Cf. *phenacetin.* (2) $C_{10}H_{11}O_4N$ = 209.2. Acetamidomethylsalicylic acid. Colorless crystals, m.205.

benzaconine $C_{32}H_{43}O_{10}N$ = 601.7. Picraconitine, 14-benzoylaconine, napelline. An alkaloid produced by the partial hydrolysis of aconitine, m.130.

benzacridine $C_{17}H_{11}N$ = 229.3. Phenonaphthacridine. 4 isomers.

benzal The benzylidene* radical.

benzalagen Analgen.

benzalcohol Benzyl alcohol*.

benzalcyanhydrin Mandelonitrile*.

benzaldehyde* $C_6H_5 \cdot CHO$ = 106.1. Phenylaldehyde, benzene carbonal, benzene methylal, benzoylhydride, artificial almond oil. Colorless liquid of bitter almond odor, b.179, sparingly soluble in water. Used in dye manufacture, perfumes, and drugs as flavoring (NF, BP); a reagent for alkaloids and fusel oil. **amino ~** $NH_2 \cdot C_6H_4CHO$ = 121.1. ***ortho-* ~** Anthranilaldehyde. Colorless leaflets, m.39 (decomp.), slightly soluble in water. ***meta-* ~** Yellow, amorphous powder. ***para-* ~** Scales, m.70, soluble in water. **α-bromo ~** Benzoyl bromide. **α-chloro ~** Benzoyl chloride. **cyano ~** Benzoyl cyanide. **diethylamino ~** See under *diethylaminobenzaldehyde.* **dihydroxy ~** $C_6H_3(OH)_2CHO$ = 162.1. **2,3- ~** Pyrocatechualdehyde. **2,4- ~** β-Resorcylaldehyde. **2,5- ~** Gentisaldehyde. **3,4- ~** Protocatechualdehyde*. **dimethoxy ~** Veratraldehyde*. **dinitro ~** $(NO_2)_2C_6H_3CHO$ = 196.1. Dinitrobenzene carbonal. Yellow crystals. **2,4- ~** m.72. **2,6- ~** m.123, slightly soluble in water. **3-ethoxy-4-hydroxy ~** Bourbonal. **hydroxy ~** ***ortho*- ~** Salicylaldehyde*. ***meta*- ~** Colorless needles, m.104, soluble in water. ***para*- ~** Colorless needles, m.115 (sublimes), sparingly soluble in water. **hydroxymethoxy ~** Vanillin*. ***p*-isopropyl ~** Cumic aldehyde. ***p*-methoxy ~** Anisaldehyde*. **methyl ~** Tolualdehyde*. **methylenedioxy ~** Piperonal*. **nitro ~** $NO_2 \cdot C_6H_4 \cdot CHO$ = 151.1. ***ortho*- ~** Yellow needles, m.44, slightly soluble in water. ***meta*- ~** Colorless needles, m.58, slightly soluble in water. ***para*- ~** Colorless prisms, m.106, insoluble in water. **3,4,5-trihydroxy ~*** $C_7H_6O_4$ = 154.1. Gallaldehyde. m.212 (decomp.); from the fungus *Boletus scaber.* **2,4,5-trimethyl ~*** $C_{10}H_{12}O$ = 148.2. Durylaldehyde. m.7, b_{11mm}122. **trinitro ~*** $(NO_2)_3C_6H_3CHO$ = 242.1. Trinitrobenzene carbonal. **2,4,6- ~** Yellow scales, m.119, insoluble in water.

b. azine Benzylideneazine. **b. cyanhydrin** Mandelonitrile*. **b. green** Malachite green. **b. oxime*** Benzaldoxime. **b. phenylhydrazone** PhCH:N·NHPh = 196.3. Benzylidene phenylhydrazine. Colorless crystals, m.155.

benzaldoxime PhCH:NOH = 121.1. Benzaldehyde oxime*. Two isomeric forms, each giving a series of derivatives:

```
        N·OH           N·OH
(E)      ‖       (Z)    ‖
       Ph·CH          HC·Ph
```

(*E*)- ~ α- or *anti*-Benzaldoxime. Benzantialdoxime. Colorless leaflets, m.35. **(*Z*)- ~** β- or *syn*-Benzaldoxime. Isobenzaldoxime. Colorless needles, m.130, insoluble in water. Both forms may give *N*- and *O*- substitution products. ***p*-methoxy ~** Anisaldoxime. **methyl ~** **(*E*)-*N*-** Benzaldoxime *N*-methyl ether. ***O*-** PhCH:NOMe = 135.2. Benzaldoxime *O*-methyl ether (1) (*E*)-*O*-Methylbenzaldoxime. Colorless liquid, b.191. (2) (*Z*)-*O*-Methylbenzaldoxime. Colorless crystals, m.82. **nitro ~** colorless crystals. (1) **(*E*)- ~** (*E*)-*meta*-Nitrobenzaldoxime. m.117. (*E*)-*para*-Nitrobenzaldoxime. m.129. (2) **(*Z*)- ~** (*Z*)-*ortho*-Nitrobenzaldoxime. m.136. (*Z*)-*meta*-Nitrobenzaldoxime. m.118. *para*-(*Z*)-Nitrobenzaldoxime. m.174. ***O*-phenyl ~** PhCH:NOPh = 197.2. Colorless crystals, m.109.

b. acetate PhCH:NO·COMe = 163.2. ***(E)*- ~** Colorless crystals, m.15, soluble in water. **b. acetic acid *(E)-O*- ~** $PhCH:NO \cdot CH_2COOH$ = 179.2. Colorless crystals, m.98. ***(E)-N*- ~** $PhCH:N(:O)CH_2COOH$. Colorless crystals, m.183. **b. carboxylic anhydride** $C_8H_5O_2N$ = 147.1. A ring compound which at 145°C becomes $C_6H_4(CN)COOH$.

benzal green Malachite green.

benzalkonium $[C_6H_5 \cdot CH_2 \cdot NMe_2R]^+$. A radical which forms antibiotics and/or surfactants with Cl, saccharinate, and phthalimidate structures. **b. chloride** $[C_6H_5CH_2N(Me)_2R]Cl$. A mixture of alkyldimethylbenzylammonium chlorides. R is a mixture of alkyls from C_8H_{17} to $C_{18}H_{37}$; an antiseptic and detergent (USP, EP, BP).

benzalphthalide $C_{15}H_{10}O_2$ = 222.2. Colorless crystals, m.108, soluble in alcohol.

benzamarone $PhCH(CHPh \cdot COPh)_2$ = 464.6. Colorless crystals, m.219, slightly soluble in hot water.

benzamide* $PhCONH_2$ = 121.1. Benzene carbon amide. Colorless, monoclinic tablets, m.128, slightly soluble in water. **amino ~** ***ortho*- ~** $NH_2 \cdot C_6H_4 \cdot CONH_2$ = 136.2. Aminobenzene carbon amide. Colorless leaflets, m.108, soluble in water. ***meta*- ~** Yellow leaflets, m.70, slightly soluble in water. ***para*- ~** Yellow crystals, m.179, slightly soluble in water. **benzoyl ~** Dibenzamide*. **bromo ~*** C_7H_6NOBr = 200.0. Colorless crystals. ***ortho*- ~** m.156. ***meta*- ~** m.150. ***para*- ~** m.190. **chloral ~** See *chloralbenzamide.* **chloro ~*** C_7H_6NOCl = 155.5. Colorless crystals. ***ortho* ~** m.141. ***meta* ~** m.135. ***para* ~** m.178. **di ~*** See *dibenzamide.* **fluoro ~*** C_7H_6NOF = 139.1. Colorless crystals. ***ortho*- ~** m.116. ***meta*- ~** m.130. ***para*- ~** m.155. **hydroxy ~*** $HO \cdot C_6H_4CONH_2$ = 137.1. ***ortho*- ~** Yellow leaflets, m.140, soluble in water. ***meta*- ~** Colorless leaflets, m.170, soluble in water. ***para*- ~** Colorless needles, m.158, sparingly soluble in water. **iodo ~*** C_7H_6NOI = 247.0. Colorless crystals. ***ortho*- ~** m.184. ***meta*- ~** m.187. ***para*- ~** m.128. **methyl ~** Toluamide*. **nitro ~*** $NO_2C_6H_4CONH_2$ = 166.1. ***ortho*- ~** White needles, m.174, soluble in water. ***meta*- ~** Yellow needles, m.141, sparingly soluble in water. ***para*- ~** Colorless needles, m.153, sparingly soluble in water. ***N*-phenyl ~** Benzanilide. **silver ~** PhCONHAg = 228.0. **sodium ~** PhCONHNa = 143.1. **tri ~** $(PhCO)_3N$ = 329.4. Tribenzoylamine. Colorless crystals, m.202. **thio ~*** See *thiobenzamide.*

b. oxime* $PHC(:NOH)NH_2$ = 136.2. Benzamidoxime. White crystals, m.80, slightly soluble in water.

benzamidine* $PhC(NH_2):NH$ = 120.2. Colorless crystals, m.75, soluble in water. **b. urethane** PHC(:NH)NHCO·OC_2H_5 = 192.2. Colorless crystals, m.38.

benzamido* Benzoylamino*. The radical PhCONH–, from benzamide. **b.acetic acid** Hippuric acid*.

benzamidoxime Benzamide oxime*.

benzaminic acid *m*-Amino*benzoic acid.*

benzamino acids Amino*benzoic acids.*

benzanalgen Analgen.

benzanilide Ph·CO·NH·Ph = 197.2. Benzoylaniline. White crystals, m.160, insoluble in water. **hydroxy ~** $HO \cdot C_6H_4 \cdot$CONH·Ph = 137.1. Colorless crystals. ***ortho*- ~** m.140. ***meta*- ~** m.170. ***para*- ~** m.162. Used in organic synthesis. **methyl ~** Benzotoluide. **nitro ~** See *nitrobenzanilide.* **thio ~** See *thiobenzanilide.*

benzanthracene(s)* $C_{18}H_{12}$ = 227.1. Generally, hydrocarbons in which a benzene and anthracene ring have a double bond in common. Cf. *benzopyrene.*

benzantialdoxime See *benzaldoxime.*

benzathene penicillin BP name for penicillin G.

benzazide Benzoylazide*.

benzazimide $C_7H_5ON_3$ = 147.1. 1,2,3-Benzotriazin-4-one. Colorless crystals, m.212.

benzazine (1) Quinoline*. (2) Iso*quinoline**.

benzazole **1- ~** Indole*. **2- ~** Iso*indole**.

benzdiazine Benzodiazine.

benzdifuran Benzodifuran.

benzdioxazine Benzodioxazine.

Benzedrine Trademark for a brand of amphetamine.

benzene* C_6H_6 = 78.1 Benzol (German), benzole (French), phenylhydride, cyclohexatriene, phene. Colorless crystals, d.0.879, m.5.4, b.80, insoluble in water, miscible with organic solvents. An intermediate for polystyrene, nylons, polyesters, alkyds. Also used as a solvent in organic synthesis, in the manufacture of dyes, in photography, as a motor fuel, and in electrotechnics; vapor inhalation may cause depression of bone marrow, convulsions, and paralysis. Cf. *b. structure.* **acetyl ~** Acetophenone*. **amino ~** Aniline*. **aminoethoxy ~** Phenetidine*. **aminoethyl ~***

$C_6H_4NH_2C_2H_5$ = 121.2 ***ortho-*** **~** Colorless liquid, b.215. ***meta-*** **~** Colorless liquid, b.214. ***para-*** **~** White leaflets, m.5. **anilino ~** Diphenylamine*. **arsenobis ~** Arsenobenzene. **azimino ~** Benzotriazole. **azo ~*** See *azobenzene.* **azoxy ~** See *azoxybenzene.* **benzoyl ~** Benzophenone*. **benzyl ~** Diphenylmethane*. **bromo ~*** Phenyl bromide. **bromochloro ~*** BrC_6H_4Cl = 191.5. **1,3- ~** Colorless liquid, b.196. **1,4- ~** White prisms, m.67. **bromofluoro ~*** BrC_6H_4F = 175.0. **1,4- ~** Colorless liquid, b.153. **bromoiodo ~*** BrC_6H_4I = 282.9. **1,2- ~** Colorless liquid, b.258. **1,3- ~** Oily liquid, b.252. **1,4- ~** White needles, m.92. **bromonitro ~*** $Br{\cdot}C_6H_4{\cdot}NO_2$ = 202.0. ***ortho-*** **~** Colorless crystals, m.38, insoluble in water. ***meta-*** **~** Colorless crystals, m.52.6, insoluble in water. ***para-*** **~** White, monoclinic crystals, m.125, insoluble in water. **chloro ~*** Phenyl chloride*. **1-chloro-x-dinitro ~*** $ClC_6H_3(NO_2)_2$ = 202.6. Characteristics vary with the positions of the 2 nitrogens (-*x*-) as follows: **-2,3-** Colorless needles, m.78, soluble in alcohol. **-2,4-** Yellow crystals, m.51, soluble in alcohol; a reagent for amines. **-2,5-** Yellow crystals, m.64, soluble in petroleum ether. **-3,4-** Colorless crystals, m.41, soluble in alcohol. **-3,5-** Colorless needles, m.53. **chloronitro ~*** $ClC_6H_4NO_2$ = 157.6. ***ortho-*** **~** Colorless needles, m.32, insoluble in water. ***meta-*** **~** Colorless rhombs, m.44, insoluble in water. ***para-*** **~** White, monoclinic prisms, m.83, insoluble in water. **chlorotrinitro ~** Picryl chloride. **cyano ~*** Benzonitrile*. **diamino ~** Phenylenediamine. **diaminoazo ~** $(NH_2)_2C_6H_3{\cdot}N{:}NPh$ = 212.3 **2,4- ~** Yellow needles, m.118, slightly soluble in water. **2,4- ~ HCl** Chrysoidine orange. **diazo ~** See *diazobenzene.* **diazoamino ~** $PhN{:}N{\cdot}NHPh$ = 197.2. Yellow leaflets, m.96, insoluble in water. **dibromo ~*** $C_6H_4Br_2$ = 235.9. ***ortho-*** **~** Colorless needles, m.−1, insoluble in water. ***meta-*** **~** Colorless crystals, m.1, insoluble in water. ***para-*** **~** White, monoclinic crystals, m. 89, insoluble in water. **dichloro ~*** $C_6H_4Cl_2$ = 147.0. ***ortho-*** **~** Colorless liquid, b.179, insoluble in water. A solvent. ***meta-*** **~** Colorless liquid, b.173, insoluble in water. ***para-*** **~** White needles, m.53, insoluble in water. **dicyano ~*** ***ortho-*** **~** Phthalonitrile*. ***meta-*** **~** Isophthalonitrile*. ***para-*** **~** Terephthalonitrile. **diethoxy ~*** $C_6H_4(OEt)_2$ = 166.2 **1,2- ~** White crystals, m.166. **1,3- ~** White prisms, m.12. **1,4- ~** Hydroquinone diethyl ether. White scales, m.71. **diethyl ~*** $C_6H_4(C_2H_5)_2$ = 134.2. ***ortho-*** **~** Colorless liquid, b. 185, insoluble in water. ***meta-*** **~** Colorless liquid, b.181, miscible with alcohol. ***para-*** **~** Colorless liquid, b.183, insoluble in water. **difluoro ~*** $C_6H_4F_2$ = 114.1. ***para-*** **~** Colorless liquid, b.89, insoluble in water. **dihydro ~** Cyclohexadiene*. **dihydrodioxo ~** Benzoquinone*. **dihydrooxo ~** Benzenone. **dihydroxy ~*** $C_6H_4(OH)_2$. **1,2 ~** Pyrocatechol*. **1,3- ~** Resorcinol*. **1,4- ~** Hydroquinone*. **diiodo ~*** $C_6H_4I_2$ = 329.9. **1,2- ~** Colorless prisms, m.23, soluble in alcohol. **1,3- ~** White rhombs, m.34, soluble in alcohol. **1,4- ~** White leaflets, m.129, soluble in alcohol. **dimethoxy ~** $C_6H_4(OMe)_2$ = 138.2. **1,2- ~** Veratrole*. **1,4- ~** Hydroquinone dimethyl ether. Colorless leaflets, m.55, insoluble in water. **dimethyl ~** Xylene*. **dinitro ~*** $C_6H_4(NO_2)_2$ = 168.1. **1,2- ~** Colorless scales, m.117, sparingly soluble in water. **1,3- ~** Yellow needles, m.90, sparingly soluble in water. Used in the manufacture of explosives. **1,4- ~** White needles, m.171, sparingly soluble in water. **diphenyl ~** $C_6H_4Ph_2$ = 230.3. ***para-*** **~** Terphenyl. White leaflets, m.205, insoluble in water. **ethenyl ~** Styrene*. **ethoxy ~** Phenetole*. **ethyl ~*** PhC_2H_5 = 106.2. Colorless liquid, b.137, insoluble in water. **ethyldimethyl ~*** $Et{\cdot}C_6H_3{\cdot}Me_2$ = 134.2. Colorless liquid, b.183, insoluble in water. **ethylmethyl ~** MeC_6H_4Et = 120.2. **1,2- ~** Colorless liquid, b.159. **1,3- ~** Colorless liquid, b.159, **1,4- ~** Colorless liquid, b.162. **ethynyl ~** Phenyl*acetylene.* **fluoro ~*** C_6H_5F = 96.1. Fluobenzene, phenyl fluoride. Colorless liquid, d.1.023, b.86, soluble in alcohol. **formamido ~** Formanilide. **hexachloro ~** See *hexachlorobenzene.* **hexaethyl ~*** $C_6(C_2H_5)_6$ = 246.4. Colorless monoclinics, m.129, insoluble in water. **hexahydro ~** Cyclohexane*. **hexahydrohexahydroxy ~** Inositols*. **hexahydropentahydroxy ~** Pinite. **hexahydroxy ~** $C_6(OH)_6$ = 174.1. B.hexol. Colorless needles, decomp. 200, slightly soluble in water. **hexaiodo ~** C_6I_6 = 833.5. Brown needles, decomp. 150. **hexamethyl ~** $C_6(CH_3)_6$ = 162.3. Mellitene. White rhombs, m.164. **hexaoxo ~** See *triquinoyls.* **hydrazino ~** Phenylhydrazine. **hydrazo ~** $PhNH{\cdot}NHPh$ = 184.2. Colorless crystals, m.131, sparingly soluble in water. **hydroxy ~** Phenol*. **hydroxyazo ~** $HO{\cdot}C_6H_4N{:}NPh$ = 198.2. **1,2- ~** White needles, m.83, slightly soluble in water. **1,4- ~** White prisms, m.152, slightly soluble in water. **inorganic ~** $B_3H_6N_3$ = 80.5. The substance formed by the action of ammonia on borane. It resembles benzene in physical properties. **iodo ~*** PhI 204.0. Colorless liquid, b.188, insoluble in water. **iodosyl ~*** PhIO = 220.0. White powder, explodes 210, soluble in water. **iodyl ~** $PhIO_2$ = 236.0. Colorless needles, explodes 230, sparingly soluble in water. **isocyano ~** Phenyl isocyanide*. **isopropyl ~** Cumene*. **isopropylmethyl ~** Cymene*. **methoxy ~** Anisole*. **methoxypropenyl ~** Anethole*. **methyl ~** Toluene*. **methylisopropyl ~** Cymene*. **nitro ~*** $PhNO_2$ = 123.1. Yellow liquid, b.210, with almond odor, slightly soluble in water. Tautomeric forms: $PhNO_2$ and PhONO. **nitroso ~** PhNO = 107.1. Colorless prisms, m.68, insoluble in water. **octa ~** Cyclooctatetraene*. **pentaamino ~*** $C_6H(NH_2)_5$ = 153.2. Colorless needles, soluble in water. **pentabromo ~*** C_6HBr_5 = 472.6. Colorless needles, m.160, insoluble in water. **pentachloro ~*** C_6HCl_5 = 250.3. Colorless needles, m.85, insoluble in water. **pentaethyl ~*** $C_6H(C_2H_5)_5$ = 218.4. Colorless liquid, b.277, insoluble in water. **pentamethyl ~*** $C_6H(CH_3)_5$ = 148.2. White crystals, m.53, insoluble in water. **pentyl ~** $Ph{\cdot}C_5H_{11}$ = 148.2. Colorless liquid, b.129, soluble in alcohol. **phenyl ~** Biphenyl.* **1-propenyl ~*** See *1-propenylbenzene.* **tetrahydro ~** Cyclohexene*. **tetramethyl ~*** $C_6H_2Me_4$ = 134.3. **1,2,3,4- ~** Prehnitene. **1,2,3,5- ~** Iso*durene.* **1,2,4,5- ~** Durene. **trihydroxy ~** $C_6H_3(OH)_3$ = 126.1. **1,2,3- ~** Pyrogallol*. **1,2,4-** See *trihydroxybenzene.* **1,3,5- ~** Phloroglucinol*. **trimethyl ~*** $C_6H_3Me_3$. **1,2,3- ~** Hemimellitene. **1,2,4- ~** Pseudocumene. **1,3,5- ~** Mesitylene*. **trinitro ~*** See *trinitrobenzene.* **trinitrotriazido ~** $C_6(NO_2)_3(N_3)_3$ = 336.1. Yellow solid, m.131, insoluble in water, soluble in acetone; a detonator and substitute for mercury fulminate. **vinyl ~** Styrene*.

b.arsonic acid $PhAsO(OH)_2$ = 202.0. A reagent for Sn, Zr, and Th. **amino ~** Arsanilic acid. **b.azoaniline** See diazoamino*benzene.* **b.azobenzene** See *azobenzene**. **b.azo-1-naphthylamine** $PhN{:}NC_{10}H_6NH_2$ = 241.3. Red needles, m.123. **b.carbinol** Benzyl alcohol*. **b.carbonal** Benzaldehyde*. **b.carbonamidine** Benzamidine*. **b.carbonitrile*** Benzonitrile*. **b.carboxylic acid** Benzoic acid*. **b.diamine** Phenylenediamine. **b.diazonium chloride ~** $Ph{\cdot}NNCl$. Formed by diazotizing aniline with acid nitrous acid; explosive when dry. **b.diazonium hydroxide** $PhN{:}N{\cdot}OH$. An intermediate in the formation of diazoaminobenzene from aniline. **b. dibromide** See

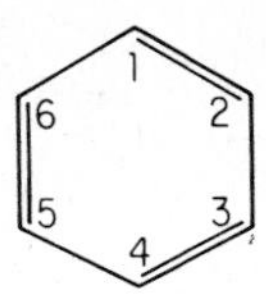

Fig. 5. The benzene ring. Substitution positions in the benzene ring are designated by: *ortho- o-* 1,2-; *meta- m-* 1,3-; *para- p-* 1,4-; *vicinal v-* 1,2,3-; *symmetric s-* 1,3,5-; *asymmetric a-* 1,2,4-.

dibromo*benzene.* **b.dicarbinol** Xylenediol. **b. dicarbonal** Phthalaldehyde*. **b.dicarbonitrile*** Phthalonitrile*. **b.dicarboxylic acid*** $C_6H_4(COOH)_2$ = 166.1. **1,2-** ~ Phthalic acid*. **1,3-** ~ Iso*phthalic acid**. **1,4-** ~ Terephthalic acid*. **b. dichloride** Dichlorobenzene.* **b.diol*** See *dihydroxybenzene* under *benzene.* **b.disulfonic acid*** $C_6H_4(SO_3H)_2$ = 238.2. Three isomers used in organic synthesis. **b.disulfonyl chloride** $C_6H_4(SO_2Cl)_2$ = 275.1. Colorless, viscous liquid, b.80, slightly soluble in water. **b.disulfoxide** PhSO·SOPh = 250.3. Colorless crystals, m.45. **b.dithiol** Phenylene dithiol. **b. hexabromide** Hexabromo*cyclohexane**. **b.hexacarboxylic acid** $C_6(COOH)_6$. Mellitic acid. **b.hexachloride** Hexachloro*cyclohexane**. **b.indenone** Aposafranone. **b.methylal** Benzaldehyde*. **b.monosulfonic acid*** B.sulfonic acid*. **b. nucleus** See *b. ring.* **b.pentaamine*** $C_6H(NH_2)_5$ = 153.2. Colorless needles, soluble in water. **b. pentacarboxylic acid*** $C_6H(COOH)_5$ = 298.2. Colorless crystals, decomp. 238, soluble in water. **b. positions** See *b. ring.* **b. ring** The graphical representation of b. structure as a hexagon with numbers representing the C atoms (see Fig. 5). **b. series** C_nH_{2n-6}. Aromatic hydrocarbons and homologs of b. See Table 13. **b.siliconic acid*** C_6H_5SiOOH = 138.2. Silicobenzoic acid. Glassy scales, m.92, insoluble in water. **b. structure** Theories of the arrangement of the C atoms in the b. molecule; i.e., plane formula (Kekulé, 1865), diagonal formula (Claus, 1867), prism formula (Ladenburg, 1869), bridge formula (Claus, 1870), centric formula (Armstrong and Baeyer, 1892), partial valence formula (Thiele, 1899); the modern concept of a dynamic formula. See *Pauling structure.* **geometrical** ~ Tetrahedra illustrating the arrangement of the atoms in the b. molecule. **b.sulfamide** See *b.sulfonamide.* **b.sulfide** Diphenyl sulfide*. **b.sulfinic acid*** PhSO·OH = 142.2. Colorless prisms, m.83, decomp. 100, soluble in water, isomeric with b.sulfonyl. **b.sulfochloride** B.sulfonic chloride. **b.sulfonamide*** $PhSO_2NH_2$ = 157.2. White plates, m.156 (decomp.), soluble in water. **b.sulfonanilide** Ph·SO_2·NH·Ph = 219.3 **b.sulfone chloride** See *b.sulfonic chloride.* **b.sulfonic acid*** $PhSO_2$·OH = 158.2. Colorless leaflets, m.65, soluble in water. **b.sulfonic chloride** $PhSO_2Cl$ = 176.6. Colorless oil, m.14, insoluble in water; a reagent for amines. **b.sulfonyl*** The acyl radical derived from b.sulfonic acid, Ph·SO_2—.

TABLE 13. BENZENE SERIES

Compound	Formula
Benzene, phenylhydride	PhH
Toluene, methylbenzene	PhMe
Xylene, dimethylbenzene	$C_6H_4Me_2$
Mesitylene, trimethylbenzene	$C_6H_3Me_3$
Durene, tetramethylbenzene	$C_6H_2Me_4$
Pentamethylbenzene	C_6HMe_5
Hexamethylbenzene	C_6Me_6

b.sulfoxide Ph_2SO_2 = 218.3. Colorless crystals, m.128, insoluble in water. **b.tetracarboxylic acid*** $C_6H_2(COOH)_4$. **1,2,3,4-** ~ Mellophanic acid. **1,2,3,5-** ~ Prehnitic acid. **1,2,4,5-** ~ Pyromellitic acid. **b.tetrol** Tetrahydroxybenzene. **b.tricarboxylic acid*** $C_6H_3(COOH)_3$. **1,2,3-** ~ Hemimellitic acid. **1,2,4-** ~ Trimellitic acid. **1,3,5-** ~ Trimesitinic acid. **b.triol*** See *trihydroxybenzene.* **b.trisulfonic acid*** $C_6H_3(SO_3H)_3$ = 318.3. Colorless, deliquescent crystals; used in organic synthesis. **b.triyl*** Phenenyl. The radical $C_6H_3\equiv$.

-benzeneazo* Suffix indicating the PhN:N— group. Cf. *phenylazo.*

benzenide* The anion $C_6H_5^-$, from benzene.

benzenium* The cation $C_6H_7^+$, from benzene.

benzeno-* Prefix indicating a —C_6H_4— bridge.

benzenoid (1) Related structurally to benzene. (2) A suggested structure for benzene involving 6 *carbonoids,* q.v.

benzenone C_6H_6O = 94.1. Dihydrooxobenzene. 1,2- and 1,4-isomers exist. **hydroxy** ~ Hydroquinone. **methyl** ~ Toluenone.

benzenyl The benzylidyne* radical.

benzenylamidine Benzamidine*.

benzenylamidoxime PhC(NH_2):NOH = 136.20. It occurs in (*E*) and (*Z*) forms. Colorless crystals, m.89. **acetyl** ~ PhC(NH_2):NO·COMe = 178.2. Colorless liquid, m.16.

benzenylaminothiophenol $C_{13}H_9NS$ = 211.3. Phenylbenzothiazole. Yellow needles, with rose odor, m.115, soluble in alcohol; a perfume.

benzenyl trichloride. Benzotrichloride.

benzfuran The ring structure.

benzhexol hydrochloride BP name for trihexylphenidyl hydrochloride.

benzhydrazoin $C_{19}H_{16}N_2$ = 272.3. Colorless crystals, m.55.

benzhydrol Benzohydrol. **b. ether** $(Ph_2CH)_2O$ = 350.5. Colorless crystals, m.109, sparingly soluble in water.

benzhydryl* Diphenylmethyl*, benzohydryl. The radical Ph_2CH—, from diphenylmethane. **b.amine** Ph_2CH·NH_2 = 138.3. Colorless liquid, b.288. **b.benzoic acid** Ph·CHOH·C_6H_4·COOH = 228.2. Crystals, m.164 (decomp.), slightly soluble in water. **b.hydroxylamine** Ph_2CH·NH·NHOH = 214.3. Colorless crystals, m.78.

benzidam Aniline*.

benzidine* $NH_2C_6H_4$·$C_6H_4NH_2$ = 184.2. *p*-Bianiline, [1,1'-biphenyl]-4,4'-diamine†. Gray scales, m.133, slightly soluble in water, soluble in boiling water. A reagent for sulfates, blood, and small quantities of higher-valency metals. **dimethoxy** ~ Bianisidine. **dimethyl** ~ Tolidine.

b. conversion The formation of b. from hydrazobenzene by boiling mineral acids. **b.dicarboxylic acid** Diaminodiphenic acid. **b. sulfate** $C_{12}H_{12}N_2H_2SO_4$ = 282.3. White crystals, soluble in water; used similarly to benzidine. **b.sulfone** $(NH_2C_6H_3)_2SO_2$ = 246.3. Colorless crystals, m.350, insoluble in water. **b. test** Adler's reaction. B. produces a purple color with an acetic acid-ether extract of blood. Certain metals also react.

benzidino* The radical *p*-NH_2·$C_6H_4C_6H_4$·NH—, from benzidine.

benzil* PhCO·COPh = 210.2. Diphenylethanedione†, bibenzoyl, diphenyldiketone. Yellow needles, m.95, insoluble in water; a reagent. **bisazo** ~ Ph·CN_2·CN_2·Ph = 234.3.

dimethoxy ~ Anisil. ***p*-hydroxy ~** Yellow needles, m.130, soluble in alkali.
benzilamphioxime (*E,Z*)-Benzil dioxime*.
benzilaniline Benzylaniline.
benzilantioxime (*E,E*)-Benzil dioxime*.
benzildianil PhC(:NPh)·(PhN:)CPh = 364.4. Colorless crystals, m.142.
benzil dioxime* $C_{14}H_{12}O_2N_2$ = 240.3.

HO·N ‖ Ph·C·C·Ph ‖ N·OH	N·OH ‖ Ph·C·C·Ph ‖ HO·N	N·OH ‖ Ph·C·C·Ph ‖ N·OH
(*E,E*) m.237	(*Z,Z*) m.207	(*E,Z*) m.165

Crystalline solids, decomp. by heat, insoluble in water: a reagent for nickel. **b. peroxide** $Ph_2C_2N_2O_2$ = 238.2. Colorless crystals, m.114.
benzilic acid* $Ph_2C(OH)COOH$ = 227.2. Diphenylglycolic acid. White, monoclinic crystals, m.150, slightly soluble in water.
benzilidene The benzylidene* radical.
benzilmonoxime $C_{14}H_{10}N_2O$ = 222.2.

C(Ph)—C(Ph)
‖ ‖
N—O—N

Crystals in isomeric forms: **(*E*)- ~** m.114; **(*Z*)- ~** m.137; insoluble in water.
benzilsynoxime (*Z,Z*)-Benzil dioxime*.
benzimidazole C_6H_4·N:CH·NH = 118.2. *o*-Phenyleneformamidine. Colorless crystals, m.167.
benzimidazolone $C_6H_4(NH)_2$:CO = 134.1. *o*-Phenyleneurea. Colorless crystals, m.308.
benzimidazolyl* 1*H*-benzimidazol-2-yl†. The radical $N_2C_7H_5$—, from benzimidazole.
benzimidoyl* Iminophenylmethyl†. The radical PhC(:NH)—.
benzin, benzine Benzoline, petroleum b. A mixture of hydrocarbons obtained from the distillation of crude petroleum, b.120–150. Clear colorless liquid, d.0.640–0.675, insoluble in water, miscible with organic solvents. A solvent for oils, resins, alkaloids, and rubber; a textile cleaner and a fuel. Cf. *benzene*.
benzindole $C_{12}H_9N$ = 167.2. Naphthazole, naphthindole.

1*H*-Benz[*g*]indole m.175

3*H*-Benz[*e*]indole m.40

benzinduline Aposafranone.
benzisosulfonazole Saccharin.
benzisoxazole C_7H_5ON = 119.1. Indoxazene, isoindoxazene.

C_6H_4<(O)(CH)>N **2,1- ~** Anthranil.

benzo- Prefix indicating (1) The phenylene* radical. (2)* The benzene ring in a polycyclic compound in addition to the prefixed base component.
benzoate* PhCOOR or PhCOOM. A salt of benzoic acid. **b. of soda** Sodium benzoate.
benzoazimidole $C_6H_6ON_3$ = 135.1. Colorless crystals, m.157.
benzobis-*m*-methylimidazole $C_{10}H_{10}N_4$ = 186.2. Colorless crystals, m.145.
benzocaine $NH_2C_6H_4COOEt$ = 165.2. Ethyl *p*-aminobenzoate*. Colorless crystals, m.90, insoluble in water; a surface anesthetic (USP, EP, BP).
benzocarbolic acid Phenylbenzoate*.
benzocinnoline 5,6-Naphthisodiazine.
benzodianthrone Helianthin.
benzodiazepines A group of drugs with tranquilizing, anticonvulsant, and sedative properties; as, chlordiazepoxide hydrochloride, diazepam.
benzodiazine $C_8H_6N_2$ = 130.1. Naphthyridine†, benzdiazine. Isomers with 2 N atoms in the following positions:

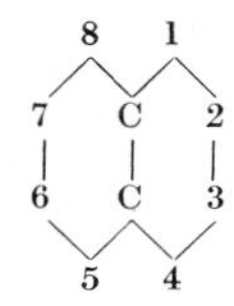

1,2- ~ Cinnoline*. **1,3- ~** Quinazoline*. **1,4- ~** Quinoxaline*. **1,5- ~** Pyridopyridine. **1,8- ~** Naphthyridine*. **2,3- ~** Phthalazine*. **2,7- ~** Copyrine. **alpha- ~** Cinnoline*. **beta- ~** Phthalazine*. **para- ~** Quinoxaline*.
benzodiazole $C_7H_6N_2$ = 118.1. **1,2- ~** Iso*indazole*. **2,1- ~** Indazole*. **1,3- ~** Benzimidazole.
benzodiazthine C_6H_4·(NH)·(S·CH):N = 150.2. Phenylsulfocarbizine. Colorless crystals, m.129.
benzo[ ]difuran $C_{10}H_6O_2$ = 158.2. Five isomers, as: **[1,2-*b*:5,4-*b*′]-**.

m.63

benzodioxazine $C_7H_5O_2N$ = 135.1. Benzodiazine derivatives containing 2 O atoms and 1 N atom.
benzodioxdiazine $C_6H_4N_2O_2$ = 136.1. Heterocyclic compounds, similar to benzodiazine, but with 2 O and 2 N atoms.
benzodioxtriazine $C_5H_3O_2N_3$ = 137.1. Heterocyclic compounds, similar to benzodiazine, but with 2 O and 3 N atoms.
benzofluorenes C_6H_4·CH_2·$C_{10}H_6$ = 216.3. A group of aromatic hydrocarbons. **1,2- ~** Chrysofluorene.
benzofuran* O·C_6H_4·CH:CH = 118.1. Coumarone. Colorless liquid, b.169, insoluble in water. **dihydro ~** Coumaran. **dihydrooxo ~** Benzofuranone. **1,2-dimethyl ~** $C_{10}H_{10}O$ = 146.2. Colorless liquid, b.210. **hydro ~** C_8H_8O = 120.2. Hydrocoumarone. Colorless, fragrant liquid, b.188. **2-methyl ~** C_9H_8O = 132.2. Colorless liquid, b.197.

oxodihydro ~ O·C_6H_4·CO·CH_2 = 134.1. Ketocoumaran. Colorless crystals, m.97.

benzofurancarboxylic acid Coumarilic acid.

benzofuranone $C_8H_6O_2$ = 134.1.
2- ~ C_6H_4·O·CO·CH_2
3- ~ 3-Coumaranone. C_6H_4·CO·CH_2·O

benzofuranyl* The radical OC_8H_5-, from benzofuran.

benzoglycolic acid Mandelic acid.

benzoglyoxaline Benzimidazole.

benzohydrol Ph_2CHOH = 184.2. Diphenylmethanol. Colorless needles, m.68, insoluble in water. Cf. *Michler's hydrol.*

benzohydroximic acid* $C_7H_7NO_2$ = 137.1. Two isomers:

N·OH	HO·N
‖	‖
HO·C·Ph	HO·C·Ph
(*E*)-~ (or α-~)	(*Z*)-~ (or β-~)

benzohydryl The benzhydryl* radical.

benzohydrylidene Diphenylmethane*.

benzoic **b. acid** See *benzoic acid.* **b. alcohol** Benzyl alcohol*. **b. aldehyde** Benzaldehyde*. **b. amide** Benzamide*. **b. anhydride** Benzoic (acid) anhydride*. **b. ether** Ethylbenzoate*. **b. sulfimide, b. sulfinide** Saccharin.

benzoic acid* Ph·COOH = 122.1. Phenylformic acid, benzenecarboxylic acid*. White needles, m.121, sparingly soluble in water, soluble in alcohol or ether; in benzoin and cranberries. Used in calico printing, in the manufacture of aniline dyes, and in standardization of bomb calorimeters; also for fungal skin diseases, as athlete's foot. **acetylamino ~** CH_3CONH·C_6H_4COOH = 179.2. Acetamidobenzoic acid. ***ortho*- ~** Colorless needles, m.185. ***meta*- ~** White crystals, decomp. 248. ***para*- ~** Colorless needles, decomp. 250. **amino ~*** NH_2·C_6H_4·COOH = 137.1. ***ortho*- ~** Anthranilic acid*. ***meta*- ~** Benzaminic acid. Yellow crystals, m.174. ***para*- ~** PABA. White crystals, m.186. Essential for the metabolic processes of bacteria. Sometimes included in the B group of vitamins. A sunscreen agent. **4-amino-3,5-dinitro ~** Chrysanisic acid. **aminohydroxy ~*** HOOC·C_6H_3·NH_2(OH) = 153.1. Aminosalicylic acid. **2,3- ~** Crystals, m.164. **2,5- ~** Violet crystals, m.252 (decomp.), soluble in water. **4,2- ~** m.150. **azo, azodi ~** Azobenzoic acid. **azoxy ~** See *azoxybenzoic acid.* **bromo ~*** Br·C_6H_4COOH = 201.0. ***ortho*- ~** Colorless needles, m.150. ***meta*- ~** White needles, m.155. ***para*- ~** Colorless prisms, m.251. **carbamoyl ~** Phthalamic acid*. **chloro ~*** ClC_6H_4COOH = 156.6. ***ortho*- ~** Rhombic crystals, m.137. ***meta*- ~** White crystals, m. 153. ***para*- ~** Colorless, monoclinic crystals, m.236. **dichloro ~*** $Cl_2C_6H_3COOH$ = 191.0. **2,5- ~** Colorless needles, m.156. **2,6- ~** White needles, m. 127. **3,4- ~** White needles, m.203. **dihydroxy ~*** $(OH)_2C_6H_3COOH$ = 154.1. **2,3- ~** (+$2H_2O$ = 190.2). Pyrocatechoic acid. Colorless needles, m.204. **2,4- ~** (+$3H_2O$ = 208.2). β-Resorcylic acid. White needles, m.206. **3,5- ~** (+1½H_2O = 181.1). α-Resorcylic acid. Colorless prisms, m.232. **2,5- ~** Gentisic acid. Colorless needles, m.199. **3,4- ~** Protocatechuic acid*. **2,6- ~** γ-Resorcylic acid. Colorless needles, m.167. **2,4-dihydroxy-6-methyl ~*** Orsellic acid. **dimethoxy ~*** Veratric acid*. **dimethoxyformyl ~** Opianic acid. **dimethoxyhydroxy ~** Syringic acid. **dimethyl ~*** $Me_2C_6H_3COOH$ = 150.2. **2,3- ~** *v*-Xylic acid. Colorless prisms, m.144. **2,4- ~** Xylic acid. Colorless, monoclinic crystals, m.126. **2,5- ~** Iso*xylic acid.* **2,6- ~** White needles, m.116. **3,4- ~** White prisms, m.163. **3,5- ~** 1,3,5-Mesitylinic acid. **dinitro ~*** $(NO_2)_2C_6H_3COOH$ = 212.1. **2,3- ~** Colorless crystals, m.201. **2,4- ~** Colorless prisms, m.179. **2,5- ~** Colorless needles, m.177. **2,6- ~** Colorless needles, m.202. **3,4- ~** Colorless needles, m.163. **3,5- ~** Colorless crystals, m.203. **ethoxy ~*** EtO·C_6H_4·COOH = 166.2. ***ortho*- ~** Colorless liquid or crystals, m.19. ***meta*- ~** Colorless needles, m.137. ***para*- ~** Colorless needles, m.195. **ethyl ~*** EtC_6H_4COOH = 150.2. ***ortho*- ~** Colorless needles, m.68. ***meta*- ~** Colorless needles, m.47. ***para*- ~** Colorless leaflets, m.112. **formyl ~** Phthalaldehydic acid*. **formyldimethoxy ~** Opianic acid. **fluoro ~*** $C_7H_5O_2F$ = 140.1. Fluobenzoic acid. Colorless rhombs, m.182, soluble in hot water. **hexahydro ~** $C_6H_{11}COOH$ = 128.2. Naphthenic acid, cyclohexane carboxylic acid. **hydrazino ~** NH_2NH·C_6H_4COOH = 152.2. ***ortho*- ~** Colorless crystals, m.155. ***meta*- ~** Colorless crystals, m.231. ***para*- ~** Colorless needles, m.258. **hydrazinohydroxy ~*** Orthin. **hydrazo ~*** $(NH\cdot C_6H_4COOH)_2$ = 272.3. ***ortho*- ~** Colorless leaflets, m.205. **hydroxy ~*** HO·C_6H_4COOH = 138.1. ***ortho*- ~** Salicylic acid*. ***meta*- ~** Colorless, rhombic crystals, m.201. ***para*- ~** Colorless, monoclinic crystals, m.213. **hydroxydimethoxy ~*** Syringic acid. **hydroxymethoxy ~** Vanillic acid*. **hydroxymethyl ~*** Cresotic acid. **isopropyl ~** Cumic acid. **methoxy ~** MeO·C_6H_4COOH = 152.2. ***ortho*- ~** White leaflets, m.99. ***meta*- ~** White needles, m.102. ***para*- ~** Anisic acid*. **methyl ~** Toluic acid*. **methylenedioxy ~*** Piperonylic acid*. **nitro ~*** $NO_2C_6H_4COOH$ = 167.1. ***ortho*- ~** Colorless needles, m.148. ***meta*- ~** Silky needles, m.141. ***para*- ~** White scales, m.242. **pentamethyl ~*** C_6Me_5COOH = 192.3. Colorless needles, m.210. **phenyl ~*** PhC_6H_4COOH = 198.3. ***ortho*- ~** Colorless needles, m.111. ***meta*- ~** White leaflets, m. 160. ***para*- ~** White needles, m.219. **sulfamine ~** $NH_2SO_2C_6H_4COOH$ = 201.2. ***ortho*- ~** Colorless crystals, m.167. The anhydride is saccharin. ***meta*- ~** Colorless crystals, m.238. ***para*- ~** Colorless crystals, decomp. 280. **sulfo ~*** HSO_3·C_6H_4COOH·$3H_2O$ = 256.2. ***ortho*- ~** Colorless crystals, m.250. ***meta*- ~** (+$2H_2O$ = 292.3). Colorless crystals, m.141. ***para*- ~** Colorless needles, m.259. **tetrahydro ~** C_6H_9COOH = 126.08. A solid, d.1.072. **tetramethyl ~*** Me_4C_6HCOOH = 178.2. **2,3,4,5- ~** Colorless crystals, m.165. **2,3,5,6- ~** Colorless crystals, m.127. **trihydroxy ~*** **2,3,4- ~** See *trihydroxybenzoic acid.* **3,4,5- ~** Gallic acid*. **trimethoxy ~** Asaronic acid. **trimethyl ~*** $Me_3C_6H_2COOH$ = 164.2. **2,3,4- ~** Prehnitilic acid. Colorless crystals, m.167. **3,4,5- ~** α-*Isodurylic acid.* **2,3,5- ~** γ-*Isodurylic acid.* **2,4,5- ~** Durylic acid. Colorless crystals, m.150. **2,4,6- ~** β-*Isodurylic acid.*

b. (acid) anhydride* PhCO·O·COPh = 226.2. Colorless crystals, m.42, b.360.

benzoin (1)* PhCO·CH(OH)Ph = 212.2. Benzoylphenylmethanol, bitter almond oil camphor. **(+)- ~, (−)- ~** White hexagons, m.132, b.343, absent from gum b.; an antiseptic for ointments. (2) Benzoinum, gum b. The balsamic resin *Styrax* species. See *b. gum.* **hydroxy ~** PhCH(OH)CH(OH)Ph = 214.3. White leaflets, m.138. **isohydro ~** Colorless, monoclinic crystals, m.120.

b. condensation A reaction between 2 aromatic aldehydes under the influence of KCN to form water and a compound R·CH(OH)·CO·R. Cf. *aldol condensation.* **b. gum** Benjamin gum benzoinam, kemenian (Malay), luban jawi (Arabic). Used as chest infection inhalants and skin antiseptic (USP, EP, BP). **Siam ~** The resin from *Styrax benzoin* and other species (chief constituents: benzoic acid, an essential oil, vanillin); used in perfumery and cosmetics. **Sumatra ~** The resin from *Styrax* species (chief constituents: cinnamic and benzoic acids); used for varnishes and in the manufacture of cinnamic acid. **b. oxime** $Ph_2 \cdot CH(OH)C:NOH$ = 227.3. Phenyl-α-hydroxybenzyl ketoxime. ***(E)*- ~, *α*- ~** Crystals, m.152; a sensitive reagent for copper or molybdenum. ***(Z)*- ~, *β*- ~** Crystals, m.99.

benzoiodohydrin $PhCOOC_3H_5ClI$ = 324.5. Benzoyl chloroiodoglycerol ether. Brown fat, soluble in alcohol.

benzol See *benzene.* **b. bichloride** See *dichlorobenzene* under *benzene.* **b. bromide** See *bromobenzene* under *benzene.* **b. chloride** See *chlorobenzene* under *benzene.*

benzoline Benzine.

benzomorpholine $C_6H_4 \cdot O \cdot NH(CH_2)_2$ = 135.2. Colorless oil, b.268.

benzonaphthacene **alpha- ~** Dibenzanthracene*. **beta- ~** $C_{22}H_{14}$ = 278.4. Pentacene. m.270.

benzonaphthene $C_{13}H_{10}$ = 166.2. 1*H*-Phenalene. m.85.

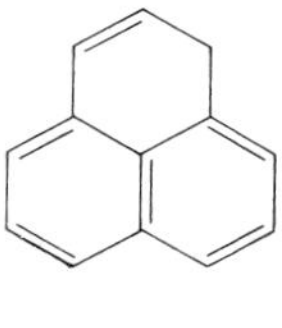

benzo-*α*-naphthindole $C_{16}H_{11}N$ = 217.3. Phenyl-α-naphthylcarbazole.

benzonaphthol Naphthylbenzoate.

benzonitrile* $C_6H_5 \cdot CN$ = 103.1. Phenyl cyanide, benzenecarbonitrile*. Colorless liquid, m. −13, b.191. It resembles bitter almond oil in odor; used in the synthesis of dyes and drugs. **amino ~*** $NH_2C_6H_4CN$ = 118.1. Anthranilo nitrile. ***meta*- ~** Colorless needles, m.53. ***para*- ~** Colorless needles, m.86. **hydroxy ~** Salicylonitrile*. **methoxy ~** Anisonitrile. **methyl ~** Cyano*toluene.* **nitro ~** $NO_2C_6H_4CN$ = 148.1. ***ortho*- ~** White, silky needles, m.109. ***meta*- ~** Colorless needles, m.117. ***para*- ~** White leaflets, m.147.

benzoparadiazine Quinoxaline*.

benzoparoxazine C_8H_7ON = 133.2.

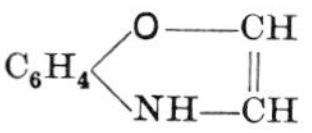

```
     O——CH
C6H4<    ||
     NH——CH
```

2-phenyl- Colorless crystals, m.103.

benzopentazole $C_4H_3N_2$ = 79.1. Heterocyclic compounds of the type:

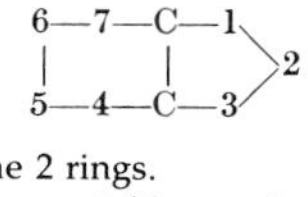

```
6—7—C—1
|   |   >2
5—4—C—3
```

with 5 N atoms in the 2 rings.

benzoperoxide Commercial benzoyl peroxide*.

benzophenanthrazine Phenanthrophenazine.

benzo[ ]phenanthrene **~[*a*]~** Chrysene*. **~[*def*]~** Pyrene. **~[*l*]~** Triphenylene.

benzo[ ]phenazine $C_{10}H_6 \cdot N:C_6H_4:N$ = 234.3. Naphthophenazine, phenonaphthazine. **~[*a*]~** Yellow needles, m.143. **~[*b*]~** Red leaflets, m.233.

benzophenid Phenyl benzoate*.

benzophenol Phenyl benzoate*.

benzophenone* $Ph \cdot CO \cdot Ph$ = 182.2. Diphenyl ketone*, diphenylmethanone†, benzoylbenzene. Colorless prisms; 2 allotropic forms: **alpha- ~** Stable, m.48; **beta- ~** Labile, m.26. A mild hypnotic; used in organic synthesis. **bis(dimethylamino) ~** Michler's ketone. **diamino ~*** See *diaminobenzophenone* under *diamino-*. **dihydroxy ~*** See *dihydroxybenzophenone* under *dihydroxy-*. **nitro ~*** $NO_2 \cdot C_6H_4COPh$ = 227.2. ***ortho*- ~** Colorless crystals, m.105. ***meta*- ~** Colorless needles, m.94. ***para*- ~** White leaflets, m.138. All soluble in alcohol. **pentahydroxy ~*** Maclurin. **thio ~*** See *thiobenzophenone.* **trihydroxy ~*** $C_{13}H_{10}O_4$ = 230.2. Several isomers.

b.carboxylic acid Benzoylbenzoic acid. **b.dicarboxylic acid** $(C_6H_4COOH)_2CO$ = 270.2. **2,2′- ~** Colorless crystals, m.150 (decomp.). **b.dicarboxylicdilactone** $C(O \cdot C_6H_4 \cdot CO)_2$ = 252.2. A spiro compound. Colorless crystals, m.212. **b. oxime*** $Ph_2C:NOH$ = 197.2. Benzophenoxime. White crystals, m.140. **b. sulfide** Thioxanthone.

benzophenoxime Benzophenone oxime*.

benzophosphinic acid $HOOC \cdot C_6H_4PHO(OH)$ = 186.1. Colorless crystals, m.300.

benzopinacol $Ph_2COH \cdot COH \cdot Ph_2$ = 366.5. Tetraphenylethyleneglycol, benzopinacone. Colorless crystals, m.185.

benzopinacone Benzopinacol.

benzopseudoxazole Anthranil.

benzopurpurine $[SO_3Na(NH_2)C_{10}H_5 \cdot N:N(Me)C_6H_4]_2$ = 726.7. C.I. Direct Red 2. Used to dye cotton red and as a biological stain and indicator, changing at pH 4.0 from purple (acid) to scarlet (alkali).

benzopyran C_9H_8O = 132.2. Heterocyclic compounds; e.g.,

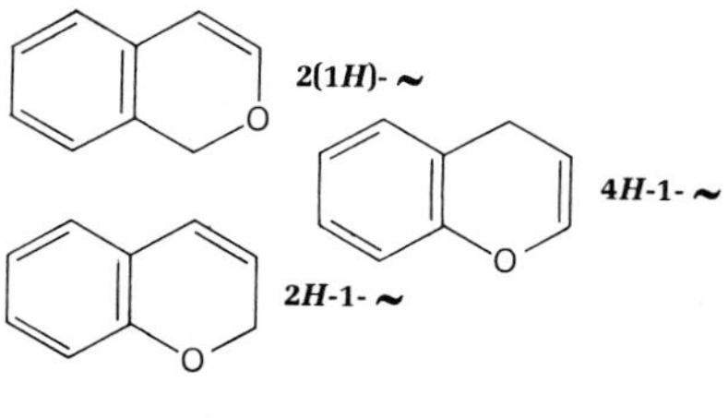

2(1*H*)- ~ **4*H*-1- ~** **2*H*-1- ~**

dihydro ~ Chroman*. **oxo ~** Benzopyrone. **1*H*-2-b.-1-one** Isocoumarin. **2*H*-1-b.-2-one** Coumarin. **4*H*-1-b.-4-one** Chromone.

benzopyranyl* The radical OC_9H_7-, from benzopyran.

benzopyrazine Quinoxaline*.

benzopyrazole Iso*indazole.*

benzopyrene $C_{20}H_{11}$ = 252.3. An anthracene ring with a naphthalene ring attached in the 9,1,2 positions; the principal carcinogen in coal tar pitch. Cf. *benzanthracenes, benzopyrine.*

benzopyridine Quinoline*.

benzopyrilium Compounds derived from cationic benzopyran.

benzopyrine $Ph \cdot COOH \cdot C_{11}H_{12}ON_2$ = 310.4. Antipyrine benzoate. Cf. *benzopyrene.*

benzopyrone Benzopyranone†. **benzo-1,2-pyrone** Coumarin*. **benzo-1,4-pyrone** Chromone. **benzo-2,1-pyrone** Iso*coumarin.* **phenyl ~** Flavone.

benzopyrrole Indole*.

benzopyrylium Benzopyrilium.

benzoquinol Hydroquinone.

benzo[ ]quinoline† $C_{10}H_6(N(CH)_3)$ = 179.2. **~[*c*]~** Phenanthridine*. **~[*f*]~** m.94. **~[*g*]~** Fluorescent solution, m.114. **~[*h*]~** m.52.

f c b g h N a

benzoquinone*

O:C·CH:CH·C:O = 108.1.
CH:CH

Quinone, 2,5-cyclohexadiene-1,4-dione†, paradioxobenzene, benzen(edi)one. Gold prisms, m.116 (sublimes), slightly soluble in water, soluble in ether; a reagent. **dihydroxydinitro ~*** Nitranilic acid. **dimethyl ~*** Xyloquinone. **3,4- ~,1,2- ~** Red leaflets, m.122 (decomp.) **2,3- ~,1,4- ~** Yellow needles, m.57. **2,5- ~,1,4- ~** Phlorone. **methyl ~*** Toluquinone. **semi ~** See *benzosemiquinone.* **tetrachloro ~*** Chloranil.

b.chlorimide C_6H_4ONCl = 141.6. m.116 (sublimes), soluble in water. **b. compounds** Organic derivatives of b.; often yellow dyestuffs. **b.dichlorimide** $C_6H_4N_2Cl_2$ = 175.0. m.164 (decomp.), soluble in water. **b.diphenylmethane** Fuchsone. **b. monoxime** O:C·CH:CH·C(NOH)·CH:CH = 123.1 Quinoxime. The major tautomer of nitrosophenol. **b. pigments** See *benzoquinones.*

benzoquinones Yellow compounds characterized by the benzoquinone grouping; as, brazilin.

benzoselenadiazole C_6H_4:(N_2Se) = 183.1. **1,2,3- ~** Isopiaselenole. m.34. **2,1,3- ~, iso ~** Piaselenole.

benzoselenazole C_6H_4·SeN·CH = 182.1.

benzoselenofuran C_6H_4·Se·CH:CH = 181.1. Benzoselenophene, selenonaphthene. **iso ~** $C_6H_4(CH)_2$Se. Benzoisoselenofuran.

benzosemiquinone ion, *p*-* The anion

·O—⟨benzene ring⟩—O⁻

benzosulfimide Saccharin.

benzosulfonazole $C_7H_5O_2NS$ = 167.2.

1- ~ C_6H_4<SO_2, CH>N; 2- ~ C_6H_4<SO_2, N>CH_2

benzosulfonazolone Saccharin.

benzotetrazine $C_6H_4\cdot(N{:}N)_2$ = 132.1.

benzotetrazole $C_5H_4N_4$ = 120.1. Heterocyclic compounds containing 4 N atoms, one of which forms NH. **1,3,5,7- ~** Purine; which is numbered 9,7,1,3. See *purine.*

7, 6, C—1, 2, 5, C—3, 4

benzothiadiazole C_6H_4:(N_2S) = 136.2. **1,2,3- ~** Isopiazthiole. m.36. **2,1,3- ~, iso ~** Benzisothiadiazole, piazthiole.

benzothiazine C_8H_7NS = 149.2. Heterocyclic compounds; as: **1,4,2- ~**

C_6H_4<S·CH_2, N>CH

benzothiazole C_6H_4(S·CH:N) = 135.1. Methenylaminothiophenol. Colorless liquid, b.234. **iso ~, phenyl ~** Benzenylaminothiophenol.

benzothiofuran Thianaphthene.

benzothiophene Thianaphthene.

benzothiopyran C_9H_8S = 148.2. A group of heterocyclic compounds; e.g.: **1,2- ~** Thiochromene.

C_6H_4<S·CH_2, CH=CH>

benzothiopyrone C_9H_6OS = 162.2. Thiochromone. A group of heterocyclic compounds; as,

C_6H_4<S·CO, CH:CH> **1,2-** Thiocoumarin m.80

C_6H_4<S·CH, CO·CH> **1,4-** Thiochromone m.78

benzotoluide PhCONH·C_6H_4Me = 211.3. Methylbenzanilide, *N*-benzoyltoluidine. ***ortho-* ~** Rhombic needles, m.146. ***meta-* ~** Monoclinic prisms, m.125. ***para-* ~** Rhombic needles, m.158, b.232.

benzotriazepine The ring system

NH—H, N=N

benzotriazine $C_7H_5N_3$ = 131.1. A group of heterocyclic compounds; as

C_6H_4<N:N, CH:N> **1,2,3-** Phentriazine

C_6H_4<N:N, N:CH> **1,2,4-** Phenpyrrodiazole

benzotriazole $C_6H_5N_3$ = 119.1. A group of heterocyclic compounds; as: **1,2,3- ~** Benzisotriazole. Colorless crystals, m.99.

C_6H_4<NH, N>N

benzotrichloride $C_6H_5\cdot CCl_3$ = 195.5. Toluene trichloride, α-trichlorotoluene, phenylchloroform. Yellowish liquid of penetrating odor, m.−5, b.214, decomp. in water; used in the synthesis of aniline dyes.

benzotrifluoride $C_6H_5CF_3$ = 146.1. Colorless liquid, b.102; an intermediate in the manufacture of dyestuffs and pharmaceuticals.

benzotrifuran $C_{12}H_6O_3$ = 198.2. **1,3,5-~** The heterocyclic compound

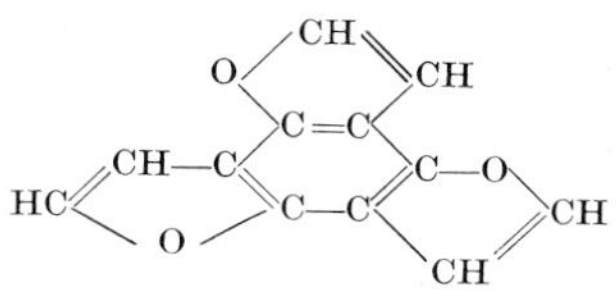

benzoxanthene $C_{10}H_6 \cdot CH_2 \cdot O \cdot C_6H_4$ = 232.3.
Naphthoxanthene, phenonaphthoxanthene.

benzoxazine* C_8H_7ON = 133.2. Heterocyclic compound; e.g.: **4-*H*-3,1-~**

$C_6H_4 \cdot N{:}CH \cdot O \cdot CH_2$

b. dione. Isatoic anhydride.

benzoxazole $C_6H_4 \cdot O \cdot CH{:}N$ = 119.1.
Methenylamidophenol. Colorless crystals, m.32. Cf. *benzisoxazole.* **amino ~** $C_7H_6ON_2$ = 134.1. Colorless crystals, m.130; 2 isomers. **anilino ~** $C_7H_4ON \cdot NHPh$ = 210.2. Colorless crystals, m.137. **phenyl ~** $C_7H_4ON \cdot Ph$ = 195.2. Colorless crystals, m.103.

benzoxazolone $C_6H_4 \cdot O \cdot CO \cdot NH$ = 135.1. ***N*-ethyl ~** $C_7H_4EtO_2N$ = 163.1. Colorless crystals, m.29.

benzoxazolyl* The radical ONC_7H_4-, derived from benzoxazole.

benzoxdiazine $C_7H_6ON_2$ = 134.1. A group of heterocyclic compounds; similar to benzodiazine, which contain 1 O and 2 N atoms in the ring.

benzoxdiazole $C_6H_4(NON)$ = 120.1. Benzofurazan. Crystals, m.55.

benzoxtetrazine $C_5H_4ON_4$ = 136.1. A group of heterocyclic compounds, similar to benzodiazine, which contain 1 O and 4 N atoms in the ring.

benzoxtriazine $C_6H_5ON_3$ = 135.1. A group of heterocyclic compounds containing 1 O and 3 N atoms in the ring; as: **1,2,3,4-~**

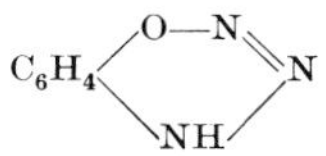

benzoxy The benzoyloxy* radical.

benzoyl* The aryl radical C_6H_5CO-, from benzoic acid. Cf. *chlorobenzoyl, nitrobenzoyl.* **b.acetaldehyde*** $PhCOCH_2 \cdot CHO$ = 148.2. Colorless crystals. **b. acetate*** $PhCO \cdot O \cdot OCMe$ = 164.2. Acetyl benzoate. **b.acetic acid*** $PhCO \cdot CH_2 \cdot COOH$ = 164.2. Phenyl ketoacetic acid. Colorless needles, m.61, decomp. 103; used in organic synthesis. **b.acetic ethyl ester** $Ph \cdot CO \cdot CH_2 \cdot COOEt$ = 192.2. Oily liquid, b.148. **b.acetone** $Ph \cdot CO \cdot CH_2 \cdot CO \cdot Me$ = 162.2. Acetylacetophenone. A homolog of b.acetaldehyde. Colorless liquid, b.260; used in organic synthesis. **b.acetonitrile** $Ph \cdot CO \cdot CH_2 \cdot CN$ = 145.2. Cyanacetophenone. Colorless crystals, m.80. **b. acetyl** $Ph \cdot CO \cdot CO \cdot Me$ = 148.2. Colorless, oily liquid, b.214. **b.amide** Benzamide*. **b.amino*** The benzamido* radical. **b.aniline** Benzanilide. **b.anthraquinone** $C_6H_4{:}(CO)_2{:}C_6H_3COPh$ = 312.3. White crystals, m.182. **b.azide*** $C_6H_5 \cdot CO \cdot N_3$ = 147.1. Azimide, benzazide, benzoyl nitride. Colorless crystals or liquid, m.20. **b. benzoate*** $C_{14}H_{12}O_2$ = 212.2. Colorless oil, d.1.118, insoluble in water. Cf. *benzyl benzoate.* **b.benzoic acid*** $PhCOC_6H_4COOH$ = 226.2. Benzophenonecarboxylic acid. Colorless crystals. ***ortho-*~** m.127. ***meta-*~** m.162. ***para-*~** m.194. **b.benzylamine** $PhCO \cdot NHCH_2Ph$ = 211.3. Colorless crystals, m.105. **b. bromide*** $C_6H_5 \cdot COBr$ = 185.0. α-Bromobenzaldehyde, benzene carbonyl bromide. Colorless liquid, b.218, decomp. in water. **b.butanol** $PhCO(CH_2)_4OH$ = 178.2. Colorless crystals, m.49. **b. chloride*** C_7H_5OCl = 140.6. α-Chlorobenzaldehyde, bénzene carbonyl chloride. Colorless liquid, b.198; decomp. with water. A reagent for alcohol and lysidine, and used in organic synthesis. **b. cyanide*** $C_6H_5 \cdot CO \cdot CN$ = 131.1. α-Cyanobenzaldehyde. Colorless scales, m.32. **b.cyclobutane** $C_6H_5 \cdot CO \cdot CH{:}(CH_2)_3$ = 160.2. Colorless liquid, b.258. **b.cyclopropane** $C_6H_5 \cdot CO \cdot CH{:}(CH_2)_2$ = 146.2. B.trimethylene. Colorless liquid, b.239. **b. disulfide** PhCOSSCOPh = 274.4. Dibenzoyl disulfide. Colorless prisms, m.128 (decomp.). **b. fluoride*** C_7H_5OF = 124.1. Benzene carbonyl fluoride. Colorless liquid, b.162. **b. formaldehyde** $C_8H_6O_2$ = 134.1. **anhydrous ~** $PhCO \cdot CHO$. Phenylglyoxal. Colorless liquid, b_{125mm} 142. **hydrated ~** $PhCO \cdot CH(OH)_2$. Colorless crystals, m.73. **b.formic acid** $C_6H_5 \cdot CO \cdot CO_2H$ = 150.1. Phenylglyoxylic acid. Colorless crystals, m.65. **b.formoxime** $C_6H_5 \cdot CO \cdot CH{:}NOH$ = 149.2. Isonitrosoacetophenone. Colorless crystals, m.127; a reagent for ferrous salts. **b.glycine, b.glycocoll** Hippuric acid*. **b.glycolic acid** $PhCO \cdot OCH_2 \cdot COOH$ = 180.2. Colorless prisms. **b.hydrazine** $PhCONHNH_2$ = 136.2. α-Hydrazinobenzaldehyde. Colorless crystals, m.112. **bi ~** $(PhCONH)_2$ = 240.3. Dibenzoylhydrazine. Colorless crystals, m.233. **b.hydride** Benzaldehyde*. **b. hydrogen peroxide** $PhCO \cdot O_2H$ = 138.1. Colorless crystals, m.41. **b. hydroxide** Benzoic acid*. **b. iodide*** C_7H_5OI = 232.0. Benzene carbonyl iodide, α-iodobenzaldehyde. Colorless leaflets, m.3 (decomp.), decomp. in water. **b.methane** Acetophenone*. **b.methanol** $PhCOCH_2OH$ = 136.2. Colorless crystals, m.83. **b.naphthol** Naphyl benzoate*. **b. oxide** Benzoic anhydride. **b. peroxide*** $(PhCO)_2O_2$ = 242.2. Dibenzoyl peroxide*, luzidol. Colorless, rhombic crystals, m.103. A flour "improver" and a bleaching agent. **b.phenylhydrazine** PhCONHNHPh = 212.3. Colorless crystals, m.145; an antiseptic. **b.propionaldehyde** $PhCO(CH_2)_2CHO$ = 162.2. Colorless liquid, b.245. **b.propionic acid*** $PhCO \cdot (CH_2)_2COOH$ = 178.2. Colorless needles, m.116. **b. pseudotropine** Tropacocaine. **b.pyrocatechol** Dihydroxybenzophenone. **b. sulfide** $PhCO \cdot S \cdot COPh$ = 242.3. Colorless crystals, m.48. **b.sulfonic imide** Saccharin. **b.thiourea** $PhCONH \cdot CS \cdot NH_2$ = 180.2. Small, colorless prisms, m.170. **b.trimethylene** B.cyclopropane. **b.urea** $PhCONH \cdot CO \cdot NH_2$ = 164.2. Crystalline solid, m.215.

benzoylation The introduction of the PhCO— radical into a molecule.

benzoylene The radical $-C_6H_4 \cdot CO-$. **b. guanidine** $C_8H_7ON_3$ = 161.2. Benzglycocyanidine. **b.urea** $C_8H_6O_2N_2$ = 162.2. Tetrahydrodioxoquinazoline.

benzoyles A group of organic, aromatic compounds containing the benzoylene radical.

benzoyloxy* Benzoxy. The radical C_6H_5COO-, from benzoic acid.

benzpinacone Benzopinacol.

benzpyrazole Iso*indazole.*

benzpyrene Benzopyrene.

benzsynaldoxime See *benzaldoxime.*

benzthiophene Thianaphthene.

benztrioxazine $C_6H_5O_3N$ = 139.1. A group of heterocyclic

compounds with 3 O and 1 N atoms in the ring; as:
8,2,3,4- ~

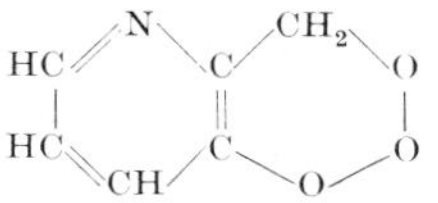

benztropine mesylate $C_{21}H_{25}ON \cdot CH_4O_3S$ = 403.5. Benztropinemethanesulfonate. Cogentin. White crystals, m.143, very soluble in water. An antiparasympathetic, used for Parkinson's disease (USP, BP).

benzyl* Phenylmethyl†. The aryl radical or $PhCH_2-$, derived from toluene. **b.acetamide** *N-* **~** $PhCH_2NHCOMe$ = 149.2. *α-* **~** Hydrocinnamamide. Colorless crystals, m.60. **b. acetate*** $MeCOOCH_2Ph$ = 150.2. Colorless liquid, b.206. Cf. *methyl benzoate.* **2-b.acetoacetic acid** $MeCOCH(COOH)CH_2Ph$ = 192.3. Colorless liquid, b_{13mm}160. **b.acetophenone** Propiophenone*. **b.acrylic acid** $CH_2:C(PhCH_2)COOH$ = 162.2. Methylenehydrocinnamic acid. Colorless crystals, m.69. **b. alcohol*** $PhCH_2OH$ = 108.1. Phenylmethanol, benzalcohol, phenmethylol, α-hydroxytoluene. Colorless, aromatic liquid with a sharp, burning taste, b.205, slightly soluble in water. Used as a local anesthetic and in perfumery. **hydroxy ~** $C_6H_4(OH)CH_2OH$ = 124.1. **(1)** ***ortho-*** **~** Salicyl alcohol. **(2)** ***meta-*** **~** Colorless needles, m.67, decomp. 300. **(3)** ***para-*** **~** Colorless needles, m.110. **hydroxymethoxy ~** Vanillic alcohol*. **isopropyl ~** Cuminol. **methoxy ~** Anisalcohol. **methylenedioxy ~** Piperonyl alcohol*. **nitro ~** $NO_2C_6H_4CH_2OH$ = 153.1. **(1)** ***ortho-*** **~** Colorless needles, m.74. **(2)** ***meta-*** **~** White, rhombic crystals, m.27. **(3)** ***para-*** **~** Colorless needles, m.93, b.179. **oxy ~** See *hydroxybenzyl alcohol.* **b.amine*** $PhCH_2NH_2$ = 107.2. Phenylmethylamine. Colorless liquid, b.184; used in organic synthesis. **di-** $(PhCH_2)_2NH$ = 197.3. Diphenylmethylimine. Colorless liquid, b.300. **imino ~** Benzamidine*. **tri ~** $(PhCH_2)_3N$ = 287.4. Colorless crystals, m.91. **b.aniline** $PhCH_2NHPh$ = 183.3. Colorless prisms, m.32; used in organic synthesis. **di ~** $(PhCH_2)_2NPh$ = 273.4. Colorless crystals, m.68. **b. azide** $PhCH_2N_3$ = 133.1. A liquid, b_{25mm}108. **b.benzene** Diphenylmethane*. **b. benzoate*** $Ph \cdot COOCH_2 \cdot Ph$ = 212.2. Colorless, oily, aromatic liquid with a sharp, burning taste, m.18; a scabicide (USP, BP). Cf. *benzoyl benzoate.* **b. bichloride** Benzylidene dichloride. **b. bromide***. $PhCH_2Br$ = 171.0. α-Bromotoluene. Colorless liquid, b.198.5. A poison gas in chemical warfare. **b. carbamate** $PhCH_2O \cdot CO \cdot NH_2$ = 151.2. Colorless liquid, b.198. **b. carbamide** B.urea. **b. chloride*** $PhCH_2Cl$ = 126.6. α-Chlorotoluene. Colorless liquid, b.178. Used in the synthesis of bitter almond oil and aniline dyes. **b. cinnamate** $C_8H_7COOCH_2Ph$ = 238.3. Cinnamein. Colorless prisms, m.39; used in perfumery. **b.cyanamide** $PhCH_2NHCN$ = 132.2. Crystalline solid, m.33. **b. cyanide*** $PhCH_2CN$ = 117.2. Phenylacetonitrile, α-tolunitrile. Colorless liquid, b. 233.5; used in organic synthesis. **nitro ~** $NO_2 \cdot C_6H_4 \cdot CH_2CN$ = 162.1. ***ortho-*** **~** Colorless needles, m.83. ***para-*** **~** Colorless prisms, m.115. **b. dichloride** Benzylidene dichloride. **b. diphenyl** $PhCH_2 \cdot C_6H_4Ph$ = 244.3. ***ortho-*** **~** Monoclinic needles, m.54. ***para-*** **~** Colorless leaflets, m.85. **b.diphenylamine** $PhCH_2 \cdot NPh_2$ = 259.1. Colorless crystals, m.87. **b. disulfide** $PhCH_2S \cdot SCH_2Ph$ = 246.4. Colorless leaflets, m.71. **b. ether** $(PhCH_2)_2O$ = 198.3. Benzyl oxide. Colorless, oily liquid, b.297. **b.ethylaniline** $Ph \cdot N \cdot Et(CH_2 \cdot Ph)$ = 211.3. Yellowish liquid, b.286 (decomp.). **b.ethylbenzene** $PhCH_2C_6H_4 \cdot Et$ = 196.3. Colorless liquid, b.194. **b. ethyl ether*** $Et \cdot OCH_2Ph$ = 136.2. Colorless liquid, b. 186, insoluble in water. **b. ethyl ketone*** $Et \cdot CO \cdot CH_2Ph$ = 148.2. 1-Phenylbutan-2-one. Colorless liquid, b.223, insoluble in water. **b.formamide** $PhCH_2 \cdot NH \cdot CHO$ = 135.2. ***o*-nitro ~** $NO_2 \cdot C_6H_4 \cdot CH_2 \cdot NHCHO$ = 180.2. Colorless crystals, m.89. **b. fumarate** $(PhCH_2OOC \cdot CH:)_2$ = 296.3. Bibenzyl fumarate. White, odorless solid; an antispasmodic. Cf. *b. succinate.* **b.hydrazine** $PhCH_2NHNH_2$ = 122.2. Colorless liquid, b_{35mm}135. **di ~** $PhCH_2NHNHCH_2Ph$ = 212.3. Colorless crystals, m.65; used in organic synthesis. **b.hydroxylamine** ***O-*** **~** $PhCH_2ONH_2$ = 123.2. Colorless liquid, b.123. ***N-*** **~** $PhCH_2NHOH$. Colorless crystals, m.57. **di ~** $PhCH_2NHOCH_2Ph$ = 213.3. **tri ~** $(PhCH_2)_2NOCH_2Ph$ = 303.4. **b. idene** Benzylidene*. **b. iodide*** $PhCH_2I$ = 218.0. α-Iodotoluene. Colorless liquid, m.24 (decomp.). Used in the manufacture of drugs and dyes. **b. isothiocyanate*** $PhCH_2NCS$ = 149.2. B. mustard oil. Colorless liquid, b.241. **b. morphine** Peronine. **b. mustard oil** Benzyl isothiocyanate*. **b.naphthalene*** $C_{10}H_7CH_2Ph$ = 218.3. Monoclinic prisms. **1- ~** m.58. **2- ~** m.35. **b. naphthyl ketone*** $PhCH_2COC_{10}H_7$ = 246.3. Colorless scales, m.57. **b. nitrile*** B. cyanide*. **b. penicillin sodium** EP, BP name for penicillin G. **b.phenanthrene*** $PhCH_2C_6H_3(CH)_2C_6H_4$ = 268.4. Colorless needles, m.155. **b.phenol** $PhCH_2C_6H_4OH$ = 184.2. ***para-*** **~** Colorless crystals, m.84. **b.phenylamine** B.aniline. **b. phthalimidine** $C_6H_4 \cdot CO \cdot N(CH_2 \cdot Ph)CH_2$ = 223.3. Colorless crystals, m.137. **b.pyridine** $PhCH_2 \cdot C_5H_4N$ = 169.2. **2- ~** b.276. **3- ~** m.34. **b. succinate** $(PhCH_2OOC \cdot CH_2)_2$ = 298.3. Bibenzyl succinate. White, odorless powder; an antispasmodic. Cf. *b. fumarate.* **b. sulfide** $(PhCH_2)_2S$ = 214.3. B. thioether. Colorless scales, m.49; used in organic synthesis. **b.sulfinic acid** $PhCH_2SO \cdot OH$ = 156.2. **b. sulfocyanide** B. thiocyanate*. **b. sulfone** $(PhCH_2)_2O_2S$ = 246.3. Colorless needles, m.150. **b.sulfonic acid** $PhCH_2SO_3H$ = 172.2. **b. sulfoxide** $(PhCH_2)_2SO$ = 230.2. Colorless leaflets, m.132. **b.tartronic acid** $PhCH_2C(OH) \cdot (COOH)_2$ = 210.2. Colorless crystals, m.143 (decomp.). **b.thiol*** $PhCH_2SH$ = 124.2. B. mercaptan, b. sulfhydrate. Colorless liquid, b.196. **b.thiourea** $PhCH_2NH \cdot CS \cdot NH_2$ = 166.2. B. sulfocarbamide. Colorless crystals, m.162. **b. thiocyanate*** $PhCH_2SCN$ = 149.2. B. sulfocyanide. Colorless prisms, m.41. **b.urea** $PhCH_2NH \cdot CO \cdot NH_2$ = 150.2. B.carbamide. Colorless needles, m.147 (decomp.).

benzylation The introduction of the benzyl radical into an organic molecule.

benzylene The phenylenemethyl* radical $-C_6H_4CH_2-$, usually in bicyclic compounds. **b. glycol** Hydrobenzoin. **b. pseudothiourea** Imido*coumothiazone.*

benzylidene* Phenylmethylene†, benzal, benzilidene, toluenyl. The PhCH= radical, from toluene. Cf. *benzylidyne.* **b.acetone** $PhCH:CHCOMe$ = 146.2. Benzalacetone, cinnamyl methyl ketone*, acetocinnamene, methyl styryl ketone. Colorless crystals with coumarin odor, m.42, soluble in alcohol; used in organic synthesis. **di ~** Styryl ketone. **b.acetophenone** $PhCH:CH \cdot CO \cdot Ph$ = 208.3. Cinnamyl phenyl ketone*. Colorless crystals, m.62, insoluble in water. **b.aniline** $PhCH:NPh$ = 181.2. Colorless crystals, m.45; used in organic synthesis. **b.azine** $PhCH:N \cdot N:CHPh$ = 208.3. Dibenzalhydrazine. Yellow prisms, m.93 (decomp.), insoluble in water. **b.coumarones** B.benzofurans. Plant pigments responsible for the golden yellow of certain plants. **b. dibromide*** $PhCHBr_2$ = 249.9. α-Dibromotoluene. A fuming oil, b.140. **b. dichloride*** $PhCHCl_2$ = 161.0. Chlorobenzal. Colorless liquid, b.206, insoluble in water. Used in chemical warfare and in dye manufacture. **b.ethylamine** $PhCH:NC_2H_5$ = 133.2. Colorless liquid, b.195. **b.hydrazine** $PhCH:N \cdot NH_2$ = 120.2. Colorless crystals, m.16.

benzylidyne* Phenylmethylidyne†, benzenyl. The radical PhC≡, from toluene.

beraunite A native iron phosphate.

Berberidaceae Barberry. Herbs, shrubs, or trees with a watery juice. Several yield drugs.

berberine $C_{20}H_{18}O_4N \cdot 6H_2O$ = 444.5. An alkaloid from the roots of *Hydrastis canadensis* (golden seal), *Berberis vulgaris*, and other Ranunculaceae. Orange crystals, m.145. Cf. *bebeerine*. **b. carbonate** $C_{20}H_{18}N \cdot H_2CO_3 \cdot 2H_2O$ = 370.4. Yellow crystals, soluble in hot water.

berberis The dried stems of *Berberis aristata* or *B. vulgaris*. The fluidextract is a purgative. See *barberry*.

berberonic acid $C_5H_2N(COOH)_3$ = 211.1. 2,4,5- Pyridinetricarboxylic acid*. Colorless, triclinic crystals, decomp. 235.

berengelite A Peruvian pitch, used for caulking.

beresovite A native mixed chromate and carbonate of lead.

Berg, Paul (1926–) American chemist; Nobel prize winner (1980). Noted for work on chemical structure of DNA.

bergamot oil A yellow-green, volatile essential oil from the rind of *Citrus bergamia* (Rutaceae), d.0.880–0.885. It contains linalyl acetate. citrene, and linalol. Used in perfumery.

bergaptol A constituent of lime oil.

bergblau A native copper carbonate.

bergenin $C_{14}H_{16}O_9$ = 328.3. Vakerin. A constituent of the barks of many tropical trees. Colorless needles, m.226.

bergenine $C_6H_3O_3 \cdot H_2O$ = 141.1. A bitter principle from *Saxifraga crassifolia;* m.140.

bergenose An ilmenite-norite (Norway).

berginization See *Bergius process (1).*

Bergius, Friedrich (1884–1949) German industrial chemist; Nobel prize winner (1931). **B. process** (1) Berginization. Production of motor fuel by the hydrogenation and liquefaction of coal at 400–450°C in hydrogen at 120–200 atm. (2) The manufacture of sugar from wood by treating sawdust with 40% hydrochloric acid and removing the acid by vapors from hot mineral oil.

Bergman, Torbern Olof (1735–1784) Swedish chemist; in 1783 adapted the alchemical symbols of circles and arcs to represent compounds by joining them together.

beriberi A disease due to deficiency of thiamin.

Berkefeld filter A porous, porcelain cylinder for the filtration of toxins and sera or the preparation of sterile solutions.

berkelium* Bk. An element, at. no. 97. Atomic weight of isotope with longest half-life (^{247}Bk) = 247.07; m.990, b.2970; forms Bk^{3+} and Bk^{4+} compounds. Prepared, by irradiation, by Cunningham and Thompson (Lawrence Radiation Laboratory) in 1959.

berkevilite Barkevite.

Berkshire sand Purified sea sand for filtration.

Berlin blue Iron(III) hexacyanoferrate(II)*.

bernstein Amber. **b. säure** Succinic acid.

Berthelot, Marcellin Pierre Eugène (1827–1907) French statesman and physical chemist: "All chemical reactions are dependent upon physical forces." Cf. *Berthollet*. **B. reaction** The determination of ammonia by reaction with an alkaline solution of phenol hypochlorite to form indophenol, which can be determined spectrophotometrically.

berthierine Iron ore (Alsace) consisting of magnetite and chamosite.

berthierite $FeSb_2S_4$. A native sulfantimonide.

Berthollet, Count Claude Louis (1748–1822) French chemist; discovered the compositions of ammonia and hydrogen sulfide and developed industrial chemistry. Cf. *Berthelot*.

berthollide Bertollide. A compound capable of variable composition; as, FeS. Cf. *daltonide*.

bertrandite $H_2Be_4Si_2O_9$. A native beryllium silicate; colorless, transparent, orthorhombic crystals.

beryl $Al_2Be_2Si_6O_{18}$. A native beryllium-aluminum silicate. Hexagonal crystals, d.2.7, hardness 7.5–8. The transparent varieties are gems; e.g., *aquamarine, emerald*.

beryllia Beryllium oxide.

beryllium* Be = 9.01218. An element of the magnesium group, at. no. 4. A hard, noncorrodible, grayish black metal in hexagons, d_{20} 1.85, m.1277, soluble in acids or alkalies. B. ores generally occur in granite rocks, e.g., beryl. B. is divalent and forms one series of compounds. It was discovered (1797) by Vauquelin in beryl, and isolated (1828) by Bussy and Woehler. Used as a moderator in nuclear reactions, in fuel "cans" for atomic fuels, in lightweight structures to impart strength and hardness to aluminum alloys, and with silver to form untarnishable alloys; also for windows in X-ray tubes. **diethyl ~** Et_2Be = 67.1. B. ethyl, b. ethide. Colorless liquid, b.186.

b. acetate* $(CH_3COO)_2Be$ = 127.1. Colorless plates, decomp. 285. **basic ~** $BeO \cdot 3Be(C_2H_3O_2)_2$ = 406.3. White octahedra, m.284, decomp. in water. **b. acetylacetonate** $Be[CH(MeCO)_2]_2$ = 207.2. Monoclinic crystals, m.108. **b. alkyls** The organic compounds of b., containing aliphatic radicals; e.g., diethylberyllium. **b. bromide*** $BeBr_2$ = 168.8. White, hygroscopic needles, m.601. **b. carbide*** Be_2C = 30.0. Yellow hexagons, decomp. by water. **b. carbonate** $BeCO_3 \cdot 4H_2O$ = 141.1. White powder, decomp. by acids; occurs also in basic salts of variable composition. **basic ~** $(BeO)_5CO_2 \cdot 5H_2O$ = 259.1. White powder, decomp. by acids. **b. chloride** (1) $BeCl_2$ = 79.9. White, deliquescent needles, m.400. (2) $BeCl_2 \cdot 4H_2O$. Yellow syrup, miscible with water. **b. dichloride oxide** Be_2OCl_2 = 104.9. White hexagons, infusible, insoluble in water. **b. fluoride*** BeF_2 = 47.0. White powder, m.800. **b. hydroxide*** $Be(OH)_2$ = 48.1. White powder, decomp. by heat, insoluble in water. **b. iodide*** BeI_2 = 262.8. White needles, m.510, decomp. by water. **b. nitrate*** $Be(NO_3)_2 \cdot 3H_2O$ = 187.1. Colorless, deliquescent crystals, soluble in water; a reagent. **b. oxalate*** $Be(C_2O_4) \cdot 3H_2O$ = 151.1. Rhombic crystals, m.($+2H_2O$)100, ($+H_2O$)220. **b. oxide*** BeO = 25.0. Beryllia. Amorphous, white powder, infusible, insoluble in water. Used to manufacture b. salts and as a refractory. **b. phosphate** Beryllonite. **b. potassium fluoride*** $BeF_2 \cdot 2KF$ = 163.2. White crystals. **b. silicate*** Be_2SiO_4. White powder, used as refractory, for spark plug porcelain, and in making b. compounds. Cf. *beryl*. **b. sodium fluoride*** $BeF_2 \cdot 2NaF$ = 131.0. Gray crystals; used to prepare metallic b. **b. sulfate*** (1) $BeSO_4 \cdot 4H_2O$ = 177.1. White tetragons; loses 2 molecules of water at 100; decomp. by further heat. (2) $BeSO_4 \cdot 7H_2O$ = 231.2. White, monoclinic crystals.

beryllonite $NaBePO_4$. Yellow, transparent crystals, d.2.845, hardness 5.5–6.

berzelianite Cu_2Se. A native selenide; thin white crusts. Cf. *crookesite, umangite*.

berzeliite $(Mg \cdot Ca \cdot Mn)_3As_2O_8$. A native arsenate of magnesium, calcium, and manganese; red, waxy masses.
berzelium A supposed element of at. wt. 212 (Baskerville).
Berzelius, Baron Jöns Jakob (1779–1848) Swedish chemist; the investigator of atomic weights, using oxygen as standard; worked on electrochemical analysis, isomerism, and the gas laws; discovered selenium and thorium and isolated silicon.
Bessemer, Sir Henry (1813–1898) English metallurgist; inventor of the B. process (1856). **B. converter** A large, egg-shaped retort used in the B. process. **B. iron** Iron made by the B. process. **B. process** Making steel by pouring molten cast iron in a specially designed converter and passing a stream of air through the molten mass to oxidize manganese, silicon, and carbon.
Best, Charles Herbert (1899–1978) Canadian codiscoverer, with Sir F. Banting, of insulin (1921).
Bestuscheff's tincture An ethereal tincture of iron chloride.
beta β. Second letter of the Greek alphabet. Symbol (italic) for phase coefficient. **b. acid** Anthraquinone-2-sulfonic acid. **b.-blocking drugs** B.-adrenoceptor blocking drugs, b.blockers. A group of drugs that prevents or blocks the effect of sympathetic nerve stimulation on the body, particularly the heart. Used to treat hypertension, abnormal heart rhythms, and in prophylaxis of myocardial infarction. **b.chlora process** A method of bleaching flour with nitrosyl chloride and chlorine. **b. gauge** A *nucleonic gage*, q.v., which uses b. rays. **b. hydrogen** See *hydrogen* (2). **b. particle** An electron or positron. **b. position** Correctly used: (1)* In conjunctive nomenclature to indicate a substituent on the C atom next but one to the principal group of a side chain of a cyclic component. (2) In sugars, to indicate a group above, rather than below (α), the plane of the ring. (3) Formerly, also used to indicate substitution of the 2d, 3rd, 6th, or 7th H atom of a bicyclic compound. **b. ray** A stream of electrons emitted from radioactive substances with the velocity of light.
Beta Beets (Chenopodiaceae). The juice of *B. vulgaris* yields sugar, betaine, and an indicator.
Betadine Trademark for povidone-*iodine*.
betaine* $Me_3N \cdot CH_2 \cdot CO \cdot O(+H_2O)$ = 135.2. Lycine, oxyneurine, dimethylsarcosine, trimethylglycine. An alkaloid, from beets, crustaceans, octopus and other cephalopods; also made synthetically. Colorless, monoclinic crystals, m.293 (decomp.), soluble in water or alcohol. **nicotine methyl ~** Trigonelline. **thio ~** Thetine. **trimethylhydroxybutyro ~** Carnitine.
betaines* Zwitterionic bases characterized by the $^-OOC \cdot CH_2 \cdot NMe_3^+$ group; e.g., *betaine*, q.v., and carnitine: $Me_3N \cdot CH_2 \cdot CHOH \cdot CH_2 \cdot CO$ (—O—). Cf. *stachydrine, trigonelline.*
beta-isoamylene Pentene*.
Betaloc Trademark for metoprolol tartrate.
betamethasone $C_{22}H_{29}O_5F$ = 392.5. White, bitter crystals, m.246 (decomp.), insoluble in water; an adrenocortical steroid (USP, EP, BP). See *corticoids.*
beta-methylethylpyridine Collidine.
beta-methylindole Skatole.
beta-naphthol See *2-naphthol.* **b. benzoate** See *2-naphthylbenzoate.* **b. orange** Tropeolin.
beta-naphthyl The 2-naphthyl* radical. **b.amine** 2-Naphthylamine*. **b. benzoate** 2-Naphthyl benzoate.
betanin The pigment of red beet *(Beta vulgaris)*; used as a food coloring.
beta-oxybutyric acid See *3-hydroxybutanoic acid.*
beta-quinine Quinidine.
beta-terpineol β-Terpineol.
betatopic Pertaining to a radioactive substance which differs from one of its isotopes by 1 electron and 1 integer in atomic number. Cf. *alphatopic.*
betatron An early form of particle *accelerator*, capable of up to 300 MeV.
betel (1) Originally the betel vine. Cf. *betel leaf.* (2) A fragment of betel nut rolled up in a betel leaf, with some lime or gambir, for chewing (southeast Asia). A stimulant. (3) Pinang. The Areca or catechu palm of southeast Asia. **b. leaf** The dried leaves of *Piper betle* (Piperaceae). **b. nut** Areca nut, buyo, semen arecae. The dried seed of *Areca catechu*, a southeast Asian palm. It contains many alkaloids (e.g., arecoline); an astringent, an anthelmintic, and a stimulant. Cf. *areca nut.* **b.phenol** Chavibetol.
bethanechol chloride $C_7H_{17}N_2O_2Cl$ = 196.7. Carbamethylcholine chloride. White crystals having a slight ammoniacal odor; a parasympathomimetic, used to produce bladder contractions (USP).
Bettendorf test A test for arsenic in presence of bismuth and antimony compounds. A freshly prepared stannous chloride solution is added to the sample; a brownish tint indicates arsenic.
Betts' process Refining lead from the hexafluorosilicate with gelatin as electrolyte.
betula A genus of the oak family (Cupuliferae). See *birch.* **b. camphor** Betulinol. **b. oil** See *birch oil.*
betulin Betulinol. **b. amaric acid** $C_{36}H_{52}O_{16}$ = 740.8. An oxidation product of betulinol.
betulinic acid $C_{30}H_{48}O_3$ = 456.7. A dibasic acid, m.195, formed by oxidation of betulin.
betulinol $C_{30}H_{50}O_2$ = 442.7. Betula camphor, betulin. An alcohol from the bark of betula species. Colorless crystals, m.258.
beudantite Biereite, corkite. A native sulfate and phosphate of iron and lead; dark green or black rhombohedra.
Beutel buret float A hydrometerlike closed glass tube, used to facilitate buret readings.
beV Billion electronvolts. The SI unit is GeV.
Bexan Trademark for a polyvinyl chloride synthetic fiber.
beyrichite Ni_3S_4. Native in metallic, gray hexagons.
bhang An intoxicating preparation made from the flowering tops of hemp *(cannabis).*
γ-BHC* See *hexachlorocyclohexane* under *cyclohexane.*
Bi Symbol for bismuth.
bi- (1) Prefix indicating 2 or double. Correctly applied to organic molecules made up of two identical halves; as, biphenyl, $(C_6H_5)_2$. Cf. *di-, bis-.* (2) Misnomer for acid salt; as, bicarbonate, which is correctly hydrogencarbonate* or hydrogen carbonate†.
biacenaphthylidene Biacene.
biacene $C_{10}H_6 \cdot CH_2 \cdot C{:}C \cdot CH_2 \cdot C_{10}H_6$ = 304.4. Red-yellow crystals, m.271, which show distinct dichroism when dissolved in concentrated sulfuric acid; indigo blue in transmitted light, red in incident light.
biacetyl* $MeCO \cdot COMe$ = 86.1. 2,3-Butanedione*, diacetyl. Colorless liquid of pungent, sweet odor, d.0.9793, b.88, slightly soluble in water. **b. dioxime** Dimethylglyoxime*. **b. monoxime** $MeCO \cdot C(NOH)Me$.
biacetylene Butadiyne*.
Bial's reagent A solution of orcinol in acidic ferric chloride, used to detect pentoses in urine.
biallyl* $CH_2{:}CH \cdot (CH_2)_2CH{:}CH_2$ = 82.1. 1,5-Hexadiene*, diallyl. Colorless liquid, b.60, insoluble in water. **b.amine**

$(CH_2{:}CH\cdot CH_2)_2NH$ = 97.2. Di-2-propenylamine. Colorless liquid, b.111.
bianiline ***N-*** ~ Hydrazobenzene. ***para-*** ~ Benzidine*.
bianisaldehyde Anisil.
bianisidine* $C_{14}H_{16}O_2N_2$ = 244.3. Dianisidine, dimethoxybenzidine. Colorless needles, m.170, slightly soluble in water. ***ortho-*** ~ A microreagent for copper (green color in presence of acetic acid and a thiocyanate; sensitivity $1{:}10^6$).
bianthryl $C_{28}H_{18}$ = 354.4.

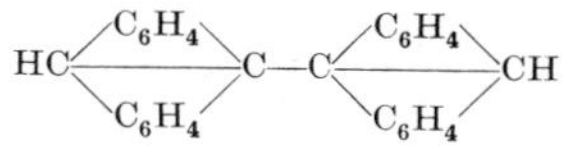

HC(C_6H_4)(C_6H_4)C—C(C_6H_4)(C_6H_4)CH

Colorless leaflets, m.300.
biarsine Diarsane*.
biaxial Having 2 axes. **b. crystals** A crystal with 2 optical axes.
bibenzal Stilbene*.
bibenzenone Diphenoquinone.
bibenzenyl Tolane.
bibenzil Bibenzyl*.
bibenzohydrol Benzopinacol.
bibenzoic acid Diphenic acid.
bibenzoyl (1) Benzil*. (2) Dibenzoyl (2).
bibenzyl* $C_{14}H_{14}$ = 182.3. Dibenzyl. **asymmetric** ~ $Me\cdot CHPh_2$. Colorless liquid, b.209. **symmetric** ~ $PhCH_2\cdot CH_2Ph$. 1,2-Diphenylethane*. Colorless needles, m.52. **b. alcohol** Hydrobenzoin.
biberine Bebeerine.
biborate An acid borate; as, sodium hydrogenborate*, $NaHBO_3$. Cf. *diborate.*
bibulous Absorbing moisture. **b. paper** Blotting or filter paper.
bicarbonate Hydrogencarbonate*, hydrogen carbonate †, dicarbonate, acid carbonate. A salt containing the $-HCO_3$ radical. **b. of potash** Potassium hydrogencarbonate*. **b. of soda** Sodium hydrogencarbonate*.
bichloride Dichloride*.
bichromate Dichromate*.
bichrome Potassium dichromate*.
biconcave See *concave lens* under *lens.*
biconvex See *convex lens* under *lens.*
bicuculline $C_{22}H_{17}O_6N$ = 391.4. An alkaloid from *Dicentra cucullaria, Corydalis sempervirens,* and *Adlumia fungosa* (Fumaraceae). Dimorphic crystals, m.177 and 196. Adlumine has two CH_3O- groups instead of the first CH_2O_2.
bicyclic* Containing 2 rings; as, naphthalene.
bicyclodecane* See *Dekalin.*
bicyclo[ ]heptane* **~[4.1.0]~** Trinorcarane. **~[3.1.1]~** Trinorpinane*. **~[2.2.1]~** Trinorbornane*.
bicyclo[ ]heptene* **~[4.1.0]~** Trinorcarene. **~[3.1.1]~** Trinorpinene*. **~[2.2.1]~** Trinorbornene*.
bicyclo numbering* In a 2-ring system, the number of C atoms in each of the three bridges connecting the tertiary C atoms is placed in square brackets in descending order; e.g., bicycloheptane.
bidesyl $PhCO\cdot CHPh\cdot CHPh\cdot COPh$ = 390.5. Dibenzoyldibenzyl. Colorless crystals, m.255.
bidiphenyleneethane $(C_6H_4)_2CH\cdot CH(C_6H_4)_2$ = 330.4. Colorless crystals, m.246. Cf. *tetraphenylene, bifluorene.*
bieberite $CoSO_4\cdot 7H_2O$. Native in red, monoclinic crystals.
biethylene 1,3-Butadiene*.
biflavonyl The structure

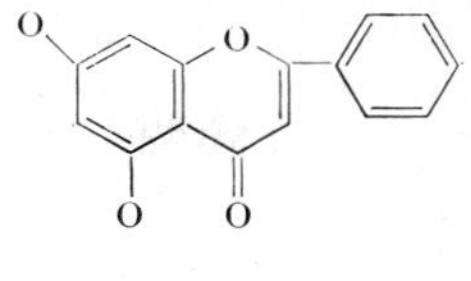

Cf. *ginkgetin.*
bifluorene $(C_6H_4)_2C{:}C(C_6H_4)_2$ = 328.4. Bidiphenylene ethylene. Red needles, m.188.
biformyl Glyoxal.
biguanide* $H_2N\cdot C({:}NH)\cdot NH\cdot C({:}NH)NH_2$ = 101.1. A condensed product of guanidine.
bihexyl Dodecane*.
bihydrazine $NH_2\cdot NH\cdot NH\cdot NH_2$ = 62.1. Buzane, dihydrobuzylene, tetrazane*.
biimino* The bridge $-NH\cdot NH-$.
biindolyl* $C_6H_4\cdot(NH\cdot CH)C\cdot C(CH\cdot NH)\cdot C_6H_4$ = 208.3. Bi-1*H*-indole.
biindoxyl Indigo white.
bikhaconitine $C_{36}H_{51}O_{11}N$ = 673.8. Bikh. An alkaloid from the root of *Aconitum ferox,* m.113.
bilateral Pertaining to 2 or both sides.
bile Gall, chola. The yellow, brown, or greenish secretion of the liver (solids 14, water 86%), d.1.026–1.032. **b. acids** Acids found in b.; e.g., glycocholic acid, $C_{26}H_{43}O_6N$. See *cholane, porphin.* **b. pigments** The coloring matter of b.; as, bilirubin (red); biliverdin (green); bilicyanin (blue); bilipurpurin (purple); bilixanthin (brown). **b. salts** The sodium salts of b. acid which aid digestion.
bilharziasis Schistosomiasis.
bilicyanin $C_{33}H_{36}O_9N_4$ = 632.7. A blue pigment obtained by oxidation of biliverdin.
biliflavin A yellow pigment derived from biliverdin.
bilifulvin Bilirubin.
bilifuscin $C_{16}H_{20}O_4N_2$ = 304.3. Brown, insoluble powder from bile pigments; used as a reagent. **meso** ~ A brown pigment formed by the decomposition of hemoglobin.
Biligrafin Trademark for iodipamide methylglucamine.
bilin A mixture of sodium taurocholate and sodium glycocholate; the main constituent of bile.
bilineurine Choline*.
bilinigrin A black oxidation product of bilirubin.
biliphain Bilirubin.
biliprasin A green pigment derived from biliverdin.
bilipurpurin Cholohematin. A purple bile pigment.
bilirubin $C_{33}H_{36}O_6N_4$ = 584.7. Bilifulvin, biliphain, hematoidin. The insoluble coloring matter of bile which is structurally related to hematoporphyrin and hematin and is derived from hemoglobin. Responsible for the color of jaundiced skin. Red-yellow needles, m.192. Cf. *porphin ring.* **hydro** ~ Urobilin.
bilirubinic acid $C_{17}H_{24}O_3N_2$ = 304.4. A monobasic acid containing 2 attached pyrrole rings, formed from bilirubin and hemin by reduction.
bilisoidanic acid $C_{24}H_{32}O_9$ = 464.5. An acid obtained from bile by treatment with nitric acid.
biliverdic acid $C_8H_9O_4N$ = 183.2. Crystals, m.114.
biliverdin $C_{33}H_{34}O_6N_4$ = 582.7. An oxidation product of bilirubin. Green, amorphous powder, soluble in alcohol. It occurs in some fish scales and fins, and yields the pigment biliflavin.

bilixanthin $C_{33}H_{36}O_{12}N_4$ = 680.7. Brown oxidation product of *bile pigments,* q.v.
billion (1) A thousand millions; 10^9 (U.S. and French usage), i.e., giga-. (2) A million millions; 10^{12} (British and German usage), i.e., tera-. Cf. *milliard.*
biloidanic acid $C_{22}H_{32}O_{12}$ = 488.5. Norsolanellic acid. An oxidation product of bile.
bimetal A sheet made of 2 metal layers and having special properties, as, corrosion resistance.
bimolecular (1) Pertaining to 2 molecules; (2) having a molecularity of 2. **b. reaction** See *reaction order (second).*
binaphthalene* $C_{10}H_7 \cdot C_{10}H_7$ = 254.3. Naphthyl naphthalene. **1,1′-~** m.160, b.365. **2,2′-~** m.187, b.452. **1,2′-~** m.70.
binaphthalenyl* The radical $C_{10}H_7 \cdot C_{10}H_6-$, from binaphthalene.
binaphthyl* Binaphthalene*.
binarite Marcasite.
binary **b. compound** A compound containing only 2 elements; e.g., NaCl, $FeCl_3$. **b. digit** Bit. The smallest unit of information in a binary system. It can take only the values 0 and 1. Employed in computers. **b. mixture** A mixture of any 2 substances. **b. salt** A salt containing 2 bases; as, $NaKSO_4$. **b. system** (1) Any combination possible with 2 metals; cf. *phase rule, alloy.* (2) See *number systems.*
binder A material used to hold solid substances together; as, bitumen.
bindheimite Native hydrous lead antimonate.
binding Holding together. **b. energy** The force that holds together the negatively and positively charged portions of an atom or molecule. Cf. *solution energy* under *energy.*
bing Spoil bank. Colliery refuse consisting of pyritic fine coal and shale.
biniodide Diiodide*.
binitro See *dinitro-*
binnite $2As_2S_3 \cdot 3Cu_2S$. A native copper-arsenic sulfide.
binoxalate Hydrogenoxalate*.
binoxide Dioxide or peroxide.
bio- Prefix (Greek *bios*) meaning life.
bioanalysis The determination of small quantities of substances by means of protozoa or bacteria.
bioassay Biological assay.
biocatalyst (1) Enzyme. (2) Ergine.
biocatalyzator A protoplasmic substance that promotes growth.
biochanin A $C_{16}H_{12}O_5$ = 284.3. 5,7-Dihydroxy-4′-methoxyisoflavone; m.215. An estrogenic flavone from red clover.
biochemical Pertaining to the matter changes within an organism. **b. oxygen demand** B.O.D. A chemical measure of the deoxygenating power of an effluent in terms of the difference between the dissolved oxygen content before and after 5 days at 20°C. Cf. *chemical oxygen demand.*
biochemistry A branch of chemistry dealing with the changes occurring in living organisms and cells.
biocolloid A gluelike organic substance; as, the glutin of bone glue.
bioctyl Hexadecane*.
biodegradable Describing a substance that can be decomposed by natural influences (e.g., biological action or sunlight); as, a soft detergent. See *sensitized photodegradation* under *photodegradation.*
biodyne A natural, cellular respiratory factor.
bioelement An element essential to life. Principally those most abundant in organisms (see Table 14). Cf. *abundance of elements.*

TABLE 14. BIOELEMENT COMPOSITION OF ORGANISMS
(Percent)

Mammals		Gymnosperms	
O	62.43	C	53.96
C	21.15	O	38.65
H	9.86	H	6.18
N	3.10	Al	0.065
Ca	1.90	Si	0.057
P	0.95	S	0.052
K	0.23	Fe	0.030
S	0.16	N	0.030
Cl	0.08	Ca	0.007
Na	0.080	K	0.006
Mg	0.027	P	0.005
I	0.014	Mg	0.003
F	0.009	Cl	0.002
Fe	0.005	Na	0.001
Br	0.002	F	0.001
Al	0.001	Mn	0.001
Si	0.001		
Mn	0.001		
	100.00		100.00

biogas A gas produced by biological means, rich in methane, as from the fermentation of agricultural or domestic waste.
Biogastrone Trademark for carbenoxolone sodium.
biogenesis The theory that all life comes only from life. Antonym: abiogenesis.
biogenic See *bioelement.*
biogeochemistry See *geochemistry.*
bioglass A glass able to form a bond between bone and orthopaedic device.
biognosis The study of life.
biological **b. assay** The determination of the active principles of a drug from the smallest quantity that will produce certain symptoms in animals or organisms. It can supplement or replace chemical methods of analysis. **b. markers** Compounds, as hydrocarbons in oil exploration, that provide clues as to the source and geological history of the chemicals sought. **b. oxygen demand** Biochemical oxygen demand.
biology The science of living matter, its forms, functions, occurrence, behavior, and evolution. See Table 15. Branches:
1. Botanical (plants)
2. Zoological (animals)
3. Anthropological (man)

bioluminescence The phosphorescence of living vegetable or animal organisms.
biomass (1) The mass of specified organisms living in a particular area. (2) Renewable biological material, as agricultural or forestry waste or energy crops, used for the production of energy.
biomechanics The study of the application of mechanics and engineering to the human body and to man-machine relationships; e.g., artificial joints. Cf. *cybernetics.*

TABLE 15. BIOLOGY-RELATED SCIENCES

Energy transformation		Biophysics
Matter changes		Biochemistry
MORPHOLOGY	Cells	Cytology
	Tissues	Histology
	Organs	Anatomy
	Organisms	Taxonomy
PHYSIOLOGY	Generative	Embryology
	Sustentative	Metabolism
	Correlative	Circulation and nerves
PATHOLOGY	Plants	Phytopathology
	Animals	Zoopathology
	Man	Pathology
ETIOLOGY	The race	Phylogeny
	The individual	Ontogeny
CHOROLOGY	Plants	Flora
	Animals	Fauna
	Plants and animals	Ecology
PALEONTOLOGY	Plants	Paleobotany
	Animals	Paleozoology
	Man	Ethnology

biometry The application of statistics to biological science.

biomolecule A molecule of protoplasm; a unit of living substance. Cf. *idioblast, protoplasm.*

bionomy The measurement of life phenomena.

bio-osmosis The osmotic pressure of living cells.

biophage A cell or organism feeding on living cells or organisms.

biophore Biomone. The smallest particle of living matter, consisting of protoplasm.

bioplasm Protoplasm.

bioplast Micelle.

biopolymer A naturally occurring macromolecule; as, produced by biosynthesis.

biorization Pasteurization at 100–300 kPa.

bios $C_5H_{11}O_3N$ = 133.1. A crystalline substance similar in character to a vitamin, which was found to be essential for the growth of certain types of yeast, m.223. Now known to consist principally of nicotinic acid and panthothenic acid.

biose A carbohydrate containing 2 carbon atoms, e.g., $HO \cdot CH_2 \cdot CHO$. Cf. *tetrose, hexose.*

biosphere The air, land, sea, and water immediately surrounding mankind.

biosterin Biosterol.

biosterol $C_{22}H_{44}O_2$ = 340.7. An alcohol resembling cholesterol.

biota The flora and fauna of a region.

biotechnology The application of living organisms, or their biological systems or processes, to the manufacture of useful products; e.g., *genetic engineering,* q.v., single-cell *protein,* q.v., biogas, q.v., drugs (as, insulin) and chemicals from biomass.

biotic Pertaining to life or living organisms.

biotin* $C_{10}H_{16}O_3N_2S$ = 244.3. A member of the vitamin B complex. Yeasts and bacteria contain or make b. Deficiency occurs only if diet consists largely of raw eggs; their white contains an antivitamin, avidin. B. is a coenzyme for carboxylases. See *vitamins,* Table 101 on p. 622.

biotite A brown-black ferrous mica.

biotoxin A toxin formed in the tissues of the living body.

bioxalate Hydrogen*oxalate**.

bioxyl Bismuthyl chloride.

biozeolite A zeolitic biological slime from sewage filters.

biperiden $C_{21}H_{29}ON$ = 311.5. Akineton. White crystals, insoluble in water; used to treat Parkinson's disease (USP, BP).

biphenyl (1)* $(C_6H_5)_2$ = 154.2. Phenylbenzene. Colorless scales, m.71, insoluble in water. (2) The biphenylyl* radical. **amino ~** $PhC_6H_4 \cdot NH_2$ = 169.2. ***ortho-* ~** Biphenylyl amine. Colorless crystals, m.45. ***para-* ~** Xenyl amine, *p*-phenylaniline, martylamine. White leaflets, m.53. **diamino ~** $NH_2C_6H_4 \cdot C_6H_4NH_2$ = 184.2. **2,2'- ~** Colorless crystals, m.81. **4,4'- ~** Benzidine*. **dimethyl ~** Ditolyl*. **methyl ~** $Ph \cdot C_6H_4 \cdot Me$ = 168.2. Phenyl tolyl. ***meta-* ~** Colorless liquid, b.275. ***para-* ~** Colorless liquid, b.265.

biphenylene (1)* The hypothetical compound $C_6H_4{:}C_6H_4$. (2) The radical $-C_6H_4 \cdot C_6H_4-$. **b.bisazo** The radical $-N{:}NC_6H_4 \cdot C_6H_4N{:}N-$. **b. oxide** Diphenylenefuran.

biphenylyl* [1,1'-Biphenyl]yl†, biphenyl, diphenyl; xenyl (*para* only). The radical $C_6H_5 \cdot C_6H_4-$, from biphenyl. **b.amine** Amino*biphenyl.* **b.diamine 2,2'- ~** Diamino*biphenyl.* **4,4'- ~** Benzidine*. **b.imide** Carbazole*. **b.mercury** $(PhC_6H_4)_2Hg$ = 507.0. White scales, m.216.

BIPP A mixture of bismuth, iodoform, and paraffin; an antiseptic paste for infected wounds.

bipropargyl 1,5-Hexadiyne*.

bipropenyl 2,4-Hexadiene*.

bipseudoindoxyl Indigo.

bipyridyl* $NH_4C_5 \cdot C_5H_4N$ = 156.2. Dipyridyl, bipyridine†.

	m.	b.	Water solubility
2,2'-	70	272	Slight
2,3'-	liquid	288	Insoluble
2,4'-	62	281	Slight
3,3'-	68	291	Very soluble
3,4'-	61	297	Very soluble
4,4'-	114	305	Hot only

biquinoline* $C_{18}H_{12}N_2$ = 256.3. Diquinoline. **2,2'- ~** Crystals, m.196. **2,3- ~** Yellow crystals, m.176. **6,6'- ~** Crystals, m.181.

biquinolyl* $C_{18}H_{12}N_2$ = 256.3. Diquinolyl, biquinoline. **2,2'- ~** m.196:

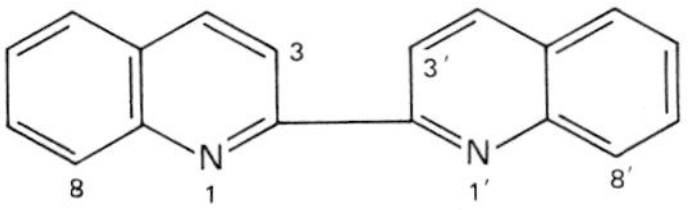

8,8'- ~ Brown crystals, m.94; reagent for cuprous ions (purple complex soluble in many solvents; sensitivity 0.2 ppm).

birch A tree of the genus Betula. **b. camphor** Betulinol. **b. oil** Sweet b. oil, betula oil. The essential oil from the bark of *Betula lenta,* black birch. Colorless oil, d.1.127–1.182, b.218–222 (chief constituent is methylsalicylate); a flavoring and liniment. **b.-tar oil** A tarry oil from the wood of *Betula alba,* white birch. Brown oil with empyreumatic odor, d.0.886–0.950, soluble in alcohol (chief constituents phenols and cresols). Used in ointments and in leather dressing. **b.wood carbon** Norit.

birdlime A viscid substance from the bark of the Ilex species and mistletoe; contains viscin, viscum, and ilicic alcohol.
birectification Analysis of fermented liquors by fractional distillation.
birefractive Doubly refracting.
birefringence Double *refraction.* **electrical ~** Kerr effect.
birotation Mutarotation.
birthwort Serpentaria.
bis-* Prefix indicating twice, correctly applied to identical radicals substituted in the same way. Cf. *bi-, di-.*
bisabolene $C_{15}H_{24}$ = 204.4. A group of monocyclic sesquiterpenes from bisabol myrrh and star anise oil.
bisacodyl $C_{22}H_{19}O_4N$ = 361.4. Biscolax, Dulcolax. White crystals, m.134, insoluble in water; a laxative (USP, BP).
bisazimethylene Ketazine.
bisazo-* Tetrazo-, disazo. Indicating 2 —N:N— groups, identically substituted. **b. compound** A compound containing 2 azo groups R·N:N·R·N:N·R, including many dyes. Cf. *diazo compounds.*
bisbenzimidazole $C_{14}H_{10}N_4$ = 234.3. A heterocyclic system related to indigo. Colorless crystals, m.305.
bischofite $MgCl_2·6H_2O$. A native magnesium chloride.
Biscolax Trademark for bisacodyl.
biscuit ware An unglazed, porous porcelain which has been fired twice.
bisdiazo- Bisazo-*. **b. amine** NH:N·NH·N:NH = 73.1. A hypothetical compound known as its derivatives. Cf. *nitrogen hydrides.* **b. hydrazine** HN:N·NH·NH·N:NH = 88.08. A hypothetical compound known as its derivatives.
bisethylxanthate Ethyl*xanthate.*
bishydrazicarbonyl Biurea*.
bismarck brown $[(NH_2)_2C_6H_3N:N]_2C_6H_4$ = 402.4. C.I.Basic Brown 1. Aniline brown, Manchester brown, triaminoazobenzene, vesuvin. Brown powder, m.143, soluble in water. Used as a dye and bacteriological stain, and to determine the decolorizing power of charcoal.
bismite $Bi_2O_3·3H_2O$. Native bismuth ocher.
bismon Colloidal bismuth metahydroxide.
bismuth* Bi = 208.9804. An element of the arsenic group, at. no. 83. Bismutum, wismuth. A pink, silvery, brittle metal resembling antimony, d_{20} 9.78, m.269, b.1570, insoluble in water, soluble in acids. It occurs native, as sulfide (bismuthinite), and in a few rare minerals; as, sulfobismuthide. B. is tri- and pentavalent and forms 2 series of compounds:

Derived from trivalent b.

Bismuthide	M_3Bi
B.(III)*, b.(3+)*, bismuthous	Bi^{3+}
Bismuthyl	BiO^+
Bismuthine	BiR_3

Derived from pentavalent b.

B.(V)*, b.(5+)*, bismuthic	Bi^{5+}
Bismuthate(V)*	BiO_3^-

B. (discoverer unknown) is mentioned by Basil Valentine (1450) as a "bastard of tin." Its name comes probably from the Arabic *wiss majat,* "a metal which easily melts," or German *wismuth, wiesen matte,* "a meadow." Used commercially in alloys of low melting point (wood metal), in type metal, and in the manufacture of b. compounds; its insoluble salts cast X-ray shadows (formerly used as b. meal); the soluble salts are very toxic. B. is very diamagnetic. Cf. *Lipowitz's alloy, rose metal, D'Arcet metal.* **trimethyl ~** See *bismuthine.*

b.(III)*, b.(3+)* See also *bismuthous.* **b.(V)*, b.(5+)*** See also *bismuthic.* **b. acetate*** $Bi(OOCMe)_3$ = 381.1. White powder, soluble in acetic acid. **b. borate*** $BiBO_3$ = 267.8 Gray powder, soluble in acids. **b. borosalicylate** Gray powder, decomp. in water. **b. camphorate** $Bi_2(C_{10}H_{14}O_4)_3$ = 1014. White powder, insoluble in water. **b. carbonate*** Bismuthyl carbonate. **b. chromate** $Bi_2O_3·2CrO_3$ = 665.9. Yellow powder, insoluble in water; a pigment. **b. dichloride*** $BiCl_2$ = 279.9. Black needles, m.163, decomp. by water. **b. gallate** $Bi(OH)_2C_7H_5O_5$ = 412.1. Dermatol, b. subgallate. Yellow powder (55% b.), insoluble in water, soluble in alkalies; an external antiseptic and astringent, used to treat hemorrhoids (BP). **b. glance** Bismuthinite. **b. gold** Au_2Bi. Occurs naturally. **b. hydroxide*** $Bi(OH)_3$ = 260.0. White powder, insoluble in water; used in the manufacture of b. salts. **b. iodate*** $Bi(IO_3)_3$ = 733.7. Heavy white powder, insoluble in water. **b. iodide** BiI_3 = 589.7. Gray crystals, m.408, insoluble in water, soluble in potassium iodide solution. **b. meal** A meal containing an insoluble bismuth salt given before an X-ray examination to render the digestive organs visible; superseded by barium meal. **b. minerals** The commonest are native b., Bi; bismuthinite (b. glance), Bi_2S_3; tetradymite, Bi_2Te_3; bismite, $Bi(OH)_3$; b. ocher (bismuthite), Bi_2O_3; bismuthite, $Bi_2H_2CO_6$; b. telluride, $Bi_2(S·Te)_2$. **b. molybdate** $Bi_2(MoO_4)_3$ = 897.8. Yellow powder, insoluble in water. **b. nickel** A native mixture of b. and nickel sulfides. **b. nitrate*** $Bi(NO_3)_3·5H_2O$ = 485.1. B. ternitrate. White, triclinic, deliquescent crystals, m.74 (decomp.), decomp. in water; an astringent, antiseptic. Cf. *b. subnitrate.* **b. ocher** See *b. minerals.* **b. organic compound** A compound with aliphatic or aromatic hydrocarbon radicals attached to b., e.g., trimethylbismuthine. **b. oxalate*** $Bi_2(COO)_6$ = 682.0. White granules, soluble in acids. **b. oxide** (1) Commonly, bismuthous oxide. (2) One of Bi_2O_3 = b. trioxide*, bismuthous oxide; Bi_2O_4 = b. tetraoxide*; Bi_2O_5 = b. pentaoxide*, bismuthic oxide. **b. oxobromide*** Bismuthyl bromide. **b. oxochloride*** Bismuthyl chloride. **b. oxofluoride*** Bismuthyl fluoride. **b. oxoiodide*** Bismuthyl iodide. **b. pentaoxide*** Bismuthic oxide. **b. permanganate*** $Bi(MnO_4)_3$ = 565.8. Black powder, insoluble in water. **b. peroxide*** Bi_2O_4 = 482.0. Brown powder, liberating oxygen at 150. **b. phosphate*** $BiPO_4$ = 304.0. White powder, decomp. by heat. **b. potassium iodide*** $BiI_3·4KI$ = 1254. Yellow crystals, soluble in water. Cf. *Dragendorff's reagent.* **b. potassium tartrate*** $BiK(C_4H_4O_6)_2$ = 544.2. Colorless crystals, soluble in water; a reagent for glucose in urine. **b. propionate*** $Bi(C_3H_5O_2)_3$ = 428.2. White powder, insoluble in water. **b. selenide*** Bi_2Se_3 = 654.8. Black crystals, decomp. by heat, insoluble in water. **b. silver** Native silver, containing 16% b. **b. subcarbonate** Bismuthyl carbonate. **b. subgallate** B. gallate. **b. subnitrate** Basic b. nitrate. A mixture of $Bi(OH)_2NO_3$ and $BiOH(NO_3)_2$ or of $(BiO)NO_3$ and (BiO)OH. An antiseptic, astringent, white powder (79–82% b.), insoluble in water, decomp. 260. Used in stomach disorders; a dusting powder and ointment (USP). Cf. *pearl white(3).* **b. sulfate*** $Bi_2(SO_4)_3$ = 706.1. White, crystalline powder, decomp. by water or heat. **b. sulfite*** $Bi_2(SO_3)_3$ = 658.1. White powder of variable composition. **b. telluride** See *b. minerals.* **b. tetraoxide** B. peroxide*. **b. tribromide*** $BiBr_3$ = 448.7. Yellow crystals, m.210, decomp. in water. **b. trichloride*** $BiCl_3$ = 315.3. White crystals, m.232, insoluble in water. **b. trimethyl** Trimethyl*bismuthine**. **b. trisulfide** Bi_2S_3 = 514.1. Brown, rhombic crystals, decomp. by heat, insoluble in water. **b. tungstate*** $BiWO_4$ = 441.5. B. wolframate*. White powder, decomp. in water. **b. violet** A triphenylmethane dye combined with b.; an antiseptic.

bismuthane* Bismuthine*.
bismuthi Official Latin for "bismuth."
bismuthic Bismuth(V)* or (5+)*. Describing compounds of pentavalent bismuth. **b. oxide** Bi_2O_5 = 498.0. Bismuth pentaoxide*. Brown powder, liberates oxygen at 150, insoluble in water.
bismuthide* Metal derivative of bismuthine, BiH_3; as, Me_3Bi. **sulfo ~** Beegerite.
bismuthine (1)* BiH_3 = 212.0. b.22. (2)* The bismuthine analogs of the *phospines,* q.v. **triethyl ~*** Et_3Bi = 296.2. Colorless liquid, b_{80mm}107. **trimethyl ~*** Me_3Bi = 254.1. A liquid, d.2.300, b.110. **triphenyl ~*** Ph_3Bi = 440.3. A solid, m.78.
bismuthinite Bi_2S_3. Bismuth glance. Native bismuth sulfide.
bismuthiol Mercaptosulfothiobiazole. A reagent for bismuth salts (red precipitate). Cf. *mercaptophenyldithiodiazolone* under *mercaptophenyl-.*
bismuthite $(BiO)_2CO_3 \cdot H_2O$. Native bismuth carbonate; green, earthy masses.
bismuthosmaltite $Co(As,Bi)_3$. Native cobalt bismuthide; or smaltite containing bismuth.
bismuthosphaerite $(BiO)_2CO_3$. Native bismuth carbonate; yellow fibrous masses.
bismuthous Bismuth(III)* or (3+)*. Describing compounds of trivalent bismuth, comprising the common bismuth compounds. **b. oxide** Bi_2O_3 = 466.0. Bismuth trioxide*. Yellow tetragons, m.850, insoluble in water. Used as bismuth subnitrate.
bismuthyl The radical Bi(O)−, from trivalent bismuth. **b. bromide** Bi(O)Br = 304.9. Bismuth oxobromide*. Brown powder, insoluble in water; used to treat dyspepsia and hysterics. **b. carbonate** $(BiO)_2CO_3$ = 510.0. Bismuth subcarbonate. White powder, insoluble in water. Used in ointments and face powders and as an antacid (EP, BP). **b. chloride** Bi(O)Cl = 260.4. Bismuth oxochloride*. White powder, decomp. at red heat, insoluble in water; an astringent and antiseptic. **b. dichromate** $(BiO)_2 \cdot Cr_2O_7$ = 665.9. Orange crystals, insoluble in water. **b. fluoride** Bi(O)F = 244.0. Bismuth oxofluoride*. White crystals, insoluble in water. **b. hydroxide** (BiO)OH. Basic bismuth hydroxide. **b. iodide** Bi(O)I = 351.9. Bismuth oxoiodide*. White powder, insoluble in water. **b. nitrate** $(BiO)NO_3 \cdot H_2O$ = 305.0. Basic bismuth nitrate. The main constituent of *bismuth subnitrate,* q.v.
bismutite Bismuthite.
bismutosmaltite Bismuthosmaltite.
bisphenol A $(HO \cdot Ph)_2CMe_2$ = 230.3. 2,2-Bis(4-hydroxyphenyl)propane. Brown crystals, m.156, insoluble in water; used in the manufacture of phenolic and epoxy resins.
bistetrazole $C_2H_2N_8$ = 138.1. Ditetrazyl:

```
N═══N          N═N
|     \       /   |
|      C──C       |
|     /       \\   |
NH──N          N──NH
```

bistort Snakeweed, adderwort. The root of *Polygonum bistorte;* an astringent.
bistriazole $C_4H_4N_6$ = 136.1. Ditriazolyl:

```
CH═N          N═CH
|    \       /    |
|     C──C        |
|    //      \\    |
NH──N          N──NH
```

Colorless liquid, b. 300.
bisulfate Hydrogensulfate*, acid sulfate. A compound containing the HSO_4- radical, from sulfuric acid. See *bi.*

bisulfide Disulfide*.
bisulfite Hydrogensulfite*, acid sulfite. A salt containing the HSO_3- radical, from sulfurous acid. See *bi.* **b. compounds** $R_2C \cdot (OH) \cdot (SO_3Na)$. Addition compounds of sodium bisulfite and an aldehyde or ketone.
bit Abbreviation for binary digit. Cf. *byte.*
bitartrate Hydrogentartrate*, acid tartrate. A salt containing the $C_4H_5O_6-$ radical, from tartaric acid. See *bi.*
bithionol $C_{12}H_6O_2SCl_4$ = 356.0. 2,2′-Thiobis(4,6-dichlorophenol). White crystals, m.186, insoluble in water.
bithiophene Dithienyl*.
bitter (1) See *bitters.* (2) An astringent taste, as of magnesium sulfate. **b. almond oil** An essential oil from the seeds of *Prunus amygdala amara,* b. almond. Pale yellow liquid, d.1.038–1.060 (chief constituents: benzaldehyde, hydrocyanic acid, phenoxyacetonitrile); a flavoring. **b. almond oil camphor** Benzoin*. **b. apple** Colocynth. **b. ash** Quassia. **b. bark** Cinchona. **b. cucumber, b. cups** Colocynth. **b. damson** Simaruba. **b. principle** Generic term for the bitter-tasting principle of a drug, e.g., due to alkaloids or glucosides. **b. root** Gentian. **b. salt** Magnesium sulfate*. **b. spar** A ferruginous dolomite. **b. stick** Chirata. **b.sweet** Dulcamara. **American ~, false ~** Waxwort. The root and bark of *Celastrus scandens* (Celastraceae). **b. wintergreen** Chimaphila. **b. wood** Quassia.
bittern Waste liquid from the solar salt industry; contains magnesium salts and bromides from seawater.
bitters (1) A group of drugs, which includes gentian, quassia, etc, that stimulates saliva and gastric juice flow. (2) Mineral waters characterized by a bitter or saline taste (e.g., due to magnesium sulfate). (3) A preparation which contains the bitter principles of plants; as, Angostura b.
bitumen(s) Native solid or semisolid hydrocarbons (naphtha or asphalt) soluble in carbon disulfide; rich in C and H. Cf. *asphaltenes, carbenes, kerotenes, protobitumen.* **albino ~** A petroleum resin. **asphaltic ~** Asphalt.
bituminous Having the qualities of bitumen. **b. coal** A coal rich in hydrocarbons (50–80% C). **b. materials** B. substances used for pavings and roofings, as, asphalt, shales, tars. **b. resins** Red, transparent fossil resins from brown coals (80% wax). Cf. *retinite resins.*
biurate Hydrogenurate*, acid urate. A salt of uric acid.
biurea* $(H_2N \cdot CO \cdot NH)_2$ = 118.1. Diurea, *p*-urazine, bishydrazicarbonyl. Colorless crystals, m.270. Cf. *urazine, biuret.* **acetylene ~** Glycol *uril.*
biuret* $NH_2 \cdot CO \cdot NH \cdot CO \cdot NH_2$ = 103.1. Allophanamide, dicarbamoylamine, carbamoylurea. Colorless needles, decomp. 190, soluble in hot water. A condensation product of urea; a reagent. **acetyl ~** $MeCO \cdot NH \cdot CO \cdot NH \cdot CONH_2$ = 145.1. Colorless needles, m.193, soluble in water.
b. reaction Pitrowsky reaction. A test for protein compounds which contain the −CO·NH− group. Drops of copper and potassium hydroxide solutions give a violet color.
bivalence, bivalent Divalence, divalent.
bivinyl 1,3-Butadiene*.
Bixaceae, Bixineae Trees and shrubs; some were formerly used to yield drugs; as, *Gynocardia odorata,* chaulmoogra oil; *Bixa orellana,* annatto.
bixin MeOOC(CH:CHCMe:CH)$_4$CH:CH·COOH = 394.5. Methyl ester of norbixin. A red, crystalline coloring matter from annatto.
Bjerrum, Niels (1879–1959) Danish physical chemist, noted for his work on the theory of electrolytic dissociation: the thermodynamic anomalies of strong electrolytes are due to interionic forces.
Bk Symbol for the element berkelium.

Black, Joseph (1728–1799) Scottish chemist and physicist, pioneer in experimental research. He introduced the term "fixed air" for the gas (carbon dioxide) given off from carbonates; opposed the *phlogiston theory,* q.v.; and developed the concept of latent heat.

black Describing a substance that reflects no colored rays of light; hence, absence of color. **chemical ~, gas ~, impingement ~** A very fine grade of carbon b. produced from the flame of natural gas; particle size, 10–30 nm. Used in rubber, paints, and lacquers.

b. alum A mixture of aluminum sulfate and activated carbon used in water treatment. **b. antimony** Antimonic sulfide. **b. balsam** Peru balsam. **b.berry** See *Rubus.* **b. body** A material that absorbs all radiant energy and transforms it into heat. Cf. *Stefan-Boltzmann equation.* **b. box** Approach to the derivation of algorithms that relates output to input and ignores the internal mechanisms of the process (i.e., "box"). **b.boy** Grass-tree gum. A resin from *Xanthorrhoea hastilis* (Liliaceae), of Australia; a varnish and sealing wax. **b. cobalt** An earthy, native cobalt. **b. cohosh** Cimicifuga. **b. copper** Copper oxide. **b. currant** The fruit of *ribes nigrum.* The syrup is a vitamin C source and flavoring (BP). **b. damp** Choke damp. **b. dogwood** Frangula. **b. drops** Opium vinegar. **b. dyes** See *nigrosines.* **b. flux** A reducing agent used in assaying; made by burning together potassium carbonate 1, argol 3 pts. **b.fish oil** Malon oil. **b. haw** *Viburnum prunifolium.* **b. henbane** Hyoscyamus. **b. Indian hemp** Apocynum. **b.jack** Sphalerite. **b. lead** (1) Plumbago. (2) Alquifou. **b.leg** Anthrax. **b. manganese** Pyrolusite. **b. metal** A black electrolytic deposit of certain metals, e.g., platinum. **b. mustard** See *mustard.* **b. pigments** See *bone b., graphite, ivory b., lampblack.* **b. potassium** Suint ash. **b. powder** An explosive: potassium nitrate 62–75, sulfur 10–19, charcoal 12–5%. **b. silver** Stephanite. **b. tellurium** (Pb, Au)(Te·S). A native telluride. **b. tin** Cassiterite. **b.top** See *tarmacadam* under *tar.*

Blackmar oil thief A device for taking oil samples from tank cars.

blackstrap An inedible grade of molasses from sugar refining; a source of alcohol.

Blagden's law The lowering of the freezing point of a solution is in proportion to the amount of dissolved substance present. Cf. *Raoult's law, Coppet's law.*

blairmorite A rock containing 71% analcite (Alberta, Canada).

Blaise reaction A Grignard reaction.

blanc fix(e) Synthetic barium sulfate produced by the action of barium chloride on aluminum sulfate; a pigment for coating paper.

blanch (1) A lead ore embedded in rocks. (2) To bleach. (3) A heat treatment of foodstuffs before preserving them, to destroy enzymes which cause deterioration. **b. liquor** A solution of calcium hypochlorite.

blancophore See *optical bleaching* under *bleaching.*

blangel Silica gel dehydrating agent, impregnated with a cobalt salt to indicate by a color change when it is hydrated.

blast (1)To smash to pieces by an explosive. (2) To subject a material to a hot firing. (3) A current of hot gases. (4) An immature blood cell. **b. burner** A large blowpipe with compressed air blown into the gas flame. **b. furnace** A smelting oven, with an air current, used in the manufacture of pig iron. **b. lamp** B. burner.

blasting The process of loosening natural deposits of rocks and other materials by explosives. **b. gelatin** A plastic high explosive; a 5–10% solution of collodion cotton in nitroglycerin. Cf. *gelignite.* **b. oil** Nitroglycerin. **b. powders** Nondetonating explosives, deflagrating powders. The black granular powders used for mining and road building; as *black powder.* Cf. *soda powder* under *soda.*

blastokolin The natural inhibitor to ripening in apples; probably associated with maleic acid.

blau gas A fuel gas produced by cracking gas oil at 550°C, comprising alkanes with some hydrogen and ethylene.

bleaching The whitening and removal of natural impurities by chemical or physical agents, e.g., chlorine or exposure to sunlight. **optical ~** The use of blancophores, certain organic compounds (e.g., derivatives of diaminostilbene-sulfonic acid), which have a short-wavelength fluorescence (blue or violet) in visible light, to enhance the whiteness of paper or textiles (e.g.) when present in very small quantities. See *optical brightening.*

b. materials Oxidizing or reducing agents, as, sulfur dioxide, sodium hydrogensulfite, hydrogen peroxide, calcium hypochlorite, chlorine water, oxides of nitrogen. **b. powder** Chloride of lime, formed by passing chlorine gas over dry slaked lime; principally calcium hypochlorite, but when it absorbs moisture, it is converted into a mixture of the chloride and hypochlorite of calcium. A disinfectant (BP).

Bleeker method The reduction of vanadium compounds by electrolysis.

blende (1) Sphalerite, zinc blende. Cf. *Sidot's b.* (2) A sulfide ore, as, antimony b. Cf. *glance.*

bleomycin sulfate A mixture of cytotoxic antibiotics, produced by *Streptomyces verticillus.* Used to treat malignant disease, particularly squamous cell carcinomas (USP).

blick The brightening of a noble metal during cupellation by adsorption on the outer layer of lead oxide.

blister steel A finely granulated steel. Cf. *cementation.*

blistering beetle Cantharides.

Blocadren Trademark for timolol maleate.

blocking The tendency of a film to adhere to itself.

block tin An alloy of tin with iron, cobalt, lead, antimony, and arsenic.

blodite, bloedite, blödit Astrakanite.

Blondlot rays n-*Rays.*

blood A red, homogeneous liquid that circulates in vertebrates through the body channels; d.1.045–1.075, pH = 7.35, constituting 70 ml/kg of the body weight, and containing solids 2.2 and water 98% in a delicately adjusted equilibrium of salts, proteins, enzymes, and organized particles (b. cells, plasma). B. carries nutritive materials from the intestines to cells and tissues, and oxygen from the lungs to the tissues; removes the waste products from the tissues and carries them to the kidneys, lungs, intestines, or skin; distributes internal secretions from one organ to another; defends against infection; and maintains an isotonic condition. B. consists of a liquid (*plasma,* q.v.) and cellular elements. The plasma contains the *serum,* q.v., and fibrinogen; the cellular elements are the red and white b. cells (corpuscles) and the b. platelets. The main constituents of the serum are albumin, globulin, glucose, and salts. The red b. cells contain the oxyhemoglobin, lecithin, and some salts. Cf. *porphin, fibrin, hemoglobin.* **arterial ~** Bright-red b., rich in oxyhemoglobin. **beaten ~** Defibrinated b. **clotted ~** A thick, semisolid mass of blood cells embedded in a network of precipitated fibrin. **concentrated ~** B. from a single donor, with some plasma and anticoagulant removed. Packed cell volume (PCV) greater than 70% (BP). **defibrinated ~** A red, homogeneous liquid, consisting of blood cells and serum. **dried ~** B. of slaughtered animals, dried and ground; a fertilizer (not less than 12% organic nitrogen). **venous ~** Dark red b., rich in carbon dioxide or reduced hemoglobin.

Cf. *b. gases.* **whole ~** Human blood. Blood drawn under aseptic conditions, containing not less than 12.5% (females) or 13.3% (males) hemoglobin. Protected from coagulation by addition of acid citrate dextrose, citrate phosphate or heparin (USP, EP, BP).

b. alkalinity The buffering property of b. B. is nearly neutral (pH = 7), but it can neutralize acidity because of its dissolved alkalies; see *buffer solution.* **b. amylase** A b. enzyme converting glycogen or starch to a reducing sugar. **b. capsules** Wright's capsules. A glass capillary tube used to take b. samples for microscopical examination. **b. cast** See *casts.* **b. clot** A clump of coagulated b. which contains the fibrin and b. cells. See *thrombin.* **b. corpuscles** B. cells. **b. count** The determination of the number of b. cells in a definite volume of b. **b. crystals** Hemin. **b. enzymes** Amylase, invertase, glycolytic enzymes, proteolytic enzymes, cholesterolesterases, and lipases. **b. gases** Carbon dioxide, oxygen, and nitrogen; e.g., b. of cats contains, the following percentages, by volume:

	CO_2	O_2	N_2
Arterial b.	25.07	13.60	1.00
Venous b.	40.83	9.93	0.77

b. glycolysis The breakdown by b. plasma of glucose into decomposition products. **b. groups** The classification of types of human b. in terms of the agglutinating behavior (clumping) of the red cells (corpuscles), in presence of the serum of a different b. The agglutinins in the b. serum of one group will not agglutinate with their own red cells, but only with the agglutinogens on the red cells of a foreign b. Principal agglutination reactions:

Serum of group (agglutinin)	Cells of blood group			
	A	B	AB	O
A	—	+	+	—
B	+	—	+	—
AB	—	—	—	—
O	+	+	+	—
% occurrence (Europeans)	42	9	3	46

Used to establish paternity, to identify bloodstains and other body fluids, e.g., saliva, and to ascertain the suitability of blood for transfusion. At present, 19 genetic markers, based on red cell antigens (6), serum proteins (4), and red cell enzymes (9) are used, giving a combined chance of exclusion of 93%. Cf. *agglutinins, haptoglobin, Rh factor* (under *Rh*). **b. hemoglobin** See *hemoglobin.* **b. hydrogen ion concentration** Mean value at 18°C: 6×10^{-8} to 2×10^{-8}; pH = 7.2–7.7. **b. osmotic pressure** 7.3 atm (freezing-point method). **b. pigment** Hemoglobin. Cf. *cytochrome.* **b. plasma** The liquid portion of the b. Composition (g per 1,000 g):

Water	901.51
Total solids	98.49
Albumin	81.92
Fibrin	8.06
Sodium chloride	5.546
Sodium carbonate	1.532
Calcium phosphate	0.298
Potassium chloride	0.359
Potassium sulfate	0.281
Sodium phosphate	0.271
Magnesium phosphate	0.218

b. platelets See *platelet.* **b. root** Sanguinaria. **b. serum** The liquid portion of the b. after removal of the fibrin, d.1.0292. Composition (g per 1,000 g):

Water	908.84
Total solids	91.16
Albumin, etc	82.59
Sodium chloride	5.591
Sodium carbonate	1.545
Potassium chloride	0.362
Calcium phosphate	0.300
Potassium sulfate	0.283
Sodium phosphate	0.273
Magnesium phosphate	0.220

b. serum, artificial Loeffler's mixture. A culture medium (250 ml glucose bouillon, 750 ml horse or beef serum). **b. sugar** The carbohydrates of b.; as, glucose. Loosely, blood glucose. **b. stone** (1) Hematite. (2) Jasper.

bloom (1) The fluorescence of lubricating oils. (2) The delicate coating which covers fresh fruit or leaves, as due to yeasts. (3) The cloudy appearance produced by aging on the surface of varnish. (4) The crystallization of a component on the surface of a material; as fat or sugar, on chocolate.

blowdown The sludge and/or concentrated feedwater removed periodically from the inside of a steam-raising boiler.

blown Describing canned foods that have expanded or burst their cans by liberation of gas (e.g., hydrogen) inside.

blown oils Polymerized oils, oxidized oils. An oil which has been oxidized by a stream of air, which converts it into a fast-drying oil; used for paints and varnishes.

blowpipe (1) A metal tube tapering to a fine point, used to blow air into a flame and to direct it as a fine conical tongue in qualitative or mineralogical analysis, soldering, melting in dentistry or jewelry manufacture. (2) A blast burner. **b. analysis** Qualitative analysis of minerals, alloys, or inorganic materials by observing their behavior in the b. flame, bead test, and reactions on charcoal or plaster of paris.

blubber oil Whale oil.

blue A spectrum of wavelength 0.000047 cm, between those of green and violet. **Chinese ~** Ferric hexacyanoferrate(II) in pigment form. **Egyptian ~** Powdered glass pigment consisting of copper oxide dissolved in a melt containing quartz sand, lime, and soda. Cf. *smalt.* **heteropoly ~** See molybdenum b.

b. cohosh Caulophyllum. **b. copper** CuS. A native, amorphous copper sulfide. **b. copperas** Copper sulfate. **b. cross** Ph_2AsCl. Diphenylchlorarsine, D.A. A nose irritant, m.39; a war poison gas. **b. dyes** See *alizarin b., alkali b., cyanin, methyl b., methylene b.* **b. ground** Kimberlite. **b. gum** Eucalyptus. **b. iron ore** Vivianite. **b. john** A native, crystalline copper sulfate with fluorspar. **b.print** A photographic copy made on ferric hexacyanoferrate(II) paper. **b. powder** Zinc dust. **b. salt** Crystallized *nickel sulfate.* **b.stone** A native, crystalline copper sulfate. Cf. *b. john.* **b. verdigris** Copper acetate. **b. verditer** Basic *copper carbonate.* **b. vitriol** (1) Chalcanthite, native copper sulfate. (2) Crystallized copper sulfate.

blushing The turbidity of lacquers and varnishes due to the precipitation of the resins by moisture or evaporation. Cf. *bloom (3).*

board A thick paper, usually taken as exceeding 0.009 in thick (U.S. usage) or 220 g/m² of weight (U.K. usage). **fiber ~** B. made by disintegrating vegetable matter into its elemental fibers and reassembling them as b. **paper ~** B.

made in a web on a continuous wire, as with paper, but with a greater thickness. **particle ~** (1) See *particleboard.* (2) Chipboard.

b. foot A unit of volume of boards sawn from logs; equal to the volume of a b. 1 in thick and 1 ft^2 in area, i.e., $\frac{1}{12}$ ft^3. Cf. *cunit, cord.*

Board of Trade unit B.T.U. Former British unit of electric energy. 1 B.T.U. = 1 kWh. Cf. *British thermal unit.*

boart Bort.

Boas reagent A solution of 5 g resorcinol and 3 g sugar in 100 g dilute alcohol. A test for hydrochloric acid in gastric juice; a rose-red color develops on boiling.

boat A small, elongated vessel of porcelain, quartz, tantalum, or platinum which can be inserted in a combustion tube. **b. conformation** See *conformation.*

bobierite $Mg_2P_2O_7 \cdot 8H_2O$. A native, crystalline phosphate in guano.

Bobina rayon Trademark for a viscose synthetic fiber.

Bobol Trademark for a viscose synthetic fiber.

B.O.D. Biochemical oxygen demand.

bodies (1) Biochemical substances of similar structure. (2) Cellular structures in protoplasm. **acetone ~** Substances, as, acetone, acetoacetic acid, or 2-hydroxybutanoic acid, in urine. **alloxur ~** A compound of uric acid and alloxan, secreted in the urine; as, the purine b. **Buchner ~** A defensive protein of the organism. **purine ~** A derivative of uric acid; as, xanthine.

body (1) The trunk of an animal or plant. (2) The largest part of an organ. (3) The consistency or viscosity of a liquid. (4) A limited portion of matter. (5) The strength of a liquid; as, wine. **black ~** See *Stefan-Boltzmann equation, black body.*

b. fluids See *blood, lymph.* **b. tube** The portion of the microscope which carries the objective, and inside which slides the draw tube.

boehmite AlO·OH. A form of bauxite.

Boerhaave, Herman (1668–1738) Dutch pioneer of modern chemistry, noted for his textbook.

bog A marsh or morass. **b. berry** The fruit of *Vaccinium oxycoccus,* cranberry. **b. butter** Butyrelite. A soft mineral occurring in marshes. **b. iron ore** Bogore. **b. manganese** Wad. **b. ore** Bogore.

Bogert, Marston Taylor (1868–1957) American chemist, noted for his work on organic synthesis.

boghead A carbonaceous rock, or cannel with a high iron carbonate content. **b. naphtha** Photogen.

bog iron ore Bogore.

bog manganese Wad.

bogore $2Fe_2O_3 \cdot 3H_2O$. Bog iron ore, marsh ore, brown iron ore, brown hematite, brown ocher, limonite. A hydrous ferric oxide with some ferrous carbonate, from marshy places; a source of iron.

Bohr, Niels (1885–1962) Danish physicist; Nobel prize (1922) for his theory of atomic structures. **B. atom** A hypothesis of *atomic* structure, q.v.; the electrons move in circular or elliptical orbits around a positive nucleus, resembling a very small solar system. **B. magneton** See *magneton.* **B. radius** The radius of the ground state orbit of the hydrogen atom, $a_0 = 5.29177 \times 10^{-11}$ m. **B. theory** Spectrum lines are produced (1) by emission of radiation (energy) when electrons drop from an orbit of greater to lower energy (energy levels); or (2) by absorption of radiation (energy) when the electrons move from an orbit of lower to higher energy. Cf. *quantum, spectrum series, Stoner quanta, correspondence principle.*

boil Quick ebullition or vaporization of a liquid by heat and/or low pressure.

boiled oil Linseed oil which has been heated to 210–260°C, and thereby hardens more readily; used in varnishes and lacquers.

boiler An open or closed vessel for evaporating liquids, cooking food, or generating steam. **vacuum ~** A closed b. in which evaporation of a liquid is caused by low pressure, with or without heat.

b. compound A substance used to prevent the formation of b. scale. Cf. *water softening.* **b. fluid** A solution which prevents the formation of a compact b. incrustation. **b. incrustation** The insoluble mass, deposited on the sides of a vessel in which hard water has been evaporated, of calcium and magnesium carbonates and sulfates. **b. mud** A loose deposit of b. incrustations. **b. scale** A compact, thick layer of successive deposits. **b. stone** B. scale.

boilers Group name for nitrocellulose and lacquer *solvents,* q.v., arranged in order of their boiling points: **low ~** B. below 100°C. **medium ~** B. near 125°C. **high ~** B. from 150–200°C.

boiling The state of ebullition; the brisk change from the liquid to the vapor state.

boiling point B.p., b. The temperature at which, under a specified pressure, a liquid is transformed into a vapor; i.e., at which the vapor pressure of the liquid equals that of the surrounding gas or vapor. Cf. *Clapeyron equation.* **b. p. apparatus** A device to determine the b.p. of a liquid under a definite pressure. **lowering of b. p** The decrease in pressure from lowering the b.p.; e.g., by lowering the pressure 10 mm, water will boil 0.37°C lower. **b. p. elevation** The raising of the b.p. of a liquid because of the presence of a dissolved substance. The rise is a function of the substance's molecular weight and may be used to determine it; 1 mole/liter in water causes a rise of 0.51°C. Cf. *Raoult's Law, Beckmann apparatus.*

Boisbaudran, Paul Émile Lecoq de (1838–1912) French scientist, discoverer of gallium, samarium, and dysprosium.

boldine $C_{19}H_{21}NO_4$ = 327.4. An alkaloid from the leaves of *Peumus boldus* (Monimiaceae). Gray powder, insoluble in water. Cf. *laurotetanine.*

boldoglucin $C_{30}H_{53}O_8$ = 541.7. A glucoside from the leaves of *Peumus boldus.* A thick syrup.

bole (1) A fine clay, colored by iron. (2) The trunk or stem of a tree. (3) A measure of corn; 6 bushels. **red ~** Ocher. **white ~** Kaolin.

boleite A native hydrous oxychloride of lead, silver, and copper; blue crystalline masses (Boleo, lower California).

Boletus A genus of edible fungi or mushrooms (Basidiomycetes).

Bologna phosphorus Luminescent barium sulfide.

bolometer A device for measuring minute quantities of radiant heat from the change in the conductivity of a black body.

Boltaflex Trademark for a mixed-polymer synthetic fiber.

bolting cloth A fabric of unsized silk, used for sieves.

boltonite Mg_2SiO_4. A native variety of fosterite.

Boltzmann **B. constant** $k = 1.3807 \times 10^{-23}$ J/K. The gas constant, R, divided by the Avogadro constant. Cf. *Maxwell-B. distribution law.* **B. equation** See *molecular free path.* **B. law** The law of equipartition of energy: The total kinetic energy of a system, due to translation, rotation, vibration, etc., is equally divided among all the degrees of freedom. The energy per degree of freedom is 0.5 RT per gram molecule.

bolus (1) Masticated food ready to be swallowed. (2) Kaolin. (3) A small, rounded mass.

bomb (1) A projectile of iron or steel filled with a nuclear device, or explosive, poisonous, or incendiary substances; may be used in chemical warfare. (2) A heavy iron tube containing

a substance which is to be oxidized for the determination of its calorific value. **atom ~** See *atom bomb.* **cobalt ~** See *cobalt bomb.* **hydrogen ~** See *hydrogen bomb.*
b. calorimeter See *bomb calorimeter* under *calorimeter.*

bombard To expose to rays, e.g., of radioactive substances, or converging cathode rays.

bombardment Exposure to a radioactive substance, e.g., to cathode rays focused on a point as in an X-ray tube; or the hitting of atomic nuclei by high-speed α particles.

bombazine A fabric having a silk warp and a cotton, linen, or woolen weft; similar to, but lighter than, poplin.

bombiosterol $C_{27}H_{46}O$ = 386.7. A sterol, m.148, in chrysalis oil.

bombykol An alcohol pheromone secreted by female moths to attract males; 500,000 scent glands yield 10 mg.

bonanza A rich vein of ore.

bond The linkage between atoms, thought to consist of an electron pair distributed between 2 nuclei and forming an electromagnetic vector along axis *ab:*

$$+\underset{a}{\text{——}}:\underset{b}{\text{——}}+$$

1. *Atomic b.* Each atom contributes one electron:
   (*a*) *Homopolar b.* (nonpolar). The electron pair is held equally by both nuclei, neither of which becomes negative with respect to the other, $a = b$; as in H_2, CH_4.
   (*b*) *Heteropolar b.* (polar). The electron pair is held unequally; hence one nucleus becomes negative and the other positive, $a > b$; as in HCl.
2. *Molecular b.* One atom contributes both electrons:
   Coordinate b. (semipolar). An unshared electron pair of an octet on a nucleus (as of N, O, F) is shared by a nucleus having an incomplete octet (generally H; also Li, Be, B, etc.).

In *electronic structure symbols,* dots represent the electrons of the particular b. However, it is more strictly correct to represent the bonds by *orbitals,* q.v. Cf. *valency, linkage, combination, compound, atomic radius, chelate.* **hydrogen ~** Weak bond between a H atom bonded to an electronegative atom, and a second electronegative atom. Both non-H atoms are usually N, O, or F. **π ~** A 3-dimensional b. produced by the overlap of 2 or more orbitals with their nodes in the same plane. See molecular *orbital.* **σ ~** A b. produced by the overlap of 2 directional orbitals. Types: bonding and antibonding, the latter having the higher energy level. Two electrons in the same state will oscillate between these states.

bonded fiber Web. Generic term to describe materials produced by assembling fibers (especially textile fibers) together without weaving. Bonding is achieved by chemical action, extraneous adhesives, or thermoplastic fibers. Used for diapers and filter cloths. See *Bonlinn.* Cf. *nonwoven.*

bonducin $C_{14}H_{15}O_5$ = 263.3. A bitter principle of bonduc seeds, *Caesalpinia crista* (Leguminosae). White, bitter powder, insoluble in water; a febrifuge.

bondur A corrosion-resisting aluminum alloy: Cu 2–4, Mn 0.3–0.6, Mg 0.5–0.9%. Cf. *acieral.*

bone The skeletal material of the vertebrates. **b. ash** Impure calcium phosphate. The remains of burnt animal bones; a fertilizer (35–38% P_2O_5). **b. black** A usually impure charcoal made from bones and blood; used to refine sugar, oil, etc. **b. earth** B. ash. **b. meal** Finely ground animal bones used as fertilizer (N 3.3–4.1, P_2O_5 20–25%). **steamed ~** Finely ground bones, previously steamed under pressure to remove the glue (N 1.6–2.5, P_2O_5 25%). **b. oil** Animal oil, Dippel's oil. A tarry oil obtained by dry distillation of bones, d.0.900–0.980, soluble in water; chief constituent, pyridine. Used as an insecticide and in organic synthesis. **b. phosphate** Calcium phosphate. **b. tallow** Soft grease obtained by boiling fresh bones; used to make cheap soaps. **b. turquoise** Fossil bones or teeth colored with $Fe_3P_2O_8$.

bongkrekik acid $C_{28}H_{38}O_7$. Produced by the action of certain bacteria on coconuts.

Bonlinn Trademark for a bonded synthetic fiber web.

Bonney's blue paint A 1% solution of brilliant green and/or crystal violet in a mixture of equal parts of rectified spirit and water. Used to sterilize skin prior to surgery.

boost See *synergist.*

Boot density bottle A specific-gravity bottle, capacity about 5 ml, having double walls with a vacuum between. Used to determine specific gravities at constant temperature.

boracic acid Boric acid.

boracite $2Mg_3B_8O_{15}\cdot MgCl_2$. White, transparent, isometric crystals (Stassfurt); hardness 7, d.2.9–3.

boracium Original name of boron (Davy).

Boraginaceae Borage family, a group of herbs, some of which are used in cooking and in traditional remedies; e.g., *Alkanna tinctoria,* alkannin; *Borago officinalis,* borage; *Pulmonaria officinalis,* lungwort.

boranate Tetrahydro*borate**.

borane(s) (1)* Borane. BH_3 = 13.8. Found as a gaseous transient, it is analogous to CH_3. (2)* Boranes. Hydroborons. Collective name for the boron hydrides, which are analogous to the alkanes and silanes. Numerous b. are known. Some have high calorific values and are used in high-energy fuels. **deca ~ (14)*** $B_{10}H_{14}$ = 122.2. m.100. **di ~ *** B_2H_6 = 27.7. Boroethane. Colorless gas, b. − 93. Unstable in air and decomp. by alcohol. **penta ~** B_5H_9 = 63.1. b.60. B_5H_{11} = 65.1. b.65. **tetra ~ (10)*** B_4H_{10} = 53.3. Borobutane. b.18. **tetradeca ~ (18)*** $B_{14}H_{18}$ = 169.5. A liquid. **triethyl ~ *** Et_3B = 98.0. Triethylborine. b.95.

boranediyl* The radical HB=.

boranetriyl* The radical B≡.

borata-* Indicating a boron anion.

borate Indicating boron as the central atom(s) in an anion. **bi ~** Dihydrogenborate*. **di ~ *** Pyroborate $M_4B_2O_5$. **dihydrogen ~ *** MH_2BO_3. Acid borate, biborate. **hydrogen ~ *** M_2HBO_3. Acid borate. **meta ~ *** A salt containing the $-BO_2$ or B_3O_6≡ radical. **ortho ~ *** Borate. A salt of boric acid. **peroxo ~ *** A salt containing the $-BO_4$ radical. Usually an oxidizing agent resembling hydrogen peroxide; used for bleaching. **pyro ~** Diborate*. **tetra ~ *** A salt containing the $=B_4O_7$ radical. **tetrahydro ~ *** Boranate, borohydride. A salt containing the $-BH_4$ radical.

boratto A silk and wool fabric.

borax $Na_2B_4O_7\cdot 10H_2O$. Zala, tinkal. A native sodium tetraborate, found in California and Asia Minor; a cleaning agent, flux, etc. (EP, BP). **b. bead** See *bead test.* **b. carmine** See *Grenacher stain, Nikiforoff stain.* **b. glass** (1) Fused borax, used as a flux. (2) See *borax glass* under *glass.* **b. methylene blue** Sahli stain.

borazine* $B_3N_3H_6$ = 80.5. Borazole. Inorganic benzene.

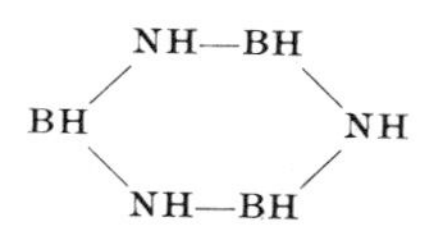

An isostere of benzene prepared by heating ammonia and diborane at 200°. Colorless mobile liquid, m.−58, b.55.

Borazole Borazine*.

bordeaux **b. B** Acid B, α-naphthalene-azo-β-naphthol-3,6-

disulfonic acid. An indicator for pH 10.5 (pink) to 12.5 (orange). **b. colors** Artificial coloring matters for foodstuffs. **b. mixture** A mixture of equal weights of copper sulfate and lime in water; a fungicide.

Bordetella Hemophilus. A genus of bacteria, small, gram-negative rods or cocci. Species of B. cause meningitis and pneumonia; *B. pertussis* causes whooping cough.

Bordet test An agglutination test to differentiate human and animal bloods.

borethyl Triethyl*borane*.

boric acid* $B(OH)_3$ = 61.8. Orthoboric acid*, boracic acid, fumarole acid. Triclinic, white crystals, m.185, soluble in water. Used as a reagent and to manufacture borates (NF, EP, BP). Found in the volcanic lagoons of Tuscany. **benzyl ~** $C_7H_7B(OH)_2$ = 136.0. Benzylboron dihydroxide. White crystals, m.161. **di ~*** $H_4B_2O_5$. **ethyl ~** $C_2H_4B(OH)_2$ = 72.9. Ethyl boron dihydroxide. Colorless crystals, sublime 40. **meta- ~*** HBO_2 or $B_3O_3(OH)_3$. **ortho- ~*** B. acid. **peroxo ~*** HBO_4. **pyro ~** Diboric acid*. **tetra ~*** $H_2B_4O_7$.

borickite Anhydrous phosphate of iron.

boride M_3B or BR_3. A binary compound of negative boron with a more positive element or radical.

boriding Boronzing.

borine BR_3. A substituted borane compound.

borium Boron.

bormethyl Trimethylborane.

bornane* $C_{10}H_{18}$ = 138.3. Camphane, bornylane. White

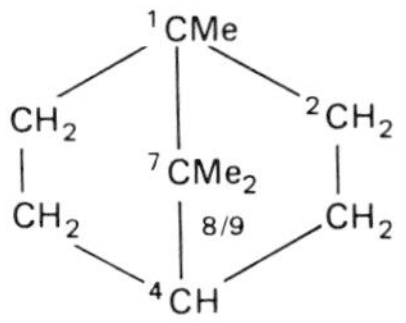

[1]CMe; CH_2, [2]CH_2; [7]CMe_2; CH_2, 8/9, CH_2; [4]CH

crystals, m.154, soluble in alcohol. **2-amino ~** Bornyl amine. **2-chloro ~** Bornyl chloride. **2,3-dihydroxy ~** Camphene glycol. **hydroxy ~** **2- ~** Borneol*. **3- ~** Epi*borneol*. **4- ~** Iso*borneol*. **oxo ~** **2- ~** Camphor. **3- ~** Epicamphor.

borneene **n- ~** 5-Methyl-2-norbornene. **iso ~** 6-Methylenenorbornene.

bornene* $C_{10}H_{16}$ = 136.2. 1,7,7-Trimethylbicyclo[2.2.1]heptene. **Δ^2- ~** Bornylene. Colorless crystals, m.110, insoluble in water.

Borneo camphor (+)-Borneol. **B. tallow** See *Borneo tallow* under *tallow*.

borneol* $C_{10}H_{18}O$ = 154.3. 1,7,7-Trimethylbicyclo[2.2.1]-heptan-2-ol. **(+)- ~** (1*R*,2*S*) form. Camphyl alcohol, Borneo camphor, baras camphor, sumatras camphor, bornyl alcohol, 2-hydroxycamphane, 2-camphanol,

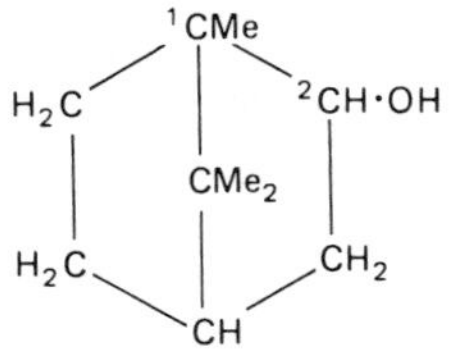

[1]CMe; H_2C, [2]CH·OH; CMe_2; H_2C, CH_2; CH

A terpene from *Dryobalanops camphora*, or prepared synthetically. Transparent leaflets, m.208, slightly soluble in water. **epi ~** The (1*S*,3*R*,4*S*) form of 1,7,7-trimethylbicyclo[2.2.1]heptan-3-ol. **iso ~** 3 enantiomers. 4-Hydroxycamphane, m.210. Used in perfumery and celluloid manufacture. **b. acetate** Bornyl acetate*. **b. salicylate** Bornyl salicylate*.

bornesitol Quebrachitol.

Born-Haber cycle The relationship of U = lattice energy of crystals, I = ionization energy, E = electron affinity, S = heat of sublimation, D = heat of dissociation, and Q = chemical heat of formation can be expressed by the diagram:

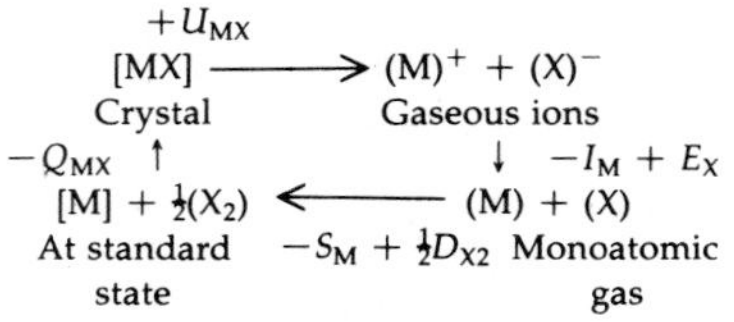

$$\begin{array}{ccc} & +U_{MX} & \\ [MX] & \longrightarrow & (M)^+ + (X)^- \\ \text{Crystal} & & \text{Gaseous ions} \\ -Q_{MX}\ \uparrow & & \downarrow\ -I_M + E_X \\ [M] + \frac{1}{2}(X_2) & \longleftarrow & (M) + (X) \\ \text{At standard state} & -S_M + \frac{1}{2}D_{X2} & \text{Monoatomic gas} \end{array}$$

bornite $FeS\cdot 2Cu_2S\cdot CuS$. Peacock copper, purple copper. A native copper iron sulfide. Red-brown, brittle, isometric crystals, d.5.0.

bornyl* Bornylyl, camphyl. The $-C_{10}H_{17}$ radical, derived from bornane. **b. acetate*** $MeCOOC_{10}H_{17}$ = 196.3. (+)-Borneol acetic ester. Colorless crystals, m.29, slightly soluble in water. **b. amine** $C_{10}H_{17}NH_2$ = 153.3. 2-Aminobornane. White crystals, m.163, soluble in water. **b. chloride** $C_{10}H_{17}Cl$ = 172.7. Pinene hydrochloride, 2-chlorobornane. Colorless crystals, m.158 decomp. by water at 49. **b. salicylate*** $C_6H_4(OH)COOC_{10}H_{17}$ = 274.4. Salit, *d*-borneol salicylic ester. Colorless solid used externally mixed with equal parts of olive oil, for rheumatism.

bornylene See *bornene*.

borobutane Tetra*borane*(10)*.

borocalcite $CaB_4O_7\cdot 4H_2O$. Bechilite. A native borate.

borofluohydric acid Tetrafluoroboric acid.

borofluoric acid Fluoroboric acid.

boroglyceride Glyceryl borate. An antiseptic paste of boric acid and glycerol.

borohydride Tetrahydro*borate**.

borol Sodium and potassium borosulfate. Colorless, transparent masses, soluble in water; a disinfectant and antiseptic.

borolon An artificial aluminum oxide; white or brown-red crystalline masses, d.3.9-4.0, obtained by fusing bauxite. An abrasive, refractory, or filtering material.

boron* B = 10.811. A metal of the aluminum group, at. no. 5, analogous to carbon; modifications:

1. *Amorphous b.* Gray powder, d.2.45, m. above 2000, b. about 3500; burns in air at about 700; insoluble in water, soluble in acids (decomp.).
2. α-*b.* Rhombohedral, dense, m.800–1200.
3. β-*b.* Rhombohedral, m.2250.
4. *Crystalline b.* Colorless tetragons of great hardness, d.2.51, sometimes variously colored by impurities; insoluble in acids and alkalies, and slowly soluble in molten alkali carbonates.

B. occurs widely distributed in small quantities in several silicates (tourmaline) and as borates (borax, ulexite), and in many alkaline lakes (California, Tibet). B. forms only one (trivalent) series of compounds.

Boride*	M_xB_y
Borate* (q.v.)	$B(OM)_3$
Borane*	B_xH_y

B. was discovered (1807) independently by Davy and Gay-Lussac and Thenard. Crystalline b. was first prepared by Woehler (1856). Metallic b. is used as an industrial catalyst; in metallurgy to give hardness; and because it absorbs neutrons, in atomic reactors. **tetraphenyl ~** A chemical reagent, particularly for the gravimetric determination of potassium. **triethoxy ~** Ethyl borate. **triethyl ~** Triethyl*borane**. **trimethoxy ~** Methyl borate.

b. alkyls See *borane.* **b. bromide*** BBr_3 = 250.5. A colorless, fuming liquid, d.2.69, b.90, decomp. by water. **b. carbide*** B_4C = 55.3. Black crystals, d.2.51; m.2350, insoluble in water. **b. chloride*** BCl_3 = 117.2. Colorless liquid, d.1.434, b.18, decomp. by water. **b. fibers** The boron analog of carbon fibers. See *synthetic graphite* under *graphite.* **b. fluoride** BF_3 = 67.8. Colorless gas, $d_{(air-1)}2.3$, b.−101. **b. hydrides** Boranes*. **b. hydroxide** Boric acid. **b. iodide*** BI_3 = 391.5. Colorless, crystalline scales, m.43, decomp. by water. **b. minerals** Chiefly borax, $Na_2B_4O_7 \cdot 10H_2O$; ulexite, $NaCaB_5O_9 \cdot 8H_2O$; borocalcite, $CaB_4O_7 \cdot 4H_2O$; sassolite, H_3BO_3; boracite, $2Mg_3B_8O_{15} \cdot MgCl_2$. **b. nitride*** BN = 24.8. Colorless, infusible, crystals, insoluble in acids or alkalies (decomp. by hydrofluoric acid). **b. oxide** B_2O_3 = 69.6. Colorless powder, m.577, slightly soluble in water or acids. **b. phosphide*** BP = 41.8. Red powder, burns at 200, insoluble in water. **b. sulfide*** B_2S_3 = 117.8. Colorless crystals, m.310, decomp. in water. **b. tribromide*** B. bromide. **b. trichloride*** B. chloride. **b. trisulfide*** B. sulfide.

boronatrocalcite Ulexite.

boronzing Boriding. The production of a hard surface layer of boride on a metal object by a process involving diffusion of boron.

borosalicylates

$$C_6H_4\langle\begin{matrix}O\\CO\cdot O\end{matrix}\rangle B-B\langle\begin{matrix}O\cdot CO\\O\end{matrix}\rangle C_6H_4$$

borosilicate Silicoborate. A salt of boric and silicic acids.

borotungstic acid $(WO_3)_9B_2O_3 \cdot 24H_2O$. Yellow liquid, d.3.0, soluble in water. Used to determine the density of minerals.

borowolframic acid Borotungstic acid.

boroxine A ring compound with alternate B and O atoms in the same plane.

borphenyl Triphenylborane*.

Borrel grinder A device for grinding organic tissues by means of flexible steel leaves which rotate at a high speed in a steel cylinder.

bort Anthracite diamond, carbonado. A dark, lustrous conglomerate of minute diamonds from Brazil; used for cutting stones and in boring machines.

boryl* Boranyl. The radical $-BH_2$.

borylia Prefix indicating a cationic boron ion.

boss Clampholder.

boswellic acid (1) $C_{30}H_{48}O_3$ = 456.7. Crystalline solid from frankincense. (2) $C_{32}H_{52}O_4$ = 500.5. An acid constituent of African olibanum, the resinous exudate of *Boswellia papyrifera* (Burseraceae).

botany (1) The science of the structure, function, occurrence, and classification of plants and vegetable organisms. (2) Waste jute cuttings for paper manufacture. Cf. *hessian.*

Botany Bay gum Acaroid resin.

botryolite Datolite.

botrytized See *wine.*

Böttcher chamber A counting apparatus for blood cells and bacteria; a microscope slide with ruled squares.

Böttger test A test for glucose in urine; a black precipitate results with sodium carbonate and bismuth subnitrate.

bottle A vessel with a narrow neck. **aspirator ~** See *aspirator.* **balsam ~** A small, wide-necked b. with a loosely fitting glass cover. **density ~** See *specific gravity.* **dropping ~** A b. with a pipet fitted into its stopper. **gas ~** A b. used for generating or washing gases, which usually has a two-hole stopper for the inlet and outlet tubing. **graduated ~** A graduated b. used for mixing liquids. **hard rubber ~** A b. made of rubber and used for certain acids. **immersion oil ~** Any of a group of b.'s of various shapes, with a glass rod attached to the stopper. **milk testing ~** Babcock b. **oil sample ~** A long, narrow b. **percolator ~** A widemouthed b., graduated in milliliters, pints, or ounces. **reagent ~** A glass b. with the name and symbol of a reagent etched on. **specific gravity ~** A small, light, accurately counterpoised and graduated b. used to determine the weight of a given volume of liquid. **specimen ~** A widemouthed b. with a closure, for holding specimens. **washing ~** A glass b. fitted with an inlet and outlet tube. The latter reaches to the bottom and at the other end has a small jet. Used for washing precipitates, or as a bubbler for washing gases. **weighing ~** A small, light, glass-stoppered container, used for weighing liquids or solids. **Woulfe ~** A glass b. with 2 or 3 necks, used as a washing or gas-generating bottle.

bottlenose oil An inferior sperm oil from the blubber of the bottlenose whale, used in soapmaking.

botulin A ptomaine produced by bacteria *(Bacillus botulinus)* and sometimes found in tinned and preserved meats.

botulism Food poisoning due to *Clostridium botulinum,* the only microorganism with spores having a high heat resistance; it occurs in canned foods.

Bouchardat, Alexander (1806–1886) French pharmacist noted for methods of urine analysis **B. reagent** A solution of 1 pt. iodine and 2 pts. potassium iodide in 20 pts. water gives a brown precipitate with alkaloids.

bougie (1) A filter cylinder made of porous porcelain. Cf. *Berkefeld filter.* (2) A taper-shaped pharmaceutical preparation for introduction into the rectum or urethra. (3) A narrow cylinder for dilation of the urethra or other body orifice. **b. unit** (French for "candle") Former French photometric standard; 0.05% of the light emitted by 1 cm^2 of platinum at its solidifying point.

bouillon Meat broth, used as food or culture medium. **glycerol ~** Koch's culture medium: 10 g Liebig meat extract, 10 g peptone, 20 ml glycerol, 1,000 ml water, and sufficient sodium carbonate solution to make alkaline to litmus. **plain ~** A culture medium: 10 g peptone; 5 g sodium chloride, and 1,000 ml bouillon stock solution, neutralized with sodium hydroxide.

b. stock A solution of 500 g lean beef in 1,000 ml water, used for culture media.

Bouin's fluid A preservative for embryological and histological material: 75 ml saturated picric acid, 25 ml 40% formaldehyde, and 5 ml glacial acetic acid (pH 1.6).

boulangerite $3PbS \cdot 2Sb_2S_3$. A native lead sulfantimonate. Gray needles or feathery masses, d.6.18. Cf. *epiboulangerite.*

bourbonal $CHO \cdot C_6H_3(OH) \cdot OEt$ = 166.2. Ethyl vanillin-3-ethoxy-4-hydroxybenzaldehyde. A synthetic substitute for vanillin, stronger in flavor.

bournonite $CuPbSbS_3$. Bluish-gray, brittle, native copper lead sulfantimonite.

boussingaultite $(NH_4)_2Mg(SO_4)_2 \cdot 6H_2O$. A native magnesium sulfate (Tuscany).

Bouveault-Blanc reaction The reduction of esters to alcohols by metallic sodium.

BOV Brown oil of vitriol. Commercial sulfuric acid (77–78% H_2SO_4 by weight).

bowenite Serpentine.

Bowen's resin A polyester, made from glycidyl methacrylate and bisphenol A; a dental filling resin.
Bowen tube Bowen potash bulb. A tube with bulbs for the absorption of gases.
Boyce burner An adjustable burner which regulates the gas and air flows.
Boyle, The Hon. Robert (1627–1691) English pioneer in the investigation of gas laws. **B.'s law** If the temperature is constant, the pressure of a given quantity of a gas is inversely proportional to the volume it occupies. pV = constant.
B.P. Abbreviation for (1) British Pharmacopoeia; (2) blood pressure; (3) the beriberi-preventing factor of the vitamin B group, e.g., thiamin.
b.p., bp Abbreviation for boiling point.
B powder Soda powder.
bpy* Indicating bipyridine as a ligand.
Bq Symbol for becquerel.
Br Symbol for bromine. Br_2 = bromine molecule. Br^- = bromide ion. Br* = excited bromine atom.
brachydome See *brachydome* under *dome.*
brackish Describing water having a chloride content exceeding about 2,000 ppm. Cf. *water.*
Bragg, Sir William Henry (1862–1942) British chemist, Nobel prize winner (1915); noted for research on crystal structure. **B. crystallogram** The photographic record obtained by B.'s method. **B. crystal model** Crystalline structure as determined by the diffraction of monochromatic X-rays. **B. method** If a crystal is placed in the path of a narrow X-ray beam, the layers of atoms in the crystal act as reflection planes for the incident ray and a series of lines, corresponding with the several orders of spectrum, will be produced on the photographic plate. $\lambda = 2d \sin \theta$, where θ is the glancing angle, d the spacings between atomic planes, and λ the wavelength. Cf. *X-ray spectrograph, crystal structure, Laue pattern.*
brain The nerve tissues in the skull which consist, chemically, of (1) white b. substance (cephalins, lecithins, paramyelins, myelins, cholesterol, phrenosterol, cerebrin, and cerebrosides); (2) buttery substance (cephaloidin, lecithin, myelin, paramyelin, aminomyelin, sphingomyelin, phrenosin, and aminolipins); (3) aqueous b. extract, containing alkaloids (hypoxanthine, etc.), amino acids, inositol, organic and inorganic acids and salts. See *cerebellum, cerebrum.* **b. death** Condition showing no spontaneous respiration, even after artificial ventilation with 5% carbon dioxide, and no brain stem reflexes. **b. sugar** Cerebrose.
bran The husk or outer covering of the grain of wheat. **b. oil** 2-Furaldehyde*.
branched chain See *branched chain* under *chain.*
Brand, Hennig An alchemist of Hamburg; discovered the first nonmetallic element (phosphorus) in 1669.
Brandt, Georg (1694–1768) Swedish mineralogist; discovered the first metallic element (cobalt) in 1735.
brandy An alcoholic beverage distilled from wine. Cf. *cognac.*
brasan Brazan.
brasileic acid $Me(CH_2)_7(CHOH)_2(CH_2)_{12}COOH$ = 386.4. Isodihydroxybehenic acid. Yellow crystals, m.99, soluble in hot alcohol. Cf. *brasilic* and *brassylic acids.*
brasileïn Brazileïn.
brasilic acid $C_{12}H_{12}O_6$ = 252.2 Brazilic acid. Colorless crystals, m.129. Cf. *brasileic, brassylic,* and *brassidic* acids.
brasilin Brazilin.
brass A copper-base alloy containing zinc. In the classics, an alloy of copper and tin. **alpha ~** B. containing less than 40% Cu. **beta ~** B. containing more than 40% Cu. **aluminum ~** An alloy: Cu 55–76, Zn 25–45, Al 1–4%. **calamine ~** Marcasite. **cartridge ~** An alloy: Cu 70, Zn 30%. **coal ~** Roman b. **iron ~** B. containing 1–9% Fe. **naval ~** An alloy: Cu 61, Zn 38, Sn 1%. **Roman ~** B. made by heating charcoal, copper, and calamine below 1000°C. **yellow ~** Muntz metal.

90% Cu, 10% Zn, red brass	m. 1040
80% Cu, 20% Zn, Dutch metal	m. 995
67% Cu, 33% Zn, yellow, ordinary b.	m. 940
60% Cu, 40% Zn, Muntz metal	m. 900

b. stone $FeCO_3$. A native carbonate of iron.
brassic See *brassidic acid.*
brassicasterol $C_{28}H_{46}O$ = 398.7. A sterol, from rape oil, m.148.
brassidic acid $Me(CH_2)_7CH{:}CH(CH_2)_{11}COOH$ = 338.6. 12-Docosenoic acid*, isoerucic acid. Colorless leaflets, m.65, slightly soluble in water. An isomer of erucic acid and cetoleic acid. Cf. *brasilic acid* and *brasileic acid.*
brassil Pyrite.
brassylic acid $HOOC(CH_2)_{11}COOH$ = 244.3. 1,11-Undecanedicarboxylic acid. Colorless crystals, m.114, slightly soluble in water. Cf. *brasileic* and *brasilic acids.*
Brauner, Bohuslav (1855–1935) Czechoslovakian chemist; noted for inorganic research and atomic weight determinations.
braunite $MnSiO_3 \cdot 3Mn_2O_3$. A native manganese silicate. Brown or gray, lustrous tetragons.
Braun tube (1)A potash bulb. (2) Former name for cathode ray tube; named for the inventor.
bravaisite Glauconite
Bravais lattice Space lattice
Bray, William Crowell (1879–1946) American chemist, noted for research in catalysis, qualitative analysis, and chemical kinetics.
Brayera A genus of Rosaceae. See *brayerin, kousso.*
brayerin $C_{31}H_{38}O_{10}$ = 570.6. A bitter, resinous principle from *Brayera* species.
brazan $C_6H_4 \cdot O \cdot C_{10}H_6$ = 218.3. Brasan, phenylenenaphthylene oxide.
braze To solder with an alloy of Cu and Zn. Cf. *solder.*
Brazil **B. gum** Angico gum. **B. nut** The edible seeds of *Bertholletia excelsa* (Myrtaceae). **b. nut oil** Castanhao oil. **b. wax** Carnauba wax. **b.wood** *See brazilwood.*
brazilein $C_{16}H_{12}O_5$ = 284.3. Dark-red crystals obtained by oxidation of brazilin; it resembles hematoxylin.
brazilianite $Na_2Al_6P_4O_{16}(OH)_8$. A monoclinic gemstone, similar to chrysoberyl in appearance (Brazil).
brazilic acid Brasilic acid.
brazilin $C_{16}H_{14}O_5 \cdot 1\frac{1}{2}H_2O$ = 313.3. Brasilin. The coloring matter of brazilwood. Colorless or yellow needles, m.250, soluble in water. Used as a dye and indicator (acids—yellow, alkalies—purple).
brazilite Baddeleyite
brazilwood (1) The heartwood of *Peltophorum dubium* (Leguminosae) of S. America. A source of brazilin. (2) A redwood of *Caesalpinia* species of S. America. **yellow ~** Morus tinctoria.
brazing Brazeing. A process for joining metals in which a molten filler metal is drawn by capillary attraction into the space between closely adjacent surfaces of the parts to be joined. The m. of the filler usually exceeds 500°C. Cf. *soldering.*
break-point See *break-point chlorination* under *chlorination.*
breath alcohol Ethanol in the breath. Legal limit in most countries falls in the range 40–150 mg/100 ml of blood.

Breath testing is based on the ratio of 1:2100 to 1:2300 for the distribution of alcohol between alveolar air and blood. Measuring methods include changes in semiconductor resistivity (Alcocheck), oxidation by fuel cell (Alcolmeter), i.r. absorption (Intoxilyzer), g.c. (Intoximeter), and color changes in acidic potassium dichromate (Breathalyzer, Ethanograph).

Breathalyzer See *breath alcohol.*

breccia Angular rock fragments in a finer matrix.

breeze Coke, 19–25 mm in size, from which no further size-graded product is removed. Used for steam raising, gas producers, briquetting, wall making, and for breaking up heavy soils.

bregenin $C_{40}H_{87}O_5N$ = 662.1. A phospholipid from brain substance.

breithauptite NiSb. A native antimonide.

Bremen **B. blue** Blue *copper carbonate.* **B. green** $Cu(OH)_2$. Verditer. A green pigment produced by the weathering of copper; or artificially, by the action of alkali with copper sulfate.

bremsstrahlung Radiation produced when β particles are stopped. They are similar in properties to low-energy X-rays, and are a body hazard.

Brenkona Trademark for a continuous-filament luster rayon yarn.

brenzcatechin Catechol.

breunnerite $MgFe(CO_3)$. A native mixture of siderite and magnesite.

brevium Early name for protactinium.

brewing The process by which malted grain is treated with hot water to produce an extract (wort). This is boiled with hops, filtered, and fermented with yeast.

brewsterite A rock-forming zeolite mineral.

Brewster's **B. angle** The particular angle of incidence p at which the reflection of approximately monochromatic, plane-polarized light from a surface is at 90° to the path of the transmitted light. Its tangent equals the refractive index n of the material. **B. law** Tan $p = n$.

Bricanyl Trademark for terbutaline sulfate.

brick A baked clay molded in various, generally rectangular, shapes. **fire ~** A b. used for furnace linings. **silica ~** A b. used as lining for high-temperature appliances. **Sil-O-Cel ~** Trademark for an insulating b.

bridge (1)* A valence bond, atom, or unbranched chain of atoms connecting 2 different parts of a molecule. Cf. *ring structures.* (2) A connecting device; as, *Kelvin b., Wheatstone b.* **b.head*** Each of the 2 atoms at which a b. contacts the molecule.

brightening Subjecting dyed materials to 35 kPa steam pressure to brighten the color.

brightness The intensity of *light* or *color,* q.v.; hence, the amount of light emitted or reflected by an object. Surface brightness is measured in candela/m^2 or lamberts. **absolute ~**, **apparent ~**, **photographic ~** See *spectral classification.*

Briglo Trademark for a viscose synthetic fiber.

brilliance Brightness.

brilliant Intensely bright. **b. cresyl blue** A dye used as a stain for blood. **b. crocein** Sodium aminoazobenzeneazo-2-naptholdisulfonate. Brown powder, soluble in water; a deep red dye. **b. green** $C_{27}H_{34}O_4N_2S$ = 482.6. Viride nitens. Bis(p-diethylamino)triphenylmethanol anhydride sulfate. Small, golden crystals, soluble in water or alcohol. A skin antiseptic (BP); also an indicator, changing at pH 2.0 from yellow (acid) to green (alkaline). **b. yellow** A dye indicator, changing at pH 0.5 from blue (acid) to yellow, and at pH 8.0 from yellow to scarlet (alkaline).

Bri-lon A brand of *Bri-Nylon,* q.v.

brimstone Sulfur.

brine (1) Water which is nearly saturated with salts, e.g., sodium chloride. (2) General term for solutions of a chloride of sodium, calcium, or magnesium, used in cooling systems.

Brinell, Johan August (1849–1925) Swedish engineer. **B. hardness** The area of indentation produced by a hardened steel ball (10 mm diameter) under a pressure of 3000 or 500 kg. B.h. = $p/\pi hd$, where p is the pressure, kg; h the depth of indentation, mm; and d is the diameter of ball, mm. **B. tester** A device for determining the B. hardness.

Brin process The preparation of oxygen gas by heating barium oxide which, at 500°, forms barium peroxide. This gives off oxygen at about 1000°C.

Brinton Reishauer bottle A specific gravity bottle which holds 100 ml of liquid.

Bri-Nylon Trademark for a polyamide synthetic fiber which has triangular facets to produce a sparkling effect.

briquet, briquette A brick of compressed solid fuel; as, coal, lignite, sawdust, or waste materials.

brisance The violence or shattering effect of an explosive, measured in terms of its detonation velocity.

britannia metal A hard silver-white alloy: Sn 80–90, Sb 10–20 pts., and small quantities of copper, and sometimes lead, bismuth, or zinc. Similar to pewter, but spun from rolled sheets instead of being cast.

Britenka Trademark for a continuous-filament rayon yarn.

britholite $(Na_2Ca_7Ce_{11})(F \cdot OH)_4[SiO_4)_9(PO_4)_3]$. A cerium silicate apatite.

British gum Dextrin.

British plate See *nickel silver* under *nickel.*

British Standards Institution See *BSI.*

British thermal unit Btu, BThU. (Cf. *Board of Trade unit,* B.T.U.). 100,000 Btu = 1 therm. A measure of energy in the former British system, which corresponds with the calories of the metric system. The heat required to raise 1 lb of water from 39 to 40°F; hence, the heat required to raise the temperature 1 lb water at its temperature of maximum density by 1°F. It varies with the temperature:

60°F	1054.68 joule
32–212°F (mean)	1055.87
39°F	1059.67
International table	1055.06

mean ~ The $\frac{1}{180}$ part of the heat required to raise the temperature of 1 lb water from 32 to 212°F = 252 cal; 1 Btu/lb = 1.8 cal/g.

brittle (1) Easily broken or pulverized. (2) In bacteriology, the dry growth of bacteria, which is friable under a platinum needle. **b. silver ore** Stephanite.

Brix **B. hydrometer** A hydrometer graduated with the Brix scale. **B. degree** The percent by weight of soluble solids in a syrup at 20°C. An arbitrary scale for the direct conversion of the saccharometer reading of a sugar solution into its specific gravity; n° Brix indicates that the solution to which it refers contains n g of sugar in 100 ml.

broach A mixture of diamond dust, bentonite, and a silicate, used for watch bearings.

brochanite $Cu_2(OH)_2SO_4$. Warringtonite. A native, basic copper sulfate, occurring in emerald-green vitreous masses. The green patina of copper.

Brodie **B. coagulometer** A device for measuring the coagulation of the blood under the microscope. **B. kymograph** An apparatus for recording blood pressure. **B. solution** A solution of salt of such specific gravity that a

column 10 m high is equivalent to a pressure of 1 atm; used in manometers.
broggerite A variety of pitchblende.
Broglie B., Maurice, Duc de (1875–1960) French physicist, noted for his work on corpuscular physics and X-ray spectra. **B., Louis Victor, Prince de (1892–)** French physicist noted for work on wave mechanics; Nobel prize winner (1929). **B. formula** An expression connecting wavelength λ (in Å) with momentum *mv* accelerated by volts *V*.

$$\lambda = h/mv = \sqrt{150/V}$$

brom- Bromo-*. **b.acetone** *Bromoacetone.*
bromacetanilide Acetbromoanilide.
bromacetate Bromoacetate.
bromacetic acid Bromoacetic acid.
bromacetol 2,2-Dibromopropane*.
bromal CBr_3CHO = 280.7. 2,2,2-Tribromoacetaldehyde*. Yellow liquid, d.2.65, b.174, decomp. in water; a hypnotic. **b. hydrate** $CBr_3 \cdot CH(OH)_2$ = 298.8. Tribomaldehyde hydrate, 2,2,2-tribromo-1,1-ethanediol*. Colorless crystals, m.53, soluble in water.
bromalin $C_6H_{12}N_4 \cdot C_2H_5Br$ = 249.2. Bromoethylformin, hexamethylenetetramine bromoethylene. Colorless crystals, m.200; soluble in water, formerly an antiepileptic and sedative.
bromalonic acid Bromomalonic acid.
bromamide Tribromoaniline*.
bromaniline Bromo*aniline**.
bromargyrite Bromyrite.
bromate* A salt of bromic acid containing the $-BrO_3$ radical, as sodium bromate, $NaBrO_3$.
bromated Brominated. (1) Combined with or containing bromine. (2) Describing a molecule undergoing *bromination (2).* **di ~** Describing a molecule containing 2 Br atoms. **mono ~** Describing a molecule containing 1 Br atom.
bromation Bromination.
bromatology The science of food and diet.
bromaurate $MAuBr_4$. Tetrabromoaurate*. A salt derived from auric bromide and a metallic bromide, MBr.
bromchlorphenol blue Bromochlorophenol blue.
bromcresol See *bromocresol.*
bromelain* A hydrolase enzyme in pineapples which converts proteins into proteoses and peptones by preferential Lys-, Ala-, Tyr-, Gly- cleavage.
bromelia Ethyl 2-naphthyl ether*.
Bromeliaceae The pineapple family.
bromeosin See *eosine.*
bromethane (1) Bromomethane*. (2) Bromoethane*.
bromethyl (1) Ethyl bromide*. (2) Bromomethyl. (3) Bromoethyl.
bromethylene Ethylene dibromide*.
bromethylformin Bromalin.
brometone $CBr_3 \cdot CMe_2OH$ = 310.8. Acetone bromoform, 1,1,1-tribromo-2-methyl-2-propanol*. White prisms, with camphorlike odor and taste, m.167, slightly soluble in water; a sedative.
bromhydrin Bromohydrin.
bromic b. acid* $HBrO_3$ = 127.9. Colorless crystals, decomp, about 100, slightly soluble in water. Its salts are bromates. **hydro ~*** See *hydrobromic acid.* **b. ether** Ethyl bromide*.
bromide* A binary salt containing negative monovalent bromine; e.g., sodium bromide, NaBr. Bromides are usually soluble in water. Formerly used as sedatives and antiepileptics. **hydro ~** An addition compound of HBr and an organic base, e.g. an alkaloid.
brominated Bromated.
bromination (1) To treat with bromine. (2) To introduce Br into an organic molecule. Bromation, bromization.
bromine* Br = 79.904. An element of the chlorine group, at. no. 35. A dark brown liquid, fuming halogen, $d_{0°}$ 3.19, m. −7.5, b.58.7, soluble in water, alcohol, chloroform, or ether. Gaseous b. has a density of 5.524 (air = 1). Elementary b. never occurs native, and is found mainly as the bromides of alkali metals, in natural waters, brines, and seawater. Solid ores are carnallite and silver bromide. B. is mainly monovalent, but in some compounds it is tri-, penta-, or heptavalent and forms the compounds:

−1 bromide*	Br^-
+1 hypobromite*	BrO^-
+3 bromate(III)*, bromite*	BrO_2^-
+5 bromate*	BrO_3^-
+7 perbromate*	BrO_4^-

B. was discovered (1826) by Balard in the mother liquors of seawater. Greek *bromos:* "stench." Liquid b. is a reagent and oxidizing agent in organic synthesis. **bi ~*** The bromine molecule, Br_2. **b. chloride*** $BrCl \cdot 10H_2O$ = 295.5. Yellow crystals or liquid, m.7, decomp. readily, verly soluble in water or ether. **b. cyanide** BrCN = 105.9. Cyanogen bromide. Colorless needles, m.52, soluble in water. **b. fluoride*** BrF = 98.9. Colorless prisms, m.50, decomp. by water. **b. hydrate** $Br \cdot 10H_2O$ = 260.1. Red octahedra, decomp. 15, soluble in water. **b. hydride** Hydrobromic acid*. **b. iodide** Iodine bromide*. **b. monochloride*** B. chloride. **b. pentachloride*** $BrCl_5$ = 257.2. Colorless liquid, used in organic synthesis. **b. sulfide** Sulfur bromide. **b. water** A saturated aqueous solution of bromine (about 3% Br); a reagent.
bromite* Bromate(III)*. A salt of bromous acid, $HBrO_2$.
bromization Bromation. To saturate with bromine. Cf. *bromination.*
bromilite Alstonite.
brommalonic acid Bromo*malonic acid.*
bromnaphthalene Bromo*naphthalene**.
bromo-* Brom-. Prefix indicating a bromine atom. **di ~** Indicating a compound or group containing 2 Br atoms. **tri ~** Indicating a compound or group containing 3 Br atoms.
bromoacetate* A salt of bromoacetic acid containing the $CH_2Br \cdot COO-$ radical.
bromoacetic acid* $CH_2Br \cdot COOH$ = 138.9. Bromacetic acid. Colorless hexagons, m.49, soluble in water.
bromoacetone* $BrCH_2COMe$ = 137.0. Monobromomethyl ketone. Colorless liquid, d.1.603, b.127; a poison gas.
bromobenzamide See *bromobenzamide* under *benzamide.*
bromobenzene* Phenyl bromide*.
bromobenzyl cyanide $C_6H_4CH(Br)CN$ = 195.0. α-Bromobenzyl cyanide, BBC. Colorless liquid, m.26; a lacrimatory poison gas.
bromocaffeine $C_8H_9O_2N_4Br$ = 273.1. White powder, m.206, slightly soluble in water.
bromocamphor Camphor, monobromated.
bromochlorophenol blue $C_{19}H_{10}Br_2Cl_2O_2S$ = 533.1. Dibromodichlorosulfonphthalein. Yellow powder, soluble in water or alcohol. An indicator, changing from yellow (pH 4.5) to blue (pH 5.5).
bromocresol b. green $C_{21}H_{14}Br_4O_5S$ = 698.0. 2,3,6,7-Tetrabromo-*m*-cresolsulfonephthalein. Gray powder, soluble in water or alcohol; a pH indicator between 4.5 (yellow) and 5.5 (blue). **b. purple** $C_{21}H_{16}Br_2C_5S$ = 520.3. Dibromo-*o*-cresolsulfonephthalein. Yellow powder; a pH indicator between pH 5.2 (yellow) and 6.8 (purple).
bromocriptine mesylate $C_{32}H_{40}O_5N_2Br \cdot CH_4O_3S$ = 750.7. Bromocriptine methanesulfonate, Parlodel. Yellow crystals. A

dopamine inhibitor used to treat prolonged lactation, acromegaly, and Parkinson's disease (USP).

bromocyanogen Cyanogen bromide.

bromoethyl (1) Ethyl bromide. (2) The radical BrC_2H_4-.

bromoethylene Ethylene dibromide*.

bromoform* $CHBr_3$ = 252.7. Tribromomethane*, formyl bromide, methenyl bromide. Colorless liquid, d.2.904, m.7.7, b.150, slightly soluble in water. Used for separation of minerals. **nitro ~** CBr_3NO_2 = 297.7. Bromopicrin, tribromonitromethane. Colorless liquid, m.10, b_{118mm}127, insoluble in water.

bromoformin Bromalin.

bromohydrin Organic compound of the type Br·R·OH. Cf. *halohydrin, chlorohydrin.* **tri ~** Allyl tribromide.

bromoketone $BrCH_2 \cdot COEt$ = 151.0. Bromomethyl ethylketone; used in chemical warfare.

bromol Tribromophenol.

bromomethane Methyl bromide. **tri ~** Bromoform.

bromomethyl (1) Methyl bromide. (2)* The radical $BrCH_2-$.

bromometry The determination of the halogen-absorbing capacity of unsaturated compounds or of materials containing them, e.g., fats; usually expressed in terms of iodine.

bromonaphthalene See *bromonaphthalene* under *naphthalene.*

bromophenol(s)* $BrC_6H_4 \cdot OH$ = 173.0. A series of compounds, formed by the action of bromine on phenol: ***ortho-* ~** A liquid, b.194, sparingly soluble in water. ***meta-* ~** A solid, m.32. ***para-* ~** A solid, m.63, sparingly soluble in water. **tri-** $C_6H_3OBr_3$ = 330.8. The solid, 1,2,4,6-$C_6H_2 \cdot OH \cdot Br_3$, m.96; sparingly soluble in water.

b. blue $C_{19}H_{10}Br_4O_5S$ = 670.0. 2,3,6,7-Tetrabromophenolsulfonphthalein. Yellow powder, soluble in water; a pH indicator from pH 3.0 (yellow) to 4.6 (blue). **b. red** $C_{19}H_{12}Br_2O_5S$ = 512.2. Dibromophenolsulfonphthalein. Yellow powder, a pH indicator from 6.0 (yellow) to 7.0 (red).

bromophosgene Carbonyl bromide.

bromopicrin Nitrobromoform.

bromoprene $CH_2{:}CBr \cdot CH{:}CH_2$ = 133.0. 2-Bromo-1,4-butadiene. An intermediate in the polymerization of synthetic rubber. Cf. *isoprene, duprene.*

bromopyrine $C_{11}H_{11}ON_2Br$ = 267.1. Monobromantipyrine. White crystals, m.114, soluble in hot water.

bromothymol blue $C_{27}H_{38}Br_2O_5S$ = 634.5. Dibromothymolsulfonphthalein. Brown crystals with green luster; a pH indicator from pH 6.0 (yellow) to 7.6 (blue).

bromotoluene* See *bromotoluene* under *toluene.*

bromoxynil* See *herbicides,* Table 42 on p. 281.

bromthymol blue Bromothymol blue.

bromum Official Latin for "bromine."

Bromwell apparatus A graduated glass cylinder of special shape for fusel oil determination.

bromyrite AgBr. Bromargyrite, bromite. An unctuous, native silver bromide. Yellow or green transparent or opaque isometric crystals.

bronchography A method of rendering the tracheobronchial tree visible by instillation of a radiopaque medium.

brongniardite $Ag_2PbSb_2S_5$. A native silver-lead sulfoantimonite which occurs in isometric crystals.

Brönner acid 2-Naphthylamine-6-sulfonic acid.

bronze A copper-base alloy usually containing up to 30% tin; e.g.:

90% Cu 10% Sn, gunmetal m.1005
80% Cu 20% Sn, bell metal .... m. 890
70% Cu 30% Sn, speculum .... m. 755

acid ~ An alloy containing Cu 82–88, Pb 2–8, Sn 8–10, Zn 0–2%. **aluminum ~** An alloy: Cu 90, Al 10%. **carbon ~** An alloy used for bearings. **coinage ~** See *coinage metals.* **cold cast ~** Misnomer for material used in sculpture that is actually powdered bronze alloy or other metal bonded by plastic resin. **jewelry ~** An alloy: Cu 87.5, Zn 12.5%. **manganese ~** An alloy: Cu 88, Sn 10, Mn 2%. **phosphor ~** See *phosphor bronze* under *phosphor.* **saffron ~** Orange *tungsten.* **silicon ~** A noncorrosive alloy of copper and tin, with 1–4% silicon. **Sillmann ~** An alloy: Cu 86, Al 10, Fe 4%. **Tobin ~** An alloy: Cu 55, Zn 43, Sn 2%. **tungsten ~** See *tungsten bronze.* **uchatius ~** A gear and bearing alloy containing Cu 92, Sn 8%.

b. blue Prussian blue. **b. disease** The formation of light green spots of basic cupric chloride, $Cu_2(OH)_3Cl$, on old b., e.g., in museums.

bronzite Schillerspar. A rhombic, native pyroxene consisting of $MgSiO_3$ with 10% $FeSiO_3$ with a bronze luster.

brookite Arkansite. A reddish-brown to black orthorhombic variety of rutile, d.4.0, hardness 6–6.5.

broom Scoparius. **Spanish ~** Spartium. **b. corn** Sorghum.

brown Describing a pigment or color made by mixing red, yellow, and black. **b. coal** Lignite. **b. dyes** See *bismark b., spirit colors.* **b. hematite** See *brown hematite* under *hematite.* **b. iron ore** See *limonite, bogore, ocher, hematite.* A hydrous iron oxide with some iron carbonate. **b. ocher** B. iron ore. **b. oil of vitriol** B.O.V. **b. pigments** See *Cassel b., ocher, sepia, van Dyck b.* **b. spar** Pearl spar. A variety of *dolomite,* q.v., with a brown tinge due to ferric or manganic oxide.

Brown B., Alexander Crum (1869–1908) Scottish chemist noted for organic research. **Crum B. rule** If the hydrogen of an aromatic compound can be converted by direct oxidation into a hydroxyl compound, substitution will take place in the *meta* position. If not, the substituting radical enters the *ortho* or *para* position. **B., Herbert Charles (1912–)** American chemist, Nobel prize winner (1979). Noted for work on organoboranes. **B., Robert (1773–1858)** British botanist, discoverer of Brownian motion.

Brownian motion Pedesis. The rapid vibratory motion of extremely small particles suspended in a liquid, caused by the bombardment of the particles by the moving molecules of the liquid. The velocity varies inversely with the size of the particles and depends also upon the viscosity of the medium. Cf. *Svedberg's equation.*

browning The formation of off-color in foods during processing.

Brownite cupel A shallow vessel used in silver analysis.

brownmillerite $4CaO \cdot Al_2O_3 \cdot Fe_2O_3$. A phase in the stabilization of dolomite by calcination with silica.

brucealin The active glycosidic principle of the seeds of *Brucea javanic,* Merr. A Chinese antiamebic drug, decomp. 140.

brucine $C_{23}H_{26}O_4N_2 \cdot 4H_2O$ = 466.5. Dimethoxystrychnine. An alkaloid from the seeds of *Strychnos* species (nux vomica, ignatia, etc.). Colorless crystals, m.178, slightly soluble in water. **b. nitrate** $C_{23}H_{26}O_4N_2 \cdot HNO_3 \cdot 2H_2O$ = 493.5. White crystals, decomp. 230, soluble in water. **b. phosphate** $(C_{23}H_{26}O_4N_2)_2H_3PO_4$ = 886.9. White crystals, soluble in water. **b. sulfate** $(C_{23}H_{26}O_4N_2)_2 \cdot H_2SO_4 \cdot 7H_2O$ = 1013. Colorless needles, soluble in water. **b. test** A solution of 0.5 g in 200 ml concentrated sulfuric acid; a test for nitric acid.

brucite $Mg(OH)_2$. Nemalite. A native magnesium hydroxide.

Bruehl receiver Receiver for a condenser in an airtight container; a number of distillates may be collected separately by revolving the apparatus so that each receiver comes under the end of the condenser in turn.

Brunck, Heinrick von (1847–1911) German chemist who developed organic chemical industry.

brunfelsia Manaca.
Bruninghaus optimum The concentration of an activator of luminescence which produces the maximum effect.
Brünner acid Brönner acid.
brunswick green (1) Green *copper carbonate.* (2) A basic copper carbonate having the same composition as *atacamite,* q.v.
brushite $CaHPO_4 \cdot 2H_2O$. A native acid calcium phosphate. Used in dentifrice.
Brussels nomenclature A widely followed customs classification of chemical and other products. Used for duties regulations and import-export records.
bryoidin $C_{20}H_{38}O_3$ = 326.5. A bitter principle of elemi gum.
bryonane $C_{20}H_{42}$ = 282.6. A saturated hydrocarbon, m.69, b.400, from the leaves of *Bryonia dioica* (Cucurbitaceae), possibly identical with laurane.
bryonia English mandrake, white bryony. The dried roots of *Brionia* species (Cucurbitaceae) containing bryonin (a glucoside), alkaloids, and resin; a cathartic. **black ~** Blackeye root. The root of *Tamus communis* (Dioscoraceae) a rubefacient. **white ~** Bryonia.
Bryophyta A division of Cryptogamia, including liverworts (Hepatica) and mosses (Musci).
BSI British Standards Institution. An organization which issues specifications, methods of testing materials, and glossaries of terms for a number of chemical products and industries. Cf. *ANSI.*
BTB Abbreviation for bromothymol blue.
B.T.U. Abbreviation for Board of Trade unit.
Btu BThU. Abbreviation for British thermal unit.
Bu Symbol for butyl. **i-Bu** Symbol for isobutyl. **BuOH** Butyl alcohol*.
bu Abbreviation for bushel.
bubble (1) A small air droplet. (2) To pass gas through a liquid. **b. counter** A device to measure the volume of a gas. A capillary tube through which the gas bubbles escape, connected with a U tube partly filled with mercury; each b. makes an electric contact. **b. gage** A small glass vessel containing a liquid through which the gas bubbles; used to indicate gas flow.
Bubblfil Trademark for a viscose synthetic fiber.
bubulin A compound of cow's dung.
bucco Buchu.
Buchner, Eduard (1860–1917) German chemist, Nobel prize winner (1907), who found that enzyme action is purely chemical. **B. funnel** A porcelain funnel with a cylindrical top, containing a perforated porcelain plate, on which a filter paper is placed. **B. number** The number of ml of 1N alcoholic potassium hydroxide solution required to neutralize 2.5 g of wax dissolved in 80% alcohol.
bucholzite A variety of fibrolite.
buchu Bucco, bucku. The dried leaves of *Barosma betulina* (Rutaceae) of S. Africa; a diuretic and anticatarrhal, used as fluid extract. Cf. *rutin, barosmin.* **b. camphor** See *diosphenol.*
buckbean Menyanthes.
Buck mortar A specially shaped mortar for grinding gold ore with mercury.
buckram A stiffening fabric, made by impregnating an open-weave cotton cloth with an adhesive.
buckthorn Rhamnus. **b. bark** Frangula. **b. berries** Rhamnus cathartica. The dried, ripe fruit of *Rhamnus cathartica* (Rhamnaceae). A cathartic. Cf. *xanthorhamnin.*
buckwheat The seeds of *Fagopyrum esculentum,* (Polygonaceae); a food.
Bucky rays Grenz rays.
Budde effect The expansion in volume of chlorine or bromine vapor when exposed to light.
buddeized Sterilized by warm hydrogen peroxide.
budding A form of cell division in which the new cell is an outgrowth of the parent cell, from which it subsequently becomes detached; as, yeast.
buddling The process of crushing an ore and washing it in a stream of water.
budiene Butadiene.
Bueb process A method for making ferrous hexacyanoferrate(II) from iron sulfate, ammonia, and hydrocyanic acid.
buergerite A species of tourmaline.
bufagin $C_{24}H_{32}O_5$ = 400.5. Crystals, m.224. A neutral, digitalislike principle in the skin-gland venom of the toad, *Bufo agua.* A Brazilian arrow poison. Cf. *sterols.*
bufanolide Bufogenan. Describing the fully saturated system of the squill-toad poison group of lactones; the configuration at the 20 position is the same as in cholesterol. Cf. *steroid.*
buffer A substance which, when added to a solution, resists a change in hydrogen ion concentration on addition of acid or alkali. **oxidation-reduction ~** A mixture of compounds which resists a change in *oxidation-reduction potential,* q.v., and so enables selective oxidation and reduction to be carried out.

b. action The capacity of a solution to neutralize, within limits, either acids or bases, without changing the original acidity or alkalinity of the solution; as, of blood, soil, plant juices. **b. capacity** (1) The millimoles of H^+ which a unit volume of the solution in question will neutralize when an excess of standard acid is added. (2) B. index. The number of moles of strong acid or base required to change the pH by one unit when added to 1 liter of the buffer solution. **b. index** B. capacity. **b. salts** Salts which behave as buffers, generally those of acids with a low dissociation; e.g., the carbonates and phosphates of the blood, which maintain a pH of 7.35, notwithstanding the absorption of carbon dioxide or the introduction of acids. **b. solution** A solution of a weak acid or base and its salts, such as acetates, borates, phosphates, phthalates, which behave as buffers. The following is a "universal" b. solution from pH 2.5 to pH 11.5 (Prideaux and Ward): A 0.1 *N* solution is made containing phosphoric, phenylacetic, and boric acids, 0.02 *N* to the hydrogen ion in each case. If *V* is the number of ml of 0.2 *N* hydrochloric acid and 0.2 *N* sodium hydroxide which is added to 100 ml of this solution with subsequent dilution to 200 ml, then $pH = 3.1 \pm 0.1185V$ (sodium hydroxide is +, hydrochloric acid is −). **b. value** The amount of standard solution (acid or alkali) to bring about a specified change in the hydrogen ion concentration of a solution.
bufogenan Bufanolide.
bufonin $C_{34}H_{54}O_2$ = 494.8. A poisonous principle in the secretion of toads and lizards, e.g., *Bufo vulgaris;* slightly soluble in water or alcohol. Cf. *phrynin.*
bufotalin $C_{24}H_{30}O_3$ = 266.2. A split product of bufotoxin, which resembles digitaligenin.
bufotanine $C_{14}H_{18}O_2N_2$ = 246.2. An alkaloid from the secretion of the parotoid gland of *Bufo vulgaris.*
bufotenine $C_{12}H_{16}ON_2$ = 204.3. 3-(2-Dimethylaminoethyl)-5-hydroxyindole. m.146. An alkaloid from the skin of the common toad and from plants.
bufotoxin $C_{34}H_{46}O_{10}$ = 614.4. A poisonous, acidic skin secretion of toads.
bugleweed The herb of *Lycopus virginicus* (Labiatae); an astringent.
bulb A spherical or bulb-shaped apparatus.
bulbocapnine $C_{19}H_{19}O_4N$ = 325.4. An alkaloid from the

bulbs of *Corydalis cava, Bulbocapnus cavus* (Papaveraceae). White crystals, m.199, soluble in alcohol. **b. hydrochloride** $C_{19}H_{19}O_4N \cdot HCl$ = 361.8. White crystals, soluble in hot water.

bulgur Dried, precooked wheat.

bulk (1) The increase in the volume of a solvent by the dissolved substance. Cf. *cut.* (2) The reciprocal of density.

bullate A bacterial growth resembling a blistered surface.

bullion (1) Coinage metal in mass form. (2) Metal money, as distinct from paper money. (3) Any metal in bar form.

bull's eye condenser A mounted concave lens for the illumination of opaque objects under the microscope.

bully tree See *chicle, balata.*

bumping Uneven boiling of a liquid, due to superheating; avoided by adding an inert solid, e.g., glass beads, pumice.

Buna Trademark for a German rubber substitute, prepared by the polymerization of butadiene.

Bundesmann test A test for the water resistance of fabrics, which are sprayed with water and rubbed on the underside.

Bunsen, Robert Wilhelm Eberhard von (1811–1899) German chemist. **b. burner** A gas burner with adjustable air supply. **B. cell** An electrolytic cell with a constant potential of 1.9 V. An anode of amalgamated zinc in 10% sulfuric acid and a cathode of carbon in conc. nitric acid. Cf. *Grove's cell.* **B. clamp** A clamp with cylindrical, rubber-covered jaws, for holding glassware. **B. eudiometer** A graduated glass tube with platinum electrodes to pass an electric discharge through a combustible gas mixture; the volume changes, and thence the composition may be determined. **B. flame** See *Bunsen flame* under *flame.* **B. funnel** A glass funnel with long stem at an angle of 60°. **B. gas bottle** A cylindrical glass bottle with outlet and inlet tubes for washing gases. **B. reactions** Flame reactions. The behavior in the Bunsen flame; e.g., substances imparting characteristic color. Cf. *analysis, bead test.* **B.-Roscoe law** Reciprocity law. The amount of substance decomposed by radiant energy is proportional to the amount of energy absorbed. **b. valve** See *bunsen valve* under *valve.*

bunsenite, bunsenine NiO. Krennerite. Native nickel oxide; green, isometric crystals.

bunt A seed-borne fungus disease of wheat.

Bunte, Hans (1848–1925) German chemist. **B. gas buret** A buret with a two-way stopcock at one end.

buoyancy Capacity to float in a liquid or gas. **b. correction** True weight = $G(1 - d + G'/\rho)$, where G is the observed weight; G' = 0.0012, the weight of 1 ml air; ρ is the density of the body being weighed; and d = 0.00014, the ratio of the densities of air and brass weights.

buphane Candelabra flower. Cape poison bulb, gifbol. *Buphane distidia* and *B. toxicaria* (Amaryllidaceae), S. Africa.

buphanine An alkaloid from the bulb of buphane.

buphanitine $C_{23}H_{24}O_6N_2$ = 424.5. An alkaloid from buphanine, m.240.

bupivicaine hydrochloride $C_{18}H_{28}ON_2 \cdot HCl \cdot H_2O$ = 342.9. Marcain(e). White crystals, soluble in water. Long-acting local anesthetic, used for nerve block and epidural anesthetics in obstetrics (USP, BP).

burbonal Bourbonal.

Burchard-Liebermann reaction Acetic anhydride produces a blue-green color with cholesterol in chloroform containing concentrated sulfuric acid.

burdock Lappa.

buret, burette A graduated glass tube with a stopcock, used in volumetric analysis to measure volumes of liquids. **automatic ~** A b. with a device for rapidly refilling to the zero point. **certified ~** A b. which has been officially tested for accuracy. **chamber ~** A b. in which drainage errors are minimized by a wide upper chamber containing the bulk of the liquid, and a small bottom graduated tube, with a tap. **gas ~** A b. used in gas analysis. **Schellbach ~** A. b. with a milk-glass scale and a central blue line which facilitates reading by giving the meniscus the appearance of a cusp.

b. cap A small glass cup that fits loosely over the top of a b. **b. clamp** A pinchcock attached to the rubber tubing outlet of a b. **b. float** A float inside the b., facilitating readings. **b. reader** A lens or device clamped on the b. to facilitate reading.

Burger's vector A measure of the magnitude and direction of slip in a crystal lattice. It is unity when it equals the interatomic distance.

burgundy mixture A fungicide. A substitute for bordeaux mixture containing copper sulfate and sodium carbonate.

burkeite $Na_2CO_3 \cdot 2Na_2SO_4$. **sesqui ~** $2Na_2CO_3 \cdot 3Na_2SO_4$. By-products from the cocrystallization of sodium sulfate and carbonate in the isolation of potassium from *trona* deposits, q.v.

Bürker chamber A counting chamber for hemocytometers; essentially an accurately ruled microscope slide.

burlap Coarse jute fabric, used in linoleum manufacture and as a wrapping.

burling A process for removing burrs and other impurities from woolen felt.

burner A device to obtain a flame by the combustion of solids, liquids, or gases. **acme ~** A b. for gas or gasoline vapor. **alcohol ~** A wick-type b. for alcohol. **Argand ~** A b. with an inner tube to supply air to the flame. **blast ~** A b. in which compressed air is blown into the flame. **Boyce ~** Acme b. **bunsen ~** See *Bunsen.* **Chaddock ~** A noncorrodible b. made from refractory. **combustion tube ~** A bunsen b. with wing tops. **evaporating ~** A round cast-iron disk with holes to fit the top of a bunsen b. **Fletcher ~** Hollow, cast-iron rings with holes or slits, for rapidly heating large vessels. **gauze top ~** A bent iron tube with gauze top. **Jansen ~** A blast b. for glass blowing with air and gas regulation. **Mekker ~** A bunsen b. with a top grid which mixes the gases and prevents "striking back." **micro ~** A very small gas b. **multiple tube ~** A gas b. with several tubes arranged in a circle or line. **pilot light ~** A b. with a small inside tube producing a small flame burning continuously. **porcelain ~** Chaddock b. **rose ~** An evaporating b. which enables very small flames to be obtained. **spectrum ~** A b. with platinum holders, for producing colored or monochromatic light from solids. **Teclu ~** See *Teclu b.*

b. attachments See *gauze top, wing top, star, tripod.*

burnettizing The preservation of wood by treatment with creosote under pressure.

burning See *calcination.*

burnt Calcined, or strongly heated. **b. alum** Anhydrous aluminum potassium sulfate. **b. lime** Calcium oxide.

Burow's solution A solution containing 4.8–5.8 g aluminum acetate per 100 ml; an external astringent (USP).

Burseraceae A group of tropical trees and shrubs that secrete resins and oils; e.g. *Commiphora myrrha,* myrrh.

bursine An alkaloid from *Capsella bursapastoris* (shepherd's purse), Cruciferae.

bursting **b. disk** A metal disk of standard composition and thickness which ruptures above a specified pressure. Used as a safety valve on pressure vessels. **b. strength tester** A device for testing the resistance of paper to bursting. A rubber diaphragm is forced through a clamped circular area of paper by means of air or glycerol, and the pressure at the bursting point is measured.

burtonization The addition of calcium salts to a water supply. Used in brewing, where a water similar to that of Burton-on-Trent (England) is required.

busbar The conducting metal rod which carries objects to be plated or otherwise treated in electrolytic deposition plants.

bushel Imperial b., Winchester b. 1 bu = 8 gal = 35.24 liters.

busulfan $Me \cdot SO_2 \cdot O(CH_2)_4O \cdot SO_2Me$ = 246.3. 1,4-Butanediol dimethanesulfonate, Myleran. White crystals, m.117, soluble in water; a cytotoxic agent used to treat leukemia (USP, BP).

butabarbital sodium $C_{10}H_{15}O_3N_2Na$ = 234.2. Sodium 5-*sec*-butyl-5-ethylbarbiturate, secbutobarbitone sodium, Butisol sodium. A hypnotic and sedative (USP). See *barbiturates.* Cf. *butobarbital.*

butacaine sulfate $(C_{18}H_{30}O_2N_2)_2 \cdot H_2SO_4$ = 711.0. Butyn. Colorless crystals, m.99, soluble in water; a surface anesthetic (USP).

butadiene* **1,3-~** $CH_2{:}CH \cdot CH{:}CH_2$ = 54.1. Bivinyl, erythrene, vinyl ethylene. Colorless gas, b. −3; a feedstock for styrene-butadiene rubber, polybutadiene and hexamethylene diamine. **bromo ~*** Bromoprene. **chloro ~*** **2-methyl ~*** Isoprene*.
 b. dicarboxylic acid 2,4-Hexadienedioic acid.

butadiyne* HC:C·C:CH = 50.1. Butadiine, biacetylene. m.−36, b. 10, readily polymerizes; forms yellow Ag, red Cu, and colorless Hg salts.

Butagas Trademark for compressed butane, used for domestic heating and lighting.

butalanin Valine*.

butalastic A general term for synthetic rubbers, in which the repeating unit of the polymer was originally a diene.

butaldehyde Butyraldehyde*.

butamben $NH_2 \cdot C_6H_4 \cdot COOC_4H_9$ = 193.2. Butyl *p*-aminobenzoate, Butesin. White crystals, m.57, insoluble in water. A local anesthetic, surface type (USP, B.P.Vet.).

Butamen process The catalytic isomerization of normal butane.

butanal* Butyraldehyde*.

butanamide* $PrCONH_2$ = 87.1. Butyramide. Colorless scales, m.115, soluble in water.

butane* $MeCH_2 \cdot CH_2Me$ = 58.1. Tetrane. Colorless gas, b.1, insoluble in water, a constituent of natural gas. **bromo ~*** Butyl bromide*. **chloro ~*** Butyl chloride*. **dibromo ~*** **2,3-~** MeCHBr·CHBrMe = 215.9. Pseudobutylene bromide. Yellow liquid, b.158, soluble in alcohol. **dihydroxy ~** Butanediol*. **dimethyl ~*** $Me_2(CH_2)_2Me_2$ = 86.2. Diisopropane. Colorless liquid, d.0.67, b.58, soluble in alcohol. **diphenyl ~** Dibenzylethane. **hydroxy ~** Butyl alcohol*. **iodo ~*** Butyl iodide*. **iso ~*** $CHMe_3$ = 34.1. 2-Methylpropane†, trimethylmethane. Colorless gas, b.−11, insoluble in water. **nitro ~*** $C_4H_9NO_2$ = 103.1. Butyl nitrite. Colorless liquid, b.151, slightly soluble in water. **phenyl ~** Butylbenzene.
 b.diamine Putrescine. **b.dicarboxylic acid** Adipic acid*.

butanedial* Succinaldehyde*.

butanediamide* Succinamide*.

butanediamine* Putrescine.

butanediol* $C_4H_{10}O_2$ = 90.1. Butylene glycol, dihydroxybutane. **1,2-~** Colorless liquid, b.191. **1,3-~** Colorless liquid, b.204. **1,4-~** Tetramethylene glycol. Colorless liquid, m.16. **2,3-~** Colorless liquid, b.184. **dimethyl ~** Pinacol*.

butanedione* $C_4H_6O_2$ = 86.1. A group of diketones: **1,3-~** $HCO \cdot CH_2 \cdot COMe$. **2,3-~** Biacetyl*.

butanediyl* 1,4-~ Butylene. The radical $-(CH_2)_4-$.

butane-ol Butyl alcohol*.

butano* Prefix indicating a $-(CH_2)_4-$ bridge.

butanoic acid* $Me \cdot CH_2 \cdot CH_2 \cdot COOH$ = 88.1. Butyric acid*, ethyl acetic acid. The 4th member of the fatty acid series. Colorless liquid of unpleasant odor, b.163, miscible with water. Produced in the decay of cheese. See *b.a. fermentation.* Used as a bacteriological reagent and a food flavor and in fruit essences. **amino ~*** **2-~** $NH_2 \cdot CHEt \cdot COOH$ = 103.1. Crystals, m.285. **3-~** $NH_2CHMe \cdot CH_2COOH$. Crystals, m.184. **4-~** $NH_2(CH_2)_3COOH$. Crystals, m.193. **2-amino-3-hydroxy ~*** $NH_2 \cdot CH(CHOH \cdot Me)COOH$. A constituent of proteins. Cf. *amino acids.* **aminomethyl ~*** Iso*valine.* **chloro ~*** **2-~** EtCHCl·COOH = 122.6. A liquid, b_{15mm}101. **3-~** $MeCHCl \cdot CH_2 \cdot COOH$. m.44, b_{13mm}100. (±)-3- m.16, 6_{22mm}116. **4-~** $Cl(CH_2)_3COOH$. m.16, b_{22mm}196. **dimethyl ~*** $Me_2CH(CH_2)_2COOH$ = 116.2. Isocaproic acid. A liquid, d.0.925, b.208. **2-ethyl-2-hydroxy ~*** $MeCH_2CEt(OH)COOH$ = 132.2. Diethoxalic acid. Colorless triclinics, m.80, soluble in water. **hydroxy ~*** EtCHOH·COOH = 104.1. Colorless crystals, m.43, b.225 (decomp.), soluble in water. **methyl ~*** **2-~** Me·$CH_2CHMe \cdot COOH$ = 102.1. Hydroangelic acid. **(*R*)-~**, **(*S*)-~**, **(±)-~**, b.177. **3-~** Iso*valeric acid.* **oxo ~** Acetoacetic acid*. **trihydroxy ~*** Erythric acid.
 b. a. anhydride* PrCO·O·OCPr = 158.2. Colorless liquid, b.192, decomp. by water to form the acid. **b. a. fermentation** The production of b. acid by fermentation of sugar or starch by *B. butyricus,* in presence of calcium carbonate.

butanol* **1-~*** Butyl alcohol*. **2-~*** *sec*-Butyl alcohol*. **1,4-dimethyl-2-~*** $Me_2C(OH)CHMe_2$ = 102.2. Colorless liquid, d.0.836, b.112, soluble in water. **3,3-dimethyl-2-~*** $Me_3C \cdot CH(OH)Me$ = 102.2. Pinacolyl alcohol. Colorless liquid, d.0.835, b.120, soluble in alcohol. **2-ethyl-2-~*** $Et_2C(OH)Me$ = 102.2. Colorless liquid, b.120, soluble in alcohol. **2-methyl-1-~*** $MeCH_2 \cdot CHMe \cdot CH_2OH$ = 88.1. A liquid, d.0.816, m.128. **3-methyl-1-~*** $MeCHMe \cdot CH_2 \cdot CH_2OH$. A liquid, d.0.812, m.−117, b.131. **3-methyl-2-~*** MeCHMe·CHMeOH. A liquid, d.0.819, b.114. **2,3,3-trimethyl-2-~*** $Me_3C \cdot COHMe_2$ = 116.2. Pentamethylethanol. Colorless liquid, b.132, soluble in alcohol.

butanone* **2-~** MeCOEt = 73.1. Ethyl methyl ketone*, MEK. Colorless, flammable liquid, d.0.808, m.−86, b.80, soluble in water; used in organic synthesis, as a solvent, and in the manufacture of colorless plastics. **dimethyl ~** Pinacolin. **hydroxy ~** Me·CO·CHOH·Me. Acetoin. An ethereal liquid formed during fermentation by yeast.

butanoyl Butyryl*.

Butaprene Trademark for a Buna-type synthetic rubber having a high flexibility at low temperatures.

butea gum Bengal kino. The dried, red, astringent juice of the dhak or pallas tree, *Butea frondosa* (Leguminosae) of India; an astringent.

butenal* Crotonaldehyde*. **2-methyl-2-~*** MeCH:CMe·CHO = 84.1. Tigl(ic)aldehyde, guaiol. Colorless liquid, b.117.

butene (1)* C_4H_8 = 56.1. Butylene. **1-~** $CH_2{:}CHCH_2Me$. Ethylethylene. b.−18. **2-~** MeCH:CHMe. Ψ-butylene, dimethylethylene. b.1. (2) Cyclobutene*.

butenic acid Butenoic acid*. **iso ~** Methyl*acrylic acid.*

butenoic acid* $C_4H_6O_2$ = 86.1. Butenic acid. **2-~** **(*E*)-~** Crotonic acid*. **(*Z*)-~** Iso*crotonic acid**. **3-** $CH_2{:}CH \cdot CH_2 \cdot$ COOH. Vinylacetic acid. Colorless liquid, b.163, soluble in water. **2-methyl-2-** ~* MeCH:CMeCOOH = 100.1. **(*E*)-** ~ Tiglic acid, cevadic acid. Colorless prisms, m.64, slightly soluble in water. Occurs with the (*Z*) form in croton and chamomile oils.

$$(E)\text{-}\sim \begin{matrix}\text{Me·C·H}\\ \|\\ \text{Me·C·COOH}\end{matrix} \qquad (Z)\text{-}\sim \begin{matrix}\text{H·C·Me}\\ \|\\ \text{Me·C·COOH}\end{matrix}$$

(*Z*)- ~ Angelic acid. Colorless, monoclinic crystals, m.46, sparingly soluble in water. A constituent of the roots of *Angelica, Chamomile,* and *Arnica* species; a flavoring. **3-methyl-2-** ~* Senecioic acid.

butenol* C_4H_8O = 72.1. **2,1-** ~ MeCH:CH·CH_2OH. Crot(on)yl alcohol. b.122. **3,1-** ~ CH_2:CH·CH_2·CH_2OH. Allylcarbinol. b.115. **3,2-** ~ MeCH(OH)CH:CH_2. **(±)-** ~ b.97. In wood spirit oil.

butenyl* Crotyl (the 2 form only). The radical C_4H_7—. Two isomers, corresponding to 1- and 2-butene.

butenylidene* Crotonal. **2-** ~ The radical MeCH:CH·CH=.

butenylidyne* **2-** ~ The radical MeCH:CH·C≡.

Butesin Trademark for butamben.

butine C_4H_6 = 54.1. Butyne*.

butobarbital $C_{10}H_{16}O_3N_2$ = 213.3. 5-*N*-Butyl-5-ethylbarbituric acid, butobarbitone. Soneryl. Fine, white, slightly bitter powder, m.123, soluble in water. A hypnotic and sedative (EP, BP). See *barbiturates.* Cf. *butabarbital sodium.*

butobarbitone BP name for butobarbital.

butopyronoxyl $C_{12}H_{18}O_4$ = 226.3. Pale brown aromatic liquid, b.258, miscible with water; an insect repellent (USP).

butoxy ***n-*** ~* The radical Me$(CH_2)_3$O—, derived from butanol. ***sec-*** ~* Et·CH(Me)O—. ***tert-*** ~* Me_3C·O—.

butter (1) A food prepared by churning cream. It contains not less than 80% of milk fat, water, vitamins, proteins, lactose, and mineral salts. (2) A soft, inorganic chloride, e.g., b. of antimony (antimonic chloride). (3) A low-melting vegetable fat, vegetable b.; as, illipé b. from *Bassia* species. **b. fat** The oily portion of mammal's milk composed of 88% of the glycerides of oleic, stearic, and palmitic acids, and 6% of the glycerides of butanoic, hexanoic, octanoic, and decanoic acids; d.0.912, m.32. **b. rock** A soft, greasy exudate of iron and aluminum from rocks. **b. yellow** (1) A yellow dye often used for tinting butter. (2) *p*-Diaminoazobenzene. (3) 4-Dimethylaminobenzene-1-azonaphthalene. (4) *N,N'*-Dimethyl-*p*-aminoazobenzene.

butternut tree A tree of the genus Bassia (Sapotaceae). The seeds yield a fat used for soap.

button A small, round lump of metal left in the crucible after fusion, or on charcoal after reduction of an ore.

Butvar Trademark for a butyryl polyvinyl plastic.

butyl Bu. **iso** ~* *i*-Bu. The radical $(CH_3)_2$CH·CH_2—. **normal** ~* The hydrocarbon radical CH_3·CH_2·CH_2·CH_2—. **secondary** ~* 1-Methylpropyl†. The radical CH_3·CH_2·CH·(Me)—. **tertiary** ~* 1,1-Dimethylethyl†. The hydrocarbon radical $(CH_3)_3$C—.

b. acetate $C_6H_{12}O_2$ = 116.2. Colorless liquid, b.125, slightly soluble in water; a solvent. **b. alcohol*** Butyl alcohol* (below). **b. aldehyde** Butyraldehyde*. **b.amine*** $C_4H_{11}N$ or BuNH_2 = 73.1. **normal** ~ Colorless liquid, b.78, soluble in alcohol. **iso** ~ Colorless liquid, b.66, miscible with water. **b.benzene** $C_{10}H_{14}$ = 134.2. **iso** ~ Colorless liquid, b.167. ***sec-*** ~ Colorless liquid, b.171. **b. benzoate*** $C_{11}H_{14}O_2$ = 178.2. PhCOOBu. Colorless oil, b.248; insoluble in water. **iso** ~ Colorless liquid, b.237, insoluble in water. **b. bromide*** C_4H_9Br = 137.0. **normal** ~ Colorless oil, b.101, insoluble in water. **iso** ~ Colorless oil, b.90, insoluble in water. **b. butyrate*** PrCOOBu = 144.2. **normal** ~ Colorless liquid, b.165, slightly soluble in water. **iso** ~ Colorless liquid, b.157, slightly soluble in water. **isobutyl isobutyrate** Colorless liquid, b.146, insoluble in water. **b. carbamate*** NH_2COOBu = 117.1. Colorless crystals. **iso** ~ Colorless crystals, m.55, insoluble in water. **b. Carbitol** See *Carbitol.* **b. Cellosolve** See *Cellosolve.* **b. chloral** MeCHClCCl_2·CHO = 175.4. Croton chloral. Colorless liquid, b.165, soluble in water. **b. chloral hydrate** $C_4H_7O_2Cl_3$ = 193.5. Colorless crystals, m.78, soluble in water. **b. chloride*** C_4H_9Cl = 92.6. **normal** ~ Colorless liquid, b.78, insoluble in water. **iso** ~ Me_2CHCH_2Cl. Colorless liquid, b.69, insoluble in water. ***tert-*** ~ Me_3C·Cl. Colorless liquid, b.51. **b. cyanide*** C_5H_9N = 83.1. Valeronitrile*. Colorless liquid, b.141, insoluble in water. **iso** ~ Me_2CHCH_2CN. Colorless liquid, b.127, slightly soluble in water. ***tert-*** ~ Me_3CCN. Crystalline solid, m.15. **b. ether** $C_8H_{18}O$ = 130.2. B. oxide. Colorless liquid, b.141, soluble in water. **iso** ~ Colorless liquid, b.122, slightly soluble in water. **b. formate*** HCOOBu = 102.1. **normal** ~ Tetrylformate. Colorless liquid, b.107, slightly soluble in water. **iso** ~ Colorless liquid, b.98, miscible with alcohol. **b. hydrate** B. alcohol*. **b. hydride** Butane*. **b. hydroxybenzoate** BP name for b. paraben. **b. iodide*** BuI = 184.0. **normal** ~ 1-Iodobutane*. Colorless liquid, b.130, insoluble in water. **iso** ~ 1-Iodo-2-methylpropane*. Colorless liquid, b.120, insoluble in water. ***sec-*** ~ Me·CH_2·CHIMe. Colorless liquid, b.118. ***tert-*** ~ Me_3CI. Colorless liquid, b.99. **b. isobutyrate** See *b. butyrate.* **b. isocyanide*** BuNC = 83.1. Butylcarbylamine. Colorless liquid, m.115, slightly soluble in water. **b. isovalerate** See *b. valerate.* **b. ketone** Bu·CO·Bu = 142.2. Colorless liquid, b.181, slightly soluble in water. **b. lithium** BuLi = 64.1. A reagent used in organic synthesis. See *Wittig reaction.* **b. mercaptan** B. thiol*. **b.methoxyphenol** Butylated hydroxyanisole. **b. mustard oil** See *b. thiocyanate.* **b. nitrite** BuNO_2 = 103.1. Colorless liquid, b.67, soluble in alcohol. **b. oxide** B. ether. **b. oxide linkage** A bond linking the first and fourth C atom of a chain through an O atom. Cf. *sugar, furanose.* **b.paraben*** $C_{11}H_{14}O_3$ = 194.2. Butyl *p*-hydroxybenzoate, butyl paraben. White powder; a pharmaceutical preservative (USP, BP). **b. phenyl ether*** $C_{10}H_{14}O$ = 150.2. Butyl phenylate. Colorless liquid, b.198, insoluble in water; an antiseptic. **b. phenyl ketone*** BuCO·Ph = 162.2. Phenylpentanone. **normal** ~ Colorless liquid, b.240. insoluble in water. **iso** ~ Colorless liquid, b. 225, insoluble in water. **b. phthalate** $C_6H_4(COOBu)_2$ = 278.3. Colorless liquid, b_{20mm}204, used instead of mercury in diffusion pumps. **b. propionate*** EtCOOBu = 130.2. Colorless aromatic liquid, b.136, soluble in alcohol. Used in fruit essences. **b. sulfide** Bu_2S = 146.3. **normal** ~ Colorless liquid, b.182, insoluble in water. **iso** ~ $(Me_2CH·CH_2)_2S$. ***sec-*** ~ $(MeCH_2CHMe)_2S$. Colorless liquid, b.165. **b. thiocyanate*** C_4H_9SNC = 115.2. B. mustard oil. Colorless liquid, b.167, insoluble in water. **iso** ~ Colorless liquid, b.162, insoluble in water. ***sec-*** ~ b.160. ***tert-*** ~ An oil, b.140. **b. thiol*** C_4H_9SH = 90.2. B. mercaptan. Colorless liquid, b.92; occurs in the odorous skunk secretion. **b. valerate*** $C_9H_{18}O_2$ = 158.2. **normal** ~ BuCOOBu. Colorless liquid, b.184, soluble in alcohol; a fruit flavor. **iso** ~ $Me_2CHCH_2COOCH_2CHMe_2$. Colorless liquid, b.169, insoluble in water.

butyl alcohol* $C_4H_{10}O$ or BuOH = 74.1. 1-Butanol*, butyl hydroxide, hydroxybutane. **normal** ~, **primary** ~

Me·CH_2·CH_2·CH_2OH. 1-Butanol*, propylcarbinol. Colorless liquid, b.117, miscible with alcohol or ether; a solvent, defrother, dehydrator, and penetrant. **iso ~** * Me_2CH·CH_2OH. 2-Methyl-1-propanol*. Colorless liquid, d_{33} 0.7980, m.−108, b.108, slightly soluble in water. It can be produced by fermentation (Fernbach's process). **secondary ~** * MeCH_2CHMeOH. 2-Butanol*. Colorless liquid, b.100, soluble in water. **tertiary ~** * Me_3COH. 2-Methyl-2-propanol*. Colorless liquid or rhombic crystals, m.25, b.83, miscible with water. **tribromo ~** Brometone. **trichloro ~** Chlorbutanol.

butylated **b. hydroxyanisole** $C_{11}H_{16}O_2$ = 180.2. 2-*tert*-Butyl-4-methoxyphenol. White crystals, m.64, insoluble in water; an antioxidant (BP). **b. hydroxytoluene** $C_{15}H_{24}O$ = 220.4. 2,6-Di-*tert*-butyl-*p*-cresol. White crystals, insoluble in water; an antioxidant (BP).

butylene (1) Butene*. **iso ~** 2-Methyl-1-*propene.* (2) The 1,4-butanediyl* radical. **diiso ~** 2,2,4-Trimethyl-1-pentene. **triiso ~** 2,2,4,6,6-Pentamethyl-3-*heptene.* **methyl ~** Trimethylethylene.

b. glycol Butanediol*. **b. oxide** C_4H_8O = 72.08. Tetrahydrofuran*. An intermediate, and a solvent for polyvinyl chloride.

butylidene* The radical, Me$(CH_2)_2$CH=. ***sec*- ~** * The radical EtC(Me)=.

butylidyne* The radical Me$(CH_2)_2$C≡.

butyl rubber (1) Generic name for vulcanizable, elastic copolymers of 2-methyl-2-propene and small amounts of dialkenes. (2) A mixture of 2-methyl-2-propene 98 and butadiene or isoprene 2% (U.S. usage).

Butyn Trademark for butacaine sulfate.

butyne* C_4H_6 = 54.1. Butine. An unsaturated hydrocarbon containing a triple bond: **1- ~** CH:C·CH_2Me. Ethylacetylene. Colorless crystals, m.18. **2- ~** MeC:CMe. Crotonylene, dimethylacetylene. Colorless mass from coal gas, m.27. **3-methyl-1- ~** * Me_2CH·C:CH = 68.1. m.−90, b.28.

butynoic acid* MeC:C·COOH = 84.1. Tetrolic acid. Colorless leaflets, m.76, soluble in water.

butyraldehyde* PrCHO = 72.1. Butanal*, butal. Colorless liquid, b.77. **iso ~** * Me_2CH·CHO. Colorless liquid, b.63, miscible with water. **hydroxy ~** Aldol. **trichloro ~** Butyl chloral.

butyramide Butanamide*. **iso ~** 2-Methyl*propanamide**.

butyrate A salt or ester of butanoic acid containing the $C_4H_7O_2$− radical.

butyrelite Bog butter. A fat in peat.

butyric acid Butanoic acid*. **iso ~** 2-Methyl*propionic acid**.

butyric aldehyde Butyraldehyde*.

butyrin $(BuCOO)_3C_3H_5$ = 344.5. Glycerol 1,2,3-tributanoate, tributyrin. Yellow liquid, b.307, insoluble in water; a constituent of butterfat. Cf. *glyceride.*

butyrobetaine The base Me_3N^+·$(CH_2)_3$CO(:O)$^-$ = 145.07. m.180. Occurs in mussels *(Arca noae),* snakes *(Python moluris),* and anthrozoans *(Actinia equina).*

butyrolactone* $C_4H_6O_2$ = 86.6. γ-B.*, butanolide. The internal anhydride of butanoic acid. Colorless liquid, b.204.

butyrometer An instrument to determine the amount of butterfat in milk.

butyrone 4-Heptanone*.

butyronitrile Propyl cyanide*.

butyrophenone Phenyl propyl ketone*. **iso ~** *Phenyl* iso*propyl ketone**.

butyrous Butterlike in consistency.

butyrum Latin for "butter." **b. antimonii** Antimonous chloride.

butyryl 1-Oxobutyl†, butanoyl. **normal ~** * The MeCH_2·CH_2·CO− radical, derived from butanoic acid. **iso ~** * The Me_2CHCO− radical, derived from isobutanoic acid.

b. chloride C_4H_7OCl = 106.6. **normal ~** Butanoyl chloride. Colorless liquid, b.101. **iso ~** Me_2CH·COCl. Colorless liquid, d.1.0174.

buxidine Colorless prisms; an alkaloid from the leaves of *Buxus sempervirens.*

buxine $C_{19}H_{21}O_3N$ = 311.4. An alkaloid from the leaves of *Buxus sempervirens,* Buxaceae. **pseudo ~** $C_{24}H_{48}ON_2$ = 380.7. Ψ-Buxine. An alkaloid from the leaves of *Buxus sempervirens.*

Buxton's fluid A mixture of 50 ml water, 20 ml glycerol, 40 g gum arabic, 50 g chloral hydrate, and 0.5 g cocaine hydrochloride; a microscopical mountant.

buyo Betel nut.

buzane Tetrazane*.

buzylene Tetrazene*. **dihydro ~** Tetrazane*.

BWR Boiling water *reactor.*

byerite A bituminous coal resembling albertite.

byerlite A substance resembling asphalt, made by heating petroleum residues with sulfur in air.

bynin Hordenin. A malt gliadin. Cf. *proteins.*

by-product A product other than the principal material from a manufacturing process.

byssinosis A pulmonary disease due to inhalation of cotton dust. Cf. *siderosis.*

byssochlamic acid $C_{18}H_{20}O_6$ = 332.4. A dianhydride produced by the action on a tetrabasic acid of *Byssochlamis fulva.*

byssus (1) Flax. (2) Lint. (3) The protein threads which attach the edible sea mussel, *Mytilus,* to rocks.

byte A sequence of adjacent bits (as, 8, 16, or 32) that can be processed by a computer as a unit.

bythium Ekatellurium. A supposed element of the sulfur group.

bytownite A triclinic anorthite feldspar. Cf. *silica minerals.*

Bz (1) The benzoyl radical, PHCO−. (2) The benzene ring.

BzH Benzaldehyde*.

BzOH Benzoic acid*.

C

C Symbol for (1) carbon; (2) coulomb. **C. acid** 2-Naphthylamine-4,8-disulfonic acid.
°C Symbol for degree Celsius and degree centrigrade.
C Symbol for (1) capacitance; (2) heat capacity.
c Symbol for (1) centi- (10^{-2}); (2) critical state or value (as subscript); (3) crystalline state (as subscript or between parentheses).
c Symbol for (1) amount of substance concentration; (2) specific heat capacity; (3) speed of *light* in free space; (4) speed of sound; (5) *cis*-.
c_p Symbol for specific heat capacity at constant pressure.
c_V Symbol for specific heat capacity at constant volume.
CA See *Chemical Abstracts Service.*
Ca Symbol for calcium.
ca Abbreviation for (1) candle. (2) approximately (circa).
cabasite Chabazite.
cabbage The vegetable *Brassica oleracea* (Cruciferae). **c. seed oil** The edible fatty oil from the seeds of *Brassica oleracea* (Cruciferae); used in soap, ointments, liniments, and to substitute for olive oil. **c. sugar** A triose from c. leaves. **c. tree bark** Yaba bark.
cabrerite $(Ni \cdot Co)_3As_2O_8 \cdot 8H_2O$. A green, fibrous, native arsenate of nickel and cobalt.
cacaine Theobromine.
cacao Cocoa. Cf. *coca.* **c. butter** Oil of theobroma. A yellow oil expressed from the seeds of *Theobroma cacao.* It has a low m. and is used for receptacles for drugs (USP, BP). **c. red** $(C_{17}H_{16}O_7)_n$. The red pigment of c., soluble in ether or alcohol. **c. syrup** A pharmaceutical flavoring containing c., sucrose, glucose, glycerol, a little sodium chloride, vanillin, and a preservative, in water (USP).
cacaorin $C_{60}H_{84}N_4O_{15}$. A glucoside from cacao beans; oxidized to cacao red.
cachet A container, usually of molded rice paste, to hold dry powders and to administer dry medicaments in a tasteless form.
cacodyl The dimethyl arsine* radical, Me_2As-, from arsine. **di~** Tetramethyl*diarsane**. **ethyl~** Tetraethyl*diarsane**.
c. chloride Me_2AsCl = 140.4. Dimethylarsenic chloride*. Colorless liquid, b.100. **c. cyanide** Me_2AsCN = 131.0. Dimethylarsenic cyanide*. Colorless crystals, m.33. **c. hydride** Me_2AsH = 106.0. Dimethylarsine*. Colorless, volatile liquid, b.36, miscible with alcohol. **c. oxide** $Me_2As \cdot O \cdot AsMe_2$ = 226.0. Alkarsine. Colorless liquid, b.120.
cacodylates Salts of cacodylic acid, containing the $Me_2AsO \cdot O-$ radical; soluble in water. **thio~** Salts of thiocacodylic acid containing the $Me_2AsO \cdot S-$ radical; insoluble in water.
cacodylic acid $Me_2AsO \cdot OH$ = 138.0. Dimethylarsinic acid*, alkargen. A monobasic acid. Colorless crystals, m.200, soluble in water; used in the manufacture of cacodylates. **ethyl~** Diethylarsinic acid.
cacotheline $C_{20}H_{22}N_2O_5(NO_2)_2$ = 462.4. A red product of the action of nitric acid on brucine; a reagent for tin and sulfur dioxide.
cacoxenite A yellow iron phosphate occurring with limonite.
cactine An alkaloid from *Cereus grandiflorus,* night-blooming cactus.
cactoid The combined principles from the leaves and stems of *Cereus grandiflorus;* a heart stimulant.
cactus Cactaceae family with succulent stems and thorny leaves, including the general *Anhalonium, Cereus,* and *Opuntia.* Cf. *prickly pear, mescal buttons, cochineal.* **c. alkaloids** See *Anhalonium, cactine, pectenine, pilocereine.*
cadalene $C_{15}H_{18}$ = 198.3. 1,6-Dimethyl-4-isopropylnaphthalene. A reduction product of cadinene and zingiberene.
cadaverine $NH_2(CH_2)_5NH_2$ = 102.2. Pentamethylene diamine, 1,5-pentanediamine*. A ptomaine formed by the hydrolysis of proteins; an isomer of sapine. Colorless liquid, m.9.
cadaverines Ptomaines.
cade oil Juniper tar oil. Yellow oil from the dry distillation of *Juniperus oxycedrus;* d.0.980–1.055, soluble in alcohol; chief constituent, cadinene. Used in soaps, ointments, and pharmaceuticals.
cadinene $C_{15}H_{24}$ = 204.4. A sesquiterpene, b.275; a constituent of essential oils from juniper species and cedars.
cadion $C_{18}H_{13}O_2N_6$ = 345.3. *p*-Nitrobenzenediazoamino-azobenzene. Brown powder, insoluble in water; a reagent for cadmium (pink) and magnesium (blue). **c. 2D** 4-Nitronaphthalenediazoaminoazobenzene. A reagent for cadmium.
cadmia An ancient name for zinc carbonate.
cadmium* Cd = 112.41. An element of the zinc family, at. no. 48. White, ductile metal, d.8.625, m.321, b.770, insoluble in water. Obtained from zinc ores, and as a by-product in zinc distillation. C. is generally divalent, but monovalent (cadmous) compounds are known. C. compounds are poisonous, and its salts are little ionized. Discovered (1817) simultaneously by Strohmeyer and Herman. Used for pigments, plastics and plating. **diethyl ~** Et_2Cd = 170.5. C. diethyl. Colorless liquid, b_{20mm}64. **dimethyl ~** Me_2Cd = 142.5. Methyl cadmium. Colorless liquid, b.106. Used in Grignard's reaction.
c. acetate* $Cd(OOCMe)_2 \cdot 3H_2O$ = 284.5. White monoclinics, soluble in water. A reagent for sulfur, selenium, or tellurium in steel. **c. alloys** A mixture of c. usually with Ag, Na, Tl, or Zn; used to reduce the m. **c. amalgam** An amalgam of c. and mercury, used in the c. cell. **c. borotungstate** $Cd_2B_2W_9O_{32} \cdot 18H_2O$ = 2737. Yellow crystals, soluble in water, used to separate minerals mechanically. See *Klein's liquid.* **c. bromate*** $Cd(BrO_3)_2$ = 368.2. Colorless crystalline powder, soluble in water. **c. bromide*** $CdBr_2$ = 272.2. Colorless crystals, m.568, soluble in water. **crystalline ~** $CdBr_2 \cdot 4H_2O$ = 344.3. White crystals, soluble in water; used in photography. **c. bromide oxide*** $CdBr_2 \cdot CdO \cdot H_2O$ = 418.6. Yellow powder, decomp. by water. **c. carbonate*** $CdCO_3$ = 172.4. Otavite. White rhombs, decomp. by heat, insoluble in water. **c. cell** C. normal element. Weston element. An H-shaped glass vessel with a c. amalgam cathode, a mercury anode, and a saturated c. sulfate solution electrode. It has a reproducible potential of 1.0186 volt, at 18°C. **c. chlorate*** $Cd(ClO_3)_2$ = 279.3. White crystals, soluble in water. **c. chloride*** $CdCl_2$ = 183.3. White

crystals, soluble in water. **crystalline ~** $CdCl_2 \cdot 2H_2O$ = 219.3. White monoclinics, soluble in water; a reagent for pyridine bases, and used in photography, the dye industry, and textile printing. **c. chloride oxide*** $CdCl_2 \cdot CdO \cdot H_2O$ = 329.7. Colorless powder, decomp. 280, slightly soluble and slowly decomp. by hot water. **c. chloroacetate*** (1) **mono ~** $(CH_2Cl \cdot COO)_2Cd \cdot 6H_2O$ = 407.5. Colorless crystals. (2) **di ~** $(CHCl_2 \cdot COO)_2Cd \cdot H_2O$ = 386.3. Colorless crystals. (3) **tri ~** $(CCl_3 \cdot COO)_2Cd \cdot 1\frac{1}{2}H_2O$ = 464.2. Rhombic crystals. **c. cinnamate*** $(C_6H_5 \cdot CH{:}CHCOO)_2Cd$ = 406.7. White powder, insoluble in water. **c. cyanide*** $Cd(CN)_2$ = 164.4. Colorless powder, soluble in water. **c. diethyl** Diethyl *c.* **c. dimethyl** Dimethyl *c.* **c. ferricyanide** C. hexacyanoferrate(III)*. **c. ferrocyanide** C. hexacyanoferrate(II)*. **c. fluoride** CdF_2 = 150.4. White crystals, m.520, soluble in water or alcohol. **c. fluosilicate** C. hexafluorosilicate*. **c. formate*** $Cd(OOCH)_2$ = 202.4. Colorless powder, soluble in water. **c. fumarate*** $Cd_2C_4H_2O_4$ = 338.9. White powder, insoluble in water. **c. hexacyanoferrate (II)*** $Cd_2[Fe(CN)_6]$ = 436.8. C. ferrocyanide. Yellow crystals, soluble in water. **c. hexacyanoferrate(III)*** $Cd_3[Fe(CN)_6]$ = 641.0. C. ferricyanide. Brown crystals, soluble in water. **c. hexafluorosilicate*** $CdSiF_6$ = 254.5. C. fluosilicate. White hexagons, soluble in water. **c. hydroxide*** $Cd(OH)_2$ = 146.4. White hexagons, slightly soluble in water. **c. iodate*** $Cd(IO_3)_2$ = 462.2. White crystals decomp. by heat, slightly soluble in water. It forms the hydrate $Cd(IO_3)_2 \cdot H_2O$ = 480.2, and the ammonate $Cd(IO_3)_2 \cdot 4NH_3$ = 530.3, which explodes on heating. **c. iodide*** CdI_2 = 366.2. Colorless scales; two water-soluble allotropes: **alpha- ~** m.388. **beta- ~** m.404. Used as a reagent for alkaloids and nitrous acid; and in photography. **c. lactate*** $Cd(C_3H_5O_3)_2$ = 290.5. Colorless powder, slightly soluble in water. **c. line** The red radiation from c. vapor. Former fundamental standard of length (1,533,164.14 waves/m). **c. lithopone** See *cadmium lithopone* under *lithopone.* **c. maleate** $Cd_2C_4H_2O_4 \cdot 2H_2O$ = 374.9. White powder, slightly soluble in water. **c. minerals** C. is associated with zinc, though in small quantities. Its chief ores are greenockite (CdS) and the carbonate and oxide. **c. nitrate*** (1)$Cd(NO_3)_2$ = 236.4. Colorless powder, m.350. (2) Hydrate: $Cd(NO_3)_2 \cdot 4H_2O$ = 308.5. Colorless prisms, m.59, soluble in water. A reagent for zinc and hexacyanoferrates(II); used in the manufacture of yellow and orange glazes. **c. normal element** C. cell. **c. oxalate*** (1) CdC_2O_4 = 200.4. White powder, m.340. (2) Hydrate: $CdC_2O_4 \cdot 3H_2O$ = 254.5. White powder, insoluble in water. **c. oxide** (1) CdO = 128.4. Amorphous brown powder, insoluble in water. (2) Cd_2O = 240.8. Cadmous oxide. Yellow powder, decomp. by heat. (3) Cd_4O = 465.6. C. suboxide. Brown amorphous powder, decomp. by water. (4) C. peroxide, insoluble in water; CdO_2, Cd_3O_5, or Cd_5O_8. **c. oxybromide** C. bromide oxide*. **c. oxychloride** C. chloride oxide*. $CdCl_2 \cdot H_2O$ = 329.7. Colorless powder, decomp. 280; slightly soluble and slowly decomp. by hot water. **c. permanganate*** $Cd(MnO_4)_2$ = 350.3. Purple crystals, soluble in water. **c. peroxide** C. oxide (4). **c. phosphate*** $Cd_3(PO_4)_2$ = 527.2. White powder, insoluble in water. **c. plating** A rustproofing coating of c. on iron or steel. **c. potassium cyanide*** $Cd(CN)_2 \cdot 2KCN$ = 294.7. White crystals, soluble in water. **c. potassium iodide*** $CdI_2 \cdot 2KI \cdot H_2O$ = 716.2. White powder, yellows on aging, soluble in water; a reagent for alkaloids. **c. red** (1) A mixture of c. selenide, sulfide, and barite. (2) A mineral pigment: Cd 60, S 25, and Se 15%. **c. selenate*** $CdSeO_4$ = 255.4. Colorless crystals, soluble in water. **c. selenide** CdSe = 191.4. Red powder, insoluble in water; used in rubber manufacture for abrasion resistance. **c. suboxide** Cd_4O = 465.6. C. oxide (3). **c. succinate*** $CdC_4H_4O_4$ = 228.5. White powder, slightly soluble in water. **c. sulfate*** $CdSO_4$ = 208.5. White rhombs, m. 1000, soluble in water. **crystalline ~** (1) $CdSO_4 \cdot 4H_2O$ = 280.5. White crystals, soluble in water. (2) $3CdSO_4 \cdot 8H_2O$ = 769.5. White monoclinics, soluble in water. Used in c. cells. **c. sulfide*** CdS = 144.5. Greenockite. (1) **Orange ~** Insoluble in water; used as a pigment and in pyrotechnics. (2) **Light yellow ~** C. yellow, jaune brilliant. Hexagonal crystals, insoluble in water; used as a pigment for coloring soaps, ceramics, rubber, and in pyrotechnics and solar cells. **c. sulfite** $CdSO_3$ = 192.5. White powder, slightly soluble in water. **c. tartrate*** $Cd(C_4H_4O_6) \cdot H_2O$ = 278.5. White, crystalline powder, slightly soluble in water. **c. telluride** CdTe = 240.0. Black crystals, m.1040; used in solar cells. **c. tungstate*** $CdWO_4$ = 360.3. Colorless powder, insoluble in water. **c. valerate*** $Cd(C_5H_9O_2)_2$ = 314.2. Colorless scales, soluble in water. **c. wolframate*** C. tungstate*. **c. yellow** C. sulfide*.

cadmous Cadmium(I)* or (1+)*. Describing a compound of monovalent cadmium.

cadoxen $[Cd(en)_3][OH]_2$. Triethylenediamine cadmium hydroxide. A stable solvent for cellulose.

Caesalpinia A genus of subtropical trees (Leguminosae); e.g.: *C. coriaria,* dividivi; *C. sappan,* brazilwood.

caesium See *cesium.*

caffalic acid $C_{34}H_{54}O_{15}$ = 702.8. An acid in coffee.

caffeic acid $C_9H_8O_4 \cdot H_2O$ = 198.2. (*E*) form of 3,4-dihydroxycinnamic acid. Yellow prisms in black fir resin, m.224 (decomp.), soluble in water. **dimethylhydro ~** $C_{11}H_{14}O_4$ = 210.2. Colorless crystals, m.96. **hydro ~** $C_9H_{10}O_4$ = 182.2. 3,4-Dihydroxyphenylpropionic acid. Colorless crystals, m.213. **iso ~** (*Z*) form of c. a.

caffeine $C_8H_{10}O_2N_4 \cdot H_2O$ = 212.2. Theine, methyltheobromine, 1,3,7-trimethylxanthine, guaranine, psoraline. A diureide in coffee (approx. 100 mg/cup of brew), tea, and cola nuts. White, bitter needles, m.230, soluble in water; a diuretic and stimulant to central nervous system and heart. **bromo ~** See *bromocaffeine.* **chloral ~** See *chloral caffeine.* **ethoxy ~** See *ethoxycaffeine.* **hydroxy ~** $C_8H_{10}O_3N_4$ = 210.2. White needles, m.345, soluble in hot water; a diuretic.

c. arsenate $C_8H_{10}O_2N_4 \cdot H_3AsO_4$ = 336.1. White powder, soluble in hot water. **c. benzoate** $C_8H_{10}N_4O_2 \cdot C_7H_6O_2$ = 316.3. White, crystalline powder, soluble in water. **c. citrate** $C_8H_{10}O_2N_4 \cdot C_6H_8O_7$ = 386.3. White crystals, soluble in water. **c. hydrate** $C_8H_{10}O_2N_4 \cdot H_2O$ = 212.2. White, bitter needles, soluble in water; a central nervous system stimulant (BP). **c. hydrobromide** $C_8H_{10}O_2N_4 \cdot HBr$ = 275.1. White crystals, soluble in water. **c. hydrochloride** $C_8H_{10}O_2N_4 \cdot HCl \cdot 2H_2O$ = 266.7. White crystals, soluble in water. **c. nitrate** $C_8H_{10}O_2N_4 \cdot HNO_3$ = 257.2. White crystals, soluble in water. **c. phenolate** $C_8H_{10}O_2N_4 \cdot C_6H_6O \cdot H_2O$ = 306.3. White crystals, soluble in water. **c. salicylate** $C_8H_{10}O_2N_4 \cdot C_7H_6O_3$ = 332.3. White crystals, soluble in water. **c. sodium benzoate** A mixture of c. and sodium benzoate, soluble in water (USP).

caffetannic acid $C_{16}H_{18}O_9$ = 354.3. Chlorogenic acid, m.208. A tannin in coffee, nux vomica, St. Ignatius beans, as its calcium or magnesium salt. Amorphous powder, soluble in water.

cahinca (1) David's root. The dried root of *Chiococca racemosa* (Rubiaceae). See *chiococcine.* (2) The root of a S. Amer. shrub *Chiococca anguifera* used as an antidote for snake poison.

Cailletet, Louis (1832–1913) French ironmaster, noted for work on liquefaction of gas.

cairngorm A variety of quartz.
cajeput Cajuput, tree tea, swamp tea. Leaves of *Melaleuca leukadendron* (Myrtaceae). E. India. **c. oil** The colorless essential oil of cajeput; chief constituents eucalyptol and terpineol. $[\alpha]_D = -10°$ to $-4°$.
cajuput Cajeput.
cajuputol Eucalyptol.
cake A solid crystalline mass. **niter ~** Sodium sulfate containing about 33% sulfuric acid. A by-product of nitric acid manufacture. **salt ~** Sodium sulfate.
caking Transformation of a powder into a solid mass by moisture, heat, or pressure.
Cal, cal Abbreviations for large (kg-cal, Calorie) and small (g-cal, calorie) calories, respectively. Replaced by joule of SI system.
calabar bean Physostigma. **false ~** See *pseudophysostigmine.*
calabarine An alkaloid from calabar bean.
Calais sand Very fine sand from Calais, France; an abrasive.
calamene $C_{15}H_{24}$ = 204.4. A sesquiterpene from calamus oil.
calamine (1) $Zn(OH)_2 \cdot Zn_3Si_2O_7 \cdot H_2O$. Native zinc silicate. (2) C. lotion. **c. lotion** A suspension of zinc carbonate colored with ferric oxide plus (sometimes) bentonite. Has astringent and soothing properties for skin conditions (USP, BP). **electric ~** See *electric calamine.* **c. brass** Marcasite.
calaminth Basil thyme, mountain minth. *Calamintha officinalis* (Labiatae); an expectorant.
calaminthone $C_{10}H_{16}O$ = 152.2. A ketone in French oil of marjoram, distilled from the leaves of *Calamintha nepeta* (Labiatae).
calamus Sweet flag, sweet grass, sweet cane. Dried rhizomes of *Acorus calamus* (Araceae); a stimulant and carminative. Cf. *acorin, acoretin.* **c. oil** The brown oil from calamus; chief constituents, asarone (2,4,5-trimethoxyphenyl-1-propene) and eugenol. Used in perfumes and flavors.
calaverite $(Au,Ag)Te_2$. Native gold telluride containing 40% Au (Colorado, California).
calc. Abbreviation for calculated.
calcareous, calcarious Containing calcium. **c. algae** Limestone from the algae of *Lithothamnium,* whose skeletons contain calcium carbonate 95% and magnesium carbonate 5%. Cf. *coccolith.* **c. sinter** Travertine, onyx. Tufa deposits of calcium carbonate formed by the evaporation of water containing calcium hydrogencarbonate.
calcarone, calcarella A Sicilian sulfur furnace.
calcein Fluorescein complexone. A fluorescent indicator for the determination of calcium by titration with EDTA; color change green to pink.
calcic Pertaining to calcium.
calciferol Common name for ergocalciferol.
calcification Hardening of tissue by calcium salt formation.
calcii Official Latin genitive of *calcium.*
calcimeter An apparatus to determine carbonates by liberating carbon dioxide with acid and absorbing it in alkali; as, Schrotter apparatus.
calcination Roasting. (1) Oxide formation by heating oxy salts; e.g., calcium oxide from calcite. (2) Expelling the volatile portions of a substance by heat. Cf. *roasting.*
calcined Heated to a high temperature. **c. phosphate** A fertilizer obtained by roasting finely ground phosphate rock with potassium salts.
calcinol Calcium iodate.
calcioferrite Iron phosphate in sedimentary beds of limonite.
calciovolborthite Native copper and calcium vanadate.
calcite $CaCO_3$. Calcspar. Crystalline: Iceland spar (birefringent), corn spar, satin spar. Amorphous: chalk, stalactite, baryte. Spongy: mountain milk. Flaky: schiefer spar. **baryto ~** Alstonite. **chloro ~** Hydrophilite.
calcitization Conversion of marble to calcite.
calcitonin A polypeptide hormone from pig thyroid gland or salmon (salcatonin); used to lower blood calcium levels in hypercalcemic states (BP).
calcium* Ca = 40.078. An element of the magnesium family, at. no. 20. White crystals, d.1.415, m.840, b. 1485, soluble in water forming the hydroxide, and in acids forming salts. The third most abundant metal, occurring as carbonates (limestone), sulfates (gypsum), fluorides (fluorspar), phosphates (apatite), arid complex silicates (feldspar). Isolated by Davy (1808) by electrolysis; made by electrolyzing a fused mixture of fluorspar and c. chloride. C. is divalent, and forms one series of compounds. Adult daily requirement 0.5–0.6 g. Used in organic synthesis, as a deoxidizer in alloys, in the preparation of metals, as reducing agent, as a purifier for argon, and, with lead, in antifriction alloys. The radioactive isotope ^{47}Ca (half-life, 4.7 days) has a great metabolic similarity to, without the undesirable qualities of, ^{90}Sr. It emits gamma radiation and is used for biological and medical purposes. **c. acetate*** $Ca(OOCMe)_2 \cdot H_2O$ = 176.2. White needles, decomp. by heat, soluble in water. Used in the manufacture of acetic acid, acetone, and dyes.; and in dialyzing solutions in hemo*dialysis,* q.v. (BP). **c. aminosalicylate** $C_{14}H_{12}N_2O_6Ca \cdot 3H_2O$ = 398.35. White crystals with bitter-sweet taste, soluble in water; a tuberculostatic agent. (BP). **c. arsenate** $Ca_3(AsO_4)_2$ = 398.1. Tricalcium orthoarsenate. White powder; slightly soluble in water. **c. arsenite*** $Ca_3(AsO_3)_2$ = 366.1. White granules, soluble in water. **c. behenate** $Ca(C_{22}H_{43}O_2)_2$ = 719.2. Colorless crystals. **c. benzoate*** $Ca(C_7H_5O_2)_2 \cdot 3H_2O$ = 336.4. Colorless crystals, soluble in water; a food preservative. **b. bisulfite** C. hydrogensulfite*. **c. bitartrate** C. hydrogentartrate*. **c. bromide*** $CaBr_2$ = 199.9. White granules, m.760, soluble in water. Used in photography. **crystalline ~** $CaBr_2 \cdot 6H_2O$ = 308.0. White needles, m.38, soluble in water. **c. butanoate*** $Ca(C_4H_7O_2)_2 \cdot H_2O$ = 232.3. Colorless crystals, soluble in water. **c. carbide*** CaC_2 = 64.1. Gray crystals, decomp. by water, soluble in alcohol; used to prepare acetylene gas. Cf. *acetylith.* **c. carbonate*** $CaCO_3$ = 100.1. A widely distributed rock (calcite). Rhombohedra, or white rhombs, or an amorphous, fine, white powder, d.2.7–2.949, decomp. 825, insoluble in water. Used medicinally as an antacid (USP, EP, BP), and in toothpaste, white paint, cleaning powder, paper fillers, and for the preparation of carbon dioxide. **gamma ~** Vaterite. **c. chlorate*** $Ca(ClO_3)_2 \cdot 2H_2O$ = 243.0. White monoclinics, decomp. by heat, soluble in water. Used in photography, pyrotechnics, manufacture of soda water, and as a weedicide. **c. chloride*** $CaCl_2$ = 111.0. White, deliquescent granules, m.774, soluble in water. Used as an additive for concrete (2% accelerates setting and imparts strength and frost resistance), as a drying agent in desiccators and in meat preservation; in freezing mixtures and for making textiles less flammable; as a dust preventer on roads (halophilite); and in the manufacture of hydrochloric acid, alizarin, and sugars. **ammonia ~** The unstable compound $CaCl_2 \cdot 8NH_3$. **crystalline ~** $CaCl_2 \cdot 6H_2O$ = 219.1. $CaCl_2 \cdot 2H_2O$ also exists. Colorless hexagons, m.29.5, soluble in water. Used to stop bleeding (USP, EP, BP), and in brewing and soda water manufacture; used to raise the boiling point of solutions. **fused ~** C. chloride. **c. tube** A glass vessel filled with c. chloride for the drying of gases or for the quantitative absorption of water. **c. chromate*** $CaCrO_4 \cdot 2H_2O$ = 192.1. Yellow crystals, slightly soluble in

water. **c. cinnamate*** $Ca(C_9H_7O_2)_2 \cdot 3H_2O$ = 388.4. White crystals, soluble in hot water. **c. citrate*** $Ca_3(C_6H_5O_7)_2 \cdot 4H_2O$ = 570.5. White crystals, slightly soluble in water. **c. cyanamide** CaNCN = 80.1. Nitrolim(e). White powder, decomp. by water; an intermediate in atmospheric nitrogen fixation; a catalyst in ammonia manufacture. **c. cyanide*** $Ca(CN)_2$ = 92.1. Calcyanide, "powdered cyanic acid." Colorless crystals, decomp. in moist air to form cyanic acid. **c. cyclamate** $C_{12}H_{24}O_6N_2S_2Ca_2H_2O$ = 454.6. White, very sweet crystals, soluble in water; a sweetening agent, but see *cyclamates.* **c. dichromate*** $CaCr_2O_7$ = 256.1. Brown, hygroscopic crystals, soluble in water. **c. disodium edetate** See *edetate.* **c. ethylsulfate** $Ca(EtSO_4)_2 \cdot H_2O$ = 308.3. C. sulfovinate. White crystals, soluble in water. **c. ferricyanide** C. hexacyanoferrate(III)*. **c. ferrocyanide** C. hexacyanoferrate(II)*. **c. ferrophospholactate** White powder, soluble in hot water; used medicinally in syrups. **c. fluoride** (1)* CaF_2 = 78.1. Colorless, regular crystals, m.1418, sparingly soluble in water, soluble in water containing carbon dioxide. Used to etch glass and in the preparation of enamels and hydrofluoric acid. (2) Fluorite. **c. fluosilicate** C. hexafluorosilicate*. **c. formate*** $Ca(OOCH)_2$ = 130.1. White crystals, soluble in water. **c. fumarate*** $CaC_4H_2O_4 \cdot 3H_2O$ = 208.2. Rhombic, white crystals, soluble in water. **c. gluconate** $(C_5H_{11}O_5COO)_2Ca$ = 430.4. White, odorless powder, soluble in water, used to treat hypocalcemia (USP, EP, BP). **c. glycerate** $Ca(C_3H_5O_4)_2 \cdot 2H_2O$ = 286.2. White powder, soluble in water. **c. glycerophosphate** $CaPO_4C_3H_5(OH)_2 \cdot 2H_2O$ = 246.2. Neurosin. White crystals, soluble in water; a "nerve" tonic. **c. glycolate** $Ca(C_2H_3O_3)_2 \cdot H_2O$ = 208.2. White crystals, slightly soluble in water. **c. hexacyanoferrate(II)*** $Ca_2[Fe(CN)_6]$ = 292.1. C. ferrocyanide. Yellow crystals, soluble in water. **c. hexacyanoferrate(III)*** $Ca_3[Fe(CN)_6]_2 \cdot H_2O$ = 442.1. Red deliquescent needles, soluble in water. **c. hexafluorosilicate*** $CaSiF_6 \cdot 2H_2O$ = 218.2. C. fluosilicate, lapis albus. Hexagons, slightly soluble in water. **c. hydride*** CaH_2 = 42.1. Hydrolith. Colorless powder, insoluble in water; used in organic synthesis. **c. hydrogensulfite*** $Ca(HSO_3)_2$ = 202.2. C. bisulfite. Used to prevent fermentation. **c. hydrogentartrate*** $CaH_2(C_4H_4O_6)_2$ = 338.2. C. bitartrate. White crystals, soluble in hot water. **c. hydroxide*** $Ca(OH)_2$ = 74.1. Hydrated lime, slaked lime. White, hexagonal crystals or powder, decomp. by heat, slightly soluble in water, soluble in ammonium chloride solution. Used in water softeners and mortar. **c. hypochlorite*** $Ca(OCl)_2 \cdot 4H_2O$ = 215.0, or $CaOCl_2 \cdot nHO$. White powder, used as disinfectant, bleaching and oxidizing agent, and to prepare chlorine. Cf. *bleaching powder.* **c. iodate*** $Ca(IO_3)_2 \cdot 6H_2O$ = 498.0. Calcinol. White prisms, decomp. by heat, slightly soluble in water; an antiseptic and iodoform substitute. **c. iodide*** CaI_2 = 293.9. A yellow powder, m.631, soluble in water; a substitute for potassium iodide. **crystalline ~** $CaI_2 \cdot 6H_2O$ = 402.0. Colorless plates, m.42, very soluble in water; used in photography. **c. isobutyrate*** $Ca(C_4H_7O_2)_2 \cdot 5H_2O$ = 304.3. Colorless needles, soluble in water. **c. lactate*** $Ca(C_3H_5O_3)_2 \cdot 5H_2O$ = 308.3. White crystals, soluble in water; used to treat rickets (USP, EP, BP). **c. larsenite** (Pb, Ca)$ZnSiO_4$. A rare mineral having a vivid yellow fluorescence (Franklin Furnace, N.J.). **c. lime** Quicklime containing 75% or less CaO. **c. linoleate** $Ca(C_{18}H_{31}O_2)_2$ = 599.0. White powder, insoluble in water. **c. magnesium chloride** Tachhydrite. **c. magnesium phosphate** Mixed c. and magnesium phosphates. White powder, insoluble in water. **c. malate*** $CaC_4H_4O_5$. (1) **active ~** $2H_2O$ = 208.2. (2) racemic ~ $3H_2O$ = 226.2. Colorless rhombs, slightly soluble in water. **c. maleate*** $CaC_4H_2O_4$ = 154.1. Colorless rhombs, soluble in water. **c. malonate*** $CaC_3H_2O_4 \cdot 4H_2O$ = 214.2. White powder, insoluble in water. **c. mandelate** $Ca(C_8H_7O_3)_2$ = 342.4. White powder, slightly soluble in water; a urinary tract anti-infective. **c. manganite** $CaO \cdot MnO_2$ = 143.0. See *Weldon process.* **c. meconate** $CaC_7H_2O_7 \cdot H_2O$ = 256.2. Yellow powder, sparingly soluble in water. **c. metaphosphate*** C. phosphate (2). **c. methylsulfate** $Ca(MeSO_4)_2$ = 262.3. White crystals, soluble in water. **c. minerals** C. is one of the most abundant metals and a main constituent of many rocks, e.g., limestone, $CaCO_3$; gypsum, $CaSO_4$; apatite, $3Ca_3(PO_4)_2 \cdot CaF_2$. **c. monophosphate** See *c. phosphate (3).* **c. nitrate*** $Ca(NO_3)_2 \cdot 4H_2O$ = 236.1. Colorless monoclinics, m.43, very soluble in water; a fertilizer (Norway saltpeter) and explosive. Cf. Baldwin's *phosphorus.* **c. nitride*** Ca_3N_2 = 148.3. Brown mass, m.900, decomp. by water. **c. nitrite*** $Ca(NO_2)_2$ = 132.1. Colorless prisms, soluble in water. **c. oleate*** $Ca(C_{18}H_{33}O_2)_2$ = 603.0. Yellow granules, soluble in alcohol. **c. oxalate*** $CaC_2O_4 \cdot H_2O$ = 146.1 or $Ca(OOC)_2$ = 128.1. White microcrystals, insoluble in water. See *raphides.* **c. oxide*** CaO = 56.1. Lime, burnt lime, quicklime, caustic lime, calx. White masses, $d_{5°}$ 3.306, m.1995, sparingly soluble in water (limewater), soluble in acids. Used as a reagent, and an an antacid and mild caustic. Commercial grades are used in mortar and fertilizers and for causticizing soda ash. **c. palmitate*** $Ca(C_{16}H_{31}O_2)_2$ = 550.9. Yellow crystals, insoluble in water. **c. pantothenate** (USP). See *pantothenic acid.* **c. permanganate*** $Ca(MnO_4)_2 \cdot 4H_2O$ = 350.0. Acerdol. Purple, hygroscopic prisms, decomp. by heat, soluble in water. An antiseptic and disinfectant; used for disinfecting drinking water. **c. peroxide*** CaO_2 = 72.1. Cream-colored powder, decomp. by water; a detergent, bactericide, and antiseptic. **c. phenolate*** $Ca(OPh)_2$ = 226.3. C. carbolate, c. phenylate, c. phenate. Pink powder, slightly soluble in water; an antiseptic. **c. phenolsulfonate** $Ca(SO_3C_6H_4OH)_2$ = 386.4. C. sulfophenate, c. sulfocarbolate. Pink powder, soluble in water; an antiseptic. **c. phosphate** Forms include:

$CaHPO_4$ = calcium hydrogenphosphate, A.C.P.
$Ca(PO_3)_2$ = calcium metaphosphate
$CaH_4(PO_4)_2$ = monocalcium phosphate as salts of phosphoric acid
$Ca_2P_2O_7$ = calcium pyrophosphate
$Ca_3(PO_4)_2$ = tricalcium phosphate

Naturally occurring phosphates include monetite, $2CaO \cdot P_2O_5 \cdot H_2O$; fluorapatite, $CaF_2 \cdot 9CaO \cdot 3P_2O_5$; dahllite, $CaCO_3 \cdot 6CaO \cdot 2P_2O_5$. (1)* **c. hydrogen orthophosphate** $CaHPO_4 \cdot 2H_2O$ = 172.1. Secondary c. p., dibasic c. p., diacid c. p.. Colorless plates, slightly soluble in water, soluble in ammonium citrate solution; used as a.c. replenisher (USP). (2)* **c. metaphosphate** $Ca(PO_3)_2$ = 198.0. White powder, insoluble in water. (3) **c. monophosphate** $CaH_4(PO_4)_2 \cdot H_2O$ = 252.1. Primary c.p., acid c.p. Colorless rhombs, soluble in water; used in baking powders. (4)* **c. pyrophosphate** $Ca_2P_2O_7 \cdot 4H_2O$ = 326.2. Colorless crystals, slightly soluble in water. (5)* **c. orthophosphate** $Ca_3(PO_4)_2 \cdot H_2O$ = 328.2. Normal c.p., tricalcium p., tertiary c.p., neutral c.p., tribasic c.p. White powder, insoluble in water. Used as c. source for bone tissues, and in the manufacture of enamels, cleansing agents, and phosphorus (NF). **c. phosphide** (1) Ca_2P_2 = 142.1. Gray granules, decomp. by water to form flammable phosphine; used in chemical warfare. (2) Ca_3P_2 = 182.2. Tricalcium diphosphide*. Yields nonflammable phosphine in water. **c. phosphinate*** $Ca(H_2PO_2)_2$ = 170.1. Monoclinic, white crystals, decomp. at red heat, soluble in water. **c. phosphonate** $2CaHPO_3 \cdot 3H_2O$. White crystals, slightly soluble in water, decomp. by heat, forming phosphine, PH_3. **c. phthalate*** $CaC_8H_4O_4 \cdot H_2O$ = 222.2. Colorless prisms, soluble in water. **c. plumbate** $Ca(PbO_3)_2$ or Ca_2PbO_4 = 351.4. Orange crystals, insoluble in cold water, decomp. by hot water. Used as an oxidizing agent; in accumulators, pyrotechnics, glass, and matches. **c. plumbite** $CaPbO_2$ = 279.3. Colorless crystals, slightly soluble in water. **c. propionate*** $Ca(C_3H_5O_2)_2$ = 186.2. White powder, soluble in water. **c. pyrophosphate*** See *c. phosphate (4).* **c. salicylate*** $Ca(OOC \cdot C_6H_4 \cdot OH)_2 \cdot 2H_2O$ = 350.3. Colorless crystals, soluble in carbonated water. **c. selenate*** $CaSeO_4$ = 183.0. White crystals, sparingly soluble in water. **c. selenite*** $CaSeO_3 \cdot 2H_2O$ = 203.1. White powder, soluble in water. **c. silicate*** $CaSiO_3$ = 116.2. Okonite. White mass, insoluble in water. **c. silicotitanite** Sphene. **c. stearate*** $Ca(C_{18}H_{35}O_2)_2$ = 607.0. White granules, insoluble in water (NF). **c. succinate*** $CaC_4H_4O_4 \cdot H_2O$ = 174.2. White crystals, soluble in water. **c. sulfate*** (1) **c.s. anhydrite** $CaSO_4$ = 136.1. White rhombs, m.1360, insoluble in water, soluble in ammonium salts; a desiccant and absorbent, for gases. (2) **c. s. sesquihydrate** $CaSO_4 \cdot \frac{1}{2}H_2O$ = 145.1. Plaster of paris. White powder containing 5% water made by heating gypsum to 120–130°C. It quickly solidifies with water; used for castings and molds and medicinally to make plaster of paris splints (EP, BP). (3) **c. s. dihydrate** $CaSO_4 \cdot 2H_2O$ = 172.1. Gypsum. White, monoclinic crystals; used as a paper filler (pearl hardening, mineral white), pigment, fertilizer, a retarder for portland cement, and a cleaning agent. **c. sulfide*** CaS = 72.1. Hepar calcis, sulfurated lime. Yellow powder, used for luminous paints. **c. sulfite*** $CaSO_3 \cdot 2H_2O$ = 156.2. White powder, soluble in dilute sulfurous acid. Used as a disinfectant in breweries, as an antichlor in bleaching, and in the manufacture of wood pulp. **c. superoxide** C. peroxide*. **c. tannate** Yellow powder, insoluble in water. **c. tartrate*** $CaC_4H_4O_6 \cdot 4H_2O$ = 260.2. White powder, sparingly soluble in water. **c. thiocyanate*** $Ca(SCN)_2$ = 156.2. C. rhodanide. White crystals, soluble in water; dissolves silk from mixed textiles. **c. thiosulfate*** $CaS_2O_3 \cdot 6H_2O$ = 260.3. White rhombs, decomp. by water and heat. **c. tungstate*** $CaWO_4$ = 287.9. C. wolframate, artificial scheelite. Tetragonal scales or white powder, insoluble in water; used in luminous pigments. **c. uranyl phosphate*** $(UO_2)_2Ca(PO_4)_2$ = 770.1. Yellow crystals, soluble in water. **c. urate** $Ca(C_5H_3O_3N_4)_2$ = 374.3. White powder, sparingly soluble in water. **c. valerate*** $Ca(C_5H_9O_2)_2 \cdot 3H_2O$ = 297.4. White crystals, soluble in water. **c. wolframate*** C. tungstate*.

calciuria An excess of calcium salts in the urine.

calcothar Colcothar.

calcspar Calcite.

calculus (1) An abnormal deposit of mineral salts in the body. **biliary ~** Gallstones. Solid deposits occurring either as: **(a) cholesterol ~**, consisting of cholesterol; **(b) pigment ~**, consisting of bile pigments, bilirubin, and calcium salts; **(c) inorganic ~**, consisting of carbonates and phosphates of calcium. **urinary ~** Solid mass of urinary sediment in the urinary tract. Commonest types: **(a) uric acid ~**, consisting of uric and urates; **(b) phosphate ~**, consisting of the triple phosphates; **(c) oxalate ~**, consisting of calcium oxalate crystals. (2) The mathematical laws of continuously varying quantities, which can be graphically expressed in curves. Cf. *diagram.*

calcyanide Calcium cyanide.

Caldwell crucible A silica crucible having an open bottom to hold a platinum or porcelain disk.

Caledon Trade name for anthraquinone vat dyestuffs.

caledonite Native basic lead and copper sulfates.

calefacient A drug applied externally to induce a sense of warmth.

calender A machine which presses moist cloth, paper, etc., between heavy rollers, to glaze its surface.

calendula Marigold. The flowering tops of *Calendula officinalis* (Compositae); a stimulant.

calendulin An amorphous substance from calendula.

caleometer An electrical instrument to measure the heat loss of a coil of wire at contant temperature.

calglucon Calcium gluconate.

Calgon Trademark for a glassy form of sodium hexametaphosphate containing a small amount of sodium pyrophosphate. It forms a soluble complex with calcium carbonate; used to soften water. See *threshold treatment.*

caliber, calibre The inside diameter or bore of a tube.

calibration (1) The graduation of a measuring instrument; (2) The determination of its error.

caliche Chile saltpeter. Crude sodium nitrate (20–50%, with sodium iodate) in the deserts of Atacama and Tarapaca (northern Chile).

calico A plain, heavy cotton cloth.

californite Massive vesuvianite, resembling jade.

californium* Cf. An element, at no. 98. Atomic weight of isotope with longest half-life (^{251}Cf) = 251.08. m.900. Forms di-, tri-, and tetravalent compounds. Prepared by irradiation, by Cunningham and Thompson (Lawrence Radiation Laboratory) in 1959.

calipers A device to measure the diameter of a tube. **micrometer ~** An instrument to measure with an accuracy up to 0.01 mm. **vernier ~** An instrument to determine the inside or outside diameter of a tube.

calisaya Cinchona.

callainite Turquoise.

callaite Turquoise.

callistephin An anthocyanin from strawberries and the purple aster *(Callistephus chinensis),* isolated as the chloride, $C_{15}H_{11}O_5Cl$. m.350+.

callitrol (*S*)-Citronellic acid. The alkali-soluble fraction, $b._{6mm}$ 118, of the wood oil of *Callitris glauca,* Australian cypress.

callophane A portable instrument utilizing daylight as a source of ultraviolet light.

calnitro A fertilizer made from ammonium nitrate and chalk containing 16–20% nitrogen.

calomel (1) Native mercurous chloride of secondary origin; a fungicide. (2) Mercurous chloride (pharmaceutical); formerly used in dusting powders and as a purgative. **c. electrode** A standard half-cell or electrode whose emf is that of mercury and calomel in contact with a solution of potassium chloride. At 20°C:

Solution strength	emf, V
Saturated potassium chloride	0.2492
1.0 *M* potassium chloride	0.2860
0.1 *M* potassium chloride	0.3379

Calor Gas Trademark for liquefied butane gas solid in small cylinders.

caloric Alchemical term for fire, which was supposed to possess weight.

calorie The metric system unit for quantity of heat. The heat required to raise the temperature of 1 g of water from $t°$ to $(t + 1)°C$. Hence the 15° cal, the 20° cal, etc., where t = 15, 20 etc. The international (SI) unit of heat is the *joule,* q.v., and the c. is defined in terms of it. **gram-~** Small c. **great ~**

Large c. **I.T. ~** International steam table c.; 1000 I.T.C. = 1/860 international kilowatt-hour. **kilo ~** Large c. **large ~** kcal, kg-cal, Cal, millithermil. The heat required to raise the temperature of 1 kg water from 14.5 to 15.5°C. **mean ~** The $\frac{1}{100}$ part of the heat required to raise 1 g water from 0 to 100°C at atmospheric pressure. **micro ~** Small c. Microthermil. **small ~** cal or gram-calorie. The quantity of heat required to heat 1 g water from 14.5 to 15.5°C. At 18°C, 1 cal = 1 therm = 0.001 kcal = 4.183 joules = 4.181×10^7 erg = 3.086 ft-lb = 0.426 kg-meter = 1.162×10^{-6} kWh = 0.003968 BThU.

Type of calorie	SI equivalent, J
I.T. c.	4.1868
15° c.	4.1858
4° c.	4.2045
Thermochemical c.	4.1840
Mean c. (0–100°C)	4.1900

calorifacient A substance which produces warmth.

calorific Carrying or holding warmth. **c. value** (1) Technical: the heat (J or kcal) obtained by the combustion of fuels or gases. (2) Physiological: the number of J or kcal derived from consuming foods or the daily diet. **gross c. value** High heat value. The number of heat units liberated when a fuel is completely burned in air, the water vapor so-formed (from the H and any moisture in the fuel) being cooled to the initial calorimeter temperature. **net c. value** Low heat value. As for gross c.v., but with the water vapor staying in the gaseous state. The two values differ by the latent heat of vaporization of the water formed (about 2.47 MJ/kg).

calorimeter An instrument to measure the amount of heat liberated or absorbed. **adiabatic ~** A c. kept at constant temperature. **bomb ~** An enclosed steel bomb for determination of the calorific value of fuels. **Emerson ~** A c. for determining the heat value of fuels. **Féry ~** A c. for determining the heat values of foods. **flame ~** A c. to measure heats of combustion of gases or vapors in oxygen at constant pressure. **isothermal ~** A c. which operates by transferring the heat of reaction to a surrounding liquid, which is simultaneously cooled by bubbling through it an inert gas, so that the heat production is just balanced. Cf. *colorimeter.*

calorimetric Pertaining to measuring heat quantities. Cf. *colorimetric analysis.*

calorimetry The measurement of heat. Cf. *colorimetry.* **differential scanning ~** A form of differential *thermal* analysis, q.v., in which energy output differences are measured instead of temperature differences.

caloriscope A device to demonstrate the release of heat in the respiration of an organism.

caloritropic Thermotropic.

calorizing A process for the production of aluminum coatings. Cf. *sheradize.*

calotropin A glucoside from *Calatropis procera* (Asclepiadaceae) which acts similarly to digitalis; an arrow poison. Cf. *akundarol.*

Calotropis floss Akund. The seed hair of *Calotropis procera* and *C. gigantea.* A fine, soft, lustrous but weak fiber, used in textiles.

calotype An early photograph (1840). Silver halide was formed on a sheet of paper by soaking it successively in appropriate reagents.

Calsolene Trademark for a highly sulfonated castor oil used as a detergent.

calumba Columba. The root of *Jateorhiza calumba (J. palmata)* (Menispermaceae), E. Africa. A former bitter stomachic tonic.

calumbic acid Colombic acid. An acid principle from calumba.

calumbin Columbin.

calutron An electromagnetic isotope separator (University of California).

calx (1) Latin for "lime" or "calcium oxide." (2) The alchemical name for an oxide or dross.

calycanthine $C_{22}H_{26}N_4$ = 346.5. An alkaloid from *Calycanthus fertilis,* Carolina allspice.

calyx The violet sheath around the flame of an air–coal gas mixture burning at a small hole.

calyxanthine Calycanthine.

cambium A layer of growing tissue between the wood and bast of a tree.

cambogia Gamboge.

cambopinic acid $C_{11}H_{18}O_2$ = 182.3. A crystalline acid from the resin of *Pinus cambodgiana.*

Cambrian See *geologic eras,* Table 38.

cambric A stiffened, plain light fabric of cottor or linen, originally from Cambrai, Belgium.

camelina The plant *C. satina* (Cruciferae), Mediterranean, which yields an oil, oil cake, and fiber.

camellin $C_{53}H_{84}O_{19}$ = 1025. A glucoside from the seeds of *Camellia japonica* (Theaceae). A red, bitter powder, soluble in water; a cardiac stimulant. Cf. *tea.*

camel's hair brush A small brush used in microoperations.

camenthol A mixture of camphor and menthol for inhalation.

camera (1) A lightproof box, compartment, or chamber. (2) An optical device for taking photographs. **chronoteine ~** A.c. taking 3,200 pictures per second (see *chronoteine*). **stereoscopic ~** A c. taking two pictures at slightly different angles simultaneously to bring out perspective.

camera lucida A prism attachment for the eyepiece of a microscope, which throws a magnified image of the object onto paper for tracing.

camomile See *anthemis, calendula, chamomile, matricaria.*

camouflet A space formed by an underground explosion; usually contains carbon monoxide.

Campden tablets A domestic food preservative, containing sodium or potassium disulfite.

campeachy wood Haematoxylon.

camphane Bornane*. **c.carboxylic acid** Camphocarboxylic acid.

camphanic acid $C_{10}H_{14}O_4$ = 198.2. α-Hydroxycamphoric acid lactone. Colorless crystals, m.201.

camphanol **2- ~** Borneol*. **3- ~** Epi*borneol.* **4- ~** Iso*borneol.*

camphanone **2- ~** Camphor. **3- ~** Epicamphor.

camphanyl The bornyl* radical.

camphene* $C_{10}H_{16}$ = 136.17. 2,2-Dimethyl-3-methylene-8,9,10-trinorbornane*.

CH; H₂C; CMe₂; CH₂; H₂C; C=CH₂; CH — 1; 6; 2; 7; 5; 3; 4

It occurs in three isomeric forms: **(±)- ~ , *i*- ~ .** Inactive c. Colorless needles, m.47, insoluble in water. **(+)- ~ , *d*- ~ .**

Dextro-c. Colorless needles, m.51, soluble in alcohol. **(−)-~, *l*-~.** Levo-c. Colorless needles, m. 52, soluble in alcohol. **chloro ~** See *chlorobornene.*

c. camphonic acid Camphoric acid*. **c. glycol** $C_{10}H_{18}O_2$ = 170.1. 2,3-Dihydroxybornane.

camphenilone $C_9H_{14}O$ = 138.2. 3,3-Dimethylbicyclo[2.2.1]heptan-2-one. **1-methyl ~** Camphenone.

camphenone $C_{10}H_{16}O$ = 152.2. 1-Methylcamphenilone. Colorless crystals, m.168.

campho acid $C_{10}H_{14}O_6$ = 230.2. Carboxylapocamphoric acid. Colorless crystals, m.196.

camphocarboxylic acid $C_{11}H_{16}O_3$ = 196.2. 2-Oxo-3-bornanecarboxylic acid. Colorless crystals, soluble in water.

camphogen Cymene*.

camphol Borneol*.

campholactone $C_9H_{14}O_2$ = 154.2. Colorless crystals, m.50.

campholenic acid $C_{10}H_{16}O_2$ = 168.2. 2,2,3-Trimethyl-*n*-cyclopenteneacetic acid. **α-~** (n = 3). A liquid. **β-~** (n = 1). Crystals, m.54.

campholic acid $C_{10}H_{18}O_2$ = 170.2. 1,2,2,3-Tetramethylcyclopentanecarboxylic acid*. Colorless prisms, m.106, slightly soluble in water.

campholide $C_{10}H_{16}O_2$ = 168.2. White crystals, m.211.

camphonanic acid $C_9H_{16}O_2$ = 156.2. 1,2,2-Trimethylcyclopentanecarboxylic acid. Colorless crystals, soluble in alcohol.

camphor (1) $C_{10}H_{16}O$ = 152.2. 1,7,7-Trimethylbicyclo[2.2.1]heptan-2-one, 2-oxobornane, 2-bornanone, 2-camphanone, Japan c., laurel c. A stearoptene from the leaves of Laurus or *Cinnamomum camphora,*

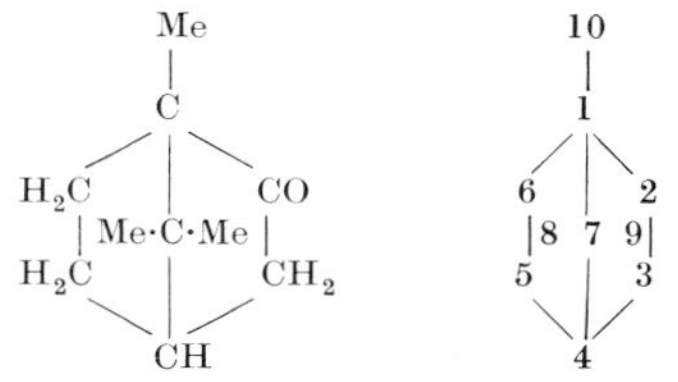

```
          Me                       10
          |                         |
          C                         1
        / | \                     / | \
    H2C   |   CO                6   |   2
     | Me·C·Me |                |8  7  9|
    H2C   |   CH2               5   |   3
        \ | /                     \ | /
          CH                        4
```

southeast Asia. (+), (−) and (±) forms. Large crystalline plates, easily broken when moistened with ether, d.0.879, m.175, $[\alpha]_D$ +44°; insoluble in water. An analgesic and rubifacient in ointments and liniments (USP, EP, BP); and a plasticizer for synthetic resins. Cf. *menthol, thymol.* (2) A group name for odorous principles of plants. See *camphors.* **alant ~** Helenin. **amido ~** $C_{10}H_{17}ON$ = 167.3. Colorless oily liquid. b.244. **anise ~** Anethole*. **artificial ~** Terpene monochlorohydrate. **azo ~** $C_{10}H_{16}N_2O$ = 180.3. Monoketazocamphorquinone. **beta-~** Epicamphor. **betula ~** Betulinol. **bitter almond ~** Benzoin*. **Borneo ~** Borneol*. **cantharides ~** Cantharidin. **champaca ~** Champacol. **cyano ~** $C_{11}H_{15}ON$ = 177.2. White crystals, m.127. **cyanomethylene ~** $C_{12}H_{15}ON$ = 189.3. Colorless crystals, m.46, b.280. **elecampane ~** Helenin. **epi ~** See *epicamphor.* **isonitroso ~** $C_{10}H_{15}O_2N$ = 181.2. Colorless crystals, m.153. **Japanese ~** Camphor. **laurel ~** Camphor. **ledum ~** Ledum camphor. **Malayan ~** Borneol*. **oxymethylene ~** $C_{11}H_{16}O_2$ = 180.2. White crystals, m.77. **parsley ~** Apiole. **peppermint ~** Menthol*. **pernitroso ~** $C_{10}H_{16}O_2N_2$ = 196.3. White crystals, m.43. **pine ~** Pinol. **Sumatra ~** Borneol*. **tar ~** Naphthalene*. **Tonka ~** Coumarin*. **thyme ~** Thymol*.

c. chlorated $C_{10}H_{15}OCl$ = 186.7. Yellow crystals, m.106, soluble in alcohol. **c.imide** $C_{10}H_{15}O_2N$ = 181.2. Camphoric imide. Colorless crystals, m. 248. **c.methylenecarboxylic acid** $C_{12}H_{16}O_3$ = 208.3. White crystals, m. 101. **c.methylimide** $C_{11}H_{17}O_2N$ = 195.3. White crystals, m.40. **c.methylisoimide** $C_{11}H_{17}O_2N$ = 195.3. White crystals, m. 134. **c. monobromated** $C_{10}H_{15}BrO$ = 231.1. Colorless crystals, m.76, soluble in alcohol; an antispasmodic, antineuralgic, and soporific. **c.nitrilo acid** $C_{10}H_{15}O_2N$ = 181.2. Cyanlauronic acid. Colorless crystals, m.152. **c. oil** The essential oil from the leaves of *Cinnamomum camphora.* Colorless liquid, d.0.87–1.04, of strong odor; an antiseptic and rubefacient. **c. oxime** $C_{10}H_{17}ON$ = 167.3. Colorless needles, m.118, insoluble in water. **c.wood oil** Oil distilled from the branches. Yellow oil of strong odor, d.1.155, containing camphor, safrole, pinene, phellandrene, cadinene; a liniment for rheumatism, bruises, and sprains. See *antimalum, camphorated oil.*

camphoramic acid $C_5H_4Me_3\cdot(CONH_2)\cdot COOH$ = 198.2. Camphoramidic acid. White crystals. **alpha-~** m. 177. **beta-~** m.183, soluble in water.

camphoranic acid $C_9H_{12}O_6$ = 216.2. Colorless crystals, m.209.

camphorated oil Camphor liniment. A rubefacient mixture of camphor 1, cottonseed or olive oil 4 pts.

camphoric acid* $C_8H_{14}(COOH)_2$ = 200.2. 1,2,2-Trimethyl-1,3-cyclopentanedicarboxylic acid*. Colorless prisms, m.187 (decomp.), soluble in water. **c. a. amide** Camphoramic acid. **c. a. diamide** $C_{10}H_{18}O_2N_2$ = 198.3. White crystals, m.197.

camphoronic acid $HOOC\cdot CH_2\cdot C(Me)COOH\cdot CMe_2\cdot COOH$ = 218.2. 2,3-Dimethyl-1,2,3-butanetricarboxylic acid*. Microcrystals.

camphoroyl* The radical $C_8H_{14}:(CO)_2=$, derived from camphoric acid.

camphors A group of odoriferous plant principles; solid volatile substances (stearoptenes), e.g., camphor, menthol, and thymol. Cf. *terpenes.*

camphoryl The monovalent radical $C_{10}H_{15}O-$, derived from camphor. **c.hydroxylamine** $C_{10}H_{15}O_3N$ = 197.2. Colorless crystals, m.225.

camphorylidene The radical $C_{10}H_{14}O=$, derived from camphor.

camphyl The bornyl* radical. **c. alcohol** Borneol*.

Camping Gaz Trademark for liquefied butane gas sold in small cylinders.

camptonide A variety of diorite.

camptonose A variety of basalt.

camwood A red dye wood from *Baphia nitida* (Leguminosae), W. Africa. Cf. *sandalwood, santalenic acid.*

canada asbestos Chrysotile. **c. balsam** A turpentinelike balsam exuded from incisions in the bark of *Abies balsamea,* of Canada and Maine. Yellow liquid, with a pleasant odor and bitter taste, dries in air to a transparent resin. Used as a microscope cement and varnish. Its refractive index equals that of glass. **c. snakeroot** Asarum. **c. turpentine** The essential oil of *Pinus maritima.* (Coniferae).

canadase Anorthosite from Maine.

canadine $C_{20}H_{21}O_4N$ = 339.4. Tetrahydroberberine. An alkaloid from *Hydrastis canadensis.* **(+)-~** Silky needles, m.133, insoluble in water.

canadol Impure hexane from petroleum; a local anesthetic.

canaigre The roots of *Rumex hymenosepalis* (Polygonaceae), Texas and Mexico; contains 25–30% tannin.

canal rays Positive rays. The positively charged molecules of gas emerging behind the perforated cathode of a high-vacuum tube of about 3.2 × 10^9 cm/s. They produce ionization, photographic action, fluorescence, and disintegration of certain substances. Cf. *ray(s), mass spectra.*

cananga Ylang-ylang.

canavalia bean The urease-rich seeds of *Canavalia* species.

canavalin A globulin of jack beans, the seed of *Canavalia* species (Leguminosae).

cancer A growth of malignant tissue.

cancrinite (1) $Na_4Al_3HCSi_3O_{15}$. Yellow, hexagonal rock, d.2.4, hardness 5–6. (2) $4Na_2O \cdot CaO \cdot 4Al_2O_3 \cdot 9SiO_2 \cdot 2(CO_2 \cdot SO_3) \cdot 3H_2O$. A constituent of steam boiler scales.

candela* Abbrev. cd. An SI base unit. 1 cd is the luminous intensity, in a given direction, of a source that emits monochromatic radiation of frequency 540 × 10^{12} Hz with a radiant intensity in that direction of 1/683 W per steradian. 1 lambert = 3183.1 cd/m^2.

candelilla wax Gama wax. Brown wax, d.0.983, m.67, from the candelilla plant (Mexico); used in candles, cements, polishes, varnishes, leather dressings, and dentistry.

candicin A mixture of heptane substances with antifungal properties from *Streptomyces griseus.* Candeptin, Vanobid. Yellow powder, insoluble in water. Used for fungal infections of skin and vagina (USP, BP).

candle Abbrev. ca. Former international unit of luminous intensity. See *foot-candle, illuminance.* **new ~** Candela*.
c. balance A balance to determine the burning rate of a c. **c.nut oil** Lumbang oil. **c.power** The luminous intensity of a standard candle. Practical standards are now a light of known luminous intensity relative to the candela. Traditional standards:

1 standard English sperm candle = 1 candle.
1 standard pentane lamp, burning pentane (International candle) = 10.0 candles.
1 standard Hefner lamp, burning pentyl acetate = 0.9 candle.
1 standard Carcel lamp, burning colza oil = 9.6 candles.

c. standard C. made of sperm wax, weight ⅙ lb, which burns 120 grains (7.776g) per hour.

candoluminescence Luminescence due to incandescent heat, i.e., temperatures exceeding 1000°C, as from a hydrogen-air flame.

canella Whitewood, cannamon bark. The bark of *Winterana canella* (Canellaceae), W. Indies; a condiment. **c. oil** The essential oil of c. Colorless liquid, d.0.920–0.935, containing eugenol, eucalyptol, and oleanolic acid.

cane sugar Sucrose made from sugar cane.

canfieldite Ag_8SnS. A rare, native sulfide.

cannabane $C_{18}H_{22}$ = 238.4. Cannabene hydride. A volatile hydrocarbon in hemp oil. Cf. *cannibene.*

cannabene $C_{18}H_{20}$ = 236.4. A hydrocarbon in hemp oil.

cannabidiol $C_{21}H_{30}O_2$ = 314.5. Crystals, m. 67. An isomer of cannabol in hemp resins.

cannabin (1) A glucoside, or (2) a resin, from *Cannabis indica.*

cannabine An alkaloid from cannabis; a hypnotic.

cannabinol $C_{21}H_{26}O_2$ = 310.4. The active principle of *Cannabis sativa.* Yellow oil, d.1.042, b_{100mm}315, insoluble in water, green fluorescence in glacial acetic acid. **acetyl ~** See *acetylcannabinol.*

cannabis Indian hemp, Indian c., bhang, ganja, hashish, marihuana, pot. The flowering tops of or resin from the female plant of *C. indica* or *C. sativa,* hemp (Urticaceae). A central nervous system stimulant producing excitement, euphoria and change of mood; also a narcotic. Dependence is psychological rather than physical; withdrawal symptoms do not seem to be produced.

cannabol $C_{21}H_{30}O_2$ = 314.5. An isomer of cannabidiol in hemp resins, m.66.5.

cannel coal A hard bituminous coal with a luminous flame, a smooth conchoidal fracture, and a high volatiles content. Used for gas production. Cf. *boghead, torbanite.*

cannibene $C_{15}H_{24}$ = 204.4. A sesquiterpene, d.0.897, b.259, from hemp oil. Cf. *cannabane.*

cannizzarization The Cannizzaro reaction.

Cannizzaro, Stanislao (1826–1910) Italian chemist noted for his work on organic chemistry and the amplification and application of Avogadro's hypothesis to the atomic theory.
C. number The mg of potassium hydroxide which react with 1 g aldehyde in C. reaction. **C. reaction** The decomposition of aromatic aldehydes by alcoholic KOH, with the formation of acids and alcohols; e.g.:

$$2RCHO + KOH \rightarrow R \cdot CH_2OH + R \cdot COOK$$

cannonite A high-explosive nitrocellulose-nitroglycerin mixture.

cannula A plastic, metal, or glass tube used to connect blood vessels; also inserted into body cavities.

cantharene C_8H_{12} = 108.2. Dihydro-*o*-xylene. Colorless liquid, b.135.

cantharides Blistering beetle, cantharis, Russian fly, Spanish fly, *Lytta* or *Cantharis vesicatoria;* a blistering agent.

cantharidic acid Cantharidin.

cantharidin $C_{10}H_{12}O_4$ = 196.2. Cantharis. Colorless crystals, slightly soluble in water, m.210; a blistering agent.

Cantharis A genus of beetles, now *Lytta.*

Canton phosphorus A luminescent mixture of oystershells 2, sulfur 1 pt.

canula Cannula.

canvas A strong, close hemp or flax fabric used in filter presses and sacking.

caoutchouc (Malaysian, "weeping tree"). Rubber. **Gaboon ~** Dambonite. **mineral ~** See *mineral caoutchouc.*

CAP Abbreviation for chloracetophenone.

capacitance* Electric c*. An isolated capacitor has unit c. when unit electrical quantity will create a unit potential difference between its plates. SI unit is the *farad,* q.v.

capacity (1) The ability to contain a force or exert energy. (2) Volume. **electrostatic ~** Capacitance*. **heat ~*** See *heat capacity.* **specific inductive ~** Relative *permittivity**. **thermal ~** Heat capacity*.

caparrosa The leaves of *Nea theifera* (Nyctaginaceae), S. America; a tea.

capers The green flower buds of *Capparis spinosa* (Capparidaceae), Mediterranean; a pickle and condiment.

capillaries The network of delicate blood vessels or other small tissue-connecting tubes.

capillarity Capillary *attraction.*

capillary A tube with a very small inside diameter. **c. analysis** (1) Early name for chromatographic analysis. (2) A filter paper dipped into a *negative* colloid absorbs both the dispersed and external phase, but in a *positive* colloid only the external phase. Cf. *adsorption.* **c. correction** A correction for the capillarity of mercury applied to mercury thermometers of diameter above 25 mm. **c. electrode** See *Lippmann electrode.* **c. electrometer** See *electrometer.* **c. pipet** A pipet for measuring fractions of a mL. **c. tubing** Glass tubing with inside diameter less than 1 mm.

capillator An apparatus for the colorimetric determination of pH values in which the solutions are compared in capillary tubes, to reduce the effect of color or turbidity.

capnometry The measurement of smoke density. Cf. *nephelometry.*

capori(e)t A disinfectant mixture of calcium hypochlorite and sodium chloride containing about 50% active chlorine.

capraldehyde Decanal*. **iso～** 4-Methyl-1-*pentanal.*
capramide (1) Octanamide*. (2) Decanamide*.
caprate Decanoate*.
capreomycin sulfate An antibiotic. A polypeptide mixture from *Streptomyces capreolus.* White solid, soluble in water. Used to treat tuberculosis resistant to drugs (USP, BP).
capric **c. acid** Decanoic acid*. **c. aldehyde** Decanal*. **c. amide** Decanamide*. **c. anhydride** Decanoic anhydride*. **c. nitrile** Decanenitrile*.
Caprifoliaceae Honeysuckle family. Shrubs of twining plants, some yielding drugs. Flowers: *Sambucus canadensis,* elder; bark: *Viburnum opulus,* cramp bark; roots: *Triosteum perfoliatum,* fever root; fruits: *Lonicera xylosteum,* xylostein.
caprilic acid Octanoic acid*.
caprin Former term for a compound of decanoic (capric) acid and glycerol.
caprine Norleucine.
caprinitrile *Decane*nitrile*.
caproaldehyde Hexanal*.
caproate Hexanoate*.
caprock A geological deposit produced by the prolonged action of heat and pressure on clays and muds.
caproic acid Hexanoic acid*. **pseudo～** *tert*-Butyl*acetic acid.* **secondary ～** Methyl propylacetic acid.
caproin A compound of hexanoic acid and glycerol.
caprolactam $(CH_2)_5C(O)N$ = 112.2. The monomer for nylon 6, produced by the photochemical reaction of nitrosyl chloride with cyclohexane.
caprone $C_{11}H_{22}O$ = 170.3. 6-Undecanone*. A volatile ketone from butter.
capronic acid Hexanoic acid*.
capronium Ytterbium.
capronyl The hexanoyl* radical.
caprophyl Dung bacteria.
caproyl The octanoyl* radical. **c. acetate** Octyl acetate*. **c. alcohol** Octanol*. **c.aldehyde** Decanal*.
capryl The hexyl* radical.
caprylate Octanoate*.
caprylene Octene*.
caprylic **c. acid** Octanoic acid*. **iso～** Triethyl*acetic acid**. **c. alcohol** Octanol*. **c. aldehyde** Octanal*.
caprylidene Octyne*.
caprylin A compound of octanoic acid and glycerol.
capryryl The octanoyl* radical.
capsaicin $C_{18}H_{27}O_3N$ = 305.4. A bitter principle, m.65, from capsicum. Cf. *zingerone.*
capsanthin $C_{40}H_{56}O_3$ = 584.9. The red carotinoid pigment of paprika, m.175.
capsic acid An active principle of pimenta.
capsicin An oleoresin from capsicum. Brown masses, soluble in alcohol; a stimulant, anodyne, and rubefacient.
capsicine An alkaloid from capsicum.
capsicol The essential oil of capsicum.
capsicum Cayenne pepper, chili, pepper red. The fruits of *Capsicum fastigiatum, C. annuum,* and other Solanaceae; a local stimulant and condiment. Cf. *paprika.* **c. resin** An oleoresin from *Capsicum;* a carminative and rubefacient.
capsularin $C_{22}H_{36}O_8$ = 428.5. A glucoside in jute leaf, *Corchorus capsularis* (Tiliaceae). Colorless needles, m.175.
capsule (1) A membranous sac enclosing a structural part of an organism. (2) A small sealed container for volatile compounds. (3) More specifically, in pharmacy, a medicament in a closed shell, e.g., methylcellulose or gelatin, which is soluble in water at 37°. **micro～** See *microcapsules.*
capsuloesic acid Aescinic acid.
captafol* See *fungicides,* Table 37 on p. 250.
captan* See *fungicides,* Table 37 on p. 250.
caput Pl., capita. A head. **c. mortuum** (1) Early name for the earthy residue from distillation or incineration. (2) Colcothar.
caracolite $PbOHCl \cdot Na_2SO_4$. An orthorhombic, native double salt.
caraguata The fiber from *Eryngium pandanifolium* (Umbelliferae), S. America; used for ropes, matting, and bags.
carajura A red pigment from *Begonia chica;* a source of carajurin.
carajuretin $C_{15}H_{10}O_5$ = 270.2. A derivative of carajurin. Scarlet needles, m.330+ (decomp.).
carajurin $C_{17}H_{14}O_5$ = 298.3. 4,5-dimethyl ether of carajuretin. m.206. The principal colored constituent of carajura.
caramel $(C_{12}H_{18}O_9)n$. Obtained by heating sugar to about 200, in presence of ammonium salts. A dark brown, soluble mass; a coloring matter for food and beverages (NF).
caramelan A brown, amorphous constituent of caramel. It forms soluble salts with lead and alkalies; is hydrolyzed by dilute acids to glucose, furaldehyde, and levulinic acid; and reduces Fehling's solution.
carana Mararo, caranna, gum carana. A gray resin, *Urotium carana* (Rutaceae), S. America.
caranda wax A wax from *Copernical australis,* Brazil; m.82, d_{25} 0.984, acid val. 2.7, iodine val. 10.5; similar to carnauba wax.
carane* $C_{10}H_{18}$ = 138.3. 3,7,7-Trimethylbicyclo[4.1.0]heptane. A constituent of essential oils. **hydroxy～** Carol. **oxo～** Carone. **c.amine** Carylamine. **c.diol** $C_{10}H_{18}O_2$ = 170.3. Dihydroxycarane, in Indian turpentine. **c. ol** Carol.
caraneol Carol.
caranna Carana.
carat Karat. A unit of (1) weight of precious stones and (2) the fineness of gold. Based on the ancient use of seeds of the locust pod tree, *Ceratoria siliqua,* which are remarkably uniform in weight. (1) When applied to a diamond, 1 c. = 200 mg. (2) When applied to the fineness of gold, 1 c. is 1/24 of an early Roman coin. As pure gold is 24 c., 15-c. gold contains 62.5% gold. **metric ～** The international c., weighing 200.000 mg = 3.086 grains. It is subdivided into points (1 point = 0.01 c.) and pearl grains (1 p.g. = 0.25 c.).
caraway The fruits of *Carum carvi* (Umbelliferae); an aromatic, carminative, and flavoring. **c. oil** The essential oil of c. seeds, d.0.907–0.915, b.180–230; used for flavoring (NF, BP). **c. water** Aqua carui: Water flavored with c. oil.
carbachol (USP) $C_6H_{15}N_2O_2Cl$ = 182.6. Carbamylcholine chloride, choryl, moryl. White, hygroscopic crystals, m.203, soluble in water; used in ophthalmology and as a bladder muscle stimulant.
carbacidometer Air tester.
carbagel Calcium chloride on porous carbon; a desiccant.
carbaldehyde* Suffix indicating an aldehyde.
carbamamidine Guanidine*.
carbamate* A salt of carbamic acid; it contains the NH_2COO- radical. **ethyl ～** Urethane.
carbamazepine $C_{15}H_{12}ON_2$ = 236.3. 5*H*-Dibenz[*b,f*]azepine-5-carboxamide, Tegretol. White crystals, m. 190, insoluble in water. Used for epilepsy and trigeminal neuralgia (USP, BP).
carbamic The radical $-NH \cdot COO-$, from carbamic acid. **c. esters** Urethanes.
carbamic acid* NH_2COOH. Amidocarbonic acid, aminoformic acid. The simplest amino acid, known only as its salts (carbamates). Cf. *thionamic acid.* **amino～** Carbazic acid*. **carbamoyl～** Allophanic acid*. **dithio～** See *dithiocarbamic acid.* **phenyl～*** Carbanilic acid*.

carbamide Urea. **c. chloride** Carbamoyl chloride. **c. peroxide** A solution in anhydrous glycerol, prepared from hydrogen peroxide and urea; an antiseptic.
carbamidine Guanidine*.
carbamido The ureido* radical. **phenyl ~** The phenylureido* radical.
carbaminate Carbamate*.
carbamonitrile Cyanamide.
carbamoyl* Aminocarbonyl†, carbamyl. The radical NH_2CO- from carbamic acid. **c.carbamic acid** Allophanic acid*. **c. chloride** NH_2COCl = 79.5. Carbamide chloride. Colorless needles, m.50, insoluble in water. Cf. *Friedel-Crafts reaction.*
carbamylic ester Phenylurethane.
carbanil aldehyde Formanilide.
carbanilic acid* PhNH·COOh = 123.1. Phenylcarbamic acid*. Known only as its derivatives.
carbanilide PhNH·CO·NHPh = 212.3. Diphenylurea*. Colorless needles, m.236, slightly soluble in water.
carbanilino- The phenylcarbamoyl* radical.
carbanilo- The anilino* radical. **c.nitrile** Cyanoaniline.
carbanions* Anions formed by removal of proton(s) from a C atom; e.g., benzenide, $C_6H_5^-$.
carbarsone $C_7H_9N_2O_4As$ = 260.1. *N*-Carbamoylarsanilic acid, *p*-carbaminophenylarsonic acid. White powder, insoluble in water; an antiprotozoan, used to treat amebiasis (USP). Cf. *acetarsol.*
carbates General name for the substituted dithiocarbamates.
carbazic acid* NH_2·NH·COOH = 76.1. Hydrazinecarboxylic acid. Powder, m.ca.90 (decomp.).
carbazide Carbonohydrazide*. **semi ~*** See *semicarbazide.* **thio ~** A compound containing the −NH·NH·CS·NH·NH− radical. **thiosemi ~** A compound containing the NH_2CS·NH·NH− radical.
carbazole* $C_{12}H_9N$ = 167.2. Dibenzopyrrole, diphenylenimide.

NH
C_6H_4———C_6H_4

Colorless crystals, m.245; an explosives stabilizer. **bi ~*** $C_{24}H_{16}N_2$ = 332.4. Dicarbazyl. Heterocyclic ring compounds, as 9,9′- ~; m. 221:

C_6H_4 C_6H_4
N——N
C_6H_4 C_6H_4

acetyl ~ See *acetylcarbazole.* ***N*-ethyl ~** $C_{14}H_{13}N$ = 196.3. Colorless crystals, m.68. **hexahydro ~** $C_{12}H_{15}N$ = 173.3. Colorless crystals, m.99. ***N*-methyl ~** $C_{13}H_{12}N$ = 182.2. Colorless crystals, m.87. **tetrahydro ~** $C_{12}H_{13}N$ = 171.2. Colorless crystals, m.119.
carbazolyl* Carbazyl. The radical $NC_{12}H_8-$ from carbazole; 5 isomers.
carbazone* A compound containing the carbazono* radical, −N:N·CO·NH·NH−. **thio ~** A compound containing the thiocarbazono* radical, −N:N·CS·NH·NH−.
carbazotate Picrate*.
carbazoyl* Hydrazinocarbonyl†. The radical NH_2·NH·CO−.
carbazylic acid An organic acid of the type $RC(:NH)NH_2$.
carbendazim* See *fungicides,* Table 37 on p. 250.
carbene (1)* Methylene*. The free radical $:CH_2$. (2) Cuprene.
carbenes (1) Divalent carbon compounds containing the free radical $:CX_2$. They are of low stability. (2) Constituents of bitumen, insoluble in carbon tetrachloride, soluble in carbon disulfide. Cf. *asphaltenes.*
carbenicillin (di)sodium $C_{17}H_{16}O_6N_2SNa_2$ = 422.4. Pyopen. White powder, soluble in water. An antibiotic used to treat infections due to gram-negative bacteria, particularly *Pseudomonas.*
carbenium* The cation CH_3^+.
carbenoid Describing the properties of a carbene.
carbenoxolone sodium $C_{34}H_{48}O_7Na_2$ = 614.7. Disodium enoxolone succinate, Biogastrone. White powder, m.122, slightly soluble in water. Used for peptic and mouth ulcers (BP).
carbetamide* See *herbicide,* Table 42.
carbethoxy The ethoxycarbonyl* radical.
carbethoxymino Urethane. The ethoxycarbonylimino* radical, −NH·CO_2Et.
carbethylic acid Ethyl carbonate.
carbide Carbonide, carburet. Cf. *acetylide.* A binary carbon compound of a metal. With water, carbides give acetylene (Li_4C); methane (Al_2C); hydrogen and methane (MgC_2); or a mixture of hydrogen, methane, and acetylene (rare-earth carbides). Carbides of the rare metals form solid, liquid, or gaseous hydrocarbons; some carbides (as SiC, B_4C) are extremely stable. Some contain the C_2^{2-} ion. Cf. *Crystolon, Carboloy.*
carbidopa $C_{10}H_{14}O_4N_2 \cdot H_2O$ = 244.3.White powder, slightly soluble in water. Inhibits the breakdown of dopamine in the body and thereby potentiates the effect of levodopa, q.v. (USP, BP).
carbimazole $C_7H_{10}O_2N_2S$ = 186.2. Ethyl 3-methyl-2-thioxo-4-imidazoline-1-carboxylate, Neo-mercazole. White, bitter crystals, with characteristic odor, m.123, slightly soluble in water; for hyperthyroidism (BP).
carbimide Iso*cyanic acid.*
carbinol Former term for (1) primary alcohols and their radical $-CH_2OH$. (2) Methanol. **acetyl ~** 1-Hydroxy-2-*propanone**. **benzyl ~** Phenethyl alcohol*. **butyl ~ primary ~** Pentyl alcohol. **secondary ~** 2-Methyl-1-*butanol.* **tertiary ~** 2-Dimethyl*propanol**. **dimethyl ~ secondary ~** Propanol*. **diphenyl ~** Benzohydrol. **ethyl ~** Propanol*. **methyl ~** Ethanol*. **phenyl ~** Benzyl alcohol. **propyl ~** Butanol*. **trimethyl ~** *tert*-Butyl alcohol.
Carbitol $EtO(CH_2)_2 \cdot O \cdot (CH_2)_2OH$ = 134.2. Trademark for diethylene glycol ethyl ether. Colorless liquid, b.195, soluble in water; used as a solvent and in cosmetics. **butyl ~** $BuO(CH_2)_2OCH_2CH_2OH$ = 162.2. Diethylene glycol butyl ether. Colorless liquid, b.222; a solvent.
carbo Charcoal. **c. animalis** Charcoal from animal matter; a decolorizing agent. **c. lignius** Wood charcoal. Used for decolorizing solutions and in blowpipe analysis. **c. sanguinarius** Blood coal. Charcoal from animal blood.
carbobenzoyl A compound containing the radical −CO·C_6H_4COOH. **c.acetic acid** Colorless, m.90. **c.formic acid** Phthalonic acid. **c.propionic acid** Colorless, m.137.
carbocinchomeronic acid Cinchomeronic acid.
carbocyanine See *cyanine dyes.*
carbocyclic* Describing a homocyclic ring compound in which all the ring atoms are carbon, e.g., benzene. Cf. *heterocyclic.*
carbodiazone* A compound containing the radical −N:N·CO·N:N−. **thio ~*** A compound containing the radical −N:N·CS·N:N−.

carbodiimide* Hypothetical compound, NH:C:NH.
Carbofrax Trademark for certain silicon carbide refractories bonded by other ceramics.
carbofuran* See *fungicides,* Table 37 on p. 250, and *insecticides,* Table 45 on p. 305.
carbohydrates Organic compounds synthesized by plants. They often fit the general formula $C_x(H_2O)_y$. *Monosaccharide* (q.v.): x and y are 2, 3, 4, 5, 6, or 7; e.g., glucose. Disaccharides: x is 12, y is 11; e.g., lactose. Trisaccharides: x is 18, y is 16; e.g., raffinose. Polysaccharides: x and y exceed 18; e.g., dextrin, cellulose. Natural c. are generally dextrorotatory, except fructose and inositol. Conjugated saccharides: (1) gums and mucilage group (saccharides and acids); (2) *glucosides,* q.v. (saccharides and another compound); (3) *tannins,* q.v. (saccharides and tannins). **c. catabolism** Achieved in animals by *glycolysis,* q.v., followed by an acetyl coenzyme A intermediate and the *citric acid cycle,* q.v.
carbohydrazide Carbonohydrazide*.
carbohydrazones Carbonohydrazides.
carbohydride Hydrocarbon.
carboids Kerotenes.
carbolate Phenolate*.
carbolfuchsin Ziehl's stain, Ziehl-Neelson. Stain for tubercle and similar bacilli, that is not removed by acid. Fuchsin 5, phenol 25, alcohol 50, water 500 pts. **c. topical solution** Castellani's paint, magenta solution. Used to treat skin infections (USP, BP).
carbolic **c. acid** Phenol*. **c. liquid** Cresylic acid. **c. oil** The phenolic fraction of coal tar, b.180–230.
carbolmethyl violet A microscope stain: 10 pts. alcoholic methyl violet 6B, 90 pts. of 5% aqueous phenol solution.
Carbolon Trademark for silicon carbide.
Carboloy Trademark for cemented tungsten carbide; used for high-speed machine tools and second in hardness to diamond.
carbolxylene A clearing solution: 3 pts. xylene, 1 pt. phenol.
carbomer A polymer of acrylic acid. Fluffy, hygroscopic powder, characteristic odor, soluble in water. A pharmaceutical gel (NF, BP).
carbometer A device to measure carbon dioxide in air.
carbomethene Ketene*.
carbomethoxy The methoxycarbonyl* radical.

carbon* C = 12.011. At. no. 6. A nonmetallic bioelement; 3 allotropes: amorphous (coal), graphite, and crystalline (diamond). m.3650 (sublimes). It occurs native as coal, graphite, and diamond; in combination with hydrogen as petroleum, with oxygen as c. dioxide. The isotope ^{14}C (half-life period, 5,730 years) is produced by irradiation of tellurium nitride, and is continuously in the atmosphere from the interaction of cosmic rays and nitrogen. See *radiocarbon dating.* Also used to label organic compounds for use as tracers; as in medicine. ^{12}C, the natural, dominant isotope of c., is the basis of the scale of atomic weights of the elements; i.e., $^{12}C = 12$. Cf. *isotopes.* C. is an element essential to vegetable and animal life. Its principal valency is 4, but some divalent c. compounds (carbenes) have been prepared. Its atoms have a greater affinity for one another than for other atoms, and give rise to numerous different (organic) compounds. The binary compounds are carbides, M_xC_y; hydrocarbons, C_xH_y; carbonyls, CO^-. **amorphous ~** C. as minute graphitelike crystallites. **asymmetric ~** See *stereoisomerism.* **fixed ~** The char remaining after removal of the *volatile matter,* q.v., from a fuel. **graphitic ~** The loss on ignition of graphite below its fusion point in air. **liquid ~** See *liquid carbon* under *liquid.* **synthetic ~** See *synthetic graphite* under *graphite.* **total organic ~** T.O.C. Measure of effluent strength involving oxidation of the organic c. to CO_2. **whetlerized ~** C. containing 5–12% Cu, to increase its absorbency. Cf. *activated c., gas c., charcoal, graphite, diamond, lampblack.*

c. apparatus An instrument to determine total c. in fuels. **c. atom** **asymmetric ~** See *asymmetric c.* **c. bisulfide** C. disulfide*. **c. black** Lampblack. **c. bond** The nonpolar electron linkage between 2 c. atoms. **c. bronze** An alloy for bearings. **c. chains** A succession of linked c. atoms in a compound. **closed ~** Aromatic compounds. **open ~** Aliphatic compounds. **c. compounds** See *organic compounds.* Characteristics: (1) nonpolarity: they do not ionize; their reactions are molecular and have a low velocity; (2) polymerism; (3) isomerism and asymmetry; (4) combustibility: all c. atoms are oxidized to c. dioxide and other products. **c. cycle** The circulation of c. between a living organism and the surrounding environments is shown in Fig. 6. **c. dating** See *radiocarbon dating.* **c. dichloride*** C_2Cl_4 = 165.8. Ethylene

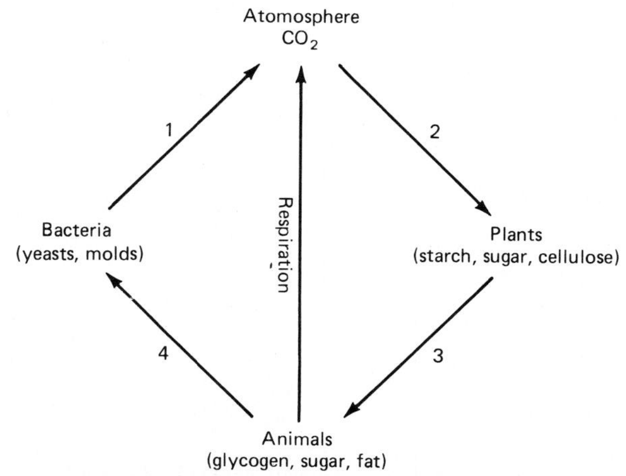

Fig. 6. The carbon cycle: (1) bacterial action, (2) photosynthesis, (3) metabolism, (4) decay.

perchloride. Colorless liquid, b.122. **c. dioxide*** CO_2 = 44.0. Carbonic acid gas, carbonic anhydride. Heavy, colorless, incombustible gas, $d_{air=1}$ 1.53, m.−65, $b._{5.3atm}$−56, soluble in water. Shipped as compressed liquid in steel tanks, and used for carbonating beverages, in refrigerators and fire extinguishers, and as a fertilizer. Formed in the body by metabolism and expired from the lungs. Used to stimulate respiration in anesthetics; solid c. d. used to treat warts and small birth marks (USP, EP, BP). See *Dry Ice* under *dry*. **c. disulfide*** CS_2 = 76.1. Colorless liquid with characteristic odor, b.46.2, slightly soluble in water; a local anesthetic and a solvent for cellulose (rayon and cellophane manufacture), sulfur, iodine, rubber. **c. fibers** See *synthetic graphite* under *graphite*. **c. group** The 4B group of the periodic table. **c. hexachloride*** C_2Cl_6 = 236.7. C. trichloride, ethyl perchloride, hexachloroethane. Colorless crystals, m.182, b.187, insoluble in water. **c. light** An electric arc light with c. electrodes. **c. monosulfide*** CS = 44.1. Colorless gas, b.−130, very unstable and polymerizes to a red solid. **c. monoxide*** CO = 28.01. Colorless, poisonous gas. b.−190, slightly soluble in water, formed during incomplete combustion of c. **c. oxysulfide** Carbonyl sulfide*. **c. paper** A tissue paper coated with a mixture of a wax and a black pigment (often c. black); used to make copies of typewriting. **c. print** A photographic process for artistic reproduction of negatives. **c. residue** Conradson c. The amount of c. produced from a lubricating oil heated in a closed crucible under standard conditions. **c. subnitride** Acetylene dinitrile. **c. suboxide** O:C:C:C:O = 68.0. Tricarbondioxide*, malonic anhydride. A pungent, lacrimatory, colorless gas, b.7, decomp. by water to malonic acid. **c. subsulfide** C_3S_2 = 100.2. Tricarbondisulfide*. Red, pungent liquid, m.−0.5, polymerized by heat. **c. tetrabromide*** CBr_4 = 331.6. Tetrabromomethane*. Colorless scales, d.3.42, m.92, insoluble in water. **c. tetrachloride** CCl_4 = 153.8. Tetrachloromethane*, phenoxin, Pyrex. Colorless liquid, b.76, slightly soluble in water. A local anesthetic, fire extinguisher, nonflammable solvent, cleaning agent, and reagent; formerly an anthelmintic (NF). **c. tetrafluoride** CF_4 = 88.0. Tetrafluoromethane*, fluoromethane. Colorless gas, b.−126, by-product in the manufacture of aluminum from cryolite. **c. tetraiodide** CI_4 = 519.6 Tetraiodomethane*. Red crystals, d_{20} 4.32, decomp. by heat, insoluble in water. **c. trichloride** Hexachlorethane*.

carbonaceous Containing carbon.

carbonado Bort. A hard, black cutting diamond.

carbonatation Formation of carbonates by carbon dioxide. Cf. *carbonation*.

carbonate A salt of the theoretical carbonic acid, containing the radical CO_3^{2-}. Carbonates are readily decomposed by acids. The carbonates of the alkali metals are water-soluble; all others are insoluble. **bi ~** Hydrogen c. **chloro ~** See *chlorocarbonate*. **hydrogen ~, bi ~** Acid c. A salt containing the ion HCO_3^-.

c. dehydratase* Carbonic anhydrase. An intracellular enzyme occurring in high concentrations in red blood cells. It catalyzes the reversal of the reaction $CO_2 + H_2O \rightleftharpoons H_2CO_3$. **c. minerals** Rock-forming minerals; as, calcite, $CaCO_3$; dolomite, $CaMg(CO_3)_2$; magnesite, $MgCO_3$; siderite, $FeCO_3$.

carbonation (1) Carbonization. (2) The precipitation of lime by carbon dioxide, e.g., in sugar refining. (3) The saturation of water with carbon dioxide, e.g., in soda water manufacture.

carbonic A compound containing tetravalent carbon. Cf. *carbonium*.

carbonic acid (1) HO·COOH. *m*-Carbonic acid, hydroxyformic acid. The hypothetical acid of carbon dioxide and water; known only as its salts (carbonates), acid salts (hydrogencarbonates), amines (carbamic acid) and acid chlorides (carbonyl chloride). (2) An old term for carboxylic acid. **ortho ~** $C(OH)_4$. Exists only as compounds, e.g., esters.

c. a. ester An organic compound in which the H of c. acid is substituted by a radical. **meta ~** Compounds of the general formula RO·CO·OR. **ortho ~** Compounds of the general formula $C(OR)_4$. **c. a. hydrate** $CO_2 \cdot 6H_2O$.

carbonic anhydrase Carbonate dehydratase*.

carbonic anhydride Carbonic acid.

carbonic ester Carbonic acid ester. **ethyl ~** $CO(OEt)_2$ = 118.1. Ethyl carbonate. Colorless liquid, b.126. **ethylmethyl ~** EtO·CO·OMe = 104.1. Colorless liquid, b.109. **ethylene ~** $CO(OC_2H_3)_2$ = 114.1. Colorless crystals, m.39. **methyl ~** $CO(OMe)_2$ = 90.1. Methyl carbonate. Colorless liquid, b.91. **methylpropyl ~** PrO·CO·OMe = 118.1. Colorless liquid, b.131.

carbonic ether Ethyl carbonate. See *ethyl carbonic ester* under *carbonic ester*.

carbonide Carbide.

carboniferous (1) Containing carbon. (2) *Geologic* period, q.v., corresponding to the Pennsylvanian and Mississippian periods.

carbonimidoyl* The radical =C:NH.

carbonite (1) Small charcoal briquettes. (2) A high explosive: nitroglycerin 17–30, sodium nitrite 24–30, flour 37–44%.

carbonitrile* Suffix indicating the −CN group (nitrile*, cyanide*), plus the systematic name ending in carboxylic acid, from which the compound's name is derived; as, cyclohexanecarbonitrile, $C_6H_{11}CN$, from cyclohexanecarboxylic acid, $C_6H_{11}COOH$.

carbonium Indicating the radical R_3C-, usually a halide.

carbonization (1) The transformation of organic matter into charcoal. (2) The distillation of coal, as in gas manufacture. **high-temperature ~** Heating coal out of air at 1000–1300°C, with the formation of gas, tar, oil, ammonia, and coke. **low-temperature ~** Heating coal at 450–700°C, with the formation of gas, petroleum (hydrocarbons from pentane to octane, and pentene to octene), and coke.

carbonize To convert to carbon by charring or burning incompletely.

carbonizer Concentrated aluminum chloride solution; removes cellulose from wool.

carbonohydrazide* $(NH_2 \cdot NH)_2C:O$ = 90.1. Carbohydrazide, carbazide. Crystals, m.154 (decomp.).

carbonohydrazido* The radical $H_2N \cdot NH \cdot CO \cdot NH \cdot NH-$, from carbonohydrazide.

carbonoid A suggested tetragonal structure of carbon, with 4 faces, one for each valency. Cf. *benzenoid*.

carbonometer A device to determine the carbonic acid content of blood. Cf. *carbometer*.

carbonothioyl† See *thiocarbonyl*.

carbonotrithioic acid Trithiocarbonic acid*.

carbonoxysulfide Carbonyl sulfide*.

carbon rheostat An electrical resistance consisting of a number of carbon plates mounted so that pressure can be placed on them by a screw and their total resistance thus altered.

carbonyl* The radical =CO. Cf. *carbonyls, sulfinyl*. **c. bromide** $COBr_2$ = 187.8. Bromophosgene. Poisonous liquid, b.64.5. **c. chloride*** $COCl_2$ = 98.9. Phosgene. Poisonous gas, b.8.2, decomp. by water; an important chemical intermediate, e.g., in the manufacture of polyurethane resins.

c.diimino† See *ureylene*. **c.dioxy*** The radical $-O \cdot CO \cdot O-$. **c.diurea** Triuret*. **c.pyrrole** $CO(C_4H_4N)_2$ = 160.2. Colorless crystals, m.63. **c. sulfide*** COS = 60.0. Carbon oxysulfide. Colorless gas, b.−50, slightly soluble in water, explosive in air. **c. thiocarbonanilide** $C_2ON_2SPh_2$ = 254.3. Colorless crystals, m.87. **thiocarbonyl thiocarboanilide** $C_2N_2S_2PH_2$ = 270.4. Colorless crystals, m.79.

carbonyls* Carboxides. Compounds of carbon monoxide and metals, some volatile; as, nickel carbonyl.

carbophenothion* See *insecticides*, Table 45 on p. 305.

carboraffin An activated charcoal, used chiefly for decolorizing sugar solutions.

Carborundum Trademark for certain silicon carbide and other abrasives.

carbosant $(C_{15}H_{23}) \cdot O \cdot COO(C_{15}H_{22})$ = 466.7. Santalyl carbonate. Carbonic acid ester of sandalwood oil. Yellow oil, insoluble in water.

carbostyril* C_9H_7ON = 145.2. 2(1*H*)-Quinolinone. Colorless prisms, m.199, slightly solube in water. **ethyl ~** See *ethylcarbostyril*. **hydro ~** C_9H_9ON = 147.2. Colorless crystals, m.163. **hydroiso ~** C_9H_9ON = 147.2. Crystals, m.71 **iso ~** C_9H_7ON = 145.2. 1(2*H*)-Isoquinolinone. Crystals, m.208. **methyl ~** Lepidone. **nitro ~** $C_9H_6O_3N_2$ = 190.2. Colorless crystals, m.168. **octahydro ~** $C_9H_{15}ON$ = 153.2. Crystals, m.151. **oxy ~** $C_9H_7O_2N$ = 161.2. Colorless crystals, m.300.

carbostyrilic acid Kynuric acid.

carboxamide* The group $-C(:O)NH_2$.

carboxamides* Amides derived from carbon acids. Cf. *sulfonamides*.

carboxamidine(s)* Amidine, q.v. *(amidines)*, whose name is derived from that of a carboxylic acid.

carboxin* See *fungicides*, Table 37.

carbox metal The alloy Pb 84, Sb 14, Fe 1, Zn 1%.

carboxy* Prefix indicating the acidic carboxyl* group, $-COOH$.

carboxyhemoglobin A compound of carbon monoxide and hemoglobin formed in the blood by carbon monoxide poisoning.

carboxyl* Oxatyl. See *carboxy*.

carboxylase* See *enzymes*, Table 30. **co ~** Thiamine pyrophosphate.

carboxylic acid* A compound of the class $R \cdot COOH$. Used both as a class name and a suffix. See *acid*, *-oic acid*. Cf. *carbylic acid*.

carboxylyase* An enzyme which splits the carboxyl group into carbon dioxide.

carboxymethylcellulose See *cellulose*.

carboy Demijohn. A 10- to 13-gal glass flask protected by wickerwork: formerly used for acids, etc. **c. inclinator** A support to enable a c. to be inclined and emptied easily.

carbro process A method of making color prints from color photographs.

Carbrosolide Trademark for silicon carbide.

carburation (1) Carbonization as applied to internal combustion engines. (2) Carburization.

carburet Carbide.

carburetor, carburettor The part of the internal combustion engine where full vaporization occurs.

carburite A mixture of equal parts carbon and iron, for recarburizing steel in the electric furnace.

carburization The dissolution of carbon in molten metals; as, steel produced by heating in a stream of carbon monoxide. **case ~** Carburization on the surface.

c. gas The production of a toughened surface layer of high-carbon steel by heating steel components in a carbon-rich gas.

carburizing Carburization.

carburolith A solid safety fuel which exudes flammable vapor under pressure. It consists of petroleum with 3% of a stabilizer (sodium silicate mixed with copper alginate and an excess of ammonia).

carbylamine (1) Isocyanide*. (2) Ethylisocyanide*.

carbylic acid An organic acid which has carbon in its acid radical; as: **ammonia ~** $R \cdot CNH \cdot NH_2$, carbazylic acid. **aquo ~** $R \cdot COOH$, carboxylic acid. **thio ~** $R \cdot CSSH$, dithionic acid. Cf. *siliconic acid, stannonic acids*.

carbynes (1) Organic compounds of doubtful existence, containing chains with $-C{:}C-$ bonds. Chaoite is said to be a naturally occurring c. (2) (sing.) The methylidyne* radical.

carcel unit The brightness of the carcel lamp, burning 42 g of colza oil per hour. 1 carcel unit = 9.6 candles = 7.5 German standard *candle*, q.v.

carcinogen A substance which produces a carcinoma in living tissues; as, benzopyrene. Cf. *neoplastigen*. **co ~** An agent that increases the effect of a c. when administered with it.

carcinoma A tumor originating from malignant epithelial cells; e.g., skin cancer (epithelioma).

carcinomic acid An unsaturated fatty acid in cancerous serum and tissue.

cardamom The seeds of *Elettaria cardamomum* (Zingiberaceae), tropical Asia; an aromatic and a spice (NF, BP). **Malabar ~** d.0.933–0.943; contains eucalyptol. **Siam ~** d.0.905; contains borneol.

c. oil The essential oil of c., d.0.895–0.905; it contains terpinene, dipentene, and citrene (NF, BP).

Cardanol $C_{15}H_{29} \cdot C_6H_4 \cdot OH$ = 302.5. 3-(8-Pentadecenyl)phenol. Trademark for a liquid obtained by the distillation of cashew nut juice, b_{10mm}225. Its esters are plasticizers.

cardenolide Cardogenan. Describing the fully saturated system of digitaloid lactones; the configuration at the 20 position is the same as in cholesterol. Cf. *steroid*.

cardiac Pertaining to the heart (χαρδια = heart). **c. glycosides** A group of glycosides, of similar chemical structure (see *digoxin*), from various plants, mainly digitalis (foxglove). Widely used since 18th century for effect on the heart, particularly the increase in force of contraction and decrease in heart rate. Used in heart failure and to control rapid heart rate. C. g. are bound to plasma proteins. See *digitalis, oubain, strophanthus*.

carding An operation in the manufacture of woolen felts which opens up the material, mixes the fibers, and removes foreign matter, by the action of wire brushes.

cardiogram, cardiograph See *electrocardiogram*.

cardioid Heart-shaped. **c. condenser** A device to concentrate light in the ultramicroscope.

cardogenan Cardenolide.

Cardol $C_{21}H_{32}O_2$ = 316.5. Trademark for an irritant phenolic oil liquid from the shell of *Anacardium occidentale*, cashew nuts.

carene* $C_{10}H_{16}$ = 136.2. **3- ~** (1*R*,6*S*)-3,7,7-Trimethylbicyclo[4.1.0]hept-2-ene. Colorless, sweet-smelling oil, d.0.8586, b.170. A terpene in essential oils and some turpentines q.v.

Carex Red couch grass. A perennial, grasslike herb (Cyperaceae).

Cargau Trademark for a protein synthetic fiber.

Carica The papaw or melon tree. *Carica papaya* (Caricaceae), S. America. Cf. *papaya*. **c. xanthin** Cryptoxanthin.

caricin (1) A glucoside from the seeds of carica. Cf. *papain*. (2) A protease, papain.

Carifil Trademark for a synthetic fiber (polypropylene with a branched fibrillated structure) used in papermaking.
cariogen(et)ic Producing dental caries. Cf. *carcinogen(et)ic.*
Carissa A genus of spiny shrubs (Apocynaceae), Asia and Australia.
caritinoid Carotenoid.
Carius, Georg Ludwig (1829–1875) German chemist. **C. furnace** A combustion furnace with 5 iron tubes.
carlic acid $C_{10}H_{10}O_6$ = 226.2. An acid from the mold *Penicillium Charlesii.* Cf. *carolic acid.*
carlosic acid $C_{10}H_{12}O_6$ = 228.2. An acid from the fungus *Penicillium Charlesii.*
Carlsbad salt (1) Natural: sal carolinum genuinum, sal thermarum carolinarum. The salts obtained by evaporation of the water of the springs at Karlovy Vary (Carlsbad), Czechoslovakia. (2) Artificial: sal carolinum factitium. Sal carlsbadense. A mixture of thenardite 44, potassium sulfate 2, sodium chloride 18, sodium carbonate 36%.
Carl's solution An insect preservative: 170 ml 95% alcohol, 60 ml 40% formaldehyde, 20 ml glacial acetic acid, 280 ml water. Cf. *fixative.*
carminative A drug relieving colic and promoting gas expulsion from the gastrointestinal tract.
carmine Coccinellin. C.I. Natural Red 4. A mixture of carminic acids, c. red, and other substances obtained by precipitating a decoction of cochineal with alum. Bright red, friable pieces, soluble in ammonia water; a microscope stain. **ammonia ~** A solution of c. in ammonia water. **blue ~** See *indigo c.* **borax ~** An alkaline c. stain containing borax, to stain the cell nuclei. **indigo ~** See *indigo c.*
c. lake A compound of c. and alumina. **c. red** $C_{11}H_{12}O_7$ = 256.2. Purple, lustrous crystals, a split product of carminic acid; a microscope stain.
carminic acid $C_{21}H_{19}O_{11}\cdot COOH$ = 492.4. Cochinilin. A glucosidal hydroxyanthrapurpurin derivative from cochineal. Purple crystals, m.136, soluble in water; a reagent for albumin and aluminum, a microscope stain, and a pH indicator changing at pH 5.5 red (acid) to magenta (alkaline).
carminite A lead ore containing phosphates, arsenates, and vanadates.
carmoisine A carmine red dye used in foodstuffs.
carnallite $KCl\cdot MgCl_2\cdot 6H_2O$ = 277.9. A native, potassium, magnesium chloride (Stassfurt, Germany). Cf. *sylvine.* **ammonium ~** The mineral $MgCl_2\cdot NH_4Cl\cdot 6H_2O$.
carnauba The Brazilian wax palm, *Copernicia ceriferai* or its root. **c. wax** Brazil wax. A wax, mainly triacontyl heptacosanoate, obtained from the young leaves of c. Yellow masses, m.83–88; a polish and coating used in pharmacy (NF, BP).
carnaubic acid $C_{23}H_{47}COOH$ = 368.6. In carnauba wax and beef kidney, m.72.
carnaubyl alcohol $C_{24}H_{49}OH$ = 354.7. A constituent of carnauba wax and wool fat, m.69, insoluble in water.
carnegieite $Na_2Al_2Si_2O_8$. A sodium feldspar.
carnegine $C_{13}H_{19}O_2N$ = 221.3. An alkaloid from Cactaceae; a dimethyl derivative of salsoline, q.v. Cf. *dioscorine.*
carnelian Cornelian.
carnine $C_7H_8O_3N_4\cdot H_2O$ = 214.2. A crystalline alkaloid from muscles, meat extract, yeast, and certain fishes; slightly soluble in water. Cf. *xanthine.*
carnitine $Me_3\cdot N\cdot CH_2\cdot CHOH\cdot CH_2\cdot CO$ = 161.2. Novain,
—————O—————┘
vitamin B_T, γ-trimethyl-β-hydroxybutyrobetaine. A base found in the muscle tissue of cephalopods. Soluble in water or alcohol. Its absence affects the development of flour beetles. Cf. *crotonbetaine.*
carnomuscarine A nitrogeneous extractive of muscles.
carnosine $C_9H_{14}O_3N_4$ = 226.2. Ignotine, β-alanylhistidine. Colorless crystals, from meat extracts, m.219 (decomp.) Cf. *anserine.*
Carnot, Sadi (1796–1832) French engineer. **C. function** The ratio between the lost and utilized heat of a living body. **C. theorem** The work obtainable from a given quantity of heat absorbed in a machine that is working in a reversible cycle depends only on the temperatures of the source of heat and refrigeration.
carnotite A radioactive mineral vanadate of uranium and potassium, (23% V).
Carnoy's fluid A mixture of absolute alcohol and glacial acetic acid used to fix animal tissues before staining.
Caro, Heinrich (1834–1910) German industrial chemist. **C.'s acid** See *Caro's acid.*
carob beans St. John's bread. The pods of *Ceratonia siliqua* (Leguminosae), rich in sugar and gum; a fodder. Its seeds were the "carats," q.v., of jewelers. Cf. *tragon.* **c. gum** The gum of c. beans; an emulsifier and tragacanth substitute.
caroba The Brazilian caraiba tree, *Jacaranda procera* (Bignoniaceae).
carobic acid A crystalline acid from caroba leaves.
carobine A caroba alkaloid.
carobone A balsamic resin from caroba leaves.
carol $C_{10}H_{18}O$ = 154.3. Caraneol, 5-hydroxycarane. A monobasic alcohol, from carane, in many essential oils.
carolic acid $C_9H_{10}O_4$ = 182.2. An acid from the mold *Penicillium Charlesii.* Cf. *carlic acid.*
carolinium A supposed element in thorium minerals.
carolite Carrollite.
carone $C_{10}H_{16}O$ = 152.2. 5-Oxocarane. A ketone of essential oils, derived from carane. Colorless oil, b.210.
carony bark Angostura bark.
Caro's acid H_2SO_5 = 114.1. Peroxomonosulfuric acid. Made by dissolving potassium peroxosulfate in concentrated sulfuric acid, or by heating the acid with an oxidizing agent, as hydrogen peroxide. A powerful oxidizing agent; converts aniline into nitroglycerin.
carotene $C_{40}H_{56}$ = 536.8. Carotin, vitamin A precursor. A lipochrome, m.174, in foodstuffs; associated with chlorophyll, animal tissues, blood serum, milk fat, etc. Isomeric forms: **alpha- ~** Optically active, m.187, maximum absorption bands at 521 and 485 nm. **beta- ~** Optically inactive, m.181, maximum absorption bands at 511 and 478 nm. A provitamin A; 2 molecules of retinol are produced from β-c. by an enzyme which splits the β molecule at C-15 (USP, EP). See *retinol, vitamin,* Table 101 on p. 622. Cf. *carotenoids.* Other associated plant pigments are neoc., ψ-c., and γ-c. m.178. **red ~** Lycopene.
carotenoids Carotinoids, polyenes. Lipochromes or carotenelike plant pigments deposited in animal tissues. They may be hydrocarbons, alcohols, ketones, or acids, with alternate double bonds; e.g., polyene hydrocarbons ($C_{40}H_{56}$), α-carotene; polyene alcohols ($C_{20}H_{29}OH$), vitamin A; polyene ketones ($C_{40}H_{50}O_2$), rhodoxanthin; polyene acids ($C_{20}H_{24}O_4$), crocetin. Cf. *androstane ring, cerebrosides.*
carotin Carotene.
carotinoids Carotenoids.
carpaine $C_{14}H_{25}O_2N$ = 239.4. An alkaloid from the leaves

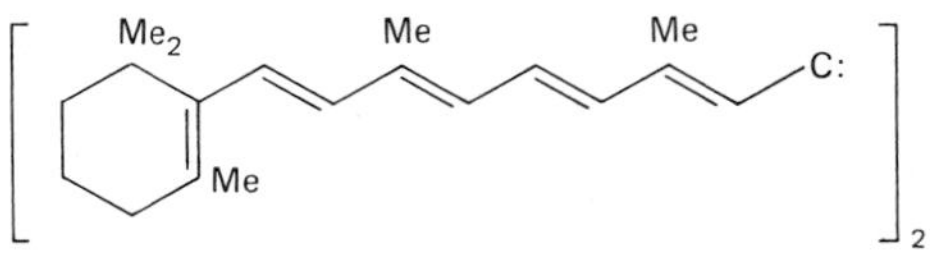

of *Carica papaya* (Caricaceae). Colorless crystals, m.119, soluble in alcohol; a diuretic and heart stimulant, resembling digitalis in action.
carpamic acid $C_{14}H_{27}O_3N$ = 257.4. An acid formed by hydrating carpaine.
carposide Caricin.
carrageen, carragheen Irish moss.
carrageenin A water-soluble extractive from carragheen. Used as a stabilizer, for suspending cocoa in chocolate manufacture, and to clarify beverages.
Carrel-Dakin solution An isotonic sodium hypochlorite solution, for bathing wounds. Cf. *Dakin solution.*
carrene Methylene dichloride.
carrollite Co_2CuS_4. A rare, native cobalt-copper sulfide.
carrotin Carotene.
carroting The preparation of hair for felting by treating the tips of the fibers on the skin with a solution of mercuric nitrate in nitric acid.
carstone A ferruginous sandstone (Norfolk, England).
carthamein An oxidation product of carthamin.
carthamic acid Carthamin.
carthamin $C_{14}H_{16}O_7$ = 296.3. The red color of safflower, *Carthamus tinctorius.* Dark red scales, insoluble in water; a dye.
carthamus African saffron. Composite plants, as *C. tinctorius,* false or American saffron or safflower, cultivated in Asia; a dye, condiment, and ingredient in making rouge.
Cartier scale A hydrometric scale for determining ethyl alcohol based on 12.5°C; used in S. America.
cartilagin Chondrogen. The protein of cartilage; the elastic substance on bone surfaces.
cartonnage A gesso made from chalk and animal glue.
carubinose D-Mannose.
carum Caraway. Cf. *ajava.*
carvacrol* $Me \cdot C_6H_3(OH)C_3H_7$ = 150.2. 2-Hydroxy-*p*-cymene, isopropyl-*o*-cresol, 2-cymophenol, oxycymol. An aromatic constituent of essential oils, camphor, origanum, caraway, savory and other Labiatae. A thick, colorless oil, d.0.978, m.0, b.236; insoluble in water. Used in perfumery and as an antiseptic.
carvacrotic acid A colorless, crystalline acid derived from carvacrol.
carvacryl The radical 2,5-$Me(Me_2CH)C_6H_3-$, from carvacrol.
carvene Citrene.
carveol $Me \cdot C_6H_7(OH)C_3H_5$ = 152.2. 5-Isopropenyl-2-methyl-cyclohexen-1-ol, 2-hydroxycitrene. Oils. Several enantiomers.
carvol Carvone.
carvomenthene $C_{10}H_{18}$ = 138.3. 1-Methyl-4-isopropylcyclohexene. Colorless liquid from carvone, d.0.806, b.175.
carvomenthol $C_{10}H_{20}O$ = 156.3. 2-*p*-Menthanol. (1*R*, 2*R*, 5*R*) form of menthol; in essential oils, d.0.904, b.222, soluble in alcohol.
carvomenthyl The radical $-C_{10}H_{17}$, from carvomenthane. **c.amine** $C_{10}H_{17}NH_2$ = 153.3. Tetrahydrocarvylamine. Colorless liquid, b.212.
carvone $C_{10}H_{14}O$ = 150.2. Mentha-1,8-dien-6-one. In the oils of caraway, cumin, and dill. Colorless liquid, b.225. **c. anilide** $C_{10}H_{13}O \cdot NHPh$ = 241.3. Yellow oil, b.180, which darkens with age.
carvylamine tetrahydro ~ Carvomenthylamine.
Carya The hickory tree, *Carya tomentosa* (Juglandaceae), N. America, cultivated for its wood and fruits (pecans).
caryin A crystalline compound from the bark of carya.
caryinite A native phosphate and arsenate of lead.
caryl-* The radical $C_{10}H_{17}-$, from carane.
caryocinesis Karyokinesis.
caryophyllin Oleanolic acid.
caryophyllinic acid Oleanolic acid.
caryophyllum Cloves. The brown, fragrant flower heads of *Eugenia caryophyllata* (Myrtaceae) which contain an oil used as a flavoring and aromatic stimulant.
CAS See *Chemical Abstracts Service.*
casca Spanish and Portuguese for "bark." **c. bark** Erythrophloeum.
cascara amarga Honduras bark. The bark of *Picramnia antidesma* (Simarubaceae), tropical America; a tonic. **c. sagrada** Sacred bark. Chittem bark, bearwood. The bark of *Rhamnus purshiana,* a shrub (western U.S.). A laxative (USP, EP, BP). See *Peristaltin, anthraglucosagradin.*
cascarilla The bark of various *Croton* and *Cinchona* species, especially *Croton eluteria.* The fluid extract or tincture is a stomachic.
cascarillic acid $C_{11}H_{20}O_2$ = 184.3. 2-Hexyl-1-cyclopropaneacetic acid. An acid from the oil of cascarilla. Colorless liquid, b.270.
cascarillin $C_{22}H_{32}O_7$ = 408.5. Colorless, bitter crystals from the bark of *Croton eluteria.*
caseate (1) Caseinate. (2) Describing pathological change in body tissues, which become cheeselike in either appearance or consistency.
caseation The curdling of milk into a cheesy mass.
case-harden To harden wrought iron by heating it in contact with carbon or potassium hexacyanoferrate(II); to produce a surface layer of steel. See *carburization, cementation.*
casein The protein of milk (1 ton from 4.2 m^3), and principal constituent of cheese. White solid, soluble in acids. Used as food; in combination with formaldehyde, as a plastic; in the leather industry; as a substitute for linseed oil in making pigments; in the manufacture of albumin, rubber, and gelatin for films; and as an adhesive when mixed with lime. The composition of a typical casein is:

Carbon	53%
Hydrogen	7%
Oxygen	22.7%
Nitrogen	15.7%
Sulfur	0.8%
Phosphorus	0.85%

Its isolectric point is pH 4.7; hence it occurs as neutral casein, metal caseinate, casein salt. **animal ~** Casein. **fibrous ~** See *Aralac, lanital.* **gluten ~** Vegetable casein. **milk ~** Casein. **saliva ~** α-Amylase*. **vegetable ~** A protein from cereal seeds, forming 10–20% of flour gluten.
caseinase Rennin.
caseinate Caseate. A compound of casein with a metal.
caseinic acid $C_{12}H_{24}O_5N_2$ = 276.3. An acid from casein.
caseinogen A compound protein of milk which yields casein by enzyme action.
Casella acid F acid. 2,7-Naphtholsulfonic acid.
Casenka Trademark for a protein synthetic fiber.
cashew The nut of *Anacardium occidentale,* tropical America. Cf. *anacardium, anacardic acid, Cardanol, Cardol.* **c. apples** Peduncles or pseudostems of the c. tree; a rich source of ascorbic acid. **c. oil** Yellow oil, used in lacquers and paints. Cf. *harvel coating.*
cashmere A fine, soft textile fiber from the coat of the Kashmir goat.
Caslen Trade name for a protein synthetic fiber.
casoid A synthetic plastic made from a casein basis.
Casolana Trademark for a protein synthetic fiber.
cassareep The evaporated juice of bitter cassava (W. Indies), a meat preservative.

cassava Manioc, mandioc, manihoi (Euphorbiaceae), S. and Central America and Africa; as, **bitter ~** *Manihot utilissima* and **sweet ~** *Manihot aipi.* Their large, tuberous roots contain starch, used to prepare tapioca and Brazilian arrowroot.

cassel **c. brown** A brown pigment found near Kassel, Germany; a fossilized, tertiary period humus. **c. yellow** Lead oxide chloride.

Casseler green Barium manganate.

casseliase A mica peridotite (Kentucky).

casselose A nephelite-melilite basalt (Texas).

casserole A porcelain laboratory disk with handle.

Cassia (1) Sweet-smelling trees (obsolete). (2) A genus of leguminous herbs and trees. (3) An inferior cinnamon from *Cinnamomum cassia,* a spice and source of oil. **purging ~** The pod pulp of *Cassia fistula.*

c. bark Chinese cinnamon. **c. buds** The unripe fruit of cinnamon species. **c. leaves** Senna. **c. oil** Chinese oil of cinnamon, from the bark of *Cassia chinensis.* It is darker, less agreeable, and heavier than true cinnamon oil. **c. seeds** The fruits of *Cassia fistula,* E. India; a laxative and poultice.

cassina Holly from coastal southern U.S.; its leaves are used in a beverage.

cassiopeium Cp. The name given to *lutetium,* q.v., by Auer von Welsbach, who separated it from ytterbium in 1905.

cassiterite SnO_2. Tinstone, block tin, tin spar, stream tin. Brown, black, red, or yellow tetragons, d.6.9, hardness 6–7, found as pebbles in streams (Malaysia).

cast (1) To form a molten substance into a definite shape by allowing it to cool in a mold. (2) See *casts.* **dip ~** A c. formed by dipping a solid mold in molten metal, and withdrawing it so that a shell of metal is left on the mold. **slush ~** A c. formed by filling a hollow mold with molten metal and then emptying it, so that a solid metal lining remains in the mold.

castable A dry, premixed blend of refractory cement (or bond) to which a liquid is added immediately before use for casting or lining plant.

Castellani's paint Trademark for carbol fuchsin solution.

castine An alkaloid from *Vitex Agnus castus* (Verbenaceae).

Castle's **C. extrinsic factor** Vitamin B_{12}. **C. intrinsic factor** A glycoprotein in gastric juice.

castonin A globulin from European chestnut.

castor (1) The beaver, *Castor fiber,* an amphibious rodent. (2) Castoreum. A brown, strong-smelling solid from the preputial follicles of the beaver. **c. beans** The seeds of *Ricinus communis.* **c. oil** Ricinus oil from the c. bean. Yellow liquid, $d_{25°}$ 0.945–0.965; a purgative (USP, EP, BP). Cf. *dericin, ricinolein.* **c. pomace** The solid residue after extraction of c. oil from c. bean. A fertilizer: nitrogen 4.1–6.6, phosphorus oxide 1–1.5, potassium oxide 1–1.5%. **c. seeds** C. beans.

castoreum Castor. **c. oil** Essential castor oil. Yellow liquid, obtained by distillation of castoreum.

casts Cylindrical structures formed in uriniferous tubules, and found in sediment of urine. **blood ~** C. consisting of red cells, indicating renal hemorrhage. **epithelial ~** C. consisting of epithelial cells, indicating inflammation. **granular ~** C. consisting of granular material, indicating an inflammatory kidney. **hyaline ~** C. consisting of material of hyaline appearance. Found in kidney disorders and sometimes in the normal state. **pus ~** C. consisting of pus cells or leucocytes, indicating renal infection.

cata- Prefix from the Greek κατα, meaning "down," "against," "back," or "of a lower order." Cf. *ana-.*

catabiotic Capable of using up or dissipating. **c. force** Energy derived by an organism from the metabolic effects of its food.

catabolism Disintegration or the breaking down of tissues; the destructive metabolism by which living matter is transformed into waste materials and eliminated from the body. Cf. *anabolism.* **carbohydrate ~** See *carbohydrate catabolism* under *carbohydrates.*

catadyn An oligodynamic catalyst; as, silver.

catalase* A hemoprotein enzyme which splits peroxides to oxygen and water; isolated from bacterial cultures as octahedra. Cf. *peroxidase.*

catalasometer An instrument to measure catalase activity from the change in level of a liquid in a graduated tube passing through the stopper of the vertical cylinder in which the reaction occurs.

catalinite Beach pebbles of Santa Catalina Island; a green, red, or brown variety of quartz.

catalysant A substance which is catalyzed. Cf. *substrate.*

catalysate The product of catalysis.

catalysis Contact action or cyclic action. The effect produced by a small quantity of a substance (catalyst) on a chemical reaction, after which the substance appears unchanged. Due to formation of an intermediate compound (chemical); adsorption (mechanical); excitation by light *(photochemical).* **heterogeneous ~** Acceleration of a chemical reaction in a heterogeneous system. **homogeneous ~** Acceleration of a chemical reaction in a homogeneous mixture. **iso ~** The coexistence of two catalytic influences in the same system. **isomeric ~** The catalytic transformation of a substance to an isomeric form. **negative ~** (1) The catalytic retardation of a reaction; e.g., mannitol reduces the oxidation of phosphorus. (2) The catalytic halting of a reaction; e.g., oxygen stops the reaction of hydrogen and chlorine. **photo ~** Acceleration of a reaction by light. **photochemical ~** The action of light as a catalyst. **pseudo ~** Acceleration of a chemical reaction by a chemical catalyst.

catalyst Catalyzator, catalyzer. A substance which effects catalysis. **chemical ~** A substance which changes the speed of a chemical reaction, but is present in its original concentration at the end of the reaction; e.g., nitric acid in the lead-chamber process for sulfuric acid. **inducing ~** Promoter. A substance which stimulates or aids catalysis. **mechanical ~** A substance which influences the speed of a chemical reaction without itself undergoing a change, e.g., the reaction of $2H_2O_2 \rightarrow 2H_2O + O_2$ is accelerated by colloidal Pt or Au. **negative ~** Retarder. A c. which retards a chemical reaction. **positive ~** Accelerator. A c. which accelerates a chemical reaction. Cf. *enzyme(s), sensitizer.* **stereoscopic ~** A c. used in polymerization processes to predetermine the final spatial configuration, and hence the physical and chemical properties, of large polymeric molecules.

catalytic Pertaining to catalysis. **c. action** Catalysis. **c. agent** Catalyst. **c. converter** Device fitted to automobiles to reduce emission of SO_2 and NO_n pollutants. **c. force** The mechanical expression of the change in the speed of a chemical reaction due to a catalyst. **c. poison** Anticatalyzers, paralyzers. A substance which counteracts the effect of a catalyst. **c. reforming** See *reforming.*

catalyzer Catalyst.

catamorphism The chemical and physical changes produced on rock by the action of wind and water. Cf. *anamorphism.*

cataphoresis Kataphoresis. The downward motion of electrically charged particles suspended in a medium under the influence of an electric field. Cf. *endosmosis.* Its antonym *anaphoresis* implies upward motion.

catapleiite A complex zirconium silicate.

Catapres Trademark for clonidine hydrochloride.

catechin $C_{15}H_{14}O_6$ = 290.3. An acid from catechu, mahogany wood, and kino. Yellow powder, soluble in water;

used for tanning and calico printing. (±)-~ $+3H_2O$. *dl*-Catechol, acacatechin. The principal constituent of *Acacia catechu.* Thin needles, m.214. (+)-~ $+4H_2O$. *d*-Catechol, gambir catechin. Thin needles, m.94, soluble in alcohol. (−)-~ $+4H_2O$. Epicatechol. Thin needles, m.240.

catechol (1) Pyrocatechol. (2) Catechin. Condensed tannin. **benzoylpyro ~** Dihydroxybenzophenone. **dimethoxy ~** Veratrole*. **methyl ~** Guaiacol*.

c. amines Physiologically important derivatives of c.; as, *epinephrine,* q.v., *norepinephrine bitartrate,* q.v., and dopamine.

catechu Cutch, cashoo, black catechu. An extract of the dark heartwood of *Acacia catechu;* an astringent. Cf. *japonic acid.* **pale ~** Gambir.

c. palm Areca.

catechuic acid Catechin.

catechuretin $C_{38}H_{28}O_{12}$ = 676.6. Produced from catechu by the action of sulfuric acid.

catechutannic acid A red, amorphous anhydride of catechin from catechu.

catena-* Prefix indicating a linear polymeric chain structure.

catenane A compound composed of large interlocking rings, which are otherwise unbonded.

catenary The curve assumed by a chain suspended at each end.

catenation The formation of polymers based on a chain or ring of atoms of one kind.

catgut An aseptic cord prepared from animal intestines; used for ligatures.

cathartic A drug which causes the evacuation of the bowels, e.g., castor oil. Cf. *purgative, laxative, aperient.*

cathartic acid $C_{180}H_{96}O_{82}N_2S$. The glucosidal and active principle of senna.

cathartin (1) A bitter principle of senna (Lassaigne and Feneulle). (2) A compound from the ripe fruits of *Rhamnus cathartica* (Winkler). (3) A compound from jalap.

cathartomannite $C_{21}H_{44}O_{19}$ = 600.6. A nonfermentable carbohydratelike substance from senna.

cathetometer A precision instrument to measure small vertical displacements. It consists usually of a microscope with cross wires and scale mounted on a rigid vertical rod.

cathidine, cathine, cathinine Three alkaloids from African tea, the leaves of *Catha edulis* (Celastracae), N. Africa. Cf. *katine.*

cathine $C_9H_{13}ON$ = 151.2. (1*R*,2*R*)-2-Amino-1-phenyl-1-propanol, D-norpseudoephedrine. A crystalline alkaloid, m.77, soluble in water, from bushman's tee (boesmanstee), the leaves of *Catha edulis* (Celastraceae), S. Africa. Cf. *celastrine, katine.*

cathode Kathode, negode, katode. The negative pole or electrode of an electric device. **c. deposit** (1) A metal precipitate produced by electrolysis. (2) A metallic mirror formed on a vacuum tube near the cathode. See *cathodic sputtering.*

cathode rays Negative rays. A stream of negatively charged electrons issuing from the cathode of a vacuum tube perpendicular to the surface, with velocities (10^9 to 10^{10} cm/s) depending upon the potential difference. When they are stopped by a solid substance, they produce heat, phosphorescence, X-rays, pressure, or photographic action. C. rays are analogous to the β rays of radioactive matter.

cathode ray tube TV table upon whose screen data are displayed.

cathodic sputtering Disintegration of the cathode in an electric discharge tube, covering the surrounding glass with a mirror of cathode metal.

catholyte The liquid close to the cathode. Cf. *anolyte.*

cation* Kathion, cathion, negion. A positively charged atom, radical, or group of atoms, which travels to the cathode or negative pole during electrolysis. Cf. *anion, ion.*

cationotropy A case of *ionotropy,* q.v., in which a H^+ or M^+ breaks off a molecule and leaves a negative ion in dynamic equilibrium. Cf. *prototropy.*

catlinite Pipestone. A red clay (Upper Missouri); used for pipes.

catoptric An optical system with metallic reflectors instead of glass mirrors.

catoptrics The study of mirrors and reflected light.

catoptron A mirror.

catoptroscope A device to examine objects by reflected light.

caulophyllin A resinous precipitate obtained by pouring concentrated tincture of caulophyllum into water.

caulophylline An alkaloid from blue cohosh, *Caulophyllum thalictroides.* Colorless crystals.

caulophylloid The combined principles from caulophyllum; an antispasmodic.

caulophyllogenin $C_{30}H_{48}O_5$ = 488.7. Crystals, m.279; from caulophyllum.

caulophyllum Squawroot, blue cohosh. The rhizome and roots of *C. thalictroides* (Beriberidaceae), N. America; a diuretic and antispasmodic.

caulosaponin $C_{54}H_{38}O_{17}$ = 958.9. Leontin. A glucoside, m.255, from caulophyllum.

caustic (1) Corrosive or burning. (2) The hydroxide of a light metal, e.g., caustic soda. (3) Describing a curve or surface of maximum brightness produced by the concentration of rays of light after reflection or refraction. (4) A drug which destroys soft body tissues, used to destroy pathological tissues. **lunar ~** Silver nitrate. **Vienna ~** Potassium hydroxide. Cf. *soda lime.* **volatile ~** Ammonium hydroxide.

c. cracking C. embrittlement. A form of metallic failure of steam-raising plant. Very fine cracks develop in the overlaps of riveted seams on the dry side of the boiler. Associated with high localized stresses and high concentrations of c. soda produced by the evaporation of alkaline waters. **c. curve** The line produced by the intersection of a number of reflected or refracted rays of light. **c. embrittlement** C. cracking (more correctly). **c. lime** Calcium oxide. **c. potash** Potassium hydroxide. **c. soda** Sodium hydroxide*.

causticity The extent to which a mixture (as, NaOH + Na_2CO_3) is caustic (as a percentage): $100(2x - y)/y$, where x and y are the mL of hydrochloric acid necessary to make a solution of the substance neutral to phenolphthalein and methyl orange, respectively. It expresses the relative proportions of carbonate and hydroxide.

causticized ash A mixture of soda ash (Na_2CO_3) and caustic soda (NaOH), containing 15–45% NaOH. Used as water softener, for cleansers, and in the manufacture of leather.

cauterize Localized treatment, as with heat, alkali, or laser beam, to coagulate tissue or blood.

Cavendish, Henry (1731–1810) English philosopher and chemist; a pioneer in studies of specific heat and eudiometry and the discoverer of hydrogen. He determined the compositions of water, air, and nitric acid.

Caventou, Joseph Bienaimé (1795–1877) French pharmacist; discoverer of brucine and strychnine, and (with Pelletier) of quinine.

cavitation The production of emulsions by disruption of a liquid into a two-phase system of liquid and gas, when the hydrodynamic pressure in the liquid is reduced to the vapor pressure.

cayaponine A purgative alkaloid from *Cayaponia globosa* (Cucurbitaceae), Brazil.

cayenne pepper Pepper made from seeds of capsicum.

cc Abbreviation for cubic centimeter. **cc test** A unit of evaluation of hydrogen peroxide. The number of cubic centimeters of 0.1 N potassium permanganate equivalent to 2 cc of sample.
CChem Abbreviation for Chartered Chemist (of the U.K.).
CCTV Closed-circuit *television.*
Cd Symbol for cadmium.
cd* Symbol for candela.
Ce Symbol for cerium.
ceara rubber The coagulated latex of *Manihot glaziovii* (Euphorbiaceae), Brazil and Argentina.
cebil gum The reddish yellow tears from *Piptadenia cebil* (Leguminosae), Brazil. Cf. *angico gum.*
cebur balsam Tagulaway.
cecidomin $C_{23}H_{28}O_{15}$ = 544.5. A glucosidal constituent of gall from *Cecidomyia tiliae.* Orange needles, m.227–231.
cecilose A pyroxenite (Maryland).
cedar A group of trees of the pine family. **red ~** Juniperus. **white ~** Thuja.
c. gum Yellow tears from *Cedrela toona,* Indian mahogany (Meliaceae), Queensland. **c.leaf oil** C. oil. **c. oil** The essential oil, d.0.870–0.890, from *Juniperus virginiania;* used in microscopy as clarifying agent and for oil-immersion lenses. It contains citrene, cadinene, and borneol. **c.wood oil** Oil from *Cedrela odorata,* W. Indies and S. America. Odorous liquid, d.0.945–0.960, soluble in alcohol; contains cedrene and cedrol.
cedarite Chemavinite.
cedrene $C_{15}H_{24}$ = 204.4. An oil from red cedar, d_{15}·0.984, b.123. **c. camphor** Cedrol.
cedrin An active principle of cedron. Colorless, bitter crystals, soluble in water.
cedrol $C_{15}H_{26}O$ = 222.4. Cedrene camphor. The crystalline portion of oil of red cedar, m.87, b.294.
cedron The seeds of *Simaba cedron* (Simarulbaceae), Central America; a febrifuge.
cedronella Lemongrass.
cedronine An alkaloid from cedron.
cefaloridine Cephaloridine.
cel The velocity imparted by 1 dyne in 1 s to 1 g.
celadonite A glauconite.
Celafibre Trademark for a cellulose acetate synthetic fiber.
Celafil Trademark for a ruptured, continuous-filament acetate yarn.
Celanide Trademark for lanatoside C.
celandin Chelidonium.
celandine Pilewort. **greater ~** Chelidonium. **lesser ~** Pilewort.
Celanese Trademark for a cellulose acetate synthetic fiber.
Celaperm Trademark for a cellulose acetate synthetic fiber.
celastin Menyanthin.
celastrine An alkaloid from the seeds of Celastraceae. White crystals; a stimulant. Cf. *cathine.*
Celcos Trademark for a cellulose acetate synthetic fiber.
celcure salts A 10% solution of chalcanthite 50, sodium dichromate 45, chromous acetate 5 pts. Used to impregnate wood to protect it against fungal and insect attack.
Celechrome Trademark for a cellulose acetate synthetic fiber.
celery The vegetable *Apium graveolens* (Umbelliferae). Cf. *apiole, apigenin.* **c. seeds** Used as a spice and for making **c. seed oil,** which contains selinene and apiole. A colorless liquid, d.0.870-0.895.
celestine, celestite $SrSO_4$. Strontium sulfate, deposited in sedimentary rocks.
celiac disease Gluten-induced enteropathy.
Celite Trademark for certain diatomaceous earth products, particularly filter aids, fillers, abrasives, heat-insulating materials, insulating plasters and cements.

cell (1) *Biological.* The anatomical unit of life from which all living matter is constituted and develops. It consists essentially of a small protoplasmic mass which is a changing colloidal chemical system of proteins, and is differentiated into a nucleus and cell body. This protoplasm is divided into cytoplasm (cell body) and nucleoplasm (cell nucleus). (2) *Galvanic.* Electrical element. A device for transforming chemical into electric energy. Its action is due to the passage of ions through an electrolyte. It usually consists of a metal in an electrolyte containing ions of that substance, connected by a tube containing an electrolyte, to a second electrode similar to but not identical with the first. Cf. *galvanic battery.* (3) *General.* A small container with two parallel transparent sides. **air ~** The air-containing chambers of vegetable tissues. **alkali ~** Photoelectric c. **alkaline ~** See *button c.* (below). **amoeboid ~** A c. resembling an amoeba, e.g., a leucocyte. **asexual ~** A c. which reproduces by simple division or fission. **bichromate ~** Dichromate c. **blood ~** See *blood.* **Bunsen ~** See *Bunsen cell.* **button ~** Small cell based on such alkaline systems as Ag_2O/Zn and MnO_2/Zn. **cadmium ~** See *cadmium cell.* **carbon-zinc ~** The familiar domestic dry cell, based on the Leclanché cell plus zinc chloride in the electrolyte. **chromatophore ~** A biological unit containing coloring matter. **Clark ~** See *Clark cell.* **columnar ~** An elongated c., a number of which form part of a type of epithelial tissue. **concentration ~** A galvanic combination of 2 electrodes of the same substance immersed in solutions containing ions of the substance differing only in the concentrations of the ions. **Daniell ~** An electrical zinc-copper c. (about 4 volts). **daughter ~** A c. produced by the divison of another (mother) c. **dead ~** A c. which has ceased to perform vital functions; it may serve for mechanical protection, e.g., horn. **dichromate ~** An electric amalgamated carbon-zinc c. (2.0 volts) in a 12% solution of potassium dichromate containing about 9% sulfuric acid. **diffusion ~** A percolator for extracting sugar from beets. **dry ~** See *dry cell.* **electrical ~** (1) An accumulator. (2) A voltaic c. (3) An electrolytic c. **electrolytic ~** (1) A c. used for electroplating. (2) Galvanic c. **electroresponsive ~** See *electroresponse.* **embryonic ~** A c. from which tissues develop. **epithelial ~** A c. of the skin, body cavity surface, or gland. **fuel ~** See *fuel cell.* **galvanic ~** See *cell* (2) and *galvanic.* **giant ~** An abnormally large c. **Leclanché ~** An electrical zinc-carbon c. (about 1.5 volts). **load ~** An electric transducer for transforming weights into electric currents; used for weighing and based on the change in electrical resistance when the internal structure of a steel billet is distorted under load. **locomotive ~** A c. capable of independent movement. **mast ~** See *mast cells.* **optical ~** See *selenium.* **oxidation-reduction ~** An electrolytic c. utilizing an oxidation-reduction reaction. **photoconductive ~** Photoelectric, photoemissive, photovoltaic. See *photoelectric cell.* **reversible ~** An electrolytic c. in which the energy spent in reversing the chemical changes in the c. is approximately equal to that given out by the direct operation of the c. **secondary ~ biological ~** A c. formed by the coalescence of other cells. **electrical ~** An accumulator. **selenium ~** See *selenium.* **solar ~** A device to convert solar energy into chemical power by the action of solar photons on a semiconductor (as, crystalline or amorphous silicon; gallium arsenide; indium phosphide) junction. Presently only about 15% efficient. **somatic ~** Body c., e.g., leucocyte or epithelial c., as distinct from gamete. When in milk, associated with udder infection. **standard ~** A c. giving a definite voltage. See *cadmium* **c.** **transition ~ biological ~** A c. whose characteristics are between 2 well-defined types of cell, and which is supposed to change from one type to the other. **electrical ~** A voltaic c. which contains an electrolyte undergoing a definite change with temperature; e.g., $ZnSO_4 \cdot 7H_2O$ changes at 39°C to $ZnSO_4 \cdot 6H_2O$, with corresponding change in the emf. **voltaic ~** See *cell (2).* **Weston ~** Cadmium c.

c. constant *K*; resistance capacity. The ratio of the measured conductivity of a solution to its known conductivity. Used to correct for peculiarities in the shape of a conductivity c. SI unit is m^{-1}. **c. division** See *karyokinesis.*.

cellase Cellulosase. An enzyme which digests cellulose.

Cellit Trade name for technical acetylcellulose.

cellobiose $C_{12}H_{22}O_{11}$ = 342.3. Cellose. A disaccharide, glucose-β-glucoside, m.225; an intermediate product in the hydrolysis of cellulose; and a bacteriological reagent.

cellodextrin Cellulose dextrin. The water-soluble constituent formed by the prolonged action of concentrated mineral acids on cellulose; precipitated by alcohol.

celloidin Collodion wool. A concentrated solution of pyroxylin. Used in microscopy for embedding specimens or section cutting.

cellon Tetrachloroethane*.

Cellophane (1) Trade name and -mark for the film obtained by the precipitation of a viscose solution with ammonium salts. Used as a wrapping. (2) (Not cap.) In the U.S., a generic name for film produced from wood pulp by the viscose process.

cellose Cellobiose.

Cellosolve $EtOCH_2CH_2OH$ = 90.1. Trademark for glycol ethyl ether, hydroxy ether, 2-ethoxyethanol*. Colorless liquid, d.0.935, b.135, soluble in water; a lacquer solvent. **butyl ~** $BuO \cdot CH_2 \cdot CH_2OH$ = 118.2 Glycol butyl ether, 2-butoxyethanol*. Colorless liquid, b.171; a solvent for brushing lacquers. **ethyl ~** Cellosolve. **methyl ~** $MeO(CH_2)_2 \cdot OH$ = 76.1. Ethylene glycol monomethyl ether, methoxyethanol*. Volatile liquid, b.124; miscible with toluene; a solvent for nitrocellulose.

c. acetate $MeCOO(CH_2)_2OH$ = 104.1. Hydroxyethyl acetate glycol monoacetate. Colorless volatile liquid, b.153, soluble in toluene; used for lacquers. **c. butylate** $C_6H_{14}O_2$ = 118.2. Colorless liquid, b.170; a solvent for nitrocellulose.

cellosteric effect Inhibition of enzyme activity due to a change in enzyme structure.

cellotriose Procellose.

cellulase* An enzyme that endohydrolyzes the 1,4-β-glucosidic linkage in cellulose.

Celluloid Zylonite, Xylonite. A malleable substance prepared from nitrocellulose (pyroxylin or nitrated cotton, usually containing 10.8-11.1% nitrogen) and camphor, rendered less flammable by the addition of ammonium phosphate and other ingredients.

cellulosans Pentosans associated with cellulose in wood.

cellulosate Sodium cellulose formed by the action of sodium in liquid ammonia, on cellulose; analogous to alcoholates.

cellulose $(C_6H_{10}O_5)n$ = $(162)n$. A carbohydrate polymer of 1,4-β-linked glucopyranose units in the walls and skeletons of vegetable cells.

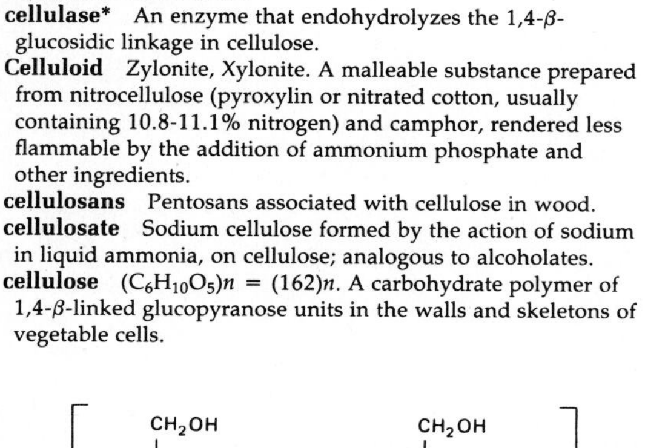

A c. *micelle* contains 1,500–2,00 such units arranged in 40–60 *cellobiose* chains, each containing 15 to 20 pairs of glucose residues. It is almost pure in absorbent cotton and filter paper, and is a colorless, transparent mass, insoluble in water or other solvents, but soluble in ammonium copper hydroxide and in sulfuric acid with the formation of amyloid (vegetable parchment). See *collodion, Celluloid, pyroxylin.* Sources:

1. Seed fibers (87–91%), as cotton.
2. Woody fibers (58–62%), as, coniferous and deciduous woods.
3. Bast fibers (32–37%), as, flax.
4. Leaf fibers, as, sisal.
5. Fruit fibers, as, coconut.

C. is divided into the following fractions by digestion with 17.5% sodium hydroxide. **alpha- ~** Long-chain c. molecules, insoluble in cold 17–18% sodium hydroxide solution under specified conditions; not a chemical entity, but a convenient measure of the "true" c. present in a plant or pulp. **beta- ~** The fraction precipitated by acid from sodium hydroxide solution. Cf. *alpha-c.* above. **gamma- ~** The fraction that is not precipitated by acid from sodium hydroxide solution. Cf. *alpha-c., beta-c.* above. **amino ~** A red, water-soluble powder formed by the action of sodium amide on c. nitrate; each NO_3 group is replaced by NH_2. **B ~** Bacterial c. C. produced by bacterial synthesis from sugars. It consists of long glucose anhydride chains. Cf. *beta-c.* **carboxymethyl ~** (Sodium) c.m.c. Carmellose sodium, sodium cellulose glycolate, cellulose gum. White, bulky solid, made by action of chloroacetic acid and alkali on wood pulp. A thickener, emulsifier, stabilizer, cation exchange substrate in chromatography (USP, BP); and used in adhesives, cosmetics and confectionery. **colloidal ~** Amyloid. **Cross and Bevan ~** C. isolated from vegetable material by successive chlorination and alkali treatments. **Guignet ~** Amyloid. **hemi ~** See *hemicellulose.* **holo ~** See holocellulose. **microcrystalline ~** Partially depolymerized c. from α-c.; used in pharmacy and as bulk-producing laxative (NF, BP). **oxidized ~** A white, gauze lint local hemostatic (USP). Soluble in dilute alkalies; contains 16–24% COOH. **pseudo ~** Hemicellulose.

c. acetate The esters of c. and acetic acid, used in industry and as rayon, imitation leather, fabrics, yarns, bristles, lacquers. **mono ~** $C_6H_9O_4 \cdot COOMe$ = 204.2. C. monoacetate. **di ~** $C_6H_8O_3(COOMe)_2$ = 246.2 C. diacetate. **tri ~** $C_6H_7O_2(COOMe)_3$ = 288.3. C. triacetate. Colorless, amorphous mass, insoluble in water. **tetra ~** $C_6H_6O(COOMe)_4$ = 330.3. C. tetraacetate. White, amorphous mass, softening 150, insoluble in ordinary solvents. **penta ~** $C_6H_5(COOMe)_5$ = 372.3. C. pentacetate. White, amorphous mass, insoluble in water. **c. dextrin** Cellodextrin. **c. gum** Carboxymethylcellulose. **c. nitrate** The esters of cellulose and nitric acid, used extensively in explosives and in lacquers. Cf. *nitrocellulose.* There exist, according to the degree of nitration: **tri ~** $C_{12}H_{17}O_7(NO_3)_3$ = 459.3. C. trinitrate. **tetra ~** $C_{12}H_{16}O_6(NO_3)_4$ = 504.3 C. tetranitrate. Tri ~ and tetra ~ are white, amorphous masses; insoluble in ordinary solvents, and are the principal constituents of collodion. **penta ~** $C_{12}H_{15}O_5(NO_3)_5$ = 549.3. C. pentanitrate. White, amorphous mass soluble in mixed alcohol and ether. **hexa ~** $C_{12}H_{14}O_4(NO_3)_6$ = 594.3. C. hexanitrate. The principal constituent of guncotton. White, amorphous mass; ignites about 160; insoluble in ordinary solvents.

Celotex Trademark for an insulating wallboard produced by pressing bagasse or other fibers.

celsian $BaO \cdot Al_2O_3 \cdot 2SiO_2$. A barium feldspar refractory.

Celsius, Anders (1701–1744) Swedish astronomer. **C. scale*** The SI temperature scale, formerly called centrigrade scale. Based on the freezing (0°) and boiling (100°) points of water. It equals the thermodynamic temperature in excess of (or below) 273.15 K. Also, 1 degree Celsius (°C) of temperature interval exactly equals 1 kelvin (K).

Celta Trademark for a cellulose acetate synthetic fiber.

celtium Early name for hafnium in zirconium minerals (Urbain and Dauvillier).

cement (1) A plastic material which hardens to form a connecting medium between solids. (2) Portland cement. A fine, gray powder (probably $3CaO \cdot SiO_2$, $3CaO \cdot Al_2O_3$, and $2CaO \cdot SiO_2$). Made by heating a calcareous material (limestone, marl, or chalk) with an argillaceous material (clay or shale, $Al_2O_3 \cdot SiO_2$) at 1350–1800°C to vitrification. The resulting clinker is mixed with 2% gypsum and ground. Composition: lime 62–67, chalcedony 18–20, alumina 4–8, iron 2–3, magnesium 1–4, potassium and sodium 0.5%, titanium and manganese traces. Used for concrete (cement 1, sand 2, gravel 4 pts, and for each 100 kg cement, 50 liters of water). **adamantine ~** Mixed powdered pumice stone with silver amalgam; used to fill teeth. **aluminous ~** High-alumina c., ciment fondu. A strong acid-resistant hydraulic c. Made by grinding the clinker of monocalcium aluminate and dicalcium silicate formed by fusing bauxite and lime. Alumina content at least 32% and ratio to lime 0.85–1.3 (U.K. standard). Has a history of failure in precast c. roof beams. **bituminous ~** C. prepared from natural pitch. **chalcedony ~** C. made from chalcedony. **clinker ~** Uncrushed portland c. **dental ~** See *dental cements.* **glass ~** A mending material for broken glass, e.g., a mixture of resins. **glycerol ~** A glue containing glycerol. **Hensler's ~** A mixture of litharge 3, quicklime 2, china clay 1 pts., ground with oil; used to fill cracks in stoneware. **hydraulic ~** A c. that sets under water; as, portland c. **portland ~** See *portland cement.* **quick-setting ~** A mixture of alkyd resin 11–20, nitrocellulose solution 35, solvent 11–21, and plasticizer 4–8%. **refractory ~** See *refractory cements.* **Roman ~** A quick-setting c. made by mixing burnt clay with lime and sand. **rubber ~** A coating mixture of unvulcanized rubber and sulfur dissolved in oil. **zinc oxide ~** See *dental cements.*

cementation (1) The setting of a plastic material. (2) Heating wrought iron in a bed of charcoal or hematite, to convert it into steel (blister steel). (3) Heating one metal in contact with another to coat it.

cementing Binding together. **c. electron** Electrons of the atomic nucleus which hold protons together.

cementite Fe_3C. Iron carbide in steel.

cementum The bonelike covering of the root *dentine*, q.v., of teeth, consisting of calcium salts deposited in a collagenous matrix.

cenospheres Hollow, spherical coal structures.

Cenozoic See *geologic eras*, Table 38.

cent Abbreviation for the Latin *centum* (a hundred) or *centesimus* (a hundredth).

cental 100-lb weight.

Centaurea A genus of herbs containing 470 species, e.g., *C. behen.* See *behenic acid.*

centaureidin $C_{18}H_{18}O_8$ = 362.3 A flavone decomposition product of centaurin.

centaurin A glucoside from the roots of *Centaurea jacea.*

centaurine An alkaloid from *Erythrea centaurium.*

centaury The *Erythraea* species, formerly used for tonic preparations. **American ~** Sabbatia.

centelloside A triterpene-sugar constituent of *Centella asiatica,* Sri Lanka. Yellow, neutral gum, soluble in water, and similar in properties to asiaticoside.

centesimal Centigrade.

centi-* Abbrev. c. Prefix denoting 1/100.

centigrade Abbrev. °C. See *Celsius scale.*

centimeter, centimetre* The one-hundredth part of a meter (0.3937 in). **cubic ~** Abbrev.: cm^3, cc. A cube whose side is 1 cm long. Since 1964 it has equaled the milliliter; thus 1/1,000 liter = 1 mL = 0.0610237 in^3. In the SI system, the cc is preferred to the mL, but the mL is acceptable for the measurement of liquids and gases. Cf. *liter.* **reciprocal ~** Kaiser.

centimeter-cube Cubic centimeter.

centinormal 0.01 N. See *normal solution* under *normal.*

centipoise cP. 0.01 *poise* q.v. A unit of viscosity (water at 20°C as unity).

centner (1) Zentner. A German standard weight of 50 kg. (2) A drachm weight divided into 100 equal portions.

centrifugal Tending to move outward, away from a center. **c. force** The tendency of one body moving round another to leave its axis of motion.

centrifugalization Centrifugation.

centrifugation Separation of solids from liquids, or liquids of different specific gravity, by subjecting them to fast rotation.

centrifuge A machine for centrifugation consisting of a rapidly rotating container, holding test tubes or bottles containing the mixture. **ultra ~** See *ultracentrifuge.* **c. tube** A heavy-walled glass container, often graduated, used in a c.

centrifuging Centrifugation.

centripetal Tending to move toward the center.

centron The nucleus of an atom.

centrosome A small, highly refractive body of protoplasm between nucleus and cell body, taking an active part in cell division. See *karyokinesis.*

cephaeline $C_{28}H_{38}O_4N_2$ = 466.6. An alkaloid from the root of *Cephaelis ipecacuanha* (Rio ipecac) and *Cephaelis acuminata* (Carthagena ipecac). White crystals, sparingly soluble in water; an emetic. **c. hydrochloride** $C_{28}H_{38}O_4N_2 \cdot HCl$ = 503.1. White crystals, soluble in water; an emetic.

Cephaëlis A genus of Rubiaceae (Central and S. America) which yields ipecacuanha.

cephalanthin A glucoside from *Cephalanthus occidentalis,* buttonbush (Rubiaceae), N. America.

cephaletin An amorphous bitter principle from swamp dogwood, *Cephalanthus occidentalis,* (Rubiaceae), N. America.

cephalexin $C_{16}H_{17}O_4N_3S \cdot H_2O$ = 365.4. Cefalexin, Ceporex(in), Keflex. White powder, slightly soluble in water. A broad-spectrum antibiotic for gram-positive and -negative organisms (USP, BP).

cephalin (1) Kephalin. (2) A white, crystalline acid principle from *Cephalanthus occidentalis.*

cephaloridine $C_{19}H_{17}O_4N_3S_2$ = 415.5. Cefaloridine, Ceporin, Loridine. White powder, freely soluble in water. Use is limited by occasional toxic effect on kidney (USP, EP, BP).

cephalosporins Bacteriocidal antibiotics derived from *Cephalosporium C* and *acremonium.* They have a similar structure to penicillins, with a β-lactam ring and 7-aminocephalosporamic acid. They destroy bacteria by bursting their cell wall. Active against gram-positive and -negative organisms. Examples: cephalexin, cephradine, cefuroxime.

Ceporex, Ceporexin Trademark for cephalexin.

Ceporin Trademark for cephaloridine.

cera Wax. **c. alba** Bleached beeswax. **c. flava** Unbleached beeswax.

ceramels Mixtures of refractory oxides or carbides embedded in a metallic base, which combine the properties of metals and ceramics.

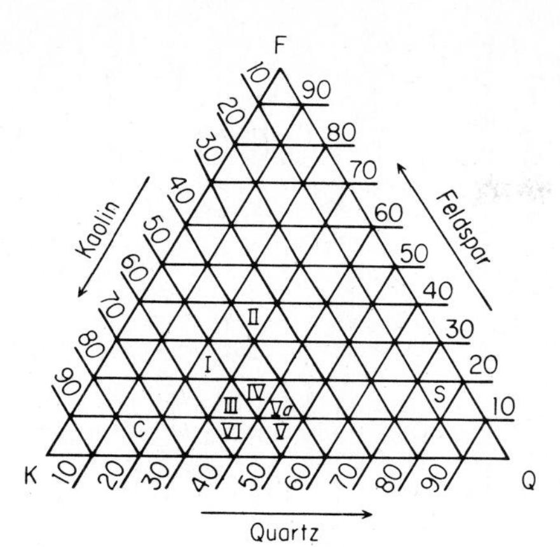

Fig. 7. Ceramic ware.

I Hard porcelain	Va German whiteware
II Soft porcelain	VI Calcareous whiteware
III Japanese porcelain	C Clay
IV Stoneware	S Sand
V Whiteware	F Feldspar

ceramic **c. alloys** Mixtures, providing important engineering materials, that are hard, strong, and light. They can be shaped by normal metal powder techniques such as extrusion and sintering. Main c. a. are silicon nitride (Syalon), silicon carbide, zirconium, and alumina. **c. ink** A ground mixture of potassium carbonate 1, borax 1, litharge 2, cobalt salt 2 pts., mixed with oil, written on glass or porcelain, and burnt in with a bunsen burner.

ceramics The art of making objects (vases, pottery) from clay, and burning metallic colors into them. See Fig. 7.

ceramide A minor constituent of animal tissues, allied to the cerebrosides: Cf. *sphingolipid.*

cerane Hexacosane*.

cerargyrite AgCl. Hornsilver, argentum cornu. A native isomer of silver chloride.

cerasein Cerasin. A resin from the bark of *Prunus (Cerasus) sarotina,* wild cherry; a diuretic and antipyretic.

cerasin (1) Kerasin. (2) Cerasein. (3) Cerasinose.

cerasinose A carbohydrate produced from cherry gum by boiling with dilute sulfuric acid.

cerasite A Japanese variety of iolite.

cerate (1) A pharmaceutical compound of oil or lard with wax, resin, or spermaceti. (2) A soap or metallic salt made from lard. (3) Heptacosanoate*.

ceratin Keratin.

ceratum A mixture of white wax 3, lard 7 pts.

cerberetin $C_{19}H_{26}O_4$ = 318.4. A decomposition product of cerberin. Yellow powder, m.85.

cerberidin A digitalis substitute from *Thevetia cerbera.* Yellow powder, soluble in water.

cerberin $C_{27}H_{40}O_8$ = 492.6. A glucoside from the seeds of *Cerbera odallam* and *Thevetia neriifolia,* (Apocynaceae) southeast Asia. A senna substitute. Cf. *odallin.*

cereal (1) An edible seed of Gramineae; e.g., wheat, rye, oats, rice, maize, barley. (2) A preparation of (1).

cerebellum A part of the brain, similar to the cerebrum, that is responsible for coordination of movement with the senses; as, ability to walk along a straight line. It lies below the cerebral hemispheres.

cerebral Pertaining to the brain. **c. depressants** Drugs which temporarily depress brain functions. **c. excitants** Drugs which stimulate brain activity.

cerebric acid A fatty acid from the white substance of brain tissue.

cerebrin $C_{17}H_{33}O_3N$ = 299.5. A colorless fatty principle, obtained from brain tissue by boiling with barium hydroxide.

cerebrine Cerebrum siccum. Dried brain substance.

cerebron $C_{48}H_{93}O_9N$ = 828.3. An amino lipotide from the white brain tissue.

cerebronic acid $C_{25}H_{50}O_3$ = 398.7 and $C_{24}H_{48}O_3$ = 384.6. A mixture of 2-hydroxypentacosanoic acid and 2-hydroxytetracosanoic acid, m.80.5, obtained by hydrolysis of the white brain substance.

cerebrose $C_6H_{12}O_6$. Brain sugar. Galactose from brain tissue.

cerebrosides Galactosides. Monoglycosyl ceramides of the type RCH(OH)CH(NHCOR′)CH_2O—sugar. Fatty substances in lipids, as, brain and nerve tissue.

cerebrovascular Pertaining to the blood vessels of the brain.

cerebrum The largest area of the brain, with mainly nerve cells (gray matter) plus nerve filaments or fibers (white matter). It is folded, convoluted, and divided into 2 sections (hemispheres); each hemisphere contains the higher centers for coordination, movement, and sensation on the opposite side of the body. Conscious thought and voluntary action originate in the c. and are coordinated with involuntary functions.

Cerenkov effect The residual fluorescence shown by certain liquids after exposure to, and then removal of, a radioactive substance.

ceresin Earth wax, ozocerite. A purified mineral wax, m.58; a beeswax substitute.

cerevisin A dried brewer's yeast derived from the yeast *Saccharomyces cerevisiae.*

cerevisterol A sterol from yeast, m.265. Cf. *ergosterol.*

ceria Ceric oxide.

ceric (1) Cerium(IV)* or (4+)*. A compound of tetravalent cerium. (2) Pertaining to wax. **c. acid** Heptacosanoic acid*. **c. ammonium nitrate** $Ce(NO_3)_4 \cdot 2NH_2NO_3$ = 548.2. Orange crystals, soluble in water, readily reduced to cerous compounds. **c. fluoride** $CeF_4 \cdot H_2O$ = 234.1. White powder, insoluble in water; used in ceramics. **c. hydroxide** $Ce(OH)_4$ = 208.1. Yellow powder , insoluble in water. **c. nitrate** $Ce(NO_3)_4$ = 388.1. Orange hygroscopic crystals, soluble in water. **c. oxide** CeO_2 = 172.1. Cerium dioxide*, ceria. Yellow powder, m.1950, insoluble in water; used in ceramics and gas mantles. **c. sulfate** $Ce(SO_4)_2 \cdot 4H_2O$ = 404.3. Yellow needles, forming a basic salt with excess water. Used as reducer in photography, and as oxidizing agent in analytical chemistry.

ceride A tertiary lipid formed from higher monovalent alcohols and fatty acids.

cerin (1) $C_{30}H_{50}O_2$ = 442.7. A sterol constituent of cork; colorless, acicular crystals. (2) Heptacosanoic acid*.

cerinic acid Heptacosanoic acid*.

ceriometry Volumetric analysis with ceric sulfate solution.

cerite A rare-earth hydrous silicate found in gneiss (Bastnaes, Sweden). **fluo~** See *fluocerite.*

cerium* Ce = 140.12. A rare-earth element, at no. 58. A gray, malleable metal, d.7.042, m.800, b.3420; burns in air like magnesium, decomp. in water. Occurs as silicate (cerite and allanite) and phosphate (monazite), with other rare-earth metals. Discovered by Klaproth (1803), independently by Berzelius and Hisinger, and first isolated by Mosander (1839). C. forms two series of compounds, with valencies 3 (c.(III)* or (3+)*, cerous, Ce^{3+}) and 4 (c.(IV)* or (4+)*, ceric, Ce^{4+}). C. is prepared by electrolysis of the fused chloride, and is used in

alloys for lighter flints (misch metal), for machine-gun tracer bullets, as a "getter" of noble gases in radio tubes, and as a scavenger in steelmaking. The compounds are poisonous.
c.(III)* or (3+)* See *cerous*. **c.(IV)* or (4+)*** See *ceric*. **c. dioxide*** Ceric oxide.

cermet A material in which a ceramic is heat-bonded to a metal; used at high temperatures.

cerolein A constituent of beeswax, soluble in cold alcohol.

ceromel Honeycomb; a mixture of wax and honey.

ceromelissic acid $C_{33}H_{66}O_2$ = 494.9. An acid, m.94, from lardacein.

ceropic acid $C_{36}H_{68}O_{30}$ = 980.9. White microcrystals from the needles of *Pinus sylvestris* (Coniferae).

ceroplastic acid $C_{33}H_{70}O_2$ = 498.9. An acid, m.97 from lardacein.

cerosate Tetracosanoate*.

cerosic acid Tetracosanoic acid*.

cerosin Tetracosanyl tetracosanoate*. A wax found as coating on sugar cane.

cerosinyl The tetracosanyl* radical, $C_{24}H_{49}-$.

cerotate Hexacosanoate*.

cerotene $C_{27}H_{54}$ = 378.7. An unsaturated hydrocarbon from wax. Cf. *cerylene, carotene.*

cerotic acid Hexacosanoic acid*.

cerotin Hexacosyl hexacosanoate*.

cerotinate Hexacosanoate*.

cerotinic acid Hexacosanoic acid*.

cerotol Hexacosyl alcohol*.

cerous Cerium(III)* or (3+)*. Containing trivalent cerium. **c. acetate** $Ce(C_2H_3O_2)_3 \cdot H_2O$ = 335.3. White crystals, m.115, soluble in water. **c. ammonium nitrate** $Ce(NO_3)_3 \cdot 3NH_4NO_3 \cdot 10H_2O$ = 746.4. Large crystals, soluble in water. **c. benzoate** $Ce(C_7H_5O_2)_3$ = 503.5. White powder, soluble in hot water. **c. bromate** $Ce_2(BrO_3)_6 \cdot 9H_2O$ = 1210. White crystals, soluble in water. **c. bromide** $CeBr_3 \cdot 7H_2O$ = 505.9. White, hygroscopic crystals, soluble in water. **c. carbonate** $Ce_2(CO_3)_3 \cdot 5H_2O$ = 550.3. White powder, insoluble in water. **c. chloride** $CeCl_3$ = 246.5. White crystals, m.848, soluble in water. **crystalline ~** $CeCl_3 \cdot 7H_2O$ = 372.6. Pink crystals, soluble in water. **c. citrate** $CeC_6H_5O_7$ = 329.2. White powder, insoluble in water. **c. fluoride** CeF_3 = 197.1. Colorless crystals, m.1324, slightly soluble in water. **c. hydroxide** $Ce(OH)_3$ = 191.1. Yellow powder, insoluble in water. **c. iodide** $CeI_3 \cdot 9H_2O$ = 683.0. Pink crystals, soluble in water. **c. lactate** $Ce(C_3H_5O_3)_3$ = 407.3. White powder, sparingly soluble in water. **c. malate** $Ce_2(C_4H_4O_5)_3$ = 676.5. White powder, soluble in water. **c. nitrate** $Ce(NO_3)_3 \cdot 6H_2O$ = 434.2. White crystals, soluble in water. **c. oxalate** $Ce_2(C_2O_4)_3 \cdot 9H_2O$ = 706.4. Pink crystals, usually containing lanthanum and praseodymium salts, insoluble in water. A raw material in the preparation of rare earths. **c. oxide** Ce_2O_3 = 328.2. Gray powder, d.6.9, insoluble in water. **c. phosphate** $CePO_4$ = 235.1. Yellow rhombs, insoluble in water. **c. salicylate** $Ce(C_7H_5O_3)_3$ = 551.5. White powder, insoluble in water. **c. sulfate** $Ce_2(SO_4)_3$ = 568.4. Monoclinic crystals, soluble in water. **crystalline ~** $Ce_2(SO_4)_3 \cdot 8H_2O$ = 712.5. Pink crystals, soluble in water; used to prepare aniline black. **c. valerate** $Ce(C_5H_9O_2)_3$ = 443.5. White powder, soluble in water.

ceroxylin $C_{20}H_{32}O$ = 288.5. Crystals in the waxy exudation of *Ceroxylon andicola,* wax palm (S. America).

Certification (Trade) Mark A trademark certified by the British Standards Institution.

Certified Reference Materials Materials of accurately ascertained properties issued by the Community Bureau of Reference (BCR) of the EEC and the International Atomic Energy Agency as analytical control standards. They include nuclear, mineral, biological, and plant materials; metals, solvents, and particles of standardized size.

cerulean blue Ceruleum.

cerulein $C_{20}H_{12}O_7$, or $C_{20}H_{10}O_6 \cdot H_2O$ = 364.3. Coerulein. An acidic or internal anhydride coloring matter from gallein.

ceruleum A blue pigment: cobaltous stannate and calcium sulfate.

cerulic acid An oxidation product of coffee.

cerulignone $C_{16}H_{16}O_6$ = 304.3. Coerulignone, cedriret. Blue needles, m.291, obtained by treating crude pyroligneous acid with potassium dichromate. **hydro ~** See *hydrocerulignone.*

cerumen Human ear wax.

cerusa, ceruse, cerussa White lead.

cerussite $PbCO_3$. White lead carbonate produced by the action of carbon dioxide and water on lead ores.

cervantite Sb_2O_4. A secondary mineral formed by the oxidation of antimony sulfide.

ceryl The hexacosyl* radical $C_{26}H_{53}-$, from hexacosyl alcohol. **c. alcohol** Hexacosyl alcohol*.

cerylate Hexacosanolate*.

cerylene $C_{27}H_{54}$ = 378.1. A hydrocarbon from Chinese wax. Cf. *cerotene.*

cerylic The hexacosyl* radical.

cesiated Coated with cesium.

cesium* Cs = 132.9054. Caesium. A rare element of the alkali family, at. no. 55. A silver, soft metal, d.1.87, m.28, b.670, which reacts violently with water. It is a highly electropositive element (valency 1). C. occurs in mineral waters and minerals and in the ashes of a few plants, and is prepared by the distillation of c. hydroxide with magnesium powder in hydrogen. Discovered spectroscopically by Kirchhoff and Bunsen (1860); isolated by Setterberg (1881). C. metal is a rubidium substitute; its salts are microchemical reagents and a sensitive coating in photoelectric cells. ^{137}Cs, in the form of needles, is used to treat malignant diseases. **c. acetate*** $CsC_2H_3O_2$ = 192.0. Deliquescent, white crystals, m.194, soluble in water. **c. alum** Aluminum c. sulfate. **c. benzoate** $CsC_7H_5O_2$ = 254.0. White powder, soluble in water. **c. bicarbonate** C. hydrogencarbonate*. **c. bisulfate** C. hydrogensulfate*. **c. bitartrate** C. hydrogentartrate*. **c. bromate*** $CsBrO_3$ = 260.8. White crystals, soluble in water. **c. bromide*** CsBr = 212.8. White crystals, soluble in water. **c. carbonate*** Cs_2CO_3 = 325.8. White, hygroscopic crystals, soluble in water; used in soda water manufacture. **c. chloride*** CsCl = 168.4. White, regular crystals, m. 646 (sublimes), soluble in water; used in soda water manufacture. **c. chromate*** Cs_2CrO_4 = 381.8. Yellow crystals, soluble in water. **c. cyanide*** CsCN = 158.9. White crystals, soluble in water. **c. dichromate*** $Cs_2Cr_2O_7$ = 481.8. Yellow crystals, soluble in water. **c. formate** $CsCHO_2$ = 177.9. Deliquescent, white powder, m.265, soluble in water. The monohydrate loses water at 41°C. **c. hexachloroplatinate*** $Cs_2[PtCl_6]$ = 673.6. Yellow octahedra, soluble in water. **c. hexafluorosilicate*** $Cs_2[SiF_6]$ = 407.9. C. fluosilicate. White crystals, soluble in water. **c. hydrogencarbonate*** $CsHCO_3$ = 193.9. C. bicarbonate. White crystals, soluble in water. **c. hydrogensulfate*** $CsHSO_4$ = 230.0. C. bisulfate, c. acid sulfate. White crystals, soluble in water. **c. hydrogentartrate*** $CsHC_4H_4O_6$ = 282.0. C. bitartrate, c. acid tartrate. White rhombs, soluble in water. **c. hydroxide*** CsOH = 149.9. A gray, deliquescent mass, m.275, soluble in water. **c. iodate*** $CsIO_3$ = 307.8. White crystals, soluble in water. **c. iodide*** CsI = 259.8. White crystals, m.621, soluble in water. **c. ion** Cs^+, present in solutions of c. salts in water. **c. manganese chloride*** $2CsCl \cdot MnCl_2 \cdot 3H_2O$ = 516.6. Pink crystals, soluble in water, **c. nitrate*** $CsNO_3$ = 194.9. Colorless prisms, m.414 (decomp.), soluble in water; used in pyrotechnics. **c. nitrite*** $CsNO_2$ = 178.9. White crystals, soluble in water. **c. oxalate*** $Cs_2C_2O_4$ = 353.8. White powder, soluble in water. **c. oxide** (1) Cs_2O = 281.8. Normal c. oxide, c. monoxide*. Orange crystals, soluble in water. (2) Cs_2O_2 = 297.8. C. dioxide*. (3) Cs_2O_3 = 313.8. C. trioxide*. (4) Cs_2O_4 = 329.8. C. tetraoxide*, c. peroxide. **c. perchlorate*** $CsClO_4$ = 232.4. White crystals, soluble in water. **c. periodate*** $CsIO_4$ = 323.8. White crystals, soluble in water. **c. permanganate*** $CsMnO_4$ = 251.8. Violet prisms, soluble in water. **c. peroxide** C. Tetraoxide*. Yellow granules, decomp. violently in water. **c. rhodium sulfate** $Cs_2SO_4 \cdot Rh_2(SO_4)_3 \cdot 24H_2O$ = 1288. Rhodium c. alum. Yellow octagons, m.110, soluble in water. **c. rubidium alum** Aluminum c. rubidium sulfate*. **c. rubidium chloride*** $CsCl \cdot RbCl$ = 289.3. Yellowish crystals, soluble in water. **c. salicylate*** $CsC_7H_5O_3$ = 270.0. White powder, soluble in warm water. **c. sulfate*** Cs_2SO_4 = 361.9. Colorless needles, soluble in water; used in soda water manufacture. **c. sulfide** (1) Cs_2S = 297.9. C. monosulfide*. (2) Cs_2S_2 = 329.9. C. disulfide*. (3) Cs_2S_3 = 362.0. C. trisulfide*. (4) Cs_2S_5 = 426.1. C. pentasulfide*. **c. sulfite*** Cs_2SO_3 = 345.9. White crystals, soluble in water. **c. tartrate, acid** C. hydrogentartrate*. **c. tetraoxide*** See *c. oxide (4).* **c. trinitride*** CsN_3 = 174.9. **c. trisulfide*** See *c. sulfide.*

cespitine $C_5H_{13}N$ = 87.2. An isomer of pentylamine from coal tar.

cetaceum Spermaceti.

cetane Hexadecane*.

Cetavlon Trademark for cetrimide.

cetene 1-Hexadecene*.

cetenylene 1-Hexadecyne*.

cetic acid $C_{15}H_{30}O_2$ = 242.4. A fatty acid from spermaceti.

cetin $C_{15}H_{31}COOC_{16}H_{33}$ = 480.0. Hexadecyl palmitate*. The principal constituent of spermaceti. Colorless crystals, m.50, insoluble in water.

cetolic acid $Me(CH_2)_9CH:CH(CH_2)_9COOH$ = 338.6. (Z)-11-Docosenoic acid*. An isomer of erucic acid in marine animal oils.

cetostearyl alcohol A mixture chiefly of hexadecyl and octadecanyl alcohols, m. below 43; used in ointments (BP).

Cetraria (1) A genus of lichens. (2) Iceland moss: *C. vulpina;* yields vulpinic acid; *C. islandica,* island moss, yields cetrarin.

cetraric acid $C_{18}H_{16}O_8$ = 360.3. A crystalline, dibasic acid from Iceland moss.

cetrarin $C_{30}H_{30}O_{12}$ = 582.6. Cetrarinic acid. A white, crystalline, bitter principle of Iceland moss, insoluble in water. A demulcent. **c. base** A brown, amorphous mixture of cetrarin, cetraric acid, stearic acid, and other substances from Iceland moss; used similarly to cetrarin.

cetrarinic acid Cetrarin.

cetrimide Cetavlon, Savlon. A mixture of tetra- (mainly), do-, and hexadecyltrimethylammonium bromides. White, bitter powder, soluble in water. An analytical spot test reagent, cationic detergent, and disinfectant. Used to clean skin, wounds, and plastic instruments; also in ointments and dusting powders (EP, BP).

cetyl The hexadecyl* radical, $C_{16}H_{32}-$. **c. alcohol** Hexadecyl alcohol* (NF). **c. cyanide** Heptadecane nitrile*. **c. ether** Dihexadecyl *ether**. **c. pyridinium chloride** $C_{21}H_{38}NCl \cdot H_2O$ = 358.0. 1-Hexadecylpyridinium chloride. White powder. Cationic disinfectant, used similarly to cetrimide (USP, BP). **c. trimethylammonium bromide** CTAB. See *cetrimide.*

cetylate (1) Palmitate*. (2) A compound derived from hexadecyl alcohol containing the radical $C_{16}H_{33}O-$.

cetylene 1-Hexadecene*.

cetylic acid Palmitic acid*.

cetylide $(C_{16}H_{31}O_2)_3C_{16}H_{18}(OH)_3$ = 1028. A hydrolysis product of cerebrin.

cevadic acid 2-Methyl-2-butenoic acid*.

cevadine Veratrine.

cevine $C_{27}H_{43}O_8N$ = 509.6. An alkaloid hydrolysis product of veratrine.

cevitamic acid Vitamin C.

cevollite A fibrous zeolite alteration product of melilite.

ceyssatite A white absorbent earth of almost pure silica (Ceyssat, France); used to treat eczema.

Cf (1) Symbol for californium. (2) (With period) abbreviation for confer or compare.

c.g. Abbreviation for center of gravity.

CGPM See *standards.*

cgs, CGS Abbreviation for centimeter-gram-second units. See *SI units* under *SI.*

ch Abbreviation for chain.

chabazite $CaAl_2Si_4O_{12} \cdot 6H_2O$. Phacolite, cabazite. A zeolite, produced artificially by the action of water on cement. Cf. *acadialite, gmelinite.*

Chaddock burner A small pottery furnace. **C. clamp** A wooden test-tube holder. **C. support** A buret stand with wire springs to hold two burets.

chaetomin An antibiotic from cultures of *Chaetomium cochliodes.*

Chaga's disease A form of trypanosomiasis in S. America.

Chain, Sir Ernst Boris (1906–1979) German-born British biochemist. Nobel prize winner (1945). Codeveloper of penicillin.

chain (1) A series of connected, successive events or substances; as in c. reaction. (2) Similar atoms linked by homopolar bonds, as a carbon c. (3) A measure of length: 1 chain = 4 rods = 100 links = 20.11684 m. **bacterial ~** A chainlike bacterial growth. **long ~** A c. of more than 8 bacteria. **short ~** A c. of 2–8 bacteria. **branched ~** See *lateral chain* and *forked chain* below. **carbon ~** A series of connected carbon atoms. **closed ~** A ring formed by a c. of atoms, e.g., the benzene ring. **forked ~** A branch of two small chains at the end atom of a long c. **lateral ~** A branch c. in the middle of a longer c. **nuclear ~** A reaction resulting from some types of nuclear fission. Cf. *atom bomb.* Thus, 1 neutron + $^{235}U \rightarrow$ fission products + 2 neutrons + gamma rays. Then, 2 neutrons + $^{235}U \rightarrow$ fission products + 4 neutrons + gamma rays. The 4 neutrons react similarly, and so on. **open ~** A number of atoms joined in a line. **periodic ~** See *periodic system.* **side ~** (1) See *side chain.* (2) An atomic group of protoplasm which reacts with a nonprotoplasmic group. **straight ~** Open c.

c. reaction A series of successive reactions, each of which depends on the preceding one. See *nuclear chain* above.

chainomatic Describing an analytical balance with no separate weights less than 0.1 g. Weights are determined by adjusting and measuring the length of the fine chain that hangs from one of the arms to balance the substance.

chair conformation See *conformation.*

chaksine $C_{11}H_{19}O_2N_3$ = 225.3. An alkaloid in *Cassia absus.*

chalcacene $C_{36}H_{18}$ = 450.5. A pyrolytic product of acenaphthene. Red needles, m.359.

chalcanthite $CuSO_4 \cdot 5H_2O$. Blue crystals, formed by the evaporation of cupriferous mine waters.

chalcedony A cryptocrystalline, amorphous quartz, which occurs as a variety of semiprecious stones: agate (banded), cornelian (opaque), chrysoprase (green opaque), bloodstone (dark opaque, red spots), jasper, silicified wood moss agate (milky with dendritic manganese oxide), onyx (agate with alternating light and dark bands), plasma (green mottled), prase (gray-green), sardonyx (golden or blood-red).

chalcedonyx White and gray layered agate.

chalcocite Cu_2S. A native copper sulfide.

chalcogens* Chalkogens. The elements oxygen, sulfur, selenium, tellurium, and polonium. Cf. *halogen.*

chalco(ge)nides* Compounds of the chalcogens.

chalcomorphite A native, hydrous aluminum silicate.

chalcone* (1) Ph·CH:CH·CO·Ph = 208.3. Phenyl styryl ketone*, 1,3-diphenyl-2-propen-1-one†. (2) Aralkylic ketones. Natural yellow and orange substituted benzylideneacetophenone derivatives of the type Ar·CO·CH:CHAr. Cf. *flavones.*

chalcophanite $ZnO \cdot 2MnO_2 \cdot 2H_2O$. A metallic, bluish black zinc manganite.

chalcophyllite $Cu_7As_2O_{12} \cdot 14H_2O$. An oxidation product of enargite; hexagonal, green, transparent crystals.

chalcopyrite $CuFeS_2$. A metallic, yellow ore.

chalcopyrrhotite $CuFe_4S_6$. A brown ore.

chalcose A hexose component of the antibiotic chalcomycin.

chalcosiderite $CuFe_6(PO_4)_4 \cdot 8H_20$. A green ore.

chalcostibite $CuSbS_2$. Wolfsbergite. A native, metallic, gray copper sulfantimonide, d.4.8–5.0.

chalcuite A bluish green turquoise (New Mexico).

chaldron An obsolete unit of dry measure; equals 36 bushels.

chalk $CaCO_3$. Creta, calcite. Amorphous shell residues, deposited from ocean oozes. See *calcium carbonate.* **French ~** A grade of talc.

chalking To treat with chalk, e.g., in dyeing by passing cotton through a 0.6% suspension of chalk.

chalkogen(ide)s Chalcogen(ide)s*.

chalmersite $CuFe_2S_3$. A native copper iron sulfide.

chalones Substances which oppose the proliferation of cancerous tissues.

chalybeate A natural water containing iron salts.

chalybite Spathic iron ore. Native ferrous carbonate. Cf. *siderite.*

chamazulene $C_{14}H_{16}$ = 184.3. 1,4-Dimethyl-7-ethylazulene. A blue hydrocarbon from chamomile oil.

chamber (1) A boxlike receptacle. **cloud ~** A c. making traveling particles visible. **fog ~** Cloud c. **ionization ~** A partly evacuated box, with galvanometer for detecting rays. (2) The leaded c. used in sulfuric acid manufacturing, q.v.

c. **acid** Impure 60–70% sulfuric acid from the c. process. c. **crystals** Nitrosyl sulfate*. c. **process** The manufacture of sulfuric acid in lead chambers from the interaction of sulfur dioxide, air, steam, and oxides of nitrogen.

Chamberland, Charles Edouard (1851–1908) French bacteriologist. **C. filter** A porous clay cylinder for filtering bacteria from solutions. **C. flask** A glass flask with side tube, for growing bacteria.

chamomile A genus of composite plants; especially *Anthemis* species. **common ~, English ~, or Roman ~** The flowers of *Anthemis nobilis;* a febrifuge used as an infusion or herb tea (EP, BP). **German ~** See *Matricaria.* **wild ~** Stinking mayweed, *Anthemis cotula.* See *anthemidine.*

c. **oil** Essential oil from the *Anthemis nobilis* flowers. Blue liquid, d.0.910; contains esters of butanoic and 2-methyl-2-butenoic acids. **German ~** The essential oil of *Matricaria,* d.0.930-0.940; it contains esters of hexanoic acid, azulene, and chamazulene.

chamomillin A bitter principle of chamomile.

chamosite A constituent of berthierine.

champacol $C_{15}H_{26}O$ = 222.4. Guaiol. Camphor from the wood of *Michelia champaca* (Magnoliaceae), Java. Colorless needles, m.87, insoluble in water.

chandoo Chundoo. The best quality of raw opium, prepared for smoking. Cf. *mudat, yenshee.* **c. dross** Yenshee.

change of state Passing from the solid to the liquid or from the liquid to the gaseous state; or the reverse. See *condensation, freezing, fusion, vaporization, sublimation.*

chaoite A rare mineral form of carbon. See *carbynes.*

Chaperon cell A voltaic cell consisting of amalgamated zinc and copper in a solution of potassium hydroxide; 0.98 volt.

char (1) To carbonize or burn incompletely. See *fixed carbon* under *carbon.* (2) Charcoal.

charas Cannabis from hemp resin.

charcoal Amorphous carbon from the incomplete combustion of animal or vegetable matter, e.g., wood. Used for the adsorption of gases or coloring matters, in blowpipe analysis, and as a pigment (c. black). **activated ~** See *activated carbon.* **animal ~** C. prepared from bone or blood. **vegetable ~** or **wood ~** C. from the incomplete combustion of wood, nutshells, or fruit stones. Used medicinally, for hyperacidity and some forms of indigestion. See *activated carbon;* industrially, as a clarifying and decolorizing agent, and in gas masks. Cf. *carbonization, Norit.*

C. Cloth Trademark for activated charcoal, in a flexible and strong form, produced from a rayon-based cloth.

Chardin filter paper A paper for the filtration of agar-agar solution for culture media.

Chardonize Trademark for a cellulose acetate synthetic fiber.

Chardonnet silk A *viscose rayon,* q.v.

charge (1) A load or burden, e.g., the c. of ore in a furnace, the c. of electricity in a capacitor. (2) In physics, a definite quantity of electricity. **atomic ~** The electricity carried by an atom or ion, which depends on the number of valence electrons. **elementary ~** The c. of a proton, $e = 1.602189 \times 10^{-19}$ C, or 1.602×10^{-20} emu, or 4.803×10^{-10} esu. **fictitious ~** The quantity of electricity contained in the material of a capacitor as distinct from the true c. of the plates. **ionic ~** See *ionic c.* **nuclear ~** See *atomic structure.* **residual ~** The quantity of electricity remaining in a capacitor after discharge. **specific ~** Electronic ratio. The ratio of electron charge to electron mass, $e/m_e = 1.7588 \times 10^{-13}$ C/kg.

charging rod (1) A piece of sealing wax, hard rubber, or glass used for charging electroscopes. (2) A device in which a Celluloid tube is rubbed over flannel to charge electroscopes.

Charles, Jacques Alex Caesar (1746–1823) French chemist. **C.'s law** Dalton's law, Gay-Lussac law. The volume of a gas at 0°C increases with each degree centigrade by $\frac{1}{273}$ if the pressure is constant. The pressure of a gas increases with each degree centigrade by $\frac{1}{273}$, if the volume is constant. Hence the ratio between the increase in pressure per degree and the pressure at 0°C (pressure coefficient) is the same for all gases.

Charleston phosphate A soft phosphate mineral (27% phosphorus pentaoxide).

charlock The seeds of *Sinapis (Brassica) arvensis,* field mustard; condiment.

Charlton white Lithopone.

charpie Lint.

Charpy test A measure of the effect of impact on sheet material. It is the dynamic analog of the *Brinell test,* q.v.

charring Carbonizing.

Chatterton's compound A cement for glass.

chaulmoogra oil A yellow oil from the seeds of *Taraktogenos Kurzii* (Bixineae), India, d.0.946-0.951, m.5-20, iodine no.

103.5. It contains chaulmoogric, gynocardic and hydnocarpic acids. Formerly used medicinally to treat leprosy and skin diseases.

chaulmoogrene $C_{18}H_{34}$ = 250.5 *n*-Docecyl-Δ^4-cyclopentane, from chaulmoogric acid.

chaulmoogric acid $C_5H_7(CH_2)_{12}COOH$ = 280.5. 2-Cyclopentene-1-Tridecanoic acid. **(S)-~** Tridecoic acid. Colorless crystals from chaulmoogra oil, m.68, soluble in alcohol.

chaulmugra Chaulmoogra.

chavibetol $C_{10}H_{12}O_2$ = 164.2. 5-Allyl-1-hydroxy-2-methoxybenzene. Colorless liquid, b. 254. **iso ~** Betelphenol, a constituent of essential oils. Cf. *eugenol.*

chavicic acid An amorphous acid from chavicine.

chavicine An alkaloid from black pepper.

chavicol $C_6H_4\cdot(OH)C_3H_5$ = 134.2. *p*·Allylphenol. A colorless constituent of betel oil, b.237. **methyl ~** Estragole.

chebulinic acid (1) $C_{28}H_{20}O_{19}$ = 660.5. A principle from the seeds of *Terminalia chebula,* southeast Asia. (2) $C_{41}H_{34}O_{28}$ = 974.7. An acid derived from tannin. (3) $C_{34}H_{30}O_{23}$ = 806.6. Eutannin. An acid from myrobalans. Rhombic prisms, decomp. 234, soluble in hot water.

checkerberry Gaultheria.

cheddite A high explosive: potassium chlorate 70–90, aromatic nitro compounds 0-20, paraffin 0–15%.

cheese A food prepared from the casein of skimmed or unskimmed milk, and flavored by the activity of certain bacteria.

cheilosis Painful reddening of lips at junction with mouth due to riboflavin deficiency.

cheiramidine $C_{22}H_{26}O_4N_2\cdot H_2O$ = 400.5. An alkaloid from *Remijia puridieana* (Rubiaceae).

cheiramine A secondary alkaloid from *Remijia purdieana.*

Cheiranthus Wallflower. The herb *C. cheiri* (Cruciferae), which yields alkaloids, glucosides, and acids.

cheirantic acid An oleic-type acid from the oil of *Cheiranthus.*

cheirantin A glucoside from *Cheiranthus.*

cheirin A digitalislike glucoside from *Cheiranthus.*

cheirinine $C_{18}H_{35}O_{17}N_3$ = 565.5. An alkaloid from the leaves of *Cheiranthus.* Colorless crystals, sparingly soluble in water.

chekan, cheken The shrub *Myrtus cheken* (Myrtaceae), Chile: its bark is an astringent.

chekenetin $C_{11}H_7O_6\cdot H_2O$ = 253.2. Olive crystals, from *Myrtus (Eugenia) cheken (Myrtaceae),* Chile.

chekenine $C_{13}H_{11}O_3$ = 215.2. A volatile alkaloid from *Myrtus cheken.* Rhombic, yellow scales.

chekenone $C_{40}H_{44}O_8$ = 652.8. A crystalline principle from *Myrtus cheken.*

chelate (Greek *chele*—a "crab's claw"). Pertaining to a molecular structure in which a ring can be formed by the residual valencies (unshared electrons) of neighboring atoms. **c. compound** An organic compound in which atoms of the same molecule are coordinated; e.g., coordinate compounds of metals with ethylenediamine

$$\begin{array}{l} CH_2\cdot NH_2 \searrow \\ | \qquad\qquad\quad M \\ CH_2\cdot NH_2 \nearrow \end{array}$$

c. groups Groups capable of bonding to a central atom by 2 or more coordinating atoms.

chelation See *sequestering agent.*

chelatometry Complexometric titration.

chelen Ethyl chloride*.

chelerythrine $C_{21}H_{17}O_4N$ = 347.4. A narcotic alkaloid related to sanguinarine from the seeds of *Chelidonium majus* (Papaveraceae). Red crystals, m.203, slightly soluble in alcohol.

chelidonate A salt of chelidonic acid containing the radical $C_7H_2O_6-$.

chelidonic acid Pyrone-2,6-dicarboxylic acid*.

chelidonine $C_{20}H_{19}O_5N$ = 353.4. An alkaloid from *Chelidonium majus* and opium. Colorless crystals,m.130, insoluble in water, a narcotic. **homo ~** See *homochelidonine.*

Chelidonium (1) Papaveraceous plants, e.g., *C. majus,* greater celandine. (2) Celandin. The leaves and stems of *C. majus,* a cathartic.

chelidonoid The combined principles of *Chelidonium majus:* chelidonine, chelerythrine, protopine, and other alkaloids.

chelidoxanthine A yellow, crystalline, bitter principle from *Chelidonium majus.*

chelonin A brown, bitter, amorphous powder from *Chelone glabra,* snakehead (Scrophulariaceae).

chemavinite Cedarite, Canadian amber. A mineral resin from Hudson Bay.

chemical (1) Pertaining to chemistry. (2) See *chemicals.* **C. Abstracts Service** CAS. Arm of the American Chemical Society, established 1907, providing in *Chemical Abstracts* about 900 abstracts/week from 12,000 scientific journals from over 150 countries. **CAS registry number** Unique number given to each definable chemical, in the format Y-XX-X, where Y is 2–6 digits and X is a single digit. Used as a basis for data retrieval; 6 million chemicals had a number in 1983. **c. action** A change in the molecular composition of a substance produced chemically or by heat, light, or electricity. See *reaction.* **c. activity** Reactivity. **c. affinity** See *affinity.* **c. antidote** A c. used to counteract the effect of a poison, e.g., sodium nitrite or thiosulfate for cyanide poisoning. **c. burns** The corrosive effect of chemicals on the skin. **c. change** Reaction. **c. compound** See *compound.* **c. constitution** See *structure.* **c. denudation** The process by which salts in the soil are dissolved by water and carried to the sea. **c. energy** The energy relations of c. reactions. Intensity is expressed by c. affinity or c. potential; capacity is expressed by the equivalent weight or concentration. **c. entities** The fundamental concepts of chemistry; as, atoms, ions, molecules, and free radicals. **c. equation** The expression of a reaction in terms of c. formulas, e.g., reacting substances ($A + B$) = reaction products ($C + D$). **ionic ~** The general statement of a c. reaction which shows only those substances actually undergoing change; thus: $Ba^{++} + SO_4^{=} = BaSO_4$. **molecular ~** The specific statement of a chemical reaction showing the relative proportions of substances involved; thus: $BaCl_2 + H_2SO_4 = BaSO_4 + 2HCl$. **c. garden** Silicate garden. Vegetationlike growths produced by dropping Group II and transition metal salts into a silicate solution. **C. Mace** See *tear gases.* **c. oxygen demand** C.O.D. A chemical measure of the deoxygenating power of an effluent in terms of the amount of oxidizing agent (e.g., boiling acidic potassium dichromate) reduced. More comprehensive oxidation is achieved than in the B.O.D. test. C.O.D. ranges 2–5 times B.O.D., depending on effluent composition. **c. pulp** Plant material treated by the sulfate or sulfite processes for papermaking or rayon manufacture. **c. reaction** See *reaction.* **C. Reference Substance** A substance of established purity prepared for use in official (e.g., BP and BP Codex) assays. **c. societies** The largest are:

Royal Society of Chemistry, London 1841
Société Chimique de France, Paris 1858
Gesellschaft Deutscher Chemiker, Frankfurt . 1867
American Chemical Society, Washington . . . 1876

c. solvents See *solvent.* **c. system** An equilibrium characterized by definite proportions of reacting substances and reaction products. Cf. *phase* rule, *system.*

chemicals Compounds or substances of definite molecular composition. Generally restricted to a single molecular species; whereas *drug* can refer to a substance derived from a vegetable or animal source or a mixture of substances. The grades of purity of chemicals are:

C.P.	Chemically pure, the highest grade
U.S.P. or B.P.	Tested to conform with the requirements of the U.S. or British Pharmacopoeia, respectively
A.R.	Analytical reagent
Pure	For general work
Tech	For technical work
Crude	An impure or unrefined grade

kinetic ~ Refrigerants.

chemiluminescence See *chemiluminescence* under *luminescence.*

chemistry The science of the fundamental structure of matter, the composition of substances, their transformation, analysis, synthesis, and manufacture. **analytical ~** C. dealing with the detection (qualitative) and determination (quantitative) of substances. **animal ~** C. dealing with the composition of animal tissues and fluids. Cf. *biochemistry* below. **applied ~** Chemical technology. C. applied to some useful end, either directly (through industry), or for the welfare of man. **astro ~** The c. of the composition of celestial objects, as, stars and nebulae. **bio ~** The c. of life; c. dealing with the composition of animal and vegetable matter, the changes occurring in the living organism, the transformation of food into living material, and the elimination of waste products. **biological ~** Biochemistry. **commercial ~** The compounding of substances for some utilitarian purpose. **electro ~** C. dealing with the relation between electrical and chemical energy and their transformation. Cf. *electrolysis.* **empirical ~** Chemical knowledge obtained by experimentation and observation, rather than from theory. **engineering ~** C. applied to the composition and properties of engineering materials. **fermentation ~** A branch of biochemistry dealing with the catalytic changes produced by enzymes. **food ~** C. dealing with the quality, composition, and examination of foods. **forensic ~** The application of chemical knowledge to legal matters. **galvano ~** Electrochemistry. **geo ~** The c. of the earth's surface and the changes occurring in the atmosphere, hydrosphere, and lithosphere. **geological ~** Geochemistry. **histo ~** The c. of the composition of and chemical changes occurring in the tissues of plants and animals. **histological ~** Histochemistry. **historical ~** The study of the evolution of chemical thought through the ages. **inorganic ~** The c. of polar compounds, usually those not containing carbon. **judicial ~** Forensic c. **legal ~** Forensic c. **manufacturing ~** C. applied to the large-scale preparation of substances. **mechano ~** A branch of physical c. dealing with the mechanical properties of substances, e.g., surface tension. **medical ~** C. applied to medicine, to diagnose and combat disease. **meta ~** The study of the subatomic or characteristically atomic properties of matter; as, adsorption. **micro ~** See *microchemistry.* **microscopical ~** See *microchemistry.* **mineral ~** Mineralogical or inorganic c. **mineralogical ~** C. dealing with the composition and formation of minerals. **organic ~** The c. of carbon compounds; generally those which are nonpolar. **pathological ~** The c. of the composition of abnormal tissues and body fluids, and the changes caused by disease. **pharmaceutical ~** C. applied to the preparation, testing, and composition of drugs. **photo ~** The study of the relations of radiant and chemical energy and their transformation. **photographic ~** C. applied to photography. **physical ~** A branch of theoretical c. which deals with the transformation of physical and chemical energies. **physiological ~** The c. of the composition of and chemical changes in the healthy animal or vegetable organism. **phyto ~** The c. of plants and plant functions. **pure ~** A branch of theoretical c. that deals with chemical forces alone. **radio ~** A branch of theoretical c. which studies the composition and structure of the atom as revealed by the radioactive elements. **stereo ~** The theoretical c. of the spatial structure of molecules. **stoichiometric ~** C. dealing with chemical force as expressed by atomic weight and valency, the proportions in which substances react, and the distribution of atoms of a molecule. **structural ~** A branch of stereochemistry dealing with the internal molecular arrangement of carbon and other compounds. **synthetic ~** The building up of compounds from simpler substances. **technical ~** C. applied to technology. **theoretical ~** The deduction of laws which govern the experimentally established facts of c. **therapeutical ~** A branch of medical c. dealing with the effect of drugs on the living organism. **thermo ~** Physical c. dealing with the relation of heat and heat radiation during chemical changes. **topo ~** See *topochemistry.* **toxicological ~** C. dealing with the composition and detection of poisons. **vegetable ~** The biochemistry of the composition of plants and the changes occurring during their normal functions. **zoö ~** The c. of animals and animal functions, excluding man.

chemoimmunology The study of the chemical changes occurring in the immunization of an organism.

chemokinesis The increase in activity of an organism due to a chemical substance.

chemology An obsolete alternative term for chemistry.

chemolysis The dissolution of organic matter during decay, due to chemical (not bacterial) action.

chemometrics The application of mathematical and statistical methods to chemical problems.

chemoresistance The specific resistance of a cell due to a chemical substance.

chemosmosis Chemical reactions taking place through an intervening semipermeable membrane.

chemostat A device for maintaining constant chemical and physical conditions; as, the composition of the medium in the continuous culture of bacteria.

chemosynthesis A chemical reaction due to oxidation by bacteria; as, $H_2S \rightarrow H_2SO_4$. Cf. *photosynthesis.*

chemotaxis Chemotropism. The attraction or repulsion of cells or organisms by chemicals.

chemotherapy Treatment of (1) infection by microorganisms, or infestation by insect or worm, with a chemical substance or drug; (2) malignant disease by cytotoxic drug or hormone.

chemotropism Chemotaxis. **negative ~** The repulsion of microorganisms by chemicals. **positive ~** The attraction of microorganisms by chemicals.

Chemstrand Trademark for a mixed-polyester synthetic fiber.

chemurgic Chemistry applied to farming. **C. Council** An organization of Dearborn, Mich., U.S.A., to advance the industrial use of American farm products through applied science.

chenopodin A bitter principle from *Chenopodium album.*

chenopodium (1) The goosefoot family. (2) American wormseed, Mexican tea. The fruit of *C. anthelminticum.* **c. oil** American wormseed oil. The essential oil from the seeds of *C. ambrosioides;* an anthelmintic. **c. seeds** The fruits of *C. ambrosioides* or *C. anthelminticum,* American wormseed; an anthelmintic.

chernawinite Canadian amber. A resinous mineral similar to amber.

chernozem A black earth from semiarid steppe lands. The top layer is rich in humus; the bottom layers are rich in calcium carbonate.

cherry A tree of the subgenus *Cerasus* of *Prunus* (Rosaceae). Cf. *cerasinose.* **wild ~** See *wild cherry.*

c. juice Succus cerasi. A sour, aromatic liquid from the ripe fruit of *Prunus cerasus.* Contains not less than 1% of malic acid; a pharmaceutical flavor (NF).

cherry laurel *Prunus laurocerasus* (Rosaceae); a sedative. **c. l. oil** The essential oil of c. l., d.1.054-1.066, soluble in alcohol; contains hydrogen cyanide and benzaldehyde.

chert (1) Petrosilex. A mainly quartz splintering rock from sedimentary strata. (2) A hard rock, of unknown composition, on the ocean floor at great depths.

Cherwell See *Lindemann.*

Cheshunt mixture A substitute for bordeaux mixture, in which the lime is replaced by ammonium carbonate.

chessylite Azurite.

chevkinite A complex mixture of hydrated oxides of K, Na, Fe, Ca, Mn, La, and Ce.

Chevreul, Michael Eugéne (1786–1889) French chemist noted for work on fats and textile dyeing.

chi The Greek letter χ.

chibou The resin of *Bursera gummifera,* Florida and tropical America; used in plasters.

chicle Gum chicle, balata, tuno gum. The dried juice of the bully tree, *Mimusops balata* (Sapotaceae), northern S. America. Soft, gray masses, tasteless, m.49; used in chewing gum.

chicory The root of *Cichorium intybus,* Europe and Asia, naturalized in the U.S.; a coffee substitute of adulterant.

chile Capsicum. **c. niter** Sodium nitrate*. **c. saltpeter** Caliche. Sodium nitrate in large deposits in Chile.

chilenite Ag_6Bi. A silver bismuthide.

chilli, chilly Capsicum.

chilte A rubber tree, *Jatropha tepiquensis,* Mexico. It yields a latex, and a chicle used as a chewing gum base.

Chimaphila Pipsissewa. The dried leaves of *C. umbellata* (Ericaceae); an astringent.

china (1) Cinchona bark. (2) Porcelain. **c. clay** The product obtained by leaching weathered deposits of granitic rocks to remove quartz and mica. Cf. *kaolin.* **c. grass** Ramie. **c. jute** The fiber from *Abutilon avicennae* (Malvaceae), China. **c. root** (1) Similax. (2) Galangal. **c. stone** A mixture of feldspathic minerals, micas, and quartz; a flux for ceramic glazes. **c. tallow tree** *Sapium sebiferum* (Euphorbiceae), whose seeds are rich in oil. **c. wood oil** Tung oil.

chinacrin Quinacrine.

chinaldine Quinaldine.

chinaphthol Quinanaphthol.

chinaroot The dried rhizome of *Smilax china,* S. China, resembling sarsaparilla.

chinazoline Quinazoline*.

chinchona Cinchona bark.

chinese **c. blue** Prussian blue. **c. cinnamon** Cassia bark. **c. green** Lokao. Green dye from the bark of Eurasian buckthorns, *Rhamnus utilis* and *R. globosa.* **c. oil** Tung oil. **c. wax** Hexacosyl heptacosanoate formed on *Fraxinus chinensis* from the secretions of *Coccus ceriferus.*

chinic acid Quinic acid.

chinidine Cinchonidine.

chinine Quinine.

chiniofon $C_9H_6ONIS + Na_2CO_3$. Yatren, a mixture of 7-iodo-8-oxyquinoline-5-sulfonic acid with its sodium salt and sodium hydrogencarbonate. Yellow crystals, effervescing in water to form sodium iodoxyquinoline sulfonate; formerly an amebicide. **c. sodium** $C_9H_5O_4NIS \cdot Na = 373.1$.

chinol Hydroquinone*.

chinoline Quinoline.

chinovin Quinovin.

chinovose A carbohydrate from cinchona bark.

chinwood Taxus.

chiococcine An alkaloid from the roots of *Chiococcoca racemosa,* cahinca, resembling emetine.

chiolite $2NaF \cdot AlF_3$. Arkustite. A mineral resembling cryolite.

chionanthin $C_{22}H_{28}O_{10} = 452.5$. A resin from *Chionanthus virginica,* the poison ash or fringe tree (N. America); a tonic.

chip See *silicon chip.*

chipboard (1) A low-grade board made from waste paper. (2) Particleboard.

chipmunk crusher A grinding machine for ores and rocks.

chiquito A fat from *Combretum butyrosum* (Combretaceae), tropical Africa.

chiral* (Greek: "hand") Dissymmetrical. Describing a molecule that, in a given configuration or conformation, is not identical with its mirror image. All c. compounds are optically active and all optically active compounds are c. Cf. *achiral, sequence rule.*

chirality* Dissymmetry, handedness. The property of being chiral.

chirata Chiretta. The dried herb of *Swertia (Ophelia) chirata* (Gentianaceae), India; a bitter tonic.

chiratin $C_{26}H_{48}O_{15} = 600.7$. A glucoside from chirata. Yellow, hygroscopic powder, hydrolyzed to ophelic acid and chiratogenin.

chitenidine $C_{19}H_{22}ON_2 = 294.4$. An oxidation product of quinine.

chitenine $C_{19}H_{22}O_4N_2 = 342.4$. An oxidation product of quinine.

chitin $(C_8H_{13}O_5N)_n = (203)n$. The horny framework of invertebrates (crabs, lobsters, beetles); the animal analog of plant cellulose, but with a $-NHAc$ group instead of an $-OH$ group at the 2-position.

chitosamin Glucosamine.

chitosan A split product of chitin, related to mucin and chondrosin.

chittim bark Cascara sagrada.

chlamydia Obligate intracellular microorganisms possessing features of both viruses and bacteria. They contain DNA and RNA and have a discrete cell wall.

chlamydozoa A minute animal organism in a rigid sheath or capsula.

chlor- See *chloro-.*

chloracne Severe rash which may follow exposure to dioxin and similar chemicals.

chloral $CCl_3 \cdot CHO = 147.4$. Trichloroacetaldehyde†, trichloroethanal*. Colorless, oily liquid, $d_{20°}$1.512, b.98, soluble in water; some derivatives are sedatives and hypnotics. **anhydrogluco ~** Chloralose. **butyl ~** See *butyl chloral* under *butyl.* **poly ~** Hydronal.

c. alcoholate $CCl_3 \cdot CHOH \cdot OEt = 193.5$. Trichloroethylate. Colorless crytstals, m.46, soluble in water; a hypnotic. **c.**

amide C. ammonia. White crystals, m.71, soluble in alcohol; a hypnotic and analgesic. **c. ammonia** C. amide. **c.benzamide** $CCl_3CHOH(C_6H_5CONH)$ = 268.5 Colorless crystals; a hypnotic. **c. betaine** $C_7H_{12}O_3NCl_3 \cdot H_2O$ = 282.5. White crystals, m.128, soluble in water; a hypnotic **c. caffeine** A molecular mixture of caffeine and chloral, soluble in water. **c. cyanohydrin** $CCl_3 \cdot CHOH \cdot CN$ = 174.4. Trichloroacetonitrile. Colorless crystals m.60, soluble in water; a substitute for bitter almond water. **c. hydrate** $CCl_3CH(OH)_2$ = 165.4. Hydrated c., crystalline c. Colorless, deliquescent needles, m.47, b.97 (decomp.), soluble in water, alcohol, or ether. A hypnotic, a reagent for ergosterol (USP,EP,BP), and a clearing agent for microscopy. **c.uric acid** See *chloraluric acid.*

chloralbine $C_6H_6Cl_2$ = 149.0. Colorless crystals from trichlorophenol.

chloralide $C_5H_2O_3Cl_6$ = 322.8. Crystals, m.116.

chloralimide $CCl_3CH{:}NH$ = 146.4. Colorless crystals, m.155, soluble in alcohol; a hypnotic.

chloralkali Collective term for chlorine and caustic soda; as in the name of the electrolytic plant, where they are produced in the fixed ratio of 1:1.1.

chlorallylene Allyl chloride*.

chloraloin $C_{34}H_{30}O_{14}Cl_6$ = 875.3. Yellow precipitate from the action of chlorine on aloes.

chloralose $C_8H_{11}O_6Cl_3$ = 309.4. Anhydroglucochloral. *α*-~ Colorless crystals, m.185, soluble in water; a hypnotic, used on laboratory animals. *β*-~ An isomeric by-product obtained by the action of chloral on glucose. Colorless crystals, m.227.

chloraluric acid $C_{14}H_{22}O_{11}N_{12}Cl_2$ = 605.3. Colorless crystals, obtained by the action of chlorous acid on uric acid.

chlorambucil $C_{14}H_{19}NO_2Cl_2$ = 304.2. 4-[Bis(2-chloroethyl)amino]phenylbutanoic acid, Leukeran. White crystals, m.66, insoluble in water; used to treat malignant diseases, as, leukemia (USP, BP).

chloramide Chloral amide.

chloramidobenzene Chloroaniline.

chloramine* NH_2Cl = 51.5. Monochloramine. Colorless, unstable, pungent liquid. An intermediate in the preparation of hydrazine from chlorine and ammonia. **chloramine-T** Chlorazene.

chlorammine process Chlorinating water by injecting chlorine and ammonia.

chloramphenicol $C_{11}H_{12}O_5N_2Cl_2$ = 323.1. Amphicol, Chloromycetin, Leukomycin. White, bitter crystals, m.151, sapringly soluble in water. The first broad-spectrum antibiotic, q.v. *(antibiotics),* discovered; produced by *Streptomyces venezuelae;* now synthetically prepared. A bacteriostatic for severe infections e.g., typhoid and typhus, but systemic use limited by toxicity on bone marrow. Used topically in eye and ear (USP, EP, BP).

chloranil $C_6Cl_4O_2$ = 245.9. Tetrachlorobenzoquinone. Yellow scales, m.290 (sublime), slightly soluble in alcohol; an oxidizing agent in the dye industry, and seed protectant.

chloranilate A salt of chloranilic acid containing the radical $C_6Cl_2O_4-$.

chloranilic acid $C_6Cl_2O_2(OH)_2$ = 209.0. 2,5-Dichloro-3,6-dihydroxy-1,4-benzoquinone. Red leaflets, m.283, insoluble in water.

chloranion The chlorate ion.

chloranol $C_6H_2O_2Cl_4$ = 247.9. 2,3,5,6-Tetrachlorohydroquinone*. Pale yellow crystals; used as a reagent and in the dye industry.

Chlorargyrite Argentum cornu.

chlorastrolite A variety of jade.

chlorate* A salt of chloric acid which contains the ClO_3^- ion.

chlorated Referring to a substance containing readily available chlorine, as in bleaching powder. Cf. *chlorinated.*

chloraurate (1) A double salt of auric chloride with another chloride. (2) A salt containing the radical $AuCl_4-$.

chlorauric acid $HAuCl_4$ = 339.8. Yellow crystals, formed on evaporation of a solution of gold in aqua regia.

chlorauride Auric chloride.

chloraurite (1) A double chloride of the type $AuCl \cdot MCl$. (2) A compound containing the radical $AuCl_2-$.

Chlorazene $MeC_6H_4SO_2Na{:}NCl \cdot 3H_2O$ = 281.7. Trademark for sodium *p*-toluenesulfochloramine, mianine, tochlorine, chloroamine, tolamine, chloramine-T. A by-product of the manufacture of saccharin. Colorless crystals, soluble in water. An antiseptic; also used for the volumetric analysis of nitrites.

Chlorazol $C_4H_3O_4NCl_3$ = 235.4. Trademark for a proprietary line of direct dyestuffs. An acrid, oily liquid obtained from proteins by distilling with nitric and hydrochloric acid.

chlorbenz- Chlorobenz-.

chlorbutanol $CCl_3 \cdot CMe_2OH$ = 177.5. Acetone chloroform, chloretone, chlorobutanol. Colorless deliquescent crystals, m.80, sparingly soluble in water; an antiseptic and hypnotic; a preservative in eye drops and injections (NF, EP, BP).

chlorcyclizine hydrochloride $C_{18}H_{21}N_2Cl \cdot HCl$ = 337.3. White crystals, m.225, soluble in water; an antihistiminic (USP, BP).

chlordane* See *insecticides,* Table 45 on p. 305.

chlordiazepoxide hydrochloride $C_{16}H_{14}ON_3Cl \cdot HCl$ = 336.2. Librium. White, bitter crystals, soluble in water; a benzodiazepine group tranquilizer, widely used to treat anxiety (USP, EP, BP).

chlorellin An antibiotic produced by unicellular species of *Chlorella.*

chlorethyl Ethyl chloride*.

chloretone Chlorbutanol.

Chlorex Trademark for a solvent consisting primarily of dichloroethyl ether. **C. process** A method of refining lubricating oils with C.

chlorfenvinphos* See *insecticides,* Table 45 on p. 305.

chlorhexidine **c. hydrochloride** $C_{22}H_{30}N_{10}Cl_2 \cdot 2HCl$ = 578.4. 1,6-Bis(4-chlorophenyldiguanido)hexane dihydrochloride. White, bitter crystals, m. 225 (decomp.), slightly soluble in water; an antiseptic (BP). **c. gluconate** $C_{22}H_{30}N_{10}Cl_2 \cdot 2C_6H_{12}O_7$ = 897.8. A more soluble form than the hydrochloride; used as an antiseptic solution both aqueous and alcoholic (BP).

chlorhydrate Hydrochloride*.

chlorhydrin A compound containing the radicals —Cl and —OH.

chloric acid* $HClO_3$. An acid existing only in solution and as salts (chlorates). **per~*** $HClO_4$. An acid existing only in dilute solution and as perchlorates.

chloride* A salt containing the Cl^- ion, usually a binary compound in which chlorine is the negative constituent. **c. of lime** Bleaching powder. **c. of soda** Sodium c. **c. ion** Chlorion. A negatively charged atom Cl^-, formed from a soluble chloride in water.

chloridion Chloride ion.

chloridization (1) Chlorination. (2) The treatment of ores with chlorine or hydrochloric acid, to produce the chloride of the principal metal present.

chlorimetry The determination of available or free chlorine in compounds.

chlorin Crude dinitroresorcinol.

chlorinate To introduce chlorine into a compound.

chlorinated Treated with chlorine. Cf. *chlorated.* **c. lime** Bleaching powder containing not less than 30% available chlorine (BP). **c. solvents** Nonflammable, stable, noncorrosive liquids formed by the action of chlorine on acetylene; e.g., dieline, $C_2H_2Cl_2$, b.52; perchloroethane, C_2Cl_6, b.185.

chlorination (1) The introduction of chlorine into a compound, especially by the substitution of H atoms. (2) The sterilization of water by chlorine. **break-point ~** The optimum chlorine content in water, which produces no residual odor. When successive doses of chlorine are added to water, the residual chlorine rises to a maximum, then falls to a minimum (the break point), and then rises linearly. **exhaustive ~** The successive substitution of all H atoms by Cl. **super ~** See *superchlorination.*

chlorine Cl = 35.453. A halogen element, at. no. 17. Greenish-yellow, poisonous, gas with suffocating odor, m. −102, b. −33.6, d. (liquid Cl) 1.4, d. (gaseous Cl_2) 71.63 (H_2 = 2), or 2.49 (air = 1), soluble in water. After fluorine, it is the most electronegative element. It is the most abundant halogen, and occurs as chlorides (seawater, salt deposits) in many minerals, and in all vegetable and animal tissues. It is produced (with caustic soda) from the electrolysis of chlorides. Discovered by Scheele (1774). It consists of two isotopes: ^{35}Cl and ^{37}Cl and is mono-, tri-, or pentavalent, giving:

−1 chloride* . . . . Cl^- (NaCl)
+1 hypochlorite* . . . . ClO^- (NaOCl)
+3 chlorite* . . . . $ClO^-{}_2$ ($NaClO_2$)
+5 chlorate* . . . . ClO_3^- ($NaClO_3$)
+7 perchlorate* . . . . ClO_4^- ($NaClO_4$)

A bleach for textiles, paper pulp, and sponges; a poison gas in chemical warfare; a gold extractant; a disinfectant, germicide, and insecticide; an oxidizing and reducing agent. **available ~** (1) The c. that can be liberated from a substance, e.g., bleaching powder, by acids. (2) The c. equivalent of the active oxygen of an oxidizing agent, e.g., a hypochlorite. **di ~ *** The chlorine molecule, Cl_2. **eu ~** A mixture of c. and c. dioxide produced by the action of hydrochloric acid on potassium chlorate.

c. cyanide Cyanogen chloride. **c. dioxide*** ClO_2 = 67.5. C. peroxide. A yellow irritating gas, b.9.9; or red liquid, d.1.5; soluble in water, decomp. in alcohol; a strong oxidizing agent forming hypochlorites or peroxides. It explodes in contact with ammonia, methane, phosphine, or hydrogen sulfide. **c. heptaoxide*** Cl_2O_7 = 182.9. The anhydride of perchloric acid. Colorless oil, b.82, readily explodes. **c. hydrate** $Cl \cdot 7.3H_2O$ = 167.0. Yellow octahedra, a clathrate, stable below 9. **c. hydride** Hydrochloric acid. **c. monoxide*** Cl_2O = 86.9. Anhydride of hypochlorous acid. Yellow liquid or gas, b.5, soluble in water, decomp. by alkalies. **c. oxides** (1) Cl_2O = c. monozide. (2) ClO_2 = c. dioxide*. (3) Cl_2O_7 = c. heptaoxide*, anhydride of perchloric acid. (4) Cl_2O_4 = c. tetraoxide*. The existence of Cl_2O_3 and Cl_2O_5 is doubtful. **c. water** A pale green liquid of strong odor, obtained by passing c. through water. It contains 0.4 g c. per 100 ml, and on keeping, it decomposes to hydrochloric and hypochlorous acid. A reagent to determine iodine, bromine, quinine, uric acid, or xanthine; an oxidizing and bleaching agent and antiseptic.

chlorinity The weight of halides (as grams of Cl) in 1 kg seawater. Salinity = 0.03 + 1.805 × chlorinity.

chloriodic acid Iodine monochloride*.

chloriodoform $CHCl_2I$ = 210.8. Formyl dichloroiodide. Yellow, aromatic liquid.

chlorion Chloride ion.

chlorisatide $C_{16}H_{10}O_4N_2Cl_2$ = 365.2. A chlorine-substituted isatin. White, insoluble powder.

chlorite (1)* A salt containing the radical ClO_2^-, from chlorous acid. (2) A group of rock-forming minerals; e.g., clinochlore.

chlorition ClO_2^- formed by ionization of chlorous acid and chlorites.

chloritoid $Al_2O_3 \cdot FeO \cdot SiO_2 \cdot H_2O$. Phyllite mineral.

chlorknallgas An explosive mixture of chlorine and hydrogen.

chloro-* Prefix indicating a chlorine atom. *Chlor-* is sometimes used (incorrectly).

chloroacetamide See *acetamide.*

chloroacetate A salt of chloroacetic acid; contains the radical $CH_2ClCOO-$.

chloroacetic acid $CH_2ClCOOH$ = 94.5. Carboxymethyl chloride. Colorless rhombs, m.63, soluble in water. **tri ~** See *trichloroacetic acid.*

chloroacetone $CH_2 \cdot Cl \cdot CO \cdot CH_3$ = 92.5. Acetyl methyl chloride. Colorless liquid, $d_{16} \cdot 1.162$, b.119, insoluble in water; liable to spontaneous combustion if not stored carefully.

chloroacetophenone $C_6H_5 \cdot COCH_2Cl$ = 154.6. C.A.P. Phenacyl chloride. Rhombic crystals, m.59; a tear gas.

chloroacetyl* Chloracetyl. The radical $CH_2Cl \cdot CO-$, from chloroacetic acid. **c. chloride** $CH_2Cl \cdot COCl$ = 112.9. Colorless liquid, b.105, decomp. by water.

chloroacid An organic acid containing chlorine.

chloroacrylate A salt of chloroacrylic acid containing the radical $C_2H_2Cl \cdot COO-$.

chloroantimoniate A double salt of a chloride with antimony trichloride.

chloroargentate A double salt of a metal chloride with silver chloride.

chloroarsine Cacodyl chloride.

chlorobenzaldehyde $Cl \cdot C_6H_4 \cdot CHO$ = 140.6. Chlorobenzene carbonal*. ***o-*** or **2-** m.11. ***m-*** or **3-** m.17. ***p-*** or **4-** m.47.

chlorobenzoate* A salt of chlorobenzoic acid containing the radical $C_6H_4Cl \cdot COO-$.

chlorobenzoic acid See *benzoic acid.*

chlorobenzoyl (1) C_7H_5OCl = 140.6. Colorless liquid. (2) The radical $Cl \cdot C_6H_4 \cdot CO-$. **c. chloride** $ClCO \cdot C_6H_4Cl$. ***ortho-*** **~** b.238. ***meta-*** **~** b.117. ***para-*** **~** b.119.

chloroboric acid Boron trichloride*.

chlorobornene $C_{10}H_{12}Cl_4$ = 276.0. Colorless liquid from the action of chlorine on turpentine.

chlorobromoacetate A salt of chlorobromoacetic acid of the type $CHClBr \cdot COOM$.

chlorobutadiene $CH_2:CH \cdot CCl:CH_2$ = 88.5. 2-Chloro-1,3-butadiene*, *chloroprene,* q.v. Colorless liquid, d.0.9583, b.59, synthesized from acetylene. It polymerizes to form: **alpha- ~** A plastic resembling unvulcanized rubber. **beta- ~** 2 volatile, fragrant products, b_{27mm}92–97 and 114–118.

chlorobutanol See *chlorbutanol.*

chlorobutyryl* The radical $C_3H_6Cl \cdot CO-$.

chlorocalcite Hydrophilite.

chlorocamphene Chlorobornene.

chlorocarbinol Chloroethanol*.

chlorocarbonate Chloroformate. A compound containing the radical $ClCO \cdot O-$.

chlorocarbonic acid Chloroformic acid.

chlorochromic acid The compound $HCrO_3Cl$. See *chromyl chloride.*

chlorocinnamoyl The radical $ClC_6H_4 \cdot CH:CH \cdot CO-$.

chlorocinnose $C_9H_4OCl_4$ = 269.9. Colorless crystals, obtained by distillation of a mixture of cinnamic acid with phosphorus pentaoxide.

chlorocitraconyl $C_5H_4O_2Cl_2$ = 167.0. Citraconyl dichloride.

chlorocosane A liquid chlorinated paraffin containing chlorine in stable combination.

chlorocresol green $C_{21}H_{14}Cl_4O_5S$ = 520.2. 2,3,6,7-Tetrachloro-*m*-cresolsulfonphthalein. Gray powder; a pH indicator between 4.0 (yellow) and 6.0 (blue).
chlorocruorin A respiratory, dichroic, red-green pigment related to hemoglobin; found in the blood of certain molluscs and especially marine worms.
chlorocyanide Chloride cyanide*. A double salt of a chloride and cyanide: MCl·MCN.
chlorocyanogen Cyanogen chloride.
chlorodifluoromethane $CHClF_2$ = 86.5. Colorless gas, b. −39.8; an anesthetic and refrigerant. Cf. *Freon.*
chlorodracylic acid *p*-Chloro*benzoic acid.*
chloroethanol* $ClCH_2 \cdot CH_2OH$ = 80.5. Chlorcarbinol. Colorless liquid, d.1.213, b.129.
chloroethylene See *vinyl chloride.*
chloroethyl ether $(ClCH_2 \cdot CH_2)_2O$ = 143.0. Chlorex. Colorless liquid, d.1.213, b.178; a solvent.
chlorofibers Synthetic polymers of vinyl chloride or vinylidene chloride.
chlorofluoride A double salt of a chloride and fluoride. MCl·MF.
chloroform* $CHCl_3$ = 119.4. Trichloromethane*. Colorless liquid, d_{15}·1.499, m.−63.2, b. 61.2, almost insoluble in water; used as a general solvent, cleanser, anesthetic, antispasmodic, and preservative. One of the first inhalant anesthetics; seldom used now because of toxicity (NF, BP). **acetone ~** Chlorbutanol. **crystal ~** Chloroformate. **deuterio ~ *** $CDCl_3$ = 120.4. Deuterochloroform, C^2HCl_3. White solid, m.64.2–64.7, formed from CCl_3CHO and D_2O. **germanium ~** See *germanium chloroform.* **methyl ~** $MeCCl_3$ = 133.4. Colorless liquid; an anesthetic. **nitro ~** Chloropicrin. **phenyl ~** Benzotrichloride.
c. of crystallization Chloroformate.
chloroformate (1) A crystal with chloroform in its crystal structure. Cf. *water of crystallization.* (2) Chlorocarbonate. An ester of chloroformic acid; as palite.
chloroformic acid Cl·COOH = 80.5. Chlorocarbonic acid. A hypothetical acid; esters are known.
chlorofumaroyl Fumaroyl chloride.
chlorogenic acid $C_{16}H_{18}O_9 \cdot H_2O$ = 372.3. An acid in coffee and other plants. A depside of caffeic and quinic acid, m.208.
chlorogriseofulvin $C_{17}H_{18}O_6$ = 318.3. A neutral, crystalline, antibiotic product of the metabolism of *Penicillium griseofulvin,* Diercke, m.180.
chlorohydrin Chlorhydrin. A compound containing the radicals −Cl and −OH.
chloroiodic acid Iodine monochloride*.
chloroiodide A double salt of a chloride and iodide.
chloromalonic acid See *malonic acid.*
chloromenthene $C_{10}H_{17}Cl$ = 172.7. A substitution product of menthol; a yellow oil.
chloromercuriphenol $C_6H_4(OH)ClHg$ = 329.1. Colorless crystals, m.152, soluble in alkalies.
chloromethane Methyl chloride*.
chloromethyl The radical $ClCH_2-$, from methyl chloride. **di ~** The radical $-CHCl_2$. **tri ~** The radical $-CCl_3$.
c.chloroformate Palite. **c.silicane** $MeSiH_2Cl$ = 84.6. Methylchlorosilane. Colorless gas, b.7.
Chloromycetin Trademark for chloramphenicol.
chloronaphthalene See *chloronaphthalene* under *naphthalene.*
chloronitric acid (1) Nitroxylchloride. (2) Aqua regia.
chloronitrobenzene See *chloronitrobenzene* under *nitrobenzene.*
chloronitrophenol See *chloronitrophenol* under *nitrophenol.*
chloronitrous acid Nitrosyl chloride*.
chloronium ion* The cation H_2Cl^+.
chlorophane CaF_2. A mineral from Virginia.
chlorophenesic acid Dichloro*phenol.*
chlorphenesin $C_9H_{11}O_3Cl$ = 202.6. 3-(4-Chlorophenoxy-propane)-1,2-diol. White, bitter crystals, m.81, soluble in water; an antiseptic.
chlorophenic acid Monochloro*phenol.*
chlorophenisic acid Trichloro*phenol.*
chlorophenol See *phenol.* **c. indophenol** An oxidation-reduction indicator. **c. red** $C_{19}H_{12}Cl_2O_9S$ = 487.3. Dichlorophenolsulfonphthalein. Yellow powder; a pH indicator between 5.5 (yellow) and 6.5 (red).
chlorophenothane $C_{14}H_9Cl_5$ = 354.5. *p,p′*-DDT, dicophane, 1,1,1-trichloro-2,2-bis(*p*-chlorophenyl)ethane. White powder, congeals at 89, soluble in water; the insect toxicant of DDT.
chlorophenusic acid *Pentachloro*phenol.
chlorophoenicite $(Zn \cdot Mn)_3As_2O_8 \cdot 7(Zn \cdot Mn)(OH)_2$. It is green by reflected, and purple by transmitted, light. **magnesium-** C. with Mg in place of Zn.
chlorophorin $C_{24}H_{28}O_4$ = 380.5. A stilbene phenolic derivative, m.158, from iroko, *Chlorophora excelsa,* a decay-resistant African tree.
chlorophyll The chromoprotein green coloring matter of plants. Soft, green mass, insoluble in water. Solutions in organic solvents fluoresce blue or green and are colloidal. C. contains

70% chlorophyll *a*, $C_{55}H_{72}MgN_4O_6$ = 909.5
30% chlorophyll *b*, $C_{55}H_{70}MgN_4O_6$ = 907.5

both of which have been synthesized (1960) from monocyclic pyrroles, by a 30-step process. Split products are phyllopyrrole (cf. *hemopyrrole*) and etiophyllin, $C_{31}H_{34}N_4Mg$. A nonpoisonous coloring matter and deodorant, for oils, cosmetics, etc.; marketed as copper or zinc compounds. **alpha- ~** The purified chlorophyll $(C_{31}H_{29}N_3Mg)NH \cdot CO(COOMe)COOC_{22}H_{33}$. Cf. *phylloerythrin, phylloporphyrin.*
c. c Porphyrin photosynthetic pigments in marine organisms; as, c_1, $C_{35}H_{28}MgN_4O_5$.
chlorophyllide An ester of the porphyrin acid chlorophyllin; e.g., chlorophyll is the c. of phytol.
chloropicrin CCl_3NO_2 = 164.4. Nitrochloroform, aquinite, trichloronitromethane*, P.S. Colorless liquid, d_9·1.692, b.112, insoluble in water; an emetic gas.
chloroplast A green protoplasmic granule containing the chlorophyll of vegetable cells; diameter 4–5μm.
chloroplatinate **hexa ~ *** See *hexachloroplatinate* under *hexachloro-.* **tetra ~ *** See *tetrachloroplatinate(II)* under *tetrachloro-.*
chloroplatinic acid Hexachloro*platinic acid**.
chloroplatinite Tetrachloroplatinate(II)*.
chloroplatinous acid Tetrachloro*platinous acid**.
chloroprene 2-Chloro-1,3-butadiene*. **micro ~** A transparent product resembling vulcanized rubber. **omega- ~** A granular balatalike amorphous product, becoming plastic at 60°C. Used to manufacture duprene.
chloroquine phosphate $C_{18}H_{26}N_3Cl \cdot 2H_3PO_4$ = 515.9. Bitter, white crystals, darkening in light, soluble in water. Two forms: m.194 or 216. An antimalarial (USP, BP).
chlororaphin An antibiotic form *Chromobacterium.*
chlorosis (1) Anemia, due to a deficiency of hemoglobin or, with plants, of magnesium. (2) The use of chlorine and chlorides in agriculture.
chlorosyl* The radical −ClO.
chlorothalonil* See *fungicides,* Table 37 on p. 250.
chlorothiazide $C_7H_6O_4N_3ClS_2$ = 295.7. White crystals, insoluble in water. A diuretic in the thiazide group. Diuresis is caused by an increase in excretion of Na^+, K^+ and Cl^- ions. Used for edema and hypertension (USP, BP).

chlorotoluene Benzyl chloride*.
chlorotrianisene $C_{23}H_{21}O_3Cl$ = 380.9. Chlorotris(*p*-methoxyphenyl)ethylene, Tace. White crystals, m.133, soluble in water. Synthetic estrogenic substance, used for menopausal symptoms (USP, BP).
chlorotrinitrobenzene Picryl chloride.
chlorous Describing a compound containing the radical Cl≡. **c. acid*** $HClO_2$ = 68.5. Known only in solution and as its salts (chlorites).
chlorovaleric acid Monochlorovaleric acid.
chlorovalerisic acid Trichlorovaleric acid.
chlorovalerosic acid Tetrachlorovaleric acid.
chloroxalovinic acid $C_4HO_4Cl_5$ = 290.3. Pentachloroethyloxalic acid. Colorless, hygroscopic crystals, soluble in water.
chloroxylene Methylbenzyl chloride*.
chloroxylenol Chlorodimethylphenol. Several isomers.
chloroxylonine $C_{14}H_{13}O_4N$ = 259.3. m.176. An alkaloid from satinwood of *Chloroxylon swietenia* (Rutaceae), southeast Asia.
chlorphenesic, chlorphenic, chlorphenisic, chlorphenusic See *chlorophen-* spellings.
chlorpheniramine maleate $C_{16}H_{19}N_2Cl \cdot C_4H_4O_4$ = 390.9. Piriton. White crystals, m.132, soluble in water; an antihistaminic. Used to treat allergic states, as, hay fever (USP, BP).
chlorphenol See *chlorophenol* under *phenol.*
chlorproguanil hydrochloride $C_{11}H_{15}N_5Cl_2 \cdot HCl$ = 324.6. White, bitter crystals, soluble in water; an antimalarial (BP).
chlorpromazine hydrochloride $C_{17}H_{19}N_2ClS \cdot HCl$ = 355.3. Largactil, 2-chloro-10-[3′-(dimethylamino)propyl]-phenothiazine hydrochloride. White, pungent crystals, discoloring in light, m. approx. 57, soluble in water; a hypnotic which lowers body temperature and a sedative used in psychiatric disorders (BP).
chlorpropamide $C_{10}H_{13}O_3N_2ClS$ = 276.7. 1-[*p*-chlorophenyl)sulfonyl]-3-propylurea. White crystals, m.128, insoluble in water; an oral hypoglycemic (BP).
chlorpropham* See *herbicides,* Table 42.
chlorpyriphos* See *insecticides,* Table 45 on p. 305.
chlortetracycline hydrochloride $C_{22}H_{23}O_8N_2Cl \cdot HCl$ = 515.3. Aureomycin hydrochloride. Bitter, yellow crystals, soluble in water; an antibiotic of the tetracyline group, used to treat respiratory infections (USP, EP, BP).
chlorthiamid* See *herbicides,* Table 42.
chlortoluron* See *herbicides,* Table 42.
chloryl (1) The radical ClO_2-, as in $Cl_2O_5 \cdot 3SO_3$. (2) Carbachol.
chlorylene Trichloroethylene.
chocolate (1) A flavored beverage made from cocoa, milk, and sugar. (2) Theobroma paste. Ground cocoa; a basis for certain drugs.
choke damp Black damp. A mixture of carbon dioxide and other gases in mines.
chola Bile.
cholagogue A substance that stimulates the flow of bile.
cholalic acid Cholic acid.
cholane $C_{24}H_{42}$ = 330.6. A hydrocarbon parent substance of steroids, hormones, bile acids, and toad poisons; related to the carotenoids and cerebrosides by ring fracture. **c. ring** See *androstane ring.*
cholanthrene* $C_{20}H_{14}$ = 254.3. A pentacyclic, strongly carcinogenic hydrocarbon.
cholate (1) A salt or ester of cholic acid, indicated by the radical $C_{23}H_{39}O_3 \cdot COO-$. (2) Taurocholate. A salt of taurocholic acid, indicated by the radical $C_{25}H_{44}O_5NS \cdot COO-$.

choleate A salt or ester of *choleic acid (2).*
cholecalciferol $C_{27}H_{44}O$ = 384.6. Vitamin D_3, activated 7-dehydrocholecholesterol. White crystals, m.89, insoluble in water. Formed from 7-dehydrocholesterol in the skin of mammals by exposure to u.v. light (sunlight).

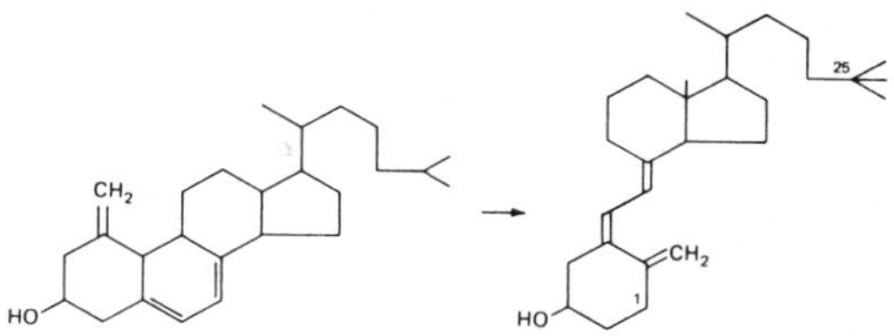

CH2 HO → CH2 HO 1 25

C. is inactive but is converted in the liver and kidneys to 25-hydroxy ~ and 1,25-dihydroxy ~ , respectively, which control calcium absorption and metabolism (USP, EP, BP). See *vitamin,* Table 101.
cholecyanin Bilicyanin.
Cholegrafin Trademark for iodipamide methylglucamine.
choleic acid (1) Taurocholic acid. (2) $C_{24}H_{40}O_4$ = 392.6. 3,12-Dihydroxy-24-cholanoic acid. From bile, m.190. **glyco ~** See *bile acids.*
cholera vaccine A suspension of killed cholera vibrios; for prophylaxis against cholera (USP, EP, BP).
cholesterilene $C_{26}H_{42}$ = 354.6. An unsaturated hydrocarbon derived from cholesterol. White crystals.
cholesterin Cholesterol.
cholesterol $C_{27}H_{45}OH$ = 386.7. Cholesterin. A sterol in blood, brain tissue, spleen, liver, bile; a principal constituent of atheromatous plaques, gallstones, and certain cysts; prepared from wool grease. Pearly scales, m.148, insoluble in water. Important in metabolic pathways of sex hormones. High levels of c. occur in some diseases, as, coronary heart disease. An emulsifier (USP). See *cholane.* **7-dehydro ~** Provitamin D_3. **iso ~** Lanosterol. **thio ~** $C_{27}H_{45}SH$ = 402.7. A solid, m.191.
c. esterase An enzyme that hydrolyzes c. esters.
cholestrophane NMe·CO·NMe·CO·CO (ring) = 142.1. 1,3-Dimethylparabanic acid. Pearly leaflets, m.156, soluble in water.
cholestyramine resin Questran. A basic anion exchange resin, consisting of a styrene-divinylbenzene copolymer with quaternary ammonium function groups. Used to remove bile salts from the intestine and treat high lipoprotein blood levels.
choletelin $C_{16}H_{18}N_2O_6$ = 334.3. A yellow oxidation product of biliverdin.
cholic acid $C_{24}H_{40}O_5$ = 408.6. Cholalic acid, 3,7,12-Trihydroxy-24-cholanoic acid; formed by the hydrolysis of bile acids. White crystals, m.198, insoluble in water. Cf. *cholane.* **deoxy ~** See *deoxycholic acid* under *deoxy-.* **glyco ~** See *glycocholic acid.* **litho ~** See *lithocholic acid.* **rhizo ~** An oxidation product of cholic acid. **tauro ~** See *taurocholic acid.*
choline* $Me_3N(OH)CH_2CH_2OH$ = 121.2. Bilineurine, sinkaline, trimethylethanolammonium hydroxide. A ptomaine in many animal and vegetable tissues. Viscous liquid, soluble in water. **acetyl ~** See *acetylcholine.* **hydroxy ~** Muscarine. **iso ~** Amanitine.
c. bases* A group of ammonium compounds, containing protonated nitrogen, and derived from ammonium hydroxide by replacing the hydrogen atoms by radicals; choline, neurine, muscarine, betaine, sinapine. **c. cation*** The ion $C_5H_{14}ON^+$.

c. (hydro)chloride $C_5H_{14}ClO_2N$ = 155.6. Colorless, hygroscopic crystals, soluble in water.
cholinesterase An enzyme in the blood and tissues, which breaks down excess *acetylcholine,* q.v.
cholohematin Bilipurpurin. A brown bile pigment.
chololic acid Cholic acid.
cholophaein $C_{16}H_{18}O_4N_2$ = 302.3. A brown biliary pigment in feces. Cf. *choletelin.*
chondrigen Cartilagin.
chondrin A mixture of mucin and gelatin from cartilage. A transparent, gelatinous mass, soluble in water.
chondrite A stone meteorite containing rounded constituents (chondrules): iron and nickel 8–16, marcasite 5%, with silicates. Cf. *achondrite.*
chondrodite Humite.
chondrodystrophy Abnormal skeleton formation in the embryo.
chondroine $C_{18}H_{21}O_4N$ = 315.4. An alkaloid from *Nectandra rodiaci* (Lauraceae), tropical America. Cf. *bebeerine.*
chondroitin $(C_{14}H_{21}O_{11}N)_n$ = (379)*n*. A mucopolysaccharide constituent of chondrin.
chondromucoid A cartilage albuminoid containing conjugated chondroitic acids.
chondroprotein A group of mucoids in the connective tissues.
chondrosamine $C_6H_{13}O_5N$ = 179.2. 2-Amino-2-deoxy-D-galactose. A widely distributed polysaccharide.
chondruel See *chondrite.*
chorion A 2-layer membrane, which covers the fertilized ovum and from which the placenta develops.
chorionic gonadotrophin A dry, sterile preparation of the gonad-stimulating substance from the urine of pregnant women. Used to induce ovulation, and treat hyponadism in males (USP, BP).
choritoid $Al_2O_3 \cdot FeO_2 \cdot SiO_2 \cdot H_2O$. A mica-type mineral.
choroid The vascular layer beneath the retina of the eye.
CHP Combined heat and *power.*
chrithmene Crithmene.
chroatol $C_{10}H_{16}2HI$ = 391.8. Terpene iodohydrate. A green oil, obtained by the action of iodine on turpentine.
chroma The degree of saturation of a color, this being the extent to which the pure color is mixed with white. High saturation means little white.
chroman* $C_6H_4 \cdot O \cdot CH_2 \cdot CH_2 \cdot CH_2$ = 134.2. Dihydrobenzopyran. Its derivatives are coronary vasodilators.
chromastrip The strip of paper used in *paper chromatography,* q.v.
chromate* Indicating chromium as the central atom(s) in an anion; as a salt of chromic acid containing the radical CrO_4^{2-}. **di ~ *** Bichromate. An acid salt derived from chromic acid containing the radical $Cr_2O_7^{2-}$, which makes aqueous solutions orange. **per ~** See *perchromate.*
c. **cell** See *dichromate cell* under *cell.* **c. green** A mixture of chrome yellow and prussian blue. Cf. *chrome green.* **c. ion** The CrO_4^{2-} ion, yellow in aqueous solutions.
chromatic Pertaining to colors. **iso ~, ortho ~** 3500–6000 Å. **tri ~** 3500–6500 Å. **pan ~** 3500–7000 Å.
c. **aberration** The refraction of the constituent rays of white light by a lens, to different extents; it produces an image fringed with color. **c. plate** A photographic plate used in conjunction with color screens to produce a contrasting colored image. Cf. *photosensitizer.*
chromatin The structural part of the cell nucleus that is most deeply stained; consists of DNA and a protein, nucleoprotein.
chromatography (chromatographic analysis) The analysis of mixtures of solutions or gases by selective adsorption on materials such as gelatin, alumina. The analytical process is based on differences in the distribution ratios of the components of mixtures between a mutually immiscible mobile and fixed phase. In particular, the formation of isolated bands which can be separated mechanically and further examined. The mobile phase (sample and carrier) can be a gas, liquid, or solid in solution, but the stationary phase (the column) can be only a liquid or solid. Hence the combinations: liquid:liquid (paper); liquid:solid (adsorption); gas:liquid (partition); gas:solid (adsorption). **affinity ~** A method of separating enzymes or proteins, in which the crude extract is passed through a column. The substrate bonds with the enzyme, which is subsequently eluted after loosening the bond chemically, as, by a pH change. **bonded phase ~** C. using a stationary phase that is chemically bonded to a support (e.g., an alkyl silyl bonded phase on a silica support), so that it is not subsequently stripped from the support. **gas ~** C. in which the mobile phase is a gas; e.g., a gas mixture is passed through the column. **gas-liquid ~** g.l.c. The substance to be analyzed is introduced into a column containing the liquid phase on an inert support, and a carrier gas is passed through the column. The emerging gas passes through a device which measures and records variations in its properties, as, a flame ionization detector. **gel ~** Isolation or detection of organic substances by the use of a gel substrate. **high performance (power, pressure) liquid ~** h.p.l.c. A refined form of c., in which the solvent is passed through the column under pressure. Detectors used to evaluate separated substances include the u.v./visible spectrophotometer; as, dyes. **normal phase ~** Column c. in which the stationary phase is more polar than the mobile phase. **pyrolysis gas ~** The use of controlled heat to decompose solids or liquids so that the gases evolved can be analyzed by gas c. **reversed phase ~** Column c. in which the stationary phase is less polar than the mobile phase. **supercritical fluid ~** C. under conditions (e.g., of temperature and pressure) close to the critical point of the mobile phase. **thin-layer ~** t.l.c, thin-film c. The use of thin films (as silica) deposited on glass for the separating column in c. **vapor-phase ~** Gas c.
chromatoplate A thin plastic or glass plate, thinly coated with a substance suitable for thin-layer chromatographic analysis.
chrome (1) Chromium. (2) Chromium ore. (3) Chromium oxide. (4) Lead chromate. **c. alum** Ammonium chromic sulfate. **c.-cobalt alloy** See *niranium.* **c. green** A mixture of chromic oxide and cobalt oxide. **c. ore** Chromite. **c. red** Lead chromate. **c. yellow** Lead chromate.
Chromel Trademark for alloys highly resistant to heat, oxidation, and acids. They contain Ni 80–90, Cr 10–20, with or without Fe 24–66%.
chromic Chromium(III)* or (3+)*. Describing a compound containing trivalent chromium. **c. acetate** $Cr(C_2H_3O_2)_3H_2O$ = 247.1. Green powder, soluble in water. **c. acid** (1)* H_2CrO_4 = 118.0. The hydrate of CrO_3; exists only in solution or as salts. (2) CrO_3. Chromium trioxide*. **c. anhydride** Chromium trioxide. **c. bromide** (1) $CrBr_3$ = 291.7. Green hexagons, soluble in water. (2) $CrBr_3 \cdot 6H_2O$ = 399.8. C. b. hexahydrate. Green scales, insoluble in water. **c. carbide** Cr_3C_2 = 180.0. Black powder, d.5.62, insoluble in water. **c. carbonate** $Cr_2(CO_3)_3$. Black powder containing some hydroxide; insoluble in water. **c. chloride** (1) $CrCl_3$ = 158.4. Purple crystals, soluble in water. (2) $CrCl_3 \cdot 6H_2O$ = 266.4. Green or purple scales, soluble in water. **c. fluoride** $CrF_3 \cdot 4H_2O$ = 181.0. Green crystals, m.1000, soluble in water; used in dyeing and printing cotton, and coloring white marble. **c. hydroxide** $Cr(OH)_3 \cdot 2H_2O$ = 139.0. A green pigment,

TABLE 16. OXIDATION STATES OF CHROMIUM

Oxidation state	Name	Ion	Color in solution
2	chromium(II)*, chromium(2+)*, chromous	Cr^{2+}	blue
3	chromium(III)*, chromium(3+)*, chromic	Cr^{3+}	green, violet
	chromite	CrO_2^-	green
6	chromyl*	CrO_2^{2+}	red
	chromate*	CrO_4^{2-}	yellow
	dichromate*	$Cr_2O_7^{2-}$	orange

	In acid solution		In alkaline solution	
Reduction	Cr^{3+} green or violet	⇌	CrO_2^- green	Oxidation
	↑		↓	
	$Cr_2O_7^{2-}$ orange	⇌	CrO_4^{2-} yellow	

insoluble in water. **c. iodide** $CrI_3 \cdot 9H_2O$ = 594.8. Black powder, soluble in water. **c. nitrate** $Cr(NO_3)_3 \cdot 9H_2O$ = 400.1. Purple prisms, m.36.5, soluble in water; a reagent. It crystallizes also with $7H_2O$ and with $6NH_3$. **c. oxide** Cr_2O_3 = 152.0. Chrome green. Green hexagons, m.2059, insoluble in water; a pigment for calico printing, ceramics, and glass. **c. oxychloride** Chromyl chloride*. **c. phosphate** $CrPO_4$ = 147.0. Plessy's green, Arnaudon's green. Green powder, insoluble in water; a pigment. It crystallizes with 3 and $6H_2O$. **c. potassium alum** Chromium potassium sulfate*. **c. potassium oxalate** $K_3Cr(C_2O_4)_3 \cdot 3H_2O$ = 487.4. Purple crystals, soluble in hot water. **c. silicide** Cr_3Si_2 = 212.2. Black crystals, insoluble in water. **c. sulfate** $Cr_2(SO_4)_3$ = 392.2. Green powder, soluble in water; used to manufacture chromium compounds. It crystallizes with $6H_2O$ (green) or $15H_2O$ (violet). **crystalline** $Cr_2(SO_4)_3 \cdot 18H_2O$ = 716.4. Violet crystals, anhydrous at 100, soluble in water; used to manufacture green inks and chromium compounds. **c. sulfide** Cr_2S_3 = 200.2. Green powder, insoluble in water. **c. sufite** $Cr_2(SO_3)_3$ = 344.2. Green crystals, d.2.2. **c. tartrate** $Cr_2(C_4H_4O_6)_3$ = 548.2. Purple scales, soluble in water.

chromicyanide Hexacyanochromate(III)*. The $Cr(CN)_6^{3-}$ ion; it imparts a yellow color to solutions.

chromising Chromizing.

chromite (1) $FeCr_2O_4$. An oxide of iron and chromium containing 68% of chromic oxides. Black streaked masses (Asia Minor), d.4.3–4.6, hardness 5.5. Used to make refractory bricks and cements, and chromium compounds or ferrochromium, and to color glass and tiles green. (2) $Cu(CrO_2)_2$ = 231.5. Chromium(III) copper (II) oxide*. **thio~** A compound of the type $M_2Cr_2S_4$.
c. ion The CrO_2^- ion, derived from chromous acid or the chromites.

chromitite The mineral Fe_2O_3, Cr_2O_3.

chromium* Cr = 51.9961. An element, at. no. 24, silver-white, hard, brittle, d.7.1,m. 1850, b.2677, insoluble in water, but dissolved rhythmically by acids (active and passive c.); discovered by Vauquelin (1797); occurs principally in chromite. C. is prepared from ores by reduction with metallic aluminum (thermit process), and is used in corrosion-resistant alloys and heavy-duty steels. C. compounds are poisonous and variously colored. They are di-, tri-, or hexavalent (see Table 16). **c. (II)***, **c.(2+)*** See also *chromous*. **c.(III)***, **c.(3+)*** See also *chromic*. **^{51}Cr** A radioactive isotope made by neutron irradiation of Cr. **c. acetate*** See *chromous*. **c. arsenide*** CrAs = 126.9. Black crystals, d.6.35, insoluble in water. **c. boride*** CrB = 62.8. Black powder, insoluble in water. **c. chloride** See *chromic, chromous, chromyl*. **c. dioxide** CrO_3 = 100.0. Black powder obtained on reducing potassium dichromate by sodium thiosulfate, and regarded as chromic chromate, $Cr_2O_3 \cdot CrO_3 \rightarrow 3CrO_2$. Used for magnetic data storage. **c. minerals** C. is common in magnesium and other rocks: e.g., chromite, $FeCr_2O_4$; crocoite, $PbCrO_4$; knoxvillite, $CrSO_4$. **c. mordants** C. used in tanning and dyeing; chrome alum. **c. oxides.** See *chromous oxide, chromic oxide (green), c. dioxide (black), c. trioxide (red)*. **c. phosphide*** CrP = 83.0. Black powder, insoluble in water. **c. plating** The electrolytic coating of metals with a layer of c. 250 μm thick, over a layer of nickel, which produces a noncorrodible surface. **c. potassium oxalate*** $K_3Cr(C_2O_4)_3 \cdot H_2O$ = 451.4. Violet crystals, soluble in water. **c.(III) potassium sulfate*** $CrK(SO_4)_2 \cdot 12H_2O$ = 499.4. Chrome alum. Red or green, cubic or octahedral crystals, m.89, soluble in water; a mordant. **c. tetrasulfide*** Cr_3S_4 = 284.2. Gray powder. **c. trioxide*** CrO_3 = 100.0. Chromic acid anhydride. Crimson needles, m.190, soluble in water, readily reduced to the green oxide. A powerful oxidizing agent. **c. tungsten steel** A heat- and chemical-resistant steel containing 5 Cr and 1% W.

chromizing Forming a protective surface layer of chromium; e.g., on steel by ion exchange through heating it in chromium vapor at 800–900°C.

chromo A paper coated with a mixture of a white pigment (e.g., china clay, blanc fixe, etc.) and an adhesive (e.g., casein), to obtain fine color prints.

chromogen (1) The parent substances of a dyestuff, or of a compound which produces a colored substance. (2) A substance in biological life which, on oxidation, forms colored compounds; as, sepia. Cf. *chromophore, auxochrome*, **c.-I** Chromotropic acid.

chromogene A light-resistant acid dye.

chromogenic Pertaining to a chromogen. **c. bacteria** Bacteria which produce colored substances.

chromoisomerism Chromotropy.

chromoisomers Differently colored modifications of a substance. Cf. *chromotropy*.

chromomere A structural unit of a chromosome containing a gene.

chromometer Colorimeter.

chromone $C_6H_4CO(CH)_2O$ = 146.1. Benzo-1,4-pyrone, 4H-1-Benzopyran-4-one†. White needles, m.59, insoluble in water; present in many vegetable pigments. Some derivatives are coronary vasodilators. **2-phenyl~** Flavone. **3-phenyl~** Is*oflavone*.

chromonucleic acid Deoxyribonucleic acid (see *nucleic acids*).

chromophilic Having an affinity for colors.

chromophore A group responsible for the color in organic substances; as, —N:N— Cf. *auxochrome*.

chromophoric Having properties of a chromophore. **c. group** The nitro, azo, quinoid, and —C:CH·CO— radicals, having the possibility of a shifting double bond. See *chromophore*.

chromophotometer Colorimeter.

chromoplast(id) A micelle of coloring matter in plant cells, usually other than chlorophyll.

chromoprotein A conjugated protein (or protein compound) and a chromophore; as, hemoglobin. Cf. *cytochrome*.

chromosan A mixture of sodium dichromate and ammonium phosphate; a dye mordant.
chromoscope A colorimeter.
chromosome Strands of chromatin in the cell nucleus of all animals and higher plants with a linear arrangement of genes along their length. The number of c. is constant for each animal species and they occur in pairs. See *meiosis, gene.*
chromosomin A principal protein of chromosomes.
chromosphere A gaseous upper "atmosphere" surrounding the sun, composed mainly of hydrogen, helium, and calcium vapors, and responsible for the Fraunhofer lines. Cf. *corona.*
chromotrope-B *p*-Nitrobenzeneazo-1,8-dihydroxynaphthalene-3,6-disulfonic acid. A colorimetric reagent for boron.
chromotropic acid $C_{10}H_4(OH)_2(SO_3H)_2$ = 288.2. Chromotrope acid, chromogen-I, 4,5-dihydroxy-2,7-naphthalenedisulfonic acid*; an intermediate and analytical reagent.
chromotropy Chromoisomerism. Change of color, especially of certain salts known in differently colored forms (chromisomers).
chromous Chromium(II)* or (2+)*. Describing a compound of divalent chromium. **c. acetate** $Cr(C_2H_3O_2)_2 \cdot 2H_2O$ = 206.1. A blue mass, soluble in water; used in calico printing and as a stain. **c. acid** $HCrO_2$ or $HO \cdot CrO$. A blue powder; a weak acid which yields chromites. **c. chloride** $CrCl_2$ = 122.9. White crystals, soluble in water to form a blue solution which absorbs oxygen. **c. hydroxide** $Cr(OH)_2$ = 86.0. Brown powder, decomp. by heat, insoluble in water. **c. iodide** CrI_2 = 305.8. White crystals, soluble in water. **c. oxalate** $Cr(OOC)_2$ = 140.0. Green scales, soluble in hot water. **c. oxide** CrO = 68.0. Chromium monoxide*. Stable only in the hydrated state. **c. sulfate** $CrSO_4 \cdot 7H_2O$ = 274.2. Blue crystals, soluble in water.
Chromspun Trademark for a cellulose acetate synthetic fiber.
chrom-X A high-carbon ferrochromium steel.
chromyl* The radical $=CrO_2$ containing hexavalent chromium. **c. amide** $CrO_2(NH_2)_2$ = 116.0. **c. chloride*** CrO_2Cl_2 = 154.9. Chromic oxychloride, chlorochromic acid. Red fuming liquid, d.1.96, b.116, decomp. in water; a powerful oxidizing agent.
chronic Long-continued. Cf. *acute.*
chronograph An instrument to record small time intervals.
chronopotentiometry Electroanalysis based on the time after which a rapid change in potential of a working electrode occurs; this is a measure of the concentration of the electroactive substance present.
chronoscope A device to measure short time intervals.
chronoteine A high-speed moving picture camera for the study of rapid-moving machinery or phenomena. Cf. *stroboscope.*
chrysalicic acid $C_7H_5O_6N_3$ = 227.1. Dinitro-*o*-aminobenzoic acid. Isomer of chrysanisic acid.
chrysamine $Na_2C_{18}H_{16}O_6N_4$ = 430.3. Flavophenine. A yellow azo dye, obtained from benzidine and toluidine.
chrysammic acid $C_{14}H_2(NO_2)_4(OH)_2O_2$ = 420.2. 1,8-Dihydroxy-2,4,5,7-tetranitroanthraquinone*, chrysamminic acid. A constituent of aloes prepared by the action of nitric acid on chrysophanic acid. A solid, insoluble in water.
chrysamminic acid Chrysammic acid.
chrysanilic acid A decompostion product of indigo blue.
chrysaniline $C_{19}H_{15}N_3$ = 291.3. *ms-p*-Aminophenyl-2-aminoacridine. Yellow crystals, m.268.
chrysanisic acid $NH_2 \cdot C_6H_2(NO_2)_2COOH$ = 227.1. 4-Amino-3,5-dinitrobenzoic acid*. Colorless crystals, m.259. Isomer of chrysalicic acid.
chrysanthemin An anthocyanin for elderberries.
chrysanthemum **chrysanthem(um monocarboxyl)ic acid** $C_{10}H_{16}O_2$ = 168.2. Colorless crystals, m.18.5. **c. dicarboxylic acid** $C_{10}H_{14}O_4$ = 198.2. Colorless crystals, m.207. See *pyrethrum.*
chrysarobin $C_{15}H_{12}O_3$ = 240.3. A neutral principle from goa powder, the exudates of *Andira (Vouacapoua) araroba* (Leguminosae), Brazil. Yellow crystals, m.207, insoluble in water. An antiseptic, and antiparasitic for skin disorders and infections. **c. triacetate** Eurobin.
chrysatropic acid Scopoletin.
chrysazin $C_{14}H_6O_2(OH)_2$ = 240.2. 1,8-Dihydroxyanthraquinone*. A solid, m.280, slightly soluble in alcohol. **tetranitro ~** Chrysammic acid.
chrysazol $C_{14}H_{10}O_2$ = 210.2. 1,8-Anthracenediol*, 1,8-anthradiol. A phenol derived from anthracene; used in the dye industry.
chryseam $C_4H_5N_3S_2$ = 159.2. White crystals; reagent for nitrites (red color).
chrysene* $C_{18}H_{12}$ = 228.3. Benzophenanthrene. Red, fluorescent scales from coal tar, m.250, slightly soluble in alcohol.
chrysin $C_{15}H_{10}O_4$ = 254.2. 5,7-Dihydroxyflavone. In poplar buds, m.275.
chrysoberyl $BeAl_2O_4$. A golden yellow gem, d.3.5–3.8. Cf. *alexandrite.*
chrysocolla $CuH_2SiO_4 \cdot H_2O$. A green mineral, d.2.0–2.2.
chrysofluorene $\underline{C_{10}H_6 \cdot CH_2 \cdot C_6H_4}$ = 204.3. Naphthylenephenylenemethane; 3 isomers, m.124–208.
chrysoidin $C_7H_{22}O_4$ = 170.3. A yellow pigment in asparagus berries.
chrysoidine $C_{12}H_{12}N_4$ = 212.3. 2,4-Diamino*azo-benzene.* **c. orange** Brown powder, soluble in water. A disinfectant, dye, and indicator at pH 7.0: orange (acid), yellow (basic).
chrysoketone $\underline{C_{10}H_6 \cdot C_6H_4 \cdot CO}$ = 230.3. Yellow crystals, m.133.
chrysolepic acid Picric acid*.
chrysolite Olivine.
chrysophanic acid $C_{15}H_{10}O_4$ = 254.2. 1,8-Dihydroxy-3-methylanthraquinone*, rheic acid, parietic acid. A constituent of rhubarb root, senna leaves, goa powder, and the wood of *Vouacapoua araroba.* Yellow crystals; a mild laxative. Cf. *rhein.*
chrysophanin $C_{20}H_{20}O_9$ = 404.4. A glucoside in rhubarb and senna.
chrysophanol Chrysophanic acid.
chrysophyscin Physcion.
chrysopicrin Vulpinic acid.
chrysoprase A green, opaque gem variety of chalcedony.
chrysoquinone $\underline{C_{10}H_6 \cdot C_6H_4 \cdot CO \cdot CO}$ = 258.3. Chrysene quinone. Red needles, m.240 (sublimes).
chrysorrhetin A yellow coloring matter of senna.
chrysotile $3MgO \cdot 2SiO_2 \cdot 2H_2O$. Canada asbestos; constitutes 95% of world asbestos output. See *asbestosis.*
chrysotoxin An active principle of ergot.
chuchuarine $C_{20}H_{15}O_2N_{12}$ = 455.4. An alkaloid from *Semecarpus anacardium* (Anacardiaceae), southeast Asia. It resembles strychnine.
chum The sediment in fatty oils.
chundoo See *chandoo.*
churning The slow stirring of milk or cream by which the fat globules aggregate to form butter.
chyazic acid Hydrocyanic acid.
chyle An emulsion of lymph and fat formed in the small intestines during digestion, and passing into the veins via the lymphatic system.
chymase Rennin.

chyme Liquid, partly digested food, passing from the stomach into the intestines.
chymia An obsolete term for chemistry.
chymification Gastric digestion.
chymogen Rennin.
chymosin* Rennin.
chymotrypsin An enzyme in the intestine that breaks down proteins. Used in eye surgery and in the reduction of secretions in chest infections (USP, EP, BP).
chymotrypsinogen A protein in beef pancreas. The precursor of chymotrypsin; it can be purified to the stage of complete homogeneity.
C.I. Abbreviation for Colour Index. See *color*.
Ci Symbol for curie (unit).
ciceric acid A mixture of oxalic and malic acid from *Cicer arietinum, Vicia sativa,* and other vetches.
Cicuta A genus of poisonous umbelliferous plants, e.g., *C. virosa,* water hemlock.
cicutoxine $C_{17}H_{22}O_2$ = 258.4. *(E)*-L-8,10,12-Heptadecatriene-4,6-diyne-1,14-diol. A conjugated alcohol, and the active principle of *Cicuta virosa* (Umbelliferae), W. Europe, m.54. A convulsant poison and isomer of enanthotoxin.
cider Fermented expressed apple juice. Cf. *perry*.
C.I.E. (Commission Internationale de l'Éclairage). Color coordinates which define any color in terms of 3 hypothetical primary colors corresponding with the wavelengths 700.0, 546.1, and 435.8 nm.
cigar burning Decomposition in a hot, defined reaction zone moving through a solid bed with characteristic velocity.
Cignolin A brand of anthrarobin.
cilia The hairlike protuberances on the surface of microorganisms, used chiefly for locomotion; also on the surface of epithelial cells in the body.
CIL-n Trademark for a polyamide synthetic fiber.
ciment fondu A hydraulic cement, formed by the complete fusion of bauxite and lime to forms monocalcium aluminate and dicalcium silicate. Cf. aluminous *cement*.
cimetidine $C_{10}H_{16}N_2S$ = 252.3. Tagamet. Crystals, m.142, sparingly soluble in water. Inhibits acid secretion in gastric juice; used for gastric and duodenal ulcers.
cimicic acid $C_{15}H_{28}O_2$ = 240.4. A monobasic acid from bedbugs, *Cimex lectularius.* Yellow crystals, m.44.2.
Cimicifuga Macrotis, black cohosh, black snakeroot. The rhizome of *C. racemosa;* an antispasmodic.
cimicifugin Macrotin. A resinoid from the rhizome of *Cimicifuga racemosa.* Brown powder, insoluble in water; an antispasmodic.
cimifuga Cimicifuga.
cimmol 1-Propenylbenzene.
cina Santonica.
cinaebene $C_{10}H_{16}$ = 136.2. A hydrocarbon from the essential oil of *Artemisia santonica.*
cinchene $C_{19}H_{24}N_2$ = 280.4. An alkaloid obtained from cinchonine by boiling with alcoholic potash.
cinchocaine hydrochloride BP name for dibucaine hydrochloride.
cinchofulvic acid Cinchona red.
cincholepidine Lepidine.
cincholine An alkaloid from cinchona bark.
cinchomeronic acid $C_7H_5O_4N$ = 167.1. 3,4-Pyridinedicarboxylic acid*. Colorless crystals, m.260 (deconp.). **carbo ~** 2,3,4-Pyridinetricarboxylic acid*. Colorless crystals, m.250. **iso ~** 2,5-Pyridinedicarboxylic acid*. Colorless leaflets, m.257, slightly soluble in water. **methyl ~** Picolinedicarboxylic acid.
cinchona (1) Quina, china, Jesuit's bark, loxa bark, huanco bark, Peruvian bark, fever bark. The bark of *Cinchona* species containing at least 3% cinchona alkaloids; a bitter and febrifuge (EP, BP). (2) A genus of trees of the Rubiaceae found in the Andes and cultivated in Sri Lanka and Java; e.g., *C. callisaya,* callisaya bark; *C. cordifolia,* Cartagena bark; *C. officinalis,* crown or loxa bark. **c. alkaloids** The more important are quinine, cinchonine, cinchonidine. **c. tannin** Quinotannic acid.
cinchonamine $C_{19}H_{24}ON_2$ = 296.4. An alkaloid from the bark of *Remijia purdieana* (Rubiaceae). Yellow crystals, m.184, insoluble in water. **c. hydrochloride** $C_{19}H_{24}ON_2 \cdot HCl \cdot H_2O$ = 350.9. Yellow crystals, soluble in water. **c. nitrate** $C_{19}H_{24}ON_2 \cdot HNO_3$ = 359.4. Yellow crystals, soluble in water. **c. sulfate** $C_{19}H_{24}ON_2 \cdot H_2SO_4$ = 394.5. Colorless crystals, soluble in water; said to have 6 times the therapeutic effect of quinine sulfate.
cinchonane $C_{19}H_{22}N_2$ = 278.4. Deoxycinchonine. Colorless crystals, m.92.
cinchonicine Cinchotoxine.
cinchonidine $C_{19}H_{22}ON_2$ = 294.3. Chinidine. The (−) isomer of cinchonine; an alkaloid from cinchona bark. White crystals, m.205, insoluble in water; has antimalarial properties.
cinchonine $C_{19}H_{22}ON_2$ = 294.4. An alkaloid from cinchona bark; the (+) isomer of cinchonidine. Colorless crystals, m.250, slightly soluble in water; a bitter with antimalarial properties and a spot reagent for bismuth (orange-red). **hydroxy ~** Cupreine.
c. benzoate $C_{19}H_{22}ON_2 \cdot C_7H_6O_2$ = 416.5. Yellow crystals, slightly soluble in water. **c. hydrochloride** $C_{19}H_{22}ON_2 \cdot HCl \cdot 2H_2O$ = 366.9. Colorless needles, soluble in water. **c. nitrate** $C_{19}H_{22}ON_2 \cdot HNO_3 \cdot H_2O$ = 375.4. White crystals, soluble in water. **c. sulfate** $(C_{19}H_{22}ON_2)_2 \cdot H_2SO_4 \cdot 2H_2O$ = 722.9. Colorless rhombs, m.198, slightly soluble in water. Used as a reagent for bismuth, hydrochloric acid, or sulfite wood pulp; and in leather manufacture.
cinchoninic acid $C_{10}H_7O_2N$ = 173.2. 4-Quinolinecarboxylic acid. Colorless crystals, obtained by oxidation of cinchonine; m.253. **methyl ~** Aniluvitonic acid.
cinchotannic acid Quinotannic acid.
cinchotenicine $C_{18}H_{20}O_3N_2$ = 312.4. An amorphous alkaloid from cinchona bark; isomeric with cinchotenine.
cinchotenine $C_{18}H_{20}O_3N_2 \cdot 3H_2O$ = 366.4. A colorless, chrystalline alkaloid produced by oxidation of cinchonine.
cinchotoxine $C_{19}H_{22}ON_2$ = 294.4. Cinchonicine. An alkaloid, m.59, obtained by heating cinchonine or cinchonidine.
cinene $C_{10}H_{16}$ = 136.2. (±)-4-Isoprenyl-1-methyl cyclohexene. Oil, b.187; used in perfumes.
cineole Eucalyptol.
cineolic acid Ascaridolic acid.
cinnabar HgS. A native, red mercuric sulfide.
cinnaldehyde Cinnamaldehyde*.
cinnamal The cinnamylidene* radical.
cinnamaldehyde* PhCH:CHCHO = 132.2. β-Phenylacrolein, benzylideneacetaldehyde, 3-phenyl-2-propenal†. Yellow, volatile liquid, d_4·1.050, b.245. Chief constituent of oil of cinnamon and cassia. Used as an itch remedy and artificial flavor. **hydroxy ~** Coumaraldehyde. **hydroxymethoxy ~** Ferulaldehyde.
cinnamamide* PhCH:CHCONH₂ = 147.2. Benzylideneacetamide. Colorless crystals, m.142, soluble in alcohol.
cinnamate* A salt of cinnamic acid containing the radical $C_9H_7O_2-$.
cinnamein $C_9H_7O_2 \cdot C_7H_7$ = 238.3. Benzylcinnamate. A colorless liquid ester from Peru balsam and tolu balsam.
cinnamene Styrene*. **aceto ~** Benzylideneacetone.

cinnamenyl The styryl* radical.

cinnamic acid* PhCH:CHCOOH = 148.2. (*E*)-3-Phenyl-2-propenoic acid*, 3-phenylacrylic acid. A constituent of balsams and storax. Colorless, monoclinic scales, m.133, slightly soluble in water; a reagent for indole. **allo ~** The (Z) form of c. a. crystals. **amino ~** $NH_2C_9H_7O_2$ = 163.2. ***ortho-* ~** Colorless needles, decomp. 160, slightly soluble in water. ***meta-* ~** Yellow needles, m.180, slightly soluble in water. ***para-* ~** Yellow needles, decomp. 175, sparingly soluble in water. **3,4-dihydroxy ~** Caffeic acid. **hydro ~** $PhCH_2CH_2COOH$ = 150.2. 3-Phenylpropionic acid. Colorless needles, m.49, sparingly soluble in water. **hydroxy ~** Coumaric and umbellic acids. **hydroxymethoxy ~** See *ferulic acid, hesperitinic acid.* **2-methoxy ~** $MeOC_9H_7O_2$ = 178.2. Colorless crystals, m.182, soluble in alcohol. **nitro ~** $NO_2C_9H_7O_2$ = 193.2. ***ortho-* ~** White needles, m.249, insoluble in water. ***meta-* ~** Yellow needles, m.197, slightly soluble in water. ***para-* ~** Colorless prisms, m.285, sparingly soluble in water.

cinnamic alcohol Cinnamyl alcohol*.

cinnamic aldehyde Cinnamaldehyde*.

cinnamic anhydride* $(PhCH:CHCO)_2O$ = 278.3. White crystals, m.127, slightly soluble in hot water.

cinnamide Cinnamamide*.

cinnamilidene (1) $PhCH:CH \cdot CH:CH_2$ = 130.2. (2) The cinnamylidene* radical. **c. acetic acid** $PhCH:CH \cdot CH:CH \cdot COOH$ = 174.2. **c. malonic acid** $PhCH:CH \cdot CH:C(COOH)_2$ = 218.2.

Cinnamomum A genus of trees (Lauraceae) which yield drugs; as, *C. camphora*, camphor; *C. cassia*, cassia bark; *C. zeylanicum*, cinnamon.

cinnamon The dried inner bark of *Cinnamomum* species, Sri Lanka, and *C. cassia*, q.v., China. A carminative, astringent, and condiment (NF, BP). **wild ~** Canella.

c. oil The essential oil from the bark of various species of *Cinnamomum*. **Cassia ~** From *C. cassia;* d.1.045–1.063, b.240–260, contains 70–85% cinnamaldehyde. **Sri Lanka ~** From *C. zeylanicum;* d.1.024–1.040, contains cinnamaldehyde and eugenol. **leaf ~** d.1.044–1.065, contains, eugenol, sapol, and cinnamaldehyde. **c. stone** A mineral of the garnet group.

cinnamone Styryl ketone.

cinnamoyl* The radical PhCH:CHCO−. **c. chloride*** PhCH:CHCOCl = 166.6. Chlorcinnamyl. Colorless crystals, m.36.

cinnamyl (1)* 3-Phenyl-2-propenyl†. The radical $Ph \cdot CH:CH \cdot CH_2-$. (2) The cinnamoyl* radical. **c. acetate** $CH_3COOC_9H_9$, in many essential oils. **c. alcohol*** $PhCH:CHCH_2OH$ = 134.2. Styryl alcohol, peruvin, styrene, styrolene alcohol, cinnamic alcohol, 3-phenyl-2-propen-1-ol*, from balsam, storax, and cinnamon bark. White needles, m.33, slightly soluble in water; an artificial flavor and antiseptic. **4-hydroxy-3-methyoxy ~*** Coniferol. **c. cinnamate*** Styracin. **c. cocaine** An alkaloid, m.121, from cocoa leaves. **c. hydride** 1-Propenylbenzene **c. methyl ketone*** Benzylideneacetone **c. phenyl ketone*** Benzylideneacetophenone.

cinnamylate Cinnamate*.

cinnamylidene* Cinnamal, cinnamilidene. The radical $PhCH:CH \cdot CH=$.

cinnoline* $C_6H_4 \cdot N:N \cdot CH:CH$ = 130.1. α-Phenol-1,2-benzodiazine. Colorless crystals, m.390. **dihydro ~** $C_8H_8N_2$ = 132.2. Colorless crystals, m.88, soluble in alcohol.

cinobufagin $C_{26}H_{34}O_6$ = 442.6. A cardiac poison from Ch'an-Su, the dried venom of the Chinese toad.

circle A ring, or a plane figure bounded by a uniformly curved line. Let *r* be the radius, *d* the diameter, and π a constant (3.14159). Then diameter = $2r$, circumference = $2\pi r$, area = πr^2. Cf. *pi.* **great ~** In crystallography, a c. which passes through 2 diametrically opposite points on the surface of a sphere. **small ~** A c. whose diameter is less than that of the great c.

circonium Zirconium*.

circuit The continuous path of an electric current.

circular Round. **c. inch** The area of a circle 1 inch in diameter = 0.7854 in^2 = 507 mm^2.

circulation A continuous movement in a regular course, as of the blood through the blood vessels.

circumference The outline of a more or less circular body. Cf. *circle.*

circumflux Flowing or winding around.

circumfusion A pouring or fusing around; as, a low-melting-point flux.

circumpolar Around a pole.

cis-*, *c- Affix indicating the *cis* position. See *stereoisomer.*

Cisalfa Trademark for a viscose-protein synthetic fiber combination.

cisoid-* *Syn-*. Affix indicating the steric relation between the nearest atoms (i.e., those linked through the smallest number of atoms) on saturated bridgeheads in a polycyclic compound. See *stereoisomer.*

cisplatin See *cisplatinum* under *platinum.*

cissampeline $C_{18}H_{21}O_3N$ = 299.4. An alkaloid from pareira root, *Cissampelos pareira* (Menispermaceae), Brazil; a diuretic.

citraconic acid* $COOH \cdot Me \cdot C:CH \cdot COOH$ = 130.1. (Z)-2-Methyl-2-butenedioic acid*, methylmaleic acid. Isomeric with itaconic acid. **c. a. anhydride*** $C_5H_4O_3$ = 112.1. An intramolecular anhydride of citraconic acid; a colorless oil.

citraconoyl* The radical $C_5H_4O_4-$, from citraconic acid.

citral* $Me_2C:CH \cdot CH_2 \cdot CH_2 \cdot CMe:CH \cdot CHO$ = 152.2. **alpha- ~** Geranial. An aldehyde of several essential oils (citron oil) in 4 isomers; *cis-* and *trans*-terpinolene and *cis-* and *trans*-citrene. Yellow volatile liquid, d.0.897, b.225, insoluble in water; used in perfumery and as a flavoring. **beta- ~** Neral. Colorless liquid, d.0.888, b_{12mm}104.

citramalic acid $HOOC \cdot CMeOH \cdot CH_2 \cdot COOH$ = 148.1. 2-Hydroxy-2-methylbutanedioic acid*, derived from citraconic acid and isomeric with itamalic acid. **(±)- ~** Monoclinic prisms, m.119.

citramide $C_6H_{11}O_4N_3$ = 189.2. Triamide of citric acid. Colorless crystals, m.215, derived from ethyl citrate by the action of ammonia.

citrate* A salt containing the radical $C_6H_5O_7\equiv$, from citric acid. **c. soluble** The phosphates in a fertilizer that are soluble in ammonium citrate solution.

Citratus A fertilizer brand of dicalcium phosphate.

citrene $C_{10}H_{16}$ = 136.2. (*R*)-4-Isopropenyl-1-methylcyclohexene, limonene. Oil, b.168. A terpene from orange and lemon rind. Used for perfumes and flavors. **c. terpin** $C_{10}H_{20}O_2$ = 172.1. Citrene dihydrate. Colorless crystals, formed by the action of water on citrene.

citresia Magnesium hydrogencitrate.

citric acid $HO \cdot C(COOH)(CH_2 \cdot COOH)_2$ = 192.1. 2-Hydroxy-1,2,3-propanetricarboxylic acid*, in fruit juices. Colorless crystals, m.153 (decomp.), soluble in water. A reagent in analysis, and constituent of soda-fountain mixtures and pharmaceuticals (USP, EP, BP). Cf. *citromyces.* **c. a. cycle** Krebs cycle. The main sequence of carbohydrate oxidation in the body: acetyl-CoA + oxaloacetate → citrate → *cis*-aconitate → isocitrate (+ CO_2) → α-oxoglutarate (+ CO_2) → succinyl-CoA → succinate → fumarate → malate →

oxaloacetate → etc. It has respiratory and biosynthetic functions, e.g., the synthesis of amino acids.

citridic acid 1,2,3-Propenetricarboxylic acid.

citrin (1) Cucurbo citrin. A hypotensor glucoside from watermelon seeds. (2) Hesperidin.

citrine A yellow variety of quartz.

citrinin $C_{13}H_{14}O_5$ = 250.3. Notalin. A yellow pigment from *Penicillium citrinum,* m.171 (decomp.); inhibits the growth of *Staphylococcus aureus.*

citrometer A hydrometer graduated in percentages of citric acid.

Citromyces A mold fungus, *C. pfefferianus,* which ferments glucose to citric acid.

citronella Lemongrass. **c. oil** The essential oil of *Andropogon* or *Cymbopogon* species (lemongrass). A perfume and mosquito repellent. **Batu ~** d.0.900–0.920. **Singapore ~** d.0.886–0.900. Both Batu and Singapore citronella contain geraniol and citronellal.

citronellal $C_{10}H_{18}O$ = 154.3. Two isomers from many essential oils: citrene form; $MeC:CH(CH_2)_3CHMeCH_2CHO$, 3,7-dimethyl-6-octenal*; terpinolene form, $CH_2:CMe(CH_2)_3CHMeCH_2CHO$. Colorless liquid, d_{17} 0.854, b.207, slightly soluble in water; used in perfumery and as a flavoring.

citronellaldehyde Citronellal.

citronellic acid $C_{10}H_{18}O_2$ = 170.3. 3,7-Dimethyl-6-octenoic acid*. **(S)-~** Callitrol.

citronellol $Me_2C:CH(CH_2)_2CHMe(CH_2)_2OH$ = 156.3. 2,6-Dimethyloct-2-ene-8-ol*. An isomer of rhodinol; two stereoisomers. An unsaturated alcohol constituent of many essential oils; colorless liquid, d.0.856, b_{17mm}118, slightly soluble in water.

citronyl Citronella oil.

citrulline (1) A resinoid of colocynth. Yellow, amorphous powder, insoluble in water; a laxative. (2)* $C_6H_{13}O_3N_3$ = 175.2. (*S*)-2-Amino-5-ureidovaleric acid. From casein and watermelon, *Citrullis vulgaris,* m.220; an intermediate in urea formation.

Citrus Aurantiaceae. A genus of trees (Rutaceae) whose fruits are edible and yield juice, rinds, oils, and acids; e.g.:

C. aurantium	Sweet orange
C. aurant. var. bigarardia	Bitter (Seville) orange
C. medica	Citron
C. med. var. *limonum*	Lemon
C. med. var. *aurantifolia*	Lime
C. decumana	Grapefruit (shaddock)

citryl Lemon oil.

civet A soft fat of strong musklike odor from the civet cat, *Viverra civetta,* southeast Asia. It contains an essential oil and ammonia. Formerly an antispasmodic; a perfume.

civetane Cycloheptadecene*.

civetone 9-Cycloheptadecen-1-one*.

C.K. wax See *C.K. wax* under *wax.*

Cl Symbol for chlorine. **Cl^-** Chloride ion. **Cl*** Excited chlorine atom. **Cl_2** Chlorine molecule. **Cl_2^+** Ionized chlorine molecule. **Cl_2*** Excited chlorine molecule.

cladding A thin sheet or veneer of metal to protect metal from corrosion.

cladonic acid An acid from *Cladonia rangiferina,* reindeer moss.

Cladothrix A quasi-branched form of Schizomycetes. **C. ochracea** A C. which oxidizes ferrous salts to ferric hydroxide and causes brown deposits in springs.

Claisen, Ludwig (1851–1930) German chemist. **C. condensation** Claisen reaction. **C. flask** A distillation flask with a U-shaped, tubulated neck. **C. reaction** A reaction of an aldehyde with an aldehyde or ketone in the presence of alkali or sodium ethanolate: $R\cdot CHO + CH_3\cdot CO\cdot R' = R\cdot CH:CH\cdot CO\cdot R' + H_2O$.

Clapeyron, Benoit-Paul Emile (1799–1864) French engineer. **C. equation** $dp/dT = \lambda/T(v_2 - v_1)$, where p = pressure, T = thermodynamic temperature, λ = heat of vaporizing 1 g of liquid, v_2 = specific volume of vapor, v_1 = specific volume of liquid. Cf. *Clausius equation.*

clarain A constituent of *coal,* q.v.

clarificant A substance for clearing a solution.

clarification A process by which a liquid is clarified, *e.g.,* filtration. Cf. *defecation.*

Clark **C. cell** A standard cell giving 1.433 volts at 15°C. It consists of a mercury anode and a cathode of amalgamated zinc in a saturated solution of zinc sulfate. **C. degree** See *hardness of water,* under *hardness,* Table 41.

Clarke, Frank Wigglesworth (1847–1931) American chemist, noted for geochemical research.

Classen, Alexander (1843–1934) German chemist. **C. switchboard** A switchboard and worktable for quantitative electrolysis.

classification **c. of compounds** See *Chemical Abstracts Service registry number, compound, chemical.* **c. of drugs** See *drug.* **c. of elements** See *periodic system.* **c. of organisms** (animals and plants). See *organism.* **c. of reactions** See *reaction.*

clathrates Substances resulting from the cagelike inclusion of gases or liquids (up to 15% by weight). The complex may be handled in solid form and the included constituent subsequently released by the action of a solvent or by melting; as, quinol and noble gases. Used to handle radioactive, gaseous isotopes.

Claude, Georges (1871–1900) French inventor. **C.'s method** Air is liquefied in stages by passage through an orifice under pressure, and the expanding gas cooled by doing work externally in a piston engine. Cf. *nitrogen fixation.*

claudetite A native arsenious oxide.

Claus, Adolf (1840–1900) German chemist noted for his benzene formula.

Clausius, Rudolf Julius (1822–1888) German physicist. **C. equation** Clausius-Clapeyron equation: $\log p = -L/4.58T + C$, where p is the vapor pressure at the thermodynamic temperature T, L the molecular heat of vaporization, and C a constant of integration. Its integrated form is

$$2.303 \log \frac{p}{760} = \frac{L}{R}\left(\frac{1}{T_{760}} - \frac{1}{T_p}\right)$$

where T_{760} and T_p are the boiling points at 760 and p mm pressure, respectively; R is the gas constant. See also *Clapeyron equation.* **C. law** The specific heat of a gas is independent of temperature at constant volume. **C.-Mosotti equation** The polarization of the dielectric per mole = $m(\epsilon_r - 1)/d(\epsilon_r + 2)$ where ϵ_r is the relative permittivity, m the molecular weight, d the density, and P the molal polarization.

clausthalite PbSe. A native selenide (Klausthal, Czechoslovakia).

clavacin An antibiotic from fungi, identical with clavatin and patulin.

clavatin Patulin from *Aspergillus clavatus.*

clavicepsin $C_{18}H_{34}O_{16}$ = 506.3. An inert glucoside, m.198, in ergot.

clavicin Patulin.

claviforme Patulin.

claviformin Patulin.
clavine $C_{11}H_{22}O_4N_2$ = 246.2. An alkaloid of ergot, m.263.
clavulinic acid $C_8H_9O_5N$ = 199.2. A β-lactamase inhibitor, produced from *Streptomyces clavuligens.* Used in combination with amoxycillin, potentiating its use. See *Augmentin.*
clay $Al_2Si_2O_5(OH)_4$. A plastic, soft, variously colored earth, formed by the decomposition of aluminum minerals. In true clay, 30% by weight of solid particles are of diameter less than 0.002 mm. Used for pottery. Cf. *ceramics.* **American ~** C. made from quartz feldspar, whitening, ball clay, and kaolin, and glazed at 1300°. **bone ~** A hard, translucent c. containing approximately 50% of kaolin and calcined animal bones (U.K. usage). **calcined ~** C. given a controlled porous structure to augment its light scattering and bulking properties. **china ~** See *china clay.* **feldspathic ~** Porcelain. **fire ~** See *fire clay.* **hydrogen ~** A c. with zeolitic properties with respect to hydrogen ions. **japan ~** Montmorillonite. **pipe ~** See *pipe clay.* **potter's ~** See *potter's clay.*
c. ironstone Siderite. **c. substance** Kaolinite.
cleaning solution Mixed concentrated sulfuric acid and sodium dichromate, used to clean chemical glassware.
clearing (1) Clarification. (2) Freeing cotton fiber from grease by boiling in a weak solution of sodium carbonate. (3) In photography, removal of fog from a plate. (4) In microscopy, removal (e.g., by chloral hydrate) of extraneous matter from plant material to render the essential plant structure visible.
cleavage (1) Separation as layers, e.g., the splitting of crystals. (2) Biology: (a) segmentation. Cf. *amitosis.* (b) division of fertilized egg.
clebrium A noncorrosive alloy of iron with Cr 13–19, Ni 2–4, Mo 0–3.2, C 0–2, Mn 0.8–2.8%.
cleiophane Sphalerite.
clematine A poisonous alkaloid from *Clematis vitalba* (Ranunculaceae).
Clematis A species of climbing, ranuculaceous plants.
Clemmensen reduction The reduction of certain aldehydes and ketones to hydrocarbons (as, acetophenone to styrene) by activated zinc and acid.
Clerget inversion The determination of sucrose polarimetrically after inversion of 100 ml solution with 5 ml strong hydrochloric acid at 69° for 7.5 min. Sucrose = $100(a - b)/(142.66 - 0.5t)$, where a and b are the polarizations originally and after inversion at t°C.
Clerici solution A molar mixture of thallium malonate and thalium formate, d.4.27; a mineralogical flotation agent for density determinations.
Cleve, Per T. (1840–1905) Swedish chemist, codiscoverer of helium and rare-earth metals. **C.'s acid** 1-Naphthylamine-6-sulfonic acid. **C.'s salts** $[Pt(NH_3)_3Cl_n]Cl$; n = 1 or 3.
Cleveland tester A flash-point apparatus.
climacteric (1) The menopause. (2) Pertaining to a critical period or stage, e.g., in the ripening of fruit.
clinical analysis The diagnostic examination of body fluids and waste products.
Clinitest tablets Trademark for cupric sulfate diagnostic tablets. Used by diabetics to test urine for sugar.
clinker The hard, partly vitrified residue after combustion of coal, also produced by volcanic action; used to manufacture cement. Cf. *slag.*
clinochlore Ripidolite. A green chlorite or aluminum magnesium silicate, d.2.65–2.78.
clinohedrite $H_2CaZnSiO_5$. An orange-colored mineral from Franklin, N.J.
clinometer A pocket device to measure the angle of a slope.
clinostat A moving-disk apparatus to determine the phototropism of a plant.
clioquinol C_9H_5ONICl = 305.5. 5-Chloro-7-iodo-8-quinolinol. Iodochlorhydroxyquin, Entero-Vioform, Vioform, Diodoquin. Yellow powder, characteristic smell, m.180, insoluble in water. For intestinal amebiasis and as cream for skin infections (USP, BP).
clofazimine $C_{27}H_{22}N_4Cl_2$ = 473.4. Lamprene. Red powder, insoluble in water. An antileprotic drug (BP).
clofibrate $C_{12}H_{15}O_3Cl$ = 242.7. Ethyl 2-(*p*-chlorophenoxy)-2-methylpropionate. Atromid. Liquid, nearly insoluble in water. Used to treat arteriosclerosis and high blood fat levels (USP, BP).
Clomid Trademark for clomiphene citrate.
clomiphene citrate $C_{26}H_{28}ONCl \cdot C_6H_8O_7$ = 598.1. (*E*)- and (*Z*)-2-[*p*-(2-chloro-1,2-diphenylvinyl)phenoxy]triethylamine citrate. Clomid. White powder, slightly soluble in water. Stimulates the production of gonadotrophins and, thus, ovulation; used to treat infertility (USP, BP).
clonazepam $C_{15}H_{10}O_3N_3Cl$ = 315.7. Clonopin, Rivotril. Yellow powder, m.239, insoluble in water. A benzodiazepine with marked anticonvulsant properties; used in epilepsy (USP).
clone (1) (Noun.) One of two or more organisms having identical genetic characteristics to the parent and formed asexually from it. (2) (Verb.) To create clones.
clonidine hydrochloride $C_9H_9N_3Cl_2 \cdot HCl$ = 266.6. Catapres, Dixarit. White, crystalline powder, soluble in water. Used for hypertension and for prevention of migraine (USP, BP).
clopidol $C_7H_7ONCl_2$ = 192.0. 3,5-Dichloro-2,6-dimethyl-4-pyridinol. m.320+. A coccidiostat used in poultry foods.
closed chain A ring compound. See *chain.*
closo*- Prefix indicating a cage or closed molecular structure; especially a boron skeleton that is a polyhedron having all triangular faces. Cf. *nido-.*
clot A solid, jellylike mass, as, of blood.
cloth A fabric woven of fiber. **cheese ~** Loosely woven, thin cotton cloth used for straining. **filter ~** A canvas used in filter presses.
clotting The coagulation of blood, lymph, or milk.
cloud point The temperature at which a solid starts to precipitate from solution in an oil when cooled under standard conditions. **c. seeding** Nucleation of rain clouds; e.g., with fine particles of silver iodide.
cloudy (1) Containing a diffused precipitate. (2) Describing a bacterial culture which does not contain pseudozoogleae.
clove(s) Caryophyllus. The dried flower buds of *Caryophyllus aromaticus* (Myrtaceae); a spice. **c. oil** Essential oil of cloves. Yellow, volatile liquid, d.1.079, b.243; consists mainly of eugenol (NF, BP).
clovene $C_{15}H_{24}$ = 204.5. A terpene from clove oil. Colorless liquid, d.0.93, b.263.
cloxacillin sodium $C_{19}H_{17}O_5N_3ClSNa \cdot H_2O$ = 475.8. White, bitter crystals, soluble in water; an antibiotic for staphylococcal infections (USP, BP).
clupanodonic acid $C_{22}H_{34}O_2$ = 330.5. Docosa-4,7,11-trien-18-ynoic acid, in fish blubber. Yellow oil, d.0.9410, b_{5mm}236, soluble in ether.
clupein A protein from herring, *Clupea harengus:* C 47.93, H 7.59, O 12.78, N 31.68%.
Cm Symbol for curium.
cm Centimeter. **cm²** Square centimeter. **cm³** Cubic centimeter = cc.
C.N. Coordination number.
cnicin $C_{20}H_{26}O_7$ = 378.4. An antibiotic from *Cnicus benedictus;* m.143. White crystals, soluble in cold water.
Co Symbol for cobalt.
CoA Symbol for coenzyme A.

coacervate The liquid product of coacervation.
coacervation, coazervation The reversible collection of emulsion particles into liquid droplets preceding flocculation. An intermediate stage between sol and gel formations. Cf. *coagulation.*
coade stone A stonelike ceramic substance made from clay and fine stones; used in sculpture.
coagel The gel formed by coagulation.
coagulability The capacity to clot.
coagulase An enzyme that clots and precipitates proteins.
coagulation The precipitation of colloids in a soft mass. **C. Factor Xa*** See *thrombokinase.*
coagulator An incubator for keeping test tubes at blood temperature.
coagulometer A device to determine the speed of coagulation of blood.
coal A native, black or brown, brittle or soft substance from the degradation of ancient forests, consisting chiefly of carbon, with hydrogen, nitrogen, oxygen, and other elements. Coals are classified according to their state of mineralization (% carbon); as, *lignite, bituminous c.,* and *anthracite;* or microscopically as:

1. *Clarain,* a lustrous mixture of leaves, wood, resinous bodies, etc., in a structureless matrix.
2. *Vitrain,* lustrous, thin bands with conchoidal fracture, in which the wood structure is destroyed.
3. *Durain,* a dull variety of c. which contains microspores and megaspores and a high ash.
4. *Fusain,* thin bands of soft, dull coal containing no volatile matter.

Calorific values in MJ/kg are:

Anthracite	31.0
Coke	27.9
Charcoal	30.2
Bituminous A	33.2
Cannel	31.8
Lignite	14.0
Peat	4.42
Wood	17.7

brown ~ Lignite. **cannel ~** A hard c. for gas production. **char ~** See *charcoal.* **hard ~** Anthracite. **mother of ~** Fusain. **soft ~** Bituminous c. **Standard C. Unit** S.C.U. The thermal energy of 1 ton of c. See *fuel equivalence.* **white ~** Waterpower.

c. brass Marcasite. **c. equivalent** See *fuel equivalence.* **c. gas** The gaseous product of the distillation of c. The average composition is:

Hydrogen	43–55%	Nonilluminating, but heating
Methane	25–45%	
Carbon monoxide	4–11%	
Olefines, acetylenes, and benzene	2– 5%	Illuminants
Nitrogen	2–12%	Impurities
Carbon dioxide	0– 3%	
Oxygen	0–1.5%	

c. liquefaction Production of fuels and chemicals from c. by 3 alternative processes: (1) Gasification (Fischer-Tropsch synthesis); (2) hydrogenation; (3) solvent extraction. **c. oil** (1) Kerosene. (2) Petroleum. **c. tar** The condensed liquid from the distillation of c., containing hydrocarbons (benzene), phenols, basic substances (pyridine), finely divided free carbon. **c. unit** See *fuel equivalence.*
Coalite Trademark for an open-structure, smokeless fuel made by low-temperature carbonization of coal at 620°C; volatile matter 10%.
Coanda effect Wall effect. The attraction of a stream of fluid to a straight object at an angle to its path.
coarse metal A fusible silicate of iron produced in the reverberatory furnace.
coazervate Coacervate.
coazervation Coacervation.
cobalamin(s) Collective name for substances possessing vitamin B_{12} activity. **5′-deoxyadenosyl ~** $C_{72}H_{100}O_{27}N_{18}PCo$ = 1740. Coenzyme B_{12}. The active form of vitamin B_{12} in the body. See *vitamin,* Table 101, and *corrin.*
cobalt* Co = 58.9332. An element of the iron group, at. no. 27, discovered by G. Brandt (1733). A steel-gray, pinkish, ductile metal, d.8.718, m.1495, b.2885, insoluble in water. It occurs as metal in meteorites, as sulfide and arsenide in smaltite and cobaltite. It is prepared by reduction of its oxides in hydrogen. C. is di- and trivalent and forms two series of compounds: c.(II)*, c.(2+)*, cobaltous, Co^{++}; c.(III)*, c.(3+)*, cobaltic, Co^{3+}. It forms many colored double salts. Its complex ions include: hexaaquacobalt(II), $[Co(H_2O)_6]^{2+}$; hexacyanocobaltate(II), $[Co(CN)_6]^{4-}$. The cobaltous compounds are the more stable; they are pink when hydrated and green when anhydrous. Metallic c. is used in alloys and in ceramics. The radioactive isotope ^{60}Co is used to detect flaws in metal articles. In medicine it is a source of γ-rays to treat malignant disease and to sterilize plastic objects, as, syringes and catheters. ^{57}Co, ^{58}Co, and ^{60}Co are used to label cobalamin. **black ~** An earthy native c. **earthy ~** Asbolite. **red ~** Erythrite. **speiss ~** An impure smaltite.

c.(II)*, c.(2+)* See also *cobaltous.* **c.(III)*, c.(3+)*** See also *cobaltic.* **c. acetate** Cobaltous acetate. **c. aluminate** Thenard's blue. **c. bloom** The mineral $Co_3(AsO_4)_2 \cdot 8H_2O$. **c. blue** A dark blue mineral pigment containing cobaltous oxide 35, zinc oxide 20, and chalcedony 25%. **c. bomb** A hydrogen bomb enclosed in c. to serve as a neutron absorber, which produces radioactivity and ^{60}Co; 500 tons deuterium reacting inside 10^5 tons c. would give sufficient ^{60}Co to kill the population of the earth in a few years. **c. carbonyl** C. tetracarbonyl*. **c. chlorides** See *cobaltichloride, cobaltic,* and *cobaltous chloride.* **c. cyanide** Cobalticyanide. **c. green** Cobalt zincate*, Rinman's green. A solid solution of cobaltous oxide and zinc oxide; a green pigment. **c. minerals** C. is usually associated with nickel and found as sulfide, or arsenide; e.g., jaipurite, CoS; linnaeite, Co_3S_4; smaltite, $CoAs_2$. **c. nitride*** Co_2N = 131.9. Black crystals. **c. nitrite** Cobaltinitrite. **c. oxides** CoO = cobaltous oxide. Co_2O_3 = cobaltic oxide, Co_3O_4 = cobaltocobaltic oxide; a blue pigment for ceramics and glass. **c. phosphide*** Co_2P = 148.8. A black, metallic substance. **c. potassium malonate*** $CoK(C_3H_2O_4)_2$ = 302.1. Pink crystals, soluble in water. **c. potassium sulfate*** $CoSO_4 \cdot K_2SO_4 \cdot 6H_2O$ = 437.3. Soluble plates. **c. sulfates** See *cobaltic sulfate, cobaltous sulfate.* **c. tetracarbonyl*** $Co(CO)_4$ = 171.0. C. carbonyl. Black crystals, m.43, insoluble in water. **c. violet** A c. phosphate pigment for oil paints. **c. yellow** Cobaltic potassium nitrite. **c. zincate*** C. green.
cobaltammine compounds Ammonates. Compounds of ammonia and cobalt, e.g., pentaamminecobalt(III) trichloride, $[Co(NH_3)_5]Cl_3$.
cobaltic Cobalt(III)* or (3+)*. Containing trivalent cobalt. **c. chloride** $CoCl_3$ = 165.3. Blue crystals, d.2.94, decomp. on heating, soluble in water. See *cobaltichloride.* **c. hydroxide** $Co(OH)_3$ = 110.0. Black powder, insoluble in water. **c. oxide** Co_2O_3 = 165.9. Cobalt sesquioxide*. Black powder, decomp. at red heat, insoluble in water. **c. potassium nitrite** $K_3[Co(NO_2)_6]nH_2O$. Potassium cobaltinitrite, cobalt yellow, potassium hexanitrocobaltate(III). Yellow precipitate obtained by adding potassium nitrite to c. nitrate acidified with acetic

acid. **c. sulfate** $Co_2(SO_4)_3$ = 406.0. Blue crystals, decomp. in water. **c. sulfide** Co_2S_3 = 214.0. Black powder, insoluble in water.

cobaltichloride **luteo ~** $[Co(NH_3)_6]Cl_3$. Hexaamminecobalt(III) trichloride*. Orange-yellow crystals, soluble in water. Reagent for pyrophosphoric acid. **praseo ~** $[Co(NH_3)_4]Cl_3$. Tetraamminecobalt(III) trichloride*. Green crystals, soluble in water. **purpureo ~** $[Co(NH_3)_5]Cl_3$. Pentaamminecobalt(III) trichloride*. Purple crystals, soluble in water. See *purpureo.* **roseo ~** $[Co(NH_3)_5 \cdot H_2O]Cl_3$. Pentaammineaquacobalt(III) trichloride*. Red crystals, decomp. by water.

cobalticyanic acid $H_3[Co(CN)_6]$ = 216.1. Hydrogenhexacyanocobaltate(III)*. Colorless needles, decomp. above 100, soluble in water.

cobalticyanide Hexacyanocobaltate(III)*. Salts of cobalticyanic acid containing the $[Co(CN)_6]^{3-}$ ion; yellow in aqueous solution.

cobaltinitrite Hexanitrocobaltate(III)*. Salts containing the $[Co(NO_2)_6]^{3-}$ ion; yellow in aqueous solution.

cobaltite The mineral (CoFe)SAs. Cf. *sehta.*

cobalto(us)cobaltic oxide Co_3O_4 = 240.8. Black crystals, insoluble in water. A spinel mixed oxide.

cobaltosic oxide Cobaltocobaltic oxide.

cobaltosulfate A salt containing the $Co(SO_4)_2^{2-}$ ion which imparts a red color to the aqueous solution.

cobaltous Co(II)* or (2+)*. Containing divalent cobalt. **c. acetate** $Co(C_2H_3O_2)_2 \cdot 4H_2O$ = 249.1. Purple crystals, soluble in water; used in invisible inks. **c. arsenate** $Co_3(AsO_4)_2 \cdot 8H_2O$ = 598.8. Erythrite, cobalt bloom. Purple, monoclinic crystals, insoluble in water; a light-blue pigment for porcelain and glass. **c. benzoate** $Co(C_6H_5COO)_2 \cdot 4H_2O$ = 373.2. Red leaflets, dehydrated at 115, soluble in water. **c. bromate** $Co(BrO_3)_2 \cdot 6H_2O$ = 422.8. Red crystals, soluble in water. **c. bromide** $CoBr_2 \cdot 6H_2O$ = 326.8. Red, hygroscopic crystals, soluble in water; used to fill hygrometers. **c. butyrate** $Co(C_4H_7O_2)_2$ = 233.1. Purple granules, soluble in water. **c. carbonate** $CoCO_3$ = 118.9. Rose powder, decomp. by heat, insoluble in water; used to prepare cobalt pigments. **c. chlorate** $Co(ClO_3)_2 \cdot 6H_2O$ = 333.9. Purple crystals, m.50, decomp. 100, very soluble in water. **c. chloride** $CoCl_2$ = 129.8. Blue crystals, sublime on heating, soluble in water. It crystallizes also as: hexahydrate, $CoCl_2 \cdot 6H_2O$; tetraammonate, $CoCl_2 \cdot 4NH_3$; pentaammonate, $CoCl_2 \cdot 5NH_3$. **crystalline** $CoCl_2 \cdot 6H_2O$ = 237.9. Ruby, monoclinic crystals, m.87, soluble in water; used in hygrometers, invisible inks, chemical barometers, and electroplating. **c. chromate** $CoCrO_4$ = 174.9. Brown powder containing some c. hydroxide, insoluble in water. **c. citrate** $Co_3(C_6H_5O_7)_2$ = 555.0. Pink powder, sparingly soluble in water. **c. cyanide** $Co(CN)_2$ and $Co(CN)_3 \cdot 3H_2O$ = 191.0. Red powder, anhydrous at 250°C, decomp. 300, insoluble in water, soluble in potassium cyanide solution. **c. fluosilicate** C. hexafluorosilicate*. **c. formate** $Co(OOCH)_2$ = 149.0. Red crystals, soluble in water. **c. hexafluorosilicate*** $CoSiF_6 \cdot 6H_2O$ = 309.1. Trigonal, pink crystals, soluble in hot water. **c. hydroxide** $Co(OH)_2$ = 93.0. Rose-red crystals, insoluble in water. **c. iodide** $CoI_2 \cdot 6H_2O$ = 420.8. Brown crystals, soluble in water. **c. linoleate** $Co(C_{18}H_{31}O_2)_2$ = 617.8. Brown powder, insoluble in water. **c. nickelous sulfate** $NiSO_4 \cdot CoSO_4$ = 309.7. A double salt of cobaltous and nickelous sulfates. **c. nitrate** $Co(NO_3)_2 \cdot 6H_2O$ = 291.0. Red crystals, m.56, decomp. at red heat, very soluble in water. Used for the preparation of cobalt pigments, for secret ink, and as a reagent in blowpipe analysis. **c. oleate** $Co(C_{17}H_{33}COO)_2$ = 621.8. Brown masses, insoluble in water. **c. oxalate** $Co(OOC)_2 \cdot H_2O$ = 165.0. Pink powder, insoluble in water. **c. oxide** CoO = 74.9. Brown powder, decomp. 2850, insoluble in water. **c. phosphate** $Co_3(PO_4)_2 \cdot 2H_2O$ = 402.8. Red powder, insoluble in water; used for light-blue pigments, in ceramics and glass. **c. propionate** $Co(C_3H_5O_2)_2 \cdot 3H_2O$ = 259.1. Red crystals, m.ca.250, soluble in water. **c. silicate** Co_2SiO_4 = 209.9. Purple crystals, insoluble in water. **c. sulfate** $CoSO_4 \cdot 7H_2O$ = 281.1. Rose vitriol. Cf. *bieberite.* Red crystals, m.97, anhydrous at 450, soluble in water. Used to prepare cobalt salts and cobalt oxides for ceramic pigments, and in galvanostegy for cobalting metals. **c. stannate** $CoSnO_3$. Blue pigment. Cf. *ceruleum.* **c. sulfide** CoS = 91.0. Brown powder, m.1100, insoluble in water. **c. tartrate** $Co(C_4H_4O_6)$ = 207.0. Pink powder, sparingly soluble in water.

Cobol *C*ommon *b*usiness *o*rientated *l*anguage. Computer language used in data processing applications.

cobra A poisonous snake of India and Africa, genus *Naja,* whose venom contains ophiotoxin and cobralysin. **c. lecithid** A hemolytic compound of lecithin and cobra toxin formed in the blood.

cobralysin The hemolytic substance of cobra venom.

cobric acid White microcrystals, from cobra venom, containing calcium sulfate.

coca Erythroxylon. The dried leaves of the S. American shrub *Erythroxylum coca;* a source of cocaine. **c. alkaloids** A group of related alkaloids; as, cocaine, $C_{17}H_{21}NO_4$; ecgonine, $C_9H_{15}NO_3$.

cocaine $C_{17}H_{21}O_4N$ = 303.4. Erythroxylon, methylbenzoylecgonine, snow. An alkaloid from the leaves of *Erythroxylum coca.* White, monoclinic scales, m.98, slightly soluble in water. A surface anesthetic for eye and nose; also a central nervous stimulant, but not used as such because can cause excitement and convulsions. **isatropyl ~** See *isatropylcocaine.* **methyl ~** Cocainidine. **tropa ~** See *tropacocaine.*

c. addiction Either inhaled as the powder, which can cause nasal septum perforation, or injected with other drugs, as heroin. **c. benzoate** $C_{17}H_{21}O_4N \cdot C_7H_6O_2$ = 425.5. White crystals, soluble in water. **c. borate** White crystals, soluble in water; used for eyewashes. **c. chloride** C. hydrochloride. **c. citrate** $(C_{17}H_{21}O_4N)_2C_6H_8O_7$ = 798.8. White crystals, soluble in water. **c. hydrobromide** $C_{17}H_{21}O_4N \cdot HBr$ = 384.3. White crystals, soluble in water. **c. hydrochloride** $C_{17}H_{21}O_4N \cdot HCl$ = 339.8. Colorless prisms, m.186, soluble in water; an anesthetic (USP, EP, BP). **c. hydroiodide** $C_{17}H_{21}O_4N \cdot HI$ = 431.3. Yellow crystals, soluble in water. **c. lactate** $C_{17}H_{21}O_4N \cdot C_3H_6O_3$ = 393.4. A syrupy mass, soluble in water; an anesthetic. **c. nitrate** $C_{17}H_{21}O_4N \cdot HNO_3$ = 366.4. White crystals, soluble in water. **c. oleate** A solution of cocaine in oleic acid, used externally. **c. salicylate** $C_{17}H_{21}O_4N \cdot C_7H_6O_3$ = 441.5. Colorless crystals, soluble in water. **c. sulfate** $C_{17}H_{21}O_4N \cdot H_2SO_4$ = 401.4. White crystals, soluble in water. **c. tannate** $C_{17}H_{21}O_4N \cdot C_{14}H_{10}O_9$ = 625.6. White powder, soluble in alcohol. **c. tartrate** $(C_{17}H_{21}O_4N)_2 \cdot C_4H_6O_6$ = 756.8. Colorless crystals, soluble in water.

cocainidine $C_{18}H_{23}O_4N$ = 317.4. Methylcocaine. An alkaloid from coca leaves, similar to cocaine.

cocatannic acid $C_{17}H_{22}O_{10} \cdot 2H_2O$(?). Yellow microcrystals, from the leaves of *Erythroxylum coca.*

Coccaceae Spherical Schizomycetes (bacteria), which includes Coccus, Diplococcus, Staphylococcus, and Streptococcus.

cocceric acid $C_{31}H_{62}(OH)COOH$ = 496.9. White, crystalline mass from cochineal wax, m.93.

coccerin $C_{30}H_{60}(C_{31}H_{61}O_3)_2$ = 1384. A wax derived from cochineal.

cocceryl The radical $C_{30}H_{60}=$ from c. alcohol. **c. alcohol**

$C_{30}H_{62}O_2$ = 454.8. A dihydric alcohol from cochineal wax. White crystals, m.102.

coccerylic acid Cocceric acid.

cocci Plural of coccus.

coccidiostat A substance that is toxic to cocci.

Coccidium A genus of sporozoa.

coccinellin Carmine.

coccognin A glucoside from the fruits of *Daphne gnidium* (Thymelaceae).

coccolith The minute calcareous skeleton of floating marine algae.

coccon Pomegranate seeds.

cocculin $C_{19}H_{26}O_{10}$ = 414.4. A crystalline constituent of *Cocculus indicus*, fishberry; a narcotic. Cf. *kukoline.*

cocculus The fruit of menispermaceous plants. **c. indicus** Fishberry, Indian berry, Oriental berry, Levant berry. The fruit of *Anamirta paniculata* (Menispermaceae), a climbing shrub, southeast Asia; a vermicide. See *cocculin, kukoline.*

coccus (1) A spherical bacterium. (2) Insects of the family Coccidae, scale insects, that yield cochineal, Chinese wax, kermes, lac, manna.

cochineal The female insect, *Coccus cacti*, raised on several species of cactus *(Opuntia)* in tropical America. Its chief constituent is carmine; used to color pharmaceutical preparations and food, and as an indicator (BP). **c. solution** An indicator prepared by macerating 3 g of unbroken c. for 4 days in 250 ml of a mixture of 1 pt. alcohol and 3 pts. water (alkalies—violet; acids—yellowish red). **c. wax** Coccerin.

cochinilin Carminic acid.

cochlear, cochleare A spoon. A pharmaceutical measure: cochlear magnum (tablespoon) = 15 mL = ½ fluid ounce; cochlear medium (dessertspoon) = 8 mL = 2 fluid drachms; cochlear parvum (teaspoon) = 4 mL.

Cochlearia (1) A genus of cruciferous plants. (2) Scurvy grass, *C. officinalis;* a diuretic and antiscorbutic. (3) *C. armoracia,* horseradish, q.v.

cochrome An alloy of Co and Cr. Cf. *Nichrome.*

cocinic acid $C_{11}H_{22}O_2$ = 186.3. A fatty acid from coconut oil.

cocinin An ester of cocinic acid; the chief constituent of coconut oil.

coclaurine An alkaloid from the leaves and bark of *Cocculus laurifolius* (Menispermaceae).

cocoa *Cacao,* q.v. (1) C. tree. The tropical shrub *Theobroma cacao* (Sterculiaceae). (2) C. seeds. (3) Brown, finely ground seeds of *Theobroma cacao,* after expression of their fat. Cf. *coca, coconut.* **c. butter** Cacao butter, oleum theobromatis, theobroma oil. Brown fat obtained by compression of cacao seeds between hot or cold plates, d.0.976–0.995, sapon. val. 192–200, iodine no. 32–37.7, m.30–35, soluble in ether (NF). **c.nut** (1) Cocoa. (2) Coconut. Cf. *coca.* **c.nut oil** See *coconut oil.* **c. oil** Cocoa butter. **c. red** $C_{17}H_{12}(OH)_{10}$. A dihydroxyhydroflavonol derivative, and the pigment of c.; 2.5–5% is formed during the drying of the white beans, by enzyme action. **c. seeds** Cacao. The dried and fermented seeds of the cocoa tree, *Theobroma cacao,* tropical and S. America. Used to make a beverage; contains cocoa butter 50, albumin 18, starch 10, theobromine 1.5%. **c. shell meal** The ground husks of c. seeds; a fertilizer: nitrogen 2.5, phosphorus pentaoxide 1, potassium oxide 2.5%.

coconut Cocoanut. The fruit of *Cocos nucifera,* a palm (Pacific and Australian islands, southeast Asia and W. Indies). Cf. *copra, coir.* **c. butter** Coconut oil. **c. cake** The material remaining after the expression of c. oil. **c. oil** A fixed oil expressed from the fruit of *Cocos nucifera* or copra (dried coconut, the coconut palm). Used in the manufacture of margarine, food products, and soap (BP). Yellow oil, d.0.9259, m.14. Sapon. val. 246–268, iodine no. 8–9.5, acid val. 5–50. Cf. *cocoa butter.*

C.O.D. Chemical oxygen demand.

cod The codfish, *Gadus morrhua.* **c. ichthulin** A vitellin from cod eggs. **c.-liver oil** See *cod-liver oil.*

codamine $C_{20}H_{25}O_4N$ = 343.4. An alkaloid of opium, isomeric with laudanidine, m.121.

CODATA See *standards.*

codeine $C_{18}H_{21}O_3N \cdot H_2O$ = 317.4. Methylmorphine. A levorotatory alkaloid from opium (0.2–0.8%). Colorless crystals, anhyd. at 155, soluble in water; a mild narcotic and analgesic. **c. citrate** $(C_{18}H_{21}O_3N)_3C_6H_8O_7$ = 1090. Colorless crystals, soluble in water. **c. hydrochloride** $C_{18}H_{21}O_3N \cdot HCl \cdot 2H_2O$ = 371.9. White needles, m.264, soluble in water. **c. phosphate** $C_{18}H_{21}O_3N \cdot H_3PO_4 \cdot 1.5H_2O$ = 424.4. White crystals, anhydrous at 235, soluble in water; an analgesic cough suppressant and control for diarrhea (USP, EP, BP). **c. salicylate** $C_{18}H_{21}O_3N \cdot C_7H_6O_3$ = 437.5. Colorless, crystals, soluble in water; used in rheumatic disorders. **c. sulfate** $(C_{18}H_{21}O_3N)_2 \cdot H_2SO_4 \cdot 5H_2O$ = 786.9. White rhombs, decomp. 278, soluble in water.

Codex Alimentarius An international compendium of food standards, compositions, descriptions, and analytical methods. Published jointly by F.A.O. and W.H.O.

cod-liver oil Oleum morrhuae, banks oil. A fatty oil from the liver of *Gadus morrhua.* Yellow oil, d_{25}-0.915–0.925, soluble in alcohol. Contains vitamins A and D, olein, morrhuic acid, phosphorus, iodine, and sulfur compounds. A source of vitamins and calories (BP). **artificial ~** An irradiated ergosterol or cholesterol.

c. o. emulsion An emulsion of c. o. with various pharmaceutical ingredients which mask its unpleasant taste (peppermint oil) or enhance its value (phosphates).

codol Retinol.

coeff. Abbreviation for coefficient.

coefficient A numerical factor by which the value of one quantity is multiplied to give the value of another; or to indicate its rate of change. **distribution ~** See *distribution coefficient.*

c. of absorption See *absorption coefficient.* **c. of expansion** A number indicating the expansion of a substance for a temperature increase of 1°C. Cf. *Charles's law.* **c. of performance** See *heat pump.*

coenzyme Cofactor, coferment. A catalyst for the activation of an enzyme. Made up of organic compounds or metal ions (e.g., K, Mg, Fe). **c. A** $C_{21}H_{36}O_{16}N_7P_3S$ = 767.5. CoA. Powder, soluble in water. In many microorganisms; important in the citric acid cycle. **acetyl ~** The thioester of CoA and acetic acid. Important in carbohydrate catabolism. **c. Q** Ubiquinone. **c. II** $C_{21}H_{28}O_{17}N_7P_3$ = 743.4. Occurs widely in animals.

coesite Synthetic crystalline silica. It is 15% denser than quartz and is very abrasive.

cofactor Coenzyme.

coferment Coenzyme.

coffee The dried seeds of the coffee tree, *Coffea arabica, C. liberica,* etc. (Rubiaceae), Asia, Africa, Central and S. America. Contains (unroasted): fat 13.27, water 11.23, albuminoids 12.07, sugar 8.55, fiber 3.92, caffeine 1.21%.

coffeic acid Caffeic acid.

coffeine Caffeine.

coffinite $U(SiO_4)_{1-n}(OH)_{4n}$. A primary constituent of certain radioactive siltstones, from South Island, New Zealand.

Cogentin Trademark for benztropine mesylate.

cognac A high-grade brandy. **c. essence** (1) Ethyl nonanoate or (2) ethyl chloride, used to imitate the c. flavor.

c. oil (1) A brandy adulterant; coconut oil, alcohol, and sulfuric acid. (2) The essential oil of cognac, containing the esters of heptanoic and octanoic acids. (3) Ethyl heptanoate; a flavoring.

Cohen **C., Ernst Julius (1869–1944)** Dutch chemist noted for research in physical chemistry. **C., Julius Berend (1859–1935)** British chemist, noted for his work on optical activity and the laws of aromatic chemistry.

cohere To hold together; as, molecules of the same type. Cf. *adhere.*

coherer An electrical resistance tube filled with a granulated conductor by which electromagnetic waves are detected from the change in resistance produced by the corresponding variations in the coherence of the particles.

cohesion The attractive force which holds the molecules of a substance together. Cf. *adhesion.*

cohesive Sticking together.

cohoba A snuff used in Haiti.

cohobation Repeated distillations, the distillate being returned each time to the residue in the distillation vessel. It was believed formerly to result in a higher degree of purity.

cohosh An Indian (Algonkian) name for various medicinal plants; e.g.: **black ~** *Cimicifuga racemosa.* **blue ~** *Caulophyllum thalictroides.* **red ~** *Actaea spicata.* **white ~** *Actaea alba.*

cohydrol A colloidal solution of graphite. Cf. *Aquadag.*

coil A loop or spiral made of wire, tubing, glass, or other material. **induction ~** A transformer for inducing an electric current, consisting of a coarse wire c. (primary c.) wound around an iron core and surrounded by a long insulated c. of fine wire in which the induced current is produced. **primary ~** See *induction coil.* **resistance ~** A series of wire coils of known electrical resistance, used to reduce the strength of an electric current or to test the resistance of an object by comparison (bridge). **Ruhmkorff ~** An induction apparatus, in which the secondary c. is a very fine, long wire permanently mounted and connected to a condenser. **secondary ~** The outer coil of an induction c. **Tesla ~** An induction c. without the iron core; used for Tesla discharges. Used to detect minute gas leaks.

coinage metals See Table 17.

coir The fibers of the coconut husk, made by retting the husks in seawater and crushing between rollers; used for cables and cordage.

coke The carbonaceous residue (70–80%) of coal after the volatile constituents have been distilled off. **native ~** Carbonite.

c. oven A retort in which coal is converted to c. **c. oven gas** Fuel gas distilled from coal.

coking Making coke by heating coal for about 12 h.

Cola A genus of sterculiaceous plants; a source of caffeine. **c. nut** Kola.

colamine $NH_2 \cdot CH_2 \cdot CH_2 \cdot OH = 61.1$. 2-Aminoethanol*. Colorless liquid, b.171.

colatannin $C_{16}H_{20}O_8 = 340.3$. Colatin. The tannin from cola; crystals, m.148, soluble in alcohol.

colatin Colatannin.

colatorium A strainer or sieve.

Colcesa Trademark for a cellulose acetate synthetic fiber.

colchiceine $C_{21}H_{23}O_6N \cdot H_2O = 403.4$. An alkaloid from colchicum. Yellow needles, m.172, soluble in water (BP).

cochicine $C_{22}H_{25}O_6N = 399.45$. Colchiceine methyl ether. An alkaloid from the seeds of *Colchicum autumnale,* meadow saffron. Yellow crystals, m.135–150, soluble in water. It arrests mitosis of the dividing cell nucleus. Used in chromosome analysis and in animal and plant breeding. A limited analgesic for gout (USP, BP). **c. salicylate** $C_{22}H_{25}O_6N \cdot C_7H_6O_3 = 537.6$. Colchisal. Yellow powder, soluble in water. **c. tannate** $C_{22}H_{25}O_6N \cdot C_{14}H_{10}O_9 = 721.7$. Yellow powder containing 38–40% colchicine; soluble in alcohol.

Colchicum (1) A genus of liliaceous plants. (2) The plant *C. autumnale,* meadow saffron, autumn crocus, wild saffron (Europe and N. Africa); a specific for gout. **c. corm** The root or bulb of *C. autumnale,* meadow saffron, which contains colchicine. **c. flowers** The blossoms of *C. autumnale.* **c. seeds** The seeds of *C. autumnale,* containing less colchicine than the corm.

colchinine $C_{22}H_{15}O_6N = 389.4$. An alkaloid, m.146.

colchisal Colchicine salicylate.

Colcord Trademark for a cellulose acetate synthetic fiber.

colcothar Prussian red, rouge. Caput mortuum, crocus martis. Red ferric oxide obtained by heating ferrous sulfate in air; a polishing material for lenses, and a pigment.

cold Relatively low degree of heat; lacking warmth. **c. cream** A white, scented ointment cosmetic, consisting of wax, spermaceti, olive oil, or other fat emulsified with rose water (USP). **c. flow** The very slow flow of an apparently solid substance, e.g., pitch. **c. storage** See *refrigeration.*

cole Rape.

colemanite $Ca_2B_6O_{11} \cdot 5H_2O$. A native calcium borate (California); used as a heat insulator.

Coleoptera An order of insects which includes the beetles. Cf. *Cantharis.*

colibacterin A vaccine from *Bacillus coli.*

colicine Colicin. A bactericidal substance produced by certain strains of Enterobacteriaceae. C. resembles bacteriophages but, unlike them, does not multiply in the cells they kill.

coliform An adjective meaning "like *Bacterium coli* in morphology and staining reactions, but not necessarily in cultural and biochemical characteristics." Incorrectly used as a noun, or to define any particular group of organisms.

colistin sulfate The sulfate of a mixture of antimicrobial peptides, produced by a strain of *Bacillus polymyxa.* White, bitter powder, soluble in water; an antibiotic for infections caused by gram-negative bacteria (USP, BP).

coliston Mixed polypeptides produced by strains of *Bacillus polymyxa* var. *colistinus.*

colla Glue. **c. animalis** Gelatin. **c. glutinum** Gluten. **c. piscium** Ichthyol. **c. taurina** Gelatin.

TABLE 17. COINAGE METALS

U.S.: "silver"	Ag 90, Cu 10
others: before 1942	Cu 75, Ni 25
1942–1945	Cu 56, Ag 35, Mn 9
present: 1 cent	Cu 95, Zn 5
5 cents	Cu 75, Ni 25
$1, 50¢, 25¢, 10¢	Cu between Cu 75, Ni 25 cladding
U.K.: bronze	Cu 95.5, Sn 3, Zn 1.5
gold (sovereign)	Au 91.66, Cu 8.34
nickel-brass (threepenny piece)	Cu 79, Zn 20, Ni 1
"silver": before 1920	Ag 92.5, Cu 7.5
1920–1927	Ag 50, Cu or Ni 50
1927–1946	Ag 50, Cu 40, Ni 5, Zn 5
1946–present	Cu 75, Ni 25 (cupronickel)

TABLE 18. COLLOIDAL SYSTEMS

	In solid	In liquid (hydrosols)	In gas (aerosols)
Solid	*Solid sols* Alloys, colored glass, certain precious stones, paper	*Suspensions* Paints, milk of magnesia, collargol	*Smokes* Iodine vapor, cement dust, hydrochloric acid, ammonia
Liquid	*Gels* Celluloid, jellies, green leaves, glue	*Emulsions* Milk, blood, liniments, crankcase oil, protoplasm	*Fogs* Sprays, mists, clouds, visible steam
Gas	*Solid foams* Rubber, pumice, plaster, fire foam, lungs, adsorbed gases, aerogels	*Foams* Lather, froths, mayonnaise, whipped cream	No example

collagen Ossein. A hydroxyproline, glycine-type protein; the chief organic constituent of connective tissue and bones: C 50.75, H 6.47, N 17.86%; yields gelatin in boiling water.

collargol (1) Colloidal silver and silver oxide formed by reduction and stabilized by egg albumen. (2) A dialyzed and evaporated alkaline solution of colloidal silver (93% Ag).

collateral Side by side.

collector A substance used in flotation; increases the capacity of the air bubbles to carry mineral particles; as, dithiocarbonates.

colletin A glucoside from *Colletia spinosa* (Rhamnaceae), S. America.

collidine $C_8H_{11}N$ = 121.2. **alpha-~** 4-Ethyl-2-methylpyridine*. Colorless liquid, in coal tar, b.179, soluble in water. **beta-~** 3-Ethyl-4-methylpyridine*. Colorless liquid, b.198, insoluble in water. Obtained from the decomposition of cinchonine or coal tar. **gamma-~** 2,4,6-Trimethylpyridine*. Colorless liquid from coal tar, b.177, sparingly soluble in water; a catalyst. **hydro ~** A ptomaine in putrefying fish.

colligate To connect together.

colligative Connected. **c. properties** Properties related by a mathematical function. **c. p. of solutions** Those properties which depend quantitatively on concentration; e.g., boiling and freezing points, osmotic and vapor pressures.

collimator A lens system to produce parallel rays.

collinsics The relationship between molecular weight M, density ρ, and refractive index n (H. Collins): $M/\Sigma V_r = \rho$, and $(n - 1)(\Sigma V_r) = \Sigma R_0$, where ΣV_r is the sum of the relative volumes, and ΣR_0 is the sum of the optical refractivities. Cf. *Lorentz-Lorenz equation.*

collinsoniod The combined principles from the root of *Collinsonia canadensis*, stoneroot (Labiatae); an antispasmodic.

collision Interaction between material systems (molecule, atom, or electron), or electromagnetic induction, resulting in a change in molecular energy. Cf. *induction, chain reaction.*

collochemistry Colloidal chemistry.

collodion A solution of pyroxylin (gun cotton, nitrated cellulose) in a mixture of alcohol and ether. Colorless, opalescent, thick liquid containing not less than 5% dissolved matter. A reagent for differentiating phenol and creosote; a protective film for wounds, burns, or ulcers; and an airtight seal (USP, BP). **c. cotton** Pyroxylin.

collodium Collodion.

colloid A state of subdivision of matter which comprises either single large molecules (*molecular c.*; as, proteins) or aggregations of smaller molecules (*association c.*; as, gold). The particles of ultramicroscopic size *(dispersed phase)* are surrounded by different matter (*dispersion medium* or external phase); both phases may be solid, liquid, or gaseous. See Table 18, above. The size and electrical charge of the particles determine the properties of the c., e.g., Brownian movement, etc. Sizes range from 1×10^{-7} to 1×10^{-5} cm (or 0.1 to 0.001 μm). The smallest particle (colloidal gold) was 1.7×10^{-7} cm (Zsigmondy). **association ~** A compound whose molecules aggregate to form colloidal particles. **dispersion ~** A dispersoid or finely divided substance. **emulsion ~** Emulsoid. A liquid dispersoid, e.g., finely divided droplets of one liquid suspended in another. **eu ~** A c. whose molecules are over 2,500 Å long. **hemi ~** A c. whose molecules are 50–250 Å long. **heteropolar ~** A c. which consists of polar molecules, as, salts. Cf. *heteropolar.* **homopolar ~** A c. of nonpolar molecules, as, hydrocarbons. **hydrophilic ~** See *hydrophile.* **hydrophobic ~** See *hydrophobe.* **irreversible ~** A c. which, once coagulated, cannot be readily returned to the colloidal state. **lyophilic ~** See *lyophile.* **lyophobic ~** See *lyophobe.* **meso ~** A c. whose molecules are 250–2,500 Å long. **molecular ~** A compound whose molecule is of colloidal size, e.g., 100–500 nm long and 0.2–1 nm thick. **protective ~** A substance which promotes stability of the colloidal state by enveloping the particles. **reversible ~** A c. which, when coagulated, can readily be converted to the colloidal state. **suspension ~** Suspensoid. A solid dispersoid, e.g., finely divided solid particles of ultramicroscopic size in a liquid.

c. equivalent The number of atoms sharing a free electric charge. **c. mill** A grinding mill for making emulsions and suspensions, or for the disintegration and dispersion of solids or liquids. **c. zone** See *orientation, zone.*

colloidal Pertaining to colloids. **c. metal** Finely divided metal particles of 1 to 100 nm. They expose a large surface and are very reactive, e.g., as industrial catalysts. **c. movement** *Brownian motion.* **c. systems** See *colloid* and Table 18.

collose An intermediate stage in liquefaction of woody tissues.

collotype Artotype. A method of printing from a gelatin surface on a glass plate sensitized with potassium dichromate and exposed under a negative.

colloxylin Collodion.

Colnova Trademark for a cellulose acetate synthetic fiber.

colocynth Bitter apple. Bitter cups. The fruit of *Citrullus colocynthis* (Cucurbitaceae), Asia Minor; a cathartic.

colocynthin $C_{56}H_{84}O_{23}$ = 1125. A glucoside from the fruit of *Citrullus colocynthis;* a strong purgative.

colog Abbreviation for cologarithm. Colog $x = -\log x = \log 1/x$.

TABLE 19. WAVELENGTH RANGE OF COLORS

Color	Wavelength, in nm
Violet	390–455
Blue	455–492
Green	492–577
Yellow	577–597
Orange	597–622
Red	622–770

Colomal Trademark for a cellulose acetate synthetic fiber.
colombic acid Calumbic acid.
colonial spirit Methanol*.
colonies Clusters of bacteria visible to the eye.
colophene $C_{20}H_{32}$ = 272.5. A turpentine hydrocarbon; a colorless liquid.
colopholic acid Colophonic acid.
colophonic acid Colopholic acid. An acid derived from turpentine; used in plasters, soaps, and cements. Cf. *abietene.*
colophonite (1) An amber-colored andralite (iron *garnet* q.v.). (2) A soil colophony. Cf. *copals.*
colophonium Rosin.
colophony BP name for rosin.
color, colour (1) The visual sensation caused by light. (2) Light of a definite wavelength or group of wavelengths which is emitted, reflected, refracted, or transmitted by an object. A c. is defined by three properties; *hue,* the wavelength of the monochromatic light, i.e., shade; *saturation,* the percentage of the light of the above wavelength present, i.e., strength; *brightness,* the amount of light reflected as compared with a standard under the same conditions, i.e., luminous flux. Cf. *C.I.E. units, dye.* **artists' ~** The finely ground pigments used by artists. **complementary ~** Any two colors of the spectrum which produce white light when blended as rays of colored light; e.g., red and green, blue and orange. **compound ~** A c. produced by mixing 2 or more primary pigment colors; as, orange, from yellow and red. **contrast ~** Complementary. **primary ~** (1) Simple colors, as red, green, and blue, combinations of which approximately produce all others possible. (2) For pigments: red, yellow and blue. See Table 19. **secondary ~** Compound c.
c. diagram A chart to illustrate the relationship of the principal colors. Colors opposite each other are complementary; if subtracted, they produce darker shades and finally black; if combined (as with beams of colored light), they produce lighter tints and finally white. See Fig. 8. **c. filter** Light filter. A solid, liquid, or gaseous layer which absorbs certain wavelengths. **C. Index** Colour Index. A numbered list of synthetic dyestuffs and inorganic pigments compiled jointly by The Society of Dyers and Colourists (Bradford, England) and the American Association of Textile Chemists and Colorists (Research Triangle Park, N.C.). It gives the scientific and commercial names, components, formulas, methods of preparation, discoverers, literature references, and descriptions of properties and methods of application. **c. photography** Photographic recording of the form and c. of an object. Negatives from exposures through differently colored screens are printed on films stained in the complementary colors. Superimposing the prints gives a transparency in natural colors. In practical c. film, 3 differently sensitized emulsions are superimposed on the same support, each producing on exposure an appropriate image in a different c. The combined effect is a c. transparency. **c. reaction** A chemical change in which a change of c. occurs. **c. scale** (1) See *c. diagram* (above), *spectrum.* (2) C. as an indication of temperature; e.g.:

Incipient red heat	500–600°C
Dark red heat	600–800°C
Bright red heat	800–1000°C
Yellowish red heat	1000–1200°C
Incipient white heat	1200–1400°C
White heat	1400–1600°C

c. screen law The photographic intensity of a c. is increased by a screen of the complementary c. and decreased by a screen of the same c. **c. theory** (1) C. of solutions of certain compounds may be due to: (*a*) *solvation,* q.v., the dehydrated compound is colorless; (*b*) association, the c. changes on dilution; (*c*) ionization, the un-ionized molecule is differently colored; (*d*) tautomerism. (2) The emission of a certain wavelength of light is due to electrons in chromophores which change their energy level.
coloradoite HgTe. A telluride from Colorado.
Coloray Trademark for a cellulose acetate synthetic fiber.
colorimeter A device to measure the intensity and shade of colored solutions or to compare standard color solutions (comparator) in quantitative analysis. Cf. *comparator, tintometer.*
colorimetric analysis The quantitative analysis of a substance by comparing the intensity of the color produced by a reagent with a standard color, e.g., produced similarly in a solution of known strength. Cf. *Nessler's test.*
colorimetry (1) Colorimetric analysis. (2) Color measurement. Cf. *calorimetry.* **adsorption ~** See *adsorption colorimetry.* **kinetic ~** C. based on the rate of development (as distinct from depth) of color in standard and sample.
coloring **c. matter** A substance which produces the sensation of color. See *color, dyes, pigment.* **c. tablet** A soluble tablet

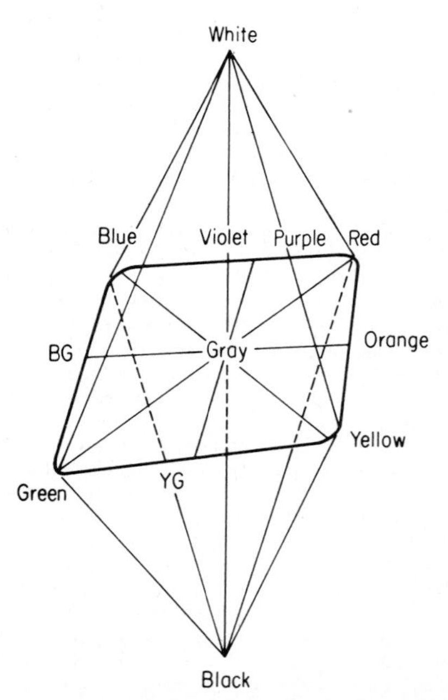

Fig. 8. Color diagram.

containing a definite amount of a pure color; to color pharmaceutical preparations.

colors *Pigments* or *dyes*, q.v. To indicate the shade of a color or dye, letters are sometimes used; thus:

G = gelb (German) = yellow
J = jaune (French) = yellow
JJ = intense yellow
O = orange
OOO = very intense orange
R = reddish tint
RR = very red
S = black or sulfonated

colostrum (1) The first secretion of the mammary glands; a thin milk. (2) Orange-colored fat, m.28–38, similar to lard, obtained from top milk by the action of 3–4 times its volume of hydrochloric acid, d.1.12, at 92°C, for 90 min.

columba Calumba.

columbate Niobate*.

columbian spirits Methanol*.

columbic Niobic.

columbin $C_{20}H_{22}O_6$ = 358.4. A glucoside from the root of *Jatrorrhiza calumba.* Yellow crystals, m.195, sparingly soluble in water; a bitter.

columbite Niobite.

columbium Cb. An element, at. no. 41. In 1949 it was officially renamed *niobium,* Nb, q.v.

columbo Calumba. **American ~** Fraserin.

columbous Niobous.

columboxy Nioboxy.

columbyl Nioboxy.

column A pillar. **c. still** A still for fractional distillation.

colza Rapeseed oil.

comanic acid Comenic acid.

combination (1) The union of 2 or more substances to form a new substance. (2) A chemical reaction in which two elements combine and form a binary compound; or two binary compounds combine and form a complex compound. **molecular ~** The union or aggregation of molecules by chemical or coordinate *bonds,* q.v.; as: *equilibrium* c. of 2 molecules of odd molecular number; *irreversible* c. or *polymerization,* q.v.; *reversible* c. or *association,* q.v. Cf. *liquid, bond.*

c. principle The *frequency,* q.v., of a radiation may be represented by the difference between two terms.

combining Uniting or joining together. **c. weight** Equivalent weight.

combustible Flammable. **c. gases** The fuel gases of industry; e.g., hydrogen, hydrocarbons, carbon monoxide, and their mixtures.

combustion Burning; a chemical change accompanied by the liberation of heat, sound, and/or light. **fractional ~** See *fractional combustion.* **heat of ~** See *heat of combustion* under *heat.* **organic ~** The decomposition of organic compounds, and collection of the products for *elementary analysis,* q.v. **slow ~** (1) The slow oxidation of a substance. (2) The combination of two gases by means of a heated wire. **spontaneous ~** (1) Accelerated c. effecting a sudden outburst of heat and light (explosion). (2) Self-ignition. **wet ~** Oxidation by strong oxidizing agents, as, sulfuric acid–dichromate mixture.

c. boat A small tray of refractory material for burning organic compounds in a c. tube. **c. capsule** A small porcelain crucible used in coal analysis. **c. furnace** A heating appliance used in the elementary analysis of organic compounds, consisting of a row of burners or electric heating units, under a gutter in which are placed the c. tube, with means for regulating the temperature. **c. train** The arrangement of apparatus for elementary organic analysis: usually hydrogen, oxygen, and air tanks; potash bulb, calcium chloride tube, c. tube and furnace, potash bulb, and calcium chloride or sulfuric acid tube. **c. tube** A wide glass, silica, or porcelain tube, resistant to high temperatures. **c. tube furnace** (1) C. furnace. (2) An electrically heated, refractory-lined, hollow spool, used for c. work.

comenic acid $C_6H_4O_5$ = 156.1. 5-Hydroxy-4-pyrone-2-carboxylic acid. Yellow crystals, m.270 (decomp.).

comfrey A perennial plant, *Symphytum officinale.* In the Middle Ages, the powder from its dried roots was used for fracture splints; contains allantoin, a former wound dressing. Also a vegetable and green manure.

commensal Living in close association with man; as, bacteria on the skin and in the intestine.

comminute To bring to a state of fine division.

comminution The process of grinding or breaking into small fragments, e.g., cutting, rasping, slicing, levigating, pulverizing, triturating.

common salt Sodium chloride.

commutator A device to interrupt or reverse an electric current. **dynamo ~** A transformer of alternating into direct current, attached to a dynamo.

comolised Highly aggregated.

comosic acid An acid substance derived from the bulbs of *Muscari comosum* (Liliaceae).

comparator (1) An electrical device to calibrate a.c. instruments. (2) See *colorimeter.*

comparison spectroscope A spectroscope in which 2 spectra can be produced simultaneously side by side for comparison.

comparison tubes A set of test tubes of the same glass and dimensions, used in colorimeters to compare colors.

comparoscope A device attached to a microscope for comparing separate slides simultaneously.

compatibility The retention of the respective properties of 2 or more drugs when administered together. **in ~** The impairing influence of 2 or more substances on one another when mixed. Cf. *incompatible.*

compatible Describing 2 drugs that do not interfere or react chemically, physically, or therapeutically with each other.

compensation apparatus Potentiometer.

complement (1) To supply a missing part. (2) Factors present in blood serum that enable antibody-antigen reactions to take place. C. is fixed or used up by the reaction, and is destroyed by heat. **c.-fixation test** The basis of antibody-antigen reactions performed in the laboratory to diagnose disease. If c. is already used up by patient's serum, it is not available to allow antibody-antigen reaction (as, hemolysis of red cells) to take place; e.g., Wassermann reaction.

complementary color See *color.* **c. c. screens** (1) Copper ruby glass (red) and copper glass (blue); (2) uranium glass (yellow) or chromium glass (greenish yellow) and cobalt glass (blue).

complex Complicated; not simple. **normal ~** A c. ion or compound which in solution dissociates reversibly into its component parts, e.g., $Cd(CN)_4^{2-}$. **penetration ~** A c. ion or compound which is sufficiently stable to retain its identity in solution, e.g., $Fe(CN)_6^{2-}$. **Π- ~** Π-adduct. Metal-unsaturated hydrocarbon complex, in which the metal atom is bonded to 2 or more contiguous atoms of the ligand; as, ferrocene.

c. acid A combination of 2 or more nonmetallic compounds or acid radicals, e.g., 2HF and SiF_4 give H_2SiF_6. **c. compound** A combinations of 2 or more compounds or ions, e.g., 4KCN and $Fe(CN)_2$ give $K_4Fe(CN)_6$. **c. ion** A c. charged radical or group of atoms, e.g., $Cu(NH_3)_2^{2+}$. **c. reaction** A chemical change in which 2 or more reactions occur simultaneously.

complexan Complexone.

compleximetry, -ometry Titration with, or of, a substance forming a slightly dissociated soluble complex.

complexing agent Chelating agent.

complexometric titration A method of volumetric analysis in which the formation of a colored complexone is used to indicate the end point; e.g., the use of *EDTA*, q.v., to determine the hardness of water.

Complexon Trade name for certain compounds of ethylenediaminetetraacetic acid. **C.-I** Trade name for trimethylaminotricarboxylic acid, a complexone.

complexone (1) The sodium salt of ethylenediaminetetra-acetic acid (EDTA). Used in *complexometric titration*, q.v. (German: *komplexon*.) (2) Complexan. In general, one of a group of polyaminopolycarboxylic acids which form anionic complexes.

component (1) An ingredient or part of a mixture (as distinct from the *constituent* of a compound). (2) The smallest number of chemical substances capable of forming all the *constituents*, q.v., of a system in whatever proportions they may be present. See *phase rule*.

Compositae Composite family. A group of herbaceous or woody plants, rarely shrubs, with many florets arranged in compact heads as a compound flower on a common receptacle. Some yield drugs; e.g., roots: *Taraxacum officinale*, dandelion; *Cichorium intybus*, chicory; *Arnica montana*, arnica root; leaves: *Ambrosia artemisiaefolia*, ragweed; herbs: *Tanacetum vulgare*, tansy; *Artemisia absinthium*, wormwood; *Senecio aureus*, ragwort; flowers: *Anthemis nobilis*, English chamomile; *Carthamus tinctorius*, safflower; seeds: *Helianthus annuus*, sunflower seeds; fruits: *Arctium lappa*, burdock fruit; juice: *Lactuca virosa*, lactucarium.

composition The elements or compounds forming a material or produced from it by analysis. Cf. *constitution*.

compost A mixed organic manure produced by the natural decomposition of waste organic matter, by means of mesophilic and thermophilic organisms.

compound (1) A substance whose molecules consist of unlike elements and whose constituents cannot be separated by physical means. A c. differs from a physical mixture by reason of the definite proportions of its constituent elements which depend on their atomic weights, by the disappearance of the properties of the constituent elements, and by entirely new properties characteristic of the c. formed. (2) To mix drugs or make up a prescription. **acyclic ~** An organic c. in which the C atoms are arranged in an open chain. **addition ~** A c. formed by the union of 2 simpler or binary compounds: $NH_3 + HCl = NH_4Cl$. **additive ~** A c. formed by the saturation of double or triple bonds. **aliphatic ~** An acyclic c. **amino ~** An organic c. containing the radical NH_2-. **aromatic ~** An organic c. in which the C atoms form one or more unsaturated rings; e.g., benzene series. **asymmetric ~** See *asymmetric compound*. **azo ~** See *azo-*. **binary ~** A c. whose molecules consist of only 2 kinds of atoms. **carbon ~** See *organic c.* **chain ~** Acyclic c. **closed chain ~** Aromatic c. **coal tar ~** A c. derived from coal tar, especially aromatic organic c. **condensation ~** A c. formed by the union of 2 or more molecules, especially organic compounds, in which one or more molecules of water are usually liberated. **coupling ~** A c. which substitutes another atom, generally hydrogen. **cyclic ~** Aromatic c. **diazo ~** See *diazo compound*. **endothermic ~** A c. formed with the absorption of heat; they are usually explosive. **exothermic ~** A c. formed with the liberation of heat; they are usually stable. **fatty ~** Aliphatic c. **homocyclic ~** An aromatic c. with only C atoms in its ring, e.g., benzene. **heterocyclic ~** An aromatic c. with atoms other than C atoms in its ring, e.g., pyridine. **index ~** A c. from which another is derived by substitution; thus methane is the i. c. for chloromethane. **inorganic ~** A c. containing no C atoms. The majority are polar. **metameric ~** Metamer. **molecular ~** (1) Addition c. (2) See *combination, association, polymerization*. **nonpolar ~** A relatively unstable c. of elements having weak electrical forces, e.g., carbon compounds. **open-chain ~** Aliphatic c. **organic ~** A c. containing C and H atoms, with or without other atoms in the molecule. The majority are nonpolar (except metallic salts of organic acids). **organometallic ~** See *organometallic compound*. **paraffin ~** A saturated aliphatic hydrocarbon. **parent ~** See *parent compound*. **polar ~** A fairly stable c. of elements of strong electromotive force, e.g., NaCl. **quaternary ~** Ammonium*. A fully substituted ammonium radical with a halogen, e.g. $N(C_2H_5)_4 \cdot I$. **ring ~** See *aromatic c.* (above). **series of c.'s** C.s containing the same positive or negative radical; e.g.: ferrous compounds, ferric compounds, ferrates; sulfides, sulfites, sulfates. **saturated ~** A c. in which all the C bonds are satisfied. **spiro ~** An organic c. in which one C atom is common to 2 rings. **substitution ~** A c. formed by the replacement of certain atoms by other atoms or groups of atoms. Cf. *index c.* (above). **sulfur ~** A c. containing sulfur, especially organic compounds containing divalent sulfur. **tertiary ~** Ternary. A c. with 4 atoms or groups, that may or may not be identical; as, Me_3N. **unsaturated ~** A carbon c. with double or triple bond(s).

c. ethers Esters. **c. microscope** See *compound microscope* under *microscope*. **c. molecule** A molecule consisting of more than one kind of atom. **c. protein** See *conjugated proteins*.

compounds, classes of **nonpolar** Do not ionize or conduct an electric current; are relatively inert; do not associate, but may polymerize; have a frame structure; exhibit isomerism; have a low relative permittivity; are generally known as organic substances. **polar** Ionize and conduct an electric current; are very reactive; form double molecules and complex ions; possess a condensed structure; exhibit tautomerism; have a high relative permittivity; are generally known as inorganic substances.

compregnation The combined application of impregnation and high pressure; e.g., in the lamination of wood veneers with plastics.

compressed gas A g. stored at high pressure in a steel cylinder for shipment; as, carbon dioxide. Fuel gases are used in cylinders under various trade names, e.g., Calor Gas.

compressibility The resistance offered by substances to high pressure. For gases: κ, the rate of change of volume with pressure, per unit of volume. Cf. *piezometer*.

Compton **C., Arthur Holly (1892–1962)** American physicist and Nobel prize winner (1927). **C., Karl Taylor (1887–1954)** American physicist. **C. effect** Free electrons scatter radiation, as X-rays, resulting in an increase in wavelength. Cf. *Raman effect*. **C. rule** Atomic weight multiplied by specific latent heat of fusion equals twice the thermodynamic melting point.

computer A device used for data handling, calculation, and control of other devices. **analog ~** A c. which represents the magnitude of variables by a continuous physical quantity such as voltage. **dedicated ~** A c. designed for a specific and limited purpose. **digital ~** A c. that uses discrete numbers to represent the magnitude of variables and other information. **host ~** A base c. used to receive information from satellite computers. **process control ~** A c. that receives readings from recording instruments and converts them into means (usually electric impulses) of operating control devices. **super ~** A c. of enormous memory and computing capability.

conalbumin A noncrystalline albumin of egg white.

conamarin A glucoside from the root of hemlock, *Conium maculatum* (Umbelliferae). Cf. *conine.*

conarachin A globulin in peanuts (8.7%), consisting principally of arginine, lysine, and cysteine.

conc., concn. Abbreviations for concentrated and concentration.

concanavalin A globulin from jack beans.

concave Presenting a depressed or hollow surface. **c. lens** See *concave lens* under *lens.* **c. mirror** See *concave mirror* under *mirror.*

concentrate, concentration (1) Chemistry: (*a*) Increasing the amount of a dissolved substance by evaporation of the solvent or by adding more of the substance. (*b*) The strength of a solution, expressed as:

Mass c., ρ, in g/liter
(Amount-of)-substance c., M, in mol/liter
Molality, m, in mol/kg of solvent
Normality, N, in (gram) equivalents/liter

(2) Physics: Gathering together that which is diffused, e.g., light, sound, or heat. **absolute ~** The c. of the active ion of a substance; e.g., the H^+ of an acid. **hydrogen-ion ~** See *hydrogen-ion concentration* under *hydrogen ion.* **ionic ~** The number of gram ions in a unit volume of solution. **molar ~, molarity ~** M. (Amount of) substance c*. **normal ~, normality ~** N. The number of equivalent moles per liter.

c. limit The minimum c. of a substance detectable by a particular method of determination. Defined as the signal equal to 3 times the noise (i.e., standard deviation of several measurements of the blank).

concentrator A mechanical device to increase concentration. Cf. *separator.*

concentric Describing a number of rings with a common center.

concha A shell.

conche Shell-like pots with granite bottoms and heavy rolls, used in chocolate manufacture.

conching The rolling of chocolate mixes, to develop flavor.

conchinine Quinidine.

conchiolin $C_{30}H_{48}O_{11}N_9$ = 710.8. A protein from mollusk shells, resembling keratin. It is the fine organic network matrix of natural pearl in which calcium carbonate crystals are distributed.

Concordin Trademark for protriptyline hydrochloride.

concrete (1) A mixture of stone (either natural gravel or crushed rock) and sand, collectively called aggregate, held together by *cement,* q.v. (2) A concentrated or clotted extract of a plant. **aerated ~** C. containing an air-entraining agent to improve its insulating power. **fiber-reinforced ~** C. whose tensile strength is raised by the addition of fibers, such as steel, glass, or polypropylene, to the mix. **poststressed ~** C. in which tension is induced in the reinforcement after hardening and transferred externally by anchorages into the structure. **prestressed ~** C. in which cracking is prevented by applying pressure to the structural member before it is loaded to neutralize the tension caused by the working load. **reinforced ~** C. formed around iron bars, rods, or wire. **spray-applied ~** Shotcrete, Gunite.

condensate In the petroleum industry, hydrocarbons that are gaseous in the subterranean reservoir and liquefy when produced.

condensation (1) Conversion to a more compact form. (2) Transformation from the gaseous to the liquid state. (3) A combination of similar molecules to form a more complex compound (polymerization, etc.). (4) A union of like or unlike molecules, usually with elimination of molecules of (e.g.) water, acid. (5) An accumulation of electrons, e.g., in a condenser. (6) The formation of a pencil of parallel or convergent light rays from divergent rays, e.g., by a convex lens. **retrograde ~** The deposition of hydrocarbon gases as liquids, from the high-pressure state, when pressure is appreciably reduced.

condensed (1) Reduced in volume. (2) Liquefied from a gas or vapor. **c. nucleus** See *condensed nucleus* under *nucleus.*

condenser (1) A device to concentrate matter or energy. (2) An apparatus for cooling vapors to liquids, e.g., Liebig ~. (3) A device for the polymerization of organic compounds. (4) A series of insulated conductors for the accumulation of electricity. (5) A system of lenses or mirrors, e.g., *Abbe condenser,* q.v. **bull's-eye ~** A convex lens c. **electrical ~** Sheets of tin foil separated by paraffin, with the alternate tin foils connected to a common terminal, for storing electrons. **high potential ~** Alternating brass sheets and glass plates in an oil-filled box. **Liebig ~** Generally 2 straight, concentric glass tubes with cold water passing through the outer, and the vapor to be condensed through the inner tube. **paraboloid ~** See *parabaloid condenser.* **reflux ~** See *reflux condenser.* **substage ~** Abbe condenser.

c. flask A large, spherical flask filled with liquid, to absorb the heat from a source of light and to concentrate its intensity. **c. tube** The inner tube of a c.

condiment A spicy or stimulating vegetable product (and common salt), used for its flavor.

condistillation The separation of organic substances by distilling with another liquid. Cf. *steam distillation.*

conductance The capacity to convey energy, as heat or electricity. **electric ~*** The reciprocal of resistance. See *siemens.* **molar ~*** Λ. The electrolytic conductivity of an electrolyte divided by its substance concentration; measured in Sm^2/mol. For an ion, its m. c. (λ_i) equals the product of its transport number and the m. c. **specific ~** Electrolytic *conductivity**. **thermal ~** See *thermal conductance.*

conduction The property of transmitting matter or energy. **air ~** C. by a current of air. **electrolytic ~** The passage of electrons by means of ions. **sound ~** The passage of sound waves in air. **thermal ~** The transmission of heat by materials.

conductivity The degree of transmitting electricity, heat, light, or sound. **electrical ~** γ, σ. The reciprocal of *resistivity,* q.v., and the analog of thermal c. Thus the current density created per unit field strength; measured in S/m. **electrolytic ~*** κ. Specific conductance. The analog of electrical c. for an electrolyte. Thus the conductance of a unit cube of electrolyte; measured in S/m. 1 mho/cm = 100 S/m. **equivalent ~** The ratio of the electrolytic c. to the number of gram equivalents of electrolyte per unit volume of solution. Cf. degree of *dissociation.* **heat ~** Thermal conductivity*. **limiting ~** The c. of a solution at infinite dilution; i.e., when dissociation is complete. **super ~** See *superconductivity.* **supra ~** Superconductivity. **thermal ~*** See *thermal conductivity.* **unilateral ~** See *valve effect.*

c. apparatus A device to determine the c. of solutions. **c. cell** A glass vessel containing 2 electrodes at a definite distance apart, and filled with a solution whose c. is to be

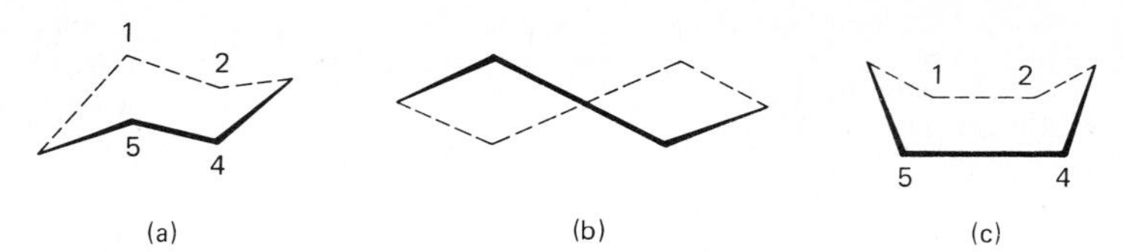

Fig. 9. Types of conformation: (*a*) chair, (*b*) twist, (*c*) boat.

determined. **c. bridge** A modified Kelvin bridge for measuring very low resistance, on which the c. is read directly from the slide wire. **c. water** A very pure water, of c. 4 × 10^{-5} S/m in a vacuum at 18°C. Cf. *water.*

conductometric Pertaining to conductivity measurements. **c. analysis** Analysis in which the discontinuity in the progressive change in conductivity determines the end point of a reaction. Cf. *potentiometric titration.*

conductor A medium which conducts. **electrical ~** Any metal, solution of salts, molten substance, or ionized gas which transmits electrons, and allows an electric current to pass. **non ~** See *insulator.*

condurangin A glucoside from condurango bark. **alpha- ~** $C_{20}H_{32}O_6$ = 368.5. White powder, m.60, insoluble in water. **beta- ~** $C_{18}H_{28}O_7$ = 356.4. Amorphous, yellow powder, sparingly soluble in water; an aid to digestion.

condurango Candurangu, condorvine. The vine, *Gonolobus condurango* (Asclepiadaceae), Colombia. **c. bark** A bitter and aid to digestion.

Condy's fluid A liquid disinfectant claimed to produce ozone by oxidation.

cone A body having a circular base and tapering point. **filtering ~** A hollow c. of porous material, e.g., alundum, which fits into a support, for rapid filtration of solutions. **measuring ~** A slightly tapered ruler to measure inside diameters. **pyrometer ~** *Seger cones.*, q.v.

conedendrin $C_{20}H_{20}O_6$ = 356.4. A constituent of many coniferous trees, m.255, insoluble in water.

conessine $C_{24}H_{40}N_2$ = 356.6. Wrightine. An alkaloid from the bark of *Wrightia zeylanica* (Apocynaceae), Sri Lanka. Colorless, deliquescent needles, m.121, slightly soluble in water; an astringent and antidiarrheal agent.

confection Pharmaceutical term for a soft paste compounded with a sweet mucilaginous liquid.

configuration The spatial arrangement of elements (as, atoms in a molecule); or of parts, in a particular form of figure. Classically in molecular c., rotation about single bonds was ignored. Now usually rotation about double, and sometimes any, bonds is ignored. Cf. *conformation.* **absolute* ~** See *absolute stereochemistry* under *stereochemistry.* **electronic ~** See *electron c.* **relative* ~** See *relative stereochemistry* under *stereochemistry.*

conformation Classically, the conformations of a molecule of defined *configuration,* q.v., are the various arrangements of its atoms in space that differ only after rotation about single bonds. Now usually extended to include double bonds and sometimes all bonds. Thus

have the same configuration but different conformations. Examples of conformations: (1) Chair* and boat*. When positions 1,2,4,5 of a 6-membered saturated ring compound lie in one plane. (2) Twist*. A median position during conversion between boat and chair, or between the two boat positions. See Fig. 9.

congealing (1) Freezing. (2) The setting of cement or solidification of a liquid. **c. point** The temperature at which a liquid freezes.

congelation Freezing, solidification. The transformation from the liquid to the solid state, by low temperature, hydration, or chemical reaction.

conglomerate An aggregation or a mass of units.

conglutin A protein from the seeds of leguminous plants.

conglutinin A protein of beef serum which clumps blood cells. See *agglutinins.*

congo **c. blue** Trypan blue. **c. paper** A filter paper stained with congo red solution. **c. red** $C_{32}H_{22}O_6N_6S_2Na_2$ = 696.7. C.I. Direct Red 28. Disodium salt of 3,3′-[(1,1′-biphenyl)-4,4′-diylbis(azo)]bis[4-amino-1-naphthalenesulfonic acid]†. A red azo dye; used as indicator (alkalies—red, acids—blue), and in detection of free mineral acids in organic acids. **c. solution** A solution of 1 g c. red in 90 ml water and 10 ml alcohol. **c. yellow** $C_{24}H_{18}O_4N_5SNa$. An orange-yellow dye.

conhydrine $C_8H_{17}ON$ = 143.2. 2-(1-Hydroxypropyl)piperidine, (hydr)oxyconine. **(+)- ~** An alkaloid from the seeds of *Conium maculatum,* poison hemlock. Colorless crystals, m.118, soluble in water. An analgesic and antispasmodic, but use limited by toxicity. **(±)- ~** Colorless crystals, m.98, soluble in water.

conic acid Coniic acid.

coniceine $C_8H_{15}N$ = 125.2. **alpha- ~** Colorless liquid, d.0.893, b.158. **beta- ~** 2-Allylpiperidine. White needles, d.0.852, m.40. **gamma- ~** 1,2,3,4-Tetrahydro-6-propylpyridine. Colorless liquid, d.0.872, b.172. **delta- ~** Piperolidine. **epsilon- ~** Methylconidine. Colorless liquid, d.0.8856, b.151.

conicic acid Coniic acid.

conicine Conine.

Coniferae Pinales; pine family. Trees or shrubs, usually with a resinous juice and awl- or needle-shaped leaves, which bear cones. It includes the old families Pinaceae and Taxaceae.

coniferaldehyde Ferulaldehyde.

coniferin $C_{16}H_{22}O_8 \cdot 2H_2O$ = 378.4. 4-*O*-*β*-D-glucosyl of coniferol. A glucoside from the cambium of coniferous trees and asparagus. Gray powder, m.185, soluble in hot water; used to manufacture vanillin.

coniferol $C_{10}H_{12}O_3$ = 180.2. 4-Hydroxy-3-methoxycinnamyl alcohol*, coniferyl alcohol. Colorless crystals, m.73, slightly soluble in water. It oxidizes to vanillin. Cf. *lignin.*

coniferyl The 4-hydroxy-3-methoxycinnamyl* radical HO·(MeO)C_6H_3CH:CH·CH_2—. **c. alcohol** Coniferol. **c. aldehyde** Ferulaldehyde. **c. benzoate** $C_{17}H_{16}O_4$. The chief constituent of Siamese gum benzoin. Colorless needles, m.73.

coniic acid Conicic acid. An acid constituent of conium.
coniine Conine.
conine $C_8H_{17}N$ = 127.2. Cicutine, 2-propylpiperidine, coniine. **(*S*)-~** An alkaloid from the seeds of *Conium maculatum,* poison hemlock (Umbelliferae). Yellow oil, d.0.862, b.166, insoluble in water; narcotic. **hydroxy ~** Conhydrine. **para ~** See *paraconine.*
c. hydrobromide $C_8H_{17}N \cdot HBr$ = 208.1. White crystals, m.210, soluble in water; an antispasmodic and antineuralgic.
c. hydrochloride $C_8H_{17}N \cdot HCl$ = 163.7. Colorless rhombs, m.211, soluble in water.
conium The seeds of *Conium maculatum,* poison hemlock (Umbelliferae); a narcotic and sedative. Cf. *conine, conamarin.*
conjugated Paired or coupled. **c. acids and bases** Chemical species that can be transformed into their respective base or acid by prototropic reaction. **c. double bonds*** Two double bonds in the relative positions indicated by the formula $-CH=CH-CH=CH-$. They form additive compounds by saturation of the 1 and 4 carbons, so that a double bond is produced between the 2 and 3 carbons. **c. foci** The anterior and posterior foci of a lens. **c. proteins** Polypeptides or proteins which contain a prosthetic group; e.g., hemoglobin which contains hematin.
conjugation (1) The linking together of centers of unsaturation. (2) The combination of large molecules, (e.g., proteins) with another compound.
conjunctive nomenclature* See *nomenclature,* Table 56.
conophor oil A substitute for linseed oil from a W. African weed.
Conradson test A destructive distillation method for the estimation of carbon residues in fuel and lubricating oils.
consecutive Following in an uninterrupted sequence or order. **c. position** Adjacent position.
conservation Protection from loss, decay, or deterioration. **c. of energy** The principle that energy is never created or destroyed but is always transformed into some other form of energy. Cf. *mass*-energy cycle. **c. of matter** The principle that matter is never created or destroyed but is always transformed into some other form of matter, e.g., the radioactive elements.
conservative Preservative. **c. motion** In metallurgy, the movement of a dislocation on the slip plane as opposed to nonconservative motion. **nonconservative motion** In metallurgy, when a dislocation moves out of the slip plane.
consistence The viscosity or solidity of a fluid.
consistency (1) The degree of solidity or fluidity. Cf. *dissipation, penetration.* (2) The percentage of solid matter in a mixture, e.g., of pulp. **c. meter** A device to determine the solidity of a semisolid or syrupy substance, e.g., gelatin; usually a disk rotating in the substance, driven by a constant force.
consistometer An apparatus to determine the hardness or consistency of semiliquids and hard, brittle, bituminous materials. Usually a plunger dropping at constant force onto the material to be tested. The depth of penetration is measured.
consolute (1) A liquid or solution which is completely miscible with another. (2) Miscible in all proportions.
const. Abbreviation for *constant.*
constant (1) That which is unchangeable, permanent, or invariable. (2) Physics: a property which remains numerically the same. (3) Mathematics: a quantity having a definite and fixed value. **basic ~** The numerical value of a permanent property which can be determined directly. **conventional ~** The numerical value of a physical property generally accepted as normal. **derived ~** The numerical value of an unchanging physical property which is indirectly determined. **experimental ~** The numerical value of a physical property determined by experiment and accepted as standard. See Table 20 on p. 152.
c. proportions Definite proportions. A fundamental law of chemistry: Every chemical compound always contains the same percentage by weight of its constituent elements (Proust).
constantan Constantin.
constantin Constantan. A thermocouple alloy: Cu 60, Ni 40%, d.8.4, m.1290.
constituent (1) An element or part of a compound. Cf. *component* of a mixture. (2) An element or compound formed from the *components,* q.v., of a system. Thus, in a system $CaCO_3 = CaO + CO_2$ there are 3 *constituents* ($CaCO_3$, CaO, and CO_2), but only 2 *components,* as any 2 will determine the amount of the third.
constitution The structure in which elements are arranged in a material. Cf. *configuration.*
constitutional formula See *constitutional formula* under *formula.*
contact (1) The touching of bodies. (2) An electric switch. **c. action** Catalysis. **c. difference, c. potential** The difference in potential of two metallic plates in close c. **c. process** The catalytic manufacture of sulfuric acid from sulfur dioxide and oxygen. **c. series** Electromotive series. **c. substance** Catalyst.
contamination Infection by accidental contact. Cf. *infection.*
Contergan Trademark for *thalidomide,* q.v.
contiguous Adjoining, as, the atoms $C-C$.
contoured Having an irregular, smooth, undulating surface, as, a relief map; used to describe bacterial cultures.
contraceptive **intrauterine ~** IUD. A device, inserted into the uterus; usually plastic, sometimes containing a copper thread or progestogen. **oral ~** A hormone used to prevent conception. (1) The pill. Combination of small doses of an estrogen, such as ethinyl estradiol or mestranol, and a progestogen, as norethisterone. Acts by inhibiting production of gonadotrophins and thus ovulation. (2) A progestogen used alone, usually in small dose taken by mouth (the minipill), as, norethisterone, or larger dose given by injection at longer intervals. Action thought to be due to changes in cervical mucus and endometrium (uterine lining). **spermicidal ~** Compound which inhibits or destroys spermatozoa. Usually used as cream; e.g., phenyl mercuric nitrate.
contractile Capable of being drawn together or shortened.
contraction (1) Chemistry: The drawing together of atoms in the molecule, depending partly upon their mutual electromotive forces = $100 - (100M/A' - A'' \ldots)$, where M is the molecular volume, A' A'', etc., the atomic volumes of the constituents. (2) Physiology: The shortening of muscles in response to a stimulus.
contrast A comparison made by accentuating points of difference. **c. stain** A double stain, in which only the special feature to be examined takes one of the colors.
contravalence Covalence.
control A blank test or standard, e.g., an experiment performed simultaneously with another to study the relative effects of different conditions. **(self-)adaptive ~** C. technique that automatically changes certain values in the control algorithm so as to adapt it to new process conditions.
conus Cone.
convallamarin $C_{23}H_{44}O_{12}$ = 512.6. A glucoside from

TABLE 20. IMPORTANT CONSTANTS
Errors Are Expressed as the Standard Deviation

Quantity	Symbol	Value (and error)
Meter	m	1,650,763.73 wavelengths in vacuo of the transition $2p_{10}$–$5d_5$ of krypton-86
Kilogram	kg	Mass of the international prototype kilogram, at Sevres, France
Second	s	Duration of 9,192,631,770 periods of radiation from the transition between two hyperfine levels of the ground state of cesium-133
Ampere	A	The constant current producing a force of 2×10^{-7} N/m length between two conductors 1 m apart under specified conditions
Kelvin	K	The fraction 1/273.16 of the thermodynamic temperature of the triple point of water
Mole	mol	Amount of substance containing the same number of elementary entities (must specify which) as there are atoms in 0.012 kg of carbon-12
Candela	cd	Luminous intensity of a monochromatic radiation source of frequency 540×10^{12} Hz with a radiant intensity of 1/683 W/sr
Acceleration of free fall (standard)	g	9.806 65 m/s^2
Avogadro constant	L, N_A	$6.022\ 045 \times 10^{23}$ mol^{-1} ± 0.000 031
Bohr magneton	μ_B	$9.274\ 078 \times 10^{-24}$ J/T ± 0.000 036
Bohr radius	a_0	$5.291\ 7706 \times 10^{-11}$ m ± 0.000 0044
Boltzmann constant	k	$1.380\ 662 \times 10^{-23}$ J/K ± 0.000 044
Electron:		
Magnetic moment	μ_e	$9.284\ 832 \times 10^{-24}$ J/T ± 0.000 036
Radius	r_e	$2.817\ 9380 \times 10^{-15}$ m ± 0.000 0070
Rest mass	m_e	$9.109\ 534 \times 10^{-31}$ kg ± 0.000 047
Faraday constant	F	$9.648\ 456 \times 10^4$ C/mol ± 0.000 027
Fine structure constant	α	$7.297\ 3506 \times 10^{-3}$ ± 0.000 0060
Gas constant	R	8.314 41 J/(K mol) ± 0.000 26
Gram molecular volume at s.t.p.	V_0	$2.241\ 383 \times 10^{-2}$ m^3/mol ± 0.000 070
Gravitational constant	G	6.6720×10^{-11} Nm^2/kg^2 ± 0.0041
Light—speed in free space	c	$2.997\ 924\ 580 \times 10^8$ m/s ± 0.000 000 012
Logarithms:		
Base of natural	e	2.718 281 828 46
Natural of 10	ln10	2.302 585 092 99
Neutron—rest mass	m_n	$1.674\ 9543 \times 10^{-27}$ kg ± 0.000 0086
Nuclear magneton	μ_N	$5.050\ 824 \times 10^{-27}$ J/T ± 0.000 020
Pi	π	3.141 592 653 59
Planck constant	h	$6.626\ 176 \times 10^{-34}$ Js ± 0.000 036
Proton:		
Charge	e	$1.602\ 1892 \times 10^{-19}$ C ± 0.000 0046
Magnetic moment	μ_p	$1.410\ 6171 \times 10^{-26}$ J/T ± 0.000 0055
Rest mass	m_p	$1.672\ 6485 \times 10^{-27}$ kg ± 0.000 0086
Rydberg constant	R_∞	$1.097\ 373\ 177 \times 10^7$ m^{-1} ± 0.000 000 083
Standard atmospheric temperature and pressure	s.t.p.	$1.013\ 25 \times 10^5$ Pa (760 mmHg) and 0°C
Stefan-Boltzmann constant	σ	$5.670\ 32 \times 10^{-8}$ W/(m^{2}K^4) ± 0.000 71
Unified atomic mass constant	m_u	$1.660\ 5655 \times 10^{-27}$ kg ± 0.000 0086
Vacuum:		
Permeability	μ_0	4×10^{-7} H/m (exactly)
Permittivity	ϵ_0	$8.854\ 187\ 82 \times 10^{-12}$ F/m ± 0.000 000 07

Principal source: *Quantities, Units, and Symbols,* The Royal Society, London.

Convallaria majalis, lily of the valley (Liliaceae). Yellow powder, soluble in water; a cardiac stimulant and diuretic, similar to digitalis.

convallaretin $C_{14}H_{26}O_3$ = 242.4. A resinoid from convallaria; an emetic.

Convallaria Lily of the valley. The flowers and roots of *C. majalis;* an antispasmodic, sternutatory, and diuretic.

convallarin $C_{34}H_{62}O_{11}$ = 646.9. A glucoside from the root of Convallaria. Colorless prisms, sparingly soluble in water: a purgative.

convallatoxin $C_{29}H_{42}O_{10}$ = 550.6. A glycoside from *C. majalis* with potent digitalislike action on the heart.

convection The transmission of heat by the rise of heated liquids or gases, and the fall of the colder parts, which, in a closed system, in turn become heated.

convergent Inclining toward one point.

conversion (1) The change from one isomer to another. (2) The change from one unit or system of measurement to another. **c. factor** A numeral by which a quantity must be multiplied in order to express it in other units. C. factors are given under the definitions of individual units.

converter A device used to change (1) energy, (2) matter, or (3) signals into another form; as, (1) phase c., (2) digester, or (3) analog-to-digital c.

convex A rounded or bulging exterior surface; opposed to concave. **c. lens** See *convex lens* under *lens*. **c. mirror** See *convex mirror* under *mirror*.

Convolvulaceae A family of twining herbs, some with a milky juice; e.g.,: *Exogonium purga*, jalap; *Convolvulus scammonia*, scammony.

convolvulin $C_{31}H_{50}O_{16}$ = 678.7. Rhodeorhetin. A glucoside from jalap. Yellow mass, soluble in alcohol; drastic purgative.

convolvulinic acid $Me(CH_2)_2 \cdot CHOH(CH_2)_9COOH$ = 244.4. 11-Hydroxytetradecanoic acid*. Colorless crystals, m.50, from jalap resin.

convolvulinolic acid Convolvulinic acid.

convolvulus The root of *C. panduratus*, manroot (Convolvulaceae), N. America; a diuretic and laxative.

convulsant A drug causing convulsions.

conydrine Conhydrine.

conyrine $C_8H_{11}N$ = 121.2. 2-Propylpyridine. Colorless crystals, m.167, from the reduction of conine by zinc dust.

Cook formula The *acetyl* number, q.v.: $(S' - S)/(1 - 0.00075S)$, where S and S' are the saponification values before and after acetylation.

Coolidge, William David (1873–1975) American physical chemist. **C. tube** (1) A modified X-ray tube; the cathode is a tungsten spiral in a molybdenum tube. (2) Electron tube. An evacuated cathode tube with a thin nickel window and a cathode consisting of a wire made incandescent by a secondary electric current. On passing a high voltage, a stream of electrons passes out of the window and produces effects similar to the β rays of radium.

cooling Depriving of heat; as, freezing. **air ~** C. by means of air currents. **water ~** C. by means of cold water.

cooperite A native sulfarsenide containing Pt 64, Pd 9%.

coordinate Related. **c. bond** See *coordinate bond* under *bond*. **c. paper** A sheet of paper ruled with horizontal and vertical lines, used to draw graphs and curves. **c. valence** See *valency*.

coordinates Means by which the position of a point, particle, object, or star is defined by reference to base lines or base planes. Cf. *diagram, abscissa, ordinate, crystal* c.

Polar c. give the latitude and longitude on a sphere of known radius.

Rectangular c. (Cartesian c.) give the distances of a point from each of 2 or 3 perpendicular lines or planes.

Vector c. Straight lines in definite directions and of definite length.

coordination The harmonious linking of parts of the same rank or order. **c. compound** One in which atoms or groups (called "coordinating" or "ligating," or ligands) are attached to a "central atom," the number of atoms or groups usually being in excess of the classical or stoichiometric valency of the central atom; as, sodium hexacyanoferrate(III), $Na_3[Fe(CN)_6]$. **c. number** (1) The c. n. of the central atom of a compound is the number of ligands directly linked to that atom. Thus with $[PtX_6]R_2$, where X may or may not be the same radical as R, the c. n. is 6. Cf.*Werner's theory*. (2) In crystallography, the c. n. of an atom or ion in a lattice is the number of near neighbors to that atom or ion.

copahene $C_{20}H_{27}Cl$ = 302.7. Colorless crystals, insoluble in water.

copahin A resinoid from copaiba balsam.

copahuvic acid Copaivic acid.

copaiba Balsam of copaiba copaiva. An oleoresin from leguminous trees of tropical America, especially *Copaifera officinalis*. Grades: **Bahia ~** $[n]_D^{20}$ 1.508, acid val. 110–150. **Maracaibo ~** $[n]_D^{20}$ 1.517, acid val. 120–160. **Maranham ~** $[n]_D^{20}$ 1.515, acid val. 140–170.

c. oil A sesquiterpene essential oil from c.; d.0.895–0.905, b.250–275. **c. resin** The resin remaining after distillation; mainly copaivic acid.

copaiva Copaiba.

copaivic acid $C_{20}H_{30}O_2$ = 302.5. Copahuvic acid. A monobasic acid from the resin of copaiba.

copalchi The barks of *Strychnos pseudoquina, Croton niveus*, S. America; used as a febrifuge.

copalin Kopalin. The resin exuded from the sweet gum tree, *Liquidambar styraciflua*.

copals Hard resin exudations from southeast Asian, S. American, and African trees used in varnish manufacture. **E. African ~** Animi resin. **fossil ~** Highgrade c. from the ground where trees yielding it have disappeared. **fresh ~** Inferior c. from living trees. **kauri ~** Kauri gum. **Manila ~** See *Manila copal*. **semifossil ~** C. collected from the ground near living trees.

Copel Trademark for an alloy: Cu 55, Ni 45%; used in thermocouples.

copernik The alloy: Fe 50, Ni 50%.

copiapite $Fe_4S_5O_{18} + H_2O$. Yellow copperas.

copis A white, transparent shell (Philippine Islands) used to manufacture screens, lamps, etc.

Coplin jar A glass box with internal perpendicular grooves to hold microscope slides during staining.

copolymer ABS ~ See *ABS copolymers*.

copper* (1) Cu = 63.546. Cuprum*. A metallic element, at. no. 29. An orange, ductile, malleable metal, d.8.90, m.1083, b.2570, insoluble in water. It occurs in nature as metal, oxide (cuprite), sulfide (c. glance, chalcopyrite), and carbonate (malachite, etc.) and was known in prehistoric times (c. age). It is prepared or refined by electrolysis of crude c. in c. sulfate solution and forms 2 series of compounds: c.(I)*, c.(1+)*, cuprous compounds, e.g., CuCl; c.(II)*, c.(2+)*, cupric compounds, e.g., $CuCl_2$. The cupric compounds are the more stable; they are blue when hydrated, and grayish when anhydrous; they are poisonous and ionize in aqueous solution. Metallic c. is used extensively for wires, sheets, coins, etc., and is a constituent of many alloys (brass, bronze, bell metal, gunmetal, etc.). (2) A general term suggested for alloys containing 98% or more of c. **black ~** Cupric oxide. **blue ~** A native c. sulfide. **indigo ~** Covelline. **rose ~** See *rosette*. **scale ~** C. in thin flakes. **silicon ~** The alloy: Cu 70–80, Si 30–20%. **wood ~** Olivenite. **yellow ~** Chalcopyrite.

c.(I)*, c.(1+)* See also *cuprous*. **c.(II)*, c.(2+)*** See also *cupric*. **c. acetoarsenite** Paris green. **c. alloys** See *aluminum bronze, bell metal, brass, constantin, Dutch metal, nickeline, Muntz metal, speculum*. **c. arsenite** See *cupric arsenite*. **c. blue** Azurite. **c. bromide** See *cuprous bromide, cupric bromide*. **c. carbonate** Verditer. **basic ~** See *cupric carbonate*. **c. chloride** See *cupric* and *cuprous chloride*. **c. dichloride oxide*** $CuCl_2 \cdot CuO$ = 214.0. Cupric oxychloride. Green powder, soluble in ammonia or acids; a green pigment. **c. froth** Tyrolite. A native copper arsenate. **c. green** A native pigment of lead chromate and c. oxide. **c.(III) hexacyanoferrate(II)*** $Cu_2[Fe(CN)_6] \cdot 7H_2O$ = 465.2. Cupric ferrocyanide. Brown powder, insoluble in water, soluble in KCN solution. **c. lazur** Azurite. **c. minerals** The chief ores of copper are native c., Cu; covellite, CuS; chalcopyrite, $CuFeS_2$: berzelianite, Cu_2Se; algodonite, Cu_6As; klaprotholite, $Cu_6Bi_4S_9$; cuprite, Cu_2O; malachite, $Cu_2(OH)_2CO_3$; azurite (chessylite), $Cu_3(OH)_2(CO_3)_2$. **c. mordant** Cupric sulfate or cupric acetate. **c. number** The number of mg of c. obtained by the reduction of Fehling's or Benedict's solution by 1 g of a carbohydrate. **c. orthosilicate** Dioptase. Chrysocolla. **c. oxychloride** (1) C. dichloride oxide*. (2) Dicopper chloride trihydroxide*. See *fungicide,* Table 37. **c. peroxide** Cu_2O_3 and $CuO_2 \cdot H_2O$. Yellow powder, decomp. in water. **c. phosphide** See *cuprous* and *cupric p.* $Cu_5P_2H_2O$ also exists. **c. shavings** Small shavings of metallic c., used as a catalyst and in preparation of cuprammonium reagent. **c. sulfate, c. vitriol** Cupric sulfate.

copperas $FeSO_4$. Native ferrous sulfate. **blue ~** Cupric sulfate. **green ~** Ferrous sulfate. **yellow ~** Copiapite. **white ~** Zinc sulfate*.

copperon Cupferron.

Coppet, Louis Cas de (1841–1911) French physicist. **C.'s law** The depression in the freezing point of a solution below 0 °C is proportional to the amount of solute. Cf. *Raoult's law.*

copra Dried coconut kernels from which oil is expressed.

copraol A fat from coconut oil.

coprastanol $C_{27}H_{48}O$ = 388.7. 5β-Cholestan-3β-ol. A sterol in the feces of humans and higher animals. It has a relatively long survival period and is used as a chemical indicator of pollution.

coprecipitation See *coprecipitation* under *precipitation.*

coprolite The fossil dung of prehistoric animals; a fertilizer (25–30% phosphorus pentaoxide).

coproporphyrin $C_{36}H_{38}O_8N_4$ = 654.7. Excreted by the liver; also the coloring matter of bottom-fermentation yeasts.

coprostanol Coprosterol.

coprosterol $C_{27}H_{48}O$ = 388.7. Dihydrocholesterol, stercorol, coprostanol. Colorless crystals, m.98, from feces.

copro-yeast See *copro-yeast* under *yeast.*

coptine A colorless, crystalline alkaloid from *Captis trifolia,* golden thread (Ranunculaceae).

copyrine $C_8H_6N_2$ = 130.1. 2,7-Pyridopyridine, 2,7-benzodiazine; m.93; an isomer of quinoxaline.

corajo Vegetable ivory. A hard, white substance from the tagud nut. Cf. *phytelephas.*

coral The solid calcareous skeletons of the coral animal *Polyps anthozoa;* chiefly calcium carbonate colored by ferric oxide. **c. bean** Sophora.

corallin (1) Peonin, aurine R. A reagent and dye from rosolic acid. (2) A pigment in *Streptothrix* species. **c. solution** A 1% solution of c. in 10% alcohol; an indicator for ammonia and weak bases (alkalies—red; acids—yellow). **c. yellow** Sodium rosolate.

corallinate Rosolate.

coralline (1) Corallin. (2) Rosolic acid.

corallite Carrollite.

Coramine Trademark for nikethamide.

corbyn A glass bottle of 1,200-ml capacity, shaped like a *Winchester,* q.v.

corchularin $C_{22}H_{36}O_8$ = 428.5. A crystalline, bitter principle from the seeds of *Corchorus capsularis,* Linn. (jute), m.174.

cord The volume occupied by an orderly pile of logs 8 ft long, 4 ft wide, and 4 ft high, the logs being laid as near parallel to the ground and to one another as possible. The gross 128 ft^3 gives 75–95 ft^3 solid wood, depending on log diameter, bark, and straightness. Cf. *cunit, board foot, festmeter.*

cordial (1) An elixir. (2) A liqueur.

cordierite Iolite.

cordite General name for smokeless powders; in particular, a c. sporting powder: nitroglycerin 30–58, nitrocellulose 65–37, mineral jelly 5–6%.

Cordura Trademark for a cellulose acetate synthetic fiber.

core The central part. **c. oil** A semisolid mixture, usually of a drying oil, a carbohydrate, and sand; used to line iron foundry molds to facilitate removal of the finished casting.

coreductase The coenzyme of a *reductase,* q.v.

Corfam Trademark for a polyester fiber with a microporous resinous binder of the urethane type. Used as a leather substitute.

corialgin $C_{27}H_{22}O_{18}$ = 634.5. 1-Galloyl-3,6-hexahydroxydiphenyl-β-D-glucopyranose. The natural coloring matter of *Caesalpina coriaria.*

coriamin A 25% aqueous solution of pryidine-β-carbonic acid diethylamide, $NC_5H_4 \cdot CON(C_2H_5)_2$.

coriamyrtin $C_{15}H_{18}O_5$ = 278.3. A glucoside from the leaves, flowers, and seeds of *Coriaria myrtifolia,* curriers sumach (Coriariaceae). Colorless crystals, m.220, sparingly soluble in water; a tetanic poison.

coriander The seeds of *Coriandrum sativum* (Umbelliferae); a carminative. **c. oil** The essential oil from coriander, d.0.863–0.875, containing linalol and pinene; a flavoring agent (NF, BP).

coriandrol Linalool.

coriarine An alkaloid from *Coriaria myrtifolia.* Cf. *coriamyrtin.*

coridine $C_{10}H_{15}N$ = 149.2. A homolog of pyridine obtained by distillation of animal matter. **hydro ~** See *hydrocoridine.*

Coriester Glucose-1-phosphoric acid. See *sugar phosphates.*

cork Suber. (1) The exterior layers of the bark of certain oaks; e.g., *Quercus suber* and *Q. occidentalis.* See *cerin, corticinic acid, suberic acid.* (2) A stopper made from c. **c. borer** A set of metallic tubes for boring holes in c. stoppers. **c. press, c. tongs** Devices for pressing and softening c. stoppers.

corm A thick underground plant stem that has the character of a tuber and bulb. See *Colchicum.*

corn (1) General term for cereals. (2) Maize. (3) A hard mass of skin over bone, caused by pressure. **Indian ~** Maize. **c. oil** Maize oil. Yellow oil, d_{25° 0.918 (NF, BP). **c. silk** The stigmas of maize. **c. spar** Crystalline calcite. **c. starch** See *corn starch* under *starch.* **c. sugar** D-Glucose. **c. syrup** A mixture of dextrins, glucose, and maltose.

cornelian A bright-red chalcedony.

cornflower Bluebottle. The blossoms of *Centaurea cyanus* (Compositae), containing cyanidin. Cf. *anthocyanins, flavones.*

cornic acid Cornin.

cornification Rendering hard and brittle by drying.

cornin Cornic acid. A crystalline substance from the bark of *Cornus florida;* dogwood (Cornaceae).

Corning filter A filter made by the Corning Glass Co. for isolating ultraviolet light.

cornstone Impure clay containing limestone.

Cornu prism A quartz prism composed of two 30°-angle right and left quartz prisms in optical contact. Cf. *Littrow prism.*

cornutol A liquid extract of ergot containing the water- and alcohol-insoluble constituents.

corona The incandescent gases surrounding the sun seen during total eclipse. Its light is partly polarized and yields unknown bright lines, and continuous (Fraunhofer) spectra. **c. discharge** The c. produced at electrodes during a high-voltage discharge. See *electrostatic precipitator.* **c. spectrum** The reversed spectrum of the sun c.

coronary arteries The a. supplying and encircling the heart.
coronene* $C_{24}H_{12}$ = 300.4.

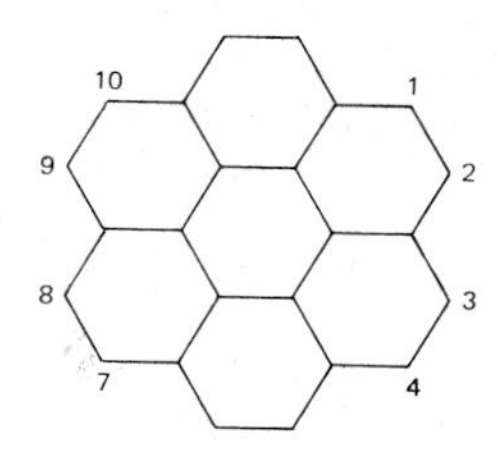

A dyestuff base, obtained from anthracene.
coronillin $C_{15}H_{21}O_{15}N_3$ = 483.4. A trinitroglucoside from the seeds of *Coronilla scorpioides* (Leguminosae). Yellow powder, soluble in water; a cardiac poison, diuretic, and heart stimulant.
coronium Protofluorine. A hypothetical element, postulated from the spectrum of the sun's corona.
corporin Progesterone.
corpse light The blue flame inside the miners' safety lamp, which indicates the presence of firedamp (methane).
corpus (1) The body. (2) The main part of an organ. **c. luteum** The yellow body that develops from the circle of cells (follicle) in the ovary after ovulation. It secretes progesterone.
corpuscle (1) A small particle. (2) A free electron. **blood ~** See *blood corpuscles* under *blood.* **light ~** See *photon.* **negative ~** See *electron.* **positive ~** See *proton.*
corr. Abbreviation for corrected.
correlation Reciprocal relationship; interdependence. **c. coefficient** r; a statistical measure of interdependence. The quantity

$$\frac{\Sigma(x - \bar{x})(y - \bar{y})}{\sqrt{\Sigma(x - \bar{x})^2\Sigma(y - \bar{y})^2}}$$

where $\bar{x}$ and $\bar{y}$ are the mean values of variables x and y. The probability of obtaining the value of r in the absence of a c. is ascertained against tables.
correlogram A graph or diagram correlating 2 or more variables.
correspondence principle The relation between the orbitals of an electron in an atom and the characteristic radiation. Cf. *Bohr theory, Pauli's principle.*
corresponding **c. state** If the pressure, volume, and temperature are expressed for each substance (whether solid, liquid, or gaseous) by the critical constants p_c, V_c, and T_c, then the (Van der Waals) equation of state will hold for all substances, in the liquid or gaseous state, for they will then correspond. **c. temperatures** Those temperatures of 2 or more substances which are equal fractions of the critical temperatures.
corrigent A drug which favorably influences the action of another drug.
corrin $C_{19}H_{22}N_4$ = 306.4. The tetracyclic porphyrin-type nucleus (without the central cobalt atom) of cobalamin compounds. See *vitamin B_{12}.* Cf. *hemoglobin.*
corrinoid Generic name for compounds containing the corrin nucleus.
corrode To disintegrate slowly by chemical action.
corronel An alloy: Ni 70, Cu 30%.
corrosion Gradual electrochemical disintegration or decomposition; e.g., of iron by acid rocks in natural water. Cf. *surrosion, erosion.* **fretting ~** C. caused by the oscillatory slipping movement of two closely fitting metal surfaces, e.g., due to vibration.
corrosive (1) A substance that destroys organic tissues by chemical means or inflammation. (2) An agent that causes corrosion. **c. poison** See *corrosive poison* under *poison.* **c. sublimate** Mercuric chloride.
corrugation Wrinkling.
corsite Napoleonite. A banded or spotted diorite (Corsica); an ornamental stone.
cortepinitannic acid $C_{32}H_{34}O_{17}$ = 690.6. Bright-red powder from the bark of Scotch fir, *Pinus sylvestris.* Cf. *pinicortannic acid.*
cortex The bark of a tree, root, or fruit.
corticin Bark tannin.
corticinic acid $C_{12}H_{10}O_6$ = 250.2. A colorless, crystalline substance from cork.
corticoids Corticosteroids. Hormones secreted by the adrenal cortex; of great physiological importance. C. all have steroid ring structures, with a $-C(O)CH_2OH$ group at the 17 position. Two main types: **gluco ~** Hormones that affect carbohydrate metabolism. Glyconeogenesis is stimulated, and glycogen storage is increased. Also influence blood pressure, healing processes, and reaction to stress. Main uses include treatment of allergic states such as asthma, and of arthritis and leukemia and control of rejection of transplants. The natural g. are cortisol, corticosterone, and cortisone. **mineralo ~** Hormones that regulate salt and water metabolism in the body. Used as replacement therapy after disease or removal of adrenal gland. Aldosterone is the chief natural hormone.
corticoles A group of lichens.
corticosteroids Corticoids.
corticosterone 11,21-Dihydroxyprogesterone. A steroid hormone from adrenal cortex extracts.
corticotrophin ACTH, adrenocorticotrophic hormone. An active hormone of the anterior lobe of the pituitary gland; controls secretion of adrenal cortex. Used to increase secretion of glucocorticoids (USP, BP).
corticrocin $C_{14}H_{14}O_4$ = 246.3. 2,4,6,8,10,12-Tetradecahexaene-1,14-dicarboxylic acid. The yellow pigment of the fungus *Corticium croceum,* red needles, decomp. 310, soluble in water.
cortisone $C_{21}H_{28}O_5$ = 360.5. Kendall compound E, 17-hydroxy-11-dehydrocorticosterone. A hormone in the adrenal gland, and in the seeds of *Strophanthus sarmentosus,* a W. African tropical vine; used to treat arthritis. It can be synthesized from ergosterol, or from hecogenin obtained from sisal waste and from ox bile. **c. acetate** $C_{23}H_{30}O_6$ = 402.5. A white, odorless, crystalline powder, m.240 (decomp.), insoluble in water; an adrenocortical hormone (USP, EP, BP).
corundum α-Alumina. Oriental topaz. A hard, native or artificial aluminum oxide, used as an abrasive and refractory. Emery is impure c. Cf. *adamantine, sapphire, oriental hyacinth.*
coruscation The emission of sparks or flashes.
corybulbine $C_{21}H_{25}O_4N$ = 355.4. An alkaloid from *Corydalis cava;* colorless crystals, insoluble in water.
corycavine $C_{23}H_{23}O_6N$ = 409.4. An alkaloid from *Corydalis cava,* m.218.
corydaline $C_{22}H_{27}O_4N$ = 369.5. An alkaloid from *Corydalis cava* and *C. tuberosa,* holewort (Fumariaceae). Colorless crystals; a diuretic.
corylin A globulin from hazel nuts.
Corynebacterium A genus of microorganisms. Irregular shape, gram-positive; as, *C. diphtheria.* Other C. cause disease in animals.

corynin $C_{50}H_{100}O_4$ = 765.3. A hydroxy acid, m.70, from the fat of the diphtheria bacillus.
corynine Yohimbine.
cosanoates A group of salts or esters derived from fatty acids, with 20–29 C atoms; as: icosanoate, $C_{19}H_{39}COO-$; pentacosanoate $C_{24}H_{49}COO-$.
cosanoic acids A group of fatty acids with 20–29 C atoms; as: icosanoic acid, $C_{20}H_{40}O_2$, m.76; tetracosanoic (cerosic) acid, $C_{24}H_{48}O_2$, m.85; heptacosanoic (cerotic) acid, $C_{27}H_{54}O_2$, m.89.
cosanols A group of aliphatic alcohols of the alkane series with 20–29 C atoms; as, iconsanol, $C_{20}H_{42}O$, m.65.
cosine The ratio of the base to the hypotenuse of a right-angled triangle is the c. of the angle subtended by these sides.
cosmetic A pharmaceutical preparation to preserve, restore, or simulate beauty.
cosmic Pertaining to the universe. **c. rays** Ultra-γ rays. A radiation of extremely short wavelength (around 10^{-15} m), frequency, and penetration, which reaches the surface of the earth from all directions of space (cosmos). First noted by Gockel (1910–11) up to 4,500 m height, and by Kohlhörster (1912–14) up to 9,200 m. Primary c. r. (mainly protons) enter the earth's atmosphere, where some collide with particles, creating secondary c. r. (include μ-mesons and neutrons).
cosmotron An early proton *accelerator,* capable of up to 3 GeV.
costunolide $C_{15}H_{20}O_2$ = 232.3. A sesquiterpene lactone having a 10-C ring; a primary constituent of costus root oil; m.106.
costus oil A perfumery oil extracted from the root of the costus, *Aplotaxis saussurea lappa.* It consists mainly of sesquiterpenes with 20% aplotaxene; $[n]_D$1.5159.
cosyl Nomenclature for radicals derived from the alkane series of hydrocarbons with 20–29 C atoms; e.g.: icosyl, $C_{20}H_{41}-$; nonacosyl, $C_{29}H_{59}-$.
cotarnic acid $C_{11}H_{12}O_5$ = 224.2. A dibasic acid oxidation product of cotarnine.
cotarnine $C_{12}H_{15}O_4N$ = 237.3. An alkaloid oxidation product of ($\pm$)-gnoscopine; an astringent. Cf. *cuprine, cupronine.* **c. phthalate** Styptol.
coto The bark from a Bolivian tree. Two varieties: Coto verum, containing cotenetin; and paracoto, containing paracotoin, cotoin, and essential oil.
cotoin $C_{14}H_{12}O_4$ = 244.2. 2,6-Dihydroxy-4-methoxybenzophenone. A constituent of paracoto bark. Yellow crystals, m.129, soluble in water; an irritant. **hydro ~** $C_{15}H_{14}O_4$ = 258.3. 6-Hydroxy-2,4-dimethoxybenzophenone. m.95. **para ~** $C_{12}H_8O_4$ = 216.2. Dihydroxymethylene phenyl coumalin. An active principle from paracoto bark. Yellow crystals, m.152, soluble in alcohol. **proto ~** $C_{16}H_{14}O_4$ = 270.3.
cotonetin $C_{20}H_{16}O_5$ = 336.3. Colorless scales, from coto bark.
co-trimoxazole Antimicrobial mixture of trimethoprim and sulfamethoxazole (1: 5 parts); used widely for urinary, respiratory, and intestinal infections (BP). See *sulfonamides.*
cotton (1) The hairs of the seeds of *Gossypium* species (Malvaceae): cellulose 91, moisture 7%. (2) A textile material spun from c. fibers. Cf. *linter.* (3) Pyroxylin or freshly nitrated c. **absorbent ~** A purified and fat-free c. (USP, EP, BP). **artificial ~** Polynosic fibers. **gun ~** Pyroxylin. **soluble ~** Nitrocellulose. **styptic ~** C. impregnated with ferric chloride and dried; used to stop bleeding.
c. gum tree See *cotton gum tree* under *gum tree.* **c. oil** C.seed oil; **c.seed meal** The residue after extracting c. oil, finely ground and used as cattle feed and fertilizer; it contains nitrogen 6.7–7.4, potassium 1.5–2, and phosphorus pentaoxide 2–3%. **c.seed oil** Oleum gossypii seminis. The yellow, viscid fixed oil expressed from the seeds of various Gossypium species. The refined oil is colorless and has a nutty odor, d_{15}·0.9264. Used extensively in pharmaceutical preparations, and as a substitute for olive oil (NF). See *Halphen test.*
cottonization The disintegration of (e.g., bast) fibers, without damage to their structural characteristics. The product is used in conjunction with cotton for spinning fabrics.
couepic acid $Me(CH_2)_3(CH{:}CH)_3(CH_2)_4CO(CH_2)_2COOH$ = 340.5. Oxoeleostearic acid. **α- ~** m.75. **β- ~** m.96. From the seed oil of *Couepia grandiflora* (Rosaceae), S. America.
Coulomb, Charles Augustin de (1736–1806) French physicist. **C. electromagnetic law** The force between 2 similar magnetic poles varies inversely as the square of the distance between them. **C. electrostatic law** The force between 2 electric charges varies: (1) inversely as the square of their distance apart; and (2) directly as the product of their electric charges. **C. unit** See *coulomb.*
coulomb* C. Coul. The SI system unit of electric charge or quantity. The quantity of electricity transported 1 c. = 2.778 $\times 10^{-4}$ ampere hour = 1.036 $\times 10^{-5}$ faraday (chemical) = 1 abcoulomb = 2.998 $\times 10^9$ statcoulombs.
coulometer Voltameter.
coulometric analysis A *conductometric* (q.v.) method, in which the change of intensity of a current passing through an electrolyzed solution is used as an end-point indicator, or is used to follow a change in composition of the solution.
coumalic acid $C_6H_4O_4$ = 140.1. 2-Oxo-2*H*-pyran-5-carboxylic acid. Colorless crystals, m.206. **dimethyl ~** $C_8H_8O_4$ = 168.2. Isodehydracetic acid. Colorless crystals, m.155.
coumalin $C_5H_4O_2$ = 96.1. 2-Pyrone*. Colorless liquid, m.5, b.207. **dimethyl ~** $C_7H_8O_2$ = 124.1. Mesitylene lactone. Colorless crystals, m.51.5. **phenyl ~** $C_2H_3PhO_2$ = 136.2. 1-Phenyl-2-pyrone. Colorless crystals, m.68. Cf. para*cotoin.*
coumaraldehyde $HO{\cdot}PhCH{:}CH{\cdot}CHO$ = 149.2. Hydroxycinnamaldehyde. **1,2-** Colorless crystals m.133, sparingly soluble in water.
coumaran $C_6H_4(CH_2)_2{\cdot}O$ = 192.1. 2,3-Dihydrobenzofuran. Colorless liquid, b.189.
coumaranone Benzofuranone.
coumaric **c. acid** $HO{\cdot}C_6H_4{\cdot}CH{:}CH{\cdot}COOH$ = 165.2. Coumarinic acid. An acid from the leaves of *Melilotus.* ***ortho-* ~** *o*-Hydroxycinnamic acid. Colorless needles, m.208, slightly soluble in water. ***meta-* ~** Colorless prisms, m.191, sparingly soluble in hot water. ***para-* ~** White needles, m.206, slightly soluble in water. **hydro ~** $HO{\cdot}C_6H_4{\cdot}CH_2{\cdot}CH_2{\cdot}COOH$ = 166.2. 3-Phenolpropionic acid. White monoclinics, m.128, soluble in water. Cf. *melilotic acid.* **hydroxy ~** Umbellic acid.
c. aldehyde Coumaraldehyde. **c. lactone** Coumarin*.
coumarilic acid $C_8H_5O{\cdot}COOH$ = 162.1. 1-Benzofurancarboxylic acid. Colorless crystals, m.190. **β-methyl-** $C_9H_7O{\cdot}COOH$. Colorless crystals, m. 189.
coumarin* $C_9H_6O_2$ = 146.1. Benzo-1,2-pyrone*, 2*H*-1-benzopyran-2-one†. The anhydride of *o*-coumaric acid in tonka beans, sweet clover, and other plants; also prepared synthetically. Colorless rhombs, m.67, soluble in hot water. **6,7-dihydroxy ~** Esculetin. **7,8-dihydroxy ~** Daphnetin. **4,7-dimethyl ~** $C_{11}H_{10}O_2$ = 174.2. Colorless crystals, m.148. **ethyl ~** $C_{11}H_{10}O_2$ = 174.2. Colorless crystals, m.71. **hydroxy ~** $C_9H_6O_3$ = 162.1. **5- ~** Colorless crystals, m.249. **7- ~** Umbelliferone. **7-hydroxy-4-methyl ~** Resocyanin. **iso ~ *** Benzo-2,1-pyrone. Colorless crystals, m.47. **isopropyl ~** $C_{12}H_{12}O_2$ = 188.3. Colorless crystals,

m.54. **7-methoxy ~** Herniarin. **methyl ~** $C_{10}H_8O_2$ = 160.2. Colorless crystals. **3- ~** m.90. **4- ~** m.82.
coumarinic acid Coumaric acid.
coumarketone A compound containing the radical *ortho*-$C_6H_4(OH)CH:CH \cdot CO-$. **methyl ~** $C_{10}H_{10}O_2$ = 162.2. *o*-Hydroxybenzylidene ketone. Colorless crystals, m.139.
coumarone Benzofuran*.
coumazonic acid $C_{10}H_{11}ON$ = 161.2. **methyl ~** $C_{11}H_{13}ON$ = 175.2. Benzotrimethylmethoxazine. Colorless crystals, m.218.
coumothiazone **imido ~** $C_8H_8N_2S$ = 164.2. Benzylene-ψ-thiourea. Colorless crystals, m.137.
counterirritant A superficial irritant, used as a distraction from the effects of other irritants or abnormal processes; e.g., rubefacients.
counterstain A microscope stain used to contrast structures already colored by another stain.
counting **c. apparatus** A device for counting bacteria or blood cells. **c. chamber** A c. apparatus microscope slide with rectangular rulings. **c. pipet** A graduated capillary glass tube, to make milk or blood smears for bacteria counts.
coupeic acid An active principle of oiticica oil.
Coupier's blue $Ph \cdot C_6H_4N:N \cdot C_6H_4 \cdot Ph$ = 334.4. Azodiphenyl, m.250.
couple (1) A pair of galvanic cells. Cf. *thermocouple, zinc-copper c.* (2) To condense or unite 2 molecules.
coupling condensation between the N of a diazo group and a C of a ring compound. **oxidative ~** Dehydrogenation (usually of a phenol) by electron transfer, to produce a free radical which dimerizes internally, couples, or reacts with another compound. Used in the synthesis of natural products.
courare Curare.
Courlene Trademark for a polyethylene synthetic fiber.
Courlose Trademark for sodium carboxymethylcellulose.
Courpleta Trademark for a cellulose acetate synthetic fiber.
Courtaulds Trademark for a protein synthetic fiber.
Courtelle Trademark for a polyacrilonitrile synthetic fiber.
Courtois, Bernard (1777–1838) French chemist; discoverer of iodine.
cousso See *kousso.*
C.O.V. Concentrated oil of vitriol; 95–96% sulfuric acid by weight.
Covadur Trademark for a viscose synthetic fiber.
covalence See *valency.* **dative ~** C. in which one of the two atoms concerned contributes both electrons. Cf. *electrovalence.* **normal ~** C. in which each of the atoms concerned contributes one electron.
coveline, covellite CuS. Native copper sulfide.
covolume The quantity *b* in *Van der Waals' equation,* q.v.
Cowper stoves Tall iron cylinders lined with firebrick to produce a hot blast for iron smelting.
coxanthin $C_{40}H_{56}O_6$ = 632.9. A carotenoid pigment of brown algae.
coyol palm A tropical palm, *Acrocomia,* yielding an oil.
CP Abbreviation for chemically pure.
Cp Abbreviation for molar heat capacity at constant pressure.
CPM Critical path method.
Cr Symbol for chromium.
cracca Tephrosia.
cracked Broken, as of a molecule split into component parts. **c. kerosine** A gasoline substitute obtained by superheating kerosine under pressure, and distilling a volatile fraction at the boiling point of gasoline.
crackene Proposed name for the hydrocarbon mixture obtained by cracking low-temperature tars.
cracking An oil refinery operation in which the feed is broken down, either by heat and pressure (thermal c.) or with a catalyst (catalytic c.), into lower-boiling-point hydrocarbons plus some bottoms. Cf. *hydrogenation, Bergius process.* **c. patterns** The characteristic spectra of hydrocarbons given by the mass spectromoter.
cradin A peptic enzyme from the leaves and twigs of the common fig, resembling papain in its action.
Crafon Trademark for bundles of fibers, each comprising a core of polymethyl methacrylate with a plastic sheath of lower refractive index. The bundles are jacketed in polyethylene, and transmit light even when flexed or bent.
Crafts, James Mason (1839–1917) American chemist, noted for organic syntheses. Cf. *Freidel-Crafts reaction.*
crateriform A round, depressed cone in a solid culture medium, due to liquefaction by bacteria.
creaming The gradual rise or fall of the disperse phase of an emulsion, depending on whether its specific gravity is less or greater than that of the continuous phase. Cf. *colloid.*
cream of tartar Potassium hydrogentartrate*.
creasote Creosote.
creatinase* An enzyme which hydrolyzes creatine to sarcosine and urea. Cf. *creatininase.*
creatine $NH_2 \cdot C(:NH) \cdot NMe \cdot CH_2 \cdot COOH$ = 131.1. Methylguanidinoacetic acid, methylglycocyamine. An amino acid from the metabolism of muscular tissue of vertebrates; excreted in urine as creatinine. Colorless monoclinics, decomp. 300, sparingly soluble in water. Cf. *alacreatine.*
creatininase* Enzyme that hydrolyzes creatinine to creatine.
creatinine $CH_2 \cdot N:C(NH_2)N(Me) \cdot C(O)$ = 113.1. (ring: CH_2 bonded to $C(O)$)
Methylglycocyamidine. An anhydride of creatine excreted by the kidney and present in urine. White prisms, decomp. 300, slightly soluble in water. Cf. *phosphagen.* **xantho ~** See *xanthocreatinine.*
creatotoxin A meat poison, or ptomaine.
creep The 3-phase, continuous deformation which occurs when a metal is subjected to a constant load: viz. primary (transient); secondary (steady state or quasi-viscous); tertiary.
creepage Tracking.
creeping Describing the behavior of (1) a precipitate rising on the walls of a wet glass container; (2) a solution which deposits crystals during crystallization on the sides and top of its container; (3) a liquid which passes through the packing of machinery; (4) flow; a nonrecoverable strain; the elongation of a metal under a stress considerably less than that required to break it.
Cremona Trademark for a polyvinyl alcohol synthetic fiber.
crenic acid $C_{24}H_{12}O_{16}$ = 556.4. An acid produced by molds in the soil.
crenilabrin A protamine from the sperm of the cunner fish, *Crenilabrus pavo.*
creosol $MeO \cdot C_6H_3(OH)Me$ = 138.2. 2-Methoxy-4-methylphenol. Colorless oil from beech wood cresols, b.220, slightly soluble in water; an antiseptic.
creosotal Creosote carbonate.
creosote An oily distillate from wood tar: chiefly cresol, hydroxycresol, and methylcresol, b.220, soluble in water; an antiseptic, local anesthetic, and caustic. **oleo ~** C. oleate
c. carbonate Creosotal. An antiseptic, oily derivative of creosote. **c. oleate** Oleocreosote. A yellow oil, insoluble in water; an antiseptic.
cresalol $C_6H_4(OH)COOC_6H_4Me$ = 228.2. A mixture of the ortho, meta, and para compounds. An antiseptic powder.
Cresatin MeC_6H_4OOCMe = 150.2. Trademark for 1,3-tolyl acetate. Colorless oil, insoluble in water; an antiseptic.
cresegol See *egols.*
cresidine $NH_2 \cdot C_6H_3Me \cdot OH$ = 123.2. Aminocresol. **5,3-** m.79.

Creslan Trademark for an acrylic synthetic fiber.
cresol* HO·C_6H_4·Me = 108.1. Methylphenol†, oxytoluol, cresylic acid. **amino ~** Cresidine. ***p*-chloro-*m*- ~** C_7H_7OCl = 142.6. Colorless crystals with a characteristic odor, m.65, slightly soluble in water. **iodo ~** Traumatol. **6-isopropyl-*m*- ~** Thymol. ***ortho*- ~** Colorless liquid, m.30, sparingly soluble in water. ***meta*- ~** Colorless liquid, m.4, slightly soluble in water. **2-methoxy-*p*- ~** Creosol. **methyl ~** Xylenol*. ***para*- ~** Colorless prisms, m.36, sparingly soluble in water. Commercial c. is a brown-red, syrupy liquid from coal tar, and contains a mixture of the three cresols. A disinfectant; toxic if ingested or large amounts absorbed through skin (BP). **triiodo ~*** Losophan. **trinitro ~*** See *trinitrocresol.*
cresolphthalein $C_6H_4(CO)_2(C_6H_3OHMe)_2$ = 346.4. Colorless crystals, m.216, slightly soluble in water; an indicator, pH 8.2 (colorless) to 9.2 (red).
cresol purple $C_{21}H_{18}O_5S$ = 382.4. *m*-Cresolsulfonphthalein. Brown powder; an indicator, pH 1.5 (red) to 2.5 (yellow). **bromo ~** See *indicator,* Table 44 on p. 302.
cresol red $C_{21}H_{18}O_5S$ = 382.4. *o*-Cresolsulfonephthalein. Brown powder; an indicator, pH 7.2 (yellow) to 8.8. (red).
cresolsulfonic acids $C_6H_3Me(OH)(SO_3H)$ = 188.2. Hydroxymethylbenzene sulfonic acids. Monobasic acids derived from the cresols; used in organic synthesis.
cresorcin $C_{22}H_{16}O_5$ = 360.4. 2,7-Dimethylfluorescein. A yellow dye indicator.
cresorcinol 2,4-Dihydroxytoluene*.
cresorcyl The dihydroxymethylphenyl† radical, $-C_6H_2(OH)_2Me$.
cresotic acids $C_6H_3Me(OH)(COOH)$ = 152.2. Isomeric acids, e.g., 2-hydroxy-3-methylbenzoic acid. m.170.
cresotinic acid Cresotic acid.
cresocy See *toloxy.*
crestmorite $2CaO \cdot 2SiO_2 \cdot 3H_2O$. A constituent of boiler scales.
cresyl (1) The hydroxymethylphenyl* radical $HOC_6H_3(CH_3)-$, from cresol. Ten isomers, derived from *o*-, *m*-, and *p*-cresols. (2) The tolyl* radical. Cf. *cresotic acids.* **c. acetate** Kresatin. **c. alcohol** Cresol*. **c. blue** An oxidation-reduction *indicator,* q.v. **c. hydrate** Cresol*. **c. hydride** Toluene*. **c. phosphate** Tricresyl phosphate. **c. violet** A stain for blood.
cresylate Homologs of the phenolates containing the radical $C_6H_4(CH_3)O-$, from cresol.
cresylic acid Mixed *o*-, *m*-, and *p*-cresols.
cresylite An explosive: picric acid 60, trinitro-*m*-cresol 40%.
creta (praeparata) Chalk.
cretaceous (1) See *geologic eras,* Table 38. (2) Describing a chalky growth of bacteria.
crill Fine fiber debris in wood pulp.
Crinovyl Trademark for a polyester synthetic fiber.
criogenine 1-Phenylsemicarbazide. A reagent for cupric ions (pink color).
cristal Crystal.
cristallisation Crystallization.
cristobalite A crystalline form of silica, formed on heating quartz to 1200. Two forms: α- and β-; transition temperature 200–275 °C. Cf. *silica.*
crit. Abbreviation for critical.
crith Krith.
crithmene $C_{10}H_{16}$ = 136.2. A terpene from samphire, *Crithmum maritimum* (Umbelliferae).
critical Pertaining to (1) a turning point or abrupt change; (2) the safe point of nuclear interaction. **c. air-blast test** A test made under standard conditions for the minimum airflow required to keep a fuel burning after ignition. Values: low-temperature coke (coalite) 0.107, anthracite 0.039, high-temperature coke 0.053. **c. angle** The angle of incidence i of a ray of light at which it is refracted through a prism so that its angle of emergence is 90°: $\sin i = 1/n$, where n is the refractive index. **c. coefficient** RTd/p, where R is the gas constant, T the c. thermodynamic temperature, d the c. density, p the c. pressure. **c. conditions** The c. temperature and c. pressure. **c. constant** A magnitude relating to the c. state. **c. density** The density of the liquid and vapor at the c. temperature and c. pressure. **c. hygrometric state** $100p/P$, where p and P are the vapor pressures of the system and of water, respectively. It determines whether a substance will deliquesce or effloresce. **c. mass** The minimum amount of fissionable material, e.g., ^{235}U or ^{239}Pu, required to sustain fission in a nuclear reactor. **c. path method** CPM. A method for controlling engineering projects, particularly those involving many concurrent and consecutive operations (as, construction). Each job is represented by an arrow on a time scale; the planned sequence of the project is denoted by joining appropriate arrows. **c. point** The conditions at which 2 phases are just about to become 1 phase. **c. pressure** The pressure necessary to condense a gas at the *c. temperature,* q.v. See Table 21. **c. solution temperature** The temperature at which a mixture of 2 liquids, immiscible at ordinary temperatures, just ceases to separate into 2 phases. It is altered considerably by impurities. **c. temperature** The temperature, T_c, at which a gas can be liquefied by the c. pressure; above this temperature the gas cannot be liquefied at any pressure. See Table 21. **c. volume** The volume of 1 g of substance at the c. temperature and c. pressure.
croceic acid 2-Naphthol-8-sulfonic acid.
crocetin $C_{20}H_{24}O_4$ = 328.4. Gardenin. **alpha- ~** The aglucone of crocin. Orange crystals, m.286, insoluble in water. Cf. *carotenoids, crocin.* **beta- ~** $C_{21}H_{26}O_4$ = 342.4. The monomethyl ester of α-c. **gamma- ~** $C_{22}H_{27}O_4$ = 355.5. The dimethyl ester of α-c.
crocic acid Croconic acid.
crocidolite $NaFe(SiO_3)_2 \cdot FeSiO_3 \cdot H_2O$. An acid-resistant asbestos. See *asbestosis.*
crocin $C_{44}H_{70}O_{28}$ = 1047. The coloring matter of saffron, *Crocus sativa.* Red powder, soluble in water; a yellow dye. Cf. *carotenoids, crocose.*
crocoisite $PbCrO_4$. Crocolite, crocoite, Siberian red lead. Red, native lead chromate.
crocoite, crocolite Crocoisite.
croconic acid $C_5H_2O_5$ = 142.1. 4,5-Dihydroxy-4-cyclo-

TABLE 21. CRITICAL TEMPERATURES AND PRESSURES OF SOME COMMON GASES

Substance	Critical temperature, °C	Critical pressure, MPa
Helium	−268	0.229
Hydrogen	−240	1.36
Oxygen	−118	5.00
Nitrogen	−147	3.41
Chlorine	144	8.07
Carbon monoxide	−140	3.51
Carbon dioxide	31	7.38
Dinitrogen oxide	36	7.26
Ammonia	133	11.4
Water	374	22.0

pentene-1,2,3-trione. Crocic acid. Yellow crystals, m.150, soluble in water. **c. acid hydride** $C_5H_5O_5$ = 145.1. A tribasic ketonic acid. Cf. *leuconic acid.*

crocose $C_6H_{12}O_6$ = 180.2. A sugar and split product of crocin. Colorless crystals, soluble in water.

crocus (1) Saffron. (2) Red ferric oxide used for polishing. **antimony ~** See *antimony crocus.* **meadow ~** Colchicum. **c. martis** Colcothar.

cromolyn sodium $C_{24}H_{14}O_{11}Na_2$ = 512.3. (Di)sodium cromoglycate, Intal, Rynacrom. White powder, soluble in water. An antiallergic drug; used to treat asthma and hay fever (USP, BP).

Cronstedt, Axel, Frederik (1722–1765) Swedish metallurgist, noted for his discovery of nickel and ore classification.

Crookes, Sir William (1832–1919) English physicist and chemist, founder of *Chemical News.* **C. glass** An optical glass which eliminates many solar ultraviolet rays. **C. radiometer** An evacuated glass bulb containing a shaft with 4 vanes; each vane, which revolves, has one side black, and the other silvered. **C. space** A dark space around the cathode of a low-pressure, high-voltage X-ray tube. **C. tube** A highly exhausted vacuum tube.

crookesite A mineral containing Tl (17%), Se, Cu, and Ag.

crore Indian unit of 10^7. Cf. *lakh.*

crotaconic acid $C_5H_6O_4$ = 130.1. An isomer of itaconic acid derived from (*E*)-2-butenoic acid.

crotaline A protein in rattlesnake venom.

crotalotoxin $C_{34}H_{54}O_{21}$ = 798.8. A crystalline principle from rattlesnake venom, *Crotalus adamanteus.*

crotamiton $C_{13}H_{17}ON$ = 203.4. *N*-ethyl-*o*-crotonotoluidide, Eurax. Colorless liquid, slightly soluble in water. An antipruritic (USP, BP).

Croton A genus of euphorbiaceous plants, yielding, e.g.: lac, *C. laciferus;* cascarilla bark, *C. cascarilla;* croton oil, *C. tiglium.* **c. oil** A yellow oil from the seeds of *C. tiglium,* d.0.940–0.955, insoluble in water; a drastic purgative and local irritant. Cf. methyl*butenoic acid, tiglium.*

crotonal The 2-butenylidene* radical.

crotonaldehyde* MeCH:CH·CHO = 70.1. Propylene aldehyde. **(*E*)-~** Colorless liquid, b.104, soluble in water; a solvent for fats and resins.

crotonarin The solid part of croton oil.

crotonbetaine $C_7H_{13}NO_2$ = 143.2. A base from beef muscle extract. Cf. *carnitine.*

crotonic acid* MeCH:CH·COOH = 86.1. (*E*)-2-Butenoic acid*, α-butenic acid. Colorless monoclinics, m.72, soluble in water. **iso ~*** (*Z*)-2-Butenoic acid*, β-crotonic acid. Colorless liquid, d.1.0252, m.15, decomp. 171. **methyl ~ 2-** See *butenoic acid.* **3- ~** Senecioic acid. **4- ~** Pentenoic acid*.

c. aldehyde Crotonaldehyde*. **c. anhydride** (MeCH:CH·CO)$_2$O = 154.2. 2-Butenoic anhydride*. Colorless liquid, b.247.

crotonoid An atomic arrangement in which an atom with free electrons (e.g., O or N) is bound by a double bond to carbon adjacent to an ethylene linkage; as R·CH:CH·HC:O. It forms coordinate compounds. Cf. *conjugated double bonds.*

crotonol Crotonolic acid.

crotonolic acid $C_9H_{14}O_2$ = 154.2. A purgative monobasic acid from croton oil.

crotonoyl* Crotonyl. The radical MeCH:CHCO−. Has (*E*) and (*Z*) forms.

crotonyl The radicals (1) 2-butenyl*, MeCH:CH·CH$_2$− and (2) crotonoyl*. **c. alcohol** 2-Buten-1-ol*.

crotonylene* 2-Butyne*.

crotoxin An active protein, containing sulfur, from snake venom. It contains 18 common amino acids.

crottel A vegetable dye resembling cudbear, from Scotland.

crotyl The 2-butenyl* radical. **c. alcohol** 2-Butenol*.

crown **c. ether** See *ethers.* **c. filler** $CaSO_4 \cdot 2H_2O$. Hydrated calcium sulfate; a paper filler. **c. glass** Glass. **c. top** Rose *burner.*

CRT Cathode ray tube.

crucible (1) A conical vessel with rounded base for fusing or incinerating. (2) The hearth of a blast furnace. **assay ~** A small procelain c. for the combustion of drugs or precipitates, in quantitative analysis. **Gooch ~** A c. with a perforated bottom for filtrations in analysis. **Hessian ~** A large clay c. for metallurgical work. **Munroe ~** A c. similar to a Gooch c. with spongy platinum deposited on filter paper as the filtering medium. **nickel ~** A fusion c. **platinum ~** A small platinum c. used in chemical analysis. **quartz ~** A c. of transparent quartz, used for high-temperature combustion. **Rose ~** A c. lid fitted with an inlet tube for burning a substance in a current of coal gas. **sillimanite ~** A superior assay c. **sintered glass ~** A c. with a base of sintered glass as filtering medium.

c. etching Diamond *ink.* **c. furnace** An electrically heated resistance wire embedded in a refractory material, which attains temperatures of 1000°C in 30 min. **c. holder** A rubber ring in a glass funnel for holding Gooch crucibles. **c. steel** Pot steel; made by the c. process. **c. tong** Scissorlike metal tongs for handling crucibles. **c. triangle** A wire or pipeclay triangle for supporting a c. over a burner.

Cruciferae The mustard family; herbs with pungent, watery juice and flowers of 4 petals and sepals, crosswise arranged; e.g.: *Brassica (Sinapis) alba,* white mustard seeds; *B. napus,* rape seed oil; *Cochlearia armoracia,* horseradish; *Isatis tinctoria,* indican, woad.

crude Unrefined or raw; e.g., c. chemicals (technical and unrefined substances, the hydrocarbons obtained from coal tar); drugs (roots, leaves, etc.). Cf. *intermediate.*

Crum Brown rule See *Crum Brown rule* under *Brown.*

cruorine Early name for hemoglobin. Cf. *erythrocruorin.*

crushing Hammering to pieces; as, opposed to *grinding,* q.v.

crutcher A mixing machine used in the soap industry.

Crylor Trademark for a polyamide synthetic fiber.

cryogenics Low-temperature operations, generally below −100°C.

cryogenin A substance producing a low temperature.

cryohydrate Cryosel. A salt that contains water of crystallization only at a low temperature; e.g., a eutectic mixture of salt and ice.

cryohydric point The temperature at which a cryohydrate crystallizes from a freezing mixture.

cryolac number The proportion of the freezing point depression of milk accounted for by the chloride and lactose present. It indicates addition of water.

cryolite Na_3AlF_6. A pale-gray mineral, d.3.0; a source of aluminum, alum, and caustic soda. Cf. *Thomsen process.* **c. glass** Milk *glass.*

cryometer A thermometer for low temperatures.

cryoscope A device to determine the freezing point of a liquid. Cf. *Hortvet cryoscope.*

cryoscopic method The determination of the molecular weight of an organic substance from the depression of the freezing point of a solution containing it. Cf. *Raoult's law.*

cryoscopy The study of physical and chemical phenomena at low temperatures; especially the depression of freezing point. Cf. *kryoscopy.*

cryosel Cryohydrate.

cryostat A low-temperature thermoregulator.

cryosurgery Surgery in which the tissue is cut, coagulated, or removed by a cryoprobe.

cryothod Podzol.

cryptal $C_{10}H_{16}O$ = 152.2. 4-Isopropylcyclohexene aldehyde. A constituent of the oil from *Eucalyptus hemiphloia.*

cryptates Cryptan(d)s. Bicyclic molecules of general formula

$$N[(CH_2)_2 \cdot O \cdot (CH_2 \cdot CH_2 \cdot O)_m \cdot (CH_2)_2]_2N$$
$$\llcorner [(CH_2)_2 \cdot O \cdot (CH_2 \cdot CH_2 \cdot O)_n \cdot (CH_2)_2] \lrcorner$$

Named cryptate-*mmn*. Form complexes with metal ions. Cf. crown *ethers.*

cryptidine $C_{11}H_{11}N$ = 157.2. An alkaloid formed by the dry distillation of quinine. Cf. *kryptidine.*

cryptocarine An alkaloid from the bark of *Cryptocarya australis* (Lauraceae), Queensland.

cryptocrystalline Microcrystalline.

cryptocyanine Kryptocyanine.

cryptogamia A division of plants characterized by having no true flowers and propagated by spores; e.g.: Thallophyta (algae, lichens, fungi), Bryophyta (mosses, liverworts), Pteridophyta (filices, ferns). Cf. *Phanerogamia.*

cryptohalite An ammonium fluorosilicate; a white efflorescence at the mouth of fumaroles and burning coal mines.

cryptometer An optical wedge to determine the covering power of paint.

cryptophanic acid $C_5H_9O_5N$ = 163.1. A dibasic acid constituent of urine.

cryptopine $C_{21}H_{23}O_5N$ = 396.4. An alkaloid of opium, m.217; an anodyne.

cryptopyrrole $C_8H_{13}N$ = 123.2. 3-Ethyl-2,4-dimethyl-pyrrole. A base derived from hemin and chlorophyllin. Colorless liquid, $b_{13mm}85$.

cryptoscope Fluoroscope.

cryptovalency Abnormal valency, e.g., tetravalent oxygen.

cryptoxanthin $C_{40}H_{55}OH$ = 552.9. Kryptoxanthin, caricaxanthin. A carotenoid, from the berries of the *Physalis* species. α-~ m.175. β-~ m.169. It is a precursor of vitamin A.

cryst. Abbreviation for crystalline, crystallization.

crystal A homogeneous and angular solid of definite form which is characterized by geometrically arranged plane surfaces (faces) and a symmetrical internal structure. See *crystal structure.* General types: homopolar, ionic, and metallic. See Fig. 10 and *crystal structure.* C. data can be set out as follows: (1) Molecular formula and formula weight. (2) Melting point. (3) System and point group. (4) Unit cell parameters (translations in A) and volume of cells, A^3. (5) Measured density d_m, number Z of molecules in unit cell, and calculated density D_c. (6) Type(s) of X-rays used, absorption coefficient *a*, and experimental methods. (7) Space group, and molecular symmetry implied. (8) Optical data. **acicular** ~ A needle-shaped c. **arborescent** ~ A slender and branching c. resembling a tree. **blood** ~ Hemin. **complex** ~ A c. with dissimilar faces. **double** ~ Twin crystals. **hemihedral** ~ A c. having half as many faces as the geometrical pattern demands. **holohedral** ~ A c. having all the faces that the geometrical pattern demands. **homopolar** ~ Nonpolar. A c. having a space lattice of atoms in which all valencies are satisfied; characteristic of organic compounds. **ionic** ~ Polar. A c. consisting of a space lattice of ions; as, Na^+ and Cl^-. Hence the entire c. is a giant molecule; characteristic of inorganic compounds. **lead chamber** ~ Nitrosyl hydrogensulfate*. **liquid** ~ A substance, usually organic with at least one polarizable group, capable of unidirectional molecular alignment in layers, giving rise to strong birefringence. L. c. are either lyotropic (solvent-induced) or thermotropic (heat-induced). A commonly used mixture for l. c. displays, E7, consists of 3 cyanobiphenyls and a cyanotriphenyl. **cholesteric** ~ L. c. in layers, with the long axes of the molecules parallel to the plane of each layer, but turning slightly from the corresponding pattern in adjacent layers. **nematic** ~ (Greek "thread".) Threadlike structures, the long axes of the l. c. molecules being parallel, but not separated into layers. **smetic** ~ A series of layers of l. c., the long axes of the molecules in each layer being perpendicular to its plane. The layers are free to slide, giving the properties of a two-dimensional fluid. **metallic** ~ Coordinate. A c. consisting of a space lattice of positive ions and electrons in which the electrons conduct an electric current. **micro** ~ A c. of microscopic size. **mixed** ~ A c. that contains 2 or more isomorphous substances, as, aluminum chromium sulfate. **nonpolar** ~ See *polar.* **polar** ~ Ionic. **racemic** ~ A c. composed of 2 optically compensating isomers. **seed** ~ Crystallon. A c. introduced into a saturated solution as a nucleus for crystallization. See *cloud seeding.* **simple** ~ A c. that belongs to a definite c. system. **Teichmann's** ~ The hemin crystals of blood smears. **twin** ~ Two crystals grown together along a common face.

c. alcohol Alcohol molecules in a c. structure. **c. ammonia** The ammonia of crystallization in the ammonates. **c. axis** An imaginary line through the center of a plane of a c. See *crystal systems.* **c. carbonate** $Na_2CO_3 \cdot H_2O$. The monohydrate of sodium carbonate. **c. chloroform** Chloroform molecules in c. structure. **c. coordinates** The designation of the axes of a c. as derived from a crystallogram. See Fig. 11 and *crystal systems.* **c. detector** A crystal which transmits electric current in one direction only; used to rectify alternating currents; as, galena. **c. ether** Ether molecules in a c. structure. **c. face** A plane surface of a c. **c. form** The external geometrical shape of a c. See *crystal systems.* **c. overgrowth** The growth of one c. around another, shown chiefly by isomorphous crystals. **c. pattern** *Space lattice,* q.v. **c. pickup** A pickup using the piezoelectric properties of a crystal to turn vibrations into an emf. **c. set** Early form of

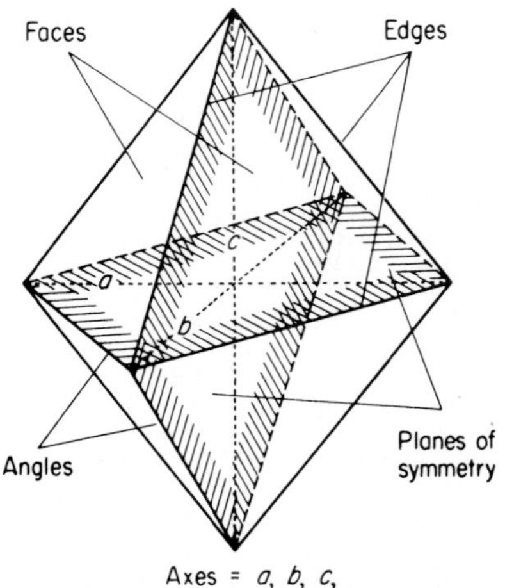

Fig. 10. Crystal.

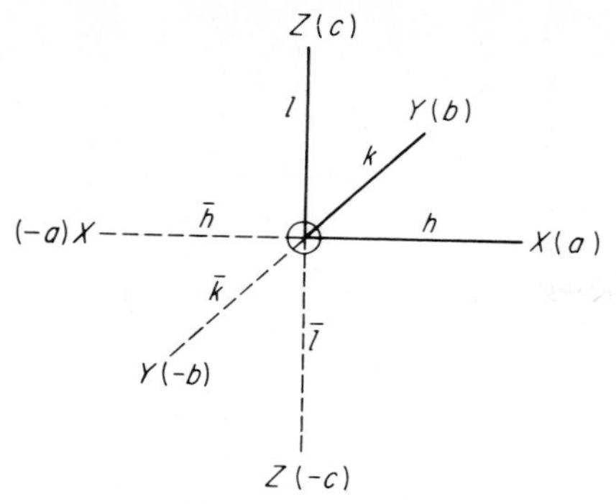

Fig. 11. Crystal coordinates. *h, k, l* are the integers (Miller indices) indicating the number of atomic planes on that axis.

radio, utilizing a crystal–cat's whisker (a pointed wire) combination. Radio signals are rectified to give an audio signal. **c structure** The internal structure of a c. as revealed by X-ray diffraction measurements. The individual atoms are arranged in a definite pattern (space lattice), and the rows of atoms act as a diffraction grating for very short X-rays. See Fig. 12. Cf. *atomic* plane. **c. systems** The 7 fundamental systems of crystallography. The simplest means of classifying crystals is in terms of their axes. See Fig. 13. **c. violet** Methyl violet. **c. water** Water of crystallization; the water molecules which are part of the crystalline structure of a substance; e.g., monohydrate, $X \cdot H_2O$; octahydrate, $X \cdot 8H_2O$.

crystalbumin An albuminoid in the crystalline lens of the eye.

crystallin (1) A solution of pyroxylin 1 pt. in methanol 4 and pentyl acetate 15 pts.; used similarly to collodion. (2) A group of proteins, found in the eye lens of vertebrates. **alpha- ~** Coagulating at 72; mol. wt. about 800,000. **beta- ~** Coagulating at 63. **gamma- ~** Mol. wt. 20,000–30,000.

crystalline Pertaining to crystals. **micro ~** Pertaining to crystals of microscopic size.

crystallite An imperfectly formed crystal.

crystallization The change from the dissolved, molten, liquid, or gaseous state to a solid state of ordered and

Fig. 12. Crystal structure (sodium chloride).

characteristic shape. **fractional ~** Repeated c. for the purification of a substance. **heat of ~** See *heat of crystallization* under *heat*. **liquid of ~** The molecules of solvent which enter into the space lattice of crystals.

crystallize (1) To assume crystalline shape. (2) To cause crystallization.

crystallized Formed into crystals.

crystallogram The photographic record obtained when X-rays are diffracted by a crystal. Cf. *X-ray spectrometer, halo, corona*. **Bragg ~** de Broglie. Siegbahn. A spectrumlike pattern produced when monochromatic X-rays pass through a slit and a rotating crystal. Cf. *Moseley spectrum*. **Clark ~** Duane. A curve obtained by plotting the ionization current at different angles when X-rays pass through a stationary crystal. **deBroglie ~** Bragg. **Debye ~** Scherer. Hull. A spectrumlike pattern of curved and concentric lines, produced on a film surrounding a crystal, which diffracts a beam of monochromatic X-rays. **Duane ~** Clark. **Hull ~** Debye. **Laue ~** The original c., in which polychromatic X-rays pass through a pinhole and a single stationary crystal, behind which is a plain photographic plate. **modified Laue ~**, **monochromatic pinhole ~** A modified Laue method using monochromatic X-rays. **Polyani ~** Siegbahn. **Scherer ~**

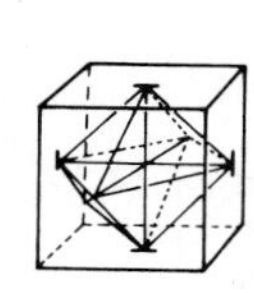

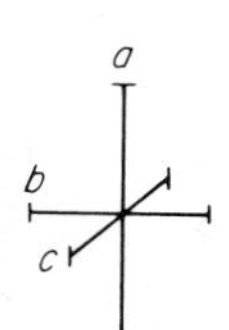

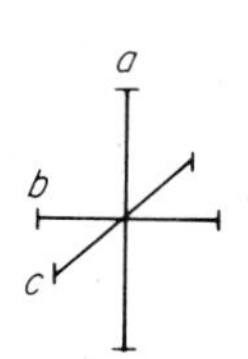

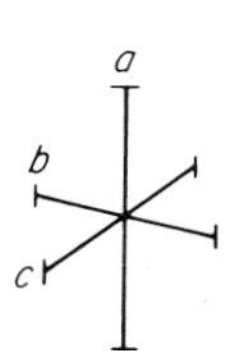

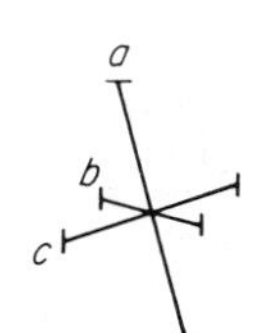

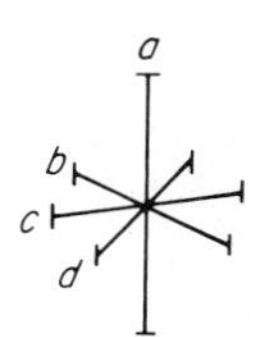

System	Interaxial angles	Length of axes	System	Interaxial angles	Length of axes
1. Isometric (cubic)	$\alpha = \beta = \gamma = 90^\circ$	a = b = c			
2. Tetragonal	$\alpha = \beta = \gamma = 90^\circ$	$a = b \lessgtr c$	5. Triclinic	$\alpha \lessgtr \beta \lessgtr \gamma \lessgtr 90^\circ$	$a \lessgtr b \lessgtr c$
3. Orthorhombic (rhombic)	$\alpha = \beta = \gamma = 90^\circ$	$a \lessgtr b \lessgtr c$	6. Hexagonal	$\alpha = \beta = 90^\circ$; $\gamma = 60^\circ$ and 120°	$a = b \lessgtr c$
4. Monoclinic	$\alpha = \gamma = 90^\circ$; $\beta \lessgtr 90^\circ$	$a \lessgtr b \lessgtr c$	7. Trigonal (rhombohedral)	$\alpha = \beta = \gamma \lessgtr 90^\circ$	a = b = c

Fig. 13. Crystal systems.

Debye. **Schiebold ~** Siegbahn. **Siegbahn ~** Polyani. Schiebold. A Laue pattern produced by monochromatic X-rays and a slowly rotating single crystal.
crystallographic apparatus A device to measure the angles and optical properties of crystals; e.g., goniometer.
crystallography The study of crystals.
crystalloids Noncolloidal substances which pass through a semipermeable membrane (obsolete).
crystalloluminescence Light emitted during crystallization, e.g., by arsenic oxide.
crystallon Seed *crystal.*
crystallose A soluble sodium salt of saccharin; a diabetic sweetening agent.
Crystolon Trademark for silicon carbide.
crystule A unit cell of a crystal.
CTAB Cetyl- (or hexadecyl-) trimethylammonium bromide. See *cetrimide.*
CTS Anhydrous aluminum sodium sulfate, a cream of tartar substitute.
cubanite $CuFe_2S_4$. A copper ore.
cube (1) A regular solid bounded by 6 equal square plane faces, the opposite faces being parallel. Cf. *space lattice.* **hydrogen ~** Hydrone. A c.-shaped alloy, Na 35, Pb 65%, for the rapid preparation of hydrogen. **oxygen ~** Cubes of sodium peroxide, with a trace of cupric oxide, for the rapid preparation of oxygen (ozone). (2) Cubé. An extract from c. root, *Lonchocarpus nicou* (Leguminosae), Peru; a fish poison and insecticide. It contains rotenone, toxicarol, and tephrosin. Cf. *derris.*
cubeb(s) The dried unripe fruit of *Piper cubeba* (Piperaceae), Java; an expectorant. **c. camphor** $C_{15}H_{20}O$ = 216.3. The solid portion of oil of c. **c. oil** Essential oil of c.; an aromatic, colorless liquid, d.0.905–0.925, b.175–180; contains cadinene, cubebene, and c. camphor.
cubebene $C_{15}H_{24}$ = 204.4. A liquid hydrocarbon from cubeb oil.
cubebic acid $C_{13}H_{14}O_7$ = 282.3. An amorphous, resinous acid from cubeb.
cubebin $C_{20}H_{20}O_6$ = 356.4. Colorless needles, m.132.
cubic Abbrev. cu. Pertaining to a cube; having three dimensions. **c. centimeter** See *cubic centimeter* under *centimeter.* **c. decimeter** 1 dm^3 = 1 liter = 1000 cm^3 = 61.0237 in^3. **c. foot** 1 ft^3 = 28,316.85 cm^3. **c. inch** 1 in^3 = 16.387 cm^3. **c. meter** 1 m^3 = 10^6 cm^3 = 1.308 yd^3 = 35.315 ft^3. **c. system** See *crystal systems.* **c. yard** 1 yd^3 = 0.765 m^3.
cubical Relating to a cube or to three dimensions. **c. atom** A theory of atomic structure (G. N. Lewis). In a stable atom or compound there are 8 valence electrons situated at the corners of a cube. **c. expansion** Volume expansion. The enlargement of a solid in 3 planes. **c. system** The isometric or regular *crystal system,* q.v.
cubit An obsolete linear measure; 1 c. = 18 in. **Bible ~** 21.8 in.
cucaivite AgCuSe. Native silver selenide.
cucoline $C_{19}H_{23}O_4N$ = 329.4. An alkaloid from the root of *Cocculus* or *Sinomenium diversifolius* (Menispermaceae). White needles, m.162, insoluble in water. **hydro ~** $C_{19}H_{25}O_4N$ = 331.4. Colorless crystals, m.198.
cucumber (1) The vine *Cucumis sativus* (Cucurbitaceae). (2) C. fruit used as food. **bitter ~** Colocynth. **oil of ~** Gourd oil.
Cucurbitaceae The gourd family of succulent herbs, generally creeping and climbing by tendrils. They contain drugs—roots: *Bryonia alba (dioica),* bryonin; fruits: *Citrullus colocynthis,* colocynth; seeds: *Cucurbita pepo,* pumpkin seed, curcurbitine; resin: *Ecballium elaterium,* elaterium.
cucurbitine An alkaloid from the seeds of *Cucurbita pepo.*
cucurbitol $C_{24}H_{40}O_4$ = 392.6. An alcohol from the seeds of the watermelon, *Cucumis citrullus* (Cucurbitaceae). Colorless crystals, m.260.
cucus oil Vetiveria oil
cudbear Persio, orchil. A purple coloring matter from lichens; also made by the action of ammoniacal substances (e.g., urine) and air on certain lichens. Known in classical times, but the name is derived from a manufacturer, Cuthbert Gordon (1758). Used as a food color and for pharmaceutical preparations.
cuinoline Cinnoline*.
cullet Crushed or broken glass.
culture (1) A growth of microorganisms. (2) C. *media,* q.v. (3) To breed, incubate, or grow microorganisms. **direct ~** A growth obtained from a natural source (tissue, sputum, etc.), and directly transferred to a c. medium. **pure ~** A growth obtained from a single species.
c. apparatus A device for growing microorganisms. **c. dish** A *petri,* q.v., or similar dish. **c. flask** A glass flask for growing microorganisms in liquid c. media. **c. media** (sing. c. medium.) The substances on which bacteria are grown; e.g.: *agar,* q.v., bouillon, gelatin glucose, blood serum. **c. slide** A microscope slide with cavities to hold a small quantity of liquid. **c. tube** A plain-rim test tube for c. media.
cumal Cumylene. The radical 4-$Me_2CH \cdot C_6H_4 \cdot CH$=, from cumaldehyde.
cumaldehyde $Me_2CH \cdot C_6H_4 \cdot CHO$ = 148.2. *p*-Isopropylbenzaldehyde. Cumic aldehyde. Colorless liquid, b.235, insoluble in water.
cumalin Coumalin.
cumaric acid Coumaric acid.
cumarin Coumarin*.
cumene* $PhCHMe_2$ = 120.2. Isopropylbenzene, (1-methylethyl)benzene†. Colorless liquid, b.153, insoluble in water; a constituent of cumin oil. **methyl ~** Cymene*. **pseudo ~** See *pseudocumene.*
cumenol Cuminol.
cumenyl* Cumyl. The radical $Me_2CH \cdot C_6H_4$—, from cumene; several isomers. **c.acrylic acid** $C_{12}H_{14}O_2$ = 190.2. A monobasic homolog of cinnamic acid. **c.amine** Cumidine. **c.sulfurous acid** $C_9H_{11} \cdot HSO_3$ = 200.3. A monobasic acid, derived from cumene by the action of sulfuric acid.
cumic Pertaining to the acid derived by oxidation from oil of *cumin,* q.v. **c. acid** $C_{10}H_{12}O_2$ = 164.2. *p*-Isopropylbenzoic acid. Colorless triclinics, m.117, sparingly soluble in water. **c. alcohol** Cuminol. **c. aldehyde** Cumaldehyde.
cumidic acid $C_{10}H_{10}O_4$ = 194.2. Dimethylphthalic acid. A dibasic acid oxidation product of durene. Colorless crystals, m. exceeds 320.
cumidine $C_9H_{13}N$ = 135.2. *p*-Isopropylaniline. Colorless liquid, b.225, insoluble in water. **pseudo ~** See *pseudocumidine.*
cumidino The radical 4-$Me_2CH \cdot C_6H_4 \cdot NH$—, from cumidine.
cumin The fruits of *Cuminum cyminum* (Umbelliferae), Europe; a stimulant, aromatic, and sedative. **c. oil.** Roman c. oil. A limpid liquid with a sharp taste, from c. seeds, d.0.900–0.930, containing cymene and cumaldehyde.
cuminal (1) Cumal. (2) Cumaldehyde. **di ~** Cuminil.
cuminalcohol Cuminol.
cuminaldehyde Cumaldehyde.

cuminic **c. acid** Cumic acid. **c. alcohol** Cuminol. **c. aldehyde** Cumaldehyde.
cuminil $C_9H_{11}CO \cdot COC_9H_{11}$ = 294.4. Dicuminoketone. Colorless crystals, m.84.
cuminol $C_3H_7 \cdot C_6H_4CH_2OH$ = 150.2. Isopropylbenzyl alcohol. Colorless liquid, from cumin seeds.
Cummings pump A mercury-vapor vacuum pump.
cumobenzyl alcohol See *phenyl-paraffin alcohols.*
cumol (1) Pseudocumene. (2) An oily mixture of trimethyl benzenes from coal tar, b.168–178.
cumulative double bonds* Bonds present in a chain with at least 3 contiguous C atoms joined by double bonds; as, $-C=C=C-$. **non ~*** All other configurations of 2 or more double bonds.
cumulenes* Generic name for compounds containing at least 3 contiguous double bonds.
cumyl Cumenyl*.
cumylene Cumal.
cumylic acid Durylic acid.
cunene Cupriethylenediamine solution; a solvent in cellulose viscosity determinations.
cunit Equal to 100 ft^3 of solid debarked wood. Cf. *unit, cord, board foot.*
cuorin $C_{71}H_{125}O_{21}NP_2$ = 1391. A phospholipid from the cow's heart.
cup A domestic measure, about 240 ml.
cupal Copper-plated aluminum.
cupaloy Trade name for a high electrical conductivity, corrosion-resistant alloy of Cu, Ag, and Cr.
cupel A flat crucible of bone ash, used in cupellation. **c. mold, mould** A brass cup with pestle for making cupels. **c. rake** An iron shovel or spatula for handling cupels. **c. tongue** A steel tong for removing cupels from the oven.
cupellation The separation of silver or gold. Unrefined metal is mixed with lead and placed in a cupel in the muffle furnace, where the impurities are volatilized or absorbed by the cupel; a button of noble metal is left.
cupferron $NH_4O \cdot NO \cdot NPh$ = 155.2. Ammonium salt of *N*-hydroxy-*N*-nitrosoaniline, copperon. Yellow crystals, m.164. Its acid solution is a quantitative precipitant reagent for aluminum, titanium, zirconium, etc. **styryl ~** 4-Stilbenzyl nitrosohydroxylamine. Used similarly to c., but gives less soluble metal complexes.
cupola A dome-shaped furnace for melting pig iron.
cupra Abbreviation for cuprammonium, e.g., c. rayon.
Cuprador Trademark for a mixed cuprammonium and polyacrylonitrile synthetic fiber.
Cupralon Trademark for a mixed cuprammonium and polyurethane synthetic fiber.
cuprammonia A solution of cupric hydroxide in ammonia water; a solvent for cellulose.
cuprammonium **c. ion** Tetraamminecopper(II) ion*. The $Cu(NH_3)_4^{++}$ ion having the characteristic deep-blue color obtained by adding excess ammonia to a cupric salt solution. **c. rayon, c. silk** See *rayon.* **c. sulfate** The solution $CuSO_4 + 4NH_3 + H_2O$. **c. viscosity** The viscosity of a solution of cellulose in c. measured under standard conditions; it measures the degree of polymerization of cellulosic materials.
cuprate* Indicating copper as the central atom(s) in an anion; as, **triethynyl c.(I)** The ion $[Cu(C_2H)_3]^{2-}$. **c. silk** Cuprammonium *rayon,* q.v.
cuprea bark The bark of *Remijia,* tropical America, which yields cinchona alkaloids.
cupreane Deoxycupreine, cupreidine, deoxycupreidine. Alkaloids from cuprea bark.
cupreidine $C_{19}H_{22}O_2N_2$ = 310.4. The (+) isomer of cupreine; a cinchona alkaloid from cuprea bark.
cupreine $C_{19}H_{22}O_2N_2$ = 310.4. (−)-Cupreidine. A cinchona alkaloid from cuprea bark. Colorless crystals, m.198, slightly soluble in water. **deoxy ~** See *cupreane.* **methyl ~** Quinine.
cuprene $(C_{11-15}H_{10})_n$. Carbene. A brown, solid polymerization product of heating acetylene in presence of copper.
cupreol $C_{20}H_{34}O$ = 290.5. A cholesterol-like substance from the bark of *Cinchona calisaya* (Rubiaceae).
Cupresa Trademark for a cuprammonium synthetic fiber.
cupressin An oil, *Cypress* species, formerly used to treat whooping cough.
cupressus Cypress.
cupric Copper(II)* or (2+)*. Cuprum(II)*. Describing compounds of divalent copper, which give the cupric Cu^{++} ion in aqueous solution. **c. abietinate** $Cu(C_{19}H_{27}O_2)_2$ = 638.4. Green scales, insoluble in water, soluble in oils (green-colored solution); a wood preservative. **c. acetate** $Cu(C_2H_3O_2)_2 \cdot H_2O$ = 199.6. Verdigris, cryst. aerugo, crystals of Venus. Bluish-green crystals, m.250 (decomp.), soluble in water; an astringent, mordant, porcelain paint, and enamel. **c. subacetate** See *c. subacetate.* **c. arsenate** $Cu_3(AsO_4)_2 \cdot 4H_2O$ = 540.5. Bluish-green crystals, insoluble in water. **c. arsenite** $CuHAsO_3$ = 187.5 or $Cu(AsO_3)_2$ = 309.4. A mixture of the neutral and acid salt. Green crystals, insoluble in water. **c. benzoate** $Cu(PhCOO)_2 \cdot 2H_2O$ = 341.8. Blue crystals, insoluble in water. **c. borate** CuB_4O_7 = 218.8. Green, crystalline powder, insoluble in water; used in ceramics and paints. **c. bromide** $CuBr_2$ = 223.4. Gray crystals, decomp. by heat, soluble in water. **c. butyrate** $Cu(C_4H_7O_2)_2 \cdot 2H_2O$ = 273.8. Green monoclinics, slightly soluble in water; a reagent for essential oils. **c. carbonate** Only basic carbonates are known, e.g., $3CuO \cdot 2CO_2 \cdot H_2O$, copper lasur, mountain blue. Used in the control of seed wheat smut, as a pigment, and in ceramics. **c. chlorate** $Cu(ClO_3)_2 \cdot 6H_2O$ = 338.5. Blue, hygroscopic crystals, soluble in water; a mordant. **c. chloride** $CuCl_2$ = 134.5. Anhydrous copper dichloride*. Brown crystals, m.498 (decomp.), soluble in water; a mordant. **crystalline ~** $CuCl_2 \cdot 2H_2O$ = 170.5. Blue rhombs, losing 2 moles water at 100°C, and decomp. at red heat; soluble in water. Used as a reagent, as a mordant, and in the manufacture of sympathetic inks. **c. chromate** $CuCrO_4$ = 179.5. Yellow liquid. **basic ~** $CuCrO_4 \cdot 2Cu(OH)_2$ = 374.7. Basic copper chromate. Brown powder, insoluble in water; a mordant. **c. citrate** $Cu_2C_6H_4O_7 \cdot 2\frac{1}{2}H_2O$ = 360.2. Green powder, soluble in water; a reagent for glucose. **c. cyanide** $Cu(CN)_2$ = 115.6. Green powder, soluble in water. Cf. *cuprocupric cyanide.* **c. dichromate** $CuCr_2O_7 \cdot 2H_2O$ = 315.6. Brown crystals, soluble in water. **c. ferrocyanide** Copper hexacyanoferrate(II)*. **c. fluoride** $CuF_2 \cdot 2H_2O$ = 137.6. Blue crystals, sparingly soluble in water. **c. formate** $Cu(HCOO)_2$ = 153.6. Blue monoclinics, soluble in water. **c. hexafluorosilicate*** C. silicofluoride. (1) *tetrahydrate.* $CuSiF_6 \cdot 4H_2O$ = 277.7. Blue prisms, soluble in water. (2) *hexahydrate.* $CuSiF_6 \cdot 6H_2O$ = 313.7. Blue octahedra, soluble in water. Used to color and harden marble; also a disinfectant. **c. hydroxide** $Cu(OH)_2$ = 97.6. Blue crystals, decomp. by heat, insoluble in water. **c. iodide** CuI_2 = 317.3. Brown powder, soluble in water. **c. ion** The Cu^{2+} ion. **c. lactate** $Cu(C_3H_5O_3)_2 \cdot 2H_2O$ = 277.7. Green crystals, slightly soluble in water. **c. nitrate** (1) $Cu(NO_3)_2 \cdot 3H_2O$ = 241.6. Blue prisms, m.115, decomp. 170, soluble in water; a reagent for detecting oxygen, and used to prepare

photosensitive papers. (2) $Cu(NO_3)_2 \cdot 6H_2O$ = 295.6. Blue crystals, m.26 (decomp.), very soluble in water. Used as (1). **c. nitrite** $Cu(NO_2)_2$ = 155.6. Green, unstable powder of variable composition, soluble in water. **c. nitroprusside** $CuFe(CN)_5(NO)$ = 279.5. Green, granular, photosensitive powder, insoluble in water. **c. oleate** $Cu(C_{18}H_{33}O_2)_2$ = 626.5. Green wax, insoluble in water. **c. oxalate** $Cu(OOC)_2$ = 151.6. Green powder, insoluble in water. **c. oxide** CuO = 79.5. Black crystals, d.6.4, m.1064, insoluble in water or potassium cyanide solution. Used as a reagent, and in ceramics and glass to produce blue and green colors. **c. oxychloride** Copper dichloride oxide*. **c. palmitate** $Cu(C_{16}H_{31}O_2)_2$ = 574.4. Blue powder, m.115, insoluble in water. **c. phosphate** $Cu_3(PO_4)_2 \cdot 3H_2O$ = 434.6. Blue rhombs, slightly soluble in water. **acid ~** $CuHPO_4$ = 159.5. Green powder, insoluble in water. A reagent for carbon dioxide in water. **c. phosphide** Cu_3P_2 = 252.6. Metallic powder, insoluble in water; used to manufacture phosphor bronze. **c. potassium chlorate** $Cu(ClO_3)_2 \cdot 2KClO_3$ = 475.5. Green crystals, soluble in water. **c. potassium chloride** $CuCl_2 \cdot 2KCl \cdot 2H_2O$ = 319.6. Green crystals, soluble in water. **c. potassium cyanide** $Cu(CN)_2 \cdot 2KCN \cdot 2H_2O$ = 281.8. Green crystals, soluble in water. **c. potassium tartrate** $K_2Cu(C_4H_4O_6)_2$ = 437.9. Blue scales, soluble in water. **c. salicylate** $Cu(C_7H_5O_3)_2 \cdot 4H_2O$ = 409.8. Green needles, soluble in water. **c. selenate** $CuSeO_4 \cdot 5H_2O$ = 296.6. Blue crystals, slightly soluble in water. **c. silicate** $CuSiO_3$ = 139.6. Green crystals, insoluble in water. **c. sodium chloride** $CuCl_2 \cdot 2NaCl \cdot 2H_2O$ = 287.4. Green crystals, soluble in water. **c. stearate** $Cu(C_{18}H_{35}O_2)_2$ = 630.5. Blue, amorphous powder, insoluble in water; a bronze for plaster. **c. subacetate** Verdigris, aerugo. Greenish-blue powder, soluble in water. The blue variety has the average composition $CuO \cdot Cu(C_2H_3O_2)_2$; the green variety $CuO \cdot 2Cu(C_2H_3O_2)_2$. Used in the manufacture of pigments (Schweinfurt green), as a mordant, and in cotton printing. **c. subcarbonate** $CuCO_3 \cdot CU(OH)_2$ = 221.1. C. carbonate. Blue monoclinics, decomp. by heat, insoluble in water. A reagent for glucose, and a pigment. **c. sulfate anhydrous ~** $CuSO_4$ = 159.6. White, amorphous powder, d.3.516, decomp. 621, soluble in water. Used to dehydrate liquids, and to detect traces of water (turns blue). **basic ~** $CuSO_4 \cdot 3Cu(OH)_2$ = 452.3. Blue powder, slightly soluble in water. **diagnostic ~** (USP) See *Clinitest tablets.* **c. s. hydrate** $CuSO_4 \cdot 5H_2O$ = 249.7. Copper sulfate, Roman vitriol, blue vitriol, bluestone. Blue, triclinic crystals, d_{16°2.286, loses 4 moles water at 100 and 5 moles at 250, soluble in 3.5 pts cold and 1 pt boiling water, insoluble in alcohol. A reagent for glucose, peptones, and picric acid; a caustic, styptic, and emetic (particularly for phosphorus poisoning). Used in the dye industry, the manufacture of green and blue pigments, electroplating, plant sprays. **c. sulfide** CuS = 95.6. Black powder or hexagons, d.3.98, insoluble in water; an antiparasitic paint for ships. **c. sulfite** $CuSO_3 \cdot H_2O$ = 161.6. Blue crystals, soluble in water. **c. tannate** Brown powder of variable composition, made by treating tannin with cupric salts; insoluble in water. **c. tartrate** $CuC_4H_4O_6 \cdot 3H_2O$ = 265.7. Blue powder, soluble in water; a reagent for glucose. **c. thiocyanate** $Cu(SCN)_2$ = 179.7. Blue powder, insoluble in water, soluble in ammonia. **c. thiosulfate** CuS_2O_3 = 175.7. Blue crystals, sparingly soluble in water.

cupricyanide Tetracyanocuprate(II)*. A compound containing the ion $[Cu(CN)_4]^{2-}$.

cupriethylenediamine $Cu[(CH_2NH_2)_2]_2(OH)_2$. A solution of ethylenediamine saturated with cupric hydroxide in which the ratio $Cu:(CH_2NH_2)_2$ is 1:2. Used as solvent for cellulose, in place of cuprammonium, for viscosity determinations.

cuprine $C_{11}H_7O_3N$ = 201.05. An alkaloid derived from cotarnine.

cupri sulphas An early name for copper sulfate.

cuprite Cu_2O. Ruberite, ruby copper, tile ore. Native cuprous oxide.

cuprocupric A complex copper salt: a mixture of cuprous and cupric salts. **c. cyanide** $Cu_3(CN)_4 \cdot 5H_2O$ = 384.8. Dicopper(I) tetracyanocuprate(II)*. Green powder; insoluble in water, soluble in potassium cyanide solution.

cuprocyanide Tetracyanocuprate(I)*. A compound containing the ion $[Cu(CN)_4]^{3-}$.

cuproine $C_{18}H_{12}N_2$ = 256.3. 2,2′-Diquinolyl. A reagent for copper (purple complex, soluble in isopentyl alcohol).

curpon α-Benzoin oxime. A quantitative precipitant for copper.

Cupro Nickel (1) Brand name for an alloy containing Cu 88.35, Ni 10, Fe 1.25, and Mn 0.40%, used for condenser plates and tubes for evaporators and heat exchangers. (2) (Not cap.) An alloy used in British *coinage*, q.v. (Table 17).

cupronine $C_{20}H_{18}O_3N_2$ = 334.4. An alkaloid derived from cotarnine.

cuprophane Transparent film made by the cuprammonium process. Cf. *rayon.*

cuprosulfate A double salt of copper sulfate and another sulfate.

cuprotungstite $CuWO_4$. A tungsten ore.

cuprous Copper(I)* or (1+)*. Cuprum(I)*. Describing a compound of monovalent copper; generally less common, and less stable than the corresponding cupric compounds. **c. acetylide** Cu_2C_2 = 151.1. Amorphous, explosive, red powder; formed from acetylene and a cupric solution. **c. bromide** Cu_2Br_2 = 286.9. Brown powder, m.484, insoluble in water. **c. carbonate** Cu_2CO_3 = 187.1. Yellow powder, decomp. by heat, insoluble in water. **c. chloride** Cu_2Cl_2 = 198.0. Resin of copper. Nantokite. Green tetrahedra, m.422, insoluble in water. A reagent in gas analysis, and for detecting arsine and stibine. **c. cupric cyanide** Cuprocupric cyanide. **c. cyanide** $Cu_2(CN)_2$ = 179.1. Colorless, amorphous powder, decomp. in red heat, insoluble in water. **c. hexafluorosilicate(IV)*** Cu_2SiF_6 = 269.2. C. silicofluoride. Red powder. **c. hydroxide** $Cu_2(OH)_2$ = 161.1. Yellow powder, insoluble in water. **c. iodide** Cu_2I_2 = 380.9. Brown crystals, m.606, insoluble in water, soluble in potassium iodide solution; when mixed with equal parts of mercuric iodide, it indicates temperature of parts of machines in frictional contact. **c. ion** The monovalent Cu^+ ion. **c. oxide** Cu_2O = 143.1. Brown, granular powder, d.5.88, insoluble in water; used in red ceramics and glass. **c. phosphide** Cu_3P = 221.6. Metallic powder insoluble in water; used to manufacture phosphorbronze. **c. potassium cyanide** $CuCN \cdot 3KCN$ = 284.9. White crystals, soluble in water. **c. sulfate** Cu_2SO_4 = 223.1. White powder, decomp. by water. **c. sulfide** Cu_2S = 159.2. Chalcocite. Black powder, m.1100, insoluble in water. **c. sulfite** $Cu_2SO_3 \cdot H_2O$ = 225.2. Brown crystals, insoluble in water. **c. thiocyanate** $Cu_2(CNS)_2$ = 243.2. C. sulfocyanate. Gray powder, m.1080, insoluble in water.

cuproxide CuO. A native copper oxide.

cuprum* Latin for "copper."

Cupuliferae A family of trees, comprising Betulaceae (birches) and Fagaceae (oaks, chestnuts, beeches,), that yield woods and drugs; e.g.: *Quercus alba,* white oak; *Q. species,* nutgalls; *Q. suber,* cork; *Betula lenta,* sweet birch oil; *Ostrya virginica,* ironwood. Cf. *quercitron.*

curangin $C_{43}H_{77}O_{20}$ = 914.0. Glucoside from *Curanga amara* (Scrophulariaceae), S. Asia. Febrifuge and vermifuge.

curare Woorara, urari, curara, curari, S. American arrow poison. A black, brittle resin of various *Strychnos* species; a motor nerve paralyzant, containing the active principles curarine and curine.

curarine $C_{40}H_{46}ON_4$ = 598.8. A crystalline alkaloid from curare. **pseudo ~** See *pseudocurarine.* **tubo ~** C., originally transported in a tube. Used, particularly in anesthesia, for its paralyzing effect; this is due to inhibition of nerve impulses to muscles by blocking the uptake of acetylcholine at the muscle nerve endings. Antidote is an anticholinesterase, e.g., neostigmine.

curcas oil An oil from the seeds of *Jatropha curcus,* Linn., physic nut (southeast Asia), d.0.919; a drying oil, emetic, and purgative.

curcin A toxic albumin of Barbados nuts, the seeds of *Curcas purgans* (Euphorbiaceae), W. Indies; resembles ricin.

curcuma Turmeric.

curcumene Terpene from the volatile oil of the rhizomes of *Curcuma aromatica.*

curcumin $[MeO(HO)C_6H_3CH:CHCO]_2CH_2$ = 368.4. 1,7-Bis(4-hydroxy-3-methoxyphenyl)-1,6-heptadiene-3,5-dione*. Orange-yellow needles, m.183, insoluble in water. Coloring matter of turmeric, the rhizome of *Curcuma longa.* An indicator, pH 7.4 (yellow) to 8.6 (brown); a reagent for beryllium; and a dye.

curdling Coagulation (of milk).

curds The precipitate obtained by curdling.

Curie **C., Irene** See *Joliot.* **C., Marie Sklodowska (1867–1934)** The French codiscoverer of radium, polonium, and radioactivity. **C., Pierre (1859–1906)** The French codiscoverer of radium, and husband of Madame C. **C. electroscope** An electroscope to detect minute amounts of radioactive substances. **C. point** The temperature above which the molecular forces of magnetism of paramagnetic bodies cease to exist. **electrical C. point** The temperature above which the increase in relative permittivity of certain crystals (e.g., phosphates) ceases. **lower C. point** The point at which permittivity measurements in an alternating field show a rapid decrease in the magnitude of the reversible polarization with decreasing temperature. **c. therapy** Radium therapy. **c. unit** Curie.

curie A unit of radioactivity; 1 Ci = 3.700×10^{10} disintegrations/s. Replaced in the SI system by the becquerel; 1 Ci = 3.700×10^{10} Bq.

curine $C_{36}H_{38}O_6N_2$ = 594.7. A paralyzant alkaloid from curare. Colorless microcrystals, m.220, soluble in water.

curite $PbO_5 \cdot UO_3 \cdot 4H_2O$. A radioactive mineral (Zaire).

curium* Cm. Element of at. no. 96. At. wt. of isotope with longest half-life (^{247}Cm) = 247.07. m.1350. Forms Cm^{3+} and Cm^{4+} compounds.

curled Describing typical growths of bacteria in parallel chains or weavy strands; as, anthrax colonies.

curling factor Griseofulvin.

current (1) A stream or flow. (2) Electric c., which moves along a conductor; its unit is the *ampere,* its quantity the *coulomb,* its potential difference the *volt.* **alternating ~** a.c. A periodically reversing electric c. **d'Arsonval ~** The high-voltage discharge of a capacitor through a wire solenoid, producing high-frequency alternations. **direct ~** d.c. A c. whose direction is always the same. **eddy ~** The c. set up around a conductor. See *eddy currents.* **Foucault ~** Electric c. induced in a mass of metal by a magnetic field of varying intensity. **high-frequency ~** An alternating c. changing its direction many times per second. **induced ~** The c. produced in an induction apparatus by a primary c. **primary ~** The c. that produces an induced c. **secondary ~** Induced c.

c. breaker A device to interrupt the electric c., e.g., a commutator or switch. **c. capacitor** A device to store electric charge. **c. changer** A device to reverse an electric c., e.g., a commutator. **c. condenser** C. capacitor. **c. density** See *density (3).* **c. regulator** A device to regulate electric c., e.g., a rheostat.

curry The powdered leaves of *Murraya koenigii* (Tutaceae), India, sometimes flavored with other spices; used to season food.

currying The incorporation of oil and grease into leather.

curtisite $C_{24}H_{18}$(?). A native, solid hydrocarbon (California).

Curtius, Theodor (1857–1928) German organic chemist. **C. reaction** The preparation of amines and urethanes by the action of water and alcohol on acid azides.

curve A continuous line that is not straight. In a graphic diagram (cf. *coordinates*) a line connecting points whose positions are defined by their abscissas and ordinates, and therefore expressing a relationship. **c. analyzer** A table with 2 moveable scales, used to measure curves.

cuscamidine Amorphous alkaloid from cusco bark.

cuscamine Crystalline alkaloid, m.218, from cusco bark.

cusco bark Red bark from *Cinchona succirubra* (Rubiaceae).

cuscohygrine Hygrine.

cusconidine Yellow, amorphous alkaloid from cusco bark.

cuscus oil Vetiveria oil.

cusp The point formed at the discontinuous join of two curves.

cusparia bark Angostura bark.

cusparine $C_{19}H_{17}O_3N$ = 307.4. 2-Homopiperonyl-4-methoxyquinoline. An alkaloid from angostura bark. Colorless crystals, m.90, insoluble in water.

cusso Kousso.

cut (1) The weight in pounds of resin added to each gallon of solvent. (2) A fraction of crude petroleum.

cutch Catechu.

cutin The waxy protective coating of plants.

cutinite The petrographic component from leaf cuticle.

cutting (1) Etching. (2) Lubricating. (3) Mixing. **c. fluid** A liquid used to keep working parts cool, e.g., in drilling. (4) Dissolving casein, e.g., in ammonia. Cf. *cut.*

cuttlefish A mollusk, genus *Sepia,* order Cephalopoda; it discharges a pigment from a gland near the liver (painter's sepia).

cwt Abbreviation for hundredweight.

Cy Abbreviation for cyanide radical, CN−.

cyamelide $(HNCO)_3$ = 129.1. 1,3,5-Trioxane-2,4,6-triimine. White, amorphous powder, insoluble in water.

cyamethine Cyanomethine.

Cyan (1) Prefix indicating the trademark of a group of blue or green phthalocyanine pigments, q.v. *(phthalocyanines).* (2) (Not cap.) Cyano-*.

Cyana Trademark for a mixed-polymer synthetic resin.

cyanacetic acid Cyanoacetic acid*.

cyanalcohol Cyanohydrin.

cyanaldehyde Cyanoaldehyde.

Cyanamid Trademark for a mixture of calcium cyanamide 65–70, calcium hydroxide 15–20%, and free carbon, used as fertilizer.

cyanamide NH:C:NH = 42.0. Urea anhydride. Colorless needles, m.46, soluble in water. **allyl ~** Sinamine. **benzyl ~** See *benzyl cyanamide.* **calcium ~** See *calcium* *cyanamide.* **diethyl ~** See *diethylcyanamide.* **diphenyl ~** Phenyl*cyanoaniline.* **phenyl ~** Cyanoaniline.

c. process A method of nitrogen fixation. Limestone and coke are heated to form calcium carbide over which (at 1000 °C) a current of nitrogen passes and produces calcium cyanamide, which is treated in autoclaves by high-pressure steam to give ammonia.

cyanamil Styracin.

cyananilide *N*-Cyanoaniline.

cyanate* A salt of cyanic acid containing the radical —OCN. **iso ~*** Carbimide. A salt of isocyanic acid containing the radical —N:C:O. **pseudo ~** Fulminate*. **sulfo ~** Thiocyanate*. **tauto ~** Fulminate*. **thio ~*** The radical —SCN.

cyanation Introduction of the radical —CN into a molecule.

cyanato-* Prefix indicating the cyanate* radical.

cyanaurite $MAu(CN)_2$. A double salt of a metal cyanide and aurous cyanide.

cyanazine* See *herbicides,* Table 42 on p. 281.

cyanethine $C_9H_{15}N_3$ = 165.2. 4-Amino-2,6-diethyl-5-methylpryimidine. Colorless crystals, m.189.

cyanetholin Ethyl cyanate*.

cyanethylamide Diethylcyanamide.

cyanic acid* HOCN = 43.0. Cyanohydroxide. Colorless, poisonous liquid, d.1.140, which polymerizes to cyamelide and fulminuric acid; its salts are cyanates. **hydro ~*** HCN = 27.03. Prussic acid. Colorless, poisonous liquid, $d_{18°}$ 0.697, m. —11, b.25, miscible with alcohol, water, or ether. Its salts are cyanides. **iso ~*** HNCO = 43.0. A monobasic acid; its salts are isocyanates. **pseudo ~, pseudoiso ~** Fulminic acid*. **sulfo ~** Thiocyanic acid*. **tauto ~** Fulminic acid*. **thio ~*** See *thiocyanic acid.* **trihydro ~** Cyanidine (1).

cyanide Nitrile*, q.v. A compound containing the radical —CN, from hydrocyanic acid. **azo ~** The radical —N:N·CN. **chloro ~** See *chlorocyanide.* **cupri ~** See *cupricyanide.* **ferri ~, ferro ~** See *hexacyanoferrate.* **hydro ~** A compound containing a HCN molecule. **iso ~** Containing the radical —NC. **sulfo ~, thio ~** Thiocyanate*.

c. poisoning C. inactivates enzymes and thus prevents oxygen uptake by all cells. Antidote: injection of dicobalt edetate plus sodium thiosulfate. **c. process** The extraction of gold from ores by leaching with potassium c. solution.

cyanidin $C_{15}H_{10}O_6 \cdot HCl$ = 322.7. An anthocyanidin, q.v. *(anthocyanidins),* from the flowers fo *Centaurea* species.

cyanidines (1) A group of compounds derived from the hypothetical trihydrocyanic acid. They contain the ring radical $C_3N_3\equiv$. (2) An anthocyanidin, q.v. *(anthocyanidins).* **trihydroxy ~** Cyanuric acid.

cyanilide Cyanoaniline.

cyanin (1) $C_{27}H_{30}O_{16}$ = 610.5. A glucoside from cornflower and other flowers, which hydrolyzes to cyanidin. (2) Anthocyanin. **phyco ~** See *phycocyanin.* **syn ~** See *syncyanin.*

cyanine (1) $C_{29}H_{35}N_2I$ = 538.5. Quinoline blue, Iodcyanin. Green metallic crystals, soluble in warm water; an indicator and photographic sensitizer. (2) Cyanin. **c. dyes** (Poly)methine dyes. A group of dyes characterized by heterocyclic rings (as, quinoline or benzothiazole) linked by one or more methine (—CH=) groups. Photographic sensitizers; e.g.: kryptocyanine (1,1′-diethyl-4,4′-carbocyanine iodide), cyanine (1,1′-diisopentylcyanine iodide), pinachrome (1,1′-diethyl-6-ethoxy-6′-methoxyisocyanine iodide). **c. hydroiodide** $C_{29}H_{35}N_2I \cdot HI$ = 666.4. Yellow needles, soluble in water.

cyanite $(AlO)_2SiO_3$. Kyanite, rhoetzite, disthene. A blue or white silicate (India), decomp. above 1100 °C to mullite and siliceous glass; a refractory for high-temperature furnace linings. Cf. *sillimanite, andalusite, mullite, fibrolite.*

cyano-* The radical N⋮C—. It acts like a halogen (forming cyanides) and like ammonia; forms many complex salts. See *cyanide.* **hexa ~** A cyclic polymer. **iso ~** The radical —NC. See *isocyanide.* **thio ~** The radical —SCN. **iso ~** The radical—NCS.

cyano salts A group of addition compounds in which the c. radical forms part of the complex ion; as, $[ZnCy_3]^-$, $[CuCy_3]^{2-}$, $[FeCy_6]^{4-}$.

cyanoacetic acid $CN \cdot CH_2 \cdot COOH$ = 85.1. Nitrilmalonic acid. Colorless crystals, m.70, soluble in water.

cyanoaldehyde $CN \cdot CH_2 \cdot CHO$ = 69.1. Nitrilmalonaldehyde. Colorless crystals, soluble in water.

cyanoaniline ***N*- ~** PhNHCN = 118.1. Cyananilide, carbanilonitrile. Colorless crystals, m.47, soluble in alcohol. ***N*-phenyl ~** $Ph_2N \cdot CN$ = 194.2. Diphenylcyanamine. Colorless crystals, m.73.

cyanobenzene See *benzonitrile.*

cyanobenziline $C_{24}H_{21}N_3$ = 351.5. Colorless crystals, m.106.

cyanobenzyl The radical $CN \cdot CH_2 \cdot C_6H_4$—, from α-benzyl cyanide. **c. cyanide** $CN \cdot CH_2 \cdot C_6H_4 \cdot CN$ = 142.2. ***ortho*- ~** α-Homophthalonitrile. Colorless crystals, m.81. ***meta*- ~** Colorless crystals, m.88. ***para*- ~** Colorless crystals, m.100.

cyanocarbonic acid Cyanoformic acid. **c. a. ester** $CN \cdot COOC_2H_5$ = 99.1. A liquid, b.115, insoluble in water.

cyanocobalamin $C_{63}H_{90}O_{14}N_{14}P \cdot Co$ = 1,357. α-(5,6-Dimethylbenzimidazoyl)cobamide cyanide. Vitamin B_{12}, antipernicious anemia vitamin, APA, Castle's extrinsic factor. Consists of a corrin nucleus bound to cobalt. The cyanide group occurs on extraction; in vivo, 5′-deoxyadenosyl occupies its place. Dark-red crystals, soluble in water. A hematopoietic used to treat pernicious anemia (USP, EP, BP). See *cobalamins.* **^{57}Co ~, ^{58}Co ~, ^{60}Co ~** Produced by the growth of certain organisms on media containing the respective cobaltous ion isotopes; used to follow cobalt absorption (USP, EP, BP).

cyanoconine $C_9H_{14}N_2$ = 150.2. 2,6-Diethyl-5-methylpyrimidine. Colorless liquid, b.205.

cyanocoumarin $C_9H_5O_2 \cdot CN$ = 171.2. 3-Cyanocoumarin. Colorless crystals, m.182.

cyanoform $HC(CN)_3$ = 91.1. Tricyanomethane*, m.93.

cyanoformate* A salt of cyanoformic acid containing the radical CN·COO—.

cyanoformic acid* CN·COOH = 71.0. A monobasic acid known only as its salts, the cyanoformates.

cyanogen (1) NC·CN = 52.0. Cyanocyanide, ethanedinitrile*, prussite, dicyanogen, oxalonitrile. Colorless, poisonous gas with odor of bitter almonds $d_{(air=1)}$ 1.806, m.—34, b.—21, soluble in water, alcohol, or ether. (2) The cyanide* (nitrile*) radical. **amino ~** Cyanamide. **bromo ~** C. bromide. **chloro ~** C. chloride. **di ~** Cyanogen. **mono ~** CN = 26.0. A compound produced by the decomposition of c. at high temperature; believed to be present in stellar bodies. **oxo ~** See *oxomonocyanogen.*

c. bands Characteristic wavelengths of emission from c. formed in arc and spark carbon electrode electrical discharges in air. **c. bromide** NCBr = 105.9. Bromine cyanide. Colorless needles, m.52, soluble in water. **c. chloride** (1) NCCl = 61.5. Chlorine cyanide. Colorless, poisonous gas or liquid, m.—5, b.15, soluble in water. (2) $C_3N_3Cl_3$ = 184.4. Solid, m.145. **c. disulfhydrate** Rubeane. **c. halides** Compounds having the general formula NCX; X is a halogen. They tend to polymerize to $N_3C_3X_3$. **c. iodide** NCI = 152.9.

Iodine cyanide. Colorless needles, soluble in water. **c. sulfide** $(CN)_2S$ = 84.1. Colorless scales, m.60, soluble in water.

cyanogenetic Yielding cyanogen; as, certain glucosides, amygdalin. Cf. *syncyanin.*

cyanohematin A compound of hematin and cyanogen.

cyanohydride Cyanohydrin.

cyanohydrin Cyanalcohol. A compound containing the radicals —CN and —OH. **c. synthesis** Addition of a carbon atom by the reaction R·HC:O + HCN → R·CHOH·CN. Cf. *Wohl's reaction.*

cyanol Aniline*.

cyanomaclurin (1) $C_{15}H_{12}O_6$ = 288.3. A tannin from the wood of *Arctocarpus integrifolia.* Colorless crystals, m.290. (2) A synthetic anthocyanidin.

cyanomethine $C_6H_9N_3$ = 123.2. 4-Amino-2,6-dimethylpyrimidine, cyamethine. Colorless crystals, m.180.

Cyanophyceae A group of blue algae. See *phycocyanin.*

cyanosis A bluish-purple color of the skin and tongue, due to lack of oxygen in the blood.

cyanotype Blueprint.

cyanoximide A compound containing the radical NC·C(:NOH)—.

cyanoximidoacetic acid NC·C(:NOH)·COOH = 114.1. Colorless crystals, m.129. **c. acetic ester.** NC·C(:NOH)COOEt = 142.1. Colorless crystals, m.127.

cyanur Indicating the trivalent cyanidine ring

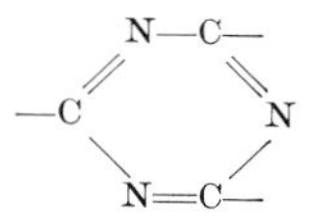

```
       N—C—
      //    \\
   —C         N
      \      /
       N=C—
```

c.amine Melamine. **c.diamine** $C_3H_5ON_5$ = 127.1. Ammeline. Colorless crystals, decomp. by heat, insoluble in water. **c.monoamine** $C_3H_4O_2N_4$ = 128.2. Ammelide. Colorless crystals, insoluble in water. **c.triamine** Melamine.

cyanuric acid HO·C:(N·C·OH)$_2$:N·$2H_2O$ = 165.1. Pyrolithic acid, trihydroxycyanidine, pyrouric acid, pyruric acid. Colorless monoclinics, sparingly soluble in water. Cf. *cyanur, melem.* **iso ~** Tricarbimide. The ketone form of cyanuric acid and main tautomer.

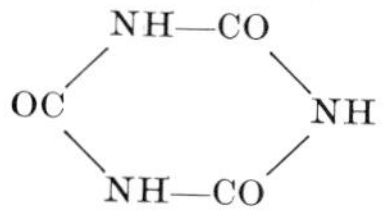

```
      NH—CO
     /     \
   OC       NH
     \     /
      NH—CO
```

thio ~ See *thiocyanuric acid.*

cyanuric azide $C_3N_3(N_3)_3$ = 204.11. Colorless crystals, m.94, insoluble in water; a detonating explosive.

cyanuric ester A derivative of cyanuric acid containing the radical $C_3N_3O_3$≡. **ethyl ~** $C_3N_3(OEt)_3$ = 213.2. Colorless crystals, m.29, b.275. **ethyliso ~** $C_3O_3N_3Et_3$ = 213.2. Ethyl tricarbimide. Colorless crystals, m.96. **iso ~** A derivative of isocyanuric acid, containing the isocyanur radical, $C_3H_3O_3N_3$≡. **methyl ~** $C_3N_3(OMe)_3$ = 171.2. Colorless crystals, m.135, soluble in alcohol. **methyliso ~** $C_3O_3N_3Me_3$ = 171.2. Methyl tricarbimide. Colorless crystals, m.175.

cyanuric trichloride Tricyanogen chloride.

cyanurin A blue compound produced in urine containing indican, on addition of an acid.

cybernetics The study of control and communications interrelationships among humans, machines, and social organization. Cf. *biomechanics.*

cybotactates Aggregates of molecules in liquids; generally oriented. Cf. *zone.*

cybotactic Pertaining to end-to-end or side-to-side arrangements of molecules. Cf. *bond, association.*

cybotaxis The cubic space arrangement of the molecules of noncrystalline substances.

cyclamates (1) Salts of cyclamic acid. (2) Sweeteners, as calcium c., about 30 times sweeter than sucrose; banned in the USA as possible carcinogens.

cyclamic acid C_6H_{11}·NH·SO_3H = 179.2. Cyclohexylsulfamic acid. White, sweet crystals, m.179, soluble in water; a sweetening agent (BP).

cyclamin Arthranitin.

cyclamiretin $C_{15}H_{22}O_2$ = 234.3. A decomposition product of arthranitin.

cycle (1) Varve. Any periodic repetition of a phenomenon; nitrogen cycle. (2) A ring or closed atomic chain; as, homocycle. **megacycles per second** 1 Mc/s = 1 megahertz. See *hertz.*

cyclic Arranged in a ring. Cf. *acyclic, aromatic.* **carbo ~** Indicating a ring of carbon atoms, e.g., benzene. **di ~** An atomic structure containing 2 rings, e.g., naphthalene. **hetero ~** A ring composed of 2 or more different kinds of atoms, e.g., pyridine. **hexa ~** An atomic structure containing 6 rings. **homo ~** A ring composed of one kind of atom. **mono ~** A molecule containing one ring only. **penta ~** An atomic structure containing 5 rings, e.g., morphine. **poly ~** An atomic structure containing 2 or more rings. **tetra ~** An atomic structure containing 4 rings. **tri ~** An atomic structure containing 3 rings.

c. action See *catalysis.* **c. compound** A compound that contains a ring of atoms, or a closed homocyclic or heterocyclic chain of atoms in its molecule. See *ring.* **c. hydrocarbons** Compounds of hydrogen and carbon, which contain a ring of carbon atoms. Cf. *benzene series, cycloparaffin.*

cyclite Benzyl bromide*.

cyclitols* Cycloalkanes with one OH group on each of 3 or more ring atoms; as, hexoses.

cyclization Ring formation.

cyclizine hydrochloride $C_{18}H_{22}N_2$·HCl = 302.8. Mar(e)zine. White, bitter crystals, m.285 (decomp.), soluble in water; an antihistamine and anti-travel-sickness remedy (USP, BP).

cyclo-* Prefix indicating a ring structure. It is italicized in inorganic names, but not in organic names.

cycloalkanes Generic name for saturated, monocyclic hydrocarbons, as, cyclohexane.

cycloalkyl Generic name for radicals derived from cycloalkanes; as, cyclohexyl.

cyclobarbitone $C_{12}H_{16}O_3N_2$ = 236.3. Phanodorm. White, odorless, slightly bitter crystals, m.172, slightly soluble in water; a sedative, used as c. calcium (EP, BP). See *barbiturates.*

cyclobutane* C_4H_8 = 56.1. *Tetramethylene,* q.v.

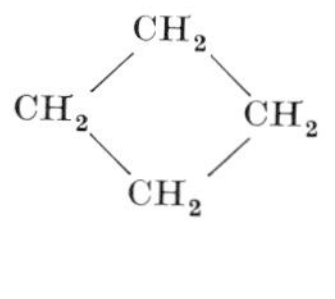

```
        CH2
       /   \
   CH2       CH2
       \   /
        CH2
```

Colorless gas, b.11. **ethyl ~** C_6H_{12} = 84.2. Colorless liquid, b.72. **methyl ~** C_5H_{10} = 70.1. Colorless liquid, b.39.
c.carboxylic acid See *alicyclic acids.*
cyclobutanol* C_4H_7OH = 72.1. Colorless liquid, b.123.
cyclobutanone* C_4H_6O = 70.1. Colorless liquid, b.99.
cyclobutene* C_4H_6 = 54.1. Colorless gas, b.3.

CH / CH   CH₂ / CH₂

cyclobutyl* The radical $CH_2 \cdot CH_2 \cdot CH_2 \cdot CH-$.
cyclobutylene Cyclobutene*.
cyclofenchene $C_{10}H_{16}$ = 136.2. A tricyclic terpene. Colorless liquid, b.144.
cyclogallipharic acid $C_{21}H_{34}OH \cdot COOH$ = 348.5. 2-Hydroxy-6-pentadecylbenzoic acid, hydroginkolic acid, m.93; from commercial tannins.
cyclogeranic acid $C_{10}H_{16}O_2$ = 168.2. Colorless crystals, m.106.
cycloheptadecene* $CH_2 \cdot CH:CH \cdot (CH_2)_{14}$ = 236.4. Civetane. Colorless crystals, m.47, b.120.
cycloheptadecenone* **9-~** $(CH_2)_7 \cdot CH:CH \cdot (CH_2)_7 \cdot CO$ = 250.4. Civetone. Extracted from glands of the civet cat.
cycloheptane* $CH_2 \cdot (CH_2)_5 \cdot CH_2$ = 98.2. Suberane, heptamethylene. Colorless liquid, b.117, soluble in alcohol.
cycloheptanol* $C_7H_{14}O$ = 114.2. Suberol. Colorless liquid, b.185.
cycloheptanone* $(CH_2)_6 \cdot CO$ = 112.2. Suberone. Colorless oil, b.180, slightly soluble in water. From distillation of calcium octanedioate.
cycloheptatriene* **1,3,5-~** Tropilidene.
cycloheptene* C_7H_{12} = 96.2. Suberene. Colorless oil, b.115, insoluble in water.
cycloheptyl* Suberyl. The cyclic radical $C_7H_{13}-$, from cycloheptane.
cyclohexadiene* C_6H_8 = 80.1. Dihydrobenzene. Any of a group of partly saturated benzenes: **1,2-~** b.78. **1,3-~** b.83; **1.4-~** b.85.
cyclohexadienylidene* Phenylidene. The radicals:

CH:CH / CH   C= / CH·CH₂ — **2,4-**   and   CH:CH / CH₂   C= / CH:CH — **2,6-**

cyclohexandiol Cyclohexanediol*.
cyclohexane* C_6H_{12} = 84.2. Hexamethylene.

CH₂—CH₂ / CH₂   CH₂ / CH₂—CH₂

Colorless hydrocarbon (1–2%) in Austrian and Caucasian petroleum, d.0.780, m.6.4, b.81, insoluble in water, soluble in alcohol or ether. Synthesized by hydrogenation of benzene. A starting point for nylon derivative manufacture. **dimethyl ~** C_8H_{16} = 112.2. Several enantiomers. **1,2-** Hexahydro-*o*-xylene. Colorless liquid, b.124. **1,3-** Hexahydro-*m*-xylene. Colorless liquid, b.120. **(*E*)-1,4-** Hexahydro-*p*-xylene. Colorless liquid, b.119. **ethyl ~** C_8H_{16} = 112.2. Ethylhexahydrobenzene. Colorless liquid, b.132. **hexabromo ~** $C_6H_6Br_6$ = 557.5. Benzene hexabromide. White monoclinics, m.212. **hexachloro ~** $C_6H_6Cl_6$ = 290.8. Benzene hexachloride, BHC. Colorless prisms, m.112, soluble in water. Like BHC, the γ-isomer (lindane, Gamene) has insecticidal properties; also a scabicide and used in veterinary medicine (USP, BP). Cf. *hexachlorobenzene.* **hexahydroxy ~** See *inositol, phenose.* **hydroxy ~** Cyclohexanol. **isopropyl-4-methyl ~** $Me_2CH \cdot C_6H_{10} \cdot Me$ = 140.3. *p*-Menthane. Colorless liquid, b.170. **methyl-** Hexahydrotoluene. **pentahydroxy ~** See *quercitol, pinite.* **propyl ~** C_9H_{18} = 126.2. Colorless liquid, b.140. **tetramethyl ~*** $C_{10}H_{20}$ = 140.3. Hexahydrodurene. Colorless liquid, b.161. **trihydroxy ~** Phloroglucitol. **trimethyl ~** C_9H_{18} = 126.2. Several enantiomers. **1,2,4-~** Hexahydro-Ψ-cumene. Colorless liquid, b.141.6. **1,3,5-~** Hexahydromesitylene. Colorless liquid, b.140.
cyclohexanediol* Each isomer has enantiomers. **1,2-~** Hexahydrocatechol. **1,3-~** Resorcitol. **1,4-~** Quinitol. **(*E*)-~** m.143. **(*Z*)-~** m.113. **isopropyl-methyl ~** Terpin.
cyclohexanehexol Inositol.
cyclohexanepentol Quercitol*.
cyclohexanol* PhH_6OH = 100.2. Hexalin. Colorless crystals, m.15; a solvent for gums, waxes, rubber, and nitrocellulose and emulsifier. **isopropyl-methyl ~** Menthanol. **trimethyl ~** $C_9H_{18}O$ = 142.2. A liquid, b_{760mm}198, soluble in most organic solvents; a mutual solvent for water-immiscible liquids.
cyclohexanone* $C_6H_{10}O$ = 98.1. Pimelinketone. Colorless liquid, b.155; a solvent. **isopropyl-methyl ~** Menthanone. **methyl ~** $C_7H_{12}O$ = 112.2. **2-~** Colorless liquid, b.166. **3-~** Colorless liquid, b.164. **4-~** Colorless liquid, b.163. **trimethyl ~** $C_9H_{16}O$ = 140.2. **2,4,5-~** Colorless liquid, b.191. **3,5,5-~** Dihydroacetophenone. Colorless liquid, b.189.
cyclohexatriene Benzene*.
cyclohexene* C_6H_{10} = 82.1. Tetrahydrobenzene. Colorless liquid, b.83. Cf. *carveol, dambonite.* **isopropyl-methyl ~** Menthene.
3-c.-1-one* $CH_2 \cdot CH:CH \cdot CH_2 \cdot CO \cdot CH_2$ = 96.1. Urinoid. A constituent of urine, contributing to its characteristic odor.
cyclohexenol* $C_6H_{10}O$ = 98.1. **(±)-3,1-** Colorless liquid, b.163.
cyclohexenyl* The radical C_6H_9-, from cyclohexene; 3 isomers.
cyclohexyl* The radical $C_6H_{11}-$, from cyclohexane. **bi ~** $C_6H_{11} \cdot C_6H_{11}$ = 166.3. Dodecahydrodiphenyl. Colorless liquid, b.240.
c.amine A toxic metabolite of cyclamate sweetening agents.
c.sulfamic acid Cyclamic acid.
cyclohexylidene* The ring radical $CH_2 \cdot CH_2 \cdot CH_2 \cdot CH_2 \cdot CH_2 \cdot C=$, from cyclohexane.
cyclol The hypothetical space-enclosing arrangement of diazine and triazine rings in globular proteins.
cyclomethycaine sulfate $C_{22}H_{33}O_3N \cdot H_2SO_4$ = 457.6.

White, bitter crystals, m.164, soluble in water; local anesthetic (USP, BP).

cyclone A static vessel of circular cross section in which fluids or gases under pressure form a vortex.

cyclonite Cyclotrimethylenetrinitramine.

cyclononane* C_9H_{18} = 126.2. Colorless liquid, b.170.

cyclooctadiene* C_8H_{12} = 108.2. A hydrocarbon constituent of rubber.

cyclooctane* $(CH_2)_8$ = 112.2. Octomethylene, m.10.

cyclooctatetraene* C_8H_8 = 104.1. Octabenzene. Yellow liquid, b. about 40.

cycloolefin(e) A ring compound with double bonds and $=CH_2$ groups. General formula, C_nH_{2n-2}.

cycloparaffin Polymethylene, naphthene. A completely saturated carbocyclic ring compound consisting of $=CH_2$ groups. General formula, C_nH_{2n}. Cf. *cycloalkane.*

cyclopentadecanone* $C_{15}H_{28}O$ = 224.4. Exaltone. Synthetic musk perfume, m.65. **3-methyl ~** Muscone.

cyclopentadiene* C_5H_6 = 66.1. Colorless liquid, b.41. **methylene ~*** Fulvene*.

cyclopentadienylide Metallocene*.

cyclopentamethylene Cyclopentane*.

cyclopentane* $(CH_2)_5$ = 70.1. A ring hydrocarbon in Caucasian petroleum. Colorless liquid, b.51. **diethyl ~** C_9H_{18} = 126.2. Colorless liquid, b.93. **dimethyl ~** C_7H_{14} = 98.2. **1,1-~** b.88. **1,3-~** b.91. **ethyl ~** C_7H_{14} = 98.2. Colorless liquid, b.101. **methyl ~** C_6H_{12} = 84.2. Colorless liquid, m.141. **pentaoxo ~** Leuconic acid. **trimethyl ~** C_8H_{16} = 112.3. Colorless liquid, b.114.
c.acetic acid $C_5H_{10} \cdot CH_2COOH$ = 129.2. Colorless liquid, b.140. **c.carboxylic acid** See *alicyclic acids.* **c.formic acid** $C_5H_{10}COOH$ = 115.2. Colorless liquid, b.214.

cyclopentanol* $C_5H_{10}O$ = 86.1. Colorless liquid, b.139.

cyclopentanone* C_5H_8O = 84.1. Adipinketone. Colorless liquid, b.130.

cyclopentene* $(CH_2)_3 \cdot CH{:}CH$ = 68.1. Δ^1-Pentamethylene. Colorless liquid, b.45. **trimethyl ~** Laurolene.
c.tridecanoic acid Chaulmoogric acid. **c.undecanoic acid** Hydnocarpic acid.

cyclopentenyl* The radical C_5H_7-, from cyclopentene.

cyclopentyl* The radical C_5H_9-, from cyclopentane.

cyclopentylene* The radical $-C_5H_8-$, from cyclopentane.

cyclophosphamide $C_7H_{15}O_2N_2Cl_2P \cdot H_2O$ = 279.1. White crystals, m.51, soluble in water; an alkylating cytotoxic agent; used to treat malignant disease, particularly lymphoma and leukemia (USP, BP).

cyclopia *C. vogelii* (Leguminosae), S. Africa. Source of bush tea, used for lung disorders. **c. fluorescin** $C_{14}H_{18}O_{12}$ = 378.3. A pigment from *C. genistoides.* **c. red** $C_{25}H_{32}O_{10}$ = 492.6. A pigment from *C. genistoides.*

cyclopin $C_{25}H_{28}O_{13}$ = 536.5. A glucoside from *Cyclopia* species.

cyclopropane* C_3H_6 = 42.1. Trimethylene, propene.

$$\begin{array}{c} CH_2 \\ / \quad \backslash \\ CH_2 - CH_2 \end{array}$$

Colorless gas, b. −35, insoluble in water; an inhalant general anesthetic (USP, BP). **acetyl ~** $C_3H_5 \cdot CO \cdot Me$ = 84.1. Trimethylene. Colorless liquid, b.113; polymerizes readily. **benzoyl ~** See *benzoylcyclopropane.* **dimethyl ~** C_5H_{10} = 70.1. Colorless liquid, b.21. ethyl ~ C_5H_{10} = 70.1. A liquid, m.21. **methyl ~** C_4H_8 = 56.1. Colorless gas, b.4. **oxo ~** Cyclobutanone*. **trimethyl ~** C_6H_{12} = 84.1. Colorless liquid, b.57.
c.carboxylic acid* See *alicyclic acids.* **c.dicarboxylic acid*** Vinaconic acid.

cyclopropyl* The radical $CH_2 \cdot CH_2{:}CH-$, from cyclopropane.

cyclopropylene Cyclo*propene**.

cyclopterin A protein from the sperm of the lumpsucker, *Cyclopterus lumpus.*

cycloserine $C_3H_6O_2N_2$ = 102.1. 4-Amino-3-isoxazolidinone. Yellow, bitter crystals produced by *Streptomyces orchidaceus,* m. 155 (decomp.), soluble in water; an antibiotic used to treat tuberculosis (USP, BP).

cyclotetraene An 8-membered ring, the vinyl analog of benzene. An intermediate in the manufacture of hydrogenated products.

cyclotrimethylenetrinitramine $(CH_2 \cdot N \cdot NO_2)_3$ = 222.1. Cyclonite, hexogen. White crystals, insoluble in water, m.204; an explosive.

cyclotron Early form of particle *accelerator,* capable of energies up to 15 MeV.

cyclural Hexobarbitone.

cydonin $C_{18}H_{25}O_{14}$ = 465.4. A gum or mucilaginous carbohydrate from quince seeds, *Cydonia vulgaris.*

cylinder A glass tube, closed at one end, used to hold liquids. **filter ~** A c. of porous material used to filter solutions. **graduated ~** A c. with a measuring scale. **steel ~** A steel c., used to ship compressed gases.

cylindrite $Pb_3FeSn_4Sb_2O_{14}$. A gray ore.

cymarigenin Apocynamarin.

cymarin $C_{30}H_{48}O_{14}$ = 632.7. A glucoside from *Apocynum cannabinum,* which hydrolyzes to cymarigenin and cymarose. Colorless prisms, m.148, slightly soluble in water; therapeutically similar to strophanthin. Cf. *apocynamarin.*

cymarose $C_7H_{14}O_4$ = 162.2. Digitose methyl ether. A sugar, m.91, from cymarin.

Cymbopogon A genus of aromatic grasses. Cf. *Andropogon.*

cymene* $C_{10}H_{14}$ = 134.2. *p*-Isopropyltoluene. A colorless liquid constituent of the oils of eucalyptus, cumin, and thyme, insoluble in water. ***ortho-* ~** b.157. ***meta-* ~** m.25, b.175. **hydroxy-2-~** Carvacrol*. **3-~** Thymol*. **c. alcohol** 7-Cymenol.

cymenol **2-~** Carvacrol*. **3-~** Thymol*. **7-~** $CH_2OH \cdot C_6H_4 \cdot CHMe_2$ = 150.2. 2-Isopropylbenzyl alcohol. b_{24}130.

cymenyl The cymyl radical.

cymol Cymenol.

cymyl Cymenyl. The methyl(1-methylethyl)phenyl† radical, $Me(Me_2CH)C_6H_3-$, from cymene. **2,5-~** Carvacryl. **5,2,-~** Thymyl.

cynapine An alkaloid from *Aethusa cynapium,* fool's parsley (Umbelliferae). Colorless crystals.

cynoctonine $C_{36}H_{34}O_{13}N_2$ = 702.7. An amorphous alkaloid, m.137, from *Aconitum septentrionale* (Ranunculaceae).

cynodine An alkaloid from *Cynodon dactylon* (Gramineae), dog's tooth, Bermuda grass or dooba grass (India).

cynoglossine An alkaloid from the root of hound's tongue. *Cynoglossum officinale* (Boraginaceae).

cynotoxin $C_{20}H_{28}O_{16}$ = 524.4. Colorless, crystalline principle from *Apocynum cannabinum.*

cynurenic acid $C_{20}H_{14}O_6N_2 \cdot 2H_2O$ = 414.4. Oxychinolin carbonic acid. A protein split product in dog's urine.

cyopin Coloring matter of blue pus. Cf. *pyocyanine.*

cypermethrin* See *insecticides,* Table 45 on p. 305.

cyperone $C_{15}H_{22}O$ = 218.3. A sesquiterpene ketone from the oil of *Cyperus rotundus.*

cypressene $C_{15}H_{24}$ = 204.4. A hydrocarbon from cypress oil, d.0.9647.
cypress oil Oleum cupressi. An essential oil from *Cupressus sempervirens.* Yellow liquid, d_{15}·0.88. Cf. *cypressene.*
cypripedoid The combined principles of the root of *Cypripedium pubescens;* a stimulant, antispasmodic, and diaphoretic.
cyprite Chalcocite.
cyproheptadine hydrochloride $C_{21}H_{21}N \cdot HCl \cdot 1.5H_2O$ = 350.9. Yellow crystals, slightly soluble in water; an antihistamine (USP, BP).
Cys* Symbol for cysteine.
cysteine* $NH_2 \cdot CH(COOH)CH_2 \cdot SH$ = 121.2. Cys*. 2-Amino-3-mercaptopropanoic acid. A cystine amino acid in proteins and urinary calculi. **acetyl ~** See *acetylcysteine.*
  c. hydrochloride $C_3H_7O_2NS \cdot HCl \cdot H_2O$ = 175.6. Used as eye drops (USP).
cystine* $(S \cdot CH_2 \cdot CH(NH_2)COOH)_2$ = 240.3. 3,3′-Dithiobialanine. An amino acid found in keratin and as minute hexagons in urine.
cytarabine $C_9H_{13}O_5N_3$ = 243.2. 1-β-D-arabinofuranosylcystosine, cytosine arabinoside, Cytosar. White, crystalline powder, soluble in water. An antimetabolic agent, used for malignant disease, especially leukemia (USP, BP).
cytase Alexin.
cytidine $C_9H_{13}N_3O_5$ = 243.2. A nucleoside from cytosine and D-ribose. Colorless needles, m.220 (decomp.).

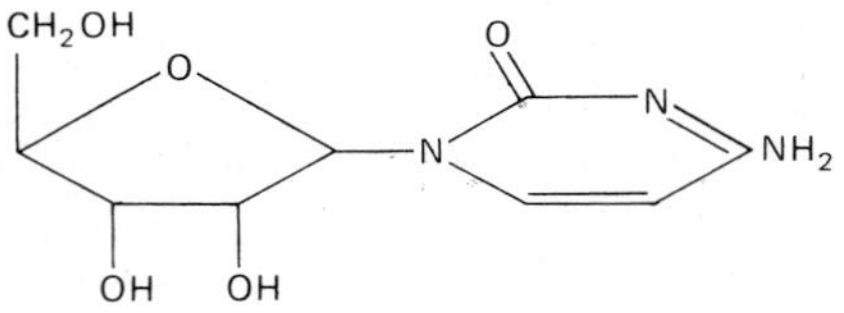

  c. phosphoric acid $C_9H_{12}N_3O_5 \cdot PO(OH)_2$. A pyrimidine nucleotide from cell nuclei.
cytidylic acid $C_9H_{14}N_3O_8P$ = 323.2. A pyrimidine mononucleotide from yeast nucleic acid, m. 240 (decomp.).
cytisine $C_{11}H_{14}ON_2$ = 190.2. Ulexine, baptitoxine, sophorine. Alkaloid from the seeds of *Cytisus, Laburnum,* and *Ulex* species (Leguminosae). Colorless crystals, m.152. soluble in water. Cf. *anagyrine, sparteine.*

TABLE 22. CYTOCHROMES

Pigment	Color	Metal, occurrence
Chlorophyll·········	Green	Mg, plants
Hemoglobin········	Red	Fe, vertebrata
Hemocyanin········	Blue	Cu, crustacea
Chlorocruorin·······	Green	Fe, worms
Achroglobin········	Colorless	Mn, mollusks

cytochemistry The study of the chemical properties of living cells.
cytochrome (1) The intracellular respiratory pigment. Cf. *chlorophyll, hemoglobin.* See Table 22. (2) A pigment of bacteria, e.g., *Mycobacterium tuberculosis,* related to chlorophyll but containing Fe and Cu. Cf. *porphin ring.*
cytoclasis The destruction of living cells.
cytoglobin A protein from white blood cells.
cytokinins A class of plant hormones promoting cell division, of which zeatin was the first isolated. Used to prolong the life of fresh vegetables and cut flowers.
cytoligneous substances The cytolytic vegetable (bacteria) or animal (erythrocytes, spermatozoa, tissue) cells.
cytology The study of living cells.
cytolysis The destruction or dissolution of living cells.
cytoplasm The protoplasm of the cell body, excluding the nucleus.
Cytosar Trademark for cytarabine.
cytosine $C_4H_5ON_3$ = 111.1. 4-Amino-2(1*H*)-pyrimidinone, derived from nucleic acid. Colorless plates, sparingly soluble in water.
cytotoxic Damaging or toxic to cells. **c. drugs** Drugs used to treat severe, life-threatening, usually malignant, disease. Three main types: alkylating agents, as cyclophosphamide and chlorambucil; antimetabolites, as methotrexate; and antibiotics, as dactinomycin.
cytotoxin An antibody having a specific action on the cells of certain organs; e.g., nephrotoxin, a toxin from bacteria *Clostridium perfringens* causing kidney damage.
cytozyme Thrombokinase.

D

D Symbol for: (1) The characteristic line of sodium. See *Fraunhofer lines.* (2) Deuterium. **2,4-~*** See *herbicides,* Table 42 on p. 281. (3) Debye units.

D Indicating right-handed chirality in carbohydrates, amino acids, peptides, cyclitols. See *sequence rule.*

D Symbol for diffusion coefficient.

D (Boldface.) Symbol for electric displacement.

d Symbol for (1) day; (2) deci- (10^{-1}); (3) density, q.v. (with period); (4) deuteron.

d Symbol for: (1) formerly, dextrorotatory; now (+). Cf. *D.* (2) Diameter. (3) Lattice plane spacing. (4) Relative density. (5) Thickness.

Δ, δ See *delta.*

da* Symbol for deka or deca*.

D.A. Blue cross.

dacite An igneous rock classed between the monazites and the quartz diorites; it consists chiefly of plagioclase, feldspar, mica, and hornblende (Yellowstone Park and California).

Dacron Trademark for a polyester synthetic fiber.

dacryagogue A substance that stimulates tears. Cf. *lacrimatory.*

dacryolin The albuminous matter of tears.

dactinomycin $C_{62}H_{86}O_{16}N_2$ = 1255. Actinomycin D. Bright-red, crystalline quinonoid-type antibiotic from cultures of *Streptomyces antibioticus;* decomp. 250. An antitumor drug (USP).

dactylin $C_{23}H_{28}O_{15}$ = 544.5. A glucoside from the pollen of orchard grass, *Dactylis glomerata.* Yellow needles, m.184, soluble in hot water.

Dag Trademark for brand of colloidal dispersions.

Daguerre, Louis Jacques Mandé (1789–1851) French physicist and painter; inventor of photography. **D. process** A polished silver plate is made light-sensitive by exposure to iodine vapor. The exposed silver iodide is developed by means of mercury vapor and thiosulfate.

daguerrotype A photograph made by Daguerre's process (1839).

dahlia Composite plants whose bulbs yield inulin and a purple color. **d. paper** A test paper impregnated with the coloring matter of d. (acids—red, alkalies—green). **d. violet** See *pyoktanin.*

dahlin (1) A purple aniline dye derived from mauveine. (2) Inulin.

dahllite $Ca_6(PO_4)_4 \cdot CaCO_3$. A white, fibrous calcium phosphate-carbonate.

Dahl's acid 2-Naphthylamine-5-sulfonic acid. **D. a. II** 1-Naphthylamine-4,6-disulfonic acid. **D. a. III** 1-Naphthylamine-4,7-disulfonic acid.

dahmenite An explosive: ammonium nitrate 91, naphthalene 6.5, potassium dichromate 2.5%.

daidzin A pigment from soybean meal.

dakerite Schroeckingerite. A fluorescent uranium mineral (Wyoming).

Dakin, Henry Drysdale (1880–1952) English-born American biochemist. **D. antiseptic** Chlorazene. **D. solution** A solution of 0.5% sodium hypochlorite with sodium hydrogencarbonate; used to treat infected wounds.

dalapon See *herbicides,* Table 42 on p. 281.

Dalen Martens machine A device to measure the tensile strength and elasticity of rubber.

Dalmatian flowers Pyrethrum.

Dalton, John (1766–1844) English chemist, mathematician, and physicist; founder of the atomic theory. **D.'s law** (1) The pressure of a gas mixture equals the sum of the partial pressures of the constituent gases. Cf. *Charles's law.* (2) Laws of proportion. See *proportion.* (3) Absorption of gases: The solubility of different gases in a mixture is unaffected by the presence of other gases so long as no chemical reaction occurs.

dalton A unit of mass in biology; 1/12 the mass of an atom of carbon-12 = $1{,}661 \times 10^{-24}$ g.

daltonide Early name for a chemical compound of constant proportions; as opposed to berthollide.

damar Dammar.

damascinine $C_{10}H_{14}O_3N_2$ = 210.2. An alkaloid from the seeds of *Nigella damascena* (Ranunculaceae).

dambonite $C_6(OH)_4(OMe)_2$ = 202.1. *p*-Dimethoxytetrahydroxybenzene. Colorless crystals from Gabon caoutchouc.

dambose *meso*-Inositol.

dammar Dammara. An oleoresin from Coniferae (Asia, Australia, S. America), resembling copal. Yellow masses, m.115–125, soluble in hot alcohol; used in varnishes and plasters. **true ~** The resin from *Dammara orientalis* (southeast Asia) or *D. australis* (New Zealand).

damourite A yellow, olivine-type muscovite.

damping Slowing-down motion, as of the swing of a balance, q.v. Cf. *Pregl.* **d. coefficient** See Table 88, group B, on p. 566. **d. period** The time needed for an excited electron to drop to the ground state and emit the characteristic radiation of the atom.

danain $C_{14}H_{14}O_5$ = 262.3. A glucoside from the root of *Danais fragrans* (Rubiaceae), Madagascar.

danaite (Fe·Co)AsS. A native sulfoarsenide.

danalite $(Zn \cdot Mn \cdot Fe \cdot Be)_7SSi_3O_{12}$. A rare native silicate and sulfide.

danburite $CaB_2Si_2O_8$. Yellow, orthorhombic calcium borosilicate, d.2.95, hardness 7.

dandelion Taraxacum.

Daniell, John Frederic (1790–1845) English chemist and physicist. **D. cell** A voltaic cell (1.08 volts) consisting of an amalgamated zinc electrode in dilute sulfuric acid (1:12) and a copper electrode in saturated copper sulfate solution.

dannemorite A rare amphibole.

dansyl The group 1-dimethylaminonaphthalene-5-sulfonyl.

Daonil Trademark for glibenclamide.

daphnandrine $C_{36}H_{38}O_6N_2$ = 594.7. An alkaloid from the bark of *Daphnandra micrantha* (Monimiaceae), Australia; m.280; a cardiac and muscular paralyzant.

daphnetin $C_9H_6O_4$ = 178.1. 7,8-Dihydroxycoumarin. Yellow crystals, m.255, soluble in water.

daphnin $C_{15}H_{17}O_9 \cdot 2H_2O$ = 377.3. A glucoside from the bark of the bay or laurel, *Daphne mezereum* (Thymelaceae). Prisms, m.223.

daphniphylline An alkaloid from *Daphniphyllum bancanum* (Euphorbiaceae), Indonesia; cardiac poison.

daphnite $H_{56}Fe_{27}Al_{29}Si_{18}O_{121}$. A hydrous iron-aluminum chlorite.

dapsone $C_{12}H_{12}O_2N_2S$ = 248.31. 4,4′-Sulfonyldianiline, diphenylsulfone. White, bitter crystals, m.180, slightly soluble in water. The first effective treatment for leprosy, widely used (USP, EP, BP).

darak Tikitiki. Philippine rice bran.

Daraprim Trademark for pyrimethamine.

darapskite $NaNO_3 \cdot Na_2SO_4 \cdot H_2O$. A native sodium nitrate and sulfate (Chile, California).

D'Arcet metal An alloy: Bi 50, Pb 25, Sn 25%; m.94; used in fire sprinklers and fusible plugs.

dari A grain similar to millet; a source of fermentation alcohol.

Darlan Trademark for a nitrile copolymer of vinyl acetate and vinylidene dinitrate. It has two CN groups directly attached to the same C atom.

dasheen Taro.

dasymeter An instrument to measure the heat loss of a furnace by analysis of the waste gases.

data bank, data base An organized collection of information.

date palm A tree whose fruit is used as food, its fiber for ropes, and its sap to make sugar.

dative bond See *covalence* under *valence.*

datolite $2CaO \cdot B_2O_3 \cdot 2SiO_2 \cdot H_2O$. A rare zeolite. Colorless or green, monoclinic masses, d.2.9, hardness 5–5.5.

Datura Thorn apple, Jamestown weed. A genus of Solanaceae; e.g., *D. stramonium.*

daturic acid Heptadecanoic acid*.

daubreeite BiOCl. Amorphous, white, native bismuthyl chloride.

daubreelite $FeCr_2S_4$. A black constituent of meteorites.

daucine An alkaloid from wild carrot, *Daucus carota* (Umbelliferae); a diuretic.

daughter cell A cell formed by division of a single cell.

daviesite A colorless, orthorhombie, native lead oxide chloride.

Davitamon-K Trade name for menadione.

Davy, Sir Humphry (1778–1829) English chemist who isolated metals by electrolysis, and devised the Davy lamp. **D. lamp** See *miner's lamp.*

Dawbarn Trademark for a mixed-polymer synthetic fiber.

dawsonite $NaAlCO_3(OH)_2$. White, monoclinic crystals.

Daxad Trademark for a group of dispersing agents.

day A unit of time. The average period from noon to noon = 1,440 min = 86,400 s. **mean solar ~** The mean time for the earth to make one complete revolution = 86,636.5 s. **sidereal ~** 86,164.09 s. The interval between successive passages of a star through the meridian.

Dayan Trademark for a mixed-polymer synthetic fiber.

dB Abbreviation for *decibel.*

d.c., dc Abbreviation for direct current.

DDNP Abbreviation for diazodinitrophenol.

DDT Common name for a mixture of insecticidal isomers with the general formula

X–⬡–CH(CCl₃)–⬡–X

The $=CH \cdot CCl_3$ group is associated with their great contact and stomach insecticidal activity. X is usually Cl. Technical DDT contains mainly *p,p′*-DDT. Yellow crystals, m. exceeds 104. Banned in some countries because of its persistence. ***o,p′*-~** Colorless plates, m.74. ***p,p′*-~** 1,1,1-Trichloro-2,2-bis(4-chlorophenyl)ethane. Colorless needles, m.109.

de-* Prefix indicating removal, e.g., of atoms, radicals, or water; as, dehydro-. Cf. *des.*

Deacon process A method of making chlorine by passing a hot mixture of hydrochloric acid gas with atmospheric oxygen over a cuprous chloride catalyst.

deactivation (1) The rendering inactive of something, e.g., a catalyst. (2) Loss of the radioactivity of a preparation.

dead In reference to a molten alloy (as steel), quiet, evolving no gas. **d. time** Time interval between the start of an input change and the start of the corresponding response.

dealcoholizing Removal of alcohol from liquids.

deamidation Deamination.

deamidizing Deaminizing.

deamination Removal of the amino group by hydrolysis:

$$R \cdot NH_2 + H_2O \rightarrow R \cdot OH + NH_3$$

oxidative ~

$$R \cdot CHNH_2 \cdot COOH \rightarrow R \cdot CO \cdot COOH + NH_3$$

reductive ~

$$R \cdot CHNH_2 \cdot COOH \rightarrow R \cdot CH_2 \cdot COOH + NH_3$$

deaminizing Splitting off ammonia from amino acids and proteins, with formation of the corresponding fatty acids.

deaquation Dehydration.

Debendox Trademark for dicyclomine hydrochloride.

Debierne, André (1874–1949) French physicist, noted for work in radioactivity.

de Broglie See *Broglie.*

debug To remove errors from a program or malfunctions from computer hardware.

Debye, Peter (1884–1966) Dutch physicist. **D. crystallogram** A spectrumlike pattern produced by refracting monochromatic X-rays by a crystal. **D.-Hückel theory** The distribution of ions in a small volume at distance l from a chosen ion is $ne(l + e\psi)kT$, where n is the number of molecules of salt in 1 ml, e the elementary electric charge, ψ the potential due to the ion, k the Boltzmann constant, and T the thermodynamic temperature.

debye D. Unit of dipole moment equal to 10^{-18} esu or 3.33564×10^{-30} coulombmeter (the SI unit).

deca- da. SI prefix denoting 10 times (U.S. standard uses deka-).

decahydronaphthalene Dekalin.

decalcify To remove lime salts from a tissue or bone.

decalescent Absorbing heat, at a certain temperature, by a bar of steel during heating, due to allotropic changes. Cf. *recalescent point.* **d. outfit** A device to determine the hardening temperature of high-carbon steel from its decalescence.

Decalin See *Dekalin.*

decal process Decalcomania. A process for transferring lithographed designs on the surface of a specially coated paper to another surface by moistening the paper.

decanal* $C_9H_{19}CHO$ = 156.3. Capraldehyde. Colorless liquid, d.0.828., b.209; from essential oils.

decanamide* $Me(CH_2)_8CONH_2$ = 171.3. Capramide. White crystals, m.108.

decane* $C_{10}H_{22}$ = 142.3. The tenth hydrocarbon of the methane series. Colorless liquid, d_{20°0.730, b.173, insoluble in water. **amino ~** $Me(CH_2)_9NH_2$ = 157.3. Decylamine. White crystals, m.17. **iodo ~*** $C_{10}H_{21}I$ = 268.2. Decyliodide. Colorless liquid, b_{15mm}132.
d.carboxylic acid* Undecanoic acid*. **d.dioic acid*.** Sebacic acid. **d.diol.*** $CH_2OH(CH_2)_8CH_2OH$ = 174.3. Decamethylene glycol. White crystals, m.71.5. **d.nitrile***

$Me(CH_2)_8CN$ = 153.3. Caprinitrile. Colorless liquid, b.244, insoluble in water.
decanoate* Caprate. A salt or ester of decanoic acid, containing the radical $C_9H_{19}COO-$.
decanoic d. acid* $Me(CH_2)_8COOH$ = 172.3. Capric acid, octylacetic acid. A fatty acid in animal fats. Colorless needles, m.31, slightly soluble in water. **d. anhydride** $[Me(CH_2)_8CO]_2O$ = 326.5. Capric anhydride. White crystals, m.239, insoluble in water.
decanol* Decyl alcohol*.
decanoyl* 1-Oxodecyl†. The radical $C_9H_{19}CO-$, from decanoic acid.
decant, decantation Separation of a sediment (liquid or solid) by pouring or syphoning off the top liquid layer.
decarbonize Remove carbon.
decarboxylase* See *enzymes,* Table 30.
decarboxylating Splitting off carbon dioxide from amino acids and proteins by bacterial action, with formation of amines.
decarboxylizing Removal of carboxyl group(s) from an organic acid; as, carbon dioxide, e.g., by yeast carboxylyase.
decarboxyrissic acid $C_{10}H_{12}O_5$ = 212.2. 3,4-Methoxy-phenoxyacetic acid. Colorless crystals, m.116.
decarburization Removal of carbon from iron and steel.
decarburize Decarbonize.
decatizing A steam or hot-water finishing process for woolen piece goods, designed to reduce and fix the luster produced on pressing, to render the fabric nonshrinking, and to produce a full and firm handle.
decatoic acid Decanoic acid*.
decatyl The decyl* radical.
decavitamin capsules (tablets) A USP formulation containing vitamins A and D, ascorbic acid, calcium pantothenate, cyanocobalamin, folic acid, nicotinamide, pyridoxine hydrochloride, riboflavin, α-tocopherol, and thiamine hydrochloride.
decay (1) The progressive decomposition of organic matter, generally due to aerobic bacteria. Cf. *putrefaction.* (2) The progressive disintegration of radioactive substances. **d. constant*** λ. Radioactive constant. The life period of a radioactive element: $n = n_0e^{-\lambda t}$, where n_0 is the number of atoms originally present, n the number of atoms present after time t, and e the base of natural logarithms.
decene* $C_{10}H_{20}$ = 140.3. Decylene, octylethylene. **1-~** $CH_2{:}CH(CH_2)_7Me$. Diamylene. Colorless liquid, b.172, insoluble in water.
Deceresol OT Trademark for the dioctyl ester of sodium sulfosuccinic acid; a wetting agent.
dechenite $Pb(VO_3)_2$. A rare mineral named after von Dechen.
Decholin Trademark for dehydrocholic acid.
deci-* Abbrev. d. SI prefix for a multiple of 10^{-1}. Cf. *deca.*
decibel Abbrev. dB. One-tenth of a *bel,* q.v.; a measure of *sound* (q.v.) *intensity.* See Table 23. Cf. *phon, barye.*

TABLE 23. RANGE OF SOUND LEVELS

0 dB =	threshold of hearing (around 1,500 Hz)
40 dB =	quiet apartment
60 dB =	typing office
70 dB =	automobile
80 dB =	motorcycle
120 dB =	jet engine
160 dB =	Pimonov siren, a noisy apparatus that can kill a small animal

decigram One-tenth of a gram.
decile The top (upper d.) and bottom (lower d.) 10% of a statistical population.
deciliter dL. One-tenth of a liter, or 100 cc, or 3.38 U.S. fluid ounces.
decilog Abbrev. dL. The expression of a ratio on a logarithmic basis.
decimal A fraction of ten; a tenth part. **d. classification** A universal system in which the whole of human knowledge, taken as unity, is divided into 10 domains represented by decimal fractions:

0.0 Generalities	0.5 Mathematics, natural science
0.1 Philosophy, etc.	0.6 Applied science, medicine
0.2 Religion, Theology	0.7 Arts, recreation, sport
0.3 Social sciences, etc.	0.8 Literature, belles lettres
0.4 Philology, etc.	0.9 History, geography, etc.

The system proceeds from the general to the particular by continued subdivision of the numbers into decimal fractions corresponding with the subdivisions of the subject. Thus, every concept in the domain of mathematics and natural science is represented by a decimal fraction greater than 0.5 or less than 0.6. Thus:

0.50 General	0.55 Geology, etc.
0.51 Mathematics	0.56 Palaeontology
0.52 Astronomy	0.57 Biology, etc.
0.53 Physics and mechanics	0.58 Botany
0.54 Chemistry	0.59 Zoology

For convenience, the first decimal point is omitted. To indicate a work dealing independently with 2 or more subjects, the numbers corresponding with those subjects are connected by a + sign. A functional relationship between 2 or more concepts is expressed by the appropriate numbers linked by a colon. Thus:

537.531	X-rays
535.4	Diffraction
548.0	Crystals

Thence: 537.531:535.4:548.0, X-ray diffraction by crystals. Further symbols indicate form, place, language, etc. See British Standard 1,000.
decimolar One-tenth of the molecular weight, in grams. **d. solution** A solution containing, in 1 liter, 1/10 g molecule; e.g., a 0.1 *M* sodium hydroxide solution contains 4.00 g/L NaOH.
decinormal One-tenth of the normal or equivalent strength. **d. solution** A solution containing, in 1 liter, 1/10 g of the equivalent weight of a substance; e.g., a 0.1 *N* solution of sulfuric acid contains 4.904 g/L H_2SO_4.
declination Deviation, bending; e.g., magnetic d., the variation of the compass needle from the true north.
decoction The solution made on boiling a solute with a solvent. Cf. *infusion.*
decoctum A soluble pharmaceutical made by boiling vegetable drugs with water, and straining.
decoic acid Decanoic acid*.
decolorant A substance which absorbs or destroys a color.
decoloration Bleaching by natural means.
decolorization, decolorizing Bleaching or destroying a color artificially.
decomp. Abbreviation for decomposed, decomposes, decomposing, or decomposition.
decompose (1) To break down, split up, or analyze a substance. (2) To rot.
decomposition The breaking down of a substance into

simpler constituents. **double ~** Metathesis. A chemical change in which 2 molecules exchange one or more of their constituents. **hydrolytic ~** Decomposition, especially of rocks, by water. Cf. *hydrolysis*. **single ~** Analysis. A chemical change breaking a molecule into its constituents.

d. apparatus A device for water electrolysis. **d. of rocks** The disintegration of rocks in successive stages: solution, hydration, disintegration, mechanical sorting.

decortication Stripping the bark, hull, or other outer layer from a plant.

decouple To treat mathematically so as to render interacting variables noninteracting.

decrepitate (1) To crackle. (2) To roast a moist material.

decrepitation Flying apart with a cracking sound when heated, e.g., crystals.

decyl* Decatyl. The radical $C_{10}H_{21}-$, from decane. **d. alcohol*** $C_{10}H_{22}O$ = 158.3. **normal ~** $Me(CH_2)_8CH_2OH$. Decatyl alcohol, nonylcarbinol, 1-decanol*. Colorless oil, b.231, soluble in alcohol. **d. aldehyde** Decanal*. **d.amine*** Amino*decane*. **d. hydride** Decane*.

decylene Decene*.

decylenic acid $C_{10}H_{18}O_2$ = 170.3. An unsaturated acid in butter; m. 0.

decylethylene Dodecene*.

decylic acid Decanoic acid*.

decyne* Decine. **1-~** $Me(CH_2)_7C{:}CH$ = 138.3. d_0·0.791, m.−36, b_{745mm}182. **3-~** b.176.

deduction (1) A conclusion drawn from established facts and data. Cf. *speculation*. (2) An inference from general to particular. Cf. *induction*.

def. See *isotopically deficient*.

defecation The industrial clarification of sugar solutions.

deferoxamine mesylate $C_{25}H_{48}O_8N_6 \cdot CH_4O_3S$ = 656.8. Desferal. A chelating compound, used in iron overdosage (USP, BP).

defervescence Cessation of boiling.

defibrinated Rendered free of fibrin. **d. blood** Blood d. by shaking with lead, beating, or chemically precipitating.

definite proportions law See *constant proportions*.

deflagration Sudden combustion, usually accompanied by a flame and crackling sound. **d. spoon** A metal spoon with long handle, used to burn substances in gases.

deflocculation Removal or reversal of flocculation.

defoamer Defrother. An agent that destroys or prevents foam. See *surfactant*.

deformability The property of a substance by which its shape, flow, or elasticity may be altered without rupture, as with pitch.

defrother Defoamer.

degasification Elimination of gases from metals before coating or plating.

degauss See *gauss*.

degeneration (1) The deterioration of cells; the loss of functional and hereditary characteristics by an organism. (2) Degeneracy. A measure of the number of equivalent levels of energy of an atomic or molecular orbital.

Degener's indicator Phenacetolin.

degradation Conversion of an organic compound to a compound containing a smaller number of carbon atoms. **energy ~** See *thermodynamics*. **mass ~** See *mass-energy cycle*. **photo ~** See *photodegradation*.

degras Wool fat. A dark-brown grease from sheep wool, used in the leather industry. A source of lanolin.

degreaser A solvent that removes fat or oil; as, dichloroethylene.

degree A position or a unit: generally a difference in temperature (as, °C), or direction (as ° angle). **d. of dissociation** See *degree of dissociation* under *dissociation*. **d. of freedom** (1) Variance. Each of the 3 variables: pressure, temperature, and specific volume, one or more of which must be fixed in order to define the state of a system. See *phase rule*. (2) The statistical analog of (1), generally equal to the number of observations minus the number of constraints imposed on the system. **d. of hardness** The *hardness*, q.v., of a mineral is indicated by its position in an arbitrary scale. See *Moh's scale of hardness*. **d. of polymerization** D.P. The number by which the simplest formula expressing the percentage composition of a polymeric substance must be multiplied in order to give the molecular weight. Usually applied to cellulose, $(C_6H_{12}O_5)_n$, which varies considerably in its d. of p., according to the state in which it occurs. Cf. *cuprammonium*. **d. of temperature** A division of the thermometer scale, usually expressed in degrees Celsius, °C.

deHaën salt $SbF_3 \cdot (NH_4)_2SO_4$ = 310.9. A double salt of antimony trifluoride and ammonium sulfate; a textile mordant.

dehydracetic acid $C_8H_8O_4$ = 168.2. Methylacetopyronone. Colorless, rhombic scales, m.108, slightly soluble in hot water. **iso ~** Dimethyl*coumalic acid*.

dehydrant A therapeutic *dehydrater*, q.v.

dehydrated food Foodstuffs from which water has been removed to reduce weight (for convenience of transport) and to ensure their keeping properties. On addition of water the original state is substantially restored.

dehydrater (1) A device to dry tissues for histological work. (2) Dehydrator. A substance that removes water; as, sulfuric acid. Cf. *desiccant*.

dehydration (1) Removal of water from compounds and crystals. (2) Removal of hydrogen from organic compounds by reducing agents (dehydrogenation).

Dehydrite $Mg(ClO_4)_2 \cdot 2H_2O$. Trademark for magnesium perchlorate; used as a desiccant. Cf. *Anhydrone*.

dehydro-* Prefix for an organic compound indicating the loss of 1 (correctly) or 2 (correctly: didehydro-) hydrogen atoms.

dehydroangustione $C_{11}H_{14}O_3$ = 194.2. A bicylic diketone from the oil of *Backhousia angustifolia*, d.1.103, b_{11mm}127. Cf. *angustione*.

dehydrocholesterol **7-~** See *cholecalciferol*.

dehydrocholic acid $C_{24}H_{34}O_5$ = 402.5. A split product of cholic acid, obtained with nitric acid. White, fluffy, bitter powder, m.231–242, slightly soluble in water; a choleretic (USP).

dehydrogenase* See *enzymes*, Table 30 on p. 214.

dehydrogenation Removal of hydrogen from organic compounds by reducing or oxidizing agents in presence of a catalyst.

dehydrolysis Removal of hydrogen and oxygen in the proportions of water from organic substances. Cf. *dehydration*.

dehydrolyzing agent A chemical which removes water from organic compounds; e.g., phosphorus pentaoxide. Cf. *desiccant*.

dehydromucic acid $C_6H_4O_5$ = 156.1 2,5-Furandicarboxylic acid. Colorless crystals from the dry distillation of mucic acid. m.320+ (sublimes).

dehydroperillic acid Thujic acid.

dehydrothiotoluidine $C_{14}H_{12}N_2S$ = 240.3. ***para-~*** 2-(4-Aminophenyl)-6-methylbenzothiazole†. Yellow needles, m.195, insoluble in water. Its solutions have a violet fluorescence; a dye.

dehydrothioxylidene $C_{16}H_{16}N_2S$ = 268.4. ***meta-~*** 2-(4-

Amino-3-methylphenyl)-4,6-dimethylbenzothiazole†. Yellow prisms, m.107, insoluble in water; a dye.

dehydroxycholic acid A constituent of bile.

deinking The removal of ink from waste paper, by chemical and mechanical means, for recycling the fiber.

deionization Demineralization.

deka da. Prefix meaning 10 times; used in U.S. metric standard. SI equivalent is deca.

Dekalin $(CH_2)_4CH(CH_2)_4CH$ = 138.3. Trademark for decahydronaphthalene. A saturated, aromatic hydrocarbon. Colorless liquid, d.0.8747, b.188, insoluble in water, miscible with ether; a cleaning fluid, solvent, and substitute for turpentine and lubricating oil. Cf. *Tetralin.*

de Khotinsky See *Khotinsky.*

Delaney Clause A U.S. Food and Drugs Administration ruling that any food additive shown to cause cancer in any person or animal at any usage level cannot be marketed for general use.

delanium A strong, compact form of carbon, made from coal by the controlled heating of a gelatinous carbonaceous mass; used to construct chemical plant.

delcosine $C_{21}H_{33}O_6N$ = 395.5. An alkaloid, m.198, isomeric and occurring with deltaline in *Delphinium consolida.*

delessite $(Mg,Fe)_4Al_4H_{10}Si_4O_{22}$. A green, hydrous silicate, related to chlorite.

deliquescence Gradual liquefaction by absorption of atmospheric moisture; as, by calcium chloride. Cf. *hygroscopic.*

deliquescent Having the property of deliquescence.

deliriant A drug that acts on the central nervous system, producing excitement, confusion, and delirium; as, belladonna.

Delphi A method of long-range technical forecasting in which participants answer carefully prepared, highly objective questionnaires, the replies being circulated until agreement is reached.

delphin blue $C_{20}H_{17}O_6N_3S$. A blue dye made by heating aniline with gallocyanin and sulfating.

delphine Delphinine.

delphinic acid An acid from the oil of *Delphinium* species (Ranunculaceae).

delphinidin $C_{15}H_{10}O_7 \cdot HCl$ = 338.7. 3,3′,4,5,5′,7-Hexahydroxyflavylium. An anthocyanidin from the flowers of *Delphinium* species.

delphinin $C_{41}H_{38}O_{21}Cl$ = 902.2. A glucoside from larkspur, *Delphinium consolida;* hydrolyzes to delphinidin and *p*-hydroxybenzoic acid.

delphinine $C_{33}H_{45}O_9N$ = 599.7. An alkaloid from the seeds of *Delphinium staphisagria* (Ranunculaceae), Europe and Asia Minor. Colorless crystals, decomp. 120, soluble in alcohol; an antispasmodic, anticonvulsive, and antineuralgic. Cf. *staphisagria alkaloids.*

Delphinium A genus of Ranunculaceae, larkspurs, which yield *staphisagria alkaloids,* q.v.

delphinoidine $C_{22}H_{35}O_6N$ = 409.5. An alkaloid from the seeds of larkspur, *Delphinium consolida* (Ranunculaceae), Europe and America.

delphisine $C_{54}H_{46}O_8N_2$ = 851.0. An alkaloid from the seeds of stavesacre, *Delphinium staphisagria* (Ranunculaceae).

delphocurarine $C_{23}H_{33}NO_7$ = 435.5. An alkaloid from the seeds of *Delphinium scopulorum.* Amorphous, white powder, m.185, resembling curare in effect. Cf. *delcosine.*

Delrin Trademark for a linear polyoxymethylene-type acetal resin, made by the polymerization of formaldehyde and having a high strength and solvent resistance. It is moldable and is used in aerosol containers; m.180.

delta Δ, δ. Greek letter. Δ is the symbol for any difference between 2 values. δ (italic) is the symbol for (1) damping coefficient; (2) a small difference between 2 values; (3) thickness. $\delta x/\delta y$ means the rate of change of x with y. *For chemical names,* used: (1)* as for *beta* position, q.v.; (2) as for beta position, but substitution on the 4th C atom from the principal group; (2)* indicating a double bond at the superscripted locant or between the superscripted locants; as, $\Delta^{2,\alpha}$, between the 2 and α C atoms. **d. acid** δ-acid. 1-Naphthylamine-4,8-disulfonic acid. **d. metal** A tough alloy: Cu 60, Zn 38.2, Fe 1.8%. **d. point** See *helium.*

deltaline $C_{21}H_{33}NO_6$ = 395.5. An alkaloid, m.180, from *Delphinium occidentale.* Cf. *delcosine.*

Delustra Trademark for a matt continuous-filament rayon yarn.

delvauxite A hydrous, ferric vanadium phosphate.

demagnetization Destruction of magnetic properties.

demargarinate To separate solid glycerides from an edible oil by cooling.

dematoid A green andralite (iron *garnet,* q.v.).

demecarium bromide $C_{32}H_{52}O_4N_4Br_2$ = 716.6. White crystals, soluble in water, m.165; a cholinesterase inhibitor, used as a miotic for glaucoma (USP).

demeclocycline $C_{21}H_{21}O_8N_2Cl \cdot HCl$ = 464.9. Demethylchlortetracycline hydrochloride, Declomycin, Ledermycin. Yellow crystals, soluble in water; an antibiotic (USP, EP, BP).

Demerol Trademark for meperidine (pethidine) hydrochloride.

demersal Describing fish and organisms living at or near the sea bottom.

demethylation Removal of a methyl group.

demethylchlortetracycline hydrochloride Demeclocycline.

demethylene Prefix used to denote the replacement of a CH_2 group.

demeton-S-methyl* See *insecticides,* Table 45 on p. 305.

demi- (French.) Half. See *semi-* (Latin) and *hemi-* (Greek). A glass carboy-shaped vessel, holding 4 gal.

demineralization Deionization. (1) Removal of inorganic constituents. (2) Removal of dissolved solids from water, e.g., boiler feed, without use of heat. Double-exchange methods are used to convert the salts into the corresponding acids, which are subsequently absorbed on synthetic substances. Cf. *base* exchange, *zeolite.* **mixed-bed ~** D. by means of an intimate mixture of cation- and anion-exchange resins, which act as numerous cation and anion-exchange columns in series.

Democritus of Miletus (Greek, *Demokritos.*) 460–370 B.C. An early Greek philosopher of Thrace, noted for his speculations on atoms and cosmology.

demolization The purely physical dispersion of a molecular system by highly superheated steam.

demulcent An oily or mucilaginous drug which soothes or protects an inflamed tissue.

demuriation The removal of hydrochloric acid from organic chlorine derivatives.

denaturation Addition of unpleasant or inert substances to a product, e.g., alcohol, to make it unfit for human consumption. (2) The production of suspensoid properties in albumins and globulins.

dendrite (1) Dentrite. A treelike or arborescent crystalline structure; as, in geology, the black designs found in the division planes of rocks (moss agate); due to infiltration and

subsequent evaporation of solutions of iron and manganese salts. (2) A protoplasmic protuberance on a nerve cell.
dendrobine $C_{16}H_{25}NO_2$ = 263.4. An alkaloid from Chin-shih-hu, *Dendrobium* (Orchideaceae), m.135; an antipyretic.
dendrology Dating trees from growth rings.
denier The unit weight of a thread or yarn expressed as the weight in grams of 9,000 m. Cf. *tex.*
Denigè's reagent A solution of 5 g yellow mercuric chloride in 20 mL concentrated sulfuric acid and 100 mL water; a test for citric acid.
denim (French: *de Nimes.*) A strong, fairly heavy, usually cotton fabric, with diagonal stripes due to 2 differently colored threads.
denitration, denitrification Removal of nitrates or nitrogen; as, the d. of soils by growing plants.
denitrify To remove nitrogen or nitrates.
Dennstedt furnace An electrically heated combustion furnace for elementary organic analysis.
densimeter A hydrometer.
densi-tensimeter An apparatus for the simultaneous determination of vapor pressure and vapor density.
densitometer An optical device to measure optical density.
density (1) ρ*,d. Weight density*, specific weight. In chemistry and physics, the weight per unit volume of a substance. For solids and liquids, d. = specific gravity. Cf. *mass density, hydrometer scales,* normal and relative *density,* (2) In photography, the blackness of the image on an exposed plate, measured photometrically. Cf. *optical* d. (3) In physics, *current density**, c.d., or *J*, the number of amperes passing per m^2 of a conductor. **electron ~** The electron probability distribution in an orbital. **limiting ~** The d. of a gas corresponding with the pressure at which it becomes an ideal gas and obeys Boyle's law. The ratio of the l. d. of 2 gases equals the ratio of their molecular weights. **normal ~** The limiting d. at atmospheric pressure; the weight in g/L of a substance measured at s.t.p., the weights being adjusted to sea level at latitude 45°. **relative ~*** *d.* Specific gravity. The ratio of the weight density of a substance to that of a reference under specified conditions. For solids and liquids, 1 mL of water at 4°C is the reference, thus giving the specific gravity. For gases and vapors, an equal volume of hydrogen. **vapor ~** The ratio of the weight of a given volume of a gas or vapor to that of the same volume of hydrogen measured at s.t.p.; 1 cc hydrogen weighs 0.00009 g. See *Avogadro law.* **weight ~*** Density (1).
d. comparator A photoelectric *densitometer,* q.v. **d. fluids** Fluids for separating minerals by flotation; as:

d.5.3 . . . . Mercury thallium nitrate solution
d.4.27 . . . . Clerici solution
d.3.5 . . . . Rohrbach solution
d.3.33 . . . . Methylene iodide
d.3.17 . . . . Toulet's solution
d.2.97 . . . . Acetylene tetrabromide
d.2.90 . . . . Bromoform
d.1.94 . . . . Ethyl iodide

densograph A graph showing the relation between the intensity of illumination and the density of the image on a photographic plate.
dental **d. alloy, d. amalgam** A mixture of mercury with either a silver-tin or copper-tin alloy. **d. cements** Luting agents. (1) ZnO + orthophosphoric acid; used to line cavities before amalgam or cement caps (crowns). (2) ZnO + eugenol. For deep cavities and temporary cementation. (3) Polycarboxylate cements. For restoration. (4) Calcium hydroxide. For pulp capping. **d. fillings** Both 2-paste systems, consisting of an organic polyester resin (as, Bowen's resin), an inorganic filler (as, quartz), and a peroxide catalyst; and 1-paste systems, that are cured by exposure to light.
dentate With the following prefixes, indicating the number of potential coordinating atoms in a ligand: for 1, uni ~; more than 1, multi ~ or poly ~; 2, bi ~; etc.
dentifrice Toothpaste.
dentin(e) The hard substance of teeth, giving elasticity. Root d. and crown d. are covered with cementum and enamel, q.v., respectively. Composition: water 5, hydroxyapatite and inorganic constituents 75, collagen 20%, with citric acid, insoluble proteins, mucopolysaccharides, and fats.
dentrite Dendrite.
denudation Stripping or making bare. Cf. *erosion.* **chemical ~** The loss of nutritive salts from agricultural land.
deodorant (1) An agent which removes, corrects, represses or masks undesirable odors. (2) An astringent preparation that reduces sweat production by its effect on sweat gland openings; as, aluminum chloride. Usually combined with an antiseptic (as, chlorhexidine hydrochloride) to reduce bacterial breakdown of sweat.
deoxidation Reduction.
deoxy- Desoxy. Prefix indicating: (1)* Replacement of an OH group by a H atom. (2) The removal of oxygen. **d.benzoin*** $PhCH_2 \cdot CO \cdot Ph$ = 196.2. Benzyl phenyl ketone*. Colorless scales, m.60, slightly soluble in water. **d.cholic acid** $C_{24}H_{40}O_4$ = 392.6. A bile acid. **d.cortone acetate** BP name for desoxycorticosterone acetate. **6-d.galactose** Fucose. **d.quinine** Quinane. **d.ribonucleic acid** See *nucleic acids.* **d.ribose** See *ribose.* **d.sugar** A group of carbohydrates, occurring in nature, which by adding water form a sugar; as digitalose.
dephlegmator Fractionating column.
dephlogisticated Without phlogiston (obsolete). **d. air** Oxygen (obsolete). **d. marine acid** Chlorine.
Dephraden Trademark for dextroamphetamine sulfate.
depilation Removal of hairs and epidermis from hides.
depilatory A substance that removes hairs from skin, e.g., barium sulfide.
depolarization The prevention or removal of polarization. **electrical ~** The prevention of electrical polarization, e.g., by separating electrodes by a porous diaphragm. **optical ~** The effect produced by placing a depolarizer between the analyzer and polarizer of a polarization apparatus.
depolarizer An optical device to refract polarized rays into ordinary and extraordinary rays.
deposit A concentration of matter at a particular place; e.g., a geologic location, test tube (precipitate), tissue, or electrode.
deposition The formation of a deposit. Cf. *electrodeposition, electroless deposition.*
depressant A drug which diminishes the functional activity of an organ or organism; e.g., alcohol and barbiturates, which are depressants of the central nervous system.
depression A lowering effect. **molecular ~** The lowering of the freezing point below 0°C of a solution containing a mole of a substance per liter. Cf. *Raoult's law.*
depressor (1) A negative *catalyst,* q.v. (2) A buffer. (3) Depressant. A substance used in flotation to reduce the tendency of gangue materials to be carried along with the froth, as, cyanides. **d. effect** The resistance of a solution to a change in pH value. Cf. *buffer.*
Depronal Trademark for propoxyphene hydrochloride.
depsides Esterlike anhydrides of phenolcarboxylic acids, from lichens or synthetic production.
depsiphore A structural arrangement of atoms common to many tanning agents. Cf. *chromophore.*
derbylite An antimoniotitanate of iron.

derby red Basic *lead* chromate.
dericin A light-colored, viscous oil from castor oil; a solvent.
derivant Derivative.
derivate Derivative.
derivation (1) The preparation of one organic substance from another. (2) The theoretical connection between the molecular structures of related organic compounds.
derivative (1) A compound, usually organic, obtained from another compound by a simple chemical process; e.g., acetic acid is a d. of alcohol. (2) An organic compound containing a structural radical similar to that from which it is derived, e.g., benzene derivatives containing the benzene ring. Cf. *anhydrosynthesis.* (3) Rate of change of a dependent variable relative to another variable. **additive ~** Additive *compound.*
d. action See *mode.*
derived units Units, q.v. *(unit),* of physical measurements which can be deduced fundamentally. Thus, the fundamental unit of length l will give the unit of area l^2 and the unit of volume l^3.
dermatol Bismuth subgallate.
dermatoscope A binocular microscope.
dermatosome The smallest fiber or fibril detected microscopically.
dermics A drug used for skin diseases.
derric acid $C_{12}H_{14}O_7$ = 270.2. 2-Carboxymethyl-4,5-dimethoxyphenoxyacetic acid. An oxidation product of derrisic acid from derris root, m.170, soluble in water. Cf. *rissic acid.*
derrid A resin from *Derris elliptica* (Leguminosae), southeast Asia; an arrow poison.
derrin Early name for rotenone.
derris (1) The leaves of *D. uliginosa* (Leguminosae), S. Sea Islands; a Fijian fish poison. It contains rotenone, toxicarol, and deguelin but is not toxic to man. Cf. oxygen *cube (2).* (2) An *insecticide,* q.v. See also Table 45 on page 305.
derritol $C_{21}H_{24}O_6$ = 372.4. An alcohol derived from rotenone. Yellow needles, m.161.
des (1) Indicating removal of the terminal ring, which is identified by a letter, from a polycyclic structure; as, des-A-androstane. (2) Alternative to "de" used in some languages; as, French and German.
desalination Potable water production from salt-containing waters. Achieved by multistage evaporation, reverse osmosis, or electrodialysis.
desamidization Deamination.
desaulesite A rare nickel-magnesium silicate.
Descartes, René (1596–1650) French philosopher, physicist, and chemist.
desclizite $Pb \cdot Zn_4V_2O_4 \cdot H_2O$. A native zinc-lead vanadate; green masses, d.6.1, hardness 3.5.
desensitization (1) Photography: rendering silver salts less sensitive to light. (2) Biochemistry: destruction of immune bodies.
desensitizer A substance which renders a photographic emulsion less sensitive; e.g., pinakryptol. Cf. *sensitizer.*
Deseril Trademark for methysergide maleate.
Desferal Trademark for deferoxamine mesylate.
desferrioxamine mesylate BP name for deferoxamine mesylate.
deshydro- Dehydro-*.
desiccant A drying agent. Weights of residual moisture in one liter of saturated air at 25°C after passing through the d. are, in mg.

P_2O_5	Practically none
$Mg(ClO_4)_2$	Practically none
KOH	0.003
H_2SO_4(95%)	0.003
NaOH	0.16
CaO	0.2
$CaCl_2$	0.36
$ZnCl_2$	0.8
$CuSO_4$	1.4

chemical ~ A d. which acts by absorption, e.g., reacts with water; as, phosphorus pentaoxide. **physical ~** A d. which acts by adsorption; as, silica gel.
desiccated Dried.
desiccation The process of drying.
desiccator A device for drying substances, e.g., a closed glass vessel containing a deliquescent substance. **vacuum ~** A d. which may be evacuated.
desicchlora An anhydrous, granulated barium perchlorate; a regenerable desiccant and absorbent for ammonia.
desivac process A process of freezing and dehydrating aqueous preparations.
deslanoside $C_{47}H_{74}O_{19}$ = 943.1. Deacetyl lanatoside C. White, hygroscopic crystals, m.220–235, slightly soluble in water; a potent cardiac glycoside from *Digitalis lanata.* Used when rapid effect required (USP, BP).
desmin $(Na_2 \cdot Ca)Al_2Si_6O_{16} \cdot 6H_2O$. An aluminum-calcium-sodium silicate; hardness 3.5–4.
desmolases Enzymes that cleave C-C bonds, as in glycolysis and respiration.
desmotrope One of a pair of tautomeric compounds. The most important types are:

1. Keto-enol tautomerism:
   =CH·CO– ⇌ =C:C(OH)–
2. Keto-cyclo tautomerism:
   –CO– . . . –C(OH)= ⇌ –C(OH). . . . O– (ring closed through –O–)
3. Imino-amino tautomerism:
   $=CH \cdot C(:NH)- \rightleftharpoons =C:C(NH_2)-$
4. Nitroso-oxime tautomerism:
   =CH·NO ⇌ =C:NOH
   nitroso- *oxime*
5. Nitro-*aci*-nitro tautomerism:
   $=CH \cdot NO_2 \rightleftharpoons =C:N(:O)OH$
6. Azo-hydrazone tautomerism:
   =CH·N:N– ⇌ =C:N·NH–

desmotropism Allelomorphism, dynamic allotropy. A form of isomerism (tautomerism) or organic compounds between 2 molecules containing the same number and kind of atoms of like valency, in the same position, but with different linkages, e.g., shifting of the double bond. See *isomerism, tautomerism.*
desorption (1) The reverse of *adsorption,* q.v. (2) The evolution or liberation of a volatile material from solution. (3) In chromatography, the removal of an adsorbate from an adsorbent.
desoxalic acid $C_5H_6O_8$ = 194.1. 1,2-Dihydroxy-1,1,2-ethanetricarboxylic acid. Colorless liquid, decomp. by heat, slightly soluble in water.
desoxy- See *deoxy-.*
desoxyalizarin Anthrarobin.
desoxycorticosterone acetate $C_{23}H_{32}O_4$ = 372.5. 11-Deoxycorticosterone acetate, DOCA. Colorless crystals, m.158, insoluble in water. Used in replacement therapy for adrenal glands (USP, EP, BP).
desoxyribonucleic acid See deoxyribo*nucleic acids.*
desoxyribose D-2-Deoxy*ribose.*
dessertspoon A measure; 2 fluid drams, about 8 ml.
destructive distillation Decomposition of organic compounds by heat out of contact with air; as, the production of wood tar.

desulfuration (1) Removal of sulfur. (2) Reduction of sulfur content; as, in steel manufacture. (3) Precipitation of sulfides with lead or mercury cyanide.

desyl PhCO·C(Ph)H−, from deoxybenzoin. **bi～** See *bidesyl.*

detaline Deltaline.

detection Qualitative identification. **d. limit** The smallest quantity detectable by a chemical reaction.

detector A device to detect electric waves; as, a vacuum tube.

detergent Abstergent. A cleansing agent. A substance that cleans. Typical constituents are surfactants, EDTA, carboxymethylcellulose, silicates, sulfates, chlorides, peroxoborates, fluorescent dyes, pigments, perfume, water. The principal classes of synthetic detergents contain as the principal constituents:

Fatty acid (natural, e.g., pine oil, or purified), fatty alcohol, rosin, and naphthenic acid hydrophobic group, an ether linkage, and an anionic hydrophilic group, e.g., carboxylate, sulfate, sulfonate.

Hydrocarbon (from petroleum, alkylbenzenes, alkylnaphthylmethylidyne compounds) as hydrophobic group, an ester linkage, and an amino or quaternary ammonium hydrophilic group.

Ether (polyoxypropylene) hydrophobic groups, with amido linkage, and an ampholytic (aminosulfonic or aminocarboxylic) hydrophilic group. **soft ～** A biodegradable d. that does not give rise to frothing in sewage disposal works. **d. builder** Compound added to a detergent to sequester Ca^{2+} and Mg^{2+} ions, and thus improve its performance; e.g., condensed polyphosphates, as, pentasodium tripolyphosphate. **d. enzymes** See *enzymes.*

determinism The principle that all properties of the atom may be related and determined.

deterministic Giving a definite output from a given set of inputs. Cf. *stochastic.*

detonation An explosion produced by a chemical change.

detonator Primer. A compound which ignites an explosive mixture.

detonics The study of detonation reactions.

detoxication The reduction in toxicity or poisons due to chemical changes caused by body metabolism.

deuteranope A person who is blind to the color green.

deuterate (1) To convert the H atom or atoms in a molecule into deuterium. (2) A substance containing heavy water of crystallization (obsolete).

deuteric acid An acid containing deuterium; as R·COOD.

deuterio-* Prefix indicating that protium has been replaced by deuterium. **d.ammonia** The compound ND_3 or N^2H_3. It differs from ammonia as follows (temperatures in K):

	NH_3	NH_2D mono ～	NHD_2 di ～	ND_3 tri ～
m.	195.2	197.9	198.6	199.0
b.	239.8	241.7	242.1	242.3

d.ammonium The ion ND_4^+. There are mono-, di-, tri-, and tetra- derivatives. **d.chloroform** See *chloroform.*

deuterium* D_2 = 4.0282. ^2H, H^b, heavy hydrogen, deuteronium, diplogen. The isotope of hydrogen, at. wt. 2.0141. **d. oxide** Heavy *water.*

deutero- Prefix meaning: (1) second in order; (2) derived from; (3) deuterio*.

deuteroalbumose An albumose derivative. Yellow powder, soluble in water.

deuteroartose $C_{156}H_{244}N_{40}O_{56}S$. A secondary artose.

deuterofibrinose A product formed from fibrin during digestion.

deuteron d; deuton, diplon. A deuterium atom nucleus; of mass 2.0141 with one positive charge. Cf. *proton, triton.*

deuteronium Deuterium*.

deuteroproteose Deuteroalbumose.

deuteroxyl The OD^- or O^2H^- ion.

deuton A deuteron.

Devarda's alloy An alloy: Cu 50, Al 45, Zn 5%; a strong reducing agent in alkaline solution.

developed dyes Colors produced from colorless substances by chemical reactions.

developer A reducing liquid to render the image on an exposed photographic plate visible, usually by formation of black silver.

development (1) Biology: the growth and differentiation in the structure of an organism, and acquisition of new, favorable characteristics. (2) Photography: the action of a developer. (3) Chromatography: separation of ions adsorbed on an adsorbent into bands, by passing a solution through the column of adsorbent. Cf. *elution.*

deviation The bending or turning of a direction or course.

devil's apple Stramonium.

devitrifaction Removal of transparency due to crystallization.

devitrification Devitrifaction.

Devonian See *geologic eras,* Table 38.

dew Moisture precipitated on a surface from an atmosphere saturated with water vapor. **d. point** The temperature of the atmosphere at which the saturation vapor pressure equals the actual (partial) vapor pressure of the water vapor in the air, and dew begins to form; an indication of the *humidity,* q.v., of the air.

Dewar, Sir James (1842–1923) British chemist. **D. flask** Vacuum flask. A silvered, double-walled, glass vessel, evacuated between the walls; used as container for liquefied gases or cold liquids.

deweylite $Mg_4Si_3O_{10}·6H_2O$. An amorphous, white silicate related to talc.

dexamethasone $C_{22}H_{29}O_5F$ = 392.5. White crystals, m.255, insoluble in water. A potent synthetic glucocorticoid, used for steroid therapy. Also as d. sodium phosphate to reduce raised intracranial pressure (USP, EP, BP).

dexamphetamine sulfate BP name for dextroamphetamine sulfate.

Dexedrine Trademark for dextroamphetamine sulfate.

dextran $(C_6H_{10}O_5)_n$ = $(162.1)n$. A gummy, fermentable polysaccharide from growths of *Leuconostoc mesenteroides* on sucrose. A polymerized glucose in which the glucose units are joined through 1,6-α-D-glucosidic links; causes ropiness in wines. **d. 40** Low-molecular-weight d. A mixture of d. in solution, average mol. wt. 40,000. Used to reduce blood viscosity. **d. 70 and d. 110** Mixtures of d.; a blood substitute.

dextranase* An enzyme that hydrolyzes 1,6-α-D-glucosidic linkages in dextran; a possible retardant for dental plaque.

dextrin $(C_6H_{10}O_5)_n$. British gum, starch gum, amylin, gummeline. A carbohydrate intermediate between starch and the sugars produced from starch by hydrolysis by dilute acids, amylase, or dry heat. An amorphous, yellow powder, soluble in water, precipitated by strong alcohol. It gives a reddish color with iodine and is not fermentable, but is converted to maltose by the action of enzymes (amylase), and to glucose by acids. Used as an adhesive, in printing, and in inks and water colors. **achroo～** See *achroodextrins.* **alpha-～** $(C_6H_{10}O_5)_6$. Schardinger d. **amylo～** See *starch soluble.*

animal ~ Glycogen. **beta-~** $(C_6H_{10}O_5)_7$. A crystalline degradation product of starch. **erythro ~** See *erythrodextrins.* **pyro ~** See *pyrodextrin.*
dextrinose Iso*maltose.*
dextro- Prefix meaning toward the right; e.g., dextrorotatory. Cf. *levo-.*
dextroamphetamine sulfate $(C_9H_{13}N)_2 \cdot H_2SO_4$ = 368.5. (+)-α-Methylphenethylamine sulfate. Dexamphetamine sulfate, Dexedrine, Durophet. White crystals, soluble in water. A sympathomimetic and stimulant, used to treat narcolepsy (a form of epilepsy). Use is limited by development of drug dependence (USP, BP).
dextrocarvol Carvone.
dextro compound A dextrorotatory compound.
dextrogyric Dextrorotatory.
dextromethorphan hydrobromide $C_{18}H_{25}ON \cdot HBr \cdot H_2O$ = 370.3. White crystals, m.127, soluble in water; a cough suppressant (USP, BP).
dextronic acid Gluconic acid.
dextropropoxyphene BP name for propoxyphene.
dextrorotary Dextrorotatory.
dextrorotatory Describing an optically active compound that rotates the plane of polarized light to the right (clockwise). Indicated by the prefix (+)-*, or *d*-. See *optical activity, asymmetric carbon.*
dextrosazone Glucosazone.
dextrose Glucose. **d. monohydrate** $C_6H_{12}O_6 \cdot H_2O$ = 198.2. Medical glucose used for intravenous feeding and hypoglycemia (USP, BP).
deyamittin $C_{18}H_{12}O_3N$. A glucoside from *Cissampelos pareira.* Colorless crystals. Cf. *dyamettin, cissampeline.*
DFP Common name for diisopropylfluorophosphonate, developed during World War II as a poison affecting the nervous system. Now used as a miotic. See *isofluorophate.*
dg Abbreviation for decigram, 0.1 g.
dhurrin $C_{14}H_{17}NO_7$ = 311.3. A glucoside from millet and sorghum, which hydrolyzes to glucose, hydrogen cyanide, and *p*-hydroxybenzaldehyde. Colorless needles, m.196.
Di Symbol for didymium, a mixture of neodymium and praseodymium.
di- Prefix meaning two, or twice. Cf. *bi-, bis-.*
dia- Prefix meaning through, or opposite.
diabantite $H_{18}(Mg, Fe)_{12}Al_4Si_9O_{45}$. A chlorite-type silicate.
diabase Dolerite. An igneous rock formed in the transition from basalt to granitoid gabbros; it consists of plagioclase, magnetite, augite, and sometimes olivine.
diabetes (1) D. mellitus. A disease characterized by high levels of glucose in the blood and urine, and reduced insulin production. (2) D. insipidus. A disease due to lack of *vasopressin,* q.v.
diabetin Fructose.
diacetamide* Me·CO·NH·CO·Me = 101.1. *N*-acetylacetamide. Colorless needles, m.78, soluble in water. **phenyl ~** Diacetanilide.
diacetanilide Me·CO·NPh·CO·Me = 177.2. *N*-Acetylacetanilide, *N*-phenyldiacetamide, acetoacetanilide. Colorless leaflets, m.37.5, soluble in water.
diacetate (1) A salt of diacetic acid. (2) A salt containing 2 acetoxy groups, CH_3COO-.
diacetenyl The 1,3-butadiyne-1,4-diyl† radical −C:C−C:C−. **d.benzene** Diphenyldiacetylene.
diacetic acid (1) $(MeCO)_2CH \cdot COOH$ = 144.1. Diacetylacetic acid. Colorless liquid in diabetic urines. (2) Acetoacetic acid. (3) Succinic acid. **acetone ~** *Hydrochelidonic acid.* **ethylidene ~** 3-Methyl*glutaric acid.* **phenylene ~** See *phenylenediacetic acid.* **propylidene ~** 3-Ethyl*glutaric acid.* **d. ester** $(MeCO)_2CHCOOEt$ = 172.1. Ethyldiacetate, ethyldiacetic ester. Colorless liquid, decomp. 200, slightly soluble in water.
diacetin $(MeCOOCH_2)_2CHOH$ = 176.2. Glyceryl diacetate. Colorless liquid, m.40, soluble in water.
diacetonamine $C_6H_{13}ON$ = 115.2. β-Aminoisopropylacetone. Formed by the action of ammonia on acetone. Yellow liquid.
diacetone Acetylacetone. **d. alcohol** $MeCOCH_2CMe_2OH$ = 116.2. Colorless liquid, b.164; a solvent for nitrocellulose and resins.
diacetyl (1) Biacetyl*. (2) Prefix indicating 2 acetyl radicals. **d. amide** Diacetamide*. **d.anilide** Diacetanilide. **d. glucose** $C_{10}H_{16}O_8$ = 264.2. Colorless crystals, m. exceeds 100, soluble in water. **d.morphine** Diamorphine. **d. peroxide** MeCO·O·O·COMe = 118.1. Yellow liquid; an antiseptic. **d.urea** MeCO·NH·CO·NH·COMe = 144.1. Colorless crystals, soluble in water.
diacetylene HC:C·C:CH = 50.1. 1,3-Butadiyne*. Colorless gas. **d. glycol** $CH_2OH \cdot C:C \cdot C:C \cdot CH_2OH$ = 110.1. Hexadiindiol. Colorless crystals, m.111.
diacetylenes C_nH_{2n-6}. Unsaturated hydrocarbons containing 2 triple bonds.
diacolation Extraction in presence of sand, sometimes used for pharmaceutical tinctures.
diad An element or radical with valency 2.
diadochite $H_{24}Fe_4P_2S_2O_{29}$. Yellow monoclinics, d.3.8, hardness 3.5–4.5.
diagenesis Geochemical activity by microbial or chemical alteration.
diagnostic Means of recognition.
diagram A graph showing the relation of one or more properties of one or more substances. Cf. *coordinates.*
Diakon Trademark for a methyl methacrylate plastic.
-dial* Suffix indicating a dialdehyde. **d.uramide** Uramil.
dialdehyde* dial. A compound containing 2 aldehyde groups. **d. starch** A product of the oxidation of starch with periodic acid.
dialin Dihydro*naphthalene.*
dialkene* Diolefin.
dialkyl A compound containing 2 alkyl radicals.
dialkylene A compound containing 2 alkene radicals.
diallag $(Mg,Fe)CaSi_2O_6$. A calcium silicate, sometimes containing alumina, d.3.2, hardness 4.
diallyl (1) Biallyl*. (2)* A compound containing 2 allyl radicals. **d. sulfide** See *allyl sulfide* under *allyl.* **d.urea** Sinapolin.
dialozite $MnCO_3$. Native manganous carbonate.
dialuramide Uramil.
dialurate A salt of dialuric acid.
dialuric acid $OC(NH \cdot CO)_2 \cdot CHOH$ = 144.1.

Tartronoylurea, 5-hydroxybarbituric acid. A monobasic heterocyclic acid from alloxan. Colorless prisms, slightly soluble in water.
dialysate Dialyzate.
dialysed Dialyzed.
dialysis Utrafiltration. Microfiltration by a semipermeable membrane, which separates ions and small molecules from large molecules and colloids. **hemo ~** Used to treat kidney failure. Blood is circulated past a semipermeable membrane which removes substances normally extracted by the kidney.
dialyzate The fraction which is dialyzed through a membrane. Cf. *diffusate.* Incorrectly used to describe the fraction which does not pass through the membrane. Cf. *tenate.*

dialyzator An apparatus for dialysis.
dialyzed Separated by dialysis. **d. iron** Colloidal ferric hydroxide solution.
dialyzer (1) The semipermeable membrane used in dialysis. Cf. *diffusion shell.* (2) Artificial kidney. The apparatus used for hemodialysis.
diamagnetic Repelled by a magnet; taking a position at right angles to the field of an electromagnet; having magnetic permeabilities less than 1. Cf. *paramagnetic.*
diamantane Adamantane.
diameter (1) dia. A straight line passing through the center of a body or figure. Cf. *circle, caliber.* (2) The length of a straight line through the center of an object. (3) The number of times the d. of a magnified object is increased.
diamide (1) Oxamide*. (2) A compound containing 2 $-CONH_2$ groups.
diamidine See *amidines.*
diamido- Prefix indicating 2 amido radicals.
diamidogen Hydrazine*.
diamine (1) Hydrazine*. (2) See *diamines.* **butane ~ *** $C_4H_{12}N_2$ = 88.2. Tetramethylenediamine*, putrescine, 1,4-diaminobutane. **diethylene ~** $NH\cdot(CH_2)_2\cdot(CH_2)_2NH$ = 86.1. Cyclobutanediamine*. **dimethylene ~ *, ethane ~ *** $NH_2CH_2CH_2NH_2$ = 60.1. Colorless liquid, b.116. **hexamethylene ~ *** Triethylenediamine*. **pentane* ~** $C_5H_{14}N_2$ = 102.2. Pentamethylenediamine*, cadaverine, 1,5-diaminopentane. Liquid, b.179; product of putrefaction. **pentamethylene ~ *** Pentanediamine*. **phenylene ~** See *phenylene*diamine. **propane ~ *** $NH_2(CH_2)_3NH_2$ = 60.1. Trimethylenediamine*. Colorless liquid, b.119. **tetramethylene ~ *** Butane*diamine**. **cyclic ~** Piperazine*. **triethylene ~ *** See *triethylenediamine.* **trimethylene ~ *** Propane*diamine**.
d. blue Trypan blue.
diamines Compounds containing 2 $-NH_2$ groups. **aliphatic ~** Compounds containing 2 amino groups attached to a carbon chain, e.g., ethylenediamine. **aromatic ~** Compounds containing 2 amino groups attached to a carbon ring, e.g., phenylenediamine. **homocyclic ~** Aromatic diamines.
diamino-* Prefix indicating 2 amino groups ($-NH_2$). **d.anthraquinone** $C_{14}H_{10}O_2N_2$ = 238.2. **1,2- ~** Blue crystals, decomp. 130, insoluble in water. **1,4- ~** Red needles, m.236, slightly soluble in water. **2,3- ~** Brown needles, sublime on heating, soluble in water. **d.azobenzene** See *2,4-diaminoazobenzene* under *azobenzene.* **d.azotoluene** $C_{14}H_{16}N_4$ = 240.3. Isomers include 2,2′-diamino-4,4′-azotoluene, m.203; 3,3′-d.-2,2′-a., m.145. **d.benzene** See *phenylenediamine.* **d.benzophenone** $NH_2\cdot C_6H_4\cdot CO\cdot C_6H_4\cdot NH_2$ = 212.3. **3,3′ ~** Yellow needles, m.237, soluble in water. **4,′4- ~** Colorless needles, m.172, soluble in water. **d.biphenyl** Benzidine*. **d.diphenic acid** $C_{14}H_{12}O_4N_2$ = 272.3. Benzidinedicarboxylic acid. Colorless crystals, insoluble in water. **d.diphenylamine** $NH_2C_6H_4\cdot NH\cdot C_6H_4NH_2$ = 199.3. Colorless scales, m.158, insoluble in water; used in organic synthesis. **d.diphenylethylene** Diaminostilbene. **d.diphenylmethane** $C_{13}H_{14}N_2$ = 198.3. **4,4′- ~** Colorless scales, m.87, soluble in water; used in organic synthesis. **d.naphthalenesulfonic acid** $C_{10}H_4(NH_2)_2(SO_3H)_2$ = 318.3. **1,5,3,7- ~** Colorless crystals, insoluble in water. **1,8,3,6- ~** Colorless prisms, soluble in water; used to make H acid. **d.naphthalenedisulfonic acid** $C_{10}H_5(NH_2)_2SO_3H$ = 238.3. **1,3,6- ~** Colorless crystals, slightly soluble in water. **1,4,2- ~** Colorless crystals, slightly soluble in water; used in organic synthesis. **2,6-diamino-1-oxohexyl†** See *lysyl.* **d.phenol*** See *aminophenols.* **d.phenol hydrochloride** Amidol. **d.stilbene** $(NH_2)_2C_6H_3\cdot CH{:}CH\cdot C_6H_5$ = 210.3. 2,4-Diaminodiphenylethylene. Yellow crystals, m.120, insoluble in water. **d.stilbene disulfonic acid** $C_{14}H_{14}N_2(SO_3)_2$ = 370.4. **1,2,4- ~** Yellow crystals, insoluble in water. **d.triphenylmethane** $C_{19}H_{18}N_2$ = 274.4. **4,4′- ~** Colorless, crystalline beads, m.139, slightly soluble in water; an intermediate.
diaminodiphosphatides Phospholipids containing 2 N atoms and 2 P atoms, e.g., assurin.
diaminophosphatides Phospholipids containing 2 N atoms to 1 P atom, e.g., aminomyeline. **diaminomonophosphatides** Diaminophosphatides.
diamminemercury(II)* See *mercur-.*
diammonium (1) $NH_4\cdot NH_4$ = 36.1. Cf. *hydrazinium.* (2) Prefix indicating the presence of 2 ammonium radicals. Cf. *ammine.*
diamond C = 12.011. Crystalline carbon, colorless or tinted isomeric crystals, d.3.53, hardness 10, insoluble and nonfusible, burning to carbon dioxide. Used as a precious stone, for cutting glass, and as bearings for delicate mechanisms. **industrial ~** (1) *Carbon, carbonado,* black Brazilian; d. in porous clusters of fine-grained microcrystals. (2) *Ballas;* nonporous, mostly round, minute d. crystal. (3) *Boart, bort;* translucent crystal which cleaves in layers. (4) Synthetic; made by heating graphite and carbon black with a catalyst (as, chromium) at 1600°C and 95,000 atm. Differs from natural diamonds in having spiral growth forms. **d. black** C.I. Natural Black 3. A black dye from hematin. (1) Waste from d. cutting and polishing. (2) Finely powdered glass, used as a polish or filter. **d. flavin** $C_6H_4(OH)\cdot C_6H_3N{:}N\cdot C_6H_3(OH)COOH$. A yellow azo dye. **d. mortar** A small, hard, steel mortar.
diamorphine (hydrochloride) $C_{21}H_{23}O_5N\cdot HCl\cdot H_2O$ = 423.9. Diacetylmorphine (hydrochloride), heroin. White crystals, m.231, soluble in water. A potent narcotic analgesic, used for severe pain and in terminal illness, being more powerful than morphia. Addictive; use and manufacture prohibited in the U.S. (BP).
diamyl (1) Decane*. (2) Dipentyl*.
diamylene (1) Decene*. (2) Indicating a compound containing 2 pentenyl radicals.
dianiline Prefix indicating 2 aniline molecules. **d.hexafluorosilicate** $(C_6H_5NH_2)_2\cdot H_2SiF_6$ = 330.3. White plates, subliming 230, soluble in water.
dianisidine Bianisidine.
dianthryl Bianthryl.
diaphanometer An instrument to measure transparency.
diaphanoscope (1) A darkened box with a source of light to view transparent objects. (2) An instrument to illuminate body cavities.
diaphoretic A drug stimulant of sweat gland secretions, e.g., pilocarpine. Cf. *sudorific.*
diaphorite $(Ag_2Pb)_5Sb_4S_{11}$. A metallic, orthorhombic sulfoantimonite.
diaphragm (1) A disk, with one or more holes, of variable size, to regulate the amount of light passing through a lens. (2) The porous wall of a galvanic cell separating the 2 liquids. (3) A semipermeable partition or wall. (4) The main muscle of respiration; the partition dividing the thorax (chest) from the abdomen.
diarabinose $C_{10}H_{18}O_9$ = 282.3. A disaccharide of arabinose, m.260.
diars* Indicating *o*-phenylenebis(dimethylarsine), $Me_2As\cdot C_6H_4\cdot AsMe_2$, as a ligand.
diarsane* $H_2As\cdot AsH_2$ = 153.9. Biarsine, diarsine, diarsyl. Cf. *arsines.* **tetraethyl ~ *** Et_4As_2 = 266.1. Ethyl cacodyl. Colorless liquid, igniting in air, b.187. **tetramethyl ~ ***

Me_4As_2 = 210.0. Dicacodyl. Colorless, poisonous liquid, b.170.
diarsenate A salt containing 2 arsenate radicals.
diarsenediyl† See *arseno.*
diarsenite A salt containing 2 arsenite radicals.
diarsenous acid Pyro*arsenic acid**.
diarsine Diarsane*.
diarsonium Cacodyl.
diarsyl Diarsane*.
diascope Portable X-ray tube for visual diagnosis by insertion into body cavities.
diaspore AlO·OH. A hydrous aluminum oxide decomposition product of rocks.
diastase α- or β-Amylase*.
diastasic action The action of amylase.
diastasimetry Measuring diastatic power.
diastatic action Diastasic action.
diathermy Slow penetration of heat.
diathesin $C_7H_8O_2$ = 124.1. Salicyl alcohol, *o*-hydroxybenzyl alcohol. Colorless crystals, m.86, soluble in water. Used to treat rheumatism.
Diatol Brand name for diethylcarbonate.
diatom A unicellular alga, usually with 2 symmetric halves. **d. ooze** Ocean sediments (about 2,750 km deep) of empty shells of diatoms (23% Ca).
diatomaceous earth Kieselguhr.
diatomic A molecule of 2 atoms, e.g., a binary compound.
diatomite Kieselguhr.
diatrizoic acid $C_{11}H_9O_4N_2I_3$ = 613.9. 3,5-Diacetamido-2,4,6-triiodobenzoic acid. A radiopaque, for X-rays of renal tract and major blood vessels (USP, BP).
diazacyclo- Prefix indicating the presence of 2 N atoms in a ring. See *aza.*
diazene Diazete.
diazepam $C_{16}H_{13}ON_2Cl$ = 284.7. 7-Chloro-1,3-dihydro-1-methyl-5-phenyl-2*H*-1,4-benzodiazepin-2-one. Valium. A drug of the benzodiazepine group; a tranquillizer, anticonvulsant, and hypnotic.
diazete $C_2H_4N_2$ = 56.1. Diazene. The heterocyclic system, $CH_2·N:N·CH_2$ (ring).
diazines (1) Hydrocarbons, consisting of a hexatomic ring with 2 N atoms and 4 C atoms. **1,2-~** Pyridazine. **1,3-~** Pyrimidine*. **1,4-~** Pyrazine*. (2) Suffix indicating a ring compound with 2 N atoms; as, benzodiazine.
diazo-* Prefix indicating the radical —N:N—, where the second bond is to an element other than C (except for cyanides). Cf. *azi, azo.* **d. compounds** Very reactive organic nitrogen compounds formed when nitrous acid acts at low temperatures on the salts of primary aromatic amines. Usually explosive; used in solution as intermediates in dyestuff manufacture. Cf. *bisazo compound.* **d. paper** See *diazo paper* under *paper.* **d. test** Pathological urine turns red on addition of diazobenzenesulfonic acid.
diazoacetate A salt of diazoacetic acid. **ethyl ~** $C_4H_6O_2N_2$ = 114.1. Colorless liquid, b_{120mm}143.
diazoamino* Azimino*, 1-triazene-1,3,-diyl†. The radical —N:N·NH—. **d.benzene*** Ph·N:N·NH·Ph = 197.2. 1,3-Diphenyltriazene*. Yellow leaflets, m.98 (decomp.); insoluble in water. **d.naphthalene*** $C_{20}H_{15}N_3$ = 297.4. 1,3-Di-1-naphthyltriazine*. Yellow leaflets, explode on heating, insoluble in water.
diazoate* Diazotate. An acidic metal salt and tautomer of diazonium hydroxide, containing the radical ArN:NO—. Isomers: (Z) (or normal) and (*E*) (or iso); either may occur ionized or un-ionized in solution, and form salts of the type ArNNOM, where M is a monovalent metal. **d. ion** The acid $PhNNO^-$ ion; exists in 2 isomers. **d. split** The decomposition of a d. to a phenol and nitrogen: ArNNOH = ArOH + N_2.
diazobenzene **d. acid** Nitranilide. **d. chloride** $C_6H_5N_2Cl$ = 140.6. Benzenediazonium chloride*. Colorless needles, decomp. by heat, soluble in water. **d. cyanide** $C_7H_5N_3$ = 131.1. Yellow prisms, m.69, slightly soluble in water. **d. hydroxide** Diazonium hydroxide. **d.imide** $C_6H_5N_3$ = 119.1. Yellow oil, d.1.098, explodes when heated, insoluble in water. **d. nitrate** $C_6H_5O_3N_3$ = 167.1. Colorless needles, explode on heating, soluble in water. **d.sulfonic acid** $C_6H_4O_3N_2S$ = 184.2. Benzeneazosulfuric acid. Red prisms, decomp. by heat, insoluble in water. Cf. *diazo test.*
diazodinitrophenol $C_6H_2N_4O_5$ = 210.1. DDNP, dinol, 4,6-dinitrobenzene-2-diazo-1-hydroxide. Yellow needles; a detonator for percussion caps.
diazoethane CH:N·N:CH (ring) = 54.1. Aziethylene, aziethane. **dimethyl ~** $C_4H_6N_2$ = 82.1. Colorless crystals, m.270.
diazohydrates Compounds containing the azoxy* group.
diazohydroxide The radical —N:NOH; 2 isomers:

```
  NOH      HON
  ||        ||
R—N      R— N
```

diazoic acid An isomer, Ph·N:NO·OH, of phenylnitroamine, from which diazoates are derived.
diazoimide Hydrogenazide*.
diazoimido **d. compounds** Compounds containing the radical $-N_3$, from hydrogenazide. **d.phenyl** Diazobenzeneimide.
diazoles Pentacyclic hydrocarbons with 2 N atoms. **1,2-~** Pyrazole*. **1,3-~** Imidazole*.
diazomethane* $N:N·CH_2$ (ring) = 42.0. Azimethylene, azimethane. Poisonous, yellow, explosive, odorless gas, used for methylation.
diazonium **d. ion** The ion PhN_2^+. **d. salts** Compounds of the type $RN_2^+X^-$, where X is an acid radical; as, benzenediazonium chloride, $PhN_2^+Cl^-$.
diazoparaffins Diazoalkanes. Aliphatic hydrocarbons containing the diazo group; as, diazomethane.
diazophenol (1) $C_6H_4ON_2$ = 120.1. *p*-Diazophenol, furo[*ab*]diazole. Colorless crystals, explode 38. (2) Internal diazo oxides formed from diazophenols by dehydration, e.g., $C_6H_4·O·N:N$ (ring).
diazosalt See *diazonium salts.*
diazosplit The decomposition of a diazonium salt and formation of an aryl compound; e.g., ArNNX = ArX + N_2, where X is an acid or halogen radical. Cf. *Sandmeyer's reaction.*
diazotate Diazoate*.
diazotetrazole CN_6 = 96.1. A heterocyclic compound. Colorless, explosive crystals.
diazotization Diazo compound formation.
diazotizing Treating primary amines with nitric oxides or nitrites (Griess, 1860); used to manufacture aniline dyes.
diazoxide $C_8H_7O_2N_2SCl$ = 230.7. 7-Chloro-3-methyl-2*H*-1,2,4-benzothiadiazine-1,1-dioxide. Eudemine, Hyperstat. White, crystalline powder, insoluble in water. Used for emergency treatment of hypertension and in chronic hypoglycemia.
diazoxy The azoxy* radical.
dibasic Describing compounds which contain 2 H atoms replaceable by a monovalent metal; or an acid which furnishes 2 H^+ ions.
dibenzacridine $C_{21}H_{13}N$ = 279.3. Six isomers composed of

an acridine molecule, plus 1 ortho-fused benzene ring on each of the outside rings.

dibenzamide (1)* $(PhCO)_2NH$ = 225.2. *N*-Benzoylbenzamide†, dibenzoylamine. Colorless crystals, m.148. (2) A compound with 2 benzamido* radicals. **ethylene ~** $(PhCONHCH_2)_2$ = 268.3. Colorless crystals, m.249. **ethylidine ~** $(PhCONH)_2CHMe$ = 268.3. Colorless crystals, m.204. **methylene ~** $C_{15}H_{14}O_2N_2$ = 254.3. Colorless crystals, m.221.

dibenzanthracene* $C_{22}H_{14}$ = 278.4. Naphthophenanthrene, benzonaphthacene. Isomers are 5 fused rings; as, bibenz[*a,j*]anthracene, m.196:

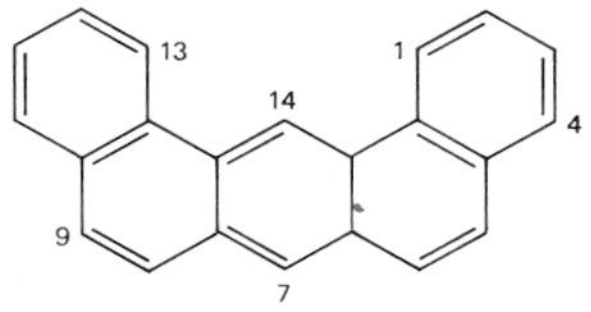

dibenzenyl (1) Tolane. (2) Dibenzylidyne*.

dibenzo- Diphenylene.

dibenzoyl (1) Benzil*. (2)* Prefix indicating 2 benzoyl* radicals, PhCO—. **d. catechol** $C_{20}H_{14}O_4$ = 318.3. Colorless crystals, m.84. **d.ethane** $PhCO \cdot CH_2 \cdot CH_2 \cdot COPh$ = 238.3. Diphenazyl. Colorless crystals, m.145. **d.furazan** PhCO·C:N·O·N:C·COPh. Colorless crystals, m.118. **d. glucoxylose** A glucoside, m.148, from *Danielia latifolia* (Bignoniaceae), Brazil. **d. ketone** Diphenyl triketone. **d.malonitrile** $(PhCO \cdot CH{:}N)_2CH_2$ = 278.3. Colorless crystals, m.130. **d.methane** $(Ph \cdot CO)_2CH_2$ = 224.3. Colorless crystals, m.81. **d. peroxide*** Benzoyl peroxide.

dibenzyl (1) Bibenzyl*. (2) Prefix indicating 2 benzyl radicals, $PhCH_2$—. **d.amine*** $C_{14}H_{15}N$ = 197.3. Colorless liquid, b.300, insoluble in water. **d. diethyl stannane*** $(C_6H_5CH_2)_2SnEt_2$ = 359.1. Colorless liquid, b.223, soluble in alcohol. **d.ethane** $Ph(CH_2)_4Ph$ = 210.3. Diphenylbutane. Colorless crystals, m.52, soluble in alcohol. **d.hydrazine** $(PhCH_2NH—)_2$ = 212.3. Colorless crystals, m.65; used in organic synthesis. **d. ketone*** $Ph \cdot CH_2 \cdot CO \cdot CH_2 \cdot Ph$ = 212.3. Colorless crystals, m.34, insoluble in water. **d.mercury** $Hg(C_6H_5CH_2)_2$ = 382.9. Colorless needles, soluble in alcohol.

dibenzylidene* A compound containing 2 $C_6H_5 \cdot CH=$ radicals. **d.acetone** Styryl ketone.

dibenzylidyne* Dibenzenyl. Prefix indicating 2 benzylidyne* radicals, ≡CPh.

diborane diphosphine $B_2H_6 \cdot 2PH_3$ = 95.7. White, crystalline solid, dissociates above −30 into diborane and phosphine.

diboran(6)yl* The radical $H_2B(H_2)BH$—.

dibromo-* Dibrom-. Prefix indicating 2 Br atoms. **d.acetic acid** $C_2H_2O_2Br_2$ = 217.8. Colorless crystals, m.48, soluble in water. **d.anthracene** $C_{14}H_8Br_2$ = 336.0. **9,10- ~** Yellow needles, m.221, insoluble in water. **d.anthraquinone** $C_{14}H_6O_2Br_2$ = 366.0. Isomers are yellow crystals, with m. ranging 195–290. **d.benzene** See *benzene*. **d.indigo** $C_{16}H_8O_2N_2Br_2$ = 420.1. Tyrian purple. Purple crystals in *Murex* and *Purpura* species. **d.isethionate** $C_{17}H_{18}O_2N_4Br_2 \cdot 2C_2H_6O_4S$ = 722.4. White, bitter crystals, m.190, soluble in water; an antiseptic. **d.quinone chloroimide** A reagent for phenols.

d.thymolsulfonphthalein An indicator, changing at pH 7.0 from yellow (acid) to blue (alkaline).

dibromide* A salt ionizing to 2 bromide ions.

dibucaine hydrochloride $C_{20}H_{29}O_2N_3 \cdot HCl$ = 379.9. Cinchocaine hydrochloride, Nupercainal. White crystals, m.98, soluble in water. Mainly used for surface anesthesia (USP, BP).

dibutyl (1)* Prefix indicating 2 butyl (C_4H_9—) radicals. (2) Octane*. **d.beryllium** $(C_4H_9)_2Be$ = 123.2. Colorless liquid, b_{25mm}170, decomp. in water. **d.cadmium** $(C_4H_9)_2Cd$ = 226.6. An oil, d.1.306, $b_{12.5mm}$103. **d. phthal(ate.)** See *phthalate*. **d.tin dibromide** $(C_4H_9)_2SnBr_2$ = 392.7. White needles, m.20. **d.tin dichloride** $(C_4H_9)_2SnCl_2$ = 303.8. Colorless needles, m.43.

di-*i*-butyl- Prefix indicating 2 isobutyl radicals. **d.cadmium** $(C_4H_9)_2Cd$ = 226.6. An oil, d.1.269, b_{20mm}91. **d.mercury** $(C_4H_9)_2Hg$ = 314.8. Colorless liquid, d.1.835, b.205, slightly soluble in water.

dibutyrin $C_3H_5(OH)(OOC \cdot C_3H_7)_2$ = 232.3. Glyceryl dibutyrate. Colorless liquid, d.1.803, b.282; formed from butyrin by steapsin.

dicacodyl (1) Tetramethyl*diarsane**. (2) Containing 2 cacodyl radicals.

dicamba* See *herbicides*, Table 42 on p. 281.

dicarbazyl Bi*carbazole**.

dicarbonate Hydrogen*carbonate**.

dicarboxyl Oxalic acid*.

dicarboxylic acid A compound with 2 —COOH groups.

Dicel Trademark for a continuous-filament cellulose acetate yarn.

dicetyl $Me(CH_2)_{30}Me$ = 450.9. Dotriacontane*, bicetyl*; m.70.

dichlofluanid* See *fungicides*, Table 37 on p. 250.

dichloricide *p*-Dichloro*benzene*.

dichloride An inorganic salt containing 2 chloride atoms.

dichloro-* Dichlor-Prefix indicating the presence of 2 Cl atoms in an organic molecule. **d.acetal** $C_6H_{12}O_2Cl_2$ = 187.1. Colorless liquid, b.180–185. **d.acetamide** See *dichloroacetamide* under *acetamide*. **d.acetic acid** See *dichloroacetic acid* under *acetic acid*. **d.acetone** See *dichloroacetone* under *acetone*. **d.acetylchloroanilide** $Cl_2CH \cdot CO \cdot NCl_2Ph$ = 274.0. A chlorinating agent. **d.acetyl chloride** C_2HOCl_3 = 147.4. Colorless liquid, b.107; hydrolyzed by water or alcohol. **d.amide T** $CH_3 \cdot C_6H_4 \cdot SO_2NCl_2$ = 240.1. *p*-Toluene dichlorosulfamine. Yellow crystals of strong chlorine odor, m.78–83, insoluble in water; an antiseptic. **d.aniline** See *dichloraniline* under *aniline*. **d.anthracene** See *dichloroanthracene* under *anthracene*. **d.benzene** See *dichlorobenzene* under *benzene*. **d.diethyl ether** 2,2′-Dichloroethyl ether. **d.diethyl sulfide** Mustard gas. **d.difluoromethane*** See *dichlorodifluoromethane* under *methane*. **d.ethyl ether** $C_4H_8OCl_2$ = 143.0. Dichloroether. **1,2,- ~** $CH_2Cl \cdot CHCl \cdot OEt$. Colorless liquid, d.1.174, b.142. soluble in alcohol. **2,2- ~** $(CH_2Cl \cdot CH_2)_2O$. Bis(2-chloroethyl)ether*. Colorless liquid, d.1.213, b.178, insoluble in water; a solvent for fats and pectins. **d.ethylene** CHCl:CHCl = 96.9. 1,2-Acetylene dichloride, dieline, dioform. Colorless liquid of pleasant odor, d.1.278, b.55. A mixture of the 2 stereoisomers; (*E*)- ~ d.1.265, b.48; (*Z*)- ~ d.1.291, b.60. Used as rubber solvent; for extractions; as a metal degreaser, anesthetic, and refrigerant. **d.hydrin** $C_3H_6OCl_2$ = 129.0 **alpha- ~** Dichloroisopropyl alcohol, 1,3-dichloro-2-propanol*. Colorless, ethereal liquid, $d_{19°}$ 1.367, b.174, miscible with alcohol; a solvent. **beta- ~** Dichloropropyl alcohol, 2,3-dichloro-1-propanol*. Colorless liquid, $d^{17°}$ 1.355, b.182. **d.methane** Methylene dichloride. **d.naphthalene** See *dichloronaphthalene* under *naphthalene*.

d.nitrohydrin $C_3H_5O_3NCl_2$ = 174.0 1,3-Dichloro-2-propanol nitrate. Colorless liquid, d.1.459, insoluble in water. **d.phenolindo-*o*-cresol** An oxidation-reduction indicator. **d.phenolindophenol** Tillman's reagent. **d.phenosulfonphthalein** A pH indicator, changing at 6.0 from yellow (acid) to red (alkaline). **d.propene*** 1,3- ~ An insecticide. **d.quinone chloroimide** A reagent for phenols.

dichlorophen(e) $C_{13}H_{10}O_2Cl_2$ = 269.1. White powder, m.175, insoluble in water; used to treat tapeworm infestation. **d.amide** $C_6H_6O_4N_2Cl_2S_2$ = 305.1. 4,5-Dichloro-*m*-benzenedisulfonamide. White crystals, m.237, slightly soluble in water; a carbonate dehydratase inhibitor, used to treat glaucoma (USP, BP).

dichlorprop* See *herbicides*, Table 42 on p. 281.

dichroine $C_{16}H_{19}O_3N_3$ = 301.3. Febrifugine. An antimalarial alkaloid from the roots and leaves of *Dichroa febrifuga*, Lour., m. 140, slightly soluble in water.

dichroism (1) The property by which certain crystals exhibit different colors when viewed in different directions, or when viewed by reflected or refracted light. (2) The property of showing different colors when viewed through different thicknesses, e.g., certain indicator solutions. **circular ~** Characteristic iridescent colors, as emitted by liquid crystals in polarized light; they vary with temperature, angle of incident light, and nature of the compound.

dichroite Iolite.

dichromat A person having reduced color discrimination; capable of recognizing only 2 out of 3 primary colors. Cf. *deuteranope, protanope.*

dichromate (1)* Bichromate, pyrochromate. A salt containing the radical $=Cr_2O_7$. (2) Sodium dichromate. **d. cell** See *dichromate cell* under *cell.*

dichromatic Having different colors according to the thickness of the layer through which the solution (e.g., dyestuffs) is viewed. Cf. *dichroism.*

dichromic acid* The hypothetical acid, $H_2Cr_2O_7$ or $2CrO_3 \cdot H_2O$, from which dichromates are derived.

dichroscope An instrument to determine the refractive power of crystals.

dick Ethyl arsinedichloride*.

dickite A form of kaolin.

Dick test An antitoxin test for scarlet fever (BP).

dicodeine $C_{72}H_{84}O_{12}N_4$. A polymer of codeine.

diconic acid $C_9H_{10}O_6$ = 214.2. A citric acid derivative.

dicophane DDT.

dicyanide* A compound containing 2 cyanide radicals.

dicyanine A $C_{21}H_{27}N_2O_2I$ = 466.4. 1,1′-Diethyl-4,2′-dimethyl-6,6′-diethoxy-2,4′-carbocyanine iodide. A quinoline dye infrared photosensitizer. Cf. *cyanine dyes.*

dicyano- (1)* Indicating a compound containing 2 cyanide radicals. (2) Cyanogen. **d.acetylene** $CN \cdot C{:}C \cdot CN$ = 76.1. Colorless crystals, m.21. **d.diamide** $NH_2 \cdot C(:NH) \cdot NHCN$ = 84.1. Cyanoguanidine, param. An isolog of amidinourea. Colorless scales, m.210, soluble in water. **d.diamidine** $NH_2 \cdot C(:NH) \cdot NH \cdot CONH_2$ = 102.1. Amidinourea. The amide of guanidinecarboxylic acid. Colorless crystals, m.105, soluble in water. Cf. *Grossmann reagent.*

dicyanoaurate(I)* See *aurocyanide.*

dicyanogen $CN \cdot CN$ = 52.0. Oxalonitrile, ethane dinitrile. Colorless, poisonous gas, b.−25. Cf. *cyanogen.*

dicyclic Bicyclic*.

dicyclomine hydrochloride $C_{19}H_{35}O_2N \cdot HCl$ = 346.0. Benacol, Debendox, Merbentyl. White, bitter crystals, m.173, soluble in water; an antispasmodic for intestinal colic and nausea (USP, BP).

didehydro- See *dehydro-.*

didiphenylamino- Prefix indicating 2 diphenylamino radicals. **d. hexafluorosilicate** $[(C_6H_5)_2NH]_2 \cdot H_2SiF_6$ = 482.5. White crystals, m.169.

didymia Didymium oxide.

didymium Di. A supposed element discovered by Mosander (1841) in the earth didymia; it is a mixture of neodymium and praseodymium. "D. salts" are mixtures of neodymium and praseodymium salts; hence, the symbol Di in the following subentries means Nd and Pr. **d. carbonate** $Di_2(CO_3)_6 \cdot 6H_2O$. Pink powder, insoluble in water. **d. chloride** $Di_2Cl_6 \cdot 12H_2O$. Purple crystals, soluble in water. **d. nitrate** $Di_2(NO_3)_6 \cdot 12H_2)$. Purple, asymmetric crystals, soluble in water. **d. oxide** Di_2O_3. Gray powder, insoluble in water. **d. sulfate** $Di_2(SO_4)_3 \cdot 6H_2O$. Red crystals, slightly soluble in water.

didymolite A dark-gray, monoclinic mineral consisting of aluminum calcium silicate, $Ca_2Al_6Si_9O_{29}$. Occurs in twinned crystals.

die casting Producing a shape by forcing a measured quantity of molten aluminum into a hardened alloy steel die (0.3–1.6 tons/cm^2).

dielectric A nonconductor of electricity. **d. constant** Relative *permittivity**. **d. strength** Electric strength. The potential (volts) at which an insulator breaks down under specified conditions. Cf. *Clausius-Mosotti equation, Helmholtz equation of inductance.*

dieline Dichloroethylene*.

Diels' hydrocarbon $C_{18}H_{16}$. The theoretical basis of the sterol molecule. **Diels-Alder reaction** Reaction for synthesizing 6-membered ring compounds by the addition of an alkene (dienophile) to the 1,4 positions of a conjugated diene; as, $CH_2{:}CH \cdot CH{:}CH_2$.

dien* Indicating diethylenetriamine, $(H_2N \cdot CH_2 \cdot CH_2)_2NH$, as a ligand.

diene* Suffix indicating 2 double bonds. **d. series** Diolefins.

dienestrol $OH \cdot Ph(C \cdot CHMe)_2PhOH$ = 266.3. 4,4′-(Diethylideneethylene)diphenol. An estrogen, m.233, for menopausal symptoms (USP, EP, BP).

dienol Catalytically dehydrated castor oil; a drying oil for paints.

diesel fuel Petroleum fraction boiling in the range 230–340°C and containing mostly C_{13} to C_{25} hydrocarbons. Combustion is started by spontaneous ignition, at higher compressions than in a gasoline engine.

diet The customary or prescribed food of an individual.

dietary A systematic diet repeated at definite time intervals. **d. standard** The amount of nourishment required per day by a man, corresponding with 3000 kcal approximately, and varying according to the work performed and by the person's weight; e.g., 120 g protein (429 kcal), 500 g carbohydrates (2050 kcal), 50 g fat (465 kcal), mineral matter, and vitamins. Cf. *pelidisi.*

Dieterici's rule A modified van der Waal equation that allows for the effect of molecules near the boundary of the gas.

$$p(V - b)e^{a/RTV} = RT$$

where a and b are constants.

diethacetic acid Diethyl*acetic acid.**

diethanolamine $NH(CH_2 \cdot CH_2OH)_2$ = 105.1. Dihydroxydiethylamine. Colorless liquid, d.1.10, b_{150mm}217; a solvent, emulsifying agent, and detergent. Cf. *triethanolamine.*

diethoxalic acid 2-Ethyl-2-hydroxy*butanoic acid**.

diethyl-* Prefix indicating 2 ethyl radicals. **d.acetal** Acetal. **d.acetic acid** See *diethylacetic acid* under *acetic acid.* **d. aldehyde** Acetal. **d.amine*** Et_2NH = 73.1. A secondary amine in putrefying fish. Colorless liquid, d_{15}·0.712, b.56, soluble in water. **d.amino*** The radical Et_2N-, from diethylamine. **d.aminobenzaldehyde** $C_{11}H_{15}ON$ = 177.2.

Colorless needles, m.78. **d.aminophenol** $C_{10}H_{15}ON$ = 165.2. Colorless crystals, m.74. **d.aniline** See *diethylaniline* under *aniline.* **d.arsine** (1)* Et_2AsH = 134.1. Ethyl cacodyl. (2) $(AsEt_2)_2$ = 266.1. Colorless liquid, b.186. **d.benzene** See *diethylbenzene* under *benzene.* **d.beryllium*** Et_2Be = 67.1. Colorless liquid, m.12. **d.cadmium*** Et_2Cd = 170.5. An oil, d.1.656, $b_{19.5mm}640$. **d.carbamazine** $C_{16}H_{29}O_8N_3$ = 391.4. Hetrazan. White, hygroscopic crystals, m.136, soluble in water; an anthelmintic, particularly for filaria infestations (USP, BP). **d. carbonate** Et_2CO_3 = 118.1. Diatol. Colorless, flammable liquid, d.0.975, b.126; a solvent for nitrocellulose. **d.carbonic ether** Ethyl carbonate. **d.cyanamide** $C_5H_{10}N_2$ = 98.1. Colorless liquid, b.186. **d. dioxide*** EtO·OEt = 90.1. Ethyl peroxide. Colorless liquid, d.0.825, b.65, slightly soluble in water. **d. disulfide*** EtS·SEt = 122.2. Ethyl dioethane, ethyl disulfide. Colorless liquid, d.0.993, b.153, soluble in water. It is toxic to *Ascaris.* Cf. *dithiene.* **d. ether*** Ether (2). **d. glycerol** $EtOCH_2 \cdot CHOH \cdot CH_2OEt$ = 148.2. Colorless liquid, b.191; a solvent. Cf. *diethyline.* **d. glycol ether** Ethylene diethanolate. **d. hydrine** Diethyline. **d. ketone*** EtCOEt = 86.1. 3-Pentanone*. Colorless liquid, d.0.814, b.103, soluble in water. **d. malate*** EtOOC·CHOH·COOEt = 176.2. Colorless liquid, b.250, soluble in water; a lacquer solvent. **d.malonic acid*** See *diethylmalonic acid* under *malonic acid.* **d.mercury*** Et_2Hg = 258.7. Colorless liquid, d.2.44, b.159, insoluble in water. **d.nitramine** See *diethylnitroamine* under *nitroamine.* **d.nitrosoamine** See *nitrosoamines.* **d.oxamide** EtNH·CO·CO·NHEt = 144.2. Oxal ethyline. A nicotine alkaloid. Colorless needles, m.175, slightly soluble in water. **d. peroxide** D. dioxide. **d. phosphate** Ethyl phosphate. **d.phosphine*** Et_2PH = 90.1. Colorless liquid, b.85. **d.phosphoric acid** $(EtO)_2PO \cdot OH$ = 154.1. Colorless liquid, soluble in water. **d. phthalate** See *phthalate.* **d.propion** $C_{13}H_{19}ON \cdot HCl$ = 241.8. 2-(Diethylamino)propiophenone hydrochloride. Tenuate. An appetite suppressant and sympathomimetic; used to treat obesity (USP, BP). **d. selenide*** Et_2Se = 137.1. Selenium diethyl. Colorless liquid, b.108, insoluble in water. **d.stilbestrol** $C_{18}H_{20}O_2$ = 268.4. α, α'-Diethyl-1-(*E*)-4,4'-stilbenediol. White crystals, m.170, insoluble in water. An estrogen for ovarian insufficiency, menopausal symptoms, prostatic cancer, and an oral postcoital contraceptive (BP). **d.s. diphosphate** Fosfesterol, Honvan. Used for prostatic cancer. **d. sulfate** Ethyl sulfate. **d. sulfide*** Et_2S = 90.2. Ethyl sulfide. d_{20}·0.837, b.94, insoluble in water. **dichloro ~** Mustard gas. **d. sulfine** The radical $=SEt_2$. **d. sulfite** Ethyl sulfite. Et_2SO_3 = 138.2. Colorless liquid, b.161, decomp. by water. **d. sulfone*** Et_2SO_2 = 122.2. Ethyl sulfone. Colorless rhombs, m.73, sparingly soluble in water. **d. sulfoxide** Et_2SO = 106.2. Colorless liquid, $b_{15mm}88$, miscible with alcohol. **d. tartrate*** $EtOOC \cdot (CHOH)_2 \cdot COOEt$ = 206.2. Colorless liquid, b.280, soluble in water; a solvent for nitrocellulose and resins. **d. telluride*** Ethyl telluride. **d.tin*** Et_2Sn = 176.8. Yellow oil d.1.654, b.150, insoluble in water. **d.tin dibromide*** Et_2SnBr_2 = 336.6. White needles, m.63. **d.tin dichloride*** Et_2SnCl_2 = 247.7. White crystals, m.84. **d.tin difluoride*** Et_2SnF_2 = 214.8. White plates, m.229. **d.tin diiodide*** Et_2SnI_2 = 430.6. White crystals, m.45. **d.tin oxide*** Et_2SnO = 192.8. White powder, insoluble in water. **d.toluamide** An insect repellent (USP, BP). **d.toluidine** $Me \cdot C_6H_4 \cdot NEt_2$ = 163.3. ***para-* ~** Colorless liquid, b.229. **d. zinc*** Et_2Zn = 123.5. Colorless liquid, b.118, ignites in air.

diethylene (1) Cyclobutane*. (2)* Prefix indicating 2 ethylene radicals. **d.diamine** Piperazine*. **d. dioxide** Dioxane. **d. disulfide** $S \cdot (CH_2)_2 \cdot S(CH_2)_2$ = 120.2. Dithiane. Colorless crystals, m.111, insoluble in water. **d. glycol** $CH_2OH \cdot CH_2 \cdot O \cdot CH_2 \cdot CH_2OH$ = 106.1. 2,2'-Oxybisethanol. Colorless liquid, d.1.1175, b.224, miscible with water or acetone. A solvent for gums and nitrocellulose; a hygroscopic agent; an antifreeze; a softener for glues, paper, and cork. **d. glycol butyl ether** Butyl*carbitol.* **d. glycol ethyl ether** Carbitol. **d. oxide** Dioxane.

diethylethylene 3-Hexene*.

diethylidene (1) 2-Butene*. (2)* Prefix indicating 2 ethylidene radicals, =CHMe.

diethylin The diethylamino* radical. **nitroso ~** $Et_2N \cdot NO$ = 102.1. Colorless liquid, d.0.951, b.177; used in organic synthesis.

diethyline $EtOCH_2 \cdot CHOEt \cdot CH_2OH$ = 148.2. 1,2-Diethylglycerinester. Colorless liquid. Cf. *diethyl* glycerol.

dietics, dietetics The science of the regulation of food.

dietzeite $7Ca(IO_3)_2 \cdot 8CaCrO_4$. A native calcium chromate and iodate, in Chilean nitrate.

differential By selective increments. **d. adsorption** The selective adsorption of dyestuffs. See adsorption *indicator.* **d. reduction** Selective reduction by metals of one component of a mixture.

differentiation (1) Mathematics: Defining an infinitesimal increment of a quantity. (2) Biology: The development of new characteristics of a cell or organism; the specialized division of labor among the various cells, tissues, or organs.

diffraction The bending of a ray of light at the edge of an object. Cf. *refraction.* Interference color fringes result when a number of parallel light waves strike a number of closely spaced, parallel edges (a grating), e.g., rows of atoms in a crystal. **crystal ~** See *X-ray spectrograph, crystal.*

d. formula The wavelength of a radiation is $\lambda = s \sin d/n$, when the angle of incidence is 90°; or $(s/n)(\sin i - \sin d)$, where i is the angle of incidence, d the angle of diffraction, s the distance between the lines of the grating, and n the order of the spectrum. **d. grating** A glass, film, or metal plate with fine rulings, used to produce a series of spectra, and to make spectroscopic measurements.

diffusate That which passes through a dialyzer.

diffuse (1) Hazy in appearance. (2) Passing through a membrane. (3) Spreading through a gas, liquid, or solid. **d. series** See *diffuse series* under *series.*

diffused (1) Widely scattered, without definite limits. (2) Spreading or passing through an object; e.g., diffused poison. **d. reflection** The ratio of reflected to incident light at a surface. See *albedo.*

diffusion The spreading or scattering of a material (gas, liquid) or energy (heat, light). **d. analysis** Determination of particle size and molecular weight by d. **d. constant** $D = ml/Act$, where m is the mass of a substance diffusing, in time t, through a cylinder of length l and cross section A, and with a concentration of c. **d. of energy** The irregular reflection of light or heat waves from a surface, part being absorbed. **d. law** The velocity of d. of 2 substances is in inverse proportion to the square root of the vapor density or molecular weight. Cf. *Graham law, atmolysis.* **d. of matter** The spreading or intermixing movement of gaseous or liquid substances, due to molecular movement. **d. shell** A membrane for dialysis (dialyzer).

diflavine The monohydrochloride of 2,7-diaminoacridine; an antiseptic.

difluoro-* Prefix indicating 2 F atoms. **d.benzene** See *difluorobenzene* under *benzene.* **d.dichloromethane** Dichlorodifluoro*methane.*

difluoroamino The radical $-NF_2$ produced in the synthesis of tetrafluorohydrazine, N_2F_4.

difluoroethylene $H_2C{:}CF_2$ = 64.0. Odorless gas.

difluoride* A compound containing 2 F atoms.
diformyl Glyoxal*.
digallic acid $C_6H_2(OH)_3 \cdot COOC_6H_2(OH)_2COOH$ = 322.2. White needles, m.275 (decomp.), slightly soluble in water; occurs in Chinese tannin and plant galls.
digalloyl See *tannin.*
digentisic acid $C_6H_3(OH)_2 \cdot COO \cdot C_6H_3(OH)_2COOH$ = 307.2. Occurs in tannins. Needles, m.204, slightly soluble in water.
digestant A digestion aid.
digester A large metal vessel with cover and safety valve, used to decompose, soften, or cook substances at high pressure and temperature. See *autoclave.*
digesting shelf A rack to hold Kjeldahl flasks.
digestion (1) In physiology, the transformation of food, mainly by enzymes, into simpler molecules (as, proteins to amino acids) that can be absorbed into the bloodstream. (1) The treatment of substances with chemicals under heat and pressure, e.g., wood to make wood pulp. (3) The disintegration of substances by strong chemical agents as in the Kjeldahl determination. **artificial ~** D. of food by enzymes outside the living body. **gastric ~** D. in the stomach, e.g., by pepsin. **intestinal ~** D. in the intestines. **peptic ~** Gastric d.
digilanid One of the 3 natural glycosides of *Digitalis lanata* (woolly foxglove), specifically designated d. A, d. B, and d. C; more properly called lanatosides, q.v. Enzyme hydrolysis yields the glycosides digitoxin, gitoxin, and digoxin, respectively, together with a molecule of glucose; the glycosides are further hydrolyzed by acid to digitoxigenin, gitoxigenin, and digoxigenin, with 3 molecules of digitoxose from each.
digipoten A mixture of glycosides from digitalis leaves.
digit (1) Three-fourths of an inch. (2) A measure of the extent of an eclipse; one-twelfth of the apparent diameter of the sun or moon. (3) An integer under 10.
digital In computers, referring to the processing of information as discrete digits (usually binary 0 and 1), rather than as continuously varying data in an analog computer.
digitalein $C_{22}H_{33}O_9$ = 441.5. A cardiac glycoside of digitalis leaves. White powder, soluble in water.
digitaligenin $C_{24}H_{32}O_3$ = 368.5. A split product from the glycoside of digitalis seeds.
digitalin $(C_5H_3O_2)_n$. A glycoside from digitalis leaves, m.244 (decomp.). **amorphous ~** Crude d. **crude ~** Amorphous d. A mixture of digitoxin, digitalin, and digitalein. Amorphous, yellow powder, soluble in water or alcohol, giving foaming solutions. **crystalline ~** Digitonin. **French ~** Homolle's d. A mixture of glycosides from digitalis leaves prepared by Homolle's method. Yellow, amorphous powder, m.100. **German ~** A mixture of glycosides from digitalis leaves, mainly digitonin. Yellow powder, soluble in water. **true ~** $C_{35}H_{56}O_{14}$ = 700.8. D. verum, Schmiedeberg's d., from the seeds and leaves of digitalis. White powder, m.217, soluble in water.
digitalis leaf The leaves of *Digitalis purpurea*, purple foxglove (Scrophulariaceae); a cardiac stimulant. Contains many glycosides and active principles. (USP, EP, BP). See *cardiac glycosides.*
digitalose A methyl pentose from digitalis.
digitogenin $C_{27}H_{44}O_5$ = 448.6. 2,3,15-Spirostanetriol. A split product of digitonin.
digitonin $C_{55}H_{90}O_{29}$ = 121.5. A glycoside from digitalis leaves. Colorless crystals, decomp. by heat, insoluble in water. Physiologically inactive; hydrolyzes to glucose, galactose, and digitogenin.
digitoxigenin $C_{23}H_{34}O_4$ = 374.5. Colorless crystals, m.253, physiologically active.
digitoxin $C_{41}H_{64}O_{13}$ = 765.0. A steroid glycoside from digitalis leaves; the chief active principle. White leaflets, with bitter taste, m.240, slightly soluble in water. A cardiac glycoside; slow to act but of long duration (USP, BP).
digitoxose $C_6H_{12}O_4$ = 148.2. 2,6-Dideoxy-*ribo*-hexose. A crystalline split product of digitoxin and gitoxin.
diglyceride A compound for the type $HO \cdot C_3H_5(OR)_2$. Cf. *glyceride.*
diglycerol $C_6H_{14}O_5$ = 166.2. L, ±, and *meso* forms of 3,3′-oxybis-1,2-propanediol. Liquids.
diglycol- Prefix indicating 2 hydroxymethyl, $-CH_2OH$, groups. **d. laurate** Odorless, yellow oil, emulsible in water, soluble in hydrocarbons.
diglycollamic acid $HOOC \cdot CH_2 \cdot NH \cdot CH_2 \cdot COOH$ = 133.1. *N*-Carboxymethylglycine. Colorless prisms, m.150 (decomp.), soluble in water.
diglycol(l)ic d.acid $(CH_2 \cdot COOH)_2O$ = 134.1. Oxybisethanoic acid. Colorless prisms, m.148, soluble in water. **d. anhydride** $O \cdot CH_2 \cdot CO \cdot O \cdot CO \cdot CH_2$ = 116.1. Colorless crystals, m.97, soluble in alcohol.
diglycolide $O \cdot CH:C \cdot OH \cdot O \cdot C \cdot OH:CH$ = 116.1. Colorless crystals, m.86.
digoxigenin $C_{23}H_{34}O_5$ = 390.5. A hydrolysis product of digoxin, m.220.
digoxin $C_{41}H_{64}O_{14}$ = 781.0. White powder, m.265 (decomp.), insoluble in water. A commonly used cardiac glycoside; intermediate in speed and duration of action (USP, EP, BP). Hydrolyzes to 3 molecules of digitoxose and digoxigenin.

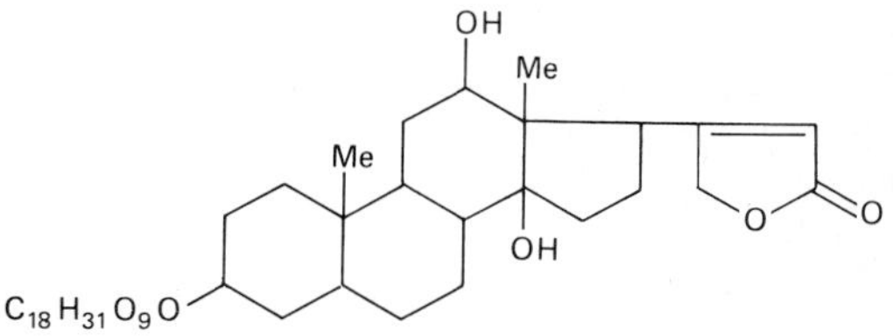

dihexyl (1)* Two hexyl radicals. (2) Dodecane*.
dihomo* See *homo(2).*
dihydracrylamic acid Dilactamic acid.
dihydracrylic acid $C_6H_{10}O_5$ = 162.1. Colorless crystals, decomp. by heat; soluble in water.
dihydrate* A substance containing 2 molecules of water.
dihydric Dibasic. Having 2 OH groups.
dihydride A binary compound containing 2 H atoms, e.g., calcium hydride.
dihydro-* Prefix indicating 2 additional H atoms in an organic compound. **d.anthracene** $CH_2 \cdot C_6H_4 \cdot CH_2 \cdot C_6H_4$ = 180.2. Diphenylene dimethylene. **d.benzene** Cyclohexadiene*. **d.bromide** An organic compound containing 2 molecules of hydrobromic acid, e.g., morphine dihydrobromide. **d.carveol** $C_{10}H_{18}O$ = 154.3. 5-Isopropenyl-2-methylcyclohexanol. Colorless liquid, b.225. **d.carvone** $C_{10}H_{16}O$ = 152.2. 5-Isopropenyl-2-methylcyclohexanone. Colorless liquid, b.222. **d.cholesterol** Formed from cholesterol in the body, and excreted in the gut. **d.cymene** $C_{10}H_{16}$ = 136.2. The dienes: citrene, phellandrene, terpinene, etc. **d.ergotamine mesylate** $C_{33}H_{37}O_5N_2 \cdot CH_4O_3S$

= 679.8. Dihydergot. Yellowish powder, sparingly soluble in water. A vasoconstrictor for migraine (USP). **d.lutidine** $C_7H_{11}N$ = 109.2. A ptomaine from cod-liver oil. **d.morphinone hydrochloride** $C_{17}H_{19}O_3N \cdot HCl$ = 321.8. White crystals, soluble in water; a more potent analgesic than morphia, with shorter action. **d.naphthalene** See *dihydronaphthalene* under *naphthalene*. **d.pentine** C_5H_8 = 68.1. Pentadiene*. **2,4-~** MeCH:CH·CH:CH$_2$. Piperylene. **2,3-~** MeCH:C:CHMe. **d.propene** $CH_2 \cdot CH_2 \cdot CH_2$ = 42.1. Cf. *cyclopropane*. **d.quinazoline** See *quinazoline*. **d.quinoline** C_9H_9N = 131.2. Colorless liquid, b.223. **d.resorcinol** $C_6H_8O_2$ = 112.1. 1,3-Cyclohexanedione. Colorless prisms, decomp. 105, soluble in water. **d.tachysterol** $C_{28}H_{46}O$ = 398.7. 9,10-Secoergosta-5,7,22-trien-3β-ol. Hytakerol, AT 10. White powder, m.128. Used to treat hypocalcemia and vitamin D–resistant rickets (USP, BP).

dihydrol A supposed polymer of water, $(H_2O)_2$, in equilibrium with the normal molecule. Cf. *trihydrol, hydrone theory*.

dihydroxy-* Prefix indicating 2 OH groups. **d.acetone** $(CH_2OH)_2CO$ = 90.1. White solid, m.70; converted by alkali into fructose. **d.acetic acid** Glyoxylic acid*. **d.amine** HO·NH·OH. A hypothetical amine. **d.anthraquinones** Dibasic phenols derived from anthraquinone: **1,2-~** Alizarin. **1,3-~** Xanthopurpurin. **1,4-~** Quinizarin. **1,5-~** Anthrarufin. **1,8-~** Chrysazin. **2,3-~** Hystazarin. **2,6-~** Anthraflavin. **2,7-~** Isoanthraflavic acid. **d.benzene** See *dihydroxybenzene* under *benzene*. **d.benzoic acid*** See *dihydroxybenzoic acid* under *benzoic acid*. **d.benzophenone** HO·C_6H_4·CO·C_6H_4·OH = 214.2. Isomers: **2,5-~** m.122; **2,2′-~** m.59, b.340; **2,3′-~** m.126; **3′,4′-~** m.205; **4,4′-~** m.210. **3,4-d.benzoyl†** See *protocatechuoyl*. **d.cholenic acid** $C_{24}H_{38}O_4$ = 390.6. An isomer of apocholic acid. Cf. *sterols*. Long needles, m.260. **d.cinnamic acid** Caffeic acid. **d.fluorescein** Gallein. **d.naphthalene** $C_{10}H_8O_2$ = 160.2. Naphthalenediol. **1,4-~** m.176. **1,5-~** m.265. **1,6-~** Colorless prisms, m.137, soluble in alcohol. **1,7-~** Colorless needles, m.178, soluble in water. **1,8-~** Colorless needles, m.140, slightly soluble in water. **2,3-~** Rhombic crystals, m.162, soluble in hot water. **2,6-~** m.218. **2,7-~** Colorless needles, m.190, slightly soluble in water. **d.palmitic acid** $C_{16}H_{32}O_4$ = 288.4. White needles, m.125; in cod-liver oil. **d.phthalophenone** Phenolphthalein. **d.propionic acid** Glyceric acid*. **d.stearic acid** $C_{18}H_{36}O_4$ = 316.5. Rhombic crystals, m.127, in castor oil; 10 isomers. **d.succinic acid** Tartaric acid*.

dihydroxyl* Indicating 2 hydroxyl groups.

diimide (1) The hypothetical compound HN:NH, known only as its organic derivatives. (2)* A compound containing 2 imide groups.

diimines* Organic compounds containing 2 imine radicals.

diimino-* Prefix indicating 2 imino groups. **d.hydrazine** The hypothetical compound HN:N·N:NH, known only as its derivatives. Cf. *nitrogen hydrides*.

diindogen Indigo.

diine Diyne*.

diiodide* A compound containing 2 I atoms.

diiodo-* Diiod-. Prefix indicating 2 I atoms in an organic compound. **d.acetic acid*** $C_2H_2O_2I_2$ = 311.8. Yellow crystals, m.110, slightly soluble in water. **d.acetylene*** C_2I_2 = 277.8. Small needles, insoluble in water. **d.hydrine** Diiodoisopropanol. **d.tyrosine** $C_9H_9NO_3I_2$ = 433.0. White crystals, m.205, slightly soluble in water. An amino acid in gorgonin, spongin, and thyroglobulin.

diiodoeosin See *rose bengal(e)*.

diiodoform $I_2C:CI_2$ = 531.6. Tetraiodoethylene*, ethylene periodide. Yellow needles, insoluble in water; decomp. by light; an antiseptic.

diisatogen $C_{16}H_8O_4N_2$ = 292.3. The heterocyclic system

C_6H_4<(CO·O)(N)>C—C<(O·CO)(N)>C_6H_4

diisoamyl Diisopentyl.

diisobutyl (1) Biisobutyl, 2,5-dimethylhexane. (2)* Prefix indicating 2 isobutyl groups. **d.amine** $C_8H_{19}N$ = 129.2. Colorless liquid, b.140, sparingly soluble in water. **d. ketone*** $Me_3 \cdot C \cdot CO \cdot CMe_3$ = 142.2. Valerone. Colorless liquid, b.181, insoluble in water.

diisopentyl Diisoamyl. (1) 2,7-Dimethyloctane*. (2)* Prefix indicating the presence of 2 isopentyl groups. **d.amine** $C_{10}H_{23}N$ = 157.3. Colorless liquid, d.0.788, b.190, slightly soluble in water. **d. ketone*** $C_{11}H_{22}O$ = 170.3. Yellow liquid, b.226, insoluble in water, miscible with alcohol or ether.

diisopropenyl (1) CH_2:CMe·CMe:CH_2 = 82.1. 2,3-Dimethylbutadiene*. Colorless liquid, b.70. Cf. *hexadiene*. (2)* Prefix indicating 2 isopropenyl groups.

diisopropyl (1) Prefix indicating 2 isopropyl radicals. (2) Biisopropyl*. (2,3-dimethylbutane). **d. fluorophosphonate** Isofluorophate. **d. ketone*** $C_7H_{14}O$ = 114.1. Colorless liquid, b.124, soluble in alcohol.

dika fat The fat of *Irvingia gabonensis* (Simarubaceae), Sierra Leone.

diketo- Dioxo*.

diketone* A compound containing 2 =CO radicals. Cf. *dione*. **α-~** A compound containing the radical —CO·CO—. **β-~** A compound containing the radical —CO·CH_2·CO—. **γ-~** A compound containing the radical —CO·CH_2·CH_2·CO—.

dil. Abbreviation for dilute.

dilactamic acid MeCHOH·COO·CHMe·CONH$_2$ = 161.2. Colorless crystals, soluble in water.

dilactic acid $C_6H_{10}O_5$ = 162.1. (1) Lactoyl lactate, lactolactic acid. MeCHOH·COO·CHMe·COOH. (2) Lactylic anhydride, lactic acid anhydride. MeCHOH·CO·O·CO·CHOHMe.

Dilantin Trademark for phenytoin sodium.

dilatancy Inverse thixotropy. A form of *thixotropy*, q.v., in which a viscous suspension sets solid under the influence of pressure. Cf. *pseudoplasticity*.

dilatation Distension or expansion.

dilatometer An instrument for measuring expansion (dilation) due either to a change in temperature or to chemical action.

dilatometry The measurement of small volume changes in liquids, due to physical or chemical actions.

dilaudid $C_{17}H_{19}O_3 \cdot HCl$ = 307.8. Dihydromorphinone hydrochloride. White powder, m.310, soluble in water; a heroin substitute.

dilituric acid 5-Nitrobarbituric acid.

dill Anetum. The dried ripe fruit of *Peucedanum graveolens* (Compositae); a spice and carminative. **d. oil** The essential oil of d., d.0.895–0.915, soluble in alcohol. It contains phellandrene, terpinene, and carvone (BP). **d. water** A filtered solution of 20 ml of d. oil in 600 ml of 90% alcohol, to which is added 1,000 ml of water and 50 g of powdered talc.

diloxanide fuorate $C_{14}H_{11}O_4NCl_2$ = 328.1. 4-(*N*-Methyl-2,2-dichloroacetamido)phenyl 2-fuorate. White crystals, m.115, slightly soluble in water; used to treat intestinal amebiasis (BP).

diluent (1) An inert solid or liquid, used to increase the bulk of another substance. (2) An inert substance used to increase the bulk of a solution; not necessarily a solvent for solute. (3) A liquid added to lacquer to increase flow and evaporation.

dilution (1) The process of diluting. (2) The state of being diluted or diffused; as, a solute in a solvent. **heat of ~** The heat, in joules (or cal), absorbed on diluting infinitely one gram molecule of a substance with water.

d. law Ostwald dilution law. **d. ratio** A measure of the solvent power of a diluent. The volume of diluent to produce incipient precipitation of a solute, divided by the total volume of solvent. **d. rule** To mix or dilute *a* with *b* to give *c*, use the diagram:

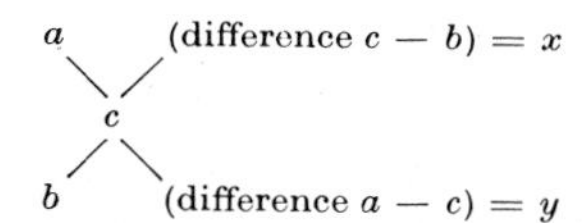

and take *x* parts of *a* and *y* parts of *b*, where *a*, *b*, and *c* are expressed in the same unit.

dimazon $C_{18}H_{19}O_4N_3$ = 341.3. Diacetylaminoazotoluene. A red dye, related to Scarlet R. Red crystals, m.75, insoluble in water; promotes growth of epithelial tissue.

dimenhydrin(ate) $C_{17}H_{21}ON \cdot C_7H_7O_2N_4Cl$ = 470.0. 8-Chlorotheophylline diphenhydramine, containing 54% 5-diphenhydramine ($C_{17}H_{21}ON$) and 46% 8-chlorotheophylline ($C_7H_7O_2N_4Cl$). Diphenhydramine theoclate, Dramamine. White, bitter crystals, m.104, soluble in water; an antihistamine; used principally as an antiemetic. (USP, BP).

dimension A magnitude in one direction. **four ~s** The concept of space and time; as, length, width, height, and duration. **three ~s** The concept of space.

dimensional equation A mathematical expression in terms of dimensions; e.g.,

$$\text{energy} = ml^2t^{-2}$$

where *m* is mass, *l* length, and *t* time.

dimer A condensation product or polymer of 2 molecules. Cf. *trimer, dimeric.*

dimercaprol $C_3H_8OS_2$ = 124.2. 2,3-Dimercapto-1-propanol, BAL. Colorless liquid with mercaptan odor, b_{15mm}122, soluble in water; antidote for poisoning from arsenic, bismuth, chromium, gold, mercury, nickel, and thallium. Also for arsenical dermatitis, reactions to gold therapy, and promotion of copper excretion from the body (USP, EP, BP).

dimercurammonium The radical HgN−. **d. chloride** HgNCl = 250.1. Yellow powder. **d. oxide** $(HgN)_2O$ = 445.2. White powder.

dimercuriammonium Dimercurammonium.

dimercurous ammonium The radical $-HgNH_2$.

dimeric Relating to a dimer.

dimethacetic acid 3-Methyl*propionic acid.*

dimethano-* Prefix indicating 2 $-CH_2-$ bridges in a ring.

dimethicone Simethicone. A silicone. Colorless liquid, insoluble in water. Used in creams for bed sores, and for defoaming properties (BP).

dimethoate* See *insecticides,* Table 45 on p. 305.

dimethoxalic acid 3-Methyl*propionic acid.*

dimethoxy-* Prefix indicating 2 −OMe radicals. **d.ethane** $MeOCH_2 \cdot CH_2OMe$ = 90.1. Glycol dimethyl ether. Colorless liquid, b.85; a lacquer solvent. **(3,4-d.phenyl)methyl†** See *veratryl.* **d.phthalaldehydic acid** Opianic acid.

dimethyl (1) Ethane*. (2)* Prefix indicating 2 Me− radicals.

d. acetal $(MeO)_2CH \cdot Me$ = 90.1. Ethylidene dimethyl ether. Colorless liquid, d_0·0.879,b.63, miscible with water.

d.acetic acid 3-Methyl*propionic acid**. **d.acetylene** 2-Butyne*. **d.amine*** $Me_2 \cdot NH$ = 45.1. Volatile liquid, b.7, soluble in water; a bait for boll weevils. **d.aminoazobenzene** $C_{14}H_{15}N_3$ = 225.3. Butter yellow. Yellow scales, soluble in water; an indicator (acids—rose-red, alkalies—yellow) and oil coloring matter. **d.aminobenzene** Dimethylaniline. **d.aminobenzylidene thiocyanate** $C_{12}H_{12}NOS_2$. A colorimetric reagent for copper, mercury, and silver. **d.aniline** $Ph \cdot NMe_2$ = 121.2. Brown oil, b.193, soluble in water. Used in organic synthesis, as reagent for nitrates, and to manufacture methyl violet. **d.anilinesulfonic acid** $C_8H_{11}O_3NS$ = 201.2. **1,2-~** Yellow masses, m.229, soluble in alcohol. **1,4-~** Brown scales, m.257, insoluble in water. **d.anthracene** $C_{16}H_{14}$ = 206.3 **1,3-~** Colorless leaflets, m.83, insoluble in water. **1,4-~** Colorless scales, m.74, insoluble in water. **2,3-~** m.252. **2,6-~** m.250. **d.arsine*** Me_2AsH = 106.0. Colorless liquid, b.36, miscible with alcohol. **d.arsinic acid** Cacodylic acid. **d.benzene** Xylene*. **d.benzoic acids** Xylic acid. **d.beryllium*** Me_2Be = 39.1. White needles, sublime 200, decomp. by water to methane. **d.cadmium*** Me_2Cd = 142.5. An oil, b.106. **d. carbonate** Methyl carbonate. See *carbonic ester.* **d.diaminotoluphenazine** Neutral red. **d. ether*** Methyl ether. **d.ethyl†** See tert-*butyl.* **d.ethylene** Butene*. **d.ethylmethane** Iso*pentane**. **d.glyoxime*** MeC(:NOH)C(:NOH)Me = 116.1. Diacetyldioxime. White needles, m.234, insoluble in water; a reagent for nickel or palladium. **d.hydantoin.** $C_5H_8N_2O_2$ = 128.1 White solid, m.178, (sublimes), very soluble in water; a plasticizer. **d. ketone** Acetone*. **d.mercury*** Me_2Hg = 230.7. Colorless liquid, b.94. **d.nitrosamine*** $Me_2N \cdot NO$ = 74.1. Nitrosodimethylin. Yellow oil, b.153, insoluble in water. **d.naphthalene** See *dimethylnaphthalene* under *naphthalene.* **3,7-d.-2,6-octadienyl†** See *geranyl.* **d.octane** $C_{10}H_{22}$ = 142.3. Liquids. **2,6-~** b.159. **2,7-~** b.160. **3,6-~** b.160. **d. oxalate** $C_4H_6O_4$ = 118.1. Colorless prisms, m.54. **d.oxamide** $C_4H_8O_2N_2$ = 116.1. **N,N-~** $NH_2 \cdot OC \cdot CO \cdot NMe_2$. Oxalmethylin. Colorless leaflets, m.104, soluble in water. **N,N′-~** $MeNH \cdot OC \cdot CO \cdot NHMe$. Colorless needles, m.209, slightly soluble in water. **2,2-d.-1-oxopropyl†** See *pivaloyl.* **d.oxyquinizine** Antipyrine. **d.phenol** Xylenol. **d.phenyl†** See *xylyl.* **d.phenylenediamine** $NH_2C_6H_4NM_2$ = 136.2. Aminodimethylaniline. ***ortho-*~** b.218. ***meta-*~** b.270. ***para-*~** Brown crystals, m.41, insoluble in water; a reagent for cellulose. **d.phenylenediamine hydrochloride** $C_8H_{12}N_2 \cdot HCl$ = 172.7. ***para-*~** Hygroscopic crystals, soluble in alcohol. **d.phenylenediamine sulfate** $C_8H_{12}N_2 \cdot H_2SO_4$ = 234.3. ***para-*~** Brown crystals, soluble in water. **d.phosphine*** Me_2PH = 62.1. Colorless liquid, b.25, insoluble in water. **d.phosphinic acid** $(CH_3)_2PO \cdot OH$ = 94.0. Colorless crystals, m.76, soluble in water. **d.phosphoric acid** $(MeO)_2PO(OH)$ = 126.0. **d. phthalate*** $C_{10}H_{10}O_4$ = 194.2. $o\text{-}C_6H_4(COOMe)_2$. Colorless, aromatic oil, d.1.190, b.280, insoluble in water; used as an insect repellent (BP). **d.propylmethane** 2-Methylpentane.* **d.pyrazine** Ketine. **d.pyrazole** $C_5H_8N_2$ = 96.1. A precipitant for cobalt salts. **d.pyridine** Lutidine. **d.pyrrole** C_6H_9N = 95.1. Colorless oil, b.165, slightly soluble in water. **d. selenide*** Me_2Se = 109.0. Selenium dimethyl. Colorless liquid, b.58, insoluble in water. **d.silane** Me_2SiH_2 = 60.2. Colorless gas, b.−20. **d. sulfate*** $(CH_3O)_2SO_2$ = 126.1. Colorless liquid, b.188, soluble in water. A poisonous gas; used for methylating, and as a reagent for coal tar oils. **d. sulfide*** Me_2S = 62.1. Methyl sulfide, methyl thiomethane. Colorless liquid, d.0.845. b.37, insoluble in water. **d. sulfoxide** $Me_2 \cdot S{:}O$ = 78.1. m.19, b.189, miscible with water. An oxidizing agent, and solvent and reagent in organic reactions. **d. telluride*** Me_2Te = 157.7. b.182. **d.thetine** $C_4H_8O_2S$ = 120.2. Colorless crystals, decomp. by heat, soluble in water. **d.thiophene** C_6H_8S = 112.2. **2,3-** ~ $HC \cdot S \cdot CMe{:}CMe \cdot CH$. Colorless liquid, b.136. **2,4-** ~ Colorless liquid, b.138. **2,5-** ~ Colorless liquid, b.135, insoluble in water. **d.xanthine** Theobromine. **d. yellow** Dimethylaminoazobenzene.

dimethylamino* Dimethyline. The radical $-NMe_2$.

dimethylene (1) Ethylene*. (2)* Prefix indicating 2 methylene radicals. **d.diamine** Ethylenediamine*. **d.imine** $CH_2 \cdot NH \cdot CH_2$ = 43.1. Ethyleneimine, vinylimine. Colorless liquid, b.55, insoluble in water. **d. oxide** Ethylene oxide*. **d. sulfide** Ethylene sulfide.

dimethyline The dimethylamino radical. **nitroso** ~ Dimethyl*nitrosamine.*

dimetric Tetragonal. Cf. *crystal* system.

dimine Cyclohexylethylamine dithiocarbamate, m.93, decomp. 100; a catalyst in rubber vulcanization.

diminution A lessening or decrease.

dimorphic Occurring in 2 crystalline forms having different melting points.

dimorphism The property of being dimorphic. Cf. *polymorphism.*

dimsylsodium Abbreviation for the sodium derivative of dimethyl sulfoxide. It forms the ion $CH_3 \cdot SO \cdot CH_2Na^+$; used in analytical chemistry for the titration of very weak acids.

DIN Deutsches Institut für Normung; responsible body for the Deutsche Industrie-Norm (standards).

dinaphthacridine $C_{14}H_8 \cdot N \cdot C_{14}H_8 \cdot CH$ = 379.5. Six isomers. Cf. *dibenzacridine.*

dinaphthanthracene Dibenzanthracene*.

dinaphtho- Dinaphthylene*.

dinaphthol $HO \cdot C_{10}H_6 \cdot C_{10}H_6OH$ = 286.3. Binaphthol. **alpha-** ~ 4,4′-Dihydroxy-1,1′-binaphthyl. Colorless, rhombic crystals, m.300, insoluble in water. **beta-** ~ 2,2′-Dihydroxy-1,1′-binaphthyl. **(+)-** ~ White needles, m.218, insoluble in water.

dinaphthoxanthene $C_{21}H_{14}O$ = 282.3. Colorless crystals, m.199.

dinaphthyl (1) Binaphthalene*. (2)* Prefix indicating 2 naphthyl* radicals, $C_{10}H_7-$. **d.mercury** $(C_{10}H_7)_2Hg$ = 454.9. Rhombic crystals, m.243, insoluble in water. **d.methane** $C_{10}H_7 \cdot CH_2 \cdot C_{10}H_7$ = 256.3. **1,1′-** ~ or **alpha-** ~ Colorless prisms, m.109, soluble in hot alcohol. **2,2′-** ~ or **beta-** ~ Small needles, m.92, soluble in alcohol. **d.tin** $(C_{10}H_7)_2Sn$ = 373.0. White powder, m.200.

dinaphthylene* Prefix indicating 2 naphthylene* groups, $C_{10}H_6=$. **d.methane** See *picene* and *fluorene.* **d.thiophene** $C_{10}H_6 \cdot S \cdot C_{10}H_6$ = 284.4. Colorless crystals, m.147.

Dindevan Trademark for phenylindanedione.

dindyl Diindolyl.

dineric Having 2 liquid layers (phases). Cf. *dimer.*

Dingler **D., Emil Maximilian (1806–1874)** German technical chemist and editor. **D., Johann Gottfried (1778–1855)** German apothecary and founder of *Polytechnischen Journal* (1820).

dinicotinic acid $C_7H_5O_4N$ = 167.1. **3,5-** ~ Pyridinedicarboxylic acid*. Colorless crystals, m.323(decomp).

dinitraniline Dinitro*aniline*.*

dinitril Zetec.

dinitro-* Prefix indicating 2 nitro groups. **d.aminobenzene** Dinitro*aniline.* **d.aniline** See *dinitroaniline* under *aniline.* **d.anthraquinone*** $C_{14}H_6O_6N_2$ = 298.2. **1,5-** ~ Yellow crystals, m.385, slightly soluble in water. **2,7-** ~ Yellow needles, m.290, sparingly soluble in water. **d.benzaldehyde** $C_7H_4O_5N_2$ = 196.1. **2,4-** ~ Yellow prisms, m.72, slightly soluble in water. **d.benzene** See *dinitrobenzene* under *benzene.* **d.benzoic acid** See *dinitrobenzoic acid* under *benzoic acid.* **d.chlorobenzene** See *1-chloro-n-dinitrobenzene* under *benzene.* **d.cresol** $C_7H_6O_5N_2$ = 198.1. Saffron substitute. An orange, poisonous coal tar dye, m.104. **d.glycerol** $C_3H_6(NO_2)_2$ = 134.1. D. glycerin, glycerol-1,3-dinitrate. An oil, d.1.47, $b_{15mm}148$; a high explosive. **d.naphthalene*** $C_{10}H_6O_4N_2$ = 218.2. Yellow crystals, m.153, insoluble in water. **1,2-** ~ m.103. **1,3-** ~ m.145. **1,4-** ~ m.129. **1,5-** ~ m.216. **1,6-** ~ m.162. **1,7-** ~ m.156. **1,8-** ~ m.170. **d.naphthol*** $C_{10}H_6O_5N_2$ = 234.2. A group of intermediates in organic synthesis: **2,4-α-** ~ m.138. **4,5-α-** ~ Decomp. 230. **4,8-α-** ~ Decomp. 235. **1,6-β-** ~ m.195. **1,8-β-** ~ m.198. **d.orthocresol** DNOC.* **d.phenamic acid** Picramic acid. **d.phenol*** $C_6H_4O_5N_2$ = 184.1. Hydroxydinitrobenzene. **2,3-** ~ Yellow needles, m.144, slightly soluble in alcohol; a pH indicator. **2,4-** ~ Yellow scales, m.114, slightly soluble in water. **2,6-** ~ Yellow needles, m.62, slightly soluble in water. **3,4-** ~ Colorless needles, m.134. **3,5-** ~ Colorless leaflets, m.122. **2,5-** ~ or **3,6-** ~ Colorless leaflets, m.104. **d.phenylhydrazine** $(NO_2)_2C_6H_3NH \cdot NH_2$ = 198.1. Red needles, m.197 (explode); a reagent for aldehydes and malic acid. **d.resorcinol** $C_6H_4O_6N_2$ = 200.1. **2,4,1,3-** ~ Yellow scales, m.142, soluble in hot water. **crude** ~ Solid green, dark green, chlorin, diquinolyldioxime. Brown mass, sparingly soluble in water; a green pigment in calico printing, and in dyeing iron-mordanted textiles. **d.salicylic acid*** $C_6H_2(OH)(NO_2)_2COOH$. A reagent for glucose. **d.toluene*** $C_7H_6O_4N_2$ = 182.1. **2,4-** ~ Colorless needles, m.71, slightly soluble in water. **2,3-** ~ m.63. **2,5-** ~ Colorless needles, m.48, soluble in alcohol. **2,6-** ~ Colorless needles, m.66, soluble in alcohol. **3,4-** ~ Colorless needles, m.59, insoluble in water. **3,5-** ~ Colorless needles, m.92, sparingly soluble in water. **d.xylene** $C_8H_8O_4N_2$ = 196.2. **2,4,1,3-** ~ Colorless prisms, m.93, soluble in hot alcohol. **2,3,1,4-** ~ Colorless needles, m.124, sparingly soluble in alcohol. **2,5,1,4-** ~ Colorless hexagons, m.93, soluble in alcohol.

dinitrogen oxide* N_2O = 44.01. Nitrous oxide, nitrogen monoxide, laughing gas, dental gas. Colorless gas, b.−88, soluble in water. An inhalation anesthetic, usually in combination with other agents and oxygen. (USP, BP). A preservative. See *Entonox.*

dinitrogen tetraoxide* See *nitrogen tetraoxide.*

dinitroso-* Prefix indicating 2 $-NO$ groups; as, **d.resorcinol** Used as reagent for iron.

dinol Diazodinitrophenol.

dinonylphthalate $C_6N_4(COOC_9H_{19})_2$ = 470.6. Bis(3,5,5-trimethylhexyl) phthalate. Pale, viscous liquid, d.0.972; a plasticizer for cellulose and polymer plastics, especially for water resistance.

dinor* As for *nor-*, but involving two such Me (def. 1) or CH_2 (def. 2) groups.

dinoseb* $C_{10}H_{12}O_5N_2$ = 240.2. DNBP, BNSBP, DNOSBP. See *herbicides,* Table 42 on p. 281.

dinucleotide A compound formed from 2 purine bases: as, adenine-uracil.

dioctyl (1) Hexadecane*. (2)* Prefix indicating 2 octyl radicals.

Diodoquin Trademark for clioquinol.

dioform Dichloro*ethylene**.
-dioic* Suffix indicating 2 —COOH groups.
diol* Indicating 2 —OH groups. Cf. *glycols*.
diolefin(e) Dialkene*, diene series. An unsaturated aliphatic hydrocarbon with the general formula C_nH_{2n-2} and 2 double bonds, e.g., propadiene, C_3H_4.
diolein $C_{39}H_{72}O_5$ = 621.0. **α,α′-~** (Z,Z)-Glycerol-1,2-di-9-octadecanoate. m27; formed from olein by action of lipase.
Diolen Trade name for a polyester synthetic fiber.
Dional Trademark for propyliodone.
-dione Suffix indicating 2 carbonyl groups, =CO; as, butadione. Cf. *diketone*. **dionedioic acids** Organic acids containing 2 carbonyl and 2 carboxyl groups; as, hexan-3,4-dione-1,6-dioic acid, HOOC·CH_2·CO·CO·CH_2·COOH.
diopside $CaMgSi_2O_6$. A monoclinic, yellow, rock-forming pyroxene, d.3.2, hardness 5–6. Cf. *baikalite*.
dioptase CuH_2SiO_4. A greenish, hexagonal, hydrous copper silicate, d.3.3, hardness 5.
diopter Dioptre. The unit of "power" of a lens. The reciprocal of the focal length in meters.
dioptra An optical device to measure heights and angles.
dioptrics The study of light refraction.
diorite An igneous rock containing quartz, plagioclase, and small amounts of femic minerals; as, malchite.
diorsellinic acid Lecanoric acid.
dioscin $C_{24}H_{38}O_9·3H_2O$ = 524.6. A saponin from the roots of *Dioscorea japonica* (Dioscoraceae). White, silky needles, m.276 (decomp.), insoluble in water.
dioscorea Yam root. Tropical shrubs with edible tubers (yams). That of *Dioscorea villosa* (wild yam) is an antispasmodic and diaphoretic. **d. sapotoxin** $C_{23}H_{28}O_{10}$ = 464.5. A toxic principle from *Dioscorea* species.
Dioscorides Greek philosopher of Anazarba, Asia Minor, who made mercury from cinnabar (A.D. 50).
dioscorine $C_{13}H_{19}O_2N$ = 221.3. An alkaloid from the rhizome of *Dioscorea hirsuto* (Dioscoreaceae), Java.
dioscoroid The combined principles from the root of *Dioscorea villosa*, wild yam. The fluid extract is an antispasmodic.
diose Biose.
diosmeleoptene $C_{10}H_{18}O$ = 154.3. A terpene of peppermintlike odor; an isomer of borneol from the essential oil *Diosma* (or *Barosma*) *betulina*. Cf. *buchu*.
diosphenol $C_{10}H_{16}O_2$ = 168.2. Buchu camphor, 2-hydroxy-6-isopropyl-3-methyl-2-cyclohexenone. Crystals, m.83; from the essential oil of *Diosma betulina*.
dioxadiene 1,4-Dioxin.
dioxane O·$(CH_2)_2$·O·$(CH_2)_2$ = 88.1. Diethylene dioxide, dihydro-*p*-dioxin. Colorless liquid, d.1.038, m.11, b.101, soluble in water. A solvent used in the silk and varnish industries and as a dehydrant in histology. Cf. *dithiane*.
dioxazole $C_2H_3O_2N$ = 73.1. Heterocyclic compounds with 1 N and 2 O atoms; as, 1,2,3-~ CH:CH·NH·O·O
dioxdiazine $C_2H_2O_2N_2$ = 86.1. Heterocyclic hydrocarbons with 2 N and 2 O atoms in the hexatomic ring. Isomers: 1,2,3,4-~; 1,2,3,5-~; 1,2,3,6-~; 1,2,4,5-~; 1,3,2,4-~; 1,3,2,5-~; 1,3,4,5-~; 1,3,4,6-~; 1,4,2,3-~; 1,4,2,5-~; 1,4,2,6-~. First 2 numbers indicate positions of O atoms; last 2 of N atoms.
dioxdiazole $CH_2O_2N_2$ = 74.0. Heterocyclic hydrocarbons with 2 N and 2 O atoms in the pentatomic ring. Isomers: 1,2,3,4-~; 1,2,3,5-~; 1,3,2,4-~; 1,3,4,5-~. The O atoms occupy the positions indicated by the first 2 numbers. Cf. *dithiodiazole*.

dioxide A compound containing 2 O atoms: (1)* normal or true dioxides in which the valency of oxygen is 2; as, manganese d., O:Mn:O; (2) abnormal dioxides or peroxides, such as barium d. (peroxide) Ba<(O—O) (ring)
The true dioxides, only, give oxygen with concentrated acids, and chlorine with concentrated hydrochloric acid. Peroxides, only, give hydrogen peroxide.
dioximes* Compounds containing 2 oxime* radicals. **alpha-~** Glyoximes. **beta-~** Glyoxime peroxides. **gamma-~** Compounds in which the 2 oxime radicals are separated by the ethylene radical; e.g., succinaldehyde dioxime, HON:$CHCH_2$·CH_2·CH:NOH.
dioxin (1) $C_{12}H_4O_2Cl_4$ = 322.0. 2,3,7,8-TCDD, 2,3,7,8-tetrachlorodibenzo-*p*-dioxin. m.295. An impurity in the manufacture of certain agricultural chemicals, as, trichlorophenoxyacetic acid; causes chloracne. (2) Heterocyclic hydrocarbons; as, **1,4-~** O·CH:CH·O·CH:CH = 84.1. **dihydro-~** *p*-Dioxane.
dioxindole C_6H_4·NH·CO·CHOH = 149.2. 1,3-Dihydro-3-hydroxy-2*H*-indol-2-one. Colorless prisms, m.180, decomp. 195, soluble in water.
dioxo-* Prefix indicating a compound containing 2 oxo groups; as, diketones. **1,4-d.-1,4-butanediyl†** See *succinyl*. **1,2-d.-1,2-ethanediyl†** See *oxalyl*. **1,6-d.-1,6-hexanediyl†** See *adipoyl*. **d.hydrindene** Indandione*. **1,5-d.-1,5-pentanediyl†** See *glutaryl*. **1,3-d.-1,3-propanediyl†** See *malonyl*. **d.purine** Xanthine. **d.triazolidene** Urazole. **d.trimethylpurine** Caffeine.
Dioxogen A brand of hydrogen peroxide.
dioxolan(e)s* $C_3H_6O_2$ = 74.1. Heterocyclic compounds with 3 C and 2 O atoms.
diotriazine CHO_2N_3 = 87.0. Heterocarbons with 3 N and 2 O atoms in a hexatomic ring. Isomers: 1,2,3,4,5-~; 1,2,3,4,6-~; 1,3,4,5,6-~; 1,3,2,4,5-~; 1,3,2,4,6-~; 1,4,2,3,5-~. The N atoms occupy the last 3 positions.
dioxy (1)* Suffix indicating a —O—R—O— radical, where R is a bivalent radical; as, carbonyldioxy, —O—CO—O—. (2) Dihydroxy (incorrect). **d.ethylene** Dioxane. **d.tetrazotic acid** $CH_2O_2N_4$. A structural arrangement, known only in combination.
dipalmitate* A compound containing 2 palmitic acid radicals.
dipalmitin $C_{35}H_{68}O_5$ = 536.9. Glycerol 1,2-dihexadecanoate, glycerol dipalmitate. Formed from palmitin by the action of lipase.
diparalene BP name for chlorcyclizine hydrochloride.
dipentene Cinene. **d. glycol** Terpin.
dipentyl* Diamyl. Indicating a compound containing 2 pentyl radicals. **d.amine** $C_{10}H_{23}N$ = 157.3. Colorless liquid, b.187, soluble in alcohol. **d. ketone** 6-Undecanone*. **d. phthalate** See *dipentyl phthalate* under *phthalate*. **d. sulfide*** $(C_5H_{11})_2S$ = 174.3. Colorless liquid; and odorant for natural gas.
dipeptide A peptide of the type NH_2RCO·NH·R·COOH; as, anserine.
diphenate (1) A salt of diphenic acid. (2) Diphenolate*.
diphenazyl Dibenzoylethane.
diphenhydramine See *dimenhydrinate*. **d. hydrochloride** $C_{17}H_{21}ON·HCl$ = 291.8. Benadryl. White, crystalline powder, m.168, soluble in water; an antihistaminic, sedative, and antiemetic (USP, BP).
diphenic **d. acid** $(C_6H_4·COOH)_2$ = 242.2. 2,2′-

Biphenyldicarboxylic acid. Colorless crystals, m.229, soluble in hot water. **d. anhydride** $C_{14}H_8O_3$ = 224.2. Colorless crystals, m.213.

diphenimide $C_{14}H_9O_2N$ = 223.2. Diphenic acid imide. Colorless crystals, m.219.

diphenol A compound containing 2 phenolic OH— groups.

diphenolate* A compound containing 2 phenoxy groups, PhO—.

diphenoquinone $O:C_6H_4:C_6H_4:O$ = 184.2. Bibenzenone. Biphenyl-4,4′-quinone. Violet-brown crystals, m.165 (decomp.).

diphenoxylate hydrochloride $C_{30}H_{32}O_2N_2 \cdot HCl$ = 489.1. Lomotil. White powder, m.222, sparingly soluble in water. Reduces intestinal motility; for diarrhea (USP, BP).

diphenyl- (1)* Prefix indicating 2 phenyl radicals. (2) The biphenyl* radical. **d.acetic acid** $C_{14}H_{12}O_2$ = 212.2. Colorless needles, m.148, soluble in hot water. **d.acetonitrile** $C_{14}H_{11}N$ = 193.2. Colorless crystals, m.72. **d.acetylene** Tolane. **d.amine** See *diphenylamine**. **d.aniline** Triphenylamine*. **d. anthrone** $Ph \cdot C_6H_3 \cdot CO \cdot C_6H_3 \cdot Ph \cdot CH_2$ = 346.4. Colorless crystals, m.192. **d.benzene** See *diphenylbenzene* under *benzene.* **d.benzidine** $C_{24}H_{20}N_2$ = 336.4. A sensitive reagent for nitrates, zinc, etc., m.151. **d.benzylsultam** $C_{19}H_{15}O_2NS$ = 321.4. Colorless crystals, m.210. **d.carbazide** $CO(NH \cdot NHPh)_2$ = 214.3. Phenylhydrazine urea. White powder, m.170. An indicator (alkalies: pink; acids: colorless), and reagent for cadmium (blue-violet), chromium (red), and mercury (blue). **d.chloroarsine** Blue cross. **d.cyanoarsine** Ph_2AsCN = 255.2. D.C. Colorless prisms, m.32, with odor of bitter almonds. A nose-irritant poison gas. **d.diacetylene** $Ph \cdot C:C \cdot C:C \cdot Ph$ = 202.3. Diacetenylbenzene. Colorless needles, m.88, soluble in alcohol. **d.dicarboxylic acid** Diphenic acid. **d.diethylene** $Ph \cdot CH:CH \cdot CH:CH \cdot Ph$ = 206.3. Colorless crystals, m.148. **d. diketone** Compounds containing 2 phenyl and 2 carbonyl groups. **d.phenyliminotriazole** Nitron. **d.ethane** Bibenzyl*. **d. ether*** Ph_2O = 170.2. Phenyl ether. Colorless monoclinics or oil, d.1.073, m.38, b.259, almost insoluble in water; a heat transfer agent. **d.ethylene** Stilbene*. **d.guanidine** $(PhNH)_2C:NH$ = 211.3. Melaniline. White needles, m.147, slightly soluble in water; a standard in acidimetry. **d.hydantoin sodium** Phenytoin sodium. **d.hydrazine** $PhNH \cdot NHPh$ = 184.2. Yellow triclinics, m.34.5, slightly soluble in water; a reagent for aldehydes and ketones. **d. ketone*** Benzophenone*. **d.methane*** $Ph \cdot CH_2 \cdot Ph$ = 168.2. Methylenediphenyl, ditan, benzylbenzene. Colorless needles, m.26.5, insoluble in water; used in organic synthesis. **d.methanol** Benzohydrol. **d. nitrogen** The free radical Ph_2N. **4,5-d.-octane-2,7-dione*** $MeCO \cdot CH_2 \cdot CHPh \cdot CHPh \cdot CH_2 \cdot COMe$ = 294.4. Colorless crystals, m.161. **d. oxide** D. ether*. **d.pentadienone** Styryl ketone. **d.phosphine** Ph_2PH = 186.2. Liquid, b.280, foul smell, readily oxidized by air. **d. sulfide*** Ph_2S = 186.3. Phenyl sulfide. Colorless liquid, d.1.119, b.296, insoluble in water. **d. sulfone*** Ph_2SO_2 = 218.3. Phenyl sulfone. White needles, m.128. **d. tetraketone** $Ph \cdot CO \cdot CO \cdot CO \cdot CO \cdot Ph$ = 266.3. Colorless crystals, m.87. **d.tetraketoxime** $C_{16}H_{11}O_4N$ = 281.3. Colorless crystals, m.176. **d.thienylmethane*** $(C_4H_3S)CHPh_2$ = 250.4. Colorless crystals, m.63, soluble in alcohol. **d.thiocarbazone** Dithizone. **d.thiourea** Thiocarbanilide. **d.tin*** Ph_2Sn = 272.9. Yellow, amorphous powder, m.226, insoluble in water. **d. triketone** $PhCO \cdot CO \cdot COPh$ = 238.2. Dibenzoyl ketone. Colorless crystals, m.70. **d.urea*** Carbanilide.

diphenylamine* Ph_2NH = 169.2. *N*-phenylbenzenamine†, phenylaniline. Colorless leaflets, m.54, sparingly soluble in water. A reagent for nitrates in water, milk, etc., and an indicator in iron dichromate titrations. **hydroxy ~** $C_{12}H_{11}ON$ = 185.2. Aminophenylphenol. **3- ~** Colorless crystals, m.82. **4- ~** Colorless crystals, m.70. **thio ~** Phenothiazine*.

d. blue Diphenylbenzidine. An oxidation product of d.; an oxidation-reduction indicator. **d. chlororasine** $NH \cdot (C_6H_4)_2 \cdot AsCl$ = 277.6. Adamsite, D.M. Yellow crystals, m.195, insoluble in water. The vapors are a nose irritant.

diphenylamino* The radical $Ph_2N—$.

diphenylene (1) The biphenylene radical $(C_6H_4)_2=$. (2)* Dibenzo ~. A compound with 2 $C_6H_4=$ radicals. **d.acetic acid** Fluorenecarboxylic acid. **d. dimethylene** Dihydroanthracene. **d.furan** $C_6H_4 \cdot O \cdot C_6H_4$ = 168.2. Dibenzofurfurane, diphenylene oxide. Colorless crystals, m.81. **d.furan-*p*-oxazine** Phenoxazine*. **d.furfurane** D.furan. **d.glycolic acid** 9-Hydroxy-9*H*-fluorene-9-carboxylic acid. Colorless crystals, m.169.

HO · C · COOH

H_4C_6 —— C_6H_4

d. imide Carbazole*. **d. ketone** Fluorenone*. **d.methane** Fluorene*. **d. oxide** D.furan. **methylene ~** Xanthene*. **d.phenylmethane** $(C_6H_4)_2 \cdot CHPh$ = 242.3. Colorless crystals, m.146. **d.pyrene** $C_{24}H_{14}$ = 302.4. Yellow crystals, m.225.

d.pyrone Xanthone. **d.pyrrole** Carbazole*. **d. sulfide** D.thiophene. **methylene ~** Thioxanthene*. **d.thiophene** $C_6H_4 \cdot S \cdot C_6H_4$ = 184.3. Diphenylene sulfide. Colorless crystals, m.97.

diphenylmethyl* The benzhydryl* radical.

diphos* Indicating ethylenebis(diphenylphosphine), $(Ph_2PCH_2—)_2$, as a ligand.

diphosgene $Cl \cdot COOCCl_3$ = 197.8. Perstoff. A lung irritant, m.−57, b.128; used in World War I.

diphosphane* $(PH_2)_2$ = 66.0. Diphosphine.

diphosphate (1)* A salt of diphosphoric acid. (2) A compound containing 2 phosphate, q.v., radicals.

diphosphenyl $Ph \cdot P:P \cdot Ph$ = 216.2. Yellow crystals.

diphosphoglyceric acid $[(HO)_2PO \cdot O]_2C_2H_3 \cdot COOH$ = 266.0. m.174. Occurs in blood.

diphosphonic acid See *diphosphonic acid* under *phosphonic acid.*

diphosphoric acid (1)* Pyro*phosphoric acid.* (2) d.(IV) a.* See *hypophosphoric acid.* (3) d.(III,V) a.* $(HO)_2PO \cdot PO(OH)_2$.

diphthaloyl $C_{16}H_8O_4$ = 264.2. A heterocyclic ketone. Colorless crystals, m.234.

dipicolinic acid $C_7H_5O_4N$ = 167.1. 2,6-Pyridinedicarboxylic acid*. Colorless needles, m.252 (decomp.), slightly soluble in water.

dipicrylamine Hexanitrodiphenylamine.

dipipanone hydrochloride $C_{24}H_{31}ON \cdot HCl \cdot H_2O$ = 404.0. 4,4-Diphenyl-6-(1-piperidinyl)-3-heptanone. White, bitter crystals, m.125, soluble in water; a powerful analgesic for severe pain, but use limited by risk of addiction (BP).
dipiperidyls Compounds formed by reduction of pyridine or bipyridyl. Soluble in water, giving strongly alkaline solutions which absorb carbon dioxide readily and regenerate it if heated.
diplococcin An antibiotic from certain strains of milk streptococci; used to treat bovine mastitis.
diplogen Deuterium*.
diploid Describing cells having a normal duplicate number of chromosomes. **d. number** The 46 chromosomes normally present in the nuclei of human cells, established at fertilization by the union of sperm and ovum, each of which carries the haploid number of 23 chromosomes. The d. n. is specific to each plant and animal species; as, dog 78, horse 66. See *haploid number.*
diplomethane Deuteriomethane.
diplon Deuteron.
dipole (1) A coordinated valence link between 2 originally neutral atoms, whereby one loses and the other gains a share of 2 electrons. Cf. *covalence.* (2) The electrical symmetry of a charge of positive electricity very close to an equal negative charge; measured by the d. moment. **d. moment** A molecular constant (p or μ) indicating the distribution of electrical charges in a neutral molecule. It is zero if they are symmetrically distributed. One coulomb-meter = 2.99793×10^{29} debye = 2.99793×10^{11} esu. Cf. *Debye-Hückel theory, association.*
Dippel, Johann Konrad (1673–1734) German alchemist. **D.'s oil** The distillation product of bones and other animal matter, chiefly containing pyridine and bases. Formerly used to denature alcohol.
dipropargyl (1) Bis(propynyl). (2) Prefix indicating two 2-propynyl* radicals.
dipropyl-* Prefix indicating 2 propyl radicals, $MeCH_2CH_2-$. **d.amine** Pr_2NH = 101.2. Colorless liquid, b.110, soluble in water. **d.beryllium*** Pr_2Be = 95.2. Colorless liquid, b.245. **d.cadmium*** Pr_2Cd = 198.6. Colorless liquid, $b_{21mm}84$. **d. ether*** Propyl ether. **d. ketone*** 4-Heptanone*. **d.methanol** Pr_2CHOH = 116.2. Heptan-4-ol, 3-propylbutanol. Colorless liquid, b.154, soluble in alcohol. **d.mercury*** Pr_2Hg = 286.8. Colorless liquid, b.190, insoluble in water. **d. sulfide*** Pr_2S = 118.2. Propyl thioether. Colorless liquid, b.141, insoluble in water. **d.tin*** Pr_2Sn = 204.9. Colorless liquid. **d.tin dibromide*** Pr_2SnBr_2 = 364.7. Yellow crystals, m.54. **d.tin dichloride*** Pr_2SnCl_2 = 275.8. White crystals, m.81.
diprotocatechuic acid $C_{14}H_{10}O_7$ = 290.2. Needles, m.237, soluble in alcohol; occurs in tannins. Cf. *diresorcylic acid.*
dipsomania Acute craving for alcohol.
dipyrazolone A compound containing 2 pyrazolone groups.
dipyridine $C_{10}H_{10}N_2$ = 158.2. Colorless needles, m.108, soluble in water. Cf. *bipyridyl**.
dipyridyl Bipyridyl*.
diquat* See *herbicides,* Table 42 on p. 281.
diquinidine Diconchinine. An alkaloid from cinchona.
diquinoline Biquinoline.
diquinolyl Biquinolyl.
Dirac constant Symbol: $\hbar$. The Planck constant divided by 2π; i.e., 1.05458×10^{-34} Js.
direct dyes Substantive *dyes.*
direct-vision spectroscope A spectroscope with prisms arranged so that the emergent rays follow the direction of the incident rays.
diresorcinol $C_{12}H_{10}O_4 \cdot 2H_2O$ = 254.2. White crystals, m.310, soluble in hot water.
diresorcylic acid $(HO)_2 \cdot C_6H_3 \cdot COO \cdot C_6H_3(OH)_2COOH$ = 307.2. An isomer of digentisic acid. Microneedles, decomp. 215, soluble in hot water; occurs in tannins.
disaccharides Carbohydrates formed from 2 simple sugars (monosaccharides) and yielding them on hydrolysis. See *carbohydrates.*
disagglomeration The chemical transformation of compact masses into a fine powder.
disalicylic acid (1) Salicylide. (2) Diplosal.
disassimilation Oxidation of assimilated material, which liberates energy.
disassociation Dissociation. **photo ~** The disarrangement of molecules under the influence of light; e.g., with silver salts.
disazo compound Bisazo compound.
disc Disk.
discharge (1) The sudden escape or liberation of stored or accumulated energy; e.g., electricity (spark discharge), or chemical energy (explosion). (2) Any waste liquid from a manufacturing plant. (3) The output of a pump. **disruptive ~** A crackling d. of electric energy. **silent ~** A gradual loss of electric energy due to the conductivity of air. Cf. *saturation current.*
Dische reaction The development of a blue color when a substance is heated with diphenylamine under standard conditions. A stain reaction for nuclear substances in cells, due to the presence of deoxyribose. Cf. *Feulgen reaction.*
Discol Trademark for an internal-combustion fuel: alcohol 50, benzene 25, hydrocarbons 25%.
discrasite Dyscrasite.
discutient A drug which dissipates morbid matter.
diselane* H_2Se_2.
diselenide A compound of the type R·Se·Se·R.
disgregation Dispersion. Cf. *aggregation.*
dish A shallow or flat glass or metal vessel. **crystallizing ~** A shallow glass d. used for evaporation and crystallization. **culture ~** A shallow flat-bottom d. of heavy glass; used to grow bacteria cultures (petri d.). **filtering ~** A d.-shaped cone of porous material. **incineration ~** See *incineration dish.* **moisture ~** A d. with a ground glass stopper.
disilane See *disilane* under *silane.*
disilanyl* The radical Si_2H_5-. Cf. *disilyl.*
disilicic acid See *silicic acid.*
disiloxane* $(SiH_3)_2O$ = 78.2. Colorless, odorless, combustible gas, b. −15.
disilyl* Indicating two silyl, SiH_3-, groups.
disinfect(ion) To free from infection, by destroying or removing harmful microorganisms. Cf. *sterilization.*
disinfectant An agent that disinfects, and usually destroys microorganisms but not bacterial spores; e.g., chlorine, phenol.
disinfest To free from infesting insects, rodents, or other small animals. Cf. *disinfect.*
disintegration (1) Decomposition. (2) See *atomic energy.* **artificial ~** See *radioelements.*
disintoxicate Detoxicate.
Disipal Trademark for orphenadrine hydrochloride.
disk A round plate. **alundum ~** A porous alundum d. used as a filter. **bursting ~** A diaphragm designed to rupture at a predetermined pressure, to safeguard against excessive pressure the equipment in which it is fitted.
d. assay The assay of antibiotics using disks dipped in different strengths of test solution and incubated in a medium with the active bacterium. The antibiotic concentration is

estimated from the size of the inhibited growth zone around each d.

dislocation The plastic deformation of metals. **edge ~** D. in which the d. line is at right angles to the direction of slip. **screw ~** D. in which the d. line is parallel to the direction of slip.

d. line A crystal defect, which marks the boundary between the regions of slip and nonslip.

dismutation The conversion of one substance into two; as, Cannizzaro's aldhyde reaction.

disodium **d. cromoglycate** Cromolyn sodium. **d. edetate** BP name for edetate disodium.

disopyramide phosphate $C_{21}H_{29}ON_3 \cdot H_3PO_4 = 437.5$. Norpace, Rhythmodan. White powder, soluble in water. Used to control arrhythmias of heart (USP, BP).

disoxidation Reduction.

dispensary A place where medicines are dispensed.

dispensing The compounding of medicines. **d. dose** See *dose.*

dispergator A peptizing agent.

dispersed Finely divided. Colloidal. **d. phase** Colloidal matter. Cf. *colloid.* **d. system** An apparently homogeneous substance which consists of a microscopically heterogeneous mixture of 2 or more finely divided phases (solid, liquid, or gaseous), e.g., liquid and liquid (milk); solid and gas (smoke). See *colloid.*

dispersion (1) Scattering of light, which depends on the size of particles present. (2) Separation by refraction of the constituent rays of a beam of nonhomogeneous light. For refractive index n and wavelength λ, d. $= dn/d\lambda$. **d. curve** A curve relating d. (2) and the wavelength of the light used. **d. medium** The material surrounding dispersed matter.

dispersivity Dispersion.

dispersoid A finely divided substance.

displacement (1) A chemical change in which one element, molecule, or radical is removed by another. (2) An ionic change in which one element exchanges charges with another element by oxidation or reduction. **electron ~** D. (2).

d. law The first enhanced spark spectrum of an element has a structure similar to that of the element preceding it in the periodic table. **d. reaction** Metals: $M + YX = MX + Y$; the metal M, being more positive than the metal Y, is oxidized. Nonmetals: $N + YX = YN + X$; the nonmetal N, being more electronegative than the nonmetal X, is reduced. **d. series** Electromotive series.

disproportionation The conversion of 2 like molecules into 2 or more unlike molecules. Cf. *dismutation.*

disruption Tearing apart suddenly.

dissection Cutting to pieces, as, the removal of tissues from an animal.

dissemination (1) Dispersion. (2) The natural scattering of seeds.

dissipation The transformation of mechanical into heat energy. **d. constant** $Q = E/\omega\mu$, where E is Young's modulus, $\omega = 2\pi\nu$, μ is the internal viscosity, and ν is the frequency. Q measures internal friction; e.g., quartz 100,000, silver 6,000, lead 30.

dissociated Split into simpler constituents.

dissociation The physical breaking apart of a molecule. **degree of ~** α. The ratio of molecules that become dissociated at a given temperature and concentration of solution to those that remain undissociated. Measured by the ratio of the equivalent conductivity under the conditions concerned to the conductivity of the solution at infinite dilution. Cf. *Ostwald law.* **electrolytic ~** Ionization. The breaking up of a molecule into 2 or more negatively and positively charged components (ions); usually when a polar compound is dissolved in water. Cf. *solvation.* **photo ~** See *disassociation.* **thermal ~** D. of a molecule of solid, liquid, or gas into simpler molecules or atoms by heat, and reversal at low temperatures.

d. constant The d. c., $k = [H^+][X^-]/[HX]$, in the application of the law of *mass* action, q.v. **d. pressure** The sum of the partial pressures of dissociated molecules in a system.

dissolution (1) Solution. (2) Hydrolysis of organic tissues.

dissolve To bring a solid into solution.

dissolved In a state of solution. **d. substance** Solute.

dissolvent Solvent.

dissonance (1) Discord. (2) A combination of sounds which produces beats.

dissymmetry Absence of complete symmetry. See *chiral.*

dist. Abbreviation for distilled.

Distalgesic Trademark for an analgesic; a combination of propoxyphene hydrochloride and acetaminophen (paracetamol).

distance The length between 2 points. Cf. *magnitude.*

Distaval Trademark for *thalidomide,* q.v.

distearin $C_{39}H_6O_5 = 554.5$. Glycerol 1,2-dioctadecanoate, glycerol distearate. m.75.

disthene Cyanite.

distibenediyl† See *antimono.*

distilland That which undergoes distillation.

distillate A liquid produced by condensation from its vapor during distillation. Cf. *tenate.*

distillation Purification of a liquid by boiling it, and condensing and collecting the vapors. **cold ~** D. at low temperatures, e.g., in a vacuum. **con ~** See *condistillation.* **destructive ~** D. of complex organic matter, e.g., wood, into a number of split and oxidation products. **fractional ~** Slow d. of a mixture and separate collection of the distillates at each boiling point, or after temperature intervals. **isothermal ~** The transfer of water vapor from a weak to a strong solution in a closed space, owing to the difference in vapor pressures. **molecular ~** D. carried out at a residual gas pressure below 1 μm, where the mean free path of the residual gas molecules is greater than the width of the distilling gap. **repeated ~** Cohobation. **steam ~** D. by the passage of steam through a liquid.

d. apparatus A device for d.; generally a closed vessel connected through a condenser to a receiver. **d. flask** A flask having a long neck with tubular outlet. **d. value** The ratio of the concentration of a substance in the vapor of its boiling solution to its concentration (c) in the liquid: $\log(1 - c)/\log(1 - V)$, where V is the volume of the distillate. Cf. *Henry's law.*

distilled Describing a liquid that has been vaporized and condensed. **d. water** Vaporized and condensed water, used extensively in the laboratory. **double ~** Water that has been distilled twice, the second time in a glass or platinum still.

distillery A place where distillation occurs; generally a plant making alcohol from fermented sugars. **d. waste** The residue from the stills of an alcohol d.; used as a fertilizer.

distribution (1) The occurrence of a substance on the earth's surface. Cf. *abundance.* (2) The assimilation or spread of a substance in the animal organism. **d. coefficient** Partition coefficient, Overton coefficient. The ratio of solubility in protoplasm to solubility in water, which measures the diffusibility of a substance into the cell protoplasm. **d. law** If a substance is dissolved in 2 immiscible liquids, a and b, then the ratio of its concentrations in each is constant. Cf.

Nernst law. **d. principle** Michael's rule. If HX adds onto an alkene linkage, where X is a halogen, it unites with the C atom having the lesser number of hydrogens. Cf. *Markovnikov rule.*

distyrene $(PhCH:CH-)_2$ = 206.3. White crystals, m.124.

disulfate (1)* Pyrosulfate, $M_2S_2O_7$ or MHS_2O_7. A salt of disulfuric acid. (2) Bisulfate.

disulfide* A compound containing 2 sulfide radicals. **alkyl ~** An organic compound containing a C chain and the −S:S− radical. **diethyl ~*** Et_2S_2 = 122.2. Colorless liquid, b.151. **dimethyl ~*** Me_2S_2 = 94.2. Colorless liquid, b.112. **d. ion*** The anion S_2^{2-}.

disulfiram $C_{10}H_{20}N_2S_4$ = 296.5. Bis(diethylthiocarbamoyl)disulfide, Antabuse. m.70, slightly soluble in water. Inhibits oxidation of acetaldehyde in the blood, causing unpleasant symptoms; used to treat alcoholism (USP, BP).

disulfite* Metasulfite, pyrosulfite. A salt of disulfurous acid, $H_2S_2O_5$; few are known.

disulfo- Prefix indicating (1) 2 sulfonic acid groups. (2) Dithio*. **d. acid** 1-Naphthylamine-4,8-disulfonic acid. **d. chloride** *Sulfur* mono*chloride.* **d. cyanate** Dithionate*. **d.cyanic acid** Dithionic acid*. **d.metholic acid** Methionic acid. **d.naphtholic acid** Naphthalenedisulfonic acid*.

disulfole Dithiole.

disulfonic acid Containing 2 sulfonic acid groups.

disulfuric acid* $H_2S_2O_7$ = 178.1. Pyrosulfuric acid. Fuming crystals obtained by freezing Nordhausen acid.

disulfuryl* Pyrosulfuryl. Indicating the radical $=S_2O_5$, from disulfuric acid. **d. dichloride*** $S_2O_5Cl_2$ = 215.0. Sulfur pentoxydichloride. A fuming liquid, d.1.844, b_{730mm}151, decomp. by water.

ditalimfos* See *fungicides,* Table 37 on p. 250.

ditan Diphenylmethane*.

ditellane* H_2Te_2.

diterpenes A group of compounds of general formula $C_{20}H_{32}$. See *terpenes.*

dithiane (1) Diethylene disulfide. (2) Heterocyclic hydrocarbons with 2 S atoms in a hexatomic ring.

dithiazine $C_3H_3NS_2$. Heterocyclic hydrocarbons with 2 S and 1 N atom in the hexatomic ring.

dithiazol $C_2H_3NS_2$. Heterocyclic hydrocarbons with 2 S and 1 N atoms in the pentatomic ring.

dithiene Heterocyclic hydrocarbons with 2 S atoms in the hexatomic ring.

dithienyl* $C_8H_6S_2$ = 166.3. Bithienyl, bithiophene. **2,2′-~** or **α-~** Colorless crystals, m.33. **d. ketone*** $(C_4H_3S)_2CO$ = 194.3. Thienone. Colorless crystals, m.88, insoluble in water. **d.methane** $(C_4H_3S)_2CH_2$ = 180.3. Colorless crystals, m.43. **d.toluene** $(C_4H_3S)_2CH\cdot Ph$ = 256.4. Colorless crystals, m.75.

dithio* Indicating 2 consecutive S atoms, −S−S−, when they replace O atoms.

dithiobisalanine Cystine.

dithiocarbamic acid $NH_2\cdot CS\cdot SH$ = 93.2. Colorless needles, decomp. by water.

dithiocarbonate* See *xanthate.*

dithiocarbonic acid $HO\cdot CS\cdot SH$ = 94.1. Theoretical acid. Cf. *xanthic acid.*

dithiocarboxy* Indicating the radical −CSSH.

dithiodiazole $CH_2N_2S_2$. Heterocyclic hydrocarbons with 2 S and 2 N atoms in the pentatomic ring.

dithioethylidene A compound containing the bivalent $-S\cdot CH(CH_3)\cdot S-$ radical. **ethylene ~** $C_4H_8S_2$ = 120.2. Colorless liquid, b.173.

dithiol (1)* Indicating 2 thiol groups. (2) Toluene-1,2-dimercapto-4-methyl-3,4-dithiolbenzene; reagent for tin, tungsten, and molybdenum. (3) British anti-*lewisite.*

dithiole (1) $C_3H_4S_2$ = 104.2. Disulfole. Erroneously called dithiazole. (2) Heterocyclic hydrocarbons with 2 S atoms in the pentatomic ring.

dithion A mixture of sodium dithiosalicylates. Yellow powder, soluble in water; used in veterinary medicine.

dithionate* Hyposulfate. A salt of the type $M_2S_2O_6$, from dithionic acid. **d. ion*** The anion $[O_3SSO_3]^{2-}$.

dithionic acid* $HO\cdot SO_2\cdot SO_2\cdot OH$ = 162.1. Known only in solution and as salts (dithionates).

dithionite* Hyposulfite. A salt of the type $M_2S_2O_4$.

dithionous acid* $HO\cdot SO\cdot SO\cdot OH$ = 130.1. Hyposulfurous acid, sulfoxylic acid. Known only in solution, and as salts (dithionites).

dithiourazole $C_2H_3S_2N_3$ = 133.2. Colorless crystals, m.245. Cf. *urazole.*

dithiourethane Ethyl dithiocarbamate.

dithiozone Dithizone.

dithizone $PhN:N\cdot CS\cdot NH\cdot NHPh$ = 256.3. Diphenylthiocarbazone, dithiozone. Blue crystals, soluble in carbon tetrachloride (green color), alkalies (red color), and sulfuric acid (blue color). A microreagent for lead (red), copper (brown), and zinc (purple); also used to separate these from other metals.

dithranol BP name for anthralin.

ditolyl* $Me\cdot C_6H_4\cdot C_6H_4\cdot Me$ = 182.3. Dimethylbiphenyl, bitolyl. **2,2′-~** Colorless liquid, b.272, insoluble in water. **2,3′-~** Colorless liquid, b.228, insoluble in water. **3,3′-~** Colorless liquid, b.286, insoluble in water. **4,4′-~** Colorless prisms, m.121. **d.amine** $(MeC_6H_4)_2NH$ = 197.3. Nine isomers; as: ***ortho*-~** Colorless liquid, b.313, sparingly soluble in water. ***meta*-~** Colorless liquid, b.319, insoluble in water. ***para*-~** Colorless needles, m.79, insoluble in water. **d.tin*** $(C_6H_4Me)_2Sn$ = 301.0. Yellow, amorphous powder, m.111, soluble in benzene.

ditophal $C_{12}H_{14}O_2S_2$ = 254.4. Diethyldithiophthalate. Yellow, viscous liquid, d.1.170–1.185; used to treat leprosy.

diurea Biurea*.

diureide A compound that contains 2 $-NH\cdot CO\cdot NH_2$ radicals. Cf. *ureylene.*

diuretic A drug that increases excretion of urine; as, chlorothiazide.

diuresis An increase in the volume of urine excreted; either physiological or after diuretics. **osmotic ~** D. due to high concentration of solute (as, glucose in diabetes) in urine.

Diuril Trademark for chlorothiazide.

divalent Bivalent. The capacity of an element to combine with 2 univalent (or 1 divalent) atoms or radicals; as, Fe(II), oxygen.

divaric acid $C_{10}H_{12}O_4$ = 196.2. 2,4-Dihydroxy-6-propylbenzoic acid. m.169 (decomp.). From the lichen *Evernia divaricata.*

divi-divi The pods of *Caesalpinia coriaria,* used for tanning.

divinyl (1) 1,3-Butadiene*. (2)* Prefix indicating 2 vinyl radicals. **d.acetylene** $CH_2:CH\cdot C:C\cdot CH:CH_2$ = 78.1. 1,5-Hexadien-3-yne*. An oil, d.0.785, b.84; basis for synthetic rubber. **d. ether*** $(CH_2:CH_2)_2O$ = 72.1. Vinyl ether, vinethine, ethenoxyethene. Colorless liquid, b.29. **d. ketone** 1,5-Pentadien-3-one*. **d. sulfide*** $(CH_2:CH)_2S$ = 86.2. Vinyl sulfide. Colorless oil, d.0.913, b.101, slightly soluble in water.

dixanthenylurea $[O(C_6H_4)_2CHNH]_2CO$. The alcohol-insoluble product of xanthydrol and urea; used to determine the latter in urine.

Dixarit Trademark for clonidine hydrochloride.

dixylyl Indicating 2 xylyl* radicals, $Me_2C_6H_3-$. **d.tin*** $(C_6H_3Me_2)_2Sn$ = 329.0. Colorless crystals, m.157.
-diyne* Suffix indicating 2 triple bonds.
djalmaite Yellow to black radioactive crystals (Brazil), containing 72% tantalum oxide.
djenkol bean See *jenkolic acid.*
djenkolic Jenkolic.
djenkolik acid See *jenkolic acid.*
dkg Symbol for decikilogram, 100g. Cf. dakg, 10 kg.
DL-, ***dl-*** Symbols formerly indicating a racemic (±) or *meso* enantiomer, and thus optically inactive.
dL Symbol for (1) deciliter; (2) decilog.
DMSO Abbreviation for dimethyl sulfoxide.
DNA* Symbol for deoxyribo*nucleic acid.* **D. ligase** See *polydeoxyribonucleotide.*
DNOC* DNC. See *insecticides,* Table 45 on p. 305.
Döbereiner, Johann Wolfgang (1780–1849) German chemist. **D.'s matchbox or lamp** A portable lamp producing a flame by passing hydrogen gas over platinum sponge in contact with air. **D.'s rule** The atomic weights of similar elements, A, B, and C, is approximately 2B = A + C. See *triad.*
Döbner's violet See *Döbner's violet* under *violet.*
DOCA Desoxycorticosterone acetate.
docosane* $Me(CH_2)_{20}Me$ = 310.6. Crystals, m.44, soluble in alcohol.
docosanoic acid* Behenic acid.
docosenoic acid* **(Z)-11-~** Cetoleic acid. **12-~** Brassidic acid. **(Z)-13-~** Erucic acid.
docosoic acid Behenic acid.
doctor solution Sodium plumbite solution containing flowers of sulfur. **d. s. test** Gasoline is mixed with d. solution; sulfur is shown by the formation of lead sulfide. **d. s. treatment** Petroleum is agitated with sodium plumbite solution and free sulfur.
dodecahedro-* Infix indicating 8 atoms bound into a dodecahedron with triangular faces.
dodecahedron A solid with 12 equal surfaces.
dodecanal* Lauraldehyde*.
dodecane* $Me(CH_2)_{10}Me$ = 170.3. Colorless liquid, b.214, insoluble in water. **1-bromo ~*** $Me(CH_2)_{11}Br$ = 249.2. Dodecyl bromide*. Colorless liquid, b_{45mm}177. **1-chloro ~*** $Me(CH_2)_{11}Cl$ = 204.8. Dodecyl chloride*. Colorless liquid, b_{18mm}145.
dodecanic acid Lauric acid*.
dodecanoic acid* Lauric acid*. **12-hydroxy ~** Sabinic acid.
dodecanol* Dodecyl alcohol*.
dodecanoyl* Indicating the $Me(CH_2)_{10}CO-$ radical, when substituted on a C atom. When unsubstituted, lauroyl* is used.
dodecenal* $Me(CH_2)_9CH:CH \cdot CHO$ = 196.3. An aldehyde in the oil from *Eryngium foetidum* (Umbelliferae).
dodecene* $CH_2:CH(CH_2)_9Me$ = 168.3. Dodecylene, decylethylene. (*E*) form is plant wound hormone.
dodecenoic acid* **(Z)-9-~** Lauroleic acid.
dodecoaldehyde Lauraldehyde*.
dodecoic acid Lauric acid*.
dodecyl* Lauryl. The radical $Me(CH_2)_{10}CH_2-$. **d. alcohol** $Me(CH_2)_{11}OH$ = 186.3. 1-Dodecanol*, lauryl alcohol. Colorless crystals, m.24. **d.amine*** $C_{12}H_{27}N$ = 185.4. Aminododecane. Colorless crystals, m.28, insoluble in water. **d. gallate*** $C_{19}H_{30}O_5$ = 338.4. 3,4,5-Trihydroxybenzoate. White powder, m.97, insoluble in water; an antioxidant (BP).
dodecylene 1-Dodecene*.
dog grass Triticum.
dogwood (1) A genus of trees and shrubs. (2) Cornus. The dried roots of *C. florida* of eastern U.S.; an astringent. **alder ~** Frangula. **Jamaica ~** Piscidia. **pond ~** or **swamp ~** See *cephaletin.*
Dolan Trademark for a polyacrylonitrile synthetic fiber.
Dolantin Trademark for meperidine hydrochloride.
dolerite Diabase. A coarse-grained basalt.
dolerophanite Cu_2SO_5. A copper persulfate in volcanic sublimates.
Dolezalek electrometer A quadrant electroscope.
dolichol A very long chain isoprenoid alcohol in tobacco; mol. wt. approx. 1,200.
Dolime Trademark for the dolomitic lime made by burning dolomite.
dolomite $CaMg(CO_3)_2$. An important calcium-magnesium carbonate, which forms mountain ranges as a white, grayish, or yellowish rock, d.2.85–2.95, hardness 3.5–4.
dolomol $Mg(C_{18}H_{35}O_2)_2$ = 591.3. Magnesium stearate, containing small amounts of the oleate and 7% magnesium oxide. A soft, white, unctuous dusting powder, insoluble in water.
Dolophine Trademark for methadone hydrochloride.
Doloxene Trademark for propoxyphene hydrochloride.
dome A d.-shaped crystal face, chiefly in the rhombic system. **d. faces** Prism faces developed parallel to a lateral axis, and intersecting the other 2 axes. **brachy ~** A d. face parallel to the brachy (shorter) axis. **macro ~** A d. face parallel to the macro (longer) axis.
domesticine $C_{19}H_{19}O_4N$ = 325.4. An alkaloid from the sacred bamboo of Japan. *Nandina domestica* (Berberidaceae). Cf. *nandinine.*
domeykite Cu_3As. A rare copper arsenide.
domingite $Pb_3Sb_4S_9$. Warrenite. A lead sulfostibate.
donarite An explosive: ammonium nitrate 70, trinitrotoluene 25, nitroglycerin 5%.
donaxine Gramine.
Donnan, Frederick George (1870–1956) English physical chemist. **D. equilibrium** If a nondiffusible substance is separated from diffusible substances by a semipermeable membrane, the ions will pass through in different amounts and establish an electrostatic difference (membrane potential), the osmotic pressure within being greater than that of the outer solution.
donor See *induced reaction.*
dopa Levodopa.
dopamine A sympathomimetic substance secreted from some areas in the brain tissue.
dopant See *dope (3).*
dope (1) A term for any drug that affects the speed of action, stamina, or courage of an animal or human being. (2) A substance, or lacquer, used to stiffen fabrics, e.g., aircraft wings. (3) The addition of a small quantity of "dopant" with the intention of changing a compound's properties; for example, adding 0.005% B to Si reduces the latter's resistivity by a factor of 10^3.
Doppler principle (1842.) Wave-type radiation emitted by a moving object decreases and increases in wavelength as the object approaches and recedes from the observer, respectively.
dopplerite $C_{12}H_{14}O_6$. Masses of humus embedded in peat: C 56.5, H 5.5%.
doron Trade name for a glass wool fiber bonded with a plastic or woven nylon; used to make "bulletproof" suits.
dorosmic acid Heptadecanoic acid*.
dose The quantity of medicine given at one time. **absorbed ~** See *gray (2).* **fatal ~** The minimum d. to cause death. **lethal ~** Fatal d. **maximum ~** The largest d. that can be

given with safety. **median lethal ~** LD_{50}. The expression of toxicity in mg/kg of body weight. The dose that kills 50% of a group of 30 or more animals receiving equal amounts of the substance. **minimum ~** The smallest d. to produce a physiological action. **poisonous ~** Toxic d. **safe ~** A d. between the minimum and maximum d. **toxic ~** The minimum d. that will produce poisonous and harmful effects.
d. equivalent See *sievert.*
dosimetry The measurement of radiation doses, given or absorbed.
dotriacontane* $Me(CH_2)_{30}Me$ = 450.9. Dicetyl. Plates, m.74; occurs in plants.
dotriacontanoic acid* Lacceroic acid.
double bond Δ. A condition that exists in unsaturated compounds where 2 single valence bonds connect 2 atoms. They are readily saturated by addition of 2 other atoms. Cf. *unsaturated.* **conjugated ~*** See *conjugated double bonds.* **(non)cumulative ~*** See *cumulative double bonds.* **isolated ~*** Bonds that are neither conjugated nor cumulative, as,

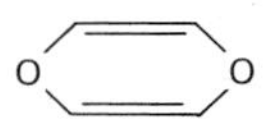

double salt A compound which crystallizes as a single substance but which, in solution, dissociates as 2 substances; as, $CaTiO_3$ is the double oxide of CaO and TiO_2.
doublet See *multiplet, duplet.*
double weighing Elimination of irregularities in the balance arm by weighing a substance first on one pan of the balance and then on the other.
doubling The conversion of spun yarn into thread, by twisting a single yarn, or by twisting together 2 or more single yarns.
douglasite $K_2FeCl_4 \cdot 2H_2O$. An ore from Stassfurt, Germany, producing miner's damp by reaction to form hydrogen.
DOV Distilled oil of vitriol (approx. 96% sulfuric acid).
Dowicide-H Trademark for sodium tetrachlorophenolate containing an excess of alkali; a wood preservative.
Down's syndrome Mongolism. A congenital chromosomal abnormality in which the person has 47 chromosomes, low mentality, and characteristic facial features.
Dowtherm Trademark for a eutectic mixture of diphenyl ether and 26.5% biphenyl, m.12, b.258; used for controlled heating.
doxepin hydrochloride $C_{19}H_{21}ON \cdot HCl$ = 315.8. Sinequan. White crystals, soluble in water. An antidepressant in the tricylcic group (USP, BP).
doxorubicin $C_{27}H_{29}ON \cdot HCl$ = 580.0. Adriamycin. Orange crystals, soluble in water. A cytotoxic drug for leukemia and similar diseases (USP).
doxycycline $C_{22}H_{24}O_3N_2 \cdot H_2O$ = 462.5. 6-Deoxy-5β-hydroxytetracycline, Vibramycin. Yellow crystals, soluble in water. A tetracycline antibiotic for many infections, including respiratory ones (USP, BP).
doxylamine succinate $C_{21}H_{28}O_5N_2$ = 388.5. Doxaminium succinate. White powder with a characteristic odor, m.102, soluble in water; an antiemetic or antihistamine (USP).
DP Abbreviation for: (1) degree of polymerization; (2) diastasic power.
dr Abbreviation for dram. **dr ap** Apothecaries' dram. **dr av** Avoirdupois dram. **dr fl** Fluidram.
dracaenic acid $C_{12}H_{12}O_3$ = 204.2. An acid from *Dracaena draco,* the dragon tree of Teneriffe.
drachm Dram.
dracilic, dracylic acid *p*-Amino*benzoic acid**.
dracoalban $C_{20}H_{40}O_4$ = 344.5. White powder, m.200, from dragon's blood.
dracone A large, flexible container towed in water, for transporting oil in bulk.
dracoresin, dracoresene $C_{26}H_{44}O_2$ = 388.6. A yellow resin from dragon's blood.
draconic acid Anisic acid*.
draconis sanguis Dragon's blood.
draconyl $C_{14}H_7$ = 175.2. A hydrocarbon distilled from dragon's blood.
draconylic acid Anisic acid*.
dracylic acid *p*-Amino*benzoic acid.*
draft See draught.
Dragendorff, Johann Georg Noel (1836–1898) German analyst. **D. reaction** The sulfates of many alkaloids give an orange-red precipitate with D.'s reagent. **D.'s reagent** Made by suspending 1.5 g bismuth subnitrate in 20 ml hot water, then adding 7 g potassium iodide and 20 drops of dilute hydrochloric acid.
dragon gum Tragacanth.
dragon's blood Sanguis draconis. The resinous exudation from the fruits of rattan palms (India, southeast Asia). Odorless, tasteless masses, insoluble in water, soluble in alcohol, giving a red solution.
Draka-Saran Trademark for a mixed-polymer synthetic fiber.
Dralon Trademark for a polyacrylonitrile synthetic fiber.
dram Drachm. A unit of apothecaries weight, ʒ. One dram = 0.16 ounce = 3 scruples = 60 grains = 3.8879351 g. **fluid ~** Fluidram. One-eighth of a fluidounce: 60 minims.
Dramamine Trademark for dimenhydrinate.
Draper D. effect Photochemical induction observed when a mixture of hydrogen and chlorine is exposed over water to diffused daylight. **D. law** Absorption law. Only rays which are absorbed by a system can produce a chemical change in it.
drastic Describing a drug having powerful irritant and purgative action.
draught A current of air or gas. **d. tube** A glass tube used in qualitative analysis to heat a substance in a current of air.
Drawinella Trademark for a cellulose acetate synthetic fiber.
draw tube A tube sliding within another; as, the d. t. of a microscope carrying the eyepiece.
drazoxolon* See *fungicides,* Table 37 on p. 250.
Drierite A brand of anhydrous calcium sulfate dessicant.
driers, dryers (1) Siccatives. Oxidizing substances which hasten the drying of varnishes, paints, etc.; as, lead resinates. (2) Usually, dryers. Mechanical devices to remove moisture by heat and/or air currents.
drift The uncertain motion of an indicating pointer, e.g., of a galvanometer.
drikold Dry Ice.
drilling mud A slurry of finely ground material, as barite or bentonite, used as a lubricant and coolant; also used to purge drillings.
drimin $C_{13}H_{14}O_4$ = 234.1. A crystalline substance, m.256, from *Drimys winteri* (Magnoliaceae), S. America. Cf. *winter's bark.*
drimol $C_{28}H_{58}O_2$ = 426.8. A wax from *Drimys granatensis* (Magnoliaceae), S. America.
Dr.Ing. See *Ing.*
drip (1) Liquid expressed from the muscle substance; in particular, the red serum that oozes from frozen meat during thawing due to the failure of the fiber to reabsorb it. (2) Infusion. A method of introducing fluid into the body continuously. **intravenous ~** Fluid fed into a vein via a

needle or cannula. **nasogastric ~** Fluid or liquid food fed into the stomach via a nasal tube.

driped Describing a hide steeped in chrome alum, dried, and soaked in melted paraffin wax.

dripping An unbleached, untreated cooking fat containing not less than 99% saponifiable matter and not more than 1.5% free fatty acids.

droperidol $C_{22}H_{22}O_2N_3F$ = 379.4. Droleptan, Inapsine. White crystals (darken on exposure), m.146, insoluble in water. A tranquilizer of the butyrophenone group.

dropping d. bottle A small glass bottle with pipet, or special stopper, enabling the contents to be delivered dropwise. **d. funnel** A separatory funnel with long stem and glass stopcock. **d. pipet** A small glass tube drawn out at one end, with a rubber bulb at the other.

drop reaction Spot test.

drops Portions of liquid, ordinarily 0.1–0.3 ml. **Prince Rupert ~** Solidified glass d. with tips, under great internal strain, which shatter to a fine powder when the tips are broken off.

dross Scum, scurf. The impurities floating on molten metals. **opium ~** Yenshee.

Druce, F. G. R. English chemist noted as codiscoverer of the element dvimanganese (rhenium).

drug A medicinal substance. Drugs are classified according to composition or constituents, structure or physical features, effect and use, origin and source. **addiction-producing ~** See *addiction-producing drugs.* **crude ~** A commercial d. before refining. **ethical ~** A drug sold only on a doctor's prescription. **inorganic ~** An inorganic salt, acid, or base used as a medicine. **organic ~** An organic compound used as a medicine.

Drummond D., Sir Jack Cecil (1891–1952) British chemist noted for his work on nutrition, especially in connection with wartime rationing. **D., Thomas (1797–1840)** English engineer who invented the calcium light.

dry (1) Free from liquid. (2) Evaporate. **air-~** Containing some equilibrium moisture content in contact with air; as, paper 5–8%. **bone-~** Completely free from water. **oven-~** D. to the extent that moisture has been removed in an oven at 100°C or other stated temperature.

d. battery Set of d. cells. **d. box** Glove box. **d. cell** An electric cell containing moist paste instead of a liquid electrolyte. See *Gassner cell.* **D. Ice** Drikold. Trademark for solid carbon dioxide used as a refrigerant. **d. method** Analysis by heat, e.g., blowpipe analysis.

dryers Driers.

drying Removal of liquid by heat, vacuum, or chemical agents. Cf. *dehydration.* **flash ~** Pneumatic d. The rapid removal of water by dispersion of moist solids in a hot gas stream. **fluidized bed ~** See *fluidized bed dryer.* **pneumatic ~** Flash d.

d. agent (1) Drier. (2) An agent that removes water; as, heat or chemicals; e.g., metallic sodium used to dry alcohol. Cf. *desiccant.* **d. oil** A liquid fat, e.g., tung oil, which absorbs oxygen and becomes hard and resinous; usually the glycerides of linoleic and linolenic acids. **d. oven** A receptacle for d. by heat. **d. tube** A U-shaped glass tube filled with a d. agent, e.g., calcium chloride, sulfuric acid; used for d. gases and vapors.

DTE DDT.

dualin An explosive: nitrogen 50, nitrated sawdust 50%.

duboisine Hyoscyamine.

Duboscq colorimeter See *colorimeter.*

Ducilon Trademark for a polyamide synthetic fiber.

ductile Capable of being drawn out into a fine wire.

ductility The extent to which a solid is ductile.

ductless glands Endocrine glands. Glands of the mammalian body that secrete hormones or enzymes into the blood. The site of action of the hormone is usually at a distance from the gland.

Dudley D. apparatus A glass apparatus to determine sulfur in iron and steel by the bromine method. **D. pipet** A pipet to deliver 100 ml water at 10°C in 35 s. Used for viscosity tests.

duff A low-grade, small-size, high-ash, anthracite.

dufrenite $H_3Fe_2PO_7$. A fibrous, green, hydrous iron phosphate.

dufrenoysite $Pb_2As_2S_5$. A gray, orthorhombic, brittle lead sulfarsenide.

dugaldine A glucoside from *Helenium (Dugaldia) hoopesii,* western sneezewood (Compositae).

dugong oil A cod-liver oil substitute from the superficial fat of the sea cow, *Halicore australis,* a herbiverous aquatic mammal.

dulcamara Bittersweet, woody nightshade, scarlet berry, felonwood, violet bloom. The stems of *Solanum dulcamara* (Solanaceae). It contains glucosides. Cf. *dulcamarin.*

dulcamarin $C_{22}H_{34}O_{10}$ = 458.5. a glucoside from *Solanum dulcamara.*

dulcamarrhetin $C_{16}H_{26}O_6$ = 314.4. A brown resin split product of dulcamarin.

dulcin (1) Sucrol. (2) Dulcitol.

dulcine Dulcitol.

dulcite Dulcitol.

dulcitol $HOCH_2(CHOH)_4CH_2OH$ = 182.2. Galactitol, euonymin, dulcin, 1,2,3,4,5,6-hexanehexol*, dulcite, dulcose, melampyrine, hexahydrohexane. ***meso-~*** From the sap of *Melampyrum, Scrophularia,* and *Euonymus* species. Colorless needles, m.189, soluble in water; a sweetening agent. **iso ~** Rhamnose.

Dulcolax Trademark for bisacodyl.

dulcose Dulcitol.

Dulin rotarex A centrifuge to determine mineral matter in asphalts.

Dulkona Trademark for a matt continuous-filament rayon yarn.

Dulong, Pierre Louis (1745–1838) French chemist. **D. and Petit law** Most elements have approximately equal molar heat capacity at room temperature, with a value for atomic weight × specific heat capacity of 25 J/K/mol at constant pressure. Certain elements of low atomic weight and high melting point obey the law only at high temperatures, e.g., beryllium.

Dumas, Jean Baptiste André (1800–1884) A French chemist noted for research on atomic weights, the gravimetric decomposition of water, and the composition of air. **D. bulb** A thin glass bulb drawn out to a fine opening; used in vapor density determinations.

dumortierite $Al_8HBSi_3O_{20}$. A dark, orthorhombic rock.

dumosa oil An oil from *Eucalyptus dumosa.* Cf. *lerp.*

dundakine An alkaloid from the bark of *Sarcocephalus esculentus,* dundaki (Rubiaceae), Cameroon.

dundasite A rare aluminum and lead carbonate.

dunder Waste saccharin liquors from sugar factories and distilleries.

dung bacteria Caprophyl. Bacteria which occur normally in manure.

dunite A green peridotite rock, consisting of chrysolite and olivine (Corundum Hill, N.C.).

Dunning colorimeter An instrument for the colorimetric estimation of phenolsulfonaphthalein excreted with urine in the renal function test.

dunninone 2-Hydroxynaphthoquinone, m.98.

duotal Guaiacol carbonate.

duplet Doublet. A pair of electrons shared by 2 atoms, corresponding with a single, nonpolar bond.

Duponal Trademark for a group of surface-active fatty alcohol sulfates.

Dupont nitrometer An arrangement of gas burets for the determination of nitrogen in explosives.

Duprene Trademark for a synthetic rubber made by polymerization of chlorobutadiene. See *elastomers,* Table 24 on p. 202.

Durabolin Trademark for nandrolone.

durain A constituent of *coal,* q.v.

duralium An alloy: Al 93–95, Cu 3.5–5.5%, and small amounts of Mg and Mn; used for chemical equipment.

Duralumin Trademark for an alloy: Mg 0.5, Mn 0.25–1.00, Cu 3.5–4.5, Al 93–95%, with traces of Fe and Si. It can be machined and resists dilute acids and seawater.

durangite $AlNaFAsO_4$. An orange, monoclinic mineral.

durene $C_6H_2Me_4$ = 134.2. Durol, 1,2,4,5-tetramethylbenzene*. Colorless monoclinics, m.79, insoluble in water. **iso ~** 1,2,3,5-Tetramethylbenzene*. Colorless liquid, b.195. **hexahydro ~** Tetramethyl*cyclohexane**.

Duriron Trademark for a resistant alloy: Fe 84.5, Si 14, other elements 1.5%.

Durite Trademark for a phenol-formaldehyde plastic.

durol Durene.

Durophet Trademark for dextroamphetamine sulfate.

durra Kaffir corn, broom corn, shallu. *Sorghum vulgare (Andropogon sorghum),* S. Africa.

durrin $C_{14}H_{17}NO_7$ = 311.3. A cyanogenetic glucoside from durra.

duryl The 2,3,5,6-tetramethylphenyl† radical Me_4C_6H-, from durene.

durylene The 2,3,5,6-tetramethyl-1,4-phenylene† radical

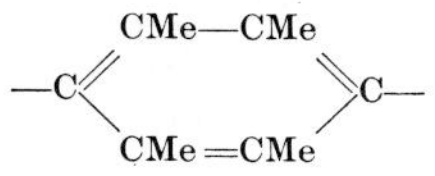

```
      CMe—CMe
 —C<          >C—
      CMe=CMe
```

durylic acid $C_6H_2Me_3 \cdot COOH$ = 164.1. 2,4,5-Trimethylbenzoic acid*, cumylic acid. Colorless crystals, m.150, soluble in water. **iso ~** Trimethylbenzoic acid. *α*- **~**, **3,4,5- ~** Needles, m.315. *β*- **~**, **2,4,6- ~** White scales, m.152. *γ*- **~**, **2,3,5- ~** White plates, m.127. All slightly soluble in water.

dust A finely powdered earth or waste material. The average size of outdoor d. is 0.5 μm. See *particulate, air pollution* under *air.* **cosmic ~** Finely divided matter in the outer atmosphere; supposed to originate from meteors and comets.
d. chamber An enlargement in a gas flue in which solid particles can collect. **d. precipitator** See *electrostatic precipitator.* **d. reticulation** See *pneumoconiosis.*

dusting Applying a fine powder. **d. powders** The fine-grained solids sprinkled over wounds or on the skin; as, talcum powder. *BP:* A white, free-flowing, absorbable powder containing up to 2.0% magnesium oxide, with maize starch which has been modified so that it will not gel on steam sterilization. *USP:* Prepared cornstarch, containing not more than 2% magnesium oxide.

Dutch **D. elm disease** A disease that often kills elm trees. Caused by the fungus *Ceratocystis ulmi* and spread by bark beetles of the genus *Scolytus.* **D. liquid** Ethylene dichloride*. **D. metal** An alloy: Cu 80, Zn 20%. **D. process** The preparation of white lead by the slow action on lead of carbon dioxide evolved from fermenting bark. **D. white** A white lead pigment containing 66% barium sulfate.

dvi Two or second (Sanskrit). Applied to the second undiscovered element of a group in the periodic system. Cf. *eka-elements.*

dvicesium Francium*.

dvimanganese Rhenium*.

dvitellurium Polonium*.

dwi Dvi.

dwt Abbreviation for (1) pennyweight; (2) deadweight (of ships); see *ton.*

Dy Symbol for dysprosium.

dyad A divalent element or radical.

dyamettin A glucoside from the root of *Cissampelos pareira.* Cf. *deyamittin.*

dydrogesterone $C_{21}H_{28}O_2$ = 312.5. 9*β*,10*α*-Pregna-4,6-diene-3,20-dione. Dehydroprogesterone, Duphastone. A progestational agent used mainly to treat menstrual disorders.

dye A natural or synthetic coloring matter used in solution to stain materials, as opposed to pigments which are used in suspension. A d. consists of a chromophore group and a salt-forming group (anchoring group). Cf. *dyes.* **d. bath** A solution of a d. for dyeing. **d.stuff** Dye. **d. wood** A wood that yields a coloring matter on extraction. Cf. *brazilwood* under *Brazil, haematoxylon.*

Dyer, Bernard (1856–1948) British chemist noted for his work on the analysis of foodstuffs and agricultural products.

dyes Dyestuffs. Synthetic d. are usually coal tar or aniline colors. Classification (based on application):

1. Substantive or direct d.: dye by immersion in *acid* or *basic* bath.
2. Adjective or mordant d.: require a fixing agent.
3. Sulfur d.: need a sodium sulfide bath followed by oxidation.
4. Vat d.: applied in their soluble, colorless, reduced (leuco) state and oxidized afterward to an insoluble color.
5. Ingrain colors: deposited on formation of an insoluble dye by chemical reaction. The fiber is immersed successively in the reagents.
6. Disperse d.: used in dispersion form.

acid ~ (1) D. which color the acidophile or basic granules of the protoplasm (generally the cytoplasm); as, eosin. (2) D. which color fibers in acid solution; e.g., nitro and azo d. **acid-mordant ~** D. which color animal fibers in acid solution with a mordant. **adjective ~** Mordant d. **artificial ~** Synthetic as opposed to natural coloring matters, e.g., coal tar. **azine ~** See *azine dyes.* **azo ~** D. containing the azo group. **azoxy ~** D. containing the $-N{:}N({:}O)-$ group. **bacteriological ~** D. used to stain bacteria; they may be acid, basic, neutral, or specific. **basic ~** (1) D. which color the basophile or acid granules of the protoplasm (generally the nucleoproteins); as, hematin. (2) D. used for dyeing in alkaline solution. **cationic ~** A generalized name for basic d. It includes fast and fugitive d. having a protonated N atom group, usually in the triphenylmethane structure

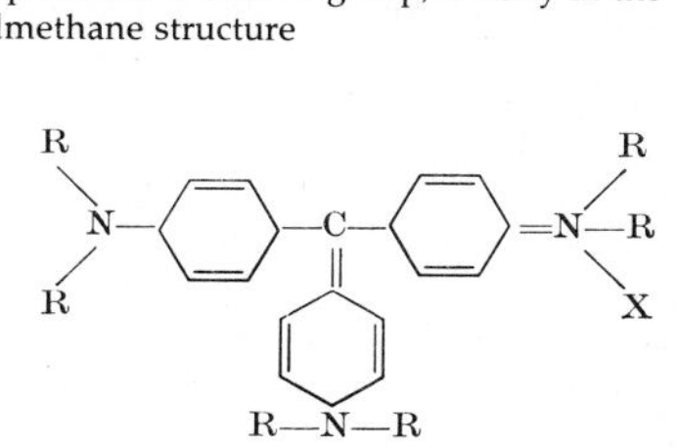

```
 R                              R
  \                            /
   N—<benzene>—C—<ring>=N—R
  /            ||              \
 R          <ring>              X
               |
             R—N—R
```

where R is an alkyl or aryl group, and X is the anion of a mineral or organic acid. Used to dye acrylic fibers with fast shades. **direct ~** Substantive dyes. **diperse ~** Used in dispersion for dyeing synthetic fibers. **hydroxyketone ~**

Aromatic compounds containing a carbonyl and hydroxyl group. **leuco ~** Vat dyes. **mineral ~** Inorganic substances used in dyeing; e.g., iron salts. **mixed ~** Polygenetic d. **monogenetic ~** Substances that dye only in one color. **mordant ~** Substances that require an additional substance for fixation on the fiber. **natural ~** Coloring matters of vegetable or animal origin, e.g., carmine. **neutral ~** D. which color both the acidophile and basophile portions of protoplasm. **nitro ~** D. containing the nitro radical in combination with OH and NH_2 groups. **nitroso ~** See *nitroso dyes.* **polygenetic ~** D. that produce 2 or more colors, e.g., alizarin. **provisional ~** See *writing ink* under *ink.* **pyronine ~** See *pyronine dyes.* **specific ~** D. which color protoplasm selectively. **Stenhouse ~** Deeply colored d. made by the interaction of furaldehyde, an aromatic amine, and the salt of a mineral acid. **substantive ~** Substances which stain fibers directly, without use of a mordant. **sulfide ~** Insoluble d. used in a sodium sulfide bath and oxidized afterward. **sulfite ~** D. used in sodium sulfite solution. **thiazine ~** See *thiazine dyes.* **triphenyl methane ~** D. derived from triphenylmethane. **vat ~** See *dyes(4).* **vegetable ~** D. derived from plants; as, alkanet, litmus, madder, rouge, turkey red (red and purple); fustic, gamboge, saffron (yellow and orange); indigo, wood (blue and green).

dyn Abbreviation for *dyne.*

dynad The intraatomic field of force.

Dynalkol Trademark for a motor fuel containing gasoline 70, alcohol 26, benzene 4%.

dynamic Describing forces not in equilibrium, and resulting in motion; opposed to static. **d. allotropy** Desmotropy. **d. formula** See *benzene ring.* **d. isomerism** Tautomerism.

dynamics The study of forces not in equilibrium. Classifications: electrodynamics (electrons), thermodynamics (atoms, molecules), gravitation mechanics (masses).

dynamites A class of explosives, usually a mixture of nitroglycerin (TNG) with an absorbing inert material. E.g.:

Kieselguhr d. . . . . TNG 72–75, kieselguhr 28–25%
Pittsburgh d. . . . TNG 40, sodium nitrate 44, wood pulp 15, calcium carbonate 1%
U.S.A. straight d. . . TNG 15–75, sodium nitrate 5–66, combustible materials 5–20, calcium carbonate 1%

dynamo A machine converting mechanical into electric energy.

dynamometer An apparatus to measure the force or power developed by an engine.

dyne Dyn. The cgs unit of force. The force which, acting for 1 s, gives to 1 g a velocity of 1 cm/s; 1 dyne = 10^{-5} newtons. **d. centimeter** The work done by a force of 1 dyne exerted along a distance of 1 cm. 1 dyne-cm = 1 erg.

Dynel Trademark for a high-melting-point synthetic fiber comprising a copolymer of vinyl chloride 60, acrylonitrile 40%, m.115.

dypnone PhCO·CH:CMePh = 222.3. 1,3-Diphenyl-2-buten-1-one. A condensation product of 2 molecules of acetophenone, $b_{22mm}225$.

dysalbumose An albumose obtained from fibrin by the action of pepsin. Brown powder, insoluble in water.

dyscrasite Ag_3Sb. A gray, mineral rhombic, d.9.6, hardness 3.5.

dyslysin (1) $C_{24}H_{36}O_3$ = 372.5. A resinous, dehydrated split product of choline acids. (2) An anhydrous decomposition product of bile acids.

dysprosium* Dy = 162.50. At. no. 66. A rare-earth metal discovered in holmia (Lecoq de Boisbaudran, 1886). It is trivalent, forms yellow or greenish salts, and occurs in small amounts in samarskite and gadolinite; m.1410, b.2580. **d. acetate*** $Dy(C_2H_3O_2)_3 \cdot 4H_2O$ = 411.7. Yellow needles, decomp. 120, soluble in water. **d. chloride*** $DyCl_3$ = 268.9. Green crystals, m.680, soluble in water. **d. nitrate*** $Dy(NO_3)_3 \cdot 5H_2O$ = 438.6. Yellow crystals, m.88, soluble in water. **d. oxalate*** $Dy_2(C_2O_4)_3 \cdot 10H_2O$ = 769.2. Yellow prisms, insoluble in water. **d. oxide*** Dy_2O_3 = 373.0. Colorless powder, d.7.81, insoluble in water. **d. sulfate** $Dy_2(SO_4)_3 \cdot 8H_2O$ = 757.3. Yellow crystals, soluble in water.

dystectic mixture An alloy or mixture containing those proportions of constituents which produce the highest constant-melting point. Cf. *eutectic mixture.*

E

E Symbol for exa- (10^{18}).
E Symbol for (1) activation energy; (2) einstein; (3) electric displacement; (4) electromotive force; (5) energy; (6) entgegen (see *stereoisomer*); (7) illuminance; (8) Young's modulus.
E (Boldface.) Symbol for electric field strength.
e Symbol for (1) base of natural logarithms = 2.718 281 828; (2) electron.
e Symbol for (1) elementary charge on an electron or proton; (2) linear strain.
ε See *epsilon.*
η See *eta.*
EAN Equivalent *atomic number.*
earth (1) Soil. (2) The solid portion of the globe, or lithosphere. See *abundance.* (3) The globe, as distinguished from other heavenly bodies. **diatomaceous ~** Kieselguhr. **fuller's ~** See *fuller's earth.* **green ~** Terra verde. **infusorial ~** Kieselguhr. **rare ~** See *rare-earth metals.* **red ~** Ocher. **siliceous ~** Kieselguhr.
e. age About 5×10^9 years. **e. alkali metals** See *alkaline.* **e. constants** See also *mass*, Table 52 on page 355.

Equatorial radius = 6,378,150 meters = 3,963.2 miles
Polar radius = 6,356,911 meters = 3,950.0 miles
One degree latitude at equator = 111.7 km
One degree latitude at pole = 110.6 km
Mean density = 5.517 g/cc
Mean density of continental surface = 2.67 g/cc
Volume = $1.083\ 160 \times 10^{12}$ km^3
Area = 510,100,934 km^2
Area of land = 148,847,000 km^2
Area of ocean = 361,254,000 km^2
Mean distance from earth to sun = 149,500,000 km = 92,900,000 miles
Mean distance from earth to moon = 384,393 km = 238,854 miles
First man-made earth satellite = 1957

---

e. flax Amianthus. **e. inductor** An electric coil, which rotates in a frame. **e.nut oil** Arachis oil. **e. oil** Petroleum or asphalt. **e. shellac** Acaroid resin. **e. wax** Ozocerite.
earthenware An opaque, porous ceramic made from kaolin, ball and flint clays, and china stone.
East Indian oil An essential oil from *Anethum sowa* (Umbelliferae), containing citrene and apiol.
Eastman Cellulose Acetate Trademark for a cellulose acetate synthetic fiber.
Easton syrup A pharmaceutical preparation containing iron, quinine, and strychnine phosphates; formerly a tonic.
eau French for "water." **e. de cologne** Cologne *water.* **e. de javelle** A bleaching solution of potassium hypochlorite. **e. de labarraque** A bleaching solution of sodium hypochlorite.
eblanin Paraxanthine.
ebonite Black, vulcanized, hard rubber. Cf. *vulcanite.*
ebony A hard, dark, heavy wood from *Diospyros* species (Ebenaceae).
ebul, ebulus The dwarf elder, *Pilea grandis* (Urticaceae), W. Indies. The berries are used for an alcoholic beverage.
ebullient Bubbling or boiling.
ebulliometry Ebullioscopy.
ebullioscope An apparatus to determine boiling points.
ebullioscopic equation Mol. wt. = $p(d/t)$, where p is weight (in g) of a substance per 100 g solvent, d the molecular rise in boiling point, and t°C the observed rise in boiling point.
ebullioscopy The determination of molecular weights from the increase in boiling point of a solution. Cf. *Raoult's law.*
ebullition Bubble formation.
ecbalin Elateric acid.
ecballium Elaterium.
ecbolics Oxytocics.
ecboline An alkaloid of ergot.
eccentric (1) Not concentric. (2) Not circular, e.g., an orbit. (3) With the axis away from the center. (4) Away from the center.
ecgonidine Anhydroecgonine.
ecgonine $C_9H_{15}O_3N \cdot H_2O$ = 203.2. Tropinecarboxylic acid. A split product of cocaine. White, monoclinic prisms, m.198, soluble in water. **anhydro ~** See *anhydroecgonine.* **methylbenzoyl ~** Cocaine.
e. hydrochloride $C_9H_{15}O_3N \cdot HCl$ = 221.7. Colorless, triclinic leaflets, m.246, soluble in water.
echelon cell A wedge-shaped glass cell used in absorption spectroscopy for different thicknesses of absorbing liquid.
echicaoutchin $C_{25}H_{40}O_2$ = 372.6. An elastic resinlike substance from dita bark.
echiine An alkaloid from *Echium vulgare* (Boraginaceae), which produces tetanic convulsions.
echinochrome A brown respiratory pigment in sea urchins. **e. A** 2-Ethylpentahydroxynaphthaquinone. A pigment derived from a leuco compound in the ripe eggs of certain marine invertebrates. Cf. *fertilizin.*
echinococcus A genus of tapeworm. The larval stage causes hydatid disease in man. The adult worm infests dogs.
echinopsine $C_{10}H_9ON$ = 159.2. 1-Methyl-4(1*H*)-quinolinone. m.152. An alkaloid from *Echinops* species, globe thistle (Compositae); similar in action to strychnine.
echinulate A growth of bacteria characterized by toothed outgrowths.
echinulin $C_{29}H_{39}O_2N_3$ = 461.6. A colorless substance formed with the pigments auroglaucin and flavoglaucin in the mycelium of the mold *Aspergillus glaucus.*
ec(h)othiopate iodide $C_9H_{23}O_3NIPS$ = 383.23. White, hygroscopic crystals, m.120 (decomp.), soluble in water; a miotic used for glaucoma (USP).
echtgelb Eger's yellow.
echugin A glucoside derivative of digitoxigenin, from *Adenium bohemianum,* an active principle of African arrow poisons, which arrests the heart in the contracted state. Colorless, rhombic scales, soluble in water.
eclogite A rock chiefly of augite and hornblende.
ECOIN European Core Inventory. EEC list of substances designating existing substances as "EINECS" (European

Inventory of Existing Chemical Substances), with others being new substances.
ecology A branch of biology dealing with the habits of organisms and relations to their surroundings.
ectohormone Pheromone.
ectoplasm The outer, compact hyaline layer of cell protoplasm.
eddy currents The electric currents set up by alternating currents in metal near or in an electric circuit. Cf. Foucault *current*.
Edecrin Trademark for ethacrynic acid.
edenite Hornblende containing iron.
Eder's solution A solution of mercuric chloride and ammonium oxalate, used to measure the intensity of X-radiation in terms of the amount of Hg precipitated.
edestan A protein from slightly denatured edestin.
edestin A globulin, mol. wt. 29,000, from hemp, rye, and cotton seeds, deficient in cystine and lysine.
edetate A salt of ethylenediaminetetraacetic acid. **e. calcium disodium** $C_{10}H_{12}O_8N_2CaNa_2 \cdot nH_2O$ (n = 2 and 3). Sodium calciumedetate. Used for lead and other heavy metal poisoning (USP, BP). **e. disodium** $C_{10}H_{14}O_8N_2Na_2 \cdot 2H_2O$ = 372.2. White crystals, soluble in water; a metal complexing agent and laboratory anticoagulant; used to treat hypercalcemia and calcium deposits in the eye (USP, BP).
edetic acid Ethylenediaminetetraacetic acid*.
edge filtration Streamline filtration.
edingtonite $BaAl_2Si_3O_{10} \cdot 4H_2O$. A gray, rhombic silicate, d.2.7, hardness 4–4.5.
edinol $C_7H_9O_2N$ = 139.2. 5-Amino-2-hydroxybenzyl alcohol, 5-aminosaligenin. m.139; A photographic developer.
Edison, Thomas Alva (1847–1931) American physicist and inventor. **E. cell** An accumulator; iron and nickel oxide electrodes in 20% potassium hydroxide solution. **E. effect** An electrode opposite a glowing filament becomes negatively charged. **E. Lelande cell** A cell (0.7 volt) consisting of an amalgamated zinc electrode and copper electrode in a solution of potassium hydroxide.
edrophonium chloride $C_{10}H_{16}ONCl$ = 201.7. Ethyl(3-hydroxyphenyl)dimethylammonium chloride, Tensilon. White, bitter, saline crystals, m.168, soluble in water; an anticholinesterase for diagnosis of myasthenia gravis (BP).
EDTA (1)*Ethylenediaminetetraacetic acid. (2)* (Not cap.) Indicating EDTA as a ligand.
effervescence The bubbling of a liquid due to escape of gas, not to boiling.
effervescent Having the property of effervescence. **e. mixture** A mixture of substances which form a gas when moistened, e.g., baking powder. **e. salts** Medicinally active substances mixed with sodium hydrogencarbonate and citric or tartaric acid.
efficiency (1) The ratio of useful work to energy supplied. (2) Statistics: The proportion of the total available data that is relevant according to the theory of probability.
efflorescence The property of crystals becoming anhydrous and crumbling when exposed to air, e.g., washing soda.
efflorescent Tending to effloresce.
effluent (1) A waste product discharged from a process. See *water pollution*. (2) Emergent, e.g., radiation.
efflux The flow of a fluid.
effuse Describing a veil-like growth of bacteria.
effusiometer An instrument (Bunsen) to determine the molecular weight of gases from effusion measurements.
effusion (1) A pouring out; a discharge. (2) The escape of a gas under pressure through an aperture. The relative e. rates of gases into air are inversely proportional to the square roots of their molecular weights. (3) In medicine, the discharge of blood or fluid into a body cavity or body tissues.
Efudix Trademark for fluorouracil.
Efylon Trade name for a polyamide synthetic fiber.
e.g. Abbreviation for exempli gratia (for example).
Eger's yellow E.Y., echtgelb. The yellow dye $HSO_3 \cdot C_6H_4 \cdot N$: $N \cdot C_6H_3 \cdot (SO_3H) \cdot NH_2$.
egg An ovum. **acid ~** A closed storage tank for strong acids. **storage ~** An e. kept in water-glass solution or in a frozen condition.
e. albumen A crystallizable protein, mol. wt. 33,800, from egg white.
eglantine $Ph \cdot CH_2COOC_4H_9$ = 192.3. Isobutyl-α-toluate. Colorless liquid, b.254, insoluble in water; a flavoring.
eglestonite Hg_4Cl_2O. A rare, native mercuric chloride oxide (Thomas Egleston, 1832–1900).
egols Antiseptic substances consisting of the *o*-nitro-*p*-sulfonates of phenol, cresol, or thymol, with mercury and potassium; e.g.: cresegol, from cresol; thymegol, from thymol.
EGTA Ethylene glycol bis(β-aminoethyl ether) *N,N,N',N'*-tetraacetic acid. A complexone, analogous to EDTA; used to determine calcium in presence of magnesium.
Ehrlich, Paul (1854–1915) German biochemist and father of immunology. **E. 606** Arsphenamine. **E. theory** Side-chain theory. An obsolete theory of the immune response involving reaction between "receptors" (antibodies) and antigenic "haptophores."
eicosa- Alternative spelling of the prefix *icosa-*, q.v.
eigenfunction Characteristic function (German). If, when an operator operates on an operand, the result can be expressed as the product of a constant and the operand, then the constant is known as the eigenvalue (eigenwert), and the operand as the eigenfunction.
eikonometer A measuring scale on a microscope eyepiece seen superimposed on the image of the object.
eikosane Icosane*.
-ein, -eine A suffix, often indicating an internal anhydride.
EINECS See *ECOIN*.
Einstein, Albert (1879–1955) German-born American physicist and mathematician, Nobel prize winner (1921). **E. equation for diffusion** The diffusion coefficient $D = RT/6\pi Lr\eta$, where L is the Avogadro constant, r the radius of the molecule, R the gas constant, T the thermodynamic temperature, and η the viscosity. **E. formula** $E = mc^2$, where m is the mass in grams, c the velocity of light in cm/s, and E the energy in ergs. Cf. *mass*-energy cycle. **E. law** In a photochemical reaction, one quantum of energy affects one molecule of matter, thus: $HI + h\nu = H + I$. **E. particle equation** The motion of a particle in a liquid is related to the radius of the particle and the viscosity of the liquid. **E. theory** See *relativity*. **E. unit** See *einstein*. **E. viscosity formula** $\eta = \eta_0(1 + k\phi)$, where η is the viscosity of a colloidal solution, η_0 the viscosity of the pure suspending medium, ϕ the total colloid per unit volume, and k a constant (1.5 to 4.75).
einstein A unit of energy, $E = 6.02 \times 10^{23}$ quanta. The amount of absorbed radiation to activate 1 gram molecule of matter. Cf. *Avogadro constant*.
einsteinium* Es. Element at. no. 99. At. wt. of isotope with longest half-life ^{252}Es = 252.08. Produced by the bombardment of californium and berkelium in a cyclotron (U. of California, 1954). It forms trivalent compounds.
eitropic Describing a fiber whose surface has been modified, as by carbon impregnation, to produce conductivity.

eka- One or first (Sanskrit). Prefix applied to the first undiscovered element in a group of the periodic system. Cf. *dvi.*

eka-elements The missing elements of the periodic table. Cf. *eka, dvi.* E.g.: gallium (formerly e.aluminum), germanium (formerly e.silicon).

elaio-, elaeo-, eleo- (Greek *elaio:* olive oil) A combining form meaning oil, oily.

elaeodic acid Ricinoleic acid.

elaeolite Nephelite.

elaeomargaric acid Eleostearic acid.

elaeometer A hydrometer for oils.

elaeostearic acid Eleostearic acid.

elaidic acid* $C_{18}H_{34}O_2$ = 282.5. (*E*)-9-Octadecenoic acid*. An isomer of oleic acid. Colorless leaflets, m.52, insoluble in water.

elaidin $C_{57}H_{104}O_6$ = 885.4. The elaidic ester of glycerol in many nondrying oils other than fish oils; an isomer of olein.

elain Ethylene*.

elaoptene The liquid part of an essential oil.

elastase* A pancreatic enzyme that hydrolyzes elastin.

elastic Describing a substance that assumes its original shape after a force, causing distortion, is removed. **e. coefficient** Young's *modulus.* **e. constants** The numerical expressions of the force to which a solid, liquid, or gas can be subjected without deformation of its shape or condition after the force has ceased to act. **e. fluid** A gas, as compared with a liquid. **e. limit** The stress or force which produces a permanent change in the length of a substance. **e. modulus** Young's *modulus.*

elastica Rubber.

elasticity The property of being elastic. **adiabatic ~** See *adiabatic elasticity.* **cubical ~** Bulk *modulus.* **limit of ~** Elastic limit. **longitudinal ~, modulus of ~** Young's *modulus.*

elasticum A carbohydrate layer between the epithelium and cortex of wool, responsible for the shrinkage of the latter.

elastin A protein of yellow, elastic tissue, partly attacked by pepsin and digestible by trypsin.

elastomer Contraction of "elastic polymer." A generic term (Fisher) for all substances having the properties of natural, reclaimed, vulcanized, or synthetic rubber, in that they stretch under tension, have a high tensile strength, retract rapidly, and recover their original dimensions fully. Typical elastomers contain long-polymer chains. Synthetic rubbers account for about 70% of world rubber production; about half of synthetics production is *SBR,* q.v. See Table 24 on p. 202.

elateric acid $C_{20}H_{28}O_5$ = 348.4. Ecbalin. An acid of elaterium, m.200.

elaterin $C_{32}H_{44}O_8$ = 556.7. Memordicin, from the juice of *Ecballium elaterium* and *Memordica balsamina* (Cucurbitaceae). White crystals. **alpha-~** Curcurbitacin E. m.232. **beta-~** m.195. Insoluble in water; a drastic purgative.

elaterite $(CH_2)_n$. Mineral caoutchuc, elastic bitumen. A flammable, elastic, brown mineral resin, d.0.8–1.23.

elaterium Ecballium. Sediment from the juice of *Ecballium elaterium* (squirting cucumber) containing chiefly elaterin; a powerful cathartic.

elaterone $C_{24}H_{30}O_5$ = 398.5. A ketone from elaterium. Colorless crystals, m.300.

elatic acid $C_8H_{12}O_2$ = 140.2. An acid from rosin.

elayl Ethylene* (2). **e. chloride** Ethylene dichloride*.

elder Sambucus. **dwarf ~** (1) Ebul(us). (2) *Aralia hispida.*

Eldred's wire Nickel-steel wire with a copper jacket and a platinum sheath, which may be sealed into glass.

eldrin Barosmin.

electret The electrical equivalent of a permanent magnet. Thus carnauba wax, when solidified in a strong electric field, acquires an orientation of molecules in the direction of this field, which can be retained for several years. The charge is on the surface and may amount to 1.7 nC/cm^2.

electric Charged with or capable of developing electricity. Cf. *electrical.* **e. arc** The luminous arc produced by the passage of electricity at high voltage from one electrode to another. **e. attraction** The force by which oppositely charged bodies are drawn together. **e. axis** The axis of a crystal that offers least resistance to the passage of an electric current. **e. battery** A series of *dry cells,* q.v., or voltaic *cells.* **e. calamine** $ZnSiO_4 \cdot H_2O$. A native zinc silicate (U.K. usage). **e. capacitance*** See *capacitance.* **e. charge density*** The electric charge per unit volume. **e. conductance*** See *electric conductance* under *conductance.* **e. conductivity** See *conductivity.* **e. current** The quantity of electricity; the number of electrons flowing per unit time. In esu units it is the amount of electricity transferred in 1 s; in emu units it is a current of such strength that 1 cm of the wire experiences a side thrust of 1 dyne, if at right angles to a magnetic field of unit intensity. The SI unit is the *ampere,* q.v.; its quantity the *coulomb,* q.v.; and its potential difference the *volt,* q.v. **e. eye** A photoelectric cell used to indicate a null point or end point, or to control a process automatically. Cf. *sectrometer.* **e. field** The forces around an electrically charged body. **e. field strength*** *E.* The force exerted on a unit charge by a field. **e. flux density** Electric charge per unit area, in C/m^2. **e. furnace** A *furnace,* q.v., for heating or processing molten electrolytes. **e. lines of forces** Imaginary curves radiating from a positive toward a negative charge. **e. potential** Electromotive force. **e. radiation** See *electromagnetic radiation* under *radiation.* **e. resistance*** See *resistance.* **e. spark** A luminous discharge produced by the disruptive passage of electricity at high voltage from one electrode to another. **e. surface density** Surface charge density. **e. tension** Electromotive force. **e. transformer** See *transformer.*

electrical Pertaining to electricity. **e. birefringence** Kerr effect. **e. capacitance** See *capacitance.* **e. capacitor** See *condenser.* **e. capacity** See *capacitance.* **e. cell** Voltaic cell. **e. condenser** See *electrical condenser* under *condenser.* **e. conductivity** See *electrical conductivity* under *conductivity.* **e. Curie point** See *electrical Curie point* under *Curie.* **e. current** Electric c. **e. elements** Voltaic cells. **e. flux** The flow of an electric current. **e. pressure** Electromotive force. **e. units** See Table 25 on p. 204. Conductance is the reciprocal of resistance. Units: siemens (SI), reciprocal ohm, mho (cgs); emu = electromagnetic units, based on the strength of magnetic poles; esu = electrostatic units, based on the strength of electric charges.

electricity A form of energy that produces magnetic, chemical, thermal, and radiant effects, generated by friction or induction; or chemically produced. (1) *Material concept:* All-pervading negative electrons; their continuous motion is a "current," their abrupt motion a "discharge," their absence a "positive charge." (2) *Dynamic concept:* A stress or strain in the ether resulting in "electric waves" and "radiation." (3) *Magnetic concept:* A field of force. **acid ~** Positive e. **atmospheric ~** The e. of the atmosphere, e.g., from charged clouds. **chemical ~, dynamic ~** Current or galvanic as opposed to static e. It indicates electrons in motion. It may be generated by chemical reaction (voltaic), induction (faradic), or magnets (magnetic). **faradic ~** Induced e. **frictional ~** Static e. obtained by friction, e.g., rubbing a glass rod with

TABLE 24. TYPES, USES, AND PROPERTIES OF ELASTOMERS

Elastomer	Trade names (examples)	Composition and manufacture	Principal uses	Properties
Natural rubber		*cis*-1,4-polyisoprene	Truck tires, off-the-road tires, dipped and proof goods, textile backing, footwear, drug sundries, mechanical goods, latex, foamed products	Abrasion resistance, resilience, good high- and low-temperature performance, tear strength
Homopolymers				
Neoprene (CR)	Perbunan C	Emulsion polymerization of chloroprene (2-chloro-1,3-butadiene)	Mechanical goods, wire coatings, heels and soles	Resistance to oil, ozone, abrasion, solvents
Polybutadiene (BR)	Buna 85, Diene, Philprene-*cis*-4	*cis*-1,4-polybutadiene	Blends with SBR in tires	Abrasion resistance, good oxidation and low-temperature resistance, low hysteresis (low heat buildup)
Polyisoprene (IR)	Coral Rubber, Ameripol SN, Natsyn	Ionic polymerization (in solution) of isoprene	Tires, adhesives, bathing caps, sneaker soles, dipped and proof goods, foamed products, rubber bands	Man-made duplicate of natural rubber
Homo- and copolymers				
Epichlorohydrin	Herclor, Hydrin	Homopolymer is CO type. Copolymer (with ethylene oxide) is ECO type	Gaskets, pump and valve parts, hose, belting	High gas impermeability; good resistance to abrasion, aging, and solvents
Copolymers				
Styrene-butadiene rubber (SBR)	Buna S, Buna Huls	Emulsion polymerization of butadiene with styrene	Tires, heels, soles, foamed products, mechanical goods, wire coatings, floorings	Resilience, tear strength, resistance to high and low temperatures and abrasion
Butyl rubber	Enjay Butyl	Ionic polymerization (in solution) of 2-methyl-2-propene with small amounts of isopropene	Auto motor parts, body mountings, tubes, tire linings, mechanical goods, cable insulation	Shock absorption, sun and ozone resistance; good containment of air; incompatible with other rubbers unless chlorinated
Nitrile rubber (NBR)	Perbunan, Butaprene, Hycar	Emulsion polymerization of butadiene with acrylonitrile	Coated paper, leather and textiles	Oil resistance; resistance to abrasion, ozone, solvents, high and low temperatures; resilience

Copolymers (continued)				
Ethylene-propene copolymer (EPM, EPDM)	C 23	Ionic polymerization (in solution) of ethylene with propene	Wire and cable coatings, weather stripping, steam hose, automotive parts	Ozone and sunlight resistance, good electrical properties, resilience, abrasion resistance; can be oil-extended
Fluoroelastomer	Viton A	Polymerization of hexafluoropropene with vinylidene fluoride	Automotive and aircraft parts	High resistance to chemicals and extremes of temperature
	Kel F	Polymerization of chlorotrifluoroethylene with vinylidene fluoride		
Polyacrylate	Lactoprene	Copolymerization of methyl or ethyl acrylate with small amounts of chloroethyl vinyl ether	Automotive transmission gaskets	Good oil resistance at high temperatures
	Acrilan	Copolymerization of acrylic ester with acrylonitrile		
Polycondensation products				
Polyurethane	Elastothane, Texin, Vulcollan, Vulcaprene	Polycondensation of diisocyanates with polyesters	Furniture, foams, linings	Good ozone and weather resistance
Silicone rubber	G.E. Silicone R, Siloprene	Polycondensation of hydrolyzed dimethyldichlorosilane	Cable insulation, coatings (as emulsion) and medical devices	Stable, good electrical properties and compatible with human tissue
Polysulfide rubber	Thiokol	Condensation of sodium polysulfides with aliphatic dihalogen compounds	Putties and hose	Good chemical resistance to organic solvents
Chemical conversion of high polymers				
Halogen-substituted rubber	Hypalon	Chlorosulfonation (in solution) of polyethylene	White tires, sealing and insulating	High resistance to chemicals, temperature and mechanical effects; good colorability

TABLE 25. ELECTRICAL UNITS

Unit of	SI system	cgs system emu	cgs system esu
Resistance	1 ohm	10^9	1.1126×10^{-12}
Current (strength)	1 ampere	10^{-1}	2.9980×10^9
Electromotive force (potential)	1 volt	10^8	0.0033357
Capacitance	1 farad	10^{-9}	8.9876×10^{11}
Quantity (charge)	1 coulomb	10^{-1}	2.998×10^9
Inductance	1 henry	10^9	1.11126×10^{-12}
Work	1 joule	10^7	10^7

fur. **galvanic ~** Dynamic. **induced ~** Faradic. A current of high voltage produced in a secondary coil when a current passes through the primary coil. **negative ~** A current of electrons passing from anode to cathode outside, and from cathode to anode inside, a galvanic cell. **photo ~** See *photoelectricity*. **positive ~** (1) An absence or deficiency of negative electrons. (2) A current in the direction opposite to negative e. According to this convention, a current flows from negative to positive. **primary ~** Power derived from nuclear or hydro sources. **static ~** Frictional, as opposed to dynamic, e. It indicates electrons accumulated at rest, and their sudden escape from the surface of an insulated conductor. **thermo ~** (1) E. produced by heat. (2) The heating effects of e. **tribo ~** E. produced by friction. **voltaic ~** A current produced by an electric battery.

electride An alkali metal salt involving a trapped electron as the anion; as, $Cs^+(18\text{-crown-6})_2e^-$. Cf. *alkalide*.

electrification Charging with electricity or electrons.

electrify To charge with electricity.

electrion Early name for electron.

electroanalysis Analytical methods based on electrolysis or conductometry; as, electrometric titration.

electrocardiogram ECG. Correctly, the tracing or record produced by an electrocardiograph and often used synonymously with it.

electrocardiograph Apparatus to record electrical changes produced by heart action transmitted through the body and picked up by electrodes usually on the chest wall and limbs. The record is made on paper or TV screen. Used to diagnose abnormalities of heart rhythm and myocardial infarct.

electrochemical Pertaining to both chemistry and electricity. **e. constant** Faraday constant*. **e. deposition** The formation of a metallic layer by electrolysis for (1) recovery of metals from ores, or refining metals; (2) electroplating, to produce a protective or ornamental coating, or for reproduction, e.g., in photoengraving. **e. equivalent** The mass in grams of any element deposited from an electrolytic cell by an electric current of 1 coulomb: $m = (A/V)0.000\ 010\ 36$, where A is the atomic weight, and V the valency of the element. Cf. *Faraday laws*. **e. machining** Finishing an article by making it an e. anode, the cathode being shaped to impart the required form when material is transferred to it from the anode by electroplating in reverse. **e. series** Electromotive series. **e. spectrum** Polarogram.

electrochemistry The science of transforming chemical into electrical energy, and vice versa.

electroconvulsive therapy ECT, electric shock treatment. Electric shock applied to the head by electrodes, causing a convulsion in the anesthetized patient; used to treat depressive illness.

electrocratic Describing a colloid stabilized by an electric charge.

electrode The device by which an electric current passes into or out of a cell, apparatus, or body. It may be a simple wire or complex device (hydrogen e.), or the container of the cell itself. **auxiliary ~** A standard e. used during electrodeposition to measure the potential at which this occurs. **calomel ~** See *calomel e., Hildebrand e.* **capillary ~** See *Lippmann e.* **dropping ~** A standard e. formed by a stream of mercury falling in fine droplets through a capillary tube into the electrolyte. A fresh surface is thus obtained continuously. Cf. *polarograph*. **gas ~** See *gas electrode*. **gas-jet ~** See *sprudel effect*. **glass ~** A thin glass membrane separating solutions of known and unknown pH value, the potential difference between the 2 sides being measured. **Hildebrand ~** See *Hildebrand electrode*. **hydrogen ~** See *hydrogen electrode*. **membrane ~** See *membrane electrode*. **negative ~** The cathode, or negatively charged pole, by which the current "passes out," and to which anions are attracted. **positive ~** The anode, or positively charged pole, by which the current "enters," and to which cations are attracted. **pX ~** Ion-selective e. that responds to the activity of an ion (X) species rather than to its concentration. Cf. *pH*, a particular case (H^+ ion). **quinhydrone ~** See *quinhydrone electrode*. **reversible ~** An e. which owes its potential to reversible ionic changes as $H_2 \rightleftharpoons 2H^+ + 2e^-$.

e. potential The potential developed by a material in equilibrium with a solution of its own ions. See Table 26, below.

electrodeposition The precipitation of a metal on an electrode. **e. analysis** The quantitative e. of an element from

TABLE 26. STANDARD ELECTRODE POTENTIALS (at 25°C)

Electrode reaction	E_0 (volt)	Electrode reaction	E_0 (volt)
$Au \rightarrow Au^+$	+1.68	$Co \rightarrow Co^{2+}$	−0.277
$Pt \rightarrow Pt^{2+}$	+1.19	$In \rightarrow In^{2+}$	−0.342
$Ir \rightarrow Ir^{3+}$	+1.000	$Cd \rightarrow Cd^{2+}$	−0.402
$Pd \rightarrow Pd^{2+}$	+0.987	$Fe \rightarrow Fe^{2+}$	−0.440
$Hg \rightarrow Hg^{2+}$	+0.857	$Cr \rightarrow Cr^{3+}$	−0.74
$Rh \rightarrow Rh^{3+}$	+0.80	$Zn \rightarrow Zn^{2+}$	−0.762
$Ag \rightarrow Ag^+$	+0.799	$V \rightarrow V^{3+}$	−0.876
$Cu \rightarrow Cu^+$	+0.521	$Nb \rightarrow Nb^{3+}$	−1.1
$Cu \rightarrow Cu^{2+}$	+0.337	$Mn \rightarrow Mn^{2+}$	−1.18
$H_2 \rightarrow H^+$	0.000	$Ti \rightarrow Ti^{3+}$	−1.21
$Fe \rightarrow Fe^{3+}$	−0.036	$U \rightarrow U^{4+}$	−1.50
$Pb \rightarrow Pb^{2+}$	−0.126	$Zr \rightarrow Zr^{4+}$	−1.53
$Sn \rightarrow Sn^{2+}$	−0.136	$Al \rightarrow Al^{3+}$	−1.66
$Mo \rightarrow Mo^{3+}$	−0.2	$Be \rightarrow Be^{2+}$	−1.85
$Ni \rightarrow Ni^{2+}$	−0.250	$Mg \rightarrow Mg^{2+}$	−2.37
		$Na \rightarrow Na^+$	−2.714

a solution. The electrode is weighed before and after deposition.

electrodynamics The study of moving charges.

electrodynamometer An instrument to measure the intensity of alternating currents.

electroencephalogram EEG. An electrical record of the waveform of electric currents developed in the human brain; used to diagnose epilepsy and pathological conditions in the brain.

electroendosmosis Electroosmosis.

electroflotation *Flotation,* q.v., produced by generation of a gas from electrodes immersed in a solution.

electrofocusing Concentration of chemical substances by utilizing their isoelectric points in a gel medium. Used to separate proteins.

electroforming The production of metallic tubes, sheets, or patterns by electrolysis.

electrofuge A leaving group that does not take away the bonding electron pair.

electrographic analysis The qualitative analysis of a metallic surface by placing it in contact with a gelatin-coated paper saturated with an electrolyte and making it the anode of an applied electric current. After removal of the electrolyte, a suitable reagent is added to the paper. Any suitable other metal may be used as electrode on the other side of the paper.

electrographite See *synthetic graphite* under *graphite.*

electroless deposition The deposition of a metal in solution on another solid metal by chemical means, instead of by means of an electric current as in electrodeposition, e.g., by the reducing action of phosphinates in nickel plating.

electroluminescence Electrophotoluminescence. The adiabatic emission of light by certain substances when placed in an electric field.

electrolysis The separation of the ions of an electrolyte; hence, the decomposition of a compound, liquid or molten, or in solution, by an electric current. **internal ~** The separation of a metal in the presence of a much more electropositive (i.e., baser) metal, by inserting an anode of the baser metal in the solution, and connecting it directly with a platinum cathode on which the metal is deposited. No external current is required; e.g., Cu on Zn immersed in $CuSO_4$ solution.

electrolyte (1) A substance that dissociates into 2 or more ions, to some extent, in water. Solutions of e. thus conduct an electric current and can be decomposed by it *(electrolysis).* (2) Sulfuric acid, d.1.150–1.835, used in batteries and accumulators. (3) The ionized salts present in the body and blood plasma; as, Na^+, K^+, Cl^-. **non ~** A substance that does not dissociate into ions. **strong ~** An e. that is highly dissociated even at moderate dilutions, and does not obey Ostwald's dilution law. **weak ~** An e. that is fully or partly dissociated only at high dilution, and obeys the Ostwald dilution law.

electrolytic Pertaining to decomposition by an electric current. **e. apparatus** An ammeter, voltmeter, rheostat, and rotating platinum anode for quantitative electrodeposition. **e. dissociation** See *electrolytic dissociation* under *dissociation.* **e. gas** A mixture of hydrogen (2 vol.) and oxygen (1 vol.) obtained by electrolysis of water. **e. separation** The graded *electrodeposition,* q.v., of metals from a solution, by varying the applied potential according to the electrode potentials of the metals. **e. solution tension theory** Nernst theory.

electrolyze, electrolyse To subject to electrolysis.

electromagnet Soft iron around which is wound an insulated wire; while an electric current passes through the wire, the iron is magnetized.

electromagnetic Pertaining to electricity and magnetism. **e. field** The area of force surrounding an electromagnet or a conductor through which a current flows. The intensity of the magnetic field at the center of a circular conductor of radius r is $2I/r$, where I is the current. **e. law** See *Coulomb.* **e. radiation** See *electromagnetic radiation* under *radiation.* **e. separation** The separation of the magnetic constituents of ores by means of an electromagnet. **e. spectrum** See *radiation* (diagram). **e. units** e.m.u., emu. A system of electrical units based on dynamics; it includes the SI units (volts, ampere, ohm, etc.) which are multiples or fractions of the cgs e. units. See *electrical units,* Table 25.

electromerism Mobile electron tautomerism within a molecule, the electron pairs of which vary in position, although the atomic nuclei do not. Cf. *chelate.*

electrometer An instrument for measuring electrostatic potential difference. **absolute ~** Galvanometer. Ammeter. **capillary ~** A null-point e. which detects 1 mA by the motion, in a capillary tube, of a dilute sulfuric acid–mercury interface. **emanation ~** See *emanation.* **photo ~** Photogalvanometer. **quadrant ~** Galvanoscope. An instrument indicating the presence and direction of an electric current, but not its strength. Essentially a magnetic needle inside a wire coil.

electromotion Mechanical action produced by electricity.

electromotive Pertaining to electromotion. **e. force** E, e.m.f., emf. Electric pressure, voltage. The energy driving a current, measured in *volts,* q.v., per unit quantity of electricity flowing through the circuit. For molar concentrations, the emf of an electrolytic cell is the algebraic difference between the electrode potentials of the ions of the metals forming the electrodes. **e. series** Displacement series, Volta series, constant series. The elements in decreasing order of their negative potentials: *(negative)* F, Cl, O, N, Br, I, S, Se, Te, P, V, W, Mo, C, B, Au, Os, Pt, Ir, Ta, Pd, Ru, Sb, Bi, As, Hg, Ag, Cu, Si, Ti, H *(zero),* Sn, Pb, Ge, Zr, Ce, Ni, Co, Tl, Nb, Cd, Fe, Cr, Zn, Mn, U, Gd, In, Ga, Al, rare earths, Be, Sc, Y, Mg, Li, Ca, Sr, Ba, Na, K, Rb, Ce *(positive).*

electron (1) An alloy: Mg 90, Al 5%, and a little Zn, Mn, or Cu. (2) Negatron, β particle. A subatomic *particle,* q.v., whose accumulation on an insolated conductor produces static electricity, and whose flow through a conductor produces an electric current. Electrons are in orbitals around the atomic nucleus, and their number and arrangement account for valency and other properties. Electrons are liberated from the atom by radioactive disintegration, and transferred from one atom to another in oxidation-reduction reactions. They are made visible by the Wilson track method, and Millikan's fog chamber. Their mass changes with their velocity. Constants are:

Rest mass . . . . 9.1095 × 10^{-28} g
Charge . . . . 1.6022 × 10^{-19} C
Spin (angular momentum) . . . 0.52729 × 10^{-34} Js

Cf. *atom.* **Auger ~** See *secondary electron* below. **binding ~** An e. which holds together the positive charges in the atomic nucleus. **free ~** See *free electron.* **heavy ~** Meson. **metastasic ~** An e. which changes its position in the atom owing to radioactive changes; generally moving from the valence orbital into the interior. **negative ~** Negatron. A normal, negatively charged electron; as distinct from a positive e. **nuclear ~** E. of the atomic nucleus. **paired ~** One of 2 electrons constituting a nonpolar *bond,* q.v. **phoretic ~** E. which conducts by passing freely from atom to atom when the atoms' outer orbitals are in contact. **photo ~** E. liberated from a surface by exposure to light.

piezo ~ A supposedly disk-shaped e. in the helium nucleus. **positive ~** Positron. **recoil ~** E. scattered by bombardment of a substance with α or β rays. **secondary ~** Auger e. E. emitted by a metal surface irradiated with X-rays of 150–200 kV. These electrons affect photographic film to extents that depend on the atomic number of the surface metal and are used in qualitative analysis. **twin ~** See *paired electrons.* **valency ~** Any of the electrons in the outer orbital of an atom which are responsible for valency. They can pass from one atom to the other (polar bond) or be held in common by 2 atoms (nonpolar bond). See *bond, valency.*

e. affinity The capture by a substance, e.g., an oxidizing agent, of the electrons of other substances. **e. beam** A stream of electrons, as in a cathode tube. **e. compounds rule** The position of the phase boundaries, at room temperature, in the equilibrium diagram of a binary alloy depends on the e. concentration. **e. configuration** The arrangement of electrons in energy levels. See Table 27 on pp. 207–208. **e. density** See *orbital.* **e. diffraction** The diffraction of a stream of electrons by a surface. Cf. electron *microscope.* **e. displacement** A shift of an e. pair held in common between 2 atomic nuclei toward one nucleus. See *Lucas theory.* **e. distribution curve** A curve showing the e. distribution among the different available energy levels. **e. exchange polymer** See *electron-exchange polymer* under *polymer.* **e. eye** Iconoscope. **e. formula** A chemical notation depicting the e. displacement in an organic compound. **e. fugacity** The tendency by an electrode in a solution to lose electrons. **e. gas** A system consisting of free electrons shared by all atoms, as in a metal; see *atomic structure.* **e. lens** The electrostatic field surrounding an aperture in a charged conductor. A circular hole will focus electrons with a focal length of $2V/(G_2 - G_1)$, where V is the energy of incident particles in volts, and G_1, G_2 are the potential gradients on the 2 sides of the plate. **e. microscope** See *electron microscope* under *microscope.* **e. optics** The control of e. motion by means of charged electric fields. Cf. lens effect on light. **e. pair** A pair of electrons which are held in common by 2 atoms. **e. probe analysis** Quantitative analysis by comparison of the characteristic X-ray intensities produced by a focused electron beam from the sample and from a standard. **e. screening effect** See *screening effect.* **e. spin resonance** e.s.r. A spectroscopic technique analogous to *nuclear* magnetic resonance, q.v., in which radiation of measurable frequency and wave length is used to supply energy to protons instead of to electrons. **e. transfer** The passage of one or more electrons from an atom or to an atom or ion in an oxidation-reduction reaction. **e. tube** A device for e. discharge; as, thermionic valve. **e. volt** eV. The energy acquired by an e. when it falls through a potential of one volt. 1 eV = 1.602 1892 $\times 10^{-19}$ joule.

electronate To cause electron transfer or reduction. **de ~** To cause oxidation.

electronation reactions Oxidation-reduction reactions.

electronegative (1) Having a negative charge or excess of electrons. (2) Capable of capturing electrons. **e. element** Elements generally located on the right side of the periodic table, especially the nonmetals. **e. ion** Anion, or negative ion. **e. radical** An acid radical or a group of atoms having a negative charge.

electronic Pertaining to electrons. **e. charge** A quantity of electricity numerically equal to the charge on a proton = 1.602 1892 $\times 10^{-19}$ C. **e. configuration of elements** See *electron configuration.* **e. formula** See *electronic formula* under *formula.* **e. mass** The mass of a negative electron moving with a velocity much less than that of light. **e. number** The number of peripheral electrons in the elements of a compound. **e. ratio** Specific *charge.* **e. structure symbol** A notation showing the distribution of electrons in the molecule. See *bond, molecular diagrams.*

electronics Radionics. The study of the applications of semiconductors and vacuum tubes (U.K.; valves) in electric circuits.

electroosmosis Electroendosmosis. The production of osmosis through a membrane by an electric current.

electropainting Electrolytic deposition of paint in a thin layer on a metal surface, which is made the anode.

electrophilic Describing preferential attraction to regions of high electron density. Cf. *nucleophilic.*

electrophoresis The migration of suspended particles in an electric field. In particular, the accelerated chromatographic separation of compounds by immersing each end of the medium in an electrolyte and applying an electric potential. Cf. *ionophoresis.*

electrophorogram A paper chromatograph produced by electrophoresis.

electrophorus An instrument consisting of insulated disks of ebonite and brass, used to produce frictional electricity.

electrophotography A method of photocopying in which a zinc oxide-coated paper is negatively charged by a corona discharge, exposed to light via the document which dissipates the charge locally, and developed by contact with a positively charged resinous toning powder, which is subsequently fixed by fusion.

electroplating The formation of a metallic coat on a baser metal by electrolysis.

electropolishing The production of a highly polished and chemically clean surface on a stainless steel object by making it the anode in an electrolyte, to reverse the process of electroplating; about 0.01 mm is removed.

electropositive (1) Having a positive charge or a deficiency of electrons. (2) Capable of losing or giving up electrons. **e. elements** The elements on the left-hand side of the periodic table, especially the light metals. **e. ion** Cation. An atom, or a group of atoms, which has lost one or more negative electrons or gained a proton, and has become positive; as, NH_4^+.

electropotential Electrode potential.

electrorefining The purification of metals by electrolysis.

electroresponse The increase in resistance of certain cells with increase of current.

electrorheological fluids Jammy fluids. Suspensions of fine, nonmetallic particles in oil that rapidly acquire solidlike properties when a voltage is applied across their flow. Response time to the voltage is less than 1 ms. Under voltage and stress, they creep rather than crack.

electroscope A device to detect electric charges or gaseous ions. **gold-leaf ~** Two strips of gold leaf suspended from an insulated conductor and in a glass vessel. **measuring ~** A gold-leaf e. which can be rotated so that an electrostatic and gravitational balance is established. The position of the leaf is read in a low-focus microscope.

electroscopy The measurement of the degree of ionization of a gas in terms of the rate of fall of the leaf of a charged gold electroscope.

electrosol A colloidal solution of a metal obtained by passing an electric discharge between metal electrodes in distilled water.

electrostatic Pertaining to electric charges at rest. **e. capacity** The ratio of quantity of electricity to difference of potential. **e. law** See *Coulomb.* **e. mixing** See *electrostatic*

TABLE 27. ELECTRON CONFIGURATION AND OXIDATION STATES OF THE ELEMENTS

At. no.	Element	Distribution of electrons									Oxidation states		
		1s	2s	2p							−	0	+
1	Hydrogen	1									1		1*
2	Helium	2											
		1s	2s	2p									
3	Lithium	2	1									.	1*
4	Beryllium	2	2										. 2*
5	Boron	2	2	1							3 . .	.	. . 3*
6	Carbon	2	2	2							4*. . 1	.	. 2 . 4
7	Nitrogen	2	2	3							4 3*. .		
8	Oxygen	2	2	4							2*.	.	. 2
9	Fluorine	2	2	5							1*		
10	Neon	2	2	6									
		1	2	3s	3p								
11	Sodium	2	8	1									1*
12	Magnesium	2	8	2									1 2*
13	Aluminum	2	8	2	1							0	1 . 3*
14	Silicon	2	8	2	2						4 . . .	0	. 2 3 4*
15	Phosphorus	2	8	2	3						3 . .	.	1 . 3*. 5*
16	Sulfur	2	8	2	4						2*.	.	. 2 . 4 . 6
17	Chlorine	2	8	2	5						1*	.	. . 3 . 5
18	Argon	2	8	2	6								
		1	2	3s	3p	3d	4s	4p					
19	Potassium	2	8	2	6		1						1*
20	Calcium	2	8	2	6		2						. 2*
21	Scandium	2	8	2	6	1	2						1 . 3*
22	Titanium	2	8	2	6	2	2				1	0	. 2 3 4*
23	Vanadium	2	8	2	6	3	2				1	0	1 2*3*4 5*
24	Chromium	2	8	2	6	5	1				2 1	0	1 2*3*4 5 6
25	Manganese	2	8	2	6	5	2				3 2 1	0	1 2*3*4 5 6 7
26	Iron	2	8	2	6	6	2				2 .	0	1 2*3*4 . 6
27	Cobalt	2	8	2	6	7	2				1	0	1 2*3*4 5
28	Nickel	2	8	2	6	8	2				1	0	1 2*3*4
29	Copper	2	8	2	6	10	1						1*2*3 4
30	Zinc	2	8	2	6	10	2						1 2*
31	Gallium	2	8	2	6	10	2	1					1 2 3*
32	Germanium	2	8	2	6	10	2	2			4 . . .	.	. 2*3 4*
33	Arsenic	2	8	2	6	10	2	3			3 . .	.	1 . 3*. 5*
34	Selenium	2	8	2	6	10	2	4			2 .	.	. 2*. 4*. 6
35	Bromine	2	8	2	6	10	2	5			1*	.	. . 3 . 5 . 7
36	Krypton	2	8	2	6	10	2	6					. 2
		1	2	3	4s	4p	4d	5s	5p				
37	Rubidium	2	8	18	2	6		1					1*
38	Strontium	2	8	18	2	6		2					. 2*
39	Yttrium	2	8	18	2	6	1	2					. . 3*
40	Zirconium	2	8	18	2	6	2	2				0	1 2 3 4*
41	Niobium	2	8	18	2	6	4	1			1	.	1 2 3 4 5*
42	Molybdenum	2	8	18	2	6	5	1			2 .	0	1 2 3*4 5 6*
43	Technetium	2	8	18	2	6	6	1			1	0	1 2 3 4*5*6 7
44	Ruthenium	2	8	18	2	6	7	1			2 .	0	1 2*3*4*5 6 7 8
45	Rhodium	2	8	18	2	6	8	1			1	0	1 2*3*4 . 6
46	Palladium	2	8	18	2	6	10					0	. 2*. 4
47	Silver	2	8	18	2	6	10	1					1*2 3
48	Cadmium	2	8	18	2	6	10	2					1 2*
49	Indium	2	8	18	2	6	10	2	1				1 2 3*
50	Tin	2	8	18	2	6	10	2	2		4 . . .	.	. 2*3 4*
51	Antimony	2	8	18	2	6	10	2	3		3 . .	.	. . 3*. 5*
52	Tellurium	2	8	18	2	6	10	2	4		2 .	.	. 2*. 4*. 6
53	Iodine	2	8	18	2	6	10	2	5		1*	.	1 . 3 . . . 7
54	Xenon	2	8	18	2	6	10	2	6				. 2 . 4 . 6*. 8

TABLE 27. ELECTRON CONFIGURATION AND OXIDATION STATES OF THE ELEMENTS (*Continued*)

At. no.	Element	Distribution of electrons												Oxidation states		
		1	2	3	4*s*	4*p*	4*d*	4*f*	5*s*	5*p*	5*d*	6*s*	6*p*	−	0	+
55	Cesium	2	8	18	2	6	10		2	6		1				1*
56	Barium	2	8	18	2	6	10		2	6		2				. 2*
57	Lanthanum	2	8	18	2	6	10		2	6	1	2				. . 3*
58	Cerium	2	8	18	2	6	10	2	2	6		2				. . 3*4
59	Praseodymium	2	8	18	2	6	10	3	2	6		2				. . 3*4
60	Neodymium	2	8	18	2	6	10	4	2	6		2				. . 3*4
61	Promethium	2	8	18	2	6	10	5	2	6		2				. . 3*
62	Samarium	2	8	18	2	6	10	6	2	6		2				. 2 3*
63	Europium	2	8	18	2	6	10	7	2	6		2				. 2 3*4
64	Gadolinium	2	8	18	2	6	10	7	2	6	1	2				. . 3*
65	Terbium	2	8	18	2	6	10	9	2	6		2				. . 3*4
66	Dysprosium	2	8	18	2	6	10	10	2	6		2				. . 3*4
67	Holmium	2	8	18	2	6	10	11	2	6		2				. . 3*
68	Erbium	2	8	18	2	6	10	12	2	6		2				. . 3*
69	Thulium	2	8	18	2	6	10	13	2	6		2				. . 3*
70	Ytterbium	2	8	18	2	6	10	14	2	6		2				. 2 3*
71	Lutetium	2	8	18	2	6	10	14	2	6	1	2				. . 3*4
72	Hafnium	2	8	18	2	6	10	14	2	6	2	2				1 . 3 4*
73	Tantalum	2	8	18	2	6	10	14	2	6	3	2		1		1 2 3 4 5*
74	Tungsten	2	8	18	2	6	10	14	2	6	4	2		2	0	1 2 3 4*5*6*
75	Rhenium	2	8	18	2	6	10	14	2	6	5	2		1	0	1 2 3 4*5*6 7
76	Osmium	2	8	18	2	6	10	14	2	6	6	2			0	. 2*3 4*5 6 7 8
77	Iridium	2	8	18	2	6	10	14	2	6	7	2		1	0	1 2 3*4*5 6
78	Platinum	2	8	18	2	6	10	14	2	6	9	1			0	. 2*. 4*5 6
79	Gold	2	8	18	2	6	10	14	2	6	10	1				1 2 3*. 5
80	Mercury	2	8	18	2	6	10	14	2	6	10	2				1 2*
81	Thallium	2	8	18	2	6	10	14	2	6	10	2	1			1*2 3
82	Lead	2	8	18	2	6	10	14	2	6	10	2	2			. 2*. 4
83	Bismuth	2	8	18	2	6	10	14	2	6	10	2	3			. . 3*. 5
84	Polonium	2	8	18	2	6	10	14	2	6	10	2	4			. . . 4*
85	Astatine	2	8	18	2	6	10	14	2	6	10	2	5	1*		1
86	Radon	2	8	18	2	6	10	14	2	6	10	2	6			
		1	2	3	4	5*s*	5*p*	5*d*	5*f*	6*s*	6*p*	6*d*	7*s*			
87	Francium	2	8	18	32	2	6	10		2	6		1			1*
88	Radium	2	8	18	32	2	6	10		2	6		2			. 2*
89	Actinium	2	8	18	32	2	6	10		2	6	1	2			. . 3*
90	Thorium	2	8	18	32	2	6	10		2	6	2	2			. . . 4*
91	Protactinium	2	8	18	32	2	6	10	2	2	6	1	2			. . 3 4 5*
92	Uranium	2	8	18	32	2	6	10	3	2	6	1	2			. . 3 4*5 6*
93	Neptunium	2	8	18	32	2	6	10	5	2	6		2			. . 3 4 5*6 7
94	Plutonium	2	8	18	32	2	6	10	6	2	6		2			. . 3 4*5 6 7
95	Americium	2	8	18	32	2	6	10	7	2	6		2			. 2 3 4*5 6
96	Curium	2	8	18	32	2	6	10	7	2	6		2			. . 3*4
97	Berkelium	2	8	18	32	2	6	10	8	2	6		2			. . 3*4
98	Californium	2	8	18	32	2	6	10	10	2	6	1	2			. 2 3*4
99	Einsteinium	2	8	18	32	2	6	10	11	2	6	1	2			. 2 3*
100	Fermium	2	8	18	32	2	6	10	12	2	6		2			. 2 3*
101	Mendelevium	2	8	18	32	2	6	10	13	2	6		2			. 2 3*
102	Nobelium	2	8	18	32	2	6	10	14	2	6		2			. 2 3*
103	Lawrencium	2	8	18	32	2	6	10	14	2	6	1	2			. 2 3

*Indicates a principal oxidation state.

PRINCIPAL SOURCE: F. A. Cotton and G. Wilkinson, *Advanced Inorganic Chemistry*, 4th ed., John Wiley, New York.

mixture under *mixture*. **e. precipitator** Equipment to collect fine particles of flue dust on a series of panels at a different electric potential from their surroundings. Used to reduce particulate pollution from boilers. **e. series** Triboelectric series. **e. units** esu, e.s.u. A cgs system of electrical units based on static data. See *electrical units*, Table 25 on p. 204.

electrostenolysis The precipitation of metals in the pores of a membrane by electrolysis.

electrostriction (1) The contraction of the solvent of a solution due to attraction of the water dipoles by the ions of the solute. (2) Mechanical deformation due to the application of an electric charge. Cf. *piezoelectricity*.

electrosynthesis Synthetic reactions produced by an electric current.

electrotaxis Electrotropism. The motion of living cells caused by an electric current.

electrotherapy The application of low-frequency or direct currents to weak or paralyzed muscles.

electrotropism Electrotaxis.

electrotyping The reproduction of type, reliefs, etc., by copper electrodeposited on the layer of graphite with which a mold or object has been covered.

electrovalence A fundamental atomic linkage due to transfer as distinct from sharing of electrons. Cf. *polar bond.*

electroviscosity The increase in the effective viscosity of a solid suspension due to the double layer on the particles.

electrowinning The separation of metals from their ores by electrolysis. Cf. *electrorefining.*

electrum (1) Amber (2) An alloy: Au 80, Ag 20%, d.13.0–16.0, hardness 2.5–3.0.

elemene $C_{15}H_{24}$ = 204.4. Oils, b. range 107–130. From elemi.

element(s) Matter consisting of atoms of one type; a substance that cannot be further decomposed by chemical means; a chemical unit; an ultimate constituent of matter. There are about 109 known elements, but doubt surrounds those above at. no. 103. The atoms of an e. may consist of a mixture of 2 or more atoms of different mass which have similar chemical properties, but different atomic weights (isotopes). The atomic weight is the mean of those of the isotopes calculated in the proportions present. The elements of higher atomic weight are radioactive. For a history of when the various elements were discovered, see Table 28, this page. **abundance of ~** See *abundance of elements.* **alkaline ~** See *alkali metals.* **basylous ~** E. of the first, second, and third groups of the periodic table, whose oxides form bases with water. **bio ~** See *bioelement.* **biogenic ~** A chemical element contained in living organisms. **Bunsen ~** See *Bunsen cell.* **chemical ~** Element. **dvi- ~** See *dvi.* **eka- ~** See *eka-elements.* **electrical ~** Galvanic, or voltaic *cell.* An electrolytic cell. **electron configuration of ~** See *electron configuration.* **electronegative ~** E. tending to gain electrons; generally on the right of the periodic table. **electropositive ~** E. tending to give up electrons; generally on the left of the periodic table. **extinct ~** Supposed e. of very low atomic weight, which may have formerly existed on the earth. See *nebulium.* **group ~** E. belonging to the same group in the periodic table; as opposed to period elements. **haloid ~** Halogens*. **heating ~** Resistance wire, used in heating apparatus. **inert ~** E. of the zero group of the periodic table; *noble gases,* q.v. **metallic ~*** E. of the lower half and left half of the periodic table, e.g., the light and heavy metals. **negative ~** Electronegative e. **nonmetallic* ~** E. in the right upper half of the periodic table. **normal ~** Standard *cell.* **period ~** An e. whose properties are closely related to e. in the same period of the periodic table, e.g., Mn, Fe, Co, Ni. Cf. group *element(s).* **primal ~** The original, nebulous form of matter from which all elements are supposed to have evolved: the protyle, pantogon, or urstoff. **principal ~** The 14 most common, abundant, and important e., viz., Al, Ca, C, Cl, H, Fe, Mg, N, O, P, K, Si, Na, S. Cf. *abundance of elements.* **radio ~** An e. that is either naturally

TABLE 28. HISTORICAL TABLE OF ELEMENTS

Date	Element	Discoverer
	Prehistoric and Archaic Time	
	Carbon	
	Sulfur	
	Gold	
	Silver	
4000 B.C.	Copper	Egypt
3500 B.C.	Iron	
?	Lead	
1600 B.C.	Tin	Chaldea
1000 B.C.	Antimony	
300 B.C.	Mercury	Theophrastus
	Alchemistic Period	
1220	Arsenic	Albertus Magnus
1450	Bismuth	Basil Valentine
1520	Zinc	Paracelsus
1669	Phosphorus	Brand
	Beginnings of Chemistry	
1733	Cobalt	Brandt
1750	Platinum	Wood
1751	Nickel	Cronstedt
1758	Sodium in salts	Marggraf
	Potassium in salts	Marggraf
	Founding of Chemistry	
1766	Hydrogen	Cavendish
1772	Nitrogen	Rutherford
1774	Oxygen	Priestley
	Chlorine	Scheele
	Manganese	Scheele, Gahn
	Barium	Scheele
1782	Tellurium	Müller
	Molybdenum	Hjelm
	Founding of Chemistry (continued)	
1783	Tungsten	d'Elhujar
1787	Strontium	Cruickshank
1789	Zirconium	Klaproth
	Uranium	Klaproth
	Titanium	Gregor
1794	Yttrium	Gadolin
1797	Chromium	Vauquelin
	Beryllium	Vauquelin
1801	Columbium	Hatchett
1802	Tantalum	Ekeberg
1803	Cerium	Klaproth
	Osmium	Tennant
	Iridium	Tennant
1804	Palladium	Wollaston
	Rhodium	Wollaston
	Beginning of Electrochemistry	
1808	Calcium	Davy
	Magnesium	Davy
	Boron	Davy
1812	Iodine	Courtois
1817	Selenium	Berzelius
	Lithium	Arfvedson
	Cadmium	Strohmeyer
1823	Silicon	Berzelius
1826	Bromine	Balard
1827	Aluminum	Wöhler
1828	Thorium	Berzelius
1830	Vanadium	Sefström
1839	Lanthanum	Mosander
1841	(Didymium)	Mosander
1843	Erbium	Mosander
1845	Ruthenium	Claus

TABLE 28. HISTORICAL TABLES OF ELEMENTS (*Continued*)

Year	Element	Discoverer
	Beginnings of Spectroscopy	
1860	Cesium	Bunsen
1861	Rubidium	Bunsen
	Thallium	Crookes
1863	Indium	Reich, Richter
	Beginnings of Systematization	
1872	Eka-elements	Mendeleyeev
1875	Gallium	Boisbaudran
1878	Terbium	Delafontaine
	Ytterbium	Marignac
	Holmium	Soret
	Thulium	Cleve
1879	Samarium	Boisbaudran
	Scandium	Nilson
1880	Gadolinium	Marignac
1885	Praseodymium	Welsbach
1886	Dysprosium	Boisbaudran
	Neodymium	Welsbach
	Germanium	Winkler
	Fluorine	Moissan
	Modern Chemistry	
1894	Argon	Ramsay, Rayleigh
1895	Helium	Clove, Ramsay
1898	Neon	Ramsay, Travers
	Krypton	Ramsay, Travers
	Xenon	Ramsay, Travers
	(radioactivity)	
	Polonium	Curie
	Radium	Curie
	Actinium	Debierne, Giesel
	Radon (niton)	Curie, Dorn
1900	Europium	Demarcay
1907	Lutetium	Urbain, Welsbach
1913	Brevium (protactinium) (mass spectrograph)	Fajans
1923	Hafnium	Coster, Hevesy
1925	Masurium (technetium)	Noddack, Tacke
	Rhenium	Noddack, Tacke
1926	Illinium (promethium) (magneto-optic method)	Hopkins
1930	Virginium (francium)	Allison, Murphy
1931	Alabamine (astatine)	Allison, etc.
1932	Deuterium	Urey, Brickwedde, Murphy
1935	Ekarhenium (neptunium)	Rolla
	Modern Chemistry (continued)	
	Ekaosmium (plutonium) (artificially produced)	Rolla
	Tritium	Rutherford, Oliphant, Harteck
1939	Francium	Perey
1940	Neptunium	Abelson, McMillon
	Plutonium	Seaborg, McMillon, Kennedy, Wahl
1944	Americium	Seaborg, James, Morgan, Ghiorso
	Curium (artificially produced)	Seaborg, James, Morgan, Ghiorso
1947	Promethium	Marinsky, Glendenin
1949	Berkelium (artificially produced)	Seaborg, Thompson, Ghiorso
	Californium	as berkelium
1952	Einsteinium	University of California, Argonne National Laboratory, Los Alamos Scientific Laboratory
1953	Fermium (from fission debris)	as einsteinium
1955	Mendelevium (artificially produced)	Ghiorso
1957	Nobelium (artificially produced)	International team (Nobel Institute for Physics, Stockholm)
1961	Lawrencium (artificially produced)	Ghiorso, Sikkeland, Larsh, and Latimer, University of California
1964 to 1983	At. nos. 104–109	American, U.S.S.R., and W. German teams

radioactive or made so by bombardment with particles; as radio*sodium*. **radioactive ~** Those e. of at. no. 82 and above. **terminal ~** E. of the zero group of the periodic table which form the terminals of each period, i.e., the noble gases. **transition ~** See *transition elements.* **transuranic ~** See *transuranic elements.* **typical ~** The e. of the third period of the periodic table, which show the typical characters of each group: Na, Mg, Al, Si, P, S, Cl, Ar.

For alphabetical list of elements giving *atomic* weights and *atomic* numbers, see Table 11 on p. 59.

For numerical list of atomic numbers including *oxidation states*, see Table 27 on p. 207. See also *ion, radioactive elements, isotopes.*

elementary (1) Ultimate, simple, fundamental. (2) Pertaining to elements. **e. analysis** The determination of carbon and hydrogen in organic compounds. Cf. *elementary analysis* under *analysis.* **e. charge** The charge of a proton, $e = 1.602\ 1892 \times 10^{-19}$ C. **e. molecule** An association of similar atoms; as, O_2. **e. particles** See *elementary particle* under *particle.* **e. space** The space surrounding the positively charged nucleus, in which the electrons are arranged in orbitals.

elemi (1) Gum elemi. A soft, yellow aromatic resin from *Canarium commune* (Burseraceae), Philippines. Used in ointments and to toughen varnishes. (2) Originally the resin of *Amyris elemi fera.* **e. oil** An essential oil, d.0.870–0.910, from e.

elemicin $C_{12}H_{16}O_3$ = 208.3. 1-Allyl-3,4,5-trimethoxybenzene. Colorless liquid, $b_{10}144$, from elemi oil.
elemin $C_{40}H_{68}O$ = 565.0. A crystalline resin in elemi.
elemol $C_{15}H_{26}O$ = 222.4. A camphor from elemi. **alpha-~** m.46. **beta-~** $b_{10mm}144$.
elemonic acid An acid, m.220, from elemi.
eleonorite $Fe_6P_4O_{19} \cdot 8H_2O$. Brown, rhombic masses, d.2.52–2.59, hardness 7.
eleoptene The liquid terpene or hydrocarbon of an essential oil. Cf. *stearoptene.*
eleostearic acid $Me(CH_2)_3(CH:CH)_3(CH_2)_7COOH$ = 278.4. 9,11,13-Octadecatrienoic acid*. From the oil of the seeds of *Elaeococca vernicia.* Colorless scales. **alpha-~** (9*Z*,11*E*,13*E*) form. m.48. **beta-~** (*E,E,E*) form. m. 71. Insoluble in water. Alpha-~ and beta-~ occur in tung oil. **oxo ~** Couepic acid.
eline Ekaiodine. Early name for astatine.
elinvar An alloy of Fe, Ni, and Cr, used to manufacture watch hairsprings (*el*asticity *invar*iable).
elixir (1) A liquid preparation for the administration of drugs in a pleasant form. Usually sweetened and aromatized alcoholic liquids or cordials. (2) The alchemical principle of everlasting life and youth, and the means of converting the baser metals into gold, (philosopher's stone). **simple, aromatic ~** A suspension of talc in an aqueous-alcoholic solution of a syrup, with orange flavoring (NF).
elkerite Pseudofucosan. A native fucose pentosan bitumen formed by the slow oxidation of crude oil. Cf. *fucosan.*
ellagene $C_6H_4 \cdot CH_2 \cdot C_6H_2 \cdot CH_2 \cdot C_6H_4$ = 254.3. A hydrocarbon produced by the reduction of ellagic acid. Small plates, m. 220, soluble in warm benzene.
ellagic acid $C_{14}H_6O_8$ = 302.2. Gallogen. Yellow crystals, decomp. by heat, sparingly soluble in hot water. It occurs or is produced by hydrolysis of many tannins; as, myrobalan.
ellagitannic acid $C_{14}H_{10}O_{10}$ = 338.2. A tanninlike substance from the pods of *Caesalpinia coriaria,* (Leguminosae) and the root bark of *Punicia granatum* (Punicaceae); slightly soluble in water.
ellagitannin A tannin giving ellagic acid on hydrolysis. Two types, derived from myrobalans and from galloyl derivatives of ellagic acid.
elliptone $C_{20}H_{16}O_6$ = 352.3. A toxic constituent of derris root, *D. elliptica.*
elm bark Ulmus, elm bark, slippery elm. The bark of *Ulmus campestri* or *U. fulva,* (Urticaceae); a demulcent, and adulterant for cassia. Cf. *ulmin, ulmic acid.*
Elon Trademark for monomethyl-*p*-amino-phenolsulfonate, a photographic developer.
Elphal process Trademark for a process for coating steel with aluminum using electrophoresis.
elpidite $Na_2ZrSi_6O_{15} \cdot 3H_2O$. A red, pearly silicate, d.2.5, hardness 7.
Eltroxin Trademark for levothyroxine sodium.
eluant A liquid used for the extraction of one solid from another, e.g., in chromatography.
eluate The solution resulting from elution.
elution A process of extracting one solid from another.
elutriation The process of washing and separating suspended particles by decantation.
elwotite A hard alloy of tungsten containing not more than 30% titanium.
Em Symbol for emanation.
emamium Early name for actinium.
eman A radioactive unit for the radon content of the atmosphere; 3.7 becquerel/dm^3.
emanation An early name for gaseous disintegration product of radioactive substances. **actinium ~** AcEm, Actinon. Early name for radon-219. **radium ~** RaEm. Early name for radon-222. **thorium ~** ThX. Thoron. Early name for radium-220.
e. electroscope An electroscope (Rutherford) to measure the emanation of radioactive bodies.
embalming fluid Solution injected into the arteries after death to preserve the tissues; as, formaldehyde, borax, phenol. Often mixed with a dye, as, eosine.
Embden E. ester A glucose-phosphoric acid ester in muscle. **E.-Meyerhof pathway** See *glycolysis.*
embedding The fixing of tissues in a solid mass so that they can be cut into thin sections. **e. bath** A device for heating paraffin wax and placing tissues in the molten liquid. **e. oven** An incubator with drawers.
embel(i)ate A salt of embelic acid.
embelic acid $C_{17}H_{26}O_4$ = 294.4. 2,5-Dihydroxy-3-undecyl-1,4-benzoquinone. Embelin. The active principle, m.145, from the berries of *Embelia ribes* (Myrsinaceae).
embelin Embelic acid.
embellic acid Embelic acid.
embolite $2AgBr \cdot 3AgCl$. A rare mineral; gray masses, d.5.8, hardness 1.5.
embolus Foreign material, usually blood clot or atheroma, blocking an artery or vein, and reducing or stopping blood flow. E. may also be air or tumor cells.
embrocation Pharmaceutical term for a preparation applied by rubbing.
embryo The developing organism; from fertilization to 8th week of intrauterine life in humans. Cf. *fetus.* **e. replacement** The replacement into the uterus and subsequent implantation of an e. fertilized in the laboratory. See *in vitro fertilization.*
embryology The study of the development of an organism.
emdite Abbreviation for ethylammonium ethyldiethylcarbamate; an alternative to hydrogen sulfide in qualitative inorganic analysis.
emerald A green variety of beryl; a precious stone. **oriental ~** A green variety of corundum.
e. green Paris green.
emergent column The portion of the thread of a thermometer which is not immersed in the substance whose temperature is being measured. Correction for its contraction (mercury in glass) $N(T - t) \times 0.000156$, where N is the number of degrees on the emergent stem, T the temperature as read, and t the temperature halfway up the emergent column given by a second similar thermometer.
emeri Emery.
emery Impure, crystalline corundum mixed with oxides of iron; a polishing, grinding, and abrasive material. **e. cloth** An abrasive fabric coated with glue and e. **e. paper** Paper coated with glue and e. used in place of sandpaper.
emetamine $C_{29}H_{36}O_4N_2$ = 476.6. An alkaloid from ipecac. Colorless crystals, m.156.
emetic A drug that causes vomiting; used especially in cases of poisoning; e.g., ipecac (uanha). **mechanical ~** Causing vomiting by tickling the throat with a feather.
emetine $C_{29}H_{40}O_4N_2$ = 480.6. An alkaloid related to cephaeline from the root of *Cephaelis ipecacuanha,* ipecac. Colorless crystals, m.74, sparingly soluble in water; an amebicide. **iso ~** $C_{29}H_{44}O_4N_2$ = 484.7. An alkaloid from ipecac, m.98.
e. hydrochloride $C_{29}H_{40}O_4N_2 \cdot 2HCl$ = 553.6. White crystals, m.53, soluble in water; used to treat amebiasis (USP, EP, BP).
emf, e.m.f. Abbreviation for electromotive force.

emission The liberation of energy, especially as radiation. **e. spectrum** See *emission spectrum* under *spectrum.*

emmenagogue A drug that produces or increases menstrual flow.

emmer A variety of wheat, formerly abundant in central Europe and the Near East.

Emmerling tube A cylinder filled with glass beads for the absorption of gases by liquids.

emodin $C_{15}H_{10}O_5$ = 270.2. 1,3,8-Trihydroxy-6-methylanthraquinone*. A purgative principle in many *Rhamnus* species. Orange prisms, m.260, sparingly soluble in water. **aloe ~** 1,8-Dihydroxy-3-hydroxymethylanthraquinone*. An isomer of emodin and constituent of aloes, cascara, senna, and frangula. Orange needles, m.222.

emollient A drug applied externally to soothe skin, e.g., talcum.

empiric(al) Describing knowledge gained by experience, without theoretical considerations. Cf. *scientific.*

empirical formula See *empirical formula* under *formula.*

emplectite $CuBiS_2$. Gray masses, d.6.4, hardness 2.5.

empyreal air See *oxygen.*

empyreumatic Any odorous substance formed by the destructive distillation of vegetable or animal matter. **e. oils** Oily e., e.g., creosote.

emu, e.m.u. Abbreviation for electromagnetic units.

emulgator Emulsifier.

emulsification Making an emulsion from 2 immiscible liquids by agitation, with addition of an emulsifier to prevent the droplets coalescing. **de ~** The breaking up of an emulsion, e.g., by adding an excess of the dispersed phase, heating, freezing, or centrifuging.

emulsifier Emulsifying agent, emulgator. A substance which makes an emulsion more stable (as, ammonium linoleate), by reducing the surface tension or protecting the droplets with a film.

emulsify To make an emulsion.

emulsifying wax (1) NF: Cetostearyl alcohol and polysorbate mixture. (2) BP: Cetostearyl alcohol 90, sodium lauryl sulfate 10, and water 4 pts., heated at 115°C.

emulsion Emulsoid. A fluid consisting of a microscopically heterogeneous mixture of 2 normally immiscible liquid phases, in which one liquid forms minute droplets suspended in the other liquid. Cf. *colloid, emulsifier, Aerosol.* **invert ~** Water in oil e.

en- (1) Greek prefix meaning "in," "on," or "at." (2)* Indicating ethylenediamine as a ligand.

enamel (1) A gloss paint. (2) A vitreous, opaque, or transparent glaze, fused over metal or pottery. **grayware ~** A mixture of feldspar 50, borax 30% with anhydrous sodium carbonate, sodium nitrate, and cryolite. **jewelry ~** A mixture of silica 14, boric acid 20, sodium nitrate 10, potassium nitrate 23%. **tooth ~** The mainly inorganic covering of the crown of teeth. When mature, it is the hardest body tissue, being as hard as sapphire. **white cover ~** A mixture of quartz 10–22, feldspar 18–33, borax 18–34, soda 3–10, sodium nitrate 2–5, cryolite 3–17%; colored with a metallic oxide, generally lead tetraoxide.

enamic form See *enimization.*

enamine A tautomeric form of Schiff's base.

enanth A synthetic polymer (U.S.S.R. origin) made by the high-pressure polymerization of ethylene and certain amino acids.

enanthaldehyde Heptanal*.

enanthic **e. acid** Heptanoic acid*. **e. aldehyde** Heptanal*. **e. ether** Ethyl hexyl ether*.

enanthine Heptyne*.

enanthol Heptanol*.

enanthotoxin $C_{17}H_{22}O_2$ = 258.4. (*E,E,E*)-2,8,10-Heptadecatriene-4,6-diyne-1,14-diol*. A poisonous resin from the rhizomes of *Oenanthe* (five-finger root); m.87.

enanthyl The hexyl* radical, $C_6H_{13}-$.

enanthylic acid Heptanoic acid*.

enantiomers* Molecules that are mirror images of one another.

enantiomorph (1) A crystal which corresponds with another crystal as its mirror image. (2) The opposite optically active substance, e.g., the dextro and levo forms.

enantiomorphous Related as enantiomorphs.

enantiotropic Existing in 2 crystal forms, one stable above, the other below, the transition-point temperature.

enantiotropy The state of being enantiotropic.

enargite Cu_3AsS_4. Gray, metallic rhombs, d.4.3–4.5, hardness 3.

encapsulation The enclosure of a reactant in a minute protective capsule, which is broken in the presence of the other reactant; as, in coatings on carbonless copy papers.

endo- Prefix indicating a "bridge" linkage joining 2 nonadjacent atoms in a ring of an organic molecule.

endocrine Pertaining to the secretions of ductless glands. See *ductless glands, hormone.*

endogenous Generated or originating within. **e. purines** The purine waste products of metabolism in excretions. Cf. *exogenous.*

endomorph A mineral enclosed within another (the perimorph).

endoscope A device to render internal organs or surfaces visible. Cf. *fiber optic.*

endosmosis The diffusion of a liquid through an organic membrane. Cf. *exosmosis.*

endosulfan* See *insecticides,* Table 45 on p. 305.

endothermic Absorbing heat. Antonym: exothermic. **e. compound** A compound absorbing heat during formation, and liberating heat on decomposition. **e. reaction** A slow chemical change which absorbs a definite number of joules; the stable exothermic compounds are transformed into unstable e. compounds.

endotoxin (1) Generally, any of the bacterial poisons set free by autolysis after cell death. (2) The O antigen from gram-negative bacteria.

endoxerosis The internal decline of a plant.

endoxy- Prefix indicating oxygen in a ring. Cf. *oxa(2).*

end point (1) The stage in titration when the reaction is complete. (2) The point of balance between 2 forces.

-ene* Suffix indicating a double bond; as, hexene. Cf. *diene, -idene.*

enema A liquid preparation for rectal injection; either a water and soap solution to cause evacuation, or a drug, as prednisolone, for inflammation.

energetics (1) The study of forces at work. (2) A philosophy which denies the existence of passive inert matter and conceives the universe as arrangements of energies in space.

e(ne)rgon See *quantum.*

energy Capacity to do work; e.g., heat, light, electricity, chemical action, or mechanical energy. The work done by the force which produces a change in the velocity of a body or a change in its shape and configuration, or both. It can be defined by *force × distance,* or *one-half mass × square of velocity.* **alternative ~** Types of energy other than those from fossil fuels; as, solar, biomass, wind, wave, tide, geothermal, and hydrogen. **atomic ~** See *atomic energy.* **available ~** Free e. **chemical ~** E. involved in chemical changes. See *thermodynamics.* **conservation of ~** E. may be transformed from one form to another, but cannot be

TABLE 29. ENERGY CONVERSION FACTORS

	joule	cal@	Btu@	kWh	erg
1 joule =	1	0.23885	0.947817×10^{-3}	2.77778×10^{-7}	10^7
1 cal@ =	4.1868	1	3.96825×10^{-3}	1.16298×10^{-6}	4.18680×10^7
1 Btu@ =	1.05506×10^3	2.52000×10^2	1	2.93071×10^{-4}	1.05506×10^{10}
1 kWh =	3.60000×10^6	8.59859×10^5	3412.14	1	3.60000×10^{13}
1 erg =	10^{-7}	2.38846×10^{-8}	9.47817×10^{-11}	2.77778×10^{-14}	1

@International Table.

destroyed or created. **creation of ~** E. is created when matter is annihilated; as, in a star. Cf. *mass-energy cycle.* **degradation of ~** The tendency for e. always to be changing to a lower form, e.g., into heat (Kelvin). **destruction of ~** The conversion of e. into matter. **dissipation of ~** Degradation of e. **dynamic ~** Kinetic e. **electric ~** Electricity. **free ~** See *Gibbs function, Helmholtz function.* **internal ~** *U*. Thermodynamic e. Changes to the total e. of a system in any form (e.g., e. taken up from its surroundings) alter its i. e. See *enthalpy, Gibbs function, Helmholtz function.* **ionization ~** See *ionization energy.* **kinetic ~** The e. of a body due to its motion = 0.5 mass × (speed)^2. **latent ~** Potential e. **lattice ~** The forces which hold the atoms together in a crystal. **mechanical ~** E. manifesting itself in mechanical work. **medial ~** The e. of a medium or solvent on which its dielectric powers and its dissociation, tautomerism, etc., depend. **potential ~** Static, or latent, e. The unreleased e. in a body due to its position, composition, or condition; as, a lifted weight, an endothermic compound, or a compressed gas. P. e. = mass × distance × acceleration due to gravity. **radiant ~** See *radiation.* **solution ~** The e. forces between the molecules of solute and solvent. Cf. *bond.* **specific ~** The power of a body due to its relative position (potential e.) and motion (kinetic e.).

e. change of atom The change of voltage, $V = 12{,}336/\lambda$, where λ is the wavelength. **e. crops** Examples are efficient sunlight converters (as, sugar cane for ethanol production), short-rotation forestry, bacteria and algae. **e. density*** The energy content per unit volume. **e. equivalence** See *fuel equivalence.* **e. level diagram** A figure giving a relative picture of the energy levels of an ion, atom, crystal, or molecule. Cf. *quantum.* **e. levels** The energy values corresponding to atomic or molecular orbitals. An electron changing from one to another will absorb (excitation) or emit (radiation) energy. **e.-mass cycle** See *mass-energy cycle.* **e. of activation** See *activation.* **e. quanta** See *quantum theory, Stoner quanta.* **e. units** Units of e. are based on the work to produce a change in speed (kinetic energy), or of condition or shape (potential energy), or both. The SI unit of e. is the *joule,* q.v. One J = 10^7 ergs = 0.101 972 kg·m = 0.737 562 ft·lb = 0.277 778 × 10^{-6} kWh. See Table 29.

enfleurage Removal of the perfume from picked flowers (e.g., jasmine) by placing them near an odorless mixture of tallow and lard, which absorbs the perfume to give a pomade from which the perfume is subsequently extracted.

enflurane $C_3H_2OClF_5$ = 184.5. 2-Chloro-1,1,2-trifluoroethyldifluoromethyl ether. Enthrane, Methylfluether. Colorless liquid; an anesthetic similar to halothane (USP).

engineering Utilizing the physical properties of matter in inventing, designing, constructing, building, and managing structures and machines. **chemical ~** The branch of e. concerned with chemical plants.

Engler, Carl (1842–1925) German chemist noted for his work on petroleum. **E. degree** A relative unit of viscosity.

enhanced Intensified, e.g., enhanced spectrum lines.

eniac. Abbreviation for electronic numerical integrator, an automatic calculator. It transforms quantities into electric pulses which are counted at 100,000 per second.

enidin, enin *Flavanols,* q.v., from grapes.

enimization An intramolecular rearrangement of amines and imines analogous to the keto-enol equilibrium: $=C{:}C(NH_2)- \rightleftharpoons CH \cdot C({:}NH)=$.

enim A glucosidal coloring matter of black grape skins.

Enka, Enkalon Trademarks for polyamide synthetic fibers.

Enkalene Trademark for a polyester-terylene synthetic fiber.

enndecane Nonadecane*.

ennea- Prefix indicating nine. Cf. *nona-.*

enneadecane Nonadecane*.

enniatin A See *lateritiin.*

enol form The alcohol form of a ketone characterized by the radical $=C{:}C(OH) \cdot R$. See *isomerism.*

enolic ester type See *ketonic ester type.*

Enovid Trademark for norethynodrel.

ensilage (1) Silage, or fodder preserved in a silo. (2) The process of making silage.

ensmartis Ammonium iron(III) chloride*.

enstatite $Mg_2Si_2O_6$. Magnesium pyroxene. A gray, rhombic silicate in meteors, d.3.1, hardness 5.5.

enterokinase Enteropeptidase*. An enzyme of the small intestine, which transforms trypsinogen to trypsin.

Entero-Vioform Trademark for clioquinol.

enthalpimetry The use of heat-change measurements to indicate a reaction end point in chemical analysis.

enthalpy* *H*. Heat content. $H = U + pV$, where U is the internal energy, p the pressure, and V the volume.

Enthrane Trademark for enflurane.

Entonox Trademark for 50% mixture of nitrous oxide and oxygen; used for self-administration analgesia, particularly in obstetrics.

entropy* *S*. In a reversible change, the ratio of the energy taken up to the thermodynamic temperature at which it is taken up. It is thus the unavailable energy of a system or a measure of the system's disorder.

-enyl Suffix indicating a univalent radical derived from an acyclic hydrocarbon with one double bond, as, butenyl.

-enyne Suffix indicating a hydrocarbon with a double and triple bond; as, butenyne.

enzyme(s) (Greek: "in yeast") Ferment. A catalyst, usually a protein, produced by living cells; it has a specific action and optimum activity under definite conditions (as, pH value). The first e. crystallized was urease (Sumner, 1926). Classification is based on the type of reaction an e. catalyzes. There are 6 classes (divisions), see Table 30 on p. 214. The publication cited at the foot of the table classifies and gives the nomenclature and characteristic reactions of over 2,100 e. **amylolytic ~** E. hydrolyzing starch to dextrin and maltose; e.g., amylase. **anti ~** A substance which inhibits e. action. See *coenzyme.* **blood ~** E. of the *blood,* q.v. **co ~** See *coenzyme.* **coagulating ~** E. changing soluble proteins to insoluble products, e.g., thrombin. **decarboxylizing ~** E.

TABLE 30. CLASSIFICATION OF ENZYMES

Division	Definition	Basis for systematic name	Recommended name	Examples
1. Oxidoreductases	E. that catalyze oxidoreduction reactions.	Donor:acceptor oxidoreductase. (Oxidized substrate is regarded as an H donor.)	Dehydrogenase or reductase. Oxidase used only where O_2 is the acceptor.	Catalase. Lipoxygenase. Glucose oxidase. Peroxidase.
2. Transferases	E. transferring a group (e.g., Me or glycosyl) from one compound (generally regarded as the donor) to another (the acceptor).	Donor:acceptor grouptransferase. (Donor is often a cofactor charged with the group to be transferred.)	Acceptor grouptransferase or donor grouptransferase.	Aspartate aminotransferase. Hexokinase.
3. Hydrolases	E. catalyzing the hydrolytic cleavage of C—O, C—N, C—C, phosphoric anhydride, and some other bonds.	Hydrolase.	Substrate, with suffix -ase.	Amylase. Cellulase. α-D-Galactosidase. Rennin. Trypsin.
4. Lyases	E. cleaving C—C, C—O, C—N, and other bonds by elimination, leaving double bonds or adding groups to double bonds.	Substrate grouplyase.	Use expressions like decarboxylase, aldolase, dehydratase (for elimination of water) or synthase (where reverse reaction is much more important).	Aspartate ammonia-lyase. Carbonate dehydratase. Oxaloacetate decarboxylase.
5. Isomerases	E. catalyze geometric or structural changes within a molecule.	According to the type of isomerism, isomerases are called: racemases, epimerases, cis-trans-isomerases, tautomerases, mutases, or cyclo-isomerases.		Glucosephosphate isomerase. Triosephosphate isomerase.
6. Ligases (synthetases)	E. catalyzing the joining together of 2 molecules coupled with the hydrolysis of pyrophosphate bond in ATP or a similar triphosphate. (Bonds are often high-energy bonds.)	X:Y ligase (ADP-forming).	Use synthetase, if no other short term (e.g., carboxylase) available.	Acetyl-CoA synthetase. Polydeoxyribonucleotide synthetase(NAD^+).

SOURCE: Abstracted from *Enzyme Nomenclature*, International Union of Biochemistry, Academic Press.

splitting carbon dioxide from organic acids; e.g., carboxylyase. **detergent ~** Mainly serine-active, alkaline, proteolytic e. aiding soil and stain removal by attacking protein and carbohydrate components in the substance to be removed. **glycolytic ~** E. hydrolyzing glucosides to glucose and residue, e.g., glucosidase. **hydrolytic ~** E. hydrolyzing carbohydrates to simple sugars, e.g., amylase. **inactive ~** A nonreactive e. **inverting ~** E. which change optically inactive sugars to optically active sugars, or vice versa; e.g., invertase. **iso ~** E. with a specific catalytic character, existing in multiple forms as aggregates of subunits; as, lactic dehydrogenase. **lipolytic ~** E. hydrolyzing esters (fats, etc.) to fatty acids and alcohols; e.g., lipase. **pro ~** Zymogen. **proteolytic ~** E. hydrolyzing proteins to albumoses and peptones, e.g., pepsin.

E. Commission numbers Numbers for classifying e. used in the publication cited at the foot of Table 30, opposite. The first number corresponds to the "division" in the table, and subsequent ones to subdivisions; as, catalase is EC 1.11.1.6. **e. unit** See *katal.*

enzymolysis A chemical change produced by enzyme action.

Eocene See *geologic eras,* Table 38.

eosine $C_{20}H_8O_5Br_4$ = 647.9. Tetrabromofluorescein, bromeosin. Red, triclinic needles, insoluble in water; a dye and indicator (alkalies—green fluorescence, acids—yellow). **soluble ~** (1) Commercial e. (2) E. Yellowish.

E. Bluish Erythrosin. **e. dyes** Dyestuffs for wool and silk derived from fluorescein, e.g., eosine. **e. soluble** E. Yellowish. **E. Yellowish** $C_{20}H_6O_5Br_4K_2$ = 724.1 or $C_{20}H_6O_5Br_4Na_2$ = 691.9. A red powder, soluble in water; a dye.

eosins A group of dyestuffs derived from triphenylmethane.

EPA Environmental Protection Agency of the U.S. government.

epanorin $C_{25}H_{25}O_6N$ = 435.5. **(S)-~** A yellow constituent of the lichen *Lecanora epanora.* m.135.

Epanutin Trademark for phenytoin sodium.

ephedrine $C_6H_5 \cdot CH(OH) \cdot CH(NH \cdot Me)Me$ = 165.2. 2-Methylamino-1-phenyl-1-propanol. An alkaloid from *Ephedra* species (Gnetaceae), Europe; also produced synthetically. **(−)-~** Levo e. (1*R*,2*S*) form. An alkaloid from the Chinese drug MaHuang from *Ephedra equisetina* (Gnetaceae). White, bitter crystals, m.41, soluble in water; used similarly to e. hydrochloride. **pseudo ~** Alkaloid enantiomers, as, (+)-(1*S*,2*S*), from *Ephedra vulgaris* (Gnetaceae), Japan. Colorless crystals, m.115, insoluble in water. **pseudoephedrine hydrochloride** See *pseudoephedrine hydrochloride.*

e. hydrochloride $C_{10}H_{15}ON \cdot HCl$ = 201.7. Odorless, bitter, white crystals, m.218, soluble in water or alcohol (EP,BP). Used as a mydriatic, and in ophthalmology as an atropine substitute. **L-~** Colorless crystals, readily soluble in water. Used as an adrenaline substitute to raise blood pressure; also in bronchial asthma and hay fever. **e. sulfate** $(C_{10}H_{15}ON)_2 \cdot H_2SO_4$ = 428.5. Fine, white, odorless crystals, darkened in light. Soluble in water or alcohol; a sympathomimetic (USP).

epi- (1) Greek prefix indicating "upon," "through," or "toward." (2)*Prefix indicating a bridge or intramolecular connection; as,

R·C·C·C·R (bridged by O)

Cf. *epoxy.* **epi-position** ε- The 1,6-position of naphthalene or other condensed rings.

Epibloc Trademark for biodegradable rat poison that renders the male sterile.

epiborneol See *epiborneol* under *borneol.*

epiboulangerite $Pb_3Sb_2S_8$. A metallic sulfantimonide of lead.

epicamphor $C_{10}H_{16}O$ = 152.2. 1,7,7-Trimethylbicyclo[2.2.1]heptan-3-one*. 3-Camphanone. An isomer of camphor.

epicatechol (2*R*,3*R*)-Catechin.

epichlorite A basic, aluminum-iron silicate of the chlorite group.

epichlorohydrin C_3H_5OCl = 92.51. Chloromethyloxirane†, 1-chloro-2,3-epoxypropane,

CH_2—O—$CH \cdot CH_2Cl$ (ring)

(±)-~ A colorless liquid, d_0 1.203, b.117, insoluble in water, miscible with alcohol or ether. Used as a solvent for resins and nitrocellulose, and in the manufacture of varnishes, lacquers, and cements for celluloid articles. Cf. *epihydrin.*

epicyanohydrin C_3H_5OCN = 83.1. Cyanopropylene oxide. Colorless prisms, m.162, soluble in hot water or alcohol.

epidemiology The scientific study of epidemics; i.e., diseases of populations rather than of individuals.

epidichlorohydrin $H_2C{:}CCl \cdot CH_2Cl$ = 111.0. 2,3-Dichloro-1-propene*. Colorless liquid, b.96, insoluble in water; a solvent.

epididymite $Na_2O \cdot 2BeO \cdot 6SiO_2 \cdot H_2O$. Rhombic crystals, d.3.55.

epidioxide Indicating a substance containing 2 O atoms attached to different C atoms of the same skeleton. Cf. *epoxide.*

epidioxy* The bridge —O·O—.

epidithio* The bridge —S·S—.

epidosite A crystalline schist derived from diabase.

epidote $HCa_2(Al \cdot Fe)_3Si_3O_{16}$. Pistacite. A crystal line schist of gneisses, garnet rock, amphibolite, etc. Green, monoclinic masses, d.3.3–3.5, hardness 6–7.

epidotization The disintegration of feldspar, hornblende, augite, and biotite into epidote.

epidoxation An oxidizing reaction in which an unsaturated alkene link is converted to a cyclic 3-membered ether.

>C:C< → >C—C< (bridged by O)

epiethylin $O \cdot CH_2 \cdot CH \cdot CH_2OC_2H_5$ = 102.1. Glycidyl ethyl ether, 2,3-epoxy-1-ethoxypropane. Colorless liquid, b.124; a solvent for varnishes.

epigenite $Cu_7As_2S_{12}$. A copper sulfoarsenide containing iron.

epiguanine $C_6H_7ON_5$ = 165.2. 2-Amino-6-methyl-8-oxopurine, methylguanine. A purine in the urine, especially during leukemia.

epihydric alcohol 2,3-Epoxy-1-*propanol**.

epihydrin $CH_2 \cdot O \cdot CHMe$ = 58.1 Propylene epoxide. Colorless liquid, d.0.859, m.35, soluble in water.

e. alcohol 2,3-Epoxy-1-*propanol**. **e. carboxylic acid** $C_4H_6O_3$ = 102.1. Colorless crystals, m.225.

epiiodohydrin C_3H_5OI = 184.0. 1,2-Epoxy-3-iodopropane*. Colorless liquid, b 160, insoluble in water; a solvent.

Epikote Tradema.κ for a class of condensation polymers prepared from epichlorohydrin and diphenyloylpropane; used in coatings and varnishes.

Epilim Trademark for sodium valproate.

epimerase* See *enzymes,* Table 30.

epimeric Having the character of epimerides.

epimerides Epimers. Isomers differing only in the arrangement of H and OH on the lowest-numbered asymmetric C atom of a chain; as, D-glucose and D-mannose.
epimers Epimerides.
epimino* Alternative name for the imino bridge group −NH−.
epinephrine $C_6H_3(OH)_2 \cdot CHOH \cdot CH_2NHMe$ = 183.2. (*R*)-1-(3,4-Dihydroxyphenyl)-2-methylaminoethanol. Adrenaline (U.K. usage), suprarenaline, Adrenine. White powder, m.205, slightly soluble in water. A neurotransmitter and hormone produced from the adrenal medula. E. increases glycogenolysis to produce glucose for energy, stimulates the heart, causes a rise in blood pressure and a relaxation of muscle in the trachea and bronchi. Used for severe allergic states and asthma. Often used as the tartrate (USP, BP). Cf. *norepinephrine bitartrate.* **e. bitartrate** $C_9H_{13}O_3N \cdot C_4H_6O_6$ = 333.30. Adrenaline bitartrate, suprarenin. Gray crystals, darkening on exposure, m.149, soluble in water; a sympathomimetic (USP, EP, BP).
epinine $(HO)_2C_6H_3(CH_2)_2NHMe$ = 167.1. 3,4-Dihydroxyphenylethylamine. Colorless crystals; synthetic substitute for epinephrine.
epiphenylin $O \cdot CH_2 \cdot CHCH_2OPh$ = 150.2. Glycidyl phenyl ether, 2,3-epoxy-1-phenoxypropane*. Colorless liquid, b_{22mm}131; a solvent for varnishes.
epiphyte An organism living on, but not feeding on, another organism, e.g., orchid.
epipolic Fluorescent.
epispastic Vesicant.
episperm The membrane between the shell and kernel of a seed.
epistilbite $Ca_2Al_4Si_{11}O_{30}$. Colorless monoclinics, d.2.24–2.36, hardness 3.5–4.
epitaxy The growth of oriented single crystals on the surface of a single crystal; as from the liquid phase (lpe) or vapor phase (vpe).
epithelium Layer(s) of cells that cover the body surface and line body cavities, organs, and glands. See *carcinoma.*
epithio* The bridge −S−.
epithioximino* The bridge −S·O·NH−.
epm Abbreviation for equivalents per million = parts per million per equivalent weight in mg.
e.p.n.s. Abbreviation for electroplate on nickel-silver.
Epontol Trademark for propanidid.
epoxide Oxirane. A product of epoxidation. E. have many uses in plastics (see *epoxy resins*), plasticizers and drugs.
epoxy* Indicating an −O− bridge in a molecule attached to different C atoms, which may or may not be otherwise united. **e.imino*** The bridge −O·NH−. **e.nitrilo*** The bridge −O·N=. **e.propanol** 2,3-Epoxy-1-propanol*. **e. resins** A group of synthetic resins containing the group $(-O \cdot C_6H_4 \cdot CMe_2 \cdot C_6H_4 \cdot O \cdot CH_2 \cdot CH(OH)CH_2-)_n$, where *n* is 0 to 9. They adhere to smooth surfaces, and resist weather and chemicals. **e.thio*** The bridge −O·S−.
EPR Trademark for ethylenepropylene rubber, which has the special qualities of resistance to heat and oxidation.
epsilon ϵ, E. Greek letter. ϵ (italic) is symbol for (1) emissivity; (2) epi-position; (3) linear strain; (4) molar (linear) absorption coefficient; (5) permittivity. **ϵ-acid** 1-Naphthylamine-3,8-disulfonic acid.
epsomite $MgSO_4$. Colorless, rhombic prisms, d.1.7, hardness 2.
Epsom salts Magnesium sulfate.
EPTC* See *herbicides,* Table 42 on p. 281.
epuration The purification of sugary liquors by defecation, etc.
epuré A bituminous constructional mixture from Trinidad.
eqn Abbreviation for equation.
Equanil Trademark for meprobamate.
equation (1) A symbolical expression of equality. (2) An expression of a chemical reaction. The formulas of the reacting substances are placed on the left, those of the reaction products on the right, with the equality sign between. This indicates that the total number of atoms of each kind on each side balance. The molecular equation $2NaOH + H_2SO_4 = Na_2SO_4 + 2H_2O$ states that 2 molecules of sodium hydroxide and one molecule of sulfuric acid give one molecule of sodium sulfate and 2 molecules of water. **analytical ~** An e. showing the decomposition of a compound into simpler constituents. See *analysis.* **balanced ~** An e. where the 2 sides are in some state of equilibrium. Indicated by ⇌ rather than =. **ionic ~** An e. expressed in terms of ions as, $2OH^- + 2H^+ = 2H_2O$. **metathetical ~** An e. expressing a metathesis. **molecular ~** An e. expressed in terms of molecules. Many molecular equations can be expressed in ionic form (see above). **numeric ~** A mathematical e., each term of which is dimensionless. **transmutation ~** An e. which indicates the disintegration or synthesis of elements. Thus, ${}^1_1H + {}^{11}_5B \rightarrow 3{}^4_2He$ means that element no. 5 (boron, at. wt. 11) is bombarded by a proton (element no. 1, hydrogen, weight 1) and forms 3 atoms of element 2 (helium, weight 4). The sum of the atomic numbers (prefixed subscripts), and also the sum of the mass numbers (prefixed superscripts), is the same on both sides.
e. of state A group of mathematical expressions derived from the equation $pV = RT$, which define the physical conditions of a homogeneous liquid or gaseous system by relating concentration or volume, pressure, and thermodynamic temperature for a given mass of substance. See *Ramsay-Young equation, van der Waals' equation, Dieterici's rule,* and *Clausius equation.*
equilibrium A condition in which contending forces are balanced. Cf. *triple point.* **chemical ~** The balanced state reached when chemical reaction apparently stops; decomposition and recombination proceed with equal speed. See *mass action.* **disturbed ~** The result of the removal of one or more reaction products or reacting substances from a chemical e., causing it to shift. **heterogeneous ~** A chemical e. between 2 (or more) phases. **homogeneous ~** A chemical e. in a single phase. **invariant ~** An e. in which the quantity of one component approaches zero. **ionic ~** An e. that involves a balanced condition of ions. **kinetic ~** The balanced state of 2 opposite reactions. Cf. *static equilibrium* (below). **molecular ~** An e. in which the components are molecules. **monophase ~** Homogeneous e. **polyphase ~** Heterogeneous e. **stable ~** A mobile condition which, after a casual displacement, is again restored. **static ~** The equilibrium attained when all reaction ceases.
e. constant Dissociation constant.
equilin $C_{18}H_{20}O_2$ = 268.4. 3-Hydroxy-17-oxoestratetraene. An estrogenic hormone produced from pregnant mares' urine.
equipartition (1) The orderly arrangement of atoms, e.g., in a crystal. (2) The condition of molecules in a gas, where the molecules keep the same average distance apart under the same pressure. **e. of energy** The total energy of a molecule is divided equally among its different degrees of freedom.
equisetic acid 1,2,3-Propenetricarboxylic acid*.
Equisetum A genus of herbaceous, spore-producing plants (horsetail, scouring rush, bottlebrush) which are rich in silica.
equiv. Abbreviation for equivalent.
equivalence The relative combining powers of a set of atoms or radicals. **e. point** Stoichiometric point. The point of a

titration where the amounts of titrant and substance being titrated are chemically equivalent.

equivalent (1) The weight in grams of an element that combines with or displaces 1 g of hydrogen. (2) The weight of a substance contained in one liter of a normal solution: determined by (*a*) dividing the atomic weight of an element by its valency; (*b*) dividing the molecular weight of a compound by the valency of its principal atom or radical; (*c*) calculating the quantity of a substance that combines with 1 g of hydrogen or with 8 g of oxygen. (3) Having the same valency. **electrochemical ~** See *electrochemical equivalent.* **gram ~** Equivalent weight. **hard coal ~** See *fuel equivalence.* **milli ~** The weight of a substance, in g, contained in 1 ml of a normal solution. **oil ~** See *fuel equivalence.* **toxic ~** See *toxic equivalent.*

e. charge The amount of electricity in a gram e. of substance, i.e., 6.022×10^{23} electron charges or 96,500 C. Cf. *faraday.* **e. conductivity** See *conductivity.* **e. weight** Gram equivalent; the equivalent of a substance in grams, calculated by dividing its molecular weight by its effective valency. With *acids,* the number of replaceable H atoms; with *bases,* the number of OH groups.

equiviscous temperature An expression of viscosity in terms of the temperature at which a substance has a standard viscosity.

Er Symbol for erbium.

era See *geologic eras.*

erbia Erbium oxide*.

erbium Er = 167.26; at. no. 68. A rare-earth metal, discovered in 1879 by Cleve in the erbia of Marignac (1878) and Berlin (1860). (The erbia discovered and named in 1843 by Mosander was actually terbium.) A metallic substance, d.4.77, m.1530, b.2880, insoluble in water, soluble in acids; trivalent and forms red salts. **e. acetate*** $Er(C_2H_3O_2)_3 \cdot 4H_2O$ = 416.5. Pink, triclinic crystals, soluble in water. **e. nitrate*** $Er(NO_3)_3 \cdot 6H_2O$ = 461.4. Pink crystals, soluble in water. **e. oxalate** $Er_2(C_2O_4)_3 \cdot 10H_2O$ = 778.7. Red powder, decomp. 575. **e. oxide*** Er_2O_3 = 382.5. Erbia. Orange powder, insoluble in water. At high temperature it glows green. **e. sulfate*** $Er_2(SO_4)_3 \cdot 8H_2O$ = 766.8. Red crystals, soluble in water.

Erdmann E., H. (1862–1910). German chemist noted for analytical methods. **E., Otto Linneé (1804–1869)** German chemist noted for atomic weight determinations. **E. float** A small, partly filled, sealed glass tube for obtaining accurate buret readings. **E. reagent** (1) Concentrated sulfuric acid containing 2 drops of conc. nitric acid per 100 ml; a reagent for alkaloids. (2) A solution of 8-amino-1-naphthol-4,6-disulfonic acid gives a red color with nitrates.

erdschreiber A culture medium made by autoclaving earth with solutions of sodium nitrate and sodium dihydrogenphosphate.

eremophilone $C_{15}H_{22}O$ = 218.3. 1(10),11-Eremophiladien-9-one. m.42, in the wood oil of *Eremophila mitchelli* (Myoporaceae), Australia.

erg The cgs unit of work or energy: the work necessary to overcome the resistance of one dyne acting through one centimeter.

$$1 \text{ joule} = 10^7 \text{ ergs}$$
$$1 \text{ foot-poundal} = 4.214 \times 10^5 \text{ ergs}$$
$$1 \text{ calorie } (20^\circ C) = 4.182 \times 10^7 \text{ ergs}$$

ergamine Histamine.

ergine (1) $C_{16}H_{17}N_3O$ = 267.3. The amide of lysergic acid. m.135 (decomp.). (2) Ergone. A categorical name for substances which exert biological catalytic effects in small quantities, e.g., vitamins.

ergobasine Ergonovine.

ergocalciferol $C_{28}H_{44}O$ = 396.7. Calciferol, vitamin D_2. White crystals, m.117, insoluble in water. Formed by irradiating ergosterol with u.v. light. Used to treat vitamin D deficiency (USP, EP, BP). See *vitamin,* Table 101 on p. 622.

ergochromes Dimeric xanthenes derived from secalonic acids, as, ergochrysin.

ergochrysin $C_{28}H_{28}O_{12}$ = 556.5. Ergochrome AC. Yellow pigment, m.285, from ergot.

ergodic Having the same average value of a specific property, whether derived from a large number of small samples or small number of large samples.

ergoflavin $C_{15}H_{14}O_7$ = 306.3. Ergochrome CC. Yellow pigment, m.350 (decomp.), from ergot.

ergometrine EP, BP name of ergonovine.

ergon The energy quantum of an oscillator, equal to the product of the Planck constant and frequency.

ergonomics The scientific study of the relationship between man and his environment; the science of fitting the occupation to the worker.

ergonovine $C_{19}H_{23}O_2N_3$ = 325.4. Ergometrine, ergobasine. An alkaloid from ergot. Colorless crystals, m.159, soluble in water with blue fluorescence. Important for controlling hemorrhage in obstetrics (USP, EP, BP).

ergosinine $C_{30}H_{35}N_5O_5$ = 545.6. An alkaloid, decomp. 228, from ergot.

ergosterin Ergosterol.

ergosterol $C_{28}H_{44}O$ = 396.7. Ergosterin. Platelets, m.168. An inert alcohol occurring in ergot, yeasts, and bacteria; the provitamin of vitamin D_2. See *sterols.* **irradiated ~** Ergocalciferol.

ergot Ergota, spurred rye, *Secale cornutum.* The dried sclerotium of *Claviceps purpurea* (Hypocreaceae), a fungus on numerous grasses, especially replacing the grain of rye, *Secale cereale* (Gramineae). Hard, black, shining, elongated mycelia. Its active principles include ergometrine, ergotoxine, histamine, tyramine, and acetylcholine. Inert constituents are ergotinine, ergothioneine, secalonic acid, schlererythrin, clavicepsin, ergosterol, fungisterol, guanosine, amino acids, and secale aminosulfonic acid.

ergotamine tartrate $(C_{33}H_{35}O_5N_5)_2C_4H_6O_6$ = 1313. Gynergen. Colorless crystals, decomp. 180, soluble in water; from ergot. Increases blood pressure; used to treat migraine (USP, EP, BP).

ergothioneine $C_9H_{15}O_2N_3S \cdot 2H_2O$ = 265.3. See *ergot.*

ergotine A brown extract of ergot.

ergotinine $C_{35}H_{39}O_5N_5$ = 609.6. The anhydride of ergotoxine and an alkaloid of ergot. Yellow needles, darkening in air, m.219–229, insoluble in water.

Ergotrate Trademark for ergonovine maleate.

Ericaceae Heather family; shrubs or trees with bell-shaped flowers. Many contain glucosides or drugs; e.g.: *Arctostaphylos uva ursi,* bearberry; *Ledum latifolium,* Labrador tea.

ercinol $C_{10}H_{16}O$ = 152.2. An alcohol of *Ledum palustris,* wild rosemary (Ericaceae). Colorless oil with a peculiar odor; darkens and yellows on exposure.

ericolin $C_{35}H_{56}O_{21}$ = 812.8. A glucoside from *Ledum palustris,* wild rosemary (Ericaceae). A brown, soft mass, soluble in water.

erigeron Fleabane, colt's-tail, prideweed. The herb or seeds of *E. canadense* (Compositae); an astringent and tonic. **e. oil** The essential oil of e., d.0.845–0.865, containing citrene.

erilite A plastic made from casein and formaldehyde. Cf. *Galalith.*

erinite $Cu_3(AsO_4)_3 \cdot 2Cu(OH)_2$. An emerald green, native copper arsenate; it is a decomposition product of enargite.

eriochrome Omega chrome. One of a class of dyestuffs used as indicators in titration with *EDTA,* q.v.

eriodictin A flavone dyestuff present with hesperidin, in citrin.
eriodictyol A chalcone of phloroglucinol occurring as glucoside in eriodictyon.
eriodictyon Mountain balsam. Yerba santa. The dried leaves of *Eriodictyon californicum* (Hydrophyllaceae); an aromatic and expectorant. **e. oil** The essential oil of e., d.0.937.
eriophyesin $C_{23}H_{18}O_{15}$ = 534.4. A glucosidal coloring matter of nail gall from *Eriophyes macrotrichus.* Brown needles, m.259.
Erlangen blue Prussian blue.
Erlenmeyer, R. A. K. E. (1825–1909) German chemist. **E. flask** or **erlenmeyer** A conical glass flask with flat bottom.
ergone Ergine (2).
erose Describing an irregularly toothed edge, as with certain bacterial growths.
erosion Denudation. The wearing away, especially of rocks, due to weather or chemicals, e.g., alkaline waters. Cf. *corrosion, surrosion.*
errhine A drug which increases nasal secretions. Cf. *sternutatory.*
error (1) Deviation from the truth. (2) The principle that the accuracy of a determination depends on the accuracy of the data on which it is based. **cavitation ~** Damage characterized by pitting in hydraulic systems under extreme conditions of high-speed liquid flow; believed to be due to the collapse of vapor-filled cavities. **protein ~** See *protein error.* **standard ~** Standard deviation.
e. curve Normal probability curve.
erucic acid $C_{22}H_{42}O_2$ = 338.6. (Z)-13-Docosenoic acid*, found in the oil of grape seeds. Colorless needles, m.34, insoluble in water. **iso ~** Brassidic acid.
erucidic acid Brassidic acid.
eryodiction Eriodictyon.
erythorbic acid $C_6H_8O_6$ = 176.1. D-Araboascorbic acid, iso*ascorbic acid,* q.v., m.168 (decomp.). An antioxidant, preventing beer haze.
erythrene Butadiene*.
erythric acid (1) $CH_2OH(CHOH)_2COOH$ = 136.1. (+)-Trihydroxybutanoic acid. (2) Erythrin. (3) Alloxan.
erythrin (1) $C_{20}H_{22}O_{10}$ = 422.4. Erythritol ester of lecanoric acid. Brown needles, m.165, sparingly soluble in water, found in *Roccella tinctoria.* (2) A native cobaltous arsenate.
erythrina (1) The coral tree (Papilionaceae), W. Indies, Central America, and southeast Asia. (2) A genus of Leguminosae.
erythrite (1) Red cobalt. A native cobalt arsenate. (2) Erythritol.
erythritic acid Erythric acid.
erythritol $H(CHOH)_4H$ = 122.1. (*R**, *S**)-1,2,3,4-butanetetrol†, erythrite, phycite, lichen sugar, from the lichen *Protococcus vulgaris.* Tetragonal prisms, m. 112, soluble in water. **penta ~** See *pentaerythritol.*
e. benzene Fuchsin. **e. tetraacetate** $C_{12}H_{18}O_8$ = 290.3. White crystals; a plasticizer for cellulose acetate, to increase u.v. light transmission. **e. tetranitrate** Erythrityl tetranitrate.
erythrityl tetranitrate $C_4H_6(NO_3)_4$ = 302.1. Tetranitrol, Cardilate. Yellow crystals, m.61 (explode), slightly soluble in water. A vasodilator, used in prophylaxis of angina (USP).
erythro- (1) Prefix meaning "red." (2) See *threo-.*
erythrocentaurin $C_{17}H_{24}O_5$ = 356.4. The red coloring matter of *Erythraea centaurium* and *Sabbatia angularis.* Needles, darkening in light, m.136, insoluble in water.
erythrocephaelin The coloring matter of ipecac.
erythrocruorin Hemoglobin from invertebrates. The hemoglobin of vertebrates has a higher molecular weight and lower isoelectric point.
erythrocyte A red blood cell, its color being due to hemoglobin.
erythrodextrin A dextrin split product of starch, giving a red color with iodine. Cf. *achroodextrins.*
erythroglucin Erythrol (2).
erythrol (1) Erythritol. (2) $CH_2OH \cdot CHOH \cdot CH:CH_2$ = 88.1. 3-Butene-1,2-diol*. **(±)-~** White crystals, m.197.
erythrolaccin The alkali-soluble coloring matter of shellac.
erythrolitmin $C_{26}H_{23}O_{13}$ = 543.5. Red crystals derived from *Corona solis.*
erythromycin $C_{37}H_{67}O_{13}N$ = 733.9. Ilotycin, Ilosone. A broad-sprectrum antibiotic from *Streptomyces erythreus,* Waksman. White, bitter, hygroscopic crystals, insoluble in water. Used particularly when penicillin allergy is present. Reported effective for legionnaire's disease (USP, EP, BP).
erythronium The name given by del Rio (1804) to vanadium; rediscovered by Sefström (1831).
erythrooxyanthraquinone $C_{14}H_8O_3$ = 224.2. 1-Hydroxy-9,10-anthraquinone*. Orange needles, m.195, soluble in alcohol.
erythrophloeine An alkaloid from the bark of *Erythrophloeum guinense* (Leguminosae), Guinea; an ordeal poison. Yellow syrup, sparingly soluble in water. Cf. *sassy bark.*
Erythrophloeum Leguminous trees (W. Africa) furnishing sassy bark; an ordeal poison.
erythroprotid $C_{13}H_8O_5N$ = 258.2. A red product obtained by boiling proteins with concentrated alkali.
erythroquine reaction A characteristic color is formed on adding to an alkaloid, in succession, chloroform, bromine water, and potassium hexacyanoferrate(II).
erythroretin $C_{38}H_{36}O_{14}$ = 716.7. A red coloring matter from rhubarb.
erythrose $CHO \cdot (CHOH)_2 \cdot CH_2OH$ = 120.1. Trihydroxybutyraldehyde. A tetrose sugar derived from erythrol, in dextro and levo forms. Isomer of threose. See *threo-.*
erythrosiderite K_2FeCl_5. A rare potassium-ferric chloride mineral.
erythrosin (1) $C_{13}H_{18}O_6N_2$ = 298.3. A red compound obtained by treating tyrosine with nitric acid. (2) $C_{20}H_8O_5I_4$ = 835.9. Tetraiodofluorescein, pyrosin. Yellow crystals, insoluble in water; a dye. (3) $C_{20}H_6O_5I_4Na_2$ = 879.8. Eosin. Bluish, sodium iodoeosin, sodium tetraiodofluorescein. Brown powder, soluble in water; a dye and hydrogen ion indicator, changing at pH 2.0 from orange (acid) to magenta (basic). Cf. *iodeosin.*
erythroxyline Cocaine.
erythroxylon Coca.
erythrytol Erythritol.
Esbach, George Hubert (1843–1890) French physician. **E. reagent** A solution of 1 g trinitrophenol and 2 g citric acid in 100 mL water, for determining albumin in urine.
esca, ESCA Electron spectroscopy for chemical analysis.
escalation A mathematical procedure used in the costing of chemical plant and production costs, which allows for increases in the cost of plant, raw materials, labor, energy, etc.
escharotic A substance that produces a crust of dead tissue.
Eschka mixture A mixture of 2 pts. magnesium oxide and 1 pt. dried sodium carbonate; a fusion reagent for determining sulfur in coal.
eschwegite $5Y_2O_3 \cdot 5(Ta, Nb)_2O_5 \cdot 10TiO_2 \cdot 7H_2O$. A reddish-gray, isotropic rare-earth mineral.
esciorcin, aesciorcin Escorcin.
escorcin $C_9H_8O_4$ = 180.2. Escorcinal, aescorcin. Brown powder derived from esculetin; used in ophthalmology to stain the cornea temporarily to detect ulceration.

esculetin $C_9H_6O_4$ = 178.1. Aesculetin, 6,7-dihydroxycoumarin. m.270; soluble in alkalies. **methoxy ~** Gelseminic acid. **methyl ~** Scopoletin.
esculetinic acid $C_9H_8O_5$ = 196.2. Aesculetinic acid. Colorless crystals, m.168.
esenbeckic acid An acid from the bark of *Esenbeckia febrifuga* (Rutaceae), tropical America.
eserine Physostigmine.
Eskornade Trademark for phenylpropanolamine hydrochloride.
Esmarch bottle A device for sampling water at any depth by removing and replacing a stopper mechanically.
esoxin A ptomaine from the sperm of white, freshwater fish.
esparto A tall grass, *Stipa tenacissima* or *Lygeum spartum* (Gramineae), N. Africa; a source of paper pulp.
esperium Early name for plutonium.
e.s.r. Electron spin resonance.
essence (1) A solution of a volatile or essential oil in alcohol. (2) The active principle of a plant. (3) A *fruit essence,* q.v. **e. niobe** Methyl benzoate.
essential Pertaining to an essence. **e. oils** Volatile ethereal oils of characteristic odors, distilled from plants. Distinguished from fatty oils by their volatility, nongreasiness, and nonsaponifying property. They exist in plants, as such, imparting the characteristic odor to flowers, leaves, or woods; as terpenes (oil of turpentine, juniper, etc.); or are developed from plant constituents by enzyme action (oil of mustard) or heat (cade). They are flammable, solube in alcohol or ether, slightly soluble in water, and can contain hydrocarbons, alcohols, phenols, ethers, aldehydes, ketones, acids, and esters. **e. salt** A salt obtained by evaporating plant juices.
essonite $6CaO \cdot 3SiO_2 + 2Al_2O_3 \cdot 3SiO_2$. Garnet lime-alumina, wolfsbergite. Calcium garnet. A pale-yellow mineral.
ester (1)* Ethereal salt. An organic salt formed from an alcohol (base) and an organic acid by elimination of water:

$$\underset{\text{alcohol}}{R \cdot OH} + \underset{\text{acid}}{R' \cdot COOH} = \underset{\text{ester}}{R'COO \cdot R} + \underset{\text{water}}{H_2O}$$

(2) Compounds of the type R_2SO_4 or $R \cdot COOM$ are erroneously called esters. (3) The ethyl esters of an acid. **acid ~** Ether acid. Compounds of polybasic acids in which not all the carboxyl hydrogen atoms are replaced by alcohol radicals, e.g., $R \cdot OOC \cdot R \cdot COOH$. **basic ~** Compounds of polyhydric alcohols in which not all the hydroxyl groups are replaced by acid radicals, e.g., $R \cdot COO \cdot R \cdot OH$. **imido ~** See *imido ester.* **mixed ~** An e. in which the radicals R of $R \cdot COOR$ are different. **rosin ~** Ester gum.
e. gum Rosin ester. A compound obtained by esterifying rosin, with a polyhydric alcohol, e.g., glycerol. **e. number** The difference between the saponification number and acid number, i.e., the mg of potassium hydroxide to saponify the neutral esters in 1 g of a fat, oil, or wax. **e. oils** Pale, edible oils produced by esterification of refinery acids with glycerol.
esterase* An enzyme acting on esters.
esterification Formation of an ester by dehydration or catalytic agents. **inter ~** The controlled rearrangement of fatty acids between the glycerol molecules of a fat, e.g., by heating at 45–95°C with a Na-K alloy. One use is to improve the creaming properties of lard.
e. law Meyer's law. Esterification of aromatic acids is retarded or prevented, respectively, if one or both *ortho*-positions are substituted.
estersil An ester of $Si(OH)_4$.
estimation An approximate evaluation.
estradiol $C_{18}H_{24}O_2$ = 272.4. Estra-1,3,5(10)-triene-3,17 β-diol, Oestradiol, Progynon. A female sex hormone produced by the ovary; used for ovarian failure and hormonal replacement. Cf. *estrone.*
estradiovalerate $C_{23}H_{32}O_3$ = 356.5. Estradiol 17-valerate, Delestrogen, Progynova. White crystals, m.147, insoluble in water; an estrogen (USP).
estragole $MeO \cdot C_6H_4 \cdot C_3H_7$ = 150.2. Methyl chavicol, *p*-allylanisole, *p*-methoxyallyl benzene. An ether of anise odor, b.216, in estragon oil.
estragon oil Tarragon oil. An essential oil, d.0.9–0.94, from *Artemisia dracunulus* (Compositae).
estrane $C_{18}H_{17}$ = 233.3. Oestrane, 2-methyl-1,2-cyclopentanoperhydrophenanthrene. A tetracyclic hydrocarbon related to the steroids. 5β form:

Me
H H
H H
H

e.diol $C_{18}H_{30}O_2$ = 278.4. 3,17-Dihydroxyestrane. Cf. *estrone.*
e.triol $C_{18}H_{30}O_3$ = 294.4. 3,16,17-Trihydroxyestrane.
estratriene $C_{18}H_{24}$ = 240.2. $\Delta^{1,3,5}$-estratriene. The parent compound of estrone. Cf. *androstane.*
estrin Estrone.
estriol $C_{18}H_{24}O_3$ = 288.4. Oestriol, Theelol, trihydroxyestrin. A trihydric alcohol from pregnancy urine. White needles, m.218, insoluble in water.
estrogen A substance, either a natural hormone or synthetic, that stimulates secondary sexual characteristics of the human adult female, causing changes in the uterus during the menstrual cycle and inhibiting production of gonadotrophin. In the human male, e. causes atrophy of sexual organs and feminization. In animals, estrus is produced, this being a period of extreme sexual excitement culminating in ovulation.
estrone $C_{18}H_{22}O_2$ = 270.4. Oestrone, Theelin, thelykinine, oestrane, Progynon, Menformon. A female sex hormone from the placenta and urine of pregnant women; produced by ovarian follicles. A monohydric keto alcohol. White, monoclinic crystals, m.260, insoluble in water.
estrus See *estrogen.*
esu, e.s.u. Abbreviation for *electrostatic units,* q.v.
Et* Abbreviation for the ethyl radical, $-C_2H_5$.
eta η. Greek letter. η (italic) is symbol for (1) dynamic viscosity. (2)* Hapto*. In chemical names, indicating that 2 or more contiguous atoms of a group are attached to the central atom (e.g., metal) of a coordination compound. (3) Overpotential.
etacrynic acid Ethacrynic acid.
Étard's reaction The oxidation of methyl to aldehyde radicals by chromyl chloride: $R \cdot CH_3 + Cr(OCl)_2 = R \cdot CHO + H_2O + CrCl_2$. **E. salt** The complex potassium trisulfatochromate(III), $K_3[Cr(SO_4)_3]$.
etch figure The characteristic pattern produced when a polished metallic surface is etched by a suitable reagent.
eteline See chlorinated *solvents.*
ethacetic acid Butanoic acid*.
ethacrynic acid $C_{13}H_{12}O_4Cl_2$ = 303.1. [2,3-Dichloro-4-(2-methylenebutyryl)phenoxy]acetic acid, etacrynic acid, Edecrin. White crystals, slightly soluble in water. A quick-acting diuretic (USP, EP, BP).
ethal Hexadecyl alcohol*.

ethaldehyde Hexadecanal*.
ethalic acid Palmitic acid*.
ethambutol hydrochloride $C_{10}H_{24}O_2N_2 \cdot 2HCl$ = 277.2. Myambutol. White crystals, m.201, soluble in water. An antituberculous drug used with others to prevent development of resistant strains of bacilli (BP).
ethamine *Ethyl*amine*.
ethamolin Ethanolamine.
ethanal* Acetaldehyde*. **hydroxy ~ *** Glycolaldehyde*. **trichloro ~ *** Chloral. **e. acid** Glyoxalic acid.
ethanamide Acetamide*.
ethanamidine Acetamidine*.
ethane* C_2H_6 = 30.07. Methylmethane, dimethyl, ethylhydride. An alkane. Colorless gas, $d_{air=1}$1.049, m.−172, b.−86, slightly soluble in water; a constituent of natural gas. The ethyl (C_2H_5−), ethylene (−$CH_2 \cdot CH_2$−), ethylidyne (≡C·Me), and ethanediylidene (=CH·CH=) radicals are derived from ethane. Unsaturated hydrocarbon derivatives include ethene* (ethylene*), CH_2:CH_2, and ethyne* (acetylene*), CH:CH. **amino ~** *Ethyl*amine*. **bromo ~ ***, **chloro ~ ***, etc. See *ethyl bromide, ethyl chloride,* etc. **dibenzyl ~** See *dibenzylethane.* **dibromo ~ ***, **dichloro ~ ***, etc. See *ethylene dibromide, ethylene dichloride,* etc. **dihydroxy ~** Glycol*. **diphenyl ~** Dibenzylethane. **hexabromo ~ *** See *hexabromoethane.* **hexachloro ~ *** See *hexachloroethane.* **hydroxy ~** Ethanol*. **nitro ~** See *nitroethane.* **nitroso ~** See *nitrosoethane.* **perchloro ~** Hexachloro e. **phenyl ~** Ethyl*benzene.* **triethoxy ~ *** Ethylidyne triethyl ether.
e.dial* Glyoxal*. **e.diamide** Oxamide*. **e.diamine*** Ethylenediamine*. **e.dinitrile*** Cyanogen. **e.dioic acid*** Oxalic acid*. **e.diol*** Glycol*. **e.dioyl** The oxalyl* radical. **e.dioyl chloride** Oxalyl dichloride*. **e.dithiol*** Dithioglycol. **1,2-e.diyl†** See *ethylene (2).* **e.nitrile*** Acetonitrile*. **e.sulfinic acid*** Ethyl*sulfinic acid.* **e.sulfonic acid*** Ethyl*sulfonic acid.* **e.thial*** Sulfaldehyde. **e.thiol*** EtSH = 62.1. Ethyl mercaptan, ethyl hydrosulfide*. Colorless liquid, b.37, slightly soluble in water; an odorous constituent of feces. **e. thioamide** Thioacetamide*. **e.thiolic acid** Thioacetic acid*.
ethanediylidene* The radical =CH·CH=.
ethanite A synthetic rubber, prepared by subjecting natural gas to a heating and cooling process.
ethano-* Prefix indicating a −$CH_2 \cdot CH_2$− bridge.
Ethanograph See *breath alcohol.*
ethanoic acid* Acetic acid*. **dihydroxy ~** Glyoxylic acid*. **e.a. anhydride** Acetic anhydride*.
ethanol* C_2H_6O = 46.07. Ethyl alcohol*, alcohol, spirit, spirit of wine, grain alcohol, absolute alcohol, ethyl hydrate, etc. (1) Et·OH. Absolute alcohol, dehydrated alcohol. Colorless liquid, d_4^{25}0.78505, m.−117.3, b.78.3, miscible with water or ether; a reagent and solvent. (2) 99% alcohol and lower concentrations. Used extensively for tinctures and pharmaceutical preparations, as a solvent and preservative, as an antiseptic, and in perfumery (USP, BP). (3) Grain alcohol, cologne spirits. Colorless liquid (ethanol 90, water 10%). (4) Diluted alcohol, proof spirit. Colorless liquid, ethanol about 49, water 51% (by weight). (5) Denatured alcohol. Alcohol made unpotable by the addition of substances such as methanol, pyridine, formaldehyde, or other denaturant. Used in industry, the arts and commerce, principally as a solvent or fuel. See also *methylated spirit.* **amino ~ *** Colamine. **butoxy ~ *** Butyl *Cellosolve.* **chloro ~ *** Ethylene chlorohydrin. **cyano ~** Ethylene cyanohydrin. **ethoxy ~ *** Cellosolve. **imino ~** See *iminoethanol.* **oxybis ~** Diethylene glycol. **phenyl ~** See *phenylethanol.* **tribromo ~ *** See *tribromoethanol.* **trichloro ~ *** See *trichloroethanol.* **trimethyl ~** *tert*-Butyl methanol.
ethanolamine $NH_2(CH_2)_2 \cdot OH$ = 61.1. 2-Aminoethanol*, β-hydroxyethylamine, monoethanolamine. Colorless liquid, d.1.04, b_{150mm}171, soluble in gasoline; used for injections and sclerosing (BP), and for dry cleaning. **di ~** See *diethanolamine.* **tri ~** See *triethanolamine.*
ethanolate* Ethylate, ethoxide. A compound derived from ethanol by replacing the OH group hydrogen by a monovalent metal (M); as, MOEt.
ethanoyl The acetyl* radical.
ethene* C_2H_4 = 28.05. (1)* *Ethylene,* q.v. olefiant gas. Colorless, flammable gas of peculiar odor, $d_{air=1}$0.978, b.−103, slightly soluble in water. From ethene the radicals vinyl, −CH:CH_2, and vinylene, −CH:CH−, are derived. An important intermediate for polyethylene, polystyrene, PVC, SBR, and polyester. (2) The radical ethylene*. **1,2-e.diyl†** See *vinylene.* **e. series** Alkenes*, olefins. The homologs of ethene; a group of *aliphatic hydrocarbons,* q.v., C_nH_{2n}.
ethenium The organic cation $MeCH_2^+$.
etheno-* Prefix indicating a −CH:CH− bridge. Cf. *ethano-.*
ethenol* Vinyl alcohol*.
ethenone* Ketene*.
ethenyl (1) The ethylidyne* radical. (2)† The vinyl* radical.
ethenylidene† The vinylidene* radical, =CH:CH_2.
ether (1)* See *ethers.* (2)* $C_4H_{10}O$ = 74.1. Ethylic ether, ethyl oxide, ethoxyethane, diethyl ether*, sulfuric ether. Colorless liquid, d.0.720, b.35, slightly soluble in water, miscible with alcohol. A reagent and solvent for fats, resins, alkaloids, and an anesthetic (USP, BP). (3) Physics: (A)ether. A hypothetical, all-pervading medium of the universe; once believed the source of radiation, light, heat, and electricity. Cf. *etheron.* **acetic ~** Ethyl acetate*. **aldehyde ~** Croton aldehyde*. **allyl ethyl ~ *** $EtOCH_2CH$:CH_2 = 86.1. 3-Ethoxypropylene. Colorless liquid, b.66, insoluble in water. **anesthetic ~** Ether (2). **anhydrous ~** Ethyl ether which has been distilled over sodium; a reagent and solvent. **butyl ethyl ~ *** BuOEt = 102.2. Colorless liquid, b.92, insoluble in water. **decachloro ~** Perchloro ether. **dichloro ~** Dichloroethyl ether. **diethyl ~ *** Ether (2). **dihexadecyl ~ *** $(C_{16}H_{33})_2O$ = 466.9. Cetyl ether. White leaflets, m.55, soluble in water. **dimethyl ~ *** See *methyl ether* under *methyl.* **ethyl heptyl ~ *** $C_7H_{15}OEt$ = 144.3. Oenanthic ether. Colorless oil, d.0.840, b.143, insoluble in water; used in flavoring extracts and in organic synthesis. **formic ~** Ethyl formate*. **hydrobromic ~** Ethyl bromide*. **hydrochloric ~** Ethyl chloride*. **hydrocyanic ~** Ethyl cyanide*. **hydroiodic ~** Ethyl iodide*. **isopropyl ~** See *propyl ether.* **methyl ~** See *methyl ether.* **methyl *n*-naphthyl ~ *** $MeOC_{10}H_7$ = 158.2. Methoxynaphthalene. **~ 1- naphthyl ~** b.265. **~ 2- naphthyl ~** Jara-jara. m.72; used in perfumes. **methyl pentyl ~ *** $MeOC_5H_{11}$ = 102.2. b. 92. **petroleum ~** See *petroleum ether.* **sulfuric ~** Ether (2).
e. acid An acid *ester,* q.v. **e. alcohol** A compound of the type R·O·R·OH; as, the diether of a dihydric alcohol. **e. of crystallization** The molecules of e. as a component part in a crystal lattice.
ethereal Resembling or made with ether. **e. fruit oil** See *ethereal fruit oil* under *oil.* **e. liquid** A highly volatile liquid. **e. oil** Essential oil. **e. salt** Ester.
etherene Ethylene*.
etheric acid Acetoacetic acid*.
etheride A compound containing the radical −COX; X is a halogen.
etherification The process of making an ether from an alcohol. Cf. *ethers.*

etherin Ethylin. **e. theory** A theory of the constitution of organic compounds (Dumas and Boullay, 1828).

etherion A supposed element, at. wt. 0.001, expelled from substances at high temperatures and low pressures.

etheron Aetheron. A supposed particle of the ether, smaller and faster than an electron; a mass of $\frac{1}{17} \times 10^9$ that of hydrogen, speed 473,000 km/s.

etherophosphoric acid Ethyl phosphate.

etherosulfuric acid Ethyl hydrogensulfate*.

ethers (1)* Compounds of general formula R-O-R. Indicated by the name ether or by the infixes -oxy- or -oxa-. (2) The halogen derivatives of alkyl and aryl radicals, as R·Cl, and the esters of inorganic or organic acids, as $R \cdot NO_2$, are both sometimes incorrectly called ethers. **complex ~, compound ~** (1) Esters*. (2) Mixed e. **crown ~** Polyethers, of a shape resembling a crown. Form complexes with metal ions, as: **15-crown-5-~** $(CH_2 \cdot CH_2 \cdot O)_5$ = 220.3. $b_2$116. **18-crown-6-~** $(CH_2 \cdot CH_2 \cdot O)_6$ = 264.3. m.39.

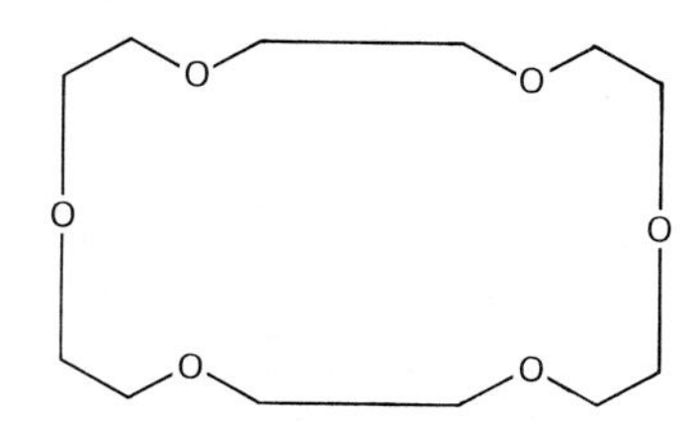

cyclic ~ E. in which the initial C atom in the series is linked directly to the oxide O; as, ethylene oxide, $CH_2 \cdot CH_2 \cdot O$.

haloid ~ Alkyl or aryl halides. **mixed ~** Compound. Alkyl or aryl ethers with 2 different radicals. **simple ~** Alkyl or aryl ethers having 2 like radicals, R·O·R. **thio ~** Alkyl or aryl sulfides in which the e. oxygen is replaced by sulfur. Cf. *thiols.*

ethide A compound of the ethyl radical and a metal; as, diethylplumbane* (lead ethide), Et_2Pb.

ethidine The ethylidene* radical.

ethine Acetylene*. **e. series** Alkynes*.

ethinyl The ethynyl* radical. **e. estradiol** $C_{20}H_{24}O_2$ = 296.4. Feminone, Lynoral. White crystals, m.144, insoluble in water. An estrogen component of many oral contraceptives. Used to treat menopausal symptoms and other conditions due to estrogen lack (USP, EP, BP).

ethiodized oil Iodized oil, Lipiodol. A sterile iodine addition product of vegetable oils, usually made by treating poppyseed oil with hydriodic acid (38–42% of organically combined iodine). A radiopaque medium (USP, BP).

ethionamide $C_8H_{10}N_2S$ = 166.2. 2-Ethylthioisonicotinamide. Yellow crystals, m.163, insoluble in water. Antituberculous agent, used when bacteria are resistant to drugs.

ethionic acid $HO \cdot SO_2 \cdot CH_2 \cdot CH_2 \cdot SO_2OH$ = 109.2. Ethylenedisulfonic acid, known only in solution. Cf. *isethionic acid.* **amino ~** Taurine*.

ethiops mineral Black mercurous sulfide.

ethisterone $C_{21}H_{28}O_2$ = 312.5. Pregneninolone, anhydrohydroxyprogesterone, Oraluton. White crystals, darkening in light, m.274 (decomp.), insoluble in water; a progestational hormone (BP).

ethocaine Procaine hydrochloride.

Ethocel Trademark for ethylcellulose.

ethohexadiol $C_8H_{18}O_2$ = 146.2. 2-Ethyl-1,3-hexanediol*. Colorless oil, soluble in water, distills 240–250; an insect repellent.

etholide A tertiary lipid formed from alcohol acids by the esterification of the hydroxyl group of one with the carboxyl group of the other molecule.

ethopropazine hydrochloride $C_{19}H_{24}N_2S \cdot HCl$ = 348.9. White, bitter crystals, slightly soluble in water. Used for Parkinson's disease (USP, BP).

ethosuximide $C_7H_{11}O_2N$ = 141.2. 2-Ethyl-2-methylsuccinimide, Zarontin. White powder, m.46, soluble in water; an anticonvulsant used for petit mal epilepsy (USP, BP).

ethoxalyl* Ethoxyoxacetyl†. The radical EtOOC·CO—.

ethoxide Ethanolate*.

ethoxy* The radical C_2H_5O—, from ethanol. **e.acetic acid** Ethylglycolic acid*. **e.acetone** $MeCOCH_2OEt$ = 102.1. Colorless liquid. b.128; a solvent. **e.aniline** $C_8H_{11}ON$ = 137.2. Colorless liquid, d_0·1.11, b.286, sparingly soluble in water. **e.butyric acid** Ethylhydroxy*butanoic acid.* **e.caffeine** $C_{10}H_{14}O_3N_4$ = 238.2. Colorless crystals, m.140. Slightly soluble in water; a narcotic. **e.carbonyl*** The radical EtOOC—. **e.catechol** $C_8H_{10}O_2$ = 138.2. A homolog of guaiacol. **e.oxoacetyl†** See *ethoxalyl.*

ethoxyl The ethoxy* radical.

Ethyl (1) (cap) Trademark for an antiknock compound to prevent or reduce knocking in internal-combustion engines. Also a trademark for other products not necessarily associated with fuels or internal-combustion engines. See *Ethyl gas.* (2)* (not cap.) The radical C_2H_5— or Et—, from ethane. ***N*-e.acetamide*** MeCONHEt = 87.1. Colorless liquid, b.200; used in organic synthesis. **e. acetate*** Me·COOEt = 88.1. Acetic ether, acetic ester, acetidin. Colorless liquid, m.—82, b.77, slightly soluble in water. Used as a reagent in organic synthesis, as a solvent for lacquers, in the separation of dyes, and as a flavoring in pharmacy (NF). **e. acetoacetate*** $MeCOCH_2COOEt$ = 130.1. Acetoacetic ester, diacetic ether. Colorless liquid, b.181, slightly soluble in water; a solvent. **e.acetylene** Butyne*. **e. acid phosphate** See *e. phosphate* below. **e. acid sulfate** E. hydrogensulfate*. **e. acrylate** $C_5H_8O_2$ = 100.1. Colorless liquid, b.99. **e. alcohol** Ethanol*. **e. aldehyde** Acetaldehyde*. **e.allyl** C_5H_{10} = 70.1. Colorless liquid, b.70 **e.amine*** $EtNH_2$ = 45.08. Ethamine, aminoethane. A ptomaine from putrefying yeast and wheat flour. Colorless liquid, b.17, miscible with water. **e.amino*** The radical EtNH—, from ethylamine. **e. aminoacetate*** NH_2CH_2COOEt = 103.1. Ethylglycine, e. glycol. **e. aminobenzoate** Benzocaine. ***N*-e.aminobenzoic acid** EtNH·C_6H_4·COOH = 165.2. Colorless prisms, m.112, slightly soluble in water. **e.aniline** See *ethylaniline* under *aniline.* **e.anthracene** $C_{16}H_{14}$ = 206.3. Colorless scales, m.60, insoluble in water. **e. dihydro ~** $C_{16}H_{16}$ = 208.3. Colorless oil, m.320, insoluble in water. **e.arsine dichloride*** $EtAsCl_2$ = 174.9. Dick. A liquid, d.1.66, b.155; a vesicant and lung irritant, formerly a war gas. **e.benzene** See *ethylbenzene* under *benzene.* **e. benzoate*** PhCOOEt = 150.2. Colorless liquid, b.213, slightly soluble in water. **e.benzoic acid** See *ethylbenzoic acid* under *benzoic acid.* **e. benzoylacetate** $PhCOCH_2COOEt$ = 192.2. Benzoyl acetic ester. Colorless liquid, b.267, insoluble in water. **e.benzylaniline** Ph·N·Et·(CH_2Ph) = 211.3. Ethylbenzylphenylamine. b_{710mm}285. **e. borate** A salt of ethanol and boric acid. **e. orthoborate** $B(OEt)_3$ = 146.0. Boron triethoxide, triethylic borate. Colorless, flammable liquid. **e. metaborate** $(EtO)_2(BO)_2$ = 143.7. Colorless, heavy liquid. **e. pyroborate** EtB_3O_5 = 141.5. E. borate. A colorless, gummy mass. **e.boric acid** $EtB(OH)_2$ = 73.9. White crystals, sublime 40, soluble in water. **e. bromide*** EtBr = 109.0. Aethylium bromatus, aether bromatus, monobromoethane, bromic ether, bromoethane*. Colorless liquid, d.1.450, m.−116, b.39, almost insoluble in water. Cf. *ethylene dibromide.* **e. bromoacetate** $C_4H_7O_2Br$ = 167.0. Colorless liquid, b.159, insoluble in water. **e. butyrate*** PrCOOEt = 116.2. Ethylbutyric ester. Colorless liquid, b.121, insoluble in water; a pineapple flavoring. **e. isobutyrate** $Me_2CHCOOEt$. Colorless liquid, b.110, sparingly soluble in water. **e. cacodyl** Diethylarsine*. **e. carbamate** Urethane. **e.carbazole** $C_{14}H_{13}N$ = 195.3. **9-~** Colorless sclaes, m.68, soluble in hot alcohol. **e. carbimide** *E. isocyanate.* **e. carbinol** Propanol*. **e. carbonate** Et_2CO_3 = 118.1. Carbethylic acid, carbonic ester, diethylcarbonic ether, diatol. Colorless liquid, d.0.978, b.126, insoluble in water. Used in organic synthesis; a solvent for nitrocellulose and resins. **e. orthocarbonate** $C(OEt)_4$ = 192.3. Tetraethyl carbonate, tetraethoxymethane. Colorless, aromatic liquid. **e.carbonic acid** Propionic acid*. **e.carbostyril** $C_{11}H_{11}ON$ = 173.2. Colorless crystals, m.168. **e.carbylamine** E. isocyanide. **e.cellulose** Ethocel, lumarith E.C. Made by heating ethyl chloride, with alkali cellulose produced by treating cellulose pulp with sodium hydroxide. It has film-forming properties, good strength, and grease resistance. **e. chaulmoograte** The e. esters of the mixed acids of chaulmoogra oil; clear, yellow liquid, d.0.904. **e. chloride*** C_2H_5Cl = 64.5. Monochloroethane, hydrochloric ether, aethylium chloratus, aether, chloratus, Kelene, chloroethane*, chelen. Colorless liquid, d_0·0.921, m.−141, b.12.2, slightly soluble in water; an anesthetic (USP, EP), constituent of cognac essence, refrigerant, and ethylating agent. **e. chloroacetate*** $CH_2Cl \cdot COOEt$ = 122.6. Colorless liquid, b.145, insoluble in water. **e. chloroacetoacetate*** MeCOCHClCOOEt = 164.6. Colorless liquid, b.197, slightly soluble in water; used in organic synthesis. **e. chlorocarbonate** E. chloroformate. **e. chloroformate*** ClCOOEt = 108.5. E. chlorocarbonate. Colorless liquid with a pungent odor, d.1.135, b.93, decomp. in water. **e. chloropropionate*** MeCHClCOOEt = 136.6. Ethylchlorpropionic ester, aether chlorpropionicus. Colorless liquid, b.146, slightly soluble in water. **e. cinnamate*** PhCH:CH·COOEt = 176.2. Ethylcinnamic ester, aether cinnamylicus. Colorless oil, m.12, insoluble in water; a strawberry flavoring. **e.crotonic acid*** MeCH:CEt·COOH = 114.1. Colorless, monoclinic crystals, m.40 (sublime), sparingly soluble in water; a peppermint flavoring. **e. cyanoacetate*** $CN \cdot CH_2 \cdot COOEt$ = 113.1. Ethylcyanacetic ester, aether cyanaceticus. Colorless liquid, d.1.066, b.207, insoluble in water. **e. cyanate*** EtOCN = 71.1. Cyanetholin. Colorless, unstable liquid, insoluble in water. **e. isocyanate*** EtNCO = 71.1 E. carbimide. Colorless liquid, b.60, soluble in alcohol. **e. cyanide*** EtCN = 55.1. Propiononitrile*, hydrocyanic ether, ether cyanatus, propanenitrile*. Colorless liquid, b.97, soluble in water. **e. isocyanide** E. isocyanide. **e. diacetoacetate** $(MeCO)_2CHCOOEt$ = 172.2. Colorless liquid, d.1.101, b.200, slightly soluble in water. **e. dibromoacetate*** $Br_2CHCOOEt$ = 245.9. Ethyldibromoacetic ester. Colorless liquid, miscible with alcohol or ether. **e. dichloroacetate*** $Cl_2CHCOOEt$ = 157.0. Ether dichloraceticus, ethyldichloracetic ester. Colorless liquid, b.156, slightly soluble in water. **e. dichloride** Ethylene dichloride*. **e. diethyl acetoacetate** $MeCO \cdot CEt_2 \cdot COOEt$ = 186.2. Diethylacetoacetic e. ester. Colorless liquid, b.218, insoluble in water. **e. diethylmalonate** $Et_2C(COOEt)_2$ = 216.3. Colorless liquid, b.223, insoluble in water. **e. diiodosalicylate** $C_2H_3I_2 \cdot COOC_6H_4 \cdot OH$ = 418.0. Ether diiodosalicylicus. Colorless crystals, m.132, sparingly soluble in water; an iodoform substitute. **e. dimethylmalonate** $Me_2 \cdot C \cdot (COOEt)_2$ = 188.2. Colorless liquid, b.197, insoluble in water. **e. dioxythiocarbonate** $C_5H_{10}O_2S$ = 134.2. Colorless, strongly refractive liquid, soluble in alcohol. Cf. *xanthic acid.* **e.diphenylamine** $EtNPh_2$ = 197.3. Diphenylethylamine. Colorless liquid, b.296, insoluble in water. **e.diphenylphosphine** $EtPPh_2$ = 214.2. Diphenylethylphosphine. Colorless liquid, b.293, insoluble in water. **e. diselenide** EtSe·SeEt = 216.0. Diethyl selenide, e. perselenide. A heavy, brown, pungent liquid, b.186. **e. disilicate** $[(EtO)_3Si]_2O$ = 246.5. Colorless, oily, flammable liquid with peppermint odor. **e. disulfide** EtS·SEt = 122.2. E. persulfide, e. dithioethane. Colorless liquid with garlic odor, d_{22}·0.993, b.153, sparingly soluble in water. **e. dithiocarbonate** $CO(SEt)_2$ = 150.3. Dithiourethane, d.1085, b.197. **e.ene** See *ethylene.* **e. ether** Et_2O = 74.1. Diethyl ether*. Colorless liquid, b.35, slightly soluble in water; a solvent. One of the first inhalation anesthetics; now largely superseded (USP, EP, BP). **e. fluoride** EtF = 48.6. Fluoroethane*. Colorless gas, b.−32, soluble in water. **e. formamide*** EtNH·CHO = 73.1. Colorless liquid, b.199, soluble in alcohol. **e. formate*** HCOOEt = 74.1. Ethylformic ester. Colorless liquid, b.54, slightly soluble in water; a flavoring. **e. furoate*** C_4H_3COOEt = 124.1. Colorless liquid, m.34; a solvent and industrial perfume. **E. gas, E. gasoline** E. petrol, "Ethyl." When used as a trademark, Ethyl designates a gasoline mixed with a fluid containing tetraethylplumbane, halogens, and other constituents of an antiknock compound. **e. glycerate*** $CH_2OH \cdot CHOH \cdot COOEt$ = 134.1. Colorless liquid, b.235, slightly soluble in water. ***N*-e.glycine** $EtNH \cdot CH_2 \cdot COOH$ = 103.1. Colorless scales, decomp. 160, soluble in water. **e. glycol ether** Cellosolve. **e. glycolate*** $HOCH_2COOEt$ = 104.1. E. hydroxyacetate. Colorless liquid, b.160, soluble in alcohol; a solvent for nitrocellulose and resins. **e.glycolic acid*** $EtO \cdot CH_2 \cdot COOH$ = 104.1. Ethoxyacetic acid. Colorless liquid, b.206, soluble in alcohol. **e. green** Ethylated *methyl* violet. **e. heptanoate*** $C_6H_{13}COOEt$ = 158.2. E. oenanthate, cognac oil. Colorless oil in old wines; a flavoring. **e. hexanoate*** $C_5H_{11}COOEt$ = 144.2. E. caproate. Colorless liquid, m.167, soluble in alcohol. **e. hydrate** Ethanol*. **e.hydrazine*** $NH_2 \cdot NHEt$ = 60.1. Colorless liquid, b.100, soluble in water. **e.hydride** Ethane*. **e.hydrine** Ethylin. **e. hydrobromide** E. bromide*. **e. hydrochloride** E. chloride*. **e.hydrogensulfate*** $EtHSO_4$ = 126.1. E.sulfuric acid, acid e. sulfate. Colorless syrup, decomp. by heat or water. Used for precipitating casein from milk. **e. hydroperoxide** EtO·OH = 62.1. Colorless liquid, b_{150mm}26, soluble in water. **e. hydrosulfide*** Ethanethiol*. **e. hydroxyacetate** E. glycolate*. **e.hydroxylamine** C_2H_7ON = 61.1. ***O*-~** NH_2OEt. Ethoxylamine. Colorless liquid, b.68, soluble in water. ***N*-~** EtNHOH. Hydroxylethylamine. Colorless leaflets, m.60, soluble in water. **e. hydroxypropionate** $C_5H_{10}O_3$ = 118.1. **2-~** E. lactate. **3-~** $CH_2OH \cdot CH_2 \cdot COOEt$. Colorless liquid, b.187, soluble in water; a solvent for resins. **e. iodide*** EtI = 156.0. Iodoethane*, hydroiodic ether, e. hydroiodic ester, ether iodatus. Colorless liquid, ethereal odor, d.1.92, b.72, slightly soluble in water. Used for mineral separation. **e. iodoacetate*** $CH_2I \cdot COOEt$ = 214.0. K.S.K. Colorless, heavy oil, d.1.8, b.179; a lacrimatory. **e. isobutyl ether** $EtOC_4H_9$ = 102.2. Colorless liquid, b.78, insoluble in water. **e. isobutyl ketone** $EtCOC_4H_9$ = 114.2. 2-Methylhexan-4-one*. Colorless liquid, b.136, insoluble in water. **e. isobutyrate*** $Me_2CH \cdot COOEt$ = 116.1. E. dimethyl acetate. Colorless liquid, d_{20}·0.869, b.110, slightly soluble in water. **e. isocyanate** See *e. cyanate* above. **e. isocyanide*** EtNC = 55.1. E.carbylamine. Colorless liquid,

d_4 0.759, b.79, soluble in water. **e. isopentyl ether*** $EtOC_5H_{11}$ = 116.2. Colorless liquid, b.112, insoluble in water. **e. isophthalate*** $C_6H_4(COOEt)_2$ = 222.2. Colorless oil, b.285. **e. isopropyl acetoacetate*** $MeCO \cdot CH(CHMe_2) \cdot COOEt$ = 172.2. Colorless liquid, b.200, slightly soluble in water. **e. isopropyl ether*** Me_2CHOEt = 88.1. Isopropyl ethoxide. Colorless liquid, b.54, soluble in water. **e. isopropyl ketone*** $Me_2CH \cdot CO \cdot Et$ = 100.2. 2-Methylpentan-3-one*. Colorless liquid, b.115, slightly soluble in water. **e. isosuccinate** $MeCH(COOEt)_2$ = 174.2. Colorless liquid, b.198, sparingly soluble in water. **e. isosulfocyanate** See *e. thiocyanate* below. **e. isothiocyanate** See *e. thiocyanate* below. **e. isovalerate*** Me_2CHCH_2COOEt = 130.2. Colorless liquid, d.0.972, b.134, insoluble in water. **e. ketone** Diethyl ketone*. **e. lactate*** MeCH(OH)COOEt = 118.1. Ethyl-2-hydroxypropionate. Colorless liquid, b.154, soluble in water; a solvent for nitrocellulose and resins. **e. malate*** $EtOOC \cdot CH_2 \cdot CHOH \cdot COOEt$ = 190.2. Colorless liquid, b.248, miscible with water; used in lacquers. **e. malonate*** $CH_2(COOEt)_2$ = 160.2. Colorless liquid, d.1.061, b.198, slightly soluble in water; a plasticizer for cellulose acetate. **e.malonic acid*** $EtCH(COOH)_2$ = 132.1. Colorless, prisms, m.111, decomp. 160, soluble in water. **e. mannitol** $C_{10}H_{10}O_5$ = 210.2. A substitution product of mannitol; a colorless syrup. **e.mercury chloride*** EtHgCl = 265.1. Silvery, iridescent leaflets, d.3.5, m.192, insoluble in water. **e.mercury hydroxide*** EtHgOH = 246.7. Silvery, iridescent leaflets, m.190, insoluble in water. **e. methyl ether*** Et·O·Me = 60.1. E. methyl oxide. Colorless liquid; an anesthetic. **e. monotartrate*** $C_6H_{10}O_6$ = 178.1. Colorless rhombs, m.90, soluble in water. **e.morphine hydrochloride** $C_{19}H_{23}O_3N \cdot HCl \cdot 2H_2O$ = 385.9. Dionine. An analgesic, principally as cough suppressant (EP, BP). **e. mustard oil** E. isothiocyanate. See *e. thiocyanate.* **e.naphthalene*** $C_{10}H_7 \cdot Et$ = 156.2. Naphthylethane. **1-~** Colorless liquid, b.256, insoluble in water. **2-~** Colorless liquid, b.251, insoluble in water. **e.naphthylamine*** $EtNH \cdot C_{10}H_7$ = 171.2. Ethylaminenaphthalene. **1-~** Colorless liquid, b.303. **2-~** Colorless liquid, b.305. **e. naphthyl ether*** $C_{10}H_7OEt$ = 172.2. Ethoxynaphthalene. **1-~** Colorless liquid, m.6, insoluble in water. **2-~** Bromelia. Colorless crystals, m.37, insoluble in water. **e.nitramine** See *nitroamine.* **e. nitrate** $EtNO_3$ = 91.1. Aether nitricus. Colorless liquid, d.1.116, m.−112, b.88, insoluble in water. **e. nitrobenzoate** $NO_2 \cdot C_6H_4COOEt$ = 195.2. **e. *ortho-*~** Colorless, triclinic crystals, m.30, soluble in alcohol. **e. *meta-*~** Colorless prisms, m.48, insoluble in water. **e. *para-*~** Colorless crystals, m.57, soluble in alcohol. **e. nitrocarbamate** Urethane. **e.nitrolic acid** Acetonitrolic acid. **e. nonanoate*** $C_8H_{17}COOEt$ = 186.3. E. perlargonate, enanthic ether. Colorless liquid with strong vinous odor and acrid taste, d_{17} 0.8635, b.277, insoluble in water; a flavoring. **e. octanoate*** $C_{10}H_{20}O_2$ = 172.3. E. caprylate. Colorless liquid, b.30, soluble in alcohol. **e. oleate*** $C_{20}H_{38}O_2$ = 310.5. Yellow oil with unpleasant taste, d_{20} 0.870, iodine val. 75–84. **e. orthophosphate*** See *e. phosphate.* **e. orthosilicate*** See *e. silicate.* **e. oxalate*** $(COOEt)_2$ = 146.1. Colorless liquid, b.186, slightly soluble in water. **e. oxide** Ether*. **e.oxydithiocarbonic acid** Xanthic acid. **e. palmitate*** $C_{15}H_{31} \cdot COOEt$ = 284.5. Colorless needles, m.24, insoluble in water. **e. pelargarate** E. nonanoate*. ***N*-e.pentylaniline** $Et \cdot N(C_5H_{11})Ph$ = 191.3. Ethylpentylphenylamine. Colorless oil, b.262, soluble in alcohol. **e. pentyl ketone*** $C_8H_{16}O$ = 128.2. Colorless liquid, sparingly soluble in water, miscible with alcohol or ether. **normal ~** $Et \cdot CO \cdot C_5H_{11}$. 3-Octanone*. d.0.825, b.165. **active ~** $MeCH_2 \cdot CHMe \cdot CH_2 \cdot COEt$. 5-Methyl-3-heptanone*. b.161. ***tert-*~** $MeCH_2 \cdot CMe_2 \cdot COEt$. 4,4-Dimethyl-3-hexanone*. d.0.825, b.151. **e. perchlorate*** $EtClO_4$ = 128.5. Perchloric ether. Colorless, explosive liquid with a sweet odor. **e. perchloride** Hexachloroethane*. **e. peroxide** Diethyl dioxide*. **e. persulfide** E. disulfide. **e.phenol*** $Et \cdot C_6H_4OH$ = 122.2. ***ortho-*~** Phlorol. Colorless liquid, d_0 1.037, b.207, soluble in alcohol. ***para-*~, 1,4-~** Colorless needles, b.214, insoluble in water. **e. phenolate** Phenetole*. **e. phenylacetate*** $PhCH_2CO \cdot OEt$ = 164.2. Colorless liquid, b.229, insoluble in water. **e.phenylacetylene** PhC:CEt = 130.2. Colorless liquid, d_{21} 0.923, b.202, soluble in alcohol. **e. phenyl carbamate** Phenylurethane. **e. phenyl ether** Phenetole*. **e.phenylhydrazine alpha-~** $PhNEt \cdot NH_2$ = 136.2. Colorless oil, b.237, soluble in alcohol. **beta-~** Colorless oil, sparingly soluble in water. **e. phenyl ketone*** PhCOEt = 134.2. Colorless leaflets, d.1.015, m.15, b.218, insoluble in water. **e.phenylmethanol** Ph(Et)CHOH = 136.2. Ethylphenylhydroxymethane. Colorless liquid, b.212, insoluble in water. **e. phenyl sulfone*** $PhSO_2Et$ = 170.2. Colorless scales, m.42, sparingly soluble in water. **e.phenylurea** PhNH·CO·NHEt = 164.2. Colorless needles, m.99, insoluble in water. **e. phosphate** E. phosphoric acids: **e. *mono*~** EtH_2PO_4 = 126.0. Colorless oil, forming metallic salts (e. phosphates). **e. di~** $Et_2 \cdot HPO_4$ = 154.1. Colorless syrup, forming salts (diethyl phosphates). **e. normal~** or **tri~** Et_3PO_4 = 182.2. Colorless, aromatic liquid, b.215, decomp. in water. **pyro** or **tetra~** $Et_4P_2O_7$ = 290.2. Colorless, odorous oil. **e.phosphine*** $EtPH_2$ = 62.1. Colorless liquid, b.25. **e.phosphonic acid, e. phosphite** E. phosphorous acids: (1) EtH_2PO_3 = 110.0. E.phosphonic acid*, acid e. phosphite. (2) $Et_2 \cdot HPO_3$ = 138.1. Diethylphosphonic acid*, diethyl phosphite. (3) Et_3PO_3 = 166.2. Normal e. phosphite, triethyl phosphite. Colorless liquid of disagreeable odor. (4) $Et_3H_2P_2O_5$ = 202.1. Diethyldiphosphonic acid*, e. pyrophosphorous acid. **e.phosphoric acid** See *e. phosphate.* **e. phthalate*** $C_6H_4(COOEt)_2$ = 222.2. Colorless liquid d_{20} 1.118, b.295, insoluble in water. **e. iso~** See *e. isophthalate* above. **e. *meta-*~** See *e. isophthalate* above. **e. *para-*~** See *e. terephthalate* below. **e. tere~** See *e. terephthalate* below. **e. propiolate*** HC:COOEt = 86.1. Colorless liquid, b.119, insoluble in water. **e. propionate*** EtCOOEt = 102.1. Colorless liquid, d.0.896, b.98, slightly soluble in water; used in perfumery. **e. propyl** Pentane*. **e. propylene** Pentene*. **e. propyl ether*** Et·O·Pr = 88.1. E. propyl oxide. Colorless liquid, b.60, slightly soluble in water. **e. propyl ketone*** Et·CO·Pr = 100.2. 3-Hexanone*. Colorless liquid, b.122, sparingly soluble in water. **e. pyruvate*** MeCO·COOEt = 116.1. Colorless liquid, d.1.060, b.144; a solvent for nitrocellulose. **e. racemate** An optically active isomer of e. tartrate. **e. red** $C_{23}H_{23}N_2I$. 1,1-Diethyl isocyanine iodide. A quinoline dye; a photosensitizer. **e. salicylate*** $HO \cdot C_6H_4 \cdot COOEt$ = 166.2. Sal ethyl. Colorless liquid of pleasant odor and taste, d_{20} 1.132, b.231, insoluble in water: an antiseptic and solvent used like methyl salicylate. **e. selenide** Et_2Se = 137.1. Diethyl selenide*, perethyl selenide. Colorless oil, b.107. **e.silanetriyl*** The radical Si≡. **e. silicate** (1) Et_4SiO_4 = 208.3. E. orthosilicate*. Colorless, flammable liquid; isolog of e. orthocarbonate. (2) Et_2SiO_3 = 134.2 E. metasilicate*. Colorless liquid, b.165, decomp. by water; isolog of e. carbonate. (3) $Et_6Si_2O_7$ = 342.5. E. disilicate. **e.stannic acid** $EtSnO_3H$ = 196.8. White powder, insoluble in water. **e. succinate*** $EtOOC(CH_2)_2COOEt$ = 174.2. Colorless liquid, d.1.044, b.217, insoluble in water; a plasticizer. **e. iso~** E. succinate. **e.succinic acid** $HOOC \cdot CH_2 \cdot CHEt \cdot COOH$ = 146.1.

Colorless prisms, m.98, soluble in water. **e. succinyl succinate*** $C_{12}H_{16}O_6$ = 256.3. Green prisms, soluble in water (blue fluorescence). **e. sulfas** E. sulfate. **e. sulfate** Et_2SO_4 = 154.2. Normal e. sulfate, diethyl sulfate*. Colorless oil with peppermint odor, d.1.184, b.208, insoluble in water. **acid ~, mono ~** E. hydrogensulfate*. **e.sulfinic acid** Et·SO·OH = 94.1. Ethanesulfinic acid*. Colorless syrup, an isolog of propionic acid. **e. sulfite** Diethyl sulfite. **e. sulfohydrate** Ethanethiol*. **e. sulfone** Diethyl sulfone*. **e.sulfonic acid** $EtSO_2$·OH = 110.1. Ethanesulfonic acid*. Colorless, deliquescent crystals, soluble in water. **e.sulfonic chloride** $EtSO_2Cl$ = 128.6. Ethanesulfonyl chloride. Colorless liquid, b.177, decomp. by water. **e.sulfonic oxide** Diethyl sulfoxide. **e. tartrate** $EtOOC(CHOH)_2COOEt$ = 206.2. E. racemate, diethyl tartrate*. Colorless liquid, b.280, slightly soluble in water; a solvent for nitrocellulose, gums and resins. **e.tartronic acid*** $C_5H_8O_5$ = 148.1. Colorless scales, m.116 (decomp.), soluble in water. **e. telluride** Et_2Te = 185.7. Tellurium ethyl. A heavy, red oil, giving off yellow fumes. **e. ditelluride** Et_2Te_2 = 313.3. Diethyl ditelluride*, e. pertelluride. A dark-red liquid decomp. by water. **e. terephthalate** $C_6H_4(COOEt)_2$ = 222.2. E. *p*-phthalate. Colorless liquid, b.119, insoluble in water. **e. thioalcohol** Ethanethiol*. **e. thiocarbonate** (1) $CS(EtO)_2$ = 134.2. A liquid, d.1.028, b.162. (2) $(EtS)_2CS$ = 166.3. Yellow oil of unpleasant odor; isolog of e. carbonate and e. sulfite. **e. thiocyanate*** EtSCN = 87.1. E. sulfocyanide. Colorless liquid, b.148, insoluble in water. **e. iso ~** EtNCS = 87.1. E. mustard oil. Colorless liquid, b_{753mm}131, insoluble in water. **e.tin tribromide*** $EtSnBr_3$ = 387.5. Colorless needles, m.310, soluble in water. **e. toluate*** MeC_6H_4COOEt = 164.2. **e. ortho- ~** Colorless liquid, b.221, insoluble in water. **e. meta- ~** Colorless liquid, b.228, insoluble in water. **e. para- ~** Colorless liquid, b.229, insoluble in water. **e.toluene** E. methyl*benzene*. **e.urethane** See *ethylurethane* under *urethane*. **e. valerate** BuCOOEt = 130.2. Colorless liquid, b.145, insoluble in water. **e. vanillate*** $C_{10}H_{12}O_4$ = 196.15. Colorless crystals, m.44, insoluble in water. **e. vanillin** See *ethyl vanillin* under *vanillin*. **e. vinyl ether*** EtO·C_2H_3 = 72.1. Colorless liquid, b.36, soluble in water. **e. violet** An indicator changing at pH 2.0 from blue-green (acid) to purple (alkali). **e. xylene** Ethyldimethyl*benzene*.

ethylal Acetaldehyde*.

ethylamine* See *ethylamine* under *ethyl*.

ethylate Ethanolate*.

ethylation The introduction of an ethyl group into a compound.

ethylene (1)* Ethene*, q.v. (2)* 1,2-Ethanediyl†, acetene, elayl. The radical $-CH_2 \cdot CH_2-$. Cf. *ethylidene*. **azi ~** Diazoethane. **bromo ~** Vinyl bromide*. **chloro ~** Vinyl chloride*. **di ~** See *diethylene*. **dichloro ~*** ClHC:CHCl = 96.6. Acetylene dichloride, dioform. ***cis*- ~** b.48. ***trans*- ~** b.60. Colorless liquids, immiscible with water. **diethyl ~** 3-Hexene*. **dimethyl ~** 2-Butene*. **diphenyl ~** Stilbene*. **oxo ~** Ketene*. **pentyl ~** Heptene*. **perchloro ~** See *tetrachloroethylene*. **phen ~, phenyl ~** Styrene*. **poly(chlorotrifluoro ~)** See *poly(chlorotrifluoroethylene)*. **tetrachloro ~*** See *tetrachloroethylene*. **tetraiodo ~*** Diiodoform. **tetraphenyl ~*** See *tetraphenylethylene* under *tetraphenyl*. **trimethyl ~** See *trimethylethylene* under *trimethyl*. **vinyl ~** 1,3-Butadiene*.

e. acetate E. diacetate*. **e. alcohol** Glycol*. **e. aldehyde** Acrylaldehyde*. **e. benzoate** Ph·COO·CH_2:CH_2·OOC·Ph = 270.3. Colorless prisms, m.73, insoluble in water. **e. bromide** E. dibromide*. **e. bromohydrin** Glycol bromohydrin. **e.carboxylic acid** Acrylic acid*. **e. chlorohydrin** $ClCH_2 \cdot CH_2OH$ = 80.5. Chloroethyl alcohol, 2-chloroethanol*, 1-hydroxy-2-chloroethane. Colorless liquid, b.128, miscible with water. Used in organic synthesis, and in forcing the sprouting of plants. **commercial ~** A 40% solution, d.1.097, b.96, used to introduce the OEt group into a molecule. **e. chloride** E. dichloride*. **e. cyanohydrin** C_3H_5ON = 71.1. 2-Cyanoethanol*, 1-hydroxy-2-cyanoethane. Colorless liquid, m.221, miscible with water. **e. cyanide** E. dicyanide*. **e. diacetate*** $(MeCOOCH_2)_2$ = 146.1. Colorless liquid, b.186, soluble in water. **e.diamine*** $NH_2CH_2 \cdot CH_2NH_2$ = 60.1. Diaminoethane, 1,2-ethanediamine*, in ptomaines. Colorless crystals, m.10, soluble in water. Cf. *sublamine*. **e. hydrate** Colorless liquid with an ammoniacal odor; used for aminophylline injections (USP, BP). **e.tetraacetic acid** $(CH_2 \cdot COOH)_2(N \cdot CH_2)_2(CH_2 \cdot COOH)_2$ = 292.3. EDTA, versenic acid. White crystals, slightly soluble in water, decomp. above 160; made from an alkali cyanide, formaldehyde, and ethylenediamine. It forms slightly ionized complexes with alkaline earths and other elements. Used as the more soluble sodium salt, as an analytical reagent, e.g., to titrate the hardness salts of water with Eriochrome Black T as indicator; also in detergents, rubber processing, and scale prevention. Cf. *sequestering agent*. See disodium *edetate*. **e. dibenzamide** See *ethylene dibenzamide* under *dibenzamide*. **e. dibromide*** $BrCH_2 \cdot CH_2Br$ = 187.9. 1,2-Dibromoethane*, e. bromide, glycol dibromide. **e. dicarbonitrile** E. dicyanide*. **e.dicarboxylic acid *cis*- ~** Maleic acid*. ***trans*- ~** Fumaric acid*. **e. dichloride*** $ClCH_2 \cdot CH_2Cl$ = 99.0. 1,2-Dichloroethane*, e. chloride, elayl chloride, vinylene chloride, Dutch liquid. Colorless liquid, b.84, slightly soluble in water. Used in organic synthesis, as a solvent for lacquers and fats, and as a textile cleanser. Cf. *dichloroethylene*. **e. dicyanide*** $C_4H_4N_2$ = 80.1. Succinonitrile*, e. cyanide, 1,2-dicyanoethane*. Colorless crystals, m.54, soluble in water. **e. diethanolate** EtO·CH_2CH_2OEt = 118.2. Diethyl glycol ether. **e. dihydrate** Glycol*. **e. diiodide*** $(CH_2I)_2$ = 281.9. Diiodoform, e. iodide, 1,2-diiodoethane*. Yellow prisms, m. 81, slightly soluble in water. **e. dinitrate** $(CN_2NO_3)_2$ = 152.1. EGDN, e. nitrate. Yellow liquid, exploded by heat or impact, insoluble in water. **e. dinitrite** $(CH_2NO_2)_2$ = 120.1. Glycol dinitrite. Colorless liquid, b.96, insoluble in water. **e. dioxide** Dioxane. **e.dioxy*** The radical $-OCH_2 \cdot CH_2O-$. **e. diphenolate** E. diphenyl ether. **e. diphenyldiamine** $C_{14}H_{16}N_2$ = 212.3. Colorless crystals, m.59, insoluble in water. **e. diphenyl ether** $C_{14}H_{14}O_2$ = 214.3. E. diphenolate. Colorless crystals, m.98, sparingly soluble in water. **e. dithiol*** HS·CH_2·CH_2·SH = 94.2. Glycol sulfohydrate, e. disulfhydrate, dithioethylene glycol. Colorless liquid, b.146, soluble in alcohol. **e. disulfhydrate** E. dithiol*. **e.disulfonic acid** Ethionic acid. **e. glycol*** Glycol*. **e. g. dinitrate** E. dinitrate. **e. g. ethyl ether** Cellosolve. **e. hydride** Ethane*. **e. monoacetate** Glycol acetate. b.187. **e. naphthalene** Acenaphthylene*. **e. oxide*** $(CH_2)_2$:O = 44.5. Oxirane*, dimethylene oxide, 1,2-epoxyethane*. Colorless gas, b_{746mm}14, soluble in water. E. oxide is used to make e. glycol and thence Terylene. **e. perchloride** Carbon dichloride*. **e. periodide** Diiodoform. **e. series** See *olefin*. **e. sulfate acid ~** $C_2H_4(HSO_4)_2$ = 222.2. E. sulfuric acid. A colorless syrup. **basic ~** $(OH)C_2H_4(HSO_4)$ = 142.1. E. hydroxysulfuric acid; known only as its compounds. **e. sulfide** $CH_2 \cdot S \cdot CH_2$ (ring) = 60.1. Thiirane, dimethylene sulfide. Liquid, b.55, polymerizes rapidly. **e.sulfonic acid** Ethionic acid. **e. sulfuric acid** See *e. sulfate* above. **e. tetrabromide*** Tetrabromoethane*.

e. tetrachloride* Tetrachloroethane. **e. thiocyanate*** $C_2H_4(SCN)_2$ = 144.21. Colorless rhombs, m.90, sparingly soluble in water. **e. trichloride*** Trichloroethylene*. **e.urea** $C_3H_6ON_2$ = 86.1. Colorless needles, m.131, soluble in water.

ethylene glycol* See *glycol.*

ethylic acid Acetic acid*.

ethylidene* Ethidene. The radical $CH_3 \cdot CH=$, from ethane; isomeric with ethylene. **e. acetone** MeCO·CH:CH·Me = 84.1. A liquid b_{741mm}122. **e. cyanohydrin** MeH·C·(OH)CN = 71.1. 1-Cyanoethanol*, lactonitrile, 1-hydroxy-1-cyanoethane. Colorless liquid, b.182 (decomp.), miscible with water. **e. dibromide*** $MeCHBr_2$ = 187.9. 1,1-Dibromoethane*. Colorless liquid, b.112, insoluble in water. **e. dichloride*** $C_2H_4Cl_2$ = 99.0. Ethylidene perchloride, 1,1-dichloroethane*, chlorinated hydrochloric ether. Colorless liquid, b.59, sparingly soluble in water. **e. diiodide*** $MeCHI_2$ = 281.9. 1,1-Diiodoethane*. Colorless liquid, b.178, insoluble in water. **e. glycol** Me·CH(OH)$_2$. Known only in its derivatives, e.g., chloral hydrate. **e.lactic acid** Lactic acid*. **e. perchloride** E. dichloride*. **e. urea** $C_3H_6ON_2$ = 86.1. Colorless needles, m.126, decomp. 165, slightly soluble in water. **e. urethane** $C_8H_{16}O_4N_2$ = 204.2. Colorless needles, m.125 (decomp.), soluble in water.

ethylidyne* Ethenyl. The radical MeC≡. **e. amide** Acetamidine*. **e. diphenylamidine** $C_{14}H_{14}N_2$ = 210.3. Colorless needles, m.131, sparingly soluble in water. **e.tricarboxylic acid** $C_5H_6O_6$ = 162.1. Colorless prisms, decomp. by heat, insoluble in water. **e. trichloride** Trichloroethane*. **e. triethyl ether** $C_8H_{18}O_3$ = 162.2. 1,1,1-Triethoxyethane*. Colorless liquid, b.145, decomp. by hot water.

ethylin A compound derived from glycerol by substituting one or more ethoxy groups for hydroxy groups: **mono ~** $C_3H_5(OH)_2OEt$ = 120.1. Colorless liquid, b.230. **di ~** $C_3H_5 \cdot OH \cdot (OEt)_2$ = 148.2. Diethylin. 1,3-Diethylglycerol. **tri ~** $C_3H_5(OEt)_3$ = 176.2. Colorless liquid, b.185.

ethylogen A complex carbide, from which water liberates ethylene slowly.

ethyloic- Prefix indicating the radical $-CH_2 \cdot COOH$ as a side chain.

ethylol Oxethyl. The hydroxyethyl radical, $CH_2OH \cdot CH_2-$.

ethyne* Acetylene*.

ethynodiol diacetate $C_{24}H_{32}O_4$ = 384.5. 19-Nor-17α-pregn-4-en-20-yne-3β,17-diol diacetate. White crystals, soluble in water. A progestational compound, used to treat menstrual disorders; also used for oral contraception (USP, BP).

ethynyl* Ethinyl, acetenyl, acetylenyl. The radical HC:C−, from acetylene. **e. bromide** Bromoacetylene.

etiline Tetrachloroethane*.

etiography The study of phenomena which destroy solid material; e.g., corrosion, thermal and mechanical stresses.

etiology The study of the origin of a disease.

etiophyllin $C_{31}H_{34}N_4Mg$ = 486.9. Aetiophyllin. The decarboxylated magnesium base of chlorophyll. Blue tablets, m.205. See *porphin.*

etioporphyrin III $C_{32}H_{38}N_4$ = 478.7. Parent structure of physiologically important tetrapyrroles (as, chlorophylls and hemes). Violet crystals, m.361. See *porphin.*

eto See *ure.*

Ettinghausen effect The galvanomagnetic change in temperature of a plate through which a current flows. Cf. *Hall, Leduc, Nernst* effects.

ettringite $Ca_{12}Al_4(OH)_{24}(SO_4)_6 \cdot 5H_2O$. A mineral found in contact zones of chalk and dolomite (Eire).

Eu Symbol for europium.

eucairite CuAgSe. A white, metallic, isometric, native selenide.

eucalin $C_{12}H_{12}O_6$ = 252.2. A nonfermentable disaccharide from the hydrolysis of melitose, resembling inositol. Cf. *eucalyptolene.*

eucalyptene $C_{10}H_{16}$ = 136.2. A terpene derived from eucalyptol. Colorless liquid, b.170, soluble in alcohol; an antiseptic.

eucalyptol $C_{10}H_{18}O$ = 154.3. Epoxymenthane, cineole, cajeputole. Colorless liquid, d_2·0.927, b.176, insoluble in water, miscible with alcohol. A chief constituent of eucalyptus and cajeput oils; an antiseptic, and expectorant.

eucalyptolene $C_{10}H_{16}$ = 136.2. An isomer of eucalyptene, in eucalyptus oil. Yellow liquid, b.300, miscible with alcohol.

eucalyptus (1) A genus of trees (Myrtaceae). Cf. *lerp, mallee bark.* (2) Blue gum tree. The dried leaves of *Eucalpytus globulus* (Myrtaceae), originally Australia. **e. oil** The essential oil of e. leaves, varying from peppermint to turpentine in odor. Chief constituents: eucalyptol, pinene, hexanal, valeraldehyde, butyraldehyde. Yellow liquid, d.0.91–0.93, miscible with alcohol; an antiseptic and flavoring (NF, EP, BP).

eucasin Ammonium caseinate.

eucasol Anytol.

eucatropine hydrochloride $C_{17}H_{25}O_3N \cdot HCl$ = 327.8. White granules, m.184. soluble in water; a mydriatic (USP).

eucazulene $C_{15}H_{18}$ = 198.3. A blue hydrocarbon from eucalyptus oil.

euchinine Quinine diethyl carbonate.

eu-chlorine, euchlorine A mixture of chlorine and chlorine dioxide obtained from potassium chlorate and concentrated hydrochloric acid.

euchroite $Ca_4As_2O_9$. Green rhombs, d.3.4, hardness 3.

eucol Guaiacol acetate.

eucolloid A colloid, with satisfied primary valencies, forming particles of chain length over 2,500 Å (mol. wt. over 1,000). Cf. *mesocolloid.*

eucryptite $LiAlSiO_4$. Transparent, native, hexagonal aluminum lithium silicate.

euctolite A basaltic rock from Italy.

eucupin $C_{24}H_{34}O_2N_2$ = 382.5. $O^{6'}$-Isopentylcupreine. One of the *quinine alkaloids,* q.v. White powder, m.152, insoluble in water.

eudalene $C_{14}H_{16}$ = 184.3. 7-Isopropyl-1-methylnaphthalene. b_{11}.140. A hydrocarbon obtained from eudesmol and selinene by treatment with sulfur.

eudeiolite A native calcium-iron-cerium niobate-titanate-thorate. Brown, shining mass, d.3.44, hardness 4.

Eudemine Trademark for diazoxide.

eudialite $Na_{13}(Ca \cdot Fe)_6(Si \cdot Zr)_{20}O_{52}Cl$. A native silicate-chloride.

eudiometer A graduated glass tube with platinum electrodes, closed at one end; used to measure volume changes during combination of gases.

eudiometry Gasometry.

euflavine $C_{14}H_{14}N_3Cl$ = 259.7. 2,8-Diamino-10-methylacridinium chloride. A dye similar to acriflavine.

eugenic acid (1) Eugetinic acid. (2) Eugenol*.

eugenics The study of the favorable alteration of the genetic endowment of individuals, populations, or mankind. Cf. *euphenics, euthenics.*

eugenol* $C_{10}H_{12}O_2$ = 164.2. 2-Methoxy-4-(2-propenyl)phenol†. 4-Allylguaiacol. In many essential oils, as, oil of cloves; isomer of chavibetol. Colorless liquid, d_{18}·1.063, b.253, slightly soluble in water. Used as a dental analgesic, as an antiseptic (USP, BP), in perfumery, and for manufacturing

vanillin. **iso ~** 5-Propenylguaiacol. An isomer with the side chain —CH:CH·Me. Colorless liquid, b.240. **methyl ~** $C_{11}H_{14}O_2$ = 178.2. Eugenol methyl ether, 4-allyl-1,2-methoxybenzene. Colorless liquid, b.244, insoluble in water.
e. acetamide $C_{12}H_{15}O_3N$ = 221.3. Colorless scales, m.110, soluble in water; an anesthetic.

eugetic acid Eugetinic acid.

eugetinic acid $C_6H_2(OH)(OMe)\cdot C_3H_5\cdot COOH$ = 208.2. Eugenic acid, eugetic acid. Colorless prisms, m.124, sparingly soluble in water.

euglenarhodon A lutein carotenoid from certain flagellates.

Euglucon Trademark for glibenclamide.

eukairit CuAgSe. A native selenide.

eulachon oil Candlefish oil. The oil of *Thaleichthys pacificus,* candlefish; a cod-liver oil substitute.

eulytin $Bi(SiO_4)_3$. A yellow silicate, d.6.1, hardness 5–6.

Eumydrin Trademark for atropine methylnitrate.

eunatrol Sodium oleate.

euonymin Euonymus. A glucoside from the bark and root of *Euonymus* species. Yellow powder; a cholagogue and cathartic.

euonymit Dulcitol.

euonymoid The combined principle from *Euonymus.* Green powder, soluble in alcohol; mild cathartic.

euonymus (1) *Euonymus.* A genus of shrubs (Celastraceae). (2) Indian arrowroot, wahoo, spindle tree, burning bush. The dried bark of the root of *E. atropurpureus* (Celastraceae); a cathartic.

euosmite $C_{34}H_{29}O_2$ = 469.6. A fossil resin.

euparin $C_{13}H_{12}O_3$ = 216.2. Yellow crystals, m.121, from the leaves of *Eupatorium purpureum* (Compositae), purple hemp; an aromatic.

eupatheoscope An instrument to measure the cooling effect of air currents. *Cf. kata*thermometer.

eupatirin Stevioside.

eupatorin A glucoside from the leaves of *Eupatorium perfoliatum,* boneset, thoroughwort; and *Eupatorium cannabinum,* hempweed (Compositae), Europe, America, Asia.

eupatorium (1) *Eupatorium.* A genus of Compositae; as, *E. cannabinum,* hemp. (2) Indian page, boneset, thoroughwort. The leaves and flowering tops of *Eupatorium perfoliatum.*

euphenics A branch of genetics concerned with the replacement of defective cells with genetically healthy cells.

euphorbain A protease in the latex of certain Euphorbiaceae, e.g., *E. lathyris.*

Euphorbia A genus of plants characterized by a milky juice. **e. gum** Potato gum. A latex gum from the W. African *E.*: rubber 10, resins 50%. Cf. *euphorbium.*

Euphorbiaceae The spurge family: a group of herbs, shrubs, and trees, often with a milky, acrid juice yielding rubber; e.g.: roots, *Euphorbia resinifera* (euphorbium); herbs, *Mercurialis perennis* (mercurialis); barks, *Croton eluteria* (cascarilla bark); seeds, *Ricinus communis* (castor oil), *Croton tiglium* (croton seed); glands, *Mallotus philippinensis* (kamala); plant products, *Hevea brasiliensis* (para rubber).

euphorbol $C_{31}H_{52}O$ = 440.8. Crystals, m.127, from the resinous juice of *Euphorbia* species.

euphorbium A resin from *Euphorbia resinifera,* Morocco; an emetic and aperient. Cf. *euphorbia* gum.

euphorine Phenylurethane.

euphthalmine $C_{17}H_{25}O_3N\cdot HCl$ = 327.8. Eucatropine. A synthetic alkaloid, the mandelic acid derivative of eucaine. White powder, m.113, soluble in water; a mydriatic.

eupion An antiseptic constituent of wood tar.

eupittone Eupittonic acid.

eupittonic acid $C_{19}H_8O_3(OMe)_6$ = 470.5. Pitacol, eupittone, hexamethoxyaurine. A hexamethoxylated amino dye. Orange needles, decomp. 200, soluble in alkalies, with blue coloration.

eupurpuroid The combined principles from *Eupatorium purpureum,* purple boneset, Joe-Pye weed; a diuretic.

eupyrine $C_{18}H_{18}O_5N$ = 328.3. Vanillin ethyl carbonate, *p*-phenetidine. Green crystals, m.87, sparingly soluble in water; an antipyretic.

euquinine Quinine diethyl carbonate.

Eurax Trademark for crotamiton.

Euresol Trademark for resorcinol monoacetate; an antiseptic.

eurobin Chrysarobin triacetate. Yellow powder; an antiseptic.

europium* Eu = 151.96; at. no. 63. A rare-earth metal discovered (1896) by Demarcay in cerium minerals. m.820. Principal valency 3; its compounds are rose-colored. **e. nitrate*** $Eu(NO_3)_3$ = 338.0. Colorless crystals, soluble in water. **e. oxide*** Eu_2O_3 = 352.00. White powder, insoluble in water. **e. sulfate*** $Eu_2(SO_4)_3\cdot 8H_2O$ = 736.2. Colorless crystals, soluble in water.

eurybin A glucoside from *Eurybia* (or *Olearia*) *moschata,* New Zealand.

euryhaline Capable of living in fresh or salt water; as, salmon.

euscopol Scopolamine hydrobromide.

eustenin Theobromine sodium iodide.

eutannin Chebulinic acid.

eutectic An alloy of metals in proportions such that it has the lowest possible melting point. **e. alloy** A mixture of metals which solidifies completely at the e. temperature.
e. mixture A mixture of substances which has the lowest possible constant melting point. Cf. *dystectic mixture.*
e. temperature The melting point of a eutectic mixture.

euthenics A branch of biomedicine that aims to provide the human organism with improved environment, feeding, exercise, and education. Cf. *eugenics.*

eutrophication Stream pollution by excessive growth of green plants (algae) due to fertilizer-type contaminants. The plants cut off sunlight from the fauna of the lower layers.

eutropic series A series in which crystalline form and physical constants show a regular variation.

eutropy The progression of the isomorphism of crystals of salts of an element, with its atomic number.

euxanthic acid $C_{19}H_{18}O_{11}\cdot H_2O$ = 440.4. Purreic acid, hamathionic acid. Yellow crystals from Indian yellow (purree), decomp. 160, soluble in hot water.

euxanthin $C_{19}H_{16}O_{10}$ = 404.3. The principal constituent of Indian yellow (purree). Yellow crystals.

euxanthinic acid Euxanthic acid.

euxanthone $(HO\cdot C_6H_3)_2CO\cdot O$ = 228.2. Purrone, purrenone, porphyric acid, 1,7-dihydroxyxanthone. Yellow crystals, m.237. **iso ~** 1,6-Dihydroxyxanthone. Yellow crystals, m.258, soluble in alkalies with fluorescence.
methoxy ~ Gentisin.

euxanthonic acid $C_{13}H_{10}O_5$ = 246.2. Yellow needles, decomp. 200; soluble in hot water.

euxantogen Mangiferin.

euxenite (Ca,Ce,Th,U,Y)(Cb,Ta,Ti)$_2$O$_6$. Polycrase. A mineral containing helium.

eV Abbreviation for electronvolt.

evacuant A cathartic.

evacuate To remove a gas from a container; to produce a vacuum.

Evans blue $C_{34}H_{24}O_{14}N_6S_4Na_4$ = 960.8. Azovan blue. Blue, hygroscopic powder; used in the determination of blood and plasma volumes (USP).

evansite $H_6Al_3PO_{10}$. A hydrous aluminum phosphate.
evaporate (1) To convert a liquid into its vapor by heat or low pressure. (2) To concentrate a solution by removing part of the solvent as gas or vapor.
evaporation The process of converting a liquid into a vapor. **direct ~** E. by means of flame. **indirect ~** E. by means of steam, water, oil bath, or other indirect source of heat. **latent heat of ~** The amount of heat energy absorbed by a unit weight of substance as it passes from the liquid to the vapor state. If T is the thermodynamic scale boiling point, and E the molecular elevation of boiling point of a solution, then the latent heat of e. $= 0.02T^2/E$. See *latent heat.* **spontaneous ~** E. without artificial heat, e.g., by sun heat or air currents. **vacuum ~** E. by exposure to low pressure.
e. burner A gas burner with a perforated disk top to heat a large area. **e. dish** A shallow dish to contain evaporating solutions.
evaporator A device to volatilize liquids. **solar ~** An appliance that heats a liquid from above, by infrared radiation.
evaporimeter An instrument to measure the rate of evaporation of a liquid in terms of the rate of fall of its level.
evaporite Natural deposit of mineral salts produced by evaporation of large volumes of water.
Everitt's salt Iron(II) potassium hexacyanoferrate(II)*.
evernesic acid Everninic acid.
evernic acid $C_{17}H_{16}O_7 = 332.3$. Colorless needles, m.170, insoluble in water. A homolog of lecanoric acid, from *Evernia prunastri,* a lichen.
everniine $C_{16}H_{14}O_7 = 318.2$. Yellow powder, from gums and lichens, soluble in water.
everninic acid $C_9H_{10}O_4 = 182.2$. Evernesic acid, 4-methoxy-6-methylsalicylic acid, from lichens. Colorless crystals, m.170, soluble in hot water. **ethyl ester of ~** Lichenol. **methyl ester of ~** Sparassol.
Evipal Trademark for hexobarbitone. **e. sodium** E. soluble; sodium hexobarbitone.
Evipan Trademark for hexobarbitone.
evodene A terpene from *Evodia rutaecarpa.*
evodiamine $C_{19}H_{17}ON_3 = 303.4$. An alkaloid from *Evodia rutaecarpa.* Colorless crystals, m.278.
evolution (1) Chemistry: the escape or liberation of a gas. (2) Biology: the gradual development of a species through many generations, by which new species are formed and develop into more specialized organisms. **atomic ~** See *spectral classification.*
evolutionary operation EVOP. Philosophy prescribing the optimization of systems through systematic and organized effort. Concerned principally with processes.
Ewens-Bassett system See *nomenclature.*
Ewer and Pick's acid Naphthalene-1,6-disulfonic acid.
ex An inscription on a pipet or measuring vessel, indicating that it is to be used to deliver (not to contain) its nominal volume. Cf. *in.*
ex- Prefix (Greek) indicating "out of" or "out from."
exa-* Abbrev. E. SI system prefix for a multiple of 10^{18}.
exaltation The amount by which the molecular refraction of a compound exceeds the sum of the refractions of its atoms; an indication of constitution.
exalton(e) Cyclopentadecanone.
exanthema (1) An infectious disease, often viral, with a characteristic rash, the exanthem; as, measles, chicken pox. (2) A copper deficiency, pathological condition of certain citrus fruits.
excelsin A globulin, mol. wt. 14,738, from brazil nuts.
exception By exception. A process control approach, whereby only conditions requiring the operator's attention are brought to notice.
excipient An inert carrier for a medicinal agent; as, starch in tablets.
excitation Activation. The passage of an electron from its normal (ground) state to one of higher energy level due to absorption of radiation. On the electron's return to the normal state, the absorbed radiation is emitted, e.g., as thermal or kinetic energy, or as fluorescence. An asterisk (as a right superscript) denotes an electronic or nuclear excited state, with m (for "metastable") an alternative for nuclear excited states; as, He^*, $^{110}Ag^m$. **photochemical ~** Excitation. **thermal ~** See *thermal excitation.*
excitomotors Drugs that excite the nerve activities.
exciton The energy of an electron, which can be passed from atom to atom. Thus describing a molecule (as found in silicon) which is a stable complex of 2 pairs of electrons and unoccupied energy levels ("holes") that electrons can fill. Applied primarily to the solid state. See *semiconductor.*
exclusion principle Pauli's principle.
excretin $C_{20}H_{36}O = 292.5$. A cholesterol-like substance in human feces. Yellow needles, m.96, insoluble in water.
exinite A perhydrous constituent of coal with preserved protective plant structures, e.g., cuticles.
exo- Prefix indicating substitution in a side chain. Cf. *endo-.*
exobiology The study of life in space.
exocondensation Ring formation.
exocyclic Pertaining to a cyclic compound with substitution or a double bond in the side chain.
exogenous Produced outside. **e. purines** The purine bodies of excretions, which have passed through the system and have their origin in food. Cf. *endogenous.*
exosmosis The diffusion of salts through the plant membrane from the protoplasm into water. Antonym: endosmosis.
exothermic Indicating liberation or escape of heat. **e. compound** A stable compound formed with liberation of heat, as result of an e. reaction. **e. reaction** A spontaneous reaction in which heat is liberated; usually rapidly and sometimes explosively. Cf. *endothermic.*
expansion An increase in dimension. **adiabatic ~** The rapid e. of a gas, with a cooling effect. **cubical ~** Volumetric e. **linear ~** Increase in length. **thermal ~** The increase in volume due to increase in temperature. It depends on the coefficient of e., which is relatively small for solids and liquids, but large for gases. **volumetric ~** Increase in 3 dimensions.
e. equations (1) *Linear* e. of solids: If l_0 is the length at 0°C, x the coefficient of e., then the length at t°C is: $l_t = l_0(1 + xt)$. The coefficient of e. is $(l_t - l_0)/t \cdot l_0$. (2) *Cubical* e. of solids; $l_t = l_0(1 + at + bt)$, where a and b are the first and second coefficients of e. For isotropic solids or liquids (approximately): $V_t = V_0(1 + 3xt)$. E. of *gases:* The volume V_t of a gas at temperature t°C (at constant pressure) for an original volume V_0 at 0°C is given by $V_t = V_0(1 + 0.00367t)$. Cf. *Charles's law, gas* laws. **e. regulator** Thermoregulator.
expectorant A drug that promotes the removal of mucus from the respiratory tract. It irritates the respiratory tract and stomach; as, ammonium chloride.
experiment A test or trial to illustrate natural phenomena, or to determine some unknown fact by careful observation and sometimes structured variation of the operating conditions. Observation, by comparison, involves acceptance of the conditions of nature.
experimental Pertaining to knowledge obtained by actual tests, or based on facts, not speculation.

experimentation The performance of experiments.

expired air Air from the lungs of living organisms; it has less O_2, but more CO_2 and moisture, than normal air.

exploding atom An atom undergoing radioactive change by the spontaneous emission of electrons or helium nuclei.

explosion A sudden, violent, and noisy exothermic chemical change in which heat, light, and gases are produced. Cf. *implosion.* **e. forming** The formation of metal shapes by the action of an e. on molten metal in a mold. **e. spectrum** See *explosion spectrum* under *spectrum.*

explosives A group of endothermic compounds or mixtures which cause explosions. They are characterized by sensitivity to ignition and detonation, and have a high detonation velocity and explosive strength. Generally applied to mixtures used for ordnance, pyrotechnics, or mining operations, approximately 1 kg of e. being used for every 8.4 tons of coal mined. Classification by composition:

1. Gunpowders (mixtures of potassium nitrate, sulfur, and charcoal).
2. Nitrate mixtures (as, class 1, with barium or sodium nitrate in place of potassium nitrate).
3. Potassium chlorate mixtures.
4. Dynamites (nitroglycerin).
5. Guncottons (nitro compounds).
6. Picric acid or its derivatives.
7. Spreng e. (not explosive, but becoming so on addition of an oxidizing substance and detonator).
8. Miscellaneous and fulminating mixtures.
9. Nuclear e. (*atom* bomb, q.v.). Cf. *propellant.*

Classification by chemical properties:

1. N—O bonds; (*a*) Nitrates and nitric esters (e.g., ammonium nitrate, nitrocellulose); (*b*) carbon-nitro groups (e.g., tetranitromethane, picric acid)
2. Peroxides and ozonides (e.g., hydrogen peroxide)
3. Chlorine compounds (e.g., methyldichloroamine)
4. Self-linked nitrogen compounds (contain N—N, N=N, and N≡N linkages; e.g., hydrogen azide, lead azide)

blasting ~ E. used in mining. **compound ~** E. that are a mixture of substances, e.g., gunpowder. **detonating ~** High e. **forbidden ~** A group of unstable chemicals, mixtures, and devices which, by U.S. law, may not normally be transported; as, nitroglycerin, diethylene glycol dinitrate, and mixtures of a chlorate with either an ammonium salt or an acidic metal salt. **high ~** Detonating e. E. that are more sudden than gunpowder. They have high brisance and shattering power, but little propelling power. **low ~** E. that can be fired by flame. **nondetonating ~** E. that explode like gunpowder. **permissible ~** E., for use in underground coal mines, that have passed the certification tests of the Mining Enforcement and Safety Administration (MESA), U.S. Dept of the Interior, and have been marked as such. **propulsive ~** E. used in shells and guns, e.g., cordite. **simple ~** Pure or single explosive substances, e.g., nitroglycerin.

exponent See *mantissa.*

exponential If the base of the napierian system of logarithms *e* (0.43429) is raised to the power indicated by the variable x, then e^x is the exponential of x.

expressed Squeezed out. **e. oil** A fatty *oil*, q.v., as compared with essential oil (distilled). See *fats.*

expression (1) The squeezing out of a liquid, e.g., oils. (2) A mathematical symbol or equation.

expt Abbreviation for experiment.

extender A material added to serve as a diluent, or to cheapen a mix, or to adjust the physical condition of a product; e.g., china clay in paints.

extinction Fading out. **e. coefficient** The quantity $1/cd$ (log I_0/I), where I_0 and I are the intensities of light, respectively, falling on and transmitted by a solution d cm thick and of molar concentration c. Characteristic curves relate the e. c. and the wavelength of light used. Cf. *absorbance.*

Exton reagent A reagent for albumin (a solution of 200 g $Na_2SO_4 \cdot 10H_2O$ and 50 g sulfosalicylic acid in 1 L water).

extract (1) A dried plant juice; e.g., of aloes, kino, opium. (2) A pharmaceutical preparation made from vegetable tissues by expression, maceration, digestion, or infusion with a solvent. **acet ~** E. prepared by macerating a drug with acetic acid. **alcoholic ~** E. prepared by exhaustively extracting a drug with alcohol. **aqueous ~** E. prepared by infusing or percolating a powdered drug with water. **dry ~** Powdered e. **fluid ~** See *fluidextract.* **nitrogen-free ~** The difference between the weight of a vegetable material and the combined weights of its moisture, ash, fat, protein, and crude fiber; e.g., mainly carbohydrates. **powdered ~** An evaporated and powdered e. **solid ~** A thick, semisolid e.

extractant The liquid phase used to remove a solute from another liquid phase.

extraction Dissolution and removal of one constituent of a mixture in a solvent. **alcohol ~** See *alcoholic extract* under *extract, alcoholic tincture* under *tincture.* **back ~** Stripping. **dilute acetic acid ~** See *vinegar.* **ether ~** See *oleoresin.* **liquid-liquid ~** The transfer of a solute from one liquid phase to another liquid (immiscible) phase. **water ~** See *decoction, infusion.*

e. apparatus Apparatus used for the extraction of fats, oils, or waxes from substances, preparatory to or as part of analysis. See *Soxhlet apparatus.* **e. coefficient** The ratio of the concentrations of a solute in the organic and aqueous phases. **e. flasks** Small round flasks of high-resistance glass. **e. thimble** A cup made of some fat-free, porous material (filter paper, Alundum) in which substances are extracted. **e. tube** A tube holding e. thimbles, so that the solution flows downward into the container, and the solvent vapors pass above the thimbles to the condenser; e.g., Soxhlet tube.

extractive Unspecified substance(s) extracted from a specific solvent in chemical analysis, e.g., ether extractives.

extrapolation The deduction, as by graphical methods, of a value that lies beyond the range of values already established. Cf. *interpolation.*

extrusion Forcing of a substance through an aperture.

exudate A material that has filtered through the walls of living cells and accumulates in the adjacent tissues. **vegetable ~** Types: (1) Soluble in water, insoluble in alcohol—gums. (2) Soluble in alcohol, insoluble in water—resins.

exude To ooze out; e.g., a soft or fusible substance from a harder or less fusible material, under heat or pressure.

E.Y. Eger's yellow.

eyebright The herb of *Euphrasia officinalis* (Scrophulariaceae); a traditional remedy, used as an eye lotion.

F

F Symbol for (1) farad; (2) fluorine. **F acid** Casella acid. 2-Naphthol-7-sulfonic acid.
F11, F12, etc. See *Freon.*
°F Symbol for degree Fahrenheit.
F Symbol for: (1) degree of freedom; (2) Faraday constant; (3) force; (4) Helmholtz function; (5) hyperfine structure quantum number.
F (Boldface.) Symbol for force (vector).
f Symbol for femto- (10^{-15}).
f Symbol for (1) frequency; (2) function; thus, $x = f(t)$, means x is a function of t.
Φ, ϕ, φ See *phi.*
faba Latin for "bean." **f. physostigma** See *physostigma.*
fabiana The dried leaves of *F. imbricata* (Solanaceae), S. America. **f. resin** $(C_{18}H_{30}O_2)_3$. F. tannoid. A crystalline resin from f. Cf. *crocin, pichi.*
fabianol $C_{54}H_{90}O_2$ = 771.3. A volatile, oily liquid from *Fabiana imbricata.*
fabric General term for a cloth material of fibrous structure. Cf. *textile, staple, fiber.* Types: (1) Felt, made by interlocking fibers by the combined action of physical entanglement, chemical action, moisture, and heat, usually from animal fibers, e.g., wool. (2) Knitted f., made by interlooping the fibers. (3) Woven f., made by interlacing spun yarns, or continuous filaments containing one or more filaments, in 2 directions at right angles. **bonded ~** F. made by applying a synthetic resin (by spraying, impregnating, or printing), in the form of fiber, powder, or liquid, to a web or sheet of fibers, and subjecting the resulting product to heat or to an organic solvent to complete the binding action; and finally to curing. The binder may be a thermoplastic fiber comprising 100% of the web. **needled ~** A f. of improved strength and closeness made by punching projecting fibers into the mesh. **nonwoven ~** See *nonwoven.*
Fabrikoid Trademark for pyroxylin-treated fabrics, used for bookbinding, etc.
fabulite Synthetic strontium nitrate used for imitation diamonds.
fac-* Affix indicating 3 groups occupying the corners of the same face of an octahedron.
face-centered cube A unit of crystal structure: a *space lattice,* q.v., in which a fifth atom is located centrally to the 4 corners of a plane.
facing The addition of a coloring matter to food to improve its appearance, as in flour.
Factice Trademark to designate a line of vulcanized vegetable oils.
factitious Made artificially.
factor (1) A reacting substance (reactant) which takes part in a chemical change. (2) When two quantities are multiplied together, each is a f. of the multiplication. **accessory food ~** Vitamin. **analytic ~** Gravimetric f. or volumetric f. **conversion ~** See *conversion factor.* **filtrate ~** Obsolete name for vitamin B_3. **gravimetric ~** A quantity which, multiplied by the weight of precipitate obtained in gravimetric analysis, gives the amount of the related substance being determined. **round ~** See *round factor.* **volumetric ~** A quantity which, multiplied by the number of mL of a standard solution, gives the amount of the corresponding substance that is being determined in volumetric analysis. **volumetric correction ~** A quantity (determined by experiment) which, multiplied by the number of mL of a volumetric solution, gives the number of mL of a normal solution.
f. quantity An aid to rapid calculation in quantitative analysis; e.g., in the determination of carbon in steel as CO_2, the f. quantity is 2.73; if 2.73 g of steel is taken for analysis, the number of decigrams of CO_2 gives the percent of carbon present.
facultative Permissive or optional. **f. aerobe** A bacterium which prefers anaerobic conditions, but can live in an atmosphere containing oxygen. **f. anaerobe** A bacterium which prefers an oxygen atmosphere, but can live in its absence.
FAD* Symbol for flavin-adenine dinucleotide.
fadeometer An instrument for comparing the fading properties of dyed fabrics, etc., which are exposed under standard conditions to a carbon arc.
faeces See *feces.*
fagacid An acid resin from beech wood, soluble in alkalies; an antiseptic in soaps.
fagarine The 3 alkaloids of *Fagara coco,* a tree of Argentina; they are differentiated by the prefixes α, β, γ. The α-fagarine $(C_{19}H_{23}NO_4)$ is a possible substitute for quinidine; β-fagarine is identical with skimmianine; γ-fagarine is $C_{13}H_{11}O_3N$.
fagopyrism Photosensitivity in animals due to eating buckwheat. Cf. *hypericism.*
fahlerz Tetrahedrite.
Fahr Abbreviation for Fahrenheit.
Fahrenheit, Gabriel Daniel (1686–1736) German physicist. **F. scale** A thermometer scale invented by F. based on the lowest temperature he could obtain by freezing mixtures. Water freezes at 32°F and boils at 212°F. See *temperature scales.*
faïence Glazed pottery.
faille A plain weave used for dress or shirt fabrics, with a pronounced rib.
Fairbanks cement testing machine A lever device resembling a balance for testing the tensile strength of cement blocks.
Fajans-Soddy law When an α particle is expelled from a radioactive substance, the product is 2 places lower in the periodic table. A β-particle change (expulsion of an electron) produces a rise of one place.
fallout The radioactive debris from a nuclear explosion. It enters the human body mainly with food. The principal radioactive elements, the foods conveying them, and the organs affected are: iodine-131 (milk, thyroid gland), strontium-90 (all foods, bones), cesium-137 and carbon-14 (all foods, reproductive cells). Believed to increase the incidence of leukemia, cancer, and gene mutation.
false body A term that describes a marked decrease in viscosity resulting from an increase in shear rate, followed by

rapid recovery on release of the shear. Applied to inks, paints, coating mixes, etc. Cf. *thixotropy.*
false hellebore Adonis.
false unicorn Aletris.
Fament's process The removal of phosphorus and sulfur from iron by a current of hydrogen in heated retorts.
family (1) A group, or a part of a period of elements that have similar properties. See *periodic table.* (2) Order. A biological division higher than genus.
fanghi di sclefani A fine, yellow volcanic powder, mainly sulfur, with small amounts of manganese, iron, and calcium.
Fansidar Trademark for antimalarial mixture of sulfadoxine and pyrimethamine.
Fansil Trademark for sulfadoxine.
farad F; the SI unit of electric capacitance; the capacitance of a capacitor charged to a potential of 1 volt by 1 coulomb of electricity; 1 farad = 10^{-9} abfarad = 10^{-9} emu = 8.988×10^{11} esu = 8.988×10^{11} statfarad. Cf. *faraday constant.* **micro ~** The one-millionth part of a f.; the practical unit.
Faraday, Michael (1791–1867) English chemist and physicist. **F. cell** A system in which 2 crystalline quartz prisms each acts as a dispenser, and as a polarizer or analyzer. **F. effect** A beam of polarized light passed through a magnetic field is rotated in the direction of the lines of magnetic force. **F. laws** (1) The weight of an ion deposited electrolytically is proportional to the strength of the current passing through the solution. (2) 1 *faraday,* q.v., liberates 1 gram equivalent of an ion by electrolysis. See *electrochemical equivalent.* **F. tube** Ammonia tube. A V-shaped tube of hard glass, which inverted, is used for distillation and crystallization of liquids under pressure. **F. unit** Farad. **F. washing bottle** Washing bottle.
faraday See *faraday constant.*
faraday constant* *F;* the quantity of electricity which liberates one gram equivalent of a metal by electrolysis. 1 faraday (based on ^{12}C) = 96,487 coulomb. 1 f. (chemical) = 96,496 C. 1 f. (physical) = 96,522 C. Cf. *farad, Faraday laws.*
faradiol $C_{30}H_{50}O_2$ = 442.7. A dihydric alcohol from colt's foot, *Tussilago farfara* (Compositae).
faradization The therapeutic use of induced high-voltage currents.
farina (1) Flour, or fine meal. (2) A starch, usually from potato.
farinaceous Containing or consisting of flour.
farinose The substance of which the cell walls of starch granules are composed.
farnesene $C_{15}H_{24}$ = 204.4. 3,7,11-Trimethyl-1,3,6,10-dodecatetrene*. A sesquiterpene from citronella oil.
farnesol $CMe_2{:}CH(CH_2)_2CMe{:}CH(CH_2)_2CMe{:}CH \cdot CH_2OH$ = 222.4. Isomer of nerolidol, d.0.895, $b_{0.2mm}$120, from the flowers of *Acacia farnesiana,* oil of cassia, ambrette seed oil, etc.; has a floral odor; used in perfumery.
Farrar's process Pig iron is treated with ammonium chloride, potassium hexacyanoferrate(II), and manganese dioxide.
fastness The extent to which a dye or dyed fabric, etc., resists change of color on exposure to light and/or air.
fast breeder reactor See *fast breeder reactor* under *reactor.*
fast red Azo-2-naphthol-1-naphthylaminosulfonic acid. A red indicator dye, pH 10.5–12.1.
fat (1) A solid or liquid oil, the glyceryl esters of the higher fatty acids, e.g., tristearin. See *fats.* (2) Abounding, rich. Cf. *lean.* **animal ~** A f. of animal origin. **hard ~** A mixture of mon-, di-, and triglycerides of saturated fatty acids. Used in pharmacy (EP, BP). **mineral ~** See *mineral oil* under *oil.* **vegetable ~** A f. of vegetable origin. **wool ~** Degras. **f. asphalt** An asphalt of low gravel content. **f. clay** A clay of good plasticity. **f. coal** Coal rich in volatile matter. **f. ore** A high-grade ore. **f. sand** A sand for molding, containing large amounts of clay and alumina. **f. soluble** A substance soluble in oils; as, vitamin A.
fatal Causing death. **f. dose** The quantity of a drug that causes death.
fathom Nautical unit of 6 feet = 1.8288 m.
fatigue The local deformation of metals by repeated stresses.
fats (1) Greasy or oily substances. (2) Fixed oils, fatty oils, expressed oils. The glyceryl ester of a fatty acid, or of a mixture of fatty acids. Generally odorless, colorless, and tasteless if pure, but they may be flavored according to origin. F. are insoluble in water, soluble in most organic solvents. They occur in animal and vegetable tissue and are generally obtained by heating or boiling or by extraction under pressure. F. are important in the diet, as a source of energy and vitamin A. Cf. *oil, wax, sterols.* **polyunsaturated ~** F. containing more than one double bond; as, from linoleic acid. **saturated ~** F. containing no double bonds; as, from palmitic acid. Dietary excess contributes to atheroma. **unsaturated ~** F. containing one or more double bonds; as, from oleic acid.
fatsin $C_{31}H_{53}O_{20}$ = 745.8. A glucoside from *Fatsia japonica* (Araliaceae), Japan.
fatty acids (1) Organic, monobasic acids derived from hydrocarbons by the equivalent of oxidation of a methyl group to an alcohol, aldehyde, and then acid:

$$R \rightarrow CH_3 \rightarrow R \cdot CH_2OH \rightarrow R \cdot CHO \rightarrow R \cdot COOH$$

(*a*) Saturated: $C_nH_{2n+1}COOH$ (as, acetic acid, CH_3COOH);
(*b*) unsaturated: $C_nH_{2n-1}COOH$ (see *acrylic acids*);
(*c*) unsaturated: $C_nH_{2n-3}COOH$ (see *acetylene acids*);
(*d*) unsaturated: $C_nH_{2n-5}COOH$ (as, linolenic acid, $C_{17}H_{29}COOH$). (2) The three acids occurring most frequently in fats as glyceryl esters: palmitic, stearic, and oleic acids.
fatty compounds Aliphatic compounds.
fault finder An instrument for measuring the resistance of conductors; used to locate faults.
fayalite (1) Fe_2SiO_4. A ferrous orthosilicate olivine (Fayal, Madeira).
FDA Food and Drug Administration of the U.S. government.
Fe Symbol for iron (ferrum).
Feather Trademark for a viscous synthetic fiber. **f. ore** Jamesonite.
febrifacient A drug that produces fever.
febrifuge A drug that reduces fever, e.g., an antipyretic.
feces, faeces Excrements or alimentary refuse.
feedback The measurement of how much a particular property (as temperature) differs from a required value, and the use of this measurement to restore the latter.
feedstock The raw material of a chemical process; as, petroleum fed to a refinery. Petroleum is the f. for 75% of W. Europe's chemical and pharmaceutical industries.
Fehling, Herman von (1812–1885) German chemist. **F. solution** A mixture for the determination of reducing sugars: solution A (copper sulfate 34.639 g in 500 mL water) and solution B (sodium potassium tartrate 173, sodium hydroxide 60 g, in 500 mL water), mixed in equal parts before use. The copper produced by reduction at the boiling point is weighed or measured volumetrically. Cf. *Benedict's solution, Pavy's solution.*
Feic Abbreviation for the ferricyanide [hexacyanoferrate(III)*] ion, $Fe(CN)_6^{3-}$.
feints The impure portion of the second distillate of fermentation alcohol.

Feiser solution A solution of sodium dithionite 16, sodium hydroxide 6.6, and sodium anthraquinone-2-sulfonate 2 g, in 100 mL water; used to absorb oxygen in gas analysis.
feldspar Felspar. (1) Igneous, crystalline rocks, chiefly silicates of alumina with soda, potash, or lime. (2) The mineral $K_2O,Al_2O_3,6SiO_2$.
felite A form of belite.
fellmongering The operation of separating the wool of sheep from the pelt.
felspar Feldspar.
felvation A technique for classifying powders; involving fluidization, eleutriation, and sieving.
Femergin Trademark for ergotamine tartrate.
femic Containing iron and magnesium. **f. minerals** Igneous rocks richer in iron than in aluminum.
Feminone Trademark for ethinyl estradiol.
femto-* Abbrev. f; SI prefix for 10^{-15}.
-fen, -phen Suffix indicating a sulfur-containing hydrocarbon.
fenchane $C_{10}H_{18}$ = 138.3. 1,3,3-Trimethylbicyclo[2.2.1]heptane†. An isomer of bornane, pinane, and carane. Cf. *terpenes.*
fenchanol Fenchyl alcohol.
fenchanone Fenchone.
fenchene $C_{10}H_{16}$ = 136.2. *x,x*-Dimethyl-*y*-methylene[2.2.1]heptane*, or *z*-trimethylbicyclo[2.2.1]hept-2-ane*. Terpene constituent of essential oils; isomers: **alpha-~** (x = 7, y = 2) d.0.866, b.155. **beta-~** (x = 2, y = 5) d.0.860, b.152. **gamma-~** (z = 2,5,5) d.0.854, b.146. **delta-~** (z = 1,5,5) b.140.
fenchol Fenchyl alcohol.
fencholic acid $C_{10}H_{18}O_2$ = 170.3. 3-Isopropyl-1-methyl-1-cyclopentanecarboxylic acid*. Colorless crystals, m.18, soluble in alcohol.
fenchone $C_{10}H_{16}O$ = 152.2. 1,3,3-Trimethylbicyclo[2.2.1]heptan-2-one*. In essential oils. Colorless liquid, d.0.9465, b.192, insoluble in water. Cf. *pinone.*
fenchoxime $C_{10}H_{17}ON$ = 167.3. Colorless crystals, m.161, insoluble in water.
fenchyl The radical $C_{10}H_{17}-$, derived from fenchane. **f. alcohol** $C_{10}H_{18}O$ = 154.3. 1-Hydroxyfenchane. Colorless crystals, m.39, in pine oil.
fenchyval The isovaleric ester of fennel oil.
fenfluramine $C_{22}H_{28}ON_2 \cdot C_6H_8O_7$ = 528.6. Ponderax. White crystals, m.149, slightly soluble in water. An appetite suppressant, used to treat obesity (BP).
fenitrothion* See *insecticides,* Table 45 on p. 305.
fennel Foeniculum. The dried fruit of *Foeniculum vulgare* (Umbelliferae); an aromatic, carminative, and spice. **water ~** The oil, d.0.85–0.89, from *Oenanthe aquatica.*
f. oil Oleum foeniculi. The essential oil from *F. vulgare.* Colorless, aromatic liquid, d.0.965–0.975, containing chiefly pinene, phellandrene, and anethole; a carminative and flavoring (NF).
fenoprofen calcium $C_{30}H_{26}O_6Ca \cdot 2H_2O$ = 558.6. Calcium (±)-*m*-phenoxyhydratropate, Fenopron, Nalfon, Progesic. White crystals. An analgesic and anti-inflammatory, used for rheumatic and arthritic conditions (USP, BP).
Fenopron Trademark for fenoprofen calcium.
fenoprop* See *fungicides,* Table 37 on p. 250.
Fentazin Trademark for perphenazine.
fentin **f. acetate*** See *fungicides,* Table 37 on p. 250. **f. hydroxide*** See *fungicides,* Table 37 on page 250.
Fenton's reagent Hydrogen peroxide containing some Fe^{2+}; used for oxidizing sugars and alcohols.
fentanyl citrate $C_{22}H_{28}ON_2 \cdot C_6H_8O_7$ = 528.6. Phentanyl citrate, Sublimaze. White crystals, m.149, sparingly soluble in water. A potent analgesic related to meperidine; used for severe pain and as an adjunct to anesthesia (USP, BP).
fenugreek The seeds of *Trigonella foenum greekum* (Leguminosae), Morocco, India; a condiment.
Feoc Abbreviation for the ferrocyanide [hexacyanoferrate(II)*] ion. $Fe(CN)_6^{4-}$.
ferberite $FeWO_4$. A metallic iron tungstate, containing less than 20% manganese tungstate. Cf. *hubnerite.*
fergusonite $Y(Nb \cdot Ta)O_4$. A native, brown yttrium niobate and tantalate, d.5.8–5.9, hardness 5.5–6.
ferment (1) Former synonym for enzyme. (2) Undergoing fermentation. **co~** Coenzyme. **organized ~** A microorganism. **pro~** Zymogen. **unorganized ~** An extracted enzyme.
fermentation The anaerobic breakdown of organic compounds, by enzymes or microorganisms, to simpler products. Industrial applications include brewing, antibiotics, fats, enzymes, vitamins, and single-cell protein. Main feedstocks are petroleum-derived *n*-alkanes, sugar, and molasses. **acetic ~** The production of vinegar from alcoholic liquids by f. **alcoholic ~** The conversion of sugar to alcohol and carbon dioxide by yeast cells. **amylolytic ~** The hydrolysis of starch to dextrins by saliva. **butyric ~** The formation of butanoic acid. **lactic ~** The souring of milk; e.g., the formation of lactic acid from sugars. **malolactic ~** The conversion of malic acid (e.g., in wines) into lactic acid by f. (e.g., by lactobacilli). **panary ~** The f. processes in bread manufacture. **vinous ~** The production of wine by f.
f. alcohol Ethanol made by f. **f. chemistry** Zymurgy. **f. tube** A bent glass tube with bulb, used in tests for f. to collect the gases evolved.
Fermi, Enrico (1902–1954) Italian-born physicist noted for his synthesis of neptunium, by bombarding uranium with neutrons. A pioneer of the first sustained nuclear reaction. Cf. *ausonium, fermium.* **F. surface** A surface separating energy states that are filled with electrons from those that are not filled.
fermium* Fm. An actinoid element, at. no. 100. Atomic weight of isotope with longest half-life (^{257}Fm) = 257.10. Discovered (1953) by ion-exchange separation of the debris from the thermonuclear explosion "Mike" (Pacific, 1952). It forms Fm^{3+} ions in solution.
ferrate (1)*Indicating iron as the central atom(s) in an anion. (2) A deep, red salt containing the radical FeO_4^{2-}, as in Na_2FeO_4, sodium ferrate(VI)*.
ferret A device that is passed through a pipeline to scrape and wash it.
ferri- Prefix indicating the presence of a ferric ion, Fe^{3+}, in a compound.
ferriammonium- Prefix indicating the presence of both trivalent iron, Fe^{3+}, and ammonium, NH_4^+; as, $NH_4 \cdot FeCl_4$, ammonium iron(III) chloride*.
ferric Iron(III)*, iron(3+)*, ferrum(III)*, ironic. A compound of trivalent iron. Usually more stable than the corresponding ferrous salt; yellow, brown, or red in color. Seldom used medicinally, as not so well absorbed from the intestine as ferrous compounds. **f. acetate** $Fe(C_2H_3O_2)_3$ = 233.0. Brown scales, soluble in water; a dye mordant. **basic ~** $Fe(OH)(C_2H_3O_2)_2$ = 190.9. Red, amorphous powder, sparingly soluble in water; used in the dye industry. **f. ammonium** Ammonium iron(III). **f. ammonium citrate** Red, deliquescent scales, prepared by evaporating together solutions of citric acid and ferric hydroxide. Soluble in water;

a hematinic. **f. arsenate** $FeAsO_4 \cdot 2H_2O$ = 230.8. Brown powder, insolube in water. See *ferrous arsenate.* **f. arsenite** $4Fe_2O_3 \cdot As_2O_3 \cdot 5H_2O$ = 926.7. Brown powder. **f. benzoate** $Fe(PhCOO)_3$ = 419.2. Brown powder, used to make ferrated cod-liver oil. **f. bromide** $FeBr_3$ = 411.0. Red crystals, decomp. by heat, soluble in water. **f. chloride** $FeCl_3$ = 162.2. Ironic chloride, iron 3-chloride. Brown crystals, m.298, soluble in water. **hydrous ~** $FeCl_3 \cdot nH_2O$. Orange crystals (n = 5, 6, or 12), soluble in water; used in the dye industry and as a styptic. **f. chromate** $Fe_2(CrO_4)_3 \cdot nH_2O$. A brown solution miscible with water; a mordant. **acid ~** $Fe_2(Cr_2O_7)_3$ = 759.6. Brown granules, soluble in water; used in pigments. **f. citrate** $FeC_6H_5O_7 \cdot 3H_2O$. Ferri citras. Red scales, slowly soluble in water. **green ~** Ammonium iron(III) citrate. **f. citrate (^{59}Fe)** An injection for investigating hematological disorders (BP). **f. ferricyanide** Iron(III) hexacyanoferrate(III)*. **f. ferrocyanide** Iron(III) hexacyanoferrate(II)*. **f. fluosilicate** Iron(III) hexafluorosilicate*. **f. formate** $Fe(HCOO)_3$ = 190.9. Red crystals, soluble in water. **f. hydroxide** $Fe(OH)_3$ = 106.9. Ironic hydroxide. Brown powder, insoluble in water; an antidote for arsenic, and used in the rubber industry. **f. iodide** FeI_3 = 436.5. Black crystals, soluble in water. **f. nitrate** $Fe(NO_3)_3 \cdot 9H_2O$ = 404.0. Colorless rhombs, m.47, soluble in water. **f. n. solution** An aqueous solution (33% f. nitrate). Red liquid; used in dyeing, calico printing, tanning, and in iron pigments. **f. oleate** $Fe(C_{18}H_{33}O_2)_3$ = 900.2. Red, soft mass, used in paints. **f. oxalate** $Fe_2(COO)_3$ = 243.7. Ironic oxalate. Yellow scales, decomp. 100, soluble in water. **f. oxide** Fe_2O_3 = 159.7. Ironic oxide, red iron oxide. Pompeian red, iron sesquioxide; in nature: red hematite, martite. Red hexagons or powder, m.1541, insoluble in water; a pigment and abrasive for polishing metals; used for coloring lotions, as, calamine (NF). **magnetic ~** Iron(II) diiron(III) oxide*. Used for data storage tapes. **f. perchlorate** $Fe(ClO_4)_3$ = 354.2. Brown crystals, soluble in water. **f. phenolate** Compounds of f. iron and phenol of variable composition; purple, deliquescent masses, soluble in water. **f. phosphate** $FePO_4 \cdot 2H_2O$ = 186.8. Ironic phosphate. Yellow rhombs, insoluble in water. **f. subcarbonate** Sesquioxide of iron, saffron of mars, red oxide of iron, hydras ferricus, crocus martis. A precipitate from iron solutions of variable mixtures of f. oxide and f. hydroxide. **f. subsulfate** $Fe_4O(SO_4)_5$ or $Fe_2(SO_4)_3 \cdot Fe_2(SO_4)_2O$ = 719.7. Monsel's salt. Brown scales, soluble in water; a styptic. **f. sulfate** $Fe_2(SO_4)_3 \cdot 9H_2O$ = 562.0. Yellow rhombs, soluble in water. **f. sulfide** Fe_2S_3 = 207.9. Yellow crystals, decomp. in heat or solvents. **f. thiocyanate** $Fe(SCN)_3 \cdot 3H_2O$ = 284.1. Ironic sulfocyanide, f. rhodanate. Brown granules, soluble in water. **f. vanadate** $Fe(VO_3)_3$ = 352.7. Iron metavanadate. Brown powder, insoluble in water.

ferricyanic acid Hydrogen hexacyanoferrate(III)*. The hypothetical acid, $H_3Fe(CN)_6$, from which the hexacyanoferrate(III) salts (ferricyanides) are derived.

ferricyanide See *ferricyanic acid.*

ferriferous (1) Containing iron in the ferric state. (2) Containing iron.

ferriferrous Ferrosoferric. Describing a compound containing both di- and trivalent iron. See *iron(II) diiron(III)* compounds under *iron.*

ferrimanganese An alloy of iron and manganese; used to deoxidize and desulfurize molten steel.

ferrimanganic A compound containing iron(III) and manganese.

ferripotassium A double salt containing potassium and trivalent iron. See *iron(III) potassium* compounds under *iron.* **f. cyanide** Potassium hexacyanoferrate(III)*.

ferripyrine $2FeCl_3 \cdot 3C_{11}H_{12}ON_2$. Contains ferric chloride 36, antipyrine 64%. Red crystals, soluble in water; a styptic and hematinic.

ferrisodium A double salt containing sodium and trivalent iron. See *iron(III) sodium* compounds under *iron.*

ferrite (1) An unstable double oxide of ferric oxide (as, $Na_2Fe_2O_4$) which exists in strongly alkaline solutions. Cf. *ferrate.* (2) An allotrope of iron, having a body-centered cubic structure; used for magnetic tape. **ferrous ~** Fe_3O_4. Magnetic iron oxide.

ferritin A protein-iron complex that stores iron in the cells of the spleen, liver, and marrow.

ferro- Prefix indicating metallic iron (as in ferroaluminum) or divalent iron (as in ferrocyanide). **f.prussiate** Potassium hexacyanoferrate(II)*.

ferroaluminum An alloy: Fe 80, Al 20%, d.6.30, m.1480.

ferroammonium Describing a compound containing divalent iron and ammonium; as, $(NH_4)_2FeBr_4$, *ammonium* iron(II) bromide*, q.v.

ferrocarbon titanium An alloy obtained by the reduction of titanium oxide with carbon; used to deoxidize molten steel.

ferrocene* $(C_5H_5)_2Fe$ = 186.0. Bis(η-cyclopentadienyl)iron*. Orange crystals, m.173, soluble in ether; notable for its stability.

ferrocerium A pyrophoric alloy made by fusing cerium chloride and iron; used as flint for lighters.

ferrochrome Ferrochromium.

ferrochromium An alloy: Fe 50, Cr 50%, d.6.9, m.1458.

ferroconcrete Reinforced concrete.

ferrocyanic acid Hydrogen hexacyanoferrate(II)*. The hypothetical acid $H_4Fe(CN)_6$ from which the hexacyanoferrate(II) salts (ferrocyanides) are derived.

ferrocyanide See *ferrocyanic acid.*

ferrodolomite The mineral $CaFe(CO_3)_2$.

ferroelectricity The property of a crystal which exhibits reversible spontaneous polarization.

ferroferric Ferriferrous.

ferroferricyanide Iron(II) hexacyanoferrate(III)*.

ferroferrocyanide Iron(II) hexacyanoferrate(II)*.

ferromagnesium (1) An alloy of magnesium and iron. (2) A compound containing divalent iron and magnesium.

ferromagnetism Magnetism due solely to magnetic iron.

ferromanganese The alloy: Fe 50, Mn 50%, d.7.0, m.1325; used for tough steels.

ferromanganic Describing a compound containing divalent iron and trivalent manganese.

ferromanganous Describing a compound containing divalent iron and divalent manganese. See *iron(II) manganese* compounds under *iron.*

ferromolybdenum A steel containing about 2% Mo; used for high-speed lathe tools.

ferron 8-Hydroxy-7-iodo-5-quinolinesulfonic acid. m.265 (decomp.). A reagent for metal ions.

ferronickel The alloy: Fe 74.2, Ni 25, C 0.8%, d.8.1, m.1500; used for steel and tools. **valve steel ~** The alloy: Fe 67.8, Ni 32, C 0.2%, d.8.0, m.1480.

ferrophosphorus A by-product of heating phosphate rock, silica, and coke; used to increase fluidity during steel casting.

ferropotassium Iron(II) potassium. Describing a compound of divalent iron and potassium with another radical. **f. cyanide** Potassium hexacyanoferrate(II)*.

ferropyrine Ferripyrine.

ferrosilicon A hard steel alloy: Fe 97.6, Si 2, C 0.4%; used in deoxidized steel. **f. zirconium** An alloy of iron, silicon, and zirconium, used to purify molten steel and improve shock resistance.

ferrosilite Clinoferrosilite. Acicular crystals of $FeSiO_3$ in obsidian.
ferrosoferric Ferriferrous. **f. oxide** Iron(II) diiron(III) oxide*.
ferrotitanium An alloy made by reducing titanium dioxide with powdered aluminum and iron; used to deoxidize molten steel.
ferrotungsten Tungsten steel. A tool alloy: Fe 94.5, W 5, C 0.5%. **high-speed ~** High-speed tool steel. The alloy: Fe 75, W 18, Cr 6, V 0.3, C 0.7%.
ferrous Iron (II)*, iron(2+)*, ferrum(II)*, ironous. Prefix indicating compounds containing divalent iron. They have generally a green color, give the f. ion Fe^{2+} in aqueous solution, and are reducing agents. **f. acetate** $Fe(CH_3COO)_2$ = 173.9. Green crystals, used in solution as iron liquor (printer's liquor), and as a mordant. **f. ammonium** Ammonium iron(II) compounds. **f. ammonium gluconate** $C_{12}H_{22}O_{14}Fe \cdot 2H_2O$ = 482.2. Yellow powder with caramel odor, soluble in water; a hematinic. **f. arsenate** $Fe_3(AsO_4)_2 \cdot 6H_2O$ = 553.5. Green, amorphous powder, insoluble in water or alcohol, soluble in ammonia or hydrochloric acid. **f. bromide** $FeBr_2 \cdot 6H_2O$ = 323.7. Red crystals, soluble in water. **f. carbonate** $FeCO_3$ = 115.9. Green rhombohedra, decomp. by heat, soluble in carbonated water; used in carbonated "ferruginous" waters. **f. chloride** $FeCl_2 \cdot 4H_2O$ = 198.8. Green, monoclinic crystals, soluble in water. Used as a reducing and adsorbing reagent, in calico printing, as a stain, and in the extraction of copper from ores. **f. ferricyanide** Iron(II) hexacyanoferrate(III)*. **f. ferrite** Fe_3O_4. Magnetic iron oxide. **f. ferrocyanide** Iron(II) hexacyanoferrate(II)*. **f. fluoride** FeF_2 = 93.8 and $FeF_2 \cdot 8H_2O$ = 238.0. White powder, sparingly soluble in water. **f. fluosilicate** Iron(II) hexafluorosilicate*. **f. fumarate** $FeC_4H_2O_4$ = 169.9. Brown powder, slightly soluble in water; a hematinic (USP, BP). **f. gluconate** $C_{12}H_{22}O_4Fe \cdot 2H_2O$ = 322.2. Fergon. Gray-green granules; used for iron deficiency anemia (USP, EP, BP). **f. hydroxide** $Fe(OH)_2$ = 89.8. White powder, rapidly oxidizes to brown; insoluble in water. **f. hypophosphite** Iron(II) phosphinate*. **f. hyposulfite** Iron(II) thiosulfate*. White powder, soluble in water; rapidly oxidizes. **f. iodide** $FeI_2 \cdot 4H_2O$ = 381.7. Green, deliquescent scales, anhydrous at 177, soluble in water. **f. nitrate** $Fe(NO_3)_2 \cdot 6H_2O$ = 287.9. Green crystals, m.61, soluble in water. **f. oxalate** $Fe(COO)_2$ = 143.9. Yellow powder, soluble in water. **f. oxide** FeO = 71.8. Black iron oxide, iron monoxide. Black powder, m.1419, insoluble in water. **f. perchlorate** $Fe(ClO_4)_2 \cdot 6H_2O$ = 362.8. Soluble green powder. **f. phosphate** $Fe_3(PO_4)_2 \cdot 8H_2O$ = 501.6. Blue, monoclinic crystals, insoluble in water; native as vivianite. **f. phosphide** FeP = 86.8. Black powder. Also: Fe_2P, Fe_2P_3, Fe_3P, Fe_3P_4. **f. structure** The crystal structure of cast iron. **f. sulfate** (1) **f.s. heptahydrate, crystallized ~** $FeSO_4 \cdot 7H_2O$ = 278.0. Iron sulfate, green vitriol, green copperas, Ferri sulphas. Blue-green, monoclinic crystals, d.1.875, m.64 (decomp.), soluble in water; native as melanterite. Used for mordanting wool, as a disinfectant, and in the manufacture of ink and prussian blue. Widely used for iron deficiency anemia (USP, EP, BP). (2) **f.s. pentahydrate** $FeSO_4 \cdot 5H_2O$ = 242.0. Green crystals, d.2.2; native as siderotilate. (3) **f.s. tetrahydrate** $FeSO_3 \cdot 4H_2O$ = 208.0. (4) **f.s. monohydrate** $FeSO_4 \cdot H_2O$ = 169.9. White powder, obtained by heating iron sulfate to 140; native as szomolnokite. (5) **anhydrous ~** $FeSO_4$ = 151.9. White powder obtained by heating iron sulfate to 300. **f. sulfide** FeS = 87.9. Pyrite. Black crystals, m.1197, insoluble in water; used for making hydrogen sulfide. **f. tantalate** See *tantalite*. **f. tartrate** $FeC_4H_4O_6$ = 203.9. White crystals, soluble in water. **f. thiocyanate** $Fe(SCN)_2 \cdot 3H_2O$ = 226.0. Green crystals, soluble in water. **f. titanate** $FeTiO_3$. Ilmenite. **f. tungstate** See *ferberite, wolframite*.
ferrovanadium An alloy of iron and vanadium, used in the manufacture of steel for automobile parts.
ferroverdin $C_{30}H_{20}O_8N_2Fe$ = 592.3. Green pigment produced by streptomycetes; insoluble in water.
ferrox process A process for dissolving hydrogen sulfide from industrial gases in a suspension of ferric hydroxide in alkali.
ferroxyl indicator A jelly of potassium hexacyanoferrate(II) and phenolphthalein in agar-agar, used to test for the corrosion of iron. Iron wire electrodes will turn it red (cathode) and blue (anode).
ferrugineous Ferruginous.
ferruginous (1) Containing iron. (2) Describing a drug whose therapeutic effect depends on the presence of iron, e.g., chalybeates.
ferrum* Official Latin for "iron." **f.(II), f.(III) ions*** Alternative names for ferrous and ferric ions, respectively. **f. reductum** Frary's metal, ferry metal. A mixture of iron and iron(II) diiron oxide obtained by the reduction of ferric oxide in hydrogen.
fertilizer Fertiliser. A plant food added to soil. Classification: (1) Primary nutrients (or elements), containing N (e.g., ammonia, urea), P (phosphates), or K (chloride). (2) Secondary nutrients, containing Ca (lime), Mg (dolomite), or S (sulfates). (3) Trace nutrients containing Fe, Mn, Zn, Cu, B or Mo. (4) Organic f. (manure, legumes). **f. grade** The minimum guaranteed plant food expressed in terms of *nitrogen*, available *phosphoric acid*, and water-soluble *potash*.
fertilizin A substance secreted by certain marine invertebrates which fertilizes their ripe eggs; related to *echinochrome A*, q.v.
Ferula A genus of Umbelliferae, yielding asafetida, galbanum, sagapenum, and sumbul.
ferulaldehyde $C_{10}H_{10}O_3$ = 178.2. *p*-Coniferaldehyde, hadromal. Colorless crystals, m.83, used in perfumery. It occurs in woody tissues, and gives a red color with phloroglucinol in hydrochloric acid.
ferulene $C_{15}H_{24}$ = 204.4 (+)-9-Aristolene. A dihydrosesquiterpene from fennel, b_{7mm}126.
ferulic acid $C_{10}H_{10}O_4$ = 194.2. Ferulaic acid, 4-hydroxy-3-methoxycinnamic acid. **(*E*)- ~** A constituent of asafetida and black fir resin. Colorless needles, m.174 (decomp.), soluble in hot water. Cf. *coniferol*. **hydro ~** $C_{10}H_{12}O_4$ = 196.2. Colorless crystals, m.89, soluble in water. **iso ~** Hesperitinic acid.
fervanite A vanadium mineral similar to steigerite.
fervorization The process of heating soil or culture media at 137° for one hour to stimulate the germination of seeds immersed therein.
Féry calorimeter A thermoelectric device for determining the calorific power of foods. **F. refractometer** An instrument for the direct reading of the refractive index of a transparent liquid.
fetron A mixture: stearic acid 3, petrolatum 97%; a base for ointments.
fetuin A globulin of low molecular weight in the fetal serum of cows, sheep, and other mammals.
fetus The unborn child or animal; from the 8th week of intrauterine life to birth in humans. Cf. *embryo*.
Feulgen reaction The restoration of color to Schiff reagent; a stain test for nuclear material in cells, due to the presence of deoxyribose. Cf. *Dische* reaction.
feverbark See *Cinchona*.
feverfew The herb of *Pyrethrum parthenium* (Compositae): a carminative. Cf. *pyrethrum camphor*.

fiber A long ribbon or threadlike cell or tissue of vegetable or animal origin, used for paper (cf. *tracheid*), textiles, cordage, wickerwork and brushes. **animal ~** A f. obtained from animals, as alpaca, silk, wool. **artificial ~** (1) F. made from mineral matter; as, spun glass, metallic threads. Cf. *mineral fiber* below. (2) F. imitating natural f.; as, rayon. **bast ~** A f. from the bast of plants; as, hemp, jute. **bonded ~** See *bonded fiber.* **chloro ~** Collective name for polymers of vinyl chloride and vinylidene chloride. **conjugate ~** Hybrid f. A synthetic f. made by injecting 2 different f.-forming solutions into the same spinneret. **cordage ~** A f. used for making ropes; e.g., hemp, sisal. **crude ~** The residue after boiling fat-free ground plant successively for ½ hour in 1.25% sulfuric acid and 1.25% caustic soda. **glass ~** See *glass fiber.* **hard ~** The leaf or structural f. from tropical and subtropical plants, used for cordage, mats, and sacking; e.g., agave, Manila hemp. **high wet-modulus ~** Synthetic cellulose f. having high tensile strength, wet strength, and dimensional stability. **horn ~** Vulcanized f., leatheroid. A hard, tough substance made by compressing layers of paper treated with acids or zinc chloride. **hybrid ~** Conjugate f. **linen ~** F. used for making threads or yarn from flax. **man-made ~** Synthetic f. **mineral ~** Asbestos f. Cf. *artificial fiber* above. **polynosic ~** See *polynosic.* **soft ~** Bast or stem f.; e.g., flax, hemp, or jute. **split ~** F. made by heating and stretching polymer sheets under conditions which split them into fibers. **staple ~** F., natural or synthetic, in comparatively short and uniform lengths, from which it is spun alone or in admixture into continuous threads. Natural fibers usually occur in this form, but f. made as a continuous filament, e.g., rayon, may be cut if used to make cotton-type fabrics. Cf. *filament* yarn. **sugarcane ~** Bagasse. **synthetic ~** See individual names and *polymers.* **vegetable ~** F. from plants: (1) Hair fibers (from seed hairs), e.g., *Gossypium,* cotton. (2) Bast fibers (from stalks and stems), e.g., *Linum* (Linaceae), flax. (3) Cordage fibers (from vascular bundles), e.g., Agave (Amaryllidaceae), sisal. (4) Paper fibers (from Gramineae and Coniferae), e.g., cereal straws, pine.

f. dimensions See Table 31. **f. fineness value** See *tex, denier, micronaire value.* **f. optic** A system for transmitting an intense beam of light through a narrow, flexible "cable" containing glass f. about 0.15 mm in diameter. Used for communications. See *endoscope.*

Fibestos Trademark for a cellulose acetate synthetic fiber.
Fibramine Trademark for a cellulose-viscose synthetic fiber.
fibre Fiber. **f. V** Dacron.
Fibrelta Trademark for a cellulose-viscose synthetic fiber.
fibrid A fluffy synthetic material having the physical properties of both fibers and films. Used for nonwoven fabrics.

TABLE 31. DIMENSIONS OF SOME COMMON FIBERS

Fiber	Length, mm	Diameter, mm
Cotton	20–30	0.019
Linen	8–70	0.020
Hemp	5–55	0.025
Jute	0.8–6	0.018
Cereal straw	0.3–2.0	0.015
Bamboo	0.8–4.0	0.012
Coniferous wood	2–7	0.040
Deciduous wood	0.7–1.8	0.025
Silk	—	0.013
Wool	20–150	0.060
Asbestos	1.0	0.003

fibril A small fiber or filament.
fibrin A protein from blood (0.1–0.4%) formed from fibrinogen by the action of an enzyme, thrombin; its formation is essential in the clotting of blood. Colorless or yellow, horny masses, insoluble in water, but swelling in dilute acid to a gel. Used technically in photography; the dye, textile, and leather industries; also in foods. **muscle ~** Syntonin. **vegetable ~** Gluten. Gluten f. A by-product in the manufacture of starch. Yellow, horny masses, insoluble in water.

f. foam Artificial foam made by clotting foamed human fibrinogen with thrombin; then freezing, drying, and heating at 130 for 3 hours. A hemostatic aid in surgery (BP).

fibrinogen A globulin of the blood plasma which produces fibrin by the action of the fibrin enzyme, thrombin, and essential for the coagulation of blood. Cf. *fibroinogen.*
fibrinolysin A substance formed in the blood that causes fibrin clots to dissolve.
fibrinolysis The hydrolysis of fibrin.
Fibro Trade name for a cellulose-viscose, synthetic staple fiber.
Fibroceta Trademark for a cellulose acetate synthetic fiber.
fibroin $(C_{15}H_{23}O_6N_5)_n$ = (369.4)*n*. A protein constituent of silk and spider webs.
fibroinogen Renatured fibroin, e.g., produced by neutralization of a solution of fibroin in cupriethylenediamine solution. Cf. *fibrinogen.*
Fibrolan Trademark for cellulose dyes.
Fibrolane Trademark for a protein synthetic fiber.
fibrolite Al_2O_3,SiO_2. Sillimanite. A yellow, native silicate.
fibroplastin Paraglobulin.
Fibrovyl Trademark for a polyvinyl chloride synthetic fiber.
fibrox A fibrous variety of *siloxicon,* q.v.; a heat insulator.
fichtelite $C_{19}H_{34}$ = 262.5. White crystals, m.46, in peat beds (Fichtelgebirge, Bavaria).
ficin* An enzyme cleaving certain peptide linkages; the active principle of oje, the milky sap of the *Ficus* species.
fiducial error limit A statistical expression of the limits of error based on a certain reference figure, e.g., as in 10 ± 0.1.
field (1) The region or space within which a phenomenon occurs. (2) In optics, the area visible at any one time through an instrument, e.g., a microscope. **atomic ~** The space surrounding an atom which cannot normally be penetrated by other atoms. **electric ~** The space surrounding an electrically charged body in which its action is perceptible. **electromagnetic ~** See *electromagnetic field.* **gravitational ~** The region surrounding the earth in which bodies are attracted toward the earth's center. **intraatomic ~** Dynad. **interatomic ~** The space between dynads. Cf. *molecular diagrams, atomic* radius. **magnetic ~** The space surrounding a magnet, in which its action is perceptible. **molecular ~** The space surrounding a molecule.
field intensity The intensity of the energy of a limited region or space. **electric ~** See *electric field strength* under *electric.* **magnetic ~** See *magnetic field strength.*
fig The dried fruit of *Ficus carica* (Urticaceae); a food and laxative (BP). Cf. *cradin.*
figwort Scrophularia.
filament A fine, threadlike body or structure. **carbon ~** A fine thread of graphite or carbon, used in electric bulbs. **nuclear ~** The threadlike chromatin in the nucleus of a cell. **protoplasmic ~** A threadlike protoplasm, as, cilia. **tungsten ~** A fine thread of tungsten, used in electric bulbs.

f. yarn Reeled filaments of synthetic fiber, e.g., rayon, as distinct from staple fiber.

filamentous (1) Having the shape of a filament. (2) A growth of bacteria composed of long, interwoven threads.

TABLE 32. CLASSIFICATION OF FILTERS

Filter	Time to pass 100 mL water	Pore sizes of materials retained (with example)
Coarse	1–10 s	0.5–3 μm (kaolin)
Medium	10–30 s	0.1–1 μm (bacteria)
Fine	30–100 s	0.05–0.5 μm (colloidal gold)
Ultrafine (colloidal):		
Fast	1–5 min	Fine colloids
Medium	6–30 min	Benzopurpurine dyes
Fine	50–150 min	Albumen
Finest	> 150 min	Congo red dyes

filar micrometer A scale attachment for microscopes; 0.001 mm can be estimated.

Filices (Sing., *filix,* q.v.) Ferns. An order of Pteridophytae or spore-bearing plants which yield aspidium, polystichum, filicin, etc.

filicic acid (1)$C_{14}H_{14}O_5$ = 262.3. Colorless crystals, decomp. 185, insoluble in water; a constituent of *Aspidium* species, and a decomposition product of filicin. (2) $C_{36}H_{42}O$ = 490.7. White crystals, m.125, from *Aspidium* species. Cf. *filixic acid.*

filicin $C_{35}H_{40}O_{12}$ = 652.7. Filicic acid anhydride. A constituent of the root of *Aspidium filixmas;* a vermifuge.

filiform Describing a uniform growth of bacteria along the line of inoculation.

filitannic acid The tannin of the male fern.

filix (Pl. *Filices,* q.v.) A fern, as filixmas, the male fern. **f. extract** A fluidextract made from the root of the male fern, *Aspidium filixmas.* It contains filicic acid, filicin, albaspidin, aspidinol, and filmaron.

filixic acid $C_{35}H_{38}O_{12}$ = 650.7. An acid, m.183, from filix extract. Cf. *filicic acid.*

fillers Materials used (1) to fill the pores of paper; (2) to increase the bulk or weight of substances (loading); (3) to modify the properties of synthetic substances or insulating compositions.

film (1) A membrane or covering layer. (2) A light-sensitive, flexible, transparent sheet coated with opaque silver salts for making photographs. (3) The transparent developed form of (2). **thin ~** A material on a substrate, with a thickness not greater than 10 μm and uniformity within 20% of its average value.

f. badge Badge worn by those working with radiation, a film inside measuring the amount they receive. **f. gage** (1) The number of 100 in^2 per 1 lb weight of film (U.S. usage). (2) The weight in g of 10 in^2 of film (U.K. usage).

filmaron $C_{47}H_{54}O_{16}$ = 874.9. An amorphous substance, m.60; the anthelmintic principle of filix extract.

filter (1) A strainer or purifier. (2) Chemistry: a porous material through which a liquid passes for the purpose of (*a*) removing a precipitate or suspended matter or (*b*) clarifying the liquid. See Table 32. Cf. *membrane.* (3) Physics: an absorbing, semitransparent substance, e.g., light f. **asbestos ~** A mixture of asbestos and glass wool used for filtration. **Berkefeld ~** A tube made of diatomaceous earth, used for filtering and sterilizing water. **Chamberland ~** A tube made of porous clay, used for the filtration and sterilization of liquids, vaccines, and serums. **folded ~** F. paper, folded in alternate directions, used for rapid filtration. **gas ~** A device for removing solid or liquid impurities from gases. **gel ~** The use of gel as a *molecular sieve,* q.v. **glass ~** (1) A glass Gooch-type crucible having a sintered glass base of appropriate pore size. Used in quantitative analysis. (2) Glass wool. **Gooch ~** A platinum, glass, or porcelain crucible with a perforated bottom covered by asbestos fibers, in which a precipitate may be heated and weighed. **Kelly ~** A leaf f. for filtering slurries under pressure. **light ~** Filter (3). **membrane ~** A disk of nitrocellulose (porosity 0.03–3 μm) used in analytical and bacteriological work. **optical ~** A device that partially absorbs specific electromagnetic radiation in the infrared to ultraviolet range, such as glass or films separated by thin layers. **paper ~** F. paper. **paper pulp ~** F. paper moistened and pulped for use similar to glass wool. **Pasteur ~** A tube of unglazed porcelain for filtration by pressure or vacuum. **streamline ~** See *streamline filtration.* **ultra ~** See *ultrafiltration, membrane filter* above. **vacuum ~** See *vacuum filtration* under *filtration.*

f. aid A powder, e.g., kieselguhr, added to the solution to be filtered to form a porous bed on the f. and facilitate filtration. **f. bag** A sack of fiber for straining liquids. **f. cloth** Strong canvas for a f. press. **f. cone** A cone of porous material, as, Alundum, for filtration. **f. crucible** Gooch crucible. **f. cylinder** A porous tube used as a f. **f. flask** A conical flask with side arm made of heavy glass, for vacuum filtration. **f. mantle** A glass or metal tube around a f. cylinder. **f. paper** Unsized, porous paper used for filtration; made in various textures and grades of purity. **qualitative ~** A common f. paper used for straining, clearing, and purifying solutions or for collecting suspended matter. **quantitative ~** A high-grade f. paper, resistant to dilute acids and consisting of pure cellulose with a small ash prepared by repeated acid and water washings of the paper pulp. **f. paper analysis** (1) Identification of substances by a spot test or *f. paper test,* q.v. (2) Capillary analysis. (3) The germination of seeds on moist f. paper and examination of the exudation from the rootlets, e.g., by fluorescence methods. **f. paper test** A color reaction made with minute quantities of materials and reagents, with which a f. paper is successively moistened. **f. press** A frame on which perpendicular metal plates are suspended and pressed together by a screw (or hydraulically). Liquid to be filtered is pumped into canvas bags between the plates, and the tightening of the screw furnishes the pressure for filtration. **f. pump** A pressure or vacuum pump used for filtration. Laboratory pumps are usually small vacuum pumps connected to the water faucet and operated by the water current drawing the air from a container. **f. tube** (1) A glass tube connecting a Gooch crucible to a filter flask. (2) Bougie.

filtrate The clear liquid that has passed through a filter. Cf. *tenate.*

filtration The process of separating a solid from a liquid using a porous substance through which only the liquid passes. **centrifugal ~** The separation of filtrate from precipitate by centrifuging. **direct ~** The usual gravity separation of a solid from a liquid through a filter. **edge ~** Streamline f. **forced ~** Centrifugal, pressure, or vacuum f. **meta ~** See *metafiltration.* **pressure ~** F. by forcing the liquid by air pressure through the filter. **streamline ~** See *streamline filtration.* **vacuum ~** F. by drawing the liquid by suction through the filter.

Filtros Brand name for a porous, acid-proof filter material.

fimbriate Describing long filaments on the borders or edges of objects or tissues; as, of bacterial growth in colonies.

fineness (1) The state of subdivision of a powder or granulated substance, determined by sieves. (2) The purity of a gold alloy, expressed in parts per 1,000. See *gold.*

finings Substances added to a fermented beverage to clear it of suspended matter, e.g., yeast, and render it brilliant, e.g., isinglass. **blue ~** Potassium hexacyanoferrate(II) used to remove iron from wine; prohibited in many countries.

finishing material A substance used in industry for the last or finishing stage of manufacture; as, of textiles: (1) Substances making fibers soft, hygroscopic, and pliable; e.g., glycerol. See *plasticizers.* (2) Coloring substances, e.g., dyes. (3) Waterproofing substances, e.g., plastics. (4) Fireproofing substances, e.g., water glass. (5) Antiseptic substances, e.g., phenol. (6) Inert substances added as loadings or fillers; e.g., barytes, gypsum, chalk, clay.

finn oil Tall oil.

Finsen lamp A mercury-vapor arc lamp in a quartz container; a source of ultraviolet rays to treat skin diseases.

Fioco Trademark for a viscose cellulose synthetic fiber.

Fiolax Trademark for an alkali-free glass resistant to sudden temperature changes; used for chemical glassware.

fir (1) A coniferous tree of the *Abies* genus. (2) Firkin. (3) A *Pinus* species. **balsam of ~** Canada balsam. **oil of ~** Pine oil.

fire A bright flame caused by combustion. **f. air** Scheele's name for oxygen. **f.brick** A fire-resistant brick for lining furnaces, containing mullite, cristobalite, and tridymite as crystalline phases. **f.clay** Stowbridge clay. A refractory clay containing more silica and alumina than basic oxides. **f.damp** An explosive mixture of methane and air in coal mines; detected by the "corpse light" in a Davy lamp. **f. extinguisher** An agent that extinguishes fires by cooling the burning substance, e.g., with water; or by covering it with a medium in which combustion cannot occur, as, carbon tetrachloride, fire foam. **f. foam** A colloidal blanket of alumina and carbon dioxide to extinguish fires, e.g., in oil tanks, produced by spraying interacting solutions containing alum, and sodium carbonate and glue. **f. point** The minimum temperature at which an oil will burn continuously. Cf. *flashpoint.* **f. polishing** Smoothing the sharp edges of glass by slightly fusing them in a flame. **f.proofing** Coating or impregnation with a substance that reduces combustibility, as, water glass for textiles.

firkin Abbrev. fir; an obsolete volumetric measurement: 1 fir = 9 U.S. gal = 34.06 L. (2) A wooden vessel.

firn Granular, compressed snow, 60 m below the surface in Arctic regions.

Fischer **F., Emil (1852–1919)** German chemist noted for his synthesis of polypeptides and carbohydrates. **F., Ernst Otto (1918–)** German chemist. Nobel prize winner (1973); noted for work on C-metal complexes. **F., Hans (1881–1945)** German biochemist noted for his work on blood chlorophyll and bile pigments and for the synthesis of hemin (Nobel prize 1940). **F. projection** See *Fischer projection* under *projection.* **F.-Tropsch process** Used in the production of liquid fuels and other synthetic hydrocarbons; as, from coal.

fisetic acid Fisetin.

fisetin $C_{15}H_{10}O_6$ = 286.2. Fisetic acid. 3,3′,4′,7-Tetrahydroxyflavone. A yellow coloring matter from the wood of *Quebracho colorado,* fustic, or *Rhus cotinus.* Yellow needles, m.330. Cf. *fustin.* **iso~** Luteolin.

fish **f. bean** See *tephrosia.* **f. berry** Cocculus indicus, Indian berry, oriental berry. The dried fruit of *Anamirta cocculus* (Menispermaceae); a narcotic poison. **f. glue** See *isinglass.* **f. guano** F. scrap, f. tankage. A fertilizer made from nonedible fish and offal by cooking, expressing the oil, drying, and grinding (nitrogen 6–10, phosphorus pentaoxide 0.4–8%). **f. oil** Liquid fats obtained from fishes, characterized by great absorption of oxygen without drying to a varnish. They are colored dark red by concentrated acids and yield little or no elaidin. **f. poison** (1) A ptomaine produced by decaying f. proteins. (2) A poison produced by certain species of fishes, as, fugin or ichthyotoxin. (3) A poison used by natives to stupefy f. before catching them; usually derived from *Derris, Cracca,* and *Lonchocarpus* species. **f. scrap** F. guano. **f. tankage** F. guano.

fissile Describing material that undergoes atomic fission.

fission A division. (1) Biology: the separation of a single cell into 2 or more equal parts capable of developing to the original size of the parent cell. Cf. *karyokinesis.* (2) Astronomy: the separation of a spherical and semiliquid body into 2 parts that revolve around each other, as, double stars formed by f. from a single star. (3) Atomic chemistry: the splitting of an atom into 2 atoms of nearly identical weight; e.g., the capture of a neutron by a nucleus of ^{235}U, forming a highly unstable ^{236}U nucleus, which at once divides into 2 fragments (^{142}Ba and ^{91}Kr) and at the same time emits 3 neutrons and gamma radiation. The energy of motion of the fragments is transformed into heat, and the reaction may be propagated from atom to atom as a chain reaction, as in the atomic bomb. Cf. nuclear *fusion.*

fistelin The aglucone of the glucoside fustin.

Fittig, Rudolf (1835–1910) German organic chemist. **F. reaction** F. synthesis. **F. synthesis** The formation of an aromatic homolog from an aryl iodide or bromide by means of an alkyl iodide or bromide and metallic sodium; e.g., RBr + R′Br + 2Na = R−R′ + 2NaBr.

five finger grass Cinquefoil. The herb of *Potentilla reptans* (Rosaceae); an astringent. **American ~** The herb of *P. canadensis* used similarly. Cf. *tormentil.*

Fixanal Trademark for an analytical chemical, accurately weighed and sealed in a glass ampul for the rapid preparation of volumetric solutions.

fixation The process of rendering permanent. (1) Photography: the dissolving of light-sensitive silver salts from plates, films, or paper, to make them insensitive to the further action of light. (2) Microscopy: the preparation of minute structures in their original form on a slide. (3) Immunology: the prevention of hemolysis by the *complement,* q.v. (4) Industry: the combining of atmospheric nitrogen in the form of a useful compound. See *nitrogen fixation* under *nitrogen.* **complement ~** See *complement-fixation test* under *complement.* **nitrogen ~** See *nitrogen fixation* under *nitrogen.*

fixative A substance used to make an object permanent, as: (1) a mordant used in dyeing; (2) a varnish for pastel paintings; (3) an agent (formalin and acetic acid) used in biology to make permanent tissues; (4) a f. used in perfumery to make an odorous substance less volatile; as, methyl anisate.

fixed Made permanent or definite. **f. air** Early name for carbon dioxide. **f. carbon** See *fixed carbon* under *carbon.* **f. oil** A liquid fat which absorbs oxygen and becomes resinous (drying oil) or remains liquid (nondrying oil) as compared with evaporating (essential or volatile) oil. **f. proportions, law of** See *constant proportions.* **f. white** Barium sulfate*.

fixing The art of rendering permanent. **f. bath** A 20% solution of sodium thiosulfate; used to fix photographic plates, films, papers.

Fizeau, Armand Hippolyte Louis (1819–1896) French physicist noted for research on the interference of light and heat, and for the determination of the velocity of light and electricity.

fl Abbreviation for "fluid." **fl dr** A fluidram or drachm. **fl oz** A fluidounce.

flagstaffite $C_{12}H_{24}O_3$. An orthorhombic mineral, m.99–105.

flame A source of heat; essentially a stream of gas or vapor heated as the result of chemical reaction, usually oxidation. Its luminosity may depend on glowing solid particles, e.g., dust, carbon, etc. **acetylene ~** The hot, ignited gases emerging from a blowpipe fed with acetylene and compressed air or oxygen. **augmented ~** A f., the energy of combustion of which has been increased by subjecting it to a diffuse electric discharge. **Bunsen ~** The f. produced by a gas bunsen burner; either nonluminous (with air) or luminous (without air). **dark ~** A nonluminous f., e.g., produced by burning pure hydrogen in oxygen. **hydrogen ~** A bluish, nonluminous f. produced by the oxidation of hydrogen in air. **luminous ~** A bright- or pale-colored f., e.g., sodium light. **nonluminous ~** Dark f. **oxidizing ~** The nonluminous f. of a gas burner, used in blowpipe analysis for oxidation. **oxyhydrogen ~** The hot gas mixture from a blowpipe fed with compressed hydrogen and oxygen. **reducing ~** The luminous blowpipe f., due to the presence of solid particles of carbon. **solar ~** Protuberances on the sun due to hydrogen flames or luminous gases.

f. coloration A qualitative analytical test performed by placing a substance, moistened with hydrochloric acid on a platinum wire, in a nonluminous bunsen flame and observing the resulting color.

Bright yellow	Sodium
Brick red	Calcium
Crimson	Strontium
Red	Lithium
Green-yellow	Molybdenum, boron
Green	Barium
Blue	Indium
Blue-white	Lead, arsenic, antimony
Purple	Potassium
Blue, changing to green	Copper

f. cutting Cutting ferrous metals by oxidation. The metal is heated to 815°C by oxyacetylene jets in the cutting torch, and a stream of oxygen is applied through a central jet. **f. hardening** The surface hardening of iron by heating a thin surface layer to the hardening temperature with an oxyacetylene f., followed by rapid cooling. **f. photometer** An instrument for the photoelectric measurement of the concentration of an element in a solution from the intensity of the characteristic wavelength of light emitted when the solution is sprayed into a f. under standard conditions. **f. reactions** See *f. tests* below. **f. spectra** Spectra produced by the vapors of elements; used for spectroscopic analysis. The characteristic lines of f. spectra are due to electrons falling back to normal orbits from the easily excited levels. Cf. arc and spark *spectrum*. **f. temperature** In degrees Celsius:

Alcohol and air	1705
Gas bunsen burner:	
No air	1710
Half air	1810
Full air	1870
Hydrogen and air	1900
Gas and oxygen	2200
Hydrogen and oxygen	2420
Acetylene and air	2460
Acetylene and oxygen	3000
Thermit (Al + Fe)	3000

f. tests Qualitative tests made with a bunsen burner, such as f. colorations, bead tests, and blowpipe tests.

flame-ionization detector A device in which the change in conductivity of a standard (usually hydrogen) flame due to the inclusion of another gas or vapor is used to detect or determine the latter, as in gas chromatography. Responds to organic, but not common inorganic (e.g., water, ammonia), compounds.

flameproof Describing that which cannot be ignited. **f. group** The maximum gap dimensions in electrical apparatus that will prevent a surrounding gas from being ignited: ammonia, methane, 1; *n*-butane, carbon monoxide, ethane, propane, vinyl chloride, 2; butene, ethylene, propene, coal gas, 3; acetylene, hydrogen, water gas, 4 (maximum gap too small to be practical).

flame-resistant Flame-retardant. Resistant to ignition, but capable of being ignited. Additives used to confer f.-resistance include antimony trioxide and zinc carbonate.

flammable (U.S. and U.K. usage) Inflammable. Combustible; able to be set on fire.

flash (1) A sudden, luminous, temporary flame. (2) A volatile mixture, thrown on the fires of a kiln to produce a colored glaze on bricks or tiles that are being baked; thus: **black ~** Containing manganese. **zinc ~** Containing zinc salts (yellow and green shades). **f. drying** See *flash drying* under *drying*.

flashbulb A mixture of combustible solids used in photography; as, magnesium wire.

flash point The lowest temperature at which the vapors of a liquid decompose to a flammable gaseous mixture. It is a constant of oils, and is below the burning point (the lowest temperature at which the gas will burn steadily). **closed ~** F. p. determined in a vessel which is exposed to air only at the moment of application of the flame. **open ~** F. p. determined with the sample exposed to air during preliminary heating.

flash spectrum The reversal of the Fraunhofer lines of the solar spectrum as a bright line spectrum immediately before a total eclipse.

flask A glass, metal, paraffin, plastic, or rubber receptacle or vessel for holding solids, liquids, or gases. **Abderhalden ~** A glass vessel intermediate in shape between a beaker and an erlenmeyer f. **acetylation ~** A small, pear-shaped f. used in menthol determinations. **assay ~** A conical glass beaker used for precipitation. **boiling ~** A spherical glass vessel with long neck. **delivery ~** A f. graduated to deliver a specified volume of liquid as distinct from containing this volume. Cf. *volumetric flask* below. **Dewar ~** See *Dewar flask*. **erlenmeyer ~** A conical-shaped f. with narrow neck and wide, flat bottom **filtering ~** An erlenmeyer f. with heavy walls and side tube. **mercury ~** A commercial unit of measurement of mercury equivalent to 76 lb. **silica ~** A f. of semiopaque silica. **vacuum ~** Dewar vessel. A double-walled glass vessel of appropriate shape for holding liquefied gases. **volumetric ~** A f. with a calibration mark on it. **Wurtz ~** Distillation f.

flavaniline $C_{16}H_{14}N_2$ = 234.3. 2-(4-Aminophenyl)-4-methylquinoline. Colorless prisms, m.97, soluble in benzene.

flavanols Vegetable dyes derived from flavonol. See Table 33, next page.

flavazine $C_{16}H_{13}N_4O_4SNa$ = 380.4. The sodium salt of 1-*p*-sulfophenylmethyl-4-phenyldiazonium-5-pyrazolone; a yellow acid dye.

flavianic acid $C_{10}H_6O_8N_2S$ = 314.2. 8-Hydroxy-5,7-dinitro-2-naphthalenesulfonic acid. Yellow crystals, m.100; used in the separation of amines and as a selective precipitant for zirconium.

TABLE 33. FLAVANOLS

	A. Anthocyanidins
Pelargonidin.	3,4′,5,7-Tetrahydroxy∼[a]
Fisetinidin	3,3′,4′,7-Tetrahydroxy∼
Cyanidin	3,3′,4′,5,7-Pentahydroxy∼
Delphinidin	3,3′,4′,5,5′,7-Hexahydroxy∼
Peonidin	3,4′,5,7-Tetrahydroxy-3′-methoxy∼
Enidin	3,4′,5,7-Tetrahydroxy-3′,5′-dimethoxy∼
Myrtillidin	3,3′,4′,5,7-Pentahydroxy-5-methoxy∼
	B. Anthocyanins
Chrysanthemin . . .	5-β-Glucoside of cyanidin
Pelargonenin	5-β-D-Glucoside of pelargonidin
Enin	3-Glucoside of enidin
Pelargonin	3,5-Diglucoside of pelargonidin

[a]∼ = flavylium

flavin(e) (1) $C_{10}H_6N_4O_2$ = 214.2. Isoalloxazine. (2) Quercetin. (3) A yellow plant pigment; as, riboflavine (the free yellow pigment), lumiflavin (a cleavage product formed on irradiation of riboflavine), protein flavin (united to proteins), and carbohydrate flavin (united to carbohydrates). **f. adenine dinucleotide** FAD. A respiratory coenzyme.

flavine (1) Acriflavine. (2) Flavin.

flavoglaucin $C_{19}H_{28}O_3$ = 304.4. A yellow pigment, m.103, isolated from growing cultures of *Aspergillus glaucus* and other molds.

flavol $C_{14}H_{10}O_2$ = 210.2. 2,6-Anthracenediol*. m.295(decomp.).

flavone $C_{15}H_{10}O_2$ = 222.2. 2-Phenyl-4*H*,1-benzopyran-4-one, 2-phenylchromone. Colorless needles, m.119, insoluble in water. In many flower colors. See *flavones.*

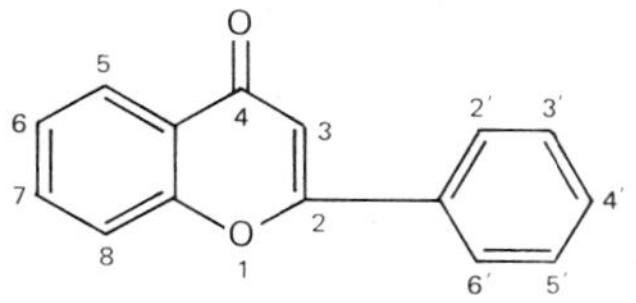

iso∼ As f., but 3-phenyl- (instead of 2-phenyl-). m.148.

flavones Vegetable coloring matters; the hydroxy or methoxy derivatives of flavone. See Table 34. Cf. *anthoxanthins.*

flavonoids A major family of natural pigments, giving blue, red, and yellow colors; it includes the flavanols and flavones.

flavonol $C_{15}H_{10}O_3$ = 238.2. 3-Hydroxyflavone. Yellow needles, m.170, soluble in alcohol.

flavopannin $C_{21}H_{26}O_7$ = 390.4. A monobasic acid from the root of *Aspidium athamanticum.*

flavophenine Chrysamine. Cf. *pannol.*

flavoprotein See *flavoprotein* under *protein.*

flavopurpurin $C_{14}H_8O_5$ = 256.2. 1,2,6-Trihydroxyanthraquinone*. C.I. Mordant Red 4. A red dye; m.330+.

flavoxanthin $C_{40}H_{56}O_3$ = 584.9. A carotinoid pigment, m.184, from the petals of *Ranunculus acer.*

flavylium* 2-Phenyl-1-benzopyrylium†. The tricyclic cation $C_6H_4{:}C_3H_2O^+ \cdot C_6H_4$, giving f. salts. See *flavanols.* Cf. *anthocyanidins.*

flax Byssus. The bast fiber of *Linum usitatissimum,* flax plant, grown in Europe and Egypt for the fiber, in the U.S., U.S.S.R., and Argentina for the seeds (linseed oil). Unbleached flax is used for ropes, twine, and coarse fabrics. Bleached flax is used for linen, lace, etc.; linen rags for high-quality paper. **earth ∼** Asbestos. **mountain ∼** (1) A fine, silky asbestos. (2) The herb of *Linum catharticum,* purging flax; a laxative. **New Zealand ∼** See *New Zealand hemp* under *hemp.* **stone ∼** Asbestos.

Flaxedil Trademark for gallamine triethiodide.

flaxseed Linseed.

fleabane The herb or seeds of *Erigeron Canadense* (Compositae); a diuretic.

Fleischl hemometer An optical instrument to determine hemoglobin in the blood by comparison with blood-colored glass wedges.

Fleming F., Sir Alexander (1881–1955) Scottish bacteriologist. Nobel prize winner (1945). Discoverer of penicillin (1928). **F., Sir Arthur (1881–1960)** British physicist, pioneer in the development of the thermionic valve, radio, and radar.

Fleming tube A glass apparatus for the absorption of carbon dioxide in the determination of carbon in steel.

Flemming's solution A fixative and preservative for small organisms: 25 mL 1% chromic acid, 10 mL 1% osmic acid, 5 mL glacial acetic acid, 60 mL water.

Fletcher furnace A laboratory gas or gasoline furnace for metals or ceramics. **F. burner** A gas ring burner.

flex Flexible insulated copper wire, for electrical connections.

flexibility (1) Ability to bend without breaking. (2) Adaptability.

flexography Relief-type printing with quick-drying inks containing volatile solvents.

flexure Any curved or bent portion or section.

flint SiO_2. Flintstone. An opaque quartz in chalkstone, resembling chalcedony. Used in the ceramic, glass, and road-making industries. **f. brick** A firebrick made of powdered f. **f. glass** Potash-lead glass. A highly refractive and easily fusible glass; used in optical and chemical apparatus **f. stone** Flint.

float A buoyant, sealed glass tube used in burets for easier reading. **f. stone** A light, porous quartz that floats on water.

floats A finely ground phosphate rock; a fertilizer.

floccose Describing a growth of bacteria in short, curved chains, resembling wool threads.

flocculant Chemical used to effect flocculation; e.g., polymers such as polyacrylamides.

flocculation Coagulation (of a finely divided precipitate).

flocculent (1) Woolly or cloudy, flakelike, and noncrystalline. (2) Describing a growth of bacteria in which small, adherent

TABLE 34. FLAVONES

Flavonol	3-∼[a]
Chrysin	5,7-Di∼
Apigeninidin	4′,5,7-Tri∼
Galangin	3,5,7-Tri∼
Fustin	3,3′,4′,7-Tetra∼
Luteolin	3,4,5,7-Tetra∼
Morin	2′,3,4′,5,7-Penta∼
Quercetin	3,3′,4′,5,7-Penta∼
Tricetin.	3′,4′,5,5′,7-Penta∼
Myricetin	3,3′,4′,5,5′,7-Hexa∼
Quercetagetin	3,3′,4′,5,6,7-Hexa∼

[a]∼ = hydroxyflavone

masses of bacteria of various shapes float in the culture medium.

Florence test The formation of brown needles or plate-shaped crystals by a solution of iodine in potassium iodide in presence of semen.

florentium Early name for promethium.

flores (1) The flowers or blossoms of a plant. (2) A chemical obtained by sublimation. See *flowers*. **f. martiales** Ammonium iron(III) chloride.

Florey, Lord Howard Walter (1897–1968) Australian-born chemist, codeveloper of penicillin. Nobel prize winner (1945).

Floridin Trademark for a variety of fuller's earth from Florida.

florigen A plant hormone associated with gibberellin.

Flory, Paul John (1910–1985) American chemist. Nobel prize winner (1974); noted for work on macromolecular chains.

floss (1) A fluffy, silky thread, e.g., *Calotropis floss*, q.v. (2) The floating scum of oxides produced in the puddling of iron; a catalyst, e.g., for the polymerization of unsaturated styrenes.

flotation The concentration of ores by grinding with a frothing agent, floating them on water, and agitating the mixture by compressed air. The wet gangue settles, and the concentrated ore is skimmed off. Cf. *density fluids, Owen process*. **f. activator** A reagent producing a metallic coat; as, sodium sulfide or copper sulfate. **f. collector** An agent that increases the carrying capacity of air bubbles; e.g., xanthates. **f. depressor** An agent preventing the gangue from being carried by the air bubble; as cyanides. **f. frother** A reagent producing a foam of stable air bubbles; as, f. oils. **f. oils** Petroleum and wood oils (pine oil, creosote) used to wet the metallic particles. **f. regulator** A reagent that controls pH value; as, lime.

flour (1) Wheat f., farina, tritici. The white starchy powder made by bolting wheat. (2) A powdered cereal or seed used for food. **baker's ~** Second-grade wheat f. **bleaching ~** See *Agene, betachlora*, and *Golo processes*. **buckwheat ~** Powder made from buckwheat. **enriched ~** Plain white f., to which vitamin concentrates and calcium salts, or a proportion of the wheat germ, have been added. **graham ~** Unbolted wheat meal. **patent ~** High-grade, white wheat meal, which has been bolted and from which all bran has been removed. **rye ~** Powdered rye.

flouve oil A mixture containing principally esters and coumarin from the sweet-scented vernal, *Anthoxanthum odoratum* L., d.1.1291; used in perfumes.

flow The motion of a fluid. Cf. *flux*. **cold ~** See *cold flow*. **molecular ~** The relative number of gas molecules which pass through a fine orifice: $Q = p_2 - p_1/(W\sqrt{\rho})$, where Q is the quantity of gas in ml/s which flows through an opening at a difference of pressure $(p_2 - p_1)$, ρ is the density of the gas at 0.100 Pa pressure, and W is the resistance overcome. **f.chart** A diagrammatic, step-by-step representation of the approach to a computer program. Uses shapes (as, diamond, rectangle) to represent a specific type of action. **f.sheet** The diagrammatic representation of an industrial process, showing the sequence and interdependence of the successive stages.

flowers (1) A chemical obtained by sublimation; usually a metallic oxide; as f. of sulfur. (2) The blossoming portion of a plant, consisting normally of a calyx (composed of sepals), corolla (composed of petals), and stamens and pistils. Many fowers contain coloring materials, essential oils and odoriferous substances.

Flox Trademark for (1) a viscose cellulose synthetic fiber. (2) Mixtures of liquid oxygen and fluorine.

floxacillin sodium Flucloxacillin.

Floxapen Trademark for flucloxacillin.

fl oz Abbreviation for fluidounce.

fluavil $C_{20}H_{32}O$ = 288.5. A resin from guttapercha, m.42, soluble in alcohol.

flucloxacillin $C_{19}H_{16}O_5N_3SClFNa \cdot H_2O$ = 493.9. Floxacillin sodium, Floxapen. White crystals, soluble in water; an antibiotic for staphlococcal and mixed infections (BP).

fluctuate To vary or move within certain limits.

fluctuation Successive rises and falls.

flucytosine $C_4H_4ON_3F$ = 129.1. 5-Fluorocytosine, Alcobon, Ancobon, Ancotil. White crystals, sparingly soluble in water. An antifungal agent for systemic infections (USP, BP).

fludrocortisone acetate $C_{23}H_{31}O_6F$ = 422.5. White crystals, m.225, soluble in water. A corticosteroid with gluco- and mineralocorticoid properties; used for adrenal failure and replacement therapy (USP, BP)

flue A channel for gases or liquids.

fluellite $AlF_3 \cdot H_2O$ = 102.0. Hydrous aluminum fluoride. Orthorhombic crystals, d.2.17, hardness 3.

fluid A form of matter that cannot permanently resist any shearing force, which causes flow. **elastic ~** A condition of matter in which the molecules flow apparently without resistance; e.g., a gas. **electrorheological ~** See *electrorheological fluids*. **inelastic ~** A condition of matter in which the molecules move freely but are restricted by gravitation; e.g., a liquid. **Newtonian ~** Describing fluids for which the shear rate and shear stress increase linearly together at constant temperature and pressure; the viscosity (shear stress/rate) is therefore constant. **non-Newtonian ~** Heterogeneous f. (as, sols, gels) and liquids of high molecular weight. **perfect ~** A hypothetical state of matter in which the molecules offer no mechanical resistance. **viscous ~** A syrup or soft mass which flows slowly.

fluid bed See *fluidized bed dryer*.

fluid dram Fluidram. A pharmaceutical measurement. 1 fl dr = 60 minims = 3.69661 mL (U.S.) or 3.55 mL (U.K.).

fluidextract An alcoholic extract of a vegetable representing the drug weight by volume; e.g., 1 g of the drug corresponds with 1 mL of fluid extract. Cf. *tincture*.

fluid friction Viscosity.

fluidics The use of fluid flow (as, water or air in a tube) to operate process controls. Cf. *Coanda effect*.

fluidity (1) The property of flowing easily. (2)* The reciprocal of dynamic velocity.

fluidization, fluidizing The suspension and maintenance, in a state of turbulent motion, of solid material in a finely divided form in a stream of gas. This increases the surface activity of the particles. Used in catalytic processes, the gasification of brown coal, and the cracking of petroleum.

fluidized bed **f. b. dryer** A dryer (as for textiles) in which the drying medium is a bed of sand particles or small glass spheres, 0.1–1.0 mm diameter. This is fluidized by passing hot air upwards through it, thus providing good heat transfer. **f. b. reactor** The reactant(s) are held in suspension by a hot-air stream during the reaction; as, in the combustion of low-grade coal for steam generation.

fluidounce Fluid ounce. A pharmaceutical measure of volume. U.S.: 1 fl oz = 29.5735 ml = 8 fl dr = $\frac{1}{128}$ gal. U.K.: 1 fl oz = 28.4131 ml = 8 fl dr = $\frac{1}{160}$ imperial gal.

fluid wax Liquid waxes obtained from the oils of marine animals. Consist of esters of monohydric alcohols, with traces of glycerides.

fluo-, fluor- (1) Fluoro-*, q.v. (2) Prefix indicating the property of fluorescence (fluo- only).

fluocerite $(Ce \cdot La \cdot Nd \cdot Pr)_2OF_4$. A mineral containing the fluorides of these four elements.
fluocinolone acetonide $C_{24}H_{30}O_6F_2$ = 452.51. Synalar. White crystals, m.275, insoluble in water; a potent glucocorticoid for skin conditions (USP, BP).
fluoflavine $C_{14}H_{10}N_4$ = 234.3. A yellow, fluorescent substance, m.360+, soluble in alcohol.
fluohydric acid Hydrofluoric acid*.
fluon Trade name for polytetrafluoroethylene (U.K. usage).
fluoracetamide Fluoroacetamide.
fluoran $O:(C_6H_4)_2:C:(O \cdot C_6H_4) \cdot CO$ = 300.3. *o*-Phenolphthalein anhydride. Colorless needes, m.182, soluble in acids; an intermediate in the manufacture of dyes.
fluorandiol Fluorescein.
fluoranthene* $C_{16}H_{10}$ = 202.3. Idryl. A hydrocarbon in coal tar. Colorless needles m.110, soluble in hot water.

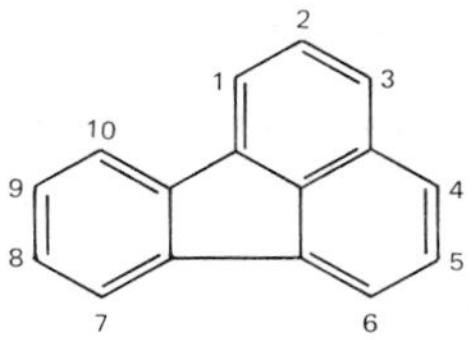

2 1 3 10 9 4 8 5 7 6

fluoranthraquinone $C_{15}H_7O_2$ = 219.1. Colorless crystals, m.188, soluble in alcohol.
fluorapatite The mineral $CaF_2 \cdot 3Ca_3(PO_4)_2$. Cf. *apatite.*
fluoration Fluorination.
fluoremetry Fluorimetry*.
fluorene* $C_6H_4 \cdot CH_2 \cdot C_6H_4$ = 166.2. Diphenylenemethane. (9)
Fluorescent, colorless scales, m.113, insoluble in water. Occurs in coal tar; used in the manufacture of dyes. Its radicals are fluorenyl and fluorenylidene. **amino ~** Fluorenylamine*. **benzo ~** Chrysofluorene. **chryso ~** See *chrysofluorene.* **naphtho ~** See *naphthofluorene.* **oxo ~** Fluorenone*.
f. alcohol $C_6H_4 \cdot CHOH \cdot C_6H_4$ = 182.2. Fluorenol*.
Colorless crystals, m.153, soluble in alcohol. **f.carboxylic acid** $C_{14}H_{10}O_2$ = 210.2. Diphenyleneacetic acid.
fluorenic acid $C_{14}H_{10}O_2$ = 210.2. Colorless crystals, soluble in water.
fluorenol* Fluorene alcohol.
fluorenone* Mol. wt. = 180.2. Diphenylene ketone,

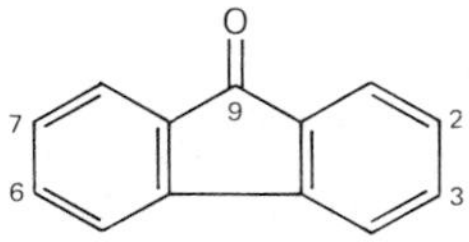

O 9 7 2 6 3

oxofluorene. An oxidation product of fluorene. Yellow prisms, m.84, soluble in alcohol. ***o*-hydroxy ~*** $C_{13}H_8O_2$ = 196.2. Oxyfluorenone. Colorless crystals, m.115.
fluorenyl* Fluoryl. The radical $C_{13}H_9-$, from fluorene; 5 isomers. **f.amine** $C_{13}H_9NH_2$ = 181.2. Aminofluorene. White crystals, insoluble in water.
fluorenylidene* Fluorylidene. The radical $C_{13}H_8=$, from fluorene.

fluores An early name for fluorite.
fluorescein $HO \cdot C_6H_3 \cdot C(C_6H_4 \cdot COOH):C_6H_3:O$ = 332.3. (—O—)
Uranine (K or Na salt), 3,6-dihydroxyfluoran, resorcinolphthalein, fluorandiol, dioxyfluoran. Orange-red powder, soluble in alkalies with orange color and green fluorescence. Used in the manufacture of eosine and other dyes and as an indicator for pH 3.6 (yellow) to pH 5.6 (fluorescent). **dihydroxy ~** Gallein. **dimethyl ~** Cresorcin. **potassium ~** See *potassium fluorescein.* **sodium ~** $C_{20}H_{10}O_5Na_2$ = 376.3. Uranine, soluble f. Hygroscopic, orange powder, soluble in water; used to diagnose corneal abrasions or ulcers, which show a green fluorescence (USP, BP). **tetrabromo ~** Eosine. **tetrabromodichloro ~** Phloxin. **tetraiodo ~** Erythrosin.
f. paper Zellner's paper. Paper impregnated with a solution of f. in alcohol; an indicator.
fluorescence The property of certain solids, liquids, and gases when illuminated to radiate unpolarized light of a different (usually greater) wavelength. Due to electrons, displaced by the radiation to excited singlet states, returning to lower singlet energy levels. F. ceases abruptly when the excitation ceases. The energy is greater than that of phosphorescence. Cf. *phosphorescence, luminescence.* **delayed ~** F. involving a circuitous route between the excited and lower electronic states, which results in a f. lifetime, after cessation of the exciting radiation, approaching that of phosphorescence.
f. analysis The examination of substances in ultraviolet light with the object of identifying or determining them or assessing their quality or purity from the wavelength and intensity of the f. produced. **f. efficiency** The intensity of f. emitted by a substance per absorbed quantum of light. **f. microscopy** Microscopical examination using ultraviolet instead of visible light; many structures fluoresce. **f. serology** The diagnosis of disease from the f. of serums. Cf. *fluorimetry.*
fluorescent Epipolic. Having the property of fluorescence. **f. screen** A glass plate covered with a f. substance, e.g., a cyanoplatinate. Used to make rays visible that are normally invisible to the eye. **f. unit** The standard luminescence produced by 1 mg of radium element on 1 cm^2 of a barium cyanoplatinate screen. See *fluorimeter.*
fluorescin $C_{20}H_{14}O_5$ = 334.3. Yellow powder, insoluble in water; a dye. Cf. *fluorescein.*
fluoric acid Hydrofluoric acid*.
fluoridation Addition of fluoride to domestic drinking water. Combats dental caries by converting outer crystals of enamel from calcium hydroxyapatite to the fluoroapatite, which is less soluble in bacterial acids. Recommended U.K. level is 1 ppm. Some natural waters contain higher levels. Cf. *fluoride toothpaste, fluorination.*
fluoride* A salt of hydrofluoric acid containing the radical F^-. **acid ~** MHF_2. Hydrogendifluoride*.
f. toothpaste A dentifrice containing fluoride, usually added in the form of sodium fluoride, stannous fluoride, or sodium monofluorophosphate. See *fluorination.*
fluorimeter Fluorometer, photofluorometer. An instrument to measure the intensity of fluorescence, especially the fluorescence caused by X-rays, cathode rays, and radium.
fluorimetry* Fluorometry. Measurement of the wavelength and intensity of *fluorescence,* q.v.
fluorination The introduction of fluorine into an organic molecule. Cf. *fluoridation.*
fluorine* F = 18.998403. A halogen element, at. no. 9. A poisonous, pale-green gas, $d_{air=1}1.31$, m. −220, b. −187,

d_{liq}1.56, soluble in water, alcohol, or ether. F. is the most negative element, and more reactive than oxygen. It has a valency of 1 and forms one series of compounds, the fluorides. Discovered (1771) by Scheele in fluorite, cryolite, and other minerals, and isolated (1886) by Moissan. Liquid f. has a bright yellow color; solid f. is colorless. F. salts are of the types MF or M_2F_2 (normal fluorides), MHF_2 (hydrogendifluorides, acid fluorides). **proto ~** Coronium.
f. hydride Hydrofluoric acid*. **f. oxide** See *oxygen fluorides.*

fluorinion The fluoride ion, F^-.

fluoriodine Iodine pentafluoride*.

fluorion The fluoride ion, F^-.

fluorite CaF_2 = 78.1. Fluorspar. Siliceous sinter. Native calcium fluoride, variously colored, brittle, d.3.18, hardness 4. Used as flux in the steel and glass industries and in the manufacture of hydrofluoric acid.

fluoro-* Prefix indicating the presence of fluorine.

fluoroacetamide* $F \cdot CH_2CONH_2$ = 77.1. White crystals, soluble in water, m.109. An arrow poison; it acts by blocking the citric acid cycle.

fluoroacetanilide $F \cdot CH_2CONHC_6H_5$ = 153.2. White solid, sparingly soluble in water, m.75.

fluoroborate Borofluoride, fluoboride. A salt of fluoroboric acid, containing the tetra ~ radical, BF_4^-.

fluoroboric acid HBF_4 = 87.8. Borofluoric acid, fluoboric acid. Colorless liquid, b.130, miscible with water.

fluorocarbon A member of the chain system $(-CF_2 \cdot CF_2-)_n$ which forms highly unreactive substances; as, $C_{10}F_{21}SO_3H$, used to reduce spray in chromium plating, being unaffected by hot chromic acid. The most important fluoropolymer is poly(tetrafluoroethylene), mol. wt. 400,000–9,000,000, used as a protective coating in the chemical and electrical industries. See *Teflon.*

fluorochromate A salt of fluorochromic acid, containing the radical $CrOF-$.

fluorochrome A substance capable of inducing fluorescence in another substance.

fluoroform* CHF_3 = 70.0. Fluoform, trifluoromethane*. Colorless gas, d.2.53, b_{40atm}20, slightly soluble in water.

fluorogen Fluorophore.

fluorogermanate M_2GeF_6. A salt of fluorogermanic acid.

fluorogermanic acid H_2GeF_6 = 188.6. (Hydro)fluogermanic acid. Obtained by passing GeF_4 into water.

fluorography The taking of miniature snapshots of a fluorescent screen in mass chest radiography; can process 200 patients per hour.

fluoroiodate A compound derived from iodates by partial replacement of the oxygen by fluorine; as, difluoroiodates, $R \cdot IO_2F_2$.

fluorometer Fluorimeter.

fluoromethane CF_4 = 88.0. Carbon tetrafluoride*. Colorless gas; a by-product in the manufacture of aluminum from cryolite.

fluorometry Fluorimetry*.

fluorones A group of compounds of the type derived from fluorene.

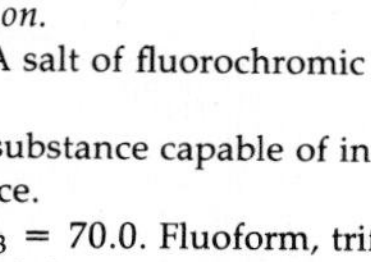

$$C_6H_4\langle{}^{O}_{C}\rangle C_6H_3{:}O \quad (\text{C bearing } R)$$

fluoronium* The cation H_2F^+.

fluorophore A group of atoms which confers fluorescence on a compound; as, the oxazine ring. Cf. *chromophore.*

fluorophosphonate A member of a series of compounds having the general formula $(RO)_2POF$, where R is an alkyl group. They have a high toxicity as lethal inhalants and were evolved during World War II as possible poison gases. They affect the eyes without producing tears, and death ensues rapidly.

fluoroplumbic acid H_2PbF_6 = 323.2. (Hydro)fluoplumbic acid. White powder obtained by passing PbF_4 into water.

fluoropolymer See *fluorocarbon.*

fluoroscope (1) A fluorescent screen used to make visible certain rays, e.g., X-rays. (2) An apparatus to determine the fluorescence of a solution by comparison with a standard.

fluorosilic acid* Hexafluorosilicic acid*.

fluorosis Disease (and in particular, brown mottling of the enamel of the teeth) due to an excess of fluorine ions in the system; usually derived from drinking water.

fluorosulfonic acid HSO_3F = 100.1. Colorless liquid, m.−87, b.163; an intermediate in fine chemical manufacture.

fluorouracil $C_4H_3O_2N_2F$ = 130.1. 5-Fluorouracil. 5FU, Efudix, Fluoro-uracil. White crystals, sparingly soluble in water. Used for malignant disease, particularly of the gastrointestinal tract, and for skin conditions (USP).

fluorspar Fluorite.

fluoryl The fluorenyl* radical.

fluorylidene The fluorenylidene* radical.

fluosilicate Hexafluorosilicate*.

fluosilicic acid Hexafluorosilic acid*.

Fluosol-DA Trademark for an artificial blood substitute used in animal experiments. Active ingredients are perfluorodecalin and fluorotripropylamine.

Fluothane Trademark for halothane.

fluoxymesterone $C_{20}H_{29}O_3F$ = 336.4. White crystals, m.278, insoluble in water; an androgen.

fluphenazine hydrochloride $C_{22}H_{26}ON_3SF_3 \cdot 2HCl$ = 510.4. Moditen, Permitil, Modecate. White crystals, soluble in water. A sedative and tranquilizer of the phenothiazine group, used for psychiatric disorders (USP, BP).

flussigas Similar to Calor Gas.

flux (1) A continuous flow or discharge. (2) A substance that causes other substances to melt more readily by dissolving their oxides or surface impurities, e.g., sodium carbonate. (3) The capacity of a nuclear reactor, in neutrons/cm^2/s.
aluminum ~ A mixture of the alkali chlorides and fluorides. **black ~** A f. and reducing agent used in metallurgy; a mixture of potassium carbonate and charcoal made by heating tartar. **radiant ~** The amount of radiant energy (in W) emitted per unit of time. **oxidizing ~** A mixture of borax glass 30, boric acid 20, silica 5, potassium chlorate 20, sodium peroxoborate 25%. **reducing ~** A mixture of borax glass 50, boric acid 15, argol 25, animal charcoal 10%. **soldering ~** A mixture of borax glass 55, boric acid 35, silica or dry sodium silicate 10%.
f. density (1) Luminous flux*. (2) Electric f. d.*

fly ash The flue dust resulting from the combustion of such fuels as pulverized coal.

foam A heterogeneous mixture of a gaseous phase in a liquid phase; finely divided gas bubbles suspended in a liquid. Cf. *flotation, colloid.*

foaming Describing a liquid which forms a surface foam on heating or on agitation.

focal Pertaining to a focus.

focus A point on which rays converge and produce an intensified effect. **acoustic ~** The meeting point of waves reflected from a concave surface. **anterior ~** A point before an optical system, which corresponds with the posterior focus. **aplanatic ~** The point from which rays pass through a lens without spherical aberration. **conjugate ~** The mutually convertible anterior and posterior foci of a lens. **posterior ~** A point behind an optical system corresponding with the anterior focus. **principal ~** The point on which parallel rays passing through a lens converge. **real ~** The point at which convergent rays intersect. **virtual ~** The point at which rays, if prolonged, would intersect one another.

focusing The adjustment of lenses or mirrors to produce a distinct image, e.g., in a microscope.

fog A heterogeneous mixture of a liquid phase in a gaseous phase; finely divided liquid droplets suspended in a gas. See *colloid.* **f. chamber** Cloud chamber. A small container in which a haze is produced by sudden pressure changes. Cf. *Wilson tracks.*

fogging (1) Producing a mist by agitating or heating a liquid in a closed container. (2) In photography, a uniform haze over the print, due to overexposure or light leakage.

foil A very thin sheet of metal, e.g., of aluminum; usually not thicker than 0.15 mm.

folacin Folic acid.

folia Latin for "leaves."

folic acid $C_{19}H_{19}O_6N_7$ = 441.4. Pteroylglutamic acid.

NH_2 (pteridine ring with N, N, N, N; OH) $—CH_2•NH•C_6H_4•CO—$
$NH•CH•CH_2•CH_2•CO_2H(CO_2H)$

Folacin, Folvite. Yellow crystals, insoluble in water. A member of the vitamin B complex. Exists in the body as tetrahydrofolate, which is essential in transfer of carbon groups (as, Me) in metabolism of purine and pyrimidine. Deficiency causes severe anemia. Tests for f. a. in fluids involve absorption by *lactobacillus casei* (USP, EP, BP).

Folin, Otto (1867–1934) American biochemical analyst, born in Sweden. **F. apparatus** A glass, specially designed absorption tube, used for the rapid determination of nitrogen, urea, or ammonia in urine or other fluids. **F.-Ciocalteu reagent** Produces a blue color with phenolic groups in drug analysis.

Folutein Trademark for chorionic gonadotrophin.

fomometer A small glass cup attached to a short length of capillary tubing, used for the rapid estimation of surface tension from the equilibrium height of the liquid in the tube when the cup is half-filled and the f. is inverted on a level surface.

fongisterol $C_{28}H_{48}O$ = 400.7. 24-Methyl-7-cholesten-3-ol, fungisterol. A sterol, q.v. *(sterols),* m.144, from ergot.

food A substance which builds up tissues and supplies living organisms with energy. The essential constituents are water, carbohydrates (starch, glucose, etc.), fats, proteins (albumins, etc.), salts, vitamins (q.v.), and minerals. See *food requirements* (below). Daily adult mineral requirements, in grams:

K	3
Na	2
Cl	2
P	1.3
Ca	0.45
Mg	0.35
I	0.015
Fe	0.014
Mn	0.003
Cu	0.002
Co	0.002
Cr	0.0001
Se	0.0001
Zn, Al, Si, F	traces

See also *food sources* (below). **animal ~** F. derived from animals; as, meat, fish, milk, eggs, liver. **flesh-forming ~** The proteins and salts necessary as tissue builders. **heat-forming ~** The carbohydrates and fats which supply body heat. **iron ~** F. containing iron, from 0.01921% (parsley) to 0.00015% (lemon). **nitrogenous ~** The proteins, lecithins, and other foods containing nitrogen. **nonnitrogenous ~** The carbohydrates and salts. **staple ~** F. which is most frequently used; as, bread, potato, maize, rice, etc. **vegetable ~** F. from plants; e.g., cereals, nuts, vegetables, fruits, sugars.

f. and drug acts Legislative measures in force: in the U.S. since Jan. 1, 1907 (now the U.S. Federal Food, Drug and Cosmetic Act, 1938), and in the U.K. since 1860 (the Food and Drugs Act, 1955), prohibiting the manufacture, sale, or transportation of adulterated, misbranded, poisonous, or deleterious foods, drugs, medicines or liquors; and designating official names for drugs. **f. requirements** Depend principally on activity, age, size (weight and height), sex, pregnancy, and lactation. Generally, 3000 kcal/day for active men, 2400 for women (plus 300 if pregnant or 500 if lactating); mainly from carbohydrates and fats. Also (for adults) about 50 g/day protein, 10–18 mg iron, 0.4–0.50 g calcium, plus vitamins and other minerals. **f. sources:**

1. CEREALS: seeds of Gramineae; rich in carbohydrates; e.g., *Avena sativa,* oats.
2. PULSES: seeds of Leguminosae; rich in proteins; e.g., *Arachis hypogaea,* peanuts.
3. NUTS: oily seeds of various plants; rich in fats; e.g., *Carya* species, pecan.
4. ROOTS and TUBERS: underground reserves of certain plants; generally rich in carbohydrates; e.g., potato.
5. GREEN VEGETABLES: leaves, inflorescences, stems, and young shoots of certain plants; generally containing vitamins and salts; e.g., *Brassica oleracea,* cabbage, cauliflower.
6. FRUITS used as VEGETABLES: fleshy fruits of various plants; e.g., *Solanum lycopersicum,* tomato.
7. FRUITS and BERRIES: generally containing acids, sugars, and salts; e.g., *Fragaria vesca,* strawberry.
8. MEAT and FISH: rich in proteins.
9. DAIRY PRODUCTS: rich in calcium and vitamins.
10. MISCELLANEOUS: mushrooms and yeast; honey, sugar, and maple syrup; sago, tapioca; spices and condiments.

Cf. *Codex Alimentaris.* **F. Standards Committee** A U.K. committee set up in 1947 to advise on the control of composition, description, labeling and advertising of f.

foot Abbrev. ft; a unit of length. 1 ft = 12 in = 0.304801 meter. **board ~** See *board foot.* **cubic ~** cu ft or ft^3. A measure of volume. 1 ft^3 = 1,728 in^3 = 2,832 cm^3. **square ~** sq ft or ft^2. A measure of area. 1 ft^2 = 144 in^2 = 929.03 cm^2.

footcandle Abbrev. fc; a measure of illuminance (illumination), now replaced in the SI system by the lux: 1 fc = 1 lumen/ft^2 = the illumination from a standard candle on a surface of 1 ft^2 1 ft distant. 1 fc = 1.076 milliphot = 10.764 lux.

foot pound Abbrev. ft·lb; a unit of work in the fps system: the work required to lift 1 lb 1 ft, where g = 32.2 ft/s^2. 1 ft·lb = 1.356 J = 0.3766 × 10^{-6} kWh. 1 hp = 33,000 ft·lb/min of work. **f.-p.-second** fps system. A system of measurements based on the foot, pound, and second. Cf. *S.I., cgs, mks system.*

foot-poundal Poundal.

foots The sediment which settles from an oil on standing: chiefly albuminous matter.
forbesite A rare, natural cobalt nickel arsenate.
forbidden **f. explosives** See *forbidden explosives* under *explosives.* **f. lines** Spectrum lines corresponding with atomic transitions not in harmony with Pauli's principle. When a gas is excited by, e.g., electric energy, certain lines only are seen in the absorption spectrum; others (f. lines), never. Thus the line may correspond with the passage of an electron from *A* to *B* and *B* to *C*, but never from *A* (the ground state) to *C*.
force The interaction between 2 bodies whereby their state of rest or motion, or their form or size, is changed. Cf. *attraction, repulsion.* The unit of force is the newton (SI system), dyne (cgs system) or poundal (fps system). Force = (mass × speed)/time = mass × acceleration. **catabiotic ~** The heat energy derived by a living organism from food. **catalytic ~** The chemical work performed by a catalyst. **centrifugal ~** See *centrifugal force.* **chemical ~** See *affinity.* **cohesive ~** See *cohesion.* **electromotive ~** See *electromotive force.* **kinetic ~** See *kinetic energy.* **latent ~** See *potential energy.* **living, vital ~** The energy obtainable from a living organism or cell.
forceps A small V-shaped instrument for grasping objects such as analytical weights.
forensic Pertaining to courts of law. **f. analysis** Chemical analysis performed for the purpose of assisting justice. **f. chemistry** Legal chemistry, judicial chemistry. The application of chemistry for purposes of criminal or civil law, e.g., f. analysis. **f. medicine** See *forensic medicine* under *medicine.*
forgenin $HCOONMe_4$ = 119.2. Tetramethylammonium formate. Colorless, hygroscopic crystals. In small doses it stimulates; in large doses it acts similarly to curare.
forked chain See *forked chain* under *chain.*
Forlion Trademark for a polyacrylonitrile synthetic fiber.
formal $CH_2(OMe)_2$ = 76.1. Methylal, dimethoxymethane*, methyl aldehyde, formyl aldehyde, oxymethylene. Colorless liquid; an anesthetic and hypnotic.
formaldehyde* H·CHO = 30.03. Methanal*, methylene oxide, formalin, Formol, Formalin. The simplest aldehyde, derived from methanol. Colorless gas, b.−21, soluble in water; marketed as an aqueous solution. **poly ~** See *Delrin.* **f. solution** A 37% aqueous solution of f. Colorless liquid with characteristic odor. Used as a reagent, preservative, sterilizing agent, and fumigant; used industrially in synthesis and in the formation of plastics (USP, BP).
formaldoxime H_2C:NOH = 45.04. The simplest oxime. Colorless liquid, b.84.
formalin (1) A 40% formaldehyde solution; may contain some methanol as a stabilizer. (2) (Cap. F.) Trademark for formaldehyde.
formamide (1)* H·CO·NH_2 = 45.04. Formylamine, methanamide*. Colorless, hygroscopic liquid, d.1.337, b_{10mm}105, soluble in water. (2) Containing the formylamino* radical. ***N*-allyl ~** C_4H_7ON = 85.1. Colorless liquid, b.109. **ethyl ~** C_3H_7ON = 73.1. Colorless liquid, b.199. **phenyl ~** Formanilide.
formamidine HC(:NH)NH_2. Known only as salts. **diphenyl ~** PhN:CH·NHPh = 196.2. Methenyldiphenyldiamine. Colorless crystals, m.135. ***o*-phenylene ~** Benzimidazole.
formamido (1) HN:CH·NH_2 = 44.6. (2) The formylamino* radical.
formamine Hexamethylenetetramine.
formamyl The carbamoyl* radical.
Formanek's indicator Alizarin green, dihydroxydinaphthazoxonium sulfonate. A green oxazine dye indicator: pH 0.3 (violet), 1.0 (pink), 12.0 (yellow), 14.0 (brown).
formanilide C_7H_7ON = 121.1. Phenylformamide, formamidobenzene, carbanil aldehyde. Yellow prisms, m.50, soluble in water. **thio ~** See *thioformanilide.*
formate* Formiate. A compound containing the radical H·COO−, from formic acid.
formation (1) The process of being made. (2) Geology: an assemblage of rocks having their origin, age, or composition in common. **heat of ~** See *heat of formation.*
formazan* NH_2·N:CH·N:NH = 72.1.
formazyl (1) Ph·N:N·CH:N·NH·Ph = 224.3. Diphenyl formazan*. Colorless crystals, m.116. (2) The radical $(PhN:N)_2H$·C−. **f.carboxylic acid** PhN:N·C(COOH):N·NHPh = 268.3. Red needles, m.162 (decomp.), insoluble in water; a colorimetric reagent for silver. **f. hydride** Formazyl. **f. methyl ketone** PhN:N·C(COMe):N·NHPh = 266.3. Colorless crystals, m.134.
formcoke A calcined agglomerate prepared from bituminous coals; used in blast furnaces.
formhydroxamic acid The theoretical compound H·CO·NH·OH.
formiate Formate*.
formic acid* H·COOH = 46.03. Formylic acid, methanoic acid*, aminic acid. The first member of the aliphatic monocarboxylic acid series. Colorless liquid, $d_{20°}$1.218, m.8.6, b.101, miscible with water. Contained in ants, spiders, and various plants; a reagent for nitrates in water and the analysis of essential oils. **acetyl ~** Pyruvic acid*. **amino ~** Carbamic acid. **aminobenzoyl ~** Isatic acid. **benzoyl ~** See *benzoylformic acid* under *benzoyl.* **carbamoyl ~** Oxamic acid*. **carbobenzoyl ~** Phthalonic acid. **formyl ~** Glyoxylic acid*. **hydrazino ~** Carbazic acid*. **hydroxy ~** Carbonic acid. **phenyl ~** Benzoic acid*. **styryl ~** Cinnamic acid*.
formic **f. aldehyde** Formaldehyde*. **f. nitrile** Hydrocyanic acid.
formimidoyl* The radical CH(:NH)−.
formin The glycerol esters of formic acid; e.g., **mono ~** $C_3H_5(OH)_2OOCH$, **tri ~** $C_3H_5(OOCH)_3$. **bromoethyl ~** Bromalin.
formine Hexamethylenetetramine.
formohydrazide Formylhydrazine*.
Formol Trademark for an antiseptic solution of formaldehyde in methanol and water. **f. nitrogen** The nitrogen of unsubstituted amino groups in the protein molecule which combines with formaldehyde. **f. titration** The determination of amino acids by titration with an alkali, with formaldehyde added to annul the alkalinity of the amino group. **f. toxoid** F.T. A preparation produced by treating a toxin with formaldehyde solution until its specific toxicity has been removed. Used to diagnose diphtheria. Cf. *Schick test.*
formolite A yellow, nonfusible, solvent-insoluble resin made by the action of aromatic compounds on formaldehyde in presence of concentrated sulfuric acid. Used in paints, inks, and adhesives.
formonitrile Hydrocyanic acid.
formonitrolic acid (NO_2)·CH:NOH = 90.01. Methylnitrolic acid, nitroformaldehyde oxime. Unstable liquid.
formothion* See *insecticides,* Table 45 on p. 305.
formoxime HCH:NOH = 45.4. Formaldoxime. Colorless liquid, b.84, decomp. in hot water. **nitro ~** Formonitrolic acid.
formoxyl Formyl*. **f. hydride** Formic acid*.

formoyl The formyloxy* radical.

formula A combination of chemical symbols expressing the composition of a molecule. Formulas indicate: (1) the kind of elements and the number of atoms (2) the weight relations of these elements and the molecular weight, (3) the percentage by weight of the elements present, (4) the valency of the elements. Formulas for gases and vapors also indicate (5) the volume relations, (6) the density, (7) the weight of 1 liter, (8) the volume occupied by 1 gram. Thus, chemical formula H_2O means: (*a*) A molecule of water consists of 2 atoms of H and 1 atom of O. (*b*) The molecular weight is the sum of the atomic weights; hence $2 \times 1.0079 + 15.999 = 18.0148$. (*c*) Water is 11.2% H and 88.8% O (by weight). (*d*) The valency of H is + 1; hence 2 atoms are the equivalent of O (valency −2). (*e*) 2 volumes H and 1 volume O combine to give 2 volumes water (vapor). (*f*) The density of water vapor is: (*i*) compared with oxygen, 1.13 (ii) compared with hydrogen, 8.9 (iii) compared with air, the molecular weight divided by 28.95; hence 0.622. (*g*) A gram molecule of any gas or vapor occupies 22.4 liters; hence 1 liter of water vapor weighs 0.804 g. (*h*) As 18 g water vapor occupies 22.4 liters, 1 g occupies 1.243 liters. Cf. *nomenclature.* See Table 35, next column. **abbreviated ~** An abbreviation of radicals; as, Am for ammonium, Ph for phenyl. **atomic ~** Structural f. **Beckmann's ~** See *Beckmann reaction.* **constitutional ~** A notation indicating constitution by valence bonds or the linkage between radicals and atoms, e.g., $C_2H_4 \cdot (COOH)_2$. **dynamic ~** See *benzene ring* under *benzene.* **electronic ~** A notation that indicates the electropositive or electronegative character of an atom in a compound. **empirical ~** A f. showing the simplest ratio between the atoms, but not the number of atoms nor the way they are linked together; e.g., CH_3 is the e. f. of ethane. **general ~** An expression representing the formulas of a series of compounds; thus, $M'_2M''(SO_4)_2 \cdot 12H_2O$ is the general f. for the alums, in which M′ is a monovalent metal, and M″ a trivalent metal. C_nH_{2n+2} is the general equation for the alkane series. **graphic ~** Diagrammatic f. A f. showing space relations; as, a tetrahedron for carbon. Cf. *stereoisomer.* **ionic ~** The symbol for an electrically charged atom or radical, as Na^+ or NH_4^+. **line ~** A f. showing molecular structure by lines joining the atoms to represent bonds; as, H−C≡C−H. **molecular ~** A formula for a complex compound indicating the participating molecules; thus $NH_3 \cdot HCl$ is the molecular f. for NH_4Cl. **octet ~** A f. showing the number of electrons; as, H:Ö:H. Cf. *Lewis*-Langmuir theory. **polar ~** Octet f. **polarity ~** A f. which indicates the relative position of the electron pairs held in common between 2 atoms. **rational ~** A f. indicating by the use of radicals the intraatomic arrangement; thus $Al_2(SO_4)_3$. **space ~, stereometric ~** See *stereoisomer, projection.* **structural ~** A f. indicating the 2- or 3-dimensional structure of a compound and the linkages of its atoms. **transmutation ~** See *transmutation equation* under *equation.*

formulary A selected list of drugs, chemicals, and medicinal preparations with descriptions; tests for their identity, purity, and strength; and formulas for preparing them; especially one issued by official authority, and accepted as a standard. Cf. *pharmacopoeia.*

formula weight The molecular weight expressed in grains.

Formvar Trademark for a polyvinyl formaldehyde plastic.

formyl* The radical HC(:O)−, from formic acid. **f.acetic acid** $CHO \cdot CH_2 \cdot COOH = 88.1$. **f. aldehyde** Formaldehyde*. **f. amide** Formamide*. ***N*-f.biphenyl-2-amine** $C_{13}H_{11}ON = 197.2$. Colorless crystals, m.73, soluble in alcohol. **f. bromide** Bromoform*. **f. chloride** Chloroform*. **f.hydrazine*** $NH_2NH \cdot CHO = 60.1$. Formohydrazide. Colorless crystals, m.54, soluble in alcohol. **di ~** $C_2H_4O_2N_2 = 88.1$. Colorless crystals, m.106, soluble in alcohol. **f. trichloride** Chloroform*. **f. triiodide** Iodoform*.

formylamino* Formamide. The radical $H \cdot CO \cdot NH-$, from formamide.

TABLE 35. FORMULA SIGNS

	Points in the center of the line separate composite groups or radicals, as, $CH_3 \cdot COOH$, or the constituents of a double salt, $M_2SO_4 \cdot M_2(SO_4)_3$.
— or = or ≡ · or : or ⁝	Lines or dots, either single, double, or triple, indicate valence bonds in a structural formula. A central, boldface dot indicates a free radical; as, $\cdot CH_3$.
:	Double dots also indicate an electron pair in an octet formula.
,	Commas indicate interchangeable elements; (Ni,Fe)AsS means FeAsS and NiAsS in variable proportion.
( )	Parentheses group elements together, or indicate radicals, as, $(NH_4)_2SO_4$.
[ ]	Square brackets are used: (1) in addition to parentheses, as $Fe_4[Fe(CN)_6]_3$, to indicate radicals; or in coordination compounds, to show the relationship to the central atom; (2) to denote an isotopically labeled atom, as, $[^{14}C]H_4$.
{[( )]}	This is the normal nesting order, but in coordination compounds, brackets [ ] enclose the complex ion or neutral coordination entity; as, $[Co\{SC(NH_2)_2\}_4][NO_3]_2$.
+	The plus sign indicates positive charge or a cation, K^+.
—	The minus sign indicates negative charge or an anion, Cl^-.
→	An arrow indicates direction of a reaction.
⇌	Reversible double arrows indicate equilibrium or tendency to react.
*	An asterisk indicates (1) an excited atom or molecule; (2) a radioactive atom.
1,2, . . .	An numeral after a symbol or parenthesis is a multiplier for that symbol; a numeral before a formula is a multiplier of the whole formula. Thus $2Al_2(SO_4)_3 \cdot 6H_2O$ means 2 molecules of crystalline aluminum sulfate, together having 6 molecules of water of crystallization.
^1A	Small superscript numerals before the symbol of an element indicate the mass number; thus $^2H_2$, ^6Li, etc. An isotopically substituted compound is written, e.g., $^{14}CH_4$. Cf. [ ] above.
$_1A$	Small subscript numerals before the symbol of an element indicate the atomic number; $_1H$, $_3Li$.
M, X, L	These letters indicate a metal, halogen, and ligand (unspecified), respectively.
R^1, R^2, R^3, etc.	These indicate radicals (or R, R′, R″ for up to 3 groups), where the groups are different. Where the groups are the same, R_2, R_3, etc., are used.

formylation Vilsmeir reaction. The introduction of the formyl radical into an organic compound.
formylene The methylidyne* radical.
formylic acid Formic acid*.
formyloxy* Formoyl, methanoyl. The radical H·CO·O−, from formic acid.
forsterite Mg_2SiO_4. A white, orthorhombic magnesium silicate of the olivine group, containing iron oxide, d.3.2–3.3, hardness 7.
Fortral Trademark for pentazocine.
Fortrel Trademark for a polyester synthetic fiber.
fortified Strengthened. **f. wine** (1) Wine to which a fermentable sugar has been added. (2) Wine to which alcohol or wine brandy has been added.
Fortisan Trademark for regenerated cellulose produced by the simultaneous stretching and deacetylation of cellulose acetate filaments.
Fortran *For*mula *tran*slation computer language; used in scientific and technical programming.
fosfesterol BP name for diethylstilbestrol diphosphate.
foshagite $5CaO \cdot 3SiO_2 \cdot 3H_2O$. A constituent of boiler scale.
fossil Remains of a prehistoric organism imprinted or entombed in geological formations.
fossil copal See *copals.*
fossilin Vaselin.
foto- Photo-.
fotosensin A condensation product of phthalic acid and resorcinol, with small amounts of copper and iron, used to sensitize plant growth.
fouadin Stibophen.
Foucault, Jean Bernard Léon (1819–1868) French physicist. **F. current** Eddy current. **F. pendulum** A Galileo pendulum. **F. prism** A polarizing prism using a thin film of air for reflection. Cf. *nicols.*
Fowler, Sir Ralph (1899–1944) British mathematical physicist.
fowlerite (Mn, Fe, Ca, Zn, Mg)SiO_3. A red, brown, or yellow, natural, triclinic silicate, d.3.4–3.7, hardness 5–6.5.
Fowler's series For lines of the helium spectrum: $1/\lambda = 4N(1/9 - 1/n^2)$, where n is 4, 5, 6; λ the wavelength; and N a constant.
foxglove See *digitalis.*
foyaite A nepheline syelite (Portugal).
f.p., fp Abbreviation for freezing point.
fps Abbreviation for foot-pound-second system.
Fr Symbol for francium.
fraction A separated portion of a whole.
fractional Pertaining to separated parts. **f. column** Dephlegmator. Same as f. distillation tubes. **f. combustion** A method of separating gas mixtures by removing one constituent by combustion. **f. condensation** A method of separating gas mixtures by lowering the temperature or increasing the pressure until one of the gases liquefies. **f. condensation tube** An air-cooled, vertical condenser tube of special shape to promote condensation. **f. crystallization** A method for separating or purifying compounds in solution by successive, slow crystallizations, the mother liquor being removed at each stage for a further crystallization. **f. distillation** A method of separating volatile substances by collecting separately the distillates evaporating at certain temperatures. **f. distillation tube** Shaped glass tubes used to collect distillates separately at different temperatures. **f. expression** The collection of plant oils or juices expressed at different temperatures. **f. filtration** The filtration of solutions consecutively through coarse, medium, and fine filters. **f. precipitation** See *salting out.* **f. weights** Analytical weights of less than 1 gram.
fractionating Separating into parts. **f. column** Dephlegmator. A device for fractional distillation. Cf. *still head.*
fractionation Fractional distillation.
fracture A sharp edge produced on breaking a solid substance. Cf. *cleavage.* **conchoidal ~** The irregular fracture of an amorphous body. **crystalline ~** A f. producing the plane faces and sharp edges of a crystal, i.e., cleavage.
fragility Brittleness. The characteristic of easily breaking apart.
francium* Fr. Alabamine. Actinium-K. The radioelement, at. no. 87, formed by the α decay of actinium. At. wt. of isotope with longest half-life (^{223}Fr) = 223.02. It is separated from actinium compounds by fractional precipitation with ammonium carbonate on a rare-earth carrier; its properties are similar to those of cesium. Discovered by Perey (1939); named for France. Valency is one.
Franck-Condon principle The change in state of an electron in a molecule takes place so rapidly in relationship to the vibrational motion that the separation and velocity of the nuclei are virtually unchanged.
franckeite $Pb_4FeSn_3Sb_2S_{14}$. A rare, native sulfostibide.
francolite A calcium phosphate containing calcium carbonate.
frangula Buckthorn bark, arrow wood. The dried bark of *Rhamnus frangula* (Rhamnaceae); a laxative (EP, BP). **f. emodin** Emodin.
frangularoside A rhamnoside from the bark of black alder.
frangulic acid A dihydroxyanthraquinone from frangula.
frangulin $C_{20}H_{20}O_9$ = 404.4. A glucoside from the bark of *Rhamnus frangula.* **A-** Yellow crystals, m.226, insoluble in water.
Frangulineae Rhamnales. Plants comprising the Rhamnaceae (buckthorn) and Vitaceae (grape) families.
frangulinic acid Frangulic acid.
frankincense Olibanum. **American ~** An oleoresin from the bark of *Pinus palustris,* common pine. Cf. *rosin, gum* thus.
Frankland **F., Sir Edward (1825–1899)** British chemist noted for research on organometallic compounds, valency, water supply, the theory of flames; a coworker of Lockyer in the discovery of helium in the sun. **F., Percy Faraday (1858–1946)** British chemist, noted for his work on stereochemistry. **F. notation** The grouping together of radicals in a formula, and the assumption that certain salts are addition compounds; thus SO_2HO_2 for H_2SO_4. **F. reaction** The synthesis of hydrocarbons by the zinc-alkyl condensation: $Zn(CH_3)_2 + 2R \cdot Br = 2R \cdot CH_3 + ZnBr_2$.
Franklin **F., Benjamin (1706–1790)** American statesman and scientist, founder of the American Philosophical Society. **F., Edward Curtis (1862–1937)** American chemist noted for his theory of the *ammonia* system, q.v.
franklinic Static.
franklinite (Fe,Zn,Mn)Fe_2O_4. A black, brittle, isometric spinel, d.5–5.1, hardness 6.1–6.5; a source of iron and zinc.
Frary's metal An alloy made by electrolytic deposition of barium (2%) and calcium (1%) in molten lead; used for bearings and shrapnel bullets.
Fras fuel Petroleum in gel form, for flamethrowers.
fras(s) Excrement or other fragments left by insects on infestation of food.
Fraude reagent Perchloric acid*.
Fraunhofer, Joseph von (1787–1826) German physicist. **F. lines** The dark lines in spectra of the sun and other stars, produced by the absorption of certain rays of the photosphere by incandescent gases of the solar atmosphere. Some 20,000 have been measured, and more than 5,000 identified with chemical elements. F. lines and the corresponding bright

emission lines indicate stellar conditions; as, composition, motion, temperature, magnitude, and mass.

fraxetin $C_{10}H_8O_5$ = 208.2. A crystalline split product of fraxin, m.227.

fraxin $C_{16}H_{18}O_{18}$ = 498.3. A glucoside from the bark of *Fraxinus* and *Aesculus* species. Yellow needles, m.205, soluble in water.

fraxinol $C_{11}H_{10}O_5$ = 222.2. Crystals, m.171; from *Fraxinus excelsior.*

Fraxinus A genus of trees, including ash. Cf. *manna.*

fraxitannic acid $C_{26}H_{32}O_{14}$(?). A tannin from the leaves of the ash, *Fraxinus excelsior.* Brown powder, soluble in water.

free Uncombined, unattached, or available. **f. acid** (1) See *degree of acidity* under *acidity.* (2) The hydrochloric acid of stomach juices. **f. charge** An electric charge on a body, e.g., f. electrons attached to an atom or molecule and not a part of its structure. **f. electron** An electron not confined to 1 atom or molecule, but free to move from one to another under the influence of an electric current. **f. energy** See *Gibbs function, Helmholtz function.* **f. path** The average distance traveled by an electron, ion, or molecule before colliding with another. **f. radical** See *free radical* under *radical.* **f. valence** An unsatisfied bond.

freedom Variance. In the phase rule: $P + F = C + 2$, where P is the number of phases, C the number of component substances reacting, F the degree of f., an integer which indicates the least number of variable factors (pressure, temperature, or volume) which must be arbitrarily fixed in order to define the state of equilibrium of a chemical system. A one-component solid-liquid-gas system is invariant, i.e., $F = 0$; a one-component liquid-gas system is univariant, $F = 1$; a one-component gas system is bivariant, $F = 2$.

freeness The extent to which a pulp for the manufacture of paper has not been hydrated by beating, and, therefore, parts with its water by simple drainage. Antonym: *wetness.*

freeze (1) To lower the temperature until a liquid solidifies. (2) The interlocking of accurately machined and polished steel parts of machinery; it is avoided by chromium-plating one part. **deep ~** Refrigeration of food at a low temperature. **quick ~** Rapid refrigeration of foods. The ice crystals formed are relatively small and do not rupture tissues, and the thawed food retains much of the flavor of the original.

f. drying A method of drying a substance (as, food) by freezing it below 0°C, and removing the water as ice, but sublimation under a high vacuum.

freezing The solidification of a liquid or solution when its temperature is lowered. **f. attachment** A device attached to microtomes for freezing tissues by liquid or solid carbon dioxide. **f. mixture** Frigorific mixture. A mixture of substances that absorbs heat and thus lowers the surrounding temperature; e.g.:

Ammonium nitrate 1, anhydrous sodium carbonate 1, water 1 pt.	−10 to −26°C
Ammonium chloride 5, powdered ice or snow 10 pts.	− 5 to −18°C
Sodium chloride 5, ammonium nitrate 5, ice or snow 12 pts.	−15 to −25°C
Calcium chloride 3, ice or snow 1 pt.	−40 to −70°C
Sulfuric acid (dilute) 10, ice or snow 8 pts.	−65 to −90°C
Solid carbon dioxide and ether or chloroform	−77°C

f. point Solidification or f. temperature. **f. point apparatus** A device for the accurate determination of f. point, used to determine the concentration of solutions, osmotic pressure, and molecular weight. See *cryoscopy.* **f. point depression** Solutions freeze at a lower temperature than the pure solvent proportionally to the concentration of the solute. See *Raoult's law.* A solution containing 1 mol of solute dissolved in 1,000 g of water lowers the f. point by 1.86°C. **f. salt** Crude sodium chloride.

freibergite $(Ag,Cu)_3Sb_2S_7 \cdot (Fe,Zn)_4SbS_7$. A gray sulfantimonide, d.4.85–5.

freierslebenite $(Pb,Ag_2)_5Sb_4S_{11}$. A black, metallic sulfantimonide, d.2–2.5.

Fremy, Edmond (1814–1894) French chemist noted for developing the manufacture of iron and steel, sulfuric acid, and artificial rubies. **F. salt** Potassium hydrogenfluoride.

French chalk A hydrated silicate of magnesium.

Freon Arcton (U.K. usage). Trademark for a group of halogenated hydrocarbons (usually based on methane), containing one or more fluorine atoms; widely used as noncorrosive and nontoxic refrigerants, propellants, and insecticide solvents. **F11** Trademark for trichlorofluoro*methane.* **F12** CCl_2F_2. Trademark for dichlorodifluoro*methane;* a refrigerant and propellant for aerosols. **F21** Trademark for dichlorofluoro*methane.* Used in heat transfer tubes. **F22** Trademark for chlorodifluoro*methane.*

frequency (1) The rapidity with which an occurrence, oscillation, or vibration is repeated. (2) The number ν of complete vibrations or waves per unit time. The reciprocal of the period. Thus $\nu = c/\lambda$, where c is the velocity of light, and λ the wavelength of the ray concerned; e.g., for green light $\nu = 6 \times 10^{14}$ hertz. F. is related to energy e and mass m by $h\nu = E = mc^2$. **high ~** A rapid alternating electric current. **molecular ~** The molecular vibrations. **radiation ~** The f. of the emitted radiation is proportional to the amount of energy radiated; $W_1 - W_2 = E = h\nu$, where W_1 and W_2 are the energy of the atom in the initial and final state, respectively, E is the energy in joules, and h is the Planck constant.

f. curve Normal probability curve.

Fresenius, Karl Remigus (1818–1897) German chemist. **F. desiccator** A desiccator with bell-shaped cover. **F. nitrogen bulb** A conical flask with side tubes having 2 bulbs near the base.

Fresnel, Augustin Jean (1788–1827) French physicist noted for experiments on light. **f** A little-used unit of wave frequency, expressed as 10^{12} s^{-1}.

Freund acid 1-Naphthylamine-3,6-disulfonic diacid.

Freundlich, Herbert (1881–1941) German-born chemist, noted for his work on colloid chemistry.

freyalite A variety of thorite.

friction The resistance offered to sliding motion by rubbing. **fluid ~** Viscosity. **internal ~** The resistance to bending or metals due to their crystalline structure. **mechanical ~** See *friction coefficient* below.

f. coefficient μ The force F necessary just to move an object along a horizontal plane divided by the normal force N; thus $F = \mu N$. It depends on the material of the substances, not on the velocity or area of the surface in contact. Cf. *lubricant.*

Friedel, Charles (1832–1899) French organic chemist.

Friedel-Crafts **F.-C. condensation** The condensation of hydrocarbons and halogen compounds in the presence of anhydrous aluminum chloride according to the F.-C. reaction; e.g., alkyl halides yield hydrocarbons. **F.-C. reaction** The synthesis of an aromatic hydrocarbon homolog by the catalytic action of aluminum chloride: $R + R' \cdot Cl + AlCl_3 = R \cdot R' + HCl + AlCl_3$.

friedelin $C_{30}H_{50}O$ = 426.7. A sterol from cork related to the hydrocarbon $C_{30}H_{52}$.
Friedrichs **F. condenser** A condensing worm surrounded by a glass or metal tube. **F. gas bottle** A glass cylinder with a spiral tube for washing gases.
frieseite $Ag_2Fe_5S_8$. A rare, native sulfide.
frigorific Describing an agent that produces coldness. **f. mixture** Freezing mixture.
Frilon Trademark for a polyamide synthetic fiber.
fringes The dark, parallel, equidistant lines observed in the interferometer.
frit (1) Enamel. A complex alkaline borosilicate glass, usually containing fluorine, produced by melting a mixture such as borax, feldspar, quartz, and cryolite. (2) To sinter.
fritted, frittered Having been heated near the melting point. **f. glass** Glass powder heated sufficiently for the particles to adhere together without coalescing completely; used in filters.
fritting Becoming pasty and beginning to melt; as, some soft coals.
Froehde reagent A reagent for alkaloids: 5 mg molybdic acid in 1 mL hot concentrated sulfuric acid.
frontier orbital See *frontier orbital* under *orbital.*
frost Dew produced in frozen form. **hoar ~** F. produced at a dew point of less than 0°C.
froth Foam. **iron ~** Spongy hematite.
frother A chemical used in the flotation process. Thus, pine oil produces a thin transient foam; cresylic acid, a heavy permanent foam.
frothing Foaming. **f. agent** A substance which produces a froth when shaken with a liquid; as, saponin.
F.R.S. Abbreviation for Fellow of the Royal Society (London).
F.R.S.C. Abbreviation for Fellow of the Royal Society of Chemistry (London).
fructofuranosidase* **β-D- ~** See *invertase.*
fructosans Sugar anhydrides which hydrolyze to fructose. Cf. *hemicellulose.*
fructose $C_6H_{12}O_6$ = 180.2. Levulose, fruit sugar, D-fructose, *l*-fructose. A carbohydrate in sweet fruits and honey. Colorless needles, m.103, soluble in water, alcohol, or ether; a preservative; an intravenous infusion in parenteral feeding (USP, EP, BP). **inactive ~** Fructose, acrose, or formose. An unfermentable carbohydrate formed by polymerization of formaldehyde in limewater. **L- ~** *d*-Fructose. m.102. **pseudo ~** Psicose.

CH_2OH | C=O | HOCH | HCOH | HCOH | CH_2OH

D-form

O, CH_2OH, HO, HO, OH, OH

D-pyranose form

fructoside A glucoside which hydrolyzes to fructose.
fruit The ripened ovary of a plant together with parts of the flower that share in its development. Some used in traditional remedies; e.g., anise, caraway, fig, hops, orange, pepper, vanilla. See also *seeds.* **f. essence** An artificial mixture imitating the taste of a fruit and consisting of esters of alkyl radicals with organic acids, usually in dilute alcoholic solution. **f. oil** See *oil.* **apple ~** Chiefly pentyl valerate. **banana ~** Chiefly pentyl acetate. **pear ~** Chiefly pentyl acetate. **pineapple ~** Chiefly ethyl butyrate. **f. sugar** Fructose.
frusemide EP, BP name for furosemide.
ft Abbreviation for foot. ft^2 = square foot. ft^3 = cubic foot.
fuberidazole* See *fungicides,* Table 37.
fuchsin $C_{20}H_{19}N_3HCl$ = 337.8. Magenta red, C.I. Basic Violet 1A, rosaniline hydrochloride, aniline red, azaleine, harmaline, rosein, erythrobenzene, rubine, solferino. A red dyestuff, a mixture of the hydrochlorides of rosaniline and pararosaniline. Red rhombs, with green fluorescence, soluble in water; a dyestuff, coloring matter for inks and microscopical stains. **acid ~** A mixture of the disulfonic and trisulfonic acids of pararosaniline; a dye and stain. **English ~** A mixture of the acetates of rosaniline and pararosaniline; a dyestuff. **German ~** Fuchsin.
fuchsite $H_2KAl_3(SiO_4)_3Cr$. A variety of muscovite.
fuchsone $O{:}C_6H_4{:}C(C_6H_5)_2$ = 258.3. 4-(Diphenylmethylene)-2,5-cyclohexadien-1-one. Brown crystals, m.167. An intermediate in the manufacture of rosaniline dyes.
fucitol $C_6H_{14}O_5$ = 166.2. **D- ~** D-1-Deoxygalactitol. A sweet alcohol, m.153, from fucus.
fucoiden A mucilaginous constituent of seaweed, $R^1-R^2-O\cdot SO_3M$, where R^1 is fucose, R^2 another carbohydrate complex, and M is Na, K, Mg, or Cu.
fucosan A fucose polymer in the cell walls of marine algae.
fucose $C_5H_9MeO_5$ = 164.2. 6-Deoxygalactose. **D- ~** From glycosides in Convolvulaceae spp., m.142. **L- ~** From frog spawn, sea urchin, milk. m.141.
fucosite The least water-soluble portion of seaweed, containing fucose, algarose, and an oil. **pseudo ~** Elkerite.
fucoxanthin $C_{42}H_{58}O_6$ = 658.9. Fucoxanthinol acetate. A carotenoid pigment of brown algae. Red crystals, m.168.
fucusamide Fucusine.
fucusine $C_{15}N_{12}O_3N_2$ = 268.3. Fucusamine. A crystalline alkaloid from fucus.
fucusol $C_5H_4O_2$ = 96.1. A colorless oil, resembling furaldehyde.
fuel A material that furnishes heat on combustion. Classification: (1) *Natural* or *solid f.;* as, *wood, peat, lignite, coal,* q.v. (2) *Prepared* or *dried f.;* as, briquets and compressed fuels. (3) *Liquid f.;* as, petroleum, gasoline. See Table 36 on p. 248. (4) *Gaseous f.;* as, coal or water gas. **fossil ~** F. derived from coal, lignite, peat, natural gas, oil, shales, or tar sands. **Fras ~** See *Fras fuel.* **metallic ~** Finely powdered magnesium or aluminum. Cf. *Thermit.* **pulverized ~** Ground furnace fuel; typically 80% below 200 μm.

f. cell A means of converting chemical energy, as, of a chemical reaction, into electrical energy by reverse electrolysis; e.g., the catalytic conversion of H and O into water. **f. equivalence** Several conventions exist for converting different fuels into common energy units. Often in the U.S., oil is expressed as "oil equivalent (o.e.)," based on 10,000 kcal/kg net, and coal as "hard coal equivalent (c.e.)," based on 7000 kcal/kg net. Thus 1 ton c.e. = 0.7 ton o.e. (See *net calorific value* under *calorific*). Other fuels, as gas, are equated to these units on the basis of their thermal content. Nuclear and hydro power are also expressed as o.e. or c.e., on the basis of the notional amount of these fossil fuels needed to produce the same power output in a conventional steam power station. **f. gases** The compressed gases used for welding and cutting metals; as, blau gas, butane, coal gas. **f. oil** Crude petroleum.
fugacity The escaping tendency of a substance in a

TABLE 36. PROPERTIES OF LIQUID FUELS

Fuel	Specific gravity at 15°C	Viscosity, Pa·s at 25°C	Boiling point, °C	Calorific value, kJ/g	Flash point, °C	Composition, %			
						C	H	O	S
Methyl alcohol	0.796	0.531	64.6	19.9	—	37.5	12.5	50.0	—
Ethyl alcohol	0.794	1.07	78.3	26.8	—	52.0	13.0	35.0	—
Benzole mixture	0.878	0.510	86.0	40.2	−9	91.6	8.1	—	0.3
Aviation petrol	0.720	0.439	84	44.0	—	85.0	15.0	—	0.01
Petrol no. 1	0.740	0.481	104	43.9	—	85.3	14.7	—	0.02
Petrol no. 3	0.745	0.538	112	43.6	—	85.5	14.5	—	0.03
Tractor oil	0.780	0.991	166	43.6	37	86.3	13.7	—	0.01
Kerosene	0.793	1.13	196	43.6	40	86.3	13.6	—	0.08
Diesel oil	0.870	3.47	300	43.1	75	86.3	12.8	—	0.9
Light fuel oil	0.895	35.4	348	42.1	80	86.2	12.4	—	1.4
Heavy fuel oil	0.949	850	360	41.3	110	86.0	11.9	—	2.1
Heptane	0.691	0.411	98	44.8	—	84.0	16.0	—	—
Methylated spirit	0.832	0.850	76	30.0	−8	59.5	11.6	28.9	—
Methylated spirit (commercial)	0.824	0.779	74	34.3	−12	70.7	11.5	17.8	—

heterogeneous mixture which enables a chemical equilibrium to respond to altered conditions. In a dilute solution obeying the gas laws, the f. equals the osmotic pressure. In others, it is the pressure for which these laws are still valid. Cf. *activity.*

fugin Fugutoxin, tetrodonine. A poisonous protein from the fish *Tetrodon* species (Japan, China).

fugitive Unstable; not fast (as a color).

fugitometer An apparatus for rapidly testing colored materials for fastness to light. Cf. *fadeometer.*

fugugetin $C_{17}H_{12}O_6 \cdot 5H_2O$ = 402.4. The coloring matter of fukugi, *Garcinia spicata* (Japan), m.288. Cf. *garcinin.*

fugutoxin Fugin.

Fujiwara reaction Test for organic halides. A red color is produced on heating with pyridine and sodium hydroxide.

Fukui, Kenichi (1918–) Japanese chemist. Nobel prize winner (1981). Noted for work on frontier orbital theory.

Fulcher spectrum That portion of the hydrogen spectrum which consists of many fine lines due to a lower excitation level of the H_2 molecule.

Fulcin Trademark for griseofulvin.

fulgenic acids The compounds $R_2C:C(COOH)\cdot C(COOH):CR_2$, in which R is a hydrogen, alkyl, or aryl radical.

fulgides Anhydrides of fulgenic acids, of the type

```
R2C:C·CO
    |     >O
R2C:C·CO
```

They are phototropic. **diphenyl ~** *A* form (yellow-green) ⇌ *B* form (blue). **triphenyl ~** *A* form (orange) ⇌ *B* form (blue).

fulgaration Melting together with an electric spark. Cf. *fritting.*

fulgurator An atomizer of solutions for producing flame spectra.

fulgurite Fritted sand produced by lightning passing through the soil.

fuller's earth Calcium montmorillonite. An impure kaolin containing magnesium and iron; used for decolorizing solutions and oils; also used as a substitute for absorbent charcoal, as a dusting powder and for paraquat poisoning.

full gas A class of combustible gases, consisting of saturated hydrocarbons, e.g., water gas.

fulling Felting (wool industry). Treatment of a textile surface so that the fibers project so as to give a bulky texture.

fulmar oil An oil from *Procellaria glacialis,* a seabird (U.K.); a substitute for cod-liver oil.

fulmargin A colloidal solution of silver.

fulminate* A salt of fulminic acid containing the radical CNO−. Used to detonate high explosives. **silver ~** AgONC = 149.9. White explosive needles.

fulminating Causing detonation or explosion. **f. caps** Small amounts of fulminates used to explode the charge of shells, etc.; usually containing mercuric fulminate. **f. gold** Aurodiamine. **f. powder** Percussion powder.

fulminic acid* HONC = 43.03. Paracyanic acid, ψ-isocyanic acid. An isomer of cyanic acid known only as its very explosive salts.

fulminurate A salt of fulminuric acid.

fulminuric acid $CN\cdot CH(NO_2)\cdot CONH_2$ = 129.1. 2-Cyano-2-nitroacetamide*. Colorless needles, m.138, explode 145, soluble in water; a trimer of cyanuric acid.

fulvalene $C_{10}H_8$ = 128.2. A compound known only in solution.

fulvene* C_6H_6 = 78.1. 5-Methylene-1,3-cyclopentadiene*. An isomer of benzene; a yellow oil.

```
CH=CH
|      >C=CH2
CH=CH
```

Fulvicin Trademark for griseofulvin.

fumaramic acid $NH_2CO\cdot CH:CH\cdot COOH$ = 115.1. 4-Amino-4-oxo-2-butenoic acid. Colorless crystals, m.220.

fumaramide* $NH_2CO\cdot CH:CH\cdot CONH_2$ = 114.1. Colorless crystals, m.266 (decomp).

fumarate hydratase* Enzyme that catalyzes the reversible elimination of water from L-malate to give fumarate.

fumarhydrazide $C_4H_6O_2N_4$ = 142.1. Colorless crystals, m.220.

Fumariaceae A family of herbs yielding alkaloids; e.g.: *Corydalis tuberosa* (corydaline), *Fumaria officinalis* (fumarine).

fumaric acid* $(H \cdot C \cdot COOH)_2$ = 116.1. (*E*)-2-Butenedioic acid*, in *Fumaria officinalis;* an isomer of maleic acid. Colorless prisms, m.286, slightly soluble in water. Used to manufacture polyester and alkyd resins. **ethyl ~** $C_6H_8O_4$ = 144.1. Ethylethylenedicarboxylic acid. Colorless crystals, m.194. **methyl ~** Mesaconic acid.

f. a. series $C_nH_{2n-2}(COOH)_2$. A group of dibasic acids with a double bond and many isomers. See *maleic, pentenedioic,* and *allylmalonic* (under *allyl*) *acids.*

fumaroid See *malenoid.*

fumarole A small hole from which volcanic gases and vapors escape. **f. acid** Boric acid*. **f. gases** The vapors escaping from fumaroles; those of Italy contain steam, boric acid, carbon dioxide, ammonia, and hydrogen sulfide.

fumaroyl* The radical $-OC \cdot CH{:}CH \cdot CO-$, from fumaric acid. **f. chloride** $C_4H_2O_2Cl_2$ = 153.0. Colorless liquid, d.1.410, b.160.

fume Visible or invisible particles of solid or liquid suspended in a gas. **f. cupboard** A safety-glass-enclosed shelf or table, with a ventilating device, for experiments involving poisonous or unpleasant fumes or gases.

fumigant A gaseous *insecticide,* q.v., or disinfectant, or a substance producing one. **food ~** A f. used to destroy insect pests and microorganisms on foodstuffs; as, ethylene oxide.

fumigate To disinfect by means of vapors or smokes.

fumigatin $C_8H_8O_4$ = 168.2. 3-Hydroxy-2-methoxy-5-methyl-1,4-benzoquinone. Maroon crystals, m.116, in cultures of *Aspergillus fumigatus;* an antibiotic. Cf. *penicillin.*

fumigation Disinfection by means of volatile substances.

fuming Emitting smoke or vapors.

function (1) Any specific power. (2) Physiology: the work or purpose of an organ. (3) Mathematics: one quantity is a function of another when for each value of the latter, y, there corresponds a definite value of the former, x. Thus $x = f(y)$. **chemical ~** (1) **simple ~** Describes a substance containing only one type of radical, which may, however, be repeated several times in the same molecule, e.g., polyhydroxy alcohol. (2) **complex ~** Describes a molecule having 2 or more different types of radical; as, amino acids.

fundament The foundation or basis of a structure, either physical or mental.

fundamental Pertaining to the basis for groundwork. **f. chain** The longest chain of a branched hydrocarbon.

fungi Plural of "fungus."

fungicide An agent that destroys fungi and their spores. See Table 37 on p. 250 for agricultural f. See text for individual medical f.; as, amphotericin B, griseofulvin. Cf. *herbicide, insecticide, rodenticide.*

Fungilin Trademark for amphotericin B.

fungisterin Fungisterol.

fungisterol $C_{28}H_{48}O$ = 400.7. 24-Methyl-7-cholesten-3-ol. Fungisterin. An inert alcohol in ergot; colorless; waxy masses, m.152.

Fungizone Trademark for amphotericin B.

fungus (Pl., fungi) (1) A main division of *Thallophyta,* or primitive plants, either parasitic or saprophytic, distinguished from *Algae* by the absence of chlorophyll; e.g.: *Eumycetes* (mushrooms), *Schizomycetes* (bacteria), *Phycomycetes* (algalike fungi), *Blastomycetes* (yeasts), *Hypomycetes* (molds). (2) Popularly, the mushrooms and toadstools. **f. dyestuff** A pigment obtained from a f.; as litmus.

funnel A glass tube with one enlarged, and usually conical, end. **Buchner ~** A porcelain f. having a flat, perforated, round bottom; used for rapid filtration by suction. **double-wall ~** A metal f. with 2 walls, between which hot water or steam is circulated. Used for hot filtration. **dropping ~** A separatory f., with long stem and glass stopcock. **Hirsch ~** A porcelain f. with a fixed porcelain plate. **hot-air ~** Double-wall f. **hot-water ~** Double-wall f. **separatory ~** A f. of varied shape with a stopcock on its stem; used to separate immiscible liquids. **tap ~** Separatory f.

funnel tube A glass tube with a conical or thistle-shaped top; used to convey liquids into a chemical apparatus.

fur Abbreviation for "furlong."

furac See *lead dithiofuroate.*

Furacin Trademark for a brand of nitrofurazone, 5-nitro-2-furaldehyde semicarbazone, an antibacterial; also used for trypanosomiasis and in veterinary medicine.

furacrolein Furfuracrolein.

furacrylic Furanacrylic.

Furadantin Trademark for nitrofurantoin.

fural The furfurylidene* radical.

furaldehyde 2- ~* $C_4H_3O \cdot CHO$ = 96.1. Furfural, 2-furancarboxaldehyde†, α-furfuraldehyde, furol, furfurol, furfuryl aldehyde, 2-furancarbonal. Colorless liquid, turning yellow on standing, $d_{22} \cdot 1.159$, m.−36, b.161, soluble in water. Used as a reagent for urea, alkaloids, santonin, cholesterol, ketones, or phenols; for the manufacture of furfuryl alcohol and lubricating oil stabilizers. Produced from pentosan-rich materials, as, corn cobs.

furan* C_4H_4O = 68.1. Furfuran, tetrol.

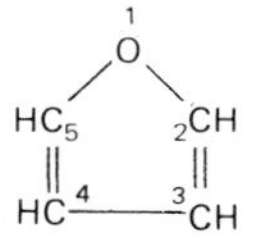

1
O
HC_5 $_2CH$
HC 4 3 CH

Colorless liquid, b.31, insoluble in water; in conifer tar. **bi ~*** $C_8H_6O_2$ = 134.2. 2,3′-Bifuryl*. Unstable yellow liquid. **dimethyl ~** C_6H_8O = 96.1. Colorless liquid, b.94, soluble in water. A tanning agent and solvent for polyvinyl plastics. **methyl ~** Sylvan. **tetrahydro ~*** Butylene oxide. **thio ~** Thiophene*.

2-f.carboxaldehyde† Furaldehyde*. **f.carboxylic acid*** **2- ~** Pyromucic acid. **3- ~** β-Furoic acid*. **f.dione** Maleic anhydride*. **f.methylamine** Furfurylamine.

furanacrylic acid $C_7H_6O_3$ = 138.1. β-2-Furylacrylic acid. White crystals, m.141, insoluble in water.

furano* The bridge $-C_4H_2O-$.

furanose The 5-membered cyclic form of sugars; as, *glucose,* q.v. Cf. *pyranose.*

furanoside A glucoside derived from pentoses, having a furan ring; as, adenosine.

furanylmethyl 2- ~† See *furfuryl.*

furanylmethylene 2- ~† See *furfurylidene.*

furazan* N:CH·CH:N (└O┘) = 70.1. Azoxazole, oxdiazole. **ethylmethyl ~** $C_6H_8ON_2$ = 124.1. Colorless liquid, b.170. **phenyl ~** $C_8H_6ON_2$ = 146.1. Colorless crystals, m.30.

f.carboxylic acid $C_3H_2O_3N_2$ = 114.1. Colorless crystals, m.107. **methyl ~** $C_4H_4O_3N_2$ = 128.1. Colorless crystals, m.74. **f.dicarboxylic acid** $C_4H_2O_5N_2$ = 158.1. Colorless crystals, m.178. **f.propionic acid** $C_5H_6O_3N_2$ = 142.1. Colorless crystals, m.86.

furazanyl* The radical C_2HON_2-, from furazan.

furfuracrolein $C_7H_6O_2$ = 122.1. Colorless crystals, m.51.

furfural Furaldehyde*.

TABLE 37. FUNGICIDES

Common name	Systemic (s)	Chemical name
Benodanil*	s	2-Iodobenzanilide
Benomyl*	s	Methyl 1-(butylcarbamoyl)benzimidazol-2-ylcarbamate
Bordeaux mixture		Combination of copper sulfate and lime
Burgundy mixture		Combination of copper sulfate and sodium carbonate
Calomel		See *mercurous chloride** below
Captafol*		*N*-(1,1,2,2-Tetrachloroethylthio)cyclohex-4-ene-1,2-dicarboximide
Captan*		1,2,3,6-Tetrahydro-*N*-(trichloromethyl)thiophthalimide
Carbendazim*	s	Methyl benzimadazol-2-ylcarbamate
Carbofuran*	s	2,3-Dihydro-2,2-dimethylbenzofuran-7-yl methylcarbamate
Carboxin*	s	5,6-Dihydro-2-methyl-1,4-oxathiin-3-carboxanilide
Chlorothalonil*		Tetrachloroisophthalonitrile
Copper oxychloride		Dicopper chloride trihydroxide
Dichlofluanid*		*N'*-Dichlorofluoromethylthio-*N',N'*-dimethyl-*N*-phenylsulfamide
Ditalimfos*		*O,O*-Diethyl phthalimidophosphonothiolate
Drazoxolon*		4-(2-Chlorophenylhydrazono)-3-methyl-5-isoxazolone
Fenoprop*	s	(±)-2-(2,4,5-Trichlorophenoxy)propionic acid
Fentin acetate*		Triphenyltin acetate
Fentin hydroxide*		Triphenyltin hydroxide
Fuberidazole*		2-(2-Furyl)benzimidazole
Guazatine*	s	bis(8-Guanidino-octyl)amine
Imazalil*	s	1-(β-Allyloxy-2,4-dichlorophenethyl)imidazole
Iprodione*		3-(3,5-Dichlorophenyl)-*N*-isopropyl-2,4-dioxoimidazolidine-1-carboxamide
Mancozeb*		Complex of zinc and maneb containing 20% manganese and 2.5% zinc
Maneb*		Manganese ethylenebis(dithiocarbamate) polymer
Mercurous chloride*		Dimercury dichloride
Nabam*		Disodium ethylenebis(dithiocarbamate)
Nuarimol*	s	2-Chloro-4'-fluoro-α-(pyrimidin-5-yl)benzhydryl alcohol
Oxamyl	s	*N,N*-Dimethyl-2-methylcarbamoyloxyimino-2-(methylthio)acetamide
Oxycarboxin*	s	2,3-Dihydro-6-methyl-5-phenylcarbamoyl-1,4-oxathiin 4,4-dioxide
Prochloraz		*N*-Propyl-*N*-[2-(2,4,6-trichlorophenoxy)ethyl]imidazole-1-carboxamide
Propiconazole	s	1-[[2-(2,4-Dichlorophenyl)-4-propyl-1,3-dioxolan-2-yl]methyl]-1*H*-1,2,4-triazole.
Propineb*		Zinc propylenebis(dithiocarbamate) polymer
Quintozene*		Pentachloronitrobenzene
Tecnazene* (TCNB)		1,2,4,5-Tetrachloro-3-nitrobenzene
Thiophanate-methyl*	s	Dimethyl 4,4'-(*o*-phenylene)bis(3-thioallophanate)
Thiram*		Tetramethylthiuram disulfide
Triadimifon*	s	1-(4-Chlorophenoxy)-3,3-dimethyl-1-(1,2,4-triazol-1-yl)butanone
Tridemorph*	s	2,6-Dimethyl-4-tridecylmorpholine
Triforine*	s	1,4-bis(2,2,2-Trichloro-1-formamidoethyl)piperazine
Zineb*		Zinc ethylenebis(dithiocarbamate) polymer

*Indicates ISO-approved name. Some countries do not use certain ISO names because of their similarity to established names.

furfuralcohol $C_4H_3O \cdot CH_2OH$ = 98.1. Furancarbinol. A solid, m.200 (sublimes), soluble in water.

furfuraldehyde Furaldehyde*.

furfuramide $C_{15}H_{12}O_3N_2$ = 268.3. Hydrofuramide. Colorless crystals, m.117, decomp. 250; an isomer of furfurine.

furfuran Furan*.

furfurine $C_{15}H_{12}O_3N_2$ = 268.3. Brown rhombs, m.116, insoluble in water. An isomer of furfuramide.

furfuroin $C_{10}H_8O_3$ = 176.2. Furfuryl-furfural, furfuryl-fural. The compound

```
CH ---- O ---- C—CO·CH2·C ---- O ---- CH
  \\ CH·CH //        \\ CH·CH //
```

furfurol(e) Furaldehyde*.

furfurostilbene $C_{10}H_8O_2$ = 160.2. Colorless crystals, m.101.

furfuryl The radical C_5H_5O-, from furaldehyde; **2-~*** 2-Furanylmethyl†, α-furfuryl. The radical

O
HC C·CH_2—
CH·CH

3-~ β-Furfuryl. The furylmethyl* radical. **f. acetate** $CH_3COO·C_5H_5O$. Colorless liquid, b.176, insoluble in water; a solvent. **f. alcohol** $C_5H_5O·CH_2OH$ = 112.1. Furylcarbinol. Colorless liquid, b.170, soluble in water; used to make adhesives and chemically resistant resin binders. **f. aldehyde** Furaldehyde*. **f.amide** Furfuramide. **f.amine** C_5H_7ON = 97.1. Colorless liquid, b.145, soluble in water. **f.fural** Furfuroin. **f. methyl ether** $C_6H_8O_2$ = 112.1. Colorless liquid, b.135.

furfurylidene **2-~*** 2-Furanylmethylene†, furfural, fural. The radical O·CH:CH·CH:C·OH=, from furaldehyde.

furil* $C_4H_3O·CO·CO·C_4H_3O$ = 190.2. Difurylglyoxal, bipyromucil. Yellow needles, m.165, insoluble in water. Cf. *furoin.* **f.dioxime** A color reagent for copper.

furlong An eighth part of 1 mile.

furnace An apparatus for heating, fusing, or hardening materials by exposing them to high temperatures. Classification:

Furnace	Fuel	Combustion products
Blast	+	+
Reverbatory	–	+
Muffle	–	–

+In contact with the charge.
–Not in contact with the charge.

arc ~ A device for obtaining high temperatures by an electric arc. **blast ~** A tall oven, in which molten iron is produced by heating a mixture of iron ore and coal by a blast of hot air. **combustion ~** An oven, elongated to take a horizontal combustion tube, and heated by a row of burners or a coil of resistance wire; used in organic analysis. **crucible ~** A device for heating crucibles by gas, oil, or electricity. **electric ~** See *arc furnace* above or *resistance furnace* below. **Héroult ~** An electric arc f. used for the reduction of iron. **induction ~** Inductance f. A f. heated from outside by electrically induced currents. **muffle ~** A f. of refractory material, and completely enclosed except for a small air inlet. **reducing ~** (1) A shaft f. in which ores are reduced to metals. Cf. *aludel.* (2) A f. in which an atmosphere of reducing gases is maintained. **resistance ~** (1) A modified arc f., in which the electrodes dip into the heated material. (2) A wire-wound electrical resistance coil embedded in a refractory material. **reverberatory ~** A f. for roasting ores, so constructed that the flame and hot gases are reflected by the curved roof into direct contact with the material to be heated, which is not contaminated by solid fuel. **revolving ~** A sloping, revolving metal cylinder lined with firebricks, down which the charge passes and up which the hot gases are driven. **roasting ~** See *roasting furnace.* **tank ~** A large oil- or gas-heated container in which glass is melted.

furoates* The esters of furoic acid, used in perfumery.

furoic acid* *α-~* Pyromucic acid. *β-~* $C_4H_3O·COOH$ = 112.1. 3-Furancarboxylic acid. Needles, m.122. **f. acid esters** See *furoates.*

furoin* $C_4H_3O·CHOH·CO·C_4H_3O$ = 192.2. A condensation product of furan. Colorless crystals, m.135.

furol Furaldehyde*.

Furon Trademark for a polyamide synthetic fiber.

furonic acid $(C_4H_5O)·CH_2·COOH$ = 140.1. 2-Furanpropanoic acid, furfuryl acetic acid. Colorless crystals, m.58, soluble in water.

furosemide $C_{12}H_{11}O_5N_2SCl$ = 330.7. 4-Chloro-*N*-furfuryl-5-sulfamoylanthranilic acid. Frusemide, Lasix. White crystals. A potent diuretic, used for edema, heart failure, hypertension; also used to cause rapid excretion of some poisons (USP).

furoyl **2-~*** Pyromucyl. From pyromucic acid. **3-~*** The radical CH:CH·O·CH:C·CO—. **2-f. chloride*** $C_4H_3O·COCl$ = 130.5. Colorless liquid, b.60.

furyl* Furanyl†. The radical $-C_4H_3O$; 2 isomers:

O | O
HC C— | HC CH
CH—CH | CH·C
2-~ | 3-~

Cf. *furfuryl.* **f.acrolein** $(C_4H_3O)CH:CH·CHO$ = 122.1. 3-(2-Furyl)propenal*. Yellow crystals, m.51, insoluble in water. **f. alcohol** Furfuralcohol.

furylidene The radicals

O | O
HC C= | HC CH_2
CH—CH_2 | CH—C
2-~ | 3-~

furylmethyl **3-~*** 3- or β-Furfuryl. The radical O·CH:CCH(:CH)CH_2—, from furaldehyde.

fusafungine Locabiotal. An antibiotic from *Fusarium lateritium* 437; a spray for respiratory tract infections.

fusain Mineral charcoal. Mother of coal. See *coal.*

fuschin Fuchsin.

fuscochlorin A dark-green pigment from algae.

fuscorhodin A dark-red pigment from algae.

fuse (1) To melt. (2) A safety device for electrical instruments; a fine wire which melts when the electric current becomes too strong. (3) Fuze. A device for igniting an explosive charge. **Abel ~** An ignition f. consisting of potassium perchlorate and copper sulfate, and ignited by an electric current. **combination ~** A military f. for shells consisting of a time f. and concussion f. **electric ~** A device for igniting an explosive charge by electric sparks, e.g., Abel f. **concussion ~** An explosive mixture which ignites by concussion, e.g., fulminates. **time ~** A slow-burning material which ignites an explosive mixture after a certain time.

fused Cooled to a compact mass after having been molten or sintered; as, slag.

fused ring See *ortho- and peri-fused.* **cis-~, trans-~** Saturated aromatic rings fused with a *cis* or *trans* configuration of the exocyclic groups at the bridgeheads. See *stereoisomer.*

fusel oil Potato spirit, fermentation pentyl alcohol. A crude mixture of isopentyl, pentyl, butyl, and propyl alcohols, obtained in spiritous fermentation; a source of pentyl acetate.

fusibility The property of becoming liquid when heated.

fusible Capable of being melted. **f. alloys** An alloy having a melting point lower than the mean melting point of the constituents, e.g., Wood's alloy. **f. metals** A metal or alloy of relatively low melting point.

fusidic acid A bacteriocidal antibiotic, from the fungus *Fusidium coccineum,* with a steroid nucleus. It is active against penicillin-resistant staphylococci.

fusing Melting. **f. point** Melting point.

fusinites Highly carbonized cuticular residues in coal. Cf. *exinite.*

fusion The act of melting or flowing together. **alkaline ~** The substitution of the $-SO_3H$ group of an organic compound by an $-OH$ group by action of concentrated caustic soda, followed by treatment with acid. **aqueous ~** The liquefaction of a substance below 100 °C, by solution in its water of crystallization. **false ~** Aqueous f. **nuclear ~** A process analogous to *fission,* q.v., in which heat is liberated by the f. of light elements by accelerating them to speeds equivalent to temperatures of over 10^6 °C. See *thermonuclear reaction.* **watery ~** Aqueous f.

f. heat See *specific latent heat of fusion* under *latent.* **f. mixture** A mixture of sodium and potassium carbonates; fused with insoluble substances of high melting points to render them soluble as carbonates. **f. point** Melting point.

fustic (1) **old ~** Yellow brazilwood, fustic wood. The wood of *Morus tinctoria* (Urticaceae), S. America; used to make morin and in drying textiles. (2) **young ~** The wood of *Rhus cotinus* (Anacardiaceae). Cf. *morin.*

fustin (1) $C_{15}H_{12}O_6$ = 288.3. (2*S*,3*S*)-3,3′,4′,7-Tetrahydroxyflavone. From the wood of *Rhus cotinus* (Anacardiaceae). Silvery needles, m.218, hydrolyzed to fisetin. (2) The coloring matter of the male fern, *Aspidium filixmas.* (3) A coloring matter from sumac. (4) The coloring matter of fustic.

fuze Fuse *(3).*

G

G Symbol for giga- (10^9). **G acid** 2-Naphthol-6,8-disulfonic acid.

G Symbol for (1) conductance; (2) Gibbs function; (3) gravitational constant; (4) shear modulus; (5) weight.

g Symbol for (1) gram; (2) gaseous state (as subscript or between parentheses).

g Symbol for acceleration of free fall.

γ See *gamma.*

Ga Symbol for gallium.

Gabbet solution A solution of 2 g methylene blue in 25 g sulfuric acid and 75 mL water; a bacteriological stain.

gabbro Igneous rocks consisting of plagioclase and pyroxenes.

gaboon Okoumé. The tree *acoumea klaineana,* exported from Africa. Used for plywood.

gadinine $C_7H_{16}NO_2$ = 146.2. A ptomaine from putrefying fish.

gadol Suggested name for vitamin A (from *Gadus,* the cod).

gadoleic acid $C_{20}H_{38}O_2$ = 310.5 (Z)-9-Icosenoic acid*. A fatty acid, m.23, from cod-liver oil.

Gadolin, Johann (1760–1852) Finnish chemist and mineralogist.

gadolinite Principally $4BeO \cdot FeO \cdot Y_2O_3 \cdot 2SiO_2$. Ytterbite. Yttria. A rare-earth silicate found at Ytterby in black, monoclinic masses, d.4–4.5, hardness 6.5–7; a source of Gd, Ho, and Re.

gadolinium* Gd = 157.25. A rare-earth metal, at. no. 64, discovered (1880) by Marignac in gadolinite, d.7.9, m.1311, b.3100. Principal valency 3. It is ferromagnetic below 16°C. **g. acetate*** $Gd(C_2H_3O_2)_3 \cdot 4H_2O$ = 406.4. Colorless, crystalline powder, d.1.611, soluble in water. **g. bromide*** $GdBr_3 \cdot 6H_2O$ = 505.1. Colorless crystals, soluble in water. **g. chloride*** $GdCl_3 \cdot 6H_2O$ = 371.7. Colorless crystals, soluble in water. **g. hydroxide*** $Gd(OH)_3$ = 208.3. White powder, insoluble in water. **g. oxalate*** $Gd_2(C_2O_4)_3 \cdot 10H_2O$ = 758.7. Monoclinic crystals, dehydrated 110, slightly soluble in water. **g. oxide*** Gd_2O_3 = 362.5. Colorless crystals, insoluble in water. **g. sulfate*** $Gd_2(SO_4)_3 \cdot 8H_2O$ = 746.8. Colorless crystals, soluble in water.

gadose A fat from cod-liver oil; a yellow, greasy mass, m.34, slightly soluble in alcohol. **anhydrous ~** Pure, anhydrous g. **glycerinated ~** G. containing 25% glycerol. **hydrous ~** G. containing 25% water.

gadusene An unsaturated hydrocarbon from cod-liver oil.

gagat A soft coal.

gage Gauge. (1) An instrument for measuring the dimension of an object or the pressure or flow of a liquid. (2) The diameter of a wire, the thickness of a sheet or plate. **film ~** See *film gage.* **high and low ~** A g. that registers both maximum and minimum dimensions. **hot-wire ~** (1) A resistance wire sealed in a vacuum tube, used to measure very low pressures by determining (*a*) change of current at constant voltage; (*b*) change of total watts at constant temperature (resistance); (*c*) change of resistance at constant current. (2) A small thermocouple. **ionization ~** A device to measure low pressures from the ionization produced in a gas by a definite electron current. **McLeod ~** See *McLeod gauge.* **nucleonic** See *nucleonic gage.* **resistance ~** Hot-wire g.

g. pressure Absolute pressure, less atmospheric pressure; as in psig.

Gahn, Johann Gottlieb (1745–1818) Swedish mineralogist and discoverer of manganese (1780).

gahnite $ZnAl_2O_4$. Automolite, zinc-spinel. A vitreous, native zinc aluminate, d.4–4.6, hardness 7.5–8.

gaidic acid $C_{16}H_{30}O_2$ = 254.4. A monobasic, unsaturated acid, and homolog of elaidic acid. Colorless crystals, m.39.

Gaillard tower An absorption tower in which a spray of lead-chamber sulfuric acid falls through rising furnace gases and is concentrated.

gain The ratio of the strength of an output signal (particularly electrical) to that of the corresponding input signal.

gaine An intermediate explosive, used to pass on the action of the detonator to the main, less sensitive explosive charge.

gaize A very friable, argillaceous sandstone, converted into a pozzolana when heated.

gal Abbreviation for gallon.

galactagogue A drug that increases the secretion of milk.

galactan $(C_6H_{10}O_5)_n$ = $(162.1)_n$. Gelose. The carbohydrate in the cell wall of algae, e.g., *Irideae, Laminarioides,* obtained as an insoluble dextrorotatory gum from agar-agar. It yields galactose on hydrolysis, and mucic acid on oxidation.

galactaric acid† Mucic acid.

galactitol Dulcitol.

galactolipids Cerebrosides. Lipid substances of brain and nerve tissues.

galactometer (1) A graduated glass funnel for determining the fat in milk. (2) A hydrometer for determining the specific gravity of milk.

galactonic acid $C_6H_{12}O_7$ = 196.2. Pentahydroxyhexanoic acid.* A monobasic acid, derived from galactose. **D-~** m.140.

galactosamine Chondrosamine.

galactosan Galactan.

galactosazone $C_{18}H_{21}O_4N_4$ = 357.4. Yellow needles, m.192–195.

galactose $C_6H_{12}O_6$ = 180.2. Dextrogalactose, cerebrose. **dextro ~** A hexose sugar, derived from milk sugar by hydrolysis. Colorless hexagons, m.168, soluble in water. **levo ~** m.162–163, soluble in water.

galactosemia An inborn error of metabolism resulting in high blood galactose level due to the absence of an enzyme.

galactosidase* **α-D-~** Melibiase. **β-D-~** Lactase.

galactoside Cerebroside. Glycolipids containing nitrogen, galactose, and a fatty acid; as, phrenosin.

galacturonic acid $C_6H_{10}O_7$ = 194.1. **D-~** A hydrolysis product of pectic substances; has furanose and pyranose forms. Structural formula follows.

COOH
O
HO
OH
OH
OH

Galalith Trademark for artificial horn prepared by the action of formaldehyde on casein.

galangal Galanga, India root, chinaroot, kaw-liang ginger, kaw-liang kiang. The dried rhizome of *Alpinia officinarum* (Zingiberaceae), Asia; an aromatic and carminative. **g. oil** The essential oil, d.0.910–0.940, of g.; it contains α-pinene and cadinene.

galangin $C_{15}H_{10}O_5$ = 270.2. 3,5,7-Trihydroxyflavone. Yellow crystals, m.215; from the root of galangal. Cf. *alpinin, kaempferide.*

galbanum A gum resin from *Ferula galbaniflua.* White, yellow, or red tears of waxy consistency, m.100; an ingredient of plasters.

galegine $Me_2C:CH \cdot CH_2 \cdot N:C(NH_2)_2$ = 127.2. α-2-Isopentenylguanidine. An alkaloid from *Galega officinalis* (Leguminosae); reduces blood pressure.

Galen (130–circ. 200 A.D.) Iatrochemist who advocated the use of vegetable in place of mineral preparations in medicine. Cf. *galenical.*

galena PbS. Galenite. A gray, isometric, native lead sulfide, d.7.3–7.6, hardness 2.5. **false ~, pseudo ~** Sphalerite.

galenical A medicine of vegetable origin, especially the liquid preparations, e.g., decoctions. Cf. *Galen.*

galenite Galena.

galenobismuthite A sulfobismuthide of lead.

Galilei, Galileo (1564–1642) Italian physicist and astronomer, who invented the thermometer and telescope.

galipeine $C_{20}H_{21}O_3N$ = 323.4. An alkaloid from angostura bark; colorless needles, insoluble in water.

galipene $C_{15}H_{24}$ = 204.4. A sesquiterpene from *Galipia officinalis,* the source of angostura bark.

galipidine $C_{19}H_{19}O_3N$ = 309.4. An alkaloid from angostura bark. Lustrous plates, m.110.

galipol $C_{15}H_{26}O$ = 222.4. A sesquiterpene alcohol in the oil of the angostura bark. Colorless crystals, m.89.

galipot resin The exudation of *Pinus maritima* (Pinaceae); a source of pimaric acid. Cf. *gallipot.*

galiquoid Proposed name for a colloidal system of a gaseous phase dispersed in a liquid phase; e.g., foams.

gall (1) Bile. (2) Nutgall. (3) A large swelling on plant tissues caused by invasion of parasites after puncture by an insect.

gallacetophenone $C_8H_8O_4$ = 168.2. 2′,3′,4′-Trihydroxyacetophenone. Alizarin Yellow C. Yellow powder, m.173, soluble in water.

gallamine triethiodide $C_{30}H_{60}O_3N_3I_3$ = 891.5. Bencurine iodide, Flaxedil. White crystals, very soluble in water. Blocks nervous impulses at neuromuscular junction causing paralysis; used to paralyze muscles during surgery (USP, EP, BP).

gallanilide $PhNHC_6H_2(OH)_3$ = 217.2. 3,4,5-trihydroxybenzanilide. Colorless crystals, m.205, soluble in hot water.

gallate (1)*$C_6H_2(OH)_3COOM$. A salt of gallic acid. (2) g.(III)*. A salt of gallic hydroxide; as, $NaGaO_2$.

gallein $C_{20}H_{12}O_7$ = 364.3. Gallin, pyrogallolphthalein, anthracene violet. Brown powder or green scales, decomp. by heat, sparingly soluble in wàter. Used as an indicator (alkalies—bright red; acids—pale brown), in the determination of phosphates in urine, and in the manufacture of dyes.

gallic (1) Gallium(III)* or (3+)*. Describing a trivalent gallium compound. (2) Pertaining to nutgall. **g. acid*** $C_7H_6O_5 \cdot H_2O$ = 188.1. 3,4,5-Trihydroxybenzoic acid*. White, triclinic crystals, m.225, slightly soluble in water. A constituent of nutgall, mangoes, and other vegetable matter; used as a reagent for detecting ferric salts and mineral acids. **g. bromide** Gallium tribromide*. **g. chloride** Gallium trichloride*. **g. compounds** See *gallium.* **g. hydroxide** Gallium hydroxide*. **g. iodide** Gallium triiodide*. **g. nitrate** Gallium nitrate*. **g. oxide** Gallium trioxide*. **g. sulfate** Gallium sulfate*.

gallicin $C_8H_8O_5$ = 184.2. Methyl gallate,* methyl gallic ester. Colorless needles, m.202, soluble in hot water.

gallin Gallein.

gallipot A plastic or steel bowl widely used in medicine and by druggists. Cf. *galipot.*

gallitannic Tannin.

gallium Ga = 69.723. Austrium. A metallic element of the third subgroup of the periodic system, at. no. 31. Gray octahedra, d.5.94, m.29.8. Discovered (1875) by Lecoq de Boisbaudran in zinc blende, after its existence had been predicted by Mendeleev from the periodic system (ekaaluminum). The valencies of g. are 2 (g.(II)*, g.(2+)*, gallous) and 3 (g.(III)*, g.(3+)*, gallic), the latter compounds being the stabler; it forms alums of general formula, $MGa(SO_4)_2 \cdot 12H_2O$. Used in quartz thermometers and for brightening optical mirrors. **g.(II)*, g.(2+)*** See also *gallous.* **g.(III)*, g.(3+)*** See also *gallic.* **g. acetate*** $4Ga(C_2H_3O_2)_3 \cdot 2GaO_3 \cdot 5H_2O$ = 1313. White crystals, decomp. 128, soluble in water. **g. acetyl acetonate** $Ga(C_5H_7O_2)_3$ = 367.0. White crystals: **alpha-~** monoclinic, d.1.42; **beta-~** rhombic, d.1.41; m.194, soluble in water. **g. arsenide** GaAs = 144.6. Gray crystals, m.1240; used in solar cells. **g. bromide** G. tribromide*. **g. dibromide*** $GaBr_2$ = 229.5. Gallous bromide. Colorless powder, decomp. by water. **g. dichloride*** $GaCl_2$ = 140.6. Gallous chloride. Colorless crystals, m.164, decomp. by water. **g. diiodide*** GaI_2 = 323.5. Gallous iodide. Colorless powder, decomp. by water. **g. hydride** Ga_2H_6 = 145.5. Digallane. A gas, decomp. 130. **g. hydroxide** $Ga(OH)_3$ = 120.7. Gallic hydroxide. White powder, insoluble in water. **g. iodide** G. triiodide*. **g. monoxide*** GaO = 85.7. Gallous oxide. Blue mass obtained by heating g. trioxide in a stream of hydrogen. **g. nitrate** $Ga(NO_3)_3$ = 255.7. Colorless crystals, soluble in water. **g. oxalate** $Ga_2(C_2O_4)_3 \cdot 4H_2O$ = 475.6. White crystals, insoluble in water. **g. phosphate** $Ga_3(PO_4)_2$ = 304.2. The white, monocrystalline form, grown from seed crystals, is used in electronics. **g. sulfate** $Ga_2(SO_4)_3$ = 427.6. Gallic sulfate. Colorless crystals, soluble in water. It forms double salts with alkali sulfates analogous to alums. **g. sulfide** Ga_2S_3 = 235.6. White powder, insoluble in water. **g.-tin alloy** A liquid metal, m.15: Ga 88, Sn 12%. **g. tribromide*** $GaBr_3$ = 309.4. Colorless, deliquescent crystals, soluble in water. **g. trichloride*** $GaCl_3$ = 176.1. Gallic chloride. Colorless, deliquescent needles, m.76, soluble in water. **g. triiodide*** GaI_3 = 450.4. White powder, soluble in water. **g. trioxide*** Ga_2O_3 = 187.4. Gallic oxide. Colorless, friable mass, insoluble in water.

gallocyanin $C_{15}H_{12}O_5N_2$ = 300.3. A purple dye from nitrosodimethylaniline and gallic acid.

gallogen $C_{14}H_6O_8$ = 302.2. Anhydrous ellagic acid. A constituent of divi-divi, the pods of *Caesalpinia coriaria.* Yellow powder, insoluble in water; an astringent.

gallols A group of compounds of pyrogallol, resorcinol, and chrysarobin with various acids; as, eugallol (pyrogallol monoacetate).

gallon A measure of volume. (1) 1 U.S. liquid g. = 3.785412 liters = 4 quarts = 8 pints = 0.833 imperial g. = 231 in^3. (2) 1 U.S. dry g. = 4.404884×10^{-3} m^3. (3) 1 imperial liquid g. = 4.546092 liters = 275 in^3 = 1.20095 U.S. g. = volume occupied by 10 lb distilled water at 62°F, bar. 30 in. **Winchester ~** See *Winchester gallon.* **wine ~** A measure of capacity used in the U.S. and U.K. (until 1826), equivalent to 231 in^3 or 5 wine bottles.

gallotannic acid Tannin.

gallotannin Tannin.

gallous Gallium(II)* or (2+)*. Describing a divalent gallium compound. **g. bromide** Gallium dibromide*. **g. oxide** Gallium monoxide*.

galloyl* 3,4,5-Trihydroxybenzoyl†. The radical $(HO)_3C_6H_2CO-$, from gallic acid.

Galvani, Luigi (1737–1798) Italian physician and anatomist, discoverer of the galvanic current.

galvanic Electric, voltaic. Pertaining to an electric current (direct current) produced by chemical action. **g. battery** A series of voltaic cells. **g. current** A stream of electrons produced by a displacement reaction. Cf. voltaic *cell.* **g. element** Voltaic *cell.* **g. pile** A pile of disks of 2 different metals, placed alternately and separated by moistened paper; used to produce a g. current.

galvanism (1) Electric currents (direct currents) produced by chemical action (as opposed to heat, friction, or induction). (2) The use of direct currents in medicine.

galvanize, galvanise To protect a metal with a layer of a less oxidizable metal.

galvanized iron (1) Iron coated with tin by electrolysis, and then immersed in a zinc bath. (2) Iron immersed in molten zinc and so coated by that metal without the aid of electricity.

galvanograph A photographic record from a galvanometer whose mirror deflects a beam of light onto a moving film or paper.

galvanolysis Electrolysis (obsolete).

galvanomagnetic effect (1) A g. difference in potential. See *Hall effect.* (2) A g. difference in temperature. See *Ettinghausen effect.*

galvanometer An instrument to detect and measure the strength of an electric current. A magnetic needle is suspended in a wire coil. The deflection of the needle produced by a current through the coil is measured by an optical system. The coil may be fixed and the magnet movable, or vice versa. **absolute ~** An instrument measuring current directly by means of 2 equally strong electromagnets. **astatic ~** An instrument with 2 magnetic needles of equal magnetic moment, suspended parallel to each other, with their poles in opposite directions to eliminate terrestrial magnetism. **ballistic ~** A g. to determine the capacitance or energy produced in the discharge of a capacitor. **D'Arsonval ~** A delicate g.: a magnet and mirror are suspended inside a coil and the deflection of the mirror is read from a reflected light beam on a scale, or the scale may be read in the mirror by means of a telescope. Cf. *galvanograph.* **differential ~** An instrument with 2 equal coils through which 2 separate currents are sent and their comparative strengths thus determined. **Einthoven ~** String or thread g.: a delicate instrument to detect minute electric currents, consisting of a silvered quartz or platinum thread stretched between the poles of a strong magnet. The magnified shadow of the thread is read on a screen. **Kelvin** See *Kelvin galvanometer.* **mirror ~** Reflecting g. **photo ~** Galvanograph. **reflecting ~** Mirror g.: a small mirror is attached to the galvanic needle. Its deflection is read on a scale by reflected light. **string ~** Einthoven g. **tangent ~** The strength of the current through a tangent g. is proportional to the tangent of the angle of deflection. **thread ~** Einthoven g.

galvanometry The measurement of electric currents.

galvanoscope Quadrant *electrometer.*

galvanostalametry A method of measuring electrolytic and time parameters by using gas formation at an electrode to indicate an electrode reaction.

galvanostegy (1) Galvanotropism. (2) Electrolytic tinning to protect against hardening by the nitrite process.

galvanotaxis The response of a living organism to an electric current.

galvanotropism Galvanostegy. The motion of living cells in an electric curernt.

gama wax Candelilla wax.

gambin $C_{10}H_7O_2N$ = 173.2. **R ~** Reddish, 3-nitroso-1-naphthol, a nitroso dye. **Y ~** Yellowish, nitrosonaphthol, 1-nitroso-2-naphthol, a nitroso dye.

gambir Pale catechu. The dried extract from a decoction of the leaves and twigs of *Ouroupariа* or *Uncaria gambir* (Rubiaceae), Asia. Brown powder, insoluble in water; an astringent and tan. **g. catecholcarboxylic acid** $C_{16}H_{14}O_8$ = 334.3. White solid, from gambir. **(+)-*β*- ~** m.259. **(−)-*β*- ~** m.261. **(±)-*β*- ~** m.252.

gamboge Camboge, cambogia, gummi guttae. A gum resin from *Garcinia hanburii* and other Guttiferae. Gray or brown cylinders forming a colloidal solution in water; a pigment.

Gamene Trademark for γ-hexachloro*cyclohexane.*

games theory The use of models to represent and provide a means of studying competitive market interactions.

gamete A sexual cell, as an ovum, capable of uniting with another sexual cell, as a spermatozoon, to form a zygote, or fertilized cell. G. cells have the haploid number of chromosomes. Cf. *somatic.*

gamma Γ, γ. Greek letter. Γ (italic) is symbol for surface concentration. γ (not italic) is symbol for photon. γ (italic) is symbol for (1) conductivity; (2) cubic expansivity; (3) nuclear gyromagnetic ratio; (4) shear strain; (5) surface tension; (6) unit of magnetic flux density = 1 nanotesla; (7) unit of weight = 10^{-6}g. *In chemical names* as for *beta position (2),* but substitution on the third C atom from the principal group. Formerly, also positions 4 and 5 on the naphthalene ring. **g. acid** γ-acid. 2-Amino-8-naphthol-6-sulfonic acid. **g.-benzene hexachloride** BP name for γ-hexachloro*cyclohexane.* **g. iron** An allotropic, nonmagnetic variety of iron existing above 860°C, and crystallizing in the cubic system. **g. particles** A misnomer for g. rays. **g. radiography** Radiography in which a small radioactive γ-ray source replaces an X-ray apparatus. **g. rays** γ-rays. Radiation similar to X-rays, but having shorter wavelengths; emitted by radioactive substances and also from secondary radiation caused by β-rays striking matter. Used in medicine to sterilize plastic materials, as, tubing, syringes.

gammagraph A γ-ray radiograph.

gamone A substance that acts as a carrier in the interactions between gametes at fertilization.

gangaleodin $C_{18}H_{14}O_7Cl_2$ = 413.2. Needles, m.214. A chlorinated depside which occurs in lichens of the *Lecanora* species.

ganglion A nerve cell or group of nerve cells.

gangue The earthy portion of an ore. It forms a fusible slag which flows away from the metallic portion on reduction. Cf. *flotation.*

ganister A fine, compact, hard sandstone, used for grinding and for furnace hearths.
ganja, ganjah Hindi term for round, compressed masses of cannabis.
ganomalite $(Ca,Mn)Pb_3Si_3O_{11}$. A rare, native silicate.
ganomatite $(Fe,As,Sb)_2O_3$. A gray or brown mineral, d.2.3.
ganophyllite $Mn_7Al_2Si_8O_{26} \cdot 6H_2O$. A brown, monoclinic silicate.
Gantanol Trademark for sulfamethoxazole.
Gantrisin Trademark for sulfisoxazole.
garage poison Petromortis. A mixture of carbon monoxide and air from the exhaust of combustion engines.
garancin A preparation of the coloring matter of madder having 3–4 times its dyeing powers.
garbage Refuse from households. **g. tankage** The dried and ground product obtained by steaming and degreasing g.; a fertilizer.
garcinin A pigment from fukugi; a dye for silk. Cf. *fugugetin.*
Gardenal Trademark for phenobarbital. Cf. *Gardinol.*
garden celandine The dried herb of *Chelidonium majus;* a cathartic and diaphoretic.
gardenic acid $C_{14}H_{10}O_6 = 274.2$. A quinone split product of gardenin.
gardenin (1) $C_{14}H_{12}O_6 = 276.2$. A yellow, crystalline principle from the resin of *Gardenia lucida* (S. Asia). (2) Crocetin.
Gardinol Trademark for detergent alcohols made by reduction of sulfonated fatty acids, e.g., $HO \cdot R \cdot HSO_3$. Cf. *Gardenal.*
gargle (1) A disinfecting solution for rinsing the throat. (2) To wash the throat.
garlic Allium. The fresh bulb of *Allium sativum* (Liliaceae); an irritant, expectorant, and condiment. **g. oil** The essential oil of g., containing diallyl sulfide and disulfide and the compounds $C_6H_{10}S_3$ and $C_6H_{10}S_4$.
garnet A red, yellow, or green, transparent silicate. General type: $A_3B_2Si_3O_{12}$, in which A is generally a divalent element, and B a trivalent metal. **aluminum ~** $A_3Al_2Si_3O_{12}$: (1) *Grossularite,* $Ca_3Al_2Si_3O_{12}$; varieties include hessonite, succinite, romanzovite, wiluite. (2) *Pyrope,* $Mg_3Al_2Si_3O_{12}$, red to black crystals, d.3.7. (3) *Almandite,* $Fe_3Al_2Si_3O_{12}$, red crystals, d.3.9–4.2. (4) *Spessartite,* $Mn_3Al_2Si_3O_{12}$, red to brown crystals, d.4.2. **calcium ~** Essonite. **chromium ~** $A_3Cr_2Si_3O_{12}$: (5) *Uvarovite,* $Ca_3Cr_2Si_3O_{12}$, emerald green crystals, d.3.5. **iron ~** $A_3Fe_2Si_3O_{12}$: (6) *Andradite,* $Ca_3Fe_2Si_3O_{12}$; varieties (black, amber, or green) include topazolite, colophonite, melanite, pyreneite, jelleteite, dematoid. (7) *Manganese g.,* $Mn_3Fe_2Si_3O_{12}$; varieties (black, brown, or green) include rothoffite, allochroite, polyadelphite, aplome. (8) *Sodium g.,* $Na_6Fe_2Si_3O_{12}$, lagoriolite. **manganese ~** See *iron garnet (7)* above. **sodium ~** See *iron garnet (8)* above. **titanium ~** $A_3Fe_2(Si,Ti)_3O_{12}$: e.g., *schorlomite,* $Ca_3(Fe,Ti)_2(Si,Ti)_3O_{12}$.
g. rock A metamorphic rock containing g. as an accessory mineral, e.g., schists.
garnierite $(Ni,Mg)SiO_3 \cdot nHO$. Noumeite. A green, amorphous silicate, d.2.5–4, hardness 7.5–8.
garryine $C_{22}H_{33}O_2N = 343.5$. An alkaloid from the leaves of *Garrya fremontii,* skunkbush (Cornaceae), California and Oregon. An oil, m.75–80 (monohydrate).
gas (1) The vapor state of matter; a nonelastic fluid, in which the molecules are in free movement and their mean positions far apart. Gases tend to expand indefinitely, to diffuse and mix readily with other gases, to have definite relations of volume, temperature, and pressure, and to condense or liquefy at low temperatures or under sufficient pressure. One cc of any g. contains under standard conditions 27×10^{18} molecules. Cf. *Avogadro, Charles's,* and *Boyle's law.* (2) Fuel g. **air ~** G. made by blowing air through hydrocarbons. **anesthetic ~** (1) The vapors of a volatile anesthetic agent. (2) Loosely, dinitrogen oxide. **blau ~** See *blau gas.* **coal ~** Fuel g. distilled from coal, consisting of aliphatic hydrocarbons, methane, ethane, etc. **coercible ~** Liquefiable g. **combustible ~** A flammable g. produced by incomplete combustion, e.g., coke-oven g. **compound ~** A gaseous compound, e.g., methane. **compressed ~** See *compressed gas.* **CS ~** *o*-Chlorobenzylidene*malononitrile.* **electrolytic ~** A mixture: 2 vol. H_2, 1 vol. O_2. **elementary ~** An element that is gaseous under ordinary conditions, e.g., hydrogen. **flammable ~** A g. that is able to burn in air, e.g., hydrogen. **forest ~** Producer g. from wood charcoal. **fuel ~** A g. used to produce heat by combustion. **full ~** A flammable g. consisting of saturated hydrocarbons, e.g., water g. **ideal ~** Perfect g. **inert ~** (1) Noble g.* (2) A g. that does not readily react chemically; e.g., nitrogen. **intestinal ~** Gases produced during digestion: nitrogen, hydrogen sulfide, etc. **lacrimatory ~, lachrymatory ~** G. producing profuse secretion of tears. **laughing ~** Dinitrogen oxide*. **liquefied natural ~** LNG. Natural gas that has been liquefied by cooling to about −140° for ease of transportation or storage. **liquefied petroleum ~** LPG. Butane and propane liquefied under pressure for use from cylinders. **marsh ~** Methane. **natural ~** (1) The flammable gas from oil wells, used as a fuel. Principally methane, with ethane, propane, butane, and higher hydrocarbons, and small amounts of hydrogen, ethylene, carbon dioxide, and carbon monoxide. Calorific value 60 GJ per ton. Estimated world reserves about 9×10^{13} m^3, of which about 40% is in the U.S.S.R. and 25% in the Middle East. See *fuel.* (2) Helium. **noble ~*** The members of the zero group of the periodic table, consisting of inert elements: He, Ne, Ar, Kr, Xe, and Rn. **noxious ~** Any poisonous g. or a g. with a strong odor. **oil ~** Natural g. A fuel g. distilled from crude petroleum. **olefiant ~** Ethylene*. **oxygen ~** Oxygen*. **perfect ~** A g. that obeys the g. laws. None is known, but it is assumed that as the pressure becomes infinitely small, the g. approaches nearer and nearer to the ideal state, where there is no internal resistance to molecular motion. **permanent ~** An obsolete term for gases which are not liquefiable. However, at sufficiently low temperature and high pressure all gases condense. **petroleum ~** Oil. **poison ~** (1) Lung irritants (phosgene). (2) Lacrimators (bromoacetone). (3) Sternutators (diphenyl cyanoarsine). (4) Vesicants (mustard g.). (5) G. which affects body enzymes. **producer ~** See *producer gas.* **propellant ~** A g. used in a pressurized dispenser to expel the contents when a valve is opened. Cf. *aerosol.* **rich ~** Full g. **rock ~** Natural g. **sewer ~** G. from the decay of organic material. **sludge ~** A fuel g. (methane 70, carbon dioxide 30%) from the activated sludge sewage treatment process. **sour ~** Natural g. containing impurities, chiefly hydrogen sulfide. **sternutatory ~** A g. that produces sneezing. **suffocating ~** A g. that is nonrespirative and smothering and that finally stops respiration. **sun ~** The gaseous constituents of the sun, e.g., hydrogen, helium, carbon dioxide, etc. **toxic ~** A g. that causes poisoning. **tracer ~** A radioactive g. used as a radioactive indicator, e.g., to detect leaks in underground piping. **two-dimensional ~** A layer of adsorbed atoms. Cf. *adatom.* **vesicant ~** A g. that blisters the skin. **volcanic ~** G. from volcanoes (carbon dioxide, nitrogen, hydrogen, sulfur dioxide, etc.). **war ~** Poison g. **water ~** A fuel prepared by passing steam over glowing coal and enriching the hydrogen produced with hydrocarbons and carbon monoxide.

g. analysis apparatus See *Orsat apparatus.* **g.bag** An oval rubber container for holding gases. **g. balance** A balance for determining the specific gravity of gases. **g. ballons** A blown, spherical glass container with one or more necks, for weighing gases. **g. bath** Air bath. **g. battery** G. cell. **g. black** Lampblack. **g. bleaching** Bleaching by sulfur dioxide or chlorine. **g. buret** A graduated glass tube with a stopcock on each end; used in gasometric analysis. **g. calorimeter** An apparatus to determine the heat value and tar content of g. **g. carbon** The amorphous, compact residual carbon remaining after distillation of g. from coal; used for electrodes. **g. cell** An electrolytic cell formed from 2 g. electrodes. **g. chromatography** See *chromatographic analysis.* **g. collecting tube** An elongated cylinder or bulb with a stopcock at each end. **g. constant** R in the g. law equation is independent of the chemical nature of a g., but depends on the units of measurement: $R = pV/T$, where p, V, and T are the pressure, volume, and thermodynamic temperature, respectively, under conditions of an ideal g. For SI units, $R = 8.314\ JK^{-1}mol^{-1}$. Cf. *equation of state.* **g. cylinder** A steel tank or iron bottle used to ship liquefied gases. **g. electrode** An electrode (usually a finely divided metal) which holds a g. on its surface, and behaves as a reversible electrode when placed in a solution. Cf. *hydrogen* electrode. **g. filter** A device to remove solid or liquid particles from gases. **g. generating bottle** A device to generate gases in the laboratory. See *Kipp generator.* **g. generator** A device to manufacture g., e.g., the retort of a g. plant. **g.holder** Gasometer. A g. storage tank, 2 overlapping halves expanding within each other and sealed by a liquid; or the g. may be displaced from a g.-tight container by allowing water to flow in. **g. integral process** G.I. A 2-stage water-g. process for producing g. from low-grade coals. Cf. *Lurgi process.* **g. laws** The combination of Boyle's, Gay-Lussac's, and Charles's laws in the equation $pV = RT$. See *g. constant* above, *equation of state.* **g. liquor** The liquor from washing g. from the distillation of coal. It contains ammonia, sulfides, and carbonates. **g. mask** Respirator. **g. pipet** A series of glass bulbs mounted on a frame; used in gasometric analysis. **g. regulator** (1) A device for regulating the pressure of the g. taken from a cylinder in which it is compressed. (2) A device for regulating temperature by controlling the g. supply. **g. thermometer** A thermometer based on the variation in pressure or volume of a g., generally hydrogen. **constant volume ~** Measures the variation in pressure of a g. confined at constant or nearly constant volume. **constant pressure ~** Measures the change in volume of a g. confined at constant pressure, generally air pressure.

gaseous Describing the third state of matter, as opposed to the solid and liquid states.

gasification The manufacture of *producer gas,* q.v., by the combustion of coal or other organic material (as, wood) under controlled temperature and air flow conditions.

gasohol Ethanol or methanol produced from crops such as sugar cane or cassava; mixed with gasoline for automotive fuel. Cf. *Discol, Dynalkol.*

gasol 3-C and 4-C mixed alkanes and alkenes produced in the synthesis of oil from carbon monoxide and hydrogen, by the Fischer-Tropsch process, using a metal catalyst.

gasoline, gasolene Petrol. The fraction of crude petroleum, 40–180°C; chiefly C_6H_{14}, C_7H_{16}, and C_8H_{18}. Used for internal combustion engines. Cf. *gasohol.* **Ethyl ~** See *Ethyl gas* under *Ethyl.*

gasoloid Proposed name for a gaseous dispersed phase in a solid surrounding phase.

gasometer See *gas*holder.

gasometric Pertaining to gas analysis.

gasometry Gas analysis.

gassed Overcome by noxious gas.

Gassner cell A voltaic dry cell (1.3 volts); zinc and carbon electrodes in a paste of zinc oxide 1,ammonium chloride 1, calcium sulfate 3, zinc chloride 2 pts., moistened with water.

gastric Pertaining to the stomach. **g. content** Semidigested food mixed with digestive enzymes. **g. digestion** The decomposition of food materials in the stomach; chiefly the hydrolysis of proteins by pepsin with hydrochloric acid as activator. **g. juice** The secretions of the stomach glands, containing the digestive enzymes. **g. lavage** Stomach pump or washout. Fluid, usually water, poured into the stomach by wide-bore tube and syphoned out with the stomach contents.

gastrin A hormone, secreted from the gastric epithelium into the blood stream, which stimulates production of the stomach's digestive enzymes.

gatsch A soft hydrocarbon wax, m.320–460, obtained in the synthesis of oil from carbon monoxide and hydrogen by the Fischer-Tropsch process; a source of edible fatty acids.

gauchamacine Guachamacine.

gauge Alternative form of *gage,* q.v.

gaultheria The dried leaves of *Gaultheria procumbens,* wintergreen, partridgeberry, or checkerberry (Ericaceae), N. America. **g. oil** Wintergreen oil.

gaultheric acid Methyl salicylate*.

gaultherilene $C_{10}H_{16} = 136.2$. A terpene in wintergreen oil.

gaultherin $C_{14}H_8O_8 = 304.2$. Monotropitoside. A glucoside from the bark of *Betula lenta,* black birch; hydrolyzes to methyl salicylate and glucose.

gaultherolin Methyl salicylate*.

Gauss, Karl Friedrich (1777–1855) German mathematician who developed the concept of the 3 fundamental units: length, mass, and time.

gauss symbol: G; the unit of flux density of a magnetic field (intensity or field strength); a force of one dyne on a unit magnetic pole: 1 gauss (emu) = $\frac{1}{3} \times 10^{-10}$ esu = 10^{-4} tesla (the SI unit). 1 γ = 0.00001 gauss. **de ~** To render iron nonmagnetic; e.g., during World War II, coils through which an electric current passed were placed in ships' hulls to protect against magnetic mines.

Gaussian curve Normal probability curve.

Gautier receiver A glass apparatus for collecting samples during *vacuum distillation,* q.v.

gauze A light, loosely woven fabric, or fine wire netting. **absorbent ~** Cotton, or cotton and rayon mixture, woven into fabric for surgical dressings (USP, EP, BP). **petrolatum ~** Absorbent g. saturated with not less than 4 times its weight of white petrolatum; a protective (USP, EP). **g. top** G. covering the top of a bunsen burner to prevent the flame from striking back.

Gay-Lussac, Joseph Louis (1778–1850) French chemist and physicist. **G.-L. hydrometer** A hydrometer used for alcoholic liquids; graduated in percentage of alcohol (by volume). **G.-L. law** The volumes of reacting gases and the volume of the reaction product are in simple proportions, and can be expressed by whole numbers. **G.-L. tower** A tower used in the chamber process for the manufacture of sulfuric acid to absorb the oxides of nitrogen from the crude acid produced, to form nitrous vitriol (mainly nitrosyl hydrogensulfate).

gaylussite $Na_2Ca(CO_3)_2 \cdot H_2O = 224.1$. A natural carbonate.

gazogene A fuel gas made by burning charcoal.

GB See *sarin.*

Gd Symbol for gadolinium.

Ge Symbol for germanium.

Geber Abu Abdallah Jaber. Arabian alchemist and writer (9th century); the discoverer of sulfuric and nitric acids. Latin writings attributed to him contain speculations on the alchemical "elements."

Gecesa Trademark for a polyamide synthetic fiber.

gedanite A fossil resin resembling amber.

gedrite A variety of anthophyllite containing alumina.

gee-lb Slug.

gehlenite $2CaO \cdot Al_2O_3 \cdot SiO_2$. A green, resinous, tetragonal silicate, d.2.9–3, hardness 5.5–6, associated with spinel.

geic acid Ulmic acid.

Geiger-Müller counter Geiger counter. A metal cylinder, with a fine axial wire, containing gas (as, argon) at low pressure. β particles, caused by radiation, cause ionization, which can be detected by amplification.

geikielite Ilmenite.

geissine $C_{40}H_{48}O_3N_4$ = 632.8. Geissospermine. m.217 (decomp). An alkaloid from the bark of *Geissospermium laeve*, pereira bark (Apocynaceae), Brazil.

Geissler Heinrich (1814–1879) German physicist who determined the coefficient of expansion of water. **G. bulb** A *potash bulb*. **G. tube** A sealed and partly evacuated glass tube, used in the study of electric discharges through gases and for spectroscopic examination.

geissospermine Geissine.

gel (1) Jel, jelly. A colloidal solution of a liquid in a solid. (2) To form a gel. Cf. *coagel*. **alco ~** A solid colloidal solution in alcohol. **hydro ~** A solid colloidal solution in water. **irreversible ~** A g. that cannot be converted into a sol. **reversible ~** A g. that becomes a liquid sol on suitable treatment, and can be gelled again. **silica ~** See *silica gel*.
g. filtration See *molecular sieve*.

gelate Gelatinize. To cause solidification of a colloidal solution.

gelatin $C_{76}H_{124}O_{29}N_{24}S$(?). An albumin usually obtained by boiling animal bones and cartilage under pressure with water. Yellow films, which soften and swell in cold water; insoluble in alcohol, soluble in hot water; coagulated by tannin and hardened by formaldehyde. G. is an amphoteric compound which combines with cations to form "gelatinates," and with anions to form "gelatin salts":

1. Nonionized (isoelectric) .... pH 4.7
2. Ionized (e.g. with sodium chloride):
   (*a*) Metal gelatinate (e.g. sodium gelatinate) ... pH $<$ 4.7
   (*b*) Gelatin salt (e.g. gelatin chloride) .... pH $>$ 4.7

G. is used as a nutrient, hemostatic, excipient, culture medium, and blood substitute; for photographic papers, films, and plates; in glues; as a clarifying agent, adhesive, sizing or stiffening agent, and colloidal protector. Cf. *glue*. **absorbable ~** G. foam. Gelfoam, Sterispon. A sponge or film of g. Used as hemostatic in surgical operations (USP, BP). **ana ~** One of the 2 constituents of gelatin which produces a gel. **animal ~** Gelatin. **blasting ~** See *blasting gelatin*. **bone ~** G. from bones. **Chinese ~** Vegetable g. **chromatized ~** A mixture: 5 pts. 10% g. solution, 1 pt. potassium dichromate. **formalin ~** Glutolin. **glycerinated ~** A mixture of equal parts of glycerol and g. **Japanese ~** Agar-agar. **nitro ~** A mixture of nitroglycerin and nitrocellulose. Cf. *gelignite*. **plain ~** A culture medium: gelatin 100, peptone 10, sodium chloride 5 g in 1,000 ml bouillon stock solution, neutralized with sodium hydroxide. **silk ~** Sericin. **vegetable ~** A gelatinous substance from vegetable tissues; as, agar.
g. culture A bacterial culture grown on a medium containing g. **g. disk** A small disk of medicated g. used to apply drugs to the eye. **g. filtration** Filtration through a gel to separate substances of different molecular size. **g. sugar** Glycine*.

gelatinate A compound of *gelatin*, q.v., with a positive ion or radical.

gelatinize (1) To gelate. (2) To convert into gelatin.

gelatinous Resembling gelatin.

gelation The formation of a gel. **con ~** See *congelation*. **re ~** See *regelation*.

Gelidium latifolium A plant (Eire), source of agar.

gelignite Blasting gelatin, gelatin dynamite. An explosive: nitroglycerin, nitrocellulose, potassium nitrate, and wood-meal. Cf. *dynamites*.

gelling point The concentration and temperature at which semiliquids become solid.

gelometer An instrument that measures gel strength in terms of the weight of shot, running into a hopper attached to a hard rubber plunger, required to force the plunger 4 mm below the gel surface; e.g., the Bloom g. Cf. *penetrometer*.

gelose Galactan.

gelsem(in)ic acid Scopoletin.

Gelsemium A genus of loganiaceous (de Candolle), apocynaceous (Decaisne), or rubiaceous (Chapman) plants.

gem A precious stone. **artificial ~, synthetic ~** A precious stone made by a chemical process.

***gem*-** Prefix indicating that the radicals in a disubstituted compound are both on the same C atom.

geminal coupling The H–C–H linkage of a CH_2 group.

gemmatin $C_{17}H_{12}O_7$ = 328.3. A coloring matter from the fungus *Lycoperdon gemmatum*.

-gen Suffix meaning "to produce" or "to bear"; as, hydrogen.

genalkaloid An alkaloid in which the amino group has been converted into an aminooxy group. A g. has the same therapeutic effect as, but is less toxic than, the parent alkaloid.

gene Part of the DNA molecule; responsible for the inherited ability to synthesize a specific protein; each g. has a specific position (locus) on the molecule. See *chromosome*.

generate To produce a gas or electric current.

generator An apparatus for producing a gas or electricity.

generator gas A fuel gas obtained by blowing air through layers of heated coal or coke; chiefly carbon monoxide and carbon dioxide.

generic name Name given to a class of similar substances; as, alkanes.

genetic engineering A branch of biotechnology. The incorporation of segments of DNA from one organism into the DNA of another organism using recombinant DNA technology. Cutting and joining can be achieved by enzymes.

Geneva nomenclature An international system of naming carbon compounds adopted in 1892. Now largely obsolete. See *nomenclature*.

genistein $C_{15}H_{10}O_5$ = 270.2. Prunetol, 4′5,7-trihydroxyisoflavone. m.301 (decomp.). A plant flavone; the aglucone of genistin.

genistin $C_{21}H_{20}O_{10}$ = 432.4. A glucoside of genistein, m.254 (decomp.), from soybean meal. Cf. *plant* pigments.

genoline oil Linseed oil polymerized by boiling; a lithographic varnish.

gentamicin sulfate A broad-spectrum antibiotic, produced by *Micromonosporia purpurea*. Used for gram-positive and -negative organisms, but limited by toxic effect on the inner ear (USP, BP).

genthite Nickel gymnite.

gentian Gentian root. The dried rhizome and roots of *Gentiana lutea* (Gentianaceae); a bitter tonic and stomachic. See *aromatin*. **g. violet** USP name for methyl violet.

Gentianaceae Gentian family, a group of herbs with a bitter juice, containing little tannin. Roots: *Gentiana lutea* (gentian); herbs: *Swertia chirayita* (chirata).

gentianic acid Gentisic acid.

gentianin (1) An extract containing the bitter principle of the root of *Gentiana* species. (2) Gentisin.

gentianine $C_{10}H_9O_2N$ = 175.2. An alkaloid from *Enicostemma littorale* (Gentianaceae), m.82.
gentianite $C_{16}H_{32}O_{16}$ = 480.4. A carbohydrate in the root of *Gentiana* species.
gentianose $C_{18}H_{32}O_{16}$ = 504.5. A trisaccharide from gentian, m.212.
gentienin $C_{14}H_{10}O_5$ = 258.2. An isomer of gentisin, m.225.
gentiin $C_{25}H_{28}O_{14}$ = 552.5. A glucoside from *Gentiana* species, m.274.
gentiobiose Iso*maltose.*
gentiopicrin $C_{16}H_{20}O_2$ = 244.3. A glucoside from the root of *Gentiana* species. Yellow crystals, m.191, soluble in water.
gentisein $C_{13}H_8O_5$ = 244.2. 1,3,7-Trihydroxyxanthone. Orange crystals, m.321.
gentisic acid $C_7H_6O_4$ = 154.1. Hydroquinonecarboxylic acid, gentisinic acid, gentianic acid, 2,5-dihydroxybenzoic acid*. Colorless crystals, m.204, soluble in water. A metabolic product of the mold *Penicillium patulum.* Cf. *patulin.*
gentisin $C_{14}H_{10}O_5$ = 258.2. Gentianin, 1,7-dihydroxy-3-methoxyxanthone. The yellow pigment of the root of *Gentiana.* Yellow needles, m.267, sublimes and decomp. 400, slightly soluble in water. **iso ~** 1,3-Dihydroxy-7-methoxyxanthone. m.241.
gentisinic acid Gentisic acid.
genus (1) A group of related organic compounds (Mulliken). (2) A group of related species of plants or animals.
geochemistry The study of the chemical changes occurring on the earth's crust. **bio ~** The study of the chemical composition of living matter in relationship to organically formed rocks.
geocoronium A hypothetical element assumed to exist in the upper layers of the atmosphere.
geocronite $Pb_5Sb_2S_8$ or $5PbS \cdot Sb_2S_3$. A metallic, rhombic sulfide, d.6.4–6.5, hardness 2–3.
geodesy The science of the form and dimensions of the earth and its surroundings.
geodynamics The study of the forces and causes which change the earth surface. Cf. *geomorphology.*
geoffroyin Rhatanin.
geologic(al) Pertaining to geology.
geologic eras The time intervals during which certain rock strata were formed on the earth's surface. Subdivided into periods and into epochs. The rocks formed during a period constitute a geologic system; those formed during an epoch constitute a series, which are subdivided into formations, many characterized by the remains of life. See Table 38.
geology The science of the physical history of the earth and its surface structure. **dynamic ~** The study of the causes of geological changes. **economic ~** The study of economically important minerals. **eolic ~** The study of changes produced by wind. **paleontologic ~** The study of remains of life in relation to rock formations. **stratigraphic ~** Historical g.
geometric(al) Pertaining to the principles of geometry. **g. conversion** The change from one g. isomer to another; as, maleic acid to fumaric acid. **g. isomer** An optically inactive compound which exists in 2 or more geometrically different atomic arrangements; as, (*E*) and (*Z*) forms. **g. progression** A series of numbers related by a constant ratio between successive terms; e.g., 1,4,16,64, . . .
geomorphology The study of the form of the earth surface. Cf. *geodynamics.*
Geon Trademark for a polyvinyl chloride and vinylvinylidene copolymer synthetic fiber.
geostatics The science of the loose sediments of the earth's crust.

TABLE 38. GEOLOGIC ERAS

ERA / *Period or system* / Epoch or series	Approximate commencement (millions of years ago)
CENOZOIC	
Quaternary	
Holocene	0.01
Pleistocene	2
Tertiary	
Pliocene	7
Miocene	26
Oligocene	38
Eocene	54
Paleocene	65
MESOZOIC	
Cretaceous	136
Jurassic	195
Triassic	225
PALEOZOIC	
Permian	280
Pennsylvanian[a]	320
Mississippian[a]	345
Devonian	395
Silurian	440
Ordovician	500
Cambrian	570
PRECAMBRIAN	4,500

[a]Collectively, the *Carboniferous* period

geostationary Describing a satellite that is stationary relative to the earth's surface.
geothermal Geothermic. Referring to heat below the earth's surface. **g. energy** Steam from wells 1,000 m deep at 300°C used in turbines to generate electricity. Artificially fractured dry rock can generate synthetic g. e. using water injection. **g. gradient** The clinal rise in temperature below the earth's crust (average 0.65°C per 20 m of depth).
geranial α-Citral*.
geranic acid $C_{10}H_{16}O_2$ = 168.2. 3,7-Dimethyl-2,6-octadienoic acid*. The oxidation product of citral; oily liquid, b_{20mm}119. **iso ~** Colorless crystals, m.103.
geraniol* $Me_2C:CHCH_2 \cdot CH_2 \cdot CMe:CH \cdot CH_2OH$ = 154.3. (*E*)-3,7-Dimethyl-2,6-octadien-1-ol†. A constituent of the oils of geranium, eucalyptus, citronella, and ylang. Colorless liquid, d.0.881, b.231, insoluble in water; an insect bait. **dihydro ~** Citronellol.
geranium Cranesbill. The dried rhizome of *Geranium maculatum.* **g. oil** Occurs in the leaves of *Pelargonium* species, d.0.889–0.906. Contains α-citral, citronellol, phellandrene, and (*E*)-2-butenoates. **Turkish ~** Palmarosa oil. The essential oil (70% geraniol) of *Andropogon (Cymbogen) martini* (Geraniaceae).
geranyl* 3,7-Dimethyl-2,6-octadienyl†. The radical $C_{10}H_{17}-$, from geraniol. **g. acetate** $CH_3 \cdot COOC_{10}H_{17}$ = 196.3. Geraniol acetate. Colorless liquid, decomp. 245, slightly soluble in water.
Gerhardt, Charles Friederich (1816–1856) French chemist. **G. test** A test for acetoacetic acid in urine with ferric chloride solution.
gerhardtite A native, basic copper nitrate.
geriatrics The study of old age.
germ A bacterium, microbe, or spore. **g. cell** An embryonic cell.

germanate Indicating germanium as the central atom(s) in an anion; as, a salt of the type M_2GeO_3, from germanium dioxide. **thio ~** A salt of the type M_2GeS_3, from germanium disulfide.

germane* GeH_4 = 76.6. Germanium hydride. Colorless gas, burning with a blue flame, b.−90. **di ~*** Ge_2H_6 = 151.2. Germanoethane. A gas, b.29. Cf. tetraethyl*germanium.* **tri ~*** Ge_3H_8 = 225.8. Germanopropane. A liquid, d.2.20, b.110.

germanic Germanium(IV)* or (4+)*. Describing a compound of tetravalent germanium. **g. chloride** Germanium tetrachloride*. **g. oxide** Germanium dioxide*.

Germanin Trademark for suramin.

germanite A copper sulfarsenide mineral from Tsumeb, Namibia, which contains Ge 5–10, Ga 0.3–1%, and is a principal source of these metals.

germanium* Ge = 72.59. A metallic element, at. no. 32, of the carbon family. d.5.36, m.937, b.2825, insoluble in water. Discovered (1886) by Winkler in argyrodite; predicted (1871) by Mendeleev (ekasilicon). G. is a rare metal occurring in argyrodite, euxenite, and germanite. Its valency is 2 or 4; hence the compounds: g.(II)*, g.(2+)*, germanous, Ge^{2+}; g.(IV)*, g.(4+)*, germanic, Ge^{4+}. G. is used in semiconductor devices; it gives strength to aluminum alloys, hardness to magnesium alloys, and refractive power to glass. **tetraethyl ~*** Et_4Ge = 188.8. G. tetraethyl, g. ethide. Colorless liquid, b.160, insoluble in water.

g.(II)*, g.(2+)* See also *germanous.* **g.(IV)*, g.(4+)*** See also *germanic.* **g. alkyls** Organometallic compounds in which tetravalent g. replaces carbon; as, tetraethyl ~. **g. chloroform** $GeHCl_3$ = 180.0. Colorless liquid, b.72, decomp. by water. **g. dibromide*** $GeBr_2$ = 232.4. Germanous bromide. Colorless crystals, decomp. by heat. **g. dichloride*** $GeCl_2$ = 143.5. Germanous chloride. Colorless liquid. **g. diiodide*** GeI_2 = 326.4. Yellow crystals. **g. dioxide*** GeO_2 = 104.6. G. oxide, germanic oxide. White powder, m.1025, slightly soluble in water. **g. disulfide*** GeS_2 = 136.7. Germanic sulfide. White powder, decomp. by heat or by water. **g. ethide** Tetraethyl ~*. **g. hydride** Germane*. **g. hydroxide*** $Ge(OH)_2$ = 106.6. Yellow powder, insoluble in water. Cf. *germanoformic acid.* **g. iodide** See *germanium tetraiodide.* **g. monosulfide*** GeS = 104.7. Germanous sulfide. Brown, metallic plates. **g. monoxide*** GeO = 88.6. Germanous oxide. Gray, volatile powder, soluble in hydrochloric acid. **g. oxide** See *germanium dioxide* or *germanium monoxide.* **g. sulfide** See *germanium disulfide* or *germanium monosulfide.* **g. tetrabromide*** $GeBr_4$ = 392.2. Germanic bromide. Colorless, fuming liquid, m.26. **g. tetrachloride*** $GeCl_4$ = 214.4. Germanic chloride. Colorless liquid, b.86, decomp. by water. **g. tetrafluoride*** GeF_4 = 148.6. Germanic fluoride. Colorless, hygroscopic crystals, soluble in water. **g. tetraiodide*** GeI_4 = 580.2. Germanic iodide. Red crystals, m.144, decomp. by water.

germanoformic acid HGeOOH. A hydrous oxide obtained by heating $Ge(OH)_2$ with alkali. A red powder and reducing agent.

germanous Germanium(II)* or (2+)*. Describing compounds of divalent germanium, which are generally less stable than the germanic compounds. **g. chloride** Germanium dichloride*. **g. oxide** Germanium monoxide*. **g. sulfide** Germanium monosulfide*.

german silver Nickel silver.

germicidal Destructive to germs.

germicide An agent that destroys microorganisms, especially disease germs. Cf. *disinfectant, Rideal-Walker test.*

germination The sprouting of a seed or spore. **g. capacity** The percentage of grain, etc., which can be made to germinate. **g. energy** The g. capacity in a specified time; e.g., for barley, 3 days at 16°C.

germinator A device to determine the germinating energy of seeds; a perforated disk holding seeds over water at a definite temperature.

geronic acid $C_9H_{16}O_3$ = 172.2. 2,2-Dimethyl-6-oxoheptanoic acid*. b.280. An oxidation product of β-carotene and β-ionone.

gerontin $C_5H_{14}N_2$ = 102.2. A leukomaine from dog liver.

gerontology The study of the symptoms and reactions of the decline in physical and organic functions.

gersdorffite (Ni,Fe)AsS or NiS_2,$NiAs_2$. Plesite. A native, metallic sulfarsenide, d.5.2–6.3, hardness 5.5.

Geryk pump A vacuum pump. Cf. *Guericke.*

gesnerin The 5-β-D-glucoside of 4′,5,7-trihydroxyflavylium chloride. An anthocyanin from the orange flowers of *Gesneria* species.

gesso A plaster (whiting and glue) base on the canvases of early paintings.

gestalt (German for "shape"). (1) A synergic mental pattern derived from many separate sense impressions. (2) Showing properties other than can be derived from the individual constituents by summation.

Gesterol Trademark for progesterone.

gestogens Compounds similar to progestogen.

Gestone Trademark for progesterone.

getter (1) A substance that "cleans" gases in vacuum tubes. **absorptive ~, chemical ~** A g. that reacts with the gas; as, Li. **adsorptive ~, physical ~** A g. that binds gases on its surface; as, zirconium. (2) A metal (e.g., thallium) coating on the filament of a tungsten lamp, to prolong its life.

gettering Obtaining and maintaining a high vacuum in a container, e.g., by adding an activated metal which absorbs gas molecules.

GeV Symbol for 10^9 electronvolts.

geyserite A hydrous silicic acid sinter produced near geysers.

ghatti gum Indian gum.

ghee Indian clarified butter.

ghetta acid $C_{34}H_{68}O_2$ = 508.9. A fatty acid from ghedda wax (southeast Asia).

Ghosh, Sir Jnan Chandra (1895–1959) Indian physical chemist, noted for his work on electrolytes and the theory of dissociation.

giallioline Lead antimonate.

gibberellins Plant hormones originally produced from *Gibberalla fujikuroi*, an organism that causes elongation of rice shoots disease; and by fermentation. They promote the stem growth of trees, cereal crops, and tobacco. White, crystalline, optically active acids. **g. A_1** $C_{19}H_{24}O_6$. **g. A_3** $C_{19}H_{22}O_6$ = 346.4. Gibberellic acid. Crystals, m.235 (decomp.). Used in horticulture to combat delayed germination and to delay the ripening of citrus fruits. **g. A_4** $C_{19}H_{24}O_5$. Cf. *auxins.*

Gibbs G., Josiah Willard (1839–1903) American mathematician and physicist noted for the development of the *phase* rule, q.v., and thermodynamics. **G. function*** G. Thermodynamic potential. G. free energy. $G = U + pV - TS$, where U is the internal energy, p the pressure, V the volume, T the thermodynamic temperature, and S the entropy. At constant temperature and pressure, the change in G is a measure of the work done. Cf. *Helmholtz function.* **G.-Helmholtz equation** The relationship between the internal energy U and Helmholtz function A of a system. $U = A - T(dA/dT)_V$. For a reversible electric cell, the relationship between the chemical energy transformed and the maximum energy obtainable electrically is

$$E = H/nF + T\ dE/dT$$

where E is the emf of the cell, H the heat equivalent of the chemical change for molar quantities expressed in electrical units, F the Faraday constant, T the thermodynamic temperature at which the cell is working, and n the valency, or the number of charges carried by a mole of the substances undergoing change; dE/dT is the rate of change in emf with temperature of the cell. **G., Oliver Wolcott (1822–1908)** American chemist noted for his work on complex compounds. **G. paradox** Work results when 2 gases of thermodynamically identical physical properties (e.g., N_2 and CO) are mixed, but not when 2 portions of the same gas are mixed. **G. phase rule** See *phase rule.*

gibbsite $Al(OH)_3$. A native aluminum hydroxide.

gibrel $C_{19}H_{21}O_6K$ = 384.5. Potassium gibberellate; used to increase the microbial activity of the soil.

Giemsa, Gustav (1867–1948) German chemotherapist. **G. stain** A staining for white blood cells and bacteria: Azur II Eosin 0.3, Azur II 0.8, glycerol 250 g; and 250 mL methanol. **G. ultrafilter** A device for sterilizing and filtering small quantities of biological liquids through a collodion membrane.

giga* G. SI prefix for a multiple of 10^9.

gigantolite A pseudomorph of iolite.

Gilbert G., Sir Joseph Henry (1817–1901) British chemist, noted for agricultural research. **G., Ludwig Wilhelm (1769–1824)** German chemist, and editor of *Annalen der Physik.* **G., Walter (1932–)** American chemist, Nobel prize winner (1980). Noted for work on chemical structure of DNA. **G. William (1540–1603)** British natural philosopher, physician to Queen Elizabeth I, and a pioneer in magnetism and electricity.

gilbert An obsolete unit of magnetic quantity. 1 gilbert = 0.795775 A (the SI unit). **pra ~** See *pragilbert.*

Gilead balm Balm of Gilead, Mecca balsam. An oleoresin from *Balsamodendron gileadense* (Burseraceae). Cf. *poplar buds.*

Giles flask A volumetric flask with long neck, graduated at x and at $(x + 10\%x)$ of its volume; used to prepare normal solutions.

gill A liquid measure: 1 U.S. gill = 118.29 mL = 0.83267 U.K. gill.

gillenia Indian physic, American ipecac. The root bark of *G. trifoliata* or *G. stipulacea* (Rosaceae); an emetic and cathartic.

gilpinite Uranvitriol.

gilsonite Uintaite. A black, brittle, lustrous hydrocarbon mineral.

gin An alcoholic beverage made by distillation of a fermented extract of grain in the presence of juniper leaves. **artificial ~** Fancy g. to which flavoring essences have been added. **fancy ~** A mixture of g. and neutral alcohol.

gingelly Sesame.

ginger Zingiber. The dried rhizome of *Zingiber officinalis* (Scitaminaceae), Asia, W. Indies, Africa; an aromatic, flavoring, and carminative (BP). **jamaica ~** The yellow roots, with the skin removed. **wild ~** Asarum.

g. oil The essential oil of g., d.0.882–0.900, b.155–300, containing phellandrene and zingiberene.

gingerin An oleoresin from ginger.

gingerol An essential oil from ginger.

ginkgetin $C_{32}H_{22}O_{10}$ = 566.5. A yellow biflavonyl pigment from the leaves of *Ginkgo biloba,* maidenhair tree, m.343.

ginkgolic acid $C_{22}H_{34}O_3$ = 346.5. (*Z*)-2-Hydroxy-6(8-pentadecenyl)benzoic acid*. An unsaturated acid from the fruit of *Ginkgo biloba.*

ginning The removal of the larger seed hairs from the cotton plant. Cf. *linter(s).*

ginseng Panax. The dried roots of *Panax quinquefolium* (Aralia); a reputed tonic that may cause hypertension.

gismondine Gismondite.

gismondite $CaAl_2Si_4O_{12}$. Gismondine, abrazite. A gray, hydrated, monoclinic zeolite, d.2.4, hardness 5–5.5.

gitalin $C_{28}H_{48}O_{10}$ = 544.7. A glucoside, m.253, from digitalis.

githagenin $C_{28}H_{44}O_4$ = 444.4. The aglycone of githagin.

githagin A saponin from corn cockel, *Agrostemma githago;* hydrolyzes to githagenin and glucuronic acid.

gitogenic (1) Having a digitalislike effect. (2) The structure of digitalis aglucones.

gitoxigenin $C_{23}H_{34}O_5$ = 390.5. 3,14,16-Trihydroxy-20(22)-cardenolide. m.222. A split product of gitoxin.

gitoxin A glucoside from the leaves of digitalis; it hydrolyzes to 1 mole gitoxigenin and 3 moles digitoxose.

glacial Describing a compound of icelike, crystalline appearance, especially the solid form of a liquid compound; as, glacial acetic acid.

gladiolic acid $C_{11}H_{10}O_5$ = 222.2. 2,3-Diformyl-6-methoxy-5-methylbenzoic acid*. From *Penicillium gladioli.* Silky needles, m.160; an antibiotic. With ammonia it gives a deep green color, changing after 12 hours to red and then orange.

glair Prepared white of egg used for tempera painting.

glance General term for minerals with a glassy luster, e.g., lead glance.

gland An organ or group of cells that secretes specific substances, e.g., enzymes, sweat, mucus.

Glanzstoff Trademark for a viscose synthetic fiber. Cf. *rayon.*

Glaser furnace A combustion furnace used for organic elementary analysis.

glaserite $Na_2SO_4.3K_2SO_4$. Aphthitalite, arcanite. A colorless, vitreous sulfate, d.2.6, hardness 3–3.5 (Stassfurt).

glass An amorphous, hard, brittle, often transparent material; a fused mixture of the silicates of the alkali and alkaline-earth or heavy metals. See Table 39. Composition: between $(K,Na)_2O$, $(Ca,Pb)O$, $6SiO_2$ and $5(K,Na)_2O$, $7(Ca,Pb)O,36SiO_2$.

TABLE 39. TYPICAL GLASS COMPOSITIONS, %

Composition	Soda, window	Flint	Bottle	Borosilicate	Lead	Aluminosilicate	Silica
(A) SiO_2	71.5	54	74	80.5	35.0	58.7	96.3
Al_2O_3	1.5	—	0.5	2.4	—	22.4	0.4
B_2O_3	—	—	—	12.9	—	3.0	2.9
(B) Na_2O	14.0	—	17	3.8	—		
K_2O	—	10	—	—	7.0	} 1.4	} 0.4
(C) CaO	13.0	—	5	0.4	—	6.0	—
PbO	—	36	—	—	58.0	—	—
MgO	—	—	3.5	—	—	8.5	—

Formula: $(K,Na)O-Si_nO_{2n-1}O(Ca,Pb)O-Si_nO_{2n-1}-O(K,Na)$.

Classification:

1. Potash-lime g.: hard, resistant to water and acids, d.2.4; used for chemical glassware.
2. Soda-lime g.: more fusible and less resistant than potash-lime g., d.2.65; used for windows.
3. Potash-lead g.: readily fusible and highly refractive; as, crystal g., d.2.9–3.6; flint g., d.3.3–3.6; paste for artificial gems and lenses; crown g. (containing barium oxide), d.1.5–1.56.
4. Bottle g. (Na, K, Ca, and Al silicates), d.2.73.
5. Opaque g.: opacified by barite, smalts, or bone ash.
6. Colored g.:
   (*a*) Yellow: antimony, iron, silver, uranium.
   (*b*) Red: gold chloride, ocher, cuprous oxide, selenium.
   (*c*) Green: ferrous sulfate, copper, chromium oxide.
   (*d*) Blue: cobalt oxide, traces of copper.
   (*e*) Iridescent: the action of vapors of metallic chlorides on the hot g.
   (*f*) Nacreous: Addition of scales of mica.

bio ~ See *bioglass*. **blown ~** A g. that is blown into shape. **bohemian ~** Potash g. **borax ~** A g. with a low expansion coefficient, which contains borax. **borosilicate ~** A heat-resistant silicate g. containing at least 5% boric acid, d.2.25, m.730. Cf. *Pyrex*. **bottle ~** A g. that is blown into shape in a mold. **bulletproof ~** Plate g. sheets cemented together by a transparent medium. **canary ~** Uranium g. **cast ~** Plate g. **chemical ~** An acid g. or alkali-resistant g. for chemical apparatus. **chromium ~** A g. colored yellow by chromium compounds. **clock ~** G. similar in shape to that used for covering clock faces; used to cover beakers, etc. **cobalt ~** A g. colored purple-blue by cobalt compounds; a light filter. **conductive ~** G. rendered electrically conductive to a desired extent by treatment with tin chloride and heating to produce a layer of tin oxide. **copper ~** G. colored blue or red by copper compounds. **cover ~** A thin g. square used to cover microscope specimens on the slide. **crown ~** A hard optical g. silicate of sodium with calcium and aluminum oxides; formerly made by blowing and spinning, to form a disk, from which small windowpanes were cut. **cryolite ~** Milk g. **crystal ~** Flint g. **electric bulb ~** A lime g. used for electric bulbs. **fiber ~** G. fiber. **flint ~** A soft optical g. made from sand, potash, and lead oxide. **float ~** G. solidified as a continuous sheet on a bath of molten metal (as, tin) at 1000 °C. It has a high surface finish, flatness, and absence of distortion. **frosted ~** Opaque g. having a roughened surface. **iron ~** G. colored yellow, olive green, or pale blue by iron compounds. **Jena ~** Optical and heat-resisting g. made at Jena. **laminated ~** Safety g. made by cementing thin sheets of g. together with a plastic at 90–130 °C and 1.7–2.4 MPa. It may crack but will not splinter under impact. **lead ~** A soft g. with a low melting point, containing lead oxide; e.g., flint g. **lime ~** G. containing calcium oxide; e.g., venetian g. **manganese ~** G. colored violet by manganese compounds. **milk ~** G. colored milky white by cryolite. **Muscovy ~** Muscovite. **normal ~** A g. of definite chemical composition. **opal ~** G. colored milky white by calcium phosphate or bone ash. **optical ~** See *crown glass, flint glass* (both above). **organic ~** Synthetic g. Synthetic resins having the appearance of g., e.g., Perspex. **plate ~** A thick g. made by pouring molten g. on iron tables, then rolling and polishing it; used for mirrors and windows. **porous ~** G. containing pores of molecular dimensions, made by leaching boric acid from heat-treated borosilicate g. (SiO_2 96, void space 25%). Used in filters and salt bridges. **potash ~** Bohemian g. G. containing more potassium than sodium, e.g. crown g. **rolled ~** Inferior plate g., made by passing molten g. between iron rolls. **ruby ~** Dark-red g. containing copper compounds or colloidal gold. **safety ~** Laminated g. **sheet ~** Flat sheets made by blowing long cylinders, splitting them longitudinally, and flattening them out. **silica ~** Though not a true g., fused silica is often used for a transparent, resistant g. **silicate-flint ~** A Jena g.: SiO_2 29–53, PbO 67–36, K_2O 3–8, Na_2O 0–1, Mn_2O_3 0.04–0.06, As_2O_3 0.2–0.3%; used for optical purposes. **soda ~** G. containing more sodium than potassium; e.g., venetian g. **sol-gel process ~** G. produced by a new method involving the collapse of a microporous structure at temperatures well below the 800–950 °C used traditionally. **soluble ~** Water g. **spun ~** G. fiber. **synthetic ~** Organic g. **thallium ~** G. containing Tl in place of Pb. **toughened ~** Heat-treated plate g. used to prevent splintering under impact. Cf. *laminated glass*. **uranium ~** A dichroic, greenish-yellow glass containing uranium compounds, used for light filters. **watch ~** A small *clock glass*, q.v. **water ~** Sodium or potassium silicate. **window ~** G. plates made by blowing the molten g. into cylinders, then slitting and flattening them out on tables. **zinc-crown ~** An optical g.: SiO_2 65.4, K_2O 15, NaO_2 5, BaO 9.6, ZnO 2.0, As_2O_3 0.4, Mn_2O_3 0.1, B_2O_3 2.5%.

g. beads Solid or hollow spheres; used to prevent excessive ebullition of heated liquids or to determine the specific gravity of liquids. **g. colors** See *glass (6)*. **g. cullet** (1) Broken g. waste. (2) Powdered waste from g. manufacture; used as abrasive in matches, primers, polishes, soaps, and cements. **g. cutters** Small, mounted diamond fragments, used to cut glass. **g. drops** Prince Rupert drops. **g. fiber** Fiberglass. Glass in filaments, of 3 types: (1) For reinforcing plastics, as GRP (see below). Has low alkali content. (2) For reinforcing cement. An alkali-lime-silica-zirconia mixture. (3) For insulation. Similar to flat glass, but with some silica replaced by ulexite or rasorite. **g. fiber-reinforced plastics** GRP. Materials giving good corrosion resistance, high strength-to-weight ratio, high thermal and electrical insulation; suitable for fabricating in complex shapes. **g. gage** A metal disk with round holes, used to measure the outside diameter of g. tubing. **g. marking** Ceramic ink. **g. of antimony** The fused mass resulting from the incomplete oxidation of antimony glance. **g. paper** Calico or paper covered with thin glue and sprinkled with powdered g.; used for polishing. **g. tubing** A hollow g. rod, used in scientific apparatus. **barometer ~** Capillary g. t. **capillary ~** A thick-walled g. tube having a bore of less than 1 mm. **g. tank** The container lined with aluminum silicate in which g. is melted. **g. wool** Glass fiber.

glassine A thin, hard, and almost transparent paper made from well-beaten chemical wood pulp.

Glauber, Johann Rudolf (1603–1668) Dutch iatrochemist who prepared many metallic salts. **g. salt** Crystalline sodium sulfate decahydrate.

glauberite $CaSO_4.Na_2SO_4$. A calcium sodium sulfate (Stassfurt).

glaucine $C_{21}H_{25}O_5N = 371.4$. An alkaloid from the sap of *Glaucium flavum*, yellow horned poppy (Papaveraceae). Yellow prisms, m.119.

glauchochroite $CaMnSiO_4$. A rare silicate of the olivine group.

glaucodot $(Fe,Co)S_2$, $(Fe,Co)As_2$. A native sulfarsenide. Cf. *alloclasite*.

glauconite Bravaisite. An amorphous, green, granular iron, potassium, aluminum, magnesium, calcium silicate formed from oceanic sediments of all ages.
glaucophane $NaAlSi_2O_6.(Fe,Mg)SiO_3$. An amphibole, rock-forming mineral. Gray, monoclinic masses, d.3–3.1.
glaucopicrine An alkaloid from the roots of *Glaucium flavum* and *Chelidonium majus.*
glaze A glassy coating. **enamel ~** A suspension of metallic oxides in a glass which is burned into pottery or ironware. **porcelain ~** A mixture of feldspar, lime, and quartz fused into ware. **salt ~** A glassy covering of a silicate of sodium and aluminum produced on earthenware by adding salt to the kiln during firing. **transparent ~** A glass covering for earthenware.
glazed Having a glossy appearance.
GLC, glc Gas-liquid chromatography. See *gas-liquid chromatography* under *chromatographic analysis.*
gleditschine Stenocarpine. An alkaloid from the leaves of *Gleditschia triacanthos,* three-thorned acacia or honey locust tree (Leguminosae), U.S.
glendonite A pseudomorphous calcite (Australia).
gliadin $C_{685}H_{1068}O_{211}N_{196}S_5$ = 15,586. Prolamin, vegetable protein. A simple protein from gluten, the protein of cereals. See *gluten.*
glibenclamide $C_{23}H_{28}O_5N_3SCl$ = 494.0. Glyburide, Euglucon, Daonil. White crystals, m.173, insoluble in water. An oral hypoglycemic agent, used for mild diabetes (BP).
glioma A malignant tumor of nerve tissue.
Gln* Symbol for glutamine.
globin $C_{700}H_{1098}O_{196}N_{184}S_2$ = 15,292. An animal protein in hemoglobin. Insoluble in water, soluble in acids or alkalies, coagulated by heat, redissolved by acids. **hemo ~** See *hemoglobin.* **oxyhemo ~** See *oxyhemoglobin* under *hemoglobin.*
globucid The antibiotic *p*-aminophenylsulfonamide ethylthiodiazole, m.184, sparingly soluble in water.
globularimin $C_{24}H_{30}O_{12}$ = 510.5. An amorphous glucoside from the leaves of *Globularia alypium,* Globulariaceae (S. Europe).
globulin A simple protein, coagulated by heat, insoluble in water, soluble in dilute solutions of salts. **acid ~** Syntonin. **crystallin ~** A g. from the eye lens. **immune ~** (Human) normal immuno ~. G. from plasma of not less than 1,000 donors, containing most of the antibodies present in blood. Used to prevent infectious diseases and rubella in pregnancy (USP, EP, BP). **$Rh_0(D)$ ~** Anti-D(Rh_0)immunoglobulin. G. from a number of donors naturally or artificially immunized against RhD antibody. Given after delivery to prevent rhesus disease of the newborn in subsequent pregnancies (USP, BP). See *Rh factor* under *Rh, immunoglobulins* under *immuno.* **serum ~** A simple blood protein. Horse g. has the proposed formula $C_{628}H_{1002}O_{209}N_{160}S_5$.
globulins Simple *proteins,* q.v., in vegetable and animal tissues.
globulol $C_{15}H_{26}O$ = 222.4. A sesquiterpene alcohol from eucalyptus oil. b.283.
globulose A split product of globulins produced by peptic digestion.
globulus A small sphere, e.g., a button of metal.
glomeruli Microscopical capillary structures in the kidney that filter urine from the blood.
Glon Trademark for a vinyl chloride–type plastic.
glonoin Nitroglycerin.
glove box Dry box. A closed box with a sloping glass front through which air can be drawn. A pair of gloves sealed into it enable operations to be carried out safely (e.g., free from noxious fumes) and under observation.
Glover tower A tower in *sulfuric acid manufacture.*
Glu* Symbol for glutamic acid.
glucagon $C_{153}H_{225}O_{49}N_{43}S$ = 3483. A polypeptide hormone secreted by the pancreas. White crystals, soluble in acids; from bovine or porcine pancreatic glands. Causes glucose production by liver. Used for hypoglycemic states in diabetes (USP, BP).
glucaric acid Saccharic acid.
glucic acid $C_3H_4O_3$ = 88.1. Acrolactic acid, β-hydroxyacrylic acid, 3-hydroxypropenoic acid. Colorless liquid, soluble in water.
glucide A group term including carbohydrates and glucosides, q.v.
glucin Sodium aminotriazine sulfonate. A sweetening agent.
glucinic acid $C_{12}H_{16}O_9 \cdot 3H_2O$ = 358.3. A hexabasic acid formed in the decomposition of glucose by acids or alkalies. Colorless crystals.
glucinum Obsolete name for beryllium.
Glucitol Sorbitol.
gluco (1) Indicating glucose. (2) Incorrect for *glyc(o)-*.
glucochloral Chloralose.
glucocholic acid $C_{24}H_{39}O_4 \cdot NHCH_2COOH$ = 465.6. Colorless needles, m.134, slightly soluble in water.
glucocorticoid See *corticoids.*
glucofurone $C_6H_{10}O_6$ = 178.1. The γ-lactone of gluconic acid.
glucogen Glycogen.
glucohydrazones Intermediate compounds of the osazone reaction (heating aromatic hydrazines and hexoses).
glucokinin Insulin.
gluconate A salt of gluconic acid containing the radical $HOCH_2(CHOH)_4COO-$. **aluminum ~** A tanning salt—partly a colloidal suspension of aluminum hydroxide in gluconic acid. **calcium ~** See *calcium gluconate.*
gluconeogenesis Formation of glucose in the liver from noncarbohydrate sources, as, proteins and amino acids.
gluconic acid $CH_2OH(CHOH)_4 \cdot COOH$ = 196.2. Pentahydroxyhexanoic acid, dextronic acid, glycogenic acid. Derived from glucose by oxidation. **D- ~** Dextronic acid, maltonic acid. White powder, m.131, pleasant, sour taste; solutions form lactones and become plastic. Used for fruit jellies, as a sequestrant in paint strippers, and in electroplating. **L- ~** White, crystalline solid, soluble in water.
glucoprotein Mucoprotein. A conjugated protein having one or more heterosaccharide prosthetic groups with few sugar residues, lacking a serially repeating unit, and bound covalently to the polypeptide chain.
glucosamine $CH_2OH(CHOH)_3CHNH_2 \cdot CHO$ = 179.2. The amine of glucose and a split product of chitin, b.110 (decomp.).
glucosazone $C_{18}H_{22}O_4N_4$ = 358.4. A reaction product of monosaccharides and aryl hydrazines. Their characteristic crystalline forms and melting points are used to separate and identify monosaccharides.
glucose $C_6H_{12}O_6$ = 180.2. Dextrose, phlorose, grape sugar, saccharum amylaceum. A monosaccharide carbohydrate constituent of many animal and vegetable fluids (blood, sweet fruits, etc.), formed by hydrolysis of starch, cane sugar, and glucosides. **D- ~** Colorless needles, d.1.562, m.147, soluble in alcohol or water. A reagent to detect carbon dioxide in blood, tellurous acid, etc., and (as glucose liquid or syrup) a nutrient; used in the manufacture of confectionery and the

production of beer and alcoholic liquors, curing of tobacco, tanning; and as reducing agent. Several isomers, as:

HCO
|
HCOH
|
HOCH
|
HCOH
|
HCOH
|
CH_2OH
D-
m.147

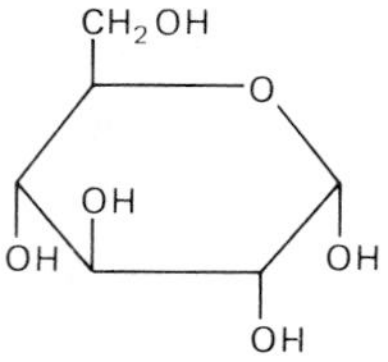

CH_2OH O OH OH OH OH

α-D-pyranose form

HOH_2C HO O OH OH OH

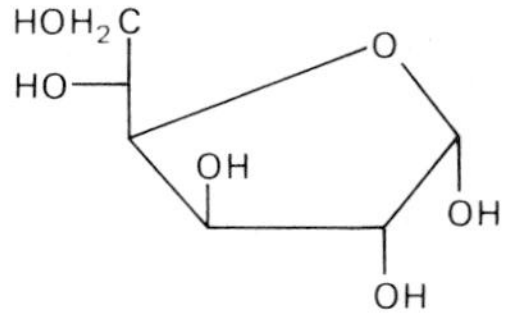

β-D-furanose form

medicinal ~ Dextrose monohydrate.

g. evolué A reducing bacteriological medium, prepared by heating a 10% solution of g. in 0.1 *N* caustic soda at 100°C for 15 min. **g. gelatin** A culture medium: glucose 10, gelatin 100, peptone 10, sodium chloride 5, bouillon stock 1,000 pts., neutralized with caustic soda. **g. imine** $C_6H_{12}O_5$:NH = 179.2. A solid, m.128. **g. isomerase** An enzyme that converts glucose into fructose. The conversion product of hydrolyzed starch, e.g., high fructose corn syrup, has a high sweetening power. **g. liquid** Glucosum. A syrup made by the incomplete hydrolysis of starch, and containing glucose and dextrins; a nutrient (USP). **g. oxidase*** An enzyme, from *Aspergillus niger,* oxidizing glucose to gluconic acid. **g. oxime** $C_6H_{12}O_5$:NOH = 195.2. A reaction product of hydroxylamine and glucose, m.138. **g.phosphate isomerase*** Enzyme catalyzing the interconversion of D-glucose and D-fructose 6-phosphates. **g. syrup** Corn syrup, liquid glucose, starch hydrolysate. A liquid hydrolysis product of edible starch (U.K. usage); solids not less than 70%, D-glucose not less than 20%, and sulfated ash not more than 1% of dry matter.

glucosidase **α-D- ~*** Maltase. An enzyme which hydrolyzes terminal 1,4-linked α-D-glucose residues, as in maltose, with release of α-glucose.

glucoside (1) A compound of glucose; as, maltose. (2) See *glycoside.* A neutral, nonnitrogenous vegetable constituent decomposed by heat, dilute acids, alkalies, enzymes, bacteria, or fungi, to form a sugar (glucose) and another compound, e.g., salicin. Glucosides are the ethers of monosaccharides; 2 types: alpha and beta: **methyl ~** $C_7H_{14}O_6$ = 194.2. **alpha- ~** Long needles, m.168, $[\alpha]_D$ +157°. **beta- ~** Rectangular prisms, m.104, $[\alpha]_D$ −33°. Natural glucosides have the suffix *-in* and can be classified as: ethylene derivatives, e.g., jalapin; benzene derivatives, e.g., arbutin; styrene derivatives, e.g., daphnin; anthracene derivatives, e.g., digitoxin; cyanogen derivatives, e.g., amygdalin.

glucosimine Obsolete term for amino sugar.

glucosin Ptomaine bases obtained by the action of ammonia on carbohydrates.

glucosone $CH_2OH\cdot(CHOH)_3COCHO$ = 178.1. An osone of glucose and aldehyde ketone; a reaction product from glucosazone.

glucotin A cement mixture of isinglass, gelatin, and acetic acid.

glucuronic acid $CHO(CHOH)_4COOH$ = 194.1. An aldehyde-hydroxy acid in urine, m.175.

glue Colla. Impure gelatin obtained from animal organs by boiling with water, straining, and drying as thin, hard, and brittle cakes; an adhesive. **albumen ~** G. obtained from flour in starch manufacture. **bone ~** Artificial isinglass, from hides and bones. **fish ~** Isinglass. **liquid ~** G. acidified with acetic or nitric acid. **marine ~** Waterproof g. made of caoutchouc or shellac in turpentine. **skin ~** G. from hides. **vegetable ~** Acacia. **waterproof ~** A fish g. dissolved in hot milk.

g. sniffing The inhalation of glues or solvents to produce an altered mental state.

Glumiflorae Monocotyledonous plants, comprising the families Cyperaceae and Gramineae.

gluside, glusidum Names for saccharin.

glutaconic acid 2-Pentenedioic acid*.

glutamic acid* $NH_2\cdot CH(COOH)CH_2\cdot CH_2\cdot COOH$ = 147.1. Glu*. 2-Aminoglutaric acid, glutanic acid. Colorless crystals, decomp. 208, soluble in water. A constituent of proteins in seeds and beets, and the Japanese flavoring, ajinimoto.

glutamine* $H_2N.CO(CH_2)_2\cdot CHNH_2\cdot COOH$ = 146.1. Gln*. 2,5-Diamino-5-oxopentanoic acid. Slightly soluble in water.

glutaminic acid Glutamic acid*.

glutamyl* The radicals: **α- ~** $-OC\cdot CHNH_2(CH_2)_2\cdot COOH$. **γ- ~** $HOOC\cdot CHNH_2(CH_2)_2\cdot CO-$.

glutamoyl* The radical $-OC\cdot CHNH_2(CH_2)_2CO-$.

glutanic acid Glutamic acid*.

glutaric acid* $COOH(CH_2)_3COOH$ = 132.1. Deoxyglutaric acid, pentanedioic acid*. A dibasic acid in sheep wool. Colorless, monoclinic crystals, m.97, soluble in water; a constituent of sheep-wool grease. **amino ~** Glutamic acid*. **dimethyl ~** $C_7H_{12}O_4$ = 160.2. Colorless crystals, m.127. **ethyl ~ 2- ~** $C_7H_{12}O_4$ = 160.1. Colorless crystals, m.60. **3- ~** Propylidenediacetic acid. Colorless crystals, m.67. **methyl ~ 2- ~** $C_6H_{10}O_4$ = 146.1. Colorless crystals, m.76. **3- ~** Ethylidenediacetic acid. Colorless crystals, m.86.

glutaronitrile* $CN(CH_2)_3CN$ = 94.1. Trimethylene cyanide, pentane dinitrile*. Colorless liquid, b.286.

glutaryl* 1,5-Dioxo-1,5-pentanediyl†. The radical $-OC(CH_2)_3CO-$.

glutathione $C_8H_{15}O_2N_3S(COOH)_2$ = 307.2. γ-Glutamylcysteylglycine. A tripeptide in blood, animal organs, and germinating plants. It plays an important part in metabolism.

glutazine $C_5H_6O_2N_2$ = 126.1. 4-Amino-2,6-pyridinedione. Colorless crystals, m.300 (decomp.).

glutelins Simple vegetable proteins, coagulated by heat, insoluble in water or dilute salts, soluble in dilute acids or alkalies; as, glutenin.

gluten A brown, sticky mixture of 2 proteins, glutelin and gliadin, in the seeds of cereals, which remains after washing the starch out of wheat flour with water. It confers the toughness on dough. The amino acids of glutens are glutamic acid (12–24%), leucine, proline, and arginine. **g.-induced enteropathy** Celiac disease. A disease of the small intestine causing failure of food absorption due to the effect of g. on its lining; cured by gluten-free diet.

glutenin A wheat protein, soluble in dilute alkalies.

glutin $C_{192}H_{294}N_{60}SO_{70}$(?). A protein in gelatin.

glutine Glue of animal origin.

glutinic acid HOOC·CH:C:CH·COOH = 128.1. 2,3-Pentadienedioic acid. White solid, m.146.
glutinosin $C_{48}H_{60}O_{16}$(?). An antibiotic from the soil fungus *Metarrhizium glutinosum;* a severe skin irritant.
glutol A reaction product of starch and formaldehyde. Cf. *amyloform.*
Glutolin $C_{204}H_{336}N_{60}SO_{70}$. Trademark for formaldehyde gelatin. A protein derived from gelatin.
glutose The unfermentable reducing portion of cane molasses. A complex mixture of anhydrofructose; predominantly the compounds formed by the condensation of amino acids and their amides with simple sugars.
Gly* Symbol for glycine.
glycal A cyclic enol ether derivative of a sugar having a double bond between C atoms 1 and 2 of the ring.
glyceraldehyde* CH_2OH·CHOH·CHO = 90.1. 2,3-Dihydroxypropanal*, glycerose. Colorless solid, m.132, soluble in water.
glycerals Compounds derived from glycerol and aldehydes, similar to the acetals.
glycerate* Glycerinate. A salt or ester of glyceric acid, containing the radical $C_3H_5O_4-$.
glyceric acid* CH_2OH·CHOH·COOH = 106.1. 2,3-Dihydroxypropanoic acid*. Occurs as (+)- and (−)-acids. Colorless syrup, soluble in water, formed during alcoholic fermentation. **2-phenyl ~** Atroglyceric acid.
glyceric aldehyde Glyceraldehyde*.
glyceride An ether or ester derived from glycerol. The fats and oils are mainly triglycerides of fatty acids, e.g., palmitin.
glycerin(e) USP name for glycerol*. **g. trinitrate** Nitroglycerin.
glycerinate Glycerate*.
glycerino Glycero.
glycerinum Glycerol*.
glycero The radicals (1) 1,2,3-propanetriyl*; (2) $-OH_2C$·CHO·$CH_2O=$, from glycerol.
glycerogen A German wartime substitute for glycerol (glycerol 40, propylene glycol 40, other higher alcohols 20%). Made by hydrogenating inverted sucrose.
glycerol* $(CH_2OH)_2CHOH$ = 92.1. Glycerin(e), glycerinum, 1,2,3-propanetriol*, propenyl hydrate. Colorless, sweet syrup, d.1.260, m.17 (solidifies at lower temperature), b.290, soluble in water, insoluble in organic solvents. Obtained by the saponification of fats in the soap industry; used as a mordant, plasticizer, solvent, and reagent, in the manufacture of printer's ink and rolls, and explosives; also used for application to the skin and as suppositories (USP, EP, BP). **absolute ~** G. free from water. **diethyl ~** See *diethyl glycerol.* **dithio ~** See *Dimercaprol.* **mesitylene ~** Mesicerin. **pentyl ~** $C_5H_9(OH)_3$. Trihydroxypentane.
g. diacetate Diacetin. **g. dinitrate** Dinitroglycerol. **g. diphenyl ether** $C_{15}H_{16}O_3$ = 244.3. 1,3-Diphenoxy-2-propanol*. White crystals, m.80; a plasticizer for nitrocellulose. **g. distearate** Distearin. **g. monochlorohydrin** $C_3H_7O_2Cl$ = 110.5. **alpha- ~** 3-Chloro-1,2-propanediol*. Colorless liquid, d.1.322, $b_{0.5mm}81$, miscible with water; used in the synthesis of 2,3-epoxy-1-propanol. **beta- ~** 2-Chloro-1,3-dihydroxypropane; $b_{14mm}124$. **g. monophenyl ether** $C_9H_{12}O_3$ = 168.2. 1-Phenoxy-2,3-propanediol*, autodyne. White solid m.53; a plasticizer. **g.phosphoric acid** $C_3H_5(OH)_2H_2PO_4$ = 172.1. An oily constituent of lecithins and nerve tissues. **g.sulfuric acid** $C_3H_5(OH)_2·HSO_4$. **g. tributyrate** Butyrin. **g. trilaurate** Laurin. **g. trinitrate*** Nitroglycerin. **g. tripalmitate** Palmitin. **g. tristearate** Stearin.

glycerolphosphate Lecithin. A salt of glycerolphosphoric acid containing the radical $= PO_4·C_3H_5(OH)_2$.
glycerophosphoric acid See *glycerolphosphoric acid* under *glycerol.*
glycerose Glyceraldehyde*.
glycerosulfuric acid Glycerosulfuric acid.
glyceroyl* The radical CH_2OH·CH(OH)·CO−.
glyceryl The 1,2,3-propanetriyl* radical. **g. aldehyde** Glyceraldehyde*. **g. chloride** Trichlorohydrin. **g. hydroxide** Glycerol*. **g. laurate** Laurin. **g. linoleate** Trilinolein. **g. monostearate** $C_{20}H_{42}O_4$ = 346.5. A commercial emulsifying and dispersing agent, used mainly in cosmetics. A hard fat, containing 30–40% of the α isomer, m.54–60, dispersible in water. **g. trinitrate** Nitroglycerin.
glycide 2,3-Epoxy-1-*propanol**.
glycidol 2,3-Epoxy-1-*propanol**.
glycin (1) Glycine*. (2) *p*-Hydroxyphenylaminoacetic acid; a developer. (3) Mannitol. (4) Beryllium. (5) Glycyrrhiza.
glycine* $NH_2·CH_2$·COOH = 75.1. Gly*. Glycocoll, aminoacetic acid (USP), aminoethanoic acid*, glycocin, gelatin sugar. Sweet, colorless, monoclinic crystals, d.1.575, m.232, slightly soluble in water. **acetyl ~** See *acetylglycine.* **amidinomethyl ~** Creatine. ***N*-benzoyl ~** Hippuric acid*. **carbamoyl ~** Hydantoic acid*. **glycyl ~** The simplest peptide, $NH_2CH_2CONHCH_2COOH$. ***N*-methyl ~** Sarcosine*. **2-methyl ~** Alanine*. **trimethyl ~** Betaine*.
g. anhydride 2,5-Piperazinedione. **g. betaine** Betaine*.
glycinin The principal protein of the soybean.
glycinium* The cation $^+H_3N·CH_2$·COOH.
glycirrhiza Glycyrrhiza.
glyc(o)- (Greek: "sweet") Indicating sugars or glycine.
glycocholate A salt of glycocholic acid.
glycocholeic acid $C_{27}H_{45}O_5N$ = 463.7. A bile acid compound of glycogen and choleic acid. Colorless prisms, m.175, slightly soluble in hot water.
glycocholic acid $C_{26}H_{43}O_6N$ = 465.6. A bile acid compound of glycine and cholic acid. Colorless needles, m.134, soluble in water.
glycoclastic Glycolytic.
glycocoll Glycine*. **g. betaine** Betaine*.
glycogen $(C_6H_{10}O_5)_n$. Animal starch, glucogen, liver sugar. A stored carbohydrate in the animal organism, especially in the liver. Colorless, tasteless powder, readily hydrolyzed to glucose (red with iodine). Acids hydrolyze it to glucose, and enzymes to maltose. **g.-storage disease** An inborn error of metabolism, in which g. hydrolysis cannot occur owing to a congenital deficiency of various enzymes.
glycogenase Amylase*.
glycogenesis The synthesis of glycogen from glucose, occurring in the liver and muscles.
glycogenolysis The successive breaking down of glycogen in animal tissues. Cf. *staircase reaction.*
glycol (1) See *glycols.* (2)* $CH_2OH·CH_2OH$ = 62.1. Ethylene g*., 1,2-ethanediol, dihydroxyethane. Colorless liquid, d.1.115, $b_{20mm}198$, miscible with water. An antifreeze (60% in water freezes at −49°C) and a solvent for cellulose esters; used to manufacture low-freezing dynamites. **benzylene ~** Hydrobenzoin. **but(yl)ene ~** Butanediol*. **diethylene ~** Carbitol. **diphenyl ~** Hydrobenzoin. **ethylidene ~** $MeCH(OH)_2$. Known only in derivatives; as, acetals. **mesitylene ~** See *mesicerin.* **phenyl ~** Cinnamyl alcohol* **propylene ~*** 1,2-Propanediol*. **tetramethylene ~** Butanediol*. **tetramethylethylene ~** Pinacol*. **tetraphenyl ~** Benzopinacol. **trimethylene ~** 1,3-Propanediol*.
g. acetate See *ethylene monoacetate, ethylene diacetate* (both

under *ethylene*). **g.aldehyde*** $CH_2OH \cdot CHO$ = 60.0. Hydroxyaldehyde† glycolal, hydroxyethanal*. Colorless plates, m.96, soluble in water. **g.amide** $CH_2OH \cdot CONH_2$ = 75.1. 2-Hydroxyacetamide*. Colorless solid, m.120, soluble in water. **g. bromohydrin** $CH_2OH \cdot CH_2Br$ = 125.0. Ethylene bromohydrin, 2-bromo-1-ethanol*. Colorless liquid, b.147, soluble in water. **g. chlorohydrin** $CH_2OH \cdot CH_2Cl$ = 80.5. Ethylene chlorohydrin, 2-chloro-1-ethanol*. Colorless liquid, b.128, miscible with water. **g. cyanohydrin** $CH_2OH \cdot CH_2CN$ = 71.1. Ethylene cyanohydrin. Colorless liquid, b.222, miscible with water. **g. diacetate** Ethylene diacetate*. **g. dibromide** Ethylene dibromide*. **g. dichloride** Ethylene dichloride*. **g. dicyanide** Ethylene dicyanide*. **g. diiodide** Ethylene diiodide*. **g. dinitrate** Ethylene dinitrate. **g. dinitrite** Ethylene dinitrite. **g. ethers** A group of compounds used as lacquer solvents; as, g. butyl ether (see *butyl Cellosolve* under *Cellosolve*). **g. leucine** Norleucine. **g. monoacetate** Ethylene monoacetate. **g. mercaptan** Ethylenethiol*. **g. sulfhydrate** Ethylenedithiol*. **g. thiourea** $C_3H_4ON_2S$ = 116.1. A solid, soluble in water, m.200 (decomp.).

glycolal Glycol aldehyde*.

glycoleucine Norleucine.

glycolic g. acid* $OHCH_2 \cdot COOH$ = 76.1. Glycollic acid. Hydroxyacetic acid†, hydroxyethanoic acid*. Colorless leaflets, m.78 (decomp.), soluble in water. **diphenyl ~** Benzilic acid*. **phenyl ~** Mandelic acid.

g. aldehyde Glycol aldehyde*. **g. amide** Glycol amide. **g. anhydride** $(CH_2OH \cdot CO)_2O$ = 134.1. 1,4-Dioxane-2,5-dione. Colorless powder, m.129, insoluble in water.

glycolide $(-C(O) \cdot CH_2 \cdot O-)_2$ = 116.0. Glycollide. Colorless leaflets, m.86, soluble in alcohol.

glycolipids Fatty substances, yielding on hydrolysis fatty acids and a carbohydrate, usually glucose. They contain no phosphorus; e.g., kerasin.

glycoloyl* Hydroxyacetyl†, glycolyl. The radical $OH \cdot CH_2 \cdot CO-$.

glycols Diols. *Term formally used for dihydric alcohols. Now limited to ethylene and propylene glycol.

glycolyl The glycoloyl* radical.

glycolysis Glycolytic or Embden-Meyerhof pathway. The decomposition of glucose into pyruvic acid (pyruvate) by enzymes; the energy source of animals and anaerobic organisms.

glycolytic See *glycolysis.*

glycophospholipids Fatty substances which yield a fatty acid, a carbohydrate, and phosphoric acid on hydrolysis.

glycoproteins Glucoproteins. Conjugated proteins containing a carbohydrate radical and a simple protein; e.g., ichthulin.

Glycosal Trademark for glycerol salicylate.

glycosamine $C_6H_{11}O_5NH_2$ = 179.2. A decomposition product of chitin. Colorless crystals, slightly soluble in water.

glycoside A natural compound of a sugar with another substance, which hydrolyzes to a sugar plus a principle: (e.g., coniferin yields glucose plus coniferyl alcohol as the principle); *glucosides* yield glucose, *fructosides* yield fructose, *galactosides* yield galactose, etc. Many pigments (as anthocyanins), saponins, and tannins are glycosides. Examples (parent compound in parentheses): sinigrin (ethylene), arbutin (benzene), daphnin (styrene), quercitrin (flavone, anthoxanthin), cyanin (anthocyanins), frangulin (anthracene), prulaurasin (cyanogen), indican (indoxyl), digitalin (cholane).

glycosidic bond The ether linkage between monosaccharide units of a polysaccharide; as, $\alpha(1,4)$ or $\beta(1,6)$, the numbers indicating the participating C atoms.

glycosuria Sugar in the urine. Usually due to diabetes, but may be physiological or due to some drugs.

glycotropin A hormone resembling *prolactin,* q.v.

glycuronate A salt of glucuronic acid.

glycuronic acids Uronic acids.

glycyl* Aminoacetyl†. The radical $NH_2 \cdot CH_2CO-$, from glycine. It occurs in peptides, e.g., glycylalanine.

glycyrrhetinic acid $C_{30}H_{46}O_4$ = 470.7. m.302. Aglycone from licorice root.

glycyrrhiza (1) Licorice. The dried, aqueous extract of licorice root. Lustrous, black, brittle mass, soluble in water (BP). (2) The rhizome and roots of *Glycyrrhiza glabra typicat,* Spanish licorice, and *G. glabra glandulifera,* Russian licorice (Leguminosae). A flavoring in medicines (NF).

glycyrrhizic acid $C_{44}H_{64}O_{19}N$ = 911.0. Crystals, m.220, from licorice.

glycyrrhizin $C_{44}H_{64}O_{19}$ (?). A glucoside of glycyrrhetinic acid. Brown scales, m.205 (decomp.), soluble in water; optically inactive.

glyoxal* $(CHO)_2$ = 58.0. Oxalaldehyde, ethanedial*, diformyl, oxal. Colorless, deliquescent powder or liquid, d.1.14, m.15, b.50, soluble in water. **difuryl ~** Furil*. **dimethyl ~** Biacetyl*. **diphenyl ~** Benzil*. **methyl ~** Pyruvaldehyde*. **phenyl ~** Benzoyl formaldehyde. **poly ~, trimeric ~** $C_{12}H_{18}O_8$ = 290.3. A threefold polymer of g., known as its acetone derivative.

glyoxalase An enzyme in all animal tissues, except pancreas and lymph glands. It converts glyoxal or its substituents (as, R is Me) into glycolic acid or its substituents:

$$R \cdot CO \cdot CHO + H_2O = R \cdot CHOHCOOH$$

glyoxalene Imidazole*.

glyoxalic acid $CHO \cdot COOH$ = 74.0. Ethanol acid, oxoethanoic acid*, oxaldehydic acid. Colorless rhombs, soluble in water, forming $(HO)_2CH \cdot COOH$. Cf. *glyoxylic acid.* **amino ~** Oxamic acid*. **aminophenyl ~** Isatic acid. **carboxyphenyl ~** ***o*- ~** Phthalonic acid. ***p*- ~** Terephthalic acid. **methyl ~** Pyruvic acid*. **phenyl ~** Benzoylformic acid.

g. a. hydrate Glyoxylic acid*.

glyoxaline Imidazole*.

glyoxime $(CH{:}NOH)_2$ = 88.1. Colorless prisms, m.178, soluble in water. Cf. *dimethylglyoxime* (under *dimethyl*), *Tschugajew's reaction.*

glyoxyl Glyoxyloyl*.

glyoxylic acid* $(HO)_2CHCOOH$ = 92.1. Dihydroxyacetic acid*. The hydrated and crystalline form of *glyoxalic acid,* q.v.; in unripe fruit.

glyoxyloyl* The radical $OHC \cdot CO-$, from glyoxylic acid.

glyphosate* See *herbicides,* Table 42 on p. 281.

Glyptal Trademark for synthetic resins and plasticizers prepared from a polyhydric alcohol and phthalic anhydride.

gm Abbreviation for gram. Correct usage is g.

Gmelin G., Christian Gottlieb (1792–1860) German chemist, noted for making artificial ultramarine. **G., Johann Friedrich (1748–1804)** German physician and writer on chemistry. **G., Leopold (1788–1853)** German chemist, noted as discoverer of potassium hexacyanoferrate(III). A prolific author. **G. test** Nitric acid is dropped on filter paper saturated with urine; concentric rings of various colors appear in presence of bile acids.

gmelinite $(Ca,Na_2)Al_2Si_3O_{12}$. A native chabazite.

gneiss Crystalline metamorphic rocks; typically of quartz or feldspar.

gnoscopine $C_{22}H_{23}O_7N$ = 413.4. (±)-Narcotine. An opium alkaloid, m.229, in the mother liquor of narceine; synthesized by dehydration of a molecular mixture of meconin and cotarnine.

goa Araroba, crude chrysarobin, Brazil powder. Yellow powder from the cavities in the trunks of *Andira araroba*, Leguminosae (Brazil). It contains 80% chrysarobin, and resin, gum, etc. Cf. *yaba bark.*

goaf The space left in a coal mine after removal of the coal.

Göckel condenser A Liebig-type condenser with a U-shaped inside tube; at the base, the tube is connected with an airtight receiver.

go-devil A cylindrical brush or scraper used to scrub the interior of pipes by the action of the liquid or gas flowing through.

goethite FeO·OH. Ruby mica. A hydrated oxide of iron. Cf. *göthite.*

Goethlin solution See *Göthlin solution.*

goiter Goitre. Any enlargement of the thyroid gland, sometimes due to iodine deficiency in the diet. **endemic ~** G. occurring in a specific geographical area, due to low iodine levels in the soil.

Golay column A 50-m-long tube, wound on a cylindrical former, to contain the filling used in gas chromatography.

gold* Au = 196.9665. Aurum*. An element, at. no. 79. Yellow, ductile, noble metal, $d_{17.5}$ 19.32, m.1064, b.3080, insoluble in acid or alkalies, soluble in aqua regia. It occurs in nature as an uncombined metal, has valencies of 1 and 3, and tends to form complex compounds:

g.(I)*, g.(1+)*, aurum(I)*, aurus ...... Au^+
g.(III)*, g.(3+)*, aurum(III)*, auric .... Au^{3+}
dicyanoaurate(III)* .... $Au(CN)_2^-$
(disulfido)thioaurate(III)* .... $AuS(S_2)^-$
tetrachloroaurate(III)* .... $AuCl_4^-$
tetrahydroxoaurate(III)* .... $Au(OH)_4^-$

coinage ~ An alloy: Au 90, Cu 10%. Cf. *coinage metals.* **colloidal ~** ^{198}Au is used to treat malignant effusions and measure liver blood flow (USP, EP, BP). **fulminating ~** Aurodiamine. **glucosylthio ~** Aurothioglucose. **hall-marked ~** Standard g. **liquid ~** A mixture of an organic g. compound with an adhesive and essential oil, or with oxides of bismuth, chromium, and rhodium, brushed onto ceramics and burned in to produce a pattern. **mosaic ~** See *mosaic gold.* **rhodium ~** Rhodite. A native alloy: Au 57–66, Rh 34–43%. **rolled ~** Describing a mechanically applied surface layer of not less than 9-karat g., at least 10 μm thick (U.K. usage). **standard ~** Pure g. (24 karats) and 4 alloys of 22, 18, 14, and 9 karats are legal U.K. standards. Four alloys are used in the U.S.: 22, 18, 14, and 10 karats. **white ~** An alloy of g. with 20% Pd; used in jewelry. **yellow ~** The alloy: Au 41.67, Cu 38.5, Ag 5.83, Zn 12.83, Ni 1.17%.

g.(I)*, g.(1+)* See also *aurus.* **g.(III)*, g.(3+)*** See also *auric.* **g. alloys** See *coinage metals, gold plate* (below), *standard gold* (above). The g. content of alloys is indicated in karats, or by the fineness: parts per 1,000. See *carat.* **g. amalgam** A fusible, crumbling amalgam of Au 40%, Ag, and Hg. **g. bromide** See *auric* or *aurous bromide.* **g. chloride** See *auric* or *aurous chloride.* **g. cyanide** See *auric* or *aurous cyanide.* **g. dichloride*** $AuCl_2$ = 267.9. Red crystals, decomp. by water. **g.-filled** Describing a g. surface finish of lower quality than rolled g. **g. foil** Thin leaves of hammered g., used for gilding or dental work. **g. (^{198}Au) injection** A sterile, gelatin-stabilized, colloidal solution of ^{198}Au; used to estimate reticuloendothelial activity (USP, BP). **g. iodide** See *auric iodide* or *aurous iodide.* **g. leaf** G. foil. **g. monobromide*** Aurous bromide. **g. monochloride*** Aurous chloride. **g. monoiodide*** Aurous iodide. **g. number** A measure of the protecting action of a colloid. The weight, in mg which when added to 10 mL of a 0.005–0.006% red g. sol, just prevents the color change to blue (due to coagulation) on addition of 1 ml of 10% sodium chloride solution. **g. pentafluoride*** AuF_5 = 292.0. Dark-red solid. **g. perchloride** Auric chloride. **g. plate** An alloy of gold, silver, and copper; e.g., 18 karats: Au 18, Ag 2, Cu 4 pts.; 20 karats: Au 20, Ag 2, Cu 2 pts. **g.-plated** Describing g. plating equivalent in quality to rolled *g.*, q.v. (U.K. usage). **g.-plating** The electrodeposition of g. from a solution of g. cyanide in potassium cyanide. **g. size** A solution of white and red lead and yellow ocher in linseed oil, used to seal permanent microscopical preparations. **g. sodium thiomalate** $C_4H_3O_4AuNa_2S$ = 390.1. A mixture of mono- and disodium salts of gold thiomalic acid. Yellow powder, soluble in water. Used to treat arthritis (USP, BP). **g. sponge** Spongy, metallic g. obtained by precipitating g. solution with oxalic acid, and drying and heating the precipitate. **g. terchloride** Auric chloride. **g. tribromide*** Auric bromide. **g. trichloride*** Auric chloride. **g. tricyanide*** Auric cyanide. **g. triodide*** Auric iodide. **g. trioxide*** Auric oxide. **g. trisulfide*** Auric sulfide. **g.-washed** Describing a g. surface finish of lower quality than rolled g.

golden **g.rod** Solidago. **g.seal** Hydrastis. **g. yellow** Naphthalene yellow.

Goldschmidt's process Thermit(e) process.

Golgi apparatus Lipochondria. A homologous cytoplasmic structure in most animal cells. Spongy structures which mobilize the fat and protein reserves.

Golo process A process of bleaching flour with nitrosyl chloride.

gomabrea An exudation from a Chilean tree; used as a substitute for gum arabic.

Gomberg, Moses (1866–1947) American chemist, noted as pioneer in the study of free radicals.

gonadorelin $C_{55}H_{75}O_3N_{17}$ = 1,182. A polypeptide hormone produced in the hypothalamus. Used in pituitary and gonadal dysfunction (BP).

gonadotrophin A hormone controlling function of the gonads, secreted by the anterior pituitary gland; in the urine of pregnant women and animals. Used to stimulate the ovaries in infertility. See *chorionic gonadotrophin.*

gonane Androstane*.

gondoic acid $C_{20}H_{38}O_2$ = 310.5. (Z)-11-Icosenoic acid*, m.24; from rape and fish oils.

goniometer An optical device for measuring angles, especially of crystals.

Gooch, Frank Austin (1852–1929) American chemist. **G. crucible** A crucible with a perforated base; used in analysis for filtering through glass or asbestos.

goosefoot Chenopodium.

Gore phenomenon The recalescence of an alloy or steel on cooling, due to transition to another crystalline form.

gorgonin A scleroprotein from the skeletal tissue of coral, *Gorgonia cavollisa* (sea fans), which contain 9% diiodotyrosine.

gorli oil The fixed oil of *Oucoba echinata* (Flacourtiaceae), S. Africa, resembling chaulomoogra oil.

gosio gas Me_3As = 120.0. Trimethylarsine*. A gas of garlic odor generated by certain molds growing in media containing carbohydrates and arsenic compounds. Discovered by Gosio (1891).

goslarite $ZnSO_4$,$7H_2O$. A mineral zinc sulfate.

gossypin The cellulose of cotton.

Gossypium (1) The cotton plant, a genus of Malvaceae. (2) Cotton, or the hairs of the seeds of *G. herbaceum.*

gossypol The toxic principle of cottonseeds. **bound ~** An ether-insoluble product formed from g. in the commercial manufacture of cottonseed meal.

göthite Fe_2O_3,H_2O. Pyrrhosiderite. A crystalline hydrated ferric oxide. Cf. *goethite.*

Göthlin solution An artificial serum: sodium chloride 6.5, sodium carbonate 1, potassium chloride 0.1, calcium chloride 0.13 g/L of water.

Göttling, Johann Friedrich August (1755–1809) German apothecary; the first in Germany to accept Lavoisier's theory.

Goulard's extract A solution of basic lead acetate, $Pb(OH)_2 \cdot Pb(OAc)_2$; a reagent for phenols.

Gouy layer A diffuse layer of positive and negative ions responsible for the stability of colloidal particles.

gr Abbreviation for grain.

Gräbe See *Graebe.*

gracilaria A seaweed from Vancouver, from which agar-agar is made.

grade Gradient. The ratio of the rise of a slope to its length; the sine of the angle of slope.

graded Differentiated. **g. potential** Analysis by electrodeposition in which metals separate at specific voltages.

grading Sorting on the commercial scale according to size, quality, rank, etc., by gravity action (air separators), centrifugal force (cyclone separators), or mechanical action (screening).

graduate A measure for liquids, generally a cone-shaped or cylindrical vessel marked with lines showing the volume.

graduated Divided into units by a series of lines, as a vessel marked for measuring liquids; e.g., thermometers.

Graebe, Carl (1841–1927) German organic chemist who established the constitution of naphthalene. Cosynthesizer (with Liebermann) of alizarin.

graebite A natural, organic mineral coloring matter; a derivative of polyhydroxyanthraquinone.

Graetz rectifier A device for converting alternating to direct current, consisting essentially of 4 electric cells with lead and aluminum plates in a solution of sodium hydrogencarbonate.

grafting An operation of organic-chemical synthesis, in which a group of elements or a radical is attached to a basic molecular chain, e.g., cellulose.

Graham, Thomas (1805–1869) Scottish chemist and pioneer in the study of colloids, who introduced bronze coinage. **G. law** The velocities of diffusion of any 2 gases are inversely proportional to the square roots of their densities. Cf. *diffusion law.* **G. salt** Calgon. Soluble sodium hexametaphosphate. Prepared by strongly heating monosodium dihydrogen orthophosphate and cooling the molten mass rapidly.

grain (1) A unit of the apothecaries, avoirdupois, and troy weights (originally that of an average wheat g.):

$$1 \text{ grain} = 64.79891 \text{ mg} = 0.06479891 \text{ g} = \tfrac{1}{20} \text{ scruple} = \tfrac{1}{60} \text{ dram} = \tfrac{1}{480} \text{ ounce}$$

(2) The seeds of Gramineae, the cereals. (3) The appearance of a heterogeneous surface. **g. alcohol** Ethanol*. **g. germinator** Germinator. **g. oil** Fusel oil. **g. tester** (1) Germinator. (2) A device for sectioning grains.

graininess Lack of homogeneity of deposits due to aggregations of particles.

gram* Abbrev. g; gramme. A unit of weight in the SI and cgs systems; equal to 0.001 kg. Also the weight of 1.000 ml water at 4°C.

$$1 \text{ g} = 1{,}000 \text{ mg} = 1{,}000{,}000\gamma = 1/1{,}000 \text{ kg} = 15.43236 \text{ grains} = 0.03527 \text{ av. oz} = 0.03220 \text{ ap. oz}$$

kilo ~ 1,000 g, 1 kg. **micro ~** The one-millionth part of a g., 1 μg (or 1 γ). **milli ~** The one-thousandth part of a g., 1 mg.

g. atom The atomic weight of an element in grams. **g. calorie** Small *calorie.* **g. equivalent** The equivalent weight of a substance in grams: *n* g of a substance, where *n* is atomic weight/valency. **g. molecular solution** Molar solution. **g. molecular volume** The volume at 0°C and 760 mm pressure occupied by 1 mole of an ideal gas = 22.41383 liters. **g. molecular weight** G. molecule. **g. molecule** 1 mole: the molecular weight of a substance in grams.

-gram Suffix indicating a mechanical record; as, spectrogram. Cf. *-graph.*

Gram, Hans, C. J. (1853–1938) Danish bacteriologist. **g.-negative** Describing bacteria that are decolorized by G. stain. **g.-positive** Describing bacteria that retain the G. stain. **G.'s iodine solution** A solution: iodine 1, potassium iodide 2 pts. in 200 pts. water; a microscopical stain. **G.'s stains** Solutions used in bacteriology, as: aniline 15, saturated alcoholic solution of methyl violet 7, absolute alcohol 10 mL in 100 mL water.

gramicidin A compound isolated from cultures of certain bacteria in phosphate-enriched soils; toxic to all gram-positive organisms; m.229. **g. S** Soviet g. An antibiotic from strains of *Bacillus brevis* from U.S.S.R. soils. Colorless needles, m.267.

Graminaceae, Gramineae The grass family, a group of plants that yield cereals, sugar, starch, and essential oils. E.g.:

Saccharum officinarum	Cane sugar
Agropyron repens	Triticum
Zea mays	Indian corn
Avena sativa	Oats
Triticum vulgare	Wheat
Oryza sativa	Rice
Hordeum distichum	Barley

gramine $C_{11}H_{14}O_2$ = 178.2. 3-(Dimethylaminomethyl)indole. m.138. An alkaloid from barley.

gramme Gram.

granatine An alkaloid from pomegranate.

granatonine Pseudo*pelletierine.*

granatotannic acid $C_{20}H_{16}O_{13}$ = 464.3. An amorphous substance from the root bark of *Punica granatum,* pomegranate.

granatum Pomegranate.

grandiflorine An alkaloid from the fruit of *Solanum grandiflorum.*

granidiorite A form of granite.

granite A crystalline, igneous rock, of quartz, orthoclase, with both muscovite and biotite, cooled slowly under great pressure.

granular Grainlike.

granulated Made of small particles.

granulation The process of converting a substance into granules, e.g., by rapidly quenching drops of a molten metal (as with granulated zinc). Cf. *slugging, spheronizing.*

granules (1) Small grains having in bulk the properties of semifluids. Their flow through an orifice is a function of the orifice area and is practically independent of the head. (2) Medicinal substances in small pellets.

granulose (1) β-Amylose. A sugar of starch plants, enclosed by an envelope of starch cellulose. It gives a blue color with iodine solution. (2) The product obtained when cotton wool is charred.

grape (1) Vine, *Vitis vinifera.* (2) The edible fruit of *Vitis* species. **mountain ~** The root of *Berberis aquifolium* (Berberidaceae). **Oregon ~** Mountain g.

g.fruit The edible fruit of *Citrus maxima,* rich in vitamin C. **g. pomace** A fertilizer. The dried cake remaining after pressing juice from grapes; contains about 1.2% N, small amounts of P and K. **g.-seed oil** An oil expressed from g. seeds; used as lubricant for watches and for coating raisins; d.0.923, sapon. no. 182. **g. sugar** Glucose.

graph (1) A record obtained by physical means. (2) A line drawing relating 2 or more variables.

-graph Suffix indicating (1) a pictorial record, e.g., photograph; (2) an instrument to make mechanical records; as, spectograph.

graphic Pertaining to diagrams. **g. formula** A spatial structural formula, or a geometrical drawing, indicating the isomeric forms of certain carbon compounds.

graphite Black lead, plumbago. A native or artificially made allotropic carbon. Shining, amorphous masses or hexagonal lamellae, d.1.9-2.3, hardness 0.5–1.0. Used as pigment; for crucibles, retorts, electrodes and pencils; and as a lubricant. **colloidal ~** Deflocculated suspensions of g. in oil (Oildag) or in water (Aquadag), used as lubricants. **pyrolytic ~** A light polycrystalline form of carbon produced at about 2000°C by gas deposition. Its thermal conductivity is high and low, parallel to and across the plane of deposition, respectively; a coating for missiles and electronic apparatus; compatible with living tissue. **synthetic ~** Artificial carbon, electrographite. A broad spectrum of industrially and medically useful substances, as carbon fibers and pyrolytic g., produced by treating carbon-containing compounds above 1000°C. These substances have good mechanical, thermal, and electrical properties that can be tailored to suit the end use; as, in aerospace, nuclear reactors, and motor brushes. **white ~** A form of boron nitride, used as a refractory and dry lubricant; d.2.2.

graphitic acid $C_{11}H_4O_5$ = 216.2. Yellow powder produced from native graphite by the action of potassium chlorate and nitric acid. **pyro ~** See *pyrographitic oxide*.

graphology The study of handwriting as a guide to character. It is widely practiced, but there is no scientific proof that its interpretations are correct.

graphon sulfate The black substance produced by the slow action on graphite of potassium chlorate and concentrated sulfuric acid.

grappa A brandy prepared by fermenting pressed pomace, adding grape residues, and distilling.

GRAS Acronym for generally recognized as safe. Used to describe food additives (U.S. usage).

grasses See *Graminaceae*.

grating A latticework or screen composed of lines, e.g., a glass, metal, or film with minute fine rulings (often 20,000 per cm); produces a series of spectra by the dispersion of a ray of light. Cf. *diffraction grating*. **concave ~** A slightly concave piece of speculum metal on which lines are ruled; it focuses the light. **plane ~** A g. which requires parallel rays of light; e.g., a slit, and collimator.

g. spectroscope A spectroscope in which the spectrum is produced by a diffraction grating, and not by a refracting prism.

gratiolin $C_{20}H_{34}O_7$ = 386.5. A glucoside from *Gratiola officinalis*, hedge hyssop (Scrophulariaceae); yellow needles.

gratiosolin $C_{46}H_{84}O_{25}$ = 1037. A glucoside from *Gratiola officinalis*, hydrolyzed by water to gratiosoletin.

graukalk Technical calcium acetate.

gravimetric Describing measurement by weight. **g. analysis** Quantitative analysis by weighing precipitates.

gravimetry Measurement by weight.

gravitation (1) The universal attraction between material bodies. Its intensity varies directly with the product of the 2 masses. See *gravitational constant*. (2) The tendency of objects to move toward the center of the earth. See *acceleration of free fall*.

gravitational Pertaining to gravitation. **g. constant*** The force F of g. attraction between 2 masses, m and m_1, separated by the distance s: Gmm_1/s^2; where G is the g. constant = 6.6720×10^{-11} Nm^2/kg^2. **g. effect** Westling effect. The loss of weight of body A weighed underneath body B on a beam balance. The effect is specific for those pairs of elements whose atomic numbers are related by $B^2/A^2 = n$, where n is an integer.

graviton Hypothetical gravitational analog of the photon, having zero mass and a spin of 2.

gravity (1) The attractive force of the earth. See *acceleration of free fall*. (2) Specific g. See *relative density* under *density*. **API ~** American Petroleum Institute g. (in degrees) = (141.5/specific gravity at 60°F) − 131.5; e.g., Arabian light oil = 34°, Arabian heavy oil = 27°. **distillation ~** The specific g. of 200 ml of distillate from 200 ml of alcoholic liquor. **original ~** O.G. The specific g. of a wort before fermentation as determined from the amount of alcohol in the fermented liquor. It is given by the residual g. plus the g. lost according to the spirit indication. **present ~** The actual specific g. of a fermented liquor. **residual ~** The specific g. of the liquid remaining after all the alcohol has been removed by distillation from 200 mL fermented liquor, and the residue is made up to 200 mL. **specific ~** See *relative density* under *density*.

gravure A process of printing from an inked metal surface which has been etched with acid in such a way that the darker the shade, the deeper the etch and the more ink it holds available for transfer to the paper.

grax Whale tissue remaining after extraction of oils, etc.; a fertilizer.

gray (1) Grey (U.K. usage). Ash color, a mixture of white and black pigments. (2)* Gy. SI unit of absorbed radiation dose equal to 1 J/kg. 1 Gy = 100 rad.

grease (1) A soft fat. (2) A dark, low-grade waste product containing lard, tallow, bone, horse or fish fat, stearins, etc. G. usually has an unpleasant odor, high unsaponifiable matter, and free fatty acids; is used as a lubricant. (3) Oil thickened with soap. Cf. *wax*. **black ~** Dark, fatty matter obtained from cottonseed oil; used in candle manufacture. **cup ~** An emulsion: mineral oil 80%, lime soaps and water 1%; a lubricant. **Yorkshire ~** Lanolin.

Greek alphabet See Table 40, next page. See individual names for their application in chemistry, etc.

green (1) Grass color: a hue obtained by mixing yellow and blue pigments. (2) Unused, raw, untreated, or incompletely treated. **brilliant ~** A derivative of malachite green; a bacterial stain.

g. oil The anthracene oil fraction of coal tar. A source of vinylcarbazole, m.64; used in the manufacture of polymers. **g. soap** Soft soap. Soap made by saponification of vegetable oils with potassium hydroxide, and then colored green. Used to clean the skin and as an enema (USP, BP). **g. vitriol** Ferrous sulfate.

greenhouse effect Warming of the earth, caused by an increase in CO_2 from the burning of fossil fuels, creating a screen that prevents the escape of radiation. Estimated at about 3°C increase by the year 2040.

greenockite CdS. A rare mineral sulfide.

greensalt A wood-preserving solution of potassium dichromate, copper sulfate, and arsenic acid.

greensand (1) A sandy deposit containing glauconite. (2) Natural sand dampened for molding.

greenstone A variety of jade.

Gregory mixture G. powder. A mixture of rhubarb rhizome, ginger, and magnesium carbonate; formerly used as a stomachic and general panacea.

TABLE 40. The Greek Alphabet

Α	α (ɑ)	alpha *(al'-fah)*	a
Β	β	beta *(bā'-tah)*	b
Γ	γ	gamma *(gam'-ah)*	g (hard)
Δ	δ (∂)	delta *(del'-tah)*	d
Ε	ϵ (ε)	epsilon *(ep'-si-lon)*	e (short)
Ζ	ζ	zeta *(zā'-tah)*	z
Η	η	eta *(āt'-ah)*	e (long)
Θ	θ (ϑ)	theta *(thā'-tah)*	th
Ι	ι	*(ī-ō'-tah)*	i
Κ	κ (ϰ)	kappa *(kap'-ah)*	k
Λ	λ	lambda *(lam'-dah)*	l
Μ	μ	mu *(mū)*	m
Ν	ν	nu *(nū)*	n
Ξ	ξ	xi *(zī)*	x
Ο	ο	omicron *(ō'-mi-kron)*	o (short)
Π	π	pi *(pī)*	p
Ρ	ρ	rho *(rō)*	r
Σ	σ (ς)	sigma *(sig'-mah)*	s
Τ	τ	tau	t
Υ	υ	upsilon *(up'-si-lon)*	u
Φ	ϕ (φ)	phi *(fī)*	ph (f)
Χ	χ	chi *(chī)*	ch (as in loch)
Ψ	ψ	psi *(psī; sī)*	ps
Ω	ω	omega *(o-me'-gah)*	o (long)

greisen A granite in which feldspar is replaced by quartz.

Grenacher stain Carmines of alum, borax, and hydrochloric acid, used to stain nucleic and muscle tissues.

Grenet battery An electrolytic carbon-zinc cell.

grenz rays Infraroentgen rays, Bucky rays, long-wave X-rays. Very soft X-rays produced at low voltages and absorbable by glass; used to treat skin diseases. Cf. *radiation.*

grey See *gray.*

GRI Abbreviation for government rubber-isobutene; a synthetic rubber. Cf. *GRS.*

Griess, Peter (1829–1888) German-born British chemist. **G. reaction** The substitution of amino radicals by hydroxy, halogen, or nitrile radicals by diazotization and treatment with water. **G.-Ilosva reagent** A solution of sulfanilic acid and 1-naphthylamine in acetic acid; a reagent for nitrites.

Griffith **G. white** Lithopone. **G. cracks** Surface flaws in glass.

Grignard, Victor (1871–1935) French chemist. Nobel prize winner (1912). **G.'s reaction** Magnesium alkyl condensation: a reaction by which a C atom is introduced into the hydrocarbon radical or a compound by G.'s reagent to pass from a lower to a higher member of a homologous series. **G.'s reagent** Compounds of the general type R·Mg·X, where R is an organic radical, and X a halogen.

Typical Grignard reactions:

1. Formation of a *hydrocarbon.*

$$\mathrm{RMgI} \xrightarrow[\text{dilute acids}]{\text{hydrolysis in}} \mathrm{Mg}\!\begin{smallmatrix}\diagup \mathrm{I}\\ \diagdown \mathrm{OH}\end{smallmatrix} + \mathrm{RH}$$

2. Preparation of an *acid.*

$$\mathrm{RMgI} \xrightarrow{CO_2} \mathrm{R\cdot COOMgI} \xrightarrow{\text{hydrolysis}} \mathrm{Mg}\!\begin{smallmatrix}\diagup \mathrm{I}\\ \diagdown \mathrm{OH}\end{smallmatrix} + \mathrm{R\cdot COOH}$$

3. Preparation of a *ketone.*

$$\mathrm{RMgI} \xrightarrow{R'CN} \begin{smallmatrix}\mathrm{R}\diagdown\\ \mathrm{R'}\diagup\end{smallmatrix}\mathrm{C{:}NMgI} \xrightarrow{\text{hydrolysis}} \mathrm{Mg}\!\begin{smallmatrix}\diagup \mathrm{I}\\ \diagdown \mathrm{OH}\end{smallmatrix} + \mathrm{NH_3} + \begin{smallmatrix}\mathrm{R}\diagdown\\ \mathrm{R'}\diagup\end{smallmatrix}\mathrm{CO}$$

4. Preparation of (*a*) a *secondary alcohol,* (*b*) a *ketone,* or (*c*) a *tertiary alcohol.* Polaises reaction:

$$\mathrm{RMgI} \xrightarrow{R'CHO} \begin{smallmatrix}\mathrm{R}\diagdown\\ \mathrm{R'}\diagup\end{smallmatrix}\mathrm{C}\begin{smallmatrix}\diagup \mathrm{OMgI}\\ \diagdown \mathrm{H}\end{smallmatrix} \xrightarrow{\text{hydrolysis}} \begin{smallmatrix}\mathrm{R}\diagdown\\ \mathrm{R'}\diagup\end{smallmatrix}\mathrm{C}\begin{smallmatrix}\diagup \mathrm{OH}\\ \diagdown \mathrm{H}\end{smallmatrix} \quad (a)$$

$$(a) \xrightarrow{\text{oxidize}} \begin{smallmatrix}\mathrm{R}\diagdown\\ \mathrm{R'}\diagup\end{smallmatrix}\mathrm{CO} \quad (b)$$

$$(b) \xrightarrow{R''MgI} \begin{smallmatrix}\mathrm{R}\diagdown\\ \mathrm{R'}\diagup\end{smallmatrix}\mathrm{C}\begin{smallmatrix}\diagup \mathrm{R''}\\ \diagdown \mathrm{OMgI}\end{smallmatrix} \xrightarrow{\text{hydrolysis}} \begin{smallmatrix}\mathrm{R}\diagdown\\ \mathrm{R'}\diagup\end{smallmatrix}\mathrm{C}\begin{smallmatrix}\diagup \mathrm{R''}\\ \diagdown \mathrm{OH}\end{smallmatrix} \quad (c)$$

Grilon Trademark for a polyamide synthetic fiber.

grinding The process of powdering a substance by lateral motion, as opposed to perpendicular motion (crushing).

griphite $8[(NaAlCaFe)_3Mn(PO_4)_2\cdot 5(OH)_2]$. A native, garnet-type hydroxylphosphate.

griseofulvin $C_{17}H_{17}O_6Cl = 352.8$. Fulcin, Fulvicin, Grisovin. Cream powder produced by *Penicillium griseofulvum,* slightly soluble in water; an antifungal for skin and nails (USP, BP).

Grisovin Trademark for griseofulvin.

grit See *particulate.*

grog Broken bricks, or burnt, ground fireclay; a refractory.

Grossmann reagent $(C_2H_6ON_4)_2H_2SO_4$. An ammoniacal solution of dicyanodiamidine sulfate; yellow precipitate with nickel.

grossularite A green calcium aluminum garnet.

Grotthus' law Radiation must be absorbed to produce a reaction.

ground (1) Powdered. (2) Earth. A conducting path between an electric circuit or equipment and the earth. (Also, to create such a path.) **g.nut** Arachis. **g.nut oil** Arachis oil. **g. state** The normal or unexcited state of an atom. **g.wood** A form of mechanical wood pulp.

group (1) A number of elements having similar properties, e.g., the alkali metals. See *periodic table* under *periodic.* (2) A number of atoms that pass through a series of reactions unseparated. See *radical.* (3) A number of elements with similar reactions. See *qualitative analysis* under *analysis.* **characteristic ~*** An atom or g. that is incorporated into a parent compound other than by a direct carbon-carbon linkage, but including groups —CN and =C:X, where X is O, S, Se, Te, NH, or substituted NH. It includes g. such as —OH, $-NH_2$, —COOH, single atoms (as, halogen), =O, N, and substituents such as piperidino and acetyl. It does not apply to substituents such as Me, Ph, 2-pyridyl. **negative ~** A negatively ionized group, as SO_4^{2-}. See substitutive *nomenclature* in Table 56 under *nomenclature.* **positive ~** A positively ionized g. of a metal or radical, as NH_4^+. **principal ~*** The characteristic group chosen for expression as suffix in a particular name.

g. precipitant A reagent that precipitates elements of the same group, e.g., hydrogen sulfide. **g. properties** The properties of elements belonging to the same g., e.g., of a vertical division of the periodic table. **g. reaction** The precipitation of elements in a definite analytical g. Cf. *precipitant.*

Grove, Sir William Robert (1811–1896) British scientist. **G.'s cell** A voltaic cell (1.91 volts) of amalgamated zinc in sulfuric acid (d.1.136) and platinum in concentrated nitric acid.

growth An increase in size. **bacterial ~** The appearance of a bacterial colony after incubation. **inorganic ~** The aggregation of solid particles, by crystallization, periodic precipitation, or colloidal growth. **g. hormone** A hormone secreted by the anterior pituitary gland, that influences growth in children and protein metabolism in adults.

GRP See *glass fiber–reinforced plastics.*

GRS Abbreviation for government rubber-styrene (U.S. usage). See *SBR.*

grumose Clotted.

grundy Granulated pig iron.

grunerite $Fe_7H_2(SiO_3)_8$, from Massachusetts.

gryolite $CaO \cdot 3SiO_2 \cdot 2H_2O$. A constituent of certain boiler scales. Cf. *cryolite.*

G salt The sodium or potassium salt of G acid.

guachamacine Gauchamacine. An alkaloid from guachamaca, the bark of *Malouetia nitida* (Apocynaceae), Venezuela; an arrow poison.

guaethol Thanatol.

guaiac A resin from *Guaiacum officinale* (Zygophyllaceae), S. America. **g. resinic acid** $C_{20}H_{26}O_4$ = 330.3. An acid of guaiac; yellow crystals. **g. wood** Lignum vitae, guaiaci lignum. The heartwood of *Guaiacum* species; a dye. **g. wood oil** The essential oil of g. d.0.965–0.975, soluble in alcohol. It contains 2-methyl-2-butenal*. **g. yellow** The coloring matter of guaiac wood; yellow crystals.

guaiacene C_5H_8O = 84.1. 2-Methyl-2-butenal*. An oil from the distillation of guaiac wood.

guaiaci lignum Guaiac wood.

guaiacin $C_{14}H_{24}O$ = 208.3. An alcohol, the odorous principle of balsam wood. Colorless crystals, m.91, soluble in alcohol; used in perfumery.

guaiacol* $C_6H_4(OH)OMe$ = 124.1. 2-Methoxyphenol†, methylpyrocatechin; in wood tar. Colorless prisms, m.32, soluble in water. A reagent to detect lignin, narceine, chelidonine, nitrous acid, and acacia. **allyl ~** **4-~** Eugenol*. **5-~** Chavibetol. **methyl ~** Creosol. **propenyl ~** Iso*eugenol.* **vinyl ~** Hesperetol.

g. acetate $C_9H_{10}O_3$ = 166.2. Eucol. A liquid, b.238. **g. carbonate** $(C_7H_7O)_2CO_3$ = 274.3. White crystals, m.87, soluble in ether; a substitute for g. oleate. **g. oleate** Oleoguaiacol. A mixture of g. and oleic acid in ether.

guaiaconic acid $C_{10}H_{24}O_5$ = 244.3. A resinous acid, from guaiac. Brown powder; used in the guaiac test for blood.

guaiacum Guaiac.

guaiacyl The 2-methoxyphenyl* radical.

guaiene $C_{15}H_{24}$ = 204.4. Oils. *α*-~ b.78. *β*-~ b.138; from guaiac.

guaiol Champacol.

guaj- Variant of *guai-.*

guajene Guaiene.

guanamine CH:N·C(NH_2):N·C(NH_2):N = 111.1. 2,3-Diamino-1,3,5-triazine. m.325. **acet ~** $C_4H_7N_5$ = 125.1. Colorless crystals, m.265.

guanase Guanine deaminase*.

guanazyl The radical —N:NC(:N·NH·CNH·NH_2)—. **g. benzene** $Ph_2C_2H_4N_6$ = 266.3. Colorless crystals, m.199.

guanethidine $C_{10}H_{22}N_4 \cdot H_2SO_4$ = 296.4. Ismelin. White crystals, m.254, soluble in water; a sympatholytic. **g. sulfate** $(C_{10}H_{22}N_4)_2 \cdot H_2SO_4$ = 494.7. White crystals, soluble in water; used as an antihypertensive and for eye conditions in thyrotoxicosis and glaucoma (USP, EP, BP).

guanidine* $(NH_2)_2C:NH$ = 59.1. Carbamidine, carbondiamide imide, uramine. Colorless crystals, soluble in water; an isolog of urea. **aminobutyl ~** Agmatine. **benzoylene ~** See *benzoylene guanidine.* **bi ~*** $(H_2N \cdot C(:NH))_2NH$ = 116.1. m.130. **carbamoyl ~** $HN:C(NH_2)NHCONH_2$ = 102.1. Prisms, m.110. Amidinourea, dicyandiamidine, param. Colorless crystals, m.205. **diphenyl ~** See *diphenylguanidine* under *diphenyl.* **guanyl ~** Bi*guanidine**. **isopentyl ~** Galegine. **lactoyl ~** Alacreatine. **nitro ~** $CH_4O_2N_4$ = 104.1. Colorless crystals, m.240.

g. phosphoric acid Phosphagen.

guanidines* Compounds derived from guanidine, e.g., containing the radical =N·C(:NH)·N=. See *creatine.*

guanidinium* Describing a salt from guanidine with the g. cation, $C(NH_2)_3^+$, or its derivatives.

guanidino* (Aminoiminomethyl)amino†, guanidino. The radical NH_2·C(:NH)·NH—, from guanidine. **g.phosphoric acid** Phosphagen. **g.propanoic acid** Alacreatine.

guanido The guanidino* radical.

guanine $C_5H_5ON_5$ = 151.1. Imidoxanthine, 2-amino-6-oxopurine, 2-aminohypoxanthine. In guano, fish scales, human liver, and spleen. Colorless needles, decomp. above 360, insoluble in water. **g. deaminase*** Guanase. An enzyme converting g. into xanthine; in adrenals and pancreas.

guanite Struvite.

guano Bird manure. The partly decomposed excrements of sea birds from the islands off the western coast of S. America, especially Peru; an excellent fertilizer and a source of guanine.

guanoline $C_4H_9O_2N_3$ = 131.1. Guanidinocarbonic ethyl ester. Colorless crystals, m.114.

guanosine $C_{10}H_{13}N_5O_5$ = 283.2. Vernine. A nucleoside in the pancreas. **g. phosphoric acid** Guanylic acid.

guanyl The amidino* radical.

guanylic acid $H_2PO_4 \cdot C_5H_7O_4 \cdot C_5H_4ON_5$ = 378.2. A nucleic acid containing guanosine; in pancreas, ox liver, and yeast.

guar *Cyamopsis tetragonoloba.* An Indian plant resembling soya, grown in the U.S. The seeds are a source of g. gum, a mannogalactan mucilage. **g. gum** An emulsifier, stabilizer, and thickening agent in pharmacy (NF).

guarana tannin Paullinia tannin.

guaranine Caffeine.

guard tube A tube which usually contains calcium chloride, to prevent access of atmospheric moisture to gas absorption bulbs during weighing. Cf. *witness.*

guavacine Guvacine.
guayule The desert shrub *Parthenium argentatum* (Compositae), Mexico; cultivated in central California. **g. rubber** Rubber formerly made from g.
guazatine* See *fungicides,* Table 37 on p. 250.
Guericke, Otto von (1602–1688) German philosopher noted for the Magdeburg hemispheres.
guhr Trade abbreviation for kieselguhr.
Guignet's green $3CrO_3 \cdot B_2O_3 \cdot 4H_2O$(?). A green pigment resulting from the fusion of potassium dichromate and crystalline boric acid.
guinea green $C_{37}H_{35}N_2O_6S_2Na$ = 690.8. A dye used as food color and indicator, changing at pH 6.0 from magenta (acid) to green (alkaline).
Guldberg, Cato (1836–1902) Norwegian chemist. **G.-Guye rule** The critical temperature of a substance is 1.4–1.9 times its boiling point (in K). **G. rule** The boiling point of a liquid is two-thirds the critical temperature of its gas (in K). **G. and Waage law** See *mass action.*
gulose $C_6H_{12}O_6$ = 180.2. A *monosaccharide,* q.v., isomeric with glucose.
gum A mucilaginous plant stem excretion; complex carbohydrates yielding sugars on hydrolysis. Gums dissolve or swell in water, and are insoluble in alcohol. See *polysaccharides.* Classification: (*a*) Arabin type: completely soluble in water, e.g., *Acacia senegal* (gum arabic). (*b*) Bassorin type: slightly soluble in water, e.g., *Astragalus gummifer* (tragacanth). (*c*) Cerasin type: swelling in water, e.g., *Prunus cerasus* (cherry gum). Individual gums are dealt with under their respective names. **g. arabic** Acacia. **g. benjamin** See *benzoin.* **g. copal** Copal **g. dragon** Tragacanth. **g. elastic** Caouthchouc. **g. resin** See *gum resins.* **g. running** The process of melting gums in varnish manufacture. **g. sugar** Arabinose. **g. thus** (1) Olibanum. (2) In naval stores, the crystalline pine oleoresin collected from the scarified faces of trees being worked for turpentine. **g. tragacanth** Tragacanth. **g. tree** See *gum tree.*
gumbotil A gray, leached, deoxidized clay containing tourmaline and epidote, from glacial formations in Kansas.
gummeline Dextrin.
Gummon Trademark for an insulating material of tar and asbestos.
gummy Sticky; resembling gum.
gum resins Oleoresina. Aromatic exudations of plants; a mixture of various substances (as essential oils) with gum. Formerly used in pharmacy: ammoniac, asafetida, myrrh, and scammony. Cf. *resin(s).*
gum tree Red gum, sweet gum. A large lumber tree, *Liquidambar styraciflua* (Mississippi swamps). **blue ~** Eucalyptus. **cotton ~** A timber tree, *Nyssa sylvatica* (Asia), which yields edible fruit.
guncotton $C_{12}H_{14}O_4(NO_3)_6$ = 594.3. Cellulose hexanitrate. A highly nitrated, sparingly soluble, explosive cellulose. **soluble ~** Pyroxylin.
Gunite (U.K. usage) Trademark for a spray-applied concrete. Cf. *Shotcrete.*
gunmetal Bronze. The alloy: Cu 86–90, Sn, Zn 10, Pb, Ni, Sb, Fe, Al, etc., approximately 2%.
gunny A jute bagging cloth.
gunpowder A granulated, explosive mixture of charcoal about 2, sulfur 3, potassium nitrate 15 pts. Its properties depend largely on the size and shape of the grains, their density and hardness, glazing, and moisture content.
Gunter's chain A measure of length, 20.1168 m.
Gunzberg **G. reagent** Phloroglucinol (1:15) and vanillin (1:15), mixed shortly before use, to detect free inorganic acids. **G. test** A drop of G. r. is evaporated; on adding the unknown and again evaporating, a pink color indicates hydrochloric acid.
guoethol Thanatol.
gur Jaggery. A crude Indian sugar obtained by evaporating unclarified cane juice in open pans.
gurjun A balsam varnish, from *Dipterocarpus* species (India). **g. oil** The essential oil of g., d.0.915–0.925, b.255, containing sesquiterpenes.
gut Intestines of sheep, cleansed, treated with alkali, and twisted to a cord; corrosion-resistant. Cf. *catgut.*
Guthrie test To detect excess phenylalanine in blood of newborn infants. Phenylalanine prevents inhibition of the growth of *Bacillus subtilis* by β-thionylalanine. See *phenylketonuria.*
gutta (Pl., guttae) Latin for "drop."
guttameter A device to measure surface tension by the number of drops formed.
gutta-percha $(C_{10}H_{16})_2$. Gummi plasticum. The purified, coagulated, milky exudate of *Palaquium* species (Sapotaceae). Yellow masses, sticks, or sheets, with red streaks, insoluble in water, partly soluble in turpentine oil. It contains fluavil, albane, and a volatile oil; softens at 65. Used for insulating, in dentistry, and as a rubber substitute (USP).
Gutzeit, Heinrich Wilhelm (1845–1888) German chemist. **G. arsenic test** Zinc and dilute sulfuric acid added to the substance in a test tube and covered with a filter paper moistened with mercuric chloride solution form a yellow spot on the paper due to $(HgCl)_2AsH$. This turns black forming $(HgCl)_3As$, then Hg_3As_2.
guvacine $C_6H_9O_2N$ = 127.1. 1,2,5,6-Tetrahydro-3-pyridinecarboxylic acid*. An alkaloid, m.271, from the betel nut, the fruit of *Areca catechu,* a southeast Asian palm; an anthelmintic.
Guyton de Morveau, Louis Bernard (1737–1816) French lawyer who introduced the first chemical nomenclature.
g.w.a. Grams of water in air. See *Mohr liter.*
gymnemic acid Four constituents of *Gymnema sylvestre* (Asclepiadaceae), Australia, India, Africa. **g. a. A_1** $C_{49}H_{74}O_{16}$ = 919.1. m.285.
gymnosperm A large group of plants in which the seeds are not enclosed in an ovary, e.g., the conifers. Cf. *angiosperm.*
Gynergen Trademark for ergotamine tartrate.
gynocardic acid (1) $C_{18}H_{34}O_2$ = 282.5. From the oils of *Gynocardia odorata* and *G. Prainii,* m.67, insoluble in water. (2) A mixture, from gynocardia and chaulmoogra oils, of hydnocarpic, taraktogenic, and gadoleic acids.
gynocardine $C_{12}H_{17}O_8N$ = 303.3. A glucoside, m.165, from the seeds of *Gynocardia odorata.* It yields on hydrolysis glucose, hydrogen cyanide, and ethylfumaric acid. Cf. *chaulmoogra oil.*
gypsum Selenite. A native hydrated *calcium* sulfate, q.v. Cf. *anhydrite, plaster of paris, phosphogypsum.*
gyration Revolution in a circle.
gyrolite $H_2Ca(SiO_3)_3,H_2O$. A mineral from Radzein, Czechoslovakia.

H

H Symbol for (1) henry; (2) hydrogen: H^+, hydrogen ion (proton); H*, excited hydrogen atom; ^2H, deuterium; ^3H, tritium. **H acid** 1-Amino-8-naphthol-3,6-disulfonic acid. **H ion** See *hydrogen ion.* **H lines** (1) Hydrogen spectrum lines in Å (Fraunhofer nomenclature in brackets):

H_α [C] .... 6573λ (red)
H_β [F] .... 4861λ (blue)
H_γ [G] .... 4340λ (blue)
H_δ [h] .... 4101λ (violet)

(2) Fraunhofer line H, 3968λ, due to calcium.

H Symbol for (1) Boltzmann function; (2) enthalpy; (3) Hamiltonian function; (4) indicated *hydrogen;* (5) light exposure.

H (Boldface.) Symbol for magnetic field strength.

h Symbol for (1) hecto- (10^2); (2) hour.

h Symbol for (1) height; (2) Planck constant.

ħ Symbol for Planck constant divided by 2π.

ha Abbreviation for hectare.

Haber, Fritz (1868–1934) German chemist. Nobel prize winner (1916). **H. process** The synthesis of ammonia from nitrogen and hydrogen in presence of a catalyst. Cf. *nitrogen fixation.*

habitat The surroundings in which a living organism is commonly found. **abyssal ~** A h. in deep sea. **alpine ~** A h. in high mountains. **fossorial ~** A h. in burrows and caves. **littoral ~** A h. near a shore. **pelagic ~** A h. in open sea.

hadal Describing ocean depths exceeding 6,000 m.

Hadfield, Sir Robert (1859–1940) British chemist, noted for development of stainless steels. **H. process** A metal oxide is reduced by heating with granulated aluminum and fluorspar.

hardromal Feruladehyde.

hadron See *subatomic particle* under *particle.*

hæ- See words beginning with *hae-, he-.*

haem Heme.

haema-, haemo- U.K. usage for *hema-, hemo-,* q.v.

haemanthine $C_{18}H_{23}O_7N$ = 365.4. An alkaloid from buphane.

haematoxylin Hematoxylin.

Haematoxylon, hematoxylon (1) A genus of leguminous trees of Central America, specifically *H. campechianum.* (2) The heartwood of *H. campechianum,* logwood, campeachy wood, which contains hematoxylon.

haeterolite The mineral $ZnO \cdot Mn_2O_3$.

hafnium* Hf = 178.49. Celtium, Oceanum. Named for Hafnia (Latin for "Copenhagen"). An element of the carbon group, at. no. 72, discovered (1924) by Coster and Hevesy in zircon and baddeleyite. Valency 4, d.13.3, m.2200. **h. carbide*** HfC = 190.5. Gray powder, m.3887. A mixture of h. carbide 25 and tantalum carbide 75% has a high melting point, 4200°C. **h. dichloride oxide*** $HfOCl_2$ = 265.4. White powder, insoluble in water, soluble in hydrochloric acid. **h. hydroxide*** $Hf(OH)_4$ = 246.5. White powder, insoluble in water. **h. oxide*** HFO_2 = 210.5. White powder, m.3025, insoluble in water, soluble in acids. **h. sulfate*** $Hf(SO_4)_2$ = 370.6. White, crystalline powder, soluble in water.

Hägglund, Erik (1888–1959) Swedish authority on the chemistry and technology of wood products.

Hahn, Otto (1879–1968) German chemist. Nobel prize winner (1947). Noted for work on radioactive elements.

hahnium Ha. Un-nil-pentium*. Unofficial name for element at. no. 105. Its isotopes are unstable with short half-lives.

hairari root The roots of *Lonchocarpus* species containing rotenone. Cf. *cube* (2).

halazone $C_6H_4(SO_2NCl_2) \cdot COOH$ = 270.1. *p*-Sulfondichloraminobenzoic acid. White powder with strong chlorine odor, m.213, soluble in water; used to sterilize water for drinking (USP).

Haldol Trademark for haloperidol.

half-life, half-period $t_{1/2}$. The period in which the activity of a radioactive substance falls to half its value. It ranges from 10^{15} y (vanadium) to 10^{-16} s (beryllium).

halide* Halogenide*. A compound of the halogens, as MX, MX_2, or MX_3, in which M is a metal, X a halogen (F, Cl, Br, I). **acid ~** A compound of the type R·COX. **alkylaryl ~** A compound of the type RX, where R is an alkyl or aryl radical. **magnesium ~** A compound of the type RMgX, e.g., Grignard's reagent.

h. lamp An alcohol torch with copper tube, which ordinarily burns without color but gives a green flame in presence of organic halides; used to detect leaks of halide refrigerants.

halite A native sodium chloride.

Hall **H., Sir Arthur Daniel (1864–1942)** British agricultural chemist. **H., Carl von (1819–1880)** Austrian chemist, noted for work on vanadium compounds. **H. Charles M. (1863–1914)** American chemist, inventor of the H. process. **H., Edwin H. (1855–1938)** American physicist, discovered H. effect (1879).

H. effect The production of a voltage across a current-carrying conductor (as indium antimonide) located at right angles to an electric field. **H. formula** If E is the difference of potential between the lower and upper edge of a metal plate, then $E = R_H jH$, where R_H is a constant specific to individual metals (H coefficient), H the magnetic field strength, and j the electric current density. **H. process** The electrolytic process by which metallic aluminum is recovered from aluminum oxide. **H. purinometer** A graduated glass tube of special construction, for the determination of purines in urine.

halloylite Halloysite.

halloysite $Al_2O_3 \cdot 2SiO_2 \cdot 4H_2O$. Halloylite. A refringent micaceous silicate constituent of clay. **meta ~** A clay mineral similar to h., but having the same formula as kaolinite.

Hallwach's effect Photoelectric effect.

halo (1) A series of luminous, concentric circles around a source of illumination, caused by the refraction of light on passing through solid or liquid particles suspended in the atmosphere. (2) A circular photographic image produced when X-rays pass through an amorphous substance. Cf. *X-ray analysis.*

halo- Prefix indicating presence of a halogen.

halochromism (1) The formation of colored salts from

colorless organic bases by addition of acids. Cf. *solvatochromism.* (2) The production of colorless solutions in some solvents and colored solutions in others.

haloform A compound of the type CHX_3, in which X is a halogen; as, chloroform. **h. reaction** A reaction analogous to that by which iodoform is made from alcohol or acetone.

halogen* The nonmetallic elements of the seventh group of the periodic table: F, Cl, Br, I, and At. Halogens are multivalent and have oxidation numbers of −1 (chlorides), 1 (hypochlorites), 3 (chlorites), 5 (chlorates), and 7 (perchlorates). **h. acids** The hydrogen compounds of the halogens: hydrofluoric, hydrochloric, hydrobromic, and hydroiodic acid.

halogenation Introduction of a halogen into an organic compound, by addition or substitution; as, chlorination.

halogenide* See *halide.*

halohydrin An organic compound of the type X−R−OH, where X is a halogen; as $Cl \cdot CH_2 \cdot CH_2 \cdot OH$.

haloid Resembling or derived from halogens. **h. acid** An inorganic acid, HX, containing a halogen but no oxygen. **h. elements** The halogens.

halonium ions* Halogen cations of the type H_2X^+; as, iodonium, H_2I^+.

haloperidol $C_{21}H_{23}O_2NClF$ = 375.9. Haldol, Serenace. White crystals, m.149, insoluble in water. A tranquilizer, used to treat anxiety and psychiatric illness, e.g., mania (USP, BP).

halophile A bacterium that can grow in saline media.

halothane $C_2HBrClF_3$ = 197.4. Colorless liquid, d.1.875, b.50, slightly soluble in water. An inhalation anesthetic; liver damage may occur after repeated exposure to h. over a short period (USP, BP).

Halothene Trade name for a chlorinated polyethylene.

Halowax $C_{10}H_7Cl$ = 162.6. Trademark for 2-chloronaphthalene. Colorless solid, m.56; used in gasoline to lubricate valve stems of internal-combustion engines.

Halphen reagent A 1% solution of sulfur in carbon disulfide. **H. test** To 1 mL oil add 1 mL H. reagent and 1 mL pentyl alcohol; heat in a brine bath for 30 min; 1% cottonseed oil gives a red color.

Hamamelidaceae Witch-hazel family. Shrubs that yield witch hazel (*Hamamelis virginiana*), storax (*Liquidambar orientalis*), sweet gum (*Liquidambar styraciflua*).

hamamelin A precipitate from the extract of the bark of *Hamamelis virginiana;* an external cooling lotion.

Hamamelis Hamamelidaceae.

hamamelitannins Tannic acids from the bark of witch hazel; 3 forms: **alpha-~** $C_{34}H_{36}O_{22}$. Hydrolyzed to gallic acid and glucose. **beta-~** $C_{20}H_{20}O_{14} \cdot 6H_2O$ = 592.5. The principal form, hydrolyzed to gallic acid and the hexose hamamelose. Fine white needles, m.115, soluble in water. **gamma-~** $C_{27}H_{32}O_{19}$. Yellow crystals, m.222 (decomp.), hydrolyzed by enzymes to gallic acid, methoxygallose, and glucose.

hamamelose See *β-hamamelitannins.*

hamartite Bastnasite.

hamathionic acid Euxanthic acid.

Hamiltonian function A special operator equation used to convert the energy equation into the wave equation.

hand Obsolete unit of length: 1 hand = 4 in.

hanksite $9Na_2SO_4 \cdot 2Na_2CO_3 \cdot KCl$. A mineral from California.

Hanovia lamp An evacuated quartz tube with 2 mercury reservoir electrodes; used to produce mercury-vapor radiation rich in ultraviolet light.

Hantzsch, Arthur (1857–1935) German organic chemist noted for studies of the stereochemistry of nitrogen compounds. **H.-Widman system** See *nomenclature.*

Hanus solution A solution of iodine monobromide in glacial acetic acid, used in the determination of iodine values of oils containing unsaturated organic compounds.

haploid Describing cells which have half the number of chromosomes in the diploid cell. **h. number** The 23 chromosomes normally present in the nucleus of the human sperm and ovum. Cf. *diploid number.*

hapten A chemical grouping on an antigen molecule that confers the antigenic property on it, but is not in itself an antigen.

hapto* (Greek: "to fasten") See *eta(2).*

haptogen The protein of the membrane of the globules of fat in milk.

haptoglobin A protein in serum which combines with free hemoglobin molecules. Used in blood grouping tests.

harbolite A carbonaceous, asphaltic deposit containing 3% bitumen and coal (Turkey).

hard A condition of water, due to the presence of calcium and magnesium salts. See *hardness.* **biologically ~** Resistant to biological decomposition. **h. salt** Hartsalz.

Harden, Sir Arthur (1865–1940) British biochemist. **H.-Young ester** Fructose diphosphate, produced in the fermentation of sugar by yeast. Cf. *sugar phosphates.*

hardening A process which makes a material more resistant to cutting, breaking, or bending. **work ~** The increase in resistance to deformation produced on cold-working a metal. **h. of fats** Hydrogenation. **h. of steel** Tempering.

hardness (1) The state or quality of being hard. Resistance to cutting, bruising, scratching, or grinding. (2) The presence of calcium and magnesium salts in water (usually carbonates and hydrogencarbonates), which incrusts boilers and impairs the lathering of soap by forming insoluble fatty acid salts. **Brinell ~** See *Brinell hardness.* **Mohs ~** See *Mohs scale of hardness.* **Shore ~** See *Shore hardness.* **h. of water** A measure of the calcium and magnesium content of water. **permanent ~** A condition of magnesium or calcium sulfate or carbonate or other calcium salts (except hydrogencarbonates) in water, which cannot be removed by simple boiling, but can be removed by chemical treatment. See *permutite.* **temporary ~** A condition of magnesium or calcium hydrogencarbonate in water; the water is softened by boiling, insoluble calcium carbonate being formed. **total ~** The total amount of calcium and magnesium salts in water. See Table 41.

hardpan (1) An accumulation of hard cementing material containing Ca or Fe, in the lower horizon of a topsoil. (2) Erroneously used to describe bedrock underlying surface deposits.

hardware The electronic, electrical, magnetic, and mechanical parts of a computer system. Cf. *software.*

hardystonite $Ca_2ZnSi_2O_7$. A mineral from Franklin, N.J.

Hare, Robert (1781–1858) American chemist noted for work on the oxyhydrogen flame, the colorimeter, gas analysis, and artificial graphite.

Harkins, William Draper (1873–1951) American chemist noted for research on atomic structure. **H. theory** With the exception of the noble gases, elements of odd atomic number are rarer than adjacent elements of even atomic number.

Harlon Trademark for a mixed-polymer synthetic fiber.

harmala red An oriental red dye from harmel.

harmaline $C_{13}H_{14}ON_2$ = 214.3. Harmine dihydride. An alkaloid from the seeds of *Peganum harmala* (Rutaceae), south U.S.S.R., Turkey. Colorless octahedra, m.238, slightly soluble in water; an anthelmintic. Cf. *turkey red.*

harman Aribine.

harmel Wild rue, *Peganum harmala,* a common weed of the

TABLE 41. UNITS OF WATER HARDNESS

Unit	ppm	Grains per U.S. gal	Clark degree	French degree	German degree
1 ppm	1.0	0.058	0.07	0.10	0.056
1 grain per U.S. gal	17.1	1.000	1.20	1.71	0.958
1 Clark degree	14.3	0.829	1.00	1.43	0.800
1 French degree	10.7	0.583	0.70	1.00	0.560
1 German degree	17.9	1.044	1.24	1.78	1.000

Russian and Turkish steppes; a vermifuge. Its seeds contain the alkaloids harmaline, harmine, and aribine. Cf. *harmala red.*

harmine $C_{13}H_{12}ON_2$ = 212.3. Yajeine. An alkaloid from the seeds of *Peganum harmala.* m.264 **3,4-dihydro ~** Harmaline.

harminic acid $C_{10}H_8O_4N_2$ = 220.2. An oxidation product of harmine or harmaline.

harmonic progression A series whose terms are the reciprocals of an arithmetic progression.

harmony The adaptation of component parts to a state of equilibrium; fitting together.

harmotome $(K,Ba)(Al_2Si_5O_{14})\cdot 5H_2O$. A zeolithic mineral having ion-exchange properties.

Harrison Narcotic Act (U.S.) An Internal Revenue regulation to govern the production, importation, manufacturing, compounding, dispensing, selling, and giving away of opium or coca leaves, or their salts, derivatives, or preparations.

hartin $C_{20}H_{34}O_4$ = 338.5. A substance in fossil wood or lignite. Cf. *hartite.*

hartite $(C_6H_{10})_n$. A hydrocarbon in lignite and fossil wood.

Hartmann's solution An isotonic solution of sodium lactate, sodium chloride, and calcium chloride for intravenous infusion.

Hartman's solution Thymol 12.5, ethyl alcohol 10, sulfuric ether 20 g; used to desensitize dentin selectively.

Hartridge unit A photoelectric measure of diesel smoke intensity: clean air 0, complete opacity 100.

hartsalz A mixture of sylvinite and kainite (16% K); a fertilizer.

hartshorn Spirit of hartshorn. Ammonium hydroxide*. **h. salt** Ammonium carbonate.

harvel coating A waterproof paint made with cashew nut oil.

hashish The dried leaves and stalks of *Cannabis indica,* q.v.

Hassel, Odd (1906–1981) Norwegian chemist. Nobel prize winner (1969). Noted for work on structural chemistry.

hatchettenine C_nH_{2n+2}. Rock tallow. A yellow, native, waxy hydrocarbon, soluble in ether.

Hatschek, Emil (1869–1944) Hungarian-born British chemist, noted for his work on colloids.

Hauptman, Herbert (1917–) American chemist; Nobel prize winner (1985) for work on crystal structures.

hausmannite A native manganese oxide, Mn_3O_4, or manganomanganite, Mn_2MnO_4; brown masses.

hauynite $3NaAlSiO_4$. A mineral from the Laacher See, Germany.

Hayem solution Sodium sulfate 5, sodium chloride 1, mercuric chloride 0.5, water 200 pts.; used in the microscopical analysis of blood.

Haynes alloy Co 45, Cr 26, W 15, Ni 10, C 0.4, B 0.4%; used for engine rotor blades operating at above 870°C.

hazardous chemicals Chemicals that may cause loss of life or property by improper handling, shipping, or storing.

hazard warning symbols A system of pictographs and color schemes, evolved by the United Nations or the EEC's Council of Europe, for display on containers with hazardous chemicals and on vehicles transporting them. See Fig. 14 on p. 276.

hazen unit A measure of the color of water.

Hb Abbreviation for hemoglobin.

H.C. A mixture of zinc and hexachloroethane, used to produce artificial smoke.

γ-HCH* See *insecticides,* Table 45 on p. 305.

HDPE High-density *polyethylene.*

He Symbol for helium.

heart **h. sugar** Inositol. **h. attack** Popular name for myocardial *infarct.* **h. cut** The middle fractions of a series of fractionations. **h.-lung machine** Total life support system that includes a blood oxygenator.

heat A form of energy that can be transmitted from one body to another (1) by radiation: electromagnetic radiation, when stopped by a substance, causes its molecules to vibrate faster, and, so, produce h.; (2) by contact: molecular vibrations are transmitted directly by conduction and convection. It is supposed that at −273.15°C, the absolute zero, these vibrations cease. **animal ~** The h. evolved during metabolism. **atomic ~** The amount of h. required to raise the temperature of a gram atom of substance from 0 to 1°C. **latent ~** See *latent heat.* **mechanical equivalent of ~** See *h. equivalent* below. **molecular ~** The amount of h. required to raise the temperature of one mole of a substance by 1°C, i.e., specific h. × molecular weight. Cf. *Kopp's* law. **radiant ~** H. waves transmitted through space. **radioactive ~** H. evolved during radioactive decomposition. **sensible ~** The h., in J/kg, that must be added to a liquid to bring it to its boiling point. **specific ~** Specific h. capacity*.

h. of absorption The quantity of h. consumed or liberated when a gas is dissolved. **h. of activation** H. involved in catalytic processes. **h. of adhesion** The quantity of h. consumed or liberated in the formation of heterogeneous mixtures. **h. of admixture** H. of mixing. **h. of adsorption** The quantity of h. liberated when a substance is adsorbed or condensed on the surface of a solid. **h. of aggregation** H. involved in the formation of aggregates; as, h. of condensation, crystallization. **h. of association** The quantity of h. absorbed on the formation of coordinate compounds. **h. capacity*** Thermal capacity. The amount of h. required to raise the temperature of a body 1 K (i.e., 1°C); usually expressed in J/K or cal/°C. **atomic ~** Heat capacity per gram atom. Cf. *Dulong and Petit law.* **molecular ~** Cf. *Kopp's* law. The molecular h. capacity is the sum of the atomic h. capacities. **specific ~*** The h.c. per unit mass. **h. of combination** H. of formation or h. of hydration. **h. of combustion** The number of joules liberated per gram atom or gram molecule when an element or compound, respectively, is completely oxidized. **h. of compression** H. produced when a gas is compressed. **h. of condensation** The reverse of h. of evaporation. **h. conductivity** Thermal conductivity*.

Corrosive substance

Oxidizing substance

Explosive substance

Radioactive substance

Flammable substance

Toxic substance

Harmful or irritating substance

Fig. 14. Hazard warning symbols.

h. content Enthalpy*. **h. of cooling** The h. liberated at a certain temperature during cooling; it indicates an allotropic rearrangement. **h. of crystallization** The quantity of h. liberated or absorbed per mole on crystallization. **h. of decomposition** The quantity of h. liberated or absorbed during the complete decomposition of a mole of substance. **h. degree** The intensity of h. See *temperature scales.* The entire absence of h. is considered the absolute zero at −273.15°C. **h. of dilution** The quantity of h. consumed or liberated when a liquid is diluted. **h. of dissociation** The h. involved in the disruption of certain bonds. **h. of dissolution** H. of solution. **h. effect** Joule effect. The h. developed by an electric current in a metallic circuit = RI^2t joules, where R is the internal resistance, I the electric current (equal to the emf divided by the total resistance), and t the time. Cf. *Peltier effect.* **h. engine** An arrangement for converting h. into work. **h. equivalent** A factor to convert h. into energy units: 1 mean calorie = 4.1900 joules; 1 calorie (15°C) = 4.1858 joules. The conversion factors for the *gas constant R,* q.v., per degree per mole: 82.07 cc·atm, 1.9885 cal, 8.314 J. **h. of evaporation** See specific *latent* h. **h. of explosion** The h. liberated from a mole of explosive. **h. flux density*** Irradiance*. **h. of foods** See *heat value* below. **h. of formation** The quantity of h. liberated or consumed when a compound is formed from its component elements; it depends on the physical condition (solid, liquid, or gaseous) of the reacting molecules: e.g., S (rhombic) = S (monoclinic) + 322 J. **h. of fuels** See *heat value* below. **h. of fusion** See specific *latent heat of fusion* under *latent.* **h. of hydration** The amount of h. consumed or liberated when a substance takes up water. **h. index** Maumené number. The temperature (in °C) produced by mixing 50 mL oil with 10 mL concentrated sulfuric acid. **h. of isomerization** The h. involved in the formation of isomers. **h. of linkage** The amount of h. required to form or disrupt certain atomic bonds. **h. of mixing** For solid and liquid, see *heat of solution;* for liquid and liquid, see *heat of dilution;* for gas and liquid, see *heat of absorption.* **h. of neutralization** The quantity of h. liberated on neutralization: $H^+ + OH^- \rightarrow H_2O$ + 57,300 J (at 18°C). **h. number** H. index. **h. of oxidation** See *heat of combustion.* **h. pump** A means of concentrating low-temperature h. into h. at higher temperatures, utilizing a principle similar to that of the domestic *refrigerator,* q.v. Effectiveness is measured by the coefficient of performance, this being the heat output divided by the electrical energy input (for the compressor). Values are normally 3–5. **h. quantity** The amount of h. energy, expressed in joules (or cal). **h. of racemization** The quantity of h. consumed or liberated on the change from one stereoisomer to the other. **h. rays** See *infrared.* **h. of reaction** The quantity of h. consumed or liberated in a chemical reaction; as, h. of neutralization. **h. regenerators** Stoves used in the blast furnace process for iron, which are heated by wastes gases, and then cooled by heating up cold gases for the blast. See *Cowper stoves.* **h. shocking** Preliminary h. treatment to ensure that only the most h.-resistant organisms in a culture survive; the culture thus has increased virility. **h. of solidification** The quantity of h. liberated on freezing or solidifying. **h. of solution** H. of dissolution. The quantity of h. liberated or consumed when a solid dissolves in a liquid. **h. of sublimation** The quantity of h. required to convert a solid into a gas at constant temperature. **h. summation** See *Hess law.* **h. of swelling** The h. evolved when a colloid, e.g., gelatin, absorbs water. **h. transfer coefficient** Thermal conductance. **h. of transition** The quantity of h. liberated or consumed at the transition temperature, when a substance passes from one allotropic form to another. **h. treatment** Exposure of manufactured parts to h. to remove internal stresses; sometimes in a vacuum for alloys. **h. value** *Foods:* The h. liberated on oxidation of 100 g food = $4.1C + 4.1P + 9.3F$ kcal/100g, where C, P, and F are the percentage of carbohydrate, protein, and fat, respectively. *Fuels:* The h. obtained on the complete combustion of fuels = $34.1C + 144H - 12.6(O - N)/100$ kJ, where C, H, O, and N are respectively the percentage of carbon, hydrogen, oxygen, and nitrogen in the fuel. **h. of vaporization** See *specific latent heat of vaporization* under *latent.*

heater A device by which temperature can be raised. **direct ~** An open fire or flame, or heater functioning by direct addition of steam. **indirect ~** A tube or plate steam heater; or in the laboratory, a water, steam, or other bath.

Heaviside layer See *Kennelly-Heaviside layer.*

heavy (1) Not light. (2) Large in quantity. **h. acids** Those used in large quantities; as, sulfuric, hydrochloric, and nitric acids. **h. chemicals** Those manufactured in large quantities; as, chloralkali and sulfuric and nitric acids. **h. hydrogen** Deuterium **h. metal** A metal of specific gravity greater than 4. The heavy metals are located in the lower half of the periodic table. They have complex spectra, form colored salts and double salts, have a low electrode potential, are mainly amphoteric, yield weak bases and weak acids, and are oxidizing or reducing agents. **h. spar** Barite. **h. water** See *heavy water* under *water, deuterium, heavy water reactor* under *reactor.*

hecogenin A steroid prepared from the juice of the sisal leaf, *Agave sisalana,* by autofermentation followed by hydrolysis; used in the synthesis of cortisone.

hectare* Abbrev. ha; a unit of area in the metric system. 1 hectare = 100 ares = 2.471 acres = 10,000 sq. meters.

hecto-* Abbrev. h; SI system prefix for a multiple of 10^2.

Hector's base $C_{14}H_{12}N_4S$ = 268.3. An oxidation product of phenylthiocarbamide; m.239.

hedenbergite $CaO \cdot FeO \cdot 2SiO_2$. A *silica* mineral, q.v., of the *pyroxene* group, q.v.

hedeoma The dried leaves and tops of *H. pulegioides,* American pennyroyal, an annual herb of U.S. and Canada. **h. oil** The essential oil of *H. pulegioides.* Colorless or yellowish liquid of pungent odor, used to repel mosquitos and fleas.

hedeomol $C_{10}H_{18}O$ = 154.3. The ketone of the oil from hedeoma. Colorless liquid, b.217. Cf. *pulegone.*

hedera Ivy.

hederagenin $C_{30}H_{45}O_4$ = 469.7. A glucoside from the seeds of *Hedera* species, ivy (Araliaceae). Cf. *hederaglucoside.*

hederaglucoside $C_{32}H_{54}O_{11}$ = 494.7. Helexin. A glucoside from *Hedera helix,* common ivy (Araliaceae). White powder, m.233, soluble in alcohol.

hedgehog crystals Crystals of ammonium urate in urinary deposits.

hedyphane $3Pb_3As_2O_8 \cdot PbCl_2 \cdot (Ca,Ba)O$. A mineral from Franklin, N.J.

Heerwagen pipet A pipet with piston, for delivering small quantities of mercury.

Hefner **H. lamp** A device for burning pentyl acetate with a flame 4 cm high. A standard for photometric measurements. **H. unit** The horizontal luminous intensity of the H. lamp burning at 760 mm in an atmosphere containing 8.8% water vapor. 1 H. unit = 0.90 candlepower.

Hehner number The percentage of water-insoluble fatty acids and unsaponifiable matter in a fat or oil.

Heilbron, Sir Ian (1886–1959) Scottish chemist, noted for his work on synthetic organic chemistry.

Heim's cage A metal box for breeding small rodents for experimental purposes.

Heisenberg, Werner (1901–1976) German physical chemist. Nobel prize winner (1933). **H. principle** The uncertainty or indeterminacy principle. The limit to observational experiments is reached when the observational or determining factors begin to interfere with normal happenings in the experiment under observation. More specifically, the product of the uncertainty of 2 such related parameters (as, position and momentum of a particle) approximately equals the Dirac constant.

helenene $C_{19}H_{26}$ = 254.4. A product of distilling crude helenin with phosphoric acid.

helenin (1) C_6H_8O = 96.1. Inula camphor. True helenin. Colorless crystals, m.72, from the roots of *Inula helenium.* (2) $C_{21}H_{28}O_{31}$ = 776.5. Crude helenin. A principle from the root of *Inula helenium:* alantol, alant camphor, and alantic anhydride. (3) Inulin. (4) $C_{20}H_{25}O_5$ = 345.4. Bitter crystals from *Helenium autumnale,* sneezewort (Compositae). (5) Helenene.

helenine An active nucleoprotein antiviral agent from the mycelia of *Penicillium funiculosum.*

helenite Mineral caoutchouc.

helexin Hederaglucoside.

helianthic acid $C_{14}H_9O_8$ = 305.2. An acid from the seeds of *Helianthus annuus,* sunflower (Compositae).

helianthin $Me_2N \cdot C_6H_4 \cdot N{:}N \cdot C_6H_4SO_3H$. Dimethylaminoazobenzenesulfonic acid. A red dye. Its sodium salt is methyl orange.

helianthus Sunflower.

helicin $C_{13}H_{16}O_7 \cdot \frac{3}{4}H_2O$ = 297.8. Salicylaldehyde glucose. An oxidation product of salicin. Colorless crystals, m.175, soluble in water.

helicoprotein A glucoprotein from the snail *Helix.*

helictite A stalactite or stalagmite of irregular shape.
helide A supposed and unproven compound of helium, e.g., $He_{10}Hg$, mercury helide.
heliostat Motorized device used for reflecting the sun's rays, as in thermal energy generation.
heliotrope A quartz, semiprecious stone with a greenish tint and red specks.
heliotropic acid Piperonylic acid*.
heliotropin (1) Piperonal*. (2) A purple diazo dye. (3) The odorous principle of *Heliotropium.*
heliotropine An alkaloid from *Heliotropium europaeum* (Boraginaceae).
heliotropism Growth or orientation movement in plants due to the stimulus of sunlight.
helisterol $C_{26}H_{44}O_2$ = 388.6. A colorless sterol from plant carotenoids.
helium* He = 4.002602. A chemically inert gas and element, at. no. 2; a constituent of the atmosphere, of radioactive minerals, and of natural gases. Discovered (1895) by Ramsay and Cleve, after its presence in the sun was indicated spectroscopically (1869) by Lockyer and Frankland. Thus, named from the Greek for "sun." Colorless gas, $d_{air=1}0.137$, m.−272, b.−270. When cooled at 2.2 K, it is transformed into liquid He_{II}, which exhibits superconductivity and very low viscosity. Cf. *lambda phenomenon, superfluid.* Used to prevent oxidation in arc welding; used in admixture with 20% O_2 as an inhalant, since requires less effort to breath than O_2-air mixture (USP, BP). Also for preventing N_2 bubbles from forming in the blood of divers. See *bends.* **ortho ~** H. in which the electron spins are parallel. **para ~** Parhelium. H. in which the electron spins are antiparallel. It gives the principal spectral line 20.528 Å.
h. compounds H. has a valency of zero and forms no compounds. It is occluded in some minerals as the result of radioactive decay. **h. nucleus** The atomic nucleus of He consists of 2 protons and 2 neutrons. Cf. *packing effect.* In *radioactive disintegration,* q.v. He nuclei (α rays) are thrown off at high speed. Their positive charge is neutralized by 2 electrons, forming He gas.
helix (1) The coil of wire in an electromagnet. (2) A spiral arrangement of the periodic system. (3) A snail.
hellebore A genus of ranunculaceous plants. **American ~** *Veratrum viride.* **black ~** The root of *Helleborus niger.* **false ~** American h. **green ~** The root of *H. viridis.* **white ~** *Veratrum album.*
Helmert's equation The acceleration of free fall (due to gravity) at sea level, $g = 978.0318(1 + 0.0053024 \sin^2\varphi - 0.0000058 \sin^2 2\varphi)$ cm/s², where φ is the latitude. Subtract $(0.30877 - 0.00044 \sin^2\varphi)h - 0.000072\, h^2$, at a height h km above the earth's surface.
Helmholtz, Herman Ludwig Ferdinand von (1821–1894) German scientist, noted for his generalizations on the conservation of force. **Gibbs- ~ equation** See *Gibbs-Helmholtz equation* under *Gibbs, Josia.*
H. equation of inductance $L = (E/R)[1 - t\exp(-R/T)]$, where E is the electromotive force, R the resistance, L the self-inductance, T the temperature, and t the time. **H. function** A, F. H. free energy, work function. $A = U - TS$, where U is the internal energy, T the thermodynamic temperature, and S the entropy. The change in A, thus, gives the maximum work obtainable from a system. **H. layer** The electrical double layer of opposite charges formed on the surface of a charged solid in contact with a liquid.
helminth A worm; some cause parasitic diseases.
helminthiasis A disease caused by parasitic worms, e.g., schistosomiasis.
helminthic Anthelmintic.
Helmont, Johann Baptist van (1577–1644) Belgian alchemist whose work represents the transition from speculative to experimental chemistry. He originated the term gas (chaos).
Helvella A family of cryptogams.
helvellic acid $C_{12}H_{20}O_7$ = 276.3. A dibasic, poisonous acid from *Helvella esculenta,* which causes hemoglobinuria.
helvetium Astatine*.
helvite $(BeMoFe)_7Si_9O_{12}S$. A brittle, lustrous, greenish mineral.
hema- Prefix denoting blood. See also *hemo-*. Cf. *haemo-*.
hemachate Blood agate. A brown agate.
hemacyanin Hematocyanin. A blue coloring matter in bile. Cf. *hemocyanin.*
hemagglutinins Substances (agglutinins) which cause the clumping of red blood corpuscles.
hemanthine An alkaloid of *Haemanthus toxicarus* (Amaryllidaceae); constituent of Australian arrow poisons, resembling scopolamine.
hematein $C_{16}H_{12}O_6$ = 300.3. An oxidation product of hematoxylin. Brown powder, insoluble in water; an indicator in alkalimetry.
hematin $C_{34}H_{32}N_4O_4FeOH$ = 633.5. Phenodin. An oxidation product of the hemoglobin of blood. Brown powder, soluble in alkalies. Structure similar to heme and hemin. See *porphin ring.* **oxy ~** $C_{34}H_{32}N_4O_7Fe$ = 664.5. The coloring matter of oxyhemoglobin. It yields, on strong oxidation, hematinic acid; and, on heating, pyrrole and pyrrole derivatives. Cf. *porphin.*
hematine **h. crystals** Hematoxylin. **h. extract, h. paste** Hematoxylin paste; more strictly applied to its oxidation product.
hematinic An agent used to treat anemia.
hematinic acids A group of di- and tribasic acids obtained by the strong oxidation of hematin.
hematite, haematite Fe_2O_3. Red iron ore. Raddle. An iron ore commonly used for the manufacture of iron and steel. **brown ~** Hydrated h. or limonite. **red ~** Kidney ore. Rhombic or reniform, native Fe_2O_3. **spicular ~** Specularite.
hematocrit A small centrifuge, used to separate blood cells from plasma, to measure their volume. **h. estimation** Packed cell volume. PCV. The percentage volume of blood cells after whole blood has been centrifuged in a h.
hematocrystallin Hemoglobin.
hematoidin Bilirubin.
hematology The study of the blood and blood diseases.
hematolysis Hemolysis.
hematoporphyrin $C_{34}H_{38}N_4O_6$ = 598.7. Cruentine. Dark-violet powder obtained by adding hemin to glacial acetic acid saturated with hydrobromic acid, and neutralizing with sodium hydroxide. Cf. *porphin.*
hematoxylic acid Hematoxylin.
hematoxylin $C_{16}H_{14}O_6$ = 302.3. Haematoxylic acid, haematine, logwood crystals, campeachy wood, Jamaica wood, steam black. The coloring principle of *Haematoxylon campechianum,* logwood. Colorless crystals, soluble in water and alkalies (purple color); with acids the color changes to yellow. On exposure it turns black with formation of h. A mordant dye, an indicator in the titration of alkaloids, a microscope stain, and a reagent for copper and iron. **h. paste** Logwood paste, h. extract, logwood extract. A technical grade of h.; a coloring material in the textile and leather industries.
hematoxylon Haematoxylon.
heme $C_{34}H_{33}O_4N_4FeOH$ = 636.9. Haem. The prosthetic group of hemoglobin and some enzymes; structure similar to hematin, but an iron(II) complex.
hemellitic acid 2,3-Xylic acid.

hemellitol Hemimellitene.
hemerythrin A relatively rare iron-protein compound in living tissue.
hemi- Prefix (Greek) indicating "half." See also *semi-* (Latin) and *demi-* (French). **h.acetal** See *acetals.*
hemialbumose Propeptone. A decomposition product of albumin, related to peptone.
hemicellulose (1) A constituent of the cell wall of bacteria. (2) Pseudocellulose. A group of gummy substances intermediate in composition between cellulose and the sugars.
hemicolloid A colloidal particle having a chain length up to 250 Å and a polymerization of 20–100 molecules. Cf. *mesocolloid.*
hemihedral Describing a crystal which has only half the number of faces that the symmetry of the system requires. Cf. *holohedral.*
hemimellitene $C_6H_3Me_3$ = 120.2. Hemimellitol, 1,2,3-trimethylbenzene*. Colorless liquid, b.175.
hemimellitic acid $C_6H_3(COOH)_3$ = 210.1. 1,2,3-Benzenetricarboxylic acid*. Colorless needles, m.196 (decomp.), slightly soluble in water. Cf. *hemellitic acid.*
hemimellitol (1) $HO \cdot C_6H_2Me_3$. (2) Hemimellitene.
hemin $C_{34}H_{32}O_4N_4FeCl$ = 652.0. Protoporphyrin iron(III) chloride complex, Teichmann's crystals, hematin chloride. The characteristic microcrystals obtained by heating a crystal of sodium chloride, a drop of glacial acetic acid, and blood on a microscope slide; an identification test for blood. See *porphin.*
hemipic acid Hemipinic acid.
hemipinic acid $C_6H_2(OMe)_2(COOH)_2$ = 226.2. Hemipic acid. 3,4-Dimethoxy-1,2-benzenedicarboxylic acid*. Colorless crystals, m.166, soluble in water; a split product of nicotine.
hemiquinonoid Describing the structure

$$R_2C\langle{}^{C=C}_{C=C}\rangle C=O$$

Cf. *quinonoid.*
hemisotonic Isotonic (2).
hemiterpenes Hydrocarbons of the general formula C_5H_8, related to the *terpenes,* q.v.; as, isoprene.
hemitrope A twin crystal.
hemlock (1) The fir tree *Tsuga canadensis* of W. and N. America. (2) The poisonous plants and shrubs of the *Conium* species (Umbelliferae). **poison ~, spotted ~** Conium. **water ~** Cicuta.
h. alkaloids Alkaloids from the seeds and bark of *Conium* species: conine, conhydrine. **h. bark** The bark of the h. fir, used in tanning. **h. fir** The tree *Tsuga canadensis* of W. and N. America. **h. spruce** H. fir. **h. tannin** $C_{20}H_{16}O_{10}$ (?). A tannin from the bark of *Tsuga* species.
hemo- Prefix denoting blood. See also *hema-.*
hemocuprein A blue copper-protein compound in the red blood cells and livers of mammals.
hemocyanin Haemocyanin. A blue coloring matter from the blood of mollusks; related to hematin but contains copper instead of iron. Cf. *hemacyanin, chromoprotein.*
hemocytometer Hemacytometer, hemameter, etc. A microscope slide with square rulings; used for counting blood cells.
hemodyn See *periston.*
hemoglobin Haemoglobin. Hb. Mol. wt. about 65,000. A pigmented protein occurring in the red blood cells of vertebrates and acting as the oxygen carrier of their blood. The h. molecule consists of 4 polypeptide chains (i.e., 2 pairs), each attached to the heme molecule. Each species of animal has a different sequence of amino acids. **abnormal ~** Either abnormal proportions of HbF (see *fetal ~* below) or HbA_2 in the adult, or the presence of HbS (sickle cell disease) or HbC, both of which have 1 or 2 different amino acids in the chains. **adult ~** H. in adults, composed of two different types: 90–98% HbA (comprising 2 α chains of 141 amino acids each, and 2 β chains of 146 each); and 2–10% HbA_2 (with 2 δ chains of 146 amino acids each, and 2 α chains). **carbamino ~** Reduced h. combined with carbon dioxide. **fetal ~** HbF (with 2 γ chains of 146 amino acids, and 2 α chains). Occurs in the fetus from about the 12th week of pregnancy until age one year, progressively diminishing from age 6 months; adult h. increases from birth. **normal ~** See *adult hemoglobin* above. **oxy ~** H. combined with O_2; present in arterial blood. **reduced ~** H. of venous blood, without oxygen.
hemoglobinometer An instrument to determine hemoglobin in the blood, originally by matching against a color chart, as in the Haldane h. Now estimated by photometer.
hemolysin A substance causing hemolysis, e.g., bacterial h. or antibodies formed in the body on its own red cells.
hemolysis Hematolysis. The destruction of red blood cells, releasing hemoglobin. (1) In vivo, occurs normally at the end of the life span of the red cell, or, pathologically, because of poisons, bacterial hemolysins, antibodies to red cells, or the presence of an abnormal hemoglobin. (2) In vitro, the dissolution of red blood cells by chemicals, heating, freezing, or biological agents, which causes the blood to become transparent and clear. Used in many laboratory tests on blood and bacteria.
hemolytic Describing an agent that destroys red cells.
hemolyzate The product of hemolysis.
hemophilia A hereditary disease in males due to absence or deficiency of blood clotting Factor VIII; characteristically, the blood fails to clot normally after surgery and minor trauma.
hemophilus See *Bordetella.*
hemoporphyrin $C_{16}H_{18}O_3N$ = 272.3. A decomposition product of hematin, closely allied to phylloporphyrin (a decomposition product of chlorophyll). Cf. *porphin.*
hemopyrrole $C_8H_{13}N$ = 123.2. A decomposition product of both hemoglobin (hemoporphyrin) and chlorophyll (phylloporphyrin).
hemostatic An agent which when applied externally checks the flow of blood; as, absorbable *gelatin.* Cf. *styptic.*
hemp The plant *Cannabis indica* or *C. sativa* (Urticaceae): (1) The flowering tops (cannabis) yield a resin (cannabin); (2) the leaves and stalks are hashish; (3) the seeds yield h. seed oil; and (4) the stems yield a fiber used for ropes or paper. **Bombay ~** Sunn h. **bow-string ~** A tough fiber from *Sanseviera Zeylanica* (Liliaceae), Sri Lanka. **Canadian ~** *Apocynum cannabinum.* **china ~** See *china jute* under *jute.* **Deccan ~** The fiber from *Hibiscus cannabinus* (Malvaceae), India. **manila ~** Abaca. **Mauritius ~** The fiber *Furcraea gigantea* (Amaryllidaceae). **New Zealand ~** N.Z. flax. The fiber from *Phormium tenax.* **sann ~** Sunn h. **sisal ~** See *sisal.* **sunn ~** Bombay h. A fiber from the stems of *Crotalaria juncea* (Leguminosae), India, Australia.
h.seed The seeds of *Cannabis* species. **h.seed oil** A green, nondrying oil, from h.seeds; chief constituent, linolein; d_{15}0.925–0.932, m. −15 to 23. Used for soap, paints, and varnishes.
Hempel, Walther (1851–1916) German analytical chemist. **H. gas buret** A glass apparatus used to absorb gases by solid or liquid reagents. The measured gas enters the absorption bulbs through a capillary tube and pushes the absorbing liquid partly into the leveling bulbs. After absorption, the residual gas is pulled back into the buret and measured.

H. palladium tube A U-shaped glass tube filled with palladium sponge for use in gas analysis to absorb hydrogen.
henbane *Hyoscyamus,* q.v.; poisonous to hens.
hendec- See words beginning with *undec-*.
Henderson, George Gerald (1862–1942) British chemist, noted for his work on terpenes. **H. process** Roasting copper ores with salt, subsequent leaching of the chlorides, and precipitation of the metals.
henequen The fiber of *Agave fourcroydes* (Mexico, W. Indies); used for ropes. Cf. *sisal.*
henicosane* $C_{21}H_{44}$ = 296.6. An alkane, m.40. **h.di(carb)o(xyl)ic acid** Japanic acid.
henicosanoic acid* $C_{21}H_{42}O_2$ = 326.6. White needles in fats, m.74.
henna The powdered leaves of *Lawsonia inermis* (Lythraceae), Turkey, Egypt, and Iran. A brown dye, especially for hair.
Henry **H., Joseph (1797–1878)** American physicist noted for his research in magnetism. Cf. *henry.* **H., William (1775–1836)** English chemist and coworker of Dalton. **H.'s law** The amount of gas dissolved in a liquid is proportional to the pressure of the gas at constant temperature.
henry* *H*; the SI system unit of inductance. If the emf induced is 1 volt when the inducing current varies uniformly at the rate of 1 amp/s, the inductance of the closed circuit is 1 h. 1 $H = 1.1126 \times 10^{-12}$ esu (or stathenry) = 10^9 emu (or abhenry) = 1 quadrant = 1 secohm.
hentriacontane* $C_{31}H_{64}$ = 436.8. An alkane, m.69, from the roots of *Oenanthe crocata* (Umbelliferae), and in beeswax.
hepar Greek for "liver." **h. antimoni** Sodium or potassium antimonate. **h. calcis** Calcium sulfide. **h. reaction** A test for sulfur. The compound is reduced with soda and carbon, and the mass moistened on a silver coin. A black stain indicates sulfur. **h. sulfuris** Liver of *sulfur.*
heparin A blood anticoagulant from the livers or lungs of domestic animals. A conjugated glucuronic acid glucoside. Used in thrombosis and embolism, where it prevents the formation of thrombin, and in the laboratory to obtain unclotted blood (USP, BP).
hepatica The dried plant *Hepatica triloba,* liverwort; a mild, mucilaginous astringent.
hepatitis B immune globulin Immune *globulin,* q.v., used against hepatitis B.
hepotic acid Heptanoic acid*.
hepta* Septa. Indicating 7 times.
heptacosane* $C_{27}H_{56}$ = 380.7. An alkane, m.59, in beeswax.
heptacosanoic acid* $C_{26}H_{53}COOH$ = 410.7. White powder, m.83, insoluble in water. Cf. *neocerotic acid.*
heptad Heptavalent.
heptadecane* $C_{17}H_{36}$ = 240.5. Dioctylmethane. **h.carboxylic acid*** Stearic acid*. **h.nitrile*** $Me(CH_2)_{15}CN$ = 251.5. Margaronitrile, cetyl cyanide. White crystals, m.53.
heptadecanoic acid* $C_{17}H_{34}O_2$ = 270.5. Margaric acid, daturic acid. Colorless mass, m.60, soluble in petroleum ether; occurs in lichens (also synthesized).
heptadecanone* $C_{17}H_{34}O$ = 254.5. **2-~** Methyl pentadecyl ketone*. m.48. **9-~** Dioctyl ketone*, nonylone. m.53.
heptadecoic acid, heptadecylic acid Heptadecanoic acid*.
heptadiene* C_7H_{12} = 96.2. **2,4-~** MeCH:CH·CH:CH·CH_2Me. Colorless liquid, b.107. **h.one** Phorone.
heptaldehyde Heptanal*.
heptamethylene Cycloheptane*.
heptanal* $Me(CH_2)_5CHO$ = 114.2. Oenanthal, enanthaldehyde, heptoic aldehyde, heptyl aldehyde; from castor oil. Colorless, fragrant liquid, d.0.850, b.155, soluble in water; used in organic synthesis.
heptane* $Me(CH_2)_5Me$ = 100.2. Methylhexane, dipropylmethane, heptyl hydride. Colorless liquid, d.0.690, b.95, insoluble in water, highly flammable; in the needles of *Pinus sabiana;* a solvent. **bicyclo ~** See *carane, fenchane.* **heptahydroxy ~** Volemitol.
heptane-1,7-dicarboxylic acid* Azelaic acid.
heptanedioic acid* Pimelic acid.
heptanedioyl* Pimeloyl*. The radical $-OC(CH_2)_5CO-$.
heptanoate* Enanthate. Salt or ester of heptanoic acid, containing the radical $C_6H_{13}COO-$.
heptanoic acid* $Me(CH_2)_5COOH$ = 130.2. Heptoic acid, heptylic acid, (o)enanthic acid. Colorless, oily liquid with unpleasant odor, d.0.9345, b.223, insoluble in water.
heptanol* **1-~** $C_7H_{15}OH$ = 116.2. Heptyl alcohol*, heptylic alcohol, (o)enanthol. Colorless liquid, d.0.830, b.176, soluble in water; used in organic synthesis. **4-~** $Me(CH_2)_2CHOH(CH_2)_2Me$. Isohexylcarbinol. Colorless liquid, d.0.814, b.149, soluble in alcohol.
heptanone* **2-~** $C_5H_{11}C(O)Me$ = 114.2. Methyl pentyl ketone*. A constituent of oil of cloves and cinnamon. **3-~** Et·CO·C_4H_9. Butyl ethyl ketone*. Colorless liquid, b.147, insoluble in water. **4-~** $(C_3H_7)_2CO$. Dipropyl ketone*. Colorless liquid, b.144; a solvent.
heptanoyl* The radical $Me(CH_2)_5CO-$.
heptene* $CH_2:CH(CH_2)_4Me$ = 98.2. Heptylene, pentylethylene. Colorless liquid, d.0.703, b.98, soluble in alcohol. **2,2,4,6,6-pentamethyl-3-~** $Me_3C·CH_2·CMe:CHCMe_3$ = 168.3. **trimethylbicyclo ~** See *bornene, carene, fenchene.*
heptenophos* See *insecticides,* Table 45 on p. 305.
heptenyl* The radical $C_7H_{13}-$. **h. methyl carbonate** See *heptyne methyl carbonate.*
heptenylene Heptyne*.
heptine Heptyne*.
heptoglobin A protein of blood serum having an inherited gene and, therefore, used to establish parenthood.
heptoic **h. acid** Heptanoic acid*. **iso ~** 5-Methyl*hexanoic acid**. **h. alcohol** Heptanol*. **h. aldehyde** Heptanal*.
heptols Pentahydric alcohols, from heptoses.
heptose A sugar having 7 C atoms. See *carbohydrates.*
hepturonic acid A pentahydroxy aldehyde acid derived from heptoses; general formula, CHO·$(CHOH)_5$COOH. Cf. *uronic acids.*
heptyl* The radical $Me(CH_2)_6-$, from heptane. **h. acetate*** $AcOC_7H_{15}$ = 158.2. Colorless liquid, d.0.874, b.190, insoluble in water; used in artificial flavorings. **h. alcohol*** Heptanol*. **h. aldehyde** Heptanal*. **h.amine*** $Me(CH_2)_6NH_2$ = 115.2. 1-Aminoheptane. Colorless liquid, d.0.78, b.155, slightly soluble in water. **h. ether** $(C_7H_{15})_2O$ = 214.3. Diheptyl ether*, enanthylic ether, heptyloxyheptane*. Colorless liquid, d.0.815, b.265, insoluble in water. **h. formate*** $HCOOC_7H_{15}$ = 144.2. Colorless, aromatic liquid, d.0.894, b.176, insoluble in water; used in organic synthesis.
heptylene Heptene*.
heptylic acid Heptanoic acid*.
heptyne* $HC:C(CH_2)_4Me$ = 96.2. Heptenylene, heptine, oenanthine. Colorless liquid, d.0.831, b.104, soluble in alcohol. **h. methyl carbonate** $Me(CH_2)_4C:C·COOMe$ = 154.2. Methyl octynoate*, from the ricinoleic acid of castor oil, d.0.930, insoluble in water; a perfume.
herapathite $4Qu·3H_2SO_4·2HIO_4·6H_2O$. Artificial tourmaline. Quinine periodate sulfate. Produced as optically active crystals when iodine vapor is passed into a solution of quinine sulfate. See *polarizing disk.*
herb The leafy, flowering, or fruiting stems of some smaller plants. Many herbs are used as teas (infusions); others, in the

TABLE 42. HERBICIDES

Common name	Type[a]	Chemical name
Atrazine[b]	s	2-Chloro-4-ethylamino-6-isopropylamino-1,3,5-triazine
Barban[b]	t	4-Chlorobut-2-ynyl 3-chlorophenylcarbamate
Benazolin[b]	t	4-Chloro-2-oxobenzothiazolin-3-ylacetic acid
Bentazone[b]	c	3-Isopropyl-(1*H*)-benzo-2,1,3-thiadiazin-4-one 2,2-dioxide
Bromoxynil[b]	c	3,5-Dibromo-4-hydroxybenzonitrile
Carbetamide[b]	s	(*R*)-(−)-1-(Ethylcarbamoyl)ethyl phenylcarbamate
Chlorpropham[b]	s	Isopropyl 3-chlorophenylcarbamate
Chlorthiamid[b]	s	2,6-Dichloro(thiobenzamide)
Chlortoluron[b]	s,t	3-(3-Chloro-*p*-tolyl)-1,1-dimethylurea
Cyanazine[b]	s	2-(4-Chloro-6-ethylamino-1,3,5-triazin-2-ylamino)-2-methylpropionitrile
2,4-D[b]	t	(2,4-Dichlorophenoxy)acetic acid
Dalapon	t	2,2-Dichloropropionic acid
Dicamba[b]	t	3,6-Dichloro-2-methoxybenzoic acid
Dichlobenil[b]	s	2,6-Dichlorobenzonitrile
Dichlorprop[b]	t	(±)-2-(2,4-Dichlorophenoxy)propionic acid
Dinoseb[b]	c,s	2-*sec*-Butyl-4,6-dinitrophenol
Diquat[b]	c	1,1′-Ethylene-2,2′-bipyridyldiylium ion
EPTC[b]	s	*S*-Ethyl dipropylthiocarbamate
Glyphosate[b]	t	*N*-(Phosphonomethyl)glycine
Lenacil[b]	s	3-Cyclohexyl-6,7-dihydro-1*H*-cyclopentapyrimidine-2,4-dione
MCPA[b]	t	(4-Chloro-*o*-tolyloxy)acetic acid
MCPB[b]	t	4-(4-Chloro-*o*-tolyloxy)butanoic acid
Mecoprop[b]	t	(±)-2-(4-Chloro-*o*-tolyloxy)propionic acid
Methabenzthiazuron[b]	s,t	1-(Benzothiazol-2-yl)-1,3-dimethylurea
Metoxuron[b]	s,t	3-(3-Chloromethoxyphenyl)-1,1-dimethylurea
Paraquat[b]	c	1,1′-Dimethyl-4,4′-bipyridyldiylium ion
Propham[b]	s	Isopropyl phenylcarbamate
Sodium chlorate	s,t	Sodium chlorate
2,4,5-T[b]	t	2,4,5-Trichlorophenoxy)acetic acid
TCA[b]	s	Trichloroacetate
Terbutryn[b]	s	2-*tert*-Butylamino-4-ethylamino-6-methylthio-1,3,5-triazine
Tri-allate[b]	s	*S*-2,3,3-Trichloroallyl diisopropylthiocarbamate
Trifluralin[b]	s	α,α,α-Trifluoro-2,6-dinitro-*N*,*N*-dipropyl-*p*-toluidine

[a]c = contact; t = translocated (i.e., systemic); s = soil-acting.
[b]ISO-approved names. Some countries do not use certain ISO names because of their similarity to established names.

manufacture of essential oils; e.g., absinthium, peppermint, tansy. **bitter ~** Snakehead. The leaves of *Chelone glabra* (Scrophulaceae); an anthelmintic. Cf. *chelonin.* **blanket ~** Mullein. **felon ~** Mugwort. **Fuller's ~** Saponaria.

herbicide Weedicide. A chemical used in agriculture to destroy unwanted plants, especially grasses. About 10% of world crops are lost to weeds. See Table 42, above. **contact ~** A h. that kills only the areas it contacts. **systemic, translocated ~** A h. that spreads through the plant via the sap system, thus killing all the plant. Cf. *fungicide.*

Hercules stone Magnetite.

hercynite $FeAl_2O_4$. Iron spinel. A black mineral, d.3.92, hardness 7.5–8.

herderite Be(OH,F)$CaPO_4$. A calcium phosphate mineral, d.3.0, hardness 5.

heretine An alkaloid from *Heritiera javanica* (Sterculiaceae), Indonesia.

hermetic Airtight. **h. art** Magic or alchemy. **h. casing** A watertight casing.

Hermite process The manufacture of hypochlorite bleaching liquor by the electrolysis of sodium chloride 0.5 and magnesium chloride 0.05% in water.

herniarin $C_{10}H_8O_3$ = 176.2. Methylumbelliferone, 7-methoxycoumarin, from *Herniaria glabra* (Carvophyllaceae).

heroin(e) Diamorphine.

herpatite Herapathite.

herrerite (Zn,Cu)CO_3. A cupriferous smithsonite.

Hertz, Heinrich Rudolf (1857–1894) German physicist noted for research in theoretical physics.

hertz* Abbrev. Hz; an SI system–derived unit. A periodic occurrence of once per second has a frequency of 1 Hz.

hertzian waves Term embracing electromagnetic radiations of wavelength 10^{-3} to 10^3 m. Cf. *radiation.*

Herzberg, Gerhard (1904–) German-born Canadian chemist. Nobel prize winner (1971). Noted for work on molecular spectroscopy.

Herzberg's stain (1) A solution of iodine, potassium iodide, and zinc chloride in water; used to stain rag fibers *red,* chemically treated paper pulp *blue,* and groundwood pulp or lignin *yellow.* (2) A solution of potassium iodide in sulfuric acid.

hesion value A measure of the combined effects of adhesion and cohesion of butter to a solid surface.

hesperetic acid Hesperitinic acid.

hesperetin $C_{16}H_{14}O_6$ = 302.3. 3′,5,7-Trihydroxy-4′-methoxyflavanone. A split product of hesperidin, and a chalcone of phloroglucinol. Yellow crystals, soluble in alcohol.

hesperetol $CH_2 \cdot CH{:}C_6H_4 \cdot (OMe)OH$ = 151.2. 5-Vinylguaiacol, 3-hydroxy-4-methoxystyrene. Colorless crystals, m.57.

hesperidene (*R*)-Citrene.

hesperidin $C_{28}H_{34}O_{15}$ = 610.6. Citrin, vitamin I. A glucoside which occurs with eriodictin in the unripe fruits of

Citrus aurantium. Yellow powder, decomp. 251, soluble in water. It splits on hydrolysis to hesperitinic acid, glucose, and phloroglucinol.
hesperidine An alkaloid from the leaves of *Peucedanum galbanum,* wild celery (Umbelliferae).
hesperitinic acid $C_{10}H_{10}O_4$ = 194.2. 3-Hydroxy-4-methoxycinnamic acid, isoferulic acid. Yellow needles, m.233.
Hess H., Germain Henri (1802–1850) German-born Russian chemist, and a founder (1840) of thermochemistry. **H., Victor Franz (1883–1964)** German physicist, noted for work on cosmic radiation. Nobel prize winner (1936). **H. law** The law of constant heat summation. The net amount of heat liberated or absorbed in a chemical reaction is the same, whether the reaction is performed in one or successive steps. **H. rays** Cosmic rays. **H. viscosimeter** A graduated capillary tube with a rubber bulb, used for determining the viscosity of biological solutions.
hessian A plain woven fabric of hemp or jute. Used for sacking, and as waste for paper manufacture. C. *botany.* **h. crucible** (1) A sand crucible. (2) A large clay crucible.
hessite Ag_2Te. Silver telluride. A black mineral, d.8.3–9, hardness 2.5–3.
hessonite $Al_2(Ca,Fe)_2Si_8O_{16}$. Cinnamon stone. A garnet, d.3.5, hardness 6.5–7.
Het Acid Trademark for 1,4,5,6,7,7-hexachlorobicyclo[2,2,1]-5-heptene-2,3-dicarboxylic acid. Unique among dibasic acids in containing over 54% by weight of stable Cl. Used to impart flame resistance to resins.
hetero- Prefix (Greek) indicating "unlikeness" or "difference."
heteroalbumose A form of albumose, insoluble in water, precipitated by saturation with sodium chloride.
heteroartose The protein $C_{74}H_{130}N_{20}O_{24}S$.
hetero atom* A heterocyclic atom.
heterobaric Possessing different mass numbers; as, isotopes.
heterocycle A ring of different types of atoms. Antonym: homocycle. See *heterocyclic compound.*
heterocyclic* Pertaining to dissimilar atoms in a ring. **h. atom** Any atom, other than carbon, C, in an atomic ring; e.g., N, O, S, Se, P, As. **h. compound** A ring compound having atoms other than C in its nucleus; as:

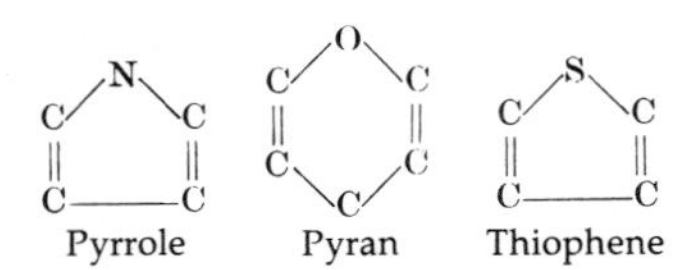

Antonym: homocyclic.
heterofil A composite filament, in which polymers of different characteristics are spun together so that the filaments coalesce longitudinally.
heterogeneity The state of being composed of particles or aggregates of different substances; hence, matter that is of dissimilar composition. Antonym: homogeneity.
heterogeneous Opposite to *homogeneous,* q.v. *(homogeneity).* Describes a substance that consists of more than one phase, and therefore is not uniform; as, colloids. **h. reaction** A chemical change in which 2 or more reactions take place simultaneously.
heterogenesis The derivation of a living thing from something unlike itself; e.g., of viruses from the complex cell.
hetero ion An adsorption complex ion whose charge is due to an adsorbed simple ion, e.g., a protein complex with adsorbed OH−.
heterolysis (1) The dissolution of a cell by an external agent. Cf. *autolysis.* (2) The hemolytic action of the blood serum of one animal species on the blood cells of another species. (3) A reaction in which a bond is severed and one fragment retains both bonding electrons: A: B→A + : B. Cf. *homolysis.*
heterolyzate The filtered liquid portion of the products of heterolysis.
heterometry A form of turbidimetric titration in which nucleating chemical systems are studied by light absorption measurements.
heterophase Forming 2 or more states of aggregation. Cf. *phase.*
heteropolar An unequal distribution of electric charges in a bond, so that one atom is more positive or negative than the other. Cf. *homopolar.*
heteropoly acids The complex acids of heavy metals with phosphoric acids; as, phosphomolybdic acid.
heteropoly blue Molybdenum blue.
heterotopes Elements having different atomic numbers and, therefore, occurring in different parts of the periodic table. Antonym: isotopes. Cf. *isobar.*
heterotype A compound which differs in properties from compounds of a similar type.
hetol Sodium cinnamate.
hetralin $C_6H_{12}N_6 \cdot C_6H_6O_2$ = 278.3. Dihydroxybenzenehexamethylenetetramine. Colorless needles, decomp. 155, soluble in water; a substitute for hexamethylenetetramine.
heulandite $CaAl_2Si_6O_{15}$. A zeolite.
heuristic Describing an approach to scientific problems involving trial-and-error procedures.
Heusler alloys See *Heusler's alloy* under *alloy.*
Hevea See *rubber.*
Hevesy, George de (von) (1885–1966) Hungarian-born German chemist, codiscoverer of hafnium. Nobel prize winner (1943).
hex Hexamethylenetetramine.
hexa- Prefix (Greek) denoting "six."
hexaammine* Indicating 6 $-NH_3$ groups. **h.cobalt (III)*** The cation $[Co(NH_3)_6]^{3+}$.
hexaaqua* Indicating 6 H_2O molecules. **h.chromium (III)*** The cation $[Cr(H_2O)_6]^{3+}$.
Hexa-Betalin Trademark for pyridoxine hydrochloride.
hexabiose Hexobiose. A carbohydrate (disaccharide) consisting of 2 hexoses; as, lactose, sucrose.
hexaborane* B_6H_{10} = 74.9. Colorless liquid, m.−65, decomp. in water.
hexabromide number An analytical value of fats indicating their content of acids with 3 (or more) double bonds; the mg of Br needed to brominate 100 g fat.
hexabromo-* Prefix indicating 6 Br atoms. **h.ethane** C_2Br_6 = 403.4. Yellow needles, decomp. 210, slightly soluble in water. **h.silicoethane** See *silicon bromides.*
hexachloro-* Prefix indicating 6 Cl atoms. **h.benzene** C_6Cl_6 = 284.8. Colorless needles, m.229, insoluble in water. Used in organic synthesis, in airfield flares, and in waterproofing of dopes. Cf. *benzene* hexachloride. **h.ethane** C_2Cl_6 = 236.7. Carbontrichloride, hexoram. White, rhombic crystals, m.184, insoluble in water. Has an inherent tendency to agglomerate; so additives used to improve flow and stability. Used in organic synthesis; the manufacture of explosives and fireworks, smoke screens, and disinfectants. Cf. *H.C.* **h.platinate(IV)*** Chloroplatinate. A salt of h. platinic acid, containing the anion $[PtCl_6]^{2-}$. **h.platinic acid*** See *hexachloroplatinic acid* under *platinic acid.*
hexachlorophane BP name for hexachlorophene.
hexachlorophene $C_{13}H_6O_2Cl_6$ = 406.93. 2,2′-Methylenebis(3,4,6-trichlorophenol). Hexachlorophane. Phisoscrub, Sterzac. White crystals with phenolic odor,

m.161–167, insoluble in water; used for hand washing prior to surgery, and as a dusting powder (USP, BP).
hexacontane* $C_{60}H_{122}$ = 843.6. An alkane. Colorless crystals, m.102.
hexacosane* $C_{26}H_{54}$ = 366.7. Cerane. An alkane. Colorless crystals, m.57.
hexacosanoic acid* $C_{25}H_{51}COOH$ = 396.7. Cerot(in)ic acid, cerinic acid. In peanut oil, vegetable waxes, and the wax from tubercle bacilli. White crystals, m.83, insoluble in water.
hexacosanol* Hexacosyl alcohol*.
hexacosyl* Ceryl. The radical $Me(CH_2)_{24}CH_2-$. **h. alcohol*** $C_{26}H_{53}OH$ = 382.7. Hexacosanol*, ceryl alcohol. Colorless crystals, m.79, insoluble in water. From Chinese wax and wool fat. **h. heptacosanoate*** $C_{26}H_{53}OOC \cdot C_{26}H_{53}$ = 775.4. Ceryl cerotate. Colorless crystals, m.84. In Chinese wax. **h. palmitate** $C_{15}H_{31}COOC_{26}H_{53}$ = 609.1. In opium wax.
hexacyano-* Indicating 6 $-CN$ groups. **h.ferrate(II)*** The ferrocyanide ion, $[Fe(CN)_6]^{4-}$. **h.ferrate(III)*** The ferricyanide ion, $[Fe(CN_6]^{3-}$.
hexacyanogen C_6N_6 = 156.1. A ring polymer, m.120.
hexad (1) In crystallography, showing 6 similar faces on rotation of the crystal around its axis of symmetry. (2) Hexavalent.
hexadecanal* $Me(CH_2)_{14}CHO$ = 240.4. Polymerizes to the trimer. m.34.
hexadecane* $C_{16}H_{34}$ = 226.4. Cetane, dioctyl. An alkane. White leaflets, m.18, isoluble in water.
hexadecanoic acid* Palmitic acid*. **hydroxy ~ 11- ~** Jalapinolic acid. **16- ~** Juniperic acid.
hexadecanol* Hexadecyl alcohol*.
hexadecanoyl* The radical $Me(CH_2)_{14}CO-$.
hexadecene* $CH_2:CH(CH_2)_{13}Me$ = 224.4. Cetene, cetylene. m.4.
hexadecenoic acid* **7- ~** $Me(CH_2)_7CH:CH(CH_2)_5COOH$ = 240.2. Hypogaeic acid, physetoleic acid. A fatty acid in tallow and some oils (notably, arachis). Colorless needles, m.33, insoluble in water. **(Z)-9- ~** Zoomaric acid, palmitoleic acid. m.40; from marine animal oils.
hexadecimal system See *number systems.*
hexadecoic acid Hexadecanoic acid*.
hexadecyl* Cetyl. The radical $-C_{16}H_{33}$, from hexadecane. **h. alcohol*** $C_{16}H_{33}OH$ = 242.4. 1-Hexadecanol*, cetyl alcohol, cetol, ethal. Colorless wax, m.40, insoluble in water; used in pharmacy (NF). A solution in kerosine forms a film on water surfaces; used to reduce losses due to evaporation. **h.amine*** $C_{16}H_{33}NH_2$ = 241.5. Cetylamine. Colorless solid, insoluble in water. **h. cyanide*** Heptadecane nitrile*. **h. ether** $(C_{16}H_{33})_2O$ = 466.9. Dihexadecyl ether*. White leaflets, m. 55, soluble in water. **h. iodide*** $C_{16}H_{33}I$ = 352.3. 1-Iodohexadecane*. Colorless scales, m.22, insoluble in water. **h.trimethylammonium bromide** Cetavlon.
hexadecylene Hexadecene*.
hexadecyne* $C_{16}H_{30}$ = 222.4. 1-Hexadecine, cetenylene, m.15.
hexadiene* C_6H_{10} = 82.1. **1,4- ~** $CH_2:CH \cdot CH_2CH:CHMe$. **1,5- ~** Biallyl. **2,4- ~** $(MeCH:CH-)_2$. Bipropenyl, dipropylene. Colorless liquid, b.82. **2,5-dimethyl-2,4- ~** $(Me_2C:CH-)_2$ = 110.2. Colorless liquid, m.15, b.135.
hexadienedioic acid* **2,4- ~** $(HOOC \cdot CH:CH-)_2$ = 142.1. Muconic acid. Colorless crystals, m.260, decomp. 272, soluble in water.
hexadienic acid Hexadienoic acid*.
hexadienoic acid* **(*E,E*)-2,4- ~** $MeCH:CH \cdot CH:CH \cdot COOH$ = 112.1. Sorbic acid. Colorless needles, m.134, b.228 (decomp.), soluble in water; from the unripe berries of mountain ash, *Sorbus.* A selective fungistatic for certain foods.
hexadiine Hexadiyne*.
hexadiyne* **1,5- ~** $CH:C \cdot CH_2 \cdot CH_2 \cdot C:CH$ = 78.1. Bi-2-propynyl, bipropargyl. Colorless liquid, b.85.
hexaethylbenzene C_6Et_6 = 246.4. Colorless monoclinics, m.129, insoluble in water.
hexafluorodisilane* See *hexafluorodisilanes* under *silanes.*
hexafluorosilicate* A salt of hexafluorosilicic acid.
hexafluorosilicic acid* H_2SiF_6 = 144.1. Fluorosilicic acid*. Gas, b.−19, decomp. in water.
hexagalloyl mannite $C_{48}H_{38}O_{30}$ = 1095. A brown tanning powder, soluble in water.
hexagon A plane figure with 6 sides and 6 angles. **h. tester** An electrically heated device for determining the flash and fire points of oils.
hexagonal crystal system A crystal system made up of 4 axes; 3 equal and intersecting in one plane at 60°, and one of a different length in a plane at right angles.
hexagonite $CaMg_3(SiO_3)_4Mn$. A mineral from Edwards, N.J.
hexahedro-* Affix indicating 8 atoms bound into a hexahedron.
hexahydrate Containing 6 molecules of water.
hexahydric Containing 6 $-OH$ groups.
hexahydro- Prefix indicating 6 more H atoms than are normally present. **h.anthracene** $C_{14}H_{16}$ = 184.3. White leaflets, m.63, insoluble in water. **1,2,3,4,5,8-h.anthraquinone** $C_{14}H_{14}O_2$ = 214.3. Yellow crystals, m.175. **h.benzene** C_6H_{12} = 84.1. Cyclohexane*, hexamethylene. Saturated benzene. A hydrocarbon in Austrian and Caucasian petroleum. Colorless liquid, d.0.76, b.79, insoluble in water. **h.benzodipyrazolone** Dipyrazolone. **h.benzoic acid** Naphthenic acid. **h.cumene** C_9H_{18} = 126.2. Trimethylcyclohexane*. Colorless liquid, d.0.787, b.139, insoluble in water. **h.cymene** $C_{10}H_{20}$ = 140.3. *p*-Hexahydrocymol. Colorless liquid, d.0.802, b.156, insoluble in water. **h.diphenyl** Phenylcyclohexane. **h.mellitic acid** $C_6H_6(COOH)_6$ = 348.2. Colorless crystals, decomp, by heat, soluble in water. **h.mesitylene** $C_6H_9Me_3$ = 126.2. 1,3,5-Trimethylhexahydrobenzene. Colorless liquid, b.136. **h.naphthalene** $C_{10}H_{14}$ = 134.2. Colorless liquid, d.0.924, b.208. **h.phenol** Cyclohexanol*. **h.pyrazine** Piperazine*. **h.pyridine** Piperidine*. **h.salicylic acid** $C_6H_{10}OH(COOH)$ = 144.2. Colorless crystals, m.110, soluble in water. **h.thymol** Menthol*. **h.toluene** $C_6H_{11}Me$ = 98.2. Colorless liquid, d.0.769, b.101, insoluble in water. **h.xylene** C_8H_{16} = 112.2. ***meta-* ~** Colorless liquid, d.0.771, b.118, insoluble in water. ***para-* ~** Colorless liquid, d.0.769, b.121, insoluble in water.
hexahydroxy- Prefix indicating 6 OH groups. **h.benzene** See *hexahydroxybenzene* under *benzene.*
hexaiodo- Prefix indicating 6 I atoms in a molecule.
hexakontaine Hexacontane*.
Hexalin Trademark for cyclohexanol*.
hexalite An explosive of high detonating velocity. Used to describe (1) Mannitol hexanitrate. (2) RBX. Cyclotrimethylene trinitroamine mixed with beeswax. (3) Hexanitrodiphenylamine. (4) Dipentaerythritol hexanitrate. Cf. *hexolite.*
hexamethylated Containing 6 Me groups.
hexamethylbenzene* C_6Me_6 = 162.3. Mellitene. Colorless rhombs, m.164, soluble in alcohol.
hexamethyldisilane* $Si_2(CH_3)_6$ = 146.4. White solid, m.13.
hexamethyleneamine Hexamethylenetetramine.
hexamethylene (1) 1,6-Hexanediyl†. The radical $-(CH_2)_6-$. (2) Cyclohexane*. **amino ~** $C_6H_{11}NH_2$ = 99.2. Aminohexahydrobenzene. Colorless liquid, b.133. **h.amine** Hexamethylenetetramine. **h.diamine** Triethylenediamine*.

hexamethylenetetramine $(CH_2)_6N_4$ = 140.2. Hexamethyleneamine, hexine, hexamine, methenamine (USP), formin, aminoform, hex, naphthamine. Colorless rhombs, m.280, soluble in water; used as a urinary antiseptic, a deodorant, a reagent for metals and alkaloids, a rubber accelerator, a solid portable fuel (hexamine), in the manufacture of synthetic resins, and as an absorbent for phosgene in gas masks.
h. alliodide A condensation product of h. and allyl iodide; a reagent for cadmium. **h. bromoethylene** Bromalin. **h. camphorate*** $(CH_2)_6N_4 \cdot C_8H_{14}(COOH)_2$. Amphotropine. A molecular combination of camphoric acid and hexamethylenamine. Colorless crystals, soluble in water. **h. methylenecitrate** $C_6H_5O_7(CH_2)_6N_4$ = 342.3. Helmitol. White crystals, m.165. Used in place of h. **h. sulfosalicylate** Hexal. **h. tetraiodide** $(CH_2)_6N_4I_4$ = 647.8. Siomine. Colorless powder, decomp. 138; used in place of iodides.

hexamethylparafuchsin Methyl violet.

hexamethylviolet Methyl violet.

hexamine Hexamethylenetetramine.

hexanal* $Me(CH_2)_4CHO$ = 100.2. Hexoic aldehyde, caproaldehyde. Colorless liquid, b.131.

hexanamide $Me(CH_2)_4CONH_2$ = 115.2. Caproamide.

hexane* $Me(CH_2)_4Me$ = 86.2. Caproyl hydride, hexyl hydride. Colorless liquid, d.0.658, b.69, highly flammable, insoluble in water; a solvent. **iso ~ *** $Me(CH_2)_2CHMe_2$. Ethylisobutane. Colorless liquid, d.0.701, b.62, soluble in alcohol. **amino ~** Hexylamine*. **bromo ~ *** Hexyl bromide*. **chloro ~ *** Hexyl chloride*. **hexachlorocyclo ~** See *hexachlorocyclohexane* under *cyclohexane*. **2-methyl ~ *** $Me_2CH(CH_2)_3Me$. Colorless liquid, d.0.680, b.90, soluble in alcohol. **neo ~** See *triptane*.
h.carboxylic acid Heptanoic acid*. **h.dial*** Adipaldehyde*. **h.diamide*** Adipamide*. **h.dicarboxylic acid*** Suberic acid. **h.dioic acid*** Adipic acid*. **h.dioyl chloride** Adipoyl chloride*. **1,6-h.diyl†** See *hexamethylene.*

hexanitrin Mannitol hexanitrate.

hexanitro- Prefix indicating 6 nitro groups. **h.diphenylamine** $[2,4,6-(NO_2)_3C_6H_2]_2NH$ = 439.2. Dipicrylamine. Yellow crystals; a high explosive. **h.mannitol** A misnomer for mannitol hexanitrate. **h.phenyl sulfide** $[C_6H_2(NO_2)_3]_2S$ = 456.3. A high explosive.

hexanoate* Caproate. A salt or ester of hexanoic acid, containing the radical $C_5H_{11}COO-$.

hexanoic acid* $Me(CH_2)_4COOH$ = 116.2. Caproic acid, hexoic acid, hexylic acid, butylacetic acid. A fatty acid in animal fats. Colorless liquid, b.206, slightly soluble in water. **2-amino ~ *** Norleucine. **2,6-diamino ~ *** Lysine*. **5-methyl ~ *** $Me_2CH(CH_2)_3COOH$ = 130.2. Isoheptoic acid, isoheptylic acid. Colorless liquid, b.210, soluble in alcohol. **4-oxo ~ *** Homo*levulinic acid.* **pentahydroxy ~ *** Galactonic acid.

hexanol* $C_6H_{14}O$ = 102.2. **1- ~** $Me(CH_2)_5OH$. Hexyl alcohol, capryl alcohol. Colorless liquid, d.0.820, b.158, slightly soluble in water. **2- ~** $MeCH(OH)(CH_2)_3Me$. Colorless liquid, d.0.833, b.137, slightly soluble in water. **3- ~** $Et(CHOH)(CH_2)_2Me$. Colorless liquid, d.0.834, b.134, slightly soluble in water.

hexanone* **2- ~** MeCOBu = 100.2. Butyl methyl ketone*. b.127. **3- ~** EtCOPr. Ethyl propyl ketone*. b.123.

hexanoyl* Capronyl. The radical $C_5H_{11}CO-$, from hexanoic acid. **h. chloride** $C_5H_{11}COCl$ = 134.6. Capronyl chloride. Colorless liquid, b.138.

hexaphenyl (1) Sexiphenyl. (2) Indicating 6 phenyl radicals. **h.ethane** $Ph_3C \cdot CPh_3$ = 486.7. The dimer of a free *radical,* q.v. **h.tin** $Ph_3Sn \cdot SnPh_3$ = 700.0. White crystals, m.233.

hexaprismo*- Affix indicating 12 atoms bound into a hexagonal prism.

hexasaccharose A polysaccharide of the general type $(C_6H_{10}O_5)_6 \cdot H_2O$.

hexavalent Sexavalent. Describing an element with a valency of 6.

hexecontane Hexacontane*.

hexenal* $PrCH:CH \cdot CHO$ = 98.1. α,β-Hexenic aldehyde, propylacrolein. An oil $b_{17mm}48$, in green leaves.

hexene* **1- ~** $CH_2:CH(CH_2)_3Me$ = 84.2. Hexylene. Colorless liquid, d.0.683, b.68, insoluble in water. **2- ~** $MeCH:CH(CH_2)_2Me$. **3- ~** $MeCH_2CH:CHCH_2Me$. Δ^2-Hexylene, diethylethylene.
h. glycol Pinaco*. **h. iodide** $C_6H_{12}I_2$ = 338.0. Diiodohexane*. Yellow liquid, decomp. by heat, d.2.024.

hexenic* Hexenoic.

hexenoic acid* $C_6H_{10}O_2$ = 114.1. Hexenic acid. **2- ~** $MeCH_2CH_2CH:CH \cdot COOH$. Propylacrylic acid. White solid, m.32. **3- ~** $MeCH_2CH:CHCH_2COOH$. Hydrosorbic acid. White solid, b.208. Cf. *pyroterebic acid.*

hexenoic aldehyde Hexenal*.

hexenyl* The unsaturated radical $C_6H_{11}-$, from hexene. **h. alcohol** $C_6H_{11}OH$ = 100.2. Colorless liquid, d.0.891, b.137, soluble in water.

hexestrol $HO \cdot C_6H_4(C_2H_5)CH \cdot CH(C_2H_5) \cdot C_6H_4 \cdot OH$. Dihydrodiethylstilbestrol. An estrogenic hormone.

hexine (1) Hexyne*. (2) Hexadiene*. (3) Hexamethylenetetramine.

hexinic acid $C_7H_{10}O_3$ = 142.2. α-Propyltetronic acid. Colorless crystals, m.126.

hexitols $C_6H_{14}O_6$ = 182.2. Hexahydric alcohols, as dulcitol, from hexoses.

hexobarbitone 5-(1-Cyclohexen-1-yl)-1,5-dimethylbarbituric acid. Evipal, Evipan, cyclural, hexobarbital; an anesthetic (BP).

hexobiose Hexabiose.

hexogen Cyclotrimethylenetrinitramine. A constituent of high explosives.

hexoic The hexanoyl* radical. **h. acid** Hexanoic acid*.

hexokinase* Enzyme that catalyzes the formation of a D-hexose 6-phosphate (as, D-glucose 6-phosphate) and ADP, from the corresponding D-hexose and ATP.

hexolite A high explosive prepared from nitrated cyclic compounds (U.S. usage). Cf. *hexalite, hexonit.*

hexone $Me_2CH \cdot CH_2 \cdot COMe$ = 100.2. Isobutyl methyl ketone*. Colorless liquid, b.118, insoluble in water; a solvent.
h. bases Histone bases. Organic bases containing 6 C atoms, formed by hydrolysis of proteins and histones, e.g., lysine.

hexonic acids Acids obtained by the oxidation of hexoses: $C_5H_5(OH)_6 \cdot COOH$. Cf. *gluconic acid.*

hexonit Nitroglycerin containing 10% hexogen; a high explosive. Cf. *hexolite.*

hexoran Hexachloroethane.

hexosans Hemicelluloses which arc hydrolyzed to hexoses.

hexose A carbohydrate (monosaccharide) containing 6 C atoms; as, glucose. Widely distributed in plants and animals; each has 3 optical isomers: D, L, and (±).

hexotriose A carbohydrate (trisaccharide) consisting of 3 hexoses, e.g., raffinose.

hexoylene 2-Hexyne*.

hexuronic acid $CHO(CHOH)_4COOH$ = 194.1. A tetrahydroxy aldehyde acid obtained by oxidation of hexoses; as, glucuronic acid.

hexyl* Enanthyl, capryl. The radical $Me(CH_2)_5-$, from *n*-hexane. **iso ~ *** The radical $Me_2CH(CH_2)_3-$, from isohexane.
h. acetate* $AcOC_6H_{13}$ = 144.2. Colorless liquid, d.0.890,

b.169, insoluble in water. **h. alcohol*** 1-Hexanol*. **h. aldehyde** Hexanal*. **h.amine*** $C_6H_{15}N$ = 101.2. Aminohexane. A poisonous base obtained by autolysis of protoplasm. **h. bromide*** $C_6H_{13}Br$ = 165.1. 1-Bromohexane*. Colorless liquid, d.1.1705, b.156. **h. chloride*** $C_6H_{13}Cl$ = 120.6. 1-Chlorohexane*. Colorless liquid, d.0.8741, b.132. **h. formate*** $HCOOC_6H_{13}$ = 130.2. Colorless liquid, d.0.898, b.153, miscible with alcohol. **h. iodide*** $C_6H_{13}I$ = 212.1. Iodohexane*. Colorless liquid, d.1.453, b.70. **h. methyl ketone*** $C_8H_{16}O$ = 128.2. 2-Octanone*. A liquid, d.0.818, b.173. **h.resorcinol** $C_6H_{13}C_6H_3(OH)_2$ = 194.3. White crystals, m.64, soluble in alcohol.

hexylene Hexene*.

hexylic acid Hexanoic acid*.

hexyne **1-~** $Me(CH_2)_3C:CH$ = 82.1. *n*-Butylacetylene, 1-hexine. Colorless liquid, d.0.712, b.72. **2-~** MeC:CPr. Methylpropylacetylene. Colorless liquid, d.0.7377, b.84.

Heyrovsky, Jaroslav (1890–1967) Czechoslovak chemist, codiscoverer of rhenium (1925) and inventor of the polarograph. Nobel prize winner (1959).

HF High frequency. See *radiation.*

Hf Symbol for hafnium.

HFCS High fructose corn syrup. See *glucose isomerase.*

Hg Symbol for mercury (hydrargyrum).

HHDN* See *insecticides,* Table 45 on p. 305.

hibiscus Musk seed.

Hickman pump A borosilicate glass diffusion pump for producing high vacuum, containing butyl phthalate instead of mercury. Cf. *vacuum pump.*

hiddenite An emerald-green gem variety of spodumene.

hidrotic A substance that causes sweating; as, a diaphoretic.

hielmite Hjelmite. A calcium, iron, manganese, and yttrium tantalate and stannate (Sweden). Black rhombs, d.58, hardness 5.

hierro Spanish for "iron."

Hi-flash Trademark for an oil having a high flash point.

high Above the average; great. **h. explosive** See *high explosives* under *explosives.* **h. flash point** Describing a substance, e.g., an oil, that ignites at high temperatures only. **h.-frequency** HF. Describing a rapidly alternating electric current or wave. **h.-frequency spectrum** An X-ray spectrum produced by a h.-f. current. **h. furnace** A blast furnace. **h.-grade** Describing (1) a pure substance; (2) a concentrated substance; (3) a rich ore. **h.-speed steel** A tool steel rich in carbon. Cf. *ferrotungsten.*

HILAC Heavy ion linear accelerator. A high-voltage bombarding device for producing isotopes.

Hildebrand **H. electrode** A platinum electrode in a bell-shaped tube, through which hydrogen passes; used as a hydrogen electrode, being partly immersed in the solution and partly exposed to the gas. **H. rule** The molar entropy of vaporization is a function of the molar concentration of the vapor involved. Cf. *Trouton's rule.*

hillclimbing Computer optimizing technique that involves "climbing" up a contoured surface (representing the parameter values) to the highest peak (representing the optimum conditions).

Hillebrand, William Francis (1853–1925) American chemist noted for work on mineral analysis.

Hilt's law In a vertical section (e.g., of coal seams) the deeper seams are of higher rank than the upper.

hindrance A retarding factor. **steric ~** See *steric hindrance* under *steric.*

hinokitiol $C_{10}H_{12}O_2$ = 164.2. A constituent of the coloring matter of certain Japanese woods. m.52.

Hinshelwood, Sir Cyril (1897–1967) British chemist. Nobel prize winner (1956). Noted for work on reaction kinetics.

Hipersil Trademark for a Fe-Si alloy having magnetic properties of special use in electric transformers.

hippurate* A salt of hippuric acid, containing the radical $PhCONHCH_2COO-$.

hippuric acid* $PhCONHCH_2COOH$ = 179.2.*N*-Benzoylglycine†, benzamidoacetic acid. Colorless crystals, d.1.371, m.188 (decomp.), soluble in hot water.

hippuroyl* The radical $PhCONHCH_2C:O-$, from hippuric acid. **h. hydrazine*** $Ph \cdot CO \cdot NH \cdot CH_2CO \cdot NHNH_2$ = 193.2. Colorless crystals, m.162.

hiptagenic acid $NO_2 \cdot (CH_2)_2COOH$ = 117.1. 3-Nitropropanoic acid*. m.67. A hydrolysis product of hiptagin.

hiptagin $C_{10}H_{14}O_9N_2$ = 306.2. A glucoside from *Hiptage madablata.* Colorless, silky needles, m.112.

hiragashira oil An oil from the livers of *Scoliodon lacticaudus,* containing scoliodonic acid.

hiragonic acid $C_{16}H_{20}O_2$ = 244.3. A liquid, unsaturated fatty acid from sardine oil.

hircine (1) A strongly odorous fossil resin. (2) The odorous principle in the suet of goats.

hirudin The active principle of a secretion from the buccal glands of leeches. It prevents the coagulation of blood.

hirudo The leech *Sanguisuga medicinalis;* formerly used to remove blood by suction, assisted by hirudin.

His* Symbol for histidine.

histamine $C_3H_3N_2 \cdot CH_2 \cdot CH_2 \cdot NH_2$ = 111.1. Ergamine, 1*H*-imidazole-4-ethanamine. m.86. An amine derived from histidine in ergot. Occurs in plant and animal tissues. A powerful uterine stimulant and vasodilator. **h. hydrochloride** The soluble hydrochloride of histamine. **h. phosphate** $C_5H_9N_3 \cdot 2H_3PO_4$ = 212.2. Colorless prisms, m.129, soluble in water. A diagnostic aid for gastric acid secretions (USP, BP).

histazarin Hystazarin.

histidine* $(C_3H_3N_2)CH_2 \cdot CH(NH_2)COOH$ = 155.2. His*. 2-Amino-3-imidazolylpropanoic acid. An amino acid derived from the protamin of fishes, or by the action of sulfuric acid on ptomaines. D-~ White scales, m.288, soluble in water. (±)-~ Tetragonal prisms, m.285, soluble in water. L-~ Colorless leaflets, m.277. **thiol ~** A constituent of proteins.

histidyl* The radical $C_6H_8O_2N_3-$, from histidine.

histo- Prefix (Greek) denoting "tissue."

histochemistry The chemistry of the histological structures of the body.

histogram A diagram derived by dividing a number of values obtained for a particular determination into classes differing by a standard amount, and plotting the number of samples in each class against the corresponding values. The probability curve (see *normal probability curve* under *normal*) can then be obtained.

histology The study of the structure of tissues. **normal ~** The study of healthy tissues. **pathological ~** The study of diseased tissues. **phyto ~** The science of plant tissues. **zoo ~** The science of animal tissues.

histolysis The disintegration or liquefaction of tissues.

histone (1) See *hexone bases.* (2) A protein from cell nuclei, soluble in water, coagulated by heat.

histoplasmin A liquid concentrate of the soluble growth products of the fungus *Histoplasma capsulatum.* A clear, amber liquid, miscible with water; a diagnostic aid for histoplasmosis, a disease due to *Histoplasma* (USP).

Hittorf, Johan Wilhelm (1824–1914) German physicist and discoverer of cathode rays. **H. cell** Transference cell. **H. number** Transport number*. **H. tube** A modified Crookes tube.

HLB Hydrophilic Lyophilic Balance –JT Davies 2nd Int Congr of Surface Activity I pp 426–439

hjelmite Hielmite.
HMT Hexamethylenetetramine.
Ho Symbol for holmium.
hoangnan The bark of *Strychnos malaccensis* (Loganiaceae); an arrow poison.
hoarhound Horehound.
Hoff, Jacobus Hendricus van't See *van't Hoff.*
Hoffman **H., Friedrich (1660–1742)** German physician and professor of chemistry. **H. clamp** A clamp with one V-shaped and one flat jaw. **H. drops** A mixture of alcohol and ether. **H. electrolytic apparatus** Two inverted burets with platinum electrodes, with an overflow tube and bulb, to demonstrate the decomposition of water. **H., Roald (1937–)** American chemist. Nobel prize winner (1981). Noted for extension of Debye-Hückel theory and Woodward-H. rules.
Hofmann, August Wilhelm von (1818–1892) German chemist, founder of the coal tar industry. **H. reaction** The shortening of C chains by addition of bromine to an alkaline solution of an acid amine. The amine of the shorter C chain is formed: $R \cdot CONH_2 + Br_2 + 3NaOH = R \cdot NH_2 + Na_2CO_3 + NaBr + H_2O$. **H. sodium press** An iron barrel, piston, and screw used to make sodium wire or ribbon.
Hofmeister series The order in which anions and cations may be arranged according to their powers of coagulation of an emulsoid, in neutral, acid, or alkaline solutions. Cf. *lyotropic series.*
hog bane Hyoscyamus.
hogbomite $MgO(AlFe)_2 \cdot TiO_2$. A mineral from Cameroon.
hoggin A natural deposit of gravel held together with a little clay.
hog gum Kuteera gum, gum hogg. A variety of Bassora gum from *Sterculia urens;* used in marbling paper.
hogshead hhd; a measure: 52.5 imperial gal (238.5 liters). (2) A cask of capacity 100–140 gal (U.S.); 50–100 gal (U.K.). **Irish ~** 52 gal. **United Kingdom ~** 54 gal.
hogweed Scoparius.
Holborn-Kurlbaum pyrometer An instrument for determining high temperatures in furnaces.
holder A device for retaining or holding objects in a definite position. **animal ~** A wire net or metal tube, used to secure small animals for biological experiments. **buret ~** A clamp to hold burets. **clamp ~** Boss. Iron screws to attach clamps to supports. **crucible ~** A wire clamp or frame for holding crucibles. **culture dish ~** A wire frame for holding petri dishes together in a sterilizer. **gas ~** See *gasholder* under *gas.* **plate ~** A lighttight box for holding photographic plates. **watch glass ~** A pair of circular wire springs for holding 2 watch glasses together.
hole See *exciton.*
hollyhock (1) Althaea. (2) The flowers of *Althaea rosea* (Malvaceae); an emollient.
holmia (1) A mixture of the oxides of holmium and dysprosium. (2) Holmic oxide.
holmic A trivalent holmium compound. **h. oxide** Ho_2O_3 = 377.9. Gray powder, insoluble in water.
holmium* Ho = 164.9304. A rare-earth metal, at. no. 67, in gadolinite, discovered (1879) by Cleve. m.1470, b.2650. Valency 3. Its salts are slightly yellow. **eka ~** Einsteinium. **h. chloride** $HoCl_3$ = 271.3. Holmic chloride. Colorless powder, soluble in water. **h. oxalate*** $Ho_2(C_2O_4)_3$ = 593.9. Colorless powder, insoluble in water. **h. oxide*** Holmic oxide.
holo- Prefix (Greek) indicating the "whole" or "entirety."
holocellulose The total carbohydrate constituents of wood.
Holocene See *geologic eras,* Table 38.
holography A photographic method giving a 3-dimensional image. It uses interference and diffraction effects to store, and subsequently release, light waves. It is therefore applicable to acoustic rays, X-rays, and visible rays.
holohedral A crystal which has the full number of faces required for the maximum and complete symmetry of the system. Cf. *hemihedral.*
holosiderite Meteoric iron.
Holtz, Wilhelm (1836–1913) German physicist. **H. machine** An induction machine.
homarine $C_7H_7NO_2$ = 137.1. *N*-Methyl-2-pyridinecarboxylic acid. In shellfish, *Arbatia pustulosa.* Cf. *trigonelline.*
homatropine $C_{16}H_{21}O_3N$ = 275.3. An alkaloid condensation product of tropine and mandelic acid: tropine mandelate. Colorless prisms, m.97, slightly soluble in water. **h. hydrobromide** $C_{16}H_{21}O_3N \cdot HBr$ = 356.3. Homotropine hydrobromide. A very poisonous, soluble form of h. Colorless prisms, m.213, soluble in water. Used in opthalmology as a mydriatic (USP, EP, BP). **h. hydrochloride** $C_{16}H_{21}O_3N \cdot HCl$ = 311.8. Homotropine hydrochloride. A soluble form of h.; white crystals, m.216, soluble in water; used in ophthalmology. **h. methyl bromide** $C_{17}H_{24}O_3NBr$ = 370.3. White powder, m.195, used similarly to h. hydrobromide (USP).
homeopathic Homoeopathic. Pertaining to homeopathy. **h. dose** An extremely small amount. **h. vial** An elongated bottle for h. tablets.
homeopathy A system of medicine employing extremely small doses of agents, which if administered in health in slightly larger quantities, would produce symptoms similar to those for the relief of which they are given.
homilite $FeCa_2B_2Si_2O_{10}$. A native borosilicate, d.3.0–3.28, hardness 5.5.
homo- (1) Prefix (Greek) indicating "similarity." (2)* Indicating the addition of a CH_2- group, particularly in sterol structures. **di ~*** Indicating the addition of 2 CH_2- groups.
homoantipyrine $C_{12}H_{14}ON_2$ = 202.3. 1-Ethyl-5-methyl-2-phenyl-3-pyrazolone. Colorless crystals, m.73.
homoarecoline $C_9H_{15}O_2N$ = 169.2. Ethoxyarecaidine. Yellow liquid, soluble in water. **h. hydrobromide** $C_7H_{10}EtO_2N \cdot HBr$. Colorless crystals, m.118, soluble in water; an anthelmintic.
homoatomic ring Homocycle.
homocamphoric acid $C_{11}H_{18}O_4$ = 214.3. An acid, similar to camphoric acid, but having a $-CH_2-$ bridge. Distillation of its calcium salts yields camphor.
homocentric Having the same center. **h. rays** Light rays that are parallel or have a common focus.
homochelidonine $C_{21}H_{23}O_5N$ = 369.4. An alkaloid from the seeds of *Chelidonium majus* and sanguinaria. Colorless crystals, slightly soluble in alcohol.
homochemical A binary compound.
homochromic Possessing the same color, but having a different molecular composition.
homochromoisomers Substances having similar absorption spectra, but different molecular compositions.
homocycle, homocyclic Pertaining to compounds which contain a closed chain or ring of atoms of the same type, usually C atoms; e.g., benzene.
homodetic Describing polypeptide rings consisting of amino acid residues.
homogeneity, homogeneous Of uniform or similar nature throughout. Antonym: heterogeneous.
homogentisic acid $(HO)_2C_6H_3 \cdot CH_2COOH$ = 168.2. 2,5-Dihydroxyphenylacetic acid. Colorless crystals, m.147; an

intermediate in the oxidation of tyrosine and phenylalanine. Excreted in the urine in alkaptonuria, an inborn error of *metabolism,* q.v.

homoisohydric Having the same ions and a constant hydrogen-ion concentration in a solution.

homolog, homologue Member of a series of compounds whose structure differs regularly by some radical, e.g., $=CH_2$, from that of its adjacent neighbors in the series.

homologous Of similar structure. **h. lines** A pair of spectrum lines, the relative intensities of which are independent of the electrical discharge producing them. **h. series** A series of organic compounds which differ by CH_2 or some similar multiple. Cf. *alkanes.*

homology The similarity of organic compounds and their gradation of properties, as shown by a homologous series.

homolysis Describing a reaction in which a bond is severed and each fragment retains a bonding electron. $A{:}B \rightarrow A\cdot + \cdot B$. Cf. *heterolysis.*

homometric Having the same X-ray pattern.

homonataloin $C_{22}H_{24}O_9$ = 432.4. A constituent of Natal aloes.

homophase Forming a single and similar state of aggregation. Cf. *phase.*

homopolar (1) Having an equal distribution of electric charges between 2 atoms (generally C), neither of which becomes negative or positive; hence, a bond in which both atoms share the electron pair equally. (2) Covalent.

homopyrocatechol $Me \cdot C_6H_3 \cdot (OH)_2$ = 124.1. 3,4-Dihydroxytoluene*. Colorless prisms, m.65, soluble in water. Cf. *orcinol.*

homosalicylic acid $C_8H_8O_3$ = 152.2. Cresotic acid. A compound containing a $-COOH$, $-OH$, and CH_3- group attached to the benzene ring; 10 possibilities.

homosaligenin Salicyl alcohol*.

homotaraxasterol $C_{25}H_{40}O$ = 356.6. A sterol, m.164, from dandelion.

homotope An element in a vertical group of the periodic table; thus Br is a h. of Cl, Tl a h. of Ga.

Honduras bark Cascara amarga.

honey Mel. **h.dew** A sugary, sticky excretion from certain leaf-sucking insects; toxic to bees and a contaminant of honey.

honeystone $Al_2C_{12}O_{18} \cdot 18H_2O$. Aluminum mellate, mellite. A natural hydrocarbon-aluminum derivative (Germany). Yellow, resinous mineral, d.1.6, hardness 2.

Hönigschmid, Otto (1878–1945) German chemist noted for inorganic work.

honing guide A device for keeping a dissecting knife at the proper angle while honing.

Honvan Trademark for diethylstilbestrol diphosphate.

hood A glass-enclosed ventilator shaft or flue in the laboratory; carries away fumes.

hoof and horn meal A fertilizer (11–15% N) made by processing, drying, and grinding hooves and horns.

Hooke, Robert (1635–1703) English philosopher. **H.'s law** The strain, or alteration in length produced, is proportional to the stress applied, so long as the elastic limit is not exceeded.

hoolamite A mixture of pumice, iodine pentaoxide, and fuming sulfuric acid used as absorbing agent and reagent for carbon monoxide. It changes from white through blue-green to violet-brown.

hopcalite (1) A mixture of cobalt, copper, silver, and manganese oxides, used in gas masks as a catalyst to oxidize carbon monoxide. (2) A mixture of manganese dioxide and cupric oxide (3:2), prepared by heating hydrated manganese oxide and copper carbonate.

Höpfner process A method of recovering copper by electrolysis.

Hopkins, Sir Frederick Gowland (1861–1947) British chemist, discoverer of the importance of vitamins in diet. Nobel prize winner (1929).

Hopogan Trademark for magnesium peroxide.

hop oil A green, odorous oil from hops. Its chief constituents are terpenes, humulene, and geraniol; a flavoring. Cf. *humulene.* **Spanish ~** Origanum oil.

hopper A funnel-shaped trough, or trap inlet.

Hoppe-Seyler, Ernst Felix Immanuel (1825–1895) German physiologist noted for the development of biochemical analysis.

hopred $C_{38}H_{26}O_{15}$ = 722.6. A hydrolytic split product of the coloring matter phlobabene, from hops.

hops Humulus. The dried spikes or strobiles of *Humulus lupulus* (Urticaceae, Moraceae). They contain tannin, humulone, and lupulin, and have a bitter, aromatic taste; used in beer as a flavoring and preservative. **h. oil** See *hop oil.* **h. substitutes** Bitter principles; as, quassia or camomile.

horbachite A native sulfide of gold, iron, and nickel.

hordein A protein of barley, the seeds of *Hordeum sativum.*

hordeine Hordenine.

hordenine $HO \cdot C_6H_4(CH_2)_2N \cdot Me_2$ = 165.2. Hordeine, ephedrine. An alkaloid, m.118, from malted barley.

hordeum Barley.

horehound The dried leaves and flowering tops of *Marrubium vulgare;* an expectorant and stimulant.

hormone A substance produced by the internal secretion of endocrine glands and certain tissues. Hormones are chemical messengers, which circulate in the bloodstream and coordinate the functions of organs and other endocrine glands by exciting them to activity; as, epinephrine, estrogen, insulin, testosterone, levothyroxine. **food ~** Vitamin. **plant ~** A h. which promotes growth of plants, the main classes being auxins, cytokinins, and gibberellins.

horn A substance composed mostly of keratin and containing insoluble mineral salts, particularly calcium phosphate.

hornblende $Ca(MgFe_3) \cdot (SiO_3)_4$. An amphibole in colors ranging from black through green and olive to white.

horn lead Phosgenite.

horn quicksilver Native mercurous chloride.

horn silver Argentum cornu.

horse chestnut The seeds of *Aesculus hippocastanum;* an astringent and febrifuge. **h. c. tannin** $C_{26}H_{24}O_{11}$ = 512.5. A tannin from the h. c.

horsehair The hair from the mane or tail of horses; used for mattresses and fabrics. **vegetable ~** Spanish moss. The fibers of *Tillandsia usneoides* (Bromeliaceae), America; used for pillows and mattresses.

horsemint oil Monarda oil. A brown essential oil, from the leaves and stems of *Monarda punctata;* used in the preparation of liniments. Cf. *monarda.*

horsenettle See *Solanum.* **h. berries** Solanum.

horse oil Yellow oil from horse fat; used in soap manufacture.

horsepower hp, HP; a unit of power or the rate of doing work. 1 hp = 33,000 ft·lb/min = 550 ft·lb/s = 76.1 kg·m/s = 745.7 W. **boiler ~** 9809.5 W. **metric ~** E.g., *chevaux* (French) or *Pferdekraft* (German): 735.5 W.

horseradish The root of *Cochlearia armoracia* (Cruciferae); a condiment. It contains potassium myronate.

horsetail See *propene*tribarboxylic acid, *Equisetum.*

horsfordite Cu_6Sb. A native copper-antimony ore.

Hortvet **H. cryoscope** An apparatus for determining the freezing point. **H. tube** A graduated glass centrifuge tube for determining the volume of a sediment.

Horwood process A method of flotation of partly roasted sulfides of Fe, Cu, Pb, and Zn, to separate the zinc.

Hoskin furnace An electric heating device for metallurgical and dental laboratories.

hot (1) Having or producing the sensation of heat. (2) A colloquialism describing material that is producing a dangerous degree of radioactivity. **h.-air sterilizer** See *sterilizer.* **h.bed** A h. mass of decomposing animal manure, used to accelerate plant growth. **h.-water funnel** See *hot-water funnel* under *funnel.* **h.-wire gage** See *hot-wire gage* under *gage.*

houdriforming reaction A dehydration or isomerization reaction which requires precious metals on an acidic support as catalyst.

hour Abbrev. h; a unit of time, 60 min, 1/24 of a day.

Howard **H. chamber** A microscope slide with a small counting chamber; used for yeast cells, molds, etc. **H. pan** A jacket and coil type of vacuum evaporating kettle, used to concentrate sugar juice.

Howe, Harrison Estell (1881–1942) American chemist, noted for his development of industrial chemistry.

howlite $2CaO \cdot 5B_2O_3 \cdot 2SiO_2 \cdot 5H_2O$. A mineral from Long Beach, Calif.

hp HP; abbreviation for "horsepower."

H salt The sodium salt of H acid.

HTGR High-temperature, gas-cooled *reactor.*

H.T.P. High-test hydrogen peroxide (more than 75% H_2O_2 by weight).

H.T.S.T. See *H.T.S.T. pasteurization* under *pasteurization.*

huantajayit A native sodium-silver chloride.

huanuco bark Cinchona bark.

Hübl **H. number** Iodine number. **H. solution** A solution of iodine and mercuric chloride used to determine iodine numbers of unsaturated compounds.

hübnerite $MnWO_4$. A native ferruginous tungstate occurring in western U.S. Red crystals, d.7.17. Cf. *ferberite.*

Hückel theory See *Debye-Hückel theory.*

Hudson, Claude Silbert (1881–1952) American chemist, noted for work on enzymes and carbohydrates. **H. rule** To assign α and β to a sugar in comparison with their molecular rotation, R_α and R_β, select names so that $R_\alpha - R_\beta$ is equal to and of the same sign as the difference for the 2 corresponding forms of glucose.

hue A property of *color,* q.v.

Huff separator An electrostatic concentrator for crushed ores.

huile French for "oil."

Hulett, George Augustus (1867–1955) American physical chemist. **H. still** A device for distilling mercury.

hülsneride $FeWO_4$. A native ferrous tungstate.

Humatin Trademark for paromycin.

humboldite A variety of melilith.

Humbold penetrometer A device for testing asphalt by measuring its resistance to the pressure of a needle.

humectant A hygroscopic substance (e.g., glycerol) used to ensure the absorption of a certain amount of atmospheric moisture by the material to which it is added. Cf. *plasticizer.*

Hume-Rothery rule The majority of elements in the B subgroups of groups IV, V, VI, and VII of the periodic table have $(8 - N)$ near neighbors in the solid state, where N is the number of the group to which the element belongs.

humic acid A general term for acids derived from humus.

humicolin An antibiotic from *Aspergillus humicola* from cave soils. A weakly acidic, yellow oil, which inhibits spore germination of many common fungi.

humidity Dampness, hygrometric state. The amount of water vapor in the air. **absolute ~** The weight of water vapor in a unit of moist air, in g/m^3. **equivalent ~** The relative h. of the air in contact with a particular substance when it neither gains nor loses moisture. **relative ~** Abbrev. r.h., R.H. The weight of water vapor contained in a given volume of air expressed as a percentage of the weight that would be contained in the same volume of saturated air at the same dry-bulb temperature. See *dew point.*

humidor A compartment whose atmosphere is kept saturated with water vapor.

huminite A native hydrocarbon from Sweden.

humite Chondrodite. A basic magnesium fluorosilicate.

hummer A microphone used in conductivity measurements to determine, by the absence of hum, a point at which no current is passing.

humoceric acid $C_{19}H_{34}O_2 = 294.5$. An acid from peat wax.

humulene $C_{15}H_{24} = 204.4$. 3 isomeric sesquiterpenes (α-, β-, and γ-) from hop oil. Colorless oils.

humulic acid $C_{15}H_{22}O_4 = 266.3$. Humulinic acid. Product of the action of alkali on humulon.

humulin Lupulin.

humulone $C_{21}H_{30}O_5 = 362.5$. A bitter constituent, m.66, of the soft resin of hop lupulin. Cf. *lupulone.*

humulo tannin $C_{25}H_{24}O_{13} = 532.5$. A white, amphoteric powder from hops. It loses water at 130 and forms a phlobaphen.

humulus Hops.

humus (1) The top layer of the soil, containing the organic decomposition products of vegetation (leaf mold, etc.). (2) The decayed products of plant life.

Hund's rules (1) Electrons tend to avoid the same orbital. (2) Two electrons, each singly occupying 2 equivalent states, tend to have parallel spins in the lowest state.

Hünefeld solution A reagent for blood: alcohol 25, glacial acetic acid 1.5, chloroform 5, turpentine 15 ml.

Huntington mill An ore crusher: steel rollers are pressed by centrifugal force against a heavy encasing steel ring.

Huppert's reagent A 10% aqueous solution of calcium chloride, used to detect biliary pigments in urine.

huttonite $ThSiO_4$. A monoclinic mineral, d.7.18, in New Zealand beach sands.

Huygens, Christian (1629–1695) Dutch physicist, noted for the development of optical instruments, and a wave theory of light. **H. ocular** Negative ocular. A magnifying eyepiece lens on a microscope or telescope: 2 convex lenses mounted convex sides down. Cf. Ramsden *ocular.*

HWR Heavy water *reactor.*

hyacinth (1) A transparent, red gem variety of zircon. (2) Erroneously applied to light-colored garnet, to red spinel from Brazil, and to red quartz. **Ceylon ~** Garnet. **false ~** Garnet. **oriental ~** Corundum (rose-colored).

hyacinthozontes A sapphire-blue gem beryl.

hyaenasic acid Pentacosanoic acid*.

Hyalase Trademark for hyaluronidase.

hyaline Resembling glass.

hyalite A clear, colorless gem opal.

hyalophane $(K_2,Ba)Al_2(SiO_3)_4$. A barium feldspar.

hyalosiderite A deep-olive-green gem olivine.

hyaluronic acid $-(C_8H_{13}O_4N)_n \cdot O \cdot (C_6H_8O_5)_nO-$ consisting of alternate 1,4-linked *N*-acetylglucosamine and glucuronic acid units. A viscous, highly polymerized mucopolysaccharide in mammalian fluids and connective tissue.

hyaluronidase Hyalase, Wydase. A mucolytic enzyme from

mammalian testes. Used to facilitate the spread of fluid in the tissues so as to promote absorption; for bruising (USP, BP).

hybrid An organism, usually a plant, obtained by crossbreeding.

hybridization (1) Producing a hybrid. (2) A process in which electrons are raised from a lower to a higher energy state, and which lowers the total energy.

Hycar Trademark for an oil-resistant synthetic rubber.

hychlorite Antiformin.

Hycoloid Trademark for a cellulose nitrate plastic.

hydantoic acid* $NH_2C(O)NHCH_2COOH$ = 118.1. Ureidoacetic acid*, *N*-(aminocarbonyl)glycine†, glycoluric acid. Colorless prisms, m.171, slightly soluble in water. **iso ~** $NH_2 \cdot C(:NH) \cdot O \cdot CH_2 \cdot COOH$.

hydantoin $\underline{CH_2 \cdot NH \cdot CO \cdot NH \cdot CO}$ = 100.1. Glycolylurea, imidazoledione. Colorless needles, m.216, soluble in hot water or alcohol. Cf. *barbituric acid.* **5-ureido ~** Allantoin.

hydnocarpic acid $(C_5H_7)(CH_2)_{10}COOH$ = 242.3. 2-Cyclopentenyl-11-undecanoic acid*. White crystals, m.60, from the oil of *Hydnocarpus* species; formerly used to treat leprosy. Cf. *chaulmoogric acid.*

Hydnocarpus A genus of trees (Bixaceae), whose seeds yield oils formerly used to treat leprosy; as, kavatel oil, maroti oil. Cf. *chaulmoogra oil.* **h. oil** An oil of *H. wightiana,* d.0.947, m.21–24. Cf. *Moogrol.*

hydr- Prefix (Greek) indicating water or hydrogen. Cf. *hydro-, hydrido.*

hydracetamide $(MeCH)_3N_2$ = 112.2. Yellow powder, soluble in water.

hydracid An acid without O atoms, as HCl.

hydracrylic acid $CH_2OH \cdot CH_2COOH$ = 90.1. 3-Hydroxypropionic acid*. Colorless crystals, decomp. by heat. **amino ~** Serine*. **phenyl ~** Tropic acid*.

hydroacrylo The 3-hydroxypropylidyne* radical $HOCH_2 \cdot CH_2 \cdot C\equiv$. **h.nitrile** $HO(CH_2)_2CN$ = 71.1. 3-Hydroxypropanenitrile*. Colorless liquid, d.1.059, b.221, soluble in water.

hydragog(ue) A cathartic producing a watery stool; e.g., jalap.

hydralazine hydrochloride $C_8H_8N_4 \cdot HCl$ = 196.61. 1-Hydrazinophthalazine hydrochloride, White, bitter powder, m.275, soluble in water; a vasodilator used to treat hypertension (USP).

Hydralo Trademark for an activated alumina desiccant.

hydramine A hydroxyalkylamine or dihydric alcohol in which an OH group is replaced by the NH_2 group: $HO \cdot R \cdot NH_2$.

hydramyl Pentane*.

hydrangea Seven barks. The dried roots of *H. arborescens* (Saxifragaceae); a cathartic and diuretic.

hydrangin $C_9H_6O_3$ = 162.1. m.230. From plants, including hydrangea.

hydranthranol Hydroanthranol.

hydrargillite $Al_2O_3 \cdot 3H_2O$. A native aluminum hydroxide. Colorless needles.

hydrargyrate Mercurate*.

hydrargyrol $C_6H_4OH \cdot SO_3Hg$ = 373.7. *p*-Phenylmercury thiosulfate; an antiseptic.

hydrargyrum Latin for "mercury."

hydrastin A concentrate containing the principles of golden seal.

hydrastine $C_{21}H_{21}O_6N$ = 383.4. An alkaloid from *Hydrastis canadensis,* golden seal. **(−)-~** White prisms, m.132, slightly soluble in water.

hydrastinine $C_{11}H_{13}O_3N$ = 207.2. A decomposition product of hydrastine. Colorless crystals, m.116, slightly soluble in water.

hydrastis Golden seal, yellow pucoon, orange root, turmeric root. The dried roots of *H. canadensis* (Ranunculaceae). **h. alkaloids** Compounds of the isoquinoline group; as, hydrastine.

hydratated Hydrated.

hydrate (1)* A crystalline substance containing one or more molecules of water of crystallization; as, $Na_2SO_4 \cdot 7H_2O$, sodium sulfate heptahydrate. (2) A substance containing water combined in the molecular form, as, $H_2SO_4 \cdot H_2O$ or $[Zn(H_2O)_6]^{2+}$, hexaaqua zinc. (3) Hydroxide. (4) A *solvate,* q.v. **carbo ~** See *carbohydrates.*

hydrated Combined with water. Cf. *hydrate, solvate.* **h. ion** An ion surrounded by oriented water molecules. **h. lime** Dry calcium hydroxide.

hydration Combination with water, but not necessarily in the form of a hydrate. **heat of ~** The energy difference between anhydrous and hydrous compounds (in joules or cal).

hydratisomery Isomerism depending on the structure of hydrated crystals. Thus, $CrCl_3 \cdot 6H_2O$, chromic chloride hexahydrate, occurs as:

Violet	$[Cr(H_2O)_6]Cl_3$
Grayish-green	$[CrCl(H_2O)_5]Cl_2 \cdot H_2O$
Green	$[CrCl_2(H_2O)_4]Cl \cdot 2H_2O$

hydratropic acid* $Ph \cdot CH \cdot MeCOOH$ = 150.2. 2-Phenylpropanoic acid*. Colorless crystals, m.265, soluble in water. **hydroxy ~** Atrolactic acid.

hydraulic Pertaining to liquids, especially water and oils. **h. cement** A cement that sets under the action of water instead of atmospheric carbon dioxide and moisture. **h. lime** Limestone which, on burning, yields a quicklime that will set or harden under water. Cf. *hydrated lime.* **h. mining** Excavation by means of a strong jet of water. **h. mortar** A mortar that hardens under water. **h. press** High pressure press with water or other fluid to transmit the pressure.

hydraulics Hydromechanics. The study of the mechanical properties of liquids.

hydrazi-* Prefix indicating the $-NH \cdot NH-$ group attached to the same atom. Cf. *hydrazo.*

hydrazid(o)-* Infix indicating an organic derivative of As or P in which an OH group has been replaced by a $-NH \cdot NH_2$ group. Cf. *hydrazono.*

hydrazide* An acyl hydrazine of the type $R \cdot CO \cdot NH \cdot NH_2$ or $R \cdot SO_2 \cdot NH \cdot NH_2$. **acet ~** See *acethydrazide.* **hydroxy ~** See *hydroxyhydrazides.*

h. ion* The anion $H_3N_2^-$.

hydrazidine (1)* A compound with the formula $RC(NH \cdot NH_2):N \cdot NH_2$.

hydrazido- See *hydrazid(o)-.*

hydrazimethylene The theoretical compound $\underline{CH_2 \cdot NH \cdot NH}$, from which the hydrazi compounds are derived. **benzoylphenyl ~** $NH \cdot NH \cdot CPh(COPh)$. Colorless crystals, m.151. **diphenyl ~** $(NH \cdot NH \cdot CPh)_2$. Colorless crystals, m.147.

hydrazine* $H_2N \cdot NH_2$ = 32.05. Diamine, diamidogen. Colorless liquid, d.1.08, m.1.4, b.113, soluble in water. Used as a reducing agent in organic synthesis; and in liquid ammonia solution, as a nitridizing agent, analogously to the oxidizing action of hydrogen peroxide in aqueous solution. It gives the radicals:

$H_2N \cdot NH-$	Hydrazino*, hydrazyl*
$H_2N \cdot N=$	Hydrazono*

−HN·NH− .... Hydrazo*, hydrazi*
=N·N= .... Azino*

Cf. *hydrazinium.*
h.carbamide Biurea*. **h.carboxamide** Semicarbazide*. **h.carboxylic acid** Carbazic acid*. **h.dicarbamide** Biurea*. **h. chloride, h. dihydrochloride** $N_2H_4 \cdot 2HCl$ = 105.0. Colorless crystals, m.198, soluble in alcohol; used in organic synthesis. **h. formate** $N_2H_4 \cdot 2HCOOH$ = 124.1. Cubic crystals, m.128, soluble in water. **h. hydrate*** $N_2H_4 \cdot H_2O$ = 50.6. Colorless, fuming liquid, b.119. **h. nitrate** $N_2H_4 \cdot NO_3$. White crystals; a high explosive. **h. sulfate** $N_2H_4 \cdot H_2SO_4$ = 130.1. Colorless scales, m.254, soluble in water; a reagent for separating copper.

hydrazines* Compounds from hydrazine: *methylhydrazine,* $MeNH \cdot NH_2$; *phenylhydrazine,* $PhNH \cdot NH_2$. They are analogous to peroxides. **acyl ~** Hydrazides*. **alkyl ~** Compounds of the type $RNH \cdot NH_2$, $R_2N \cdot NH_2$, or $RNH \cdot NHR$, in which the H is replaced by an alkyl radical. **aryl ~** Organic compounds from hydrazine, in which the H is replaced by an aryl radical.

hydrazinium* Hydrazonium. Salts of hydrazine and an acid, with cations of the type $R \cdot NH \cdot NH_3^+$, $RNH_2^+ \cdot NH_2$, $RNH_2^+ \cdot NH_2^+R$. h.(1+) and h.(2+) mean, respectively, that 1 or both N atoms bear a charge.

hydrazino-* Infix for the radical $H_2N \cdot NH-$, from hydrazine. **h. acids** Organic compounds, derived from hydrazine, of the general type $H_2N \cdot NH \cdot R \cdot COOH$. **h.carbonyl**† See *carbazoyl.*

hydrazo- (1)* Infix indicating the bridge $-NH \cdot NH-$ attached to different atoms. (2) The azino* group. Cf. *hydrazi-, biimino.* **h.amine** Triazane*. **h.benzene** $(Ph \cdot NH)_2$ = 184.2. **1,1′-~** 1,1′-Diphenylhydrazine*. **1,2′-~** 1,2′-Diphenylhydrazine*, *N*-bianiline. Colorless scales, m.132, slightly soluble in water; an intermediate in the manufacture of benzidine. **h.benzoic acid** See *hydrazobenzoic acid* under *benzoic acid.* **h. compounds** Compounds of the general formula $RNH \cdot NHR$. **h.dicarbonamide** $(NH_2 \cdot CO \cdot NH)_2$ = 86.1. Hydrazoformamide. Colorless leaflets, m.245, soluble in water. **h.dicarbonimide** Urazole. **h.formamide** H. dicarbonamide. **h.toluene** $MeC_6H_4NH \cdot NHC_6H_4Me$ = 212.3. Ditoluylhydrazine. **1,2-~** Colorless leaflets, m.156 (decomp.), soluble in water. **1,3-~** Colorless crystals, soluble in alcohol. **1,4-~** Monoclinic, colorless crystals, m.130, insoluble in water.

hyrazoate A compound from hydrogen azide, $NH \cdot N:N$.

hydrazoic acid Hydrogen azide*.

hydrazoin A compound containing the radical $-CH:(N_2)=$. See *hydrazone**.

hydrazone* A compound of general formula $R:N \cdot NR_2$, as resulting from the action of hydrazines with aldehydes or ketones. **diphenyl ~** Osazone.

hydrazonic acid A compound of the general type $R \cdot C(OH):N \cdot NH_2$. Metameric with amide oximes and tautomeric with acethydrazides. **sulfino ~*** Suffix indicating the modified sulfinic acid group, $-S(OH):N \cdot NH_2$. **sulfono ~*** Suffix indicating the modified sulfonic acid group, $-S(O)(OH):N \cdot NH_2$.

hydrazonium Hydrazinium*.

hydrazono* The radical $=N \cdot NH_2$, from hydrazine. Cf. *hydrazid(o).*

hydrazyl* A free radical of the type $R_2N \cdot N \cdot$, from the hydrazino group.

-hydric Suffix indicating the hydroxyl group; as, dihydric.

hydride (1)*A compound of hydrogen with a more positive element containing the anion H^-; as, sodium hydride, NaH; (2) Formerly, also hydrogen with a member of the phosphorus group or a radical, RH; as, methylhydride (methane). **antimony ~** Stibine*. **arsenic ~** Arsine*. **phosphorus ~** Phosphine*.

hydrido-* Indicating the H atom, usually in other than carbon and boron compounds. Cf. *hydro-.*

hydrin The hydrogen acid ester of a polyhydric alcohol (as, glycerol) of the type $HO-R-X$.

hydrindene Indan*.

hydrindone Indenone*.

hydrine Hydrin.

hydriodic acid HI = 127.9. A 57% solution of hydrogen iodide in water; a reagent for nitrites, methoxy, and general reducing purposes.

hydriodide Hydroiodide.

hydrion (1) Hydrogen ion. (2) A proton: the positive nucleus of an atom (Soddy).

hydro- (1)* Indicating the H atom, usually in carbon and boron compounds. Cf. *hydrido.* (2) Prefix (Greek) indicating "water."

hydroacridine $C_{13}H_{11}N$ = 181.2. Colorless crystals, m.169, insoluble in water.

hydroalium An alloy resistant to alkalies: Al 90–92, Mg 7–9, Si and Mn 0.2–0.6%.

hydroangelic acid 2-Methyl*butanoic acid**.

hydroanthracene Dihydroanthracene.

hydroanthranol $C_{14}H_{12}O$ = 196.2. Hydranthranol. Colorless needles, m.76, soluble in hot water. Cf. *oxanthranol.*

hydroaromatics Naphthenes.

hydroatropic acid MeCHPhCOOH = 150.2. 2-Phenylpropionic acid*. Colorless liquid, b.264, soluble in water; an isomer of hydrocinnamic acid.

hydrobenzoin $(Ph \cdot CHOH)_2$ = 214.3. Benzylene glycol, diphenyldihydroxyethane. Colorless leaflets, m.136, slightly soluble in water.

hydroberberine $C_{20}H_{21}NO_4$ = 339.4. White crystals, m.167, insoluble in water.

hydrobilirubin Urobilin.

hydroboration The addition of a compound containing a $B-H$ bond to an organic compound containing C with multiple bonds; e.g., the reaction of a carbonyl with diborane.

hydroboron See *boranes.*

hydrobromic acid* HBr = 80.9. A 40% solution of hydrogen bromide in water. Colorless liquid, d.1.38; a reagent.

hydrobromide Hydrogen bromide. A salt of an organic base (e.g., an alkaloid) and hydrobromic acid. Cf. *bromide.*

hydrocaffeic acid See *hydrocaffeic acid* under *caffeic acid.*

hydrocarbobase See *hydrocarbobase* under *base.*

hydrocarbon A compound consisting of only C and H. The number of hydrocarbons is very large. **aliphatic ~** A compound consisting principally of C atoms in chains. **aromatic ~** A compound consisting principally of C atoms in 1 or more rings. **cyclic ~** Aromatic h. **normal ~** A h. without side chains. **saturated ~** A h. in which all 4 valencies of the C atoms are satisfied. **unsaturated ~** A h. with one or more double or triple bonds between the C atoms.
h. black Lampblack. **h. burner** An oil stove with vaporized kerosene as fuel. **h. radical** A group of atoms of C and H, with one or more free bonds. Cf. *organic radicals.* **h. series** A group of hydrocarbons arranged as homologs, which generally differ by $=CH_2$. Cf. *nitrogen hydrides.* See Table 43.

hydrocarbostyril C_9H_9ON = 147.2. Colorless prisms, m.163, soluble in water.

TABLE 43. HYDROCARBON SERIES

	Formula
Alkanes*, paraffins, or methane series	C_nH_{2n+2}
Alkenes*, olefins, or ethylene series	C_nH_{2n}
Alkynes*, acetylenes, or ethyne* series	C_nH_{2n-2}
Terpenes* or alkones	C_nH_{2n-4}
Benzenes and diacetylenes	C_nH_{2n-6}
Phenylene series	C_nH_{2n-8}
Indene series	C_nH_{2n-10}
Naphthalene series	C_nH_{2n-12}
Biphenyl series	C_nH_{2n-14}
Stilbene series	C_nH_{2n-16}
Anthracene series	C_nH_{2n-18}
Fluoranthene series	C_nH_{2n-20}
Pyrene series	C_nH_{2n-22}
Chrysene series	C_nH_{2n-24}
Binaphthyl series	C_nH_{2n-26}
Perylene series	C_nH_{2n-28}

hydrocarpic Misnomer for hydnocarpic.

hydrocellulose $C_{12}H_{22}O_{11}$ = 342.3. A compound obtained from cellulose by prolonged treatment with concentrated acids.

hydroceruliguone $C_{16}H_{18}O_4$ = 274.3. A solid, m.190, soluble in water.

hydrochelidonic acid $CO(CH_2 \cdot CH_2COOH)_2$ = 174.2. Acetonediacetic acid, 4-oxoheptanedioic acid*. Crystals, m.142, soluble in water.

hydrochinone Hydroquinone*.

hydrochlorate Hydrochloride.

hydrochloric acid HCl = 36.46. Muriatic acid. A solution of hydrogen chloride gas in water (NF, EP, BP). **concentrated ~** Not less than 35% HCl. A clear, colorless, fuming liquid, d.1.18, used extensively as a reagent and in organic synthesis. **dilute ~** A solution of about 20% HCl. **fuming ~** A solution of about 37% HCl, d.1.19. **nitro ~** Aqua regia.

hydrochloride A salt of hydrochloric acid and an organic base, especially an alkaloid, usually more soluble than the base. It differs from chlorides in retaining the H atom; as, *Alk*-HCl.

hydrochlorothiazide $C_7H_8O_4N_3ClS_2$ = 297.7. White powder, m.267 (decomp.), insoluble in water; a diuretic (USP, BP). See *chlorothiazide.*

hydrocinnamic acid $Ph(CH_2)_2COOH$ = 150.2. 3-Phenylpropionic acid*. Colorless needles, m.49, slightly soluble in water. **amino ~** Phenylalanine. **aminohydroxy ~** Tyrosine*. **hydroxy ~** ***ortho-*~** Melilotic acid. ***para-*~** Phloretinic acid. **methylene ~** Benzylacrylic acid.

hydrocinnamic aldehyde $C_9H_{10}O_3$ = 166.2. 3-Phenylpropanal*. Colorless liquid, b.222, insoluble in water; used in organic synthesis.

hydrocollidine A ptomaine from putrefying fish.

hydrocoridine $C_{10}H_{17}N$ = 151.3. A ptomaine produced by *Bacillus allii* or *Bacterium album* in agar cultures.

hydrocortisone $C_{21}H_{30}O_5$ = 362.5. White crystals, m.214, insoluble in water. A hormone secreted by the adrenal cortex; used widely as the h. acetate and h. sodium succinate for intravenous injections and for skin conditions (USP). **h. acetate** $C_{23}H_{32}O_6$ = 404.5. White crystals, m.220, insoluble in water; used similarly to h. (USP, BP). See *corticoids.*

hydrocotarnine $C_{12}H_{15}O_3N$ = 221.3. An alkaloid from opium. Colorless crystals, m.53, insoluble in water; a hypnotic.

hydrocotoin See *hydrocotoin* under *cotoin.*

hydrocoumaric acid $C_9H_{10}O_3$ = 166.2. (Hydroxybenzene)propanoic acid. ***ortho-*~** Melilotic acid. ***para-*~** Phloret(in)ic acid.

hydrocoumarone Hydro*benzofuran.*

hydrocracking Conversion of inferior fuel oil fractions to more valuable fuels (as, jet and diesel) in the presence of catalysts and hydrogen.

hydrocupreine $C_{19}H_{24}N_2O_2$ = 312.4. An alkaloid from cuprea bark. **methyl ~** Hydroquinine*.

hydrocyanic acid HCN = 27.03. Prussic acid, hydrogen cyanide*, formonitrile. Colorless, very poisonous gas, with an almond odor; d.0.697, b.25, soluble in water. Used as a poison gas, in metallurgy and mining (cyanide process), and in organic synthesis.

hydrocyanide The salt of an organic base with hydrocyanic acid, containing the HCN molecule.

hydrodiffusion Diffusion into water.

hydrodynamics The study of the mechanical properties of liquids, especially water.

hydrodynamometer An instrument for measuring the velocity of a fluid in motion.

hydroextractor Hydro. A rapid centrifuge for drying or dehydrating crystals, textiles, etc.

hydroferricyanic acid Ferricyanic acid.

hydrofluorogermanic acid Fluorogermanic acid.

hydrofluoric acid* HF = 20.00. Phthoric acid. A solution of hydrogen fluoride in water. A colorless liquid which must be kept in paraffin, rubber, or plastic bottles. Used to etch glass; and as a reagent. Its salts are the fluorides.

hydrofluoride A salt of hydrogen fluoride and an organic base, usually an alkaloid. Cf. *fluoride.*

hydrofluosilicic acid Hexafluorosilicic acid*.

hydroforming The cold forming of metal objects by pressure.

hydrofranklinite Chalcophanite.

hydrogel Water-swollen, rigid, 3-dimensional network of cross-linked, hydrophilic macromolecules (20–95% water). Used in paints, printing inks, foodstuffs, pharmaceuticals, and cosmetics. Gelatin or starch can be used as a base.

hydrogen* (1) History: H was probably first discovered in the 16th century by Paracelsus, and first investigated in 1766 by H. Cavendish, who later showed that water is produced when this gas burns (Greek ὕδωρ, "water," and γενναω, "to produce"). Liquid and solid H were first prepared by Dewar in 1898. (2) H_2 = 2.016. Hydrogen gas. Colorless, inflammable gas, m.−259, b.−253, slightly soluble in water, alcohol, or ether. With the 3 isotopes H, D, and T, there are 6 possible molecular forms: HH, DD, TT, HD, HT, DT. H_2 consists of ortho (symmetrical, alpha) and para (antisymmetrical, beta) forms, which differ slightly in physical properties owing to the different spin of their atomic nuclei. The nuclei in the ortho form have parallel spins, and those in the para form have antiparallel spins. At 20°C, h. consists of 75% ortho and 25% para forms. Used as fuel in torches for cutting metals, welding, and melting; in the production of synthetic stones or gems, the annealing of steel, the hydrogenation of oils, the cracking of hydrocarbons, and the production of synthetic ammonia. (3) H = 1.00794. The simplest element, at. no. 1. Hydrogen atoms, the basis of the valence system, being taken as unity. Elements combining directly with H atoms have a negative oxidation number; elements replacing H have a positive oxidation number.
(4) Constants of H atom and molecule:

Mass of atom = 1.662×10^{-24} g
Radius of molecule = 10^{-8} cm

Mean free path of molecules at 760 mm pressure and 0°C = 1.6×10^{-5} cm/s
Average velocity at 760 mm and 0°C = 1.70×10^{-6} cm/s
1 m^3 weighs 89.97 g (Regnault value)
1 liter at 760 mm and 0°C weighs 0.0899 g

(5) Isotopes of H are: protium*, ^1H, mass 1.00783; deuterium* (diplogen), ^2H or D, mass 2.0141; tritium*, ^3H or T, mass 3.0161. See *h. (2)*. Also: ^4H, a transient that decays to tritium and a neutron; ^5H, a short-lived β emitter; muonium, which behaves as a light h. isotope. (6) Compounds of H: The 3 types include (*a*) nonmetallic, e.g., SbH_3, in which H is the positive part, often gaseous and volatile; (*b*) salts or hydrides, e.g., NaH, in which the H is the negative part; transparent crystals; (*c*) metallic (H alloys), e.g., PdH_n, in which the H is alloyed with, occluded in, or absorbed in the metal. For other types of H compounds see *hydrocarbon, nitrogen hydrides, silanes, stannanes, boranes*. **activated ~** Atomic hydrides, **arseniuretted ~** Arsine*. **atomic ~** H gas subjected to a strong electromagnetic field, the molecules being broken into single atoms or ions. Used in certain blowpipes to obtain high temperatures. **excited ~** H gas subjected to a high potential at low pressure to emit characteristic radiation. **heavy ~** Deuterium*. **indicated ~*** Prefix *H* (with locant) used to distinguish between isomers; as, *pyran*, q.v. **labeled ~** Deuterium*. **nascent ~** Freshly generated H, which owes its greater chemical activity to its atomic form. **ortho-~**, **para-~** See *hydrogen (2)*. **phosphoretted ~** Phosphine*. **proto ~** See *protohydrogen*. **sulfuretted ~** H. sulfide*. **telluretted ~** H. telluride*. **triatomic ~** H_3 = 3.024. A modification of H obtained by exposure to (1) α rays, (2) vacuum discharge, (3) corona discharge, (4) a glass-tube ozonizer, (5) high-frequency Tesla discharge. It is very unstable; reduces S, As, Hg, and N; can be condensed by liquid air and decomposed (to H_2) by Pt, Ni, Co, etc. **trivalent ~** Misnomer for triatomic h.

h. acid Hydracid. An acid containing no oxygen; as, HCl. **h. azide*** NH·N:N = 43.03. Hydrazoic acid, hydronitric acid, (di)azoimide; from which the unstable azides are derived. Explosive liquid, b.37; a strong protoplasmic poison. **h. bomb** Deuterium bomb (more correctly). An atomic bomb in which a powerful explosion results from the 3 reactions of 2 atoms of deuterium; thus (n = neutron):

$$^2H + {}^2H \rightarrow {}^3H + {}^1H + 4.0 \text{ MeV}$$
$$^2H + {}^2H \rightarrow {}^4He + {}^1n + 3.3 \text{ MeV}$$
$$^2H + {}^3H \rightarrow {}^4He + {}^1n + 17.6 \text{ MeV}$$

The fuse for the reaction is a powerful fission bomb. **dry h. b.** A h. b. in which the main charge is solid lithium deuteride made from ^6Li. First exploded 1954. **wet h. b.** A h. b. in which the main fuel is liquefied deuterium. First exploded 1952. **h. bond** See *hydrogen bond* under *bond*. **h. bromide** HBr = 80.9. (1)* Hydrobromic acid gas. Colorless gas, $d_{air=1}2.71$, b.−68.7, soluble in water. Used as a reagent, in organic synthesis, and forms salts (bromides). (2) Hydrobromide. Cf. *bromide*. **h. calomel cell** An electrolytic cell comprising a h. gas electrode and calomel electrode, for conductivity and pH measurements. **h. cell** An electrolytic cell with h. gas electrodes. **h. chloride** HCl = 36.46. (1)* H. chloride gas, hydrochloric acid gas. Colorless gas, $d_{air=1}1.268$, b.−83, very soluble in water; used as a reagent and in organic synthesis. (2) Hydrochloride. **h. clay** See *hydrogen clay* under *clay*. **h. cyanide*** Hydrocyanic acid. **h. difluoride ion*** The anion HF_2^-. **h. dioxide** H. peroxide*. **h. electrode** A gas electrode whose potential is set up between finely divided metal saturated with h., and the h. ions of the solution in which it is placed. It depends on the pH value of the solution. **h. equivalent** The number of replaceable H atoms in a molecule of an acid; or the number of replaceable OH groups in the molecule of a base. **h. fluoride*** HF = 20.01. Hydrofluoric acid gas, phthoric acid. Colorless, poisonous gas, $d_{air=1}0.713$, b.19.4, soluble in water; used to etch glass. **h. iodide*** HI = 127.9. Hydroiodic acid gas. Colorless gas, $d_{air=1}4.38$, b.34, soluble in water; used as a reagent and in organic synthesis. **h. ion** See *hydrogen ion*. **h. line** The spectrum lines due to h. Cf. *Bohr atom*. **h. molecule** See *hydrogen (2), (4)* and *(5)*. **h. overvoltage** The excess emf needed to liberate h. at a metallic surface forming one of a pair of electrodes in a solution; e.g.: Pt 0, Ag 0.33, Pb 0.45, Zn 0.70. **h. oxide** Water. **h. peroxide** (1)* H_2O_2 = 34.01. H. dioxide, oxygen, auricome, perhydrol, peroxide of h. Clear, colorless liquid with a faint odor of nitric acid, d.1.458, m.−2, b.85, soluble in water. Marketed as a 3–90% solution, designated according to the number of volumes of oxygen it evolves, as, "10 volumes" (cf. *perhydrol, c.c. test*). In the solid state it is explosive; in solution it decomposes into water and oxygen; this action is accelerated by alkalies and retarded by acids. It can act as a reducing or oxidizing agent and as a fuel, reagent, disinfectant, antiseptic, antichlor, and bleach (USP, EP, BP). It occurs in traces in natural waters exposed to the sun. 100 vol. h. p. is 30.36% (by vol.), or 27.52% (by wt.). H. p. forms 2 eutectics with water: 45% H_2O_2 at −53.5, and 60% at −55°C. During World War II, stable 410-vol. (90%) h. p. was produced in the U.K. The Germans used 85% h. p. in their V weapons, as a high-energy propellant. Made by oxidizing an alkyl anthraquinol or isopropanol. (2)* h. p. ion. The anion O_2^{2-}. Cf. *hydroperoxide*. **h. p. hydrates** $H_2O_2 \cdot H_2O$, $H_2O_2 \cdot 2H_2O$. They are stable if pure, but otherwise decompose explosively. **h. p. of crystallization** Compounds which crystallize with one or more molecules of H_2O_2. **HTP** See *HTP*. **h. persulfide** H_2S_2 = 66.1. Yellow oil, decomp. by alcohol into sulfur and h. sulfide; bleaches litmus. **h. phosphide** Phosphine*. **h. selenide*** H_2Se = 81.0. Colorless, poisonous gas, m.−60, b.−42; very soluble in water. **h. sulfate** Sulfuric acid*. **h. sulfide*** H_2S = 34.08. Sulfuretted h., hydrosulfuric acid, hepatic gas. Colorless gas having the odor of rotten eggs, b.−59.6, $d_{air=1}1.189$; very soluble in water; a reagent for precipitation of metals. **h. s. ion** The anion HS^- in inorganic compounds. **h. s. water** An aqueous solution of H_2S. Colorless liquid of characteristic odor; used as a reagent in qualitative analysis, for the precipitation of heavy metals, and as an antichlor. **h. telluride*** H_2Te = 129.6. A combustible gas or liquid, m.−50, b.−2; soluble in water with decomposition.

hydrogenate (1) To introduce H into a molecule; as, the saturation of unsaturated compounds. (2) To reduce. Cf. *oxidation*.

hydrogenated Treated or saturated with H.

hydrogenation The process of causing combination with H; e.g., saturation of aliphatic unsaturated compounds with H in the presence of Ni or Pd as catalyst.

hydrogen ion H^+. Proton. A positively charged H atom constituent of aqueous solutions of acids. All acids dissociate, to some extent, into this ion, HCl = $H^+ + Cl^-$; and the strength or "acid" character (sour taste, change of color of indicator, reaction with metals, etc.) depends on the extent of this dissociation, which varies inversely with the concentration of the acid within certain limits. **h.-i. concentration** $[H^+]$, c_{H^+}. Potential of hydrogen; "momentary," true, actual, or active acidity. The amount of H^+ per unit volume (mole per liter, $[H^+]$ or c_{H^+}) of an aqueous solution. It denotes the true acidity or alkalinity of such solutions, and is expressed by the pH value (potential of hydrogen), which is the reciprocal logarithm of the number of gram ions of hydrogen per liter, pH = $\log_{10} 1/c_{H^+}$. Pure water contains 10^{-7} gram ion of H per liter. Since there are an equal number of OH^- ions, pH(7.0) + pOH(7.0) = constant (14.00). Accordingly, pH values 0–7 indicate an acid solution; pH 7 neutrality; pH 7–14 an alkaline solution. **h.-i. conversion** If h.-i. concentration $c_{H+} = a \times 10^{-b}$, pH = $-b \log a$. If pH = $x \cdot yz$, c_{H+} = [antilog $(1 - yz/100)] \times 10^{-(x+1)}$. **h.-i. determination apparatus** (1) Electrometric: Measures the potential of a H electrode (which depends on the pH of the solution) against a standard calomel electrode. (2) Colorimetric: Compares the color of an indicator added to the solution with its color in a solution of known pH. **h.-i. indicator** A dye that has definite colors at different pH values. See *indicator.* **h.-i. recorder** An automatic potentiometer for recording and controlling the acidity and alkalinity of solutions; used in industry.

hydrogenite A mixture: silicon 25, sodium hydroxide 60, slaked lime 15%; ignites on burning to give 270–370 L/kg of hydrogen gas.

hydrogenium A volatile, metallic element of which H is the supposed vapor.

hydrogenize To hydrogenate.

hydrogenolysis The cleavage of a C—C or C—O bond accompanied by the addition of H_2; as, $R \cdot R' + H_2 \rightarrow RH + R'H$. Cf. *hydrogenation.*

hydrogenomonas A genus of bacteria occurring in the soil and oxidizing hydrogen gas to form water by catalytically reducing carbon dioxide:

$$CO_2 + H_2 \rightarrow HCHO + O$$
$$HCHO + O_2 \rightarrow H_2O + CO_2$$

hydroginkolic acid Cyclogallipharic acid.

hydrohaeterolite The mineral $2ZnO \cdot 2Mn_2O_3 \cdot H_2O$.

hydrohalic Composed of hydrogen and halogens.

hydrohematite $2Fe_2O_3 \cdot H_2O$. A crystalline, hydrated mineral.

hydrohydrastine $C_{11}H_{13}O_2N$ = 191.2. An alkaloid, from hydrastine, m.66, soluble in alcohol.

hydroiodide Hydriodide. A compound, usually an alkaloid, combined with hydrogen iodide.

hydrokinetics The science of the motion of fluids under a force.

hydrol See *hydrone theory.*

hydrolapachol A hydroxynaphthoquinone, m.93, used as an acid-base indicator (colorless to red at pH 5 to 6).

hydrolase* See *enzymes,* Table 30.

hydrolith Calcium hydride*.

hydrolysis A decomposition reaction caused by water, $AB + H_2O = AOH + HB$, which, in its ionic form, $H_2O = H^+ + OH^-$, is the reverse reaction of neutralization. Cf. *hydrogenolysis.*

hydrolyst A catalyst causing hydrolysis; as, a hydrolase.

hydrolyte A substance that undergoes hydrolysis.

hydrolytic Pertaining to hydrolysis. **h. condensation** An erroneous term applied to condensations in which water is eliminated. **h. dissociation** (1) See *dissociation.* (2) Hydrolysis. **h. enzymes** See *hydrolytic enzymes* under *enzymes.*

hydrolyze To cause hydrolysis.

hydromagnesite $3MgCO_3 \cdot Mg(OH_2) \cdot 3H_2O$. A chalklike magnesium carbonate from Lodi, N.J.

hydromechanics Hydraulics.

hydromel A fermented (mead) or unfermented mixture of water and honey.

hydrometallurgy The recovery of metals from primary ores, concentrates, or secondary materials by processes based on aqueous solution chemistry. See *leaching.*

hydrometer Aerometer. A device to measure the specific gravity of liquids. Usually a graduated, hollow, weighted glass tube which sinks in the liquid to a certain depth which, read on the scale, indicates the specific gravity of the liquid. **Sikes ~** A hydrometer in which 1 degree equals a mean specific gravity interval of 0.002.

h. scales The graduations on a h.; as, Baumé, Twaddle, Beck, Brix, Balling, and Sikes. Conversion of:

$$°\text{Bé to } d: \quad d = \frac{144.3}{144.3 \pm \text{Bé}}$$

$$°\text{Tw to } d: \quad d = 1 + \frac{\text{Tw}}{200} = 1 + 0.005\text{Tw}$$

$$°\text{Brix to } d: \quad d = 1 \pm \frac{400}{\text{Brix}} \text{ at } 15.6°\text{C}$$

$$°\text{Balling to } d: \quad d = 1 \pm \frac{200}{\text{Balling}} \text{ at } 17.5°\text{C}$$

(d = specific gravity)

See *API gravity* under *gravity.*

Hydron Trademark for polyhydroxyethylmethacrylate. Used for contact lenses and artificial skin.

hydronal Polychloral, viferral. A polymerized product of pyridine and chloral; a hypnotic.

hydronaphthoquinone $C_{10}H_8O_2$ = 160.2. **1,2-~** Colorless leaflets, m.60, soluble in water. **1,4-~** Colorless needles, m.175, soluble in water.

hydrone (1) An alloy: Na 35, Pb 65%; used to make hydrogen gas by the action of water. (2) The active molecule H_2O. **h. theory** Water is a complex mixture of active molecules: hydrone, H_2O; hydrol, H_4O_2; and inactive or associated molecules, polyhydrones, $H_{2n}O_n$.

hydronitric acid Hydrogen azide*.

hydronitrogens Nitrogen hydrides.

hydronitrous acid Nitroxylic acid*.

hydronium ion The protonated water molecule, or solvated hydrogen ion, H_3O^+. Cf. *protophilic.*

hydroperoxide* An organic compound containing an —OOH group, e.g., as formed in the oxidation of rubber. Hydroperoxides have oxidizing properties.

hydrophane A transparent opal.

hydrophile Lyophile. A substance, usually a colloid or emulsion, which is wetted by water.

hydrophilic (1) Lyophilic. Describing a substance that readily associates with water. Antonym: hydrophobic. (2) Protophilic. **h. colloid** Finely divided particles forming stable suspensions in water. Antonym: hydrophobic colloid.

hydrophilite $CaCl_2$. Native calcium chloride (chlorocalcite), occurring as white incrustations on Mt. Vesuvius.

hydrophobe Lyophobe. A substance, usually colloidal, which is not wetted by water.

hydrophobic Describing a substance that does not adsorb or absorb water. Antonym: hydrophilic. **h. colloid** Finely divided, suspended particles in water which precipitate readily.

hydropirin Sodium acetyl salicylate.

hydropolysulfide* Polysulfane*. Compounds of the type $R \cdot S_2H$, $R \cdot S_3H$, $R \cdot S_nH$; as, ethyl hydrodisulfide or ethyl disulfane, EtSSH.

hydroponics Tank culture. The cultivation of plants in aqueous solutions of inorganic salts, without soil. See *water culture.*

hydroquinine $C_{20}H_{26}O_{22} \cdot 2H_2O$ = 654.5. Methylhydrocupreine. White crystals; a quinine substitute in malaria, developer in photography, and reducer in chemical analysis.

hydroquinol Hydroquinone*.

hydroquinone* $C_6H_4(OH)_2$ = 110.1. *p*-Dihydroxybenzene*, 1,4-benzenediol†, (hydro)quinol. White needles, m.174,

soluble in water. Forms clathrate compounds. A photographic developer. **ethyl ~** H. ethyl ether. **hydroxy ~** 1,2,4-Trihydroxybenzene*. **tetrachloro ~** Chloranol.
h.carboxylic acid Gentisic acid. **h. dimethyl ether** 1,4-Dimethoxy*benzene**. **h. ethyl ether** HOC_6H_4OEt = 138.2. Ethylhydrochinone, *p*-ethoxyphenol*. Colorless leaflets, m.66, soluble in water; a reducing agent.

hydroscopic Hygroscopic.

hydroseleno-* Prefix indicating the —SeH group.

hydrosilicofluoric acid Hexafluorosilicic acid*.

hydrosilicon See *silanes.*

hydrosol A colloidal suspension in water.

hydrosorbic acid Hexenoic acid*.

hydrosphere The liquid portion of the earth's surface, as the oceans, lakes, rivers, etc. Cf. *lithosphere, atmosphere.* Principal constituents: oxygen 85.8, hydrogen 10.7, chlorine 2.1, sodium 1.1%. Distribution, in Mkm^3: oceans 1,330, lakes 0.25, rivers 0.02, ice 4.0, groundwater 0.25.

hydrostatics The study of liquids in equilibrium.

hydrosulfate An addition combination of an organic base, usually an alkaloid, with sulfuric acid, without replacement of the hydrogen of the acid.

hydrosulfide (1) Thiol*, in organic compounds. (2) Hydrogensulfide*, in inorganic compounds.

hydrosulfite Dithionite*.

hydrosulfuric acid (1) Hydrogen sulfide*. (2) Dithionic acid*.

hydrosulfurous acid Dithionous acid*.

hydrotaxis The motion of organisms or cells toward water.

hydrotetrazone An aromatic compound containing 4 consecutive N atoms in the molecule; e.g., dibenzylidenediphenyldihydrotetrazone, PhCH:N·NPh·NPh·N:CHPh. Cf. *tetrazone.*

hydrotherapy The treatment of disease by water; particularly, the exercising of arthritic joints and paralyzed limbs in warm water.

hydroumbellic acid $C_9H_{10}O_4$ = 182.2. 3-(2,4-Dihydroxyphenyl)propanoic acid*. m.165.

hydrous Containing water. Cf. *anhydrous.* **h. salt** A salt containing water of crystallization.

hydroxamic acid* An organic compound containing the radical —C(:O)·NH·OH. **iso ~** Hydroximic acid*.

hydroxamino The hydroxyamino* radical.

hydroxamphetamine hydrobromide $C_9H_{13}ON \cdot HBr$ = 232.2. White crystals, m.191, soluble in water; an adrenergic used for vasoconstrictor effect (USP).

hydroxides* Compounds containing the OH^- ion. In general, the h. of metals (M) are bases; those of nonmetals (N) are acids.

MOH	Bases
NOH	Acids
ROH	Alcohols, phenols
RCO·OH	Organic acids

alkyl ~ Alcohols*. **aryl ~** Phenols*. **inorganic ~** Bases.

hydroxidion Hydroxyl ion*.

hydroximic acid* An organic compound of the type R·C(:NOH)·OH, isomeric with hydroxamic acids. **acet ~** $CH_3C(OH)$:NOH. Colorless crystals, m.59. **di ~** HON:C(OH)—C(OH):NOH. **sulfino ~*** Suffix indicating the modified sulfinic acid group, —S(:NOH)OH. **sulfono ~*** Suffix indicating the modified sulfonic acid group, —S(:NOH)(O)(OH).

hydroximino† Oxime*.

hydroxo-* Indicating the anionic ligand group OH^-.

hydroxocobalamin $C_{62}H_{89}O_{15}N_{13}CoP$ = 1,346. The —CN in vitamin B_{12} is replaced by —OH. Red crystals, soluble in water. Produced from *Streptomyces grisens* during production of streptomycin. Used to treat pernicious anemia (USP, BP). See *cyanocobalamin, vitamin,* Table 101.

hydroxonium Hydronium.

hydroxy-* Oxy-. Prefix indicating the —OH group in an organic compound. Cf. *hydroxyl, hydroxides, hydroxo.*
h.acetic acid Glycolic acid*. **h.acetophenone** $C_6H_4(OH)COMe$ = 136.2. ***ortho-* ~** b_{10mm} 97. ***meta-* ~** m.95. ***para-* ~** m.110. **h.acetyl†** See *glycoloyl.* **h. acid*** An organic compound containing both the h. and carboxyl radicals: HO·R·COOH. See *lactic acid series.* **h.amides** Oxyamides. Compounds containing the radicals $-CONH_2$ and —OH; as, $CHOH \cdot CONH_2$, glycol amide. **h.amino*** The radical —NH·OH, from hydroxylamine. **h.anthracene** Anthrol*. **h.anthraquinone** $C_{14}H_8O_3$ = 224.2. **1- ~** m.190. **2- ~** Yellow leaflets, m.302, slightly soluble in water. **h.apatite** Compounds of the type $M_{10}(PO_4)_6(OH)_2$, where M is Ba, Sr, or Ca. The principal mineral in phosphorite deposits, biological tissue, human bones and teeth. An anticaking agent and polymer catalyst. **h.azobenzene** $C_{12}H_{10}ON_2$ = 198.2. ***ortho-* ~** Colorless needles, m.83, slightly soluble in water. ***para-* ~** Colorless prisms, m.152, slightly soluble in water. **h.azobenzene compounds** $R \cdot N{:}N \cdot C_6H_4OH$. Obtained by the action of diazo compounds on phenols in alkaline solution. They form dyes. **h.benzaldehyde** $C_7H_6O_2$ = 122.1. ***ortho-* ~** Colorless liquid, d.1.159, b.197, slightly soluble in water. ***meta-* ~** Colorless needles, m.104, soluble in water. ***para-* ~** Colorless needles, m.116, soluble in water. **h.benzamide** $C_7H_7O_2N$ = 137.1. ***ortho-* ~** Yellow leaflets, m.140, soluble in water. ***meta-* ~** Colorless leaflets, m.167, soluble in water. ***para-* ~** Colorless needles, m.162, soluble in water. **h.benzene** Phenol*. **h.benzoic acid** $C_7H_6O_3$ = 138.1. ***ortho-* ~** Colorless needles, m.158, slightly soluble in water. ***meta-* ~** Rhombic crystals, m.200, slightly soluble in water. ***para-* ~** Colorless, monoclinic crystals, m.201, slightly soluble in water. **h.benzyl alcohol** $C_7H_8O_2$ = 124.1. ***ortho-* ~** Salicyl alcohol*. ***meta-* ~** Colorless needles, m.67, slightly soluble in water. ***para-* ~** Colorless needles, m.120, soluble in water. **h.butanoic acid** $C_4H_8O_3$ = 104.1. **2- ~** Colorless crystals, m.43, soluble in water. **3- ~** $CH_3CHOHCH_2COOH$. **4- ~** $CH_2OH(CH_2)_2COOH$. **h.caffeine** See *hydroxycaffeine* under *caffeine.* **h.chloroquinone sulfate** $C_{18}H_{26}ON_3Cl \cdot H_2SO_4$ = 433.9. White, bitter crystals, m.198 or 240, soluble in water; an antimalarial and antiarthritic (USP, BP). **h.choline** Muscarine. **h.cinnamic acid** Coumaric acid. **h.citric acid** $C_6H_8O_8$ = 208.1. Colorless liquid, soluble in water; found in sugar beets. **h.coniine** Conhydrine. **10-h.-2-decenoic acid*** $C_{11}H_{20}O_3$ = 200.3. An optically inactive acid constituting the major portion of the ether-soluble fraction of royal *jelly,* q.v. **h.ethylamine** $NH_2(CH_2)_2OH$ = 61.1. Colorless liquid, d.1.022, b.171, produced by the putrefaction of kephalin and serine. **h.formic acid** Carbonic acid. **h.glutamic acid** $NH_2(OH)C_3H_4(COOH)_2$ = 163.1. **3- ~** Obtained by extraction of protein hydrolysate in butane. **h.hexanoic acid** $C_6H_{12}O_3$ = 132.2. 2-Hydroxycaproic acid. Colorless crystals, m.60, slightly soluble in water. **h.hydrazides*** Oxyhydrazides. Compounds containing the OH and hydrazide groups; as, $HOCH_2CONHNH_2$, glycolhydrazide. **h.imino** The oxime* radical. **h.isobutryic acid** Acetonic acid. **h.isophthalic acid** $HO \cdot C_6H_3(COOH)_2$ = 182.1. Hydroxy-1,3-benzenedicarboxylic acid*. Colorless needles, slightly soluble in water. **2- ~** m.234. **4- ~** m.305. **5- ~**

m.288. **h.malonic acid** Tartronic acid*. **h.methyl*** Methylol. The radical HO·CH_2—. **h.peucedanin** $C_{15}H_{14}O_5$ = 274.3. A lactone, m.142, from *Peucedanum officinale* and *Imperatoria osthruthium.* **h.phenylacetic acid** *p*-HO·C_6H_4·CH_2COOH = 152.2. m.148; produced from tyrosine by intestinal putrefaction. **h.phthalic acid** HO·$C_6H_4(COOH)_2$ = 183.1. Hydroxy-1,2-benzenedicarboxylic acid*. **3-~** Colorless prisms, decomp. by heat, soluble in water. **4-~** Colorless rosettes, decomp. 181, soluble in water. **1-h.-2-propanone** See *1-hydroxy-2-propanone* under *propanone.* **h.propionic acid** Lactic acid*. **h.quinol** 1,2,4-Trihydroxybenzene*. **h.quinoline** C_9H_7NO = 145.2. **2-~** Carbostyril. **4-~** Kynurine. **8-~** A precipitant for aluminum, magnesium, and zinc. **h.quinolinecarboxylic acid 4-~** Kynurenic acid. **h.succinic acid** Malic acid*. **h.terephthalic acid** $C_8H_6O_5$ = 182.1. 2-Hydroxy-1,4-benzenedicarboxylic acid*. Colorless powder, slightly soluble in water. **h.toluene** Cresol*. **h.toluic acid** $C_8H_8O_3$ = 152.2. Methylhydroxybenzoic acid, cresotic acids: 10 possibilities, according to the positions of the Me, COOH, and OH groups. Colorless needles, soluble in alcohol; some are used in organic synthesis. **h.urea** $NH_2CONHOH$ = 76.1. Colorless needles,m.130, soluble in water. **h.valeric acid 2-~** C_3H_7·CHOH·COOH = 118.1. Colorless needles, m.31, soluble in water.

hydroxyl* The —OH group, in inorganic names, but not as the anion. Cf. *hydroxides.* Its H is replaceable by positive elements (K, Na, etc.); the entire group, by halogens. Cf. *hydroxy-, hydroxo-.*

hydroxylamine NH_2OH = 33.0. Oxyammonia. Colorless crystals, decomp. 130 (explode), m.33, soluble in water. Used as a reducing agent and in the manufacture of synthetics. **aminonitrosophenyl ~** Cupferron. **di~** The hypothetical compound HO·NH·OH.

h. hydrochloride $NH_2OH·HCl$ = 69.5. Oxammonium hydrochloride. Colorless crystals, m.151 (decomp.), soluble in water. Used as a reagent in determining gold, silver, copper, acetone, glucose, and colchicine; in organic synthesis; and as a reducing agent (developer). **h. hydrosulfate** $(H_2NOH)_2H_2SO_4$ = 164.2. Oxammonium sulfate, hydroxylaminosulfate. Colorless crystals, m.140, soluble in water; a reagent and reducing agent.

hydroxylamines* Organic compounds containing the —NHOH group from hydroxylamine and its derivatives.

hydroxylamino See *hydroxy*amino.

hydroxylation Oxidation as opposed to hydrolysis, by which hydroxy groups are formed in an organic molecule.

hydroxylimide A metamer of amideoximes of the type (R·NH)(OH)·C:NH.

hydroxylin A product of the acid hydrolysis of lignin.

hydroxynaphthalene Naphthol*.

hydroxyphenyl-* Prefix indicating the HOC_6H_4— group. **h.methyl†** See *salicyl.*

hydroxytryptamine A substance of potent biological activity in animals and plants, e.g., in venoms.

hydrozincite $ZnCO_3·2Zn(OH)_2$. Zinc in bloom. Massive or fibrous incrustations in zinc mines.

hyenanchin $C_{15}H_{18}O_7$ = 310.3. Mellitoxin. A crystalline poison from the seeds of *Hyaenanche globosa (Toxicodendron capense),* hyena poison, boesmansgif (Euphorbiaceae), S. Africa; decomp. 234. **iso~** Decomp. 299.

hyenic acid Pentacosanoic acid*.

hygric acid* $C_6H_{11}O_2N$ = 129.2. 1-Methyl-2-pyrrolidenecarboxylic acid, 1-methylproline. **4-hydroxy ~** $C_6H_{11}O_3N·H_2O$ = 163.2. A crystalline, toxic principle from *Croton goubuga,* Transvaal croton bark (Euphorbiaceae), S. Africa.

hygrine $C_8H_{15}ON$ = 141.2. **(±)-~** An alkaloid from coca leaves. **(S)-~** A liquid, d.0.935, b.195. **cusco ~** $C_{13}H_{24}ON_2$ = 224.3. An alkaloid, b_{23mm}170, from coca leaves.

hygrol Colloidal mercury.

hygrometer Psychrometer. A device for measuring the amount of moisture in the atmosphere. **chemical ~** A hygroscopic mixture of chemicals which indicates atmospheric moisture by a change of color (cobalt salts) or the formation of crystalline precipitate (camphor solutions). **physical ~** A dry-and wet-bulb thermometer. The difference between the readings gives the humidity from tables. **whirling ~** A physical h. attached to a handle so that it may be whirled around rapidly, thereby producing rapidly circulating air around the thermometer bulbs.

hygrometric Pertaining to humidity. **h. paper** A filter paper impregnated with a solution of cobalt chloride 4, sodium chloride 2, acacia 1, water 11, glycerol 1 pts. The amount of moisture is indicated by colors ranging from red (moist) to blue (dry). **h. scale** Erroneous term for hydrometric scale. **h. state** Humidity.

hygrometry The measurement of the moisture content of the atmosphere.

hygroscopic Becoming moist; as of a substance that absorbs water from the atmosphere; e.g., calcium chloride or phosphorus pentachloride. See *desiccant, deliquescence.*

hygroscopy Hygrometry.

hygrosterol A dextrorotary phytosterol from the roots of *Hygrophylia spinosa,* m.194.

hylergography The study of the influence of foreign matter on living cells.

hylogenesis The theory of the formation of matter (ὕλη = *hyle,* Greek for "matter").

hylon The positive nucleus of the atom (obsolete).

hylotropic Describing a substance that can undergo a change in phase (e.g., be melted) without change of composition.

hylotropy The property of having a constant melting or boiling point. Cf. *azeotropy.*

hymolal salts The salts of sulfuric esters of monohydric alcohols of high molecular weight, e.g., sodium dodecyl sulfate; detergents.

hyoscine BP name for scopolamine.

hyoscyamine $C_{17}H_{23}O_3N$ = 289.4. (*S*)-Tropine tropate, duboisine. An alkaloid from *Hyoscyamus niger* and other solanaceous plants, isomeric with atropine. Colorless needles, m.107, slightly soluble in water; a hypnotic, sedative, and antispasmodic (USP). **h. hydrobromide** $C_{17}H_{23}O_3N·HBr$ = 370.3. Colorless prisms, m.152, soluble in water; a hypnotic (USP). **h. hydrosulfate** $(C_{17}H_{23}O_3N)_2·H_2SO_4$ = 676.8. H. sulfate. Colorless crystals, m.199, soluble in water (USP, EP, BP).

hyoscyamus Henbane, poison tobacco, hog bane. The dried leaves and tops of *H. niger* (Solanaceae), which should contain not less than 0.065% alkaloids (hyoscine, hyoscyamine, etc.); a sedative, analgesic, and antispasmodic (EP, BP).

Hypalon Trademark for a rubbery material obtained by the chlorination and sulfonation of polyethylenes.

hypaphorine $C_{13}H_{17}O_2N$ = 219.3. Trimethyltryptophane, an alkaloid from the seeds and bark of *Hypaphorus subrumbrans* (Solanaceae). Colorless crystals, soluble in water.

hyper- Prefix (Greek) indicating an "excess"; e.g., hyperoxide.

hyperchromic Describing a radical that increases the intensity of a coloring material.
hyperconjugation The interaction of a σ bond with a π bond.
hypergolic Describing propellants that are self-igniting, usually comprising 2 constituents, e.g., hydrogen peroxide with kerosene or hydrazine.
hypericin $C_{30}H_{16}O_8$. The red coloring matter from St. John's wort. Causes light sensitivity.
hypericism Sensitivity to light in animals caused by eating plants of the genus *Hypericum.* Restlessness and irritation result.
hyperon See subatomic *particle* under *particle.*
hyperoxide ion* Superoxide. The anion O_2^-.
hypersonic Exceeding 5 times the speed of sound. Cf. *supersonic.*
Hyperstat Trademark for diazoxide.
hypersthene $(FeMg)O \cdot SiO_2$. A brown or green, ferruginous, orthorhombic, pyroxene mineral resembling enstatite.
hypertensin Renin.
hypertonic Describing a solution having a higher osmotic pressure than blood, or another solution with which it is compared. Cf. *isotonic, hypotonic.*
hypervitaminosis Condition resulting from excess levels of fat-soluble vitamins in the body.
hypnone Acetophenone*.
hypnotic An agent that produces sleep, e.g., barbiturates. Cf. *soporific, somnifacient.*
hypo- (1) Prefix (Greek) indicating "below" or "under" or "too little"; as, a lower oxidation state in hypochlorous acid, HClO. (2) Common name for sodium dithionite (hyposulfite), used as a photographic fixing agent.
hypobromite* A compound derived from hypobromous acid containing the radical —BrO. **h. nitrogen** The nitrogen which can be liberated from organic compounds by hypobromites.
hypobromous acid* HBrO = 96.9. An unstable compound of monovalent bromine, b.40 (in vacuo).
hypochlorite (1)* A compound containing the radical —ClO. (2) The hypochlorites of sodium, potassium, calcium, or magnesium, used in bleaching.
hypochlorous acid* HClO = 52.5. An oxyacid of chlorine containing monovalent chlorine; readily oxidized to chlorous acid or reduced to free chlorine.
hypochlorous ion* The ClO^- ion.
hypodermic Subcutaneous. Beneath the skin; as, an injection.
hypogaeic acid 7-Hexadecenoic acid*.
hypoglycemia Low sugar or glucose level in the blood.
hypoglycin A $C_7H_{11}O_2N$ = 141.2. The amino acid

$$CH_2{=}C\overset{CH_2}{\frown}CH \cdot CH_2 \cdot CH \cdot (NH_2)COOH$$

A hypoglycemic agent from unripe "ackee" fruit of the W. Indies, which prevents gluconeogenesis in the liver.
hypoiodous acid* HIO = 143.9. A hypothetical acid. **ammonio ~** H_2IN = 142.9. A solution of iodine in liquid ammonia; a nitridizing agent.
hypokalemia Low level of potassium ion in the blood.
hypomagnesemia Grass staggers. A magnesium deficiency disease of crops and cattle.
hypon A hypothetical noble gas, at. no. 118.
hyponitrite* A compound containing the radical $N_2O_2^{2-}$; general type $M_2N_2O_2$. Used in chemical synthesis.
hyponitrous acid* $H_2N_2O_2$ = 62.0. An oxyacid of monovalent nitrogen; decomp. to dinitrogen oxide. An active reducing and oxidizing agent.
hypophamine **alpha-~** Oxytocin. **beta-~** Vasopressin.
hypophorine $C_{14}H_{18}N_2O_2$ = 246.3. Trimethyltryptophan. An amino acid, decomp. 255, from the proteins of erythrina seeds; it causes tetanus in frogs.
hypophosphate* A salt of hypophosphoric acid*.
hypophosphite See *phosphinic acid.*
hypophosphoric acid* $(HO)_2OP \cdot PO(OH)_2$ = 162.0. Diphosphoric(IV) acid*. m.17, decomp. to phosphine and phosphoric acid.
hypophosphorous acid Phosphinic acid*.
hyposochromic Hypsochromic.
hyposulfate (1) Dithionate*. (2) Thiosulfate*.
hyposulfite (1) Dithionite*. (2) Hypo or antichlor. Sodium thiosulfate used in bleaching and as a photographic fixing bath. (3) Thiosulfates* generally. **h. process** Extraction of roasted ores with sodium dithionite solution and precipitation of the silver with sodium sulfide.
hyposulfuric acid Dithionic acid*.
hyposulfurous acid Sulfoxylic acid*.
hypothesis A theory which has not been fully proved by experiment. Cf. *theory, law.*
Hypovase Trademark for prazosin hydrochloride.
hypotonic Describing (1) a solution having a lower osmotic pressure than blood; Cf. *hypertonic, isotonic.* (2) weakness or less than usual strength of muscles.
hypoxanthine $C_5H_4ON_4$ = 136.1. Sarkine, 6-oxopurine, xanthoglobulin. Colorless needles, decomp. 150, insoluble in water. Cf. *purine.*
hypsochrome An atom or group that when introduced into a compound (such as a dye) causes a shift in color toward the red end of the spectrum; the opposite of a *bathochrome.* Cf. *hyperchromic.*
hyptolide $C_{18}H_{24}O_8$ = 368.4. The bitter principle of the leaves of *Hyptis pectinata* (Labiataceae). Colorless needles, soluble in hot water.
Hyraldite Trademark for a bleaching preparation consisting principally of sodium thiosulfate and formaldehyde.
hyrax A synthetic resin mounting agent for microscopic work (refractive index 1.75).
hyssop The dried leaves of *Hyssopus officinalis;* an aromatic carminative. **hedge ~** See *gratiolin.* **wild ~** Verbena. Cf. *ysopol.*
h. oil Colorless oil from hyssop, d.0.932, soluble in alcohol; a flavoring.
hystazarin $C_{14}H_8O_4$ = 240.2. 2,3-Dihydroxyanthraquinone*. Orange needles, m.260, soluble in alcohol; a dye.
hystarzine Hystazarin.
hysteresis The lag or retardation of an effect behind its cause, as: (1) the magnetic lag, or retention of the magnetic state of iron in a changing magnetic field; (2) the retardation of a chemical system from reaching equilibrium.
Hytor compressor A rotary pump or blower employing a centrifuged liquid to obtain suction or pressure.
hyzone Triatomic *hydrogen.* Cf. *ozone.*

I

I Symbol for iodine.
I Symbol for (1) electric current; (2) intensity: radiant (I_e), luminous (I_v), sound (I); (3) ionic strength; (4) moment of inertia; (5) nuclear spin quantum number.
i Symbol (as subscript) referring to a typical ionic species i.
i- Symbol for (1) iso; (2) optically inactive; now, *meso-* or (±)-.
iatrochemists A 16th-century school of medicine based on the principles of Paracelsus that there should be a chemical balance in the body.
ibogaine $C_{20}H_{26}ON_2$ = 310.4. An alkaloid, m.152, from the roots of *Tabernanthe iboga,* a narcotic arrow poison (Zaire).
ibuprofen $C_{13}H_{18}O_2$ = 206.3. 2-(*p*-Isobutylphenyl)propionic acid, Brufen, Nurofen. White crystals, insoluble in water. An analgesic and anti-inflammatory agent, used principally for rheumatic pains and arthritis (USP, BP).
-ic (1) Suffix indicating a higher valency, as compared with -ous. Thus, ferrous 2 and ferric 3. *Allowed only for elements exhibiting 2 valencies; usage is discouraged. (2) A termination of acids generally. Cf. *-ate* (the salt).
icaroscope An instrument for observing the sun from the afterglow of its image projected on a phosphorescent screen.
ice Frozen or solid water. Transparent, colorless solid, d.0.92, m.0. Several structural forms known. **Dry ~** Trademark for solid carbon dioxide; a refrigerating packing material. **salt ~** Frozen brine, m.−21. A 1.35 kg eutectic mixture has the same cooling effect as 0.45 kg solid carbon dioxide (Dry Ice).
i. flowers Water flowers. Negative i. crystals produced in a slab of i. by exposure to heat rays. **i. point** The m. of i. on the Kelvin temperature scale: 273.15 K. **i. ton** The theoretical number of heat units required to melt one ton of i. at 0°C to water of 0°C; 660 gigajoule/ton. **i. water mixture** A mixture of pure water and crushed i.; used to maintain a constant temperature of 0°C.
Iceland **I. agate** An obsidian from Iceland. **I. moss** Cetraria. The dried lichens *Cetraria islandica;* gray, white, brown, or red plant bodies. Cf. *cetraric acid, cetrarin, lichen, stearic acid.* **I. spar** A transparent, double-refracting calcite used in nicol prisms.
ichthammol The ammonium salts of the sulfonated oily substance resulting from the destructive distillation of bituminous schist or shale. Black viscous liquid with a strong odor, soluble in water; used for treatment of eczema (BP). Cf. *ichthyol.*
ichthulin A glycoprotein, q.v. *(glycoproteins),* from the eggs of the carp.
ichthylepidin A protein of fish scales, intermediate between collagen and keratin.
ichthyocolla Isinglass.
ichthyol $C_{28}H_{36}S_3O_6(NH_4)_2 \cdot 2H_2O$ = 634.9. Ammonium ichthyol sulfonate, anysin. Brown syrup of empyreumatic odor and burning taste, obtained by distillation of bituminous shale; soluble in water; an antiseptic and astringent. Cf. *ichthammol.* **i. sulfonic acid** $C_{28}H_{38}O_6S_3$. The dibasic acid, from ichthyol. Cf. *sulfoichthyolic acid.*
ichthyolate A compound containing the radical $=C_{28}H_{36}S_3O_6$.
ichthyophthalmite A gem variety of apophyllite.
I.C.I. Imperial Chemical Industries, PLC.
iconoscope Electric eye emitron. An instrument in which invisible radiation, e.g., X-rays, is rendered visible by impact on a luminescent screen.
icosa-* Eicosa. Prefix for 20.
icosahedro-* Affix indicating 12 atoms bound into a triangular icosahedron.
icosane* $C_{20}H_{42}$ = 282.6. An alkane in petroleum. Colorless liquid, m.37, insoluble in water. Cf. *laurane.*
icosanic acid Icosanoic acid*.
icosanoic acid* $Me(CH_2)_{18}COOH$ = 312.5. Eicos(an)oic acid, arach(id)ic acid. White leaflets, m.77, insoluble in water; a constituent of butter and Japan wax.
icosanol* Icosyl alcohol*.
icosenoic acid* $C_{20}H_{38}O_2$ = 310.5. **(Z)-9-~** Gadoleic acid. **(Z)-11-~** Gondoic acid.
icosinene $C_{26}H_{38}$ = 350.6. A liquid hydrocarbon from ozocerite.
icosyl* The radical $C_{20}H_{41}-$, from icosane. **i. alcohol** $Me(CH_2)_{19}OH$ = 298.6. Icosanol*, arachidic alcohol. Colorless solid, m.71, in Malagasy Rep. palm wax.
I.C.T. coefficient A numerical expression of the activity of an insecticide. The ratio of the average percent paralysis of a test insect over a given insecticide concentration range to that of a standard. I = insect, C = carrier, T = toxicant.
Id* Alternative symbol for iodine, where I is inconvenient.
-ide Suffix indicating, generally, a monoatomic (as, chloride) or homopolyatomic (as, disulfide), electronegative constituent of a molecule. A few specified heteropolyatomic groups also have this ending (as, cyanide), as well as organic anions formed by removal of a proton from a carbon atom (as, benzenide, $C_6H_5^-$). See *radicals,* Table 70 on p. 494. Cf. *-ate.*
idene* Suffix indicating a bivalent radical derived from a univalent radical; thus, MeCH=, ethylidene. Cf. *idyne.*
idioblast A biophore.
idiosyncrasy Abnormal, constitutional, or personal reaction to the effects of certain substances. Cf. *allergy.*
iditol $C_6H_{14}O_6$ = 182.2. Idite. A hexahydric (+)- and (−)-alcohol.
idocrase A gem variety of vesuvianite.
idonic acid $C_6H_{12}O_7$ = 196.2. A monobasic, pentahydroxy acid, from iditol.
idoplatinic acid Platinic acid.
idosaccharic acid $C_6H_{10}O_8$ = 210.1. A dibasic, tetrahydroxy acid, from iditol.
idose $C_6H_{12}O_6$ = 180.2. A hexose of aldose sugar, isomeric with glucose; m.156.
idoxuridene $C_9H_{11}O_5N_2I$ = 354.1. 2′-Deoxy-5-iodouridine, Herplex, Herpid. White crystals, slightly soluble in water. An antiviral that acts by preventing the formation of DNA; used for herpetic eye and skin infections (USP, BP).
idrialene $C_{22}H_{14}$ = 278.4. A hydrocarbon from asphalt.
idryl Fluoranthene*.
-idyne* Suffix indicating a trivalent radical derived from a univalent radical; as, MeC≡, ethylidyne. Cf. *idene.*
IEC See *standards.*

Igamid Trademark for superpolyamide synthetic fibers.
Igepon Trademark for a series of anionic surfactants used as detergents, wetting agents, emulsifiers, dispersants, and foaming agents; including esters of sodium isethionate and oleic acid, and sulfonamides derived from *N*-methyltaurine or *n*-cyclohexyltaurine and fatty acids.
ignatia Saint-Ignatius's-bean. The dried, ripe seeds of *Strychnos ignatia* (Loganiaceae). It contains 2% alkaloids, mainly strychnine and brucine.
igneous Plutonic. Describing rocks formed from a molten state. Cf. *sedimentary rock.*
ignis A fire.
ignite (1) To heat a substance at a high temperature until no more loss in weight occurs. (2) To set fire to a reaction mixture.
ignition Combustion burning, or setting on fire. In analysis: (1) Complete oxidation of an organic compound by heating in oxygen gas. (2) Heating an inorganic compound until all volatile matter has been driven off. (3) Placing a flame directly or indirectly in contact with a reaction mixture until the reaction starts and continues to completion. **pre ~** See *knock.*
i. point Kindling temperature. The temperature at which a substance begins to burn. Cf. *flash point.*
ihlenite $Fe_2(SO_4)_3 \cdot 12H_2O$. A native sulfate.
ilang-ilang Ylang-ylang.
Ile* Symbol for isovaline.
Iletin Trademark for a brand of insulin.
Ilex A genus of shrubs and trees (Aquifoliaceae), including the hollies. Cf. *ilexanthin.*
ilexanthin $C_{17}H_{23}O_{11}$ = 403.4. The yellow coloring matter of *Ilex aquifolium,* holly. Yellow needles, m.198, insoluble in cold water.
ilicic alcohol $C_{30}H_{50}O$ = 426.7. α-Amyrin, m.185, prepared from *Ilex* species; a constituent of birdlime.
ilicin A bitter principle from holly, *Ilex aquifolium.*
ilicyl alcohol $C_{22}H_{38}O$ = 318.5. A wax, m.139.
illicium Star *anise.*
illinium Early name for promethium.
illinum An acid-resistant alloy of Ni, Cr, Co, W, Al, Mn, Ti, B, and Si.
illipé (1) The fat of *Bassia latifolia* or *B. longifolia.* (2) Borneo tallow (a misnomer). See *tallow.*
illipene $C_{64}H_{106}$ = 875.5. An unsaturated hydrocarbon from the unsaponifiable matter of illipé.
illuminance* The quantity of light thrown on an object: $E = \Phi/A$, where E is the i., Φ the luminous flux, and A the surface area; measured in lux*, meter-candles, footcandles, or phot.
illuminant An agent that produces light.
illumination (1) The act of lighting up. (2) Illuminance*. **axial ~** Light passing in the direction of the axis of a miroscope. **dark-ground ~** Light passing at right angles to the direction of the axis of a microscope; permits some objects, as, spirochaetes, to be viewed more clearly. **direct ~** Light falling directly on an object on the stage of a microscope from above. **indirect ~** Dark-ground i. Cf. *ultramicroscope.*
ilmenite $FeTiO_3$. Menacantite, geikielite. A native, black titanate; a gem.
ilmenium A supposed element, which proved to be a mixture of niobium and tantalum.
Ilosone Trademark for erythromycin.
Ilotycin Trademark for erythromycin.
ilvaite $CaFe_2Fe(OH)(SiO_4)_2$. Yenite. A native silicate, sometimes used as a gem.
im- Prefix indicating the imino* group.
image (1) The likeness or a reproduction of an object. (2) The picture of an object formed by rays of light after passing through an optical system. **real ~** An image formed where rays of light meet or converge. **virtual ~** An apparent image, formed in the direction from which rays enter the eye. The rays do not converge where the image is seen, but would do so if extended backward.
image stone A gem variety of pyrophillite.
imasatic acid Isamic acid.
imasatin The lactam of isamic acid, q.v.
imazalil* See *fungicides,* Table 37 on p. 250.
imazine An organic compound containing the radical $=C:N \cdot CH:N-$.
imbibition Absorption of a liquid by a solid or gel.
Imelon Trademark for a polyamide synthetic fiber.
Imhoff sludge A fertilizer made from sewage sludge settled with the aid of anaerobic bacteria: N 2–3.3, P 1%.
imid Imide*.
imidazole* $C_3H_4N_2$ = 68.1. 1,3-Diazole, glyoxaline. Colorless prisms, m.88, soluble in water. The ring compound

```
      N:CH
HC<        >NH
 4    CH       1
```

2,4,5-triphenyl ~ Lophine.
imidazoledione Hydantoin.
imidazoleethylamine Histamine.
imidazoletrione Parabanic acid.
imidazolium* The cation $C_3H_5N_2^+$.
imidazolone $C_3H_4ON_2$ = 84.1. 2(3*H*)-Imidazolinone, glyoxalone. Colorless needles, m.250 (decomp.). Cf. *creatinine.*
imidazolyl* The radical $C_3H_3N_2-$, from imidazole; 4 isomers. **i.ethylamine** Histamine **i.thiol** $C_3H_3N_2SH$ = 100.1 ***meta-* ~** Colorless crystals, m.222.
imide (1)* A compound from a dicarboxylic acid of general formula $-C(O) \cdot NH \cdot (O)C-$, with or without the H substituted; as, $C(O)(CH_2)_2(O)C \cdot NH$, succinimide. Cf. *imine.* (2)* A compound containing the =NH group. **cyclic ~** A ring formed by replacing 2 —OH groups by =NH. **di ~** See *diimide.*
i. chloride Imine chloride.
imidic acid Acid in which the =O atom of a carboxylic acid group has been replaced by =NH; as, heptanimidic acid, $Me(CH_2)_5C(:NH)OH$. **sulfin ~*** Suffix indicating the modified sulfinic acid group, $-S(:NH)OH$. **sulfon ~*** Suffix indicating the modified sulfonic acid group, $-S(:NH)(O)OH$.
imido* Radical formed by removal of the H from the corresponding imide compound; as, succinimido, $(CH_2CO)_2N-$. **acet ~** The acetimidoyl* radical.
i.carbamide Guanidine*. **i.dicarbonic acid** $HOOC \cdot NH \cdot COOH$ = 105.1. **i. esters** Compounds of the type $R \cdot C(:NH) \cdot OR$, obtained as hydrochloride by the action of hydrochloric acid on a mixture of a nitrile and alcohol. **i. ethers** A group of compounds, from i.carbonic acid; of the general type $R \cdot C(:NOH) \cdot OR$. **i. hydrogen** The H of the NH group, which is replaceable by metals, such as K. **i.urea** Guanidine*.
imidodiphenyl Carbazole*.
imidoxanthin Guanine.
-imidoyl* Suffix indicating the $RC(:NH)-$ radical from the corresponding imidic acid, $RC(:NH)OH$; as, formidoyl, $CH(:NH)-$.
iminazolone Imidazolone.
imine* A compound containing the group =C:NH, as, in the imino ethers, $R_1 \cdot C(:NH) \cdot OR_2$. **i. chloride** A compound

containing the radical $-C(:NH)Cl$; formed by the action of hydrochloric acid on nitriles.

imineazole Imidazole*.

iminio-* Prefix indicating the imino group in a configuration involving quadricovalent nitrogen, $R_4N^+X^-$.

imino-* Prefix indicating the $=NH$ group attached to 1 or 2 C atoms; as, $=C:NH$ or $-C \cdot NH \cdot C-$. Cf. *carbonimidoyl, epimino.* **hydroxy ~** † Oximido. The oxime* radical, $=NOH$.

i.acetic acid $NH(CH_2COOH)_2$ = 133.1. Colorless rhombs, m.225, soluble in water. **i.acetonitrile** $NH(CH_2CN)_2$ = 95.1. Colorless leaflets, m.75, soluble in water. **i. bases** Compounds containing the $=C:NH$ group, as guanidine. **i.ethanol** $NH(CH_2CH_2OH)_2$ = 105.1. Colorless crystals, m.28, soluble in water. **i.methylamino*** The radical $HN:CH \cdot NH-$. **i. nitrogen** See *ammonia nitrogen* under *ammonia.* **i.phenylmethyl**† See *benzimidoyl.* **i.urea** Guanidine*.

imipramine $C_{19}H_{24}N_2 \cdot HCl$ = 316.9. Tofranil. A tranquilizer and antidepressant. **i. hydrochloride** $C_{19}H_{24}N_2 \cdot HCl$ = 316.9. White crystals,burning taste, m.170, soluble in water; used to treat depression (USP, EP, BP).

immersion Submersion in a liquid. **oil ~** Connecting an object and objective of a microscope with oil. **water ~** Connecting an object and objective of a microscope with water.

i. electrode An electrode that can be lifted from the liquid and immersed at will. **i. objective** See *immersion objective* under *objective.*

immiscible Describing liquids that will not mix. **i. solvent** A liquid that dissolves a solute from a solution with which it does not mix. See *distribution coefficient.*

immune Completely resistant to a disease **i. globulin** See *immune globulin* under *globulin.* **i. serum** The serum from the blood of an actively immunized animal, containing the antibodies for a certain disease.

immunity The resistance of an individual or organism to infection. **acquired ~** I. acquired by a previous attack of the disease, or by inoculation with bacterial preparations. **active ~** I. in which the cells of the organism manufacture antibodies, stimulated by bacterial preparations or a slight attack of the disease. **natural ~** I. with which an individual is born. **passive ~** I. that depends solely on inoculated immunizing sera.

immunization The process of enabling an organism to withstand the harmful effects of microorganisms, to endure the metabolic products of the invader without injury, and to destroy the parasite.

immuno- **i.assay** A method of protein measurement by antibody-antigen reaction. A specific antibody is formed by injecting the protein into a laboratory animal. The result of the antibody-antigen reaction can be measured by electrophoresis or filtration. **radio ~** Immunoassay in which either the antibody or the antigen is labeled by a radioactive isotope. **time-resolving fluoro ~** As *radioimmunoassay,* but labeled with a rare-earth metal (as, terbium) that fluoresces in u.v. light. **i.chemistry** The study of the chemical phenomena of immunity. **i.compromised** Characterized by suppression of the immune system: (1) by cytotoxic drugs for treatment of cancer, or prevention of transplant rejection (i.suppression); (2) by a congenital or acquired disease, as AIDS. **i.gen** Antigen. **i.globulins** Antibodies. Proteins produced by modified lymphocytes (plasma cells). Structurally, 2 chains, heavy and light (according to molecular weight), joined by disulfide bonds. Every plasma cell can react (one at a time) to all antigens to produce immunoglobulins. See *immune globulin* under *globulin.* **i.suppression** See *i.compromised.* **i.therapy** Treatment stimulating the body to form antibodies; used in cancer chemotherapy.

immunology The study of immunity.

impact Sudden collision. **atomic ~** Collisions between electrons and protons; as, in the bombardment of a gas with cathode rays. **molecular ~** Collisions between molecules, which are essential for the progress of a reaction.

imperatorin $C_{16}H_{14}O_4$ = 270.3. A crystalline principle from masterwort, the root of *Imperatoria ostruthium* (Umbelliferae). Colorless prisms, m.98, soluble in alkalies. Cf. *osthruthin.*

imperialine $C_{27}H_{43}O_3N$ = 429.6. A colorless, crystalline alkaloid, m.267, from *Fritillaria imperialis* (Liliaceae).

imperial jade A green aventurine gem quartz (China).

imperial yu stone Imperial jade.

impermeable Not permitting passage.

impervious Impenetrable, nonabsorbent.

impinger An apparatus for sampling dust in air by drawing it at high velocity through a glass tube onto a wet glass plate on which microscopic counts are made.

implant A foreign substance (e.g., organ, tissue, or drug) which is introduced surgically (e.g., subcutaneously or intramuscularly) into the tissues of the body. In drug implants the i. is in a sparingly soluble form. The slow release of the drug obviates repeated injections.

implosion Explosion inward, as, the collapse of the walls of a vessel under internal vacuum.

impregnate To saturate or charge with a gas or liquid.

impregnation Saturation with a material having special properties; as, waterproofing. Cf. *introfaction.*

improver (1) A mixture of starch and salts (as, phosphates) added to flour to stimulate the yeast and improve the rising properties. (2) Bleaching or whitening agent (as, peroxodisulfates) added to flour to remove or mask the color due to carotene.

I.M.S. Industrial methylated spirit.

Imuran Trademark for azothioprine.

in (1) An inscription on a pipet or measuring vessel indicating that it is to be used to contain (not to deliver) its nominal volume. Cf. *ex.* (2) Abbreviations for *inch,* q.v. **in²** Square inch. **in³** Cubic inch. **in-** Prefix indicating (*a*) within; (*b*) not, e.g., *in*organic. **-in** Suffix indicating (*a*) a neutral carbohydrate, as, insulin; (*b*) a glucoside, as, amygdalin; (*c*) a protein, as albumin; (*d*) a glyceride, as, palmitin.

In Symbol for indium.

inactivate To destroy activity.

inactivation Destruction of activity (as, of a catalyst or a serum) by chemical or physical means.

inactive *i-*. Describing a compound having an asymmetric C atom but no optical activity. It is thus either racemic or *meso.* **divisibly ~** Capable of being resolved into 2 optically active substances or racemic compounds. **indivisibly** *meso-**. Incapable of being resolved into its optically active components.

incandescence A state of glowing with intense brilliance.

incandescent Emitting heat or light or both by virtue of being at a high temperature. **i. light** An electric light bulb producing light by the passage of an electric current at a low pressure through a fine metallic element.

incaparina A protein-rich food for children, developed under the auspices of the World Health Organization in Central America. Composition: cottonseed flour 38, ground corn 29, sorghum 29, *Torula* yeast 3, calcium carbonate 1, protein 27.5%; vitamin A 45 units per gram.

Inca stone A gem pyrite.

incendiary An agent that causes combustion. Classes: (1) Spontaneously inflammable solids, as, phosphorus; (2)

metallic powders, as, thermit; (3) oxidizing combustible mixtures, as, potassium nitrate; (4) flammable material, as, carbon disulfide.

inch A unit of length in the fps system. 1 inch = 1/12 foot = 1/36 yard = 1/63,360 mile = ¼ hand. In accordance with an agreement between English-speaking countries (1959), 1 in = 2.54000 cm. **cubic ~** in^3 or cu in. A unit of volume in the fps system. 1 in^3 = 0.0005787 ft^3 = 16.387 cm^3. **micro ~** 10^{-6} in = 254 Å. **square ~** in^2 or sq in. A unit of area in the fps system. 1 in^2 = 0.006944 ft^2 = 645.2 mm^2.

inchi grass oil An oil from *Cymbopogon caesius* containing borneol, terpineol, camphene, and citrene; a substitute for palmarosa oil.

inch-pennyweight The product of the gold content of a gold reef (in pennyweights per ton) and the width of the reef (in inches). A measure of the value of a reef in the S. African mining industry; 500 is a high figure.

incidence The striking contact of one body with another. **angle of ~** The angle made with the normal by a beam of light striking a surface.

incineration Cremation. The process of burning to ashes. **i. dish** A flat dish for reducing substances to ash in analysis.

incipient Beginning. **i. red heat** The point at which a substance begins to glow. Cf. *color* scale.

inclination Deviation. The angle of an object above the horizon.

inclinator A stand for large bottles or carboys so that they can easily be tipped for emptying.

inclusion A state of being enclosed in or surrounded by a substance; as, suspended foreign matter in a crystal. See *clathrates.*

incompatibility Inability to be mixed without impairing the original properties. **chemical ~** Describing substances which, when mixed, react with each other. **physical ~** The property of repellent substances; as, water and oil. **physiologic ~** The i. of drugs that have a mutually antagonistic effect. **therapeutic ~** The i. of drugs that have opposite therapeutic effects.

incompatible Applied to a substance which for chemical, physical, or physiological reasons cannot be mixed with another without a change in the nature or effect of either.

incomplete Not carried to its greatest possible extent. **i. equilibrium** Equilibrium that has not reached a balance. **i. reaction** Reversible *reaction.*

incompressibility The property of not being compressible. Ability to resist pressure without change of form or volume.

incompressible volume That part of a gas which is not uniformly compressed according to the gas laws; the quantity *b* of *van der Waals' equation,* q.v.

Inconel Trademark for a corrosion-resisting alloy containing 76 Ni, 15 Cr, and 9% Fe; m.400, d.8.51.

increment The augmentation of the quantity of substance, e.g., during crystallization.

incrustation Formation of a crust or scale.

incubation time (1) The period between implanting an infection in a culture medium or organism and the first signs of growth. (2) The period between a person becoming infected and the first signs of disease.

incubator A chamber or box at a definite temperature (usually 37°C), in which bacterial cultures are grown.

indaconine $C_{27}H_{47}NO_9$ = 529.7. An alkaloid, m.94, from aconite.

indaconitine $C_{34}HN_{47}O_{10}$ = 1228. Acetylbenzoyl-ψ-aconine. An alkaloid from aconite, m.202.

indamine $ClH_2N:C_6H_4:N \cdot C_6H_4NH_2$ = 197.2. Phenylene blue. Obtained by oxidation of *p*-phenylenediamine and aniline.

indamines Dyestuffs with a quinonoid nucleus, except that N atoms are in the place of both O atoms.

indan* $C_6H_4:(CH_2)_3$ = 118.2. Hydrindene, 2,3-dihydroindene. Colorless liquid, b.176, insoluble in water.

indandione* $C_9H_6O_2$ = 146.1. Dioxohydrindene. **1,2- ~** White crystals, m.107. **1,3- ~** White crystals, m.130.

indanone C_9H_8O = 132.2. Oxohydridene. **1- ~** m.42. **2- ~** m.60.

indanyl* The radical C_9H_9-, from indan; 4 isomers.

indazole (1)* $C_7H_6N_2$ = 118.1. Benzopyrazole, 2,1-benzodiazole. **1*H*- ~** Colorless crystals, m.146. **2*H*- ~** Is(o)indazole.

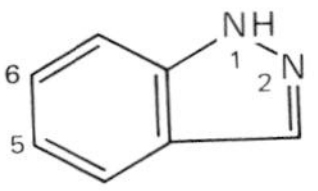

1*H* form 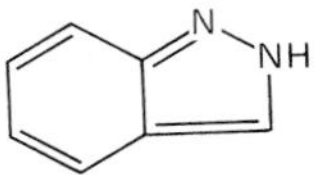   2*H* form

(2) A compound containing the i. nucleus; as: **2-phenyl-2*H*- ~** $C_6H_4(N \cdot Ph)(CH):N$. Colorless crystals m.142.

indene* C_9H_8 = 116.2.

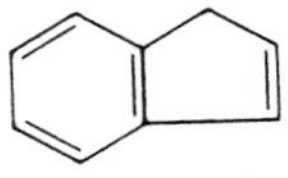

Colorless liquid, d.1.040, b.188, soluble in alcohol. **dihydro ~** Indan*.

indenone* C_9H_6O = 130.1. Indone, hydrindone. Yellow liquid.

indenyl* The radical C_9H_7-, from indene; 7 isomers.

indeterminate That which cannot be predicted, but which (when it happens) can be determined.

indeterminism See *Heisenberg principle.*

index (Pl., indexes or indices) (1) Mathematical: exponent. See *mantissa.* (2) Physical: a numerical ratio of measurement in comparison with a fixed standard. (3) Bibliographical: a classified list; as subjects or patent numbers. **i. compound** The *parent compound,* q.v., under which derivatives are listed.

Indian **I. agate** A gem moss agate. **I. arrowroot** Euonymus. **I. balsam** Peru balsam. **I. barley** Sabadilla. **I. bel** Bael. **I. cannabis** Cannabis. **I. corn** Maize. **I. fig** Prickly pear. **I. ginger** Asarum. **I. gum** (1) Ghatti gum, gummi indicum. The exudation of *Anogeissus latifolia* (India). Yellow tears, soluble in water; a mucilage. (2) Sterculia gum. **I. hemp** Cannabis. **I. hippo** Gillenia. **I. laburnum** Cassia. **I. licorice** Abrus root. **I. ocher** A native ferric oxide war paint (N. American Indians). **I. physic** Gillenia. **I. pink** Spigelia. **I. poke** Veratrum. **I. red** (1) Red ocher from Ormus (Persian Gulf); an early coloring material. (2) I. ocher. **i. rubber** Rubber. **I. saffron** Turmeric. **I. sage** Eupatorium. **I. shot** Cannabin. **I. tobacco** Lobelia. **I. topaz** A saffron-yellow topaz. **I. turnip** The corn of *Arisaema triphyllum.* **I. yellow** (1) Purree. A yellow I. pigment, containing the magnesium salt of euxanthic acid. (2) Cobalt and potassium nitrite. (3) Euxanthone.

india rubber Rubber.

indican (1) $C_{14}H_{17}O_6N$ = 295.3. A glucoside from woad, *Isatis tinctoria* (Cruciferae); also from indigo, *Indigofera* species (Leguminosae). Colorless leaflets, m.57, soluble in water; hydrolyzing to glucose and indoxyl; also formed from indole in the intestine during putrefaction. (2) $C_8H_6N \cdot SO_4K$ = 251.3. I. of urine, indoxyl sulfate. A normal constituent of urine from metabolism of tryptophan. **i. meter** A device for estimating the quantity of i. in urine. **i. test** To 5 mL urine add a few drops of *Obermayer's reagent* (q.v.), shake, and add chloroform. The chloroform separates with a blue color if i. is present.

indicarmine Indigo carmine.

indicator (1) A substance that changes in physical appearance, e.g., color, at or approaching the end point of a chemical titration, e.g., on the passage between acidity and alkalinity. See Table 44 on p. 302. **adsorption ~** A substance that indicates the end point of a precipitation reaction by being released from the adsorbed state; as, rhodamine on silver chloride. **achromatic ~** A mixture of 2 indicators or of an i. and dye which produces at the end point a color complementary to that of the i. at its transition point. The mixture thus appears colorless or gray at its end point; e.g., methyl red 0.125, methylene blue 0.0825%. **acid-base ~** Hydrogen ion. **ammonio system ~** An i. that indicates the presence of an ammonio acid or ammonio base in a liquid ammonia solution; e.g., hydrazobenzene, yellow and red, respectively. **aquo system ~** Hydrogen ion. **chelatometric ~** Metal i. **Clark and Lubs ~** Phthalein *indicator*, q.v., covering the pH range 1.0–9.0. **complexometric ~** Metal i. **compound ~** A mixture of indicators; as, universal i. **external ~** Outside i. **fluorescent ~** A substance that indicates an end point or pH value by a change of fluorescence, intensity, or color; as, quinine sulfate. **hydrogen-ion ~** A substance that indicates by its color the approximate hydrogen-ion concentration of a solution. **inorganic ~** A metallic salt used in titrations; as, potassium chromate or hexacyanoferrate(II). **inside ~, internal ~** An i. added to a liquid to be titrated; as, litmus in neutralization titrations. **metal ~** A chelating dyestuff which changes color with increase or decrease in metal ion concentration by forming metal-dye complexes; e.g., Eriochrome Black-T, used in complexometric titrations. **metallochromic ~** An i. that is a complexing agent for the metal ion being treated, and changes in color at a titration end point. **neutralization ~** An i. that shows the end point of an acid-alkali neutralization reaction. **one-color ~** An i. that changes from colored to colorless, e.g., phenolphthalein. **outside ~** External. An i. to which a drop of the titrated liquid is added on a porcelain plate. **oxidation-reduction ~, redox ~** A substance that indicates the state of oxidation by its color; as, compounds of Mn. E.g.: At pH 7.0 the potentials are:

Indigo disulfonate	−0.121
Methylene blue	+0.011
o-Chlorophenol indophenol	+0.233
Hexacyanoferrate(II)	+0.40

phthalein ~ Synthetic phthalein dye, used as an i. See Table 44. **radioactive ~** See *radioactive tracer*. **redox ~** Oxidation-reduction i. **screened ~** An i. mixed with another coloring matter (not necessarily an i.) to make the color change sharper. Cf. achromatic *indicator* (above). **turbidity ~** A semicolloid which flocculates at a certain pH value (isoelectric point) and so indicates when this point has been reached in volumetric analysis. **two-color ~** An i. that changes from one color to another. **universal ~** A mixture of indicators covering a wider pH range than each individual i. E.g.: a solution of methyl orange 0.1, methyl red 0.4, bromothymol blue 0.4, 1-naphtholphthalein 0.32, phenolphthalein 0.5, and cresolphthalein 1.6 g in 100 ml 70% alcohol.

Red	3.0	Greenish blue	9.0
Yellow-orange	5.0	Blue	10.0
Yellow	6.5	Reddish violet	12.0
Green	8.0		

vegetable ~ An i. coloring matter, from plants.

i. exponent The pH value at which the color change of an i. is most rapid; theoretically the midpoint of the i. range. **i. paper** A paper impregnanted with an i. and dried. See *test paper*. **i. range** The pH values over which the color of an i. changes. See Table 44. **i. yellow** The chromophore of rhodopsin, q.v.

indicolite A blue gem tourmaline.

indigo $(C_6H_4 \cdot C(O) \cdot C{:})_2$ (rings closed by —NH—) = 262.2. I. tin†, synthetic i. Dark-blue rhombs, d.1.35, sublime 300, decomp. 390, insoluble in water, alcohol, or ether; soluble in hot aniline or hot chloroform; used in dyeing and as a reagent. **chinese green ~** A dye from the bark of *Rhamnus chlorophora* (China); used to dye silk. **dibromo ~** Murex. **leuco ~** I. white. **natural ~** I. blue. **soluble ~** I. white or I. carmine.

i. blue The blue color obtained by fermentation from various species of *Indigofera* (Leguminosae); used in dyeing and printing inks. **i. carmine** $C_{16}H_8N_2O_2(SO_3Na)_2$ = 466.3. Soluble i., sodium indigotin disulfonate, sodium coerulin sulfate, i. extract. Blue powder or paste, soluble in water. Used as a dye, and clinically in a function test of the kidneys (USP, BP). **i. copper** $CuSO_3$. Covellite. A native copper sulfite. **i. disulfonate** An oxidation-reduction and pH indicator, changing at 12.5 from blue (acid) to yellow (alkali). **i. extract** I. carmine. **i. red** Indirubin. **i. white** $C_{16}H_{12}O_2N_2$ = 264.2. Biindoxyl, leuco indigo, soluble indigo. Colorless powder, insoluble in water, oxidized to i. blue; used for vat dyeing of textiles with indigo.

indigoid dyes Dyes having the $-C(:O) \cdot C(-):C(-) \cdot C(:O)-$ group in combination with NH or S groups; as, indigo.

indin $C_{16}H_{10}O_2N_2$ = 262.3. An isomer of indigo, soluble in water.

indirubin $C_{16}H_{10}O_2N_2$ = 262.3. Indigo red, oxindole-$\Delta^{3,2}$-ψ-indoxyl. A red isomer of indigo in urine.

indium* In = 114.82. A ductile metal element of the aluminum subgroup, at. no. 49. Silver, crystalline masses, d.7.362, m.156, b.2080, insoluble in water, soluble in acids. Discovered by Reich and Richter (1863), and named from the indigo blue lines of its spectrum. It is usually trivalent but may be di- or monovalent; it forms low-melting alloys. **i. bromide** $InBr_3$ = 354.5. I. tribromide*. Yellow powder, soluble in water. **i. chloride** InCl. I.monochloride*. $InCl_2$. I.dichloride*. $InCl_3$. I.trichloride*. **i. cyanide** $In(CN)_3$ = 192.9. Colorless, poisonous powder, soluble in water. **i. dichloride*** $InCl_2$ = 185.7. Yellow liquid, decomp. by water into the trichloride and i. **i. hydroxide*** $In(OH)_3$ = 165.8. Colorless powder, insoluble in water. **i. iodide** InI_3 = 495.5. Yellow, hygroscopic crystals, soluble in water. **i. nitrate*** $In(NO_3)_3 \cdot 3H_2O$ = 354.9. White crystals, soluble in water. **i. oxide*** In_2O_3 = 277.6. Yellow powder, insoluble in water. **i. phosphide** InP = 145.8. Metallic, m.1070; used in solar cells. **i. sulfate** $In_2(SO_4)_3$ = 517.8. Gray powder, hygroscopic, poisonous, and soluble in water. **i. hydrogensulfide*** $In(SH)_3$ = 214.0. Yellow powder, precipitated from aqueous i. salt solutions by hydrogen sulfide. **i. sulfide*** In_2S_3 = 325.8. Red powder, insoluble in water. **i. trichloride*** $InCl_3$ = 221.2. I. chloride. White, hygroscopic, poisonous crystals, sublime 500, soluble in water.

Indocid Trademark for indomethacin.

indogen The radical $C_6H_4(NH) \cdot CO \cdot C=$. **di ~** Indigo. **pseudo ~** Isatin.

indogenide A compound containing the indogen radical.

indole* $C_6H_4 \cdot (NH) \cdot CH{:}CH$ (ring closed) = 117.2. 1*H*-1-Benzazole, ketole, benzopyrrole. Colorless leaflets, m.52, soluble in water. Occurs in oil of jasmine, clove oil, and in intestinal

TABLE 44. PROPERTIES OF COMMON INDICATORS

Name	Common name	pH range	Titration end point, pK_{Ind}	Color change: Acid	Color change: Alkaline	Solvent	Concentration, %	Volume of 0.05 NaOH per 100 mg	Drops added to 10 ml of sample
Thymolsulfonephthalein	Thymol blue	1.2–2.8	2.3	Rose	Yellow	Water	0.04	4.3	4
		8.0–9.6	8.8	Yellow	Blue				
Tetrabromophenolsulfonephthalein	Bromophenol blue	3.0–4.6	4.1	Yellow	Purple	Water	0.04	3.0	4
Dibromo-*o*-cresolsulfonephthalein	Bromocresol purple	5.2–6.8	6.3	Yellow	Purple	Water	0.02	3.7	4
Dibromothymolsulfonephthalein	Bromothymol blue	6.0–7.6	6.9	Yellow	Blue	Water	0.04	3.2	6
Phenolsulfonephthalein	Phenol red	6.8–8.4	7.7	Yellow	Purple	Water	0.02	5.7	4
o-Cresolsulfonephthalein	Cresol red	7.2–8.8	8.2	Yellow	Purple	Water	0.02	5.3	4
m-Cresolsulfonephthalein	*m*-Cresol purple	0.5–2.5	1.5	Red	Yellow	95% alcohol	0.02	—	4
		7.6–9.2	8.4	Yellow	Purple				
Tetrabromo-*m*-cresolsulfonephthalein	Bromocresol green	3.2–5.8	4.8	Yellow	Blue	Water	0.02	—	4
Dichlorophenolsulfonephthalein	Chlorophenol red	5.0–6.0	5.5	Yellow	Red	Water	0.04	—	4
Dibromophenolsulfonephthalein	Bromophenol red	5.4–7.0	6.2	Yellow	Red	Water	0.04	—	4
2,6-Dinitrophenol	—	1.7–4.4	—	Colorless	Yellow	Water	0.10		
2,4-Dinitrophenol	—	2.0–4.7	—	Colorless	Yellow	Water	0.10		
2,5-Dinitrophenol	—	4.0–6.0	—	Colorless	Yellow	Water	0.10		
p-Nitrophenol	—	5.0–7.6	—	Colorless	Yellow	Water	0.50		
m-Nitrophenol	—	6.5–8.5	—	Colorless	Yellow-orange	Water	0.50		
	Alizarin yellow	10.0–12.0	—	Pale-yellow	Orange	Water	0.10		
Tropeolin 00	Orange IV	1.3–3.2	—	Red	Yellow	Dilute alcohol	0.5	—	2
	Methyl orange	3.1–4.8	—	Red	Orange-yellow	Water	0.2	—	6
	Methyl red	4.2–6.3	5.1	Red	Yellow	60% alcohol	2.0	—	3
	Neutral red	6.8–8.0	—	Red	Yellow	Water	1.0	—	3
	Litmus	5.0–8.0	—	Red	Blue	60% alcohol	2.0	—	15
Turmeric	Curcumin	7.8–9.2	—	Yellow	Brown	Dilute alcohol	1.0	—	4
Phenolphthalein	—	8.2–10.0	9.7	Colorless	Red	70% alcohol	1.0	—	10
Tropeolin 0	—	11.0–13.0	—	Yellow	Orange	Dilute alcohol	1.0	—	7

Other common indicators

Alizarin blue
Alizarin red
Alkali blue
Aminoazobenzene
Amphomagenta
Aurin
Azolitmin
Benzeneazobenzylaniline
Benzeneazonaphthylamine
Benzopurpurin
Brilliant green
Brilliant yellow

Bromophenol purple
Carminic acid
Cochineal
Congo red
o-Cresolphthalein
Crystal violet
Curcumin
Dimethylaminoazobenzene
Dinitrophenol
Diphenylaminoazobenzene
Erythrosin
Ethyl violet

Fuchsin, basic
Hematoxylin
Hydrolapachol
Iodine green
Lacmoid
Lophine
Malachite green
Metanil yellow
Methyl blue
Methyl green
Methyl violet
α-Naphtholphthalein

α-Naphthylamineazosulfanilic acid
Nile blue
Nitrophenolsulfophthalein
Phenacetolin
Poirrier's blue
Propyl red
Purpurin
Rosolic acid
Thymolphthalein
Trinitrobenzene
Tropeolin 000
Xylenol blue

putrefaction, and has a fecal odor. Used as a microchemical reagent for cellulose and, diluted, in orange blossom perfume. Cf. indolyl. **diketo ~** Isatin. **dihydrooxo ~** Oxindole. **hydroxy ~** Indoxyl. **iso ~*** $C_6H_4 \cdot CH_2 \cdot N{:}CH =$ $C_6H_4 \cdot CH \cdot NH \cdot CH$. 2-Benzazole. Stable below −196° and in solution under N_2. **2-methyl-1*H*- ~** C_9H_9N = 131.2. β-Methylindole, methyl ketol. Colorless crystals, m.59, soluble in water. **3-methyl-1*H*- ~** Skatole. **nitro ~** $C_8H_6O_2N_2$ = 162.1. **3- ~** Yellow needles, m.213.

indolol Indoxyl.

indolone **1-** Phthalimidine. **2- ~** Oxindole. **3- ~** ψ-Indoxyl.

indolyl* The radical NC_8H_6-, from indole. **iso ~*** The radical NC_8H_6-, from iso*indole.* **3-i.acetic acid*** $NC_8H_6 \cdot CH_2 \cdot COOH$ = 175.2. b.197, soluble in acetone. An auxin plant hormone with cell-enlargement properties.

indomethacin $C_{19}H_{16}O_4NCl$ = 357.8. Indocid. White crystals, m.160, insoluble in water. An analgesic and anti-inflammatory, used to treat arthritis (USP, BP).

indone (1) Indenone*. (2) 1-Indanone.

indophenine $C_{24}H_{14}O_2N_2S_2$ = 426.5. Colorless powder, insoluble in water.

indophenol $CO(CH{:}CH)_2 \cdot CN \cdot C_6H_4 \cdot OH$ = 199.2. Hydroxyphenyliminobenzenone. Used to synthesize sulfur dyes.

Indopol Trademark for a range of moisture-resistant polybutenes, mol. wt. 300–1,900.

indoxyl $C_6H_4 \cdot NH \cdot CH{:}C(OH)$ = 133.2. **alpha- ~** 3-Hydroxyindole. Yellow crystals, m.85, soluble in water; used in organic synthesis. Also in keto (pseudo) form. **i. potassium sulfate** Indican.

indoxylic acid $C_9H_7O_3N$ = 177.2. An oxidation product of indoxyl, b.122 (sublimes and decomp.), soluble in water.

induced Caused or produced indirectly. **i. current** A high-frequency current produced by an induction coil. **i. radioactivity** Radioactivity produced by bombardment with neutrons, protons, or other particles. See *radioelements.* **i. reaction** Sympathetic reaction. If a slow reaction between substances *A* and *C* is hastened by promoting a fast reaction between *A* and *B*, then *A* is the *actor* or *donor* (usually an oxidizing or reducing agent), *B* the *inductor*, and *C* the *acceptor.*

inducer Inductor.

inductance* Induction. The extent to which a magnetic field is created as a result of a variation in current. Measured in henrys. **self- ~** Resistance to a change in a current by the creation of a back emf. **mutual ~** Creation of an emf in one circuit as a result of variation in the current of another; as in a transformer.

induction (1) A process of inference by which one passes from particular data to general principles. Cf. *deduction.* (2) Inductance*. (3) See *induced reaction.* (4) A change (produced by radiation) in the energy of a molecule, due to interaction with another molecule, which is at a distance from it greater than the diameter of the first molecule. Cf. *collision.* **chemical ~** See *induction (4).* **electromagnetic ~** Inductance*. **mutual ~** See *mutual inductance* under *inductance.* **photochemical ~** See *photochemical induction.* **self- ~** See *self-inductance* under *inductance.* **i. coil** Electric transformer. A wire spool inside another, used to obtain high-frequency alternating currents from a continuous current passed through the primary (inner) coil. **i. furnace** See *induction furnace* under *furnace.*

inductive capacity Relative *permittivity*.*

inductivity Relative *permittivity*.*

inductor See *induced reaction.*

indulines Blue or black *azine dyes,* q.v., with aryl substitution at all N atoms.

indurated Hardened, as in the firing of clays.

indyl The indolyl radical.

-ine Suffix indicating (1) a halogen, as, chlorine; (2) an alkaloid or nitrogen base, as, morphine. Cf. *-in* (under *in*).

inert Sluggish; having little or no chemical action. **i. elements** The *noble gases**, q.v., so called because of their low reactivity with other elements. **i. substance** A substance that is resistant to chemical or physical action.

inertia The tendency of a physical body to remain in an unchanged condition, either in a state of uniform motion, or at rest. **moment of ~** A factor equal to Σmr^2 in the mathematic treatment of a rotating body, where m is the mass of each unit a distance r from the axis of rotation. Cf. *momentum.*

infarct An area of dead tissue in the body due to blockage of arterial blood supply. **myocardial ~** Heart attack. An i. of heart muscle, or myocardium, due to blockage of an artery in the heart (coronary artery).

infection (1) Disease due to successful invasion and growth of microorganisms or protozoans in tissues of an organism, human or animal. (2) Transmission of infection. Cf. *contamination.* **airborne ~** Aerial i. I. caused by inhalation of dust particles or droplets containing microorganisms. **droplet ~** I. caused by inhalation of droplets from mouth and nose containing viruses or bacteria; e.g., measles, common cold. **focal ~** I. in which the bacterial growth is restricted to a small area of the organism. **mixed ~** I. caused by more than one kind of bacterium.

infectious disease An infection due to bacteria or viruses spread between humans or animals by direct contact or by airborne route.

infiltration (1) The deposition of minerals from solution in the pores of a rock. (2) The spread of a foreign substance in the body tissues, as, an injected solution or spread of malignant cells.

infinitesimal Smaller than any assigned quantity. Negligible.

inflammable Flammable. **i. air** The original name for hydrogen.

infra Beyond. **i.luminescence** Luminescence whose wavelengths are in the infrared region. **i.phonic** Infrasonic. **i.photic** Pertaining to radiation of a wavelength too long to be visible; as, i.red. **i.red** i.r. Electromagnetic radiation in the wavelength range 10^{-3} to 10^{-6} m (10^7–10^4 Å), which overlaps a portion of the visible spectrum. See the accompanying table. Cf. *radiation.* **i.röntgen rays** Grenz rays. **i.sonic** Pertaining to sound whose frequency is too slow to be perceived by the human ear (below 16–20 Hz). See *frequency, sound.*

	Infrared rays, %	Visible rays,
Sunlight	60	34
Incandescent lamp	95	4.8
Carbon arc	80	15
Resistance wire	99	0.5

infundibuliform A funnel-shaped bacterial growth.

infusible Not capable of being fused. **i. white precipitate** *Mercuri*diammonium chloride.

infusion Infusum. A solution obtained by steeping vegetable drugs in water at or below its boiling point, and straining. Cf. *decoction.*
infusoria A class of protozoa. Erroneously applied to diatoms (protophyta).
infusorial earth Diatomaceous earth, tripolite, kieselguhr. A light, earthy sedimentary rock consisting of empty shells of diatoms and other protophyta. Used as a filtration aid, and adsorbent.
infusum Infusion.
Ing. (French: *ingénieur;* German: *Ingenieur*) Title indicating a scientific degree at graduate level. **Dr.-Ing.** Title indicating a higher scientific degree, similar to a Ph.D.
ingredient Any constituent of a mixture. Cf. *constituent.*
inhaler (1) A device to administer vapors, gases, fine powders, or droplets. (2) A device to filter dust from air to be breathed. Cf. *respirator.*
inhibition A restraint or encumbrance.
inhibitor A substance that slows down a chemical reaction. **vapor-phase ~** An organic compound which is solid at ordinary temperatures, and evolves a vapor which surrounds a metal article in a closed container and produces on its surface an invisible protective film; e.g., nitrites of nitrogen bases. Cf. *vapor.*
inhibitory phase Protective *colloid.*
initiator Trigger. Cf. *promoter.*
injection The administration of a substance into a part of an organism: intravenous (into the bloodstream via a vein), intramuscular (into muscular tissue), subcutaneous (under the skin), intra-articular (into a joint cavity), intraarterial (into an artery), or epidural (near the nerves emerging from the spinal chord).
ink (1) A colored liquid, used for writing. (2) A colored paste or liquid used for printing. **aniline ~** A solution of an aniline dye in a volatile solvent or dilute gum; used for printing in bright colors or at high speeds, e.g., by the gravure process. Cf. *flexography.* **canceling ~** A suspension of lampblack in oil, used for stamp pads. **Chinese ~** India i. **copying ~** An iron-tannin i. **diamond ~** A mixture of barium sulfate and hydrofluoric acid, used for writing on glass. **fugitive ~** An i. that disappears on treatment with water or bleaching chemicals. Used for printing checks. **india ~** Finely divided lampblack suspended in water or gum. **invisible ~** Secret i., sympathetic i. An i. normally invisible to the human eye, but rendered visible by heat (lemon juice), light (soap solution in ultraviolet light), water (cobalt salts), or chemicals (iodine vapor on a starch i.). **long ~** Free-flowing (printing) i. **marking ~** A solution of silver nitrate, used to write indelibly on paper, textiles, laundry, etc. **printing ~** Five types:(1) A suspension of a pigment (usually carbon black) in a mineral oil, which dries by absorption of oil into the paper; used for high-speed letterpress printing. (2) A suspension of a pigment in a drying oil (e.g., linseed oil), which dries by formation of a protective layer of hard linseed oil varnish over the pigment; used for general lithographic and letterpress work. (3) An aniline *ink,* q.v., which dries mainly by evaporation. (4) A warm, molten resinous pigment which dries by simple solidification. (5) Special formulations, such as those cured by u.v. light. **secret ~** Invisible i. **short ~** Tacky (printing) i. **sympathetic ~** Invisible i. **writing ~** (1) Blue-black i. Normally, a slightly acidic solution containing principally an iron salt, a tannin, and a blue aniline dye (provisional color) to render the i. visible while it is being used. Oxidation of the iron-tannin compound produces a permanent blue-black color. (2) A solution of an aniline dye in a dilute gum; used for fountain pens, and for colored inks. (3) Ball-point i. A solution of a dye in a waxy medium.
i. transfer coefficient The ratio of the amount of i. on the paper to that on the printing surface in a printing process.
INN Abbreviation for International Nonproprietary Name, as given to drugs by the World Health Organization. May be prefixed by "p" (proposed) or "r" (recognized).
innocuous Describing a harmless substance.
innoxious Harmless. Antonym: toxic or noxious.
-ino A suffix denoting (not exclusively) substitution by one of the groups NH_2, NHR, NR_2, NH, or NR. Cf. *amino.*
inoculation Innoculation (U.K. usage). (1) The injection or insertion of microorganisms or their products into the body. Prophylactically, i. of attenuated organisms, or antigenic material from them, stimulates the production of antibodies. (2) The planting of bacteria on a culture medium.
inoculum Innoculum (U.K. usage). The substance to be inoculated.
inorganic (1) Unorganic. Pertaining to chemicals that do not contain carbon (carbonates and cyanides excepted). Cf. *organic.* (2) Devoid of an organized structure. **i. chemistry** Chemistry which deals with inorganic or polar compounds, which do not contain carbon as a principal element.
inosamines* Generic name for compounds of general name *n*-amino-*n*-deoxyinositol.
inoses* Generic name for the 2,3,4,5,6-pentahydroxy-cyclohexanones.
inosine $C_{10}H_{12}N_4O_5$ = 268.2. Hypoxanthin riboside. White needles, m.218 (decomp.), soluble in hot water.
inosinic acid $C_{10}H_{13}N_4O_8P$ = 348.2. Inosinphosphoric acid. A nucleotide from adenylic acid of nucleoproteins.
inosite Inositol*.
inositol* $C_6H_6(OH)_6 \cdot 2H_2O$ = 216.2. (1)* Generic name for 9 stereoisomeric cyclohexanehexols. Configuration indicated by a prefix or locants above/below ring. (2) Muscle sugar, 1,2,3,5/4,6-cyclohexanehexol*, dambose, inose, hexahydroxybenzene, Bios I, nucite, phaseomannite. Found in barley, peas, beans, and animal flesh, in the form of its phosphoric acid ester. Optically inactive *(meso)* white crystals, m.200 (decomp.), soluble in water. **dextro ~** Colorless crystals, m.247. **levo ~** Colorless crystals m.246.
i.hexaphosphoric acid Phytic acid.
inquartation Quartation.
insecticide An agent used to destroy insects, generally by dusting or spraying on plants. About 10% of world crops are lost because of insect pests. Cf. *fumigant.* Classes: (1) Contact i., which corrodes the surface of soft-bodied insects; as, kerosene, (2) Stomach i., which poisons through the intestinal tract, as chlordane. (3) Systemic i., carried in the sap stream of plants, thereby rendering such plants poisonous to insects; as, aldicarb. See Table 45, opposite.
insect wax Chinese wax.
insemination The introduction of semen into the vagina.
insipid Tasteless.
insipin $(C_{20}H_{23}O_2N_2)OCH_2CO \cdot H_2SO_4 \cdot 3H_2O$. Quinine diglycol sulfate. An almost tasteless quinine substitute.
in situ In the normal or natural place or position.
insol. Abbreviation for "insoluble."
insolation Solarization, irradiation. Exposure to sunrays.
insolubility The quality of being immiscible with, or insoluble in, a liquid.
insoluble Incapable of dissolving in a liquid.
inspirator (1) Respirator. (2) A device for controlling automatically the proportions of the constituents of a mixture of gases.
inspissation Thickening a liquid by evaporation.

TABLE 45. INSECTICIDES

Common name	Systemic (s)	Chemical name
Aldicarb[a]	s	2-Methyl-2-(methylthio)propionaldehyde *O*-methylcarbamoyloxime
Aldrin[a]		Contains not less than 95% of HHDN (see below)
Alphametrin		α-Cyano-3-phenoxybenzyl-3-(2,2-dichlorovinyl)- 2,2-dimethylcyclopropanecarboxylate
Azinphos-methyl[a]		*S*-(3,4-Dihydro-4-oxobenzo[*d*]-[1,2,3]-triazin-3-ylmethyl) *O,O*-dimethyl phosphorodithioate
γ-BHC[a](γ-HCH[a])		γ-isomer of 1,2,3,4,5,6-Hexachlorocyclohexane
Carbofuran*	s	2,3-Dihydro-2,2-dimethylbenzofuran-7-yl methylcarbamate
Carbophenothion[a]		*S*-(4-Chlorophenylthio)methyl-*O,O*-diethyl phosphorodithioate
Chlordane[a]		1,2,4,5,6,7,8,8-Octachloro-3a,4,7,7a-tetrahydro-4,7-methanoindane
Chlorfenvinphos[a]		2-Chloro-1-(2,4-dichlorophenyl)vinyl diethyl phosphate
Chlorpyriphos[a]		*O,O*,-Diethyl *O*-(3,5,6-trichloro-2-pyridyl) phosphorothioate
Cypermethrin		(*R,S*)-α-Cyano-3-phenoxybenzyl (1*R,S*)-*cis, trans*-3-(2,2-dichlorovinyl)-2,2-dimethylcyclopropanecarboxylate
DDT		A complex mixture in which *pp'*-DDT predominates
pp'-DDT		1,1,1-Trichloro-2,2-bis(4-chlorophenyl)ethane
Demeton-S-methyl[a]	s	*S*-2-Ethylthioethyl *O,O*-dimethyl phosphorothioate
Derris		A mixture of rotenone and related compounds extracted from roots of *Derris, Lonchocarpus*, and *Tephrosia* spp.
1,3-Dichloropropene[a]		1,3-Dichloropropene
Dimethoate[a]	s	*O,O*-Dimethyl *S*-methylcarbamoylmethyl phosphorodithioate
DNOC[a]		4,6-Dinitro-*o*-cresol
Endosulfan[a]		1,4,5,6,7,7-Hexachloro-8,9,10-trinorborn-5-en-2,3-ylenedimethyl sulfite
Fenitrothion[a]		*O,O*-Dimethyl *O*-4-nitro-*m*-tolyl phosphorothioate
Formothion[a]	s	*S*-(*N*-Formyl-*N*-methylcarbamoylmethyl) *O,O*-dimethyl phosphorodithioate
γ-HCH[a]		See *γ-BHC* (above)
Heptenophos[a]	s	7-Chlorobicyclo[3.2.0]-hepta-2,6-dien-6-yl dimethyl phosphate
HHDN[a]		A polychlorinated dimethanonaphthalene
Malathion[a]		*S*-1,2-bis(Ethoxycarbonoyl)ethyl *O,O*-dimethyl phosphorodithioate
Metaldehyde[a]		2,4,6,8-Tetramethyl-1,3,5,7-tetraoxacyclooctane
Methiocarb[a] (Mercaptodimethur[a])		3,5-Dimethyl-4-(methylthio)phenyl methylcarbamate
Nicotine[a]		(*S*)-3-(1-Methylpyrrolidin-2-yl)pyridine
Omethoate[a]	s	*O,O*-Dimethyl *S*-methylcarbamoylmethyl phosphorothioate
Oxamyl	s	*N,N*-Dimethyl-2-methylcarbamoyloxyimino-2-(methylthio)acetamide
Oxydemeton-methyl[a]	s	*S*-2-Ethylsulfinylethyl *O,O*-dimethyl phosphorothioate
Permethrin[a]		3-Phenoxybenzyl (1*R,S*)-*cis,trans*-3-(2,2-dichlorovinyl)-2,2-dimethylcyclopropanecarboxylate
Petroleum oil		Aliphatic hydrocarbons from highly refined petroleum oil
Phorate[a]	s	*O,O*-Diethyl *S*-ethylthiomethyl phosphorodithioate
Phosalone[a]	s	*S*-6-Chloro-2,3-dihydro-2-oxobenzoxazol-3-ylmethyl *O,O*-diethyl phosphorothioate
Pirimicarb[a]		2-Dimethylamino-5,6-dimethylpyrimidin-4-yl dimethylcarbamate
Pirimiphos-methyl[a]		*O*-2-Diethylamino-6-methylpyrimidin-4-yl *O,O*-dimethyl phosphorothioate
Pyrethrum[a]		Mixture of esters of chrysanthemic and pyrethric acids from pyrethrolone, cinerolone, and jasmolone alcohols
Tar oil		Aromatic hydrocarbons from coal tar distillates
Thiofanox[a]	s	3,3-Dimethyl-1-(methylthio)butanone *O*-methylcarbamoyloxime
Thiometon[a]		*S*-2-Ethylthioethyl *O,O*-dimethyl phosphorothioate
Triazophos[a]		*O,O*-Diethyl *O*-1-phenyl-1,2,4-triazol-3-yl phosphorothioate

[a]ISO-approved names. Some countries do not use certain ISO names because of their similarity to established names.

inspissator An evaporator.
instrument A mechanical or electrical device or appliance usually involved in measurement.
insuccation Soaking a material with water.
insufflation The introduction of fine powder, gas, or air into body cavities, openings, or wounds.
insulation The mechanical placing apart or separation of a physical system. **electrical ~** The prevention of the passage of electricity. **heat ~** The prevention of the passage of heat. See *thermal resistance, thermal conductivity*.
insulator A protective and separating agent; a nonconductor. **electrical ~** A device to prevent the passage of electricity from a conductor. **thermal ~** A packing that is nonconductive to heat.
insulin $C_{257}H_{387}N_{65}O_{66}S_6$ = 5634. Iletin, glucokinin. White, levorotatory crystals, m.233. A protein hormone secreted by the islets of Langerhans in the pancreas; deficiency results in diabetes. I. for treatment of diabetes is primarily obtained from bovine or porcine pancreas. Zinc is added to the extract to produce a purer crystalline form. The solubility and pH affect the onset and duration of action. **human ~** I. from either the human pancreas or a biosynthetic modification of porcine i. The chemical structure of biosynthetic h. i. is identical to that from the human pancreas, apart from one

different amino acid; it causes less allergic response than porcine i. **h. i. crb** (Chain *r*ecombinant DNA *b*acteria.) A biosynthetic form. **h. i. emp** (*E*nzyme-*m*odified *p*orcine.) A biosynthetic form. **protamine zinc ~** Protamine, being relatively insoluble, causes a longer duration of action of 24–36 h (USP, EP, BP).

i. injection Soluble i., regular i. Added zinc 40 μg/100 units; pH 3–3.5; onset of action, 30 min; duration, 6–12 h (USP, EP, BP). **i. units** One International Unit is the activity in 0.04167 mg of Fourth International Standard Preparation (1958); this is a mixture of 52% bovine and 48% porcine insulin containing 24 units/mg. **i. zinc ultralente suspension** Extended i. I. buffered with acetate so that more zinc combines with it. Duration of action, 30–36 h (USP, EP, BP).

intaglio A process of printing from plates which have been etched slightly in recess; the ink filling these is absorbed by the paper. See *rotogravure.*

Intal Trademark for cromolyn sodium.

integration (1) Assimilation or synthesis, as opposed to disintegration. (2) The summation of a series of values of a continuously varying quantity. Cf. *calculus.*

intensification (1) A process of concentrating force. (2) In photography, to increase the density of a photographic image.

intensity The strength or amount of energy per unit space, area, or time. **acid ~** Hydrogen-ion concentration. **color ~** The (1) brilliance or (2) saturation of a color. Cf. *Beer's law.* **electric ~, electric field ~** Electric field strength*. **heat ~** Temperature. **light ~** Luminance*. **magnetic field ~** Cgs system concept of a magnetic field which exerts a force of 1 dyne on a unit magnetic pole (gauss). **magnetization ~** The magnetic moment per unit volume. **sound ~** Energy transfer per unit area and time.

i. factor Of acidity, pH; of redox, rH,

inter- Prefix (Latin) indicating "between."

interdecolation The forcing apart of materials with a layered structure (as, mica) by energy released by chemical action; as between titanium disulfide and lithium.

interface Interphase. The boundary between 2 phases. Cf. *zone.*

interference A conflict between 2 agencies which produces a retardation effect, or a waste of energy. **chemical ~** In analysis, i. by a material, such as another chemical species, that causes an error in the results. **light ~** The effect produced by 2 sets of light waves that offset each other to cause diminished intensity, such as darkness. **sound ~** The effect produced by 2 sets of sound waves that offset each other to cause diminished intensity, such as silence. **spectroscopic ~** I. caused by emission or absorption bands (lines) of another species that overlap those of the substance of interest.

i. colors Complementary *colors.*

interferometer An instrument to determine the wavelength of light from interference by waves of known lengths. Cf. *fringes.*

interferons Glycoproteins, liberated by cells infected by actively dividing viruses, that reduce the activity of the virus. Types: *i. α.* Produced from leucocytes. *i. β.* Produced from fibroblasts. *i γ.* Produced from T lymphocytes. Used for herpes and hepatitis B infection. They retard the growth of tumors in animals, and, thus, are thought of as a treatment for cancer. **i. induction** Stimulation of the body's natural synthesis of i. by injection of certain agents. **i. units** 1 ampoule of International Reference Preparation contains 5,000 units of i. α.

intermediate A chemical used in organic synthesis; also in the production of pharmaceuticals, dyes, and other artificial products; usually a derivative of "crudes" or raw materials. Important i. are ethylene, propylene, benzene, phenol, and butadiene.

intermetallic Describing compounds of 2 or more metals (as distinct from alloys); e.g., NiAl or $CrBe_2$. **i. compound** A compound of metals in stoichiometric proportions. Cf. *alloy.*

intermolecular Referring to action between molecules. Cf. *intramolecular.*

internal Pertaining to the inside. **i. anhydride** A compound formed by elimination of water from the atoms of a molecule. **i. compensation** The property of an optically inactive molecule that contains 2 asymmetric C atoms, one dextro-, the other levorotatory. **i. reaction** A reaction within a molecule due to atomic rearrangement. **i. salt** An organic compound formed by the union of a basic and acid radical within the molecule. **i. standard** A substance added to a sample for analysis which calibrates the assay. The results are usually ratioed, in that any changes in the sample are reflected by both the i.s. and the sample.

international Agreed upon between nations. **i. atomic weights** Values for *atomic* weights, q.v., selected by the I. Union of Pure and Applied Chemistry. **I. Organization for Standardization** I.S.O. An association of many countries, concerned with the standardization of technical data, nomenclature, specifications, and testing methods. See *standards.* **I. Practical Temperature Scale** See *temperature.* **I. System of Units** See *SI.* **I. Union of Pure and Applied Chemistry** IUPAC. An i. organization that standardizes chemical *nomenclature,* q.v., notation, symbols, data, atomic weights, etc. Its principal publication is *Pure and Applied Chemistry.* A similar organization exists for biochemists (IUB) and physicists (IUP). **i. unit** I.U. A measure of the potency of a substance. See *vitamin units, penicillin units* (under *penicillin*).

interphase Interfacial *zone.*

interpolation The deduction, as by graphical methods, of a value that lies within the range of values already established. Cf. *extrapolation.*

interruptor A device for breaking an electric current.

interstice A small space or capillary in a structure or tissue. **atomic ~** The distance between the atoms in a molecule.

intertraction Barophoresis. The increase in density of a colloidal solution (e.g., albumin) placed on a salt solution of nearly equal density, due to the rapid diffusion of the solute.

intolerance Inability to withstand the effects of a drug, or digest or metabolize certain foods or parts of foods.

intoxication Poisoning by a drug.

Intoxilyzer, Intoximeter See *breath alcohol.*

intra- Prefix (Latin) indicating "within."

intraannular Within the ring. **i. tautomerism** The redistribution of double bonds within a ring. Cf. *intranuclear tautomerism.*

intraatomic Pertaining to atomic structure. **i. matter** Matter from which atoms are assumed to be constructed, e.g., electrons and positive nuclei.

intramolecular Pertaining to different parts of the same molecule. Cf. *intermolecular.* **i. action** A reaction occurring within the individual molecule. **i. condensation** Ring formation. A reaction in which the atoms of an organic compound combine or rearrange and form a condensation (usually a ring) compound and another (usually binary) compound. **i. oxidation and reduction** An internal oxidation and reduction reaction; as, $C_6H_4 \cdot CH_3 \cdot NO_2 \rightarrow C_6H_4COOH \cdot NH_2$.

intramuscular Inside muscular tissue; as, of an injection.
intranuclear (1) Within an atomic nucleus. (2) Within a molecular ring system. **i. tautomerism** The shifting of a double bond within one or more rings.
Intraval sodium Trademark for thiopental sodium.
intravenous Within veins, e.g., an injection.
intravital (1) Within the living organism. (2) Within a lifetime.
introduction Causing the entry of a different type of atom into an organic molecule, e.g., chlorination.
introfaction A change in the fluidity and specific wetting properties of an impregnating material, due to an introfier.
introfier Impregnation accelerator. A substance that speeds up the penetrating power of fluids.
intrusion Forcing a material into the cavities or pores of a substance.
intumescence (1) Swelling up, especially of certain crystals on heating. (2) Popping, puffing. The violent escape of moisture on heating.
inula camphor Helenin.
inulenin $(C_6H_{10}O_5)_n \cdot 2H_2O$. A carbohydrate associated with inulin. Colorless needles, soluble in water.
inulic acid Alantic acid.
inulin $C_6H_{11}O_5(C_6H_{10}O_5)_nOH$. Alantin, alant starch, dahlin, sinistrin. A polysaccharide from the rhizome of *Inula helenium* or *Dahlia variabilis.* White powder, m.160 (decomp.), soluble in hot water. Used to measure glomerular filtration rate of kidneys.
inulinase* An enzyme that endohydrolyzes 2,1-β-D-fructosidic linkages in inulin.
in vacuo In a *vacuum,* q.v.
Invar Trademark for the ferronickel: Ni 36, steel 64% (carbon content 0.2%), d.8.0, m.1500. It has a low coefficient of heat expansion; used for precision instruments.
inversion (1) The turning of a levo to a dextro compound, or vice versa. (2) The change of an isomeric compound to its opposite, as a *cis* to a *trans* compound. (3) The hydrolysis of an optically active disaccharide to 2 optically active monosaccharides; e.g., the hydrolysis of cane sugar to glucose and fructose by dilute acids, alkalies, or enzymes, resulting in a change in the direction and degree of rotation of polarized light. Cf. *Walden inversion, Clerget inversion.* (4) In an emulsion of 2 immiscible liquids, the interchange of the internal and external phases. **dipole ~** Symmetrization. The reversal of the normal activity of functional groups in organic chemistry.
i. point The temperature at which i. takes place.
invertase β-D-Fructofuranosidase*, saccharase, invertin. An enzyme of the pancreatic juice and of yeast, which hydrolyzes terminal, nonreducing β-D-fructofuranoside residues in β-D-fructofuranosides; converts cane sugar into invert sugar.
invertin Invertase.
invert soap A cationic, surface-active detergent, so called because it ionizes oppositely to soap; e.g., quaternary ammonium or sulfonium compounds.
invert sugar Approximately 50% glucose and 50% fructose, obtained by the acid hydrolysis of cane sugar. It is slightly levorotatory, fermentable; it reduces Fehling's solution and is used in brewing. **i. s. solution** A partially inverted solution of sucrose containing at least 62% solids, 3–50% i. s., and equal weights of fructose and glucose.
in vitro Describing a biological reaction which can be performed outside the living organism in the laboratory; as, in a test tube or petri dish, on a microscope slide, etc. Cf. *in vivo.* **i. v. fertilization** I.V.F. Fertilization in the laboratory of a (human) ovum, removed from an ovary, by sperm. (Used in conception of "test tube babies.") See *embryo replacement.*
in vivo Describing a reaction which takes place within the living organism. Cf. *in vitro.*
inyoite $2CaO \cdot 3B_2O_3 \cdot 13H_2O$. A native borate (S. California).
iod- See *iodo-.*
iodal $CI_3 \cdot CHO$ = 421.7. A liquid resembling chloral.
iodalbin A red compound of blood albumin and iodine, of molasseslike odor.
iodaniline Iodoaniline*.
iodate* A salt of iodic acid, containing the radical IO_3^-.
iodeosin $C_{20}H_8O_5I_4$ = 835.9. Erythrosin, tetraiodofluorescein. A red indicator powder, soluble in alcohol (alkalies—rose-red, acids—yellow). **i. solution** A 0.0002% solution of iodeosin in ether. This is added to dilute alkali and titrated until the rose tint passes from the ether into the aqueous solution.
iodi- See *iodo-.*
iodic **i. acid*** HIO_3 = 175.9. Metaiodic acid. Colorless rhombs, m.110, soluble in water. Used as an oxidizing agent; as a reagent for alkaloids, biliary pigments, naphthol, thiocyanates, and guaiacol; in organic synthesis, and for volumetric solutions **per ~** See *periodic acid.* **i. anhydride** Iodine pentaoxide*.
iodide* MI_n. A binary compound of iodine with a metal. **i. ion*** The I^- ion.
iodimetry Iodometry.
iodinated (^{131}I) serum A sterile solution of human serum albumin, treated with ^{131}I and freed from iodide; used to diagnose lung conditions; as, small tumors or emboli, and to estimate blood volume.
iodine* I = 126.9045. Id* (if I* is inconvenient). Iodum. A nonmetallic element, at. no. 53, of the halogen group. Rhombic, bluish-black, lustrous plates or scales, d.4.948, m.114, b.184, slightly soluble in water, soluble in alcohol or iodide solutions. Discovered by Courtois (1811) and named after its purple vapors (Greek: *iodes,* the "violet" and *ion,* "similar"). Obtained from the mother liquor of Chile saltpeter and seaweed ash, and widespread in nature. Valency: usually 1 (iodides*), or 3 (iodonium*), or 5 (iodates*). Used as a reagent in volumetric analysis; in organic synthesis; in the manufacture of iodides, iodates, and iodine preparations; and as an antiseptic and caustic. Used medically (^{125}I and ^{131}I) as sodium iodide and iodinated albumin. I. is also an essential trace element, present in thyroid hormones; deficiency in diet leads to goiter and hypothyroidism. Recommended daily intake 150 μg. **eka ~** Early name for astatine. **povidone- ~** $(C_6H_9ON)_nI$. 1-Vinyl-2-pyrrolidinone polymer with iodine. Betadine. An antiseptic (USP, EP, BP). **solution of ~** (1) Lugol solution. (2) Colorless Lugol solution; decolorized with sodium thiosulfate. (3) *Iodine water,* q.v. **tincture of ~** An alcoholic 7% iodine solution in 5% potassium iodide solution; an antiseptic (USP).
i. acetate* $IC_2H_3O_2$ = 185.9. A solid prepared from chlorine dioxide and i. in glacial acetic acid. **i. bromides** IBr, i. monobromide; IBr_3, i. tribromide; IBr_5, i. pentabromide. **i. chlorides** ICl, i. monochloride; ICl_3, i. trichloride. **i. cyanide*** ICN = 152.9. Cyanogen i. Colorless crystals, m.146, soluble in water. **i. cycle** See Fig. 15. **i. dioxide*** IO_2 = 158.9 or I_2O_4 = 317.8. Yellow powder, decomp. into its elements at 130. **i. disulfide** Sulfur iodide. **i. green** A phenolphthalein dye pH indicator, changing at 1.0 from yellow (acid) to blue-green (alkaline); also stains liquefied xylem in plant tissues. **i. fluoride** See *iodine pentafluoride.* **i. monobromide*** IBr = 206.8. Purple crystals, m.36, soluble in water (decomp.). Used

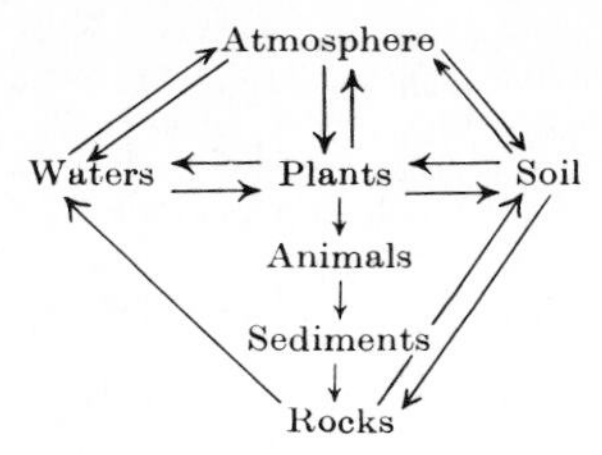

Fig. 15. The iodine cycle.

in analysis (iodine numbers) and in organic synthesis. **i. monochloride*** ICl = 162.4. Brown oil, d.3.182, b.101 (decomp.), decomp. by water; used in analysis and organic synthesis. **i. number** I. absorption value, Hübl number, Wijs number. The quantity of i., in mg, absorbed by 1 g fat or oil under specified conditions; it indicates the amount of unsaturated acids present **hydro ~** The number of parts of hydrogen (calculated as the equivalent amount of I) absorbed by 100 parts of a fat. **i. oxide** (1) I. dioxide. (2) I. pentaoxide. (3) I_4O_9 = 651.6. Green crystals, m.100 (decomp.).A green supposed oxide of i. **i. pentabromide*** IBr_5 = 526.4. Brown liquid, decomp. by water; a reagent. **i. pentafluoride*** IF_5 = 221.9. Fluoriodine. Colorless, fuming liquid, b.97, highly reactive, decomp. by water. **i. pentaoxide*** I_2O_5 = 333.8. Iodic anhydride. White crystals, soluble in water; an oxidizing agent. **i. sulfate*** $I_2(SO_4)_3$ = 542.0. Yellow crystals, soluble in water. **i. tincture** See tincture of *iodine*. **i. tribromide*** IBr_3 = 366.6. Tribromoiodine. Brown liquid, soluble in water. **i. trichloride*** ICl_3 = 233.3. Yellow, deliquescent crystals, decomp. 25, soluble in water, an antiseptic and disinfectant. **i. value** I. number. **i. water** An aqueous 0.02% solution of i.; a reagent.

iodinin A purple-bronze antibiotic pigment from *Chromobacterium iodinum.*

iodipamide methylglucamine $C_{20}H_{14}O_6N_2I_6 \cdot 2C_7H_{17}O_5N$ = 1,530. Biligrafin, Cholegrafin. Yellow liquid; a radiopaque, used for gall bladder and bile ducts (USP).

iodite* A salt of the hypothetical iodous acid containing the IO_2^{2-} ion. **hypo ~ *** MOI. A salt of hypoiodous acid.

iodized Admixed with iodine or an iodide. **i. oil** BP name for ethiodized oil. **i. salt** Common salt containing about 0.3% of potassium iodide.

iodo-* Prefix indicating an iodine atom. Cf. *iodosyl.*

iodoacetic acid* CH_2ICOOH = 185.9. Carboxymethyl iodide. Yellow crystals, m.82 (decomp.), insoluble in water.

iodoaniline* $C_6H_4INH_2$ = 219.0 Aminoiodobenzene. ***ortho-* ~** Colorless needles, m.57. ***meta-* ~** Colorless leaflets, m.25. ***para-* ~** Colorless needles, m.63. Insoluble in water.

iodobenzene* C_6H_5I = 204.0. Colorless liquid, b.188, insoluble in water.

iodocasein A compound of milk casein and 18% iodine. Brown powder, insoluble in water.

iodochlorhydroxyquin Clioquinol.

iodocrase $Ca_6[Al(OH \cdot F)],Al_2(SiO_4)_3$. Vesuvianite.

iodocresine, iodocresol Traumatol.

iodoethylene* $CH_2{:}CHI$ = 153.9. Vinyl iodide*. Colorless liquid, b.56, slightly soluble in water.

iodoform* CHI_3 = 393.7. Triiodomethane*, methenyl iodide, formyl triiodide. Iodoformum. Yellow hexagons, sublime 119 (decomp.), slightly soluble in water or alcohol; an antiseptic.

iodogorgoic acid $HO \cdot C_6H_2I_2 \cdot CH_2CH(NH_2)COOH$ = 433.0. 3,5-Diiodotyrosine*. **(±)- ~** Rectangular prisms, m.204. **D- ~** White needles, m.194 (decomp.). **L- ~** From thyroid gland and marine organisms, m.200.

iodomethane* Methyl iodide*.

iodometric Iodimetric. Pertaining to iodometry. **i. acid value** Acidity determined in terms of the amount of iodine liberated from an iodide-iodate mixture.

iodometry Iodimetry. A method of volumetric analysis, by titration of a standard iodine solution and standard sodium thiosulfate solution: $I_2 + Na_2S_2O_3 \rightleftharpoons 2NaI + Na_2S_4O_6$. Starch solution is indicator. Used to determine halogens, halides, hydrogen sulfide, sulfur dioxide, arsenic salts, manganese dioxide, ferrous salts, etc.

iodonium* The cation H_2I^+.

iodopanoic acid Iopanoic acid.

iodophor A bactericidal complex of iodine and a nonionic surface-active agent, which releases iodine in water; as, povidone-*iodine.*

iodophosphonium Phosphonium iodide*.

iodopropionic acid* $C_3H_5O_2I$ = 200.0. **2- ~** $MeCH_2ICOOH$. Colorless prisms, m.45, soluble in water. **3- ~** CH_2ICH_2COOH. Colorless leaflets, m.82, soluble in water.

iodoso The iodosyl* radical.

iodosyl* Iodoso. The radical $-IO$, **i.benzene*** C_6H_5IO = 220.0. Colorless powder, explodes 210, soluble in water.

iodous acid The hypothetical acid H_2IO_2, known as its salts. **hypo ~ *** The unstable acid HIO; decomposes to HI and HIO_3.

iodothyrin $C_{11}H_{10}O_3NI_3$ = 584.9. The iodine compound of the thyroid glands.

iodotoluene C_7H_7I = 218.0. Methyliodobenzene. ***ortho-* ~** Colorless liquid, b.211. ***meta-* ~** Colorless liquid, d.1.70, b.204. ***para-* ~** Colorless leaflets, m.35. All insoluble in water.

iodoxy The iodyl* radical.

iodoxyl $C_8H_3O_5NI_2Na_2$ = 492.9. Neiopax, uroselectan-B, disodium iodomethamate. White powder, m. (free acid) 171 (decomp.), soluble in water; an injection radiopaque for urinary tract.

iodstarin $Me(CH_2)_{10} \cdot CI_2(CH_2)_4COOH$ = 522.2. 6,6-Diiodoheptadecanoic acid*, decomp. by light to *tariric acid,* q.v.

iodum Iodine.

iodyl* Iodoxy. The radical $-IO_2$. **i.benzene*** $C_6H_5IO_2$ = 236.0. Colorless needles, m.167, explode 230, soluble in water.

iodyrite AgI. A native silver iodide.

iolanthite A jasper from Oregon.

iolite $Mg_4Al_8O_6(SiO_3)_{10}$. Cordierite, dichroite, water sapphire. A blue, orthorhombic magnesium aluminum metasilicate, hardness 7, d.2.65; a gem.

ion See Table 70 on p. 493 for a list of ions. An electrically charged (1) atom, (2) radical, or (3) molecule. If positively charged, it is called a *cation;* if negative, an *anion.* Ions travel in solution to the cathode (cations) and anode (anions). Cf. *ionic theory.* Classification: (1) Positive atoms (K^+). (2) Negative atoms (Cl^-). (3) Positive radicals (NH_4^+). (4) Negative radicals (NO_3^-). (5) Positive molecules. (6) Negative molecules. **acid ~** Anion. **amphoteric ~** Zwitter i. **aquated ~** See *solvated ion* below. **basic ~** Cation. **colloidal ~** Micelle. **hydrogen ~** See *hydrogen ion.* **hydroxyl ~** See *hydroxyl.* **molecular ~** A positive or negative, gaseous i., e.g., one produced under the influence of radiation. Cf. *activation, excitation.* **negative ~** Anion. **positive ~** Cation. **solvated ~** An i. surrounded by oriented molecules of solvent. **zwitter ~** Amphoteric i. An

i. having both a negative and positive charge; as, $^{+}NH_3RCCO^{-}$.

i. burn I. spot. A defect produced in cathode ray tubes, believed to be due to ions issuing from the thermionic cathode. **i.-exchange column** A vertical tube filled with an i.-exchange resin. See *ion-exchange resin* (under *resin*), *zeolite*. Used to separate mixtures into their constituents, e.g., by percolating a solution down the column. **i. implantation** Altering the properties and composition of solids by embedding accelerated particles in them. **i. pair** Two ions, held together by the attraction of their opposite charges. They are either adjacent (tight i. p.) or separated by 1 or more solvent molecules (loose i. p.). **i. product** Solubility product. **i. retardation** The slowing down of the passage of ions through a gel or resin in ion exchange, due to weak ionic associations. **i. spot** I. burn. **i. states** (1) Solid i., surrounded by oppositely charged ions, as in a polar crystal. (2) Liquid i., surrounded by oriented molecules of the solvent, as in a solution. (3) Gaseous i., with kinetic motion overcoming electrostatic charge, as in an ionized gas.

ionene $Me \cdot C_6H_3 \cdot C_4H_6 \cdot Me_2$ = 174.3. 1,2,3,4-Tetrahydro-1,1,6-trimethylnaphthalene. A hydrocarbon formed on dehydration of ionone. b.240. Cf. *irene*.

ionic Pertaining to electrically charged atoms, radicals, or molecules. **i. charge*** The unit electric charge carried by the hydrogen ion, $e = 1.602 \times 10^{-19}$ C $= 4.803 \times 10^{-10}$ esu $= 1.602 \times 10^{-20}$ emu. The charge of an ion is denoted by a suffix superscript; as, S^{2+}. **i. conductivity** The passage of a curent in solution by means of ions. **i. migration** The passage of ions toward the anode or cathode; as, in electrolysis. **i. mobility** (1) The motion of ions in a solution, expressed numerically by the velocity under a gradient of 1 volt/cm. It is a periodic function of the atomic weight: H^+, 3.25×10^{-4}; OH^-, 1.78×10^{-4}; K^+, 6.5×10^{-4} cm/s. (2) The velocity of ions under the influence of an electric current. **i. number** The number of extranuclear electrons of an ion. It equals the atomic number plus or minus the valence electrons. **i. potential** Assuming the ion in a crystal to be a charged sphere of definite extension, then the i. p. = valence/radius of ion. Cf. *ionization energy*. **i. radius** The distance, in Å, of the periphery of the effective sphere from the center of a charged atom or group of atoms. **i. reactions** Reactions characteristic of an ion; as, the precipitation of Cl^- by Ag^+. **i. theory** Electrolytes dissociate into component ions when dissolved in water. **i. velocity** The product of the mobility of an ion and the actual potential gradient. Cf. *migration*.

ionidine $C_{19}H_{25}O_4N_4$ = 373.4. An alkaloid from the California poppy, *Eschscholtzia californica*.

ionise Ionize.

ionium The isotope thorium-230.

ionization, ionisation (1) The removal of one or more outer electrons from an atom or molecule, by which gases become electrically charged under the influence of a strong electrostatic field or radioactive rays: $H_2 \rightarrow H_2^+ + e$. Cf. *excitation, ionization energy*. (2) Electrolytic *dissociation*. **degree of ~** D. of *dissociation*.

i. constant Dissociation constant. **i. energy** I. potential. The energy, in V, to remove an electron from an atom in its ground state $= 12.345/\lambda$, where λ is the wavelength of a radiation in Å. Cf. *orbital, quantum*. **second i. energy** Second i. potential. The energy to remove a second electron completely. It is greater because of the positive charge on the ion. **i. gage** See *ionization gage* under *gage*. **i. potential** I. energy.

ionize (1) To become electrically charged. (2) To dissociate.

ionized, ionised (1) Describing a molecule that is separated into oppositely charged atoms or radicals, the sum of positive and negative charges being zero. (2) Describing a molecule or atom from which electrons have been removed, whereby it becomes positively charged. **non ~, un-~** Remaining in the molecular condition.

ionizing The process of *ionization*, q.v. **i. potential** Ionization energy. **i. solvent** A liquid that facilitates dissociation, e.g., water or liquid sulfur dioxide. Organic liquids are generally nondissociating.

ionogenic A little-used term for ionic.

ionomer A polymer having covalent bonds between the constituents of the long-chain molecules, and ionic bonds between the chains.

ionone $Me_3C_6H_6CH{:}CHCOMe$ = 192.3. A ketone terpene, and isomer of irone. **alpha-~** Colorless liquid, d.0.934, b.120. **beta-~** Colorless liquid, d.0.949, b.135. Both forms slightly soluble in water; used in artificial extract of violets.

ionophoresis The separation of ions by *electrophoresis*, q.v.

ionosphere The radio-reflecting part of the Kennelly-Heaviside layer of the upper atmosphere, consisting of changing areas of ionized molecules.

iontophoresis Migration of ionic medication through unbroken skin under the influence of a direct electric current.

ionotropy The existence of tautomeric ions, a positive or negative atom or group becoming detached from an unsaturated molecule, thus leaving a − or + charge on the residual ion resulting in isomeric ions in dynamic equilibrium. When the detached atom is positive (H^+), the phenomenon is *cationotropy;* when it is negative (OH^-), *anionotropy;* in the special case of H^+, *prototropy* (cf. *pseudo acid*), and in the case of OH^-, pseudobasicity (cf. *pseudo base*).

iopanoic acid $C_{11}H_{12}O_2NI_3$ = 570.9. 2-(3-Amino-2,4,6-triiodobenzyl)butanoic acid. Telepaque. Creamy powder, insoluble in water, m.156 (decomp.); radiopaque medium for the gall bladder (USP, BP).

iophendylate $C_{19}H_{29}O_2I$ = 416.3. A sterile mixture of the isomers of ethyliodophenylundecanoate. Myodil, Pantopaque. Yellow liquid with ethereal odor, slightly soluble in water; a radiopaque medium for spinal chord and brain (USP, BP).

Iosol Trademark for a series of organic solvent–soluble dyes for plastics and lacquers.

iothalamic acid $C_{11}H_9O_4N_2I_3$ = 613.9. White powder, slightly soluble in water. A radiopaque medium for arteries and renal tract (BP). Also used as the meglumine, q.v. *(meglumine iothelamate),* and sodium salts (USP, BP).

Ipatiev reaction The catalytic reduction of ketones to alcohols, with nickel oxide under pressure.

Ipatiew Ipatiev.

ipecac Ipecacuanha, ipecac root. The dried root of *Cephaelis ipecacuanha,* Rio ipecac, or *Cephaelis acuminata,* Carthagena ipecac (Rubiaceae). It contains 1.75% alkaloids (emetine, cephaeline, etc.); an emetic, and expectorant; used for certain cases of poisoning in children (USP, EP, BP). **American ~** Gillenia. **Goanese ~** Naregamine.

ipecacuanhic acid $C_{14}H_{18}O_7$ = 298.3. Brown powder from the roots of *Psychotria ipecacuanha.*

I.P.K. International prototype *kilogram.*

I.P.M. International prototype *meter.* Former international standard for the meter.

ipoh Arrow poisons from Strychnos species (Malaysia). Cf. *upas.*

ipomic acid Sebacic acid.

ipomoea (1) Mexican scammony root. The dried root of *I. orizabensis* (Convolvulaceae). (2) A genus of Convolvulaceae, yielding jalap and turpeth; a cathartic.

TABLE 46. OXIDATION STATES OF IRON

Oxidation state	In acid solution	In alkaline solution
2	Fe^{2+} i.(II)*, i.(2+)*, ferrous (green)	$HFeO_2^-$ hydrogenferrate(II)*, hypoferrite
3	Fe^{3+} i.(III)*, i.(3+)*, ferric (brown or yellow)	FeO_2^- ferrate(III)*, ferrite
4	——	$HFeO_3^-$ hydrogenferrate(IV)*, hypoferrate
6	——	FeO_4^{2-} ferrate(VI)* (purple)

ipomoein A globulin from the sweet potato, *Ipomoea batatas.*
iprodione* See *fungicides,* Table 37 on p. 250.
ipuanine Artificial *emetine.*
ipuranol $C_{23}H_{38}O_2(OH)_2$ = 380.6. An alcohol from buphane, m.290.
I.R. Insoluble residue (in analysis).
Ir Symbol for iridium.
i.r., IR Infrared.
iradiation Irradiation.
iregenone $MeEtC_6H_3 \cdot CMe_3$ = 176.3. A terpenelike hydrocarbon in the roots of Iris species.
irene $C_{14}H_{20}$ = 188.3. 1,2,3,4-Tetrahydro-1,1,2,6-tetramethylnaphthalene. (±)-~ b_{10mm}123. Cf. *irone, ionene.*
iretol $MeO \cdot C_6H_2(OH)_3$ = 156.1. 1,2,3-Trihydroxy-5-methoxybenzene, 5-methoxypyrogallol. Colorless crystals, m.186; used in perfumery.
Iridaceae A family of plants that yield drugs; e.g., *Iris florentina* (orris), *I. versicolor* (iris).
iridane 1,2-Dimethyl-3-isopropylcyclopentane. A derivative of linalool.
Iridaz Trademark for a fluorescent, white rayon yarn.
iridescence Rainbow colors on a surface; as seen in opals.
iridescent Describing a surface colored like that of nacre (mother of pearl); usually due to very thin films of air or other material which cause light ray interference. **i. quartz** A gem rock crystal with fine interstices containing air films which produce combinations of colors.
iridic Iridium(IV)* or (4+)*. Describing a compound of tetravalent iridium.
iridious Iridous.
iridium* Ir = 192.22. A metallic element, at. no. 77, of the platinum family; discovered by Tennant (1804). White, brittle, lustrous metal, d.22.42, m.2410, b.4250, soluble in aqua regia. Used mainly, in alloys with the noble metals, to coat hydrogen electrodes and harden platinum for jewelry (10% Ir); also as a catalyst, and, with Os, for pen tips and compass bearings. **i. compounds** Valency 2: i.(II)*, i.(2+)*, irido-, as IrO. Valency 3: i.(III)*, i.(3+)*, iridous, as $IrCl_3$. Valency 4: i.(IV)*, i.(4+)*, iridic, as IrO_2. **i. disulfide*** IrS_2 = 256.3. Black powder, insoluble in water. **i. sodium chloride*** $Na_2IrCl_6 \cdot 12H_2O$ = 667.1. Sodium iridichloride. Yellow crystals, m.50. **i. tetrabromide*** $IrBr_4$ = 511.8. Hygroscopic, brown powder or blue crystals, soluble in water. **i. tetrachloride*** $IrCl_4$ = 334.0. Red crystals, soluble in water. **i. tetraiodide*** IrI_4 = 699.8. Hygroscopic, black powder, soluble in water. **i. trioxide*** Ir_2O_3 = 432.4. Iridium black. Black powder, soluble in hydrochloric acid; a ceramic pigment.
irido- See *iridium compounds.*
iridosmine Osmiridium.
iridous Iridium(III)* or (3+). Describing a compound of trivalent iridium.
iris (1) An iridescent quartz or other mineral. (2) A genus of plants *(Iridaceae),* q.v.
Irish **I. diamond** A gem quartz crystal from Ireland. **I. moss** Carragheen, killeen. The red seaweeds *Chondrus crispus* and *Gigartina stellata.* The dried thallus of kelps (Ireland, N. America); a demulcent, and a clarifying agent in brewing.
Irium Trade name for sodium dodecyl sulfate; a detergent.
iroko African teak from *Chlorophora excelsa.*
iron* Fe = 55.847. Ferrum*. A metallic element, at. no. 26, in the eighth group of the periodic table. Black, lustrous, cubic, magnetic metal, d.7.85, m.1535, b.2750. Compounds: see Table 46. The many modifications of metallic i. (wrought, cast, steel, reduced i.) consist chiefly of its allotropes. **alibated ~** I. covered with a protective layer of aluminum. **bis(η-cyclopentadienyl) ~** Ferrocene*. **cast ~** The molten i. from blast furnaces. Gray masses, m.1275–1505. **ingot ~** The malleable i. from the Bessemer process. **malleable ~** Wrought i. **meteoric ~** See *meteoric iron.* **native ~** Meteorites. **passive ~** I. rendered insoluble in dilute acids, e.g., by immersing it in fuming nitric acid or hydrogen peroxide, or by making it the anode in electrolysis. This property is lost by mechanical shock. **pig ~** Cast i. **Quevennes ~** Reduced i. **reduced ~** Finely powdered metallic i. obtained by heating ferric oxide in hydrogen. **specular ~** Specularite. **steel ~** A gray metal, m.1375. See *steel.* **white ~** (1) Cast i. containing only combined carbon. (2) Marcasite. **wrought ~** Cast i. heated and hammered. A gray metal, d.7.86, m.1505. Pure i. melts at 1535, and pure cementite, Fe_3C, containing 6.67% carbon at 1850. The solid solutions and compounds formed depend on the carbon content and are: α = ferrite; β = beta iron; γ = austenite, a solid solution of carbon in gamma iron; δ = delta iron (stable between 1400 and 1539); Fe_3C = cementite; α + Fe_3C = pearlite. Ferrite is the chief constituent of wrought i., stable below 723°C. Irons contain up to 0.005, steels 0.008–1.7, and cast irons 1.7–5% carbon.
i.(II)*, i.(2+)* See also *ferrous.* **i.(III)*, i.(3+)*** See also *ferric.* **i.(III) acetate*** Ferric acetate. **i. alloys** See *steel, ferrimanganese, silicon alloys.* **i. alum** See *iron alum* under *alum.* **i. arsenate** (1) Ferric arsenate. (2) Ferrous arsenate. **i.(III) benzoate*** Ferric benzoate. **i. black** Finely divided antimony. **i. brass** A brass containing 1–9% i. **i. bromide** (1) Ferric bromide. (2) Ferrous bromide. **i. buff** Nanking yellow, ferric hydroxide. **i. carbide** (1) Fe_3C = 179.6 Regular, gray crystals, d.7.07, insoluble in water. (2) FeC_4 = 103.9. A gray, crystalline mass. **i. carbonate** (1) Ferric carbonate. (2) Ferrous carbonate. **i. carbonyl** (1) $Fe(CO)_5$ = 195.9. I. pentacarbonyl*. Yellow, viscous liquid, b.103, decomp. 180. (2) $Fe_2(CO)_9$ = 363.8. Diferrocarbonyl. Orange crystals, decomp. by heat. (3) $Fe(CO)_4$ = 167.9. I. tetracarbonyl*. Green crystals. **i. chloride** (1) Ferric chloride. (2) Ferrous chloride. **i.(III) chromate*** Ferric chromate.

i.(III) citrate* Ferric citrate. **i. citrate green** Ammonium iron citrate*. **i. cyanides** See *i.(II)* and *i.(III) hexacyanoferrates* below. **i. dextran** Imferon. A sterile, aqueous, colloidal solution of a complex of ferric hydroxide with partly hydrolyzed dextran. An injectable form of iron, used to treat iron-deficiency anemia (USP, BP). **i.(II) diiron(III) chloride*** Fe_3Cl_8 or $FeCl_2 \cdot 2FeCl_3$ = 451.2. Yellow deliquescent crystals soluble in water. **i.(II) diiron(III) hydroxide*** $Fe_3(OH)_8$ = 303.6. Black powder, soluble in hydrochloric acid. **i.(II) diiron(III) oxide*** Fe_3O_4 or $FeO \cdot Fe_2O_3$ = 231.5. Magnetic iron oxide, martial ethiops, black iron oxide. Black, regular crystals, m.1537, insoluble in water. **i.(II) diiron(III) sulfate*** $Fe_3(SO_4)_4$ = 551.8. Several derivatives; some occur as minerals. **i.(II) diiron(III) sulfide*** Fe_3S_4 or $FeS \cdot Fe_2S_3$ = 295.8. Black powder, insoluble in water. **i. dinitrosothiosulfates** $M[Fe(NO_2)S_2O_3]H_2O$, where M is K or Rb. **i.(II) disulfate*** FeS_2O_7 = 232.0. White monocrystals. **i. disulfide*** FeS_2 = 120.0. Yellow rhombs, m.1171, produced by the precipitation of ferric salts with hydrogen sulfide. **i. family** See *iron period* below. **i. flint** An opaque variety of quartz, containing i. **i.(II) fluoride*** Ferrous fluoride. **i.(III) formate*** Ferric formate. **i. founding** Making a facsimile of a pattern by running molten i. into a mold of the pattern. **i. froth** A fine, spongy variety of hematite. **i. garnet** See *iron garnet* under *garnet*. **i. glance** Hematite. **i. hexachloroplatinate(IV)*** $Fe[PtCl_6]$ = 463.8. Yellow crystals. **i.(II) hexacyanoferrate(II)*** $Fe_2[Fe(CN)_6]$ = 323.6. Ferrous ferrocyanide. White powder, insoluble in water. **i.(II) hexacyanoferrate(III)*** $Fe_3[Fe(CN)_6]_2$ = 471.3. Ferrous ferricyanide. Blue powder, decomp. by heat, insoluble in water; a pigment. **i.(III) hexacyanoferrate(II)*** $Fe_4[Fe(CN)_6]_3$ = 859.3. Ferric ferrocyanide, ironic ferrocyanide, insoluble *Prussian blue*, q.v., Turnbull's blue. Dark-blue crystals, insoluble in water. A pigment (Prussian blue, Berlin blue, Paris blue) and blue ink (with oxalic acid). **i.(III) hexacyanoferrate(III)*** $Fe[Fe(CN)_6]$ = 267.8. Ferric ferricyanide. Red solid, soluble in water. **i.(II) hexafluorosilicate*** $Fe[SiF_6] \cdot 6H_2O$ = 306.0. Ferrous fluo(ro)silicate. Colorless crystals, soluble in water. **i.(III) hexafluorosilicate*** $Fe_2[SiF_6]$ = 537.9. Ferric fluo(ro)silicate. A fresh-colored gel, soluble in water. **i. hydroxide** (1) Ferric hydroxide. (2) Ferrous hydroxide. **i. iodide** (1) Ferrous iodide. (2) Ferric iodide. **i. liquor** A solution of ferrous acetate. **i.(II) manganese(II) chloride*** $FeMnCl_4$ = 252.6. Orange crystals, soluble in water. **i.(II) manganese(II) iodide*** $FeMnI_4$ = 618.4. Brown prisms, soluble in water. **i. minerals** I. is, after aluminum, the most abundant metal in minerals: meteorites, Fe; hematite, Fe_2O_3; pyrite, FeS_2; siderite, $FeCO_3$. **i. mordant** (1) Ferric sulfate. (2) Ferrous nitrate. **i. nitrate** (1) Ferric nitrate. (2) Ferrous nitrate. **i. nitride** Fe_4N. A catalyst for the synthesis of ammonia. **i. oleate** (1) Ferric oleate. (2) Ferrous oleate. **i.(III) oxalate*** Ferric oxalate. **i. oxide** (1) FeO. Ferrous oxide. (2) Fe_2O_3. Ferric oxide. (3) Fe_3O_4. i.(II) diiron(III) oxide*. **black ~** Ferric oxide. **magnetic ~** Ferrous oxide. **red ~** Ferric subcarbonate. **i. paranucleinate** Triferrin. **i. pentacarbonyl*** See *iron carbonyl* above. **i.(III) perchlorate*** Ferric perchlorate. **i. period** The central part of the fourth period in the periodic table, consisting of the elements Cr, Mn, Fe, Co, Ni, and Cu. All form colored salts and ions and 2 or more series of compounds. **i.(III) phenolate*** Ferric phenolate. **i. phosphates** (1) Ferric phosphate. (2) Ferrous phosphate. **i. phosphide*** FeP = 86.8. Black powder, d.5.2. Also: Fe_2P, Fe_2P_3, Fe_3P, Fe_3P_4. **i.(II) phosphinate*** $Fe(H_2PO_2)_2$ = 185.8. Ferrous phosphinate. White powder, rapidly oxidizes. **i.(III) potassium hexacyanoferrate(II)*** $FeK[Fe(CN)_6] \cdot H_2O$. Blue, insoluble precipitate from ferric and hexacyanoferrate(II) ions. **i.(III) potassium oxalate*** $FeK_3(C_2O_4)_3 \cdot 3H_2O$ = 491.2. Brown crystals, decomp. 230, slightly soluble in water. **i.(III) potassium sulfate*** $Fe_2(SO_4)_3 \cdot K_2SO_4 \cdot 24H_2O$ = 1006. Iron alum. Violet crystals, soluble in water. **i. protocarbonate** Ferrous carbonate. **i. protochloride** Ferrous chloride. **i. protosulfide** Ferrous sulfide. **i. putty** A mixture of ferric oxide and boiled linseed oil, used for pipe joints. **i. pyrite** Pyrites. **i. rust** $2Fe_2O_3 \cdot 3H_2O$. Hydrated ferric oxide. **i. salts** See also *ferrous, ferric*. **i. sand** See *iron sand* under *sand*. **i.(III) sodium oxalate*** $FeNa_3(COO)_6 \cdot 4\frac{1}{2}H_2O$ = 469.9. Green crystals, soluble in water; used in photography. **i.(III) sodium sulfate*** $FeNa(SO_4)_2 \cdot 12H_2O$. Sodium iron alum. Brown octahedra, soluble in water. **i. spar** Siderite. **i. sponge** (1) Ferric oxide. (2) I. produced as a honeycomb lump by direct reduction of the ore at 1100°C in a reducing atomosphere produced by burning fuel oil. Preferred for steel manufacture as it is uniform in composition, low in foreign constituents, and easily handled. **i. stone** Siderite. **i. subcarbonate** Ferric subcarbonate. **i. subsulfate** Ferric subsulfate. **i. sulfate** (1) Ferrous sulfate. (2) Ferric sulfate. **i. sulfide** (1) FeS Ferrous sulfide. (2) FeS_2. I. disulfide*. See *pyrite*. (3) Fe_2S_3. Ferric sulfide. (4) Fe_3S_4. I.(II) diiron(III) sulfide*. **i.(II) sulfite*** Ferrous sulfite. **i. tantalate** Tantalite. **i. tartrate** (1) Ferric tartrate. (2) Ferrous tartrate. **i. tersulfate** Ferric sulfate. **i.(III) thiocyanate*** Ferric thiocyanate. **i.(II) thiosulfate*** FeS_2O_3 = 168.0. Ferrous hyposulfite. White powder, soluble in water, rapidly oxidizes. **i. trichloride*** Ferric chloride. **i. trioxide*** Ferric oxide. **i. tungstate*** See *ferberite, reinite, wolframite*. **i. vanadate** Ferric vanadate.

irone $C_{14}H_{22}O$ = 206.3. Iron. A terpene from orris root, and isomer of ionone. Colorless liquid, d.0.939, b_{16mm}144, slightly soluble in water; used in violet perfume.

irradiance* Heat flux density*. Radiation power on a surface, measured in J/m^2.

irradiation (1) Exposure to radiation. (2) More specifically, exposure to ultraviolet radiation, especially wavelengths of 250–300 nm. I. is measured by *Eder's solution*, q.v.

irregular Not according to rule.

irreversible Describing a reaction that cannot be reversed, and usually proceeds to completion in one direction.

irritant (1) An agent that produces inflammation or irritation; as, histamine. (2) The metals Ni, Mn, Cr, W, and Si, which in iron alloys are an i. to carbon.

Irvine, Sir James Colquhoun (1887–1952) British chemist noted for research on carbohydrates.

isabellin An alloy of Al, Mn, and Cu; used for standard electrical resistances.

isaconic acid Itaconic acid.

isamatic Isamic.

isamic acid $C_{16}H_{11}O_3N_3$ = 293.3. Red prisms, m.164, slightly soluble in water; obtained from isatin.

isanic acid $C_{17}H_{25}COOH$ = 274.4. 17-Octadecene-9,11-diynoic acid*. Colorless crystals, m.41; from the oil of tsano nuts.

isano oil Tsano oil.

isaphenic acid $C_{17}H_{11}NO_3$ = 277.3. White leaflets, m.295, insoluble in benzene.

isapiol See *isoapiol* under *apiol(e)*.

isatic acid $NH_2 \cdot C_6H_4CO \cdot COOH$ = 165.2. Isatinic acid, (2-aminophenyl)-2-oxoethanoic acid. White powder, decomp. by heat, soluble in water.

isatide $C_{16}H_{12}O_4N_2$ = 296.3. Dihydroxybioxindol, isatin-3,3′-pinacol. Colorless crystals, m.237, insoluble in water.

isatin $C_6H_4 \cdot NH \cdot C{:}O \cdot CO$ = 147.1. 1*H*-indole-2,3-dione, lactam of isatic acid. Red needles, m.198, soluble in hot water; used in the manufacture of dyestuffs. **i. anilide** $C_{14}H_{10}ON_2$

= 222.2. m.120. **i. chloride** C_8H_4ONCl = 165.6. Brown needles, decomp. 180, insoluble in water, used in the manufacture of dyestuffs.
isatinic acid Isatic acid.
isatoic acid $C_8H_7O_4N$ = 181.2. 2-Carboxyaminobenzoic acid*.
isatoxime $C_8H_6O_2N_2$ = 162.1. Nitrosoindoxyl. Yellow needles, m.202 (decomp.), slightly soluble in water.
isatropic acid $C_{18}H_{16}O_4$ = 296.3. 1,2,3,4-Tetrahydro-1-phenyl-1,4-naphthalenedicarboxylic acid. Several isomers. Cf. *atropic acid.*
isatropylcocaine $C_{19}H_{22}O_4N$ = 328.4. An alkaloid from coca leaves.
isazol CH:N·O·CH:CH = 69.1. An isomer of oxazole.
ischidrotic An agent that causes retention or suppression of perspiration; as, aluminum chloride.
isentropic Involving no change in entropy.
iserite A sand rich in titanium and iron.
isethionate $CH_2OH \cdot CH_2 \cdot SO_3M$. A salt of 2-hydroxyethane sulfonate; used in surfactant synthesis.
isethionic acid $CH_2OH \cdot CH_2 \cdot SO_3H$ = 126.1. 2-hydroxy-1-ethanesulfonic acid*, from which taurine is derived. A syrupy liquid. With oleic acid it forms detergents (Igepons).
ishkyldite $H_{20}Mg_{15}Si_{11}O_{47}$. A structural variety of chrysotile (Volga district).
isindazole 2*H*-Indazole*.
isinglass (1) Fish glue, ichthyocolla. A pure gelatin from the swimming bladders of fishes *(Acipenser)*; an adhesive and clarifying agent. (2) Mica in thin sheets, used for windows in ovens and goggles. **Japanese ~** Agar. **vegetable ~** Agar.
Isle Royal greenstone Chlorastrolite.
Ismelin Trademark for guanethidine.
ISO International Organization for Standardization. See *standards.*
iso- *i*-. Prefix (Greek "equal"), indicating a similarity (usually isomeric) between molecules. For iso- compounds not listed below, look under the corresponding normal (root) compound.
isoallyl The radicals (1) iso*propenyl**; (2) 1-propenyl†.
isoamoxy The isopentyloxy* radical.
isoamyl The isopentyl* radical.
isoamylene Trimethylethylene.
isoanthraflavic acid $C_{14}H_8O_4$ = 240.2. 2,7-Dihydroxyanthraquinone*. Yellow crystals, m.353, soluble in alcohol.
isoantibody An antibody, synthesized by an individual, that can act against cells of other individuals of the same species.
isobar (1) A line drawn through points on a chart which have the same barometric pressure at a given time. (2) Two or more atomic species having the same mass number, but different atomic numbers.
isobaric Pertaining to isobar (2). Cf. *isotopes, nuclide.*
isobenzofurandione Phthalic anhydride*.
isobenzofuranone Phthalide*.
isobestic point The wavelength at which the molar absorption (extinction) coefficients of 2 substances, one of which can be converted into the other, are equal.
isobutenyl 2-Methyl-1-propenyl†. The radical $Me_2C{:}CH-$, from isobutene.
isobutoxy* 2-Methylpropoxy†. The radical $Me_2 \cdot CH \cdot CH_2O-$, from isobutyl alcohol.
isobutyl* 2-Methylpropyl†. The radical $Me_2 \cdot CH \cdot CH_2-$, from isobutane.
isobutyryl* 2-Methyl-1-oxopropyl†. The radical $Me_2C \cdot CH \cdot CO-$, from isobutyric acid.
isocarb A line in a diagram showing equal carbon contents.
isocatalysis See *isomeric catalysis* under *catalysis.*
isocetic acid Pentadecanoic acid*.
isocholine Amanitine (2).
isochore The curve obtained by plotting the pressure and temperature of a gas at constant volume. See *reaction isochore.*
isochromatic Having the same color throughout.
isochrone A line joining points of equal stability (expressed as log gelation time) on triangular diagrams; used to express the stability of colloids.
isochronism (1) The condition of occurring at equal intervals of time. (2) The condition of lasting for equal periods of time.
isochronous Occurring in the same period of time at equal intervals.
isocorybulbine $C_{21}H_{21}O_4N$ = 351.4. An alkaloid from *Corydalis cava.*
isocrotonic acid* See *crotonic acid.*
isocyanate* Carbimide. A compound containing the radical $-N{:}C{:}O$.
isocyanide* Isonitrile, carbylamine. A compound containing the radical $-NC$.
isocyanine A member of a group of dyes used as photographic sensitizers. See *cyanine dyes.*
isocyano-* Prefix indicating the radical $-NC$. Cf. *isocyanide.*
isocyanuric acid See *cyanuric acid.*
isocyclic (1) Describing a closed-chain compound containing the same number of atoms as the compound with which it is i.; e.g.: the benzene ring (6 C atoms) is i. with the pyridine ring (5 C atoms and 1 N atom). (2) Homocyclic (erroneous).
isodecanol A mixture of branched-chain primary alcohols, used in the manufacture of plasticizers.
isodigitoxigenin $C_{23}H_{34}O_4$ = 374.2. Lustrous, white crystals, m.271, from digitalis. See *cardiac glycosides.*
isodimorphism The condition in which isomorphs have 2 crystalline forms in common. Cf. *isomorphism.*
isodisperse Dispersible in solutions having the same pH value.
isodulcite Rhamnose.
isodurenol C_6HMe_4OH = 150.2. 2,3,4,6-Tetramethylphenol*. White crystals, m.80.
isoduridine $Me_4C_6H \cdot NH_2$ = 149.2. 2,3,4,6-Tetramethylaniline*. White crystals, m.23, b.259.
isodurylic acid See *durylic acid.*
isodynamic (1) Having equal force. (2) Generating or liberating the same amount of energy.
isoelectric point The point of electric neutrality; the pH value at which a substance (protein, etc.) is neutral. At a lower or higher pH value it acts either as a base or acid, respectively. Coagulation of colloids occurs at or near the i. point. For typical values, see Table 47.
isoelectronic Describing species having the same electronic configuration, but not necessarily the same mass and nuclear charge. Cf. *isostere.*
isofluorophate $C_6H_{14}O_3FP$ = 184.1. DFP, Dyflos. Yellow, irritant liquid, d.105, insoluble in water; a parasympathomimetic and miotic (USP).
isogam A map that shows contours of equal differences in gravity.
isogamous Having morphologically identical male and female gametes.
isohemagglutinin Isonin. A constituent of blood serum responsible for agglutination in blood-group tests.

For other iso- compounds (or radicals), look under the root compound (or radical) that is prefixed by the iso-.

TABLE 47. ISOELECTRIC POINTS OF VARIOUS PROTEINS

pH	Protein
4.4	Serum globulin
4.6	Ovalbumin
4.7	Gelatin, casein, serum albumin
5.3	Fibrinogen
5.5	Edestin, pseudoglobulin
6.8	Homoglobin, oxyhemoglobin
8.1	Globin

isohydric Describing a neutralization, during which the pH value does not change. Cf. *buffer.* Hence a set of solutions of similar H^+ concentration. **homo ~** See *homoisohydric.*
isoindoledione Phthalimide*.
isokom A line drawn through a diagram to connect points of equal viscosity.
isolate (1) To separate or prepare an element or compound. (2) That which is isolated.
isoleucine* Et·CHMe·$CHNH_2$·COOH = 131.2. Ile*. 2-Amino-3-methylpentanoic acid*. **(±)-~** Rhombic or monoclinic plates, m.292, slightly soluble in water. **D-~** Greasy leaflets, m.283.
isolog, isologue A member of a series of compounds of similar structure, but having different atoms of the same valency and usually of the same periodic group.
isologous series A set of isologs; e.g.: water—H_2O, M_2O, R_2O, M_2S, M_2Se, M_2Te; hydroxylamine—R_2NOH, R_2POH, R_2IOH, R_2TlOH; methane—CH_4, CR_4, SiR_4, SnR_4, PbR_4.
isomer* A member of a group of compounds having identical molecular formulas, but differing in the nature or sequence of bonding of their atoms (i.e., constitution) or in the arrangement of their atoms in space. Cf. *metamers.* **conformational ~** One of a group of molecules that differ in *conformation,* q.v. **constitutional ~*** One of a group of isomers, whose atoms differ in their nature or sequence of bonding. **stereo ~** See *stereoisomer.*
isomerase* See *enzymes,* Table 30.
isomeric Pertaining to isomerism.
isomeride One of a set of compounds that have similar structural groups but not necessarily the same number of atoms; thus, anabasine is an *isomer* of nicotine, but nornicotine is an *isomeride.*
isomerism The phenomenon of isomers. See Table 48 on p. 314. Four factors determine it:

P = arrangement of the atom in the molecule
T = type of the resulting compound
V = the valency of the principal atom
L = the linkage between the atoms

An organic molecule may show the following differences (*d*) or similarities (—):

P	*T*	*V*	*L*	Type of isomerism	
d	—	—	—	= isomerism and stereoisomerism	
d	*d*	—	—	= metamerism	dynamic isomerism
—	*d*	—	*d*	= desmotropism	dynamic isomerism
d	*d*	—	*d*	= tautomerism	dynamic isomerism
—	—	*d*	—	= pseudomerism	

isomerization A reaction that produces an isomer of the main reactant.
isomery Isomerism.
isometric Of the same dimension.
isometry Isomerism.
isomorph One of a set of similarly shaped crystals having different compositions; e.g., Na_2SO_4, Na_2SeO_4.
isomorphic Pertaining to similar crystalline forms.
isomorphism The crystallization of different compounds in the same form. These elements and radicals can replace one another without causing any essential alteration in crystalline form. See *isomorphs.*
isomorphous mixture A mixture of isomorphs usually found in minerals and represented in parentheses in the formula; as, $(Fe,Mn)CO_3$, which indicates that Mn and Fe are interchangeable and that each behaves as the equivalent weight of either element.
isomorphs The elements arranged in the order of periodic groups and according to their isomorphism.

1. Cl, Br, I, F, Mn; as $KMnO_4$ and $KClO_4$.
2. S, Se, Te, as sulfides and tellurides; Cr, Mn, Te, as K_2RO_4; As, Sb, as MR_2 (glances).
3. As, Sb, Bi; Te (as element), P, V (in salts); N, P (in organic compounds).
4. K, Na, Rb, Cs, Li; Tl, Ag.
5. Ca, Sr, Ba, Pb; Fe, Mn, Zn, Mg; Ni, Co, Cu; Ce, La, Pr, Nd, Er, Y, with Ca; Cu, Hg, with Pb; Cd, Be, In, with Zn; Tl, with Pb.
6. Al, Fe, Cr, Mn; Ce, U, in oxides M_2O_3.
7. Cu, Ag (as monovalent compounds), Au.
8. Pt, Ir, Pd, Rh, Ru, Os; Au, Fe, Ni; Sn, Te.
9. Cd, Si, Ti, Zr, Th, Sn; Fe, Ti.
10. Ta, Nb.
11. Cr, Mo, W.

isoniazid $C_6H_7NO_3$ = 141.1. Isonicotinic acid hydrazide, Nydrazid, Rimifon. White, crystalline powder, m.172, slightly soluble in water. A bacteriostatic antituberculous drug, used with similar drugs to prevent development of drug-resistant bacilli (USP, EP, BP).
isonicotinhydrazine Isoniazid.
isonin Isohemagglutinin.
isonitrile Isocyanide*.
isonitro *aci*-nitro*. See *aci-nitro* under *nitro.*
isonitroso Oxime*.
isooctenyl alcohol A mixture of 2 isomeric primary allyl alcohols, used as a solvent and source of plasticizers.
isopentyl* Isoamyl. The $Me_2CH·CH_2CH_2-$ radical.
isopentyloxy* Isoamoxy. The $Me_2CH(CH_2)_2O-$ radical.
isophane A sterile, aqueous suspension of crystals of protamine and zinc, used as an injection (USP), or as i. insulin for diabetes (BP).
isopiestic Isotonic.
isoporous Describing the equally spaced, cross-linked chemical structure of certain polymers; as, polystyrene; responsible for ion-exchange properties.
isopral CCl_3·CMeH·OH = 163.4. Trichloroisopropanol. White prisms, m.49, soluble in water; a hypnotic.
isoprenaline hydrochloride EP, BP name for isoproterenol hydrochloride.
isoprene* CH_2:CMe·CH:CH_2 = 68.1. 2-Methyl-1,3-butadiene*. A distillation product of india rubber, d.0.679, b.34; the unit structure of terpenes, carotenoids, phytol, and rubber. Cf. *duprene.* **di ~** The geraniol and linalool chain. **tri ~** The farnesol and nerolidol chain. Cf. *vinyl* compounds.
isoprenoid Having isoprene as a basis of its structure.
isopropanol Isopropyl alcohol*.

For other iso- compounds (or radicals), look under the root compound (or radical) that is prefixed by the iso-.

TABLE 48. CLASSIFICATION AND EXAMPLES OF ISOMERISM

A. Isomerism

1. Constitutional isomerism:

$CH_3 \cdot CH_2 \cdot CH_2 \cdot CH_3$ and $(CH_3)(CH_3) > CH \cdot CH_3$

Chain isomerism

$CH_3 \cdot CH_2 \cdot CH_2Cl$ and $CH_3CHCl \cdot CH_3$

Position isomerism

$CH_3 \cdot CH_2 \cdot CH{:}CH_2$ and $CH_3 \cdot CH{:}CH \cdot CH_3$

Metamerism

2. Stereoisomerism:

$(H)(R) > CH{:}CH < (H)(R)$ and $(R)(H) > CH{:}CH < (H)(R)$

(*E*) and (*Z*) (or *cis* and *trans*)

isomerism

$(D)(E) > C < (A)(B)$ and $(D)(E) > C < (B)(A)$

(*R*) and (*S*) isomerism
A, B, D, and E are different
atoms or radicals

B. Metamerism

$(R)(H) > C{:}NOH \rightleftharpoons (R)(HO) > C{:}NH$

Oxime Imide

C. Desmotropism

$R \cdot C(=O)(CH_2R) \rightleftharpoons R \cdot C(OH)(=CH \cdot R)$

Keto(ne) form Enol form

$R \cdot C(=O)(NH_2) \rightleftharpoons R \cdot C(OH)(=NH)$

Acid amide Acid Imide

D. Tautomerism

$R \cdot CN \rightleftharpoons R \cdot NC$

Cyanide Isocyanide

E. Pseudomerism

$R \cdot OC{:}N$ $R \cdot N{:}C{:}O$

Cyanate Isocyanate

F. Coordination Isomerism

The different arrangement of radicals around the 2 central atoms in combination:

$[Cr(NH_3)_6] \cdot [Cr(SCN)_6]$ and

$[Cr(NH_3)_4(SCN)_2] \cdot [Cr(NH_3)_2(SCN)_4]$

isopropyl* 1-Methylethyl†. The radical Me_2CH-. **i. alcohol*** $Me_2CH \cdot OH$ = 60.1. Isopropanol, 1-methyl ethanol. Colorless liquid, d.0.780, b.82, soluble in water; a germicide (BP), denaturant, and perfume additive. **i.benzene** Cumene*. **i. ether** $(Me_2CH)_2O$ = 102.2. Colorless liquid, d.0.7247, b.69; a solvent for waxes, fats, and resins. **i.metacresol** Thymol*. **i.toluene** Cymene*.

isopropylbenzyl* The radical $Me_2CH \cdot C_6H_4 \cdot CH_2-$, from cymene.

isoproterenol hydrochloride $C_{11}H_{17}O_3N \cdot HCl$ = 247.7. Isoprenaline hydrochloride, Isuprel, Medihaler-iso, Proternol. Bitter, white crystals, m.169, soluble in water; an adrenergenic used to treat asthma and raise blood pressure and heart rate (USP, EP, BP).

isopurinol Allopurinol.

isopyknoscopy Determination of the end point of a volumetric reaction from the specific gravity of the solution containing the reactants. Cf. *pyknometer.*

isopyre An impure gem variety of opal.

isoquinoline* See *isoquinoline* under *quinoline.*

isorhamnetin A glucoside from the pollen of ragweed and bulrush.

isorheic Describing liquids that have equal viscosity.

isosmotic Isotonic.

isostere One of 2 or more molecules or atomic groups having an analogous arrangement of electrons and similar physical properties. Cf. *ionic number, isoelectronic.*

isosteric Pertaining to similar electronic arrangements. **i. properties** The physical properties due to electronic arrangements. See *isosterism.*

isosterism Describing the similarity in physical properties of elements, ions, or compounds due to their similar or identical electronic arrangements.

isotachophoresis An electrophoretic method of separating ion species of the same charge type. They are given the same migration velocity in an electric field and are separated according to their net mobilities.

isotactic See *isotactic polymers* (under *polymer*), *polyvinyl chain.*

isoteniscope An instrument for the static determination of

For other iso- compounds (or radicals), look under the root compound (or radical) that is prefixed by the iso-.

TABLE 49. NATURAL ISOTOPES

Atomic number		Mass number
1	Hydrogen	1, 2
2	Helium	4, 3
3	Lithium	7, 6
5	Boron	11,10
6	Carbon	12, 13
7	Nitrogen	14, 15
8	Oxygen	16, 18, 17
10	Neon	20, 22, 21
12	Magnesium	24, 26, 25
14	Silicon	28, 29, 30
16	Sulfur	32, 34, 33, 36
17	Chlorine	35, 37
18	Argon	40, 36, 38
19	Potassium	39, 41, 40
20	Calcium	40, 44, 42, 48, 43
22	Titanium	48, 46, 47, 49, 50
23	Vanadium	51, 50
24	Chromium	52, 53, 50, 54
26	Iron	56, 54, 57, 58
28	Nickel	58, 60, 62, 61, 64
29	Copper	63, 65
30	Zinc	64, 66, 68, 67, 70
31	Gallium	69, 71
32	Germanium	74, 72, 70, 73, 76
34	Selenium	80, 78, 82, 76, 77, 74
35	Bromine	79, 81
36	Krypton	84, 86, 82, 83, 80, 78
37	Rubidium	85, 87
38	Strontium	88, 86, 87, 84
40	Zirconium	90, 94, 92, 91, 96
42	Molydenum	98, 96, 95, 92, 97, 100, 94
44	Ruthenium	102, 104, 101, 100, 99, 96, 98
46	Palladium	106, 108, 105, 110, 104, 102
47	Silver	107, 109
48	Cadmium	114, 112, 111, 110, 113, 116, 106, 108
49	Indium	115, 113
50	Tin	120, 118, 116, 119, 117, 124, 122, 112, 114, 115
51	Antimony	121, 123
52	Tellurium	130, 128, 126, 125, 124, 122, 123
54	Xenon	129, 132, 131, 134, 136, 130, 128, 124, 126
56	Barium	138, 137, 136, 135, 134, 132, 130
57	Lanthanum	139, 138
58	Cerium	140, 142, 138, 136
60	Neodymium	142, 144, 146, 143, 145, 148, 150
62	Samarium	155, 154, 147, 149, 148, 150, 144
63	Europium	153, 151
64	Gadolinium	158, 160, 156, 157, 155, 154, 152
66	Dysprosium	164, 162, 163, 161, 160, 158, 156
68	Erbium	166, 168, 167, 170, 164, 162
70	Ytterbium	174, 172, 173, 171, 176, 170, 168
71	Lutetium	175, 176
72	Hafnium	180, 178, 177, 179, 176, 174
73	Tantalum	181, 180
74	Tungsten	184, 186, 182, 183, 180
75	Rhenium	187, 185
76	Osmium	192, 190, 189, 188, 186, 187, 184
77	Iridium	193, 191
78	Platinum	195, 194, 196, 198, 192, 190
80	Mercury	202, 200, 199, 201, 198, 204, 196
81	Tellurium	205, 203
82	Lead	208, 206, 207, 204
84–91	Radioactive series	
92	Uranium	238, 235, 234

NOTE: The isotopes are arranged in approximate order of natural abundance. Elements not shown have no natural isotopes.

For other iso- compounds (or radicals), look under the root compound (or radical) that is prefixed by the iso-.

vapor pressure from the change in level of a liquid in a U-tube.

isotherm An expression of equal temperature for a number of points, systems, or phases. Cf. *reaction isotherm.*

isothermal Having a uniform or constant temperature. **i. change** The change in volume of a gas under such conditions that the temperature remains constant. Cf. *adiabatic.* **i. distillation** See *microdiffusion analysis.* **i. line** Isotherm.

isothermals A group of curves that pass through regions having the same temperature at a given time; as, on a climatic chart.

isothermic Isothermal.

isothiocyanate* Sulfocarbimide. A compound of the type R·NCS. Cf. *mustard oil.*

isothiocyanato-* Isothiocyano-. Prefix for the radical −N:C:S.

isothiocyanic acid* HNCS = 59.1. Sulfocarbimide. It yields isothiocyanates.

isotoma The dried herb *Isotoma longiflora* (Lobeliaceae), W. Indies and southeast Asia.

isotomine An alkaloid from isotoma.

isotone One of a number of atomic species having an equal number of neutrons in their nuclei.

isotonic (1) Isosmotic. (2) Hemisotonic. Describing a solution that has the same osmotic pressure as blood serum. Cf. *hypertonic, hypotonic.* **i. salt solution** Normal *saline.* **i. water** A natural water: the sum of its mineral constituents amounts to about 300 millimols per liter, which corresponds with an osmotic pressure of 770 kPa, and a freezing-point depression of −0.57 °C. If lower, the water is *hypotonic;* if higher, *hypertonic.*

isotopes Atomic species differing in mass number but having the same atomic number. The mass number is shown as a prefix thus: ^{35}Cl, ^{37}Cl. It is written out or spoken as chlorine-35, etc. (1) Originally i. indicated radioactive disintegration products which occupied an identical place in the periodic table. (2) The term now applies to all other elements that consist of 2 or more atomic species as revealed by their high-frequency or mass spectra; e.g., chlorine is made up of the isotopes ^{35}Cl and ^{37}Cl, and its atomic weight of 35.5 is the resultant of the isotopic weights 35 and 37. See Table 49 on p. 315 for isotopes occurring naturally. Certain i. are produced by gamma-ray irradiation; e.g., ^{131}I from ^{130}Te, and ^{42}K from ^{41}K. Used in medicine (e.g.; ^{131}I, ^{14}C) and *radioactive tracer elements,* q.v. **heterobaric ~** I. of different atomic weights: as, the products of different disintegration series. Cf. *radioactive elements.* **isobaric ~** I. of the same mass number but different atomic number; as, Pb-210 and Po-210. Cf. *isobar.* **metastable ~** Represented, e.g., ^{137m}Ba. See *metastable.* **radio(active) ~** I., with unstable nuclei, that emit radiation; used for labeling. **stable ~** I., with stable nuclei, that do not emit radiation.

isotopic Pertaining to isotopes. **i. nuclides** Isotopes. **i. reaction** Chemical change involving isotopes, as $^1H^2HO + N^1H_3 = {}^1H_2O + N^1H_2{}^2H$. **i. weight** I.W. Term formerly used for the relative atomic mass (atomic weight) of an isotope.

isotopically deficient* Abbrev. *def;* an isotopically labeled compound with one or more nuclides present in less than the natural ratio; e.g., [*def*^{13}C]$CHCl_3$.

isotropic Having similar properties in every direction. Cf. *anisotropic.*

isovol A line joining points corresponding with the same content of volatile matter; used on coal survey maps.

Isovyl Trademark for a polyvinyl chloride synthetic fiber.

isoxazolyl* The radical C_3H_2ON-, from is*oxazole;* 5 isomers.

isoxsuprine hydrochloride $C_{18}H_{23}O_3N \cdot HCl$ = 337.9. Duvadilan, Vasodilan. White crystals, slightly soluble in water. A vasodilator. Also inhibits uterine contractions and is used to prevent premature labor (USP, BP).

Issoglio test A test for water-soluble or steam-volatile substances characteristic of rancid fats.

Isuprel Trademark for isoproterenol hydrochloride.

istizin $C_{14}H_8O_4$ = 240.2. 1,8-Dihydroxyanthraquinone*. A spot reagent for Li, Mg, and Al (red color).

itaconic acid $CH_2{:}C(COOH)CH_2COOH$ = 130.1. 2,3-Dicarboxypropene, methylenebutanedioic acid, m.161 (decomp.). Cf. *fumaric acid.*

itakolumite An elastic sandstone from Brazil; used in refractories to minimize thermal fracture.

itamalic acid $CH_2OH \cdot CH \cdot COOH \cdot CH_2COOH$ = 148.1. Not known in free state. Cf. *citramalic acid.*

-ite Suffix: (1) Formerly used for the salt of an -ous acid; e.g., sulfite, from sulfurous acid. (2)* Now limited to a few acceptable such anions, as: nitrite, arsenite, hypochlorite. See *radicals,* Table 70 on p. 493. (3) A mineral.

iterative Descriptive of the repeated execution of a series of steps, as in a computer program loop.

I.U. International unit.

IUB International Union of Biochemistry. See *International Union of Pure and Applied Chemistry.*

-ium Suffix denoting: (1) The presence of a metal, nonmetal, or compound in a salt-forming (cationic) capacity. Cf. *onium.* (2) Many elements, generally metals; as, sodium, vanadium.

IUPAC Abbreviation for International Union of Pure and Applied Chemistry.

ivaine $C_{24}H_{42}O_5$ = 410.6. An alkaloid obtained from *Achillea moschata* (Compositae).

ivanuol $C_{24}H_{32}O_5$ = 400.5. An alcohol in iva oil. Yellow, bitter oil of pleasant odor.

iva oil A green oil, distilled from the flowering tops of *Achillea* species; used in flavoring extracts.

ivory The bonelike substance from the tusks of animals. **vegetable ~** Corajo.
i. black Animal charcoal prepared by calcining the refuse from i. working; a decolorizing agent and filtering medium.

ivy The climbing plant *Hedera helix* (Araliaceae); the leaves and berries are a stimulant. **American ~** Virginia creeper, wood vine. The bark and twigs of *Vitis hederacea* (Vitaceae); an astringent. **ground ~** Gill-go-over-the-ground. The herb of *Glechoma hederacea* (Labiatae); a diuretic. **poison ~** *Rhus toxicodendron.*

I.W. Abbreviation for *isotopic* weight, q.v.

iztac chalchihuitl A white or green gem variety of Mexican onyx.

For other iso- compounds (or radicals), look under the root compound (or radical) that is prefixed by the iso-.

J

J Symbol for joule. **J acid** 6-Amino-5-naphthol-7-sulfonic acid. **J phenomenon** The absorption of X-rays by a substance varies discontinuously with a change in wavelength λ; thus certain λ excite the characteristic radiation of the absorber.

J Symbol for (1) moment of inertia; (2) nuclear spin quantum number; (3) rate of reaction; (4) total angular momentum quantum number; (5) sound intensity.

J (Boldface.) Symbol for (1) electric current density; (2) magnetic polarization.

jaborandi See *Pilocarpus*. **j. oil** Yellow oil distilled from j. leaves, d.0.865, b.220, insoluble in water; used in pharmacy and as a source of pilocarpine.

jaboridine $C_{10}H_{12}O_3N_2$ = 208.2. An alkaloid from jaborandi leaves.

jaborine $C_{22}H_{32}O_4N_4$ = 416.5. An alkaloid from jaborandi leaves. Yellow syrup, insoluble in water; a miotic.

jaboty fat White tallow from Brazilian kernels of *Erisma calcaratum* and *E. uncinatum* (Vochysiaceae), m.42; a substitute for cacao butter.

jacinth Hyacinth.

jack Sphalerite.

jacket (1) The false or double wall of a container by means of which it may be cooled or heated. (2) The iron sheath of a furnace. (3) The brick covering of a boiler.

Jackson, Sir Herbert (1863–1936) British chemist noted for chemical research applied to industry.

jacobine $C_{18}H_{25}O_6N$ = 351.4. An alkaloid from *Senecio jacobaea*, ragwort (Compositae); the cause of Winton's disease of grazing animals.

jaconecic acid $C_{10}H_{16}O_6$ = 232.2. The acid hydrolysis product of jacobine.

jacutinga A Brazilian ore containing approximately: hematite 98, chalcedony 0.9, manganous oxide 0.25%; a source of iron.

jade (1) $3MgO \cdot CaO \cdot 2SiO_2$. True j., nephrite. A *silica* mineral, q.v., of the amphibole group. A fusible green or flaked gem (China, Siberia), d.3.0, hardness 6.0. See *imperial j.* (2) $NaAlO_2 \cdot 2SiO_2$. Jadeite, greenstone. A *silica* mineral, q.v., of the pyroxene group resembling j.

jadeite Jade (2).

jaggery sugar Gur.

jaipurite CoS. A native cobaltous sulfite.

jalap The dried root of *Exogonium* or *Ipomoea purga* (Convolvulaceae), Mexico and W. Indies. It should contain not less than 7% of resins, and is a hydragogue cathartic.

jalapic acid $C_{17}H_{30}O_9$ = 402.4. An acid, m.120, from jalap.

jalapin $C_{34}H_{56}O_{16}$ = 720.8. Orizabin. A resinous glucoside, m.150, from jalap. Cf. *convolvulin.*

jalapinolic acid $Me(CH_2)_4CHOH(CH_2)_9COOH$ = 272.4. (+)-11-Hydroxyhexadecanoic acid*, from jalap resin.

jalapoid The combined principles from extracted jalap. A cathartic.

jamaica **j. dogwood** The bark of *Ichthyomethia piscipula* or *Piscidia erythrina*. **j. wood** Haematoxylon.

jamaicine A bitter principle from the bark of the cabbage tree, *Andira inermis* (Leguminosae), Jamaica. It resembles berberine.

jamboo (1) Jambu. The seeds of *Piper jaborandi* (Piperaceae). (2) Jambul.

jambosine $C_{10}H_{15}O_3N$ = 197.2. A crystalline alkaloid from the root of *Jambosa vulgaris* (Myrtaceae), the rose apple of the tropics.

jambul The bark of *Eugenia jambolanum*, the Java plum tree (Myrtaceae), southeast Asia. **j. seeds** The seeds of *E. jambolanum;* a carminative.

jambulol $C_{16}H_8O_9$ = 344.2. A dibasic, pentahydroxy acid found in the seeds of *Eugenia jambolanum* (Myrtaceae), Java. It resembles ellagic acid.

James **J. powder** Pulvis antimonialis. Antimoniated calcium phosphate. **J. tea** Labrador tea.

jamesonite $Pb_2Sb_2S_5$. Feather ore. A native sulfide.

Jamin effect A capillary tube filled with alternate air and water bubbles sustains a finite pressure, owing to oily contamination on the glass surface, which prevents complete wetting.

jammy fluids See *electrorheological fluids.*

Janus green A green dye intravital stain.

japaconine $C_{26}H_{41}O_{10}N$ = 527.6. An amorphous alkaloid, m.97, from Japanese aconite.

japaconitine $C_{34}H_{49}NO_{11}$ = 633.8. Acetylbenzoyljapaconine. An alkaloid, m.204, from Japanese aconite.

japan A varnish for metallic and wooden articles. Cf. *stove.* **j. camphor** Camphor. **j. clay** Montmorillonite. **j. tallow** J. wax. **j. wax** A glyceride containing palmitic acid and small amounts of icosanoic and henicosanoic acids; from the berries of *Rhus succedanea* and other species of sumac tree. Yellow wax, d.0.970, m.53, soluble in benzene; used in the manufacture of candles, wax, and leather dressing.

japanic acid $C_{21}H_{40}O_4$ = 356.5. Henicosanedioic acid*. m.112, soluble in alcohol. From japan wax.

japanning Varnishing by successive applications of a lacquer and heating in an oven.

japonic acid A tannin from catechu.

japopinic acid $C_{14}H_{22}O_2$ = 222.3. A monobasic acid from Japanese turpentine.

jar A small earthen vessel, without spout or handle. **bell ~** Bell glass. **Leyden ~** An electric condenser; a glass vessel lined internally and externally with tinfoil. **Naples ~** See *Naples jar.*

jara jara Methyl 2-naphthyl *ether**.

jargon, jargoon Zircon.

jargonia Zirconium oxide.

jasmine oil A colorless essential oil from the flowers of *Jasminum grandiflorum* (Oleaceae). It contains benzyl acetate, linalol, linalyl acetate, and indole, d.1.008, soluble in alcohol; used in perfumery.

jasper Touchstone, Indian stone, bloodstone, Lydian stone. A gem variety of chalcedony. Formerly used to estimate the gold content of an alloy by comparing the color of the streak made on it with streaks made by standard alloys. **porcelain ~** Porcellanite.

jateorhiza Calumba.

jatrophone $C_{20}H_{24}O_3$ = 312.4. From *Jatropha gossipifolia.*

Yellow powder, m.152, soluble in cold water; acid solutions are fluorescent; used to treat stomach disorders.
jaune French for "yellow." **j. brillant** Cadmium sulfide. **j. d'or** Martius yellow.
Java **J. pepper** Cubeb. **J. plum** Jambul. **J. tea** Orthosiphonin.
javanicin $C_{15}H_{14}O_6$ = 290.3. An antibiotic, weakly acidic pigment from cultures of *Fusarium javanicum.* Red plates, m.212.
javelle Eau de javelle.
javellization The sterilization of water supplies by hypochlorites.
jaw oil An oil from the jaws of the black fish, *Globicephalus melas;* a lubricant for fine machinery.
Jeans, Sir James (1877–1946) British mathematician, astronomer, and physicist, noted for his correlation of the physical theories of the structure of matter with cosmic science.
jecolein A glyceride from cod-liver oil.
jecorin $C_{105}H_{186}O_{46}N_5SP_3$. A protein in the liver, spleen, and brain.
jel See *gel.*
jelletite A green variety of andradite. See *garnet.*
jellose A carbohydrate plant constituent, related and having similar properties to fruit pectins, and composed of glucose, galactose, and xylose.
jelly A soft, usually transparent *colloid* system, q.v.: liquid suspended in a solid; as, water in gelatin. **mineral ~**, **petroleum ~** See *vaselin.* **royal ~** A secretion of the pharyngeal glands of nursing bees, *Apis mellifica,* and the sole food of queen larvae.
jellying power The capacity of substances to solidify in solution; as, gelatin.
jelutong The exudate from *Dyzra costulata* (Apocynaceae); used in chewing gum. Cf. *pontianac.*
jenkolic acid $CH_2[SCH_2CH(NH_2)COOH]_2$ = 254.3. Djenkolic acid, m.300+ (decomp.), from djenkol beans, *Pithecalobium lobatum* (Leguminosae).
Jenner stain A 2-solution microscope stain for white blood cells: (1) 0.5% eosin in methanol; (2) 0.5% methylene blue in methanol.
jeppel oil Bone oil.
jequiritin Abrin.
jervine $C_{27}H_{39}O_3N \cdot 2H_2O$ = 461.6. An alkaloid from *Veratrum* species (Liliaceae). White prisms, m.240, insoluble in water.
jesaconitine $C_{40}H_{51}NO_{12}$ = 737.8. An *aconite* alkaloid, q.v. m.130 (decomp.).
Jesuits' balsam Copaiba balsam. **Jesuit's bark** Cinchona bark.
JET See *thermonuclear reaction.*
jet A dense, black, polished lignite, or cannel coal rendered black by fossilization; sometimes used for jewelry. **j. engine** A turbine that compresses air into its combustion chamber, the reaction products of which drive the compressor turbine and provide the j. propulsion.
Jetspun Trademark for a viscose synthetic fiber.
Jew's pitch Asphalt.
jig A vibrating screen submerged in water for ore concentration.
jigger Jig.
Jimson weed Stramonium.
johannite Uranvitriol.
johimbine $C_{23}H_{32}O_4N_2$ = 400.5. An alkaloid from the bark of *Corynanthe yohimbehe,* an African tree. Cf. *yohimbine.*
Joint Codex Alimentarius Commission A United Nations organization formulating health and purity criteria; and standards of quality, appearance, and packaging of food.
jojoba The shrub *Simmondsia Californica* (Buxaceae), California and Mexico. **j. oil** The oil from the nuts of j., resembling whale sperm oil.
Joliot-Curie **J.-C., Frédéric (1900–1958)** French physical chemist. **J.-C., Irene (1897–1956)** French physical chemist, wife of Frédéric J.-C., and daughter of Pierre and Marie Curie. Frédéric and Irene J.-C. won the Nobel prize (1935) for their work on induced radioactivity.
Jolly balance A spring balance for determining the specific gravity of a solid by weighing it alternately in air and water.
Jones **J. reagent** A bath of molten zinc in which a metal to be etched is immersed. **J. reductor** A glass tube filled with amalgamated, granulated zinc or zinc wire spirals, used for reducing solutions of Fe, Ti, Mo, V, Cr, and U before their volumetric determination by oxidation-reduction methods.
jordanite $AsS_3 \cdot 4PbS$. A native sulfide.
Jorissen test A pinch of phloroglucinol is added to a solution of the sample in hydrochloric acid. If the solution is made alkaline, a red or orange color results in presence of formaldehyde or acetaldehyde.
joseite Bi_3Te. A bismuth tellurite.
josephinite A native iron and nickel alloy from Oregon.
Joule, James Prescott (1818–1889) (Pronounced "jool") English physicist noted for work on energy transformations. Cf. *joule.* **J. effect** (1) The change in length of a ferromagnetic material along the axis of the applied magnetic field, when the field strength is changed. (2) Heat effect. **J. equivalent** The mechanical equivalent of heat; or the quantity of energy which is equivalent to unit quantity of heat: 4.186 $\times 10^7$ ergs = 1 calorie (20°) = 4.186 J/cal. **J.-Kelvin effect** The fall in temperature of a gas forced at high pressure through a small orifice which is proportional to the difference in pressure on the two sides; used in the Linde-Hampson process of liquefying gases. **J.-Thomson effect** J.-Kelvin effect. **J.'s law** (1) The heat produced by a current I is equal to RI^2t, where t is the time during which the current flows and R the resistance of the circuit. (2) The internal energy of a volume of gas does not vary with temperature, if the volume remains constant. (3) The molecular heat of a solid compound is the sum of the atomic heats of its constituents. **J. unit** J. equivalent.
joule* J; the SI system unit of work, energy, and quantity of heat. It is the work done by a force of 1 newton when displaced 1 m. One j. is practically equivalent to the energy expended in one second by one ampere passing through one ohm of resistance. Hence: 1 joule = amp^2 × ohm × second = watt × second = (volt^2 × second)/ohm = 10^7 erg = 0.2391 calorie (20°) = 0.1020 kg·m = 0.7376 ft·lb = 6.241 $\times 10^{18}$ electronvolt = 0.2778 $\times 10^{-6}$ kWh = 9.471 $\times 10^{-4}$ Btu (mean).
juar Sorghum.
Juerst ebullioscope A device for determining small percentages of alcohol from the boiling point of the liquid, which varies directly with the percentage of alcohol.
juglansin A globulin from walnuts and butternuts.
juglone $C_{10}H_6O_3$ = 174.2. 5-Hydroxy-1,4-naphthoquinone, nucin. Brown prisms, m.153, insoluble in water; from the bark of the European walnut, *Juglans regia.*
juice (1) A fluid from a vegetable or animal tissue. (2) A colloquial term for electric current.
jumble beads Abrus.
Jungius, Ioachim (1587–1657) German alchemist and

forerunner of Boyle in showing the importance of observation and experiment.

juniper The tree *Juniperus communis* (Coniferae) of temperate zones. **gum ~** Sandarac.

j. berries The dried, ripe frutis of j.; a diuretic. **j.-berry oil** An essential oil distilled from j. berries. Yellow liquid, d.0.865, b.120, soluble in alcohol; a carminative and flavoring. It contains pinene, cadinene, and j. camphor. **j. tar** Cade oil.

juniperic acid $C_{16}H_{32}O_3$ = 272.4. 16-Hydroxyhexadecanoic acid*. m.95, from savine, *Juniperus sabina,* and arborvitae.

juniperin A bitter principle from juniper, *Juniperus communis.*

juniperus (1) The dried tops, wood, and berries of juniper; used as a source of an essential oil, and in fumigation. (2) A genus of Coniferae yielding berries, oils, and wood: *J. communis,* juniper; *J. sabina,* savine.

junket Curds and whey, a food prepared by coagulating milk with rennet.

Jurassic See *geologic eras,* Table 38.

jurubeba *Solanum insidiosum.*

jute The bast fiber of *Corchorus* species (Tiliaceae), southeast Asia and S. America; used in gunnysacks and twine. **China ~** The fiber of *Abutilon avicenna* (Malvaceae), China.

juvabione An insect-growth regulator derived from softwoods.

juvenile water A water of magmatic or deep-seated origin, supposed to come for the first time to the earth surface; e.g., artesian wells. Antonym: vadose water.

K

(See also under C)

K Symbol for (1) computer memory, in approximate number of thousands (or 1,024) of bytes or words; (2) kelvin; (3) potassium (kalium). **K acid** 1-Amino-8-naphthol-4,6-disulfonic acid. **K electrons** The 2 electrons in the innermost shell of the atom. **K line** (1) One of a group of lines having the shortest wavelength that are visible when K radiation passes through an X-ray spectrograph. (2) The Fraunhofer line, λ 3933.7, due to calcium. **K radiation** The homogeneous X-rays emitted by metals used as an anticathode in an X-ray tube; assumed to originate from the excitation of the K electrons. **K spectrum** The spectrum produced by K radiation. The frequency of any line is approximately proportional to $c(Z - b)^2$, where Z is the atomic number, c and b are constants. The atomic number Z $= Q_K + 1 = Q_L + 7.4$, where Q_K and Q_L express the frequencies of the K and L series of radiation, respectively.

K Symbol for (1) equilibrium constant (see *mass action*); (2) kinetic energy; (3) luminous efficacy.

k Symbol for kilo- (10^3).

k Symbol for (1) Boltzmann constant; (2) circular wave number; (3) rate constant of a reaction; (4) thermal conductivity.

K, κ See *kappa.*

K.A. An aluminum alloy similar to Duralumin.

kabaite A native hydrocarbon in meteorites.

kæ- See words beginning with *kae-, ke-.*

kaempferide $C_{16}H_{12}O_6$ = 300.3. 4′-Methyl ether of kampferol. A constituent of galangal, m. 228, soluble in sulfuric acid (blue fluorescence) and in alkali (yellow color).

ka(e)mpferol $C_{15}H_{10}O_6$ = 286.2. 3,4′,5,7-Tetrahydroxyflavone. m.274, from *Indigofera erecta.*

kaf(f)ir Sorghum. **k. corn** Durra, the grain of sorghum.

kafirin A protein from sorghum.

Kahle's solution See *fixative.*

kahweol $C_{20}H_{26}O_3$ = 314.4. A highly unsaturated alcohol, m.90, from coffee bean oil.

kainit Synthetic kainite, containing 30% potassium sulfate.

kainite $MgSO_4 \cdot KCl \cdot 3H_2O$. A native potassium magnesium sulfate associated with carnallite, from Stassfurt (Germany) and Poland; made synthetically; used as a potash fertilizer and in water treatment. The AOAC defines it as potassium and sodium chloride sometimes with magnesium sulfate, containing not less than 12% potassium oxide.

kairine $C_{10}H_{13}ON$ = 163.2. 1,2,3,4-Tetrahydro-8-hydroxyquinoline. b.280. **k. hydrochloride** $C_{10}H_{13}ON \cdot HCl$ = 199.7. Gray crystals, soluble in water; an antipyretic. Cf. *thalline.*

kairoline $C_{10}H_{13}N$ = 147.2. 1,2,3,4-Tetrahydro-1-methylquinoline. Colorless liquid, b.248, soluble in alcohol; an antipyretic.

kaiser K; reciprocal centimeter, cm^{-1}. A little-used unit of length.

Kaiserling solution A solution for preserving tissues, consisting of 3 g potassium acetate, 1 g potassium nitrate, 75 ml water, and 30 ml formaldehyde.

kakodyl Cacodyl.

kakoxene The mineral $2Fe_2O_3 \cdot P_2O_5 \cdot 12H_2O$.

kaladana The dried seeds of *Ipomoea hederacea;* a purgative. **k. resin** A purgative resin from kaladana.

kalaite A turquoise containing copper and iron. Cf. *callaite.*

kali German for "potassium hydroxide." **k. ammonsalpeter** A mixture of potassium chloride and ammonium nitrate used as fertilizer (nitrogen 16, potassium oxide 27%).

kalicrete A portland cement, containing iron, which is resistant to alkali soils.

kalimeter Calcimeter.

kalinite $K_2Al_2(SO_4)_4 \cdot 24H_2O$. A native sulfate.

kaliophylite $KAlSiO_4$. A nephelite *silica* mineral, q.v.

kalium Latin and German for "potassium."

kalk German for "lime." **k. ammon** A mixture of ammonium chloride and calcium chloride used as fertilizer. **k. ammonsalpeter** Calnitro.

kallaite Kalaite.

Kalle acid 1-Naphthylamine-2,7-disulfonic acid.

kallekrein Padutin. An enzyme from salivary glands, which reacts with a plasma protein to form bradykinin, a vasodilator.

kalmia Mountain laurel, sheep laurel, lambkill. The leaves of *K. latifolia* (Ericaceae); a cardiac sedative.

kalsilite $2KAlSiO_4$. A polymorph mineral resembling nepheline; minute grains in volcanic rocks of high potassium oxide content (Uganda).

kamacite A component of meteoric iron-nickel alloys, present only when the nickel content is less than 6.5%.

kamala Glandulae Rottlerae. A light, granular powder which consists of the glands and hairs from the capsules of *Mallotus philippinensis* (Euphorbiaceae); a purgative.

kamalin $C_{30}H_{28}O_8$ = 516.6. Rottlerin. The bitter principle of kamala. Brown powder, m.200, soluble in alcohol; an anthelmintic and purgative.

kamarezite A native basic copper sulfate, in green compact masses.

kambi An aromatic gum, resembling elemi, from *Gardenai lucida* (Rubiaceae), India.

Kamerlingh-Onnes, Heika (1853–1926) Dutch physicist and Nobel prize winner (1913), noted for experiments at very low temperatures.

kampometer An instrument for measuring heat radiation. Cf. *kapnometer.*

kanamycin A basic antibiotic from *Streptomyces kanamyceticus,* consisting of two amino sugars linked glucosidally to 2-deoxystreptamine; related chemically to neomycin. **k. sulfate** $C_{18}H_{36}O_{11}N_4 \cdot H_2SO_4$ = 582.6. White crystals, soluble in water; a bacteriocidal, effective against most gram-negative organisms (USP, EP, BP).

Kanebain Trademark for a polyvinyl alcohol synthetic fiber.

kanerol $C_{20}H_{50}O$ = 426.7. An alcohol from the roots of *Kanher nerium* (India). Colorless crystals, m.185, insoluble in water.

kanirin Trimethylamine oxide.

kanyl alcohol $C_{10}H_{18}O_2$ = 170.3. An alcohol from the liver oil of tarabakani, *Paralithodes camtschatica* (Japan Sea).
kaoliang oil Koryan oil.
kaolin $Al_2O_3 \cdot 2SiO_2 \cdot 2H_2O$. China clay, porcelain clay, bolus alba, terra alba, white bole, argilla. Derived from Kao-ling (Chinese: "high ridge"), the original site. A gray, fine, inert powder used in ceramics, as filler for paper and textiles, and for pencils. Three types: kaolinite (by weathering), dickite (by moderate heating), nakrite (by hypogene processes). Cf. *atalpo clay, fuller's earth, kaolinite.* **light ~** K. containing not more than 0.5% of coarse particles; an absorbent for diarrhea (USP, BP).
kaolinite $Al_2O_3 \cdot 2SiO_2 \cdot 2H_2O$. Clay substance. The chief constituent of kaolin or clay; heated above 1100°C, it breaks down to mullite and cristobalite. White, fine powder, used medicinally as an absorbent and dusting powder, and as a clarifying agent.
kaolinization Natural rock disintegration and formation of clay from the feldspars of decomposed granite (orthoclase).
kaon A subatomic *particle.*
Kapilon Trademark for acetonaphthone.
kapnometer An instrument for measuring the density of smokes. Cf. *kampometer.*
kapok The Malay name for the cottonlike down from the seedpods of *Eriodendron anfractuosum,* (Sterculiaceae); used in mattresses. **k. oil** Yellow oil, expressed from k. seeds, d.0.923, soluble in alcohol; used in soap and edible fats manufacture.
kappa K, κ. Greek letter. κ (not italic) is the symbol for a kaon. κ (italic) is the symbol for (1) compressibility; (2) electrolytic conductivity; (3) magnetic susceptibility; (4) thermal diffusivity (*a* is preferred). **k. number** The mL of 0.1 *N* potassium permanganate consumed by 1 g of dry cellulose pulp under specified conditions. It measures bleachability.
Kapron Trademark for a polyamide synthetic fiber.
karabin $C_{21}H_{49}O_6$ = 397.6. A resinous substance from *Nerium oleander* (Apocynaceae).
karakin $C_{10}H_{14}O_9N_2$ = 306.2. A crystalline glycoside from the berries of the karaka tree, *Corynocarpus laevigata,* m.122.
karat U.S. usage for *carat,* q.v.; the unit of gold fineness, as distinct from the unit of gem weight.
karaya gum An Indian gum similar to tragacanth, used in toothpastes.
Karbogel Trademark for a porous, solid, granular desiccant made from coal. It can be reactivated at 100°C, and does not readily disintegrate.
karitene $C_{32}H_{56}$ = 440.8. A solid hydrocarbon, m.64, from shea fat.
Karle, Jerome (1918–) American chemist. Nobel prize winner (1985) for work on crystal structures.
Karl Fischer reagent A solution of iodine and sulfur dioxide in methyl alcohol and pyridine, used to determine water by titration. In presence of water, the iodine passes into the combined state, and the end point is shown by the presence of excess iodine.
karpholite The mineral $(Mn,Fe)O \cdot (Fe,Al)_2O_3 \cdot SiO_2 \cdot 2H_2O$.
karyokinesis Division of the cell nucleus and its chromosomes, the chromosome number remaining diploid. Cf. *meiosis, haploid.*
karyotype The picture and arrangement of the chromosomes of the somatic, diploid cell. Cell division is arrested when the chromosomes are most readily visible, a photomicrograph taken, and the chromosomes arranged in pairs. See *colchicine, somatic, diploid.*
Kastle-Meyer reagent (For blood) see *phenolphthalin(e).*
kata- Cata-. (1) Greek prefix indicating "down" or "below." (2) Kappa. **k.thermometer** An instrument that measures the cooling effect of airflow as distinct from the temperature of the air: used to evaluate ventilation conditions.
katabolism Catabolism.
katal The unit of enzyme activity. **micro ~** μkat; the unit of production (or consumption) of 1 μmol/s of reaction product (or reaction substrate).
katalysis Catalysis.
kataphoresis Cataphoresis.
katchung Chinese for "peanuts." (Kachang = Malay.) **oil of ~** Peanut oil.
katharometer An instrument for the analysis of a gas mixture by the changes in its thermal conductivity.
kathode Cathode.
katine $C_{10}H_{18}ON_2$ = 182.3. An alkaloid from the leaves of African tea, *Catha edulis* (Celastraceae). Cf. *celastrine, cathine.*
kation Cation.
katode Cathode.
kauri Kauri gum, kauri resin. A resin exuded from the kauri tree, *Agathis australis,* a conifer (New Zealand), and dug from the soil. It contains a parent substance of the catechins, 1-leucomaclurinyl glycol. Amberlike resin, d.1.05, m.180–230, insoluble in water; used in varnishes. **k. gum** Kauri. **k. oil** An oil obtained by distilling peat from prehistoric kauri forests.
kaurinic acid $C_{10}H_{16}O_2$ = 168.2. An acid constituent (about 1.5%) of kauri.
kaurinolic acid $C_{17}H_{34}O_2$ = 270.5. An acid constituent (about 15%) of kauri.
kaurolic acid $C_{12}H_{20}O_2$ = 196.3. The principal constituent (50%) of kauri.
kauronolic acid $C_{12}H_{34}O_2$ = 210.4. A constituent (10%) of kauri.
kava Kavakava, kawa. The dried rhizome of *Piper methysticum* (Piperaceae), Polynesian islands; mild diuretic and stimulant. **k. resin** A resin derived from kava, containing kavaic acid.
kavaic acid $C_{13}H_{12}O_3$ = 216.2. An acid derived kava. Yellow crystals, m.165.
kavain $C_{14}H_{14}O_3$ = 230.3. Methysticin, kawain. **(*R*)-(*E*)- ~** A principle from kava. White needles, m.102, soluble in alcohol.
kavatel oil An oil obtained from the seeds of *Hydnocarpus Wightiana;* used in place of chaulmoogra oil.
kawa Kava.
Kayexalate Trademark for sodium polystyrenesulfonate.
Keen tester An instrument to determine hardness of metals by the impact-ball method.
kefir Kephir, kephyr. A fermentation product of goat's milk produced by the *Bacillus caucasicus;* a nutritious food. Cf. *koumiss.* **k. milk** A beverage made from cow milk and kefir powder. **k. powder** White, irregular bodies containing a yeast fungus and several bacteria; used to make kefir and kefir milk.
Keflex Trademark for cephalexin.
keilhauite A native titanate and silicate of aluminum, calcium, iron, and the rare-earth metals. Cf. *sphene.*
Kekulé, Friedrich August (1829–1896) German chemist noted for work on the constitution of carbon compounds. **K. ring** See *benzene ring.*
Kelene Trademark for ethyl chloride.
Kelheim Trademark for a viscose synthetic fiber.
kellin A glucoside from the fruit of *Ammi visnaga.*
kelp (1) Varec. The large brown seaweeds of the Pacific and

Atlantic shores (Laminaria). (U.S. usage.) (2) The ash of seaweeds, formally a source of alkali and iodine (U.K. usage). **dried ~** A fertilizer containing nitrogen 1.6–3.3, phosphorous pentaoxide 1–2, potassium oxide 15–20%. Cf. *algin(ate).* **giant ~** *Macrocystis pyrifera.* A k. 60 m long.

Kelvin, William Thompson (Lord Kelvin) (1824–1907) British physicist, pioneer in electrical science. **K. bridge** A set of coils for accurate resistance measurements. **K. galvanometer** An elliptical coil round a magnetic needle suspended on a fine wire with mirror attached, to detect small electric currents. **K. scale** See *temperature.*

Kemadrin Trademark for procyclidine hydrochloride.

kemfert Trona potash. A potassium chloride from the waters of Searles Lake, California.

kemp A coarse, low-grade wool fiber, which resists dyeing and lacks the characteristic epidermal scales of the normal wool fiber.

kenaf *Hibiscus cannabinus.* A S. African jute substitute.

Kendall K., Edward Calvin (1886–1972) American chemist; isolated levothyroxine from the thyroid gland. **K., James (1889–1978)** British inorganic and physical chemist.

Kennelly-Heaviside layer A stratum of electrons, 160–320 km above the earth's surface, created by solar rays. It is impenetrable by radio waves and is the seat of the aurora. Cf. *ionosphere.*

kepayang oil An oil resembling chaulmoogra oil, from the seeds of Pokok kepayang.

kephalin (1) Cephalin, brain lipoid. A lecithin phosphatide from spinal and brain tissue of mammals. The constituent choline is replaced by cholamine. Yellow solid, of characteristic odor, slightly soluble in alcohol. Cf. *cuorin.* (2) A group of phosphatides from white brain substance resembling *lecithin* (q.v.) in structure, but with aminoethyl alcohol in place of the choline residue. White, brittle, very hygroscopic solids, forming colloidal solutions in water; hydrolyzed slowly to stearic acid and a mixture of unsaturated fatty acids.

kephaloidin A constituent of the buttery substance of *brain,* q.v.

kephyr Kefir.

ker A Polish synthetic rubber, made from alcohol.

keracyanin An anthocyanin from cherries.

kerasin $C_{48}H_{93}O_8N$ = 812.3. A cerebroside associated with phrenosin in the brain tissue. White powder, m.180 (decomp.), insoluble in water.

kerasol $C_{20}H_{10}O_4I_4$ = 821.9. Tetraiodophenolphthalein. The soluble sodium salt renders the gallbladder visible in X-rays.

keratin $C_{41}H_{71}O_{14}N_{12}S$ = 988.2. A protein in horns, feathers, shells, fingernails; used to coat pills; resistant to digestion by pepsin and trypsin and insoluble in water. Believed to exist in a folded chain form, converted into a straight chain by simultaneous steaming and stretching. **meta ~** The compound formed when the $-SH$ groups in k. are oxidized to the $-S-S-$ form.

keratoelastin A protein from the eggshells of fishes, reptiles, and monotremes.

keratolytic A substance that dissolves keratin or skin, e.g., salicylic acid for the treatment of corns.

kerenes That portion of kerotenes which is insoluble in organic solvents.

kerites The naturally occurring kerotene bitumens and protobitumens.

kermes The dried insects, *Coccus ilicis,* from the leaves of Oriental oak species; an ancient red textile dye. **k. mineral** Kermesite.

kermesite Sb_2S_2O. Lyrostibnite. A native, red antimony sulfide pigment.

kern German for "ring" and "nucleus," especially the benzene ring.

kernel Atomic *nucleus.*

kernite $Na_2B_4O_7 \cdot 4H_2O$. A sodium borate from Kern County, Calif. Colorless, monoclinic crystals, d.1.908, hardness 2.5.

kerogen Organic matter in hydrocarbon source rock, originating from biological material. It is insoluble in organic solvents and helps to characterize the oil and gas potential of the rock.

keroles That portion of kerotenes which is soluble in pyridine, but insoluble in chloroform.

kerols That portion of kerotenes which is soluble in both chloroform and pyridine.

kerones That portion of kerotenes which is insoluble in organic solvents.

kerosine (Official U.S. and U.K. usage) Coal oil, astral oil, kerosene, paraffin (common U.K. usage). A mixture of hydrocarbons, b.150–280; the fraction in the distillation of petroleum between gasoline and the oils; a fuel and cleanser.

kerotenes That portion of bitumen which is insoluble in carbon disulfide. Cf. *kerenes.*

Kerr K. constant If L is the difference between the extraordinary and ordinary refractive indexes, λ the wavelength of the light, E the electric field strength, then $L = k\lambda E^2$, where k is the Kerr constant. **K. effect** Electrical birefringence. The electrooptical double refraction of light by liquids in an electric field.

kesso oil Japanese valerian oil. An essential oil from the roots of *Valeriana officinalis.* Green liquid, d.0.996, insoluble in water.

kessyl alcohol $C_{15}H_{26}O_2$ = 238.4. Kersyl alcohol, *α*-~ m.85, b_{11mm}156, from kesso oil.

ketal Obsolete term for: (1) the carbonyl*, =CO, group; as, dimethyl ketal (= acetone). (2) $R_1 \cdot R_2 \cdot C \cdot (OR_3)(OR_4)$. Ketone acetates. See *acetals.*

Ketalar Trademark for ketamine hydrochloride.

ketamine hydrochloride $C_{13}H_{16}ONCl \cdot HCl$ = 274.2. (±)-2-(*o*-Chlorophenyl)-2-(methylamino)cyclohexanone hydrochloride. Ketalar. An anesthetic.

ketazin $C_6H_{12}N_2$ = 112.2. Colorless liquid, d.0.836, b.131, soluble in water.

ketazine Bisazimethylene. An azine compound containing the radical =C:N·N:C=.

ketene* $CH_2:CO$ = 42.04. Ketoethylene, ethenone, carbomethene. The simplest ketone. Colorless gas of penetrating odor, b. −56, decomp. by water; an acetylation agent. **k. diacetal** $CH_2:C(OEt)_2$ = 116.2. 1,1-Diethoxyethylene*. Colorless liquid, b.125; reacts with water to form ethyl acetate, and polymerizes to a white solid.

ketenes* $R_2C:CO$. A group of reactive organic compounds; 2 types: **aldo ~** R·H·C:CO. Colorless; polymerized by pyridine and incapable of autooxidation. **keto ~** $R_2C:CO$. Colored; readily autooxidized and form additive compounds. Both aldo ~ and keto ~ are homologs of ketene. Cf. *ketines.*

ketimide $R_2:C:NX$ where X is an acyl radical. Cf. *imine.*

ketimine Imine*.

ketine $C_6H_8N_2$ = 108.1. 2,5-Dimethylpyrazine. Colorless liquid, d.0.990, b.153, soluble in water.

ketines R·C:CH·N:CR·CH:N, derived from ketine. (ring closure drawn from first C to final N)

Cf. *ketenes.*

keto- (1) Prefix formerly indicating the carbonyl group, =C:O, joined to 2 carbon atoms, as, in a *ketone,* q.v. (2)* Used generically, as in keto-enol. **k. acid** See *ketone acid.* **k.-enol** See *isomerism,* Table 48 on p. 314.

ketoamine An organic compound containing the carbonyl

and amino group, formed by the action of ammonia on ketones. Cf. poly*peptide.*

ketocoumaran Oxodihydro*benzofuran**.

ketodestrin Estrone.

ketohexose* A monosaccharide of 6 C atoms, with a ketone group rather than an aldehyde group; e.g., fructose. Cf. *aldohexose.*

ketohydroxyestrin $C_{18}H_{22}O_2$ = 270.4. A hormone from the urine of pregnant women and mares; an anhydride of estriol. Cf. *sterols.*

ketoimine A compound containing an imino and carbonyl group.

ketoindole Oxindole.

ketoketenes See *ketenes.*

ketol Ketone alcohols. A compound containing a carbonyl and hydroxy group. **alpha-~** A compound containing the R—CO·CH_2OH group. **beta-~** A compound containing the R—CO·CH_2·CH_2OH group. **saturated ~** An α- or β-ketone alcohol. **unsaturated ~** Acivinyl alcohols. A compound containing the unsaturated R—CO·CH:CHOH group.

ketole Indole*.

ketone* R·CO·R. An organic compound containing the carbonyl group, =C:O, joined to 2 C atoms. Nomenclature: naming the 2 radicals before the term *ketone* or attaching the suffix *-one* to the hydrocarbon; CH_3·CO·CH_3 is dimethyl ketone or propanone (acetone); or attaching the prefix *oxo-*, or the suffix *-quinone.* Classification: (1) Aliphatic, saturated: acetone*, propanone*, Me·CO·Me. (2) Aliphatic, unsaturated; 3-buten-2-one*, MeCOCH:CH_2. (3) Aliphatic, diketones*: biacetyl*, 2,3-butanedione*, MeCOCOMe. (4) Cyclic: cyclobutanone*, CO·CH_2·CH_2·CH_2. (5) Quinones: benzoquinone*, CO·CH:CH·CO·CH:CH. (6) Aromatic: acetophenone, Ph·COMe. **acid ~** Ketone acid*. **aldehyde ~** A compound containing the =CO and —CHO groups. **amino ~** Ketoamine. **benzyl phenethyl ~*** PhCH_2·CH_2·CO·CH_2·Ph = 224.3 Colorless liquid, m. 324, soluble in alcohol. **butyl methyl ~*** 2-Hexanone*. **di ~*** See *diketone.* **dibutyl ~*** 5-Nonanone*. **diethyl ~*** 3-Pentanone*. **diheptyl ~*** 8-Pentadecanone*. **dimethyl ~*** Acetone*. **dipropyl ~*** 4-Heptanone*. **ethyl methyl ~*** 2-Butanone*. **ethyl pentyl ~*** 3-Octanone*. **ethyl propyl ~*** 3-Hexanone*. **heptyl methyl ~*** 2-Nonanone*. **methyl naphthyl ~*** $C_{10}H_7$·CO·Me = 170.2. **1-~** Acetonaphthone. Colorless crystals, m.34, soluble in alcohol. **methyl pentyl ~*** 2-Heptanone*. **methyl propyl ~*** 2-Pentanone*. **methyl thienyl ~*** (C_4H_3S)COMe = 126.2. Colorless liquid, b.213. **mixed ~** A k. with 2 different radicals attached to the carbonyl group; as, Me·CO·Et, ethyl methyl k. **nitroso ~** A compound containing the =CO and nitroso groups. **olefin ~** A k. of the alkene series. **paraffin ~** A k. of the alkane series. **phenyl thienyl ~*** (C_4H_3S)COPh = 188.2. Colorless crystals, m.55, soluble in alcohol. **simple ~** A k. with the same 2 radicals attached to the carbonyl group. **tri ~** A compound containing 3 carbonyl groups.

k. acid* Oxo acid. A compound containing the radicals =CO and —COOH. **alpha-~** A compound containing the radical —CO·COOH, e.g., pyruvic acid or 2-oxopropanoic acid. **beta-~** A compound containing the radical —COCH_2· COOH, e.g., acetoacetic acid. **gamma-~** A compound containing the radical —CO·CH_2·CH_2·COOH, e.g., levulinic acid. **delta-~** A compound containing the radical —CO·CH_2·CH_2·CH_2·COOH, e.g., acetylbutanoic acid. **k. alcohol** See *ketol.* **k. base** Michler's k. **k. bodies** Acetone bodies. Collective term for acetone, acetoacetic acid, and 3-hydroxybutanoic acid, which accumulate in the blood in diabetes, starvation, and vomiting. See *ketosis.* **k. color** An artificial color containing the carbonyl group, e.g., alizarin. **k. form** See *ketonic ester type.* **k. group** The carbonyl, =C:O, group attached to 2 C atoms; it usually confers reducing powers.

ketonic Pertaining to a ketone. **k. ester type** An isomer of an enolic ester–type compound:

```
 R·C = O              R·C—OR″
     |                    ||
   HCR″       ⇌          HC
     |                    |
   COOR′                COOR′
Ketonic ester type   Enolic ester type
```

Cf. *desmotropism.*

ketonuria Acetonuria. The excretion of acetone in urine, as occurs in ketosis.

ketose* A sugar containing a ketone group. Cf. *aldose.*

ketoside A glucoside which yields a ketose on hydrolysis.

ketosis The presence of ketone bodies in the blood.

ketotriazole Triazolone.

ketoxime (1)* Acetoxime. A compound containing the =C:NOH group, e.g. Me_2C:N·OH, acetoxime. (2) A compound containing the —HC·NO— group. Cf. *Beckmann rearrangement.* **tetra ~** See *diphenyltetraketoxime.*

Kevadon Trademark for thalidomide.

Kevlar Trademark for a high-strength aramid.

key atom (1) An atom in a chain whose change in electronic structure induces corresponding changes in the other atoms of the chain. (2) An atom in a ring whose oscillations cause a shift of bonds. Cf. *porphin ring.*

kg Abbreviation for kilogram.

kgf Abbreviation for kilogram force.

khat Cafta, Arabian tea. The dried leaves of *Catha edulis;* a tea.

khelin Khellin. A synthetic dimethoxymethylfuranochromone derivative, used for its specific coronary vasodilatory activity. Also obtained from the seeds of the wild Mediterranean plant *Ammi visnaga,* Lam.

Khotinsky, Achilles de (1850–1933) Russian-born American instrument designer. **de K. cement** A cement for glass and porcelain; insulating, covering, and connecting electric wires, glass, rubber, wood, etc.; resistant to ordinary solvents.

kibbled Broken up into small lumps of about 1 cm diameter.

kidney A paired mammalian organ that eliminates nitrogenous waste from the bloodstream; also controls the water concentration and electrolyte content, and, in part, the acid-base equilibrium. Much of the filtrate is reabsorbed, together with useful substances, as, amino acids, glucose, salt, proteins. See *glomeruli.*

kidney ore Red *hematite.*

kies General term for sulfide ores.

kieselguhr Diatomite. Tripoli powder. Guhr. A diatomaceous or infusorial earth. Used as an absorbent for nitroglycerin (dynamite), in chromatography, for filtration and insulation, and as an abrasive in soaps.

kieserite $MgSO_4$·H_2O. A native magnesium sulfate. White, compact masses in the Stassfurt salt beds.

Kikuchi lines The black and white lines which appear when a stream of electrons is scattered by a crystal surface. Cf. electron *microscope.*

killeen Irish moss.

Killiani reaction The synthesis of a higher homolog by forming a nitrile, followed by hydrolysis; e.g., pentose to hexose: R·CHO → R·CHOH·CN → R·CHOH·COOH → R·CHOH·CHO.
kiln (1) A potter's oven for baking bricks. (2) A furnace for calcining or drying coarsely broken ore or stone. (3) A building for drying malt.
kilo-, kilo Abbrev. k; (1)* SI prefix indicating 1,000 units; (2) kilogram.
kilocalorie Abbrev. kcal; a great (or large) calorie: 1 kcal = 1,000 calories.
kilocycle A frequency of 1,000 periodic cycles per second = 1 kHz.
kilogram* Abbrev. kg; the base unit of mass in the SI system. It is defined as the mass of the international prototype kilogram and approximately equals the mass of 1 dm^3 of water at 4°C. 1 kg = 1,000 g = 2.2046 lb = 15,432 grains. Cf. *liter.* **international prototype ~** A cylinder of platinum-iridium alloy; the standard k. weight.
kilogram-meter A metric system unit of energy; the energy to raise 1 kg by 1 m. 1 kgf·m = 7.233 ft·lb = 9.80665 newton-meters (N·m).
kilometer* Abbrev. km; a unit of length in the metric system. 1 km = 1,000 meters = 3280.84 ft = 1093.61 yd = 0.62137 mile. (1 mile = 1.609 km.)
kilonem A unit of nutrition equivalent to 667 kcal; supplied by a liter of milk.
kilowatt* kW*. A unit of power. 1 kW = 1,000 W = 1,000 J/s = 1.340 hp (electric) = 238.7 cal/s (thermochemical). (1 hp = 746 W). **k.hour** A commercial unit of electric energy. 1 kW·h = 1,000 W·h = 3,600,000 J = 3,412 Btu (international table).
kilurane A suggested unit of radioactivity. 1 ku = 1,000 uranium units (uranes).
kimberlite Blue ground. An igneous mineral in which diamonds occur.
kinase (1)* Suffix of a group of transferase enzymes; as, hexokinase. (2) Zymoexcitor. Formerly the term for a substance in living tissues, which activated or transformed a zymogen into an active enzyme. Cf. *activation (2).* **entero ~** See *enterokinase.*
kinematics The science of motion, apart from the forces that produce it.
kinetic (1) Pertaining to motion. (2) Dealing with forces that influence the motion of bodies. **k. chemicals** Gases used as refrigerants. **k. energy** The force possessed by a body due to its motion = $mv^2/2$, where m is the mass and v the velocity. **k. theory** The hypothesis that all molecules are in motion, which is most rapid in gases, less rapid in liquids, and very slow in solids, and is a function of temperature. $pV = \frac{1}{3} nmv^2 = RT$, where n is the number of molecules. See *Brownian movement, gas laws.*
kinetin Zeatin.
king's yellow Orpiment.
kinic acid Quinic acid.
kino Kino gum. The dried juice of *Pterocarpus marsupium* (Leguminosae), tropical Africa. Brown, brittle fragments, soluble in water; used as an intestinal astringent, and in the tanning and textile industries. **African ~** Kino. **American ~** The dried juice of *Coccoloba uvifera;* a k. substitute. **marri ~** A red gum from *Eucalyptus calophylla* (Myrtaceae).
kinoin $C_{14}H_{12}O_6$ = 276.2. A resin from kino.
kinovin Quinovin.
Kipp generator A 3-compartment apparatus for generating gases, generally hydrogen sulfide, from a solid reagent in the middle and a liquid reagent in the lower compartment.
Kirchhoff, Gustav Robert (1824–1887) German physical chemist noted for development of the spectroscope. **K. equation** $\log p = A + B/T + C \log T$, where p is the vapor pressure of a gas at the thermodynamic temperature T; and A, B, and C are constants.
kish Crystalline graphite deposited in iron furnaces from molten iron.
kisidwe oil A hard, white fat from the nuts of *Allanblackia* species (Ghana); used in soap manufacture.
kitol A substance containing vitamin A; found in whale liver.
kittel plates A device used in a fractionating column to ensure rapid renewal of the contacting surface between the liquid being distilled and its vapor.
kittool fiber Kittul. A fiber from the leaves of a Sri Lanka palm, *Caryota urens;* used in brushes.
kittul Kittool.
Kjeldahl, Johan (1849–1900) Danish analytical chemist. **K. apparatus** An arrangement for distilling ammonia from an organic compound. **K. flask** A pear-shaped flask of heat- and acid-resistant glass with a long neck. **K. method** A method to determine nitrogen in an organic compound by digesting with concentrated sulfuric acid and distilling the ammonia from the ammonium sulfate formed into a measured quantity of standard sulfuric acid.
Klaproth, Martin Heinrich (1743–1817) German chemist, who discovered uranium, titanium, and zirconium.
klaprotholite $Cu_6Bi_4S_9$. A native copper bismuth sulfide.
kleinite $HgCl_2 \cdot 3HgO$. A native mercury oxide and chloride.
Klein's liquid A saturated solution of cadmium borotungstate, d.3.28; used to separate minerals.
Klug, Aaron (1926–) Lithuanian-born British chemist. Nobel prize winner (1982). Noted for work on molecular biology.
klydonograph A device to record automatically a temporary excess voltage by a spark passing through a moving film. The resulting Lichtenberg diagram indicates the nature of the current.
klystron An electronic device for producing oscillations of frequency 300–400,000 MHz. A beam of electrons is shot through a cavity, and repelled back through it by a reflector electrode.
km Abbreviation for kilometer = 1,000 meters.
k.m.f. Abbreviation for the keratin-myosin-fibrinogen system, a major group of proteins.
knallgas The mixture of hydrogen, 2 vol., and oxygen, 1 vol., produced by electrolysis of water.
knap To break up, e.g., lumps of ore.
knead To work together a number of ingredients, usually by hand.
knock The nearly instantaneous and high-pressure explosion of a compressed mixture of fuel and air in an internal-combustion engine. Much of the energy is absorbed by the walls of the cylinder as radiant energy, thus reducing the available mechanical energy. **anti ~** A substance added to a liquid fuel which slightly retards the explosion and thereby reduces the energy wasted. Antiknock gasoline contains tetraethylplumbane. See *Ethyl gasoline.*
knock compound Antiknock. A substance which, when added to gasoline, reduces the knock.
Knop, Johann Ludwig Wilhelm (1817–1871) German agricultural chemist. **K.'s solution** A nutrient for plants: potassium nitrate 1, potassium dihydrogenphosphate 1, magnesium sulfate 1, calcium nitrate 4 pts., and a trace of ferric phosphate in 1,000 pts. water.
Knorr, Ludwig (1859–1921) German chemist, discoverer of

antipyrine. **K. alkalimeter** A device for determining carbon dioxide.

knot kn; the speed of 1 nautical mile/h = 0.51444 m/s.

knotgrass **English ~** The herb of *Polygonum aviculare* (Polygonaceae); an astringent. **Russian ~** The herb of *P. erectum;* an astringent.

knoxvillite $CrSO_4$. A native chromous sulfate.

kobold Early name for cobalt.

Koch, Robert (1843–1910) German bacteriologist, discoverer of the tubercle bacillus. **K.'s acid** 1-Naphthylamine-3,6,8-trisulfonic acid. **K. bacillus** Mycobacterium tuberculosis. The tuberculosis bacillus. **K. flask** A pear-shaped flask used for growing bacterial cultures.

koechlinite $Bi_2O_3 \cdot MoO_3$. A native molybdenum oxide.

kogi An enzyme used to produce *miso,* q.v.

kohl A mixture of carbon black and galena; used as a black pigment and cosmetic (eye shadow); of Eastern origin.

Kohlrausch, Friedrich (1840–1910) German physicist. **K. law** The conductivity of an electrolyte is the sum of the conductivities of its component ions when complete ionization occurs. **K. bridge** A Wheatstone bridge–type instrument to measure the conductivity of an electrolyte.

koilin A scleroprotein lining of birds' gizzards.

koilonychia Flattening and concavity of the nails, due to iron deficiency.

kojic acid $C_6H_6O_4$ = 142.1. 5-Hydroxy-2-hydroxymethyl-4*H*-pyran-4-one. m.153; formed from glucose by certain molds, e.g., *Aspergillus oryzae.* Cf. *ascorbic acid.*

kok-saghyz Russian dandelion from the Ukraine; the roots yield 10–20% of a rubber constituent.

kola Cola, kola nuts. The dried seeds of *Cola* species (Sterculiaceae) of Africa; it contains 2–3% caffeine. Used as a stimulant in beverages.

kolanin $C_{40}H_{56}O_{21}N_4$ = 928.9. A glucoside from kola, which hydrolyzes to caffeine, glucose, and kolared.

kolared $C_{14}H_{13}(OH)_5$ = 266.3. A red coloring matter from kola nuts.

kollagraph An instrument to measure the jointing capacity of a soldering system in terms of the surface and interfacial characteristics of the fluxed joint.

Kollidon Trademark for povidone.

kolm A radioactive asphaltic mineral: 0.45–0.027% Pb.

komplexon German for "complexone."

konel The alloy ferrotitanium 8, cobalt 17, nickel 73%; a platinum substitute for radiotube filaments.

koniogravimeter An instrument to determine dust in air.

Kopp, Hermann (1817–1892) German physical chemist. **K.'s law** Every element has the same specific heat in its solid free state as in its solid compounds. Thus, the molecular heat is the sum of the atomic heats of the component atoms.

Koppeschaar solution A 0.1 *N* bromine solution. Cf. *potassium bromate.*

koppite $(Ca,Ce,Fe,Hg,Na_2)O \cdot NbO_2 \cdot H_2O$. A vitreous, brown mineral.

koprosterol Coprosterol.

Koresin Trademark for a plastic made by the reaction of acetylene with *p-tert*-butylphenol in presence of zinc naphthenate. Used to render synthetic rubber tacky.

kornerupine A rare Indian aluminum magnesium borosilicate.

Koroseal Trademark for a vinyl chloride plastic; proof against water, but not against water vapor.

koryan oil Kaoliang oil. An oil from *Andropogon sorghum,* koryan corn (Manchukuo), d.0.926, iodine no. 121.

kosam seeds The fruits of *Brucea sumatrana* (Simarubaceae), China; an astringent.

kosin Koussin.

koso Kousso.

kosotoxin $C_{26}H_{34}O_{10}$ = 506.6. A yellow, amorphous principle from kousso.

Kossel **K. lines** Diffracted X-ray used in X-ray analysis. **K. press** A metal syringe for making sodium wire.

kotoin Cotoin.

Köttstorfer number Saponification number.

koumiss Kumyss. A beverage prepared by fermentation of mare's milk with kefir yeast.

kounidine $C_{21}H_{24}N_2O_5$ = 384.4 An alkaloid from Chinese *Gelsemium* species; produces muscular and respiratory weakness.

Kourbatoff's reagents Four etching agents for steel: A. 4% nitric acid in isopentyl alcohol. B. Equal parts of 20% hydrochloric acid in isopentyl alcohol and a saturated solution of nitroaniline in alcohol. C. Equal parts of 4% nitric acid in acetic acid, methanol, ethanol, and isopentyl alcohol. D. 3 pts. of saturated nitrophenol in alcohol and 1 pt. 4% nitric acid in alcohol.

koussein A yellow, anthelmintic, amorphous principle from kousso.

koussin $C_{25}H_{32}O_8$ = 460.5. Kosin. Yellow needles, m. 161. A resin from kousso; an anthelmintic.

kousso Cusso, brayera. The dried flowers of *Brayer anthelmintica (Hagenia Abyssinica),* Rosaceae; an anthelmintic. Cf. *brayerin.*

kovar A group of alloys (e.g., Fe 53.8, Ni 29, Co 17, Mn 0.2%) which show a sharp change in coefficient of expansion at certain temperatures; used for thermostats and to cement glass in vacuum apparatus.

Kr Symbol for krypton.

K radiation See under *K.*

kraft paper A strong (German; *kraft* = "strength") wrapping paper made from unbleached wood pulp prepared by the sulfate process.

krameria Rhatany root, payta. The dried root of *K. triandra,* Peru, or *K. argentia,* Brazil (Polygalaceae); an astringent (EP, BP).

krantzite A variety of retinite.

Kraut's reagent A microchemical reagent for ephedrine made by mixing *A* and *B* and diluting to 100 mL. *A*: 8 g bismuth nitrate in 20 mL concentrated nitric acid. *B*: 27.2 g potassium iodide in 50 mL water.

kreatine, kreatinine See *creatine, creatinine.*

Krebs cycle See *citric acid cycle.*

Kreis test A test for 3 C-chain, unsaturated aldehydes, characteristic of rancid fats. A red color develops in the aqueous layer after an ethereal solution of fat has been treated with hydrochloric acid and phloroglucinol solution.

kremersite $KCl \cdot NH_4Cl \cdot FeCl_3 \cdot 1\frac{1}{2}H_2O$. A native iron chloride.

krennerite (1) $(Ag, Au)Te_2$. A native telluride. (2) Bunsenite.

kreosol Cresol*.

kreotoxin A meat ptomaine formed by bacteria.

kresatin $MeCOOCH_2C_6H_4OH$ = 166.2 ***m*-~** Cresyl acetate. Colorless oil; an antiseptic.

krilium The sodium salt of hydrolyzed polyacrylonitrile; it stimulates the effect of humus in aggregating soil particles and improving soil structure.

krinosin $C_{38}H_{79}O_5N$(?). A lipid from brain substance.

krith Crith. The weight of 1 liter of hydrogen, at 0°C and 760 mm = 0.0899 g. A term formerly used to express the density of a gas with reference to hydrogen; thus the density of chlorine is 35.5 krith.

krocodylite Crocidolite.
krügite $MgSO_4 \cdot K_2SO_4 \cdot 4CaSO_4 \cdot 2H_2O$. A native sulfate.
Krupp's disease Fragility shown by steels after tempering and reheating.
kryogenin (1) See *cryogenin.* (2) $NH_2CO \cdot C_6H_4NH \cdot NH \cdot CONH_2$ = 194.2. Colorless, bitter powder, soluble in water; an antipyretic and antiseptic.
kryoscopy (1) Determination of the molecular weight of a substance from the lowering of the freezing point of its solutions. See *cryoscopy.* (2) In general, the production of low temperatures. (3) The study of the phenomena occurring during the cooling of molten alloys.
kryptidine $C_{11}H_{11}N$ = 157.2. 2,4-Dimethylquinoline. A homolog of quinoline.
krypto- Also *crypto-*. Hidden; invisible; latent.
kryptocyanine $C_{25}H_{25}N_2I$ = 480.4. 1,1′-Diethyl 4,4′-carbocyanine iodide. A sensitizer for the infrared; used for photography through haze.
kryptol A granular mixture of graphite, Carborundum, and clay; used as a resistance in electric furnaces.
krypton* Kr = 83.80. Colorless, inert, noble gas and element, at. no. 36, in the atmosphere (1 per 1,000,000); $d_{air=1}3.708$, m. −157, b. −153. Only compounds with fluorine are known.
kryptoxanthin Cryptoxanthin.
K.S.K. Ethyl iodoacetate.
Ku Symbol for the element name kurchatovium.
Kufasa Trademark for a cuprammonium synthetic fiber.
kuhseng A Chinese drug, the dried roots of *Sophora flavescens* (Leguminosae). Its chief alkaloid is matrine.
kukersite An Esthonian shale oil containing 43% organic matter (kerogen).
kukoline An alkaloid from *Cucullus diversifolius* (Asclepiadaceae).
kumyss Koumiss.
Kundt, August (1839–1894) German physicist. **K.'s constant** (1) See *Kundt's rule.* (2) A value obtained by dividing Verdet's constant by the magnetic susceptibility of the substance. **K. effect** The rotation of the plane of polarized light in certain vapors and gases under the influence of magnetic forces. **K.'s rule** An increase in refractive index produces a shift in the absorption bands of a solution toward the red, to an extent defined by K.'s constant. **K.'s tube** A horizontal tube containing a light powder. On sounding a note at the open end, the sand assumes heaps whose distance apart measures the velocity of sound in the gas and, hence, its specific heat.
Kunkel, Johann (1638–1703) German alchemist who published, under the pseudonym "Baron von Lowenstjern," a chemical textbook, *Laboratorium Chymicum.*
kunzite $LiAl(SiO_3)_2$. A native, pink variety of spodumen.
kupfernickel Niccolite.
kupferron Cupferron.
kupramite A preparation used in gas masks to absorb ammonia.
Kuralon, Kuravilon Trademarks for polyvinyl alcohol synthetic fibers.
kurchatovium Ku. Suggested name for element of atomic number 104. Cf. *Rutherfordium.*
kurchi The root of *Holarrhena antidysenterica* (Apocynaceae); a febrifuge and antidysenteric.
kurchine An alkaloid obtained from *kurchi,* q.v.
Kurehalon Trademark for a mixed-polymer synthetic fiber.
Kuremona Trademark for a polyvinyl alcohol synthetic fiber.
kurrajong oil A thick, red oil from the seeds of *Brachychiton populenum.*
Kurrol's salt $(KPO_3)_n$. Potassium polyphosphate, having ion-exchange properties. A long-chain metaphosphate made by slowly cooling and seeding molten potassium metaphosphate; mol. wt. exceeds 100,000, m.838.
kusamba A narcotic obtained by macerating opium with rose water.
kussin Koussin.
kuteera gum Hog gum.
kVAr Wattless component. Abbreviation for kilovar; equal to 1,000 volt-ampere reactive.
Kwells Trademark for scopolamine hydrobromide.
kW·h Abbreviation for kilowatthour.
kX A unit for the wavelength of X-rays. $kX \times 1.00202$ = 1 angstrom.
kyanite Cyanite.
kyanizing The preservation of wood by mercuric chloride solution.
kyanol Early name for aniline.
kymograph An instrument to record variations of blood pressure.
kynurenic acid $C_6H_4 \cdot C \cdot OH \cdot C \cdot COOH \cdot CH{:}N$ = 189.2. (ring closure between C_6H_4 and N)
4-Hydroxy-2-quinolinecarboxylic acid. Minor tautomer of 4(1*H*)-quinolinone-2-carboxylic acid.
kynuric acid $C_9H_7O_5N$ = 209.2. Oxalylanthranilic acid, carbostyrilic acid. Colorless crystals, m.288.
kynurine C_9H_7ON = 145.2. 4(1*H*)-Quinolinone. Colorless crystals, m.201, soluble in water.
kyrine A basic substance from the hydrolysis of proteins.
kytoplasm Cytoplasm.

L

L Symbol for liter. **L acid** 1-Naphthol-5-sulfonic acid. **L electrons** The electrons in the second shell of the atom. **L lines** (1) The group of lines following the *K* lines, q.v. (2) The Fraunhofer lines λ 3,820 to λ 3,858, due to iron. **L particles** Leptons. **L radiation** The second series of homogeneous X-rays characteristic of the metal used as anticathode, and analogous to *K* radiation, q.v. **L spectrum** The spectrum due to L radiation. Cf. *K* spectrum, *Moseley spectrum.*

L Indicating left-handed chirality in carbohydrates, amino acids, peptides, cyclitols. See *sequence rule.*

L Symbol for (1) Avogadro constant; (2) Lagrangian function; (3) a ligand (unspecified); (4) luminance; (5) orbital angular momentum quantum number; (6) radiance; (7) self-inductance.

L (Boldface.) Symbol for angular momentum.

l Symbol for (1) liter; (2) liquid state (as subscript or between parentheses).

l Symbol for: (1) length. (2) Formerly, levorotatory; now (−). Cf. *L*. (3) Mean free path.

Λ, λ See *lambda.*

La Symbol for lanthanum.

lab Rennin, lab ferment. The enzyme of rennet; it coagulates milk.

laballenic acid $Me(CH_2)_{10}CH{:}C{:}CH(CH_2)_3COOH = 277.2$. Octadeca-5,6-dienoic acid*. A constituent of seed oils, especially of *Leucascephalotes,* Spreng.

Labarraque, Antoine Germaine (1777–1850) French apothecary. **L. solution** A solution of sodium hypochlorite (minimum 2.4% available Cl); a disinfectant and deodorant.

labeling Rendering a substance identifiable by means of a radioactive isotope or other means (e.g., fluorescent material). Thus, ^{14}C is used to identify many organic compounds into which it is introduced. **specific ~** L. of one particular atom in a molecule; e.g., ^{14}C in $Me[^{14}C]OOH$. Compare this with the isotopically substituted compound $Me^{14}COOH$. **uniform ~** L. of all atoms of the same kind in a molecule; as, $[^{14}C]H_3[^{14}C]OOH$.

LaBel tube A distilling tube with 2 or more bulbs and return tubes, used for fractional distillation.

Labiate Mint family, aromatic herbs that yield important essential oils; e.g., herbs: *Mentha piperita,* peppermint; *Lavandula spica,* lavender; *Thymus serpyllum,* wild thyme; *Monarda punctata,* horsemint; leaves: *Salvia officinalis,* sage; *Rosmarinus officinalis,* rosemary; *Thymus vulgaris,* thyme; *Ocimum basilicum,* sweet basil; flowers: *Lavandula vera,* lavender; rhizomes: *Collinsonia canadensis,* stoneroot.

labile Relatively unstable. **l. acid** See *labile acid* under *acid.* **l. state** Temporary stability.

Labrador tea James tea, marsh tea, wild rosemary. The leaves of *Ledum palustre* (Ericaceae) used as tea in Labrador; an expectorant. It contains ledum camphor, ericinol, ericolin.

labradorite Saussurite. An iridescent lime-soda feldspar.

laburnine An alkaloid from *Cytisus laburnum* (Leguminosae). Cf. *cytisine.*

lac (1) Lacca, resina lacca, lacca gum. A resin exuding around twigs of *Croton* species (Euphorbiaceae) and *Ficus* species (Moraceae), southeast Asia; caused by the bite of the female insect *Coccus lacca.* (2) The bodies of *Coccus lac;* a red dye. (3) Milk, as, *lac sulfuris,* milk of sulfur. **alpha- ~** Suggested name for the insoluble portion of shellac. **seed ~** L. broken from the twigs. **shel ~** See *shellac.* Prepared by melting and straining seed l. **stick ~** The crude red l. as taken from the tree.

lacca See *lac, shellac.* **l. coerula** Litmus.

laccaic acid B $C_{24}H_{16}O_{12} = 496.4$. Laccainic acid. Red crystals in lac, decomp. 180.

laccase* An oxidoreductase enzyme, which oxidizes a benzenediol to a benzosemiquinone.

laccer(o)ic acid $C_{31}H_{63}COOH = 480.9$. Dotriacontanoic acid*. A fatty acid, m.95, in waxes. Cf. *tritriacontanoic acid.*

lachry- See *lacri-.*

lacmoid $C_{12}H_9O_4N = 231.2$. Resorcinol blue. Violet scales, soluble in water; an indicator: blue (pH 6.4) in alkaline, red (pH 4.4) in acid solution. **l. tincture** A solution of 0.5 g lacmoid in 100 mL each of water and alcohol; an indicator.

lacmus Litmus.

lacquer (1) A varnish, natural or containing shellac. (2) A solution of a resin which dries by evaporation of the solvent and leaves a protective covering. (3) A solution of a cellulose ester in a solvent. **acetate ~** A relatively nonflammable solution of cellulose acetate in carbon tetrachloride. **bronzing ~** A solution of nitrocellulose in pentyl acetate with suspended aluminum or bronze powder. **brush ~** A l. applicable with brushes. **Burmese ~** A natural l. exuding from the stems of *Melanorrhea usitata* (Anacardiaceae), which blackens on exposure. **cellulose ~** Lacquer (3). **Chinese ~** Japanese l. **collodion ~** See *collodion.* **dip ~** A l. for coating by immersion. **glyptal ~** A solution of the synthetic resins made from glycerol and organic acids. **Japanese ~** The resinous sap of *Rhus vernicifera* (Anacardiaceae), Japan and China; often colored by pigments and thinned with camphor oil or turpentine. **nitrocellulose ~** A l. containing cellulose nitrate in an organic solvent.
l. solvent Organic liquids which dissolve resins, gums, or nitrocellulose; used in the manufacture of lacquers, varnishes, and rayon. See *solvent.*

lacri- Prefix indicating "tears."

lacrimation Secretion of tears.

lacrimator(y) A tear-producing substance; as, ethyl iodoacetate. Cf. *dacryagogue.* **l. gas** See *tear gases.*

lacry- Lacri-.

lactalbumin The albumin of milk.

lactam* Lactan. An organic compound containing the −NH−CO− group in a ring formed by the elimination of water from −COOH and $-NH_2$ groups. Cf. *lactim.* **β-l. antibiotics** Penicillins and cephalosporins containing a 4-membered β-lactam ring. Bacteria often produce β-lactamase, destroying penicillin. See *clavulinic acid.*

lactamic acid Alanine*.

lactamide* $Me \cdot CHOH \cdot CONH_2 = 89.1$. Lactic acid amide, 2-hydroxypropanamide*. Colorless crystals, m.74, soluble in water.

lactamine Alanine*.

lactan Lactam*.

lactase β-D-Galactosidase*. An enzyme in intestinal juice, which hydrolyzes terminal, nonreducing β-D-galactose residues in β-D-galactosides; as, lactose to glucose and galactose.
lactate (1)* A salt of lactic acid containing the radical Me·CHOH·COO−; used as flour conditioners and food emulsifiers. (2) To produce milk.
lactation The secretion of milk.
lactazam A compound containing the −NH·NH·CO− group in its ring; as, 2-pyrazolone. Cf. *phenyllactazam.*
lactazone Lactoxime.
lactic acid* $C_3H_6O_3$ = 90.1. 2-Hydroxypropanoic acid*, ethylidenelactic acid, acid of milk, fermentation l. acid. A fermentation acid from milk or carbohydrates. Colorless liquid, d.1.240, m.18, b_{12mm}119, soluble in water. Used as a reagent to detect glucose and pyrogallol; in organic synthesis; in the leather, textile, and tanning industries; in pharmacy (USP, EP, BP); also to substitute citric and tartaric acids.

```
    COOH            COOH
     ⋮               ⋮
HO ▶ C ◀ H       H ▶ C ◀ OH
     ⋮               ⋮
     Me              Me
(S), L or levo   (R), D or dextro
```

dextro ~ L. acid produced in muscle tissues by metabolism of glycogen for energy, or by the action of *Micrococcus acidi paralactici.* A solid, m.26. **levo ~** L. acid formed by the action of *Bacillus acidi levolactica.* **para ~** See *dextro ~*. **(±)- ~** Occurs in potatoes, starch, molasses. m.17. **sarco ~** See *levo ~*.
l. acid series $C_nH_{2n}(OH)COOH$. Monobasic hydroxy acids; e.g.: carbonic acid, HO·COOH; lactic acid, hydroxypropionic acid, HO·C_2H_4·COOH.
lactic anhydride MeCH(OH)CO·O·HCMe(COOH) = 162.1. 2-Hydroxypropanoic anhydride. White powder, m.250, soluble in water.
lactide O·CO·MeCH·O·CO·CHMe = 144.1. Dilactide*. 3,6-Dimethyl-1,4-dioxacyclohexane-2,5-dione. Monoclinic crystals, m.128, slightly soluble in water. Cf. *lactides.*
lactides* R_2C·O·CO·CR_2·CO·O. Intermolecular cyclic esters formed by self-esterification from 2 or more molecules of a hydroxy acid. Cf. *lactone.*
lactim* Lactin. An organic compound containing the −N:C(OH)− group in its ring. Lactims are isomeric with lactams, but differ in that they are formed by the elimination of the 2 H of the NH_2 group and the O of the CO group:

```
  −NH           −N
   |            ||
  −CO    and  −C·OH
 Lactam        Lactim
(keto type)  (enol type)
```

lactin (1) Lactose. (2) Lactim*.
Lactobacillus A bacillus which causes lactic acid fermentation of milk; present in the intestine. ***L. bulgaricus*** Causes formation of yoghurt.
lactobiose Lactose.
lactochrome $C_6H_{18}O_6N$ = 200.2. A coloring matter isolated from milk.
lactocrit Lactokrit. An instrument to determine the amount of fat in milk.
Lactofil Trademark for a synthetic protein fiber.
lactoflavin Riboflavine.
lactoglobulin $C_{1864}H_{3012}N_{468}O_{576}S_{21}$(?). A protein from milk.
lactol The cyclic form of hydroxy aldehydes (aldolactols) and hydroxy ketones (ketolactols); as, γ-valerolactol.
lactolactic acid Dilactic acid.
lactolide* Names for *lactones*, q.v., based on the systematic name of the nonhydroxylated hydrocarbon with the same number of C atoms as the acid. As: 5-pentanolide CH_2·$(CH_2)_3$·CO (ring closed through O)
lactometer A hydrometer to determine the specific gravity of milk.
lactone* An anhydro cyclic ester produced by intramolecular condensation of a hydroxy acid with the elimination of water. **γ- ~** or **gamma- ~** the commonest and most stable type. The pentaatomic ring, R·CH·CO·O·CH_2·CH_2. The nomenclature of lactones is derived by adding (1) -olide to the hydrocarbon, or (2) -olactone to the acid; as, butanolide, or γ-butyrolactone. Cf. *lactide.* **δ- or delta- ~** Analog of γ-lactone, but based on a pyran rather than a furan ring. **l. isomerism** The shift from lactone to aldehyde acid; as,

$$C_6H_4\langle^{CO}_{CHOH}\rangle O \rightleftharpoons C_6H_4\langle^{COOH}_{CHO}$$

lactonic acid Galactonic acid.
lactonitrile Me·CHOH·CN = 71.1. 2-Hydroxypropanenitrile*. Colorless liquid, d.0.992, b.183.
lactoprene A copolymer of acrylic acid containing small proportions of other monomers. A rubber substitute which resists oils, oxygen, and aging.
lactose $C_{12}H_{22}O_{11}\cdot H_2O$ = 360.3. Milk sugar, sugar of milk, lactobiose, glucose galactoside. A disaccharide in human and cow milk. Colorless, rhombic crystals, d.1.525, m.203 (decomp.), slightly soluble in water. Used in pharmacy (for tablets), medicine (as nutrient) (USP, BP), and industry. **l. litmus agar** A culture medium similar to litmus milk *agar.*
lactoxime Lactazone. An unsaturated lactam containing the −CH=N−O−CO− group in the ring.
lactoyl The radical CH_3·CHOH·CO−, from lactic acid. **l. lactate** Dilactic acid. **l. urea** $C_4H_6O_2N_2\cdot H_2O$ = 132.1. Colorless rhombs, m.145, soluble in water.
lactucarium The dried juice of *Lactuca virosa*, wild lettuce (Compositae); a mild narcotic and sedative.
lactucerol $C_{30}H_{50}O$ = 426.7. α-Lactuc(er)ol, taraxasterol. An alcohol from *Lactuca* species, m.226.
lactulose A synthetic disaccharide which is not absorbed in the small intestine. It causes an osmotic catharsis.
lactyl The lactoyl radical.
ladanum Labdanum. **l. oil** An essential oil from the resin of *Cistus creticus* (Cistaceae), Mediterranean islands. Yellow oil, d.1.01, insoluble in water; used in perfumery.
Ladenburg, Albert (1842–1911) German chemist. **L. flask** A distilling flask, with several bulbs in its long neck, used for fractional distillation. **L. formula** See *benzene structure.* **L. law** The velocity of a photoelectron is proportional to the square root of the voltage exciting it.
ladle A vessel or pot with handle, for holding, transporting, and pouring molten metal.
læ- See words beginning with *lae-*, *le-*.
laetrile Amygdalin. A substance from apricot pits (stones) said by some to have anticancer properties. Can be metabolized to cyanide and cause poisoning.
laevo- Levo-.
laevulinic acid Levulinic acid.
laevulose Fructose.

lag To surround with insulation so as to keep heat in (or out). **magnetic ~** See *hysteresis.* **nitrogen ~** See *nitrogen lag* under *nitrogen.*

lagam balsam A balsam resembling copaiba.

lagoriolite $3Na_2O \cdot Al_2O_3 \cdot 3SiO_2$. A mineral of the garnet group.

lake An adsorption compound of a coloring matter with a metallic oxide (mordant) produced by coprecipitation. Lakes are usually bright in color and insoluble, but the same dye can give different shades with different mordants.

lakh Indian unit of 10^5. Cf. *crore.*

laking Hemolysis.

lamb Abbreviation for lambert.

Lamb, Arthur Becket (1880–1952) American chemist, noted for research in inorganic chemistry.

lambda Λ, λ. Greek letter. Λ is symbol for: (1) (Not italic) l. *particle* (below). (2) (Italic) molar conductance of an electrolyte. λ (italic) is symbol for (1) absolute activity; (2) decay constant; (3) linear expansivity; (4) mean free path; (5) microliter; (6) thermal conductivity; (7) wavelength. *In chemical names,* its superscript signifies the bonding number; i.e., the sum of the number of skeletal bonds and the number of H atoms associated with an atom in a parent compound. **l. particle** See *subatomic particle* under *particle.* **l. phenomenon** The rapid loss of entropy of helium at 2.2 K, which accompanies its acquisition of unusual properties in the He_{II} liquid state.

lambert Abbrev. lamb; unit of brightness of a perfectly diffusing surface, which radiates or reflects one lumen/cm². 1 lamb = 1 phot = 1 lumen/cm² = 0.3183 candle/cm² = 2.054 candles/in² = 929 footlamberts = 3,183.10 candelas/m² (the SI unit).

Lambert's law $I = I_0 e^{ad}$, where I_0 and I are the intensities of the incident and emergent radiation, respectively, of rays of light of wavelength λ; d is the thickness of the transmitting medium in cm; and a is the linear absorption coefficient of the medium for the wavelength λ.

Lambrecht's polymeter A meteorological instrument; a hygrometer and thermometer with humidity conversion scales.

lamb's wool A fabric made from the wool of sheep of up to 8 months old.

lamella (1) A medicated disk or wafer. (2) Lamina.

lamina A thin, flat plate or scale; as in mica.

Laminaria Tangle, sea girdle. A genus of *Phaeophyaceae* or algae of the brown seaweed type (Laminariaceae). Cf. *kelp.* **L. tents** The dried stalks of *L. digitata;* used in medicine to dilate cavities and orifices. They swell as they absorb water.

laminaribiose A glucose disaccharide derived from laminarin.

laminarin $(C_6H_{10}O_5)_n$. A β-D-glucose polymer in the form of a spiral chain of 1,3-glucopyranose units typical of polysaccharides in seaweeds. It comprises approximately 33% of the fronds of brown seaweed and certain algae.

laminated Split, or in thin layers.

lamine An alkaloid from the blossoms of *Lamium album,* dead nettle (Labiatae).

laminography The radiography of a thin layer of a thick specimen. The radiation source and recording film are rotated, during exposure, around the plane of the specimen under investigation.

lamp A device for obtaining light or heat. **alcohol ~** A vessel with wick, for burning alcohol. **arc ~** See *arc lamp.* **blast ~** See *blast burner* under *burner.* **carcel ~** See *carcel unit.* **electric ~** A device for transforming an electric current into light. **filament ~** A device for heating a thread of carbon, tantalum, or tungsten to incandescence in an evacuated glass bulb. **gas-filled ~** (1) Vacuum l. (2) A filament l. which contains a small quantity of gas, e.g., N_2 or A. **halide ~** See *halide lamp.* **Harcourt ~** A l. to burn a definite quantity of pentane at a definite rate; a standard of brightness. **Hefner ~** A standard l. used in photometry. **mercury ~** An evacuated quartz envelope containing mercury vapor. On the passage of an electric current it emits an intense bluish light rich in ultraviolet rays; used in spectroscopy, photography, and fluorescence analysis. **microscope ~** A point source of light for microscopes. **miner's ~** See *miner's lamp.* **neon ~** A vacuum type l. containing a trace of neon. **Nernst ~** An electric l. with metal oxide filament, which becomes a conductor when heated, and emits infrared rays on passage of an electric current. **pentane ~** Harcourt l. **quartz ~** Mercury l. **spectrum ~** A device for coloring a nonluminous flame with vapors, sprays, or solid particles; used in spectroscopy of metals. **vacuum ~** A vacuum tube, variously shaped, through which passes an electric current. Different residual gases in the tubes produce different colors.

lampblack Carbon black. Paris black. Finely divided carbon obtained by burning gas or oil under a slowly rotating metal cylinder; a paint, ink, and paper pigment. Cf. *soot.*

Lamprene Trademark for clofazimine.

lamprobolite A basaltic hornblende.

lana Wool. Flannel.

lanain, lanalin Lanolin.

lanarkite $PbSO_4 \cdot PbCO_3$, derived from $PbSO_4 \cdot PbO$. A native lead sulfate.

lanatoside See *digilanid.* **l. C** $C_{49}H_{76}O_{20} = 985.1$. Cedilanid, Celanide. White crystals, insoluble in water. A cardiac glycoside, less cumulative than digoxin; used similarly.

lanaurin A porphin-type coloring matter in suint.

Landmark process The synthesis of ammonia by passing steam over TiN_2 and carbon, and the recovery of TiN_2:

$$(1)\ 2TiN_2 + 3C + 3H_2O \rightarrow 2TiN + 3CO + 2NH_3$$
$$(2)\ N_2 + 2TiN \rightarrow 2TiN_2$$

Landolt, Hans Heinrich (1831–1910) Swiss physical chemist, noted for work on optical refraction.

land plaster Gypsum

Landsberger apparatus An apparatus to determine molecular weights from the rise in boiling point of a solution.

landsbergite Ag_3Hg_4. Argental. A native silver amalgam.

langbeinite $K_2SO_4 \cdot 2MgSO_4$. A native sulfate.

Lange solution A colloidal solution of gold.

Langevin, Paul (1872–1946) French physical chemist. **L. formula** The activity coefficient of ions is: $\alpha = \Lambda_0/\kappa$, where Λ_0 is the equivalent conductivity and κ the relative permittivity per volt.

langley A unit for the timing of exposure tests on fabrics or plastics; 1 calorie of solar radiant energy/cm² of exposed area = 41,840 J/m².

Langmuir, Irving (1881–1957) American physical chemist noted for experiments on atomic hydrogen, low-pressure and high-vacuum reactions, and development of theories of isosterism and atomic structure. Nobel prize winner (1932). Cf. *Lewis*-Langmuir theory. **L.-Blodgett film** Monolayer coating produced by pulling a solid through water on whose surface is an organic liquid with both hydrophilic and hydrophobic ends. **L. theory** (1) Electrons occupy positions in imaginary shells around the atom. See *orbit, electron configuration.* (2) In the solid state, secondary valencies supplant the effect of primary valences.

lanigerin $C_{17}H_{14}O_5$ = 298.3. A red coloring matter from insects of the Coccidae group. Cf. *cochineal.*

lanital An artificial wool formerly produced from casein. Cf. *Aralac.*

lanoceric acid $C_{30}H_{60}O_4$ = 484.8. A dibasic fatty acid, m.105, from wool grease.

lanolin Lanum, lanain, lanalin, lanesin, lanichol, laniol, adeps lanae hydrous, hydrous wool fat. Used as an external excipient and in cosmetics. **anhydrous ~** Adeps lanae, wool fat. Yellow fat from sheep's wool, consisting of cholesterol esters of higher fatty acids, m.38–42, soluble in alcohol. Used in pharmacy (USP, BP).

Lanon Trademark for a polyester synthetic fiber.

lanopalmic acid $C_{15}H_{30}O_3$ = 258.4. A fatty acid from wool grease.

lanosterol $C_{30}H_{50}O$ = 426.7. Isocholesterol. Lanostadienol. A triterpenoid. An unsaturated sterol closely related to cholesterol, m.140 (approximately 25% in wool wax).

Lansil Trademark for a continuous-filament acetate yarn.

lantanine An alkaloid from *Lantana brasiliensis* (Verbenaceae), Brazil; an antipyretic. Cf. *yerba sagrada.*

lanthanides Lanthanoids*.

lanthanite $La_2(CO_3)_3 \cdot 9H_2O$. A dull, infusible mineral.

lanthanoids* Lanthanides. The series of elements La through Lu, at. nos. 57–71. Cf. *rare-earth metals.*

lanthanum* La = 138.9055. A rare-earth metal, valency 3, at. no. 57, discovered (1839) by Mosander. A gray, lustrous metal, d.6.15, m.921, b. 3465, in rare-earth minerals (cerite, samarskite, lanthanite, and gadolinite). It decomposes in cold water slowly. Forms pyrophoric alloys and is a catalyst in ashing biological material. **l. acetate*** $La(CH_3COO)_3 \cdot 1\frac{1}{2}H_2O$ = 343.1. White crystals, soluble in water. **l. carbonate*** $La_2(CO_3)_3 \cdot 8H_2O$ = 602.0. White crystals, insoluble in water. **l. chloride*** $LaCl_3 \cdot 7H_2O$ = 371.4. White, triclinic crystals, soluble in water. **l. nitrate*** $La(NO_3)_3 \cdot 6H_2O$ = 433.0. White, deliquescent crystals, m.40, soluble in water. **l. oxalate*** $La_2(C_2O_4)_3$ = 541.9. White crystals, insoluble in water. **l. oxide*** La_2O_3 = 325.8. Infusible, white powder, d.6.41, insoluble in water. **l. sulfate*** $La_2(SO_4)_3 \cdot 9H_2O$ = 728.1. Colorless needles, d.2.81, decomp. by heat, slightly soluble in water. **l. sulfide*** La_2S_3 = 374.0. Yellow powder, d.4.997, m.2300–2350, insoluble in water.

lanthionine* $S(CH_2 \cdot CH \cdot NH_2 \cdot COOH)_2$ = 208.2. From wool, hair, and certain antibiotics, e.g., subtilin. White crystals, decomp. 294, slightly soluble in water.

lanthopine $C_{23}H_{25}O_4N$ = 379.5. Lantol. A colorless, crystalline alkaloid of opium, m.200.

lantol Lanthopine.

Lanum Trademark for lanolin.

Lanusa Trademark for a viscose synthetic fiber.

Lanvis Trademark for thioguanine.

lapacho **l. bark** The bark of *Avicennia nitida,* black mangrove; a tan. **l. wood** The heartwood of *A. tomentosa* (Verbenaceae), the tropics.

lapachoic acid Lapachol.

lapachol $C_{15}H_{14}O_3$ = 242.3. Lapachoic acid, targusic acid, 2-hydroxy-(3-methyl-2-butenyl)-1,4-naphthoquinone, tecomin. m.141. Occurs widely in plants, such as lapacho wood.

laparoscopy Inspection of the inside of the abdominal cavity via a laparoscope (an instrument with fiber-optic illumination) for diagnosis and some surgical procedures.

lapis (1) Latin for "stone." (2) An alchemical term for a nonvolatile substance. **l. albus** A native calcium hexafluorosilicate. **l. amiridis** Emery. **l. calaminaris** Calamine. **l. causticus** Fused sodium or potassium hydroxide. **l. imperalis** Silver nitrate. **l. lazuli** (1) A mixture of minerals, usually calcite, pyrite, lazulite, sodalite, and häyenite; the ancient source of ultramarine. (2) A native copper silicate. Blue, green, or purple compact masses; a semiprecious stone. **l. lunaris** Fused silver nitrate.

Laplace transform Technique for changing differential equations into a normal algebraic form. Cf. *z-transform.*

lappa Burdock, clotbur, bardana. The dried roots of *Arctium lappa* (Compositae); a diuretic and diaphoretic.

lappaconitine $C_{32}H_{44}O_8N_2$ = 584.7. An alkaloid of *Aconitum septentrionale* (Ranunculaceae), m.223.

larch The bark of *Larix europaea (Pinus larix),* Coniferae; an astringent. **American ~, black ~** Tamarac. **European ~** Larch.O

l. agaric Agaric. **l. extract** An extract of the bark of *Pinus larix;* a tan.

lard Adeps. The purified abdominal fat from the pig (excluding unbleached pig grease). White, granular mass, m. not below 25, insoluble in water; used in cooking and pharmaceutical preparations. **l. benzoinated** Adeps benzoinatus. L. containing 1% benzoin; an antiseptic. **l. oil** A colorless oil, chiefly triglycerides of oleic acid, pressed from l., d.0.915, soluble in alcohol; a lubricant.

lardacein A wax, d.0.6969, m.55, from the scale of *Ceroplastens rubens,* an insect from tea and citrus trees.

lardine A substitute for lard prepared by hydrogenating cottonseed oil.

Largactil Trademark for chlorpromazine hydrochloride.

large calorie See *large calorie* under *calorie.*

Larix A genus of Coniferae yielding timber, oils, and resins; e.g., *L. americana,* tamarac; *L. europaea,* larch.

larixine Maltol.

larkspur Delphinium.

Larmor, Sir Joseph (1857–1942) British physicist. **L. theorem** The angular velocity of an electrified particle moving at right angles to the direction of a magnetic field of constant strength is independent of the linear speed of the particle.

larvicide An agent that destroys larva, the first stage of development of insects.

Lasché unit A unit of diastatic activity: the time in minutes required by an infusion of 0.55 g malt to convert 5 g soluble starch in 100 mL at 62.5°C.

laser Soft X-ray. Abbreviation for *l*ight *a*mplification by *s*timulated *e*mission of *r*adiation. A beam of light of uniform wavelength, traveling in phase in a series of parallel lines. It can be focused accurately to produce a high-energy concentration. On exposure for short intervals to ordinary light, ruby emits l. Used to study giant molecules and crystal structure. Used in medicine to coagulate small areas of tissue without affecting intervening structures; as, in retina surgery. Cf. *maser.*

Lasix Trademark for furosemide.

latensification The intensification of a latent photographic image by preexposure; due to the fact that most photographic materials disobey the Bunsen-Roscoe law of photochemical equivalence.

latent Not manifest or readily available. **l. energy, l. force** Potential energy. **l. heat** The amount of heat required to change the state of a body from solid to liquid at its melting point, or from liquid to gas at its boiling point. Likewise the heat liberated if a gas changes to a liquid, or a liquid to a solid, at the condensation or freezing point, respectively. **molar l. h.** The l. h. per mole. **specific l. h. of fusion*** The number of joules required to convert 1 gram of a substance from solid into liquid, without a change of temperature. For water it is 331 J. **specific l. h. of vaporization*** The number of joules required to convert 1 gram of substance from liquid into vapor without a change of temperature. For water at 100°C, it is 2,257 J. **l. image** The change produced on a photographic plate by exposure to light, made visible by developing.

lateral chain See *lateral chain* under *chain.*

laterite Murram. A natural hydrated oxide of iron and aluminum, and sometimes titanium (cf. *bauxite*), which occurs as a brown surface earth and is used for building roads.

lateritiin $C_{36}H_{63}O_9N_3$ = 681.9. Enniatin A. An antibiotic pigment produced by the growth of *Fusaria,* m.121.

latex The milky juice or exudation of plants obtained by tapping the trunk. **rubber ~** A colloidal emulsion, d.0.983, pH 7.0–7.2, containing: water 60, rubber 36, sterols 1, proteins 0.3, quebracho 1.5, sugars 0.25, ash 0.5%. Cf. *lutoid, rubber.*

lather A foam or froth of soap and water.

lat. ht. Abbreviation for latent heat.

latices Plural of *latex.*

lattice See *space lattice.*

laudanidine $C_{20}H_{25}NO_4$ = 343.3. Tritopine, laudanine. **(−)-~** m.184. **(+)-~** m.181. **(±)-~** m.166.

laudanine Laudanidine.

laudanosine $C_{21}H_{27}O_4N$ = 357.5. *N*-Methyltetrahydropapaverine. An alkaloid from opium. Yellow crystals, m.89, insoluble in water.

laudanum Tincture of opium. A brown solution of opium in alcohol; a powerful anodyne, formerly used as a sedative and hypnotic.

Laue, Max Theodor Felix von (1880–1960) German physicist, noted for research on subatomic phenomena. Nobel prize winner (1914). **L. diagram** L. pattern. **L. equation** An expression of the conditions to be obeyed simultaneously for a crystal to show diffraction. **L. method** The diffraction of X-rays by means of a crystal. **L. pattern** The photographic record produced when X-rays are diffracted by a crystal.

laughing gas Dinitrogen oxide*.

laumonite $CaOAl_2O_3 \cdot 4SiO_2 \cdot 2H_2O$. A vitreous mineral. Cf. *lawsonite.*

Lauraceae Laurel family: aromatic shrubs or trees, all parts of which yield an essential oil; e.g.: *Cinnamomum camphora,* camphor; *C. cassia,* cassia; *Laurus nobilis,* sweet bay.

lauraldehyde* $C_{12}H_{24}O$ = 184.3. Dodecanal*, lauric aldehyde. Colorless leaflets, m.44, insoluble in water.

laurane $C_{20}H_{42}$ = 282.6. An isomer of icosane in laurel fat, m.69.

laurel Laurus, sweet bay, bayberry, noble berry. The dried leaves and berries of *Laurus nobilis,* used for making laurel oil. **cherry ~** See *prune.* **ground ~** Arbutus. **mountain ~** Kalmia. **New Zealand ~** See *pukateine.*

l. berries oil An essential oil from the fruits of l., soluble in alcohol, and containing pinene, eucalyptol, eugenol, and estragole. **l. camphor** Camphor. **l. fat** Bay oil, l. oil. A buttery fat, m.32–36, obtained by pressing l. leaves; used in veterinary work. **l. leaves oil.** Bay leaves oil. Yellow oil distilled from laurel, d.0.924, insoluble in water; a flavoring. **l. oil** (1) L. fat. (2) L. leaves oil. **l. tallow** Undung. **l. wax** (1) Myristin. (2) Bayberry wax.

laureline $C_{19}H_{19}O_3N$ = 309.4. An apomorphine alkaloid from *Laurelia novae zealandicae* (Monimiaceae).

laurene Pinene*.

Laurent, Auguste (1807–1853) French organic chemist, discoverer of aniline, caffeine, naphthalene, and their derivatives. Noted for his theory of types and radicals. **L. acid** 1-Naphthol-5-sulfonic acid.

lauric acid* $Me(CH_2)_{10}COOH$ = 200.3. Decylacetic acid, dodecanoic acid*. Colorless needles, m.47, insoluble in water. **12-hydroxy ~** Sabinic acid.

lauric aldehyde Lauraldehyde*.

laurin $C_3H_5 \cdot (Me(CH_2)_{10}O_2)_3$ = 603.0. Trilaurin, laurostearin, glycerol trilaurate; in the seeds of laurel, palms, coconuts. White needles, m.47; used in medicinal soaps.

laurite RuS_2. A native sulfide.

lauroleic acid $C_{12}H_{22}O_2$ = 198.3. (Z)-9-Dodecenoic acid*. An unsaturated acid, $b_{2mm}90$, from the head oil of the sperm whale.

laurolene $CMe{:}CMe \cdot CHMe \cdot CH_2 \cdot CH_2$ (ring closed) = 110.2. 1,2,3-Trimethyl-1-cyclopentene*. Colorless liquid, d.0.800, b.121. **iso ~** $CMe_2 \cdot CMe{:}CH \cdot CH_2 \cdot CH_2$ (ring closed). 1,1,2-Trimethyl-2-cyclopentene*. Colorless liquid, d.0.791, b.109.

laurone $(C_{11}H_{23})_2CO$ = 338.6. 12-Tricosanone*. Colorless crystals, d.0.789, m.69.

lauronolic acid $C_9H_{14}O_2$ = 154.2. 1,2,3-Trimethyl-2-cyclopentene-1-carboxylic acid. Laurolenic acid. **(*R*)-~** m.13.

laurostearin Laurin.

laurotetanine $C_{19}H_{21}O_4N$ = 327.4. A strychninelike alkaloid from the bark of *Litsea latifolia* (Lauraceae), S. Asia. Cf. *glaucine.*

lauroyl* Indicating the $Me(CH_2)_{10}CO-$ radical.

Laurus Laurel.

lauryl The radicals (1) dodecanoyl*; (2) dodecyl*; (3) lauroyl*. **l. alcohol** Dodecyl alcohol*.

laurylene (1) $C_{12}H_{24}$ = 168.3. An unsaturated hydrocarbon in mineral oils. (2) Pinene*.

lautal A hard aluminum alloy containing Cu 4–5, Si 1.5–2, Fe 0.4–0.7%.

Lauth's violet Thionine.

lava Liquid rock which has reached the earth's surface and is ejected from volcanoes. Cf. *magma.*

lavandula Lavender.

lavender The dried flowering tops of *Lavandula latifolia;* a carminative. **l. flower oil** Colorless essential oil from lavender flowers, d.0.885, insoluble in water; used in perfumery (NF, BP). **l. spike oil** An essential oil distilled from the herb *L. officinalis.* Colorless liquid, d.0.905, insoluble in water; used in pharmaceutical preparations.

Lavoisier, Antoine Laurent (1743–1794) French chemist, regarded as the founder of modern chemistry by virtue of his study of the role of oxygen in combustion. He formulated the theory of the conservation of matter, and laid the basis for chemical nomenclature. He was unjustly accused and executed during the French revolution.

law A generalized statement of facts or principles. **empirical ~** A l. which is the result of experience, without theoretical considerations. **natural ~** The formulation of systematized experience on the workings of nature. **periodic ~** (1) A l. that expresses a periodic variation. (2) See *periodic system.*

lawang oil An oil from the bark of a southeast Asian plant, resembling mace oil.

Law cell An electric cell (1.37 volts): a zinc anode, and carbon cathode in a 15% solution of ammonium chloride.

lawrencium* Lr. Element of at. no. 103. At. wt. of isotope with longest half-life (^{260}Lr) = 260.11. The natural element decayed out of existence shortly after the birth of the universe. Formed by bombarding californium with ^{10}B nuclei in a heavy-ion linear accelerator by Ghiorso and coworkers, University of California (1961). Gives Lr^{3+} ions.

lawsone $C_{10}H_6O_3$ = 174.2. 2-Hydroxy-1,4-naphthoquinone.

The coloring matter of henna leaves, *Lawsonia alba.* Red needles, m.194.

lawsonite $CaO \cdot Al_2O_3 \cdot 2SiO_2 \cdot 2H_2O$. A vitreous, colorless mineral. Cf. *laumonite.*

laxative A mild cathartic; as, bisacodyl.

layer A mass of uniform thickness covering an area. **molecular ~** A film of single molecules alongside one another over a surface. Cf. *zone.*

lazulite $(MgFeCa) \cdot (AlOH)_2P_2O_8$. A blue, vitreous mineral.

lazurite $2NaAlSiO_4 \cdot Na_2S$. A mineral from Ovalle, Chile.

lb Abbreviation for pound(s). **lb ap** Abbreviation for apothecary's pound. **lb av** Abbreviation for avoirdupois pound. **lbf** Abbreviation for pound-force. **lb t** Abbreviation for troy pound.

LB film See *Langmuir.*

LC Abbreviation for "lethal concentration."

LCAO Abbreviation for "linear combination of atomic *orbitals.*"

LCD Abbreviation for "liquid crystal display."

LD Abbreviation for "lethal dose."

LD 50 See *median lethal dose* under *dose.*

ld- Indicating a racemic compound; now (±)-.

LDPE Abbreviation for "low-density *polyethylene.*"

leaching Washing or extracting the soluble constituents from insoluble materials.

lead* Pb = 207.2. Plumbum*. A metallic element, at. no. 82. A gray, soft, monoclinic or regular crystalline metal, d.11.34, m.327, b.1740, insoluble in water, soluble in nitric acid; principal ore, galena. L. was known to the Romans (plumbum nigrum). Used in electrotechnics, as a solder, and in low-melting-point alloys. It forms 3 series of compounds and has valencies of 2 [l.(II)*, l.(2+)*, plumbum(II)*, plumbous] and 4 [l.(IV)*, l.(4+)*, plumbum(IV)*, plumbic]. L. salts are poisonous. Metallic l. consists mainly of ^{208}Pb (52%), ^{206}Pb (24%) and ^{207}Pb (22%). See *radioactive elements.* **actinium ~** The isotope Pb-207. **antimonial ~** See *antimonial lead.* **argentiferous ~** (1) PbAgS. (2) An alloy of silver and lead obtained in the cupellation process. **black ~** Graphite. **brown ~** Vanadinite. **horn ~** Phosgenite. **pyrophoric ~** Finely divided l. which oxidizes spontaneously on exposure to air. **radio ~** The natural radioactive mixture of l. isotopes. **red ~** L. tetraoxide. **Siberian red ~** Crocoisite. **secondary ~** L. recovered from scrap. **thorium ~** The isotope Pb-208. **toxic ~** (1) U.S. pollution limit is 1.5 $\mu g/m^3$ air (3-month av.) based on keeping children's blood level below 30 $\mu g/dl$. Electroencephalogram differences have been detected in children down to blood Pb levels of 7 $\mu g/dl$. (2) U.K. limit (1986) is 0.15 mg Pb/m^3 air (8-h av.), but 0.10 mg/m^3 for Et_4Pb. Temporary suspension from work if blood level greater than 80 $\mu g/100$ ml, or 40 $\mu g/100$ ml for pregnant women. In gasoline, a limit of 0.15 g/l. See *paint.* **uranium ~** (1) The isotope Pb-206. (2) A conduit or vein which is being followed. **white ~** L. carbonate*.

l. accumulator A reversible voltaic storage cell (2.2 volts): an anode of l. and a cathode of l. dioxide, suspended in sulfuric acid (d.1.1). **l. acetate*** $Pb(C_2H_3O_2)_2 \cdot 3H_2O$ = 379.3. L. sugar, sugar of l., plumbous acetate. Colorless, efflorescent crystals, m.75, soluble in water. Used as a reagent, in manufacture of l. salts, as a mordant in dyeing, in the tanning industry, and as an astringent lotion for bruises (BP). **basic ~** L. subacetate. **l. alkyls** Organic compounds of tetravalent l. and alkyl radicals; as, tetraethylplumbane, Et_4Pb. Some are antiknock compounds. **l. alloys** Usually have low melting points. Cf. *type* metal, *solder, carbox metal, pewter, linotype metal.* **l. antimonate*** $Pb_3(SbO_4)_2$ = 993.1. Naples yellow, plumbous stibnate, giallioline. Orange powder, insoluble in water; a pigment for paint, ceramics, and glass. **l. arsenate*** $Pb_3(AsO_4)_2$ = 899.4. Colorless, poisonous crystals, insoluble in water. **l. aryls** Organic compounds of tetravalent l. and aryl radicals; as, tetraphenylplumbane, Ph_4Pb. **l. azide*** $Pb(N_3)_2$ = 291.2. L. azoimide, l. nitride. Colorless crystals made from sodium azide with l. acetate, explodes 350; a detonating agent. **l. azoimide** L. azide*. **l. benzoate*** $Pb(C_7H_5O_2)_2$ = 449.4. Colorless crystals, soluble in water. **l. bioxide** L. dioxide*. **l. black** Graphite. **l. borate** $Pb(BO_2)_2 \cdot H_2O$ = 310.8. L. drier. White powder, insoluble in water; a paint drier. **l. borosilicate** A mixture of l. borate and l. silicate; used in optical glass. **l. bromate*** $Pb(BrO_3)_2$ = 463.0. Colorless crystals, slightly soluble in hot water. **l. bromide*** $PbBr_2$ = 367.0. Colorless rhombs, m.380, insoluble in water. **l. butyrate*** $Pb(C_4H_7O_2)_2$ = 381.4. Colorless scales, m.90, insoluble in water. **l. carbonate*** $PbCO_3$ = 267.2 White powder, decomp. by heat, insoluble in water; a pigment. Occurs naturally as cerrusite. Cf. *Dutch process.* **basic ~** $2PbCO_3 \cdot Pb(OH)_2$ = 775.6. Plumbous subcarbonate, basic l. carbonate, white l., ceruse, cerussa, l. flakes. White, heavy powder, decomp. by heat, insoluble in water; a pigment. **l. chamber** See *sulfuric acid manufacture.* **l. chlorate*** $Pb(ClO_3)_2$ = 374.1. White, monoclinic crystals, decomp. 230, soluble in water. Cf. *Thiel-Stoll solution.* **l. chloride*** $PbCl_2$ = 278.1. Colorless rhombs, m.501, insoluble in water; a reagent for silver, alkaloids, and carbonates; and, mixed with l. oxides, a pigment. **l. chloride oxide** (1) $PbCl_2 \cdot PbO$ = 501.3. White tetragons, insoluble in water. (2) $PbCl_2 \cdot 2PbO$ = 708.5. Brown solid, m.693. **l. chromate*** $PbCrO_4$ = 323.2. Leipzig yellow, paris yellow, "chrome" yellow. Fusible, yellow, monoclinic crystals, insoluble in water; a reagent and pigment. **basic ~** $PbCrO_4 \cdot PbO$ = 546.4. Plumbous subchromate. Red crystals, insoluble in water; a pigment. **l. citrate*** $Pb_3(C_6H_5O_7)_2$ = 999.8. Colorless crystals, soluble in water. **l. cyanate*** $Pb(OCN)_2$ = 291.2. Colorless crystals, decomp. by heat, insoluble in water. **l. cyanide*** $Pb(CN)_2$ = 259.2. Colorless crystals, insoluble in water, soluble in potassium cyanide solution; used in metallurgy. **l. dichromate*** $PbCr_2O_7$ = 423.2. L. bichromate. Red powder, insoluble in water; a pigment. **l. dioxide*** L. oxide (6). **l. dish** A container for corrosive liquids and hydrofluoric acid; as, in glass etching. **l. dithiofuroate** $(C_4H_3O \cdot CS_2)_2Pb$ = 493.6. Furac III. Red powder; a vulcanizing accelerator. **l. dithionate*** PbS_2O_6 = 367.3. Colorless crystals, soluble in water. **l. drier** L. borate. **l. ethyl sulfate** $Pb(C_2H_5SO_4)_2 \cdot H_2O$ = 475.4. Colorless liquid, soluble in water; used in organic synthesis. **l. ferrocyanide** L. hexacyanoferrate(II)*. **l. flake** L. carbonate basic. **l. fluoride*** PbF_6 = 321.2. Fusible, white powder, insoluble in water. **l. fluo(ro)silicate** L. hexafluorosilicate*. **l. flux** A reducing and desulfurizing agent used in the assay of gold and silver, and consisting of sodium hydrogencarbonate 16, potassium carbonate 16, flour 8, borax glass 4 pts. **l. formate*** $Pb(CHO_2)_2$ = 297.2. Plumbous formate. Lustrous, rhombic needles, d.4.63, decomp. 190, soluble in water; a reagent. **l. gleth** Lead oxide (2) **l. hexacyanoferrate(II)*** $Pb_2[Fe(CN)_6]$ = 626.4. Yellow powder, insoluble in water. **l. hexafluorosilicate*** $PbSiF_6$ = 349.3. L. fluorosilicate, l. silicofluoride. **monohydrate ~** (H_2O). Colorless crystals, soluble in water. **dihydrate ~** $(2H_2O)$. White, monoclinic prisms, soluble in water. **tetrahydrate ~** $(4H_2O)$. Monoclinic crystals, soluble in water. **l. hydrate** See *l. hydroxide* (next). **l. hydroxide** (1) $Pb(OH)_2$ = 241.2. L. hydrate. White powder, m.145 (decomp.), soluble in water. (2) $2PbO \cdot H_2O$ = 464.4. White, cubic crystals. **l. iodate*** $Pb(IO_3)_2$ = 557.0. Colorless crystals, insoluble in water. **l. iodide*** PbI_2 = 461.0. Plumbi iodidium. Yellow hexagons, m.375, insoluble in water. Used for bronzes, mosaic gold, printing, and photography. **l. lactate*** $Pb(C_3H_5O_3)_2$ = 385.3. White powder, soluble in water. **l. laurate*** $(C_{12}H_{23}O_2)_2Pb$ = 605.8. White powder, m.105, insoluble in water. **l. linoleate** $Pb(C_{18}H_{31}O_2)_2$ = 766.1. Yellow paste, insoluble in water; used in varnishes. **l. malate*** $Pb(C_4H_4O_5)\cdot 3H_2O$ = 393.3. Plumbous malate. White powder, slightly soluble in water. **l. minerals** See *galena, massicot, cerussite, anglesite, crocoisite, wulfenite.* Cf. *brongniardite, carminite, caryinite, lanarkite, plagionite, zinkenite.* **l. molybdate*** $PbMoO_4$ = 367.1. Yellow powder, insoluble in water; a reagent. **l. monosulfide*** L. sulfide (1). **l. monoxide*** L. oxide (yellow). See also *massicot.* **l. myristate** $(C_{14}H_{27}O_2)_2Pb$ = 661.9. White powder, m.109, insoluble in water. **l. naphthalenesulfonate*** $Pb(C_{10}H_7SO_3)_2$ = 621.6. White crystals, insoluble in water. **l. nitrate*** $Pb(NO_3)_2$ = 331.2. Colorless octahedra, decomp. 223, soluble in water. Used as a reagent, a mordant in the dye and textile industries, an oxidizing agent in organic synthesis, a sensitizer in photography and used in the manufacture of l. salts, matches, and pyrotechnics. **l. nitride** L. azide*. **l. nitrite*** $Pb(NO_2)_2$ = 299.2. Yellow crystals, soluble in acids. **l. oleate*** $Pb(C_{18}H_{33}O_2)_2$ = 770.1. White, fatty granules, insoluble in water; used in varnishes. **l. oxalate*** PbC_2O_4 = 295.2. White crystals, decomp. 300, insoluble in water. **l. oxide** (1) *L. suboxide*, Pb_2O = 430.4. Black powder, decomp. by heat, insoluble in water. (2) *Yellow monoxide*, PbO = 223.2. Litharge, massicot, l. protoxide. Yellow rhombs, m.888, insoluble in water. Used as a reagent, pigment, and rubber filler; in ointments and manufacture of glass and pottery; and for acid-resisting cements and putty. (3) *Red monoxide*, PbO = 223.2. Plumbous oxide. Red hexagons, insoluble in water. (4) *L. sesquioxide*, Pb_2O_3 = 462.4. L. trioxide, plumbous plumbite, l. metaplumbate. Red powder; a pigment. Decomposed by acids into PbO + PbO_2. (5) *L. oxide red*, Pb_3O_4 = 685.6. Plumbous plumbate, l. orthoplumbate, red l. oxide, minium, sandix, orthoplumbate, l. tetraoxide*, plumbous-plumbic oxide. Heavy, orange powder, d.9.07, decomp. 500, insoluble in water. Used as a reagent and pigment, for cements, and in ceramics. (6) *L. dioxide**, PbO_2 = 239.2. Plumbous peroxide, brown l. oxide, l. superoxide. Brown hexagons, d.8.91, decomp. by heat, insoluble in water; an oxidizing agent. It forms salts, PbX_4, and plumbates. **l. palmitate*** $(C_{16}H_{31}O_2)_2Pb$ = 718.0. White powder, m.112, insoluble in water. **l. pentasulfide*** See *l. sulfide* below. **l. peroxide** L. oxide (6). **l. peroxonitrate** Colorless, hygroscopic crystals, soluble in water. **l. phosphate** (1) $Pb_3(PO_4)_2$ = 811.5. L. orthophosphate*. White powder, insoluble in water. (2) $Pb(PO_3)_2$ = 365.1. L. metaphosphate*. White crystals. (3) $Pb_2P_2O_7$. L. pyrophosphate*, q.v. (4) $PbHPO_4$ = 303.2. L. hydrogenphosphate. White crystals. **l. phosphinate*** $Pb(H_2PO_2)_2$ = 337.2. L. hypophosphite. White powder, decomp. by heat, insoluble in water. **l. phosphonate** $PbHPO_3$ = 287.2. White powder, decomp. by heat, insoluble in water. **l. picrate*** A very sensitive explosive. **l. plumbate l. orthoplumbate** L. oxide (5). **l. metaplumbate** L. oxide (4). **l. poisoning** Plumbism, saturnism. Anemia, paralysis, mental changes, and colic produced by frequent contact with l. See *toxic lead* above. **l. propionate*** $Pb(C_3H_5O_2)_2$ = 353.3. Colorless crystals, soluble in water. **l. protoxide** L. oxide (2). **l. pyrophosphate*** $Pb_2P_2O_7\cdot H_2O$ = 606.4. Plumbous pyrophosphate. Colorless rhombs, decomp. by heat, insoluble in water. **l. salicylate*** $Pb(C_7H_5O_3)_2\cdot H_2O$ = 499.4. White crystals, soluble in water. **l. selenate*** $PbSeO_4$ = 350.2. White powder, insoluble in water. **l. selenide*** PbSe = 286.2. Brown powder. **l. sesquioxide** L. oxide (4). **l. silicate*** $PbSiO_3$ = 283.3. White crystals, insoluble in water; used in ceramics, l. glass, paints, enamels and for fireproofing fabrics. **l. stearate*** $Pb(C_{18}H_{35}O_2)_2$ = 774.1. Yellow fat, m.116, insoluble in water; used in paint driers and varnishes. **l. styphnate** The l. salt of trinitroresorcinol; used in primings for gun ammunition. **l. subacetate** $2Pb(C_2H_3O_2)_2\cdot Pb(OH)_2$ = 891.8. Basic acetate of l. White powder, soluble in water. **l. subacetate solution** Colorless liquid (189.5 g/l), d.1.24; a reagent and clarifier. Cf. *l. water* (below), *Goulard's extract.* **l. subcarbonate** L. carbonate, basic. **l. suboxide** L. oxide (1). **l. sugar** L. acetate*. **l. sulfate*** $PbSO_4$ = 303.3. White rhombs, m.1170, insoluble in water; a pigment. **acidic ~** $Pb(HSO_4)_2\cdot H_2O$ = 419.3. Crystals, decomp. by heat. **basic ~** $PbSO_4\cdot PbO$ = 526.5. White powder, insoluble in water, decomp. by heat. **l. sulfide** (1)* PbS = 239.3. Galena, galenite. Dark-gray crystals, m.1120, insoluble in water; used in ceramics and as a decolorizer. (2) PbS_5 = 367.5. L. pentasulfide*. An unstable yellow powder. **l. sulfite** $PbSO_3$ = 287.3. White granules, insoluble in water. **l. superoxide** L. oxide (6). **l. tartrate*** $PbC_4H_4O_6$ = 355.3. White powder, insoluble in water. **l. tetrachloride*** $PbCl_4$ = 349.0. Plumbic chloride. Yellow oil, $d_0$3.18, decomp. by water and at 105. **l. tetraethide, l. tetraethyl** Tetraethyl*plumbane**. **l. thiocyanate*** $Pb(SCN)_2$ = 323.4. L. rhodanate. Yellow, monoclinic crystals, soluble in water. **l. thiosulfate*** PbS_2O_3 = 319.3. L. hyposulfite. White powder, decomp. with age, insoluble in water. **l. tree** The branching crystalline growth of l. precipitated from its solution by more positive metals. **l. tungstate*** $PbWO_4$ = 455.0. L. wolframate*, raspite, stolzite. Yellow powder, insoluble in water; a pigment. **l. vanadate*** $Pb(VO_3)_2$ = 405.1. Yellow powder, insoluble in water; a pigment. **l. water** A 1% solution of l. subacetate. **l. white** L. carbonate, basic. **l. wire** (1) A wire made from l. (2) A chromel-alum or other alloy wire of a thermocouple of a pyrometer. **l. wolframate*** L. tungstate*. **l. yellow** L. chromate*.

Leadate Trademark for dimethyl dithiocarbamate, a rubber accelerator.

leadhillite $Pb(OH)_2\cdot PbSO_4\cdot 2PbCO_3$. A native sulfate of lead, from Leadhill, Scotland.

leaf (1) See *leaves.* (2) A thin metal sheet or foil, as, gold l.; used for ornamenting.

lean Deficient. Cf. *fat.* **l. clay** A clay of poor plasticity. **l. coal** A low-gas-content coal. **l. ore** A low-metal-content ore.

least squares Approach used in *regression analysis,* q.v., that minimizes the sum of the squares of the deviations of individual values from the resulting line of best fit.

leather A tanned skin. **mountain ~** Paligorskite. **l. tankage** A fertilizer made from l. scraps by digestion with steam, drying, and grinding.

leaves The stem appendages of a plant, containing chlorophyll and possessing respiratory openings. Some are used in pharmacy; e.g., belladonna, digitalis.

LaBel, Joseph Achille (1847–1930) Alsatian who discovered the asymmetric carbon atom, (independently of van't Hoff) in 1874; prepared the first optically active compound of asymmetric nitrogen.

LeBlanc, Nicolas (1742–1806) French chemist, and founder of the alkali industry. **L. soda process** The manufacture of sodium carbonate by treatment of salt cake (Na_2SO_4) with carbon and limestone.

lecanoric acid $C_{16}H_{14}O_7$ = 318.3. Diorsellinic acid. Colorless crystals, m.175, decomp. by hot water to orsellic acid. **l. a. monomethyl ether** Evernic acid.

LeChâtelier, Henry Louis (1850–1936) French chemist,

noted for the law of chemical equilibrium and work on metallurgy.

lecithin Monoaminomonophospholipid. A group of substances of general composition:

$$\begin{array}{l} CH_2\cdot OOCFA \\ | \\ CH\cdot OOCFA \\ | \\ CH_2O\cdot PO_4\cdot N\cdot R_3 \end{array}$$

FA is a fatty acid; R an alkyl radical. Lecithins are the esters of oleic, stearic, palmitic, or other fatty acids with glycerolphosphoric acid and choline. Cf. *kephalin.* **animal ~** Contains P 3.9–4.0, N 1.8–2.0%. **plant ~** Contains P 3.3–3.7, N 1.5–1.7%. **soya ~** L. derived from soybean; equal parts of true l. and kephalin.

Leclanché cell A voltaic cell (1.46 volts); an anode of amalgamated zinc, and a cathode of carbon suspended in a solution of ammonium chloride, with manganese dioxide as a depolarizer.

lectins Proteins from plant seeds that agglutinate erythrocytes. Used to study the structure and functions of animal surface membranes.

Ledermycin Trademark for demeclocycline.

leditannic acid $C_{15}H_{20}O_8$ = 328.3. A tannin from *Ledum* species (Ericaceae).

ledixanthin $C_{33}H_{34}O_{13}$ = 638.6. A red coloring matter from *Ledum* species.

Leduc effect Thermomagnetic difference of temperature: the effect of a magnetic field on the distribution of heat.

ledum camphor A stearoptene from *Ledum palustre,* wild rosemary (Ericaceae). Cf. *Labrador tea.*

lees The albuminoid sediment of fermented liquids, e.g., wine.

Leeuwenhoek, Anton van (1632–1723) A lens maker of Delft, Holland; discovered bacteria microscopically (1675).

legume The pod of a leguminous plant.

legumelin An albumin from peas and beans.

legumin A globulin from leguminous plants.

Leguminosae Pulse family: a group of plants with edible seeds, rich in protein; some contain drugs. Roots: *Glycyrrhiza glabra,* Spanish licorice; barks: *Acacia mimosa,* mimosa bark; woods: *Haematoxylon campechianum,* logwood; leaves: *Cassia acutifolia,* senna; herbs: *Megicago sativa,* alfalfa; fruits: *Cassia fistula,* purging cassia; seeds: *Arachis hypogoea,* peanut; plant products: *Acacia senegal,* gum arabic.

lehrbachite HgPbSe. A mercury ore.

Leibnitz, Gottfried Wilhelm Freiherr von (1646–1716) German mathematician and philosopher.

Leipzig yellow Lead chromate pigment.

leishmaniasis Diseases due to protozoan (genus *Leishmania*) infections, e.g., kala azar.

Leloir, Luis Federico (1906–) French-born Argentine chemist. Nobel prize winner (1970), noted for work on sugar nucleotides.

Lémery, Nicolas (1645–1715) French physician and chemist, noted for his chemical textbook.

lemon The ripe fruit of *Citrus limonum* (Rutaceae). Cf. *lime.* **l. chrome** Barium chromate. **l.grass** See *lemongrass oil.* **l. juice** Lime juice. **l. oil** Oleum limonis, oil of l. A fragrant, yellow oil, from fresh l. peel, d.0.849–0.855, $[\alpha]_D$ +57–65, insoluble in water; a flavoring (NF, BP). **dried l. peel** Cortex limonis. The outer rind of fresh ripe fruit of *Citrus medica,* lemon (Rutaceae); a flavoring (BP). **l. yellow** (1) Lead chromate. (2) Barium chromate.

lemongrass oil Indian mellisa oil. Yellow oil from *Cymbopogon* or *Andropogon citratus,* a grass of southeast Asia; d.0.89, insoluble in water; a flavoring and perfume. It contains citral and methyl heptenone. Cf. *citronella oil, verbena oil.*

lenacil* See *herbicides,* Table 42 on p. 281.

Lenard, Philipp (1862–1947) German physicist. Nobel prize winner (1905). **L. rays** Long, blue streamers produced when cathode rays penetrate thin sheets of aluminum or gold.

length The shortest distance along a defined path, measured in the SI system by the meter. For relative magnitudes of length, see Table 50, opposite.

lens A piece of glass or other transparent material with one or both faces curved, which converges or diffuses light. The principal focus of a l. = $1/f_1 + 1/f_2$, where f_1 and f_2 are conjugate focal distances, or the distances of the object and of the image. **achromatic ~** A l. corrected for chromatic aberration to bring different spectral rays to one focus. **aplanatic ~** A l. corrected for spherical aberration to bring rays into focus in the same plane. **apochromatic ~** A system of lenses which corrects for chromatic and spherical aberration by bringing the 3 principal spectral regions to one focus. **biconcave ~** A l. with a concave surface on each side:) (. **biconvex ~** A l. with a convex surface on each side: ( ). **bifocal ~** A double l., each portion of which has its own focus. **compound ~** Two or more lenses combined in series. **concave ~** A l. in which the center is thinner than the periphery:) |. **contact ~** L. placed below the eyelid, generally made from materials based on polyhydroxyethylmethacrylate. **convex ~** A l. in which the center is thicker than the periphery: (|. **convex-concave ~** A l. with both a concave and a convex surface:)). **electron ~** See *electron optics.* **planoconcave ~** A l. with both a plane and concave surface: | (. **planoconvex ~** A l. with both a plane and convex surface: (|. **quartz ~** A l. made of quartz, used with ultraviolet rays, to which glass is relatively opaque. **rocksalt ~** A l. made of salt, which permits the passage of many ultraviolet rays to which glass and quartz are opaque.

lenticular Having the shape of a lentil or lens.

lentine $C_6H_4(NH_2)_2\cdot 2HCl$ = 181.1. *m*-Diaminobenzene hydrochloride. Colorless crystals which turn pink with age; used as a reagent for nitrites in water.

leonardite A naturally oxidized lignitic coal.

leonite $MgSO_4\cdot K_4SO_4\cdot 4H_2O$. A Stassfurt salt.

leontin Caulosaponin.

leonurus Motherwort.

lepargylic acid Azelaic acid.

lepathinic acid $C_{20}H_{18}O_{14}$ = 482.4. A crystalline principle from the roots of *Rumex* species (Polygoniaceae).

lepidene $C\cdot Ph(CPh)_3\cdot O$ = 372.5 Tetraphenylfuran. Colorless crystals, m. 175.

lepidine $C_{10}H_9N$ = 143.2. 4-Methylquinoline, cincholepidine. An alkaloid from cinchona bark. Colorless oil, d. 1.086, b.266, slightly soluble in water; used in organic synthesis. **aminophenyl ~** Flavaniline. **oxo ~** Dibenzoyl*stilbene.*

lepidocrocite FeO·OH. A scaly form of goethite, but having a different X-ray structure.

lepidolite $(Li,Na,K)_2Al_2(SiO_3)_3(F,OH)_2$. Lithium mica.

lepidometer An instrument to measure the scaliness of animal fibers in terms of the maximum tension developed when the fibers are suspended, roots downward, and rubbed between rubber surfaces.

lepidone $C_{10}H_9ON$ = 159.2. 4-Hydroxylepidine, methyl-2(1*H*)-quinolinone. Colorless crystals, m.223, slightly soluble in water.

TABLE 50. VARIOUS MAGNITUDES OF LENGTH

Object or distance	Magnitude, units	Magnitude, meters
	Light years	
Einstein universe	2,000,000,000	2.0×10^{25}
Distance to farthest spiral nebula	140,000,000	1.4×10^{24}
Milky Way diameter	50,000	5.0×10^{20}
Distance to Sirius from sun	8.9	8.9×10^{16}
	Kilometers	
Solar system diameter	10,000,000,000	1×10^{13}
Distance to sun from Earth	150,000,000	1.5×10^{11}
Sun radius	696,000	7.0×10^{8}
Distance to moon from Earth	384,000	3.8×10^{8}
Earth circumference at equator	40,080	4×10^{7}
Earth radius	6,380	6.4×10^{6}
Moon radius	1,740	1.7×10^{6}
	Meters	
Mount Everest, elevation	8,840	8.8×10^{3}
Large tanker, length	310	3.1×10^{2}
Lowest sound, wavelength	16	1.6×10^{1}
	Centimeters	
Man, average height	172	1.72
	Millimeters	
Highest sound, wavelength	17	1.7×10^{-2}
	Micrometers	
Smallest visible particle	50	5×10^{-5}
Cells of Drosophila	7–25	2.5×10^{-5}
Coccus bacteria	2	2×10^{-6}
	Nanometers	
Red blood cell	800	8×10^{-7}
Red light, wavelength	770	7.7×10^{-7}
Smallest microscopic particle	100	1×10^{-7}
Short ultraviolet wavelength	13	1.3×10^{-8}
Thickness of oil film	5	5×10^{-9}
Ultramicroscopic particle	4	4×10^{-9}
Molecules	0.2–5	5×10^{-9}
Gas molecule, distance	1	1×10^{-9}
	Picometers	
Atoms, interbond length	100–600	6×10^{-10}
Electron orbit in hydrogen	53	5.3×10^{-11}
Hard X-ray wave	19	1.9×10^{-11}
Electron diameter	0.038	3.8×10^{-15}
Nucleus of gold atom	0.0004	4×10^{-17}
Nucleus of hydrogen atom	0.00002	2×10^{-18}

leptandra Culver's root. The dried rhizomes and roots of *Veronica (Leptandra) virginica* (Schrophylariaceae); a cathartic.

leptandrin Yellow crystals from the roots of *Veronica (Leptandra) virginica;* purgative glucoside.

leptandroid An extract containing the combined principles of *Leptandra;* a laxative.

leptokurtosis Describing a *histogram,* q.v., which is more peaked in the middle and rather longer in the tail than corresponds with a normal probability curve.

leptology The study of the fine structure of matter.

leptometer A device to compare the viscosity of 2 liquids simultaneously.

leptons Together with quarks, believed to be the only elementary *particles,* q.v., of matter.

leptospirosis A disease of humans and animals, caused by the organism leptospira, producing liver damage and meningitis.

lerp A variety of manna found on the Australian shrub, mallee, *Eucalyptus dumosa.*

Lessing, Rudolf (1878–1964) English fuel chemist. **L. ring** Contact ring: a 6-mm metal tube, split and bent; used as packing material in gas-absorption towers and fractionating columns.

lethal Fatal or deadly. **l. dose** The minimum quantity of a substance that causes death. **l. gas** Hydrocyanic acid.

letterpress Printing by direct impression from inked projecting type.

Leu* Symbol for leucine.

leucacene $C_{54}H_{32}$ = 680.8. A pyrolytic product of acenaphthene containing 12 hexatomic and 4 pentatomic rings. Silky crystals, m.250, insoluble in water.
leucanea Leucophala. A S. American plant; used as a fuel, food, and fertilizer. A leguminosae having a high N-fixation capacity.
leucaniline $CH(C_6H_4 \cdot NH_2)_3$ = 289.4. Tris(4-aminophenyl)methane, methenyltrianiline; 3 isomers exist. ***para-*** **~** Colorless leaflets, m.148, insoluble in water; used in organic synthesis. Cf. *leucoaniline.* **tetramethyl ~** See *tetramethyl leucaniline* under *tetramethyl.*
leucaurin $CH(C_6H_4OH)_3$ = 292.3. Tris(4-hydroxyphenyl)-methane*, triphenylolmethane, derived from aurin. ***o*-methyl ~** Leucorosolic acid.
leucic acid Leucinic acid.
leucicidin A toxin formed in cultures of *Staphylococci*, very similar in properties to the lysins.
leucine* $C_6H_{13}O_2N$ = 131.2. Leu*. 2-Amino-4-methylpentanoic acid*. A constituent of many proteins (27% in zein). Colorless leaflets, decomp. 283, slightly soluble in water; used in organic synthesis; 2 optically active isomers exist. Cf. *leucyl.* **iso ~*** Et·CHMe·CHNH$_2$·COOH. Ile*. 2-Amino-3-methylpentanoic acid*. **D- ~** Greasy leaflets, m.283. **(±)- ~** Rhombic or monoclinic plates, m.292, slightly soluble in water. **nor ~** See *norleucine.*
leuc(in)ic acid $C_5H_{10}(OH)COOH$ = 132.2. 2-Hydroxy-4-methylpentanoic acid*; m.77 (racemic), m.82 (active), soluble in alcohol.
leucinimide $C_{12}H_{22}O_2N_2$ = 226.3. The imide of leucylleucine.
Leucippus of Elea A contemporary of Zeno, Empedocles, and Anaxagoras (about 450 B.C.); the founder of the ancient Greek atomistic theory, later developed by Democritus.
leucite $K_2O \cdot Al_2O_3, 1 \cdot 4SiO_2$. A vitreous, white or gray, crystalline mineral in lava. Cf. *Lucite.*
leuco- Prefix meaning "white," and indicating the presence of a triphenylmethane group.
leucoaniline $(NH_2C_6H_4)_2 \cdot CH \cdot C_6H_3(NH_2)Me$ = 303.4. Colorless crystals, m.100, insoluble in water; used in organic synthesis. Cf. *leucaniline.*
leuco bases Leuko bases. A group of colorless derivatives of triphenylmethane produced by the reduction of dyes which, on oxidation, are reconverted into the dye.
leucocyte Leukocyte. White blood cell (corpuscle).
leucocytosis An increase in the number of white blood cells in circulating blood.
leucodendron Kienabossie, langbeen. The plant *L. concinnum* (Proteaceae), S. Africa; malaria remedy.
leucodrin $C_{15}H_{16}O_8$ = 324.3. Proteacin. A crystalline, bitter principle, m.212, from leucodendron.
leucoglycodrin $C_{27}H_{44}O_{10}$ = 528.6. An amorphous, white glucoside from leucodrin.
leucoindigo Indigo white.
leucoline Quinoline*.
leucomaclurin glycol The parent substance of the catechins, from kauri gum. It forms acacin and anacardin by removal of water.
leucomaines A group of poisonous nitrogen compounds, formed in animal tissues by metabolic activities; generally of the uric acid or creatinine type.
leucomalachite green $Ph \cdot CH \cdot (C_6H_4NMe_2)_2$ = 330.5. 4,4′-Bis(dimethylamino)triphenylmethane. Colorless crystals, m.93, insoluble in water, which give malachite green on oxidation.
leuconic acid $CO \cdot (CO)_3 \cdot CO$ = 140.1. Pentaoxocyclopentane*. Colorless needles, produced by strong oxidation of croconic acid.
leucopyrites $FeAs_2$. A silvery mineral.
leucorosolic acid $C_{20}H_{19}O_3$ = 307.4. *o*-Methylleucarin. Colorless crystals, from rosolic acid.
leucoscope (1) An optical pyrometer. (2) A photometer for the study of colored light.
leucosin An albumin from wheat, rye, and barley embryos.
leucotaxine A polypeptide of low molecular weight formed by the action of vesicants on the skin. It produces increased permeability and accumulation of leucocytes.
leukotrienes Biologically active compounds formed from polyunsaturated fatty acids; as arachidonic acid.
leucotrope $C_{15}H_{18}NCl$ = 247.8. Benzyldimethylphenyl-ammonium chloride. m.116. Used in dyeing.
leucyl* 2-Amino-4-methyl-1-oxopentyl†. The radical $C_4H_9 \cdot CH(NH_2) \cdot CO-$, from leucine. **iso ~*** The radical $MeCH_2 \cdot CHMe \cdot CH(NH_2) \cdot CO-$, from isoleucine.
leukemia A malignant disease of the blood produced inter alia by exposure to radiation. It is characterized by excessive production of white cells in the bone marrow, increased numbers of abnormal such cells in circulation, and anemia.
Leukeran Trademark for chlorambucil.
leuko- Leuco-.
leukomaines Leucomaines.
Leukomycin Trademark for chloramphenicol.
leukonic acid Leuconic acid.
leukotine $C_{21}H_{20}O_6$ = 368.4. A crystalline substance in paracoto bark.
leunaphos A fertilizer: diammonium phosphate and ammonium sulfate (N 20, P_2O_5 15%).
leunaphoska A fertilizer: diammonium phosphate, ammonium sulfate, and potassium chloride (N 13, P_2O_5 10, K_2O 13%).
leuna saltpeter A fertilizer from Leuna, Germany: ammonium sulfate 165, ammonium nitrate 100 pts.
levallorphan tartrate $C_{19}H_{25}ON \cdot C_4H_6O_6$ = 433.5. *N*-Allyl-3-hydroxymorphinan tartrate. White, bitter crystals, m.176 soluble in water; an antagonist to morphia (BP).
levan A soluble polysaccharide in the leaves of certain grasses, made up of repeating units of fructofuranose.
Levant wormseed Santonica.
levarterenol bitartrate Norepinephrine bitartrate.
leveler A liquid added to a lacquer solvent to adjust volatility and/or viscosity.
leveling bulb A glass vessel connected by rubber tubing to a buret; used in gas analysis to equalize the pressure in both containers.
leverierite $Al_2O_3 \cdot 2SiO_2 \cdot nH_2O$. A pearly, brown clay, resembling kaolin.
levigation The reduction of a substance to a powder by grinding in water, followed by fractional sedimentation, to separate the coarse particles.
levo-, laevo- Prefix meaning "toward the left"; as, levorotatory.
levodopa $C_9H_{11}O_4N$ = 197.2. (*S*)-3-(3,4-Dihydroxyphenyl)-L-alanine. Dopa, *l*-dopa. Larodopa, Dopar. Levorotatory. Used in Parkinson's disease and said to replace the low levels of dopamine in the brain (USP, EP, BP).
levogyric Levorotatory.
Levol's alloy The alloy: Ag 71.9, Cu 28.1%.
levonorgestrel The (−) form of norgestrel and more active than it. The progestogen component of some oral contraceptives.
Levophed Trademark for norepinephrine bitartrate.
levorotatory Describing an optically active compound that

rotates the plane of polarized light to the left (counterclockwise). Indicated by the prefix (−)-, or *l*-.
levorphanol tartrate $C_{17}H_{23}ON \cdot C_4H_6O_6 \cdot 2H_2O = 443.5$. (−)-3-Hydroxy-*N*-methylmorphinan. Levo-Dromoran, Dromovan. White crystals, m.116, sparingly soluble in water. A potent analgesic (USP, BP).
Levothyroid Trademark for levothyroxine sodium.
levothyroxine sodium $C_{15}H_{10}O_4NI_4Na \cdot nH_2O = 798.9$. Eltroxin, Levothyroid, Synthyroid. Yellowish powder, slightly soluble in water. A hormone for the thyroid gland; used to treat hypothyroidism (USP, EP, BP).
levulic acid Levulinic acid.
levulin Synanthrose.
levulinamide $MeCOCH_2CH_2CONH_2 = 115.1$. Colorless crystals, m.107.
levulinic l. acid $MeCOCH_2CH_2 \cdot COOH = 116.1$. Levulic acid, 3-acetylpropionic acid. Colorless leaflets, m.33, soluble in water. **homo ~** $C_6H_{10}O_3 = 130.1$. 4-Oxohexanoic acid. m.40. **methyl ~** Homolevulinic acid.
l. aldehyde $C_5H_8O_2 = 100.1$. 3-Acetylpropionaldehyde. Colorless liquid, d.1.016, decomp. 187, soluble in water. **l. hydrazide** $MeCO \cdot CH_2 \cdot CH_2 \cdot CO \cdot NHNH_2 = 130.1$. Colorless crystals, m.82. **l. imine** $MeC(:NH)CH_2 \cdot CH_2 \cdot COOH = 115.1$. Colorless crystals, m.95.
levulosans See *fructosans.*
levulose Fructose.
levymite $Ca(Al_2Si_4O_{12}) \cdot 6H_2O$. A zeolitic ion-exchange material.
Lewis, Gilbert Newton (1875–1946) American chemist, noted for thermodynamic theories applied to chemistry and atomic structure. **L. acid** See *Lewis acid* under *acid.* **L.'s atom** See *atomic structure.* **L. base.** See *Lewis acid* under *acid.* **L. color theory** Color is produced by the absorption of certain rays by those electrons of a molecule which vibrate with the same frequency. **L.-Langmuir theory** The atom is built up of successive shells which hold 2, 8, 18, 32, 18, and 8 electrons as their maximum capacities. **L. symbols** Bonds illustrated by indicating electrons as dots. Cf. *octet, formula, bond.* **L. theory** A chemical bond is *polar* when an electron passes from one atom to another; *nonpolar* when 2 atoms share a pair of electrons equally.
lewisite British anti ~ $C_3H_8OS_2 = 124.2$. Dithiol (U.S. usage). BAL. (±)-3-Hydroxy-1,2-propanedithiol. An antidote for l. and other metal poisons, e.g., bismuth, mercury, gold.
l. I $ClH:CH \cdot AsCl_2 = 207.3$. (*E*)-2-Chlorovinyldichloroarsine. An irritant liquid, d.1.89, b.190 (decomp.); a vesicant poison. Cf. *leucite, Lucite.* **l. II** $(ClCH:CH)_2AsCl$.
Lexan Trademark for a thermoplastic polycarbonate condensation product of bisphenol-A and phosgene.
ley (1) The mixture of salts 10–20 and glycerol 6–8% formed by saponification of crude fats by sodium hydroxide in soap manufacture. (2) Describing a system of farming of crops in rotation, interspersed with ploughing. Cf. *lye.*
Leyden jar See *Leyden jar* under *jar.*
li. Abbreviation for link.
Li Symbol for lithium.
liatris Deer's-tongue. The dried leaves of *Liatris odoratissima* (Compositae); used for flavoring and in tobacco.
Libavius, Andreas (1540–1616) German alchemical writer and pioneer in blowpipe analysis.
liberation The act of setting free, as the formation of carbon dioxide from chalk.
libration A real or apparent oscillating motion. Usually applied to the movement of the moon relative to the earth.
Librium Trademark for chlordiazepoxide hydrochloride.
licanic acid A gelling fatty acid in drying oils used for paints.
lichen Algae and fungi which live symbiotically, i.e., 2 primitive plants, one with, the other without, chlorophyll, which live together; e.g., Iceland moss. They yield coloring matter (litmus, orchil, zearin), acids (e.g., orsellic acid), carbohydrates, and depsides. **l. starch** Lichenin. **l. sugar** Erythritol.
licheniformin An antibiotic from *Bacillus licheniformis.*
lichenin $(C_6H_{10}O_5)_n = (162.1)_n$. [$n$ = 80–160.] Lichen starch, moss starch. A β-glucopyranose carbohydrate derived from Iceland moss, *Cetraria islandicus,* which is digested by invertebrates only. White powder, soluble in hot water, m.10.
lichenol $C_9H_9O_4 \cdot C_2H_5 = 210.2$. The ethyl ester of everninic acid, from the oil of oak moss, *Evernia prunastri.* Cf. *sparassol.*
lichesteric acid $C_{18}H_{31}O_2 \cdot COOH = 324.5$. m.125, insoluble in water. From Iceland moss.
Lichtenberg figures The pattern formed by an electric spark passing through a thin layer of insulator, as, sulfur. Cf. *klydonograph.*
licorice Glycyrrhizia (U.S. usage). Cf. *liquorice.* **Indian ~** Abrus. **Spanish ~** Glycyrrhiza.
lidocaine hydrochloride $C_{14}H_{22}ON_2 \cdot HCl \cdot H_2O = 288.8$. Lignocaine hydrochloride. Xylocaine. White, bitter crystals, m.77, soluble in water. A local anesthetic, and given intravenously to control heart rhythm after myocardial infarction (USP, EP, BP).
Lieben solution A solution of iodine in potassium iodide.
Liebermann, Carl (1842–1914) German chemist noted for the synthesis of alizarin (with Gräbe). **L. reaction** Sodium nitrite in concentrated sulfuric acid gives a brown color, changing to blue, in presence of a phenol. The mixture poured into water gives a red solution, which changes to blue on addition of alkali (formation of *p*-nitrosophenol).
Liebig, Justus Freiherr von (1803–1873) German chemist, founder of agricultural chemistry. **L. condenser** A glass tube, surrounded by a wider tube through which water circulates. **L. extract** A meat extract used as a nutrient for making biological bouillon. **L. potash bulb** A triangularly bent glass tube with 2 or more bulbs filled with potassium hydroxide; used in gas analysis.
Liesegang, Raphael Edward (1869–1947) German chemist. **L. rings** A periodic precipitation, formed as bands in gelatin, by the gradual diffusion toward one another of 2 mutually precipitating ions.
life (1) The vital force: the principle underlying the phenomena of organized beings. It depends on the *protoplasm,* which exercises the function of *metabolism, growth,* reproduction, *adaptation,* and *evolution,* q.v. (2) A synonym for "time" or "time period"; as: **damping ~** See *damping period.* **half- ~** See *half-life.* (3) Colloquially, the period of usefulness of a machine or other inanimate object.
l. elements The *bioelements,* q.v., necessary for an organism. **l. everlasting** The dried herb of *Gnaphalium obtusifolium (Antennaria dioica),* Compositae. **l.root** Senecio.
ligancy* The number of neighboring atoms bonded to an atom, irrespective of the nature of the bonds.
ligand A group of atoms around a central atom in a complex compound; e.g., CN^- and F^- are the ligands in $[Fe(CN)_6]^{4-}$ and $[SiF_6]^{2-}$, respectively. See *radicals,* Table 70 on p. 494. **l. field theory** The use of light to study the effect on the energy levels of a metal ion when a l. approaches it to form a complex.
ligase* See *enzymes,* Table 30.
ligasoid A disperse colloidal system, consisting of a liquid phase suspended in a gaseous phase; as, a fog.

ligating See *coordination compound.*

light (1) A form of radiant energy. Cf. *radiation, spectrum.* In particular, electromagnetic waves of the visible spectrum: red, orange, yellow, green, blue, violet; or radiation which, when stopped by an object, renders it visible by transformation into visible rays. The average velocity of l. in air is 299,711 km/s; in free space (i.e., a vacuum) 299,792.4 km/s. Ratios (velocity in vacuo)/(velocity in medium) are shown in the following table. The illuminance (intensity) of l. is measured in lux, the

Medium	Violet, λ 4200	Red, λ 6500
Air	1.000297	1.000292
Water	1.3420	1.3320
Diamond	2.4570	2.4108

quantity in lumens, the quality as colors. (2) Not heavy; as, l. metals. **absorbed ~** Those radiations which are influenced by matter. **actinic ~** Radiation rich in ultraviolet rays and that affects a photographic plate. Cf. *irradiation.* **artificial ~** A gas-filled lamp operated at a color temperature of 2848 K; adopted as the normal artificial l. by an international commission (1931). **axial ~** The rays passing through the optic axis of a system. **diffracted ~** L. that has undergone *diffraction,* q.v. **diffused ~** L. that has been scattered or reflected. **Finsen ~** Sunlight passed through a copper sulfate solution, which absorbs the yellow, red, and infrared rays. **monochromatic ~** L. consisting of apparently one wavelength only, as the sodium flame, which, however, is made up of several wavelengths. Absolutely monochromatic l. is obtained by screening a spectrum, by using interference filters, or by producing lasers. **polarized ~** L. whose vibrations are all in the same plane, are parallel to one another, and change uniformly; as, circularly or elliptically polarized l. Polarized light can be obtained by reflection, or by passage of light through certain minerals, as, Iceland spar. **reflected ~** L. that is turned back from a smooth surface such as a mirror. **refracted ~** L. that passes at an angle through a transparent medium—gas, liquid, or solid—and thereby changes its direction, as in lenses. **scattered ~** See *scattering, Raman effect, Tyndall cone effect.* **transmitted ~** L. that has passed through a medium, without absorption. **ultraviolet ~** See *radiation, ultraviolet ray* under *ray.* **Wood's ~** See *Wood's light.*

l. filter Color screen. **l. metal** A metallic element with a density below 4. In particular, the alkali, alkaline-earth, and rare-earth metals; characterized by a single valence (1, 2, or 3), a simple spectrum, strong electromotive force (positive), and colorless compounds. Cf. *periodic* table. **l. oils** Fractional distillates from coal tar, b.110–210; a source of benzene, toluene, and xylenes. **l. pipe** A tube or rod of plastic which transmits l. internally around corners. **l. scattering** See *light scattering* under *scattering.* **l. standard** Nonluminous samples are viewed normally and illuminated at 45° with the appropriate standard sources having color temperatures of 2848, 4800, and 6600 K. **l. year** A unit of astronomical distance: the distance that l. travels in 1 year at the rate of 299,992 km/s = 9.46055×10^{12} km = 5.87850×10^{12} mile.

lightning The electric flash associated with thunderstorms; or artificially produced. **l. jar** Leyden jar.

lignin Generic name for an amorphous, highly polymerized product which forms the middle lamella of many plant fibers (especially woods), and cements the fibers together by means of an intercellular layer surrounding them; insoluble in water; soluble in sodium sulfite solution, or dioxane. It contains OMe groups in proportions depending on the plant species (e.g., there are more in hardwoods than in softwoods), and 4 condensed molecules of coniferol. It occurs in wood-pulp digester liquor wastes; a source of vanillin and plastics. **alkali ~** L. obtained by acidification of an alkaline extract of wood. **Klason ~** L. obtained from wood after the wood has been extracted with 72% sulfuric acid. **native ~** L. obtained by direct extraction of wood with ethanol. **Willstätter ~** L. obtained from wood after it has been extracted with fuming hydrochloric acid. **l. compounds** See *lignosulfonates.*

lignite A brown coal, in which the original structure of the wood is still recognizable (25–45% C). See *jet, hartite.*

lignocaine hydrochloride BP name for lidocaine hydrochloride.

lignoceric acid Tetracosanoic acid*.

lignoceryl alcohol $C_{24}H_{49}OH$ = 354.7. An isomer of carnaubyl alcohol in tall oil.

lignosulfonates Lignin compounds with a wide variety of uses (dispersants, stabilizers, emulsifiers, complexing agents, grinding aids); made from the liquor from the sulfate pulping of wood.

lignum Latin for "wood." **l. benedictum** The wood of guaiac. **l. cedrum** Cedarwood. **l. nephriticum** Wood from *Pterocarpus indicus,* Philippines, or *Eysenhardtia polystachia,* Mexico. Gives highly fluorescent infusions. **l. rhodium** The wood of *Amyris balsamifera,* tropical America. **l. sanctum, l. vitae** Guaiac.

ligroin Ligroine. A petroleum distillation product. d.0.707–0.722, b.90–120; a solvent.

ligustrin Syringin.

ligustron A crystalline principle from the bark of *Ligustrum vulgare.* Colorless crystals, m.105.

Ligustrum A genus of oleaceous shrubs, the privets.

lilacin Syringin.

Liliaceae Lily family, a group of herbs with flowering stems, from bulbs or corms, with parallel-veined leaves. Some yield drugs; e.g.: *Smilax china,* chinaroot; *Schoenocaulon sabadilla,* sabadilla; *Allium sativum,* garlic; *Aloe vera, ferox,* etc., aloes.

Liliiflorae An order of phanerogamous plants (Junaceae, Liliaceae, Amaryllidaceae, Iridaceae, Dioscoraceae, and Bromeliaceae).

Lilion Trademark for a polyamide synthetic fiber.

liliquoid A dispersed or colloidal 2-liquid-phase system, as, emulsions.

lilolidine $C_{11}H_{13}N$ = 159.2. 1,2,5,6-Tetrahydro-4*H*-pyrroloquinoline. Colorless liquid, $b_{15mm}156$.

lily of the valley Convallaria.

limanol A preparation of salt-marsh mud, formerly used to treat rheumatism.

lime (1) Calcium oxide, calx, quicklime, burnt l. (USP). (2) The fruit of *Citrus aurantifolia* (Rutaceae); an antiscorbutic. (3) The linden or l. tree, *Tilia Europaea* (Tiliaceae). Cf. *linden.* **burnt ~** Calcium oxide. **chloride of ~, chlorinated ~** Bleaching powder. **fat ~** Calcium oxide. **hydrated ~** Calcium hydroxide. **quick ~** Calcium oxide. **slaked ~** A dry, white powder, chiefly calcium hydroxide, obtained by treating quicklime with sufficient water to satisfy its chemical affinity. **sulfurated ~** Calcium sulfide (USP). **unslaked ~** Calcium oxide.

l. juice The expressed sap from limes, the ripe fruits of

Citrus aurantifolia (Rutaceae); an antiscorbutic. **l. nitrate** Commercial calcium nitrate as fertilizer. **l. oil** An essential oil expressed from the rinds of the fruit of *Citrus limetta* (Rutaceae), containing citral and citrene; a flavoring and perfume. **l.water** Liquor calcis. A saturated solution of calcium hydroxide in water; a reagent.

limelight Illumination produced by heating lime to brilliant white heat with an oxyhydrogen flame.

limene $C_{12}H_{24}$ = 168.3. A sesquiterpene from lime oil.

limestone A pulverized rock consisting mainly of calcium carbonate; used in fertilizers, concrete mixtures, and in metallurgy. **magnesian ~** See *magnesian limestone.*

limettin A constituent of lime oil.

liminal (1) Barely perceptible by the senses. (2) The lowest or minimal quantity.

liming material A medium for the application of lime or limestone in agriculture, e.g., burnt lime mixed with marl.

limit (1) A border or boundary. (2) In mathematics, the value upon which an infinite series converges. Thus, the series $1 + \frac{1}{2} + \frac{1}{4} + \frac{1}{8} + \cdots$ has a limit of 2. **elastic ~** See *elastic limit.*

l. of elasticity See *elastic limit.*

limiting curve A line on a graph which represents a boundary between 2 phases.

limnology The study of lakes and inland seas.

limonene Citrene.

limonite Bog ore. A hydrated ferric oxide, 60–70%, with some carbonate, from Rotorua, New Zealand; a source of iron.

Linaceae Flax family: a group of herbaceous (sometimes woody) plants, yielding drugs and fibers; as, *Linum usitatissimum,* yielding linen, flaxseed, linseed oil.

linaloe oil An essential oil from a Mexican wood. Colorless, fragrant liquid, d.0.88, insoluble in water; used in the manufacture of linalool.

linalo(o)l $Me_2C:CH(CH_2)_2CMe(OH)CH:CH_2$ = 154.3. 3,7-Dimethyl-6-octadien-3-ol. A terpene of bergamotlike odor in coriander and linaloe oils. Colorless liquid. **(*R*)- ~** d.0.875, b.198. **(*S*)- ~** Coriandrol. d.0.866, b.199, slightly soluble in water. Both used in perfumery.

linalyl* The radical $C_{10}H_{17}-$. 2,6-Dimethylocta-2,7-dienyl, from linalool. **l. acetate** $Me_2C:CH(CH_2)_2CMe(OCOMe)\cdot CH:CH_2$ = 196.3. Linaloolacetic ester. Colorless liquid, d.0.91, b.220, slightly soluble in water; used in perfumery.

linamarin $C_{10}H_{17}O_6N$ = 247.3. An acetone cyanohydrin glucoside, m.141, from flax and *Phaseolus lunatus.*

linarin A glucoside from toadflax, *Linaria vulgaris* (Scrophulariaceae).

linarite $PbSO_4\cdot Cu(OH)_2$. A basic lead sulfate.

lindane Hexachloro*cyclo*hexane*.

Lindemann, Frederick Alexander (Lord Cherwell) (1886–1957) British physicist noted for work on aeronautics and atomic energy; personal scientific adviser to Sir Winston Churchill.

linden A tree of genus *Tilia* (Tiliaceae.) **American ~** Basswood, American lime, *T. americana,* which yields l. oil. **European ~** *T. europaea,* yielding tilia and lime (3). **l. flowers** See *lime (3).* **l. oil** Basswood oil. An oil from l. flowers and seeds; resembles cottonseed oil.

Linderström-Lang, Kaj Ulrik (1896–1959) Danish biochemist, from the Carlsberg Laboratory, Copenhagen.

lindgrenite $2CuMoO_4\cdot Cu(OH)_2$. A Chilean mineral. Transparent, green, biaxial holohedra.

Lindol Trademark for tricresyl phosphate.

line, lines The dimension of extension which possesses length but neither breadth nor thickness. **absorption ~** A l. corresponding with light which has been absorbed on its passage through a medium. **bright ~** The emission l. characteristic of an element. **C ~** The dark Fraunhofer l. of λ 6562.8 which corresponds with the hydrogen l., H_α. **calcium ~** The characteristic l. of calcium, especially the Fraunhofer l. G, H, and K. **D_1 ~, D_2 ~** The yellow light of the sodium flame; λ 5895.9, λ 5890.0, respectively. **dark ~** Absorption or Fraunhofer l. **emission ~** The bright l. characteristic of an element. **Fraunhofer ~** A dark l. in the solar spectrum caused by the absorption of light passing through incandescent vapor. Cf. *solar* spectrum. **hydrogen ~** See *H lines* under H. **hyperfine ~** L. close together, due to nuclear moments. **K ~** See *K line* under K. **metallic ~** L. due to metallic elements. **nebular ~** L. characteristic of nebulae. Cf. *nebulium.* **off- ~** Mode of using a computer in which there is no physical link between the computer and the system it is being used to guide. **on- ~** Mode of using a process control computer in which output from the system goes directly to the computer. **solar ~** See *solar spectrum.*

l. of force The direction in which a force acts; as: **electric ~** The curves radiating from a positive toward a negative charge. **magnetic ~** The curves or l. made visible by small iron filings in the field of force of a magnet. **l. spectrum** A graph of intensity (or intensities), or lack thereof, of wavelengths emitted or absorbed by an entity, e.g., light bulb. L. are caused by excited atoms and ions; bands are due to molecules.

linear Pertaining to one dimension: length. **l. absorption coefficient** See *absorption coefficient.* **l. expansion** The lengthwise expansion of materials under the influence of heat. Cf. *cubical expansion.* **l. programming** Technique for determining the optimum outcome from situations represented by linear equations and constraints.

linen A textile manufactured from the fibers of the flax stem.

linhay A concrete platform on which china clay is stored before shipment.

liniment An oily liquid used medicinally for external application.

linin (1) Oxychromatin. A morphological or structural element of the cell (linin network). (2) The active principle of purging flax, *Linum catharticum* (Linaceae), Europe. Colorless, bitter crystals; a purgative.

lining (1) The inside portion of a furnace. (2) A thin coating. **acid ~** L. of silica bricks. **basic ~** L. of magnesite bricks. **neutral ~** L. of coal or chrome bricks.

link Abbrev. li; an obsolete linear measure used in surveying. 1 li = 0.66 ft.

linkage Lines used in structural formulas to represent valency connections between atoms. Produced by a pair of electrons, one from each atom. See *bond, polar.* **cross- ~** Networks of bonds between linear polymers; associated in plastics with reduced solubility and high softening temperature.

linnaeite $(Co,Ni)_3S_4$. Linnaetite. Cobalt-nickel pyrite. A steel-gray mineral with a coppery-red tarnish. Occurs in isometric crystals and also massive.

linnaetite Linnaeite.

linoleate $C_{17}H_{31}COOM$. Octadecadienate*. A compound of a metal (M) and linoleic acid.

linoleic acid $Me(CH_2)_4CH:CH\cdot CH_2CH:CH\cdot(CH_2)_7COOH$ = 280.5. Linolic acid, *cis,cis*-9,12-octadecadienoic acid*. A fatty acid glyceride in all drying oils. An important dietary fatty acid. Yellow oil, d.0.921, b.230, insoluble in water.

linolenic acid $C_{17}H_{29}COOH$ = 278.4 α- ~ (Z,Z,Z)-9,12,15-Octadecatrienoic acid*. Colorless liquid, d.0.922, insoluble in water. γ- ~ (Z,Z,Z)-6,9,12-Octadecatrienoic acid*.
linoleum Canvas coated with a mixture of linseed oil, powdered cork, and pigment. Cf. *oilcloth.*
linolic acid Linoleic acid.
linotype metal The alloy Pb 83.5, Sb 13.5, Sn 3.0%; used for printing type.
linoxyn Solid, oxidized linseed oil, used in the manufacture of linoleum.
linron Trade name for bleached, stabilized linen fiber.
linseed The dried seeds of flax, *Linum usitatissimum* (Linaceae). **l. cake, l. meal** The solid residue of l. after removal of the oil; a cattle feed. **l. oil** Flaxseed oil, oleum lini. Yellow oil from linseed, d.0.932, m.−27, sapon. val. 188–195, iodine val. 170–192, insoluble in water, soluble in organic solvents. Used in paints, varnishes, lacquers, rubber substitutes, linoleum, and leather. A demulcent and emollient (EP, BP). **boiled ~** A l. oil which has been thickened by boiling, and dries rapidly on exposure to air; used in varnishes and driers.
lint Charpie byssus. A soft and flexible linen for dressing wounds. **cotton ~** An inferior short fiber from the cotton plant.
linter A machine for removing cotton linters from the seed.
linters The short cotton fibers left after ginning; a source of cellulose for explosives or paper.
linum Latin for linseed (flax).
lionite Impure, native tellurium.
lion's-tooth Taraxacum.
liothryonine sodium $C_{15}H_{11}O_4NI_3Na$ = 672.9. White solid, insoluble in water; a thyroid hormone, used for urgent treatment of severe hypothyroidism (BP).
lipase An enzyme in the liver, the pancreas, and oil seeds, which hydrolyzes the esters of glycerol and fatty acids; as, triacylglycerol lipase, steapsin. **gastric ~** A l. of the stomach. **pancreatic ~** Steapsin. **vegetable ~** A l. in many plants.
lipid* Lipide, lipin. A generic term for fats and lipoids, the alcohol-ether-soluble constituents of protoplasm, which are insoluble in water. They comprise the fats, fatty oils, essential oils, waxes, sterols, phospholipids, glycolipids, sulfolipids, aminolipids, chromolipids (lipochromes), and fatty acids. **complex ~** A l. that contains P, N, or both. **tertiary ~** A l. that contains neither P nor N.
Lipiodol Trademark for an injection of ethiodized oil.
lipochondria Golgi apparatus.
lipochrome A fatty pigment or coloring matter in natural fats, such as egg yolk and butter. See *carotenoids.*
lipoclastic Lipolytic.
lipoid, lipoidic Having the character of a lipid.
lipoids A group of nitrogenous fats (lecithins, cholesterol, and phospholipids).
lipolysis The decompositon or dissolution of a fat; the reverse of saponification.
lipolytic Lipoclastic. An agent that decomposes a fat into its alcohol (glycerol) and fatty acid.
lipophilic Dissolving in fatlike solvents.
lipoprotein A complex of a simple protein with a higher fatty acid.
Lipowitz's alloy The alloy Bi 50, Pb 27, Sn 13, Cd 10%; m.72; used in automatic fire sprinklers.
lipoxygenase* Enzyme that oxidizes linoleates; used to whiten bread.
lippia Yerba dulce. The dried leaves and inflorescence of *Lippia dulcis* (Verbenaceae); a demulcent or expectorant. **l. citriodora** Lemon-scented verbena. The dried leaves of *Aloysia citriodora* (Verbenaceae); a sedative.
lippianol A monohydric alcohol from the essential oil of *Lippia species* (Verbenaceae).
Lippmann, Edmond Oskar von (1857–1940) German organic chemist, noted for work in the sugar industry. **L. electrode** Capillary electrode. An early form of standard mercury dropping *electrode,* q.v. **L. electrometer** Capillary *electrometer.*
Lipscomb, William Nunn (1919–) American chemist. Nobel prize winner (1976), noted for work on boron hydrides.
liq Abbreviation for "liquid."
liquation The extraction of metals from ores by heating on an inclined hearth and collecting the molten metal.
liquefaction The change to liquid form, especially the condensation of gases to a liquid. **coal ~** See *coal liquefaction.*
liquefon A unit of the starch-liquefying power of enzymes: $\log_{10} L = (S - 1078)0.000565$, where L = liquefons per 10 cc infusion, S = mg of starch liquefied in 1 hour.
liquescent Tending to become fluid or liquid.
liqueur A strongly flavored and sweetened alcoholic beverage.
liquid A state of matter intermediate between a solid and a gas, shapeless and fluid, taking the shape of the container and seeking the lowest level. Cf. *solvent, parachor.* **anomalous ~** A fluid that is not a true liquid; e.g., paste. Cf. anomalous *viscosity.* **associated ~** Polar l. A l. in which the molecules form groups. **Newtonian ~** A true l. which does not alter in viscosity on stirring. Cf. *thixotropy.* **nonassociated ~** Normal l. **normal ~** Nonpolar l. A l. consisting of independent molecules, with no coordinate bonds or unshared electrons in the octet. **polar ~** Associated l. **semipolar ~** A l. intermediate beween associated and normal; as, alcohol.

l. acetylene Acetylene gas compressed into steel cylinders containing infusorial earth and acetone; used in welding. **l. air** See *liquid air* under *air.* **l. ammonia** Ammonia gas liquefied under pressure in steel cylinders. **l. carbon** A mixture of crushed coal, water, and a viscosity stabilizer; can be fired in boilers like a liquid. **l. carbon dioxide** Carbon dioxide gas compressed in steel cylinders. Used as a refrigerant; also for producing pressure in carbonated drinks. **l. chlorine** Chlorine gas compressed in steel cylinders; used in bleaching . **l. crystal** See *liquid crystal* under *crystal.* **l. ethylene** Ethylene gas compressed in steel cylinders; used in welding, ripening citrus fruits, and synthesis of ethanol. **l. gold** A solution of gold sulforesinate in essential oils, sometimes containing Bi and Rh, used to give a gold surface glaze to ceramics after firing. **l. helium** Helium gas compressed in steel cylinders; used for filling balloons. **l. hydrogen** Hydrogen gas compressed in steel cylinders; used in welding, production of high temperatures, hydrogenation of oils, and cracking of petroleum. **l. hydrogen sulfide** Hydrogen sulfide gas compressed in steel cylinders; used in chemical industry and as a reagent. **l. nitrogen** See *liquid nitrogen* under *nitrogen.* **l. nitrous oxide** Dinitrogen oxide gas compressed in steel cylinders; used as an anesthetic, as a substitute for oxygen in jewelers' blowpipes, and as a preservative. **l. oxygen** Oxygen gas compressed in steel cylinders; used in welding, anesthesia, mine rescue work, oxygen therapy, and for blasting. **l. silver** A mixture of silver powder and platinic chloride suspended in essential oils, used to give a silver surface glaze to ceramics after firing. **l. smoke** The first distillate in the fractional distillation of wood, which contains acetic acid, tar, and phenol compounds (pyroligneous acid); used in the preservation or "smoking" of meat. **l. sulfur dioxide** Sulfur dioxide gas compressed in steel cylinders; a reducing agent and reagent.

liquidambar Copal balsam, copalin. The balsamic exudates of *Liquidambar styraciflua* (Hamamelidaceae), N. America. Cf. *storax.*

liquidated Describing the surface of an ingot or casting which shows exudations or protuberances due to inverse segregation.

liquidus A curve relating to a liquid phase of a 2-component solution. See *solidus.*

liquor An aqueous solution. **l. ammoniae** Ammonium hydroxide. **l. calcis** Limewater. **l. of flints** A solution of silica in potash (potassium silicate). **l. trinitrin** A 1% solution of nitroglycerin.

liquorice EP, BP name for glycyrrhiza.

liroconite $Cu_9Al_4(OH)_{15}(AsO_4)_5 \cdot 20H_2O$. A native hydrated arsenate.

lisoloid A disperse or colloidal system consisting of a liquid phase surrounded by a solid phase; as, a jelly.

Lissajou's figure The pattern produced by a spot of light reflected from 2 mirrors, mounted on the ends of two tuning forks vibrating at right angles.

Lissapol C Trademark for sodium oleoylsulfate; a wetting agent. **L. NX** Trademark for an octyl cresol condensation product with ethylene oxide; an anionic surfactant.

Lister, Baron Joseph (1827–1912) English surgeon who founded antiseptic surgery.

Listerine Trademark for an antiseptic solution containing boric acid, benzoic acid, thymol, and essential oils of *Eucalyptus, Gaultheria,* etc.

liter, litre* Abbrev. L*, l*; the SI and metric unit of volume or capacity. (1) In the SI system, the l. exactly equals the cubic decimeter (the correct SI unit), and is acceptable usage for liquids and gases. 1 liter = 1,000 mL = 0.264172 U.S. gal = 33.8140 U.S. fl oz. (2) Formerly (1901–1964), the volume occupied by 1 kg of pure water at 4°C and 760 mm pressure. Although intended to be 1,000 cc, it proved to be 1000.028 cc. **micro ~** μL (or λ, lambda); one-millionth part of a liter. **milli ~** mL; one-thousandth part of a liter; 1 mL = 10^{-3} liter = exactly 1 cc. **Mohr ~** An obsolete unit of volume: the volume of 1 kg of water at 15°C weighed in vacuo (1,000.91 cc).

litharge Lead oxide, yellow.

lithate Urate.

lithia Lithium oxide. **l. mica** Lepidolite. **l. water** A solution of lithium hydrogencarbonate.

lithic acid Uric acid.

lithii Latin designation for salts of lithium.

lithionite A lepidolite containing ferrous iron. Silvery scales from Cornwall, England.

lithiophyllite A native lithium-manganese phosphate.

lithium* Li = 6.941. An element, at. no. 3, the first member of the alkali metals in group 1 of the periodic table; valency 1. Silver-gray metal, d.0.524, m.180, b.1340, reacting with water (stored in kerosene). L. is widely distributed, in small quantities; e.g., in amblygonite, lepidolite, petalite, spodumene, and mineral waters. Discovered (1817) by Arfvedson, and named from the Greek *lithos* ("stone"). It forms one series of compounds, all soluble in water. Used as "scavenger" in purifying metals; as deoxidizer in copper alloys. Traces of Li added to Al increase hardness; with Pb it produces a bearing metal. L. salts are used in soaps, lubricants, and medicine; as rubber stabilizers; and in Ni-Co batteries; l. deuteride and l. tritide as hydrogen bomb boosters. **l. acetate*** $Li(C_2H_3O_2) \cdot 2H_2O$ = 102.0. White rhombs, m.70, soluble in water. **l. aluminum hydride** $LiAlH_4$ = 37.95. Crystals, m.125 (decomp.), explodes in water. Powerful reducing agent. **l. amide*** $LiNH_2$ = 22.96. Colorless cubes, m.390, decomp. in water; used as catalyst. **l. arsenate*** $Li_3AsO_4 \cdot \frac{1}{2}H_2O$ = 168.7. White powder, soluble in water. **l. benzoate*** $LiC_7H_5O_2$ = 128.1. White powder of cool taste, soluble in water. **l. bicarbonate** L. hydrogencarbonate*. **l. borate** **l. metaborate** See *l. metaborate* below. **l. tetraborate** $Li_2B_4O_7 \cdot 5H_2O$ = 259.2. White powder, soluble in water. **l. bromide*** LiBr = 86.8. Colorless, hygroscopic powder, m.442, soluble in water. **l. carbide** Li_2C_2 = 37.90. White powder. **l. carbonate*** Li_2CO_3 = 73.9. Lithonate, Priadel. Colorless prisms, m.700, slightly soluble in water. Used in psychiatry to treat mania, manic depression, and depression; its effect is thought to be due to l. ions (USP, EP, BP). **acid ~** L. hydrogencarbonate*. **l. chlorate** $2LiClO_3 \cdot H_2O$ = 198.8. A deliquescent solid, m.50. **l. perchlorate** See *l. perchlorate** below. **l. chloride*** $LiCl \cdot 2H_2O$ = 78.4. Colorless octahedra, m.606, soluble in water, a pyrotechnic (red fires). **l. chromate*** $Li_2CrO_4 \cdot 2H_2O$ = 165.9. Yellow, hygroscopic crystals, soluble in water. **l. dichromate** $Li_2Cr_2O_7$ = 229.9. Orange crystals, soluble in water. **l. fluosilicate** L. hexafluorosilicate*. **l. fluoride*** LiF = 25.94. White crystals, soluble in water; a ceramic enamel. **l. hexachloroplatinate(II)*** $Li_2[PtCl_6] \cdot 6H_2O$ = 529.8. L. platinichloride. Yellow crystals; used in analysis. **l. hexafluorosilicate*** $Li_2[SiF_6] \cdot 2H_2O$ = 192.0. L. fluo(ro)silicate. White monoclinics, d.2.33, soluble in water. **l. germanate*** Li_2GeO_3 = 134.5. Colorless powder, m.1239, soluble in water. **l. hydride*** LiH = 7.95. White solid, m.680, decomp. in water to H_2 and l. hydroxide. **l. hydrogencarbonate*** $LiHCO_3$ = 68.0. L. bicarbonate. White powder, slightly soluble in water (lithia water). **l. hydroxide*** LiOH = 23.95. White crystals, soluble in water; a reagent. **l. iodate*** $LiIO_3$ = 181.8. White powder, soluble in water. **l. iodide*** $LiI \cdot 3H_2O$ = 187.9. White, hygroscopic crystals, m.720, soluble in water. **l. metaborate*** $LiBO_2$ = 49.75. White powder, soluble in water. **l. myristate*** $C_{14}H_{27}O_2Li$ = 234.3. White solid, m.224. **l. nitrate*** $LiNO_3$ = 68.9. Colorless rhombohedra, m.253, soluble in water; used in pyrotechnics (red fires). **l. nitride*** Li_3N = 34.83. A catalyst in ammonia synthesis. **l. nitrite*** $LiNO_2 \cdot H_2O$ = 70.96. Colorless needles, soluble in water. **l. oxalate*** $Li_2C_2O_4$ = 101.9. Colorless crystals, soluble in water. **l. oxide*** Li_2O = 29.88. Lithia. Colorless, caustic powder, sublimes 600, soluble in water; used to manufacture l. salts. **l. palmitate*** $C_{16}H_{31}O_2Li$ = 262.4. White solid, m.225. **l. perchlorate*** $LiClO_4 \cdot 3H_2O$ = 160.4. Deliquescent solid. **l. phosphate*** $Li_3PO_4 \cdot H_2O$ = 133.8. Colorless rhomboids, m.857, soluble in water. **l. platinichloride** L. hexachloroplatinate(II)*. **l. silicate** **l. metasilicate** Li_2SiO_3 = 90.0. White rhombs, m.1201, insoluble in water. **l. orthosilicate** Li_4SiO_4 = 119.8. Basic l. s. White crystals, m.1256, insoluble in water. **l. silicide** Li_6Si_2 = 97.8. Blue crystals, decomp. by heat or water. **l. stearate*** $C_{18}H_{35}O_2Li$ = 290.4. White solid, m.221. **l. sulfate*** $Li_2SO_4(+H_2O)$ = 109.9. Colorless, monoclinic crystals, m.843, soluble in water. Used for red fires, and sometimes in psychiatric disorders, similarly to l. carbonate. **l. sulfide*** Li_2S = 45.94. Colorless powder, soluble in water. **l. sulfite*** $Li_2SO_3 \cdot H_2O$ = 112.0. Soluble needles. **l. thallium tartrate*** $LiTlC_4H_4O_6 \cdot 2H_2O$ = 395.4. Triclinic crystals. **l. thiocyanate*** LiSCN = 65.0. L. rhodanate. Colorless, hygroscopic crystals, soluble in water; a reagent. **l. vanadate*** $LiVO_3 \cdot H_2O$ = 123.9. Yellow crystals, soluble in water.

litho- Prefix (Greek "stone"); as, lithography.

lithocholic acid $C_{24}H_{40}O_3$ = 376.6. An acid from bile. Cf. *choline, sterols.*
lithofellic acid $C_{20}H_{36}O_4$ = 340.5. Microscopic crystals, m.206, insoluble in water.
lithoform A zinc phosphate coating for protecting zinc from corrosion.
lithography Offset. A method of printing in which the design is produced photographically on a zinc plate, from which it is transferred on a rotary machine to a rubber blanket, which is brought into contact with the paper.
lithology Petrology.
lithol red Sodium 2-hydroxynaphthaleneazonaphthalene-1-sulfonate. A red dye used in cosmetics.
lithomarge $K_2O \cdot Al_2O_3 \cdot 6SiO_2 \cdot nH_2O$. A mottled, hydrated aluminosilicate clay from Germany.
lithopone Lithophone, Griffith white, Orr white, Charlton white. A stoichiometric mixture of zinc sulfide and barium sulfate; used as a white pigment in paints, in the rubber industry, for oilcloth manufacture, and as a filler. **cadmium ~** A red or yellow pigment analogous to l., in which the zinc sulfide is replaced by cadmium sulfide.
lithosphere The solid earth crust as compared with the liquid (hydrosphere) and gaseous (atmosphere) layers of the earth surface. See *abundance of elements,* Table 2.
lithuric acid $C_{15}H_{19}NO_9$ = 357.3. White crystals, m.205, from bladder stones.
litidionite $NaKCuSi_4O_{10}$. Mineral, containing a ladder silicate ion, found in the crater of Mt. Vesuvius.
litmin **azo ~** See *azolitmin.* **erythro ~** See *erythrolitmin.*
litmocydin An antibiotic from *Proactinomyces cyaneus.*
litmus Lacmus, turnsole, lacca coerulea. A purple coloring matter from lichens (*Roccella* and *Dendrographa* species). Blue powder, usually mixed with calcium carbonate and compressed into small cubes; it contains azolitmin and lecanoric acid. An indicator for volumetric analysis. **l. milk** Milk colored with l. and used as a culture medium to detect production of acidity. See *litmus milk agar* under *agar.* **l. paper** Filter paper impregnated with l. solution—neutral (pH 7.0), purple; alkaline (pH 8.0), blue; acid (pH 6.0), red. **l. tincture** A saturated solution of l. in alcohol, water, or both; used as an indicator and for culture media.
Litol process Trademark for a process to produce benzene from coke-oven light oil by a single catalytic hydrogenation step (to remove most impurities), followed by dealkylation.
litre* See *liter.*
Little, Arthur Dehon (1863–1935) American chemist noted for developments in chemical engineering.
littoral Pertaining to the shore. Cf. *pelagic.*
Littrow prism A glass spectrograph prism with 90°, 60°, and 30° angles, which reflects light internally from one surface. Cf. *Cornu prism.*
liver An organ responsible for maintaining balanced metabolism of carbohydrates, fats, proteins, hormones, and enzymes. It also regulates and stores vitamins, and detoxifies drugs and poisons. **l. of sulfur** The result of fusing together potassium carbonate and sulfur. It contains chiefly potassium sulfide and polysulfides; used to treat skin diseases. **l. ore** Cinnabar. **l. sugar** Glycogen.
Liverpool test A method for evaluating commercial caustic soda by titration. The at. wt. of Na is taken as 24; on this basis a sodium hydroxide content of 97.5% is reported as 100%.
liverwort The dried herb of *Hepatica triloba* (Ranunculaceae); a mild astringent. **English ~** Lichen caninus, ground liverwort. The lichen *Peltigera canina;* a mild purgative.
livetin A protein from egg yolk.
livingstoneite $2Sb_2O_3 \cdot HgO$. Antimonite containing mercury.
lixiviation The extraction and separation of a soluble substance from insoluble matter.
lixivium The extract obtained by lixiviation.
llama An animal textile fiber similar to, but coarser than, alpaca.
LLDPE Linear low-density *polyethylene.*
lm Abbreviation for lumen.
ln Abbreviation for natural logarithm.
LNG Liquefied natural *gas.*
loading A heavy substance, usually of mineral nature (clay, gypsum, etc.), added to textiles, papers, rubber, etc., to give weight or smoothness. Cf. *fillers.*
loadstone, lodestone Magnetite.
Loalin Trademark for a polystyrene plastic.
loam Clay mixed with sand in nature.
loban Indian name for gum *benzoin.*
lobate Describing a growth of bacteria in which the borders of the cultures show lobes, deep undulations, or fissures.
lobelia Indian tobacco. The dried leaves and flowering tops of *Lobelia inflata* (Lobeliaceae); an antispasmodic and expectorant.
lobeline $C_{22}H_{27}O_2N$ = 337.5. Inflatine. An alkaloid from lobelia seeds. **(−)-~** Yellow syrup, m.130, insoluble in water; an antispasmodic and sedative (BP). **(±)-~** m.110. **l. sulfate** $(C_{18}H_{23}O_2N)_2 \cdot H_2SO_4$ = 668.8. Yellow, hygroscopic crystals, soluble in water. A stimulant; has been used to treat the tobacco habit.
lobeloid The combined principles of Indian tobacco; an expectorant.
lobinol The principal skin irritant of poison oak, *Rhus diversiloba.*
locabiotal Fusafungine.
locant Number (as, 1-) or letter (as, *m*-) giving the position of an atom or a group in a molecule. **superscript ~** Generally used in special cases to supplement ordinary l. As, $\Delta^{2,\alpha}$ (see *delta*).
Lockyer, Sir Joseph Norman (1836–1920) English astronomer who discovered (with Frankland) helium in the sun's chromosphere.
locust **l. bean** Carob bean. **l. tree** The tree *Robinia pseudacacia* of semidesert N. America, which yields robin.
lode A vein of metallic deposit in a rock fissure. **mother ~** The great gold vein of Central California.
lodestone Magnetite.
Loeb, Jacques (1859–1924) American physiologist, noted for his work on the effects of ions on protoplasm. **L. collection** A collection of elements and compounds in the United States National Museum.
Loeffler See *Löffler.*
Loew theory All substances acting on aldehyde or amino groups are poisonous to the living tissue, because they change the dynamic equilibrium of the protoplasm.
Löffler, Friedrich A. J. (1852–1915) German bacteriologist noted for his bacterial stains. **L. methylene blue** A bacteriological dye prepared by dissolving 0.5 g methylene blue in 40 ml alcohol, 2 ml *N*/10 potassium hydroxide, and 98 mL water. **L. mixture** A culture medium for the growth of bacteria: glucose bouillon 250, horse- or beef-blood serum 750 mL.
log Abbreviation for logarithm. **$\log_{10}$** Common logarithm. **$\log_e$, ln** Natural logarithm.
Loganiaceae Logania family. A group of herbs, shrubs, or trees, many poisonous; e.g., seeds: *Strychnos nux vomica,* nux vomica; bark: *Strychnos malaccensis,* hoang-nan; rhizomes:

Gelsemium sempervirens, gelsemine; extractive: *Strychnos castelnaeana,* curare.

loganin $C_{17}H_{26}O_{10}$ = 390.4. A glucoside from nux vomica, m.215.

logarithm The logarithm (log) of a number n is the power x to which the logarithmic base a must be raised to give that number, $n = a^x$; $x = \log_a n$. **common ~** L. whose base is 10. Abbreviated log or $\log_{10}$. **Napierian ~** Natural l. **natural ~** Napierian l. L. whose base is e or 2.718281828. Abbreviated $\log_e$ or ln.

l. conversion: ln ( ) = 2.3025850930 log ( ), where ln is the natural, and log the common, logarithm.

logarithmic Pertaining to logarithms. **l. sector method** Quantitative spectroscopy based on the difference in length between 2 tapering wedges, produced by photographing the 2 spectra concerned after passage of the incident light through a wedge-shaped aperture in a disk which rotates in front of the spectrograph.

logwood Haematoxylon. **l. crystals** Hematoxylin. **l. extract** Reddish-blue paste made by extracting l.; used in dyeing textiles and leather and for the manufacture of hematoxylin **l. paste** Hematoxylin paste. Cf. *brazilwood, fustic.*

-logy Suffix derived from the Greek λογος ("word"), indicating a science or doctrine; as, pathology.

loiponic acid $C_7H_{11}NO_4$ = 173.2. (3*RS*, 4*RS*)-3,4-Piperidinedicarboxylic acid. m.257. An oxidation product of quinine.

lokav Prussian blue.

loliin A volatile constituent of the seeds of *Lolium* species.

lolium Poisonous darnel. A grass, *L. temulentum,* found in wheat and oat fields during wet seasons.

Lomonósov, Michájlo Vasilievič (1711–1765) Russian scientist noted for theories of molecular structure and conservation of energy and matter.

London **L. clay** A tough, compact lower Eocene clay formation, red-brown at the surface and blue-gray below. **L. paste** A mixture of quicklime and caustic soda moistened with alcohol.

longitudinal Lengthwise, or parallel to the longer axis of a body. Antonym: latitudinal.

longitude See *coordinates.*

lookafor An instrument to locate faults in telephone lines. It measures electrically the time for an electric impulse to travel at a known speed from a given point to the fault, and return.

looking-glass ore Fe_2O_3. A lustrous variety of hematite.

loomite A short-fibered variety of talc.

loose ion pair See *ion pair.*

looseness The extent to which a dyestuff may be removed by friction.

lophine $C_{21}H_{16}N_2$ = 296.4. 2,4,5-Triphenyl-1*H*-imidazole. Colorless needles, m.275, insoluble in water; a fluorescent neutralization indicator.

lophophorine $C_{13}H_{17}O_3N$ = 235.3. Methoxyanhalonine. An alkaloid from mescal buttons, the buds of *Anhalonium lewinii,* a cactus of Mexico. It is similar to anhalonine and mescaline and produces hallucinations..

Lopressor Trademark for metoprolol tartrate.

Loramine Trademark for a class of foaming agents based on fatty acid ethanolamines.

lorandite $TlAsS_2$. A native sulfoarsenide.

lorazepam A tranquilizing drug of the benzodiazepine group; used in anxiety states.

Lorentz, Hendrik Antoon (1853–1928) Dutch pioneer of the electron theory. **L.-Lorenz equation*** The molar refraction $R_m = M(n^2 - 1)/\rho(n^2 + 2)$, where M is the molecular weight, ρ the density, and n the refractive index.

loretin $C_9H_4IN \cdot OH \cdot SO_3H$ = 351.1. 8-Hydroxy-7-iodo-5-quinolinesulfonic acid, yatren. Yellow crystals, m.265 (decomp.), soluble in water; a reagent for copper.

loretinates The metal salts of loretin, e.g., bismuth loretinate (loretin bismuth).

Lorexane Trademark for γ-hexachloro*cyclohexane.*

loriodendrin A bitter principle from the bark of the tulip tree, *Liriodendron tulipifera* (Magnoliaceae). White scales, m.82, insoluble in water. Cf. *tulipiferine.*

Lorol Trademark for (1) a surfactant containing principally sodium dodecyl sulfate; (2) a mixture of aliphatic alcohols formed by the high-pressure hydrogenation of coconut oil.

Loschmidt number The number of gas molecules per cm^3 at 0°C and 760 mm pressure = 2.687×10^{19}. Cf. *Avogadro constant.*

losophan $C_6HI_3(OH)Me$ = 485.8. Triiodometacresol. Colorless crystals, insoluble in water; an external antiseptic.

lotase An enzyme of *Lotus arabicus,* converting lotusin into hydrocyanic acid and lotoflavin.

lotion A liquid preparation of a medicine for external use; usually of an antiseptic or soothing nature. **yellow ~** A solution of corrosive sublimate in limewater.

lotoflavin $C_{15}H_{10}O_6$ = 286.2. A yellow coloring matter from *Lotus arabicus.* Cf. *lotusin.*

lotusin $C_{28}H_{31}NO_{16}$ = 637.6. A glucoside from the leaves of *Lotus arabicus* (Leguminosae). Yellow crystals, hydrolyzed to lotoflavin.

lotus metal The bearing alloy Pb 75, Sb 15, Sn 10%.

loudness The magnitude of the physiological sensation produced by a sound. See also *sound loudness.*

loup A simple magnifying lens.

louver, louvre A ventilator of sloping boards which keep out rain, but not air.

lovage Sea parsley, levisticum, ligusticum, Chinese tang-kui, man-mu. The dried herb of *Levisticum officinale* (Umbelliferae); an aromatic and carminative. Cf. *eumenol.* **Scotch ~** The root of *Ligusticum scoticum,* a chewing tobacco. **l. oil ~** A colorless, fragrant oil from l., insoluble in water; used in perfumery.

Lovibond tintometer A colorimeter for comparing the color of a liquid with a standard series of tinted glass slides.

low Weak or poor. **l.-carbon steel** A steel of low carbon content. **l. explosives** Unstable chemicals and mixtures (carbon, sulfur, and nitrates) such as black gunpowder, blasting powder. Antonym: high *explosives.*

löweite $2MgSO_4 \cdot 2Na_2SO_4 \cdot 5H_2O$. A vitreous, yellow or white, fusible mineral.

Lowenstein-Jensen medium A culture medium for growing *Mycobacteria,* containing malachite green and beaten eggs.

Lowenstjern See *Kunkel.*

Lowig process The production of caustic soda from soda ash by heating at low red heat with ferric oxide, and hydrolyzing the resulting sodium ferrite with water.

lox A mining explosive consisting of liquid oxygen. With fuses it explodes like gunpowder; with detonators it detonates like dynamite.

loxa bark Cinchona bark.

lozenge A medicated tablet, usually for throat diseases.

LPOxo Trademark for a process that treats an alkene, such as ethylene, with carbon monoxide and hydrogen in the presence of a rhodium catalyst, to produce *n* and iso aldehydes, e.g., butyraldehyde, that may subsequently be hydrogenated to alcohols.

LPG Liquefied petroleum *gas.*

LSD $C_{20}H_{25}ON_3$ = 307.4. *N,N*-Diethyllyserganide, lysergide. (+)-~ A potent, addictive, psychedelic drug, m.83; used to treat psychiatric illness.
Lu Symbol for lutecium.
lubanol Coniferol.
luboil Lubricating oil.
lubricant An agent that reduces friction between moving surfaces. **liquid ~** An oil or semiliquid grease. **solid ~** A l. such as graphite, talc, soap, or sulfur.
lubricating oil A heavy distillate of petroleum used for lubricating machinery.
lubrication Making smooth or slippery. **boundary ~** The type of l. which occurs when the frictional effect is influenced by the nature of the underlying surface as well as by the chemical constitution of the lubricant; e.g., with thin l. films. **dry ~** A solid (e.g., polytetrafluoroethylene) impregnant for the metal surface of valve plugs. See *tribology.*
lubricator Lubricant.
lucalox A strong, transparent refractory, made by the high-temperature firing of compressed fine alumina. Also used for gem bearings and as an electrical insulator.
Lucas, Howard Johnson (1885–) American chemist. **L. theory** In an organic compound, the substituting radical affects the electronic orbits of a carbon atom by pulling the electronic pair toward the substituting radical if it has a high electron attraction, or vice versa.
lucerne Alfalfa.
Lucidol Trademark for organic peroxides, especially benzoyl peroxide.
luciferase An enzyme in a luminous mollusk, *Pholas dactylas,* which is luminescent in cold, but not hot, aqueous solutions.
luciferin A water-soluble protein from a luminous mollusk, *Pholas dactylus,* or the firefly, *Cypridina hilgendorfi.*
lucinite $Al_2O_3 \cdot P_2O_5 \cdot 4H_2O$. A native phosphate.
Lucite Trademark for plastic based on polymerized methyl methacrylate resin, widely used for the enclosures of airplanes. Cf. *leucite, lewisite.*
Lucitone Trademark for methyl methacrylate resins used in making dentures.
lucium A supposed chemical element, discovered in 1896, which proved to be a mixture of rare-earth metals.
Lucretius (98–55 B.C.) The earliest scientific writer to conceive an atomic theory.
lucumin A glucoside from the bark of *Lucuma glycyphloea.* Colorless needles. See *monesia* bark.
ludlamite $Fe_3(PO_4)_2$. A green mineral.
Ludolf's number The mathematical constant π.
Ludwig effect Soret effect.
ludwigite $3MgO \cdot B_2O_3 \cdot Fe_3O_4$. A blue or green mineral.
luffa Vegatable sponge, gourd towel, washrag sponge, loofah. The fibrous skeleton of the fruits of *Luffa cylindrica* (Cucurbitaceae).
Lugol solution An aqueous solution containing iodine 5 and potassium iodide 10 g per 100 mL. Used to treat overactive thyroid glands.
lukabro oil An oil from *Hydnocarpus anthelmintica,* Thailand, similar to hydnocarpus oil, d_{30}·0.946.
Lully, Raymond (1235–1315) Ramon Lull, Raymundus Lullus. Spanish alchemist who prepared nitric acid, aqua regia, alcohol, and potassium carbonate.
Lumarith **L. CA** Trademark for a cellulose acetate plastic. **L. EC** Ethocel. **L. Vn** Koroseal.
lumbang oil Candlenut oil. A colorless oil expressed from the seeds of *Aleurites moluccana,* candlenut, d.0.923, iodine val. 155-160, insoluble in water; used in soap manufacture and paints.
lumen (1) Abbrev. lm; the SI-system unit of luminous flux (luminosity). The light emitted by a point source, with a uniform intensity of 1 candela, in a solid angle of 1 steradian: 1 lumen = 0.001496 watt = 0.07958 spherical candlepower. 1 lumen emitted per square foot has a brightness of 1.076 millilamberts. (2) The bore of a small tube; e.g., the cavity of a thermometer tube, blood vessel, etc.
lumiere Wood An early international term for Wood's light.
Luminal Trademark for phenobarbital. **L. sodium** Sodium ethyl phenyl barbitrate. Cf. *luminol.*
luminance* Brightness of a surface or point source, in cd/m^2.
luminescence Emission of light under the influence of various physical agents; as mechanical (tribo ~), electrical (electro ~), radiant (photo ~), thermal (thermo ~), or chemical (chemi ~) means. Cf. *baroluminescence, candoluminescence, fluorescence, phosphorescence.* **cathodo ~** Emission of light due to electron or ion bombardment. **chemi ~** Emission of light by chemical reaction without appreciable temperature increase; as by luminol. Cf. *phosphorescence.* **crystallo ~** Emission of light during crystallization; as, arsenous acid from hydrochloric acid solution. **electro ~** (1) Emission of light due to passage of electricity through gases at low pressure and temperature; as, in vacuum tubes. (2) L. produced by a phosphor-treated surface in contact with an electrically conducting surface, e.g., conductive glass, to which an electric potential is applied. Used for electric signs. **photo ~** Emission of light on exposure to invisible radiation, by transfer from one wavelength into another; as, ultraviolet into visible rays. **radio ~** Emission of light by radioactive substances. **thermo ~** Emission of light after slight heating; as, by chlorophane. **tribo ~** Emission of light by friction or other mechanical means without temperature rise; as, with quartz.
luminiferous Giving off light without a rise in temperature. Cf. *incandescent.*
luminizing Rendering luminous by treatment with radioactive substances.
luminoflavin A luminescent substance in urine and plant products having a blue fluorescence produced by the irradiation of flavin with ultraviolet light.
luminol $NH_2 \cdot C_6H_3 \cdot (CO \cdot NH)_2$ = 177.2. 3-Aminophthalic hydrazide. White crystals. m.330, soluble in water. The alkaline solution becomes brightly luminescent when treated with hydrogen peroxide and potassium hexacyanoferrate(III). Cf. *Luminal.* **l. reaction** An analytical test in which l. in alkaline solution is used to produce a fluorescence.
luminometry The analytical applications of bio- and chemi*luminescence,* q.v.
luminophore (1) A substance which emits light. (2) A species, such as an organic radical or complex, which produces or increases the luminescence of a compound.
luminosity Luminous flux*.
luminous Pertaining to luminescence. **l. flux*** The radiant power of a source in lumens. **l. heater** A device to burn coal gas without a supply of excess air. It burns silently, without soot formation, and is hotter than ordinary gas burners. **l. intensity*** The l. flux per unit solid angle. Measured in candelas. **l. paint** A pigment which glows in the dark after exposure to light; usually sulfides of calcium, barium, and zinc, with a radioactive substance; used for watch dials. Cf. *phosphor.*
lumisterol $C_{28}H_{44}O$. Irradiated ergosterol. m.118. Cf. *ergocalciferol.*
lumophore Luminophore.
lunar **l. caustic** Fused silver nitrate. **l. cornea** Fused silver chloride.

Lunge, Georg (1839–1923) German chemist, noted for work on technological analytical methods. **L. nitrometer** Nitrometer.
lupanine $C_{15}H_{24}ON_2$ = 248.4. An alkaloid from the seeds of *Lupinus angustifolius* (Leguminosae). Yellow syrup with green fluorescence in water. **iso ~** Matrine.
luparenol $C_{15}H_{24}O$ = 220.4. An unsaturated alcohol from the higher-boiling fraction of hop oil, b_{3mm}125–128.
luparol $C_{16}H_{26}O_2$ = 250.4. A phenolic ether derived from luparone. Colorless liquid, d.0.9170, b_{2mm}123.
luparone $C_{13}H_{22}O$ = 194.3. A ketone derived from the higher-boiling fractions of hop oil. Colorless liquid, d.0.8861,b_{3mm}75.
lupetazin $C_6H_{14}N_2$ = 114.2. Dimethylpiperazine. White crystals, insoluble in water.
lupetidine $C_7H_{15}N$ = 113.2. 2,6-Dimethylpiperidine. Colorless oil.
lupin(e) *Lupinus.* A genus of leguminous plants; some are poisonous, others are forage plants. **l. alkaloids** The alkaloids from *Lupinus* species; as, lupinine.
lupinidine (1) $C_8H_{15}O_2N$ = 157.2. An alkaloid from the seeds of *Lupinus luteus* and *L. niger.* Yellow syrup. (2) Sparteine.
lupinin $C_{29}H_{32}O_{16}$ = 636.6. A glucoside from *Lupinus* species. Yellow crystals, hydrolyzed to glucose and lupigenin. Cf. *lupinine.*
lupinine $C_{10}H_{19}ON$ = 169.3. An alkaloid from the seeds of *Lupinus* species (±)-**~** Colorless crystals, m.67, insoluble in water. Cf. *lupinin.* **anhydro ~** $C_{10}H_{17}N$. **dimethyl ~** $C_{10}H_{17}ONMe_2$.
lupulin Humulin. Brown granules, of hop taste, consisting of the glandular trichomes from the strobiles of *Humulus lupulus,* hops; said to have sedative properties, as in hop pillows.
lupulinic acid *α*-**~** Humulone. *β*-**~** Lupulone.
lupulone $C_{26}H_{38}O_4$ = 414.6. A constituent of the soft resin of lupulin. m.93. Cf. *humulone.*
lupulus Hops.
lupus lapidis Stone disease. Deterioration of stonework attributed to microorganisms.
Lurgi process The complete gasification of coal of high ash and moisture content, in producers at 2.5 MPa in a continuous stream of oxygen and superheated steam. Benzene, tar, oils, and ammonia are by-products.
luster The reflection from the fractured surface of a rock, metal, or crystal.
Lustron Trademark for a polystyrene plastic.
lustrus Tygan.
lute A mixture of fireclay and water used to seal cracks in crucibles.
lutein (1) $C_{40}H_{56}O_2$ = 568.9. The carotenoid, q.v. *(carotenoids),* coloring matter of egg yolk: l. 70, zeaxanthin 30%, and in leaves, m.196. Related to carotene. (2) A yellow pigment from the fully developed corpora lutea.
luteo- Prefix (Latin), indicating "orange-yellow."
luteo compounds $[M(NH_3)_6]X_3$ and $[M(NH_3)_6]X_2$. Yellow hexaamminecobalt compounds, e.g., $[Co(NH_3)_6]Cl_3$, hexaamminecobalt(III) chloride.
luteol Oxychlordiphenylquinoxaline. A sensitive indicator and reagent for ammonia; yellow in alkali and colorless in acid solutions.
luteole A carotene in yellow corn grain.
luteolin $C_{15}H_{10}O_6$ = 286.2. Isofisetin. 3′,4′,5,7-Tetrahydroxyflavone. A flavone-derivative coloring matter of weld, *Reseda luteola,* an ancient dye. Yellow crystals, m.328, insoluble in water; a dye.
luteo salts See *luteo compounds.*
lutetium* Lu = 174.967. Lutecium. A rare-earth metal, at. no. 71, valencies 3 and 4. Discovered (1905) by Auer von Welsbach and named cassiopeium; and (1907) by Urbain and named after Lutetia (= Paris). Metallic, m.1660, b.3360. **l. chloride** $LuCl_3$ = 281.3. White crystals, soluble in water. **l. oxide** Lu_2O_3 = 397.9. White powder, insoluble in water.
lutidine $C_5H_3NMe_2$ = 107.2. Dimethylpyridine*. **2,4-~** Colorless liquid, b.157, soluble in water. **2,6-~** The commonest lutidine, obtained from tar and bone oils. Colorless liquid, d.0.942, b.142, soluble in water. **3,6-~** Colorless liquid, b.163, soluble in water. **dihydro ~** See *dihydro*lutidine. **hexahydro ~** Lupetidine.
lutidinic acid $C_7H_5O_4N \cdot H_2O$ = 185.1. 2,4-Pyridinedicarboxylic acid*. Colorless crystals, m.249, soluble in water.
lutidone C_7H_9ON = 123.2. 2,4-Dimethyl-4(1*H*)-pyridinone. Colorless crystals, m.225. Major tautomer of 4-hydroxy-2,6-dimethylpyridine.
lutoid A yellow, nonrubbery constituent of rubber latex, constituting 6–8% of the total solids. L. contains 85% water with proteins, salts, and lipoids; it makes the latex resistant to photooxidation and mechanically stable, but reduces thermal stability.
luvitherm A German thermoplastic film of high softening point, made from unplasticized polyvinyl chloride.
lux Abbrev. lx; the SI-system unit of illuminance, as produced by a luminous flux of 1 lumen uniformly distributed over 1 m^2. 1 lux = 1 lumen/m^2 = 0.0001000 phot = 0.09290 footcandle.

l. hour The light emitted by a Hefner standard candle in 1 hour at 1 m distance. A measure of the intensity of light exposure.

Clear sun	100,000 lux
Dull sun	300
Average room lighting	70
Bright moonlight	0.3
Lower limit of vision	0.000006

LWR Light water *reactor.*
lx Abbreviation for lux.
lyaconine Acolytine.
lyase* See *enzymes,* Table 30.
lyate ion A solvent molecule minus a proton, as OH^- from water.
lychnin A poisonous glucoside from *Lychnis* species (Caryophyllaceae).
lycine $C_5H_{11}O_2N$ = 117.1. An alkaloid from the leaves of *Lycium halimifolium* (Solanaceae). White, hygroscopic crystals, soluble in hot water. Cf. *lysine.*
lycoctonine $C_{25}H_{41}O_7N$ = 476.6. An alkaloid from the roots of *Aconitum lycoctonum* (Ranunculaceae). White crystals, m.100, soluble in water.
lycopene $C_{40}H_{56}$ = 536.9. Licopin. An unsaturated hydrocarbon and carotenoid, q.v. *(carotenoids),* pigment from many plants, as, tomato. Red crystals, m.168, isomeric with carotene.
lycopersicin A fungistatic principle from the tomato plant.
lycopodium Vegetable sulfur. Yellow powder, the flammable spores of *Lycopodium clavatum,* a club moss. A dusting powder, microscope reagent, and pill coating.
lycorine $C_{16}H_{17}O_4N$ = 287.3. Narcissine. (−)-**~** An alkaloid from the bulbs of *Lycoris radiata* (Amaryllidaceae). Colorless polyhedra, m.280, slightly soluble in water. **hydroxydimethyl ~** Sekisanine.
Lycra Trademark for an elastomeric synthetic fiber.
lyddite An explosive containing principally picric acid.
lydian stone Jasper.
lye (1) The alkaline solution obtained by leaching wood

ashes. (2) A solution of sodium or potassium hydroxide. Cf. *ley.*

Lyman series The first group of spectrum lines of hydrogen. Cf. *Balmer series, Paschen series.*

lymph A transparent, slightly alkaline liquid, permeating the animal organism and resembling blood serum. **l. glands** Collections of lymphoid tissue in a capsule, usually arranged in groups in the body. Part of the defense mechanism against infection.

lymphocyte A white blood cell, formed in the bone marrow and in lymphoid tissue. Found in circulating blood, lymphoid tissue, glands, and spleen. Two types: **T ~** L. dependent on a factor from the thymus gland to develop immune capability; responsible for cellular immune response. **B ~** L. often smaller than T l. Produces immunoglobulins in response to antigenic stimulus.

lymphoid tissue Collection of lymphocytes in the body, other than in circulating blood.

lynoestranol $C_{20}H_{28}O$ = 284.4. White crystals, m.162, insoluble in water. A progesterone similar to norethindrone. Used with estrogen for oral contraception.

Lynoral Trademark for ethinylestradiol.

lyochromes A group of natural, water-soluble plant pigments. Cf. *carotene.*

lyogel A gel which contains a large proportion of the dispersion medium used.

lyonium ion A solvent molecule plus a proton, as H_3O^+ from water. Cf. *hydronium ion.*

lyophil(e), lyophilic Attracting liquids. Describing a colloidal system in which the dispersed phase is a liquid and attracts the dispersing meidum. Cf. *hydrophile.*

lyophilized biological A biological substance (as blood plasma) prepared in dry form by rapid freezing and dehydration, in the frozen state under high vacuum. It is made ready for use by addition of sterile, distilled water. Cf. *desivac process, Lyovac process.*

lyophobe, lyophobic Repelling liquids. A colloidal system in which the dispersed solid phase has no attraction for the dispersion medium; hence it tends to separate. Cf. *hydrophobe.*

lyosorption The adhesion of a liquid to a solid as the adsorption of a solvent film on suspended particles. Cf. *adsorption.*

lyotrope (1) An ion or radical of a lyotropic series. (2) A readily soluble substance.

lyotropic series The arrangement of ions, radicals, or salts in the decreasing order in which they salt out or coagulate colloids by their dehydrating effect. The following salts usually appear in the same order: Potassium hexacyanoferrate(II) > sodium citrate > sodium hydrogenphosphate > sodium fluoride > sodium sulfate > sodium tartrate > sodium thiosulfate > sodium acetate > sodium formate. Cf. *Hofmeister series.*

Lyovac process A process of freezing and deyhdrating aqueous preparations.

Lys* Symbol for lysine.

lysatine $C_6H_{13}O_2N_3$ = 159.2. A crystalline alkaloid from casein.

lyse To break down body cells; as by the action of antibodies on blood. Cf. *hydrolyze.*

lysergic acid $C_{16}H_{16}O_2N_2$ = 268.3. A monobasic acid from ergot. **l. a. diethylamide** (1) LSD. (2) LSD 25.

lysidine N:CMe·NH·$(CH_2)_2$ = 84.1. Ethylene ethenyldiamine, 2-methyl-2-imidazoline. Red, hygroscopic crystals, m.105, soluble in water. **l. solution** A 50% aqueous solution of l.; a solvent for uric acid. **l. tartrate** Colorless crystals, soluble in water.

lysimeter A device for determining approximate solubility; or the water content of soils.

lysin An antibody which dissolves cells formed in *Staphylococci* cultures; as, hemolysin.

lysine* $NH_2\cdot(CH_2)_4\cdot CHNH_2\cdot COOH$ = 146.2. Lys*. 2,6-Diaminohexanoic acid*. An isolog of ornithine in many proteins. White needles, m.224, soluble in water; from casein. Cf. *lycine, lysin.*

lysis (1) The dissolution of a substance by the action of a lysin; as, hemolysis. (2) The decompositon of a substance; as, electrolysis. (3) The cleavage of a bond with addition of: H−OH, hydrolysis; $H-NH_2$, ammonolysis; $HO-C_2H_5$, alcoholysis; H−H, hydrogenation.

Lysivane Trademark for ethopropazine hydrochloride.

lysogen A substance that produces or generates a lysin.

Lysol Trademark for a cresylic disinfectant and antiseptic.

lysozyme* An enzyme in body secretions which hydrolyzes 1,4-β linkages in mucopolysaccharides; it dissolves certain airborne bacteria by lysis.

lysyl* 2,6-Diamino-1-oxohexyl†. The radical $NH_2(CH_2)_4CHNH_2\cdot CO-$, from lysine.

lytic Pertaining to (1) lysins, e.g., hemolytic; (2) lysis, e.g., hydrolytic.

lyxose $CH_2OH(CHOH)_3CHO$ = 150.1. A *monosaccharide,* q.v. (+) and (−) forms.

M

M Symbol for (1) mega or million (10^6); (2) metal (unspecified). **M acid** 1-Amino-5-naphthol-7-sulfonic acid. **M radiation** A series of homogeneous X-rays characteristic of the metal used as anticathode, and fainter than the K and L series. **M series** The spectral lines produced by M radiation on diffraction through a crystal grating. Cf. *Moseley spectrum.* **M shell** The third layer of electrons in an atom.

M Symbol for (1) bending moment; (2) magnetic quantum number; (3) molar mass; (4) molar solution; (5) mutual inductance.

M (Boldface.) Symbol for (1) magnetization; (2) moment of force (vector).

M_r Symbol for relative molecular mass (molecular weight).

m Symbol for: (1) metastable atomic nucleus. (2) Meter. **m²** Square meter. **m³** Cubic meter. (3) Milli- (10^{-3}).

m Symbol for (1) mass; (2) meta position; (3) molality.

m (Boldface.) Symbol for magnetic moment.

♏ Symbol for minim.

μ See *mu.*

mA Abbreviation for milliampere.

Mac See also *Mc.*

macassar oil Yellow fat from the seeds of *Schleichera trijuga,* India and Malaysia.

MacConkey's medium A solid bacterial culture medium, particularly for growing *Salmonella* bacteria. Sodium taurocholate is added to plain *agar,* q.v., to prevent overgrowth of other intestinal bacteria.

mace (1) Macis. The dried covering tissues of the seeds of *Myristica fragrans;* a condiment. (2) (Cap.) See *tear gases.* **m. oil** An essential oil from mace. Colorless liquid, d.0.91; a flavoring.

macene $C_{10}H_{18}$ = 138.3. A terpene from mace oil.

maceral General name for the microscopic structures of the mineral constituents of coals.

macerate To break up a solid by soaking in a liquid.

Mache unit M.E.; the quantity of radiation which produces a saturation current of 10^{-3} esu. 1 mache = 13.2 Bq.

machine steel A steel containing less than 0.3% carbon; easily machined.

macht metal A forging alloy containing Cu 60, Zn 38, Fe 2%.

Mach number A unit of speed, expressed as a multiple of the local speed of sound.

mackay bean The dried seeds of *Entada scandens* (Leguminosae), Queensland; a coffee substitute.

mackenite metals A group of heat-resisting Ni—Cr or Ni—Cr—Fe alloys.

Mackenzie amalgam An amalgam made by grinding together the solid alloys Hg—Bi and Pb—Hg.

Mackey test A test of the autoxidation fire hazards of oils.

macle (1) A variety of andalusite. (2) A twin crystal.

MacLeod, John James Rickard (1876–1935) Scottish-Canadian biochemist, awarded Nobel prize (with Banting) in 1923 for share in discovery of insulin.

maclurin $C_6H_3(OH)_2CO \cdot C_6H_2(OH)_3$ = 262.2. 2,3′,4,4′,6-Pentahydroxybenzophenone*, *osage orange* (q.v.), moringatannic acid. Yellow crystals from the wood of *Maclura aurantiaca,* m.200, soluble in hot water; a dye.

macro- Prefix (Greek μακρός = "broad"), indicating "large."

macroaxis The long axis in orthorhombic or triclinic crystals.

macrobacterium A large bacterium.

macrocarpine An alkaloid from *Thalictrum macrocarpum* (Ranunculaceae). Yellow crystals, soluble in water.

macrochemistry (1) The chemistry of reactions that are visible to the unaided eye. Cf. *microchemistry.* (2) Chemical operations on a large scale.

macrocyclic Containing rings of more than 7 C atoms.

macrodome See *dome.*

Macrogol 300 Polyethylene glycol.

macrograph Photomacrograph.

macrolides A group of broad-spectrum antibiotics containing a macrocyclic lactone ring; as, spiramycin and erythromycin.

macromolecular chemistry The study of the preparation, properties, and uses of substances containing large and complex molecules: i.e., mol. wt. exceeding 1,000. Cf. *polymer.*

macroscopic Describing objects visible to the naked eye. Cf. *microscopic.*

macrotin Cimicifugin.

macrotoid The combined principles from the root of *Cimicifuga racemosa;* an antispasmodic.

macrotys Cimicifuga.

maculanin Potassium *amylate.*

madder *Turkey red,* q.v. Garance. The root of *Rubia tinctorum* species. It contains glucosides which yield, on fermentation, alizarin and purpurin; a dye and pigment in lakes.

Maddrell salt A long-chain, high-molecular-weight sodium metaphosphate, made by heating sodium metaphosphate at 300; soluble in potassium salt solutions.

mafenide acetate $C_7H_{10}O_2N_2S \cdot C_2H_4O_2$ = 246.3. α-Aminotoluene-*p*-sulfonamide acetate. Sulfabenzamine, Sulfamylon. White powder, m.166, freely soluble in water. A cream for burns (USP).

mafic A rock-forming material, mainly magnesium and iron silicates.

mafurite A mineral association of kieserite and *augite,* q.v.

magdala red $C_{30}H_{21}N_4Cl$ = 473.0. Naphthalene red. A safranine dye of the naphthalene series.

magenta Fuchsin.

magic acid Trade name for a superacid, being a mixture of HSO_3F and SbF_5.

magma (1) Geology: a liquid, molten rock from which igneous rocks are formed. It is known as m. only while it is below the earth's surface. Cf. *lava, migma.* (2) Pharmacy: a suspension of a fine precipitate in water.

magnalite An aluminum piston alloy containing Cu 4, Ni 2, Mg 1.5%.

magnalium An alloy of Al with Mg 2–10 and sometimes Cu 1.5–2.0%; used in aircraft, balances, and, automobiles.

magnefen A dead-burned dolomite.

magneform A cold-forming technique, especially for aluminum. The workpiece is placed on or around a fixed induction coil, in which a sudden intense magnetic field is produced, that drives the workpiece into a metal die or deforms certain areas of it.

magnesia Magnesium oxide. **calcined ~** Magnesium oxide obtained by heating the carbonate. **cream of ~** Magnesium hydroxide mixture. **fluid ~** A solution of magnesium hydrogencarbonate: 2.65 g/100 ml. **milk of ~** (1) A suspension containing 7–8.5% magnesium hydroxide; an antacid and cathartic (USP). (2) Trademark for a suspension of magnesium hydroxide. **ponderous ~** Heavy magnesium oxide.

m. alba A hydrated magnesium carbonate. Cf. milk of *magnesia*. **m. a. levis** $4MgCO_3 \cdot Mg(OH)_2 \cdot 5H_2O$. **m. a. ponderosa** $MgCO_3 \cdot Mg(OH)_2 \cdot 4H_2O$. **m. cement** A cement made by heating hydrated magnesium chloride to redness and adding water. **m. glass** Glass containing 3–4% magnesium oxide, used for electric light bulbs. **m. mixture** Colorless liquid made by dissolving 55 g hydrated magnesium chloride and 70 g ammonium chloride in 650 ml water and adding 350 ml 10% ammonium hydroxide. Used in analysis to precipitate phosphates and arsenates. **m. niger** An early name for pyrolusite. **m. usta** Magnesium oxide, made by prolonged calcining of magnesium carbonate at a low temperature.

magnesian limestone Limestone containing about 10% magnesite.

magnesioferrite $MgFe_2O_4$. A spinel.

magnesite $MgCO_3$. A refractory lining for furnaces and substitute for plaster of paris.

magnesium* Mg = 24.305. An alkaline-earth metal element, at. no. 12, valency 2. Silver metal, d.1.74, m.650, b.1110, insoluble in water, decomp. by acids. Occurs native in magnesite and dolomite (carbonates); in serpentine, asbestos, talcum, biotite and meerschaum (silicates); in Stassfurt salts and mineral waters; and in seawater. M. was first obtained by electrolysis (1830) by Liebig and Bunsen; it forms only one series of compounds yielding the Mg^{++} ion in aqueous solutions. M. in bars, sheets, ingots, ribbons, wire, and powder is used in electric batteries, aircraft construction, flashbulbs, and pyrotechnics; as a deoxidizer in making brass, and for Thermit. **ethyl ~*** Et_2Mg = 82.4. M. ethide. Colorless liquid.

m. acetate* $Mg(C_2H_3O_2)_2 \cdot 4H_2O$ = 214.5. Colorless, monoclinic crystals, soluble in water. **m. acid citrate** M. hydrogencitrate*. **m. alkyl compounds** R_2Mg, where R is an alkyl radical. **m. alkyl condensation** Grignard reaction. **m. aluminate*** $MgAl_2O_4$ = 142.3. Spinel. Colorless cubes, m.2135, insoluble in water; a refractory. **m. arsenide** Mg_3As_2 = 222.8. Decomp. by water. **m. aryl compounds** R_2Mg, where R is an aryl radical. **m. benzoate*** $Mg(C_7H_5O_2)_2$ = 266.5. White powder, soluble in water. **m. bicarbonate** M. hydrogencarbonate*. **m. bisulfate** M. dihydrogensulfate. **m. borate** $Mg(BO_2)_2$ = 109.9. M. metaborate*. White powder, soluble in water. **m. bromate*** $Mg(BrO_3)_2 \cdot 6H_2O$ = 388.2. Colorless crystals, soluble in water, decomp. by heat. **m. bromide*** $MgBr_2 \cdot 6H_2O$ = 292.2. Colorless hexagons, decomp. by heat, soluble in water; used for the electrolytic preparation of m. **ethyl ~*** EtMgBr = 133.3. Grignard's reagent; used in organic synthesis. **m. butyrate*** $Mg(C_4H_7O_2)_2$ = 198.5. Colorless, hygroscopic crystals, soluble in water. **m. cacodylate** $Mg(Me_2AsO_2)_2$ = 298.3. White powder, soluble in water. **m. carbide** (1) MgC_2 = 48.33. (2) MgC_3 = 60.34. **m. carbonate*** $MgCO_3$ = 84.3. Magnesite, dolomite, heavy m. carbonate. Colorless rhombs, decomp. 350, soluble in water. Used to prepare m. salts, fireproofing compositions, and toothpaste. **basic ~** $4MgCO_3 \cdot Mg(OH)_2 \cdot 5H_2O$ = 485.6. Magnesia alba levis, light m. carbonate, magnesii carbonas (USP, EP, BP). White powder, slightly soluble in water, soluble in ammonia; a dusting powder, antacid, and laxative. **heavy ~** Magnesia subcarbonas ponderosa. A gastric antacid (EP, BP). **light ~** Basic m. carbonate (EP, BP). **m. chlorate*** $Mg(ClO_3)_2 \cdot 6H_2O$ = 299.3. Hygroscopic, colorless crystals, m.40, soluble in water. **m. chloride*** $MgCl_2$ = 95.2. Fused m. chloride. White crystals, m.708, soluble in water. Used in the manufacture of metallic magnesium; for fireproofing; for magnesia cements, composition flooring, and artificial stones. **hydrated ~** $MgCl_2 \cdot 6H_2O$ = 203.3. Crystallized m. chloride. Colorless, bitter, hygroscopic crystals, m.100, soluble in water. Used as a saline cathartic, for low m. blood levels (USP, EP, BP), and as a dialysis fluid; also for disinfecting, fireproofing, and dressing fabrics. **m. chromate*** $MgCrO_4 \cdot 7H_2O$ = 266.4. Orange crystals, m.100, soluble in water. **m. citrate*** $Mg_3(C_6H_5O_7)_2 \cdot 14H_2O$ = 703.3. Colorless scales, soluble in water; used in pharmacy. **effervescent ~** A granular mixture of m. citrate, sodium hydrogencarbonate, citric acid, and sugar; a saline laxative (USP). **m. copper alloy** An activator for the Grignard reaction. **m. dichromate*** $MgCr_2O_7$ = 240.3. Brown, hygroscopic crystals, soluble in water. **m. dihydrogensulfate*** $MgH_2(SO_4)_2$ = 218.4. M. bisulfate. Colorless crystals, soluble in water; a cathartic. **m. dioxide*** M. peroxide*. **m. dust** Finely powdered m. metal used in pyrotechnics, for photographic flashbulbs, and as a chemical reagent. **m. fluoride*** MgF_2 = 62.3. White powder, m.1396, insoluble in water; used in ceramics and glass. **m. fluosilicate** M. hexafluorosilicate*. **m. formate*** $Mg(CHO_2)_2 \cdot 2H_2O$ = 150.4. Colorless prisms, soluble in water. **m. halides** RMgX, where R is an aryl or alkyl radical and X a halogen. **m. hexafluorosilicate*** $MgSiF_6$ = 166.4. M. silicofluoride. White powder; used in ceramics. **hexahydrate ~** $6H_2O$ = 274.5. Trigonal crystals, soluble in water. **m. hydrate** M. hydroxide*. **m. hydrogencarbonate*** M. bicarbonate. A solution of m. carbonate in carbonated water. **m. hydrogensulfide*** $Mg(HS)_2$ = 90.4. A compound which, on warming, yields pure hydrogen sulfide. **m. hydroxide*** $Mg(OH)_2$ = 58.3. M. hydrate. Brucite. White rhombohedra, decomp. by heat, insoluble in water; used to manufacture m. oxide, milk of magnesia, and sugar (USP, EP, BP). **m. hydroxide phosphate** $Mg(OH)_2 \cdot Mg_3(PO_4)$ = 226.2. A constituent of boiler scale. **m. h. p. mixture** Cream of magnesia. An aqueous suspension (≡ 8.25% m. hydroxide), made from a mixture of m. sulfate, sodium hydroxide, and light m. oxide. An antacid (BP). **m. iodate*** $Mg(IO_3)_2 \cdot 4H_2O$ = 446.2. Crystals, decomp. by heat, soluble in water. **m. iodide*** MgI_2 = 278.1. Colorless powder, decomp. by heat, soluble in water. **hydrated ~** $MgI_2 \cdot 8H_2O$ = 422.2. Colorless, hygroscopic crystals, soluble in water. **m. laurate*** $Mg(C_{12}H_{23}O_2)_2$ = 422.9. Colorless crystals, m. 150. **m. lime** Quicklime containing less than 20% m. oxide. **m. magma** An aqueous suspension of 7–8.5% m. hydroxide, with a flavoring agent. Cf. milk of *magnesia*. **m. minerals** M. is abundant in rocks (olivines, micas, pyroxenes, silicates, and amphiboles). **m. molybdate*** $MgMoO_4$ = 184.2. Colorless crystals, soluble in water. **m. myristate*** $(C_{14}H_{27}O_2)_2Mg$ = 479.0. White powder, m.132. **m. nitrate*** $Mg(NO_3)_2 \cdot 6H_2O$ = 256.4. Colorless, monoclinic or triclinic crystals, m. 90, soluble in water; used as a reagent and in pyrotechnics. **m. nitride*** Mg_3N_2 = 100.9. Yellow, amorphous mass, decomp. by water. **m. nitrite*** $Mg(NO_2)_2 \cdot 2H_2O$ = 152.3. Colorless, hygroscopic crystals, soluble in water; a reagent. **m. oleate*** $Mg(C_{18}H_{33}O_2)_2$ = 587.2. M. oleinate. Yellow oil, insoluble in water; a varnish drier. **m. oxalate*** $Mg(OOC)_2 \cdot 2H_2O$ = 148.4. Colorless crystals, soluble in water. **m. oxide*** MgO

= 40.30. Magnesia, calcined magnesia, periclase, ponderous magnesia, magnesia usta. White hexagons, m.1900, insoluble in water. Used commercially in heat insulations, refractories, rubber manufacture; as a dusting powder, antacid, and laxative. **heavy ~** Amorphous m. oxide, obtained by heating basic m. carbonate, d.3.58. (USP, EP, BP). **light ~** Crystalline m. oxide obtained by heating m. carbonate d.3.36 (EP, BP). See also *magnesium peroxide.* **m. palmitate*** $Mg(C_{16}H_{31}O_2)_2$ = 535.1. White soap, m.120, insoluble in water; a varnish drier. **m. perchlorate*** $Mg(ClO_4)_2$ = 223.2. Anhydrone. White granules; a regenerable desiccant. **m. permanganate*** $Mg(MnO_4)_2 \cdot 6H_2O$ = 370.3. Blue granules, soluble in water; an antiseptic. **m. peroxide*** MgO_2 = 56.3. M. superoxide, m. dioxide. White powder containing 25% MgO_2 and 75% MgO, insoluble in water. Used to detect bilirubin; also used for bleaching silk and wool. **m. phosphate** $Mg_3(PO_4)_2 \cdot 4H_2O$ = 334.9. M. orthophosphate*. Monoclinic crystals, insoluble in water. **acid ~** $MgHPO_4 \cdot 7H_2O$ = 246.4. M. hydrogenorthophosphate*. Colorless hexagons, insoluble in water. **dibasic ~** Acid m. phosphate. **monobasic ~** $Mg(H_2PO_4)_2$ = 218.3. M. dihydrogenphosphate*, m. biphosphate. Yellow crystals, insoluble in water; a laxative. **m. pyrophosphate*** $Mg_2P_2O_7 \cdot 3H_2O$ = 276.6. Colorless crystals, insoluble in water. **tribasic ~** M. phosphate. **m. phosphide*** Mg_3P_2 = 134.9. Black powder, decomp. by water. **m. phosphinate*** $Mg(H_2PO_2)_2 \cdot 6H_2O$ = 262.4. Colorless crystals, insoluble in water. **m. phosphonate*** $MgHPO_3$ = 104.3. M. phosphite. Colorless crystals, soluble in water. **m. potassium chloride** $MgCl_2 \cdot KCl \cdot 6H_2O$ = 283.9. Colorless solid, decomp. by heat. **m. propionate*** $Mg(C_3H_5O_2)_2$ = 170.4. White powder, soluble in water. **m. pyrophosphate*** See *magnesium phosphate.* **m. silicate*** $MgSiO_4$ = 116.4. White powder, insoluble in water; used in rubber. **m. silicides** (1) Mg_2Si = 76.7. Gray leaflets, decomp. by water to silane. (2) MgSi = 52.4. **m. stearate*** $(C_{18}H_{35}O_2)_2Mg$ = 591.3. Dolomol. White, soapy powder, m. 132. M. stearate containing the equivalent of 7–8% m. oxide is used for tablet making (NF, BP). **m. succinate*** $MgC_4H_4O_4$ = 140.4. White powder, soluble in water. **m. sulfate*** (1) $MgSO_4$ = 120.4. Anhydrous m. sulfate. White powder, soluble in water. (2) **m. s. monohydrate** $MgSO_4 \cdot H_2O$ = 138.4. Kieserite. d.2.3. (3) **m. s. dihydrate** $MgSO_4 \cdot 2H_2O$ = 156.4. Desiccated m. sulfate. White powder, soluble in water. (4) **m. s. heptahydrate** $MgSO_4 \cdot 7H_2O$ = 246.5. Crystallized m. sulfate, Epsom salt (USP), Epsom salts (EP, BP). Magnesii sulfas, bitter salt. Colorless, tetragonal or monoclinic crystals, decomp. by heat, soluble in water. Used as a reagent, a constituent of bleaching solutions, and a refrigerant; as a saline cathartic and anticonvulsant; for loading and warp-sizing cotton goods and textiles, and for fireproofing; in mineral waters, in the leather industry, and in electric batteries. **m. sulfide*** MgS = 56.4. Brown cubes, decomp. by heat, soluble in water. **m. sulfite*** $MgSO_3 \cdot 6H_2O$ = 212.5. White, crystalline powder, m. 260, soluble in water. **m. tartrate*** $C_4H_4O_6Mg \cdot 5H_2O$ = 262.4. White, monoclinic crystals, soluble in water. **acid ~** $MgH_2(C_4H_4O_6)_2 \cdot 4H_2O$ = 394.5. White rhombs, soluble in water. **m. tetraborate** MgB_4O_7 = 179.5. White powder, soluble in water; used in driers. **m. thiocyanate*** $Mg(SCN)_2$ = 140.5. M. rhodanate. Colorless, hygroscopic crystals, soluble in water; a reagent. **m. thiosulfate*** $MgS_2O_3 \cdot 6H_2O$ = 244.5. M. hyposulfite. Colorless crystals, soluble in water. **m. trisilicate** $2MgO \cdot 3SiO_2 \cdot nH_2O$ = 260.9. White, hygroscopic powder. An antacid (USP, BP). **m. tungstate*** $MgWO_4$ = 272.2. M. wolframate*. Colorless crystals, insoluble in water; used for fluorescent screens and luminous paints. **m. urate*** $MgC_5H_2O_3N_4$ = 190.4. White powder, insoluble in water. **m. wolframate*** M. tungstate.*

Magnesol Trademark for an acid silicate of magnesium, used in chromatographic columns.

magneson *p*-Nitrobenzeneazoresorcinol. **m. II** *p*-Nitrobenzeneazo-1-naphthol.

magnesyl The radical —MgX. Cf. *Grignard's reagent.*

magnet A lodestone: iron that attracts iron. **bar ~** A bar of magnetized soft iron. **electro ~** Iron rendered temporarily magnetic by an electric current passing through a coil around it. **horseshoe ~** A magnetized bar of iron bent into a U shape.

magnetic Pertaining to or possessing magnetism. **m. declination** The deviation of the compass needle from the true axis of the earth, due to the m. pole not being coincident with the geographic poles. **m. deflection** The deflection of radioactive rays or particles by a field, according to the sign and magnitude of their charges. **m. elements** The elements of the iron family, all of which are m.; other elements have only slight magnetism. **m. field** The lines of force in the space around the m. poles of a magnet. **m. f. strength** Symbol: *H;* the interaction between magnetic elements, measured in A/m. **m. flux*** The magnitude of a m. field, as given by the product of m. flux density and area involved. **m. f. density*** M. induction*. **m. guard** A mask of magnetized steel wire gauze, used to protect workers from iron dust. **m. induction*** Symbol: *B;* m. flux density*. The m. flux per unit area taken perpendicularly to the direction of the m. flux. Measured in tesla (SI system) or gauss (cgs system). **m. ink character recognition** MICR. Method of reading documents that uses the automatic recognition of characters printed with magnetic ink, as checks. **m. intensity** M. field strength*. **m. iron ore** Magnetite. **m. meridian** The direction registered by a compass needle at any place. **m. moment*** The ratio of the torque of interaction to the magnetic flux density; in joules/tesla. **m. optic** See *magneto-optic effect.* **m. ore** (1) Magnetite. (2) A m. ore. **m. permeability*** μ; the total m. induction in a m. field of unit strength. **relative ~** μ_r; the ratio of the m. p. to that in a vacuum. **m. polarization*** The optical activity acquired by a substance in a m. field. **m. pole** The point on which m. lines of force converge. **m. potential** Magnetomotive force*. **m. pyrite** Pyrrhotite. **m. reluctance** The ratio of magnetic potential difference to magnetic flux. **m. rotation** M. polarization. **specific ~** The ratio of the m. rotation of a substance to that of water under the same conditions. **m. separator** A device, usually a powerful electromagnet, for separating magnetic from nonmagnetic minerals. **m. spectrum** The pattern (lines of force) produced by iron filings scattered on a plane surface in a m. field. See *magnetic field.* **m. susceptibility*** $\mu_r - 1$, where μ_r is the relative permeability. **m. units** See *weber, tesla* (SI); *maxwell, gauss, oersted* (cgs).

magnetism (1) The property of substances (as iron) which under certain conditions attract or repel each other or a like substance. (2) The science of magnetic phenomena. **electro ~** M. due to induction currents. **ferro ~** M. due to iron and independent of an electric current. **meta ~** The property of loss of m. at high and low temperatures. **photo ~** See *photomagnetism.*

magnetite $FeO \cdot Fe_2O_3$. Magnetic iron ore, hercules stone, lodestone. A black, dense magnetic mass. Cf. *iron(II)diiron(III) oxide.*

magnetization (1) The act of rendering magnetic. (2)* $(B/\mu_0) - H$, where B is the magnetic flux density, μ_0 the magnetic constant, and H the magnetic field strength.

magnetochemistry The application of magnetic susceptibilities to chemical problems.

magnetoelectricity A current of electricity produced by magnetism. Cf. electro*magnetism.*

magnetometer A device to measure magnetic force.

magnetometric titration Volumetric analysis in which changes in paramagnetism of an ion on addition of a complexing agent are followed by means of a magnetic-resonance spectrometer.

magnetomotive force* F_m, mmf. The integral of magnetic field strength round a closed path.

magneton* μ; the unit of magnetic moment of a subatomic particle. **Bohr ~** $\mu_B = eh/4\pi m_e = 9.2741 \times 10^{-24}$ J/T, where e is the electron charge, m_e its mass, and h the Planck constant. **nuclear ~*** μ_N; the magnetic moment per unit electron mass, i.e., $\mu/1{,}836$.

m. theory A theory of atomic structure (Parsons), in which magnetons form octets.

magneto-optic **m. effect** A characteristic time lag in m. rotation used in qualitative analysis. **m. rotation** Magnetic rotation, magnetic polarization. The rotation of polarized light passing through a magnetic field depends on the strength of the field, the wavelength of light, and the nature of the substance. Cf. *Verdet's constant.*

magnetophone An instrument that records sound on a magnetic film; as, a tape recorder.

magnetostriction The reversible change in dimensions of certain ferromagnetic materials in a magnetic field. Cf. *Joule effect, Villari effect.*

magnetotactic Sensitive to magnetic fields.

magnification An apparent increase in size, e.g., produced by a microscope.

magnifier A lens used to read scales on instruments.

magnifying power The ratio of the actual size of an object to its amplified image. Cf. *auxiometer.*

magnitude A measurement of an object. Cf. *unit, length, mass, time.* **astronomical ~** The size of a star, measured in terms of its brightness.

magnochromite A variety of chromite containing magnesium.

Magnoliaceae A family of trees and shrubs including *Illicium,* star anise, and *Magnolia,* magnolin.

magnolin A crystalline glucoside from the fruit of *Magnolia tripetala* (Magnoliaceae); insoluble in water.

magnolite Hg_2TeO_4. A white mineral.

magnolium The alloy Pb 90, Sb 10%.

Magnox Magnesium alloy; used to clad nuclear reactor fuel elements.

Magnus, Albertus (1193–1280) Albrecht, Graf von Bollstädt. German philosopher, the founder of the European school of alchemists.

Magnus **M., Heinrich Gustav (1802–1870)** German chemist. **M. rule** Each metal has a specific voltage at which it is deposited from a solution containing a mixture of metallic salts. **M. salt** $[Pt(NH_3)_4]PtCl_4$. Tetraammineplatinum(II) tetrachloroplatinate(II)*. Green needles, soluble in water (Magnus, 1828).

Maillard reaction The reaction between amino acids and reducing sugars which occurs in foods in hot conditions; a brown color, as in toasted cereals.

main cell An amalgamated zinc cathode and a lead dioxide anode in sulfuric acid (2.5 volts).

maisin A protein from maize.

maize Indian corn. The seeds of *Zea mays,* a cereal. Cf. *zea,* corn *starch.* **m. oil** BP name for corn oil.

maizolith An insulating material: cornstalks and corncobs pressed into sheets.

majolica Lustrous pottery enameled with tin oxide.

Makrolon Trademark for a polycarbonate (mol. wt. exceeds 200,000), made by reacting carbonyl chloride with bisphenol.

malabar tallow Fat from the seeds of *Valeria indica,* used in chocolate manufacture and in sizing yarn.

malachite $Cu_2(OH)_2CO_3$. Dense smaragd or emerald-green masses, which may be polished. **azur ~** Bluish-green m. from Arizona. **blue ~** Azurite. **pseudo ~** $Cu_3(PO_4)_2 \cdot H_2O$. Phosphochalcite. A green, native copper phosphate.

m. green (1) Pulverized malachite; a pigment. (2) $C_{23}H_{25}N_2Cl$ = 364.9. Victoria green B, benzal green. A triphenylmethane dye. Green crystals, soluble in water. A reagent for detecting sulfites in presence of thiosulfates; bacteriological stain. It is a pH indicator, changing at 1.0 from yellow (acid) to blue-green (alkaline). **leuco ~** See *leucomalachite.*

malacolac An ether-soluble, soft lac resin from shellac.

malacolite Augite.

malacon An impure zircon.

Malaguti, Faustino Jovita (1802–1878) Italian-born French chemist, noted for his work on the mass action of salts.

malakograph An apparatus to measure the rate of softening of wax by means of a falling weight attached to an indicator arm.

malamide* $NH_2CO \cdot CH_2 \cdot CHOH \cdot CONH_2$ = 132.2. 2-Hydroxybutanediamide*, malic amide. Colorless crystals, m.156. Cf. *malonamide.*

malaria A disease, due to a protozoan plasmodium which spends part of its life cycle in the anopheles mosquito. M. is transmitted to humans by the bite of an infected mosquito.

malate* A salt of malic acid, which contains the radical $-OCO \cdot CH_2 \cdot CHOH \cdot COO-$.

malathion* See *insecticides,* Table 45 on p. 305.

Malay camphor (+)-Borneol.

malayite A pigment made by heating silica, calcium carbonate, and tin oxide at 1400°C for 6 h. Used to make cloudy glass.

malchite A diorite containing quartz feldspar, hornblende, and biotite.

maldonite Au_2Bi. A pink, native alloy.

maleamic acid $NH_2CO \cdot CH{:}CH \cdot COOH$ = 115.1. Aminomaleic acid. Colorless crystals, m. 152.

maleate* A salt of maleic acid which contains the radical $-OCO \cdot CH{:}CH \cdot COO-$.

male hormone See *androgen, androsterone.*

maleic **m. acid*** $HOOC \cdot CH{:}CH \cdot COOH$ = 116.1. (Z)-2-Butenedioic acid*, ethylenedicarboxylic acid. Colorless prisms, m. 130 (decomp.), soluble in water; an isomer of fumaric acid. See *malenoid.* Used to adjust acidity in pharmaceutical preparations (BP). **amino ~** Maleamic acid. **methyl ~** Citraconic acid*. **m. anhydride*** $(HC \cdot CO)_2{:}O$ = 98.1. *cis-* **~** Butenedioic anhydride*. Colorless, trimetrical crystals, m.56, insoluble in water; used to manufacture polyester resins, alkyd resins, fungicides, and plasticizers. **m. hydrazide** $C_4H_4O_2N_2$ = 112.1. Colorless crystals, m. 250. A growth retarder in agriculture.

maleinamic acid Meleamic acid.

malenoid The (*Z*) form of geometrical isomerism as compared with the (*E*) form (fumaroid):

HC·CR	RC·CH
‖	‖
HC·CR	HC·CR
(*Z*) or *cis*	(*E*) or *trans*

maleoyl* Maleyl. The radical $C_4H_2O_2=$, from maleic acid.

maletto tannin $(C_{19}H_{20}O_9)_n$. Brown powder, from the bark of *Eucalyptus occidentalis,* soluble in water.

malic acid* $HOOC \cdot (OH)HC \cdot CH_2 \cdot COOH$ = 134.1.

Oxyethylenesuccinic acid, hydroxybutanedioic acid*. An intermediate in metabolism which exists in unripe fruit; 3 isomers. **dextro ~, (*R*)- ~** Colorless needles, m. 133, soluble in water. **inactive ~, (±)- ~** Colorless crystals, m.129, decomp. by heat, soluble in water. **iso ~** Methyltartronic acid. White solid, m. 160 (decomp.), soluble in water. **levo ~, (*S*)- ~** Common m. a. Colorless crystals, m.100, decomp.140, Cf. *aceric acid.* **3-hydroxy ~** Tartaric acid. **2-methyl ~** Citramalic acid.

malic amide Malamide*.

mallardite $MnSO_4 \cdot 7H_2O$. A vitreous, fusible, yellow or white mineral.

malleability The ability to withstand hammering or rolling without fracture or return to the original shape.

malleable Having malleability. **m. casting** A small iron casting, made m. by heating.

mallee bark The bark of *Eucalyptus occidentalis,* source of a commercial grade of eucalyptus oil.

Mallet, John William (1832–1912) Irish-born American chemist, noted for atomic weight determinations.

mallophene $Ph \cdot N{:}N \cdot C_5H_2N(NH_2)_2 \cdot HCl$ = 249.7. Phenylazo-α-diaminopyridine hydrochloride. Red powder, soluble in water; an antiseptic. Cf. *mellophanic acid.*

mallotoxin Rottlerin.

mallow Malva. The dried leaves of *Malva sylvestris* and *M.* species; herb teas. **marsh ~** Althaea.

malol Ursolic acid.

malonaldehydic acid* $CHO \cdot CH_2 \cdot COOH$ = 88.1. Formylacetic acid, decomp. 50; in tea.

malonamic acid $COOH \cdot CH_2 \cdot CONH_2$ = 103.1. The half-amide of malonic acid.

malonamide* $NH_2CO \cdot CH_2 \cdot CONH_2$ = 102.1. Propanediamide*, malonic diamide. Colorless crystals, m. 170. Cf. *malamide.*

malonate* $M_2C_3H_2O_4$. A salt of malonic acid.

malonic **m. amide** Malonamide*. **m. anhydride** Carbon suboxide. **m. dinitrile** Malononitrile*. **m. ester** $CH_2(COOEt)_2$ = 160.2. Diethyl malonate*. Colorless liquid, d.1.055, b.198. Its sodium compounds react with alkyl halides, yielding homologs of m. ester (m. e. synthesis).

malonic acid* $CH_2(COOH)_2$ = 104.1. Propanedioic acid*, methanedicarboxylic acid, occurring in many plants. Colorless, triclinic crystals, m. 132 (decomp.), soluble in water. **allyl ~** See *allylmalonic acid.* **bromo* ~** $CHBr(COOH)_2$ = 183.0. Colorless needles, decomp. 112, soluble in water. **butyl ~*** $CHBu(COOH)_2$ = 160.2. ***n*- ~** White crystals, m.102. **iso ~** m. 107. ***sec*- ~** m.76. **chloro ~*** $CHCl(COOH)_2$ = 138.5. Colorless crystals, m.133. **diethyl ~*** $CEt_2(COOH)_2$ = 160.2. White powder, m.121. **dimethyl ~*** $CMe_2(COOH)_2$ = 132.1. Colorless crystals, m.193. **ethyl ~*** $CHEt(COOH)_2$ = 132.1. White crystals, m.112, decomp. 160. Cf. *pyrotartaric, glutaric acid.* **ethylene ~** Vinaconic acid. **hexadecyl ~** $C_{19}H_{36}O_4$ = 328.5. White solid, m.121. **hydroxy ~** Tartronic acid*. **keto ~** Mesoxalic acid*. **methyl ~** Iso*succinic acid.* **nitrile ~** Cyanoacetic acid*. **oxo ~** Mesoxalic acid*. **oxy ~** Tartronic acid*.

malon oil Blackfish oil. An oil from the pilot whale, *Globicephalus melas.*

malononitrile* $CH_2(CN)_2$ = 66.1. Propanedinitrile*, methylene dicyanide. White powder, m.30, soluble in water; used in organic synthesis. **o-chlorobenzylidene ~** $Cl \cdot C_6H_4 \cdot CH{:}C(CN)_2$ = 188.6. CS gas. White powder, m. 52. A tear gas that also causes respiratory irritation.

malonurea Barbitone. Cf. *malonyl urea.*

malonyl* 1,3-Dioxo-1,3-Propanediyl†. The radical $-OC \cdot CH_2 \cdot CO-$, from malonic acid. **m. urea** Barbituric acid.

maloyl* The radical $-CO \cdot CH_2 \cdot CH(OH) \cdot CO-$, from malic acid.

malt Maltum. The grain of *Hordeum distichum* or *H. sativum* (barley), partly germinated, then dried. Yellow grains of biscuit odor and taste; used as a nutritive and digestant, and in the manufacture of malt extract, and beer. **m. extract** Extractum malti. A dark syrup obtained by evaporating an infusion of m.; used as a starch digestant, emulsifying agent, and vehicle for cod-liver oil. **m. liquor** An alcoholic beverage derived from fermented infusions of m.; as, beer.

maltase α-D-Glucosidase*.

maltha A tar from the oxidation of petroleum. Dark, asphaltlike masses, insoluble in water.

malthene Petrolene.

maltobionic acid $C_{12}H_{22}O_{12}$ = 358.3. An oxidation product of maltose.

maltoboise Maltose.

maltodextrin Amyloin. A polysaccharide, constitutionally between dextrin and maltose, and produced from the starch in barley during the manufacture of malt. Its composition depends on the relative amounts of maltose and dextrin.

maltol $C_6H_6O_3$ = 126.1. 3-Hydroxy-2-methyl-4*H*-pyran-4-one. Larixine, larixinic acid. A principle from the bark of the larch, *Larix europaea.* White crystals, m.161, soluble in water. A microchemical reagent, especially for vanadium.

maltonic acid Dextro*gluconic* acid.

maltosazone $C_{12}H_{14}O_7(:N \cdot NHPh)_4$. The osazone of maltose, used in the synthesis of $C_{220}H_{142}O_{58}N_4I_2$ = 4023. Heptatribenzoylgalloyl-*p*-iodophenylmaltosazone; one of the largest molecules synthesized (Emil Fischer).

maltose $C_{12}H_{22}O_{11} \cdot H_2O$ = 360.3. Malt sugar, 4-*O*-α-D-glucopyranosyl-D-glucose†, glucose-α-glucoside. A dextrodisaccharide from malt and starch. Colorless crystals, $+[\alpha]_D 138°$, soluble in water. A sweetening agent and fermentable intermediate in brewing. **iso ~** $C_{12}H_{22}O_{11}$ = 342.3. Gentiobiose. A disaccharide, from the hydrolysis of starch; a sweetener.

malt sugar Maltose.

maltum Malt.

malva Mallow.

Malvaceae Mallow family, a group of mucilaginous plants, e.g., *Malva sylvestris,* mallow.

malvidin The blue aglucone of the glucoside malvin in wild mallow.

malvin A glucoside and anthocyan pigment from mallow.

malvon $C_{29}H_{36}O_{20}$ = 704.6. An oxidation product of malvin and the glucoside of a didepside sugar ester.

man The highest living organism of the third group of living beings. Elementary composition of the human body: O 66.0, C 17.6, H 10.1, N 2.5, Ca 1.5%.

manaca The dried root of *Brunfelsia hopeana* (Solanaceae), Brazil; a diuretic and diaphoretic.

management information system MIS. A computer-organized version of the normal financial, sales, production, etc., reporting systems used by management.

Manchester brown Bismark brown.

mancona bark Sassy bark.

mancophalic acid $C_{10}H_{30}O_2$ = 182.3. An amorphous resin from manila copal.

mancozeb* See *fungicides,* Table 37 on p. 250.

mandarine (1) The reddish-yellow fruit of *Citrus nobilis.* (2) $C_{10}H_6(OH)N{:}NC_6H_4SO_3H$. Orange II, 2-naphtholorange. A monoazo dye. **m. oil** The essential oil from m., b. 175–179.

M. and B. 693 The 693d experiment of May and Baker, which produced the first of the sulfapyridine drugs (in 1936).

mandelic acid $Ph \cdot CH(OH)COOH$ = 152.2. Phenylglycolic acid, amygdalic acid, α-hydroxybenzeneacetic acid†, three

TABLE 51. OXIDATION STATES OF MANGANESE

Oxidation state	Name	Ion	Color of compounds
2	M.(II)*, m.(2+)*, manganous	Mn^{2+}	Slightly pink
3	M.(III)*, m.(3+)*, manganic	Mn^{3+}	Slightly green
5	Manganate(V)*, manganite	MnO_4^{3-}	Blue
6	Manganate(VI)*	MnO_4^{2-}	Dark green
7	Permanganate*	MnO_4^-	Dark purple

isomers. **inactive ~**, **(±)-~**. Colorless rhombic crystals, m.118, soluble in water. **levo ~**, **(S)-**. The natural form, from amygdala, m.133. **para ~** Inactive. **methyl ~** Atrolactic acid. **phenyl ~** Benzilic acid*.

Mandelin's reagent A reagent for alkaloids: 0.5 g vanadium chloride in 100 mL concentrated sulfuric acid.

mandelonitrile* Ph·CH(OH)CN = 133.2. **(±)-~** Benzaldehyde cyanohydrin. Yellow oil, d.1.124, b.170 (decomp.), insoluble in water; used in organic synthesis.

mandioc Tapioca.

mandrake Podophyllum.

mandrel A handle or shaft in which a rotating tool is held.

maneb* See *fungicides,* Table 37 on p. 250.

manelemic acid A constituent of elemi. **alpha-~** $C_{37}H_{56}O_4$. **beta-~** $C_{44}H_{80}O_4$.

mangabeira *Hancornia speciosa,* Gom. A plant from Bahia, the latex of which is a source of rubber.

mangal A noncorrosive aluminum alloy containing 1.5% manganese.

mangan Manganese. **m. blende** MnS. A native manganese sulfide.

manganate* Indicating manganese as the central atom(s) in an anion. **m.(V)*** Manganite. A salt containing the radical MnO_4^{3-}. **m.(IV)*** A salt containing the radical MnO_4^{2-}. **per ~*** A salt containing the radical MnO_4^-.

manganese* Mn = 54.9380. A metallic element, at. no. 25. A grayish-pink, lustrous, brittle metal, d.7.2, m.1240, b.1960, reacts with boiling water, soluble in acids. Discovered (1774) by Scheele, isolated (1789) by Gahn, and named from the Greek *manganidso* ("to purify") in allusion to its use to neutralize the green iron color in the manufacture of glass. Valencies: see Table 51. **black ~** Pyrolusite. **dvi ~** Rhenium*. **eka ~** Technetium*. **red ~** (1) Rhodonite. (2) Rhodochrosite.

m.(II)*, m.(2+)* See also *manganous.* **m.(III)*, m.(3+)*** See also *manganic.* **m. acetate** Manganous acetate. **m. blende** MnS. A native m. sulfide. **m. boride*** MnB_2 = 76.6. A black powder, insoluble in water. **m. boron** A manganese bronze: Cu 88, Sn 10, Mn 2%. **m. carbide*** Mn_3C = 176.8. Black crystals. **m. chloride** See *manganic, manganous.* **m. copper** M. *bronze.* **m. dioxide*** MnO_2 = 86.9. M. peroxide, pyrolusite, battery m. Black powder, decomp. 390, insoluble in water. Used as an oxidizing agent; in halogen manufacture; in electric dry cells; in paints and varnishes; as a black or purple color for glass and ceramics; for making manganese compounds; in the rubber industry. **m. dithionate*** MnS_2O_6 = 215.1. Manganous hyposulfate. Colorless needles, soluble in water. **m. green** Barium manganate*. **m. heptoxide*** Mn_2O_7 = 221.9. Permanganic acid anhydride. Green liquid, rapidly decomp. to m. dioxide; a powerful oxidizing agent. **m.(II) hexacyanoferrate(II)*** $Mn_2[Fe(CN)_6]$ = 321.8. Manganous ferrocyanide. Green powder, soluble in cyanide solutions. **m.(II) hexafluorosilicate*** $Mn[SiF_6]\cdot 6H_2O$ = 305.1. Manganous fluo(ro)silicate. Pink, hexagonal prisms, soluble in water. **m.(II) m.(III) oxide*** Mn_3O_4 = 228.8. Manganic manganous oxide. Red crystals, insoluble in water. **m. minerals** Principal ores: pyrolusite, MnO_2; braunite, Mn_2O_3; manganite (manganese spar rhodochrosite), MnO(OH); mangan blende, MnS. **m. nitrides** Mn_5N_2 = 302.7. Mn_3N_2 = 192.8. **m. nodules** Concentric encrustations of m. and iron oxides and other metals, occurring alternately around a clay nucleus in tropical Pacific waters. **m. oxides** (1) MnO. Manganous oxide. (2) Mn_3O_4. M.(II) m.(III) oxide*. (3) Mn_2O_3. Manganic oxide. (4) MnO_2. M. dioxide*. (5) MnO_3. M. trioxide*. (6) Mn_2O_7. M. heptoxide*. **m. peroxide** M. dioxide*. **m.(II) phosphinate*** $Mn(H_2PO_2)\cdot H_2O$ = 202.9. Manganous hypophosphite. Pink crystals, soluble in water. **m. potassium sulfate*** $Mn_2(SO_4)_3\cdot K_2SO_4\cdot 24H_2O$ = 100.5. Mangan alum. Green crystals, soluble in water. **m. protoxide** Manganous oxide. **m. sesquioxide** M.(II) m.(III) oxide*. **m. silicate** Mn_2SiO_4. See *braunite, rhodonite, tephroite.* **m. spar** (1) Rhodonite. (2) Rhodochrosite. **m. steel** An extremely hard and ductile steel containing 12% Mn. **m. sulfate** See *manganic sulfate, manganous sulfate.* **m. tetrachloride*** $MnCl_4$ = 196.8. Green solid. **m. tetrafluoride*** MnF_4 = 130.9. Brown solid. **m. titanium** An alloy of m. and titanium, used in the steel industry. **m. trioxide*** MnO_3 = 102.9. An acidic oxide, which forms manganates. **m. tungstate*** See *hübnerite.*

manganesium Manganesum. An early name for manganese.

manganic Manganese(III)* or (3+)*. Describing compounds of trivalent manganese; generally unstable and yield the green Mn^{3+} ion, which readily decomposes to the stable Mn^{2+} state (manganous). **m.(VI) acid*** H_2MnO_4 = 120.9. An acid known only as salts of the type M_2MnO_4, manganate(VI). **per ~*** $HMnO_4$ = 119.9. Red, unstable liquid; forms salts of the type $MMnO_4$, permanganates. **m. chloride** $MnCl_3$ = 161.3. An unstable compound which forms more stable double salts; soluble in water, and decomposes to $MnCl_2$ and Cl_2. **m. hydroxide** $Mn(OH)_3$ = 106.0. An unstable hydroxide, which forms brown MnO(OH); a pigment for textiles. **m. manganous oxide** Manganese(II) manganese(III) oxide*. **m. metaphosphate** $Mn_2(PO_3)_6\cdot 2H_2O$ = 619.7. Pink crystals. **m. oxide** Mn_2O_3 = 157.9. Manganese trioxide*, m. sesquioxide, black m. oxide, braunite. Black powder, insoluble in water. **hydrated ~** $Mn_2O_2(OH)_2$ = 175.9. Black powder. **m. sulfate** $Mn_2(SO_4)_3$ = 398.0. Green crystals, decomp. by water or at 160; a powerful oxidizing agent.

manganiferous Containing or carrying manganese.

manganin An alloy: Mn 12, Cu 84, Ni 4%, m.910; used in electric heating elements.

manganite (1) $Mn_2O_3\cdot H_2O$. Acerdese. A native manganic oxide hydrate. (2) Manganate(V)*.

manganomanganic oxide Manganese(II) manganese(III) oxide*.

manganosite MnO. A native, emerald-green manganous oxide.

manganostilbite A mineral: manganous oxide with orpiment and antimonous sulfide.

manganotantalite A mineral: manganous oxide, and tantalum oxide, tin, and tungsten.

manganous Manganese(II)* or (2+)*. Describing the manganese salts containing the Mn^{2+} ion; usually colorless or slightly pink, soluble in water. **m. acetate** $Mn(C_2H_3O_2)_2\cdot 4H_2O$ = 245.1. Pink, monoclinic crystals, soluble in water.

m. arsenate $MnHAsO_4$ = 194.9. Manganese hydrogenarsenate*. Pink powder, soluble in water. **m. benzoate** $Mn(C_7H_5O_2)_2$ = 297.2. Manganese benzoate*. Colorless scales, soluble in water. **m. borate** MnB_4O_7 = 210.2. Manganese borate, manganese siccative. White powder, soluble in water; a paint drier. **m. bromide** $MnBr_2$ = 214.7. Manganese dibromide*. Pink crystals, soluble in water. **m. carbonate** $MnCO_3$ = 114.9. Manganese carbonate*, native as dialozite and rhodochrosite. Pink rhombs, decomp. by heat, insoluble in water. **m. chloride anhydrous ~** $MnCl_2$ = 125.8. Pink crystals, m.650, soluble in water. **m. c. hydrate** $MnCl_2 \cdot 4H_2O$ = 197.9. Pink, monoclinic crystals, m.87, soluble in water. Used as a mordant and disinfectant; in the manufacture of manganese salts; in the glass industry. **m. chromate** $MnCrO_4$ = 170.9. Brown powder, partly soluble in hot water. **m. ferrocyanide** Manganese hexacyanoferrate*. **m. fluoride** MnF_2 = 92.9. Pink powder, soluble in water. **m. fluosilicate** Manganese hexafluorosilicate*. **m. formate** $Mn(OOCH)_2 \cdot 2H_2O$ = 181.0. Rhombic, red crystals, soluble in water. **m. hydrate** M. hydroxide. **m. hydroxide** $Mn(OH)_2$ = 89.0. Native as pyrochroite. White hexagons, decomp. by heat, insoluble in water. **m. iodide** $MnI_2 \cdot 4H_2O$ = 380.8. Pink, monoclinic crystals, decomp. by heat, soluble in water. **m. linoleate** $Mn(C_{18}H_{31}O_2)_2$ = 613.8. Brown, fatty mass, insoluble in water; used in paint driers. **m. nitrate** $Mn(NO_3)_2 \cdot 6H_2O$ = 287.0. Pink, monoclinic crystals, m.26, soluble in water; a reagent. **m. oleate** $Mn(C_{18}H_{33}O_2)_2$ = 617.9. Brown, granular, fatty mass, insoluble in water; used in ointments, and as a drier for varnishes. **m. oxalate** $Mn(OOC)_2 \cdot 2\frac{1}{2}H_2O$ = 188.0. White, crystalline powder, decomp. 150; slightly soluble in water; a drier. **m. oxide** MnO = 70.9. Manganese protoxide. Green powder, insoluble in water, soluble in acids. **manganic ~** Manganese(II) manganese(III) oxide*. **m. phosphate** (1) **m. orthophosphate** $Mn_3(PO_4)_2 \cdot 7H_2O$ = 480.9. Manganese orthophosphate*. Red powder, insoluble in water; A reagent. (2) $MnHPO_4$ = 150.9. Manganese hydrogenphosphate*. Pink crystals, soluble in water; used in analysis. (3) **m. pyrophosphate** $Mn_2P_2O_7$ = 283.8. White powder, d.3.58, insoluble in water. **m. propionate** $Mn(C_3H_5O_2)_2$ = 201.1. Pink powder, slightly soluble in water. **m. silicate** $MnSiO_3$ = 131.0. Pink crystals, m.1218, insoluble in water; used in ceramics and glass. **m. succinate** $MnC_4H_4O_4$ = 171.0. White crystals, soluble in water. **m. sulfate** (1) **anhydrous ~** $MnSO_4$ = 151.0. Native as mallardite. Red crystals, soluble in water. (2) **m. s. tetrahydrate** $MnSO_4 \cdot 4H_2O$ = 223.1. Labile, pink prisms, slightly hygroscopic, m.30, soluble in water. Used in the ceramic and glass industries; as a mordant in the textile industry. (3) **m. s. pentahydrate** $MnSO_4 \cdot 5H_2O$ = 241.1. Stable at 8–27. (4) **m. s. heptahydrate** $MnSO_4 \cdot 7H_2O$ = 277.1. Red prisms, d.3.1, m.280 (decomp.). **m. sulfide** MnS = 87.0, or $MnS \cdot H_2O$ = 105.0. Manganese sulfide*, native as alamandite and mangan blende. Gray, pink, or brown fusible powder, insoluble in water; a pigment. **m. sulfite** $MnSO_3$ = 135.0. Gray crystals, insoluble in water. **m. tartrate** $MnC_4H_4O_6$ = 203.0. White crystals, slightly soluble in water. **m. valerate** $Mn(C_5H_9O_2)_2 \cdot 2H_2O$ = 293.2. Brown powder, slightly soluble in water.

manganum Latin for "manganese."

mangiferin $C_{19}H_{18}O_{11}$ = 422.4. Euxanthogen. A principle from the leaves of *Mangifera indica;* thin, yellow needles, m.271.

mango The tree *Mangifera indica* (Anacardiaceae), India. **m. gum** An amber-colored or red-yellow resin from the m. tree.

mangrove An extract from the bark of *Rhizophora mucronata,* containing 30–35% tannin; used in tanning.

Manihot A group of S. American shrubs and herbs (Euphorbiaceae) which yield cassava, Brazilian arrowroot, cassareep, and ceara rubber.

Manila M. copal A resin from *Agathis dammara,* a conifer of the Philippine Islands; contains 80% mancophalic acid. **M. hemp** Abaca. **M. paper** A strong paper, originally made from old M. rope (M. fibers).

manioc Cassava.

manioca Tapioca.

manna The dried saccharine exudation of *Fraxinus ornus* (Oleaceae), the Orient. It contains mannitol, and forms a yellowish-white mass of sweet, slightly acrid taste. It can be compressed into tacky lumps for storage and subsequent purification and consumption, and is said to have been the food of the Israelites; a mild cathartic. **Armenian ~** Glucose. A m. from oak trees, containing glucose. **Australian ~** A m. from the eucalyptus species, containing melitose. Cf. *lerp.* **yeast ~** Yeast gum.

mannan A glucoside constituent of yeast gum and manna, analogous to araban.

Mannich reaction The condensation of an amine and an aldehyde with a compound containing an acidic H atom attached to a C or N atom.

mannide $C_6H_{10}O_4$ = 146.1. An anhydride of mannitol, b.317. **iso ~** A solid, m.87, b.274.

mannitan $C_6H_{12}O_5$ = 164.2. D-1,4-Anhydromannitol. Crystals, m.146.

mannite Mannitol.

mannitol $C_6H_{14}O_6$ = 182.2. Mannite. A hexahydric alcohol from manna and many plants (larch, sugarcane, *Viburnum, Syringa,* and *Fraxinus* species); or, by electrolysis, from glucose. 3 isomers: D, L, and (±). Colorless needles, m.166, soluble in water. A mild laxative when given by mouth. An osmotic diuretic when given intravenously; used to reduce intracranial pressure and to counteract poisoning (USP, BP). A reagent for detecting glucose. **D-m. hexanitrate** $C_6H_8(NO_3)_6$ = 452.2. Hexanitrin, MHN, nitromannite, m.108; a substitute for mercury fulminate in high explosives.

mannitose Mannose.

mannoheptitol Perseitol.

mannoheptonic acid $C_7H_{14}O_8$ = 226.2. **D- ~** White solid, decomp. 175, soluble in water.

mannoheptose $C_7H_{14}O_7$ = 210.2. **α-D- ~** A heptose, m.134, in the avocado fruit of *Persea gratissima* (Lauraceae).

mannolite Chlorazene.

mannosans $(C_6H_{10}O_5)_n$. Polysaccharides, which hydrolyze to mannose.

mannose $CH_2OH(CHOH)_4CHO$ = 180.2. A hexose or fermentable monosaccharide and isomer of glucose from manna; 2 optically active forms. **D- ~** Seminose. Colorless prisms, m.132, soluble in water.

mannoside A glucoside which yields mannose. Cf. *rhamnomannoside.*

mannotriose $C_{18}H_{32}O_{16}$ = 504.5. Glucose galactose galactoside. A trisaccharide from manna, indigestible by man.

manocryometer A device to determine the melting point under pressure.

manometer An instrument to measure the pressure of gases or liquids. **gas ~** See *gage, McLeod gauge.* **mercury ~** A U-tube filled with mercury; the difference in the heights of the arms indicates the pressure. **sphygmo ~** An instrument for measuring blood pressure. **spring ~** An instrument constructed from a coiled tube into which the gas or steam passes and, according to its pressure, uncoils and records on the dial. Cf. *McLeod gauge.* **water ~** A U-tube filled with water, with one end open and the other connected to a gas container. The difference in the heights of the water columns indicates the gas pressure. Classification of m.:

Ionization gage	0.0001–0.1 μm Hg
Pyrometric gage	0.001–1 μm Hg
McLeod gauge	0.01–1,000 μm Hg
Micromanometer	0.01–10 mm Hg
U-tube	1–1,000 mm Hg
Single tube	2–2,000 mm Hg
Spring manometer	3–70 kPa (gage)
Diaphragm	3–1,400 kPa (gage)
Carbon pile	0.3–14 MPa (gage)
Crystals	0.7–140 Mpa (gage)
Steel tube spiral	1–10 MPa (gage)

manool The starting point for the synthesis of ambergris perfumes. A diterpene alcohol from the oil of the New Zealand yellow pine, *Dacrydium biforme.*
manoscopy Gas-volumetric analysis.
mantissa As an example, the number 8.32 in the number 8.32×10^4; 4 is the exponent.
mantle The outer wall of a furnace. **filter ~** Berkefeld filter. **gas ~** Welsbach mantle.
Manucol Trademark for an alginate thickening agent.
manure Refuse, e.g., excreta, straw, etc.; used as fertilizer. Average composition: nitrogen 0.6, phosphorus pentaoxide 0.3, potassium oxide 0.6%. **m. salt** A potassium salt (chiefly chloride) containing 20–30% potash.
manzoul A narcotic mixture of hashish and muscat nut.
maple The tree *Acer saccharum* (U.S.) or *A. campestris* (U.K.). **red ~** The bark of *A. rubrum,* swamp m.; used by American Indians to cure sore eyes. **m. sugar** Brown mass, chiefly sucrose with glucose, coloring matter, and proteins, from evaporated m. syrup. **m. syrup** The concentrated sap of *A. Saccharum* (1–2 kg per tree).
mapp gas Generic name for a stabilized industrial fuel gas (methylacetylene and propadiene) which combines the versatility of acetylene with the safety of propane.
MAR Abbreviation for microanalytical reagent.
mar-aging Maraging. Modification of the martensitic structure of steel by heating at 450–500°C for 3 hours, followed by air cooling without quenching. It produces high strength, impact, and ductility values.
Maranta Arrowroot.
marble $CaCO_3$. Native limestone recrystallized under the influence of heat, pressure, or both, in many forms and colors.
marc (1) The residual vegetable tissue and mucilage after expression of oil from a plant or nut kernel. (2) The cellular tissue left after complete extraction of the juice from sugar beet or sugarcane. Cf. *bagasse.*
Marcain(e) Trademark for bupivicaine hydrochloride.
marcasite FeS_2. Coal brass, white iron, spear, binarite, coxcomb, radiated pyrites. Yellow, orthorhombic crystals.
Marcet **M., Alexander (1770–1822)** Swiss physician, who became an English chemist and noted lecturer-demonstrator. **M., Jane Haldemand (1769–1858)** Swiss exponent of popular science *("Conversations on Chemistry").*
Marchand **M., Richard Felix (1813–1850)** German chemist, noted for atomic weight determinations. **M. tube** A U-shaped calcium chloride tube, with bulb and side tube attached.
marcitine $C_8H_{19}N_3$ = 157.3. A basic substance from putrid pancreas.
marennin Green pigment in certain French oysters, derived from the chlorophyll of microorganisms present.
maretine $MeC_6H_4NH \cdot NH \cdot CONH_2$ = 165.2. 1-(3-oyl) semicarbazide. Colorless crystals, m.184, insoluble in water.
Marezine Trademark for cyclizine hydrochloride.
marfacing Mar-aging, q.v., to produce a hard-faced deposit.
Marfanil Trademark for mafenide acetate.
margaric acid Heptadecanoic acid*.
margarine A butter substitute; a solid emulsion of fats in milk serum. Named after its inventor, Mège-Mouries, 1870. **oleo ~** The liquid fat from which m. is made by hydrogenation, which saturates the double bond.
margarite $CaAl_4H_2Si_2O_{12}$. Lustrous, pearly, monoclinic masses of various shades.
margaron $(C_{16}H_{33})_2O$ = 466.9. Dihexadecyl ether*. White powder from beef suet; an ointment base.
margaronitrile Heptadecane nitrile*.
Marggraf, Andreas Sigismund (1709–1782) German chemist, founder of the beet sugar industry.
margosa oil Neem oil, veepa oil, veppam fat, oil of azedarach, from the seeds of *Melia azedarach,* the bead tree, Indian lilac, cape syringa or china tree (Meliaceae), Asia and Africa.
margosic acid $C_{22}H_{40}O_2$ = 336.6. A fatty acid from margosa oil; probably impure oleic acid.
marialite $2NaCl.2Na_2O.3Al_2O_3 \cdot 18SiO_2$. A vitreous, green mineral.
Mariana's trench The ocean's deepest area (11,000 m).
Marie Davy cell Amalgamated zinc in dilute sulfuric acid, as anode; carbon, in a paste of mercurous sulfate, as cathode (1.5 volts).
Marignac **M., Jean Charles Galissard de (1817–1894)** Swiss chemist, noted for determinations of atomic weights and investigations of rare earth metals. **M. salt** Potassium stannum sulfate.
marigold Calendula.
marihuana Cannabis.
marinating Picklingin brine.
Mariotte **M., Edme (1629–1684)** A French prior. **M. law** The product of volume and pressure of a gas is constant (1676). Cf. *Boyle's law.*
marjoram The herbs *Origanum* (Labiatae). **sweet ~** The herb and leaves of *O. major.* **wild ~** The herb of *O. vulgare.* It yields origanum oil.
m. oil An essential oil from *Origanum majorana* (Labiatae). Odorless liquid, d.0.9, insoluble in water; a perfume. Cf. *origanum oil.* **French ~** The essential oil of *Calamintha nepeta* (Labiatae) containing calaminthone.
marking Branding. **m. apparatus** A microscope objective, used to make small circles on the cover glass, marking fields for reference. **m. ink** A solution of silver nitrate; used to mark textiles in laundries. **m. nut** Semecarpus.
Markovnikov rule In the addition reaction of a polar molecule to an alkene or alkyne, the least hydrogenated C atom of the latter will combine with the more electronegative atom of the former. Thus, when HCl is added to a double bond, the Cl atom combines with the C atom having the lowest number of H atoms.
marl (1) A soil consisting of clay, sand, and chalk; used as a fertilizer. (2) An earthy or soft rock deposit rich in calcium carbonate.
Marlex Trademark for a high-density polyethylene synthetic fiber.
Marlspun Trademark for a cellulose acetate synthetic fiber.
marmatite A native zinc, iron, and manganese sulfide.
marmelide $C_{16}H_{14}O_4$ = 270.3. A coumarin derivative, which accelerates melanin formation by enzyme action.
Marme's reagent A solution of potassium iodide 6 and cadmium iodide 3 g, in 18 ml water. A reagent for alkaloids (white or yellow precipitate).
Marmite (1) Trade name for a food prepared by treating

dried yeast with an acid under pressure, and neutralizing the product. (2) (Not cap.) A tall, lidded vessel for boiling large volumes of bouillon or media.

Marne, N. H. Johann Bernhard Herrmann. German phlogistic chemist who attempted the first classification of elements (1786).

maroti oil Fatty oil, d.0.96, expressed from the seeds of *Hydnocarpus wightiana* (Flacourtiaceae).

marrubium Horehound.

Marsh **M., James (1789–1846)** English chemist. **M. test** Marsh-Berzelius test for arsenic. The substance is added to pure zinc, and pure hydrochloric acid is slowly added; the evolved hydrogen with any arsine, AsH_3, is passed through a long tube which is heated at the end so that arsenic is deposited; or the escaping hydrogen is ignited, and a cold porcelain dish is held above the flame. The arsenic deposits on it as a black mirror that can be dissolved in potassium hypochlorite solution, while any antimony deposit does not dissolve.

Marshall apparatus A device for the determination of urea in blood.

marsh **m. gas** The gaseous products, chiefly methane, formed from decaying, moist organic matter in marshes. Cf. *firedamp.* **m.mallow** Althaea. **m. mint** Wild mint. The herb *Mentha sativa* (Labiatae). **m. ore** Bogore. **m. tea** Labrador tea.

Martens **M. densitometer** An optical device to measure the density of the silver deposit on photographic plates. **M. illuminator** A photometer to determine illumination efficiency. **M. spectroscope** A direct-vision spectroscope. **M. test** The determination of the temperature at which the free end of a testpiece, subjected to a specified bending stress at a gradually increasing temperature, first shows signs of deflection.

martensite Normally, a solid solution of 2% carbon in iron, present in quenched steel. On slow cooling it decomposes into iron and iron carbide. **alpha-∼** The tetragonal form of normal (beta-) martensite.

martial ethiops Magnetic ferric oxide.

Martin **M.'s flask** A culture flask, consisting of a glass bulb with 3 long necks; used to manufacture toxins. **M.'s centrifuge** A small laboratory centrifuge driven by waterpower. **M.'s filter** A Berkefeld filter with a funnel, for filtering toxins.

martite A native ferric oxide. See *hematite.*

Martius yellow The calcium salt of naphthalene yellow.

Marvinol Trademark for a polyvinyl rubber substitute.

Marzine Trademark for cyclizine hydrochloride.

mascagnine A mineral ammonium sulfate.

maser Abbreviation for *m*icrowave *a*mplification by *s*timulated *e*mission of *r*adiation. An ultrasensitive amplifier based on the use of a synthetic crystal at −269°C, so that the thermal motion of the atoms present does not interfere with reception of very weak radio signals. Cf. *laser.*

mash A warm mixture of malted barley and water used to prepare brewer's wort.

masked Hidden; concealed. **m. element** An element combined in an organic compound so that its properties are subdued or hidden, and it does not give the usual reactions; e.g., iron is masked in hemoglobin. **m. radical** An atomic group present in an organic compound, but combined so that its usual properties are subdued.

maslin A mixture of grains, especially wheat and rye.

Masonite Gun fiber. Trademark for a wood fiber constructional and insulating material. **M. process** Chips of wood are placed in autoclaves (guns) and subjected to high-pressure steam. Upon sudden release, the chips explode, and are then recombined as slabs, etc.

mass Abbrev. *m.* A definite quantity of *matter,* q.v., which offers resistance to change of motion. The physical quantity of an electron, atom, molecule, or an assembly of these, e.g., mole. Unit: kilogram. Unlike *weight,* q.v., mass is unchangeable; e.g., 1 g water always contains the same number of molecules. For various magnitudes of mass, see Table 52. **active ∼** The number of moles (gram molecules) in unit volume (1 liter). Cf. *unified atomic mass constant.*

TABLE 52. VARIOUS MAGNITUDES OF MASS

	Kilograms
Einstein universe	8×10^{75}
Galaxy (Milky Way system)	2.3×10^{52}
Large stars	1×10^{33}
Sun (331,950 × earth)	1.99×10^{30}
Earth	5.98×10^{24}
Mercury (1/19 × earth)	3.2×10^{23}
Moon (1/81 × earth)	7.4×10^{22}
Hydrosphere (1,660,000,000,000 million tons)	1.7×10^{21}
Atmosphere (5,140,000,000 million tons)	5.1×10^{18}
Average asteroid (1,000,000,000 million tons)	1×10^{18}
Large tanker (250,000 tons)	2.5×10^{8}
Meteorites, annual fall (40,000 tons)	4×10^{7}
1 ton (2,205 lb)	1×10^{3}
Man, average weight (166 lb)	7.6×10^{1}
U.S. quarter (25¢)	5.6×10^{-3}
Microbalance sensitivity (0.4 μg)	4×10^{-13}
Sodium by flame test (0.07 μg)	7×10^{-14}
Thiols by odor (0.002 μg)	2×10^{-15}
Colloid particle	2×10^{-21}
Protein molecule (1,000,000 × H)	1×10^{-21}
Oil film (0.000 000 2 cm square)	2×10^{-24}
Cane sugar molecule (342 × H)	5.7×10^{-25}
Water molecule (18 × H)	3×10^{-26}
Hydrogen atom	1.663×10^{-27}
Electron (1/1,836 × H)	9.11×10^{-31}

relative atomic ~* See *atomic weight.* **relative molecular ~*** See *molecular weight.*
m. action A law of chemical reaction: In a homogeneous system, the product of the substance concentrations of the participating substances on one side of the equation, when divided by the product of the substance concentrations of the substances on the other, is constant (*K*) for each temperature. It applies to both direct and reversible reactions (Guldberg and Waage). Hence, the velocity of a chemical reaction is proportional to the substance concentrations of the reacting substances. **m. a. constant** The constant *K* in the equation, which applies to other equilibria, e.g., dissociation. **m. conservation** A law of physics: matter cannot be destroyed or created; all changes of matter are transformations (analogous to the law of energy conservation). Cf. *energy, matter.* **m. density*** The mass of a body divided by its volume. Cf. weight *density.* **m. energy cycle** The transformation of matter into energy and vice versa. Cf. *relativity, materialization.* **m. fraction*** The mass of the component divided by the mass of the system. Its correct unit is unity; but g/kg, ppm, %, w/w, etc. are in use. **m. law** (1) M. conservation. (2) M. action. **m. number*** Nucleon number*. The nearest integer to the number expressing the m. of the corresponding neutral atom. It equals the total number of protons and neutrons. Dehoted by a prefix superscript; as, ^{32}S. Cf. *isobar.* **m. spectra** Aston spectra. Spectra of isotopes. The separation of an element into its isotopes by making it the anode in a vacuum discharge tube. The canal rays (formed behind the perforated cathode), when exposed to a magnetic field, are deviated from a straight path proportionally to their mass, and made visible by a fluorescent screen or photographic plate. **m. spectrogram** The photographic images produced when positive rays, deflected by a magnetic or electric field, fall on a photographic plate. **m. spectrograph*** An apparatus that separates beams of ions according to their mass-to-charge ratio, the deflection and intensity of the beams being recorded on photographic film. **m. spectrometer*** An apparatus that separates ions according to their mass-to-charge ratio, the ions being measured electrically. **m. spectroscope*** Term embracing both m. spectrograph and m. spectrometer. **quadrupole ~** A device in which the ions to be analyzed are separated by superimposed radio frequency and d.c. electric fields.

massecuites A mixture of syrup and crystals of cane sugar, used in the sugar industry.

massicot A native lead monoxide, d.9.3, m.600.

mast cells Tissue cells, which provide the histamine contributing to many symptoms of allergy.

mastic (1) Mastiche. The concrete, resinous exudations from *Pistacia lentiscus* (Anacardiaceae), Mediterranean. Yellow or green, transparent resin, insoluble in water, soluble in alcohol. Used in lacquers, incense, plasters, chewing gum; and as a wound covering (BP). (2) A mixture of finely powdered rock and bituminous material, used for highway construction. (3) A mortar for plastering walls; finely ground limestone, sand, litharge, and linseed oil. **m. oil** An essential oil from m., d.0.863, b.157, containing α-pinene.

mastic acids A group of resinous acids obtained from mastic; e.g., mastichic acid, $C_{20}H_{32}O_2$ (α-resin); masticin, $C_{20}H_{31}O$ (β-resin).

mastication The stage of chewing food in which it is mixed with the salivary enzymes, and in which amylolytic changes begin. Cf. *digestion.*

masticatory An agent to increase the secretion of saliva; e.g., chewing gum.

mastix Mastic (1).

masurium Early name for technetium.

masut Mazout. The residue remaining after the distillation of benzine and kerosene from Russian petroleum: C 87, H 12, O 1%; a fuel oil; burns at about 100°C.

mat Matte (2).

match A small strip of wood, paper, or wax tipped with a pyrophoric mixture. **lucifer ~** A m. tipped with a paste of scarlet or yellow phosphorus or P_4S_3, gum, red lead, and sometimes potassium chlorate; ignites when rubbed on sandpaper. **safety ~** A m. tipped with a mixture of antimony sulfide, potassium chlorate, potassium dichromate, and gum and coloring pigments; ignites when rubbed on paper coated with red phosphorus, glass powder, and gum.

maté Paraguay tea, yerba, Jesuit tea, Brazil tea, yerba maté. The dried leaves of *Ilex paraguayensis* (Aquifoliaceae), S. America, containing caffeine. Its infusion is a beverage. **Zapek ~** Maté leaves which have been heated.

materialization The production of matter, e.g., electrons and positrons, by the transformation of γ rays. Cf. *mass*-energy cycle.

Materia Medica (1) Knowledge of the natural history, physical characters, and chemical properties of drugs. (2) Pharmacy. (3) Pharmacology. (4) Therapeutics.

matesterin $C_{23}H_{40}O_3$(?). A dihydroxy sterol from maté. White needles, m. 270.

matico The dried leaves of *Piper angustifolium* (Piperaceae). **m. camphor** $C_{12}H_{16}O$ = 176.3. A terpene from matico. **m. oil** An essential oil from m., d.0.930–1.130, soluble in alcohol.

matildite $AgBiS_2$. A silver ore.

matlockite $PbO \cdot PbCl_2$. A lead mineral from Matlock, England.

matrass Distilling flask (obsolete).

matricaria German chamomile. The dried flower heads of *M. chamomilla* (Compositae); a febrifuge (EP, BP).

matrine (1) $C_{15}H_{24}ON_2$ = 248.4. Isolupanine. The chief alkaloid of kuh-seng; 4 isomers: **alpha-~** Needles, m.77. **beta-~** Rhombic prisms, m.87. **gamma-~** A liquid, d.1.088, b.223. **delta-~** Prisms, m.84. All soluble in water.

matrix (1) Groundmass; rock or earth which contains a metallic ore or mineral. (2) The impression left in a rock by a fossil or crystal. (3) The material surrounding a precious stone. (4) A conventional arrangement of numbers in horizontal rows and vertical columns, the latter being the more numerous; used to interpret quantum numbers. Cf. *Pauli's* principle. **m. theory** The spectral line which corresponds with a transition from one quantum state to another is expressed by an amplitude or intensity factor and a frequency or energy factor. Cf. *quantum theory.*

matte The crude metal, obtained by smelting sulfide ores, which still contains some sulfur.

matter Any body subject to gravitation; hence any substance that occupies space. Cf. *mass.* **annihilation of ~** The theory that m. is destroyed in the interior of a star by the transformation of mass into radiation. Cf. *energy, mass*-energy cycle. **conservation of ~** See *mass conservation.* **creation of ~** The assumption that the diffuse gaseous nebula absorb radiation and transform it into m. Cf. *energy, mass*-energy cycle. **destruction of ~** See *annihilation of matter* (above), *disintegration of matter* (following). **disintegration of ~** The radioactive transformations by which m. of one kind is transformed into m. of another with liberation of energy. Cf. *radioactivity.* **transformation of ~** Chemical changes; reactions. **volatile ~** See *volatile matter.*

Matthiessen's rule The product of resistivity and

temperature coefficient of resistance is constant irrespective of a metal's purity.

matt salt Ammonium hydrogendifluoride.

maturation Geochemical changes, as in oil, gas, and coal fields, resulting in the loss of volatile compounds and increasing aromaticity of the residue.

Maumené **M., Edme Jules (1818–1891)** French chemist. **M. number** The rise in temperature occurring in the M. test. **M. test** 50 g of oil is added to 10 mL of concentrated sulfuric acid; if the rise in temperature exceeds 70 °C, drying oils are present.

mauvein $C_{27}H_{24}N_4$ = 404.5. Aniline purple, Perkin's mauve. A violet dye of the phenyl safranine group; the first aniline dye (Perkin, 1856).

max. Abbreviation for maximum.

maximal Having attained the greatest value. **m. work** The greatest amount of energy obtainable from a process or reaction.

maximum The largest quantity or value. **m. boiling-point mixture** That mixture of 2 or more liquids which has the highest boiling point. **m. temperature** The temperature above which the growth of bacteria does not take place.

maxite Leadhillite.

maxivalence The highest valency of an element; for some elements it is related to the group number in the periodic table.

Maxolon Trademark for metoclopramide.

Maxton screen A rotating screen.

Maxwell **M., James Clerk (1831–1879)** Scottish physicist. **M.-Boltzmann distribution law** An expression, based on the theory of probability, for the number of moles per unit volume of gas at equilibrium n_i which have a certain energy E. $n_i = n_0 \cdot e^{-E}/kT$, where n_0 is the number per unit volume for $E = 0$ and k is the Boltzmann constant. **M. demon** A hypothetical creature able to violate the second law of *thermodynamics,* q.v. **M. electromagnetic law** If n is the refractive index of a medium, and ϵ_r the relative permittivity of a medium: $\epsilon_r = n^2$, provided the frequencies of the electrical and light vibrations are the same.

maxwell Abbrev. Mx; the cgs unit of magnetic flux: 1 maxwell = 10^{-8} *webers* (the SI unit), q.v.

may apple Podophyllum.

Mayer, Julius Robert von (1814–1878) German physicist, who originated the mechanical theory of heat and conservation of energy and matter. Cf. *Meyer.*

mayer Abbrev. My; a unit of heat capacity. The heat capacity of a body which is raised 1 °C by 1 joule (4.19 My for 1 g water at 20 °C).

Mayer's hemalum A histological stain: 1 g hematein, 50 ml 90% alcohol, 50 g alum, 0.5 g thymol, and 1,000 ml water. **M.'s reagent** A solution of mercuric chloride 1.35 and potassium iodide 5 g, in 100 ml water; gives a white precipitate with alkaloids.

maysin A globulin in corn meal (0.25%), which coagulates at 70 °C.

Mayow, John (1643–1679) English chemist who discovered that the atmosphere consists of gases, one supporting life and combustion.

mazout (French: "fuel oil") Masut.

McBain **McB., James William (1882–1953)** Canadian-born American physical chemist, noted for his concept of the colloid micelle. **McB.-Baker balance** A silica spring balance enclosed in heavy glass tubing, for the study of sorption under pressure. **McB. centrifuge** An ultracentrifuge consisting of a spinning rotor (350,000 r/min) driven by and floating on air.

McCance reagent A solution containing silver nitrate and gelatin acidified with sulfuric acid; an etch to detect sulfur in iron or steel.

mcg Abbreviation for micro*gram,* q.v.

McGill metals A group of Al bronze casting alloys containing 2% Fe.

McLaurin process A low-temperature carbonization method for coal.

McLeod gauge A device for measuring low gas pressures (0.01–1000 μm Hg) in a high-vacuum system by trapping a known volume of gas and compressing it to a measurable pressure.

MCPA* Methoxone. See *herbicides,* Table 42 on p. 281.

MCPB* See *herbicides,* Table 42 on p. 281.

Mc/s Abbreviation for megacycles per second.

Me Abbreviation for methyl, CH_3-.

mead Hydromel.

meadow **m. crocus** Colchicum. **m. lily** The bulb of *Silium candidum* (Liliaceae); a mucilaginous emulsion. **m. saffron** *Colchicum autumnale.* **m.sweet** Queen-of-the-meadow. The herb of *Spireae ulmaria* (Rosaceae).

mean Arithmetic mean. **arithmetic ~** Average, mean. The quotient obtained by dividing the sum of n numbers by n; hence, $(a + b + c + d)/4$ = arithmetic mean of a, b, c, and d. **geometric ~** The nth root of n numbers multiplied by one another; hence, $\sqrt[4]{abcd}$ = geometric mean of these 4 numbers. **harmonic ~** The quotient obtained by dividing n by the sum of the reciprocals of n numbers. **logarithmic ~** The logarithm of the geometric m. **proportional ~** Weighted mean. The quotient obtained by dividing $ma + nb + oc + \cdots$ by $m + n + o + \cdots$; thus the atomic weight is the proportional mean of the isotopic atomic weights (multiplied by the percentages present). **quadratic ~** The square root of the quotient obtained by dividing the sum of n-squared numbers by the number of added numbers; as, $\sqrt{(a^2 + b^2 + c^2 + d^2)/4}$. Numerical values follow the order: quadratic (highest), arithmetic, geometric, harmonic (lowest). **weighted ~** Proportional m.

m. free path The average distance molecules are supposed to travel without collision. **m. refractive index** The average refractive index of a substance for the extreme red and violet rays. **m. time between failures** MTBF. The total number of functioning hours of one or more similar pieces of equipment (e.g., printer) divided by the number of failures occurring during those hours.

measles vaccine A live preparation of measles virus grown on chick embryo cells (USP, BP).

measure A device to determine a physical quantity, generally length, diameter, volume, and capacity; e.g., rules, calipers, graduates, etc.

meat The edible portion of animal flesh, excluding fish. **m. bases** An arbitrary analytical number of m. foodstuffs: Total nitrogen (= insoluble and coagulable nitrogens, proteoses, peptones, and gelatin) × 3.12. **m. extract** A partly evaporated bouillon, used for culture media. **m. meal** A fertilizer consisting of cooked, dried, and powdered m., with little bone. It contains nitrogen 10–11.5, phosphorus pentaoxide 1–5%. **m. sugar** Inositol.

Mecca balsam Balm of Gilead.

mechanical Pertaining to the physical forces of masses and their control. **m. analysis** Analysis by mechanical as distinct from physical or chemical means, e.g., sedimentation. **m. antidote** The use of the stomach pump to remove a poison from an organ. **m. equivalent** Joule. The quantity of energy which, transformed into heat, yields 1 calorie of heat = 4.182 J (at 20 °C). **m. pulp** A pulp obtained by the wet grinding of logs (stone groundwood) or mechanical reduction of wood chips (refiner, thermomechanical, or chemithermomechanical pulp). Used to manufacture cheap grades of paper, e.g., newsprint.

mechanics The study of forces or bodies (solid, liquid, or gaseous) which involve no change in state or composition: e.g., machinery (levers, wheels, screws), hydraulics, and pneumatics. **quantum ~** See *quantum theory*. **soil ~** The application of statistical methods to data deduced from the measured properties of geological sediments. **wave ~** See *wave mechanics*.

mechanism (1) A machine or instrument which transforms or transmits mechanical force. (2) A description of the steps by which a reaction proceeds.

mechlorethamine hydrochloride $MeN(C_2H_4Cl)_2 \cdot HCl$ = 192.5. 2,2′-Dichloro-*N*-methyldiethylamine hydrochloride. Mustine hydrochloride. Nitrogen mustard, Mustargen. White, hygroscopic, vessicant crystals, soluble in water. A cytotoxic agent used to treat leukemia and other malignant diseases (USP, BP).

meclozine hydrochloride $C_{25}H_{27}N_2Cl \cdot 2HCl \cdot H_20$ = 481.9. White crystals, m.224 (decomp.); an antihistaminic, used to prevent nausea and vomiting (BP).

mecocyanin $C_{27}H_{30}O_{16}Cl$ = 646.0. An anthocyanin from the poppy.

meconate A salt of meconic acid, containing the radical $C_5H_2O_3(COO)_2=$.

meconic acid $C_7H_4O_7 \cdot 3H_2O$ = 254.1. 3-Hydroxy-4-pyrone-2,6-dicarboxylic acid. From opium. White crystals, soluble in water.

meconidin $C_{21}H_{23}NO_4$ = 353.4. Yellow, amorphous powder, m.58, insoluble in water.

meconin $C_{10}H_{10}O_4$ = 194.2. Opianyl. 6,7-Dimethoxyphthalide. The lactone of meconinic acid, derived from opium. Colorless crystals, m.102, soluble in water.

meconinic acid $C_6H_2(OMe)_2(CH_2OH)COOH$ = 212.2. 1,2-Methoxy-3-carboxyl-4-methanolbenzene. It exists only as its salts and its lactone, meconin.

mecoprop* See *herbicides*, Table 42 on p. 281.

media Plural of medium.

median* The middle reading of a range of readings listed in order of magnitude. See *normal probability curve* under *normal*.

medical Pertaining to the diagnosis and treatment of disease. **m. jurisprudence** Forensic *medicine*.

medicine (1) The science and art of healing. (2) A drug or substance, often liquid, administered to the body to correct a disturbance of its normal function. **clinical ~** The study of disease by practical methods, in particular, by studying the patient. **forensic ~** Legal medicine. Science applied to the detection of crime. **patent ~** A m. or drug protected by letters patent. **preventive ~** A branch of medical knowledge which aims to prevent disease. **veterinary ~** The application of medical knowledge to the treatment of diseases of animals.

Mediolanum Trademark for a urea-casein dithiocarbonate synthetic fiber.

medium (1) A substance, such as a solvent, that contains something else or that acts as the transmitter of a force. (2) See *culture medium*. (3) An average or mean.

Medrol Trademark for methylprednisolone.

medroxyprogesterone acetate $C_{24}H_{34}O_4$ = 386.5. Depo-Provera. White crystals, insoluble in water, m.204; used for uterine bleeding, endometriosis, breast cancer, and contraception (USP).

meerschaum $H_2Mg_2(SiO_3)_3 \cdot H_2O$. Sepiolite. A common, porous rock-forming silicate, d.2.

Mees, Charles Edward Kenneth (1882–1960) British-born American chemist, noted for his pioneer work in photographic science.

mega- Prefix (Greek) for: (1) "Large." (2)* One million times. (3) A unit of activity of penicillin preparations; 1 million units = 1 megaunit = 600 mg.

megapascal Abbrev. MPa; 1 megapascal = 10^6 Pa = 9.87 atmos.

megarrhizin A glucoside from the root of *Megarrhiza californica* (Cucurbitaceae).

megaton A measure of explosive power, equivalent to 1 million tons of TNT.

megavolt One million volts.

meglumine iothalamate $C_{18}H_{26}I_3N_3O_9$ = 809.1. Yellow, viscous liquid; a radiopaque (USP, BP).

meiler A pit or heap of wood covered with soil, for the manufacture of charcoal.

meinorite A vitreous, white, translucent, native aluminum calcium silicate.

meiosis The process of division of a gamete, in which the number of chromosomes is halved and chromosomal material exchanged. See *haploid*. Cf. *mitosis*.

Meissner, Paul Traugott (1778–1850) Austrian pharmaceutical chemist, noted for his "Handbuch."

MEK Abbreviation for methyl ethyl ketone. See 2-*butanone*.

Meker burner A bunsen burner with metal screen in an enlarged opening. The gas and air are intimately mixed and thus produce a high temperature (about 1700°C).

mekonine Meconin.

Mekralon Trademark for a polypropylene synthetic fiber.

mel Honey. The saccharine substance deposited by the honeybee, *Apis mellifera*, rich in fructose; contains invert sugar. Cf. *nectar, ceromel, hydromel, oxymel*.

melaconite CuO. A native black copper oxide.

melam $C_6H_8N_{11}$ = 234.2. White powder, obtained with melamine by heating potassium thiocyanate and ammonium chloride in intimate contact; insoluble in water. See *albene, melem*.

Melamac Trademark for a melamine-formaldehyde resin.

melamine $N{:}C(NH_2){\cdot}N{:}C(NH_2){\cdot}N{:}C(NH_2)$ = 126.1. (ring closure bracket from first N to last C)

Cyanurotriamine, cyanuramide, 2,4,6-triamino-1,3,5-triazine*. Colorless crystals, slightly soluble in water, insoluble in organic solvents, sublime 350; form a plastic with formaldehyde, used to give wet strength to paper.

melaminylphenylarsonic acid $C_9H_{11}O_3N_6As$ = 326.1. Used, as its disodium salt (melarsen), to treat sleeping sickness. See *melarsoprol*.

melampsorin $C_{23}H_{28}O_{15}$ = 544.5. A glucosidal coloring matter of galls, from *Melampsora goepporriana*. Yellow crystals, m.235.

melampyrine Dulcitol.

melampyrite Dulcitol.

melaniline Diphenylguanidine.

melanin(e) $C_{77}H_{98}O_{33}N_{14}S$ = 1780. Black coloring matter (chromoprotein) from certain insects, hair, and dark skins; soluble only in alkali.

melanite A black andradite. See *iron garnet* under *garnet*.

melanoidins Dark-colored substances formed by the interaction of reducing sugars and amino acids when heated; largely responsible for the color of molasses.

melanosis An abnormal diffuse pigmentation of the skin.

melanterite $FeSO_4 \cdot 7H_2O$. A native ferrous sulfate.

melarsen See *melaminylphenylarsonic acid*.

melarsoprol $C_{12}H_{15}ON_6AsS_2$ = 398.31. Creamy, bitter powder, m.217 (decomp.), insoluble in water; used to treat sleeping sickness (BP).

melded Generic term describing fabrics made by melting and welding; as, bonded fabrics.
meldola blue New blue. Naphthol blue. A methylene blue–type dye.
melee A diamond for glass cutting, weighing less than a quarter carat.
melem $(C_6H_6N_{10})_n$ = $(218.2)n$. An amine of cyanuric acid, obtained with melam by heating ammonium thiocyanate.
melene $C_{30}H_{60}$ = 420.8. Triacontene*. An unsaturated hydrocarbon from beeswax. Colorless fat, d.0.89, m.62, insoluble in water.
meletin Quercetin.
melezitose $C_{18}H_{32}O_{16}$ = 504.5. Melicitose. A trisaccharide from manna or the sap of conifers and poplars, which hydrolyzes to glucose and turanose (glucose-fructose). Cf. *melizitose.*
melibiase α-D-Galactosidase*. An enzyme in low (but not high) yeasts, which hydrolyzes raffinose.
melibiose $C_{12}H_{22}O_{11}$ = 342.3. 6-*O*-α-D-Galactopyranosyl-D-glucose†. A disaccharide, $[\alpha]_D$ +143°, from Australian manna, yellow mallow, also by hydrolysis of raffinose.
melicitose Melezitose.
melilite (1) Anhydrous aluminum-calcium silicates in igneous rocks. (2) $(Al, Fe)_2(Ca, Mg)_3Si_2O_{10}$.
melilith $Ca_4Si_3O_{10}$. A melilite.
melilot Melilotus, sweet yellow clover. The leaves of *Melilotus officinalis* (Leguminosae), which contain coumarin and coumaric and melilotic acid.
melilotic m. acid $HO \cdot C_6H_5(CH_2)_2COOH$ = 167.2. 2-Hydroxybenzenepropanoic acid†. From *Melilotus* species. Colorless crystals, m.81 (forms a lactone). **m. lactone** $C_9H_8O_2$ = 148.2. m.25.
Melinex Trademark for a polyester synthetic film; it is stretched and heated during manufacture to improve strength.
melinite A high explosive of the lyddite type.
melissic m. acid $C_{30}H_{60}O_2$ = 452.8. Triacontanoic acid*. In plant waxes and beeswax. Colorless scales, m. 94, insoluble in water. **m. alcohol** 1-Triacontanol*. **m. palmitate** Triacontyl palmitate*.
Melissos of Samos (470–410 B.C.) Greek philosopher famous for the statement "Nothing can come from nothing."
melissyl alcohol 1-Triacontanol*.
melitic Mellitic.
melitose $C_{12}H_{22}O_{11}$ = 342.3. A disaccharide from Australian manna.
melitriose Raffinose.
melizitose A sugar from *Alhagi maurorum* (Leguminosae), yielding manna. Cf. *melezitose.*
Mellaril, Melleril Trademarks for thioridazine hydrochloride.
mellisic Melissic.
mellispolynology The microscopical analysis of honey pollen.
mellite (1) Pharmacy: medicated honey. (2) Mineralogy: honeystone.
mellitene Hexamethylbenzene*.
mellitic acid $C_6(COOH)_6$ = 342.2. Hexacarboxylbenzene. Colorless needles, m.287, soluble in water. Its aluminum salt occurs in peat as honeystone. **pyro ~** See *pyromellitic acid.*
mellitoxin (1) Hyenanchin. (2) A poisonous constituent of certain honeys (New Zealand).
melloene Describing a process for increasing the gluten tenacity of flour by treatment with sulfur dioxide and water vapor.
mellon(e) C_9H_{13} = 121.2. A hydrocarbon obtained, with melam and melem, on igniting mercuric thiocyanate. See *pharaoh's serpents.*
mellophanic acid $C_{19}H_6O_8$ = 254.2. 1,2,3,4-Benzenetetracarboxylic acid*. Colorless crystals, m.238. Cf. *mallophene, pyromellitic acid.*
Mellor, Joseph William (1869–1938) English chemist, noted for his handbook of inorganic chemistry.
mellorite A complex garnet-type lime-ferric oxide silicate.
Mellot's metal D'Arcet metal.
melon $(C_6H_3N_9)_n$. An amine of cyanuric acid, obtained with and similar to melem.
melonite (1) Ni_2Te_3. Tellurnickel. Red granules. (2) $Ca_4Al_6Si_6O_{25}$. A silica mineral of the scapolite group.
melphalan $C_{13}H_{18}O_2N_2Cl_2$ = 305.2. L-3-[*p*-[Bis(2-chloroethyl)amino]phenyl]alanine. Nitrogen mustard, Alkeran. White crystals, m.177 (decomp.), insoluble in water. An antineoplastic agent, used for myelomatosis and other cancers (USP, BP).
melting Fusing. The transformation of a solid into a liquid by means of heat. **m. point** The temperature at which a solid changes to a liquid, and the liquid and solid phases are in equilibrium under a pressure of 760 mmHg. The highest known m. p. is 4200°C (a mixture of hafnium carbide 25 and tantalum carbide 75%). The lowest is that of helium. **American m. p.** See *American melting point* under *American.* **m. p. tube** A capillary tube attached to a thermometer bulb and heated until the contents fuse. Cf. *Thiele tube.* **m. salt** See *melting salt* under *salt.*
membrane A thin, enveloping or lining substance which divides a space or an organ. **animal ~** A skinlike tissue obtained from animal tissues (parchment), used for dialyzing. **hollow fiber ~** Industrial process involving selective gas permeation through hollow polysulfone fiber membranes. **semipermeable ~** A tissue that permits the passage of certain substances, e.g., ions, but prevents the passage of others, e.g., colloids. Classification:

1. Sieves: as such, or coarse filter paper
2. Cell filters: very fine filter paper
3. Bacterial filters: kieselguhr
4. Colloidal filters: parchment or collodion
5. Molecular sieves: copper hexacyanoferrate(II).

m. electrode E. protected from interfering constituents by a membrane through which the gas to be measured can diffuse.
menachanite Menaccanite, menacanite, menacconite. A titaniferous magnetic iron oxide from Cornwall. Cf. *ilmenite.*
menadiol sodium diphosphate $C_{11}H_8O_8Na_4P_2 \cdot 6H_2O$ = 654.1. Pink, hygroscopic powder with characteristic odor, soluble in water; used to prevent hemorrhage due to low vitamin K level (USP).
menadione $C_{11}H_8O_2$ = 172.2. Menaphthone. 2-Methyl-1,4-naphthoquinone. Davitamon-K. Vitamin K.

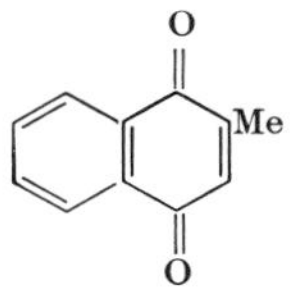

O
Me
O

Yellow crystals, m.106, insoluble in water. A synthetic vitamin K; used to prevent hemorrhage due to low vitamin K levels in the newborn (USP). Also used as its sodium bisulfite compound (USP).
menaphthone Menadione.
menaphthyl The naphthylmethyl* radical, $C_{10}H_7 \cdot CH_2-$.
Mendele(y)ev M., Dmitri Ivanovitch (1834–1907) Mendeléeff, Mendelejeff. Russian chemist, one of the discoverers of the periodic law (see *Meyer, Lothar*) and predictor of several eka-elements. **M. chart** Periodic table.

M. group A vertical group of the periodic table. **M. law** Periodic law. **M. system** Periodic system.

mendelevium* Md. An actinoid element, at. no. 101. At. wt. of isotope with longest half-life (^{256}Md) = 256.10. Discovered (1952) by Ghiorso and coworkers by bombarding ^{253}Es with helium accelerated in a cyclotron, to give ^{256}Md. Named for Mendelyeev. It forms Md^{3+} ions.

mendozite White, fibrous sodium-aluminum sulfate.

meneghinite $4Pb_2SbS_2$. A native sulfide.

menhaden The fish *Brevoortia tyrannis,* used to make oil and fertilizer.

meniscus The flat or crescent-shaped surface of a liquid in a tube, either concave (when the liquid wets the material of the container, as water and glass), or convex (when liquid does not wet, as mercury and glass). **m. reader** (1) A colored streak placed behind a buret to enable the m. to be read more exactly. It is customary to read the lowest point. (2) A lens or clamp and card attached to the buret.

Menispermaceae Moonseed family, a group of woody, climbing tropical plants; as, *Anamirta paniculata,* cocculus indicus. Cf. *cucoline, deyamittin, sinomenine.*

menispermine $C_{18}H_{24}N_2O_2$ = 300.4. An alkaloid from *Cocculus indicus (Anamirta paniculata)* and *Menispermum canadense.* Colorless crystals.

menispermoid The combined principles from *Menispermum canadense.*

menotrophin BP name for menotropin.

menotropin An extract from human postmenopausal urine containing a mixture of follicle stimulating and luteinizing hormones. Menotrophin, FSH. Yellowish powder, soluble in water. Used for infertility due to failure to ovulate (USP, BP).

menstruum Obsolete term for a solvent for the extraction of drugs.

mensuration The act of measuring. **m. formula(e,s)** The mathematical equations by which plain cubical, or spherical, figures or bodies are measured.

mentha (1) Mint, e.g., peppermint. (2) A genus of Labiatae; e.g.: *M. crispa,* spearmint oil; *M. piperita,* peppermint oil; *M. spicata,* spearmint. **m. camphor** 1-Menthol*. **m. viridis** Spearmint.

menthadiene* A group of terpenes with 2 double bonds, derived from menthane, e.g.: **1,3-*p*- ~** α-Terpinene. **1,4(8)-*p*- ~** Terpinolene **1,5-*p*- ~** α-Phellandrene. **1(7)2-*p*- ~** β-Phellandrene. **1,8-*p*- ~** Citrene.
m.dione Thymoquinone. **m.one** Carvone.

menthane* $C_{10}H_{20}$ = 140.3. Terpane. 4-Isopropyl-1-methylcyclohexane*, hexahydrocymene, menthonaphthene. A saturated hydrocarbon, parent substance of many terpenes:

```
              CH3 (7)
               |
              CH
            /     \
(6) H2C    (1)    CH2 (2)
     |             |
(5) H2C    (4)    CH2 (3)
            \     /
              CH
               |
       CH3—CH—CH3
       (9)   (8)  (10)
```

Colorless liquid, d.0.807, b.169, insoluble in water. **amino ~** Menthylamine. **dihydroxy ~** Terpin. **epoxy ~** Eucalyptol. **3-hydroxy ~** Menthol*.
m.diol Terpin.

menthanol* A hydroxy derivative of menthane. **2- ~** Carvomenthol. **3- ~** Menthol*.

menthanone A ketone derivative of menthane, e.g.: **3- ~** Menthone*.

menthe A peppermint liqueur (créme de menthe), prepared from menthol and alcohol.

menthene* $C_{10}H_{18}$ = 138.3. **3- ~** Colorless liquid, d.0.814, b.167, insoluble in water.

menthenol* A hydroxy derivative of menthene. **3- ~** Pulegol.

menthenone $C_{10}H_{16}O$ = 152.2. A ketone derivative of menthene. **3-$\Delta^{4,8}$- ~** Pulegone; **3-$\Delta^{8,9}$- ~** Isopulegone.

menthenyl* The radical $C_{10}H_{17}-$, from methene.

menthol* $C_{10}H_{20}O$ = 156.3. 5-Methyl-2-(1-methylethyl)cyclohexanol†. Peppermint camphor, 3-hydroxymenthane*, menthacamphor. A terpene alcohol, in many essential oils. **(−)- ~** Colorless crystals, m.42, slightly soluble in water. Used in perfumery, flavoring extracts, confectionery; an inhalant and rubifacient ointment (USP, BP).

menthonaphthene Menthane*.

menthone* $C_{10}H_{18}O$ = 154.3. 2-Isopropyl-5-methylcyclohexanone*. 3-Terpanone. **(+)- ~** A colorless liquid, d.0.896, b.207, soluble in water.

menthyl* The radical $C_{10}H_{19}-$, from menthane. **m.amine** $C_{10}H_{19}NH_2$ = 155.3. 3-Aminomenthane. Colorless liquid, b.205. **carvo ~** See *carvomenthylamine.*

menyanthes Marsh trefoil, buckbean. The dried leaves of *M. trifoliata* (Gentianaceae); an aromatic bitter.

menyanthin $C_{33}H_{50}O_{14}$ = 670.8. Celastin. A bitter glucoside from the leaves of the buckbean; soluble in water.

menyanthol $C_7H_{11}O_2$ = 127.2. A split product of menyanthin.

mepacrine Quinacrine.

meperidine hydrochloride $C_{15}H_{21}O_2N \cdot HCl$ = 283.8. Ethyl-l-methyl-4-phenylisonipecotate hydrochloride. Pethidine hydrochloride, Demerol, Dolantin. White powder, m.188, soluble in water. A potent analgesic and adjunct to anesthesia. Addictive (USP, EP, BP).

mephitic Obsolete term for foul, noxious, or poisonous. **m. air** Black damp, choke damp. Obsolete name for (1) carbon dioxide; (2) nitrogen.

meprobamate $C_9H_{18}O_4N_2$ = 218.3. 2-Methyl-2-propyl-1,3-propanediol dicarbamate. Equanil, Miltown. White crystals, m.105, soluble in water; a sedative causing muscular relaxation (USP, EP, BP).

mepyramine maleate BP name for pyrilamine maleate.

mer A monomeric unit.

mer-, mere-, meri- (1) Prefix (Greek) indicating a "part." (2) See *meridional.*

meranti A tree, *Shorea spp.,* exported from southeast Asia. Used for plywood and general purposes.

Merbentyl Trademark for dicyclomine hydrochloride.

Mercadium Trademark for an orange- to maroon-colored pigment: principally mercury and cadmium sulfides in a common crystal lattice.

mercaptal Thioacetal*.

mercaptan (1) A thiol*. (2) Ethanethiol*.

mercaptide A metal thiolate* or sulfide*.

mercapto-* Prefix indicating a thiol group, −SH. **m.dimethur*** See *insecticides,* Table 45 on p. 305. **m.succinic acid** Thiomalic acid. **m.sulfothiobiazole** Bismuthiol.

mercaptol Thioacetal* produced from a thiol and ketone in the presence of acid.

mercaptophenyl- Prefix indicating the presence of the radical $HS \cdot C_6H_4-$. **m.dithiodiazolone** $C_8H_6N_2S_3$ = 226.3. White crystals. A reagent for bismuth (red precipitate). Cf. *bismuthiol.*

mercaptopurine $C_5H_4N_4S \cdot H_2O$ = 170.2. Purine-6-thiol, 6MP, Puri-Nethol. Yellow crystals, m.300 (decomp.), insoluble in water; an antimetabolite for leukemia (USP, BP).
mercaptothiazole **2-~** The hypothetical ring compound CH:CH·N·C(SH)·S.

Mercer **M., John (1791–1866)** English cotton printer who invented the mercerization process (1850); developed practically (1889) by Horace Arthur Lowe. **M. process** Mercerization.
mercerization Treatment of cotton with 25% caustic soda, causing it to shrink and become stronger and denser and to acquire a milky luster; it then becomes unshrinkable and easily dyed.
mercur- Prefix indicating a mercury compound. **m.diammonium chloride** $[Hg(NH_3)_2]Cl_2$ = 305.6. Diamminemercury(II) chloride*. An infusible, white precipitate obtained from Hg^{++} and Cl^- in the presence of concentrated NH_4^+.
mercurammonium- Prefix indicating NHg_2X, where X is a halogen; as, m. bromide, NHg_2Br. Cf. *mercuriammonium, mercurdiammonium chloride.*
mercuration Mercurization.
mercurial A drug containing mercury.
mercurialine A supposed alkaloid of *Mercurialis annua* (Euphorbiaceae); probably methylamine.
mercurialis The dried herb of *M. perennis.*
mercuriammonium Mercuric ammonium. The salt formed in the presence of dilute NH_4^+ ion; as, $HgNH_2Cl$. Cf. *mercurammonium.*
mercuric Mercury(II)* or (2+)*. A compound of mercury, containing divalent Hg. **m. acetate** $Hg(C_2H_3O_2)_2$ = 318.7. White crystals, soluble in water; a reagent. **m. acetylide** $3C_2Hg \cdot H_2O$ = 691.9. Explosive, white powder, insoluble in water. **m. arsenate** $Hg_3(AsO_4)_2$ = 879.6. Yellow powder, insoluble in water. **m. borate** M. pyroborate. **m. bromide** $HgBr_2$ = 360.4. Mercury dibromide*. Colorless rhombs, m.235, soluble in water. **m. carbonate** $HgCO_3$ = 260.6. White powder, insoluble in water. Known chiefly as its basic salts, $HgCO_3 \cdot 2HgO$ and $HgCO_3 \cdot 3HgO$. **m. chloride** $HgCl_2$ = 271.5. Corrosive sublimate, sublimate, corrosive mercury chloride, mercury dichloride. Poisonous, white rhombs (antidote: 10% sodium thiosulfate solution), m.287, soluble in water. Used in preservation of wood and museum specimens; as a mordant, caustic, and reagent; and in purification of gold, photography, textile printing, etching of steel and iron, and dyeing of furs. **m. chromate** $HgCrO_4$ = 316.6. Yellow crystals, insoluble in water. **m. cyanide** $Hg(CN)_2$ = 252.6. Colorless tetragons, decomp. by heat, slightly soluble in water. Used in photography and in manufacturing cyanogen; a reagent for palladium. **m. ferrocyanide** Mercury hexacyanoferrate(II)*. **m. fluosilicate** Mercury(II) hexafluorosilicate*. **m. fulminate** $Hg(ONC)_2 \cdot \frac{1}{2}H_2O$ = 293.6. Colorless rhombs, explode on detonating or at 175°C, slightly soluble in water; a detonator. **m. halides** The divalent halogen compounds of mercury, HgX_2. **alkyl ~** RHgX, where R is an aliphatic, or alkyl, radical. **aryl ~** RHgX, where R is an aryl, or aromatic, radical. **m. hydroxide** $Hg(OH)_2$ = 234.6. White powder, insoluble in water. **m. hydroxides** Compounds of the type RHgOH. **m. iodate** $Hg(IO_3)_2$ = 550.4. White powder, insoluble in water. **m. iodide** HgI_2 = 454.4. Mercury diiodide*. **red ~** Red tetragons, m.241, insoluble in water. **yellow ~** Yellow rhombs, m.241, insoluble in water. **m. nitrate** $Hg(NO_3)_2$ = 324.6. White, hygroscopic powder, m.79 (decomp.), soluble in water; a reagent. **basic ~** (1)$Hg(NO_3)_2 \cdot HgO$ = 541.2. Deliquescent, colorless crystals. (2) $2Hg(OH) \cdot NO_3 \cdot H_2O$ = 577.2. (3) $Hg(NO_3)_2 \cdot 2HgO \cdot H_2O$ = 775.8. White, unstable powder. **m. oxalate** HgC_2O_4 = 288.6. White powder, insoluble in water. **m. oxide** HgO = 216.6. **red ~** Hydrargyri oxydum rubrum, red mercury oxide. Red, monoclinic prisms, decomp. by heat, insoluble in water. Used in the manufacture of mercury salts and ceramics, and as a caustic. **yellow ~** Hydrargyri oxydum flavum, yellow precipitate. Orange, tetragonal crystals, insoluble in water. **m. phosphate** $Hg_3(PO_4)_2$ = 791.7. White powder, insoluble in water. **acid ~** $HgHPO_4$ = 296.6. White powder, insoluble in water. **m. potassium cyanide** $HgK(CN)_3$ = 317.7. Mercury potassium tricyanide*. Colorless crystals, soluble in water; used in the manufacture of mirrors. **m. potassium iodide** $HI_2 \cdot 2KI$ = 586.8. Mayer's reagent. Yellow, deliquescent crystals, soluble in water; a reagent for alkaloids. Cf. *Toulet's solution.* **m. subsulfate** $HgSO_4 \cdot 2HgO$ = 729.8. Basic m. sulfate, turpeth mineral, Queen's yellow; powder, insoluble in water. **m. succinate** $HgC_4H_4O_4$ = 316.7. Colorless crystals, insoluble in water. **m. sulfate** $HgSO_4$ = 296.6. Mercury(II) sulfate*, m. bisulfate, m. persulfate. Yellow powder, decomp. by heat, soluble in water. Used to manufacture calomel; in the extraction of gold and silver, in electric batteries. **basic ~** M. subsulfate. **hydrous ~** (1) $HgSO_4 \cdot 2H_2O$ = 332.7. (Erroneously called "basic.") Yellow crystals, soluble in water. (2) $3HgO \cdot SO_3 \cdot 4H_2O$ = 962.0. Yellow powder. **m. sulfide** HgS = 232.7. **black ~** Black mercury sulfide, metacinnabarite. Black powder, insoluble in water; a pigment. **red ~** Red mercury sulfide, artificial cinnabar, vermilion, cinnabar. Red rhombohedra, sublimes 446, insoluble in water; a pigment. **m. sulfite** $HgSO_3$ = 280.6. White powder, becomes pink with age, soluble in water. **m. thiocyanate** $Hg(SCN)_2$ = 316.7. M. rhodanate, m. sulfocyanide. White powder, usually pressed into sticks, decomp. by heat. When ignited, it glows and forms a voluminous, cohesive, and light ash (*pharaoh's serpents,* q.v.), insoluble in water.
mercuricide Lithium mercury(II) iodide*.
mercurides R_2Hg. R is a hydrocarbon radical; as, *diethylmercury.*
mercurification Amalgamation. Cf. *mercurization.*
mercurimetry The determination of a substance by precipitating it with a mercury salt and ascertaining the mercury in the precipitate.
mercuriovegetal Manaca.
mercurius Mercury. **m. praecipitatus** The "red precipitate," HgO, of the Latin alchemist Geber.
mercurization The introduction of mercury into the formula of an organic compound.
Mercurochrome $C_{20}H_8O_6Na_2Br_2Hg$. Trademark for disodium dibromohydroxymercurifluorescein. Iridescent, green crystals, soluble in water (red color); antiseptic, less irritating than tincture of iodine.
mercurothiolate Thimerosal.
mercurous Mercury(I)* or (1+)*. A compound of monovalent mercury. **m. arsenite** Hg_3AsO_3 = 724.7. Brown powder, insoluble in water. **m. bromide** HgBr = 280.5. Yellow tetragons, sublime 350, insoluble in water. **m. carbonate** Hg_2CO_3 = 461.2. Yellow powder, slowly decomp., insoluble in water. **m. chloride** HgCl = 236.0. Hydrargyri chloridum mite, calomel, m. monochloride, mild m. chloride, m. subchloride, m. protochloride; native as horn quicksilver. Colorless rhombs or tetragons, sublimes 303, insoluble in water. Used as a reagent, fungicide, green pyrotechnic; and in ceramics, for gold colors. **m. chromate** Hg_2CrO_4 = 517.2. Red powder, insoluble in water, decomp. by heat. **m. citrate**

$Hg_3C_6H_5O_7$ = 790.9. Colorless powder, slightly soluble in water. **m. fluosilicate** Mercury(I) hexafluorosilicate*. **m. formate** H·COOHg = 245.6. White scales, soluble in water. **m. iodide** HgI = 327.5. Yellow m. iodide, hydrargyri iodidum flavum, m. protiodide, m. monoiodide. Yellow tetragons, m.290, insoluble in water. **m. nitrate** $HgNO_3 \cdot 2H_2O$ = 298.6. Colorless, monoclinic crystals, decomp. by heat, soluble in water; a reagent and caustic. **m. oxalate** HgC_2O_4 = 288.6. White powder, insoluble in water. **m. oxide** Hg_2O = 417.2. Black powder, insoluble in water. **m. phosphate** Hg_3PO_4 = 696.7. Colorless powder, insoluble in water. **m. potassium tartrate** $HgKC_4H_4O_6$ = 387.8. White crystals, insoluble in water. **m. protoxide** Black m. oxide. Native as montroydite. **m. sulfate** Hg_2SO_4 = 497.2. White, monoclinic crystals, decomp. by heat, soluble in water; used for batteries and standard cells. **m. sulfide** Hg_2S = 433.3. Black powder, insoluble in water. **m. tartrate** $Hg_2C_4H_4O_6$ = 549.2. Yellow crystals, insoluble in water. **acid ~** $HgHC_4H_4O_6$ = 349.7. M. bitartrate. White crystals, insoluble in water.

mercury* Hg = 200.59. Hydrargyrum. Quicksilver, mercurius, liquid silver. A liquid metal element, at. no. 80. Silver-white, metallic liquid, d.13.595, freezing at −38.8, b.357, insoluble in water, soluble in nitric acid. M. occurs in nature chiefly as sulfide (cinnabar). It forms monovalent [m.(I)*, m.(1+)*, mercurous] and divalent [m.(II)*, m.(2+)*, mercuric] compounds. Used as a catalyst (Kjeldahl nitrogen determination); for filling thermometers and apparatus and for silvering mirrors; in electric cells, amalgams with gold and silver, electric rectifiers, and vacuum-tube lights; and in explosives manufacture. **ammoniated ~** NH_2HgCl = 252.1. Sal alembroth, white precipitate, hydrargyrum ammoniatum, mercury and ammonium chloride. White powder, insoluble in water (USP). **dibenzyl ~*** $(C_7H_7)_2Hg$ = 382.9. Long needles, soluble in alcohol. **dibiphenylyl ~*** $(C_6H_4Ph)_2Hg$ = 507.0. Colorless crystals, m.216. **dibutyl ~*** $(C_4H_9)_2Hg$ = 314.8. Colorless liquid, d.1.835, b.205. **diethyl ~*** Et_2Hg = 258.7. M. ethide. Colorless liquid, b.159, insoluble in water. **dimethyl ~*** Me_2Hg = 230.7. M. methide. Colorless liquid, b.96, insoluble in water. **dinaphthyl ~*** $(C_{10}H_7)_2Hg$ = 454.9. M. naphthide. White powder, m.188. **diphenyl ~*** Ph_2Hg = 354.8. M. phenide. White powder, m.120, insoluble in water. **dipropyl ~*** $(C_3H_7)_2Hg$ = 286.8. Colorless liquid, b.190, insoluble in water. **ditolyl ~*** $(C_7H_7)_2Hg$ = 382.9. White powder, m.107. **repurified ~** Redistilled m. used for dental amalgams or electrodes. **soluble ~** See *soluble mercury.*

m.(I)*, m.(1+)* See also *mercurous.* **m.(II)*, m.(2+)*** See also *mercuric.* **m. alkylides** R_2Hg. R is a monovalent alkyl radical, e.g., diethyl ~. **m. alloys** Mixtures of m. with other metals. If liquid or semiliquid, they are termed amalgams. **m. amalgams** See *mercury alloys, amalgam.* **m. arc** An electric arc between m. electrodes. **m. arylides** Organic compounds of m. and an aryl radical, e.g., diphenyl ~ **m. cathode** (1) See *dropping electrode* under *electrode.* (2) The positive electrode in a rectifier or a m. vacuum-tube lamp. Cf. *polarograph.* **m. c. cup** A glass cylinder with fused-in platinum electrode; used for the electrolytic determination of m. **m. chloride** See *mercuric chloride* (sublimate), *mercurous chloride* (calomel). **corrosive ~** Mercuric chloride. **ethyl ~** C_2H_5HgCl = 265.1. Ethylmercuric chloride. White, iridescent crystals, m.193. **methyl ~*** CH_3HgCl = 251.2. Silver crystals, m.170. **mild ~** Mercurous chloride. **phenyl ~*** C_6H_5HgCl = 313.1. Chloromercury benzene. White leaflets, m.251. ***p*-tolyl ~*** $Me \cdot C_6H_4HgCl$ = 327.2. *p*-Chloromercury toluene. Silky crystals, m.233, insoluble in water. **m. cup** (1) A small tray containing m. into which wires are dipped to make electric contact. (2) The glass bulb of a thermometer containing m. **m.(II) dichromate*** $HgCr_2O_7$ = 416.6. Red crystals, insoluble in water. **m.(II) dicyanide oxide*** $Hg(CN)_2 \cdot HgO$ = 469.2. White crystals, soluble in water. **m.(II) dithiocarbonates*** Mercuric xanthates. R·HgS·CS·OR. R is an aliphatic or aromatic radical. **m. dropping electrode** See *dropping electrode* under *electrode.* **m. ethanethiolate*** $(EtS)_2Hg$ = 322.8. M. mercaptide. Colorless leaflets, m.86, insoluble in water. **m. furnace** A retort for distilling m. from cinnabar. **m.(II) hexacyanoferrate(II)*** $Hg_2[Fe(CN)_6]$ = 613.1. Mercuric ferroycanide. Brown powder, insoluble in water. **m.(I) hexafluorosilicate*** $Hg_2SiF_6 \cdot 2H_2O$ = 579.3. Mercurous fluo(ro)silicate. Colorless prisms. **m.(II) hexafluorosilicate*** $HgSiF_6 \cdot 6H_2O$ = 450.8. Rhombohedra. **m. minerals** The chief ores of m. are native m. and its sulfide cinnabar, HgS; and montroydite, HgO. **m. nitrate** See *mercuric nitrate, mercurous nitrate.* **m. ore** (1) Native m. (2) Cinnabar. **m. nitride** Hg_3N_2 = 629.8. Explosive brown powder. **m. oxide** See *mercury peroxide, mercuric oxide, mercurous oxide.* **black ~** Mercurous oxide. **red ~** Mercuric oxide. **yellow ~** Mercuric oxide. **m. peroxide*** HgO_2 = 232.6. Red powder, stable in absence of water. **m. phosphate** See *mercuric phosphate, mercurous phosphate.* **m. potassium iodide solution** See *Toulet's solution.* **m. protiodide** Mercurous iodide. **m. protochloride** Mercurous chloride. **m. protoxide** Mercurous oxide. **m. pump** Sprengel pump. **m. subchloride** Mercurous chloride. **m. subsulfate** Mercuric subsulfate. **m. sulfuret** See *mercuric sulfide, mercurous sulfide.* **black ~** Mercurous sulfide. **red ~** Mercuric sulfide. **m.(II) tetraborate*** HgB_4O_7 = 355.8. Brown powder, insoluble in water. **m. thiocyanate reagent** A solution of mercuric chloride and ammonium thiocyanate in water gives crystals of characteristic shape with copper, cobalt, and zinc solutions. **m. trap** M. well. A box used in amalgamators to prevent the escape of m. **m. vapor lamp** An evacuated glass tube containing some m. which is vaporized and gives an intense blue light in an electric discharge; used in photography and as an ultraviolet light source.

meridian A geographical unit: 4 quadrants or 40,000,000 meters.

meridional* *mer**. Indicating 3 groups on an octahedron, with 1 *cis* to the 2 others which are themselves *trans.*

merino Botany; a fine animal textile fiber from m. sheep.

Merinova Trademark for a synthetic protein fiber.

meriquinone A compound whose electronic configuration resembles that of a quinone but which contains no oxygen; as, *Wurster's red.*

merit number A numerical expression of the quality of a steel: ultimate strength, psi × elongation, in.

merochrome A chromoisomeric crystal having 2 isomeric forms.

merotropy Desmotropy. Cf. *tautomerism.*

Merrifield, Robert Bruce (1921–) American biochemist. Nobel prize winner (1984) for the development of solid-phase peptide synthesis.

merron Proton.

mersalyl $C_{13}H_{16}O_6NHg$ = 482.9. White, bitter crystals, soluble in water. Formerly a diuretic; now superseded by thiazide (as, bendroflumethiazide) and furosemide diuretics.

Mersolate Trademark for alkyl sulfonate detergents prepared by the sulfochlorination of Fischer-Tropsch alkanes.

Mersolite Trademark for mercury phenyl salicylate; a disinfectant.

Merthiolate Trademark for thimerosal.

merwinite $3CaO \cdot MgO \cdot 2SiO_2$. An orthosilicate in refractory bricks and blast furnace slags, m.1590 (approx.); formed when the mixed oxides of calcium, magnesium, and silica are heated at 1500.

mes- Prefix (Greek) meaning "middle" or "intermediate." See *meso.*

mesaconic acid $C_5H_6O_4$ = 130.1. Methylfumaric acid, *trans*-methylbutenedioic acid*. Unsaturated, dibasic acid and isomer of citraconic acid. Colorless needles, m.202, soluble in water. Cf. *furmaric acid.*

mesate *Methane*sulfonate*.

mescal An intoxicant spirit distilled from pulque, the fermented juice of Agave (Mexico).

mescal buttons The dried buds or young leaves of *Anhalonium* lewinii, q.v. Cf. *lophophorine, pellotine.*

mescaline $MeO \cdot C_6H_2 \cdot CH_2CH_2NH_2$ = 149.2. 3,4,5-Trimethoxyphenylethylamine. An alkaloid from *Anhalonium lewinii*, a cactus species of Central America; related to adrenaline. Has hallucinatory properties.

mesembrene $C_{28}H_{56}$ = 392.8. An unsaturated hydrocarbon from *Mesembryanthemum expansum* (Ficoidaceae or Aizoaceae).

mesembrine $C_{17}H_{23}O_3N$ = 289.4. An alkaloid from *Mesembryanthemum tortuosum*, the kougoed of channa (Aizoaceae), S. Africa.

mesh The number of openings per unit area in a sieve.

mesicerin $C_6H_3(CH_2OH)_3$ = 168.2. Mesityleneglycerol. Colorless, viscous liquid.

mesidine $C_9H_{13}N$ = 135.2. 2,4,6-Trimethylaniline. Colorless liquid, d.0.963, b.233.

mesitene lactone Dimethyl*coumalin.*

mesitic acid Uvitic acid.

mesitilol Mesitylene*.

mesitine spar Mesitite.

mesitite $2MgCO_3 \cdot FeCO_3$. Mesitine spar. A native carbonate.

mesitol $Me_3C_6H_2OH$ = 136.2. Mesityl alcohol*, 2,4,6-trimethylphenol*. Colorless crystals, m.69.

mesityl* 2,4,6-Trimethylphenyl†. The radical $Me_3C_6H_2-$, from mesitylene. **m. alcohol** Mesitol. **m. oxide** $MeCO \cdot CH:CMe_2$ = 98.11. 4-Methyl-3-penten-2-one*. Colorless liquid, d.0.858, b.131.

mesitylene* $C_6H_3Me_3$ = 120.2. Mesitylol, 1,3,5-trimethylbenzene*. Colorless liquid, d.0.86, b.164, insoluble in water. **dihydroxy ~** Mesorcinol. **hexahydro ~** 1,3,5-Hexahydrocumene. **hydroxy ~** Mesitol.
m. alcohol Mesitol. **m.carboxylic acid** 2,4,6-Iso*durylic acid.* **m.glycerol** Mesicerin. **m. lactone** Dimethyl*coumalin.*

mesitylenic acid Mesitylinic acid.

mesitylinic acid $Me_2C_6H_3COOH$ = 150.2. Mesitylenic acid, 3,5-dimethylbenzoic acid*. White, monoclinic crystals, m.166, slightly soluble in water.

mesitylol Mesitylene*.

meso- Prefix indicating between. **m.-compound*** Compound whose individual molecules contain equal numbers of enantiomeric groups, identically linked, but no other chiral group. Cf. *racemic compound.* **m. position** See *meso position.*

mesocolloid Particles 250 to 2,500 Å long, consisting of 100 to 1,000 molecules; between hemicolloids and eucolloids.

mesoform See *meso*-compound under *meso-*.

mesohydry A form of tautomerism which assumes divided valencies or oscillating bonds, especially between hydrogen and other atoms.

mesomer See *meso*-compound under *meso-*.

mesomeric Pertaining to (1) the *meso* form; (2) mesomerism.

mesomerism* Resonance*.

mesomethylene carbon The 7th C atom in the bornane or pinane structure, which forms a bridge in the ring system.

mesomorphic The anisotropic liquid crystal shape, intermediate in properties between the true liquid and the crystal states.

mesomorphous See *turbostratic.* Cf. *mesomorphic.*

meson (1) Any of a group of subatomic *particles*, q.v. **π- ~** Pion. Intermediate in the formation of muons. **μ- ~** Muon.

mesonin A protein constituent, 25% of wheat gluten.

mesophilic Describing organisms of optimum growth at temperature 25–40°C. Cf. *psychrophilic, thermophilic.*

mesophyll The cellular structure of a leaf, through which water passes.

meso position Formerly: (1) that of a substituting radical attached to a C atom between 2 hetero atoms in a ring; indicated by the Greek letter μ; (2) the 9 or 10 position in the anthracene ring. (3) *ms-*. Central ring fusion.

mesorcin Mesorcinol.

mesorcinol $C_9H_{12}O_2$ = 152.2. 2,4-Dihydroxy-1,3,5-trimethylbenzene*, mesorcin. Colorless, lustrous scales, m.150, insoluble in water.

mesotartaric acid See *tartaric acid.*

mesothelioma A malignant disease, usually of the pleura (the lung covering), associated with asbestos dust particles, particularly those 5–15 μm long.

mesothorium **m.-1** Radium-228. **m.-2** Actinium-228.

mesotomy The separation of optically inactive, isomers into equal parts of dextro- and levorotatory compounds. Cf. *inversion, resolution.*

mesoxalate A salt of mesoxalic acid.

mesoxalic acid* $COOH \cdot C(OH)_2 \cdot COOH$ = 136.1. Dihydroxymalonic acid, 2-oxopropanedioic acid*. Hygroscopic, colorless needles, m.120, soluble in water. **anhydrous ~** $COOH \cdot CO \cdot COOH$, from which the mesoxalyl radical is derived.

mesoxalo* The radical $HOOC(CO)_2-$.

mesoxalyl* The radical $-(CO)_3-$, from anhydrous mesoxalic acid.

mesoxalylurea Alloxan.

Mesozoic An era of geologic time, q.v. *(geologic eras),* between the Paleozoic and Cenozoic eras.

mesquite gum Brown resin from *Prosopis juliflora* (Leguminosae), New Mexico, Texas; resembling gum arabic.

Mestinon Trademark for pyridostigmine bromide.

mestranol $C_{21}H_{26}O_2$ = 310.4. White crystals, m.152, insoluble in water. A potent synthetic estrogen, used for replacement therapy, menstrual disorders and oral contraception.

mesyl* Methylsulfonyl†. The radical $MeSO_2-$.

mesylate *Methane*sulfonate*.

met-, meta- Prefix (Greek: "beyond," "over," or "after"), indicating: (1)* (ital.) the 1,3 position of benzene (cf. *ortho-*); (2) a transformation; as, in petrology; (3) a polymeric compound, e.g., metaldehyde; (4) a less hydrous acid, e.g., metaphosphoric acid; (5) a derivative of a complex compound, e.g., a metaprotein. (6) Met*. Symbol for methionine.

metaacetaldehyde Metaldehyde.

metaacetone Diethyl ketone*.

metaaluminate See *aluminate.*

mesaarsenate See *arsenate.*

metaarsenic acid See *arsenic acid.*

metaarsenite See *arsenite.*

metabisulfite Disulfite*.

metabolic Pertaining to metabolism.

metabolism The chemical reactions in a living cell or organism, by which food is transformed into living protoplasm, reserve materials are stored up, and waste materials are eliminated. **basal ~** The energy m. of an individual at rest. **constructive ~** Anabolism. The processes which build complex from simpler compounds for growth and replacement of tissues. **destructive ~** See *catabolism.* **energy ~** The heat liberated by a living organism. **inborn error of ~** A congenital disease due to absence or deficiency of an essential factor (usually an enzyme) in metabolism; as, absence of tyrosinase leads to the lack of melanin in albinism. See *phenylketonuria, galactosemia.* **synthetic ~** Constructive m.

metabolite A breakdown product of a physiologically active substance (as, a drug) produced by body metabolism; usually having biological activity. **anti ~** A substance that opposes metabolic reactions.

metaborate* See *metaborate* under *borate.*

metaboric acid* See *metaboric acid* under *boric acid.*

metacasein An intermediate protein in the digestion of caseinogen to casein by pancreatic juice.

metacellulose An isomer of cellulose (fungi and lichens), insoluble in cuprammonium.

metacenter The intersection of a vertical line through the center of buoyancy of a floating body, slightly displaced from its equilibrium position, with a line connecting the center of gravity and the equilibrium center of buoyancy. For stable flotation is should be above the center of gravity.

metacetaldehyde Metaldehyde.

metacetone Diethyl ketone*.

metacetonic acid Propionic acid*.

metachemistry See *metachemistry* under *chemistry.*

metachromatic Describing the property of certain substances which appear in different colors according to the wavelength of the light in which they are viewed.

metacinnabarite HgS. A black, native sulfide.

metacompound A derivative of benzene obtained by substitution of the first and third atoms.

meta-cresol See *meta-cresol* under *cresol.* **m. purple** Cresol purple. **m.sulfonephthalein** A pH indicator, changing at pH 2 from red (acid) to yellow (alkaline), and at pH 8.5 from yellow (acid) to purple (alkaline).

metacrolein $(CH_2{:}CHO)_3$ = 129.1. Colorless crystals, m.45.

metacryotic Describing the liquid which separates gradually from frozen fruit juices.

metadiazine Pyrimidine*.

metalement A hypothetical substance, intermediate between an element and a protyle.

metalfiltration Edge filtration through superimposed metallic strips with beveled edges, involving a change from coarse filtration (due to the strips) to fine filtration (due to the filter bed formed in their interstices).

metaformaldehyde 1,3,5-Trioxane.

metaiodate Iodate*.

metaiodic acid Iodic acid.

metaisocymophenol Carvacrol*.

metakliny An intramolecular transfer of groups. See *pinacol conversion.*

metal (1) An electropositive chemical element characterized by ductility, malleability, luster, conductance of heat and electricity, which can replace the hydrogen of an acid and forms bases with the hydroxyl radical. Metals and nonmetals differ in lattice structure, each atom being surrounded by 8–12 or 1–4 other atoms, respectively. Cf. *nonmetallic, periodic table.* (2) An alloy. **alkali ~** See *alkali metals.* **alkaline-earth ~** See *alkaline-earth metals.* **basic ~** Base m. A m. that is readily oxidized. **bell ~** An alloy of copper with 20–25% tin, used for casting bells. It has changed little over the centuries. **Deva ~** Trademark for a porous metal, impregnated with graphite; a solid lubricant. **fine ~** White m. **fusible ~** A m. or alloy of relatively low melting point, e.g., Na, Pb, Sn. **heavy ~** A m. with a density above 4, located in the lower half of the *periodic table,* q.v. **light ~** A m. with a density below 4, located in the upper part of the periodic table. **noble ~** A m. that is not readily oxidized or dissolved in acid, e.g., Au or Pt. **precious ~** Silver, gold, and platinum; they are rare in nature and difficult to isolate. **primary ~** A m. used for the first time. Cf. secondary *metal* below. **rare ~** An element that occurs only in small quantities. **rare-earth ~*** See *rare-earth metals.* **secondary ~** A m. recovered from waste or scrap. **sensitized ~** M. treated with light-sensitive material, so that designs can be photographed on it directly. Used in the mass production of metallic articles. **type ~** See *type metal.* **virgin ~** Primary m. **white ~** (1) Fine m. The almost pure cuprous sulfide obtained in the Welsh process for smelting Cu. (2) Alloys containing large proportions of Pb or Sn, e.g., pewter.

m. bath A fusible metal (as lead), used to obtain a high temperature. **m. compounds** Intermetallic compounds, usually present in alloys. **m. detector** Device for detecting concealed m. (as, guns) or inadvertently added m. (as, in industrial processes); generally based on the change produced in a high-frequency electromagnetic field.

metalammine An ammine, q.v. *(ammines),* of a metal.

metalammonia compound Metalammine.

metalbumin Paralbumin. A protein from ovarian cysts.

metalceramics Powder *metallurgy.*

metaldehyde (1) $(O{\cdot}CH{\cdot}Me)_n$ = $(44.05)n$. Metacetaldehyde. Consists mainly of the tetramer, with higher cyclic oligomers. Colorless needles, sublime 112, insoluble in water; a slugicide. Cf. *aldol, paraldehyde.*

metalepsis An early term (Dumas, 1834) to indicate a substitution.

metalepsy Substitution.

metalignitious Noncaking; as, of coals.

metallic (1) Pertaining to metals in their uncombined forms. (2)* Describing elements with the characteristics of a metal. Cf. *semimetallic, nonmetallic.* **m. carbonyls** Compounds of carbon monoxide with metals; as, nickel carbonyl $Ni(CO)_4$. **m. soap** See *soap.*

metalliferous Describing an ore that contains a metal.

metallify (1) To convert into a metal. (2) To extract a metal from its ore. (3) To give metallic properties.

metalline Resembling a metal.

metallization A process by which a surface is coated with a metal.

metallocene* Cyclopentadienylide. A metal derivative of bis(η-cyclopentadienyl) existing uncharged (as ferrocene) or as charged ions; used in organic synthesis.

metallochrome A tint imparted to metal surfaces by metallic salts.

metallogenic map A map showing the distribution of mineral deposits in relationship to geological formation and tectonic features.

metallography (1) The science of metals and of their ores, production, properties, and uses. (2) The microscopic study of the etched surfaces of metals and alloys. Cf. *mineralography.*

metalloid Semimetallic*.

metallurgy The science of preparing metals from their ores. Cf. *siderurgy.* **electro ~** The electrical preparation of metals. **hydro ~** The preparation of metals by leaching processes. **powder ~** The working of compressed metal powders, e.g., obtained by reducing the corresponding oxides. They are thus formed into solid masses by heat and pressure, and forged or drawn into wire. **pyro ~** The preparation of metals by smelting, roasting, or furnace methods.

metalorganic Pertaining to a metal in organic combination. **m. compound** R_nM. Organometallic compound. A compound of organic radicals, R, with a metal, M; e.g., dimethylzinc.

metamerism Isomerism between 2 compounds which contain the same number and kind of atoms, but with the radicals in different positions; as; $CH_3 \cdot CH_2 \cdot CH_2CHO$, an aldehyde, and $CH_3 \cdot CH_2 \cdot CO \cdot CH_3$, a ketone. Cf. *isomerism.*

metamers Metameric compounds. Substances that exhibit *metamerism*, q.v.

metamorphism In geology, a change in the texture and composition of a rock due to the external agencies (heat, wetness, etc.).

metamorphosis (1) Biology: a change of form or structure during embryonic development, in which the intermediate or larval form leads an independent existence. (2) Geology: a change in the crystalline structure of a mineral. **thermo ~** Thermometamorphism.

metanilic acid $NH_2 \cdot C_6H_4 \cdot HSO_3$ = 173.2. *m*-Aminobenzene-3-sulfonic acid. An intermediate in dyestuff manufacture, decomp. 280.

metaniline yellow Metanil yellow.

metanil yellow Metaniline yellow, sodium phenylaminobenzene metasulfonate. A yellow dye used for wool and paper, for counterstaining tissues, and as an indicator, changing at pH 2.5 from red (acid) to yellow (alkaline).

metantimonate A salt of the type $MSbO_3$.

metantimonic acid The acid $HSbO_3$, from which the metantimonates are derived.

metapeptone A digestive product of peptone.

metaperiodic acid Periodic acid*.

metaphenylene The 1,3-phenylene* radical $C_6H_4=$, from benzene. **m.diamine** $C_6H_4(NH_2)_2$ = 108.1. 1,3-Diaminobenzene. **m.diamine hydrochloride** $C_6H_4(NH_2)_2 \cdot 2HCl$ = 181.1. Metadiaminobenzene hydrochloride. White crystals, soluble in water; a reagent to detect nitrites in water.

metaphosphate A salt of the type MPO_3. See *metaphosphate* under *phosphate.*

metaphosphoric acid The monobasic acid, HPO_3. See *metaphosphoric acid* under *phosphoric acid.*

***meta* position** The 1 and 3 positions in the benzene ring.

metaprotein A hydrolytic split product of proteins.

metarchon A substance used to mask odor.

metaraminol (bi)tartrate $C_9H_{13}NO_2 \cdot C_4H_6O_6$ = 317.3. Aramine. White crystals, soluble in water, m.173; an adrenergic used to raise blood pressure (USP, BP).

metargon An isotope of argon, mass number 38.

metarsenic acid Meta*arsenic acid.*

metartrose $C_{315}H_{504}O_{106}N_{90}S$. A product obtained on digestion of the proteins of wheat.

metasilicic acid* The dibasic acid H_2SiO_3, from silicic acid.

metasomatism Natural enrichment of ores by chemical reaction with external substances.

metasomatosis Chemical alteration of a mineral, to form a new mineral.

metasome An individual mineral which has developed within another mineral.

metastable An unstable condition which changes readily, to a more or a less stable condition. Indicated by the prefix m in the case of *isotopes*, q.v. **m. electron** An electron moving in an excited orbit. **m. phase** The existence of a substance as a solid, liquid, or vapor under conditions in which it is normally unstable in that state.

metastasic electron (1) An electron which transfers from one atom to another. (2) An electron which changes its position in the atom during a radioactive change.

metastasis (1) Radioactive disintegration, in which an α particle is thrown off and 2 electrons pass from the valence into inner orbitals of the atom. (2) An area of secondary disease. See *metastasize.*

metastasize The spreading of disease, usually malignant, from the original or primary site to a distant or secondary site, by the blood or lymphatic system.

metastructure A structure having a dimension between that of the molecule and the smallest structure visible microscopically. Cf. *colloid.*

metastyrene $(C_8H_8)_n$ = $(104.2)n$. Metastyrolene. Fatty liquid, d.1.054, b.320 (decomp.), insoluble in water.

metastyrolene Metastyrene.

metasulfite Disulfite*.

metathesis A chemical reaction (as, neutralization) in which there is an exchange of elements or radicals according to the general equation: AB + CD = AD + BC.

metathiazole Thiazole*.

metatorbernite $Ca(UO_2)_2P_2O_8 \cdot 8H_2O$. A natural alteration product of *torbernite*, q.v.

metazoa A multicellular animal. Cf. *protozoon.*

Metchnikoff See *Metshnikoff.*

meteoric Pertaining to meteorites. **m. iron** The metallic iron in meteorites; usually contains nickel. **m. stone** A meteorite, mainly of aluminum silicates. **m. water** Water reprecipitated as rain, snow, etc., that has entered the lithosphere from the earth's surface.

meteorite Aerolite. A stony or metallic body that has fallen to the earth from outer space. Cf. *chondrite.* Three types: siderite (meteoric iron), siderolite (meteoric iron and stone), aerolite (meteoric stone). See *abundance.* **micro ~** A cosmic particle of diameter less than 1 mm, which loses heat by radiation and does not burn up. The earth collects 10 million tons per year.

meteorograph An apparatus that automatically records atmospheric pressure, temperature, humidity, and wind velocity.

meteorology The study of climatic conditions.

meter, metre* (1) Abbrev. m; the SI system unit of length. Originally supposed to be 1/10,000,000th of the distance from the pole to the equator. Now defined as 1,650,763.73 wavelengths in vacuo of the orange-red radiation corresponding with the transition between the energy levels $2p_{10}$ and $5d_5$ of the ^{86}Kr atom. 1 meter = 100 cm = 1,000 mm = 0.001 km = 39.37008 in. **atom ~** Angström unit. **centi ~** One-hundredth of a meter. **fest ~** German unit equal to 1 m^3 of solid wood. Cf. *cord.* **international ~** Prototype meter. The platinum-iridium bar that was formerly (pre-1961) taken as the standard meter. **kilo ~** One thousand meters. **micro ~** μm. Micron, 10^{-6} m. **milli ~** One-thousandth of a meter. **tenth ~** Angstrom unit, i.e. 10^{-10} meter.

m. angle The angle of vision on viewing a point 1 meter distant. **m. bridge** A slide-wire resistance 1 meter long, used in electrical resistance measurements. **m.-candle** See *illuminance.* **m.-kilogram** The force necessary to lift 1 kg 1 meter.

meter (2) A measuring device for determining a quantity, e.g., of matter, flow, or force. **electro ~** See *electrometer.* **gas ~** A device for determining the quantity of gas passing through a pipeline. **gaso ~** Gasholder. **photo ~** See *photometer.* **urea ~** See *ureameter.* **venturi ~** See *venturi meter.*

meth- Prefix indicating methyl, e.g., methoxy.

methabenzthiazuron* See *herbicides*, Table 42 on p. 281.
methacrylic acid* CH_2:CMeCOOH = 86.1. 2-Methylpropenoic acid*. d.1.015, b.160. Isomeric with 3-butenoic, crotonic, and isocrotonic acids.
methadone hydrochloride $C_{21}H_{27}ON \cdot HCl$ = 345.9. 6-Diamino-4,4-diphenyl-3-heptanone hydrochloride. Amidone, Dolophine, Physeptone. Colorless crystals, m.235, soluble in water; a potent analgesic, also used as a substitute to treat heroin addiction (USP, BP).
methal Myristic alcohol*.
methanal* Formaldehyde*.
methanamide* Formamide*.
methane* CH_4 = 16.04. Methyl hydride. Cf. *biogas, marsh gas, firedamp.* The simplest saturated hydrocarbon. Colorless, flammable gas, $d_{air=1}$0.558, m. −184, b. −161, slightly soluble in water. One of the chief constituents of natural gas, and formed in the decomposition of organic matter. Pure m. is obtained from aluminum carbide and water; used in the manufacture of formaldehyde and in organic synthesis. For most of its compounds see; *methyl-*, CH_3-; *methylene*, $-CH_2-$; *methylidyne*, ≡CH. **m. d** Deuteromethane, diplogen m. The isotopic compounds CH_3D, CH_2D_2, CHD_3, and CD_4. **azi ~** Diazomethane*. **bromo ~*** Methyl bromide*. **chloro ~*** Methyl chloride*. **chlorodifluoro ~*** $CHClF_2$ = 86.5. Freon 22. Gas, m. −146, b. −41. **cyano ~** Acetonitrile*. **diazo ~*** See *diazomethane.* **dibromo ~*** Methylene dibromide. **dichloro ~*** Methylene dichloride. **dichlorodifluoro ~*** CCl_2F_2 = 120.9. Freon 12. Colorless gas, d.1.40, b. −30; a refrigerant. **dichlorofluoro ~** $CHCl_2F$ = 103.9. Freon 21. A refrigerant. **dimethoxy ~*** Formal. **dimethyl ~** Propane*. **diphenyl ~** See *diphenylmethane.* **diphenylene ~** Fluorene*. **fluoro ~*** Methyl fluoride*. **hydroxy ~** Methanol*. **iodo ~*** Methyl iodide*. **methoxy ~*** Methyl ether. **methyldithio ~** Methyl disulfide. **methyl thio ~** Dimethyl sulfide*. **nitro ~*** CH_3NO_2 = 61.0. Colorless liquid, d.1.130, b.102, soluble in water. **phenyl ~** Toluene*. **tetrabromo ~*** Carbon tetrabromide*. **tetrachloro ~*** Carbon tetrachloride*. **tetrahydroxy ~** *o*-Carbonic acid. **tetramethyl ~*** Neo*pentane**. **tribromo ~*** Bromoform*. **trichloro ~*** Chloroform*. **trichlorofluoro ~*** CCl_3F = 137.4. Arcton 9, Freon 11. Liquid, b.24; used as refrigerant and for dry cleaning. **trichloronitro ~*** Chloropicrin. **triethyl ~*** Et_3CH = 100.2. 2-Ethylpentane*. Colorless liquid, d.0.689, b.96, insoluble in water. **trifluoro ~*** Fluoroform*. **triiodo ~*** Iodoform*. **trimethyl ~** Iso*butane**.
m. acid Formic acid*. **m. alcohol** Methanol*. **m. aldehyde** Formaldehyde*. **m. amide** Formamide*. **m.arsonic acid** Methylarsinic acid. **m. base** Leucomalachite green. **m. chloride** Methyl chloride*. **m.dicarboxylic acid** Malonic acid*. **m.disulfonic acid** Methionic acid. **m.phosphonic acid** Methylphosphonic acid*. **m. series** Alkanes*. **m.siliconic acid** MeSiOOH = 76.1. Silicoacetic acid. White powder, insoluble in water. **m.stannonic acid** MeSnOOH = 166.7. Methylstannic acid, stannoacetic acid. White, infusible powder, insoluble in water. **m.sulfonate*** Mes(yl)ate. An ester of m.sulfonic acid, $Me \cdot SO_2 \cdot OR$. **m.sulfonic acid*** $Me \cdot SO_3H$ = 96.1. Methylsulfonic acid. A syrup, decomp. 130. **m.sulfonyl chloride*** $MeSO_2Cl$ = 114.5. Colorless liquid, d.1.51, b.160. **m.thial*** H_2C:S = 46.1. Thioformaldehyde. Unstable. Polymerizing readily. **m.thiol*** CH_3SH = 48.09. Methyl mercaptan. Colorless liquid or gas, d.0.868, b.7.6.
methano-* Prefix indicating a $-CH_2-$ bridge in a ring compound. Cf. *methylene.*
methanoic acid* Formic acid*.
methanol* CH_3OH = 32.04. Methyl alcohol, carbinol, wood alcohol, pyroxylic spirit, wood spirit, wood naphtha, columbian spirit, colonial spirit, methyl hydroxide. Colorless liquid, d.0.810, b.64.7, flammable, soluble in water or ether; a solvent for varnishes, paints, organic compounds; a fuel; used to manufacture formaldehyde; used in organic synthesis and for denaturing.
methanolate* Methoxide. CH_3OM. A compound of a metal with the methoxy group; as sodium m., $NaOCH_3$.
methanoyl The formyloxy* radical.
methemoglobin A product derived from oxyhemoglobin, having the same composition as hemoglobin but with its oxygen more firmly bound. It contains Fe^{3+}; occurs in transudates containing blood, in blood after overdose of some drugs (as, prilocaine), and in urine after hematuria. Mol. wt. 16,666. Cf. *porphin.*
methenamine USP name for hexamethylenetetramine.
methene Methylene*. Cf. *methano-.* **m.disulfonic acid** Methionic acid.
methenyl The radicals (1) menthenyl*; (2) methine*; (3) methylidyne*. **di ~** Acetylene*.
m. bromide Bromoform*.
methide A methyl compound of a metal; as, Me_2Mg, dimethyl magnesium* (magnesium methide).
methimazole $C_4H_6N_2S$ = 114.2. Yellow powder m.145, soluble in water; antithyroid drug, used to treat hyperactive thyroid gland (USP).
methine* The group =CH−. Cf. *methylidyne.* **m. dyes** See *cyanine dyes.*
methiocarb* See *insecticides*, Table 45 on p. 305.
methiodal sodium $NaCH_2O_3 \cdot NaIS$ = 267.0. Sodium iodomethane sulfonate. White crystals with a saline taste, soluble in water; a radiopaque (USP).
methionic acid $CH_2(SO_3H)_2$ = 176.3. Methylenedisulfonic acid. Colorless, hygroscopic crystals.
methionine* $MeS \cdot CH_2 \cdot CH_2 \cdot CHNH_2COOH$ = 149.2. Met*. 2-Amino-4-methylthiobutanoic acid*, from many proteins, e.g., casein; m.283 (decomp.).
methionyl* The radical $C_4H_{10}NS \cdot CO-$, from methionine.
methohexital sodium $C_{14}H_{17}N_2NaO_3$ = 284.3. Brevital sodium. White powder, soluble in water. A short-acting barbiturate, used for anesthesia.
methose $C_6H_{12}O_6$ = 180.2. A carbohydrate synthesized by the polymerization of formaldehyde in the presence of magnesia.
methotrexate $C_{20}H_{22}O_5N_8$ = 454.4. An analog of folic acid, with an NH_2- and OMe− group replacing an −OH and H atom, respectively. A folic acid antagonist, it inhibits the growth of malignant cells, as, leukemia (USP, EP, BP).
methoxalyl* The radical $MeOOC \cdot CO-$.
methoxide Methanolate*.
Methoxone Trademark for MCPA.
methoxsalen $C_{12}H_8O_4$ = 216.2. 9-Methoxy-7*H*-furo[3,2-*g*][1]benzopyran-7-one. Xanthotoxin, Maladinine, Oxsoralen. White crystals, m.145, insoluble in water. Increases melanin pigmentation of skin; used in depigmentation conditions; as vitiligo (USP).
methoxy-* Prefix indicating a methoxy group, $-OCH_3$. **m.benzoyl†** See *anisoyl.* **m.carbonyl*** The radical MeOOC−.
methoxybenzoic acid* $C_8H_8O_3$ = 152.2. **2- ~** Colorless, monoclinic scales, m.98, soluble in water. **3- ~** Colorless needles, m.167, soluble in water. **4- ~** Anisic acid*.
methoxyl The methoxy* radical.
methoxyphenyl* Anisyl. The radical $MeO \cdot C_6H_4-$, derived from anethole. 3 isomers: *o*- ~ , *m*- ~ and *p*- ~ . **m.acetic**

acid $C_9H_{10}O_3$ = 166.2. A gravimetric reagent for Na, with which it forms an insoluble acid salt. **m.acetone** Anisacetone.

methyl* Me. The CH_3- radical. ***N*-m.acetamide** Me·CO·NHMe = 73.1. White needles, m.28, soluble in water. **m. acetate*** MeCOOMe = 74.1. Colorless liquid with apple odor, d.0.924, b.45, soluble in water; a solvent and flavoring. **m.acetic acid** Propionic acid*. **m. acetoacetate*** $MeCOCH_2COOMe$ = 116.1. Colorless liquid, b.170. **m. acetone** A mixture of methyl acetate and acetone; a rubber solvent. **m.acetyl** Acetone*. **m.acetylene** Propyne*. **m. acetylsalicylate** $C_6H_4(O·CO·Me)_2$ = 194.2. Colorless crystals, m.54, isomeric with dimethyl phthalate. **m.acrylic acid** See *methylacrylic acid* under *acrylic acid.* **m. alcohol** Methanol*. **m. aldehyde** Formaldehyde*. **m.aminophenol** Anisidine*. **m.amine*** CH_3NH_2 = 31.06. Aminomethane. Colorless gas, soluble in water. Formed on distillation of wood and bones, putrefaction of fats and fish; a constituent of mercurialis. **m.-*p*-aminophenol** $MeNH·C_6H_4OH$ = 123.2. Rhodol. Used as its hydrochloride or hydrosulfate, as a photographic developer. ***N*-m.aniline** C_6H_5NHMe = 107.2. d.0.986, b.194. **m. anthranilate*** $NH_2·C_6H_4·COOMe$ = 151.2. M. *o*-aminobenzoate. Colorless crystals, m.25, soluble in alcohol; a perfume (orange). **m.arsenious oxide** CH_3AsO = 106.0. The anhydride of arrhenic acid, Me·As:O, m.95. **m.arsine** See *methylarsine* under *arsine.* **m.arsine dichloride** CH_3AsCl_2 = 160.9. Volatile liquid, d.1.858, b.136. **m.arsinic acid mono ~** Arrhenic acid. **di ~** Cacodylic acid. **m. azide** CH_3N_3 = 57.1. Azoimidemethane, methyl azoimide. A hypothetical compound, known from its derivatives. **m.benzene** Toluene*. **m. benzoate*** PhCOOMe = 136.1. Essence niobe. Colorless, fragrant liquid, d.1.094, b.198, insoluble in water; used in perfumery. **m.benzoic acid** Toluic acid*. **m.benzyl*** See *methylbenzyl.* **m.bismuthine*** $MeBiH_2$ = 226.0. Colorless liquid, d.2.30, b.110, insoluble in water. **m. blue** A pH indicator changing at 11 from blue (acid) to brown (alkali). **m. borate** $B(OMe)_3$ = 103.9. Trimethoxyboron. It imparts a green color to a flame, and is used to test for borates. **m. bromide*** CH_3Br = 94.9. Bromomethane*. Colorless liquid, d.1.732, b.4.5, insoluble in water. Used in organic synthesis in refrigerators, and as a fumigant for foods. **2-m.-2-butenal*** MeCH:CMe·CHO = 84.1. Tiglaldehyde, tiglic aldehyde, guaiol. Colorless liquid, b.117. **m. butex** Methyl hydroxybenzoate. **m.butyl** Pentane*. **m. butyrate*** $C_3H_7COOCH_3$ = 102.1. Used in perfumes. **m. Capri blue** An oxidation-reduction indicator. **m. carbamate** Urethylan. **m.carbinol** Ethanol*. **m. carbonate** See *carbonic acid ester.* **m.carbylamine** Methyl isocyanide. **m.catechol** Guaiacol*. **m. Cellosolve** See *methyl Cellosolve* under *Cellosolve.* **m. cellulose** A cellulose m. ether (26–29% —OMe). White, fibrous powder, swelling in water to a viscous colloidal solution, insoluble in alcohol; a bulk laxative (USP, BP). **m. chloride*** CH_3Cl = 50.49. Chloromethane*. Artic, arctic. Colorless gas, b.−24, soluble in water; a refrigerant. **m. chlorofluoride** Dichlorodifluoro*methane**. **m. chloroform** $MeCCl_3$ = 133.4. Trichloroethane*, ethenyl chloride. Colorless liquid of pungent odor, d.1.346, b.74. **m. cinnamate** C_8H_7COOMe = 162.2. White crystals, m.36, insoluble in water; an insect bait. **m.cocaine** Cocainidine. **m.crotonic acids** See *butenoic acid.* **m. cyanate*** $N:COCH_3$ = 57.1. Cf. *methyl isocyanate* below. **m. cyanide** Acetonitrile*. **m.cyclohexane** Hexahydrotoluene. **m.diphenyl** Phenyltoluene. **m. disulfide** MeS·SMe = 94.2. Methyldithiomethane. **m.dopa** $C_{10}H_{13}O_4N·1½H_2O$ = 283.2. L-3-(3,4-Dihydroxyphenyl)-2-methylalanine. Metildopa, Aldomet, Dopamet. White crystals, sparingly soluble in water. An antihypertensive and inhibitor of dopamine formation (USP, EP, BP). **m.ene** Methylene*. Colorless liquid, b.112. **m.ergometrine (maleate)** BP name for m.ergonovine maleate. **m.ergonovine maleate** $C_{20}H_{25}N_3O_2·C_4H_4O_4$ = 455.5. M.ergometrine maleate. Pink, bitter crystals, slightly soluble in water; an oxytocic (USP, BP). See *ergonovine.* **1-m.ethenyl†** See *isopropenyl* under *propenyl.* **m. ether** MeOMe = 46.1. Dimethyl ether*, methoxymethane*. A gas, b.−25, soluble in alcohol; a refrigerant. **m. ethyl ketone*** 2-Butanone*. **1-m.ethyl†** See *isopropyl* under *propyl.* **m.fluether** Enflurane. **m. fluoride*** CH_3F = 34.03. Fluoromethane*. Colorless gas, b.−78. **m. fluorosulfonate** $MeFOSO_2$ = 114.1. Volatile liquid, b.92. A powerful methylating agent. **m. formate*** HCOOMe = 60.1. Methylformic acid ester. Colorless liquid, d.0.973. b.32, soluble in water; used to manufacture cellulose acetate. **m. gadenine** $C_8H_{19}NO_2$ = 161.2. A poisonous, oxygenated ptomaine from fish. **m. gallate*** Gallicin. **m.gallium dichloride** $GaMeCl_2$ = 155.7. White crystals, m.75. decomp. in water. **m.glycine** Sarcosine*. **m.glyoxal** Pyruvaldehyde*. **m.glyoxalidine** Lysidine. **m. green** $[(Me_2N·C_6H_4)Me_2\overset{+}{N}:C_6H_4:C(C_6H_4\overset{+}{N}Me_2Et)]BrCl$. A triphenylmethane dye; used to dye silk and stain mitochondria; and as a pH indicator changing from yellow (acid) through blue-green to colorless (alkaline). **m.guanidine** $MeN:C(NH_2)_2$ = 73.1. A ptomaine formed from creatine or arginine. **6-m.-5-hepten-2-one*** $Me_2C:CH(CH_2)_2COMe$ = 126.2. Colorless liquid, b.173, in essential oils. **m.heptylamine*** $MeCH(NH_2)C_6H_{13}$ = 192.2. b.174. **m. heptyne carbonate** Heptyne methyl carbonate. **m. Hexalin** Trademark for m.*cyclohexanol.* **m.hexane** Heptane*. **m. hydrate** Methanol*. **m.hydrazine*** $MeNHNH_2$ = 46.07. A liquid, b_{745mm}87. **m.hydrazone*** $MeCH:NNH_2$. **m. hydroxide** Methanol*. **m.hydroxybenzene** (1) Cresol*. (2) Salicyl alcohol*. **m. hydroxybenzoate** $C_8H_8O_3$ = 152.2. M. butex. M.paraben (N.F.). White crystals with a burning taste, m.126, slightly soluble in water; a preservative used in pharmacy (BP). **m.indole** Skatole. **m. iodide*** CH_3I = 141.9. Iodomethane*. Colorless liquid, d.2.28, b.44, insoluble in water. **m. isocyanate** MeCNO = 57.1. b.45. **m. isocyanide** CH_3NC = 41.5. M.carbylamine, m. isonitrile. Colorless liquid d.0.756, b.60. **m.isophthalic acid 4- ~** Xylidic acid. **5- ~** Uvitic acid. **m. isothiocyanate** M. *mustard oils.* **m. mercaptan** Methanethiol*. **m.mercuric chloride** MeHgCl = 251.1. White crystals with a disagreeable odor, m.170. **m.mercuric iodide** MeHgI = 342.5. Pearly leaflets, m.145, insoluble in water. **m. methacrylate** $CH_2:CMe·COOMe$ = 100.1. A source of polymers used for transparent sheet material, laminated safety glass and dentures. **m.morphine** Codeine. **m.nitramine** $Me·NH·NO_2$ = 76.1. m.38. **m. nitrate** $Me·O·NO_2$ = 77.04. Explosive liquid, d_5·1.2, soluble in water. **m. nitrite** $MeNO_2$ = 61.0. A gas, d.0.99, b.−12. **m.nitrobenzene** Nitrotoluene*. **m. nonyl ketone*** 2-Undecanone*. **m. orange** $Me_2NC_6H_4N:NC_6H_4SO_3Na$. The sodium salt of *p*-dimethylaminobenzenesulfonic acid. Yellow powder, soluble in water; an indicator (alkalies—yellow, acids—red; pH 3.1–4.4). **m. oxide** M. ether. **3-m.-1-oxobutyl†** See iso*valeryl.* **m.oxyaniline** Anisidine*. **m.paraben** M. hydroxybenzoate. **m.pentose** $C_6H_{12}O_5$. A sugar containing 6 carbons but only 5 hydroxy groups; as, fucose. **m.phenidine** See *phenacetin.* **m.phenolate** Anisole*. **(m.phenyl)amino†** See *toluidino.* **m.phenylene†** See *tolylidene.* **m. phenyl ether*** Anisole*. **methyl phenyl ketone*** Acetophenone*. **m. phosphate** $MePO_2(OH)_2$ = 112.0. m.105. **m.phosphine*** $MePH_2$ = 48.0. A gas, b.−14, soluble in water. **m.phosphonic acid*** $CH_3PO(OH)_2$ = 96.0. Colorless crystals, m.105. **m.prednisolone** $C_{22}H_{30}O_5$ = 374.5. A glucocorticoid. Medrol. White, bitter crystals, m.243 (decomp.), insoluble in water (BP). See *corticoids.* **m. propionate*** EtCOOMe = 88.1. M. propanoate*. Colorless liquid, d.0.9148, b.80, soluble in water; used in perfumes. **2-m.propoxy†** See *isobutoxy.* **2-m.propyl†** See *isobutyl* under *butyl.* **m. propyl ether*** MeOPr = 74.1. Colorless liquid, d.0.738, b.39. **m. propyl ketone*** 2-Pentanone*. **m.propylphenol** Thymol*. **m.pyridine** Picoline. **m. pyruvate*** MeCOCOOMe = 102.1. A liquid, d.1.154, b.137; a solvent for resins. **m.quinoline 2-~** Quinaldine. **4-~** Lepidine. **m. red** $Me_2NC_6H_4N{:}NC_6H_4COOH$ = 269.3. *p*-Dimethylaminoazobenzenecarboxylic acid. Red powder, insoluble in water; an indicator (alkalies—yellow, acids—violet-red; pH 3–6). **m.resorcinol** Orcinol. **m.rosaniline hydrochloride** See *methyl violet* below. **m. rubber** Early name for synthetic rubber made by polymerization of dimethylbutadiene. **m. salicylate*** $C_6H_4\cdot(OH)\cdot COOMe$ = 152.2. Artificial wintergreen oil, methylic salicylas, betula oil, gaultheria oil, sweet birch oil. Colorless liquid, d.1.183, b.222, insoluble in water. Used as a flavoring, antipyretic, antiseptic; and in antirheumatic liniments (NF, EP, BP). **m. silane** $SiH_3\cdot CH_3$ = 46.11. Colorless gas, b.−57. **m.stannic acid** MeSnOOH = 166.7. White powder, insoluble in water. **m. styryl ketone*** Benzylidene acetone. **m.succinic acid** Pyrotartaric acid. **m. sulfate** Dimethyl sulfate*. **m. sulfide** Dimethyl sulfide*. **m.sulfonic acid** Methanesulfonic acid*. **m.sulfonyl†** See *mesyl.* **m.tartronic acid** Iso*malic acid.* **m. telluride** $(CH_3)_2Te$ = 157.7. Yellow liquid with garlic odor, b.82. **m.testosterone** $C_{20}H_{30}O_2$ = 302.5. 17β-Hydroxy-17-methylandrost-4-en-3-one. Metandren, Neo-Hombreol-M, Oreton-M. White hygroscopic crystals, m.164, insoluble in water; an androgenic hormone. Used for testicular insufficiency and breast cancer (USP, EP, BP). **m. theobromine** Caffeine. **m. thiocyanate** MeSCN = 73.1. M. rhodanate. Colorless liquid, d.1.088, b.133, soluble in alcohol. **m.thionine chloride** Methylene blue. **m.thiophene** Thiotolene. **m.tin tribromide*** CH_3SnBr_3 = 373.4. White needles, m.54, soluble in water. **m.tin trichloride*** CH_3SnCl_3 = 240.1. Colorless crystals, m.43, soluble in water. **m.tin triiodide*** CH_3SnI_3 = 514.4. Yellow needles, m.87, soluble in water. **m.toluidine** Xylidine*. **m. urea** See *methyl urea* under *urea.* **m. urethane** Urethylan. **m. violet** Crystal violet, gentian violet, pyoktanin blue. A mixture of the hydrochlorides of pentamethyl-*p*-rosaniline and hexamethyl-*p*-rosaniline. Green crystals, soluble in water; a reagent, indicator and textile dye (alkalies—violet, acids—yellow; pH 2.0–3.1).

methylal Formal.

methylamino-* Prefix indicating the radical −NHMe.

methylate (1) The substitution of a methyl group for an atom or radical. (2) Denaturate. To add methanol to alcohol to render it unpotable. (3) Methanolate*.

methylated **m. ether** Ethyl ether made from m. spirit instead of from pure ethanol. **m. spirit** *Rectified spirit,* q.v., denatured by addition of 2 or more of: naphtha, mineral naphtha, pyridine, methanol and a dye.

methylbenzyl* Xylyl. The radical $Me\cdot C_6H_4\cdot CH_2-$, from xylene. **m. bromide** $MeC_6H_4CH_2Br$ = 185.1. ***ortho-~*** m.21. ***meta-~*** b.215. ***para-~*** m.38. The mixed isomers are a lachrymatory poison gas (T-stoff). Cf. bromo*xylene.* **m. chloride** $CH_3\cdot C_6H_4\cdot CH_2Cl$ = 140.6. Monochloroxylene, tolyl chloride. Colorless liquids, insoluble in water. ***ortho-~*** b.197. ***meta-~*** b.195. ***para-~*** b.192.

methylene* Carbene*, methene. The groups $-CH_2-$ and $=CH_2$, and the free radical $:CH_2$. Cf. *di-, tri-,* etc., *methylene.* **meso ~** See *mesomethylene carbon.* **trioxy ~** 1,3,5-Trioxane.

m.bis(oxy)† See *methylenedioxy.* **m. blue** $C_{16}H_{18}N_3SCl\cdot 3H_2O$ = 373.9. Tetramethylamidophenthiazinium chloride, methylthionine chloride, methylthioninae chloridum (USP); a dye of the thiazine group. Green crystals, soluble in water. Used as a redox indicator, a bacteriological stain, a diagnostic aid, and a textile dye. **alkaline ~** A stain: m. blue 5, sodium peroxodicarbonate 5 g/l in water. **m. dibromide** CH_2Br_2 = 173.8. Dibromomethane*. Yellow liquid, d.2.59, b.98, insoluble in water. **m. dichloride** CH_2Cl_2 = 84.9. Dichloromethane*, methylbichloride, carrene. Colorless liquid, d.1.377, b.41, soluble in alcohol; solvent, degreaser, and refrigerant. **m. dicyanide** Malononitrile*. **m. diiodide*** CH_2I_2 = 267.8. Diiodomethane*. Yellow liquid, d.3.335, b.180, insoluble in water; used to determine density of mineral mixtures and water-soluble substances. **m. diol** $CH_2(OH)_2$ = 48.04. Hydrated formaldehyde in its aqueous solutions. **m.dioxy*** Methylenebis(oxy)†. The radical $-OCH_2O-$. **m.disulfonic acid** Methionic acid. **m.ditannin** Tannoform. **m.imine** See *methylenimine.* **m.triol** Phloroglucitol.

methylenimine $H_2C{:}NH$ = 29.04. Azomethine.

methylic The methyl* radical. **m. acid** Formic acid*. **m. alcohol** Methanol*.

methylidyne* Carbyne. The radical CH≡. Cf. *methine.*

methylin A lignin extracted from plants by ethylene glycol monomethyl ether.

methyloic- Prefix indicating a carboxyl group as a side chain, e.g., $Et_2CH\cdot COOH$, pentane-3-methyloic acid. Cf. *ethyloic.*

methylol The hydroxymethyl* radical.

methyne The methine* radical.

methysergide maleate $C_{21}H_{27}O_2N_3\cdot C_4H_4O_4$ = 469.5. Deseril, Sansert. White crystals, slightly soluble in water. Used in prophylaxis of migraine (USP, BP).

methysticin Kavain.

methysticum Kava.

metioscope A photoemission electron microscope; the image is formed directly by electrons emitted by the object after u.v. light irradiation.

metoprolol tartrate $(C_{15}H_{25}O_3N)_2C_4H_6O_6$ = 684.8. Betaloc, Lopressor. White crystals, soluble in water. A beta-adrenergic blocking drug; used to control cardiac arrhythmias and treat angina (USP).

metoclopramide hydrochloride $C_{14}H_{22}O_2N_3Cl\cdot HCl\cdot H_2O$ = 354.3. Maxolon, Primperan, Reglan. White crystals, soluble in water. An antiemetic that also promotes gastric emptying (BP).

Metol Trademark for 4-(methylamino)phenol sulfate. A photographic developer.

metopon Methyldihydromorphinone. A narcotic and analgesic.

metoxuron* See *herbicides,* Table 42 on p. 281.

metre* See *meter.*

metric (1) Pertaining to measure. **gravi ~** Relating to analysis involving the use of the balance. **volu ~** Relating to analysis carried out by measuring volumes with pipet and buret. (2) Pertaining to the *m. system* (see below).

m. carat See *metric carat* under *carat.* **m. count** A measure of the fineness of a fiber; the length (in meters) of 1 gram. **m. slug** See *metric slug* under *slug.* **m. system** Weights and measures based on the meter, from which other scientific units are derived. The multiples and fractions of units are uniformly prefixed with the principal Greek and Latin terms used in the *SI* system, q.v., which has now become the official system of many countries and for international usage. Units used:

Length = meter = m; km, cm, mm, μ (10^{-6} m), mμ (10^{-9}m), $\mu\mu$ (10^{-12} m)
Area = square meter = m^2; are = 100 m^2
Volume = liter = 1,000 cc (cubic centimeter) and λ (10^{-6} l)
Mass = gram = g; kg, mg, and γ (10^{-6} g)

m. ton 1,000 kilograms = 2204.6 pounds.
metrication Conversion to the metric system.
metronidazole $C_6H_9O_3N_3$ = 171.2. 5-Methyl-5-nitroimidazole-1-ethanol. Flagyl. White, bitter crystals, m.161, soluble in water; an antiprotozoal and antibacterial compound used to treat trichomonoiasis, amebiasis, and anaerobic bacterial infections (USP, BP).
metronome An instrument to denote short time intervals; a mechanically driven pendulum or an electronic clicker.
-metry Suffix indicating measurement and measuring.
Metshnikoff, Elie (1845–1918) Metchnikoff. Russian physiologist, discoverer of phagocytosis.
metso Sodium metasilicate; a scouring agent.
metyrapone $C_{14}H_{14}N_2O$ = 226.3. 2-Methyl-5-nitroimidazole-1-ethanol. Metopiron(e). White crystals, m.52, sparingly soluble in water. Inhibits an enzyme necessary for production of glucocorticoids. A test drug for pituitary function (USP, BP).
Mev Abbreviation for million electronvolts.
mevalonic acid $C_6H_{12}O_4$ = 148.2. 3,5-Dihydroxy-3-methylvaleric acid, involved in the biosynthesis of cholesterol.
Mewlon Trademark for a polyvinyl alcohol synthetic fiber.
Mexican **M. poppy oil** Brown oil from the seeds of prickly poppy, *Argemone mexicana;* used in soap manufacture. **M. onyx** A variety of calcite, used in interior decorations. **M. scammony root** Ipomoea.
Meyer **M., Lothar Julius (1830–1895)** German chemist, discoverer of the periodic system. Cf. *Mendeleev.* **M., Victor (1848–1897)** German chemist. **M.'s formula** (1) See *molecular free path.* (2) An equation connecting viscosity and temperature. **M.'s law** Law of *esterification.* **M.'s tube** An absorption tube for carbon dioxide, filled with barium hydroxide; used in steel analysis.
meyerhoffite $2CaO\cdot3B_2O_3\cdot7H_2O$. A native borate.
meymacite $WO_3\cdot H_2O$. Native, hydrated, brown, resinous masses.
mezcal Mescal.
mezcaline Mescaline.
mezquit *Prosopis juliflora* of Mexico and the S.W. United States. Its gum resembles gum arabic.
MF Medium frequency. See *radiation,* Table 69.
MF resin Melamine-formaldehyde resin.
Mg Symbol for magnesium.
mg Abbreviation for milligram. **mg%** mg/100 ml.
mho A unit of electrical conductance. 1 mho = 1 ohm^{-1} = 1 siemens (the SI unit).
miamine Chlorazene.
miargyrite $AgSbS_2$. A silver sulfide ore.
miasma, miasm Noxious vapors from swamps.
miazines Metadiazines or pyrimidines. Heterocyclic compounds having 2 N atoms in the meta position; as, pyrimidine. Cf. *piazines, oiazines.*
micas (1) A group of laminated *silica* minerals, q.v.; e.g., biotite, muscovite, phlogopite, zinnwaldite. None has fixed properties. (2) $3Al_2O_3\cdot K_2O\cdot6SiO_2\cdot2H_2O$. Isinglass, muscovy glass. A native, hydrous silicate, which can be split into very thin transparent sheets. Used as an electrical insulator; as windows in furnaces and refracting instruments; and, ground, as a lubricant. (3) Geology: Prefix to describe rocks containing m; as mica basalt. **amber ~** Phlogopite. **lithia ~, lithium ~** Lepidolite. **potash ~** $KH_2Al_3(SiO_4)_3$. Potassium metasilicate. **ruby ~** Muscovite.
micelle (1) An electrically charged, colloidal particle or ion, consisting of oriented molecules. Cf. *zone.* (2) An oriented arrangement of a number of molecules; as in cellulose. Cf. *liquid, association.* (3) An aggregate of a number of molecules held loosely together by secondary bonds. Cf. *bond.*
Michael's reaction An organic addition reaction in which the sodium salts of acetoacetic acid or malonic esters disrupt the double bond to form unsaturated compounds of the type, R—CH:CH—X, where X is a carbonyl or cyanide radical.
Michler **M.'s hydrol** $(Me_2NC_6H_4)_2CHOH$ = 256.4. Bis(4-dimethylaminophenyl)methanol. White crystals, m.96; used in organic synthesis. **M.'s ketone** $(Me_2NC_6H_4)_2CO$ = 254.4. 4,4′-Bis(dimethylamino)benzophenone. Colorless plates, m.172, insoluble in water; used in the synthesis of dyestuffs and auramine derivatives and as a microreagent for metals (blue color with Hg).
miconazole nitrate $C_{18}H_{14}ON_2Cl_4$ = 416.1. Daktarin, Monistat. White crystals, m.182, very slightly soluble in water. An antifungal, in the imidazole group, used for systemic skin and vaginal infections (USP, BP).
MICR Magnetic ink character recognition.
micrinite An opaque material in *coal* durains, q.v.; and intermediate between fusain and vitrain.
micro- (1) Prefix (Greek,) indicating "small." (2)* μ. SI prefix for meter. **sub ~** See *analysis,* Table 9 on p. 38.
microanalysis See *microanalysis* under *analysis.*
microbalance A balance to weigh micro quantities. See *McBain-Baker balance.*
microbe A microorganism of animal or vegetable nature; generally causing disease; as, bacteria and pathological protozoans.
microbic Pertaining to microorganisms.
microbicide An agent that destroys microorganisms.
microbiology (1) The study of microorganisms. (2) Synonym for bacteriology.
microburner A small bunsen burner.
microcapsules Capsules whose walls are made of a material such as insolubilized gelatin and which contain a dye, fragrance, pesticide, etc. Their diameter range exceeds 1–2,000 μm; as, on carbonless copy *paper.*
microchemical (1) Pertaining to reactions observed under the microscope. (2) Pertaining to chemical reactions in miniature apparatus with small quantities. See *microanalysis* under *analysis.*
microchemistry (1) Chemical investigation by means of the microscope, especially the performance of chemical reactions on a microscope slide requiring only minute quantities of substances. (2) Qualitative and quantitative reactions performed with small quantities (μg and μl), using miniature apparatus.
microcline Amazonite.
micrococcus A minute, spherical or round bacterium.
microcosmic salt $NaNH_4HPO_4\cdot4H_2O$ = 209.1. Phosphor salt. Ammonium sodium hydrogenphosphate in blood and natural waters; used in blowpipe analysis for *bead tests,* q.v.
microcrith The weight of a hydrogen atom (obsolete).
microcrystalline Cryptocrystalline. Crystallizing in minute

crystals. **m. wax** A mixture of solid, mineral-origin hydrocarbons, e.g., precipitated during the deoiling of petroleum crude oil distillates and fractionally crystallized. White to pale amber, melting point not less than 71, iodine val. not exceeding 4.0. It should conform to the BP test for sulfur compounds in liquid paraffin. Used in chewing gum and to make paper water- and vaporproof.

microdiffusion analysis Isothermal distillation. An analytical method (milligram scale) based on the gaseous diffusion of a volatile substance from sample to reagent; e.g., of ammonia liberated from an ammonium salt to a standard acid. It is usually effected by the use of 2 petri dishes, one inside the other and both covered, each containing a reactant.

microfilm A photographic film reproducing printed matter on a greatly reduced scale and read by projection on a screen.

microfractography The microscopical study of the fracture surfaces of metals.

micrography (1) Photomicrography. (2) The measurement of physical properties with the microscope.

microlamp (1) An illuminator lamp for microscopes. (2) A small source of artificial light.

microline $K_2O,Al_2O_6,6SiO_2$. A vitreous, yellow mineral.

microliths Very small crystals, microscopic sections of rocks and slags.

micromanipulator Attachments to the microscope stage (controls and levers) to manipulate an object under observation, e.g., for dissections.

micromerol $C_{33}H_{52}O_2$ = 480.8. A monobasic alcohol from *Micromeria chamissonis* (Labiatae).

Micromet Trademark for a mixed sodium and calcium phosphate boiler-water conditioner.

micrometer (1)* μm; micron. SI system unit for one-millionth of a meter. (2) An instrument for measuring small lengths under the microscope. **m. caliper** An instrument for measuring with an accuracy of 0.01 mm.

micromicron $\mu\mu$ (mu-mu). Metric system unit for one-millionth of a micron = 10^{-12} m = 10λ. SI equivalent is the picometer, pm.

micromillimeter Nanometer*, nm.

micromonosporin An antibiotic produced by *Micromonospora* species of actinomycetes.

micron (1) Micrometer*. **milli ~** See *millimicron.* (2) A colloidal particle:

Micron	10 to 0.2 μm; 10^{-3} to 2×10^{-5} cm
Submicron	0.2 μm to 5 nm; 2×10^{-5} to 5×10^{-7} cm
Amicron	Less than: 5 nm; 5×10^{-7} cm

m. of mercury The pressure exerted by a column of Hg 1 μm high; 1 μmHg = 0.001 mmHg.

micronaire value A measure of the fineness and general quality of a fiber. A known weight of fiber is compressed to a plug of known volume, and the flow of air forced through it is measured.

micronize To reduce particles to a size below 5 μm.

microorganism A minute animal or plant, visible only through a microscope. Often used to describe bacteria and viruses.

microphone An electrical instrument to intensify or transmit sound.

microphotogram The record made by a microphotometer.

microphotograph (1) Photomicrograph. (2) Microphotogram.

microphotometer An instrument to measure and record the intensity of spectral lines by determining the density of their photographic images over small areas by means of a photoelectric cell.

microporous Having openings or cavities of microscopic size. **m. rubber** See *mipor rubber.*

micropolariscope A microscope with polariscope attached; used to study minerals and crystals.

microprocessor A single chip containing several main electronic components, as, ROM, RAM, registers, and I/O control.

microreaction A qualitative chemical reaction performed under the microscope with minute reagents Cf. *spot analysis.*

microsal A disinfectant mixture of copper carbonate and crude sulfonephenolic acids.

microscope An optical instrument, consisting of objectives and eyepiece, that magnifies minute objects for visual inspection or photographic record by direct illumination. Normal lower limit of visibility 0.10 μm. **binocular ~** A m. having two eyepieces; produces a perspective effect. **compound ~** An ordinary m., enlarging 30 to 1,000 diameters. **electron ~** A device analogous to an ordinary m., in which a beam of electrons replaces the source of light and magnetic condensers replace the lenses. The image is rendered visible by projection on a fluorescent screen. Magnifications of up to about 200,000 are obtainable. Cf. *scanning electron microscope* below. **fluorescence ~** A m. in which the illumination is filtered ultraviolet light; used to study fluorescence (q.v.) phenomena. Cf. *ultraviolet microscope* below. **ion ~** A powerful m. (magnification $2 \times 10^6 \times$). The image is produced on a fluorescent screen by the ions accelerated from a metal specimen. **photoemission ~** See *photoemission* m. **polarizing ~** A m. in which the object is on a rotating stage between crossed nicols. **scanning electron ~** A microscope that uses an electron beam to sweep to and fro over a specimen in a fixed pattern. A detector and amplifier transmit signals from the reflected and secondary electrons to an oscillotype image screen, which shows the surface ultrastructure in depth, as distinct from the conventional transmission electron m. Magnification up to 50,000×. **scanning tunneling ~** Microscope utilizing vacuum tunneling to give a 3-dimensional image of a solid surface down to the atomic level. **ultra ~** A m. in which the object is indirectly illuminated; e.g., a thin layer of a colloidal solution is illuminated at right angles to the line of sight, and the colloidal particles appear as bright points on a dark field. Lower limit of visibility 5 nm. **ultraviolet ~** A high power m., in which an almost monochromatic ultraviolet ray is the illuminator. Objects may be enlarged by 1,000 to 6,000 diameters. Lower limit 30 pm. Cf. *fluorescence microscopy.* **m. test** Microreaction.

microscopic Visible only under the microscope. Lower limit about 0.10 μm. **a ~** Invisible under the ordinary microscope. **sub ~** Amicroscopic. **ultra ~** Visible under the ultramicroscope. Lower limit about 5 nm.

microscopy (1) The study of the optical enlargement of objects, and their photography. (2) The application of the microscope to useful ends. **fluorescence ~** See *fluorescence microscopy.* **phase contrast ~** See *phase contrast microscopy.*

micro-silica Fine powder, added to fresh cement to increase its strength and impermeability.

microspectroscope A microscope with spectroscope attached; used to study spectral phenomena, such as fluorescence, polarizing and Raman absorption spectra, and the structure of spectral lines.

TABLE 53. CONTENTS ANALYSIS OF MILK FROM VARIOUS MAMMALS
(Parts per Thousand)

Origin	Water	Proteins	Fat	Sugar	Salts
Dog	754.4	99.1	95.7	31.9	7.3
Cat	816.3	90.8	33.3	49.1	5.8
Goat	869.1	36.9	40.9	44.5	8.6
Sheep	835.0	57.4	61.4	39.6	6.6
Human	875.5	12.5	35.0	75.0	2.0
Cow	871.7	35.5	36.9	48.8	7.1
Mare	900.6	18.9	10.9	66.5	3.1
Ass	900.0	21.0	13.0	63.0	3.0
Pig	823.7	60.9	64.4	40.4	10.6
Elephant	678.5	30.9	195.7	88.5	6.5

microtome An instrument for cutting thin sections of materials for microscopic examination.
microtopography The study of the fine structure of surfaces.
micro unit A unit of small measurement, usually one-millionth part; as, γ, λ, σ for microgram, microliter, microsecond, respectively.
microwave Radiation of wavelength 1 mm to 15 cm; generated by radio-frequency power tubes from high-voltage direct current. Used industrially, as, to rapidly cure plastics, and domestically to produce rapid internal heating by intramolecular friction and agitation.
micrurgy The study of surgery by microscopical methods.
miemite $CaCO_3 \cdot MgCO_3$. A vitreous, brown mineral.
Miers, Sir Henry (1858–1942) British chemist, noted for his work on mineralogy and crystallography.
Miescher pipet A small tube for diluting blood specimens for hemocytometers.
migma A stage in the formation of granites, when the rock material is fluid. Cf. *magma.*
mignonette oil Reseda oil.
migration A change of position, as the m. of ions in an electric cell. **atomic ~** See *rearrangement.*
m. tube An H-shaped glass vessel with electrodes containing a salt solution and indicator. On passing an electric current, the m. of the ions is illustrated by the color. **m. velocity** The velocity with which ions move through a solution during electrolysis. See *transport lag number.* Absolute velocity = $10^8 \times$ migration velocity/96,000. Cf. *mobility.*
mikro- Micro-.
mikrobe See *microbe.*
mil (1) A measure of thickness, especially of wire: 1 mil = 1/1,000 in = 25.400 μm. (2) Former symbol for milliliter.
milarite $HKCa_2Al_2(SiO_3)_{12}$. Green, brittle, hexagonal prisms.
mile mi. A measure of length: 1 international mile = 1.609344 km = 5,280 ft = 80 chains. **geographical ~** International nautical mile. **international nautical ~** 1.85200 km = 6076.1 ft. **U.S. statute ~** 1.609347 km.
milk (1) The opaque secretion of the mammary glands. A white emulsion, d.1.029–1.039. For contents analyses, see Table 53. **acid of ~** Lactic acid. **butter ~** Milk from which the fat has been removed. **certified ~** M. which has been chemically and bacteriologically tested, and contains few bacteria. **condensed ~** Milk from which some water has been removed and sugar added. **dried ~** M. from which water has been removed to produce a dry and relatively stable powder. **fermented ~** Koumiss. **homogenized ~** M. that has been reemulsified, e.g., after pasteurization, or treated so that the fats do not separate into cream. **modified ~** A cow's milk diluted with lactose solution for infant feeding. **pasteurized ~** M. heated at 60°C for 30 min. **reconstituted ~** The m. obtained by adding the appropriate quantity of water to dry m. **skimmed ~** M. from which the separated cream has been removed. **sterilized ~** M. that has been heated at 100°C for 45–60 min. **UHT ~** Ultra-heat-treated. M. clarified, heated to 132–138°C for 2 s, homogenized, and cooled. It keeps enchanged for considerable periods in sterile containers. **vegetable ~** An emulsion of fats from the soybean, used extensively in China for bean cakes or cheeselike foods.
m. extract Condensed m. whose casein has been partly peptonized. **m. fat** The total fats from milk. **m. powder** A milk that has been evaporated and powdered. **m. scale** A white, opal strip along the length of a buret, to facilitate accurate meniscus readings. **m. serum** Whey. **m.stone** (1) A flint, whitened by fire. (2) A hard casein-containing scale produced in dairies, due to hard water for cleaning. **m. sugar** Lactose. **m. test bottle** A graduated centrifuge tube, used to determine the fat content of m.
milk (2) Magma. An emulsion or suspension. **m. of almonds** An emulsion of 6% almond oil with acacia, in water; used in pharmaceutical preparations and cosmetics. **m. of asafetida** An emulsion of 4% asafetida in water; a sedative and carminative. **m. of barium** A suspension of barium hydroxide in water. **m. of lime** A suspension of calcium hydroxide in water. **m. of magnesia** See *milk of magnesia* under *magnesia.* **m. of sulfur** Precipitated sulfur.
milky Having a flat-white or opaque appearance. **m. quartz** Quartz with a milklike color and greasy luster.
mill (1) A crushing, grinding, or pulverizing apparatus. (2) An establishment for reducing ores by mechanical means. (3) An establishment for grinding, crushing, powdering, or other processing, as, paper mill or flour mill. (4) The equipment of a rolling (steel) mill. **assay ~** A small mechanical laboratory crusher. **ball ~** A grinding apparatus with iron or quartz balls for powdering. **drug ~** A laboratory apparatus for grinding drugs or seeds. **pebble ~** Ball mill. **porcelain ~** A laboratory machine for grinding wet or dry chemicals or bacteriological materials.
m. iron A pig iron suitable for puddling or for the basic open-hearth process.
millboard Board made from wastepaper. Cf. *chipboard.*
Miller M., William Lash (1866–1940) Canadian chemist, noted for his study of bios and physical chemistry. **M. indices** See *atomic plane.*
millerite NiS. Nickel pyrites. A native sulfide.
millet A small-grain, edible cereal, cultivated on dry, sandy soils, *Setaria italica* (Gramineae).
milli-* Abbrev. m; SI system prefix indicating one-thousandth.
milliammeter An ammeter for measuring around the 0.001 Å level.
milliard (French and U.S.) One thousand million, 10^9. Cf. *billion.*
milligram Abbrev. mg; a unit of weight: 1 mg = 1/1,000 g = 0.01543 grain. **m. atom** Abbrev. mg. at. The mg of an element, divided by its atomic weight. **m. percent** Symbol: mg %. The concentration of a solution expressed in mg per 100 mL.
milligramage The amount of radioactive exposure produced by 1 mg of radium in 1 hour.
milligram-hour Milligramage.
Millikan M., Robert Andrews (1868–1954) American physicist, noted for his research on electrical phenomena.

Nobel prize winner (1923). **M.'s rays** Cosmic rays. A high-frequency radiation from interstellar space, which penetrates the atmosphere and upper crust of the earth.

milliliter* Abbrev. ml*, mil; 10^{-3} of a liter. In the SI system, since 1964, exactly equal to 1 cc. See *centimeter.*

millimeter* Abbrev. mm; 10^{-3} of a meter.

millimicron Abbrev. mμ; a metric measure, replaced by the nanometer* of the SI system; equal to 10^{-9} m.

millimole* Abbrev. mmol; molecular weight expressed in mg; 1/1,000 g molecule (mole).

milling The operation of crushing, grinding, or powdering. **chemical ~** The removal of metal (especially aluminum) by controlled chemical reaction to produce a given shape or finish, e.g., by acid-etching unmasked areas.

millinormal Describing a solution having 1/1,000 of the concentration of a normal solution.

million 10^6 or 1,000,000.

Millon M.'s base $HO(Hg_2O)NH_2 \cdot H_2O$ or $(HOHg)_2NH_2OH$. Yellow powder, produced from a solution of mercuric oxide in ammonium hydroxide. **M.'s reagent and test** A reagent for the detection of proteins made by dissolving mercury in twice its weight of concentrated nitric acid and diluting with twice the volume of water (1849). Red color with proteins.

millstone Buhrstone. A hard stone used for grinding cereals; it usually consists of a coarse sandstone with fine quartz inclusions.

milo A cattle food containing crude protein.

milone A beverage obtained by the fermentation of whey (about 0.8% vol. of alcohol).

milorganite An organic fertilizer prepared in Milwaukee by the dehydration of sewage. Brown granules, free from bacteria and seeds: nitrogen 5.4, phosphoric acid 3%.

Milori blue A pigment similar to prussian blue, but having a red tint; prepared by the oxidation of a paste of potassium hexacyanoferrate(II) and ferrous sulfate.

Miltown Trademark for meprobamate.

mimetite $PbCl_2 \cdot 3Pb_3(AsO_4)_2$. Mimetisite. A native lead arsenate.

mimosa bark The dried bark of *Acacia mimosa* (Leguminosae); a tan.

mimosine A toxic principle in the tropical leguminous shrub *Leucaena glauca,* which causes cattle to lose hair.

min Abbreviation for (1) minute; (2) minim. ♏.

mina The Sumerian unit of weight; one imperial pound.

mine Subterranean workings for minerals, coal, or ores. Cf. *outcrop.*

mineral A native, inorganic or fossilized organic substance having a definite chemical composition and formed by inorganic reactions. **Ethiops ~** Black mercuric sulfide with some free Hg and S.

m. acid An inorganic acid. **m. adhesive** Sodium silicate. **m. alkali** An inorganic base, e.g., sodium hydroxide. **m. blue** (1) A mixture of iron(III) hexacyannoferrate(II) with calcium sulfate or barium sulfate. (2) A blue copper or tungsten ore. **m. butter** Antimonous chloride. **m. caoutchouc** Alaterite, helenite, bitumen elastic. A plastic bitumen. **m. carbon** Graphite. **m. chameleon** Potassium permanganate*. **m. charcoal** Amorphous coal with a vegetable structure, as thin layers in bituminous coal. See *coal* (fusain). **m. coal** Fusain. **m. cotton** M. wool. **m. dye** An inorganic pigment. **m. fat** Petrolatum. **m. green** Copper carbonate. **m. jelly** A semisolid mixture of hydrocarbons; as, petrolatum. **m. oil** See *mineral oil (1)* under *oil.* **m. paint** A pigment derived from a colored m. **m. pigment** A native colored ore; or an artificial inorganic color. **m. pitch** Asphalt. **m. purple** A red, iron oxide pigment, or ocher. **m. resin** A hydrocarbon mineral; as, asphalt, copal. **m. rubber** Gilsonite. **m. separating fluid** See *density fluids.* **m. spirit** White *spirit.* **m. streak** The characteristic colored streak produced when certain minerals are rubbed on a porcelain plate. **m. tallow** Hatchettenine. **m. water** A natural water containing sufficient salts or gases in solution to give it certain properties and taste. See *mineral water* under *water.* **artificial ~** A solution of certain salts in carbonated or distilled water. **m. wax** Ozocerite. **m. white** Pearl hardening. Pure natural calcium sulfate; used as a loading in paper, etc. **m. wool** M. cotton. Finely interlaced filament produced by suddenly cooling molten slag; an insulation. **m. yeast** A nonsporing yeast of the *Torula utilis* type; a contaminant of pressed yeast. On fermentation it gives high protein and low alcohol yields. **m. yellow** Lead chloride oxide.

mineralization The replacement of organic constituents by inorganic matter, e.g., in fossilized plants. Cf. *petrifaction.*

mineralize Petrify.

Minerallac Trademark for an asphalt solution; used to insulate cable joints.

mineralography (1) The descriptive branch of mineralogy. (2) The study of minerals by microscopic methods, and the photography of thin sections of the polished and etched minerals.

mineralogy The study of the occurrence, description, mode of formation, and uses of minerals. **topo ~** The m. of a particular region.

miner's inch The quantity of water that flows through 1 in^2 in a 2-in plank, the water standing 6 in above the top of the hole = 2,274 ft^3/24 h = 1.58 ft^3/min.

miners' lamp Davy lamp. An oil lamp enclosed in wire gauze which passes sufficient air for combustion, but conducts the heat of the flame away, thereby preventing explosion.

minetisite Mimetite.

minim ♏ = min; a unit of volume in the former English system: 1 minim = 0.0616 ml = $\frac{1}{60}$ fluid dram. **U.S. ~** 0.9606 imperial m. = 0.9483 grain water at 17°C.

minimal Smallest quantity.

minimum The smallest amount or lowest value.

mining The processes by which useful minerals are obtained from the earth's surface (quarries) and underground (mines). **m. engineering** The study of excavating, working, and controlling the technical processes of mines.

minioluteic acid $C_{16}H_{26}O_7$ = 330.4. A dibasic acid from the mold fungus *Penicillium minioluteum.*

Minipress Trademark for prazosin hydrochloride.

minium Pb_3O_4. Red *lead oxide.* Originally cinnabar; now applied to its chief adulterant, red lead. Cf. *sandix.*

minivalence The lowest valency of an element.

mint Mentha. **horse ~** Monarda. **marsh ~** See *marsh mint.* **mountain ~** Calaminth. **pepper ~** See *peppermint.* **spear ~** See *spearmint.*

minulite $KAl_2(OH,F)(PO_4)_2 \cdot 3.5H_2O$. A mineral from W. Australia.

minus (−)*; prefix indicating a levorotatory compound.

minute (1) Describing a particularly small object. (2)* Abbrev. min; a unit of time: 1/60th hour; denoted ′. **m. glass** An 8-shaped sealed glass vessel with fine sand which flows, in a given time, from the upper to the lower compartment.

Miocene See *geologic eras,* Table 38.

miosis Constriction of the eye pupil.

miotic Myotic. An agent that contracts the eye pupil; as, pilocarpine.

Mipolam Tradename for a copolymer of vinyl chloride and acrylonitrile; low-voltage dielectric.

mipor Microporous. **m. rubber** A soft rubber, with pores of about 0.0004 mm average diameter. **m. scheider** A diaphragm of m. rubber used in accumulators.
mirabilite $Na_2SO_4 \cdot H_2O$. A native sulfate.
miramint A tungsten-molybdenum alloy, used in cutting tools.
mirbane oil Nitrobenzene*.
Mirlon Trademark for a synthetic polyamide fiber.
mirror A highly polished surface that reflects light; made of polished metal or glass. **concave ~** A)-shaped mirror. **convex ~** A (-shaped mirror. **plane ~** A flat mirror.
mirrorstone (1) Mica. (2) Muscovite.
MIS Management information system.
misce Latin for "mix."
mischmetal (1) A mixture of rare-earth metals. (2) Commercial cerium (40–75% Ce) with La, Nd, Pr, etc., and sometimes 1–5% Fe; used for pyrophoric alloys. Cf. *Auer metal.*
mischzinn (German: "mixed tin") The alloy Sn 54.4, Pb 41.9, Sb 3.6%; used to prepare solders.
miscibility The ability of certain liquids to mix in all proportions. **m. gap** The temperature range in which certain normally miscible liquids will not mix.
miscible Capable of mixing or dissolving in all proportions. **im ~** Not able to mix.
miso An edible fermented soybean paste. Cf. *kogi.*
mispickel $FeS_2 \cdot FeAs_2$. A native iron ore.
Mississippian See *geologic eras,* Table 38.
mist (1) Fog. Cf. *colloidal systems.* (2) Pharmaceutical abbreviation for mixture.
mistletoe The leaves and young twigs of *Phoradendron flavescens;* an antispasmodic and narcotic. Cf. *viscum.*
mistura Mist. Latin for "mixture"; used in pharmacy.
Mitchell, Peter Dennis (1920–) British chemist. Nobel prize winner (1978), noted for work on chemiosmotic reactions.
mitochondrion A double-membrane structure in the living cell, which plays a role in the chemical changes involved in respiration.
mitosis Division of somatic cells, as part of cell regeneration and growth. The number of chromosomes remains the same. See *diploid, karyokinesis.* Cf. *meiosis.*
mitragynine $C_{23}H_{30}O_4N_2 = 398.5$. Mitragyne. An alkaloid, m.106, from *Mitragyna speciosa* (Rubiaceae).
Mitscherlich M., Eilhardt (1794–1863) German chemist. **M. desiccator** A desiccator, with side tubes for evacuation. **M. eudiometer** A closed glass buret, with platinum electrodes at one end and a glass stopcock at the other. **M. law** (1) The law of *isomorphism,* q.v., which is not rigidly correct: The same number of atoms of similar elements combined in the same way produce an identical crystalline structure. (2) The spectra of isomorphous substances are similar.
mitsubaene $C_{15}H_{24} = 204.4$. A sesquiterpene for *Cryptotaenia japonica,* mitsuba-zeri (Umbelliferae), Japan.
mix (1) To intermingle. (2) A physical mixture of substances, applied to rubber, etc.
mixed **m. crystal** A crystal of 2 isomorphous substances, which crystallize in the same system. **m. ester** An ester $R-COO-R'$, in which the 2 radicals, R and R′, are different. **m. ether** An $R-O-R'$ ether, in which the radicals, R and R′, are different. **m. infection** The invasion by and growth of 2 or more microorganisms in the animal body. **m. ketones** A ketone of the type $R-CO-R'$. **m. salt** A salt derived from a polyvalent acid, in which the H atoms are replaced by different metals, as $KNaNH_4PO_4$.
mixer Equipment for incorporating one or more materials into another; a steel bowl, with revolving mixing arms moving in opposite directions. Cf. *mill.* **static ~** A tubular m. with helical elements giving alternating left- and right-hand twists; designed to mix by a fluid's motion.
mixite $Cu_2O \cdot As_2O_3 \cdot nH_2O$ with 13% Bi_2O_3. An emerald mineral.
mixture (1) Substances that are mixed, but not chemically combined. **constant boiling ~** A m. of 2 liquids which, at a given pressure, distills unchanged, the boiling point remaining constant. Cf. *azeotropy.* **electrostatic ~** A m. obtained by using electric energy to accelerate conducting particles or ions in a nonconducting medium, and so to impart rapid and violent motion to the dispersed particles. Used to desulfurize fuel oils. **freezing ~** A m. of salts with water or ice which produces low temperatures. **law of ~** Law of *alligation.*
mixture (2) Mistura. A pharmaceutical preparation.
mks system Meter-kilogram-second system. A technical system of measurements recommended by the International Electrotechnical Commission (1938) as simpler than the cgs system. Subsequently rationalized and expanded to become the internationally used SI system.
mL*, ml* Abbreviation for milliliter.
mm Abbreviation for millimeter = 1/1,000 m. **mm²** Abbreviation for square millimeter. **mm³** Abbreviation for cubic millimeter.
mμ Former symbol for millimicron, 10^{-9} m; superseded (SI system) by nm.
μμ Former symbol for micromicron, 10^{-12} m; superseded (SI system) by pm.
mmf Abbreviation for magnetomotive force.
mmm Former symbol for millimicron; superseded (SI system) by nm.
Mn Symbol for manganese.
Mo Symbol for molybdenum.
m.o., MO Abbreviation for molecular *orbital.*
mobile Changing position; moving.
mobility (1) The motion of atoms, molecules, ions, or colloidal particles. The mobility, α, of an ion in a liquid; $\alpha = 1.037 \times 10^{-5}\lambda t$, where λ is the equivalent conductivity, and t the transport number of the ion. (2) The visible motion of colloidal particles and microorganisms. Cf. *Brownian motion.*
mobilometer A viscometer in which the time is noted for a disk to fall through a column of the liquid under investigation; used for oils and liquid foods.
mocha See *coffee.* **m.stone** Moss agate.
mochyl alcohol $C_{26}H_{46}O = 374.6$. An alcohol, m.234, from mochi (Japanese birdlime).
mock **m. gold** Pyrites. **m. lead** Sphalerite. **m. ore** Sphalerite. **m. silver** Britannia metal. **m. vermilion** Lead chromate.
mock-up A nonworking model of an apparatus or plant intended to show the layout and method of operation.
mode (1) The actual composition of a substance, e.g., rock, as compared with its norm, q.v. (2) Term. One of three basic control methods used by conventional instrumentation: *proportional control* (corrective action is proportional to the difference between desired and actual values, that is, the error); *reset action* (correction is proportional to both the magnitude and duration of the error); and *derivative action* (correction is proportional to the rate of change of the error). (3) In statistics, the value of highest frequency, corresponding to the peak value of a normal distribution curve.
Modecate Trademark for fluphenazine hydrochloride.
modeccin A toxin from the passion flower plant.
model (1) A geometrical arrangement by which an idea or

concept may be visualized. (2) Mathematical description of process, financial, or economic interactions. **space lattice ~** A group of wire nets and balls to show the arrangement of atoms in a crystal or molecule. Cf. *space lattice.*

moderator A substance used in a nuclear reactor (as, heavy water or graphite) to reduce the energy of the electrons.

modification (1) A slight alteration or change. (2) The conversion of cereal starch into a form in which it is readily acted on by enzymes, as in malting barley, or treating with heat.

modified soda A mixture of sodium carbonate and hydrogencarbonate; a cleanser.

Moditen Trademark for fluphenazine hydrochloride.

Modrella Trademark for a continuous-filament semimatt rayon yarn.

modular Describing a system of plant or apparatus construction or assembly, comprising a number of units which permit rapid erection and easy modification.

modulus The measure of a force or properties of mass or their effects. Often used as an abbreviation for Young's m.; as, high m. fiber. **bulk ~** Compression modulus. The volumetric m. of elasticity, being the compressive force per unit cross section, divided by the change per unit volume. **compression ~** Bulk m. **shear ~** Shear stress divided by shear angle. **wet ~** See *high wet-modulus fiber* under *fiber.* **Young's ~** Symbol: E; modulus of elasticity, m. of rigidity, longitudinal elasticity. The normal stress divided by the linear strain (relative elongation). Thus $E = Fl/\pi r^2 s$, where s is the elongation produced by a force F in a sample of length l and cross-sectional radius r; measured in MN/m^2.
m. of elasticity, m. of rigidity Young's *modulus.*

Mogadon Trademark for nitrazepam.

mohair A long, lustrous textile fiber from the Angora goat.

moho Mohorovicic discontinuity. The boundary between the earth's inner mantle and the assorted surface rocks, 16–32 km beneath the surface.

Mohr **M., Karl Friedrich (1806–1879)** German chemist and physicist. **M. condenser** A modified Liebig condenser. **M. liter** Abbrev. g.w.a.; The space occupied by an amount of water at 17.5°C having an apparent weight in air (brass weights) of 1,000 g. 1,000 g.w.a. = 1,002 ml. **M. pipet** A small buret with tap, used as a pipet. **M. salt** $(NH_4)_2Fe(SO_4)_2 \cdot 6H_2O$. Ammonium iron(II) sulfate; a standard in volumetric analysis.

Mohs **M., Friedrich (1773–1839)** German mineralogist. **M. scale of hardness** The hardness of a mineral is gauged by its ability to scratch or be scratched by one of ten standard minerals:

1. Talc	6. Orthoclase
2. Gypsum	7. Quartz
3. Calcite	8. Topaz
4. Fluorite	9. Corundum
5. Apatite	10. Diamond

Each mineral is scratchable by all below it.

Moissan **M., Ferdinand Frédéric Henri (1852–1907)** French chemist, noted for the production of artificial diamonds and the isolation of fluorine. Nobel prize winner (1906). **M. furnace** A high-temperature electric furnace. **M. process** The reduction of chromic oxide with carbon in an electric furnace lined with calcium chromite.

moistness The amount of liquid, generally water, held by a solid or gas. Cf. *wetting, absorption.*

moisture The wetness or dampness of a substance; the percentage of water contained in a substance.

mol* Abbreviation for mole. **mol. wt.** Abbreviation for molecular weight.

molal Moles per weight, as in m. solution. **m. conductivity** Molar *conductance**. **m. solution** *Concentration,* q.v., expressed as molality; e.g., a 0.5 molal solution contains 0.5 mol/kg of solvent. Cf. *molar* solution. **m. volume** See *molar volume* under *volume.*

molality* Symbol: m; concentration expressed in moles per kg of solvent. Cf. *molal* solution, *molar* solution.

molar (1)* Before the name of an extensive quantity, means divided by the amount of substance; e.g., molar *volume.* Cf. *specific volume.* (2) Amount-of-substance *concentration**. Mole per volume. See *molar solution* below. **m. conductivity** M. *conductance**. **m. latent heat** Molecular heat of vaporization. The quantity of heat (J or cal) required per mole to transform a substance from the liquid to the gaseous state. **m. solution** A solution that contains one mole per liter of a substance. Thus a 1.0 M NaCl solution contains 58.5 g/liter. **m. surface** The area of a sphere of one mole of a substance. **m. volume*** See *molar volume* under *volume.* **m. weight** The molecular weight expressed in grams; a mole.

molarity (Amount-of-) substance *concentration**. Cf. *molar* solution, *molal* solution.

molasses Treacle. The uncrystallizable syrup obtained on boiling down raw cane or beet sugar solution (70% of sugars). Cf. *affination.*

mold Mould. (1) A receptacle in which a molten or liquid mass solidifies. (2) To shape or form. (3) The loose earth on the upper surface of cultivated soil. (4) A variety of fungoid growth, usually filamentous, which grows Hypomycetes, found on damp vegetable material; e.g., *Penicillium.*

mole* Also mol*. An SI base unit. The amount of substance of a system containing the same number (i.e., the Avogadro constant) of elementary entities (must specify which; e.g., atoms, molecules, ions, electrons, or other particles) as there are atoms in exactly 0.012 kg of carbon-12. Thus, 1 mole (gram molecule) of HgCl has a mass of 236.04 g.

molecular Pertaining to single molecules. Cf. *macromolecular chemistry.* **m. association** The state in which two or more molecules are held by coordinate bonds. **m. colloid** See *molecular colloid* under *colloid.* **m. combination** See *molecular combination* under *combination.* **m. compound** Double salt. **m. conductivity** Molar *conductance**. **m. conversion** See *rearrangement.* **m. depression** The lowering of the freezing point of a solution. See *Raoult's law.* **m. diagrams** Drawings to scale of a view of the m. model. They resemble structure symbols, but show the ionic and effective radii and the shape of the molecule. **m. diameter** The diameter of a molecule calculated from (1) Sutherland's equation, (2) van der Waals' equation, (3) the thermal conductivity, (4) the specific heat capacity at constant volume; e.g., in angstrom units:

	(1)	(2)	(3)
Hydrogen	2.40	2.34	2.32
Helium	1.90	2.65	2.30
Oxygen	2.98	2.92	
Nitrogen	3.18		

m. dispersion M. rotation. **m. elevation** The raising of the boiling point of a solution. See *Raoult's law.* **m. equation** See *chemical equation.* **m. field** See *molecular field* under *field.* **m. film** A monomolecular layer; as produced by adsorption. **m. flow** See *molecular flow* under *flow.* **m. formula** A combination of chemical symbols from which the m. weight of a substance is obtained by addition of the atomic weights of the constituents. Cf. molecular *formula*. **m. free path** The mean free path of a molecule in a solution or gas, calculated from (1) Boltzmann's equation $\eta(0.3592\rho c)$ or (2) Meyer's formula, $\eta(0.3097\rho c)$, where η is the viscosity, ρ the density of the medium, and c the molecular velocity. **m. frequency** The ratio $v = k(T_s/MV^{2/3})^{1/2}$, where v is the molecular frequency, T_s the melting point of the substance (in K), M the molecular weight, V the molar volume, and k a constant whose empirical value (Nernst) is 3.08×12^{12}. **m. heat** Specific heat capacity × molecular weight. **m. h. of vaporization** See *molar latent heat* under *latent*. **m. number** (1) A number, analogous to the atomic number, obtained by arranging molecules according to their molecular frequencies. (2) The sum of the atomic numbers of the elements of a molecule; in a compound it is *even*, in a free radical it is *odd*. Cf. molecular *combination*. **m. orbital** See *molecular orbital* under *orbital*. **m. rays** A stream of molecules moving uniformly in one direction, obtained by the escape of vapor through an orifice into a vacuum, screening, and condensing the vapor on the wall of the vessel. **m. rearrangement** See *rearrangement*. **m. rotation** Specific rotation × molecular weight. **m. sieve** A zeolite having an open-network structure, used to separate hydrocarbon and other mixtures by selective occlusion of one or more of the constituents; e.g., gmelinite adsorbs methane but not iso*paraffins*. **m. solution** A true solution, in which single molecules of the solute move in the solvent. **m. s. volume** The difference between the volume of a solution containing 1 mole/l of substance and that of 1 liter of solvent. **m. velocity** The mean velocity with which molecules move, proportional to the mean kinetic energy: $19{,}300\sqrt{(T/M)}$ cm/s, where T is the thermodynamic temperature, and M the molecular weight. **m. volume** See *parachor, volume*. **m. weight** Symbol: M_r; common name for relative molecular mass*. The ratio of the average mass per formula unit of a substance to 1/12 of the mass of an atom of the nuclide ^{12}C. Obtained by adding together the atomic weights indicated by the formula of the substance; or determined by chemical or physical methods; as, lowering of freezing point, vapor pressure, or vapor density.

molecularity The number of reactant particles (as, atoms, molecules, or ions) taking part in a reaction, including any transition state.

molecule The chemical combination of 2 or more like or unlike atoms. The smallest quantity of matter that can exist in the free state and retain all its properties. In noble gases the m. is essentially monoatomic. Other gaseous m. usually consist of 2 atoms; as, O_2, Cl_2. **activated ~** A m. with one or more electrons at a higher energy level than their ground state. **biatomic ~** Diatomic m. **compound ~** A m. consisting of different atoms. **elementary ~** A m. consisting of one type of atom. **excited ~** Activated m. **gram ~** Mole. **homopolar ~** Homonuclear m. A diatomic m. composed of 2 similar atoms, e.g., H_2. **isosteric ~** One of a group of molecules having the same number of electrons, the same sum of atomic numbers, and, sometimes, the same molecular weight; as, CO_2 and N_2O. **nonpolar ~** See *polar compound*. **oriented ~** A m. having directional properties. Cf. *anisotropic*. **saturated ~** A m. in which all valencies are satisfied. **symmetric top ~** A m. with an n-fold axis of symmetry, where n exceeds 2; e.g., $CHCl_3$. **spherical top ~** A m. having two axes of symmetry, e.g., CH_4. **tetraatomic ~** A m. having 4 atoms. **tie ~** A m. in which parts of the same m. participate in more than 1 crystal lamella. **triatomic ~** A m. having 3 atoms. **unsaturated ~** A m. in which there are double or triple bonds, or both, between certain of the atoms.

molecules (constants) (1) Avogadro constant. The number of molecules per gram molecule (mole) is $L = 6.022045 \times 10^{23}$. (2) Loschmidt number. The number of molecules of a gas per mL at 0°C and 760 mm pressure is $n = 2.687 \times 10^{19}$. (3) Calculated constants:

Constant	H_2	O_2
Molecular weight	2.016	31.998
Velocity, m/s at 0°C	1,859	465
Mean free path, m $\times 10^{-9}$	965	560
Collisions, million/s	17,750	7,646
Diameter, m $\times 10^{-10}$	2.35	2.95
Mass, g $\times 10^{-25}$	46	736
Number per ml, $\times 10^{19}$	3.8	3.8

Typical molecular weights:

Insulin	5,634
Hemoglobin	65,000
Tobacco seed globulin	300,000
Hemocyanin	6,800,000
Tomato bushy stunt virus	10,000,000

molions The supposed negatively charged atomic groups of an ionized inert gas.

molybdaenum, molybdan Early names for both native molybdenum sulfide and graphite, which were frequently confused.

molybdate* Indicating molybdenum as the central atom(s) in an anion. *Simple* salts: $M_2(MoO_4)$ or $M_2Mo_2O_7$, corresponding with the chromates and dichromates, respectively. *Complex:* polymolybdate(VI) acids and salts, analogous to the poly*tungstate*s, q.v.

molybdenic Molybdic.

molybdenite MoS_2. A white or green mineral.

molybdenous Molybdenum(II)*, Mo(II)*, Mo(2+)*. A salt of divalent molybdenum.

molybdenum* Mo = 95.94. A heavy metal, at. no. 42, of the chromium group of the periodic table. Gray metal, d.10.2, m.2620, b.5560, insoluble in water or alkalies. It occurs in molybdenite, molybdite, wulfenite, and other rare minerals; and its presence in soil is important for the growth of grasses and vegetables. The metal is cast with difficulty and is used for crankshafts and connecting rods; as a resistor in heating devices and radios; as wire for vacuum tubes and contacts. Valency 2, 3, 4, 5, or 6; but the commoner compounds are derived from divalent [Mo(II)*, Mo(2+)*, molybdenous], trivalent [Mo(III)*, Mo(3+)*, molybdic], and hexavalent [molybdic, molybdate(VI)*] molybdenum. **m. blue** $MoO_2 \cdot 4MoO_3 \cdot H_2O$. Heteropoly blue. A mixture of m. dioxide and trioxide. **m. chlorides** **m. dichloride*** $MoCl_2$ = 166.8. Molybdenous chloride. Yellow, insoluble powder. **m. trichloride*** $MoCl_3$ = 202.3. Molybdic chloride. Red needles. **m. tetrachloride*** $MoCl_4$ = 237.8. Brown crystals. **m. pentachloride*** $MoCl_5$ = 273.2. Black crystals, m.194. **m. hexacarbonyl*** $Mo(CO)_6$ = 264.0. Colorless crystals. **m. hydroxide** (1) $Mo(OH)_3$ = 147.0. Black, insoluble powder. (2) $Mo(OH)_5$ = 181.0. Brown, insoluble powder. **m. minerals** Principal ores: molybdenite, MoS_2; molybdite, MoO_3; wulfenite, $PbMoO_4$. **m. orange** Pigments formed by the coprecipitation of lead molybdate, lead chomate, and lead sulfate in various proportions. **m. oxides** **m. dioxide*** MoO_2. Blue prisms; a textile pigment. **m. sesquioxide*** Mo_2O_3 = 239.9. Yellow to black mass, soluble in acid. **m. trioxide***

MoO_3 = 143.9. Molybdic anhydride. The most common oxide; white rhombs, soluble in alkalies; a reducing agent, and reagent for phosphorus pentaoxide, arsenic oxide, hydrogen peroxide and organic hydroxy compounds. **m. sulfides** **m. disulfide*** MoS_2 = 160.1. Molybdenite. Black crystals, insoluble in water. A dry lubricant for metal bearings. See *molyoil*. **m. trisulfide*** MoS_3 = 192.2. Red crystals. **m. tetrasulfide*** MoS_4 = 224.3. Brown crystals.

molybdenyl (1) The radical $MoO_2{=}$. (2) The radical $MoO{\equiv}$. **m. dichloride** $MoO(OH)_2Cl_2$ or $MoO_2Cl_2 \cdot H_2O$. **m. trichloride** $MoOCl_3$. Green, soluble crystals.

molybdic Molybdenum(III)*, (3+)*, (VI)*, or (6+)*. Describing a salt of trivalent or hexavalent molybdenum. **m. acid*** H_2MoO_4 = 162.0. M. hydroxide. Colorless needles. **m. a. dihydrate** H_4MoO_5. Yellow, monoclinic crystals, soluble in ammonia; a reagent. **m. anhydride** MoO_3. See *molybdenum oxides*. **m. ocher** Molybdite.

molybdite MoO_3. Molybdenum trioxide, molybdic ocher. Native as yellow, earthy, or capillary tufted forms.

molybdoena An obsolete term applied in confusion to both graphite and molybdenum sulfide.

molybdyl Molybdenyl.

molyoil Oil containing about 1% MoS_2; said to reduce fuel consumption.

molysite An incrustation of ferric chloride in lava and near volcanoes.

moment The power to overcome resistance. **magnetic ~** See *magnetic moment*. **m. of force** Torque. The effectiveness of a force in producing rotation around a center, in N·m (force × distance from center). **m. of inertia** See *moment of inertia* under *inertia, momentum*.

momentum The force effect of a moving body = mu = mlt^{-1}, where m is mass, u velocity, l length, and t time. **angular ~** Spin.

momordicin Elaterin.

mon- See *mono-*.

monacetin See *monoacetin* under *acetin*.

Monacrin Trademark for 3-aminoacridine.

monad A monovalent element, radical, atom, or atomic group; e.g.: Na—, NH_4—, CH_2—, or —COOH. Cf.*dyad*.

monarda American horsemint, bee balm. The dried herb of *Monarda punctata* (Labiatae). **m. oil** An essential oil from m., of thyme odor, d.0.930–0.940, containing thymol.

monardin A thymol-like terpene from horsemint oil.

monascin $C_{21}H_{26}O_5$ = 358.4. A red pigment produced by the growth of the fungus *Monascus purpureus* on rice, m.144, insoluble in water.

Monastral blue A *phthalocyanine*, q.v., containing a Cu atom. A blue pigment which has a great resistance to light, heat, and reagents.

monatomic Monoatomic*.

mona wax Peat wax. Wax extracted from peat and used in making emulsions.

monazite A native rare-earth phosphate sand, especially containing Ce, La, and Th (India, Brazil). Used in the manufacture of Welsbach burners and pyrophoric alloys.

Mond **M., Ludwig (1839–1909)** German-born English chemist. **M. gas** Fuel gas made by the passage of superheated steam over coal. **M. process** Separation of nickel and copper by carbon monoxide. Volatile nickel carbonyl, $Ni(CO)_4$, is formed and subsequently decomposed by heat. Cf. *Oxford process*.

Monel metal Trademark for a native alloy containing normally Ni 67, Cu 28, Mn 1–2, Fe 1.9–2.5%; d.8.82, m.1160–1360. Very resistant to corrosion and used in chemical plant.

monesia bark The bark of *Chrysophyllum glyciphloeum* (Brazil); a tonic and astringent.

monetite $CaHPO_4$. An apatite. Used in dentifrices.

mongolism See *Down's syndrome*.

monistic Pertaining to singleness. **m. compound** A substance that does not dissociate in solution; as, sugar. See *polar compound*.

monitor To guide or give warning. In particular, to follow the course of a process, e.g., by radioactivity tests.

monitron A device for the automatic monitoring of a radioactive area; it sounds an alarm when the intensity of the radiation exceeds the tolerance value.

monium Victorium. A supposed rare-earth element (discovered 1898) which proved to be a mixture of rare-earth metals.

monkshood Aconite.

mono- Mon-. Prefix (Greek) indicating "one."

monoacetin See *monoacetin* under *acetin*.

monoamine oxidase Amine oxidase, M.A.O. A body enzyme, particularly in the brain, that breaks down norepinephrine and affects sympathetic nerve transmission. **m. o. inhibitors** M.A.O.I. A group of drugs, related chemically to isoniazid, which inhibit the action of m. o. Used to treat depression. With some other drugs (as, sympathomimetics) or foods containing tyramine (as, cheese) can cause a dangerous rise in blood pressure.

monoamino acid An organic acid of the type, $NH_2 \cdot R \cdot COOH$.

monoatomic* Describing a molecule consisting of 1 atom.

monobasic Describing an acid having one H atom replaceable by a metal or positive radical.

monobenzone $C_{13}H_{12}O_2$ = 200.2. *p*-Benzyloxyphenol. Monobenzyl ether of hydroquinone, MBEH. An inhibitor of tyrosinase, preventing formation of melanin. An ointment for pigmented skin conditions.

monobromated Describing a compound having one Br atom. **m. camphor.** See *camphor*.

monobromethane (1) Ethyl bromide. (2) Methyl bromide*.

monobromo-* Bromo-*. Describing a compound which contains one Br atom.

monobutyrin $C_3H_7COOC_3H_5(OH)_2$ = 162.2. Glycerol monobutyrate. Colorless liquid, d.1.008, b.271, produced by lipolysis.

monochloro-* Chloro-*. Describing a compound having one Cl atom. **m.amine** NH_2Cl = 51.5. An intermediate product in the preparation of hydrazine from chlorine and ammonia; an unstable, pungent liquid. **m.ethane** Ethyl chloride*. **m.methane** Methyl chloride*.

monochroic Monochromatic.

monochromatic Monochroic. Describing a substance having one color, represented by one wavelength only. **m. analysis** The measurement of color by mixing white light with light of a pure spectral hue; only the hue and whiteness vary. **m. illuminator** M. lamp. **m. lamp** (1) A gas flame colored yellow with sodium compounds. (2) A spectrum apparatus with a narrow slit for isolating a radiation of one wavelength; a source of light for polarimetry, spectroscopy, or irradiation.

monochromatism A rare form of color vision abnormality which enables the subject to match 2 colors merely by adjusting their intensities. Its frequency of occurrence in males is 0.003%.

monclinic See *crystal systems*.

monocon An electrostatically focused cathode ray tube used as an electron gun.

monoethan Ethanolamine.

monoethyl A compound containing one ethyl radical.

monoethylin See *monoethylin* under *ethylin*.

Monofil Trademark for single-thread nylon, used in toothbrushes.
monofilm Monolayer.
monoformin $C_3H_5(OH)_2 \cdot COOH$ = 120.1. Glyceryl formate. An oil, b.165.
monogenetic (1) Biology: pertaining to nonsexual reproduction. (2) Industry: pertaining to dyestuffs which produce only one color on textiles. Cf. *polygenetic dye.*
monoglyceride A *glyceride,* q.v., containing one acid molecule.
monohydrate A compound containing one molecule of water of crystallization. **m. crystals** $Na_2CO_3 \cdot H_2O$. Sodium carbonate; a cleansing agent.
monohydric Describing a compound containing one OH group.
monolayer Monofilm. A monomolecular surface film, e.g., of octadecanol; used to retard the evaporation of large areas of water. Cf. *adsorption.*
monoleate A combination of a base with one oleic acid radical.
monolupine $C_{16}H_{22}N_2O$ = 258.4. An alkaloid from *Lupinus caudatus* (0.45%). Yellow glass. $b_{4mm}257$, soluble in alcohol; related to anagyrine.
monomer A substance composed of molecules which can polymerize with like or unlike molecules.
monometric Isometric.
monomolecular Pertaining to one molecule. **m. layer** A layer one molecule thick. Cf. *adsorption.* **m. reaction** A unimolecular reaction; having a molecularity of one. See *reaction order.* **m. zone** See *zone.*
monomorphous Occurring in one crystal form only. Cf. *dimorphic.*
mononitraniline See *nitroaniline.*
mononuclear Describing (1) an aromatic compound having one ring of atoms; (2) a cell having one nucleus.
monoolein $C_{17}H_{31}COOCH_2 \cdot CHOH \cdot CH_2OH$ = 354.5. Glycerol 1-oleate. White crystals, m.35, insoluble in water; synthesized from fats by steapsin.
monopalmitate An ester, especially of glycerol, containing one palmitic acid radical.
monopalmitin **alpha-~** $C_{15}H_{31}COOCH_2 \cdot CHOH \cdot CH_2OH$ = 330.5. Glycerol 1-palmitate. White leaflets, m.77, soluble in alcohol. **beta-~** Glycerol 2-palmitate.
monophosphate A salt containing one phosphate radical.
monopole soap A soap of highly sulfonated fatty acids.
Monoprim Trademark for trimethoprim.
monorefringent Describing an isotropic solid or mineral.
monosaccharide In general, a polyhydroxy aldehyde or ketone. See Fig. 16. Cf. *carbohydrates.*
monose Ose. A hexose or pentose. See *saccharide.*
monosilane Silane*.
monostearin **alpha-~** $C_{17}H_{35}COOCH_2 \cdot CHOH \cdot CH_2OH$ = 358.6. Glycerol 1-stearate. White needles, m.74, insoluble in water. **beta-~** Glycerol 2-stearate. White solid, m.80.
monosulfide Containing one divalent S atom, e.g., FeS. **m. equivalent** A value of lime-sulfur solutions determined by titration with iodine to disappearance of the yellow color, giving the S of calcium sulfide, as distinct from polysulfides.
monotropic Describing a substance occurring in one crystalline form. **pseudo ~** See *pseudomonotropy.*
monotropitoside A primeveroside of methyl salicylate from *Monotropa hypopitys,* yellow bird's nest (Pryolaceae).
monotype metal The alloy Pb 80, Sb 15, Sn 5%; used for printing type.
monovalent See *univalent.*
monovinylacetylene Vinylacetylene.
Monox Trademark for a mixture of silicon monoxide and dioxide; a thermal and electrical insulator, scale preventive, and corrosion inhibitor.
monoxide A binary compound containing one oxygen atom; as, PbO.
monoxy Misnomer for (mono)hydroxy.
Monsel **M. salt** Ferric subsulfate. **M. solution** Liquor ferri subsulfatis. A brown, styptic liquid containing about 15% iron.
montanic acid Octacosanoic acid*.
montanin A disinfectant consisting chiefly of hexafluorosilicic acid.
montanite $Bi_2O_3 \cdot TeO_3 \cdot 2H_2O$. A rare bismuth tellurate from Montana.
montan wax A native hydrocarbon extracted from lignites. Brown or white masses, soluble in chloroform or benzene; used as a substitute for carnauba wax.
montanyl alcohol 1-Nonacosanol*.
monte-acid An acid elevator or acid pump for raising acids to the tops of towers by means of air pressure.
Monte Carlo method Mathematical device for determining the most probable outcome from a stochastic system.
montejus An apparatus for raising liquids by air pressure (acid egg or monte-acid).
month (1) Calendar m. The twelfth part of a year. (2) Lunar m. The period of revolution of the moon around the earth, measured variously as: **anomalistic ~** = 27.5546 days, or the time of revolution of the moon from one perihel to another. **draconic, nodical ~** = 27.2122 days, or the time

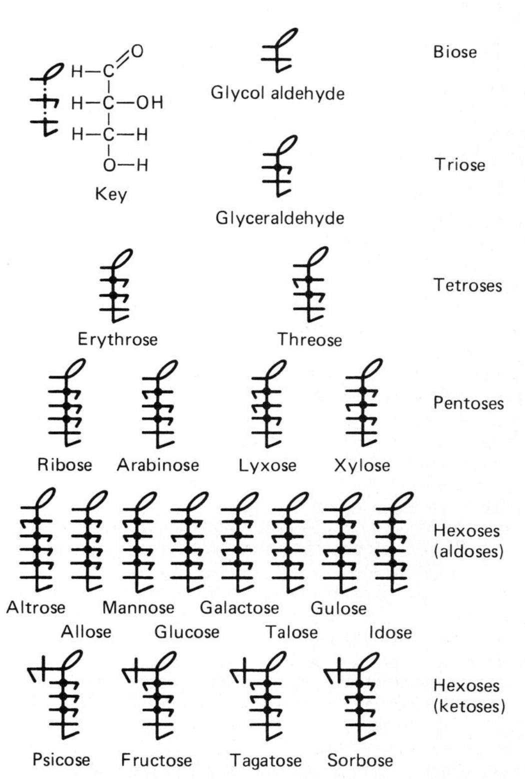

Fig. 16. Classification of monosaccharides. Only the D isomers are shown, the OH group on the highest-numbered chiral carbon atom being on the right.

of revolution of the moon from a node to the same node again. **sidereal ~** = 27.32166 days, or the time of revolution of the moon from a distant star back to that star. **synodic ~** = 29.5306 days, or the time elapsing from new moon to new moon. The ordinary month.

Monthier blue $FeNH_4[Fe(CN)_6]$. A blue pigment formed by the reaction of a hexacyanoferrate(II) and ammonium iron(II) sulfate.

monticellite $MgO \cdot CaO \cdot SiO_2$. A native calcium magnesium silicate, belonging to the olivine group, and occurring in limestone in colorless or gray crystalline masses.

montmorillonite $(Mg \cdot Ca)O \cdot Al_2O_3 \cdot 4SiO_2 \cdot nH_2O$. Japan clay. A rose-red mineral and acidic clay constituent active in base exchange. When activated by treatment with acids, m. is used to catalyze the polymerization of unsaturated styrenes; for cracking petroleum, separating hydrocarbons, and adsorbing dyestuffs. The action is believed to be linked with the formation of colloidal silica by the acid treatment. **calcium ~** Fuller's earth. **sodium ~** Bentonite.

montroydite HgO. Native mercury(I) oxide.

Moore, Stanford (1913–1982) American chemist. Nobel prize winner (1972), noted for work on the structure of pancreatic ribonuclease.

Moogrol Trademark for the ethyl esters of hydnocarpus oil.

moonmilch A pasty calcite formation found in caves where the air is saturated with moisutre.

moonstone A gem variety of feldspar with delicate, opalescent colors.

mora Mowrah.

Moracaea *Urticaceae*, q.v., yielding rubber, lac, hemp, timber, etc.

morantel tartrate $C_{16}H_{22}O_6N_2S$ = 370.4. An anthelmintic used for domestic animals.

mordant A chemical used for fixing colors on textiles by adsorption; as, soluble salts of aluminum, chromium, iron, tin, antimony. **m. dye** An artificial or natural color for fibers which usually forms an insoluble metal compound (lake) with metallic salts (mordant). **m. rouge** Aluminum acetate.

morenosite $NiSO_4 \cdot 7H_2O$. Nickel vitriol.

Morgan M., Sir Gilbert Thomas (1872–1940) British industrial and research chemist; first director of the Chemical Research Laboratory. **M., John Livingston Rutgers (1872–1935)** American physical chemist, noted for research on the liquid state. **M. equation** A modification of the Ramsay-Shields equation: $G(M/\rho) = k(t_c - t - 6)$, where G is the weight of the drop of liquid, M molecular weight, ρ density, and t_c critical temperature. Used to determine M. Cf. *stalagmometer.*

morin $C_{15}H_{10}O_7$ = 302.2. 2′,3,4′,5,7-Pentahydroxyflavone. Yellow coloring matter of various woods, as *Morus tinctoria,* m.303. A wool dye and reagent for aluminum (green fluorescence).

morindone $C_{15}H_{10}O_5$ = 270.2. 1,2,5-Trihydroxy-6-methylanthraquinone*. m.284. A red coloring matter, similar to morin, from morus.

Morely, Edward W. (1838–1923) American chemist, noted for work on water and gases.

morograph An instrument that records the death point of vegetable tissues.

-morph, -morphous Suffix (Greek), meaning "form." See *amorphous, isomorph, monomorphous, polymorph.*

morphia Morphine.

morphina Morphine.

morphine (*Mo.*) $C_{17}H_{19}O_3N \cdot H_2O$ = 303.4. Morphina, morphinum, morphia, morphium. An alkaloid derived from opium. Colorless crystals, decomp. 250, slightly soluble in water; a narcotic and potent analgesic. Used to reduce severe pain and to allay associated anxiety. Addiction occurs easily. **apo ~** See *apomorphine.* **benzyl ~** Peronine. **dia ~** See *diamorphine.* **dimethyl ~** Thebaine. **para ~** Thebaine. **methyl ~** Codeine.

m. acetate $(Mo)C_2H_4O_2 \cdot 3H_2O$ = 399.4. Yellow powder, soluble in water; a narcotic. **m. anisate** $(Mo)C_8H_8O_3$ = 437.5. The salt of morphine and anisic acid. Colorless crystals, soluble in water; a narcotic. **m. benzoate** $(Mo)C_7H_6O_2$ = 407.4. The salt of morphine and benzoic acid. White prisms, soluble in water; a narcotic. **m. hydrochloride** $(Mo)HCl \cdot 3H_2O$ = 375.8. Colorless needles, soluble in water; a narcotic and analgesic (EP, BP). **m. sulfate** $(Mo)_2H_2SO_4 \cdot 5H_2O$ = 758.8. Colorless needles, decomp. 250, soluble in water. A narcotic; used like the hydrochloride (USP, BP).

morphinone $C_{17}H_{17}NO_3$ = 283.3. The keto form of morphine. **dihydro ~** Dilaudid.

morphinum Morphine.

morphol $C_{14}H_{10}O_2$ = 210.2. 3,4-Dihydroxyanthracene. White needles, m.143, insoluble in water.

morpholine* $O \cdot (CH_2)_2 \cdot NH \cdot (CH_2)_2$ = 87.1. Tetrahydro-*p*-oxazine. Colorless liquid, d.0.9998, b.128, soluble in water; a reagent for Zn or Cu, and a corrosion inhibitor.

morphology The study of the form and structure of the living organism:

Cytology—the structure of cells
Histology—the structure of cell aggregates (tissues)
Anatomy—the structure of tissue aggregates (organs)

-morphous See *-morph.*

morrenine An alkaloid from *Morrenia brachystephana,* Argentine milkweed. Colorless crystals, m.106.

morrhua The codfish, *Gadus morrhua;* source of cod-liver oil.

morrhuic acid An acid from the saponification of cod-liver oil. Its salts are used to treat varicose veins.

morrhuin $C_{19}H_{27}N_3$ = 297.4. A liquid ptomaine in some cod-liver oils.

morrhuol The active principle of cod-liver oil: contains S, I, and P. Brown liquid, d.0.94, insoluble in water; a cod-liver oil substitute.

Morse M., Harmon Northrup (1848–1920) American chemist noted for osmosis experiments. **M. buret** A capillary buret holding 1 ml. **M. equation** An approximate formula for the potential energy of a diatomic molecule as a function of nuclear separation.

mortar (1) A cavity shaped container of glass, iron, agate, or brass; used for powdering materials with a pestle. (2) A building material: slaked lime and sand, sometimes mixed with plaster of paris or cement. Cf. *pozzolana.* **ballistic** A m. closed with a tightly fitting steel projectile, used to test explosives by measuring the angle of recoil when the explosive is fired inside the m.

morus Latin for "mulberry." **m. tinctoria** Yellow brazilwood, old fustic. The yellow wood from *M. tinctoria* (Moraceae); used to make morin and dye textiles.

Morveau See *Guyton.*

moryl Carbachol.

MOS Metal-oxide-semiconductor. An electronic device consisting of a metal electrode separated from a semiconductor by an insulator such as silicon dioxide.

mosaic m. gold (1) SnS_2. Tin bronze. A pigment stannic sulfide. (2) A yellow alloy of copper and zinc. **m. silver** An amalgam of tin and bismuth.

Mosander, Carl Gustav (1797–1858) Swedish discoverer of some rare-earth metals.

mosandrium A rare-earth metal isolated from samarskite and later separated into samarium and gadolinium.

Moseley **M., Henry Gwyn Jeffreys (1887–1915)** English physicist, noted for his work on X-ray spectra. **M. formula** The frequency ν is related to Rydberg's constant R_∞ and atomic number Z by $\nu = \frac{3}{4}R_\infty(Z - 1)^2$. **M. law** All elements can be arranged according to the frequencies of their X-ray spectra, in one continuous series which corresponds generally with the order of their atomic weights. **M. number** Atomic number. **M. series** An arrangement of elements according to increasing atomic numbers, as determined from the square root of the frequency of the principal line in their X-ray spectra. Cf. *electron* configuration. **M. spectrum** The characteristic lines produced when the X-rays from an anticathode of the metal under examination are diffracted through a crystal. Cf. *crystallogram.*

moslene $C_{10}H_{16}$ = 136.2. A terpene from *Mosla japonica* (Labiatae).

moss A cryptogamic plant, order *Musci.* **Ceylon ~** See *agar.* **Iceland ~** Cetraria. **Irish ~** Chondrus.
m. agate Mocha stone. A gem variety of chalcedony, or an agate containing dark, mossy, dendritic forms due to infiltration of iron and manganese salts. **m. gold** Native gold in mosslike form. **m. silver** Native silver in mosslike form. **m. starch** Lichenin.

Mössbauer effect The recoilless emission and resonant reabsorption of γ rays arising from nuclear excited states (lifetime $10^{-6}-10^{-10}$ s); used to study chemical phenomena spectroscopically from the resulting M. spectrum.

mote A solid particle, e.g., dust, in a gas, which acts as a nucleus of condensation, causing the fog of the modern city.

mother A progenitor. **m. cell** A single cell that divides into 2 daughter cells. **m. of coal** Fusain. **m. liquor** The residual saturated liquid which remains after the crystallization of a liquid. **m. lode** The principal vein of a metallic deposit passing through a district. **m. of pearl** Nacre.

motherwort The herb of *Leonurus cardiaca;* a bitter and antispasmodic.

motile Possessing motion.

motility The phenomena of motion, especially as observed under the microscope (Brownian motion) or detected by chemical or physical means (ionic motion).

motion Change of position. **laws of ~** (Newton). (1) A body left to itself will continue in its state of rest or of motion. (2) The rate of change of momentum of a body is proportional to the impressed force acting on it.

motor An agent that produces motion or mechanical power (muscle or machine). **hot-air ~** An apparatus in which a current of hot air drives a set of propellers. **stepping ~** A m. whose shaft rotates through a fixed angle in response to each input pulse.

mould Mold.

mountain **m. ash** Sorbus. **m. balm** Eriodictyon. **m. blue** Azurite. **m. butter** A hydrated fibrous aluminum sulfate. **m. cork** An elastic form of asbestos. **m. crystal** Rock crystal. **m. flax** A silky variety of asbestos. **m. grape** See *mountain grape* under *grape.* **m. green** Malachite. **m. laurel** Kalmia. **m. leather** A tough asbestos, occurring in thin, flexible sheets. **m. milk** A soft, spongy calcite. **m. mint** Calaminth. **m. soap** An unctuous halloysite. **m. tallow** Hatchettenine. A waxy hydrocarbon. **m. tobacco** See *Arnica.* **m. wood** A brown, compact, fibrous asbestos.

mounting The preparation of specimens for microscopic study.

Moureu, Charles Léon Francois (1863–1929) French chemist noted for research in catalysis, rubber chemistry, and autoxidation.

mouse jar A glass jar with an iron screen top, to hold small animals.

Movil, Movyl Trademark for a polyvinyl chloride synthetic fiber.

moving bed process A process for the recovery of uranium and other metals by ion exchange, in which the separate stages of absorption, backwash, and elution are carried out in separate columns, the resin moving from one stage to the next.

mowrah Mora, mowa. The seeds of *Bassia butyraceae* (Sapotaceae), India. **m. fat** A soft mass expressed from m., used in soapmaking. **m. meal** The dried and ground residue of m. seeds, after expression of the fat. It contains saponin and is unfit for animal food, but is used as fertilizer (2.7%N).

m.p., mp Abbreviation for melting point.

M radiation See M.

mRNA* See *ribonucleic acid* under *nucleic acids.*

ms Abbreviation for *meso-*, when indicating central fusion.

M.S., M.Sc. Abbreviation for Master of Science.

MTBF Mean time between failures.

MTS Trade abbreviation for sodium ethylmercuriosalicylate, a germicide.

mu μ. Greek letter. μ (not italic) is symbol for muon (μ-meson). μ (italic) is symbol for (1) chemical potential; (2) coefficient of friction; (3) Joule-Thomson coefficient; (4) linear extinction coefficient; (5) magnetic moment of particle; (6) micro- (10^{-6}); (7) micron; (8) permeability; (9) Poisson ratio. *In chemical names:* (1)* the group so-designated bridges 2 centers of coordination; (2) indicates the *meso* position.

muava Muavi. The dried bark from a leguminous tree of Madagascar; an arrow poison.

mucic acid COOH(CHOH)$_4$COOH = 210.1. Galactaric acid†. Colorless crystals, decomp. 224, insoluble in water. A *meso*compound, formed by the oxidation of D-galactose. Stereoisomeric with saccharic acid.

mucicarmine A stain for mucin: carmine 2, aluminum chloride 1, distilled water 4 pts.

mucilage (1) A paste usually prepared from dextrin or gum. (2) In pharmacy, a solution of acacia, chondrus (Irish moss), tragacanth, or starch.

mucilaginous Slimy or adhesive.

mucin A glycoprotein; the chief constituent of mucus, the slimy secretion of organs or organisms. Soluble in water, precipitated by alcohol and acid. Cf. *chitosan, mucoid.* **frog egg ~** M. from frog eggs; it yields chondrosamine on hydrolysis. **m. sugar** Fructose.

muck (1) A soil composed largely of highly decomposed organic matter, showing little of the original plant structure. Cf. *humus.* (2) Peat. (3) The miners' term for loose rock and ore.

muckite $C_{20}H_{28}O_6$. A resin found in some coal beds.

mucoid Any of a group of glycoproteins, differing from the mucins in solubility; in bone, tendon, cartilage, the cornea, white of egg, and ascitic fluids. **chondro ~** M. from cartilage yielding chondroitin sulfuric acid. **cornea ~** M. from the cornea. **ovo ~** M. that does not coagulate; forming 10% of the solids of egg white. **serum ~** M. from blood serum, forming 0.5–1% of its proteins.

mucolactonic acid $C_6H_6O_4$ = 142.0. Colorless crystals, m.122, soluble in water.

mucolipoid A compound that contains a fatty acid residue bound to a carbohydrate residue.

muconic acid 2,4-Hexadienedioic acid*.

mucopolysaccharide A compound of low protein content,

whose chemical reactions are primarily those of a carbohydrate.

mucopolysaccharidosis, (-es) A congenital enzyme deficiency disease. See *inborn error of metabolism* under *metabolism.*

mucoprotein (1) A compound of relatively high protein or peptide content, whose chemical reactions are predominantly those of a protein; e.g., plant globulins. (2) Glucoprotein (obsolete).

mucorin A protein from plant molds.

mucous Adjective of the noun "mucus."

mucus A slimy secretion from an organ or organism. **vegetable ~** Tragacanthin.

mud drilling ~ See *drilling mud.* **red ~** Bauxite sludge from the Bayer process for the manufacture of aluminum.

mudaric acid $C_{30}H_{46}O_3$ = 454.7. An acid from mudar, *Calotropis gigantea* (Asclepiadaceae).

mudat A low-grade opium; the refuse from *yenshee,* q.v.

Muencke pump A glass filter pump attached to a water faucet.

muffle A semicylindrical container of Alundun, fireclay, porcelain, or silica that protects substances placed in it from fuel gases and sudden temperature changes when heated. **m. furnace** A fireclay container with resistance wires or other means of heating, which produces high temperatures rapidly, owing to reflection of the heat from the walls.

mugwort Felon herb. The leaves of *Artemisia vulgaris* (Compositae).

muka Air-dried foliage and small branches of conifers; used in animal feeds.

mullein The dried flowers or leaves of *Verbascum thapsus* (Scrophulariaceae); a demulcent.

mullen An instrument used to evaluate the bursting strength of paper from the pressure required to force a glycerol-inflated rubber diaphragm through the clamped sample.

muller (1) A stone pestle for grinding pigments. (2) An instrument for grinding glass.

Müller's glass Hyalite.

Mulliken M., Samuel Parsons (1864–1934) American chemist. **M.'s classification** An arrangement of organic compounds based on their qualitative *(genus)* and quantitative *(order)* reactions.

mullite $3Al_2O_3 \cdot 2SiO_2$. An orthorhombic, homogeneous solid solution of alumina in sillimanite, from the Island of Mull, or artificially made by heating andalusite, sillimanite, or cyanite. The only compound of Al and Si that is stable at high temperatures. Formed in fireclay above 1060°C, does not deform under loads at 1800°C, is resistant to corrosion, and has a low coefficient of expansion. Used as a refractory.

mulls Pharmaceutical ointments of high melting point, spread on soft muslin ("mull").

mulser An emulsifying machine.

multi- Prefix (Latin), indicating "many." Cf. *poly-.*

multifrequent Describing different wavelengths, e.g., those comprising a heterogeneous beam of light.

multiple m. proportions Dalton's law. The chemical elements always combine in a definite ratio by weight, or in multiples of that ratio. **m.-tube burner** Three or more bunsen burners mounted close together, and supplied from one connection.

multiplet A spectral line, which on close examination is found to consist of 2 (doublet), 3 (triplet), or more single lines close together.

multiplying affixes* See *nomenclature.*

multipolar Having more than 2 poles.

multivalent Polyvalent. **m. vaccine** A suspension of 2 or more species or varieties of the same microorganism.

mumeson Muon*.

mundic Pyrites. **white ~** Arsenical iron pyrites.

mung Green gram. The mung bean, *Phaseolus mungo (Ph. aureus),* E. Asia; it is rich in vitamins of the B group.

mungo A low-grade shoddy.

Munktell paper A brand of filter paper.

Muntz metal Yellow brass. The alloy Cu 60, Zn 40%. It can be rolled hot or cold, and corrodes less then copper.

muon* μ. Mumeson. A subatomic *particle,* q.v., discovered in cosmic rays (1936). Produced by (1) transitions between electronic states differing in combined orbital and spin angular momentum values; (2) decay of a pion to either a positron, or an electron and 2 neutrinos. **m. spin rotation** μ.s.r. The analog of n.m.r. spectroscopy, utilizing the effect of a magnetic field on muon spin.

muonium Mu. The bound state of a muon and an electron, which behaves like a light isotope of hydrogen.

murex $C_{16}H_8O_2N_2Br_2$ = 420.1. Tyrian purple. 6,6′-Dibromoindigotin. A dyestuff used by the aristocracy of the ancient world. A secretion from *M. brandaris,* a Mediterranean rock whelk, which develops color on exposure to light.

murexan Uramil.

murexide $C_8H_4O_6N_5NH_4 \cdot H_2O$ = 302.2. Ammonium purpurate. Purple carmine. A purple coloring matter produced from uric acid by the action of nitric acid, followed by neutralization with ammonia. Brown crystals, green in reflected and red in transmitted light, soluble in water; a dye and indicator.

muriacite Anhydrite.

muriate Obsolete term for chloride. **m. of potash** Potassium chloride*. **m. of soda** Sodium chloride*.

muriatic acid Obsolete term for hydrochloric acid. **oxygenated ~** Early name for chlorine.

murram See *laterite.*

muscarine ion $MeCH_2 \cdot CHOH \cdot CH_2 \cdot CH(CH_2 \cdot NMe_3^+)O$ = 175.3. Hydroxycholine. A very poisonous alkaloid resembling pilocarpine, from mushrooms, *Amanita muscaria,* and a ptomaine of decaying fish. Cf. *amanitine.*

muscarufin (1) $C_{25}H_{16}O_9$ = 460.4. An orange dyestuff, m.276, from *Amanita muscaria.* (2) $C_{25}H_{16}O_5$ = 396.4. A glucoside from *Amanita muscaria.*

musci Mosses. A division of Bryophyta.

muscicide An agent that destroys flies.

muscle m. fibrin Syntonin. **m. sugar** Inositol.

muscone $C_{16}H_{30}O$ = 238.4. Muskine. 3-Methylcyclopentadecanone. **(*R*)-~** From musk. A thick liquid, b.328. **(±)-~** Used in perfumery.

muscovite $KH_2Al_2(SiO_4)_3$. Phengite. A potash-mica constituent of granite.

musenine Mussanine.

musennin An acid resin from the bark of mussena, *Albizzia anthelmintica* (Leguminosae); a tenifuge.

mushroom (1) A toadstool *fungus* (q.v.), e.g., *Amanita, Boletus.* (2) The edible m., *Agaricus campestris.*

musical m. notes A succession of sound impulses of definite pitch and vibration. See *sound.* **m. scale** Sound vibrations in a harmonic series, the frequency of vibration of successive notes being fixed ratios for a particular scale. The scientific diatonic scale is based upon C_3 = 256 vibrations/s; the musical even-tempered chromatic scale, on A_3 = 435 vibrations/s.

musk Moschus, deer musk, tonquin musk. The dried secretion from the follicles of *Moschus moschiferous,* the musk deer of central Asia. Dark, shiny grains of penetrating odor, soluble in water; used in perfumery. **artificial ~** $Me_3C \cdot C_6H(NO_2)_2 \cdot OMe(Me)$ = 240.3. 3-*tert*-Butyl-2,6-dinitro-3-

methoxytoluene. M. ambrette. A m. substitute. Colorless prisms, m.96, soluble in water. **vegetable ~** M. seed.
m. seed The seeds of *Hibiscus abelmoschus* (Malvaceae); used to clarify sugar.
muskeg The vegetable cover over water-bearing soils.
muskine Muscone.
Muspratt, James Sheridan (1821–1871) Irish chemist, and founder (1848) of the College of Chemistry of Liverpool.
mussanine Musenine. An alkaloid from *Acacia anthelmintica.*
must (1) Most. The expressed, unfermented juice of grapes. (2) Any juice freshly expressed from fruits.
mustard A condiment made by grinding the seeds of *Brassica (Sinapis)* species. **black ~** The finely ground seeds of *B. nigra.* It contains a fixed oil; a pungent, irritant, essential oil (chiefly allyl thiocyanate); an enzyme, myrosin; and a glucoside potassium myronate. Cf. *sinamine.* **brown ~** M. from the seeds of *B. nigra,* L. (Koch), or *B. juncea,* L. (Czernj and Cosson), or both. **French ~** Table m. **table ~** A semiliquid mixture of white mustard, salt, sugar, and vinegar. **white ~** Yellow m. English m. The finely ground seeds of *Sinapis alba,* having the same constituents as black m. except that sinalbin is present. A condiment and counterirritant.
m. gas $(CH_2Cl \cdot CH_2)_2S$ = 159.1. 2,2′-Dichlorodiethyl sulfide, yperite. An oil, d.1.28, m.14, b.215; used in World War I as a vesicant; it blisters the skin by penetrating the tissues and forming hydrochloric acid. **m. meal** The dried and ground residue of m. seeds after extraction of the oil: nitrogen 5, phosphorus pentaoxide 1, potassium oxide 1%; a fertilizer. **m. oil** (1) Oleum sinapis volatile. A volatile or essential oil obtained by distillation of macerated m. seeds. Its chief constituent is allyl isothiocyante. Colorless, pungent liquid, d.1.018, b.148, soluble in alcohol; a rubefacient. (2) Acrinyl isothiocyanate. **m. seed** The dried, ripe seed of *Brassica nigra;* a source of m. oil.
mustard oils Isothiocyanates*, isosulfocyanic esters, sulfocarbimides. Compounds containing the radical —NCS. (*a*) MeNCS. Methyl m. oil, m. isothiocyanate. Colorless crystals, m.34, b.119. (*b*) EtNCS. Ethyl m. oil. Colorless liquid, b.133. (*c*) PrNCS. Propyl m. oil. Colorless liquid, b.153. (*d*) Me_2CHNCS. Isopropyl m. oil. Colorless liquid, b.137. (*e*) Allyl isothiocyanate. See *mustard oil.*
Mustargen Trademark for mechlorethamine hydrochloride.
mustine hydrochloride BP name for mechlorethamine hydrochloride.
mutamer A compound showing dynamic isomerism. See *mutarotation.*
mutamerism. The phenomenon of changing from one newly dissolved isomer to another, and establishing an equilibrium between the two. See *mutarotation.*
mutarotation Birotation. A change of optical rotation occurring in newly prepared solutions of reducing sugars. Thus α-D-glucose solution has at first $[\alpha]_D$ +110° which drops to +87° in 25 minutes, to +52.5° in 6 hours, and then remains constant (β-D-glucose).
mutase* See *enzymes,* Table 30.
Muthman liquid Acetylene tetrabromide.
mutterlauge German for *"mother liquor."*
mutual inductance*, induction See *mutual inductance* under *inductance.*
m. wt., M.W. Abbreviations for molecular weight.
Myambutol Trademark for ethambutol hydrochloride.
mycelioid A bacterial growth that resembles the radiated filamentous appearance of molds.
mycelium A mass of protoplasmic, colorless threads, constituting the plant body of mushrooms, toadstools, and certain molds (spawn).
Mycoban Trademark for a product containing sodium propionate; used to prevent bread mold.
mycodextran $C_6H_{10}O_5$(?). An unbranched polyglucose produced by certain molds, e.g., *Aspergillus niger.* It reduces Fehling's solution only after acid hydrolysis, and gives a light-blue color with iodine.
mycogalactan $(C_6H_{10}O_5)_n$. A polysaccharide from the fungus *Aspergillus niger.*
mycophenolic acid $C_{17}H_{20}O_6$ = 320.3. An acid antibiotic, antifungal, and antibacterial from the mold fungus *Penicillium brevicompactum.*
mycoporphyrin Penicilliopsin.
mycoprotein The albuminous material of bacteria.
mycose $C_{12}H_{22}O_{11} \cdot 2H_2O$ = 378.3. α, α-Trehalose. A carbohydrate occurring in manna, several fungi, and ergot of rye.
mycostatin BP name for nystatin.
mycosterol $C_{30}H_{48}O_2$ = 440.7. A sterol, m.160, from fungi and lichens.
Mycota Trademark for undecylenic acid.
mycotoxin A toxin from fungi.
mycoxanthin A biologically active carotenoid, from algae.
mydatoxin $C_6H_{13}O_2N$ = 131.2. A ptomaine from decaying flesh.
mydine $C_9H_{11}O_2N$ = 165.2. A ptomaine in cultures of typhoid bacillus and putrefied flesh.
mydriatic An agent that causes dilation of the eye pupil, as, atropine.
myelin (1) A soft, yellow kaolin. (2) A doubly refractive lipid from organic tissues. It forms a semifluid, viscous covering around nerves. **m. forms** The cylindrical excrescences which develop when lecithin swells in water.
mykonucleic acid $C_{36}H_{52}O_{14}N_{14} \cdot 2P_2O_5$ = 1189. A nucleic acid from yeast.
Myleran Trademark for busulfan.
mylonite A rock deformed by earth movements, so as to lose all its original structure.
myocrisin Gold sodium thiomalate.
Myodil Trademark for iophendylate.
myogen An albumin in muscle juice, soluble in water and salt solutions.
myoglobin The principal red pigment of meat, which stores oxygen in muscles; similar to hemoglobin.
myology The study of muscles.
myosin The thick protein filaments of muscle that slide over the thin filaments (actin) to produce muscle contraction. Enzyme-catalyzed hydrolysis of ATP to ADP provides the energy.
myotic Miotic. An agent that contracts the eye pupil; as, pilocarpine.
myoxanthim A lipochrome, m.118, from the unsaponifiable matter of *Rivularia nitida.*
myrbane oil Nitrobenzene*.
myrcene $C_{10}H_{16}$ = 136.2. **α-~** 2-Methyl-6-methylene-1,7-octadiene*. Used in perfumery. **β-~** 7-Methyl-3-methylene-1,6-octadiene*. A constituent of bay oil and other essential oils. Colorless liquid, d.1.467, b.116.
Myrcia Tropical trees (Myrtaceae), related to the tree myrtles. **m. oil** The essential oil of *Pimenta acris* (Myrtaceae), W. Indies.
myria- Prefix (Greek) indicating 10,000.
myrica Bayberry, wax myrtle, candleberry. The dried bark of *M. cerifera,* U.S. **m. leaves** The leaves of *M. acris,* from which bay rum is prepared. **m. oil** Bay oil.
myricetin $C_{15}H_{10}O_8$ = 318.2. A ketone and yellow pigment, m.358 (decomp.), from the bark of Myrica species. See *flavones.*

myricetrin A glucoside resembling quercetin.
myricin Triacontyl palmitate*.
myricyl The triacontyl* radical.
myriocarpin A neutral, resinous substance, insoluble in water, from the fruit of *Cucumis myriocarpus,* wild cucumber (Cucurbitaceae).
myristamide $Me(CH_2)_{12}CONH_2$ = 227.4. Tetradecanamide*. White leaflets, m.103.
myristica Nutmeg, nux moschata. The kernels of the ripe fruits of *M. fragrans;* a condiment. Cf. *mace, otoba.* **m. oil** Oleum myristicae, oil of nutmeg. The essential oil of m.; d.0.880–0.910, n_D1.4690–1.4880 according to origin (southeast Asia or W. Indies) (NF).
myristic acid* $C_{13}H_{27}COOH$ = 228.4. Tetradecanoic acid*, from human subcutaneous fat (1%). Colorless leaflets, m.54, insoluble in water.
myristicene $C_{10}H_{14}$ = 134.2. An eleoptene from the essential oil of nutmeg.
myristicin $C_{11}H_{12}O_3$ = 192.2. 1-Allyl-3-methoxy-4,5-methylenedioxybenzene. Yellow liquid, d.1.1425, b_{15mm}149; from myristica and other essential oils.
myristicol $C_{10}H_{16}O$ = 152.2. A stearoptene from the essential oil of nutmeg.
myristin $(Me(CH_2)_{12}CO_2)_3C_3H_5$ = 723.2. Trimyristin, glycerol tritetradecanoate, laurel wax, myrtle wax. A fat in nutmeg butter, spermaceti, and other fats. Colorless needles, m.59, insoluble in water.
myristoleic acid (Z)-9-Tetradecenoic acid*.
myristone $(C_{13}H_{27})_2CO$ = 394.7. Tridecyl ketone*, myristic ketone. A solid, m.75.
myristoyl* The tetradecanoyl* radical, $Me(CH_2)_{12}CO-$, from myristic acid.
myrobalan Terminalia. The dried fruits of *Terminalia (Myrobalanus) chebula* (Combrataceae), Asia, containing 30% tannin. Cf. *beda nuts, chebulinic acid.*
Myrol Trademark for a mixture of methyl nitrate with 33 vol. % of alcohol. An explosive.
myronic acid $C_{10}H_{19}NS_2O_{10}$ = 377.4. An acid in potassium myronate, the glucoside of black mustard, which is hydrolyzed to mustard oil.
myroxin $C_{23}H_{36}O$ = 238.5. A principle from Peru balsam.
myrrh A resinous gum exuded from the stems of *Commiphora myrrha* (Burseraceae), Ethiopia. It contains the resin myrrhin, a volatile oil, and a gum (brown tears of balsamic odor). Used medicinally as an astringent or carminative. **m. oil** An essential oil distilled from myrrh. Colorless liquid, d.0.998, b.220–235, insoluble in water; used in perfumery. Cf. *myrrholic acid.*
myrrhin A resin from myrrh.
myrrholic acid $C_{17}H_{22}O_5$ = 306.4. An acid from myrrh.
Myrtaceae The bay of eucalyptus family, which usually have pungent and fragrant leaves, fruits, and seeds and which yield essential oils and spices; e.g., roots: *Eucalyptus globulus,* (eucalyptus); leaves: *Pimenta (Myrcia) acris* (bay leaves); flowers: *Eugenia aromatica* (cloves); fruits and seeds: *Pimenta officinalis* (pimenta).
myrtenal $C_{10}H_{14}O$ = 150.2. 2-Pinen-10-al, myrtenic aldehyde. Colorless liquid, d.0.998, b_{10mm}90.
myrtenic acid $C_{10}H_{14}O_2$ = 166.2. 2-Pinen-10-oic acid. From myrtenal. m.54.
myrtenol $C_{10}H_{16}O$ = 152.2. An alcohol, the chief constituent of myrtol. 2 isomers, (+) and (±); oily liquids.
myrtillidin $C_{16}H_{13}O_7^+$. 3,3′4′,5,7-Pentahydroxy-5-methoxyflavilium. From roses, bilberries, whortleberries, cactus.
myrtillin chloride $C_{22}H_{23}O_{12}Cl$ = 514.9. Galactoside chloride of myrtillidin.
myrtle The dried leaves of *Myrtus communis;* an antiseptic. **m. oil** (1) Yellow oil from m. leaves, which contains eucalyptol, α-pinene, and cinene, d.0.90, insoluble in water. (2) Myrtol. **m. wax** Bayberry wax.
myrtol The essential oil of *Myrtus communis;* an antiseptic. Cf. *myrtenal, myrtenol.*
mystin A preservative mixture of sodium nitrite and formaldehyde.
mytilite Mytilitol.
mytilitol $C_6H_5(OH)_6Me$ = 194.2. Mytilite, pentahydroxymethoxybenzene, C-methylinositol, from the mussell *Mytilus edulis.* Rhombohedra, m.267. Cf. *inositol.*
mytilotoxin $C_6H_{15}O_2N$ = 133.2. A leucomaine from mussels, rendered harmless by canning alkaline.
mytolin $C_{234}H_{360}O_{70}N_{60}S$(?). A protein obtained from muscles.
myxo- Prefix indicating "mucus" or "mucoid." **m.xanth(in)ophyll** $C_{46}H_{66}O_7$ = 731.0. The characteristic pigment of certain freshwater algae.
myxobacteria Slimy, moldlike organisms having an ameboid stage, as distinct from the common bacteria.

N

N Symbol for (1) newton; (2) nitrogen. **N shell** The fourth layer of electrons in an atom, which contains 32 electrons when complete.

N (1) Symbol for neutron number. (2) In chemical names, the radical prefixed by *N*- is attached to the nitrogen atom. (3) Normal solution.

n Symbol for (1) nano- (10^{-9}); (2) neutron.

n Symbol for (1) amount of substance; (2) in chemical names, normal; as distinct from iso(meric); thus, *n*-butane; (3) principal quantum number; (4) refractive index; (5) rotational frequency; (6) unspecified number; as in C_nH_{2n}.

ν, η See *nu, eta,* respectively.

Na Symbol for sodium (natrium).

naal oil An essential oil distilled from naal grass, *Cymbopogon nervatus,* a Sudanese grass. Yellow liquid, d.0.954, insoluble in water. Chief constituents: citrene and perilla alcohol.

nabam* See *fungicides,* Table 37 on p. 250.

Nacconol Trademark for a group of surface-active sodium alkyl aryl sulfonates.

nacre Mother of pearl. The hard, iridescent inside layer of oyster- and other seashells.

NAD⁺ Symbol for oxidized nicotinamide adenine dinucleotide.

NADH Symbol for reduced nicotinamide adenine dinucleotide.

nadi reagent A reagent for cytochrome *c* oxidase, e.g., in milk powders: dimethyl-*p*-phenylenediamine hydrochloride 0.02 and 1-naphthol 0.1 g are dissolved successively in 1 mL water. The blue color developed is a rough measure of the degree of oxidation.

nadorite $PbClSbO_3$. An Algerian antimonate. Brown, orthorhombic crystals.

NADP⁺ Symbol for the phosphate of NAD^+.

naehrsalz (German for "nutrient salt.") A mixture of sodium and ammonium phosphates.

naftolens $(C_3H_4)_n$. A group of unsaturated, vulcanizable hydrocarbons, b.200–380, from acid-tar by-products of mineral oil refining; extenders for rubber.

nagelschmidtite CaO 7–9, P_2O_5, SiO_2 2–3. The principal phosphatic phase in open-hearth steel furnace slags.

nagyagite Tellurium glance. Lead sulfotelluride which contains gold and antimony.

nahcolite Native sodium hydrogencarbonate.

Nailon Trademark for a polyamide synthetic fiber.

NaK An alloy of sodium with 40–90% potassium; a light liquid used as a heat-exchange fluid.

nakrite $Al_2O_3 \cdot 2SiO_2 \cdot H_2O$. A gray clay.

Nalfon Trademark for nalidixic acid.

nalidixic acid $C_{12}H_{12}O_3N_2$ = 232.2. Nalfon, Negram, NegGram. White crystals, m. 228, insoluble in water. An antibacterial agent; used for urinary infections (USP, BP).

nalorphine hydrochloride $C_{19}H_{21}O_3N \cdot HCl$ = 347.8. *N*-Allylnormorphine hydrochloride. White crystals, darkening on exposure, m.262, soluble in water; respiratory stimulant and antidote to morphine (BP).

naloxone hydrochloride $C_{19}H_{21}O_4N \cdot HCl$ = 363.8. Narcan. White powder, soluble in water. An antidote to morphine and pentazocine; used to treat narcotics overdose and addiction to morphine (USP, BP).

name See *nomenclature.* **fusion ~*** N. for a cyclic system, formed by use of a linking "o" between the names of two ring systems, denoting that the 2 systems are fused by 2 or more common atoms; as, benzofuran. **semisystematic ~*** Semitrivial n*. N., of which only a part is used in the systematic sense; as, methane (-ane), butene (-ene), calciferol (-ol). Most names in organic chemistry belong to this class. **systematic ~*** N. composed wholly of specially coined or selected syllables, with or without numerical prefixes; as, pentane, oxazole. **trivial ~*** N., no part of which is used in a systematic sense; as, xanthophyll.

nandinine $C_{19}H_{19}O_3N$ = 309.4. A diisoquinoline alkaloid from the root bark of nanten, *Nandina domestica* (Berberidaceae), Japan. Cf. *domesticine.*

nandrolone phen(yl)propionate $C_{27}H_{34}O_3$ = 406.6. Durabolin. White crystals, m.97, insoluble in water; an androgen with marked anabolic effect; used for breast cancer and muscle wasting (USP, BP).

nano* Symbol: n; SI system prefix for 10^{-9}.

nanometer* Abbrev. nm; millimicron, mμ. 10^{-9} m.

nantokite CuCl. A native cuprous chloride.

napalin An aluminum soap of dodecanoic and naphthenic or oleic acids. A thickening agent for gelled gasoline fuels for incendiary bombs.

napalite C_6H_4. Red wax, m.42, from near Napa, Calif.

napalm Gasoline or oil gelatinized with aluminum salts of higher carboxylic acids. Used in n. bombs and flamethrowers.

napelline $C_{22}H_{30}N(OH)_3$ = 359.5. Benzaconine. An alkaloid from aconite.

naphazoline **n. hydrochloride** Privine. Used as the nitrate (USP). **n. nitrate** $C_{14}H_{14}N_2 \cdot HNO_3$ = 273.3. White, bitter crystals, m.168, soluble in water; a vasoconstrictor, used for eye and nose allergies, as, hay fever (BP).

naphsultam acid $O_2S—NH—C_{10}H_6$ = 205.2. Cf. *naphsultone.* (ring: O_2S bonded to $C_{10}H_6$)

naphsultone $C_{10}H_6O_3S$ = 206.2. Naphthosultone. Colorless crystals, m.154, soluble in benzene.

naphthalin Naphthalene*.

naphtha (1) Hydrocarbons, C_6H_{14} to C_7H_{18}, b.70–90, from the distillation of petroleum, coal tar, and shale oil. (2) Gasoline. (3) A *bitumen,* q.v. **boghead ~** Photogen. **coal tar ~** Mainly benzene and its homologs. **petroleum ~** Mainly alkanes and naphthenes distilled from crude oil. **shale ~** Ligroin. It contains alkenes and alkanes. **solvent ~** A coal tar distillate. **wood ~** Mainly methyl alcohol and acetone, obtained by the distillation of wood; a solvent and denaturant for alcohol.

naphtha aceti Ethyl acetate*.

naphthacene $C_{18}H_{12}$ = 228.3. Rubene. Four linearly fused benzene rings. Orange-red leaflets, m.341.

naphthacetol 4-Acetamido-1-naphthol.

naphthacridine $C_{21}H_{13}N$ = 279.3. Dibenzacridine; 6 isomers. **di~** See *dinaphthacridine.* **pheno~** Benzacridine.

naphthal The naphthylmethylene* radical.
naphthaldehyde* $C_{10}H_7CHO$ = 156.2. Naphthalene carbonal. **1-~** Colorless liquid, b.291, insoluble in water. **2-~** Colorless crystals, m.59, soluble in water.
naphthalene* $C_{10}H_8$ = 128.2. Naphthalin, naphthene, tar camphor. A hydrocarbon from coal tar distillates:

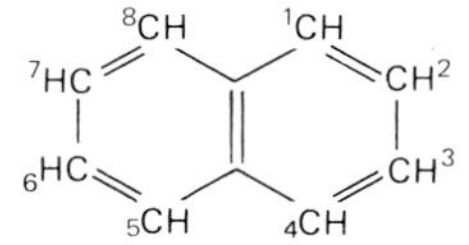

^{8}CH ^{1}CH
^{7}HC CH^2
$_6$HC CH^3
$_5$CH $_4$CH

Former nomenclature called the 1 and 8 positions α; the 2,3,6,7 positions β; and the 4,5 positions γ. Colorless, monoclinic crystals, d_4. 1.152, m.80, b.218, insoluble in water, soluble in alcohol or ether. Used as raw material for aniline dyes; as insecticide (mothballs) and introfier; for the enriching of gasoline (albocarbon); and in the manufacture of indigo and lampblack. **acetyl ~** Methyl naphthyl *ketone**. **amino ~** Naphthylamine*. **azo ~** See *azonaphthalene.* **azoxy ~** See *azoxynaphthalene.* **benzyl ~** See *benzylnaphthalene.* **bi ~*** See *binaphthalene.* **bromo ~*** $C_{10}H_7Br$ = 207.1. **1-~** Colorless crystals, m.5. **2-~** Colorless crystals, m.59. **chloro ~*** $C_{10}H_7Cl$ = 162.6. **1-~** Colorless liquid, b. 258, immiscible with water. **2-~** Colorless crystals, m.56. Used in organic synthesis and added to gasoline to lubricate the valve stems of internal-combustion engines (Halowax). **chlorinated ~** C.K. *wax.* **decahydro ~** Dekalin. **di ~** Binaphthalene*. **diamine ~** Naphthylenediamine*. **diazoamino ~** See *diazoaminonaphthalene.* **dichloro ~*** $C_{10}H_6Cl_2$ = 197.1. Colorless solids, insoluble in water. **1,2-~** m.37. **1,3-~** m.61. **1,4-~** m.68 **1,5-~** m.107. **1,6-~** m.48. **1,7-~** m.62 **1,8-~** m.88. **2,3-~** m.120. **2,6-~** m.135. **2,7-~** m.114. **dihydro ~*** $C_{10}H_{10}$ = 130.2. Dialine. Colorless liquid. **1,2-~** b_{16mm}85. **1,4-~** b.212, m.25. **dihydrodioxo ~** Naphthoquinone*. **dihydroxy ~*** Naphthalenediol*. **dimethyl ~*** $C_{12}H_{12}$ = 156.2 **1,4-~** Colorless solid, b.264. **2,3-~** b.266. **2,6-~** m.111. **dinitro ~*** See *dinitronaphthalene* under *dinitro-*. **ethoxy ~** Ethyl naphthyl ether*. **ethyl ~*** $C_{12}H_{12}$ = 156.2. **1-~** Colorless liquid, d.1.064, b.258. **2-~** b.251. **ethylene ~** Acenaphthene*. **fluoro ~*** $C_{10}H_7F$ = 146.2. **1-~** b.216. **2-~** m.59. **hexahydro ~*** $C_{10}H_{14}$ = 134.2. Colorless liquid, b.205. **hydroxy ~** Naphthol*. **isopropylmethyl ~** Eudalene. **methyl ~*** $C_{11}H_{10}$ = 142.2. **1-~** Colorless liquid, d.1.025, b.243. **2-~** Colorless crystals, m.35. **nitro ~*** See *nitronaphthalene.* **para ~** Anthracene*. **phenyl ~** See *phenylnaphthalene.* **phenyldihydro ~** Atronene. **tetrahydro ~*** Tetralin. **tetrahydrotetramethyl ~** Irene. **tetrahydrotrimethyl ~** Ionene. **tetranitro ~** See *tetranitronaphthalene* under *tetranitro-*. **trimethyl ~** Sapotalene. **trinitro ~** See *trinitronaphthalene.*

n. aldehyde Naphthaldehyde*. **n.carboxylic acid*** Naphthoic acid*. **n.diamine*** Naphthylenediamine. **n.diazo oxide** $C_{10}H_6N_2O$ = 170.2. Naphthooxadiazole. Colorless crystals, m.76. **n.dicarboxylic acid*** $C_{10}H_6(COOH)_2$ = 216.2. **1,5-~** Colorless crystals, m. exceeds 286. **1,8-~** Naphthalic acid. **n.diol** See *naphthalenediol.* **n.disulfonic acid*** $C_{10}H_6(SO_3H)_2$ = 288.3. Armstrong acid. **2,6-~** Colorless needles, soluble in water; used in organic synthesis. **2,7-~** Colorless leaflets, soluble in water. **n.diyl†** See *naphthylene.* **n. picrate** m.153. **n. red** Magdala red. **n.sulfinic acid*** $C_{10}H_7SO_2H$ = 192.2. Colorless solid. **1-~** m.85. **2-~** m.105. **n.sulfonic acid*** $C_{10}H_7SO_3H \cdot H_2O$ = 226.2. **1-~** Colorless crystals, m.90, soluble in water. **2-~** Colorless crystals, m.102, soluble in water. **n.sulfonyl chloride** $C_{10}H_7SO_2Cl$ = 226.7. **1-~** Colorless tablets, m.67, insoluble in water. **2-~** Colorless tablets, m.77, insoluble in water. **n.thiol*** Thio*naphthol.* **n. yellow** $C_{10}H_5(NO_2)_2OH$ = 234.2. 2,4-Dinitro-1-naphthol. Its calcium, sodium, or ammonium salt is Martius yellow.
naphthalenediol* $C_{10}H_8O_2$ = 160.2. Dihydroxynaphthalene. **1,2-~** m.60. **1,3-~** m.125. **1,4-~** m.176. **1,5-~** m.258. **1,6-~** m.138. **1,7-~** m.178. **1,8-~** m.140. **2,3-~** m.159. **2,6-~** m.216. **2,7-~** m.190.
naphthalenedione Naphthoquinone*.
naphthalenol† Naphthol*.
naphthalenyl† See *naphthyl.* **n.carbonyl†** See *naphthoyl.*
naphthalic acid **1,8-~** (1) $C_{10}H_6(COOH)_2$ = 216.2. 1,8-Naphthalenedicarboxylic acid*. Colorless needles, m.270, soluble in water; used in organic synthesis. (2) $C_{10}H_6O_3$ = 174.2. Yellow needles, m.190 (anhydride).
naphthalic acid lactone $C_{10}H_6(CO \cdot O)_2$ = 214.2. Colorless crystals, m.169, soluble in water.
naphthalide (1) A compound of the type $C_{10}H_7M$, as, $C_{10}H_7 \cdot Hg$, naphthyl mercury. (2) Former name for substituted naphthylamine compounds, as, acetnaphthalide.
naphthalidine Naphthylamine*.
naphthalimido The radical $C_{10}H_6(CO)_2 \cdot N-$, from naphthalic acid imide.
naphthalin Naphthalene*.
naphthalize To enrich a gas or liquid with naphthalene.
naphthamide $C_{10}H_7CONH_2$ = 171.2. Naphthalene carbonamide. **1-~** Colorless leaflets, m.202, soluble in water. **2-~** Colorless crystals, m.192, soluble in alcohol.
naphthamine Hexamethylenetetramine.
naphthane Dekalin.
naphthaquinone Naphthoquinone*.
naphthathiourea See *antu.*
naphthazin(e) $C_{10}H_6{:}N \cdot N{:}C_{10}H_6$ = 280.3. Azinodinaphthylene. Anthrapyridine. Colorless crystals, m.273.
naphthazole Benzindole.
naphthenes (1) C_nH_{2n}. Cyclic hydrocarbons, also termed hydroaromatics, cycloparaffins, or hydrogenated benzenes, from petroleum. Polycyclic n. occur in the higher-boiling fractions. (2) The naphthalene ring system. **mentho ~** Menthane*.
naphthenic acid (1) $C_6H_{11}COOH$ = 128.2. Hexahydrobenzoic acid. Colorless crystals, m.38, soluble in water. (2) $R \cdot (CH_2)_nCOOH$. Monocarboxylic acids of the naphthene hydrocarbons; R is a cyclic nucleus of 1 or more rings. They occur in crude mineral oils and are used in lubricants, paints, preservatives, and as industrial catalysts.
naphthenics The undesirable constituents of lubricating oil, which have steep viscosity-temperature curves.
naphthenyl The naphthylmethylidyne* radical.
naphthieno The radical $-C_{10}H_6 \cdot S-$, attached to 2 C atoms of different rings.
naphthindene $C_{13}H_{10}$ = 166.2. The aromatic hydrocarbon

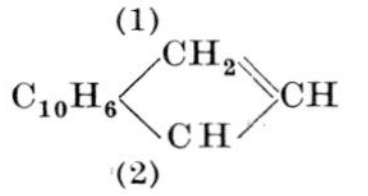

(1) CH_2 — $C_{10}H_6$ — CH — CH (2) — alpha-

(2) CH_2 — $C_{10}H_6$ — CH — CH (3) — beta-

Cf. *indene.* **peri ~** Benzonaphthene.

naphthinduline $C_{27}H_{18}N_2$ = 370.5. A crystalline coloring material, m.250. Cf. *indulines.*

naphthionic acid* $NH_2 \cdot C_{10}H_6 \cdot SO_3H \cdot \frac{1}{2}H_2O$ = 232.3. 1-Naphthylamine-4-sulfonic acid*. Colorless needles, decomp. by heat, soluble in water; used in the synthesis of dyestuffs and pharmaceuticals.

naphthisodiazine $C_{12}H_8N_2$ = 180.2. Ring compounds derived from phenanthrene by replacement of 2 C atoms by 2 N atoms. The 10 different structural possibilities are named according to the positions of the 2 N atoms in the ring:

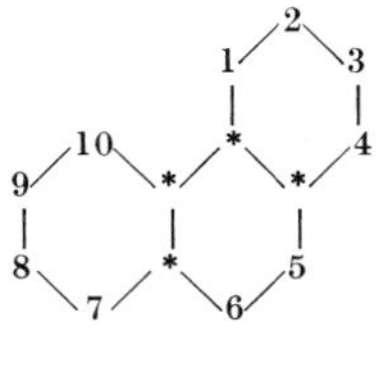

naphthisotetrazine $C_{10}H_6N_4$ = 182.2. The ring compound derived hypothetically from phenanthrene (see *naphthisodiazine*) by replacing 4 C atoms by 4 N atoms.

naphthisotriazine $C_{11}H_7N_3$ = 181.2. The ring compounds derived from phenanthrene by replacing 3 C atoms by 3 N atoms.

naphtho-* Prefix indicating the naphthalene ring in a polycyclic compound in addition to the prefixed component.

naphthodianthrene **meso ~** $C_{28}H_{14}$ = 350.4. An octocyclic hydrocarbon,

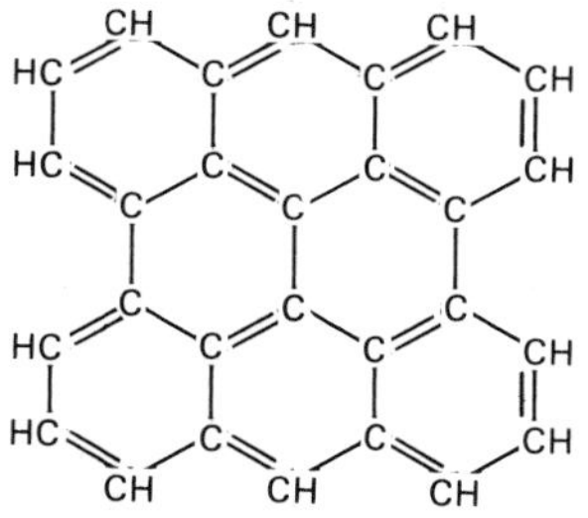

naphthodiazine $C_{12}H_8N_2$ = 180.2. The ring structures derived hypothetically from anthracene by replacing 2 C atoms by 2 N atoms.

naphthofluorene $C_{21}H_{14}$ = 266.3. Pentacyclic compounds, as: $C_{14}H_8 \cdot CH_2 \cdot C_6H_4$.

naphthoic acid* $C_{10}H_7COOH$ = 172.2. Naphthalenecarboxylic acid*. **1- ~** Colorless crystals, m.160, soluble in water. **2- ~** Colorless crystals, m.184, soluble in water. **hydroxy ~*** $C_{10}H_6(OH)COOH$ = 188.2. 1-Naphtholcarbolic acid. Colorless crystals, m.186, soluble in water. **nitro ~** See *nitronaphthoic acid.*

naphthoic aldehyde Naphthaldehyde*.

naphthol* $C_{10}H_7 \cdot OH$ = 144.2. Naphthalenol†, hydroxynaphthalene, naphthyl hydroxide or alcohol. **1- ~** α-n. Colorless, monoclinic crystals, m.94, soluble in water; an antiseptic. **2- ~** β-n. Colorless leaflets, m.122, soluble in water. **acetamido ~** See *acetamidonaphthol.* **amino ~*** $NH_2 \cdot C_{10}H_6 \cdot OH$ = 159.2. **3- ~** White crystals, m.234. **7-2- ~** m.163. **aminoazobenzene ~** Sudan G. **anilineazo ~** Sudan yellow. **benzoyl ~** N. benzoate. **dinitro ~*** See *dinitronaphthol.* **nitro ~** See *nitronaphthol.* **nitroso ~** See *nitrosonaphthol.* **thio ~** $C_{10}H_7SH$ = 160.2. 2-Naphthalenethiol*, 2-naphthyl mercaptan. White scales, m.81, soluble in water.

n. aristol Diiodo-2-naphthol. **n. AS** 2-Hydroxynaphthoic acid anilide. **n. benzoate** Naphthyl benzoate*. **n. blue** Meldola blue. **n.disulfonic acids** See *sulfonic acid.* **n. ethyl ether** Nerolin. **n. green B** $C_{20}H_{10}O_{10}N_2S_2FeNa_2$ = 604.3. The disodium ferrous salt of nitroso-2-naphtholsulfonic acid. Green powder, soluble in water. **n. orange** Tropeolin. **n.phthalein** A pH indicator changing from slight pink (pH 7.3) to green (pH 8.7), for titrating weak acids in alcoholic solution. **n.sulfonic acid** $HO \cdot C_{10}H_6 \cdot SO_3H$ = 224.2. Hydroxynaphthalenesulfonic acid†. Acid intermediates for dyestuffs. **1,2- ~** Colorless tablets, m.250, soluble in water. **2,6- ~** Schäffer's acid. Colorless leaflets, m.122, soluble in water **1,4- ~** Neville-Winther's acid. Colorless crystals, m.170 (decomp.), soluble in water. **1,5- ~** Colorless, hygroscopic crystals, soluble in water. **2,7- ~** Casella acid. Colorless crystals, m.89, soluble in water **n. yellow** $(NO_2)_2C_{10}H_4(OH)SO_3H$ = 314.2. **2,4- ~** Dinitro-1-naphtholsulfonic acid, S yellow. Yellow crystals. Disodium salt is: **n. y. S** $C_{10}H_4O_8N_2SNa_2$ = 358.2. Citronin. The potassium salt is a dye for wool, silk, and food and an oxidation-reduction and pH indicator, changing from colorless (acid) to yellow (alkali).

naphtholate* A compound derived from naphthol by replacing the OH hydrogen by a base or metal.

naphtholbenzein $C_{54}H_{38}O_5$ = 766.9. Brown powder, soluble in alcohol; an indicator in volumetric analysis (alkalies—green, acids—orange).

naphthonitrile* $C_{10}H_7CN$ = 153.2. Naphthyl cyanide*, **1- ~** Colorless needles, m.34, insoluble in water. **2- ~** Colorless leaflets, m.66, insoluble in water.

naphthophenanthrene Dibenzanthracene*.

naphthoquinaldine $C_{13}H_8N \cdot CH_3$ = 193.2. Methylbenzoquinoline. **1- ~** A liquid, b. exceeds 300. **2- ~** m.82, soluble in water.

naphtho[ ]quinoline† $C_{13}H_9N$ = 179.2. **~[2,3-*f*]~** m.170. **~[2,3-*g*]~** Orange crystals, m.245.

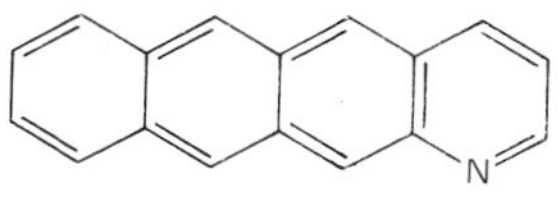

~[2,3-*h*]~ m.127.

naphthoquinone* $C_{10}H_6O_2$ = 158.2. Dihydrodioxonaphthalene. **1,2- ~** Red needles, decomp. 115, soluble in water; a reagent for resorcinol and thalline. **1,4- ~** Yellow crystals, m.125, soluble in water. **2,6- ~** Orange crystals, changing to gray at 130–135.

n. dioxime $C_{10}H_8O_2N$ = 174.2. Derivatives of naphthoquinone formed by replacing the =CO by =CN·OH groups. **1,4- ~** m.207.

naphthoquinoxaline 1,4-Naphthisodiazine.

naphthoresorcinol $C_{10}H_8O_2$ = 160.2. 1,3-Dihydroxynaphthalene*, naphthalenediol*. Colorless crystals, m.124, soluble in water; a reagent for aldehydes, sugars, and uronic acids.

naphthosultone See *naphsultone.*

naphthotetrazines $C_{10}H_6N_4$ = 182.2. Ring structures derived theoretically from anthracene by replacing 4 C atoms by 4 N atoms.

naphthothiazoles $C_{11}H_7NS$ = 185.2. Ring compounds similar in structure to the corresponding naphthoxazoles, whose O is replaced by an S atom.

naphthotriazines $C_{11}H_7N_3$ = 181.2. Ring compounds

similar to the naphthodiazines, but having one more N atom in place of a C atom.

naphthoxanthene Benzoxanthene.

naphthoxazine Phenoxazine*.

naphthoxy The naphthyloxy* radical.

naphthoyl* Naphthalenylcarbonyl†. The radical $C_{10}H_7CO-$, from naphthoic acid. **n.oxy** Naphthyloxy*.

naphthyl* Naphthalenyl†. The radical $C_{10}H_7-$, from naphthalene. **di ~** See *dinaphthyl.*

n. acetate* $CH_3COOC_{10}H_7$ = 186.2. **1-~** Colorless needles, m.130, soluble in water. **2-~** Colorless needles, m.142, insoluble in water. **n.acetic acid*** $C_{10}H_7CH_2COOH$ = 186.2. **1-~** Colorless crystals, m.131. **2-~** Colorless crystals, m.139. **n. alcohol** Naphthol*. **n.acetylene** $C_{12}H_8$ = 152.2. **1-~** b_{25mm}143. **2-~** White crystals, m.36, soluble in alcohol. **n. aldehyde** Naphthaldehyde*. **n.amine** See *naphthylamine.* **n.benzene*** An indicator, changing at pH 8.0, colorless (acid) to blue (alkaline). **n. benzoate*** $PhCOOC_{10}H_7$ = 248.3. Naphthol benzoate, benzo(yl)naphthol. White crystals. **1-~** m.56. **2-~** m.110. Soluble in water. **n. cyanide*** Naphthonitrile*. **n. ether** $(C_{10}H_7)_2O$ = 270.3. N. oxide, naphthyloxynaphthalene*. **1,1-~** Colorless crystals, m.110, insoluble in water. **2,2-~** Colorless crystals, m.105, insoluble in water. **n.hydrazine*** $C_{10}H_7\cdot NHNH_2$ = 158.2. **1-~** Colorless leaflets, m.117. **2-~** Colorless leaflets, m.125. Both soluble in water. The hydrochlorides are reagents for sugars. **n. hydroxide** Naphthol*. **n. isocyanate*** $C_{10}H_7\cdot NCO$ = 169.2. Colorless crystals; a reagent for hydroxy and amino compounds. **n. ketone** $(C_{10}H_7)_2CO$ = 282.3. **1,2-~** Colorless needles, m.135, insoluble in water. **n. mercaptan** Thio*naphthol.* **n.methanol** $C_{10}H_7\cdot CH_2OH$ = 158.2. Naphthobenzyl alcohol. Colorless crystals. **1-~** m.60, soluble in alcohol. **2-~** m.80. **n. methyl ether** $C_{10}H_7\cdot O\cdot CH_3$ = 158.2. Methoxynaphthalene*. **1-~** Colorless liquid, b.258. **2-~** Yara-yara, nerolin. Colorless powder of fruity odor, m.78, insoluble in water; used in perfumery. **n. phenyl ketone*** $C_{10}H_7\cdot CO\cdot Ph$ = 232.3. Insoluble in water. **1-~** Colorless rhombs, m.76. **2-~** White needles, m.82.

naphthylamine* $C_{10}H_7NH_2$ = 142.2. Naphthalidine. **1-~** Colorless needles, m.50, soluble in water. Used in the manufacture of Martius yellow, magdala red, and other dyestuffs; and as a reagent for nitrites and nitrates. **2-~** Colorless leaflets, m.111, soluble in water; used to manufacture azo dyes. **acet ~** Acetnaphthalide. **dimethyl-2-~** $C_{10}H_7NMe_2$ = 171.2. Colorless crystals, m.46. **ethyl-1-~** $C_{10}H_7NHEt$ = 171.2. Colorless liquid, b.303. **methyl-1-~** $C_{10}H_7NHMe$ = 157.2. Colorless liquid, b.293. **nitro ~** See *nitronaphthylamine.* **nitroso ~** See *nitrosonaphthylamine.* **phenyl ~** See *phenylnaphthylamine.* **tetrahydro ~** See *tetrahydronaphthylamine.*

n. brown $HO\cdot C_{10}H_6\cdot N:N\cdot C_{10}H_6\cdot SO_3Na$ = 400.4. C.I. Acid Brown 6. An indicator for the pH range 6.0 (orange) to 8.4 (pink). **n. hydrochloride** $C_{10}H_7N_2H\cdot HCl$ = 192.6. **1-~** Colorless needles, soluble in water; an intermediate in organic synthesis. **2-~** Colorless leaflets, soluble in water; used in organic synthesis. **n.sulfonic acid** A compound derived from naphthalene by substitution of one H by $—NH_2$, and another H by $—SO_3H$. There are 13 isomeric monosulfonic acids, $C_{10}H_6\cdot(NH_2)SO_3H$; 20 isomeric disulfonic acids, $C_{10}H_5(NH_2)(SO_3H)_2$; and 10 isomeric trisulfonic acids, $C_{10}H_4(NH_2)(SO_3H)_3$. See *naphthionic, Baden, Brönner, Cleve's, Dahl's* acids.

naphthylene (1)* Naphthalenediyl†. The radical $C_{10}H_6=$, from naphthalene. (2) Obsolete name for cyclohexene. **di ~** Perylene*. **periethylene ~** Acenaphthene*.

n.diamine* $C_{10}H_6(NH_2)_2$ = 158.2. Diaminonaphthalene. Colorless crystals. **1,2-~** m.98. **1,3-~** m.95. **1,4-~** m.120. **2,3-~** m.191.

naphthylidene* The radical $CH:CH\cdot CH_2\cdot C_6H_4\cdot C=$.

naphthylmercuric The radical $C_{10}H_7Hg-$. **n. acetate** $CH_3COOHgC_{10}H_7$ = 386.8. Colorless needles, m.154, insoluble in water. **n. chloride** $C_{10}H_7HgCl$ = 363.2. Silky, quadratic crystals, m.188, insoluble in water.

naphthylmethylene* The radical $C_{10}H_7OH=$.

naphthylmethylidene* Naphthenyl. The radical $C_{10}H_7C\equiv$.

naphthylnaphthyl* The radical $C_{10}H_7\cdot C_{10}H_6-$.

naphthyloxy* Naphtho(ylo)xy. The radical $C_{10}H_7O-$, from naphthol.

naphthyridine* $C_8H_6N_2$ = 130.1. (1)† Benzodiazine compound. (2) 1,8-Benzodiazine,

octohydro ~ $C_8H_{14}N_2$ = 138.2. Colorless crystals, m.227.

naphtol Naphthol*.

Napier, John (1550–1600) Scottish inventor of logarithms.

Napierian logarithms The exponent of the power to which the quantity e (= 2.718281828) must be raised in order to produce a given number. See *logarithm,* e.

napiform Turnip-shaped; such as cavities in culture media liquefied by bacteria.

Naples **N. jar** A glass jar for staining microscope slides. **N. yellow** Lead antimonate.

napoleonite Corsite.

Naprosyn Trademark for naproxen.

naproxen $C_{14}H_{14}O_3$ = 230.3. (+)-6-Methoxy-α-methyl-2-naphthaleneacetic acid. Naprosyn. White crystals, m.156, insoluble in water. An analgesic anti-inflammatory agent; used to treat pain, rheumatic conditions, and arthritis (USP, BP).

naptha Naphtha.

Narcan Trademark for naloxone hydrochloride.

narceine $C_{23}H_{27}O_8N\cdot 3H_2O$ = 499.5. An alkaloid from opium. Colorless prisms, m.170, soluble in water; a hypnotic and sedative. **ethyl ~** Narcyl.

n. hydrochloride $C_{23}H_{27}O_8N\cdot HCl$ = 481.9. White granules, m.190, soluble in water; a hypnotic. **n. meconate** $C_{23}H_{27}O_8N\cdot C_7H_4O_7$ = 645.6. Yellow needles, m.126, soluble in hot water. **n. sulfate** $C_{23}H_{27}O_8N\cdot H_2SO_4$ = 543.5. Yellow crystals, soluble in water. **n. valerate** $C_{23}H_{27}O_8N\cdot C_5H_{10}O_2$ = 547.6. Green powder, decomp. on ageing, soluble in hot water.

narcine Narceine.

narcissine Lycorine.

narcotic A drug that produces stupor, complete insensibility, or sleep; as, the opium group, producing sleep. **n. poison** A n. that produces stupor.

Narcotic Act See *Harrison Narcotic Act.*

narcotine Noscapine. **(±)-~** Gnoscopine.

narcyl $C_{25}H_{31}O_8N\cdot HCl$ = 510.0. Ethylnarceine hydrochloride. Colorless crystals, slightly soluble in water.

naregamine An alkaloid from *Naregamia alata,* Goanese ipecac; used as an emetic.

naringenin $C_{15}H_{12}O_5$ = 272.3. 4′,5,7-Trihydroxyflavanone. A synthetic flavone.

naringin $C_{21}H_{26}O_{11}$ = 454.4. Aurantium. A crystalline glycoside of (S)-naringenin from the flowers of *Citrus decumana*, the grapefruit or pomelo tree. Yellow prisms, soluble in hot water.

narki metal An acid-resistant alloy of iron and silicon.

NAS National Academy of Sciences.

nascency The nascent state.

nascent Describing a chemical substance, especially a gas, at the moment of its formation when it is most chemically active. See *status nascendi*.

naszogen process The production of oxygen from a mixture of a chlorate, an inert substance (as kieselguhr), and an exothermic compound.

natalite An alcohol and ether mixture; fuel for internal-combustion engines.

nataloin $C_{34}H_{38}O_{15}$ = 686.7. A principle derived from Natal aloes. Cf. *aloin*.

National Bureau of Standards NBS. The U.S. official institution for maintaining standards, including those for the quality of chemicals and apparatus.

National Physical Laboratory A British government institution which certifies scientific glassware, weights, and measuring instruments having a specified high degree of accuracy, and carries out physical research. Certified articles bear a monogram NPL.

National Research Council Established in 1916 by the National Academy of Sciences in Washington, D.C., for the better coordination of research in industrial problems.

native Describing a substance occurring in nature. Used (1) to indicate an uncombined element; as, native mercury; (2) to distinguish natural from artificial substances. **n. coke** Carbonite. **n. compound** A chemical compound that occurs in nature. **n. element** A chemical element that occurs uncombined in nature. **n. metal** A metal that occurs uncombined in nature. **n. paraffin** Ozocerite. **n. prussian blue** Vivianite. **n. soda** Natron.

natrine An alkaloid from *Solanum tomatillo*, (Solanaceae).

natrium Latin and German for "sodium."

natrocalcite Gaylussite.

natrolite $Na_2O \cdot Al_2O_3 \cdot 3SiO_2 \cdot 2H_2O$. Needle stone. Sodium aluminum silicate. A yellow zeolite. Cf. *radiolite*.

natron $Na_2CO_3 \cdot 10H_2O$. A native sodium carbonate from the Egyptian desert; hence the term *natrium* (sodium). Cf. *trona, urao*.

natronkalk Soda lime.

natrum Early (Arabic) name for trona. Cf. *natrium*.

Natta, Giulio (1903–1979) Italian chemist. Nobel prize winner (1963). Noted for work on synthesis of high polymers.

natural Not artificial or synthetic. **n. base** Alkaloid. **n. changes** (1) Reactions produced by decay, fermentation, or putrefaction of organic compounds. (2) Reactions produced by decomposition, corrosion, oxidation, or hydrolysis of inorganic compounds. **n. dyes** Coloring materials derived from the vegetable or animal world; as, carminic acid. **n. gas** (1) A mixture of hydrogen 20, methane gas 70%; or a mixture of simple hydrocarbon homologs found near oil fields. It is collected at the oil fields, and piped to the cities to be used as fuel. See *natural gas* under *gas*. (2) Helium. **n. immunity** Inherited immunity. **n. logarithm** Napierian logarithm. **n. philosophy** The interpretation of natural phenomena by theories and abstract concepts. **n. science** Knowledge of the phenomena of nature, as opposed to abstract or philosophical science. **n. ventilation** The ventilation of a mine by natural means.

Naturetin Trademark for bendroflumethiazide.

naucleine $C_{21}H_{26}O_4N_2$ = 370.5. An alkaloid obtained from *Nauclea excelsa*, Japan.

nauli **n. gum** An oleoresin from *Canarium commune*, Solomon Islands. **n. oil** An essential oil of anise odor, distilled from n. gum. Cf. *elemi*.

naumannite $(Ag_2Pb)Se$. Large cubes with metallic luster.

nauseant An agent that produces sickness.

nautical **n. mile** See *international nautical mile* under *mile*. **n. speed** See *knot*.

naval **n. brass** The alloy Cu 61, Zn 38, Sn 1%. **n. stores** (1) Group name for the materials originally used by builders of wooden ships; as, tar, rosin, turpentine, asphalt, pine oil. (2) Products obtained from rosin; e.g., turpentine, pine oil.

Navashin's solution A fixative for plant tissues: equal parts of (1) chromium trioxide 1.5, acetic acid 10, water 80%; (2) formaldehyde 40, water 60 ml.

Nb Symbol for niobium.

NBS National Bureau of Standards.

NCR No carbon required. Trademark for a coated paper containing chlorinated diphenyl, which produces copies without the use of an interleaving carbon paper.

Nd Symbol for neodymium.

Ne Symbol for neon.

neat's-foot oil Oleum bubulum. An oil obtained by boiling calves' and sheep's feet and shinbones. Yellow oil, d.0.916, soluble in alcohol; a lubricant and leather dressing.

nebulium A hypothetical light element, at. wt. 1.31. It is assumed (from spectroscopical evidence) to exist in nebulas, but its characteristic lines are due to doubly and triply ionized oxygen and doubly ionized nitrogen. Cf. *coronium, aurorium*.

nebulization The transformation of a liquid into a fine spray.

nebulizer An instrument for nebulization.

necic acids The acid products of hydrolysis of senecio alkaloids.

necines The basic products of hydrolysis of senecio alkaloids.

necrocryptoxanthol A provitamin A in yellow corn grain.

nectandrine $C_{20}H_{23}O_4N$ = 341.4. An alkaloid from the bark of *Nectandra rodiaei* (Lauraceae). White powder.

nectar The sugary juice of flowers; the source of honey.

needle (1) A sharp, pointed rod of metal for puncturing or sewing. (2) A n.-shaped crystal. (3) A magnetic n. **astatic ~** See *astatic needle*. **compass ~** A magnetized n., mounted to move freely horizontally. **dipping ~** A magnetized n., mounted to move freely vertically.

n. ore (1) An iron ore which occurs in long fibrous filaments. (2) Aikenite. A native sulfide of lead, bismuth, and copper. **n. spar** Aragonite. **n. stone** Natrolite. **n. valve** A screw valve with a long tapering point, for high-pressure gas cylinders.

neem **n. bark** Azadarach. **n. oil** Margosa oil, nim oil. An oil from the seeds of *Melia azadirachta*, containing sulfur compounds; an alcohol denaturant. Cf. *margosa oil, margosic acid*.

Nefa, Nefalon Trademarks for a polyamide synthetic fiber.

negative (1) Having a charge due to electrons, e.g., a n. ion. (2) The absence of a reaction, property, or phenomenon, e.g., n. catalysis. (3) The opposite of positive; as, a photographic n. **n. catalysis** The retardation of a chemical reaction by means of a catalyst. **n. cotton** A low-nitrated cellulose, soluble in ether or alcohol, used for photographic plates. **n. crystal** A birefractive crystal in which the refractive index of the ordinary ray exceeds that of the extraordinary ray. **n. electrode** Cathode. **n. element** An acid-forming element or an atom with 4–7 valency electrons, as C = 4, N = 5, O = 6, Cl = 7. It tends to attract additional electrons and form a stable system of 8. **n. group** (1) An acid radical. (2) A radical which causes an organic compound to become more n., and so enables the H atoms to be replaceable by metals; as, carboxyl, phenyl, and thionyl groups. **n. ion** An atom charged with one or more n. electrons; an anion. **n. plate** A photographic plate whose image is inverted: light appears black, and shadows white. **n. radical** An acid radical. **n. test** A reaction or test which indicates the absence of the substance sought.

negatron Suggested name for a negative electron. Cf. *positron.*

negion Cation. Cf. *posion.*

negode Cathode.

Negram, Neg Gram Trademarks for nalidixic acid.

neiopax Iodoxyl.

nemalite Brucite.

nematic Pertaining to threadlike liquid crystals. Cf. *smectic.*

nematicide A substance that kills worms.

Nembutal Trademark for pentobarbital sodium.

neo- Prefix (Greek) indicating (1) "new" or "recent"; (2) An isomer of the prefixed compound, of unknown structure.

neoarsphenamine $C_{12}H_{11}O_2N_2As_2(CH_2)O \cdot SONa$ = 466.1. Neoarsaminol, Neosalvarsan, neodiarsenol, Ehrlich no. 914, sodium 3,3′-diamino-4,4′-dihydroxyarsenobenzolmethanal sulfoxylate. A product of salvarsan and formaldehyde sodium sulfoxylate discovered by Ehrlich and Bertheim (1912). Orange powder, with characteristic odor, very soluble in water. Formerly used for syphilis; superseded by penicillin. Cf. *melarsoprol.*

neobrucine $C_{23}H_{26}O_4N_2$ = 394.5. An isomer of brucine.

neocaine Procaine hydrochloride.

neocerotic acid $C_{25}H_{50}O_2$ = 382.7. A fatty acid from beeswax, m.78.

neocianite The blue mineral CuO,SiO_2.

neocupferron The homologous naphthyl derivative of *cupferron,* q.v.; a reagent for iron.

neocuproine $C_{14}H_{12}N_2$ = 208.3. 2,9-Dimethyl-1,10-phenanthroline. m.160. A specific reagent for copper; it forms a colored compound extractable in isopentyl alcohol.

neocyanine A dyestuff derivative of cyanine; a photographic sensitizer for infrared rays.

neo-DHC Trademark for neohesparidin dihydrochalcone. A sweetening agent from naringin in waste grapefruit peel: 2,400 times as sweet as sucrose.

neodymia Neodymium oxide*. See also *didymia.*

neodymium* Nd = 144.24. A rare-earth metal, at. no. 60, discovered by Auer v. Welsbach (1885). Cf. *rare earths.* Yellow metal, d.6.96, m.1025, tarnishes slowly. It occurs in cerium and lanthanum minerals, and forms salts that are generally purple-colored and fluorescent. Valency 3. See *didymium.* **n. acetate*** $(CH_3COO)_3Nd \cdot H_2O$ = 339.4. Pink crystals, soluble in water. **n. acetylacetate*** $Nd(MeCOCH_2COO)_3$ = 447.5. Violet crystals, m.122, soluble in water. **n. chloride*** $NdCl_3$ = 250.6. Rose crystals, soluble in water. **n. hydroxide*** $Nd(OH)_3$ = 195.3. Pink powder, insoluble in water. **n. iodide*** NdI_3 = 524.9. Purple crystals, soluble in water. **n. oxalate*** $Nd_2(C_2O_4)_3$ = 552.5. Rose crystals, insoluble in water. **n. oxide*** Nd_2O_3 = 336.5. Blue powder, insoluble in water. **n. phosphate*** $NdPO_4$ = 239.2. Red powder; an amethyst coloring for porcelain. **n. sulfide*** Nd_2S_3 = 384.7. Green powder, m.2200.

neoepinene Isoprenaline sulfate.

neogen A silver-white alloy: Cu 58, Zn 27, Ni 12, Sn 2, Al 0.5, Bi 0.5%.

neohexane Triptane.

Neo-Hombreol Trademark for testosterone propionate. **N.-H.M.** Methyltestosterone.

neolactose An isomer of lactose.

neoline $C_{23}H_{39}O_6N$ = 425.6. An alkaloid obtained by hydrolysis of neopelline.

neolite A green, fibrous aluminum-magnesium silicate.

Neolithic Late Stone Age. Pertaining to the epoch of the beginning of agriculture, discovery of fire, and better implements of stone and bone. It was preceded by the Paleolithic age and followed by the Copper and Bronze Ages.

Neo-mercazole Trademark for carbimazole.

neomycin An antibiotic from the genus *Streptomyces,* similar in effect to streptomycin. Commercial n. contains 2 isomers, n. A (predominating) and n. B. **n. sulfate** White, hygroscopic crystals, soluble in water; an antibiotic used for nose, eye, ear, and skin infections. (USP, EP, BP).

neon* Ne = 20.179. A noble gas element, at. no. 10, discovered in the atmosphere (1898) by Ramsay and Travers. Colorless gas, $d_{H=1}9.96$, m.−249, b.−246. It occurs in the air (1 in 55,000) and consists of 3 isotopes (at. no. 20, 21, and 22). It forms no chemical compounds, and is used as a residual gas in vacuum tubes and incandescent lamps. **n. lamp, n. light, n. tube** A vacuum tube containing a trace of n. It emits an intense red light of great fog-penetrating power; used for advertising signs and beacons.

neonicotine ±-Anabasine.

neopelline $C_{32}H_{45}O_8N$ = 571.7. An alkaloid from *Aconitum napellus,* aconite.

neopentane* Me_4C = 72.1. 2,2-Dimethylpropane. Gas, m.−17.

Neophrin Trademark for phenylephrine hydrochloride.

neoplasm A new growth of an abnormal character, e.g., cancer.

neoplastigen A substance that produces a benign tumor. Cf. *carcinogen.*

Neoplatin Trademark for *cis*-platinum (see under *platinum*).

neoprene $[-CH_2 \cdot CH{:}C(Cl)CH_2-]_n$. Polychloroprene. Generic name for synthetic rubber made by polymerization of 2-chloro-1,3-butadiene (prepared by the action of hydrogen chloride on monovinylacetylene). Neoprene vulcanizates are resistant to oils, chemicals, sunlight, ozone, and heat.

Neosalvarsan Trademark for neoarsphenamine.

neosine A nitrogenous base from muscle.

neostigmine **n. bromide** $C_{12}H_{19}O_2N_2Br$ = 303.2. Prostigmin. White, bitter, crystalline powder, m.167 (decomp.), soluble in water; a parasympathomimetic; also used to reverse paralyzing drugs. (USP, BP). **n. methyl sulfate** m.143; used similarly (USP, EP, BP).

neostrychnine $C_{21}H_{22}N_2O_2$ = 334.4. An isomer of strychnine.

neotype Alstonite.

neoytterbium Ytterbium.

Neozone Trademark for phenylnaphthylamine derivatives; used as antioxidants in rubber manufacture.

nepalin $C_{17}H_{14}O_4$ = 282.3. A crystalline constituent from the *Rumex* species (Polygonaceae). Cf. *nepodin.*

nepheline A granular rock, which consists chiefly of nephelite and pyroxene.

nephelite (1) An orthosilicate rock of sodium, potassium, and aluminum: aluminum potassium silicate 20, aluminum sodium silicate 75, albite 5%. (2) $NaAlSiO_4$. Elaeolite. A *silica* mineral, q.v., typical of a group.

nephelometer A photometric or other optical device to determine the amount of suspended matter in a solution by comparison of the amount of light scattered by the suspended particles with that scattered by a standard suspension. Used to determine any substance that can be precipitated in a finely divided form. Cf. *turbidimetry.*

nephelometry Quantitative analysis by determining the

degree of light scattered from a fog or suspension. Cf. *turbidimetry.*

nephrite Jade.

nephritis Inflammation of the kidney.

nepodin $C_{13}H_{12}O_3$ = 216.2. 2-Acetyl-1, 8-dihydroxy-3-methylnaphthalene. Yellow needles, m.165. A principle from *Rumex nepalensis* (Polygoniaceae). Cf. *nepalin.*

neptunium* Np. An element, at. no. 93. Atomic weight of the isotope with the longest half-life (^{237}Np) = 237.05. Obtained by bombarding uranium with neutrons (Abelson and McMillon, 1940); also as an impurity in plutonium formed by double-electron capture in ^{235}U. m.635. It has oxidation states of 3, 4, 5, 6, and 7. It transforms into plutonium.

neptunyl* The NpO_2^{2+} radical.

Neradol Trademark for a synthetic tanning agent.

Neral Trademark for β-citral.

neriantin A crystalline glucoside from the leaves of *Nerium oleander.*

neriin A glucoside from the leaves of *Nerium oleander* (Apocynaceae), soluble in water; it resembles strophanthin and has a digitalislike action. Cf. *oleandrin.*

neriodorin A glucoside from the bark of *Nerium odorum.*

nerium Oleander.

Nernst, Walther Hermann (1864–1941) German physical chemist. Nobel prize (1921). **N. effect** The thermomagnetic difference in potential observed on passing a current through a metallic plate placed between 2 magnetic poles. **N. heat theorem** The entropy of a condensed, chemically homogeneous substance vanishes at the zero of thermodynamic temperature. **N. lamp** See *Nernst lamp* under *lamp.* **N. law** Partition law. Distribution law, Nernst theorem. A substance in contact with 2 immiscible liquids and soluble in both, will dissolve in the 2 liquids in fixed proportions, providing that no molecular association or electrolytic dissociation occurs. **N. theory** (1) The electric stimulus to the tissues of an organism is due to the dissociation of the salts surrounding the cell membranes. (2) The theory of electrolytic solution *pressure,* q.v., which is set up by the tendency of a metal to dissolve and form ions. **N. unit** A measure of flow, in L/s.

nerol $Me_2C:CH(CH_2)_2CMe:CH \cdot CH_2OH$ = 154.2. A primary alcohol derived from neroli oil. Colorless liquid, d.0.881, b.226.

nerolidol $C_{15}H_{26}O$ = 222.4. Peruviol, 3,7,11-trimethyl-1,6,10-dodecatrien-3-ol. From neroli oil and Peru balsam, d.0.880, b.277; used in perfumery.

nerolin (1) $C_{10}H_7 \cdot O \cdot C_2H_5$ = 172.2. Ethyl 2-naphthyl ether*. (2) $C_{10}H_7 \cdot O \cdot CH_3$ = 158.2. Methyl 2-naphthyl ether*. Colorless crystals with a fruity odor, m.72, insoluble in water; a substitute for nerol.

neroli oil A brown, essential oil distilled from orange blossoms, *Citrus bigardia* and *C. aurantium,* d.0.87, containing nerol, geraniol, and citrene; used in perfumery.

nerve A fibrous, whitish cord, which transmits impulses to and from the central *nervous system,* q.v.

nervon(e) $C_{48}H_{91}NO_8$ = 810.3. *N*-15-Tetracosenoylgalactocerebroside. A brain galactoside, m.155, containing nervonic acid, sphingosine, and galactose. Cf. *cerebrosides.*

nervonic acid $Me(CH_2)_7CH:CH(CH_2)_{13}COOH$ = 366.6. Selacholeic acid, (Z)-15-tetracosenoic acid*. White crystals, m.41, insoluble in water; in fish liver oils and sphingomyelin.

nervous system The means by which animals, other than the very primitive, respond to environment and stimuli, both voluntarily and involuntarily. **autonomic ~** The n. s. of automatic functions (as, secretions of glands, contraction of the intestines and blood vessels), which cannot normally be voluntarily controlled. **central ~** The brain and spinal cord. The part of the n. s. where coordination of response takes place. See *cerebrum.* **parasympathetic ~** The part of the autonomic n. s. controlling the bladder, sexual function, secretion of saliva and tears, and curvature of the eye lens. **sympathetic ~** Part of the n. s. controlling contraction of blood vessels and heart, dilation of bronchi, size of the eye pupil, sweating, secretion of epinephrine, and motility of the intestines.

neryl* The radical $C_{10}H_{17}-$, from nerol.

nesquehonite $MgCO_3 \cdot 3H_2O$. A native magnesium carbonate.

Nessler **N., Julius (1827–1905)** German analytical chemist. **N. reagent** Dissolve (A) 3.5 g potassium iodide in 10 mL water and (B) 1.7 g mercuric chloride in 30 mL water. Slowly add B to A until a permanent precipitate occurs; then dilute with 20% sodium hydroxide to 100 mL. Again add B, until a permanent precipitate forms, settle, decant, and keep solution in the dark. **N.'s test** A delicate test for detecting ammonia, aldehydes, and hexamethylenetetramine. N. reagent gives a brown precipitate with ammonia; or with traces, a yellow tint due to Hg_2NI. Used for colorimetric determination: 1 pt. per 10^8. **N. tube** A glass cylinder with a flat base; used to compare colors in colorimetric analysis.

nesslerize To treat with Nessler reagent.

nether An early (Egyptian) name for trona.

netoric acid $C_{12}H_{14}O_5$ = 238.2. A monocarboxylic acid, m.92, derived from rotenol.

netropsin $C_{18}H_{26}O_3N_{10}$ = 430.5. An antibiotic from the genus *Streptomyces.*

nettle Urtica. **horse ~** *Solanum carolinense,* q.v.

Neuberg ester $C_6H_{13}O_9P$ = 260.1. Fructose-6-dihydrogen phosphate. In animal tissue and produced in the fermentation of sugar by yeast. Cf. *Robison ester, sugar phosphates.*

neudorfite $C_{18}H_{28}O_2$ = 276.2. A resinous, oxidized hydrocarbon in Bavarian coal beds.

neumandin Isoniazid.

neuraminic acid $C_9H_{17}O_8N$ = 267.2. 5-Amino-3,5-dideoxy-D-glycero-D-galactonon-2-ulosonic acid. A lipid component.

neurine $Me_3N(OH) \cdot CH:CH_2$ = 103.2. Amatine. Trimethylvinylammonium hydroxide. Poisonous; in decaying proteins from fishes, fungi, and brain substances. **oxy ~** Betaine*.

neurodine $C_5H_{19}N_2$ = 107.2. A ptomaine in decomposing flesh.

neuron A nerve cell.

neusilber Nickel silver.

neuton The theoretical zero element, consisting of a neutron, $_0^1n$, having at. no. 0 and mass no. 1. Cf. *neutron.*

neutral (1) Neither acid nor basic. Cf. *amphoteric.* (2) Having no free electric charge. Cf. *neutron.* **n. atom** An atom in which the nuclear positive charge is balanced by negative electrons, in excited or ground state orbitals. **n. compound** A compound that has neither acid nor basic reaction. **n. element** Noble gas*. **n. molecule** A system of 2 ions (cation and anion) in a solvent. **n. oil** Light petroleum, d.0.82, flash point, 143–160, sometimes mixed with animal or vegetable oils. **n. point** The point at which H^+ ions from an acid and OH^- ions from a base give practically undissociated water, i.e., pH = 7.00. **n. principle** The nonacid and nonbasic constituents of plants; as, glucosides. **n. reaction** A reaction that is neither acid nor basic. **n. red** $C_{14}H_{16}N_4$ = 240.3. Toluylene red, dimethyldiaminotoluphenazine hydrochloride. Green powder, soluble in water (red color). Used as a dye, and as an indicator, changing at pH 7.5 from blue (acid) to magenta (alkaline), and at pH 8 to orange-yellow. **n. salt** A salt that does not show acidic or basic reaction. It differs from the normal salt which may be neutral, acid, or basic. **n. salt effect** The reduction of the dissociation of a weak acid or base by addition of an ionizable salt containing one of the ions already present. Cf. *buffer solution.* **n. solution** A solution at the *neutral* point, q.v. **n. violet** $C_{14}H_{15}N_4Cl$. A compound similar to n. red, with H in place of a methyl group.

neutrality The state of being neither acid nor basic in reaction.

neutralization (1) The process of making a solution neutral by adding a base to an acid solution, or an acid to an alkaline, or basic solution. The fundamental reaction of n. is: acid + base $\rightleftharpoons H_2O$ + salt. The reverse is hydrolysis. **heat of ~** The amount of heat liberated during the n. of a strong acid and a strong base in equivalent quantities of dilute solutions equals about 57.4 kJ/mol. (2) Proton reaction. The transfer of a proton, H^+, from an acid to a base, producing a weaker acid and a weaker base; as, $HCl + H_2O \rightleftharpoons H_3O^+ + Cl^-$; $H_3O^+ + Ac^- \rightleftharpoons HAc + H_2O$.

n. ratio The ratio of the concentration of the anion of an acid to the total concentration of the undissociated acid.

neutralize To make neutral.

neutralizer Any agent that produces neutrality.

neutrino A subatomic *particle,* q.v. Suggested by Pauli (1927) to satisfy the conservation laws of spin, momentum, and energy. Cf. *neutron.*

neutron *n* or ${}^1_0n$. A subatomic *particle,* q.v. A "building stone" of the atomic nucleus. Each isotope contains in its atom ($A - Z$) neutrons, where A is the mass number, and Z the atomic number. An electrically neutral particle, with mass 1.008665 m_u, spin $\frac{1}{2}$, which does not ionize air; produced from elements by bombardment with α particles. Is stable in an atomic nucleus, but otherwise decays with a half-life of about 800 s. Cf. *nuclear reactions, neutrino.* **anti ~** An electrically neutral particle differing from a n. in that its magnetic moment is parallel with its spin angular momentum; antiproton + proton = neutron + antineutron.

n. activation analysis The detection of traces of elements by activation with high-flux n. bombardment, and measurement of the resulting emission-energy decay rate.

Nevile and Winther's acid NW acid. 1-Naphthol-4-sulfonic acid.

nevyanskite A native alloy of iridium and osmium with allied metals.

new **n. blue** (1) Meldola blue. (2) Prussian blue. **n. fuchsin** $C_{22}H_{23}N_3 \cdot HCl$ = 365.9. Soluble fuchsin. A bluish-red dye obtained by oxidation of diamino-*o*-ditolylmethane.

Newlands, John Alexander Reina (1838–1898) British chemist; pioneer in the development of the *periodic* system, q.v., from his concept of "octaves."

Newton **N., Sir Isaac (1642–1727)** English mathematician and philosopher, originator of the law of universal gravitation. **N.'s alloy** Bi 20, Sn 30, Pb 50%; m. 94. **N.'s law** The attractive force between 2 bodies is proportional to their masses. **N.'s rings** Colored rings produced when plane and convex glass surfaces are pressed together, due to interference between the beams of light reflected from the two surfaces.

newton* Abbrev. N; the SI system unit of force; the force required to accelerate 1 kg mass at 1 m/s^2. 1 N = 10^6 dynes = 0.101972 kg-force = 7.233011 poundals = 0.224809 lb-force.

newtonian fluid See *Newtonian fluid* under *fluid.*

newtonium Name given by Mendelyeev to his hypothetical protoelement.

ngai camphor Borneol*.

Ni Symbol for nickel.

niacin USP name for nicotinic acid*.

niacinamide USP name for nicotinamide*.

Niagara blue Trypan blue.

nialamide $C_{16}H_{18}N_4O_2$ = 298.3. Bitter, white crystals, slightly soluble in water, m.152; an antidepressant in the monoamine oxidase inhibitor group (BP).

niccolate* Indicating nickel as the central atom(s) in an anion.

niccolic Nickelic.

niccolite NiAs. Copper nickel, arsenical nickel. A red, native nickel arsenide.

niccolous Nickelous.

niccolox $C_4H_{12}O_4N_8$(?). An oxamide dioxime reagent for nickel.

niccolum Latin for "nickel."

Nichols, William Henry (1852–1930) American industrial chemist. **N. medal** A prize for achievement in industrial chemistry.

Nicholson, William (1753–1815) English physicist and chemist, a pioneer in electrolyzing water.

nicholsonite $CaCO_3$. Aragonite, containing zinc.

Nichrome Trademark for a high-melting-point alloy: Ni 60, Fe 25, Cr 15%; used in electric heaters and acid-resisting apparatus. **N. wire** A wire of composition Ni 80, Cr 20%; used as a platinum substitute (for bead and flame tests), and for heating elements and electrical resistances.

nickel* Ni = 58.69. An element of the iron group, at. no. 28. A silver-white metal, d.8.9, m.1455, b.2730, insoluble in water, dilute acids or alkalies. It occurs in niccolite, Monel metal, kupfernickel, nickel glance, and nickel blende. Isolated (1751) by Cronstedt. N. has the common valencies of 2 [n.(II)*, n.(2+)*, nickelous] and 3 [n.(III)*, n.(3+)*, nickelic]; but it can possess all 6 oxidation states from −1 to +4; e.g., $[Ni_2(CO)_6]^{2-}$ to K_2NiF_6. Used as a hydrogenation catalyst (margarine manufacture); for galvanic coating of other metals (stainless steel); for acid- resisting alloys, chemical apparatus, coins and medals, electric apparatus, surgical and dental instruments; and in manufacture of n. salts. **Admiralty ~** Adnic. **arsenical ~** Niccolite. **emerald ~** Zaratite. **Raney ~** *See Raney's alloy.* **super ~** See *supernickel.*

n.(II)*, n.(2+)* See also *nickelous.* **n.(III)*, n.(3+)*** See also *nickelic.* **n. acetate*** Nickelous acetate. **n. alloys** Noncorroding alloys; as, nickeline, ferronickel. Some are harder than steel and are used for surgical instruments. Melting points (in °C) are:

With	90% Ni 10%	80% Ni 20%	70% Ni 30%	60% Ni 40%	50% Ni 50%
Cu	1440	1430[a]	1410[b]	1380[c]	1335[d]
Sn	1380	1290	1200	1235	1290

[a]Nickeline. [b]Corronel, Monel.
[c]Constantin. [d]Alpro, Copel.

See also *Numetal* (74% Ni), *permalloy* (30–80% Ni), *Nichrome* (60–80% Ni), *Monel* (67% Ni), *permivar* (45% Ni), *nickel* silver (6–30% Ni), and Table 54. **n. alumide** High-purity aluminum powder particles with a n. coating, m. exceeds 1,500. Used for spray-coating ceramics. Cf. *cermet.*

n. arsenate* Nickelous arsenate. **n. benzoate*** Nickelous benzoate. **n. black** N. peroxide (See *n. oxides* below). **n. blende** NiS. A native n. sulfide. **n. bloom** Annabergite.

TABLE 54. COMPOSITION AND HARDNESS OF SOME NICKEL ALLOYS

Alloy	Composition, percent								Brinell hardness
	Ni	Cu	Cr	Fe	Si	Mn	C	Other elements	
A nickel	99.6	—	—	0.05	0.02	0.25	0.07	—	196—cold-drawn
Duranickel	93.8	—	—	0.12	0.58	0.38	0.20	Al 4.5, Ti 0.4	235—cold-drawn
Monel	66.0	31.2	—	1.41	0.14	1.05	0.19	—	235—cold-drawn
K Monel	65.5	29.1	—	0.96	0.12	0.60	0.15	Al 3.0, Ti 0.5	340—aged
Inconel	78.5	—	14.5	6.40	0.20	0.23	0.07	—	262—cold-drawn
Inconel X	72.9	—	15.0	6.77	0.42	0.59	0.04	Al 0.9, Ti 2.4, Nb 1.0	340—aged
17-7 P.H. stainless	6.7	—	17.2	bal.	0.36	0.64	0.07	Al 1.0, P 0.02	436—hardened
G nickel	94.2	—	—	0.5	1.5	0.80	1.5	—	131
S nickel	90.0	—	—	2.0	6.0	0.78	0.68	—	241
Colmonoy 6	65–75	—	13–25	—	—	—	—	B 2.7—4.7	—
Waukesha 88	bal.	—	12.0	—	0.20	0.80	—	Sn 4, Mo 3, Bi 3.5	127
Leaded bronze	0.46	79.9	—	—	—	—	—	Sn 6.3, Pb 10.1, Zn 2.1	69
Neveroil 21	32.3	64.5	—	—	—	—	—	Sn present	17

n. borate Nickelous borate. **n.-brass** An alloy used in British coinage, q.v. *(coinage metals).* **n. bromide** Nickelous bromide. **n. carbide** Ni_3C = 188.1. A black solid. **n. carbonate** Nickelous carbonate. **n. carbonyl** $Ni(CO)_4$ = 170.7. N. carbon oxide. A poisonous gas. Colorless liquid, b.43, insoluble in water. Used in the Mond process for the isolation of n. If pure, it explodes at 60: $Ni(CO)_4 = Ni + 2CO_2 + 2C$. **n. chloride** Nickelous chloride. **n.-chrome steel** An alloy: Fe 95, Ni 3, Cr 1.5, C 0.5%; used for heat- and acid-resistant machinery. **n.-chromium triangle** A triangle made of Nichrome wire; used to heat small laboratory utensils. **n. citrate** Nickelous citrate. **n. cyanide** Nickelous cyanide. **n. dimethyl glyoxime** $Ni(C_4H_7N_2O_2)_2$ = 288.9. Red crystals, used to determine n.; sublimes at 120. **n. fluosilicate** N. hexafluorosilicate*. **n. glance** Ni_2AsS. A native arsenide and sulfide of n. **n. gymnite** Genthite. A gymnite, in which part of the magnesium is replaced by n. **n. hexafluorosilicate*** $NiSiF_6 \cdot 6H_2O$ = 308.9. N. flu(or)osilicate. Green, trigonal crystals, d.2.109, soluble in water. **n. iodide** Nickelous iodide. **n. minerals** N. is associated with cobalt and iron in meteorites; with chromium in olivine; and as sulfide, arsenide, and silicates. E.g.: bunsenite, NiO; polydymite, Ni_4S_5; niccolite, NiAs; smaltite, $(NiCoFe)As_2$; white nickel ore, $NiAs_2$; nickel ocher, $Ni_3(AsO_4)_2 \cdot 8H_2O$. **n. monoxide*** Nickelous oxide. **n. nitrate** Nickelous nitrate. **diaquatetraammine ~** $[Ni(H_2O)_2(NH_3)_4](NO_3)_2$ = 286.9. Nickeloaminnitrate. Green crystals, soluble in water; used in nickel plating. **n. ocher (ochre)** Annabergite. **n. oleate** $Ni(C_{18}H_{33}O_2)_2$. A waxy n. soap, a compound of n. with oleic acid; used in ointments. **n. oxalate** Nickelous oxalate. **n. oxides** NiO, nickelous oxide; Ni_2O_3, nickelic oxide; Ni_3O_4, nickelous nickelic oxide; NiO_2, nickel peroxide; NiO_4, nickel superoxide. **n. phosphate** Nickelous phosphate. **n. plating** The deposition, usually of pure nickel on copper, or on steel coated with copper, from saturated ammonium n. sulfate. **n. pyrites** Millerite. **n. sesquioxide** Nickelic oxide. **n. silver** German silver, neusilber, argentan, albata, British plate (U.K. usage). White alloys: copper 46–63, nickel 6–30, zinc 18–36%; used for resistance coils and cutlery. **n. steel** A very tough alloy: Fe 96.5, Ni 3.5%. **n. stibine** Ullmannite. **n. sulfate** Nickelous sulfate. **n. sulfide** Nickelous sulfide. **n. superoxide** NiO_4 = 122.7. A black oxide produced by electrolysis. **n. tetracarbonyl*** N. carbonyl. **n. vitriol** Morenosite. **n. yellow** Nickelous phosphate.

nickelic Nickel(III)* or (3+)*, niccolic. Describing compounds of trivalent nickel, which are all strong oxidizing agents, and readily reduced to the more stable nickelous compounds. **n. hydroxide** $Ni_2(OH)_6$ = 219.4. Niccolic hydroxide. A black powder, rapidly decomp. by heat; used in the manufacture of n. compounds. **n. nickelous sulfide** Ni_2S_4. Native as polydymite. **n. oxide** Ni_2O_3 = 165.4 Nickel sesquioxide*, niccolic oxide, black nickel oxide. Black powder, reduced to nickelous oxide at 600°C, insoluble in water; used in storage batteries. **n. sulfide** Ni_2S_3. Native as *beyrichite,* q.v.

nickeline (1) An alloy: Cu 80, Ni 20%. (2) An alloy: Cu 56, Ni 31, Zn 13%; used in high-resistance apparatus.

nickelous Nickel(II)* or (2+)*. Describing compounds containing divalent nickel. The anhydrous salts are yellow, the hydrous salts green. **n. acetate** $Ni(C_2H_3O_2)_2 \cdot 4H_2O$ = 248.8. Green crystals, soluble in water. **n. arsenate** $Ni_3(AsO_4)_2 \cdot 8H_2O$ = 598.0. Green powder, insoluble in water. **n. benzoate** $Ni(C_7H_5O_2)_2 \cdot H_2O$ = 318.9. Green powder, soluble in ammonium salt solutions. **n. borate** $Ni(BO_2)_2 \cdot 2H_2O$ = 180.3. Green powder, insoluble in water. **n. bromide** $NiBr_2$ = 218.5. Yellow scales, decomp. by heat, soluble in water. **ammonia ~** $NiBr_2 \cdot 6NH_3$ = 320.7. Hexaamminenickel(II) chloride*. Purple crystals, soluble in water. **hydrous ~** $NiBr_2 \cdot 3H_2O$ = 272.5. Green, deliquescent needles, soluble in water. **n. carbonate** $NiCO_3$ = 118.7. Green rhombs, insoluble in water; used in electroplating. **basic ~** $2NiCO_3 \cdot 3Ni(OH)_2 \cdot 4H_2O$ = 583.5. Green powder, insoluble in water; used in nickel plating. **n. chloride** $NiCl_2$ = 129.6. Yellow scales, soluble in water or ammonium hydroxide. It sublimes, and is used as a reagent for thiocarbonates, as an absorbent in gas masks, for nickel plating, and in sympathetic ink. **ammonia ~** $NiCl_2 \cdot 6NH_3$ = 231.8. Nickeloammonia chloride. Purple crystals, soluble in water. Cf. *ammonium nickel(II) chloride.* **hydrous ~** $NiCl_2 \cdot 6H_2O$ = 237.7. Green hexagons, soluble in water or alcohol. **n. citrate** $Ni_3(C_6H_5O_7)_2 \cdot 8H_2O$ = 698.4. Green, hygroscopic crystals, soluble in water; used in nickel plating. **n. cyanide** $Ni(CN)_2 \cdot 4H_2O$ = 182.8. Apple-green powder, insoluble in water, soluble in cyanide solutions; used in electroplating and metallurgy. **n. formate** $(HCOO)_2Ni \cdot 2H_2O$ = 184.8. Green crystals, soluble in water. **n. hydroxide** $Ni(OH)_2$ = 92.7. White powder, insoluble in water. **hydrous ~** $4Ni(OH)_2 \cdot H_2O$ = 388.8. Green powder, insoluble in water; used in the manufacture of nickelous salts. **n. iodide** NiI_2 = 312.5. Black scales, m.57, soluble in water. **n. nickelic oxide** Ni_3O_4

= 240.1. Nickel tetraoxide*. Gray powder, insoluble in water. **n. nitrate** $Ni(NO_3)_2$ = 182.7. Yellow powder, soluble in water (green color). **ammonia ~** $[Ni(NH_3)_6](NO_3)_2$ = 284.9. Hexaamminenickel(II) dinitrate*. Blue crystals, soluble in water; used with tannic acid to stain hair and fur. **hydrous ~** $Ni(NO_3)_2 \cdot 6H_2O$ = 290.8. Nickeloaquanitrate. Monoclinic, deliquescent crystals, m.57, soluble in water; used as a reagent and in nickel plating. **n. oxalate** NiC_2O_4 = 146.7. Green powder, insoluble in water. **n. oxide** NiO = 74.7. Nickel monoxide*, nickel protoxide, green nickel oxide, insoluble in water; used in green ceramic pigments. **n. phosphate** $Ni_3(PO_4)_2 \cdot 7H_2O$ = 492.1. Nickel (ortho)phosphate*. Green powder, insoluble in water. Used in nickel plating and in the manufacture of nickel yellow. **n. silver** Nickel silver. **n. sulfate** $NiSO_4$ = 154.7. Regular, yellow crystals, decomp. 830, soluble in water. **hydrous ~** (1) $NiSO_4 \cdot 7H_2O$ = 280.8. Nickeloaquasulfate. Green rhombs, dehydrated at 100, soluble in water. Used in nickel plating, for blackening brass and zinc, as a textile mordant, and as a reagent for glucose and albumose. (2) $NiSO_4 \cdot 6H_2O$ = 262.8. Green crystals, soluble in ammonia. **n. sulfide** NiS = 90.8. Black hexagons, m.797, slightly soluble in water; used in ceramics. **n. tartrate** $NiC_4H_4O_6 \cdot 5H_2O$ = 296.8. Green powder, insoluble in water. **n. thallous sulfate** $NiSO_4 \cdot Tl_2SO_4 \cdot 6H_2O$ = 767.6. A double salt; green crystals, soluble in water.

niclosamide $C_{13}H_8O_4N_2Cl_2$ = 327.1. Yomesan. A teniacide for tapeworm (BP).

Nicol **N., William (1766–1851)** British physicist noted for optical devices. **N.'s prism** See *nicols.*

nicoline C_3H_4O = 56.1 A constituent of *Lonchocarpus rufescens* (Leguminosae), Guyana; used to stupefy fish.

nicols Nicol's prism. Two prisms of Iceland spar cemented together; used as polarizer and analyzer in polariscopes. Cf. *polaroid.*

nicopyrite A pyrite containing nickel.

Nicorette Trademark for sugar-free chewing gum preparation containing nicotine; used to treat the smoking habit.

nicoteine $C_{10}H_{12}N_2$ = 160.2. A mixture of nornicotine and anabasine from Kentucky tobacco.

nicotiana Tobacco.

nicotianine A volatile, fragrant principle from tobacco.

nicotinamide* $C_5H_4N \cdot CONH_2$ = 122.1. Niacinamide. Nicotinic acid amide, vitamin PP. White crystals, m.129, soluble in water. A B-complex vitamin (USP, EP, BP); used to treat pellagra.

nicotine $C_{10}H_{14}N_2$ = 162.2. (*S*)-3-Methylpyrrolidin-2-yl)pyridine*. A diacid alkaloid from the leaves of *Nicotiana tabacum.* Yellow liquid, d.1.01, b.247, soluble in water. **dihydroxy ~** Pilocarpidine. **neo ~** Anabasine. **nor ~** See *nornicotine.*

n. group Alkaloids derived from pyridine and pyrrolidine: (1) Nicotine and pilocarpine and derivatives. (2) sparteine, diethyloxamide, and lobeline; (3) conine; (4) pyridine bases. **n. hydrochloride** $C_{10}H_{14}N_2 \cdot 2HCl$ = 235.2. Colorless, hygroscopic crystals, soluble in water. **n. salicylate** $C_{10}H_{14}N_2 \cdot C_7H_6O_3$ = 300.4. Colorless plates, m.117, soluble in water; used in skin ointments. **n. tartrate** $C_{10}H_{14}N_2 \cdot (C_4H_4O_6)_2 \cdot 2H_2O$ = 494.4. Red crystals, soluble in water.

nicotinic acid* C_5H_4NCOOH = 123.1. Niacin. Pellagra-preventing vitamin. 3-Pyridinecarboxylic acid*. Colorless, bitter needles, m.236 (sublimes), slightly soluble in water; a peripheral vasodilator and a vitamin of the B complex. Synthesized (as nicotinamide) in the body from tryptophan; is part of coenzyme NAD (USP, EP, BP). Occurs in maize as niacytin. See *vitamin,* Table 101. **hydroxy ~** $C_5H_3N \cdot OH \cdot COOH$ = 139.1. 2-Hydroxypyridine-5-carboxylic acid. Colorless crystals, m.303. Cf. hydroxy*quinolinic acid.* **iso ~*** 4-Pyridinecarboxylic acid*. Colorless crystals, m.304, soluble in water. **tetrahydro-1-methyl ~** Arecaidine.

nicotinoyl* 3-Pyridinylcarbonyl†. The radical C_5H_4NCO-, from nicotinic acid.

nicouic acid $2,4-(COOH)_2 \cdot 3-(OH) \cdot C_6H_2OCMe_2COOH$ = 284.2. From rotenone and deguelin.

nicoulin A resin from nekoe or stinkwood, *Gustavia ocotea;* a Malaysian arrow poison.

nicouline Rotenone.

nido-* Affix indicating a nestlike molecular structure; especially a boron skeleton that is very close to a closed *(closo)* structure.

niello silver Russian tula, blue silver. A bluish alloy of silver, copper, lead, and bismuth.

Nierenstein reaction The synthesis of ketones by diazomethane: $R \cdot COCl + CH_2N_2 \rightarrow R \cdot COCH_2Cl + N_2$. Cf. *Schlotterbeck reaction.*

Nieuwland, Father Julius Arthur (1878–1936) A Jesuit American chemist, noted for the synthesis of acetaldehyde, vinylacetylene, and synthetic rubber from acetylene.

nifurtimox $C_{10}H_{13}O_5N_3S$ = 287.3. Lampit. Yellow powder, soluble in water. A trypanocide, used for S. American trypanosomiasis.

nigella Black caraway, black cumin, small fennel. The dried seeds of *Nigella sativa;* a condiment.

nigelline An alkaloid in the seeds of *Nigella sativa* (Ranunculaceae).

nightshade Plants of the Solanaceae. **deadly ~** Belladonna. **garden ~** *Solanum nigrum.* **woody ~** Dulcamara.

nigraniline Nigrosine.

nigre A dark-colored layer, formed in the soap pan, between the neat soap and the lye; an isotropic solution containing a higher concentration of soap than lye, a fairly high concentration of salts, and colored impurities.

nigrite A variety of asphalt.

nigrometer An instrument to evaluate the color of carbon blacks.

nigrosine $C_{38}H_{27}N_3$ = 525.7. Aniline black. Pure nigrosine obtained by oxidation of aniline; a microscopic stain.

nigrosines Black or deep-blue azine dyes obtained by oxidation of aniline or its homologs; used in the manufacture of inks and shoe polishes, and in dyeing.

nigrotic acid 3,6-Dihydroxy-2-naphthol-7-sulfonate.

Ni-Hard Trademark for an abrasion-resistant cast iron containing Ni 4.5, Cr 1.5, Mn 1.5%.

nihil album Zinc oxide*.

nikethamide $C_{10}H_{14}ON_2$ = 178.2. Coramine. Nicotinic acid diethylamide. *N,N*-Diethyl-3-pyridinecarboxamide. Yellow liquid, congeals 22–24, miscible with water; a respiratory and cardiac stimulant (EP, BP).

Nikiforoff stain (1) A borax-carmine solution acidified with acetic acid; used to stain nuclei. (2) An alkaline concentrated methylene blue counterstain.

nile blue $C_{20}H_{19}ON_3$ = 317.4. A blue aniline dye, insoluble in water. Used as a pH indicator, changing from yellow (acid) through blue to magenta (alkaline). **n. blue sulfate** An oxidation-reduction and pH indicator, and vital stain.

nilvar An alloy similar to Invar: 36% Ni. It has a low temperature coefficient, below 100°C.

nimbicetin The tetraacetate of kaempferol, from *Melia azadirachta,* m.180 (decomp).

nimbin $C_{30}H_{36}O_9$ = 540.6. A carboxymethyl steroid from the bark of *Melia azadirachta,* m.204.

nimbiol $C_{18}H_{24}O_2$ = 272.4. An oxophenolphenanthrene derivative from *Melia azadirachta.*

nimonic alloys A range of alloys having good resistance to oxidation and to creep at high temperatures. They contain Ni, with Cr 18–21, Fe 5–11, Si 1.0–1.5, Ti 0.2–2.0, C 0.08–0.15%, and sometimes Al, Cu, or both.

Ninhydrin $C_6H_4 \cdot CO \cdot C(OH)_2 \cdot CO$ = 178.1. Trademark for 2,2′-dihydroxy-1(*H*)-indene-1,3(2*H*)-dione. Ninidrine. Colorless crystals; a reagent in the Abderhalden test for proteins, and for amino acids (blue color) in developing fingerprints.

ninidrine Ninhydrin.

niobate (1)* Indicating niobium as the central atom(s) in an anion. (2) Columbate. A salt (mixed oxide) of the type $MNbO_3$ or $M_2O \cdot Nb_2O_5$; e.g. $NaNbO_3$, sodium n.; they often form complexes, as, hexa ~, $M_8[Nb_6O_{19}]$.

niobe oil Methyl benzoate*.

niobic Niobium(V)* or (5+)*, columbic. Describing a compound containing pentavalent niobium, **n. acid** (1)* $HNbO_3$ = 141.9. An insoluble white powder derived from Nb_2O_5; soluble in alkalies to form complex niobates. (2) The compound $3Nb_2O_6 \cdot 4H_2O$, which gives the acid $H_8Nb_6O_{19}$. **n. anhydride** Nb_2O_5. Columbic anhydride.

niobite $(Fe,Mn)O \cdot (Nb,Ta)_2O_3$. Native in black crystals.

niobium* Nb = 92.9064. A metallic element, at. no. 41, discovered (1801) by C. Hachette in the mineral columbite; named in honor of America. In 1802, A. G. Ekeberg isolated a new metal from yttrotantalite whose elemental nature was not realized until 1844. H. Rose (1844) isolated from Bavarian columbite 2 metals, niobium (named after Niobe, daughter of Tantalus, since it resembled Ta) and peloponium. Niobium proved to be an element, and peloponium a mixture of tantalum and niobium. Niobium was known as columbium until 1944, when its name was changed officially. A gray, shining metal, d.8.4, m.2470, b.4800, made by the hydrogen reduction of the pentachloride. It forms trivalent [n.(III)*, n.(3+)*, niobus] and pentavalent [n.(V)*, n.(5+)*, niobic] compounds and belongs to the 5A group of the periodic table. N. is used as a reactor fuel element and in heat-resistant structures. **n. bromide** $NbBr_5$ = 492.4. Purple crystals, decomp. by water. **n. chloride** (1) $NbCl_3$ = 199.3. Niobus chloride. Violet crystals, soluble in water. (2) $NbCl_5$ = 270.2. Niobic chloride. Yellow, hygroscopic crystals, m.194, decomp. in air or water, to form hydrochloric acid. **n. fluoride** NbF_5 = 187.9. Niobic fluoride. Colorless, monoclinic crystals, m.72, hydrolyzed by water. **n. hydride** NbH = 93.9. Black powder, decomp. by heat. **n. minerals** N. occurs with Ti and usually the rare-earth metals, e.g., *niobite* (q.v.), tantalite, samarskite. **n. oxalate** $Nb(HC_2O_4)_5$ = 538.0. Colorless, monoclinic crystals, decomp. by water. **n. oxides** (1) NbO = 108.9. Insoluble in water. (2) NbO_2 = 124.9. Black, insoluble powder. (3) Nb_2O_5 = 265.8. See *niobium pentaoxide.* **n. oxybromide** $NbOBr_3$ = 348.6 Nioboxy bromide. Yellow crystals, hydrolyzed by water. **n. oxychloride** $NbOCl_3$ = 215.3. Niobyl or nioboxy chloride. Colorless needles, sublime 400, hydrolyzed by water. **n. pentaoxide*** Nb_2O_5 = 265.8. Niobic acid, niobic oxide. White crystals, m.1520 (yellow when hot), insoluble in water. **n. potassium fluoride oxide** $NbOF_3 \cdot 2KF \cdot H_2O$ = 300.1. Nioboxy potassium fluoride. White scales, soluble in water.

nioboxy Columboxy. Niobyl. Describing a compound of pentavalent niobium, containing the radical NbO≡.

niobus Niobium(III)* or (3+)*, columbus. Describing a compound of trivalent niobium.

niobyl Nioboxy.

Ni-O-Nel Trademark for an alloy resistant to acids; Ni 40, Fe 31, Cr 21, Mo 3, Cu 1.75%, and traces of Mn, Si, and C.

nioxime 1,2-Cyclohexanedione dioxime, m.188 (decomp.), slightly soluble in water. A reagent for nickel (1 pt. in 5×10^6 gives a purple-red color).

Nipagin M $HO \cdot C_6H_4COOMe$ = 152.2. Trademark for methyl *p*-hydroxybenzoate*. Solbrol. A food preservative.

Nipasol M $C_{10}H_{12}O_3$ = 180.2 Trademark for propyl *p*-hydroxybenzoate. White powder; a preservative.

nipecotic acid $C_5H_9 \cdot NH \cdot COOH$ = 129.2. 3-Piperidinecarboxylic acid. m.260.

Niplon Trademark for a Japanese polyamide synthetic fiber.

nipponium Early name for technetium.

Nipride Trademark for sodium nitroprusside.

niranium Chrome-cobalt alloy. An alloy for dental castings: Cr 64, Co 29, Ni 4.3, W 2, C 0.2%, with traces of Si and Al.

ni-resist A heat-resistant, nonmagnetic, weldable alloy of Fe, with Ni 12–15, Cu 5–7, Cr 1.5–4%; used in glass-annealing furnaces and in manifold valves of automobile engines.

Nisalpin Trademark of commercial nisin, an antibiotic produced by strains of *Streptococcus lactis;* occurs in milk and inhibits food spoilage bacteria.

nisin A polypeptide antibiotic produced by *Streptococcus lactis;* a permitted preservative for canned food (U.K. usage).

nisinic acid $C_{24}H_{36}O_2$ = 356.5. A highly unsaturated acid from fish liver oils.

nital A solution of 1–5 ml nitric acid (d.1.42) in 100 ml of 95% alcohol. An etching reagent for iron and steel.

niter, nitre Potassium nitrate. **Chile ~, cubic ~** Sodium nitrate. **Norwegian ~** Calcium nitrate. **rough ~** Magnesium chloride.

n. air See *oxygen.* **n. cake** Crude sodium sulfate containing some hydrogensulfate and nitrate; a by-product in the manufacture of nitric acid by the retort process.

nitinol Generic name for nickel-titanium intermetallic compounds used in spacecraft construction, e.g., Ti-Ni. They have high ductility and impact resistance at low temperatures.

Nitionol Trademark of a series of Ti-Ni alloys, used in nonmagnetic tools.

niton Radon*.

nitr- A combining form that means containing nitrogen.

nitracetanilide Nitroacetanilide.

nitracidium ion $H_2NO_3{}^+$. Nitrate acidium ion*, nitriacidium ion. Responsible for the nitration of organic compounds by nitric acid according to the equilibria: $2HNO_3 \rightleftharpoons H_2NO_3^+ + NO_3^-$ and $H_2NO_3^+ \rightleftharpoons NO_2^+ + H_2O$.

nitralizing A process for the preparation of steel sheets for enameling by degreasing, pickling, and immersion in fused potassium nitrate at 500°C. Blister formation is thereby minimized.

nitralloy Cr-Al steels containing 0.2–0.6% C, surface-hardened by nitridation.

Nitram Trademark for prills of ammonium nitrate fertilizer, with a deliquescence-preventing additive.

nitramide* NO_2NH_2 = 62.6. Colorless crystals, m.75. **phenyl ~** $NO_2 \cdot NH \cdot C_6H_5$ = 138.1. Colorless crystals, m.46, soluble in water.

nitramides A group of compounds derived from nitramide and differing from nitroamines by the presence of a radical —COO—; as, $NO_2 \cdot NH \cdot COOH$, nitrocarbamic acid.

nitramine (1) Picryl methyl n. An indicator, changing at pH 10.5 from colorless (weakly alkaline) to brown (strongly alkaline). (2) See *nitroamine.*

nitranilic acid $C_6H_2O_8N_2$ = 230.1. 2,5-Dihydroxy-3,6-dinitro-1,4-benzoquinone. Yellow plates, m.100 (decomp.), soluble in water.

nitranilide $C_6H_5N:NO \cdot OH$ = 138.1. Diazobenzene acid. Phenylisonitramine. Colorless crystals, m.46, soluble in water.

nitrate (1) A salt of nitric acid, or compound containing the radical $-NO_3$. (2) Nitration. **n. acidium*** See *nitracidium ion.* **n. ion** The NO_3^- ion, colorless, and forming no insoluble precipitates with metallic ions. **n. of lime** Calcium n. **n. of potash** Potassium n. **n. of soda** Sodium n. **n. of soda-potash** A crude Chilean saltpeter: sodium nitrate 75, potassium nitrate 25%; a fertilizer.

nitrated Describing an organic compound containing the $-NO_2$ group.

nitratine A mineral form of sodium nitrate.

nitration The introduction of the NO_2 group into an organic compound, usually by means of a mixture of sulfuric and nitric acids.

nitrato-* Prefix indicating the ligand $-NO_3$. Cf. *nitrito-*.

nitrator A vessel, usually double-jacketed, with heating or cooling coils and stirring device, used for nitration.

nitrazepam $C_{15}H_{11}O_3N_3$ = 281.3. Mogadon. Yellow crystals, m.228, insoluble in water. A hypnotic drug of the benzodiazepine group (EP, BP).

Nitrazine **N. Paper** Trademark for a filter paper, impregnated with sodium dinitrophenyl azonaphthol disulfonate; used to indicate pH values: yellow 4.5, olive green 6.2, blue 7.0. **N. yellow** An indicator dye; pH 6.5: yellow (acid) to blue-green (alkaline).

nitre Niter. See *oxygen.*

nitrene The $-NH$ group; analog of carbene.

nitriacidium ion Nitracidium ion.

nitric acid HNO_3 = 63. Colorless liquid, d_0 1.53, m.−40.3, b.86, soluble in water; used extensively as its aqueous solutions: (1) Fuming: 86% HNO_3 with some N_2O_4. Brown-red fuming liquid, d.1.48–1.5; an energetic oxidizing agent in chemical analysis and synthesis. (2) 70% HNO_3. NF, BP strength. (3) Concentrated: 65% HNO_3. Aqua fortis, azotic acid. Faintly yellow liquid, d.1.40–1.42. Used as a solvent for metals and an oxidizing agent; in etching and many chemical operations; and in nitrating organic compounds. (4) 32–34% HNO_3. d.1.20. (5) Dilute: 10% HNO_3. Colorless liquid, d.1.06; a reagent, solvent, and acidifying agent. **chloro ~** See *chloronitric acid.* **peroxo ~*** HNO_4. An acid of doubtful existence.

n. a. anhydride Nitrogen pentaoxide*. **n. a. hydrate** HNO_3 + 32% H_2O. $d_{15.5}$ 1.414, b.121.

nitric ether Ethyl nitrate.

nitric oxide Nitrogen oxide*.

nitridation (1) Formation of metallic nitrides by heating metals in nitrogen to increase hardness. Cf. *nitration.* (2) De-electronation in the ammonia system, analogous to oxidation in the water system. Cf. *nitridizing agent.*

nitride A binary compound of nitrogen and a metal. The alkali and alkaline-earth nitrides are readily hydrolyzed: $Mg_3N_2 + 6H_2O = 3Mg(OH)_2 + 2NH_3$.

nitridizing agent A substance that furnishes nitrogen or causes an exchange of electrons in liquid ammonia; as, hydrogenazide, HN_3; analogous to nitric acid, HNO_3, as oxidizing agent.

nitrifiable Describing a nitrogen compound that can be transformed into nitrates by soil bacteria.

nitrification Oxidation of the nitrogen in ammonia to nitrous and nitric acid or salts.

nitrifiers Soil bacteria which oxidize ammonia and its derivatives to nitrites (as nitromonas) or to nitrates (as nitrobacter).

nitrifying To cause the oxidation of ammonia or atmosphere nitrogen to nitrites and nitrates, e.g., by n. bacteria and n. catalysts.

nitrilase* A hydrolase enzyme that converts a nitrile to a carboxylate and ammonia.

nitrile* See *nitriles.* **iso ~** Isocyanide*.

n. oxides* Compounds of formula RC:NO. **n. rubber** See *elastomer.*

nitriles* Nitrile is the term in substitutive *nomenclature,* q.v., and cyanide the term in radicofunctional nomenclature for the $-CN$ radical; as, hexanenitrile and pentyl cyanide, respectively, for $C_5H_{11}CN$. **carbo ~** See *carbonitrile.* **di ~*** Dicyanides*. Compounds containing 2 $-CN$ radicals; as, hexanedinitrile, $NC(CH_2)_4CN$.

nitrilo- Prefix indicating: (1)* The bridge $-N=$. (2)* A nitrogen atom attached to 3 identical radicals. Cf. *nitrile.*

Nitrilon Trademark for a polyacrylonitrile synthetic fiber.

nitrine N_3 = 42.02. A hypothetical allotropic form of nitrogen analogous to ozone, O_3. See *active nitrogen* under *nitrogen.*

nitrite* A salt of nitrous acid, or a compound containing the group NO_2^-. The inorganic nitrites of the type MNO_2 are all insoluble, except the alkali nitrites. The organic nitrites may be isomeric, but not identical with the corresponding nitro compounds.

nitrito-* The $-NO_2$ group as a ligand, either as nitrito-*N*, $-NO_2$; or as nitrito-*O*, $-O \cdot N:O$.

nitro- (1)* A prefix denoting the radical $-NO_2$. Nitroxyl is sometimes used when the element is strongly electronegative or is a metal. Nitro compounds are usually yellowish in color, and differ from the less stable, isomeric nitrosooxy compounds. Cf. *nitroxyl, nitrite, nitrito.* (2) A misnomer for nitrate; as, nitroglycerin (trinitroglycerol). ***aci-* ~*** Isonitro-. The radical $HO(O)N=$. **iso ~** See *aci-*.

nitroacetanilide $C_8H_9O_3N_2$ = 181.2. Nitracetanilide. Colorless crystals **1,2- ~** m.92. **1,3- ~** m.151. **1,4- ~** m.210.

nitroacid A compound containing both the radicals $-COOH$ and $-NO_2$; as, $NO_2 \cdot CH_2 \cdot COOH$, nitroacetic acid.

nitroalizarin $C_{14}H_5O_2(OH)_2NO_2$ = 285.2. **1,2,3- ~** Alizarin orange. 1,2-Dihydroxy-3-nitroanthroquinone*. Orange-yellow crystals, decomp. 244, slightly soluble in water, soluble in alcohol; used as dye, and as an intermediate in organic synthesis. **1,2,4- ~** Yellow crystals, decomp. 290.

nitroamine An organic compound containing the radical $-NH \cdot NO_2$ or $=N \cdot NO_2$. **diethyl ~** $Et_2N \cdot NO_2$ = 118.1. Colorless liquid, b.206. **dimethyl ~** $Me_2N \cdot NO_2$ = 90.1. Colorless crystals, m.58, soluble in water. **ethyl ~** $EtNH \cdot NO_2$ = 90.1. Colorless liquid, m.3. **iso ~** A compound containing the radical $-N-O-N \cdot OH$ (with a ring bond joining the two N atoms). **methylphenyl ~** $MeNPh \cdot NO_2$ = 152.2. Colorless crystals, m.39, soluble in water. **phenyl ~** $PhNH \cdot NO_2$ = 138.1. Colorless crystals, m.46, soluble in water. **propyl ~** $PrNH \cdot NO_2$. Colorless liquid, b.140.

nitroamino The radical NO_2NH—. **n.acetic acid** $C_2H_4O_4N_2$ = 120.1. A homolog of nitrourethane. Colorless crystals, m.103, soluble in water (strongly acid).

nitroaniline $NH_2 \cdot C_6H_4 \cdot NO_2$ = 138.1. **1,2- ~** Colorless needles, m.71, soluble in water. **1,3- ~** Yellow needles, m.114, slightly soluble in water. **1,4- ~** Yellow needles, m.146, soluble in water. All used in organic synthesis and as indicators for strong acids. **di ~** See *dinitroaniline* under *aniline.*

nitroanilines Compounds derived from benzene by the substitution of 2 or more H atoms by one or more NH_2— and NO_2— radicals. The higher-nitrated anilines are powerful explosives.

nitroanisole $C_6H_4(OMe)NO_2$ = 153.1. ***ortho-* ~** 1-Methoxy-2-nitrobenzene*. Yellow liquid, d.1.268. m.9, b.265.

meta-~ m.38, b.258. **para-~** Colorless or yellowish plates, d.1.233, m.54, b.258. Insoluble in water, soluble in alcohol or ether.

nitroanthracene* $C_{14}H_9NO_2$ = 223.2. Nitrosoanthrone. **1-~** Yellow needles, m.146, insoluble in water, soluble in benzene or chloroform.

nitroanthraquinone* $C_6H_4(CO)_2C_6H_3NO_2$ = 253.2. **1-~** Yellow needles, m.228, subliming when heated, insoluble in water, soluble in alcohol or ether. **2-~** Yellow needles, m.184, subliming when heated, insoluble in water, soluble in alcohol or ether.

n. sulfonic acid A reagent for sugars.

Nitrobacter A soil bacterium or other microorganism that oxidizes ammonia and its derivatives, or atmospheric nitrogen, to nitrites or nitrates.

nitrobacteria Soil bacteria; as, Nitrobacter, Nitrosococcus, or Nitrosomonas.

nitrobarite $Ba(NO_3)_2$. A native barium nitrate.

nitrobenzaldehyde* $C_6H_4(NO_2)CHO$ = 151.1. **ortho-~** Yellow needles, m.44, slightly soluble in water. **meta-~** Colorless needles, m.58. **para-~** Colorless prisms, m.106, soluble in water; used in indigo synthesis.

nitrobenzamide* $NO_2 \cdot C_6H_4 \cdot CONH_2$ = 166.1. **ortho-~** Colorless needles, m.174, soluble in water. **meta-~** Yellow needles, m.140. **para-~** Colorless needles, m.198, soluble in water.

nitrobenzanilide $NO_2 \cdot C_6H_4 \cdot CONHPh$ = 242.2. **meta-~** Colorless leaflets, m.143, insoluble in water.

nitrobenzene* $C_6H_5NO_2$ = 123.1. Nitrobenzol, oil of mirbane, phenyl nitrite, essence of mirbane, artificial oil of bitter almonds. Yellow liquid, $d_{20°}$1.198, b.210, slightly soluble in water. A reagent for sulfur, calcium oxide, and glucose; a substitute for almond oil; and a raw material for aniline and derivatives. **amino ~** Nitroaniline*. **chloro ~*** $C_6H_4ClNO_2$ = 157.6. Colorless crystals. **ortho-~** m.33. **meta-~** m.44. **para-~** m.84. **dimethyl ~*** Nitroxylene. **methyl ~** Nitrotoluene*. **tri ~*** See *trinitrobenzene.*

n. azoresorcinol Magneson; a reagent for magnesium (blue color). **n. reduction** The reduction of nitrobenzene to aminobenzene (aniline) in alkaline solutions. The following are produced: nitrobenzene (2 moles), azoxybenzene, azobenzene, hydrazobenzene, aniline (2 moles). **n.sulfonyl chloride** A reagent for amines.

nitrobenzoic acid* $NO_2 \cdot C_6H_4 \cdot COOH$ = 167.1. **ortho-~** Yellow crystals, m.148, soluble in water; a reagent. **meta-~** Yellow leaflets, m.141, soluble in water. **para-~** Nitrodracylic acid. Yellow crystals, m.242, soluble in water; used in organic synthesis.

nitrobenzol Nitrobenzene*.

nitrobenzonitrile* $NO_2 \cdot C_6H_4 \cdot CN$ = 148.1. **ortho-~** 1-Cyano-2-nitrobenzene*. Colorless needles, m.109, soluble in hot water. **meta-~** Colorless needles, m.115, soluble in water. **para-~** Colorless leaflets, m.149, insoluble in water.

nitrobenzophenone See *nitrobenzophenone* under *benzophenone.*

nitrobenzoquinone $C_6H_3O_2NO_2$ = 153.1. Yellow crystals, decomp. 206, slightly soluble in water.

nitrobenzoyl* The radical $NO_2 \cdot C_6H_4 \cdot CO-$, from nitrobenzoic acid. **n. chloride** $C_7H_4O_3NCl$ = 185.6. **ortho-~** A solid, m.75. **meta-~** m.34. **para-~** m.72. **n.formic acid** $C_6H_4O_2N \cdot CO \cdot COOH$ = 195.1. **ortho-~** A solid, m.123, soluble in warm water.

nitrobenzyl* The radical $NO_2 \cdot C_6H_4 \cdot CH_2-$, from nitrotoluene; 3 isomers. **n. alcohol*** $NO_2C_6H_4CH_2OH$ = 153.1. **ortho-~** Colorless needles, m.74, soluble in water. **meta-~** Rhombic crystals, m.27. **para-~** Colorless needles, m.93. **n. bromide*** $NO_3C_6H_4CH_2Br$ = 232.0. A reagent for hydroxy compounds. **n. chloride*** $NO_2C_6H_4CH_2Cl$ = 171.6. **ortho-~** m.48. **meta-~** m.46. **para-~** m.71. **n. cyanide*** $NO_2C_6H_4CH_2CN$ = 162.1. **ortho-~** Colorless needles, m.83, soluble in hot water. **meta-~** Colorless prisms, m.115, insoluble in water. **para-~** Colorless crystals, m.110, soluble in alcohol.

nitrobiphenyl* $NO_2C_6H_4Ph$ = 199.2. **ortho-~** Colorless leaflets, m.37, insoluble in water. **meta-~** Yellow solid, m.61. **para-~** Colorless needles, m.114, insoluble in water.

nitrobromoform NO_2CBr_3 = 297.7. Bromopicrin. Colorless liquid, d.2.81, m.10, insoluble in water.

nitrocaptax 2-Mercapto-6-nitrobenzothiazole. A reagent for alkyl halides.

nitrocarbamate A salt or ester of nitrocarbamic acid or a compound containing the radical $NO_2 \cdot NH \cdot COO-$. **ethyl ~** Urethane.

nitrocarbol Nitromethane*.

nitrocarbonitrate Ammonium nitrate sensitized with carbonaceous materials, e.g., diesel oil; an explosive.

nitrocellulose $C_6H_7O_5(NO_2)_3$ = 297.1. (1) An ester of nitric acid and cellulose; as, pyroxylin (11.2–12.4% N) and guncotton (12.4–13% N). Cf. *rayon.* Yellow granules used in Celluloid and collodion; mixed with picrates as an explosive; and used in lacquers. (2) Also applied to other nitrates of cellulose.

nitrochalk A fertilizer (10% available N); ammonium nitrate and calcium carbonate. Cf. *calnitro.*

nitrochlorobenzol Chloro*nitrobenzene**.

nitrochloroform Chloropicrin.

nitrocinnamic acid* $NO_2C_6H_4CH{:}CHCOOH$ = 193.2. **ortho-~** Colorless scales, m.240, insoluble in water. **meta-~** Yellow needles, m.196, soluble in water. **para-~** Colorless prisms, m.285, insoluble in water.

nitrococcus Nitrosococcus.

nitrocolors Colored compounds containing the nitro group; as, picric acid.

nitrocompound A compound containing the $-NO_2$ group.

nitrocotton Guncotton.

nitrocresol $CH_3 \cdot C_6H_3 \cdot NO_2 \cdot OH$ = 153.1. Methylnitrophenol. **5-nitro-*m*-cresol** Yellow solid, m.54, soluble in benzene.

nitrocumene $C_6H_4 \cdot NO_2 \cdot CHMe$ = 165.2. Colorless liquid, b.224 (decomp.).

nitrodiphenyl Nitrobiphenyl*.

nitrodracylic acid *p*-Nitrobenzoic acid*.

nitrodye See *nitrodyes* under *dyes.*

nitroerythrite Nitroerythrol.

nitroerythrol $C_4H_6(NO_3)_4$ = 302.1. Butine tetranitrate, erythrol tetranitrate, tetranitroerythritol. Large plates, m.61, exploding on percussion or heating, slightly soluble in water; explosives.

nitroethane* $MeCH{:}N^+(OH)O^- \rightleftharpoons MeCH_2NO_2$ = 75.1. A colorless liquid, b.114, slightly miscible with water.

nitroform $CH(NO_2)_3$ = 151.0. Trinitromethane. Colorless crystals, m.15 (explode), soluble in water.

nitrofurantoin $C_8H_6O_5N_4$ = 238.1. Yellow, bitter crystals, slightly soluble in water; a urinary antiseptic (BP).

nitrogelatin Dynamite.

nitrogen* N = 14.0067. A gaseous element in the atmosphere, at. no. 7. Colorless gas, $d_{air=1}$0.967, m.−210, b.−196, slightly soluble in water or alcohol. First produced by Rutherford (1772) from air ("mephitic air"). Priestley (1780) called it "phlogisticated air," and Cavendish (1785) produced nitric acid from air and called it nitrogen (meaning "niter-producing"). N. occurs in many minerals, e.g., saltpeter, KNO_3. Its principal valency is 3, but it forms several oxides and series of compounds, all fairly unstable (explosives, proteins). Used to manufacture nitric acids, nitrates, cyanamides, ammonia, nitrides, and cyanides (nitriles) (NF). **active ~** Electrically excited n. It has a characteristic yellow afterglow. **alloxuric ~** N. in organic tissues bound in purine bases. **alpha-~** See *nitrogen molecule* below. **ammonia ~** The n. of ammonium salts. **atmospheric ~** See *atmospheric gases.* **beta-~** See *nitrogen molecule* below. **di~*** The n. molecule, N_2. **fixed ~** See *nitrogen fixation* below. **formol ~** The n. of amino acids. See *formol titration* under *Formol.* **ionized ~** N. atoms or molecules ionized at low pressures; they produce lines in the spectra of auroras, coronas, nebulas, and stars. **liquid ~** Misnomer for frozen n.; used to treat skin conditions, e.g., warts. **ortho-~** See *nitrogen molecule* below. **oxides of ~** See *dinitrogen oxide* (N_2O), *nitrogen oxide* (NO) below, *nitrogen dioxide* (NO_2) below, *dinitrogen tetraoxide* (N_2O_4) below (under *nitrogen tetraoxide*), *nitrogen pentaoxide* (N_2O_5) below. **para-~** See *nitrogen molecule* below. **radio ~** A short-lived isotope (mass 13) obtained by bombardment of n. with α particles. **soil ~** The n. in soil nitrates or nitrites. **urea ~** N. eliminated as urea.

n. balance The ratio of n. intake (proteins) to n. output (urea, uric acid) of the human body. In the normal adult it is unity. **n. benzide** Azobenzene*. **n. cycle** The passage of n. from the atmosphere to the soil by means of gaseous n. compounds, e.g., produced by thunderstorms; through bacteria to the plant; from the plant to the animal body; and back to the soil as excreta. **n. degradation** The annual loss of n. from the soil by crop production. **n. dioxide*** NO_2 = 46.01. Colorless crystals, b.22; or brown, irritant gas. The name is applied to the equilibrium mixture of NO_2 and N_2O_4. Cf. d*initrogen tetraoxide* below. **n. equilibrium** See *nitrogen balance* above. **n. equivalent** 1 g nitrogen ≡ 6.25 g protein. **n. fixation** The conversion of atmospheric n. into a compound:

1. Natural processes:
   (*a*) Nonsymbiotic fixation: direct conversion of atmospheric nitrogen into nitrates by soil bacteria
   (*b*) Symbiotic fixation: gradual conversion of atmospheric nitrogen into nitrates by the intermediate formation of ammonia and nitrites
2. Artificial processes:
   (*a*) The electric arc at 2800–3300°C
   (*b*) Furnace process (Hausser process)
   (*c*) Direct catalytic union of nitrogen and hydrogen to form ammonia at 20–90 MPa pressure at 600°C
   (*d*) Formation of metallic nitrides: $Al_2O_3 + 3C + N_2 = 2AlN + 3CO$ at 1750°C
   (*e*) Formation of calcium cyanamide from calcium carbide and nitrogen at 1100°C

n. hydrides Hydronitrogens. A compound of n. and hydrogen in which the H is generally replaceable by a hydrocarbon radical. For example, saturated n.h.: N_2H_4 (hydrazine); unsaturated n.h.: N_3H_5 (triazene), HN_3 (hydrogenazide). **n. iodide** N_3I = 168.9. Iodine azide. Red explosive compound, soluble in water. **n. lag** The time between protein intake and appearance of an equivalent amount of nitrogen compounds in the urine. **n. molecule** Dinitrogen*. Normal N_2 is a mixture of 1 pt. ortho-N (or β-N) with parallel atomic nuclear spins and 2 pts. para-N (or α-N) with antiparallel nuclear spins. **n. monoxide** Dinitrogen oxide*. **n. oxide*** NO = 30.01. Nitric oxide. Colorless gas, $d_{air=1}$1.0366, b.−153, soluble in water. Formed in an electric arc from air; oxidizes readily to nitrogen dioxide. **n. oxychloride** Nitrosyl chloride*. **n. partition** The distribution of the total nitrogen in urea, ammonia, uric acid, etc. **n. pentaoxide*** N_2O_5 = 108.0. Nitric acid anhydride. Colorless crystals, b.46 (decomp.), soluble in water (nitric acid). **n. peroxide** N. dioxide*. **n. sulfides** (1) N_4S_4 = 184.3. Orange crystals, m.179, decomp. 185 or by water, sublimes 135. (2) N_2S_5 = 188.3. Dinitrogen pentasulfide*. Red liquid, m.10–11, decomp. by heat. **n. sulfochloride** N_3S_4Cl. Thiazyl chloride. **n. tetraoxide** N_2O_4 = 92.0. Dinitrogen tetraoxide*. Colorless crystals, m.−9. **n. tribromide*** NBr_3 = 253.7. Red, volatile, explosive oil. **n. trichloride*** NCl_3 = 120.4. Agene. Dark, liquid, explodes 95. **n. trifluoride** NF_3 = 71.01. A liquid, m.−200. **n. trioxide** N_2O_3 = 76.0. Dinitrogen trioxide*. Green liquid, b.3.5, soluble in water.

nitrogenated Containing nitrogen. **n. oil** An oil containing N compounds, e.g., oil of bitter almonds.

nitrogenium Latin for "nitrogen."

nitrogenization The operation of (1) combining with nitrogen; or (2) impregnation with nitrogenous compounds.

nitrogenous Describing a compound containing N; or N compounds. **n. tankage** Process tankage. A fertilizer made by digesting n. waste (wood, felt, leather, hair, feathers) with steam under pressure, sometimes with sulfuric acid, and then drying and grinding.

nitroglycerin $C_3H_5(NO_3)_3$ = 227.1. Glycerol trinitrate*, glyceryl nitrate, nitroglycerol, trinitrin, nitric ester of glycerin, glonoin, trinitroglycerin, propenyl trinitrate, nitroleum, blasting oil, piroglycerina (early name). Yellow oil, d.1.60, m.13, explodes 260, insoluble in water. Used in the preparation of explosives; and as a vasodilator in angina (USP, BP).

nitroglycerol Nitroglycerin.

nitroglycol Ethylene dinitrate.

nitro group* Nitryl. The $-NO_2$ group.

nitroguanidine* $NO_2 \cdot NH \cdot CNH \cdot NH_2$ = 104.1. Colorless needles, m.230 (decomp.), slightly soluble in water.

nitrohydrochloric acid Aqua regia.

nitroic acids Compounds containing the radical $-NO(OH)_2$.

nitroleum Nitroglycerin.

nitrolevulose Fructose nitrate, dextrose nitrate. Yellow crystals, explode when heated.

nitrolic acid A compound of the type $R \cdot C(NOH)(NO_2)$. The solutions are deep red.

nitrolim(e) Calcium cyanamide.

nitrols Compounds of the type: (1) trinitrates of 1,2,3-propanetriol. (2) $R \cdot C(NO)NO_2$ or $R \cdot CH(NO)_2$. The solutions are deep blue.

nitromagnesite $Mg(NO_3)_2 \cdot nH_2O$. A native nitrate.

nitromannite Mannitol hexanitrate.

nitrometal An addition compound (nitro) of nitrogen oxide and metallic oxides, e.g., Cu_2NO_2. Nitrometals form nitrites and nitric oxide with water.

nitrometer A glass apparatus for measuring gases evolved by a chemical reaction; essentially a eudiometer with a 2-way stopcock connecting the air or a funnel to a buret in which the gas is liberated.

nitromethane* CH_3NO_2 = 61.0. Nitrocarbol. Colorless liquid, d.1.14, b.101, slightly soluble in water.

Nitromonas A soil bacterium that converts ammonium salts into nitrites and nitrates, but will not grow in organic media.

nitromuriatic acid Aqua regia.

nitron (1) $N{:}C(NPh)_2 \cdot CH \cdot NPh$ = 312.4. 1,4-Diphenyl-3,5-phenylimino-1,2,4-triazole. Yellow needles, m.190, insoluble in water; reagent for nitrates. (2) Trona; early (Greek) name. Cf. *nitrone.* **n. nitrate** $C_{20}H_{16}N_4 \cdot HNO_3$ = 375.3. An insoluble precipitate formed by nitrates and nitron.

nitronaphthalene* $C_{10}H_7NO_2$ = 173.2. **1-~** Yellow crystals, m.61, insoluble in water. Used in organic synthesis and in the removal of fluorescence from oils. **2-~** Colorless, needles, m.79, insoluble in water; used in organic synthesis.

nitronaphthoic acid $C_{10}H_6(NO_2)COOH$ = 217.2. Colorless prisms, slightly soluble in water. **1,2-~** 1-Nitro-2-naphthoic acid. m.240. **3,1-~** m.270. **4,1-~** m.222. **5,1-~** m.242. **5,2-~** m.295. **8,1-~** m.98. **8,2-~** m.300.

nitronaphthol* $C_{10}H_6(NO_2)OH$ = 189.2. **1,2-~** 1-Nitro-2-naphthol. Yellow crystals, m.103. **2,1-~** Colorless leaflets, m.128. **3,1-~** m.168. **4,1-~** Yellow needles, m.164. **5,1-~** Yellow needles, m.171. **5,2-~** Yellow needles, m.147. **6,2-~** m.158. **8,2-~** Yellow needles, m.144. Used in organic synthesis.

nitronaphthylamine* $C_{10}H_6(NO_2)NH_2$ = 188.2. **1,2-~** 1-Nitro-2-naphthylamine. Orange needles, m.126. **1,4-~** m.191. **1,5-~** m.118. **1,8-~** m.97. **2,1-~** Yellow prisms, m.144. **5,2-~** Red needles, m.142. **8,2-~** Red needles, m.103. Used in organic synthesis.

nitrone* Compounds containing the radical =C:NO−. Cf. *nitron.*

nitronic acid A group of compounds containing the radical = N(:O)·OR.

nitronium ion The nitryl* cation, NO_2^+.

nitropentaerythrol $C_5H_{11}O_4 \cdot NO_3$ = 197.1. An explosive.

nitrophenetole* $C_2H_5O \cdot C_6H_4NO_2$ = 167.2. ***ortho-~*** Yellow oil, m.2, b.267. ***meta-~*** Yellow crystals, m.34. ***para-~*** Yellow crystals, m.58.

nitrophenol* $C_6H_4(NO_2)OH$ = 139.1. ***ortho-~*** Yellow prisms, m.45, soluble in water. ***meta-~*** Yellow crystals, m.96, slightly soluble in water. ***para-~*** Colorless, monoclinic crystals, m.114, slightly soluble in water. Used to manufacture rhodamine dyestuffs, phenolphthalein, etc. *p*-Nitrophenol is an indicator for alkalimetry, and for blood and seawater; changing at pH 6 from colorless (acid) to yellow (alkaline). **chloro~*** $Cl \cdot C_6H_3(OH)NO_2$ = 173.6. In the following, the first number indicates Cl, the second NO_2; OH is the 1 position: **2,3-~** m.120. **2,4-~** m.111. **3,4-~** m.133. **4,2-~** m.87. **4,3-~** m.127. **5,2-~** m.39. **5,3-~** m.147. **6,2-~** m.70. **6,3-~** m.118.

nitrophenols Compounds derived from phenol by the substitution of one or more nuclear H atoms by the nitro group.

Nitrophoska Trademark for a fused, ternary mixture of KNO_3, NH_4Cl, and $(NH_4)_3PO_4$; a fertilizer.

nitrophthalic acid* $C_6H_3(NO_2)(COOH)_2$ = 211.1. **3-~** Yellow, monoclinic crystals, m.220, slightly soluble in water. **4-~** Colorless needles, m.161, soluble in water.

nitrophthalide $C_6H_4 \cdot CO \cdot O \cdot CHNO_2$ = 179.1. Colorless needles, m.141, insoluble in water.

nitropropane* $C_3H_7NO_2$ = 89.1. Colorless liquid, d.1.01, b.131, slightly soluble in water.

nitroprussiate Nitroprusside.

nitroprusside Containing the pentacyanonitrosylferrate(III)* radical, $[Fe(CN)_5NO]^{2-}$.

nitroquinoline* $C_9H_6N \cdot NO_2$ = 174.2. Colorless needles. **5-~** m.72, soluble in water. **6-~** m.150, soluble in water. **7-~** m.132. **8-~** m.88, insoluble in water; a reagent for palladium.

nitrosalicylic acid* $C_6H_3(NO_2)(OH)COOH$ = 183.1. **3,2,1-~ (or 3-~)** White needles, m.144, slightly soluble in water. **5,2,1-~ (or 5-~)** Colorless needles, m.230, slightly soluble in water. Used in organic synthesis.

nitrosalol $C_6H_4(OH)COOC_6H_4NO_2$ = 259.2. Yellow powder, m.148, insoluble in water.

nitrosamines Nitrosoamines (2).

nitrosates Compounds containing the $=C(ONO_2)\cdot C(:NOH)-$ group.

nitrose A solution of nitrosyl hydrogensulfate in sulfuric acid, made by passing NO_2 and NO into sulfuric acid.

nitrosites Compounds containing the $=C(ONO)\cdot C(:NOH)-$ group.

nitroso* Oximido, hydroximino. The radical −NO in connection with organic compounds. Cf. *nitrosyl.* **iso~** Oxime*. **oxy~** Nitro*.

n.-R salt $C_{10}H_4OH \cdot NO(SO_3Na)_2$; gives a blue color with cobalt and iron only.

nitrosoacetophenone **iso~** Benzoylformoxime.

nitrosoamines (1) Compounds containing both the radicals $-NH_2$ and −NO; as nitrosoaniline. (2)Yellow organic compounds containing the radical =N·NO, obtained by the action of nitrous acid on secondary amines. E.g. **diethyl~** $Et_2N \cdot NO$ = 102.1. Nitrosodiethyline, diethylnitrosamine. Yellow oil, d.0.951, b.177, soluble in water. **dimethyl~** $Me_2N \cdot NO$ = 74.1. Nitrosodimethyline, dimethylnitrosamine. Yellow oil, b.148, soluble in water. **diphenyl~** $Ph_2N \cdot NO$. **phenylethyl~** PhEtN·NO. **phenylmethyl~** PhMeN·NO.

nitrosoaniline $C_6H_4(NO)NH_2$ = 122.1. ***para-~*** Steel-blue needles, m.174, insoluble in water.

Nitrosobacter A rodlike nitrifying bacterium.

nitrosobacteria Nitrifying bacteria. See *Nitromonas.*

nitrosobenzene* C_6H_5NO = 107.1. Blue, monoclinic needles, m.68.

nitrosobenzoic acid* $C_6H_4(NO)COOH$ = 151.1. ***ortho-~*** Colorless crystals, decomp. 210, soluble in water.

Nitrosococcus A round, nitrifying soil bacterium; converts ammonia to nitrites. Cf. *Nitrosomonas.*

nitroso dyes Dyestuffs derived from benzoquinone monoxime, which contains the chromophore

```
=C—O·.
 |      ·H
=C=N·O/
```

nitrosoethane* C_2H_5NO = 59.1.

Nitrosomonas A soil bacterium of Europe, Asia, and Africa; converts ammonia into nitrites. Cf. *Nitrobacter, Nitrosococcus.*

nitrosonaphthol $C_{10}H_6(NO)OH$ = 173.2. **1,2-~** Yellow prisms, m.110, soluble in water. **2,1-~** Yellow needles, m.162, soluble in water; used in organic synthesis, and as reagent for cobalt. **4,1-~** Yellow crystals, m.194, insoluble in water.

nitrosonaphthylamine* $C_{10}H_6(NO)NH_2$ = 172.2. **1,2-~** Green needles, m.150, soluble in hot water.

nitrosonitric acid Fuming nitric acid.

nitrosooxy* The radical −O·N:O in an organic compound.

nitrosonium The nitrosyl* cation.

nitrosophenol* $OH \cdot C_6H_4 \cdot NO$ = 123.1. A solid, m.140 (decomp.), soluble in water. Cf. *benzoquinone monoxime,* its tautomeric form.

nitrososulfuric acid Nitrosyl hydrogensulfate*.

nitrosotoluene* $C_6H_4(NO)CH_3$ = 121.1. ***ortho-~*** Yellow needles, decomp. 120, soluble in water.

nitrostarch $[C_{12}H_{12}O_{10}(NO_2)_3]_n$. Starch nitrate. Orange powder, soluble in alcohol or ether. A nitrated starch used in the manufacture of explosives.

nitrostyrene $NO_2C_6H_4 \cdot CH:CH_2$ = 149.2. Nitrostyrolene.

***ortho-*~** Colorless oil, m.12 (decomp.), insoluble in water. ***meta-*~** Colorless liquid, b.235, soluble in ether. ***para-*~** Colorless prisms, m.58, soluble in hot ether.
nitrostyrolene Nitrostyrene.
nitrosubstitution The result of nitration.
nitrosulfamide $H_2N \cdot SO_2:NHNO$ = 93.0. Colorless liquid; the silver salt is a detonator.
nitrosulfonic acid Nitrosyl hydrogensulfate.
nitrosulfuric acid Nitric acid 1, sulfuric acid 2 vol.; a nitrating agent. Cf. *nitrosyl hydrogensulfate.*
nitrosyl* The radical ON— in connection with inorganic compounds, including the cation NO^+ and the ligand. Cf. *nitroso.* **n. bromide*** NOBr = 109.9. Brown liquid, decomp. −2. **n. chloride*** NOCl = 65.5. Nitrogen oxychloride. Yellow or red gas, b.−5, decomp. by water. A constituent of aqua regia. **n. fluoride*** NOF = 49.00. A gas, b.−56. **n. perchlorate*** $NOClO_4$ = 129.5. An unstable liquid. **n. hydrogensulfate*** $NO \cdot HSO_4$ = 127.1. Nitrososulfuric acid, chamber crystals, nitrosulfonic acid, n.sulfuric acid, nitrose. Colorless crystals, m.75 (decomp.); formed in the lead chambers during the manufacture of sulfuric acid. **n.sulfuric acid** N. hydrogensulfate*. **n. sulfuryl chloride** $ClSO_2NO$ = 177.5. Crystals, produced from sulfur trioxide and nitrosyl chloride.
nitrothiophene* $NO_2 \cdot C_4H_3S$ = 129.1. Colorless crystals, m.44, insoluble in water.
nitrotoluene* $Me \cdot C_6H_4 \cdot NO_2$ = 137.1. ***ortho-*~** Yellow liquid, d.1.168, b.218, insoluble in water. ***meta-*~** Yellow crystals, m.16, insoluble in water. ***para-*~** Yellow crystals, m.54, insoluble in water. All are used in organic synthesis. **di~** See *dinitrotoluene* under *dinitro.* **tri~** See *trinitrotoluene.*
nitrotoluidine* $C_6H_3(NH_2)CH_3(NO_2)$ = 152.2. A series of aminomethylnitrobenzenes, used in the synthesis of dyestuffs. They are usually slightly soluble in water; soluble in alcohol, ether, or benzene. Isomers are divided according to the positions of the 3 groups in the benzene ring (see Table 55).
nitrourea $NH_2 \cdot CO \cdot NHNO_2$ = 105.1. White crystals, decomp. by heat, slightly soluble in water.
nitrourethane $NO_2 \cdot NH \cdot COO \cdot C_2H_5$ = 134.1. Colorless leaflets, m.64, b.140 (decomp.), soluble in water.
nitrous **n. acid*** HNO_2 = 47.01. An aqueous solution of nitrogen trioxide, N_2O_3. **n. anhydride** Nitrogen trioxide. **n. ether** $C_2H_5NO_2$ = 75.1. Nitroethane*. **n. oxide** Dinitrogen oxide*. **n. vitriol** A solution of oxides of nitrogen and sometimes of sulfur dioxide in strong sulfuric acid formed during the manufacture of *sulfuric acid,* q.v.

TABLE 55. ISOMERS OF NITROTOLUIDINE

NH_2	CH_3	NO_2	Description
			ortho-Nitrotoluidine
2	1		Orange prisms, m.92
2	1	4	Yellow, monoclinic crystals, m.104
2	1	5	Yellow needles, m.127
2	1	6	Yellow leaflets, m.92
			meta-Nitrotoluidine
3	1	2	Yellow needles, m.53
3	1	4	Yellow leaflets, m.109
3	1	5	Orange needles, m.98
3	1	6	Yellow needles, m.138
			para-Nitrotoluidine
4	1	2	Yellow, monoclinic crystals, m.77
4	1	3	Red prisms, m.114

nitroxanthic acid Picric acid*.
nitroxyl The nitryl* radical.
nitroxylene $Me_2 \cdot C_6H_3 \cdot NO_2$ = 151.2. **1,2,3-~** 2,3-Dimethyl-1-nitrobenzene*. Yellow liquid, m.15, b.245. **1,2,4-~** Yellow crystals, m.29. **1,3,2-~** Yellow liquid, b.225. **1,3,4-~** A liquid, b.238. **1,3,5-~** m.74. All are insoluble in water and used in organic synthesis.
nitroxylic acid* H_2NO_2 = 48.02. Hydronitrous acid.
nitrum An early (Latin) name for trona.
nitryl (1) Nitroxyl. The inorganic radical $-NO_2$, from nitrous acid, HNO_2, when attached to a strongly electronegative group. (2) Vinylidene dinitrile. See *plastics.* **n. chloride*** NO_2Cl = 81.5. Yellow liquid. **n. fluoride*** NO_2F = 65.0. Colorless gas, b.−64, decomp. by water.
nivalic acid $C_{20}H_{26}O_6$ = 362.4. An acid from the lichen *Cetraria nivalis.*
nivenite Uranite with a large proportion of lanthanoids.
Nixie tube Trade name for a digital readout tube.
nm Abbreviation for nanometer.
n.m.r Symbol for nuclear magnetic resonance.
No Symbol for nobelium. **No. 606** Arsphenamine. **No. 914** Neoarsphenamine.
Nobel **N., Alfred Bernhard (1833–1896)** Swedish chemist, the inventor of several explosives and artificial rubber; founder of the Nobel prize. **N.'s explosive** A high explosive: nitroglycerin 25–80, guncotton 0.5–7, liquid nitro-body 0.4–9, wood meal 0.9–10, potassium or sodium nitrate 6–45%. **N. oil** Nitroglycerin. **N. prize** An annual award given to those who have made outstanding contributions to physics, chemistry, medicine, economics, literature, and peace.
nobelium* No. Element of atomic number 102. The at. wt. of the isotope with longest half-life (^{259}No) = 259.10. Discovered (1957) by joint research, in the U.S., U.K., and Sweden, on exposing ^{244}Cm on thin foil to cyclotron bombardment with accelerated carbon ions. Forms No^{2+} ions.
Noble, Alfred (1844–1914) American civil engineer who founded the Noble prize for the best technical engineering paper. Cf. *Nobel prize.*
noble **n. gas*** Inert gas, rare gas. A member of the zero group of the periodic table: He, Ne, Ar, Kr, Xe, and Rn. Only Kr, Xe, and Rn form compounds, and then only with Cl, F, N, and O atoms. Cf. *atmospheric gases.* **n. laurel** Laurel. **n. liverwort** Liverwort. **n. metal** A metal that is not readily oxidized; as, the gold, platinum, and palladium family of the periodic table.
nodakenin $C_{20}H_{24}O_9$ = 408.4. A glucoside from nodake, *Pellecedanum decursivium* (Umbelliferae), Japan.
node (1) The point at which a curve or wave motion intersects a fixed plane. (2) A knot or protuberance.
nodule A small, round lump, as, of a mineral.
noil Short-staple, wool-fiber combings used in worsteds.
noise (1) Sound which is unwanted because of its irritating effect on humans. See *sound.* **perceived ~** A measure of aircraft n. (in PNdB) that takes into account the most annoying frequencies. (2) Unwanted background signals that interfere with the wanted signal.
n. ratio See *signal-to-noise ratio.*
Nolvadex Trademark for tamoxifen citrate.
nomenclature The terminology of chemical compounds. Rules of chemical n. are formulated by the International Union of Pure and Applied Chemistry (IUPAC), sometimes in collaboration with the international unions of biochemistry (IUB) and physics (IUPAP). The two principal IUPAC source information books are *Nomenclature of Inorganic Chemistry,*

Butterworth, London, 1970, and *Nomenclature of Organic Chemistry,* Pergamon Press, Oxford, 1979. Subsequent rules are published in the journal "Pure and Applied Chemistry." The other system that is widely accepted is that of the American Chemical Society's "Chemical Abstracts" (CA). Generally, it agrees with IUPAC n., except where its computerized information-retrieval system requires a more systematic approach. IUPAC rules are more flexible in that they use more trivial and semisystematic names. CA n. is explained in the society's current "Index Guide," which also assists in converting from IUPAC to CA n. In this dictionary, IUPAC names are indicated by * and CA names (when different from or less common forms of IUPAC names) by †. See Table 56; for more details (e.g., exceptions, seniority, numbering, etc.), see the above references; also *acids, name, radical, stereoisomer, symbol.*

Nomex Trademark for a high-temperature-resistant aramid.

nomogram Nomograph.

nomograph Nomogram. Alignment chart. Graphical scales, used for the rapid calculation and solution of complicated equations.

nomography The representation of analytical or other correlations by means of charts and graphs, which eliminate calculations.

non- Prefix (Latin) indicating "not."

nona-* Prefix for nine. Cf. *ennea-.*

TABLE 56. BASICS OF IUPAC NOMENCLATURE

GENERAL

Multiplying Affixes

The three following series of multiplying affixes are used. They indicate the number of:

1. Identical unsubstituted radicals or parent compounds in a set (in organic chemistry). Greek: di, tri, tetra, penta, etc. (See *alkanes* below.) This series, together with hemi ($\frac{1}{2}$), mono (1), sesqui ($1\frac{1}{2}$), is also put to broad use in inorganic chemistry; as, 1,2-ethanediol $(CH_2OH)_2$; dinitrogen tetraoxide, N_2O_4.
2. Identical radicals or parent compounds in a set, each substituted in the same way: bis (2), tris, tetrakis, pentakis, etc. "Chemical Abstracts" (CA) uses this series for complex expressions; as, bis(2-chloroethyl)amine $(CH_2Cl \cdot CH_2)_2NH$; pentacalcium fluoride tris(phosphate), $Ca_5F(PO_4)_3$.
3. Identical rings joined to one another by a link (single or double). Latin: bi, ter, quater, quinque, sexi, septi, octi, novi, deci; as, biphenyl, $C_6H_5 \cdot C_6H_5$.

Elements

Where an approved Latin name exists, it should be used when forming names derived from that element; as, ferrum, wolfram. (CA uses iron, tungsten).

Elements above Atomic No. 103

Named by stringing together the following names and adding the suffix -ium. Thus 104 is Un-nil-quadium.

0	nil	4	quad	7	sept
1	un	5	pent	8	oct
2	bi	6	hex	9	non
3	tri				

INORGANIC

General

1. In binary compounds, that constituent is placed first which appears earlier in the sequence: Rn, Xe, Kr, B, Si, C, Sb, As, P, N, H, Te, Se, S, At, I, Br, Cl, O, F. Thus, H_2S, Cl_2O, OF_2.
   For other elements, the sequence is obtained by tracing a path through the *periodic* table, q.v., namely: F–Cl . . . At–O . . . Po–N . . . Tl–Zn . . . La–Lu–Ac–Lr–Be . . . Ra–Li . . . Fr–H . . . Rn. The element occurring last in this sequence is placed first; as, aluminum boride.
   If just one element occurs in the first sequence (Rn . . . F), it is given the ending -ide; as, nickel arsenide.
2. Where there is more than one electronegative or electropositive constituent, they are cited in the unprefixed alphabetical order; as, dialuminum tetracalcium heptaoxide, $Al_2Ca_4O_7$.

Electropositive Constituents

1. They are not modified, other than by multiplying and ligand prefixes; as, triiron tetraoxide, Fe_3O_4.
2. The suffixes -ous and -ic to indicate valency may be retained for elements exhibiting not more than two valencies, but their use is discouraged. (Such entries are therefore not marked * in this dictionary.)
3. In coordinated cations, the element is prefixed by the ligand names; as, pentaamminechlorocobalt(III) chloride, $[CoCl(NH_3)_5]Cl_2$.
4. Certain special names are allowed for polyatomic cationic and neutral radicals; as, chromyl, CrO_2; nitrosyl, NO. See *radicals,* Table 70 on p. 494.
5. Protonated cations have the ending -onium on the root of the name of the anion; as, phosphonium, PH_4^+.

Electronegative Constituents

1. If a constituent is monoatomic or homopolyatomic, its name is modified to end in -ide; as, hydrogen chloride; sodium plumbide.
2. Volatile hydrides, except those of Group 7, N and O, may be named from the root name of the element plus -ane; as, diarsane, As_2H_4. Allowed exceptions are: arsine, ammonia, bismuthine, hydrazine, phosphine, stibine, water.
3. There are some exceptional anions that are allowed to end in -ide or -ite; as, hydroxide, OH^-; arsenite, AsO_3^{3-} See *radicals,* Table 70.
4. Other electronegative constituents end in -ate. This includes polyatomic atoms, whose name is made from the root of the central atom; as, hexahydroxoantimonate(V) ion, $[Sb(OH)_6]^-$.
5. Salts. See *acids,* Table 3. Unreplaced "acid" H atoms are written out and joined to the name of the anion; as, sodium hydrogencarbonate. ("Chemical Abstracts" leaves a space between the words.)

Proportions

The proportion of constituents may be denoted by:

1. Multiplying affixes (see above); as, tricalcium bis(orthophosphate), $Ca_3(PO_4)_2$.
2. Stock system, where the oxidation number of the element is placed in roman numerals after the element; as, dilead(II)lead(IV) oxide, $Pb_2^{II}Pb^{IV}O_4$; potassium hexacyanoferrate(II), $K_4[Fe(CN)_6]$.
3. Ewens-Bassett system, where the ionic charge is placed in arabic numerals after the name of the ion; as, uranyl(2+) sulfate, UO_2SO_4; potassium tetracyanoniccolate(4−), $K_4[Ni(CN)_4]$.

TABLE 56. BASICS OF IUPAC NOMENCLATURE (*Continued*)

ORGANIC

Alkanes, Alcohols, Aldehydes, Acids, etc.
After trivial names for the first four, alkanes generally follow the Greek multiplying affixes plus the suffix -ane (n = total number of C atoms).

n		n		n	
1	Methane	15	Pentadecane	29	Nonacosane
2	Ethane	16	Hexadecane	30	Triacontane
3	Propane	17	Heptadecane	31	Hentriacontane
4	Butane	18	Octadecane	32	Dotriacontane
5	Pentane	19	Nonadecane	33	Tritriacontane
6	Hexane	20	Icosane	40	Tetracontane
7	Heptane	21	Henicosane	50	Pentacontane
8	Octane	22	Docosane	60	Hexacontane
9	Nonane	23	Tricosane	70	Heptacontane
10	Decane	24	Tetracosane	80	Octacontane
11	Undecane	25	Pentacosane	90	Nonacontane
12	Dodecane	26	Hexacosane	100	Hectane
13	Tridecane	27	Heptacosane	132	Dotriacontahectane
14	Tetradecane	28	Octacosane		

See the appropriate entry in text for other classes of compounds.

Substitutive Nomenclature
Names involving replacement of hydrogen by a group or by another element; as, 1-methylnaphthalene; 1-pentanol. Some characteristic *groups*, q.v., are cited only as prefixes to the name of the *parent compound*, q.v. They are those marked [a] in the *radical*s table, plus: IX_2 (where X is a halogen, OH, or other radical); *aci*-nitro; R-oxy (−OR); −SR, −SeR, and −TeR. When there is more than one group not listed above, the principal *group*, q.v. (cited as suffix to the parent compound) is that highest (i.e., having highest "seniority") in the following list. All other groups are cited as prefixes (in alphabetical order); as, 3-chloro-6-hydroxy-3-hexen-2-one, $HO \cdot (CH_2)_2 \cdot CH{:}CCl \cdot CO \cdot CH_3$.

1. Anionic centers in the order of corresponding acids.
2. "Onium" and similar cations.
3. Acids: in the order COOH, CO·OOH, then successively their S and Se derivatives, followed by sulfonic, sulfinic acids . . . selenonic . . . phosphonic, etc.
4. Acid derivatives of inorganic acids in which the central atom of the acid residue is linked through one or more heteroatoms to the carbon skeleton of the inorganic part of the molecule.
5. Derivatives of acids belonging to class 3 above, in the order: anhydrides, esters, acyl halides, amides, hydrazides, imides, amidines, etc.
6. Nitriles (cyanides), then isocyanides.
7. Aldehydes; then successively their S and Se analogs; then their derivatives.
8. Ketones; then their analogs and derivatives in the same order as for aldehydes.
9. Alcohols and phenols; then their S, Se, and Te analogs; then neutral esters of alcohols and phenols with inorganic acids except hydrogen halides, in the same order.
10. Hydroperoxides.
11. Amines, imines, hydrazines, etc.; then phosphines, arsines, etc.
12. Other derivatives of inorganic acids which cannot be considered as class 4 or 9 above.
13. Ethers; then, successively, their S and Se analogs.
14. Peroxides.

Radicofunctional Nomenclature
Names formed from the name of a radical and the name of a functional class; as, acetyl chloride; ethyl methyl ketone. (Substitutive names are preferred to these.) Rules are identical to substitutive nomenclature, except that suffixes are never used. Instead of the principal group being named as suffix, the "functional class name" of the compound is expressed as one word and the remainder of the molecule as another. Some functional class names used in r. n., in order of decreasing priority for choice as such:

Group	Functional class name
X in acid derivatives RCO−X, RSO_2−X, etc.	Name of X; in the order fluoride, chloride, bromide, iodide; cyanide, azide, etc.; then their S, followed by their Se, analogs
−CN, −NC	Cyanide, isocyanide
>CO	Ketone, then S, then Se analogs
−OH	Alcohol; followed by S and then Se analogs
−O−OH	Hydroperoxide
>O	Ether or oxide
>S, >SO, >SO_2	Sulfide, sulfoxide, sulfone
>Se, >SeO, >SeO_2	Selenide, selenoxide, selenone
−F, −Cl, −Br, −I	Fluoride, chloride, bromide, iodide
−N_3	Azide

Additive Nomenclature
Names signifying addition between molecules, atoms, or both; styrene oxide. "Hydro-," to denote added H atoms, is the only case where addition is expressed by a prefix; as, 1,2,3,4-tetrahydronaphthalene. Otherwise, the name of the added atom is placed, in its ion form, after the name of the parent compound. (Generally discouraged, except for a few special cases.) See also *Conjunctive N.* below.

Subtractive Nomenclature
Names involving removal of specified atoms; e.g., -ene, -yne, anhydro-, dehydro-, deoxy-; e.g., 6-deoxy-α-D-glucopyranose; 2,3-anhydro-D-gulonic acid.

Conjunctive Nomenclature
Names formed by placing together the names of two molecules, it being understood that the molecules are linked by loss of one hydrogen atom from each; as, naphthaleneacetic acid. (Especially used in "Chemical Abstracts," where it was formerly termed "additive nomenclature.") May be applied when a principal group is attached to an acyclic component that is directly attached by a C−C bond to a cyclic component.

Replacement Nomenclature
Names where C, CH, or CH_2 are replaced by a hetero atom; as, 2,7,9-triazaphenanthrene; silabenzene. Also certain names involving thio-, seleno-, telluro-, to indicate replacement of O by S, Se, or Te. R. n. is an extension of "a" nomenclature. This is based on the Hantzsch-Widman system, in which monocyclic compounds containing one or more hetero atoms is a 3- to 10-membered ring are named by combining the appropriate prefix(es) to indicate the hetero atom(s), with a suffix (as, -ole, as in triazole) indicating the number of ring members and saturation. All the prefixes end in "a"; as, O, oxa; S, thia; N, aza; Si, sila; Se, selena. Applied also to acyclic structures, but intended for use only when other n. systems are difficult to apply.

SOURCES: See *nomenclature*.

nonacosane* $C_{29}H_{60}$ = 408.8. A hydrocarbon in beeswax and *brassica.* Colorless wax, m.63.
nonacosanol* $C_{29}H_{59}OH$ = 424.8. **1-~** Montanyl alcohol. White crystals, m.85, from beeswax and *Parosela barbata* (Leguminosae). **9-~** White solid, m.75. **10-~** m.75. **12-~** m.74. **14-~** m.79. **15-~** m.84.
nonacyclic (1) Having 9 rings; as violanthrone. (2) Not acyclic.
nonadecane* $C_{19}H_{40}$ = 268.5. Enneadecane, enndecane. A solid, m.32.
nonadecanoic acid* $C_{18}H_{37}COOH$ = 298.5. Nonadecylic acid. White leaflets, m.67.
nonadecanol* $C_{19}H_{39}OH$ = 284.5. Nonadecyl alcohol. Opaque crystals, m.62.
nonadecanone* $(C_9H_{19})_2CO$ = 282.5. **10-~** Caprinone, dinonyl ketone. Colorless leaflets, m.58, insoluble in water.
nonadecyl alcohol Nonadecanol*.
nonadecylic acid Nonadecanoic acid*.
nonaldehyde Nonanal*.
nonanal* $C_8H_{17}CHO$ = 142.1. Pelargonaldehyde. b.191. In citron oil; used in perfumes.
nonane* C_9H_{20} = 128.3. Colorless liquid, d.0.718, b.149, insoluble in water. **n.carboxylic acid** Decanoic acid*. **n.dioic acid*** Azelaic acid.
nonanediol* $CH_2OH \cdot (CH_2)_7CH_2OH$ = 160.3. Nonamethylene glycol. Colorless liquid, b.149, slightly soluble in water.
nonanenitrile* $Me(CH_2)_7CN$ = 139.2. Pelargononitrile. Colorless liquid, d.0.8331, b.224, insoluble in water.
nonanoate* A salt or ester of nonanoic acid, containing the radical $C_8H_{17}COO-$.
nonanoic acid* $C_8H_{17}COOH$ = 158.2. Pelargonic acid, octanecarboxylic acid. An oxidation product of oleic acid, and a constituent of oil of *Pelargonium roseum.* Colorless leaflets, m.12, soluble in water; a flavoring.
nonanol* $C_9H_{20}O$ = 144.13. **1-~** *n*-Nonyl alcohol. **2-~** $Me(CH_2)_6CHOHMe$. Heptylmethylcarbinol. Colorless liquid, b.193, insoluble in water. **3-~** $Me(CH_2)_5CHOHEt$. Ethylhexylcarbinol. Colorless liquid, b.194. **4-~** $Me(CH_2)_4CHOHPr$. Amylpropylcarbinol. Colorless liquid, b.192. **5-~** $[Me(CH_2)_3]_2CHOH$. Dibutylcarbinol. Oily liquid, b.194.
nonanone* $C_9H_{18}O$ = 142.14. **2-~** $Me \cdot CO(CH_2)_6Me$. Heptyl methyl ketone*. A constituent of rue and clove oils. Colorless liquid, b.195. **3-~** $Et \cdot CO(CH_2)_5Me$. Ethyl hexyl ketone*, b.190. **5-~** $[Me(CH_2)_3]_2CO$. Dibutyl ketone*. Colorless liquid, b.187.
nonanoyl* Pelargonyl. The radical $Me(CH_2)_7CO-$, from nonanoic acid. **n. chloride*** $C_9H_{17}OCl$ = 176.7. Pelargonyl chloride. Colorless liquid, b.215, decomp. by water.
nonconductor A substance that does not transmit electricity, heat, or light. Cf. *insulator.*
nondecylic acid Nonadecanoic acid*.
nondrying oil A liquid fat that remains fluid on exposure; contains olein and glycerides of unsaturated fatty acids; e.g., castor oil.
nonene* C_9H_{18} = 126.2. Nonylene. **1-~** b.147; from tars, plants, fruit, animals. **2-~** b.144. **3-~** b.125.
nonferrous Other than iron.
nonine Nonyne*.
nonmetallic Describing an electronegative element in the upper right half of the periodic table. Nonmetals are generally polyvalent (except O and H), and exist in several stages of oxidation. Their oxides form acids. Cf. *metallic, semimetallic.*
nonoic acid Nonanoic acid*.
Nonox Trademark for aldolnaphthylamine derivatives; rubber antioxidants.
nonvalent Inert, having zero valency; as, neon.
nonwoven Describing any manufactured sheet, web or batt, of directional or random fibers, held together through mechanical, chemical, or physical methods, or any combination of these; but excluding weaving, knitting, stitch-bonding, traditional felting, as well as conventionally formed paper.
nonyl* The radical $-C_9H_{19}$, from nonane. **n. alcohol*** $C_9H_{19}OH$ = 144.3. Colorless liquid, b.213, insoluble in water. For other isomers see *nonanol.* **n.aldehyde** Nonanal*. **n.amine** $C_9H_{19}NH_2$ = 143.3. Colorless liquid, b.195. **n. cyanide** Decanenitrile*.
nonylene Nonene*.
nonylic acid Nonanoic acid*.
nonylone 9-Heptadecanone*.
nonyne* $Me(CH_2)_6C{:}CH$ = 124.2. Heptylacetylene, nonine. Colorless liquid, b.160, insoluble in water.
nootkatin $C_{15}H_{20}O_2$ = 232.3. A cycloheptatrienone, m.95.
nootkatone $C_{15}H_{22}O$ = 218.3. A sesquiterpene ketone flavoring in grapefruit. m.36.
nopinene β-Pinene.
nor- Prefix indicating: (1)* Replacement by H of a Me group attached to a ring system; used particularly for terpenes. (2)* Elimination of one CH_2 group from a chain or ring; used particularly for higher terpenes and steroids. Cf. *dinor, trinor.* (3) Usage to denote replacement by H of all Me groups in a monoterpene ring system has been abolished by IUPAC.
noradrenaline acid tartrate EP, BP name for norepinephrine bitartrate.
noratropine An alkaloid from various Solanaceae species, allied to atropine.
Norbide B_6C. Trademark for boron carbide, formed by heating coke and boric acid in an electric furnace. Second to diamond in hardness; used for grinding, deoxidizing steel; also as an abrasion-resistant.
norbixin See *bixin.*
norcamphane Trinorbornane*. **dimethylmethylene ~** Camphene*. **trimethyl ~** Bornane*.
norcamphanyl The trinorbornyl* radical.
nordenskioldine (Swedish). $CaSnB_2O_6$. A tin ore.
Nordhausen acid Oleum.
nordihydroguairetic acid A natural antioxidant for fats, from *Larrea divaricata,* the creosote bush of Mexico.
nordmarkite A ferromagnesian-feldspathic soda syenitic rock, from Norway.
norephedrene $Ph \cdot CH(OH) \cdot CH(NH_2)Me$ = 151.2. 2-Amino-1-phenyl-1-propanol*. Used to control allergic conditions of the nose, e.g., hay fever.
norepinephrine bitartrate $C_8H_{11}O_3N \cdot C_4H_6O_6 \cdot H_2O$ = 337.3. α-(Aminomethyl)-3,4-dihydroxybenzylalcohol tartrate. Noradrenaline acid tartrate, leventerenol. Levophed. White crystals, m.102, soluble in water. A hormone produced in the adrenal medulla, and also present at the sympathetic nerve endings, where it is a chemical transmitter of impulses. It has a constricting effect on blood vessels and is used to raise blood pressure (USP, EP, BP).
norepol Agripol.
Noreseal Trademark for a whipped-up mixture of glue, glycerol, glucose, peanut hulls, saponin, water, and formaldehyde; a cork substitute.
norethindrone $C_{20}H_{26}O_2$ = 298.4. 17-Hydroxy-19-nor-17α-pregn-4-en-20-yn-3-one. Norethisterone. White crystals, m.203, insoluble in water. A progestational steroid. Used for menstrual abnormalities and in oral contraceptives (USP, BP).

norethisterone BP name for norethindrone.
norethynodrel $C_{19}H_{12}O_2$ = 272.3. 17-Ethynyl-17-hydroxy-5(10)-estren-3-one. Enovid. A steroid, similar to progesterone, used in oral contraceptives.
Norge saltpeter A Norwegian calcium nitrate fertilizer.
norgestrel $C_{21}H_{28}O_2$ = 312.5. DL-norgestrel. White powder, m.210, insoluble in water. A progesterone used in oral contraceptives (USP).
Norgine Algin. Trademark for an adhesive made by boiling seaweed with alkali and precipitating the filtered solution with acid.
noric The division of the Triassic system containing the limestones and dolomites with some clays and sandstones.
Norit Trademark for purified charcoal made from birch; used to decolorize and deodorize syrup, oils, and pharmaceutical products.
norium A supposed rare-earth metal from zircon; actually a mixture of rare-earth metals.
norleucine $C_6H_{13}O_2N$ = 131.2. Caprine, glycoleucine, α-amino-n-caproic acid, α-amino-α-butylacetic acid, 2-aminohexanoic acid*. An amino acid from the leucine fractions of proteins of the brain and of casein.
norm A theoretical standard. **n. system** In mineralogy, a system of classification of rocks based on their theoretical chemical compositions. Cf. *mode.*
normal (1) One plane or line perpendicular to another. (2) Describing a fixed standard or an established type, as a solution. (3) Prefix indicating: (*a*) a salt with all the available H atoms replaced; (*b*) an organic compound in its "normal" (as distinct from its "iso") form. (4)* Describing a material containing as a major constituent a specified element with an atomic weight value that is not significantly different from its accepted value. Differences can arise from the source, artificial alteration, mutation, or rare geological occurrence of the material. (5) Average or mean. **n. distribution** See *normal probability curve* below. **n. element** A cell, used as standard of electromotive force. See *cadmium cell.* **n. glass** A glass whose composition can be represented by a definite chemical formula; as, $6SiO_2 \cdot CaO \cdot Na_2O$. **n. hydrocarbon** An aliphatic hydrocarbon with a straight carbon chain in its molecule and no side chains; as, n-pentane. **n. material** See *normal (4) above.* **n. pressure** Standard pressure. See *s.t.p.* **n. probability curve** N. distribution. Gaussian, error, bell, or frequency curve. Bell-shaped curve showing the probability of a value differing from the mean under normal (i.e., unbiased) conditions. Values of 1, 2, and 3 standard deviations from the mean value encompass 68.3, 95.5, and 99.7% of the results of a large population, respectively. The median, mean, and mode are identical for a symmetrical distribution. See *histogram, leptokurtosis.* **n. saline** See *normal saline* under *saline.* **n. salt** A salt in which all the hydrogen atoms of the acid have been replaced by a metal, or all the hydroxide radicals of a base replaced by an acid radical; e.g., Na_2CO_3 (normal salt), $NaHCO_3$ (acid salt). **n. solution** A solution that contains one equivalent of the active reagent in grams, in one liter of the solution. The equivalent in grams is that quantity of the active reagent which contains, replaces, unites with, or in any way, directly or indirectly, brings into reaction 1 g of H.

$$1\ N\ HCl = 36.5\ g\ HCl/liter = 1\ M\ HCl$$

$$1\ N\ H_2SO_4 = 49\ g\ H_2SO_4/liter = 0.5\ M\ H_2SO_4$$

Weaker or stronger solutions are indicated thus: 0.5 *N*, 2 *N*, etc., the prefix being the normality. Cf. *centinormal, millinormal, supernormal.* **n. state** Ground state. **n. temperature** Room temperature: 20°C. Cf. *standard temperature* (0°C). **n. thermometer** A standardized thermometer.
normality See *normal solution* under *normal.*
normalization (1) The process of restoring normality. (2) The treatment of milk after pasteurization to disperse the fat homogeneously throughout it.
normalizing The treatment of metals (especially aluminum alloys) with a molten mixture of sodium and potassium nitrates at 500°C.
normenthane Isopropylcyclohexane.
nornicotine $C_9H_{12}N_2$ = 148.2. 3-(2-Pyrrolidinyl)pyridine. **(S)-~** In Kentucky tobacco.
noropianic acid $C_8H_6O_5$ = 182.1. 5,6-Dihydroxy-1,2-phthalaldehydic acid. Colorless crystals, m.171, soluble in alcohol.
norpinane Trinorpinane*. **dimethylmethylene ~** β-Pinene*.
norpinic acid $C_8H_{10}O_4$ = 170.2. 2,2-Dimethyl-1,3-cyclobutanedicarboxylic acid*. Cf. *truxillic acid.*
Norris, Flack James (1871–1940) American chemist, noted for organic research.
Norrish, Ronald George Wreyford (1897–1978) British chemist, Nobel prize winner (1967). Noted for work on reaction mechanisms.
norsolanellic acid Biloidanic acid.
Northylen Trademark for a polythene synthetic fiber.
nortriptyline hydrochloride $C_{19}H_{21}N \cdot HCl$ = 299.8. Aventyl. Bitter, white crystals, soluble in water, m.218; an antidepressant in the tricyclic group (USP, BP).
nortropinone $C_7H_{11}ON$ = 125.2. 8-Azabicyclo[3.2.1]octan-3-one. A ketone derived from tropine by oxidation. Colorless powder, m.70.
norvaline $C_5H_{11}O_2N$ = 117.1. 2-Aminopentanoic acid*. A protein amino acid, m.291.
Norwegian saltpeter Norge saltpeter.
noscapine $C_{22}H_{23}O_7N$ = 413.4. Narcotine, Tusscapine. An alkaloid from opium. Colorless needles, m.176, insoluble in water. A cough suppressant (USP, EP, BP). Cf. *cotarnine.*
nosean An aluminum sodium silicate containing gypsum.
noselite $Al_3Na_5SSi_3O_{16}$. A *silica* mineral, q.v., of the sodalite group.
nosepiece A revolving disk on a microscope, into which the objectives are screwed.
nosology The classification of disease.
nostrum A cure-all, or quack medicine.
notalin Citrinin.
notation A system of numerals and symbols which indicate the structural characteristic of an organic compound. See under individual symbols and *formula, nomenclature.* **International Chemical ~, Dysonian ~** A tentative method for delineating the structures of organic compounds independently of national languages; published by IUPAC, 1958. It used a cipher or line of capital letters, numbers, stops, or symbols, each feature of the formulas being described as a distinct operation separated from the others by stops. Thus, unsaturation was denoted by *E*, with (e.g.) *E*3 for a triple bond; methane was *C*. **Rosanoff ~** A n. for representing sugars. The $-CHO$ group is a circle, the main C chain a vertical line, each $-OH$ group a horizontal line, and the 'tail' the $-CH_2OH$ group. Thus, xylose:

noumeite Garnierite.
novaculite Razor stone. A fine-grained, abrasive rock.
Novadelox Trademark for a mixture: benzoyl chloride 1, calcium phosphate 3 or 5 pts.; formerly used to bleach flour.
novain Carnitine.
novobiocin $C_{31}H_{36}O_{11}N_2$ = 612.6. An antibiotic from *Streptomyces niveus;* used as the sodium or calcium salt (BP).
Novocain Trademark for procaine hydrochloride.
Novodur Trademark for acrylonitrile/butadiene/styrene copolymers and graft polymers.
novolak A plastic of the *resole* type, q.v., but formed under acid conditions. Novolaks are fusible and soluble and, unlike the resoles, will not condense with like molecules without addition of a hardening agent.
noxious Harmful. **n. gases** Poisonous gases or gases of strong odor.
Noyes **N., Arthur Amos (1866–1936)** American chemist, noted for development of thermodynamics. **N., William Albert (1859–1941)** American chemist, noted for organic synthesis and atomic weight determinations.
Np Symbol for neptunium.
NPK Abbreviated description of a fertilizer containing nitrogen, phosphates, and potassium.
NPL Abbreviation for National Physical Laboratory, Teddington, U.K.
NSAID Abbreviation for nonsteroidal anti-inflammatory drug; as, aspirin, naproxen.
NSM New smoking material. A tobacco substitute.
NTP See *s.t.p.*
nu ν. Greek letter. ν (not italic) is symbol for neutrino. ν (italic) is symbol for (1) frequency; (2) Poisson ratio; (3) kinematic viscosity. ν_i is symbol for velocity of ion i.
nuarimol* See *fungicides,* Table 37 on p. 250.
nuces The Latin plural of "nut"; as, nuces nucistae = nutmeg.
nucin Juglone.
nucite Inositol.
nuclear Pertaining to a *nucleus,* q.v. **n. activation analysis** Analysis for minute quantities of metals in very small samples, from the nature and magnitude of the radioactivity of the isotopes produced on irradiation with neutrons in a nuclear reactor. **n. chemistry** The branch of chemical physics dealing with changes in the atomic nucleus. **n. equation** An expression indicating the changes occurring in the nucleus during a n. reaction. The sum of the atomic numbers (= *positive charges*) and atomic weights (= *mass*) must be equal on both sides of the equation. **n. halogen** A term erroneously applied to a halogen attached to a ring. **n. magnetic resonance** Abbrev. n.m.r. The absorption of electromagnetic radiation by atomic nuclei which do not have even numbers of protons and neutrons, under the influence of a magnetic field. The variations in line width, magnitude, and spectral position of the absorption are used as a sensitive, nondestructive method of structural analysis. Cf. *electron spin resonance.* **n. particles** The building units of the atomic nucleus. See *atomic structure.* **n. reactions** The disruption of an atomic nucleus by bombardment with protons, deuterons, or α particles, accompanied by the emission of neutrons, protons, α particles, positive or negative electrons and formation of excited states or radioelements. **n. structure** See *atomic nucleus* under *nucleus.* **n. symbols** See *atomic symbols.*
nuclease An enzyme that digests nucleoproteins. N. is present in blood and liver.
nucleate A salt of nucleic acid.
nuclei Plural of nucleus.
nucleic acid(s) Nucleinic acids. A group of compounds in which one or more molecules of phosphoric acid are combined with carbohydrate (pentose or hexose) molecules, which are in turn combined with bases derived from purine (as adenine) and from pyrimidine (as thymine). They commonly occur, conjugated with proteins, as nucleoproteins, and are found in cell nuclei and protoplasm. The molecules of the most common nucleic acids comprise 4 phosphoric acid, 4 carbohydrate, and 2 basic molecules each from purine and pyrimidine and are therefore tetranucleotides. Two main types: deoxyribo ~ (DNA) and ribo ~ (RNA). E.g.; from salmon ($C_{40}H_{56}N_{14}O_{26}P_4$), from yeast ($C_{38}H_{49}N_{15}O_{29}P_4$). **deoxyribo ~** DNA*, chromosomic acid, thymus nucleic acid. Mol. wt. > 10^6. A constituent of nuclear cell material, associated with the transmission of the characteristics of the cell and largely responsible for cell division. Structure similar to ribo ~ with thymine in the place of uracil. **recombinant ~** See *genetic engineering.* **ribo ~** RNA*. Polynucleotide of adenosine, guanosine, cytidine, and uridine linked by 3′,5′-phosphodiester bonds. Important component of living cells. **messenger ~** mRNA*. Form of RNA containing a sequence of amino acids, copied from DNA, that is used by ribosomes for protein synthesis. **ribosomal ~** rRNA*. A part of the ribosome structure, of unknown function. **transfer ~** tRNA. Form of RNA that brings amino acids to the ribosome site for protein synthesis. **syn ~** See *polynucleotide* under *nucleotide.* **thymus ~** Deoxyribo*nucleic acid.*
nucleid A metal compound of nucleic acid.
nucleinate A salt of nucleic acid and a base.
nucleinic acid Nucleic acid.
nucleoalbumin Paranuclein, phosphoglobulin, pseudonuclein. A combination of nucleic acid and albumin.
nucleolus Plasmosone. A spot in the center of the nucleus of a cell which contains RNA.
nucleon See *subatomic particle* under *particle.* **n. number*** Mass number*.
nucleonic(s) The study of the phenomena associated with atomic nuclei. **n. gage** A device for measuring and recording automatically and continuously variations in thickness of a moving web, e.g., of paper, which passes between a radioactive source and a detector, e.g., an ionization chamber.
nucleophilic Describing preferential attraction to regions of low electron density.
nucleoplasm The protoplasm of the cell nucleus.
nucleoprotein A compound of a simple protein with nuclein and a hexose (in animals) or a pentose (in plants), containing 0.5–3.0% P. White powders, soluble in weak alkalies and precipitated by acids. **ribo ~** RNP. A complex of nucleic acid and protein involved in the synthesis of proteins in the living cell, and in the production of antibodies associated with the protein portion of the ribonucleoprotein released after the nucleic acid portion has been destroyed.
nucleoside A compound containing a nitrogen base (cytosine, etc.) bonded to a carbohydrate (as, a pentose); e.g., cytidine. A *nucleotide,* q.v., without the phosphate group.
nucleotide The phosphate of a nucleoside. Thus the n. of the nucleoside cytidine is cytidine 5′-monophosphate. **poly ~** Molecule comprising several n.; as, ribonucleic acid.
nucleus A central part. **acyclic stem ~** A hydrocarbon chain, thus $CH_3 \cdot CH_2 \cdot CH_3$ is the a. s. n. of $ClCH_2 \cdot CH_2 \cdot CONH_2$. **alicyclic ~** A saturated carbon ring. **atomic ~** The positively charged center of an atom, with Z positive charges (protons) and $(A - Z)$ neutrons, where A is the mass number. The n. of an atom can be adequately specified by the

TABLE 57. NUMBER SYSTEMS

Decimal, base 10	Binary, base 2	Octal, base 8	Hexadecimal, base 16
1	1	1	1
2	10	2	2
3	11	3	3
4	100	4	4
5	101	5	5
6	110	6	6
7	111	7	7
8	1000	10	8
9	1001	11	9
10	1010	12	A
11	1011	13	B
12	1100	14	C
13	1101	15	D
14	1110	16	E
15	1111	17	F
16	10000	20	10
17	10001	21	11

2 parameters atomic and mass number. Suggested names: proton (Rutherford), hydrion (Soddy), centron (Lodge), merron, hylon, and prouton. Cf. *nuclear reactions, particle.* **benzene ~** The 6-C ring of benzene. **cell ~** The central part of protoplasm which is darker than the remainder, and contains genetic material composed of DNA (chromosomes). **condensed ~** A carbocyclic or heterocyclic compound; 2 or more rings joined by at least 2 C atoms, as in naphthalene. **heterocyclic ~** A ring compound with an atom other than C in the ring. **homocyclic ~** A ring compound of C atoms only.

nuclide* A species of atom, characterized by its atomic number (proton number) and mass number (nucleon number). **isobaric ~** Isobar. N. having the same mass number. **isotopic ~** Isotope.

nuclidic* Having the properties of a nuclide. Cf. *nucleic acid(s).*

nugget A water-worn piece of native gold.

null instrument An instrument indicating an end point, or point of balance between 2 effects, because it is not actuated at that point; e.g., the galvanometer in a Wheatstone bridge circuit.

number A definite value. **oxidation ~** See *oxidation number.* **preferred ~** (Renard). An industrial standard for the increase in size of a commodity; as, *paper sizes.* **wave ~** See *wave number.*

n. systems Progressions derived from a stated base (or radix). Thus 7 in the decimal system (i.e., 7×10^0), is 111 in binary $[(1 \times 2^2) + (1 \times 2^1) + (1 \times 2^0)]$, 7 in octal (7×8^0), and 7 in hexadecimal (7×16^0). See Table 57.

Numetal Trademark for an alloy: approximately Ni 74, Fe 18, Cu 5.3, Cr 2, Mn 0.7%. It has a high magnetic permeability and low hysteresis; used for electric cables.

Nupercaine Trademark for dibucaine hydrochloride.

nupharine $C_{18}H_{24}O_2N_2$ = 300.4. An alkaloid from the bulbs of (pond lily) *Nuphar luteum* (Nymphaeaceae).

Nurofen Trademark for ibuprofen.

nutgall Galls, galla. The excrescence of *Quercus infectoria* (Cupuliferae) and allied species, produced by the desposited eggs of *Cynips tinctoria,* a gallfly. Nutgalls contain tannic acids and are astringents.

nutmeg Myristica. **American ~** Otoba wax. **n. butter** Myristica oil. **n. oil** Myristica oil.

nutrient A food for a body, organism, or cell. See *fertilizer.*

nutrilite A nutrient.

nutriment Nourishment.

nutrition The study of food in relation to the metabolism of animals and plants.

nux A nut. **n. moschata** Myristica. **n. vomica** Poison nut, quaker nuts, quaker buttons. The dried seeds of *Strychnos nux vomica* (Loganiaceae), Sri Lanka, India, N. Australia; 25% alkaloids, principally strychnine and brucine (BP).

NVM Abbreviation for "nonvolatile matter."

NW acid Nevile and Winther's acid.

nyctanthin $C_{28}H_{27}O_4$ = 427.5. A crystalline principle from the tree of sadness, *Nyctanthes arbortristis* (Oleaceae).

Nydrazid Trademark for isoniazid.

Nylenka Trademark for a polyamide synthetic fiber.

nylon Generic term for any long-chain, synthetic polyamide which has recurring amide groups as an integral part of the main polymer chain, and can be formed into a filament whose structural elements are oriented in the axis direction. N. is characterized by high strength, elasticity, and resistance to water and chemicals. A single-figure suffix (x) denotes derivation of the repeating unit from a C_x aminocarboxylic acid; a double-figure suffix, (xy) describes the combination of a linear (C_x) diamine and a linear (C_y) carboxylic acid. Cf. *aramid, CIL-n, Monofil.* **n. 6** Polycaprolactam. **n. 11** Rilsan. **n. 12** Vestamid. **n. 66** Polymer of hexamethylene diamine and adipic acid.

Nymo Trademark for a polyamide synthetic fiber.

nystagmus An involuntary oscillatory movement of the eyeballs. A symptom of some eye or cerebellar disorders.

nystatin Mycostatin. An antifungal antibiotic produced by *Streptomyces noursei;* used to treat infections due to the fungus *Candida albicans.* (BP).

nytril A copolymer of vinyl acetate and vinylidene dinitrile. Cf. *Darlan.*

O

O Symbol for oxygen. **O shell** The 5th layer of electrons in an atom, which contains 32 electrons when complete.
O In chemical names, the radical prefixed by *O*- is attached to the oxygen atom.
o Symbol for the *ortho* position.
ö See words beginning with *e*- and *oe*-.
Ω, ω See *omega.*
oak A genus of trees, *Quercus* (Cupuliferae), which yield cork, nutgall, quercin, and tans. **poison ~** See *poison oak.* **white ~** Quercus.
o. bark The bark of oak, *Q. robur,* containing 25% tannin; a tan and astringent. **o. red** $C_{28}H_{22}O_{11}$ = 534.5. A coloring matter obtained by hydrolysis of quercitannic acid. **o. tannin** (1) Quercinic acid. (2) Quercitannic acid.
oakum Fiber made by untwisting old hemp ropes. Formerly used as a medicinal dressing and padding, and (impregnated with tar) for caulking.
oat Avena.
oatmeal Coarse flour from oats.
OBA See *optical brightening.*
obeche Wawa, ayous, samba. A tree, *triplochiton scleroxylon,* exported from Africa. Used for furniture.
Obermayer, Friedrich (1861–1925) Austrian physician. **O.'s reagent** A 0.4% solution of ferric chloride in dilute hydrochloric acid. **O's test** Equal parts of urine and chloroform are shaken with 3 drops of O.'s reagent. The intensity of the blue color of the chloroform indicates the amount of indican present in the urine.
Obermüller's test (For cholesterol.) The substance is melted in a test tube and 3 drops of propionic aldehyde added; on cooling, the mass becomes successively blue, green, orange, and brown-red.
Oberphos A dry, granular superphosphate fertilizer.
objective (1) The lens of the microscope nearest to the object. (2) The lens of a photographic camera. **achromatic ~** A compound lens to correct chromatic aberration. **immersion ~** An o. dipped into a drop of cedar oil or similar liquid covering the object, to establish a continuous film between object and microscope.
obs Abbreviation for "observed."
obsidian A volcanic glass rock used by Indians for spearheads and implements.
occlusion (1) The retention of a gas or liquid on solid particles or inside a solid mass; as, hydrogen on palladium, solvent on precipitates, or air in crystals. (2) Blockage; as, arteries by atheroma.
oceanic sediments (1) Mechanical deposits from rivers. (2) Remains of marine animals. (3) Substances formed by chemical action on the ocean floor. See the following table:

Classification	Depth, m	% calcium carbonate
Red clay	5,100	6.7
Radiolarian ooze	5,300	4.0
Diatom ooze	2,700	23.0
Globigerina ooze	3,750	64.5
Ptoropod ooze	1,900	79.2

oceanium Hafnium*.
ocher Paint rock, mineral purple, yellow earth. Powdered iron oxide, usually with clay. A yellow (limonite), red (hematite), or brown pigment in paints, varnishes, linoleum, papermaking, and oilcloth manufacture. **antimony ~** Stibiconite. **bismuth ~** Bismite. **brown ~** Bogore. **Indian ~** See *Indian ocher.* **molybdic ~** Molybdite. **nickel ~** Annabergite. **plumbic ~** Brown *lead oxide (6).* **red ~** A hematite. **Roman ~** A deep orange o. **synthetic ~** An artificial o. obtained by precipitating ferrous sulfate with soda and lime. **telluric ~** Tellurite. **tungstic ~** Tungstite. **uranic ~** See *uranic ocher.* **yellow ~** Selwynite. A mixture of ferric oxide and clay.
ocherous deposit The accumulation of precipitated ferric hydroxide and calcium carbonate from mineral waters, due to escape of carbon dioxide.
ochre Ocher.
ocimene $C_{10}H_{16}$ = 136.2. **β-~** 3,7-Dimethyl-1,3,6-octatriene*. A terpene, d.0.799, b_{21mm}74, from many essential oils.
OCR Optical character recognition.
octa-* Octo-. Prefix indicating 8.
octacosane* $C_{28}H_{58}$ = 394.8. Crystals, m.65, soluble in acetone.
octacosanoic acid* $C_{27}H_{55}\cdot COOH$ = 424.8. Montanic acid. m.90; from hardwoods and montan wax.
octadeca-* Prefix indicating 18. **o. peptide** An artificial protein containing 19 amino acids or 18 peptide groups. Cf. poly*peptide.*
octadecanal* Stearaldehyde*.
octadecane* $C_{18}H_{38}$ = 254.5. d_{20}·0.7668, m.30, b.306. Cf. *anthemane.*
octadecanoic acid* Stearic acid*.
octadecanol* $Me(CH_2)_{16}CH_2OH$ = 270.5. Octadecyl alcohol*. White leaflets, m.59, in spermaceti, whale, and linseed oils.
octadecatrienoic acid* See *linolenic acid, eleostearic acid*.*
octadecenoic acid* **6-~** Petroselinic acid. **(*E*)-9-~** Elaidic acid*. **(*Z*)-9-~** Oleic acid*. **11-~** Vaccenic acid.
octadecyl* The radical $Me(CH_2)_{17}-$. **o. alcohol** Octadecanol*.
octadecylic acid Stearic acid*.
octadecynoic acid* **6-~** Tariric acid. **9-~** Stearolic acid.
octadiene* C_8H_{14} = 110.2. Conylene. **2,3-~** Colorless liquid, d.0.770, b.130. **dimethylcyclo~** $C_{10}H_{16}$ = 136.2. 1,5-Dimethyl-1,5-cyclooctadiene*. A major constituent of rubber. **methylmethylene~** Myrcene.
octahedral Describing a crystal with 8 surfaces. **o. iron ore** Magnetite.
octahedrite Anatase.
octahedro-* Affix indicating 6 atoms bound into an octahedron.
octahedron An isometric crystal with 8 surfaces, each having equal intercepts on all 3 axes; i.e., the faces are equilateral triangles.
octal system See *number systems.*
octanal* $C_7H_{15}CHO$ = 128.2. Octylaldehyde, caprylic aldehyde. Colorless liquid, d.0.821, insoluble in water.

octane* C_8H_{18} = 114.2. Dibutyl. A colorless oil in petroleum, d.0.706, b.125, insoluble in water. **amino ~** Octylamine*. **chloro ~***. **1- ~** $C_8H_{17}Cl$ = 148.7. Octyl chloride*. b.180. **2- ~** b.175. **2,7-dimethyl ~*** $[Me_2CH(CH_2)_2-]_2$ = 142.2. Biisopentyl. Colorless liquid, b.100. **iodo ~** **2- ~** $MeCHI \cdot C_6H_{13}$ = 240.1. Colorless oil, d.1.31, decomp. 200, insoluble in water. **iso ~** 2,2,4-Trimethylpentane. See *octane number.* **100- ~** The standard against which antiknock values of fuels are evaluated; mainly iso ~. See *octane number.*

o.carboxylic acid* Nonanoic acid*. **o.dioic acid*** Suberic acid*. **o. number** Research octane number. The percentage of 2,2,4-trimethylpentane, by volume, which must be mixed with *n*-heptane to reproduce the knocking characteristics of the gasoline being tested. See *reforming, plumbane.*

octanoate* Caprylate. A salt or ester of octanoic acid, containing the radical $C_7H_{15}COOH-$.

octanoic acid* $Me(CH_2)_6COOH$ = 144.2. Caprylic acid, hexylacetic acid, octoic acid. A fatty acid in butter. Colorless leaflets, m.16, slightly soluble in water.

octanol* $C_8H_{17}OH$ = 130.2. Octyl alcohol*, octylic alcohol, caprylic alcohol. **1- ~** $Me(CH_2)_7OH$. 1-Hydroxyoctane. Colorless liquid, b.195, soluble in water. **2- ~, (+)- ~** Colorless liquid, $b_{20mm}86$. **(±)- ~** Used in perfumes, soaps, antifoams.

octanone* **2- ~** $C_6H_{13}COMe$ = 128.2. Hexyl methyl ketone*. m.−16, b.173. Perfume and flavoring. **3- ~** Ethyl pentyl ketone*. Occurs in fragrant oils. b.166. A food flavoring.

octanoyl* Caproyl, capryryl. The radical $C_7H_{15}CO-$, from octanoic acid. **o. chloride** $C_7H_{15}COCl$ = 162.7. Colorless liquid, b.196.

octapeptide An artificial peptide with 9 combined amino acids and 8 peptide groups.

octavalent* Having a valency of 8.

octazone HN:N·NH·N:N·NH·N:NH. A hypothetical compound known only in its derivatives. Cf. *nitrogen hydrides.*

Octel Trademark for an antiknock compound containing tetraethylplumbane.

octene* C_8H_{16} = 112.2. Octylene, caprylene. Alkenes in bergamot and lemon oils, as: **1- ~** $Me(CH_2)_5CH:CH_2$. Colorless liquid, d.0.722, b.123. **2- ~** $Me(CH_2)_4CH:CHMe$. **3- ~** $Me(CH_2)_3CH:CHEt$.

octet A chemically inert group of 8 valence electrons. Cf. *polar bond.* **o. theory** Lewis theory.

octine Octyne*.

octivalent Octavalent*.

octo- See *octa-*.

octoic **o. acid** Octanoic acid*. **o. alcohol** Octyl alcohol*.

octoil Ethyl hexyl phthalate, used for *vacuum* pumps, q.v.

octosan Acetylated cane sugar. A bitter denaturant.

octoxynols $C_8H_{17} \cdot C_6H_4O(CH_2CH_2O)_nH$. Tetramethylbutylphenyl ethers of polyethylene glycols. Used as surfactants and spermicides. **o.-9** *n* ranges 5–15 and averages 9; a spermicide.

octyl* Capryl. The radical $-C_8H_{17}$, from octane. **o. acetate*** $CH_3COOC_8H_{17}$ = 172.3. Capryl acetate. Colorless liquid, d.0.885, b.210. **o. alcohol*** Octanol*. **o.aldehyde** Octanal*. **o.amine*** $C_8H_{17}NH_2$ = 129.2. Colorless liquid, b.186, soluble in water or alcohol. **o. chloride*** Chloro*octane**. **o.cresol** A mixture containing principally *m*- and *p*-cresols; a germicide. **o. formate*** $HCOOC_8H_{17}$ = 158.2. The formic acid ester of octane. Colorless liquid, d.0.893, b.198, insoluble in water. **o. gallate*** $C_{15}H_{22}O_5$ = 282.3. Octyl-3,4,5-trihydroxybenzoate. White powder, m.101, insoluble in water; an antioxidant (BP). **o. iodide*** Iodo*octane**.

octylene Octene*.

octylic **o. acid** Octanoic acid*. **o. alcohol** Octanol*.

octyne* $CH:C(CH_2)_5Me$ = 110.2. Octine, caprilydene, hexylacetylene. Colorless liquid, b.125.

ocular (1) The eyepiece of an optical instrument. (2) Pertaining to the eye. **compensating ~** An eyepiece that corrects the axial aberration of the objective. **Huygens ~** See *Huygens ocular.* **Ramsden ~** An eyepiece to increase definition; 2 planoconvex lenses, with their convex surfaces facing. **stereoscopic ~** Two eyepieces giving a stereoscopic effect.

ocymene Ocimene.

od (1) Od-rays. The luminescence of living organisms. Cf. *scotography.* (2) Abbreviation for outside diameter.

odallin A glucoside from *Cerebera odollam.* Cf. *cerberin.*

odor, odour The volatile portion of a substance perceptible by the sense of smell. Classification: (1) Aromatic (eugenol), (2) sweet (nitrobenzene), (3) citric (citral), (4) floral (citronella), (5) fragrant (coumarin), (6) pungent (heliotropine), (7) ethereal (ether), (8) esteric (pentyl acetate), (9) resinous (turpentine), (10) camphoric (terpineol), (11) heavy (chloroform), (12) empyreumatic (aniline), (13) tarry (phenol), (14) narcotic (pulegone), (15) fecal (skatole), (16) caprylic (valeric acid), (17) fishy (trimethylamine), (18) garlic (phosphine). **ammoniacal ~** An o. resembling ammonia. **aromatic-** See *(1).* **balsamic ~** An aromatic o. resembling balsam. **burnt ~** An o. resembling tar or burnt organic matter. **camphorous ~** An o. resembling camphor (10). **caprylic ~** See *(16).* **citric ~** See *(3).* **delicate ~** A faint, agreeable o. **empyreumatic ~** An o. resembling burnt animal or vegetable matter (12). **esteric ~** See *(8).* **ethereal ~** A delicate, fruity or flowery o. (7). **fecal ~** See *(15).* **fishy ~** See *(17).* **floral ~** See *(4).* **flowery ~** The o. of flowers and blossoms. **foul ~** See *(15–18).* **fragrant ~** See *(5).* **garlic ~** See *(18).* **heavy ~** See *(11).* **narcotic ~** See *(14).* **pungent ~** A strong o. (6). **putrefactive ~** The o. of decaying animal matter. **resinous ~** An o. resembling rosin or turpentine (9). **spicy ~** An o. resembling spices, like anise, cinnamon, cloves (1–3). **sweet ~** See *(2).* **tarry ~** An o. resembling tar (13).

o. intensity The strength of an odor. Cf. *olfactory.* Odors are used as a warning for the detection of leaks, as, in some household gas. The o. intensity is determined (cf. *olfactory*) in terms of the minimum perceptible concentration, as shown in Table 58. **o. permanence** The persistence of an o., measured by placing a definite quantity of the oil on a filter paper,

TABLE 58. MINIMUM PERCEPTIBLE CONCENTRATION OF VARIOUS ODORS

Odor	Mg/liter of air
Musk	0.00004
Iodoform	0.018
Dipentyl sulfide	0.001
Propylthiol	0.006
Butanoic acid	0.009
Pentyl isovalerate	0.012
Valeric acid	0.029
Oil of peppermint	0.03
Methyl salicylate	0.1
Ethyl acetate	0.68
Chloroform	3.3

exposing it to air, and comparing its o. after certain time intervals. Some of the most persistent odors, in decreasing order of permanence, are: patchouli oil, sandalwood oil, cinnamon oil, cassia oil, citronella oil, origanum oil, thyme oil, neroli oil. **o. theory** Several theories exist, none of which are proven.

odorator An atomizer or nebulizer for diffusing liquid perfumes.

odoriferous Giving off odor; fragrant.

odorimetry Olfactometry. The measurement of the intensity and permanence of odors.

odorometer An apparatus for measuring the intensity of odors and stenches for industrial purposes.

odorous Having an odor. **o. principle** A terpene, essential oil, or balsam.

oe-, ö- See also words beginning with *e-*.

oenanthal Heptanal*.

oenanthe Five-finger root, water dropwort, dead tongue. The dried herb of *O. crocata,* a highly poisonous umbelliferous plant.

oenanthic acid Heptanoic acid*.

oenanthotoxin $C_{17}H_{22}O_2$ = 258.4. A conjugated trienediyne alcohol, m.87. A convulsant poison in *Oenanthe crocata* (Umbelliferae), W. Europe.

oenanthyl The hexyl* radical.

oenology The study of wine.

oenotannin A tannin in wine.

Oersted, Hans Christian (1777–1851) Danish physicist who discovered electromagnetism (1819) by passing a current of electricity through a wire in the same plane as a suspended magnetic needle.

oersted Abbrev. Oe; the cgs system unit of magnetic field strength. 1 Oe = 79.58 A/m (the SI equivalent).

oerstedite Hydrated zircon containing titanium dioxide.

oestradiol Estradiol.

oestrane Estrane.

oestrin Estrone.

oestriol Estriol.

oestrone Estrone.

official Pertaining to or listed in a pharmacopoeia.

officinal (1) For sale in an apothecary shop or drugstore. (2) Official.

offset printing See *lithography.*

O.G. Abbreviation for original *gravity.*

Ohm, Georg Simon (1787–1854) German physicist. **O's law** $I = E/R$. The strength of an unvarying electric current I is directly proportional to the electromotive force E, and inversely proportional to the resistance R of the circuit concerned.

ohm* Ω. The SI system unit of electric resistance; a primary practical unit of electricity. Equal to the resistance between 2 points of a conductor when a potential difference of 1 V produces a current of 1 A. 1 ohm = 10^9 emu (or abohm) = 1.11265×10^{-12} esu (or statohm). **international ~** Pre-SI unit, now obsolete. 1 mean international o. = 1.00049 ohms. 1 U.S. international o. = 1.000495 ohms. **mega ~** A million ohms. **micro ~** A millionth of an ohm. **reciprocal ~** Mho, ohm^{-1}. A unit of electrical conductance; the reciprocal of resistance. **true ~** The resistance of a column of mercury 106.25 cm long at 0°C, of 1 mm^2 cross section; experimentally determined (10^9 emu).

ohm^{-1} See *reciprocal ohm* under *ohm.*

ohmammeter A combined ohm- and amperemeter.

ohmmeter An instrument to measure electric resistance in ohms.

ohmoil A mineral oil with agerite (aldol-1-naphthylamine) to prevent decrease in electrical resistance on prolonged heating.

oiazines Orthodiazines. A group of heterocyclic compounds having 2 nitrogens in the *ortho* position; as, pyridazine. Cf. *miazines.*

-oic acid* Suffix indicating a carboxylic acid; as, benzoic acid.

-oid Suffix indicating resemblance or likeness, as, alkaloid.

oil (1) A liquid not miscible with water, generally combustible, and soluble in ether:

a. Fixed oils—fatty substances of vegetable and animal organisms—contain esters (usually glycerol esters) of fatty acids.

b. Volatile or essential oils—odorous principles of vegetable organisms—contain terpenes, camphors, and related compounds.

c. Mineral oils, fuel oils, and lubricants—hydrocarbons derived from petroleum and its products.

(2) Natural *oil,* q.v. (below). **aniline ~** Crude aniline. **acetone ~** See *acetone oil.* **blown ~** A fixed o. oxidized by a current of air. **boiled ~** See *linseed oil.* **compounded ~** A mixture of essential oils for flower perfumes. **crude ~** Petroleum. **distilled ~** Essential o. **drying ~** See *drying oil.* **edible ~** A fatty oil food or food accessory. **essential ~** See *essential oils.* **ethereal fruit ~** A mixture of aromatic substances resembling fruit in odor. **expressed ~** Fatty o. **fatty ~** The nonvolatile oils of plants and animals; mixtures of fatty acids and their esters (usually triglycerides), subdivided into solid (mainly stearin), semisolid (mainly palmitin), and liquid (mainly olein). Cf. *fats.* **fish ~** See *fish oil.* **fixed ~** Fatty o. **flower ~** See *essential oil* above. **liquid ~** A fatty o.; contains chiefly olein and is solidified by hydrogenation. **lubricating ~** A mineral lubrication o. **mineral ~** (1) An o. composed mainly of hydrocarbons derived from inorganic matter; as petroleum. Primeval biological processes are partly responsible for their formation. (2) Liquid paraffin, (heavy) liquid petrolatum. Colorless, nonfluorescent, odorless, tasteless mixture of liquid hydrocarbons from petroleum. d.0.860–0.905. An internal lubricant (USP, BP). **natural ~** A mixture of hydrocarbon oils with their oxidation products; as, crude petroleum. **nondrying ~** See *nondrying oil.* **quintessential ~** A highly aromatic substance from a natural essential o.; as, anethole from anise o. **red ~** See *red oil.* **residual ~** An o. that does not distill in refining processes. **semisolid ~** A fatty o., m. ca. 20, consisting chiefly of palmitin. **solid ~** A fat that consists chiefly of stearin. **stand ~** A drying o. which has been polymerized, i.e., thickened by heating in an inert atmosphere, without addition of a drier. **straight-cut ~** A mineral o. fraction having a relatively small difference between the initial and final boiling points. **synthetic drying ~** SDO. An amber, viscous liquid which polymerizes on drying to a hard, chemically resistant, protective coating. **tall ~** See *tall oil.* **volatile ~** Essential oil.

o. equivalent See *fuel equivalence.* **o. of absinthe** Wormwood o. **o. of ajava** Ajowan o. **o. of albahaca** Tolu o. **o. of almond artificial ~** Benzaldehyde*. **o. analysis** The identification and evaluation of an o. in terms of moisture content, foots content, color, specific gravity, boiling point, optical rotation, refraction, congealing or melting point, flash point, solubility, unsaponifiable matter, saponification value, ester value, iodine value, acid value. **o. of anthos** Rosemary o. **o. of ants** 2-Furaldehyde*. **o. of apple** Pentyl valerate. **o. of arachis** Peanut o. **o. of aspic** Spike o. **o. of badian** Star *anise* o. **o. of bananas** Pentyl acetate. **o. bath** A metal container filled with o. (rapeseed o.); used to heat glass apparatus to 100–200°C. **o. of bay** Laurel o. **o. of bayberry**

Myrcia o. **o. of benne** Sesame o. **o. of brazil nuts** Castanhao o. **o. cake** A compact mass of crushed seeds, from which o. has been expressed or extracted. It contains proteins 20–40; carbohydrates, salts, and residual oil 7–10%; a cattle food and fertilizer. **o. of candlenuts** Lumbang o. **o. of checkerberry** Gaultheria o. **o. of chinawood** Tung o. **o. of chinese beans** Soybean o. **o. of chinese cinnamon** Cassia o. **o.cloth** Linoleum. Canvas coated with a mixture of linseed o. and pigments. **o. of cognac** Ethyl hexyl ether*. **o. of cuscus** Vetiveria o. **o. of dolphin** Porpoise o. **o. of earthnuts** Peanut o. **o. of fir** Pine o. **o. of flaxseed** Linseed o. **o. of Florence** Olive o. **o. of garlic** Allyl sulfide. **o. gas** Combustible hydrocarbon gases; as, Blau gas. **o. of gingelly** Sesame o. **o. of glonoin** Nitroglycerin. **o. of goosefoot** Chenopodium o. **o. of gourd** Cucumber o. **o. of groundnuts** Peanut o. **o. of gynocardia** Chaulmoogra o. **o. of hoofs** Neat's foot o. **o. of illicium** Star *anise* o. **o. of katchung** Peanut o. **o. of lemon** Lemon o. **o. length** An indication of the oil-resin ratio in a varnish medium; 1:1 is a short, 3:1 a long, oil varnish. **o. of maize** Corn o. **o. of melissa** Balm o. **East Indian ~** Lemongrass o. **o. of mignonette** Reseda o. **o. of mirbane** Nitrobenzene. **o. of monarda** Horsemint o. **o. of mosoi flowers** Ylang-ylang o. **o. of mustard** See *allyl mustard oil.* **o. of mirbane** Neroli o. **o. of orange** Orange o. **o. of origanum** Wild *marjoram oil.* **o. of palmarosa** Turkish *geranium oil.* **o. of pears** Pentyl acetate*. **o. of pennyroyal** Hedeoma o. **o. of pineapple** Ethyl butanoate*. **o. of ricinus** Castor o. **o. sands** See *tar sands.* **o.seed cake** O. cake. **o. shale** Compact, sedimentary rock yielding 45–230 liters of oil per ton on distillation, and much ash (silica 30–50, ferric oxide 12–67%). Occurs in Scotland, Australia, Nevada, Colorado, Wyoming. Cf. *kerogen.* **o. of snakeroot** Asarum o. **o. of sperm** Whale o. **o. spills** Caused by cleaning oil tanks or by tanker collisions. Treated with dispersants with an aromatic content below 3% to minimize toxicity; these reduce the surface tension and allow the oil to be assimilated by the sea. **o.stone** A fine-grained stone for sharpening knives and scalpels in o. **o. sugars** Oleosacchara. A trituration of 2 mL of essential oil with 100 g of sugar; used for flavoring. **o. of tar** An empyreumatic o. obtained by distillation of pine tar. **o. of theobroma** Cacao butter. **o. thief** A tubular apparatus for taking samples of o. from the top, center, and bottom of tank cars or storage vessels. **o. of verbena** **East Indian ~** Lemongrass o. **Singapore ~** Citronella o. **o. of vitriol** Concentrated commercial sulfuric acid. **o. of wintergreen** Wintergreen o. **artificial ~** Methyl salicylate. **synthetic ~** Methyl salicylate.

Oildag Trademark for a colloidal dispersion of graphite in petroleum oil. Cf. *graphite, Dag, Aquadag.*

ointment Unguentum. A salve or fatty preparation for external medicinal or cosmetic use. **basilcon ~** 26% resin with beeswax, olive oil, and lard. **brown ~** Mother's salve: a mixture of camphorated brown plaster, suet, and olive oil. **camphor ~** A mixture of 22% camphor with lard and white wax; a rubefacient. **carbolized ~** Phenol o. **iodine ~** A mixture of iodine 4 and potassium iodide 4%, with glycerol and benzoinated lard. **iodoform ~** A mixture of 10% iodoform with benzoinated lard. **simple ~** White o. **sulfur ~** A mixture of sublimed sulfur 20 and potassium carbonate 10% with benzoinated lard; an antiparasitic. **white ~** Unguentum: a mixture of white wax 5 and white petrolatum 95%; a pharmaceutical base (USP). **yellow ~** Unguentum flavum: a mixture of yellow wax 5 and petrolatum 95% (USP). **zinc oxide ~** A mixture of 20% zinc oxide with white petrolatum or benzoinated lard; an astringent and protective.

oiticica oil A white fat from the nuts of the Brazilian oiticica tree *Licania rigida* (Rosaceae). It resembles Chinese wood oil, and contains coupeic acid.

Oklo phenomenon The presence of natural uranium, containing 0.7171% U-235 instead of the usual 0.7202%, plus fission fragments. Both suggest natural chain reactions 1,800 million years old, at Oklo, Gabon.

okonite (1) $CaH_2Si_2O_6$. A native calcium silicate in compact fibrous masses. (2) An insulating material obtained by vulcanizing a mixture of ozokerite and resin with rubber and sulfur.

okoumé See *gaboon.*

-ol* Suffix indicating a hydroxyl group in alcohols, phenols, and heterocyclics where the OH is attached to a C atom. Cf. *-ole.*

-ole Suffix denoting (1) a substance, not an alcohol, to which an ending -ol was originally given; as, indole; (2) an aromatic ether; as, anisole.

Oleaceae Olive family of trees and shrubs: *Olea europaea,* yielding olive oil; *Fraxinus ornus,* yielding manna; *Jasminium grandiflorum,* yielding jasmine oil; *Syringa vulgaris,* yielding syringin.

oleaginous Oily or greasy.

oleander Nerium. Subtropical, poisonous shrubs (Apocyanaceae), which yield essential oils, karabin, neriin, pseudocurarine, and oleandrin.

oleandrin $C_{31}H_{48}O_9$ = 564.7. A glucoside of gitoxigenin. Yellow powder, m.250, soluble in alcohol. Cf. *neriin.*

oleanol $C_{29}H_{48}O$ = 412.7. Needles, m.216–230, derived from oleanolic acid, and sometimes used as its synonym.

oleanolic acid $C_{29}H_{47}O \cdot COOH$ = 456.7. Caryophyllin. From oil of cloves and sugar beet; m.307, soluble in alcohol.

oleastene $C_{21}H_{36}$ = 288.5. A hydrocarbon from olive oil.

oleasterol $C_{20}H_{34}O$ = 290.5. A phytosterol from olive oil.

oleate* A compound of an alkaloid or a metal with oleic acid. Used medicinally for external applications; technically, in soaps and paints.

olefiant gas Ethylene*.

olefin(e) Alkene*, ethylene series. An unsaturated hydrocarbon of the type C_nH_{2n}, indicated by the suffix *-ene;* as, ethylene. **o. acid** See *acrylic acids.* **o. alcohol** An unsaturated alcohol of the type $C_nH_{2n-1}OH$; as, $CH_2{:}CH \cdot OH$, vinyl alcohol. **o. aldehyde** An unsaturated aldehyde of the type $C_nH_{2n-1}CHO$; as, $CH_2{:}CHCHO$, acrylaldehyde. **o. ketone** An unsaturated ketone of the type $C_nH_{2n}CO$: as, $MeCH{:}CH \cdot CO \cdot Me$, ethylidene acetone.

oleic **o. acid*** $Me(CH_2)_7 \cdot CH{:}CH(CH_2)_7COOH$ = 282.5. Oleinic acid, (Z)-9-octadecenoic acid*, red oil. Colorless needles, m.14, insoluble in water. It occurs in many (nondrying) oils, and is hydrogenated to stearic acid (by catalysis with nickel); used to prepare soap (NF, BP). **12-hydroxy ~** Ricinoleic acid. **iso ~** Elaidic acid* [(*E*) form of oleic acid].

o. series Acrylic acids.

olein (1) $(C_{18}H_{33}O_2)_3C_3H_5$ = 885.4. Triolein. (9Z)-Glycerol tri-9-octadecanoate. The glyceride of oleic acid. Colorless oil, m.6, insoluble in water; the main constituent of olive and many other oils. (2) Oleine. A mixture of the fatty acids obtained by steam or vacuum distillation of the products of acid hydrolysis of fats. (3) A glyceride of oleic acid, as, monoolein.

oleinic acid Oleic acid*.

oleo- Prefix indicating oil. **o.creosote** Creosote oleate. **o.guaiacol** Guaiacol oleate. **o. oil** A yellow oil of olein expressed from tallow; used to make oleomargarine.

oleomargarine (1) Margarine. A butter substitute made from a mixture of hydrogenated fatty oils; colored with aniline

dyes. (2) The liquid fat from which margarine is made by hydrogenation.
oleometer A hydrometer for oils.
oleoptene See *stearoptene.*
oleoresin (1) A natural combination of resins and essential oils occurring in or exuded from plants; e.g., benzoin. (2) A pharmaceutical preparation of ethereal extract of drugs. **o. aspidium** Oleoresina aspidii, o. of male fern. A thick, green liquid from the male fern, which contains filicic acid; formerly an anthelmintic. **o. capsicum** Oleoresina capsica, o. of red pepper, cayenne pepper. Brown, soft mass from acetone extraction of the fruit of capsicum species; a gastric stimulant and spice. **o. giner** Oleoresina zingiberis. An acetone extract from Jamaica ginger, the rhizome of *Zingiber officinale;* a carminative. **o. male fern** O. aspidium.
oleoresina Latin for "oleoresin" (2).
oleosacchara Oil sugar.
oleostearin Beef stearin. An edible solid fat from fatty tissues of the cow.
oleoyl* 1-Oxo-9-octadecenyl†. The radical $C_{17}H_{33}CO-$, from oleic acid.
oleum (1) Latin for "oil." (2) Nordhausen acid, fuming sulfuric acid. A solution of sulfur trioxide in conc. (97–99%) sulfuric acid; a reagent in the chemical industry. **o. arachis** Peanut oil. **o. lini** Linseed oil. **o. menthae viridis** Spearmint oil. **o. morrhuae** Cod-liver oil. **o. myristicae** Nutmeg oil. **o. ricini** Castor oil. **o. santali** Sandalwood oil. **o. sinapis volatile** Mustard oil. **o. terebinthinae** Turpentine oil. **o. theobromatis** Cacao butter.
oleyl alcohol $C_{18}H_{35}OH$ = 268.5. (Z)-9-Octadecen-1-ol. b_{15mm}207; in fish oils.
olfactie The maximum distance from an odorometer that an odor can be recognized.
olfactometry Odorimetry.
olfactory Pertaining to the sense of smell.
olibanoresin $C_{14}H_{22}O$ = 206.1. A neutral resin from olibanum.
olibanum Frankincense, gum thus. A gum resin from incisions in trees, e.g., *Boswellia carteri* and other species from Sudan. Yellow tears or red fragments; perfume and incense. Cf. *boswellic acid.* **o. oil** An essential oil from olibanum containing pinene, cinene, and phellandrene, d.0.895, insoluble in water.
-olide* Suffix indicating a lactone.
olifiant gas Ethylene*.
oligo- Prefix, meaning "few."
Oligocene See *geologic eras,* Table 38.
oligoclase A silica mineral of the feldspar group.
oligodynamic (1) Small but powerful. (2) Describing the inhibition of fermentation or growth of bacteria by metals, e.g., iron containers.
oligodynamics The bactericidal action of metals; as, Ag and Cu.
oligomer A polymer whose properties change with the addition or removal of one or a few repeating units. The properties of a true polymer do not change markedly with such modification.
oligonite $FeMn(CO_3)$. A native carbonate.
oligosaccharide A carbohydrate that is made up of 2–10 monosaccharide units, which may be identical or different.
oligotrophic Describing a water that is not polluted eutrophically.
olistomerism Describing reactions in which the same substrates yield the same final products, but via different intermediate stages.
olive The fruit of *Olea europaea* (Oleaceae), a food and source of olive oil. **o. oil** A fixed oil expressed from the olive. Pale yellow liquid, d.0.91, m.−6, insoluble in water; contains olein and palmitin. Used as a food, in pharmaceutical preparations as an emollient, and in the manufacture of soap (NF, BP). **o. kernel oil** Yellow oil extracted or expressed from olive kernels, d.0.918, insoluble in water; used as food, as a lubricant, and in soap manufacture.
olivenite $CuO \cdot As_2O_3 \cdot H_2O$ Wood copper. A native copper arsenate.
olivine (1) See *olivines.* (2) $(Mg,Fe)_2SiO_4$. Peridot, chrysolite. Green, orthorhombic crystals; a gem and refractory. Cf. *chrysotile, peridotites.* **o. diabase** A diabase containing o. crystals.
olivines A group of silica minerals, generally the sulfates of divalent metals.
Olsen's testing machine An apparatus to measure the tensile strength of cement. An arrangement of levers and 2 heavy forceps, between which the specimen is placed and torn apart.
oly The scum on molten metal.
O.M. Abbreviation for "organic matter."
omal Trichlorophenol*.
omega Ω, ω. Greek letter. Ω (not italic) is symbol for (1) ohm; (2) o. *particle.* Ω (italic) is symbol for solid angle. ω (italic) is symbol for (1) angular speed; (2) circular frequency; (3) solid angle. **o. chrome** Eriochrome. **o. particle** See *particle.*
omethoate* See *insecticides,* Table 45 on p. 305.
ommochrome An animal pigment produced by the metabolism of tryptophan; 180,000 silkworm eyes yielded 100 mg.
Omnopon Trademark for papaveretum (mixed opium alkaloids).
onchocerciasis A disease due to filaria (*Onchocircia* volvulous), a worm of W. and Central Africa, which causes large swellings and blindness.
oncogene A cancer-causing gene.
Oncovin Trademark for vincristine sulfate.
-one* Suffix indicating a ketone.
onion The edible bulb of *Allium cepa* (Lilaceae); a condiment and food. **o. oil** Yellow oil of penetrating odor (mainly allyl propyl disulfide), d.1.04, insoluble in water; a flavoring.
-onium* Suffix indicating a cation,, with one more proton than is required to make a neutral molecule of the central atom; e.g., ammonium, NH_4^+; phosphonium, PH_4^+; oxonium, H_3O^+. **o. compound** One containing an onium ion.
onocerin $C_{30}H_{48}(OH)_2$ = 442.7. Onocol. A dihydric alcohol from the root of *Onomis spinosa* (Leguminosae). β-∼ m.234. γ-∼ m.335.
onofrite Hg(S, Se). A native mercury sulfoselenide.
ononid $C_{18}H_{22}O_8$ = 366.4. A neutral principle from the root of *Ononis spinosa* (Leguminosae), Europe. Yellow powder, soluble in water.
ononin $C_{25}H_{26}O_{11}$ = 502.5. A glucoside of 7-hydroxy-4′-methoxyisoflavone from the root of *Ononis spinosa.* White crystals, m.218, slightly soluble in hot water.
onychograph A sphygmograph to record variations in blood pressure of the fingertips.
onyx Calcareous sinter. A gem variety of agate (chalcedony quartz) in white and black strata. **o. marble** Mexican o. A calcite resembling o.; an ornamental stone.
oocyanin Blue coloring matter from certain bird eggs.
oocyte A female egg before maturation.
oogenesis The formation and maturation of an egg.
oolite Peastone. Small, round, granular concretions, usually of limestone containing silica or iron oxide, cemented together to form a compact mass. **o. limestone** A calcareous o. used for building purposes.
oometer An instrument which grades eggs by size.

ooplasm The cytoplasm of an egg.
ooporphyrin The pale, brown pigment of eggshells.
oosperm A fertilized egg.
oospora Molds of the order Moniliales (Fungi Imperfecti).
ooze (1) The slime at the bottom of lakes and oceans. Cf. *oceanic sediments, diatom.* (2) Slow exudation of liquid. **o. leather** Leather made from calfskins by vegetable tanning, with a suede finish on the flesh side.
O.P. Abbreviation for overproof; see *proof spirit.*
Opaceta Trademark for an acetate synthetic fiber.
opacifier A substance which when added to a transparent substance makes the latter opaque, e.g., titanium dioxide when added to paper.
opacity The property of being impervious to light rays; not transparent or translucent. Cf. *opaque.*
opal $SiO_2 \cdot nH_2O$. An amorphous form of silica containing a variable amount of water. Vitreous masses, of rainbow color, ranging from transparent to translucent, which reflect light after refracting it. **agate ~** An o. of agatelike structure. **cacholong ~** An opaque, blue-white or pale-yellow o. **fire ~** A red to yellow o. with red reflections. **girasol ~** A blue-white, translucent o. with red reflections in strong light. **harlequin ~** An o. having a variegated play of colors on a red background. **hydrophan ~** A white, opaque o. that becomes transparent in water. **jasp ~** O. containing iron. **lechosos ~** O. with deep-green flashes of color. **moss ~** O. with mosslike inclusions of manganese oxide. **pearl ~** An opaque, bluish-white, lustrous o. **semi ~** A native form of silica. **tabasheer ~** An amorphous, opal-like silica in the joints of bamboo. Cf. *hydrophane, hyalite, isopyre.* **wood ~** Wood that has become silicified into o.
o. flashes The bright colors of an o. in certain positions, due to the refraction of light by thin layers. **o. glass** A milky-white glass. **o. jasper** A form of silica resembling jasper.
opalescence Milky, iridescent light reflected from a mineral or colloidal solution.
opalescent (1) Resembling an opal in appearance. (2) Having a milky turbidity.
opalescin A milk albumin which forms opalescent solutions.
opalized wood A petrified wood resembling opal.
opaque Nontransparent and nontranslucent. Cf. *opacity.*
open-chain Noncyclic. Describing a carbon chain that does not close and form a cyclic compound.
open-hearth furnace A reverberatory furnace used in the manufacture of steel; hot producer gases are passed over a large open crucible containing iron ore, scrap iron, and pig iron.
operational research OR; a philosophy prescribing the optimization of systems through systematic and organized effort. Concerned principally with nonprocess systems.
ophelic acid $C_{13}H_{20}O_{10}$ = 336.3. An amorphous split product of chiratin in *Swertia (Ophelia) chirata* (Gentianaceae); soluble in water.
ophiolite Serpentine containing calcium carbonate.
ophiotoxin $C_{17}H_{26}O_{10}$ = 390.4. A poison from the venom of the southeast Asian cobra, *Naja tripudians.*
ophioxylin $C_{16}H_{13}O_6$ = 301.3. A glucoside from the root of *Ophioxylon serpentinum* (Apocynaceae), India. Yellow crystals, m.72; an anthelmintic.
ophitron A small, lightweight microwave generator, based on an electron stream with an undulating path.
ophthalmo- Prefix (Greek) indicating "eye."
ophthalmoleukoscope An instrument to test color sensitivity by polarized colors; to detect color blindness.
ophthalmoscope An instrument to examine the interior of the eye.
opianic acid $C_6H_2(OMe)_2 \cdot CHO \cdot (COOH)$ = 210.2. 6-Formyl-2,3-dimethoxybenzoic acid*. Colorless prisms, m.150, soluble in hot water.
opianyl Meconin.
opiate A narcotic drug from opium.
opium The air-dried, milky juice of the unripe capsules of the poppy *Papaver somniferum.* Brown masses, containing inert matter and alkaloids (see below). Used as a narcotic and analgesic, and to prepare o. alkaloids. See *kusamba, meconin, chandoo, yenshee, mudat.* Pharmaceutical o. (USP, BP) contains not less than 9.5% morphine. **camphorated ~** Paregoric. A tincture containing 0.05% wt./vol. of morphine, with benzoic acid, camphor, and anise oil. Used to treat coughs and morphine dependence of newborn babies (USP, BP).
o. alkaloids A series of alkaloids obtained from or related to opium, e.g., (1) morphine group: morphine, chelidonine; (2) codeine group: papaverine, codeine, laudanosine, noscapine, hydrocotarnine, thebaine, hydrastine; (3) miscellaneous: chelerythrine. **o. vinegar** Black drops. A preparation containing 10% o.
opopanax An oleoresin from the roots of *Pastinaca opopanax* (Umbelliferae), the Orient; an incense and perfume. **o. oil** An oil, d.0.87–0.90, b.250–300, from *Commiphora katof* (Burseraceae). Cf. *balm* of Gilead.
oppanol $[Me_2C{:}CH_2]n$. A polymer of 2-methyl-1-propene; a rubber substitute.
opsogens See *opsonigenous substances.*
opsonic index The ratio of the opsonin content of the blood of a diseased person to that of a normal individual.
opsonigenous substances Aggressins, opsogens. Substances produced by the metabolism of bacteria. Cf. *endotoxin.*
opsonin A substance of the blood serum that renders invading microorganisms more susceptible to digestion by the phagocytes of the body. **common ~** Normal o. **immune ~** Specific opsonins. Substances formed in the blood upon stimulation by bacterial poisons (opsonigenous substances). **normal ~** The protective substances of the blood present in normal uninfected individuals. **specific ~** Immune o.
optical (1) Pertaining to sight. (2) Pertaining to light. **o. activity** O. rotation, opticity. The power of a substance or solution to rotate the plane of vibration of polarized light to the left (−) or right (+). It is characteristic of compounds containing an asymmetric atom (usually C); measured by the polariscope and used in quantitative analysis. Cf. *enantiomorph, meso* compounds, *racemic compound, mutarotation.* **o. bench** A rail with holders to carry lenses and mirrors, whose distances apart may be varied and measured, and whose o. axes are in the same straight line. Used for o. experiments. **o. bistability** Having 2 stable states for 1 input state. **o. brightening** The use of optical brightening agents (OBA). See *optical bleaching* under *bleaching.* **o. character recognition** OCR. Method of reading documents that uses the automatic recognition of characters by optical means. **o. contact** Close contact between two plane surfaces, such that no interference lines are produced. **o. density** Absorbance*. **o. fibers** Glass fibers used in *fiber optic* systems. **o. isomerism** See *stereoisomer.* **o. pyrometer** An instrument for measuring high temperatures optically. A calibrated filament is heated by an electric current of known strength until the glow matches that of the furnace. **o. rotation*** See *optical activity* above. **o. telephone** A telephone, the receiver diaphragm of which is connected to a mirror, so that its vibrations are magnified as bands of light of varying size.

opticity Optical activity. A term used in the sugar and brewing industries.
optics Photics. The study of visible radiation (light). Cf. *dioptrics.*
optimal Describing the most favorable, most suitable, or best factor or condition.
optimum Best. Most favorable overall, often involving a compromise. **sub ~** Less than o.
o. temperature The temperature at which an enzyme or bacterium is most active.
optoacoustic Photoacoustic. Describing a spectrometric analytical method in which the absorption of incident electromagnetic radiation by a sample in a container of constant volume produces an intermittent temperature change. This is translated into a pressure change by a transducer, which can be monitored by a microphone.
optophone An instrument to enable ordinary print to be read by the blind. A selenium cell, telephone, and illuminating device transform light waves into sound waves. Cf. *photophone.*
OR Operational research.
orange (1) A color intermediate between red and yellow. (2) The edible fruit of *Citrus* species *(Rutaceae),* q.v. (3) A group of pH indicators changing from purple or magenta (acid) to orange (alkaline). **bitter ~** Curacao or Seville o. The fruit of *Citrus bigarardia.* Cf. *orange peel* below. **sweet ~** China or Portugal o. The fruit of *Citrus aurantium.*
o.-blossom odor See *indole.* **o. dye** Tropeolin. **o.-flower oil** Neroli oil (NF). **o. oil** Oleum aurantii (USP). An essential oil expressed from the peel or rind of oranges. Yellow liquid, d.0.84–0.85, insoluble in water; a flavoring, perfume, and pharmaceutical additive (NF, BP). **bitter ~** Curacao o. oil, Seville o. oil. The oil from the rinds of o. peels of *Aurantii amara (Citrus vulgaris).* **sweet ~** The oil from sweet o. peels, *Aurantii dulcis (Citrus bigarardia).* **o. mineral** Sandix. **o. peel** The dried rinds from oranges (USP, BP). **bitter ~** From the fruits of *Citrus vulgaris* = Seville o., curacao o.; or *Citrus aurantium* (BP). **sweet ~** From the fruits of *Citrus bigarardia* = Portugal o., China o. **o.-peel oil** O. oil. **o. red** Sandix. **o.root** Hydrastis. **o. syrup** A solution of 60 ml o. tincture in 940 ml syrup (BP). **o. tincture** A tincture prepared from 50% o. peel in alcohol; a flavoring (NF, BP).
orangite An orange variety of thorite.
Orbenin See *penicillin.*
orbit The circular path of an electron moving around the atomic nucleus (Bohr's dynamic concept of the atom). Orbits correspond with the shells of the static atom (Lewis' model), in which the electrons are assumed to oscillate about certain positions at different distances from the nucleus. See *orbital (2).* (2) The part of the skull containing the eye.
orbital (1) Pertaining to motion around a center. (2) Term used, rather than 2-dimensional *orbits*, q.v., to describe the quantum mechanical probability density of electronic motion. O. shape increases in complexity, as: *s*- ~, spherical; *p*- ~, dumbbell; *d*- ~, 2 dumbbells at 90°. See *quantum.* **frontier ~** The molecular orbitals that influence the nature of molecular interactions, being the highest-energy occupied, lowest-energy unoccupied, and next-to-highest occupied (subjacent) orbitals. **linear combination of atomic ~** LCAO. An approximate approach for deriving molecular orbitals that utilizes the atomic orbitals of the bonding electrons. **molecular ~** Abbrev. m.o. The region, for a given electronic state, over which the electrons in a molecule are distributed. Usually depicted by the envelope of the region, with the signs (+ and −) of the wave function shown. For benzene, the π-bond o. has the shape of a doughnut above and below the C-atom ring. **subjacent ~** See *frontier orbital* above.
o. state See *orbital state* under *state.*
orcein $C_{28}H_{24}O_7N_2$ = 500.5. A coloring matter of orchil, synthesized from orcinol and ammonia. Brown crystals, insoluble in water, soluble in acid (red color) and alkalies (violet color); a microscope stain, antiseptic, and reagent.
Orchidaceae Orchis family, a group of perennial herbs, some epiphytic, with showy flowers; e.g., *Orchis* species, salep.
orchil Archil, orseille extract, cudbear, persio. Purple liquid with ammoniacal odor. A coloring matter obtained from lichens by treatment with ammonia and exposure to air. Contains orcein, orcinol, and litmus; a coloring for pharmaceutical preparations. Cf. *orcein.*
orcin Orcinol.
orcinaurine Homofluorescein. Red dyestuff with green fluorescence, obtained by the action of chloroform and alkali on orcinol; a fluorescent indicator for pH values 6.5–8.0.
orcinol $C_6H_3(CH_3)(OH)_2$ = 124.1. Orcin. 5-methyl-1,3-benzenediol*, methylresorcinol. White crystals, redden with age, m.107, soluble in water. It occurs in lichens and forms orcein with ammonia; a reagent for pentoses, and stain. **methyl ~** Xylorcinol. **trinitro ~** See *trinitroorcinol.*
ordeal **o. bark** (1) Poisonous bark used in trial by ordeal. (2) Casca bark. See *Erythrophloeum.* **o. bean** Poisonous fruits used in ordeals; e.g., *physostigma, tanghinia.*
order A grade in scientific classification, particularly zoology. **reaction ~** See *reaction order.*
o. of compounds (1) An arrangement of types of compounds according to their complexity:

1st order: binary compounds; as, NaCl.
2d order: tertiary compounds; as, $Al(OH)_3$.
3d order: addition or coordinate compounds; as, $NiSO_4 \cdot 6NH_3$.

(2) See *Mulliken's classification.*
ordinal Pertaining to a sequence or order. **o. number** Rydberg's number, indicating position in the periodic table; superseded by atomic number. **o. test** A reaction which determines qualitatively the elements of an organic compound.
ordinary Common. **o. iron** Cast *iron.* **o. ray** The ray of light normally reflected or refracted in a polariscope; the extraordinary ray is not.
ordinate The *y* coordinate in the cartesian system of graphical representation. Cf. *coordinates.*
ordination number Atomic number*.
Ordovician See *geologic eras,* Table 38.
ore A natural mineral from which useful substances are obtained; found mixed with earthy matter (matrix or gangue). The principal ores are oxides, carbonates, silicates, sulfides, arsenides, antimonides, and halides usually of metals. Cf. *flotation, roasting, smelt.* **mock ~** Sphalerite. **pay ~** An o. that can be worked with profit. **positive ~** See *positive ore.* **possible ~** See *possible ore.* **raw ~** An o. in its natural, but crushed, untreated condition.
o. band A zone of rock rich in ores; a vein. **o. bed** A zone of o. between sedimentary deposits. **o. body** A solid, continuous mass of o. **o. crusher** A machine for the preliminary disintegration of ores. **o. dressing** The refining or cleaning of o. by mechanical means; as, jigging, cobbing, etc. **o. mill** A concentrator or stamp mill. **o. pocket** An isolated occurrence of a rich deposit of o. **o. separator** A mechanical device in which o. is separated from rock, as a cradle or jigging machine.

oreodaphene A hydrocarbon oil of California laurel, *Umbellularia californica* (Lauraceae). Colorless oil, d.0.894, b.175.
oreodaphnol An alcohol in the oil of California laurel. Colorless, pungent liquid.
Oreton Trademark for testosterone propionate. **O.-M** Methyltestosterone.
Orfila, Mathieu Joseph Bonaventura (1787–1853) Spanish-born Frenchman; founder of toxicology.
organ Any tissue or part of an organism that has a distinct function.
organellae Organized structures within the cytoplasm of a cell with a definite function; as, mitochondria.
organic Pertaining to an organ or a substance derived from an organism. **o. acid** A compound containing one or more carboxyl radicals, —COOH; e.g.: $C_nH_{2n+1} \cdot COOH$, acetic acid series; $C_nH_{2n}(COOH)_2$, oxalic acid series; $C_nH_{2n-8}(COOH)_2$, phthalic acid series. **o. analysis** The qualitative or quantitative determination of o. compounds. See *organic analysis* under *analysis.* **o. bases** The amines and alkaloids. **o. chemistry** The study of carbon compounds. Originally restricted to compounds from organisms, but the synthesis of many of these compounds makes division into organic and inorganic chemistry of convenience only. Elements having o. combinations, in order of importance, are: C, H, O, N, Cl, Br, I, F, S, P, B, some metals. Cf. *inorganic chemistry.* **o. combustion** See *organic combustion* under *combustion.* **o. compounds** Nonpolar compounds, which generally consist of carbon and hydrogen, with or without oxygen, nitrogen, or other elements, except those in which carbon plays no important part, e.g., carbonates. Many may be classified into *aliphatic* and *aromatic* compounds, q.v. **Heilbron's Dictionary of ~** A multivolume compendium of information on o. c. **o. radicals** A group of atoms that normally passes unchanged from one molecule of a carbon compound to another (defined under their respective headings). See *radical, nomenclature, formula.*
organism A living complex, undergoing dynamic changes, which consists of protoplasm and has a definite pattern and function. Any animal, plant, or human body. Cf. *life.* System of classification originated by Linnaeus (1735), with each species of plant and animal having 2 Latin names on the pattern *genus species.* For humans, the classification would be:

Kingdom	Animal(ia)
Subkingdom	Metazoa
Phylum	Chordata
Subphylum	Vertebrata
Class	Mammalia
Order	Primate
Family	Hominidae
Genus	*Homo*
Species	*sapiens*

organoferric Describing masked iron.
organoleptic Referring to sensation; as, smell.
organolite An organic base-exchange material.
organomagnesium halides See *Grignard's reagent.*
organometallic Pertaining to the carbon-metal linkage. **o. compounds*** A class of compounds of the type R—M, where a C atom is joined directly to any other element except H, C, N, O, F, Cl, Br, I, or At; e.g., $PbEt_4$, tetraethylplumbane. Cf. *Grignard reagent, organolite.*
organosol A sol whose essential constituent is organic.
organotin compounds Used as preservatives and plastics stabilizers, and in lubricants.
organotropic A substance that acts specifically on an organism, and not on parasites.
organzine Warp silk. The reeled-off fibers from a number of silk cocoons.
oriental **o. agate** A translucent gem variety of agate. **o. amethyst** A native purple alumina. **o. cashew nut** Semecarpus. **o. emerald** A green corundum. **o. garnet** Garnet. **o. hyacinth** A rose-colored corundum. **o. powder** A mixture of gamboge with potassium nitrate; a fireworks explosive. **o. ruby** A red corundum. **o. sapphire** A blue corundum. **o. sweet gum** Styrax. **o. topaz** A yellow corundum.
orientation (1) The structural arrangement of radicals in a compound in relation to one another and to the parent compound. (2) The determination of crystal structure. (3) The direction or position assumed by a molecule, due to an electric charge, adsorption, or other cause. Cf. *zone.* **preferred ~** The principal orientation which the crystal units of a metal assume when the metal is deformed.
origanum oil (1) Spanish hop oil. An essential oil from *Origanum vulgare* (Labiatae), wild marjoram; used in veterinary medicine and in liniments. (The oil of *O. marjorana* is marjoram oil.) (2) An essential oil distilled from *Thymus vulgaris* (French usage).
orizabin Jalapin.
orlean Annatto.
Orlon Trademark for a polyacrylonitrile (polyvinyl cyanide) continuous synthetic filament.
Orlovius flask A flask similar to a wash bottle, used for handling blood samples under sterile conditions.
ormolu Mosaic gold. An alloy of equal parts of copper and zinc, used for cheap jewelry and ornaments.
ormosinine $C_{20}H_{33}N_2$ = 301.5. m.205.
ornithine* $C_5H_{12}O_2N_2$ = 132.2. 2,5-Diaminopentanoic acid*. An amino acid from the excrements of birds. ***N^δ*-amidino ~** Arginine.
o. cycle The theory that o., citrulline, and arginine are intermediate stages in the synthesis of urea in the mammalian liver; the o. behaves as a catalyst.
oroberol $C_{18}H_{14}O_8$ = 358.3. A chromogen from the leaves of *Orobus tuberosus,* m.290.
orobol $C_{15}H_{10}O_6$ = 286.2. 3′,4′,5,7-Tetrahydroxyisoflavone, m.212, obtained by hydrolysis of oroboside.
oroboside A glucoside from the leaves of *Orobus tuberosus (Lathyrus)* (Leguminosae), yielding orobol. White crystals, m.250.
orogen The mobile belt of the earth's crust, in which mountain chains are formed.
orogenic Describing the large-scale, tangential, compressive forces responsible for geological fractures.
Oropon Trademark for a tryptic puering material.
orotic acid $C_5H_4O_4N_2$ = 156.1. 2,6-Dihydroxypyrimidine-4-carboxylic acid*. White crystals, m.324; from milk.
oroxylin A $C_{16}H_{12}O_5$ = 284.3. 5,7-Dihydroxy-6-methoxyflavone. Yellow crystals, m.220, soluble in water; from *Oroxylon indicum* bark.
orphenadrine **o. citrate** $C_{18}H_{23}ON \cdot C_6H_8O_7$ = 467.5. Norflex. White crystals, m.136, sparingly soluble in water. An anticholinergic drug, used to treat Parkinson's disease (BP). **o. hydrochloride** Disipal. Used like the citrate (USP, BP).
orpiment As_2S_3. Kings' yellow, auripigment. Yellow, crystalline masses; a pigment. **red ~** Arsenic tetrasulfide*.
orris The dried rhizome of Florentine iris, *Iris florentina.* Creamy powder, insoluble in water; a dentifrice and perfume. **o. oil** An essential oil from the rhizome of *Iris florentina.* Yellow oil, m.44, insoluble in water; used in cosmetics and perfumery.
Orr white Lithopone.
Orsat apparatus A portable gas analysis apparatus: a

measuring buret and 3 or 4 gas-absorption pipets connected by a manifold.

orseille, orselle Orchil.

orseillic acid, orseillinic acid Orsellic acid.

orsellic acid $MeC_6H_2(OH)_2COOH$ = 168.2. Orsellinic acid, orseillic acid, 2,4-dihydroxy-6-methylbenzoic acid*. A split product of lecanoric acid and a constituent of many lichens, m.176 (decomp.).

orsellin, orseillin Roccellin. A constituent of orchil.

orsellinic acid Orsellic acid.

orthanilic acid $C_6H_7O_3NS$ = 173.2. *o*-Aminobenzenesulfonic acid. An isomer of sulfanilic acid, used in organic synthesis.

orthin $C_6H_3(OH)COOH(NH \cdot NH_2)$ = 168.2. 2-Hydrazino-4-hydroxybenzoic acid. Colorless crystals.

orthite Allanite.

ortho- (1)* *o*- (italic). Prefix indicating the neighboring, or 1,2 position. Cf. *meta-*, *para-*. (2) Prefix indicating the most hydrous acid known in the free state or as salts or esters, e.g., *orthosilicic acid.* Cf. *meta.* (3) Prefix indicating parallel spins; as, orthohydrogen. **o.acid** (1) An organic acid having the carboxyl group in the *ortho* position. (2) An organic acid containing one additional molecule of water in chemical combination, as $H \cdot COOH$, formic acid, and $HC(OH)_3$, orthoformic acid. (3) An inorganic acid containing the stoichiometric amount of water; as, H_3PO_4, orthophosphoric acid. ***o.* compound** A benzene derivative containing 2 substitution radicals in neighboring positions. See under the name of the compound. ***o.*-fused*** Referring to polycyclic compounds in which 2 rings have 2 (only) atoms in common; as (a) below. ***o.*- and *peri*-fused*** Referring to polycyclic compounds in which one ring contains only 2 atoms in common with each of 2 or more rings of a contiguous series of rings; as (b) below. **o.hydrogen** See *hydrogen (2).* **o.nitrogen** See *nitrogen molecule.* ***o.* position** See *ortho compound* above.

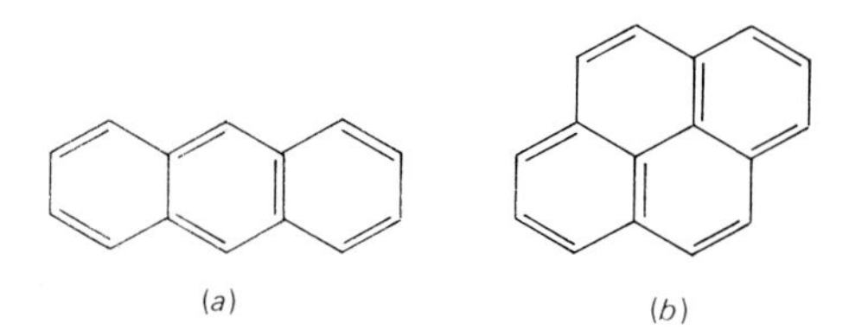

orthoacetic acid* $CH_3 \cdot C(OH)_3$, known only in its esters.

orthoaluminic acid The tautomer of aluminum hydroxide: $Al(OH)_3 \rightleftharpoons H_3AlO_3$. It forms aluminates.

orthocarbonic acid* $C(OH)_4$ or $H_2O \cdot H_2CO_3$. Tetrahydroxymethane, esters of which are known; as, $C(OEt)_4$.

orthoclase $KAlSi_3O_8$. Sunstone, potash feldspar. The commonest silicate, a constituent of many rocks. White, gray, or pink, monoclinic crystals.

orthoformic acid $HC(OH)_3$, known only as its esters, $H(COEt)_3$.

orthohydrogen See *hydrogen (2).*

orthokinetic coagulation Coagulation due to motion of particles in one direction. Cf. *perikinetic.*

orthonitric acid The hypothetical compound $O{:}N(OH)_3$.

orthophosphate* A salt of the type M_3PO_4.

orthophosphoric acid* See *orthophosphoric acid* under *phosphoric acid.*

orthorhombic system A prismatic, rhombic, or trimetric system. Crystal forms derived from a prism having 3 axes of different lengths intersecting at right angles (see Fig. 17).

orthosilicate* A salt containing the radical $\equiv SiO_4$.

orthosilicic acid* Silicic acid.

orthosiphonin A glucoside from the leaves of Java tea, *Orthosiphon stamineus* (Labiatae); a diuretic.

orthotaxy Crystalline structure forming long, parallel columns.

Ortix Trademark for a nonwoven fiber base; a blend of Terylene and polypropylene fibers, cemented with a rubber-type resin, and coated with a porous polyurethane layer under a nonporous one.

Ortizon $CO(NH_2)_2 \cdot H_2O_2$ = 94.1. A colorless, crystalline compound of hydrogen peroxide and urea; an antiseptic.

ortnithine Ornithine*.

oryza Rice.

oryzamin An extract from rice bran.

Os Symbol for osmium.

osage orange The bark of *Maclura aurantiaca* (N. America). It contains a yellow coloring material and tannin; used in the textile and leather industries. **o. o. wood** The wood of *Maclura pomiferum;* a yellow coloring matter and fustic substitute.

osamine A compound derived from sugars by replacing an OH group by NH_2.

Osann, Gottfried Wilhelm (1797–1866) German chemist noted for his work on platinum metals.

osazone $R \cdot C{:}N \cdot NHPh$. A yellow or orange, crystalline, insoluble compound obtained by heating a polyhydroxy aldehyde or ketone (aldose or ketose sugar) with phenylhydrazine hydrochloride and sodium acetate. **o. test** The identification of sugars by the microscopical examination and melting-point determination of their o. precipitates.

Osborne, Thomas Burr (1859–1935) American biochemist, noted for work on vegetable proteins.

oscillation A vibratory to-and-fro motion.

oscillograph Oscillometer. A device for recording the waveforms of high-frequency currents, e.g., by cathode ray discharge of frequency 10^{-5} s.

oscillometer Oscillograph.

oscillometry See *high-frequency titration* under *titration.*

oscine Scopoline.

-ose Suffix indicating (1) a carbohydrate, particularly a monose; (2) the substance produced by enzymic digestion of a protein, e.g., albumose.

osmate* Indicating osmium as the central atom(s) in an anion, as, a salt containing the radical OsO_4^{2-}.

osmic Osmium(IV)*, (4+)*, (VIII)*, or (8+)*. A compound of tetravalent or octavalent osmium. **o. acid** OsO_4 = 254.2. O. anhydride, osmium tetraoxide*, perosmic anhydride, perosmic oxide. Yellow crystals of irritating odor, m.20, soluble in water, depositing black o. hydroxide. An octavalent compound. Used as a reagent for adrenaline and indican, as a histological stain for fat, and in photography. **o. acid anhydride** O. acid. **o. anhydride** O. acid.

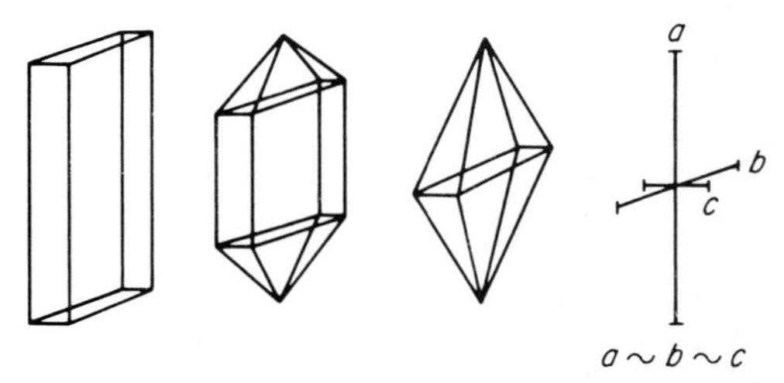

Fig. 17. The orthorhombic crystal system.

osmics The study of odors.
osmiridium Iridosmine. An alloy: osmium 17–50, iridium 77%, with other platinum metals; occurs native in platinum deposits. Insoluble in aqua regia; used to prepare osmium and its salts, and for pen nib points.
osmium Os = 190.2. A hard, white, metallic element of the platinum group, d.22.48, m.2700, b.5400, insoluble in water or acids; at. no. 76. Discovered (1803) by Smithson-Tennant. Used as a catalyst, in platinum alloys, and in pen points. Valencies: 2 (green); 3, 4 (orange); 8 (orange). It is found native as an alloy: nevyanskite. **o. dichloride*** $OsCl_2$ = 261.1. Green, hygroscopic needles, soluble in water. **o. dioxide*** OsO_2 = 222.2. Copper-red powder, insoluble in water. **o. disulfide*** OsS_2 = 254.3. Yellow powder, slightly soluble in water. **o. fluorides** (1) OsF_8 = 342.2. O. octafluoride*. Colorless crystals, m.34. (2) OsF_6 = 304.2. O. hexafluoride*. b.203. **o. hydroxide** (1) $Os(OH)_3$. Orange, insoluble powder. (2) $Os(OH)_4$. Black, insoluble powder. **o. monoxide*** OsO = 206.2. Black powder, insoluble in water. **o. potassium chloride*** OsK_2Cl_6 = 481.1. Red octahedra, soluble in water. **o. sesquioxide** Os_2O_3 = 428.4. Black powder, insoluble in water. **o. sodium chloride*** $OsNa_2Cl_6$ = 448.9. Red rhombs, soluble in water. **o. tetrachloride*** $OsCl_4$ = 332.0. Yellow needles, slightly soluble in water. **o. tetrasulfide*** OsS_4 = 318.4. Black powder, insoluble in water. **o. tetraoxide*** Osmic acid. **o. trichloride*** $OsCl_3$ = 296.6. Brown crystals, soluble in water.
osmometer A device for measuring osmotic pressure.
osmondite A solid solution of iron carbide in α iron.
osmoscope An osmometer, or device to demonstrate osmosis.
osmose Osmosis.
osmosis The diffusion of a liquid or gas through a semipermeable membrane, due to *osmotic pressure,* q.v. **electro ~** See *electroosmosis.* **reverse ~** Reversal of the o. flow of a solvent by applying a hydrostatic force exceeding the o. pressure of the solution; as in a desalination process; seawater is forced under pressure through a thin plastic membrane, which is then permeable to water but not salt.
osmotaxis The movement of cells due to osmotic pressure.
osmotic Pertaining to osmosis. **o. cell** A container separated from another by a semipermeable membrane or a finely porous wall. **o. equivalent** The ratio between the amount of water and the amount of solute passing in opposite directions through a semipermeable membrane. **o. pressure** The force exerted by dissolved substances on a semipermeable membrane which separates two solutions of different concentrations. The force of the molecular attraction between solute and solvent. It is proportional to the number of molecules in solution (concentration). Phenomena due to o. pressure are the rise of sap in plants, the expanding force of roots, and the absorption and secretion of food and waste materials.
Osmund **O. furnace** A high forge or primitive blast furnace. **O. iron** High-grade iron from an O. furnace.
Osnager, Lars (1903–) Norwegian chemist. Nobel prize winner (1968), noted for work on colloids and ionic motion.
osone A ketone-aldehyde formed from an osazone by hydrolysis with concentrated hydrochloric acid. Reduction forms a ketone-alcohol, so that the original aldose can be changed to an isomeric ketose: R·CH(OH)·CHO → R·C(:N·NHPh)·CH:N·HNPh; R·CO·CH_2OH ← R·CO·CH:O.
osotriazole N:N·NH·CH:CH (ring) = 69.1. 1,2,3-Triazole. The heterocyclic compound. **1*H*- ~** Colorless liquid, b.204, soluble in alcohol. ***N*-phenyl ~** 1-Phenylosotriazole. Colorless liquid, b.224.
osram An alloy of osmium and tungsten. **O. lamp** Trade name for an incandescent lamp having filaments of tungsten coated with osmium, cadmium, etc.
ossein Collagen. The albumenoid material (bone marrow) which remains after treating bones with dilute hydrochloric acid; it is hydrolyzed by boiling water to glue and gelatin.
ossification Formation of, or conversion into, bones, e.g., as of cartilage.
osso-albumin A protein from ossein.
osteolite An impure apatite or calcium phosphate.
osthole A methoxycoumarin derivative, the active principle of Hseh tsuang (Umbelliferae), China.
ostranite Zircon.
ostrole $C_{15}H_{16}O_3$ = 244.3. A substance like coumarin, m.84, from masterwort, the rhizome of *Imperatoria ostruthium.*
ostruthin $C_{19}H_{22}O_3$ = 298.4. A crystalline substance from the root of masterwort, *Imperatoria ostruthium* (Umbelliferae). Yellow crystals, m.119, insoluble in water. Cf. *imperatorin.*
Ostwald **O., Wilhelm (1853–1932)** German chemist. Nobel prize winner (1909). **O. indicator theory** The color changes of indicators are due to their existence as weak acids or bases, one ionic radical having a different color from the undissociated molecule. **O.'s dilution law** In a weak solution of an electrolyte AB, the dissociation constant $K = [A^+][B^-]/[AB] = \alpha^2/[(1-\alpha)V]$, where $[A^+]$ is the concentration of A ions, etc., α is the degree of dissociation of AB, and V is the volume. Cf. *activity.* **O. rule** If a substance can exist in more than one modification, the least stable is formed first, and changes ultimately into the more stable.
osyritin $C_{27}H_{30}O_{17}$ = 626.5. Yellow glucoside from *Osyris abyssinica,* Cape sumach or bergbas (Santalaceae), S. Africa; a tan.
otavite A native cadmium carbonate, usually hydrated.
otoba **o. butter** O. wax. **o. wax** O. butter, American nutmeg butter, American mace butter. The fat from the fruits of *Myristica otoba* or *M. fragrans,* m.38, having a nutmeg odor. Cf. *mace, myristica.*
otobain $C_{20}H_{20}O_4$ = 342.4. Otobite. The active constituent of otoba wax, m.137, soluble in hot alcohol.
otolith Ear stone. A concretion of calcium carbonate present in the ear.
otto of roses Attar of roses.
ouabain $C_{29}H_{44}O_{12}\cdot 8H_2O$ = 728.8. Uabain. Strophanthin G. A very poisonous glucoside from the seeds of *Strophanthus gratus.* Colorless, quadratic crystals, m. approx. 200, soluble in hot water; a rapid-acting cardiac glycoside (USP, EP, BP); constituent of Zulu arrow poison. Cf. *acocantherin, wabain.*
ounce A measure of weight in the English system. **apothecary's ~** 1 ap oz = ℥ = 1 troy oz = $\frac{1}{12}$ lb = 8 drams (ʒ) = 24 scruples (℈) = 480 grains (gr). 1 ap oz = 31.10348 g. **avoirdupois ~** 1 av. oz = $\frac{1}{16}$ lb = 7.2916 drams = 18.229 pennyweights (dwt) = 21.875 scruples = 437.5 grains = 28.3495 g = 138.449 carats = 142.045 metric carats. **fluid ~** 1 fl oz = $\frac{1}{16}$ lb = 0.05 pint = 0.25 gill = 480 minims (♏) = volume of 1 av oz distilled water at 62°F = 29.5735 ml (U.S. usage), 28.4131 ml (U.K. usage). **troy ~** 1 oz t. = 20 pennyweights = 31.10348 g = 1 ap oz.
ouroboros A symbol of Greek alchemy; viz., a snake seizing its own tail.
-ous See *-ic.*
outcrop A mineral deposit that comes to the surface of the earth.
output The total amount produced in a given time.
ouvarovite Uvarovite.

ovalbumin An albumin from egg white.
ovalene* $C_{32}H_{14}$ = 398.5.

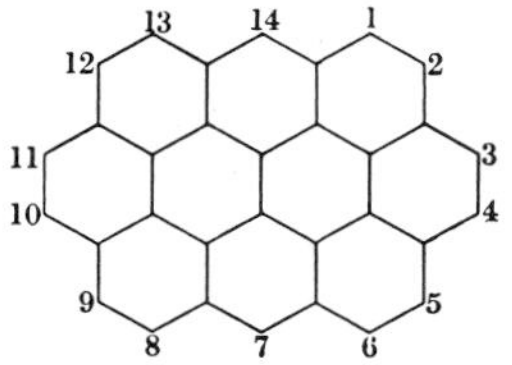

Orange needles, m.473 (sublimes), insoluble in water; the most compact configuration of benzene rings.
oven A compartment in which substances are heated. Cf. *furnace.*
overburden The top layer of soil or waste which has first to be removed in mining or quarrying.
overcooled Supercooled.
overgrowth The growth of a crystal over the surface of a crystal of a different, but usually isomorphous, substance.
overheated (1) Superheated. (2) Heated excessively.
overpotential The difference in volts between the back potential of an electrode and that of a saturated calomel electrode, immediately after electrolysis. Cf. *overvoltage, undervoltage.*
oversaturated Supersaturated.
Overton coefficient Distribution coefficient.
overvoltage (1) Overpotential. (2) Supertension. The excess voltage above the normal reversible electrode potential of a metal electrode required to decompose a solution or cause deposition on the electrode; used for the graded electrolytic deposition of metals from a mixed solution. Cf. *polarization.*
ovo An egg (Latin: *ovum*). **o.globulin** A protein of egg white precipitated by dialysis. **o.keratin** The membranous lining of birds' and sharks' eggs.
Ovoco classifier A double-screw, continuous conveyor to separate ores.
ovomucin A water-insoluble globulin from egg white.
ovomucoid A non-heat-coagulable glycoprotein in egg white.
Owen process A flotation process, in which the ores are agitated in water containing 60 g eucalyptus oil per ton.
oxa- Prefix indicating: (1) An oxygen bridge; as, $-CH_2 \cdot O \cdot CH_2-$. (2)* Oxygen in a monocyclic compound containing 3–10 members.
oxacid Oxyacid.
oxalacetic **o. acid*** $COOH \cdot CO \cdot CH_2 \cdot COOH$ = 132.1. 2-Oxosuccinic acid, 2-oxobutanedioic acid*. An unstable liquid whose methyl ester forms colorless crystals, m.74. **o. (ethyl) ester** $EtOOC \cdot CO \cdot CH_2 \cdot COOEt$ = 188.2. Colorless liquid, b.131.
oxalaldehyde Glyoxal*.
oxalamide Oxamide*.
oxalanilide Oxanilide.
oxalate* A salt of oxalic acid containing the $(COO)_2=$ radical. Only those of the alkali metals and magnesium are soluble in water. **hydrogen ~*** Acid o., bi(n) ~. A salt containing the radical HC_2O_4-. **neutral ~** Oxalate.
oxaldehyde Glyoxal*.
oxalene The tetravalent group $-N:C-C:N-$.
oxalethyline Diethyl oxamide.
oxalic **o. acid*** $HOOC \cdot COOH$ = 90.0. Dicarboxylic acid, ethanedioic acid*. Colorless, monoclinic, poisonous crystals, m.99, soluble in water. Occurs in many plants; excess intake of plants containing o. a., such as spinach, may cause o. a. kidney stones. It is prepared by passing carbon monoxide into concentrated sodium hydroxide or by heating cellulose (sawdust) with sodium hydroxide. Used as a reagent, in synthesis, as a precipitant in purifying glycerol and an ink eradicator, and in bleaching, photography, and the dye and textile industries. **chloro ~** C_2HO_4Cl = 124.5.
o. acid series General formula: $(CH_2)_n(COOH)_2$. **o. aldehyde** Glyoxal*. **o. dianilide** Oxanilide. **o. acid monoamide** Oxamic acid*. **o. a. monoanilide** Oxanilic acid.
oxalium Potassium hydrogenoxalate*.
oxalmethylin $(CONHMe)_2$ = 116.1. Dimethyloxamide. White crystals, m.210.
oxalo* The radical $HOOC \cdot CO-$. **o.acetate decarboxylase*** Enzyme catalyzing the splitting of CO_2 from oxaloacetate to give pyruvate.
oxaluramide $NH_2 \cdot CO \cdot NH \cdot CO \cdot CO \cdot NH_2$ = 131.1. Oxalan, oxamic acid ureide, m. exceeds 310.
oxalyl* 1,2-Dioxo-1,2-ethanediyl†. Ethanedioyl. The radical $-CO \cdot CO-$. **o. dichloride*** $COCl \cdot COCl$ = 126.9. Colorless liquid, b.64, decomp. by water. **o.diacetophenone** $PhCOCH_2CO \cdot COCH_2COPh$ = 294.3. Colorless crystals, m.180. **o.urea** Parabanic acid.
oxam Oxomonocyanogen.
oxamethane $C_2O_2NH_2(C_2H_5O)$ = 117.1. Acetyloxamide, ethyl oxamate. Colorless crystals, m.115.
oxamic acid* $HOOC \cdot CONH_2$ = 89.1. Aminooxoacetic acid†, oxaminic acid, oxalic acid monoamide. Colorless crystals, decomp. 210, slightly soluble in water. **phenyl ~** Oxanilic acid.
oxamide* $(CONH_2)_2$ = 88.1. Ethane diamide. Colorless crystals, decomp. 417, insoluble in water.
oxamides Compounds derived from oxamide by replacing the hydrogen atoms by alkyl or aryl radicals; e.g., $(CONHMe)_2$, dimethyl oxamide, m.210.
oxamidine Amide oxime.
oxamido The (aminooxoacetyl)amino† radical, $H_2NCO \cdot CONH-$.
oxaminic acid Oxamic acid*.
oxamyl A *fungicide* and *insecticide,* q.v.
oxamoyl* The radical $H_2NCO \cdot CO-$, from oxamic acid.
oxane Ethylene oxide*.
oxanilic acid $HOOC \cdot CONHPh$ = 165.2. Phenyloxamic acid, oxalic acid monoanilide. Colorless rhombs, m.149, slightly soluble in hot water.
oxanilide $(CONHPh)_2$ = 240.3. Colorless scales, m.245, insoluble in water; used in organic synthesis.
oxanthranol $OC \cdot C_6H_4(COH)C_6H_4$ = 209.2.

Anthrahydroquinone. Yellow needles, m.167 (decomp.).
oxatyl The carboxyl* radical.
oxazetidine An organic fluorine compound containing the $-N-O-C-C-$ group; e.g., $CF_3 \cdot NO \cdot CFCl \cdot CF_2$. The $-N:O-$ group behaves like a $=C:C=$ group in the formation of polymers.
oxazine A series of heterocyclic compounds. Designated according to the position of the indicated H, O, and N atoms, as:

4*H*-1,2-~ $O \cdot N:CH \cdot CH_2 \cdot CH:CH$

Cf. *thiazine, benzoxazine.* **naphth ~** Phenoxazine*. **par ~** 2*H*-1,4-Oxazine. **phen ~** See *phenoxazine.*
o. dyes The o. analogs of thiazine dyes.
oxazole* C_3H_3ON = 69.1. Liquid, b.70. **is ~** b.95.

oxethyl Ethylol. The 2-hydroxyethyl radical, $CH_2OH \cdot CH_2-$. Cf. *ethoxy.*

oxetone $C_7H_{12}O_2$ = 128.2. 1,6-Dioxaspiro[4.4]nonane. The heterocyclic spiro compound

$$\begin{array}{ccccc} H_2C\text{—}O & & O\text{—}CH_2 \\ | & \searrow C \swarrow & | \\ H_2C\text{—}CH_2 & & CH_2\text{—}CH_2 \end{array}$$

Oxford process The separation of nickel from copper by means of sodium sulfide. Cf. *Mond process.*

oxgall Bile from the gallbladder of oxen; used in the textile and printing industries.

oxid Oxide*.

oxidase* See *enzymes,* Table 30.

oxidation Originally, o. meant combining with oxygen; later it also indicated combination with electronegative elements. Now it has a broader meaning: an augmentation of the valence number of an ion or atom as the result of the loss of one or more electrons, thereby making it more electropositive. Cf. *hydroxylation, reduction.* **o. base** A dye produced by oxidative means; as, nigrosine. **o. number*** The o. n. of an element in any chemical entity is the charge which would be present on an atom of the element if the electrons in each bond to that atom were assigned to the more electronegative atom. Thus, for MnO_4^-, the o. n. of Mn is VII; that of O is −II. For CH_4, the o. n. of C is −IV; that of H is I. Rules consider: (1) H as positive in combination with nonmetals; (2) organic radicals as anions; (3) the groups NO and CO as neutral; (4) a bond between atoms of the same element to make no contribution to the o. n. Where doubt may exist, the o. n. should be shown by Stock *nomenclature,* q.v. For o. n. (or o. states) of the elements, see *electron configuration* (under *electron*), Table 27. Cf. *valency.* **o. process** A reaction that increases the proportion of oxygen or acid-forming elements or radicals in a compound. **o. reaction** Electronation reaction. A reaction accompanied by a correlated reduction in the valence number of another element. **o.-reduction indicators** See *oxidation-reduction indicator* under *indicator.* **o.-reduction potential** r_H, rH. Redox potential. The potential acquired by an inert electrode, e.g., platinum, immersed in a reversible oxidation-reduction system, e.g., Fe^{++}/Fe^{3+}; measured by the ratio of the oxidized and reduced forms. rH = log $1/pH_2$, where pH_2 is the hydrogen gas pressure. **o. state** The degree of oxidation corresponding to a given *o. number,* q.v. **o. value** A constant of oils. The degree of oxidation (as grams of I per 100 g sample) when a fat dissolved in carbon tetrachloride is oxidized by potassium dichromate in glacial acetic acid.

oxidative coupling The formation of a high-molecular-weight polymer when an organic compound with activated hydrogens reacts catalytically with an oxidizing agent.

oxide (1) A binary compound of oxygen generally with a metal, M_2O (basic), or nonmetal, NO_n (acidic), containing the anion $O^=$. (2) Used instead of "ether" in some languages. **acid ~** An oxygen compound of nonmetals; as, SO_2, P_2O_5, which give oxyacids with water. **amphoteric ~** An oxygen compound of the heavy metals; as, ZnO, Fe_2O_3, which may form weak acids and weak bases. **basic ~** An oxygen compound of metals; as, Na_2O, Al_2O_3, which give bases with water. **hydrous ~** An amorphous colloidal substance, which is neither a definite hydroxide nor a definite crystalline hydrate. **inert ~** An oxygen compound which forms neither acid nor basic compounds, as CO, N_2O. **metal-modified ~** A refractory made by adding small amounts of refractory metals to refractory oxides. **per ~** See *peroxide.* **primary ~** See *primary oxide.* **sub ~** See *suboxide.*

oxidimetry The use of an oxidizing agent in volumetric analysis.

oxidize To cause to unite with oxygen; to increase the proportion of electronegative elements or radicals.

oxidizer Oxidizing agent.

oxidizing The act of oxidation. **o. agent** A substance that (1) yields oxygen readily, (2) removes hydrogen from a compound, or (3) attracts negative electrons; e.g.: the common oxidizing agents are O_2, O_3, Cl_2, $KMnO_4$, $K_2Cr_2O_7$, $KClO_3$, HNO_3, H_2O_2. **o. flame** The outer zone of a gas flame containing an excess of air. Cf. *reducing flame.* **o. reaction** See *oxidation.*

oxidoreductases* See *enzymes,* Table 30 on p. 214.

oxidoreduction See *oxidation-reduction potential.*

oxime* Hydroxyimino†. A compound (*cis* or *trans*) containing the o. radical, =C:N·OH; a condensation product of aldehydes or ketones with hydroxylamine. See *aldoximes.* **acet ~** See *acetoxime.* **ald ~** See *aldoxime.* **amid ~** Amide oxime*. A compound containing 2 o. radicals. **dimethyldi ~** Dimethylglyoxime*. **form ~** See *nitrolic acid.* **glucose ~** See *glucose oxime.* **lact ~** See *lactoxime.*

oximide $(CO)_2$:NH = 71.0. The imide of oxalic acid. Colorless prisms obtained from oxamic acid by dehydration. **cyan ~** See *cyanoximide.*

oximido The oxime* group.

oximinoketone A compound of the type R·CO·C:NOH, which gives a blue color with ferrous iron.

oxine 8-Quinolinol*. **thio ~** 8-Mercaptoquinoline.

oxindole $\underline{NH \cdot C_6H_4 \cdot CH_2 \cdot CO}$ = 133.2. 2-Indolinone, 2-oxoindoline. Colorless needles, m.126, soluble in hot water.

oxirane (1)* Ethylene oxide*. (2) Epoxide. Describing the oxygen atom of the epoxide ring.

$$\begin{array}{c} | \quad\quad | \\ \text{—}C\text{—}C\text{—} \\ \searrow O \swarrow \end{array}$$

methyl ~ Epoxypropane.

oxirene $\underline{CH:CH \cdot O}$. **methyl ~** 1,2-Epoxy*propene**.

oxo (1)* Prefix indicating the =O group; as, in aldeydes and 2-oxopropanoic acid, MeCO·COOH. (2)* Infix indicating the anionic ligand O^{2-}. **o. acids** (1)* Acids containing an o. group, as, phosphonic acid, $HP(:O)(OH)_2$. (2) Ketone acids. **1-o.butyl†** See *butyryl.* **o. compounds** Compounds having an o. group, excluding carboxylic acids. See under element; as, oxo*vanadium* ions. **1-o.decyl†** See *decanoyl.* **1-o.-9-octadecenyl†** See *oleoyl.* **1-o.pentyl†** See *valeryl.* **o. process, o. reaction** The manufacture of alcohols by catalytically reacting an olefin with water-gas under pressure and reducing the resulting aldehyde: $C_2H_4 \rightarrow CH_2 \cdot CHO \rightarrow CH_3 \cdot CH_2 \cdot CH_2OH$. If carbon monoxide and water are used, an acid results. **1-o.-2-propenyl†** See *acryloyl.*

oxomalonic acid Mesoxalic acid*.

oxomonocyanogen CNO = 42.2. Oxam. A gas prepared by heating cyanogen in oxygen.

Oxone Trademark for a bleaching preparation, whose active constituent is potassium peroxosulfate.

oxonite An explosive: picric acid dissolved in nitric acid.

oxonium **o. compounds** An addition or double compound of an organic oxide with strong acids or their salts; as, $[Me_2OH]Cl$, dimethyl oxonium chloride. **o. ion*** H_3O^+. The monohydrated proton.

oxophenic acid Pyrocatechol*.

Oxsoralen Trademark for methoxsalen.

oxozone O_4. A supposed modification of oxygen. Cf. *ozone.*

oxozonide $R_2C(O{:}O){\cdot}(O{:}O)CR_2$. Any of the unstable compounds formed by the addition of 4 O atoms to the double bonds of an unsaturated organic molecule. Cf. *ozonides.*

oxy- (1)* Prefix, infix, or suffix indicating an oxygen bridge —O—. Cf. *epoxy, ether, methoxy.* (2) Misnomer for *hydroxy-,* q.v. **o.acetylene** A mixture of oxygen and acetylene gases used in a blowpipe to obtain high temperatures (3300°C); used to cut armor plate and weld. **o.hydrogen** See *oxyhydrogen.* **o.muriatic acid** Obsolete term for hydrochloric acid.

oxyacanthine $C_{35}H_{40}O_6N_2$ = 584.7. Vinetine. An alkaloid in *Berberis* spp. White crystals, m.216, soluble in alcohol.

oxyacetone 1-Hydroxy-2-*propanone**.

oxyacid An acid containing oxygen. **inorganic ~** A tertiary compound of an acid radical with hydrogen; as, H_3PO_4.

oxyalizarin Purpurin.

oxyamides Hydroxyamides. Compounds containing the radicals $-CONH_2$ and $-OH$; as, $CHOH{\cdot}CONH_2$, glycol amide.

oxyammonia Hydroxylamine.

oxyammonium compounds Hydroxylammonium compounds.

oxyanthracene Anthrol*.

oxyazo- Hydroxyazo-.

oxybenzene Misnomer for phenol.

oxybenzoic Misnomer for hydroxybenzoic (salicylic).

oxybenzyl Misnomer for hydroxybenzyl. **o. alcohol** Misnomer for salicylic alcohol.

oxybromide Misnomer for (1) a compound containing the =O and —Br groups; (2) the radical —OBr (hypobromite*).

oxybutyric o. acid See *hydroxybutanoic acid** under butanoic acid. **o. aldehyde** Aldo.

oxycanthine Oxyacanthine.

oxycarbonyl-* Prefix, R-oxycarbonyl, indicating an ester with the —COOR group.

oxycarboxin* See *fungicides,* Table 37 on p. 250.

oxychloride Misnomer for (1) a compound that contains the radicals =O and —Cl. (2) the radical —OCl (hypochlorite*).

oxychromatin Linin (1).

oxydase Oxidase*.

oxydation Oxidation.

oxydemeton-methyl* See *insecticides,* Table 45 on p. 305.

oxydizing Oxidizing.

Oxygal Trade name for sodium peroxide.

oxygen* O = 15.9994. Oxygenium, vital air, sauerstoff. A gaseous element, at. no. 8, free in the atmosphere as O_2 (23.2% by weight) and combined in many substances (50% of the earth crust). It was Hooke's *nitre air,* isolated 1774 by J. Priestley *(dephlogisticated air),* and in 1773 (but not published until 1777) by K. W. Scheele (fine air). Its name (*oxy,* "acid"; *gennao,* "I form"—Greek) is due to A. L. Lavoisier, who developed the theory of combustion. Colorless gas, $d_{air=1}$1.1053, m.—218, b.—183, slightly soluble in water. It occurs as 3 isotopes: masses 16, 17, and 18, in the proportions of 10,000:1:8. Prepared by fractional distillation of liquid air, or by electrolysis of water, and shipped in steel cylinders. Used medicinally (USP, EP, BP) for severe oxygen lack, as in respiratory diseases and heart failure; also during resuscitation from injury and for premature babies; and with dinitrogen oxide or other anesthetics; and with hydrogen or acetylene for producing high temperatures for welding and for melting metals. **active ~** Ozone, O_3. **allotropic ~** Ozone, oxozone. **di ~*** The oxygen molecule, O_2. **dissolved ~** A measure of the purity of a river and thus its suitability for fish. Commonly determined by the Winkler method or by membrane electrode. **hydroxyl ~** The oxygen of the hydroxyl group. **ketonic ~** The oxygen of the >CO group. **liquid ~** See *liquid oxygen.* **radio ~** A short-lived isotope obtained by bombardment of o. with α particles. See *radioelements.*

o. absorbent A substance that removes gaseous oxygen from a gas mixture; as, an alkaline solution of pyrogallol or chromous sulfate. **o. carrier** A catalytic substance that absorbs O_2 molecules and splits off O as atoms. **o. fluorides** (1) OF = 35.00. O. fluoride*. Colorless gas. (2) OF_2 = 54.00. O. difluoride*. Yellow liquid, b.147. Less reactive than fluorine, having the same odor, sparingly soluble in water. (3) O_2F_2 = 70.0. Dioxygen difluoride*. Brown gas.

oxygenate To enrich with oxygen.

oxygenated water Water saturated with oxygen gas.

oxygenation Saturation with oxygen.

oxygenium Latin for "oxygen."

oxygenize (1) Oxygenate. (2) Oxidize.

oxyhematoporphyrin A pigment from oxyhemoglobin; closely allied to urohematoporphyrin sometimes found in urine.

oxyhemoglobin, oxyhaemoglobin Hematoglobulin, hematocrystallin. Mol. wt. 65,000. Hemoglobin combined with oxygen.

oxyhydrogen An explosive mixture of oxygen and hydrogen. **o. blowpipe** A burner with a jet of oxygen inside a jet of hydrogen, ensuring complete combustion of both gases and an extremely hot flame. **o. flame** A blowpipe flame obtained by burning hydrogen with oxygen (2760°C). **o. light** The "lime" light produced by heating calcium oxide with the o. blowpipe.

oxylepidine Dibenzoyl*stilbene.*

oxyleucotin $C_{34}H_{32}O_{12}$ = 632.2. A crystalline principle from paracoto bark.

oxylith A mixture of sodium peroxide and bleaching powder, used to generate oxygen for welding purposes.

oxymalonic acid Tartronic acid*.

oxymel A medicated honey: clarified honey 80, acetic acid 10, water 10%; used in cough medicines.

oxymethylconiferin Syringenin.

oxymuriate Chlorate*.

oxymuriatic acid Chloric acid*.

oxymyoglobin The red coloring matter of muscle, i.e., meat.

oxyn A crystalline dihydroxy peroxide glyceride in tung oil.

oxynaphthoic acid Hydroxy*naphthoic acid.*

oxynitroso The nitrosooxy* radical.

oxyphile (1) Describing an element which does not occur in the native state but which forms minerals containing oxygen (e.g., mineral oxides); as uranium. (2) A cell structure that can be stained with acid dyes.

oxyphor Hydroxycamphor.

oxypyridine Pyridone*.

oxyquercetin Myricetrin.

oxyquinazoline Quinazoline.

oxyquinones Derivatives of $CO{\cdot}CH{:}CH{\cdot}CO{\cdot}CH{:}CH$, in which one or more H atoms are replaced by an OH, OMe, or OEt group.

oxysalt A salt of an oxyacid.

oxytetracycline $C_{22}H_{24}O_9N_2{\cdot}HCl$ = 496.9. 2-Acetyl-2-decarboxamidooxytetracycline hydrochloride. Terramycin. Yellow crystals, darkening in light, insoluble in water; an antibiotic (USP, EP, BP).

oxy-Tobias acid 2-Naphthol-1-sulfonic acid.

oxytocic A drug that increases the expulsive power of the uterus, as, oxytocin.
oxytocin $C_{43}H_{66}O_{12}N_{12}S_2$ = 1007. α-Hypophamine. A hormone produced by the posterior lobe of the pituitary body of animals (USP, EP, BP). The first polypeptide hormone to be synthesized. Used to induce labor and to control bleeding after delivery. Cf. *vasopressin.*
oxytoxin An oxidation product of a toxin. Cf. *oxytocin.*
oxytropism The response of living cells to oxygen.
oz Abbreviation for ounce(s). **oz ap** Abbreviation for apothecaries' ounces = ℥. **oz av** Abbreviation for avoirdupois ounce. **oz fl** Abbreviation for fluidounce. **oz t** Abbreviation for troy ounce.
ozamin Benzopurpurine.
ozocerite Mineral wax, native paraffin, fossil wax. A native mixture of hydrocarbons. Brown to green mass, soluble in carbon disulfide. Used as an insulator, for paints, polishes, candles. **purified ~** Ceresin.
ozogen Hydrogen peroxide.
ozokerite Ozocerite.
ozonation Impregnation or saturation with ozone. Cf. *ozonidation.*
ozonator Ozonizer. A device to generate ozone.
ozone O_3 = 48.00. Trioxygen*. A modification of oxygen gas produced usually by a silent electric discharge in air or oxygen. Faint blue gas, of intense odor, $d_{air=1}$1.658, b.−110, decomp. 270, soluble in water. Used as a bactericide, and oxidizing agent; and for bleaching oil, fats, textiles, and sugar solutions. Cf. *oxozone.* **o. paper** A filter paper impregnated with potassium iodide and starch, or with indigo solution, which turns blue on exposure to ozone. Cf. *thallium ozone paper.* **o. pollution** O. contributes to photochemical smog. The World Health Organization has proposed an air quality standard of 60 ppb o. (hourly average), while the EPA has set 120 ppb for human health and 80 ppb for welfare (visibility, avoidance of plant damage, etc.). **o. shield** O. in the stratosphere that shields the earth from much of the sun's u.v. radiation.
ozonidation Conversion into an ozonide. Cf. *ozonization.*
ozonides* Thick, oily, unstable compounds of ozone with unsaturated organic compounds containing a double bond e.g.:

$$=C\underset{O_3}{\diagdown\!\!\diagup}C=$$

Cf. *oxozonide.* **o. ion*** The anion O_3^-.
ozonization Treatment or sterilization with ozone. Cf. *ozonidation.*
ozonizer Ozonator.
ozonolysis The treatment of hydrocarbons with ozone.
ozotetrazone Vicinal *tetrazine.*

P

P Symbol for (1) peta- (10^{15}); (2) phosphorus.
P Symbol for (1) power; (2) probability density; (3) sound energy flux.
P (Boldface) Symbol for electric polarization.
p Symbol for (1) pico- (10^{-12}); (2) proton.
p Symbol for (1) dipole moment of molecule; (2) momentum; (3) *para* position; (4) pressure; (5) at constant pressure (as subscript).
p_0 Symbol for standard atmospheric pressure.
Π, π See *pi.*
Φ, φ, φ See *phi.*
Ψ, ψ See *psi.*
Pa Symbol for protoactinium.
PABA See *p-aminobenzoic acid, sulfa drugs.*
pacemaker (1) A specialized tissue area of the heart, from which originates the heartbeat. (2) An electrical device which stimulates the heart to beat at a regular normal rate. It is either implanted under the skin or worn externally.
pachnolite $NaF \cdot CaF_2 \cdot AlF_3 \cdot H_2O$. A mineral.
pachyrhizid $C_{30}H_{24}O_{10}$ = 544.5. A glucoside from *Pachyrhizus angulatus.*
pacite Arsenopyrite.
pack fong Nickel silver.
packing (1) A filling material. (2) Crowding together. **atomic ~** The crowding of atoms which have lost electrons and can approach one another more closely; as in certain stars, which have a density of up to 50,000. Cf. *spectral classification.* **nuclear ~** The crowding of hydrogen and helium nuclei and electrons within the atomic nucleus.
p. effect Loss of mass due to crowding of protons and electrons in atomic nuclei; as, in the hypothetical case 4H → He, where the atomic weight drops from 4.032, to 4.003. Cf. *mass*-energy cycle. **p. fraction** The deviation per unit mass of the atomic weight from whole numbers, in parts per 10,000. It measures the forces holding electrons and protons together.
paddy Threshed rice.
padutin Kallekrein.
pae, pæ- See *pe-.*
pagodite $2Al_2O_3 \cdot K_2O \cdot 3H_2O \cdot 5SiO_2$. Pinite. A mineral.
paint A suspension of finely ground, white pigment (as, zinc oxide, titanium dioxide, or kaolin) with added colored pigments if required, in a vehicle (linseed oil, varnish, turpentine, and organic constituents). Lead level limit in paint accessible to children: U.S.A.—0.06%; EEC—0.5%. **luminous ~** A suspension of zinc sulfide or barium sulfide in nitrocellulose lacquer. **water ~** An aqueous mixture of pigment and adhesive, e.g., casein.
p. mill A mill that grinds to 200-mesh. **p. oil** An oil for thinning paints; as, linseed oil. **p. rock** Ocher. **p. thinner** Turpentine or its substitutes.
paired electrons An electron couple. A nonpolar bond between 2 atoms, each atom furnishing one of the electrons of the pair. Cf. *twin electrons.*
palaeontology Paleontology.
palaeopathology The study of ancient disease from nonliterature sources, as, mummies, bones.
palau General name for Pd-Au alloys. A platinum substitute where resistance to heat and chemicals is not important.
Paleocene See *geologic eras,* Table 38.
paleontology The science of prehistoric life as revealed by fossil remains.
Paleozoic See *geologic eras,* Table 38.
palicourine A crystalline alkaloid from *Pali courea* species (Rubiaceae).
paligorskite Mountain leather. An asbestos substitute from the blue limestone fractures of Alaska. Average composition: MgO 8.1, Al_2O_3 14.3, SiO_2 49.5, Fe_2O_3 2.6, CaO 3.3%.
Palissy, Bernard (1499–1589) French leader in chemical technology and forerunner of Boyle.
palite $ClCOOCH_2Cl$ = 128.9. Chloromethyl chloroformate. A poison gas used during World War I. **super ~** See *superpalite.*
palladate* Indicating palladium as the central atom(s) in an anion; as, hexachloropalladate(IV)*, $M_2[PdCl_6]$.
palladic Palladium(IV)* or (4+)*. Describing a compound containing tetravalent palladium.
palladichloride M_2PdCl_6. Hexachloropalladate(IV)*.
palladious Palladous.
palladium* Pd = 106.42. A noble metal of the platinum group, at. no. 46, discovered (1803) by Wollaston. A silver-white, ductile, malleable metal, d.12.16, m.1553, b.3140, insoluble in water or acids; occurs native and alloyed with platinum metals. Used in alloys with gold or platinum in dentistry and jewelry; and with copper or silver in watches, pens, mirrors, and surgical instruments; as Pd sponge as a catalyst and hydrogen absorbent. Compounds: divalent [p.(II)*, p.(2+)*, palladous] and tetravalent [p.(IV)*, p.(4+)*, palladic]. **allo ~** A hexagonal or cubic variety of p., formerly confused with potarite.
p. asbestos Asbestos coated with metallic p.; used to absorb hydrogen. **p. black** Black, finely divided p.; a catalyst in oil hydrogenation. **p. dibromide*** $PdBr_2$ = 266.2. Brown powder, insoluble in water. **p. dichloride*** $PdCl_2 \cdot 2H_2O$ = 213.4. Brown, hygroscopic crystals, soluble in water; used as a reagent for malic acid, cocaine, iodine, mercury vapor, and carbon monoxide; and for photographic toning, porcelain prints, marking inks, and coating metals. **p. dicyanide*** $Pd(CN)_2$ = 158.5. Yellow solid, unstable to heat. **p. diiodide*** PdI_2 = 360.2. Black powder, m.100, insoluble in water. **p. dinitrate*** $Pd(NO_3)_2$ = 230.4. Brown rhombs, hydrolyzed in dilute solutions; a reagent for separating Cl and I. **p. dioxide*** PdO_2 = 138.4. Palladic oxide. Black powder, decomp. 200, insoluble in water. **p. disulfide*** PdS_2 = 170.5. Brown powder, insoluble in water. **p. family** The elements Ru, Rh, and Pd of group 8 of the 5th period of the periodic table. **p. gold** Porpezite. Native gold containing up to 10% Pd. **p. hydroxide*** $Pd(OH)_2$ = 140.4. Brown powder, insoluble in water. **p. monosulfide*** PdS = 138.5. Black powder, insoluble in water. **p. monoxide*** PdO = 122.4. Black powder, decomp. 875, insoluble in water. **p. oxides** See *palladium monoxide, palladium dioxide* above. **p. potassium chloride** See *potassium tetra-* and *hexachloropalladate.* **p. sponge** A gray spongy mass of finely

divided p.; a gas absorbent and catalyst for hydrogenation. **p. subsulfide** Pd_2S = 244.9. Insoluble, gray solid. **p. sulfate** $Pd(SO_4) \cdot 2H_2O$ = 238.5. Brown crystals, soluble in water. **p. tube** A glass tube filled with p. asbestos or p. sponge.

pallado Palladous.

palladous Palladium(II)* or (2+)*. Palladious. Describing compounds containing divalent palladium.

pallamine Colloidal palladium.

pallas An alloy of gold, palladium, and platinum; harder than platinum.

pallasite A meteorite: Ni-Fe sponge enclosing olivine.

palm See *Palmae.* **p. butter** P. oil. **p. grease** P. oil. **p.-kernel cake** The residue after expressing p.-nut oil; a cattle food. **p.-kernel oil** P.-nut oil. **p.-nut oil** P.-kernel oil. Yellow oil from the crushed fruits of W. African oil palm, d.0.952, m.26–36; consists of the glycerides of palmitic, oleic, and stearic acids. Used to manufacture soap, chocolate, and pharmaceuticals. **p. oil** Oleum palmae, p. butter, p. grease, expressed from the crushed and fermented fruit of the W. African oil palm, *Elaeis guineensis.* It consists of the glycerides of palmitic, oleic, and stearic acids and free palmitic acid; d.0.859–0.870, sapon. val. 196–205, m.33–46. Used to manufacture soap, candles, and lubricants. **red ~** A source of carotene and, thus, retinol. **p. wax** A yellow wax from *Ceroxylin andicola,* a palm of Ecuador; a beeswax substitute.

Palmae Palm family of tropical trees, e.g.: *Areca catechu,* yielding areca nut; *Copernicia cerifera,* yielding carnauba; *Elaeis guineensis,* yielding palm oil; *Cocos nucifera,* yielding coconut oil. See also, *raphia, coyol, palm, date palm, sago.*

palmarosa oil Pamorusa oil. An essential oil d.0.885–0.896, from ginger grass, *Andropogon schoenanthus,* S. Africa; an adulterant for oil of rose. It contains perilla alcohol, geraniol, and esters.

palmellin The red coloring, resembling hemoglobin, from the freshwater algae, *Palmella cruenta.*

palmetto The fruit of *Serenoa serrulata* (Florida); a leather tan.

palmic Palmitic.

palmierite $K_2Pb(SO_4)_2$. A mineral from Italy.

palmin Purified coconut fat for the manufacture of butter substitutes.

palmitamide $C_{16}H_{33}ON$ = 255.4. Palmitic acid amide, hexadecanamide*. Colorless solid, m.93.

palmitate* Hexadecanoate*, cetylate. A salt of palmitic acid, which contains the radical $C_{15}H_{31}COO-$.

palmitic p. acid* $C_{15}H_{31} \cdot COOH$ = 256.4. Hexadecanoic acid*, palm(itin)ic acid, cetylic acid, ethalic acid. A saturated fatty acid in many vegetable fats and oils. Colorless needles, m.62 (decomp.), insoluble in water; used in soap manufacture. **hydroxy ~** Juniperic acid.
p. cyanide Palmitonitrile*.

palmitin (1) $C_3H_5(C_{15}H_{31}COO)_3$ = 807.3. Tripalmitin. Tripalmitic acid ester of glycerol, which occurs in many vegetable fats and oils. Colorless fat, m.61, insoluble in water; used in soap manufacture. (2) A glyceride of palmitic acid; as, monopalmitin.

palmitinic acid Palmitic acid*.

palmitoleic acid (Z)-9-Hexadecenoic acid*.

palmitolic acid $C_{15}H_{27}COOH$ = 252.4. 7-Hexadecynoic acid*. An unsaturated fatty acid, in oils and japan wax. Colorless needles, m.47, insoluble in water.

palmitone $C_{31}H_{62}O$ = 450.8. 16-Hentriacontanone*. The ketone of palmitic acid; m.82.

palmitonitrile* $C_{15}H_{31}CN$ = 237.4. Palmitic cyanide, hexadecanenitrile*. Colorless scales, m.30, insoluble in water.

palmitoyl* The radical $C_{15}H_{31}CO-$, from palmitic acid. **p. chloride** $Me(CH_2)_{14}COCl$ = 274.9. Hexadecanoyl chloride*. Colorless liquid, m.11.

palmityl alcohol Hexadecyl alcohol*.

palmoil Palm oil.

Palmquist apparatus A portable gas analysis apparatus for the determination of carbon dioxide in air.

palmyra The palm *Borassus flabellifer* (S. Asia). The wood is used for structures, the leaves for thatch and mats, the fruits for food, the sap for toddy and jaggery sugar, the fiber for brushes; the young seedlings are ground to edible flour.

Paludrine Trademark for proguanil hydrochloride.

palustric acid $C_{20}H_{30}O_2$ = 302.5. 8,13-Abietadien-18-oic acid*. m.164, soluble in EtOH. A resin acid derived from coniferous wood.

palustrol $C_{15}H_{26}O$ = 222.4. A sesquiterpene oil from *Ledum palustre,* d.0.9544.

palygorskite Attapulgite (U.S.S.R. usage).

palynology The study of spores and pollen.

PAN* Symbol for polyacrylonitrile.

pan- Prefix indicating "all," or the "whole."

panabase Tetrahedrite.

panacea A universal remedy, sometimes applied to a quack medicine.

panacon $C_{22}H_{19}O_8$ = 411.4. Colorless crystals from ginseng, the roots of *Aralia* or *Panax quinquefolium.*

Panadol Trademark for acetaminophen (paracetamol).

Panama bark Quillaia.

panaquinol $C_{12}H_{25}O_9$ = 313.3. A bitter principle from ginseng, the root of *Panax quinquefolium.* Yellow powder, soluble in water. Cf. *panacon.*

panax (1) Ginseng. (2) A genus of araliaceus plants.

panchromatic Sensitive to light of all colors. Cf. *pantachromatic.*

panchromium Early name for vanadium.

panclastite Nitrogen tetraoxide dissolved in a combustible liquid; as, carbon disulfide. An explosive.

pancreas An endocrine gland in the abdomen behind the stomach. It secretes insulin into the bloodstream, and digestive juices containing enzymes (as, trypsin) into the intestine.

pancreatic juice The secretions of the pancreas containing the digestive enzymes: pancreatin, trypsin, steapsin, rennin, and invertase.

pancreatin Pancreatinum. A mixture of enzymes (including trypsin) from the fresh pancreas of the hog or ox. Cream powder of meatlike taste, slowly and partly soluble in water; it should convert 25 times its weight of starch into water-soluble products. Used for pancreatic insufficiency; as, fibrocystic disease (USP, BP).

pandermite $Ca_2B_6O_{11} \cdot 3H_2O$. A native source of boric acid.

Paneth **P., Friedrich Adolf (1887–1958)** German physicist. **P.'s rule** A radioelement will be adsorbed by a solid substance if its electronegative radical can form a relatively insoluble compound with the adsorbing substance.

pannic acid $C_{11}H_{14}O_4$ = 210.2. A constituent of rhizoma pannae. Colorless crystals, m.192. Cf. *pannol.*

pannol $C_{11}H_{14}O_4$ = 210.2. A constituent of *Aspidium athamanticum* (Filiceae). Cf. *pannic acid.*

panose A nonfermentable trisaccharide produced from maltose by an enzyme from *Aspergillus niger.*

panscale Calcium sulfate produced in salt manufacture by evaporation.

pansupari A mixture of betel and areca nuts and lime, used for chewing in India.

pansy The dried herb of *Viola tricolor.*

pantachromatic Existing in 2 or more colored forms. Cf. *panchromatic.*
pantal A corrosion-resistant aluminum alloy containing Cu 4.2, Mn 0.3–0.6, Mg 0.5–0.9%.
Pantocaine Trademark for tetracaine hydrochloride.
pantochromic Pantachromatic.
pantogen Protyle.
pantograph A device for copying diagrams, etc., to any scale: a set of adjustable levers connect a pencil with a tracing guide or stylus.
pantomorphism The perfect symmetry displayed by crystals.
Pantopaque Trademark for iophendylate.
Pantopon Trademark for papaveretum (mixed opium alkaloids).
pantothenic acid $OH \cdot CH_2 \cdot CMe_2 \cdot CHOH \cdot CO \cdot NH \cdot CH_2 \cdot CH_2 \cdot COOH$ = 219.2. A member of the vitamin B complex and part of coenzyme A. Colorless liquid, soluble in water. Occurs widely in all living things and tissues.
pantothenol A derivative of pantothenic acid in which an Me group is substituted by OH; activity 86% (pantothenic acid 100%).
papain* Papayotin, caricin, carase, vegetable pepsin, from the fruit of the papaw, *Carica papaya* (S. America). Gray powder, soluble in water; an enzyme with preferential cleavage at Arg-, Lys-, and Ph-X-(the bond next but one to the COOH group of phenylalanine); used to tenderize meat.
Papase Trademark for a preparation of papain.
Papaveraceae Poppy family: herbs with a milky juice, containing narcotic alkaloids; e.g.: *Papaver somniferum,* opium; *Papaver rhoeas,* rhoeadine; *Glaucium flavum,* glaucine.
papaveraldine $C_{20}H_{19}O_5N$ = 353.4. An alkaloid, m.210, from opium.
papaveric acid $C_{16}H_{13}NO_7$ = 331.3. Rhoeadic acid, m.233 (decomp.).
papaverine $C_{20}H_{21}O_4N$ = 339.4. Tetramethoxybenzyliso-quinoline. An alkaloid from opium. White, rhombic prisms or needles, m.148, insoluble in cold water; an antispasmodic. **p. hydrochloride** $C_{20}H_{21}O_4N \cdot HCl$ = 375.8. Colorless crystals, soluble in water; an antispasmodic (USP, EP, BP).
papaveroline $C_{16}H_{13}O_4N$ = 283.3. Tetrahydroxybenzyl-isoquinoline. A derivative of papaverine: the MeO groups are replaced by OH groups.
papaw (1) The edible fruit of papaya. (2) Pawpaw.
papaya Papaw, melon tree. The tropical *Carica papaya* (Passifloraceae), which yields fruit and juice containing proteolytic enzymes (papain), a milk-curdling enzyme, caricin, and carpaine. Used locally to tenderize meat. Cf. *papayotin.*
papayotin The dried milky juice of papaya; contains papain.
paper A sheet or continuous web of material formed by the deposition of vegetable, mineral, animal, or synthetic fibers, or their mixtures, with or without the addition of other substances, from aqueous suspension in such a way that the fibers are intermeshed together to form a thin but compact whole. P. may be coated, impregnated, printed, or otherwise converted without necessarily losing its identity. The usual raw materials are wood pulp, waste paper, rag, straw, esparto grass, and bagasse. Cf. *nonwoven.* **art ~** A p. coated on one or both sides with a mixture of a white or colored pigment (e.g., satin white) and an adhesive (e.g., casein), and dried and calendered to a high finish. Used for fine-screen printing. Cf. *chromo paper* below. **asbestos ~** An asbestos board or a tissue made from asbestos fibers. **azolitmin ~** A substitute for litmus p. **bank ~, bond ~** In typing, the bond is usually for the top copy, and the bank the thinner, lower-quality p. used for carbon copies. **blueprint ~** See *blueprint.* **carbon ~** See *carbon paper.* **carbonless copy ~** NCR. Multicopy system based on breaking microcapsules (on the sheet underside) containing a dye that reacts with a phenolic-clay coating (on the top side of the sheet below). **chromo ~** An art p., usually with a duller finish. **coated ~** P. with a surface coat of pigment (as, kaolin) bound by an adhesive (as, starch). **congo blue ~** See *congo blue paper.* **coordinate ~** See *coordinate paper.* **dahlia ~** See *dahlia paper.* **dialyzing ~** A parchment p. used for dialysis. **diazo ~** P. coated with diazonium salts, that on exposure to strong light and development in an alkali (as, ammonia) form diazo dyes. Used to make copies, particularly in drawing offices. **drying ~** An absorbent p. **emery ~** See *emery paper.* **filter ~** A porous, unsized p. used as an absorbent and filter. **glass ~** P. coated with fine glass powder; an abrasive. **imitation art ~** An uncoated p. with a high finish, obtained by addition of a high proportion of loading and calendering. Cf. *art paper* above. **linen ~** A p. made from linen rags. **linen-faced ~** A p. embossed with linen, an impression of which is retained on its surface. **litmus ~** A filter p. impregnated with red or blue litmus solution used as a test paper for acids and alkalies. **mechanical ~** P., as, for newsprint and magazines, containing a high proportion of *mechanical pulp,* q.v. **niter ~** A p. impregnated with saltpeter. **ozone ~** See *ozone paper.* **paraffined ~** A p. saturated with hot paraffin, used for electrical insulation and waterproofing. **parchment ~** P. prepared by intensive beating of the fibers to imitate the characteristics of parchment. Cf. *parchment.* **sand ~** See *sandpaper* under *sand.* **silver nitrate ~** See *silver nitrate paper.* **test ~** A filter p. impregnated with an indicator solution. **turmeric ~** A filter p. dipped in turmeric solution; used to identify boric acid (red color). **vegetable parchment ~** See *vegetable parchment* under *parchment.* **wax ~** A p. made waterproof by treating it with molten paraffin wax. **woodfree ~** Freesheet. P., as, for writing or printing, containing a high proportion of *chemical pulp,* q.v.
  p. analysis See *Herzberg's stain.* **p.board** A thick sheet of p. pulp mixed with size and filling materials. Cf. *board, pasteboard, millboard.* **p. chromatography** *Chromatographic analysis,* q.v., with p. as the selective absorbent, e.g., as strips hung in the solution of the sample, or as circles in the center of which is placed the solution to be tested. **p. coal** Brown coal in thin layers. **p. colors** Aniline dyes used for coloring p. **p. filter** A filter pulp cone or thick filter p. thimble. **p. pulp** Wood pulp or other fibers from which p. is deposited by drainage on a mesh. Cf. *chemical pulp,* dissolving *pulp, mechanical pulp, soda pulp, sulfite pulp, sulfate pulp, Herzberg's stain.* **p. sizes** International series. Writing and general printing. Based on A0 = 841 × 1,189 mm (= 1 m^2); A1 is $\frac{1}{2}$ A0, or 594 × 841 mm; and A2, etc., each $\frac{1}{2}$ the longest side of its predecessor. Also B (large printings) and C (envelope) series. **p. spar** Calcite, in thin plates.
papier maché Molded articles, made by boiling old paper with water and glue and soaking the semidried product in linseed oil.
Papreg Trademark for paper which has been impregnated with a plastic, e.g., for lamination.
paprika Ground red pepper used as a spice, rich in ascorbic acid.
para- Prefix (Greek "beyond" or "opposite") indicating (1)* (italic) the 1,4-position (*para* compounds) of the benzene ring; (2) nuclei with antiparallel spins as in parahydrogen; (3) polymerization, as, paraldehyde; (4) a relationship, as, paracasein. **p. compounds** See under the parent compound.
para arrowroot Tapioca.

parabanic acid CO·NH·CO·NH·CO = 114.1. Oxalylurea, imidazolidinetrione, oxalic acid ureid. Colorless plates, m.243 (decomp.), soluble in water. **dimethyl ~** Cholestrophane.
parabituminous Describing a good caking gas-coal.
parabola A plane curve, each point of which is equidistant from a straight line (axis) and a central point (focus). It resembles a circle at some points, a straight line at others.
paraboloid The solid shape traced by a parabola when rotated about the axis containing the focus. **p. condenser** A spherical microscope mirror having an elongated focus.
parabuxin $C_{24}H_{48}ON_2$ = 380.7. An alkaloid in common garden box, *Buxus sempervirens* (Euphorbiaceae).
paracasein Casein digested with rennin.
paracellulose Cellulose from the parenchyma or pith of plants (obsolete).
Paracelsus (1493–1541) Philippus Aurelius Theophrastus Paracelsus Bombastus von Hohenheim. Swiss physician and alchemist, advocate of chemical as opposed to vegetable remedies. Cf. *iatrochemists.*
paracetamol EP, BP name for acetaminophen.
parachor $P = M\gamma^{1/4}/(D - d) = 0.78 \times V$, where V is the critical volume, γ surface tension, M molecular weight, and D and d densities of a compound in the liquid and vapor state, respectively, at the same temperature. A comparison of the P of liquids is equivalent to a comparison of their molecular volumes at temperatures of equal surface tension. P is an additive constant for saturated compounds and is used to determine chemical constitution.
Paracon Trademark for an oil- and heat-resistant synthetic rubber consisting of chain esters of sebacic or succinic acid and ethylene or propylene glycols.
paraconic acid $O\cdot CH_2\cdot CH(COOH)\cdot CH_2\cdot CO$ = 130.1. Itamalic acid γ-lactone, tetrahydro-5-oxo-3-furancarboxylic acid. Colorless crystals, m.58. Cf. *citraconic acid.* **dimethyl ~** Terebic acid. **phenyl ~** See *phenylparaconic acid.*
paraconine $C_8H_{15}N$ = 125.2. An alkaloid obtained by heating butyraldehyde with ammonia. Colorless liquid with stupefying odor, b.170.
paracoto The dried bark of an unidentified tree of N. Bolivia; a substitute for *coto* bark, q.v.
paracotoin $C_{12}H_8O_4$ = 216.2. An active principle from paracoto. Yellow crystals, m.150.
paracyanogen $(CN)_5$ = 130.1. (1) An insoluble solid, sublimes if heated. (2) More correctly, the water-insoluble polymer, $(CN)_n$, of unknown molecular weight. Brown powder converted into cyanogen when heated above 860° in absence of air. Produced by prolonged pyrolysis of cyanogen at 300°.
paradiazine Pyrazine*.
paradimethylaminobenzaldehyde $C_6H_4(CHO)NMe_2$ = 149.2. A reagent for indole, skatole, pyrrole.
paradioxybenzene Hydroquinone*.
paraffin (1) See *alkanes.* (2) Hard p. (BP). White wax, d.0.890, m.47–65, insoluble in water, soluble in organic solvents. P. is a mixture of hydrocarbons occurring native in ozocerite, peat, and bituminous coal, and is a constituent of petroleum from which it is distilled. Used in the manufacture of ointments (NF, BP), waxed paper, matches, lubricants, oil crayons; and for waterproofing wood and cork. (3) See *kerosine.* **liquid ~** BP name for mineral oil. **white soft ~**, **yellow soft ~** BP names for petrolatum.
p. bath Molten paraffin. **p. oil** Petrolatum. **p. scale** A crude paraffin. **p. wax** Paraffin (2).
paraffins Alkanes*. **iso ~** Aliphatic, saturated hydrocarbons containing one −CHMe− group or a side chain. **normal ~** Aliphatic, saturated hydrocarbons containing only CH_3- and $-CH_2-$ groups.
paraffinum Paraffin.
paraform 1,3,5-Trioxane.
paraformaldehyde 1,3,5-Trioxane. Cf. *paraldehyde.*
parafuchsin $(C_6H_4NH_2)_2C(OH)C_6H_4\cdot NH\cdot HCl$. Pararosaniline chloride. A dye. Cf. *pararosaniline.*
paragenesis The passage of minerals through successive stages of chemical composition during the cooling of the earth's crust.
paraglobulin Fibr(in)oplastin. A globulin from blood serum and lymph.
paragonite $Al_3NaH_2Si_3O_{12}$. A mica-type *silica* mineral, q.v.
Paraguay tea Maté.
parahydrogen See *hydrogen (2).*
paralactic acid (*S*)-Lactic acid.
paralbumin A protein from ovarian cysts.
paraldehyde $(O\cdot CH\cdot Me)_3$ = 132.2. 2,4,6-Trimethyl-1,3,5-trioxane. A polymer of acetaldehyde. Colorless liquid of pungent odor, d.0.992, m.11, b.128, soluble in water; a reagent for alkaloids and fuchsin, and a hypnotic and sedative (USP, EP, BP). Cf. *metaldehyde, aldol.*
paraldol $C_8H_{16}O_4$ = 176.2. The dimer of aldol, m.96.
parallax The apparent displacement of an object due to a change in the position of the observer, e.g., errors in buret meniscus readings.
parallel (1) Having the same direction, but separated by equal distances. (2) Describing electric connections such that like poles of a number of units are connected to one another. Cf. *series.*
parallelosterism The relationship between isomorphous groups and their chemical compositions or physical properties.
paralyser Paralyzer.
paralysol $Me\cdot C_6H_4\cdot OK$. A mixture of cresol and potassium cresolate. Colorless crystals, m.146, insoluble in water; an antiseptic.
paralyst Paralyzer.
paralyzant A substance that causes paralysis.
paralyzer (1) An agent that prevents a chemical reaction; a catalytic poison. (2) Paralyzant.
param $N{:}C\cdot NHC({:}NH)NH_2$ = 84.1. Cyanoguanidine. A condensation product of cyanamide, formed at 150, m.204, soluble in water.
paramagnetic A substance that has magnetic properties stronger than those of air (as, iron); i.e., a magnetic permeability over 1. Cf. *diamagnetic.*
paramagnetism The property of being attracted by a magnet.
paramecium A genus of unicellular animals, or protozoans.
parameter A quantity that can be varied, but that is defined for a specific case; as, pressure.
paramisan sodium Aminosalicylate sodium.
paramorph A crystal that has undergone paramorphism.
paramorphine Thebaine.
paramorphism (1) The physical change of a mineral from one modification to another, without a change of chemical composition. (2) A rearrangement of molecular structure.
paramucic acid $C_6H_{10}O_8$ = 210.1. An isomer of mucic acid.
paramyelin $C_{35}H_{75}O_9NP$ = 721.0. White solid from brain and nerve substance.
paranaphthalene Anthracene (obsolete).
paranitraniline $NO_2\cdot C_6H_4\cdot NH_2$ = 138.1. Yellow crystals, m.148, soluble in alcohol; a reagent and intermediate.
parapectic acid $C_{24}H_{34}O_{23}$ = 690.5. An oxyacid produced from pectose by the ripening of fruits.
parapeptone Syntonin.

paraplasm The fluid portion of cell protoplasm.
paraquat See *herbicides,* Table 42 on p. 281.
paraquinoid Quinonoid.
para red A red aniline dye obtained from *p*-nitroaniline.
pararosaniline $(NH_2 \cdot C_6H_4)_2C{:}C_6H_4 \cdot NH_2Cl$ = 323.8. C.I. Basic Red 9. An organic base forming red salts. Colorless leaflets, m.188, insoluble in water; a dye. **hexamethyl ~** Methyl violet. **p. dyes** Dyestuffs derived from pararosolic acid; e.g., pararosaniline.
pararosolic acid Aurin.
parartrose $C_{120}H_{192}O_{40}N_{30}S$. A proteose obtained by digestion of wheat.
parasite An organism that obtains its nourishment from another living organism. Cf. *saprophyte, epiphyte.* Classification: Phytoparasites, vegetable parasites (bacteria and fungi); Zoöparasites, animal parasites (protozoans and metazoans, worms, etc.). Each of these is divided into occasional, temporary, and obligate; and constant, stationary, and facultative.
parasiticide An agent that destroys parasites.
parasorbic acid $C_6H_8O_2$ = 112.1. Hepenolactone. An acid from the berries of mountain ash, sorbus.
Parasporin Trademark for a bacterial insecticide, the principal active ingredients of which are spore and paraspores of *Bacillus thurungiensis,* Berliner. Used to combat alfalfa caterpillars.
paratartarics Racemic forms.
parathormone The hormone of the parathyroid. It maintains the normal level of calcium in the blood.
parathyroid Small glands near the thyroid which secrete parathormone.
parawolframate* See *tungstate.*
paraxanthine $C_7H_8O_2N_4$ = 180.2. 1,7-Dimethylxanthine, urotheobromine, a leucomaine in urine. Colorless crystals, m.298, soluble in water; an isomer of theobromine.
paraxylene 1,4-Xylene*.
parchment A specially prepared animal skin. **vegetable ~** P. paper. A paper treated with concentrated sulfuric acid to produce grease resistance, high wet strength, and some water resistance. Used to wrap fatty foods, and as a dialysis membrane.
p. paper Vegetable parchment.
paregoric A flavored camphorated tincture of *opium,* q.v., for cough mixtures.
pareira Perieira. The dried root of *Chondrodendron tomentosum* (Menispermaceae), S. America, which contains the alkaloid bebeerine.
parenchyma (1) The principal constituent of the thin cell-wall tissue of vegetable matter. (2) The main cellular component of the functional tissues of the body (not including the framework).
parent compound* P. molecule. The principal chain or ring system from which a name is derived by substitution of hydrogen with other atoms or groups; as, ethane is the p. c. of ethanol.
parenteral Describing a route of administration of drugs or food, other than by mouth or into the intestine.
parhelium See para*helium.*
parianite An asphalt from the pitch lake in Trinidad.
Paricin Trademark for a group of alkylhydroxy and acetoxy stearates.
paridin $C_{16}H_{28}O_7$ = 332.4. A glucoside from *Paris quadrifolia* (Liliaceae). Cf. *paristyphnin.*
parietic acid Chrysophanic acid.
parietin Physcion.
pariglin Smilacin.
parillic acid Parillin.
parillin $C_{40}H_{70}O_{18}$ = 839.0. Parillic acid. Salseparisin. A glucoside from the sarsaparilla root. Colorless needles or scales. Cf. *smilacin.*
parinaric acid $C_{18}H_{28}O_2$ = 276.4. 9,11,13,15-Octadecatetraenoic acid*. From the kernel fat of *Parinarium laurinum* (Rosaceae).
paris **plaster of ~** See *plaster of paris.*
p. black Lampblack. **p. blue** Iron(III) hexacyanoferrate(II)*. **p. green** $Cu(C_2H_3O_2)_2 \cdot 3Cu(AsO_2)_2$. Emerald green, Schweinfürth green, cooper acetoarsenite. **p. red** (1) Colcothar. (2) Minium. **p. violet** Methyl violet. **p. yellow** Lead chromate.
parisite A native fluoride and carbonate of the cerium metals. Brown, hexagonal pyramids.
paristyphnin $C_{38}H_{64}O_{18}$ = 808.9. A glucoside in the root of *Paris quadrifolia,* one-berry. Cf. *paridin.*
Parkerizing Trademark for a process of forming, by chemical reaction, a protective phosphate film on metal (especially cadmium) surfaces.
Parkes **P., Samuel (1761–1825)** English technical chemist. **P. process** The refining of argentiferous lead by liquation, followed by addition of zinc to the molten mass, and skimming of the surface crust (silver and zinc). The silver is isolated by distilling the zinc, and the lead purified by electrolysis.
parkesine An early plastic made by Alexander Parkes (1813–1890), from a "dough" of nitrocellulose in mixed alcohol and ether.
parkine An alkaloid from the seeds of nitta tree, *Parkia biglandalosa* (Leguminosae), Africa.
Parkinson's disease A disease of the nervous system characterized by tremor, and paucity and rigidity of movement. Associated with disease, damage, or low levels of dopamine in the brain. See *levodopa.*
Parlodel Trademark for bromocriptine mesylate.
Parma blue A triphenylrosaniline dye.
Parnate Trademark for tranylcypromine sulfate.
paromomycin An antibiotic produced by *Streptomyces rimosus* forma *paromomycinus.* **p. sulfate** Humatin. Mixed sulfates of the antibiotics from *Streptomyces rimosus paromomycinus.* Yellow powder; an antibiotic for dysentery and amebiasis (USP, BP).
paroxazine 2*H*-1,4-Oxazine.
Parr **P., Samuel Wilson (1857–1931)** American chemist. **P. apparatus** A device to determine the total carbon in fuels by means of a calorimeter and gas burets.
parsec Symbol: pc; the distance of a star whose annual parallax is 1 second of an arc. 1 pc = 3.0857×10^{13} km = 19.2×10^{12} miles = 3.26 light years = 206,266 astronomical units (distance of earth to sun).
parsley The umbelliferous plant *Carum (Apium) petroselinum* or *Petroselinum sativum.* It contains an essential oil and a camphor (apiole). Cf. *petroselinum.* **p. camphor** Apiole. **p. fruit** P. seeds. **p. leaves oil** An essential oil from the leaves of p., d.0.900–0.925, greenish yellow and of strong parsley odor; it contains apiole. **p. oil** An oil distilled from p. seeds. Colorless liquid, insoluble in water.
Parsons, Charles Lathrop (1867–1954) American chemist noted for his work on education, uranium minerals, and nitrogen fixation.
Parstelin Trademark for tranylcypromine sulfate.
parthenicine Parthenine.
parthenine An alkaloid from *Chrysanthemum (Parthenium) hysterophorum* (Compositae), W. Indies; an antipyretic.
parthenogenesis Development of the egg without previous fertilization. Cf. *agamy.*
partial Fractional. A proportion of the whole. **p. pressure**

TABLE 59. SUBATOMIC PARTICLES

Particle	Symbol	Life, s	Mass, MeV	Charge	Spin
Leptons					
Electron	e^-	Stable	0.51	−1	½
Neutrino	ν	Stable	0	0	½
Muon	$\mu^{\pm}$	2.2×10^{-6}	106	±1	½
Tau particle	τ	2×10^{-12}	1,800	−1	½
Hadrons (Resonances)[a]					
Nucleons					
Proton	p	Stable	938	+1	½
Neutron	n	Stable/800[b]	940	0	½
Hyperons					
Lambda particle	Λ	3×10^{-10}	1,116	−1	½
Xi particles	Ξ^0	2×10^{-10}	1,315	0	½
	Ξ^-	2×10^{-10}	1,321	−1	½
Sigma particles	Σ^+	8×10^{-10}	1,189	+1	½
	Σ^0	10^{-14}	1,193	0	½
	Σ^-	10^{-10}	1,197	−1	½
Omega particle	Ω^-	10^{-10}	1,672	−1	$\frac{3}{2}$
Mesons					
Kaons	$\kappa^{\pm}$	10^{-8}	494	±1	0
Pions	$\pi^{\pm}$	3×10^{-8}	140	±1	0
	π^0	10^{-16}	135	0	0

[a]Numerous other hadrons have been detected.
[b]Stable in an atomic nucleus, but otherwise decays with a half-life of about 800 s.

The fraction of the total pressure due to each constituent of a gas mixture. The p. pressures are proportional to the concentrations of the individual gases in a mixture. See *Dalton's law.*

particle A very small quantity of matter. **alpha ~** See *alpha particles.* **anti ~** An identical p., but of opposite sign; as, a positron. **beta ~** Electron. **colloidal ~** See *colloid.* **elementary ~** Fundamental p. The simplest unit of matter not as yet shown to be subdivisible. See *lepton, quark.* **gamma ~** A misnomer for cathode rays. **L ~** Lepton. **nuclear ~** See *subatomic particle* (following). **subatomic ~** Nuclear p. A "building stone" of matter. Fundamental characteristics: mass (MeV or units of proton mass), charge (electron units), spin (unit, $h/2\pi$), magnetic moment (Bohr magnetons), process of generation, destruction, and interconversion. S. particles are divided into those that interact weakly (leptons) and those that interact strongly (hadrons, or resonances). Each lepton is an elementary particle, while hadrons are composed of different combinations of *quarks,* q.v. Hadrons are subdivided into those that decay into protons (baryons) and those that decay into leptons or photons (mesons). Baryons can be further subdivided into those present in the atomic nucleus (nucleons) and others (hyperons). See Table 59.

p. accelerator See *particle accelerator* under *accelerator.* **p.board** A panel material made essentially from particles of wood, other lignocellulosic material (as, sawdust, flax shives), or both bonded with an organic binder. **p. distribution** See *Perrin equation.*

particulate Solid matter, as from boilers, contributing to *air pollution,* q.v. British Standard distinguishes between grit (above 75 μm) and dust (under 75 μm).

parting In assaying, the dissolution of silver from gold by means of nitric acid. **p. acids** Nitric acid of graded strengths used in the stages of parting.

partinium An alloy of tungsten and aluminum.

partition The distribution of a substance or ions between 2 immiscible liquids, or between a liquid and a gas. **p. chromatography** See *chromatographic analysis.* **p. coefficient** Distribution coefficient. See *Perrin equation.* **p. function*** A thermodynamic function expressing the sum of the energy levels of an atomic system; $Q = \Sigma g_n e^{-\epsilon_n/kt}$, where ϵ_n is the energy of the nth energy level and g_n its statistical weight, k the Boltzmann constant, and t the thermodynamic temperature. **p. law** *Nernst law.*

partridge berry (1) Squaw vine. (2) Gaultheria.

Partz cell An anode of amalgamated zinc in a solution of magnesium sulfate, and a cathode of carbon in a solution of potassium dichromate (2.06 volts).

parvoline $C_9H_{13}N$ = 135.2. **alpha- ~** 2-Ethyl-3,5-dimethylpyridine. A ptomaine from decaying fish or meat. Amber oil, with odor of hawthorn blossoms, b.188. **beta- ~** See *parvuline.*

parvuline $C_9H_{13}N$ = 135.2. 2,3,4,5-Tetramethylpyridine, β-parvoline. In coal tar. Amber liquid, b.228.

PAS *p-Aminosalicylic acid.*

Pascal **P., Blaise (1623–1662)** French scientist and philosopher. **P.'s law** The pressure applied to a liquid at any point is transmitted equally in all directions.

pascal* Symbol: Pa; the SI system unit of pressure or stress, equal to 1 N/m^2. 1 Pa = 1.020×10^{-5} atmos (technical) = 1.000×10^{-5} bar = 10.00 dynes/cm^2 = 7.500×10^{-3} mmHg (at 0 °C) = 1.450×10^{-4} psi = 4.019×10^{-3} in H_2O (at 15.6 °C).

Paschen **P., F. (1865–1947)** German physicist. **P. galvanoscope** A sensitive milliammeter. **P. series** The spectrum lines produced when electrons fall from an outer orbit to the third ring. Cf. *Balmer* and *Lyman series, energy levels, Bohr theory.*

passiflora Passionflower.

Passiflorinae A plant order comprising the families *Passifloraceae, Caricaceae,* and *Begoniaceae.*
passionflower Passiflora. The dried herb of *Passiflora incarnata* (Passifloraceae), N. America. The fruit juice is a beverage.
passive (1) Not active. (2) In electronics, a component that neither generates power nor amplifies a signal. Cf. *active.* **p. immunization** The process by which the blood serum of an actively immunized animal is injected into another animal, for protection against bacterial infection or toxin produced by the bacteria; as, antitetanus serum. **p. metal** Metal rendered noncorrodible by treatment with heat or strong acids. This property is sometimes lost by mechanical shock. **p. state** See *passivity.*
passivity The inertness of certain substances under conditions in which chemical activity is expected; e.g., certain metals dissolve in acids of low, but not high, concentration, owing to the formation of a thin layer of peroxide, oxygen, or salt, which prevents direct contact of acid and metal. Mechanical shock can destroy this layer. Cf. *zone.*
paste (1) A tenacious cementing substance. (2) A pharmaceutical preparation for external use. (3) A matrix in which minerals are embedded. **London ~** A mixture of equal weights sodium hydroxide and slaked lime, moistened with alcohol. **vienna ~** A mixture of sodium hydroxide and slaked lime, moistened with water.
p. blue Prussian blue. **p.board** Cardboard formed by pasting high-grade outer layers of paper on to a lower-grade middle layer or layers. It differs from ordinary board, which is formed directly without an adhesive.
Pasteur **P., Louis (1822–1895)** French chemist who established the connection between bacterial growth and disease, and laid the foundation of immunology. **P. effect** P. reaction. Cells of yeast and other organisms which can ferment sugar, irrespective of the presence of oxygen, are more active in this respect in the absence of oxygen. **P. filter** See *Pasteur filter* under *filter.* **P. flask** A glass flask for bacterial cultures with a long neck bent downward and upward with constrictions and expansions. Opposite the neck is a tubular outlet. **P. reaction** P. effect.
pasteuring Pasteurizing.
pasteurization Partial sterilization of organic fluids by heating at 65°C for not less than 30 min. Cf. *stassanization.* **H.T.S.T. ~** Flash p. Abbreviation for high-temperature, short-time p. The liquid, e.g., milk, is held at not less than 73°C for not less than 15 s, and then cooled at once to 12°C. Cf. UHT *milk.*
pasteurized milk Milk that has been partly sterilized by heating at 63 to 66°C for at least 30 min, and then immediately cooled to 12°C.
pasteurizer A machine for pasteurization.
pastil, pastille A lozenge, sugared confection, or aromatic mass burnt as incense or fumigant.
patchoulene $C_{15}H_{24}$ = 204.4. A sesquiterpene, d.0.930, b.256, from patchouli oil.
patchouli Patchouly. The herb *Pogostemon patchouli* (Labiatae), India, used in perfumery. **p. oil** An essential oil from the leaves of patchouli. Yellow, aromatic oil, d.0.970, insoluble in water. It contains cadinene, eugenol, cinnamaldehyde; used in perfumery.
patchouli alcohol $C_{15}H_{25}OH$ = 222.4. From patchouli oil. Colorless crystals, m.56.
patchouly Patchouli.
patent (1) A grant conveying public lands from the government. (2) Letters (of) patent. Any process, method, or device that has been accepted as new by the patent office and is, thereby, protected under the patent laws for a number of years. (3) Evident, not hidden. **provisional ~** An official application for a patent, which gives temporary protection until the final patent is granted. **p. blue** A phenolated rosaniline disulfonic acid; a redox indicator. **p. yellow** $PbO \cdot PbCl_2$. Mineral yellow. A yellow pigment.
Patera process A method of producing silver by chloridizing the ore, roasting, and leaching successively with water and sodium thiosulfate solution, which dissolves the silver; precipitation with sodium sulfide follows and, finally, heating the resulting silver sulfide.
path The course or track along which objects (as, electrons, ions, or molecules) move. See *Wilson fog method.* **free ~** The average distance between collisions in the travel of a molecule in a gas or liquid.
pathochemistry Chemical pathology. The study of the chemical changes of the living organism in a diseased condition.
pathogen Any living organism, as, microorganism, bacterium, or protozoan, that produces disease.
pathogenic Describing an agent that produces disease.
pathologic(al) Pertaining to a diseased condition.
pathology The study of the nature of disease; especially the functional (physiological) and structural (morphological) changes produced. **experimental ~** The study of artificially produced disease. **phyto ~** The study of plant diseases.
patina The thin and often multicolored coat of oxides formed on metallic surfaces. Cf. *brochanite.*
patronite V_2S_3. A native sulfide.
pattern (1) A design or arrangement of symbols or figures. (2) A model from which an object is made. **crystal ~** A space lattice of *crystal* structure, q.v. **Laue ~** See *Laue.*
Pattinson process Separation of silver from lead, by fractional crystallization and removal of the molten, silver-free lead crystals. Superseded by the *Parkes* process, q.v.
patulin $C_7H_6O_4$ = 154.1. Clavacin, clavatin, calviformin, penicidin. An antibiotic from several molds, e.g., *Penicillium patulum* and *Aspergillus clavatus,* m.111. A suggested cure for the common cold. Active against gram-positive and gram-negative organisms.
paucine $C_{27}H_{38}O_5N_5$ = 512.6. An alkaloid from the seeds of pauco nuts. Yellow scales, m.126, insoluble in ether. **p. hydrochloride** $C_{27}H_{39}O_5N_5 \cdot 2HCl \cdot 6H_2O$ = 693.6. Colorless needles, m.245, slightly soluble in water.
pauco nuts Graine d'owala. The fruits of *Pentaclethra macrophylla* (Leguminosae), Africa, which contain paucine.
Pauli **P., Wolfgang (1900–1958)** Austrian physicist. Nobel prize winner (1945). **P.'s principle** Exclusion principle. The number of electrons in a shell or orbit is limited. No 2 electrons can have, simultaneously, all 4 quantum numbers the same. Cf. *correspondence principle.* **P. rule** The atomic nucleus has alternately odd and even numbers of α particles.
Pauling **P., Linus C. (1901–)** American chemist. Nobel prize winner (1954 and 1963). **P. structure** A structure of the benzene molecule based on an electronic concept of interatomic bonds. Each bond results from 2 electrons moving in orbits around the 2 nuclei of the atoms they join.
paullinia (paullinio) tannin $C_{38}H_{36}O_{20}$ = 812.7. Guaranatannin. White crystals from the seeds of *Paullinia cupana* (Sapindaceae), Brazil, used similarly to cacao.
pavine $C_{20}H_{23}O_4N$ = 341.4. 2,4-Dihydropapaverine. **(±)-~** White needles, m.201, soluble in chloroform.
Pavy's solution (*A*) 4.158 g copper sulfate in 500 mL water. (*B*) 20.4 g sodium potassium tartrate, 20.4 g potassium hydroxide, 300 mL strong ammonia, and water to make 500 mL. Mix in equal parts. To determine sugars, titrate into

boiling P. solution until the blue color vanishes. Cuprous oxide is not precipitated as with ordinary *Fehling solution*, q.v.
pawpaw The seeds of the edible fruit of *Asimina triloba* (Anonaceae), E. United States; an emetic, Cf. *papaw.*
Payen, Anselme (1795–1871) French chemist noted for industrial processes (decolorizing with charcoal).
Payne's process Paynization. Making wood fireproof by successive treatments with ferrous sulfate and calcium chloride solutions.
paynize To treat wood by Payne's process.
pay ore An ore, rock, earth, or gravel that can be worked with profit.
payta Krameria.
paytine $C_{21}H_{24}ON_2$ = 320.4. An alkaloid, m.156, from cinchona bark.
Pb Symbol for lead.
PCB Polychlorinated biphenyls.
p.c.e. Abbreviation for *pyrometric* (q.v.) cone equivalent.
PCTFE* Symbol for poly(chlorotrifluoroethylene).
p.d. Abbreviation for potential difference.
Pd Symbol for palladium.
pdl Symbol for poundal.
PE* Symbol for polyethylene.
pe- A syllable indicating a higher degree of saturation or of hydrogenation; as, pyridine and pi*pe*ridine.
peach The edible fruit of *Prunus persica* (Rosaceae). **p. aldehyde** γ-Undecalactone. **p.-kernel oil** Persic oil (NF). Yellow liquid, d.0.915, m.−15, sapon. val. 189–192, iodine val. 93.5, insoluble in water; an adulterant for almond oil, and flavoring.
peacock copper Bornite.
peanut The edible seeds of *Arachis hypogoea* (Leguminosae), of the temperate zones. It contains the globulins *arachin* and *conarachin,* q.v. **p. hull meal** Ground p. shells; a fertilizer (1.5–2.5% nitrogen). **p. oil** Arachis oil, groundnut oil, earthnut oil, oil of katchung. Yellow oil expressed from peanuts, d.0.916, sapon. val. 188–196, iodine val. 103–104, insoluble in water. Used as a vehicle in pharmacy (NF), as an adulterant of olive oil, and in poor-grade soap. Cf. *Bellier's test.* **p. ore** Wolframate.
pearl A calcareous secretion from various species of mollusks, chiefly oyster. **artificial ~** (1) Culture p. (2) Synthetic p. **culture ~** A natural p. produced by artificial stimulation of the oyster; as, insertion of a grain of sand. **imitation ~** An imitation of natural pearls that produces only the outside appearance; as, alabaster coated with p. essence. **synthetic ~** A p. made by slow precipitation from a gelatinous solution of certain salts.
p. alum Aluminum sulfate for paper manufacture. **p. ash** Impure calcined potassium carbonate, made from potash. **p. essence** A product obtained from fish scales, e.g., of European minnow, the principal constituent being guanine; used to obtain a highly lustrous coating, as, on imitation pearls. **p. grain** 0.25 meter-carat. A unit of weight for pearls. **p. hardening** Gypsum used as a paper filler. **p. mica** Margarite. **p. opal** A bluish-white, lustrous opal. **p. powder** A form of bismuthyl chloride; a cosmetic. **p. sinter** A modification of silica. **p.spar** Brown spar. A dolomite with pearly luster. **p.stone** Perlite. **p. white** (1) Lithopone. (2) Calcium sulfate for the paper industry. (3) $Bi(OH)_2 \cdot NO_3$. A form of bismuth subnitrate used formerly as a cosmetic.
pearlite (1) A eutectic mixture: soft α-ferrite 87, hard cementite 13%, containing 0.9% carbon, formed during the slow cooling of molten steel. (2) *Perlite,* q.v.
pear oil An alcoholic flavoring solution of pentyl acetate.
peastone Oolite.
peat Dark soil produced by the decomposition of plants in moist places: water 25, ash 3, woody fiber 50, humus acids 22, nitrogen 1.5–2.5%. A fuel and fertilizer. Cf. *dopplerite, ulmic acid.* **p. coal** A soft coal, intermediate between p. and lignite. **p. coke** A carbonized p. produced by destructive distillation. **p. gas** A hydrocarbon gas obtained by distilling p. **p. tar** A tar obtained by distilling p. **p. wax** Mona wax. A black wax extracted by alcohol from p.; a substitute for montan wax.
pebble A small stone, worn round by the action of water. **Brazilian ~** A rock crystal or quartz, from which lenses are cut.
p. mill A power-driven, rotating steel cylinder with a porcelain lining, or steel, and partly filled with flint pebbles, porcelain, or metal balls for pulverizing or mixing materials. Cf. *ball mill, agate.* **p. powder** Gunpowder pressed into large cubic grains.
peck Abbrev. pk; a dry measure or unit of capacity in the U.S. and U.K. systems: 1 peck = 0.25 bushel = 2 gallons = 8.81 liters.
pectate A salt of pectic acid. **p. lyase*** An enzyme that eliminates Δ-4,5-D-galacturonate residues from pectates, thus causing depolymerization.
pectenine A tetanic poison alkaloid from the cactus *Cereus caespitosus* or *C. pecten* (Mexico).
pectic acid $C_{17}H_{24}O_{16}$ = 484.4. A dibasic acid obtained from ripe fruit or vegetable pectin by enzyme action or long boiling with alkali. It forms a jelly with calcium salts (as in the setting of jams and fruit preserves). **para ~** See *parapectic acid.*
pectinase Polygalacturonase*.
pectinose Arabinose.
pectins Compounds formed from the protopectin of unripe fruits, whose function in ripe fruits is the cementation of individual cells. P. are esters (usually methyl) of polygalacturonic acid, and on hydrolysis form pectic acid in overripe fruits, which gives the juice the property of jelling.
pectization Gelatinization.
pectocellulose A substance from raw flax, which yields pectic acid and cellulose.
pectograph The pattern obtained by drying a film of colloidal solution on a glass plate.
pectolinarigenin $C_{17}H_{14}O_6$ = 314.3. 5,7-Dihydroxy-4′,6-dimethoxyflavone. Yellow needles, m.215, slightly soluble in alcohol. Cf. *pectolinarin.*
pectolinarin $C_{29}H_{34}O_{15}$ = 622.6. A glucoside from the flowers of *Linaria vulgaris,* L., m.240 (decomp.), hydrolyzed to pectolinarigenin.
pectolite $4CaO \cdot Na_2O \cdot 6SiO_2 \cdot H_2O$. Ratholite. A native hydrated acid silicate.
pectolysis The clearing of fruit juices by the decomposition of pectin by polygalacturonase.
pectose A polysaccharide in fruits and vegetables, from which pectins are formed on ripening.
pectosinic acid $C_{32}H_{23}O_{31}$ = 903.2. An amorphous acid derived from pectose by treatment with alkali. It forms gelatinous salts.
pederin Potent defensive toxin found in the E. African beetle *Paederus fuscipes.*
pedesis Brownian motion.
Pediculus The louse, a genus of insect, which can infest man (e.g., head louse) and carry disease, e.g., typhus.
pedology The science of the soil.
Peek Trademark for a polyetherketone resin that, when reinforced with 60% carbon fiber, makes a composite material stronger and lighter than aluminum alloys.
peganine Vasicinone.

peganite Variscite.
Peganum harmala The commonest weed of the Russian and Siberian steppes. Its seeds contain harmaline and harmine, used in the orient to prepare turkey-red dye.
pegmatite Giant granite. An igneous rock of coarse-grained quartz, feldspar, muscovite, tourmaline, and biotite. Some pegmatites are a source of lithia minerals, rare-earth, tin, tungsten, tantalum, or uranium minerals.
pegu catechu Catechu.
pelagic Pertaining to the deep sea. Cf. *littoral.*
pelagite $xMnO_2 \cdot yFe_2O_3 \cdot 2H_2O$. A mineral.
pelargonaldehyde Nonanal*.
pelargone (1) $C_{15}H_{22}O_2$ = 234.3. (1*R*)-Furopelargone. Oil. (2) 9-Heptadecanone*.
pelargonic acid Nonanoic acid*.
pelargonidin $C_{15}H_{10}O_5 \cdot HCl$ = 306.7. 3,4′,5,7-Tetrahydroxyflavilium chloride. An anthocyanidin from the flowers of many plants, as, *Pelargonium* species (Geraniaceae.)
pelargonin Pelargonidin glucoside. An anthocyan from dahlia, geranium, and other flowers. Cf. *callistephin.* **p. chloride** $C_{27}H_{31}O_{15}Cl \cdot 4H_2O$ = 703.0. m.180 (decomp.). A pigment closely allied to that from scarlet pelargonium.
pelargononitrile Nonanenitrile*.
pelargonyl The nonanoyl* radical.
pelidisi $10W/cm^3$, where *W* is the weight in grams and cm the sitting height in centimeters of a person; used in the calculation of normal diets.
Péligot P., Eugène Melchior (1811–1872) French chemist. **P. blue** A hydrated copper oxide pigment. **P. salt** Probably the potassium salt of the unknown chlorochromic acid, $KCrO_3Cl$, obtained by heating potassium dichromate with hydrochloric acid. **P. tube** A calcium chloride tube or U-tube with a bulb in each arm, and in the bend.
pelitic Describing rocks essentially of clay origin, e.g., slate.
pellagra A deficiency disease due to lack of nicotinic acid.
Pelletan, Pierre (1782–1846) French chemist noted for his *Dictionnaire de Chimie* (1821).
Pelletier, Pierre Joseph (1788–1842) French pharmacist who discovered toluene and several alkaloids (with Caventou). Cf. *Peltier.*
pelletierine $C_8H_{15}ON$ = 141.2. Punicine. A pyrrolidine alkaloid obtained from the root bark of pomegranate, *Punica granatum.* Brown oil, insoluble in water; an anthelmintic. **iso~** (±)-Piperidinylpropanone. **pseudo~** $C_9H_{15}ON$ = 153.2. Yellow crystals, soluble in water.
p. sulfate Brown syrup, soluble in water; a teniacide. **p. tannate** Brown mass, slightly soluble in water; an anthelmintic.
pellicle (1) A thin skin. (2) The crust forming on the surface of a saturated solution during evaporation.
pellitory root Pyrethrum.
pellote (Mexican "peyotl.") Mescal buttons.
pellotine $C_{13}H_{19}O_3N$ = 237.3. *N*-Me derivative of (±)-anhalonidine. An alkaloid from pellote, the dried cactus *Anhalonium williamsi* (Mexico). Colorless crystals, m.110, slightly soluble in water. Cf. *mescaline.*
pelopium Impure niobium. A supposed element, isolated by Rose (1846) from tantalite.
peloponium Niobium.
pelosine Bebeerine.
pelotherapy Treatment by external application of natural products, as, mud.
Peltier P., Jean Charles Athanase (1785–1845) French watchmaker and experimenter. Cf. *Pelletier.* **coefficient of ~** The ratio of the quantity of heat applied to a thermocouple to the quantity of electricity obtained from it.
P. effect If a current passes through a circuit containing a thermocouple, heat is evolved at one junction and adsorbed at the other. Cf. *heat* effect.
pemphigus alcohol $C_{34}H_{70}O_2$ = 510.9. A solid, m.100–105, from the wax of the insect *Pemphigus xylostei.*
pemphigic acid $C_{33}H_{66}(OH)COOH$ = 524.9. From the wax of the insect *Pemphigus xylostei;* m. 101.
Penbritin See *penicillin.*
pencil (1) A roll or stick that contains an active substance in its center; as, a litmus p., wax p. (2) An aggregation of light rays meeting at a point.
pencil stone Pyrophyllite.
penetration (1) Entering or piercing. (2) The hardness or consistency of a material expressed as the distance that a standard needle passes vertically into it under conditions of loading (100 g), time (5 s), and temperature (25 °C). (3) The focal distance or depth of a lens. (4) The passage of radiation through materials.
penetrometer A device to measure *penetration* (2) or (4), q.v. Cf. *gelometer.*
Penex process The catalytic isomerization of light naphthas.
penicidin Patulin.
penicillamine $HS(Me_2)C \cdot CH(NH_2)COOH$ = 149.2. 2-Amino-3-mercapto-3-methylbutanoic acid*. **(*S*)-~** Cuprimine. m.200, freely soluble in water. A degradation product of penicillins; and chelating agent, used to treat a copper overload disease, lead poisoning, and arthritis.
penicillanic acid **6-amino~** $C_8H_{12}O_3N_2S$ = 216.3. An antibiotic less potent than penicillin, isolated (1959) from strains of *Penicillium chrysogenum.* White crystals, m.208 (decomp.).
penicillic acid $C_8H_{10}O_4$ = 170.2. 3-Methoxy-5-methyl-4-oxo-2,5-hexadienoic acid†. An antibiotic from *Penicillium puberulum.*
penicillin The first antibiotic discovered. Originally produced by the mold *Penicillium notatum* (A. Fleming 1928, H. W. Florey 1936). Now produced from the mold *P. crysogenum,* and semisynthetically by adding substances to this mold culture, including precursors of the side chain, R. Penicillins have a common structure, this being a fused thiazolidine and *β*-lactam ring, with different radicals R on the side chain:

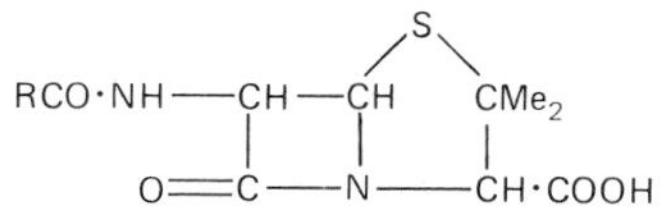

```
                          S
                        /   \
RCO·NH——CH——CH       CMe2
         |    |         |
      O==C———N————CH·COOH
```

Some principal penicillins:

Penicillin G. $C_{16}H_{18}O_4N_2S$ = 334.4. The R group is $C_6H_5 \cdot CH_2-$. Benzylpenicillin. Pfizerpen G, Crystapen G. Usually used as the sodium or potassium salt. Active against gram-positive organisms; destroyed by *β*-lactamase (USP, EP, BP).
Penicillin V. $C_{16}H_{18}O_5N_2S$ = 350.4. The R group is $C_6H_5 \cdot OCH_2-$. Phenoxymethylpenicillin. Distaquaine VK, Uticillin VK. Active against gram-positive organisms; destroyed by *β*-lactamase (USP, EP, BP).
Ampicillin. $C_{16}H_{19}O_4N_3S \cdot 3H_2O$ = 403.4. The R group is $C_6H_5 \cdot CH(NH_2)-$. Penbritin, Totacillin. Active against gram-positive and -negative organisms; destroyed by *β*-lactamase (USP, BP).
Amoxicillin. $C_{16}H_{19}O_5N_3S \cdot 3H_2O$ = 419.4. The R group is $OH \cdot C_6H_4 \cdot CH(NH_2)-$. Amoxycillin, Amoxil. Active against gram-positive and -negative organisms; destroyed by *β*-lactamase (USP, BP).

Cloxacillin sodium. $C_{19}H_{17}O_5N_3SClNa \cdot H_2O$ = 475.9. The R group is

```
o-Cl·C6H4·C ─── C
          ‖     ‖
          N    CMe
           \   /
             O
```

Cloxapen, Orbenin, Active against β-lactamase-producing staphylococci (USP, BP).

Carbenicillin disodium. $C_{17}H_{16}O_6N_2SNa_2$ = 422.4. Pyopen. Active against *pseudomonas* and other gram-positive organisms; destroyed by β-lactamase (USP, BP).

p. units 1 unit of p. is contained in 0.5988 μg of the Second International Preparation, although the international standard was discontinued in 1968 in favor of mg doses.

penicillinase* An enzyme that hydrolyzes penicillin to penicilloate. Used to treat allergic reactions to penicillin.

penicilliopsin $C_{30}H_{22}O_8$ = 510.5. Mycoporphyrin. A pigment from the mycelia of molds. Orange needles, m.330 (decomp.).

Penicillium A genus of Ascomycetes, fungi of the mildew group. ***P. glaucum*** Blue mold. The common mold, e.g., of bread. It excites alcoholic fermentation and separates (+)- and (−)-tartrates from racemic mixtures, since the (+) isomer is acted on faster than the (−) one.

penillic acid $C_{14}H_{22}O_4N_2S$ = 314.4. A product of the catalytic hydrogenation of penicillin. Thick, (+)-rotatory rods, m.182, with blue fluorescence.

pennine Penninite.

penninite Pennine. A green, crystalline chlorite from the Pennines, England.

pennone $CMe_3 \cdot CMe_2 \cdot CO \cdot Me$ = 142.2. 3,3,4,4-Tetramethylpentan-2-one*. Colorless crystals, m.63.

Pennsylvanian See *geologic eras,* Table 38.

pennyroyal (1) *Hedeoma pulegoides* of N. America; (2) *Mentha pulegium* of Europe. **p. oil American ~** The essential oil of *Hedeoma* species (Labiatae), d.0.920–0.935, containing pulegone and hedeomol. **European ~** The essential oil of *Mentha pulegium,* d.0.930–0.960.

pennyweight Symbol: dwt; a unit of weight in the English system: 1 pennyweight = 24 grains = 0.05 troy ounce = 1.5552 grams.

Pensky-Martens apparatus An instrument for determining flash points.

pent- See *penta-.*

penta-* Quinque-, quinqui-. Prefix (Greek) indicating five. Cf. *pentakis.*

pentaamino-* Prefix indicating 5 amino groups in an organic compound.

pentabasic Describing a compound that has 5 H atoms replaceable by bases or metals.

pentaborane* B_5H_9 or B_5H_{11}. See *boranes.*

pentabromo-* Describing a compound that has 5 Br atoms in its molecule. **p.benzene*** C_6HBr_5 = 472.6. Colorless needles, m.159, insoluble in water.

pentacarboxylic* Describing a compound containing 5 carboxyl groups.

pentacetate A mixture of pentyl alcohol 20 and pentyl acetate 80%, b.128–148.

pentachloro-* Indicating a compound containing 5 Cl atoms. **p.aniline*** $C_6Cl_5NH_2$ = 265.4. Colorless needles, m.232; soluble in alcohol. **p.benzene*** C_6HCl_5 = 250.3. Colorless needles, d.0.769, m.85, insoluble in water. **p.ethane*** CCl_3CHCl_2 = 202.3. Colorless liquid, d.1.834, b.161, insoluble in water; a solvent. **p.phenol*** C_6Cl_5OH = 266.3. Gray flakes, with phenolic odor, m.191.

pentacosamic acid Cerebronic acid.

pentacosane* $C_{25}H_{52}$ = 352.7. A hydrocarbon from beeswax, m.54, insoluble in water.

pentacosanoic acid* $Me(CH_2)_{23}COOH$ = 382.7. Hyenic acid. m.78.

pentacyanonitrosylferrate(III)* See *nitroprusside.*

pentacyclic Describing a molecule with 5 rings.

pentad An element or radical having a valency of 5.

pentadecane* $C_{15}H_{32}$ = 212.4. In the oil from the rhizomes of *Kampheria galanga.* Colorless liquid, d.769, b.270, insoluble in water. **p.carboxylic acid*** Palmitic acid*.

pentadecanoic acid* $C_{14}H_{29}COOH$ = 242.4. Isocetic acid, *n*-pentadecylic acid. Colorless solid, m.54, d_{100m}257, occurring in agaricus. **carboxymethyl ~** Roccellic acid.

pentadecanol* $C_{15}H_{31}OH$ = 228.4. Pentadecyl alcohol*. Colorless crystals, m.44.

pentadecanone* $C_{15}H_{30}O$ = 226.4. **2-~** Tridecyl methyl ketone*. m.39, b.294. **3-~** Dodecyl ethyl ketone*. m.38, b_{20mm}174. **8-~** $[Me(CH_2)_6]_2CO$. Dipheptyl ketone*, caprylone. Colorless crystals, m.43, b.178, soluble in alcohol.

pentadecyl* The radical $C_{15}H_{31}-$ from pentadecane. **p. alcohol** Pentadecanol*.

pentadiene* C_5H_8 = 68.1. **1,2-~** $CH_2{:}C{:}CH \cdot CH_2 \cdot CH_3$. Ethylallene. Colorless liquid, b.45. **1,3-~** $CH_3{:}CH \cdot CH{:}CH \cdot CH_3$. α-Methyl bivinyl, piperylene. Colorless liquid, d.0.696, b.43. **1,4-~** $(CH_2{:}CH)_2CH_2$. Colorless liquid, d.0.6594, b.26. **p.carboxylic acid** Hexadienoic acid*. **p.dioic acid*** Glutinic acid.

pentadienone* **1,4-p.-3-** $(CH_2{:}CH)_2CO$ = 82.1. Divinyl ketone*. b_{50mm}38. **1,5-diphenyl-1,4-pentadien-3-one** $(PhCH{:}CH)_2CO$ = 234.2. Distyryl ketone*. m.189.

pentadigalloylglucose $C_{72}H_{52}O_{46}$ = 1653. A tannin from Chinese nutgalls. **alpha-~** Brown mass, soluble in water. **beta-~** Less soluble in water than alpha-p.

pentaerythritol* $C(CH_2OH)_4$ = 136.1. Colorless solid, m.261. It forms polyesters with organic acids; used in plastics manufacture. **p. tetranitrate** $C(CH_2NO_3)_4$ = 316.1. Peritrate. Colorless crystals, m.141; a powerful explosive, and vasodilator for angina.

pentaethylbenzene* C_6HEt_5 = 218.4. Colorless liquid, d.0.89, b.277, insoluble in water.

pentagalloyl glucose $C_{41}H_{32}O_{26}$ = 940.7. A yellow tannin soluble in water.

pentaglucose Pentose.

pentaglycol $Me_2C \cdot (CH_2OH)_2$ = 104.1. A solid, m.129, soluble in water.

pentahydro- Prefix indicating 5 H atoms.

pentahydroxy-* Prefix indicating 5 hydroxy groups. **p.pentane** $C_5H_7(OH)_5$ = 152.1. A liquid, b.102.

pentaiodo-* Prefix indicating 5 iodine atoms in the molecule.

pentakis-* Prefix indicating 5 times. Generally applied to identical radicals substituted in the same way. Cf. *penta-.*

Pental Trademark for trimethylethylene.

pentaline Pentachloroethane*.

pentamethyl-* Prefix indicating 5 methyl groups in the molecule. **p.benzene*** C_6HMe_5 = 148.2. Colorless crystals, m.56, insoluble in water. **p.benzoic acid*** Me_5C_6COOH = 192.3. Colorless needles, m.210, soluble in water. **p.phenol*** Me_5C_6OH = 164.2. Colorless needles, m.125, insoluble in water.

pentamethylen- Prefix indicating 5 methylene groups in a molecule.

pentamethylene (1)* 1,5-Pentanediyl†. The radical

$-CH_2(CH_2)_3CH_2-$. (2) $(CH_2)_5$. Cyclopentane*. **hydroxy ~** Cyclopentanol*. **oxo ~** Cyclopentanone*.

p. bromide 1,5-Dibromopentane*. **p.diamine*** Cadaverine. **p.imine** Piperidine*. **p. oxide** Tetrahydropyran.

pentamidine isethionate $C_{19}H_{24}O_2N_4 \cdot 2C_2H_6O_4S$ = 592.7. White, hygroscopic crystals, m.190, soluble in water; used to treat trypanosomiasis.

pentamino- Pentaamino-*.

Pentamul Trademark for a group of emulsifiers.

pentanal* Valeraldehyde*. **4-methyl-1- ~ *** Isocapraldehyde. A liquid, d.0.830, b.170.

pentane* C_5H_{12} = 72.1. Pentylhydride. A saturated methane hydrocarbon. Colorless liquid, d.0.634, b.36, insoluble in water; an anesthetic, refrigerant, and thermometer filling. **amino ~ *** Pentylamine*. **bromo ~ *** Pentyl bromide*. **chloro ~ *** Pentylchloride*. **dimethyl ~ *** Et_2CMe_2 = 100.2. Colorless liquid, d.0.711, b.86, insoluble in water. **ethoxy ~ *** Ethyl pentyl ether*. **iodo ~ *** Pentyl iodide*. **iso ~ *** $Me_2CH \cdot CH_2Me$. *sec*-P. Colorless liquid, d.0.622, b.30. **methoxy ~ *** Methyl pentyl *ether**. **neo ~ *** CMe_4. *tert*-P. Colorless, b.10. A constituent of coal oil and gas. **secondary ~** Iso*pentane**. **tertiary ~** Neo*pentane**. **tetrahydro ~** Cyclopentane*.

p.carboxylic acid* Hexanoic acid*. **p.diamine*** **1,5- ~** Cadaverine. **p.dicarboxylic acid*** Pimelic acid. **p.dioic acid*** Glutaric acid*. **p.dione*** Acetylacetone. **p. lamp** A photometric source of illumination. Cf. Harcourt *lamp*. **p. thermometer** A low-temperature thermometer filled with colored p.

pentanediol* $C_5H_{12}O_2$ = 104.1. **1,2- ~** $Me(CH_2)_2CHOH \cdot CH_2OH$. α-n-Amylene glycol. Colorless liquid, d.0.980, b.212. **1,4- ~** $MeCHOH(CH_2)_2CH_2OH$. γ-Pentylene glycol. Colorless liquid, d.0.9954, b_{20mm}131. **1,5- ~** $CH_2OH(CH_2)_3CH_2OH$. Pentamethylene glycol. An oil, d.0.994, b.239. **2,3- ~** $MeCH_2(CHOH)_2 \cdot Me$. Ethyl methyl ethylene glycol, β-n-amylene glycol. Colorless liquid, d.0.9945, b.187.

pentanediyl† **1,5- ~** See *pentamethylene.*

pentanethiol* **1- ~** $Me(CH_2)_4SH$ = 104.2. Pentyl mercaptan. Colorless liquid, d.0.857, b.126, insoluble in water; used as an odorant for natural gas and in synthesis.

pentanoic acid* Valeric acid*. **4-oxo ~** Levulinic acid.

pentanol **1- ~ *** Pentyl alcohol*. **2- ~ *** *sec*-Pentyl alcohol*. **3- ~ *** $MeCH_2 \cdot CHOH \cdot CH_2 \cdot Me$. d.0.815, b.116. **2,5-dimethyl-2- ~** $Me_2COH \cdot CH_2 \cdot CHMe_2$ = 116.2. Colorless liquid, b.130, soluble in alcohol. **1,5-dimethyl-3- ~ *** $Me_2CH \cdot CHOH \cdot CH \cdot Me_2$. Colorless liquid, d.0.832, b.131, soluble in water. **methyl ~** $Me_2C(OH)(CH_2)_2Me$ = 102.2. Colorless liquid, d.0.830, b.115, soluble in alcohol.

pentanone* A ketone derived from pentane. **2- ~** $Me \cdot CO \cdot Pr$ = 86.1. Methyl propyl ketone*, acetyl propane. Colorless liquid, d.0.812, b.102. **3- ~** $Et \cdot CO \cdot Et$. Diethyl ketone*. m.−42, b.102, soluble in water. **phenyl ~** Butyl phenyl ketone*. **tetramethyl ~** Pennone.

pentaoxide* Pentoxide. A binary compound containing 5 atoms of oxygen; as, nitrogen pentaoxide, N_2O_5.

pentaprismo-* Affix indicating 10 atoms bound into a pentagonal prism.

Pentasol Trademark for a mixture of pentyl alcohols, b.116–136.

pentasulfide A compound containing 5 sulfur atoms; as, K_2S_5.

pentatriacontane* $Me(CH_2)_{33}Me$ = 493.0. White crystals, m.75.

pentavalent* An atom or group of atoms of valency 5.

pentazane Pyrrolidine.*

pentazdiene A compound derived from $HN{:}N \cdot NH \cdot N{:}NH$.

pentazocine $C_{19}H_{27}ON$ = 285.4. Fortral, Talwin. An analgesic.

pentazolyl* The radical $-N \cdot N{:}N \cdot N{:}N$ (ring closed between first and last N).

pentels* The Group 5B elements of the periodic table, N, P, As, Sb, Bi.

pentene* **1- ~** $Me(CH_2)_2CH{:}CH_2$ = 70.1. n-Amylene, propylethylene. Colorless flammable liquid, b.40, insoluble in water. **2- ~** Isoamylene, ethylmethylethylene. Colorless, flammable liquid, b.37, insoluble in water. Three forms:

Et·CH ‖ HC·Me	Et·CH ‖ Me·CH	Et·C·Me ‖ HCH
trans	*cis*	*asym*

pentenedioic acid* **2- ~** $HOOC \cdot CH_2CH{:}CH \cdot COOH$ = 130.1. Propenedicarboxylic acid, glutaconic acid. Colorless crystals, m.134.

p. a. anhydride* $O \cdot CO \cdot CH_2 \cdot CH{:}CH \cdot CO$ (ring) = 112.1. Glutaconic acid anhydride. Colorless crystals, m.87.

pentenic acid Pentenoic acid*.

pentenoic acid* $C_5H_8O_2$ = 100.1. **2- ~** $EtCH{:}CHCOOH$. Propylideneacetic acid, γ-methylcrotonic acid. Colorless liquid, d.0.990. b.201. **3- ~** $MeCH{:}CH \cdot CH_2COOH$. Ethylidenepropionic acid. Colorless liquid, d.0.987, b.194. **4- ~** $CH_2{:}CH(CH_2)_2COOH$. Allylacetic acid. Colorless liquid, d.0.984, b.189. **3,4-dimethyl-3- ~ *** $C_7H_{12}O_2$ = 128.2. Teracrylic acid. A liquid, b.218. **iso ~** See *butenoic acid.* **methyl ~ *** Pyroterebic acid.

pentenol* $C_5H_{10}O$ = 86.1. **1-penten-3-ol*** $CH_2{:}CH \cdot CHOH \cdot CH_2 \cdot Me$. Ethylvinylcarbinol. Colorless liquid, d.0.840, b.114. **3-penten-2-ol** $Me \cdot CHOH \cdot CH{:}CHMe$. Dimethylpropenylcarbinol. Colorless liquid, d.0.834, b.112. **4-penten-1-ol** $CH_2OH(CH_2)_2CHCH_2$. β-Allyl ethyl alcohol. A liquid, d.0.863, b.140. **4-penten-2-ol** $Me \cdot CHOH \cdot CH_2CH{:}CH_2$. Allylmethylcarbinol. A liquid, d.0.834, b.116.

pentenyl* Describing a series of radicals derived from pentene; as, **1- ~** $-CHCH(CH_2)_2Me$.

pentevalent Pentavalent*.

pentine Pentyne*.

pentinic acid $C_6H_8O_3$ = 128.1. α-Ethyltetronic acid. Colorless crystals, m.128.

pentite Pentitol.

pentitol $CH_2OH(CHOH)_3CH_2OH$. Pentite: 5 isomers derived from the pentoses; 2 are optically active. See *arabitol, xylitol.* **methyl ~** Rhamnitol.

pentlandite (Fe, Ni)S. A native sulfide.

pentobarbital $C_{11}H_{18}N_2O_3$ = 226.3. 5-Ethyl-5-(1-methylbutyl)barbituric acid. Nembutal, pentabarbitone. White crystals, m.129; a hypnotic (USP). **p. sodium** Used similarly to p. (USP). See *barbiturates.*

pentobarbitone BAN for pentobarbital.

pentoic acid Valeric acid*.

pentol $HC{:}C \cdot CH{:}CH \cdot CH_2OH$ = 82.1. 2-Penten-4-yn-1-ol. b_{10mm}66.

pentonic acid $CH_2OH(CHOH)_3COOH$ = 166.1. A series of pentavalent monobasic acids. See *arabonic acid.*

pentosans A group of gums or resins (hemicelluloses) which hydrolyze to pentoses; as, araban. They are constituents of cell membranes of plants. **methyl ~** Gums that yield methyl pentoses on hydrolysis.

pentose Pentaglucose. A monosaccharide sugar containing 5 C atoms, as: arabinose, xylose. **methyl ~** See *rhamnose.*
pentoside A compound containing a pentose, with a single N—C bond to a purine or pyrimidine base (nucleoside).
Pentostam Trademark for sodium stibogluconate.
Pentothal Trademark for thiopental sodium.
pentoxazol 4*H*-1,3-Oxazine.
pentoxide Pentaoxide*.
pentrite $C(CH_2O \cdot NO_2)_4$ = 316.1. Tetranitroerythrite. A high explosive.
pentyl* Amyl. The radical $Me(CH_2)_4$—, from pentane. **iso ~*** 3-Methylbutyl†, isoamyl. The radical $Me_2CH(CH_2)_2$—. ***tert*-~*** 1,1-Dimethylpropyl†. *tert*-Amyl. The radical $Et \cdot C(Me)_2$—.
p. acetate* $MeCOOC_5H_{11}$ = 130.2. **iso ~*** Isoamyl acetate, isoamylacetic ester, banana oil, pear oil, amylacetic ester. Colorless, flammable liquid, b.148, sparingly soluble in water, miscible with alcohol. Used in making fruit flavors; as a solvent for nitrocellulose, varnishes, and lacquers; for waterproofing materials, and for metallic paints and liquid bronzes. **p.acetic ether** Iso*pentyl acetate**. **p. alcohol*** $CH_3(CH_2)_4OH$ = 88.1. 1-Pentanol*, fusel oil, grain oil, potato spirit, amyl alcohol. Colorless liquid, d.0.817, m.—79, b.138, miscible with alcohol. Used as a solvent and reagent, and in pharmaceuticals, artificial silk, varnishes, lacquers, mercury fulminate, and photography. **active ~** See iso*pentyl alcohol* or secondary *pentyl alcohol* below. **alpha-~** Secondary p.a. **beta-~** Tertiary p.a. **dextro-~** See iso*pentyl alcohol* or secondary *pentyl alcohol* below. **iso ~*** $Me_2CH \cdot CH_2 \cdot CH_2OH$. Isobutylcarbinol, 2-methylbutanol, isoamyl alcohol. Occurs in (+) and (—) forms. Colorless liquid, d.0.825, b.132, the chief constituent of fusel oil. **levo-~** See iso*pentyl alcohol* above or secondary *pentyl alcohol** below. **neo ~** 2,2-Dimethyl*propanol**. **normal ~, primary ~** P. alcohol*. **secondary ~*** $Me(CH_2)_2CH(Me)OH$. α-Amyl alcohol, secondary butylcarbinol, 1-methylbutanol*, 2-pentanol*; occurs in (+) and (—) forms. Colorless, flammable liquid, d.0.809, b.118, miscible with water; used in organic synthesis and as a solvent. **tertiary ~*** $MeCH_2C(Me)_2OH$. Amylene hydrate, dimethylethylcarbinol, 1-dimethylpropyl alcohol. Colorless, flammable liquid, d.0.809, m.—12, b.102, soluble in water; used in fruit ethers. **p. aldehyde** Valeraldehyde*. **p.amine*** $C_5H_{11}NH_2$ = 87.2. ***n*-~** Normal amylamine, 1-aminopentane. Colorless liquid, b.104, miscible with water. **iso ~** Colorless liquid, b.95, soluble in water. Used as an emulsifier and in dye manufacture. Cf. *dipentylamine, tripentylamine.* **p.benzene*** $C_5H_{11} \cdot C_6H_5$ = 148.2. Phenylpentane. Colorless liquid, b.201, soluble in alcohol. **p. benzoate*** $C_6H_5COOC_5H_{11}$ = 192.3. ***n*-~** Colorless liquid, b.260, soluble in alcohol. **iso ~** Colorless liquid, b.262, insoluble in water. **p. bromide*** $C_5H_{11}Br$ = 151.0. **n-~** 1-Bromopentane*. Colorless liquid, b_{740mm}129, soluble in alcohol. **iso ~** β-Bromopentane. Colorless liquid, b.120, insoluble in water. **p. butanoate*** $C_3H_7COOC_5H_{11}$ = 158.2. **isopentyl butanoate** Isoamylbutyric ester. Colorless liquid, b.179, slightly soluble in water; used in fruit ethers. **pentyl isobutanoate** Colorless liquid, b.154, slightly soluble in water. Used in organic synthesis and in artificial fruit essences. **isopentyl isobutanoate** Colorless liquid, b.169, miscible with alcohol. **p. carbamate*** $C_5H_{11}CO_2NH_2$ = 131.2. Amylcarbamic ester, amyl urethane. Colorless crystals, m.60, soluble in alcohol. Cf. *aponal.* **p. chlorocarbonate** $ClCOOC_5H_{11}$ = 150.6. **iso ~** Colorless liquid, b.100, insoluble in water. **p. chloride*** $C_5H_{11}Cl$ = 106.6. ***n*-~** 1-Chloropentane*. Colorless liquid, b_{740mm}107, miscible with alcohol. **iso ~** β-Chloropentane. Colorless liquid, b.100, insoluble in water. **p. cyanide*** $C_5H_{11}CN$ = 97.2. Capronitrile. Colorless liquid, b.163, insoluble in water. **iso ~** Isocapronitrile. Colorless liquid, b.155. **p. ether** $(C_5H_{11})_2O$ = 158.3. ***n*-~** Pentyloxypentane. Yellow liquid, b.169, insoluble in water. **iso ~** Isoamyl oxide. Colorless liquid, b.173, insoluble in water; a solvent. **p. formate*** $HCOOC_5H_{11}$ = 116.2. ***n*-~** Colorless liquid, b.123, slightly soluble in water. **iso ~** Colorless liquid, b.130, slightly soluble in water; used in fruit essences. **p. furoate*** $C_4H_3O \cdot COOC_5H_{11}$ = 182.2. ***n*-~** Colorless liquid, b.233; used in perfumery and lacquers. **iso ~** b_{25mm}136. **p. hydrate** See *pentyl alcohol.* **p. hydride** Pentane*. **p. hydrosulfide** 1-Pentanethiol*. **p. iodide*** $C_5H_{11}I$ = 198.0. **n-~** 1-Iodopentane*. Colorless liquid, b_{739mm}155, miscible with alcohol. **iso ~** Isoiodopentane. Colorless liquid, b.148, insoluble in water; an antiseptic. **p. isocyanide*** $Me(CH_2)_4NC$ = 97.2. Pentylcarbylamine. Colorless liquid, b.155. **p. ketone** 6-Undecanone*. **p. mercaptan** 1-Pentanethiol*. **p. mustard oil** $C_5H_{11}NCS$ = 129.2. **iso ~** Isopentyl isothiocyanate. Colorless liquid, d.0.942, b.184, sparingly soluble in water. **p. nitrate** $C_5H_{11}NO_3$ = 133.1. **iso ~** Colorless liquid, b.147, slightly soluble in water. **p. nitrite*** $C_5H_{11}NO_2$ = 117.1. ***n*-~** Yellow, flammable liquid, b.96, insoluble in water. **iso ~** Colorless liquid, b.94, slightly soluble in water. Used as an antidote for cyanide poisoning, an antispasmodic, as a reagent for phenols and wormseed oil and in perfumes. **p. oxide** Pentyl ether. **p. oxyhydrate** Pentyl alcohol*. **p.phenol** $C_5H_{11} \cdot C_6H_4OH$ = 164.2. Isopentyl-*p*-phenol. White needles, m.94, slightly soluble in water. Used in the manufacture of synthetic resins, varnishes, and antiseptic emulsions. **p. phenylhydrazine** $C_5H_{11}(NH_2)NPh$ = 178.3. Colorless liquid, b.175; a reagent for aldehydes. **p. phenyl ketone*** $C_5H_{11}COC_6H_5$ = 176.3. Colorless liquid, b.242, insoluble in water. **p. phthalate** $C_6H_4(COOC_5H_{11})_2$ = 306.4. Amoil. Colorless crystals, b_{11mm}205, used in vacuum pumps, instead of mercury. **p. propionate*** $C_2H_5COOC_5H_{11}$ = 144.2. Colorless liquid, b.160, sparingly soluble in water. **p. salicylate*** $C_6H_4(OH)COOC_5H_{11}$ = 208.3. ***n*-~** Colorless liquid, b.277, soluble in water. **iso ~** Colorless liquid, b.270, insoluble in water; an antirheumatic and fruit flavor. **p. sulfate** $(C_5H_{11})_2SO_4$ = 238.3. Colorless liquid, $b_{2.5mm}$117, decomp. by water. **p. sulfide** $(C_5H_{11})_2S$ = 174.3. ***n*-~** Colorless liquid, b.204. **iso ~** Colorless liquid, b.215, insoluble in water. **p. thiocyanate** $C_5H_{11}SCN$ = 129.2. Amyl mustard oil. Colorless liquid, b.197, miscible with alcohol or ether. **p.urea** $NH_2CONHC_5H_{11}$ = 130.2. White crystals, m.90, slightly soluble in water. **p.urethane** $C_5H_{11} \cdot OCONH_2$ = 131.2. Amyl carbamate. Crystals, m.65, soluble in water. **p. valerate*** $C_4H_9COOC_5H_{11}$ = 172.3. **iso ~** Isoamylvaleric ester, apple essence. Colorless liquid, b.196, insoluble in water. Used in fruit flavorings, and as a sedative. **p. xanthate** The pentyl ester of xanthic acid; a flotation reagent.
pentylene Pentadiene*.
pentylidene* Amylidene. The radical $Me(CH_2)_3CH$=.
pentylidyne* The radical $Me(CH_2)_3C$≡.
pentyloxy* The radical $Me(CH_2)_4O$—.
pentyne* C_5H_8 = 68.1 **1-~** HC:C$(CH_2)_2$Me. *n*-Pentine, propylacetylene. Colorless liquid, d.0.722, b.40. **2-~** Me·C: C·Et. Ethylmethylacetylene, 2-pentine. A liquid, d.0.687, b.56. **p.dioic acid** Glutinic acid.
PEO* Symbol for poly(ethylene oxide). See *polyethylene glycol.*
peonol $C_9H_{10}O_3$ = 166.2. 2-Hydroxy-4-methoxyacetophene. From the root of *Paeonia montana* (Japan). Colorless needles, of aromatic odor, m.50.
pepo Pumpkin seed.

pepper Piper. The dried fruit of tropical *Piper* species. **black ~** The dried, unripe fruits of *Piper nigrum* (Piperaceae), containing piperine, chavicine, essential oils, and a resin; a condiment and source of p. oil. **cayenne ~** Red p. **Jamaica ~** Pimenta. **long ~** The dried, unripe fruit of *P. longum* a carminative. **red ~** The dried, ground, unripe fruits of *Capsicum frutescens* (Solanaceae); a condiment. **Spanish ~** Red p. **white ~** Black p. without the outer skin; milder than black p.

p. oil An oil from the fruits of *Piper nigrum*. Yellow liquid, d.0.87–0.91, insoluble in water; contains cadinene, phellandrene, and dipentene.

peppermint Mentha piperita. The dried leaves and flowering tops of *Mentha piperita* (Labiatae). Used for flavoring (NF, EP, BP), and for manufacturing oil and menthol. **p. camphor** Menthol*. **p. oil** An essential oil distilled from p. Colorless liquid of strong odor, d.0.91, insoluble in water (50–90% menthol). Grades: American, Chinese, English, Italian, Japanese, and Spanish; a flavoring, antispasmodic (NF, EP, BP), and perfume. **p. water** A saturated solution of p. oil in water; a carminative (NF).

pepsin* An enzyme of gastric juice that hydrolyzes proteins to proteases, peptones, and peptides, preferably in acid solution. Obtained from pigs' stomachs, as fine, white grains, slightly hygroscopic and soluble in water or dilute acid. **vegetable ~** Papain*.

pepsinogen The precursor of pepsin in the cells of the stomach glands.

peptide A compound of 2–10 amino acids joined through the main (not side) chain by the p. amide bond, $-C(:O)NH-$; e.g.: dipeptides, $NH_2 \cdot R \cdot CO \cdot NH \cdot R \cdot COOH$ (as, carnosine); tripeptides, $NH_2 \cdot R \cdot CO \cdot NH \cdot R \cdot CO \cdot NH \cdot R \cdot COOH$ (as, glutathione); tetrapeptides, $NH_2 \cdot R \cdot CO \cdot NH \cdot R \cdot CO \cdot NH \cdot R \cdot CO \cdot NH \cdot R \cdot COOH$ (as, triglycylglycine). Peptides use the same symbols as amino acids. **poly ~** Contains 10–100 amino acid residues, and resembles the peptones and proteins.

peptization The change from a jelly to a liquid other than by melting.

peptolysis The hydrolysis of peptone to amino acids.

peptolytic An agent that splits peptones.

peptones Simple mixtures of proteoses and amino acids, soluble in water, diffusible through parchment, not coagulated by heat; formed by the action of pepsin on albuminous bodies. Cf. *gastric digestion.* **beef ~** Meat p. **gelatin ~** P. obtained by digestion of gelatin. **meat ~** An extract from fresh, lean beef treated with the digestive juices of the pancreas. Brown powder, soluble in water; a nutrient and culture medium. **meta ~** A digestion product of p. **para ~** Syntonin. **silk ~** P. from silk; used in biochemical tests. **true ~** Tryptone.

peptonization The conversion of proteins into peptones.

peptonize To change into peptone.

peptonizing tube A parchment tube for dialyzing proteins.

peptotoxine A ptomaine from decomposing proteins.

per-* Prefix (Latin *per* = "through") indicating: (1) very or more than ordinary; as, indicating that all the H atoms, except those whose replacement would affect the nature of the characteristic groups present, have been replaced by halogen atoms of the same kind; e.g., perfluoropentane, $CF_3(CF_2)_3CF_3$; perbromo-, perchloro-, periodo-. (2) above or beyond; as, the higher oxidation state of the peracids of Group 7 elements.

peracetic acid* $Me \cdot CO \cdot O \cdot OH = 76.1$. Ethaneperoxoic acid†. An isomer of methyl hydrogencarbonate, made by the action of hydrogen peroxide on acetic acid. Used to bleach textiles and as a selective oxidant in the formation of epoxides from olefins.

peracid, per- acid (1)* Group 7 acids of higher oxidation state, as perchloric acid, $HClO_4$. (2) A peroxo-* acid. (3) A peroxy-* acid.

peracidity Excessive acidity.

Peractivin Trademark for *p*-toluene sulfomonochloride; mild oxidant in textile bleaching.

per(oxo)borate See *borate.*

perbromo-* See *per- (1).*

Perbunan Trademark for a copolymer of butadiene and styrene. Cf. *buna.*

Perbutan Trademark for a copolymer of butadiene and acrylic nitrile. Cf. *buna.*

percarbide A binary compound of carbon containing excess of carbon.

percarbonate Peroxodicarbonate*. A salt of the hypothetical peroxodicarbonic acids containing the radicals CO_4^{2-} or $C_2O_6^{2-}$; decomp. in aqueous solution to hydrogen peroxide and carbonates.

percarbonic acid H_2CO_4 or $H_2C_2O_6$. Carbonoperoxoic acid†. The hypothetical peroxodicarbonic acid, source of the per(oxodi)carbonates.

percentage* Parts in 100 parts by weight or volume.

percentile* A division of the area under a probability curve (see *normal probability curve*) giving upper and lower quartiles and deciles, etc.

perch Rod.

perchlorate* A salt of perchloric acid containing the radical ClO_4^-.

perchloric **p. acid*** $HClO_4 = 100.5$. Colorless liquid, d.1.12, $b_{60mm}39$; contains 20% $HClO_4$ in water. Used as an oxidizing agent, in electroanalysis, for the destruction of organic material; and reagent for potassium. **p. a. anhydride** Cl_2O_7. Chlorine heptaoxide. **p. a. dihydrate** $HClO_4 \cdot 2H_2O = 136.5$. A liquid oxidizing agent (72.4% p. acid), b.203. **p. a. hydrate** $HClO_4 \cdot H_2O = 118.5$. Explosive solid, m.50. **p. ether** Ethylperchlorate*.

perchloride A chloride that contains more chlorine than the corresponding normal chloride; e.g., $FeCl_2 \cdot 2FeCl_3$, iron(II) diiron(III) chloride*.

perchloro-* See *per- (1).*

perchloroethane Hexachloroethane*.

perchloroether $(Cl_5C_2)_2O = 418.6$. Decachloroethyl ether. Colorless scales, m.69 (decomp.).

perchloroethylene* Tetrachloroethylene*.

perchloryl* The radical $-ClO_3$.

perchromate A salt from perchromic acid; M_3CrO_8 (deep orange-brown); rapidly decomposed in aqueous solution.

perchromic acid $H_3CrO_8 = 183.0$. Contains tri- and hexavalent chromium.

percin A ptomaine from perch and pike (78% arginine).

percolate (1) To strain, or pass through fine interstices. (2) To extract soluble matter by the passage of water through a powder. (3) The solution obtained by percolation.

percolater Percolator.

percolator A conical, long glass vessel, with tubulated bottom, for the extraction of drugs. **p. bottle** A wide-mouthed, graduated bottle in which the percolate is collected.

Percorten Trademark for desoxycorticosterone acetate.

percrystallization The crystallization of a solute from a solution dialyzing through a membrane.

percussion The striking of a sharp blow. **p. cap** A detonator or primer. **p. figure** Strike figure. The radiating lines formed in certain minerals by striking them with a sharp

hammer. **p. powder** Fulminating powder. An explosive that ignites by p.
perdistillation Distillation through a dialyzing membrane. Cf. *pervaporation.*
pereira Pareira.
pereiro bark The bark of *Geissospermum laeve* (tropical America); an antipyretic.
perezol A 0.5% alcoholic solution of pipitzahoic acid; an indicator (acids—colorless, alkalies—deep orange).
perezone Pipitzahoic acid.
perferrate Misnomer for ferrate(VI)*, M_2FeO_4.
perfluoro-* See *per- (1).*
perforated plate A porcelain plate with small holes; usually inserted in a funnel and covered with filter paper for rapid filtration.
perfume A volatile, fragrant substance resembling a natural, odoriferous substance in odor: (1) *natural,* if obtained by extraction of flower, herb, blossom, or plant; (2) *artificial,* if a mixture of natural oils or oil constituents; (3) *synthetic,* if a mixture of synthetically produced substances. Cf. *essential oils, odor, terpenes.*
perfusion The passage of a fluid through spaces.
pergamyn An artificial parchment paper (German).
pergenol Solid hydrogen peroxide. A mixture of sodium peroxoborate and hydrogentartrate, which, with water, gives hydrogen peroxide.
perhydrate Ortizon.
perhydro-* Prefix indicating a fully hydrogenated hydrocarbon. Cf. *dihydro.*
perhydrol (1) A 30% solution of hydrogen peroxide. Known as "100 volumes" or (inaccurately) "100%" *hydrogen* peroxide, q.v., because it evolves 100 times its volume of O_2. (2) A supposed combination of 2 hydroxyl groups with an inert element or compound; as, $N{:}N(OH)_2$.
peri- Prefix (Greek) meaning "around." **p. acid** 1-Naphthylamine-8-sulfonic acid. ***p.*-fused*** See *ortho-.*
periclase MgO. Native in aggregates of small cubes.
pericline Albite.
pericyclic reaction See *pericyclic reaction* under *reaction.*
pericyclivine $C_{20}H_{22}N_2O_2$ = 322.4. An alkaloid from *Gabunia odoratissimi* (Apocyanoaceae).
pericyclo- Prefix describing a bond or valency that extends partly around a ring; as,

```
O————CH
|   HC    CH
|   ||    ||
|   HC    CH
O————CH
```

peridote A gem variety of olivine.
peridotites A group of crystalline, igneous rocks; chiefly olivine.
perigee The point in the moon's orbit at which it is nearest the earth. Cf. *apogee.*
perihelion The point of the orbit of a heavenly body at which it is nearest to the sun. Cf. *aphelion.*
perikinetic Concerning Brownian motion. Cf. *orthokinetic.*
perilla alcohol $C_{10}H_{16}O_4$ = 200.2. Colorless solid from palmarosa and naal oils.
perilogic series The allologic series of pyrene: $C_{10}H_8$, $C_{16}H_{10}$, $C_{22}H_{12}$, etc.
perimeter The sum of the lengths of the bounding lines of a figure.
perimidine* $C_{11}H_8N_2$ = 168.2. Green crystals, m.227, soluble in alcohol.
perimidinyl Any of eight isomeric radicals, $C_{11}H_7N_2-$, from perimidine.
perimorph A mineral that encloses another. Cf. *endomorph.*
perinaphthodiazine $C_{12}H_8N_2$ = 180.2. *N*-naphthodiazine. A group of isomers of naphthodiazine and naphthisodiazine (1,2-, 1,3-, 1,4-).
perinaphthotriazole $C_{10}H_7N_3$ = 169.2. π-Naphthotriazole. A series of compounds isomeric with naphthotriazole and naphthisotriazole (1,2,3-; 2,1,3-).
period (1) A regular interval between recurring phenomena. (2) A set of elements in the periodic table. (3) A family of closely related elements with consecutive atomic numbers; as, 24Cr—29Cu. (4) A geologic measure. See *geologic eras.* **damping ~** See *damping period.* **decay ~** Half-life. **gold ~** The elements Os, Ir, Pt, Au. **half ~, half-life** Half-life. **incubating ~** The time necessary for the full development of a bacterial culture. **iron ~** Iron family. The elements Mn, Fe, Co, Ni, forming the central portion of the 4th p. of the periodic table. **life ~** Half-life. **long ~** The 4th, 5th, or 6th p. of the periodic table. **platinum ~** The elements Os, Ir, Pt, Au, and Hg; part of the 6th p. of the periodic table. **rare earth ~** See *rare-earth metals.* **short ~** Either of periods 2 or 3 of the periodic table. **silver ~** The elements Ru, Rh, Pd, Ag, and Cd; the center of the fourth p. of the periodic table. **transitional ~** See *transition elements.* **uranium ~** The 7th p. of the periodic table (the radioactive elements).
periodate* A salt of periodic acid, containing the IO_4^- radical.
periodic (1) Pertaining to a regularly recurring event, phenomenon, or characteristic. (2) Pertaining to the highest valency of iodine, as *p. acid.* **p. acid*** $HIO_4 \cdot 2H_2O$ or H_5IO_6 = 227.9. Orthoperiodic acid*. Colorless, monoclinic crystals, m.140 (decomp.), soluble in water. Can be dehydrated to $H_4I_2O_9$. **p. chain** See *electron configuration.* **p. law** (1) Mendeleev law. The properties of the elements change periodically when the elements are arranged in increasing order of atomic number. (2) Moseley's law: The properties of the elements are a p. function of their atomic number. **p. precipitation** See *Liesegang rings.* **p. properties** Those properties of elements, which when plotted against atomic weight, show the relationship exemplified in the periodic table; as, valency, atomic volumes, atomic heats, wavelengths of X-ray spectra, compressibilities. **p. spiral** Helix Chemica. A graphical representation of the periodic system. **p. system** A classification of elements into one system, from which their properties could be deduced, and unknown elements and their properties predicted. Represented as a chain, table, q.v. (*periodic* table), spiral, or helix, so that elements with similar properties come together. **p. table** An arrangement of the p. system, developed independently (1869) by D. Mendeleev and (1870) by Lothar Meyer, and, in a crude form, by Newlands. See Table 60 on p. 433.
periodicity (1) The occurrence of a phenomenon at regularly recurring intervals. See *periodic properties.* (2) The occurrence of similar properties in a group of chemical elements in the periodic table.
periodo-* See *per- (1).*
peripheral Situated near or at the surface, or on the circumference of a curvilinear figure. **p. speed** The velocity of a point on the circumference of a rotating circular object: p. s. = $n \cdot \pi \cdot d$ m/min, where n is the rate of rotation in rpm, and d the diameter in meters.
periphery The surface or outer part; in particular, a circumference.
periplocin $C_{30}H_{48}O_{12}$ = 600.7. A glucoside from the bark of

TABLE 60. PERIODIC TABLE OF THE ELEMENTS

Group	1A	2A	3A	4A	5A	6A	7A	8			1B	2B	3B	4B	5B	6B	7B	0
ACS[d] format	1	2	3d	4d	5d	6d	7d	8d	9d	10d	11d	12d	13	14	15	16	17	18
Name	Alkali metals[a]	Alkaline-earth metals[b]	Rare-earth metals	Transition elements[c]									Triels	Tetrels	Pentels	Chalcogens	Halogens	Noble gases
Period																		
1	H 1																	He 2
2	Li 3	Be 4											B 5	C 6	N 7	O 8	F 9	Ne 10
3	Na 11	Mg 12											Al 13	Si 14	P 15	S 16	Cl 17	Ar 18
4	K 19	Ca 20	Sc 21	Ti 22	V 23	Cr 24	Mn 25	Fe 26	Co 27	Ni 28	Cu 29	Zn 30	Ga 31	Ge 32	As 33	Se 34	Br 35	Kr 36
5	Rb 37	Sr 38	Y 39	Zr 40	Nb 41	Mo 42	Tc 43	Ru 44	Rh 45	Pd 46	Ag 47	Cd 48	In 49	Sn 50	Sb 51	Te 52	I 53	Xe 54
6	Cs 55	Ba 56	La 57	Hf 72	Ta 73	W 74	Re 75	Os 76	Ir 77	Pt 78	Au 79	Hg 80	Tl 81	Pb 82	Bi 83	Po 84	At 85	Rn 86
7	Fr 87	Ra 88	Ac 89															

Lanthanoids[e] (Lanthanides)	Ce 58	Pr 59	Nd 60	Pm 61	Sm 62	Eu 63	Gd 64	Tb 65	Dy 66	Ho 67	Er 68	Tm 69	Yb 70	Lu 71
Actinoids[f] (Actinides)	Th 90	Pa 91	U 92	Np 93	Pu 94	Am 95	Cm 96	Bk 97	Cf 98	Es 99	Fm 100	Md 101	No 102	Lr 103

[a] H is not included in the alkali metals.
[b] Be and Mg are not included in the alkaline-earth metals.
[c] The transition elements commence with the Group 3A elements.
[d] ACS American Chemical Society.
[e] Lanthanoids include lanthanum, La.
[f] Actinoids include actinium, Ac.

Periploca graeca (Asclepiadaceae). Yellow powder, m.205, soluble in water; a cardiac glycoside.

periscope Altiscope. An arrangement of lenses and mirrors enabling an observer to see over intervening obstacles or from a submarine.

periscopic Applied to lenses having a concave-convex surface.

perisphere The "atmosphere" around a molecule, ion, or radical in which its influence is felt. Cf. *molecular diagrams.*

peristalsis The rhythmic contractions of the intestines by which their contents are propelled through and mixed with digestive enzymes.

peristaltic pump A device with rollers on a flexible plastic tube which propel the contents by p. motion. Used in automatic chemical analysis to take successive samples of liquid of determined quantity.

Peristaltin $C_{14}H_{18}O_8$ = 314.3. Trademark for a glucoside from the bark of *Rhamnus purshiana,* cascara sagrada. Brown, hygroscopic crystals, soluble in water; a cathartic.

periston A colloidal solution containing polymerized vinylpyrrolidone (hemodyn) and inorganic salts; used (in Germany during World War II) as a blood substitute. Cf. *povidone.*

peritoneum A smooth, transparent membrane lining the abdominal cavity and covering the surface of the intestines. See peritoneal *dialysis.*

Peritrate Trademark for pentaerythritol tetranitrate.

Perkin, Sir William Henry (1838–1907) English chemist, who made the first synthetic dye, mauvein (1856). **P.'s mauve** Mauvein. **P.'s reaction** The formation of unsaturated acids of the cinnamic type by the condensation of aromatic aldehydes with fatty acids in presence of acetic anhydride: $Ar \cdot CHO + CH_3COONa = Ar \cdot CH{:}CH \cdot COONa + H_2O$. **P.'s violet** Mauvein.

perlite A complex volcanic silicate of aluminum. At 900° it is softened and expanded to a foam, by its own steam, of 20 times its volume; used in fire extinguishers and for insulation. Cf. *pearlite.*

Perlofil Trademark for a polyamide synthetic fiber.

perloline $C_{36}H_{22}O_3N_4$ = 558.6. An alkaloid from perennial ryegrass (*Lolium perenne* L.), soluble in water. Solutions in chloroform are golden, with a green fluorescence visible in concentration of 1 in 5×10^6.

Perlon Trademark for a polyamide synthetic fiber. **P.-U** Trademark for a polyurethane synthetic fiber.

permafrost Describing areas of permanently frozen ground, e.g., 33% of Canada.

permalloy A magnetic alloy of nickel (30–80%) and iron.

Permalon Trademark for a mixed-polymer synthetic fiber.

permanent Everlasting, fixed, enduring. **p. gas** A gas that cannot be liquefied or condensed by pressure alone. **p. set** The deformation that an object retains after removal of the deforming force. **p. white** Precipitated barium sulfate.

permanganate* $MMnO_4$. A salt of permanganic acid. Purple in color. Good oxidizing agents, many permanganates are disinfectants. **p. ion*** The negatively charged ion MnO_4^-, which produces a purple color in solution.

permanganic acid* $HMnO_4$ = 119.9. An acid derived from heptavalent manganese, stable only in dilute solutions; decomp. to manganese dioxide and oxygen.

permanganyl The radical MnO_3-. **p. chloride** MnO_3Cl = 138.4. The green-brown acid chloride of permanganic acid which emits purple fumes in moist air and explodes if heated. **p. fluoride** MnO_3F = 121.9.

permeability (1) The ability to pass or penetrate a substance or membrane. (2) The quantity of air flowing through a body in unit time under standard conditions of area, thickness, and pressure. Cf. *porosity, osmosis.* **magnetic ~*** See *magnetic permeability.*

permeable Pervious or porous. **semi ~** Allowing the passage of some substances but not others; as, a semipermeable *membrane.*

permeate To pass through the pores of a body without rupture of its parts.

permethrin* See *insecticides,* Table 45 on p. 305.

Permian See *geologic eras,* Table 38.

Perminal Trademark for a sodium alkylnaphthalene sulfonate wetting agent.

Permitil Trademark for fluphenazine hydrochloride.

permittivity* ϵ. Absolute p. The value of the electric displacement divided by electric field strength. Measured in F/m. **relative ~*** ϵ_r. Dielectric constant, inductivity, specific capacitance. The p. of a dielectric divided by that of a vacuum. For practical purposes, it is the ratio of the capacitance of a condenser, when the dielectric is the substance under investigation, to that with a vacuum.

permivar An alloy: Ni 45, Co 25, Fe 30%, which has a high magnetic permeability and low hysteresis loss; used in the cores of loading coils for telephone circuits.

permonosulfuric acid Caro's acid.

permutation (1) Substitution, e.g., of radicals. (2) Transmutation.

Permutit Trademark for a group of cation and anion exchangers.

permutite $Na_2Al_2H_6Si_2O_8$. An artificial zeolite obtained by melting aluminum silicate, sodium carbonate, and sand together; used for water softening (p. process). Regenerated by a strong solution of sodium chloride: (1) hard water + sodium-permutite = soft water + calcium-permutite. (2) Calcium-permutite + sodium chloride = calcium chloride solution + sodium-permutite. **natural ~** A zeolite mineral, boronite or refinite, used for water softening. Also used for estimating ammonia in blood and urine.

permutoid reaction A reaction of double decomposition between a soluble and insoluble substance. Cf. *permutite.*

pernambuco Lima wood, Nicaragua wood. The red wood from *Caesalpinia echinata* (Leguminosae); a dye.

pernitric acid Peroxonitric acid*.

Pernot furnace A reverberatory puddling furnace with a circular inclined hearth, used in making steel.

peroffskite $CaTiO_3$. Perovskite.

peronine $C_{24}H_{25}O_3N \cdot HCl$ = 411.9. Benzylmorphine hydrochloride. Colorless powder, soluble in water; a narcotic.

perosis Displacement of the ankle joint of chickens, due to manganese deficiency.

perosmic A compound of octavalent osmium. **p. acid** See *osmic acid.*

Pérot lamp A mercury-vapor lamp light source for optical instruments.

peroxidase* A hemoprotein enzyme that catalyzes the oxidation of a donor by H_2O_2. Cf. *catalase.*

peroxide (1)* Superoxide. A derivative of hydrogen peroxide or a compound containing the $-O-O-$ or $=O_2$ group, in which 2 O atoms are singly linked. They liberate hydrogen peroxide with acids and are strong oxidizing agents; as, H_2O_2. (2) Used loosely for *dioxides,* q.v., which liberate oxygen with acids. **acid ~** See *peroxo-, peroxy-.* **hydro ~*** See *hydroperoxide.*
p. ion* The anion O_2^{2-}. **p. of hydrogen** Hydrogen peroxide*. **p. value** A measure of the degree of deterioration by oxidation of oils and fats, determined analytically.

peroxo-* Prefix indicating the inorganic acid of the prefixed trivial name, but with substitution of $-O-$ by $-O-O-$. Cf. *peroxy-.* **p.diphosphoric acid** $H_4P_2O_8$ = 194.0. Crystalline solid. **p.disulfuric acid** $H_2S_2O_8$ = 194.1. Persulfuric acid. An acid in lead accumulators obtained by electrolyzing sulfuric acid; a strong oxidizing agent. **p.monosulfuric acid*** Caro's acid. **p.nitric acid** HNO_4 = 79.0. **p.nitrite ion** The anion NOO_2^-.

peroxy-* Prefix indicating the organic acid of the prefixed name, but with substitution of $-O-$ by $-O-O-$; as, peroxypropionic acid, EtC(O)OOH. Cf. *peroxo.*

peroxydol *Sodium* peroxoborate*.

perparaldehyde CHMe·O·CHMe·O·O·CHMe·O = 148.2. A thin oil, b_{12mm}45, insoluble in water.

perphenazine $C_{21}H_{26}ON_3ClS$ = 404.0. Fentazin, Trilafon. White, bitter crystals, m.98, insoluble in water; a tranquilizer and antiemetic (BP). See *phenothiazine.*

perphosphoric acid Peroxodiphosphoric acid*.

perrhenate* $MReO_4$. Colorless, and (except those of Ag, Tl, K, Rb, and Cs) soluble in water.

perrhenic acid* $HReO_4$ = 251.2. Colorless, stable powder.

Perrin **P., Jean B (1870–1942)** French chemist and physicist. Nobel prize winner (1926). **P. equation** The distribution of particles in a colloidal system $= 1.33\pi r^3 g(D - d)h = (RT/L) \ln (n_1/n_2)$, where r is the radius of the particles, D their density, d the density of the medium, g the acceleration of free fall, n_1 and n_2 the number of particles per unit volume in 2 layers of liquid a distance h apart, L the Avogadro constant, R the gas constant, and T the thermodynamic temperature.

perry A fermented beverage made from pear juice. Cf. *cider.*

persalt A salt of a peracid (1).

perseite Perseitol.

perseitol $CH_2OH(CHOH)_5CH_2OH$ = 212.2. Perseite, from *Laurus persea* (Lauraceae), a tropical tree. Colorless needles, m.188, soluble in water; occurs as (+) and (−) compounds.

perseulose $C_7H_{14}O_7$ = 210.2. *galacto*-Heptulose. A ketoheptose. L-~ m.110, produced by sorbose bacteria from perseitol.

Pershbecker furnace A rotating furnace, heated by wood, for roasting mercury ores.

Persian **P. flowers** Pyrethrum. **P. red** Indian red.

persic oil Peach kernel nut oil (USP).

persilicic Describing an acid or acidic rock containing more than 60% Si.

persimmon The fruit-bearing trees *Diospyros virginiana,* N. America; and *D. kaki,* China, Japan (Ebenaceae).

persio Orchil.

personal equation Differences in reading instruments or in making chemical operations, due to temperamental or other inherent qualities or habits of the operator.

persorption The intimate and almost molecular mixture of a gas and solid due to absorption. Cf. *sorption.*
Persoz **P., Jean François (1805–1869)** French technical chemist. **P.'s solution** 10 g zinc chloride, 10 mL water, 2 g zinc oxide; dissolves silk but not wool.
Perspex Trademark for a transparent polymerized methyl methacrylate plastic; made from acetone, methyl alcohol, hydrogen cyanide, and sulfuric acid. It is light in weight, a good dielectric, and can "pipe" light round bends.
perstoff Diphosgene.
persulfate A salt derived from per(oxodi)sulfuric acid, which contains the radical $S_2O_8^{2-}$; made by the electrolysis of sulfate solutions.
persulfide A sulfide containing more S than is required by the normal valency of the element; as, Na_2S_2.
persulfuric acid Peroxodisulfuric acid*.
perthio The radical =S=S.
perthiocarbonates M_2CS_4. Salts formed from a solution of carbon disulfide in alkali disulfides.
pertusarene $C_{60}H_{100}$ = 821.5. A solid hydrocarbon from the lichen *Pertusaria communis.*
pertussis vaccine A sterile suspension of killed p. bacilli in normal saline. An immunizing agent for whooping cough (USP, EP, BP).
Peru **P. apple** Stramonium. **P. balsam** Indian balsam. A reddish-brown balsam from *Myroxylon (Toluifera) pereirae* (Leguminosae), tropical America; a local irritant (USP).
peruol Benzyl benzoate*.
peruscabin Benzyl benzoate*.
Peruvian **P. balsam** Peru balsam. **P. bark** Cinchona.
peruvin Cinnamyl alcohol*.
peruviol Nerolidol.
pervaporation The evaporation of a liquid through a dialyzing membrane; as, parchment.
pervesterol A sterol, q.v. *(sterols),* isolated from the fat of algae.
pervious Allowing the passage of fluids.
perylene* $C_{10}H_6{:}C_{10}H_6$ = 252.3. 1,1′,8,8′-Binaphthylene*. m.274.
peryallartine $C_6H_8 \cdot CMe \cdot CH_2 \cdot CHNOH$ = 165.2. Perillaldehyde-α-antialdoxime; 2.0 times sweeter than sucrose.
pessary Pharmaceutical term for a medicated solid body for vaginal insertion.
pesticide General term for chemicals to combat pests; thus, including *insecticide*s, rodenticides, *fungicides*, *herbicides*, weedicides. The world loss of crops due to pests is estimated at about 35%. For ISO nomenclature, see the tables accompanying the specific pesticide types just given in italics above.
pestle A blunt, rounded instrument, for pounding drugs or chemicals in a mortar, or ores in a stamp mill.
PET Symbol for poly(ethylene terephthalate).
peta* Symbol: P; SI system prefix for a multiple of 10^{15}.
petalite $Li_2O \cdot Al_2O_3 \cdot 8SiO_2$. Found in S.W. Africa and used in ceramics to produce resistance to high, varying temperatures.
petalon The hypothetical disk-shaped nucleus of helium.
pethidine EP, BP name for meperidine.
Petit, Alexis Therèse (1791–1820) French physicist. Cf. *Dulong and Petit law.*
petitgrain **p. oil** An essential oil distilled from the leaves and fruits of *Citrus bigardia.* Yellow liquid, d.0.887, insoluble in water. Used in perfumery to adulterate neroli oil; contains linalool, citrene, and esters. **p. citronier oil** The essential oil from the unripe fruits of *Citrus medica,* d.0.869–0.878; contains citral and esters of linalool.
PETP* Symbol for poly(ethylene terephthalate).
petrichor An oil believed to be responsible for the odor of damp earth.
petri dish A flat, shallow, circular glass dish for bacterial cultures.
petrifaction Mineralization, silification. The process of changing organic matter into stonelike substances by the gradual infiltration and replacement of the tissues by mineral matter.
petrified wood Fossil wood that has been gradually changed into stone by the slow infiltration of silica, details of its structure being preserved.
petrochemical A chemical product derived from petroleum; e.g.: Chemical and hydrocarbon solvents, synthetic detergents, plastics and resins, agricultural chemicals (pesticides), ammonia and nitrogenous fertilizers, synthetic rubbers, synthetic fibers, glycerol, glycols, and other esters.
Petroff **P. equation** The coefficient of friction f of a shaft revolving in a lubricating oil of dynamic viscosity μ, at n rpm and p N/m^2 of projected area = $\pi^2 \mu n R / 230 p C$, where R is the shaft radius, and C the radial clearance. **P. reagent** A sulfonic acid of nonalkane hydrocarbons. A by-product of the refining of petroleum oils with fuming sulfuric acid; a catalyst in fat manufacture (Twitchell's method).
petrograd standard A measure of the volume of sawn softwood: 165 ft^3 = 4.67 m^3.
petrographical Pertaining to the description of rocks and stones.
petrography The study of rocks and stones as aggregates of minerals. Cf. *petrology.*
Petrohol Trademark for isopropyl alcohol synthesized by hydration of propene from petroleum cracking stills.
petrol U.K. usage for gasoline.
petrolate A general term for products derived from petroleum.
petrolatum Petroleum jelly, Vaseline, paraffin ointment, cosmolin, fossolin, adeps mineralis, yellow petrolatum. A purified mixture of semisolid hydrocarbons from petroleum; a yellow jelly, d.0.82–0.88, m.38–60, soluble in alcohol or chloroform; an ointment base, lubricant, cleanser, rust preventive, and leather dressing (USP, BP). **heavy ~** Mineral *oil* (2). **hydrophilic ~** A mixture of p. with cholesterol, octadecanol, and white wax; used in ointments. **light ~** d.0.830–0.870. **liquid ~** Mineral *oil* (2). **white ~** Albolene. A decolorized p. ointment base (USP). **yellow ~** Petrolatum.
p. albumin White p. **p. jelly** Petrolatum. **p. liquidum** Mineral *oil* (2).
Petrolene (1) Trademark for malthene, the oily or soft constituents of bitumen, soluble in petroleum spirit. (2) (Not cap.) Asphalt.
petroleum (1) Mineral oil, rock oil, coal oil, earth oil, seneca oil, crude oil, naphtha. A native mixture of gaseous, liquid, and solid hydrocarbons. Thick, brown or yellow oil, obtained from wells, springs, and lakes, d.0.78–0.97 (extreme limits 0.65–1.07); insoluble in water, soluble in organic solvents. (2) The fraction of crude p. distilling at 150–300°C. Classification:

A. Paraffin base: mainly C_nH_{2n+2} (C_4H_{10} to $C_{35}H_{72}$); small amounts of C_nH_{2n} ($C_{21}H_{42}$ to $C_{26}H_{52}$) and C_nH_{2n-x}. Sulfur occurs as thiophanes, $C_nH_{2n}S$.
B. Naphthene base (or aslphalt base): mainly C_nH_{2n} (naphthenes and alkenes); moderate amounts of C_nH_{2n-x}, where x is 2, 4, 6, 8, etc., up to 20; also some aromatic hydrocarbons, little or no C_nH_{2n-2}.

Estimated world reserves, about 7 × 10^{11} bbl, of which about 55% is in the Middle East and 17% in the Western Hemisphere. **p. asphalt** The residues from Trinidad p. **p. coke** The residue from the distillation of p.; used in

TABLE 61. PETROLEUM FRACTIONS

Product	Temperature, °C	Composition
	Boiling point	
Cymogen	0	C_4H_{10}
Rhigolene	18–21	C_4H_{10}, C_5H_{12}
Petroleum ether	40–60	C_5H_{12}, C_6H_{14}
Gasoline	40–180	C_6H_{14} to $C_{10}H_{22}$
Naphtha	70–90	C_6H_{14} to C_7H_{18}
Ligroin	90–120	C_7H_{16} to C_8H_{18}
Benzine	120–150	C_8H_{18} to C_9H_{20}
Kerosine, coal oil, photogene	150–300	C_9H_{20} to $C_{16}H_{34}$
Lubricating oils	Over 300	
Light		C_{12} to C_{20}
Medium		C_{16} to C_{22}
Heavy		C_{18} to C_{26}
Paraffin oil, liquid petrolatum, Albolene, Stanolax, Nujol, etc.	300	C_{10} to C_{18}
	Melting point	
Vaseline petrolatum, petroleum jelly	38–50	C_{20} to C_{22}
Paraffin, Parowax, Cerolene	45–65	C_{23} to C_{25}
Paraffin wax	50–80	C_{27} to C_{30}

In practice, however, the fractions ("cuts") are:

Fraction	%
1. Below 95°C at 1 atm, gasoline or naphtha	2–80
2. 95–135°C at 1 atm, kerosine or coal oil	3–25
3. All vacuum fractions to 40 mm, gas oil (η below 50 s)[a]	7–50
4. Light lubricating oil (η = 55–99)	2–15
5. Medium lubricating oil (η = 100–199)	3–20
6. Viscous lubricating oil (η exceeds 199)	0–22

[a]η = Saybolt viscosity in seconds.

metallurgical processes, carbon pencils, and dry batteries. It may contain petrolatum or paraffin wax. **p. ether** See *petroleum fractions.* A solvent. **p. fractions** See Table 61. **p. furnace** An oil burner. **p. jelly** Petrolatum. **p. naphtha** Benzine. **p. oil** See *insecticides,* Table 45 on p. 305. **p. ointment** Petrolatum. **p. pitch** Asphalt. **p. refining** The industrial separation of the constituents of p., including cracking, decolorizing, distilling, filtering, and skimming. **p. spirit** Ligroin.

petroline A paraffin obtained by distillation of Indian petroleum.

petrology Lithology. A branch of science dealing with the origin (petrogeny), structure, and composition (petrography) of rocks.

petromortis Garage poison.

petroselic acid Petroselinic acid.

petroselinic acid $Me(CH_2)_{10}CH{:}CH(CH_2)_4COOH$ = 282.6. 6-Octadecenoic acid*. An isomer of oleic acid, in parsley seeds.

petroselinum (1) Parsley. The dried herb of *Petroselinum sativum* (Umbelliferae); used medicinally, and in cooking. (2) Parsley seeds.

petrosilane $C_{20}H_{42}$ = 282.6. A saturated hydrocarbon, m.69, in the unsaponifiable matter of parsley oil.

petrosilex Chert. A hard, siliceous rock or flint; used for grinding.

petrous Hard or stonelike.

petrox Petroxolin.

petroxolin Petroxolinum liquidum, liquid petrox, petrolatum saponatum liquidum. Paraffin oil saponified with ammonium oleate. Brown liquid, insoluble in water; used medicinally in external preparations.

Pettenkofer P., Max von (1818–1901) German chemist. **P. test** (1) On adding urine to a mixture of sugar and sulfuric acid, a crimson color indicates the presence of bile acids. (2) The determination of carbon dioxide in air by absorption in a known volume of standard baryta solution, and titration of the excess with alkali.

petzite $(Ag, Au)_2Te$. A native telluride.

pewter (1) A gray alloy: Sn 83–75, Pb 0–20, with sometimes Sb 0–7 and Cu 0–4%. The lead is omitted from p. household utensils. (2) Calcined tin used for polishing marble. (3) A p. pot. **British ~** Contains at least 90% Sn, hardened with Cu and Sb or Bi, or both, with not more than 0.5% Pb.

PF Abbreviation for phenol-formaldehyde plastic.

Pfeilring reagent A fat-splitting catalyst made by the action of sulfuric acid on aromatic hydrocarbons and castor oil. Cf. *Twitchell reagent.*

Pferdekraft See *metric horsepower* under *horsepower.*

Pfeufer's green Blue-green dye from the fungus *Chlorospenium aeruginosum.*

Pfund German for "pound."

pH *p*H, pH, *pH*, p_H. Symbol for the logarithm of the reciprocal of the hydrogen-ion concentration, c_H or $[H^+]$; hence pH = log $(1/c_{H^+})$. **international standard scale ~** pH = pH (standard) + $FE/2.3RT$, where F is the Faraday constant and E the e.m.f. of the cell.

p-H Parahydrogen. See *hydrogen* (2).

Ph Abbreviation for phenyl, C_6H_5-.

phacolite Chabazite.

phacometer An instrument to measure the refractive index of a lens.

phaeo-, phæo- See words beginning with *pheo-*.

phage A virus which destroys bacteria. Often used synonymously with *bacteriophage.*

phagocytes Cells that envelop and digest invading microorganisms, harmful cells, or foreign matter. **fixed ~** P. located in the connective tissue (endothelial cells). **motile ~** P. that move in the blood and lymph (leukocytes).

phagocytolysis The destruction of phagocytes by microorganisms.

phagocytosis The destruction of microorganisms, by phagocytes. See *Metshnikoff.*

phagolysis The destruction of phagocytes.

phallin A toxic hemolytic protein from *Amanita phalloides,* a poisonous mushroom.

phanerogamia The large group of flowering plants which produce seeds.

phanquone $C_{12}H_6O_2N_2$ = 210.2. 7-Phenanthroline-5,6-quinone. Orange crystals, slightly soluble in water; an antiamebiatic (BP).

pharaoh's serpents A stick of *mercuric* thiocyanate, q.v. When ignited, it glows and swells during its formation to a voluminous ash that resembles a moving serpent. It liberates N_2, CS_2, and Hg vapor and leaves a gray residue of *mellon,* q.v.

Phar. D. Abbreviation for Doctor of Pharmacy.

Phar. M. Abbreviation for Master of Pharmacy.

pharmaceutic(al) Pertaining to drugs. **p. chemistry** The analysis of drugs and isolation of their active constituents.

pharmacist Apothecary. A druggist (U.S.) or chemist (U.K.).

pharmacodynamics The study of the effects of drugs on living organisms.

pharmacognosy The study of the identification, properties, and quality of crude drugs.

pharmacokinetics The study of the time course of drug and metabolite concentrations in body fluids and excreta.

pharmacolite $CaHAsO_4 \cdot 2H_2O$. A native arsenate.

pharmacology The study of drugs, their origin and composition *(pharmacy),* identification *(pharmacognosy),* and effects on living organisms *(pharmacodynamics).*

pharmacopoeia, pharmacopeia Official lists of drugs and chemicals issued by many countries. A p. contains a description of each drug, its composition, tests for identification and purity, and its medicinal doses. Substances listed are called "official" or "officinal," and must have the specified purity for medical use.

Aust.P.: *Oesterreichisches Arzneibuch*
BP: *British Pharmacopoeia*
Fr.P.: *Pharmacopée Française*
Ger.P.: *Deutsches Arzneibuch*
It.P.: *Farmacopea Ufficiale della Repubblica Italiana*
Jap.P.: *The Pharmacopoeia of Japan*
Neth.P.: *Nederlandse Farmacopee*
Nord.P.: *Nordic Pharmacopoeia*
Rus.P.: *State Pharmacopoeia of the USSR*
Span.P.: *Farmacopea Espanola*
Swiss.P.: *Pharmacopoeia Helvetica*
USP: *United States Pharmacopeia*
USNF: *National Formularly* (USA)

In addition, there are in use:

Eur.P.: *European Pharmacopoeia* (EP is used in this dictionary.)
The Extra Pharmacopoeia (Martindale)
Cf. *formulary.*

pharmacosiderite A native arsenate of iron.

pharmacotherapy The treatment of disease with drugs.

pharmacy The art of preparing drugs for medicinal use.

phase (1) A solid, liquid, or gaseous, homogeneous substance, that exists as a distinct and mechanically separate portion in a heterogeneous system. Cf. *colloid, zone, micelle.* (2) The succession of electrical impulses of an alternating current. (3) A stage in the growth of microorganisms. (4) A subdivision of the changes occurring in protoplasm during *karyokinesis,* q.v. **activatory ~** The active stage or rapid growth of organisms, especially bacteria. **continuous ~** External or enclosing p. The surrounding (dispersion) medium in a heterogeneous mixture. See *colloid.* **discontinuous ~** Dispersed p. **dispersed ~** Internal or enclosed p. The solute or insoluble part of a colloidal solution, as distinct from the solvent. **dispersion ~** Continuous p. **enclosed ~** The discontinuous or separated medium in a heterogeneous mixture. **enclosing ~** Continuous p. **inhibitory ~** The passive stage or slow growth of an organism. **oriented ~** Misnomer for *zone,* q.v. **suspended ~** Enclosed p.

p. coefficient See *symbols,* Table 88, Group *B,* on p. 566. **p. contrast microscopy** When light waves pass through an object whose refractive index is greater than that of its surroundings, they are retarded and emerge out of p. with those forming the background. If the p. difference is half the wavelength, the two sets of waves will cancel each other, and the object will appear dark. Used to improve visibility in microscopical work. **p. converter ~** A device for changing the phases of an alternating electric current. **p. reversal** The change of the components of an emulsion; thus, an emulsion of oil in water, converted into an emulsion of water in oil. **p. rule** Gibbs: A mathematical generalization of systems in equilibrium: $F = C + 2 - P$, where P is the number of phases, F the degrees of freedom, C the number of components. $F = 0$ is invariant (a *point* on a diagram), $F = 1$ is monovariant (a *line* on a diagram), $F = 2$ is divariant (an *area* on a diagram). Thus, for water:

Solid ⇌ liquid	$C = 1, P = 2, F = 1$
Solid ⇌ liquid ⇌ vapor	$C = 1, P = 3, F = 0$

phaselin An enzyme from the bean of *Dilkas mexicana,* resembling papain.

phaseolin The chief protein of the navy bean, *Phaseolus vulgaris.*

phaseoline An alkaloid obtained from string beans, *Phaseolus vulgaris.*

phaseolunatin $C_{10}H_{17}NO_6 = 247.3$. A cyanogenetic glucoside, m.144, from *Phaseolus lunatus,* lima beam (Leguminosae).

phaseomannite Inositol.

phasine A group of vegetable proteins from seeds, that agglutinate the red blood cells.

phasotropy Dynamic isomerism in which the H atom of amidines and formazyl derivatives oscillates from one nitrogen to the other:

```
H—NR'          H   NR'
|              |   ||
R·N=CH         R·N—CH
```

Ph.C. Abbreviation for Pharmaceutical Chemist.

Ph.D. Abbreviation for Doctor of Philosophy.

Phe* Symbol for phenylalanine.

phellandrene $C_{10}H_{16} = 136.2$. **α-~** 1,5-*p*-Menthadiene. 5-Isopropyl-2-methyl-1,3-cyclohexadiene. A terpene from the seeds of water fennel, *Phellandrium aquitanium* (Umbelliferae); constituent of certain eucalyptus oils, elemi oil, and oil of water hemlock. Colorless, (+)- and (−)-rotatory liquid, b.176.

phen- (1)* Indicating 1,10 phenanthroline as a ligand. (2) Prefix derived from phenyl, indicating a benzene derivative. (3) A suffix. See *-fen.*

phenacetein Phenacetolin.

phenacetin $C_{10}H_{13}O_2N = 179.2$. Acetophenetidin(e), acetophenetidide. White, bitter scales, m.135, insoluble in water; an analgesic and antipyretic. Usually used with aspirin and codeine as APC; use is limited by toxic effect on kidney (USP).

phenacetol Phenoxy acetone.

phenacetolin $C_{16}H_{12}O_2 = 236.3$. Phenacetein. An indicator (alkalies—red; acids—yellow).

phenacite Be_2SiO_4. A native gem silicate.

phenacyl* 2-Oxo-2-phenylethyl†. The radical $PhCOCH_2-$. **p. alcohol** See *hydroxyacetophenone.* **p. bromide** $PhCOCH_2Br$. White powder, a reagent for hydroxy compounds.

phenacylidene* The radical $PhCOCH=$.

phenacylidin $C_6H_4(OMe)NH \cdot CH_2COPh = 241.3$. Colorless powder, insoluble in water; an antipyretic in veterinary medicine.

Phenamine Trademark for direct dyestuffs, for cotton and paper.

phenanthrahydroquinone $C_{14}H_8(OH)_2 = 210.2$. 9,10-Dihyroxyphenanthrene*. A solid, m.146, soluble in water.

phenanthraquinone Phenanthrenequinone.
phenanthrene* $C_{14}H_{10}$ = 178.2. Phenanthrine. An isomer of anthracene from coal tar:

Colorless scales or leaflets, d.1.063, m.99, insoluble in water; used in the synthesis of dyes and drugs. **benzo ~** Chrysene*. **dihydroxy ~** Phenanthrahydroquinone. **hydroxy ~** Phenanthrol*. **methylisopropyl ~** Retene. **p.dione** Phenanthrenequinone. **p.hydroquinone** Phenanthrenequinone.
phenanthrenequinone $C_{14}H_8O_2$ = 208.2. 9,10-Phenanthrenedione†. Phenanth-9-quinone. An oxidation product of phenanthrene. Orange needles, m.209, insoluble in water. **p. dioxime** $C_{14}H_{10}O_2N_2$ = 238.2. A compound derived from p. by replacement of the two =C:O groups by two =C:N·OH radicals. **p. dioxime anhydride** $C_{14}H_8ON_2$ = 220.2. A furazan derivative. Colorless crystals, m.181. **p. monoxime** $C_{14}H_9O_2N$ = 223.2. A compound derived from p. by the replacement of one =C:O group by =C:N·OH. Colorless crystals, m.158; used in organic synthesis.
phenanthrenol Phenanthrol*.
phenanthrenone Phenanthrone*.
phenanthrenyl† See *phenanthryl.*
phenanthridine* $C_{13}H_9N$ = 179.2. Benzo[*c*]quinoline. Colorless crystals, m.104, soluble in alcohol.
phenanthridinone $C_{13}H_9ON$ = 195.2. The heterocyclic compound 6-oxophenanthridine. Colorless crystals, m.293, soluble in alcohol; used in organic synthesis.
phenanthrine Phenanthrene*.
phenanthrol* $C_{14}H_9OH$ = 194.2. Hydroxypheanthrene. Crystals. **1- ~** m.156 (subl.). **2- ~** m.169.
phenanthroline(s)* $C_{12}H_8N_2$ = 180.2. Heterocyclic compounds derived from phenanthrene by substituting 2 N atoms for 2 CH groups in the ring. **batho ~** 4,7-Diphenyl-1,10-phenanthroline. A sensitive, specific reagent for iron (red color). **dihydroxy ~** Snyder reagent. Cf. *bathocuproine, neocuproine.* ***ortho*- ~** 1,10-Compound; an oxidation-reduction indicator, and reagent for copper (brown color, soluble in isopentyl alcohol), iron, and cobalt. ***meta*- ~**, **pseudo ~** 2,9-Compound. ***para*- ~**, **iso ~** 3,8-Compound.
phenanthrone* $C_6H_4·C_6H_4·CH_2·CO$ = 194.2. Solid, m.148, soluble in water.
phenanthrophenazine $C_{20}H_{12}N_2$ = 280.3. 4 isomers, as: α,γ-Dibenzophenazine.

Colorless crystals, m.217, soluble in alcohol; used in organic synthesis.
phenanthryl* Phenanthrenyl†. Describing a group of 5 isomeric radicals, $C_{14}H_9-$, from phenanthrene.
phenanthrylene* The radical $-C_{14}H_8-$, from phenanthrene.
phenate Phenolate*.
phenazine* $C_6H_4·N·C_6H_4·N$ = 180.2. Dibenzoparadiazine, azophenylene. Yellow needles, m.170, soluble in water; used in the manufacture of dyes. **benzo ~** See *benzophenazine.* **dibenzo ~** $C_{20}H_{12}N_2$ = 280.3. Yellow crystals. **naphtho ~** Benzophenazine. **phenanthro ~** See *phenanthrophenazine.*
phenazone (1) $C_{12}H_8N_2$ = 180.2. Benzocinnoline. A heterocyclic compound. Yellow prisms, m.156, soluble in alcohol. (2) A former name for antipyrine.
phenazonium Derivatives of phenazone in which one N atom is pentavalent.
phene Synonym for the benzene ring.
phenedin Phenacetin.
phenelzine sulfate $C_6H_5·(CH_2)_2·NH·NH_2·H_2SO_4$ = 234.3. Phenethylhydrazine sulfate. Naidil. Pearly plates, m.166, soluble in water. An antidepressant in the monoamine oxidase inhibitor group; used to treat psychotic depression (USP, BP).
phenenyl The benzenetriyl* radical.
Phenergan Trademark for promethazine hydrochloride.
phenethicillin Common name for 6-phenoxypropionamidepenicillin. An antibiotic similar in origin and properties to penicillin. **p. potassium** $C_{17}H_{19}O_5N_2KS$ = 402.5. Bitter, white powder, soluble in water; an antibiotic (BP).
phenethyl* Phenyl ethyl. The radical $PhCH_2CH_2-$. **p. alcohol*** $PhCH_2·CH_2OH$ = 122.2. Colorless liquid, b.212. A constituent of oil of rose, geranium, etc.; used in perfumes.
phenethylamine $PhCH_2CH_2NH_2$ = 121.2. **1- ~**, **α- ~**. Colorless liquid, d.0.9395, b.182, insoluble in water, alcohol, or ether. **2- ~**, **β- ~**. Colorless liquid, d.0.9580, b.198, soluble in water. Base structural unit of the catecholamines. ***p*-hydroxy ~** Tyramine.
phenethylene Styrene*.
phenetidine* $NH_2·C_6H_4·OEt$ = 137.2. *ar*-Ethoxybenzenamine†, aminoethoxybenzene, ethoxyaniline, aminophenetole. **aceto ~** Phenacetin. ***N*-acetyl ~** Phenacetin. **ethyl carbonate ~** Eupyrine. ***meta*- ~** Colorless liquid, b_{100mm} 180, soluble in water. ***ortho*- ~** Colorless liquid, b.228, soluble in water. ***para*- ~** Colorless liquid, b.224, soluble in water; used to manufacture phenacetin. **phenylglycol ~** Amygdophenin.
phenetidines Synthetic drugs derived from phenetidine.
phenetidino* The radical $EtOC_6H_4NH-$.
phenetole* Ph·O·Et = 122.2. Ethyl phenyl ether*, ethoxybenzene*. Colorless liquid, d.0.892, b.172, insoluble in water. **acetamido ~**, **acetylamino ~** Phenacetin. **amino ~** Phenetidine.* **azo ~** See *azophenetole.*
phenetyl The ethoxyphenyl* radical. **p.urea** Sucrol.
phengite Muscovite.
phenic acid Phenol. **di ~** See *diphenic acid.*
phenicate (1) Phenolate*. (2) To sterilize or disinfect with phenol.
phenil Phenyl*.
phenindione $C_{15}H_{10}O_2$ = 222.2. 2-Phenyl-1,3-indandione. Dindevan, Hedulin. White crystals, m.150, soluble in water; an anticoagulant used to treat vascular thrombosis. (BP).
phenixin Carbon tetrachloride*.
phenmethyl The benzyl* radical.

phenmethyltriazine $C_6H_4 \cdot N:N \cdot C(Me):N$ = 145.2. Colorless crystals, m.89, soluble in alcohol.
phenmiazin Quinazoline*.
phenobarbital $C_{12}H_{12}O_3N_2$ = 232.2. 5-Ethyl-5-phenylbarbituric acid. Phenobarbitone. Gardenal, Luminal. White crystals, m.176, slightly soluble in water. A sedative and anticonvulsant; used to treat epilepsy (USP, EP, BP).
phenobarbitone EP, BP name for phenobarbital.
phenodiazine **alpha-~** Cinnoline*. **beta-~** Phthalazine*.
phenodin Hematin.
phenol* (1) See *phenols.* (2) $C_6H_5 \cdot OH$ = 94.1. Carbolic acid, hydroxybenzene, phenyl hydroxide, phenic acid, phenylic acid. Colorless needles, d.1.072, m.42, b.182, slightly soluble in water, soluble in alcohol or glycerol. Used as an antiseptic and disinfectant (USP, BP) and in the manufacture of dyes, synthetic drugs, plastics, and insulating materials. **aceto ~** Hydroxyacetophenone. **allyl ~** Chavicol. **allylmethoxy ~** Eugenol*. **amino ~** See *aminophenol* under *amino.* **aminodinitro ~*** Picramic acid. **aminoethyl ~** Tyramine. **arsenobis ~** Arsenophenol. **benzoylamino ~** Hydroxy*benzanilide.* **betel ~** Chavicol. **bis ~** See *bisphenol.* **bromo ~*** See *bromophenols.* **chloro ~** A halogenated derivative of p. E.g.: (1) $OH \cdot C_6H_4 \cdot Cl$. Chlorophenic acid, monochlorophenol*. ***ortho-* ~** m.9. ***meta-* ~** m.33. ***para-* ~** m.37. (2) $HO \cdot C_6H_3 \cdot Cl_2$. Chlorophenesic acid, dichlorophenol*. (3) $HO \cdot C_6H_2 \cdot Cl_3$. Chlorophenisic acid, trichlorophenol*. (4) $HO \cdot C_6H \cdot Cl_4$. Chlorophenosic acid, tetrachlorophenol*. (5) $HO \cdot C_6Cl_5$. Chlorophenasic acid, pentachlorophenol*. **diamino ~** See *aminophenol* under *amino.* **dichloro ~** See *chlorophenol* above. **dimethyl ~** Xylenol*. **(dimethylamino)ethyl ~** Hordenine. **dinitro ~*** See *dinitrophenol* under *dinitro-.* **ethyl ~*** Phlorol. **hexahydro ~** Cyclohexanol*. **hydroxy ~** Dihydroxy*benzene.* **iodized ~** A mixture of iodine 20, phenol 60, glycerol 20%; an antiseptic. **iodo ~** An iodized derivative of p. E.g.: (1) $OH \cdot C_6H_4 \cdot I$. Iodophenic acid, monoiodophenol*. ***ortho-* ~** m.40. ***meta-* ~** m.40. ***para-* ~** m.94. (2) $OH \cdot C_6H_3 \cdot I_2$. Iodophenesic acid, diiodophenol*. (3) $OH \cdot C_6H_2 \cdot I_3$. Iodophenisic acid, triiodophenol*. **2,4,6- ~** m.156. **isopentyl ~** See *pentylphenol* under *pentyl.* **isopropylmethyl ~** Thymol. **liquefied ~, liquid ~** P. containing 10% (USP), or 20% (BP) water. **methoxy ~** Guaiacol*. **methoxymethyl ~** Creosol. **methoxypropenyl ~** Anethole*. **methyl ~** Cresol*. **methylamino ~** Metol. **monobromo ~*** C_6H_4BrOH = 173.0. Brown oil; an antiseptic. **monochloro ~*** C_6H_4ClOH = 128.6. Colorless liquid; an antiseptic. **nitro ~** See *nitrophenol.* **propenyl ~** Anol. **propyl ~** Thymol*. **seleno ~** See *selenophenol.* **sulfo ~** Aseptol. **thio ~** See *thiophenol.* **triamino ~*** See *aminophenols.* **trichloro ~*** See *chlorophenol* above. **trimethyl ~*** **2,3,5- ~** Isopseudocuminol. **2,4,6- ~** Mesitol. **trinitro ~** Picric acid*.

p. acids Compounds having a hydroxy and a carboxyl group attached to a ring. E.g., Hydroxybenzoic acids (salicylic acid, $HO \cdot C_6H_4 \cdot COOH$); hydroxyphthalic acids, $HO \cdot C_6H_3(COOH)_2$. **p. bismuth** Bismuth phenolate*. **p. blue** $C_{14}H_{14}N_2O$ = 226.3. Dimethylaminophenylimide. A blue indicator dye. **p. camphor** A germicide mixture of phenol and camphor. **p. coefficient** Antiseptic power compared with phenol as unity. Cf. *Rideal-Walker test.* **p. derivatives** See *phenols.* **p. ethers** Ph—O—R. **p. formaldehyde resin** A synthetic resin formed by the condensation of p., or a p. with formaldehyde. Cf. *plastics.* **p. glycerol** A 16% solution of p. in glycerol (BP). **p.phthalein** See *phenolphthalein.* **p. red** Phenolsulfonephthalein. **p.tricarboxylic acid** $OH \cdot C_6H_2 \cdot (COOH)_3$ = 226.1. **2,4,5- ~** A solid, decomp. 245.
phenolate (1) To disinfect with phenol. (2)* Phenate, carbolate, phenylate. A salts of a metal containing the phenoxy radical; as, sodium p., PhONa.
phenolic Pertaining to phenol.
phenology The study of bird migrations, with special reference to climate.
phenoloid Algoid and similar plant substances, which contain compounds similar to phenol.
phenolphthalein $C_{20}H_{14}O_4$ = 318.3. 3,3-bis(*p*-Hydroxyphenyl)phthalide. Phenothalin. White, triclinic crystals, m.258, slightly soluble in water, soluble in alcohol; used as an indicator (acids—colorless; alkalies—deep red), a cathartic (USP, BP), and in dye synthesis. Cf. *phenolphthaline.* **tetrabromo ~** An indicator.
phenolphthalide $Ph \cdot CO \cdot OCR_2$, in which R is an aromatic radical. Cf. *fluorescein.*
phenolphthalin(e) $(C_6H_4OH)_2:CH \cdot C_6H_4 \cdot COOH$ = 320.3. Phthalin. Kastle-Meyer reagent (when in solution). 4,4′-Dihydroxytriphenylmethanol-2″-carboxylic acid. m.237. Made by reducing phenolphthalein with zinc and alkali. Reagent for oxidases, blood, peroxides, cyanides, and copper. Cf. *phenolphthalein.*
phenols* Aryl hydroxides; as, phenol and derivatives. In particular, hydroxybenzenes. Compounds containing one or more hydroxy groups attached to an aromatic or carbon ring. Classification:

1. Monohydroxybenzenes; e.g.: phenol, C_6H_5OH.
2. Dihydroxy benzenes (diphenols); e.g., catechol, resorcinol, hydroquinone, $C_6H_4(OH)_2$.
3. Trihydroxy benzenes (triphenols); e.g., pyrogallol, phloroglucinol, $C_6H_3(OH)_3$.
4. Polyhydroxy benzenes (polyphenols); e.g., pentahydroxybenzene, quercitol, $C_6H(OH)_5$.

Cf. *phenol acids* **di ~** P. containing two —OH groups. **poly ~** P. containing four or more —OH groups. **sulfurized ~** Syntans. **tri ~** P. containing three —OH groups.
phenolsulfonate* $HO \cdot C_6H_4 \cdot SO_3M$. Sulfophenylate, sulfocarbolate. A salt of phenosulfonic acid.
phenolsulfonephthalein $(OH \cdot C_6H_4)_2:C \cdot C_6H_4 \cdot SO_2 \cdot O$ = 354.4. Phenol red. A sulfuric acid derivative of phenolphthalein, generally marketed as the monosodium salt. Bright-red crystals, slightly soluble in water. Used as a test for the secreting power of the kidneys (USP), and as a pH indicator changing at 7.7 from yellow (acid) to red (alkaline).
phenolsulfonic acid* $C_6H_4(OH)SO_3H$ = 174.2. Three isomers, soluble in water, obtained by heating phenol with concentrated sulfuric acid. ***ortho-* ~** Aseptol.
phenolsulfuric acid Phenylsulfuric acid.
phenomena Plural of *phenomenon.*
phenomenon An event or manifestation; that which is apparent, as distinct from that which merely exists.
phenonaphthazine Benzophenazine.
phenones $Ph \cdot CO \cdot R$. A series of ketones; as, propiophenone, $Ph \cdot CO \cdot Et$.
phenopiazine Quinoxaline*.
phenoquinone $C_6H_4O_2 \cdot 2C_6H_6O$ = 296.3. A solid, m.71, soluble in water.
phenoresorcin Resorcinol phenolate*. A mixture of phenol and resorcinol; used to treat skin diseases.
phenosafranine $C_{18}H_{15}N_4Cl$ = 322.8. 3,6-Diaminophenyl phenazine chloride. An aniline dye used in photography to

prevent fogging. Green needles, soluble in water with a red color; an oxidation-reduction indictor. Cf. *aposafranone.*

phenosalyl An antiseptic mixture of phenol, salicylic acid, menthol, and lactic acid.

phenose $C_6H_6(OH)_6$ = 180.2. Hexahydroxycyclohexane*. Colorless, deliquescent powder, soluble in water. Cf. *inositol.*

phenosolvan A mixture of esters of higher aliphatic alcohols, b.130. A scrubbing agent in the recovery of phenols from effluents.

phenostal $(COOPh)_2$ = 242.2. Diphenyl oxalate, diphenyloxalic ester; a germicide.

phenothalin Phenolphthalein.

phenothiazine* $C_6H_4 \cdot S \cdot C_6H_4 \cdot NH$ = 199.3. Thiodiphenylamine, phenthiazine. Colorless rhombs, m.180, soluble in water; used in the manufacture of dyes, and as a veterinary anthelmintic. (2) Group of drugs derived from p. by substitution at the 10 (i.e., N atom) and 2 positions (S = 5 position).

phenotyping Identification of certain blood groups, especially for haptoglobin.

phenoxarsine $C_{12}H_9AsO$ = 244.1. A compound similar to phenoxazine, but with AsH in place of NH.

phenoxaselenin $C_{12}H_8OSe$ = 247.2. A compound analogous to phenoxazine, with Se in place of S.

phenoxatellurin $C_{12}H_8OTe$ = 295.8. A compound analogous to phenoxazine, with Te in place of S.

phenoxazine* $C_6H_4 \cdot O \cdot C_6H_4 \cdot NH$ = 183.2. Naphthoxazine. Colorless leaflets, m.150, soluble in alcohol.

phenoxetol Trademark for phenoxyethanol.

phenoxide Phenolate*.

phenoxin Carbon tetrachloride*.

phenoxy (1)* The radical PhO—, from phenol. (2) Generic name for a polyhydroxy ether synthetic thermoplastic. **p. acetaldehyde** $PhOCH_2CHO$ = 136.2. m.38. Unstable. **p.acetone** $PhO \cdot CH_2 \cdot COMe$ = 150.2. Phenacetol. b.230. **p.benzamine hydrochloride** $C_{18}H_{22}ONCl \cdot HCl$ = 340.3. White crystals, m.139, slightly soluble in water. A sympatholytic drug; used for hypertension and epinephrine-secreting tumors (USP, BP). **p.benzene*** Diphenyl ether*. **p.ethanol** $C_8H_{10}O_2$ = 138.2. Phenoxetol. Colorless liquid. Used as a preservative and topically for infected wounds. **p.methylpenicillin potassium** $C_{16}H_{17}O_5N_2KS$ = 388.5. Penicillin V. White crystals, soluble in water. A synthetic penicillin, which is destroyed by penicillinase (EP, BP). Also used as phenoxymethylpenicillin calcium.

phenpiazine Quinoxaline*.

phenthiazine Phenothiazine*.

phenetetrol Apionol.

phenthiol Thiophenol*.

phentolamine mesylate $C_{17}H_{19}ON_3 \cdot CH_4O_3S$ = 377.5. Regitine, Rogitine. White crystals, m.240, soluble in water. A sympatholytic drug; used for hypertension and epinephrine-secreting tumors (USP, BP).

phentriazine $C_6H_4 \cdot N{:}N \cdot N{:}CH$ = 131.1. 1,2,3-Benzotriazine. Yellow crystals, m.120, soluble in alcohol; used in organic synthesis.

phenyl* Ph. The radical C_6H_5—, from benzene. **hydroxy ~*** The radical $-C_6H_4OH$, from phenol. Cf. *phenoxy.*

phenylacetaldehyde* $PhCH_2CHO$ = 120.2. α-Tolualdehyde. Colorless liquid, b.194, soluble in water; a perfume (hyacinth odor).

phenylacetamide $PhCH_2 \cdot CONH_2$ = 135.2. Crystals, m.157, slightly soluble in water.

phenylacetanilide See *phenylacetanilide* under *acetanilide.*

phenyl acetate MeCOOPh = 136.2. Colorless liquid, d.1.093, b.196, slightly soluble in water; used in organic synthesis.

phenylacetic p. acid* $PhCH_2COOH$ = 136.2. α-Toluic acid. Crystals, m.77. **dihydroxy ~** Homogentisic acid. **2-hydroxy ~** Mandelic acid. **2-methyl ~** Hydratropic acid*. **p. anhydride** $(Ph \cdot CH_2CO)_2O$ = 254.3. A solid, m.73. **p. nitrile** Benzyl cyanide*.

phenylacetyl* The radical $Ph \cdot CH_2CO$—, from phenylacetic acid. **p. chloride*** C_8H_7OCl = 154.6. Colorless, fuming liquid.

phenylacetylene PhC:CH = 102.1. Colorless liquid, d.0.937, b.139; insoluble in water.

phenylalanine* $C_9H_{11}O_2N$ = 165.2. Phe*. Aminohydrocinnamic acid, a constituent of proteins.

phenylaldehyde Benzaldehyde*.

phenylamine Aniline*.

phenyl amines A group of aromatic amino compounds: primary phenyl amines $(Ar \cdot NH_2)$, e.g., aniline; secondary phenyl amines (Ar_2NH), e.g., diphenylamine; tertiary phenyl amines (Ar_3N), e.g., triphenylamine.

phenylaniline ***ar*- ~** Amino*biphenyl.* ***N*- ~** Diphenylamine*.

phenylate Phenate*.

phenylation Introduction of the phenyl group into a molecule.

phenyl azide* $C_6H_5N_3$ = 119.1. Triazobenzene. Colorless liquid, d.1.098, $b_{24mm}74$.

phenylazo* Infix indicating the radical PhN:N—. Cf. *azobenzene.* **p.aniline** See *aminoazobenzene* under *amino.* **p.ethane** PhN:NEt = 134.2. Benzeneazoethane. Colorless liquid, b.280. **p.methane** PhN:NMe = 120.2. Colorless liquid, b.150.

phenylbenzamide Benzanilide.

phenylbenzene Biphenyl*.

phenylbenzhydryl The benzhydryl* radical.

phenyl benzoate* PhCOOPh = 198.2. Benzophenid, phenol benzoate, benzocarbolic acid. Colorless, monoclinic crystals, m.68, soluble in water.

phenylbenzoyl* The radical $Ph \cdot C_6H_4CO$—. **p.benzoic acid** $PhC_6H_4COC_6H_4COOH$ = 302.3. Diphenylphthaloic acid, 4-phenylbenzophenone-2-carboxylic acid; used in the synthesis of dyes.

phenylbenzyl (1) The radical $Ph \cdot C_6H_4 \cdot CH_2$—. (2) Indicating a phenyl and benzyl group. **p.amine** $PhNHCH_2Ph$ = 183.3. *N*-Benzylaniline. Colorless crystals, m.38, soluble in alcohol; used in organic synthesis. **phenyl benzyl ketone** See *phenylpropiophenone* under *propiophenone.* **p.tin chloride** $Ph \cdot C_6H_4CH_2SnCl_2$ = 356.8. Colorless needles, m.83.

phenylborane $PhBH_2$ = 89.9. An isolog of aniline.

phenyl bromide* C_6H_5Br = 157.0. Bromobenzene*. Colorless liquid, d.1.497, b.156.

phenylbutanoic acid* $C_{10}H_{12}O_2$ = 164.2. Phenylbutyric acid. Colorless crystals. **3- ~** $MeCHPh \cdot CH_2 \cdot COOH$. **(+)- ~** $b_{12mm}157$. **(±)- ~** m.47. Soluble in water. Used in organic synthesis. **4- ~** $Ph(CH_2)_3COOH$. m.52, soluble in water.

phenylbutazone $C_{19}H_{20}O_2N_2$ = 308.4. White crystals, m.107, insoluble in water; an analgesic and antipyretic.

phenylbutyric acid Phenylbutanoic acid*.

phenyl carbamate* NH_2COOPh = 137.1. Colorless leaflets, m.141, slightly soluble in water.

phenylcarbamido The phenylureido* radical.

phenylcarbamoyl* The radical $Ph \cdot NH \cdot CO$—, from phenylcarbamic acid.

phenylcarbylamine Phenyl isocyanide*.

phenyl chloride C_6H_5Cl = 112.6. Chlorobenzene*. Colorless liquid, d.1.107, b.132.
phenyl chloroform $C_6H_5CCl_3$ = 195.5. Benzotrichloride. Colorless liquid.
phenylcyanamide PhNH·NC = 118.1. Cyanoanilide. A solid, m.47, soluble in water.
phenyl cyanide Benzonitrile*.
phenylcyclohexane $C_{12}H_{16}$ = 160.3. Hexahydrodiphenyl. Colorless oil, b_{13mm}107.
phenyldicarbinol Xylenediol*.
phenyl diethyl aminoethyl nitrobenzoate A precipitant for nitrates and perchlorates.
phenyldihydronaphthalene Atronene.
phenyldimethylpyrazolone Antipyrine.
phenyl disulfide Ph·S·S·Ph = 218.3. Phenyldithiobenzene*. Colorless needles, m.60, insoluble in water.
phenylene* Benz, benzo. The radical C_6H_4=, derived from benzene by replacement of 2 H atoms; 3 isomers: *ortho, meta,* and *para* compounds. **p.bisazo*** The radical −N:N·C_6H_4·N:N−. **p. blue** Indamine. **p.diacetic acid** $C_6H_4(CH_2COOH)_2$ = 194.2. ***ortho-*** ~ m.150. ***meta-*** ~ m.190. ***para-*** ~ m.244. **p.diamine** $C_6H_4(NH_2)_2$ = 108.1. Diaminobenzene, benzenediamine. **1,2-** ~ m.104. **1,3-** ~ m.63. **1,4-** ~ Ursol. m.140. Used as a dye for hairs and furs, as an indicator, and as a reagent for nitrogen. **methyl** ~ Diamino*toluene*. **phenylazo** ~ Chrysoidine. **p.dithiol** $C_6H_4(SH)_2$ = 142.2. Benzenedithiol*. **1,3-** ~ Dithio*resorcinol*. **1,4-** ~ Dithiohydroquinone. Colorless crystals, m.98. **p.disazo** P.bisazo*. **p.thiourea** C_6H_4·NH·C:S·NH = 150.2. Thiobenzimidazoline. **p.urea** $C_7H_6ON_2$ = 134.1. Colorless crystals, m.312, soluble in alcohol; used in organic synthesis.
phenylephrine hydrochloride $C_9H_{13}NO_2$·HCl = 203.7. Bitter, white crystals, soluble in water, m.142; an adrenergic. Used to treat hypotension; also used in inhalers for hay fever and asthma (USP).
phenylethane Ethyl*benzene.*
phenylethanol Phenethyl alcohol*.
phenylethenyl† **2-** ~ See *styryl.*
phenyl ether Diphenyl ether*.
phenylethyl (1) The phenethyl* radical. (2) Indicating the phenyl and ethyl radicals together as separate entities. **p. alcohol** Phenethyl alcohol*. **p.barbituric acid** Luminal. **phenyl ethyl ether** Phenetole*. **phenyl ethyl ketone** Ethyl phenyl ketone*.
phenylethylene Styrene*.
phenylformamide Formanilide.
phenylformic acid Benzoic acid*.
phenyl gamma acid 2-Phenylamino-8-naphthol-6-sulfonic acid.
phenylglucosazone $C_{18}H_{22}O_4N_4$ = 358.4. A condensation product of phenylhydrazine and glucose. Yellow needles, soluble in water. **alpha-** ~ m.205. **beta-** ~ m.145.
phenylglycine $PhNHCH_2COOH$ = 151.2. Phenylglycocoll, m.127, soluble in water; used in the synthesis of indigo.
phenylglycocoll Phenylglycine.
phenylglycolic acid Mandelic acid.
phenylglycol Cinnamyl alcohol*.
phenylglyoxal Benzoyl formaldehyde.
phenylglyoxylic acid Benzoylformic acid.
phenyl hydrate Phenol*.
phenylhydrazide A derivative of phenylhydrazine of the type Ph·NH·NH·COR or Ph·N(COR)NH_2: β (symmetric) or α (asymmetric).
phenylhydrazine $PhNHNH_2 \cdot \frac{1}{2}H_2O$ = 117.2. Hydrazobenzene. Colorless liquid, d.1.097. b.243, slightly soluble in water; a reagent for aldehydes and ketones. **ethyl** ~ See *ethylphenylhydrazine* under *ethyl.* **phthaloyl** ~ See *phthaloylphenylhydrazine.*
p. hydrochloride Ph·NH·NH_2·HCl = 144.6. Colorless crystals, m.200, soluble in water; a reagent for glucose, formaldehyde, and urea. **p. urea** Diphenylcarbazide.
phenylhydrazone Ph·NH·N:CH_2 = 120.2. **acetone** ~ See *acetone phenylhydrazone.*
phenylhydrazones Ph·NH·N:CHR or Ph·NH·N:CR_2. Formed by the action of phenylhydrazine on aldehydes or ketones.
phenyl hydroxide Phenol*.
phenylhydroxylamine* PhNHOH = 109.1. Colorless needles, m.80, soluble in water; used in organic synthesis.
phenylic Pertaining to phenol or the phenyl radical. **p. acid** Phenol*. **p. alcohol** Phenol*.
phenylid Aniline*.
phenylidene The cyclohexadienylidene* radical.
phenylindanedione $C_{15}H_{10}O_2$ = 222.2. Dindevan. 2-Phenylindane-1,3-dione. White crystals, m.150, soluble in water; an anticoagulant (USP). See *phenindione.*
phenylindazole See *indazole.*
phenyl isocyanate PhNCO = 119.1. A lachrymatory, d.1.096, b.163; a reagent for hydroxy and amino compounds.
phenyl isocyanide* PhNC = 103.1. Phenylcarbylamine, isocyanobenzene. Colorless liquid, d.0.978, b.165, decomp. by water.
phenylisonitramine Nitranilide.
phenyl isothiocyanate* See *phenyl mustard oil.*
phenyl ketone Benzophenone*.
phenylketonuria An inborn error of metabolism due to lack of phenylalanine hydroxylase. Phenylalanine accumulates, and phenylpyruvic acid is excreted in the urine. See *Guthrie test.*
phenyllactazam C_6H_4·CPh:N·NPh·CO = 298.3. Colorless crystals, m.181, soluble in alcohol.
phenyl mercaptan Thiophenol*.
phenylmercuric The phenylmercury(II) radical, C_6H_5Hg−. **p. acetate** $CH_3COOHgPh$ = 336.7. Rhombic, colorless crystals, m.149. A preservative and spermicide (NF). **p. chloride** PhHgCl = 313.1. Leaflets, m.251. **p. cyanide** PhHgCN = 303.7. White prisms, m.204. **p. nitrate** $PhHgNO_3$ = 339.7. Rhombic scales, m.180, soluble in water. The basic compound (m.190, decomp.) is a nonirritant, low-toxicity antibacteriostatic; a preservative and spermicide (NF, BP).
phenylmethane Toluene*.
phenylmethyl† See *benzyl.*
phenylmethylene† See *benzylidene.*
phenylmethylidyne† See *benzylidyne.*
phenyl mustard oil C_6H_5NCS = 135.2. Phenyl isothiocyanate*, thiocarbanil, phenylthiocarbonimide. Colorless liquid, d.1.138, b.221, insoluble in water; used in organic synthesis.
phenylnaphthalene* $C_{10}H_7Ph$ = 204.3. **1-** ~ Colorless liquid, b.324, soluble in alcohol. **2-** ~ Colorless scales, m.102, soluble in alcohol.
phenylnaphthylamine $C_{10}H_7$·NH·Ph = 219.3. **1-** ~ Colorless needles, m.60, insoluble in water. **2-** ~ Colorless leaflets, m.107, insoluble in water; an antiaging agent for rubber.
phenylnaphthylcarbazole Benzonaphthindole.
phenylnitroamine Ph·NH·NO_2 = 138.1. An isomer of diazoic acid. Colorless leaflets, m.46, explodes 98, soluble in water.

phenylo Phenyl*.
phenylogic series The hydrocarbon series; C_6H_6, $C_{12}H_{10}$, $C_{18}H_{24}$.
phenylon Antipyrine.
phenyloxydisulfide $Ph_2S_2O_2$ = 250.3. A solid, m.45, soluble in ether.
phenylparaconic acid $C_{11}H_{10}O_4$ = 206.2. An aromatic lactone. (+)-~ Crystals, m.134, soluble in water.
phenyl-paraffin alcohols Aromatic alcohols with an −OH radical in the side chain; as, $Ph \cdot CH_2OH$, benzyl alcohol.
phenyl-peri acid Phenyl-1-naphthylamine-8-sulfonic acid.
phenylphenol $Ph \cdot C_6H_4 \cdot OH$ = 170.2. ***meta-*** **~** m.78. ***ortho-*** **~** White crystals, m.56, soluble in water; phenol coefficient 38. A preservative for glues, and (with certain legal limitations) for noncitrus fruits. ***para-*** **~** m.165.
phenylphosphine $PhPH_2$ = 110.1. Phosphaniline. Colorless liquid, b.160, obtained by the action of hydriodic acid and phosphorus on phenyl chloride.
phenylphosphinic acid* $Ph \cdot (H)PO \cdot OH$ = 142.1. A solid, m.158, soluble in water.
phenylpropanolamine hydrochloride $C_9H_{13}ON \cdot HCl$ = 187.7. (±)-Norephedrine hydrochloride. Eskornade, Propadrine. White crystals, m.102, freely soluble in water. An adrenergic vasoconstrictor; used for hay fever and incontinence (USP, BP).
phenyl-2-propenyl† **3-~** See *cinnamyl.*
phenylpropiolic acid PhC:C·COOH = 146.1. Needles, m.136, soluble in water; used in organic synthesis.
phenylpropionic acid **2-~** Hydratropic acid*. **3-~** Hydrocinnamic acid.
phenylpropiophenone See *phenylpropiophenone* under *propiophenone.*
phenyl propyl ketone* $PhCOCH_2CH_2Me$ = 148.2. Butyrophenone. Colorless liquid, b.233, used in organic synthesis. **phenyl isopropyl ketone** b.217.
phenylpyridine $C_6H_5 \cdot C_5H_4N$ = 155.2. **2-~** Colorless liquid, b.269, insoluble in water. **3-~** Colorless oil, b.270, insoluble in water. **4-~** Colorless scales, m.77, soluble in water.
phenyl salicylate* $C_6H_4(OH)COOPh$ = 214.2. Salol, phenylis salicylas. Colorless crystals, m.42, soluble in water.
phenylsalicylic acid* $HO \cdot C_6H_4 \cdot C_6H_4 \cdot COOH$ = 214.2. Hydroxybiphenylcarboxylic acid. Several isomers.
phenyl semicarbazide **1-~** $Ph \cdot NH \cdot NH \cdot CO \cdot NH_2$ = 151.2. White scales, m.172, soluble in water; used in organic synthesis.
phenyl stannane $Ph_{2n+2}Sn_n$. An organic compound of tetravalent tin; $Ph_{12}Sn_5$.
phenyl sulfate $PhOSO_3M$. A salt of phenylsulfuric acid.
phenyl sulfhydrate Thiophenol*.
phenyl sulfide Diphenyl sulfide*.
phenyl sulfone Diphenyl sulfone*.
phenylsulfonyl* The radical $PhSO_2$−. **p.benzene** Diphenyl sulfone*.
phenylsulfuric acid $C_6H_5OSO_3H$ = 174.2. A phenylester of sulfuric acid. An unstable acid whose salts are sometimes found in urine.
phenylthiocarbonimide Phenyl mustard oil.
phenylthiohydantoic acid $PhN:C(NH_2)S \cdot CH_2 \cdot COOH$ = 210.3. Phenyliminocarbaminthioglycolic acid. White powder, m.150, insoluble in cold water, soluble in hot water; a reagent for copper and cobalt. ***ortho-*** **~** $PhNH \cdot C(:NH)S \cdot CH_2 \cdot COOH$. Insoluble in water.
phenylthiourea $H_2N \cdot C(:S)NHPh$ = 152.2. m.154, soluble in water.
phenyltin The radical $C_6H_5Sn\equiv$. **p. tribenzyl** $PhSn(CH_2Ph)_3$ = 469.2. Colorless liquid, b_{5mm}290; soluble in organic solvents except alcohol. **p. tribromide** $PhSnBr_3$ = 435.5. Colorless liquid, b_{25mm}185. **p. trichloride** $PhSnCl_3$ = 302.2. Colorless liquid, b_{25mm}143.
phenyltoluene* $Ph \cdot C_6H_4 \cdot Me$ = 168.2. Diphenylenemethane, methyldiphenyl, phenyltolyl; 3 isomers; ***meta-*** **~** Colorless liquid, d. 1.031, b.272, insoluble in water. ***ortho-*** **~** Colorless liquid, b.260, insoluble in water. ***para-*** **~** Colorless liquid, d.1.015, b.263, insoluble in water.
phenyltolyl Phenyltoluene*.
phenyl tolyl ketone* $Ph \cdot CO \cdot C_6H_4 \cdot Me$ = 196.2. ***meta-*** **~** Colorless liquid, d.1.088, b.314, insoluble in water. ***ortho-*** **~** Colorless liquid, b.315, soluble in alcohol. ***para-*** **~** Colorless, monoclinic crystals, m.59, insoluble in water.
phenylurea $PhNH \cdot CO \cdot NH_2$ = 136.2. Colorless, monoclinic crystals, m.146, slightly soluble in water.
phenylureido* Phenylcarbamido. The radical $PhNH \cdot CO \cdot NH$−.
phenylurethrane $Ph \cdot NH \cdot COO \cdot Et$ = 165.2. Euphorin, ethyl phenyl carbamate, carbamilic ether. White crystals of clove odor and taste, m.51, slightly soluble in water.
phenyl vinyl ketone* $CH_2:CH \cdot COPh$ = 132.2. Acrylophenone.
phenytoin sodium $C_{15}H_{11}O_2N_2Na$ = 274.3. 5,5-Diphenylhydantoin sodium salt. Dilantin, Epanutin. White crystals, m.295, soluble in water. An anticonvulsant; used to treat epilepsy (USP, EP, BP).
pheochrome Cellular tissue stained dark by chromium salts.
pheophorbide $C_{31}H_{31}N_4(COOH)_2COOMe$ = 608.7. A split product of chlorophyll obtained by saponification of pheophytin.
pheophytin $C_{31}H_{31}N_4(COOH)COOMe(COOC_{20}H_{39})$ = 887.2. A split product of chlorophyll obtained by treatment with oxalic acid.
pheoretin $C_{14}H_8O_7$ = 288.2. A resinous principle from rhubarb root. Brown powder, soluble in alkalies.
pheromone Ectohormone. A chemical substance which acts as a medium of communication between living bodies; e.g., a scent, the external secretion of termites, or bombykol from moths. An active area of research, some pheromones being used in pest control. Cf. *allomone.*
Ph.G. Abbreviation for Graduate of Pharmacy.
phi Φ, ϕ, φ. Greek letter. Φ (italic) is symbol for (1) heat flow rate; (2) luminous flux; (3) potential energy; (4) radiant flux (power). ϕ (italic) is symbol for (1) cylindrical and spherical coordinates; (2) electric potential; (3) fluidity; (4) osmotic coefficient; (5) phase displacement; (6) scattering angle; (7) volume fraction. φ is symbol for phenyl.
phial A vial or small bottle.
philippium A supposed rare-earth metal obtained from samarskite, shown to be mixed terbium and yttrium.
phillipite $Fe_2Cu(SO_4)_4 \cdot 12H_2O$. A native copper sulfate. Blue masses.
phillipsite A complex clay containing Na, K, and Ca.
philosopher's **p. stone** An imaginary substance supposed by the alchemists to transform base metals into gold, and, if taken internally, to prolong life; identical with the "elixir" of the iatrochemists and with the alkahest. **p. wool** Zinc oxide.
philosophy An appreciation of the many branches of science in one system of correlated and generalized knowledge. It differs from true *science,* q.v., in its more imaginary and speculative viewpoint.
phlobabene $C_{50}H_{46}O_{25}$ = 1047. A glucoside from various tree barks and hops; yields glucose and hopred on hydrolysis.
phlobaphen A series of resinous oxidation products of tannin materials, formed by boiling them with acids. Yellow or brown substances, soluble in dilute alkali.

phlobatannin Catechol tannin. The coloring matter in the rinds of many fruits.

phlogistic (1) Inflammatory. (2) Pertaining to burning or fire.

phlogisticated air Priestley's term for nitrogen. **de ~** Oxygen.

phlogiston ϕ. The hypothetical component of flammable substances (Stahl). **p. theory** (Becher, 1669.) An all-pervading substance, phlogiston was supposed to be the combustible principle and to escape during combustion or oxidation in the form of heat and smoke. Cf. *phlogisticated* air.

phlogopite $K_2(Fe^{++}Mg)_6 \cdot Al_2Si_6O_{20} \cdot (OH,F)_4$. Rhombic mica. A natural mica, used as an electrical insulator. **fluoro ~** $K_2Mg_6Al_2Si_6O_{20}F_4$. A synthetic industrial mica.

phlogosine A ptomaine in cultures of staphylococcus.

phlolaphenes Pyrocatechol esters of aliphatic C_{16-22} acids made by pyrolysis of pine bark; used as plasticizers.

phloret(in)ic acid $OH \cdot C_6H_4(CH_2)_2COOH = 164.2$. 3-(4-Hydroxyphenyl)propanoic acid. m.128, soluble in alcohol.

phloretin $(OH)_3C_6H_2 \cdot CO(CH_2)_2C_6H_4 \cdot OH = 274.3$. Colorless crystals, in fruit seeds, m. ca. 263 (decomp.), soluble in ether.

phloroglucin(e) Phloroglucinol*.

phloroglucinol* $C_6H_3(OH)_3 \cdot 2H_2O = 162.1$. Phloroglucine, 1,3,5-trihydroxybenzene, 1,3,5-benzenetriol*, 3,5-dihydroxyphenol. Yellow crystals, m.217, soluble in water. A reagent for hydrochloric acid in gastric juice; pentoses, lignin, and pentosans; hydrated chloral, etc.; and lignin in mechanical wood pulp. Tautomeric forms:

COH, HC, CH, HOC, COH, CH ⇌ CO, H_2C, CH_2, OC, CO, CH_2

phenol form — ketone form

hexahydro ~ Phloroglucitol. **triamino ~*** $C_6H_9O_3N_3 = 171.2$. 2,4,6-Triamino-1,3,5-trihydroxybenzene. **trinitro ~*** $C_6H_3O_9N_3 = 261.1$. 2,4,6-Trinitro-1,3,5-trihydroxybenzene. **p.phthalein** Gallein. **p. test** Add hydrochloric acid and a fragment of phloroglucinol and heat; a cherry-red color and precipitate indicate pentoses or lignified fibers.

phloroglucite Phloroglucitol.

phloroglucitol $C_6H_{12}O_3 = 132.2$. Methylenetriol, 1,3,5-cyclohexanetriol, hexahydrophloroglucinol, phloroglucite. Colorless crystals, m.184, soluble in water.

phlorol $C_6H_4(OH)Et = 122.2$. *o*-Ethylphenol*. Colorless oil obtained from cresols, b.211, soluble in alcohol.

phlorone $(CH{:}CMe \cdot CO)_2 = 136.2$. 2,5-Dimethyl-1,4-benzoquinone. Yellow crystals, m.125. Cf. *phorone.*

phlorose α-D-Glucose.

phloxin Tetrabromodichlorofluorescein. An aniline-dye pH indicator changing at 3.5 from colorless (acid) to magenta (alkaline).

phocenic acid Valeric acid*.

phocenin $C_3H_5(OOC \cdot C_4H_9)_3 = 344.5$. Trivalerin. The valeric acid ester of glycerol.

phoe- See words beginning with *pho-*.

phoenicochroite $PbCrO_4$. A native lead chromate.

pholcodine $C_{23}H_{30}O_4N_2 \cdot H_2O = 416.5$. *O*-(2-morpholinoethyl)morphine. White, bitter crystals, m.99, soluble in water; a cough suppressant (EP, BP).

phon A unit of loudness. When a sound is *n* decibels greater than the reference tone (20 μPa at 1 kHz), then the loudness is *n* phons; 1 p. is approximately the minimum loudness change that the human ear can detect. Cf. *decibel.*

phonene A group of similar sounds.

phonic Sonic. **infra ~** Infrasonic. **ultra ~** Ultrasonic.

phonochemical Describing a reaction induced or produced by sound. Cf. *ultrasonic.*

phonolite A variety of feldspar.

phonometer Sonometer.

phonosensitive Affected by sound.

phorate* See *insecticides,* Table 45 on p. 305.

-phore Suffix indicating to carry, bear, or bring forth, as in chromo ~. Cf. *chromogen.*

phoretic See *phoretic electron* under *electron.*

phorone $Me_2C{:}CH \cdot CO \cdot CH{:}CMe_2 = 138.2$. Diisopropylideneacetone. Yellow prisms, m.28, insoluble in water. Cf. *phlorone.*

phoronomy The study of motion, time, and relativity.

phoryl resins Resins made by the reaction of phenols with an excess of phosphoryl chloride (with calcium chloride as catalyst); the resulting arylphosphoryl chloride is reacted with a dihydroxyphenol. Colorless, glassy plastics, $[n]_D 1.55-1.63$, mol. wt approx. 15,000, resistant to mineral acids and many chemical reagents. Used as paint undercoats, to ensure adhesion to metals.

phosalone* See *insecticides,* Table 45 on p. 305.

phosgene* Carbonyl dichloride*. **di ~** Trichloromethyl chloroformate. **thio ~*** See *thiophosgene.*

phosgenite $PbCl_2 \cdot PbCO_3$. Horn lead. An ore.

phospha-* Phospho-. Prefix indicating the presence of phosphorus.

phosphagen Guanidinophosphoric acid, phosphocreatinine. A compound of phosphorus and creatinine, involved in muscle contraction.

phospham $(PN_2H)_n = (60.0)n$. A white, infusible polymer.

phosphamic acid Aminophosphoric acid.

phosphamide $PO \cdot NH \cdot NH_2 = 78.0$. White powder, insoluble in water.

phosphane* Phosphine*.

phosphaniline Phenylphosphine.

phosphatase **acid ~*, alkaline ~*** Two enzymes that (1) hydrolyze orthophosphoric monoester to an alcohol and orthophospate; (2) catalyze transphosphorylations. Alkaline p. occurs in tissues throughout the body, and acid p. in the prostate gland.

phosphate Indicating phosphorus as the central atom(s) in an anion, as: (1) Salt of (ortho)phosphoric acid containing the radical PO_4:

M_3PO_4	(Ortho)phosphate*; normal or tertiary phosphate
M_2HPO_4	Hydrogen(ortho)phosphate*; monoacid, monohydric, dibasic, or secondary phosphate
MH_2PO_4	Dihydrogen(ortho)phosphate*; diacid, dihydric, monobasic, or primary phosphate
$(M,M')PO_4$	Double phosphate
$(M,M',M'')PO_4$.	Triple phosphate

(2) Salts of other phosphoric acids; as: $M_4P_2O_6$, di(IV) ~* or hypophosphate*; MPO_3, metaphosphate*; $M_4P_2O_7$, di ~* or pyrophosphate*. **acid ~** A mono- or dihydrogen p.; as, M_2HPO_4, MH_2PO_4, $M_2H_2P_2O_7$. **alkaline ~** A p. of sodium or potassium. **bi ~** Dihydrogenphosphate*. **bone ~** Calcium p. **di ~*** A salt containing two phosphate groups. **(III,V) ~** A salt of $(HO)_2P \cdot O \cdot PO(OH)_2$. **(IV) ~** A salt of $(HO)_2OP \cdot PO(OH)_2$. **dibasic ~** A compound of the type M_2HPO_4. **dihydrogen ~*** A compound of the type MH_2PO_4. **earthy ~** A p. of the alkaline-earth metals.

hypo ~* A salt of diphosphoric(IV) acid, $M_4P_2O_6$. **meta ~*** A salt of metaphosphoric acid, MPO_3. **monobasic ~** An acid phosphate, MH_2PO_4. **monohydrogen ~** An acid phosphate, M_2HPO_4. **normal ~** A salt in which all H atoms of the acid have been displaced; as, Na_3PO_4. **ortho ~*** A salt of orthophosphoric acid, M_3PO_4. **poly ~** A compound having the general formula $M_{n+2}P_nO_{3n+1}$, where M is an alkali metal. Made by fusing together $NaPO_3$ and $Na_4P_2O_7$ in the desired proportions; used for water softening by double decomposition. **pyro ~*** A salt of pyrophosphoric acid, $M_4P_2O_7$. **stellar ~** Calcium p. occurring in star-shaped crystals. **super ~** See *fertilizer, superphosphate.* **triple ~** (1) A salt of the type MM′M″PO_4; as, NH_4KNaPO_4. (2) A calcium, magnesium, and ammonium p. sometimes occurring in urine.

p. ester Any of the compounds of general formula $ROP(O)(OH)_2$. **di ~** Any of the compounds of general formula $ROP(O)OH \cdot O \cdot P(O)(OH)_2$. **p. of lime** Apatite. **p. rock** A sedimentary rock containing calcium p.

phosphated, phosphatic Containing phosphates.

phosphatides Phospholipids.

phosphatins Organic phosphates in animal tissues.

phosphazide A compound of the type $R_3P{:}N \cdot N{:}NR$.

phosphazine A compound of the type $R_3P{:}N \cdot N{:}CR_2$.

phosphazo The radical $-N{:}P-$.

phosphene $CH{:}(CH)_2{:}CH \cdot PH$ = 84.1. Phosphurane. (ring: CH ... PH joined) Cf. *pyrrole.*

phosphenyl The radical $=PC_6H_5$. **p. chloride** $C_6H_5PCl_2$ = 179.0. An unstable liquid d.1.319, b.225. **p. oxychloride** $C_6H_5OPCl_2$ = 195.0. A liquid, d.1.375, b.258.

phosphenylic acid $C_6H_5 \cdot H_2PO_3$ = 158.1. Phenylphosphorous acid. A crystalline solid, m.158 (decomp.), soluble in alcohol.

phosphide* A metal derivative of phosphine, PH_3; as, Na_3P.

phosphinate* See *phosphinic acid (2).*

phosphine* **gaseous ~** PH_3 = 34.00. Phosphoreted hydrogen, hydrogen phosphide. Colorless, poisonous gas of characteristic odor, b.−86, slightly soluble in water. **liquid ~** P_2H_4 = 66.1. Diphosphane*. Colorless liquid, d.1.01, b.51, insoluble in water. **solid ~** $P_{12}H_6$ = 378.5. Colorless crystals, decomp. by heat, insoluble in water.

p. oxide An organic derivative containing the ≡PO group; as, Et_3PO. **p. sulfide** An organic derivative containing the ≡PS group; as, Et_3PS.

phosphines* Compounds derived from phosphine by the replacement of H atoms: RPH_2 = primary p., e.g., $MePH_2$, methylphosphine. R_2PH = secondary p., e.g. Me_2PH, dimethylphosphine. R_3P = tertiary p., e.g., Me_3P, trimethylphosphine. Cf. *phosphonium compounds.*

phosphinic acid (1)* $H_2PO(OH)$ = 66.0. Hypophosphorous acid. Colorless liquid, d.1.493, m.27 (decomp.), soluble in water. (2)* Phosphinate, hypophosphite. Derivatives of (1); as, $Me \cdot HPO \cdot OH$, methylphosphinic acid*; $Me_2 \cdot PO \cdot OH$, dimethylphosphinic acid*. Cf. *phosphonic acid.*

phosphinico* Phosphonoso. The $=PO(OH)$ radical.

phosphinimine A compound of the type $R_3P{:}NR$.

phosphino* The H_2P- radical.

phosphinoborine $R_2BPR'_2$. An inorganic *polymer,* q.v.

phosphinoso The $HO \cdot P=$ group. Cf. *phosphonoso.*

phosphinous acid* H_2POH.

phosphinoyl* Indicating the $H_2P(O)-$ radical.

phosphite* A salt of phosphorous acid containing the radical $\equiv PO_3$: e.g., Na_3PO_3, sodium phosphite. **meta ~** A phosphite containing the radical $-PO_2$.

p. esters Compounds of the type $P(OR)_3$, where R is an aryl or alkyl group.

phospho (1) Phospha*. (2) In biochemistry, the $-PO(OH)_2$ group (phosphono*) when linked to a heteroatom. **p.albumin** An albuminous substance containing phosphorus. **p.globulin** Nucleoalbumin. **p.lipid** An ester of a fatty acid containing nitrogen and phosphorus radicals. **p.nitrogen** Phosphazote. A fertilizer: urea solution and phosphate rocks. **p.protein** See *phosphoprotein.*

phosphoaminolipid A complex lipid containing phosphorus and amino nitrogen. Cf. *phospholipids.*

phosphobenzene **p. A** P_5Ph_5 = 540.4. Phosphor(o)benzene. Yellow solid, m.155. **p. B** P_6Ph_6 = 648.5. m.183–195.

phosphocalcite Pseudo*malachite.*

phosphocerite Rhabdophane.

phosphochalcite Pseudo*malachite.*

phosphodiester Compound containing the p. bridge, $-OP(O)(OH)O-$.

phosphoglobulin Nucleoalbumin.

phosphogypsum A gypsum produced in the manufacture of phosphoric acid from rock calcium phosphate; used in cement.

phosphoinositide A lipid containing radicals derived from inositol and phosphoric acid.

phospholeum A mixture of a phosphate or a phosphoric acid with soap, used to prevent precipitation of the latter in hard water.

phospholipids Phosphatides, phospholipins, phospholipoids. Lipoid substances that occur in cellular structures and contain esters of phosphoric acid; as, sphingomyelins. **amino ~** Lecithins.

phospholipins, phospholipoids Phospholipids.

phosphomolybdic acid $H_3[PMo_{14}O_{40}] \cdot 12H_2O$ = 2233. Trihydrogen dodecamolybdophosphate. Yellow solid; a reagent.

phosphonate* See *phosphonic acid (2).*

phosphonic acid (1)* $HPO(OH)_2$ = 82.0. Tautomer of phosphorous acid, $P(OH)_3$. (2)* Phosphonate*. Generally, derivatives of p. a.; as, $C_6H_5PO(OH)_2$, phenylphosphonic acid. Cf. *phosphinic acid, phosphono.* **di ~*** $[(OH)HP(O)]_2O$ = 146.0. Pyrophosphorous acid. A tetrabasic acid of trivalent phosphorous. Colorless needles, m.38, decomp. in water. **ethyl ~** See *ethylphosphonic acid* under *ethyl.*

phosphonitrile* The radical $=PN$. **p. dibromide*** $PNBr_2$ = 204.8. Phosphorus bromonitride. Rhombic crystals, m.190. **p. dichloride*** $PNCl_2$ = 115.9. Formed by the action of ammonium chloride on phosphorus chloride. It readily polymerizes at 260°C to a substance resembling rubber.

phosphonium* The ion PH_4^+, analogous to ammonium. Cf. *-onium.* **p. bromide*** PH_4Br = 114.9. Bromophosphonium. **p. chloride*** PH_4Cl = 70.5. Chlorophosphonium. **p. compounds** Quaternary phosphines. Compounds derived from phosphonium hydroxide by replacement of its H atoms: e.g., $R_4P(OH)$, tetra-*R*-phosphonium hydroxide. **p. hydroxide*** PH_4OH = 52.0. **p. iodide*** PH_4I = 161.9. Iodophosphonium.

phosphono* The radical $(HO)_2PO-$, from phosphonic acid. Cf. *phospho.*

phosphonoso (1) The phosphinate* group. (2) The phosphinico* radical. Cf. *phosphinoso.*

phosphonous acid* $HP(OH)_2$.

phosphonoyl* Phosphinylidene†. Indicating the $HP(O)=$ radical.

phosphoprotein A group of conjugated proteins consisting of a simple protein combined with phosphoric acid by a peptide or ester bond. Occur in egg yolk and milk.

phosphor (1) German for "phosphorus." (2) A substance which phosphoresces when stimulated by an external energy source, which is usually radiation. Thus, sodium iodide activated with thallium iodide is used to detect and measure γ radiation from the fluorescence produced. Used on TV screens and scintillation counters. **p. bronze** Cu 82–94, Sn 9.5–10.8, Zn 1.0–2.0, P 0.01–0.1, impurities 0.30%. Used for suspension threads for galvanometer mirrors. **p. hydrogen compounds** See *phosphines*. **p. salt** Microcosmic salt. **p. tin** A white alloy of phosphorus and tin, m.370.

phosphorate An organic compound, analogous to a nitrosate, containing the radical $=P_2O_4$; as, $C_6H_{10}P_2O_4$, cyclohexane phosphorate.

phosphorated Containing phosphorus; as, p. oil.

phosphorescence (1) The continuous emission of light from an electronic triplet state to a lower singlet state of a substance, without apparent temperature rise, upon continuous exposure to heat, light, or electric discharge. Since two unpaired spins (triplet state) is essentially forbidden, p. continues after cessation of the exciting radiation. (2) The luminosity of a living organism; as, glowworms. (3) In particular, the faint green glow of white phosphorus in air, due to slow oxidation. Cf. *luminescence, fluorescence, scattering*.

phosphorescent paint A luminous paint.

phosphoret(t)ed hydrogen Phosphines*.

phosphoric Containing pentavalent phosphorus. **p. acid*** H_3PO_4 = 98.0. Orthophosphoric acid*. Colorless crystals, m.38, soluble in water. Marketed as: 85%—colorless oil, d.1.7, miscible with water; 10%—dilute p. acid, colorless liquid, d.1.057. A reagent, and stimulant of secretion of gastric juices (NF, BP). **di ~*** See *pyrophosphoric acid* below. **hypo ~*** See *hypophosphoric acid*. **meta ~*** HPO_3 = 80.0. Clear, viscous liquid. "Glacial metaphosphoric acid, sticks" contains 17–18% sodium oxide. **ortho ~*** P. acid*. **poly ~** Acids of general formula $H_{n+2}P_nO_{3n+1}$, as pyrophosphoric acid. Condensation products of orthophosphoric acid above 400 °C, to which they revert on dilution. **pyro ~*** $H_4P_2O_7$ = 178.0. Disphosphoric acid*. Colorless crystals, m.61, soluble in water. The p. acids are derived from phosphorus pentaoxide by addition of water.

$P_2O_5 + H_2O = 2HPO_3$, metaphosphoric acid.
$P_2O_5 + 2H_2O = H_4P_2O_7$, pyrophosphoric acid.
$P_2O_5 + 3H_2O = 2H_3PO_4$, orthophosphoric acid.

p. anhydride Phosphorus pentaoxide*.

phosphorimetry Analysis based on the spectrophotometric measurement of the phosphorescence from a sample excited by ultraviolet or visible light.

phosphorite (1) Apatite. (2) An organic compound analogous to a nitrosite, containing the radical $=P_2O_3$; as, $C_6H_{10}P_2O_3$, cyclohexane phosphorite.

phosphoro The —P:P— group.

phosphoroscope An instrument in which the duration of a phosphorescence phenomenon is measured. Also used to discriminate against interfering fluorescence radiation simultaneously emitted from the same sample.

phosphoroso The OP— group.

phosphorous (1) Phosphorescent. (2) Describing a compound of trivalent phosphorus. **p. acid*** $P(OH)_3$ = 82.0. Orthophosphorous acid. Tautomer of phosphonic acid, $OPH(OH)_2$. Yellow crystals, m.70, soluble in water; a reagent and reducing agent. **hypo ~** Phosphinic acid*.

phosphorus* P = 30.97376. A nonmetallic element of the nitrogen group, at. no. 15. Main allotropes, in decreasing order of reactivity:

1. *White* (*ordinary, yellow,* or *regular*) p. Regular, white crystals, usually compressed into yellow, waxlike sticks, d.1.82, m.44, b.280 (P_4 and P_2 molecules in the vapor), insoluble in and stored under water protected from light (which changes it into the red form).
2. *Red (amorphous)* p. Red rhombohedra, d.2.2, b.280, insoluble in water or carbon disulfide. Formed by heating white p. to 400; nonpoisonous and nonluminous.
3. *Black,* or *β-black,* p. Black powder, d.2.7, m.588; conducts electricity.

P. (Greek: "bringing light") was discovered in 1669 by the alchemist H. Brandt of Hamburg, and was prepared independently by Boyle and Kunkel. It forms 2 main series of compounds, derived from tri- and pentavalent p. **amorphous ~** See (2). **Baldwin's ~** A phosphorescent form of commercial fused calcium nitrate. **black ~** See (3). **Bologna ~** Luminescent barium sulfide. **Canton ~** See *Canton phosphorus*. **Homberg's ~** Phosphorescent calcium chloride, made by heating lime and ammonium chloride. **ordinary ~** See (1). **radio ~** The isotope of mass 32; a radioactive indicator. Used, as sodium phosphate, to treat malignant red cells and in detection of some cancers (USP, EP, BP). **red ~** See (2). **regular ~** See (1). **salt of ~** Acid ammonium phosphate; used for bead tests. **scarlet ~** Scarlet powder, intermediate in activity between red and yellow p.; used for matches. **vitreous ~** See (1). **white ~** See (1). **yellow ~** See (1).

p. bromides PBr_3, p. tribromide; PBr_4, p. tetrabromide; PBr_5, p. pentabromide. **p. bromonitride** Phosphonitrile dibromide*. **p. chloride** See *phosphorous dichloride, trichloride, pentachloride*. **p. dichloride** P_2Cl_4 = 203.8. Diphosphorus tetrachloride*. Oily, fuming liquid, decomp. by water. **p. diiodide** P_2I_4 = 569.5. Red solid, m.110, soluble in carbon disulfide. **p. group** Group 5B of the periodic table, composed of the elements N, P, As, Sb, and Bi. **p. halides** The halogen compounds of p. **p. hydrides** See *phosphines*. **p. oxides** P_2O, p. monoxide; P_4O_6, p. trioxide; P_4O_8, p. tetraoxide; P_4O_{10}, p. pentaoxide. **p. oxychloride** Phosphoryl chloride*. **p. pentabromide*** PBr_5 = 430.5. Yellow crystals, m.100, decomp. by water; used in organic synthesis. **p. pentachloride*** PCl_5 = 208.2. More correctly, $[PCl_4]^+[PCl_6]^-$. Yellow rhombs, fuming in air, b.160 (sublimes), decomp. by water; a reagent and chlorinating agent. **p. pentafluoride*** PF_5 = 126.0. Colorless, irritant gas, b.−75. **p. pentaoxide** P_4O_{10} = 283.9. White powder, d.2.38, soluble in water (forms phosphoric acids); a dehydrating agent for gases, and a reagent. **p. pentaselenide*** P_2Se_5 = 456.7. Black solid, decomp. by heat. **p. pentasulfide*** P_2S_5 = 222.2. Yellow or gray crystals, m.270, decomp. by water. **p. tetraoxide** P_4O_8 = 251.9. Colorless, orthorhombic crystals, m.100, soluble in water. **p. tribromide*** PBr_3 = 270.7. Colorless, fuming liquid, d.2.88, b.175, decomp. by water. **p. trichloride*** PCl_3 = 137.3. Clear, fuming, poisonous liquid, d.1.613, b.76, decomp by water. **p. triiodide*** PI_3 = 411.7. Red prisms, m.61, decomp. by heat or water. **p. trioxide*** P_4O_6 = 219.9. Phosphorous anhydride. Colorless, monoclinic crystals, m.23, soluble in water. **p. trisulfide*** P_2S_3 = 158.1. Yellow crystals, m.290, decomp. by water.

phosphoryl* The radical ≡P:O. **thio ~*** The radical ≡PS. **p. bromide*** $POBr_3$ = 286.7. m.55, decomp. by water. **p. chloride*** $POCl_3$ = 153.3. Phosphorous oxychloride. Colorless, fuming liquid, d.1.711, b.110; a reagent and catalyst in chlorination and dehydration. **p. fluoride*** POF_3 = 104.0. Colorless, fuming liquid, b.−40. **p. nitride*** PON = 61.0. Insoluble, white solid. **p. triamine*** $PO(NH_2)_3$ = 95.0. Insoluble, white solid, decomp. by heat.

phosphorylation The introduction of a phosphate group into an organic molecule in biochemical systems; as achieved by ATP.

phosphotungstate Phosphowolframate. A salt of phosphotungstic acid.

phosphotungstic acid $H_3[PW_{12}O_{40}]$ = 2880. Trihydrogen dodecatungstophosphate. Phosphowolframic acid. Green crystals, soluble in water; a reagent for alkaloids (Scheibler's reagent), and for albumin (Salkowski's reagent).

phosphovitin A nonlipoid phosphoprotein, N 12, P 10%, in egg yolk.

phosphowolframate Phosphotungstate.

phosphowolframic acid Phosphotungstic acid.

phosphurane Phosphene.

phosphuranolite $(UPb)O \cdot P_2O_5 \cdot nH_2O$. A mineral.

phosphuret(t)ed Containing phosphorus in its lowest state of oxidation. **p. hydrogen** See *phosphines.*

phot The cgs unit of illuminance (light intensity): 1 phot = 1 lumen/cm^2 = 10^4 lux.

photic Pertaining to radiation.

photics Optics.

photo- Pertaining to light.

photoactinic Emitting visible and ultraviolet rays having photochemical activity.

photobacteria A light-producing or phosphorescent vegetable microorganism.

photobiotic Pertaining to organisms that live habitually in the light.

photocatalysis The acceleration of certain reactions by light.

photocatalyst A substance that aids photochemical reactions; as, chlorophyll.

photochemical (1) Pertaining to the chemical effects of light. (2) A chemical used in photography. **p. activation** See *irradiation.* **p. catalysis** The acceleration of a reaction by light. Cf. *photochemical reaction.* **p. dissociation** The splitting of a molecule by the influence of light, low pressure, and low temperature; as, $H_2 \rightarrow 2H$. Cf. *photochemical excitation.* **p. effect** Chemical changes produced by light. **p. equivalent** The one quantum of energy, $h\nu$, which excites one molecule; hence, total radiation $E = nh\nu$, for n activated molecules. **p. excitation** Dissociation of excited atoms; as, $H_2 \rightarrow 2H^*$. Cf. *photochemical dissociation.* **p. induction** Draper effect. The period between exposure to light and the p. reaction which results. **p. processes** A reaction in which light energy is stored as chemical energy; as, in photosynthesis. **p. reaction** A reaction influenced by light. **p. smog** Combination of smoke and fog made more acidic by reactions initiated by u.v. light. See *ozone pollution.*

photochemistry The study of the relations between radiant and chemical energy. Cf. *photosynthesis, irradiation.*

photochromism The property of changing color reversibly on exposure to u.v. or visible radiation.

photoconductivity The property of a substance to conduct electricity when illuminated.

photodegradation Decomposition due to the action of light on a substance in the absence of oxygen. **sensitized ~** Inbuilt degradation, particularly in plastics, to reduce litter pollution. Achieved by (1) making the plastic photosensitive by copolymerization with carbon monoxide or a vinyl ketone monomer; (2) using photoactive additives (prodegradants); e.g., a carbonyl, as benzophenone, or transition metal compounds.

photodeposition The formation of a thin polymer film on a surface by exposure to ultraviolet light in a monomer gas which reacts with the surface.

photodynamic Describing a substance that fluoresces in light; as, chlorophyll.

photoelectric **p. cell** Generic term for a device which produces changes in an electric circuit by the action of light. (1) Photoconductive cell (see *selenium cell*). (2) Photoemissive cell (e.g., the alkali cell), in which an emission of electrons occurs in a vacuum or across a gas-filled space. Used by photomultipliers. (3) Photovoltaic cell (e.g., the rectifier cell) which depends on contact between a metal and a semiconductor. Used for solar cells. Favored materials are gallium arsenide, silicon, indium phosphide. **p. effect** Hallwach's effect. The discharge of electrons from the surface of a metal under the influence of light, leaving the metal positively charged.

photoelectricity The transformation of light into electricity, e.g., the photoelectric effect.

photoelectrometer Photogalvanometer.

photoelectrons Electrons emitted from a surface under the influence of light.

photoemission A microscopical method of examining surface details by heating the surface to the point where electrons are emitted or by bombarding it with u.v. photons.

photofabrication Chemical machining, chemical milling. Etching away unwanted metal to obtain a desired shape, which has been imposed on the surface photographically.

photoflavins A group of 9-alkylated alloxazines.

photofluorimeter Fluorimeter.

photofluoroscope A fluorescent screen used to make X-rays visible, e.g., coated with zinc sulfide.

photogalvanometer A recording mirror galvanometer to measure minute quantities of light falling on a photosensitive cell. Cf. *microphotometer.*

photogen Boghead naphtha. A constituent of vegetable or animal organisms that causes their luminescence.

photograph A picture or image produced and fixed on a chemically sensitive surface. **micro ~** A p. that shows the object in a size much smaller than normal. Cf. *photomicrograph.*

photographic Pertaining to methods of recording or reproducing images by the action of light on light-sensitive surfaces. **p. brightness** See *spectral classification.* **p. chemicals** Chemicals for developing, fixing, toning, and sensitizing. **p. developer** An organic reducing agent, as pyrocatechol or Metol, used to develop an image by the reduction of light-exposed silver salts to metallic silver. **p. film** A transparent, flexible sheet of cellulose coated with a light-sensitive emulsion. **p. filter** See *color filter.* **p. intensifier** A substance used to intensify an image; as, mercuric chloride. **p. negative** An image in which black appears white, and white black. **p. paper** Paper coated with a light-sensitive chemical; as, silver salts. **p. plate** A glass plate coated with a light-sensitive emulsion. See *chromatic plate.* **p. positive** An image copied from a p. negative, in which the light values appear in their true shades. **p. reducer** A substance that reduces or bleaches the density of an image; as hexacyanoferrate(II). **p. screens** See *color filter.* **p. sensitizer** A dye which increases sensitivity toward certain wavelengths of light. See *photosensitizer.* **p. spectrum** The range of wavelengths, in 10^{-7} m, that can be recorded by photography. With quartz lenses and plates or films sensitized by dyes, extends from extreme ultraviolet (1.9) to infrared (9.2).

photography The study of recording visible objects by light-sensitive plates or films, and copying them on light-sensitive papers or other media. **color ~** The production of colored images by physical or chemical techniques.

photogravure See *gravure.*

photohalide A halogen salt sensitive to light.

photohyalography Etching glass by a photomechanical process.

photoisomeric change The transformation of one isomer into another by light.

photolysis Decomposition or chemical action due to the action of light on a substance in solution: (1) photolyte alone is active: e.g., decomposition of hydrogen sulfide in hexane;

(2) photolyte reacts with solvent: e.g., decomposition of hydrogen sulfide in water.

photolyte A substance decomposed by light.

photolytic Pertaining to the decomposition or dissociation of a substance by radiant energy.

photomacrograph Macrograph. A magnified photograph. Cf. *photomicrograph.*

photomagnetism Magnetic phenomena produced by light.

photometer A device to measure luminous intensity at discrete or several wavelengths. Cf. *nephelometer, colorimeter.* **micro ~** See *microphotometer.* **spectro ~** See *spectrophotometer.*

photometric Pertaining to the measurement of luminous intensity. **p. curve** Graph of quantities related to luminous intensity versus quantities related to electromagnetic energy. **p. standards** See *candle.*

photometry (1) The measurement of luminous intensity. (2) The measurement of brightness from the density of photographic images.

photomicrograph (1) Micrograph. The magnified photograph of a microscopic object obtained by means of a camera attached to a microscope. (2) Photometric curve. Cf. *photomacrograph.*

photomultiplier Device that amplifies light by photon-induced electron ejection with subsequent amplification. Commonly used as a detector in spectroscopes, spectrometers, and spectrophotometers. See *photoelectric cell.*

photon γ. One quantum of electromagnetic radiation, of rest mass 0; energy given by $h\nu$, where h is the Planck constant and ν the frequency.

photoperiodism The rhythmic changes in the composition of plant sap, due to light.

photophone An instrument that transforms radiant energy into sound waves by means of selenium, which alters in electrical resistance on exposure to light. Cf. *optophone.*

photophoresis The motion of small particles under the influence of light. Cf. *radiation pressure.*

photopic See *photopic spectrum* under *spectrum.*

photopolymer A material which, on exposure to radiation (especially u.v. light) increases in molecular weight, in particular by polymerization. Used to produce surface coatings (as polystyrene on wood), printing inks, and printed circuits.

photopolymerization A condensation reaction of molecules due to light.

photoproduct A substance synthesized in a living organism by light. Cf. *precursor.*

photoradiation therapy Technique utilizing porphin dyes to detect tumors and render them sensitive to radiation.

photoreaction A chemical reaction that occurs under the influence of light.

photosensitivity (1) The capacity of an organ or organism to be stimulated by light. (2) The absorption of a certain portion of the spectrum by a chemical system. Cf. *irradiation.* (3) Extreme sensitivity of skin to sunlight, due to disease or drugs (as, sulfonamide reaction).

photosensitizer A dyestuff which, when added to the substance of a photographic plate, increases the plate sensitivity by absorbing the light of the wavelengths to which it best responds.

1. Halogenated fluoresceins:
   (*a*) Eosin .... Yellow-green
   (*b*) Erythrosin .... Yellow-green
2. Derivatives of pyridine, quinoline, and acridine:
   (*a*) Isocyanines:
      (1) Pinaflavole .... Green
      (2) Pinachrome .... Orange
      (3) Ethyl red .... Orange-red, λ 6,500 Å
   (*b*) Carbocyaninse:
      (1) Pinacyanole .... Red, 6,800
      (2) Naphthocyanole .... Deep red, 7,500
      (3) Kryptocyanine .... Extreme red, 8,000
      (4) Dicyanine .... Infrared, 9,000
      (5) Neocyanine .... Infrared, 10,000

Cf. *photographic spectrum, chromatic plate, cyanin.*

photosphere The outer radiating surface of the sun, probably composed of incandescent clouds in a less luminous medium (cf. *chromosphere*): the source of the luminous portion of the solar spectrum.

photostabilization Reduction of the photodegradation processes. Approaches: (1) u.v. screener; (2) u.v. absorber; (3) excited state quencher; (4) free radical scavenger, hydroperoxide decomposer, or both.

photosynthesis (1) Synthesis caused by light. (2) The important reaction of all green plants in synthesizing glucose, and thence starch, from carbon dioxide and water by the catalytic action of chlorophyll and absorption of light and heat.

photosyntometer A device to demonstrate photosynthesis of plants.

phototaxis Phototropism. The response of cells or organisms to light; as, phototropism.

phototronic cell Photoelectric cell.

phototropism (1) The color change undergone by certain compounds, e.g., titanium dioxide, on exposure to light of certain wavelengths, followed by reversion to the original color in the dark or on irradiation with light of a different wavelength; attributed to impurities present. (2) The movement of living organs toward light.

phototropy (1) A reversible change of color induced by colored light; as, in fulgides. (2) The change in or loss of color of dyestuffs in light of a specific wavelength.

photovoltaic effect A change in potential of an electrode due to light. See *solar cell* under *cell.*

phrenosin $C_{48}H_{93}NO_9$ = 828.3. A cerebroside from brain substance, m.200; hydrolyzed to sphingosine, phrenosinic acid, and galactose.

phrenosinic acid $Me(CH_2)_{21}\cdot CHOH\cdot COOH$ = 384.6. Cerebronic acid, neuro stearic acid. An acid in phrenosin, which combines with bacterial toxins.

Phrilon Trademark for a polyamide synthetic fiber.

Phrix Trademark for a viscose cellulose synthetic fiber.

phrynin A poisonous protein from the skin secretions of toads. See *bufonin.*

phthalal The 1,2-phenylenedimethylidyne† radical, $=CH\cdot C_6H_4\cdot CH=$.

phthalaldehyde $C_6H_4(CHO)_2$ = 134.1. Phthalic aldehyde. **iso ~*** m.89. ***ortho-* ~*** m.52. ***para-* ~***, **tere ~*** m.116.

phthalaldehydic acid* $C_6H_4(CHO)COOH$ = 150.1. **1,2- ~** *o*-Aldehydebenzoic *o*-formylbenzoic acid. Colorless crystals, m.97. **1,3- ~, iso ~*** White needles, m.164. **1,4- ~, tere ~*** Colorless needles, m.246. **dimethoxy ~** Opianic acid. **dihydroxy ~** Noropianic acid.

phthalamic acid* $C_6H_4(CONH_2)COOH$ = 165.2. Phthalaldehyde acidamide, *o*-carbamoylbenzoic acid. Colorless crystals, m.148, soluble in alcohol.

phthalamide Phthaldiamide*.

α-phthalamidoglutarimide Thalidomide.

phthalanil $OC\cdot C_6H_4\cdot C(O)N\cdot Ph$ = 223.2.

N-Phenylphthalimide, m.210.

phthalanone Phthalide*.

phthalate* A salt of phthalic acid containing the radical

$C_6H_4(COO)_2$=; used for buffers, q.v., and standard solutions, and in vacuum pumps. **dibutyl ~*** $C_{16}H_{22}O_4$ = 278.3. Colorless liquid, d.1.046, b.340; a plasticizer. **diethyl ~*** $C_{12}H_{14}O_4$ = 222.2. Colorless liquid, d.1.119, b.290; a plasticizer. **dipentyl ~*** $C_6H_4(COOC_5H_{11})_2$ = 306.4. Colorless liquid, d.1.023, b.340, insoluble in water; a plasticizer.

phthalazine* $C_6H_4 \cdot CH{:}N \cdot N{:}CH$ = 130.1. β-Benzo-*o*-diazine, β-phenodiazine, 2,3-benzodiazine. Colorless crystals, m.91, soluble in alcohol. **dihydroxyoxo ~** Phthalazinone. **tetrahydro ~** $C_8H_{10}N_2$ = 134.2.

phthalazinone $C_8H_6ON_2$ = 146.1. **1(2*H*)- ~** 1-Hydroxyphthalazine. Colorless crystals, m.183, soluble in water.

phthaldiamide* $C_6H_4(CONH_2)_2$ = 164.2. Phthalamide, phthalic diamide. Crystals, m.220 (decomp.), insoluble in water.

phthaleins Colored compounds derived from phthalophenone by the substitution of its H atoms by OH or NH_2 groups; related to the colorless leuko compounds (phthalines). Cf. *indicator.* **cresolsulfone ~** See *cresol red.* **phenol ~** See *phenolphthalein.* **phenolsulfone ~** Phenol red. **phloroglucinol ~** Gallein. **resorcinol ~** Fluorescein. **thymolsulfone ~** Thymol blue.

phthalic p. acid* $C_6H_4(COOH)_2$ = 166.1. Alizarinic acid, 1,2-benzenedicarboxylic acid*. Colorless rhombs, m.213, soluble in water; used in dye manufacture. **dimethoxy ~** Hemipinic acid. **dimethyl ~** Cumidic acid. **hydroxy ~** See *hydroxyphthalic acid* under *hydroxy.* **iso ~*** 1,3-p. a. or *meta*-p. a. Colorless needles, m.300 (sublimes). ***meta*- ~** Iso*phthalic acid**. **methyl ~** See *xylidic acid, uvitic acid.* ***ortho*- ~** Ordinary p. a. ***para*- ~** Terephthalic acid*. **tere ~*** See *terephthalic acid.*

p. acid series $C_nH_{2n-8}(COOH)_2$. Cf. *aromatic acids.* **p. aldehyde** Phthalaldehyde*. **p. amide** Phthaldiamide*. **p. anhydride*** $C_6H_4{:}(CO)_2{:}O$ = 148.1. Phthalandione, obtained by heating p. acid. White prisms, m.128, slightly soluble in water; used extensively in organic synthesis. **p. diamide** Phthaldiamide*. **p. nitrile** Phthalonitrile*.

phthalide* $C_6H_4 \cdot CO \cdot O \cdot CH_2$ = 134.1. Isobenzofuranone, 1-phthalanone. Colorless crystals, m.68, soluble in alcohol. **bis(hydroxyphenyl) ~** Phenolphthalein. **bis(hydroxytolyl) ~** Cresolphthalein. **dimethoxy ~** Meconin. **xanthilidene ~** Fluoran.

phthalidene The phthalidylidene* radical.

phthalidyl* The radical $C_6H_4 \cdot CO \cdot O \cdot CH-$.

phthalidylidene* The radical $C_6H_4 \cdot CO \cdot O \cdot C=$.

phthalimide* $C_6H_4 \cdot CO \cdot NH \cdot CO$ = 147.1. *o*-Phthalic imide, 1,3-isoindoledione. Colorless crystals, m.228 (sublimes). **iso ~** A tautomer of p.

phthalimidine $C_6H_4 \cdot CO \cdot NH \cdot CH_2$ = 133.2. 1-Isoindolinone. Colorless crystals, m.150, soluble in alcohol. **benzyl ~** See *benzylphthalimidine.*

phthalimido* The radical $CO \cdot C_6H_4 \cdot CO \cdot N-$.

phthalines $C_6H_4 \cdot (CHR_2) \cdot COOH$. Colorless compounds which, on reduction, form phthaleins.

phthalizine Phthalazine*.

phthalocyanines $C_{32}H_{18}N_8M$. Colored compounds containing 4 iso*indole* rings linked in a 16-membered ring of alternate C and N atoms around a central atom, usually a metal (Cu or Fe). Blue to green pigment dyestuffs, very stable to light, and to dilute acids and alkalies. Cf. *porphyrins, Monastral blue.*

phthalonic p. acid $COOH \cdot C_6H_4CO \cdot COOH \cdot 2H_2O$ = 230.2. Carbobenzoylformic acid, *o*-carboxyphenylglyoxylic acid. White crystals, m.140, soluble in water. **p. anhydride** $C_9H_4O_4$ = 176.1. A solid, m.186.

phthalonitrile* $C_6H_4(CN)_2$ = 128.1. 1,2-Dicyanobenzene. m.141. **homo ~** Cyanobenzyl cyanide. **iso ~, *meta*- ~** Isophthalonitrile*. Colorless crystals, m.161. ***para*- ~, tere ~** Terephthalonitrile*. Colorless crystals, m.222.

phthaloperine The ring structure.

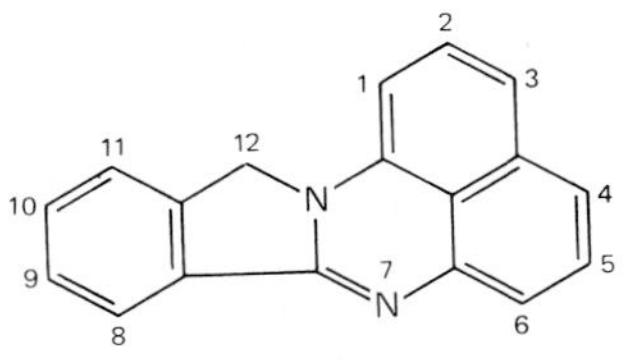

phthalophenone $C_6H_4 \cdot C(Ph)_2 \cdot O \cdot CO$ = 286.3. 3,3-Diphenyl-1(3*H*)-isobenzofuranone. The anhydride of the hypothetical triphenylmethanol-*o*-carboxylic acid, and parent substance of the phthaleins and phthalins. Colorless leaflets, m.115, slightly soluble in water; used in organic synthesis. **dihydroxy ~** Phenolphthalein.

phthaloyl* The radical *(ortho)* $-OC \cdot C_6H_4 \cdot CO-$, from phthalic acid. **di ~** See *diphthaloyl.* **iso ~*** *(meta)* $-C \cdot C_6H_4 \cdot CO-$.

p. alcohol 1,2-Xylenediol*. **p. chloride** $C_6H_4 \cdot (COCl)_2$ = 203.0. 1,2-Benzenedicarbonyl chloride*. ***ortho*- ~** Colorless oil, b.276, soluble in ether. ***meta*- ~** m.41. ***para*- ~** m.77. **p.hydrazine** $C_6H_4 \cdot CO \cdot (NH)_2 \cdot CO$ = 162.1. Colorless crystals, m.200. **p.hydroxamic acid** $C_6H_4 \cdot (CO)_2 \cdot N \cdot OH$ = 163.1. Colorless crystals, m.230, soluble in water. **p.phenylhydrazide** $C_6H_4(CO \cdot NH \cdot NHPh)_2$ = 346.4. Colorless crystals, m.161. **p.phenylhydrazine** $C_6H_4 \cdot (CO)_2 \cdot N \cdot NHPh$ = 238.2. **alpha- ~** Colorless crystals, m.178, soluble in alcohol. **beta- ~** Colorless crystals, m.210.

phthaluric acid $C_{10}H_7O_4N$ = 205.2. Colorless crystals, m.192.

phthalyl The phthaloyl* radical.

phthioic acid $C_{26}H_{52}O_2$ = 396.7. Phthoic acid, 3,13,19-trimethyltricosanoic acid*. A saturated, branched-chain fatty acid, m.48, from the lipoids.

phthoic acid Phthioic acid.

phthoric acid Obsolete name for hydrofluoric acid.

phtiocol A pigment associated with *carotene,* q.v.

phycinic acid An acid from the alga *Protococcus vulgaris.* White needles, m.136, insoluble in water.

phycite Erythritol.

phycobilins Photosynthetically active, blue or red plant pigments having a tetrapyrrole structure; as, seaweeds.

phycocerythrin A globulin.

phycochrome A chlorophyll-like pigment in freshwater algae.

phycocyanin A blue pigment in blue-green algae (Cyanophyceae); active in photosynthesis.

phycoerythrin The red pigment of brown algae (Florideae); active in photosynthesis.

phycology The study of seaweeds or algae.

Phycomycetes An order of thallophytes, algal fungi.

phycophaein The brown pigment of certain algae.

phyll-, phyllo- Prefix evoking a leaf or chlorophyll.
phyllanthin $C_{30}H_{37}O_8$ = 525.6. A glucoside from *Phyllanthus niruri* (Euphorbiaceae). Colorless needles, insoluble in water.
phyllins The derivatives of chlorophyll.
phyllite Chloritoid.
Phyllocontin Trademark for aminophylline.
phylloerythrin A split product of chlorophyll in the stomachs of herbiverous animals; identical with the porphyrin of human feces. Cf. *chlorophyll.*
phylloporphyrin $C_{32}H_{36}O_2N_4$ = 508.7. A split product of chlorophyll. See *porphin.*
phyllopyrrole $C_9H_{15}N$ = 137.2. 4-Ethyl-2,3,5-trimethylpyrrole. A pyrrole fragment of the *porphin ring,* q.v. White plates, m.69.
phylloquinone Phytonadione. See Table 101 on p. 622.
phylology The study of seaweeds.
phylum A primary division of the animal or plant kingdoms, which have 12 or 4 phyla, respectively.
physalin B $C_{28}H_{30}O_9$ = 510.5. A carotinoid pigment found in the sepals of *Physalis alkekengi* and *Ph. franchetti,* winter cherry. Yellow powder, soluble in water.
physalite The mineral $Al_2O_3 \cdot SiO_2$.
physcic acid Physcion.
physcion $C_{16}H_{12}O_5$ = 284.3. 1,8-Dihydroxy-3-methoxy-6-methylanthraquinone*. Physcic acid, chrysophyscin, parietin. An orange, crystalline principle, m.209, from the lichen *Parmelia (Physcia) parietina.*
Physeptone Trademark for methadone hydrochloride.
physeteric acid $Me(CH)_2)_7CH:CH(CH_2)_3COOH$ = 226.4. 5-Tetradecenoic acid*. **(Z)-** ~ m.20. A constituent of sardine and whale oils.
physetoleic acid 7-Hexadecenoic acid*.
physic (1) Medicine. (2) Cathartic. Purgative. (3) Puddling or working molten iron in order to remove impurities. **p. nut** *Jatrophan ureas,* L. (Malagasy Rep.). Its oil, d_{15}·0.9228, $[n]_D$ 1.4733, iodine value 93–109, is used for soap manufacture and has purgative properties and fuel value.
physical Pertaining to the energy relations of substances. **p. analysis** Testing or determining the p. properties of a material. **p. chemistry** A branch of science that employs experimental or theoretical p. methods to solve chemical problems. Subdivisions: electrochemistry, thermochemistry, and thermodynamics; optical methods; radiochemistry and photochemistry. **p. properties** See *physical properties* under *properties.* **p. solution** A solution from which the solute can be recovered chemically unchanged. **p. test** A test to determine physical, as distinct from chemical or other, properties; as, hardness.
physicochemical Pertaining to physical chemistry or physical chemical properties.
physics The science of energy. The study of the phenomena due to forces acting on matter, and changes that do not involve a change in the composition of the material. Subdivisions: mechanics, heat, electricity, magnetism, light, and radioactivity. **classical** ~ The branch of p. dealing with the mechanical, thermal, optical, and electrical properties of matter. **modern** ~ The branch of p. dealing with the relations among atoms, electrons, photons, quanta, and radiation.
physiochemical Biochemical.
physiography Physical geography. The study of the general properties of the earth and its atmosphere.
physiological Pertaining to the functions and activity of living organisms (plant, animal, and human), especially in their normal condition, as opposed to their diseased (pathological) state. **p. action** The effect of a substance on living organisms. **p. chemistry** A branch of science that studies the chemical changes in normal living organisms. **p. salt solution** (1) Normal saline. An isotonic solution of sodium chloride in water, usually 0.9%. (2) A solution resembling the salts of normal blood serum: calcium chloride 0.25, potassium chloride 0.10, sodium chloride 9.00 g; water 1,000 ml.
physiology The study of the functions and biological activities of living organisms; as, respiration. **animal** ~ The functions of the animal organism and its parts. **human** ~ The functions of the human organism, apart from its mental aspect. **phyto** ~ Plant p. **plant** ~ The functions of plants and their organs. **zoo** ~ Animal p.
physostigma Calabar bean. The seeds of *Physostigma venenosum* (Leguminosae), Africa, which contain the alkaloids physostigmine, eseridine, and calabarine.
physostigmine $C_{15}H_{21}O_2N_3$ = 275.4. White crystals, turning pink, slightly soluble in water; a parasympathomimetic; used as a miotic (USP). **p. salicylate** $C_{15}H_{21}O_2N_3 \cdot C_7H_6O_3$ = 413.5. Eserine salicylate. The common form of p. Yellow crystals, m.186, soluble in water; a miotic (USP, EP, BP). **p. sulfate** $(C_{15}H_{21}O_2N_3)_2 \cdot H_2SO_4$ = 648.8. Eserine sulfate, physostigminae sulfas. Yellow crystals, m.140, soluble in water; a miotic (USP, EP).
phyt-, phyto- Prefix (Greek), indicating plant or vegetable.
phytalbumin A vegetable albumin.
phytane $C_{20}H_{42}$ = 282.6. A saturated hydrocarbon obtained by reduction of phytol. Occurs in oil shale. Colorless liquid, b_{10mm}169, insoluble in water.
phytate The inositol hexaphosphate anion. It is absorbed by tooth enamel and is said to be cariostatic.
phytelephas The negrito palm. *P. macrocarpa* (Ecuador), whose fruit yields corajo or vegetable ivory. Cf. *tagud nut.*
phyterythrin The red coloring matter of plants, especially of leaves in autumn.
phytic acid $C_6H_6O_6(H_2PO_3)_6$ = 660.0. Inositolhexaphosphoric acid. A powder, m.214, slightly soluble in water. A metabolic intermediate in plant seeds. See *sugar phosphates.*
phytochemistry The study of chemical changes occurring in plants.
phytochrome General name for the coloring matters of plants necessary for their synthetic metabolism; as, chlorophyll.
phytohormone Any of the auxins.
phytol $C_{20}H_{39}OH$ = 296.5. 3,7,11,15-Tetramethyl-2-hexadecen-1-ol*. A colorless oil from chlorophyll, d.0.864, b_{10mm}203; a polymer of isoprene.
phytolacca Pokeroot. The direct root of *P. decandra* (Phytolaccaceae); an emetic. **p. berries** Pokeberries.
phytolaccagenin $C_{31}H_{48}O_7$ = 532.7. A glycone, m.317 (decomp.), from *Phytolacca americana,* S. Africa and America.
phytolaccic acid An acid from pokeberries. Brown gum, soluble in water.
phytolaccin A neutral principle from the seeds of *Phytolacca* species. Needles, insoluble in water; a laxative.
phytolaccine An alkaloid obtained from the roots of *Phytolacca* species.
phytomenadione BP name for phytonadione.
phytonadione $C_{31}H_{46}O_2$ = 450.7. Vitamin K_1. Phytomenadione, phylloquinone, antihemorrhagic vitamin. Yellow liquid, d.0.963, insoluble in water. Essential for prothrombin synthesis in the liver. An antidote to the coumarin group of anticoagulants (USP, BP).
phytopathology The study of diseased plants.
phytopharmacy The study of fungicides and insecticides.

phytopyrrole See *chlorophyll.*
phytosterin Phytosterol.
phytosterol (1) $C_{26}H_{44}O \cdot H_2O$ = 390.6. Phytosterin. An isomer of cholesterol, m.135–144. Its presence in all vegetable fats (0.5–1%) distinguishes them from animal fats which contain cholesterol. (2) Any sterol derived from plants. See *cholane.* **p. test** To distinguish vegetable from animal fats: Treat the ether extract of the saponified fats with glacial acetic acid. The melting points: phytosterol acetate, m.125–127; cholesterol acetate, m.114.
phytosterolin $C_{34}H_{56}O_6$ = 560.8. A glucoside, m.275–288, that yields phytosterol on hydrolysis.
phytyl* The 3,7,11,15-tetramethyl-2-hexadecenyl radical, $C_{20}H_{39}-$.
phytylmenaquinone Phytonadione.
pi Π, π. Greek letter. π (not italic) is the symbol for (1) mathematical constant equal to the ratio of the circumference of a circle to its diameter: 3.141 592 653 59; (2) pion (π meson). *π* (italic) is the symbol for osmotic pressure. **p. adduct** See Π-*complex* under *complex.* **p. bond** See *π-bond* under *bond.* **p. complex** See Π-*complex* under *complex.*
piaselenole $C_6H_4 \cdot N_2 \cdot Se$ = 183.1. 2,1,3-Benzoselenadiazole, m.76. **iso ~** 1,2,3-Benzoselenadiazole. **tolu ~** See *tolupiaselenole.*
piassaba A fiber from *Leopoldina piasoaba,* Brazil; used for ropes. **Bahia** A fiber from *Attalea funifera,* Brazil.
piazines *p*-Diazines. Heterocyclic compounds having N atoms in the *para* position; as, pyrazine. Cf. *miazines, oiazines.*
piazthiole $C_6H_4 \cdot N_2S$ = 136.2. 2,1,3-Benzothiadiazole. Colorless crystals, m.44. **iso ~** 1,2,3-Benzothiadiazole.
PIB* Symbol for polyisobutylene.
picamar $C_6H_2(OMe)_2(OH)(C_3H_7)$ = 196.2. 2-Hydroxy-3,4-dimethoxy-1-propylbenzene. Colorless liquid, b.245, used in perfumery.
picein $C_{14}H_{18}O_7$ = 298.3. The β-D-glucoside of 4′-hydroxyacetophenone from the leaves of the Norway spruce, *Picea excelsa,* and bark of *Salix nigra,* willow (Salicaceae). White powder, m.194, soluble in water. Cf. *piceoside.*
picene* $C_{22}H_{14}$ = 278.4. Dibenzphenanthrene. An aromatic, high-melting hydrocarbon from coal tar. Blue, fluorescent leaflets, m.364, insoluble in water. **p. perhydride**

$$C_{10}H_{16}\begin{matrix} CH_2-CH_2 \\ \hline \end{matrix}C_{10}H_{16}$$

Mol. wt. = 300.5. Docosahydropicene. Colorless crystals, m.175, insoluble in water.
picenic acid $C_{21}H_{14}O_4$ = 330.3. Colorless solid, m.201.
piceoside See *salinigrin.* From species of *Coniferae, Rosaceae,* and *Salicaceae.* Cf. *picein.*
pichi The woody and resinous branches of *Fabiana imbricata* (Solanaceae), Chile; a diuretic. Cf. *fabiana.*
pichurim beans Sassafras nuts. The seeds of *Nectandra puchury* (Lauraceae), Brazil and Venezuela; an aromatic. **p. camphor** $C_{12}H_{24}O_2$ = 200.2. An aromatic resembling laurel camphor, from p. beans. **p. fat** The fatty matter of p. beans (30% laurin and p. camphor); a flavoring.
pickeringite $MgAl_2(SO_4)_4 \cdot 22H_2O$. A native magnesia alum. Long, fibrous masses.
pickle (1) Dilute acids, used to remove oxides, carbonates, or other scales. (2) A fruit or vegetable preserved in spiced vinegar. **p. inhibitor** A substance added to p. (1) to restrain its corrosive action.
pickling To clean metal by immersion in a pickle containing an inhibitor.
pico-* SI prefix for the multiple 10^{-12}.
picoampere 0.001 milliampere.

picoline $C_5H_4N \cdot Me$ = 93.1. Methylpyridine†. Member of a group of homologs of pyridine, from the dry distillation of bones and coal. **alpha- ~** Colorless liquid, d.0.952, b.128,

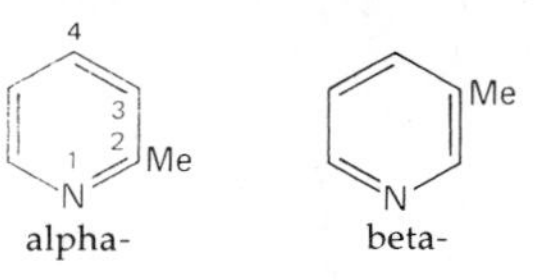

soluble in water; a nerve sedative. **beta- ~** (1) Colorless liquid, b.144, soluble in alcohol; used to produce nicotinic acid. (2) A coal tar base fraction containing 2,6-lutidine and β- and γ-picolines. **gamma- ~** 4-Methylpyridine. Colorless liquid, d.0.974, b.143, soluble in water. **tetrahydro ~** $C_5H_8N \cdot CH_3$ = 97.2. Colorless liquid, b.132.
picolinic acid $C_5H_4N \cdot COOH$ = 123.1. 2-Pyridinecarboxylic acid*. Colorless needles, m.136, soluble in water. **3-hydroxy ~** $C_6H_5O_3N$ = 139.1. Oxypicolinic acid. Colorless crystals, m.250. See *hydroxynicotinic acid* under *nicotinic acid.*
picolyl The pyridylmethyl* radical, $C_5H_4NCH_2-$, from picoline.
picotite The mineral $MgO \cdot Al_2O_2$, with iron and 7% chromium trioxide.
picraconitine Picroaconitine.
picradonidin A digitalislike glucoside from *Adonis* species (Ranunculaceae).
picramic acid $(NO_2)_2(NH_2)C_6H_2(OH)$ = 199.1. Dinitrophenamic acid, 2-amino-4,6-dinitrophenol*. Dark-red crystals, m.168, insoluble in water.
picramide* $C_6H_2(NH_2)(NO_2)_3$ = 228.1. 2,4,6-Trinitroaniline*. Yellow leaflets, m.193, soluble in acetic acid. **methylnitro ~** Tetryl.
picramnine An alkaloid from the bark of *Picramnia antidesma* (Simarubaceae), Honduras bark.
picranisic acid Picric acid*.
picrasmine $C_{35}H_{46}O_{10}$ = 626.7. From the wood of *Picrasma quassioides* (Simarubaceae), Himalayas.
picrate* $(NO_2)_3C_6H_2OM$. Explosive salts of picric acid. Carbazotate.
picric acid* $\underline{CNO_2{:}CH \cdot CNO_2{:}C(OH) \cdot CNO_2{:}CH}$ = 229.1.
2,4,6-Trinitrophenol*, picranisic acid, picronitric acid, chrysolepic acid. Yellow leaflets, m.122 (explode), soluble in water. Used as a reagent and in the manufacture of explosives and dyes. Cf. *lyddite.*
picrin A bitter principle from *Digitalis purpurea* (Scrophylariaceae). **bromo ~** Nitrobromoform.
picrite Describing propellant action based on picric acid.
picro- Prefix (Greek) indicating "bitter."
picroaconitine $C_{31}H_{45}O_{10}N$ = 591.7. 14-Benzoylaconine, picraconitine, m.130. A bitter principle from the bulbs of *Aconitum napellus* (Ranunculaceae).
picroadonidine A bitter principle from *Adonis* species.
picrocarmine A microscope stain: carmine 1, ammonia 5, water 50, saturated picric acid solution 50 pts.
picrocrocin (1) $C_{16}H_{26}O_7$ = 330.4. Saffron bitter, m. 155. A glucoside from saffron. (2) $C_{10}H_{14}O$ = 150.2. A ketone, b.209, from saffron involved in the action of the hormone *fertilizin,* q.v. Cf. *pikrococin.*
picroerythrin $C_{12}H_{16}O_7 \cdot 3H_2O$ = 326.3. Bitter crystals, m.158, soluble in water, insoluble in alcohol.
picrolonic acid $C_{10}H_8N_4O_5$ = 264.2. 3-Methyl-4-nitro-1-(*p*-nitrophenyl)-5-pyrazolone, m.116. A microreagent for calcium (rectangular crystals) and alkaloids.
picromerite The mineral $(K_2Mg)SO_4 \cdot 6H_2O$.

picronigrosine An alcoholic solution of picric acid and nigrosine; a microscope stain.
picronitric acid Picric acid*.
picropodophyllin A crystalline principle from *Podophyllum* species. Cf. *podophyllin.*
picroroccellin $C_{20}H_{22}O_4N_2$ = 354.4. Bitter crystals from *Rocella tinctoria.*
picrosclerotine An alkaloid from ergot.
picrotin $C_{15}H_{18}O_7$ = 310.3. A decomposition product of picrotoxin, m.240.
picrotone $C_{14}H_{16}O_3$ = 232.3. A ketone derived from picrotoxin.
picrotonol $C_{14}H_{16}O_4$ = 248.3. An α-ketol, degradation product of picrotoxin.
picrotoxin $C_{30}H_{34}O_{13}$ = 602.6. A neutral principle from the fruit of *Anamirta paniculata* (or *Cocculus indicus*), fishberries. White needles, soluble in water.
picrotoxinin $C_{15}H_{16}O_6 \cdot H_2O$ = 310.3. A decomposition product of picrotoxin. Colorless crystals, m.201, soluble in water.
picryl* 2,4,6-Trinitrophenyl†. The radical $(NO_2)_3C_6H_2-$, from picric acid. **p.amine** Picramide*. **p. chloride*** $ClC_6H_2(NO_2)_3$ = 247.5. (1) 2-Chloro-1,3,5-trinitrobenzene*. Yellow prisms, m.84, soluble in chloroform; used to manufacture explosives.
Pictet **P., Amé (1857–1937)** Swiss chemist noted for his work on vegetable alkaloids. **P., Raoul (1842–1929)** Swiss chemist who liquefied oxygen, nitrogen, hydrogen, and carbon dioxide at −140°C.
P. crystals $SO_2 \cdot 7H_2O$ = 190.2. White crystals formed when liquid sulfur dioxide evaporates.
Pictol Trademark for monomethyl-*p*-aminophenol sulfate; a photographic developer.
piedmontite The mineral $4CaO \cdot 3(AlMn)_2 \cdot O_3 \cdot 6SiO_8 \cdot H_2O$.
pieso- Piezo-.
pieze The uniform pressure (equal to 1 kPa) which, spread over a surface of 1 m^2, produces a force of 1 sthene (French usage). **hecto ~** Bar.
piezo- Prefix (Greek) indicating "pressure."
piezochemistry The study of chemical reactions under high pressure.
piezocontrol The maintenance of a definite radio frequency with a quartz oscillator. Cf. *piezoelectricity.*
piezocrystallization Crystallization under great pressure. See *diamond.*
piezoelectricity An electric current produced by pressure exerted on certain crystals; as, quartz. Cf. *electrostriction, quartz* oscillator.
piezoelectron A supposedly disk-shaped electron pressed between 2 petalons.
piezometer An instrument to determine compressibility.
pig A cast metal bar or brick. **p. iron** An iron p., molded in sand.
pigment (1) A fine, insoluble white, black, or colored material; used, suspended in a vehicle, as a paint, or ink. E.g.: mineral pigments, as, ocher; animal pigments, as, carmine; vegetable pigments, as, madder lake; synthetic or artificial pigments, as, phthalocyanines. (2) A coloring matter in the tissues of plants or animals; as, the chromoproteins. **carotenoid ~** See *carotenoids.*
p. dye An insoluble dye, which does not form lakes. **p. green** A green color prepared from an iron salt and nitroso-2-naphthol.
pigmentation A deposit of coloring matter in a living organism.
pigmentolysis The dissolution or destruction of pigmentation.

pikrococin $(C_6H_{10}O_5) \cdot CH \cdot CH_2 \cdot CMe_2 \cdot C(CHO){:}CMe \cdot CH_2$ (ring closed between CH and CH_2) = 313.4. A glucoside of safranal in saffron. Cf. *picrocrocin.*
pilchardine A commercial blend of oils from pilchard, sardine, and grayfish.
pile A bundle or stack. **galvanic ~** Voltaic p. **thermo ~** Sheets or bars, of 2 or more metals, that produce an electric current when heated at their junctions of contact. **uranium ~** Uranium-graphite containing ^{235}U in a thick ferroconcrete casing. It emits heat by producing plutonium by radioactive decomposition of the ^{235}U. The graphite damps down the action and enables the heat to be utilized. Cf. *reactor, chain* reaction, *fission, moderator.* **voltaic ~** A series of metallic disks, forming a galvanic battery.
pilewort Lesser celandine, *Ranunculus ficaria;* used as ointment.
piliganine $C_{15}H_{24}ON_2$ = 248.4. An alkaloid from piligan, *Lycopodium saururus,* S. American club moss. Yellow powder, insoluble in water.
pill (1) A small, round mass of a mixture of an active drug and inert material; used in medicine. Cf. *contraceptive tablet.* (2) A pellet of cesium "flashed" in a vacuum tube to remove all oxygen.
pilling The formation of small, tight balls of fluff on the surface of a fabric.
pilocarpidine $C_5H_4N \cdot CMe(NMe_2) \cdot COOH$ = 194.2. An alkaloid from the leaves of *Pilocarpus.* Colorless syrup, insoluble in water. **p. nitrate** $C_{10}H_{14}N_2O_2 \cdot HNO_3$ = 257.2. Colorless crystals, soluble in water.
pilocarpine $C_{11}H_{16}O_2N_2$ = 208.2. An alkaloid from jaborandi. See *pilocarpus.* Colorless needles, m.34, soluble in water; it increases saliva and sweat, constricts pupils, and decreases intraocular tension. Used to treat glaucoma. **p. hydrochloride** $C_{11}H_{16}O_2N_2HCl$ = 244.7. Colorless, hygroscopic needles, m.202, soluble in water (USP, BP). **p. nitrate** $C_{11}H_{16}O_2N_2 \cdot HNO_3$ = 271.3. Pilocarpinae nitras. Colorless crystals, m.178, soluble in water (USP, EP, BP). **p. phenate** $C_{11}H_{16}O_2N_2 \cdot C_6H_5OH$ = 302.4. Aseptolin. Colorless oil, soluble in water. **p. salicylate** $C_{11}H_{16}N_2O_2 \cdot C_7H_6O_3$ = 346.4. White crystals, m.120, soluble in water. **p. sulfate** $(C_{11}H_{16}N_2O_2)_2 \cdot H_2SO_4$ = 514.6. Colorless crystals, m.133.
pilocarpus Jaborandi. The dried leaflets of *Pilocarpus pennatifolius* (Rutaceae), tropical America; a diaphoretic (0.6% alkaloids).
pilocereine $C_{45}H_{65}O_6N_3$ = 744.0. Pilocerine. An alkaloid from *Lophocereus shotti.*
pilosine $C_{16}H_{18}N_2O_3$ = 286.3. An alkaloid from pilocarpus. White crystals, m.171.
pilosinine $C_9H_{12}N_2O_2$ = 180.2. An alkaloid from pilocarpus. White crystals, m.79.
pilot burner A small burner, permanently alight, attached to a larger burner for relighting purposes.
pilot plant An experimental assembly of manufacturing equipment to develop and test a new process on a reduced scale.
pimanthrene $C_{16}H_{14}$ = 206.3. 1,7-Dimethylphenanthrene. A hydrocarbon from copal and pimaric acid.
pimaric acid $C_{20}H_{30}O_2$ = 302.5. 8(14),15-Pimaradien-18-oic acid. A dextrorotatory acid from burgundy pitch and galipot resin. Crystals, m.218, soluble in hot alcohol. See *abietic acid.*
pimelic acid $CH_2(CH_2CH_2 \cdot COOH)_2$ = 160.2. 1,5-Pentanedicarboxylic acid*, heptanedioic acid*. The 6th member of the oxalic series, m.105, soluble in alcohol.
pimelinketone Cyclohexanone*.
pimelite A native, green nickel-iron silicate, similar to meerschaum.

pimenta Allspice, Jamaica pepper. The fruit of *Pimenta officinalis* (Myrtaceae), the tropics. A condiment and stimulant. **p. oil** Allspice oil. Yellow oil from the pimenta of the W. Indies, d.1.05, insoluble in water; a perfume and flavoring.
pimpinella The roots of *Pimpinella saxifraga* (Umbelliferae); a diuretic.
pimpinellin $C_{13}H_{10}O_5$ = 246.2. A bitter, crystalline principle from pimpinella. Colorless needles, insoluble in water. **iso ~** A constituent of lime juice.
Pinaceae Pines: trees and shrubs with a resinous juice and awl- or needle-shaped leaves. Some yield drugs; e.g.: *Abies balsamea* (Canada fir), Canada balsam; *Juniperus oxycedrus,* cade oil; *Pinus sylvestris* (Scotch fir), wood tar; *Thuja occidentalis* (arborvitae), thuja. Cf. *Abies.*
pinachrome $C_{26}H_{29}IN_2O_2$ = 528.4. *p*-Ethoxyquinaldine *p*-ethoxyquinoline ethyl cyanine. An indicator, pH 5.8–7.8, (acids—colorless, alkalies—red violet).
pinacoid See *pinakoid.*
pinacol $Me_2C(OH) \cdot C(OH)Me_2$ = 118.2. Tetramethylethylene glycol, pinacone, 2,3-dimethyl-2,3-butanediol*. Colorless crystals, m.38, soluble in water. **p. condensation** Two aldehydes or ketones are reduced and linked together; as in the formation of pinacol from acetone: $2Me_2CO + H_2O = Me_2COH-COHMe_2$. **p. conversion** An intramolecular transfer of a CH_3- group from one C atom to another; as in the change from pinacol to pinacolin: $Me_2C(OH)-C(OH)Me_2 \rightarrow Me_3C \cdot COCH_3$.
pinacolin $Me \cdot CO \cdot CMe_3$ = 100.2. *tert*-Butyl methyl ketone*, 3,3-dimethyl-2-butanone*. Colorless oil of peppermint odor, b.106, slightly soluble in water.
pinacolines Ketones containing a tertiary alkyl group, $R \cdot CO \cdot CR_3$.
pinacolone Pinacolin.
pinacols $R_2C(OH) \cdot C(OH)R_2$. Dihydric alcohols.
pinacolyl alcohol $CMe_3 \cdot CHOH \cdot Me$ = 102.2. 3,3-Dimethyl-2-butanol*. **(±)- ~** Colorless liquid, d.0.812, b.121, soluble in water.
pinacone* Pinacol*. **p. rearrangement** See *pinacol conversion.*
pinacones See *pinacols.*
pinacyanol $C_{25}H_{25}N_2I$ = 480.4. A cyanine dye for sensitizing photographic plates to red; a histological stain.
pinaflavole An isocyanine dye; a photosensitizer for green.
pinakoid A prism crystal face intersecting one axis of the system and parallel to the other 2. **brachy ~** A p. intersecting the brachy (broad) axis. **macro ~** A p. intersecting the macro (long) axis.
pinakryptol A green dye; a photographic desensitizer.
pinalic acid Valeric acid*.
pinane* $C_{10}H_{18}$ = 138.3. 2,6,6-Trimethylbicyclo[3.1.1]heptane*. A terpene hydrocarbon in many essential oils. Cf. *pinene.*

CH(Me)
2 10
1CH 3CH2
7CH2
CMe2 CH2
CH

pinang *Areca* nut.
pinanyl* The radical $C_{10}H_{17}-$, from pinane.
pinaverdol $C_{22}H_{21}IN_2$ = 440.3. 1,6,1′-Trimethyl isocyanine iodide. An isocyanine dye for sensitizing plates to the orange of the spectrum.

pinch An approximate measure: 1–2 g.
pinchbeck A yellow alloy: Cu 83, zinc 17%; imitation gold for jewelry.
pinckneyin A glucoside from the bark of *Pinckneya pubens* (Rubiaceae).
pine A general name for coniferous trees that yield turpentine, resin, tar, pitch, sawn timber, and pulp (especially kraft). See *Pinaceae.* **p. camphor** Pinol. **p.cone oil** Turpentine. **p. leaf oil** An essential oil distilled from p. needles. **p. needle oil** Oleum pini. An essential oil distilled from fresh p. needles, the leaves of *Pinus pumilio.* Yellow liquid, d.0.865–0.875, b.165, containing pinene, citrene, and bornyl acetate; an inhalant (NF). **p. oil** Crude turpentine from distillation of pinewood: chiefly γ-terpinene, eucalyptol, fenchyl alcohol, borneol, and α-terpineol. **p. tar** Wood *tar.* **p. tar oil** Red distillate from p. tar, d.0.97, insoluble in water; used in ore flotation.
pineapple The fruit of *Ananas sativus* (Bromeliaceae). Cf. *bromelain.*
pinene* $C_{10}H_{16}$ = 136.2. **alpha- ~** 2,6,6-Trimethylbicyclo[3.1.1]hept-2-ene†. Australene, laurene. A terpene constituent in oils of turpentine, savine, and fir. Colorless, aromatic liquid, d.0.859, b.155. slightly soluble in water. **beta- ~** Nopinene. Pseudopinene. **(1*S*,5*S*)- ~** Occurs in *turpentine,* q.v. Oil, b.164. Intermediate in the manufacture of aroma compounds, and a flavoring ingredient.

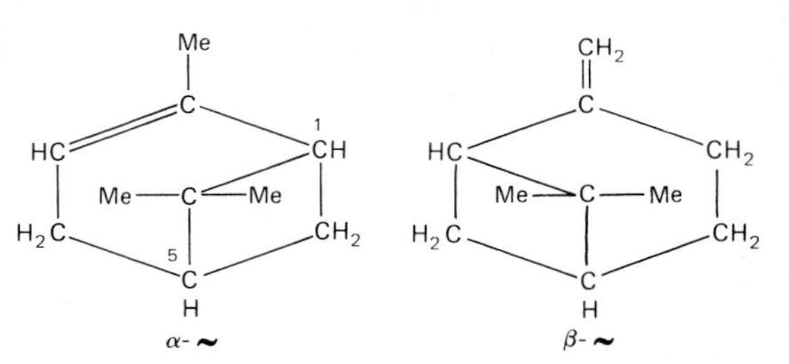

p. hydrochloride $C_{10}H_{16}HCl$ = 172.7. Insoluble solid, m.134.
pinicortannic acid $(C_{16}H_{18}O_{11})_2 \cdot H_2O$ = 790.7. A brown tannin from the bark of Scotch fir, *Pinus sylvestris.* Cf. *cortepinitannic acid.*
pinipicrin $C_{22}H_{36}O_{11}$ = 476.5. A bitter principle from the needles of *Thuja occidentalis* and *Pinus sylvestris* (Pinaceae).
pinitannic acid $C_{14}H_{16}O_8$ = 312.3. A brown tannic acid from the wood of *Pinus sylvestris.*
pinite (1) $C_6H_{12}O_5$ = 164.2. Hexahydropentahydroxy-benzene. Colorless solid, m.150, in the resin from *Pinus lambertiana.* Cf. *quercitol.* (2) Pagodite. **methoxy ~** Quebrachitol.
pinking See *knock.*
pinkroot Spigelia.
pink salt $(NH_4)_2SnCl_6$ = 367.5. Tin-ammonium chloride, diammonium hexachlorostannate(IV)*. Pink crystals; a textile mordant.
pINN See *INN.*
pinnoite MgB_2O_4. A native borate.
pinol $C_{10}H_{16}O$ = 152.2. Sobrerone, pine camphor. A terpene, from pine needles: (+) and (−) forms. **(1*R*,5*R*)- ~** Colorless liquid, d.0.952, b.183, insoluble in water. **p. hydrate** $C_{10}H_{16}O \cdot H_2O$ = 170.3. Sobrerol. m.150, soluble in water.
pinoline Rosin spirit.
pinone *β*- ~ $C_9H_{16}O$ = 140.2. 7,7-Dimethylbicy-

clo[3.1.1]heptan-2-one*. A terpene constituent of many essential oils.

pinonic acid $C_{10}H_{16}O_3$ = 184.2. 3-Acetyl-2,2-dimethylcyclobutane acetic acid. **(+)-~** An oxidation product of pinene, m.69. **(±)-~** m.105. A wetting agent for mercerizing processes.

pinosylvin $C_{14}H_{12}O_2$ = 212.2. (*Z*)-3,5,-Dihydroxystilbene. A constituent of the heartwood of the *Pinus* spp., m.156, slightly soluble in water, soluble in alcohol (violet fluorescence).

pint pt; a measure of volume, dry and liquid. **U.S. liquid ~** 1 pint = 0.5 quart = 0.125 gal = 4 gill = 16 fl oz = 28.875 in^3 = 128 fl drams = 7,680 minims = 0.473177 liter. **British ~** Imperial p. **dry ~** 1 dry pint = 0.5506 liter. **imperial ~** British p. 1 pint = 20 fl oz = 34.6593 in^3 = 568.25 mL.

pinte An early French measure of capacity; renamed *liter,* without change of volume.

Pintsch gas A fuel gas made by spraying oil into a hot retort.

Pinus An important genus of *Coniferae,* q.v., which yields turpentine, galipot resin, pitch, tar, combopinic acid, ceropic acid, pinite, pinipicrin. Cf. *Pinaceae, Abies.*

Piobert effect The surface markings on polycrystalline iron and soft steel at or near the yield point.

pion A subatomic *particle.*

Pipanol Trademark for trihexylphenidyl hydrochloride.

pipe A 115-gallon cask of wine (former U.K. usage).

pipe clay A fine, grayish-white clay, similar to kaolin, used for heat-resisting apparatus, and as a whitening.

pipecoline $C_5H_{10}N \cdot Me$ = 99.2. Methylpiperidine. **N-~, 1-~** Colorless liquid, d.0.818, b.107. **α-~, 2-~** and **β-~, 3-~** have (*R*), (*S*) and (±) forms. **γ-~, 4-~** Colorless liquid, d.0.867, b.133.

pipecolinic acid $C_5H_9NH \cdot COOH$ = 129.2. 2-Piperidinecarboxylic acid. **(*S*)-~** Colorless crystals, m.261, soluble in alcohol; occurs widely in legumes.

piper (1) Pepper. (2) A genus of *Piperaceae,* q.v.

Piperaceae The pepper family, a group of shrubs, or climbing tropical plants, containing aromatic substances; e.g.: *Piper betle,* betel leaf; *Piper nigrum,* black, white pepper; *Houttuynia californica,* yerba mansa.

piperazidine Piperazine*.

piperazine* $HN \cdot (CH_2)_2NH(CH_2)_2 \cdot 6H_2O$ = 194.2. Diethylenediamine, piperazidine, hexahydropyrazine, dispermin(e). Glassy leaflets, m.104, soluble in water; used, as the adipate, citrate, and phosphate, for threadworm and roundworm. **dimethyl ~** Lupetazin.

p. adipate $C_{10}H_{20}O_4N_2$ = 232.3. White crystals, m.250 (decomp.), soluble in water (EP, BP). **p. citrate** See *piperazine.* (USP, EP, BP). **p.dione** $NH \cdot (CH_2 \cdot CO)_2 \cdot NH$ = 114.1. Dioxopiperazine, glycine anhydride. The anhydride of the dipeptide glycylglycine, from silk fibroin. **p. phosphate** $C_4H_{10}N_2 \cdot H_3PO_4 \cdot H_2O$ = 202.1. White powder, soluble in water. (USP, BP).

piperic acid $C_7H_5O_2 \cdot CH{:}(CH)_2{:}CH \cdot COOH$ = 218.2. Piperinic acid, 5-(1,3-benzodioxol-5-yl)-2,4-pentadienoic acid†. An unsaturated monobasic acid, from piperonal. Yellow needles, m.216, slightly soluble in water.

piperidic acid $NH_2(CH_2)_3COOH$ = 103.1. 4-Aminobutyric acid*. Colorless crystals, m.183, soluble in water. **homo ~** $NH_2(CH_2)_4COOH$ = 117.1. 5-Aminovaleric acid*. Colorless crystals, m.158, soluble in water.

piperidine* $CH_2 \cdot (CH_2)_2 \cdot NH \cdot CH_2 \cdot CH_2$ = 85.1. Hexahydropyridine. Pentamethylene imine. Colorless liquid, d.0.862, b.106, soluble in water; a vasodilator and a part of the structure of many alkaloids. **dimethyl ~** Lupetidine. **methyl ~** Pipecoline. **piperyl ~** Piperine. **propyl ~** Conine.

p.carboxylic acid **2-~** Pipecolinic acid. **3-~** Nipecotic acid.

piperidineethanol $C_7H_{15}ON$ = 129.2. Yellow liquid, d.0.973, $b_{45mm}116$; an intermediate in the manufacture of drugs and plastics.

piperidinium* The cation $C_5H_{10}NH_2^+$, from piperidine. **p. compounds** Derivatives of p.; as, $[C_5H_{10}NH(Me)]I$.

piperidino* 1-Piperidinyl†. The piperidyl radical, but with the external bond in the 1 position instead of the 2 position.

piperidinyl **1-~†** See *piperidino.*

piperidyl **2-~*** The radical $C_5H_{10}N-$, from piperidine. Cf. *piperidino.* **p.urethane** $C_5H_{10}N \cdot CO \cdot OC_2H_5$ = 157.2. Colorless liquid, b.211.

piperine $C_{17}H_{19}O_3N$ = 285.3. Piperylpiperidine. An alkaloid from black pepper. Colorless, monoclinic crystals, m.129, slightly soluble in water, soluble in alcohol or ether.

piperinic acid Piperic acid.

piperolidine $C_8H_{15}N$ = 125.2. Octahydropyrrocoline, δ-coniceine. Colorless liquid, d.0.904. **(±)-~** b.161. **(−)-~** b.158.

piperonal* $(CH_2{:}O_2) \cdot C_6H_3 \cdot CHO$ = 150.1. Heliotropin, piperonylaldehyde; from piperine, having a heliotrope odor. Colorless needles, m.37, slightly soluble in water.

piperonoyl The piperonyloyl* radical.

piperonyl* The radical 3,4-$(CH_2O_2){:}C_6H_3 \cdot CH_2-$, from piperine. **p. alcohol*** $C_8H_8O_3$ = 152.2. 3,4-Methylenedioxybenzyl alcohol. A solid, m.51, slightly soluble in water. **p.aldehyde** Piperonal*.

piperonylic acid* $C_8H_6O_4$ = 166.1. Methyleneprotocatechuic acid, 3,4-methylenedioxybenzoic acid, heliotropic acid, from paracoto bark, m.228.

piperonylidene* The radical 3,4-$(CH_2O_2){:}C_6H_3 \cdot CH=$, from piperonal. Cf. *piperic acid.*

piperonyloyl* Piperonoyl. The radical 3,4-$(CH_2O_2){:}C_6H_3 \cdot CO-$, from piperonylic acid.

piperovatine $C_{17}H_{23}O_2N$ = 273.4. An alkaloid from the fruits of *Piper ovatum* (Piperaceae), Trinidad. Colorless crystals, insoluble in water.

piperyl The radical $(CH_2{:}O_2){:}C_6H_3 \cdot CH{:}CH \cdot CH{:}CH \cdot CO-$, from piperic acid.

piperylene 1,3-Pentadiene*.

piperylhydrazine $C_5H_{12}N_2$ = 100.2. Piperidylamine*. Colorless liquid, b.146, soluble in alcohol.

pipestone Catlinite.

pipet A graduated, open glass tube used for measuring or transferring definite quantities of liquids. See *pipet* under *automatic, Babcock, capillary, counting, gas, Mohr.* **auswaschen ~** A micropipet which *contains* a specified volume of liquid, as distinct from *delivering* it; and must be washed out with water after it has drained. **solid-interface ~** A p. with no dead-air space between the plunger and liquid.

pipette Pipet.

pipitzahoac The dried roots of *Perezia* species (Compositae), Mexico.

pipitzahoic acid $C_{15}H_{20}O_3$ = 248.3. Perezone. Aurum vegetable. Yellow needles from pipitzahoac, m.103, soluble in water. See *perezol.*

pipitzahoin A colored principle from the roots of *Perezia adnate* (Compositae); an indicator.

pipsissewa Chimaphila.

piquia fat A fat from the kernels of *Caryocar villosum*

(Caryocaraceae), Brazil (souari or butternut); it resembles palm oil.
pirimicarb* See *insecticides,* Table 45 on p. 305.
pirimiphos-methyl* See *insecticides,* Table 45 on p. 305.
Piriton Trademark for chlorpheniramine maleate.
piroglycerina An early name (due to the discoverer, Soluero) for nitroglycerin.
pisang Malay for "banana." **p. wax** A wax from the leaves of the Java banana tree, *Musa paradisiaca* (Musaceae), containing pisangceryl alcohol.
pisangceric acid Pisangcerylic acid.
pisangceryl alcohol $C_{13}H_{28}O$ = 200.4. A saturated alcohol, m.78.
pisangcerylic acid $C_{24}H_{48}O_2$ = 368.6. A monobasic acid, m.71, from pisang wax.
pisanite (Fe, Cu)$SO_4 \cdot 7H_2O$. A native sulfate.
piscidia Jamaica dogwood. The dried bark of *P. erythrina;* a W. Indies fish poison, and narcotic.
piscidic acid $C_{11}H_{12}O_7$ = 256.2. 4-Hydroxybenzyltartaric acid. Obtained from piscidia. Colorless needles, m.183, soluble in water.
piscidin $C_{29}H_{24}O_8$ = 500.5. A neutral principle from piscidia (dogwood); an anodyne.
pisolite A hard, compact form of aragonite.
pistacite Epidote.
pistil Modified leaves forming the central part of a flower.
pistomesite $MgCO_3 \cdot FeCO_3$. A native carbonate.
pitayamine An alkaloid from the bark of *Cinchona pitayensis* (Rubiaceae).
pitch (1) A heavy, liquid or dark residue obtained by distillation of tar. See *rosin.* (2) To add yeast, with or without sugar, in order to start fermentation. (3) The distance between the threads of a screw. (4) The vibration frequency of the keynote of a tune. **archangel ~** (1) Originally, pine tar p. from Archangel. (2) A blend of pine pitch with various oils, used to caulk boats (U.S. usage). **black ~** Naval p. **Burgundy ~** Principally the solid resin obtained by heating and straining the air-dry solid resin exuded by Norway spruce *Picea excelsa,* and European silver fir, *Abies pectinata.* **Canada ~** The resin from *Tsuga* species. **earth ~** Asphalt. **Jew's ~** Asphalt. **mineral ~** (1) Asphalt. (2) Bitumen. **naval ~** The dark, solid residue from the distillation of various tars. **petroleum ~** Asphalt. **Trinidad ~** Asphalt.
p.blende Impure *uranite,* q.v., in which radium was discovered. Cf. *broggerite.* **p. coal** Specular coal. **p.stone** A dark-colored igneous rock, similar to obsidian.
pithecolobine An alkaloid from the bark of *Pithecolobium saman* (Leguminosae), southeast Asia. Brown oil, insoluble in water.
pitot tube A vertical U-tube with a movable scale, closed at one end and containing a liquid to record differences in pressure; an anemometer.
Pitressin Trademark for vasopressin.
Pitrowsky test Biuret reaction.
pittacol Eupittonic acid.
pitticite Scorodite.
pituitary Hypophysis sicca. The dried, cleaned, and powdered posterior lobe of the pituitary gland of the ox; formerly used in medicine.
pituri The powdered leaves and twigs of the p. plant, *Duboisia hopwoodii* (Solanaceae), Australia. Used locally as a narcotic stimulant.
piturine $C_{10}H_{16}N_2$ = 164.3. An alkaloid from pituri. Brown oil resembling nicotine.
piuri Indian yellow.
pivaldehyde* Me_3CCHO = 86.1. Trimethylacetaldehyde, 2,2-dimethylpropanal*. Colorless liquid, d.0.793, m.3, b.75; used in organic synthesis.
pivalic acid* CMe_3COOH = 102.1. Trimethylacetic acid, 2,2-dimethylpropanoic acid*. Colorless crystals, m.35, soluble in water. Cf. *valeric acid.*
pivaloyl* 2,2-Dimethyl-1-oxypropyl†. The radical $CMe_3 \cdot CO-$.
pix Latin for "pitch"; as, pix burgundica (Burgundy pitch). **p. liquida** Wood *tar.*
p*K* Symbol for the logarithm of the reciprocal of the dissociation constant of an electrolyte: $pK = \log(1/K)$.
pk Abbreviation for peck.
placebo An inert pharmaceutical preparation which is prescribed for psychological rather than for therapeutical reasons, or as a control in drug trials.
place isomerism The isomerism of chemical substances of similar compositions, which differ in structure by the positions of radicals; as, *ortho, meta,* and *para.*
placement A method of applying fertilizer in pockets or continuous narrow bands near to the seed at sowing time.
placer An alluvial or glacial deposit of sand or gravel containing gold or other precious minerals and metals. **p. mining** The extraction of precious metals from sand by washing.
plagioclase General name for triclinic feldspars. Cf. *gabbro.*
plagionite $5PbS \cdot 4Sb_2S_3$. Cf. *lead* minerals.
plague vaccine A sterile suspension in isotonic salt solution of killed *Yersinia pestis;* a specific plague vaccine (USP).
Planck **P., Max Karl Ernst Ludwig (1858–1947)** German physicist. Nobel prize winner (1918). **P. constant*** The fundamental constant, $h = 6.62617 \times 10^{-34}$ Js. Thence, energy = $h\nu$, where ν is frequency. See *quantum theory.* **P.'s element of action** P. constant. **P.'s formula** The energy radiated from a black body at wavelength λ micrometers = $C_1\lambda^{-5}/[e^{C2/\lambda T} - 1]$, where e is the base of napierian logarithms; $C_1 = 2\pi hc^2$, $C_2 = hc/k$, where h is the Planck constant, k the Boltzmann constant, c the velocity of light, and T the thermodynamic temperature. **P.'s unit** P. constant.
plane A surface, imaginary or real, which is level with itself in all directions. **chiral ~** See *sequence rule.* **stereoisometric ~** See *stereoisomer.*
p. of symmetry An imaginary p. passing through a crystal so that, for each face or angle of the crystal, there is a similar face or angle on the opposite side of the p., a line joining the two faces being perpendicular to the p.
planet A celestial body moving around the sun in a nearly circular orbit. Cf. *meteoric.*
planetary Pertaining to a planet. **p. atmosphere** The gases surrounding a planet: Mercury, none; Venus, CO_2; Earth, N_2; Mars, H_2O.
plangi Tie dye. A form of reserve dyeing used in the Far East, in which patterned areas of cloth are brushed up and tied in small bundles, so that the cloth inside is unaffected by the dye bath.
plant (1) The machinery and equipment used in manufacturing processes. (2) A living organism of the vegetable kingdom, which generally contains chlorophyll and photosynthesizes food; hence (unless a parasite or saprophyte) it requires for life only inorganic substances. Cf. *animal.* Plants are grouped into 4 phyla and consist of 2 systems:

1. Protective system:
   (*a*) The surface—epidermis, cork, bark

(*b*) The skeleton—bast fibers, collenchyma, sclerotic parenchyma
2. Nutritive system:
(*a*) Absorbing tissues—epithelium of roots, root hairs, etc.
(*b*) Assimilating tissue—chlorophyll parenchyma
(*c*) Conducting tissue—conducting parenchyma, vascular bundles, latex cells
(*d*) Storage tissue—reserve tissue of seeds, bulbs, tubers, and water tissues
(*e*) Aerating system—intercellular spaces, stomata, and lenticels
(*f*) Receptacles for secretions and excretions—glands, oil, resin and mucus canals, crystal sacs

p. acids The organic acids in vegetable organisms; as, citric acid (lemons). **p. elements** The elements known to be essential to p. growth: C, H, O, N, S, P, K, Ca, and Mg; traces of Fe, Na, Si, Al, Cl, Cu, Mo, Mn, Zn, B, and F may be essential. **p. food** See *fertilizer.* **p. pigments** The coloring matter of plants, chiefly:
1. *Chlorophyll pigments:* the green and reddish colors of leaves. Cf. *porphin ring.*
2. *Carotenoids,* q.v.: the lipochromes or fatty pigments of plants.
3. *Flavones* and *flavanols,* q.v.: the fairly soluble pigments of blossoms and fruits. Cf. *anthocyanins.*

plantain The herbaceous tree *Musa sapientum paradisiaca* (Scitamineae); its fruit is the Adam's apple. Cf. *banana.*

plantose An albuminous substance from rapeseed.

plaque (1) A high-molecular-weight polysaccharide film formed on teeth by cariogenic bacteria. (2) Clear zone in a bacterial culture on an agar plate caused by lysis of bacterial cells due to activity of bacteriophage.

Plaskon Trademark for a urea- or melamine-formaldehyde plastic.

plasma (1) A green mottled variety of chalcedony. (2) Generally applied to a gas that is sufficiently ionized for its properties to depend on the ionization. It contains approximately equal numbers of positive ions and electrons, so the mixture is electrically neutral, highly conductive, and affected by magnetic fields. A plasma is produced by temperatures above 20,000°C in controlled thermonuclear fusion reactors, such as the tokamak device; it is produced naturally in St. Elmo's fire and in the sun and other stars. (3) The liquid part of the blood, containing p. proteins, fibrinogen, and dissolved metabolites (as, glucose). **dried ~** Sterile p. obtained by pooling approximately equal volumes of the liquid portions of citrated whole blood from not more than 12 donors; a blood volume replenisher (EP, BP). **proto ~** Protoplasm.

p. protein fraction A sterile preparation of serum albumin and globulin from healthy donors; used as a dried plasma, particularly for burns (USP, BP). **p. torch** A device in which temperatures up to 35,000°C are obtained by injecting a gas jet (as, argon or nitrogen) tangentially into an electric arc formed between a rod and a nozzle electrode in a chamber. The resulting p. jet of hot gases can be used for welding, cutting hard rock or hard metal, or spraying metal.

plasmid A small ring of DNA coated with protein. Plasmids occur in some microorganisms and can confer drug resistance on other cells, but are not part of the hereditary system of the mother cell.

plasmin* An active fibrinolytic and proteolytic enzyme in human blood plasma.

plasminogen The inactive precursor of plasmin.

plasmolysis Dissolution of the protoplasm of a cell when it is bathed in water or a salt solution.

plasmosome Nucleolus.

plastein A substance formed when peptic digests of certain proteins (e.g., insulin) are concentrated and treated with more pepsin under slightly acid conditions.

plaster (1) In pharmacy, a preparation spread on fabric for application to the skin. (2) In general, a paste for coating surfaces or making molds. **adhesive ~** A mixture of rosin and wax, for coating textiles. **hard-burnt ~** An insoluble anhydrite.

p. of paris A zeolitic type of hydrated calcium sulfate, made by heating gypsum. Dissolved as $CaSO_4 \cdot \frac{1}{2}H_2O$, it quickly solidifies in the presence of water to $CaSO_4 \cdot 2H_2O$; used to make molds for taking impressions of objects, and to make splints for limbs.

plastic Soft or moldable, pliable. Cf. *plastics.*

plasticity Capability of being formed or shaped in any desired way.

plasticization The conversion of hard, glassy polymers into a soft, rubbery solid by the action of an organic compound, usually an ester.

plasticizer A liquid having a low vapor pressure at room temperatures. Used to (1) modify flow properties, as, of synthetic resins; (2) reduce evaporation rate, as, of a paint solvent; (3) impart flexibility and toughness to a plastic, paint, or varnish film, e.g., phthalates in lacquers. Cf. lacquer *solvent.*

plastics A group of organic materials which, though stable in use at ordinary temperatures, are plastic at some stage of manufacture and then can be shaped by application of heat, pressure, or both. Synthetic rubber and certain inorganic materials, e.g., glass, comply with this definition but are not usually regarded as p. Cf. *elastomer, polymer.* See Table 62 on p. 456. **ABS ~** P. of good impact strength obtained by dispersing an elastomer into a rigid acrylonitrile-butadiene-styrene copolymer. **casein ~** P. made from milk; as, Galalith. **cellulose ~** P. made from nitrocellulose and camphor; as, Celluloid. **contact-pressure ~** P. that form laminates at a pressure of 100 kPa or less. **ethenoid ~** P. comprising the acrylic, vinyl, and styrene types. **phenol ~** P. made by condensation of phenol and formaldehyde; as, Bakelite. **rosin ~** Phenol p. **thermoplastic ~** P. that become moldable when heated; as, vinyl polymers. **thermosetting ~** P. that harden irreversibly when heated; as, phenol-aldehyde (Bakelite).

plastisol (1) A plastic used as a solution or emulsion, e.g., for coating. (2) The product resulting from plasticization.

Plastofilm Trademark for a plastic made from reclaimed Pliofilm.

plastometer (1) An instrument to measure the hardness of rubber from the depth of indentation of a hard body. Cf. *Brinell tester.* (2) A device to measure the plasticity of a material by timing its flow through successive increments of length of a capillary tube.

plate A thin sheet of metal, glass, etc., with a flat surface, e.g., silver p. or p. glass. **black ~** P. for tinning, in the untinned state. **photographic ~** A glass p. coated with an emulsion containing light-sensitive silver salts. Cf. *chromatic plate.*

p. amalgamation A method of extracting gold from finely crushed ore by floating it over a copper surface coated with mercury.

platelet Blood p. The smallest cell in the blood, diameter 2–4 μm, nonnucleated, containing only dark-staining granules. Platelets are important in normal blood clotting mechanisms.

Platforming Process Patented process for the catalytic

TABLE 62.
CLASSIFICATION OF PLASTICS

Some of the principal trade names are given in parentheses. In some cases there are separate definitions under the name headings.

ABS (Abson, Novodur).
Acrylic (Acrilan, Lucite, Orlon, Perspex, Plexiglas.) Glass substitutes.
Alkyd Polyalcohol-phthalic anhydride esters, resinoids, resins and resin mixtures (Avlin, Dacron, Dulux, Duraplex, Encron, Esterol, Fortrel, Glyptal, Mylar, Terylene). Used in hardened forms; as, for electrical insulation and cements.
Allyl (Allymer). Similar to alkyds.
Aniline-formaldehyde (Cibanite). Heat-hardening.
Aramid (Kevlar, Nomex). Aromatic analog of polyamides.
Bituminous (Cetec, Thermoplax). Cold molding.
Caffelite From coffee beans.
Casein (Galalith). Billiard balls and ivory substitute.
Cellulose (Avril, Cellophane, Zantrel). Transparent film and fibers.
Cellulose acetate (Bakelite C.A.I., Lumarith, Textolite).
C.a.-butyrate (Hercose C).
C.a.-propionate (Hercose A.P.).
Cellulose nitrate (pyroxylin) (Celluloid, Fiberlac, Nitron). Flammable film and solid articles.
Cellulose propionate (Forticel).
Chlorinated diphenyl (Arochlor).
Chlorinated rubber (Parlon). Acid-resistant coatings.
Copal ester (Kopal).
Coumarone (Cumarone)-Indene (Brofo, Cumar, Paradene).
Cyclohexanone-formaldehyde
Epoxy, epichlorhydrin-bisphenol (Epikote, Araldite). Coatings and varnishes.
Ethyl cellulose (Ethocel, Lumarith E.C.).
Formaldehyde-sulfonamide (Santolite).
Furane (Duralon).
Fluorocarbon Tetrafluoroethylenes. (Gore-Tex, Teflon, Tetran).
Hydrogenated rosin (Staybelite).
Lignin (Benalite, Lignolite, Meadol). Mainly an extender for other plastics.
Melamine (Catalin, Melamine). Wet-strengthening agent for paper.
Methylcellulose (Methocel).
Phenol-aldehyde (Aerolite, Albertol, Amberlite, Bakelite, Catalin, Formalite, Formica, Indurite, Micarta, Mouldrite, Phenolite, Textolite). Heat-hardening moldings or coatings; oil-soluble resins for use in varnishes, enamels, paints, and lacquers; adhesives for lamination.
Phenol-copal (Beckopol).
Polycarbonate (Lexan, Makrolon, Merlon).
Polyamide (Capron, Nylon, Quina, Tactel, Vydyne). Principally fibers, filaments, and bristles.
Polyamide-aldehyde (Melopar).
Polyester (Actol, Polylite).
Polyethylene (Epolene, polythene, Tyvek) and *Polypropylene* (Propathene). *Polyolefins.* Transparent film, and moldings and extrusion material.
Polystyrene (Lustron, Styron).
Rubber hydrochloride (Pliofilm, Plioform).
Silicone (Silastic).
Sulfonamide-aldehyde (Sanolite).
Terpene (Piccolyte, Rezinel).
Urea-(form)aldehyde (Aldur, Beatl, Beetle, Melamac, Plaskon, Pollopas, Uformite). Heat-hardened moldings.
Urethane (Estane, Hetrofoam, Lycra, Numa).
Vinyl polymers (P. = Polyvinyl).
P. acetate, PVAC (Alvar, Vinylite).
P. acetate and vinylidene dinitrile, or nytril (Darvan).
P. alcohol, PVAL (Vinal).
P. aldehyde (Formvar).
P. chloride, PVC (Chloroprene, Geon, Korolac, Plioflex, Tygon).
P. chloride and acetate (Elastiglas, Vinyon).
Polyvinylidene (Saran).

upgrading of low-octane gasoline from natural gas or crude petroleum.

platina (1) "Little silver." An early name for platinum. (2) Native platinum, often containing Ir, Rh, Ru, Os, Au, and Ag.

platinammines Compounds containing the tetrammine*platinum*(IV)* cation.

platinammonium The diammine*platinum*(IV)* cation.

platinate* Indicating platinum as the principal atom(s) in an anion. **hexabromo ~** Platinibromide, bromoplatinate. A salt containing the radical $[PtBr_6]^{2-}$. **hexachloro ~** Platinichloride, chloroplatinate. A salt containing the radical $[PtCl_6]^{2-}$. **tetrachloro ~** Platinochloride, chloroplatinite. A salt containing the radical $[PtCl_4]^{2-}$. **tetracyano ~** Platinocyanide, cyanoplatinite. A salt containing the radical $[Pt(CN)_4]^{2-}$.

plating (1) A process by which a surface is coated with a metal; as, *silver* plating. (2) Transferring a bacterial suspension to a culture medium. **close ~** A nonelectrolytic process in which sheets of metal are soldered to the surface to be plated. **electro ~** Coating with a metal by electrodeposition.

platinibromide A salt containing the hexabromo*platinate*(IV)* anion.

platinic Platinum(IV)* or (4+)*. Describing a substance containing tetravalent platinum. **p. acid** H_2PtO_3 = 245.1. Trioxoplatinic(IV) acid*. White powder, soluble in alkalies (forming platinates) and acids (forming platinic compounds). **hexachloro ~** $H_2PtCl_6 \cdot 6H_2O$ = 517.9. Brown crystals, soluble in water. Used as a reagent for separation of potassium from sodium; for platinization; in photographic toning; in ceramics, for metallic lusters; and in catalysts. **p. bromide** $PtBr_4$ = 514.7. Platinum tetrabromide*. Brown crystals, soluble in water. **p. chloride** (1) $PtCl_4$ = 336.9. Brown crystals, soluble in water. (2) Hexachloro*platinic acid.* **p. hydroxide** $Pt(OH)_4$ = 263.1. Brown powder, decomp. by heat, insoluble in water. **p. iodide** PtI_4 = 702.7. Platinum tetraiodide*. Brown powder, insoluble in water, soluble in iodide solutions. **p. oxide** PtO_2 = 227.1. Platinum dioxide*. Black powder, m.430 (decomp.), insoluble in water. **p. sulfate** $Pt(SO_4)_2$ = 387.2. Platinum sulfate. Green-black, deliquescent crystals, soluble in water; a microchemical reagent. **p. sulfide** PtS_2 = 259.2. Platinum disulfide*. Black needles, decomp. by heat, insoluble in water, soluble in ammonium sulfide solution.

platinichloride A salt containing the hexachloro*platinate*(IV)* anion.

platiniferous An ore or substance containing platinum.

platiniridium A native alloy of Pt and Ir, often containing Rh, Ru, and Cu.

platinite Anion containing divalent platinum; as, tetrachloro*platinate*(II)*.

platinize To coat with metallic platinum.
platinized asbestos Asbestos impregnated with a solution of a platinum salt and ignited, so that metallic platinum results; a catalyst.
platinochloride A salt containing the hexachloro*platinate*(IV)* anion.
platinocyanide A salt containing the tetracyano*platinate*(II)* anion.
platinoid The alloy Cu 61, Zn 24, Ni 14, W 1–2%; used for electrical-resistance coils.
platinous Platinum(II)* or (2)*. Describing a compound containing divalent platinum. **p. acid** H_2PtCl_4 = 338.9. Tetrachloroplatinous acid. **p. bromide** $PtBr_2$ = 354.9. Platinum dibromide*. Brown, deliquescent crystals, decomp. 200, insoluble in water. **p. chloride** $PtCl_2$ = 266.0. Platinum dichloride*. Brown crystals, insoluble in water, soluble in chloride solutions. **p. cyanide** $Pt(CN)_2$ = 247.1. Platinum dicyanide*. Yellow powder, insoluble in water, soluble in cyanide solutions. **p. hydroxide** $Pt(OH)_2$ = 229.1. Black powder, decomp. by heat, insoluble in water. **p. iodide** PtI_2 = 448.9. Platinum diiodide*. Black powder, decomp. 325, insoluble in water, soluble in acids. **p. oxide** PtO = 211.1. Platinum monoxide*. Violet to black powder, decomp. by heat, insoluble in water. **p. sodium chloride** Sodium tetrachloroplatinate(II)*. **p. sulfide** PtS = 227.1. Platinum monosulfide*. Black powder, decomp. by heat, insoluble in water, soluble in ammonium sulfide solution.
platinum* Pt = 195.08. A noble metal of the fourth period, at. no. 78, described (1741) by Wood. Silver-gray, d.21.37, m.1773, b. 3800, insoluble in water, alcohol, acids, or bases; soluble in aqua regia. Used as a catalyst (as, to purify car exhausts and as a fuel cell catalyst), foil, wire powder (p. black), gauze, and gray sponge; and for acid-proof containers, electrodes, and jewelry. P. forms 2 series of compounds:divalent [Pt(II)*, Pt(2+)*, platinous] and tetravalent [Pt(IV)*, Pt(4+)*, platinic], each having many double salts. ***cis*-~** $Pt(NH_3)_2Cl_2$ = 300.0. *cis*-Diamminedichloroplatinum, cisplatin, DDP. An antineoplastic drug, used to treat tumors of testis and ovary.
diammine ~ (IV)* The cation, $[Pt(NH_3)_2]^{4+}$.
tetraammine ~ (IV)* The cation, $[Pt(NH_3)_4]^{4+}$.
p.(II)*, p.(2+)* See also *platinous.* **p.(IV)*, p.(4+)*** See also *platinic.* **p. alloy** An alloy of p. with another noble metal; as, gold. **p. black** A black powder of finely divided p.; a catalyst. **p. chloride** (1) Commercial hexachloroplatinic acid. See *platinic acid.* (2) Platinic chloride. (3) Platinous chloride. **p. dibromide*** Platinous bromide. **p. dichloride*** Platinous chloride **p. dicyanide*** Platinous cyanide. **p. diiodide*** Platinous iodide. **p. dioxide*** Platinic oxide. **p. disulfide*** Platinic sulfide. **p. iridium** An alloy: Pt 90, Ir 10%. **p. metals** A group of noble metals that occur together in nature, and form 2 series in the periodic table: platinum series—Re, Os, Ir, Pt; and the palladium series—Tc, Ru, Rh, Pd. **p. minerals** P. occurs in nature alone or associated with the other metals of the group, e.g., platiniridium, Pt·Ir. See also *cooperite, platina.* **p. monosulfide*** Platinous sulfide. **p. monoxide*** Platinous oxide. **p. pentafluoride*** $[PtF_5]_4$ = 1160. A pentavalent p. compound. Red solid, reacts with water to form the fluoroplatinate. **p. sponge** Gray, spongy, porous, metallic p. mass, obtained by reduction of hexachloroplatinic acid; a catalyst. **p. sulfate*** Platinic sulfate. **p. tetrachloride*** Platinic chloride. **p. tetrafluoride oxide*** $PtOF_4$ = 287.1. Red solid, sublimes 150 (orange vapor). **p. yellow** An alkaline chloroplatinate coating for fluorescent X-ray screens.
platinum cladding The bonding of a layer of platinum to a bar of other metal, and the working down of the result to a desired thickness; for laboratory ware.
platize Platinize.
platosammine A compound containing the diammineplatinum(II)* cation, $[Pt(NH_3)_2]^{2+}$.
Plato unit The weight (grams) of total solids in 100 g of wort.
Plattner **P., Karl Friedrich (1800–1858)** German mineralogist. **P.'s process** The extraction of gold as trichloride by passing chlorine gas through the gold-bearing pulp.
plattnerite PbO_2. A native oxide.
Plausen mill A colloid mill.
Plavia Trademark for a viscose cellulose synthetic fiber.
plazolite $3CaO \cdot Al_2O_3 \cdot 2(SiO_2 \cdot CO_2) \cdot 2H_2O$. A Californian silicate. Colorless crystals.
pleiad A group of isotopes (obsolete).
pleio- See words beginning with *pleo.*
Pleistocene See *geologic eras,* Table 38.
pleochroic Pleochromatic.
pleochroism Pleochromatism.
pleochromatic Pleochroic. Showing more than one color; as, a fluorescent solution.
pleochromatism Pleochroism. The capacity of certain optically biaxial crystals to transmit polarized light, so that a complementary color is seen at right angles to the direction of the ray. Cf. *dichroism.*
pleomorphic Occurring in more than one form.
pleomorphism The capacity to crystallize in 2 or more different crystal systems.
pleonast The mineral $(Mg \cdot Fe)O \cdot (Al \cdot Fe)_2O_3$.
plessite Gersdorffite.
Plessy's green Chromic phosphate.
pleurisy root Asclepias.
Plexiglas Trademark for a methyl acrylate plastic.
pliers Pincers with long jaws for holding, bending, or cutting. **button ~** A circular disk with a round hole, used to hold assay buttons during polishing.
plinol A hydroxy derivative of iridane.
Pliny the Elder Cajus Secundus Plinius, 23–79. Roman soldier noted for his scientific observations and writings *(Historia naturalis).*
Pliocene See *geologic eras,* Table 38.
Pliofilm Trademark for rubber hydrochloride. Cf. *Plastofilm.*
Pliolite Trademark for resinous materials obtained when rubber is cyclized with agents such as tin tetrachloride. Used to produce water resistance, e.g., coated paper.
pliowax *Pliolite,* q.v., containing 60% paraffin wax; a hot-melt coating, e.g., for paper.
plodding Compression of warm soap into shape in a screw compressor.
Plotnikow effect The longitudinal scattering of rays (especially infrared) by solid objects or fluids. Cf. *Raman effect.*
plotting Making a graph describing a relationship between 2 unknowns, by determining a number of pairs of values that represent the relation, and depicting them in terms of the lengths of 2 lines at right angles (coordinates).
Plucker tube A glass tube with 2 electrodes, containing a gas under reduced pressure; used in spectroscopy.
plumbagin $C_{11}H_8O_3$ = 188.2. Methyljuglone, 5-hydroxy-2-methyl-1,4-naphthoquinone*. Yellow crystals, m.79, from Plumbaginaceae spp. Exhibits bactericidal activity.
plumbago Native graphite. Black lead. Used in the manufacture of "lead" pencils, crucibles, etc., and as a lubricant.
plumbane (1)* PbH_4 = 211.2. Lead tetrahydride. A gas formed in the cathodic reduction of finely divided lead. (2)*

R_4Pb. An organic compound of quadrivalent lead, as: **tetraethyl ~*** Et_4Pb = 323.4. Tetraethyllead, lead tetraethide. Colorless liquid; an antiknock compound for gasoline, which increases its octane number by up to 8 units. See *toxic lead* (under *lead*), *ethyl gasoline* (under *Ethyl*).

plumbate* Indicating lead as the central atom(s) in an anion. (1) Salt derived from lead dioxide: metaplumbate(IV)*, M_2PbO_3; orthoplumbate(IV)*, M_4PbO_4. (2) Salt derived from lead hydroxide, $Pb(OH)_2$: plumbate(II)*, plumbite ($MHPbO_2$ and M_2PbO_2).

plumbic Lead(IV)* or (4)*, plumbum(IV)*. Describing a compound of tetravalent lead. **p. ocher** Brown *lead* oxide (6).

plumbiferous Containing lead.

plumbism Poisoning by lead. See *toxic lead* under *lead.*

plumbite See *plumbate(II).*

plumbocalcite Lead containing calcite.

plumbogummite A native aluminum and lead phosphate.

plumbosolvency The degree of solubility of lead in a liquid.

plumbous Lead(II)* or (2+)*, plumbum(II)*. Describing a compound of divalent lead. **p. compounds** See *lead.*

plumbum* Latin for *"lead,"* q.v. **p. candidum** Early name for tin. **p. cinereum** Early name for bismuth. **p. nigrum** Early name for lead.

plumiera Sucuuba bark. The dried bark of *Plumiera sucuuba* (Apocynaceae); an anthelmintic.

plumose Having a fleecy or feathery appearance. **p. growth** A fleecy, feathery growth of bacteria on a culture medium. **p. mica** A muscovite resembling asbestos.

plural gel A gel formed by the simultaneous gelification of a mixture of 2 or more sols.

pluranium An alleged new element discovered by Osann (1828) in a platinum ore: probably mixed titanium dioxide, silica, and zirconium oxide.

plus (+). Prefix indicating a dextrorotatory compound. Cf. D. **p. or minus** (±). Prefix indicating a racemic mixture (or *meso-*compound when the cause of optical inactivity is unknown).

plutonic Igneous. General name for rocks that have crystallized below the earth's surface; as, granite.

plutonium* Pu. An element, at. no. 94. At. wt. of isotope with longest half-life ^{244}Pu = 244.06. Obtained by bombarding uranium with neutrons (Seaborg, etc., 1940); named for the planet Pluto. ^{239}Pu is isolated from the uranium fuel elements of nuclear reactors. It is also a transformation product of Np. Exhibits oxidation states 3–7. m.640, b.3230. Used in nuclear reactors and weapons. **trans ~** Early name for native barium oxide.

p. enrichment. The separation of ^{239}Pu from its other isotopes; possibly by lasers.

plutonyl* The cation PuO_2^{2+}.

pluviometer An instrument to measure rainfall.

PMMA* Abbreviation for poly(methyl methacrylate).

pn* Indicating propylenediamine, $H_2N \cdot CH(Me)CH_2 \cdot NH_2$, as a ligand.

pneumatic (1) Pertaining to air. (2) Pertaining to gases. Cf. *aerodynamics.* **p. drill** A drill operated by compressed air. **p. jig** A device for separating minerals by an air blast. **p. trough** A vessel of water containing an inverted cylinder, filled with water, for the collection of gases.

pneumatics The study of the mechanical properties of gases.

pneumatology (1) The science of respiration. (2) The science of gases.

pneumatolysis The production of ore deposits by liberation of gases or vapors during solidification of igneous rocks.

pneumoconiosis General name for pulmonary diseases due to inhalation of dust. Cf. *silicosis, asbestosis.*

pneumokoniosis Pneumoconiosis.

pockeling The development of circular patches on fresh surfaces of certain solutions, e.g., soap.

pocket (1) A sac-shaped cavity or hole. (2) A small body of ore. (3) An enlargement of an ore vein. (4) A unit of measure, e.g., for hops.

poco oil The essential oil from *Mentha aquatica,* rich in linalool acetate.

podophyllic acid $C_{22}H_{24}O_9$ = 432.4. A monobasic acid derived from podophyllin by hydrolysis.

podophyllin A resin obtained from podophyllum; contains podophyllic acid, podophyllotoxin, and picropodophyllin. Yellow powder, insoluble in water, soluble in alcohol. Used in treating warts (USP, BP).

podophyllotoxin $C_{22}H_{22}O_8$ = 414.4. The active principle of podophyllin. Yellow powder, insoluble in water.

podophyllum Mandrake, May apple. The dried rhizome of *Podophyllum peltatum* (Berberidaceae) which contains 3% podophyllin. Used medicinally as the fluidextract (USP). **p. resin** Podophyllin.

podzol, podsol A soil that has undergone podzolization.

podzolization, podsolization The division of soil into 3 layers by continual leaching with rainwater. The top layer contains decaying organic matter; the center layer the material out of which silica, iron oxide, and alumina have been washed; and the bottom layer redeposited extractives.

Poggendorff P., Johann Christian (1796–1877) German physicist. **P. cell** An amalgamated zinc anode and carbon cathode in a solution of potassium dichromate 12, sulfuric acid 25, water 100 pts. (2.01 volts.) **P. compensation method** The determination of the potential of a cell by comparing the potential required to balance it with that required to balance a standard, in terms of the length of a resistance wire. Cf. *Wheatstone bridge.*

Pohl's commutator An electrical-contact key that enables the direction of current to be reversed instantaneously; used in physiology.

poi See *taro.*

poidometer A rapid-action industrial weighing machine.

poikilothermism The ability of living organisms, especially lower forms, to adapt their body temperature to that of their environment.

point (1) That which has position but not magnitude. (2) A minute spot. (3) A numerical value on a scale. (4) A unit of weight used by jewelers 1 p. = 0.01 carat. (5) A printer's unit, equal to 0.3514 mm. **boiling ~** The minimum temperature at which the vapor pressure of a liquid equals that of the external pressure (such as atmospheric pressure, under vacuum distillation, at high or low altitude). **condensation ~** The maximum temperature at which a vapor changes to the liquid state. **critical ~** The temperature above which a gas cannot be liquefied by pressure only. **dew ~** The temperature at which the moisture of the air condenses. **end ~** In titration: the stage at which one drop of solution completes the reaction. **freezing ~** The maximum temperature at which a liquid changes to the solid state. **isoelectric ~** See *isoelectric point.* **liquefaction ~** (1) Melting p. (2) Condensation p. (3) Critical p. **melting ~** The minimum temperature at which a solid becomes liquid. **quadruple ~** The temperature at which all 4 phases in a 2-component system can exist. **refraction ~** The spot on a surface at which a ray of light is refracted. **set ~** See *set point.* **slip ~** The minimum temperature at which a fat or wax slips downward in a vertical capillary tube, as distinct from melting completely. **triple ~** See *triple point.*

pointage test Pointage titration. The control of zinc phosphating baths from the volume of 0.1 *N* sodium hydroxide to neutralize (to phenolphthalein) a 10 mL aliquot of the bath.

poise Symbol: p; (1) the cgs unit of dynamic viscosity, 1 p = 1 dyne·s/cm^2 = 0.1000 Pa·s (the SI unit) = 1.45×10^{-5} lb·s/ft^2; or stokes × density. Cf. *reyn, stoke.* (2) To balance or maintain an oxidation-reduction equilibrium; as, a poising agent.

poised Describing a system that resists oxidation or reduction. Cf. *buffer.*

Poiseuille's law See *dynamic viscosity* under *viscosity.*

poising agent Poiser. A substance that stabilizes an oxidation-reduction equilibrium. Cf. *buffer.*

poison (1) A substance that causes the disturbance, disease, or death of an organism. Classification:

1. Chemical (according to composition):
   (*a*) Inorganic (alkalies, acids, metals, volatile nonmetals)
   (*b*) Organic (acids, glucosides, alkaloids, volatile substances, bacteria, animals)
2. Physiological (according to effect):
   (*a*) General: metabolic poisons, irritants, and corrosives (as, cyanides, insecticides)
   (*b*) Narcotic: depressor of brain centers (morphine, alcohol, sedatives)
   (*c*) Convulsant: stimulator of brain centers (strychnine, cocaine, atropine, caffeine)
   (*d*) Cardiac (digoxin, emetine)
   (*e*) Renal or kidney (mercury)
   (*f*) Liver (carbon tetrachloride)
   (*g*) Nerve: paralyzer (some snake venoms, lead)
   (*h*) Bone marrow (cytotoxic drugs)

(2) A substance that impairs the quality of a metal or alloy. (3) A substance that destroys or diminishes the action of a catalyst or enzyme. **acrid ~** Irritant. **arrow ~** See *arrow poison.* q.v. **bacterial ~** P. produced by bacteria; as, toxins. **corrosive ~** A substance that destroys tissue locally; as, acids. **cumulative ~** A p. retained and gradually accumulating in the body. **fish ~** A substance used to stupefy fish. **narcotic ~** A p. that produces stupor or unconsciousness.

p. ash Chionanthus. **p. effect** See Table 63 and *toxic.* **p. gas** See Table 63. **p. ivy** *Rhus toxicodendron.* **p. nut** Nux vomica. **p. oak** (1) The fresh leaflets of *Rhus toxicodendron* (Anacardiaceae). U.S. Atlantic states. Cf. *poison ivy* (above). (2) *Rhus diversiloba,* U.S. Pacific coast. Both produce highly irritant skin inflammations. **p. vapor** See Table 63. **p. vine** *Rhus toxicodendron.*

poisoning The diseased condition produced by a poison. **acute ~** A morbid condition caused by a single large dose of poison. **chronic ~** A morbid condition caused by many small doses of poison. **garage ~** A morbid condition due to inhalation of carbon monoxide from an internal-combustion engine.

poisonous dose The amount that produces marked pathological conditions.

Poisson P. distribution The probability distribution of *n* random events occurring in a given time period, $P(n) = (\mu^n - e^{-\mu})/n!$, where μ is the mean number of events in that time. Followed by many phenomena, e.g., photomultiplier. **P. ratio** The ratio of lateral strain to longitudinal strain for a bar under stress parallel to its length.

poivrette Ground olive stones; an adulterant for pepper.

TABLE 63. FATAL PERCENTAGES OF POISON VAPORS

Poison vapor	Fatal percent
Acrylaldehyde	0.001[a]
Ammonia	0.3
Arsine	0.05[a]
Bromine	0.1[a]
Carbon dioxide	30.0[a]
Carbon disulfide	0.001
Carbon monoxide	0.5[a]
Carbon tetrachloride	0.03[a]
Chlorine	0.10[a]
Chloroform	0.03[a]
Chloropicrin	0.05[a]
Dinitrogen monoxide	0.07[a]
Hydrochloric acid	0.5[a]
Hydrocyanic acid	0.048[a]
Hydrogen sulfide	0.06[a]
Phosgene	0.02[a]
Phosphine	0.2[a]
Phosphorus trichloride	0.00035
Sulfur dioxide	0.2[a]
Sulfur trioxide	0.001[a]

[a]Percentage in air fatal in 30 min. Others dangerous after 30 min.

poke Phytolacca. **Indian ~** Veratrum. **p.berries** The fresh fruits of *Phytolacca decandra;* an emetic, narcotic, and indicator. **p.root** Phytolacca.

Polaises reaction See *Grignard's reaction.*

Polan Trademark for a polyamide synthetic fiber.

polar Pertaining to a pole. **p. bond** The electrostatic union of 2 atoms established by the passage of one or more electrons from one to the other. **non ~** The union of 2 atoms established by sharing one or more pairs of electrons. **p. compound** An electrolyte: a compound that can dissociate when dissolved or fused; as, inorganic acids, bases, and salts. The atoms are held in electrostatic union. **nonpolar compound** A nonelectrolyte: an organic compound the atoms of which are held in union by sharing a pair of electrons. **p. formula** See *polar formula* under *formula.* **p. zone** See *polar zone* under *zone.*

polarimeter Polariscope. A device for measuring the rotation of polarized light. Between 2 nicol prisms is placed a column of polarizing liquid. By rotating the prism nearest to the eye until the intensity of the light passing through the liquid equals the intensity of the comparison field, the new plane of vibration of the light can be read from the scale in degrees. See *rotation.*

polarimetry Polariscopy. Measurement of the rotation of polarized light by a polarimeter.

polarisation Polarization.

polariscope Polarimeter.

polariscopy Polarimetry.

polariser Polarizer.

polarising Polarizing.

polarity The condition in which a body has 2 poles, or different properties at terminal points. **atomic ~** The loss or gain of one or more electrons by an atom. See *polar bond.* **chemical ~** The condition in which a molecule has an acid and a basic radical; as, an amino acid. **electrical ~** The condition of having a positive and negative terminal, pole, or electrode in an electrical device. **magnetic ~** The condition of having north or south pole in a magnet. **molecular ~** The distribution of electric charges in a molecule. See *polarization (3).*

p. formula A formula which shows the distribution of electric charges in the molecule. Cf. *formula, molecular diagrams.* **p. paper** A paper used to distinguish the positive and negative poles of a direct electric current. E.g., a filter paper saturated with sodium chloride and phenolphthalein is colored red at the positive terminal because of the liberation of hydroxyl ions.

polarization (1) The ability of certain substances to polarize light passing through them. (2) The stoppage or reversal of the voltaic current from an electrolytic cell due to the accumulation of dissociation products at the electrodes. See *overvoltage.* Cf. *p. potential.* (3) The orientation of a molecule in an electric field: e.g., the positive nucleus toward the negative pole, the electron cloud toward the positive pole. P. $= 3R/4N$, where R is molecular refraction and N the Avogadro constant. P. $= el/I$, where e is the charge, l the distance, and I the intensity of the field. Cf. molar *refraction.* **cell ~** The accumulation of hydrogen bubbles on the negative electrode of a battery. **circular ~** Polarized light in circular vibrations. **elliptic ~** Polarized light in elliptical vibrations. **plane ~** Polarized light, the vibrations of which are parallel and in one plane.

p. curve (1) The current-voltage curve obtained when the intensity of an electrolytic current is plotted against the polarizing electromotive force. (2) Polarogram. **p. potential** The reverse potential of an electrolytic cell, tending to oppose the direct potential effecting electrolytic decomposition in the cell; due to p. (2) Cf. *overvoltage.*

polarize To produce polarization.

polarized light A composite of 2 types of rays whose vibrations are in 2 directions at right angles: the ordinary and extraordinary rays. The ordinary ray is "lost" in the polarimeter by total reflection, and the extraordinary ray emerges as light polarized in one plane only. Optically active substances rotate this plane. Cf. *radiation, Verdet's constant.*

polarizer A device for polarizing light; as, a nicol prism through which light vibrating only in one plane passes. The p. acts as a filter by transmitting only parallel vibrations, thus ||||| , which, on passing through an optically active substance, are rotated either ///// or \\\\\ . Cf. *analyzer.*

polarizing Causing polarization. **p. angle** See *Brewster's law.* **p. disk** Polaroid. A cellulose film containing oriented iodoquinine sulfate crystals, mounted between 2 glass plates; a substitute for nicol prisms in polarimeters and used in microscopes, reading glasses, and windshields to prevent glare.

polar number Oxidation number*.

polarogram Polarization curve. The current-voltage curve produced by a polarograph. It indicates the presence of substances that are oxidizable or reducible in the voltage range scanned.

polarograph An instrument that records photographically, minute changes in the intensity of a current resulting from a gradually increasing applied voltage, in electrolysis with a dropping mercury cathode. Used to measure deposition or reduction of cations and anions, overvoltage, ionic complexity and equilibria, solubility; and for qualitative and quantitative microanalysis. Cf. *Heyrovsky, tastpolarograph.*

polaroid A *polarizing disk,* q.v. **p. vectograph** Aerial photography in which a p. is used to obtain stereoscopic effects.

polaron An electron produced in an aqueous medium by radiation. It is associated with one or more water molecules, and it polarizes the surrounding medium.

Polathene Trademark for a polythene synthetic fiber.

poldine methylsulfate $C_{22}H_{29}O_7NS$ = 451.5. White, bitter crystals, m.139, soluble in water; an antispasmodic (USP, BP).

pole(s) (1) The points at the opposite ends of an axis; 2 points that have opposite physical properties. (2) Rod. **negative ~** The cathode: an electric terminal charged with electrons. **(magnetic) north ~** That point toward which a freely suspended magnetic needle will point. **positive ~** The anode: an electric terminal that becomes positively charged by loss of electrons. **unit ~** A basis of cgs system magnetic units. A u. p. repels another u. p. with a force of 1 dyne when 1 cm from it in vacuo. 1 u. p. $= 1.257 \times 10^{-7}$ Wb.

Polenské number The number of mL of 0.1 *N* alkali (less the blank) needed to neutralize an alcoholic solution of the water-insoluble volatile fatty acids liberated on acidification of the soap made by saponification of 5 g of a fat. Cf. *Reichert number.*

polianite Pyrolusite.

policeman A device to remove precipitates from the walls of glass vessels in quantitative analysis. **platinum ~** A platinum-iridium alloy claw that fits over a glass rod, to hold a quantitative filter during ignition. **rubber ~** A small piece of rubber tubing fitting snugly over the end of a glass rod.

poliomyelitis vaccine A mixture of 3 types of live attenuated virus (USP, EP, BP).

polishing Rubbing or smoothing metal or glass surfaces. **electrolytic ~** Smoothing a metal surface by making it the anode in a suitable electrolyte (as, phosphoric acid and glycerol). The high current density produced on the small projecting portions results in their preferential dissolution.

polishing slate Gray or yellow slate used for polishing.

Pollack's cement A stiff paste of equal weights of red lead and litharge, in gelatin, for jointing metal, glass, or both: slow-setting but strong.

pollantin An antitoxin obtained by inoculating horses with pollen extract.

pollen The male sex cells or fertilizing grains of a flowering plant, which contain glucosides; e.g.: ragweed p., quercitin; timothy p. *(Phleum pratense),* dactylin. Cf. *stamen.* **p. extract** A solution of proteins from the pollens of plants believed to cause hay fever; used for immunization. **fall ~** An extract of the proteins of ragweed, goldenrod, and maize. **spring ~** An extract of the proteins of rye, timothy, orchard grass, redtop grass, and sweet vernal grass; used for immunizing to hay fever.

Pollopas Trademark for a glasslike, transparent plastic; a glass substitute.

pollucite $Cs_2O \cdot Al_2O_3 \cdot 5SiO_2 \cdot H_2O$. Pollux. A rare, native silicate, in pegmatite (island of Elba).

pollution (1) Contamination. (2) The introduction of a deleterious substance into the air or a water supply. **air ~** See *air pollution* under *air.* **water ~** See *water pollution* under *water.*

pollux Pollucite.

polonium Po. Radiotellurium, dvitellurium, RaF. An element, at. no. 84. At. wt. of isotope with longest half-life (^{209}Po) = 208.98. m.250, b.962. Formed by disintegration of radium; occurs in uranium minerals, e.g., pitchblende. The first radioactive decomposition product discovered by Madame Curie (1898).

poly- Prefix (Greek) meaning "many." (Latin: multi-.)

polyacetylenes Unbranched carbon chain compounds,

sometimes part of a ring structure, C_{10} to C_{13} predominating; formed by plants (hydrocarbons) and microorganisms (alcohols and acids).

polyacid Acid derived by condensation of molecules of the same monoacid.

polyacrylamide $(-CH_2 \cdot CH \cdot CONH_2-)_n$. White solid; used as a flocculant.

polyacrylonitrile* PAN*, poly(1-cyanoethylene). The polymer $(-CHCN \cdot CH_2-)_n$.

polyactivation Activation, e.g., of fluorescence, by more than one substance.

polyad An element or radical with a valency greater than 2, e.g., triad.

polyadelphite A brown to green manganese garnet.

polyalkane A hydrocarbon polymer of long-chain molecules containing only saturated atoms in the main chain. See *alkane.*

polyallomers Plastics having a highly crystalline and stereoregular structure, and consisting of chains of polymerized crystalline segments of each of the constituent alkene monomers. Specially suitable for blow-molding and extrusion processes; e.g., propylene-ethylene p.

polyamide A polymer, usually of a carboxylic acid (e.g., adipic acid) and its aminated derivative, in which the structural units are linked by amide or thioamide groupings; many have fiber-forming properties. Cf. *nylon.*

polyargyrite A native silver-antimony sulfide.

polyatomic Describing a molecule with 3 or more atoms.

polybasic acid A compound that yields 2 or more H ions per molecule in aqueous solution; as, H_2SO_4.

polybasite Ag_9SbS_6. A native sulfide, containing copper.

polybutenes Polybutylenes. Polymers of butene, ranging from viscous to rubbery substances.

polycaprolactam Nylon 6. A polyamide synthetic fiber.

polycarbonates Thermoplastic linear polyesters of carbonic acid, made by the polymeric condensation of bisphenols with a phosgene or its derivatives. Used for injection molding, especially where clarity is important.

Polycarpeae A group of families of Phanerograms (Nymphaeaceae, Ranunculaceae, Magnoliaceae, Myristicaceae, Menispermaceae, Berberidaceae, Lauraceae).

polychloral Hydronal.

polychlorinated biphenyls PCB. Compounds whose toxic effect is compounded by their resistance to biodegradation; as, DDT.

poly(chlorotrifluoroethylene)* PCTFE*. The polymer $(-CFCl \cdot CF_2-)_n$.

Polychol Trademark for surfactant condensation products of poly(oxyethylene) and wood alcohols.

polychromatic Showing more than one color, particularly if viewed by polarized light. **pseudo ~** Pseudodichro(mat)ic. Showing more than one color when viewed by polarized light, but otherwise colorless.

polychromatophil(e) A cell or tissue that can be stained differentially with dyes.

polycondensation A form of polymerization in which recurring structural units are formed from simpler molecules by elimination of a simple substance, e.g., water.

polycrase Euxenite.

polycrystalline Describing a polymer having simple crystal lamellas of folded chain molecules surrounded by noncrystalline polymers.

polycyclic* A molecule containing 3 or more rings; as, anthracene. Cf. *ring* system.

polydeoxyribonucleotide synthetase (NAD^+)* DNA ligase. An enzyme that joins the ends of two different, or the same, DNA chain(s).

polydymite $(Ni, Co)_4S_5$. A native sulfide.

polyelectrolyte A polymer producing large chain-type ions in solutions, that can carry positive or negative groups along the polymer chain; used as industrial flocculants.

polyene A compound containing many double bonds; as, the carotenoids. **p. grouping** A system of double bonds associated with color reactions; as, in carotene.

polyester A polymer having structural units linked by ester groupings; obtained by condensation of carboxylic acids with polyhydric alcohols.

polyether A polymer containing the $-(CH_2-CHR-O-)_n$ linkage in the main chain or side chain.

polyethylene* (U.S. usage.) $(-CH_2 \cdot CH_2-)_n$. PE*, poly(methylene)*, polythene (U.K. usage). A member of a series of straight-chain hydrocarbons of high molecular weight (18,000–20,000), made by polymerizing ethylene at very high pressures, e.g., 200 MPa, under controlled conditions, m.110–115. Polyethylenes are thermoplastic and can be extruded or molded by injection or compression. **high-density ~** HDPE. P. with specific gravity above 0.941; used for injection and blow molding. **linear low-density ~** LLDPE. A more linear structure (than LDPE), giving it better chemical resistance and strength. **low-density ~** LDPE. P. with specific gravity 0.910–0.925; the film is used for packaging.

p. glycol Poly(ethylene oxide)*, PEO*, poly(oxyethylene), Macrogol. A polyglycol derived from ethylene glycol. Used for conservation of wood and leather objects long immersed in seawater. **p. g.-400** $H(OCH_2CH_2)_nOH$. A condensation product of ethylene oxide and water, where n is 8–10. Colorless, hygroscopic liquid, miscible with water; used in ointments (NF). **p. g.-4000** Similar to p.g.-400, where n is 70–85. A wax, m.54; used in ointments (NF). **p. oxide** See *polyethylene glycol.* **poly(ethylene terephthalate)*** PETP*, PET. A polymer, as used in *Terylene,* q.v.

polygalacturonase* Pectinase. An enzyme that hydrolyzes 1,4-δ-D-galactosiduronic linkages in pectate and other galacturonans.

polygalic acid $C_{27}H_{42}O_2(COOH)_2 = 488.7$. Polygalin. An active principle from *Polygala senega.* Cf. *senega.*

polygalin Polygalic acid.

polygamarin A crystalline, bitter principle from *Polygala amara* (Polygalaceae).

polygarskite Attapulgite (U.S. usage). A hydrated, aluminum-magnesium silicate from Attapulgus, Decatur, Ga., and Ukraine; a drilling mud, fungicide base, and filler.

polygen An element that forms 2 or more series of compounds; as, chlorine (chlorides, chlorites, and chlorates).

polygenetic Producing more than one phenomenon. **p. dye** A coloring material that gives different shades with different mordants. Cf. *monogenetic.*

polyglycol A dihydroxy ether formed by dehydration of 2 or more glycol molecules, e.g., diethylene glycol.

polygon A plane figure bounded by 3 or more sides.

Polygonaceae The buckwheat family of herbs or woody plants; e.g.: *Rheum* species, rhubarb; *Rumex crispus,* rumex; *Polygonum bistoria,* bistort.

polygonin $C_{21}H_{20}O_{10} = 432.4$. A glucoside from *Polygonatum cuspidatum* (Liliaceae), Japan.

polygraph A device to record arterial and venous pulse waves simultaneously; used as a lie detector.

polyhalite $K_2SO_4 \cdot MgSO_4 \cdot 2CaSO_4 \cdot 2H_2O$. A native hydrated sulfate.

polyhydrate A compound containing more than 2 molecules of water.

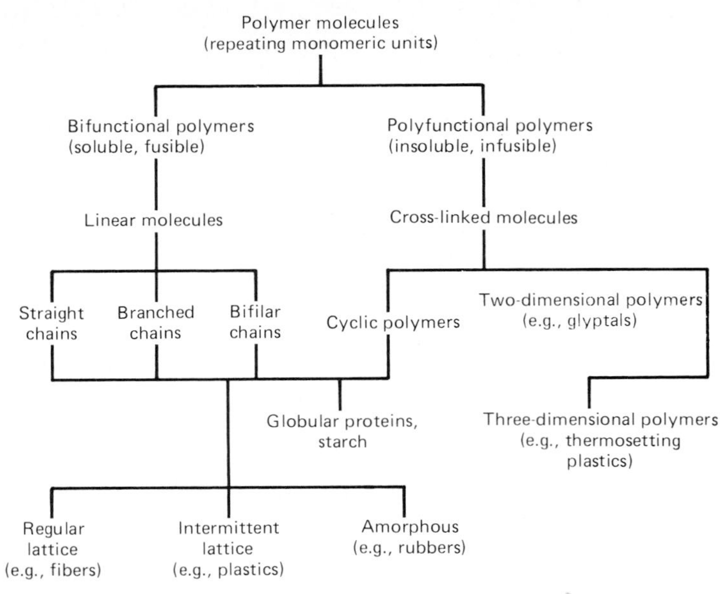

Fig. 18. Classification of polymer molecules.

polyhydric Polyol. A compound containing more than 2 hydroxyl groups.

polyhydrone $(H_2O)_n$. A polymer of *hydrone,* q.v.

polyisobutylene* PIB*. Poly(1,1-dimethylethylene). The polymer $(-CMe_2 \cdot CH_2-)_n$.

polymer Polymere, polymeride (obsolete). A member of a series of polymeric compounds. A substance composed of very large molecules consisting essentially of recurring long-chain structural units that distinguish polymers from other types of organic molecules, and confer on them tensile strength, deformability, elasticity, and hardness. Monomers, largely derived from coal and oil, are used to build up such polymers. Considerable modification of properties results on introducing a second type of monomer (B) into the main structure (monomer A), producing a *copolymer,* in which the units A and B are arranged completely at random. Alternatively, the A and B units may be arranged in order of long segments, e.g., ~A—A—A—A—B—B—B—B—B—A—A—A—A~(*block* p.). There are also *branched* polymers, in which the B units branch from the A units; and *cross-linked* polymers, in which 2 A chains are joined by one or a block of B units. Polymeric molecules are classified above in Fig. 18 (after Pinner). Examples of *high* polymers are plastics, fibers, elastomers, human tissue. Cf. *macromolecular chemistry.*

alloy ~ A p. produced by the simultaneous polymerization of 2 substances. Cf. *silicone alloy.* **blocked ~** See above. **branched-chain ~** See above. **co ~** A composite p. prepared by the polymerization of a mixture of 2 or more monomers, or of a monomer and p. of low molecular weight. Cf. alloy *polymer.* **block ~** A p. built of linearly linked polymeric units. **random ~** A p. having 2 or more types of units combined in random succession in a linear-chain structure. **cross-linked ~** See above. **electron-exchange ~** Redox p. A polymeric structure having several sites capable of accepting or donating electrons. Thus, modified cellulose with redox properties is used as a catalyst to remove oxygen from water to obtain anaerobic conditions. **graft ~** A p. produced by grafting a monomer onto a straight-chain p. to produce a branched-chain p. Thus, a fluorocarbon p. is heated sufficiently to form free radicals on its surface and then dipped into a monomer, e.g., styrene, to produce a graft p. having a printable surface. **high ~** A p. of high molecular weight, e.g., containing a large number of structural units. **high-trans ~** A rubbery p. in which a large proportion of the C atoms are arranged in a definite pattern that repeats itself consistently in the chain; as, natural rubber. **homo ~** See *tactic polymer* below. **inorganic ~** Inorganic p. structures formed on heating or by catalytic action; as, mica, silicones, inorganic rubber. **irregular ~** A p. with more than one type of repeating unit. **isotactic ~** A crystalline p. made from 1-alkenes, in which the substituents in the asymmetric C atoms all have the same configuration relative to the main chain. **linear ~** A p. in which the molecules are essentially in the form of long chains. **organized ~** A p. having a regular macroscopic structure, without necessarily showing microcrystallinity. Cf. *polyallomers.* **orientated ~** A p. film that has been stretched mechanically in 2 directions at right angles to improve its strength properties. **redox ~** Electron-exchange p. **regular ~** Tactic p. **super ~** A p. in which the polymerized molecules have an average molecular weight exceeding 10,000. **tactic ~** A p. with only one type of repeating unit. See *tacticity.*

P.R. Trade name for a polyamide synthetic fiber.

polymeric Related molecularly to an isomeric compound, but having a multiple of its molecular weight; as, acetylene and benzene. See *polymerism.* **p. dialdehyde** See *starch dialdehyde.*

polymericular weight The molecular weight of a polymer.

polymeride Polymer.

polymerisation Polymerization.

polymerism The property of certain organic compounds which have the same percentage composition, but different molecular weights, the heavier being multiples of the lighter.

Thus, C_2H_2, C_4H_4, C_6H_6, C_8H_8 are polymeric compounds. See *polymerization.*

polymerization Describing a reaction in which 2 or more molecules of the same substance combine to form a compound, from which the original substance may or may not be regenerated. Cf. *molecular association, hydrone.* **aromatic ~** The formation of an aromatic compound from two or more molecules of an aliphatic compound; as, benzene from acetylene. **carbohydrate ~** The formation of monosaccharides from formaldehyde: $6HCHO = C_6H_{12}O_6$. See *photosynthesis.* **co ~** The structural arrangement, e.g., of rubber, in which 2 or more different monomers or types of group are present in alternate sequence in a chain. **condensed ~** P. in which atomic displacement occurs. See *aldol condensation.* **degree of ~** (1) The number of times a structural unit occurs in the molecule of a polymer. (2). D.P. A measure of the chain length and molecular weight of cellulose derivatives; determined from the viscosity of the cellulose in cuprammonium solution; e.g.: cellulose acetate 150–250, regenerated cellulose 100–250, sulfite wood pulp 230–310, ramie pulp 1,000, cotton 750. **group-transfer ~** A process that permits the sequential addition of monomers during the creation of polymers. **photo ~** See *photopolymerization.* **true ~** P. in which the atoms remain in similar relative positions; as, hexaphenylethane from triphenyl methyl.

polymers Compounds having the same percentage composition, but containing different numbers of the same atoms.

polymeter (1) A device to measure 2 or more different physical properties simultaneously. (2) A hygrometer, thermometer, and barometer mounted together.

polymethylene See *cycloparaffin, polyethylene.* **p. glycols** A group of polyglycols derived from methylene glycol, $CH_2(OH)_2$, or from its anhydride (formaldehyde); as, dimethylene glycol. **p. tetrasulfide** Thiokol.

poly(methyl methacrylate)* PMMA*. The polymer $(-CMe(COOMe)CH_2-)_n$. Used in dentistry.

polymignite A native lime-niobium oxide containing numerous metallic oxides.

polymyxin B An antibacterial polypeptide from *Bacillus polymyxa,* having a specificity for gram-negative bacteria. It contains threonine and a branched C_9 fatty acid. Used to treat infection by gram-negative organisms (USP, EP, BP).

polymorph A substance that occurs in 2 or more different forms.

polymorphism Ability to crystallize in 2 or more different systems. See *dimorphism, isomorphism.*

polynosic Describing: (1) A manufactured cellulosic fiber with a fine and stable microfibrillar structure that is resistant to the action of 8% sodium hydroxide solution down to 0°; minimum wet strength 2.2 g/denier, wet elongation less than 3.5% at a stress of 0.5 g/denier (U.S. usage). (2) A regenerated cellulose fiber characterized by a high initial wet modulus of elasticity and a relatively low degree of swelling in sodium hydroxide solution (U.K. usage). (3) A fiber or filament having high tenacity and modulus of elasticity in the wet state (French usage).

polynucleated Polycyclic.

polynucleotide See *nucleotide.*

polynuclidic* Describing an element that exists in several isotopic forms.

polyol See *polyhydric.*

polyoxy- Prefix indicating more than 3 oxygen atoms.

poly(oxyethylene) See *polyethylene glycols.*

poly(oxymethylene)* $(-CH_2O-)_n$. POM*. A condensation product of formaldehyde. Cf. *trioxane.*

polyoxyl-40-stearate Polyoxyethylene stearate. The monostearate of the condensation product, $H(OCH_2 \cdot CH_2)_n \cdot O \cdot CO \cdot C_{16}H_{32}Me$, where n is 40. Waxy solid, m.40, soluble in water; used in ointments (USP).

polypeptide See *peptide.*

polyphase Having more than one phase; as, an alternating electric current.

polyphosphides M_2P_n. Compounds of phosphorus and univalent metals; as, K_2P_n.

Polyporaceae A genus of 2,000 species of spore-forming fungi; as, the edible *Boletus.*

polypropylene* $(-CH(Me) \cdot CH_2-)_n$. PP*. Poly(propylene)*, propathene. A polymer derived from propene with the aid of an organometallic catalyst. Similar in chemical and electrical properties to polyethylene, but lighter, d.0.90, and stronger, m. above 150; it retains its shape in boiling water. **p. tetramer** Has detergent action.

polypyknotic Consisting of 2 independent components of different densities.

polyquinoyl Describing an organic compound containing 4 or more =CO groups; e.g., diquinoyl, $C_2H_2(CO)_4$. Cf. *quinolyl.*

polyric oil A mixture of polymerized α-methylstyrene and castor oil in proportions having a refractive index 1.515 at 20°C; less viscous than cedarwood oil, and used to substitute it for oil-immersion lenses.

polysaccharides Polysaccharoses. Carbohydrates containing more than 10 molecules of simple sugars; as, $(C_6H_{10}O_5)_n$, polyhexoses, and $(C_5H_8O_4)_n$, polypentoses. They are theoretically derived from mono-, di-, or trisaccharides by abstraction of water. E.g., starch, dextrin, glycogen, and inulin, which hydrolyze to monosaccharides in steps (staircase reaction). See Table 64, below, and *sugar, carbohydrates.*

polysilicate See *silicate.*

polysorbate P. 80. Sorethytan. A complex mixture of polyoxyethylene ethers of mixed partial oleic esters of sorbitol anhydrides. Yellow, bitter oil, d.1.08, soluble in water; used in ointments (NF, EP, BP).

TABLE 64. WATER-SOLUBLE POLYSACCHARIDES

Source	Polysaccharide
Natural gums	
Plants	Gum arabic, gum karaya, gum tragacanth
Seeds	Locust bean gum, guar gum, psyllium gum, quince gum
Corn, wheat, maize	Cereal starches
Sorghum, potato, arrowroot, tapioca	Tuber starches
Citrus fruits	Pectin
Seaweeds	Agar, alginate, carrageenan, furcellaran
Microorganisms	Dextran, xantham gum
Chemically modified natural gums	
Plants	Carboxymethyl cellulose
Grasses	Methyl cellulose
Cotton	Hydroxymethyl cellulose
Cereals	Dextrins
Tubers	Carboxymethyl starch, hydroxypropyl starch
Citrus fruits	Low methoxy pectin
Seaweeds	Propylene glycol alginate
Microorganisms	Diethylaminoethyldextran

polyspiro* Of or relating to compounds consisting of a linear assembly of 3 or more spiro-united cyclic systems.
polystichoflavin $C_{24}H_{30}O_{11}$ = 494.2. A constituent of the rhizome of *Aspidium spinulosum*.
polystyrene* $(-CH(Ph)CH_2-)_n$. PS*. Poly(1-phenylethylene). Polymerized *styrene*, q.v., derived from petroleum. White solid; a basis for synthetic rubber of good insulating properties. P. resins have base-exchange properties.
polysulfide* (1) A binary sulfur compound containing more sulfur than is required by the normal valency of the metal; Na_2S is the normal sulfide, and Na_2S_2, Na_2S_3, Na_2S_4 and Na_2S_5 the polysulfides of sodium. (2) In organic compounds, a straight chain of 2 or more S atoms; as, diphenyl trisulfide, $Ph \cdot S_3 \cdot Ph$. **p. ion** A ion that contains 2 or more S atoms; as, AsS_2^-.
polyterpenes $(C_{10}H_{16})_n$. Compounds containing 2 or more terpene molecules.
poly(tetrafluoroethylene)* $(-CF_2 \cdot CF_2-)_n$. PTFE*. Teflon (U.S.). Fluon (U.K.). A plastic in which the hydrogen of polyethylene is replaced by fluorine; insoluble in all mineral or organic solvents, heat-resisting (up to 350); has a low power factor and relative permittivity. Used for corrosion-resistant gaskets, and on cooking vessels, nonstick rolls, and dryers.
polythene U.K. usage for polyethylene*.
polythionate* $M_2S_nO_6$. A salt of a polythionic acid in which n is 3–6 . . . (some only) . . . 20.
polythionic acids* $H_2S_nO_6$. Thionic acids. Sulfur acids in which n is 3–6 . . . (some only) . . . 20; as, trithionic etc.
polytrophic Describing a bacterium or microorganism that produces 2 or more different types of fermentation, or derives its nourishment from more than 1 organic substance.
polytropy A form of *polymorphism*, q.v.
polyurethanes A group of synthetic polymers formed by reacting hydroxyl and isocyanate groups to give the urethane group $-NH \cdot CO \cdot O-$; used to produce a wide range of fibers, lacquers, adhesives, rigid materials, and foams.
polyuronide A substance that yields one or more uronic acids on hydrolysis.
polyvalent Multivalent. Having more than one valency. **p. vaccine** A suspension of 2 or more species of the same microorganism in a liquid.
polyvinyl Describing a compound containing a number of vinyl, $-CH{:}CH_2$, groups in a polymerized form. **poly(vinyl acetate)*** $[-CH_2 \cdot CH(OOCMe)-]_n$. PVAC*. A transparent, thermoplastic solid, insoluble in water or mineral oils, soluble in most organic solvents; a heat-sealable adhesive, binder, plastic, and size. **poly(vinyl alcohol)*** $[-CH_2 \cdot CH(OH)-]_n$. PVAL*. Cream-colored powder. Made by the hydrolysis of vinyl acetate, and containing 1–22% acetyl groups. Soluble in water, insoluble in most organic solvents; an adhesive, emulsifier, and sizing agent (USP). **p. chain** The structure $-(CH_2 \cdot CHX)_n-$. It is isotactic, syndiotactic, or atactic, according to whether the configuration of the CHX group is identical, regularly alternating, or random. **poly(vinyl chloride)*** $[-CH_2 \cdot CH(Cl)-]_n$. PVC*. First suggested for the manufacture of synthetic fibers in 1913. A finely divided powder made by heating vinyl chloride under pressure with azodiisopropylnitrile. P. chloride fibers and film are resistant to chemicals, microorganisms, water, and ignition, but are decomp. by heat. **poly(vinyl fluoride)*** PVF*. The fluorine analog of p. chloride. **poly(vinylidene dichloride)*** $(-CHF_2 \cdot CH_2-)_n$. PVDC*. *Saran*, q.v. The copolymer of polyvinyl chloride and vinylidene chloride. **poly(vinylidene difluoride)*** PVDF*. The fluorine analog of poly(vinylidene dichloride).
Polyzime Trademark for a malt extract containing the enzyme constituents in an active state.
POM* Symbol for poly(oxy methylene).
pomace The pressed residue after the extraction of apple juice in cider manufacture; a cattle food.
pomade (1) A perfumed ointment, especially for the hair. (2) See *enfleurage*.
pomegranate Granatum. The dried bark of the stems and roots of *Punica granatum* (Punicaceae), containing pelletierine alkaloids; an anthelmintic and teniafuge. Cf. *coccon, granatine*. **p. tannin** Ellagitannic acid.
Pompey red Ferric oxide.
Ponderax Trademark for fenfluramine.
Ponder's stain A solution of 0.02g toluidine blue in glacial acetic acid 1, absolute alcohol 2, and water 97 mL.
ponite Rhodochrosite containing iron.
pontanin $C_{23}H_{18}O_5$ = 374.4. A glycosidal coloring matter of galls, produced by *Pontiania proxima*. Orange needles, m.284.
pontianac A resin of the copal type, used in paint. Cf. *jelutong*.
Pontocaine Trademark for tetracaine hydrochloride.
pontol A mixture of secondary and tertiary alcohols; a denaturant for ethanol.
poonac Solvent-extracted coconut meal.
poonahlite A hydrated calcium-aluminum silicate (Poonah, India).
poor gas A group of combustible gases from the blast furnace (Siemens-Martin furnace); also Mond gas and regenerator gas.
Pope, Sir William Jackson (1870–1939) English chemist noted for his work on crystallography and organic chemistry.
poplar buds Balm of Gilead buds. The air-dried, winter-leaf buds of *Populus nigra* (Salicaceae); an aromatic **p. b. oil** An essential oil from p. b., d.0.900, b.255–265, soluble in alcohol, and containing humulene, sesquiterpenes, and an alkane.
poplox A sodium silicate. See *intumescence*.
poppy Maw seed. The plant *Papaver somniferum* (Papaveraceae). **California ~** Eschscholtzia. **red ~** Corn p., corn rose. The flower petals of *Papaver rhoeas*; an anodyne expectorant. **p. capsules** Papaveris fructus. The fruit of *Papaver somniferum* (Papaveraceae). **p. seed oil** Yellow oil from *Papaver* species, m.5, insoluble in water; lubricant for fine machinery.
populoid The combined principles from the bark of *Populus tremuloides* (American aspen) or *P. tremula* (European aspen); an antipyretic.
porcelain A white, semiopaque, dense, waterproof substance obtained by strongly heating and sintering a mixture of kaolin, feldspar, and quartz; m. 850–1400. Used for domestic and laboratory utensils, and in dentistry (bonding to gold). Cf. *sillimanite*. **high-fusing ~** m.1340. **low-fusing ~** m.930. **medium-fusing ~** m.1260. (All melting points are approximate.) **p. clay** Kaolin. **p. color** A pigment used to color p.; as a metal oxide. **p. glaze** See *porcelain glaze* under *glaze*. **p. jasper** Porcellanite. **p. mill** A mill made of p., for wet or dry grinding. **p. utensils** See *casserole, crucible, combustion boat, evaporation dish, funnel, mortar, retort, spatula*.
porcellanite Porcelain jasper. A sintered clay and shale, on the borders of burned coal seams.
pore A minute opening on a surface.
poromer A permeable synthetic material resembling leather; used for footwear.
porosimeter An instrument to determine porosity from the volume of liquid absorbed, or air transmitted, in a given time.
porosity The state of a solid body penetrated by minute open spaces filled with liquid or gas. P. is expressed as the percentage of open space in the total volume. Cf. *permeability*.

apparent ~ The volume of open-pore space per unit total volume. **true ~** The volume of both open and sealed pore spaces per unit total volume.

porous Penetrated by small open spaces.

porpezite Palladium gold.

porphin(e) $C_{20}H_{14}N_4$ = 310.4. Purple solid, m.360+. A compound of 4 pyrrole rings united by methylene groups, in the center of which may be a metal (Fe or Mg). It is a structural part of chlorophyll and hemoglobin. **aza ~** See *tetraarylazadipyrromethines.*

p. derivatives

Chlorophyll group:

$C_{31}H_{34}N_4Mg$, etiophyllin
$C_{31}H_{34}N_4O_2$, pyrroporphyrin
$C_{32}H_{34}N_4O_4$, rhodoporphyrin
$C_{32}H_{36}N_4O_2$, phylloporphyrin
$C_{33}H_{34}N_4O_3$, phylloerythrin
$C_{55}H_{72}N_4O_6Mg$, chlorophyll *a*
$C_{55}H_{70}N_4O_6Mg$, chlorophyll *b*

Hemoglobin group:

$C_{30}H_{30}N_4O_4$, deuteroporphyrin
$C_{30}H_{28}N_4O_4FeCl$, deuterohemin
$C_{34}H_{32}N_4O_4FeCl$, hemin
$C_{34}H_{32}N_4O_4FeOH$, hematin
$C_{34}H_{34}N_4O_4$, protoporphyrin
$C_{34}H_{36}N_4O_4FeCl$, mesohemin
$C_{34}H_{38}N_4O_6$, hematoporphyrin
$C_{34}H_{38}N_4O_4$, mesoporphyrin
$C_{36}H_{38}N_4O_8$, coproporphyrin
$C_{40}H_{38}N_4O_{16}$, uroporphyrin
$C_{38}H_{38}N_4O_{16}Cu$, turacin

Bile pigments:

$C_{33}H_{36}N_4O_6$, bilirubin (protoporphyrin in which the alpha $-CH=$ group is replaced by $-OH$ and $O=$).

p. ring The structure:

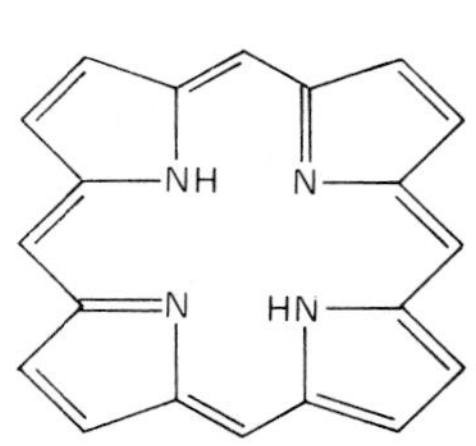

porphyric acid Euxanthone.

porphyrilic acid $C_{16}H_{10}O_7$ = 314.3. A sparingly soluble constituent of crustaceous lichens, m.274–278 (decomp.).

porphyrins A class of natural compounds incorporating the *porphin ring,* q.v., or its derivatives.

porphyrine $C_{21}H_{25}O_2N_3$ = 351.4. An alkaloid from the bark of *Alstonia constricta* (Apocynaceae). Colorless substance, m.97, soluble in acids (blue fluorescence).

porphyrite A coarse-grained igneous rock.

porphyrization Pulverization.

porphyropsin The pigment of the eyes of freshwater fish.

porphyry An igneous rock with large, red crystals set in a finer-grained or glassy, dark-red mass.

porpoise oil A yellow fatty oil from porpoises.

porporino A decorative imitation of gold; an alloy of mercury, tin, and sulfur.

Porter, Sir George (1920–) British chemist. Nobel prize winner (1967). Noted for work on reaction mechanisms.

portland cement A hydraulic cement, obtained by burning a mixture of limestone or chalk with clay and pulverizing the clinker. A greenish-gray powder of basic calcium silicates, calcium aluminates, and calcium ferrites. When mixed with water, it solidifies to an artificial rock, similar to portland stone. See *cement.*

portland stone Yellowish limestone from the Isle of Portland, Dorset, England.

Portsmouth accelerator POTG. Phenyl-*o*-tolylguanidine; an accelerator for the vulcanization of rubber.

posion Anion. Cf. *negion.*

positive (1) Opposite of negative. (2) Greater than zero. **p. crystal** A crystal whose refractive index for the extraordinary ray is greater than that for the ordinary ray. **p. electron** Positron. **p. element** An element of the upper left of the periodic table, which can yield valence electrons to another element and become positively charged. **p. group** A positively charged and base-forming group of atoms; as, NH_4^+. **p. nucleus** See *atomic nucleus.* **p. ore** An ore that has been exposed and properly worked on four sides. Cf. *possible ore.* **p. radical** An atom or group of atoms that has lost one or more electrons and so become positively charged. **p.-ray analysis** See *mass spectrograph* under *mass.* **p. rays** A stream of positively charged molecules shooting from the anode to the cathode of a discharge tube. If the cathode is perforated, the rays are made visible on a fluorescent screen behind it. See *canal rays.* **p. reaction** A reaction that produces the effect sought.

positon Positron.

positron Positon, oreston, positive electron. The antiparticle of the electron, of mass and charge equal to it, but the charge being of opposite sign. It does not decay, but is annihilated upon collision with an electron; magnetic moment + 1 magneton; spin $\frac{1}{2}$. Cf. *particle, nucleus.*

positronium A temporary electron-positron system. An unstable substance, made in the laboratory by eliminating the atomic nucleus and retaining the bonding electrons. Breaking length 16.7 Mkm. **ortho ~** A p. in which the spins of the electron and positron are parallel; mean life 1.4×10^{-7} s. **para ~** A p. in which the spins of the electron and positron are antiparallel; mean life 10^{-10} s. See *spin.*

posode Suggested name for anode. Cf. *negode.*

posologic Pertaining to medicinal doses.

posology The study of the form, quantity, and frequency of administering medicines.

possible ore An ore that may exist out of sight, or below the lowest workings of a mine. Cf. *positive ore.*

postulate An assumption not capable of proof: (1) An indisputable prerequisite; (2) a stipulated condition; (3) a demand. Cf. *axiom.*

pot Cannabis.

potable Drinkable without injury to health. Cf. *brackish,* potable *water.*

potarite A native compound of palladium and mercury. Cf. allo*palladium.*

potash (1) Potassium hydroxide*. (2) Potassium carbonate*. (3) An early name (pot ash) for wood ash used as a source of potassium carbonate. **black ~** A commerical grade of caustic soda: 5% NaOH with iron oxide and sodium carbonate impurities. **caustic ~** Potassium hydroxide*. **sulfurated ~** A mixture of potassium thiosulfate and potassium polysulfides containing not less than 12.8% sulfur as sulfide. Brown lumps, changing to green-yellow, with odor of hydrogen sulfide and acrid taste, soluble in alcohol; an astringent lotion used for acne (USP).

p. alum Kalinite. **p. bulb** Any of variously shaped glass bulbs, filled with potassium hydroxide solution; for the absorption of carbon dioxide in chemical analysis. **p. feldspar** Orthoclase, **p. glass** See *potash glass* under *glass*. **p. mica** Muscovite. **p. value** Incorrect synonym for saponification value or acid value. **p. water** Potassic water. **p. water glass** See *water glass* under *glass*.

potassa Potassium hydroxide*. **p. sulfurata** Potassium sulfide*.

potassamide NH_2K = 55.2. White (if pure) or brown (if impure), flammable solid, m.330, sublimes 400.

potassic Containing potassium. **p. water** Potash water. An aerated mineral water in which the potassium ions exceed the sodium ions.

potassii Official Latin for "of potash."

potassium K = 39.0983. Kalium. An alkali-metal element, at. no. 19. Silver-white, soft substance, d.0.87, m.62, b.760, which reacts violently with water, and rapidly oxidizes in air; stored under kerosine. It occurs abundantly in its salts, as sylvenite, kainite, and carnallite in Stassfurt (Germany) and Alsace and in smaller quantities in feldspars, rocks, and soils. First prepared (electrolytically) in 1807 by Davy. Its salts are all soluble in water and yield the colorless p. ion, K^+; valency 1. Used as a reagent and reducing agent in many chemical reactions, and in the metallic state emits weak β radiation. **p. acetate*** CH_3COOK = 98.1. Potassii acetas, diuretic salt. White, deliquescent powder, m.292, soluble in water; used as a reagent and in buffer solutions; also used to treat hypokalemia and to reduce acidity of urine (USP, BP). **acid ~** P. hydrogenacetate*. **p. acetate tungstate** A double salt of p. acetate and tungstate; a photographic toner. **p. acetylsalicylate*** $KC_9H_7O_4 \cdot 2H_2O$ = 254.3. White crystals, m.65. **p. acid sulphate** P. hydrogensulfate*. **p. alloys** Liquid or semisolid mixtures of potassium with sodium:

K, %	Na, %	m.
90	10	17
80	20	−10
70	30	−3
60	40	5
50	50	11

p. aluminate $K_2Al_2O_4 \cdot 3H_2O$ = 250.2. Colorless crystals, soluble in water. **p. aminochromate** $CrO_2(OK) \cdot NH_2$ = 155.1. Red crystals. **p. anthranilate*** $C_6H_4(NH_2)COOK$ = 175.2. White crystals, soluble in water. **p. antimonate*** $KSbO_3$ = 208.8. P. stibnate. White crystals, soluble in water. **p. pyroantimonate*** $K_2H_2Sb_2O_7 \cdot 6H_2O$ = 543.8. P. stibnate. White granules, slightly soluble in water. A reagent for sodium salts. **p. argentum cyanide** P. dicyanoargentate(I)*. **p. arsenate*** K_3AsO_4 = 256.2. Colorless crystals, soluble in water; a reagent for tannin and opium alkaloids. **acid ~** KH_2AsO_4 = 180.0. Colorless crystals, m.228, soluble in water. **p. arsenite** $KAsO_2$ = 146.0. Gray powder, soluble in water; a reagent for Ceylon cinnamon oil, and reducing agent in the manufacture of mirrors. **p. aurate** $KAuO_2 \cdot 3H_2O$ = 322.1. Soluble, yellow needles. **p. aurichloride** P. tetrachloroaurate(III)*. **p. auricyanide** P. tetracyanoaurate(III)*. **p. aurocyanide** P. hexacyanoaurate(II)*. **p. benzoate*** $C_6H_5 \cdot COOK \cdot 3H_2O$ = 214.3. White crystals, soluble in water; a food preservative. **p. benzenedisulfonate*** $C_6H_4(SO_3K)_2$ = 314.4. Colorless crystals, soluble in water; used in organic synthesis. **p. biborate** P. tetraborate*. **p. bicarbonate** P. hydrogencarbonate*. **p. bioxalate** P. hydrogenoxalate*. **p. bisaccharate** P. hydrogensaccharate. **p. bisulfate** P. hydrogensulfate*. **p. bisulfite** P. hydrogensulfite*. **p. bitartrate** P. hydrogentartrate*. **p. black** Suint ash. **p. borates** $K_2B_2O_4$, p. metaborate; $K_2B_4O_7$, p. tetraborate. **p. borofluoride** P. tetrafluoroborate*. **p. borohydride** P. tetrahydroborate*. **p. borate tartrate** A mixture of p. metaborate and p. bitartrate. White crystals, soluble in water; an antiseptic and photographic-developer retardant. **p. bromate*** $KBrO_3$ = 167.0. Colorless rhombohedra, m.434, soluble in water; a reagent (Koppeschaar solution) for titrating phenol and oxalic acid, and standard in iodometry. **p. bromide*** KBr = 119.0. Potassii bromidum. White, regular crystals, m.740, soluble in water; a reagent and photographic chemical (EP, BP). **p. butanoate*** C_3H_7COOK = 126.2. White, deliquescent crystals, soluble in water. **p. carbonate*** K_2CO_3 = 138.2. Potassii carbonas, sal tartar, potash. Colorless, monoclinic crystals or hygroscopic powder, d.2.2, m.900 (decomp.), soluble in water, insoluble in alcohol; a reagent, flux for silicates and insoluble sulfates, and neutralizing agent. **acid ~** See *potassium hydrogencarbonate*. **p. carbonyl** P. hexacarbonyl*. **p. chlorate*** $KClO_3$ = 122.5. Potassii chloras, p. oxymuriate. Colorless, monoclinic crystals, m.357, decomp. 400, soluble in water, slightly soluble in alcohol. Used as a reagent for alkaloids, phenols, indican; and in mouthwashes, and the manufacture of explosives, fulminators, and pyrotechnics. **p. chloride*** KCl = 74.6. Potassii chloridum. White, regular crystals, m.776 (sublimes), soluble in water. Used as a reagent (determination of hexafluorosilicic acid), in fertilizers and explosives, and in the treatment of hypokalemia (USP, EP, BP). **p. chloride chromate** $KClCrO_3$ = 174.7. Red crystals, decomp, by water; an oxidizing agent. **p. chromate*** K_2CrO_4 = 194.2. Yellow rhombohedra, m.971, soluble in water; a reagent and volumetric indicator, mordant, and oxidizing agent. **acid ~** P. dichromate. **p. chlorochromate** P. chloride chromate. **p. chromicyanide** P. hexacyanochromate(III)*. **P. cinnamate*** $KC_9H_7O_2$ = 186.3. White crystals, soluble in water. **p. citrate*** $K_3C_6H_5O_7 \cdot H_2O$ = 324.4. Potassii citras. Colorless crystals, decomp. 230, soluble in water; a diuretic and treatment used in reducing acidity of urine (USP, EP, BP). **dihydrogen ~, monobasic ~** $KH_2(C_6H_5O_7)$ = 230.2. White crystals, soluble in water. **p. cobalticyanide** P. hexacyanocobaltate(III)*. **p. cobaltinitrite** P. hexanitrocobaltate(III)*. **p. cyanate*** KOCN = 81.1. Colorless needles, soluble in water. **p. cyanide*** KCN = 65.1. White, regular crystals or sticks, soluble in water. Used as a reagent, for gold extraction *(cyanide process)*, and in photography. **p. dichloroaurate(I)*** $KAuCl_2$ = 307.0. P. chloraurite. **p. dichromate*** $K_2Cr_2O_7$ = 294.2. P. bichromate, p. acid chromate. Yellowish-red; triclinic or monoclinic crystals, d.2.692, m.396 (decomp.), soluble in water, insoluble in alcohol. Used as a reagent, general oxidizing agent, bactericide and caustic; in cleansing solutions with sulfuric acid, in electric batteries and photography, and in the dye, textile, tanning, printing, bleaching, and oil industries. **p. dicyanoargentate(I)*** $KAg(CN)_2$ = 199.0. P. argentocyanide. A constituent of solutions of silver cyanide in potassium cyanide solution; used for silver plating. **p. dicyanoaurate(I)*** $KAu(CN)_6$ = 288.1. P. aurous cyanide. White crystals, soluble in water; used in electroplating. **p. dicyanocuprate(I)*** $KCu(CN)_2$ = 154.7. A solution of copper cyanide in p. cyanide solution; used for copper plating. **p. dihydrogenphosphate** KH_2PO_4. See *potassium phosphates*. **p. disulfate*** $K_2S_2O_7$ = 254.3. White needles, m.210 (decomp.), soluble in water. **p. disulfite*** $K_2S_2O_5$ = 222.3. White plates, decomp. by heat, soluble in water. **p. dithionate*** $K_2S_2O_6$ = 238.3. P. hyposulfate. Colorless crystals, decomp. by heat, soluble in water. **p. dithiooxalate** $K_2C_2O_2S_2$ = 198.3. A solid reagent for nickel. **p. ethyl dithiocarbonate*** SC(SK)OEt= 160.3. P. xanthate. Yellow prisms, decomp. 200, soluble in water, decomp. by acids to alcohol and carbon disulfide; a flotation agent, and reagent for separating cobalt and nickel. **p. ferrate** K_2FeO_4 = 198.0. P. perferrate. Purple solid, rapidly decomp. in acid solution. **p. ferric ferrocyanide** Iron(III) potassium hexacyanoferrate(II)*. **p. ferricyanide** P. hexacyanoferrate(III)*. **p. ferrite** $K_2Fe_2O_4$ = 253.9. Yellow solid. **p. ferrocyanide** P. hexacyanoferrate(II)*. **p. fluorescein** $K_2C_{20}H_{10}O_5$ = 408.5. Yellow-red powder, soluble in water (pink solution, green fluorescence). **p. fluoride*** $KF \cdot 2H_2O$ = 94.1. Colorless, hygroscopic crystals, m.860, soluble in water. Used in etching glass. **acid ~** P. hydrogenfluoride*. **p. fluosilicate** P. hexafluorosilicate*. **p. formate*** H·COOK = 84.1. Colorless, deliquescent crystals, m.197, soluble in water. **p. glycerolphosphate** $K_2C_3H_7O_3PO_3$ = 264.4. White, hygroscopic crystals, soluble in water, marketed as a 50–75% solution; a reagent. **p. hexacarbonyl*** $K_6(CO)_6$ = 402.6. Gray or red, explosive crystals. **p. hexachloropalladate(IV)*** K_2PdCl_6 = 397.3. P. palladichloride, palladiopotassium chloride. Red, soluble solid, decomp. by heat. **p. hexachloroplatinate(IV)*** K_2PtCl_6 = 486.0. Platinic p. chloride. Yellow, regular crystals, decomp. by heat, slightly soluble in water; a reagent. **p. hexacyanochromate(III)*** $K_3[Cr(CN)_6]$ = 325.4. P. chromicyanide. Yellow solid, insoluble in water. **p. hexacyanocobaltate(III)*** $K_3[Co(CN)_6]$ = 332.3. P. cobalticyanide. A solid, decomp. by heat, soluble in water. **p. hexacyanoferrate(II)*** $K_4[Fe(CN)_6] \cdot 3H_2O$ = 422.4. P. ferrocyanide, yellow prussiate of potash, ferrous p. cyanide. Yellow, monoclinic crystals, soluble in water. Used as a reagent and as an anticaking agent (for deicing salt), for hardening of steel, in electroplating, photography, the textile industry, and in the manufacture of prussian blue. **p. hexacyanoferrate(III)*** $K_3[Fe(CN)_6]$ = 329.2. P. ferricyanide, red prussiate of potash, ferric p. cyanide. Red, monoclinic crystals, decomp. by heat, soluble in water or alcohol. Used as a reagent for ferrous salts, a reducing agent, an indicator and in light-sensitive papers, the textile industry, and the manufacture of prussian blue. **p. hexafluorosilicate*** K_2SiF_6 = 220.3. Colorless hexagons, insoluble in water. **p. hexanitrocobaltate(III)*** $K_3[Co(NO_2)_6]$ = 452.3. P. cobaltinitrite, cobaltic p. nitrite. Yellow tetragons, decomp. 200, soluble in water; a reagent and pigment. Cf. *aureolin yellow.* **p. hippurate*** $KC_9H_8O_3N \cdot H_2O$ = 235.3. Colorless crystals, soluble in water. **p. hydrate** P. hydroxide*. **p. hydride*** KH = 40.11. White, volatile needles, from p. heated in hydrogen. **p. hydrogenacetate*** $KH(C_2H_3O_2)_2$ = 158.2. Acid p. acetate. Colorless needles, m.148. **p. hydrogencarbonate*** $KHCO_3$ = 100.1. P. bicarbonate, acid p. carbonate, potassii bicarbonas, saleratus (*sal* aeratus). Colorless, transparent, monoclinic crystals, decomp. by heat, soluble in water. Used as a reagent in titrating arsenous and antimonous oxides; in the manufacture of potassium salts and inorganic synthesis; in foam-type fire extinguishers; in treatment of hypokalemia; and as an antacid (USP). **p. hydrogenfluoride*** KHF_2 = 78.1. P. bifluoride, acid p. fluoride, Fremy's salt. Colorless crystals, soluble in water; used for etching glass. **p. hydrogeniodate*** KHI_2O_2 = 325.9. P. biiodate, acid p. iodate. Colorless crystals, soluble in water; used in volumetric analysis. **p. hydrogenoxalate*** KHC_2O_2 = 96.1. P. bi(n)oxalate, salt of sorrel, salt of lemons, salt acetosella. White crystals, soluble in water. Used as a reagent, in stain removers, and in photography. **p. hydrogenphosphate** See *potassium phosphates.* **p. hydrogensaccharate** $(CHOH)_4(COOH)COOK$ = 248.2. Yellow crystals, soluble in water. **p. hydrogensulfate** $KHSO_4$ = 136.2. P. bisulfate, potassii bisulfas, p. acid sulfate, sal enixum. Colorless crystals, m.197 (decomp.), soluble in water; used as a reagent. **p. hydrogensulfide*** KSH = 72.2. P. hydrosulfide. Colorless, hygroscopic crystals, m.455, soluble in water. **p. hydrogensulfite** $KHSO_3$ = 120.2. P. bisulfite, p. acid sulfite. White crystals, decomp. 190, soluble in water. **p. hydrogentartrate** $KHC_4H_4O_6$ = 188.2. P. bitartrate, potassii bitartras, cream of tartar, depurated tartar. Colorless crystals, soluble in water; a reagent, refrigerant, and diuretic (BP). **p. hydroxide*** KOH = 56.1. Potassii hydroxidum, potash, p. hydrate, caustic potash, kalilauge. White, rhombohedral crystals or deliquescent sticks, d.2.044, m.360 (sublimes), soluble in water, alcohol, or ether; absorbs carbon dioxide from air. Used as a reagent; for neutralization; and in the manufacture of soft soap (NF, BP), oxalic acid, and glass. Cf. *lapis causticus, vienna caustic.* **p. hypochlorite*** KClO = 90.6. Colorless crystals, decomp. by heat, soluble in water. Cf. *eau de javelle.* **p. hyposulfate** P. dithionate*. **p. hyposulfite** P. thiosulfate*. **p. indigosulfate** $K_2C_{16}H_8O_3N_2(SO_3)_2$ = 514.6. Blue powder, soluble in water; used in dyeing textiles. **p. indoxylsulfonate** Indican (2). **p. iodate*** KIO_3 = 214.0. Colorless, regular, deliquescent crystals, m.560 (decomp.), soluble in water, a volumetric reagent and antiseptic. **acid ~** P. hydrogeniodate*. **p. iodide*** KI = 166.0. Potassii iodidum. Colorless, regular crystals, m.680, soluble in water; a volumetric reagent, solvent for iodine and iodides, antiseptic, treatment for overactive thyroid gland (USP, BP), and photographic chemical. See *Lugol solution.* **p. triiodide** See *potassium triiodide.* **p. iodohydrargyrate** Mercuric p. iodide. **p. ion*** K^+. The positively charged p. atom. **p. manganate*** K_2MnO_4 = 197.1. Green rhombohedra, decomp. 190, soluble in water. **p. permanganate** See *potassium permanganate.* **p. metaphosphate*** KPO_3. See *potassium phosphates.* **p. metasilicate*** K_2SiO_3. See *potassium silicate.* **p. molybdate*** $K_2MoO_4 \cdot 5H_2O$ = 328.2. White microcrystals, soluble in water. **p. monosulfide*** P. sulfide*. **p. myronate** $KC_{10}H_{18}O_{10}NS_2$ = 415.5. Sinigrin. Colorless, crystalline, glucoside salt from black mustard seeds, decomp. by myrosin into glucose, p. hydrogensulfate, and allyl thiocyanate; soluble in water. **p. nitranilate** $K_2C_6O_8N_2$ = 306.3. P. dihydroxydinitrobenzoquinone. Yellow crystals, slightly soluble in water. **p. nitrate*** KNO_3 = 101.1. Saltpeter, potassii nitras, niter, sal prunella. Colorless prisms or rhombohedra, d.2.109, m.374, soluble in water, insoluble in alcohol or ether. Used as a reagent, oxidizing flux, diaphoretic, and diuretic (USP, BP); and in pyrotechnics, gunpowder, fertilizers, and preservatives. **p. nitrite*** KNO_2 = 85.1. Colorless, deliquescent prisms or sticks, d.1.915, m.440, soluble in water. Used as a reagent (e.g., for cobalt, amino acids, phenols, and iodine), reducing agent, and in the manufacture of dyes. Cf. *Indian yellow.* **p. oleate*** $KC_{18}H_{33}O_2$ = 320.6. Yellow mass, soluble in water; a soap. **p. osmate** $K_2OsO_4 \cdot 2H_2O$ = 368.4. P. perosmate. Purple crystals, decomp. in warm, moist air, soluble in water; a reagent for nitrogen. **p. oxalate*** $K_2C_2O_4 \cdot H_2O$ = 184.2. Colorless, monoclinic crystals, decomp. by heat, soluble in water; used as a reagent, and in photography and stain removers. **acid ~** KHC_2O_4 = 128.1. Monoclinic crystals, slightly soluble in water. **acid ~** (hemihydrate) $KHC_2O_4 \cdot \frac{1}{2}H_2O$ = 137.1. Trimetric crystals. **acid ~** (monohydrate) $KHC_2O_4 \cdot H_2O$ = 146.1. Rhombic crystals. **p. dioxalate** $KH_3(C_2O_4)_2 \cdot 2H_2O$ = 254.2. P. trihydrogendioxalate*.

Triclinic crystals, d.1.836. **p. oxide*** K_2O = 94.2. Burnt potash, calcined potash. Colorless octahedra, decomp. 350, soluble in water; used extensively as a reagent and in the manufacture of p. salts. Cf. *potassium peroxide.* **p. palladichloride** P. hexachloropalladate(IV)*. **p. palladochloride** P. tetrachloropalladate(II)*. **p. pentasulfide*** K_2S_5 = 238.5. Yellow granules, m.220, soluble in water. **p. perborate** P. peroxoborate*. **p. percarbonate** $K_2C_2O_6 \cdot H_2O$ = 216.2. Colorless powder, decomp. by water (to hydrogencarbonate and oxygen) and acids (to hydrogen peroxide and carbon dioxide). Used as a reagent, in microscopy, photography (antihypo), and bleaching. **p. perchlorate*** $KClO_4$ = 138.5. Colorless rhombs, m.610, slightly soluble in water. Used as a reagent, oxidizing agent, pyrotechnic, antithyroid agent, and source of oxygen. **p. periodate*** KIO_4 = 230.0. Colorless rhombs, m.582, slightly soluble in water; a volumetric reagent. **p. permanganate*** $KMnO_4$ = 158.0. Potassii permanganas, p. hypermanganate. Purple, rhombic needles, d.2.703, decomp. 240, soluble in water. Used as a volumetric and oxidizing agent, antiseptic and disinfectant (USP, EP, BP); for mordants; and in bleaching and photography. **p. peroxide** K_2O_2 = 110.2. Yellow powder, decomp. in air and moisture, soluble in alcohol; an oxidizing agent. **p. peroxoborate*** KBO_3 = 97.9. A soluble solid. **p. peroxosulfate*** $K_2S_2O_8$ = 270.3. Anthion. Colorless prisms, decomp. by heat, soluble in water. Used as a strong oxidizing agent and in photography (anthion). **p. phosphates** (1) K_3PO_4 = 212.3. P. orthophosphate*. Colorless rhombs, soluble in water. (2) K_2HPO_4 = 174.2. P. hydrogenphosphate, dipotassium phosphate, monoacid phosphate. Colorless crystals, soluble in water. (3) KH_2PO_4 = 136.1. P. dihydrogenphosphate*, monobasic p. phosphate, Sörensen's p. phosphate. Colorless tetragons, d.2.338, m.96, decomp. by heat, soluble in water; a reagent and pH buffer. (4) $K_2P_2O_7 \cdot 3H_2O$ = 306.2. P. pyrophosphate*. Colorless crystals, soluble in water. (5) KPO_3 = 118.1. P. metaphosphate*. Colorless crystals, insoluble in water (NF). **p. phosphide*** K_2P_5 = 233.1. Unstable, yellow solid. **p. phosphinate*** KH_2PO_2 = 104.1. P. hypophosphite, p. hypophosphis, diacid p. phosphate. White, opaque plates or deliquescent powder, soluble in water. **p. phosphite** K_3PO_3. Normal phosphite. **p. phosphonate*** K_2HPO_3 = 158.2. White power, decomp. by heat. **p. picrate*** $C_6H_2(NO_2)_3OK$ = 267.2. P. trinitrophenate. Yellow crystals, soluble in water; used in explosives. **p. plumbate** $K_2PbO_3 \cdot 3H_2O$ = 387.4. Colorless crystals; soluble in water. **p. pyroantimonate*** $K_2H_2Sb_2O_7 \cdot 6H_2O$ = 543.8. P. stibnate. White granules, slightly soluble in water; a reagent for sodium salts. **p. pyrophosphate** See *potassium phosphates (4).* **p. salicylate*** $KC_7H_5O_3$ = 176.2. White crystals, soluble in water; an antirheumatic and antipyretic. **p. silicate** K_2SiO_3 = 154.3. P. metasilicate*. Potash-water glass. Glassy mass, soluble in water. **p. silicate solution** Potash-water glass. An aqueous syrup of p. silicate 30–38 Bé, miscible with water. Used as a cement, for fireproofing, and in bleaching and soap manufacture. **p. silver cyanide** P. dicyanoargentate(I)*. **p. sodium tartrate*** $KNaC_4H_4O_6$ = 210.2 $+4H_2O$ = rochelle salt; $+3H_2O$ = Seignette salt. Colorless crystals, soluble in water; a reagent (in Fehling's solution), depilatory, and purgative (NF). **p. sorbate** P. hexadienoate*. A water-soluble fungistatic for food, similar in action to sorbic acid (NF). **p. stannate*** K_2SnO_3 = 244.9. Colorless crystals, soluble in water; a mordant. **p. stannum sulfate** $K_2Sn(SO_4)_2$ = 389.0. P. stannous sulfate, Marignac salt. Colorless crystals, soluble in water; a reagent for mercury and bismuth salts. **p. stibnate** P. pyroantimonate*. **p. sulfate*** K_2SO_4 = 174.3. Potassii sulfas, normal p. sulfate, tarcanum, sal de duobus, tartarus vitriolatus. White rhombs or hexagons, d.2.663, m.1076 (sublimes), soluble in water. Used as a reagent, and in fertilizers and artificial mineral waters. **acid ~** P. hydrogensulfate*. **p. disulfate** See *p. disulfate.* **p. sulfide*** K_2S = 110.3. P. monosulfide*. Brown crystals, soluble in water. **p. sulfides** K_2S, K_2S_2, K_2S_3, K_2S_4, K_2S_5. Commercial "p. sulfide" contains some or all of these. **p. sulfite*** $K_2SO_3 \cdot 2H_2O$ = 194.3. White crystals, decomp. by heat, soluble in water; a reagent and mordant. **acid ~** P. hydrogensulfite*. **p. tartrate*** **D-~** $K_2C_4H_4O_6 \cdot \frac{1}{2}H_2O$ = 245.3. Tartarus, soluble tartar, sal vegetal, normal p. tartrate. Colorless, monoclinic crystals, d.1.9715, soluble in water; a reagent and laxative. **D-p. hydrogen ~** $KHC_4H_4O_6$ = 188.2. Rhombic crystals, d.1.956. **(±)-~** $K_2C_4H_4O_6$ = 226.3. Monoclinic crystals, d.1.984. **(±)-hydrogen-acid ~** Monoclinic crystals, d.1.954. **p. tellurite*** K_2TeO_3 = 253.8. Colorless powder, soluble in water; used in bacteriology culture media. **p. tetraborate*** $K_2B_4O_7 \cdot 5H_2O$ = 323.5. P. borate. Colorless hexagons, soluble in water. **p. tetrabromoplatinate(II)*** K_2PtBr_4 = 592.9. P. platinobromide An unstable solid. **p. tetrachloride iodide*** KCl_4I = 307.8. **p. tetrachloroaurate(III)*** $KAuCl_4 \cdot 2H_2O$ = 413.9. P. aurichloride. Yellow needles, soluble in water; a ceramic pigment. **p. tetrachloropalladate(II)*** K_2PdCl_4 = 326.4. P. palladochloride, palladous p. chloride. Brown prisms, soluble in water, decomp. by heat. **p. tetrachloroplatinate(II)*** K_2PtCl_4 = 415.1. P. chloroplatinite, p. platinous chloride. Red crystals, soluble in water. **p. tetracyanoaurate(III)*** $K[Au(CN)_4] \cdot H_2O$ = 358.2. P. auricyanide. Colorless solid, soluble in water. **p. tetracyanoplatinate(II)*** $K_2[Pt(CN)_4] \cdot 3H_2O$ = 431.5. P. platinocyanide. An unstable solid. **p. tetrafluoroborate*** KBF_4 = 125.9. P. borofluoride. White crystals, soluble in alcohol. **p. tetrahydroborate*** KBH_4 = 53.9. P. borohydride. White crystals, stable to heat, soluble in water; a strong reducing agent, and non-pyrophoric-specific hydrogenation catalyst. **p. tetraiodoaurate(III)*** $KAuI_4$ = 743.7. Auric p. iodide. Black crystals, decomp. by heat, soluble in water. **p. tetraoxalate** $(COOH)_3COOK \cdot 2H_2O$ = 254.2. Colorless prisms, soluble in water; a volumetric standard. **p. thiocyanate*** $KSCN$ = 97.2. P. sulfocyanide, p. rhodanate. Colorless, deliquescent prisms, m.162, soluble in water. Used as a reagent for ferric (red), copper (blue), and silver (white) salts. **p. thiosulfate*** $K_2S_2O_3$ = 190.3. P. hyposulfite. Colorless, hygroscopic crystals. **p. triiodide*** KI_3 = 419.8. Prismatic crystals, m.45 (decomp.), soluble in p. iodide solution. **p. tungstate*** $K_2WO_4 \cdot 5H_2O$ = 416.4. P. wolframate*. White, hygroscopic crystals, soluble in water; used for bronzing solutions. **p. uranyl nitrate** $2KNO_3 \cdot UO_2(NO_3)_2$ = 596.2. Yellow crystals, soluble in water. **p. uranyl sulfate*** $K_2SO_4 \cdot UO_2SO_4$ = 540.3. Yellow powder, soluble in water. **p. xanthate** P. ethyl dithiocarbonate*. **p. zincate** K_2ZnO_2 = 175.6. Colorless powder, soluble in alkalies. **p. zinc iodide** K_2ZnI_4 = 651.2. Colorless crystals, soluble in water; a reagent.

potato The tuber of *Solanum tuberosum.* **sweet ~** The tuber of *Ipoemoea batatus.*

p. culture A slice of p. used as culture medium for bacteria. **p. gum** Euphorbia gum. **p. spirit** Fusel oil. **p. starch** See *starch.*

potency (1) Mathematical: magnitude—the number of times a number is multiplied by itself; e.g., $10^3 = 1,000$; $10^{-2} = 0.01$. (2) Therapeutic: the activity of a drug. (3) Homeopathic: the degree of dilution of a remedy by a neutral medium, corresponding with the mathematical potency.

potential (1) Stored-up energy capable of performing work. (2) Voltage: low, below 301; medium, 301–651; high, above 651. **chemical ~** A measure of the tendency of a chemical reaction to take place. The increase in internal energy on adding an infinitesimally small quantity of substance to a system, the entropy and volume being constant. **electric ~** Electromotive force. **electrokinetic ~** Zeta-p. **gravitational ~** See *potential energy (1).* **half-wave ~** The p. of a standardized dropping mercury electrode at the point on the current-voltage curve where the current is 50% of its limiting value. It is a characteristic property of electroreducible substances, and is independent of their concentration. Cf. *polarograph.* **hydrogen ~** See *pH.* **ionic ~** See *ionic potential.* **magnetic ~** Magnetomotive force*. **oxidation-reduction-** See *rH.* **pseudo ~** Zeta-p. **sedimentation ~** The vertical electric field set up when suspended particles settle. **streaming ~** The p. set up between the ends of a capillary when an electrolyte is forced through it. **super ~** Overvoltage. **zeta- ~** ζ. Electrokinetic p.* The potential difference produced at a solid-liquid interface due to ions absorbed from the moving solution. It can be controlled by addition of suitable ions so as to produce or prevent flocculation (as in a fiber suspension); measured by electrophoresis.

p. alcohol The amount of alcohol that can be produced in practice by fermentation of a sugar solution of definite specific gravity. **p. difference** The difference in voltage between the electrodes of a battery, vacuum tube, or thermocouple. See *electrode potential.* **p. energy** (1) The energy of a body due to its position (as water in a high tank). (2) The heat capacity of a compound. **p. mediator** A substance used to accelerate equilibrium in measurements of oxidation-reduction potentials. Cf. *poised.*

potentiation The activity induced in one insecticide by another used with it.

potentiometer A low-resistance instrument for the accurate determination of small differences in electric potentials by the *Poggendorff compensation method,* q.v. Cf. *galvanometer, hydrogen-ion recorder.*

potentiometric titration Volumetric analysis in which the potential of an electrode immersed in the solution to be titrated is continually determined; a rapid change corresponds with the end of the reaction. Cf. *conductometric analysis.* **chrono ~** The analysis of a solution of several electrolytes having different reduction potentials, by means of a stepped potential-time curve.

potentiostat An electrical device to control the potential of an electrochemical system, e.g., in electrodeposition.

P.O.T.G. Portsmouth accelerator.

pothole A deep, natural cavity in the earth surface, often filled with a deposit of salts, as, sodium carbonate; it may be formed by the grinding action of pebbles in a stream.

potstone Talc.

potter's **p. clay** A pure, plastic clay, free from iron. Cf. *pipe clay.* **p. lead** Alquifou. **p. ore** Alquifou.

pottery Ware made from clay, molded while soft and moist, and hardened by heat; e.g.: (1) earthenware—relatively soft and fusible, and (*a*) unglazed, (*b*) glazed, (*c*) lustrous, or (*d*) enameled; (2) stoneware—hard, infusible, and containing more silica. See *porcleain.*

pounce Powdered chalk, charcoal, or (more usually) cuttlefish bones; formerly used to dry ink.

pound lb, #. A unit of weight in the fps system. The weight in vacuo of a platinum cylinder known as the imperial standard pound. Cf. *pfund.* **apothecaries' ~** = 12 oz = 96 drams = 288 scruples = 5,760 grains = 0.373242 kg. **avoirdupois ~** = 16 oz = 7,000 grains = 0.45359237 kg = 16 fl oz. **gee ~** Slug. **troy ~** (for gold and silver, etc.) = 5,760 grains = 12 oz.

p. per square inch psi. A unit of pressure. See *atmosphere.* **p.-mole** The number of pounds of a gas numerically equal to its molecular weight.

poundal Abbrev. pdl; the unit of force in the fps system. The force to accelerate 1 pound 1 foot per second each second = 0.138255 N.

pour The flow of a liquid under gravity. **p. point** (1) The lowest temperature of flow under standard conditions. (2) The temperature at which an alloy is cast.

povidone $(C_6H_9ON)_n$. 1-Vinyl-2-pyrrolidinone polymer. PVP. Kollidon. White crystals, soluble in water. Used in pharmacy as suspending and stabilizing agent, and in eye drops. Cf. *periston.* **p.-iodine** See *povidone-iodine* under *iodine.*

powder (1) An aggregation of loose, small, solid particles. (2) Discrete particles of dry material, the maximum dimension of which is less than 1,000 μm (British Standards Institution, 1958). (3) An explosive used in blasting or gunnery. **algaroth ~** Precipitated antimonous oxychloride. **baking ~** See *baking powder.* **bleaching ~** See *bleaching powder.* **dusting ~** See *dusting powders.* **effervescent ~** A mixture of salts that develops carbon dioxide in water. **flameless ~** An explosive that produces little or no muzzle flash. **Seidlitz ~** A mixture of rochelle salt, sodium hydrogencarbonate, and tartaric acid, used to make an effervescent saline water. **smokeless ~** An explosive producing little or no smoke

p. cutting The use of pyrophoric iron in a flame for cutting metals to increase temperature and eroding power.

powellite $CaMoO_4 \cdot CaWO_4$. Yellow tetragons.

powellizing Hardening wood by impregnation with a saccharin solution.

power (1) Potency. (2) The time rate of doing work, the unit of which is the *watt,* q.v. (3) See *diopter.* **combined heat and ~** CHP. The use of energy wasted during p. generation for providing residential heating, greenhouse, etc, needs. **wind ~** See *wind power.*

p. factor That proportion of the total electric power flowing in an a-c circuit which is actually delivered to the load. It measures the ratio of watts dissipated to volt-amperes used. If ϕ is the phase displacement, cos ϕ is the p. f. **p. spectrum** Plot showing the amount of total variation in a time series attributed to individual frequencies or other variables. **p. transmission** Methods: (1) mechanical, or direct; (2) hydraulic, by oil or water; (3) pneumatic or compressed air; (4) steam in pipes; (5) electricity, through wires. **p. units** See *watt, joule, horsepower.*

poyok oil A drying oil from a Nigerian tree, whose fatty acid contains licanic acid 41, eleostearic acid 31%; used in paints.

pozzolana (1) *Puzzolane,* q.v. (2) A substance mixed with lime mortar to increase its strength. **artificial ~** Burnt clay, granulated slag, certain clinkers, and burnt oil shale. **natural ~** Volcanic ash and celite.

pozzolanic action Chemical action which forms insoluble compounds in cement.

PP* Abbreviation for polypropylene.

PPD Symbol for (1) purified protein derivative; (2) piperidine pentamethylene dithiocarbonate; a rubber vulcanization accelerator.

ppm Abbreviation for parts per million. Preferred form is $/10^6$. Cf. *epm.*

ppt(n) Abbreviation for precipita(te)-tion.

Pr (1) Symbol for praseodymium. (2) Abbreviation for propyl; also: *n*-Pr for 1-propyl (normal propyl); *i*-Pr for 2-propyl (isopropyl). (3) (Italic.) Symbol for Prandtl number.

pragilbert The ratio of watts to intensity of magnetic current. Cf. *gilbert.*

pralidoxime chloride $C_7H_9ON_2Cl$ = 172.6. 2-Formyl-1-methylpyridinium chloride oxime. Protopam. White crystals,

m.220, soluble in water. An antidote to organophosphorus insecticides (USP).
Prandtl number *Pr.* Specific heat capacity at constant pressure × kinematic viscosity/thermal conductivity. Cf. *Reynolds number.*
prase (1) Greenish. (2) A gray-green chalcedony.
praseodymia The earth corresponding with the element praseodymium.
praseodymium* Pr = 140.9077. A rare-earth metal, at. no. 59. Green metal, d.6.48, m.930, slowly decomp. in water. Separated (1885) by Auer von Welsbach from the earth didymia, and occurs in cerite and rare-earth minerals. Principal valency 3. See *didymium.* **p. acetate*** $Pr(C_2H_3O_2)_3 \cdot 3H_2O$ = 372.1. Green needles, soluble in water. **p. chloride*** $PrCl_3$ = 247.3. Green needles, m.818; soluble in water. **p. oxalate*** $Pr_2(C_2O_4)_3 \cdot 10H_3O$ = 736.1. Green crystals, insoluble in water. **p. oxides** (1)Pr_2O_3 = 329.8. P. trioxide*. Yellow-green powder. (2) Pr_2O_4 = 345.8. P. tetraoxide*. Black powder. (3)Pr_2O_5 = 361.8. P. pentaoxide*. **p. phosphate*** $PrPO_4$ = 235.9. Green powder for coloring ceramics. **p. sulfate*** $Pr_2(SO_4)_3 \cdot 8H_2O$ = 714.1. Green crystals, soluble in water. **p. sulfide*** Pr_2S_3 = 378.0. Brown powder, decomp. by heat, insoluble in water.
praseolite A green alteration product of iolite.
prazosin hydrochloride $C_{19}H_{21}O_4N_5 \cdot HCl$ = 419.9. Hypovase, Minipress. White crystals, 264 (decomp.). Used for hypertension and heart failure (USP).
preboarding The setting of hosiery fibers by the combined actions of steam and pressure.
Precambrian See *geologic eras,* Table 38 on p. 259.
precancerous Describing a growth which can or may develop into a cancer.
precious Valuable, rare. **p. garnet** Almandite. **p. metals** The noble metals: gold, platinum, and silver. **p. opal** An opal exhibiting a play of delicate colors. **p. stone** A mineral used as a gem.
precipitable Describing that which can be precipitated.
precipitant A substance which, when added to a solution, causes the formation of an insoluble substance. **group ~** A reagent that will precipitate several related substances; as, hydrogen sulfide or ammonia. See *qualitative analysis* under *analysis.*
precipitate Abbrev. ppt (no period). (1) To cause a substance to be precipitated. (2) The deposit of an insoluble substance in a solution as a result of a chemical reaction after the addition of a precipitating reagent. Cf. *schwellenwert.* **banded ~** Periodic p. **black ~** Mercurous oxide. **group ~** The p. formed by a group precipitant consisting of substances of related properties. See *qualitative analysis* under *analysis.* **periodic ~** See *Liesegang rings.* **red ~** Red *mercuric* oxide. **white ~** Ammoniated *mercury.* **yellow ~** Yellow *mercuric* oxide.
precipitated Settled out; rendered insoluble. **p. bone** A by-product in the manufacture of glue from bones; chiefly calcium hydrogenphosphate. **p. chalk** Calcium carbonate produced by precipitation. **p. phosphate** Calcium hydrogenphosphate obtained from phosphate rock or processed bone. **p. vapor** The deposit of solid particles from gases or vapors on the walls of a container.
precipitation The process of producing a precipitate. **co ~** The simultaneous p. of more than one substance. Cf. group *precipitate.* **electrostatic ~** See *electrostatic precipitator.* **fractional ~** The separation of substances by precipitating them in increasing order of solubility.
precipitin An antibody formed in the serum of animals or humans that precipitates antigens, as, bacteria.
precipitinogen A substance which, on injection, causes the formation of precipitins.
precipitum The deposit formed by the action of precipitins.
precision The degree of mutual agreement between individual measurements, such as the *standard deviation,* q.v., as distinct from their accuracy. **p. instrument** An instrument capable of precise measurements.
precursor (1) A substance synthesized in the dark by an organism, and decomposed by light. Cf. *photoproduct.* (2) A substance that forms the raw material for the synthesis of protoplasm in the living animal body. (3) A substance that precedes the formation of another compound. Cf. *provitamin.*
predissociation A spectral phenomenon by which a molecule dissociates at a lower level than its dissociation energy.
prednisolone $C_{21}H_{28}O_5$ = 360.5. Sterane. Delta Cortef. White, bitter crystals, m.229 (decomp.), soluble in water; a cortisone substitute, as it has fewer side effects (USP, EP, BP). **p. sodium phosphate** $C_{21}H_{27}O_8Na_2P$ = 484.4. Bitter, white powder, soluble in water; a synthetic glucocorticoid. Used for its anti-inflammatory effect in blood diseases and allergic states (USP, BP).
prednisone $C_{21}H_{26}O_5$ = 358.4. White, bitter crystals, insoluble in water; used similarly to prednisolone (USP, EP, BP).
preform Material produced in a state ready for molding to a desired shape, e.g., plastic-impregnated wood pulp for molded panels.
Pregl, Fritz (1868–1930) Austrian chemist, noted for his development of quantitative microanalysis. Nobel prize winner (1923).
pregnancy test Detection of chorionic gonadotrophin in urine by antibody-antigen reaction, or by radioimmunoassay of blood serum.
pregnane $C_{21}H_{36}$ = 288.5. A tetracyclic hydrocarbon, androstane derivative. Parent compound of some natural steroids, including mammalian hormones. **p.diol** $C_{21}H_{36}O_2$ = 320.5. A sterol, m.233, from the urine of pregnant women. **p.dione** $C_{21}H_{32}O_2$ = 316.5. A ketone derivative of p. diol.
pregnanolone $C_{21}H_{34}O_2$ = 318.5. A metabolite of progesterone.
pregnene $C_{21}H_{34}$ = 286.2. $\Delta^{5:6}$-pregnane. An androstane derivative.
pregneninolone Ethisterone.
pregrattite A variety of muscovite from Tyrol.
prehnite $H_2Ca_2Al_2(SiO_4)_3$. A hydrous silicate.
prehnitene $C_{10}H_{14}$ = 134.2. Prenitol, 1,2,3,4-Tetramethylbenzene*. Colorless liquid, b.204, insoluble in water.
prehnitic acid $C_{10}H_6O_8$ = 254.2. 1,2,3,5-Benzenetetracarboxylic acid*. Colorless crystals, m.252. Cf. *mellophanic acid.*
prehnitilic acid 2,3,4-Trimethylbenzoic acid*.
preignition See *knock.*
preimpregnate (1) To impregnate prior to a subsequent stage in a process. (2) A substance used to hold the ingredients of a mix together, before resin impregnation and molding; e.g., polyester resins.
Prelog, Vladimir (1906–) Bosnian-born Swiss chemist. Nobel prize winner (1975), noted for work on the chirality of organic compounds.
premier alloy A heat-resisting alloy: Ni 61, Fe 25, Cr 11, Mn 3%.
prenyl The 3-methyl-2-butenyl* radical, $Me_2C{:}CH \cdot CH_2-$.
prep Abbreviation for "preparation."
preparation (1) A chemical process for the production of a

chemical compound. (2) A chemical compound. (3) The treatment of ores; as, dressing. (4) A pharmaceutical product. **p. dish** A glass dish with glass cover, for microscope slides or cover glasses. **p. jar** A large glass jar with glass cover, for museum specimens. **p. of salts** See *salts.*

prescription A written direction for compounding or administering a drug. **p. balance** A delicate scale used to weigh small quantities: less sensitive than an analytical balance.

preservative A substance that prevents decay and decomposition of organic liquids or foods, including sulfites, fluorides, benzoates, salicylates, borates, and formaldehyde. Use of p. in food is limited or prohibited in many countries. **histological ~** P. used for biological specimens; as, Zenker's solution. **wood ~** P. used for lumber; as, creosote.

press A device that applies pressure. **cork ~** A corrugated wheel rotating in a corrugated elliptical frame, for softening corks by pressure. **filter ~** See *filter press.* **plant ~** A piston in a perforated cylinder, to express juices from plants.

pressing paper A coarse filter paper for drying specimens.

presspahn Wallboard, made by hydraulic pressure, from wood pulp.

pressure *p**, P. Stress produced by the meeting of opposite forces; as, compression. Units of p. are: SI unit: 1 pascal = 1 $N/m^2 = 0.1450 \times 10^{-3}$ psi. cgs unit: 1 dyne/cm^2. 1 millibar (mb) = 1,000 dynes/cm^2. 1 mmHg at 0°C and standard gravity = 133.3 Pa. **atmospheric ~** See *atmosphere.* **critical ~** The p. required to condense a gas at the critical temperature. **disruptive ~** See *disruption.* **electrical ~** Electromotive force. **fugitive ~** The variations of atmospheric p. in the vicinity of explosions. **low ~** A p. below one atmosphere. **negative ~** Suction. A p. less than one atmosphere or the reference p. **normal ~** Standard p. **osmotic ~** See *osmosis.* **partial ~** The p. exerted by a single gaseous constituent of a gas mixture. See *Dalton's law.* **positive ~** A p. slightly higher than that of the reference p., which is sometimes atmospheric p. **radiation ~** See *radiation pressure.* **solution ~** The molecular force of a solid that tends to dissolve; a measure of the attractive forces (coordinate bonds) between the molecules of solute and solvent. **standard ~** 760 mmHg. See *s.t.p.* **total ~** The sum of the partial pressures of the constituents of a gas mixture. **ultrahigh ~** A p. exceeding 10 GPa. Produced by the force exerted on a tetrahedral crystal of pyrophyllite by 4 tapered anvils, one on each face. P. on the anvils is transmitted hydrostatically to a specimen container midway between 2 opposite edges of the tetrahedron.

p. blower A device to produce a current of air. **p. bottle** A heavy-walled glass container. **p. filter** (1) A porous porcelain cylinder through which the solution to be filtered is pumped. (2) A filter press. **p. gage** A device to measure the p. of gases or liquids; as, mechanical instruments (coiled tubes or diaphragms) or hydrostatic devices (mercury columns). Cf. *manometer.* **p. packing** See *Aerosol.* **p. tubing** Thick-walled rubber tubing used for vacuum pumps.

Prestone Trademark for ethylene glycol antifreeze.

Prezenta Trademark for a viscose cellulose synthetic fiber.

pribramite (1) A variety of sphalerite from Czechoslovakia. (2) A variety of göthite.

priceite $3CaO \cdot 4BO_3 \cdot 6H_2O$. A mineral from Death Valley, Calif.

prickly pear Opuntia, Indian fig. The fruit of a cactus species, *Opuntia;* a source of alcohol for motor fuel.

Priestley, Joseph (1733–1804) English-born American chemist, noted for his discovery of oxygen and gas experiments.

Prigogine, Ilya (1917–) Russian-born Belgian chemist. Nobel prize winner (1977), noted for work on thermodynamics.

prill A small globule or button of material, e.g., of metal formed in assays.

prilocaine hydrochloride $C_{13}H_{20}ON_2 \cdot HCl$ = 256.8. Citanest. White crystals, m.168, freely soluble in water. A local anesthetic (USP, BP).

primaquine phosphate $C_{15}H_{21}ON_3 \cdot 2H_3PO_4$ = 455.3. Bitter, orange crystals, m.203, soluble in water; an antimalarial (USP, BP).

primary The first, simplest form. **p. alcohol** An organic compound characterized by the $-CH_2OH$ group. **p. amine** An organic compound characterized by the $-CH_2NH_2$ group. **p. battery** A voltaic cell producing a potential as a result of chemical changes undergone by its constituents. **p. carbon atom** A C atom attached to one other C atom only. **p. color** See *primary color* under *color.* **p. current** The inducing current of an induction coil. **p. electricity** See *primary electricity* under *electricity.* **p. flash distillate** A very light petroleum oil fraction, which can be hydrolyzed to produce benzene, town gas, or both. **p. nucleus** An organic, cyclic compound with only H atoms attached to the ring. **p. oxide** A hypothetical unstable oxide formed during an oxidation reaction, and having the characteristics of a peroxide which loses oxygen. See *induced reaction.* **p. reaction** The principal or fastest reaction in a composite system of reactions. **p. valency** See principal *valency.*

primer (1) Detonator, percussion cap. (2) An explosive cartridge containing the detonator.

primeverose $C_{11}H_{20}O_{10}$ = 312.3. A disaccharide from the glucosides of cowslips and primrose.

primidone $C_{12}H_{14}O_2N_2$ = 218.3. Mysoline. Bitter, white crystals, m.281, slightly soluble in water or organic solvents; used for grand mal epilepsy (BP).

priming (1) Escape into the condenser of a liquid being distilled, due to splashing, bumping, etc. (2) P. sugar. **p. powder** Percussion powder. **p. sugar** Glucose or invert sugar added to beer to impart body or briskness as a result of afterfermentation. **p. tube** Primer.

primula Cowslip. The dried herb of *Primula officinalis* (Primulaceae).

primulaverin A glucoside from the root of primula. Cf. *primverin.*

primulin (1) A crystalline substance from the root of *Primula officinalis.* (2) A primrose-colored azo dye; the alkali salts of p. bases. **p. bases** A group of sulfur dyes obtained by heating *p*-toluidine and sulfur with amines.

primulite Volemitol. A crystalline sugar from the roots of *Primula officinalis.*

primverin A glucoside from the root of *Primula grandiflora.* Cf. *primulaverin.*

primycin $C_{55}H_{106}O_{17}N_3$ = 1081. An actinomycetic antibotic from the larvae of the wax moth *Galeria melonella.* White crystals, m.194 (decomp.), sparingly soluble in water.

Prince Rupert drops Pear-shaped droplets of glass, formed when molten glass falls into water. They shatter explosively into fine powder when the tips are broken.

Prince's metal A brass: Cu 75, Zn 25%.

principal The main or chief function; leading or first; as, p. valency. Cf. *principle.* **p. axis.** (1) The optical axis of a crystal. (2) The longest axis of a crystal. **p. group*** See *principal group* under *group.* **p. series** A type of spectrum that contains the strongest lines. **p. valency** That valency of an element for which it has the largest number of stable compounds.

principle (1) A theory or assumption; a fundamental concept, as, the p. of LeChâtelier. Cf. *principal.* (2) A substance on which the characteristic effect of a vegetable drug or mixture depends. **acid ~** The organic acids of a vegetable drug. **active ~** The substance on which the physiological effect of a drug depends. **basic ~** Alkaloid. **bitter ~** Amaroids. **neutral ~** A glucoside, salt, ester, or essential oil. **odorous ~** Essential oil. **proximate ~** Active p. **resinous ~** Resin. **sweet ~** Glucoside. **toxic ~** The poisonous constituent of a drug.
printer's ink See *printing ink* under *ink.*
printer's liquor A solution of ferrous acetate.
printing The mechanical reproduction of reading matter or designs. See *gravure, letterpress, lithography.*
Priscol, Priscoline Trademarks for tolazoline hydrochloride.
prism (1) A crystal or solid figure whose faces are parallelograms parallel to the axis, and whose ends are triangular or polygonal faces parallel and similar to one another. They may belong to the tetragonal, hexagonal, orthorhombic, monoclinic, or triclinic systems. (2) A triangular glass rod or hollow glass vessel, used to produce a spectrum. **hollow ~** A p.-shaped glass or quartz container for measuring the refractive index of or obtaining a more dispersed spectrum from highly refractive liquids. **nicol ~** Two similar triangular pieces of Iceland spar cemented together to form a prism. It splits a ray of light into 2 portions, polarized and reflected, and is used in polarimeters to obtain polarized light. **spectrum ~** A polished triangular glass rod used in spectroscopes.
prismatic Formed or shaped like a prism. **p. colors** Rainbow colors. The monochromatic tints produced by the unequal refraction of the constituent rays of light passing through a prism. See *spectrum.*
pristane $C_{19}H_{40}$ = 268.5. 2,6,10,14-Tetramethylpentadecane*. b.296. From shark liver and herring oil.
Privine Trademark for benzindole.
Pro* Symbol for proline.
proactinomycin An antibiotic substance from *Nocardia gardneri* (Actinomycetes).
proagglutinoid An agglutinoid that has greater attraction for the agglutinogen than the agglutinin.
probability A mathematical treatment of data to find the limits or error. Let M be the total number of cases in a series, m the number in one group, and n the number in another. Then $n + m = M$, and $(m/M)\cdot(n/M)$ is the proportion of the parts to the whole. The probability $= m/M + 2\sqrt{2mn/M^3}$; e.g.: if in 100 observed cases (M), there are 25 cases (m) of one type, and 75 cases (n) of the other, then in any other series of similar experiments there may be observed as many as 37 or as few as 13 cases of m. Correspondingly, there will be 63 to 87 cases of n.
probe electron ~ An electron beam less than 1 μm diameter. It excites the X-ray characteristics (photons) of the elements in the area on which it is focused, enabling them to be determined.
probenecid $C_{13}H_{19}O_4NS$ = 285.4. *p*-(Dipropylsulfamoyl)benzoic acid. Benamid. White crystals, m.199, insoluble in water; used to increase secretion of uric acid, as in gout, or to decrease excretion of penicillin (BP).
procainamide hydrochloride $C_{13}H_{21}ON_2\cdot HCl$ = 271.8. Pronestyl. White crystals, m.167, soluble in water; used for irregular heart rhythm (USP, BP).
procaine hydrochloride $NH_2\cdot C_6H_4\cdot COOCH_2\cdot CH_2\cdot NEt_2\cdot HCl$ = 272.8. 2-(Diethylamino)ethyl *p*-aminobenzoate. Ethocaine, Novocain, Syncaine. Colorless crystals, m.155, soluble in water; a local anesthetic (USP, EP, BP). **p. h. base** $C_{13}H_{20}N_2O_2$ = 236.3. Novocaine base. *N*-diethylaminoethyl *p*-aminobenzoate. White granules, m.61, insoluble in water. **p. h. nitrate** Colorless crystals, m.100, soluble in water. **p. h. penicillin** The monohydrate of the p. salt of benzyl penicillin. White powder, soluble in water. Administered intramuscularly to create a depot from which penicillin is released slowly into the blood (BP).
procellose $C_{18}H_{32}O_{16}$ = 504.5. Procellulose. Cellotriose. A trisaccharide from cellulose, m.208.
process A method used in the manufacture or treatment of substances. **type ~** The principal operations involved in a chemical plant. **unit ~** (1) A single chemical operation carried out on a nonlaboratory scale. (2) Unit operation. One of the successive individual operations necessary for the production of a definite compound; as, crushing, grinding, separating, leaching, dissolving, concentrating, evaporating, distilling, mixing.
p. tankage A fertilizer made from nitrogenous waste material by the action of acid or alkali and pressure. Cf. *nitrogenous tankage.*
prochiral* Describing an achiral molecule having at least one pair of features that can be distinguished only by reference to a chiral object or to a chiral reference frame.
prochirality* The quality exhibited by prochiral molecules.
prochloraz See *fungicides,* Table 37 on p. 250.
prochlorite A mineral containing penninite and chlorite.
prochlorperazine maleate $C_{20}H_{24}N_3ClS\cdot 2C_4H_4O_4$ = 606.1. Stemetil. A phenothiazine with piperazine side chain. Yellow crystals, m.201, insoluble in water; a tranquilizer and antiemetic (USP, BP). Also used as prochlorperazine methanesulfonate (BP).
Procion Trademark for a class of fast, soluble azo dyestuffs for cellulose fibers, of the dichlorotriazinyl type. Procions are fixed to the fibers by esterification of one or more OH groups of the cellulose molecule.
procyclidene hydrochloride $C_{19}H_{29}OH\cdot HCl$ = 323.9. Kemadrin. Bitter, white crystals, m.226, slightly soluble in water; used to treat Parkinson's disease (USP, BP).
Prodag Trademark for a semicolloidal suspension of graphite in water; a mold wash in metal casting and rubber curing. Cf. *Aquadag.*
prodegradant See *sensitized photodegradation* under *photodegradation.*
prodorite An acid-resisting concrete which contains pitch.
producer An apparatus producing fuel gas, by passing air over red-hot coke. **p. gas** A combustible mixture of nitrogen, carbon monoxide, and small quantities of carbon dioxide, hydrogen, and methane obtained by passing steam and air over coal at 1000°C.
product The substance manufactured. **reaction ~** Reactant. The compound formed by a reaction. **split ~** A decomposition product. **substitution ~** A *derivative,* q.v.
Profax Trademark for polypropylene.
profile paper Coordinate paper.
profilogram A diagram showing the variation of a property over a relatively flat surface; as the roughness of paper.
proflavine $C_{13}H_{11}N_3SO_4$ = 305.3. 3,6-Diaminoacridine. Brown crystals, soluble (with fluorescence) in water; an antiseptic.
progallin-P Propyl gallate.
Progesic Trademark for fenoprofen calcium.
progesterone $C_{21}H_{30}O_2$ = 314.5. **(±)-~** 4-Pregnene-3,20-dione. α-P. Gesterol, Gestone, Proluton. White crystals, m.130, insoluble in water. A hormone secreted by the corpus luteum and placenta; it produces progestational effects on

uterus and mammary gland and maintains the developing placenta. Used for menstrual abnormalities, endometrial carcinoma, and contraception (USP, EP, BP). **(±)-~** β-P. m.184.

progestogen Generic name for drugs used to maintain pregnancy, treat uterine bleeding, and effect contraception.

progression Series. A series of related increasing or decreasing numbers; e.g.: **arithmetic ~** 1, 2, 3, 4,. . . . **geometric ~** 1, 2, 4, 8,. . . . **harmonic ~** ½, ⅓, ¼,. . . .

proguanil hydrochloride $C_{11}H_{16}N_5Cl \cdot HCl$ = 290.2. Paludrine. Bitter, white crystals, m.245, soluble in water; an antimalarial, used for prophylaxis (BP).

Progynon Trademark for estrone.

proidonite SiF_4. A mineral.

projection (1) Convention showing the spatial arrangement of a three-dimensional object on paper. (2)* For molecules, a broken line represents a bond projecting behind the plane of the paper; a thick line, one projecting out of the paper; a normal line, one in the plane of the paper; and a wavy line, one that cannot (or is not to) be specified. **Fischer ~** A p. with the main chain vertical and the principal functional group at the top. Atoms or groups, attached to a tetrahedral center, that lie behind the central atom's plane appear vertically above and below it, while those in front of its plane appear to the right and left; as shown in (a), D-glyceraldehyde. Cf. *monosaccharide.* **Haworth ~** A p. used for cyclic carbohydrates, showing the hydroxyl group on the hemiacetal carbon above the plane (α form) or below it (β form); as shown in (b), α-D-glucopyranose. **Newman ~** Two atoms and their common bond are represented by a circle. Radial lines stopping at the circumference represent bonds to atoms farthest away; lines to the center represent bonds to the nearest atoms. Coincident projected bonds are drawn at a small angle to each other; as shown in (c), 1,2-dichloroethane. **Rosanoff ~** See *Rosanoff notation* under *notation.*

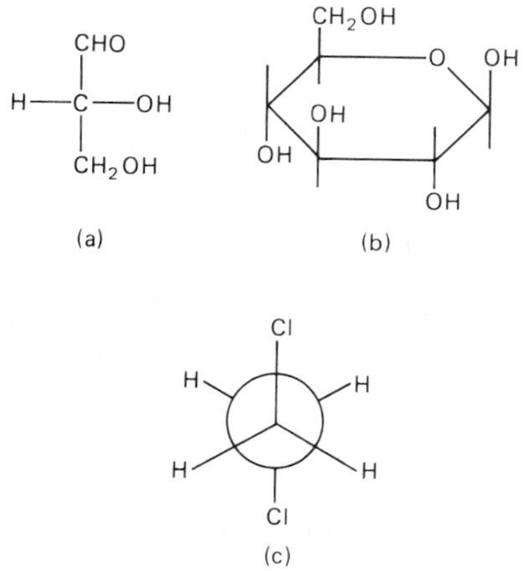

CHO
H—C—OH
CH_2OH

(a)

CH_2OH
O OH
OH
OH
OH

(b)

Cl
H H
H H
Cl

(c)

pro-knock A substance that induces knocking in an internal-combustion engine. Cf. anti*knock.*

prolactin A hormone from the anterior pituitary which causes growth of mammary glands of mammals and secretion of milk.

prolamines A group of proteins from cereals; insoluble in water, soluble in dilute alcohol. **prolamine from oats** Contains cysteine and histidine. **prolamine from sorghum** Contains no tryptophanol.

Prolene Trademark for a polypropylene synthetic textile fiber.

proline* $NH \cdot (CH_2)_3 \cdot CH \cdot (COOH)$ = 115.1. Pro*. 2-Pyrrolidinecarboxylic acid*. An amino acid, protein split product, which does not react with nitrous acid or ninhydrin. **4-hydroxy ~** $C_5H_9O_3N$ = 131.1. 4-Hydroxy-2-pyrrolidenecarboxylic acid. Oxyproline. Several enantiomers. An amino acid from proteins.

prolon A synthetic protein fiber, analogous to rayon.

Proluton Trademark for progesterone.

prolyl* The radical $NH \cdot (CH_2)_3 \cdot CH \cdot (COO-)$, from proline.

promazine hydrochloride $C_{17}H_{20}N_2S \cdot HCl$ = 320.9. 10-[3-(Dimethylamino)propyl]phenothiazine hydrochloride. Sparine. White, hygroscopic crystals, m.179; a tranquilizer, sedative, and antiemetic (USP, BP). See *phenothiazine.*

promethazine **p. hydrochloride** $C_{17}H_{20}N_2S \cdot HCl$ = 320.9. 10-[2-(Dimethylamino)propyl]phenothiazine hydrochloride. Phenergan. Yellow crystals, m.223, slightly soluble in water. An antihistamine; used for allergies (USP, EP, BP). **p. theoclate** $C_{17}H_{20}N_2S \cdot C_7H_7O_2N_4Cl$ = 499.0. Avomine. Prevents travel sickness (BP).

promethium* Pm. Illinium, florentium. Rare-earth element of at. no. 61. At. wt. of isotope with longest half-life (^{145}Pm) = 144.91. A product of the fission of ^{238}U. m.1080.

Promin Trademark for *p,p′*-diaminodiphenylsulfone *n*-didextrose sulfonate. Bitter, yellow, hygroscopic solid, soluble in water; used to treat leprosy.

promoter (1) Catalyst accelerator. (2) In flotation: See *collector.*

Pronestyl Trademark for procainamide hydrochloride.

prontosil *p*-[(2,4-Diaminophenyl)azo]benzene sulfonamide. Prontosil rubrum. The first chemotherapeutic agent active against streptococci. Developed from a dye. A precursor of the sulfa drugs.

proof **over ~** *n* overproof describes a spirit, 100 volumes of which when diluted to (100 + *n*) volumes, gives proof spirit. **under ~** A spirit that is *n* underproof contains in 100 volumes (100 − *n*) volumes of proof spirit.

p. gallon U.S.: 1.8927 liters at 60°F. U.K.: 2.5926 liters at 15°C, of ethyl alcohol. **p. spirit** Double the ethyl alcohol content (thus 50% alcohol is 100 proof); defined as follows: (1) Dilute aqueous ethyl alcohol containing 50% EtOH by volume at 60°F (U.S. usage). (2) Dilute aqueous ethyl alcohol which at 51°F weighs exactly 12/13 of an equal measure of distilled water. At 60°F it contains 49.28% by weight or 57.10% by volume of EtOH; d.0.91976. (U.K. usage).

propadiene* $H_2C:C:CH_2$ = 40.06. Allene*, dimethylenemethane. **dioxo ~** Carbon suboxide.

Propadrine Trademark for phenylpropanolamine hydrochloride.

propalanine Amino*butanoic acid**.

propaldehyde Propionaldehyde*.

propamidine $NH:C(NH_2) \cdot C_6H_4O \cdot (CH_2)_3 \cdot O \cdot C_6H_4 \cdot (NH_2)C:NH$ = 312.4. 4,4′-Trismethylene dioxydibenzamidine. An antiseptic for eye infections.

propanal* Propionaldehyde*. Cf *propenal, proponal.* **2,2-dimethyl ~*** Pivaldehyde*. **2-methyl ~*** Iso*butyraldehyde**.

propanamide* $EtCONH_2$ = 73.1. Propionamide, propionic amide, propionylamine. Colorless leaflets, m.79, soluble in water. **2-hydroxy ~*** Lactamide*. **methyl ~*** **2- ~** $Me_2CH \cdot CONH_2$ = 87.1. Isobutyramide. Crystals, m.118, soluble in water. ***N*- ~** Colorless liquid, b.193, with high relative permittivity (340 at −40°C) for a nonionic liquid; used in electrochemical research.

propane* C_3H_8 = 44.10. The third member of the alkane series. Colorless gas, $d_{air=1}$ 1.558, b. −39, slightly soluble in water; a fuel for internal-combustion engines. **acetyl ~** 2-Pentanone*. **bromo ~*** Propyl bromide*. **chloro ~*** Propyl chloride*. **diamino ~** P. diamine*. **1,2-dibromo ~** $C_3H_6Br_2$ = 201.9. Propylene bromide. Colorless liquid, b.131, soluble in water. **1,2-dichloro ~** $C_3H_6Cl_2$ = 113.0. Propylene chloride. Colorless liquid, d.1.166, b.97, soluble in water. **dihydroxy ~** Propanediol*. **epoxy ~*** Propylene oxide, propene oxide. **1,2- ~** $MeCH \cdot CH_2 \cdot O$ = 58.1. Colorless liquid, m.35, soluble in water. **1,3- ~** $CH_2 \cdot CH_2 \cdot CH_2 \cdot O$ Trimethylene oxide. **hydroxy ~** Propanol*. **iodo ~*** Propyl iodide*. **methoxy ~*** Methyl propyl ether*. **phenyl ~** Cumene*. **1,2,3-trichloro ~*** $CH_2 \cdot Cl \cdot CHCl \cdot CH_2Cl$ = 147.4. Trichlorohydrin. Allyl trichloride. Colorless liquid, d.1.417, b.158, insoluble in water. **trihydroxy ~** Glycerol*.

p.carboxylic acid* Butanoic acid*. **p.diamide*** Malonamide*. **p.diamine*** $NH_2(CH_2)_3NH_2$ = 74.1. Trimethylenediamine. A ptomaine from beef broth cultures of coma bacillus. Colorless liquid, b.136; a reagent for mercury. **p.dicarboxylic acid*** Glutaric acid*. **p.dinitrile*** Malononitrile*. **p.dioic acid*** Malonic acid*. **p.diol*** Dihydroxypropane. **1,2- ~** $MeCHOH \cdot CH_2OH$ = 76.1. Propylene glycol*. Colorless, hygroscopic, viscous liquid, d.1.036, b.188, soluble in water; a solvent and vehicle for drugs (NF, BP). **1,3- ~** $CH_2OH \cdot CH_2 \cdot CH_2OH$. Trimethylene glycol. A liquid, b.215, soluble in water. **p.dione*** 1,2-Pyruvaldehyde*. **1,2,3-p.tricarboxylic acid*** $CH \cdot COOH(CH_2COOH)_2$ = 176.1. Tricarballylic acid. Colorless rhombs, m.166 (decomp.), soluble in water; in beet molasses. **p.triol*** Glycerol*. **1,2,3-p.triyl*** Glyceryl, propenyl. Prefix indicating the $-CH_2 \cdot CH(-) \cdot CH_2-$ radical, from glycerol.

propanidid $C_{18}H_{27}O_5N$ = 337.4. Epontol. Pale, greenish-yellow liquid, very slightly soluble in water. An intravenous anesthetic (BP).

propano-* Prefix indicating a $-(CH_2)_2-$ bridge.

propanoic acid* Propionic acid*.

propanol* C_3H_8O = 60.1. **1- ~** $CH_3 \cdot CH_2 \cdot CH_2OH$. *n*-Propyl alcohol*. Colorless liquid, d.0.799, b.97, soluble in water. **2- ~** $CH_3 \cdot CHOH \cdot CH_3$. Isopropyl alcohol*. Colorless liquid, d.0.780, b.82, soluble in water. **2-chloro-1- ~*** $MeCHCl \cdot CH_2OH$ = 94.5. Propylene chlorohydrin. Colorless liquid, b.134. **1-chloro-2- ~*** $MeCHOH \cdot CH_2Cl$ = 94.5. Propylene chlorohydrin, chloroisopropyl alcohol. Colorless liquid, b.126. **diethoxy ~*** **2,3-d.-1-p.** Diethyline. **1,3-d.-2-p.** Diethylglycerol. **1,1-diethyl-1- ~*** Et_3COH = 116.2. Colorless liquid, d.860, b.140, soluble in water. **1,1-dimethyl-1- ~*** $MeCH_2 \cdot CMe_2OH$ = 88.1. *tert*-Pentyl alcohol. d.0.809, m.−12, b.102. **2,2-dimethyl-1- ~*** $Me_3C \cdot CH_2OH$. Neopentyl alcohol*. m.53, b.114, slightly soluble in water. **2,5-dimethyl-2- ~*** $Me_2COH \cdot CH_2 \cdot CHMe_2$. Colorless liquid, b.130, soluble in alcohol. **1,5-dimethyl-3- ~*** $Me_2CH \cdot CHOH \cdot CH \cdot Me_2$. Colorless liquid, d.0.832, b.131, soluble in water. **2,3-epoxy-1- ~*** $O \cdot CH_2 \cdot CH \cdot CH_2OH$ = 74.1. Glycidol, glycide, epihydrin alcohol. Colorless liquid, b_{751mm}162, miscible with water; used in organic synthesis, as a solvent and textile processing aid, and in plastics manufacture. **methyl ~** Butyl alcohol*. **trichloromethyl ~** Chlorbutanol.

propanone* Acetone*. **dihydroxy ~** Dihydroxyacetone. **diphenyl ~** Dibenzyl ketone*. **1-hydroxy-2- ~*** $MeCOCH_2OH$ = 74.1. Acetol. Colorless liquid, b.145, soluble in water; a reducing agent.

propantheline bromide $C_{20}H_{30}O_3NBr$ = 412.4. White crystals, m.158, soluble in water. An antispasmodic, used for indigestion and peptic ulceration (USP, BP).

propargyl The 2-propynyl* radical. **bi ~, di ~** Hexadiyne*.

propargylic acid Propiolic acid*.

Propathene Trademark for polypropylene.

propellant An explosive material which generates a large volume of hot gas at a predetermined rate. Propellants may be solid (plasticized cellulose nitrate) or liquid (concentrated hydrogen peroxide). **bi ~** A p. whose action depends on the interaction of the fuel and an oxidant. **mono ~** A p. whose action is due to the fuel alone. See Table 65.

propenal* Acrylaldehyde*. Cf. *propanal, proponal.*

propene **1- ~*** $MeCH:CH_2$ = 42.08. Propylene. Colorless gas, soluble in water. **1,3- ~** Cyclopropene*. Reactive gas, b.−36 (decomp.). **bromo ~*** Allyl bromide*. **chloro ~*** Allyl chloride*. **1,2-dichloro ~*** $MeCCl:CHCl$ = 111.0. Allylene dichloride. b.55. **1,2-epoxy ~*** $MeC(O):CH$ = 96.1. Allylene oxide. Colorless liquid, b.62, slightly soluble in water. **2-methyl-1- ~*** $Me_2C:CH_2$ = 56.1. Isobutylene. Gas, b.−6.

p.dicarboxylic acid Pentenedioic acid*. **1-p.-1,3-diyl†** See *propenylene.* **p. oxide** Epoxy*propane**. **2-p.-1-thiol*** $CH_2:CH \cdot CH_2SH$ = 74.1. Allyl mercaptan. Yellow liquid, b.90, miscible with alcohol. **1,2,3-p.tricarboxylic acid*** $COOH \cdot CH_2 \cdot C(COOH):CH \cdot COOH$ = 174.1. Aconitic acid, citridic acid, achilleic acid, adonic acid. Colorless, crystalline leaflets, m.190, obtained from *Aconitum, Equisetum* (horsetail), *Adonis* and *Achillea* species, or made from citric acid. It occurs also in beets and sugarcane.

propenol* Allyl alcohol*. **phenyl ~*** Cinnamyl alcohol*.

propenyl (1)* 1-Propenyl. Propylen. The radical $MeCH:CH-$, from propene. **iso ~*** 1-methylethenyl†. The radical $CH_2:CMe-$.

1-p.benzene $MeCH:CH \cdot C_6H_5$ = 118.2. Phenylallylene. Colorless liquid, b.160, soluble in alcohol. **p.phenol** Anol. (2) 2-Propenyl†. See *allyl* radical. (3) The 1,2,3-propanetriyl* radical. **p. hydrate** Glycerol*. **p. trinitrate** Nitroglycerin.

TABLE 65. SPECIFIC IMPULSES OF VARIOUS PROPELLANTS

Propellant	Specific impulse[a] (s)
Hydrogen peroxide, nitric acid, or dinitrogen tetraoxide with alcohols, petroleum fuels, or hydrazine hydrate	210–250
Liquid oxygen with ethanol, kerosene, liquid ammonia, liquid acetylene, or liquid hydrogen	240–345
Liquid ozone with kerosine or liquid hydrogen	275–370
Liquid fluorine with kerosine, liquid ammonia, hydrazine, or liquid hydrogen . .	265–355
Perchlorates with organic polymers	180–245
Cordite	190–235
Hydrogen peroxide (90%), ethylene oxide, hydrazine, propyl nitrate, nitromethane . .	135–220

[a]Specific impulse is the thrust divided by the mass flow rate.

propenylene* 1-Propene-1,3-diyl†. The radical $-CH_2 \cdot CH:CH-$.
propenylidene* The radical $CH_3 \cdot CH:C=$, from propene.
propeptone Hemialbumose.
properties The characteristics of a substance. **additive ~** (1) P. that depend on the quantity of matter; as, mass. (2) P. of a molecule which can be calculated from those of constituent atoms; as, molecular weight. Cf. *constitutive p.* below. **atomic ~** P. depending on atomic characteristics; as, spectrum lines. Cf. *molecular p.* below. **chemical ~** The chemical reactions of a substance. Cf. *physical p.* below. **constitutive ~** (1) P. that depend on the quality of matter; as, melting point. (2) P. due to constitution; as, optical rotation. Cf. *additive p.* above. **extensive ~** P. depending on quantity; as, momentum. Cf. *intensive p.* below. **general ~** P. inherent in all matter; as, gravity. Cf. *special p.* below. **intensive ~** P. depending on quality; as, solubility. Cf. *extensive p.* above. **mechanical ~** Physical p. of a substance; as, tensile and tear strength, but not brightness. **molar ~** P. depending on the amount of substance; as, m. boiling-point elevation. **molecular ~** P. depending on molecules; as, isomerism. Cf. *atomic p., molar p.* (both above). **physical ~** Physical phenomena which remain unchanged so long as there is no change in molecular composition. Cf. *mechanical p.* above. **radiant ~** The color and intensity per mol of a substance. **special ~** P. exhibited to varying extents by some but not all matter; as, color. Cf. *general p.* above. **thermal ~** The p. of a substance due to absorption or liberation of heat, such as melting and boiling.
propham* IP(P)C. See *herbicides,* Table 42 on p. 281.
prophetin $C_{20}H_{36}O_7$ = 388.5. A glucoside from *Ecballium officinale* and *Cucumis prophetarum* (Cucurbitaceae), soluble in water.
prophoretin $C_{20}H_{30}O_4$ = 334.5. An amorphous, resinous split product of prophetin.
prophylactic A preventive, particularly of disease.
prophylaxis Preventive treatment.
propiconazole See *fungicides,* Table 37 on p. 250.
propilidene The propylidene* radical.
propine Propyne*.
propineb* See *fungicides,* Table 37 on p. 250.
propinol 2-Propynyl alcohol*.
propinyl The propynyl* radicals.
propiolic acid* $CH:C \cdot COOH$ = 70.0. Carboxyacetylene, propargylic acid, acetylenecarboxylic acid, propynoic acid*. Colorless liquid, b.144, soluble in water. **methyl ~** Butynoic acid.* **phenyl ~** See *phenylpropiolic acid.*
p. acid series See *acetylene acids.*
propioloyl* The radical $HC:C \cdot CO-$, from propiolic acid.
propion The propionyl* radical. **p. aldoxime** Propionaldehyde oxime*. **p.amide** Propanamide*. **p.anilide** Propionylaniline.
propionaldehyde* $MeCH_2CHO$ = 58.1. Propanal*, propaldehyde, propyl aldehyde. Colorless liquid, d.0.807, b.48, soluble in water. **dihydroxy ~** Glyceraldehyde*. **dimethyl ~** Pivaldehyde*.
p. oxime* $EtCH:NOH$ = 73.1. Propion(ic) aldoxime. Colorless crystals, m.21.
propionamide Propanamide*.
propionate* Propanoate*. C_2H_5COOM. A salt or ester of propionic acid.
propionic p. aldehyde Propionaldehyde*. **p. anhydride*** $(MeCH_2CO)_2O$ = 130.1. Propionyl oxide, propanoic anhydride*. Colorless liquid, d.1.017, b.168, decomp. by water.
propionic acid* $MeCH_2COOH$ = 74.1. Carboxyethane, propanoic acid*, pseudoacetic acid. Colorless liquid, d.0.987, b.141, soluble in water. **acetyl ~** Levulinic acid. **amino ~** Alanine*. **aminohydroxy ~** Serine*. **aminohydroxyphenyl ~** Tyrosine*. **aminoindyl ~** Tryptophan*. **benzoyl ~*** See *benzoylpropionic acid* under *benzoyl.* **carbamoyl ~*** Succinamic acid*. **dihydroxy ~*** Glyceric acid*. **dihydroxyphenyl ~** Hydro*caffeic acid.* **dimethyl ~*** Pivalic acid*. **hydroxy ~*** **2- ~** Lactic acid*. **3- ~** Hydracrylic acid. **2-methyl ~*** $Me_2CH \cdot COOH$ = 88.1. Isobutyric acid. Colorless liquid, m.80, slightly soluble in water. **3-nitro ~*** Hiptagenic acid. **oxo ~*** Pyruvic acid*. **phenol ~** Hydrocoumaric acid. **phenyl ~** **2- ~** Hydratropic acid. **3- ~** Hydrocinnamic acid.
propiononitrile* Ethyl cyanide*.
propiono The propionyl* radical.
propionyl* Propion(o), propanoyl. The radical $MeCH_2CO-$, from propionic acid. ***N*-p.aniline** $EtCONHPh$ = 149.2. Propionanilide. Colorless leaflets, m.104, soluble in water. **p. chloride*** $C_2H_5 \cdot COCl$ = 92.5. b.80. **p.salicylic acid*** $C_6H_4(OOC \cdot CH_2Me)COOH$ = 194.2. ***o*- ~** White scales, m.95, insoluble in water.
propiophenone* $MeCH_2COPh$ = 134.2. m.20, b.218. **β-phenyl ~** $PhCH_2COPh$ = 196.2. Benzyl phenyl ketone*, α-phenylacetophenone. Colorless solid, m.60; used in organic synthesis.
proponal $(C_3H_7)_2C:(CONH)_2:CO$ = 212.3. Diisopropylbarbituric acid. Colorless crystals, m.145, soluble in water; a hypnotic. Cf. *propanol.*
proportion A ratio; a numerical comparison of a part to the whole. **definite ~** Proust's law of definite (constant) proportions: the same compound always consists of the same elements combined in the same proportions by weight. **equivalent ~** Compounds of elements *A* and *C* contain *m* atoms of *A* and *n* atoms of *C*. Compounds of elements *B* and *C* contain *x* atoms of *B* and *y* atoms of *C*. Compounds of elements *A* and *B* contain *p* atoms of *A* and *q* atoms of *B*. *m, n; x, y;* and *p, q* are whole numbers, usually small. Hence *p, q* equal either *m, x* or whole multiples of them, usually small. **multiple ~** Law of multiple proportions. If 2 elements combine in more than one p., the molecules of one compound must be formed by adding a whole number of atoms of one or both elements to one or more molecules of the other compound.
p. limit The least p. of a substance that can be detected in another specified substance.
proportional control See *mode.*
propoxy* The radical $MeCH_2 \cdot CH_2O-$, from propanol. **iso ~*** The radical Me_2CHO-, from isopropyl alcohol.
propoxyphene hydrochloride $C_{22}H_{29}O_2N \cdot NCl$ = 375.9. Dextropropoxyphene hydrochloride, Distalgesic, Doloxene. White crystals, m.163, soluble in water. An analgesic, related to methadone (USP, BP).
propranolol hydrochloride $C_{16}H_{21}O_2N \cdot HCl$ = 295.8. 1-(Isopropylamino)-3-(1-naphthyloxy)-2-propanol hydrochloride. Inderal. White crystals, m.94, soluble in water. A betadrenergic blocking agent; used to slow rapid heart rate and to prevent irregular cardiac rhythms (USP, BP).
propyl* Pr. $MeCH_2CH_2-$. α-Propyl, normal propyl, *n*-Pr. **iso ~*** The radical Me_2CH-, from isopropane. 1-Methylethyl*. β-Propyl, *i*-Pr.
p. acetate* $MeCOOPr$ = 102.1. Colorless liquid, d.0.891, b.102; soluble in water. **p.acetylene** $PrC:CH$ = 68.1. Colorless liquid, b.48, insoluble in water. **p. alcohol*** Propanol*. **p. aldehyde** Propionaldehyde*. **p.amine*** $PrNH_2$ = 59.1. **normal ~*** $PrNH_2$. Propaneamine. Colorless liquid, b.49, soluble in water. **iso ~*** Me_2CHNH_2.

b.32, soluble in water. **p.aniline*** PhNHPr = 135.2. Colorless liquid, b.222, soluble in water. **p.benzene*** PhPr = 120.2. Phenylpropane*. Colorless liquid, b.158, insoluble in water. **p. benzoate*** PhCOOPr = 164.2. Colorless liquid, b.229, sparingly soluble in water. **p.benzoic acid*** $C_6H_4PrCOOH$ = 164.2. ***ortho-*** **~** Colorless leaflets, m.58, soluble in water. ***para-*** **~** Colorless crystals, m.140, soluble in water. **p. bromide*** PrBr = 123.0.1-Bromopropane*. Colorless liquid, b.71, soluble in water. **p. butex** P. paraben. **p. butanoate*** PrCOOPr = 130.2. Colorless liquid, b.143, soluble in alcohol. **p. carbamate*** NH_2COOPr = 103.1. Colorless prisms, m.53, soluble in water. **p. chloride*** **normal ~** PrCl = 78.5. 1-Chloropropane*. Colorless liquid, b.46, soluble in water. **iso ~** MeCHClMe. 2-Chloropropane*, secondary, or propyl 2-chloride. Colorless liquid, b.37, soluble in water. **p. cyanide*** **normal ~** PrCN = 69.1. Butyronitrile*. Colorless liquid, b.118, soluble in water. **iso ~** Me_2CHCN = 69.1. A liquid, b.107. **p. ether** Pr_2O = 102.2. Dipropyl ether*, propoxypropane*. Colorless liquid, b.91, soluble in water. **iso ~** Diisopropyl ether*. Colorless liquid, b.68, insoluble in water; a solvent for gums, waxes, and asphalt. **p. formate*** HCOOPr = 88.1. Colorless liquid, b.81, soluble in water. **p. gallate** $C_{10}H_{12}O_5$ = 212.2. 3,4,5-Trihydroxybenzoate. White crystals, m.147, slightly soluble in water; a preservative (NF, BP) **p. 4-hydroxybenzoate** P. paraben. **p.hydroxylamine** **alpha- ~** $PrONH_2$ = 75.1. ***N-*** **~** PrNHOH. m.45. **p. iodide*** PrI = 170.0. 1-Iodopropane*. Colorless liquid, d.1.748, b.102, sparingly soluble in water; used in organic synthesis. **p. isocyanide** **normal ~*** PrNC. **iso ~*** Me_2CHNC = 69.1. Liquid, b.87. **p. isothiocyanate*** PrNCS = 101.2. P. mustard oil, b.137. **p. ketone** 4-Heptanone*. **p. mercaptan** P. thiol*. **p.nitroamine** $PrNHNO_2$ = 104.1. Colorless liquid, b.130, soluble in water. **p. nitrate*** $PrNO_3$ = 105.1. Colorless liquid, b.119, soluble in alcohol. **p. nitrite*** $PrNO_2$ = 89.1. Colorless liquid, b.57, soluble in alcohol. **p.nitrolic acid** $PrC(NO_2)\cdot NOH$ = 132.1. m.74, soluble in water. **p.paraben** $C_{10}H_{12}O_3$ = 180.2. P. *p*-hydroxybenzoate. White crystals, m.90, slightly soluble in water; a fungistatic and preservative (NF, BP). **p.pyridine** PrC_5H_4N = 121.2. **2- ~** b.166. **3- ~** b.158. **4- ~** b.177. **p. sulfide** Dipropyl sulfide*. **p. thiol*** PrSH = 76.2. P. mercaptan, p. hydrosulfide, hydrosulfopropane. Colorless liquid, b.67, soluble in water. **p.urea*** $NH_2\cdot CO\cdot NHPr$ = 102.1. Colorless crystals, m.107, soluble in water.

propylen The 1-propenyl* radical.

propylene (1)* The radical $-CHMe\cdot CH_2-$. Cf. *propylidene.* (2) Propene*. (3) The propenylene* radical. Cf. 1-*propenyl.* **poly ~** See *polypropylene.*
p. aldehyde Crotonaldehyde*. **p. bromide** 1,2-Dibromo*propane**. **p. chloride** 1,2-Dichloro*propane**. **p. chlorohydrin** Chloro*propanol.* **p.diamine*** Propanediamine*. **p. dichloride** 1,2-Dichloropropane*. **p. glycol** 1,2-Propanediol*.

propylidene* Propilidene. The radical $MeCH_2CH=$, from propane. **p. bromide*** 1,1-Dibromopropane*. **p. chloride*** 1,1-Dichloropropane*.

propyliodone $C_{10}H_{11}O_3NI_2$ = 447.0. Propyl-3,5-diiodo-4-oxo-1(4*H*)-pyridineacetate. Dionosil. White crystals, m.188, insoluble in water; a contrast medium for bronchography (USP, BP).

propyloic- Prefix indicating the radical $-CH_2\cdot CH_2\cdot COOH$ on a side chain. Cf. *methyloic-.*

propylthiouracil $C_7H_{10}ON_2S$ = 170.2. Bitter powder, m.220, slightly soluble in water; an antithyroid drug used for thyrotoxicosis (USP, BP).

propynal* CH:C·CHO = 54.0. Propiolaldehyde. Oily liquid, b.61.

propyne* MeC:CH = 40.1. Allylene, methylacetylene. Colorless gas, b.−24, soluble in ether. Cf. *propadiene.* **3-bromo ~*** $HC:CCH_2Br$ = 119.0. Propargyl bromide. b.89. Lachrimatory liquid. **dimethyl ~** Methylbutyne*. **methyl ~** Butyne*. **phenyl ~** Methylphenyl*acetylene.*

propynoic acid* Propiolic acid*.

propynol* 2-Propynyl alcohol*.

propynyl **1-p*** The radical MeC:C−. **2-p*** Propargyl, protyl, archyl. The radical $HC:C\cdot CH_2-$. **2-p. acetate*** $MeCOOC_3H_3$ = 98.1. Colorless liquid, d.1.005, b.124. **2-p. alcohol*** $CH:C\cdot CH_2OH$ = 56.1. Propinol. Colorless liquid, d.0.972, b.114, soluble in water.

prorennin Rennin*.

pros- Prefix (Greek) indicating "by," "near," or "at." **p. position** The 2,3-position of the naphthalene ring.

prosapogenin $C_{36}H_{56}O_{10}$ = 648.8. The 3-β-D-glucopyranoside of quillaic acid; decomp. 220.

prosopite A native, hydrated calcium aluminum fluoride.

prospecting Searching for ore deposits.

prostaglandins Hormonelike, unsaturated hydroxy fatty acids, first found in the prostate gland, but occurring in most animal tissues. Synthesized in the body from linear fatty acids with 20 C atoms (as, arachidonic acid). Related structurally to prostanoic acid. They influence uterine contractions and blood pressure. Used to produce therapeutic abortion and induce labor. Examples are: **p. E_2** Also called dinoprostone, Prostin E_2. **p. $F_{2\alpha}$** Also called dinoprost, Prostin $F_{2\alpha}$:

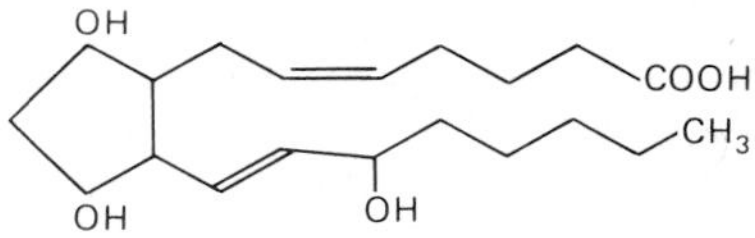

prostanoic acid The precursor of prostaglandins, having the structure:

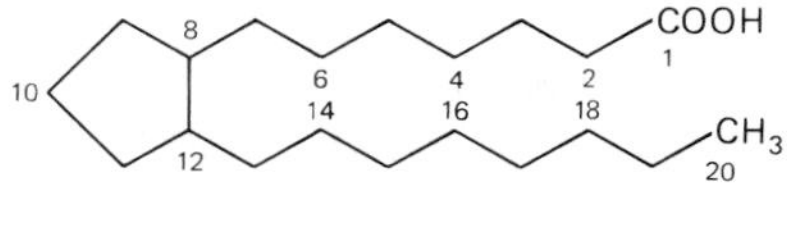

prostate A gland at the neck of the male bladder.

prosthesis (1) Replacement or substitution. (2) An artificial limb or part of body.

prosthetic group A group, being either a metal ion or organic radical not derived from an amino acid, in the complex molecule of a conjugated protein; as, heme.

Prostigmin Trademark for neostigmine.

Prostin Trademark for prostaglandins.

protactinium* Pa. Protoactinium, brevium, ekatantalum. Element, at. no. 91. At. wt. of isotope with longest half-life (^{231}Pa) = 231.04. Discovered (1918) by Hahn and Meitner. m.ca. 1227. A radioactive disintegration product of uranium.

protamin Any of the simple proteins in the sex cells of fishes; as, salmine.

protamine $C_{16}H_{32}O_2N_9$ = 382.5. An amine from

TABLE 66.
CLASSIFICATION OF PROTEINS

I. Method 1
  A. Simple proteins—substances that yield only α-amino acids or their derivatives on hydrolysis:
    1. Albumins—soluble in water or dilute salt solution, and coagulable by heat; as, ovalbumin
    2. Globulins—insoluble in water, soluble in salt solutions, and coagulated by heat; as, edestin from hemp seed
    3. Glutelins—insoluble in neutral solvents, soluble in dilute acids or bases, and coagulated by heat; as, glutenin
    4. Prolamines (gliadins)—soluble in 80% alcohol, insoluble in water, absolute alcohol, or neutral solvents; as, zein
    5. Albuminoids—insoluble in all neutral solvents; as, elastin
    6. Histones—soluble in water, or dilute acids, precipitated by ammonia, not coagulated by heat; as, globin
    7. Protamins—soluble in water, uncoagulable by heat; as, salmine
  B. Conjugated proteins—protein combined with another (prosthetic) substance:
    1. Nucleoproteins—compounds of one or more proteins with a nucleic acid; as, nucleohistones
    2. Glycoproteins—compounds of protein with carbohydrates; as, mucins
    3. Phosphoproteins—compounds of proteins with a phosphorus compound other than lecithin or nucleic acids; as, casein
    4. Chromoproteins (hemoglobins)—compounds of proteins with a chromophoric group; as, hemoglobin
    5. Lecithoproteins—compounds of proteins with lecithins
    6. Lipoproteins—compounds of proteins with fatty acids
  C. Derived proteins—primary and secondary split products of proteins:
    1. Proteans—insoluble products from the action of water or enzymes; as myosan
    2. Metaproteins—from the action of acids or bases
    3. Coagulated proteins—from the action of heat or alcohol
    4. Proteoses—from further hydrolysis; soluble in water, not coagulated by heat, and precipitated in saturated solution by ammonium or zinc sulfate
    5. Peptones—soluble in water, noncoagulable by heat, not precipitated by ammonium sulfate
    6. Peptides—compounds of amino acids containing peptide groups
II. Method 2 (References in parentheses are to Method 1.)
  A. Simple proteins:
    1. Protamins (*A7*)
    2. Histones (*A6*)
    3. Albumins (*A1*)
    4. Globulins (*A2*)
    5. Glutelins (*A3*)
    6. Alcohol-soluble proteins (*A4*)
    7. Scleroproteins (*A5*)
    8. Phosphoproteins (*B3*)
  B. Conjugated proteins:
    1. Glucoproteins (*B2*)
    2. Nucleoproteins (*B1*)
    3. Chromoproteins (*B4*)
  C. Hydrolysis products:
    1. Infraproteins (*C2*)
    2. Proteoses (*C4*)
    3. Peptones (*C5*)
    4. Polypeptides (*C6*)
III. Method 3 (According to physiological role.)
  A. All types of proteins
    1. Transport and storage proteins; as, hemoglobin, myoglobin.
    2. Enzymes, q.v.; as, trypsin.
    3. Structural proteins, present in hair, wool, bone, skin, teeth, nail, horn; as, α- and β-keratin, collagen.
    4. Contractile proteins, being involved in muscle movement; as, myosin, actin.
    5. Information proteins; as, hormones (insulin) and proteins involved in cell growth and photosynthesis.

spermatozoa and fish spawn. **p. sulfate** A heparin antagonist (USP), used as an injection and prepared from the sperm of certain fishes.

Protan Trademark for sodium formate. A masking agent, neutralizer in wool dyeing, and solubilizer and buffer in the chemical industry.

protanope Describing a person who is color blind to red.

protargyl Protyle.

proteacin Leucodrin.

protean A group of derived *proteins,* q.v.

protease An enzyme that splits proteins to proteoses and peptones. **gastric ~** Pepsin*. **vegetable ~** Bromelain*. Papain*.

protective colloid A covering envelope for colloidal particles, stabilizing them against coagulation by electrolytes. Cf. *cone.*

proteid Protein (obsolete).

proteidin An immunizing bacterial solvent developed from complex proteins and bacteriolytic enzymes.

protein(s) Nitrogenous organic compounds, containing more than about 100 amino acid residues, mol. wt. 8,000–200,000, in vegetable and animal matter. P. yield amino acids on hydrolysis and are foods assimilated as amino acids and reconstructed in the protoplasm. Cf. poly*peptide, proteinate, protein salt* (below). See Table 66. **alcohol-soluble ~** Prolamines. Gliadins. **bacterial ~** A toxic p. formed by bacteria. **bitterness ~** A muco ~ containing 33% carbohydrate and forming complexes with sugars, mol. wt. 166,000–175,000; responsible for bitter flavors. **coagulated ~** A simple p. rendered insoluble by heat or chemical agents. **compound ~** Conjugated p. **conjugated ~** Complex p. containing a protein and a nonprotein molecule. **derived ~** A decomposition product of p., in complexity of structure between proteins and amino acids. **flavo ~** P. with flavine adenine dinucleotide or mononucleotide as the prosthetic group. They are enzymes in oxidation and reduction reactions. **gamma ~** A mixture: approx. protein 50 (often from soybeans), starch 50%; a paper size. **hemo ~** Hemoglobin p. **purified p. derivative** P.P.D. See *tuberculin.* **single-cell ~** P. produced by microorganisms; as, by the action of yeasts on hydrocarbon oils. See *Pruteen.* **synthetic ~** See poly*peptide.* **textured vegetable ~** P. (as, from soya bean, oil feedstock, or other high-energy substrate) that has

been given the texture of lean meat; for human consumption or spinning.
p. error A change in the color of an acid-base indicator due to the amphoteric nature of p. present. **p. salt** A compound of a p. with an acid; as, casein hydrochloride. Cf. *proteinate.*

proteinase An enzyme that hydrolyzes proteins. **acid ~** Pepsin*. **alkali ~** Trypsin*.

proteinate A compound formed from a protein and a base; as, sodium caseinate. Cf. *protein* salt.

proteoclastic Proteolytic.

proteolysis The conversion of proteins into soluble peptones by decomposition or hydrolysis.

proteolytic Proteoclastic. Indicating proteolysis.

proteoses A group of derived *proteins,* q.v.

protheite A variety of pyroxene.

prothesis Prosthesis.

prothrombin A protein in blood plasma essential for the clotting of blood. Cf. *menadione.*

protium* The hydrogen isotope, mass 1.00783. Cf. *deuterium, tritium.*

proto- Prefix (Greek) indicating "first." **protosalt** Obsolete term for a compound that contains the smallest amount of the negative radicals or atoms of the same positive element; as, HgI (mercurous iodide), mercury protoiodide; protosulfates, protoxides.

protoactinium Protactinium*.

protobastite A variety of enstatite.

protobitumen A partially reduced carbohydrate, which on further reduction yields oil; as, algarose (synthetic) and algarite (natural).

protocatechol (1) Catechol. (2) Pyrocatechol.

protocatechualdehyde* $C_6H_3(OH)_2CHO = 138.1$. 3,4-Dihydroxybenzaldehyde*, dihydroxybenzenecarbonal*, protocatechoic aldehyde. Colorless leaflets, m.153, soluble in water. **methyl ~** Vanillin*. **methylene ~** Piperonylic acid*.

protocatechuic **p. acid*** $C_6H_3(OH)_2COOH = 154.1$. 3,4-Dihydroxybenzoic acid*, from the oil of shepherd's purse. Colorless crystals, decomp. 199, soluble in water; a photographic developer. **p. aldehyde** Protocatechualdehyde*.

protocatechuoyl* 3,4-Dihydroxybenzoyl†. The radical $(OH)_2 \cdot C_6H_3 \cdot CO-$, from protocatechuic acid.

protocotoin $C_{16}H_{14}O_6 = 302.3$. m.141. A principle from paracoto bark.

protocurarine An alkaloid from curare.

protofluorine An element, at. wt. 2.1, assumed to exist in nebulae, and between hydrogen and helium in the periodic table. Cf. *coronium, nebulium.*

protogenic Yielding up a proton. Cf. *protophilic.*

protohydrogen An element, assumed to exist in bright stars and nebulae, of atomic weight less than that of hydrogen.

protolysis Proton transfer. A reaction in which a proton is transferred.

protomorph A particulate substance in cellular juices; an intermediate state between the organized cell and the disorganized juice.

proton Prouton, uron. (1) Originally the positive *nucleus,* q.v., of an atom (Rutherford, Bohr). (2) The nucleus of the H atom, e.g., the hydrogen ion, H^+. A subatomic *particle,* q.v., of mass 1.67264×10^{-27} kg and one positive charge; spin $\frac{1}{2}$; magnetic moment 1.41061×10^{-26} J/T; a "building block" of atomic nuclei. **anti ~** Negative p. The analog of the positron, produced by collision of high-velocity p. with a copper target.
p. acceptor A substance that gains a hydrogen ion; as, a base. **p. bombardment** The production by fast-moving protons of an excited isotope which yield γ rays. Cf. *radioactivity.* **p. donor** A substance yielding a hydrogen ion; as, an acid. **p. number*** Atomic number*. **p. reaction** (1) Neutralization. Cf. *prototropy.* (2) Nuclear reactions.

protonated Given a proton; as, *onium compounds.*

protones A group of hydrolyzed products of protamins.

protophile A weak base which unites with a proton. Cf. *acid (3).*

protophilic Hydrophilic. The tendency of a molecule to unite with hydrogen ions (protons). See *hydronium ion.* Cf. *protogenic.*

protophyllin Chlorophyll hydride, a colorless substance that changes to green chlorophyll by the action of carbon dioxide.

protophyta A single-celled plant; as, yeasts. Cf. *diatom.*

protoplasm Protoplasma bioplasm. The jellylike colloidal material forming the basis of living cells—both plant and animal. Contains proteins, carbohydrates, and fats, and includes both cytoplasm and nucleus. See *cell, paraplasm.*

protoplasma Protoplasm.

protoporphyrin $C_{34}H_{34}O_4N_4 = 562.7$. A constituent of hemoglobin and the brown pigment of eggshells, having a nucleus of 4 pyrrole rings. Cf. *phylloporphyrin.*

protosulfate A sulfate containing the smallest amount of SO_4 radical.

prototropic Pertaining to reactions influenced by protons.

prototropy Pseudoacidity, mobile H-tautomerism, mobile proton tautomerism. Ionotropy in which a detached H atom causes a molecule to exist as tautomeric ions. Cf. *aci-, anionotropy.*

prototype An original type, model, or measure, e.g., the international *kilogram.*

protoveratrine $C_{32}H_{51}O_{11}N = 625.8$. An alkaloid, m.270, from the rhizome of *Veratrum album* (Liliaceae); a hypotensive drug.

protovitamin Provitamin.

protozoan See *protozoon.*

protozoon (Pl., protozoa.) Protozoan. A unicellular animal: the lowest class of the animal kingdom; as, ameba. Cf. *protophyta.*

protractor An instrument to measure angles.

protriptyline hydrochloride $C_{19}H_{21}N \cdot HCl = 299.8$. Concordin. White crystals, soluble in water. An antidepressant of the tricyclic group (USP, BP).

protuberance A projecting part. **solar ~** The giant streamers of incandescent gases on the sun's surface. Cf. *coronium, corona.*

protyl The 2-propynyl* radical.

protyle Archyl, protargyl, pantogen, urstoff. The hypothetical substance from which all chemical elements are derived.

Proust **P., Louis Joseph (1755–1826)** French chemist. **P.'s law** The law of definite (constant) *proportions.*

proustite Ag_3AsS_3. Ruby silver. Native, red hexagons.

Prout **P., William (1785–1850)** English physician. **P. hypothesis** (1815.) All elements are multiples of a protyle or of hydrogen. Cf. *atomic* structure.

prouton Suggested name for proton.

provitamin A precursor to a vitamin. **p. A** Carotene. See *β-carotene.* **p. D₂** Sterols, e.g., ergosterol, converted into vitamin D_2 by exposure to ultraviolet or cathode rays. **p. D₃** See 7-dehydro*cholesterol.* Cf. *β-carotene.*

proximate The nearest approach. **p. analysis** See *proximate* *analysis* under *analysis.* **p. principle** The active principle of a drug; as, an alkaloid.

prozane Triazane*.

prulaurasin $C_{14}H_{17}NO_6$ = 295.3. A racemic mandelonitrile glucoside, m.122, from cherry laurel, *Prunus laurocerasus.*

prunasin $C_{14}H_{17}NO_6$ = 295.3. A (+)-mandelonitrile glucoside, m.147, from *Prunus padus.*

prune Prunus. The partly dried fruit of *Prunus domestica* (Rosaceae); a food and mild laxative.

prunetol Genistein.

prunicyanin An anthocyanin from plums.

prunol Ursolic acid.

prunus Prune. **p. amygdalus** Almond. **p. domestica** Prune. **p. spinosa** Blackthorn. **p. virginiana** Wild cherry.

prussian blue $Fe_4[Fe(CN)_6]_3$ = 859.3. Iron(III) hexacyanoferrate(II)*. *Turnbull's blue,* q.v. Blue precipitate formed from solutions of ferric salts and potassium hexacyanoferrate(II). **native ~** Vivianite. Cf. *Milori blue.*

prussian red Colcothar.

prussiate (1) Cyanide*. (2) Hexacyanoferrate(II)* or -(III)*. **red ~** Potassium hexacyanoferrate(III)*. **yellow ~** Potassium hexacyanoferrate(II)*.

prussic acid Hydrocyanic acid.

prussite Cyanogen.

Pruteen Trademark for granular, single-cell protein produced from *Methylophilus methylotrophus* feeding on methanol. It contains 72% protein (from atmospheric nitrogen), and is rich in lysine and methionine; an animal food.

PS* Abbreviation for polystyrene.

ps Abbreviation for "pseudo."

pseud-, pseudo- *ps*- or ψ-. Prefix (Greek) indicating "false" or "similar to." Indicating derivation from a hypothetical parent substance or tautomeric form (Baeyer).

pseudacetic acid Propionic acid*.

pseudaconitine $C_{36}H_{51}O_{12}N$ = 689.8. Acetylveratryl-ψ-aconine. A crystalline alkaloid, m.211, from *Aconitum ferox.*

pseudo acid An organic compound that forms salts, but has no carboxyl radical. The acid character is due to a hydroxyl group attached to an N atom; as, oxime compounds. Cf. *prototropy.*

pseudoaconitine Pseudaconitine.

pseudoallyl The iso*propenyl** radical.

pseudoalum $MAl_2(SO_4)_4 \cdot 24H_2O$, where M is divalent as Mn, Fe^{2+}, Mg. Cf. *true alum* under *alum.*

pseudoasymmetric Indicating an atom that is bonded tetrahedrally to 1 pair of enantiomeric groups, and also to 2 atoms or achiral groups that are different from each other; as:

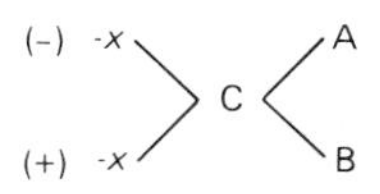

pseudo base An organic compound that becomes basic in presence of acids. Cf. *amphoteric.*

pseudobrookite $2Fe_2O_3 \cdot 3TiO_3$. A vitreous mineral.

pseudobutylene 2-Butene*.

pseudocellulose Hemicellulose.

pseudocumene $C_6H_3Me_3$ = 120.2. Cumol, 1,2,4-trimethylbenzene*. Colorless liquid, b.170, insoluble in water.

pseudocumidine $C_6H_2(NH_2)Me_3$ = 135.2. 2,4,5-Trimethylaniline, 1-amino-2,4,5-trimethylbenzene*. Colorless needles, m.66, soluble in water.

pseudocumyl The trimethylphenyl* radical, $Me_3C_6H_2-$, from pseudocumene; as, 2,3,5- or *asym*-; 2,4,5 or *sym*-; 2,3,6- or *vic*-.

pseudocurarine A nonpoisonous alkaloid in *Nerium oleander* (Apocynaceae).

pseudocyanate Fulminate*.

pseudocyanic acid Fulminic acid*.

pseudodichroism See *polychromatic.*

pseudoephedrine hydrochloride $C_{10}H_{15}ON \cdot HCl$ = 201.7. Sinufed, Sudafed. The (+)-(1*S*,2*S*) enantiomer of ephedrine. A sympathomimetic, used for asthma and hay fever (USP, BP).

pseudogalena Sphalerite.

pseudoindoxyl See *indoxyl.*

pseudoisatin Isatin.

pseudoisocyanic acid Fulminic acid*.

pseudo(iso)merism Isomerism between 2 molecules containing the same number and kind of atoms, but having different atomic linkages and valencies; as, cyanide and isocyanide. See *isomerism.*

pseudoisotope A radioactive element resembling an isotope in its reactions, but having a different atomic number.

pseudomalachite See *malachite.*

pseudomerism See *pseudoisomerism.*

Pseudomonas A genus of motile gram-negative rod organisms from soil and food, and often forming greenish-blue pigments. They decompose food and pharmaceuticals, and can cause infection.

pseudomonotropy Allotropy in which the transition temperature is below the melting point.

pseudomorph A crystal having the general outline of one crystal system, but actually an aggregation of minute crystals of another system. **chemical ~** A p. produced by chemical substitution or alteration. **physical ~** A p. produced by a change in allotropic form.

pseudonarcissine An alkaloid from the bulbs of *Narcissus pseudonarcissus* (Amarylladaceae).

pseudonuclein Nucleoalbumin.

pseudopelletierine $C_9H_{15}NO \cdot H_2O$ = 171.2. 9-Methyl-9-azabicyclo[3.3.1]nonan-3-one. An alkaloid from the root of the pomegranate. Prisms, m.63, soluble in water. Cf. *pelletierine.*

pseudophenanthroline 4,7-Naphthisodiazine.

pseudophysostigmine $C_{15}H_{21}O_3N_2$ = 277.3. An alkaloid from calinuts or false calabar beans. White crystals, soluble in alcohol.

pseudoplasticity The condition in which a fluid's viscosity decreases with increasing shear rate under steady shear flow. Cf. *dilatancy.*

pseudopolychroism See *polychromatic.*

pseudosolution A colloidal suspension or emulsion.

pseudotannin Tannin split products unable to convert hide into leather.

pseudoviscosity See *viscosity.*

pseudowavellite A hydrous aluminum-calcium phosphate (Florida) containing 8–9% phosphorus pentaoxide and sand; a fertilizer.

pseudoxanthine A leucomaine from muscle tissue, resembling xanthine in properties but not composition.

pseudozoogloea A clump of living bacteria resembling a protozoon microscopically.

psi Ψ, ψ. Greek letter. Ψ (italic) is the symbol for (1) electric flux; (2) the wave function in the *Schrödinger equation.* ψ is the symbol for pseudo in chemical names.

psia Abbreviation for pounds per square inch absolute; a unit of pressure.

psicose $CH_2OH \cdot CO \cdot (CHOH)_3 \cdot CH_2OH$ = 180.0. Allulose, D-ribo-2-oxohexose. An epimer of (+)-fructose. A reducing, nonfermentable constituent of cane sugar molasses. Cf. *glutose.*
psig Abbreviation for pounds per square inch gage; a unit of pressure.
psilocin A hallucinogenic indole occurring in some mushroom species.
psilomelane $(Ba \cdot H_2O)_2Mn_5O_{10}$. A hydrated pyrolusite (Saxony, Germany).
psophometer A device to measure signal-to-noise ratios in communications systems.
psoralens Substances found in plants, as bergamot and parsley, that have a photosensitizing effect. They have a coumarin molecule with a furan ring. See *PUVA treatment.*
psoraline Caffeine.
psychedelic A mind-expanding drug; as LSD.
psychomimetic Describing drugs that produce hallucinations and psychotic states.
psychotrine $C_{28}H_{36}O_4N_2$ = 464.6. An alkaloid in ipecac.
psychrometer See *hygrometer.* **sling ~** Whirling *hygrometer.*
psychrometric chart A chart showing the drying temperature plotted against the weight of water vapor removed per unit weight of dry air.
psychrophilic Describing organisms of optimum growth at temperature 4–10°C. Cf. *mesophilic.*
psylla **p. alcohol** Tritriacontanol*. **p. wax** A solid wax from the leaf house, *Psylla alni.*
psyllic **p. acid** Tritriacontanoic acid*. **p. alcohol** Tritriacontanol*.
psyllostearyl alcohol Tritriacontanol*.
psyllostearylic acid Tritriacontanoic acid*.
pt Abbreviation for (1) pint; (2) part.
pteridine* $C_6H_4N_4$ = 132.1. A diaminopyrimidine derivative. Yellow plates, m.140, soluble in water (violet fluorescence).
Pteridophyta A main division of Cryptogamia, as Filices (ferns), Equisetaceae (horsetails), Lycopodiaceae (club mosses).
pterin A yellow purine pigment from mammalian tissues. Cf. *rhodopterin.*
pterocarpine An alkaloid from red sandalwood, *Pterocarpus santalidus* (Leguminosae).
pterygospermin The antibacterial principle from *Moringa pterygosperma,* Gaertn.
pteryolglutamic acid Folic acid.
PTFE* Symbol for poly(tetrafluoroethylene)*.
ptomaine Animal, cadaveric, or putrefactive alkaloid. Any of the amino compounds that result from the decomposition of proteins by microorganisms; e.g.: aminovaleric acid (meat), cadaverine (animal tissues), diethylamine (fish), morrhuic acid (cod-liver oil), mydine (human tissue), tetanine (cultures of tetanus bacillus), tyrotoxine (dairy foods).
ptyalase α-Amylase*.
ptyalin α-Amylase*.
Pu Symbol for plutonium.
pucherite $BiVO_4$. A mineral.
puddle, puddling (1) The conversion of cast iron into wrought iron by fusion in a reverberatory furnace, in contact with the hematite lining of the furnace, where oxidation of carbon to carbon monoxide occurs. (2) Clay, moistened and well-worked.
puering Bating. The cleaning of depilated leather hides, by the action of tryptic enzymes.
pukateine $C_{17}H_{17}O_3N$ = 283.3. An apomorphine alkaloid found in *Laurelia novae zealandiae.* Cf. *laureline.*
pulegium The Labiatae *Hedeoma pulegoides* (American p.) and *Mentha pulegium* (European p.).
pulegol $C_{10}H_{18}O$ = 154.3. 3-Menthenol. 2-Isopropylidene-5-methylcyclohexanol. From the essential oils of *Mentha pulegium* (Labiatae).
pulegone $C_{10}H_{16}O$ = 152.2. 3-$\Delta^{4(8)}$-Menthenone. An aromatic ketone in oil of hedeoma. Colorless liquid, d.0.932, b.221, insoluble in water.
Pulfrich refractometer A refractometer used especially for oils and fats.
pullulan A poly-α-1,6-maltotriose.
pullulanase* An enzyme which hydrolyzes 1,6-α-D-glucosidic linkages in pullulan, amylopectin, and glycogen.
pulp Any soft mixture of solid particles and liquids as, *paper pulp.* **dissolving ~** A p. similar to, but purer than, that for paper; used to produce viscose rayon and high-tenacity tire cord.
pulpwood A raw material for paper pulp manufacture, as, conifer trunks.
pulque The fermented sap of an *Agave* species (Mexico). Cf. *mescal.*
pulsatance See Table 88, Group *B,* footnote *b,* on p. 266.
pulse The edible seeds of leguminous plants; as, peas, lentils.
pulverization The reduction of a substance to a powder.
pulverizing Powdering.
pulvinate (1) Convex or cushion-shaped, as a colony of bacteria. (2) A salt of pulvinic acid.
pulvinic acid $C_{18}H_{12}O_5$ = 308.3. Orange pigment from lichen spp. ***p,p'*-dihydroxy ~** Atromentin.
pulvis Latin for "powder."
pumice (stone) A light, porous stone of volcanic origin, which consists of the silicates of aluminum, sodium, and potassium; an abrasive and catalyst base (USP).
pump A machine for drawing or forcing liquids or gases from one container or level into another. **acid ~** See *acid pump.* **air ~** See *air pump.* **backing ~** A low-power p. to produce a low pressure or partial vacuum in preparation for a high-power booster p. **filter ~** A low-power p. operated by a water faucet. **heat ~** See *heat pump.* **Hickmann ~** See *vacuum pump.* **mercury ~** See *Sprengel pump.* **metering ~** Type of p. used to add chemicals to a process at a controllable level; as, mono and Moyno p. **suction ~** Filter p. **Töpler ~** A p. that removes air by entrainment between drops of mercury falling in a tube. **vacuum ~** See *vacuum pump.*
pumpkin seed Pepo. The seeds of varieties of *Cucurbita pepo;* an anthelmintic.
punctiform Describing a bacterial colony, near the limit of natural vision.
pungent Sharp or biting, as a p. odor.
punicic acid $C_{18}H_{30}O_2$ = 278.4. (9*Z*,11*E*,13*Z*)-9,11,13-Octadecatrienoic acid*. m.44, from the seed oil of *Punica granatium,* pomegranate (Punicaceae).
punicine (1) Pelletierine. (2) A purple oxidation product of colorless shellfish juices.
pure Free from contamination. **bacteriologically ~** Containing no live bacteria. **chemically ~** Containing no other substance. Cf. *chemicals.*
purgatin Purgatol.
purgative An agent that causes evacuation of the bowels; as, castor oil.
purgatol $C_{14}H_5O_2(OH)(C_2H_3O_2)_2$ = 340.1. Purgatin, anthrapurpurin diacetate. Orange crystals, m.177, insoluble in water; a purgative.

purine* $C_5H_4N_4$ = 120.1.

```
            1       2
           N——————CH
         //          \\
     6 HC              N 3
         \ 5       4 /
          C=========C
         /           \
     7 HN             N 9
          \         //
             CH
             8
```

Colorless needles, m.212, soluble in water. **endogenous ~** See *endogenous purines.* **exogenous ~** See *exogenous purines.*

p. alkaloids Purine bases. **p. bases, p. bodies** The alkaloids derived from purine; as theobromine, caffeine, 2,6-dioxopurine (xanthine), 2,6,8-trioxopurine (uric acid). Many p. b. are hydrolytic products of nucleoproteins, and occur in animal waste products. **p.dione** Xanthine. **p. ring** The heterocyclic arrangement of the atoms in the molecule of purine. **p. skeleton** The p. ring. **p.trione** Uric acid.

PuriNethol Trademark for mercaptopurine.

purinometer A small buret for estimating purine bases in urine.

purinone Hypoxanthine.

purity See *chemicals.*

Purkinje (Purkyně), Jan Evangelista (1787–1869) Czech physiologist. **P. effect** Optical sensation increases with increasing intensity of light more rapidly for the red than for other spectral colors.

purone $C_5H_8O_2N_4$ = 156.1. 2,8-Dioxo-1,4,5,6-tetrahydropurine. A reduction product of uric acid produced by electrolysis.

purple A reddish-blue coloring matter of the purple snail. **antique ~** Dibromoindigo. **tyrian ~** Murex. **visual ~** The photosensitive material of the retinal rods of the eye. It sensitizes the eye to dim light, and contains vitamin A. Cf. *rhodopsin.*

p. carmine Murexide. **p. of cassius** Red, colloidal tin oxide with adsorbed gold, formed on adding alkali to a mixture of solutions of stannous, stannic, and gold chlorides; used in the manufacture of ruby glass and red enamels. **p. copper** Bornite.

purpureo Describing metal ammines of tetra-, tri-, or bivalent metals with NH_3 molecules and negative radicals; as, pentaammine, $[Co(NH_3)_5]X_3$; and tetraammine, $[Co(NH_3)_4]X_3$, where X is a halogen or other negative radical. There are also salts of the type $[Co_2(NH_3)_{10}Cl_2](NO_3)_4$.

purpuric acid $C_8H_5N_5O_6$ = 267.2. An oxidation product of uric acid, related to alloxantin. Cf. *murexide.*

purpurin $C_{14}H_8O_5$ = 256.2. 1,2,4-Trihydroxyanthraquinone*, oxyalizarin 6. From the glucoside of madder; also obtained synthetically. Red needles, m.256 (decomp.), soluble in water. **anthra ~** See *anthrapurpurin.*

purpurogallin $C_{11}H_8O_5$ = 220.2. A tetrahydroxy constituent or hydrolysis product (with glucose) of many galls. See *eriophyesin, pontanin.*

purpuroxanthene Xanthopurpurin.

purpuroxanthin Xanthopurpurin.

purree Indian yellow.

purreic acid Euxanthic acid.

purr(en)one Euxanthone.

pus A liquid from infected wounds, consisting of dead cells and leukocytes, with microorganisms and serum. **blue ~** Blue p. produced by *Pseudomonas pyocyaneus.* Cf. *cyopin.*

pustulant An agent or microorganism that produces small inflammations of the skin (pustules); as, chicken pox virus. Cf. *vesicant.*

putrefaction Putrescence. The progressive chemical decomposition of organic matter, especially proteins, generally by anaerobic bacteria. See *decay.*

putrefactive alkaloids Ptomaines.

putrescence Putrefaction.

putrescent Undergoing putrefaction.

putrescine $NH_2(CH_2)_4NH_2$ = 88.2. Tetramethylenediamine*. 1,4-Butanediamine*. A ptomaine from the decay of animal tissues and the action of certain bacteria. Colorless liquid with unpleasant odor.

putrid Putrefactive.

putty A mixture of chalk 11, raw linseed oil 3 parts; used for setting glass and filling holes and cracks. **iron ~** See *iron putty.*

p. powder A polishing powder made by heating tin in air, removing the dross, and igniting the product of the action of nitric acid on the residue; used in white enamel and opal glass.

PUVA treatment Therapy for certain skin conditions (as, psoriasis) by a *p*soralen (as, methoxypsoralen), together with *u.v. A* radiation.

puzzolane A native, silicate-rich, lime-poor cement (Puzzuoli, Italy); used by the Romans. Cf. *pozzolana.*

PVAC* Abbreviation for poly(vinyl acetate).

PVAL* Abbreviation for poly(vinyl alcohol).

PVC* Abbreviation for poly(vinyl chloride).

PVDC* Abbreviation for poly(vinylidene dichloride).

PVDF* Abbreviation for poly(vinylidene difluoride).

PVF* Abbreviation for poly(vinyl fluoride).

PVP Providone.

PWR Pressurized-water *reactor.*

pwt Abbreviation for pennyweight. See *troy.*

py* Indicating pyridine as a ligand.

pycnometer Pyknometer.

pycnosis Pyknosis. The concentration of chromatin material into an area of the nucleus, which becomes small and deeply stained; usually a result of cell degeneration.

pydine $C_7H_{13}ON$ = 127.2. 3-Oxa-7-azabicyclo[3.3.1]nonane. A bicyclic combined piperidine and pyran ring:

```
            1
  8CH2—CH  —CH22
   |     |     |
  7NH   CH2    O   3
   |     |     |
  6CH2—CH  —CH24
            5
```

pyelogram Pyelograph. X-ray examination of the kidneys and urinary tract by means of radiopaque injection.

pyknometer A graduated glass vessel of definite volume with a glass stopcock, with or without a thermometer, used to weigh a definite volume of liquid in order to determine its specific gravity, W/V, where W = weight, V = volume.

pyo- Prefix (Greek) indicating "pus."

pyoctanin Pyoktanin.

pyocyanase An enzyme produced by the growth of *Pseudomonas pyocyaneus.*

pyocyanine $C_{13}H_{10}ON_2$ = 210.2. A blue antibiotic pigment from *Pseudomonas aeruginosa.*

pyogenic A microorganism that produces pus, especially in wounds.

pyoktanin Dahlia violet. Methyl violet, methyl aniline violet. **p. blue** Methyl violet. **p. yellow** Auramine.

Pyopen Trademark for carbenicillin disodium.

pyr- See *pyro-*.
pyracetic acid Pyroligneous acid.
pyracin A lactone of an acid produced from pyridoxine, in α and β forms. It prevents anemia in chicks.
pyraconitine $C_{32}H_{41}NO_9$ = 583.7. An alkaloid, m.171, from aconite.
pyramid A polyhedron whose base is a polygon and whose faces are triangles, with a common vertex and the sides of the polygon as bases. Cf. *crystal systems*.
pyran* C_5H_6O = 82.1. The compounds

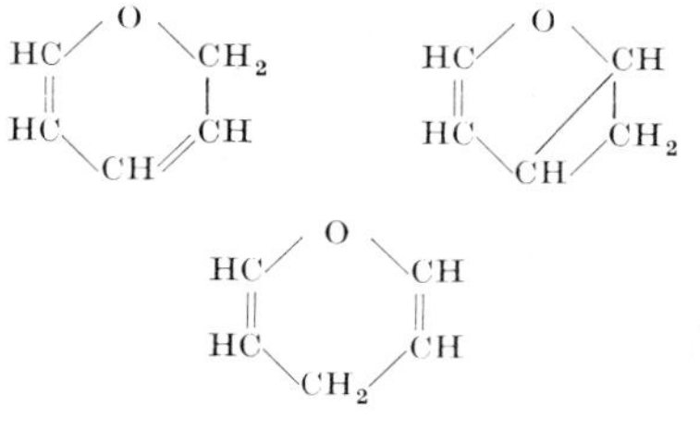

benzo ~ See *benzopyran*. **dioxo ~** Pentenedioic anhydride*. **oxo ~** Pyrone*.
p.dione Pentenedioic anhydride*.
pyranil black A black sulfur dye.
Pyranol (1) Trademark for dielectric material, principally of the askarel type. (2) Sodium acetyl salicylate. (3) (Not cap.) Chlorinated biphenyl.
pyranone* Pyrone*.
pyranose The 6-membered cyclic form of sugars; e.g., glucose (glucopyranose).

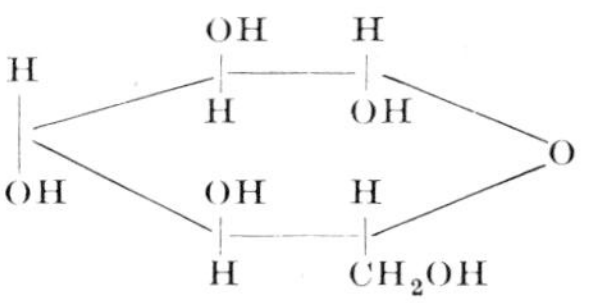

Cf. *furanose*.
pyranoside A glucoside having a pyran ring; as, adenosine. Cf. *furanoside*.
pyranthrene* $C_{30}H_{16}$ = 376.5.

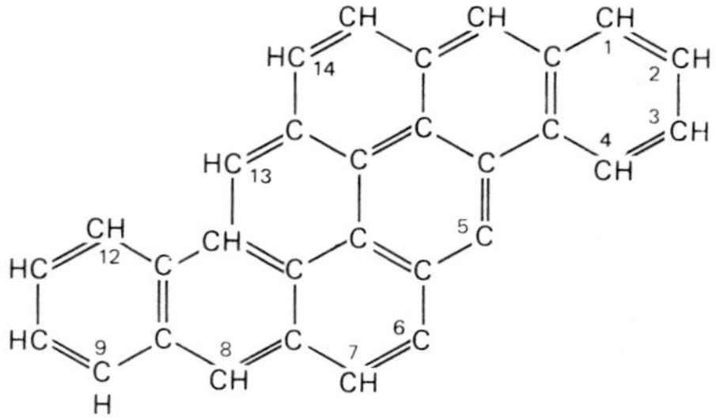

pyranyl* A group of radicals: C_5H_5O-, from pyran.
pyrargyrite $3Ag_2S \cdot Sb_2S_3$. Acrosite, aerosite. A red ore.
pyrazinamide $C_5H_5ON_3$ = 123.1. Pyrazinecarboxamide. Zinamide. An antituberculous drug.
pyrazine* CH:CH·N:CH·CH:N = 80.1. Paradiazine, piazine, 1,4-diazine. An isomer of pyrimidine and pyridazine. Colorless crystals, m.47, soluble in water. **dimethyl ~** Ketine. **hexahydro ~** Piperazine*. **tetraphenyl ~** Amaron.
pyrazole* CH:CH·NH·N:CH = 68.1. α-Pyrromonazole-1,2-diazole. Colorless needles, m.70, soluble in water.
dihydro ~ Pyrazoline*. **tetrahydro ~** Pyrazolidine*.
pyrazolidine* $C_3H_8N_2$ = 72.1. Tetrahydropyrazole. b.140.
phenyl ~ *N*-Phenylpyrazolidine. Colorless liquid, b.160.
pyrazoline **2- ~*** CH:N·NH·CH_2·CH_2 = 70.1. Dihydropyrazole. Colorless liquid, b.144, soluble in water.
oxo ~ Pyrazolone*.
pyrazolinium A pyrazoline compound with protonated nitrogen.
pyrazolone(s)* $C_3H_4ON_2$ = 84.1. Oxopyrazoline(s).

3- ~ NH·NH·CO·CH:CH
4- ~ NH·N:CH·CO·CH_2
5- ~ NH·N:CH·CH_2·CO

5- ~ Solid, m.165 (sublimes), soluble in water.
dimethylphenyl ~ Antipyrine. **3-methyl ~** Butyrolactazam.
pyrazolyl* The radical $C_3H_3N_2-$, from pyrazole; 4 isomers.
Pyrene (1) Trademark for a carbon tetrachloride fire extinguisher. Cf. *Pyrex*. (2)* (Not cap.) $C_{16}H_{10}$ = 202.3. Benzo[*def*]phenanthrene. A tetracyclic hydrocarbon from coal tar:

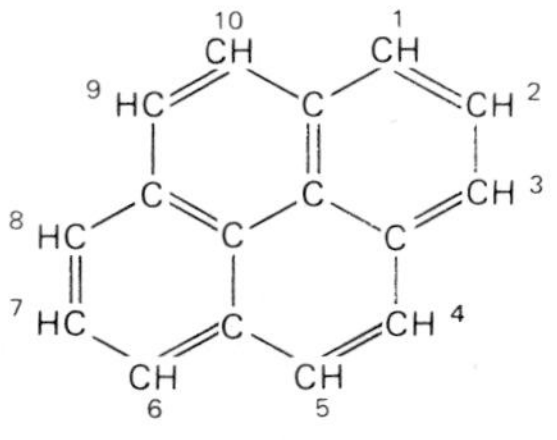

Colorless, monoclinic crystals, m.148, insoluble in water.
pyreneite A grayish-black andralite. Cf. iron *garnet*.
pyrenite Tetryl.
pyreno-* Prefix indicating the pyrene ring in a polycyclic compound in addition to the prefixed base component.
pyrenyl* The radical $C_{16}H_9-$, from pyrene.
pyrethric acid MeOOC·C_8H_{12}·COOH = 212.2. An oil. See *pyrethrum*.
pyrenoids Large, spherical structures in the chloroplasts of certain algae.
pyrethrin Dalmatian flowers. Insect flowers. Persian flowers. An active principle of *pyrethrum*, q.v.
pyrethrol $C_{21}H_{34}O$ = 302.5. Pyretol. An alcohol from the leaves of *Chrysanthemum cinerariaefolium* (insect powder). White needles, m.222, insoluble in water.
pyrethrolone A complex mixture of keto alcohols, whose esters are the active principles of pyrethrum flowers.
pyrethrone A ketone from pyrethrum which forms pyrethrol.
pyrethrum (1)* An insecticide. A mixture of esters of chrysanthemic and pyrethric acids from pyrethrolone, cinerolone, and jasmolone alcohols. (2) Pellitory root. The dried root of *Anacyclus pyrethrum* (Compositae); local irritant.
p. camphor $C_{10}H_{16}O$ = 152.1. A terpene from the essential oil of *Chrysanthemum parthenium* (Compositae). **p. extract**

An extract of p. containing 25% by weight of pyrethrin, with petroleum as diluent.

pyretol Pyrethrol.

Pyrex (1) Trademark for a variety of chemical, cooking, and other glassware, including low-expansion, heat-resistant, or chemically resistant glasses. **P. EC** P. glass permanently bonded with a thin, transparent, electrically conducting metal film, used to heat the glass. (2) Trademark for a carbon tetrachloride fire extinguisher. Cf. *Pyrene.*

pyridazine* CH:CHN:NCH:CH = 80.1. 1,2-Diazine. Colorless liquid, b.206, soluble in water.

pyridazinone $(CH_2)_2\cdot CH:N\cdot NH\cdot CO$ = 98.1. A dihydrooxopyridazine. Colorless liquid, b.170, soluble in alcohol.

pyridinol Pyridol. Hydroxypyridine. The hydroxy form of pyridone: $HO\cdot C_5H_4N \rightleftharpoons O:C_5H_5N$.

pyridinole $C_6H_4\cdot(NH)\cdot C_5H_3N$ = 168.2. A group of heterocyclic compounds.

pyridine* C_5H_5N = 79.1. The heterocyclic compound

6 HC—N(1)=CH 2; 5 HC—CH(4)—CH 3

in coal tar, bone oil, and vegetable distillation products (tobacco smoke). Colorless liquid, d_{15} 0.9893, m.−42, b.115, soluble in water or ether; an antiseptic, denaturant for alcohol, solvent for rubber and paint, and reagent for acetone and blood. **allyl ~** See *allyl pyridine.* **benzyl ~** See *benzylpyridine* under *benzyl.* **dihydrooxo ~** Pyridone*. **dimethyl ~*** Lutidine. **ethyldimethyl ~*** Parvoline. **ethylmethyl ~*** **2,5- ~** Aldehydine. **4,2- ~, 3,4- ~** Collidine. **hexahydro ~** Piperidine*. **hydroxy ~*** Pyridinol. **methyl ~*** Picoline. **oxo ~** Pyridone*. **propyl ~*** Conyrine. **tetramethyl ~*** Parvuline. **trimethyl ~*** γ-Collidine.

p. bases The homologs of pyridine, $C_nH_{2n-5}N$; as: picolines, lutidines, collidines, parvulines, rubidine, viridine. **p.carboxylic acid*** **2- ~** Picolinic acid. **3- ~** Nicotinic acid*. **4- ~** Isonicotinic acid*. **p.dicarboxylic acid*** **2,3- ~** Quinolinic acid. **2,4- ~** Lutidinic acid. **2,5- ~** Iso*cinchomeronic acid.* **2,6- ~** Dipicolinic acid. **3,4- ~** Cinchomeronic acid. **3,5- ~** Dinicotinic acid. **p.pentacarboxylic acid*** $C_5N(COOH)_5$ = 299.2. (+2- or $3H_2O$.) Soluble acid, decomp. 220. **p.sulfonic acid*** $C_5H_4NSO_3H$ = 159.2. Colorless needles, soluble in water; used in organic synthesis. **p. thiocyanate reaction** The precipitation of certain metals, e.g., Zn, Ni, Co,, by pyridine and ammonium thiocyanate; used in analysis. **p.tricarboxylic acid*** $C_5H_2N(COOH)_3\cdot 2H_2O$ = 247.2. **1,2,3,4- ~** Carbo*cinchomeronic acid.* **1,2,3,5- ~** m.323. **1,2,3,6- ~** m.71, b.130. **1,2,4,5- ~** m.235. **1,2,4,6- ~** m.277 (decomp.). **1,3,4,5- ~** Berberonic acid.

pyridinium The cation $C_5H_5N^+$, from pyridine. Cf. *piperidinium.*

pyridinyl† See *pyridyl.* **3-p.carbonyl†** See *nicotinoyl.*

Pyridium Trademark for phenylazodiaminopyridine; a urinary antiseptic.

pyridol Pyridinol.

pyridone* C_5H_5ON = 95.1. Pyridinone†, ketopyridine, oxopyridine*; a group of heterocyclic ketones or quinones; e.g., 2-oxopyridine, m.106.

pyridopyridine $C_8H_6N_2$ = 130.1. (1) Benzodiazine compounds. (2) 1,5-Naphthyridine, 1,5-benzodiazine. Yellow needles, m.75.

pyridoquinoline $C_{12}H_8N_2$ = 180.2; and $C_{13}H_9N$ = 179.2. = 169.1. Heterocyclic compounds containing 1 (as, benzoquinolines) or 2 (as, diazaphenanthrenes) N atoms in a quinoline and pyridine ring fused together.

pyridostigmine 3-[[(Dimethylamino)carbonyl]oxy]-1-methylpyridinum. **p. bromide** $C_9H_{13}O_2N_2Br$ = 261.1. Mestinon. White, bitter, deliquescent crystals, m.154, soluble in water; an anticholinesterase. Used for myasthenia gravis, a disease of muscle weakness (USP, BP). See *acetylcholine.*

pyridotropolone $C_{10}H_7O_2N$ = 173.2. Pale-yellow needles, m.168.

pyridoxal $C_8H_9O_3N$ = 167.2. Vitamin B_6. An oxidation product of pyridoxine. Colorless crystals, m.170 (decomp.), soluble in water.

pyridoxine hydrochloride $C_8H_{11}NO_3\cdot HCl$ = 205.6. Vitamin B_6 hydrochloride. 5-Hydroxy-6-methyl-3,4-pyridinedimethanol hydrochloride. Benadon, Hexa-Betalin. White crystals, darkening in light, m.206 (decomp.), soluble in water. Artificially produced. Deficiency in humans is rare. Some drugs are antagonists, and convulsions occur in babies fed on p. h.–deficient milk powders (USP, EP, BP).

pyridyl* Pyridinyl†. The radical $-C_5H_4N$, from pyridine. ***N*-pyridyl** 1-Pyridyl*. P. with substitution on the N atom of the ring. **ter ~** See *terpyridyl.*

p.amine* Aminopyridine*.

pyridylidene* The radical $C_5H_5N=$, from pyridine.

pyrilamine maleate $C_{17}H_{23}N_3O\cdot C_4H_4O_4$ = 401.5. Mepyramine hydrochloride, Anthisan. White crystals, m.100, soluble in water; an antihistamine used for allergies, as hay fever (USP, BP).

pyrimethamine $C_{12}H_{13}N_4Cl$ = 248.7. 2,4-Diamino-5-(*p*-chlorophenyl)-6-ethylpyrimadine. Daraprim. White crystals, m.240, insoluble in water. An antimalarial (USP). See *Fansidar.*

pyrimidine* CH:CH·CH:N·CH:N = 80.1. **1,3- ~** *m*-Diazine. Tautomer of pyrimidinone. Colorless crystals of pungent odor, m.22, soluble in water. **4-amino-2-oxo ~** Cytosine. **2,6-dioxo ~** Uracil. **methyldioxo ~** Thymine. **2,4,6-trioxo ~** See *barbituric acid.*

p. bases Compounds in nucleoproteins: **dioxo ~** Uracil. **trioxo ~** Barbituric acid. **p.dione** Uracil. **p.tetrone** Alloxan. **p.trione** Barbituric acid.

pyrimidinone $C_4H_4N_2O$ = 98.1. 1*H*-4, 3*H*-4, and 2(1*H*) are tautomers of 4-hydroxypyrimidine. m.179. **4-amino ~** Cytosine. Cf. *primidone.*

pyrimidinyl* The radical $C_4H_3N_2-$, from pyrimidine.

pyrite (1) FeS_2. Brassil, fool's gold. Yellow, shining crystals. (2) Native iron sulfide.

pyrites Generic name for sulfide minerals; as, tin pyrite (stannite). **arsenical ~** Mispickel. **arseno ~** See *arsenopyrite.* **auriferous ~** A p. containing gold. **capillary ~** Millerite. **cobalt ~** Smaltite. **copper ~** The yellow ore $CuFeS_2$. **coxcomb ~** Marcasite. **iron ~** Pyrite. **magnetic ~** Pyrrhotite. **nickel ~** Millerite. **radiated ~** Marcasite. **spear ~** Marcasite. **tin ~** Stannite. **white iron ~** Marcasite.

pyro(-) (1) Pyr-. Prefix (Greek) indicating "heat" or "fire"; as, pyrometer. (2) Pyroxylin.

pyroacetic **p. acid** Crude acetic acid from wood distillation. **p. spirit** Acetone*.

pyroacid Diacid*. An acid produced by the loss of 1 mole of water from 2 moles of an *orthoacid*, q.v.; e.g., by heat. Cf. pyro*phosphoric acid.*
pyroantimonate See *antimonate.*
pyroantimonic acid See *antimonic acid.*
pyroarsenate A salt of pyroarsenic acid.
pyroarsenic acid See *arsenic acid.*
pyrobitumen A constituent of shale oil; converted to bitumen by heat.
pyroboric acid Di*boric acid**.
pyrocatechin Pyrocatechol*.
pyrocatechoic acid 2,3-Dihydroxy*benzoic acid**.
pyrocatechol* (1) $C_6H_4(OH)_2$ = 110.1. Catechol, pyrocatechin, *o*-dihydroxybenzene, 2-hydroxyphenol, 1,2-benzenediol*. Colorless leaflets, m.105, soluble in water; a photographic developer. Cf. *hydroquinone.* (2) Protocatechols. Dihydric phenols derived from catechol; as, methylpyrocatechol, $C_6H_3Me(OH)_2$.
pyrocatechualdehyde $(HO)_2C_6H_3 \cdot CHO$ = 138.1. **ortho-~** 2,3-Dihydroxybenzaldehyde*. White crystals, m.108.
Pyoceram Trademark for a hard, thermal-shock-resistant, homogeneous composite having the properties of both glass and a ceramic.
pyrochlore (Ca, Fe, Ce)O·(Nb, Ti, Th)$O_2 \cdot H_2O$. A blue, vitreous mineral.
pyrochroite $Mn(OH)_2$. An amorphous mineral.
pyrochromate Dichromate*.
pyroclastic Describing volcanic ash deposits that are partly igneous and partly sedimentary in origin.
pyrocoll $C_{10}H_6O_2N_2$ = 186.2. Pyrrolecarboxylic acid anhydride. Yellow leaflets, m.269, insoluble in water.
pyrocomane 1,4-Pyrone.
pyrocondensation A union of molecules due to heat; as, formation of biuret from urea.
pyrodextrin A brown, tasteless heat-decomposition product of starch, involving the linkage of linear chains from amylose with branched chains from amylopectin.
pyroelectricity Thermal deformation. The property of crystals which are electrically charged by heat.
pyrogallate* A salt of pyrogallol.
pyrogallic acid Pyrogallol*.
pyrogallol* $C_6H_6O_3$ = 126.1. Pyrogallic acid, 1,2,3-trihydroxybenzene, 1,2,3-benzenetriol*. White needles or leaflets, m.133, decomp. 293, very soluble in water or ether; a reagent (absorbs oxygen), weak reducing agent, and photographic developer. **acetyl ~** Gallacetophenone. **trimethoxy ~** $C_6H_3(OMe)_3$ = 168.2. Pyrogallol trimethyl ether. Colorless crystals, m.47, soluble in alcohol. **p.carboxylic acid*** 2,3,4-Trihydroxybenzoic acid*. **p.phthalein** Gallein. **p. red** P. sulfonephthalein. An indicator for chelatometric titration.
pyrogen An impurity in a drug which when injected into the bloodstream produces fever.
Pyrogen(e) dye Trademark for a sulfur dye.
pyrogenic Describing a reaction induced by heat.
pyrogram The graph obtained by applying gas chromatography to the pyrolysis products of a compound.
pyrographitic oxide $C_{22}H_2O_4$ = 330.3. Gray-black powder produced by heating graphitic acid; used for crucibles, and formerly to coat electric bulb filaments.
pyrokomane Pyrone*.
pyroligneous Pertaining to wood distillation. **p. acid** Wood vinegar. Pyracetic acid. Impure acetic acid from the destructive distillation of pine tar and wood. **p. alcohol** Methanol*. **p. spirit** Methanol*.
pyroluminescence Luminescence produced by low-temperature flames. Cf. *candoluminescence.*
pyrolusite MnO_2. Polianite. Cf. *psilomelane.*
pyrolysis Thermolysis. Decomposition of organic substances by heat.
pyrolythic acid Cyanuric acid.
pyromagnetic Pertaining to heat and magnetism.
pyromellitic acid $C_6H_2 \cdot (COOH)_4 \cdot H_2O$ = 272.2. 1,2,4,5-Benzenetetracarboxylic acid*. Colorless prisms, m.275, soluble in water. Cf. *mellophanic acid.*
pyrometallurgy The study of metallurgical heat treatment.
pyrometer An instrument to measure high temperatures. **electrical ~** A thermocouple to measure the current produced at a metal-metal junction. **mechanical ~** A thermocouple (lever system) that magnifies the heat distortion of 2 joined metal strips. **optical ~** A p. to measure the intensity of light emitted by a hot body by comparison with a standard. **radiation ~** A p. to measure heat radiated by a hot body, e.g., by a thermocouple. **resistance ~** A thermometer to measure the change in resistance of a heated conductor.
pyrometric Pertaining to high temperature. **p. cone** Seger cone. **p. c. equivalent** A measure of refractoriness in terms of the softening temperature of a Seger cone.
pyrometry The study of the measurement of high temperatures.
pyromorphite $3Pb_3(PO_4)_2 \cdot PbCl_2$. Green lead ore (apatite group).
pyromucic acid $C_4H_3O \cdot COOH$ = 112.1. **2-~** 2-Furancarboxylic acid*, α-furoic acid. Colorless, monoclinic crystals, m.132, soluble in water. Cf. *furoates.*
pyromucyl The 2-furoyl* radical. **bi ~** Furil*.
pyrone* $C_5H_4O_2$ = 96.1. Pyranone*, pyrokomane. (Numbers give O (= 1), then CO position.) **1,2-~, α-~** O·CO·CH:CH·CH:CH. b.207. **1,4-~, γ-~** Colorless crystals, m.32. **benzo ~* 1,2-~** Coumarin. **1,4-~** Chromone. **p.carboxylic acid** $C_5H_3O_2 \cdot COOH$ = 140.1. **2-p.-5-carboxylic acid** Prisms, 207 (decomp.). **p.dicarboxylic acid* 2,6-~** $C_7H_4O_6$ = 184.1. Chelidonic acid. From *Chelidonium majus*, resembling meconic acid.
pyrones Heterocyclic compounds derived from 1,2- or 1,4-pyrone; in numerous natural coloring materials, and synthetic dyes and drugs. See *coumalin.* **benzo ~** Chromones.
pyronine dyes Xanthene dyes containing the chromophore

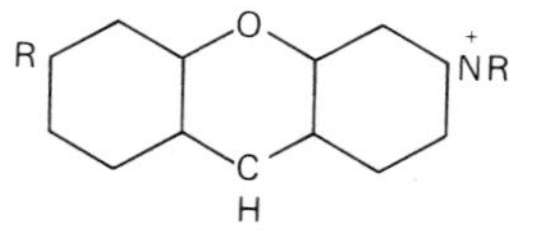

pyrope Aluminum *garnet.*
pyrophore A pyrophoric substance.
pyrophoric Producing sparks when rubbed, or burning spontaneously in air; e.g., finely divided metals. **p. alloy** A Cr—Ce—Fe alloy, used in gas lighters. **p. lead** The p. Pb—C mixture obtained on heating lead tartrate. **p. reaction** A reaction that produces flame.
pyrophosphorus Any of several substances or mixtures that ignite spontaneously on exposure to air.
pyrophosphate* Diphosphate*. A salt of pyrophosphoric acid; as, $M_4P_2O_7$ and $M_2H_2P_2O_7$.

pyrophosphite Diphosphonate*. $M_4P_2O_5$ and $M_2H_2P_2O_5$.
pyrophosphoric acid* See *phosphoric acid.*
pyrophosphorous acid See di*phosphonic acid.*
pyrophosphoryl The radical $\equiv P_2O_3$. **p. chloride** $(Cl_2PO)_2O$ = 251.8. Colorless, fuming liquid, b.250, hydrolyzed by water to orthophosphoric and hydrochloric acids.
pyrophyllite $Al_2O_3 \cdot 4SiO_2 \cdot H_2O$, resembling talc and montmorillonite; its melting point is raised from 1315 to 2093 by pressure. Used as filler for paint, rubber, and textiles; as a polishing agent for foods; in lubricants and toilet preparations; and in the production of ultrahigh *pressures,* q.v. Cf. *image stone.*
pyroracemamide $MeCO \cdot CO \cdot NH_2$ = 87.1. Colorless crystals, m.124.
pyroracemic acid Pyruvic acid*.
pyroracemic aldehyde Pyruvaldehyde*.
pyrosine Erythrosin.
pyrostilbnite Kermesite.
pyrosulfate Disulfate*.
pyrosulfite Disulfite*.
pyrosulfuric acid Disulfuric acid*.
pyrosulfuryl The disulfuryl* radical.
pyrotartaric acid $MeCH(COOH)CH_2COOH$ = 132.1. Methylsuccinic acid, 2-methylbutanedioic acid*, pyrovinic acid. Colorless, triclinic crystals, m.113, soluble in water. Cf. *ethyl*malonic acid. **hydroxy ~** Citramalic acid.
pyrotechnics (1) The science of explosives which are burned rather than detonated (U.S. and U.K. usage), including cool-burning propellants used as gas generators. (2) The study of all explosive substances (European usage).
pyroterebic acid $Me_2C{:}CH \cdot CH_2COOH$ = 114.1. 4-Methyl-3-pentenoic acid*. Colorless crystals, m.207.
pyrothioarsenate $M_4As_2S_7$.
pyrotritaric acid Uvic acid.
pyrouric acid Cyanuric acid.
pyrovanadic acid See *vanadic acid.*
pyrovinic acid Pyrotartaric acid.
pyroxene $CaO \cdot MgO \cdot 2SiO_2$. A white, green, or black silicate.
pyroxenite Websterite.
pyroxylic spirit Methanol*.
pyroxylin Pyroxylinum, soluble guncotton, collodion cotton, trinitrocellulose, collodium, colloxylin. A solution of nitrated cellulose in a solvent of high boiling point; it consists of cellulose trinitrate, $C_6H_7O_5(NO_3)_3$, and tetranitrates, $C_6H_6O_5(NO_3)_4$. Used as an artificial skin collodion (USP, BP), and in the manufacture of artificial silk, leather, cloth, and lacquers. See *rayon.*
pyrrhosiderite Göthite.
pyrrhotine Pyrrhotite.
pyrrhotite Fe_6S_7 or $Fe_{11}S_{12}$. Pyrrhotine. Natural or synthetic magnetic pyrite; the former usually contains nickel.
pyrrilium Pyrylium.
pyrrole* ***1H*-~** $CH{:}CH \cdot NH \cdot CH{:}CH$ (ring) = 67.1. Azole. Colorless liquid from bone oil and coal tar, $d_{20} 0.9481$, b.130, soluble in alcohol or ether; its derivatives are antiseptics. Cf. *porphin ring.* **dibenzo ~*** Carbazole*. **dihydro ~** Pyrroline*. **ethyldimethyl ~** Cryptopyrrole. **tetrahydro ~** Pyrrolidine*.
p.-1-carboxylic acid* $C_4H_4N \cdot COOH$ = 111.1. A solid, decomp. 191, soluble in water.
pyrrolidine* $CH_2 \cdot CH_2 \cdot NH \cdot CH_2 \cdot CH_2$ (ring) = 71.1. Pentazane, butene imide, tetramethyleneimine, tetrahydropyrrole. Colorless liquid, b.88, soluble in water. **dioxo ~** Succinimide*. **methyl pyridyl ~** Nicotine.
p. alkaloids See *nicotine group.* **p.carboxylic acid*** Proline*. **p.dione*** Succinimide*.
pyrrolidinium A derivative of cationic pyrrolidine. Cf. *piperidinium.*
pyrrolidinyl* The radical C_4H_8N-, from pyrrolidine.
pyrroline* C_4H_7N = 69.1. **3-~** 2,5-Dihydro-1*H*-pyrrole. Colorless liquid, b.90, soluble in water.
pyrrolinium A cationic (N^+) derivative of pyrroline. Cf. *piperidinium.*
pyrroloindole $C_{10}H_7N$ = 141.2. A tricyclic compound, m.35.
pyrrolopyridine $C_7H_5N_2$ = 117.1. A compound with a fused hexa- and pentaatomic ring, each having 1 N.
pyrroloquinoline $C_{12}H_9N$ = 167.2. A tricyclic system. Crystals, m.108.
pyrrolyl* The radical C_4H_4N-, from pyrrole. ***N*-pyrrolyl** The substituting radical is attached to the N atom. **alpha-~** $C_4H_4N \cdot CO \cdot Me$. Colorless crystals, m.90.
pyrrolylene 1,3-Butadiene*.
pyrromonazole **alpha-~** Pyrazole*. **beta-~** 1*H*-Imidazole*.
pyrrotriazole 1*H*-1,2,3,4-Tetrazole.
pyrroyl The pyrrolylcarbonyl† radical, $CH{:}CH \cdot CH{:}CH \cdot N \cdot$ (ring) $CO-$, from pyrrolecarboxylic acid.
pyrryl The pyrrolyl* radical.
pyruric acid Cyanuric acid.
pyruvaldehyde* $Me \cdot CO \cdot CHO$ = 72.1. Pyroracemic aldehyde, pyruvic aldehyde, ketoacetaldehyde, methyl glyoxal, 2-oxopropanal*. Yellow, pungent liquid, b.72, polymerizes to a glassy mass.
pyruvic acid* $Me \cdot CO \cdot COOH$ = 88.1. Pyroracemic acid, oxoacetic acid, acetylformic acid, 2-oxopropanoic acid*. Colorless crystals, or liquid, d.1.288, m.14, b.165 (decomp.), soluble in water; an intermediate product in the metabolism of proteins, fats, and carbohydrates.
pyruvic aldehyde Pyruvaldehyde*.
pyruvonitrile* CH_3COCN = 69.1. Acetyl cyanide, 2-oxopropanenitrile*. Colorless liquid, b.93, decomp. in water.
pyrivinium pamoate $C_{52}H_{56}N_6 \cdot C_{23}H_{14}O_6$ = 1,151. Viprynium embonate. Orange crystals, insoluble in water; an anthelmintic (USP, BP).
pyrvolidine An alkaloid from the leaves of the wild carrot, *Daucus carota* (Umbelliferae).
pyrylium $CH{:}CH \cdot CH{:}O(R) \cdot CH{:}CH$ (ring). Describing cationic compounds derived from pyran.
pyx (1) Trial of the pyx. The annual independent testing of coins issued by the British Mint, for composition and weight, since the 13th century. (2) Pix. **p. liquida** Wood *tar.*

Q

Q Symbol for (1) nuclear quadrupole moment; (2) partition coefficient; (3) quantity of electricity (charge); (4) quantity of heat; (5) quantity of light (Q_v); (6) reactive power.
q Symbol for (1) a generalized coordinate; (2) heat flow rate (Φ is preferred).
qt Abbreviation for quart.
Qu Abbreviation for quinine.
quad A unit of energy; 10^{15} Btu (U.S. usage). U.S. consumption about 80 q annually.
quadrant (1) The quarter of a circle. (2) The distance from the pole to the equator = 10^7 meters = 0.25 meridian. (3) See *henry*.
quadratic A cubical tetragon. Cf. *crystal* systems.
quadri- Tetra-*.
quadrilateral Describing a 4-sided figure.
quadrimolecular Tetramolecular*.
quadro-* Affix indicating 4 atoms bound into a quadrangle (e.g., square).
quadroxalate Hydrogen*oxalate*.
quadroxide Tetraoxide*.
quadruple point The temperature at which 4 phases are in equilibrium, e.g., a saturated solution containing excess of solute. Cf. *triple point*.
quaker buttons Nux vomica.
qualitative Pertaining to the kind or type. **q. analysis** The methods by which the constituents of a substance are detected. **q. reaction** A reaction that detects one substance in a mixture.
quality control Ensuring that specific q. levels are maintained by a manufacturing process, as by obtaining and analyzing quality data. Action must be taken if quality deviates by more than a specified amount.
quanta Plural of "quantum."
quantitative Pertaining to an amount. **q. analysis** The methods by which the amount of a constituent is determined. **q. reaction** (1) A reaction of q. analysis. (2) A reaction that proceeds wholly or almost to completion.
quantity The amount, active mass, or concentration of a substance. **physical ~** The numerical value of a phenomenon: *Extensive* when the quantity has an additive effect; as, mass. *Intensive* when the quantities added increase the intensity of a property; as, temperature.
q. of electricity See *coulomb*.
quantivalence Valency.
quantization A change, as from a normal to an excited atom, in which electrons are raised to a higher energy level by the absorption of energy. Equally, from an excited to a normal level, when energy is emitted.
quantum (Pl. quanta.) E(ne)rgon. The unit amount of energy *E* set free or bound during the emission or absorption of radiation: $E = h\nu$, where ν is the atomic frequency number of the radiation, and *h* is the Planck constant. Cf. *Compton effect, Raman effect*. **q. group** A number that defines spectroscopic terms as a resultant of the orbital numbers of an atom. Cf. *Stoner quanta*. **q. mechanics** A departure from classical Newtonian mechanics, q. m. was developed by Planck from his discovery that black bodies radiate in discrete quanta, equal to $h\nu$. It has wide application in that it applies to electromagnetic radiation, serves to explain atomic spectra, and was subsequently developed into wave mechanics. **q. number** A number describing a q. state. See *symbol*, Table 88—groups *J* and *N*. $^2P_{3/2}$ indicates a q. state of $L = 1$, total angular momentum quantum number $\frac{3}{2}$, and spin multiplicity $(2S + 1)$ equal to 2. Uppercase letters are used for systems; lowercase, for single electrons. **magnetic ~** *M,m*. Describes states in an external magnetic field. For a given value of *l*, it can have values $-l \ldots 0 \ldots +l$. **orbital angular momentum ~** *L, l*. Azimuthal q. n. For an s electron, $l = 0$; for a p electron, $l = 1$, etc. **principal ~** *n* corresponds to the shell or energy level of an electron in the Bohr atom. $n = 1 = K$, $n = 2 = L$, etc. **spin ~** Can take values of $\pm\frac{1}{2}$. *S* is the electronic and *I* (or *J*) the nuclear spin quantum number. **q. relation** See Table 67. **q. state** See *energy levels*. **q. theory** Energy changes occur in continuous pulsations, not gradually, and always in multiples of a definite quantity. See *q. mechanics* above. **q. unit** Planck constant. **q. yield** In photochemistry, the ratio of the number of product molecules formed to the number of protons required.
quark Together with leptons, believed to be the only elementary *particles*, q.v., of matter. Quarks are the constituents of the hadrons. There are 4 q. types (flavors): u (up), d (down), s (strange), and c (charm); each q. comes in 3 primary colors. The proton consists of 2 u quarks and 1 d quark.
quarry An open or surface working for minerals, as limestone. Cf. *mine, outcrop*.
quart Qt. 0.25 gallon. **U.S. liquid ~** 0.946353 liter. **U.S. dry ~** 1.1012 liters. **imperial ~** 1.13652 liters, or 69.355 in^3; 2.5 lb distilled water occupies 1 qt. **Winchester ~** See *Winchester quart*.
quartation Inquartation. Separation (parting) of silver from gold by dissolving the silver in nitric acid; it is quantitative for Au 25, Ag 75%, and, if necessary, additional silver must be added for separation.
quarter (1) One-fourth part. (2) An old English measure of capacity, generally for grain: 8 bushels. **q. wave plate** A transparent plate, e.g., mica, used in petrological determinations of refractive index; it produces a phase

TABLE 67. QUANTUM RELATION OF RADIATION AND ENERGY

Type	λ	ν	hν	Q
Red	7,500	400	2.65	159
Yellow	5,900	508	3.37	203
Blue	4,900	612	4.05	244
Violet	4,550	659	4.37	263
Ultraviolet	3,950	759	5.03	303

λ = wavelength, angström
ν = frequency ($\times 10^{12}$ per s)
$h\nu$ = quantum per molecule ($\times 10^{-31}$ J)
Q = quantum energy of 1 mole of matter, nJ (6.022×10^{23} $h\nu$)

difference of ¼ wavelength of light between 2 emergent beams.

quartering A method of sampling, e.g., coal, by dividing a heap into approximately equal quarters, mixing opposite quarters, redividing the mixture into quarters, and continuing these operations until a sample of suitable size results.

quartile The top (upper q.) and bottom (lower q.) 25% of a statistical population.

quartz SiO_2. Silicon dioxide, silica. (1) Glassy or crystalline, native silica; or massive in veins; as, rose or milky q. (2) Prepared artificially by seeding an aqueous solution of sodium hydroxide with small q. crystals, and autoclaving. Pure q. occurs in colorless hexagons, d.2.66, hardness 7, mean refractive index 1.55. It occurs in α and β forms (transition point 573), is stable below 870, and changes to cristobalite above 1,200. See *silica.* Many colored varieties are semiprecious stones. Native varieties: amethyst (red-violet, transparent), catalinite (green, red, and brown mottled), cat's-eye (green and brown), citrine (yellow), milky quartz (opalescent), rock crystal (colorless hexagons), rose quartz (pale rose), smoky quartz (gray). Q. is very resistant to acids, m.1715, and is used for chemical apparatus: rock crystals (transparent to ultraviolet rays) for optical apparatus; threads for delicate suspensions. **q. apparatus** Chemical utensils, transparent and highly resistant to sudden temperature changes, made from fused rock crystal. **q. clock** Clock utilizing the piezoelectric properties of a q. crystal. **q. lamp** Mercury-vapor lamp. **q. lens** (1) A rock crystal lens transparent to ultraviolet rays. (2) A fused quartz lens. **q. oscillator** A section cut from a q. crystal; used to tune radio circuits, as it shows sharply tuned resonance. **q. resonator** Q. oscillator. **q. rock** Quartzite.

quartzite A yellow, impure quartz.

quasar Quasi-stellar radio source. A distant astronomical object that emits strong radiofrequency radiation and shows a high red shift.

quasi-chemical Having the attributes of a chemical compound, but not conforming to the law of constant proportions.

quassia Bitterwood, bitter ash. The dried wood of *Picrasma excelsa,* Jamaica q., or *Quassia amara,* Suriname q. (Simarubaceae); a hop substitute. **q. wood** Quassia.

quassic acid $C_{30}H_{38}O_{10}$ = 558.6. A glucoside from quassia.

quassin $C_{22}H_{28}O_6$ = 388.5. A bitter principle from quassia (0.03%). Colorless crystals, m.220, slightly soluble in water.

quassoid The total bitter principles of quassia; a febrifuge and bitter.

quaternary (1) The last *geologic* period, q.v. (2) A compound containing 4 different elements (types of atoms); as, $NaHSO_4$. **ammonium compounds** Q. amines. Organic derivatives of NH_4OH in which the hydroxyl group and the 4 H atoms are replaced by radicals; as, NMe_4I. Used as surfactants, antiseptics and disinfectants; as benzylkonium chloride. See *ammonium, -onium.* **q. carbon atom** A C atom linked to 4 other C atoms.

quebrachamine An alkaloid from quebracho. Colorless crystals, m.142, insoluble in water.

quebrachine Yohimbine.

quebrachite Quebrachitol.

quebrachitol $C_7H_{14}O_6$ = 194.2. 2-Methoxy derivative of L-*chiro*-inositol. L-inosite methyl ether, quebrachite, bornesitol. Colorless crystals, m.191. Occurs widely in plants. Cf. *pinite.*

quebracho, quebracho blanco Aspidosperma. The dried bark of *Aspidosperma quebracho blanco,* (S. America). It contains tannins and a number of alkaloids. **q. colorado** The dried wood of *Schinopsis* species, which contains fisetin. **q. extract** A tanning extract from the heartwood of q. (35–65% tannin). **q. gum** The dried juice of *Schinopsis lorentzii* (Anacardiaceae), Argentina; a tan.

quebrachomine An alkaloid from quebracho.

queen of the meadow (1) The dried root of *Eupatorium purpureum* (Compositae), an astringent and diuretic. Cf. *euparin.* (2) Meadowsweet.

queen's **q. delight** Stillingia. **q. root** Stillingia. **q. yellow** Mercuric subsulfate.

queen substance A substance secreted by queen honeybees, which inhibits queen rearing and the development of ovaries in workers. Its principal constituent is related to the 10-hydroxy-2-decenoic acid present in royal jelly.

quenching (1) Cooling suddenly, as in tempering steel. (2) Producing a diminution, e.g., of fluorescence.

quercetagetin $C_{15}H_{10}O_6$ = 286.2. Yellow crystals, m.318, from African marigold. See *flavones.*

quercetin $C_{15}H_{10}O_7$ = 302.2. Meletin, sophoretin, quercetinic acid, flavin. Yellow dye obtained by decomposition of q., m.312, soluble in water. See *flavones.*

quercetinic acid Quercetin.

quercic Quercinic acid.

quercimetin Quercitrin.

quercin $C_6H_{12}O_6$ = 180.2. Bitter crystals, from oak bark and acorns.

quercinic acid $C_{31}H_{50}O_4$ = 486.7. Quercic acid, oaktannin. Brown crystals, m.204; from the wood of *Quercus* species.

quercitannic acid $C_{28}H_{28}O_{14}$ = 588.5. A tannin from oak bark; hydrolyzed to oak red.

quercite Quercitol*.

quercitin (1) A glucoside in ragweed pollen. (2) Quercetin.

quercitol* $CHOH\cdot(CHOH)_2\cdot CH_2\cdot(CHOH)_2$ = 164.2. (ring closed from first to last CHOH)

Cyclohexanepentol*, quercite, acorn sugar, pentahydroxycyclohexane. Colorless, sweet crystals in oak bark, m.224 (decomp.), soluble in water.

quercitrin $C_{21}H_{22}O_{12}\cdot 2H_2O$ = 502.4. Quercimetin, quercitrinic acid. A rhamnoside of quercitron bark, oak bark, tea leaves, hops, horse chestnut. Yellow crystals, m.160–200, insoluble in water; hydrolyzed by acids to rhamnose and quercetin, and used in their manufacture.

quercitrinic acid Quercitrin.

quercitron The coarse, powdered bark (25% tannin) of *Quercus tinctoria* (Cupuliferae), N. America; a tan and dye.

quercus The dried bark of *Quercus alba,* white oak (Cupuliferae); an astringent and tan.

Questran Trademark for cholestyramine resin.

Quevenne's iron Reduced iron.

quickening liquid A solution of mercuric nitrate or cyanide, used in electroplating.

quicklime Calcined limestone; chiefly calcium oxide in natural association with magnesium oxide.

quicksilver Mercury. **horn ~** A native mercurous chloride.

quick-vinegar process Vinegar manufacture by passing weak alcohol slowly through wood shavings covered with *Bacterium aceti.*

quillaia Quillaya, quillaja, soapbark, Panama bark. The bark of *Quillaja saponaria* (Rosaceae), S. America, containing saponins; a foam producer, emulsifier, and sternutatory (BP).

quillaic acid $C_{30}H_{46}O_5$ = 486.7. A saponin from the bark of *Quillaja saponaria.* Crystals, m.294.

quina Cinchona bark.

quinacetine $C_{27}H_{31}O_2N_3$ = 429.6. An alkaloid from cinchona bark. **q. sulfate** $(C_{27}H_{31}O_2N_3)_2H_2SO_4 \cdot H_2O$ = 975.2. Colorless powder, soluble in water; an antipyretic.

quinacridine $C_{20}H_{12}N_2$ = 280.3. *Ortho*-fused pentacyclic compounds of 3 benzene rings and 2 pyridine rings; as, dibenzophenanthrolines.

quinacridone Quin(2,3-b)-acridone-7,14(5,12)dione, C.I. Pigment Violet 19. A red pigment which can be converted into other red and a blue pigment.

quinacrine $C_{23}H_{30}ClON_3$ = 400.0. Mepacrine, Atabrin(e). **q. hydrochloride** $C_{23}H_{30}ClN_3O \cdot 2HCl \cdot 2H_2O$ = 508.9. Bitter, yellow crystals, soluble in water; used for tapeworm and formerly for malaria (USP, BP).

quinalbarbitone sodium EP, BP name for secobarbital sodium.

quinaldic acid 2-*Quinoline*carboxylic *acid;* used in analysis. **4-hydroxy ~** Kynurenic acid.

quinaldine $C_6H_4 \cdot CH:CH \cdot CMe:N$ = 143.2. 2-Methylquinoline*. Colorless liquid, d.1.186, b.244, soluble in water; used medicinally. **naphtho ~** See *naphthoquinaldine.*

quinaldinium compounds Derivatives of cationic (N^+) quinaldine.

quinalizarin $C_{14}H_8O_6$ = 212.2. 1,2,5,8-Tetrahydroxyanthraquinone*. A reagent for beryllium; the violet alkaline solution becomes blue.

quinamine $C_{19}H_{24}N_2O_2$ = 312.4. A crystalline alkaloid from cinchona bark. Cf. *apoquinamine.*

quinanaphthol $C_{20}H_{24}N_3O_2(OHC_{10}H_6SO_3H)_2$ = 786.9. Quinaphthol, chinaphthol, quinine 2-naphthol monosulfate. Yellow crystals, m.185, soluble in water.

quinane $C_{20}H_{24}N_2$ = 292.4. Deoxyquinine. The heterocyclic reduction product of quinine.

quinaphthol Quinanaphthol.

quinate A salt of quinic acid.

quinazerin Quinizarin.

quinazine Quinoxaline*.

quinazoline* $C_6H_4 \cdot N:CH \cdot N:CH$ = 130.1. Phenmiazine, benzopyrimidine, 1,3-benzodiazine. Colorless crystals, m.48. **dihydrooxo ~** Quinazolone. **methyl ~** $C_9H_8N_2$ = 144.3. **2- ~** Colorless crystals, m.35. **oxo ~** Quinazolone. **tetrahydrodioxo ~** Benzoyleneurea.

quinazolone $C_8H_6ON_2$ = 146.1. Dihydrooxoquinazoline. Cf. *quinoxaline.*

quince seeds The dried seeds of *Pyrus cydonia* (Rosaceae). Used medicinally as a demulcent (as the mucilage).

quinene $C_{20}H_{22}ON_2$ = 306.4. Koenig's term for the heterocyclic parent compound of the quinine alkaloids.

quinetum A mixture of the alkaloids of cinchona bark in the proportions in which they occur in nature. Cf. *quinium.*

quinhydrone $C_6H_4O_2 \cdot C_6H_4(OH)_2$ = 218.2. A compound of benzoquinone and hydroquinone dissociating in solution. Green powder, m.171 (sublimes), soluble in hot water. **q. electrode** A platinum wire in a saturated solution of q.; used as a reversible electrode standard in pH determinations: pH = 2.03 + $E/0.0577$, where E is the electromotive force against a saturated calomel electrode at 18°C.

quinhydrones Intermediate reduction products of benzoquinones.

quinia Quinine.

quinic acid $C_6H_7(OH)_4 \cdot COOH$ = 192.2. 1,3,4,5-Tetrahydroxycyclohexanecarboxylic acid*. Kinic acid. **(−)- ~** In cinchona bark. Colorless, monoclinic crystals, m.162 (decomp.), soluble in water.

quinicine $C_{20}H_{24}O_2N_2$ = 324.4. An amorphous cinchona bark alkaloid, m.60, isomeric with quinine.

TABLE 68. QUININE-RELATED COMPOUNDS

$R^1(R^2)$[a]	(−)-isomer	(+)-isomer
H (*1*)	Cinchonidine	Cinchonine
OH (*1*)	Cupreine	Cupreidine
OH (*2*)	Hydrocupreine	Hydrocupreidine
OCH_3 (*1*)	Quinine	Quinidine
OCH_3 (*2*)	Hydroquinine	Hydroquinidine
OCH_3 (*3*)	Ethylquitenine	Ethylquitenidine
OC_2H_5 (*2*)	Optochine	Optochinidine
OC_5H_{11} (*2*)	Eucupin	(Eucupidine)
OC_8H_{17} (*2*)	Vuzine	(Vuzidine)

[a] (*1*) is $-CH:CH_2$; (*2*) is $-CH_2 \cdot CH_3$; (*3*) is $-COOC_2H_5$

quinidine $C_{20}H_{24}O_2N_2$ = 324.4. An alkaloid from cinchona bark: the (+) isomer of quinine. Colorless prisms, m.168, insoluble in water, soluble in alcohol, ether, or chloroform. Used as a cardiac rhythm regulator. **q. gluconate** $C_{20}H_{24}O_2N_2 \cdot C_6H_{12}O_7$ = 520.6. Bitter, white powder, soluble in water; a cardiac rhythm regulator (USP). **q. sulfate** $(C_{20}H_{24}O_2N_2)_2 \cdot H_2SO_4 \cdot 2H_2O$ = 783.0. Bitter, colorless needles, soluble in water; a cardiac depressant and rhythm regulator. (USP, EP, BP).

quinine $C_{20}H_{24}O_2N_2$ = 324.4. (8α, 9*R*)-6′-Methoxycinchonan-9-ol†. Quinina, quinia, chinine. *Qu.* A levorotatory alkaloid from the bark of *Cinchona* species. Colorless needles, m.175, slightly soluble in water, soluble in alcohol or ether. Its (+) isomer is quinidine (see Table 68). An antimalarial, usually as its salts. Q. kills the parasite after it has entered the red blood cells, but not when in the liver; also used for night cramps. **deoxy ~** Quinane. **eu ~** Q. diethyl carbonate.

q. alkaloids Alkaloids from cinchona bark, related to q. Q. contains 4 asymmetric C atoms.

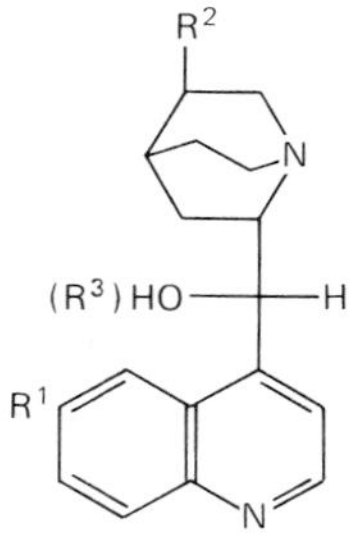

q. diethylcarbonate $QuEt_2CO_3$ = 442.3. Euquinine. Needles, m.89, slightly soluble in water; a tasteless q. substitute. **q. hydrochloride** $QuHCl \cdot 2H_2O$. Needles, m.156, soluble in water (EP, BP). **q. sulfate** $Qu_2H_2SO_4 \cdot 2H_2O$ = 872.8. Colorless, silky needles, m.205, slightly soluble in water; soluble in alcohol, ether, or very dilute acids; an antipyretic and antimalarial (USP, EP, BP). **q. synthesis** Skraup's reaction.

quininic acid $MeO \cdot C_9H_5N \cdot COOH$ = 203.2. Yellow prisms, m.285 (decomp.), slightly soluble in water; used in organic synthesis.

quininone $C_{20}H_{22}N_2O_2$ = 322.4. The ketone of *quinine,* q.v.; R^3 is CO, R^2 is $CH:CH_2$, and R^1 is OCH_3.

quinisatin $C_9H_5NO_3$ = 175.1. 2,3,4(1*H*)-Quinolinetrione.

The heterocyclic lactam or lactim of quinasitinic acid:

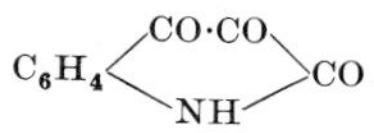

$$C_6H_4 \langle \begin{smallmatrix} CO\cdot CO \\ NH \end{smallmatrix} \rangle CO$$

Red crystals, m.272 (decomp).
quinite Quinitol.
quinitol 1,4-Cyclohexanediol*.
quinium Crude quinine containing alkaloids of cinchona bark. Amorphous, white mass; used to prepare quinine and its salts. Cf. *quinetum.*
quinizarin $C_{14}H_8O_4$ = 240.2. 1,4-Dihydroxyanthraquinone*, quinazerin. Red crystals, m.200, insoluble in water.
quinizine Antipyrine.
quinoa Dried seeds of *Chenopodium quinoa* (Chenopodiaceae), Peru and Chile; an edible, starchy flour, containing 12–19% protein.
quinochromes Blue, fluorescent products of the oxidation of vitamin B_1. Cf. *thiochrome.*
quinogen $R\cdot CO\cdot CO\cdot CH_2\cdot C\cdot R\cdot(OH)\cdot CO\cdot CH_3$. An intermediate compound in the condensation of α-diketones to quinones.
quinoid Quinonoid.
quinol Hydroquinone*.
quinoline* C_9H_7N = 129.2. 1-Benzazine, chinoline. A decomposition product of quinine; a distillation product from bone oil and petroleum.

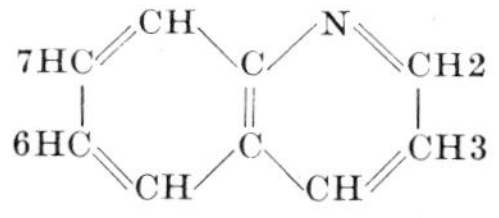

7HC, CH, N, CH2, 6HC, CH, C, C, CH, CH3

Colorless, refractive oil, d_{20}·1.0944, m.−20, b.238, aromatic odor, soluble in hot water or benzene; a solvent for resins, camphor, and terpenes. **amino ~ *** See *aminoquinoline* under *amino.* **(aminophenyl)methyl ~** Flavaniline. **benzo ~ †** See *benzo[ ]quinoline.* **dimethyl ~ *** Kryptidine. **hydroxy ~ *** Quinolinol*. **iso ~ *** $C_6H_4\cdot CH{:}N\cdot CH{:}CH$ = 129.2. Leucoline, 2-benzazine. Colorless crystals, m.23, slightly soluble in water. **i. alkaloids** *Alkaloids*, q.v., from 1-benzyl-*N*-methyltetrahydroisoquinoline; e.g., the papaverine, corydaline, morphine, hydrastine, cryptopine, berberine, and fumarine groups. **methyl ~ *** **2- ~** Quinaldine. **4- ~** Lepidine. **nitro ~ *** See *nitroquinoline.* **2(1*H*)-oxo ~** Carbostyril*. **tetrahydro ~** See *tetrahydroquinoline* under *tetrahydro-.* **tetrahydrohydroxy ~** Kairine.
q. acids $C_9H_6N\cdot COOH$ = 173.2. Q.carboxylic acids. **2- ~** Quinaldic acid, m.156. **3- ~** m.275. **4- ~** Cinchoninic acid. m.254. **5- ~** m.342. **6- ~** m.291. **7- ~** m.249. **8- ~** m.187. **q. aldehyde** $C_9H_6N\cdot CHO$ = 157.2. **2- ~** Colorless crystals, m.71, soluble in water. **q. alkaloids** See *alkaloid.* **q. blue** Cyanine. **q. dyes** Dyestuffs, photographic sensitizers, and indicators containing the configuration

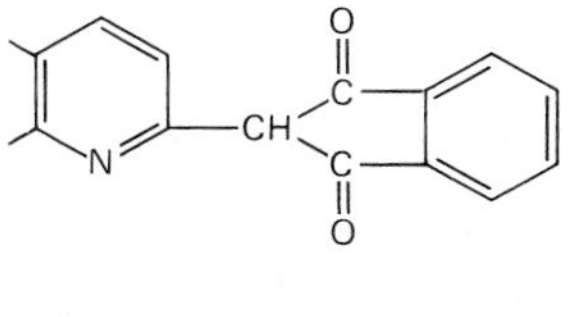

O, C, CH, N, C, O

q. hydrochloride $C_9H_7N\cdot HCl$ = 165.6. Colorless crystals, m.94, soluble in water. **q. red** $C_{26}H_{19}N_2Cl$. 1,1′-Benzilidene 2,2′-quinocyanine chloride. A q. dye.
quinolinic acid $C_5H_3N(COOH)_2$ = 167.1. 2,3-Pyridinedicarboxylic acid*. Colorless crystals, m.190 (decomp.), insoluble in water. **hydroxy ~** $C_7H_5O_5N$ = 183.1. 2-Hydroxypyridine-5,6-dicarboxylic acid*.
quinolinium compounds Cationic (N^+) quinoline derivatives, analogous to pyridinium and quinaldinium compounds.
quinolinol* C_9H_7NO = 145.2. Hydroxyquinoline. **2- ~** Minor tautomer of carbostyril. **4- ~** Minor tautomer of kynurine. **5- ~** m.224. **6- ~** m.193, b.360. **7- ~** decomp. 238. **8- ~** Oxine, m.76; a precipitant for aluminum.
quinolinone **2(1*H*)- ~** Carbostyril*. **4(1*H*)- ~** Kynurine*.
quinolizine* C_9H_9N = 131.2. Heterocyclic compounds of 2 pyridine rings, with a common N atom and one saturated C atom.
quinolone* Carbostyril*.
quinolyl* Quinolinyl†. Quinoyl. The radical C_9H_6N-, from quinoline; 7 possibilities of substitution. See *quinoline.*
quinone Benzoquinone*. **anthra ~** See *anthraquinone.* **benzo ~** See *benzoquinone.* **hydro ~ *** See *hydroquinone.* **meri ~** See *meriquinone.* **naphtho ~ *** See *naphthoquinone.* **semi ~** The anion *(para-)*

·O, O⁻

quinones Benzoquinones*.
quinonoid† Quinoid, paraquinoid. Relating to the benzoquinone structure. Cf. *hemiquinonoid.* **q. dyes** See *alizarin.*
quinonyl The dioxocyclohexadienyl† radical, $C_6H_3O_2-$, from benzoquinone.
quinoquinoline 1,10-Naphthodiazine.
quinosol Potassium oxyquinoline sulfate. Chinosol. White powder, soluble in water. A preservative for anatomical specimens and a keratolytic for skin conditions.
quinotannic acid $C_{14}H_{16}O_9$ = 328.3. Cinchonatannin. A tannin from cinchona bark. Yellow powder, soluble in water.
quinovic acid $C_{30}H_{46}O_5$ = 486.7. White crystals, m.302, soluble in chloroform.
quinovin $C_{30}H_{48}O_8$ = 536.7. Kinocin, chinovin. A glucoside from cinchona bark, m.235
quinovose A glucoside in cinchona bark; a stereoisomer of rhamnose.
quinoxaline* $C_8H_6N_2$ = 130.1. 1,4-Benzodiazine, quinazine, phenpiazine. Colorless crystals, m.305, slightly soluble in water; used in organic synthesis. **naphtho ~** Naphthisodiazine. **1,2,3,4-tetrahydro ~** $C_8H_{10}N_2$ = 134.2. Colorless crystals, m.97, soluble in alcohol. **q.dicarboxylic acid** $C_8H_4N_2(COOH)_2$ = 218.2. **2,3- ~** Colorless crystals, m.190, soluble in water.
quinoxalinyl* The radical $C_8H_5N_2-$, from quinoxaline.
quinoxime Benzoquinone monoxime.
quinoyl The quinolyl* radical.
quinque-, quinqui- Penta-*.
quintal (1) A unit of weight in the avoirdupois system, equal to 100 lb. (2) A unit of weight in the metric system equal to 100 kg.
quintavalent Pentavalent.

quintenyl Pentyl*.
quintessence A concentrated extract; as, an essential oil.
quintozene* See *fungicides,* Table 37 on p. 250.
quintuple point The temperature of a system at which 5 phases exist in equilibrium. Cf. *triple point, quadruple point.*
quire A number of sheets of paper (usually 144) of the same dimensions and weight per unit area.
quitenidine $C_{19}H_{22}O_4N_2$ = 342.4. An oxidation product of quinidine. Cf. *chitenine.*
quotient The numerical result of division. **albumen ~** The ratio of blood albumen to total albumen. D- **~** The ratio of glucose to nitrogen in urine.
q.v. Abbreviation for (1) quod vide—"which see"; (2) quantum viz—"as much as you wish."

R

R Symbol for (1) roentgen; (2) organic *radical.* **R acid** 2-Naphthol-3,6-disulfonic acid. **2R acid** 2-Amino-8-naphthol-3,6-disulfonic acid. **R salt** Sodium salt of R acid.

R Symbol for (1) (molar) gas constant; (2) rectus, see *sequence rule;* (3) refraction; (4) reluctance (R_m); (5) resistance.

R_∞ Symbol for Rydberg constant.

r Symbol for (1) formerly, racemic, now (±); (2) radius.

ρ See *rho.*

Ra Symbol for radium.

ra Abbreviation for "radioactive"; as *ra*Cl, radiochlorine.

rabies vaccine Inactivated virus grown in human diploid cells; used for prevention of r. either before or after exposure to the virus.

rabble An iron stirrer for molten metal; as, rabbling a charge of ore in a reverberatory furnace.

rabelaisin A poisonous glucoside from *Rabelaisa philippinensis* (Philippines); an arrow poison.

***ra*C** Radioactive carbon.

RaC′ Radium C′. Early name for polonium-214.

racahout Meal prepared from the edible acorn; used, with sugar and flavoring, as a food.

racemase* See *enzymes,* Table 30.

racemate (1)* Any homogeneous phase containing equimolar amounts of enantiotropic molecules. (2) A salt of a racemic acid, generally of (±)-tartaric acid.

racemation Racemization.

raceme A racemic compound.

racemic* (±)-*, *r-*, *dl-*. Indicating that equal amounts of enantiomeric molecules are present together, irrespective of whether in crystalline, liquid, or gaseous form. Inactive optically, but separable into dextro- and levorotatory forms. Cf. *meso.* **r. acid** (±)-Tartaric acid. **r. compound*** A homogeneous solid phase composed of equimolar amounts of enantiomeric molecules. **r. mixture*** A mixture of equimolar amounts of enantiomeric molecules present as separate solid phases.

racemization Racemation, racemisation. The transformation of optically active substances into optically inactive substances or mixtures. **auto ~** Spontaneous r. **partial ~** R. that affects only a few asymmetric groups.

racephedrine Racemic ephedrine.

rackarock An explosive that is made, as required, from a mixture of potassium chlorate and nitrobenzene, sometimes with picric acid.

racking (1) Separation of ores by washing on an inclined plane (rack). (2) The final stage in brewing, when the liquor is clarified. (3) The almost complete separation of solid glycerides from cod-liver oil on cooling.

rad (1) A unit of *r*adiation *a*bsorbed *d*ose; an energy absorption of 100 erg/g of tissue. 1 rad = 10^{-2} gray (the SI unit). Cf. *rem.* (2) Radian.

radar *Ra*dio *d*etection *a*nd *r*anging. Radiolocation. The location of an object by means of radiation, of wavelength 10^{-2} to 10 m, reflected from it.

raddle Hematite.

radian rad. (1) An arc of a circle that is as long as the radius. (2)* The angle subtended at the center by the arc of a circle equal to the radius: $180°/\pi = 57.29578° = 57°17'44.8062''$ = 1 radian. The SI unit for plane angles.

radiance* Symbol: *L;* radiant flux per unit solid angle per projected unit area, in $W/m^2 \cdot sr$.

radiant Diverging from a common center in all directions. **r. flux** See *radiant flux* under *flux.* **r. heat** Heat waves. **r. matter** (1) The residual gas in a luminous vacuum tube. (2) Radioactive matter. **r. state** (1) The condition of emitting light; as, incandescence, luminescence, fluorescence. (2) Crooke's fourth state of matter.

radiated Describing a rosette-shaped arrangement of crystals. **r. pyrite** Marcasite.

radiation (1) Transmission of energy through space, unassociated with motion of material particles, and without loss or change; *electromagnetic r.* accounts for interference, diffraction, refraction, and polarization. (2) Emission of material particles moving at high velocity; *corpuscular r.* accounts for the photoelectric and Compton effects. (3) Sometimes, transfer or diffusion of energy through matter, as heat waves. Cf. *ray, irradiation.* **background ~** R. from natural sources (cosmic and γ rays; equal to a human dose equivalent of about 1.2 mSv/year) plus any man-made radiation. **corpuscular ~** A stream of particles; as, α rays, positively charged atomic nuclei; or β rays or cathode rays, negatively charged particles or electrons moving with the velocity of light. **cosmic ~** See *cosmic rays.* **electromagnetic ~** Term embracing r. in the electromagnetic spectrum (see Table 69 on p. 492), all having the velocity of light. Consists of an electric field oscillating sinusoidally at right angles to the direction of propagation, accompanied by a similar magnetic field at right angles to the direction of both. Emitted discontinuously in photons. See *Planck constant.* **hard ~** R. of short wavelength and thus high penetrating power. **hertzian ~** See *hertzian waves.* **immaterial ~** The vibratory disturbance of a medium; as, sound waves in air. **K ~, L ~, M ~** See under *K, L, M,* etc. **material ~** Corpuscular r. **mechanical ~** Those material radiations which cause a vibration of molecules; as, heat waves. **monochromatic ~** A single-colored r., which consists of waves of equal or approximately equal wavelength. **permissible ~** The maximum dose of r. regarded as safe for occupational exposure in proximity to X-rays, γ-rays, β-rays, electrons, or positrons; i.e., generally (for persons over 18 years) 0.05–0.30 Sv/year, depending on the part of the body exposed (U.S. National Committee on Radiation Protection). **photic ~** See *electromagnetic radiation* above. See also *light.* **polarized ~** An electromagnetic r. in which the direction of the electromagnetic field, i.e., the plane of the waves, is parallel. In *plane*-polarized r. the plane remains constant; in *circular*-polarized r. the plane rotates; in *elliptical*-polarized r. the plane rotates and also changes in quantity. **soft ~** Opposite of hard *radiation,* q.v. **solar ~** (1) See *solar constant.* (2) The total r. from the sun's surface: 3.79×10^{26} J/s.

r. constants See *Planck's formula, Stefan-Boltzmann equation.* **r. dose** See *gray* (absorbed), *sievert* (equivalent). **r. effect** The phenomena resulting when r. falls on matter or is

TABLE 69. ELECTROMAGNETIC RADIATION

Frequency, kHz	Wavelength Å	Wavelength m	Type[a]	Source
3×10^{-1}	10^{16}	10^{6}		
				AC power line
3	10^{15}	10^{5}		
			Very low f. VLF	
3×10^{1}	10^{14}	10^{4}		Longwave radio
			Low f. LF	
3×10^{2}	10^{13}	10^{3}		
			Medium f. MF	AM radio
3×10^{3}	10^{12}	10^{2}		
			High f. HF	
3×10^{4}	10^{11}	10		
			Very high f. VHF	FM radio, TV
3×10^{5}	10^{10}	1		
			Ultrahigh f. UHF	
3×10^{6}	10^{9}	10^{-1}		
			Superhigh f. SHF	Radar
3×10^{7}	10^{8}	10^{-2}		
			Extremely high f. EHF	Microwave generator
3×10^{8}	10^{7}	10^{-3}		
3×10^{9}	10^{6}	10^{-4}		
			Infrared	Hot bodies
3×10^{10}	10^{5}	10^{-5}		
3×10^{11}	10^{4}	10^{-6}		
			Visible region	Sun and electric light
3×10^{12}	10^{3}	10^{-7}		
			Ultraviolet	
3×10^{13}	10^{2}	10^{-8}		
			(Soft)	
3×10^{14}	10	10^{-9}		
			X-rays	X-ray tube
3×10^{15}	1	10^{-10}		
			(Hard)	
3×10^{16}	10^{-1}	10^{-11}		
3×10^{17}	10^{-2}	10^{-12}		
			γ-rays	Radioactive emission
3×10^{18}	10^{-3}	10^{-13}		
3×10^{19}	10^{-4}	10^{-14}		
			Cosmic rays	Cosmos
3×10^{20}	10^{-5}	10^{-15}		

[a]f. = frequency. Note that divisions are not to scale.

intercepted by a body. The effects of r. may be physical (thermionic, photoelectric, fluorescent), chemical (photographic), or biological (photosynthetic). **r. hypothesis** See *r. theory of chemical reaction.* **r. intensity*** Radiant flux per unit solid angle, in W/sr. **r. pressure** The force exerted by light on particles that it strikes; as, radiometer. Sunlight has a r. pressure of 4.5×10^{-10} N. **r. temperature** See *radiator.* **r. theory of chemical reaction** The application of the quantum theory to chemical reactions. Before a molecule can react, the electrons of its constituent atoms become excited. The energy of activation may be calculated from the temperature coefficient of the reaction rate.

radiator A body that emits *radiation,* q.v. **perfect ~** Black body: a material absorbing all radiant energy (reflecting no light) and transforming it into heat. See *Stefan-Boltzmann equation, Wien law.*

radical (1) A group of atoms that behaves as a single atom in a chemical reaction. See Table 70 on p. 493, *acid.* Cf. *nomenclature, ion.* (2) A free *radical,* q.v. **acid ~** An electronegative group of atoms that remains intact in ordinary chemical reactions. **acyl ~*** Ac. An acid organic r. of the general type $R(CO-)_n$. **alkyl ~*** Al. An aliphatic organic r.; as, Me. **aryl ~*** Ar. An aromatic organic r.; as, Ph or Bz. **bi ~** A molecule having an unsatisfied valency on 2 different atoms. **free ~** A r., usually organic, highly reactive and short-lived, with an unsatisfied electron valence pair;

TABLE 70. NAMES FOR RADICALS AND IONS
See also *acids*, Table 3

Atom or group	As uncharged atom, molecule, or radical	As cation or cationic radical	As anion	As ligand	As prefix for substituent in organic compounds
H	(mono)hydrogen	hydrogen	hydride	hydrido	
F	(mono)fluorine	fluorine	fluoride	fluoro	fluoro[a]
OF	oxygen (mono)fluoride				fluorooxy, F—O—
Cl	(mono)chlorine	chlorine	chloride	chloro	chloro[a]
ClO		chlorosyl	hypochlorite	hypochlorito	chlorosyl,[a] O=Cl—
ClO_2	chlorine dioxide	chloryl	chlorite	chlorito	chloryl[a]
ClO_3		perchloryl	chlorate	chlorato	perchloryl[a]
ClO_4			perchlorate	perchlorato	
ClS		chlorosulfanyl			chlorothio, Cl—S—
ClF_2	chlorine difluoride		difluorochlorate(I)		difluorochloro
Br	(mono)bromine	bromine	bromide	bromo	bromo[a]
I	(mono)iodine	iodine	iodide	iodo	iodo[a]
IO		iodosyl	hypoiodite		iodosyl[a]
IO_2		iodyl			iodyl[a]
ICl_2			dichloroiodate(I)		dichloroiodo
O	(mono)oxygen		oxide	oxo	oxo, O=; oxy, —O—; oxido, $-O^-$
O_2	dioxygen	dioxygen(1+), O_2^+	peroxide, O_2^{2-}; hyperoxide, O_2^-	peroxo dioxygen	dioxy, —O—O—
O_3	trioxygen (ozone)		ozonide		trioxy, —O—O—O—
H_2O	water			aqua	oxonio, H_2O^+-
H_3O		oxonium			
HO	hydroxyl		hydroxide	hydroxo	hydroxy
HO_2	perhydroxyl		hydrogenperoxide	hydrogenperoxo	hydroperoxy
S	(mono)sulfur		sulfide	thio, sulfido	thio, —S—; sulfido, $-S^-$; thioxo, S=
HS	sulfhydryl		hydrogensulfide	mercapto	mercapto
S_2	disulfur	disulfur(1+)	disulfide	disulfido	dithio, —S—S—
SO	sulfur monoxide	sulfinyl (thionyl)			sulfinyl
SO_2	sulfur dioxide	sulfonyl (sulfuryl)	sulfoxylate	sulfur dioxide	sulfonyl
SO_3	sulfur trioxide		sulfite	sulfito	sulfonato, $-SO_3^-$
HSO_3			hydrogensulfite	hydrogensulfite	sulfo, $(HO)O_2S-$
H_2S	dihydrogen sulfide				sulfonio, H_2S^+-
H_3S		sulfonium			
S_2O_3			thiosulfate	thiosulfato	
SO_4			sulfate	sulfato	sulfonyldioxy, $-O-SO_2-O-$
Se	(mono)selenium		selenide	seleno	seleno, —Se—; selenoxo, Se=
SeO		seleninyl	selenoxide		seleninyl
SeO_2	selenium dioxide	selenonyl			selenonyl
SeO_3	selenium trioxide		selenite	selenito	
SeO_4			selenate	selenato	
Te	(mono)tellurium		telluride	telluro	telluro
CrO_2	chromium dioxide	chromyl			
UO_2	uranium dioxide	uranyl			
N	(mono)nitrogen		nitride	nitrido	nitrilo, N≡
N_2	dinitrogen	dinitrogen(1+), N_2^+		dinitrogen	azo, —N=N—; azino, =N—N=; diazo[a], $=N_2$; diazonio, $-N_2^+$
N_3			azide	azido	azido[a]
NH	aminylene	aminylene	imide	imido	imino
NH_2	aminyl	aminyl	amide	amido	amino
NH_3	ammonia			ammine	ammonio, H_3N^+-
NH_4		ammonium			
NH_2O			hydroxylamide	hydroxylamido-*O*; hydroxylamido-*N*	aminooxy, H_2NO-; hydroxyamino, HONH—
N_2H_3	hydrazyl	hydrazyl	hydrazide	hydrazido	hydrazino
N_2H_4	hydrazine			hydrazine	
N_2H_5		hydrazinium(1+)		hydrazinium(1+)	
N_2H_6		hydrazinium(2+)			
NO	nitrogen oxide	nitrosyl		nitrosyl	nitroso[a]
N_2O	dinitrogen oxide			dinitrogen oxide	azoxy

TABLE 70. NAMES FOR RADICALS AND IONS (*Continued*)
See also *acids*, Table 3

Atom or group	As uncharged atom, molecule, or radical	As cation or cationic radical	As anion	As ligand	As prefix for substituent in organic compounds
NO_2	nitrogen dioxide	nitryl	nitrite	nitro (nitrito-*N*); nitrito-*O*	nitro[a], $-NO_2$; nitrosooxy, $-O-N{=}O$
NS		thionitrosyl			
NO_3			nitrate	nitrato	
N_2O_2			hyponitrite	hyponitrito	
P	(mono)phosphorus		phosphide	phosphido	phosphinetriyl
H_2P			dihydrogenphosphide	dihydrogenphosphido	phosphino
PH_3	phosphine			phosphine	phosphonio, H_3P^+-
PH_4		phosphonium			
PO		phosphoryl			phosphoroso, $OP-$; phosphoryl, $OP{\equiv}$
PS		thiophosphoryl			thiophosphoryl
PH_2O_2			phosphinate	phosphinato	
PHO_3			phosphonate	phosphonato	
PO_4			phosphate	phosphato	
$P_2H_2O_5$			diphosphonate	diphosphonato	
P_2O_7			diphosphate	diphosphato	
AsO_4			arsenate	arsenato	
CO	carbon monoxide	carbonyl		carbonyl	carbonyl
CS		thiocarbonyl		thiocarbonyl	thiocarbonyl
HO_2C	carboxyl			carboxyl	carboxy
CO_2	carbon dioxide			carbon dioxide	carboxylato
CS_2	carbon disulfide			carbon disulfide	dithiocarboxylato
ClCO	chloroformyl			chloroformyl	chloroformyl
H_2NCO	carbamoyl			carbamoyl	carbamoyl
H_2NCO_2			carbamate	carbamato	carbamoyloxy
CH_3O	methoxyl		methoxide or methanolate	methoxo or methanolato	methoxy
C_2H_5O	ethoxyl		ethoxide or ethanolate	ethoxo or ethanolato	ethoxy
CH_3S	methylsulfanyl		methanethiolate	methylthio or methanethiolato	methylthio
C_2H_5S	ethylsulfanyl		ethanethiolate	ethylthio or ethanethiolato	ethylthio
CN		cyanogen	cyanide	cyano	cyano, $-CN$; isocyano, $-NC$
OCN			cyanate	cyanato; isocyanato	cyanato, $-OCN$; isocyanato, $-NCO$
ONC			fulminate	fulminato	
SCN		thiocyanogen	thiocyanate	thiocyanato; isothiocyanato	thiocyanato, $-SCN$; isothiocyanato, $-NCS$
SeCN			selenocyanate	selenocyanato; isoselenocyanato	selenocyanato, $-SeCN$; isoselenocyanato, $-NCSe$
CO_3			carbonate	carbonato	carbonyldioxy, $-O-CO-O-$
HCO_3			hydrogencarbonate	hydrogencarbonato	
CH_3CO_2		acetoxyl	acetate	acetato	acetoxy
CH_3CO	acetyl	acetyl		acetyl	acetyl
C_2O_4			oxalate	oxalato	

SOURCE: *Nomenclature of Inorganic Chemistry*, IUPAC, Butterworth, London.
[a] Used only as prefixes in substitutive *nomenclature*.

as, $\cdot CH_3$. **organic ~** A group of atoms which confers characteristic properties on a compound containing it, or which remains unchanged during a series of reactions; as, $-COOH$, carboxyl. Indicated by R for a single monovalent radical. When more than one radical, by R^1, R^2, R^3, etc. (R, R′, R″ for up to 3 groups) where the groups are different; and by R, R_2, R_3, etc., where the groups are the same.

r. weight The sum of the atomic weights of the elements in the r.

radicofunctional nomenclature* See *nomenclature*.

radio- Prefix (Latin) indicating "rays" or "radiation."

radioactive Able to give off rays. **r. constant** Decay constant*. **r. decay** The speed of disintegration: $n/n_0 = e^{-\gamma t}$. **r. disintegration** The breaking up of an atom of a r. element. **artificial ~, synthetic ~** The breaking up of an element by bombardment with particles (protons, neutrons, deuterons, etc.). **r. elements** The elements of high atomic weight, which disintegrate spontaneously with emission of rays. There are 4 series (where n is an integer): (1) Actinouranium ($4n + 3$); ${}^{235}_{92}U \rightarrow {}^{207}_{82}Pb$. (2) Uranium ($4n + 2$);

$^{238}_{92}U \rightarrow ^{206}_{82}Pb$. See *radon* decay chain, Table 71. (3) Thorium ($4n$); $^{232}_{90}Th \rightarrow ^{208}_{82}Pb$. (4) Neptunium ($4n + 1$); $^{237}_{93}Np \rightarrow ^{209}_{83}Bi$. There are 3 types of radiation: (1) The emission of α particles, or helium nuclei, lowers the mass number by 4 and the atomic number by 2. (2) β particles, or negative electrons, increase the atomic number by 1. (3) γ rays, or X-rays, are a secondary phenomenon caused by β rays striking an obstacle. **r. equilibrium** An equilibrium in a mixture of elements whose decay is balanced by the formation of fresh products, as in the transformation of radon into helium. **r. fallout** See *fallout.* **r. indicator** R. tracer. **r. material** (U.K. usage.) A substance containing one of the following elements in such proportion that the Bq/g of substance is greater than the quantity specified (for solids): Ac, Pb, Po, Pa, Ra, Th, and U, all 370; K, 33,000; Ra (as gas or vapor), 37. **r. tracer** R. indicator. Radiothor. A r. substance mixed in minute amounts with an isotope; the mixture behaves as a single chemical substance, although the indicator may always be detected by its radioactivity. See *labeling.* **r. series** See *radioactive elements.* **r. units** See *radioactivity units.*

radioactivity The spontaneous disintegration of elements with emission of rays. It is shown usually by elements of atomic weight over 207, and is unaffected by chemical or physical influences. **artificial ~** Induced r. **atmospheric ~** The r. of the atmosphere, which comprises r. due to natural sources (e.g., cosmic rays), medical and industrial uses of r., and radioactive fallout. **induced ~** Artificial r., synthetic r. Temporary r. produced in an element by bombardment with particles. Cf. *radioelements, nuclear reactions.* **synthetic ~** Induced r.
r. units See *becquerel, gray, sievert.*

radioassay Chemical analysis by means of radioactive indicators.

radiobiology The study of the effects of radiation (especially light) on living organisms.

radiocarbon dating Technique used to date archaeological plant finds that measures the amount of ^{14}C remaining that was originally fixed from the atmosphere. Accuracy dependent on age; ± 150 y for a 2,000-year-old object. Suitable for objects up to 80,000 years old. Cf. *aminostratigraphy.*

radiochemistry The study of radioactive elements and their reactions.

radiochromatography Chromatography in which identification of constituents of a separated mixture is aided by adding radioactive isotopes.

radioelements (1) Elements of at. no. 82 (Pb) and above. Other elements may be slightly radioactive; as, K. (2) Synthetic r., artificial r. Elements of low atomic number made temporarily radioactive by exposure to high-velocity protons, deuterons, neutrons, or α particles. Cf. *nuclear reactions.*

radiogenic Produced by radioactive action; as, radiolead.

radiogram An X-ray photograph.

radiograph Roentgenogram. (1) X-ray photograph. (2) Gammagraphs, taken with γ rays. **auto ~** A r. produced by direct contact of a radioactive sample with a photographic negative.

radiography X-ray photography. **mass ~** MMR. Routine examination by the reproduction in miniature of an X-ray image on a fluorescent screen or photographic film.

radioimmunoassay See *radioimmunoassay* under *immuno-.*

radiolead (1) The natural radioactive mixture of lead isotopes in minerals. (2) Radium G. ^{206}Pb. A radioactive disintegration product of polonium-210.

radiolite A variety of natrolite.

radiolocation Radar.

radiology The study of radioactivity and radioactive elements.

radiolucent Radio translucent. Offering very little or no resistance to the passage of X-rays. Cf. *radiopaque.*

radioluminescence Fluorescence caused by radioactive rays striking a screen treated with a suitable substance.

radiolysis Decomposition by high-energy radiation, e.g., formation of formaldehyde from methanol.

radiometallography The examination of the structure of metals by photography with radioactive substances and fluorescent screens.

radiometer (1) An instrument measuring the penetrating power of radioactive rays. (2) An apparatus demonstrating the mechanical effect of light. (3) A bolometer. (4) A thermocouple. Cf. *radiator.*

radiometric titration The use of a radioactive indicator to follow the transfer of material between 2 phases in equilibrium; e.g., the titration of $^{110}AgNO_3$ against potassium chloride.

radiomicrometer A delicate thermopile, to detect small changes in radiation intensity.

radiomimetic Producing effects similar to those of ionizing radiation.

radion A particle thrown off by a radioactive element (obsolete). Cf. *radon.*

radionitrogen See *radionitrogen* under *nitrogen.*

radionuclide An isotope that undergoes radioactive decay.

radiopaque Impervious to the passage of X-rays. Cf. *radiolucent, radioparent.*

radioparent Allowing the free passage of X-rays. Cf. *radiopaque, radiolucent.*

radioscope An electroscope to detect radioactive substances.

radiosodium See *radiosodium* under *sodium.*

radiotellurium Polonium*.

radiotherapy Treatment of disease, usually malignant, with ionizing radiation.

radiothor Radioactive indicator.

radiothorium Thorium-228.

radish The root of *Raphanus sativus* (Cruciferae); a vegetable and source of an indicator color.

radium* Ra. A radioactive element, at. no. 88, a disintegration product of thorium. At. wt. of isotope with longest half-life (^{226}Ra) = 226.03. It disintegrates to radon and helium. Discovered (1898) by Dr. and Mme. Curie. Gray powder, m.700, decomp. by water, and forms a series of salts; extremely toxic, and deposits in the bones of organisms. **RaA, RaB, etc.** See *radon.* **r. carbonate*** $RaCO_3$ = 286.0. White powder, insoluble in water; marketed mixed with barium carbonate. **r. emanation** Radon. **r.F** Polonium-210. **r.G** See *radiolead.* **r. sulfate*** $RaSO_4$ = 322.1. White powder, insoluble in water; used mixed with barium sulfate. **r. therapy** Curie therapy. The use of ^{226}Ra to treat disease. **r. units** (1) Micrograms of Ra (0.001 mg). (2) *Curie*s, q.v., millicuries, and microcuries; the emanation from 1 g, 1 mg, and 1 μg Ra, respectively. (3) Mache units, a concentration of Ra emanation corresponding with 1/2,500 microcurie. Cf. *urane.*

radius Any line from the center to the circumference of a circle: radius = 0.5 diameter = circumference/2π.

radix Latin for "root." See *base (4).* **r. sarsale** Sarsaparilla.

radome A solid shape produced by the electrophoretic deposition of a ceramic powder on an electrode, from which it is subsequently removed and consolidated by sintering.

radon* Rn. Niton, actinon, radium emanation. The element of at. no. 86; the last of the series of noble gases. At. wt. of isotope with longest half-life (^{222}Rn) = 222.02. It is produced

TABLE 71. RADON DECAY CHAIN

Early name	Isotope	Half-life	Radiation
Radon	Radon-222	3.8 days	α
Radium A	Polonium-218	3.1 min	α, β
Radium B	Lead-214	26.8 min	β
Radium C	Bismuth-214	19.7 min	α, β
Radium C′	Polonium-214	160 μs	α
Radium D	Lead-210	21 yr	α, β
Radium E	Bismuth-210	5.0 days	α, β
Radium F	Polonium-210	138.4 days	α
Radium G	Lead-206	Stable	

in radioactive minerals by radioactive disintegration. See Table 71. Discovered by Mme. Curie in 1898 (niton), and isolated by Dorn (1901). A colorless liquid, m. −71, b. −62, d.9.97; glows with a blue light which turns orange at lower temperatures.

rador An early name for *radar.*

raestelin $C_{23}H_{18}O_{15}$ = 534.4. A glucoside, the coloring matter of galls produced by *Raestelia lucerta.* Green needles, m.243.

raffia Raphia.

raffinase Melibiase.

raffinate The purified product obtained by a refining operation. In particular, a refined oil from the fractionation of crude lubricating oils.

raffinose $C_{18}H_{22}O_{16} \cdot 5H_2O$ = 584.5. Melitriose, melitose. A carbohydrate from sugar beet, cottonseed, or eucalyptus; hydrolyzed to fructose, glucose, and galactose. Colorless needles, m.119, soluble in water.

Ra G See *radiolead.*

Ragsky test The toxicological detection of chloroform by the reaction $6CHCl_3 \rightarrow C_6Cl_5 + 3Cl_2 + 6HCl$, whose products are deposited as white needles, detected by iodide-starch paper, and absorbed by silver nitrate solution, respectively.

ragweed The dried leaves and flowers of *Ambrosia artemisiaefolia;* an astringent and styptic.

ragwort Senecio.

raies ultimes Ultimate lines. The strongest and most persistent lines of a spectrum from a given element, and used to identify it.

rainfall The depth of water precipitated annually (or as otherwise specified) from the atmosphere in a liquid or solid state: 1 inch r. = 5.61 U.S. gal/yd^2. 1 cm r. = 10.0 L/m^2.

raised growth A heavy bacterial growth, with abrupt or terraced edges.

raisin The dried ripe fruit of grapes, *Vitis vinifera* (Vitaceae). **r.-seed oil** Grape-seed oil. Yellow oil, d.0.92–0.93. Used in foods, as a lubricant, and in soapmaking.

ralstonite $3Al(OH,F)_3,(Na_2,Mg) \cdot F_2 \cdot 2H_2O$. A mineral.

RAM Random access memory.

Raman **R., Sir Chandrasekhara Venkata (1888–1970)** Indian physicist, Nobel Prize winner (1930). **R. effect** The scattering of monochromatic light into different wavelengths due to interaction of the chemical species with the incident energy. Cf. *scattering, luminescence.* **R. lines** The shift in wavelengths of the R. effect, indicating molecular structure and the type of atomic bonds. **R. spectrum** The characteristic line patterns on a spectrophotometer, usually taken at right angles through a substance illuminated with a laser. The intensity of R. scattering of monochromatic light is analyzed as a function of frequency of the scattered light. Provides information on molecular vibrations and crystalline structure.

ramie Rhea fiber. The bast fiber from the nettle *Boehmeria tenacissima;* used for weaving grass cloth or Canton linen, and for making Bible papers. **false ~** China grass from *B. nivea.*

ramigenic acid $C_{16}H_{20}O_6$ = 308.3. An acid produced by the mold fungus, *Penicillium Charelsii.*

rammelsbergite $NiAs_2$, in nature.

Ramsay **R., Sir William (1852–1916)** British chemist, Nobel prize winner (1904); discovered the noble gases and some radioactive elements. **R.-Young equation** $p = kT - c$, where p is the pressure; T, the thermodynamic temperature of a gas or homogeneous liquid, the volume being constant; and k and c, constants. **R.-Young law** $A/B = A'/B'$, where A is the boiling point of compound a, B the boiling point of b at pressure x, A' the boiling point of a, and B' the boiling point of b at pressure y.

Ramsden ocular See *ocular.*

random access memory RAM. Computer memory space available for the programmer's use. The time taken to arrive at an address location is independent of the position of the location previously addressed. Cf. *ROM.*

rancid Having the peculiar tainted smell of oily substances that have begun to spoil, owing to formation of free fatty acids; as, rancid butter.

Raney's **R. alloy** The alloy Ni 30, Al 70%; used in place of *Devarda's alloy,* q.v., for the determination of nitrates by reduction to ammonia. **R. catalyst** R. nickel. A highly active, finely divided nickel catalyst, prepared by dissolving the aluminum out of R. alloy by alkali; used for hydrogenating organic compounds.

Rankine scale A thermometer scale based on absolute zero of the Fahrenheit scale: −460°F = 0°R. Cf. Kelvin scale of *temperature.*

rankinite $3CaO \cdot 2SiO_2$. A natural silicate (Ireland); a constituent of high-lime blast-furnace slags.

Ranunculaceae The crowfoot family; e.g.: *Helleborus niger,* black hellebore; *Anemone hepatica,* liverwort. Cf. *aconite alkaloids.*

ranunculine An alkaloid cardiac poison from *Ranuncula* species.

Raoult **R., François Marie (1830–1901)** French chemist. **R.'s law** The lowering of the freezing point or of the vapor pressure p of a solution is proportional to the amount of substance dissolved in the solution: $(p_0 - p)/p_0 = aM/bm$, where p_0 is the vapor pressure of the pure solvent, a the weight in grams of the solute of molecular weight m, and b the weight in grams of the solvent of molecular weight M. It is used to determine molecular weight. Cf. *Blagden's law, Coppet's law, freezing point depression.*

rapeseed oil Colza oil, cole oil, rape oil. Brown oil from the seeds of rape, *Brassica napus* (Cruciferae). Unpleasant odor, d_{25} 0.906–0.910, m.20, soluble in alcohol; used as a lubricant, and in the heat treatment of steel and the manufacture of rubber substitutes.

raphanin $C_{17}H_{26}N_3O_4S_5$(?). An antibiotic anthocyanin from radishes, which prevents the germination of seeds. Yellow syrup, $b_{0.06mm}$135, soluble in water.

raphia Raffia. A fiber from the leaves of the palm *Raphia pedunculata* (Malagasy Rep.), used for mats and basketry. **r. wax** A wax from palm leaves; little used, as it has poor solubility in solvents.

raphides Small crystals of calcium oxalate in plants.

rare Not common. **r.-earth metals*** The elements Sc and Y, plus the lanthanoids (La through Lu; at. no. 57–71). Two classes: ceria earths and yttria or gadolinite earths, which are similar in physical and chemical properties. **r.-earth minerals** The r.-earth metals generally occur associated with one another as phosphates and silicates; e.g.: zircon, $ZrSiO_4$; monazite, $CePO_4$. **r. earths** (1) The oxides of the r.-earth metals. (2) In a wider sense, the oxides of Sc, Y, La and even of Hf, Zr, Th. **r. gases** Noble gases*.

rarefaction Making less dense by increasing the volume without changing the amount of a gas; or decreasing the amount of gas in a certain volume.

rarefied Describing a gas at pressure less than atmospheric.

rasmosin A resin from the root of *Cimicifuga racemosa* (Ranunculaceae).

Rasorite Trademark for sodium borate ore concentrates from California.

raspberries The fruits of *Rubus idaeus* (Rosaceae); used as food and in pharmaceutical preparations.

raspberry leaves The dried leaves of *Rubus* species. Used in pharmacy as a flavoring vehicle (USP) and astringent.

raster (1) A grid of narrow parallel beams used in the photoelectric scanning of letters and figures in electronic "readers." (2) A grid on a video display terminal.

rat White rat, *Rattus norvegicus*, used in animal experiments. **r. unit** The minimum quantity of pituitary hormone injected subcutaneously that will cause the formation of one or more corpora lutea in the female white r.

ratafia A cordial prepared by steeping crushed fruit kernels.

ratany Krameria.

rate The relative degree of speed. **r. constant*** Symbol: k; velocity constant. See *reaction order*. **r. of decomposition, r. of formation** The velocity of a chemical reaction expressed in mol/s/mL.

ratholite Pectolite.

raticide A rat-killing chemical.

ratio Proportion.

rational Based on reasoning and not on direct experience. **r. analysis** The expression of analytical results so that the combination of elements present is indicated; e.g., in water analysis. Thus, clay would be expressed in terms of calcium silicate, rather than as Ca, Si, and O. Cf. *ultimate analysis*. **r. formula** A combination of chemical symbols indicating the probable atomic links in the molecule. **r. units** See *ultimate rational units*.

rattlesnake Common name for various species of snakes of the genera Sistrurus and Crotalus. **r. root** Senega. **r. venom** The poisonous proteins from the fangs of *Crotalus* genus.

Raulin's solution A culture fluid for fungi.

rauwolfia alkaloids Alkaloids from a variety of *Rauwolfia* species, particularly *R. serpentina*, Benth; e.g.: reserpine, rauwolfine, δ-yohimbine.

rauwolfine $C_{19}H_{26}O_2N_2$ = 314.4. An alkaloid, decomp. 235, from the bark of *Rauwolfia caffra* (Apocynaceae).

raw Crude, unrefined, or unfinished. **r. material** A crude material from which useful substances can be made. **r. ore** An ore in its natural state.

ray(s) A beam of light, heat, or rapidly moving particles. Cf. *radiation*. **absorbed ~** A light r. transformed into heat (molecular motion) on passing through matter. **actinic ~** Chemically active (especially ultraviolet) light. **alpha ~** α particles. Positively charged, high-velocity helium nuclei produced by the disintegration of radioactive elements. Cf. *canal ray* below. **Becquerel ~** X-rays produced by disintegration of radioactive elements. **beta ~** Negatively charged, high-velocity electrons produced by the disintegration of radioactive elements. Cf. *cathode ray* below. **Blondlot ~** n-rays. **canal ~** Positively charged particles produced by the electric discharge in an evacuated tube having a perforated cathode through which the particles pass. Cf. *alpha ray* above. **cathode ~** High-velocity electrons moving at right angles to the cathode of a vacuum tube and producing X-rays on striking a solid. **chemical ~** Actinic r. **cosmic ~** See *cosmic rays*. **extraordinary ~** See *ordinary ray*. **gamma ~** See *gamma rays*. **grenz ~** See *grenz rays*. **hard ~** A ray of short wavelength and high penetrating power. **heat ~** See *infrared ray* following. **infrared ~** Heat radiation beyond the red portion of the visible spectrum. **K ~** See *K radiation* under *K*. **L ~** See *L radiation* under *L*. **Lenard ~** Residual cathode r. that pass through windows of thin metal foil in a vacuum tube. **M ~** See *M radiation* under *M*. **Millikan ~** See *cosmic rays*. **molecular ~** See *molecular rays*. **n- ~** Blondlot r. A nonluminous radiation from certain flames that produces fluorescence on striking certain substances. **negative ~** Cathode or β rays. **od- ~** The radiations of living organisms. Cf. *scotography*. **ordinary ~** See *ordinary ray*. **phonic ~** Sound waves. **photic ~** Visible light r. **positive ~** Canal and α rays. **roentgen ~** X-rays. **secondary ~** Radiation produced after a primary r. strikes matter. **soft ~** A ray of long wavelength that has low penetrating power. **ultimate ~** See *raies ultimes*. **ultra-γ ~** See *cosmic rays*. **ultraviolet ~** The radiation from beyond the violet end of the visible spectrum; they produce fluorescence. **u.v. ~** Ultraviolet r. **W ~** Rays between u.v. and X-rays. **X- ~** See *X-rays*.

Rây, Sir Prafulla Chandra (1861–1944) Indian chemist, noted for his work on the organic nitrites and history of chemistry.

Rayleigh **R., (Lord) John William Strutt (1842–1919)** English physicist, Nobel prize winner (1904); noted for the discovery of the noble gases, and determinations of the densities of gases. **R., (Lord) Robert John (1875–1947)** English physicist, noted for his work on radioactivity and optics.

Rayolande Trademark for a chemically modified viscose rayon synthetic fiber.

rayon Generic term for synthetic textile fibers whose chief ingredient is cellulose or one of its derivatives. **acetate ~** The acetic ester of cellulose, prepared by treating the cellulose of cotton or wood pulp with acetic anhydride, acetic acid, and concentrated sulfuric acid in presence of a catalyst; precipitating with water; and dissolving in acetone, from which it is spun. Trademarks: Celanese and Lustron. **cuprammonium ~** Cupra ~, r. glanzstoff, Pauly r. Made by dissolving cotton or wood pulp in an ammoniacal copper solution, which is forced through fine orifices into a setting bath of dilute sulfuric acid. **nitrocellulose ~** (Chardonnet) Prepared by treating cotton with nitric and sulfuric acid, dissolving the resulting trinitrocellulose in alcohol and ether (collodion solution), and forcing it through fine nozzles into water or warm air. The flammable filaments are denitrated with sodium hydrogen sulfide. **seaweed ~** R. made by extruding an extract of seaweed in an aqueous alkali into acid. The algin is precipitated as an alkali-insoluble filament, e.g., calcium alginate. **tang ~** A Norwegian textile made from staple fiber derived from r. and treated seaweed. **viscose ~** The commonest process of making r. involves soaking wood pulp or cotton linters in 18% caustic soda solution, treating the soda cellulose with carbon disulfide (xanthation), dissolving in caustic soda (viscose), and forcing the viscose through fine outlets into a setting and bleaching bath.

razor stone Novaculite.

Rb Symbol for rubidium.
R.B.C. Abbreviation for "red blood cells" (corpuscles).
RBX See *hexalite.*
rd. Abbreviation for: (1) rod, (2) rutherford.
RDX A high explosive of the tetranitrocellulose type.
Re Symbol for rhenium.
Ré Abbreviation for Réaumur.
react To enter into a chemical combination.
reactants Molecules that act with one another to form a new set of molecules (resultants).
reacting Undergoing a reaction. **r. weight** Equivalent weight.
reaction (1) That force which tends to oppose a given force (Newton's laws). (2) The acidity or alkalinity of a solution. (3) A chemical change: the transformation of one or more molecules (reactants) into others. **acid ~** A positive test for the presence of hydrogen ions (acidity); as, blue litmus turning red. **addition ~** Molecules combine to form a more complex molecule. **aldol ~** See *aldol condensation.* **alkaline ~** A positive test for the presence of hydroxyl ions (alkalinity); as, red litmus turning blue. **amphoteric ~** The r. of a substance that has both acid and alkaline properties. **analytic ~** Decomposition r. **analytical ~** A r. used to determine the quantity or quality of matter, or both. **balanced ~** A reversible r. that does not go to completion in either direction. **bimolecular ~** A r. of the second order. **Bunsen ~** See *Bunsen reactions* under *Bunsen.* **catalytic ~** A r. whose rate is accelerated by a catalyst. Cf. *induced reaction.* **chain ~** See *chain reaction, trigger.* **color ~** A r. involving a change in color. **combination ~** Elements unite to form a compound. **complete ~** A chemical change that proceeds to completion; as, precipitation. **complex ~, composite ~** A chemical change in which more than one r. occurs simultaneously. **concurrent ~** A chemical change consisting of a series of connected reactions that could not occur separately. **condensation ~** A r. in which atoms are removed from 2 or more molecules, the residues combining to form a single molecule. **counter ~** A reversible r. **coupled ~** A concurrent r. **decomposition ~** Molecules decompose into simpler molecules, atoms, or both. **diazo ~** See *diazotization.* **displacement ~** See *displacement reaction.* **dissociation ~** A compound molecule splits into element molecules. **electron ~** See *oxidation reaction.* **endothermic ~** A r. in which heat is consumed. **esterification ~** $R \cdot OH + HA = RA$ and H_2O, where R is an aryl or alkyl radical, and HA an acid; a catalyst (as zinc chloride) is required. **etherification ~** $R \cdot OH + HOR' = R \cdot OR' + H_2O$, where R and R′ are aryl and/or alkyl radicals. **exothermic ~** A r. in which heat is liberated. **flame ~** See *Bunsen reactions.* **group ~** A r. typical of a certain group of elements, as in qualitative analysis. **heat of ~** See *heat of reaction* under *heat.* **heterogen(e)ous ~** A composite r. **hydrolysis ~** The reversal of a neutralization r. by the action of water. **incomplete ~** A balanced r. **induced ~** See *induced reaction.* **ionic ~** An instantaneous r. between ions in solution. **irreversible ~** A r. that proceeds to completion and cannot be reversed. **main ~** The principal r. **metathetic ~** R. involving interaction of molecules. **micro ~** See *microreaction.* **molecular ~** In general, a slow r. between molecules and not ions. **monomolecular ~** A r. of the first order. **negative ~** The absence of r., e.g., in a colorimetric, qualitative analytical test. **neutral ~** A r. that is neither acid nor basic. **neutralization ~** See *neutralization.* **nuclear ~** See *nuclear reactions.* **opposing ~** The reverse r. in a balanced r. **oxidation-reduction ~, oxidoreduction ~** See *oxidation reaction.* **pericyclic ~** R. involving bond reorganization in a cyclic compound. **peritectic ~** A r. between a solid phase (α) and a liquid phase producing a second solid phase (β). **photo ~** See *photoreaction.* **positive ~** A r. that is definite, e.g., in a colorimetric, qualitative analytical test. **primary ~** The principal or main r. that occurs in a composite system of reactions. **principal ~** Primary r. **proton ~** See *neutralization.* **pyrogenic ~** See *pyrogenic.* **qualitative ~** See *qualitative reaction.* **quantitative ~** See *quantitative reaction.* **restitution ~** The reverse of substitution r. **reversible ~** An incomplete r. A r. that under suitable conditions can proceed from right to left or left to right. **secondary ~** Subsidiary r. A r. between the resultants of a r. **side ~** A r. simultaneous with the principal r., occurring between the same reactants or their products, and usually forming different products. **simultaneous ~** Side r. **staircase ~** See *staircase reaction.* **subsidiary ~** Secondary r. **substitution ~** An element molecule substitutes one of the elements in a compound molecule. **successive ~** A r. made up of a number of component reactions which occur in succession, the reactants being in turn the resultants of the preceding r. **sympathetic ~** See *induced reaction.* **synthetic ~** Combination and addition r. **topochemical ~** See *topochemistry.* **trigger ~** See *trigger reaction.* **unimolecular ~** See *reaction order.*
r. constant* Symbol: k; velocity constant. See *reaction order.* **r. control** (1) Speed of r. depends on amounts of substances present and energy (potential energy, electromotive force, and free energy). A r. can be accelerated mechanically (as by increasing the number of molecules coming in contact with one another, by increasing the surface by subdivision); thermally (by increasing the velocity of the molecules by heating); electrically (by electrolysis); optically (by irradiation). (2) R. acidity is controlled by buffer solutions. **r. equation** See *equation.* **r. isochore** An equation relating temperature and the equilibrium constant K at constant volume (or pressure) of a gas: $d(\log_e K)/dT = -U/RT^2$; where T is thermodynamic temperature, U the decrease in total energy, and R the gas constant. By integration, the maximum work at any temperature can be determined. **r. isotherm** An equation for ideal gases indicating the maximum external work or diminution in free energy obtainable from a chemical reaction at constant temperature and volume: $A = RT \log K - RT\Sigma_v \log C$, where A is the decrease in free energy (affinity), and $\Sigma_v \log C$ represents the concentration. **r. law** If a system in equilibrium undergoes a restraint, a change tends to take place, and partly annul the constraint. **r. order** A classification of reactions:

First order: monomolecular r. The r. rate depends on the concentration of only 1 reactant. Velocity $= dx/dt = k(a - x)$, where a is the initial amount of substance, x the amount that has changed after time t, and k the rate constant.
Second order: bimolecular r. The rate depends on 2 concentrations: $2A$ or $A + B = 1$ or more products. Velocity $= k(a - x)(b - x)$.
Third order: trimolecular r. The rate depends on 3 concentrations: $3A$ or $2A + B$ or $A + B + C = 1$ or more products. Velocity $= k(a - x)(b - x)(c - x)$.

r. promoter Promoter. **r. sensitivity** The dilution at which a r. still gives identifiable end products. **r. velocity** Rate constant*. See *reaction order.*
reactivation The rendering active again of a catalyst or serum which has become inactivated. Cf. *revivification.*
reactor Nuclear r., thermal r., atomic pile. A plant that generates steam by controlled nuclear fission, the steam then

being used to produce electric power. The fuel is usually uranium-235, plutonium-239, or thorium-232; the neutrons released maintain a chain reaction. **boiling water ~** BWR; a type of light water reactor in which steam is generated in the reactor. **fast breeder ~** A r. that produces its own fuel. Liquid sodium is the coolant. **heavy water ~** HWR; a r. that uses heavy water as the moderator to slow down the neutrons. **high-temperature, gas cooled ~** HTGR; a r. that uses graphite as the moderator and helium as the coolant. **light water ~** LWR; a r. with ordinary water as the coolant. **pressurized-water ~** PWR; a type of light water reactor in which pressurized water acts as the coolant and moderator, and the steam is generated after a heat exchanger. **RBMK ~** Russian light water-cooled r. with graphite moderator.

read-only memory ROM. Memory from which a computer can read only, the writing in of the programs having been unalterably performed during manufacture. Cf. *RAM.*

reagent (1) A chemical substance that reacts or participates in a reaction. (2) A substance used for the detection or determination of another substance by chemical or microscopical means; especially analysis. Types:

1. Precipitants—produce an insoluble compound.
2. Solvents—used to dissolve water-insoluble materials.
3. Oxidizers—used in oxidation.
4. Reducers—used in reduction.
5. Fluxes—used to lower melting points.
6. Colorimetric reagents—analogous to (1), but produce colored soluble compounds.

See *volumetric solutions.* **r. solution** The aqueous solution of a r. used in the laboratory; usually 10% concentration.

realgar* Arsenic tetrasulfide*.

real time (1) The actual time during which a process is proceeding. (2) The ability to compute outputs quickly enough for them to be effective in running the operation.

ream A number of sheets of paper of the same kind, size, and weight per unit area; usually 500, but also 480, 516, 1,000, etc., according to trade custom.

rearrangement Intramolecular conversion, migration, or transposition, in which the atoms or atomic groups redistribute or arrange themselves in a different manner. Cf. *pinacol conversion, enol form.*

reastiness The flavor produced in salt-preserved meat due to oxidation of the fats present.

Réaumur R., René Antoine Ferchault de (1683–1757) French natural philosopher, noted for his invention of a thermometer scale based on the freezing point (0°R) and the boiling point (80°R) of water. **R. degree** °R or °Ré. 80°R = 100°C. Cf. *Rankine scale.*

recalescence An increase in the emission of visible light during the cooling of molten metals; explained by thermally excited electrons falling back to a lower energy level in the atom. Cf. *luminescence.*

recalescent point The temperature at which evolution of heat occurs during the cooling of steel. It is lower than the decalescent point by an amount that is a measure of the hysteresis of steel.

recarbonize To restore the carbon content of steel after decarbonizing.

recarburize Recarbonize.

receiver A vessel in which the products of a distillation are collected.

receptor (1) A radical or atom, on the surface of a cell, that reacts with drugs, hormones, poisons, etc. Receptors are associated with the active group of an enzyme, and can be "blocked" (inactivated) by drugs. (2) A term used in Ehrlich's side-chain theory.

reciprocal The quotient *r* obtained by dividing unity by a number *n*: $r = 1/n$. **r. ohm** mho. Former unit of conductance; replaced in the SI system by the siemens. **r. salt pair** Two salts which may act either as the reactants or resultants of a reaction; as, $NaCl + KNO_3 \rightleftharpoons NaNO_3 + KCl$.

reciprocating motion Motion to and fro in a straight line; as, a piston.

reciprocity failure See *latensification.*

reciprocity law Bunsen-Roscoe law.

recoil Return motion. **r. atom** The atom of a radioactive substance in the process of expelling an α particle; the high velocity causes a recoil. **r. radiation** Radiation produced by bombarding gaseous atoms with α particles; its source is the action of the r. atom.

recovery The extraction of a valuable constituent from a raw material, by-product, or waste product; as, silver from photographic wastes.

recryst. Abbreviation for "recrystallization."

recrystallization Repeated crystallizations for the purpose of purification.

recrystallize To purify a substance by repeated crystallization.

rectification (1) The redistillation of a liquid for purification. (2) The transformation of an alternating into a direct current. (3) The evaluation of the length of a curved line.

rectified spirit Spirits of wine. Redistilled alcohol; 84% by weight of absolute alcohol, 50° over proof (O.P.), and a specific gravity 0.8382 at 60°F. Cf. *proof spirit.*

rectifier A device to convert alternating to direct current, e.g., quartz r. **Dryr ~** An asymmetric conductor consisting of a disk of lead sulfide waxed on one side, with tinfoil on both sides.

r. cell Photoelectric cell.

rectorite A white aluminum silicate, similar to kaolinite.

recuperator Regenerator.

red The least refracted portion of the visible spectral colors. **natural ~** See *carmine.*

r. acid 1,5-Dihydroxynaphthalene-3,7-disulfonic acid. **r. antimony** Kermesite. **r. arsenic** Realgar. **r. bole** R. ocher. **r. brass** An alloy: Cu 78–83, Zn 7–9, Pb 6–10, Sn 4–2%. **r. cake** Commercial sodium vanadate, used in the manufacture of ceramic colors and catalysts. **r. cedar** Juniper. **r. chalk** R. ocher mixed with clay. See *red hematite* under *hematite.* **r. chrome** Lead chromate. **r. cobalt** Erythrite. **r. copper ore** Cuprite. **r. couch grass** Carex. **r. hematite** See *hematite.* **r. iron ore** Hematite. **r. lead** Lead tetraoxide. **r. lead ore** Crocoisite. **r. liquor** A solution of aluminum acetate, used in dyeing. **r. manganese** Rhodonite. Rhodochrosite. **r. mercury iodide** Mercuric iodide. **r. mercury oxide** Mercuric oxide. **r. metal** A copper alloy. **r. mustard** Black *mustard.* **r. ocher** Reddle, r. bole. A r., often impure, hematite; a pigment. **r. oil** Commercial oleic acid. **r. orpiment** Realgar. **r. oxide** Ferric sesquioxide. **r. oxide of zinc** Zincite. **r. pepper** Capsicum. **r. phosphorus** See *phosphorus (2).* **r. precipitate** Mercuric oxide. **r. prussiate** Hexacyanoferrate(III)*. **r. p. of potash.** Potassium hexacyanoferrate(III)*. **r. p. of soda.** Sodium hexacyanoferrate(III)*. **r.root** Sanguinaria. **r.-sensitive plate** A photographic plate with the emulsion made orthochromatic by addition of a sensitizer; as, neocyanine. **r. silver ore dark ~** Pyrargyrite. **light ~** Proustite. **r. stone** Ferric oxide. **r. vitriol** Bieberite. **r. zinc ore** Zincite.

reddingite A native iron and manganese phosphate from Redding, California.

reddingtonite A native chromium sulfate.

reddle Hematite.

redistillation Repeated distillation of a liquid, usually to purify it.
Redon Trademark for a polyamide synthetic fiber.
Redonda phosphate A phosphate ore (40% phosphorus pentaoxide), W. Indies.
redox Reduction-oxidation. See *oxidation-reduction potential.* **r. equilibrium** The state in which a reaction is poised at a definite rH value. Cf. *oxidation-reduction potential.* **r. indicator** See *redox indicator* under *indicator.* **r. polymer** See *electron-exchange polymer* under *polymer.* **r. series** Oxidizing and reducing reactions arranged in order of intensity.
redoxokinetic effect The current-rectifying power of certain reversible electrodes. It depends on the kinetics of their redox reactions. Used for electrometric titration.
redruthite Cu_2S. A black mineral.
reduce To add one or more electrons to an atom. Hence: (1) to decrease the oxidation number of an atom with a corresponding increase in the oxidation number of another atom (which is oxidized); (2) to deprive of oxygen; (3) to add hydrogen. Antonym: *oxidize.*
reduced (1) Brought or restored to metallic form. (2) Brought to a lower stage of oxidation. **r. iron** Finely powdered iron obtained by heating ferric oxide in a current of hydrogen. **r. oil** Crude petroleum from which hydrocarbons of low boiling point have been removed by distillation or evaporation.
reducer (1) Reducing agent. (2) A solution that decreases the intensity of a photographic image; as, equal volumes of sodium thiosulfate 10 and potassium hexacyanoferrate(III)* 3% solutions. **Jones ~** An apparatus used in chemical analysis to reduce ferric iron to ferrous iron by passing the solution through granulated zinc. Cf. *reductor.*
reducible Capable of being reduced.
reducing **r. agent** Reducer. A substance that is readily oxidized by reducing another substance; as, nascent hydrogen, stannous chloride, formaldehyde, and zinc dust. Cf. *oxidizing agent.* **r. flame** The luminous portion or inner cone of the blowpipe or bunsen burner where an excess of unburnt carbon or hydrocarbon gas acts as a reducing agent on substances in it. **r. sugar** A mono- or disaccharide (as, glucose or fructose) that reduces copper or silver salts in alkaline solutions. Cf. *Fehling solution.*
reductase* See *enzymes,* Table 30.
reductic acid See *reductones.*
reduction (1) Making less or smaller. Converting to a fine state. Reproducing on a smaller scale. (2) A chemical reducing reaction, being the antonym of *oxidation,* q.v.; e.g.: (*a*) removal of oxygen; as, $CuO \rightarrow Cu$; (*b*) addition of hydrogen; as, $Mg \rightarrow MgH_2$; (*c*) change from a higher to lower oxidation number; as, from the ferric to the ferrous state, viz.: $Fe^{3+} + e^- = Fe^{2+}$, where the atom gaining the electron (e^-) is reduced, and that losing it is oxidized. **Birch ~** R. of organic compounds with sodium and liquid ammonia. **r. intensity, r. oxidation** See *oxidation-reduction potential.* **r. potential** The potential of a platinum electrode immersed in a solution containing an ion being reduced.
reductones Apparent vitamin C. Reductic acid. Reducing substances present in certain processed foods which react similarly to ascorbic acid, and interfere with its determination. Do not prevent scurvy.
reductor (1) An apparatus for a reducing reaction, e.g., for determination of phosphorus in steel. See *Jones reductor.* (2) A metal or metal amalgam which acts as a reducing agent.
Redwood number A measure of oil viscosity in terms of rate of flow from a standard Redwood viscometer (former U.K. standard).
Reech's theorem The ratio of the adiabatic to the isothermal elasticity of a fluid equals the ratio of their specific heat capacity at constant pressure and constant volume; equals 1.66 (for monatomic gases).
Reevon Trademark for a polyethylene synthetic fiber.
refikite $C_{20}H_{16}O_2$. A white resin, in lignite.
refine To purify.
refinery (1) A building or apparatus for refining. (2) A shallow hearth furnace for making wrought iron.
refining heat The temperature 655°C, that imparts fineness of grain and toughness to steel.
reflection The rebounding of an incident ray of light, heat, or sound from a surface, as a reflected ray having the same form, quality, and intensity.
reflux Backward flow, return. **r. condenser** A vertical or inclined condenser, from which the condensed liquid flows back into the distilling vessel. **r. valve** A check valve.
Reformatzky reaction A condensation reaction between ketones and 2-bromoaliphatic acids in the presence of zinc or magnesium: $R_2CO + BrCH_2COOR + Zn \rightarrow (ZnO \cdot HBr) + R_2C(OH)CH_2COOR$.
reforming Oil refinery process that increases the octane number of the naphtha feed from 20–50 up to the 85–95 required in high-octane gasoline, at about 510°C and 2.5 MPa in the presence of a $Pt\text{-}Al_2O_3\text{-}Cl_2$ catalyst.
refract To change the direction, deviate, or bend (usually rays).
refraction* Symbol: *R;* the change in direction of rays of light obliquely incident on, or traversing a boundary between, 2 transparent media, or a medium of varying density. Cf. *light, diffraction.* **atomic ~** The specific r. of an element multiplied by its atomic weight. **double ~** Birefringence. Having more than one refractive index, according to the direction of the traversing light. Occurs in all except isometric crystals; transparent substances with internal strains, e.g., glass; and substances with different structures in different directions, e.g., fibers. **electrical ~** See *Kerr effect.* **index of ~** See *refractive index.* **molar ~** The sum of the atomic r. of the atoms in a molecule. For molecules of normal structure, it equals specific r. times molecular weight. **specific ~** Symbol: *r;* equal to $(n - 1)/\rho$ for a given wavelength, where n is the refractive index and ρ is density; r can vary with ρ.
refractive Refringent. Pertaining to refraction. **r. constant** The sum of the refractivities of the pure constituents of a solution, each multiplied by the ratio of its mass per unit volume of solution to its own density when pure. **r. index*** n. The ratio of the velocities of light in a medium and in air under the same conditions. Measured by the ratio of the sines of the angles of incidence and refraction. Determined by a refractometer. See Table 72. **absolute ~** The r. index in a vacuum: $n \times 1.0029$.
refractivity (1)* Refractive index minus 1. (2) Refraction*.
refractometer An instrument to determine the refractive index of a substance at various temperatures.
refractoriness Resistance to softening on application of heat. Cf. *Seger cones.* **r. under load** The resistance to deformation of a refractory subjected to stress at a definite temperature.
refractory A material that is slow to soften and resists heat; as, brick. See Table 73. **r. cements** Materials used to hold r. bricks together. **straight ~** Fireclay, silica, magnesite, etc., which have high refractoriness (>1650°C) but poor bonding strength. **synthetic ~** (1) Vitrifying at about 9000°C, as portland cement, glass, fusible clays, which have a good bonding strength. (2) Air-setting mixtures of refractories and fluxes of good bonding strength.

TABLE 72. REFRACTIVE INDEX n_D OF VARIOUS COMPOUNDS AT 23°C

Compound	n_D
Methyl alcohol	1.3279
Water	1.3328
Acetone	1.3598
Hexane	1.3750
Heptane	1.3868
1,2-Dichloroethane	1.4417
Chloroform	1.4430
Eucalyptol	1.4552
Glycerol	1.4671
Decahydronaphthalene	1.4750
Isopentyl phthalate	1.4870
Xylene	1.4957
Pentachloroethane	1.5006
Anisole	1.5150
Chlorobenzene	1.5218
1,3-Dibromopropane	1.5220
1,2-Dibromoethane	1.5353
2-Nitrotoluene	1.5440
Nitrobenzene	1.5506
Tri-*o*-cresyl phosphate	1.5560
2-Toluidine	1.5700
Aniline	1.5840
Bromoform	1.5940
Quinaldine	1.6088
Iodobenzene	1.6168
Quinoline	1.6239
1-Chloronaphthalene	1.6318
1-Bromonaphthalene	1.6569
Diiodomethane	1.7400

refrigerant An agent that produces the sensation of cold or used to obtain a low temperature. See *freezing mixture.* **mechanical ~** A noncorrosive liquid of suitable vapor pressure, used in refrigeration; it should be nontoxic and nonflammable; e.g.: carbon dioxide, ammonia, dichlorodifluoromethane.

refrigeration The production of cold; the lowering of the temperature of a body by conducting away its heat.

refrigerator A machine to produce refrigeration by compressing a gas, cooling it, and expanding it into a low-pressure pipe system, where it absorbs the heat, and is pumped out and compressed again.

refringent Refractive.

refuse Waste, garbage, or sewage.

regelation The fusion of ice under pressure, followed by solidification of ice particles into a solid mass.

regeneration The repair or renewal to the original state; as, of an electric battery, or catalyst. Cf. *revivification, Weldon process.*

regenerative Capable of being utilized anew. **r. furnace** A furnace in which the incoming fuel gases are preheated by the waste gases.

regenerator A series of checkerwork brick chambers through which the flue gases and fuel gases of a furnace are alternately directed.

Regitine Trademark for phentolamine mesylate.

Regnault R., Henri Victor (1810–1878) French chemist, noted for research in physical chemistry. **R. cell** An anode of amalgamated zinc and a cathode of cadmium in a solution of sulfuric acid 1, calcium sulfate 1, water 12 pts; yields 0.34 volt. **R. value** The weight of 1 m^3 of hydrogen: 89.99 g. Cf. *krith.*

regression analysis Mathematical method for determining the best line or curve through (usually scattered) data relating a dependent variable to an independent variable. Thus, the line of *least squares* (q.v.) fit, $y = ax + b$, is found by solving the 2 equations: $\Sigma y = a\Sigma x + nb$ and $\Sigma xy = a\Sigma x^2 + b\Sigma x$, where n is the total number of readings of individual values x and y. **multi ~** As for r. a., but involving 2 or more independent variables.

Reguir cell An anode of copper and cathode of lead oxide in a solution of copper sulfate and sulfuric acid; yields 1.4 volts.

regular system The cubic system of crystal forms.

regulator (1) Buffer solution. (2) A substance used in flotation processes for pH control; as, lime.

regulus (1) A small, compact mass of metal in the bottom of the crucible on reduction of ores. (2) A compound of one or more metals with sulfur; usually brittle and crystalline with a dull greasy luster.

Reich, Ferdinand (1799–1882) German mining engineer; discoverer of indium (with Richter, 1863).

Reichert R. number The number of mL of 0.1 *N* alkali required to neutralize the volatile, water-soluble acids liberated on acidification of the soap produced by saponification of 10 g fat. **R.-Meissl number, R.-Wollny number** Similar to the R. number, referred to 5 g fat and determined according to specified procedures. Cf. *Polenské number.*

Reimer's reaction A reaction of phenols and chloroform in alkaline solutions to form phenol aldehydes.

TABLE 73. FORMULAS AND PROPERTIES OF SOME REFRACTORY MATERIALS

Refractory	Formula	Temperature of failure under load, °C	Bulk density, g/cc	Porosity, %	Fusion point, °C
Alundum	Al_2O_3	1550[a]	2.6	—	1750–2000
Bauxite brick	$Al_2O_3 \cdot TiO_3$	1350[b]	1.6	46–50	1750–2000
Chrome brick	$Cr_2O_3 \cdot Al_2O_3 \cdot MgO$	1450[c]	2.8–3.2	—	1850–2050
Fireclay brick	$SiO_2 \cdot Al_2O_3$	1500[b]	1.7–2.1	20–30	1500–1750
Magnesite brick	MgO	1550[c]	2–2.8	24–40	2150–2615
Silica brick	SiO_2	1600[c]	2–2.2	18–43	1685–1800
Silicon carbide	SiC	1650[a]	2.0–2.6	17–34	2200–2240
Zirconia brick	ZrO_2	1510[b]	3.4–4	19	2000–2600

[a]No failure under a load of 170 kPa
[b]Softens
[c]Shears

Reinecke's acid $[Cr\cdot(NH_3)_2\cdot(SCN)_4]H$. Diammonotetrathiocyanochromic acid. A reagent in the isolation of organic bases, e.g., choline from proteins. **R. salt** $[Cr(NH_3)_2(SCN)_4]NH_4\cdot H_2O)$. Used similarly to R.'s acid; also as reagent for mercury (a red color or precipitate).

reinforced Made stronger. **r. concrete** Concrete with steel lengths inserted during laying, for added strength. **r. wine** Wine to which alcohol has been added.

reinite $FeWO_4$. The native tungstate.

Reinsch test The detection of a small quantity of arsenic by depositing it from solution as a black stain on metallic copper.

Rekylon Trademark for a polyamide synthetic fiber.

relative Dependent upon or connected with some other phenomenon. **r. analgesia** See *relative analgesia* under *analgesia.* **r. atomic mass*** See *atomic weight.* **r. density*** See *relative density* under *density.* **r. molecular mass*** See *molecular weight.* **r. permittivity** See *relative permittivity* under *permittivity.* **r. weight** Atomic weight.

relativity Einstein's theory: the concept that matter, space, and time are relative and not absolute; that all physical phenomena depend upon position in or relation to the universe. **general ~** Mass exerts an attraction on light; hence: (1) large bodies, as the sun, deflect light waves, and (2) spectral lines of atoms on the surface of very massive stars are shifted toward the red. **special ~** Energy *E* has mass; hence mass *m* increases slightly when energy content is increased. $m = E/c^2$, where *c* is the velocity of light, an absolute constant of nature. Cf. *mass*-energy cycle.

relaxation **r. effect** The time lag of the response of a chemical equilibrium reaction to a change in external conditions. **r. time*** See Table 88—Group *B*, on p. 566.

relay (1) A sensitive electromechanical device which, when operated by a comparatively weak current (as from a photoelectric cell) will cause corresponding action in a more powerful circuit. (2) A substance of natural origin used as an intermediate in the course of a long synthesis.

relict(a) Residue(s).

rem The *r*oentgen *e*quivalent *m*an dose of an ionizing radiation; has the same biological effect as 1 rad of X-radiation. 1 rem = 10^{-2} sievert (the SI unit).

remanence The permanence of a permanent magnet.

remedy An agent that cures or alleviates disease.

remolinite Atacamite.

Remsen, Ira (1846–1927) American chemist noted as teacher and writer.

renal Pertaining to the kidneys.

Renard number Preferred *number.*

renature To restore to a natural or original condition. Cf. *denaturation, fibroinogen.*

render (1) To melt down; to clarify. (2) To remove fat from animal tissues by heat.

renes An extract from the kidneys of sheep, pigs, or calves.

renin Hypertensin. A hormone secreted by the kidneys. It helps to maintain normal blood pressure. **anti ~** A substance of the same origin as, but opposite in effect to, r. Cf. *rennin.*

rennase Rennin*.

rennet A preparation of the lining of the stomach of the calf; a source of rennin.

rennin* Rennase, prorennin, chymogen, caseinase, lab ferment, chymosin*. An enzyme of the gastric juice from the fourth stomach of calves. It hydrolyzes caseinogen, coagulating milk with formation of curd and whey; used in cheese manufacture. Cf. *renin.*

renosine A nucleoprotein in the kidney.

rensselaerite A variety of talc.

rep Abbreviation for *r*oentgen *e*quivalent *p*hysical. A measure of absorbed radiation dose; equal to 9.3×10^{-3} Gy.

repand Wrinkled; as, a bacterial culture with an uneven surface.

repeatability Describing the extent to which the same result can be obtained over and over again using the same apparatus. Cf. *reproducibility.*

repeating unit The original structural unit of a polymer, which is repeated to form the polymer, e.g., the diene unit in a butadiene rubber.

repercolation A repeated percolation.

replacement (1) The substitution of an atom or radical by another atom or radical: X + HA = HX + A. (2) The change from one mineral to another by gradual substitution during geological ages. **r. nomenclature** See *nomenclature.*

replica (1) A copy or reproduction taken from an original. (2) In microscopy, a technique in which a mold is made of a specimen, e.g., a fiber, in moist cellulose acetate, and then coated with a 0.001-mm layer of silver by evaporation in a high vacuum, as the basis for building up a solid copper replica by electrodeposition.

replicable Describing results that are always the same, within experimental error, when obtained by tests on the replicate samples of the same substance.

repp A plain weave, in which one of the elements (usually the finer) covers and conceals the other.

reproducibility Describing the extent to which results are always the same, within experimental error, when obtained by different operators using different sets of apparatus. Cf. *repeatability.*

reprography Collective term for all processes for reproducing texts or illustrations.

repulsion The tendency of 2 bodies to move away from each other.

resacetophenone $MeCO\cdot C_6H_3(OH)_2 = 152.2$. 2,4-Dihydroxyacetophenone. White needles, m.147.

resazurin $C_{12}H_7O_4N = 229.2$. Blue dyestuff used to test milk. Reduction to red resorufin, and eventually to colorless dihydroresorufin, occurs in presence of mastitic cells or of milk more than 24 hours old.

research (1) Scientific investigation directed to the discovery or examination of a new or existing fact or phenomenon. (2) A literary search, followed by planned experimental investigation or verification. **operational ~** Abbrev. OR; the use of scientific methods, and particularly applied statistics, to provide a quantitative basis for policy decisions. **r. octane number** See *octane number.*

reseda oil Mignonette oil. Fragrant oil, used as a flavoring and scent, extracted from plants of the *Reseda* genus, such as mignonette.

resenes The constituents of resins, insoluble in alkalies. They contain O; but not as OH, COOH, lactone, or ester. The most valuable varnish resins, as, dammar, copal, dragon's blood, are rich in r.

reserpine $C_{33}H_{40}O_9N_2 = 608.7$. An alkaloid from the trees *Rauwolfia serpentina* (India), and *Alstonia conesticta* (Australia). Brown crystals, m.270 (decomp.), insoluble in water; used as the vegetable extract to treat hypertension (USP, EP, BP).

reserve Resist. A substance that will resist the dyeing of those portions of a fabric onto which it has been printed, e.g., wax. Cf. *batik.*

reservoir A storage receptacle.

reset action See *mode.*

residual That which remains. **r. gas** The small amount of

TABLE 74. MECHANISMS OF ION-EXCHANGE RESINS

Cation-exchange resins						
(1)	R^-Na^+	$+ M^+A^-$	$\rightleftharpoons$	R^-M^+	$+ Na^+A^-$	(Water softening)
(2)	R^-H^+	$+ M^+A^-$	$\rightleftharpoons$	R^-M^+	$+ H^+A^-$	(Salt splitting)
(3)	R^-H^+	$+ M^+(OH)^-$	$\rightarrow$	R^-M^+	$+ H_2O$	(Neutralization)
Anion-exchange resins						
(4)	$R^+(OH)^-$	$+ M^+A^-$	$\rightleftharpoons$	R^+A^-	$+ M^+(OH)^-$	(Salt splitting)
(5)	$R^+(OH)^-$	$+ H^+A^-$	$\rightarrow$	R^+A^-	$+ H_2O$	(Neutralization)

gas that remains in an evacuated apparatus, e.g., a vacuum tube.

residue That which remains as the ash after an ignition.

resilience The property of returning to the original shape after distortion within elastic limits. The energy stored upon distortion is measured by the work done per unit volume.

resilient Elastic, rebounding.

resin(s) (1) Natural r. Flammable, amorphous, vegetable products of secretion or disintegration, usually formed in special cavities of plants. Generally insoluble in water and soluble in alcohol, fusible, and having a conchoidal fracture. They are the oxidation or polymerization products of the terpenes, and are mixtures of aromatic acids and esters. Some are official, as, *benzoin;* others are nonofficial, as, *copals, sandarac.* (2) Raw materials for *plastics* (q.v.) fabrication. See *synthetic resins* below. Cf. *resinoid, rosin.* **alkyd ~** R. made from phthalic anhydride and glycerol. **artificial ~** Resin (2). **Bowen's ~** See *Bowen's resin.* **gum ~** R. containing gum which softens in water; as, olibanum, myrrh, gamboge, scammony. **ion-exchange ~** A solid solution of an active electrolyte in a highly stable insoluble matrix (usually a synthetic r.). It exists in cation-active and anion-active forms, and can react with the ions present in (e.g.) water. See Table 74. $R^{\pm}$ is the active r., M^+ a given metallic ion, and A^- an anion. The squared groups are insoluble r. complexes. The functional groups are:

Cation exchanges:	Strong—sulfonic acid
	Weak—carboxylic acid
Anionic exchanges:	Strong—quaternary ammonium
	Weak—amino

Used to demineralize water. Also in medicine, for high levels of blood potassium and for renal failure; as, sodium polystyrene sulfonate. **modified ~** A mixture of a synthetic and natural r., e.g., colophony, which may be an extender or may confer special properties. **oil-reactive ~** A synthetic phenol-formaldehyde r., soluble in oil to form a quick-hardening varnish. **oil-soluble ~** R. not reacting with an oil but soluble in it. **oleo ~** Balsams. A viscous mixture of r. with essential oils; as, Canada balsam, turpentine, pitch, storax, white dammar. **single-stage ~** Resol. **synthetic ~** A heterogeneous group of compounds produced synthetically from simpler compounds by polymerization, condensation, or both. Upon them are based many *plastics,* q.v. The term was originally used to describe such synthetic substances having the properties of r., but is now used in a wider sense. **true ~** A natural r. that is neither a gum nor oleo r.; as, colophony, kauri, sandarac, mastic, shellac, jalap, copal. **vinyl ~** A synthetic polymerization product of vinyl compounds. Cf. *Vinylite.*

r. acids Organic acids derived from r.; as, pimaric acid. **r. of copper** Cuprous chloride (obsolete).

resina (1) Latin for "resin." (2) Rosin.

resin amines Insoluble, amorphous, solid bases from low-temperature tar, which fuse to glassy masses and are deposited from solution as tough films.

resinate A salt of a resin acid mixture.

resinene Generic name for neutral resins.

resineon An essential oil, distilled from black tar oil, b.148.

resinification (1) The process of oxidation or polymerization of essential oils by which they become solid resins. (2) Artificial condensation polymerization, or similar processes, by which resinlike substances are formed. Cf. *plastics.*

resinite A resinous constituent of coal.

resinoic acid Generic name for acidic resins.

resinoid A substance resembling a resin in physical properties, except that it changes to an insoluble and infusible solid on heating. Cf. *plastics.*

resinol (1) Resole. (2) The noncrystalline constituents of tar, soluble in sodium hydroxide solution and precipitated from solution in an organic solvent by light petroleum. Cf. *resinotannol.*

resinotannol A colored alcohol of resin esters which gives the tannin reaction. Cf. *resinol.*

resinous Having properties of a resin.

resist Reserve.

resistance (1) Opposition to force or external conditions. (2)* The opposition that a conductor offers to the passage of an electric current; measured in *ohms,* q.v. The reciprocal of electrical conductance. **body ~** See *immunity.* **external ~** The opposition to the passage of an electric current outside the source of current; as in wires. **internal ~** Electrical resistance within the generating device. **specific ~** Resistivity*. **thermal ~** See *thermal resistance.*

r. box An arrangement of rheostat coils in mounted rows in a container. By turning dials or inserting plugs, degrees of resistance may be obtained. **r. capacity** See *cell constant.* **r. coil** A spool of wire for increasing the r. in a circuit. **r. gage** See *resistance gage* under *gage.* **r. to infection** See *immunity.* **r. thermometer** See *pyrometer.* **r. wire** Nichrome wire for heating units or r. devices having a definite r.

resistivity* Specific resistance, $\rho = RA/l$, where R is the electrical resistance, A the cross section, and l the length of the material. The reciprocal of conductivity. See Table 75 on p. 504.

resistor A metal (as chromium) which becomes hot on the passage of an electric current; used in electric heaters. **age ~** A rubber additive which inhibits oxidation; e.g., aromatic diamine antioxidants.

resocyanin $C_{10}H_8O_3$ = 176.2. 7-Hydroxy-4-methylcoumarin, 4-methylumbelliferone. Colorless crystals, m.185; used in perfumery.

resoflavin $C_{14}H_3O_4(OH)_3$ = 286.2. Yellow dye, from resorcinol.

resol A single-stage synthetic resin produced from a phenol and an aldehyde. Cf. *resole.*

TABLE 75. RESISTIVITY OF VARIOUS ELEMENTS AT 0°C

Element	Resistivity, $\Omega \cdot m \times 10^{-8}$
Silver	1.47
Copper	1.60
Aluminum	2.50
Zinc	5.5
Iron	8.9
Lead	19.2
Mercury	94

resole Resinol. A compound formed by the condensation of a phenol with an aldehyde in presence of a catalyst and alkali, and hardened by heat. Cf. *Bakelite, novolak.*

resolution (1) Mesotomy. The separation of a racemic mixture into its optically active components. Cf. *racemization, inversion.* (2) The separation of spectral lines into a number of component lines by intense electric or magnetic fields. Cf. *Stark effect.*

resolving power The power of a lens to produce a detailed and distinct image of an object at a certain distance.

resonance (1) The vibrations set up by sound or electromagnetic waves in a material that is capable of vibrating with the same or a multiple frequency. (2) Rays emitted by an atom, of the same wavelength as those previously absorbed. (3)* Mesomerism*. The representation of molecular electronic structure in terms of the wave functions of contributing structures. (4) Subatomic *particles*.

resonant Producing sound.

resonator An instrument used to intensify sound waves or electromagnetic oscillations by resonance. Cf. *quartz* oscillator.

Resonium A Trademark for sodium polystyrene sulfonate.

resorcin Resorcinol*. **r. blue** A microchemical stain.

resorcinol* $C_6H_4(OH)_2$ = 110.1. Resorcin, 1,3-benzenediol, metadihydroxybenzene. Colorless crystals, m.116, very soluble in water or ether. Used in the manufacture of fluorescein, eosin, and other dyes, synthetic drugs, and photographic developers; and as a reagent, reducing agent, and keratolytic for skin conditions (USP, BP). **dimethyl ~** Xylorcinol. **dithio ~** $C_6H_4(SH)_2$ = 142.2. 1,3-Phenylenedithiol, m.26, b.245. **hexahydro ~** Resorcitol. **hexyl ~** See *hexylresorcinol* under *hexyl.* **methyl ~** Orcinol. ***p*-nitrobenzeneazo ~** $(HO)_2C_6H_3N{:}N \cdot C_6H_4 \cdot NO_2$ = 259.2. A reagent for magnesium (0.1 mg in 5 ml gives a blue color). **trimethyl ~** Mesorcinol. **trinitro ~** Styphnic acid*.
 r. monoacetate $C_8H_8O_3$ = 152.2. Euresol. Yellow oil, b.283, insoluble in water; a skin antiseptic. **r. phenolate** Phenoresorcin. **r.phthalein** See *fluorescein.* **r. yellow** Tropeolin.

resorcitol $C_6H_{10}(OH)_2$ = 116.2. 1,3-Cyclohexanediol*. **(1*RS*,3*RS*)- ~** m.117. **(1*RS*,3*SR*)- ~** m.86.

resorcyl The 1,3-dihydroxyphenyl* radical, $-C_6H_4(OH)_2$, from resorcinol.

resorcylate A salt of dihydroxy*benzoic acid.*

resorcylic acid Dihydroxy*benzoic acid**.

resorption The absorption of superfluous material; as, the products of inflammation.

resorufin $C_{12}H_7O_3N$ = 213.2. 7-Hydroxy-2-phenoxazone. A heterocyclic compound; used to detect halogens, which destroy its intense fluorescence in alkaline solution. See *resazurin.*

respiration The process of breathing, by which air is inhaled (inspiration) and expelled (expiration) by a living organism. In mammals r. depends partly on the chemical equilibrium between oxygen, carbon dioxide, and H^+ ions, and partly on nervous impulses from the respiratory area in the brain. In the lungs the hemoglobin is oxidized to oxyhemoglobin. In the tissues (where an excess of carbon dioxide exists) the oxyhemoglobin releases the oxygen and removes the carbon dioxide.

respirator (1) Inspirator, inhaler. A gas mask or screen of fine wire or gauze with or without adsorbing or chemical reagents, worn over the mouth or nose to protect the wearer from dust, smoke, poisonous or irritating gases. (2) A mask to enable drugs to be inhaled. (3) A machine to perform movements of respiration mechanically.

respiratory Pertaining to respiration. **r. quotient.** Abbrev. r.q.; the ratio of the volumes of inhaled oxygen to expelled carbon dioxide.

response time The time it takes an output to reach a specified percentage of the ultimate steady-state value.

restitution The reverse reaction of *substitution,* q.v.: an element is either oxidized or reduced to the free state.

restorative An agent that aids, renews, or promotes a healthy physical condition.

resublimed Purified by repeated sublimation.

resultants The products of a chemical reaction: reaction products. Cf. *reactants.*

resuscitation The restoration of consciousness or life in one apparently dead or very gravely ill.

retamine $C_{15}H_{26}ON_2$ = 250.4. An alkaloid from *Retama sphaerocarpa;* long needles, m.166, soluble in EtOH.

retardation Slowing up of a chemical reaction; negative catalysis.

retarder A substance added to a reacting mixture to prevent the reaction from becoming too vigorous; as, potassium bromide in photographic developers to prevent overdevelopment. Cf. *accelerator, catalytic poison.*

retene $C_{18}H_{18}$ = 234.3. 1-Methyl-7-isopropylphenanthrene. A hydrocarbon from pine tar. Lustrous leaflets, m.98, slightly soluble in water.

retentate Suggested term for the material retained by a semipermeable membrane. Cf. *dialyzate.*

retention The holding or retaining of a substance or property.

Retger's law The physical properties of mixed crystals vary in proportion to their percentage compositions.

reticulated Retiform. Resembling a network.

reticulin The protein of the fibers of reticular tissue.

retiform Reticulated.

Retin A Trademark for tretinoin.

retina The portion of the eye on which the image is focused by the lens; bundles of sensitive rods and cones, which carry the sensation of vision to the brain.

retinene A yellow protein pigment formed in the eye retina by the action of light on rhodopsin. It subsequently forms vitamin A, which forms more rhodopsin.

retinite $C_{12}H_{18}O$ = 178.3. An amorphous, gray, native substance. **r. resins** Hard, brittle, brown resins from brown coal. They contain no wax and are derived from amber. Cf. *bituminous resins.*

retinol (1) $C_{20}H_{30}O$ = 286.5. Vitamin A (formerly A_1).

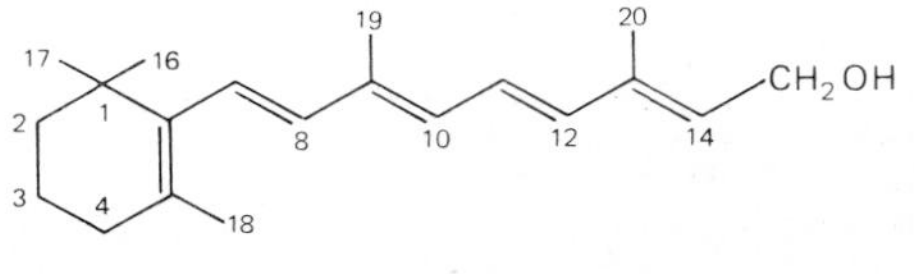

Green, fluorescent, readily oxidized fat, soluble in water. Derived from β-carotene. Essential for night vision, being part of visual purple or rhodopsin in the retina of the eye (USP, BP). (2) $C_{32}H_{16}$ = 400.5. Resinol, codol, resinoil, obtained by distilling pitch or rosin. Yellow liquid, b.280, insoluble in water; an antiseptic.

retonation A wave propagated backward through the burned gases from the starting point of an explosion.

retort A distilling vessel. Originally a flask with a bent neck (alembic).

retro- Prefix (Latin) indicating "backward."

retrogression A reversal; a reverse reaction.

retronecic acid $C_{10}H_{16}O_6$ = 232.2. An acid hydrolysis product of retrorsine, m.181.

retronecine $C_8H_{13}NO_2$ = 155.2. A basic hydrolysis product of retrorsine, m.121.

retrorsine $C_{18}H_{25}NO_6$ = 351.4. An alkaloid from *Senecio retrorsus* (Compositae), m.207.

retting Loosening the fiber of vegetable materials by the action of moisture, enzymes, and/or bacteria. Cf. *coir, flax.*

retzbanyite Native bismuth lead sulfide.

reussinite A resinlike, reddish-brown hydrocarbon in certain coal deposits.

reverberatory Flickering, blowing downward. **r. furnace** See *reverberatory furnace* under *furnace.*

reversible Capable of being restored to an original condition. **r. action** The reestablishment of an original condition. **r. cell** An electric cell which after discharge can be activated by an external current; as, an accumulator. **r. colloid** See *reversible colloid* under *colloid.* **r. electrode** An electrode which owes its potential to ionic changes of a r. nature. See *reversible electrode* under *electrode.* **r. reaction** A reaction that establishes an equilibrium, and can proceed from right to left or from left to right: $A + B \rightleftharpoons C + D$.

revertose $C_{12}H_{22}O_{11}$ = 342.3. A disaccharide produced by the action of α-D-glucosidase on glucose solution.

revive (1) To restore original activity, e.g., of a catalyst. (2) To reduce a metallic ore or oxide to the metal.

revivification The restoration of an active condition. **r. of carbon** R. of spent carbon black by (1) burning in the absence of air, (2) washing with acids to remove its ash, (3) washing with an alkaline solution.

Rey, Jean (1630) An alchemist who experimented with metallic oxides.

reyn A unit of dynamic viscosity: 1 r. = centipoise $\times$ 6.9 $\times$ 10^6. Cf. *Reynolds number.*

Reynolds number An expression of fluid flow in a pipe: *Re* = (density of fluid $\times$ velocity of flow $\times$ diameter of pipe)/ dynamic viscosity of the liquid. Different fluids and gases show similar flow patterns at the same R. n.

Rf Symbol for the element name rutherfordium.

Rf value In paper-strip chromatography, the proportion of the total length of climb of a solution that is reached by a spot characteristic of one of the constituents present.

RG acid 1-Naphthol-3,6-disulfonic acid.

Rh Symbol for rhodium. **Rh factor** Rh antigen. A type of antigen on red blood cells, first discovered in *Rhesus* monkeys. There are 3 Rh factors (C,D,E), of which D is the most important. About 85% of the Caucasian population has the D antigen and is called Rh-positive. Isoantibodies can occur after pregnancy and blood transfusion, and may cause anemia in the fetus. Cf. *blood groups.*

rH, r_H Symbol for oxidation-reduction potential: rH = log (1/pH_2), where pH_2 is the hydrogen pressure, in atmospheres. Cf. *pH.*

r.h., R.H. Abbreviation for relative humidity.

rhabdophane $(Ce,Nd,Pr,La)_3(PO_4)_2$. Phosphocerite. A mineral.

Rhamnaceae Buckthorn family; shrubs from which drugs are derived; as, *Rhamnus frangula,* frangula; *R. purshiana,* cascara sagrada. Cf. *cathartin.*

rhamnegin $C_{12}H_{10}O_5$ = 234.2. A glucoside from buckthorn berries.

rhamnetin $C_{16}H_{12}O_7$ = 316.3. A methyl ester of quercetin and a split product of xanthorhamnin; a yellow color.

rhamnicogenol A pentahydroxymethylanthranol occurring as primeveroside in purgative buckthorn and other *Rhamnus* species.

rhamnicoside $C_{26}H_{30}O_{15}\cdot 4H_2O$ = 654.6. A glucoside from the stem bark of purgative buckthorn, *Rhamnus cathartica* (Rhamnaceae). Cf. *Chinese green.*

rhamnin A fluidextract of *Rhamnus frangula* containing rhamnetin. **xantho ~** See *xanthorhamnin.*

rhamninose $C_{18}H_{32}O_{14}$ = 472.4. A trisaccharide from Persian berry, *Rhamnus infectoria,* which hydrolyzes to 2 rhamnose and 1 glucose residues.

rhamnite Rhamnitol.

rhamnitol $Me(CHOH)_4CH_2OH$ = 166.2. A reduction product of rhamnose. Colorless crystals, m.121, soluble in water.

rhamnofluorin A constituent of *Rhamnus* species.

rhamnogalactoside Glucosides that are hydrolyzed to rhamnose and galactose; as, robinin.

rhamnoglucoside A glucoside that is hydrolyzed to rhamnose and glucose; as, rutin.

rhamnol $C_{20}H_{36}O$ = 292.5. An alcohol, m.132, from cascara sagrada.

rhamnomannoside A glucoside that is hydrolyzed to rhamnose and mannose; as, baptisin.

rhamnose $Me(CHOH)_4CHO\cdot H_2O$ = 182.2. 6-Deoxymannose†, isodulcite. A methyl pentose from rhamninose and various glucosides. L-**~** Colorless crystals, m.93, soluble in water. Cf. *quinovose.*

rhamnoside (1) A glucoside that is hydrolyzed to rhamnose. (2) $C_{21}H_{20}O_9$. A split product of rhamnoxanthin; occurs in *Rhamnus* species.

rhamnoxanthin A crystalline glucoside from *Rhamnus frangula.*

Rhamnus A genus of trees and shrubs; many have a purgative bark or fruit. **R. cathartica** Buckthorn berries. The dried fruits of *Rhamnus cathartica.* **R. frangula** See *frangula.* **R. purshiana** See *cascara sagrada.*

rhatanin $C_{10}H_{13}O_3N$ = 195.2. *N*-Methyltyrosine, geoffroyin. A glucoside of rhatany root; needles, m.280.

rhatany Krameria.

rhe A unit of fluidity; the reciprocal of a unit of viscosity (poise). 1 rhe = 10 $Pa^{-1}s^{-1}$.

rheadine $C_{21}H_{21}O_6N$ = 383.4. An alkaloid from opium; white prisms, m.246.

rhea fiber Ramie.

rheic acid Chrysophanic acid.

rhein $C_{15}H_8O_6$ = 284.2. Rheinic acid. 4,5-Dihydroxyanthraquinone-2-carboxylic acid. From senna leaves, rhubarb, and lichens. Golden prisms, m.321.

rheinic acid Rhein.

rhenate* Indicating rhenium as the central atom(s) in an anion. M_2ReO_4. Rhenates are colorless, soluble in water (except the Ag, Tl, K, Rb, and Ce salts). **per ~*** $MReO_4$.

rhenic acid* H_2ReO_4. A dibasic acid which changes to $HReO_4$. **per ~*** $HReO_4$ formed on dissolving Re_2O_7 in water.

rhenium* Re = 186.207. Dvimanganese. Bohemium,

uralium. The element of at. no. 75, a noble metal of Group 7A, discovered by Noddack (1924) by means of its X-ray spectrum, in platinum ores. Named after the Latin for "Rhine." Obtained as a by-product in the roasting of molybdenum sulfide. Silvery, d.21.0, m.3180 (the highest of the elements, except W and C), b. ca. 5400. Used in high-temperature technology, for electric contacts and acid-resistant chemical plant, and as a catalyst to dehydrogenate alcohols to aldehydes or ketones: 2 isotopes, mass 185 and 187; and oxidation states −1 through 7. **r. black** A catalyst for selective hydrogenation reactions, prepared by the hydrogenation reduction of r. heptaoxide in a solvent. **r. chlorides** (1) $ReCl_7$ = 434.4. R. heptachloride*. Green, volatile crystals, hydrolyzed in water. (2) $ReCl_6$ = 898.9. R. hexachloride*. Brown, volatile crystals. (3) $ReCl_4$ = 328.0. R. tetrachloride*. Brown crystals. (4) $ReCl_3$ = 292.6. Hexagonal, red crystals, soluble in water. **r. fluoride** ReF_6 = 300.2. Yellow crystals, m.26. **r. oxides** (1) Re_2O_8. R. peroxide. Bluish-white solid, soluble in water. (2) Re_2O_7. R. heptaoxide*. Yellow solid, m.200, soluble in water forming perrhenic acid. (3) ReO_3. Red solid, changes to Re_2O_7. (4) ReO_2. R. dioxide*. Black powder formed by heating $NaReO_4$ in hydrogen. (5) Re_3O_8. Blue solid. **r. sulfide** $Re_2S_7 \cdot H_2O$. R. heptasulfide*. Black powder, insoluble in acids or alkalies, oxidized by nitric acid.

rheochrysin $C_{22}H_{22}O_{10}$ = 446.4. A glucoside from rhubarb. Cf. *rheopurgin.*

rheology The science of the deformation and flow of matter. Cf. *softener.*

rheometer (1) A galvanometer. (2) An instrument to measure the velocity of the blood current.

rheopexy A form of thixotropy, in which solidification occurs as the result of a regular, gentle motion instead of vigorous shaking.

rheopurgin A constituent of rhubarb which decomposes into 4 glucosides. Cf. *rhein.*

rheoscope An instrument for detecting an electric current.

rheostan The alloy Cu 52, Zn 18, Ni 25, Re 5%; used for rheostats.

rheostat An instrument to regulate electrical resistance.

rheotannic acid $C_{26}H_{26}O_{14}$ = 562.5. Rhubarb tannin. Yellow powder, soluble in water.

rheotome A current breaker.

Rheovot Trademark for a mixed-polymer synthetic fiber.

rhesus See *Rh.*

rheum Rhubarb.

rhigolene The first condensation product of the fractional distillation of petroleum, b.21; consists mainly of butane and pentane. Used to produce coldness of the skin before surgical operations; also as a solvent.

rhinanthin A glucoside from the seeds of *Alectorolophus hirsutus* or *Rhinanthus major* (Scrophulariaceae).

rhine metal An alloy: Sn 97, Cu 3%; d.7.35, m.300.

rhinestone Colorless, highly refractive glass; a semigem.

rhizobia Soil bacteria which live symbiotically with legumes, forming nodules that enable the plant to utilize atmospheric nitrogen.

rhizocarpic acid $C_{28}H_{23}O_6N$ = 469.5. A yellow constituent of the lichen *Lecanora epanora.* Cf. *epanorin.*

rhizocholic acid An oxidation product of cholic acid.

rhizoid Describing an irregularly branched bacterial growth.

rhizome An underground, swollen plant stem, characterized by the presence of leaf bases. Some are used medicinally.

rhizonic acid $C_9H_{12}O_4$ = 184.2. 2-Hydroxy-4-methoxy-3,6-dimethylbenzoic acid. White crystals, m.232. **iso ~** 4-Hydroxy-2-methoxy-3,6-dimethylbenzoic acid. White crystals, decomp. 156, decomp. by boiling water.

rhizoplane Soil particles adhering to roots.

rhizosphere The zone of soil subject to the influence of plant roots; it supports a high biological activity.

rho ρ. Greek letter. ρ (italic) is symbol for (1) charge density; (2) (mass) density; (3) reflection coefficient; (4) reflection factor (reflectance); (5) resistivity. **r. 2** Progesterone.

rhodacene $C_{30}H_{20}$ = 380.5. A pyrolytic product from acenaphthene. Violet crystals with green metallic luster, m.339.

rhodalline Allyl thiourea.

rhodamic acid Rhodanine*.

rhodamines Basic red and violet dyes closely allied to the fluoresceins; obtained by condensation of phthalic anhydride with *p*-alkylated aminophenols. **tetraethyl ~** $C_{28}H_{30}N_2O_3$ = 442.6. Colorless, but its salts are red dyes with an oxonium or quinonoid structure.

rhodan The thiocyanate* radical.

rhodanate Thiocyanate*.

rhodanic acid (1) Rhodanine*. (2) Thiocyanic acid*.

rhodanide Thiocyanate*.

rhodanine* $\underline{CH_2 \cdot CO \cdot NH \cdot CS \cdot S}$ = 133.2. Rhodanic acid. 2-Thioxo-4-thiazolidinone†. ***p*-dimethylaminobenzylidene ~** $(C_3HNOS_2):CH \cdot C_6H_4 \cdot NMe_2$ = 264.4. A reagent for silver (flocculent, red precipitate).

rhodanizing Plating with rhodium (especially on silver) to prevent tarnishing.

rhodanometry (1) The use of free cyanogen to determine the absorption values of oils. Cf. *iodine number.* (2) Thiocyanometry. Titration with thiocyanate solutions to determine silver, mercury, etc.

rhodeol $CH_3(CHOH)_4CH_2OH$ = 166.2. D-1-Deoxygalactitol, rhodeite, rhodeitol; oxidized to D-fucose.

rhodeoretin Convolvulin.

rhodeose D-Fucose.

Rhode test Concentrated sulfuric acid and dimethyl aminobenzaldehyde give a purple color with proteins.

Rhodia, Rhodiafil, Rhodialin Trademarks for cellulose acetate synthetic fibers.

rhodiene $(C_{10}H_{16})_n$. A hydrocarbon from *Rhodium lignum,* rosewood.

rhodinol $Me_2C:CH \cdot CH_2 \cdot CH_2 \cdot CHMe \cdot CH_2 \cdot CH_2OH$ = 156.3. (*S*)-3,7-Dimethyl-7-octen-1-ol*. A constituent of citronella and geranium oils; used in perfumes.

rhodite Rhodium gold.

rhodium* Rh = 102.9055. A metal in Group 8, at. no. 45; usual valency 2 or 3. Grayish-white, ductile metal, d.12.44, m.1970, b.3730, insoluble in water. Occurs in the platinum ores and gold gravels of S. America. Discovered by Wollaston (1804). Used for plating (superior to chromium plating); and in alloys, thermoelements, and astronomical measuring instruments. **r. black** Finely divided r. metal obtained by precipitation of r. salt solutions by formaldehyde; a catalyst. **r. cesium alum** Cesium rhodium sulfate. **r. chloride** $RhCl_3 \cdot 4H_2O$ = 281.3. R. sesquichloride. Brownish, deliquescent powder, decomp. 475, soluble in water. **r. gold** Rhodite. A native alloy: Au 57–66, Rh 34–43%. **r. hydroxide*** $Rh(OH)_3$ = 153.9. Black powder, decomp. by heat, insoluble in water. **r. nitrate*** $Rh(NO_3)_3 \cdot 2H_2O$ = 325.0. Red, deliquescent crystals, soluble in water. **r. oxides** (1) RhO = 118.9. R. monoxide*. Gray powder, insoluble in water. (2) Rh_2O_3 = 253.8. R. sesquioxide. Gray powder obtained by heating r. with barium peroxide; insoluble in water. (3) RhO_2 = 134.9.

R. dioxide*. Brown powder, insoluble in water. **r. sulfate*** $Rh_2(SO_4)_3 \cdot 12H_2O$ = 710.2. Yellow crystals, soluble in water. **r. sulfide*** RhS = 135.0. Blue powder, decomp. by heat, insoluble in water.

rhodizite White or green, native calcium borate.

rhodizonic acid $\underline{C \cdot OH:C \cdot OH \cdot (CO)_3 \cdot CO}$ = 170.1. A dihydroxytetrone that acts as a dibasic acid. Its sodium salt is a reagent for barium and strontium (brown color).

rhodochrosite $MnCO_3$. Red manganese. A native carbonate.

rhododendrin $C_{16}H_{22}O_7$ = 326.3. A constituent of *Rhododendron chrysanthemum* (Ericaceae). Colorless crystals, m.187, soluble in hot water.

rhododendrol $C_{10}H_{12}O_2$ = 164.2. A constituent of *Rhododendron chrysanthemum.* Colorless crystals, m.80.

rhodol Methyl-*p*-aminophenol. Used as a photographic developer.

rhodolite A variety of garnet.

rhodonite $MnSiO_3$. Red manganese, manganese spar. A native manganese silicate.

rhodopsin The protein pigment of the visual purple in the retina of the eye, associated with the mechanism of vision, particularly night vision. R. is changed into retinene and opsin by the action of light.

rhodopterin A purple oxidation product of pterin.

rhodopurpurin Rhodovibrene.

rhodotannic acid $C_{14}H_6O_7$ = 286.2. A tannin from the leaves of *Rhododendron ferrugineum.*

rhodovibrene Rhodopurpurin. A purple carotenoid pigment produced by bacteria.

rhodoviolascin $C_{42}H_{60}O_2$ = 596.9. Methoxybacteriopurpurin. m.216. Main carotenoid in Athiorhodaceae spp.

rhodoxanthin $C_{40}H_{50}O_2$ = 562.8. Thujorhodin. A red carotenoid in the red berries of *Taxus baccata,* yew.

Rhoduline Trademark for aniline dyes of the safranine type; used in textile printing.

Rhoeadales An order of plants comprising the families: *Papaveraceae, Cruciferae, Resedaceae.*

rhoeadic acid Papaveric acid.

rhoeadine $C_{21}H_{21}O_6N$ = 383.4. (+)-*O*-Merhoeagenine. A crysalline alkaloid from *Papaver rhoeas* (Papaveraceae).

rhoeagenine $C_{20}H_{19}O_6N$ = 369.4. **(+)~** An alkaloid from *Paparverrheos.*

rhoetzite Cyanite.

Rhofil Trademark for a polyamide synthetic fiber.

rhomb A rhombic crystal. **r. spar** A rhombic dolomite.

rhombic Describing a crystal with 3 unequal axes, all at right angles; as, iodine. **r. dodecahedron** A crystal of the isometric system with 12 faces; each face is parallel to 1 axis and has equal intercepts on the other 2 neighboring faces. **r. mica** Phlogopite. **r. quartz** Feldspar (obsolete). **r. spar** A form of dolomite. **r. system** The orthorhombic crystal system.

rhombohedral Describing a crystal of the trigonal system with a vertical axis of 3-fold symmetry and 3 horizontal axes of 2-fold symmetry. **r. system** A modified trigonal system of crystals with 3 horizontal axes of 2-fold symmetry in place of the 3-fold symmetry of the hexagonal system.

rhombohedron A crystal form bounded by 6 rhombic faces.

rhometer An instrument to measure resistivity.

rhotanium An alloy of palladium and rhodium.

Rhovyl Trademark for a polyamide synthetic fiber.

rhubarb Rheum. The dried rhizomes and roots of *Rheum palmatum,* and *Rheum* species (Polygonaceae). It contains rheopurgin, chrysophanic acid, rhein, and tannin and is a purgative (BP). Cf. *aporetin.* **r. tannin** A rheotannic acid.

Rhus A genus of trees and shrubs of the cashew family (Anacardiacea). ***R. aromatica*** Yields a stimulant. ***R. cotinus*** Fustic. ***R. coriaria*** Sumac. ***R. diversiloba*** The poison oak of the Pacific coast. ***R. glabra*** Sumac berries. An astringent. ***R. toxicodendron*** Poison oak, poison ivy.

rhythmic Occurring at regular intervals. **r. deposition** Precipitation or condensation in gas or vapor mixtures at periodic intervals; as produced from containers of hydrochloric acid and ammonia in a closed bell jar. **r. precipitation** Periodic precipitation. Precipitation in bands or zones in a colloid. See *Liesegang rings.*

ribichloric acid $C_{14}H_8O_9$ = 320.2. An acid constituent from *Galium aparine,* goose grass (Rubiaceae).

riboflavine $C_{17}H_{20}N_4O_6$ = 376.4. Vitamin B_2, lactoflavin. Bitter, orange crystals, m.280 (decomp.), slightly soluble in water. Part of the molecule of FAD; essential in oxidation and reduction in the tissues. Deficiency causes minor disorders in humans, but failure to grow in animals (USP, EP, BP).

ribonic acid $CH_2OH(CHOH)_3COOH$ = 166.1. Derived from ribose. **D-~** m.112. **L-~** m.105.

ribonucleic acid See *ribonucleic acids* under *nucleic acids.*

ribonucleoprotein See *ribonucleoprotein* under *nucleoprotein.*

ribose $CH_2OH \cdot (CHOH)_3 \cdot CHO$ = 150.1. **D-~** A pentose sugar constituent of some nucleic acids and coenzymes, existing in the furanoside and pyranoside forms. **D-2-deoxy~** $HOH_2C(CHOH)_2 \cdot CH_2 \cdot CHO$ = 134.1. A combined sugar in deoxyribonucleic acid and cellular nuclear material. Cf. *Feulgen reaction, Dische reaction.*

r. ketohose Psicose.

ribosome Ribo*nucleoprotein* particles; the main constituent of cytoplasm, providing sites for protein synthesis.

rice The seeds or grains of *Oryza sativa;* mainly starch (rice flour). A food, and dusting powder.

rich High in content; having a great percentage. **r. gas** The second group of combustible gases obtained by the distillation of coal and coke; as, coke-oven gas. **r. iron** Iron having a high silicon content.

Richards, Theodore William (1868–1928) American chemist, Nobel prize winner (1914); noted for his atomic weight and physical chemistry experiments.

richellite A native iron, calcium phosphate from Richelle, Belgium.

richmondite A heterogeneous mineral associated with galena, and containing galena 36, antimonous sulfide 22, chalcocite 19, ferrous sulfide 14, sphalerite 6, argentite 2, manganous sulfide 0.5%, and bismuth sulfide traces.

Richter R., Jeremias Benjamin (1762–1807) German chemist noted for his work on volumetric analysis and stoichiometry. **R.'s law** Wenzel's law. Each equivalent weight of an acid will completely neutralize an equivalent weight of a base. **R. scale** Logarithmic scale used to measure the strength of earthquakes. Zero on the scale corresponds to an earthquake causing a reading of 1 μm on a seismometer 100 km from the epicenter of the earthquake. Recordings go up to about 9. **R., Victor von (1841–1891)** German chemist, author of textbooks.

ricin An albuminous toxin of castor-oil beans, *Ricinus communis;* it agglutinates red blood cells.

ricinate Ricinoleate. A salt of ricinoleic acid.

ricinic acid Ricinoleic acid.

ricinine $C_8H_8O_2N_2$ = 164.2. An alkaloid from the castor-oil plant, *Ricinus communis.*

ricinoleate Ricinate. A salt of ricinoleic acid.

ricinoleic acid $Me(CH_2)_5 \cdot CHOH \cdot CH_2CH:CH(CH_2)_7COOH$

= 298.5. Ricinic acid, (*R*)-(*Z*)-12-hydroxy-9-octadecenoic acid*, elaeodic acid. Yellow fatty acid from castor oil, m.6, b_{10mm}245, insoluble in water. Its alkali salts detoxify antigens.
ricinolein $C_{57}H_{104}O_9$ = 933.4. Glyceryl triricinoleate, triricinolein, in castor oil (80%).
Ricinus The castor-oil plant, *R. communis* (Euphorbiaceae), whose seeds yield castor oil and ricinine.
rickardite Cu_3Te_3. A copper mineral.
Rickettsiae Microorganisms, larger than viruses, but, like them, inhabit living cells; many cause disease, as, typhus, Rocky mountain spotted fever. R. are transmitted by ticks and lice.
Rideal, Sir Eric (1890–1975) British chemist, noted for work on colloids.
Rideal-Walker **R-W test** The evaluation of an antiseptic from the amount required to prevent the growth of *Salmonelli typhi* under standard working conditions. Phenol is the standard of comparison. **R-W value** Phenol coefficient.
riebeckite A blue amphibole asbestos in ironstone seams (W. Australia); it contains about 6% crocidolite.
Rifadin Trademark for rifampin.
rifampicin EP, BP name for rifampin.
rifampin $C_{43}H_{58}O_{12}N_4$ = 823.0. Rifampicin, Rifadin. Reddish-brown crystals, slightly soluble in water. An antibiotic, from *Streptomyces mediterranei,* for gram-positive bacteria and myocobacteria (USP, EP, BP).
riffles (1) Small waves. (2) A corrugated surface. (3) A *sampler,* q.v.
rigidity The state of being inflexible or stiff. **r. modulus** Young's *modulus.*
Rilsan, Rilson Trademark for polyamide synthetic materials, including nylon 11.
Rimifon Trademark for isoniazid.
rimnic acid The chief constituent of rimu resin.
rimose Describing a bacterial culture which shows fissures and cracks.
rimu resin Red resin from the rimu (red pine), *Dacrydium cupressinum* (Coniferae), New Zealand.
ring (1) A circular structure. (2) A closed chain of atoms. (3) A system of rings; as, the porphin r. See *ring structures* below. **bacterial ~** A growth of bacteria on the surface of a medium in a petri dish, which adheres to the glass forming a rim. **benzene ~** See *benzene ring.* **fused ~** See *ortho- (and peri-) fused* under *ortho-*. **heterocyclic ~** See *heterocycle.* **homocyclic ~** See *homocycle.* **Liesegang's ~** See *Liesegang ring, rhythmic precipitation.* **Newton's ~** See *Newton's rings* under *Newton.*

r. assembly* Two or more cyclic systems (single rings or fused systems) directly joined to each other, where the number of direct junctions is one less than the number of cyclic systems. **r. breakage** Disruption of an atomic r., changing an aromatic to an aliphatic compound. **r. closure** R. formation. **r. compound** A compound, the atoms of which form a r. or closed chain; as, a cyclic compound. **r. formation** Exocondensation. The closing of an atomic chain; as, the change from an aliphatic to an aromatic compound. **r. reaction** R. test. A chemical reaction (precipitate or color formation) at the boundary between one liquid superimposed as a layer on another. **r. structures** The atomic arrangements in aromatic molecules which may contain one or more rings connected or joined by simple or double bonds or by a single atom (spiro compound). See *carbocyclic.* **r. symbol** A graphical representation of an atomic r., as skeleton symbols. **r. system** A diagrammatic representation on a plane surface of the geometrical arrangement of atoms. **r. test** Ring reaction.
Ringelmann smoke chart A set of 6 cards, printed with straight lines at right angles, which are viewed at 15 m from a chimney. The cards show progressive darkening from white to black and can be matched against the shades of the smoke.
Ringer **R., Sidney (1835–1910)** English physiologist. **R. injection, R. solution** An isotonic solution: sodium chloride 0.86, potassium chloride 0.03, calcium chloride 0.033%, in water; used in physiological experiments, and given intravenously to correct severe fluid loss (USP). Cf. *saline.* **lactated ~** R. solution containing not more than 330 mg sodium lactate per 100 mL. Used to correct acidosis and fluid loss (USP).
rinkite $3NaF \cdot 4CaO \cdot 6(Ti,Si)O_2 \cdot Ce_2O_3$. A mineral.
Rinmann's green $CoZnO_2$(?). Green mass obtained by melting zinc compounds with a trace of cobalt salts. A blowpipe identification test for zinc compounds.
rINN See *INN.*
rinneite $FeCl_2 \cdot 3KCl \cdot NaCl$. A Stassfurt salt.
rio arrowroot Tapioca.
Rio Tinto process The extraction of copper from its ores after atmospheric oxidation.
ripener Ethylene gas used to ripen imported fruit (dates, persimmons, bananas, avocados) which are exposed to the gas (1:1,000).
ripening (1) Progressive, enzymatic hydrolytic changes in fruit: the fruit becomes sweeter, tannins and organic acids disappear, proteins are hydrolyzed, starch changes to sugar and some sugars form esters, and characteristic flavors develop. The cell wall changes, and the color generally deepens. The term is applied similarly to cheese. (2) The gradual formation of light-sensitive centers in the emulsion of a photographic plate. (3) Fermentation, as of tobacco, in which organic catalysts produce maturity. (4) A stage in the manufacture of rayon in which the solution attains the necessary properties for successful spinning. **artificial ~** Ripener.
ripidolite $(Al,Cr)_2O_3 \cdot 5(Mg,Fe)O \cdot 3SiO_2 \cdot 4H_2O$. A green or white mica.
riptography Analysis in which a solution is titrated with a precipitating agent, and the properties of the clear layer are measured.
rissic acid $C_6H_2(OMe)_2COOH \cdot OCH_2COOH$ = 256.2. 2-(Carboxymethoxy)-4,5-dimethoxybenzoic acid†. Risic acid. A split product of rotenone; a lower homolog of derric acid. White crystals, m.262.
ristocetin An antibiotic produced by *Nocardia lurida* (USP).
Ritz formula The wavenumber of a line in a series of the spectrum: $A - R/[m + \mu + (d/m)^2]^2$, where R is the Rydberg constant; m is an integer; and A, μ, and d are constants.
rivotite A native, basic antimony copper carbonate.
rms Abbreviation for root mean square.
Rn Symbol for radon.
RNA* Symbol for ribo*nucleic acid.*
RNP Abbreviation for ribo*nucleoprotein.*
roan Sheepskin tanned with sumach. Cf. *skiver.*
roast (1) To heat in air. (2) The product obtained by heating substances, especially ores, in air. **r. gases** The gases formed on roasting ores, especially sulfides; mainly sulfur dioxide, arsenous fumes, air, and the gases of the fuel used.
roaster Roasting furnace.
roasting Oxidation of ores by heat in a current of air, to remove sulfur and arsenic, and make the ores more porous for chlorination or hyposulfitization. **r. furnace** A long firebrick furnace; ore is fed into one end, raked toward the firebox, and removed after about 3 hours. **r. temperature** See Table 76.
robin A poisonous nucleoprotein from the bark of the locust

TABLE 76. EFFECTS ON VARIOUS COMPOUNDS OF ROASTING AT SELECTED TEMPERATURES

Temperature	Effect
100	Water removed
250	Ag_2O decomposes
350	Sulfides begin to burn; $CuSO_4$ dehydrated
600	$FeSO_4$ decomposes to Fe_2O_3
653	$CuSO_4$ becomes basic
655	Ag_2SO_4 melts
702	Basic $CuSO_4$ decomposes to CuO
807	Ag_2SO_4 decomposes
1050	CuO forms Cu_2O
1100	Fe_2O_3 forms Fe_3O_4

tree, *Robinia pseudacacia* (Leguminosae), N. America; an emetic and purgative.

robinin A coloring material from the heartwood of *Robinia pseudacacia,* locust.

robinose $C_{18}H_{32}O_{14}$ = 472.5. A trisaccharide in *Robinia* species; it hydrolyzes to 2 molecules of rhamnose and 1 of glucose.

Robinson, Sir Robert (1886–1975) British chemist, Nobel prize winner (1947); noted for organic research on plant constituents.

Robiquet **R., Henri Edme (1822–1860)** French chemist noted for work on fermentation and photography. The son of **R., Pierre Jean (1780–1840)** French chemist, noted as an analyst of vegetable materials.

Robison **R., Robert (1884–1941)** British biochemist noted for his work on yeast and fermentation. **R. ester** Glucose 6-phosphoric acid, produced in the fermentation of sugar by yeast. Cf. *Neuberg ester, sugar phosphates.*

roburite An explosive: ammonium nitrate 87, dinitrotoluene 11, chloronaphthalene 2%.

Roccal Trakemark for benzalkonium chloride.

Roccella A genus of lichens, as *R. tinctoria,* litmus plant.

roccellic acid $C_{17}H_{32}O_4$ = 300.4. 3-Carboxy-2-methylpentadecanoic acid. A dibasic acid from *Roccella fuciformis* and other fungi or lichens. Colorless scales, m.131, soluble in petroleum ether.

roccellin Orsellin.

rochelle salt Potassium sodium tartrate*.

Rochleder, Friedrich (1819–1874) Austrian chemist noted for his plant analyses.

rock A natural aggregation of mineral matter in the earth crust. **basic ~** See *basic rocks.* **igneous ~** See *igneous rock.* **primary ~** R. that has not undergone weathering and disintegration. **secondary ~** R. that has been altered by disintegration. **sedimentary ~** See *sedimentary rock.*
r. asphalt Sandstone or limestone impregnated with asphalt. **r. breaker** A machine that crushes rock. **r. candy** Crystallized sucrose. **r. cork** A variety of asbestos. **r. crystal** A transparent, colorless variety of quartz. **r. gas** Natural gas. **r. meal** Calcium carbonate deposited from water. **r. milk** A fine-quality r. meal. **r. oil** Petroleum. **r. quartz** Quartz. **r. ruby** A red variety of garnet. **r. salt** Common table salt. Sodium chloride obtained solid by mining. Cf. *sea salt.* **r. sand** The debris of abraded r. **r. silk** A variety of fine asbestos. **r. tallow** Hatchettenine. **r. tar** A crude petroleum. **r. wool** A fibrous substance made by blowing a jet of steam against a small stream of molten lime and siliceous r.; a heat insulator. Cf. *slag* wool.

Rocky Mountain spotted fever vaccine A suspension of inactivated *Rickettsia rickettsii,* used for prophylaxis.

rod (1) Pole, perch. A unit of length in surveying: 1 rod = 25 links = $5\frac{1}{2}$ yards = $\frac{1}{4}$ chain = 5.02921 meters. Cf. *rood.* (2) A stick-shaped bacillus. **square r** A unit of area in surveying (normally 1/160 acre).

rodenticide A chemical substance used to destroy rodent pests; as, zinc phosphide, barium carbonate, 1-naphthylurea (ANTU), warfarin, sodium fluoroacetate. Cf. *insecticide.*

Rodinal $NH_2C_6H_4 \cdot OH$ = 109.1. Trademark for *p*-aminophenol; a photographic developer.

Roentgen See *Röntgen.*

roentgen R. A unit of X- or γ-ray exposure dose. 1 R produces in 1 kg of air a charge of 2.58×10^{-4} C. See *rem.* **r. photograph** X-ray photograph. **r. rays** X-rays. **r. ray analysis** X-ray analysis. **r. tube** See *X-ray tube.*

roentgenogram X-ray photograph.

roentgenography X-ray photography.

roentgenology The study of X-rays.

Roese-Gottlieb method The determination of fat in dairy products. The moist food is extracted with a mixture of ethyl ether and petroleum ether, and the separated ethereal layer evaporated in a weighed vessel.

Roesler's process The separation of copper and silver from gold by fusion with sulfur or antimony sulfide to obtain copper and silver sulfides.

rogascope An electromagnetic device to detect cracks in, or variations of the diameter of, wire threads.

Rogitine Trademark for phentolamine mesylate.

Rohrbach's solution An aqueous solution, d.3.58, of barium and mercuric chlorides, for determining the density of minerals by the suspension method.

Röhrig tube A separating device for fat extraction.

ROM Read-only memory.

Roman **R. cement** A cement made by heating clay and limestone. **R. ocher** A native, orange variety of ocher. **r. vitriol** Cupric sulfate.

romanium An alloy of aluminum with small amounts of tungsten, copper, and nickel.

romeite $5CaO \cdot 3Sb_2O_5$. A native antimonate.

RON See *octane number.*

Rongalite $CH_2O \cdot NaHSO_2 \cdot H_2O$ = 136.1. Trademark for an addition product of formaldehyde; a substitute for sodium dithionite.

Röntgen, Wilhelm Konrad (1845–1923) German physicist, Nobel prize winner (1901); discoverer of X-rays (1895).

röntgen See *roentgen.*

rood A survey measure: 1 rood = 1,210 yd^2 = 925 m^2 = 0.025 acre. Cf. *rod.*

root An underground portion of a plant that conducts moisture and salts to the leaves. It maintains the plant in position, and stores reserve materials. Some roots are used medicinally; as, glycyrrhiza, rhubarb. Cf. *rhizome.*

Roozeboom, Hendrick Willem Bakhuis (1854–1907) Dutch chemist noted for his practical application of the phase rule.

ropiness Stickiness or stringiness of an organic liquid (also beer, milk, or bread) caused by microorganisms of the *B. mesentericus* group.

Rosaceae Rose family of herbs, shrubs, and trees; several yield drugs and edible fruits; e.g.: *Prunus domestica,* prune; *P. amygdalus amara,* bitter almonds; *Quillaja saponaria,* soapbark.

rosaniline Methylpararosaniline. A triphenylmethane dye. Red needles, soluble in water. **r. hydrochloride** Fuchsin.

roscoelite A vanadium-mica ore (California).

Rose **R., Friedrich (1839–1899)** German chemist, noted for work on cobalt-ammonia compounds. **R., Gustav (1798–1873)** German chemist noted for his system of crystallography. **R., Heinrich (1795–1864)** German chemist; originator of the use of hydrogen sulfide in qualitative analysis. Gustav and Heinrich R. were sons of **R., Valentin (the younger) (1762–1807)** German chemist, noted for analytical methods; son of **R., Valentin (the elder) (1736–1771)** German chemist who first prepared *rose* metal.

rose The dried petals of *Rosa gallica* (Rosaceae); a perfume. **cabbage ~** The petals of *Rosa centifolia*, used for rose water. **Christmas ~** Black hellebore. **corn ~** Red *poppy*. **dog ~** The ripe fruit of *Rosa canina*, used for making pill masses. It contains invert sugar, citric and malic acids.

r. bengal(e) Diiodoeosin. A red adsorption indicator in the titration of chlorides by silver nitrate. **r. bengal sodium ^{131}I injection** A sterile solution of r. bengal with ^{131}I in the molecular structure; a diagnostic aid in liver disease (USP). **r. copper** Rosette copper. **r. metal** Rose's metal. An alloy: Bi 2, Sn 1, Pb 1 pts., m.94. **r. oil** (1) An essential oil perfume obtained by cold extraction of r. petals. (2) The volatile oil from flowers of Rosaceae; e.g.: *Rosa gallica* and *R. alba*. Yellow liquid with a r. odor, d_{30}·0.848–0.863; a perfume and emollient (NF). **r. oxide** $Me_2C{:}CH(C_5OH_8)Me$ = 154.3. Terpenoid ether in r. oil. Perfume used in soap manufacture. **r. process** Separation of gold from its ores by fusion of the zinc precipitates and aeration to oxidize the baser metals to removable borax-silica flux. **r. quartz** A pink gem variety of quartz. **r. vitriol** Cobaltous sulfate. **r. water** Water saturated with the odoriferous principles of r. oil (NF).

rosein Fuchsin.

roselite A native, hydrated oxide containing Ca, Co, Mg, and As.

rosellane Rosite.

rosemary Rosmarinus. **wild ~** Labrador tea.

r. oil The volatile oil from the flowering tops of *Rosmarinus officinalis* (Labiatae). Yellow liquid, with characteristic odor and warm camphoraceous taste, d.0.895–0.910, insoluble in water; a flavoring.

Rosenstein process A method of making hydrochloric acid from chlorine and water gas.

Rosensthiel's green Barium manganate.

roseo compound $[Co(NH_3)_5]X_3 \cdot H_2O$. A red cobalt pentaammine.

rosette A disk with a centrally arranged pattern of crystals. **r. copper** A disk formed on the surface of molten copper by sudden cooling with water.

rosin Colophony. The resin after distilling turpentine from the exudation of species of pine, e.g., *Pinus palustris*. It contains abietic acid (80–90%) and its anhydride. Yellow, brittle mass, insoluble in water, soluble in alcohol; used as an insulator, for sizing paper, and in plastics. **r. acetal** A highly polymerized compound of repeating oxymethylene units. **r. ester** Ester gum. **r. grades** A is darkest; color decreases progressively through the alphabet to X. WG = window glass. WW = water white. **r. jack** A yellow variety of sphalerite. **r. oil** Blue, fluorescent oil obtained by the destructive distillation of rosin and pitch. It contains complex hydrogenated retenes, phenols, and resin acids; a lubricant. Cf. *retinol*. **r. spirit** A hydrocarbon mixture from the destructive distillation of rosin. **r. tin** A red variety of cassiterite.

rosindone $C_{22}H_{14}N_2O$ = 322.4. Rosindulone. Red crystals, m.224, soluble in water; a dye.

rosinduline $C_{22}H_{15}N_3$ = 321.4. An aniline dye, m.199, soluble in water. **r. scarlet** A red oxidation-reduction indicator.

rosindulone Rosindone.

rosinoil Retinol (2). Cf. *rosin* oil.

rosinol Retinol (2).

rosite Decomposed anorthite.

rosmarinus The dried leaves and plants of *Rosmarinus officinalis*, rosemary (Labiatae); an aromatic.

rosolic acid $C_{20}H_{16}O_3$ = 304.3. Corallin. Red scales with a green luster, m.270 (decomp.), slightly soluble in water; an indicator and dyestuff intermediate. Cf. *aurin*.

rosterite A light-red variety of beryl.

rosthornite $C_{24}H_{40}O$ = 344.6. A native, brown resin, d.1.076 (Carinthia).

rostone A synthetic stone molded from finely ground shale mixed with slaked lime and moistened with 18–22% water.

rot **dry ~** Dry wood decay due to *Merulius lacrymans*. **wet ~** Wet wood decay due to organisms other than *M. lacrymans*; commonly *Coniphora cerebella*.

rotamerism Geometric *isomerism*.

rotameter An indicator of fluid flow.

rotary (1) Rotatory. (2) Revolving. **r. kiln** A long drum, usually inclined, which rotates and can be heated; used, e.g., in the production of lime from limestone. **r. movement** A circling or turning motion.

rotate To turn or twist.

rotation Turning around an axis, or with a circular motion. **magnetic ~** The optical activity of a liquid between magnetic poles. **molecular ~** The product of the molecular weight (divided by 100) and the specific rotation. **specific ~** $[\alpha]_D^t = \alpha/l \cdot c$, where α is the observed angle of rotation, l the length of the layer of liquid in decimeters, c the g/l of substance, and $[\alpha]_D$ the specific rotation when the examination is made with the sodium flame (D line), at t°C. Cf. *polarimeter*.

rotational Pertaining to rotation.

rotatory Optically active; capable of turning the plane of polarized light. **r. dispersion** The ratio of the specific rotations of a substance with lights of 2 different wavelengths. **r. power** Specific optical rotary power*.

rotaversion The rearrangement of a *cis* to a *trans* isomer, or vice versa.

rotenic acid $C_{12}H_{12}O_4$ = 220.2. Isotubaic acid. White crystals, m.182, derived from rotenone. Cf. *tubaic acid*.

rotenoid Describing substances related structurally to rotenone, in leguminous fish-poison plants.

rotenone $C_{23}H_{22}O_6$ = 394.4. Derrin. Tubatoxin, nicouline. A crystalline, insecticidal principle, m.163, insoluble in water; from *derris* root, q.v., *cube* root, q.v., and *Cracca (Tephrosia) vogelii*.

rothic acid $C_{14}H_{12}O_7$ = 292.2. A split product of nucitannin.

rothoffite A native, brown garnet, which contains magnesium, iron, and calcium.

rotogravure A high-speed printing process, in which a rotating metal cylinder is etched to various depths, corresponding with the shade gradations of an illustration. An ink, prepared with a volatile solvent, transfers from the etching to the paper, in amounts corresponding with the depths. Cf. *intaglio*.

rotor (1) The rotating portion of a piece of machinery, as distinct from the stationary stator around or inside which it revolves. (2) The spinning top of an ultracentrifuge.

rotoscope A stroboscope for the observation of rapid mechanical motion. Cf. *chronoteine*.

rotten The state resulting from natural decomposition. **r. stone** (1) Terra cariosa. Light, friable, fine grains of silica formed by the decomposition of siliceous limestone (Derbyshire); a polishing material. (2) Tripoli.

rottisite A native, hydrated nickel silicate.
rottlera Kamala.
rottlerin $C_{30}H_{28}O_8$ = 516.6. Mallotoxin, kamalin. Yellow leaflets from kamala, m.206, soluble in ether or alkalies (red color).
Rotwyla Trademark for a viscose synthetic fiber.
Rouelle, Guillaume François (1703–1770) French chemist; the first physiological chemist.
rouge (1) A cosmetic dye made from safflower mixed with talc. (2) A cosmetic that simulates flushed skin. (3) Colcothar. **jeweler's ~** Colcothar.
r. flambé A glaze containing copper, of Chinese origin.
roughage Fiber. Fibrous food material (usually cellulose), which stimulates the large intestine by distension with bulky, nonabsorbable material.
round factor 1000 in = 25.4 m.
Roussin's salt $K[Fe_4(NO)_7S_3]$ = 568.7. Dark-colored salt formed when nitrogen oxide is passed through a suspension of precipitated ferrous sulfide in potassium sulfide solution.
ROV Refined oil of vitriol: 95–96% sulfuric acid.
Rowland, Henry Augustus (1848–1901) American physicist. **R.'s value** The wavelength of a Fraunhofer line in the solar spectrum.
rowlandite A native silicate of yttrium, with cerium, lanthanum, and thorium.
Royal Society, The A society, in London, founded in 1662 for furthering the natural and physical sciences.
Roylon Trademark for a polyamide synthetic fiber.
Ru Symbol for ruthenium.
rubber Caoutchouc, elastica, india r. An elastic substance obtained from the coagulated milky juice (latex) of *Hevea* (para r.) and *Ficus* (india r.) species. It contains the hydrocarbon unit $-CH_2 \cdot CMe:CH \cdot CH_2-$, and is a linear polymer: mol. wt. approx. 300,000. The main constituent is *cis*-1,4-polyisoprene, d.0.92, with varying quantities of oxidation products, resins 2–60, water and impurities 3–15%. Soluble in chloroform, carbon disulfide, ether, or benzene. On destructive distillation it yields isoprene, dipentene, and other hydrocarbons. See also Table 24 on p. 202. **butyl ~** See *elastomer.* **cold ~** Synthetic r. made by a cold process, e.g., at 5–15°C instead of 50°C, as with SBR. It has improved wearing properties. **cyclized ~** A finely divided powder produced by heating a solution of pure natural r. with, e.g., stannic chloride, and evaporating the solvent. The r. molecules are cross-linked to give a ring structure and good plastic properties. Cf. *Pliolite.* **deuterio ~** A synthetic r. of high strength and specific gravity made from a polyisoprene derivative in which all the H atoms are replaced by heavy H atoms. **fluorine ~** See *elastomer.* **guayule ~** See *guayule rubber.* **hard ~** Ebonite, vulcanite. **inorganic ~** $(NPCl_2)_n$. Polymerized cyclic chlorophosphazine. An inorganic polymer. **nitrile ~** See *elastomer.* **silicone ~** See *elastomer, silicone rubber.* **stero ~** Synthetic r. made from stereopolymers. **synthetic ~** Elastomer.
r. accelerators See *accelerator.* **r. acid** Sulfuric acid produced by the oxidizing action of water organisms on sulfur in rubber. **r. base** Generic name for dental-impression material; either silicone elastomer or polysulfide r. **r. goods** Articles prepared by mixing various substances with r. and subjecting to vulcanization, acceleration, and aging. **r. graft** A blend of natural r. and a polymer, as, styrene. **r. hydrochloride** $C_{10}H_{18}Cl_2$. A reaction product of r. with dry hydrogen chloride at below 10°; used as an adhesive and waterproof wrapping (Plioform), and in paints. **r. sources:**

Euphorbiaceae:	
Heavea brasiliensis	Para r.
Manihot	Ceara r., q.v.
Micrandra	Venezuelan r.
Sapium	Bolivian r., Colombian r.
Moraceae:	
Ficus elastica	India r.
Castilloa	Mexican r., W. Indian r.
Apocynaceae:	
Funtumia	African r.
Landolphia	Malagasy r.
Clitandra	Central African r.
Willoughbeia	Indonesian r.
Hancornia	Mancabeira r.
Asclepiadaceae:	
Asclepias	Milkweed r.
Compositae:	
Parthenium argentatum	Guayule r., q.v.

r. sulfide A reinforcing substance formed in r. during vulcanization. **r. uses** See *elastomers,* Table 24 on p. 202.
Rubbone $C_{10}H_{16}O$ (approx.). Brand name for an orange, viscous gum, produced by atmospheric oxidation of rubber. Used to accelerate the oxidation of polymerized linseed oil.
rubeane $(NH_2 \cdot C:S)_2$ = 120.2. Rubeanhydride, ethane dithioamide. Obtained by heating hydrogen sulfide and cyanogen. Red crystals, soluble in water; a microreagent for copper.
rubeanhydride Rubeane.
rubeanic acid Rubeane.
rubefacient An agent that produces redness of the skin, with pain alleviation due to local warmth; as, mustard.
rubella German measles. **r. vaccine** A vaccine of attenuated live virus; used to prevent r. in early pregnancy.
rubellite Tourmaline.
rubene (1) Naphthacene. (2) $C_{18}H_8R_4$. A compound that absorbs and regenerates oxygen; it contains the dibenzofulvene group. **diphenylditolyl ~** $C_{44}H_{30}$ = 558.7. White crystals, m.375, violet fluorescence. **tetraphenyl ~** Rubrene.
ruberite Cuprite.
ruberythric acid Rubianic acid.
ruberythrinic acid Rubianic acid.
rubia Madder. The root of dyer's madder, *Rubia tinctorium* (Rubiaceae).
Rubiaceae The madder family of herbs, shrubs, and trees that yield drugs; e.g.: bark: *Cinchona* species, quinine alkaloids; herbs: *Mitchella repens,* squaw vine; roots: *Rubia tinctorum,* madder, alizarin; seeds: *Coffea arabica,* coffee; extract: *Oruouiarpa gambir,* gambir.
rubianic acid $C_{26}H_{28}O_{14}$ = 564.5. Ruberythr(in)ic acid. A glucoside from rubia, hydrolyzed to glucose and alizarin. Yellow prisms, m.259.
rubicelle Spinel.
rubicene* $C_{26}H_{14}$ = 326.4. Red needles, m.305, insoluble in water.
rubidine $C_{11}H_{17}N$ = 163.3. A *pyridine base,* q.v.
rubidium* Rb = 85.4678. An alkali metal element, at. no. 37, discovered by Bunsen (1861). Named from the Latin for the "red" of its spectral lines. A soft, white metal, d.1.52, m.38, b.690, decomp. in water or alcohol, soluble in ether; occurs in Stassfurt salts, lepidolite, leucite, and mineral waters. **r. acetate*** $RbC_2H_3O_2$ = 144.5. Colorless crystals, soluble in water. **r. alum** Aluminum rubidium sulfate*. **r. bromide*** RbBr = 165.4. Colorless, regular crystals, m.683, soluble in water. **r. carbonate*** Rb_2CO_3 = 230.9. Colorless, deliquescent crystals, m.837, soluble in water. **r. chloride*** RbCl = 120.9. Colorless crystals, m.710, soluble in water.

r. chromate* Rb_2CrO_4 = 286.9. Yellow crystals, soluble in water. **r. dichromate*** $Rb_2Cr_2O_7$ = 386.9. Orange crystals, soluble in water. **r. fluoride*** RbF = 104.5. White crystals, soluble in water. **r. hexachloroplatinate(IV)*** $Rb_2[PtCl_6]$ = 578.7. R. platinichloride. Yellow crystals, soluble in water. **r. hexafluorosilicate*** $Rb_2[SiF_6]$ = 313.0. White octahedra, soluble in water. **r. hydride*** RbH = 86.5. Colorless needles, decomp. 300. **r. hydroxide*** RbOH = 102.5. Gray, deliquescent powder, soluble in water; used in the manufacture of glass and rubidium salts. **r. iodate*** $RbIO_3$ = 260.4. Colorless prisms, soluble in water. **r. iodide*** RbI = 212.4. Colorless cubes, soluble in water. **r. ion** The cation, Rb^+. **r. nitrate*** $RbNO_3$ = 147.5. Colorless hexagons, soluble in water. **r. oxide*** Rb_2O = 186.9. Colorless octahedra, soluble in water. **r. perchlorate*** $RbClO_4$ = 184.9. Colorless crystals. **r. peroxide*** Rb_2O_2 = 202.9. Yellow needles, m.600, decomp. by water. **r. platinichloride** R. hexachloroplatinate(IV)*. **r. sulfate*** Rb_2SO_4 = 267.0. Colorless rhombs, m.1051, soluble in water. **r. sulfide*** $Rb_2S \cdot 4H_2O$ = 275.1. Colorless crystals, soluble in water. **r. sulfite*** Rb_2SO_3 = 251.0. Colorless crystals, soluble in water. **r. tartrate*** $Rb_2C_4H_4O_6$ = 319.0. Colorless crystals, soluble in water. **r. hydrogentartrate** $RbHC_4H_4O_6$ = 234.5. White prisms, soluble in water.

rubijervine $C_{27}H_{43}O_2N$ = 413.6. An alkaloid from *Veratrum album* and *V. viride* (Liliaceae), m.240.

rubin An anthocyanin from radishes.

rubine Fuchsin.

rubitannic acid $C_{14}H_{22}O_{12} \cdot \frac{1}{2}H_2O$ = 391.3. A tannin from the leaves of *Rubia tinctorium*.

rubixanthin $C_{40}H_{56}OH$ = 553.9. An orange carotinoid pigment from *Rosa rubiginosa*.

rubrene $C_{42}H_{28}$ = 532.7. 5,6,11,12-Tetraphenylnaphthacene. A red, fluorescent, solid hydrocarbon, m.331. When illuminated in air it forms oxyrubrene, $C_{42}H_{28}O_2$, which dissociates when heated, to $C_{42}H_{28}$ and O_2. This reversible oxidation is rare. Cf. *hemoglobin, rubene*. **r. bromide** A solid; m.500, among the highest of known organic compounds.

rubrocyanine Ruhrgasol.

rubrofusarin An antibacterial, red pigment from the mycelia of *Fusaria* cultures.

rubroglaucine $C_{16}H_{12}O_5$ = 284.3. A red pigment related to auroglaucin.

rubrones Substances prepared by bubbling air through a solution of rubber with cobalt linoleate as catalyst. Used in paints and varnishes, and for molding.

rubrum scarlatinum Scarlet R.

Rubus (1) A genus of shrubs and herbs of the family *Rosaceae*, q.v. (2) Blackberry, cloudberry, dewberry, fingerberry. The dried bark of the rhizomes of *Rubus* species, which contain tannin; used as an astringent; e.g., *Rubus idaeus*, raspberry.

ruby Al_2O_3. A red, transparent corundum gem, similar to sapphire. **artificial ~** A r. made by fusing chromium sesquioxide and powdered alumina.

r. alamandine Spinel. **r. arsenic** Arsenic tetrasulfide*. **r. balas** Spinel. **r. blende** A red variety of sphalerite. **r. copper** Cuprite. **r. glass** A dark-red glass, colored by colloidal gold. Ordinary red glass is colored by selenium. **r. mica** Goethite. **r. silver** Proustite. **r. spinel** Spinel. **r. sulfur** Arsenic tetrasulfide*.

Rudolfi's equation The degree of dissociation α, related to the dilution v: $\alpha^2/(1 - \alpha)\sqrt{v}$ = constant. See *Ostwald's dilution law*.

rue Ruta. The dried, shrubby *Ruta graveolens;* a condiment containing rutic and rutinic acids. **r. oil** Oleum rutae. An essential oil from the leaves of *Ruta graveolens*. Green liquid, d.0.837, b.230, insoluble in water.

rufianic acid Quinizarinsulfonic acid, 1,4-dihydroxy-anthraquinone-2-sulfonic acid. White crystals; a reagent for the separation of amino acids.

rufigallic acid $C_{14}H_8O_8 \cdot 2H_2O$ = 340.2. 1,2,3,5,6,7-Hexahydroxy-9,10-anthraquinone. Orange needles, sublime if heated, insoluble in water; a dye.

rufigallol $C_{14}H_8O_8$ = 304.2. 1,2,3,5,6,7,-Hexahydroxyanthraquinone. Red crystals.

rufiopin $C_{14}H_8O_6$ = 272.2. 1,2,5,6-Tetrahydroxyanthraquinone. Orange cystals, soluble in alcohol.

rufol $C_{14}H_{10}O_2$ = 210.2. 1,5-Dihydroxyanthracene, 1,5-anthra(cene)diol*. Yellow needles, decomp. 265, soluble in alcohol (blue fluorescence).

ruggedized Rendered resistant to the deteriorating influences (as, heat, humidity, dust, vibration, and corrosive gases) of a working environment.

rugosimeter An instrument to measure the roughness of a flat surface in terms of the resistance to airflow between it and a plane surface resting on it, under a standard pressure.

rugosity Roughness, angularity. The ratio of measured specific surface to hypothetical surface area, of a particle, assumed to be spherical.

Ruhmkorff R., Heinrich Daniel (1803–1877) German electrical researcher. **R. coil** An induction or spark coil.

ruhrgasol A mixture of CO_2, CO, H_2, N_2, and C_3 and C_4 aliphatic hydrocarbons (35% by vol. of C_3H_6) recovered from coke-oven gas; an auto fuel in Germany.

rule (1) An empirical relationship of physical and chemical properties; as, Crum *Brown* rule. Cf. *theory, law, hypothesis*. (2) A ruler. **phase ~** See *phase rule*. **slide ~** See *slide rule*.

rum A distillation product from fermented molasses.

Rumford, Count (Benjamin Thompson) (1753–1814) American scientist noted for his work on the nature of heat, and cofounder of the Royal Institution, London.

rumicin $C_{15}H_{10}O_4$ = 254.2. Yellow crystals, m.182, from the roots of *Rumex crispus*. It resembles chrysophanic acid.

Runge, Friedlieb Ferdinand (1795–1867) German chemist, discoverer of aniline. See *aniline*.

Rupert's drops See *Prince Rupert drops*.

Rush, Benjamin (1745–1813) American physician and pioneer in chemical education, noted as the first American author and professor of chemistry.

russellite $Bi_2O_3 \cdot WO_3$ = 697.8. A mineral that occurs sparingly, as yellow pellets, in concentrates of tungsten ores (Cornwall, England).

russium Early name for francium.

rust (1) Iron oxide mixed with hydroxides and carbonates, formed on the surface of iron exposed to moisture and air. (2) A red fungus on cereal grain. **white ~** A form of corrosion of steel due to excess chloride from the galvanizing operation.

r.proofing Plating a metal with a less corrodible metal; as, tin, cadmium, or zinc. See *galvanize, sheradize*.

ruta Rue.

Rutaceae Rue family, yielding acrid, resinous principles and essential oils; e.g.: *Galipea (Cusparia) officinalis*, angostura bark; *Pilocarpus jaborandi*, jaborandi leaves; *Aegle marmelos*, bael fruit.

ruthenate Indicating ruthenium as the central atom(s) in an anion, as: M_2RuO_4. M is a monovalent metal. Red salts, soluble in water, derived from hexavalent ruthenium. **hexachloro ~*** $M_2[RuCl_6]$. **per ~** $MRuO_4$, e.g., $NaRuO_4$. Dark-green crystals, soluble in water.

TABLE 77. OXIDATION STATES OF RUTHENIUM

Oxidation state	Name	Ion	Color in solution
2	r.(II)*, r.(2+)*, ruthenious	Ru^{2+}	blue
3	r.(III)*, r.(3+)*, ruthenic	Ru^{3+}	gray
4	ruthenite	RuO_3^{2-}	yellow
6	ruthenate(VI)*	RuO_4^{2-}	orange-red
7	perruthenate	RuO_4^-	green

ruthenic Ruthenium(III)* or (3+)*. Describing a compound derived from trivalent ruthenium; as, $RuCl_3$.

ruthenious Ruthenium(II)* or (2+)*, ruthenous. Describing a compound with divalent ruthenium.

ruthenium* Ru = 101.07. A rare metallic element of the platinum group, at. no. 44. Discovered (1845) by Claus. Gray or silvery, brittle metal, d.12.06, m.2310, b. ca. 4000, insoluble in water or acids. It can have every positive oxidation state from 1 to 8, and forms coordination compounds and double salts. See Table 27 on p. 207 and Table 77 on p. 513. **r. bromide** $RuBr_3$ = 340.8. Dark, hygroscopic crystals; soluble in water. **r. carbonyl** $Ru(CO)_5$ = 241.1. Orange crystals. **r. chloride** (1) **r. dichloride*** $(RuCl_2)_n$ = $(172.0)n$. Ruthenious chloride. Blue solid, insoluble in water. (2) **r. trichloride*** $RuCl_3$ = 207.4. Ruthenic chloride. α-~ Black crystals. β-~ Deliquescent, brown crystals, soluble in water, decomp. by alcohol; a catalyst. **r. chloride hydroxide** $Ru(OH)Cl_2$ = 189.0. R. red. Red powder. **ammoniated ~** $[Ru_3(NH_3)_{14}O_2]Cl_6 \cdot 4H_2O$ = 858.4. Brown powder, soluble in water; a reagent for pectin, plant mucin, gums; also a microscope stain. **r. fluoride** RuF_5 = 196.1. Green crystals, m.87, decomp. in water. When heated with iodine, it is converted into r. trifluoride. **r. hydrochloride** H_2RuCl_6 = 315.8. Hexachlororuthenic acid*. Orange crystals, soluble in water. **r. hydroxide** (1) $Ru(OH)_3$ = 152.1. Black powder, insoluble in water. (2) $Ru_2O_3 \cdot nH_2O$. Ruthenic hydroxide. Yellow powder, soluble in water. (3) $RuO_2 \cdot nH_2O$. R. tetrahydroxide. Black powder, insoluble in water. **r. minerals** R. is associated with the platinum metals, and occurs as laurite, RuS_2, and the element. **r. nitrosotrinitrate** $Ru(NO)(NO_3)_3 \cdot 4H_2O$ = 389.1. Deliquescent, red solid, decomp. by water (to nitric acid) and heat. **r. oxide** (1) RuO_2 = 133.1. R. dioxide*. Violet crystals, insoluble in water, formed by the action of oxygen on hot r. (2) Ru_2O_3 = 250.1. R. sesquioxide. Brown powder, insoluble in water. (3) RuO_4 = 165.1. R. tetraoxide*. An octavalent compound. Yellow crystals, m.25, insoluble in water. (4) Ru_2O_5 = 282.1. R. pentaoxide*. Black powder. **r. oxychloride** R. chloride hydroxide. **r. red** See *ruthenium chloride hydroxide*. **r. sesquioxide** R. oxide (2). **r. silicide** RuSi = 129.2. Metallic prisms, insoluble in water. **r. sulfide** RuS_2 = 165.2. Laurite. Gray cubes, insoluble in water. **r. tetraoxide*** R. oxide (3). **r. trifluoride*** RuF_3 = 158.1. Brown crystals, insoluble in water.

ruthenous Ruthenious.

Rutherford **R., Daniel (1749–1819)** Scottish botanist who discovered nitrogen (1772). **R., Lord Ernest (1871–1937)** New Zealand physicist, Nobel prize winner (1908); noted for research on the structure of the atom. **R. atomic theory** An atom consists of a small positive nucleus with *A* free positive charges in its nucleus, surrounded by a system of *A* electrons (*A* is the atomic number). The nucleus contains equal numbers of positive and negative charges which balance.

rutherford Symbol: rd; a unit of radioactivity equal to 10^6 disintegrations/s. 1 rd = 10^6 Bq.

rutherfordite A mixed, native phosphate of cerium, neodymium, praseodymium, and lanthanum.

rutherfordium Rf. Un-nil-quadium*. Ekahafnium. Suggested name for element at. no. 104. Its isotopes are unstable, with short half-lives. Cf. *Kurchatovium.*

rutic acid (1) $C_{10}H_{20}O_2$ = 172.3. A monobasic acid from rue. (2) Decanoic acid*.

rutile TiO_2. Dark, tetragonal crystals; a source of titanium compounds.

rutin (1) $C_{27}H_{30}O_{16}$ = 610.5. A hydroxyflavone glucorhamnoside from cowslip and other plants. Yellow needles, m.214; reputedly effective for capillary disorders. (2) Barosmin.

rutinic acid $C_{25}H_{28}O_{15}$ = 568.5. The coloring material of rue, *Ruta graveolens.*

rutonal $C_{11}H_{10}N_2O_3$ = 218.2. Methylphenylbarbituric acid. White crystals, m.227.

Rydberg **R., Johannes Robert (1854–1919)** Swedish physicist. **R. constant*** A constant connecting the wave number of atomic spectrum lines. $R_\infty = 2\pi^2 m_e e^4/h^3 c$, where m_e is the mass of an electron and e its charge, h is the Planck constant, and c the velocity of light. $R_\infty = 1.097\,3732 \times 10^7\ m^{-1}$. Cf. *Ritz formula.* **R.'s formula** The wavenumber of a spectrum radiation line = $R(1/n^2 - 1/n'^2)$, where n is the number of the inner, and n' that of the outer, orbit in a given jump, and R is the R. constant.

rye The grain of *Secale cereale*, used in making bread. **spurred ~** Ergot.

Rynacrom Trademark for cromolyn sodium.

S

S Symbol for (1) sulfur; (2) siemens. **S acid** 1-Amino-8-naphthol-4-sulfonic acid. **2S acid** 1-Amino-8-naphthol-2,4-disulfonic acid. **S yellow** Naphthol yellow.

S Symbol for (1) apparent power; (2) entropy; (3) sinister (see *sequence rule*); (4) spin angular momentum quantum number. In chemical names, the radical prefixed by *S*- is attached to the sulfur atom.

s Symbol for (1) second; (2) solid state (as subscript or between parentheses).

s (1) Symbol for distance along a path. (2) In chemical names, symmetrical or secondary.

Σ, *σ* See *sigma.*

sabadilla Cevadilla, Indian barley. The seeds of *Veratrum sabadilla* or *Schoenocaulon officinale* (Liliaceae), Mexico and Central America; containing veratrine and sabadine. Formerly used to treat pediculosis.

sabadine $C_{29}H_{51}O_8N$ = 541.7. An alkaloid obtained from sabadilla and veratrum. Colorless needles, m.238.

sabal Saw palmetto berries. The partly dried, ripe fruit of *Serenoa serrulata* or *Sabal serrulata* (Palmacea); a sedative. **s. fiber** The split leaves from *Sabal palmetto,* thatch palm; used for matting.

sabalol The active principle of sabal.

Sabatier, Paul (1854–1941) French chemist, Nobel prize winner (1912); pioneer in the hydrogenation of vegetable oils.

sabatrine $C_{51}H_{86}O_{17}N$ = 985.2. An alkaloid from sabadilla seeds.

sabbatia American centaury. The dried herb of *Sabbatia angularis* (Gentianaceae); a febrifuge.

sabbatin A glucoside from *Sabbatia elliottii,* quinine flower.

sabin Absorption unit (in acoustics), after W. C. Sabine; equivalent to that of 1 ft^2 of open window.

sabina Savine.

sabina oil Savine oil.

sabinane Thujane*.

sabinene Thujene*.

sabinic acid $HO(CH_2)_{11}COOH$ = 216.3. λ-Hydroxylauric acid, 12-hydroxydodecanoic acid*, m. 78. From savine and juniper oils.

sabinol $C_{10}H_{16}O$ = 152.2. **(±)-~** 4(10)-Thujen-3-ol. b.208. An alcohol from savine oil.

saccharase Invertase.

saccharate Sucrate. (1) A salt of saccharic acid. (2) A compound of a saccharide and a metallic oxide; as, $CaO \cdot C_{12}H_{22}O_{11} \cdot H_2O$.

saccharetin $(C_5H_7O_0)_n$. The yellow crust of sugarcane; probably a phlobaphen.

saccharic acid $(CHOH)_4 \cdot (COOH)_2$ = 210.1. Glucaric acid. Obtained on oxidation of hexoses. **D-~** m.125, very soluble in water. Cf. *saccharinic acid.*

saccharide (1) A compound of an organic base with sugar; as, casein saccharide. (2) See *polysaccharides, sugar, carbohydrates.*

saccharification (1) Conversion into sugar. (2) Impregnation with sugar solutions, as, in malting.

saccharify To convert starches into sugar.

saccharimeter A device to determine sugar in solution; as, a polarimeter. Cf. *saccharometer.*

saccharimetry The determination of the sugar content of a solution from its optical activity.

saccharin, saccharine $C_6H_4(SO_2) \cdot NH \cdot CO$ = 183.2. 1,2-Benzisothiazolin-3-one-1,1-dioxide. Gluside, sykose, benzosulfimide, benzoylsulfonic imide, saccharinose, sulfobenzoic acid imide, saccharinol, glucid, saccharol, sulfinid, guarantose, Saxin, agucarina. White crystals, m.228 (decomp.), slightly soluble in water, and 400 times sweeter than sucrose; a sugar substitute, viewed as a potential carcinogen. World Health Organization suggests a limit of 2.5 mg/day/kg of body weight. **soluble ~** $C_7H_4N \cdot NaO_3S \cdot 2H_2O$ = 241.2. The sodium salt of s.; often s. mixed with $NaHCO_3$.

s. sodium (1) Soluble s. (USP, BP). (2) A lactone of a saccharic acid.

saccharinic acid $C_6H_{12}O_6$ = 180.2. Acids obtained by oxidation of sugars; 24 possible isomers. **gluco ~** $CH_2OH \cdot (CHOH)_2 \cdot COH(Me) \cdot COOH$. Obtained from glucose and fructose; 8 isomers. **iso ~** $CH_2OH \cdot CHOH \cdot CH_2 \cdot COH(CH_2OH) \cdot COOH$, from lactose and cellulose; 4 isomers. **meta ~** $CH_2OH(CHOH)_2CH_2 \cdot CHOH \cdot COOH$; 8 isomers. **para ~** $CH_2OH \cdot CH_2 \cdot COH(COOH) \cdot CHOH \cdot CH_2OH$, from galactose and milk sugar; 4 isomers. All form lactones when their aqueous solutions are evaporated. Cf. *saccharonic acid.*

saccharobiose Sucrose.

saccharol Saccharin.

saccharolactic acid Mucic acid.

saccharometer A fermentation tube.

Saccharomyces Yeasts, q.v., which ferment sugar.

saccharon $C_6H_8O_6$ = 176.1. The lactone of saccharonic acid.

saccharonic acid $HOOC \cdot CMeOH \cdot (CHOH)_2 \cdot COOH$ = 194.1. Obtained by oxidation of sucrose. Colorless crystals, soluble in water.

saccharose (1) The $-CH(OH) \cdot C(:O)-$ group; the basic building block of carbohydrates. (2) Sucrose.

Saccharum S. officinarum Sugarcane. **S. saturnii** Early name for an aqueous lead acetate solution.

SAE number (Society of Automotive Engineers.) A measure of the relative viscosity of lubricating oils, related to the Saybolt universal viscosity.

Safa Trademark for a polyamide synthetic fiber.

safety Avoidance of hazards. **s. lamp** Miner's lamp. **s. tube** A device to oppose the effects of sudden pressures or sudden flows of liquids.

safflorite (Fe, Co)As_2. A native, rhombic arsenide.

safflower Carthamus.

saffron Crocus. The stigmas of *Crocus sativus* (Iridaceae), containing the glucoside crocetin. A yellow food coloring, carminative. **American ~, bastard ~, false ~** Carthamus. **Indian ~** Turmeric. **meadow ~** Colchicum.

s. bronze Orange *tungsten.* **s. glucoside** Crocetin. **s. of mars** Ferric subcarbonate. **s. substitute** Dinitrocresol.

safranal $C_{10}H_{14}O$ = 150.2. 2,6,6-Trimethyl-1,3-cyclohexadienal. An aldehyde occurring in saffron as the glucoside pikrococin: $CHO \cdot C{:}CMe \cdot CH{:}CH \cdot CH_2 \cdot CMe_2$.

safranine (1) $C_{18}H_{14}N_4Cl$ = 321.8. Phenosafranine. A

phenazine dye used as a stain in microscopy and the textile industry. (2) A class of azine dyes.
safraninol $C_{18}H_{13}ON_3$ = 287.3. A derivative of safranine, in which the :NH group is replaced by an oxo, :O.
safrene $C_{10}H_{16}$ = 136.2. A terpene from sassafras oil.
safrole* $C_6H_3(OCH_2O)\cdot CH_2\cdot CH:CH_2$ = 162.2. Shikimol. 4-Allyl-1,2-(methylenedioxy)benzene; in sassafras oil. Colorless oil, m.10, b.232, insoluble in water; an anodyne and perfume. **iso ~*** $C_6H_3(OCH_2O)\cdot CH:CH\cdot CH_3$. A liquid, b.251, insoluble in water.
safron Saffron.
safrosin $C_{22}H_8O_5Br(NO_2)_2$ = 524.2. A scarlet textile dye, chiefly dibromodinitrofluorescein.
sagapenum A gum resin from *Ferula persica* (Umbelliferae).
sage Salvia. **s.brush** Artemisia. **s. oil** The essential oil of *Salvia officinalis*, d.0.915–0.925, soluble in alcohol; contains eucalyptol.
sagenite Silica containing rutile.
sago (1) An edible starch from the pith of palms, chiefly *Metroxylon rumphii* (India and southeast Asian islands). (2) The pith of several palms and tree ferns; as, Palmaceae; Cyatheaceae, tree ferns; and Cycadaceae.
Sahli's stain A solution of borax and methylene blue in water; a stain for blood cells.
sahlite Augite containing iron.
Sainte-Claire Deville, Etienne Henri (1818–1881) French chemist noted for mineralogical and inorganic research.
Saint Ignatius bean Ignatia.
Saint John's bread Carob beans.
saké, saki Japanese beer prepared from rice, water, tané-koji, and saké yeast; a yellow, aromatic liquid with pleasant taste: ethyl alcohol 13–14, sugar 0.9%.
Sakurai S., Joji (1858–1939) Japanese chemist noted for organic research. **S.-Landsberger apparatus** A device for rapidly determining approximate molecular weights by the vapor-pressure method.
sakuranin $C_{16}H_{14}O_5$ = 286.3. (*S*)-4′,5-Dihydroxy-7-methoxyflavanone. A glucoside, m.212, from sakura, the Japanese cherry tree.
sal Latin for "salt." **s. acetosella** Potassium hydrogenoxalate*. **s. aeratus** Potassium hydrogencarbonate*. **s. alembroth** Ammoniated *mercury.* **s. amarum** Magnesium sulfate*. **s. ammoniac** Ammonium chloride*. **s. carolinum facticum** An artificial Carlsbad salt. **s. communis** Sodium chloride*. **s. de duobus** Potassium sulfate*. **s. enixum** Potassium hydrogensulfate*. **s. epsom** Magnesium sulfate*. **s. ethyl** Ethyl salicylate*. **s. fossile** Sodium chloride*. **s. glauberi** Sodium sulfate*. **s. marinum** Sodium chloride*. **s. mirabile** Sodium sulfate*. **s. perlatum** Sodium phosphate. **s. prunella** A concentrated, refined saltpeter, made by fusion to remove moisture and molded into a cake; used for meat curing. **s. rupium** Rock salt. **s. sedatirum** Borax. **s. sedative** Boric acid. **s. soda** Sodium carbonate*. **s. tartari** Potassium carbonate*. **s. volatile** Ammonium carbonate*.
salacetol $C_6H_4(OH)COOCH_2COMe$ = 194.2. Salantol, salicylacetol, acetosalicyclic ester. Colorless needles, m.71, soluble in water.
salamandaridine An alkaloid from the poisonous skin secretion of the salamander species. Cf. *samandaridine.*
salamander A truncated cone of plumbago, for heating a crucible uniformly.
salamanderine $C_{34}H_{60}O_5N_2$ = 576.5. A poisonous alkaloid from the skin of the salamander species. Cf. *samandarine.*
salamide Salicylamide*.
salantol Salacetol.
Salazopyrin Trademark for sulfasalazine.
salbutamol $C_{13}H_{21}O_3N$ = 239.3. Albuterol, Ventolin. White crystals, m.156, sparingly soluble in water. A sympathomimetic, used for asthma (BP).
salcatonin Calcitonin.
saldanine An alkaloid from *Datura arborea*, a Mexican shrub; a local anesthetic.
salep The tubers of *Orchis mascula, O. latifolia*; demulcents and a food. Cf. *arrowroot.*
saleratus Sal aeratus.
salic Containing alumina. **s. minerals** Igneous rocks that contain more alumina than iron. Cf. *alferric.*
salicin $C_{13}H_{18}O_7$ = 286.3. Saligenin. A glucoside from the bark of *Populus tremula*, American aspen; *Spiraea* and *Salix* species, willows. Colorless leaflets, m.201 (decomp.), soluble in water; an antipyretic, and reagent for nitric acid.
salicoside Salicin.
salicoyl The salicyloyl* radical.
salicyl* (2-Hydroxyphenyl)methyl†. The radical *o*-HO·$C_6H_4\cdot CH_2-$, from salicylic acid. Cf. *salicyloyl.* **s.fluorone** 2,6,7-Trihydroxy-9-(2-hydroxyphenyl)-3*H*-xanthen-3-one. A color reagent for rare-earth elements.
salicylacetol Salacetol.
salicylal (1) Salicylaldehyde*. (2) The salicylidene* radical, 1,2-$C_6H_4(OH)CH=$.
salicyl alcohol* $C_6H_4(OH)CH_2OH$ = 124.1. Saligenin, saligenol, *o*-hydroxybenzyl alcohol*. A hydrolysis product of salicin. Colorless needles, m.86, soluble in water; an antipyretic. ***p*-amino ~** Edinol.
salicylaldehyde* $C_6H_4(OH)CHO$ = 122.1. Salicylic aldehyde, salicylal, *o*-hydroxybenzaldehyde. Colorless, aromatic liquid, b.197, slightly soluble in water; a reagent for acetone, and used in perfumery. **methoxy ~** Vanillin*. **s. glucose** Helicin.
salicylamide* $C_6H_4(OH)CONH_2$ = 137.1. Salamide, *o*-hydroxybenzamide. Colorless leaflets, m.138, slightly soluble in water; an antipyretic and analgesic.
salicylate* $C_6H_4(OH)COO\cdot M$. A salt of salicylic acid: M is a monovalent metal.
salicylic acid* $C_6H_4(OH)COOH$ = 138.1. *o*-Hydroxybenzoic acid*. Colorless needles, m.158, slightly soluble in water. A reagent for ferric salts, nitrites, formaldehyde; a preservative; used in skin diseases and for softening warts (USP, EP, BP). **acetamidoethyl ~** Benzacetin (1). **acetamidomethyl ~** Benzacetin (2). **acetyl ~** See *aspirin.* ***p*-amino ~** $C_7H_7NO_3$ = 153.1. PAS. Bulky powder, darkening in air, with acetous odor, slightly soluble in water; a tuberculostatic antibacterial (USP). **homo ~** Cresotic acid. **5-hydroxy ~** Gentisic acid. **methoxymethyl ~** Everninic acid. **methyl ~** Cresotic acid. **nitro*** See *nitrosalicylic acid.* **phenyl ~*** See *phenylsalicylic acid.* **propionyl ~*** See *propionylsalicylic acid.* **sulfo ~** See *sulfosalicylic acid.*
s. a. anhydride Salicylide. **s. a. collodion** A mixture of 100 grams s. a. with sufficient flexible collodion to make 1 liter; a keratolytic (USP).
salicylic aldehyde Salicylaldehyde*.
salicylic amide Salicylamide*.
salicylide $C_{28}H_{16}O_8$ = 480.4. Tetrasalicylide, an anhydride of salicylic acid. Colorless crystals, m.260.
salicylidene* Salicylal. The radical 1,2-$C_6H_4(OH)CH=$.
salicylol $C_7H_8O_2$ = 124.06. Colorless, fragrant liquid from various plants; used in perfumery.
salicylonitrile* $C_6H_4(OH)CN$ = 119.1. *o*-Hydroxybenzylnitrile. Colorless crystals, m.98, soluble in water.
salicyloyl Salicoyl. The radical *o*-HO·$C_6H_4\cdot CO-$.
salicylresorcinol $C_6H_4(OH)\cdot CO\cdot C_6H_3(OH)_2$ = 230.2. The

ketone of salicylic acid and resorcinol, 2,2′4-trihydroxybenzophenone. Colorless leaflets, soluble in water.

salicylyloyl The salicyloyl* radical.

salify To form a salt.

saligenin (1) Salicyl alcohol*. (2) Salicin. **homo ~** Salicyl alcohol*.

saligenol Salicyl alcohol*.

salimeter A hydrometer to determine the density of salt solutions. Cf. *salinimeter.*

saline (1) Saltlike. (2) A salt spring or well. (3) Describing the taste of common salt. (4) Containing sodium chloride. **physiologically normal ~** A sterilized 0.9% solution of common salt in water. Isotonic with blood and tissue fluids. Chemically normal is 5.85%.

salines Salt springs; salt lands.

salinigrin $C_{13}H_{16}O_7$ = 284.3. Piceoside. The β-D-glucoside of 4′-hydroxyacetophenone, m.195. From the bark of *Salix nigra,* willow.

salinimeter A hydrometer for determining the salt content of brine or seawater. Cf. *salimeter.*

salinity (1) A comparative indication of the concentration of salts in natural waters. (2) The number of grams of salt in 1 kg seawater, when bromides and iodides are converted to chlorides, the carbonates to oxides, organic matter destroyed, and the mass heated at 450°C for 72 hours. $S = 0.03 + 1.805$ Cl content.

salinometer An instrument which uses the electrical conductivity of water to control salt content. Cf. *salinimeter.*

saliretin $C_{14}H_{14}O_3$ = 230.3. A yellow resin from salicin.

saliseparin Smilacin.

salit Bornyl salicylate*.

saliter, salitre Sodium nitrate*.

salithymol $C_6H_4(OH)COOC_{10}H_{13}$ = 270.3. Thymol salicylate. Colorless crystals, insoluble in water; an antiseptic.

saliva The alkaline secretion of the salivary glands; contains digestive enzymes (α-amylase), salts (potassium thiocyanate), proteins (albumin). The composition depends on the diet and can cause tartar formation on teeth. Cf. *sputum.*

Salix The willows (Salicaceae) whose bark yields salicin. **S. alba** The European or white willow. **S. fragilis** The brittle willow, snap willow. Its bark is an astringent and febrifuge. **S. nigra** The pussy willow (American, black, or swamp willow). Its bark is an antipyretic and sedative.

Salkowski's solution A solution of phosphotungstic acid; used to test for albumose in urine.

salmak Ammonium chloride.

salmiac Ammonium chloride.

salmine $C_{30}H_{57}O_6N_{14}$ = 709.9. A protamine from salmon spermatozoa.

salmonellosis Food poisoning, which can be fatal, due to organisms mainly of the *Salmonella typhimurium* type, especially from poultry.

salol Phenylsalicylate*. **nitro ~** See *nitrosalol.*

salseparin Smilacin.

salseparisin Parillin.

salsoline $C_{11}H_{15}NO_2$ = 193.2. (*S*)-1,2,3,4-Tetrahydro-7-methoxy-1-methyl-6-isoquinolinol. m.220. An alkaloid from *Salsola Richteri* (Cactaceae). Cf. *carnegine.*

salt (1) See *salts.* (2) Common s., halite, or sodium chloride. **air ~** See *air salt.* **baker's ~** Ammonium carbonate. **bay ~** Sodium chloride from seawater. **bitter ~** Magnesium sulfate*. **Carlsbad ~** A mixture of sodium and potassium sulfates, sodium hydrogencarbonate, and sodium chloride. **common ~** Sodium chloride*. **diuretic ~** Potassium acetate*. **Epsom ~** Magnesium sulfate*. **Everitt's ~** Potassium hexacyanoferrate(III)*. **Glauber ~** Sodium sulfate*. **green ~** Uranium tetrafluoride*. **Homberg's ~** Boric acid. **iodized ~** Sodium chloride, with a trace of iodide, for table use. **melting ~** A s., e.g., the carbonate and phosphate of sodium, melted with cheese during processing to improve emulsification and texture. **microcosmic ~** See *microcosmic salt.* **Mohr ~** See *Mohr salt.* **Monsel's ~** Ferric subsulfate. **pepetic ~** A mixture of sodium chloride and pepsin. **phosphor ~** Microcosmic s. **Plimmer's ~** Antimony sodium tartrate. **Preston's ~** An aromatized ammonium carbonate; a smelling s. **rochelle ~** Potassium sodium tartrate*. **rock ~** Sodium thioantimonate*. **sea ~** Sodium chloride from seawater. **Seignette's ~** Potassium sodium tartrate*. **solar ~** S. produced by evaporation of seawater by the sun. **Sorrel ~** Potassium hydrogenoxalate*. **spirits of ~** Commercial hydrochloric acid. **Stassfurt ~** Stassfurt salts. **sweet ~** Sodium chlorite*. **table ~** Sodium chloride*.

s. of amber Succinic acid*. **s. cake** (1) Impure sodium sulfate, e.g., as a by-product of the Leblanc soda process, or from natural Glauber salt. (2) A synthetic s. cake made by fusing sulfur and sodium carbonate together in the correct proportions. **s. deposits** Saline residues. The accumulation of salts from the evaporation of natural waters, as at Strassfurt, q.v. *(Stassfurt salts),* and in the desert. Chiefly carbonates, chlorides, sulfates, and borates of sodium, potassium, calcium, and magnesium. **s. glaze** See *salt glaze* under *glaze.* **s. hydrates** The solid phases, salt and water; hence any crystal with one or more molecules of water of crystallization. **s. ice** Frozen brine, m.−21. **s. of lemon** Potassium hydrogenoxalate*. **s.peter** See *saltpeter.* **s. of phosphorus** Ammonium sodium hydrogenphosphate. **s. solution** Saline solution. **s. of sorrel** Potassium hydrogenoxalate*. **s. of tartar** Potassium hydrogentartrate*. **s. of tin** Stannous chloride. **s. of vitriol** Zinc sulfate*. **s. of wormwood** Potassium carbonate*.

salting Treating with salt. **s. in** The mutual increase in the solubilities of an electrolyte and an organic compound added to the same solvent. **s. out** Aiding liquid–liquid extraction by addition of an electrolyte. Separation of a substance from its solution by adding soluble salts; as, precipitation of proteins by salts.

saltpeter, saltpetre Potassium nitrate*. **Chile ~** Sodium nitrate*. **German ~** Ammonium nitrate*. **Norge ~, Norway ~** Calcium nitrate*.

salts Substances produced from the reaction between acids and bases; a compound of a metal (positive) and nonmetal (negative) radical: M·OH (base) + HX (acid = MX (salt) + H_2O (water). **acid ~** S. containing unreplaced H atoms from the acid; as, $NaHSO_4$. **acidic ~** S. having an acid reaction. **alkaline ~** S. having a basic reaction. **amphoteric ~** S. having both acid and basic reactions. **basic ~** S. containing unreplaced hydroxyl radicals of the base; as $Bi(OH)Cl_2$. **binary ~** Compounds of 2 bases and one acid radical; as, $NaKSO_4$. **complex ~** S. made up of more than one simple acid or metallic radical, but which ionize in solution into only 2 types of ions. Thus potassium hexacyanoferrate(II): $K_4Fe(CN)_6 \rightleftharpoons 4K^+ + Fe(CN)_6^{4-}$. Cf. *Werner's theory.* **double ~** A molecular combination of 2 s.; as, alums: $M_2SO_4 \cdot M_2(SO_4)_3 \cdot 24H_2O$. Cf. complex *salts.* **ethereal ~** An ester. **mixed ~** S. of 2 or more metals; as, $NaKSO_4$. **neutral ~** S. having a neutral reaction, as potassium chloride. **normal ~** Compounds of a base and acid that have completely neutralized each other. **oxy ~** Compounds of a base with an oxyacid radical. **triple ~** S. containing 3 metals; as, triple chloride.

salufer Sodium hexafluorosilicate*.

salumin Aluminum salicylate.

Saluric Trademark for chlorothiazide.

salvarsan Arsphenamine.
salve See *ointment.*
salvia Sage, save. The dried leaves of *Salvia officinalis* (Labiatae). It contains an essential oil, resin, tannin, and bitter principles; a spice. Cf. *sclareol.* **s. oil** Sage oil.
salvianin Monardein. A coloring matter from *Salvia coccinea.*
salviol $C_{20}H_{30}O_2$ = 302.5. 8,11,13-Abietatriene-2α,12-diol, m.108. From the roots of *Salvia multiorrhiza.*
sama condition Sama-zustand. A temperature difference in complete equilibrium; as, gases at low pressures in a gravitational field.
samandaridine $C_{21}H_{31}O_3N$ = 345.5. An alkaloid from *Salamandra maculosa,* a salamander, m.287. Cf. *salamandaridine.*
samandarine $C_{19}H_{31}O_2N$ = 305.5. An alkaloid from the skin secretion of salamanders, m.187. Cf. *salamanderine.*
samaric Samarium(III)* or (3+)*. Pertaining to trivalent samarium. **s. bromide** $SmBr_3 \cdot 6H_2O$ = 498.2. Green crystals, soluble in water. **s. chloride** $SmCl_3$ = 256.7. Green crystals, m.686, soluble in water. **hydrate ~** $SmCl_3 \cdot 6H_2O$ = 364.8. Green trigonal crystals, soluble in water. **s. hydroxide** $Sm(OH)_3$ = 201.4. Colorless powder, insoluble in water. **s. nitrate** $Sm(NO_3)_3 \cdot 6H_2O$ = 444.5. Yellow prisms, soluble in water. **s. oxalate** $Sm_2(C_2O_4)_3 \cdot 10H_2O$ = 744.9. Colorless crystals, insoluble in water. **s. sulfate** $Sm_2(SO_4)_3 \cdot 8H_2O$ = 733.0. Yellow, monoclinic crystals, slightly soluble in water. **s. sulfide** Sm_2S_3 = 396.9. Yellow powder, m.1900.
samarium* Sm = 150.36. A rare-earth metal and element, at. no. 62, discovered by Boisbaudran (1879). Gray metal, d.7.7, m.1080, b.1800, soluble in acids. Valency 2 [s.(II)*, s.(2+)*, samarous] and 3 [s.(III)*, s.(3+)*, samaric). Green and pink salts. Cf. *samaric, samarous.* **s. trioxide*** Sm_2O_3 = 348.7. White powder, insoluble in water.
samarous Samarium(II)* or (2+)*. Pertaining to bivalent samarium. **s. chloride** $SmCl_2$ = 221.3. Brown needles, m.740; soluble in water, evolving hydrogen. **s. sulfate** $SmSO_4$ = 246.4. Orange powder, insoluble in water.
samarsiite Black, native niobate and tantalate of uranium, cerium, and yttrium metals, including samarium.
sambucinin $C_{24}H_{42}O_7N_2$ = 470.6. An antibiotic from sambucus.
sambucus Elder flowers. The dried flowers of *Sambucus* species, elder.
sambunigrin $C_{14}H_{17}NO_6$ = 295.3. (*S*)-α-(β-D-glucopyranosyloxy)benzeneacetonitrile. m.152, from the leaves of *Sambucus nigra;* hydrolyzes to glucose and (−)-mandelonitrile.
samin $C_{13}H_{14}O_5$ = 250.1. A hydrolysis product of sesamolin. Colorless needles, m.103.
samneh Rendered butterfat (Israel).
samphire Common name for *Crithmum maritimium* (Umbelliferae), an English herbal; a pickle and a vegetable.
sample A representative portion of a substance, systematically taken for the purpose of judging its quality by analysis. See *quartering.*
sampler Riffles. A device for automatically splitting aggregates of ore, coal, cement, etc., for analysis. **air ~** A device for continuously collecting air pollutants, e.g., by drawing atmospheric air through a filter to collect particulate matter for subsequent analysis.
sand Particles of disintegrated siliceous rock; quartz. **black ~** Ilmenite. **Calais ~** An extremely fine s. (Calais), used to polish platinum ware. **iron ~** Titanomagnetite: Fe 60, Ti 5%. **oil ~** See *tar sands.* **silver ~** Fine s., washed with hot acids and water, for grinding substances before their extraction. **tar ~** See *tar sands.*
s. bath A heating vessel filled with s., to obtain a uniform distribution of heat; used similarly to a water bath. **s.blast** A stream of s. projected by compressed air or steam; used as an abrasive and metal cutter, and for frosting glass. **s.paper** An abrasive made by coating stout paper or thin cloth with glass, silicon carbide, garnet, alumina, or zirconia.
sandalwood White s., santalum, white sanders. The heartwood of *Santalum album* (Santalaceae); a source of s. oil and incense. **red ~** Red sanders, ruby wood. The heartwood of *Pterocarpus santalinus* (leguminosae); a coloring. See *santalin.* **yellow ~** Yellow sanders. A yellow s.; a coloring.
s. oil **East Indian ~** The essential oil of white s., d.970–0.985, containing santalol, santalenic acid, and its esters. **West Indian ~** The essential oil from *Amyris balsamifera* (Rutaceae).
sandarac(h) Gum juniper, sandarach. The resin from *Callitris quadrivalvis (Thuja articulata),* a pine of N.W. Africa.
Sanderit Trademark for a polyamide synthetic fiber.
sandix Orange mineral. A pale-orange, native lead oxide; a pigment.
Sandmeyer's reaction The transformation of diazo compounds into halogen compounds in presence of cuprous halogen salts: $Ph \cdot N_2Cl \rightarrow PhCl + N_2$.
sandstone A sedimentary rock, of coherent grains of sand.
sang de boeuf Red pottery glaze produced by reduction of copper oxide.
Sanger, Frederick (1918–) British biochemist. Nobel prize winner (1958 and 1980). Noted for work on insulin and genetic manipulation of DNA.
sanguinaria (1) Bloodroot, redroot, tellerwort. The dried rhizomes of *Sanguinaria canadensis* (Papaveraceae). It contains the alkaloids sanguinarine, homochelidonine, fumarine, and chelerythrine (see *opium alkaloids*); an emetic and expectorant. (2) A green bloodstone with red spots. (3) Hematite.
sanguinarine $C_{20}H_{15}O_4N$ = 333.3. An alkaloid from the root of *Sanguinaria canadensis* and *Stylophorum diphyllum.* White needles, m.213, soluble in alcohol (red color).
sanguis draconis Dragon's blood.
sanidine A glassy orthoclase.
Sankey diagram A diagram depicting the flows in a process. The distance between the double lines is proportional to the flow.
sansa The residue after pressing oil from olives.
Sansert Trademark for methysergide maleate.
santal Santalenic acid. **oil of ~** Sandalwood oil.
santalene $C_{15}H_{24}$ = 204.4. A terpene from sandalwood oil. **alpha-~** b.252. **beta-~** $b_{7mm}126$. **gamma-~** $b_{10mm}120$.
santalenic acid $C_{15}H_{24}O_5$ = 284.4. The coloring of red sandalwood, m.104, insoluble in water.
santalin A $C_{33}H_{26}O_{10}$ = 582.6. Red needles, m.302. Pigment from red sandalwood.
santalol $C_{15}H_{24}O$ = 220.4. Terpenes from sandalwood. Colorless liquids. Used in perfumes.

(R is $-CH_2 \cdot CH_2 \cdot CH{:}CHMe \cdot CH_2OH$)

alpha-~ Arheol, d.0.979. **beta-~** d.0.973, $b_{17mm}177$.
s. methyl ester Thyresol.
santalum Sandalwood.
santalyl The radical $C_{15}H_{23}-$. **s. carbonate** Carbosant. **s. chloride** $C_{15}H_{23}Cl$ = 238.8. Colorless liquid, $b_{10mm}155$.
santene C_9H_{14} = 122.3. 2,3-Dimethylbicyclo[2.2.1]hept-2-ene. A liquid, b.142.
santenol $C_9H_{16}O$ = 140.2. 1,7-Dimethylbicyclo[2.2.1]octan-2-ol. *α*-~ m.84 (subl.). *β*-~ m.108 (subl.).
santiaguine $C_{38}H_{43}O_2N_4$ = 296.4. An alkaloid from *Adenocarpus*. Cf. *adenocarpine*. **s. hydrochloride** m.241.
Santochlor Trademark for *p*-dichlorobenzene.
santol $C_8H_6O_3$ = 150.1. Colorless crystals, from red sandalwood.
Santomerse Trademark for an alkylated aryl sulfonate preparation having surface-active properties.
santonica Levant worm seed, cina, xantholine, semen cinae. The dried flower heads of *Artemisia* (Compositae); contains santonin, artemisin, essential oils, resins, and gums.
santonic acid $C_{15}H_{22}O_4$ = 266.3. An acid from santonica, m.171. **apo~** $C_{14}H_{20}O_3$ = 236.3. White crystals, m.164. **hydroxy~** $C_{14}H_{20}O_6$ = 284.3. Colorless crystals, decomp. 215.
santonin $C_{15}H_{18}O_3$ = 246.3. Santoninic acid lactone. A neutral, nonglucosidal bitter principle from santonica. Colorless leaflets, m.172 (sublimes and decomp.), insoluble in water; formerly an anthelmintic. **hydroxy~** Artemisin.
santoninic acid $C_{15}H_{20}O_4$ = 264.3. Colorless crystals, m.179.
s.ap. Abbreviation for apothecaries' scruple = ℈.
sap (1) The circulating plant juices that assist growth. (2) The surface of a rock softened by weathering.
Sapamine Trademark for trimethyl-*β*-oleoaminoethylammonium sulfate, a wetting agent.
Sapindales An order of plants (families Anacardiaceae, Aquifoliaceae, Aceraceae, Sapindaceae).
sapine $C_5H_{14}N_2$ = 102.2. A nontoxic isomer of cadaverine.
sapo A soap from olive oil and sodium hydroxide.
sapogenin $C_{14}H_{22}O_2$ = 222.3. Sapogenol. A decomposition product of saponin. Colorless needles, m.257, insoluble in water.
sapogenol Sapogenin.
saponaretin Vitexin.
saponaria Soaproot, soapwort, bruisewort, *S. officinalis* (Caryophyllaceae) which contains saponin and sapotoxin.
saponarin $C_{21}H_{24}O_{12}$ = 468.4. A glucoside, m.232, from saponaria.
saponification The conversion of an ester into an alcohol and acid salt; as, fats into soaps by an alkali: $(R \cdot COO)_3G + 3NaOH = 3R \cdot COONa + G(OH)_3$. G = glycerol.
s. equivalent The quantity of fat in grams saponified by one liter of normal alkalies = 56,108/sap. no. **s. number, s. value** The quantity of potassium hydroxide in mg to saponify 1 g of fat.
saponin $C_{32}H_{54}O_{18}$ = 726.5. A glucoside from soapwort, quillaia, and especially the heartwood of *Mora excelsa* (Guyana), which contains 4–5%. White powder, soluble in water; a toxic foam producer. **sasanqua ~** $C_{73}H_{11}O_{22} \cdot 3H_2O$. A glucoside from the seeds of *Camelia sasanqua* (Theaceae), decomp. 222; hydrolyzes to pentose, galactose, and prosapogenin.
saponins Amorphous glucosides that produce foaming solutions in *Saponaria* species (Hippocastanaceae).
saponite A native, hydrous silicate of magnesium and aluminum.
Sapotaceae Tropical shrubs, many of which bear edible berries, and are sources of saponin; as, *Mimuseps globosa*, balata, chicle. Cf. *butternut, illipé, sapotin.*
sapotalene $C_{13}H_{14}$ = 170.3. 1,2,7-Trimethylnaphthalene*. Colorless crystals, insoluble in water.
sapotin $C_{29}H_{52}O_{20}$ = 720.7. A glucoside from the seeds of *Achras* or *Sapota sapotilla*. Colorless crystals, m.240.
sapotoxin $C_{17}H_{26}O_{10}$ = 390.4. A toxin from the bark of quillaia.
sappanwood Sibucao. The wood of *Caesalpini sappan* (Leguminosae); a dye.
sapphire Al_2O_3. A native, blue gem corundum. **Brazilian ~** Tourmaline. **water ~** Iolite.
saprine $C_5H_{14}N_2$ = 102.2. A ptomaine from decaying meat.
sapropel A submarine deposit formed by the sedimentation of dead algal colonies.
sapropelic coal A coal that contains microscopic oil-bearing algae.
sapropelitic Resembling coal or asphalt.
saprophyte A vegetable microorganism in air, soil, or water, which feeds on dead or decaying animals or plants. Cf. *parasite*.
Saran Trademark for a vinyl chloride-vinylidene chloride copolymer; used as a chemical-resistant lining for piping; as film, for food wrap; in filament form, used to make rotproof fabrics, and to render fabrics flameproof and water-vapor-resistant.
Sarcina (Pl. Sarcinae) A genus of *Coccaceae* (Schizomycetes) which forms balelike packs. Cf. *bacteria*.
sarcine Hypoxanthine.
sarcocol A gum resin from *Penaea sarcocolla* (Penaeaceae), Africa.
sarcolactate A salt of (*S*)-lactic acid.
sarcolactic acid (*S*)-Lactic acid.
sarcolite A melilitic aluminum silicate, which contains lime and sodium.
sarcoma A malignant tumor formed by cells of connective tissue, muscle, or bone. **Rous ~** A s. produced by the injection of cell-free filtrates of tumor tissues into chickens; the filtrates contain the causative viruses.
sarcosine* $MeNH \cdot CH_2 \cdot COOH$ = 89.1. Methylglycine, sarkosine, methylaminoacetic acid. Colorless rhombs, m.210, soluble in water. **dimethyl~** Betaine*.
sard A brown sardonyx.
sardinianite $PbSO_4$. A monoclinic anglesite.
sardonyx A brown, translucent chalcedony; a semiprecious stone. Cf. *sard*.
Sarelon Trademark for a protein synthetic fiber.
sarin $(Me_2CHO)MeFPO$ = 140.1. GB. 1-Methylethyl ester of methylphosphonofluoridic acid. A nerve gas; inhibits choline esterase.
sarkine Hypoxanthine.
sarkokaulin $C_{13}H_{24}O_2$ = 212.3. An alcohol, m.78, from the wax of the candle bush, *Sarcocaulon burmain* (Leguminosae), S. Africa.
sarmentogenin $C_{23}H_{34}O_5$ = 390.5. A trihydroxycardenolide from the seeds of *Strophanthus sarmentosus;* a heart poison used in the synthesis of cortisone.
sarracenine An alkaloid from the roots of *Sarracenia* species, flytrap or pitcher plant. White needles, soluble in alcohol.
sarsaparilla Radix sarsae. The dried root of *Smilax medica* (Liliaceae). It contains glucosides (smilacin, parillin), resin, saponins, and essential oils. **American ~, false ~** The root of *Aralia nudicaulus* (Araliaceae). **Indian ~** The root of *Hemidesmus indicus* (Asclepiadaceae). **Jamaica ~** Sarsaparilla.
sarsasaponin A glucoside from sarsaparilla or smilax; an emetic.
sarverogenin $C_{23}H_{32}O_7$ = 420.5. A cardenolide found with and similar to sarmentogenin.

SAS (1) Society for Applied Spectroscopy. (2) Abbreviated trade name for the sodium salt of an alkane sulfonic acid. The sulfonate groups are randomly distributed along a straight hydrocarbon chain. Used in the manufacture of soft detergents.

sasanqua See *saponin.*

Sasol South African plant producing from coal: liquid fuels, styrene, butadiene, and ammonia. See *coal liquefaction* under *coal.*

sassafras The dried bark of the root of *Sassafras variifolium* (lauraceae), N. America. It contains an essential oil, resin, tannin, and wax. **Australian ~** Atherospermine.
s. nuts Pichurim beans. **s. oil** Oleum sassafras. An essential oil from the bark of *S. officinalis,* d.1.065–1.095, containing safrole, eugenol, camphor, pinene, and phellandrene.

sassa gum A red gum from *Albizzia fastigiata* (Leguminosae), Ethiopia.

sassoline Sassolite.

sassolite $B(OH)_3$. Sassoline. A native boric acid; occurs in triclinic scales at the fumaroles.

sassy bark Mancona bark, casca bark, doom bark. The bark of *Erythrofloeum guinense* (Leguminosae), W. Africa; an ordeal poison.

satellite Natural or artificial body orbiting around a planet. **remote sensing ~** S. orbiting about 800 km above the earth that collects data about weather, mineral deposits, fish stocks, and crops.

satin **s. spar** A smooth compact variety of calcite. **s. white** A mixture of calcium sulfate and aluminum hydroxide, produced by the coprecipitation of lime and aluminum sulfate in presence of water; a pigment for coating paper.

saturate (1) To link up all the atomic bonds in a molecule so that only single bonds exist. (2) To dissolve sufficient substance in a solution, so that no more can be dissolved.

saturated Completely satisfied. **super ~** See *supersaturated.* **s. compound** An organic compound with neither double nor triple bonds. **s. hydrocarbons** Alkanes*. **s. solution** A solution that contains so much dissolved substance that no more will dissolve at a given temperature.

saturation (1) Complete neutralization of an acid or base. (2) Complete or maximum absorption of a substance by a solvent. (3) Complete satisfaction of the valency bonds in a molecule. (4) A property of *color,* q.v. **super ~** See *supersaturated.*
s. current The maximum current that can pass as a silent discharge through a gas or vapor without decomposing it. **s. isomerism** Isomerism between 2 compounds, one of which is saturated and the other unsaturated; as, acetone, $CH_3 \cdot CO \cdot CH_3$, and allyl alcohol, $CH_2{:}CH \cdot CH_2OH$. **s. point** (1) The concentration at which a solution is saturated with a particular substance. (2) In color printing, the stage at which one color becomes dominant at the expense of the others.

saturnism Lead poisoning.

saturnus The alchemical name for lead.

saunders Sandalwood.

Säure German for "acid."

saussurite An impure labradorite.

save Salvia.

savine Sabina. The fresh tops of *Juniperus sabina,* containing essential oil, tannin, and resin. Cf. *juniperic acid.* **s. oil** Oleum sabinae. An essential oil from *Juniperus sabina.* Yellow liquid, d.0.903, which contains pinene, cadinene, thujene, and sabinol.

savory The herb *Satureia hortensis* (summer s.) and *S. montana* (winter s.); an aromatic and carminative.

saw palmetto berries Sabal.

saxicoles A group of lichens.

saxifrage Pimpinella.

Saxin A brand of saccharin.

Saybolt seconds A relative unit of viscosity; the time necessary for a specified volume of a liquid to flow through the orifice of a Saybolt viscosimeter at a definite temperature. Cf. *SAE number.*

Sb Symbol for antimony (from Latin, *stibium*).

SBR GRS. A synthetic rubber produced by polymerization at 50°C, or as cold rubber at 5–15°C.

Sc Symbol for scandium.

scagliola A 19th-century imitation marble made from colored plaster.

scalar Describing a quantity that has magnitude but no direction, e.g., density. Cf. *vector.*

scale (1) A thin, flaky leaflet. (2) A crust of oxides formed on the surface of metals. (3) Boilerstone. The incrustation of insoluble salts formed by the evaporation of water. (4) Markings at regular intervals, on instruments, drawings, and graphs. (5) A balance used for relative rough weighings. **conversion ~** A graph of 2 or more parallel scales, used for the rapid solution of proportional problems. Cf. *nomograph.*
s. copper Copper in thin flakes. **s. stone** Wollastonite.

scaling index A measure of the degree of corrosion of a metal in a liquid; the gain in weight in mg/cm^2 under specified conditions.

scalpel A small, curved dissecting or surgeon's knife.

scammonin $C_{34}H_{56}O_{16}$ = 720.8. A glucoside derived from scammony.

scammony The dried root of *Convolvulus scammonia* (Convolvulaceae), Asia Minor. It contains the glucoside scammonin, a gum, and a resin; a cathartic. **Mexican ~** Ipomoea.

scandia Scandium oxide*.

scandium* Sc = 44.95591. A rare-earth metal, at. no. 21. Predicted by Mendeleev (as ekaboron), and discovered (1879) by Nilson; obtained from thortveite. Gray metal with pink tinge, m.1540, b. ca. 2780, soluble in acids; valency 3. **s. acetyl acetonate** $Sc(MeCOCHCOMe)_3$ = 342.3. White plates, m.187, soluble in water. **s. chloride** $ScCl_3$ = 151.3. Colorless flakes, sublimes 800, soluble in water. **s. hydroxide*** $Sc(OH)_3$ = 96.0. Colorless powder, insoluble in water. **s. oxalate*** $Sc_2(C_2O_4)_3 \cdot H_2O$ = 372.0. White crystals, m.140, insoluble in water. **s. oxide*** Sc_2O_3 = 137.9. Scandia. Colorless powder, insoluble in water. **s. sulfate*** $Sc_2(SO_4)_3$ = 378.1. Colorless crystals, soluble in water.

scapolite Wernerite. A powder or green calcium, aluminum, sodium silicate containing chlorine.

scarlet A group of dyes, as: **s. R (Michaelis)** $MeC_6H_4N{:}NC_6H_3MeN{:}NC_{10}H_6OH$ = 380.5. Sudan Red, C.I. Solvent Red 24. Brown powder, insoluble in water; used as a dye.

scatol Skatole.

scattering (1) Dispersing. (2) The splitting of molten metals on pouring. **elastic ~** The condition where the energy of the scattered light equals that of the incident light, as in the Tyndall effect. **inelastic ~** The condition where the energy of the scattered light is different from that of the incident light. Fluorescence and Raman phenomena contain both elastic and inelastic light. **light ~** The emission of light from a particle or molecule under illumination, due to resonance and excitation; e.g.: *Tyndall effect,* where the initial and final state of the scattering medium and the incoming and scattered quantum remain unchanged, except for a new direction of motion; *fluorescence,* where the scattered light results from excitation and subsequent emission between discrete energy states; *Raman effect,* where the scattered light results from a continuum of virtual energy states; *Compton effect,* where a quantum of high frequency (X-rays) dislodges an electron from the scattering substance. Cf. *luminescence.*

scavenger A purifying substance; as, metallic lithium which removes impurities from alloys.

Schäffer's acid $HSO_3 \cdot C_{10}H_6 \cdot OH$ = 224.2. Armstrong's acid. β-Naphtholsulfonic acid, 2-hydroxynaphthalene-6-sulfonic acid; used in organic synthesis.

schapbachite $PbS \cdot Ag_2S \cdot Bi_2S_3$. A native sulfide.

schappe Silk waste.

Schardinger dextrin α-Dextrin.

Scheele **S., Carl Wilhelm (1742–1786)** Swedish apothecary noted for his discovery of oxygen, chlorine, ammonia, manganese, and barium. **S.'s green** $CuHAsO_3$. An acid copper arsenite used as pigment.

scheelite $CaWO_4$. A native calcium tungstate.

scheelium Tungsten (obsolete).

scheererite A mineral hydrocarbon, m.45, b.92.

Scheibler **S., Carl (1827–1899)** German chemist noted for developments in the sugar industry. **S.'s reagent** A solution of phosphotungstic acid; yellow precipitate with sulfates of the alkaloids.

Schick test A test for susceptibility to diphtheria in which intracutaneous injection of toxin, from *Corynebacterium diphtheriae,* produces a local reaction.

schieferspar A flaky variety of calcite.

Schiff **S., Hugo (1834–1915)** German organic chemist. **S. bases*** $R \cdot N:CHR$. Condensation products of aromatic amines and aliphatic aldehydes forming azomethines substituted on the N atom.: $PhNH_2 + OCH \cdot Ph = PhN:CHPh + H_2O$. **S. reagent** Thioacetic acid*. **S. solution** A solution of 0.2 g rosaniline and 15 mL sulfurous acid in 200 mL water; a test for aldehydes which restore the red color.

schiller spar Bronzite.

schinus oil An essential oil from the pepper tree, *Schinus molle* (Anacardiaceae), N. America, d.0.850, containing phellandrene, pinene, and carvacrol.

schist A crystalline rock that can be split into scales or flakes.

schistic Not aschistic, q.v. *(aschistic process).*

schistosomiasis Bilharziasis. A helminth disease in tropical countries.

Schizomycetes *Schizophyta,* fission fungi, *bacteria,* q.v. Plant microorganisms of the chlorophyll-free, fungi class.

Family 1: Coccaceae, round or spherical in shape.
Genus I: Streptococci, beadlike chains.
Genus II: Micrococci, grapelike clusters.
Genus III: Sarcina, balelike packs.
Genus IV: Planococci, like II but mobile.
Genus V: Planosarcina, like III but mobile.
Family 2: Bacteriaceae, cylindrical or rodlike in shape.
Family 3: Spirillaceae, curved or S-like in shape.

Schizophyta Schizomycetes.

schlempe Vinasse.

schlieren Describing the region of changing refraction in an otherwise optically homogeneous medium, e.g., heat waves seen over a hot surface.

Schlippe **S., Carl Friedrich von (1799–1874)** German-born Russian chemist. **S.'s salt** $Na_3SbS_4 \cdot 9H_2O$. Sodium tetrathioantimonate*.

Schlotterbeck reaction The synthesis of ketones from aldehydes: $R \cdot CHO + CH_2N_2 \rightarrow R \cdot CO \cdot CH_3 + N_2$. Cf. *Nierenstein reaction.*

Schmidt test A test for glue; white precipitate with a solution of ammonium molybdate.

Schmoluchowski's equation The average path length, in μm, of a particle in a dispersed system: $2.37\sqrt{K} \cdot RT \cdot t/L\eta r$, where R = gas content, L = Avogadro constant, T = thermodynamic temperature, t = period of vibration of the particle, η = viscosity of the medium, r = radius of the particle.

schneebergite $CaSbO_3$. A native antimonite.

Schneider's furnace A retort for the distillation of zinc from zinc-lead ores.

Schoenbein, Christian Friedrich (1799–1868) German chemist noted as discoverer of ozone and for work on catalysis.

Schoenherr process A nitrogen fixation method in which the air circulates spirally around a 6-meter electric arc.

schöenite $K_2SO_4 \cdot MgSO_4 \cdot 9H_2O$. A stassfurt salt.

Schöllkopf's acid (1) 1-Naphthol-4,8-disulfonic acid. (2) 1-Naphthylamine-4,8-disulfonic acid. (3) 1-Naphthylamine-8-sulfonic acid.

schorl Tourmaline.

Schörlemmer, Carl (1834–1900) German chemist, noted for his textbooks.

schorlomite A titanium *garnet,* q.v.

Schötten **S., Carl (1853–1910)** German organic chemist noted for his organic synthesis methods. **S. reaction** Acylation in alkaline solution with benzoyl chloride.

schou oil A gelatinous product of the oxidation of soybean oil; an emulsifying agent in the margarine industry.

schraufite $C_{11}H_{16}O_2$. A fossil resin in Carpathian sandstone.

schreibersite $(FeNiCo)_3P$. A mixed phosphide, in certain meteorites.

schreinering Reduction of the fiber interstices of a knitted fabric to give a tighter structure and higher density.

Schrödinger **S., Erwin (1887–1961)** Austrian physicist, Nobel prize winner (1933); noted for his atomic concepts. **S. atom** Pulsating or fluctuating atom. The atom is regarded as a sphere of electricity which may vary in its density, but which may pulsate, with absorption or liberation of radiation. **S. equation** Wave equation. An equation, based on de Broglie's equation, in which ψ^2 determines the statistical charge density. In 3 dimensions it is: $\nabla^2\psi + (8\pi^2 m/h^2)(E - U)\psi = 0$, where ∇^2 is the Laplace operator, ψ the wave function, E the total energy, U the potential energy, and m the mass of the particle. Cf. *Heisenberg principle.*

schroeckingerite Dakerite.

Schroeder's paradox Polymers swell more in a liquid than in its vapor, owing to small temperature differences.

Schrötter apparatus Calcimeter.

Schultz number The classification number of a dyestuff as given in "Farbstofftabellen," by Gustav Schultz. Cf. *Color* Index.

Schulze's rule The precipitating effect of an ion varies with its valency.

Schumann rays The extreme ultraviolet portion of the spectrum which affects a photographic plate. Cf. *ultraviolet.*

Schütz-Borrisow rule Enzyme activity; $x = tK\sqrt{c}$, where x is the amount of substance digested, t the reaction time, e.g., 24 hours, K a constant, and c the concentration of the enzyme.

Schwarza Trademark for a viscous synthetic fiber.

schwatzite A tetrahedrite containing mercury.

Schweinfurt green Cupric subacetate.

Schweitzer **S., Mathias E. (1818–1860)** German chemist. **S.'s reagent** An ammoniacal solution of cupric hydroxide, which dissolves cellulose.

schwellenwert Liminal value, threshold value. The minimum quantity of electrolyte required to precipitate a colloidal solution.

sciadopitene $C_{20}H_{32}$ = 272.5. A diterpene, m.96, from the

wood oil of *Sciadopitys verticillata*, the parasol pine, or umbrella fir of Japan.
sciagraph, sciagram X-ray photograph.
science Systematized and verifiable knowledge reached by observation, measurement, and experiment. Science describes, measures, and coordinates facts, and attempts to explain their ultimate cause. Cf. *philosophy*.

Formal science:
Logic—ideas and concepts
Mathematics—numbers and magnitudes
Geometry—space and extension
Phoronomy—motion, time, and relativity
Natural science:
Physics—energy transformations
Chemistry—matter transformations
Astronomy—the universe
Geology—the earth
Biological science:
Botany—structure and functions of plants
Zoology—structure and functions of animals
Anthropology—man
Psychology—human behavior
Economics—practical applications
Sociology—human society

scientific Based on systematized and verifiable facts or experience. Cf. *empiric*.
scilla Squill.
scillain (1) An amorphous glucoside obtained from the bulbs of *Scilla maritima*, squill. (2) Scillipicrin.
scillarabiose $C_{12}H_{22}O_{10}$ = 326.3. A disaccharide hydrolyzed to rhamnose and D-glucose. Cf. *scillaren*.
scillaren **s. A** $C_{37}H_{54}O_{13}$ = 706.8. A glucoside from squill, *Scilla maritima*. Crystals, m.270, hydrolyzed to scillaridin and scillarabiose; a cardiac stimulant and diuretic. **s. B** A mixture of glucosides of greater physiological activity than s. A.
scillaridin $C_{24}H_{30}O_3$ = 366.5. **A-~** A bufatetraenolide.
scillin A yellow, crystalline glucoside from the bulbs of *Scilla* species.
scillipicrin Scillain. Yellow, amorphous glucoside from the bulbs of *Scilla maritima*; a diuretic.
scintillation (1) Burning with brilliant sparks, as an iron wire in oxygen. (2) The emission or production of sparks. **s. counter** Detector that measures radiation (e.g., γ or X) by the light scintillations produced on its phosphor coating.
scission (1) The splitting of a molecule. (2) Ring breakage. The opening of an atomic ring. (3) Fission, the division of a living cell. See *karyokinesis*.
Scitaminaceae The Musaceae or banana family, a group of tropical plants from which drugs are obtained: as, *Zingiber officinale*, ginger; *Musa sapientum*, banana.
sclareol $C_{20}H_{36}O_2$ = 308.5. m.106. An unsaturated, dihydric alcohol, the principal constituent of oil of sage.
sclerethryrin A red coloring matter in ergot.
sclero-, sklero- Prefix (Greek) indicating "hard."
sclerolac An ether-soluble, hard lac resin from shellac.
sclerometer An instrument for determining the hardness of materials from the pressure on a moving diamond point necessary to produce a scratch.
scleron A light, noncorrodible alloy of Al with Si, Cu, Fe, Mn, Zn, and Li.
scleroproteins A group of proteins in animal skeletons.
scleroscope An instrument for determining the hardness of substances from the extent to which a steel ball rebounds on being dropped from a certain height. Cf. *Shore hardness*.
sclerotic acid Sclerotinic acid, ergotic acid. A brown substance from ergot.
sclerotin An early name for pectin.
scolecite $CaAl_2Si_3O_{10} \cdot 3H_2O$. A white or yellow mineral.
Scoline Trademark for suxamethonium chloride.
scoliodonic acid $C_{24}H_{38}O_2$ = 358.6. A highly unsaturated acid from hiragashira oil.
scombrin A protamine from mackerel sperm: 88.8% arginine.
scoparin $C_{22}H_{22}O_{11}$ = 462.4. A yellow, crystalline principle from scoparius. Crystals, m.253 (decomp.).
scoparius Spartium. Broom tops. The dried tops of *Cytisus scoparius* (Leguminosae). It contains sparteine and scoparin; a diuretic.
scopine $C_8H_{13}NO_2$ = 155.2. 2,3-Epoxytropan. A product of hydrolysis of scopolamine, and isomer of scopoline. Needles, m.76.
scopola The dried rhizome of *Scopola carniolica* (Solanaceae). It contains the belladonna alkaloids. Cf. *atroscine*.
scopolamine $C_{17}H_{21}O_4N$ = 303.4. Hyoscine. A levorotatory alkaloid from Solanaceae. Colorless syrup, soluble in alcohol. **inactive ~** Atroscine.
s. hydrobromide $C_{17}H_{21}O_4N \cdot HBr \cdot 3H_2O$ = 438.3. Kwells, Colorless crystals, m.196 (anhydrous), soluble in water. A hypnotic and anticholinergic; used for preoperative sedation, for travel sickness, and as a mydriatic. (USP, EP, BP).
scopoleine $C_{17}H_{21}O_4N$ = 303.4. A crystalline alkaloid from several *Scopola*, *Duboisia*, and *Atropa* species.
scopoletin $C_{10}H_8O_4$ = 192.2. Chrysatropic acid, gelsem(in)ic acid, 7-hydroxy-6-methoxycoumarin, β-methylesculetin. Colorless crystals, m.204. Occurs widely in plants.
scopoline $C_8H_{13}O_2N$ = 155.2. Oscine. A decomposition product of scopolamine. Colorless crystals, m.110, soluble in water.
scopometer An instrument with an optical wedge for visual measurement of turbidity by observing the disappearance of an illuminated target.
scopometry A branch of *nephelometry*, q.v.; matching colors or turbidities by comparing an illuminated line against a field of constant intensity.
scorification The assay of ores by roasting, fusion, and oxidation of gold and silver ores with lead and borax glass in a shallow clay vessel in a muffle.
scorifier A vessel for scorification.
scorodite $Fe_2O_3 \cdot As_2O_5 \cdot 4H_2O$. Pitticite. A native hydrated ferrous arsenate.
scorodites $M_2O_3 \cdot N_2O_5 \cdot nH_2O$. *M* is ferric iron or aluminum, and *N* is arsenic or phosphorus.
scotography Skiagraphy. The study of human radiation, radioactivity, aura, or od-*rays*, q.v.
scotoma Blind spot. A spot or area in the visual field where there is no vision.
scouring (1) Corroding; as by certain ores that attack furnaces. (2) Cleaning; as, removing the grease or stain from a vessel. (3) Diarrhea in calves. **s. cinder** A basic slag that attacks the lining of a shaft furnace. **s. rush** Equisetum.
SCR See *silicon-controlled rectifier*.
screen (1) A sieve of wire cloth, textile, or perforated metal plates, used to sort particles according to size. (2) A prepared surface on which light or images are projected. (3) An apparatus with *circular* apertures as compared with a sieve (*square* apertures). **fluorescent ~** A plate coated with calcium tungstate or barium thiocyanate; used to make ultraviolet rays, X-rays, etc., visible to the eye. **revolving ~** A steel cylinder, usually inclined, with round holes.
s. analysis The separation of a material into particles of definite sizes, by screens of graded sizes.

screening effect In any atom the inner-orbital electrons act as screens between the nucleus and the outer orbital electrons, and thus decrease the effective nuclear charge on the latter. Cf. *Pauling structure, Lucas theory.*
Scrophularia Figwort, rose noble. The herb of *S. nodosa* (Scrophylariaceae). **water ~** Bishop's leaves. The leaves of *S. aquatica;* used externally for poultices.
Scrophulariaceae Figwort family. Herbs and shrubs that contain glucosides and drugs; e.g., leaves: *Digitalis purpurea,* digitalin; herbs: *Veronica officinalis* (speedwell), veronica; rhizomes: *Veronica (Leptandra) virginica,* leptandra.
scrubber A device for washing or absorbing gases; used in chemical plants for purification, dissolving, or reacting gases in or with liquids.
scrubbing Removal of impurities by extraction from the separated phase in liquid–liquid extraction.
scruff Surface dirt or impurities. Cf. *scurf.*
scruple ℈. A unit of apothecaries' weight: 1 scruple = 20 grains = 1.295978 grams.
scullcap Scutellaria.
scum The impurities on the surface of molten materials. Cf. *oly.*
scurf Material that flakes off; dross. Cf. *scruff.*
Scutellaria Scullcap, helmet flower. The dried plant of *S. lateriflora* (Labiatae), N. America. It contains scutellarin and an essential oil; an antispasmodic.
scutellarin $C_{10}H_8O_3$ = 176.1. A nontoxic, crystalline principle from the leaves of *Scutellaria.* Yellow needles, m.199, insoluble in water.
scyllitol $C_6H_6(OH)_6$ = 180.2. *scyllo*-Inositol. 1,3,5/2,4,6-Inositol. Occurs in fish and plants.
SDO Synthetic drying oil.
Se Symbol for selenium.
sea **s. salt** (1) Commercial sodium chloride from evaporated seawater. (2) The residual mixture of salts on evaporating seawater. **s.water** See *hydrosphere, water.* **s.weed** Kelp.
seal Water, mercury, wax, oil, or other material placed around joints to prevent passage of liquids or gases.
sealing wax A colored, scented mixture of resins and shellac; used for sealing.
seam A stratum or bed of a mineral or ore.
sebacic acid $COOH(CH_2)_8COOH$ = 202.3. Ipomic acid, decanedioic acid*. Colorless leaflets, m.133, soluble in water.
sec (1) Abbreviation for (a)* (italic) secondary; (b) second, the unit of time, formerly; now s. (2) Dry (French).
Secale **S. cereale** Rye. **S. cornutum** Ergot.
secaline Trimethylamine*.
secalonic acid $C_{14}H_{14}O_6$ = 278.3. Yellow crystals, in ergot. See *ergochromes.*
secalose A carbohydrate from rye. White, hygroscopic powder. Cf. *trifructosan.*
secbutobarbitone sodium Butabarbital sodium.
Secchi, Angelo (1818–1876) Italian Jesuit astronomer, noted for spectrum analysis and polaroscopic experiments.
sechometer A hand-driven induction apparatus.
seco-* Indicating that a compound is formed by breaking ring(s) of the parent-named compound.
secobarbital sodium $C_{12}H_{17}N_2NaO_3$ = 260.3. Sodium 5-allyl-5-(1-methylbutyl)barbiturate. Secobarbitone sodium. Quinalbarbitone sodium. Seconal sodium. Bitter, white powder, soluble in water; a barbiturate (USP). Cf. *secbutobarbitone sodium.*
secohm A unit of self-inductance: 1 ohm/s.
Seconal sodium Trademark for secobarbital sodium.
second* s*, sec, ″. (1) A basic SI system unit. The duration of 9,192,631,770 periods of the radiation corresponding to the transition between the two hyperfine levels of the ground state of the cesium-133 atom. (2) More commonly, the $\frac{1}{60}$ part of a minute: 1/86,164.09 of a sidereal day; 1/(24 × 60 × 60) of a mean solar day. **s. ionization constant** See *ionization energy.*
secondary (1) Second in order. (2) Next in importance. **s. alcohol** An organic compound containing the radical =CHOH. **s. amine** An organic compound containing the radical =NH. **s. carbon atom** A carbon atom directly attached to 2 others. **s. metal** Metal recovered from scrap, sweepings, skimmings, drosses, etc. Cf. primary *metal.* **s. reaction** See *secondary reaction* under *reaction.* **s. X-rays** The characteristic scattered radiation emitted from a substance exposed to X-rays; used for analytical purposes.
secretin A hormone, secreted by the epithelial lining of the first part of the intestine (duodenum), stimulating the pancreas to secrete pancreatic juice.
secretion The separation of a substance, other than a waste material, from a living organism; as, resins from plants, serum from wounds and products from glands in the body.
secretor A person whose ABO blood group substances exist also, in a soluble form, in other body fluids, such as semen, saliva. **non ~** In general, a person whose secreted body fluids (as, semen, saliva) do not contain certain blood grouping substances.
section A thinly cut piece of a substance for microscopic study. **histological ~** A thin cut of a plant or animal tissue. **metallographic ~** A thin cut of a metal.
sectrometer A vacuum-tube Titrimeter for potentiometric titrations; a cathode ray tube replaces the microammeter, and the end point is a sudden permanent change in the shadow angle on a fluorescent screen. Cf. *Titrimeter.*
securite A mine explosive not ignited by firedamp; contains ammonium nitrate and oxalate, and dinitrobenzene.
sedanolid $C_{12}H_{18}O_2$ = 194.3. The lactone of sedanonic acid, in celery seeds, *Apium graveolens.*
sedative A calming agent that counteracts stimulation, irritation, or excitement; as, barbiturates. Cf. *stimulant.*
sediment A deposit of an insoluble material, especially if settled by gravitation. Cf. *precipitate.*
sedimentary rock A rock formed by the accumulation of grains or fragments of rock carried by water or air.
sedimentation The precipitation or settling of insoluble materials from a suspension, either naturally (by gravity) or artificially (by a centrifuge). **free ~** S. in which the particles exert no mutual interference. Stokes law then applies. **hindered ~** The opposite of free s. **rate of ~** See *settling, Stokes law.*
see Salt. **s. mixte** A natural mixture of sodium chloride and magnesium sulfate ($7H_2O$), deposited in the salt lakes of the Volga regions, U.S.S.R.
Seebeck, Thomas Johann (1770–1831) German physicist; discoverer of thermoelectricity and the magnetism of cobalt and nickel.
seed lac See *lac.*
seeds The product of fertilized and developed ovules of plants; usually rich in proteins, carbohydrates, and oils, and an important source of food; e.g., those of Gramineae (grains) and Leguminosae (peas and beans). See *fruit.*
Seekay wax C.K. *wax.*
seepage (1) The percolation of a fluid through a porous material. (2) The fluid that results from s. (3) The separation of 2 phases.
Sefström, Nils Grabriel (1787–1845) Swedish chemist and mineralogist, discoverer of vanadium.

Seger **S., Hermann A. (1839–1893)** German technologist noted for ceramic research. **S. cones** Pyrometric cones. Small pyramids of various clay and salt mixtures; used to indicate the temperature of a furnace. Each cone softens at a definite temperature, ranging from 500 to 2000°C.
seggar Clay boxes in which ceramics are kilned.
segregate To separate.
sehta CoAsS. Indian name for cobaltite, as used to make blue-enameled metalware.
Seidlitz powder An effervescent mixture of potassium sodium tartrate, sodium hydrogencarbonate, and tartaric acid; used in alkaline mineral waters.
seifert solder See *seifert solder* under *solder.*
Seignette's salt Potassium sodium tartrate*.
seismometer An instrument for recording earth shocks. See *Richter scale.*
sekisanine $C_{34}H_{34}O_9N_2$ = 614.7. Hydroxydimethyllycorine. A physiologically inactive alkaloid from the bulbs of *Lycoris radiata* (Amaryllidaceae). Colorless prisms. Cf. *lycorine.*
selacholeic acid $Me(CH_2)_7CH:CH(CH_2)_{13}COOH$ = 366.6. (Z)-15-Tetracosenoic acid*, nervonic acid, m.41. From shark-liver oil.
selachyl alcohol $C_{18}H_{35}OC_3H_5(OH)_2$ = 342.6. 2,3-Dihydroxy-1-octadec-9′-enyloxypropane. **(S, Z)-∼** A liquid glyceryl ether from shark liver. Cf. *batyl.*
selection rules Important rules of wave mechanics, which control the possible transitions that an electron can make between states.
selena-* Prefix indicating =Se in a ring compound. Cf. *seleno.*
selenate* Indicating selenium as the central atom(s) in an anion; as, M_2SeO_4. A salt of selenic acid.
selenic Selenium(IV)* or (4+)*; (VI)* or (6+)*. A compound of tetra- or hexavalent selenium. **s. acid*** $H_2SeO_4 \cdot nH_2O$ = 145.0. An isolog of sulfuric acid. Colorless prisms, m.58, soluble in water; forms selenates.
selenide* A binary, inorganic (m_2Se) or organic compound of divalent selenium. **di∼** R·Se·Se·R. Cf. *disulfide.* **hydro∼** The hydroseleno* radical.
-seleninic acid* Suffix indicating the $-SeO_2H$ radical; as, $PhSeO_2H$, benzeneseleninic acid. Cf. *selenonic acid.*
selenino-* Prefix indicating the radical (HO)OSe−. Cf. *seleninic acid, selenono.*
seleninyl* The radical =SeO. **s. chloride*** $SeOCl_2$ = 165.9. Selenium oxychloride. Colorless liquid, m.10, decomp. by water.
selenious Selenium(II)* or (2+)*; (IV)* or (4+)*; selenous. A compound containing divalent or tetravalent selenium; as, $SeCl_2$, SeO_2. **s. acid*** H_2SeO_3 = 129.0. Colorless crystals, decomp. by heat, soluble in water; forms selenites. **s. oxide** Selenium dioxide*.
selenite (1)* M_2SeO_3. A salt of selenious acid. (2) $CaSO_4 \cdot 2H_2O$. A native gypsum.
selenium* Se = 78.96. A nonmetal element of the sulfur group, at. no. 34. Modifications: (1) *Metallic.* Gray hexagons, m.217. b.685, insoluble in water, soluble in ether. (2) *Crystalline.* Red, monoclinic crystals, m.175, soluble in carbon disulfide. (3) *Amorphous.* Red powder, obtained by precipitation, m.100, soluble in carbon disulfide. (4) *Colloidal.* Red solution, slowly depositing amorphous s., m.70.
S. was discovered (1817) by Berzelius in the lead chambers of a sulfuric acid plant. It burns with a blue flame and garliclike odor to its dioxide, SeO_2. It has valencies of 2, 4, and 6, and forms ions: Se^{2-}, selenides; SeO_3^{2-}, selenites; SeO_4^{2-}, selenates. The electrical resistance of metallic s. decreases with increase in intensity of illumination, and it is used for s. cells, for the optophone, and for making red glasses, enamels, and glazes. **cycloocta∼** The ring structure Se_8, analog of S_8.
s. bromide (1) Se_2Br_2 = 317.7. S. monobromide*. Red liquid, m.−46. (2) $SeBr_2$ = 238.8. S. dibromide*. Brown liquid. (3) $SeBr_4$ = 398.6. S. tetrabromide*. Orange crystals, soluble in carbon disulfide. **s. cell** An arrangement of metallic s. plates enabling electricity or sound to be transmitted by means of the variation of electrical resistance of the cell with light intensity. **s. chloride** (1) Se_2Cl_2 = 228.2. S. monochloride*. Brown crystals. (2) $SeCl_2$ = 149.9. S. dichloride*, selenous chloride. Brown oil. (3) $SeCl_4$ = 220.8. S. tetrachloride*, selenic chloride. Yellow crystals. **s. dibromide*** See *selenium bromide.* **s. dichloride*** See *selenium chloride.* **s. diethyl** Diethyl selenide*. **s. dimethyl** Dimethyl selenide*. **s. dioxide*** See *selenium oxide.* **s. hydride** Hydrogen selenide*. **s. monobromide*** See *selenium bromide.* **s. monochloride*** See *selenium chloride.* **s. nitride** N_2Se_2 = 185.9. Yellow solid, explodes 200, insoluble in water. **s. oxide** SeO_2 = 111.0. S. dioxide*, selenious acid anhydride. Colorless crystals, sublimes 310, soluble in water. **s. oxides** Organic compounds that contain the seleninyl* radical =SeO; as, Me_2SeO. **s. oxychloride** Seleninyl chloride*. **s. sulfide** (1) SeS = 111.0. Yellow solid, m.118, insoluble in water. (2) SeS_2 = 143.1. S. disulfide*. Orange powder. Used to treat dandruff (USP, BP).
seleno- (1)* Prefix indicating the divalent atom =Se. Cf. *sulfo-.* (2) Prefix indicating an organic compound containing selenium in place of oxygen. **s.naphthene** Benzoselenofuran.
selenocyanate* The radical NCSe−.
selenofuran Se·CH:CH·CH:CH = 131.0. Selenophene. Colorless liquid, b.110, insoluble in water; resembles thiophene and burns in air with a blue flame, forming selenium.
selenoid Solenoid. A hollow cylinder, wound with resistance wire, used to produce fields of electric force.
-selenol* Suffix indicating the −SeH group.
selenole Selenofuran.
selenomercaptan R·SeH. See *hydroseleno* and *selenol.*
-selenonic acid* Suffix indicating the radical $-SeO_3H$, analogous to sulfonic acid. Cf. *seleninic acid.*
selenone* Compound containing the group $=SeO_2$.
selenonium* The cation SeH_3^+.
selenono-* Prefix indicating the radical HO_3Se-. Cf. *selenonic acid, selenino.*
selenonyl-* Prefix indicating a selenone containing the radical $-SeO_2-$. Cf. *sulfonyl.*
selenophene Selenofuran.
selenophenol C_6H_6Se = 157.1. PhSeH. Colorless liquid, b.183.
selenophthalide $C_6H_4 \cdot CO \cdot Se \cdot CH_2$ = 197.1. Colorless crystals, m.58.
selenopyronine Selenoxanthene.
selenotungstate A salt containing the green radical $=WSe_4$. **di∼** A salt containing the red radical $=WSe_2O_2$.
selenourea* $NH_2 \cdot CSe \cdot NH_2$ = 123.0. m.200 (decomp.). Air- and light-sensitive.
selenous Selenious.
selenoxanthone 9-Oxoselenoxanthene.
selenoxide* The group =SeO.
selenuretted A substance impregnated or combined with hydrogen selenide.
selenyl The seleninyl* radical.

self-inductance*, self-induction See *self-inductance* under *inductance.*
selinene $C_{15}H_{24}$ = 204.4. A sesquiterpene. Oils. α-~ b.270. β-D-~ From celery seed oil. b_{16mm}135, soluble in alcohol.
sellaite MgF_2. A native magnesium fluoride.
selwynite Yellow *ocher.*
semecarpus Marking nut, Oriental cashew nut, acajou nut. The fruit of *S. anacardium* (Anacardiaceae), southeast Asia; a black stain.
semen The fecundating fluid of the male containing sperm. See *Florence test.*
semi- Prefix (Latin) indicating "half;" *hemi-* (Greek); *demi* (French); *halb-* (German).
semicarbazide* $NH_2 \cdot NH \cdot CO \cdot NH_2$ = 75.1. Hydrazine carboxamide†, aminourea, carbamoylhydrazine. An amide and hydrazide of carbonic acid. Colorless prisms, m.96, soluble in water; a reagent for aldehydes and ketones. **4-amino ~** Carbonohydrazide*. **s. hydrochloride** $CH_5ON_3 \cdot HCl$ = 111.5. Aminourea hydrochloride. Colorless prisms, m.175, soluble in water; a reagent for aldehydes and ketones.
semicarbazido* 2-(Aminocarbonyl)hydrazino†. The group $NH_2 \cdot CO \cdot NH \cdot NH \cdot -$.
semicarbazino The semicarbazono* radical.
semicarbazone* $R_2C{:}N \cdot NH \cdot CO \cdot NH_2$, where one R can be a H atom. A condensation product of aldehydes or ketones and semicarbazide.
semicarbazono-* (Aminocarbonyl)hydrazono†, semicarbazino. Prefix indicating the radical $=N \cdot NH \cdot CO \cdot NH_2$.
semicoke Fuel made from coal by low-temperature carbonization at 594; smokeless, with little ash.
semiconductor A substance with electrical characteristics that enable it to act as an amplifier (in transistors), rectifier, thermistor, or switch. Button-size semiconductors are made from germanium or silicon (plus added impurities). **cadmium mercury telluride ~** Used for infrared sensors. **n-type ~** A s. that conducts electricity by means of "spare" electrons. **p-type ~** A s. that conducts electricity by means of electron "holes."
semidines $R \cdot C_6H_4 \cdot NH \cdot C_6H_4 \cdot NH_2$. Aromatic amines, *ortho* or *para* according to the position of the NH_2 group. **s. rearrangement** A special type of benzidine rearrangement, in which only half the molecule rotates; as, $R \cdot C_6H_4 \cdot NH \cdot NH \cdot C_6H_4 \cdot R \rightarrow R \cdot C_6H_4 \cdot NH \cdot C_6H_3RNH_3$.
semidrying oils Fatty oils that thicken slowly on exposure to light and air.
semimetallic* Describing an element midway in properties between metals and nonmetals, as, arsenic. Cf. *metallic, nonmetallic.*
seminose D-Mannose.
semiopal A native silica.
semipermeable Permitting the passage of certain molecules, and hindering others. **s. membrane** A diaphragm through which certain substances pass, while others are retained; as, a cell membrane. See *osmosis.*
semipervine $C_{19}H_{16}N_2$ = 272.3. A brown alkaloid from Carolina jasmine, *Gelsemium sempervireus.*
semiprecious Describing a decorative gem or metal which is inferior to the precious grades.
semisilica brick A firebrick made from a siliceous clay, or a mixture of fireclay and ganister (80–92% silica).
semisolid Soft and slowly flowing; as, asphalt.
semivalence A monoelectronic link between 2 rigid systems, characteristic of unstable, intermediate addition compounds and less stable than the ordinary nonpolar bond (bielectronic link).
Semmler, Friedrich Wilhelm (1860–1931) German organic chemist noted for work on the essential oils.
senarmontite Sb_2O_3. A native antimony trioxide.
seneca oil Petroleum.
senecifolidine $C_{18}H_{25}NO_7$ = 367.4. An alkaloid from *Senecio latifolius.* Rhombic plates, m.212.
senecifoline $C_{18}H_{27}NO_8$ = 385.4. An alkaloid from *Senecio latifolius* (Compositae), S. Africa. Colorless plates, m.194, soluble in ether. **s. hydrochloride** $C_{18}H_{27}NO_8 \cdot HCl$ = 421.9. White crystals, m.260, soluble in water.
senecine An amorphous alkaloid from *Senecio vulgaris.*
senecio Liferoot, ragwort, squaw-weed. The dried herb of *Senecio aureus.* **s. alkaloids** Alkaloids from s.; as, jacobine, necine, retronecine, retrorsine, senecifoline, senecifolidine.
senecioic acid $Me_2C{:}CH \cdot COOH$ = 100.1. 3-Methyl-2-butenoic acid*.
senega Senega snakeroot. Rattlesnake root. The dried root of *Polygala senega* (Polygalaceae); contains senegin and polygalic acid (BP).
senegenin $C_{30}H_{45}O_6Cl$ = 537.1. Senegeninic acid. A dibasic acid-hydrolysis product of senegin. Colorless powder, m.290.
senegeninic acid Senegenin.
senegin (1) $C_{32}H_{52}O_{17}$ = 708.8. A saponin derived from senega. (2) $C_{20}H_{32}O_7$ = 384.5. A hydrolysis product of (1). Cf. *senegenin.*
seneski A natural coke from the intrusion of igneous basaltic rock into a coal seam (20% ash); used to produce water gas.
seniority The priority, in a prescribed order, in which names are used. See *nomenclature.*
senna The dried leaflets of *Cassia acutifolia,* Alexandria s.; or *C. angustifolia,* India s., Tinnevelly s. (Leguminosae); contains glucosides, acids, and resins; a cathartic (USP, EP, BP). **American ~** The leaves of *C. marilandica.*
sensibilizer An agent that renders an enzyme active.
sensitive (1) Responding readily to a test or force. (2) See *anaphylaxis.*
sensitiveness, sensitivity (1) The degree of accuracy of a test or instrument. (2) The speed with which light acts on a photographic plate. (3) The property of exploding by mechanical shock.
sensitization (1) Biochemistry: Rendering a cell sensitive to the action of a complement by treating it with a specific antibody. (2) The process by which a person is sensitized to a certain antigen by a second exposure, as with pollens and hay fever. (3) Photography: (*a*) coating a surface with light-sensitive emulsions; as silver salts; (*b*) rendering the photographic emulsion more sensitive by addition of dyes which absorb certain portions of the spectrum. (4) Treatment of paper with chemicals so that ink writing cannot be eradicated without producing a telltale stain.
sensitizer (1) In chemistry, a species other than a catalyst which induces changes in another species; for instance, the transfer of energy from an excited molecule (sensitizer) to a different species that subsequently emits the excess energy as radiation. (2) Biology: A specific substance that occurs in small quantities in serum, and in larger quantities during immunization. (3) Photography. See *photosensitizer.*
sensor A device that indicates a change in physical or chemical conditions (as, of a gas mixture) in a manner that can be translated into data readings; as, a thermistor.
separator A device or machine for separating materials of different densities by the aid of air or water. See *centrifuge.*
separatory funnel Separating funnel. A tap funnel or device for separating 2 immiscible liquids.

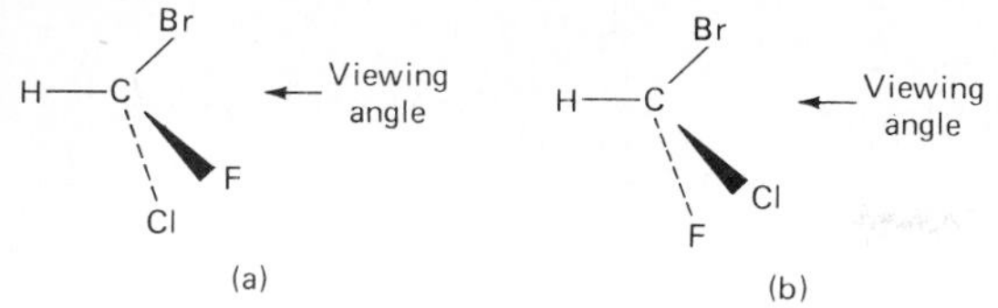

Fig. 19. Enantiomers: (*a*) rectus, or (*R*), and (*b*) sinister, or (*S*), according to the sequence rule.

Sephadex Trademark for a hydrophilic, insoluble, molecular-sieve chromatographic medium, made by cross-linking dextran.
sepia The dried, inky juice of a cuttlefish or squid; a dye.
sepiolite $2MgO \cdot 3SiO_2 \cdot 4H_2O$. A very absorptive, native magnesium silicate; similar to meerschaum.
sepsine $C_5H_{14}N_2O_2$ = 134.2. A ptomaine from decaying yeast.
sepsis Poisoning produced by microorganisms or putrefaction. Cf. *asepsis.*
septa-, septi- Hepta-.
septic Pertaining to putrefaction.
septicemia Blood poisoning. A morbid condition caused by the multiplication of pathogenic bacteria in the blood.
Septra, Septrin Trademarks for co-trimoxazole.
sequence rule* The procedure for determining the absolute molecular chirality (handedness) of a compound; i.e., it specifies in which of two enantiomeric forms (denoted *R* and *S*) each chiral element of a molecule exists. The three types of chiral element are chiral center (as, an asymmetric C atom); chiral plane (as, the benzene ring plane); chiral axis (as in the chiral allenes, $CH_2{=}C{=}CH_2$). The s. r. arranges atoms and groups in an order of precedence (also: "preference"). In the simplest of cases, i.e., a C atom with 4 different tetrahedrally arranged ligands (see Fig. 19), the s. r. gives the precedence order: Br > Cl > F > H. Viewing the molecule from the side remote from the least preferred ligand (H), the path from the most preferred (Br) to Cl to F is clockwise and is symbolized *R* (Latin *rectus:* "right"); thus, (*R*)-bromochlorofluoromethane. Equally, D-glyceraldehyde is (*R*)-glyceraldehyde. (Carbohydrates, amino acids, peptides, and cyclitols may use D and L terminology.) Similarly the other enantiomer is symbolized *S* (Latin *sinister:* "left"). Other cases are more complex. *R** and *S** indicate chiral centers where the relative, but not absolute, configuration of the chiral centers is known. Order of precedence of some common groups:

$$I > Br > Cl > HSO_3 > F > OH > NO_2 > NH_2 > COOH > CHO > Ph > pentyl > Bu > Pr > Et > Me > H$$

For a full summary of the s. r. see *Nomenclature of Organic Chemistry,* IUPAC, Pergamon Press. See *stereoisomer.*
sequestering The removal of a metal ion from a system by forming a complex ion which does not have the chemical reactions of the ion removed; e.g., the removal of Ca^{2+} ions from water by means of Graham's salt. **s. agent** A substance added to a system to preclude the normal ionic effects of the metals present.
sequestration Chelation, complexing. The reversible reaction of a metallic ion with a molecule or ion to form a complex molecule which does not have all or most of the characteristics of the original metallic ion.
sequestric acid Ethylenediaminetetraacetic acid*.
sequiatannic acid Sequoia tannin.
sequoia tannin $C_{21}H_{20}O_{10}$ = 432.4. A tannin from the cones of *Sequoia gigantea,* the mammoth tree of California. Brown powder, soluble in water.
Ser* Symbol for serine.
Seraceta Trademark for an acetate synthetic fiber.
seralbumin The albumin of the blood.
Serenace Trademark for haloperidol.
serenoa Sabal.
serge blue Methylene blue.
sericin $C_{15}H_{25}O_3N_5$ = 323.4. Silk gelatin, silk glue. An amorphous substance from silk, q.v.
sericite A flaky muscovite, causing silicosis.
series A succession of compounds, objects, or numbers, arranged systematically according to a rule. See *progression, hydrocarbon series.* **aliphatic ~** See *aliphatic.* **alkane ~** See *alkanes.* **alkene ~** See *olefin.* **alkyne ~** See *alkynes.* **analogous ~** See *analogs.* **aromatic ~** See *aromatic.* **Balmer ~** The hydrogen lines, H_α, H_β, H_γ . . . , which correspond with an electron transition from superior orbits to the second orbit. Cf. *hydrogen, Bohr theory.* **benzene ~** See *benzene series.* **chemical ~** See *series of compounds.* **diffuse ~** The spectrum lines resulting from transition from the *p* state to the *d* state. **displacement ~** Electromotive series. **electrical ~** (1) See *electromotive force.* (2) See *series of cells.* **ethylene ~** See *olefin.* **fatty ~** See *alkanes.* **fuzzy ~** The spectrum lines caused by transit from the outermost *orbit,* q.v. **galvanic ~** See *galvanic battery.* **geologic ~** See *geologic eras.* **homologous ~** Compounds differing by a definite radical or atomic group; as, CH_2. **homotopic ~** The elements in a group or family of the periodic table. **hydrocarbon ~** See *hydrocarbon series.* **isologous ~** See *isologous series.* **isotopic ~** The isotopes, q.v., of an element. **isomorphous ~** See *isomorphism.* **isosteric ~** See *isostere.* **K ~** See K *radiation.* **Lyman ~** The hydrogen lines in the ultraviolet spectrum, due to transition of electrons from superior orbits to the first orbit. Cf. *energy levels.* **methane ~** Alkanes*. **Paschen ~** The hydrogen lines in the infrared spectrum, due to transition of electrons from superior orbits to the third orbit. Cf. *energy levels.* **periodic ~** See *periodic system.* **principal ~** The spectrum lines caused by transition from the *p* state to the lowest or *s* state. Cf. *Rydberg's formula.* **radioactive ~** See *radioactive elements.* **sharp ~** The spectrum lines produced by an electron transition from the *p* state to the *s* state.
s. of cells Electric cells arranged in s. with the anode of one connected to the cathode of another. Cf. *parallel.* **s. of compounds** Compounds of an element whose valency is the same throughout: e.g., iron forms the ferrous and ferric s. of salts. **s. of lines** See *Balmer, Lyman, Paschen series* (all above). **s. notation** See *quantum number.*
serine* $CH_2OH \cdot CHNH_2 \cdot COOH$ = 105.1. Ser*. 2-Amino-3-hydroxypropanoic acid*. Hydroxylalanine. Colorless crystals from sericin and horn. A constituent of many proteins, m.246 (decomp.), insoluble in alcohol.
seriplane test A test for the evenness of yarn. The sample is wound on an inspection board in uniformly spaced panels, and assessed by comparison with photographs of standard yarns similarly wound.
Serogan Serum gonadotrophin. Trademark for the follicle-stimulating sex hormone from serum of pregnant mares.
serology The study of reactions in or of serum.
seroreaction A reaction that occurs in a serum as a result of immunization.
serotonin $C_{10}H_{12}ON_2$ = 176.2. 5-HT. 5-Hydroxytryptamine. A protein in human blood; possibly a neurotransmitter.
serpentaria Virginia snakeroot. The dried rhizome and roots of *Aristolochia serpentaria* (Aristolochiaceae).

serpentine $Mg_3Si_2O_7 \cdot 2H_2O$. Green, massive or lamellar oxides in rocks, often containing ferrous masses. Cf. *ophiolite.*

serum (Pl. sera.) (1) The clear, liquid portion of a body fluid. (2) The clear, amber, alkaline fluid of the blood from which the cellular elements and fibrinogen have been removed by clotting. It contains the salts, soluble proteins, and carbohydrates; used in biochemical and therapeutic work. See *immunity.* **milk ~** Whey.
s. albumin A protein, mol. wt. 67,000, from s. **s. globulin** A protein from blood serum.

servo A general term describing mechanisms which control automatically a changing physical condition, e.g., temperature. **s. mechanism** An amplifying actuator mechanism, whose output is compared with its input in order that the difference between the two be maintained as close to null as possible. The compensation necessary is a measure of changes in the input; e.g., speed-control systems. **s. recorder** The compensation required to maintain a null between the input and output is used to deflect a pen on the recorder paper, the extent of the deflection being proportional to the change in input signal.

sesame oil Ben(n)e, gingelly, simsim, til, or ufuta oil. The oil extracted from the seeds of *Sesamum indicum* (Pedaliaceae), $d_{25°}$ 0.918; an olive oil substitute. Used in pharmacy in preparation of drugs, and as a food (NF, EP, BP). Cf. *Villavecchia test.*

sesamin $C_{20}H_{18}O_6$ = 354.4. D- ~ In sesame oil. An aromatic ether. Colorless crystals, m.123, slightly soluble in alcohol.

sesamol $HO \cdot C_6H_3 \cdot O \cdot CH_2 \cdot O$ = 138.1. A hydrolysis product of sesamolin; responsible for the Baudouin color test.

sesamolin $C_{20}H_{18}O_7$ = 370.4. A substance in sesame oil, m.94; hydrolyzes to sesamol and samin.

sesqui- Prefix (Latin) indicating 1½, or proportion 3:2.

sesquicarbonate A compound of carbonic acid and a base in the proportion 3:2. **s. of soda** $NaHCO_3 \cdot Na_2CO_3 \cdot 2H_2O$. Sodium s. Snowflake crystals; used as a neutralizing agent, and in the manufacture of soap, glass, and cleansers.

sesquichloride A compound of chlorine and metal in the proportion 3:2; as, Fe_2Cl_3.

sesquioxide A compound of oxygen and a metal in the proportion 3:2; as, Al_2O_3, Fe_2O_3.

sesquisalt A compound of an acid and base in the proportion 3:2; as, $Fe_2(SO_4)_3$.

sesquisoda A molecular mixture of $NaHCO_3$ and Na_2CO_3.

sesquiterpenes $C_{15}H_{24}$. Terpenes formed by the theoretical polymerization of 3 isoprene units; as, cadinene, clovene, santalene.

Setilose Trademark for an acetate synthetic fiber.

setoff Transfer of ink from a printed to an unprinted surface by direct contact, due to slow drying of the ink. Cf. *offset printing.*

set point The value at which an instrument is required to control.

setting The hardening of semiliquid mixtures on crystallization (as cement) or organic condensation (as polymers).

settling The precipitation of insoluble materials from suspension in a liquid, and their gradual sinking by gravitation. **hindered ~** S. prevented by some factor other than specific gravity. **rate of ~** The velocity of fall of particles in a liquid. If the particle is large and causes eddies, velocity = $k\sqrt{d(S - s)/s}$, where d is the diameter of the particle, S and s are the specific gravity of solid and liquid, respectively, and k is a constant (9.3 for spheres, 9.0 for irregular particles). Otherwise, Stokes law applies.

sewer gas The gases from decomposition of sewage. See *biogas.*

sexiphenyl $Ph(C_6H_4)_4Ph$ = 458.6. Hexaphenyl, a hydrocarbon chain of 6 benzene rings. 9 isomers; solids, m.146–465.

sexivalent Hexavalent.

Sextol Trademark for methylcyclohexanol.

seybertite A complex, hydrated, native iron, calcium, aluminum silicate.

sfax A variety of *esparto,* q.v., from near Sfax, N. Africa.

shadowgram, shadowgraph X-ray photograph.

shadowing A process to give an electron photomicrograph a 3-dimensional appearance by depositing on the specimen an opaque substance; e.g., metal is evaporated from an electrically heated filament at such a distance from the specimen that the metal ions reach it in almost parallel straight lines.

shale A fine-grained sedimentary rock, with splintery uneven fractures. Cf. *slate.* **alum ~** See *alum shale.* **Esthonian ~** Kukersite. **oil ~** See *oil shale.*
s. naphtha A petroleum from shale. **s. oil** A crude oil from bituminous shales by destructive distillation; chief constituent, *kerogen.* Cf. *oil shale.* **s. spirit** The lower-boiling fractions from distilling oil.

shallu Durra.

shearing Side cutting or a lateral motion; as in grinding. Cf. *crushing.*

sheave A grooved pulley wheel.

sheep The herbiverous mammal *Ovis aries,* providing meat, wool, leather, and endocrine glands. **s. laurel** Kalmia. **s. oil** Lanolin. **s. sorrel** The dried herb of *Rumex acetosella.*

sheerness The combined qualities of transparency, surface gloss, and smoothness, as of nylon hose.

Sheffield plate Copper with a fused-on layer of sheet silver, rolled out and worked into articles of desired form; displaced by electroplating (1837).

shell (1) The husk of a fruit. (2) The calcareous or siliceous covering of marine invertebrates. (3) A projectile filled with explosives. (4) An energy level of an atomic nucleus containing electrons, itself divided into subshells, *s, p, d, f,* etc. See *atomic structure.* **s. lime** A fertilizer made by grinding mollusks, containing 90% calcium (or magnesium) carbonate. **s. marl** A fertilizer made by grinding natural deposits of shells, containing not less than 80% calcium (or magnesium) carbonate.

shellac The purified resin *lac,* q.v., obtained from plants by the incisions of an insect, *Laccifer lacca (Coccus lacca).* Brown leaflets, insoluble in water; used in varnishes, polishing materials, sealing wax, and pyrotechnics. It contains aleuritic acid 30, resin acid mixture 35–38%. **earth ~** Acaroid resin.

shelloic acid An acidic constituent of shellac, in which it occurs to the extent of less than 1%.

shellolic acid $C_{15}H_{20}O_6$ = 296.3. 10-Hydroxyshellene-1,12-dicarboxylic acid. Colorless crystals, m.200, soluble in hot water.

shepherd's purse The freshly gathered green herb of *Capsella bursa-pastoris;* a stimulant.

sherardize To galvanize articles by covering them with zinc dust and heating in a tightly closed retort. Cf. *calorizing.*

sherbet Sorbet. (1) An effervescing drink, sold in powder form, consisting of sugar, sodium hydrogencarbonate, tartaric acid, and flavoring materials. (2) A frozen fruit juice used as a dessert.

shibuol $C_{14}H_{20}O_9$ = 332.3. A phenol from kaki shibu, the unripe kaki fruit of Japan.

shift (1) A slight change in the wavelength of a spectral line caused by (*a*) density (cf. *pressure*), (*b*) mass (cf. *relativity*), (*c*)

TABLE 78. BASE SI UNITS

Quantity	Name	Symbol
length	meter	m
mass	kilogram	kg
time	second	s
electric current	ampere	A
thermodynamic temperature	kelvin	K
amount of substance	mole	mol
luminous intensity	candela	cd

motion (cf. *Doppler* principle), (*d*) absorption (cf. *Compton effect*), (*e*) scattering (cf. *Raman effect*), (*f*) emission (cf. *luminescence*). (2) A change of workers. **antistokes ~** A change to higher energy. **stokes ~** A change to lower energy.

shikimol Safrole*.

shikonin $C_{16}H_{16}O_5$ = 288.3. **(*R*)-~** A principle from shikon, the dried root of *Lithospermum erythrorhizon* (Boraginaceae), Japan.

shilajatu A mineral gum from India.

shirlacrol A solution of phenolic tars and sodium hydroxide; used as a textile industry wetting agent.

Shirlan Trademark for a mildew preventive for textiles; 30 times as powerful as zinc chloride.

Shirlastain Trademark for a stain evolved by the *Shirley Institute*, q.v., for distinguishing various textile fabrics.

Shirley Institute Headquarters of the U.K. Cotton, Silk, and Man-Made Fibres Research Association, Manchester.

shock (1) A violent collision between bodies. (2) The concussion a collision occasions. (3) The effect of an electric discharge on the animal body.

shoddy Wool waste recovered for reuse from knitted fabrics; a better grade than mungo.

shogaol $C_{17}H_{24}O_3$ = 276.4. 1-(4-Hydroxy-3-methoxyphenyl)-4-decen-3-one*. A pungent constituent of ginger, resembling zingerone in having OH, OMe, and CO groups. Colorless liquid, b_{15mm}235, soluble in water.

Shore hardness The height of rebound of a diamond-pointed hammer falling under gravity on an object; a high-carbon steel is taken as 100.

short circuit An electric current that passes directly between leads which touch at a point between the source of current and its destination.

shortite $Na_2CO_3 \cdot 2CaCO_3$. A pyroelectric, crystalline mineral from Wyoming.

Shotcrete (U.S. usage) Trademark for a spray-applied concrete. Cf. *Gunite.*

shotgun pattern Irregular points on a graph which do not coincide with a theoretical curve.

shunt An alterable resistance in parallel with a galvanometer; used to control the current passing through it by diverting or "shunting."

SI Abbreviation for Système International d'Unités, an extension and refinement of the metric system accepted by most countries as the only international system. Main features are 7 base units, the meter and kilogram in place of the centimeter and gram; the unit of force, the newton is independent of the earth's gravitation; introduction of g into equations is unnecessary; the unit of all forms of energy is the joule (newton × meter), and of power the joule per second (watt) in place of calories, kilowatthour, Btu, and horsepower; electrostatic and electromagnetic units are replaced by SI electrical units; multiples of units are mainly restricted to steps of a thousand or a thousandth. **SI units** See Tables 78 to 82 on the next four pages (from *Quantities, Units, and Symbols,* The Royal Society, London).

Si Symbol for silicon.

sial (Derived from Si and Al.) A hypothetical solid or semisolid rock substance on which the land masses of the earth are assumed to be supported. Cf. *sima.*

sialagogue An agent that increases the flow of saliva; as, lemon juice.

sairesinolic acid $C_{29}H_{47}O_2 \cdot COOH$ = 472.7. A resin acid, from Thai benzoin gum, m.274.

sib(ling)s Progeny having one or both parents in common.

sibucao Sappanwood.

siccative Drier. A solution of manganese or zinc salts of resin acids; a drying accelerator for varnish or paint.

TABLE 79. DERIVED SI UNITS WITH SPECIAL NAMES

Quantity	Unit	Symbol	Definition[a]
plane angle	radian	rad	dimensionless
solid angle	steradian	sr	dimensionless
energy	joule	J	$kg\,m^2\,s^{-2} = N\,m$
force	newton	N	$kg\,m\,s^{-2} = J\,m^{-1}$
pressure	pascal	Pa	$kg\,m^{-1}\,s^{-2}$, $N\,m^{-2}$, $J\,m^{-3}$
power	watt	W	$kg\,m^2\,s^{-3} = J\,s^{-1}$
electric charge	coulomb	C	A s
electric potential difference	volt	V	$kg\,m^2\,s^{-3}\,A^{-1} = J\,A^{-1}\,s^{-1}$, JC^{-1}
electric resistance	ohm	Ω	$kg\,m^2\,s^{-3}\,A^{-2} = V\,A^{-1}$
electric conductance	siemens	S	$kg^{-1}\,m^{-2}\,s^3\,A^2$, Ω^{-1}, AV^{-1}
electric capacitance	farad	F	$A^2\,s^4\,kg^{-1}\,m^{-2} = A\,s\,V^{-1}$, CV^{-1}
magnetic flux	weber	Wb	$kg\,m^2\,s^{-2}\,A^{-1} = V\,s$
inductance	henry	H	$kg\,m^2\,s^{-2}\,A^{-2} = V\,s\,A^{-1}$
magnetic flux density	tesla	T	$kg\,s^{-2}\,A^{-1} = V\,s\,m^{-2}$, $Wb\,m^{-2}$
luminous flux	lumen	lm	cd sr
illuminance	lux	lx	$cd\,sr\,m^{-2}$
frequency	hertz	Hz	s^{-1}
activity (of a radioactive source)	becquerel	Bq	s^{-1}
absorbed dose (of ionizing radiation)	gray	Gy	$J\,kg^{-1}$
dose equivalent (of ionizing radiation)	sievert	Sv	$J\,kg^{-1}$

[a]Either a raised dot (U.S. standard) or space between the product of two units is acceptable.

TABLE 80. MULTIPLES OF SI UNITS

Multiple	Prefix	Symbol	Multiple	Prefix	Symbol
10^{-1}	deci	d	10	deca[a]	da
10^{-2}	centi	c	10^2	hecto	h
10^{-3}	milli	m	10^3	kilo	k
10^{-6}	micro	μ	10^6	mega	M
10^{-9}	nano	n	10^9	giga	G
10^{-12}	pico	p	10^{12}	tera	T
10^{-15}	femto	f	10^{15}	peta	P
10^{-18}	atto	a	10^{18}	exa	E

[a]U.S. standard uses deka.

side chain A group of 2 or more similar atoms, generally C atoms, that branch off from a ring of atoms or a longer chain of atoms: 2 types:

```
                CH3
                |
                CH2
                |
—CH2—CH2—CH—CH2—CH2—
```

Branched side chain

```
      CH
  HC/    \\C—CH2—CH3
  ||       |
  HC\     /CH
      CH//
```

side chain

s.-c. isomery The isomery of molecules that differ in structure by the arrangement of the side chain atoms; as:

$C_6H_5 \cdot CH_2 \cdot CH_2 \cdot CH_3$ — Propylbenzene

$C_6H_5 \cdot CH(CH_3)_2$ — Isopropylbenzene

s. c. substitution A reaction in which substitution takes place in the side chain of a molecule.

side cut A distillate obtained by fractional distillation.

siderazote Fe_5N_2. A volcanic incrustation.

side reaction A subsidiary reaction which occurs simultaneously with the main reaction.

sidereal Pertaining to the fixed stars. Cf. *solar.* **s. day** 86,164.09 seconds. **s. year** See *sidereal year* under *year.*

siderite (1) $FeCO_3$. Clay ironstone, chalybite, spathose, spathic iron ore, a native iron carbonate. (2) An iron meteorite. A body of metallic iron with nickel, cobalt, etc., from outer space.

siderocyte A blood cell containing iron.

siderography (1) The study of the natural surface condition of siderites. (2) The etching of steel and iron, and its microscopic study.

siderolite Mesoderite. A meteorite of spongy meteoric iron, with embedded grains of silicate minerals; as, olivite.

siderology Siderurgy.

sideroplesite A form of bruennerite.

siderosis A pulmonary disease due to inhalation of iron dust. Cf. *byssinosis.*

siderostat An instrument to transmit a beam of light along the optical axis of a fixed horizontal telescope.

siderotilate $FeSO_4 \cdot 5H_2O$. A native sulfate.

siderurgy Siderology. A branch of science that deals with the metallurgy of iron.

Sidgwick, Nevil Vincent (1873–1952) British physical chemist, noted for his electronic theory of valency.

Sidot's blends An artificial zinc sulfide, which contains traces of copper; used in fluorescent screens for X-rays or radioactive rays.

Siegbahn, Karl Manne Georg (1886–1978) Swedish physicist, Nobel prize winner (1924); noted for his work on crystal structure.

Siemens **S., Carl Friedrich von (1872–1941)** German industrialist. **S., Karl Wilhelm (Sir Charles William) (1823–1883)** German-born British chemist, inventor of the S. Process. **S. furnace** A reverberatory furnace, heated by gas. **S.-Halske process** A method of dissolving copper sulfides in a solution of ferrous sulfate and sulfuric acid, and obtaining metallic copper by electrolysis. **S.-Martin process** A method for producing steel in a reverberatory furnace by adding scrap iron to iron ores. **S. ozonizer** Two concentric glass tubes, the outer covered and the inner lined with tinfoil, which act as electrodes for a silent discharge passed through oxygen flowing between them. **S. process** A method for making wrought iron directly from iron ores. **S. producer** A furnace used to manufacture producer gas.

siemens Symbol: S; an SI-derived unit. The electric conductance of a conductor in which a current of 1 amp is produced by a potential difference of 1 volt. 1 S = 1 mho.

sienna Raw sienna. Brown-yellow clay; a permanent pigment. It contains hydrated ferric oxide and manganic oxide. **burnt ~** A burnt form of s., richer and brighter in color than raw s.

sieve An apparatus with square apertures, to separate particles according to size. Cf. *screen* (round holes), *fineness.* See also *molecular sieve.*

sievert* An SI-derived unit. 1 Sv is the dose equivalent when the absorbed dose of ionizing radiation multiplied by the dimensionless factors Q (quality factor) and N (product of any other multiplying factors) stipulated by the International Commission on Radiological Protection is 1 J/kg. 1 Sv = 100 rem.

sigma Σ, σ. Greek letter. Σ (not italic) is the symbol for (1) s. *particle;* (2) "the sum of," in mathematical expressions. σ (not italic), in chemical names, signifies that one atom of a group is attached to a metal by a s. bond. σ (italic) is the symbol for conductivity. **s. bond** See *bond.* **s. particle** See *particle.* **s. phenomenon** Anomalous *viscosity.* **s. reaction** Σ test: the Wassermann test for syphilis.

sigmatropic Describing a unimolecular, one-step rearrangement in which a sigma bond migrates to a new position.

signal-to-noise ratio Measure of the relative influence of noise on a control signal. Usually taken as the magnitude of the signal divided by the standard deviation of the background signal. The smallest signal that can be measured reliably equals 3 times the standard deviation of the background.

Sikes hydrometer See *Sikes hydrometer* under *hydrometer.*

sikimin $C_{10}H_{16}$ = 136.2. A terpene in the leaves of the sikimi plant *Illicium religiosum* (Magnoliaceae), Japan. Cf. star *anise.*

silage A fodder made of finely cut green plants packed tightly in tanks (silos) and fermented.

silal A heat-resisting iron containing 5% Si.

silane* (1) SiH_4 = 32.12. Monosilane, silicomethane, silicohydride. Colorless gas, b. −112, strong odor, decomp. in water. (2) See *silanes.* **bromo ~*** SiH_3Br = 111.0. Colorless

TABLE 81. TRADITIONAL UNITS WITH SI EQUIVALENTS

Quantity	Unit	Equivalent
length	angstrom	10^{-10} m
	inch	0.0254 m
	foot	0.3048 m
	yard	0.9144 m
	mile	1.609 34 km
	nautical mile	1.852 00 km
area	square inch	645.16 mm^2
	square foot	0.092 903 m^2
	square yard	0.836 127 m^2
	square mile	2.589 99 km^2
volume	cubic inch	$1.638\ 71 \times 10^{-5}$ m^3
	cubic foot	0.028 316 9 m^3
	U.S. gallon	0.003 785 412 m^3
	U.K. gallon	0.004 546 090 m^3
mass	pound	0.453 592 37 kg
density	pound/cubic inch	$2.767\ 99 \times 10^4$ kg m^{-3}
	pound/cubic foot	16.0185 kg m^{-3}
force	dyne	10^{-5} N
	poundal	0.138 255 N
	pound-force	4.448 22 N
	kilogramme-force	9.806 65 N
pressure	atmosphere (standard)	101.325 kPa
	torr (1 mmHg at 0°C)	133.322 Pa
	pound (f)/sq in. (psi)	6894.76 Pa
energy	erg	10^{-7} J
	calorie (I.T.)	4.1868 J
	calorie (15°C)	4.1858 J
	calorie (thermochemical)	4.184 J
	Btu (I.T.)	1055.06 J
	foot poundal	0.042 1401 J
	foot pound (f)	1.355 82 J
power	horsepower	745.70 W
temperature	degree Fahrenheit	(°F − 32) + 273.15 K

gas, b.1.8. **chloro ~*** SiH_3Cl = 66.6. Colorless gas, b.−30. **chloromethyl ~*** $MeClSiH_2$ = 80.4. A volatile liquid, decomp. by water to silica; used to make textiles water-repellent. **di ~*** Si_2H_6 = 62.2. Silicoethane, a gas, m.−132, b.−15. **dibromo ~*** SiH_2Br_2 = 189.9. Colorless liquid, d.2.17, b.66. **dichloro ~*** SiH_2Cl_2 = 101.0. Colorless gas, b.8.3. **dimethyl ~*** Me_2SiH_2 = 60.2. Colorless gas, b.−20. **ether ~** $(SiH_3)_2O$ = 78.2. Disilane oxide. Colorless gas, b.15. **ethoxytriethyl ~** Et_3SiOEt = 160.3. Triethylsilane ethyl oxide, triethyl silicol ethyl ether. Colorless liquid, b.153, insoluble in water. **hexafluorodi ~*** Si_2F_6 = 170.2. A gas, m.−19. **hydroxy ~** Silicol. **methyl ~*** $MeSiH_3$ = 46.14. Methylmonosilane. Colorless gas, b.−57. **tetra ~** Si_4H_{10} = 122.4. Silicobutane. Liquid, m.−88, b.107. **tetrabromo ~*** Silicon bromide (1). **tetrachloro ~*** Silicon chloride (1). **tetraethyl ~*** Et_4Si = 144.3. Silicon tetraethyl, siliconomane. Colorless liquid, d.0.7682, b.153. **tetrafluoro ~*** Silicon fluoride (1). **tetraiodo ~*** Silicon iodide (1). **tetramethyl ~*** Me_4Si = 88.2. Silicon tetramethyl. Colorless liquid, d.0.645, b.27. **tetraphenyl ~*** Ph_4Si = 336.5. Silicon tetraphenyl, tetraphenyl silicon. Colorless crystals, m. 233. **tri ~** Si_3H_8 = 92.3. Silicopropane. A gas, m.−117. **tribromo ~*** $SiHBr_3$ = 268.8. Silicobromoform. Colorless liquid, d.2.7, b.109. **trichloro ~*** Silicochloroform. **trichlorethyl ~*** $EtSiCl_3$ = 163.5. Colorless liquid, d.1.239. **trichlorophenyl ~*** $PhSiCl_3$ = 211.6. Colorless liquid, d.1.326, b.197, decomp. in water. **triethyl ~*** Et_3SiH = 116.3. Triethyl silicon, silicoheptane. Colorless liquid, d.0.751, b.107, insoluble in water. **trifluoro ~*** $SiHF_3$ = 86.1. Silicofluoroform. Colorless gas, b.−80. **triiodo ~*** SiI_3H = 409.8. Silicoiodoform. Red liquid, d.3.314, b.220.

s.diol A disubstituted chlorosilane of the type $R_2Si(OH)_2$. Silanediols condense to form chain or ring structures. **s.diyl*** Silylene†, silicylene. The radical $-SiH_2-$, from silane. **s.triol** A hydrolysis product of a monosubstituted chlorosilane of the type $R \cdot Si(OH)_3$. Silanetriols condense to form 3-dimensional polymeric resins. **s.triyl*** The radical RSi≡; as, methyl silanetriyl.

silanes* Silican(e)s, silicohydrides, hydrosilicons. The branched or unbranched silicon hydrides. Compounds similar to hydrocarbons, in which tetravalent Si replaces the C atom; as, SiH_4, silane. S. are very reactive, ignite in air, and form derivatives. See *silane*.

silanol Silicol. The trivalent group ≡SiOH.

Silastic Trademark for a heat-stable silicone.

silavans Group name for colorless, high-melting-point, strong polymers, containing silicon, carbon, and nitrogen.

silbamin Silver fluoride.

Silberrad, Oswald John (1878–1960) British chemist, noted for his work on explosives.

Silesia explosive A high explosive: potassium chlorate 75, nitrated resin 25%.

silex A heat- and shock-resistant glass (98% quartz). **liquid ~** Water glass.

Sil-Fos Trademark for an alloy, m.625–705: Cu 80, Ag 15, P 3%; used for brazing alloys containing copper.

silica SiO_2 = 60.1. Silicon dioxide*, silicic acid anhydride. Occurs abundantly in nature (12% of all rocks), and exists in

TABLE 82. UNITS ALLOWED IN CONJUNCTION WITH SI SYSTEM

Quantity	Unit	Symbol	Definition
angle	degree	°	$(\pi/180)$ rad
	minute	′	$(\pi/10{,}800)$ rad
	second	″	$(\pi/648{,}000)$ rad
area	barn[a]	b	10^{-28} m^2
	are[a]		10^2 m^2
concentration (amount of substance)	—	M	10^3 mol/m^3 = mol/dm^3
energy	electronvolt	eV	$1.602\ 1892 \times 10^{-19}$ J
	erg[a]	erg	10^{-7} J
	kilowatthour[a]	kWh	3.6 MJ
force	dyne[a]	dyn	10^{-5} N
illuminance	phot[a]	ph	10^4 lx
length	angstrom[a]	Å	10^{-10} m = 10^{-1} nm
	astronomical unit	AU	$149{,}597.9 \times 10^6$ m
	micron[a]	μm	10^{-6} m
	parsec	pc	30.857×10^{15} m
magnetic flux density	gamma[a]	γ	10^{-9} T
mass	ton	t	10^3 kg = Mg
	unified atomic mass	u	$1.660\ 5655 \times 10^{-27}$ kg
pressure	bar[a]	bar	10^5 Pa
radiation dose:			
exposure	roentgen	R[a]	2.58×10^{-4} C/kg
absorbed	rad	rd[a]	0.01 Gy
radioactivity	curie	Ci[a]	3.7×10^{10} Bq
temperature	degree Celsius	°C	K
time	minute	min	60 s
	hour	h	60 min = 3600 s
	day	d	24 h = 86,400 s
	year	a	see *year*
viscosity:			
dynamic	poise[a]	P	10^{-1} Pa s
kinematic	stokes[a]	St	10^{-4} m^2/s
volume	liter, litre	l, L	10^{-3} m^3 = dm^3

[a]Indicates units to be abandoned as quickly as possible.

6 crystalline forms. Classification: (1) Phenocrystalline or vitreous minerals; see *quartz, cristobalite.* (2) Cryptocrystalline and amorphous minerals; see *chalcedony.* (3) Amorphous and colloidal minerals; see *opal.* **amorphous ~** Colorless powder, m.1650, insoluble in water, soluble in hot alkalies or hydrofluoric acid; used for chemical glassware. **colloidal ~** See *colloidal silicon dioxide* under *silicon dioxide.* **crystalline ~** Colorless, transparent prisms, m.1760, insoluble in water, soluble in hydrofluoric acid. Used in optical instruments, kitchenware, and chemical plant. The main crystalline forms (quartz, tridymite, and cristobalite) have definite transition points (870 and 1470 °C, respectively).

s. brick A firebrick containing over 92% s.; its crystalline phase is cristobalite and tridymite. **s. gel** Gelatinous s. which, if activated, absorbs water. Used to dry blast-furnace gases, air, and other gases; also in pharmacy (NF). **s. minerals** Rock-forming minerals comprising the groups: amphiboles, andalusite, cancrinite, sodalite, chlorite, feldspar, garnet, iolite, leucite, melilite, mica, nephelite, olivine, pyroxene, scapolite, topaz, tourmaline, zeolite, zoisite; also beryl, quartz, serpentine, talc. **s. rock** Hard, compact, quartzitic sandstones and quartzite, used for refractories. **s. sand** A commercial source of silica produced from sand and weakly cemented sandstone deposits (Carboniferous onwards). Used for foundry molding and glass manufacture.

silicam $Si(NH)_2$ = 58.1. Silicon diimide. White powder, insoluble in water. Forms silicon nitride, Si_3N_4, when heated.

silicane See *silane, silanes.*

silicate* Indicating silicon as the principal atom(s) in an anion, as, a salt derived from silica or the silicic acids. Silicates form the largest group of minerals (see *silica*), and are derived from M_4SiO_4, *ortho*silicate*, and M_2SiO_3, *meta*silicate*, which may combine to form *poly*silicates. Except for the alkali silicates, they are insoluble in water. See *silica minerals.* **fibrous ~** **natural f. s.** Asbestos. **man-made f. s.** Glass, silica, and aluminosilicate fibers, rock wool, slag wool.

s. garden See *chemical garden.* **s. of soda** Sodium silicate.

siliceous Containing silica. **s. algae** See *siliceous alga* under *alga.* **s. deposit** S. sinter. The solid accumulation of silica deposited from hot mineral springs. Cf. *geyserite.* **s. earth** Silica of diatomite origin, purified by boiling with dilute acid, washing, and calcining; a filter medium and component of dusting powders (NF). **s. sinter** S. deposit.

silicic (1) Containing silicon. (2) Containing silicic acid. **s. acid** See Table 83. H_4SiO_4 = 96.1. Orthosilicic acid*. White powder, slightly soluble in water. **di ~*** $H_6Si_2O_7$. Pyro s. a. White, insoluble powder. **meta ~*** $(H_2SiO_3)_n$ = $(78.1)n$. Hypothetical acid corresponding to long-chain anions.

TABLE 83.
SILICIC ACIDS

$H_4SiO_4 = SiO_2 \cdot 2H_2O$, ortho~*
$(H_2SiO_3)_n = nSiO_2 \cdot nH_2O$, meta~*
$H_6Si_2O_7 = 2SiO_2 \cdot 3H_2O$, di~*, pyro~
$H_6Si_3O_{10} = 3SiO_2 \cdot 4H_2O$, tri~*
$H_{2n}Si_nO_{3n}$ = cyclic ~

tri ~* $H_6Si_3O_{10}$ = 250.3. White, insoluble powder. **tetrahydrogen decawolframo ~*** $SiO_2 \cdot 10WO_3 \cdot 2H_2O$ = 2415. Silico(deci)tungstic acid. White powder; a reagent for cesium (insoluble salts).

silicide Compounds of the type M_xSi_y, as, Mg_2Si, $CaSi_2$, Fe_3Si.

silicification The gradual replacement of rocks or fossils by silica. Cf. *petrifaction.*

silicified Describing an organic material, e.g., wood, that has been petrified.

silicium Silicon.

silico- Prefix indicating silicon, generally in organic compounds. **s.benzoic acid** PhSiOOH = 138.2, m.92, insoluble in water. **s.bromoform** $SiHBr_3$ = 268.8. Heavy, colorless liquid, d.2.7, b.116, decomp. by water. **s.butane** See *silanes.* **s.calcium** A product of the electric furnace used to deoxidize steel. **s.chloroform** $SiHCl_3$ = 135.5. Colorless liquid, d.1.34, b.34, decomp. by water. **s.decitungstic acid** Tetrahydrogen decawolframo*silicic acid.* **s.ethane** See *silanes.* **s.fluoride** Hexafluorosilicate*. **s.fluoric acid** Hexafluorosilic acid*. **s.heptane** Triethyl silane*. **s.hydrides** Silanes*. **s.iodoform** $SiHI_3$ = 409.8. Heavy, colorless liquid, d.3.4, b.220, decomp. by water. **s.methane** Silane*. **s.oxalic acid** HOOSi·SiOOH = 122.2. White, unstable solid.

silicol R_3SiOH. Hydroxysilane. **triethyl ~** Et_3SiOH = 132.3. Silicoheptyl alcohol. Colorless liquid, b.154, insoluble in water.

silicon* Si = 28.0855. Silicium. A nonmetallic element of the carbon group, at. no. 14. Allotropic modifications: (1) *Amorphous:* Brown powder, d.2.35. (2) *Crystalline:* Gray crystals, m.1412, b. ca. 2480, insoluble in water. (3) *Graphitoidal:* Dense crystals, or graphitelike masses deposited from molten s. (4) *Adamantine:* Hard needles. Principal valency 4. S. forms many complex compounds on the earth surface (rocks). Used in alloys to impart hardness, and in semiconductors. See *silica minerals.* **ethyl ~** The radical ≡SiEt. Cr. *silanes.* **methyl ~** The radical ≡SiMe. **radio ~** A s. isotope, mass 27. Cf. *radioelements.*

s. alkyls (1) Hydrogen compounds of s. corresponding with hydrocarbons; as, SiH_4, silane. (2) Organic compounds of s. and alkyl radicals; as, Me_4Si. See *silanes.* **s. alloys** Noncorrodible alloys of s. with metals; as, Duriron. Cf. *silicon copper.* **s. borides** SiB_3, SiB_4, and SiB_6 exist. Black, irregular crystals, of high m.; very hard, and good conductors of electricity. **s. bromides** (1) $SiBr_4$ = 347.7. S. tetrabromide*. Colorless, fuming liquid, b.154, decomp. by water to silicic acid. (2) Si_2Br_6 = 535.6. S. tribromide*. Colorless solid, b.240, decomp. by water. **s. bronze** A noncorrodible alloy: Cu, Sn, with 1–4% Si. **s. carbide*** SiC = 40.10. Colorless plates, dissociates 2250; used in refractories and abrasives. **s. chip** A wafer of pure s. printed with alternate insulating and semiconducting layers, on which the pattern of an electric circuit is etched. Wafers fused together can contain thousands of circuits. **s. chlorides** (1) $SiCl_4$ = 169.9. S. tetrachloride*. Colorless, fuming liquid, d.1.524, b.58, decomp. by water to silicic acid. Used in electrotechnics, and mixed with ammonia vapors, in smoke screens. (2) Si_2Cl_6 = 268.9. S. trichloride, b.146, decomp. by water. (3) Si_3Cl_8 = 367.9. S. octachloride*. White powder. **s. controlled rectifier** SCR. Thyristor. A fast-acting switching device made from 4 alternate layers of n- and p-type silicon. **s. copper** An alloy: Si 20–30, Cu 70–80%, used in metallurgy. **s. dioxide*** Silica. **colloidal ~** Used in pharmacy as a suspending agent and stabilizer (NF). **s. disulfide*** SiS_2 = 92.2. White needles, sublime when heated, decomp. by water. **s. ethane** See *silanes.* **s. ethyl** Tetraethyl*silane**. **s. fluorides** (1) SiF_4 = 104.1. S. Tetrafluoride*. Colorless, suffocating gas, b_{1810mm} −65, decomp. by water to hexafluorosilicic acid, soluble in alcohol. (2) Si_2F_6 = 170.2. S. subfluoride. White powder. **s. hydrides** Silanes*. **s. iodides** (1) SiI_4 = 535.7. S. tetraiodide*. Colorless solid, m.121, insoluble in water. (2) Si_2I_6 = 817.6. S. subiodide. Colorless solid, m.250 (in vacuo), decomp. by water. **s. iron** Ferrosilicon. Iron contianing 2–15% Si; used in metallurgy. **s. magnesium** See *magnesium silicides.* **s. methane** Silane*. **s. methyl** Tetramethyl*silane**. **s. nitride** Si_3N_4 = 140.3. White powder insoluble in water, existing in 2 hexagonal phases stable below and above 1400–1450°C, respectively. Very resistant to thermal shock and chemical reagents; used as a support for catalysts and in stator blades of high-temperature gas turbines. **s. octachloride*** See *silicon chlorides.* **s. oxide** Silica. **s. oxychlorides** Si_2OCl_6, b.137; $Si_4O_4Cl_8$, b.200; $Si_4O_3Cl_{10}$, b.153; also $(SiOCl_2)_n \cdot O(SiCl_3)_2$, where n = 1 to 4. **s. steel** Steel containing 2–3% Si; hard and brittle. **s. sulfide** S. disulfide*. **s. tetrabromide*** See *silicon bromides.* **s. tetrachloride*** See *silicon chlorides.* **s. tetrafluoride*** See *silicon fluorides.* **s. tetraiodide*** See *silicon iodides.* **s. tetraphenyl** Tetraphenyl*silane**. **s. tungstic acid** Silicotungstic acid. **s. zirconium** An alloy used to purify molten steel.

silicone (1) Contraction of silicoketone. A polymer containing −Si(R_2)O− groups. Lower molecular weight compounds are oils (used as lubricants and in polishes); higher are inert solids with good electrical insulation properties. (2) $H_3Si_3O_2$ = 119.3. Yellow solid. **s. alloy** A compound produced by the simultaneous polymerization of 2 silicones; e.g., tetravinyl s. and methyl hydrogen siloxane give a s. alloy of high water repellency. **s. release paper** Protective backing paper that is easily removed when required, as on self-adhesive labels. **s. rubber** A s. that retains its elastic properties between −50 and +291, and can be kneaded; used for protective coatings on wires and for high-temperature lubricants.

siliconic acid R·SiOOH, analogous to organic acids. Cf. *carbylic acid.*

silicono The radical (HO)OSi−, derived from metasilicic acid.

Silicool Trademark for a protein synthetic fiber.

silicosis A form of pneumoconiosis due to silica dust less than 10 μm in diameter. U.K. limit is 0.1 mg/m^3 of respirable air.

silicotungstate A salt of silicotungstic acid, especially with the alkaloids.

silicotungstic acid $H_4[SiW_{12}O_{40}]$ = 2878. Tetrahydrogen dodecawolframosilicate*. Dodecawolframosilicic acid*. Yellow crystals, soluble in water; used in alkaloid analysis.

silicyl The silyl* radical. **di ~** The disilanyl* radical.

s. oxide $(R_3Si)_2O$; as **hexaethyl ~** $(Et_3Si)_2O$ = 246.5. Colorless liquid, b.231.

silicylene The silanediyl* radical.

silk (1) Fibroin, sericin. The fibrous envelope of the silkworm before the chrysalis state (cocoon). It consists of fibroin (the fiber protein) and sericin (the gummy protein). (2) A sieve for grading flour: no. 5 = 0.270, no. 8 = 0.190 mm aperture. (3) A series of parallel fine-line inclusions in certain gems (e.g., rubies). Cf. *asterism.* **"all- ~"** S. containing fillers, but no other fibers. **artificial ~** Rayon. **net ~** S. fabric made from yarns of continuous s. filament. **pure ~** S. fibers without fillers. **schappe ~, spun ~** Describing a fabric made from silk-waste staple fiber. **vegetable ~** (1) The floss from the seeds of *Calotropis gigantea* (Asclepiadaceae), Asia. (2) Kapok.

silk warp Organzine.
Silliman, Benjamin (1779–1864) American chemist and geologist, who founded the American Journal of Science and Arts.
sillimanite Al_2SiO_5. (1) Rhombic aluminum silicate, m.1820; used for porcelain. Above 1545 it forms mullite and siliceous glass. See *cyanite.* (2) Fibrolite. **s. ware** Laboratory utensils resistant to mechanical and thermal shock.
Sillman bronze An alloy: Cu 86, Al 10, Fe 4%.
silo (1) A tank or channel for conveying solid material in small pieces. (2) A pit or chamber in which fodder is fermented. See *silage.*
Silon Trademark for a polyamide synthetic fiber.
siloxanes* Compounds whose skeleton is made up of alternate Si and O atoms; as, trisiloxane, $H_3Si \cdot O \cdot SiH_2 \cdot O \cdot SiH_3$.
siloxicon Si_2OC_2. A refractory obtained by heating quartz, carbon, and sawdust in the electric furnace. Cf. *fibrox.*
siloxy* The radical H_3SiO-.
Silphenylene Trademark for heat-resistant, laminating resins derived from silicones.
silumin A noncorrodible alloy: Si and Al, sometimes containing 4–5% Cu.
silundum Silicon carbide*.
Silurian See *geologic eras,* Table 38.
silva Describing a lithographic printing process with zinc plates coated with silver halide emulsion instead of a dichromate. The silver image is converted into an ink image, enabling large-scale printing plates to be made from miniature negatives.
silvan Sylvan.
silver* Ag = 107.8682. Latin: *argentum**, which may be used, instead of s., in naming its compounds. A metal of the gold family; element of at. no. 47. White, lustrous metal of regular crystalline structure, d.10.50, m.961, b.2210, soluble in nitric acid or hot conc. sulfuric acid, insoluble in hydrochloric acid, water, or cold sulfuric acid. Valency 1. Used in jewelry, coins, instruments, and in the manufacture of s. salts for photography and pharmacy. **antimonial ~** Dyscrasite. **black ~** Stephanite. **blue ~** Niello. **coinage ~** See *silver alloys, coinage metals.* **colloidal ~** Collargol. **800- ~** The alloy Ag 80, Cu 20%. **fulminating ~** S. nitride. **german ~** Nickel s. **horn ~** Argentum cornu. **liquid ~** Mercury*. **moss ~** Native s. in mosslike form. **nickel ~** See *nickel silver.* **niello ~** See *niello silver.* **quick ~** Mercury*. **ruby ~** Proustite. **sterling ~** The alloy Ag 92.5, Cu 7.5%.
s. acetate $AgC_2H_3O_2$ = 166.9. Colorless crystals, decomp. by heat, soluble in water. **s. acetylide*** Ag_2C_2 = 239.8. S. carbide. An explosive white powder. **s. alloys** Principally:

Coinage:
(See *coinage metals.*)
Jewelry:
Ordinary: Ag 80, Cu 20%
Hall-marked: Ag 92.5, Cu and/or Cd 7.5%

s. amalgam $AgHg_2$ = 509.1. A silvery, brittle solid. **s. ammonium nitrate solution** Ammonium silver nitrate*. **s. antimonide*** Ag_3Sb = 445.4. S. stibide. Black rhombs, insoluble in water. **s. arsenate*** Ag_3AsO_4 = 462.5. Red powder, insoluble in water. **s. arsenide*** Ag_3As = 398.5. Black precipitate, decomp. when dried. **s. arsenite*** Ag_3AsO_3 = 446.5. Yellow powder, insoluble in water. **s. azide*** AgN_3 = 149.9. A curdy, white, explosive powder. **s. benzamide** See *silver benzamide* under *benzamide.* **s. benzoate*** $AgC_7H_5O_2$ = 229.0. Colorless powder, soluble in hot water. **s. borate** $Ag_2B_4O_7$ = 371.0. S. tetraborate*. Unstable, white powder, soluble in cyanide solutions. **s. bromate*** $AgBrO_3$ = 235.8. Colorless tetragons, decomp. by heat, soluble in hot water. **s. bromide*** AgBr = 187.8. Native as bromyrite and embolite. Yellow, regular cyrstals, m.427, decomp. 700, insoluble in water, soluble in thiosulfate solution; a light-sensitive coating in photography. **s. carbonate*** Ag_2CO_3 = 275.7. Heavy, yellow powder, d.6.0, decomp. 200, insoluble in water, soluble in cyanide solutions. **s. chlorate*** $AgClO_3$ = 191.3. Colorless tetragons, m.230, decomp. 270, soluble in water. **s. chloride*** AgCl = 143.3. Lunar cornea. Native as horn s., cerargyrite, embolite. Colorless, regular crystals, d.5.553, m.455, insoluble in water, soluble in ammonium hydroxide. Used in the manufacture of pure s. and s. salts; for s. plating; and in photography and photometry. **s. chlorite*** $AgClO_2$ = 175.3. Yellow powder, slightly soluble in water. **s. chromate*** Ag_2CrO_4 = 331.7. Red crystals, d.5.623, insoluble in water, soluble in ammonium hydroxide. **s. colloidal** Collargol, argentum, colloidale. An allotrope of s., with a small percentage of albumin; black scales giving a fairly stable colloidal suspension with water. **s. cyanate*** AgOCN = 149.9. Colorless powder, decomp. by heat, slightly soluble in water. **s. cyanide*** AgCN = 133.9. White crystals, decomp. by heat, insoluble in water, soluble in cyanide solutions. **s. dichromate*** $Ag_2Cr_2O_7$ = 431.7. Purple, triclinic crystals, decomp. by heat or alcohol, slightly soluble in water. **s. dithionate*** $Ag_2S_2O_6$ = 375.9. **s. ferricyanide** S. hexacyanoferrate(III)*. **s. ferrocyanide** S. hexacyanoferrate(II)*. **s. fluoride*** $AgF \cdot H_2O$ = 144.9. Tachiol. Silbamin. Yellow, deliquescent tetragons, m.435, soluble in water. **s. fluosilicate** S. hexafluorosilicate*. **s. fulminate*** $Ag_2O_2C_2N_2$ = 299.8. Colorless needles, explode when heated, slightly soluble in water; used in detonators. **s. glance** Argentite. **s. hexacyanoferrate(II)*** $Ag_4[Fe(CN)_6] \cdot H_2O$ = 661.4. S. ferrocyanide. Yellow crystals, insoluble in water. **s. hexacyanoferrate(III)*** $Ag_3[Fe(CN)_6]$ = 535.6. S. ferricyanide. Orange crystals, slightly soluble in water, soluble in ammonium hydroxide. **s. hexafluorosilicate*** Ag_2SiF_6 = 357.8. S. fluo(ro)silicate. White powder. **tetrahydrate ~** Globular granules, slightly soluble in water. **s. hypochlorite*** AgOCl = 159.3. An unstable bleaching agent. **s. hypophosphate*** See *silver phosphates* below. **s. iodate*** $AgIO_3$ = 282.8. Colorless, monoclinic crystals, decomp. by heat, insoluble in water. **s. iodide*** AgI = 234.8. Native as iodyrite. Yellow hexagons, d.5.67, m.536, insoluble in water; soluble in cyanide, iodide, or thiosulfate solutions; used in photography and in cloud seeding. **s. laurate*** $AgC_{12}H_{23}O_2$ = 307.2. Colorless powder, m.213. **s. leaf** Stillingia. **s. myristate*** $AgC_{14}H_{27}O_2$ = 335.2. White powder, m.211, insoluble in water. **s. nitrate*** $AgNO_3$ = 169.9. Argenti nitras. Colorless hexagons or rhombs, d.4.352, m.209 (decomp.), soluble in water. Used as a reagent (especially for halogens); has antiseptic properties (USP, BP); used for s.-plating, permanent marking of laundry, and dyeing hair and fur; and in the manufacture of s. salts and in photography. **fused ~** Lunar caustic, molded s. nitrate. White, hard sticks. **toughened ~** White, molded crystal masses, made by fusing s. nitrate 95 and potassium nitrate 5 pts., soluble in water; a caustic used for warts. (USP, EP, BP). **s. nitrate paper** A filter paper impregnated with a solution of nitrate and dried in the dark. A test for arsenic (yellow), phosphorus (black), chromates (red), and uric acid (brown). **s. nitride*** Ag_3N = 337.6. Fulminating s. Gray, explosive solid, insoluble in water. **s. nitrite*** $AgNO_2$ = 153.9. Colorless crystals, d.4.453, decomp. 150, slightly soluble in water; a reagent for standardizing permanganate solutions, determining nitrites, and differentiating alcohols. **s. ores** Silver is associated in nature with copper and gold minerals generally in binary compounds, as: native s., Ag; argentite, Ag_2S; proustite, Ag_3AsS_3; stromeyerite, AgCuS; bromyrite, AgBr; argentiferous lead, AgPbS. **s. oxalate*** $Ag_2C_2O_4$ = 303.8. White crystals, detonated by heat, insoluble in water. **s. oxide*** Ag_2O = 231.7. Brownish powder, d.7.521, decomp. 330, insoluble in water. **s. palmitate*** $AgC_{16}H_{31}O_2$ = 363.3. White powder, m.209, insoluble in water. **s. perchlorate*** $AgClO_4$ = 207.3. White crystals, m.486, soluble in water. **s. period** Heavy metals of the 5th period in the periodic table: Ru, Rh, Pd, Ag, Cd, In. **s. permanganate*** $AgMnO_4$ = 226.8. Violet, monoclinic crystals, decomp. by heat, slightly soluble in water. **s. peroxides** (1) Ag_2O_2 = 247.7. Black, insoluble solid, d.7.44, decomp. 110. (2) Ag_2O_4 = 279.7. An unstable solid. **s. phosphates** (1) Ag_3PO_4 = 418.6. S. orthophosphate*. Yellow powder, m.849, insoluble in water; used in photography. (2) $Ag_4P_2O_7$ = 605.4. S. pyrophosphate*. White, insoluble solid, m.585. (3) Ag_2PO_3 = 294.7. S. hypophosphate*. **s. phosphide*** AgP_2 = 169.8. Gray powder. **s. plate** A metal article plated or covered with metallic s. See *Sheffield plate.* **s.-plating** The electrolytic deposition of s. on another metal. **s. salicylate*** $C_6H_4(OH)COOAg$ = 245.0. Pink crystals, soluble in water; an antiseptic. **s. salt** Sodium anthraquinone 2-sulfonate; used in stripping dyed rags by reducing agents. **s. selenide*** Ag_2Se = 294.7. Gray powder, insoluble in water. **s. sodium chloride*** $NaAgCl_2$ = 201.8. Colorless crystals, decomp. by water. **s. sodium cyanide** Sodium argentum cyanide. **s. sodium thiosulfate*** $Na_4Ag_2(S_2O_3)_3$ = 644.1. White crystals, soluble in water. **s. solder** An alloy of 40, 50, or 60% Ag with Cu, Zn, and Cd. **s. stearate*** $AgC_{18}H_{35}O_2$ = 391.3. White powder, m.205, soluble in water. **s. stibide** S. antimonide*. **s. sulfate*** Ag_2SO_4 = 311.8. Colorless, triclinic or rhombic crystals, m.651, soluble in water; a reagent and electroplating agent. **s. sulfides** (1) Ag_2S = 247.8. Native as argentite and acanthite. Gray, regular crystals, m.676, insoluble in water; used in ceramic pigments. (2) Ag_2S_2 = 279.9. Black solid. **s. sulfite*** Ag_2SO_3 = 295.8. Colorless crystals, decomp. 100, slightly soluble in water. **s. tartrate*** $Ag_2C_4H_4O_6$ = 363.8. White crystals, soluble in water. **s. telluride*** Ag_2Te = 343.3. Native as hessite. Black powder, insoluble in water. **s. thiocyanate*** AgSCN = 165.9. White powder, decomp. by heat, insoluble in water, soluble in cyanide or ammonia solutions. **s. thiosulfate*** $Ag_2S_2O_3$ = 327.9. White solid, soluble in water, decomp. by heat. **s. tree** Arbor Dianae. **s. vanadate** Ag_3VO_4 and $Ag_4V_2O_7$, m.385; a catalyst. **s. vitellin** Argyrol.

silvering Coating with metallic silver, chemically (reduction of silver salts) or electrolytically.

silvestrene $C_{10}H_{16}$ = 136.2. 3-Isopropenyl-1-methylcyclohexene, sylvestrene. ***(R)*-~** A terpene from oils of *Pinus* spp. Colorless liquid, d.0.863, b.177.

silvinate Abietadienoate*. A salt of 7,13-abietadien-18-oic acid.

silyl* Silicyl*. The radical $-SiH_3$, analogous to the methyl group $-CH_3$. **di~** Indicating 2 s. groups.
s.ene† See *silanediyl* **s.thio*** The radical H_3SiS-.

silylation Replacement of an active H in an organic molecule by a silyl group; used to protect active sites in organic synthesis.

sima A contraction of silica and magnesium magma. Semiliquid rock on which the sial floats.

simaruba Bitter damson. The dried root bark of *Simaruba amaris* or *S. officinalis* (Simarubaceae), tropical America; an astringent (contains quassin).

Simarubaceae Tropical shrubs and trees; e.g.: *Picrasma excelsa,* Jamaica quassia; *Ailanthus glandulosa,* Chinese sumac; *Picramnaea antidroma,* cascara amarga.

simethicone A mixture of fully methylated siloxane polymers, $Me_3Si(-OSiMe_2-)_nMe$. Dimethicone. Colorless, insoluble liquid. Used in creams; also used for its defoaming properties, and with antacids (BP).

simmer Boil gently.

Simon's test Acetaldehyde and sodium nitroprusside, added together, give a red and blue color with primary and secondary amines, respectively.

simple Not complex, as, s. spectrum; not mixed, as, s. ether; not double, as, s. salt.

simsin Sesame.

simulator A device that provides a model, usually mathematical and computerized, to reproduce the functioning of a process, equipment, etc.

simultaneous reaction (1) Side reaction. (2) Secondary reaction. (3) One of 2 or more reactions that occur at the same time in the same reacting system.

sinactine $C_{20}H_{21}O_4N$ = 339.4. L-Tetrahydroepiberberine. A diisoquinoline alkaloid from *Sinomenium actum.*

sinalbin $C_{30}H_{42}O_{15}N_2S_2$ = 734.8. A glucoside from the seeds of *Brassica (Sinapsis) alba,* white mustard seed (Cruciferae). Colorless crystals, m.83, hydrolyzed to sinapine, *p*-hydroxybenzyl mustard oil, and glucose.

sinamine $C_4H_6N_2$ = 82.1. Allylcyanamide. An amine from black mustard seed. **thio~** Allyl thiourea.

sinapic acid $C_{11}H_{12}O_5$ = 224.2. 3-(4-Hydroxy-3,5-dimethoxyphenyl)-2-propenoic acid. Derived from sinapine.

sinapine $C_{16}H_{24}O_5N^+$ = 310.4. An alkaloid from sinalbin, the glucoside of white mustard; hydrolyzes to sinapic acid and choline.

sinapis **s. alba** White *mustard.* **s. nigra** Black *mustard.*

sinapolin $C_{14}H_{12}O_2N_2 \cdot C_7H_{12}O_2N$ = 382.4. Diallylurea, in mustard oil.

sine The ratio of the length of the side opposite an angle of a right-angled triangle to that of the hypotenuse. Cf. *cosine.*

singular solution A solution with a maximum or minimum on its vapor-pressure curve.

sinigrin Potassium myronate.

sinistrin $C_6H_{10}O_5$ = 162.1. A levorotatory carbohydrate from squill.

sinkaline Choline*.

sinomenine $C_{19}H_{23}O_4N$ = 329.4. An isoquinoline alkaloid from *Sinomenium acutum* (Menispermaceae). Colorless needles, m.182.

sinter (1) Saline incrustations formed around mineral springs. (2) See *sintering.* **calcareous ~** Tufa, travertine, or onyx. **iron ~** Amorphous scorodite. **pearl ~** A modification of silica. **siliceous ~** (1) Geyserite. (2) Fluorite.

sintering (1) The coalescence by heat of crystalline or amorphous particles into a solid mass, due to the formation of allotropic crystals. Cf. *fritted.* (2) Coalescence, in a boiler, of the fuel ash at a temperature below the fusion point.

Sinufed Trademark for pseudoephedrine.

siomine Hexamethylenetetramine tetraiodide.

sipalin A plasticizing mixture of the cyclohexyl and methylcyclohexyl esters of adipic acid.

siphon A ∩ shaped tube with one short leg which takes up liquid and delivers it to a lower level. The s. must be primed to start the flow, which is then maintained by hydrostatic head.

sipylite A negative niobate of erbium and other rare-earth metals.
Sirius Trademark for a viscose synthetic fiber.
sirup Syrup.
sisal S. hemp. A fiber from the leaves of *Agave sisalana,* a cultivated plant of Mexico and E. Africa; used in making rope, twine, and sacking. Cf. *henequen.* **s. wax** A hard wax from s. waste, m.63, decomp. 95, d.1.007, sapon. val. 55, I. val. 26.
sitostane See *stigmastane.*
sitosterol $C_{29}H_{47}OH$ = 412.7. From wheat, corn, bran, and calabar beans; occurs in cigarette smoke. Crystals, m.163. **beta- ~** 22-Dihydrostigmasterol. Crystals, m.136.
six hundred six (606) Salvarsan. The 606th compound tested by Ehrlich; the first effective treatment for syphilis.
Six Hundred Sixty-Six (666) Trademark for γ-hexachlorocyclohexane, an insecticide.
sizing The dressing and preparation of: (1) textiles for printing, (2) surfaces to receive paint, (3) paper to control water or ink absorption due to capillary attraction. **fortified ~** Rosin s. whose effect is enchanced by reaction with maleic anhydride, which produces 2 extra carboxyl groups.
s. materials Starch, gums, gelatin, rosin, tragacanth, albumin, casein, and plastics; used to size textiles or paper.
skatole C_9H_9N = 131.2. 3-Methyl-1*H*-indole, β-methylindole. Colorless leaflets with strong fecal odor, m.95, insoluble in water. A protein decomposition product; used as odor enhancer.
skatoxyl The radical C_9H_8ON-, from skatole.
skelgas Pentane*.
skep A heat-resistant, Russian, synthetic rubber copolymer of ethylene and propene.
skiadin viscous Injection of iodized oil.
skiagenol A vegetable radiopaque oil (20% iodine).
skiagram, skiagraph (1) Radiograph. Skiogram. A photograph made by X-rays. (2) Scotograph.
skiameter A device to measure the intensity of X-rays preparatory to a photographic exposure.
skimmianine $C_{14}H_{13}O_4N$ = 259.3. β-Fagarine. m.176. An alkaloid from the Japanese plant, *Skimmia japonica* (Rutaceae).
skimming Removing floating matter from the surface of a liquid.
skimmiol Taraxerol.
skin effect The phenomenon of an alternating current flowing near the surface of a conductor, thus increasing its resistance.
skiogram Skiagram.
skiver A sheepskin, split, and tanned with sumach, Cf. *roan.*
sklero- See *sclero.*
skleron An aluminum alloy containing Li, Cu, Zn, and Mn.
sklodowskite $MgO.2UO_3.2SiO_2.7H_2O$. A radioactive mineral (Congo). Named for Mme. Curie (née Sklodowska).
skotography Scotography.
Skraup **S., Zdenko Hans (1850–1910)** Polish chemist. **S. synthesis** Quinoline synthesis. The ring formation $C_6H_4\cdot N{:}CH\cdot CH{:}CH$ (ring closed), obtained by heating an aromatic amine with a free *ortho* position (as aniline) with glycerol and concentrated sulfuric acid in presence of an aromatic nitro compound (as nitrobenzene).
skullcap Scutellaria.
skunk The mammal *Mephitis mephitis;* it has an offensively odorous secretion. See *butyl thiol.* **s.bush** Feverbush. The leaves of *Garrya Fremontii* (Cornaceae), California; containing garryine. **s. cabbage** The rhizomes of *Symplocarpus foetidus* (Araceae).
skutterudite $CoAs_3$. A native arsenide.
slack (1) Slake. (2) Lumpy and damp, as lime exposed to air. (3) Loose. (4) Slow.
slacken To mix ores with slag to prevent fusion of the nonmetallic portions.
slag The vitreous mass which separates from fused metals during the melting of ores. **basic ~** Thomas s. A slag of calcium phosphate and free lime, produced in the manufacture of steel by the basic hearth process; a fertilizer. **electro ~** A homogeneous ingot produced from a consumable electrode of the material immersed in a pool of molten slag in a water-cooled mold which forms the other electrode. **soda ~** See *soda slag.* **Thomas ~** Basic s.
s. wool A fibrous packing material made by pouring molten s. into a pan with steam injection.
Slagceram Trademark for a strong, chemical-resistant, glassy material produced by heating blast-furnace slag with sand, followed by addition of metal oxides.
slake Slack. To quench.
slaked lime Calcium hydroxide.
slashing The stiffening of warp yarn, before spinning, with a solution containing an oil and starch. It reduces the abrasive action of the shuttle.
slat A thin, flat piece of solid material.
slate A dense, fine-textured rock whose mineral constituents are indistinguishable to the unaided eye. It has parallel cleavage planes and breaks into thin plates. Cf. *shale.* **polishing ~** Gray or yellow shale, used for polishing.
slide A plane glass plate. **lantern ~** 80 × 80 mm for projection. **microscope ~** 25 × 75 mm for observation. **projector ~** 23 × 35 mm.
s. rule A ruler with a central s. between 2 scales graduated logarithmically; used widely before the introduction of the electronic calculator. **chemical ~** A s. rule graduated for rapid chemical calculations.
slime A fine powder suspension; as, ore crusher mud. **s. molds** Myxomycetes.
slip (1) A fluid suspension of clay, fluxing material, and water, used to coat ceramics before final heating to give a glaze; as, zinc oxide and clay. (2) The sliding of atoms over one another in a crystal. **s. direction** The direction in which crystal slip occurs. **s. plane** The plane in which crystal slip occurs.
sludge A soft mud. **activated ~** S. produced by bubbling air through sewage, to promote growth of aerobic bacteria; a fertilizer: nitrogen 4–6, phosphorus pentaoxide 2.5–4%, dry basis. **Imhoff ~** S. produced by the action of anaerobic bacteria; a fertilizer: nitrogen 1.5–2.5, phosphorus pentaoxide 1%, dry basis.
s. acid The tarry sediment in oil refining tanks; impurities from the oil mixed with the strong sulfuric acid refining agent. **s. acid phosphate** A superphosphate manufactured with s. acid. **s. gas** See *sludge gas* under *gas.*
slug (1) The mass that acquires an acceleration of 1 ft/s^2 when acted upon by a force of 1 lb. Thus 1 s. = 32.17 lb. (2) A large tablet. **metric ~** The mass that acquires an acceleration of 1 m/s^2 when acted upon by a force of 1 kg.
slugging The mechanical compression of powders to form oversize tablets, often with a binder; often as a preliminary to granulation. Cf. *spheronizing.*
sluice A long, inclined trough with baffles for washing gold-bearing earth. **s.box** A wooden box in which the gold accumulates on washing auriferous gravel.
slum The insoluble oxidation products deposited from lubricating oil; eliminated by solvent extraction of the oil.
Sm Symbol for samarium.
smallpox vaccine Prepared from living vaccina virus from a

calf. Following worldwide eradication of smallpox, used for special risks; as laboratory workers.
smalls Slack (of coal).
smalt A blue glass or pigment of cobalt, potash, and silica.
smaltine Smaltite.
smaltite $CoAs_2$. Cobalt pyrites. Smaltine. A native diarsenide. Cf. *bismuthosmaltite.*
smaragd A green gem variety of beryl.
smectic Pertaining to liquid crystals of the soap type, having disk-shaped molecules.
smell See *odor.*
smelt (1) To obtain metals from their ores by a process that involves (*a*) roasting to remove volatile constituents; (*b*) reduction (smelting proper) in which the fused metals are separated from gangue; (*c*) purification of the metals. (2) The material obtained in (*b*).
smilacin (1) $C_{18}H_{30}O_6$ = 342.4. Salseparin, pariglin. A glucoside from sarsaparilla. (2) $C_{26}H_{42}O_3$ = 402.3. A solid, decomp. 160.
Smilax A genus of climbing plants (Liliaceae); as, sarsaparilla and chinaroot.
Smith, Edgar Fahs (1854–1929) American chemist, noted for his writings and educational methods.
smithite $AgAsS_2$. A native sulfide.
Smithson, James (1765–1829) English chemist noted for his bequest for the foundation of the Smithsonian Institution.
Smithsonian Institution A government establishment in Washington, D.C., created in 1846 according to the will of James Smithson "to increase and diffuse knowledge among men."
smithsonite $ZnCo_3$. Calamine. A native, yellow carbonate.
smog A toxic haze or fog, resulting from geographical location and temperature inversions that prevent normal diffusion of air pollutants. Principal ingredients: carbon dioxide, carbon monoxide, sulfur dioxide and sulfuric acid, nitrogen oxides, ethylene and other alkenes, formaldehyde, acrylaldehyde, and soot.
smoke (1) The dispersed system of solid carbon in air escaping or expelled from a burning substance. (2) A colloidal solid phase suspended in a gaseous phase; also termed sogasoid, from so(lid) gas(eous). **dark ~** See *Ringelmann smoke chart.*
s. screen A s. or fog produced by chemical reaction, to screen objects from observation; e.g., vapors of silicon tetrachloride and ammonia. **s. stone** Smoky quartz.
smokeless powder An explosive consisting mainly of nitrocellulose.
smoky **s. quartz** A smoky, gray or brown variety of quartz. **s. topaz** S. quartz used for jewelry.
Sn Symbol for tin (stannum).
snake One of a large class of reptiles. **s.head** Balmony, turtlebloom. The leaves of *Chelone glabra* (Scrophulariaceae); an anthelmintic and detergent. Cf. *chelonin.* **s. lily** Iris. **s. poison** S. venom. **s.root** Snakeroot. **s. venom** The poisons secreted by certain snakes, which hemolyze the blood, or cause paralysis or necrosis of muscles. **s.weed** Bistort.
snakeroot **black ~** Cimicifuga. **Canada ~** Asarum. **Seneca ~** Senega. **Texas ~, Virginia ~** Serpentaria.
sniol Trade name for a polyvinyl chloride synthetic fiber.
snow (1) A crystalline, finely divided form of water. **carbon dioxide ~** Dry Ice. Frozen carbon dioxide obtained by rapid evaporation of liquid carbon dioxide; temperature −110; a refrigerant, sometimes mixed with ether. Used to treat warts and other skin conditions. **dinitrogen oxide ~** The s. formed by the rapid evaporation of liquid N_2O. (2) Cocaine (slang).
Snyder reagent 4,7-Dihydroxy-1,10-phenanthroline. A reagent for ferrous iron (stable red compound).
soap A salt of a higher fatty acid with an alkali or metal. Soaps exist in 2 microcrystalline forms, viz., hexagonal plates and curd fibers, and in 3 types of solution, viz., isotropic solutions (including lyes and nigre), and neat and middle soaps, the 2 latter being conic, anisotropic "liquid crystal" forms. **castile ~** A s. made from sodium carbonate and olive oil. **essence of ~** An alcoholic s. solution, used in pharmacy. **green ~** S. liniment. **hard ~** An ordinary s., made with soda, giving a poor lather. **invert ~** See *invert soap.* **marine ~** Saltwater s. **medicinal ~** Sapo mollis, green s. A soft s. that yields not less than 44.0% fatty acids (USP, BP). **metallic ~** The salts of heavy metals with oleic, stearic, palmitic, erucic, and lauric acids. Used as paint and ink driers (Co, Mn), also for decolorizing varnish (Zn, Fe, Ni, Co, Cr), and waterproofing textiles (Al, Mg) and leather. **middle ~** A phase sometimes formed in s. boiling at concentrations intermediate between those of neat s. and isotropic solutions. A conic, anisotropic, plastic solution, darker in color than neat s. **neat ~** The upper layer in the s. pan; an anisotropic solution (63% fatty acid for sodium, and 40% fatty acid for potassium, soaps). **potash ~** A soft s. made with potassium hydroxide. **saltwater ~** S. containing hexanoic, octanoic, decanoic, and myristic acids. Not readily precipitated by Ca^{2+} and Mg^{2+} ions; made from coconut oil. **soda ~** A hard s. made with sodium hydroxide. **soft ~** Potash s. **toilet ~** S. containing 70% or more of fatty and resin acids. **transparent ~** S. made transparent by adding methyl alcohol. **white curd ~** S. made from tallow.
s.bark Quillaia. **s. liniment** Green s. A solution of soft s. in 70% alcohol, containing camphor and rosemary oil or lavender oil (USP). **s.root** Saponaria. **s. tree** Quillaia. **s.wort** Saponaria.
soapstone Talc.
Sobrero, Ascanio (1812–1888) Italian discoverer of nitroglycerin (1847). Cf. *Nobel.*
sobrerol Pinol hydrate.
sobrerone Pinol.
soda Sodium carbonate*. **baking ~** Sodium hydrogencarbonate. **caustic ~** Sodium hydroxide* (solution). **chlorinated ~** Sodium hypochlorite*. Sal-sodium carbonate. **scotch ~** An impure grade of sodium carbonate. **sesqui ~** A molecular mixture of $NaHCO_3$ and Na_2CO_2. **washing ~** Sodium carbonate*.
s. alum A double salt of aluminum and sodium sulfates. **s. ash** Commercial anhydrous sodium carbonate (99% Na_2CO_3). Used widely in industry. **s. feldspar** Albite. **s. lime** A mixture of calcium and sodium hydroxides, and sometimes potassium hydroxide. Used in anesthetics to absorb expired CO_2 (NF). **s. mint** Compound *sodium hydrogencarbonate.* **s. niter** Native sodium nitrate. **s. powder** B-powder. An early blasting powder made from Chile saltpeter glazed with graphite to prevent deliquescence. **s. process** (1) A method of manufacturing sodium carbonate. See *Le Blanc soda process.* (2) See *soda pulp* following. **s. pulp** Paper pulp obtained by digesting straw, bagasse, etc., with sodium hydroxide at about 70 kPa pressure. **s. slag** A slag obtained in the desulfurization of pig iron: chalcedony 35, sodium oxide 22, sulfur 7 pts.; used in bottle glass melts to oxidize the sulfides. **s. water** A beverage make by injecting carbon dioxide into water, sometimes containing sodium hydrogencarbonate. Cf. aerated *water.*
sodalite $Na_4Al_3Si_3O_{12}Cl$. A silicate that contains salt.
sodamide Sodium amide*.
Soddy, Frederick (1877–1956) British chemist, Nobel prize

winner (1921); noted for his research on radioactive elements.

sodic Containing sodium (obsolete).

sodii Official Latin for "of sodium."

sodiomalonic Sodium malonic.

sodion Sodium ion: Na^+.

sodium Na = 22.98977. Natrium. An alkali-metal element, at. no. 11. A tetragonal, crystalline, soft metal, silvery-white when freshly cut; rapidly dulling in air; stored under kerosine. Becomes brittle at low temperature, d_{15}·0.9732, m.97, b.883, decomp. by water, insoluble in alcohol or ether. Isolated by Davy (1807). Used as a dehydrating agent, flux, reactor coolant, reducing agent, conductor in cables; and in organic synthesis. **radio ~** The isotope of mass 24, half-life 15.5 hours, formed from s. by bombardment with deuterons; decomposes to magnesium with emission of β rays.

s. abietate $C_{20}H_{29}O_2Na$ = 324.4. The s. salt of abietic acid, produced when rosin is saponified for use as a size for paper. **s. acetate*** CH_3COONa = 82.0. Colorless, monoclinic crystals, m.58, soluble in water. Used as a mordant, reagent for alkaloids; for filling thermophores; and in photography, and the manufacture of acetic acid, ethyl acetate, and pigments. **hydrated ~** $NaC_2H_3O_2 \cdot 3H_2O$ = 136.1. Colorless, monoclinic crystals, m.58, soluble in water. Used as a Na^+-ion source in dialyzing solutions. **s. acetate tungstate*** $Na_2(CH_3CO)WO_4$ = 336.9. S. acetwolframate. White crystals, soluble in water; a microscope reagent. **s. acetyl arsanilate** $NaAsO_2 \cdot C_6H_4 \cdot NHCOCH_3$ = 264.1. Yellow crystals. **s. acetyl salicylate** $C_2H_3O \cdot OC_6H_4COONa$ = 202.1. Hydropirin, Pyranol; used similarly to aspirin. **s. alginate** The sodium salt of algin; a protective colloid for pharmaceuticals and cosmetics (NF). **s. alizarin sulfonate** $C_{14}H_5O_2(OH)_2SO_2Na$ = 342.3. Alizarin carmine. Orange powder, soluble in water; a dye, and indicator for strong acids (yellow) and strong alkalies (violet), except carbonates. **s. alum** Aluminum s. sulfate*. **s. aluminate** $Na_2Al_2O_4$ = 163.9. Colorless powder, m.1850, soluble in water. **s. amalgam** A mercury amalgam (2–10% Na); a reducing agent. **s. amide*** $NaNH_2$ = 39.1. Sodamide. Colorless crystals, m.208, decomp. by water. Used similarly to s., but forms explosive products when exposed to air. **s. aminohippurate** See *aminohippuric acid.* **s. aminosalicylate** EP, BP name for aminosalicylate sodium. **s. ammonium acid phosphate** Ammonium sodium hydrogenphosphate*. **s. amytal** Trademark for amobarbital sodium. **s. anoxynaphthonate** $C_{26}H_{16}O(SO_3Na)_3$ = 653.6. Blue, hygroscopic powder; a dye used to investigate cardiac disease. **s. antimonate*** $2NaSbO_3 \cdot 7H_2O$ = 511.6. Colorless octahedra, slightly soluble in water. See *sodium pyroantimonate.* **s. argentum cyanide** $NaAg(CN)_2$ = 182.9. Silver sodium cyanide. Yellow, soluble solid produced in the cyanide process for silver extraction. **s. arsenates normal ~** $Na_3AsO_4 \cdot 12H_2O$ = 424.1. Colorless crystals, m.85, soluble in water; a reagent. **acid ~** (1) $Na_2HAsO_4 \cdot 12H_2O$ = 402.1. Dodecahydrate. Colorless, rhombic crystals, m.28, soluble in water. (2) $Na_2HAsO_4 \cdot 7H_2O$ = 312.0. Heptahydrate. Colorless prisms, m.57, soluble in water; a reagent. (3) Na_2HAsO_4 = 185.9. Anhydrous. White powder, soluble in water; used in dyeing and printing textiles. **s. arsenide*** Na_3As = 143.9. Black solid; evolves arsine in presence of water. **s. arsenite*** $NaAsO_2$ = 129.9. Gray powder, soluble in water. **s. hydrogenarsenite** Na_2HAsO_3 = 169.9. **s. arsphenamine** The s. salt of Salvarsan. **s. ascorbate** $C_6H_7O_6Na$ = 198.1. White crystals, soluble in water; used similarly to ascorbic acid (USP). **s. aspartate*** $HOOC \cdot CH_2 \cdot CHNH_2 \cdot COONa$ = 155.1. S. aminosuccinate. Colorless needles, soluble in water.

s. auribromide S. tetrabromoaurate(III)*. **s. aurichloride** S. tetrachloroaurate(III)*. **s. aurothiomalate** BP name for gold sodium thiomalate. **s. azide*** NaN_3 = 65.0. A poisonous, crystalline salt used to make explosives. **s. azo-1-naphthol sulfanilate** Tropeolin. **s. barbitone** See *sodium barbitone* under *barbitone.* **s. benzenesulfonate*** $C_6H_5SO_3Na$ = 180.2. S. sulfobenzene. Colorless crystals, soluble in water. **s. benzoate*** $C_7H_5O_2Na$ = 144.1. White crystals, soluble in water; a preservative and antiseptic (NF, BP). **s. bicarbonate** S. hydrogencarbonate*. **s. binoxalate** S. hydrogenoxalate*. **s. biphosphate** See *sodium phosphates.* **s. bismuthate** $NaBiO_3$ = 280.0. Fawn powder, used in manganese determinations. **s. bisulfate** S. hydrogensulfate*. **s. bisulfite** (1) S. hydrogensulfite*. (2) A mixture of (1) and s. disulfite (NF). **s. bitartrate** S. hydrogentartrate*. **s. borates** (1) **s. metaborate** $NaBO_2$ = 65.8. Colorless prisms, m.966, soluble in water. (*a*) *tetrahydrate* ~ $Na_2B_2O_4 \cdot 4H_2O$ = 203.7. Colorless prisms, m.57, soluble in water. (2) **s. peroxoborate*** $NaBO_4 \cdot 4H_2O$ = 169.9. S. perborate, peroxydol. White crystals, soluble in water. Used as reagent, oxidizing agent, antiseptic, deodorant, bleach; and in dentifrices and detergents (BP). (3) **s. tetraborate*** (*a*) *calcined* ~ $Na_2B_4O_7 \cdot 10H_2O$ = 381.4. Borax. White powder, soluble in water; a reagent. (*b*) *fused* ~ $Na_2B_4O_7$ = 201.2. Borax glass, anhydrous borax. Colorless, vitreous mass, m.741, slightly soluble in water, soluble in alcohol; used as a reagent and antacid and in ointments. (*c*) *s. b. pentahydrate* $Na_2B_4O_7 \cdot 5H_2O$ = 291.3. Colorless octahedra, slightly soluble in water or alcohol; a reagent. (*d*) *s. b. decahydrate* $Na_2B_4O_7 \cdot 10H_2O$ = 381.4. Borax, sodii boras (NF), s. biborate, s. pyroborate. Colorless, monoclinic crystals, d_{17}·1.72, m. red heat, slightly soluble in water, insoluble in alcohol; used as a reagent, flux, and preservative; and for borax beads. (4) **s. tetrahydroborate*** $NaBH_4$ = 37.83. S. borohydride. White crystals, m.400 (decomp.), soluble in water. Much used mild reducing agent. **s. bromate*** $NaBrO_3$ = 150.9. Colorless crystals, m.384, soluble in water; an oxidizing agent. **s. bromide*** (1) NaBr = 102.9. Colorless cubes, m.768, soluble in water; a reagent. (2) $NaBr \cdot 2H_2O$ = 138.9. Colorless, monoclinic crystals, soluble in water. **s. butyrate*** C_3H_7COONa = 110.06. Colorless, deliquescent crystals, soluble in water. **s. calciumedetate** BP name for edetate calcium disodium. **s. carbide*** Na_2C = 58.0. Gray powder, decomp. by water. **s. carbonate*** Soda, washing soda. Cf. *Solvay process, ammonia* soda process. (1) **anhydrous ~** Na_2CO_3 = 106.0. White powder, d.2.476, m.852 (decomp.), soluble in water; used as a reagent, in freezing mixtures, and in pharmacy (NF, EP, BP). Cf. *sesquisoda, soda ash.* (2) **s. c. monohydrate** $Na_2CO_3 \cdot H_2O$ = 124.0. White crystals, soluble in water. Reagent and photographic chemical (NF, EP, BP). (3) **s. c. decahydrate** $Na_2CO_3 \cdot 10H_2O$ = 286.1. Colorless, monoclinic crystals, d.1.458, m.34, soluble in water; a reagent, precipitant, neutralizer; used in alkaline baths and in pharmacy (EP, BP). See *soda ash.* **s. carboxymethylcellulose** See carboxymethyl*cellulose.* **s. carminate** $C_{22}H_{19}O_{13}Na$ = 514.4. A red dye in microscopy. **s. chlorate*** $NaClO_3$ = 106.5. Colorless cubes or tetragons, m.250 (decomp.), soluble in water; a reagent, explosive (in pyrotechnics), herbicide; used for chlorine dioxide production. **s. chloride*** NaCl = 58.4. Table salt, common salt, rock salt, sea salt. Colorless cubes, d.2.176, m.800, soluble in water, insoluble in alcohol. Used as a reagent and condiment; for freezing mixtures and production of chloralkali; and widely in industry. The principal constituent, sometimes with glucose, of physiological salt solutions for fluid and sodium depletion (USP, EP, BP). **s. chlorite*** $NaClO_2$ = 90.4. Colorless crystals, soluble in water; an oxidizing agent. **s. 6-chloro-5-nitrotoluene-3-sulfonate** White powder; reagent for potassium. **s. chloroplatinate** S. hexachloroplatinate(IV)*. **s. chloroplatinite** S. tetrachloroplatinate(II)*. **s. chlorosulfonate** $NaClSO_3$ = 138.5. Fine crystals, readily hydrolyzed to hydrochloric acid; a reagent for sulfonation and chlorination. **s. cholate** Bile salts. **s. chromate*** $Na_2CrO_4 \cdot 10H_2O$ = 342.1. Yellow, triclinic crystals, m.20, soluble in water; a reagent and mordant. **Cr-51** A sterile injection of radioactive s. chromate made from ^{51}Cr (USP, EP, BP). **s. chromite** Na_2CrO_3 = 146.0. Green needles, soluble in water. **s. cinnamate*** $NaC_9H_7O_2$ = 170.1. Hetol. White crystals, soluble in water. **s. citrate*** $C_6H_5O_7Na_3 \cdot 2H_2O$ = 294.11. Sodii citras. Colorless crystals, m.150, soluble in water; a reagent, refrigerant, and, with glucose, a blood anticoagulant (USP, EP, BP). **s. acid citrate** $C_6H_6O_7Na_2 \cdot 1\frac{1}{2}H_2O$ = 263.1. White powder, soluble in water; a blood anticoagulant (BP). **s. cobaltic nitrite, s. cobaltinitrite** S. hexanitrocobaltate(III)*. **s. coerulin sulfate** Indigo carmine. **s. cromoglycate** BP name for cromolyn sodium. **s. cyanamide** Na_2NCN = 86.0. Hygroscopic, white powder. **s. cyanide*** NaCN = 49.01. White, deliquescent crystals, soluble in water. Used as a reagent, and in electroplating, case hardening, flotation, metal extraction (cyanide process), and in rubber accelerators. **s. cyclamate** $C_6H_{12}O_3NSNa$ = 201.2. White, sweet crystals, soluble in water; a sweetening agent. **s. dichromate*** $Na_2Cr_2O_7 \cdot 2H_2O$ = 298.0. S. bichromate. Red, triclinic crystals, m.320, soluble in water; an oxidizing agent. **s. dicyanoaurate(I)*** $NaAu(CN)_2$ = 272.0. Aurous sodium cyanide. White crystals, soluble in water; used in electroplating. **s. diethyl barbiturate** $NaC_8H_{11}O_3N_2$ = 206.1. Barbital sodium. White crystals, soluble in water; a hypnotic. **s. diethyl dithiocarbamate*** $Et_2N \cdot CS \cdot SNa$ = 171.3. Sensitive reagent for copper (brown color). **s. dihydrogenorthoperiodate*** $Na_3H_2IO_6$ = 293.9. S. paraperiodate. An oxidizing agent. **s. dihydrogenphosphate** See *sodium phosphates*. **s. dimethylaminoazobenzene sulfonate** Methyl orange. **s. dinitrocresolate** Antinonnin. **s. dioxide** S. peroxide*. **s. diphenylhydantoin** Phenytoin sodium. **s. disulfate*** $Na_2S_2O_7$ = 222.1. S. pyrosulfate. Colorless crystals, soluble in water. **s. disulfite*** $Na_2S_2O_5$ = 190.1. S. pyrosulfite. Colorless crystals, soluble in water. **s. dithionate*** $Na_2S_2O_6 \cdot 2H_2O$ = 242.1. S. hyposulfate. Transparent prisms, soluble in water. **s. dithionite*** $Na_2S_2O_4 \cdot 2H_2O$ = 210.1. S. hydrosulfite. Colorless crystals, decomp. red heat, soluble in water. **s. dithiosalicylate** Dithion. **s. divanadate** $Na_2V_4O_{11} \cdot 9H_2O$ = 587.9. Orange crystals, slightly soluble in water; used in the manufacture of dyes and inks. **s. ethanolate*** C_2H_5ONa = 68.1. S. ethoxide. White, hygroscopic powder; a reagent and escharotic. **s. ethylmercurithiosalicylate** Thimerosal. **s. ethyl sulfate** Ethyl s. sulfate*. **s. ferricyanide** S. hexacyanoferrate(III)*. **s. ferric oxalate** S. trioxalatoferrate(III)*. **s. ferrite** $Na_2Fe_2O_4$ = 221.7. Decomp. by water. **s. ferrocyanide** S. hexacyanoferrate(II)*. **s. fluorescein** See *sodium fluorescein* under *fluorescein*. **s. fluoride*** NaF = 41.99. Colorless cubes, m.982, soluble in water; a reagent for blood, an antiseptic and dental prophylactic (USP, BP). **s. fluosilicate** S. hexafluorosilicate*. **s. formaldehyde hydrogensulfoxylate** $NaHSO_2 \cdot HCHO \cdot 2H_2O$. A reducing agent in dyeing and a preservative (NF). **s. formate*** HCOONa = 68.0. Colorless rhombs, decomp. by heat, soluble in water; a reagent for arsenic and phosphorus. **s. fusidate** $C_{31}H_{47}O_6Na \cdot \frac{1}{2}H_2O$ = 547.7. The s. salt of fusidic acid; an antibiotic from *Fusidium coccineum* used for staphylococcal infections (BP). **s. germanate** Na_2GeO_3 = 166.0. Colorless crystals, soluble in water. **s. halides** The sodium salts of the halogen acids. **s. hexachloroplatinate(IV)*** $Na_2[PtCl_6] \cdot 6H_2O$ = 561.9. S. platinichloride, s. platinic chloride. Red solid, m.100, soluble in water. **s. hexacyanoferrate(II)*** $Na_4[Fe(CN)_6] \cdot 12H_2O$ = 520.1. S. ferrocyanide, yellow prussiate of soda. Yellow prisms, soluble in water. Used as a reagent, in photography, and in pigment manufacture. **s. hexacyanoferrate(III)*** $Na_3[Fe(CN)_6] \cdot H_2O$ = 298.9. S. ferricyanide, red prussiate of soda. Red, hygroscopic crystals, soluble in water. Used as a reagent, in photographic papers, and in the manufacture of pigments. **s. hexafluorosilicate*** Na_2SiF_6 = 188.1. S. fluo(ro)silicate, s. silicofluoride, salufer. White crystals, slightly soluble in water; a by-product of superphosphate manufacture. Used as a reagent, antiseptic, and enameling and laundry chemical. **s. hexanitrocobaltate(III)*** $Na_3[Co(NO_2)_6] \cdot \frac{1}{2}H_2O$ = 412.9. P. cobaltinitrite, s. cobaltic nitrite. Purple, hygroscopic crystals, soluble in water; reagent for potassium. **s. hydrate** S. hydroxide*. **s. hydride*** NaH = 24.00. Colorless crystals, decomp. by water or heat. **s. hydrogencarbonate*** $NaHCO_3$ = 84.0. S. bicarbonate, acid s. carbonate, baking soda. White powder, d.2.206, decomp. 270, soluble in water, insoluble in alcohol. Used as an antacid (USP, EP, BP) and reagent; in baking powders and pharmaceutically in effervescent mixtures; and to make volumetric solutions. **compound ~** Soda mint. S. hydrogencarbonate tablets containing peppermint oil; a flavoring (BP). **s. hydrogendifluoride*** $NaF \cdot HF$ = 62.0. S. bifluoride. Colorless crystals, soluble in water; used in ant powders, and for etching glass. **s. hydrogenlactate*** $NaH(C_3H_5O_3)_2$ = 202.1. S. bilactate. Colorless liquid, soluble in water. **s. hydrogenoxalate*** $NaHC_2O_4$ = 112.0. S. binoxalate. Colorless, monoclinic crystals, soluble in water; a stain remover. **s. hydrogen peroxide** (1) Sodyl hydroxide. (2) $2NaOH \cdot H_2O_2$ = 114.0. A white solid. **s. hydrogenphosphate** See *sodium phosphates*. **s. hydrogensulfate*** $NaHSO_4 \cdot H_2O$ = 138.1. S. bisulfate. Colorless crystals, m.300, soluble in water; a reagent and flux. **s. hydrogensulfide*** $NaHS \cdot 2H_2O$ = 92.1. S. hydrosulfide. Deliquescent, colorless crystals, decomp. by heat. **s. hydrogensulfite*** $NaHSO_3$ = 104.1. S. bisulfite. White crystals, soluble in water; a pharmaceutical (NF), reagent, disinfectant, bleach, and preservative. **s. hydrogentartrate*** $NaHC_4H_4O_6 \cdot H_2O$ = 190.1. S. bitartrate. Colorless crystals, soluble in water. Used as a reagent and in effervescent mixtures. **s. hydrosulfite** S. dithionite*. **s. hydroxide*** NaOH = 40.00. S. hydrate, soda, caustic soda, sodii hydroxidum. White, hygroscopic powder, or white flakes, plates, pellets, or sticks, d.2.13, m.318; very soluble in water, alcohol, or ether. Used extensively in chemistry, the chemical industry, metallurgy, photography (NF, BP). Commercial grades of purity: (1) reagent from sodium (for special analytical work); (2) reagent from alcohol (for general analytical work); (3) reagent, purified (for general chemical work); (4) pharmaceutical; (5) technical (fused or in flakes for industrial purposes). **s. hydroxide solution** Commercial grades: 95%; 40%, d.1.43; 31%, d.1.34; 15%, d.1.17; 5%, d.1.06. **s. hydroxylamine sulfonates** **mono ~** $HO \cdot NH(SO_3Na)$ = 135.1. **di ~** $HO \cdot N \cdot (SO_3Na)_2$ = 237.1. **s. hypobromite*** NaOBr = 118.9. Colorless powder, soluble in water; an oxidizing agent. **s. hypochlorite*** NaOCl = 74.4. Colorless powder, decomp. by heat, soluble in water, decomp. by ether; an oxidizing and bleaching agent and disinfectant (USP). Cf. *eau de javelle*. **s. hyponitrite*** $Na_2N_2O_2$ = 106.0. Colorless crystals, soluble in water. **s. hypophosphite** S. phosphinate*. **s. hyposulfate** (1) S. dithionate*. (2) S.

thiosulfate*. **s. hyposulfite** (1) S. dithionite*. (2) S. thiosulfate*. (3) S. hydrogensulfite*. **s. indigotin sulfonate** Indigo carmine. **s. iodate*** $NaIO_3$ = 197.9. White powder, decomp. by heat, soluble in water; a reagent. **s. iodide*** NaI = 149.9. Colorless cubes, m.664, soluble in water; a reagent, and used in medicine like potassium iodide (USP, EP, BP). $Na^{131}I$, in sterile solution, is a diagnostic and therapeutic agent (USP, BP). **s. i. dihydrate** $NaI \cdot 2H_2O$ = 185.9. Colorless crystals, soluble in water. **s. iodoeosin** Erythrosin. **s. iodohippurate I-131** A diagnostic aid containing radioactive ^{131}I (EP, BP). **s. iodotheobromate** Theobromine s. iodide. **s. ion** The cation Na^+. **s. iothalamate** See *iothalamic acid.* **s. iridichloride** Iridium s. chloride*. **s. lactate*** $NaC_3H_5O_3$ = 112.1. A thick syrup, soluble in water; an electrolyte replenisher (USP, BP), sometimes compounded with Ringer's solution. **s. lauryl sulfate** $NaC_{12}H_{25}SO_4$ = 288.4. S. dodecyl sulfate. Irium. White powder; a detergent, and emulsifier in pharmacy (NF, BP) and toiletries. **s. malonic ester** EtO(NaO)C:CH·COOEt = 182.1. Sodiomalonic ester. White needles. Formed, but not usually isolated, in the malonic ester synthesis. **s. manganate*** $Na_2MnO_4 \cdot 10H_2O$ = 345.1. Green, monoclinic crystals, decomp. by heat, soluble in water. **s. metabisulfite** S. disulfite*. **s. metaborate** See *sodium borates.* **s. metaphosphate*** $NaPO_3$ = 102.0. Two crystalline forms; one soluble in water (s. trimetaphosphate), the other insoluble. Melting either form and quickly cooling gives a water-soluble glass (Graham's salt). **s. metasilicate*** See *sodium silicate.* **s. metavanadate*** $NaVO_3$ = 121.9. Green crystals, soluble in hot water; a reagent. **s. methyl arsenite** $CH_3Na_2AsO_3$ = 183.9. Disodium methyl arsonate, monomethyl disodium arsenite*, arrhenal, arsynal, neoarsycodyl, stenosine. Colorless crystals, m.135, soluble in water; a cacodylate substitute. **s. methanolate*** CH_3ONa = 54.0. White powder, decomp. by water. **s. molybdate*** $Na_2MoO_4 \cdot 2H_2O$ = 241.9. Colorless leaflets with pearly luster, soluble in water; a reagent. **s. molybdophosphate** $Na_3[PMo_{12}O_{40}]$ = 2083. Trisodium dodecamolybdophosphate. S. phosphomolybdate. Yellow crystals, soluble in water; used as a reagent for alkaloids and vegetable fats, and in microscopy. **s. monosulfide*** S. sulfide. **s. monoxide*** S. oxide*. **s. morrhuate** The sodium salt of a fraction of the fatty acids of cod-liver oil having a high iodine content; a sclerosing compound, used to treat varicose veins (USP). **s. naphthionate*** $NaC_{10}H_6 \cdot NH_2SO_4$ = 261.2. S. naphthylaminesulfonate. Colorless prisms, soluble in water; reagent for nitrous acid. **s. 2-naphthol-1-sulfonate*** $C_{10}H_7(OH)SO_3Na$ = 247.2. White tablets, soluble in water. **s. naphthoquinone sulfonate** White powder; a reagent for nitrogen in amino acids. **s. naphthylamine sulfonate** S. naphthionate*. **s. nipagin-M** Soluble methylhydroxybenzoate. **s. nipagol-M** Soluble propylhydroxybenzoate. **s. nitranilate** $C_6(NO_2)_2O_2(ONa)_2$ = 274.1. S. dinitrodihydroxyquinonate. Brown powder, soluble in water. **s. nitrate*** $NaNO_3$ = 85.0. Saliter. Soda niter, cubic niter, caliche, Chile saltpeter. Colorless rhombohedra, m.307, soluble in water. Used as a flux in metallurgy, as an oxidant; in the manufacture of acids, fertilizers, explosives, and glass. **s. nitride*** Na_3N = 83.0. An unstable, explosive compound formed by passing an electric arc between a platinum cathode and a sodium anode in liquid nitrogen. **s. nitrite*** $NaNO_2$ = 69.0. Pale-yellow crystals or yellow sticks, d.2.167, m.271, soluble in water; a reagent in the manufacture of azo dyes and synthetics; used for cyanide poisoning (USP). Cf. *mystin.*
s. nitroferricyanide, s. nitroprussiate, s. nitroprusside S. pentacyanonitrosylferrate(III)*. **s. nitrosohydroxylamine sulfonate** $ON \cdot N(ONa)SO_3Na$ = 186.0. **s. oleate*** $C_{18}H_{33}O_2Na$ = 304.4. Yellow, unctuous granules, soluble in water. **s. orthovanadate** See *sodium vanadate.* **s. oxalate*** $Na_2C_2O_4$ = 134.0. White crystals, soluble in water; a reagent and stain remover. **acid ~** $NaOOC \cdot COOH \cdot H_2O$ = 130.0. Colorless, monoclinic crystals, soluble in water. **s. oxide*** Na_2O = 62.0. S. monoxide*. Gray mass, sublimes and m. at red heat, decomp. by water (to s. hydroxide); a reagent and strong base. **s. paraperiodate** S. dihydrogenorthoperiodate*. **s. pentacyanonitrosylferrate(III)*** $Na_2[Fe(CN)_5NO] \cdot 2H_2O$ = 298.0. S. nitroferricyanide, s. nitroprusside. Nipride. Red, transparent crystals, soluble in water; reagent for alkali sulfides, acetone, formaldehyde, amino acids, and alkaloids. A hypotensive agent (USP, BP). **s. pentobarbital** See *pentobarbital.* **s. perborate** See *sodium borates.* **s. percarbonates** S. peroxodicarbonates*. **s. perchlorate*** $NaClO_4$ = 122.4. Colorless rhombohedra, m.482 (decomp.), soluble in water; a reagent and explosive. **s. periodate*** $NaIO_4 \cdot 3H_2O$ = 267.9. Efflorescent hemihedra, dehydrated at 300. **s. permanganate*** $NaMnO_4 \cdot 3H_2O$ = 196.0. Purple crystals, decomp. by heat, soluble in water; an oxidizing agent and disinfectant. **s. peroxide*** Na_2O_2 = 78.0. S. dioxide, s. superoxide. Pale-yellow powder, d.2.805, decomp. by heat; in water produces heat, s. hydroxide, and hydrogen peroxide. Used as a flux for minerals; in bleaching and the preparation of calcium peroxide, s. peroxoborate, and ferrate; and in the oxidation of organic matter in analysis. *Use with caution!* (it can ignite organic matter). **s. peroxodicarbonates*** S. percarbonates. (1) **normal ~** $Na_2C_2O_6$ = 166.0. (2) **s. monocarbonate** Na_2CO_4 = 122.0. Used in detergents, photography, and for bleaching. **s. peroxodisulfate** $Na_2S_2O_8$ = 238.1. S. persulfate. White crystals, soluble in water; a bleaching agent, and reagent for indican and adrenaline. **s. pertechnate** $Na^{99}TcO_4$. Used for tumor scanning in various organs, including brain, lung, heart, thyroid. **s. phenolphthaleinate** $C_{20}H_{12}O_4Na_2$ = 362.3. Red syrup, soluble in water. **s. phenol sulfonate*** $C_6H_4(OH)SO_3Na$ = 196.2. S. sulfocarbolate. Colorless crystals, soluble in water. **s. phenyl ethyl barbiturate** Luminal sodium, phenobarbital s. White, hygroscopic powder, m.147; a hypnotic. **s. *o*-phenylphenolate*** $Ph \cdot C_6H_4ONa$ = 192.2. White powder; a glue preservative. Cf. *phenylphenol.* **s. phosphates** (1) $NaH_2PO_4 \cdot H_2O$ = 138.0. Monosodium dihydrogen(ortho)phosphate*, s. biphosphate. Transparent rhombs, soluble in water; used (as $NaH_2PO_4 \cdot 2H_2O$) as a purgative (BP). (2) **anhydrous ~** Na_2HPO_4 = 142.0. Sörensen's s. phosphate, disodium hydrogen(ortho)phosphate*. White, hygroscopic powder; a reagent and buffer. **dibasic ~, hydrous ~** $Na_2HPO_4 \cdot 12H_2O$ = 358.1. Disodium orthophosphate, monohydrogen s. phosphate. Transparent rhombs, m.35, soluble in water; a reagent, purgative, and treatment for calcium and phosphorus disorders (USP, EP, BP). (3) $Na_3PO_4 \cdot 12H_2O$ = 380.1. S. orthophosphate*. Colorless hexagons, m.77, soluble in water. (4) **s. diphosphate** See *diphosphate.* (5) **s. metaphosphate*** $NaPO_3$ = 102.0. Insoluble form, colorless crystals; soluble form, colorless crystals; m.627 (glass obtained by quenching melt). (6) **s. pyrophosphate*** $Na_4P_2O_7 \cdot 10H_2O$ = 446.0. Colorless, monoclinic crystals, m.988, soluble in water; a reagent. (7) **s. tetraphosphate** $Na_6P_4O_{13}$ = 469.8. (8) **radioactive ~** Sterile phosphate solution (^{32}P); a diagnostic agent (BP). **s. phosphide*** Na_3P = 99.9. Red solid, evolves phosphine with water, decomp. by heat. **s. phosphinate*** $NaH_2PO_2 \cdot H_2O$ = 106.0. S. hypophosphite.

Colorless, deliquescent prisms or white granules, soluble in water; a reagent in gas analysis. **s. phosphite*** Na_3PO_3 = 147.9. **s. phosphonate** (1) $Na_2HPO_3 \cdot 5H_2O$ = 216.0. Colorless rhombohedra, m.53, soluble in water. (2) NaH_2PO_3 = 104.0. **s. phosphomolybdate** S. molybdophosphate. **s. phosphotungstate** S. tungstophosphate. **s. phosphovanadate** S. vanadophosphate. **s. phthalate*** $C_6H_4(COONa)_2$ = 210.1. White powder, slightly soluble in water. **s. platinichloride** S. hexachloroplatinate(IV)*. **s. platinochloride** S. tetrachloroplatinate(II)*. **s. plumbate** $Na_2PbO_3 \cdot 3H_2O$ = 355.2. Yellow powder, decomp. by water. **s. polystyrene sulfonate** Kayexalate, Resonium A. Brown powder, insoluble in water; a cation-exchange resin for excess blood potassium level (USP). **s. propionate*** Et·COONa = 96.1. White granules, soluble in water. An antifungal agent (NF). **s. pyroantimonate** $H_2Na_2Sb_2O_7 \cdot 6H_2O$ = 511.6. Colorless powder, slightly soluble in water. **s. pyrophosphate***. See *sodium phosphates.* **s. pyrosulfate** S. disulfate*. **s. pyrosulfite** S. disulfite*. **s. pyrovanadate** $Na_4V_2O_7$ = 305.8. Gray crystals, slightly soluble in water. **s. rosolate** $C_{20}H_{15}O_3Na$ = 326.3. S. corallinate. Red powder with green luster, soluble in water; a stain in microscopy. **s. salicylate*** $NaC_7H_5O_3$ = 160.1. White scales, soluble in water; an antirheumatic and antipyretic (USP, EP, BP); and a reagent for free acid in gastric juice. **s. selenate*** $Na_2SeO_4 \cdot 10H_2O$ = 369.1. White crystals, soluble in water; a reagent. **s. selinite*** Na_2SeO_3 = 172.9. Colorless powder, soluble in water; used as a reagent in bacteriology, and in red glass. **s. sesquicarbonate** See *sesquicarbonate of soda.* **s. silicate liquid ~** Water glass, soluble glass. An amorphous powder or heavy, liquid, aqueous solution, soluble in water. **solid ~** Na_2SiO_3 = 122.1. S. metasilicate*. Colorless, monoclinic crystals, m.1056, soluble in water. Used for detergents, pigments, cracking catalysts, gels and sols, foundry binders, adhesives and cements. **s. silver cyanide** S. argentumcyanide. **s. stannate** $Na_2SnO_3 \cdot 3H_2O$ = 266.7. White hexagons, soluble in water; a dye mordant. **s. stannite** HSnOONa = 174.7. Known only in solution; analogous to s. formate. **s. stearate** $C_{18}H_{35}O_2Na$ = 306.5. White, unctuous powder, soluble in water; used (with s. palmitate) for suppositories (NF). **s. stibogluconate** Pentostam. Colorless powder made by the action of antimony trichloride and gluconic acid in presence of alkali; it contains 30.0–34.0% Sb. Soluble in water, insoluble in alcohol or ether; used to treat leishmaniasis (BP). **s. succinate*** (1) $Na_2C_4H_4O_4$ = 162.1. Colorless powder, soluble in water. (2) $Na_2C_4H_4O_4 \cdot 6H_2O$ = 270.1. Colorless prisms, soluble in water. **s. sulfalizarate** Alizarin red. **s. sulfanilate*** $C_6H_4(NH_2)SO_3Na \cdot 2H_2O$ = 231.2. S. aniline sulfonate, s. *p*-aminobenzene sulfonate. Colorless flakes, soluble in water. **s. sulfantimonate** S. thioantimonate*. **s. sulfates anhydrous ~** Na_2SO_4 = 142.0. Salt cake, niter cake, native as thenardite. Colorless powder, d.2.673, m.884, soluble in water (EP, BP). **s. s. monohydrate** $Na_2SO_4 \cdot H_2O$ = 160.1. Native as mirabilite. White powder, soluble in water. **s. s. heptahydrate** $Na_2SO_4 \cdot 7H_2O$ = 268.1. Colorless rhombs or tetragons, soluble in water. **s. s. decahydrate** $Na_2SO_4 \cdot 10H_2O$ = 322.2. Glauber's salt, ordinary s. sulfate. Colorless, monoclinic crystals, d.1.462, m.38, soluble in water, a reagent, precipitant, cathartic (USP, EP, BP). Cf. *sodium hydrogensulfate.* **s. sulfides** (1) Na_2S = 78.1. Anhydrous s. monosulfide*. Amorphous, pink powder, decomp. by heat, soluble in water. (2) $Na_2S \cdot 9H_2O$ = 240.2. Colorless, hygroscopic crystals, soluble in water; a reagent, instead of hydrogen sulfide. (3) Principal polysulfides: (*a*) Na_2S_2 = 110.1, disulfide*; (*b*) Na_2S_3 = 142.2, trisulfide*; (*c*) Na_2S_4 = 174.2, tetrasulfide*; (*d*) Na_2S_5 (+ $8H_2O$) = 206.4, pentasulfide*. **s. sulfites** (1) Na_2SO_3 = 126.0. **anhydrous ~** Colorless prisms, m.150 (decomp.), soluble in water; a reagent and reducing agent. (2) $Na_2SO_3 \cdot 7H_2O$ = 252.1. **s. s. heptahydrate** Colorless prisms, m.100 (decomp.), soluble in water; a reagent and preservative. **s. sulfocarbonate** S. trithiocarbonate*. **s. sulfoxylate formaldehyde** $NaHSO_2 \cdot CH_2O \cdot 2H_2O$ = 154.1. White prisms, m.64, soluble in water; a reagent. **s. superoxide** S. peroxide*. **s. tartrate** $Na_2C_4H_4O_6 \cdot 2H_2O$ = 230.1. White, trimetric crystals, soluble in water; a reagent, cathartic, and refrigerant. **s. taurocholate** (1) $NaC_{26}H_{44}NSO_7$ = 537.7. Yellow powder, soluble in water, from the bile of carnivorous animals; a supplement for bile, to aid digestion. **s. tellurate*** $Na_2TeO_4 \cdot 5H_2O$ = 327.7. White powders, soluble in water. **s. tellurite*** Na_2TeO_3 = 221.6. White powder, soluble in water; used in bacteriology. **s. tetraborate***. See *sodium borates.* **s. tetrabromoaurate(III)*** $NaAuBr_4 \cdot 2H_2O$ = 575.6. Auric s. bromide. Black crystals, soluble in water. **s. tetrachloroaurate(III)*** $NaAuCl_4$ = 361.8. Auric s. chloride. Yellow crystals, soluble in water. **s. tetrachloropalladate(II)*** Na_2PdCl_4 = 294.2. Red, hygroscopic crystals, soluble in water; a reagent for gases (methane, carbon monoxide). **s. tetrachloroplatinate(II)*** Na_2PtCl_4 = 382.9. S. platinochloride. Brown crystals. **s. thioantimonate*** $Na_3SbS_4 \cdot 9H_2O$ = 481.1. Schlippe's salt, s. sulfantimonate. Colorless tetrahedra, decomp. by heat, soluble in water; a reagent for alkaloids. **s. thioaurate** $NaAuS \cdot 4H_2O$ = 324.1. S. aurosulfide. Colorless crystals. **s. thiocarbonate** S. trithiocarbonate*. **s. thiocyanate*** NaSCN = 81.1. S. sulfocyanate, s. sulfocyanide, s. rhodanate. White, deliquescent rhombs, m.287, soluble in water; a reagent, especially for ferric ions. **s. thiopentone** See *thiopental sodium.* **s. thiosulfate*** $Na_2S_2O_3 \cdot 5H_2O$ = 248.2. S. hyposulfite, s. subsulfite, antichlor. Colorless, monoclinic crystals, d.1.729, m.40–45, decomp. 48, soluble in water. A volumetric reagent, group precipitant instead of hydrogen sulfide, "antichlor" in bleaching, mordant, photographic fixing salt, solvent for lead and silver, and antidote for cyanide poisoning (USP, EP, BP). **s. thiotetraphosphate** $Na_6P_4O_{10}S$ = 453.9. A detergent. **s. trioxalatoferrate(III)*** $2Na_3[Fe(C_2O_4)_3] \cdot 10H_2O$ = 957.9. S. ferric oxalate. Green, monoclinic crystals, dehydrated at 100 ($4H_2O$) and above, soluble in water. **s. tripolyphosphate** $Na_5P_3O_{10}$ = 367.9. STPP. A builder in synthetic detergents. **s. trithiocarbonate*** Na_2CS_3 = 154.2. S. sulfocarbonate. Brown crystals, soluble in water. An antiseptic, and reagent for nickel and cobalt. **s. trititanate** $Na_2Ti_3O_7$ = 301.6. White needles, m.1128, insoluble in water. **s. truxillate** $Na_2C_{18}H_{14}O_4 \cdot 10H_2O$ = 520.4. White powder, soluble in water. **s. tungstate*** $Na_2WO_4 \cdot 2H_2O$ = 329.9. S. wolframate*. Colorless prisms, m.100, soluble in water; a reagent for alkaloids, bile pigments, and tannins; and waterproofing and fireproofing agents. **s. tungstophosphate** $Na_3[PW_{12}O_{40}] \cdot 18H_2O$ = 3270. Trisodium dodecatungstophosphate. S. phosphotungstate. White granules, soluble in water; a reagent for alkaloids, potassium, ferrous salts, and uric acid. **s. uranate** Na_2UO_4 = 348.0. Uranium yellow. Orange rhombs, soluble in water; used as a reagent, and in the manufacture of green glass and ceramic pigments. **s. uranyl acetate** $UO_2(C_2H_3O_2)_2 \cdot 2NaC_2H_3O_2$ = 552.2. Yellow crystals, soluble in water; a reagent. **s. valproate** $C_8H_{15}O_2Na$ = 166.2. 2-Propylpentanoate. Epilim. White crystals, freely soluble in water. An anticonvulsant, for epilepsy (BP). **s. vanadate** Na_3VO_4 = 183.9. S. orthovanadate*. Colorless crystals, soluble in water; used as a reagent and in the manufacture of inks and dyes. **s. metavanadate** See *sodium metavanadate.* **s. pyrovanadate** See *sodium pyrovanadate.* **s. vanadophosphate** $Na_3[PV_{12}O_{40}] \cdot 21H_2O$ = 1730. Trisodiumdodecavanadophosphate. S. phosphovanadate. Yellow crystals, soluble in water; a microscope reagent. **s. versenate** Versene. **s. warfarin** See *warfarin sodium.* **s. wolframate*** S. tungstate*. **s. zincate** Na_2ZnO_2 = 143.4. White powder, decomp. by water into zinc and sodium hydroxides.

sodyl The radical NaO—. **s. hydroxide** NaOOH = 56.0. White powder, explodes if heated.

soft biologically ~ Having low resistance to biological decomposition.

s.-sized paper Bibulous or blotting paper. **s. soap** Green soap.

softener (1) A substance that increases plasticity: either a true s. or a lubricant; as, stearic acid or pine tar, respectively, for rubber. (2) See *soft water* under *water, permutite.* (3) A substance that improves pliability and softness. **anti ~** A substance that stiffens; as, benzidine in rubber.

softening (1) Making plastic. (2) Removing salts. **water ~** See *soft water* under *water, permutite.*

s. temperature The point at which substances without a sharp melting point change from viscous to plastic flow.

software The programs, associated documentation, and programming know-how for computers. Cf. *hardware.*

sogasoid A dispersion of a solid phase in a gaseous phase; as, smoke.

soil The surface layer of the earth; the weathered mineral and rock fragments with decomposed vegetable and animal matter. Of the s. of the world 1,000 million hectares are used for agriculture. Cf. *humus, subsoil.* See Table 84. **s. amendment** A material added to s. to improve it other than by plant nutrients, e.g., sand added to prevent hardening of clay. **s. bacteria** Protophyta that enrich the s. by ammonification and nitrification of nitrogen compounds; e.g., *Bacillus mycoides.* Cf. *Azobacter, Nitrobacter.* **s. horizon** Layers of a s. profile which have become differentiated as a result of processes occurring in the s. mass. **s. mechanics** See *soil mechanics* under *mechanics.* **s. profile** A section of s. showing the layers produced at different depths by geological or other causes. Cf. *soil horizon.* **s. science** Pedology. The study of the earth surface layer.

TABLE 84. COMPOSITION OF SOILS

Lime, %	Sand, %	Clay, %	Type
Lime-poor:			
0–15	80–100	0	Flying sand
		10	Loose sand
		20	Clay sand
	30–50	30	Sandy loam
		40	Mild loam
		55	Strong loam
	0–30	65	Mild clay
		75	Common clay
		90	Strong clay
Lime-rich:			
15–20	25–50	20–50	Loamy soil
	0–25	50–75	Clayey soil
15–60	40–80		Sandy soil
50–75	Little		Calcareous soil
75–95	Little		Lime soil

Type	Particles, mm diameter
Gravel	Over 2
Coarse sand	0.2–2.0
Fine sand	0.02–0.20
Silt	0.002–0.02
Clay	Below 0.002

soja Soybean.

sol (1) A colloidal solution. (2) The liquid phase of a colloidal solution. (3) Abbreviation for soluble (with period). **aero ~** A colloidal system in which the surrounding phase is a gas; e.g., fog. **electro ~** See *electrosol.* **hydro ~** A colloidal suspension in a water-liquid phase. See *gel.* **sulfo ~** See *sulfosol.*

Solanaceae Nightshade family. Herbs or shrubs with rank-scented, often poisonous foliage and a colorless juice containing alkaloids; e.g., roots: *Atropa belladonna,* atropine; leaves: *Datura stramonium,* stramonium; branches: *Fabiana imbricata,* pichi; fruits: *Capsicum frutescens,* cayenne pepper. **S. alkaloids** The alkaloids obtained from various species of s.; as: atropine, hyoscyamine, belladonine. Cf. *lycine, solandrine, solanine.*

solandrine An alkaloid from *Solandra lavis.* It resembles scopolamine.

solanellic acid $C_{23}H_{34}O_{12}$ = 502.5. A hexabasic acid from oxidation of bile acids.

solanesol A long-chain, isoprenoid alcohol from tobacco.

solanidine $C_{27}H_{42}N \cdot OH$ = 397.6. A decomposition product of solanine. Colorless crystals, m.219, soluble in alcohol.

solanin $C_{45}H_{73}O_{15}N$ = 868.1. A glycoside from *Solanum nigrum,* potato and other species. Colorless microcrystals, m.280, soluble in water; a poison.

solanine Any of the alkaloids in Solanaceae. Potato s. can cause poisoning.

Solanum Herbs and shrubs of the family Solanaceae; includes nightshades and potatoes. ***S. carolinense*** Solanum, horse nettle, poison potato, sand brier, bull nettle. Air-dry ripe fruits (Southern states); a sedative. ***S. dulcamara*** Bittersweet. ***S. grandiflora*** S. yielding *grandiflorine,* q.v. ***S. insidiosum*** Jurubeba. The root (Brazil) is a diuretic. ***S. melongena*** Eggplant. ***S. nigrum*** The common garden nightshade. ***S. tomatillo*** S. yielding *natrine,* q.v. ***S. tuberosum*** The common potato.

solar Pertaining to the sun. **s. cell** See *solar cell* under *cell.* **s. constant** The amount of s. energy falling at normal incidence on a body outside the earth's atmosphere at the earth's mean distance from the sun; 8.4 $J/cm^2/min$. **s. pan, s. pond** Flat areas surrounded by low dykes in which seawater is evaporated for salt. **s. radiation** See *solar radiation* under *radiation.* **s. rays** The visible and invisible radiation of the sun. **s. power** Production of power using s. cells. Installations up to 1 MW exist (California). Smaller units are used for domestic power, irrigation pumps, and instruments in isolated areas. See *solar cell.* **s. salt** See *solar salt* under *salt.* **s. spectrum** The spectrum produced when sunlight is refracted by a prism or grating; characterized by Fraunhofer lines. **s. wind** Plasma radiating from the sun. **s. year** See *solar year* under *year.*

solarization (1) Exposure to the sun; as in accelerated aging. Cf. *irradiation.* (2) A decrease in starch content following long exposure of plant leaves to light. (3) The partial inversion of a photographic negative into a positive by exposure to light during development; used to enhance shading effects.

solate A liquefied gel.

solation Liquefaction of a gel; the reverse of gelation.
solbrol Nipagin M.
solder Braze. A fusing metal or alloy used to unite adjacent surfaces of less fusible metals. **brass ~** Copper s. **copper ~** An alloy: Sn 5, Pb 2 pts., with zinc chloride as flux. **fine ~** Soft s. **fusible ~** An alloy of Pb, Sn, and Bi, which melts in water; used in spray fire extinguishers. **gold ~** An alloy: Au 10, Ag 6, Cu 4 pts. **hard ~** A high-melting-point alloy used as s.; it fuses at red heat: e.g., Cu + Zn + Ag. **lead ~** An alloy of equal parts of Pb and Sn, used for soldering lead. **plumber's ~** An alloy usually containing approx. Pb 65, Sn 30%, with some Sb. **seifert ~** A s. for aluminum, containing Sn 73, Zn 21, Pb 5%. **silver ~** See *silver solder.* **soft ~** A s. that fuses below red heat; as, Sn + Pb; *lead s.* (above), *fusible s.* **zinc ~** An alloy: Sn 5, Pb 3 pts.
soldering (1) Uniting metallic pieces by heat with or without an alloy (solder) and flux (borax). (2) In commerce, soft (as distinct from hard) solders. S. differs from *brazing* and fusion *welding,* q.v. **autogenous ~** Uniting metal surfaces by interfusion, without a more fusible alloy. **fusing ~** Uniting metal surfaces by filling all intervening space with a completely fused solder. **sweating ~** S. in which the solder is heated near its melting point and adheres.
solenhofen stone A fine-grained, porous limestone; contains clay.
solenoid A hollow cylinder, wound with resistance wire; used to produce fields of electric force, as to operate a valve.
solfatara A volcanic vent from which sulfur is obtained.
solferino Fuchsin.
solid (1) A substance of definite shape, and relatively great density, low internal enthalpy, and great cohesion of its molecules. It may be homogeneous (as crystals and solid solutions) or hetergeneous (as amorphous and colloidal substances). **s. solution** (1) Sosoloid. A homogeneous, s. mixture of substance; as, glass. (2) A s. solution of a solid, liquid, or gas in a solid. **s. state** Describing electronic components that utilize electronic and magnetic properties of solids.
solidago Goldenrod. The dried herb of *Solidago odora* (Compositae); a carminative.
solidify To change into the solid state.
solidifying point Freezing point.
solidus In a temperature-concentration diagram for both solid and liquid solutions whose concentrations differ, the s. curve relates to the solid phase, and the *liquidus* to the liquid phase.
soliquoid Suspension. A dispersed system of a solid phase in a liquid phase.
soln. Abbreviation for solution.
solodization Dealkalization. Removal of alkali from soils by degradation.
Solozone Trademark for a brand of hydrogen peroxide.
solubility The extent to which a substance (solute) mixes with a liquid (solvent) to produce a homogeneous system (solution). The classification used by the United States Pharmacopeia is shown in Table 85. **apparent ~** The total amount of undissociated and dissociated portions of a substance dissolved in a liquid. **degree of ~** The concentration of a saturated solution at a given temperature. S. generally increases with increase in temperature. **molar ~** c/M, where c is the g/L and M the molecular weight. **real ~** The amount of undissociated solute in a liquid. **s. curve** A graph obtained by plotting the amount of dissolved substance in a saturated solution against the temperature. **s. exponent** p or $p_s = \log 1/S$. Cf. *pH.* **s. product** $S = [M^+] \times [X^-]/[MX]$, where the brackets indicate the concentrations of the components of the dissociation equilibrium: $MX \rightleftharpoons M^+ + X^-$. If $[M^+] \times [X^-]$ exceeds S, MX will precipitate; and vice versa. E.g., NaCl is precipitated from concentrated solutions by HCl gas.

TABLE 85. USP SOLUBILITY CLASSIFICATION

Description	Parts of solvent required for 1 part of solute
Very soluble	Less than 1
Freely soluble	1–10
Soluble	10–30
Sparingly soluble	30–100
Slightly soluble	100–1,000
Very slightly soluble	1,000–10,000
Practically insoluble or insoluble	10,000+

soluble Capable of mixing with a liquid (dissolving) to form a homogeneous mixture (solution). Cf. *solubility, solution.* **s. barbital** Sodium *barbitone.* **s. cotton** Nitrocellulose. **s. glass** Sodium silicate. **s. mercury** $NH_2Hg_2NO_3 = 479.2$. Hahnemann's mercury. Black precipitate on adding ammonia to mercurous nitrate. **s. starch** See *starch soluble.* **s. tartar** Ammonium potassium tartrate*. **s. tartrate** Potassium tartrate.
solum A damp-resisting layer of material installed on the ground under a floor, e.g., bitumen.
solute A substance that mixes with or dissolves in a solvent to produce a solution.
solution (1) Dissolution. The mixing of a solid, liquid, or gaseous substance (solute) with a liquid (the solvent), forming a homogeneous mixture from which the dissolved substance can be recovered by physical processes. (2) The homogeneous mixture formed by the operation of s. **anisotonic ~** Any nonisotonic s.; as, a hypotonic or hypertonic s. **aqueous ~** A s. in which water is the main solvent. **buffer ~** A s. of acid or basic salts that can neutralize either acids or bases without appreciable change in hydrogen-ion concentration. **centinormal ~** A s. containing 0.01 equivalents per liter. **chemical ~** A s. in which solute and solvent react to form a compound that dissolves in the solvent and cannot be recovered by distillation. Cf. *physical solution.* **colloidal ~** A macroscopically homogeneous, microscopically heterogeneous, system of minute particles (colloid, dispersed phase) suspended in a liquid (continuous phase, medium). Cf. *colloid.* **concentrated ~** A s. in which the solute content is relatively great. **decinormal ~** A s. that contains 0.1 equivalents per liter. **dilute ~** A s. in which the solute is relatively small in quantity. **gram molecular ~** Molar s. **heat of ~** See *heat of solution.* **hypertonic ~** A s. whose osmotic pressure is greater than that of blood serum. **hypotonic ~** A s. whose osmotic pressure is less than that of blood serum. **ionic ~** A s. whose ions of the solute are surrounded by oriented molecules of the solvent. **isotonic ~** A s. having an osmotic pressure equal to that of blood serum; as, 0.9% w/v sodium chloride s. **molal ~** A s. containing 1 g molecule (mole) of substance per 1,000 g of s. **molar ~** A s. containing 1 g molecule of substance per liter. Cf. *normal solution.* **molecular ~** A true s. in which the molecules of solute are surrounded by molecules of solvent. Cf. *colloidal solution, ionic solution.* **normal ~** A s. containing 1 gram equivalent per liter. **normal salt ~** A s. containing 1 mole sodium chloride per liter. Cf. *isotonic*

solution. **physical ~** A s. in which solute and solvent mix but do not react chemically; the solute can be recovered on evaporation, the solvent by distillation. Cf. *chemical solution.* **physiological ~** Isotonic s. **saturated ~** A s. that normally contains the maximum amount of substance able to be dissolved. **solid ~** See *solid solution, sosoloid.* **standard ~** A s. that contains a definite amount of substance dissolved; as, a molar s. **standardized ~** A s. adjusted to a known concentration. **supersaturated ~** A s. that contains a greater quantity of solid than can normally be dissolved at a given temperature; on slow cooling, the excess precipitates under suitable conditions. **test ~** T.S. A reagent s. **volumetric ~** V.S. A standard analytical s., usually containing 1, $\frac{1}{2}$, or $\frac{1}{10}$ mole of a substance dissolved in 1 liter of water.

s. mining Winning soluble salts (as potassium chloride) by pumping water into the formation and evaporating the resulting solution. E.g., Frasch process. **s. pressure** The tendency of atoms or molecules to mix with a liquid, or to dissolve in it; measured by the osmotic pressure. **s. tension** The tendency of atoms or molecules to dissolve in a liquid and form ions; measured by the electromotive force. See *Nernst theory.* **s. theory** See *Nernst theory, Arrhenius theory.*

solvate A molecular or ionic complex of molecules or ions of solvent with those of solute; as $Cl(H_2O)_n^-$. The ions are surrounded by a zone of oriented water molecules. **crystalline ~** A crystal containing solvent as part of its lattice. **s. theory** The abnormalities of solutions are due to the formation of complexes between the ions or molecules of the solute and solvent. Cf. *hydration.*

solvation Any stabilizing interaction between solute and solvent; if the latter is water, hydrates or hydrated ions are formed, e.g., $M(H_2O)_n$.

solvatochromism The formation, by molecular addition, of a colored complex (solvate) between colorless molecules of organic compounds and those of other compounds.

Solvay **S., Ernst (1838–1922)** Belgian industrial chemist. **S. process** Making sodium carbonate and calcium chloride by treating sodium chloride with ammonia and carbon dioxide. The sodium hydrogencarbonate produced is heated and some carbon dioxide recovered; the ammonia is recovered by lime or magnesia.

solvent (1) That component of a homogeneous mixture which is excess. (2) A liquid which dissolves another substance (solute), generally a solid, without any change in chemical composition; as, water containing sugar. (3) A liquid that dissolves a substance by chemical reaction; as, acids and metals. **acid ~** A s. that acts as an acid by losing a proton to the solute. **aqueous ~** Water. **associating ~** A s. whose molecules form complexes; as, water. Cf. *bond.* **basic ~** A s. that acts as a base by gaining a proton from the solute. **chemical ~** See (3). **ionizing ~** See *polar solvent.* **lacquer ~** Organic liquids used to dissolve resins and nitrocellulose: **low-boiling ~** b. below 100 (alcohol). **medium-boiling ~** b. near 125 (toluene). **high-boiling ~** b. 150–200 (xylene). **plasticizer and softener ~** b. near 300 (camphor). **molten ~** Flux. **nonaqueous ~** A solvent other than water. **nonassociating ~** A s. that does not form complexes between its molecules or ions and the solute; as, benzene. **nonionizing ~** Nonpolar. **nonpolar ~** A s. that does not conduct an electric current; as, hydrocarbons. **normal ~** Nonassociating. **physical ~** A s. that does not react chemically with the solute. **polar ~** A s. that produces electrically conducting solutions (as, water), and causes dissociation of the solute into ions. **two-type ~** A s. having 2 groups which confer s. properties; as alcohol-ethers, HO·R·O·R; e.g., Cellosolve. **universal ~** Aqua regia.

s. action A process of making substances water-soluble.

solvolysis The effect of the nucleophilic character of a solvent on the reactions of the solute dissolved in it.

solvolytic Pertaining to solvation. **s. dissociation** Ion formation in a nonaqueous solution. Cf. *solvate theory.*

somatic Pertaining to the body; usually to cells other than gametogenic and gemete cells. S. cells have the diploid number of chromosomes.

sombrerite A "hard" mineral phosphate (35% phosphorus pentaoxide); a source of phosphorus.

Sommelet reaction The production of benzaldehyde by reaction between benzylamine and formaldehyde, preferably in presence of hexamine.

Sommerfeld **S., Arnold (1868–1951)** German physicist; developed quantum theory of atomic structure. **S. notation** See *quantum number.*

somnifacient A *hypnotic,* q.v.

somnirol $C_{32}H_{44}O_7$ = 540.7. A monohydric alcohol of *Withania* species (Solanaceae).

Somnitol $C_{33}H_{46}O_7$ = 554.7. Trademark for an alcohol from *Withania* species (Solanaceae).

Somophyllin Trademark for aminophylline.

Soneryl Trademark for butobarbital.

sonic Phonic. Pertaining to sound which is audible to the human ear. See *sound frequency.* Cf. *infrasonic, ultrasonic.*

Sonnenschein **S., Franz Leopold (1819–1879)** German forensic analyst. **S.'s reagent** A solution of phosphomolybdic acid forms a yellow precipitate with the sulfates of alkaloids.

sonochemistry The use of high-intensity ultrasound radiation to induce chemical reactions. Acoustic cavitation causes localized areas of high temperature and pressure.

sonoluminescence Luminescence induced by sound waves.

sonometer Phonometer. An instrument to measure sound vibrations.

sonora gum The exudiations of the creosote bush, *Covillea tridentata* (Mexico.).

soot An impure black carbon containing oily and empyreumatic compounds from the incomplete combustion of resinous materials or wood. It contains hydrocarbons, and if derived from coal, ammonium sulfate. Cf. *lampblack.*

sophora Coral bean. The poisonous seeds of *Sophora* species (Leguminosae), India.

sophorine An alkaloid from *Sophora* species. Colorless liquid resembling cystine and matrine. Cf. *kuhseng.*

soporific An agent that produces sleep. Cf. *hypnotic.*

sorbic acid (*E,E*)-2,4-Hexadienoic acid*. **hydro ~** Hexenoic acid*. **para ~** A lactonelike compound forming sorbic acid when heated with acid or alkali.

sorbide Sorbitan with one further water molecule removed. Any of a group of surfactants, used as emulsifiers.

sorbin, sorbinose Sorbose.

sorbitan $C_6H_8O(OH)_4$ = 164.1. Sorbitol anhydride. Generic name for anhydrides of sorbitol; derived by removal of 1 molecule of water. With fatty acids, as oleic and stearic, sorbitans form nonionic surface-active agents that are used as emulsifiers. Cf. *sorbide.*

sorbite (1) Sorbitol. (2) A mixture of ferrite and cementite, with conglomerations of carbon in steel; a transition form between pearlite and troostite.

sorbitol $HO{\cdot}H_2C(CHOH)_4CH_4OH$ = 182.2. Glucitol†, sorbite. **D-~** Occurs in many plants. Colorless crystals, m.111, soluble in water. Used chiefly for the preparation of ascorbic acid and in parenteral feeding (NF, BP). Also a sweetener and humectant and used in surfactants, pharmaceuticals, foods, and rigid polyurethane foams. **s. anhydride** See *sorbitan.*

Sorbol $C_{34}H_{70}O$ = 494.9. Trademark for an alcohol, m.78, from the wax of the berries of *Sorbus aucuparia.*

sorbose $HOH_2C(CHOH)_3 \cdot CO \cdot CH_2OH$ = 180.2. Sorbin(ose). An optically active carbohydrate from the fruit of mountain ash, *Sorbus.* Colorless rhombs, m.165, slightly soluble in water.

Sorbothane Trademark for a polyurethane-based quasi liquid that behaves as a molecular spring. It absorbs energy upon impact and dissipates it over a period of time.

sorbus Rowan tree, mountain ash. The tree *Pyrus (Sorbus* or *Mespilus) aucuparia* (Rosaceae). A decoction of the bark contains sorbitol and sorbose.

Sorel S. cement $MgO \cdot MgCl_2 \cdot 11H_2O$. A hard, quick-setting mixture of magnesium oxide and a concentrated solution of magnesium chloride. **S. floor cement** A mixture of magnesium oxide, zinc chloride, and portland cement, used for floors; 10% copper powder makes it waterproof.

Sörensen S., Sören P. L. (1868–1939) Danish chemist. **S. indicators** A group of hydrogen-ion-concentration *indicators,* q.v. **S. phosphate** Di*sodium* hydrogen *phosphate**. **S. symbols** See *pH.* **S. value** Hydrogen-ion concentration.

Soret effect, Soret principle Ludwig phenomenon. When differences of temperature are maintained in a salt solution, the solute will concentrate in the coolest parts.

sorethytan 20-monooleate Polysorbate.

sorghum A cane, *Andropogon sorghum* or *Sorghum vulgare,* from which a sugar and Indian millet (African durra) are obtained.

sorgo *Sorghum vulgare.*

sorption A reaction on a surface, especially *absorption,* q.v., or *adsorption,* q.v., and *persorption* (permeation into a very porous solid.). Cf. *monomolecular layer.* **ab~** See *absorption.* **ad~** See *adsorption.* **re~** See *resorption.* Cf. *zone.*

sorrel The leaves of *Rumex acetosa* (Polygonaceae); a refrigerant. **s. salt** Potassium hydrogenoxalate*.

sosoloid Solid solution. One solid phase dispersed in another. See *colloid.*

Soubeiran, Eugène (1797–1858) French apothecary; discoverer of chloroform.

sound Waves of compression and rarefaction moving outward through a medium, set in motion by a vibrating object. Cf. *musical notes, noise, phon.* Sound waves are described by three different properties: pitch (i.e., frequency or the related property wavelength), amplitude (i.e., intensity or loudness), and waveform (i.e., quality or timbre). At their simplest, sound waveforms can be represented as a sine curve centered on the x axis, in which the perpendicular distance from $y = 0$ to a peak represents amplitude and the horizontal distance between peaks (i.e., across one valley) represents wavelength. **speed of ~** c. Sound travels fastest through solids and slowest through gases. The speed of sound through air (at 0°C) is about 330 m/s, water 1,500 m/s, iron 5,000 m/s. The velocity of sound varies with temperature. For every 1°C rise in temperature, there is an increase in sound velocity through air of about 0.6 m/s. Cf. *Mach number.*

s. amplitude The maximum (or minimum) value of a varying quantity (e.g., sine curve). See *sound intensity* below. **s. frequency** Number of vibrations (wavelengths) per second; measured in hertz. Frequency is perceived as pitch. Frequencies between roughly 20 and 20,000 Hz are audible to the human ear. **s. intensity** I, J. The power transmitted in unit time by a sound wave across a unit area perpendicular to the wave. It is proportional to the square of the amplitude of the vibration causing the s. Measured in W/m^2 or bels (or more conveniently, decibels dB). The difference between two sounds in *bels,* q.v., is related logarithmically to the ratio of their power. Thus two sounds of relative power 1 and 1,000 will have a difference of 30 dB. S. greater than 90 dB may cause damage to hearing, depending on the time of exposure. Cf. *sound loudness.* **s. loudness** L_N. The perceived effect of sound, the magnitude of which depends on the hearer. Measured in phon. Loudness can be related to *sound intensity,* q.v., by using a standard source of frequency (e.g., 1,000 Hz) and sound intensity (e.g., 10^{-12} W/m^2). **s. wavelength** The distance between two identical phases (such as two areas of greatest compression) of a sound wave. Wavelength = velocity ÷ frequency.

sourwood The leaves of *Oxydendrum arboreum,* N. America.

southern wood *Abrotanum.*

sovprene A polymerized 2-chloro-1,3-butadiene.

Soxhlet S., Franz (1848–1913) German food analyst. **S. apparatus** A flask and condenser for the continuous extraction of alcohol- or ether-soluble materials.

soy, soya Soybean.

soybean Soja, soy, soya. The bean of *Soja hispida* (Leguminosae), China. An important local food, and source of many preparations; as, flour, sizing materials, bean cake, sauce, oil, cheese. Average composition: proteins 40, oils 18% (phospholipids 2%), urease, raffinose, stachyose, saponins, phytosterols, and isoflavone. **s. oil** Colorless, liquid oil expressed from s., d_{15} 0.925, free fatty acids (as oleic acid) 0.46%, sapon. val. 191, iodine val. 129; used in cooking (USP).

Soylon Trademark for a protein synthetic fiber.

sozolic acid Aseptol

sp. Abbreviation for: (1) spirit, (2) specific.

space The three-dimensional concept of length, width, and height: l^3, where l is a unit of length. **Crookes' ~** Dark s. **dark ~** A nonluminous region near the cathode of a vacuum tube, through which a high-frequency current is passing. **interatomic ~** The region between the outermost orbitals of 2 atoms. **intraatomic ~** The region within the outermost orbital of an electron and the nucleus of an atom. It consists of the nucleus and the orbitals of the valence electrons.

s. group A characteristic arrangement of atoms in a crystal. **s. lattice** Bravais lattice. The characteristic pattern formed by the spatial distribution of atoms or radicals in a crystal. In noncrystalline solids the s. l. is distorted, and in truly amorphous solids there is no order. **homopolar ~** A s. l. in which the constituents are neutral atoms linked to a number of adjacent similar atoms by chemical valencies (electron sharing); e.g., diamond. **ionic ~** A s. l. in which each ion of a given charge is equidistant from a small number of ions of opposite charge, arranged equidistantly around it; e.g., the sodium chloride crystal. **metallic ~** A s. l. composed of atoms of a metal which have lost one or more valency electrons; these are thus free to produce conductivity. **molecular ~** A s.l. composed of a regular arrangement of molecules held together by van der Waals forces, the constituent atoms being held by valencies.

spalling (1) The failure of a refractory material under stresses induced by temperature fluctuations. (2) The cracking or flaking of particles from a metal surface, e.g., wheels. (3) In general, splintering, cracking or breaking due to heat.

Span (1) Trademark for a group of sorbitan ester emulsifiers.

(2) (Not cap.) 9 inches. A little-used unit of length in the old English system.

spandex Generic name (U.S. Textile Fiber Products Identification Act) for stretch fibers based on synthetic, elastomeric long-chain polymers. S. comprises at least 85% polyurethane (U.S. Federal Trade Commission), e.g., Vyrene. Monofilament s. fiber returns to its original length after being stretched several times that length.

Spanish **S. broom** Spartium. **S. fly** Cantharides. **S. hops** Origanum. **S. moss** Vegetable *horsehair.*

spar A transparent or translucent, readily cleavable, crystalline mineral of vitreous luster; as, fluorspar.

sparassol $C_6H_2(COOMe)Me(OMe)OH$ = 196.2. Methyl everninate. **1,2,4,6-~** An ester and ether from the oil of *Sparassis ramona,* a lichen. Colorless powder, m.68. Cf. *lichenol.*

Sparine Trademark for promazine hydrochloride.

spark A flash of light produced chemically or physically; as, burning iron wire in oxygen or an electric discharge in air. **s. spectrum** An emission spectrum characteristic of the electrode metal, obtained from a high-voltage discharge between metallic electrodes. It is characterized by enhanced lines due to ionized atoms. Cf. *arc spectrum, isostere.*

sparking Producing electric sparks. Cf. *arcing.* **s. potential** The electric potential necessary to produce a spark in vapor or gas at ordinary temperature. It depends on the distance apart, shape, and size of the electrodes.

sparklet Device containing (1) compressed carbon dioxide, or (2) a powder (similar to baking powder) for the generation of small quantities of carbon dioxide in the laboratory or household (for beverages).

spartalite Native zinc oxide, usually pink in color due to Fe or Mn; found in New Jersey.

sparteine $C_{15}H_{26}N_2$ = 234.4. Lupinidine. An alkaloid from *Leguminosae genera.* Cf. *scoparius.* **(−)-~** Colorless oil, d.1.02, b_{18mm}180, soluble in water; a heart stimulant. Cf. *cytisine, anagyrine.*

spartium (1) Scoparius. (2) The fiber from *S. junceum,* Spanish broom (Leguminosae). Cf. *esparto.* **s. alkaloids** A group of alkaloids derived from scoparius; e.g., sparteine, $C_{15}H_{26}N_2$.

spasmotin $C_{20}H_{21}O_9$ = 405.4. Sphacelotoxin. A poisonous principle from ergot. Yellow powder, soluble in alcohol.

spathic Describing a lamellar or foliated structure. **s. iron ore** Siderite.

spathose Siderite.

spatial Pertaining to space.

spatula A blunt knife for mixing or transferring small quantities of powders.

spavin Warrant. A fireclay found underneath coal deposits; used for refractories.

spearmint Mentha veridis. The dried leaves of *Mentha spicata* (Labiatae); a carminative and flavoring (NF). **s. oil** Oleum menthae viridis, from the flowering plant *Mentha spicata.* Colorless oil, d.0.917–0.934, soluble in alcohol; it contains carvone, linalol, pinene (NF, BP).

spear pyrites Marcasite.

species (1) Subdivision of a genus of plants or animals. (2) A type; as, atomic s. **aromatic ~** A mixture of thyme, pepper mint, lavender, and cloves. **laxative ~** St. Germain tea. A mixture of senna, elder flowers, fennel, anise, and potassium hydrogentartrate. **pectoral ~** Breast tea. A mixture of althaea, coltsfoot, licorice, anise, mullein, and orris.

specific (1) Pertaining to a particular kind of matter, microorganism, plant, or animal. (2)* Before an extensive physical quantity, meaning that quantity divided by the mass. Cf. *molar.* **s. charge** The charge per unit mass of an elementary particle. For a slow electron, 1.758805 × 10^{11} C/kg. **s. conductance** Electrolytic *conductivity**. **s. conductivity** Conductivity*. **s. energy** See *specific energy* under *energy.* **s. gravity** Relative *density**. **s. heat capacity*** S. heat. The number of J (or cal) required to raise the temperature of 1 g of material by 1°C. For gases, it may be at constant pressure (c_p) or constant volume (c_V). **s. impulse** See *propellant.* **s. inductive capacity** Relative *permittivity.* **s. reaction rate** Rate constant*. **s. refraction*** See *specific refraction* under *refraction.* **s. resistance** Resistivity*. **s. rotation** See *specific rotation* under *rotation.* **s. optical rotatory power*** $\alpha_m = \alpha V/ml$, where α is the angle of rotation, l the length of the light path, and m the mass of substance in volume V. **s. surface** A/V. The ratio of the area A per unit volume V of a system. **s. volume*** Bulk. The volume occupied by a unit mass of material; the reciprocal of density. **s. weight** Weight *density**.

specpure Spectroscopically pure; e.g., as, a substance giving a pure spectrum, characteristic of itself only.

spectinomycin $C_{14}H_{24}O_7N_2 \cdot 2HCl \cdot 5H_2O$ = 495.4. Trobicin. An antibiotic produced by *Streptomyces spectabilis.* White crystals, soluble in water. Used for gonorrhea (USP, BP).

spectra Plural of spectrum. Classification:

Source of light:

Arc s.	Solar s.	Vacuum arc s.
Explosion s.	Spark s.	Vacuum flash s.
Flame s.	Stellar s.	Vacuum spark s.
Flash s.		X-ray s.

Kind of spectrum produced (See *spectrum types*):
- Emission s. or absorption s.:
  - Continuous s.:
    - A hot solid body
  - Discontinuous s.:
    - Line s.—atoms
    - Band s.—molecules

Mode of production:

Crystal-diffracted s.	Grating-refracted s.:
Grating-diffracted s.:	Primary
Primary	Secondary
Secondary	etc.
etc.	Prism-refracted s.

spectral Pertaining to a spectrum. **s. analysis** See *spectrum analysis.* **s. classification** S. types. Harvard Star classification. A systematic arrangement of the stellar spectra in a continuous sequence of types. Based on the largest and hottest (80 kK) stars emitting characteristically gaseous spectra at the blue end, through to the smaller, cooler (2.5 kK) stars emitting molecular spectra at the red end.

W. Carbon and nitrogen lines.
O. Helium lines.
B. Hydrogen and helium lines.
A. Hydrogen lines.
F. Hydrogen and faint metallic lines.
G. Many metallic lines, but no compounds.
K. Strong metallic lines, very weak bands.
M and S. Banded spectra of metallic oxides.
R and N. Increasing molecular bands.

s. tube An evacuated glass tube containing rarefied traces of gas, which produces light of characteristic spectrum when an electric current is passed between 2 electrodes. **s. types** (1) S. classification. (2) See *spectra.*

spectrochemical analysis Spectroscopic analysis.

spectrochemistry A branch of science that utilizes electromagnetic radiation for chemical analysis. See *spectrum analysis.*

spectrogram The photographic record of a spectrum with a standard comparison spectrum.

spectrograph (1) An instrument to produce a spectrogram, consisting of the slit, the lenses (collimator or camera), the dispersing system (prism, grating), and the detector (photographic, thermal, or ionic). (2) Misnomer for a device to convert speech sound waves into electrical impulses which produce a graph (voiceprint) said to be characteristic of the individual. **quartz ~** A s. used for wavelengths of 2000–8000 Å. **X-ray ~** A device for obtaining crystallograms.

spectroheliograph A device for photographing the sun's surface by means of a spectroscope employing a specific spectrum line.

spectrometer A spectroscope, but with some form of electrical measuring detection device, such as a servo recorder. **constant deviation ~** A s. in which the collimator and telescope are fixed permanently at right angles, and the prism is rotated.

spectrometry The measurement of the wavelengths and intensity of the lines or bands in a spectrum and their identification with the atoms, ions, or molecules producing them.

spectrophotometer (1) A device to measure photometrically the quantity of light of any particular wavelength range absorbed by a solution. Also measures the electromagnetic radiation emitted or scattered from solutions, gases, or solids. (2) A device to measure the intensity of the photographic image of a spectral line. **atomic absorption ~** A s. for determination of elements from their capacity to absorb light of characteristic frequency in the atomic state.

spectropolarimeter An instrument to measure optical rotation for different wavelengths of light. Cf. *dispersion.*

spectroscope (1) An instrument for analyzing light by separating it into its component rays. It consists essentially of a device for making the rays parallel (collimator), a dispersive device (prism that refracts, or a grating that diffracts, the light), and a detector such as a photomultiplier or the human eye. (2) A device by which radiations are separated into component parts; as, mass s. **abridged ~** An absorptiometer used for restricted spectral ranges. **Auger ~** A s. using secondary electrons ejected from a sample on bombardment with electrons or X-ray photons. **comparison ~** A s. for comparing 2 spectra side by side. **direct-vision ~** A low-dispersive s., with a crown-glass collimator along the axis of a single tube with the train of prisms and eyepiece. **grating ~** A s. of high dispersive power in which a series of spectra is formed by either (1) a reflection grating, a finely ruled piece of plane or concave (self-focusing) speculum metal or (2) a cast of the original transmission grating. **inelastic electron tunnel ~** A s. for investigating thin-surface adsorbates. The energy exchange between thin-film electrodes, separated by an oxide layer bearing the adsorbate, is measured when a potential is applied to the electrodes. **mass ~*** See *mass spectroscope.* **measuring ~** A s. adapted to the measurement of the wavelengths of the component rays. **micro ~** See *microspectroscope, spectrometer.* **nuclear-magnetic-resonance ~** A s. that measures the absorption of energy by spinning nuclei in a strong magnetic field. The sample is subjected to radio-frequency radiation from a transmitter at a fixed frequency (thus, 60-, 100-, 300-MHz instruments), and the energy absorption spectrum is obtained by varying the applied magnetic field. **optoacoustic ~** See *optoacoustic.* **photoacoustic ~** See *optoacoustic.* **photographic ~** A s. with the viewing telescope replaced by a camera; a *spectrograph.* **prism ~** A s. whose dispersive power depends on one or more prisms of glass, quartz, or rock salt. **reflectance ~** A s. measuring the light reflected diffusely or specularly, as radiant energy, from a solid; as, with inks, plastics, dyestuffs. **reversion ~** A s. that enables the same spectrum to be reproduced twice, but in reverse directions, thus doubling any displacement of absorption bands; used to determine carbon monoxide in blood.

spectroscopic Associated with the spectroscope. **s. analysis** Spectrum analysis, spectrochemical analysis. Minute quantities of elements may be detected by their characteristic spectral lines or bands. Cf. *flame photometer.* Methods:

1. A flame (bunsen burner) colored with a small quantity of the substance (in nm):
   Barium, green line 553, bands 534–524, etc.
   Calcium orange band 620–618
   Cesium, blue line 455, 459
   Copper, green and blue lines and bands
   Indium, blue line 451, purple line 410
   Lithium, red line 671
   Potassium, red lines 766, 769; purple line 404
   Radium, red band 670–653
   Rubidium, purple lines 420, 421
   Sodium, yellow line 589
   Strontium, red bands 686–674–662, 606
   Thallium, green line 535
2. An electric spark passing through a cup containing a solution of the substance: for Mg, Fe, Mn, Zn, Co, Ni, Cr.
3. Examination of the spectrum after electromagnetic radiation has passed through a solution of the substance. The absorption spectrum is used for molecular (infrared, n.m.r, etc.) and atomic species. The emission spectrum is also used for molecular (luminescence, Raman) and atomic (flame photometry, inductively coupled plasmas) species.

spectroscopically Pertaining to spectroscopy. **s. pure** Specpure. Showing a degree of purity enabling a substance to be identified s., the essential lines being apparent without interference by impurities.

spectroscopy The study of the properties of light by means of the spectroscope.

spectrum (1) A variously colored band of light showing in succession the rainbow colors or isolated lines or bands of colors; produced by refraction through a prism, or by diffraction by a grating. Cf. *spectra.* (2) A similar band of electromagnetic radiation, invisible to the eye and extending beyond the violet (ultraviolet) or red (infrared) portions of the visible spectrum into, respectively, the X-ray and microwave regions. See Table 69 under *radiation.* **absorption ~** The visible or invisible s. produced by a composite ray of light after it has passed through a solution or through a layer of vapor or gas, which absorbs one or more of the constituent rays. **arc ~** A s. from a substance placed between the carbon poles of an arc. **Aston ~** See *mass spectra.* **Auger ~** See secondary *electron.* **band ~** Lines so close together that they appear as a continuous band; due to molecular vibrations and rotations. Cf. *line spectrum* below **bright-line ~** See *line spectrum, flash spectrum* (both below). **chemical ~** That portion of the s. which contains the most chemically active wavelengths, as, ultraviolet rays. **comparison ~** A s. having sharp lines, as standard; usually photographed on the same plate above and below the spectrum of the sample, e.g., the arc or spark spectrum of titanium or iron. **continuous ~** A s. in which no Fraunhofer lines are visible; there is an uninterrupted change from one color to another. **dark-line ~** Reversal s. A s. containing Fraunhofer lines. Cf. *absorption spectrum* above. **diffraction** A s. produced by means of a grating. **discontinuous ~** A combined line and bands. **electrochemical ~** See *electrochemical spectrum.* **electromagnetic ~** See *radiation.* **electronic ~** Visible and ultraviolet spectra produced by changes in the electronic state of molecules. **electron spin resonance ~** See *electron spin resonance.* **emission ~** Bright lines from a source. **explosion ~** A s. produced by exploding a metallic wire or a solution on an asbestos fiber by means of an electric current. It shows lines from excitation states above the spark s. **flame ~** A s. with a bunsen flame as the source of excitation. **flash ~** (1) Explosion s. (2) A s. showing bright Fraunhofer lines on a dark background; seen during a solar eclipse. **fluorescence ~** See *fluorescence.* **furnace ~** A s. obtained with an electric furnace (3000°C) source; intermediate between the flame and arc s. **high-frequency ~** An X-ray s. produced by high-frequency currents. **hyperfine ~** Extremely thin lines close together; due to nuclear vibrations. Cf. *band spectrum* above. **infrared ~** The thermal-ray region beyond the red end of the s. **invisible ~** See *ultraviolet spectrum* below, *infrared spectrum* above. **line ~** Colored or bright lines on a dark background, or dark lines on a light background; due to atomic species. Cf. *hyperfine spectrum* above. **magnetic ~** See *magnetic spectrum.* **mass ~** Aston s. The images produced when canal rays are subjected to electric and magnetic fields, which separate the ions according to their mass. **microwave ~** S. in the microwave region; used to determine molecular geometry and to detect free radicals. **molecular ~** See *band spectrum* above, *molecular rays.* **Moseley ~** See *X-ray spectrum* below. **nebular ~** A s. obtained by photographing gaseous, planetary, or spiral nebulae with telescope and spectrograph. **normal ~** Diffraction s. A s. produced by a grating; it shows the rays of different wavelengths in proper relationship to one another, and less distorted than with a prism. **phosphorescence ~** See *phosphorescence (1).* **photoelectron ~** A s. showing the kinetic energy distribution of photoelectrons ejected from a solid. Consists of a series of peak values that permit identification of the atom involved in the ionization process. **photographic ~** A s. that affects photographic emulsions. **photopic ~** A s. bright enough to arouse color sensations in the human eye. Cf. *scotopic spectrum* below. **planetary ~** A polarized solar s., produced by light reflected from planets. **primary ~** The most prominent s. produced by a grating. **Raman ~** See *scattering, Raman spectrum.* **reversal ~** See *dark-line spectrum* above. **roentgen ~** X-ray s. **rotational ~** A s. in the microwave region produced by changes in the rotational states of nonhomopolar molecules. **scotopic ~** The s. seen by the dark-adapted human eye. Cf. *photopic spectrum* above. **secondary ~** The second most prominent s. produced by a grating. **secondary-electron ~** A spectrum used in very sensitive and nondestructive detection of substances on a surface by bombardment with low-energy electrons and measuring the energies of the secondary (Auger) electrons emitted which are characteristic of the atoms from which they originate. **solar ~** A s. produced by the light of the sun; it shows Fraunhofer lines. **spark ~** A s. produced by the excitation of a vapor by electric sparks; a characteristic s. is obtained for each electrode metal. **stellar ~** A s. produced by the light of stars. **sunspot ~** A s. obtained by passing the light from sunspots through a quarter wave plate or nicol prism; it shows the Zeeman effect. **ultraviolet ~** The dark portion beyond the violet end of the s.; chiefly chemically active rays. Cf. *irradiation.* **vibrational ~** Infrared s. produced by changes in the vibrational state of nonhomopolar molecules. **X-ray ~** Moseley s. Lines characteristic of the metal used as anticathode in an X-ray tube. The X-rays are diffracted by a crystal acting as a grating. The frequencies are proportional to the atomic number of the element used as anticathode.

s. analysis (1) The measurement of the intensity and frequency of s. lines. (2) Spectroscopic analysis. (3) The analysis of the structure of a spectrum, e.g., resolution into series or multiplets. **s. classes** See (1) *spectra,* (2) *spectral classification.* **s. lamp** See *spectrum lamp* under *lamp.* **s. lines, s. series** A mathematical relationship existing between the lines or group of lines in the s. of an element: $\nu = 1/\lambda = L + BR/(m + \alpha + \beta/m^2)^2$, where

ν = wavenumber in vacuo, waves/cm
L = wavenumber of the limit of the series
R = Rydberg constant
m = a variable integer corresponding with a definite line
α, β, B = constants

s. types

1. Continuous; as, from an incandescent lamp, carbon arc, or hot, glowing body.
2. Discontinuous:
   (a) Bright-line s., as, incandescent gas, flame of sodium, arc of iron, or spark spectra
   (b) Dark-line s.; as, solar s., or continuous s. absorbed by a hot, gaseous envelope
   (c) Bright-band s.; as, comet s.
   (d) Dark-band s.; as the absorption s. of dye solutions.

specular Mirrorlike. **s. coal** Pitch coal. A shining variety of coal. **s. hematite, s. iron** Specularite. **s. metal** Speculum metal.

specularite Fe_2O_3. Specular hematite, specular iron, gray hematite. Native, disklike crystals, with metallic luster.

speculation A conclusion drawn from incomplete knowledge of facts. Cf. *deduction, hypothesis.*

speculum metal An alloy: Cu 66, Sn 33%, As trace; used in making mirrors.

speed Popular term for amphetamine.

speise A usually fusible and brittle, native arsenide.

speisequark A German soft-curd cheese.

speiss cobalt $(FeNiCO)As_2$. An impure smaltite.

speleology The study of cave formation.

spelter Commercial zinc used for galvanizing. **hard ~** S. recovered from galvanizing-bath dross (10% iron).

Spencer, Leonard James (1870–1959) British mineralogist.

Spergon Trademark for tetrachloro-*p*-benzoquinone; a seed dressing to protect against smut disease.

sperm A male reproductive cell. **s. oil** The oil of the s. whale. **s. whale** The mammal *Physeter macrocephalus,* a source of spermaceti and ambergris.

spermaceti Cetaceum. The solid fat from the head of the sperm whale, *Physeter macrocephalus,* Linné; chiefly hexadecyl palmitate. White, unctuous mass of faint odor, d.0.94, m.45; used in ointments (USP), and standard candles.

spermicide See *contraceptive.*

sperrschicht cell A photoelectric cell sensitive over the visual spectrum.

sperrylite $PtAs_2$. A native arsenide.

spessartite $3MgO.Al_2O_3.3SiO_2$. A red aluminum garnet (Germany).

sp. gr. Abbreviation for specific gravity.

sphacelic acid An acid from ergot.

sphacelotoxin Spasmotin.

sphaerite A hydrous phosphate of aluminum.

sphalerite ZnS. Pseudogalena. Zinc blende, cleiophane blende, rosin jack, blackjack, mock ore. An isometric, native zinc sulfide. Cf. *wurtzite.*

sphene $CaTiSiO_5$. Titanite. A native, lustrous, brown calcium silicotitanite, found in rocks.

sphenoid A hemihedral crystal or half-crystal.

spherical Bell-shaped or globular.
spherocobaltite $CoCO_3$. A native carbonate.
spheroidal Sphere-shaped.
spheroidization The formation of rounded grains in alloys, generally during annealing.
spherometer An instrument to determine surface curvature.
spheronizing A process for converting fine particles of powder into pellets by maintaining them in turbulent motion against baffle bars and so forming agglomerates, e.g., for increasing the apparent density of carbon black.
spherulite A spherical aggregation of outward-radiating fibers, as from a solidifying polymer melt; gives a maltese cross in polarized light.
sphingoin $C_{17}H_{35}O_2N$ = 285.3. A brain leucomaine.
sphingolipid A lipid containing a long-chain base. General name for ceramides and cerebrosides from animal tissue.
sphingomyelins A class of sphingolipids; as, $C_{46}H_{96}O_6N_2P$ = 804.2. A phospholipid from brain, hydrolyzing to phosphoric acid, choline, sphingosine, and cerebronic acid.
sphingosine $C_{18}H_{37}O_2N$ = 299.5. (2*S*,3*R*,4*E*) = 2-Amino-4-octadecene-1,3-diol. m.80. A split product of cerebrosides.
sp. ht. Abbreviaton for specific heat.
sphygmograph A device for recording the pulse.
sphygmomanometer A device for recording blood pressure.
spice A condiment or substance used to give flavor and distinctive taste to food; the majority of spices contain essential oils.
spicular Needle-shaped.
spider poison Arachnolysin.
spiegel (1) Spiegeleisen. (2) German for "mirror."
spiegeleisen Spiegel. A white cast iron (Mn 5–20%), obtained in the blast furnace; used to make manganese steel.
Spiegler Jolle's reagent A solution of mercuric chloride 2, succinic acid 4, sodium chloride 4 g in 100 mL water; a reagent for albumin in urine.
spigelia Pinkroot, Indian pink. The dried roots of *Spigelia marilandica* (Loganiaceae); contains spigeline; a teniafuge.
spigeline An alkaloid from *Spigelia* species.
spike (1) Plantain. (2) Pepper. **s.nard.** See *spikenard.*
spikenard The aromatic root of an herb, *Nardostachys Jatamansi* (Valerianaceae). **American ~** Aralia.
spilanthol $C_{14}H_{23}ON$ = 221.3. (2*E*,6*Z*,8*E*)-2,6,8-Decatrienoic isobutylamide, isisilli (local name), m.23. From *Spilanthus acmella* (Compositae), S. Africa; a toothache remedy.
spin Angular momentum; rotation around an axis. **antiparallel ~** The s. of 2 particles in opposite directions. **electron ~** The quantized spin angular momentum, equal to $Sh/2\pi$; *S* being the spin quantum number of $+\frac{1}{2}$ (clockwise) or $-\frac{1}{2}$, and *h* the Planck constant. **parallel ~** The s. of 2 particles in the same direction.
spinacene Squalene.
spinasterol $C_{29}H_{48}O$ = 412.7. (3*β*, 22*E*,24*S*)-7,22-Stigmastadien-3-ol. A phytosterol from spinach.
spindel oil A mixture of fuel oil about 95 and wool fat 5%; used in the textile industry to preserve and lubricate raw fibers.
spindel A hydrometer.
spindle tree Euonymus.
spinel $MgAl_2O_4$. Rubicelle, ruby alamandine, ruby balas. Variously colored, isometric crystals, some used as gems.
spinels $M''M'''_2O_4$. Rock-forming aluminates or ferrates. M″ is magnesium, zinc, manganese, or ferrous iron; M‴ is aluminum, chromium, ferric iron, or manganic manganese. They have a high hardness and refractive index; e.g.; gaehnite, $ZnAl_2O_4$; franklinite, (Fe, Zn, Mn)Fe_2O_4; chromite, $FeCr_2O_4$.
spinneret (1) A platinum thimble with a flat base containing minute holes, through which rayon spinning solution is forced to form filaments. (2) The spinning organ of spiders.
spinning (1) Revolving around an axis. (2) The production of a synthetic filament by extrusion of a liquid under pressure through fine holes in a spinneret. **dispersion ~** S. in which a dispersion of fine particles of the material is made in a viscous medium. **dry ~** S. in which a hot gas evaporates the solvent of the liquid. **melt ~** S. in which the liquid is molten and solidifies into a filament on cooling. **reaction ~** S. in which solidification is effected by polymerization. **wet ~** S. in which the liquid is coagulated by a chemical precipitant.
s. electron The fourth motion of an electron in an atom. Cf. *Pauli's principle.* **s. power** The property of a fluid which enables it to be drawn out into threads; as, egg white.
spinulosin $C_8H_8O_5$ = 184.2. 2,5-Dihydroxy-3-methoxy-6-methyl-1,4-benzoquinone. Purple-bronze plates, m.201; an antibiotic from cultures of the mold *Penicillium spinulosum* (cf. *penicillin*), and *Aspergillus fumigatus.*
spiraeic acid Salicylic acid*.
spirans A spiro compound.
Spirillaceae The third family of the Schizomycetes bacteria; with wave-shaped cells: *Spirosoma,* no organs of locomotion; *Microspira,* rigid cells, few polar flagella; *Spirillum,* rigid cells, many flagella; *Spirochaeta,* flexible and mobile cells.
spirit (1) Any distilled liquid. (2) A solution of a volatile substance in alcohol. (3) Ethanol*. **cologne ~** Ethanol*. **colonial ~**, **Columbian ~** Methanol*. **Libavius ~** Stannic chloride. **methylated ~** See *methylated spirit.* **mineral ~** White s. **motor ~** Gasoline (U.S.) or petrol (U.K.). **potato ~** A whisky distilled from fermented potatoes. **proof ~** See *proof spirit.* **pyroacetic ~** Acetone*. **pyroligneous ~**, **pyroxylic ~** Methanol*. **rectified ~** 90% ethyl alcohol. **silent ~** The alcoholic s. distillate from a spiritous liquor, before addition of denaturants. **white ~** Mineral s. A solvent composed entirely of petroleum products, b.150–190, Abel's closed flash point 78 or over; a paint thinner and turpentine substitute. **wood ~** Methanol*.
s. acid Concentrated acetic acid obtained by distillation of 12% vinegar. **s. of alum** Sulfuric acid*. **s. colors** Aniline dyes insoluble in water, soluble in alcohol; used as stains and silk dyes. **s. of copper** Acetic acid obtained from copper acetate. **s. of hartshorn** Ammonium hydroxide*. **s. of salt** Hydrochloric acid*. **s. of tin** Stannic chloride. **s. of vitriol** Concentrated sulfuric acid. **s. of wine** Ethyl alcohol*. **s. of wood** Methanol*.
spirit(u)ous Having the character of spirit.
spiro **s. atom** The common atom in a spiro union. **s. compounds** Spirans, spirocyclans. Substances containing a s. union. **s. union** The union formed by a single atom which is the only common member of 2 rings; as, spiro[3.3]heptane.

CH_2 CH_2
CH_2 C CH_2
CH_2 CH_2

A free s. u.* is one constituting the only union, direct or indirect, between rings. According to the number of s. atoms present, compounds are distinguished as monospiro, dispiro, etc.
spirocyclans Spiro compounds (Baeyer).
spirodienone A derivative of cyclohexa-2,4- and -2,5-dienone; spirodienones occur widely in nature and are used in biosynthesis.

spirogyra A filamentous alga: a green feltlike mass or scum in freshwater ponds or tanks.
spironolactone $C_{24}H_{32}O_4S$ = 416.6. Aldactone. Yellow, crystals, insoluble in water, m.200; a diuretic (USP).
spitting Small explosions and scattering of materials when certain substances are dried and heated or brought together; as, with sulfuric acid and water. Cf. *decrepitation.*
splash head A device between the flask and condenser of a distillation apparatus to prevent liquid from splashing over into the latter. Usually an open glass tube in a bulb, bent away from the direction of the flask.
split product A decomposition product, e.g., of a hydrolyzed glucoside.
splitting The breaking of a molecule into 2 or more individual atoms.
spodumene $(Li,Na)_2Al_2Si_4O_{12}$. Triphane. The pink (kuntzite) and green (hiddenite) are used as gems. Some spodumene crystals are 14 m long.
spoilage Any detrimental change due to physical or chemical action.
spoil bank Bing.
sponge (1) A marine animal, *Euspongia officinalis* (Poriferae). (2) The flexible, fibrous skeleton of the animal, cut and dried. Yellow, porous masses of various shapes. (3) A metal in porous form, as platinum s. **gelatin ~** A spongy, water-insoluble form of gelatin, solubilized by pepsin; a local hemostatic (USP). **iron ~** See *iron sponge.* **platinum ~** Platinized asbestos. **vegetable ~** Tufa.
spongin A scleroprotein from bath sponge; yields diiodotyrosine and bromine.
spontaneous Sudden or voluntary, and without apparent external cause or incitement. **s. combustion** Self-ignition of flammable material, caused by the accumulation of heat on slow oxidation.
sporangia Cells containing spores.
spores The resting state of microorganisms. Single cells capable of growth and reproduciton.
spot analysis (1) A microchemical identification test, made on a porcelain plate. (2) A reaction on impregnated filter paper.
spout (1) A slightly projecting depression on the rim of a vessel through which the contents are poured. (2) A trough to conduct molten metals.
spray A stream of mixed air and finely divided liquid produced with an atomizer. **s. drying** Rapid evaporation, in which a solution is heated in the atomized state so that the dissolved substance falls out of the spray in the solid state.
sprays (1) Pharmaceutical preparations in spray form; as, nasal s. for hay fever. The active substances are dissolved, usually in liquid. (2) Technical preparations; as, pesticides.
Sprengel pump A high-vacuum mercury pump; capable of 0.001 mmHg (absolute).
spring A source of natural water. **hot ~** Water emerging from the soil at a temperature above 50°C.
spring balance A device for measuring force by the extension it produces in a spring. See *Hooke's law.*
sprinkler (1) A perforated plate near the top of a gas-scrubbing tower. It produces a rain of gas-absorbing liquid. (2) A bushing filled with a low-melting alloy, e.g., D'Arcet metal, which when heated, will release water from a pipe system.
sprudel effect Emerging gas bubbles rising in a continuous stream, or s., from gas-jet electrodes.
sprue (1) A projection left on a casting, to be broken off for testing purposes. (2) Tropical intestinal disease.
Sprunstron Trademark for a polypropylene staple fiber used for ropes.

spur feterita Sorghum.
spurred rye Ergot.
sputter To produce finely divided metal or a thin film of metal by passing a high-potential discharge between 2 electrodes of the metal in a dielectric liquid or gas; used to make magnetic data storage devices.
sputum Saliva mixed with mucus and other secretions of the mouth, air passages, and lungs. Typical analysis, percent:

Organic constituents		4.1–6.9
Fatty acids	0.02–0.97	
Soaps	Traces–0.40	
Cholesterol	Traces–0.16	
Lecithin	Traces–0.15	
Nuclein	Traces–0.48	
Proteins	0.90–0.52	
Inorganic constituents		0.3–0.9
Water		93.0–95.0

sp. vol. Abbreviation for specific volume.
sq Abbreviation for square. **sq cm** Square centimeter; cm^2. **sq in** Square inch; in^2. **sq ft** Square foot; ft^2. **sq m** Square meter; m^2. **sq mi** Square mile. **sq yd** Square yard.
squalane $C_{30}H_{62}$ = 422.8. 2,6,10,15,19,23-Hexamethyltetracosane. Colorless oil, b. ca. 340. Used for lubrication and perfumes.
squalene $C_{30}H_{50}$ = 410.7. Spinacene. 2,6,10,15,19,23-Hexamethyl-2,6,10,14,18,22-tetracosanehexaene. An unsaturated hydrocarbon in the oils of the elasmobranchs (shark family) and, as its homologs, in many marine oils. Colorless oil, d_{15} 0.8610, b_{25mm}284. Cf. *carotene.* **perhydro ~** Squalane.
square A four-sided figure, whose sides are equal and whose angles are right angles. **s. centimeter** cm^2. A unit of area 1/10,000 m^2: 1 cm^2 = 0.155 sq in. **s. foot** ft^2 or sq ft. A unit of area in the fps system = 929 cm^2. **s. inch** in^2 or sq in. A unit of area in the fps system = 6.452 cm^2. **s. meter** m^2. A unit of area in the metric system: 1 m^2 = 10,000 cm^2 = 10.7639 ft^2. **s. root** That quantity which, when multiplied by itself, gives the original quantity. **s. yard** yd^2 or sq yard. A unit of area in the fps system = 0.836 m^2.
squawroot Caulophyllum.
squaw vine Partridgeberry. The dried herb of the evergreen *Mitchella repens.*
squaw-weed Senecio.
squill Scilla. The fleshy inner scales of the bulb of *Urginea maritima* (Liliaceae); contains scillin, scillitoxin, sinistrin; an emetic and expectorant (BP).
Sr Symbol for strontium.
St Abbreviaton for stoke.
S.T. Abbreviation for surface tension.
S.T.A. Abbreviation for softening temperature of ash; used in coal analysis.
stabile Stable.
stabilizer (1) A retarding agent, for a vigorous accelerator which preserves chemical equilibrium. (2) A substance added to a solution to render it more stable; as, acetanilide to hydrogen peroxide. (3) A substance that enables a plastic material to resist chemical changes that can affect its properties; e.g., organotin s. in polyvinyl chloride absorbs hydrochloric acid and inhibits free radical formation.
stable Describing a balanced condition not readily destroyed; as, photo ~.
stachydrine $C_7H_{13}O_2N$ = 143.2. A heterocyclic amino acid in many plant juices (as *Stachys* species) and in mussels (as *Arca noae*). Levorotatory, m.117. Cf. *trigonelline.*
stachyose $C_{24}H_{42}O_{21}$ = 666.6. A nonreducing

tetrasaccharide from the roots of soybean and *Stachy stuberifera* (Labiatae), m.168. Hydrolyzes to fructose, glucose, and galactose.

stacking The process whereby typically planar, hydrophobic portions of molecules come into face-to-face contact. **intermolecular ~** S. between identical molecules; i.e., aggregation.

stagonometer Stalagmometer.

Stahl, Georg Ernst (1660–1734) German physician and chemist; developed the phlogiston theory (1697).

stain Dyes in solution for coloring materials such as wood, tissues, textiles. Cf. *paint.* **contrast ~** See *contrast stain.* **counter ~** See *counterstain.* **microscope ~** A dye used to stain preparations for microscopic examination; used for differentiating structural elements. **negative ~** A s. that stains only the surroundings of a structure and so renders it more easily visible. Cf. *positive stain* following. **positive ~** A s. that stains the structure itself. Cf. *negative stain* preceding.

s. jar A small, square glass vessel used to stain microscopic slides.

staircase reaction Chemical changes that proceed stepwise; as, the hydrolysis of starch to glucose, via α and β dextrins and di- and trisaccharides.

stalactite A hanging column of calcite formed by the slow evaporation of carbonated mineral solutions dripping from the roof of a cavern.

stalagmite A standing column of calcite formed by the slow evaporation of carbonated mineral solutions dripping onto the floor of a cavern.

stalagmometer A device to obtain drops of liquid at definite intervals; e.g., Traube's s., used to calculate surface tension from the number of liquid drops passing an orifice in a given time; also used to determine molecular weight (cf. *Morgan equation*) and the degree of association of liquids.

stalagmometry Measuring the progress of chemical reactions (as esterifications) from the change in surface tension.

Stalloy Trademark for high-silicon steel sheets and laminations used in electrical transformers.

stamen The male organ of a plant, in which the pollen is prepared. Cf. *pistil.*

stamp (1) To break up or crush ore and rock by machinery. (2) A heavy mechanical pestle for crushing ore. **s. battery** A group of pestles working mechanically in an iron mortar. **s. mill** An establishment for crushing ores by s. batteries.

standard (1) An established form of quality or quantity. (2) A substance used to establish the strength of volumetric solutions. **s. candle.** See *candle standard.* **s. cell** An electrolytic cell having a definite voltage; as, the cadmium cell. **s. conditions** In gas analysis, an atmospheric pressure of 760 mmHg, and a temperature of 0°C; sometimes abbreviated s.t.p. (standard temperature and pressure). **s. deviation** S. error. The quantity $\sqrt{\Sigma d^2/(n-1)}$; where Σd^2 is the sum of the squares of the deviations of individual values from the mean, n the number of determinations. See *normal probability curve.* **s. error** S. deviation **s. lamp** A source of light of a known light intensity at each wavelength for photometric calibration; as: pentane 10.00, Hefner 0.9, Carcel 9.6 standard candles. **s. meter.** See *meter.* **s. pressure** Pressure equal to a column of 760 mmHg at sea level in latitude 45°. **s. solution** A solution of definite concentration. See *normal solution.* **s. substance** A substance used to standardize volumetric solutions. It should be easily obtainable, pure, unaltered in air and at moderate temperatures, neither hygroscopic nor efflorescent, readily soluble in water or alcohol and should have a high molecular weight (to reduce effects of error in weighing), produce no interfering product on titration, and when using visual indicators be free from color before and after titration. **s. temperature** 0°C (273.15 K). **s. t. and pressure** See *standard conditions.* **s. volume** The normal volume occupied by 1 mole of a gaseous substance: 22.4138 liters. **s. wavelength** The wavelength of the red cadmium line, observed in air at 15°C and 760 mmHg pressure. It is equal to 6438.4696 Å. Formerly used to define the meter.

standardization The procedure of bringing a preparation to an established quality. **physiological ~** Testing of drugs or biological products (which cannot be chemically analyzed) by their pharmacological action on a normal animal. Thus: digitalis, strophanthum—frog heart; ergot—rooster; aconite—guinea pig; cannabis, adrenaline—dog.

standardized Describing a utensil, device, or preparation that has been tested, measured, and compared with a standard. See *National Bureau of Standards* (NBS, U.S.), *National Physical Laboratory* (NPL, U.K.), *volumetric glassware.* **s. deviate** For a reading, the difference between that reading and the mean of the set of readings, divided by the standard deviation of the set of readings.

standards The agencies responsible are:

(1) International s. (with their fields of coverage):

ISO	International Organization for Standardization, Geneva (highest international authority)
CGPM	General Conference on Weights and Measures (names, definitions, and symbols of units)
CODATA	International Council of Scientific Unions (values of fundamental physical constants)
IEC	International Electrotechnical Commission (electrical and magnetic subjects)

(2) National s. (The letters denote the prefix to their s.):

ANSI	American National Standards Institute
AS	Standards Association of Australia
BS	British Standards Institution
CAN2	Canadian General Standards Board
DIN	Deutsches Institut für Normung
GOST	USSR State Committee of Standards
JSA	Japanese Standards Association
NBN	Institut Belge de Normalisation
NEN	Nederlands Normalisatie Instituut
NF	Association Francaise de Normalisation
OE-NORM	Oesterreichisches Normungsinstitut (Austria)
SN	Schweizerische Normenvereinigung (Switzerland)
UNE	Instituto Espanol de Normalizacio
UNI	Ente Nazionale Italiano di Unificazione

stand oil Litho oil. An oil heated at 250–300 without addition of oxygen or driers, and allowed to settle to remove coagulated "mucilage" (foot).

stannane* SnH_4 = 122.7. Tin tetrahydride. Colorless gas, b. −52.

stannanes* A group of organic compounds containing tetravalent tin; as: tetramethyltin, Me_4Sn; tristannane, $H_3Sn \cdot SnH_2 \cdot SnH_3$.

stannate* Indicating tin as the central atom(s) in an anion, as: (1) s.(II)*. Stannite. M_2SnO_2. Salts of stannous hydroxide, possibly containing the $[Sn(OH)_6]^{4-}$ ion. (2) s.(IV)*. M_2SnO_3. A salt of stannic acid. **sulfato ~** $R_2H_2Sn(SO_4)_3$. Cf. *stannosulfate.* **thio ~** M_2SnS_3. A salt.

stannated Treated with tin salts. See *Gutzeit arsenic test.*

stannekite $C_{20}H_{22}O_3$. A resinous hydrocarbon in coal deposits. W. Czechoslovakia.

stannic Tin(IV)* or (4+)*, stannum(IV)* or (4+)*. Describing compounds of tetravalent tin. **s. acid** H_2SnO_3 = 168.7. Tin hydroxide oxide*. White, amorphous powder, insoluble in water, soluble in alkalies. **s. bromide** $SnBr_4$ = 438.3. Colorless, fuming, caustic liquid, d.3.349, m.31, soluble in water, decomp. by alcohol; used as a mordant and in tinning. **s. chloride** $SnCl_4$ = 260.5. Tin bichloride. Colorless, fuming, caustic liquid, d.2.2738, b.114, soluble in water; a mordant, reagent, and tinning agent. The vapor is used to coat glass with tin oxide, which renders it very strong, even when thin (as in bottles); also used to manufacture organotin compounds. **s. chromate** $Sn(CrO_4)_2$ = 350.7. Yellow crystals, soluble in water; a ceramic pigment. **s. ethide** Tetraethyltin*. **s. ethyl hydroxide** Triethyl*tin hydroxide**. **s. fluoride** SnF_4 = 194.7. White, deliquescent crystals, b.705. **s. hydroxide** $Sn(OH)_4$ = 186.7. White powder, insoluble in water. **s. ion** The tetravalent cation Sn^{4+}. **s. iodide** SnI_4 = 626.3. Tin tetraiodide*. Red crystals, m.144, decomp. by water. **s. methide** Tetramethyltin*. **s. oxide** SnO_2 = 150.7. Tin dioxide*, tin ash, flowers of tin, stannic anhydride. White powder, d.6.95, m.1197, insoluble in water, soluble in alkalies. Used as a pigment, and polishing material for steel, glass; and in opal glass. **s. phenide** Tetraphenyltin*. **s. sulfate** $Sn(SO_4)_2 \cdot 2H_2O$ = 346.8. Colorless rhombs, soluble in water. **s. sulfide** SnS_2 = 182.8. Tin disulfide*, mosaic gold, tin bronze. Yellow hexagons, decomp. red heat, insoluble in water; used for bronzing and gilding.

Stannine (1) Trademark for restrainers for the acid pickling of iron, steel, and ferrous alloys. (2) (Not cap.) Stannite.

stannising A process for tin-coating metal objects in a mixture of vapors of hydrogen and stannous chloride at 500–600°C.

stannite $SnS_2 \cdot Cu_2S \cdot FeS$. Stannine, tin pyrites.

stannites Stannate(II)* compounds.

stannonate RSnOOM.

stannonic **s. acids** $RHSnO_2$ or $R \cdot SnOOH$. **s. ester** $RR' \cdot SnO_2$, or RSnOOR′.

stannonium An organic compound of general formula $RSnH_3$. Cf. *-onium, stannane, stannyl.*

stannosulfate $M_2Sn(SO_4)_3$. A salt. Cf. *sulfatostannate.*

stannous Tin(II)* or (2+)*, stannum(II)* or (2+)*. Describing a compound containing divalent tin. **s. bromide** $SnBr_2$ = 278.5. Tin protobromide. Yellow crystals, m.215, soluble in water, decomp. by alcohol. **s. chloride** $SnCl_2$ = 189.6. Tin salt, tin protochloride. White crystals, m.247, soluble in water. It absorbs oxygen to form an insoluble oxychloride; a reagent, reducing agent and mordant. **hydrous ~** $SnCl_2 \cdot 2H_20$ = 225.6. Colorless, triclinic crystals, m.38 (decomp.), soluble in water; a reagent. **s. chromate** $SnCrO_4$ = 234.7. Brown powder; slightly soluble in water; a ceramic pigment. **s. citrate** $SnC_6H_6O_7$ = 308.8. Tin citrate. White powder, soluble in water. **s. fluoride** SnF_2 = 156.7. Tin difluoride*. White crystals, m.213, soluble in water. Prevents dental caries (USP). **s. hexafluorosilicate** $Sn[SiF_6]$ = 260.8. S. silicofluoride. Colorless prisms, soluble in water. **s. hydroxide** $Sn(OH)_2$ = 152.7. White powder, insoluble in water, soluble in fused alkalies [forms stannate(II) compounds]. **s. iodide** SnI_2 = 372.5. Red needles, m.316, soluble in water. **s. ion** The cation Sn^{2+}. **s. oxalate** SnC_2O_4 = 206.7. White crystals, insoluble in water; used in dyeing. **s. oxide** SnO = 134.7. Tin monoxide*, t. protoxide. Brown powder, d.6.3, burns on heating, insoluble in water; a reducing agent. **s. sulfate** $SnSO_4$ = 214.7. White crystals, soluble in water; a mordant. **s. sulfide** SnS = 150.8. Brown crystals, m.882, insoluble in water. **s. tartrate** $SnC_4H_4O_6$ = 266.8. White crystals; a mordant.

stannum* Latin for "tin." Used for naming its compounds.

stannyl* The radical H_3Sn-.

stanozolol $C_{21}H_{32}ON_2$ = 328.5. 17-Methyl-2′*H*-5α-androst-2-eno[3,2-*c*]pyrazol-17β-ol. Stromba. White crystals, insoluble in water. An androgen with anabolic effects; used for wasting diseases and anemias (USP).

staphisagria Stavesacre seed. The ripe seeds of *Delphinium staphisagria* (Ranunculaceae). **s. alkaloids** See *delphinine, delphinoidine, delphisine staphisagrine.*

staphisagrine $C_{43}H_{60}O_2N_2$ = 637.0. Staphisaine. Crystals, m.204. An alkaloid from *Delphinium staphisagria.*

staphisagroine $C_{20}H_{24}O_4N$ = 342.20. An alkaloid from staphisagria.

staphisaine Staphisagrine.

Staphylococcus (Pl., -cocci) A genus of Coccaceae (Schizomycetes) forming grapelike clusters. Many are pathogenic and produce pus. See *bacteria.*

staphylotoxin A poison from *Staphylococcus* cultures.

staple The average length of the majority of the fibers in a textile. **s. fiber.** See *fiber.*

star (1) Needle-shaped crystals that radiate from a common center. (2) A sunlike body outside the solar system. See *spectral classification.* (3) A burner attachment for holding small vessels. **blazing ~** Aletris.

s. anise See *star anise* under *anise.* **s. grass** Aletris.

starch $(C_6H_{10}O_5)_n$ = $(162.1)n$. Amylum. Carbohydrates or polysaccharides in many plant cells, and serving as their chief carbohydrate reserve. As extracted, s. is a white, amorphous powder, d.1.5, insoluble in cold water, alcohol, or ether; partly soluble in hot water; hydrolyzed to several forms of dextrin and to glucose. It consists of 2 fractions: (1) Amylose or α-amylose, a straight chain of 1,4,α-glucopyranose units, soluble in water without forming a paste, but reverting to a soluble form on storage; stained deep blue by iodine. (2) Amylopectin or β-amylose. A 1,6,α-branched form, which gels with water, and has little affinity for iodine with which it gives a violet color. It contains esterified phosphoric acid and may be associated with fatty acids. Waxy starches are almost pure amylopectin; starches from pea and lily contain up to 75% amylose. Amylose contents, percent:

Corn (maize)	23–29
Potato	18–27
Rice	14–17
Tapioca	17–21
Wheat	24–32

alant ~ Inulin. **allyl ~** A soft, gummy mass prepared by the action of s. with allyl chloride in presence of strong alkali; a coating for wood or metal. **animal ~** Glycogen. **cassava ~** Tapioca. **converted ~** S. treated by heat, by enzymes, or chemically to change its properties (e.g., solubility, viscosity); used industrially, as, for coating paper. **corn ~** Amylum. S. granules from the fruit of *Zea mays,* Indian corn (USP, BP). Used to starch fabrics, and as a dusting powder, food, antidote for iodine, reagent, and indicator for iodine. **lichen ~** Lichenin. **nitro ~** See *nitrostarch.* **rice ~** Rice flour. The s. granules from the seeds of *Oryza sativa,* rice. Used as a nutrient and to starch fabrics. **soluble ~** See *starch soluble.* **sterilizable ~** Treated maize starch that can be steam-sterilized without forming a gel. A powder for dusting and for surgeon's gloves (BP). **waxy ~** S. containing amylopectins with little or no amylose; a thickener in the food, textile, and adhesive industries. **wheat ~** The s. granules from the seeds of *Triticum vulgare,* wheat; used as a reagent and to starch fabrics (see Fig. 20 on p. 551).

s. dialdehyde Polymeric dialdehyde. The product of the periodate oxidation of s. This breaks the C_2-C_3 bond of the glucose units in the molecule, forming dialdehyde units. Used

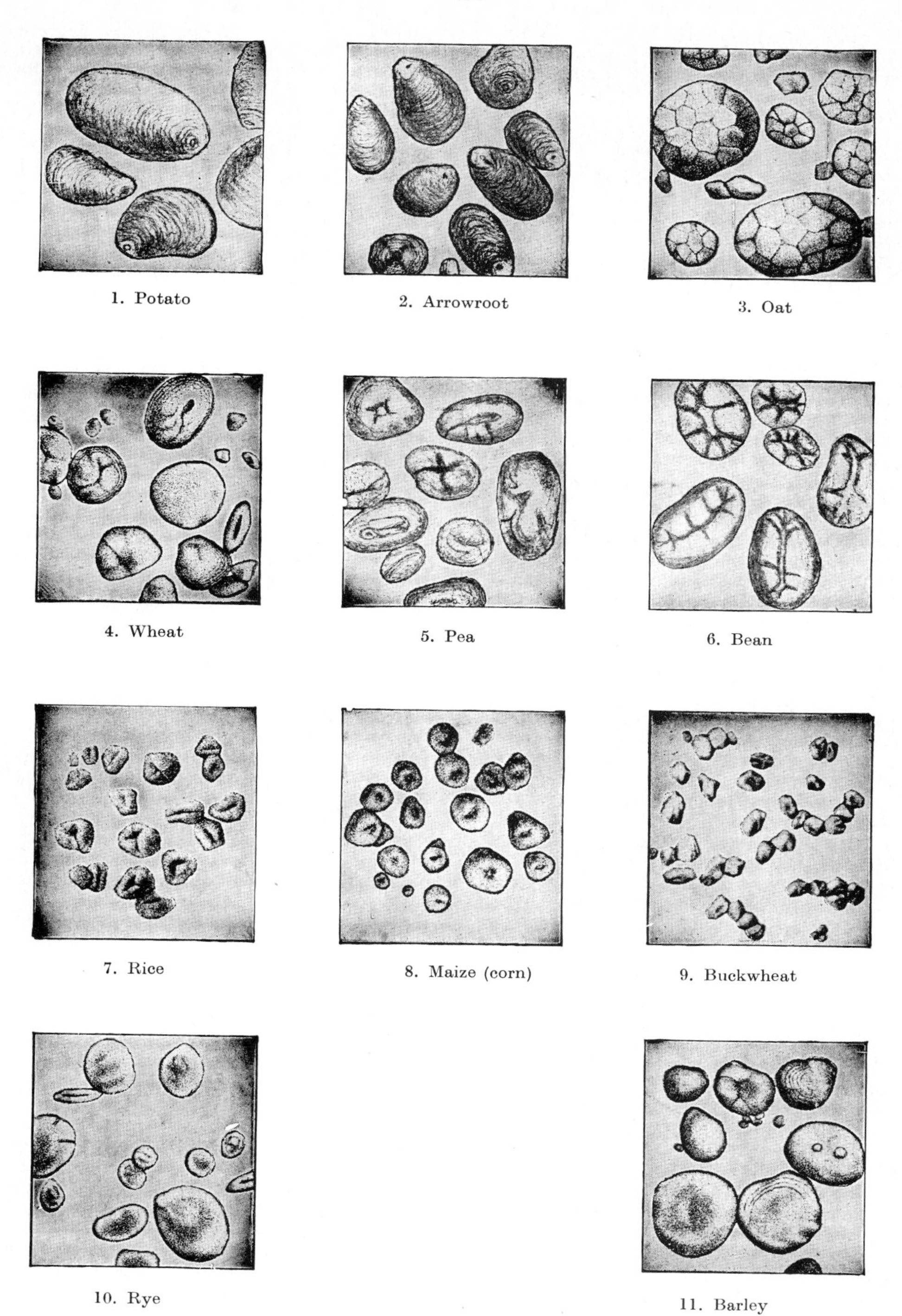

Fig. 20. Starches. (From Hawk and Bergeim, *Practical Physiological Chemistry*.

in adhesives and coatings, for textiles and wet-strengthening papers, and as a pretan for leather. **s. glycerin, s. glycerite** A protective colloid freshly prepared by heating 1 pt starch with glycerol and diluting to 100 pts with water; an emollient (NF). **s. gum** Dextrin. **s. iodide, s. iodized** Dark-blue, antiseptic powder (2% iodine), insoluble in water. Produced by the reaction of s. and iodine solution, and a test for either. **s. nitrate** Nitrostarch. **s. phosphates** Reaction products of s. and alkali metal phosphates; thickeners or binders for foods, pharmaceuticals, and adhesives. **s. soluble** $C_{36}H_{62}O_{31} \cdot H_2O$. Amylodextrin. A hexasaccharide. White powder obtained by heating s. with glycerol; yellow color with iodine solution; soluble in water; an indicator, emulsifying agent, and textile dressing. **s. sugar** Glucose obtained by heating s. with dilute sulfuric acid; used as a syrup.

Stark effect The separation of the lines of a positive-ray spectrum by action of an intense electric field.

starlite A brilliant, light-green or blue artificial gem. Blue zircon of Thailand is burned 7 hours and treated with cobalt nitrate or potassium hexacyanoferrate(II). See *zircon.*

starter A mixture of a culture of a suitable organism and nutrient medium, used to initiate large-scale fermentation; e.g.: lactic acid-producing streptococci are a s. for cheese making.

Stas **S., Jean Servais (1813–1891)** Belgian chemist, noted for atomic weight determinations. **S. pipet** A pipet whose jet has parallel walls so that the level after free drainage is always at the same point.

stassanization A method of sterilization. Milk is heated at 75°C for 15–20 s and then cooled rapidly to 4–7°C. Cf. *pasteurization.*

stassfurtite A double salt of magnesium borate and chloride form Stassfurt.

Stassfurt salts A large deposit of oceanic salts, mainly chlorides and sulfates, found in Magdeburg Halberstadt, E. Germany; principally rock salt and anhydrite, in layers under a top alluvial deposit and sandstone. Also present are gypsum, glauberite, kainite, sylvite, sylvine, and carnallite. Cf. *abraum salts.*

state (1) A condition. (2) A form of aggregation or dispersion; as colloidal s. (3) Orbital, energy level. The condition of an electron in a neutral or ionized atom. **activated ~** See *excitation.* **amorphous ~** Noncrystalline. Describing a homogeneous solid whose molecules are not symmetrically oriented. Cf. *crystal.* **change of ~** See *change of state.* **colloidal ~** Finely divided particles, surrounded by another medium. See *colloid.* **crystalline ~** A homogeneous, anisotropic solid, whose atoms are symmetrically oriented. ***d* ~** See *orbital state* below. **dissociated ~** See *dissociation.* **equation of ~** See *equation of state, gas laws.* **excited ~** See *excitation.* ***f* ~** See *orbital state* below. **fluid ~** Molten, liquid, or gaseous s. **gaseous ~** A condition in which the molecules are in rapid and irregular motion. See *gas.* **irradiated ~** See *irradiation.* **liquid ~** A condition between the solid and gaseous states. **luminescent ~** See *luminescence.* **molten ~** A liquid s. above melting-point temperature. **nascent ~** The s. of a newly formed atom or molecule. Cf. *status nascendi.* **orbital ~** The energy levels between which electrons move during a quantum transition; final states; *s, p, d, f.* Cf. *orbital series.* ***p* ~** See *orbital state* preceding. **quantum ~** See *energy levels.* **radiant ~** See *incandescence, luminescence, fluorescence.* ***s* ~** See *orbital state* above. **solid ~** (1) A condition in which the molecules are in slow motion as compared with the liquid s.; they are in relatively rigid positions, which may be symmetrical (crystalline s.) or nonsymmetrical (amorphous s.) (2) In electronics: see *solid state.* **steady ~** (1) A balance between the formation and decomposition of a substance. (2) If, in a chain of reactions, $A \rightarrow B \rightarrow C$, the concentration of B remains constant, then B is in the steady s.

static (1) The condition of being at rest or equilibrium as opposed to being in motion (dynamic). (2) The interference with radio radiation due to atmospheric electric charges or other sources of electricity. (3) In medicine: indicating the inhibition or control of growth, as, bacteriostatic; or the stopping or control of flow, hemostatic.

s. atom See *atomic structure.* **s. electricity** An electric charge at rest; often produced by friction.

statics The study of matter and forces in equilibrium.

statistics (1) The science of classifying numerical facts, judging collective numerical data, and determining probabilities from values obtained by enumeration or estimation. (2) In particular, the consideration of conditions and properties of matter (as the *mean* value of a group of units rather than the *actual* values of individual units).

stator The stationary, as distinct from the revolving portion of a machine. Cf. *rotor.*

status nascendi Nascent *state.*

staubosphere The dust of the atmosphere. (German *staub:* "dust").

Staudinger, Hermann (1881–1965) German chemist, Nobel prize winner (1953); pioneer in the basic chemistry of synthetic plastics.

staurolite Staurotide. An aluminum iron silicate gem.

staurotide Staurolite.

stavesacre seed Staphisagria.

steam Water vapor. **dry ~** S. which is free from any drops of water. **flash ~** S. at low pressure obtained from hot condensate. **superheated ~** Water vapor heated under pressure to a temperature above 100°C. **wet ~** S. containing drops of liquid water in suspension.

s. bath A vessel surrounded by s. **s. black** Hematoxylin. **s. distillation** Distillation by blowing s. through the liquid, to obtain a distillate consisting of water and dissolved substance. **s. gage.** See *pressure gage.*

steapsin See *lipase.*

stearaldehyde* $C_{17}H_{35}CHO$ = 268.5. Octadecanal*. White scales, m.64.

stearamide* $Me(CH_2)_{16}CONH_2$ = 283.5. Octadecanamide*. White leaflets, m.109.

stearate* $C_{17}H_{35}COOM$. A salt or ester of stearic acid; M is a monovalent radical.

stearic **s. acid*** $CH_3(CH_2)_{16}COOH$ = 284.5. Octadecanoic acid*, *n*-octodecylic acid; in many vegetable and animal fats. Colorless leaflets, m.69, insoluble in water. NF. s. a. contains palmitic acid; used in pharmacy. **s. aldehyde** Stearaldehyde*. **s. amide** Stearamide*.

stearin (1) $C_3H_5(C_{17}H_{35}COO)_3$ = 891.5. Tristearin, glycerol tristearate. Colorless crystals, the chief constituent of many fats, d_{65} 0.943, m.71, insoluble in water. (2) **commercial ~** A mixture of fatty acids, prepared by hydrolysis of fats; used in the manufacture of candles and solid alcohol. **lauro ~** Laurin.

stearolic acid $Me(CH_2)_7C{:}C(CH_2)_7COOH$ = 280.5. 9-Octadecynoic acid*. Colorless prisms, m.48, soluble in EtOH.

Stearone $(C_{17}H_{35})_2CO$ = 506.9. Trademark for 18-pentatriacontanone*. A solid, m.88, insoluble in water.

stearonitrile* $C_{17}H_{35}CN$ = 265.5. Octadecane nitrile*. Colorless crystals, m.41.

stearoptene Oleoptene. The oxygenated portions of an

essential oil consisting chiefly of the solid part (as, camphor) as opposed to the liquid part (as, eleoptene).

stearoxylic acid $Me(CH_2)_7CO \cdot CO(CH_2)_7COOH = 312.5$. 9,10-Dioxooctadecanoic acid*. A solid, m.86, insoluble in water.

stearoyl Octadecanoyl*. The radical $C_{17}H_{35}CO-$, from stearic acid.

stearyl The octadecanyl* radical. **s. alcohol** $C_{18}H_{38}O = 270.5$. White flakes, m.58, insoluble in water; used in ointments (NF).

steatite A variety of talc.

stechiometry Stoichiometry.

steel (1) Carbon s. A tough, elastic alloy of iron containing small quantities of carbon:

Mild or soft steel	Less than 0.15% C
Medium steel	0.15–0.30% C
Hard steel	More than 0.30% C

See *iron.* (2) Alloy steel. An alloy of iron whose properties are due to an element or elements other than carbon; as, Cr, Mn, Ni, W, Si. **alloy ~** Iron and other metal, fused together and cooled rapidly. **alloy-treated ~** S. containing metals added during manufacture for curative purposes. Cf. *alloy.* **austenitic ~** Acid-resisting s. containing Cr 18, Ni 8, Mo 1–4%. **bulletproof ~** S. containing Mn 12, C 1, P or S less than 0.02%. **carbon ~** An alloy of iron and carbon without the addition of other metals. **chrome-molybdenum ~** A light alloy of iron, with Cr 0.8–1.1, Mn 0.4–0.6. Mo 0.15–0.25, C 0.25–0.35%. **chromium-nickel ~** See *stainless steel* below. **high-speed ~** Iron alloys containing approximately C 0.65, Mn 0.2, Cr 4.7, Mo 8.5 (or W 17), V 5%; used for cutting tools, and does not lose temper if heated. **nickel ~** See *nickel steel.* **nickel-zirconium ~** A tough alloy of iron with Ni 2, Zr 0.34, Mn 1, Si 15, C 0.4%; used for armor plate and helmets. **stainless ~** A s. containing Ni, Cr, or both; does not tarnish on exposure; used in corrosive environments. **super ~** A high-speed s. containing 4% Co. Cf. *Carboloy.* **tool ~** Similar to high-speed s., but with lower alloy content. Used for blanking and forming tools; also, as constituent of hard tools.

Steele S. acid An oxidizable form of abietic acid comprising 60–90% of the resin acids in tall oil. **S. microbalance** A quartz beam with a small quartz ball, whose buoyancy is changed by increasing or decreasing the air pressure. Sensitivity 4×10^{-6} mg. Used to measure the density of 0.1 mm^3 of radon (Ramsay and Gray).

Steelon Trademark for a polyamide synthetic fiber.

Stefan-Boltzmann S.-B. equation S-B law. The radiation per m^2 per second from a black body (which is a perfect radiator) at thermodynamic temperature T to surroundings at thermodynamic temperature $t_0 = \sigma(T^4 - t_0^4)$. Cf. *Wien equation.* **S.-B. constant*** The constant σ (sigma) of the S.-B. equation: $\sigma = 5.67032 \times 10^{-8}$ $W/m^2/K^4$.

Steffen waste A by-product in the manufacture of sugar from beet; a source of amino acids, particularly L-glutamic acid.

steigerite $4AlVO_4 \cdot 13H_2O$. A yellow mineral in Colorado uranium deposits.

Stein, William Howard (1911–1980) American biochemist. Nobel prize winner (1972). Noted for work on the structure of pancreatic ribonuclease.

Stelazine Trademark for trifluoperazine hydrochloride.

stellar (1) Pertaining to stars. (2) Star-shaped; as crystals. **s. evolution** See *spectral classification.* **s. spectra** See *spectral classification.*

stellate crystals Stellar- or star-shaped crystals; as, phenylglucosazone.

Stellite Trademark for nonferrous alloys of cobalt, chromium, and tungsten; used for metal-cutting steels, wear-resistant castings, and hard-facing welding rods.

stem correction The correction to a thermometer reading for the portion of the mercury column not in the liquid.

Stemetil Trademark for prochlorperazine maleate.

stench Malodorous gases used industrially to detect gas leaks.

stenocarpine Gleditschine.

stenosation A process for increasing tensile strength, e.g., of viscose fibers, by treatment with formaldehyde.

stenosine Sodium methyl arsenite.

stephanite $5Ag_2S \cdot Sb_2S$. A native sulfide.

stepwise s. decomposition Staircase reaction. **s. dissociation** A gradual dissociation.

steradian* Symbol: sr; the SI unit for solid angle. The solid angle subtended by an area on the surface of a sphere equal to the square of the sphere's radius.

Sterane (1) Trademark for prednisolone. (2) (Not cap.) Androstane*.

stercobilin $C_{33}H_{46}N_4O_6 = 594.8$. The normal pigment of feces, m.236; allied to urobilin.

stercorite $Na(NH_4)HPO_4$. Native microcosmic salt.

stercorol Coprosterol.

Sterculia A genus of tropical plants, used for edible seeds and barks, cordage, and mats. **S. gum** Indian tragacanth, Indian gum. A tragacanthlike exudation from *S.* species (India); absorbs water. Used as a laxative and in pharmacy (BP). Also a filler for ice cream. Cf. *hog gum.*

Sterculiaceae Softwood trees or shrubs; as, *Cola acuminata,* kola nut; *Theobroma cacao,* cocoa.

stere A French unit of wood measurement equal to 1 m^3 of stacked logs. Cf. *cord.*

stereo- (1) Prefix (Greek) indicating "solid" in structure or "three-dimensional." (2) Abbreviation for stereotype.

stereochemistry Spatial or configurative chemistry. The study of the spatial arrangement of the atoms in a molecule. **absolute ~*** Absolute configuration*. Term used to describe the 3-dimensional arrangement of substituents around a chiral element. **realtive ~*** Relative configuration*. Term used to describe the positions of substituents on different atoms in a molecule relative to one another.

stereoisomers* Stereomers. Isomers that differ only in the arrangement of their atoms in space. Thus, glyceraldehyde

CHO, H, HO, C, CH_2OH and CHO, HO, H, C, CH_2OH

has s., where (conventionally) broken lines represent a bond projecting behind the plane of the paper, a thick line one projecting out of the paper, and a normal line one in the plane of the paper. Steric relationships are described by *cis-trans* (*Z,E*), see below, and *sequence rule,* q.v., nomenclature. For additional details see *Nomenclature of Organic Chemistry,* IUPAC, Pergamon Press. See *isomer.* ***cis-trans* ~*** Stereoisomers that differ only in the positions of atoms (or groups) relative to a specified plane in cases where these atoms are, or are considered as if they were, parts of a rigid structure. Atoms are termed *cis* (*c*) or *trans* (*t*) to one another when they lie, respectively, on the same or on opposite sides of the reference plane; e.g.,

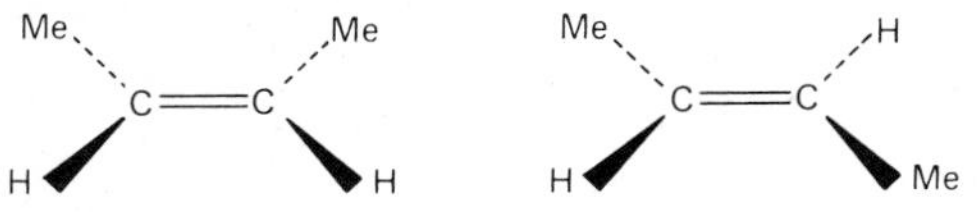

cis *trans*

The reference plane in this case is the plane of the paper. Customarily, the reference plane is perpendicular to the plane containing the double-bonded atoms; for cyclic compounds, it is the plane of the cyclic structure. Classical nomenclature used *cis* and *trans,* and they still serve relatively simple cases. For more complex compounds, Z (German *zusammen:* "together") and *E* (*entgegen:* "opposite"), respectively, are used. Thus, for a double bond, the sequence rule–preferred group attached to one of the doubly bonded atoms is compared with that of the other doubly bonded atom. If the pair are on the same side of the reference plane, they are prefixed Z; if not, *E.* Thus

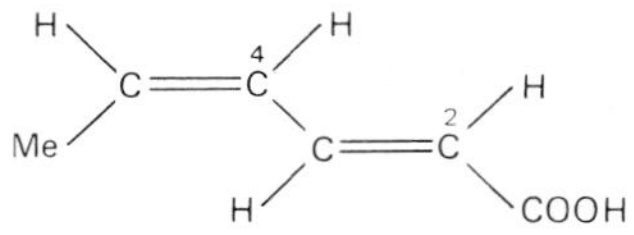

is designated (2*E*,4Z)-2,4-hexadienoic acid.
dia ~* Stereoisomers that are not enantiomeric.
stereoisomerism The phenomenon of *stereoisomers,* q.v.
stereomer See *stereoisomer.*
stereoscope A device for viewing a pair of photographs simultaneously, each with one eye, to obtain a perspective effect.
stereoscopic photographs Photographs providing a three-dimensional effect, achieved using 2 lenses separated roughly by the distance between the human eyes; used in a stereoscope.
stereoselectivity Reaction forming preferentially 1 of 2 or more stereoisomers.
stereoskiagraphy The production of stereoscopic X-ray pictures to locate an opaque object in the body.
stereotyping A printing operation in which the original type is used to emboss a plastic paper sheet (flong), which is then hardened and used to cast metal type for high-speed rotary letterpress machines.
steric Pertaining to the spatial arrangement of atoms in the molecule. **s. hindrance** The nonoccurrence of an expected chemical reaction, due to inhibition by a particular atomic grouping.
sterid Steroid.
steride A tertiary lipid in which the alcohol is a sterol.
sterile Aseptic or free from microorganisms. **s. solution** A solution made aseptic by the destruction or removal of all microorganisms.
sterilization The destruction or removal of all living bacteria, microorganisms, and spores: physically (heat, sound, light, or other radiations; adsorption; filtration) or chemically (antiseptics). **cold ~** S. by γ rays. They penetrate deeply, so that large packages may be treated intact without temperature rise.
sterilizer An autoclave, steam or bake oven, for sterilizing.
sterilizing Rendering free from all living microorganisms and their spores. Glass and metal utensils are heated in a hot-air oven or autoclave, in steam or boiling water. Solutions are treated similarly or are filtered through a stone, porcelain, or Alundum filter. **s. area** A pressure-temperature-time diagram on which the thermal death points of all bacteria occur in a wedge-shaped area.
sternbergite $Ag_2S \cdot Fe_4S_5$. A native sulfide.
sternutative An agent that produces sneezing. **poison gas ~** A s. used in explosive shells; on liberation, causes violent sneezing; as, diphenylchloroarsine.
sternutatory Able to produce sneezing.
steroid Sterid. Generic name for a family of lipid compounds comprising the sterols, bile acids, cardiac glycosides, saponins, and sex hormones. Basic structure, including ring letters:

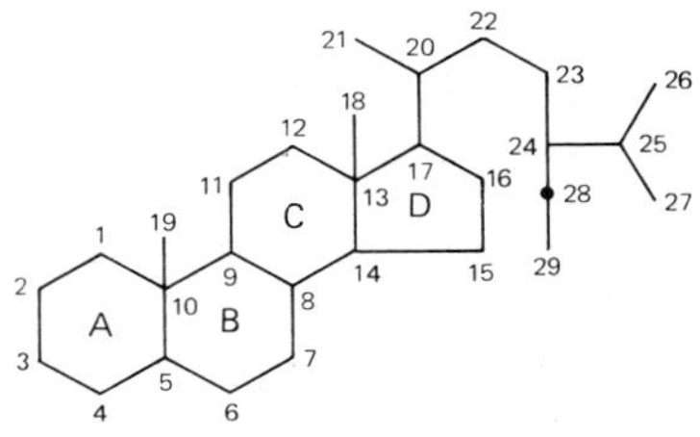

Cf. *androstane, bufanolide, cardenolide, estrane, digitoxin.*
sterols Steroid alcohols of vegetable and animal origin; consist chiefly of solid unsaturated alcohols; e.g.; farnesol. $C_{15}H_{26}O$ (acacia); ergosterol, $C_{28}H_{44}O$ (yeast, animal tissues); cholesterol, $C_{27}H_{46}O$ (animal and plant tissues); sitosterol $C_{29}H_{48}O$ (wheat).
sterro metal The alloy Cu 56.5, Zn 40, Fe 1.5, Sn 1%; stronger than gunmetal or bronze.
sterule An ampul or glass container for a sterile solution.
Stetefeldt furnace A furnace for roasting silver and copper ores in chlorine gas.
steviol $C_{20}H_{30}O_3$ = 318.5, m.215. The nonsugar component of stevioside.
stevioside $C_{38}H_{60}O_{18}$ = 804.9. A glucoside, 300 times as sweet as sucrose, from the leaves of Kaa-ha-e, *Stevia rebaudiana* (Compositae), S. America; long prisms, m.238.
Stewart-Kirchhoff law The light waves emitted by a hot gas can be absorbed by the same substance at a lower temperature. Cf. *Fraunhofer lines.*
sthene The force which in 1 second communicates to a mass equal to 1 metric ton an increase in velocity of 1 m/s (French usage).
sthenosage The waterproofing of, e.g., fabrics by means of gelatin treated with formaldehyde (Eschalier process).
stibane* Stibine*.
stibate, stibiate Antimonate.
stibic Antimonic.
stibiconite $Sb_2O_3(OH)_2$. Antimony ocher. Stibite. A native hydroxide of antimony.
stibide Antimonide*.
stibine (1)* SbH_3 = 124.8. Antimonous hydride. Colorless, poisonous gas, b. −18, sparingly soluble in water. (2)* The s. (1) analogs of *phosphines,* q.v. **pentamethyl ~*** Me_5Sb = 196.9. Pentamethylantimony. Colorless liquid, b.97, insoluble in water. **triethyl ~*** Et_3Sb = 208.9. Triethylantimony. Colorless liquid, b.159, insoluble in water. **trimethyl ~***

Me_3Sb = 166.9. Colorless liquid, b.81, insoluble in water. **triphenyl ~*** Triphenylantimony. Colorless crystals, m.48.
s. hydroxide R_2SbOH. An organic compound.
stibino* The radical H_2Sb-.
stibinoso (1) The group $(OH)_2Sb-$. (2) The group $(OH)Sb=$.
stibious Antimonous.
stibite Stibiconite.
stibium Latin for "antimony."
stibnate Antimonate*.
stibnic Antimonic.
stibnide Antimonide*.
stibnite Antimony glance.
stibnous Antimonous.
stibo The group O_2Sb-.
stibonic acid R_2SbO_2H. An organic compound.
stibonium* The radical SbH_4^+. analogous to ammonium. **s. compounds** R_4SbX. Organic compounds derived from stibine and stibonium. **s. hydroxide** R_4SbOH. **s. iodide** R_4SbI.
stibono The radical $(HO)_2OSb-$.
stibonoso (1) The group $HO \cdot OSb \cdot H-$. (2) The group $HO \cdot OSb=$.
stibophen $(NaO_3S)_2C_6H_2O_2{:}SbOC_6H_2(ONa)(SO_3Na)_2 \cdot 7H_2O$ = 847.2. Neoantimosan. Sodium antimony-bispyrocatechol-3,5-sodium disulfonate. White crystals, darkening in light, soluble in water; formerly used for schistosomiasis.
stiboso The group $OSb-$.
stick (1) A rod or cylinder, e.g., of dynamite. (2) A thick syrup obtained by evaporation of the water used in making tankage (1.5–10% N).
stick lac See *stick lac* under *lac*.
stictography The photographic fixation on paper of an image left by radio waves.
Stieglitz, Julius Oscar (1867–1937) American chemist noted for organic research.
stigmastane $C_{29}H_{52}$ = 400.7. **(5α,24R) ~** Sitostane. Derived from stigmastanol and oil shale; m.85.
stigmastanol $C_{29}H_{52}O$ = 416.7. A hydrogenation product of stigmasterol.
stigmasterol $C_{29}H_{48}O$ = 412.7. A sterol, m.170, from calabar beans, soybeans, and vegetable oils.
stilbene* $Ph \cdot CH{:}CH \cdot Ph$ = 180.2. Diphenylethylene, toluylene, bibenzylidene. Colorless leaflets, m.124, insoluble in water. **diamino ~** See *diaminostilbene* under *diamino-*. **dibenzoyl ~** $(Ph \cdot CH \cdot COPh)_2$ = 390.5. Acicular oxolepidene. **phenyl ~** Triphenylethylene*.
s. dyes Derivatives of disulfonic acids. **s. hydrate** $Ph \cdot CHOH \cdot CH_2 \cdot Ph$ = 198.2. Toluylene hydrate. Colorless crystals, m.62, soluble in alcohol.
stilboestrol EP, BP name for diethylstilbestrol.
stilbite $(Na_2 \cdot Ca)O \cdot Al_2O_3 \cdot 6SiO_2 \cdot 6H_2O$. A silicate of the zeolite group; pearly prisms.
still An apparatus in which a substance is heated to the gaseous state and then condensed. See *distillation*. **column ~** A s. for fractional distillation. **pot ~** A distillation apparatus for potable spirits. **vacuum ~** A device for distilling at a low temperature, under a reduced pressure. **water ~** A s. for water.
s. head Fractionating column. A device for selectively refluxing the less volatile components of mixtures under distillation, thus improving fractionation.
stillingia Queen's-root, silver leaf, yawroot. The dried roots of *Stillingia sylvatica* (Euphorbiaceae), N. America; an expectorant. S. oil has drying properties.
stillingine An alkaloid from stillingia.
stilpnomelane $FeSiO_3$. A native metasilicate which contains Al, Ca, and Mg.
stimmi Sb_2S_3. A black, native sulfide.
stimulant An agent that excites or increases a functional activity. Cf. *sedative*.
stimulation The acceleration of crystallization by addition of another crystal.
stinging nettles Urtica.
stinkstone Anthraconite.
stinkweed Stramonium.
stipitatic acid $C_8H_6O_5$ = 182.1. An acid metabolite in cultures of the mold *Penicillium stipitatum;* its sodium salt gives a deep-yellow solution.
Stirling's approximation A relationship of importance in statistical mechanics; viz., if N is large, then: $\log_e N! = N \cdot \log_e N - N$.
stirrer A device for agitating liquids. **magnetic ~** A piece of magnetic steel, in the liquid to be stirred, rotated from outside by a rotating electromagnet.
stirring rod (1) A glass rod suitably bent and mechanically driven, for agitating liquids. (2) A glass rod used for stirring. Cf. *policeman*.
stizolobin A globulin from Chinese velvet beans, *Mucuna (Stizolobium) pruriens* (Leguminosae).
stochastic Giving an output whose value is dependent on probability rather than rigidly determined by the input values. Cf. *deterministic*.
Stock system See *nomenclature*.
stoechiometry Stoichiometry.
stoichiometric point Equivalence point.
stoich(e)iometry The study of the numerical relationship between elements or compounds (atomic weights), the determination of the proportions in which the elements combine (formulas) and the weight relations of reactions (equations).
stok(e) Abbrev. St; the cgs unit of kinematic viscosity ν as determined from Stokes' law. It equals dynamic viscosity in poise/density at the same temperature. Viscosity of water at 20°C = 1.0018. 1 cSt = 10^{-6} m^2/s, the SI unit.
Stokes S., George Gabriel (1819–1903) English mathematician. **S. law** The rate of fall ν of a spherical body of radius r under gravity $g = 2gr^2(\sigma - \rho)/9\eta$, where η is the viscosity of the medium, and σ and ρ are the density of the substance and of the medium, respectively; used to calculate the charge on an electron (Millikan's method). For *sedimentation*, q.v., the formula is $\nu = gd(\sigma - \rho)/18\eta$, where d is the effective diameter of the particle of density σ in a liquid of density ρ. Cf. *settling*. **S. shift** See *stokes shift* under *shift*.
stolzite $PbWO_4$. Native lead tungstate.
stomach A pouch between the esophagus and the intestines, in which food is partly digested. **powdered ~** The dried, defatted powdered wall of hog s., containing intrinsic factor, which is essential for absorption of vitamin B_{12}; formerly used in cases of pernicious anemia. **s. enzyme** Pepsin. **s. juice** Gastric juice. **s. pump, s. tube** Gastric lavage. A rubber syphon tube used to wash out or sample s. contents.
stomachic Obsolete term for an agent that is said to stimulate the appetite or increase gastric secretion.
stone (1) A standard British weight: 14 pounds. (2) Concrete mineral matter or pieces of rock. (3) A concretion or *calculus*, q.v. **blue ~** Copper sulfate. **brass ~** See *brass stone* under *brass*. **gall ~** Calculus. **Lydian ~** Jasper. **precious ~** Gem. **rotten ~** See *rotten stone*. **touch ~** See *touchstone*.
s. flax Asbestos. **s. groundwood** See *mechanical pulp*. **s. root** Collinsonoid. **s. ware.** See *ceramics, earthenware, pottery*.
Stoner S. energy levels, S. quanta A modified Bohr theory,

which deals with the distribution of electrons in different energy levels. The quantum number n of Bohr is divided into functions k_1 and k_2, from which a systematic relationship of the lines of X-ray spectra is found. Cf. *Pauli's principle.*

stool The discharge from the bowels; *feces,* q.v.

stopping power The capacity of substances to absorb α particles, compared with air = $k \cdot N^{2/3}$, where N is the atomic number, and k a constant.

storax Styrax. A semiliquid balsam from the wood of *Liquidambar orientalis* (Hamamelidiaceae), W. Asia. It contains styrene, styracin, and cinnamic acid; insoluble in water. **American ~** Solid mass, softening when warmed (USP). **Levant ~** Red semiliquid. Both American s. and Levant s. are used in benzoin tinctures (USP, BP). **s. oil** The volatile oil of s., d.0.890–1.100, b.150–300, containing styrene and cinnamic esters. Cf. *liquidambar.*

storesin $C_{36}H_{55}(OH)_3$ = 538.9. A hard resin from storax. **alpha-~** An amorphous mass, m.160–168. **beta-~** A white, flocculent mass, m.140–145.

Stovarsol Trademark for acetarsol.

stove (1) A heating device whose flame is covered. (2) To heat a paint or varnish film on metal to convert it into a protective or decorative coating, or both.

s.t.p. STP, NTP (normal t. and p.). Abbreviation for standard temperature and pressure: O°C and 101.325 kPa (760 mmHg).

STPP Sodium tripolyphosphate.

strain (1) A deformation resulting from stress; the increase in length per unit original length. (2) To pass through a coarse filter. **s.-aging** The increase in resistance to deformation which occurs in time after a metal has been deformed. It increases hardness and tensile strength. **s. theory** Baeyer's theory. The assumption that the normal lines of force of the valency bonds of a C atom run from the center of a tetrahedron to each of the 4 corners.

stramonium Thorn apple leaves, Jamestown weed, jimsonweed, stinkweed, apple of Peru. The dried leaves of *Datura stramonium* (Solanaceae), containing alkaloids; an antispasmodic (EP, BP). **s. seeds** The seeds of stramonium, containing atropine, hyosycamine, oil, resin, and proteins; a hypnotic and antispasmodic.

strata Pl. of *stratum,* q.v. (1) The geological layers of rock on the earth surface. (2) The several layers of the atmosphere. (3) The seams of minerals.

stratification The formation of layers, as in Liesegang rings.

stratiform Describing a bacterial growth in a solid culture medium, characterized by liquefaction near the walls of the tube and downward.

stratometer An instrument to determine the hardness of a soil from the distance it sinks under a given impact.

stratosphere The atmosphere above approximately 12 km; temperature about −55°C.

stratum A geological layer. See *geologic eras.*

streak The colored line produced when a mineral is rubbed across unglazed porcelain. **s. test** The approximate gold content of an alloy is determined by streaking a fine-grained silica stone, covered with charcoal, with the alloy, and comparing the streak with those produced by alloys of known compositions, before and after treatment with dilute aqua regia. Cf. *touchstone.*

streamline filtration Edge filtration. Filtration with the filter in a plane perpendicular to the direction of motion (streamline) of the liquid; e.g., a pack of stout paper sheets is held together in a press; the filtrate passes by pressure or suction between the sheets and emerges through holes running along the length of the pack. Cf. *filter press.*

Strecker reaction Formation of amino acids by the action of ammonia on cyanohydrins.

strengite $FePO_4 \cdot 2H_2O$. An ore.

streptococcus (Pl. -cocci) A group of round or spherical *bacteria,* q.v., arranged in strings or rows.

streptokinase A mixture of filtrates of cultures of certain strains of hemolytic streptococci. S. causes activation of plasmin. Used for severe thrombosis or embolism (BP).

streptolin An anticoliform antibiotic from species of *Streptomyces.*

streptolysin A toxin produced by streptococci, which hemolyzes red blood cells.

streptomycin $C_{21}H_{39}O_7N_{12}$ = 571.6. An antibiotic from *Streptomyces griseus,* especially from the soil. A bitter aminoglycoside. Active against gram-positive and -negative organisms. Used mainly in tuberculosis to prevent drug resistance. **s. sulfate** $(C_{21}H_{39}O_{12}N_7)_2 \cdot 3H_2SO_4$ = 1457. White powder, soluble in water. An antibiotic, the usual preparation of s. (USP, EP, BP).

stress The force per unit area that tends to deform a body. Cf. *strain.*

striate Striped; provided with lines.

striga Parasitic plant species that reduces the yield of tropical crops such as sorghum, maize, and millet.

string A gold-plated quartz thread, 1.5–6 μm thick, for Einthoven galvanometers.

stripped atom An atom from which one or more electrons have been removed; hence, the gaseous ion of an element. Cf. *ionization energy.*

stripping (1) Back-extraction. Extracting a solute from a solvent which has already been used to extract it from another. (2) Removal of light fractions, e.g., of lubricating-oil distillates, by distillation with superheated steam. (3) Removal of dyestuffs from textiles by bleaching agents.

strobinin $C_{30}H_{24}O_8$ = 512.5. A red pigment from black aphids.

stroboscope Rotoscope. A mercury arc or neon lamp coupled with a periodically discharging capacitor. Through regulation of the frequency of the flashes (up to 11,000/s) fast-moving machinery can be observed in apparent slow motion. **s. photography** A moving picture of quick-moving objects illuminated by the stroboscope to reduce it to slow motion. Cf. *chronoteine.*

stroma (1) The granular background matrix of chloroplasts. (2) The protein background structure of an organ.

stromatin A constituent of blood cells; forms a loose compound with hemoglobin.

Stromba Trademark for stanozolol.

Stromeyer, Friedrich (1786–1835) German apothecary, discoverer of cadmium (1817).

stromeyerite $(Ag,Cu)_2S$. A native sulfide.

strontia Strontium oxide.

strontianite $SrCO_3$. A native strontium carbonate.

strontium* Sr = 87.62. An alkaline-earth metal and element, at. no. 38. Named after Strontian, Scotland. Silvery crystals, d.2.54, m.770, b.1380, slowly soluble in water forming the hydroxide. Occurs native as strontianite and celestite; forms one series of compounds: valency 2. The isotope ^{90}Sr is a constituent of radioactive fallout from nuclear explosions and has a half-life of 28 years. It is absorbed by foodstuffs, soil, vegetables, and especially milk; occurs in human bone, where an excess may cause leukemia. Cf. *calcium* (^{47}Ca), *strontium unit.* **common ~** Sr as found in minerals having a high Sr:Rb ratio; e.g., celestite.

s. arsenite $Sr(AsO_2)_2 \cdot 4H_2O$ = 373.5. White powder, soluble in water. **s. bromate*** $Sr(BrO_3)_2 \cdot H_2O$ = 361.4.

Colorless, deliquescent crystals, decomp. 240, soluble in water. **s. bromide*** $SrBr_2$ = 247.4. Colorless needles, m.630, soluble in water. **hydrous ~** $SrBr_2 \cdot 6H_2O$ = 355.5. Colorless, hygroscopic crystals, soluble in water. **s. carbonate** $SrCO_3$ = 147.6. Native as strontianite. Colorless rhombs, d.3.62, decomp. 1155, insoluble in water; a source of s. salts. **s. chlorate*** $Sr(ClO_3)_2$ = 254.5. Colorless rhombic or monoclinic crystals, m.290 (decomp.), soluble in water; a pyrotechnic for red fires. **s. chloride*** $SrCl_2$ = 158.5. Colorless crystals m.872, soluble in water **hydrous ~** $SrCl_2 \cdot 6H_2O$ = 266.6. Colorless needles, m.112 soluble in water; a reagent, dentin desensitizer, and pyrotechnic. **s. chromate*** $SrCrO_4$ = 203.6. Yellow, monoclinic crystals, insoluble in water, soluble in acetic acid. **s. citrate*** $Sr_3(C_6H_5O_7)_2$ = 641.1. Colorless crystals, soluble in water. **s. dioxide*** S. peroxide*. **s. dithionate*** $SrS_2O_6 \cdot 4H_2O$ = 319.8. Plates, insoluble in alcohol. **s. fluoride*** SrF_2 = 125.6. Colorless octahedra, m.1470, insoluble in water. **s. formate*** $Sr(OOCH)_2$ = 177.7. Colorless rhombs, m. 72, soluble in water. **hydrated ~** $Sr(OOCH)_2 \cdot 2H_2O$ = 213.7. White powder, soluble in water. **s. hexafluorosilicate*** $SrSiF_6 \cdot 2H_2O$ = 265.7. S. silicofluoride. White, monoclinic crystals, soluble in water. **s. hydroxide*** $Sr(OH)_2$ = 121.6. S. hydrate. Colorless powder, m.375, slightly soluble in water. **hydrous ~** $Sr(OH)_2 \cdot 8H_2O$ = 265.8. Colorless, deliquescent tetragons, slightly soluble in water; used to crystallize sugar from beet molasses. **s. hyposulfite** S. thiosulfate*. **s. iodide*** SrI_2 = 341.4. Colorless, hygroscopic plates, m.402, slightly soluble in water. **hydrous ~** $SrI_2 \cdot 6H_2O$ = 449.5. Colorless crystals, slightly soluble in water; a potassium iodide substitute. **s. monoxide*** S. oxide*. **s. nitrate*** $Sr(NO_3)_2$ = 211.6. Colorless cubes or octahedra, m.645 (decomp.), soluble in water. The synthetic crystals are used as imitation diamonds. **hydrous ~** $Sr(NO_3)_2 \cdot 4H_2O$ = 283.7. Colorless, triclinic crystals, soluble in water; a pyrotechnic for red fires. **s. nitrite*** $Sr(NO_2)_2$ = 179.6. White powder, soluble in water. **s. oxalate*** $SrC_2O_4 \cdot H_2O$ = 193.7. White crystals, decomp. by heat, insoluble in water; a pyrotechnic. **s. oxide*** SrO = 103.6. Strontia, s. monoxide*. Colorless rhombs, m.3,000, decomp. by water; used in beet-sugar manufacture. **s. peroxide*** SrO_2 = 119.6. White powder, soluble in water; bleaching agent. **hydrous ~** $SrO_2 \cdot 8H_2O$ = 263.7. White crystals, decomp. red heat. **s. phosphate*** $Sr_3(PO_4)_2$ = 452.8. White powder, insoluble in water. **s. hydrogenphosphate** $SrHPO_4$ = 183.6. Colorless rhombs, insoluble in water. **s. saccharate** $C_{12}H_{22}O_{11} \cdot 2SrO$ = 549.5. An almost insoluble salt used to separate sugar from molasses. The sugar is isolated by precipitation of the s. as carbonate. **s. salicylate*** $Sr(C_7H_5O_3)_2 \cdot 2H_2O$ = 397.9. Colorless crystals, soluble in water; used medicinally. **s. sulfate*** $SrSO_4$ = 183.7. Native as celestine. Colorless rhombs, decomp. white heat, insoluble in water; used in pyrotechnics. **s. sulfide*** SrS = 119.7. Colorless, regular crystals, soluble in water; a constituent of luminous paints. **s. sulfite*** $SrSO_3$ = 167.7. Colorless crystals, decomp. by heat, insoluble in water, soluble in sulfurous acid. **s. tartrate*** $SrC_4H_4O_6 \cdot 4H_2O$ = 307.7. White crystals, soluble in water. **s. thiosulfate*** $SrS_2O_3 \cdot 5H_2O$ = 289.8. S. hyposulfite. Colorless needles, soluble in water. **s. titanate** $SrTiO_3$ = 183.5. Fabulite. An imitation diamond **s. unit** S.U. A measure of the ^{90}Sr content of materials, as milk or bones, relative to their calcium content. 1 S.U. = 0.037 Bq of $^{90}Sr/g$ of Ca. Cf. *fallout, strontium.*

strophanthic acid $C_{23}H_{30}O_8$ = 434.5. A dibasic acid from strophanthin, decomp. 270, insoluble in water.

strophanthidin $C_{23}H_{32}O_6$ = 404.5. (3β, 5β, 14β)-3,5,14-Trihydroxy-19-oxo-20(22)-cardenolide. An aglucone from strophanthus, m.170.

strophanthin $C_{30}H_{47}O_{12}$ = 599.7. Methylouabain, chorchorin strophanthum. A mixture of glucosides (strophanthidin, strophanthobiose methyl ether, etc.) from the ripe seeds of strophanthus. Pale-yellow, poisonous crystals, m.179, soluble in water; a cardiac glycoside. **G ~** Ouabain. **K ~** The glucosides from *Strophanthus kombé.*

strophanthobiose $C_{13}H_{24}O_9$ = 324.3. A disaccharide obtained by partial hydrolysis of strophanthin, m. 145. Hydrolyzes to mannose and rhamnose.

strophanthum Strophanthin.

strophanthus The dried ripe seeds of *Strophanthus kombé* (Apocynaceae), Africa and Asia; contain cardiac glycosides.

structural Pertaining to the arrangement of atoms in a molecule. **s. formula** A plane representation of the atomic arrangement of a molecule.

structure Broadly, the organization of matter; as *atomic* structure, q.v. See *stereoisomer, projection, formula.*

Strutt, John William See *Rayleigh.*

struvite $Mg_2(NH_4)_2(PO_4)_2 \cdot 12H_2O$. Guanite. White crystals; in canned fish.

strychnine $C_{21}H_{22}O_2N_2$ = 334.4. Vauqueline. An alkaloid from various species of *Strychnos ignatia* and *S. nux vomica.* Colorless tetragons, m.268, slightly soluble in water; formerly used as a tonic and to improve appetite. **dimethoxy ~** Brucine. **methyl ~** Yellow powder, soluble in water. **s. hydrochloride** $(Str)HCl \cdot 2H_2O$ = 406.9. Colorless trimetric crystals, soluble in water. **s. sulfate** $(Str)_2H_2SO_4 \cdot 5H_2O$ = 856.9. Colorless prisms, anhydrous at 200, soluble in water.

strychninium The strychnine cation (N^+). Cf. *piperidinium.*

Strychnos A genus of tropical trees (Loganiaceae). They yield strychnine, nux vomica, ignatia, curare, and ipoh.

stucco A mixture of calcium sulfate, sand, and lime; a decorative building material.

Stuffer law Sulfones with 2 $=SO_2$ groups on adjacent C atoms are readily saponified.

stuffing A dyeing process; textiles are passed through the color and then through a mordant. **s. box** A box containing packing which fits around and seals the point where a rotating shaft passes through a vessel wall.

stumpage (1) The fee paid for the right to cut and remove standing timber. (2) Merchantable standing timber.

sturin $C_{36}H_{69}O_7N_{10}$ = 754.0. A protein from the sperm of the sturgeon.

sturnutatory Sternutatory.

Stutzer's reagent A suspension of cupric hydroxide in aqueous glycerol used to separate proteins from other nitrogenous constituents of plants.

stylophorum The herb of *Stylophorum diphyllum* (Papaveraceae). It contains the alkaloids, chelidonine, fumarine, sanguinarine, and chelerythrine.

styphnic acid* $C_6H_3O_8N_3$ = 245.1. 1,3-Dihydroxy-2,4,6-trinitrobenzene*. Trinitroresorcinol. Yellow crystals, m.180; used in explosives.

styptic An astringent that contracts blood vessels and stops hemorrhage locally; as, alum.

stypol Cotarnine phthalate. Yellow crystals, soluble in water; a styptic.

styracin $PhCH:CHCOOCH_2CH:CHPh$ = 264.3. Cinnamyl cinnamate*. A solid constituent of liquidambar or styrax, m.44; used in perfumery.

Styrax (1) A genus of trees and shrubs (Styraceae) that yield balsams; as, storax, benzoin. (2) Storax.

styrene* (1) $Ph \cdot CH:CH_2$ = 104.2. Ethenylbenzene†,

styrolene, phenethylene, styrol, vinylbenzene, cinnamene, phenylethylene. A constituent of storax, essential oils, and coal tar. Colorless, aromatic liquid, d.0.925, b.145, soluble in alcohol. Used in organic synthesis, and forms 2 types of derivatives; as *o*-, *m*-, and *p*-aminovinylbenzene = $NH_2C_6H_4CH:CH_2$. *p*-phenylvinylamine = $C_6H_5CH:CHNH_2$. (2) The radical $-CHPh \cdot CH_2-$.
styrilic acid Cinnamyl alcohol.
Styroflex Trademark for a polystyrene synthetic fiber.
styrol (1) Styrene*. (2) Colloidal silver.
styrolene Styrene*. **s. alcohol** Cinnamyl alcohol*.
Styron Trademark for polymerized styrene.
styrone Cinnamyl alcohol*.
styryl* 2-Phenylethenyl†. Cinnamenyl. The radical PhCH:CH−, from styrene. **s. alcohol** Cinnamyl alcohol*. **s.amine*** $PhCH:CH \cdot NH_2$ = 119.2. Colorless liquid, b.236, insoluble in water. **s. ketone** $(PhCH:CH)_2CO$ = 234.3. Dibenzylideneacetone. 1,5-Diphenyl-1,4-pentadien-3-one*. Colorless crystals, m.112. **3-s.-2-propenoic acid*** $C_8H_7CH:CH \cdot COOH$ = 174.2. Colorless crystals, m.165.
S.U. Strontium unit.
sub- Prefix (Latin), indicating "below," "almost," "under," or "near." Formerly designating a lower form of oxidation or a basic compound, and a deficiency of the substance or radical it described. Cf. *per-*.
subacetate A basic acetate; as, lead subacetate.
subatomic Pertaining to the structure of actual atoms as distinct from their function as parts of a molecule. **s. decomposition** Radioactive disintegration. **s. particle** See *subatomic particle* under *particle*. **s. reaction** A change in which an atom is disintegrated or transformed. See *nuclear chemistry*.
subatomics The study of the structure of atoms and the role of electrons and nuclei in subatomic changes.
subcarbonate A basic carbonate.
subcutaneous Located beneath the skin. **s. injection** The administration of a drug by injection under the skin.
suber Cork.
suberane Cycloheptane*.
suberic acid $(CH_2)_6 \cdot (COOH)_2$ = 174.2. Octanedioic acid*, 1,6-hexanedicarboxylic acid*. A homolog of oxalic acid, obtained by oxidation of cork. Colorless needles, m.140, soluble in water.
suberin A polysaccharide constituent of wood bark.
suberol Cycloheptanol*.
suberone Cycloheptanone*.
suberyl The cycloheptyl* radical. **s. alcohol** Cycloheptanol*.
subhalide A compound of the type $X_2M \cdot MX_2$; e.g., B_2Cl_4.
sublamine $HgSO_4 \cdot 2C_2H_4(NH_2)_2 \cdot 2H_2O$ = 424.8. Mercuric sulfate ethylenediamine. White crystals, soluble in water.
sublation A flotation process in which material absorbed on the surface of gas bubbles is collected on a layer of immiscible liquid, instead of as a foam over a liquid aqueous phase.
sublethal Not quite fatal. **s. dose** A quantity of drug below the fatal dose.
sublimate (1) The deposit formed on heating substances which pass directly from the solid to the vapor phase and then back to the solid state. Cf. *distillate*. (2) Mercuric chloride. **corrosive ~** Mercuric chloride.
sublimation The production of a sublimate; used to purify substances; as, iodine.
Sublimaze Trademark for fentanyl citrate.
submicron See *micron*.
subnitrate A basic nitrate, as, bismuth subnitrate.
subnormal Below normal.
suboxide That oxide of an element which contains the lowest proportion of oxygen.
subshell. See *shell (4)*.
subsoil The layer below the surface soil. It contains the rain-soluble organic portion of the soil.
subsonic Describing a velocity less than that of sound.
substance The material of which a body is composed; as a chemical compound. **s. concentration** See *concentration*.
substantive dyeing The coloring of fabrics with dyestuffs, without mordants.
substantive dyes A group of coal tar colors, chiefly for cotton, that dye without mordants; as benzidine dyes.
substituent* Any atom or group replacing the hydrogen of a parent compound.
substitute To replace one element or radical in a compound by a substituent.
substituted Pertaining to a compound which has undergone substitution. **s. compound** A compound obtained by substitution; a *derivative*, q.v.
substitution A reaction in which an atom or group of atoms in a (usually organic) molecule is exchanged for another. **cine ~** Reaction in which the entering group occupies the position next to that of the group being substituted.
substitutive nomenclature See *nomenclature*.
substrate The material upon which an enzyme acts.
substructure searching Computerized approach to searching a data base of chemical information for substances that contain particular combinations and arrangements of atoms and bonds.
subsubmicron Amicron.
subsulfate A basic sulfate.
subtilin An antibiotic polypeptide produced by *Bacillus subtilis*, especially from the fermentation of asparagus canning waste.
subtilisin* A proteolytic enzyme that degrades tissue proteins. Used to isolate drugs from tissues before analysis.
subtractive nomenclature See *nomenclature*.
subultramicroscopic Invisible in the ultramicroscope. Cf. *amicron*.
Sucaryl Trademark for sodium cyclohexylsulfamate, a nonsugar sweetening agent.
succinaldehyde* $C_2H_4(CHO)_2$ = 86.1. Butanedial*, b.170 (decomp.). **polymeric ~** m.65.
succinamic acid* $H_2N \cdot C(O) \cdot CH_2 \cdot CH_2 \cdot COOH$ = 117.1. 3-Carbamoylpropionic acid*. Amidosuccinic acid. White powder, soluble in water. **amino ~** Asparagine*.
succinamide* $(NH_2COCH_2)_2$ = 116.1. Butanediamide*. Colorless needles, m.242 (decomp.), soluble in water. **hydroxy ~** Malamide*.
succinamoyl The radical $-OC(CH_2)_2CONH_2$.
succinamyl The succinamoyl* radical.
succinate* $C_2H_4(COOM)_2$. A salt of succinic acid.
succinelite Succinic acid from amber.
succinic **s. acid*** $HOOC \cdot CH_2 \cdot CH_2 \cdot COOH$ = 118.1. Butanedioic acid*, ethylenedicarboxylic acid, amber acid. Occurs in amber and other resins as colorless, monoclinic prisms, m.184, slightly soluble in water; a reagent. **iso ~** $CH_3 \cdot CH(COOH)_2$. A solid, m.130 (decomp.), soluble in alcohol. **2-amino ~** Aspartic acid*. **diamido ~** Succinamide*. **dihydroxy ~** Tartaric acid*. **ethyl ~** See *ethylsuccinic acid* under *ethyl*. **formyl ~** Aconic acid. **hydroxy ~** Malic acid*. **mercapto ~ *** Thiomalic acid. **methyl ~** Pyrotartaric acid. **methylene ~** Itaconic acid. **oxo ~** Oxalacetic acid*.

s. aldehyde Succinaldehyde*. **s. anhydride*** $CH_2 \cdot CO \cdot O \cdot CO \cdot CH_2$ = 100.1. Succinyl oxide, butanedioic anhydride*. Colorless needles, d.1.104, m.120, insoluble in water. **s. peroxide** $(HOOC \cdot CH_2 \cdot CH_2 \cdot CO)_2O_2$ = 234.2. Alphozon, alphogen. White, fluffy powder, soluble in water; a germicide and mouthwash.

succinimide* $CH_2 \cdot CO \cdot NH \cdot CO \cdot CH_2$ = 99.1. Butanedioneimide. Colorless octahedra, m.124, soluble in water.

succinimido* The radical $(CH_2CO)_2N-$, from succinimide.

succinite (1) An amber-colored grossularite or aluminum *garnet*, q.v. (2) Baltic amber (4% succinic acid).

succinol An oil obtained by distilling amber.

succinonitrile* Ethylene dicyanide*.

succinoresinol $C_{12}H_{20}O$ = 180.3. An alcohol from amber.

succinyl* 1,4-Dioxo-1,4-butanediyl†. Butanedioyl. The radical $-OC \cdot CH_2CH_2 \cdot CO-$, from succinic acid. **s. chloride** $C_2H_4(COCl)_2$ = 155.0. S. dichloride*, butanedioyl chloride. Colorless, fuming, highly refractive liquid, d.1.412, b.190; used in organic synthesis. **s. choline chloride** $C_{14}H_{30}O_4N_2Cl_2$ = 361.3. Choline chloride succinate. Suxamethonium chloride. Scoline. White crystals, soluble in water, m.161; a muscle relaxant used in anesthesia (USP, BP). **s. oxide** Succinic anhydride*. **s. sulfathiazole** $C_{13}H_{13}O_5N_3S_2 \cdot H_2O$ = 373.4. Yellow crystals, slightly soluble in water; an antibacterial (USP).

succus cerasi *Cherry juice.*

sucramine The ammonium salt of saccharin.

sucrate A compound of sucrose.

sucroclastic Describing a glycolytic enzyme.

sucrol $NH_2 \cdot CO \cdot NH \cdot C_6H_4OEt$ = 180.2. Dulcin, valzin, *p*-ethoxyphenylurea, *p*-phenetolcarbamide. Colorless needles, m.173, slightly soluble in water, soluble in alcohol or ether; a sweetening 200 times sweeter than sucrose.

sucrolite A formaldehyde-urea-type plastic in which most or all of the former is replaced by sugar or molasses; some are transparent.

sucrose $C_{12}H_{22}O_{11}$ = 342.3. β-D-Fructofuranosyl α-D-glucopyranoside†. Cane sugar, saccharobiose, beet sugar, saccharose. A disacharide, hydrolyzing to D-fructose and D-glucose, $[\alpha]_D^{20}$ +66.5°. Colorless, monoclinic crystals, d.1.588, m.160 (decomp.), soluble in water, slightly soluble in alcohol. Cf. *sugar refining*. Used as a food sweetening, and in pharmacy (USP, EP, BP) and explosives. **s. octaacetate** White, bitter powder, m.69–72; a denaturant for rubbing alcohol (75 g/L).

suction The effect of sucking. Drawing a fluid into a pipe. **s. gas** The coal gas in producer gas when coal replaces coke. **s. pump** See *filter pump*.

sucuuba bark Plumiera.

Suda Trademark for a viscose synthetic fiber.

Sudafed Trademark for pseudoephedrine hydrochloride.

Sudan Trade name for a series of dyestuffs. **S. II** 2,4-Xylidineazo-2-naphthol. A reddish-orange dye. **S. IV** Scarlet Red. **S. brown** $C_{10}H_7N{:}NC_{10}H_7 \cdot OH$ = 229.4. C.I. Solvent Brown 5. A fat-soluble diazo dye, soluble in alcohol. **S. G** $Ph \cdot N_2 \cdot C_6H_4 \cdot N_2 \cdot C_{10}H_6OH$ = 352.4. C.I. Solvent Red 23. Brown powder, insoluble in water; a microscope stain and fat coloring. **S. yellow** $C_{18}H_{12}ON_2$ = 272.3. C.I. Solvent Yellow 14. Aniline azo-2-naphthol. Red powder, insoluble in water; used to color oils and varnishes.

sudanite rock A mixture of talc and magnesite.

sud cake A fertilizer made from sewage sludge and waste textile fibers (10–15% oil).

sudorific A diaphoretic; produces sweat.

suet Tallow. (1) The fat from the abdominal cavity of sheep: chiefly stearates and palmitates, with some oleates of glycerol. Colorless, unctuous mass, m.45–50, soluble in alcohol; used in ointments, cerates, and cooking. (2) The fat from beef. (3) The solid fat from any animal.

suffocating gases *Poison gases*, q.v., that stop respiration; as, Cl_2 or $COCl_2$.

sugar (1) $C_nH_{2n}O_n$ or $C_nH_{2n-2}O_{n-1}$. A sweet carbohydrate. Cf. *carbohydrates, monosaccharide, furanoside, pyranoside, sweetness.* (2) Generally sucrose. **acorn ~** Quercitol*. **amorphous ~** A hygroscopic form of s. produced by spray drying. Each particle is a hollow sphere. **avocado ~** A mannoketoheptose from the avocado, or alligator pear. **beechwood ~** Xylose. **beet ~** Sucrose from the s. beet, *Beta vulgaris.* **blood ~** Glucose. **brain ~** Cerebrose. **cabbage ~** A triose from cabbage leaves. **cane ~** A sucrose from sugarcane, *Saccharum officinalis.* **collagen ~** Glycine*. **corn ~** Glucose. **diabetic ~** Misnomer for saccharin. **fruit ~** Fructose. **gelatin ~** Glycine*. **grape ~** Glucose. **gum ~** Arabinose. **heart ~** Inositol. **invert ~** See *invert sugar.* **larch ~** Melezitose. **liver ~** Glycogen. **malt ~** Maltose. **maple ~** A crude s., chiefly sucrose, of agreeable flavor made from the sap of certain maples (*Acer saccharum, A. nigrum,* etc.), native in eastern Canada and northeastern U.S. **meat ~** Inositol. **milk ~** Lactose. **muscle ~** Inositol. **palm ~** A sucrose from the toddy palm, *Caryota urens,* or the coconut palm, *Cocos nucifera.* Cf. *gur.* **priming ~** See *priming sugar.* **reducing ~** A s. that will reduce Fehling's solution at the boiling point; indicative of free aldehyde or carbonyl groups. **simple ~** Monosaccharide. **sorghum ~** A sucrose from s. millet, *Sorghum saccharatum.* **stonecrop ~** A heptose in *Sedum spectabile.* **table ~** Sucrose. **wood ~** Xylose. **yeast ~** A sulfur-containing s., found in small traces in yeast.

s.cane *Saccharum officinarum,* the source of cane sugar. **s. carbonate** A carbonic ester of sugars, e.g., glucopyranose carbonate. **s. chemicals** Chemical products made commercially from cane s.; e.g., detergents, cosmetics, pesticides, alcohol, s.-formaldehyde plastics, and plasticizers (e.g., sucrose benzoate). **s. granulation** Crystallization in jams, due usually to sucrose or glucose, caused by under- or overinversion, respectively. **s. of lead** Lead acetate. **s. phosphates** Metabolic intermediates formed in the assimilation and utilization of carbohydrates by plants, animals, or microorganisms; e.g.: fructose 6-phosphoric acid, $C_6H_{11}O_5 \cdot PO_4H_2$ = 260.1; fructose 1,6-diphosphoric acid, $C_6H_{14}O_{12}P_2$ = 340.1; galactose 6-phosphoric acid, $C_6H_{11}O_5 \cdot PO_4H_2$ = 260.1; glucose phosphoric acids, $C_6H_{11}O_5 \cdot PO_4H_2$ = 260.1; inositol hexaphosphoric acid, $C_6H_{18}O_{24} \cdot P_6$ = 660.0 (cf. *phytic acid*); mannose 6-phosphoric acid, $C_6H_{11}O_5 \cdot PO_4H_2$ = 260.1; 6-phosphogluconic acid, $C_6H_{13}O_{10}P$ = 276.1; ribose 5-phosphoric acid, $C_5H_9O_4 \cdot PO_4H_2$ = 230.1. **s. refining** The purification of beet or cane sugar by removal of vegetable proteins and salts, followed by decolorization and crystallization of pure sucrose.

suint Greasy potassium salts of organic acids derived from dry sheep perspiration. **s. ash** Potassium black. A source of potassium salts and fertilizer.

sulf-, sulph- Prefix indicating the presence of sulfur. See *thio-, thia-, sulfo-.*

sulfa drugs $H_2N \cdot C_6H_4 \cdot SO_2 \cdot NHR$. Sulfonamides. A group of antimicrobial compounds, derivatives of *p*-aminobenzene

sulfonamide, that act by inhibiting uptake of *p*-amino*benzoic acid,* q.v.; as, sulfamethazine, sulfamethoxazole. Formerly widely used to treat infections, but now largely superseded by antibiotics.
sulfabenzamine Mafenide acetate.
sulfacetamide sodium $C_8H_9O_3N_2SNa \cdot H_2O$ = 254.2. *N*-Sulfanilylacetamide monosodium salt. Albucid, Cetamide. White crystals, freely soluble in water. Used for eye infections (USP, BP).
sulfacid (1) A thio acid. (2) A sulfonic acid.
sulfadiazine $NH_2 \cdot C_6H_4 \cdot SO_2 \cdot NH \cdot C_4H_3N_2$ = 250.3. *N'*-2-Pyrimidinylsulfanilamide. White powder, m.254, slightly soluble in water; a sulfa drug. (USP, EP, BP).
sulfadimethylpyrimidine Sulfamethazine.
sulfadimidine BP name for sulfamethazine.
sulfadoxine $C_{12}H_{14}O_4N_4S$ = 310.3. N^1-(5,6-Dimethoxy-4-pryimidinyl) sulfanilamide. Fansil. White crystals, m.198, slightly soluble in water. A long-acting sulfa drug; used for malaria and leprosy (USP, BP). See *pryimethamine.* Cf. *Fansidar.*
sulfafurazole BP name for sulfisoxazole.
sulfaldehyde Me·CS·H = 60.1. Ethanethial*, thioaldehyde. Oily liquid.
sulfamethazine $C_{12}H_{14}O_2N_4S$ = 278.3. *N'*-(4,6-Dimethyl-2-pyrimidinyl)sulfanilamide. Sulfadimidine, Sulfamezathine. White crystals, very soluble in water; a sulfa drug (USP, BP).
sulfamethizole $C_9H_{10}O_2N_4S$ = 270.3. Thiosulfil, Urolucasil. White crystals, m.209, slightly soluble in water; a sulfa drug, used for urinary infections (USP, BP).
sulfamethoxazole $C_{10}H_{11}O_3N_3S$ = 253.3. *N'*-(5-Methyl-3-isoxazolyl)sulfanilamide. Gantanol. White crystals, m.170, very slightly soluble in water. A sulfa drug, usually combined with trimethoprim (USP, BP). See *co-trimoxazole.*
Sulfamezathine Trademark for sulfamethazine.
sulfamic The radical $=N \cdot SO_2-$. Correctly designated as *N*-substituted sulfamoyl* radical. **s. acid*** $HSO_3 \cdot NH_2$ = 97.1. Sulfamidic acid*, amidosulfuric acid*, sulfaminic acid, sulfonamic acid. White crystals, soluble in water, m. ca. 250; a cleaning agent.
sulfamide Sulfonyldiamine*.
sulfamidic acid* Sulfamic acid*.
sulfamine Sulfonamide*.
sulfaminic acid Sulfamic acid*.
sulfamino- See *sulfoamino-*
sulfamoyl* The radical NH_2SO_2-.
sulfamyl The sulfamoyl* radical.
sulfane* Compounds of the type H_2S_n, as H_2S_5. **poly ~** See *hydropolysulfide.*
sulfanilamide* $H_2N \cdot C_6H_4 \cdot SO_2 \cdot NH_2$ = 172.2. Prontosil album. The antimicrobial part of the compound prontosil rubrum; the first sulfa drug (1936) and first effective treatment for streptococcal infections.
sulfanilate* A salt of sulfanilic acid.
sulfanilic acid* $NH_2 \cdot C_6H_4 \cdot SO_3H$ = 173.2. Sulphanilic acid, *p*-aminobenzenesulfonic acid. Colorless, efflorescent rhombs, decomp. 280, soluble in water. Used in microscopy, organic synthesis, the manufacture of azo dyes, and detection of nitrous acid and bile pigments.
sulfantimonate Thioantimonate.
sulfantimonide MS·SbS. A double salt of sulfide and antimonide.
sulfantimonite Thioantimonite.
sulfapyridine $H_2N \cdot C_6H_4 \cdot SO_2 \cdot NH \cdot C_5H_4N$ = 249.3. N^1-2-Pyridylsulfanilamide. M&B 693. White crystals, darkening in light, m.192, slightly soluble in water; one of the first sulfa drugs (1936), now used for dermatitis herpetiformis (USP, BP).
sulfarsenate Thioarsenate.
sulfarsenide $MAsS_2$. A double salt of sulfide and arsenide.
sulfarsenite Thioarsenite.
sulfasalazine $C_{18}H_{14}O_5N_4S$ = 398.4. 5-[[*p*-(Pyridylsulfamoyl)phenyl]azo]salicylic acid. Salazopyrin. Yellow powder, m.255, insoluble in water. A long-acting sulfa drug, used for ulcerative colitis (USP).
sulfate (1)* Indicating sulfur as the central atom(s) in an anion. (2)* M_2SO_4. A salt of sulfuric acid. **acid ~** Hydrogen ~*. **basic ~** Hydroxide s.* **bi ~** Hydrogen ~*. **di ~*** $M_2S_2O_7$. **hydrogen ~*** Acid s.*, bi ~. $MHSO_4$. **hydroxide ~*** Basic s.* $M(OH)SO_4$. **hypo ~** Dithionate*. **neutral, normal ~** Sulfate*. **peroxodi ~*** $M_2S_2O_8$. **peroxomono ~*** M_2SO_5. **pyro ~** Disulfate*. **thio ~*** Thionate. $M_2S_2O_3$.
s. ion The anion SO_4^{2-}. **s. of lime** Calcium s. **s. of potash** Potassium s. **s. pulp** Pulp obtained by digestion of wood in a solution of sodium hydroxide, containing some sodium sulfide.
sulfatide Lipoid substances containing sulfuric acid esters.
sulfation Conversion into sulfates; as, in the determination of ash content. Cf. *sulfonation.*
sulfatostannate. $R_2H_2Sn(SO_4)_3$. An organic compound. Cf. *stannosulfate.*
sulfazide $R-NH \cdot NH \cdot SO_2-R$. An organic compound.
-sulfenic* Suffix indicating the radical $-S \cdot OH$.
sulfeno-* Prefix corresponding to sulfenic.
sulfhydrate, sulfhydryl Hydrogensulfide*.
Sulfidal Trademark for colloidal sulfur.
sulfide (1)* M_2S. M is a monovalent metal. (2)* R_2S. Thioether, alkyl (or aryl) sulfide, sulfur ether. R is a monovalent organic radical. Obtained from organic halides and alkali metal sulfides. Forms colorless, volatile liquids that can be oxidized to sulfones. **di ~*** M_2S_2 or RS_2R.
s. dye A dye used in a sodium s. bath. **s. ion** The ion S^{2-}. **s. sulfur** The divalent, negative sulfur atom in sulfides, as distinct from tetra- or hexavalent, positive sulfur.
sulfidion See *sulfide ion.*
sulfilimine A compound containing the sulfinimidoyl† radical, HN:S=.
sulfime A compound containing a =C:N·S– group; as, R·CH(:NSH). Cf. *oxime.*
sulfimide (1)* The hypothetical compound $H_2S(NH)$. (2) A compound containing the radical $-SO_2 \cdot NH-$. (3) $(SO_2NH)_3$ = 237.2. A sulfuric acid imide.
sulfinate (1)* R·OH·S:O. A salt of sulfinic acid. (2) $R_2S \cdot CH_2 \cdot CO \cdot O$. Thetines. (3) A compound having the grouping R_2SX_2.
sulfindigotic acid $C_8H_5NOSO_3$ = 211.2. Obtained by the action of sulfuric acid on indigo. Cf. *thioindigo.*
sulfine Sulfonium compound. An organic compound of the type R_3SX. See *sulfonium.*
sulfinic acid* $R \cdot SO_2H$; as, ethanesulfinic acid*, $EtSO_2H$. The prefix is sulfino-*.
sulfinid Saccharin.
sulfinimidic acid* See *imidic acids.*
sulfino-* Prefix indicating the radical $-SO_2H$, from sulfinic acid. **s.hydrazonic acid*** See *sulfinohydrazonic acid* under *hydrazonic acid.* **s.hydroximic acid*** See *sulfinohydroximic acid* under *hydroximic acid.*
sulfinoxide $R_3 \cdot S \cdot OH$. An organic compound.
sulfinpyrazone $C_{23}H_{20}O_3N_2S$ = 404.5. Anturan. White

crystals, insoluble in water, m.133; a uricosuric, used to treat gout (USP, BP).

sulfinyl* Thionyl*. The inorganic group =SO. Cf. *sulfoxide, thienyl.* **s.amines*** R·N:SO. E.g.: sulfinylmethylamine, MeN:SO, b.58; sulfinylethylamine, EtN:SO, b.75. **s.aniline*** PhN:SO = 139.2. Colorless liquid, b.200; used in organic synthesis. **s. bromide*** $SOBr_2$ = 207.9. Red liquid, $b_{40mm}68$. **s. bromide chloride*** SOBrCl = 163.4. Yellow liquid, b.115. **s. chloride*** $SOCl_2$ = 119.0. Colorless, pungent liquid, b.79; used in organic synthesis. **s.dialkylamine** $R_2N \cdot SO \cdot NR_2$; as sulfinyldiethylamine, $(Et_2N)_2SO$, b.118. S. dialkyl amines are carbamides (ureas) with the C atom replaced by a tetravalent S atom. **s. fluoride*** SOF_2 = 86.1. Colorless gas, b.−32; forms, with ammonia, the compounds $2SOF_2 \cdot 5NH_3$ and $2SOF_2 \cdot 7NH_3$. **s.hydrazine** $R_2N \cdot N{:}SO$. Sulfinyldiethylhydrazine, $Et_2N \cdot NSO$. **s.imide*** SONH = 63.1. Colorless liquid, m.−85, from the action of ammonia on s. chloride; polymerizes −70 to a yellow, transparent resin, soluble in alcohol, insoluble in water. **s.toluidines** $Me \cdot C_6H_4N{:}S{:}O$ = 153.2. ***ortho-*** ~ b.184. ***meta-*** ~ b.220. ***para-*** ~ m.7, b.224. Used in organic synthesis.

sulfion The sulfide ion, S^{2-}.

sulfisoxazole $C_{11}H_{13}O_3N_3S$ = 267.3. N^1-(Dimethyl-5-isoxazolyl)sulfanilamide. Sulphafurazole, Gantrisin. White crystals, m.197, insoluble in water. A sulfa drug, used for urinary infections (USP, BP).

sulfite* M_2SO_3. A salt of sulfurous acid. **acid** ~, **bi** ~, **hydrogen** ~* $MHSO_3$. **hypo** ~ (1) Dithionite*. (2) Thiosulfate*. **sub** ~ Thiosulfate*.
s. ion The anion SO_3^{2-}. **s. pulp** Paper pulp obtained by the digestion of wood chips (generally coniferous) with ammonium, calcium, magnesium, or sodium sulfite as the main chemical.

sulfivinyl The sulfinyl* radical.

sulfo- Sulpho-. Prefix indicating: (1)* the sulfonic acid group, $-SO_3H$; (2) the presence of divalent sulfur. Cf. replacement *nomenclature.* **s. acid** Sulfonic acid*. **s. group** Sulfonic acid group. **s. salt** A salt of an acid containing sulfur.

sulfoacetic acid* $HSO_3 \cdot CH_2 \cdot COOH \cdot H_2O$ = 158.1. Colorless tablets, m.86 (sublimes), soluble in water.

sulfoamino-* Sulfamino-. Prefix indicating the radical $HO_3S \cdot NH-$ or $=NSO_3H$, from sulfamic acid.

sulfobenzide $(C_6H_5)_2SO_2$ = 218.3. Diphenylsulfone. Colorless scales, m.123, insoluble in water.

sulfobenzoic acid* See *sulfobenzoic acid* under *benzoic acid.*

sulfobromophthalein sodium $C_{20}H_8O_{10}Br_4Na_2S_2$ = 838.0. Bromsulfophthalein sodium, BSP. White, hygroscopic, bitter crystals; a diagnostic aid for liver function (USP, BP).

sulfocarbamide Thiourea*.

sulfocarbanilide Thiocarbanilide.

sulfocarbazide Thiocarbonohydrazide*.

sulfocarbimide Isothiocyanic acid*.

sulfocarbodiazone Thio*carbodiazone**.

sulfocarbolate Phenol sulfonate*.

sulfocarbolic acid **2-** ~ Aseptol.

sulfocarbonate Thiocarbonate*.

sulfocarbonic acid Trithiocarbonic acid*.

sulfochloride Sulfonyl chloride*.

sulfocyan Thiocyanate*.

sulfocyanate Thiocyanate*.

sulfocyanic acid Thiocyanic acid*.

sulfocyanide Thiocyanate*.

sulfoform $(C_6H_5)_3 \cdot SbS$ = 385.1. Triphenylstibine sulfide. Colorless needles, m.120; used to treat skin diseases. **crude** ~ Yellow liquid; a powerful reducing agent.

sulfohydrate Thiol*.

sulfoichthyolic acid A compound obtained from bituminous shales; contains sulfur as sulfonate, sulfone, or sulfide.

sulfoid Colloidal sulfur.

sulfolane $C_4H_8O_2S$ = 120.2. Tetramethylene sulfone. Tetrahydrothiophene-1,1-dioxide. Colorless solid, $b_{18mm}153$, miscible with water; a selective solvent for hydrocarbons.

sulfolene $C_4H_6O_2S$ = 118.2. **3-** ~ 2,5-Dihydrothiophene-1,1-dioxide. m.65. Used as a solvent in the petroleum industry.

sulfoleate A salt of sulfoleic acid.

sulfoleic acid Obtained by mixing sulfuric acid with oils containing oleic acid.

sulfolipids Fatty substances containing sulfur.

sulfonamic Sulfamic.

sulfonamides (1)* Sulfamine. Amides derived from sulfonic acids, containing the $-SO_2 \cdot NH_2$ radical. Cf. *carboxamides.* (2) Sulfa drugs.

sulfonamido Indicating an *N*-substituted sulfamoyl group.

sulfonaphthol Naphthol sulfonic acid.

sulfonate (1) To treat an aromatic hydrocarbon with fuming sulfuric acid. (2) A sulfuric acid derivative. (3)* A sulfonic acid ester or salt.

sulfonation Substitution of H atom(s) by $-SO_3H$ group(s). **direct** ~ Treatment of an organic compound with fuming sulfuric acid. **indirect** ~ Treatment with hydrogensulfites.

sulfonator A double-walled, cast-iron vessel with power-driven stirrer, for large-scale sulfonation.

sulfone* R_2SO_2, obtained by oxidation of sulfides. In derivatives its prefix is sulfonyl*, Et_2SO_2.

sulfonephthalein A compound similar to the phthaleins, but with a sulfone, instead of a carbonyl, group. Made by condensation of *o*-sulfobenzoic acid anhydride with phenols. Many are important *indicators*, q.v., and dyes; as, *o*-cresol ~, cresol red.

sulfonic acid* An organic compound containing the radical $-SO_3H$, derived from sulfuric acid by replacement of an −OH group. Used also as a suffix. Sulfonic acids are soluble in water and yield phenols when heated with potassium hydroxide. Used in the manufacture of dyes and synthetic drugs. **amino** ~ Sulfamic acid*. **diamino** ~ Sulfonyl diamine*. **nitro** ~ Nitrosyl hydrogensulfate*.

sulfonic anhydride* Suffix indicating the group $-SO_2 \cdot O \cdot SO_2-$.

sulfonimidic acids See *sulfonimidic acids* under *imidic acids.*

sulfonium* Sulfine. The cation H_3S^+. **s. compound** R_3SX, where X is an electronegative element or radical. **s. hydroxides** $R_3S \cdot OH$. Strong bases ionizing to R_3S^+ and OH^-.

sulfonohydrazonic acid* See *sulfonohydrazonic acid* under *hydrazonic acid.*

sulfonohydroximic acid* See *sulfonohydroximic acid* under *hydroximic acid.*

sulfonyl* Sulfuryl. The radical $-SO_2-$, derived from sulfuric acid. **s. chloride*** SO_2Cl_2 = 135.0. Sulfuric chloride, chlorsulfuric acid. Colorless, pungent liquid, b.69, decomp. by water; a reagent. **s.diamine*** $SO_2(NH_2)_2$ = 96.1. Sulfamide, m.92, soluble in water. Cf. *sulfamic acid.* **s. fluoride*** SO_2F_2 = 102.1. Sulfur oxyfluoride. An irritating vapor.

sulfoparaldehyde $(SCHMe)_3$ = 180.3. Trithioacetaldehyde. Colorless crystals, insoluble in water.

sulfophenol Aseptol.

sulfophen(yl)ate (1) Phenolsulfonate*. (2) Phenylsulfonate.

sulforhodanide Thiocyanate*.

sulfosalicylic acid $C_6H_3(OH) \cdot (COOH) \cdot SO_3H$ = 218.2. A soluble solid, m.120, a reagent for albumin.

sulfosalt (1) A salt of a thioacid. (2) A sulfonic acid ester.

sulfoselenide M_4SSe. A double salt.
sulfosemicarbazide Thiosemicarbazide*.
sulfosol A colloidal system with sulfuric acid as the disperse phase.
sulfostannite Any of the thiostannate(II) compounds, M_2SnS_2.
sulfourea Thiourea*.
sulfovinic acid Ethylhydrogensulfate*.
sulfoxide* The radical =SO.
sulfoxides* R·SO·R. Also indicated by the prefix sulfinyl*. Obtained by oxidation of thiols (analogous to the ketones). **diphenyl ~*** Ph_2SO = 202.3. Colorless crystals, m.70.
sulfoxone sodium $C_{14}H_{14}O_6N_2Na_2S_3$ = 448.4. White powder, with characteristic odor, soluble in water; an antileprotic (USP).
sulfoxylate* M_2SO_2, derived from sulfoxylic acid.
sulfoxylic acid* The hypothetical acid $S(OH)_2$, containing divalent sulfur; known as its salts, Na_2SO_2 and $NaHSO_2$. See *Rongalite.*
sulfur* Sulphur. S = 32.066. Brimstone. Solid, nonmetal element, at. no. 16. A yellow, brittle mass or transparent, monoclinic or rhombic crystals; but existing in a number of modifications that are either ring cyclosulfurs (λ-sulfur) (see definitions 1–3), or chain catenasulfurs (μ-sulfur): (1) Orthorhombic. S_α, containing S_8. Stable at ordinary temperatures, d.2.07, m.113, b.445. (2) Monoclinic, S_β. Stable above 96, forms transparent, yellow, monoclinic prisms, d.1.957, m.119, b.445. (3) Cyclosulfurs, S_n, where *n* ranges 7–20. (4) Catenasulfur, plastic sulfur. From liquid s. quenched in cold water. Changes to S_α. See Table 86. S. is found native in Sicily and U.S., but its principal source is natural gas and oil. It is insoluble in water; slightly soluble in alcohol or ether; soluble in carbon disulfide: solubility varies with the modification. Used in the manufacture of gunpowder, disinfectants, vulcanizing agents, sulfuric acid, sulfides, and pharmaceuticals. See *sulfur compounds* below. **colloidal ~** Sulfidal, sulfoid. Finely divided s. obtained by passing hydrogen sulfide through a solution of a sulfite. **elastic ~** S. mu. Amorphous, chemically pure s., 90–95% soluble in carbon disulfide; used in the rubber industry. **flowers of ~** Sublimed s. **lac ~** Precipitated s. obtained by adding sulfuric acid to a solution of a polysulfide (contains up to 45% calcium sulfate), used in pharmacy. **liver of ~** A mixture of potassium sulfide and polysulfides. **milk of ~, precipitated ~** A suspension of finely divided s. obtained by precipitation of calcium pentasulfide or a thiosulfate with an acid. Used for acne (USP, BP). **ruby ~** Arsenic tetrasulfide*. **sublimed ~** Fine, yellow powder obtained by distilling s. (USP). **vegetable ~** Lycopodium. **washed ~** S. washed with ammonium hydroxide and water.

TABLE 86. EFFECT OF HEAT ON SULFUR

Temperature, °C	Effect or form
20	Orthorhombic, S_α, containing S_8.
96	Monoclinic, S_β, containing S_8.
119	Melts to thin, yellow liquid, S_8.
160	Forms thick, brown liquid containing S_8.
220	Becomes dark brown and plastic; progressive degradation of S_8.
250	Begins to burn if heated in air.
440	Forms thin, brown liquid.
447	Boils; reddish-brown vapor; becomes lighter in color and larger in volume with higher temperatures.
860	Colorless vapor, mainly of S_2, d.2.23.

s. acids See *sulfur acids, sulfuric acid.* **s. alcohols** Thiols*. **s. anhydride** S. trioxide*. **s. bromide** S_2Br_2 = 223.9. Disulfur dibromide*, s. monobromide. Yellow liquid, d.2.63, b.195, soluble in carbon disulfide. **s. chlorides** (1) S_2Cl_2 = 135.0. Disulfur dichloride*. S. monochloride. Brown, fuming liquid, b.138, decomp. by water, soluble in carbon disulfide. Used as a reagent; for s. solvents, war gases, vulcanized oils; and in vulcanizing rubber and purifying sugar. (2) SCl_2. S. dichloride*. Dark-red liquid, changing to S_2Cl_2. (3) SCl_4. S. tetrachloride*. Brown liquid, decomp. −30; used for synthesis; and in the manufacture of varnishes, rubber pigments, and poison gas. **s. compounds** S. has a valency of 2, 4, or 6; and forms 4 series of compounds; as, hydrogen sulfide (valency −2), s. dichloride (valency 2), s. dioxide (valency 4), s. trioxide (valency 6). **s. cycle** The transformation of s. compounds by living organisms:

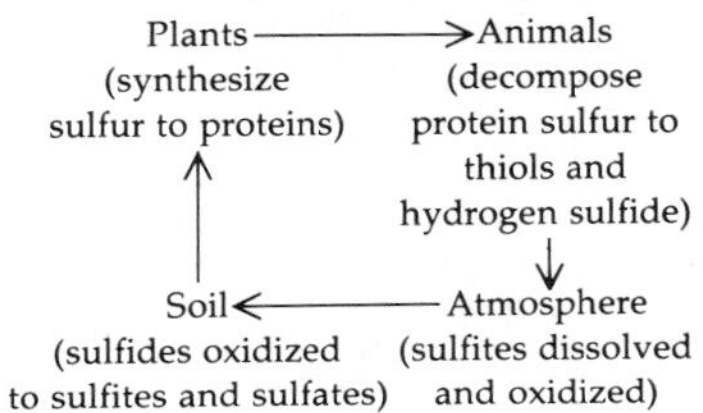

s. dichloride* SCl_2 = 103.0. Brown liquid, b.69, decomp. by water or heat to S_2Cl_2. Used as a s. solvent and for vulcanizing rubber and purifying sugar juices. **s. dioxide*** SO_2 = 64.1. Sulfurous acid anhydride. Colorless gas, b.−10, soluble in water or ether. Marketed as liquid under pressure; a reagent, bleach, and preservative (NF). Continual exposure to concentrations above 100 mg/m^3 air is regarded as undesirable. **s. dyes** Pyrogenic dyes. Organic dyes containing s., obtained by heating amino or nitro compounds with sulfur or alkali polysulfides. Their color arises from the unsaturated heterocyclic nuclei so formed. Insoluble in water, soluble in sodium sulfide solutions; permanent textile dyes. **s. ether** R_2S. See *sulfide.* **s. fluoride** S_2F_2 = 102.1. S. monofluoride. Colorless gas, readily hydrolyzed to hdyrofluoric acid, s. dioxide, and s. **s. hexafluoride*** SF_6 = 146.0. A gas of high relative permittivity which captures electrons in an electric field; used in high-voltage switches and stabilizers. **s. hexaiodide*** SI_6 = 793.5. Green crystals, soluble in carbon disulfide, decomp. by water. **s. iodide** S_2I_2 = 317.9. Iodine disulfide, s. subiodide. Gray, lustrous, brittle crystals, insoluble in water. **s. monobromide** S. bromide. **s. monochloride** S. chloride. **s. monofluoride** S. fluoride. **s. mu** Elastic s. **s. oxides** See *sulfur dioxide* above, *sulfur trioxide* below. **s. oxychlorides** (1) S_2OCl_4 = 221.9. Red liquid, b.100, decomp. by water. (2) $S_2O_3Cl_4$ = 253.9. White needles, m.57 (sublime), decomp. by water. (3) Sulfonyl chloride. **s. oxyfluoride** Sulfonyl fluoride*. **s. subbromide** S. bromide. **s. subchloride** S. chloride. **s. subiodide** S. iodide. **s. tetrachloride*** SCl_4 = 173.9. Yellow liquid, m.−31; unstable, but forms stable compounds; as, $2AlI_3 \cdot SCl_4$. **s. tetrafluoride*** SF_4 = 108.1. A toxic, gaseous fluorinating agent. **s. trioxide** SO_3 = 80.1. Sulfuric acid anhydride. **gamma-~** Liquid, b.45, obtained by distillation of oleum; rapidly polymerizes in moist air to the *beta* (straight-chain) form, and then the *alpha* (cross-linked) form, m.100. On heating, these revert to the gamma form, which may be stabilized by addition of 0.2% of boric acid. **solid ~** $(SO_3)_2$. Colorless prisms, m.15, b.46, rapidly decomp. by water to form sulfuric acid; soluble in concentrated sulfuric acid. **s. yellow** Naphthol yellow.

sulfur acids The following are known:

H_2S	Hydrogen sulfide*
H_2S_5	Dihydrogen pentasulfide*
H_2SO_3	Sulfurous acid*
H_2SO_4	Sulfuric acid*
$H_2SO_4 + SO_3$	Oleum
$H_2S_2O_3$	Thiosulfuric acid*
$H_2S_2O_4$	Dithionous acid*
$H_2S_2O_5$	Disulfurous acid*
$H_2S_2O_6$	Dithionic acid*
$H_2S_2O_7$	Disulfuric acid*
$H_2S_2O_8$	Peroxodisulfuric acid*
$H_2S_3O_6$	Trithionic acid*
$H_2S_4O_6$	Tetrathionic acid*
$H_2S_5O_6$	Pentathionic acid*
$H_2S_6O_6$	Hexathionic acid*

sulfurated, sulfuretted, sulphuretted Combined with sulfur. **s. oils** Essential oils containing sulfur compounds; as, mustard oil.

sulfuret Sulfide*.

sulfuretted hydrogen Hydrogen sulfide*.

sulfuric Pertaining to hexavalent sulfur. **s. acid*** See *sulfuracids, sulfuric acid.* **s. anhydride** Sulfur trioxide*. **s. chloride** Sulfonyl chloride*.

sulfuric acid $H_2SO_4 = 98.1$. Oil (spirit) of vitriol, hydrogen sulfate. A dibasic acid from sulfur dioxide. See *sulfuric acid manufacture.*

1. $H_2SO_4 = 98.1$. d.1.8542, m.10, b.316.
2. $H_2SO_4 \cdot H_2O = 116.1$. Colorless prisms, d.1.788, m.8, b.210, soluble in water, decomp. by alcohol.
3. $H_2SO_4 \cdot 2H_2O = 134.1$. Colorless liquid, d.1.665, m.−38, b.170; soluble in water, alcohol, or ether.
4. $H_2SO_4 \cdot nHO$. Concentrated acid, 94%, d.1.84.

4*a*. Concentrated: NF, BP: 95% w/w min.

5. $H_2SO_4 \cdot nSO_3$. Oleum (above 100%).

Cf. *sulfamic acid, sulfur acids, BOV, DOV, ROV.* **concentrated ~** See (4), (4*a*) in the preceding list. **ethyl ~** Ethylhydrogensulfate*. **fuming ~*** See (5) in the preceding list. **glycerol ~** See *glycerolsulfuric acid* under *glycerol.* **nitro ~** See *nitrosulfuric acid.* **nitroso ~** Nitrosyl hydrogensulfate*. **Nordhausen ~** Oleum. **peroxodi ~*** $H_2S_2O_8$. **peroxomono ~*** $H_2S_2O_5$. See *Caro's acid.* **phenol ~** Phenylsulfuric acid.

sulfuric acid manufacture (1) *Lead-chamber process:* Sulfur or pyrites are burned in furnaces, and the sulfur dioxide formed passes through the Glover tower, where it comes in contact with 65% s. acid from the lead chamber and the acid containing oxides of nitrogen from the Gay-Lussac tower. A portion of the sulfur dioxide is oxidized and dissolved. The remaining mixture of gases, consisting now principally of sulfur dioxide and nitrogen oxide, passes into the lead chamber, where the principal reaction occurs:

$$SO_2 + O_2 + NO \rightarrow SO_3 + NO_2 \rightarrow H_2SO_4\ (65\%)$$

The escaping nitrogen dioxide is passed through the Gay-Lussac tower, in which it is dissolved in strong sulfuric acid from the Glover tower. The acids collecting at the bottom of the lead chambers are finally concentrated in pans by heat. (2) *Contact process:* Sulfur or pyrites are burned in furnaces, and the sulfur dioxide formed passes through scrubbers. In some the gas is washed with water; in others it is dried with sulfuric acid. The dry gas enters the contact chamber in which a catalyst (generally platinum or iron oxide) oxidizes it to sulfur trioxide which, in turn, is passed through water and absorbed as sulfuric acid.

sulfuric ester Sulfuric acid ester. A R_2SO_4 compound.

sulfuric ether Commercial ether (a misnomer).

sulfuring (1) Burning sulfur in barrels to disinfect them for wines, etc. (2) Dressing growing crops with flowers of sulfur to arrest mold growth. (3) Adding sulfur to hop kiln fires to liberate sulfur dioxide as preservative.

sulfurize To combine with sulfur.

sulfurous Pertaining to tetravalent sulfur. **s. acid*** $SO(OH)_2 = 82.1$. An aqueous solution of sulfur dioxide (6% SO_2). Colorless liquid, d.1.03; a reducing agent and bleach. **di ~*** $H_2S_2O_5$. See *Caro's acid.* **ethyl ~** Ethylsulfonic acid*. **s. anhydride** Sulfur dioxide*. **s. esters** R_2SO_3.

sulfuryl The sulfonyl* radical.

sulph-, sulpho-, sulphur, sulphuretted, sulphuric, etc. (U.K. usage.) See *sulf-, sulfo-,* etc. (international usage).

sulphafurazole BP name for sulfisoxazole.

sultams* Compounds containing the group $-SO_2 \cdot O-$ as part of a ring.

sultones* Compounds containing the group $-SO_2 \cdot N=$ as part of a ring.

sulvanite Cu_3VS_4. A vanadium mineral.

sumac(h) (1) Sicilian s., the shrub *Rhus coriaria* (Anacardiaceae), S. Europe; leaves used for tanning and dyeing. (2) Any of the *Rhus* species. (3) Sumac. A tan (18–25% tannin) from the dried leaves of *Rhus* species. Cf. *fustic.* **American ~** *R. copellina* and *R. metopium.* **Chinese ~** Tree of heaven. The bark of *Ailanthus glandulosa* (Simarubiaceae); used for tanning. **curriers ~** The plant *Coriaria myrtifolia* (Coraniaceae), containing coria myrtin. **Indian ~** *R. succedania.* **smooth ~** S. berries. The dried, ripe fruit of *R. glabra.* **sweet ~** Fragrant s. The bark of the root of *R. aromatica.* **Tizra ~** *R. pentaphylla.* **Venetian ~** *R. cotinus.* **White ~** *R. glabra.*
s. bark Sweet s. **s. berries** Smooth s.

sumaresinol $C_{30}H_{48}O_4 = 472.7$. Sumaresinolic acid, m.298, from Sumatra gum benzoin.

sump An open channel or pit for collecting drainings for removal by pumping.

sun The star nearest to the earth; mass 1.99×10^{33} g, d.5.00, surface (photosphere) temperature 6000°C; corona temperature $>10^6$°C. Cf. *spectral classification, earth constants.* **s. atmosphere** H 92, He 6, metal vapors (mainly Mg, Fe, Si, Na) 2 vol. %. **s.light** Intensity on a clear day at zenith: 100,000–125,000 lux.

sunflower Helianthus. The dried seeds of *Helianthus annuus* (Compositae). **s. oil** A slow-drying oil from the seeds of s. (contain 32–45%). Consists mostly of the glycerides of oleic, linoleic, and palmitic acids; used for margarine manufacture and cooking.

sunstone Orthoclase.

sunt An aqueous extract of *Acacia arabica* (Sudan); a brown preservative for tent canvas.

S.U.P. *p*-Benzoyl-*p*-aminobenzoyl-1-amino-8-naphthol-3,8-sodium sulfonate. A substance that supplies stabilizing electrons to colloidal particles.

super- Prefix (Latin) indicating "above," "beyond," or "higher."

superacid Excessively acid in reaction; as, magic acid. **s. solution** A solution in acetic or phosphoric acid.

superacidity Increased acidity of the gastric juice.

Superayflex Trademark for a viscose synthetic fiber.

superbine A poisonous, bitter principle from *Gloriosa superba* (Liliaceae), India and S. Africa; resembles the bitter principle of squill.

supercalender A type of *calender,* q.v., used to obtain a high finish, e.g., on paper.

supercarbonate Hydrogencarbonate*.
superchlorination Sterilizing water by chlorinating in excess, and then adding an antichlor.
superconductivity Supraconductivity. The property of some elements of becoming particularly good electrical conductors at temperatures below about 11 K; e.g.: Pb 7.2, Hg 4.2, Sn 3.7, In 3.4, Tl 2.4 K. **s. motor** A device utilizing the s. of copper to produce considerable reductions in the size and weight of conventional electric motors.
supercooled Over- or undercooled. Cooled below the freezing point of a liquid without the separation of solid matter.
superfluid Supafluid. A fluid having exceptionally good heat-transfer and penetrating properties, e.g., liquid *helium*$_{II}$.
Supergas Trademark for triptane.
superheated Overheated. Describing a liquid or gas heated above its boiling point in the liquid state; as superheated steam (water vapor at above 100°C).
supermalloy The alloy Ni 75, Mo 5, Fe + Mn 20%. It has a high magnetic permeability; used in transformer cores.
supernatant Describing the liquid above a sediment or precipitate.
supernickel The alloy Cu 70, Ni 30%; retains its strength at high temperatures.
supernormal A volumetric solution of concentration greater than normal.
superoxide Hyperoxide*.
superpalite A poison gas. Cf. *palite.*
superphosphate (1) An acid phosphate. (2) A fertilizer mixture of calcium phosphate and calcium sulfate obtained by the action of concentrated sulfuric acid on phosphate minerals (apatite, phosphorite). **ammoniated ~** See *ammoniated superphosphate.* **double ~** A fertilizer made by the action of phosphoric acid on rock phosphate (40–50% phosphorus pentaoxide).
superpolymer A polymer whose molecular weight exceeds 10,000.
superpotential Overvoltage.
supersaturated More than saturated. **s. solution** A solution containing an excess of dissolved substance over that normally required for saturation at a particular temperature, obtainable by slowly cooling a saturated solution.
supersolubility Supersaturation. **s. curve** The curve relating the concentration of a supersaturated solution with the temperature; analogous to and parallel with the *solubility curve,* q.v.
supersonic Describing speed or velocity greater than that of sound. Cf. *ultrasonic.*
supersteel A high-speed *steel,* q.v.
supertension Overvoltage.
suppository Pharmaceutical term for a medicated, conical body for insertion into the rectum.
suppurative An agent that produces pus.
supra- Prefix (Latin) indicating "above."
supraconductivity Superconductivity.
Supramid Trademark for a polyamide synthetic fiber.
suprarenal gland Adrenal gland. A gland above the kidney.
suprarenaline Adrenaline.
suprasterols Sterols produced by irradiation of lumisterol. **s. II** $C_{28}H_{44}O$ = 396.7. m.110.
Suprema Trademark for a viscose synthetic fiber.
Supron Trademark for a polyamide synthetic fiber.
suramin sodium ~ $C_{51}H_{34}O_{23}N_6S_6$ = 1291. Bayer 205. Germanin. Bitter, hygroscopic powder, insoluble in ether; a urea complex, used to treat African trypanosomiasis.
surcharge The sum of the errors involved in an assay.
surface The outer part of a body having length and breadth but not thickness. **s.-active agent** See *surfactant.* **s. charge density*** The electric charge per unit surface area. Cf. current *density.* **s. energy** The product obtained by multiplying surface tension by the two-thirds power of the molecular weight and specific volume. **s. tension*** The contractile surface force of a liquid which makes it tend to assume a spherical form, e.g., to form a meniscus. It also exists at the junction of 2 liquids. It is measured directly (1 mN/m = 1 dyne/cm) or indirectly by determining the capillarity. The s. t. (at constant temperature) γ is in a constant ratio to the 4th power of the orthobaric densities of liquid D and gas d; hence $\gamma = K(D-d)^4$. Cf. *parachor.* **s. tension apparatus** Tensiometer. A device for determining the s. t., based upon the flexibility of a wire.
surfactant A surface-active agent; i.e., one that modifies the nature of surfaces, this often involving reducing the surface tension of water. Widely used as wetting agents, detergents, emulsifiers, dispersing agents, penetrants, and antifoams. Four types: cationic (as, modified onium salts); anionic (alkylarylsulfonates); nonionic (polyethylene oxides), and ampholytic (dodecyl-β-alanine).
surfusion The unstable condition of a liquid cooled below its freezing point without solidifying. Cf. *supercooled.*
surpalite Diphosgene.
surrogate A substitute for another substance; as, margarine for butter.
surrosion The increase in weight of a substance due to corrosion.
susceptible Sensitive. Readily capable of responding to an action or force; as, magnetic susceptibility. Cf. *immune.*
susotoxin $C_{10}H_{26}N_2$ = 174.3. Sustoxin. A ptomaine from cultures of hog cholera bacillus.
suspension (1) Suspensoid. (2) A thin thread on which the mirror and magnet of a galvanometer hang. **s. method** The determination of the density of a solid by placing it in a solution of known density. See *density fluids.*
suspensoid Suspension, soliquoid. Finely divided colloidal particles floating in a liquid, too small to settle, but kept in motion by Brownian motion.
sussexite A native, hydrated magnesium manganese diborate.
sustoxin Susotoxin.
Sutherland S.-Einstein equation $D = RT/Lf$, where D is the diffusion constant, R the gas constant, T the thermodynamic temperature, L the Avogadro constant, and f the frictional force on each molecule having unit velocity. **S.'s formula** The viscosity of a gas (η) at thermodynamic temperature T is given from that at 0°C (η_0) by: $\eta = \eta_0(T/273.15)^{3/2} \times (k + 273.15)/(k + T)$, where k is the S. constant.
suture A stitch, or the operation of stitching, or the s. material, used in surgery. **absorbable ~** A sterile s. prepared from the intestine of healthy animals (USP). **nonabsorbable ~** A s. that resists enzyme digestion in living animal tissues, e.g., of stainless steel.
suxamethonium chloride BP name for succinylcholine chloride.
Sved A unit of sedimentation rate.
Svedberg S., The(odor) (1884–1971) Swedish physical chemist, Nobel prize winner (1926); noted for research on colloids. **S.'s equation** The amplitude A of Brownian movement of a particle is proportional to its vibration period t. **S. unit** Abbrev. *S;* used in sedimentation; equal to 10^{-13} seconds.
swage To fashion metal, particularly iron, by drawing it into a groove, mold, or die having a desired shape.

TABLE 87. SWEETNESS FACTORS OF SOME COMPOUNDS RELATIVE TO SUCROSE = 1.0

Sugar	Sweetness factor
Neohesparidin dihydrochalcone	2400
Thaumatin	2000
Saccharin	400
Stevioside	300
Acesulfame K	200
Sucrol	200
Aspartame	180
Glycyrrhizins	50
Cyclamates	30
Peryllartine	20
Fructose	1.7
Invert sugar	1.2
Sucrose, xylitol	1.0
Glucose	0.74
Xylose	0.40
Maltose, rhamnose, galactose	0.32
Raffinose	0.23
Lactose	0.16

Swan, Sir Joseph Wilson (1828–1914) English chemist, pioneer in photography, electric carbon filament lamps, and electrodeposition of metals.

swan neck A tube bent into the shape of an upright U joined to an inverted U to create an S-shaped curve.

swarf The raw edge of a metallic object produced in casting metals, and removed for reuse.

Swedenborg, Emanuel (1688–1772) Swedish mineralogist (and later, theosophist) who invented mercury pumps.

sweetbirch oil Methyl salicylate*.

sweeten (1) To remove substances of unpleasant odor (as thiols) from spirits, petroleum, etc., usually by oxidation. (2) To purify. (3) To add a sugar.

sweet flag Calamus.

sweetness The characteristic taste of sugars. See Table 87. Cf. *sapiphore.*

swell The swelling of a canned food container due to the liberation of gas, e.g., hydrogen from the action of acidic fruit juices on the metal.

swelling The adsorption of water by an amorphous substance, to form a jelly.

SWU See *uranium enrichment.*

Syalon Trademark for a ceramic alloy.

sycoceryl alcohol $C_{18}H_{30}O$ = 262.4. Colorless crystals, m.90.

sydnone An aromatic compound having a ring structure of the type

```
       CR——CO
      /      |
R'—N        |
      \      |
       N———O
```

e.g., the anhydro derivative of a *N*-nitrose-*N*-arylglycine.

syenite A granular, igneous rock, chiefly orthoclase with or without albite, biotite, hornblende, microcline, or corundum.

sylvan CH:CH·O·CMe:CH = 82.1. 2-Methylfuran. A constituent of wood tar. Yellow liquid, b.63, soluble in alcohol.

sylvanite (Au,Ag)Te_2. Graphic tellurium.

sylvanium Tellurium*.

sylvate A salt of abietadienoic acid.

sylvic acid 7,13-Abietadien-18-oic acid*.

sylvic oil Tall oil.

sylviculture The study of forestry.

sylvine An isomorphous, native mixture of sodium and potassium chlorides.

sylvinite A sylvite, with rock salt (about 16% potassium oxide); a fertilizer.

sylvite KCl. A native potassium chloride fertilizer from Stassfurt.

Sylvius, François (1614–1672) (Dubois de la Boë) German-born Dutch physician who considered that combustion, respiration, and similar physiological functions are based on chemical reactions.

sym-* Prefix meaning "symmetrical."

symbiosis A partnership of 2 different living organisms, of mutual benefit to both. **antagonistic ~** S. in which one species derives more benefit than the other. **conjunctive ~** The living together of 2 organisms with bodily union, as, lichens, which consist of algae and fungi. **disjunctive ~** An association of 2 living organisms that are not actually united.

symbiotic fixation The conversion of atmospheric nitrogen to nitrites or nitrates by soil bacteria and leguminous plants. Cf. *nitrogen* fixation.

symbol A mark which represents a substance, quality, or relation. International standardizing bodies follow the principle that symbols neither carry a period not add an s when in the plural. See Table 88, on p. 566, and *SI units.* **chemical ~** A letter or a combination of letters that represents an atom of an element and its relative mass. Chemical symbols are used to indicate: (1) an atom, as, H; (2) an ion, as, H^+; (3) a molecule, as H_2; (4) a gram atom or gram molecule: thus H = 1.0079 g hydrogen (see *mole*); (5) an excited atom (indicated by an asterisk), as, H*; (6) an isotope by its mass number, as, ^{35}Cl; (7) an atomic nucleus by prefixing its atomic number as subscript, and its mass number as superscript; and suffixing its number of atoms per molecule as subscript, and ionic charge as superscript. Thus, $^{32}_{16}S_2^{2+}$. See *ion, nomenclature, radical, formula, alchemistic symbols.* **mathematical ~** A letter that indicates quantity; or a mark that indicates or forms part of an algebraic operation. **structure ~** A diagram that indicates the arrangement of atoms in an organic compound.

symmetric, symmetrical Having constituent parts arranged in a definite pattern, and repeated in a definite direction of space; as atoms in a crystal. See *stereoisomer.* **s. carbon atom** See *asymmetric carbon.* **s. compound** A benzene derivative with substitution of the 1-, 3-, and 5-hydrogen atoms.

symmetrization See *dipole inversion* under *inversion.*

symmetry Being symmetrical. The s. of a crystal is determined by regularities in the positions of the similar faces or edges, and is defined by the number of elements of symmetry. See *crystal systems.* **axis of ~** An imaginary axis through a symmetrical body. If this body is rotated, it occupies the same position in space more than once in a 360° turn. **center of ~** A central point of a symmetrical body, around which like faces are arranged in opposite pairs. **elements of ~** The number of planes, axes, and centers of s. of a symmetrical body. A cube has the largest possible number; viz.: 9 planes, 13 axes, and one center. **external ~** The s. of the outside crystal form. Cf. *internal symmetry* below. **geometrical ~** The external shape of a crystal, which resembles that of a geometrical figure. **internal ~** The s. of the arrangement of the atoms inside a crystal. Cf. *external symmetry* above. **plane of ~** An imaginary plane through a symmetrical body, which divides it into mirror-image halves.

TABLE 88. INTERNATIONALLY APPROVED SYMBOLS FOR PHYSICAL QUANTITIES

Where two or more symbols separated by commas are given for a quantity, these symbols are to be regarded as alternatives for which no preference is expressed; where they are separated by a dotted line, the first is preferred. Letters of the Latin alphabet are generally printed in italic type. In the United States, letters of the Greek alphabet are not italicized. Vector quantities should be in boldface type, except that the boldface is optional when the directional character of such quantities is not to be emphasized.

A. Space and time

Quantity	Symbol	Quantity	Symbol
angle (plane angle)	$\alpha, \beta, \gamma, \theta, \phi$, etc.	spherical coordinates	r, θ, ϕ
solid angle	Ω, ω	position vector; radius vector	$\mathbf{r}$
length	l	area	$A \ldots S$
breadth	b	volume	$V \ldots v$
height	h	time	t
thickness	d, δ	angular speed: $d\theta/dt$	ω
radius	r	angular acceleration: $d\omega/dt$	α
diameter: $2r$	d	speed: ds/dt	u, v, w
distance along path	s, L	acceleration: du/dt	a
generalized coordinate	q	acceleration of free fall	g
rectangular coordinates	x, y, z	speed of light in free space	c_0, c
cylindrical coordinates	r, ϕ, z	Mach number	Ma

B. Periodic and related phenomena

Quantity	Symbol	Quantity	Symbol
period	T	circular wavenumber: $2\pi\sigma$	k
relaxation time[a]	τ	circular wavevector	$\mathbf{k}$
frequency: $1/T$	ν, f	damping coefficient[c]	δ
rotational frequency	n	logarithmic decrement[c]: δ/ν	Λ
circular frequency[b]: $2\pi\nu$	ω	attenuation coefficient[d]	α
wavelength	λ	phase coefficient[d]	β
wavenumber: $1/\lambda$	$\sigma \ldots \tilde{\nu}$	propagation coefficient[d]: $\alpha + i\beta$	γ
wavevector	$\boldsymbol{\sigma}$		

C. Mechanics

Quantity	Symbol	Quantity	Symbol
mass	m	volume strain[h]: $\Delta V/V_0$	θ
density (mass density): m/V	ρ	Young modulus: σ/ϵ	E
relative density: ρ_2/ρ_1	d	shear modulus: τ/γ	G
specific volume: V/m	v	bulk modulus: $-p/\theta$	K
reduced mass: $m_1 m_2/(m_1 + m_2)$	μ	Poisson ratio	μ, ν
momentum: mu	p	compressibility: $-V^{-1}dV/dp$	κ
momentum (vector): $m\mathbf{u}$	$\mathbf{p}$	section modulus	Z, W
angular momentum	b, p_θ	coefficient of friction	$\mu \ldots f$
angular momentum (vector): $\mathbf{r} \times \mathbf{p}$	$\mathbf{L}$	viscosity (dynamic viscosity)	$\eta \ldots \mu$
moment of inertia[e]	I, J	fluidity: $1/\eta$	ϕ
second moment of area[f]	I_a	kinematic viscosity: η/ρ	ν
second polar moment of area[g]	I_p	diffusion coefficient	D
force	F	surface tension	γ, σ
force (vector)	$\mathbf{F}$	angle of contact	θ
weight	$G \ldots P, W$	work	$W \ldots A$
bending moment	M	energy	$E \ldots W$
moment of force (vector): $\mathbf{r} \times \mathbf{F}$	$\mathbf{M}$	potential energy	E_p, V, Φ
torque; moment of a couple	$\mathbf{T}$	kinetic energy	E_k, T, K
pressure	$p \ldots P$	power	P
normal stress[h]	σ	Hamiltonian function	H
shear stress[h]	τ	Lagrangian function	L
linear strain[h]: $\Delta l/l_0$	ϵ, e	gravitational constant	G
shear strain[h]: $\Delta x/d$	γ	Reynolds number: $\rho u l/\eta$	Re

D. Thermodynamics

Quantity	Symbol	Quantity	Symbol
thermodynamic temperature	$T \ldots \Theta$	heat; quantity of heat	$Q \ldots q$
common temperature	t, θ	work; quantity of work	$W \ldots w$
linear expansivity: $l^{-1}dl/dT$	α, λ	heat flow rate	$\Phi \ldots q$
cubic expansivity: $V^{-1}dV/dT$	α, γ	thermal conductivity	$\lambda \ldots k$

[a] When F is a function of time t given by $F(t) = A + B \exp(-t/\tau)$; τ is also called "time constant."

[b] Also called "pulsatance."

[c] When F is a function of time t given by $F(t) = A \exp(-\delta t) \sin[2\pi\nu(t - t_0)]$.

[d] When F is a function of distance x given by $F(x) = A \exp(-\alpha x) \cos[\beta(x - x_0)]$.

[e] $I_z = \int (x^2 + y^2)\, dm$.

[f] $I_{a,y} = \iint x^2\, dx\, dy$.

[g] $I_p = \iint (x^2 + y^2)\, dx\, dy$.

[h] More generally, stress and strain are each treated as a tensor, and a distinct notation is used.

TABLE 88. INTERNATIONALLY APPROVED SYMBOLS FOR PHYSICAL QUANTITIES (*Continued*)

D. Thermodynamics (continued)

Quantity	Symbol	Quantity	Symbol
thermal diffusivity: $\lambda/\rho c_p$	$a \ldots \alpha, \kappa$	specific internal energy: U/m	$u \ldots e$
heat capacity	C	specific enthalpy[i]: H/m	h
specific heat capacity: C/m	c	specific Helmholtz function: A/m	a, f
specific heat capacity at constant pressure	c_p	specific Gibbs function: G/m	g
specific heat capacity at constant volume	c_V	Joule-Thomson coefficient: $(\partial T/\partial P)_H$	μ, μ_{JT}
ratio c_p/c_V	γ, κ	isothermal compressibility: $-V^{-1}(\partial V/\partial p)_T$	κ, κ_T
entropy	S	isentropic compressibility: $-V^{-1}(\partial V/\partial p)_S$	κ_S
internal energy	$U \ldots E$	isobaric expansivity: $V^{-1}(\partial V/\partial T)_p$	α
enthalpy: $U + pV$	H	thermal diffusion ratio	k_T
Helmholtz function: $U - TS$	A, F	thermal diffusion factor	α_T
Gibbs function: $U + pV - TS$	G	thermal diffusion coefficient	D_T
Massieu function: $-A/T$	J		
Planck function: $-G/T$	Y		
specific entropy: S/m	s		

E. Electricity and magnetism[j]

Quantity	Symbol	Quantity	Symbol
electric charge; quantity of electricity	Q	relative permeability: μ/μ_0	μ_r
electric current: dQ/dt	I	magnetic susceptibility: $\mu_r - 1$	$\kappa \ldots \chi_m$
charge density: Q/V	ρ	magnetic moment ($\boldsymbol{T} = \boldsymbol{m} \times \boldsymbol{B}$)	$\boldsymbol{m}$
surface charge density: Q/A	σ	magnetization: $(\boldsymbol{B}/\mu_0) - \boldsymbol{H}$	$\boldsymbol{M}$
electric field strength	$\boldsymbol{E}$	mangetic polarization: $\boldsymbol{B} - \mu_0\boldsymbol{H}$	$\boldsymbol{J}$
electric potential	V, ϕ	electromagnetic energy density	w
electric potential difference	$U \ldots V$	Poynting vector: $\boldsymbol{E} \times \boldsymbol{H}$	$\boldsymbol{S}$
electromotive force	E	speed of propagation of electromagnetic waves in free space	c_0, c
electric displacement	$\boldsymbol{D}$	resistance	R
electric flux	Ψ	resistivity ($\boldsymbol{E} = \rho\boldsymbol{J}$)	ρ
capacitance	C	conductivity: $1/\rho$	γ, σ
permittivity ($\boldsymbol{D} = \epsilon\boldsymbol{E}$)	ϵ	reluctance: U_m/Φ	R, R_m
electric constant; permittivity of a vacuum	ϵ_0	permeance: $1/R_m$	Λ
relative permittivity[k]: ϵ/ϵ_0	ϵ_r	number of turns	N
electric susceptibility: $\epsilon_r - 1$	χ_e	number of phases	m
electric polarization: $\boldsymbol{D} - \epsilon_0\boldsymbol{E}$	$\boldsymbol{P}$	number of pairs of poles	p
electric dipole moment	$\boldsymbol{p} \ldots \boldsymbol{\mu}$	loss angle	δ
electric current density	$\boldsymbol{J}, \boldsymbol{j}$	phase displacement	ϕ
magnetic field strength	$\boldsymbol{H}$	impedance: $R + iX$	Z
magnetic potential difference	U_m	reactance: Im Z	X
magnetomotive force: $\oint H_s ds$	F_m	resistance: Re Z	R
magnetic flux	Φ	quality factor: $\|X\|/R$	Q
magnetic flux density; magnetic induction	$\boldsymbol{B}$	admittance: $1/Z$	Y
magnetic vector potential	$\boldsymbol{A}$	susceptance: Im Y	B
self inductance	L	conductance: Re Y	G
mutual inductance	M, L	power, active	P
coupling coefficient: $L_{12}/(L_1L_2)^{1/2}$	k	power, reactive	Q
leakage coefficient: $1 - k^2$	σ	power, apparent	S
permeability ($\boldsymbol{B} = \mu\boldsymbol{H}$)	μ		
magnetic constant; permeability of a vacuum	μ_0		

F. Light and related electromagnetic radiations

The same symbol is often used for a pair of corresponding radiant and luminous quantities. Subscripts e for radiant and v for luminous may be used when necessary to distinguish these quantities.

Quantity	Symbol	Quantity	Symbol
radiant energy	Q, Q_e	illuminance: $d\Phi/dA$	E, E_v
radiant flux; radiant power	$\Phi, \Phi_e \ldots P$	light exposure: $\int E dt$	H
radiant intensity: $d\Phi/d\Omega$	I, I_e	luminous efficacy: Φ_v/Φ_e	K
radiance: (dI/dA) sec θ	L, L_e	absorption factor; absorptance: Φ_a/Φ_0	α
radiant exitance: $d\Phi/dA$	M, M_e	reflection factor; reflectance: Φ_r/Φ_0	ρ
irradiance: $d\Phi/dA$	E, E_e	transmission factor; transmittance: Φ_{tr}/Φ_0	τ
emissivity	ϵ	linear extinction coefficient	μ
quantity of light	Q, Q_v	linear absorption coefficient	a
luminous flux	Φ, Φ_v	refractive index	n
luminous intensity: $d\Phi/d\Omega$	I, I_v	refraction: $(n^2 - 1)V/(n^2 + 2)$	R
luminance: $(dI/dA$ sec θ	L, L_v	angle of optical rotation	α
luminous exitance: $d\Phi/dA$	M, M_v		

[i] For the specific enthalpy change resulting from phase transitions the term "specific latent heat" is still used.
[j] Some of the quantities appearing as scalars in this table are treated as tensors.
[k] Also called "dielectric constant" when it is independent of $\boldsymbol{E}$.

TABLE 88. INTERNATIONALLY APPROVED SYMBOLS FOR PHYSICAL QUANTITIES (*Continued*)

G. Acoustics

Quantity	Symbol	Quantity	Symbol
speed of sound	c	reflection coefficient: P_r/P_0	ρ
speed of longitudinal waves	c_l	accoustic absorption coefficient: $1 - \rho$	$\alpha \ldots \alpha_a$
speed of transverse waves	c_t	transmission coefficient: P_{tr}/P_0	τ
group speed	c_g	dissipation coefficient: $\alpha - \tau$	δ
sound energy flux	P	loudness level	L_N
sound intensity	I, J		

H. Physical chemistry

Quantity	Symbol	Quantity	Symbol
relative atomic mass of an element		osmotic coefficient	$\phi \ldots g$
(atomic weight)	A_r	osmotic pressure	Π
relative molecular mass of a substance		surface concentration	Γ
(molecular weight)	M_r	electromotive force	E
amount of substance	n	Faraday constant	F
molar mass: m/n	M	charge number of ion i	z_i
molar volume: V/n	V_m	ionic strength: $\frac{1}{2}\Sigma_i m_i z_i^2$	I
molar internal energy: U/n	U_m	velocity of ion i	ν_i
molar enthalpy: H/n	H_m	electric mobility of ion i ($\nu_i = u_i \boldsymbol{E}$)	u_i
molar heat capacity: C/n	C_m	electrolytic conductivity[m] ($\boldsymbol{J} = \kappa \boldsymbol{E}$)	κ
at constant pressure: C_p/n	$C_{p,m}$	molar conductance of electrolyte: κ/c	Λ
at constant volume: C_V/n	$C_{V,m}$	transport number of ion i	t_i
molar entropy: S/n	S_m	molar conductance of ion i: $t_i\Lambda$	λ_i
molar Helmholtz function: A/n	A_m	overpotential	η
molar Gibbs function: G/n	G_m	exchange current density	j_0
(molar) gas constant	R	electrokinetic potential	ζ
compression factor: pV_m/RT	Z	intensity of light	I
mole fraction of substance B	x_B	transmittance: I/I_0	T
mass fraction of substance B	w_B	absorbance[n]: $-\lg T$	A
volume fraction of substance B	ϕ_B	(linear) absorption coefficient: A/l	a
molality of solute B:		molar (linear) absorption coefficient: A/lc_B	ϵ
(n_B divided by mass of solvent)	b_B, m_B	angle of optical rotation	α
amount-of-substance concentration[l]		specific optical rotatory power: $\alpha V/ml$	α_m
of solute B: n_B/V	c_B, [B]	molar optical rotatory power: $\alpha/c_B l$	α_n
chemical potential of substance B:		molar refraction: $(n^2 - 1)V_m/(n^2 + 2)$	R_m
$(\partial G/\partial n_B)_{T,p,n_C} \ldots$	μ_B	stoichiometric coefficient of molecules B	
absolute activity of substance B:		(negative for reactants, positive for	
$\exp(\mu_B/RT)$	λ_B	products: the general equation for a	
partial pressure of substance B		chemical reaction is $0 = \Sigma_B \nu_B B$)	ν_B
in a gas mixture: $x_B^g p$	p_B	extent of reaction ($d\xi = dn_B/\nu_B$)	ξ
fugacity of substance B in a gas		affinity of a reaction[o]: $-(\partial G/\partial \xi)_{T,p}$	
mixture: $\lambda_B \lim_{p \to 0} (x_B^g p/\lambda_B)$	f_B, p_B^*	$= -\Sigma_B \nu_B \mu_B$	$\boldsymbol{A} \ldots \mathcal{A}$
		equilibrium constant[p]	K
relative activity of substance B	a_B	degree of dissociation	α
activity coefficient (mole fraction basis)	f_B	rate of reaction: $d\xi/dt$	$\dot{\xi}, J$
activity coefficient (molality basis)	γ_B	rate constant of a reaction	k
activity coefficient (concentration basis)	y_B	activation energy of a reaction	E

I. Molecular physics

Quantity	Symbol	Quantity	Symbol
Avogadro constant	L, N_A	distribution function of speeds:	
number of molecules	N	$N/V = \int f dc_x dc_y dc_z$	$f(c)$
number density of molecules: N/V	n	Boltzmann function	H
molecular mass	m	generalized coordinate	q
molecular velocity	$\boldsymbol{c}, (c_x, c_y, c_z)$;	generalized momentum	p
	$\boldsymbol{u}, (u_x, u_y, u_z)$	volume in phase space	Ω
molecular position	$\boldsymbol{r}, (x, y, z)$	Boltzmann constant	k
molecular momentum	$\boldsymbol{p}, (p_x, p_y, p_z)$	$1/kT$ in exponential functions	β
average velocity	$\langle \boldsymbol{c} \rangle, \langle \boldsymbol{u} \rangle, \boldsymbol{c}_0, \boldsymbol{u}_0$	partition function	Q, Z
average speed	$\langle c \rangle, \langle u \rangle, \bar{c}, \bar{u}$	grand partition function	Ξ
most probable speed	$\hat{c}, \hat{u}$	statistical weight	g
mean free path	l, λ	symmetry number	σ, s
molecular attraction energy	ϵ	dipole moment of molecule	p, μ
interaction energy between		quadrupole moment of molecule	Θ
molecules i and j	ϕ_{ij}, V_{ij}	polarizability of molecule	α

[l]Formerly called "molarity."

[m]Formerly called "specific conductance."

[n]Formerly called "optical density."

[o]The symbol ΔG for this quantity is inappropriate since Δ implies a finite increment.

[p]For qualifying subscripts, see Group O below.

TABLE 88. INTERNATIONALLY APPROVED SYMBOLS FOR PHYSICAL QUANTITIES (*Continued*)

Quantity	Symbol	Quantity	Symbol
	I. Molecular physics (continued)		
Planck constant	h	vibrational temperature: $h\nu/k$	Θ_v
Planck constant divided by 2π	$\hbar$	Stefan-Boltzmann constant: $2\pi^5k^4/15h^3c^2$	σ
characteristic temperature	Θ	first radiation constant: $2\pi hc^2$	c_1
Debye temperature: $h\nu_D/k$	Θ_D	second radiation constant: hc/k	c_2
Einstein temperature: $h\nu_E/k$	Θ_E	rotational quantum number	J, K
rotational temperature: $h^2/8\pi^2Ik$	Θ_r	vibrational quantum number	v
	J. Atomic and nuclear physics		
nucleon number; mass number	A	nuclear quadrupole moment	Q
atomic number; proton number	Z	nuclear radius	R
neutron number: $A - Z$	N	orbital angular momentum	
(rest) mass of atom	m_a	quantum number	L, l_i
unified atomic mass constant:		spin angular momentum quantum number	S, s_i
$m_a(^{12}C)/12$	m_u	total angular momentum quantum number	J, j_i
(rest) mass of electron	m_e	nuclear spin quantum number	I, J
(rest) mass of proton	m_p	hyperfine structure quantum number	F
(rest) mass of neutron	m_n	principal quantum number	n, n_i
elementary charge (of proton)	e	magnetic quantum number	M, m_i
Planck constant	h	fine structure constant: $\mu_0e^2c/2h$	α
Planck constant divided by 2π	$\hbar$	electron radius: $\mu_0e^2/4\pi m_e$	r_e
Bohr radius $h^2/\pi\mu_0c^2m_ee^2$	a_0	Compton wavelength: h/m_ec	λ_C
Rydberg constant: $\mu_0^2m_ee^4c^3/8h^3$	R_∞	mass excess: $m_a - Am_u$	Δ
magnetic moment of particle	μ	packing fraction: Δ/Am_u	f
Bohr magneton: $eh/4\pi m_e$	μ_B	mean life	τ
Bohr magneton number[q]: μ/μ_B		level width: $h/2\pi\tau$	Γ
nuclear magneton: $(m_e/m_p)\mu_B$	μ_N	activity: $-dN/dt$	A
nuclear gyromagnetic ratio[r]: $2\pi\mu/Ih$	γ	specific activity: A/m	a
g-factor	g	decay constant: A/N	λ
Larmor (angular) frequency: $e\boldsymbol{B}/2m_e$	ω_L	half-life: $(\ln 2)/\lambda$	$T_{1/2}, t_{1/2}$
nuclear angular precession frequency:		disintegration energy	Q
$ge\boldsymbol{B}/2m_p = \gamma\boldsymbol{B}$	ω_N	spin-lattice relaxation time	T_1
cyclotron angular frequency of electron:		spin-spin relaxation time	T_2
$e\boldsymbol{B}/m_e$	ω_c	indirect spin-spin coupling	J
	K. Nuclear reactions and ionizing radiations		
reaction energy	Q	atomic attenuation coefficient	μ
cross section	σ	mass attenuation coefficient	μ_m
macroscopic cross section	Σ	linear stopping power	S, S_l
impact parameter	b	atomic stopping power	S_a
scattering angle	θ, ϕ	linear range	R, R_l
internal conversion coefficient	α	recombination coefficient	α
linear attenuation ocefficient	μ, μ_l		
	L. Quantum mechanics		
complex conjugate of Ψ	Ψ^*	anticommutator of A and B: $AB + BA$	$[A, B]_+$
probability density: $\Psi^*\Psi$	P	matrix element: $\int\phi_i^*(A\phi_j)d\tau$	A_{ij}
probability current density:		Hermitian conjugate of	
$(h/4\pi im_e)(\Psi^*\nabla\Psi - \Psi\nabla\Psi^*)$	$\boldsymbol{S}$	operator A	A^H
charge density of electrons: $-eP$	ρ	momentum operator in coordinate	
electric current density of electrons: $-e\boldsymbol{S}$	$\boldsymbol{j}$	representation	$+(h/2\pi i)\nabla$
expectation value of A	$\langle A\rangle, \bar{A}$	annihilation operators	a, b, α, β
commutator A and B: $AB - BA$	$[A, B], [A, B]_-$	creation operators	$a^\dagger, b^\dagger, \alpha^\dagger, \beta^\dagger$
	M. Solid-state physics		
fundamental translations for lattice	$\boldsymbol{a}, \boldsymbol{b}, \boldsymbol{c}$;	order of reflexion	n
	$\boldsymbol{a}_1, \boldsymbol{a}_2, \boldsymbol{a}_3$	short-range order parameter	σ
Miller indices	$h, k, l; h_1, h_2, h_3$	long-range order parameter	s
plane in lattice[s]	$(hkl); (h_1h_2h_3)$	Burgers vector	$\boldsymbol{b}$
direction in lattice[s]	$[u, v, w]$	circular wave vector; propagation	
fundamental translations	$\boldsymbol{a}^*, \boldsymbol{b}^*, \boldsymbol{c}^*$;	vector (of phonons)	$\boldsymbol{q}$
in reciprocal lattice	$\boldsymbol{b}_1, \boldsymbol{b}_2, \boldsymbol{b}_3$	circular wave vector; propagation	
lattice vector	$\boldsymbol{R}$	vector (of particles)	$\boldsymbol{k}$
lattice plane spacing	d	effective mass of electron	m^*, m_{eff}
Bragg angle	θ	Fermi energy	E_F, ϵ_F

[q]No internationally agreed symbol has yet been recommended, but $p \ldots n$ are in use.

[r]More logically called "nuclear magnetogyric ratio."

[s]Braces { } and angle brackets ⟨ ⟩ are used to enclose symmetry-related sets (forms) of planes and directions, respectively.

TABLE 88. INTERNATIONALLY APPROVED SYMBOLS FOR PHYSICAL QUANTITIES (*Continued*)

M. Solid-state physics (continued)

Fermi circular wavenumber	$\boldsymbol{k}_F$	piezoelectric coefficient (polarization/stress)	d_{mn}
work function	Φ	characteristic (Weiss) temperature	Θ, Θ_W
differential thermoelectric power	$S \ldots \Sigma$	Curie temperature	T_C
Peltier coefficient	Π	Néel temperature	T_N
Thomson coefficient	μ	Hall coefficient	R_H

N. Molecular spectroscopy

quantum number of:	
component of electronic orbital angular momentum vector along symmetry axis	Λ, λ_i
component of electronic spin along symmetry axis	Σ, σ_i
total electronic angular momentum vector along symmetry axis	Ω, ω_i
electronic spin	S
nuclear spin	I
vibrational mode	v
vibrational angular momentum (linear molecules)	l
total angular momentum (excluding nuclear spin)	J
component of $\boldsymbol{J}$ in direction of external field	M, M_J
component of $\boldsymbol{S}$ in direction of external field	M_S
total angular momentum (including nuclear spin: $\boldsymbol{F} = \boldsymbol{J} + \boldsymbol{L}$)	F
component of $\boldsymbol{F}$ in direction of external field	M_F
component of $\boldsymbol{I}$ in direction of external field	M_I
component of angular momentum along axis (linear and symmetric top molecules; excluding electron and nuclear spin; for linear molecules $K = \|\Lambda + l\|$)	K
total angular momentum (linear and symmetric top molecules; excluding electron and nuclear spin: $\boldsymbol{J} = \boldsymbol{N} + \boldsymbol{S}$[t]	N
component of angular momentum along symmetry axis (linear and symmetric top molecules; excluding nuclear spin; for linear molecules: $P = \|K + \Sigma\|$[u]	P
degeneracy of vibrational mode	d
electronic term: E_e/hc[v]	T_e
vibrational term: E_{vib}/hc	G
coefficients in expression for vibrational term for diatomic molecule: $G = \sigma_e(v + \frac{1}{2}) - x\sigma_e(v + \frac{1}{2})^2$	σ_e and $x\sigma_e$
coefficients in expression for vibrational term for polyatomic molecule: $G = \Sigma_j \sigma_j(v_j + \frac{1}{2}d_j) + \frac{1}{2}\Sigma_j \Sigma_k x_{jk}(v_j + \frac{1}{2}d_j)(v_k + \frac{1}{2}d_k)$	σ_j and x_{jk}
rotational term: E_{rot}/hc	F
moment of inertia of diatomic molecule	I
rotational constant of diatomic molecule: $h/8\pi^2 cI$	B
principal moments of inertia of polyatomic molecule ($I_A \leqslant I_B \leqslant I_C$)	I_A, I_B, I_C
rotational constants of polyatomic molecule: $A = h/8\pi^2 cI_A$, etc.	A, B, C
total term: $T_e + G + F$	T

O. Modifying signs to be used with the symbols

(*a*) Subscripts

I, II . . . / 1,2 . . .	*Especially with symbols for thermodynamic functions,* referring to different systems or different states of a system
A, B . . .	Referring to molecular species A, B . . .
i	Referring to a typical ionic species i
u	Referring to an undissociated molecule
p, V, T, S	Indicating constant pressure, volume, temperature, entropy
p, m, c, a	*With symbol for an equilibrium constant,* indicating that it is expressed in terms of pressure, molality, concentration, or relative activity
g, l, s, c	Referring to gas, liquid, solid, and crystalline states, respectively
f, e, s, t, d	Referring to fusion, evaporation, sublimation, transition, and dissolution or dilution, respectively
c	Referring to the critical state or indicating a critical value
C, D, F	*With symbols for optical properties,* referring to particular wavelengths
+, −	Referring to a positive or negative ion, or to a positive or negative electrode
∞	Indicating limiting value at infinite dilution

Some of the above subscripts may sometimes be more conveniently used as superscripts.

(*b*) Superscripts

$\ominus$	Standard (in chemical thermodynamics)
*	Indicating a pure substance
id	Ideal
E	Excess

[t]System of loosely coupled electrons.
[u]System of tightly coupled electrons.
[v]All energies are taken here with respect to the ground state as reference level.
SOURCE: *Quantities, Units, and Symbols,* The Royal Society, London.

s. groups The grouping of crystals according to the number and nature of their elements of s.; 32 groups exist, but 11 cover most common substances. See *crystal systems* under *crystal.*

sympathetic Pertaining to or related to a mutual relationship. **s. ink** An ink, visible only if heated (as cobaltous chloride solution) or when treated with a specific reagent. **s. reaction** Induced reaction.

sympathin Norepinephrine bitartrate.

sympathomimetic action Results of stimulation of the sympathetic *nervous system,* q.v., mimicked by drugs.

symphytum Bruisewort, comfrey, blackwort. The dried root of *Symphytum officinale* (Boraginaceae), Europe and Asia; a demulcent.

symplesite $Fe(AsO_4)_2 \cdot 8H_2O$. A mineral.

symplocarpus Skunk cabbage.

syn- Prefix (Greek "with" or "together") signifying union, association, or building up. ***s.* position** Former term for *cis, cisoid,* and Z. See *stereoisomer.*

synaldoxime The *trans* form of an aldoxime.

synanthrose $C_6H_{10}O_5$ = 162.1. Levulin. A carbohydrate; from the rhizomes of *Helianthus tuberosus* (Compositae).

synartesis Stabilization of the transition state of an ionizing substance.

Syncaine Trademark for procaine hydrochloride.

synchronal Occurring at the same time.

synchronism The occurrence of 2 or more phenomena at the same time.

synchronizing To produce synchronism; as, a speed controller for a number of current generators.

synchrotron Particle *accelerator,* q.v., where the tube, and thus the path of the particles, is circular. The CERN s., at Geneva, can give particles an energy of up to 500 GeV. Cf. *cyclotron.*

syncrude Oil produced from coal.

syncyanin A blue pigment produced by the bacteria *B. syncyanus* and *B. cyanogenus.*

Syndet (1) Trademark for a detergent (U.K. usage). (2) (Not cap.) Abbreviation for "synthetic detergent" (USP).

syndiazotate (Z) isomer of a diazoate.

syndiotactic See *polyvinyl chain.*

syndrome A group of signs and symptoms that occur together and characterize a disease.

syneresis The contraction of a clot or gel; as of blood.

synergism The combined action of 2 distinct agencies, the sum of the effects of which is greater than the sum of the effects of each taken separately; e.g., a mixture of trimethoprim and sulfamethoxazole has greater microbial effect than the sum of the effects of the two ingredients used separately.

synergist Booster. One component of a system exhibiting synergism.

synergy Synergism.

syngas Gas produced from coal. Used as a feedstock for chemicals production.

syngenite $CaSO_4 \cdot K_2SO_4 \cdot H_2O$. A native calcium-potassium sulfate.

synonym One of several names given to the same substance; e.g.: wood alcohol, methyl alcohol, and wood spirit are synonyms for methanol.

synourin oil A paint-drying oil produced by the chemical dehydration of certain glyceride molecules of castor oil.

syntactic Describing a cellular polymer produced by dispersing rigid microscopic particles in a fluid polymer, and then stabilizing the system.

syntan Synthetic tanning compounds. (1) A sulfurized phenol. (2) **alpha-~** An artificial tan made by the sulfonation of hydrocarbons, e.g., naphtha. **beta-~** A s. prepared by condensing pyrogallol or catechol with formaldehyde in presence of hydrochloric acid.

synthesis A reaction, or series of reactions, in which a complex compound is obtained from elements or simple compounds. Cf. *analysis.* **electro ~** See *electrosynthesis.* **organic ~** The production of dyestuffs and medicines; the artificial production of a naturally occurring substance, e.g., indigo. **photo ~** See *photosynthesis.*

synthesize To produce a complex compound from simpler compounds.

synthetase* See *enzymes,* Table 30.

synthetic (1) Produced synthetically. (2) Produced by artificial means. **s. drying oil** See *synthetic drying oil* under *oil.* **s. resins** (1) *Plastics,* q.v., produced by (*a*) polymerization (polyalkenes), (*b*) condensation (phenol-aldehydes). (2) Polymers of organic compounds; as, (*a*) association polymers (viscose silk), (*b*) hemicolloidal polymers (warm styrene), (*c*) eucolloidal polymers (cold styrene).

Synthofil Trademark for a polyvinyl synthetic fiber.

synthol A synthetic motor fuel obtained from carbon monoxide and hydrogen (water gas) at 7.5–15 MPa and 400–435°C with an iron and sodium carbonate catalyst. A mixture of higher alcohols, aldehydes, ketones, and higher fatty acids with aliphatic hydrocarbons; 34.3 kJ.

synthovo Hexestrol.

Synthyroid Trademark for levothyroxine sodium.

syntonin Muscle fibrin, parapeptone, acid albumin. An acid albumin from albumose. Yellow powder, insoluble in water, produced in the body as an intermediate product of the gastric digestion of proteins, which eventually become peptones.

syphon See *siphon.*

syr. Abbreviation for "syrup."

syringenin $C_{11}H_{14}O_4$ = 210.2. Oxymethyl coniferin. Red mass from syringa, insoluble in water; formed by hydrolysis of syringin.

syringetin $C_{17}H_{14}O_8$ = 346.3. 3,4′,5,7-Tetrahydroxy-3′,5′-dimethoxyflavone, from lilac. Yellow needles, m.288.

syringic acid $C_9H_{10}O_5$ = 198.2. 4-Hydroxy-3,5-dimethoxybenzoic acid*. m.210; obtained by hydrolysis of syringin.

syringin $C_{17}H_{24}O_9 \cdot H_2O$ = 390.3. Lilacin, ligustrin. A glucoside from *Syringa vulgaris,* lilac (Oleaceae), *Ligsturum vulgare,* privet; *Robinia pseud-acacia.* White needles, m.191, sparingly soluble in water. Cf. *syringenin.*

syrup Sirup. A concentrated aqueous solution of a carbohydrate (as, cane sugar), with or without drugs. USP s. contains 85.0% and BP s. 66.7% w/w of sucrose in water. **simple ~** Syrupus simplex. An aqueous solution of cane sugar (83%).

s. acacia A 10% solution of acacia in sucrose s.

system (1) A combination of matter containing one or more phases. (2) An organized and related group of facts, phenomena, or ideas. **binary ~** (1) A s. involving 2 components (elements or compounds). (2) See *number systems.* **condensed ~** A s. with no gaseous phase. **divariant ~** A s. with 2 degrees of freedom (cf. *phase rule*), represented by an area on a diagram. **geological ~** See *geologic eras.* **heterogeneous ~** A s. containing 2 or more phases or distinct regions, separated by definite boundaries. **homogeneous ~** A s. with no definite boundaries; its properties are the same at all parts, or vary gradually. **mobile ~** A s. that responds readily to external conditions, as pressure. **monovariant ~** A s. with one degree of freedom. Cf. *phase rule.* **nonvariant ~** A s. with no degree of freedom (cf. *phase rule*), represented by a point on a diagram. **periodic ~** See *periodic system* under *periodic.* **quaternary ~** A 4-component s. **ring ~** See *ring system* under *ring.* **stable ~** A s. that does not respond readily to a changing environment. **tertiary ~** A 3-component s. **thermodynamic ~** A s. which may consist of matter, energy, or both, and is limited by a physical or imaginary boundary. **unstable ~** See *mobile system* above.

s. of compounds See *nomenclature.* **s. of elements** See *periodic system.* **s. of stars** See *spectral classification.*

systematic name A name based on the systematic structure of a compound, e.g., pentane. Cf. *trivial name.*

Système International d'Unités See *SI.*

systemic (1) See *systemic herbicide* under *herbicide.* (2) Describing treatment, given either orally or by injection, that affects the body as a whole as well as the diseased part.

Szent-Györgyi hypothesis A mechanism permits the energy of absorbed light or of a chemical reaction occurring in one part of a living system to be available, without degradation or distortion, for reactions in other parts of the system.

szomolnokite $FeSO_4 \cdot H_2O$. A mineral.

T

T Symbol for (1) tera (10^{12}); (2) tesla; (3) tritium. **2,4,5-~** Trichlorophenoxyacetic acid.
T Symbol for (1) kinetic energy; (2) light transmittance; (3) period; (4) thermodynamic temperature; (5) at constant temperature (as subscript).
T (Boldface.) Symbol for torque.
t Symbol for (metric) ton.
t Symbol for (1) common temperature (cf. *T*); (2) tertiary (in chemical names); (3) time; (4) transport number.
$t_{1/2}$ Symbol for half-life.
T, τ See *tau.*
Θ, θ See *theta.*
Ta Symbol for tantalum.
tabacin The toxic principle of tobacco. A yellow, waxy, hygroscopic, nitrogenous acid glucoside, decomp. 110.
tabacum Tobacco.
tabashir Tabashis.
tabashis A secretion of bamboo containing lime and silica; an Indian treatment for tuberculosis.
table (1) A lamella, or flat, scalelike crystal. (2) The flat surface of a precious stone. **t. salt** Sodium chloride. **iodized ~** Sodium chloride with 0.1% of potassium or sodium iodide. **t.spoon** A domestic measure, about 15 mL.
tablet Tabloid. A medicated disk made from a drug incorporated into a mixture of absorbents or adhesives. Moistening agents—water or alcohol. Absorbents—starch, milk sugar, magnesium carbonate, magnesium oxide, or licorice root. Adhesives—cane sugar, tragacanth, acacia, glucose, gelatin, dextrin, flour, boric acid. **effervescent ~** A t. with an effervescent mixture incorporated to cause rapid disintegration in water. Cf. *pill.*
tabular **t. crystal** A table-shaped or flattened crystal. **t. spar** Wollastonite.
tabun $C_2H_5O \cdot P(:O) \cdot CN \cdot NMe_2$ = 162.1. Ethyl dimethylphosphoramidocyanidic acid. A tasteless, odorless gas that affects the nerves with rapidly fatal results; antidote: atropine.
tacamahac (1) A resin from *Calophyllum tacamahaca* (Guttiferae). (2) A resin from *Populus balsamifera,* balsam poplar (Salicaceae).
Tace Trademark for chlorotrianisene.
Tachenius, Otto 17th-century German physician and chemist, the first to use distilled water and to suspect a hidden acid in oils and fats; author of *Epistola de famoso liquore alcahest* (1655) and *Hippocrates chymicus* (1674).
tachhydrite $CaCl_2 \cdot 2MgCl_2 \cdot 12H_2O$. Yellow, very hygroscopic, native chloride.
tachiol Silver fluoride. **iso ~** Silver hexafluorosilicate*.
tachogram A curve indicating the speed of the blood current.
tachometer (1) A device to record the speed of the blood current. (2) An instrument to record the angular speed of a revolving shaft.
tachylite A dark, basic, volcanic glass.
tachymeter A speed recorder.
tachyol Tachiol.
tachysterol An isomer of ergosterol; yields vitamin D on irradiation. Cf. *ergocalciferol.*
tackiness The property of stickiness; as, of resins.
taconite A low-grade ferruginous ore from Labrador and Venezuela; a source of iron if upgraded.
Tacryl Trademark for an acrylic synthetic fiber.
Tactel Trademark for a polyamide fiber with a cottonlike feel.
tacticity* A measure of the degree of order in the succession of repeating units in the main chain of a polymer.
tactile Pertaining to touch.
tactole A bird repellent. Strips of soft plastic which prevent birds from settling and break their roosting pattern.
tæ- Alternative for words beginning with *tae-, te-.*
taeniacide Teniacide.
taeniafuge Teniafuge.
taenite A constituent of iron-nickel meteoric alloys (35–48% Ni).
tagatose $C_6H_{12}O_6$ = 180.2. D-**~** m.134. An unfermentable ketohexose.
tagetone $C_{10}H_{16}O$ = 152.2. 2,6-Dimethyl-5,7-octadien-4-one. Related to myrcene, in the volatile oil of *Tagetes glandulifera* (Compositae).
taggant A substance that can be easily detected and retrieved. Added in minute amounts to a product which has to be identified subsequently. Cf. *labeling.*
tagmanone(ol) A β-triketone in the essential oil of *Eucalyptus risdoni.*
tagud nut Vegetable ivory. The dried fruit of the negrito palm, *Phytelephas macrocarpa.*
tagulaway Cebur. The dried bark of *Parameria vulneraria* (Apocynaceae), Philippine Islands; contains a resin and caoutchouc.
Talbot, William Henry Fox (1799–1877) British archaeologist and inventor of photographic paper (1840) and plates.
talc (1) $3MgO \cdot 4SiO_2 \cdot H_2O$. Talcum, soapstone, rensselaerite, potstone, steatite, French chalk. Purified t. is used as a dusting powder, filtering material (USP, BP), and in crayons. (2) Mica used for glazing (UK usage).
talcum Talc (1).
Talin Trademark of a low-calorie sweetener (2,000 times as sweet as sucrose), from the seeds of the *Thaumatococcus danielli* fruit (W. Africa).
talisai oil An oil from the seeds of *Terminalia catappa,* country almond (Combretaceae).
talite Talitol.
talitol $CH_2OH(CHOH)_4CH_2OH$ = 182.2. Talite, obtained by reduction of talose. D-**~** and L-**~** m.88.
tallate A salt of the fatty acids of tall oil; used in paints.
tall oil A by-product from sulfate wood pulp digestion, mainly resin acids and fatty acids; as linoleic, abietic (Steele acid); linolenic, and some oleic acid, with 2,2′-dihydrostigmasterol and lignoceryl alcohol. Used in soaps, varnishes, and fruit sprays.
tallois-desmi Talmi.
tallol Floating soap, finn oil, sylvic oil. An emulsifying agent from the black liquor of sulfate wood pulp manufacture. Cf. *tall oil.*

tallow A solid fat; as *suet,* q.v. **bone ~** See *bone tallow.* **Borneo ~** Illipé butter (misnomer). A fat used in the chocolate industry. **japan ~** See *japan wax.* **Malabar** See *Malabar tallow.* **mineral ~** Hatchettenine.

Tallquist scale Perforated sections of blotting paper with 10 printed shades of hemoglobin (10–100%). Used to estimate blood counts by matching a blot of blood with the color scale.

talmi Talmi gold, tallois-desmi. Gold-plated brass used in jewelry.

talomucic acid $COOH(CHOH)_4COOH$ = 210.1. Derived from talose and altrose. White crystals, m.158 (decomp.), soluble in water; D and L forms. Cf. *mucic acid.*

talose $CH_2OH(CHOH)_4CHO$ = 180.2. An aldose and hexose; D and L forms.

Talwin Trademark for pentazocine.

tamaid See *bach.*

tamarac The dried bark of the hackmatack, *Larix laricina* or *L. americana,* larch, N. America; an astringent and stimulant.

tamarind The preserved pulp of the fruit of *Tamarindus indica* (Leguminosae), India; a mild laxative.

tamoxifen citrate $C_{26}H_{29}ON \cdot C_6H_8O_7$ = 563.7. Nolvadex. White crystals, slightly soluble in water. An antiestrogenic agent, used for breast cancer (USP, BP).

tampicin $C_{34}H_{54}O_{14}$ = 686.8. An amorphous resin from tampico jalap, *Ipomoea simulans* (Convolvulaceae).

tanacetin $C_{11}H_{16}O_4$ = 212.2. A bitter principle from tanacetum. A brown, amorphous, hygroscopic mass, soluble in water or alcohol, insoluble in ether.

tanacetone $C_{10}H_{16}O$ = 152.2. A ketone obtained from tanacetum oil, and similar to thujone. Colorless liquid, b.195.

tanacetum Tansy. The dried leaves and tops of *Tanacetum vulgare* (Compositae), Asia, Europe, and N. America. **t. balsamita** Costmary. The dried leaves of *T. balsamita;* a vermifuge. **t. oil** An essential oil from t. It contains tanacetene, tanacetone, and borneol; an anthelmintic.

tanekaha The bark of *Phyllocladus trichomanoides* (Coniferae), pine tree of New Zealand; used in tanning and dyeing leather.

tané-koji Brown powder from the mold *Aspergillus oryzae;* used to make saké.

tangent The ratio of the side opposite the angle concerned to the base of the right-angled triangle in which it occurs.

tanghinia The fruit of *Tanghinia venenifera* (Apocynaceae), Malagasy Rep.; an ordeal bean containing a cardiac poison, tanghinin.

tanghinin $C_{10}H_{16}N$ = 150.2. An alkaloid from tanghinia. Colorless scales, soluble in alcohol; a cardiac poison.

tang-kui (1) Eumenol. (2) Lovage.

tangle Laminaria.

tankage The rendered, dried, ground by-products from animal carcasses; a fertilizer and feeding stuff: nitrogen 5–10, tricalcium phosphate 8–30%. **fish ~** Fish guano. **garbage ~** The rendered, dried, ground product of waste household food materials: nitrogen 2.5–3.5, phosphorus pentaoxide 2–5, potassium oxide 0.5–1.0%. **gashouse ~** A misnomer for "spent oxide," consisting of iron oxide and substances taken out in the purification of gas (N 5–10%). **hynite ~** A t. with high N content. **process ~** Hynite t.

tank culture Hydroponics.

tannase* An enzyme in tannin-bearing plants, it hydrolyzes tannins to gallic acid.

tannate A salt of tannin.

tannic acid Tannin.

tannin $C_{76}H_{52}O_{46}$ approx.; i.e: $\underline{CHOR(CHOR)_2CHOR \cdot CHO} \cdot CH_2OR$, where R is the digalloyl group, $-CO \cdot C_6H_2(OH)_2 \cdot COO \cdot C_6H_2(OH)_3$. Tannic acid, gallotannic acid, digallic acid. The glucoside pentadigalloyl glucose, from nutgalls; m.212 (decomp.), soluble in water. Used in tanning, inks, dyes (USP). See *tannins.* **acetyl ~, condensed ~** Catechol. **ellagi ~** See *ellagitannin.* **white ~** The mixture of polyphenols extracted in ethyl acetate from aqueous infusions of green leaves.

TABLE 89. PERCENT OF TANNIN IN VARIOUS TANNING MATERIALS

Material	Form[a]	Tannin, %
Chinese galls, *Rhus semialata*	g.	70
Quebracho, *Schinopsis* species	e.	35–65
Valonia, *Quercus aegilops*	e.	30–65
Red oak, *Qu. rubra*	g.	35
Cutch, *Acacia catechu*	e.	60
Wattle, *A. decurrens*	b.	20–51
Golden wattle, *A. pycnantha*	b.	40–50
Bengal kino, *Butea frontosa*	e.	30–40
Algarobilla, *Caesalpinia brevifolia*	p.	43–67
Cascalote, *C. cacolaco*	p.	40–55
Divi-divi, *C. coriaria*	p.	30–50
Tari, *C. digyna*	p.	40–50
Kino, *Pterocarpus* species	e.	45–60
Piagao, *Xylocarpus granatum*	b.	21–48
Ribbon gum, *Eucalyptus amygdalina*	s.	58–65
Spotted gum, *E. maculata*	s.	45
Black mallet, *E. occidentalis*	b.	40–50
Mallet, *E. occidentalis*	b.	30–50
Messmate, *E. piperita*	s.	33–62
Red ironbark, *E. siderophloia*	s.	35–73
Yhva, *Eugenia* species	b.	44
Oak gum, *Spermolepsis* species	s.	43–80
Mangrove, *Rhizophora* species	b.	21–58
Guara, *Paullinia sorbilis*	f.	43–55
Tamarisk, *Tamarix* species	g.	26–58

[a]b. = Bark; e. = Extract; f. = Fruit; g. = Galls; p. = Pods; s. = Secretion or sap

tanning Converting skins or hides into leather. **t. materials** Vegetable preparations containing tannin. See Table 89.

tannins Astringent, aromatic, acidic glucosides, in various plants and trees. They precipitate alkaloids, mercuric chloride, and heavy metals; they form blue-black solutions (ink) with ferric solutions; and their strongly alkaline solutions absorb oxygen rapidly. See *tannin.* Classification:

1. Hydrolyzable tannins, ester type, RCOOR:
   (*a*) Depsides or gallotannins, $(HO)_xRCOOR'(OH)_y \cdot COOH$; as, di(2,4-dihydroxybenzoic acid), $C_{14}H_{10}O_7$.
   (*b*) Glusides or galloyl sugars, $(HO)_xRCOOR'(OH)_y$; as, digalloyl-L-glucosan.
   (*c*) Ellagitannins or diphenylmethylolids, $R(CO \cdot O)(O \cdot CO)R'$; as, ellagic acid, $C_{14}H_6O_8$.
2. Condensed tannins, keto type, $R \cdot CO \cdot R$:
   (*a*) Ketones, $HOR \cdot CO \cdot R'$; as, hydroxybenzophenone, $C_{13}H_{10}O_2$.
   (*b*) Catecholtannins or phlobatannins, $(HO)_xRCOOR'(OH)_y \cdot (COOH)$; as, the catechols.

tannoform $(C_{14}H_9O_9)_2CH_2$ = 656.5. Methyleneditannin. A condensation product of formaldehyde and tannin. Pink powder, insoluble in water.

tannyl The radical $-C_{14}H_9O_9$, from tannin.

tansy Tanacetum. **t. oil** Tanacetum oil.

tantalate (1)* Indicating tantalum as the central atom(s) in an anion. (2) $MTaO_3$. A salt of tantalic acid; mixed oxides. **hexa ~** $M_8Ta_6O_{19}$. **peroxy ~** M_8TaO_8.

tantalic acid* $HTaO_3$ = 230.0. Colorless crystals, insoluble in water; forms complex salts, as, $Na_8Ta_6O_{19}$.

tantalite $(FeMn)O \cdot Ta_2O_4$. Black crystals, often found with niobium.

tantalites $M(TaO_3)_2$. M is divalent iron, or manganese. Minerals usually mixed with the corresponding niobites.

tantalous Tantalum(III)* or (3+)*. A compound of trivalent tantalum. **t. bromide** See *tantalum bromides (1).* **t. chloride** See *tantalum chlorides (1).*

tantalum* Ta = 180.9479. A rare metal of the vanadium family, and element, at. no. 73. Gray metal, d.16.6, m.2996, b. ca. 5560, insoluble in alkalies or acids (except hydrofluoric acid). Occurs in tantalite and niobite; discovered (1802) by Ekeberg. Valencies: 5 [principal valency; t.(V)*, t.(5+)*, tantalic] and 3 [t.(III)*, t.(3+)*, tantalous]. Resistant to corrosion, and a substitute for platinum; used for electric-light filaments, surgical instruments (it may be heat-sterilized without losing hardness), rayon spinnerets, jewelry, laboratory ware, bone repair; and as an electrode in current rectifiers, and a catalyst for producing synthetic diamonds. **beta- ~** A possible allotrope of Ta formed as a film by cathode sputtering; it has a higher resistivity and lower temperature coefficient of resistance, and becomes a superconductor at much lower temperature.

t. bromides (1) $TaBr_3$ = 420.7. Tantalous bromide. Yellow crystals, m.240, decomp. by water. (2) $TaBr_5$ = 580.5. Tantalic bromide. Colorless powder, decomp. by water. **t. carbide** TaC = 193.0. A hard mixture: t. carbide 75, hafnium carbide 25%, m. 4000. **t. chlorides** (1) $TaCl_3$ = 287.3. Tantalous chloride. Yellow prisms, decomp. by water. (2) $TaCl_5$ = 358.2. Tantalic chloride. Yellow needles, m.211, decomp. in moist air to tantalic acid. **t. fluorides** TaF_5; also $MTaF_6$, M_2TaF_7, M_3TaF_8, where M is a monovalent metal. **t. minerals** Ta occurs in nature associated with Nb. Its principal ore is tantalite, $FeTa_2O_6$. **t. nitride*** TaN = 195.0. Colorless powder, soluble in mixed hydrofluoric and nitric acid. **t. oxide** Ta_2O_5 = 441.9. Tantalic acid anhydride, t. pentaoxide*. White rhombs, d.7.53, insoluble in water, soluble in fused potassium hydrogensulfate. **t. pentabromide*** See *tantalum bromides (2).* **t. pentachloride*** See *tantalum chlorides (2).* **t. pentaoxide*** See *tantalum oxide.* **t. tribromide*** See *tantalum bromides (1).* **t. trichloride*.** See *tantalum chlorides (1).*

tantcopper A copper alloy, analogous to tantiron.

tantiron An acid-resistant alloy of iron containing silica, used for chemical equipment.

tantnickel A nickel alloy analogous to tantiron.

tapioca Cassava, manioca starch, Bahia arrowroot, Rio arrowroot, mandioc, from the root of cassava, *Jatropha manihot* (Euphorbiaceae), of Brazil; a food.

tar A thick, brown to black liquid mixture of hydrocarbons and their derivatives with distinctive odor; obtained by distillation of wood, peat, coal, shale, or other vegetable or mineral materials. **coal ~** A t. from destructive distillation of bituminous coal or crude petroleum. It contains naphthalene, toluene, quinoline, aniline, cresols. Cf. *coal.* **prepared ~** Pix carbonis prep. A product made by heating commercial c. t. at 50 °C for 1 hour, with stirring. Used to treat skin diseases (USP, BP). **oil of ~** Tar oil. **pine ~** Wood t. **rock ~** Crude petroleum. **Schroeter ~** A "tar" produced by the action of aluminum chloride on tetrahydronaphthalene, followed by distillation. It contains 3,4-benzopyrene, and is carcinogenic. **Stockholm ~** A wood t. **wood ~** Pix liquida, pine t. The empyreumatic syrup from the destructive distillation of the *Pinus* species: contains resins, turpentine, and oils; used as a disinfectant and antiseptic and in the treatment of forms of eczema (USP, BP). On fractional distillation it yields an acid liquor (pyroligneous acid), empyreumatic oil (oil of t.), and a black residue (pitch).

t. camphor Naphthalene*. **t.mac(adam)** Blacktop. Road surface, consisting of 90% gravel or crushed rock, plus a bitumen-, asphalt-, or crude oil–based binder. **t. oil** Oleum picis rectificatum. The volatile oil from pine t. rectified by steam distillation, d. 0.862–0.872. An insecticide. **t. sands** Oil sands. A natural sand formation containing about 10% bitumen.

tarapacaite Native potassium chromate.

taraxacum Dandelion, lion's tooth. The dried roots of *T. officinale* (Compositae); a laxative.

taraxanthin $C_{40}H_{56}O_4$ = 600.9. A carotenoid from dandelion flowers, m.184; isomer of violaxanthin.

taraxasterol $C_{30}H_{50}O$ = 426.7. Anthesterol. A triterpene alcohol from chamomile.

taraxerol $C_{30}H_{50}O$ = 426.7. 3β-14-Taraxeren-3-ol. Skimmiol. An alkaloid from *Taraxacum officinale,* m.282.

tarchonyl alcohol $C_{50}H_{102}O$ = 719.4. A methanol alcohol, m.82, insoluble in water, from *Tarchonanthus camphoratus* (Compositae), Africa.

tare (1) The weight of an empty container. (2) A counterweight used to balance a container. (3) A fodder plant of the vetch family. (4) Any weed among corn.

target (1) Anticathode. (2) A substance exposed to bombardment by particles Cf. *ray.*

targusic acid Lapachol.

tariric acid $Me(CH_2)_{10}C:C(CH_2)_4COOH$ = 280.5. 6-Octadecynoic acid*. m.51; a glyceride in Guatemalan tarira, *Picramia* species.

tarnish A surface film of contrasting color formed on an exposed surface of a metal or mineral; usually the oxide or sulfide.

taro Dasheen. The rhizome of *Colocasia esculenta* (Araceae) of the tropics, whose poison is destroyed by fermenting, and then boiling to remove the hydrogen cyanide so liberated; a food (poi).

tarpaulin Canvas rendered waterproof by tar.

tarragon The plant *Artemisia tranunguloides* (Compositae); a spice. **t. oil** Estragon oil.

tartar (1) Tatar. A crude potassium sediment in wine casks. (2) Calculus. A deposit of saliva proteins and calcium phosphate on teeth, providing sites for pyorrhea. **cream of ~** Potassium hydrogentartrate*. **oil of ~** A saturated solution of potassium carbonate. **salt of ~** Potassium carbonate*.

tartaric acid* $C_4H_6O_6$ = 150.1. Dihydroxysuccinic acid, 2,3-dihydroxybutanedioic acid*.

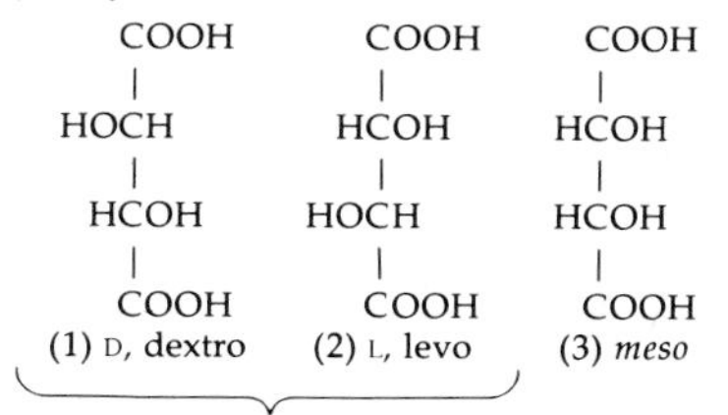

The D acid occurs in many vegetable tissues and fruits. ***dextro-* ~** (2*S*,3*S*)-(+)- ~, or ***levo-* ~** (2*R*,3*R*)-(−)- ~.

Colorless, monoclinic crystals, m.168, soluble in water; used as a reagent, and in pharmacy to produce effervescence (NF, EP, BP). **inactive ~** *Meso-* **~**. **levo- ~** See *levo-tartaric acid* above. **meso- ~** (2*RS*,3*SR*)- **~** + $1H_2O$. Colorless scales, m.110, soluble in water, no optical activity. **pyro ~** See *pyrotartaric acid.* **racemic ~** (2*RS*,3*RS*)-(±)- **~** + $1H_2O$. A mixture of D and L forms, m.205.

tartaroyl* The radical $-CO(CHOH)_2CO-$, from tartaric acid.

tartarus Potassium hydrogentartrate*. **t. stibiatus** Antimony potassium tartrate*. **t. tartarisatus** Potassium tartrate*. **t. vitriolatus** Potassium sulfate*.

tartrate* $M_2C_4H_4O_6$. A salt of tartaric acid. **acid ~, bi ~, hydrogen ~*** $MC_4H_5O_6$. **pyro ~** A salt of pyrotartaric acid.

tartrazine $Ph·N·CO·C(:N·NPh)·C(COOH):N$ = 307.3. A yellow pyrazolone dye. Often used in foodstuffs.

tartronic acid* $CHOH·(COOH)_2·\frac{1}{2}H_2O$ = 129.1. Hydroxymalonic acid, propanoldiacid, hydroxypropanedioic acid*. Colorless prisms, sublime 110, m.185 (decomp.), soluble in water. **benzyl ~** See *benzyltartronic acid.* **methyl ~** Iso*malic acid.*

t. a. series Polyhydric acids formed from carbohydrates.

tartronoyl* The radical $-CO·CHOH·CO-$. **t.urea** Dialuric acid.

taste The sensation caused by a soluble substance on the nerves of the tongue. Threshold values of taste stimulus, grams per 100 mL water: salty, 0.25 sodium chloride; sweet, 0.5 sugar; acidic, 0.007 hydrochloric acid; vanilla (coumarin), 0.0002; bitter, 0.00005 quinine.

tastpolarograph A polarograph in which the current is measured only during a short time interval before the fall of each mercury drop, and at the same time the cell voltage is altered and the chart paper advanced. It eliminates distortion of the polarogram.

tatar Tartar.

tau T, τ Greek letter. T (not italic) is symbol for (1) t. *particle;* (2) triple bond. *In chemical names,* T (not italic) indicates a nonbridging substituent. *T* (italic) is symbol for (1) mean life; (2) relaxation time (time constant); (3) shear stress; (4) transmission factor (transmittance). **t. particle** See *tau particle* under *particle.*

Taube, Henry (1915–) American chemist. Nobel prize winner (1983); for work on mechanisms of electron-transfer reactions involving metal complexes.

taurine* $NH_2·CH_2·CH_2·SO_3H$ = 125.1. Aminoethionic acid, 2-aminoethanesulfonic acid†. Prisms, m.240 (decomp.), soluble in water; formed in bile by the hydrolysis of taurocholic acid.

taurocholic acid $C_{26}H_{45}O_7NS$ = 515.7. Choleic acid, in bile. Colorless needles, m.125 (decomp.), soluble in water; hydrolyzes to taurine and cholic acid.

tauryl* (2-Aminoethyl)sulfonyl†. The radical $NH_2·CH_2·CH_2SO_2-$, from taurine.

tauto- Prefix indicating a tautomeric form.

tautocyanate Isocyanate*.

tautomeric Capable of tautomerism.

tautomerase* See *enzymes,* Table 30.

tautomerism Dynamic allotropy. Existing in a state of equilibrium between 2 isomeric forms, and able to react according to either; as, R·CN and R·NC. The molecules may differ in the linkage, bond or connections between the atoms, and the position or distribution of these atoms in the molecule. Cf. *isomerism.* **electrolytic ~** The property of an amphoteric substance in solution which produces hydroxyl ions in presence of acids, and hydrogen ions in presence of alkalies. **mesohydric ~** T. involving the intramolecular sharing of H atom(s). **protropic ~** T. of the keto-enol type.

tautourea $NH_2·C:NH(OH)$; as distinct from $NH_2·CO·NH_2$ (urea).

tawing Tanning of hides with mineral substances; as, alum.

taxicatin $C_{13}H_{22}O_7·2H_2O$ = 326.3. A crystalline glucoside from *Taxus.*

taxine $C_{37}H_{49}O_{10}N$ = 667.8. An alkaloid from *Taxus baccata.* White scales, m.80, slightly soluble in water.

taxis The movement of a cell in response to an external stimulus; as, chemotaxis.

taxometer An instrument to measure the feel of sheet materials, e.g., leather, in terms of bending length and flexural rigidity.

taxus Yew, chinawood. The poisonous dried seeds of *Taxus baccata* (Taxaceae), Africa.

Taylor, Sir Hugh Stott (1890–1974) English-born American chemist, noted for work in photochemistry and thermodynamics.

Taylor-White process Toughening steel by heating almost to fusion, successively cooling in molten lead and hot oil, reheating to 400–600°C, and cooling in air.

tazettine $C_{18}H_{21}NO_5$ = 331.4. Base VIII, ungernine. An alkaloid, m.210, from the dried corm of *Narcissus tazetta,* related to sekisanine.

TB Abbreviation for (1) tuberculosis; (2) tuberculin.

Tb Symbol for terbium.

Tc Symbol for technetium.

TCA* See *herbicides,* Table 42 on p. 281.

TCDD See *dioxin.*

TCNB See *fungicides,* Table 37 on p. 250.

Te Symbol for tellurium.

tea (1) A medicinal or beverage plant decoction or infusion. (2) The dried leaves of *Thea sinensis* or *Camellia sinensis* and other species of China, Japan, India. Average composition: water 10, extractives 32.7, tannin 11.4, caffeine 1.9, ash 6.23, fibers 30–60%.

Paronychia species	Algerian t.
Melaleuca species	Australian t.
Chenopodium anthelminticum	Mexican t.
Monarda species	Oswego t.
Gaultheria species	Salvador t.
Capraria species	West Indian t.
Thea sinensis	Chinese t. (Pekoe, souchong, congou)
Catha edulis	Arabian t. (kat)
Stachytarpheta species	Jamaican t.
Neea theifera	Caparrosa t.

Brazil ~ Maté. **James ~** See *Labrador tea.* **Jesuit's ~, Paraguay ~** Maté.

t.berries Gaultheria. **t. root** The root of *Ceanothus Americanus* (Celastraceae), an astringent. **t.spoon** About 5 mL. **t. tree** Cajeput.

teak The extremely hard wood of *Tectona grandis* (Verbenaceae), S. Asia.

tear gases Volatile lachrymatory compounds, usually halogenated and organic; as, chloroacetophenone (e.g., Mace) or halogenated cyanides, as, BrPhCHCN, bromobenzyl cyanide. Cf. CS *gas,* poison *gas.*

Teb-x-cel Trademark for fabrics treated with a sodium sulfatoethyl sulfonium compound to enable them to dry smooth, without drip drying.

Teca Trademark for an acetate synthetic fiber.

technetium* Tc. Masurium. Nipponium. The element of at.

no. 43. At. wt. of isotope with longest half-life (^{98}Tc) = 97.907. A fission product of uranium, occurring in traces in nature; in compounds main oxidation states are 4 and 5. Obtained by the extraction of neutron-irradiated molybdenum in ethyl methyl ketone; m.2200 ± 50, b.4567. Used medically, as ^{99}Tc, for scanning bones and organs (USP, EP, BP).

technician A technically qualified worker whose skill is based on some knowledge of science and mathematics, and who works under the general direction of a technologist. See *technology.*

technology The application of science to manufacturing methods.

Teclu burner A modified bunsen burner, with finely adjustable air and gas streams.

tecnazene* See *fungicides,* Table 37 on p. 250.

tecomin Lapachol.

Tecsol Trademark for a product of wood distillation; an alcohol denaturant.

tectite Tektite.

tectonics The study of the deformation of the structure of the earth's crust.

Teepol Trademark for a detergent based on higher sodium alkyl sulfates.

Teflon Trademark for polytetrafluoroethylene. **T.-100** A fully fluorinated copolymer of hexafluoropropene and tetrafluoroethylene. It resists chemical attack and has good electrical insulating properties.

Tegretol Trademark for carbamazepine.

Teichmann's crystals See *hemin.*

teilungs koeffizient Distribution coefficient.

tektite Tectite. A mineral of high silica and alumina content, often glassy and always associated with sediments; produced by weathering of granite and basalt, followed by removal of iron and the alkaline-earths by leaching.

telcomer The product of a telomerization reaction.

tele- Prefix (Greek, "far off") indicating distance.

teleidoscope A television device in which a modulated cathode ray scans an image on a uniform liquid layer spread on a revolving glass disk. Each point takes the form of a minute lens, the curvature of which depends on the instantaneous change in that area produced by the cathode ray.

telekinesy Action at a distance without contact.

telemetry Technique of transmitting data over long distances; as, by radio.

Telepaque Trademark for iopanoic acid.

telephotography Transmission of black and white pictures by telegraphy or wireless.

telescope An optical instrument for observing distant objects, and for reading measuring scales at a distance.

television TV. Receiver in which signals are converted into visible form by electrons striking a phosphor (as, zinc sulfide) on the tube screen. Picture and sound transmissions at approx. 1 m wavelength, using mainly 624 or 525 lines/screen. **closed-circuit ~** CCTV. TV camera-screen combination that permits remote visual monitoring of part of a process or process records.

telfairic acid Linoleic acid.

tellerwort Sanguinaria.

tellur* Indicating Te analogs of Se compounds, with tellur- in the place of selen-.

tellurate* Indicating tellurium as the central atom(s) in an anion, as: (1) M_2TeO_3. Tellurate(IV)*, tellurite. A salt of tellurous acid. (2) M_6TeO_6 or M_2TeO_4. Tellurate(VI)*. A salt of telluric acid.

tellur(o)- Prefix for tellurium. **t.bismuth** Tetradymite. **t.nickel** Melonite. **t.ocher** Tellurium dioxide.

telluretted Describing a compound containing divalent tellurium (obsolete). **t. hydrogen** Hydrogen telluride*.

telluric (1) Of earthly (as opposed to meteoric, or solar) origin. (2) Tellurium(VI)* or (6+)*. Referring to hexavalent tellurium. **t. acid*** H_6TeO_6 or $Te(OH)_6$ = 229.6. Orthotelluric acid*. Allotelluric acid. White crystals, m.136, soluble in water. Obtained by oxidation of tellurium dioxide. **t. bismuth** Tetradymite. **t. lead** Nagyagite. **t. lines** The lines in the solar spectrum due to absorption of rays by the atmosphere; as, Fraunhofer lines *A* and *B*. **t. ocher** Tellurite. **t. silver** Hessite.

telluride (1)* M_2Te. A compound of divalent tellurium; as, hessite, Ag_2Te; tetradymite, Bi_2Te_3. (2) Weissite. (3)* Organic compounds of the type R_2Te. **di ~, per ~** $R_2Te \cdot Te \cdot R$.

tellurinic acid* $RTeO_2H$; as, methyltellurinic acid, $MeTeO_2H$. Cf. *telluronic acid.*

tellurinyl-* Telluryl-. Prefix indicating the radical =TeO, from tellurous acid.

tellurious Tellurous.

tellurite (1) Tellurate(IV)*. (2) TeO_2. Telluric ocher. Native tellurium dioxide.

tellurium* Te = 127.60. Sylvanium. A nonmetal and element, at. no. 52. Rhombic crystals, m.451, b. ca. 1200, insoluble in water; a homolog of sulfur and selenium. Discovered by Müller (1782); occurs native as lionite in tellurite and tellurides. Inorganic compounds: see Table 90. Organic compounds: R_2Te, R-telluride; R_2TeO, R-telluroxide. **black ~** (Pb,Au)(Te,S). A lead-gold mineral. **foliated ~** Nagyagite. **graphic ~** Sylvanite. **radio ~** Polonium.

t. bromides (1) $TeBr_2$ = 287.4. Tellurous bromide. Gray needles, m.280, decomp. by water. (2) $TeBr_4$ = 447.2. Telluric bromide. White powder, hydrolyzed by water. **t. chlorides** (1) $TeCl_2$ = 198.5. Tellurous chloride. Black crystals, m.175, decomp. by water. (2) $TeCl_4$ = 269.4. Telluric chloride. White crystals, m.224, decomp. by water. **t. dibromide*** T. bromide (1). **t. dichloride*** T. chloride (1). **t. diethyl** See *ethyl telluride.* **t. dimethyl** Dimethyl telluride*. **t. dioxide*** See *tellurium oxide.* **t. disulfide*** T. sulfide. **t. glance** Nagyagite. **t. graphite** $(AgAu)Te_2$. A native telluride. Cf. *sylvanite.* **t. hydride** Hydrogen telluride*. **t. hydroxide** R_3TeOH. **t. iodide** TeI_2 = 381.4. Black crystals, insoluble in water. **t. lead** Lead containing 0.05% Te. It resists mechanical shock and stress reversal better than pure lead, and may be used with 77% or less sulfuric acid. **t. monoxide*** See *tellurium oxides (1).* **t. nitrate** $Te_2O_3(OH)NO_3$, derived from tellurous acid, $TeO(OH)_2$. **t. nitrates** $R_2Te(NO_3)_2$. **t. oxides** (1) TeO = 143.6. T. monoxide*, a rare compound. (2) TeO_2 = 159.6. T. dioxide*. Yellow octahedra, m.700, slightly soluble in water. (3) TeO_3 = 175.6. T. trioxide. Orange powder, decomp. red heat, insoluble in water. (4) R_2TeO; as,

TABLE 90. OXIDATION STATES OF TELLURIUM

Oxidation state	Name	Ion
−2	telluride*	Te^{2-}
+2, +4	t.(II)*, t.(2+)*, etc., tellurous	Te^{2+}, Te^{4+}
+4	tellurate(IV)*, tellurite	TeO_3^{2-}
+6	t.(VI)*, t.(6+)*, telluric	Te^{6+}
+6	tellurate(VI)*	TeO_4^{2-}

dimethyltelluroxide*. **t. oxychloride** $TeOCl_2$, derived from tellurous acid. **t. sulfate** $Te_2O_3SO_4$, derived from tellurous acid. **t. sulfide** TeS_2 = 191.7. Black powder, insoluble in water. **t. sulfite** $(TeO_2)_2SO_3$, derived from tellurous acid. **t. trioxide***. See *tellurium oxides.*

telluro- Prefix indicating divalent tellurium. Cf. *sulfo-.*

telluronic acid* $RTeO_3H$; as, methyl telluronic acid*. Cf. *tellurinic acid.*

telluronium* The cation TeH_3^+. Cf. *selenonium.*

tellurous Tellurium(II)*, (2+)*; (IV)*, or (4+)*. Tellurious. Describing a compound of di- or tetravelent tellurium. **t. acid** H_2TeO_3 = 177.6. Tellurious acid. Colorless octahedra, decomp. 40, slightly soluble in water. **t. bromide** See *tellurium bromides (1).* **t. chloride** See *tellurium chlorides (1).*

telluryl The tellurinyl* radical.

telomerization A polymerization reaction between 2 substances providing respectively the terminal groups and internal linkages of the resulting telomer molecule; e.g., a saturated compound (carbon tetrachloride) and a polymerizable substance (ethylene) produce low-molecular-weight vinyl polymers; as, tetrachloroalkanes, $Cl(CH_2 \cdot CH_2)_n \cdot CCl_3$.

telopsis Early name for television.

TEM Tretamine.

temp. Abbreviation for temperature.

temperature The International Practical Temperature Scale is based on a number of reference temperatures, some of which are given in Table 91. For the range of temperatures see Table 92. **absolute ~** T. on the former absolute scale, in °A, which begins at −273°C. Now effectively, and correctly, the thermodynamic t. scale. See *Kelvin scale of temperature* below. **color scale of ~** See Table 93. **critical ~** The t. above which a gas cannot be condensed to a liquid by pressure alone. **critical solution ~** See *critical solution temperature.* **flame ~** See *flame temperature.* **kelvin scale of ~** K*. The thermodynamic temperature* scale and a base unit (K) of the SI system. The freezing point of water is 273.15 K and the boiling point 373.15 K. Hence K reading = °C + 273.15. For temperature intervals, 1 kelvin = 1°C exactly. Cf. *absolute temperature* above. **maximum ~** The t. above which life or growth of bacteria ceases. **normal ~** (1) Room t. (2) T. measured on the hydrogen thermometer. (3) Standard t. **optimum ~** The t. most favorable for action or growth. **room ~** About 20°C (U.S.), or 15.5°C (U.K.). **salt ~** T. at which crystals begin to separate when a solution is concentrated by boiling. **standard ~** T. of 0°C. **thermodynamic ~*** Kelvin scale of t. **transition** See *transition temperature.*

t. coefficient A factor that indicates quantitatively the effect of t. on a property of matter; as, electrical conductivity, 2.5% per °C. **t. conversion** (1) °C to °F: multiply the °C by 9, divide by 5, and add 32. (2) °F to °C: subtract 32 from the °F, multiply by 5, and divide by 9. **t. reaction.** See *Manumené number.* **t. recorder** An electric device for recording temperature. **t. regulator** Thermoregulator. **t. scales**

1. K = kelvin (SI)
   t K = t + 273.15°C
2. °C = degree centigrade or degree Celsius
   t°C = $\frac{9}{5}t$ + 32° F = $\frac{4}{5}t$°R = t + 273.15 K
3. °F = degree Fahrenheit
   t°F = $\frac{5}{9}(t - 32)$°C = $\frac{4}{9}(t - 32)$°R
4. °R = degree Réaumur
   t°R = $\frac{5}{4}t$°C = $\frac{9}{4}t$ + 32°F = $\frac{5}{4}t$ + 273.15 K

tempered Subjected to successive heatings, of decreasing intensity. **t. steel** A hardened steel that has been reheated at a lower temperature.

tempering (1) The reheating and cooling of metal; especially hardening steel. (2) Rendering plastic materials (as clay) homogeneous. Cf. *annealing.* **t. oil** A heavy, viscous oil used for cooling metals during tempering.

template An outline (e.g., or wood, paper, or metal) of an article to be constructed; used as a pattern.

tempolabile Describing a substance that changes with time.

temulentine $C_{12}H_{42}O_{19}N$ = 504.5. An alkaloid from the seeds of *Lolium temulentum* the darnel grass associated with wheat.

temuline $C_7H_{12}ON_2$ = 140.2. An alkaloid from *Lolium.*

tenacity Ability to hold fast. Cf. *tensile strength.*

tenaculum A hook-shaped dissecting needle.

Tenasco Trademark for a viscose synthetic fiber.

tenate That portion of a liquid which is not separated by dialysis, diffusion, filtration, or distillation.

tendering unit A measure of the tendering of fabrics by

TABLE 91. INTERNATIONAL REFERENCE TEMPERATURES

	Temperature	
	K	°C
Cadmium, superconducting transition	0.519	−272.631
Lead, superconducting transition	7.199	−265.951
Hydrogen, liquid-gas equilibrium	20.28	−252.87
Oxygen, liquid-gas equilibrium	90.188	−182.962
Water, solid-liquid-gas equilibrium	273.16	0.01
Water, liquid-gas equilibrium	373.15	100.00
Zinc, solid-liquid equilibrium	692.73	419.58
Gold, solid-liquid equilibrium	1337.58	1064.43

TABLE 92. TEMPERATURE MAGNITUDES, °C

Stellar interior (hottest)	1,000,000,000
Hydrogen bomb	100,000,000
Solar corona	1,000,000
Mercury lamp	14,000
Solar surface (G star)	6,000
Atomic hydrogen torch	4,200
Stellar surface (M star)	3,000
Tungsten, melting point	3,410
Gas burner	1,700
Water, boiling point	100
Hottest climate (Tripoli)	58
Human body temperature	37
Water, freezing point	0
Coldest climate (Alaska)	−63
Liquid air	−190
Lowest experimental temperature	10^{-6} above abs. zero
Absolute zero	−273.15

TABLE 93. COLOR SCALE OF TEMPERATURES

Color	Temperature, °C
Red, visible in dark	400
Red, visible in daylight	525
Dark-red	700
Bright cherry	1,000
Orange-red	1,100
Yellow	1,200
White	1,450
Blue, dazzling	1,600

chemical attack (e.g., in laundering) in terms of the cuprammonium viscosity in reciprocal poises.

tenderometer An instrument for measuring the ripeness of peas in terms of the force required to push them through a standard grid.

tengujo Yoshino.

teniacide An agent that kills tapeworms; as, niclosamide.

teniafuge An agent that expels tapeworms.

Tenite Trademark for a cellulose acetobutyrate plastic.

Tenith Trademark for a cellulose acetate plastic.

Tennant **T., Charles (1768–1838)** Scottish industrial chemist; the original manufacturer of bleaching powder. **T., Smithson (1761–1815)** English chemist, discoverer of iridium and osmium.

tennantite $3Cu_2S \cdot As_2S_3$. A native sulfarsenide, associated with iron.

tenorite CuO. A native black copper oxide.

tenside Surface-active substance (collective name).

tensile Rigid. **t. strength** Tenacity. Resistance to pulling action, measured by the breaking stress in kg/mm^2; e.g.: steel, 50–100; ramie fibers, 70–80; rayon, 60; copper, 20–50; silk fibers, 35–44; cotton fibers, 28–44; rubber, 15–20.

tensimeter An instrument used to determine transition points from small vapor-pressure changes.

tensiometer An apparatus for measuring surface tension.

tension The stress caused by pulling; as, of rubber. **adhesion ~** The degree of wetting of a solid by a liquid. **electric ~** Electromotive force. **gaseous ~** The elasticity of a gas. **surface ~** The force exerted on the surface of a liquid, responsible for the formation of a meniscus or froth. **vapor ~** See *vapor pressure*.

tenter A machine for drying woolen felts.

tenth-meter 10^{-10} m; 1 angstrom unit.

tenth-normal solution $N/10$ or 0.1 N, decinormal. A solution that contains 0.1 of the equivalent weight of a substance per liter.

Tenuate Trademark for diethylpropion.

tephigram An entropy-temperature diagram, used for weather forecasting.

tephroite Mn_2SiO_4. A native silicate.

tephrosia (1) Devil's shoestring, Turkey pea. The dried leaves of *T. virginiana* (Leguminosae); a cathartic. (2) Fish bean, uwuwa, ombwe. The leaves of *T. vogelii* (Leguminosae), Zambia; a fish poison and parasiticide.

tephrosin $C_{23}H_{22}O_7$ = 410.4. Hydroxydequelin. **(±)-~** A crystalline principle from tephrosia, derris, and cube. Transparent prisms, m.198, insoluble in water.

ter-* Prefix (Latin) indicating "3 times." Use is limited in chemical names, in favor of *tri-* and *tris-*, q.v., to assemblies of 3 hydrocarbon ring systems; as, $(C_6H_5)_2 \cdot C_6H_4$, terphenyl.

tera* Abbrev. T; the SI system prefix for 10^{12}.

Teracol Trademark for a polymethylene oxide having the properties of a rubber substitute.

teraconic acid $Me_2C{:}C(COOH) \cdot CH_2COOH$ = 158.2. (1-Methylethylidene)butanedioic acid. Crystals, m.162 (decomp.), soluble in water.

teracrylic acid $C_7H_{12}O_2$ = 128.2. 3,4-Dimethyl-3-pentenoic acid*. A liquid, b.218.

teratogenic Describing a substance that can cause abnormalities in the fetus; as, thalidomide.

teratolite A mixed, native oxide of iron and manganese, with decomposed feldspar.

terbium* Tb = 158.9254. A rare-earth metal and element, at. no. 65; discovered by Mosander in gadolinite. m.1360, b. ca. 2960. **t. chloride*** $TbCl_3$ = 265.3. Colorless crystals, soluble in water. **t. hydroxide*** $Tb(OH)_3$ = 210.0. Colorless powder, insoluble in water. **t. nitrate*** $Tb(NO_3)_3$ = 344.9. White powder, soluble in water, decomp. by heat to Tb_2O_3. **t. oxide*** Tb_2O_3 = 365.9. White powder, insoluble in water. **t. sulfate*** $Tb_2(SO_4)_3$ = 606.0. Colorless crystals, soluble in water.

terbutaline sulfate $(C_{12}H_{19}O_3N)_2H_2SO_4$ = 548.7. Bricanyl. White crystals, freely soluble in water. A sympathomimetic, used to treat asthma (USP, BP).

terbutryn* See *herbicides*, Table 42 on p. 281.

terchloride Trichloride*.

terchoic acid Any of the polymers of phosphates of glycerol or ribitol in which the units are joined through phosphodiester linkages; occur in the walls of certain bacteria.

terebene $C_{10}H_{16}$ = 136.2. A mixture of terpenes, made by the action of sulfuric acid on turpentine. Yellowish, fragrant liquid, b.160–172, insoluble in water. Used in disinfectant soaps. Cf. *terebine*.

terebenthene Turpentine.

terebentylic acid $C_8H_{10}O_2$ = 138.2. A liquid, b.160.

terebic acid $O \cdot CMe_2 \cdot CH(COOH) \cdot CH_2CO$ = 158.2. Terebinic acid, 2,2-dimethylparaconic acid. An oxidation product of turpentine. **(±)-~** Colorless crystals, m.175. Cf. *terpenylic acid*. **pyro ~** See *pyroterebic acid*.

terebine A paint dryer, made by fusing rosin with a metallic acetate or oxide and thinning with white spirit.

terebinic acid Terebic acid.

terebinthina Turpentine.

terephthalal The 1,4-phenylenedimethylidyne† radical, $=CH \cdot C_6H_4 \cdot CH=$.

terephthalaldehyde* $C_6H_4(CHO)_2$ = 134.1. Benzene-*p*-dialdehyde. Colorless needles, m.116, soluble in water or alcohol.

terephthalic Pertaining to *p*-phthalic acid. **t. acid*** $C_6H_4(COOH)_2$ = 166.1. 1,4-Benzenedicarboxylic acid. White powder, sublimes when heated, insoluble in water; used as a reagent for alkali in wool, and in the manufacture of plasticizers, and polyester fibers and films.

terephthalonic acid $C_9H_6O_5$ = 194.1. 4-Carboxy-α-oxobenzeneacetic acid.

terephthalonitrile* $C_6H_4(CN)_2$ = 128.1 *p*-Dicyanobenzene. Colorless crystals, m.215, insoluble in water.

terephthalyl alcohol 1,4-Xylenediol*.

teresantalic acid $C_9H_{13} \cdot COOH$ = 166.2. A terpene acid in sandalwood oil. m.158.

Tergal Trademark for a polyester synthetic fiber.

Tergitol Trademark for a group of detergent sodium or amine salts of higher primary or secondary alkyl sulfates.

Terital Trademark for a polyester synthetic fiber.

Terlenka Trademark for a polyester synthetic fiber.

terlinguaite Hg_2ClO. A native oxychloride from Terlingua, Tex.

term (1) A mathematical relation. (2) See *mode*. **D ~** The diffuse spectrum lines. **P ~** The principal spectrum lines. **S ~** The sharp spectrum lines. See *series*.

terminal The end of an electric junction.

terminalia Myrobalan.

terminology (1) The study of the construction, definition, and arrangments of names and terms. (2) A nomenclature. **Geneva ~** See *Geneva nomenclature*. **(in)organic ~** See *nomenclature*.

termolecular Trimolecular.

ternary (1) Consisting of three. (2) Tertiary. **t. compound** A molecule consisting of 3 different types of atom. **t. steel** A steel consisting of iron, carbon, and one other metal. **t. system** A system of 3 components.

terne plate (French *terne:* dull) (1) Originally (1720) made in S. Wales; then (1830) widely used in the U.S. as a durable roofing material (Thomas Jefferson's house, Monticello). Made by coating tin plate with molten lead. (2) Steel sheet coated with an alloy of Pb 80–90, Sn 7–25%, sometimes with Sb. Used in the automobile industry and for roofing.

ternitrate Trinitrate*.

terotechnology A combination of management, financial, engineering, and other relevant practices applied to optimizing the economic usage of physical assets.

teroxide Trioxide*.

terpadien(on)e Menthane derivatives containing 2 double bonds.

terpane Menthane*.

terpanone A saturated ketone derived from terpenes; as, menthone.

terpene(s) (1)* Hydrocarbons in essential oils, resins, and other vegetable aromatic products. Classification: monocyclic (as, menthane, cyclohexane derivatives), bicyclic (pinane, bicyclohexane derivatives), acyclic (aliphatic terpenes). Related to the terpenes are hermiterpenes, C_5H_8 (isoprene); sesquiterpenes, $C_{15}H_{24}$ (clovene); diterpenes, $C_{20}H_{32}$; polyterpenes, $n(C_{10}H_{16})$. (2) A derivative of the hydrocarbons, $C_{10}H_{14}$, $C_{10}H_{18}$, and $C_{10}H_{20}$; as, pinol, menthol, ionone. **t. dihydrochloride** $C_{10}H_{16}\cdot 2HCl$ = 209.2. Dipentene dihydrochloride. Colorless crystals, m.50, insoluble in water. **t. hydriodide** $C_{10}H_{16}\cdot HI$ = 264.1. Dipentene iodide, terpene iodide. Brown liquid, insoluble in water. **t. hydrochloride** $C_{10}H_{16}\cdot HCl$ = 172.7. Dipentene hydrochloride, pinene hydrochloride, artificial camphor, turpentine camphor. Colorless crystals, m.125, insoluble in water.

terpenoids Derivatives of terpenes, occurring widely in plants; used as perfumes, antibiotics, and cytotoxic agents.

terpenol $C_{10}H_{17}OH$. Cyclic alcohols derived from the terpenes; as, pulegol.

terpenone Unsaturated ketone derived from terpenes.

terpenylic acid $C_8H_{12}O_4$ = 172.2. An oxidation product of turpentine. Colorless crystals, m. 90. Cf. *terebic acid.*

terpilene $C_{10}H_{16}$ = 136.2. Terpinylene, obtained by dehydrating a terpene dihydrochloride solution, b.176. **t. dihydrochloride** Eucalyptol.

terpilenol Terpineol.

terpin $Me_2C(OH)\cdot C_6H_9\cdot Me(OH)$ = 172.3. Terpine, dihydroxymenthane. ***cis-*** ~ m.157. ***trans-*** ~ m.104.
 t. hydrate $C_{10}H_{20}O_2\cdot H_2O$ = 190.3. Colorless rhombs, m.106, soluble in water, obtained by oxidation of turpentine; an expectorant (USP).

terpinene $C_{10}H_{16}$ = 136.2. Isopropylmethylcyclohexadiene. A hydrocarbon from turpentine in α and β forms. Colorless liquid, b.179, insoluble in water.

terpineol $C_{10}H_{18}O$ = 154.3. Lilacin, terpilenol. Colorless liquid. **alpha-** ~ d.0.936, m.35. Used in perfumes. **beta-** ~ d.0.923, m.33. **gamma-** ~ d.0.936, m.70. Insoluble in water; lilac odor. **pharmaceutical** ~ A mixture of isomers in which the (±)-α-t. predominates. Colorless, viscous liquid, d.0.933, b.214–224, slightly soluble in water; a disinfectant (BP).

terpinolene $C_{10}H_{18}$ = 138.3. 4-Isopropylidene-1-methylcyclohexene. In many essential oils. Colorless liquid, b.186, soluble in alcohol. Used in perfumes and flavors. Cf. *citral.*

terpinylene Terpilene.

Terposol Trademark for solvents made by the catalytic etherification of pinene.

terpyridyl $C_{15}H_{11}N_3$ = 233.3. 2,2′:6′,2″-Terpyridine. m.88. A solution in hydrochloric acid is a reagent for cobalt (orange color).

terra- Prefix indicating an earth or earthly origin. **t. alba** White clay; kaolin. **t. cariosa** Rottenstone. **t.-cotta** A coarse clay, baked and used for decorative purposes or utensils. **t. japonica** Pale catechu or gambir. **t. rossa** A fossil red earth. **t. verde** Green earth. A green disintegration product of hornblende minerals; used to fix basic dyes on textiles.

terracoles A group of lichens.

Terramycin Trademark for oxytetracycline.

terrein $C_8H_{10}O_3$ = 154.2. A dihydroxycyclopentanone. D- ~ m.123. Produced by the mold *Aspergillus terreus.*

tert-* Abbreviation and prefix indicating "tertiary."

tertiary Third in order or type. **di** ~ See *arsine.*
 t. alcohol $R_3C\cdot OH$. **t. amine** R_3N; as, trimethylamine. **t. carbon atom** A C atom attached to 3 others. **t. lipid** See *lipid.* **t. period** A *geologic* period, q.v. *(geologic eras),* and system of strata deposited during the Cenozoic era. **t. phosphate.** M_3PO_4. The normal orthophosphate.

tervalent Trivalent*.

Terylene Trademark for a synthetic fiber made from a terephthalic acid polyester and ethylene glycol; very elastic, stable, and resistant to moisture and chemicals, and twice as strong as cotton. Structural unit:

$$\left[-OC-C_6H_4-CO-O-(CH_2)_2-O-\right]_n$$

tesla* Symbol: T; the SI unit of magnetic flux density, equal to that given by a magnetic flux of 1 Wb/m^2. **T. coil** See *Tesla coil* under *coil.*

tesseral Isometric. See *crystal systems.*

test (1) Qualitative trial, or reaction. (2) An experiment to determine a property. **hot toddy** ~ An evaluation of: (*a*) *aroma:* treat 10 mL liquor with 40 mL hot water and smell in comparison with a standard; (*b*) *flavor:* sip and roll the liquid over the tongue without swallowing. **layer** ~ Ring t. **load** ~ Subjection to different pressures or weights and temperatures, and measurement of the resulting deformation. **negative** ~ See *negative test.* **ordinal** ~ See *ordinal test.* **ring** ~ A chemical reaction between 2 liquid layers, so that the reaction products form a zone between them. **slag** ~ An examination of refractories for their behavior toward slag penetration. **spalling** ~ The behavior of heated materials after repeated quenching. **spot** ~ See *spot analysis.* **streak** ~ See *streak test.*
 t. glass T. tube. **t. meal** A specially prepared food that is eaten, withdrawn from the stomach after a certain time, and analyzed. **t. paper** A filter paper impregnated with a *reagent* or *indicator,* q.v. Used with solutions that must not be contaminated by the reagent. **t. solution** T.S. A solution of reagent. **t. tube** A small cylindrical glass vessel of resistant glass; used for chemical reactions. **t.-tube baby** See *in vitro fertilization.*

testa A shell; as, testa ovi (eggshell).

Testoral Trademark for testosterone.

testosterone $C_{19}H_{28}O_2$ = 288.4. 17β-Hydroxy-4-androsten-3-one. Oreton, Testoral. A male sex hormone, secreted by the testes or synthesized. White crystals, m.154, soluble in acetone. Cf. *androsterone.* **methyl** ~ See *methyltestosterone.* **t. cypionate** $C_{27}H_{40}O_3$ = 412.6. White crystals, insoluble in water, m.101; an androgen (USP). **t. propionate** $C_{22}H_{32}O_3$ = 344.5. Testoviron. White crystals, m.120, insoluble in water; used for testicular failure (USP, EP, BP). See *androgen.*

tetan Tetranitromethane*.

tetanic A spinal cord poison producing spasmodic contractions of the muscles; as, nux vomica.
tetanine $C_{13}H_{30}O_4N_2$ = 278.4. A ptomaine from cultures of the tetanus bacillus.
tetanotoxin $C_5H_{11}N$ = 85.1. A toxin from cultures of the tetanus bacillus.
tetanthrene $C_{14}H_{14}$ = 182.3. Tetrahydrophenanthrene, an aromatic hydrocarbon.
tetanus **t. antitoxin** A sterile, nonpyrogenic solution of antibodies to the toxin produced by *Clostridium tetani.* Obtained from horses (USP, EP, BP). **t. toxoid** (USP), **t. vaccine** (BP) A sterile solution of the formaldehyde-treated growth products of the tetanus bacillus, *Clostridium tetani;* used adsorbed on aluminum hydroxide or after precipitation with alum. A prophylactic to prevent tetanus.
tetartohedral Describing crystal form with only one-quarter of the full number of faces required by the symmetry of the system.
tethelin A principle obtained from the anterior lobe of the pituitary body; supposed to control growth.
tetra-* Quadri-. Prefix (Greek) indicating "four." **t.ammine copper(II)** See *cuprammonium.* **t.arylazadipyrromethines** Stable dyestuffs made by the progressive replacement of methine bridges in porphyrins to form azaporphyrins and finally phthalocyanines.
tetrabasic Describing an acid that has 4 replaceable H atoms; as, H_4SiO_4, orthosilicic acid.
tetrabenzyl- Prefix indicating 4 benzyl radicals. **t.tin** $(CH_2Ph)_4Sn$ = 483.2. Colorless prisms, m.43, insoluble in water.
tetraborane* See *tetraborane* under *boranes.*
tetraborate $M_2B_4O_7$. A salt of tetraboric acid. See *borax.*
tetraboric acid* See *boric acid.*
tetrabromo-* Tetrabrom-. Prefix describing a compound containing 4 Br atoms. **t.benzene** $C_6H_2Br_4$ = 393.7. **1,2,3,5-~** m.99. **1,2,4,5-~** m.180. Insoluble in alcohol. **t.cresolsulfonephthalein** Bromocresol green. **t.ethane*** $CHBr_2 \cdot CHBr_2$ = 345.7. Ethylene tetrabromide. Colorless liquid, d.2.972, m. −20, insoluble in water. **t.fluorescein** Eosine. **t.phenolsulfonephthalein** Bromophenol blue.
tetracaine hydrochloride* $C_{15}H_{24}O_2N_2 \cdot HCl$ = 300.8. 2-(Dimethylamino)ethyl *p*-(butylamino)benzoate hydrochloride. Amethocaine hydrochloride. Pontocaine. Bitter crystals, m.148, soluble in water; a local anesthetic (USP, EP, BP).
tetracarboxylic acid A tetrabasic organic acid; as, ethanetetracarboxylic acid, $(CH)_2(COOH)_4$.
tetracarp Tetrachloroethylene*.
tetracene Tetrazene (2).
tetrachloride* A compound with 4 Cl atoms in its molecule.
tetrachloro-* Tetrachlor-. Prefix describing a compound with 4 Cl atoms. **t.acetone** $(CHCl_2)_2CO$ = 195.9. Bis(dichloromethyl) ketone*. Colorless crystals, m.48. **t.aniline*** $C_6HCl_4NH_2$ = 230.9. **2,3,4,5-~** 1-Amino-tetrachlorobenzene. Colorless crystals, m.118, insoluble in water. **2,3,5,6-~** m.60. **t.aurate(III)*** A salt containing the $[AuCl_4]^-$ anion. **t.benzene*** $C_6H_2Cl_4$ = 215.9. **1,2,3,4-~** Colorless needles, m.45, insoluble in water. **1,2,4,5-~** Colorless, monoclinic crystals, m.140, slightly soluble in water. **1,2,3,5-~** White needles, m.50, insoluble in water. **t.ethane*** $C_2H_2Cl_4$ = 167.8. Acetylene tetrachloride, ethylene tetrachloride*. **1,1,2,2-~** $CH_2Cl_2 \cdot CH_2Cl_2$. Colorless liquid, d.1.600, b.146, insoluble in water; a fat solvent, airplane dope, paint remover, spotting agent. **1,1,1,2-~** $CH_2Cl \cdot CCl_3$. Colorless liquid, b.129. **t.ethylene*** $CCl_2{:}CCl_2$ = 165.8. Perchloroethylene. Colorless liquid, d.1.608, b.119, insoluble in water. A spotting and drying agent, soap and fat solvent, and anthelmintic; used for hookworm (USP, BP). **t.platinate(II)*** Chloroplatinite, platinochloride. A salt of tetrachloroplatinous acid, containing the radical $PtCl_4^{2-}$. **t.methane*** Carbon tetrachloride*. **t. quinol** Chloranil.
tetracid A base or alcohol with 4 OH radicals.
Tetracol Trademark for carbon tetrachloride.
tetracosane* $C_{24}H_{50}$ = 338.7. An alkane. Colorless crystals, m.51, insoluble in water. **iso ~** 12-Methyl*tricosane.*
tetracosanoate* Cerosate, lignocerate. A compound containing the radical $C_{23}H_{47}COO-$.
tetracosanoic acid* $C_{23}H_{47}COOH$ = 368.6. Cerosic acid, Lignoceric acid. A fatty acid, m.84, from peanut oil and wood. **iso ~** 12-Methyl*tricosanoic acid.*
tetracosenoic acid* **(Z)-15-** Selacholeic acid.
tetracosyl* The radical $Me(CH_2)_{22} \cdot CH_2-$.
tetracyano-* Prefix indicating a compound with 4 −CN groups. **t.aurate*** A salt containing the $AuCN_4^-$ anion.
tetracycline $C_{22}H_{24}O_8N_2$ = 444.4. Yellow crystals, insoluble in water; an antibiotic active against gram-positive and -negative organisms (USP). Also, as the hydrochloride. Uses include respiratory infections, rickettsial diseases, and dysentery (USP, EP, BP).
tetrad (1) An atom or group of atoms of valency 4. (2) A crystal showing 4 similar faces when rotated 360° about its axis of symmetry.
tetradecanamide* $Me(CH_2)_{12}CONH_2$ = 227.4. Myristamide*. White leaflets, m.103.
tetradecane* $C_{14}H_{30}$ = 198.4. An alkane. Colorless liquid, d.0.765, b.253, insoluble in water.
tetradecanoic acid* Myristic acid*.
tetradecenoic acid* **5-~** Physeteric acid. **(Z)-9-~** $C_{14}H_{26}O_2$ = 226.4. Myristoleic acid. Colorless liquid, $b_{2mm}135$; from fish oils.
tetradecoic acid Myristic acid*.
tetradecyl* The radical $-C_{14}H_{29}$, from tetradecane.
tetradymite Bi_2Te_3. Tellurbismuth. A native telluride; usually contains sulfide.
tetraedrite A native arsenic and copper sulfide, intermediate in composition between tennantite and tetrahedrite.
tetraethyl- Prefix describing a compound with 4 Et groups. **lead ~** See *tetraethylplumbane* under *plumbane, tetraethyl gas.*
t.ammonium hydroxide Et_4NOH = 147.3. Colorless, hygroscopic needles, m.50, soluble in water. **t.ammonium iodide** NEt_4I = 257.2. Yellow crystals, soluble in water or alcohol. **t.benzene*** $C_6H_2Et_4$ = 190.3. An aliphatic, aromatic hydrocarbon. **1,2,3,4-~** Colorless liquid, d.0.888, b.254, insoluble in water. **1,2,4,5-~** Colorless liquid, d.0.887, b.250. **t. gas** "Knockless" gasoline; contains tetra-ethylplumbane. **t.germanium*** See *tetraethylgermanium* under *germanium.* **t.lead** See *tetraethylplumbane* under *plumbane.* **t.tin*** Et_4Sn = 234.9. Ethylstannane. Colorless liquid, b.181, insoluble in water. **t.urea** $Et_2N \cdot CO \cdot NEt_2$ = 172.3. Colorless liquid, b.210, soluble in water.
tetrafluoroboric acid HBF_4 = 87.8. Fluoboric acid. The hypothetical parent acid of the fluoroborates.
tetragalloyl-* Prefix indicating 4 $C_6H_2(OH)_3CO-$ radicals, from gallic acid. **t. erythrite** $C_4H_6O_4[C_6H_2(OH)_3CO]_4$ = 730.6. White crystals, decomp. 308, soluble in water.
tetragonal Describing a crystal having its 3 axes at right angles to one another: 2 of equal length, the third shorter or longer.
tetrahedrite Cu_2SbS_3, or $3Cu_2S \cdot Sb_2O_3$. Panabase, fahlerz, gray copper ore. Cf. *tetraedrite.*
tetrahedro-* Affix indicating 4 atoms bound into a tetrahedron.

tetrahedron A structure with 4 equal faces; as, a pyramid.

tetrahedronal Tetrahedral. Resembling a tetrahedron. **t. atom** A static atom in which the electron octet is divided into 4 pairs of electrons which oscillate around centers located at the 4 corners of a tetrahedron. Cf. *atomic structure.* **t. carbon** T. models of the C atom, to illustrate the bonds; as: single bond, touching on corners; double bond, touching on edge; triple bond, touching on face.

tetrahexahedron A crystal structure with 24 equal faces, 4 on each face of the cube.

tetrahydro- Prefix describing a compound with 4 H atoms in excess of the formula indicated by its suffix name. **t.benzene** C_6H_{10} = 82.1. Colorless liquid, b.82. **t.benzoic acid** See *tetrahydrobenzoic acid* under *benzoic acid.* **t.butene** Tetramethylene. **t.furan*** Butylene oxide. **t.naphthalene** Tetralin. **t.naphthol** $C_{10}H_{11}OH$ = 148.2. Forms 1,2,3,4- and 5,6,7,8-tetrahydro isomers of 1- and 2-naphthol. **t.naphthylamine** $C_{10}H_{11}\cdot NH_2$ = 147.2. **1-~** Colorless oil, b.227, soluble in alcohol. **2-~** Colorless liquid, b.251, soluble in alcohol. **t.phenol** Cyclohexenol*. **t.quinoline** $C_6H_4\cdot NH\cdot(CH_2)_3$ = 133.2. Hydronated quinoline. **1,2,3,4-~** Colorless liquid, b.250. **5,6,7,8-~** b.222. **t.thiophene** $CH_2\cdot CH_2\cdot CH_2\cdot CH_2\cdot S$ = 88.2. A liquid, b.121, immiscible with water; an odorant for hydrocarbon gases other than coal gas.

tetrahydroxy-* Prefix indicating 4 OH groups. **t.benzene*** $C_6H_2(OH)_4$ = 142.1. **1,2,3,4-~** Apionol. **1,2,4,5-~** Colorless leaflets, m.215, soluble in water. **t.benzoic acid*** $C_6H(OH)_4COOH$ = 186.1. **2,3,4,5-~** m.218 (decomp.). **2,3,4,6-~** m.309. **t. stearic acid** $C_{18}H_{36}O_6$ = 348.5. Sativic acid. A white purgative powder in strophanthus oil.

tetraiodo-* Prefix indicating 4 I atoms. **t.aurate(III)*** A salt containing the $AuI^-{}_4$ anion. **t.ethylene** $CI_2{:}CI_2$ = 531.6. Colorless crystals, m.187. **t.phenolphthalein** $C_{20}H_8I_4N_4Na$ = 834.9. Bluish powder, soluble in water; formerly a radiopaque medium.

-tetraketone* Suffix indicating 4 contiguous =CO groups. Cf. *-tetrone.*

tetrakis-* Multiplying affix analogous to *bis,* q.v., but meaning 4 times.

tetrakis hexahedron A form of the regular crystal system (4-faced cube).

Tetralin $C_6H_4\cdot C_4H_8$ = 132.2. Trademark for tetrahydronaphthalene*. Colorless liquid, d.0.967, b.205, insoluble in water; a solvent and substitute for turpentine. Cf. *Dekalin.*

tetralite Tetryl.

tetramethyl-* Prefix indicating 4 Me groups. **t.ammonium** $(CH_3)_4NX$. A group of organic compounds; as, $(CH_3)_4NBr$, t.a. bromide. **t.ammonium formate** Forgenin. **t.ammonium hydroxide** Me_4NOH = 91.2. Soluble solid, decomp. by heat. **t.benzene** $C_6H_2Me_4$ = 134.2. **1,2,3,4-~** Prehnitene, prehnitol. Colorless liquid, d.0.882, b.204, insoluble in water. **1,2,3,5-~** Iso*durene.* **1,2,4,5-~** Durene. **t.diaminobenzhydrol** See *Michler's hydrol.* **t.diaminobenzophenone** See *Michler's ketone.* **t.diaminotriphenylmethane** Leucomalachite green. **t.lead** T.plumbane*. **t. leucaniline** $(Me_2N\cdot C_6H_4)_2CH\cdot C_6H_4\cdot NH_2$ = 345.5. Lustrous crystals, m.151, soluble in alcohol; used in organic synthesis. **t.methane*** CMe_4 = 72.1. *tert*-Pentane, 2,2-dimethylpropane. Colorless liquid. **2,3,5,6-t.phenyl†** See *duryl.* **t.-*p*-phenylenediamine** Wurster's blue. **t.plumbane*** Me_4Pb = 267.3. Colorless liquid, d.1.995, b.110, insoluble in water. **t.tin*** Me_4Sn = 178.8. Colorless liquid, d.1.314, b.78, insoluble in water.

tetramethylene (1) $CH_2\cdot CH_2\cdot CH_2\cdot CH_2$ = 56.1. Tetrahydrobutene. Cyclobutane*, cyclotetramethylene. **amino ~** $C_4H_7\cdot NH_2$, b.81. **hydroxy ~** $C_4H_7\cdot OH$, b.123. **methyl ~** $C_4H_7\cdot Me$, b.39. (2)* 1,4-Butylene. The radical $-(CH_2)_4-$; as, in t.amine, $H(CH_2)_4NH_2$. (3) A compound containing 4 $CH_2=$ radicals. **t.diguanidine** Arcaine. **t.glycol** 1,4-Butanediol*. **t.imine** Pyrrolidine*. **t. oxide** Butylene oxide.

tetramine An organic compound containing 4 amino nitrogens; as, hexamethylenetetramine, $(CH_2)_6N_4$.

tetramolecular* Describing a reaction of the fourth order. See *reaction order.*

tetramorphism The property of crystallizing in 4 different crystal systems; as, phosphorus.

tetrandrine (1) $C_{19}H_{23}NO_3$ = 313.4. An alkaloid, m.217, from *Stephania tetrandra* (Menispermaceae). (2) $C_{38}H_{42}O_6N_2$ = 622.8. An alkaloid from the Chinese drug hanfengchi, m.216.

tetrane Butane*.

tetranitrate A compound containing 4 $-NO_3$ radicals; as, erythritoltetranitrate.

tetranitro-* Prefix indicating 4 $-NO_2$ radicals. **t.aniline*** $NH_2\cdot C_6H(NO_2)_4$ = 273.1. **2,3,4,6-~** TNA. A high explosive, m.220, explodes 237. **t.anthraquinone** Aloetic acid. **t.biphenyl*** $C_{12}H_6(NO_2)_4$ = 334.2. 9 isomers. m.165–262. **t.chrysazin** Chrysammic acid. **t.diphenylmethane** $C_{13}H_8(NO_2)_4$ = 348.2. Yellow prisms, m.172, insoluble in alcohol; used in organic synthesis. **t.methane*** $C(NO_2)_4$ = 196.0. Tetan. Colorless liquid, d.1.650, m.13, b.126 (decomp.), insoluble in water, soluble in alcohol; an explosive, and color reagent for unsaturation in organic compounds. **t.methylaniline** Tetryl. **t.naphthalene*** $C_{10}H_4(NO_2)_4$ = 308.2. **1,3,5,8-~** Yellow tetragons, m.194, soluble in water. **1,3,6,8-~** Long needles, m.203 (explodes), insoluble in water. **1,4,5,8-~** Yellow rhombs, m.257 (explodes), slightly soluble in water. **t.phenol*** $C_6H(OH)(NO_2)_4$ = 274.1. 1-Hydroxy-2,3,4,6-tetranitrobenzene. Yellow needles, m.130 (explodes), soluble in water. **t.phenolsulfonephthalein** An indicator (acids, yellow; alkalies, magenta).

tetranitrol Erythritoltetranitrate.

tetranthera The bark of *Tetranthera citrata* (Lauraceae).

tetraoxide* Tetroxide. A binary compound containing 4 oxygen atoms.

tetraphenyl* Describing a compound containing 4 phenyl groups. **t.enyl*** The radical $C_{24}H_{15}-$, from tetraphenylene (1). **1,1′,2,2′-t.ethane*** $CHPh_2\cdot CHPh_2$ = 334.5. Colorless needles, m.209, insoluble in water. **t.ethylene*** $Ph_2C{:}CPh_2$ = 332.4. Colorless, monoclinic crystals, m.221, insoluble in water. **t.lead** T.plumbane*. **t.methane*** CPh_4 = 320.4. Colorless crystals, m.285. **t.plumbane*** Ph_4Pb = 515.6. White needles, m.228, soluble in benzene. **t.silane*** Ph_4Si = 336.5. Colorless crystals, m.238, insoluble in water. **t.succinic acid** $(COOH)Ph_2C\cdot C\cdot Ph_2(COOH)$ = 422.5. Tetraphenylethanedicarboxylic acid. Colorless crystals, m.261. **t.tin*** Ph_4Sn = 427.1. Tetragonal crystals, m.226, insoluble in water. **t.urea*** $Ph_2N\cdot CO\cdot NPh_2$ = 364.4. Colorless crystals, m.183, insoluble in water.

tetraphenylene (1)* $C_{24}H_{16}$ = 304.4. Crystals, m.230. (2) Describing a compound with 4 phenylene groups.

tetraphosphorus monoselenide* P_4Se = 202.9. In soluble solid, m.166.

tetrasaccharide A carbohydrate that can be hydrolyzed to 4 monosaccharides; as, lupeose.

tetrasilane* $SiH_3\cdot SiH_2\cdot SiH_2\cdot SiH_3$ = 122.4. Unstable liquid, b.90.

tetrathioarsenate* M_3AsS_4.

tetrathioarsenite* M_3AsS_3.
tetrathionate* $M_2S_4O_6$, soluble in water.
tetrathionic acid* $H_2S_4O_6$ = 226.3. A polythionic acid obtained by the action of iodine on thiosulfates. When heated, it decomposes into sulfuric acid, sulfur dioxide, and sulfur; exists only in dilute solutions.
tetravalent Indicating a valency of 4.
tetrazane* $NH_2 \cdot NH \cdot NH \cdot NH_2$ = 62.1. Buzane, bihydrazine. See *tetrazone, tetrazo-*.
tetrazene (1) **1-~*** HN:N·NH·NH_2 = 60.1. Buzylene, diazohydrazine. Cf. *tetrazone*. **dihydro~** Tetrazane*. (2) $(N_4)CH \cdot N{:}N \cdot NH \cdot NH \cdot C({:}NH)NH_2 \cdot H_2O$ = 188.2. Tetracene. Yellow crystals, insoluble in water; a percussion primer.
tetrazine $C_2H_2N_4$ = 82.1. Heterocyclic compounds. **asymmetric ~** 1,2,3,5-t. **symmetric ~** 1,2,4,5-t. Red prisms, m.99, insoluble in water. **vicinal ~** 1,2,3,4-t. Ozotetrazone. Known only in its derivatives.
tetrazo- Bisazo*.
tetrazole* CH_2N_4 = 70.1. ***1H*-1,2,3,4-~** N:CH·NH·N:N (ring closure 1–2) m.156 (sublimes), soluble in water. Derivatives: **C-~** When the substituting radical is on the C atom. **N-~** When the substituting radical is on the N atom. **benzo~** See *benzotetrazole*. **bis~** See *bistetrazole*. **diazo~** See *diazotetrazole*.
tetrazolium A cationic (N^+) tetrazole derivative.
tetrazone $R_2N \cdot N{:}N \cdot NR_2$. Obtained by the action of yellow mercuric oxide on asymmetric dialkyl hydrazines. Tetrazones are strong reducing agents in acid solution. Cf. vicinal *tetrazine*.
tetrazotic acid $(CH_2)N_4$. Known only in derivatives. Cf. *tetrazole, tetrazolyl*.
tetrazolyl* The radical $-HCN_4$, from tetrazole. **t. hydrazine** $CN_4 \cdot NH \cdot NH_2$ = 99.1. Yellow crystals, m.199.
tetrels* The Group 4B elements of the periodic table: C, Si, Ge, Sn, Pb.
tetrinic acid $CH_2 \cdot CO \cdot CHMe \cdot CO \cdot O$ (ring) = 114.1. α-Methyltetronic acid. Colorless crystals, m.189. Cf. *pentinic acid*.
tetrodonine A curarelike poisonous principle, fugin, from the roe of Japanese fishes: genus *Tetrodon*.
tetrole Furan*.
tetrolic acid Butynoic acid*. **t. a. series** See *acetylene acids*.
-tetrone* Suffix indicating 4 =CO groups, not necessarily contiguous. Cf. *-tetraketone*.
tetronerythrin Red pigment from the feathers of birds.
tetronic A class of detergents made by condensing ethylene oxide and ethylene diamines. **t. acid** $CH_2 \cdot CO \cdot CH_2 \cdot CO \cdot O$ (ring) = 100.1. Colorless crystals, m.141, soluble in alcohol. Keto and enol forms in equilibrium. **ethyl~** Pentinic acid. **methyl~** Tetrinic acid. **propyl~** Hexinic acid.
tetrose Any of a class of monosaccharides with 4 C atoms; as, erythrose, threose.
tetroxide Tetraoxide*.
tetryl $(NO_2)_3C_6H_2N(Me)NO_2$ = 287.2. Tetranitromethylaniline, methylnitropicramide, pyrenite, tetralite, 2,4,6-trinitrophenylmethylnitroamine. Yellow powder, m.130, explodes 187; an explosive in detonators, and primer for less sensitive explosives.
teucrin $C_{19}H_{20}O_5$ = 337.4. A glucoside from germander, *Teucrium fruticans* (Labiatae).
teucrium (1) Cat thyme, Syrian mastic. The dried herb of *Teucrium marum* (Labiatae); contains an essential oil, camphor, resin, and glucoside. (2) *T. chamaedris*, germander. (3) *T. scordium*, water germander.
tex A unit expressing the linear weight of textile filaments: 1 tex = 1 g/km. Cf. *denier*.
textile (1) A natural or man-made fibrous material suitable to be spun and made into a yarn. Cf. *fiber*. (2) An assembly of interlacing yarns or fibers in the form of woven, knitted, or other structures. Cf. *fabric*.
textryl Generic name for sheet structures prepared by papermaking processes from synthetic fibers, which are held together at their points of contact by fibrillated, synthetic, thermoplastic polymers (fibrids). Textryls are intermediate in character between paper and nonwoven fabrics.
texture Coarse structure, e.g., the arrangement of fibers in textiles.
textured vegetable protein See *textured vegetable protein* under *protein*.
tfol An argillaceous earth containing gelatinous silica (N. Africa); a substitute for soap. **t. ointment** A mixture of tfol 20, tar 100 pts.; an antiseptic paste.
Th Symbol for thorium. **ThEm** Radium-220. **ThX** Radium-224.
thalenite $2Y_2O_3 \cdot 4SiO_2 \cdot H_2O$. A mineral.
Thales of Miletus (640–546 B.C.) Greek philosopher noted for his attempt to find a single material cause for all things: "Moisture is the principle of all life."
thalictrine $C_{20}H_{27}O_4N^+$ = 345.4. An alkaloid obtained from *Thalictrum macrocarpum*, or *T. foliolosum* (Ranunculaceae). Colorless needles, m.208, insoluble in water.
thalidomide $C_{13}H_{10}N_2O_4$ = 258.2. 2-Phthalimidoglutarimide. Contergan, Distaval, Kevadon. A hypnotic, withdrawn because it can produce severe fetal malformations when taken by women early in pregnancy.
thallasotherapy Treatment involving seawater.
thalleoquine reaction A characteristic color is obtained with alkaloids on adding, in succession, chloroform, bromine water, and sodium hydroxide.
thallic Thallium(III)* or (3+)*. Describing a compound containing trivalent thallium. **t. chloride** Thallium chloride (2).
thalline $C_9H_6N(OMe) \cdot H_4$ = 163.2. Tetrahydro-*p*-quinanisol, tetrahydro-*p*-methoxyquinoline. Colorless rhombs, m.40, soluble in water. It has a coumarinlike odor, and gives a deep-green color with ferric chloride. **t. sulfate** $(C_{10}H_{13}ON)_2 \cdot H_2SO_4 \cdot 2H_2O$ = 460.5. Colorless crystals, m.110, soluble in water.
thallium* Tl = 204.83. A rare metal of the gallium-indium family, and element, at no. 81. Bluish white, d.11.862, m.304, b.1460, insoluble in water, soluble in nitric acid. Discovered (1861) by Crookes and named from its green spectral line (Greek *thallein:* "green,"). Widely distributed in nature and obtained from lead-chamber sludge by precipitation with zinc. T. forms 2 series of compounds, and has valencies of 1 [t.(I)*, t.(1+)*, thallous] and 3 [t.(III)*, t.(3+)*, thallic]. Its salts are all cumulative poisons. Chemically it is analogous to the alkali metals, lead, and aluminum. Used as a catalyst in the manufacture of azobenzene, as an antiknock compound, and in optical glass. Principal sources: lorandite (59–60% Tl), $TlAsS_2$; vrbaite (29–32% Tl), $TlAs_2S_5$. **triethyl~*** Et_3Tl = 291.6. Thallium ethyl. An antiknock compound.
t. acetate* $TlC_2H_3O_3$ = 263.4. Thallous acetate. Colorless, hygroscopic crystals, m.110, soluble in water. **t. alkyls** R_2TlX; as, diethyl t. chloride, Et_2TlCl. **t. alum** $Tl_2SO_4 \cdot Al_2(SO_4)_3 \cdot 24H_2O$ = 1279. Soluble white cyrstals. **t. amalgam** An alloy: Tl 8.5, Hg 91.5%; used in thermometers for low temperatures (−60°C). **t. bromides** (1) TlBr = 284.3. Thallous bromide. Colorless, regular crystals, m.450, insoluble in water. (2) $TlBr_3$ = 444.1. Thallic bromide. Yellow needles, decomp. by heat, soluble in water. **t. carbonate***

Tl_2CO_3 = 468.8. Thallous carbonate. Colorless, monoclinic crystals, d.7.11, m.272 (decomp.), soluble in water. **t. chlorides** (1) TlCl = 239.8. Thallous chloride. Colorless cubes d.7.02, m.429, soluble in water: a *getter*, q.v., in tungsten lamps. (2) $TlCl_3 \cdot H_2O$ = 328.8. Thallic chloride. Thallium sesquichloride. Colorless, deliquescent crystals, decomp. 100, occurring anhydrous or with 1, 4, and 7.5 moles of water. **t. chloroplatinate** T. hexachloroplatinate*. **t. ethanolate*** EtOTl = 249.4. T. alcoholate. A saturated alcoholic solution of thallous oxide, having a high density of 3.55. **t. ethyl** Triethyl ~*. **t. fluosilicate** T. hexafluorosilicate*. **t. formate*** HCO_2Tl = 249.4. Thallous formate. Colorless liquid, used for mineralogical solutions. **t. formate-malonate** A double salt, m.60, miscible with water. Cf. *Clerici solution*. **t. glass** A high-refractive-index flint glass in which lead is replaced by thallium. **t. hexachloroplatinate.** Tl_2PtCl_6 = 816.6. T. chloroplatinate. Yellow solid, sparingly soluble in water. **t. hexafluorosilicate*** $Tl_2SiF_6 \cdot 2H_2O$ = 586.9. Hexagonal plates, soluble in water. **t. hydroxides** (1) TlOH = 221.4 (+H_2O). Thallous hydroxide. Yellow powder, decomp. 100, soluble in water; contains 1 H_2O and is a strong base. (2) $Tl(OH)_3$ = 255.4, or TlO·(OH) = 237.4. Thallic hydroxide. Orange powder, decomp. 100, insoluble in water or alkalies, soluble in acids. **t. iodides** (1) TlI = 331.3. Thallous iodide. Yellow crystals, d.7.072, m.432, insoluble in water. (2) TlI_3 = 585.1. Thallic iodide. Yellow powder, slowly decomp. in air, soluble in alcohol. **t. ions** The Tl+ (thallous) or Tl^{3+} (thallic) ion. **t. mercurous nitrate** A saturated solution, d.5.3, used for the separation of minerals. **t. nitrates** (1) $TlNO_3$ = 266.4. Thallous nitrate. Colorless rhombs, m.205, soluble in water; a reagent, indicator, and pyrotechnic for green fires. (2) $Tl(NO_3)_3 \cdot 3H_2O$ = 444.4. Thallic nitrate. Colorless, deliquescent crystals, decomp. 100. **t. oxides** (1) Tl_2O = 424.8. Thallous oxide. Black powder, m.870, soluble in water; used in the manufacture of flint glass and artificial gems. (2) Tl_2O_3 = 456.8. Thallic oxide. Brown hexagons, m.759, insoluble in water. (3) TlO or Tl·O·Tl:O and (4) Tl_3O_5 also exist. **t. oxysulfide** T. sulfoxide*. **t. ozone paper** A filter paper impregnated with t. hydroxide; colored brown by ozone. **t. peroxide** T. oxides (3). **t. phosphate** Tl_3PO_4 = 708.1. Thallous phosphate. Colorless needles, soluble in water. **t. sulfates** (1) Tl_2SO_4 = 504.8. Thallous sulfate. Colorless prisms, m.632 (decomp.), soluble in water; a reagent. (2) $Tl_2(SO_4)_3 \cdot 7H_2O$ = 823.0. Thallic sulfate. Colorless crystals, decomp. by heat, soluble in water; form double salts: $MTl(SO_4)_2$, not isomorphous with $TlAl(SO_4)_2$. **t. sulfides** (1) Tl_2S = 440.8. Thallous sulfide. Black tetragons, m.448 (decomp.), insoluble in water. (2) Tl_2S_3 = 504.9. Thallic sulfide. Black mass, decomp. by heat, insoluble in water. **t. sulfoxide*** Tl_2SO = 456.8. T. oxysulfide. A light-sensitive substance used in photoelectric (thalofide) cells. **t. trioxide*** T. oxides (2).

thallochlor The green coloring material of Iceland moss. Cf. *chlorophyll*.

thallofide cell Thalofide cell.

thallophytes Cryptogamous plants:

Peridineae	Flagellates
Cyanophyceae	Blue-green algae
Chlorophyceae	Green algae
Rhodophyceae	Red algae
Phaeophycetes	Brown algae
Diatomae	Diatoms
Schizomycetes	Bacteria
Phycomycetes	Algal fungi
Myxomycetes	Slime molds
Eumycetes	Fungi
Characeae	Stoneworts

thallous Thallium(I)* or (1+)*. Describing a compound of monovalent thallium. **t. compounds** See *thallium*.

thalofide cell Thallofide c. A photoelectric cell in which thallium sulfoxide is the light-sensitive material; the resistance decreases on illumination.

thalviol Thujone.

thanatol $C_6H_4(OEt)OH$ = 138.2. Guaethol, ajacol, pyrocatechin monoethyl ether, ethoxyhydroxybenzene. An oil, b.215, insoluble in water; used medicinally.

thanmasite $CaSiO_3 \cdot CaCO_3 \cdot CaSO_4 \cdot 15H_2O$. A mineral from Paterson, N.J.

thanomin Ethanolamine.

thapsic acid A fatty acid, m.124, from *Thapsia garganica* (Umbelliferae), insoluble in water.

thaumatin A protein from *Thaumatococcus danielli* (W. Africa); 2,000 times as sweet a sucrose.

thawing (1) The liquefaction of ice by heat. (2) The softening of frozen dynamite.

thea The Latin (pharmaceutical) term for "tea."

theaflavin, thearubigin Compounds formed by the oxidation of polyphenols in tea during fermentation; largely responsible for the characteristic flavor.

theamin $C_7H_8N_4O_4 \cdot NH \cdot C_2H_4OH$. Theophylline ethanolamine. White powder; a diuretic and bronchial dilator.

thebaine $C_{19}H_{21}O_3N$ = 311.4. **(−)-~** Paramorphine, dimethylmorphine. An alkaloid from opium. Very poisonous, colorless prisms, m.193, slightly soluble in water. **methyl ~** $C_{20}H_{23}O_3N$ = 325.4. An alkaloid from opium. Colorless crystals, insoluble in water.
t. tartrate $(Tbn)C_4H_6O_6 \cdot H_2O$ = 479.5. White crystals, soluble in water.

thebaol $C_{16}H_{13}O_3$ = 253.3. 3,6-Dimethoxy-4-phenanthrenol. m.94. From thebaine.

thebenidine $C_{15}H_9N$ = 203.2. Azapyrene. The ring structure; as, 4-t.

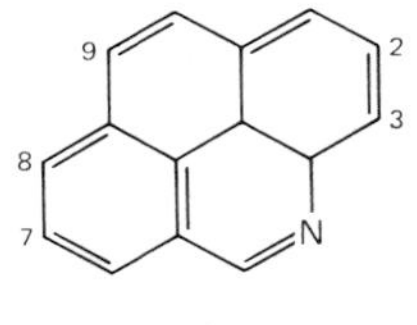

2-~ Yellow crystals, m.163 (decomp.). **4-~** m.159.

theetsee A black varnish obtained by tapping the stems or trunks of *Melanorrhoea usitata* (Anacardiaceae), Malaysia.

theine Caffeine.

thelephoric acid $C_{18}H_8O_8$ = 352.3. Black powder, from lichens, as *Thelophora*, and fungi. m.300+, soluble in pyridine.

Thénard, Louis Jacques (1777–1857) French chemist, noted for the isolation of elementary boron. **T.'s blue** $Co(AlO_2)_2$. The blue cobalt aluminate obtained by heating alum moistened with cobalt nitrate, in the blowpipe test.

thenardite Na_2SO_4. A native sodium sulfate.

thenyl alcohol 2-Hydroxymethyl*thiophene*.

thenoic acid* Thiophenecarboxylic acid*.

theobroma (1) Cocoa. (2) Plants (Sterculiaceae) that yield cacao. **oil of ~** Cocoa butter (BP).

theobromine $C_7H_8O_2N_4$ = 180.2. Cacaine, 3,7-dimethylxanthine. An alkaloid from the leaves and seeds of *Theobroma* species; an isomer of theophylline and paraxanthine. Microcrystals, m.350, soluble in water; a diuretic and stimulant (EP, BP). **iodo ~** T. sodium iodide. **methyl ~** Caffeine. **uro ~** Paraxanthine.
t. acetyl salicylate $(Tbr)C_9H_8O_4$ = 360.4. White crystals, insoluble in water. **t. hydrochloride** (Tbr)HCl = 216.7.

White crystals, soluble in water; a weak nerve stimulant. **t. sodium benzoate** $C_7H_7NaO_2N_4 \cdot NaC_7H_5O_2$ = 346.3. White crystals, soluble in water. **t. sodium citrate** $C_7H_7NaO_2N_4 \cdot Na_3C_6H_5O_7$ = 460.2. Urocitral. White powder, soluble in water. **t. sodium iodide** $C_7H_7NaO_2N_4 \cdot NaI$ = 352.0. Eustenin, iodotheobromine, sodium iodotheobromate. White, hygroscopic powder, soluble in water.

Theocine A brand of synthetic theophylline.

theogallin $C_{14}H_{16}O_{10} \cdot 2H_2O$ = 380.3. A polyphenol in unprocessed tea leaves.

theoline C_7H_{16} = 100.2. An aromatic hydrocarbon from petroleum.

theophylline $C_7H_8O_2H_4 \cdot H_2O$ = 198.2. 1,3-Dimethylxanthine. An isomer of theobromine; an alkaloid from tea leaves. Colorless needles, m.269, slightly soluble in water; a cardiac stimulant and diuretic (USP, EP, BP). **t. ethylenediamine** Aminophylline.

theorem Theory.

theory The reduction of data or facts to a principle, and the demonstration of their interrelations. **atomic ~** See *atomic theory.* **ionic ~** See *ion, ionization.* **phlogiston ~** See *phlogiston theory.* **quantum ~** See *quantum theory, energy levels.* **valency ~** See *valency.*

theotannin A tannin prepared from freshly plucked green tea leaves.

therapeutic Pertaining to the art of healing. **t. agent** A remedy or substance used to alleviate disease, pain, or injury. **t. index** The ratio: minimum fatal dose/minimum curative dose. **chemical ~** The ratio: curative dose/tolerated dose.

therapeutics The study of remedial measures.

therapy A system of treatment, as: pharmacotherapy (drugs); specific chemotherapy (chemicals); immunotherapy (vaccines–in prevention of infectious disease and in disseminated cancer); ray therapy (radiation), e.g, X-rays, ultraviolet rays, infrared rays, radioactivity. Cf. *aerotherapeutics, atmotherapy, hydrotherapy, radium therapy, photoradiation.*

therm (1) A unit of energy: 1 t. = 100,000 Btu = 1.0051×10^8 J. (2) British thermal unit. (3) Large calorie. (4) Small calorie.

thermae Natural warm springs.

thermal Pertaining to heat or temperature. **t. analysis*** Analysis depending on a change in weight or condition due to heat. Cf. *thermometric titrimeter.* **t. capacity** Heat capacity*. **t. conductance** *C*. Heat transfer coefficient. The heat passing, in unit time, through unit cross-sectional area of a substance when there is unit temperature difference between the opposite faces. The SI unit of measurement is $W/m^2 \cdot K$. The non-SI units are $Btu/h \cdot ft^2 \cdot {}^\circ F$ or $Btu/s \cdot ft^2 \cdot {}^\circ F$, or similar units. 1 $Btu/h \cdot ft^2 \cdot {}^\circ F$ = 5.67 $W/m^2 \cdot K$. The reciprocal quantity is t. resistance. Cf. *thermal conductivity.* **overall t. c.** *U*. Thermal transmittance. T. c. when applied to a composite system, as a wall or roof. Now generally replaced by t. resistance. **t. conductivity*** Symbols: λ*, *k;* heat conductivity. The heat passing, in unit time, through unit cross-sectional area of a substance when there is unit temperature gradient between the opposite faces. See Table 94. The SI unit of measurement is $W/m \cdot K$. The non-SI units are $Btu \cdot in/h \cdot ft^2 \cdot {}^\circ F$ or $Btu \cdot ft/h \cdot ft^2 \cdot {}^\circ F$, or similar units. 1 $Btu \cdot in/h \cdot ft^2 \cdot {}^\circ F$ = 0.0832 $Btu \cdot ft/h \cdot ft^2 \cdot {}^\circ F$ = 3.44×10^{-4} cal(thermochemical)/$cm \cdot s \cdot {}^\circ C$ = 0.144 $W/m \cdot K$. The reciprocal quantity is t. resistivity. Cf. *thermal conductance.* **t. constant** The heat (in J or cal) evolved during a particular reaction. **t. death point** The temperature required to kill bacterial cultures in 10 min. **t. deformation** Pyroelectricity. **t. energy** Heat energy. **t. excitation** The transition of an electron to a higher energy level or of a molecule to a higher vibrational energy level, by an increase in temperature. The fraction *Y* of a gas which is excited at the thermodynamic temperature *T* is $\log Y/(1 - Y) = -5{,}048E$, where *E* is the ionization energy (Saha). **t. expansion** The increase of volume due to heat; small for solids and liquids, large for gases. **t. intensity** Temperature. **t. ionization** The loss of an electron by an atom due to an increase in temperature. **t. resistance** Symbol: *R;* the reciprocal of t. conductance. Thus, the temperature difference which causes unit heat flow rate per unit area. Used as a measure of insulation effectiveness. Glass fiber, mineral wool, and expanded polystyrene have a value of 2.3 $m^2 \cdot K/W$ when 80 mm thick and 4.3 when 160 mm thick. Building regulations in countries with temperate climates generally require 1.4–5 $m^2 \cdot K/W$ for roofs, walls, and floors, depending on local ambient temperatures. Cf. overall *thermal conductance.* **t. radiation** See *radiation.* **t. reactor** See *reactor.* **t. resistivity** The reciprocal of t. conductivity. **t. transmittance** See *overall thermal conductance* above under *thermal conductance.*

thermel Thermocouple.

thermic Thermal.

thermil The quantity of heat required to raise, by 1°C, the termperature of a mass of 1 metric ton of a body having a specific heat equal to that of water at 15°C, at the standard atmospheric pressure of 101.3 kPa.

thermionic effect The loss of electrons from a heated body maintained at an electric potential; the basis of the thermionic valve. **t. tube, t. valve** A glass or metal container, either evacuated or containing gas, in which electrons emitted from the cathode pass to the anode under the control of 1 or more other electrodes.

thermions The ions that carry current through a vacuum in a thermionic tube; positively charged atoms, or electrons.

thermistor A sensory material or device which changes rapidly in electrical resistance with change in temperature.

Thermit Trademark for a mixture of aluminum and ferric

TABLE 94. THERMAL CONDUCTIVITY λ OF VARIOUS MATERIALS AT 25°C

Material	λ, W/m·K	Material	λ, W/m·K
Silver	430	Asbestos cement sheets	0.30
Copper	400	Paper	0.05–0.20
Aluminum	230	Wood	0.05–0.16
Zinc	115	Vermiculite	0.065
Iron and steel	80	Glass fiber	0.043
Lead	35	Cork slab	0.040
Mercury	8.3	Foam plastics	0.03–0.04
Brick	0.4–1.3	Oxygen	0.022
Concrete	1.0	Air	0.022
Glass	0.80	Sulfur dioxide	0.0072
Water	0.60		

oxide (Fe_3O_4). If ignited by a primer (magnesium powder), it liberates heat and forms aluminum oxide and molten iron at about 3000 °C. **T(e). process** Goldschmidt's process. Thermoreduction. (1) Obtaining a high temperature and molten iron by Thermit; used in welding steel, rails. (2) Obtaining metallic chromium or manganese by mixing their oxides with powdered aluminum and igniting the mixture with a primer.

thermo- Prefix (Greek) indicating "heat."

thermobalance A balance for weighing substances while they are being heated, e.g., during drying.

thermochemical standard A substance of known heat of combustion used to standardize bomb calorimeters. Benzoic acid for this purpose is certificated by the National Bureau of Standards, Washington, D.C., and by the National Physical Laboratory, Teddington, England.

thermochemistry The study of the relations of heat and chemical reactions. See *thermodynamics.*

thermochor An expression of the relationship between the molecular volume and the temperature of a substance at the boiling point. Cf. *parachor.*

thermocouple Thermel, thermoelement, thermojunction. A device for measuring temperature by the production of a thermoelectric emf at the junction of 2 different metallic wires. One wire is kept at constant low temperature (e.g., 0 °C), the other at the temperature to be measured. E.g.: copper-constantan, 25.54 mV (at maximum 500 °C); nickel-nickel chrome, 41.80 mV (at maximum 1090 °C). Cf. *pyrometer.*

thermocroic Changing color with temperature.

thermocross Thermocouple.

thermoduric Describing organisms that resist the conditions used for milk pasteurization.

thermodynamic **t. concentration** Activity. **t. potential** Gibbs' function*.

thermodynamics The study of the empirical relations between heat energy and other forms of energy. **first law of ~** Conservation of energy: "Heat and work are equivalent." A definite quantity of heat gives a fixed amount of mechanical energy and vice versa. Since any change or transformation produced in a body is proportional to the *heat equivalent,* the change in energy content of a body depends only on the difference between the original and final states. The amount of work obtained from a given process, changing at constant temperature and pressure from a higher to a lower state, is the *free* energy. **second law of ~** Degradation of energy: It is impossible, when unaided, to convey heat from one body to another at a higher temperature; therefore, in all changes, the *entropy,* or amount of irreversible energy of the participating bodies, increases. **third law of ~** Every substance has a finite positive entropy, and, at absolute zero temperature, the entropy value is zero for pure crystalline substances. Cf. *entropy.*

thermoelectric Pertaining to electricity produced by heat. **t. current** An electric current produced from a thermocouple. **t. power** The thermoelectromotive force produced at the junction of 2 metals by a temperature difference of 1 °C.

thermoelectromotive force Thermoelectric power.

thermoelement (1) Thermocouple. (2) Thermopile.

thermogenesis The generation of heat by living bodies; e.g., by organisms in compost.

thermography The detection of temperature differences and its depiction. Used in industrial testing and medical diagnostics. **pulse video ~** A technique of subjecting a material to an intense heat for a short period, then recording temperature gradients by i.r. camera or video. Any defects, as in laminates and composites, affect heat diffusion and thus temperature gradient.

thermogravimetric analysis Chemical analysis by measuring the weight changes of a system as a function of increasing temperature.

thermojunction Thermocouple.

thermokalite A mixture of trona, thenardite, thermonatrite, and sodium hydrogencarbonate.

thermolabile Decomposed or destroyed by heat.

thermoluminescence Luminescence caused by a slight increase in the temperature of a body, without production of incandescence.

thermolysis Pyrolysis. Dissociation or decomposition produced by heat.

thermomagnetic effect A difference in magnetic properties caused by heat. See *Leduc effect, Nernst effect.*

thermomechanical pulp See *mechanical pulp.*

thermometamorphism A change in allotropic forms produced by heat.

thermometer A device for determining temperature. (1) *Mechanical:* A substance that expands and contracts with temperature changes. (2) *Electrical:* A device that measures the change in or production of an electric current or resistance due to change of temperature; as, resistance thermometers. (3) *Optical:* A device that evaluates the light of an incandescent body; as, optical pyrometer. **angle ~** An L-shaped t. for insertion in the vertical side of a vessel. **armored ~** A t. surrounded by a metal casing. **Beckmann ~** See *Beckmann thermometer.* **kata ~** See *katathermometer* under *kata-.* **ultra ~** Beckmann t.

t. scale conversion See *temperature conversion.*

thermometric titrimetry Enthalpimetry. The use of temperature changes to indicate the end point of a volumetric reaction.

thermonatrite $Na_2CO_3 \cdot H_2O$. A native carbonate.

thermoneutrality The absence of a heat change when dilute solutions of netural salts are mixed and no precipitate forms.

thermonuclear reaction A fusion reaction that proceeds at high temperatures, e.g., in a star, and transmutes elements by bombardment of particles, etc. The Joint European Torus (JET) project is an attempt to simulate and control a t. r. It is based on the Russian tokamak system, which controls a ring of plasma at temperatures similar to those on the sun using magnetic fields.

thermophilic Describing organisms of optimum growth at temperature 60–80 °C. Cf. *psychrophilic.*

thermophone A device for converting a.c. electricity into thermal (and some sound) waves, by supplying it to a platinum strip immersed in a true fluid.

thermophore A device for retaining or holding heat.

thermopile Thermoelement. Soldered metal plates or bars arranged in series so as to produce a multiple thermocouple and thus a cumulative effect; on heating, a potential change is produced.

thermoplastic Rendered soft and moldable by heat. Cf. *plastics.*

thermopotential Gibb's function.

thermoredress The latent power of recovery from a strain applied to a solid, due to the action of heat.

thermoreduction Thermit process.

thermoregulator A device for mechanically regulating temperature by controlling the source of heat. See *thermostat.*

thermosetting Rendered hard by heat; as, certain *plastics,* q.v.

thermostabile, thermostable (1) Refractory, heat-resisting. (2) In biochemistry: not affected by temperatures above 55 or below 100 °C.

thermostat A device for regulating the heat of an apparatus automatically.

thermotaxy (1) A form of *orthotaxy,* q.v., produced by heat. (2) The directional tendency of regular groupings in crystalline substances, due to heat.
thermotension The subjection of a red-hot metal to high tensile stress during cooling.
thermotropic Caloritropic. Stimulated by or responding to a change of temperature.
Thermovyl Trademark for a polyvinyl chloride synthetic fiber.
thermsilid An acid-resisting alloy: Fe 84, Si 16%.
theta Θ, *θ* Greek letter. *θ* (italic) is symbol for (1) angle; (2) common temperature; (3) scattering angle; (4) volume strain.
thetine Thiobetaine, sulfinate. Organic compounds derived from the heterocyclic compound

CO<(O)(CH₂)>S<(R)(R)

e.g., dimethylthetine, where R is $-CH_3$.
theveresin $C_{48}H_{70}O_{17}$ = 919.1. White powder; a split product of thevetin.
thevetin Any of a group of crystalline glucosides from the seeds of yellow oleander, *Thevetia neriifolia* (Apocynaceae), Central America. **t. A** Colorless crystals, m.210, soluble in water. **t. B** m.200. Cf. *theveresin.*
THF Abbreviation for tetrahydrofuran.
thia-* Infix indicating S when it is replacing C, as in a heteroatomic compound. Cf. *thio-.*
thiabendazole $C_{10}H_7N_3S$ = 201.3. 2-(4-Thiazolyl)benzimidazole. Mintezol. White crystals, insoluble in water. An anthelmintic.
thiacetamide Thioacetamide*.
thiacetazone $C_{10}H_{12}ON_4S$ = 236.3. Amithiozone. Yellow, bitter crystals, soluble in water; used to treat tuberculosis and leprosy.
thiacetic Thioacetic acid*.
-thial* Suffix for a thioaldehyde; as, ethanethial*, CH_3CHS.
thialdine $C_6H_{13}NS_2$ = 163.3. Colorless crystals, m.43 (decomp.), soluble in water.
thiambutosine $C_{19}H_{25}ON_3S$ = 343.5. White, bitter crystals, m.125, insoluble in water; used to treat leprosy (BP).
thiamide Thioamide*. **iso ~** R−C:NH·(SH).
thiamin Thiamine.
thiamin(e) Vitamin B_1. **t. hydrochloride** $C_{12}H_{17}ON_4SCl \cdot HCl$ = 337.3. White, hygroscopic needles with a yeasty odor and a salty, nutty taste, m.247 (decomp.), soluble in water, insoluble in ether. A compound of the vitamin B complex (USP). **t. mononitrate** $C_{12}H_{17}O_4N_5S$ = 327.4. White crystals, soluble in water; used similarly to t. hydrochloride (USP).
thianapthene $C_6H_4 \cdot S \cdot CH{:}CH$ = 134.2. Benzothiophene, benzothiofuran. Colorless leaflets, m.31, soluble in alcohol. In lignite tar.
thianthrene* $C_6H_4{:}(S_2){:}C_6H_4$ = 216.3. Diphenylene disulfide, dibenzo-*p*-dithiin, m.159, insoluble in water.
thiazine Sulfur-containing heterocyclic compounds. Cf. *benzothiazine, phenothiazine.* **t. dyes** Dyes derived from thiazine containing 3 rings resulting from all 4 C atoms becoming part of an aryl ring; as, in thionine, toluidine blue.
thiazole* C_3H_3NS = 85.1. Metathiazole, thio-[*o*]-monazole, vitamin T. Colorless liquid, b.117, insoluble in alcohol. **dihydro ~** Thiazoline. **2-hydroxy-1,3- ~** C_3H_3ONS = 101.1. Oxythiazole. **2-hydroxy-4-methyl-1,3- ~** Colorless crystals, m.160, soluble in water. **2-hydroxy-5-phenyl-1,3- ~** Colorless crystals, m.204, soluble in water. **iso ~** b.113. **t.-azo dyes** Sensitive color reagents for cations. **t. purple** $C_{19}H_{17}BrN_2S_2$. 1,1′-Dimethylthiocarbocyanine bromide. A carbocyanine dye photographic sensitizer.

1,3-Thiazole	Isothiazole
(1) S; (5) HC, CH (2); (4) HC—N (3)	(1) S; (5) HC, N (2); (4) HC—CH (3)

thiazoline C_3H_5NS = 87.1. Member of the dihydrothiazoles. Organic compounds derived from the heterocyclic compound $CH_2 \cdot CH_2 \cdot S \cdot CH{:}N$; as:

2-Methylthiazoline, C_3H_4NSMe b.145
2-Phenylthiazoline, C_3H_4NSPh b.276

thiazolyl* The radical C_3H_2NS-, from thiazole. **iso ~** The radical C_3H_2NS-, from isothiazole.
thicken (1) To evaporate to a high viscosity. (2) To expand the ends of a metal rod.
thickness (1) The degree of viscosity or fluidity. (2) The width of a plate. (3) The radius or gauge of a wire.
Thiele tube A specially shaped test tube (like an inverted P) which, if heated at the base, ensures the circulation of the liquid in it, and thus, even distribution of heat. Used in melting-point determinations.
Thiel-Stoll solution A saturated solution of lead perchlorate, d.2.6; used for determining specific gravity by the suspension method.
thienone Dithienyl ketone*. **aceto ~** Methyl thienyl *ketone**.
thienyl* The radical $-C_4H_3S$, from thiophene. **t.diphenylmethane** Diphenylthienylmethane*. **t. ketone** Dithienyl ketone*.
Thies process Extraction of gold from crushed ores by adding bleaching powder and sulfuric acid.
thimble A porous cup of filter paper, fritted glass, or Alundum, used as a container for materials being extracted; e.g., in a Soxhlet apparatus.
thimerosal $C_9H_9O_2SHgNa$ = 404.8. Thiomersal. Ethylsodium *o*-mercaptobenzoato)mercury. Mercurothiolate. Creamy crystals, soluble in water. A disinfectant for skin sterilization (USP, BP).
thinner A liquid used to dilute a paint. Cf. *vehicle.*
thinolite Tufa or native calcium carbonate, forming layers of interlaced crystals (Nevada and California).
thio-* Prefix (Greek *theion:* "sulfur"). It indicates the replacement of oxygen by divalent sulfur. Cf. *sulfo, thia-.* **t. acid*** An organic compound in which divalent sulfur has replaced some or all of the oxygen atoms of the carboxyl groups; as,

R·CO·SH . . . .*S*-Thioic acids* (carbothioic acid*)
R·CS·OH . . . .*O*-Thioic acids* (carbothioic acid*)
R·CS·SH . . . . .Dithioic acids* (carbodithioic acid*)

See polythionic acids.
thioacetal* Mercaptal. Sulfur analogs of acetals, containing the $=C(SR')SR^2$ or $=C(OR')SR^2$ groups; e.g., as formed from thiols and aldehydes in the presence of hydrochloric acid: $R' \cdot CHO + 2HSR^2 = H_2O + R' \cdot CH(SR^2)_2$.

thioacetaldehyde* $(CH_3 \cdot CHS)_3$ = 180.3. A solid, m. 45, insoluble in water.
thioacetamide* $Me \cdot C(:S)NH_2$ = 75.1. Thiacetamide. Colorless leaflets, m.109, soluble in water. Its derivatives are thioamides.
thioacetanilide* $Me \cdot CS \cdot NHPh$ = 151.2. A thioamide. Colorless needles, m.75 (decomp.), insoluble in water; used in organic synthesis.
thioacetic acid* MeCOSH = 76.1. Thiacetic acid, thiolacetic acid. Colorless liquid, b.93, soluble in water.
thioacetin $C_5H_{10}O_3S$ = 150.2. Akcethin.
thioalcohol Thiol*.
thioaldehyde* Organic compounds containing the −CHS radical.
thioamides* $R \cdot CS \cdot NH_2$.
thioanhydride* An anhydride of a thiocarboxylic acid:

$R\langle^{CO}_{CO}\rangle S$	$R\langle^{CS}_{CS}\rangle O$	$R\langle^{CS}_{CS}\rangle S$
Thioic anhydride	Thioic anhydride	Dithioic anhydride

thioanilide $C_{10}H_6 \cdot (SH)NH_2$ = 175.2. Aminonaphthalenethiol. A precipitation reagent for Cu, Ag, Hg, and Bi.
thioaniline $(C_6H_4NH_2)_2S$ = 216.3. 4,4′-Diaminodiphenyl sulfide, m.108. ***N*-~** PhN:S = 123.2. A reactive intermediate.
thioantimonate Sulfantimonate. $M_3SbO_{(4-n)}S_n$. **mono ~ ***, **di ~ ***, etc., according to the value of *n*. See *Schlippe's salt.*
thioantimonite Sulfantimonite. R_3SbS_3 and $R_4Sb_2S_5$. Salts known only in solution.
thioarsenate $M_3AsO_{(4-n)}S_n$. **mono ~ ***, **di ~ ***, etc., according to the value of *n*.
thioarsenite $M_3SbO_{(3-n)}S_n$. **mono ~ ***, **di ~ ***, etc., according to the value of *n*.
-thioate* Suffix indicating the ester or salt of a thiocarboxylic acid; as, *S*-ethyl hexanethioate, $C_5H_{11}COSEt$.
thiobacilli Bacilli which oxidize many inorganic sulfur compounds to sulfuric acid. They require no organic food, and some tolerate 10% sulfuric acid. T. occur in soil and in deteriorating iron and concrete structures.
thiobacteria Bacteria that reduce or oxidize sulfur compounds.
thiobenzaldehyde* PhCHS = 122.2. **alpha-~** A solid, m.160 (decomp.), insoluble in water. **beta-~** A solid, m.225.
thiobenzamide* $Ph \cdot CS \cdot NH_2$ = 137.2. Colorless crystals, m.116.
thiobenzanilide $Ph \cdot CS \cdot NHPh$ = 213.3. Colorless crystals, m.98.
thiobenzimidazolone Phenylenethiourea.
thiobenzoic acid* $Ph \cdot COSH \cdot \frac{1}{2}H_2O$ = 147.2. Colorless crystals, m.24, insoluble in water. Cf. sulfo*benzoic acid.*
thiobenzophenone* $Ph \cdot CS \cdot Ph$ = 198.3. Colorless crystals, m.147.
thiobetaine Thetine.
thiocacodylate A salt of thiocacodylic acid containing the radical Me_2AsOS−.
thiocarbamic t. acid* $CS(NH_2)SH$ = 93.2. Soluble in water. **t. ester** $R_2N \cdot CS \cdot OR$. E.g., Diphenylthiocarbamic phenyl ester, $Ph_2N \cdot CS \cdot OPh$.
thiocarbamide Thiourea*.
thiocarbamoyl* The radical $-SCNH_2$.
thiocarbanilide $Ph \cdot NH \cdot CS \cdot NH \cdot Ph$ = 228.3. Sulfocarbanilide, diphenylthiourea. Colorless leaflets, m.153, insoluble in water.
thiocarbazide Thiocarbonohydrazide*.
thiocarbimide Isothiocyanic acid.
thiocarbin A thiol prepared by boiling glycerol with sodium thiosulfate; a photographic emulsion ripening accelerator, and analytical reagent.
thiocarbonate (1) A salt of a thiocarbonic acid. (2) M_2CS_3. Trithiocarbonate*. A salt of trithiocarbonic acid.
thiocarbonic t. acids* Thio acids derived from carbonic acid by substitution of oxygen atoms (see *thiocarboxy*):

$HO \cdot C\{^O_S\}H$	Monothiocarbonic acid (thiocarbonic acid)
$HO \cdot CO \cdot SH$	*S*-Monothiocarbonic acid
$HO \cdot CS \cdot OH$	*O*-Monothiocarbonic acid
$HS \cdot C\{^O_S\}H$	Dithiocarbonic acid (e.g., xanthic acid)
$CS(SH)_2$	Trithiocarbonic acid

t. acid ester An ester of a thiocarbonic acid; as, diphenylthiocarbonic ester, $PhO \cdot CS \cdot OPh$.
thiocarbonohydrazide* $(NH_2 \cdot NH)_2CS$ = 106.1.
thiocarbonyl* Carbonothioyl†. The radical =CS (analogous to =CO). **t. chloride** Thiophosgene*.
thiocarboxy-* Prefix for the HSOC− radical. Qualified by *O*− or *S*−, corresponding to the −HO(S:)C− and −HS(O:)C− forms, respectively. If it is not desired to indicate to which atom the acid H is attached, it is written $C\{^O_S\}H$.
thiocarboxylic acids* Thio acids*.
thiocarburyl chloride Thiophosgene*.
thiochrome A fluorescent quinochrome, q.v. *(quinochromes)*, which contains sulfur; an oxidation product of thiamine.
thiochromene 1,2-Benzothiopyran.
thiochromone 1,4-Benzothiopyrone.
thiocoumarin 1,2-Benzothiopyrone.
thiocresol* MeC_6H_4SH = 124.2. **ortho-~** m.15. **meta-~** A liquid. **para-~** m.43.
thioctic acid $C_3H_5S_2(CH_2)_4 \cdot COOH$ = 206.3. **(*R*)-~** Growth factor for bacteria, m.47.
thiocumazone C_6H_7ONS = 141.2. Benzodihydrothiometoxazine. Colorless crystals, m.142.
thiocumothiazone $C_8H_7NS_2$ = 181.3. Benzodihydrothiothiazine. Colorless crystals, m.166.
thiocyanate* Thiocyanide, sulfocyanate, sulfocyanide, rhodanate, rhodanide. A salt of thiocyanic acid, which contains the radical −SCN.
thiocyanic acid* HS·C:N = 59.1. Sulfocyanic acid, rhodanic acid. Colorless liquid, decomp., 200, soluble in water. **t. a. ester** R−SCN. An ester of thiocyanic acid, e.g., methyl thiocyanate, MeSCN.
thiocyanide Thiocyanate*.
thiocyanato-* Thiocyano-. Prefix for the radical NCS−. **t. dyestuffs** Dyestuffs produced from nuclear-substituted t. derivatives or aromatic amines and phenols; deeper in shade than the corresponding nonsubstituted compound.
thiocyanogen NCS·SCN = 116.2. White rhombs, m.−2, unstable, particularly in light. **t. value** An analog of the *iodine number*, q.v., in which the thiocyanate radical replaces iodine. Cf. *thiocyanometry.*
thiocyanometry Rhodanometry. Volumetric analytical methods using thiocyanates; as in the determination of silver.
thiocyanuric acid $C_3N_3S_3H_3$ = 177.3. Yellow needles, decomp. 200, soluble in alcohol.
thiodialkylamine R_2NSNR_2; as, thiodiethylamine, $Et_2NSN \cdot Et_2$.

thiodiazolidine $C_2H_6N_2S$ = 90.1. Tetrahydrothiodiazole. Heterocyclic compounds derived by saturating thiodiazoles.
thiodiazoline $C_2H_4N_2S$ = 88.1. Dihydrothiodiazole. Heterocyclic compounds derived by partially saturating thiodiazoles.
thiodiglycolic acid $S(CH_2 \cdot COOH)_2$ = 150.2. 2,2′-Thiobisacetic acid. Crystals, m.129.
thiodiphenylamine Phenothiazine*.
thioethers Alkyl and aryl *sulfides*.
thiofanox* See *insecticides*, Table 45 on p. 305.
thioformanilide Ph·NH·CSH = 137.2. Colorless crystals, m.137.
thiofuran Thiophene.
thiogenic dyes Sulfur dyes.
thioglycerol* $C_3H_8O_2S$ = 108.2. **1-~** $CH_2SH \cdot CHOH \cdot CH_2OH$. Colorless liquid, decomp. by heat. **2-~** $(CH_2OH)_2CHSH$. **1,2-~ 1,2-dithioglycerol** $C_3H_8OS_2$ = 124.2. Thick liquid, decomp. 130. **1,2-trithioglycerol** $C_3H_8S_3$ = 140.3. Heavy, odorous liquid, d.1.391.
thioglycolic acid* $HS \cdot CH_2COOH$ = 92.1. White crystals; a reagent for iron (1:10 million), and a setting agent in hair waving.
thioguanine $C_5H_5N_5S$ = 167.2. 2-Aminopurine-6(1*H*)-thione. Lanvis. Pale-yellow crystals, insoluble in water. A cytotoxic drug, used to treat leukemia (USP, BP).
thiohydroquinone* $C_6H_4(SH)_2$ = 142.2, m.98.
thiohydroxy Thiol*.
thioic t. acids* Organic compounds containing the radical —CS·OH or —CO·SH. See *thio acid, polythionic acids.* **t. anhydride** R·CS·O·CS (ring)
thioindigo $C_{16}H_8O_2S_2$ = 296.4. A permanent purple dye. m.280+ (subl.).
thioindigotic acid Sulfindigotic acid.
thioketone* R_2CS; as, 2-butanethione*, MeCSEt. Suffix: -thione*.
Thiokol Trademark for an oil-resistant, synthetic, rubberlike substance made from ethylene dichloride and sodium polysulfide.
thiol (1)* See *thiols.* (2) (Cap.) Trademark for perchlorothiol, used in the production of lubricant additives. Cf. *Thiovanic Acid.*
thiolacetic acid (1) Thioacetic acid*. (2) Thioglycolic acid*.
-thiolate* Suffix indicating a thiol salt with the S hydrogen replaced by a metal; as, sodium ethanethiolate, EtSNa.
thiole* A 5-membered ring containing a S atom. Cf. *thiols.*
thiolic t. acid *S*-thioic acid*. **t. anhydride** A compound of the type R·CO·S·CO·R (ring)
thiols* Hydrosulfides, mercaptans. Organic compounds containing the —SH group; as, methanethiol, MeSH. Prefix is mercapto-, suffix is -thiol. Cf. *hydrogen* sulfide, *thiole.* **de ~** Made commercially from the action of hydrogen sulfide on alcohols and alkenes. Used to odorize natural gas, and as catalysts for the polymerization of rubber and plastics.
thiomalic acid $HOOC \cdot CH(SH)CH_2 \cdot COOH$ = 150.2. Mercaptosuccinic acid*. A reagent for molybdenum (yellow complex).
thiomersal BP name for thimerosal.
thiometon* See *insecticides*, Table 45 on p. 305.
thion Pertaining to divalent sulfur. Cf. *-thione.* **t. dyes** Sulfur dyes. **t. kudor** A red-yellow solution of sulfur in boiling milk of lime; contains the calcium polysulfides CaS_2 to CaS_7.
thionamic acids R·NH·S(:O)OH, in which sulfur replaces the carbon of the COOH group, e.g., ethylthionamic acid, EtNH·SOOH. Cf. *carbamic acid.*
thionate Thiosulfate*. **di ~*** $M_2S_2O_6$. **penta ~*** $M_2S_5O_6$. **tetra ~*** $M_2S_4O_6$. **tri ~*** $M_2S_3O_6$.
-thione* The suffix for =S at a nonterminal C atom and thus for a thioketone; as, propanethione, $CH_3 \cdot CS \cdot CH_3$. Cf. *thioxo, thion.*
thioneine $C_9H_{15}N_3O_2S$ = 229.3. Thionene, thiazine, ergothioneine, sympectothiene; in blood and certain plants, m.290 (decomp.). Cf. *histidine.*
thionic acids Thioic acids*. See *thio acid, polythionic acids.*
thionine $C_{12}H_9N_3SCl$ = 262.7. Lauth's violet, aminophenothiazine. Greenish-black powder of metallic luster, soluble in water with violet color; a nuclear stain in microscopy, and hydrogen-ion indicator. Cf. *methylene blue.*
thiono- Thioxo-*. Cf. *thiophosgene, thione.*
thionyl* The sulfinyl* radical. Cf. *thienyl.*
thiooxamide* $(CSNH_2)_2$ = 120.2. A solid, decomp. by heat, soluble in alcohol.
thiooxybiazoline The heterocyclic compound O·CS·NH·N:CH (ring), known only as its derivatives.
thiopental sodium $C_{11}H_{17}O_2N_2NaS$ = 264.3. Thiopentone sodium. Pentothal. Yellow crystals with characteristic odor, soluble in water. A barbiturate-type, short-acting general anesthetic (USP), sometimes used with calcium carbonate (BP).
thiopentone sodium BP name for thiopental sodium.
thiophanate-methyl* See *fungicides*, Table 37 on p. 250.
thiophanes $C_nH_{2n}S$. Sulfurated hydrocarbons from crude petroleum.
thiophene* CH:CH·S·CH:CH (ring) = 84.1. Thiofuran. Colorless, benzenelike liquid, $d_{15} \cdot 1.071$, m.−37, b.84, insoluble in water; used in organic synthesis. Cf. *thienyl.* **amino ~** Thiophenine. **benzo ~** Thianaphthene. **dimethyl ~** Thioxene. **hydroxymethyl ~** $C_4H_3S \cdot CH_2OH$ = 114.2. **2-~** Thenyl alcohol. $b_{12mm}95$. **methyl ~** Thiotolene. **nitro ~** See *nitrothiophene.* **tetrahydro ~** Butene sulfide. **t. alcohol** $C_4H_3S \cdot CH_2OH$ = 114.2. b.207. **t. diiodide** $C_4H_2I_2S$ = 335.9. T. biniodide. Yellow leaflets, m.40, insoluble in water; an antiseptic. **t.carboxylic acids*** Acids derived from thiophene. (1) $C_4H_3S \cdot COOH$ = 128.1. 2-Thiophenecarboxylic acid*, thenoic acid*. m.126. (2) $C_4H_2S(COOH)_2$ = 172.2. 2,4-Thiophenedicarboxylic acid*, m.118. **t.methanol** T. alcohol. **t. sulfonate** $C_4H_3S \cdot HSO_3$ = 164.2. colorless crystals, insoluble in water. The sodium salt is an antiseptic. **t. tetrabromide*** C_4Br_4S = 399.7. Yellow crystals, m.112, insoluble in water, soluble in alcohol.
thiophenine $C_4H_3S \cdot NH_2$ = 99.2. Aminothiophene. Yellow liquid, insoluble in water.
thiophenol* PhSH = 110.2. Phenthiol, phenylmarcaptan, phenylthiol*, phenyl sulfhydrate. Colorless liquid, b.168, insoluble in water; used in organic synthesis. **aminovinyl ~** C_8H_7SN = 149.2. Colorless liquid, b.238, insoluble in water.
thiophenyl The thiophenoxy* radical, PhS—, from thiophenol. Cf. *phenoxy.* **t.acetone** Colorless liquid, b.266, soluble in alcohol.
thiophosgene* $CSCl_2$ = 115.0. Thiocarbonyl dichloride*, carbon thionyl chloride. Red liquid, b.73, insoluble in water.
thiophosphates Compounds derived from phosphoric acids by substituting divalent sulfur for one or more oxygen atoms; as, monothiophosphates, M_3PO_3S.
thiophosphoric acid* $PS(OH)_3$ = 114.1. A solid, decomp. by water, soluble in alcohol.
thiophosphorous anhydride Phosphorus trisulfide*.
thiophosphoryl* The radical PS≡, derived from

thiophosphoric acid. **t. bromide*** $PSBr_3$ = 302.7. Yellow solid, m.38, decomp. by water. **t. chloride*** $PSCl_3$ = 169.4. Colorless liquid, b.126, decomp. by water. **t. triamide*** $PS(NH_2)_3$ = 111.1. White solid, decomp. by heat or water.

thiophthalide* $C_6H_4 \cdot CO \cdot S \cdot CH_2$ = 150.2. Colorless crystals, m.60.

thiophthene $C_6H_4S_2$ = 140.2. Thienothiophene. Colorless liquid, b.225, insoluble in water, soluble in alcohol.

thiopyrophosphoryl bromide $P_2S_3Br_4$ = 477.7. Yellow liquid, decomp. by water or heat.

thioresorcinol* $C_6H_4(SH)_2$ = 142.2. Yellow powder, m.27, insoluble in water.

thioridazine hydrochloride $C_{21}H_{26}N_2S_2 \cdot HCl$ = 407.0. Mellaril, Melleril. White, bitter crystals, m.161, soluble in water; a tranquilizer of the phenothiazine group, with the piperidine group at 10 and the S·Me at 2. Used for schizophrenia and other psychiatric illness (BP).

Thiosa Trademark for tetramethylthiuram disulfide; a dip solution to control potato scurf.

thiosemicarbazide* $H_2N \cdot CS \cdot NH \cdot NH_2$ = 91.1. Colorless needles, m.181, soluble in water; used in organic synthesis and as reagent for aldehydes and ketones.

thiosinamine $C_4H_8N_2S$ = 116.2. Prisms, m.78. **t. ethyl iodide** $C_4H_8N_2S \cdot C_2H_5I$ = 272.1. Tiodine. Colorless crystals, soluble in water; used in medicine to soften scar tissue.

thiostannates M_2SnS_3. A salt of thiostannic acid.

thiostannic acid H_2SnS_3 = 216.9. Yellow, unstable crystals.

thiosulfate* $M_2S_2O_3$. A salt of thiosulfuric acid*. Cf. *sodium thiosulfate, thionate.*

Thiosulfil Trademark for sulfamethizole.

thiosulfite* A salt, $M_2S_2O_2$, of thiosulfurous acid.

thiosulfuric acid $H_2S_2O_3$ = 114.0. An unstable acid, decomp. readily to sulfur and sulfurous acid.

thiotepa $P \cdot S \cdot [N(CH_2)_2]_3$ = 189.2. Tris(1-aziridinyl)phosphine sulfide. Highly poisonous, white flakes, soluble in water, m.55; an alkylating agent used for disseminated malignant disease (USP, BP).

thiotolene $C_4H_3S \cdot Me$ = 98.2. Methylthiophene. **2-~**, *α*-**~** b.111. **3-~**, *β*-**~** b.115.

thiouracil $NH \cdot CS \cdot NH \cdot CO \cdot CH{:}CH$ = 128.1. Parent compound of antithyroid drugs; as, propylthiouracil.

thiouramil $NH \cdot CO \cdot NH \cdot C \cdot SH{:}CH \cdot CO$ = 144.1.

thiourazole $NH \cdot CO \cdot NH \cdot NH \cdot CS$ = 117.1. Colorless crystals, m.177, soluble in alcohol.

thiourea* $NH_2 \cdot CS \cdot NH_2$ = 76.1. Sulfourea, sulfocarbamide, thiocarbamide. Colorless prisms, m.180, slightly soluble in water or ether. Used in organic synthesis and as reagent for bismuth. **allyl ~** See *allyl thiourea.* **benzyl ~** See *benzylthiourea.* **diphenyl ~** Thiocarbanilide. **glycol ~** See *glycolthiourea* under *glycol.* **iso ~*** $NH_2 \cdot C{:}NH(SH)$. Pseudo ~. The tautomer of t. **phenyl ~** See *phenylthiourea.* **pseudo ~** Iso ~*.

thioureido* The radical $NH_2 \cdot CS \cdot NH-$, from thiourea.

thiourethane (1) $NH_2 \cdot CO \cdot SEt$ = 105.2, m.108 (sublimes), insoluble in water. (2) $NH_2 \cdot CS \cdot OEt$. m.40.

Thiovanic Acid Trademark for vacuum-distilled thioglycolic acid (approximately 75% with some dithioglycolic acid). Colorless liquid with faint sulfur odor, soluble in water; used as an analytical reagent, a depilatory, and in hair waving.

thioxanthene* $C_6H_4 \cdot S \cdot C_6H_4 \cdot CH_2$ = 198.3. Methylene diphenylene sulfide. Colorless crystals, m.128; used in organic synthesis.

thioxanthone $C_6H_4 \cdot S \cdot C_6H_4 \cdot CO$ = 212.3. 9-Oxothioxanthene. Colorless crystals, m.207.

thioxene $C_4H_2Me_2S$ = 112.2. Dimethylthiophene. **2,3-~** b.139. **2,4-~** b.139. **2,5-~** b.137. **3,4-~** b.145.

thioxo*- Thiono-. Prefix for =S in a thioketone.

Thiozell Trademark for a protein synthetic fiber.

thiozon Former name for the allotrope of sulfur, S_3.

thiram* Tetramethyl*thiuram* disulfide; a fungicide.

third law of thermodynamics See *thermodynamics, entropy.*

third order See *reaction order, tertiary.*

thiuram (1) The radical $R_2N \cdot CS \cdot -$. (2) $(Me_2N \cdot CS \cdot S)_2$ = 240.4. Tetramethylthiuram disulfide*. Thiram, TMT, TMTD. Bis(dimethylthiocarbamoyl) disulfide. Yellow crystals, m.155. **ethyl ~** $(Et_2N \cdot CS \cdot S)_2$ = 296.5. Bis(diethylthiocarbamoyl) disulfide, tetraethylthiuram disulfide*. Yellow crystals, m.70.
t. disulfides $R_2N \cdot CS \cdot S \cdot S \cdot CS \cdot NR_2$. **t. monosulfides** $R_2N \cdot CS \cdot S \cdot CS \cdot NR_2$. Cf. *xanthate.*

thiurea Thiourea*.

thiuret $C_6H_7N_3S_2$ = 185.3. Colorless crystals, insoluble in water.

thixotrope A colloid whose properties are changed by mechanical treatment; as, clay.

thixotropic Pertaining to thixotropy. **t. viscosity** The anomalous viscosity of sols which are about to gel.

thixotropy The property of certain gels of becoming fluid on agitation and coagulating again when at rest (as, a suspension of ferrous hydroxide); due to the mechanical destruction of the *zones*, q.v., of oriented moleules. **inverse ~, negative ~** Dilatancy.

thiylation Introduction of S-containing groups into organic compounds.

Thomas **T., Sidney Gilchrist (1850–1885)** British technologist noted for his improvement of the Bessemer process. **T. meal** Basic slag. **T. process** The use of burned dolomite as a converter lining, which reacts with the phosphorus of pig iron. **T. slag** The finely powdered, phosphatic basic slag obtained in the T. process; a fertilizer.

Thompson, Benjamin See *Rumford.*

Thompson process Electric-arc welding with the metal to be welded as electrode.

Thomsen **T., Hans Peter Jürgen Julius (1826–1909)** Danish chemist, noted for his work in thermochemistry. **T. process** The manufacture of sodium carbonate and aluminum oxide by heating powdered cryolite with lime, leaching out the sodium aluminate, and decomposing it into aluminum hydroxide and sodium carbonate by means of carbon dioxide.

thomsenolite $NaCaAlF_6 \cdot H_2O$. A native fluoride.

Thomson **T., Sir Joseph John (1856–1940)** British physicist, Nobel prize winner (1906); noted for work in theoretical physics, discoverer of the electron. **T., Thomas (1773–1852)** British chemist, noted for his textbooks. **T., William** Lord *Kelvin.*

thomsonite A gem variety of zeolite; a hydrous aluminum calcium sodium silicate.

thoria Thorium oxide.

thorianite A complex mineral; thorium oxide 70, uranium oxide 10-12%, and rare earths.

thorin Thoron(ol). APNS. Disodium salt of 4-[(2-arsonophenyl)azo]-3-hydroxy-2,7-naphthalenedisulfonic acid. A reagent for thorium (red precipitate in presence of dilute hydrochloric acid).

Thorite (1) Trademark for a nonshrink patching mortar. (2) (Not cap.) $ThSiO_4$. Freyalite. Orangite. A native thorium silicate, containing Ca, Fe, Mn, and V.

thorium* Th. A radioactive metal and element, at no. 90, in monazite and thorite; discovered (1828) by Berzelius. At. wt.

of isotope with longest half-life (^{232}Th) = 232.04. A gray, amorphous or crystalline, soft mass, readily burning in air to thorium oxide, d.11.3, m.1845, b.4787.; insoluble in water, alcohol, alkalies, or acids, soluble in aqua regia. Used to produce Th-Mg alloys. Cf. *radioactive elements.* **meso ~** See *mesothorium.* **radio ~** See *radiothorium.*

t. anhydride T. oxide. **t. chloride** $ThCl_4$ = 373.9. Colorless, deliquescent plates, m.820 (sublime), soluble in water, used in incandescent burners. **t. dioxide*** T. oxide. **t. emanation** See *thoron.* **t. hydroxide*** $Th(OH)_4$ = 300.4. Colorless, gelatinous substance, insoluble in water. **t. lead** ThD. The isotope ^{208}Pb. **t. nitrate*** $Th(NO_3)_4 \cdot 3H_2O$ = 534.1. White, hygroscopic crystals or granules, decomp. to t. oxide by heat. **t. oxalate*** $Th(C_2O_4)_2$ = 408.1. White crystals, insoluble in water. **t. o. hexahydrate** White powder. **t. oxide*** ThO_2 = 264.0. Thoria, t. dioxide. White, infusible powder, insoluble in water; main constituent of the ash of incandescent gas mantles. **t. picrate*** $Th(C_6H_2N_3O_7)_4 \cdot 10H_2O$ = 1325. Yellow powder, highly explosive. **t. series** See *radioactive elements.* **t. sulfate*** $Th(SO_4)_2 \cdot 4H_2O$ = 496.2. Colorless crystals, slightly soluble in water. **t. X.** Early name for radium-224.

thorn apple Stramonium.

Thornel Trademark for continuous filaments of graphite. See *whiskers.*

thorogummite $UO_2 \cdot 3ThO_2 \cdot 3SiO_2 \cdot 6H_2O$. A native silicate.

thoron (1) Tn, ThEm. Early name for radium-220. (2) Thorin.

thoronol Thorin.

thoroughwort Eupatorium.

Thorpe T., Jocelyn Field (1872–1940) British chemist noted for organic synthesis; coeditor of *Thorpe's Dictionary of Applied Chemistry,* the original author of which was: **T., Thomas Edward (1845–1925)** British chemist noted for atomic weight determinations and studies on gold.

thortveitite A mineral, chiefly scandium silicate (37–42% Sc_2O_3).

Thoulet's solution A concentrated solution of potassium and mercuric iodides in water, d.3.17; used to determine density by the suspension method.

thread (1) A *string,* q.v. (2) A unit of worsted yarn measure, 36 inches.

threo- Describing isomers having groups on opposite sides (e.g., D-threose), as distinct from *erythro-* on the same side (e.g., D-erythrose).

```
      CHO                 CHO
HO ----+---- H      H ----+---- OH
       |                  |
H  ----+---- OH     H ----+---- OH
      CH2OH              CH2OH
   D-Threose          D-Erythrose
```

threonine* $Me \cdot CHOH \cdot CH(NH_2)COOH$ = 119.1. Thr*. 2-Amino-3-hydroxybutanoic acid*. An amino acid essential for growth. See *amino acids.*

threose $CHO \cdot (CHOH)_2 \cdot CH_2OH$ = 120.1 A tetrose and isomer of erythrose, in D and L forms.

threshold Schwellenwert. **t. limit value** See *TLV.* **t. treatment** The process of stopping precipitation at the threshold of its occurrence. Thus certain phosphates, e.g., Graham salt, do not stop the formation of nuclei of calcium carbonate in water, but do inhibit growth of the nuclei. Used in water treatment.

thrombin* The enzyme of the blood that activates fibrinogen to fibrin, thus causing coagulation. T. from bovine blood is a local hemostatic (USP). **dried human ~** T. prepared from pooled human plasma; used as a hemostatic (BP).

thrombocytes Blood *platelets.*

thrombokinase Coagulation factor Xa*. A blood enzyme that activates prothrombin to *thrombin,* q.v.

thromboplastin Factor III. Blood substances that take part in the blood clotting mechanism. Used, in physiological salt solutions, as a hemostatic.

thrombosis The formation of clots in the blood vessels. Cf. *prothrombin.*

throwing power The efficiency of an apparatus or electrolyte for the electrodeposition of metals. $(L - M)100/(L + M - 2)$; where L is the ratio of the distance between the anode and 2 half-cathodes in the liquid, and M is the ratio of the weights of metal deposited on the 2 half-cathodes.

THT Tetrahydrothiophene.

thuja Thuya, thuva. Arbor vitae, white cedar, tree of life. The dried, young tops of *Thuja occidentalis* (Pinaceae), N. America and Europe. It contains an essential oil and tannin. **t. oil** Oil of white cedar. Sharp, camphorlike odor and taste, d.0.925, b.190, soluble in alcohol. It contains pinene, fenchene, thujone, and carvone.

thujane* $MeC_6H_8 \cdot CHMe_2$ = 138.3. 4-Methyl-1-(1-methylethyl)bicyclo[3.1.0]hexane. Sabinane. Mobile liquid, b.157.

thujene* $CH_2{:}C_6H_7 \cdot CHMe_2$ = 136.2. 4(10)-Thujene, sabinene, tanacetene. **(1*R*,5*R*)- ~** $b_{160mm}164$. A terpene from thuja and other essential oils. **hydroxy ~** Thujenol.

thujenol* $C_{10}H_{16}O$ = 152.2. 4(10)Thujen-3-ol, absinthol, thujol. b.208; in *Juniperus* and *Sabina* spp.

thujenyl* The radical $C_{10}H_{15}-$, from thujene.

thujetic acid $C_{28}H_{22}O_{13}$ = 566.5. Obtained by heating thujin with barium hydroxide.

thujetin $C_{14}H_{14}O_8$ = 310.3. Colorless crystals formed by hydration of thujigenin.

thujic acid $HOOC \cdot C_7H_5 \cdot Me_2$ = 164.2. 5,5-Dimethyl-1,3,6-cycloheptatriene-l-carboxylic acid. m.89. From the heartwood of western red cedar, *Thuja plicata.*

thujigenin $C_{14}H_{12}O_7$ = 292.2. A decomposition product of thujin. Cf. *thujetin.*

thujin $C_{20}H_{22}O_{12}$ = 454.4. A glucoside and coloring matter of thuja. Yellow crystals, insoluble in water.

thujol Thujenol*.

thujone $C_{10}H_{16}O$ = 152.2. Thalviol, 3-oxosabinane. A dextrorotatory, colorless liquid, d.0.913, b.203, from the essential oil of *Thuja* and *Salvia* species, insoluble in water.

thujorhodin Rhodoxanthin.

thujyl* The radical $C_{10}H_{17}-$, from thujane.

thulite Zoisite.

thulium* Tm = 168.9342. A rare-earth metal and element, at. no. 69, valency 3. Discovered by Cleve (1879) and independently by Soret. m.1545, b. ca. 1720. **t. chloride*** $TmCl_3$ = 275.3. Colorless crystals, soluble in water. **t. oxalate*** $Tm_2(C_2O_4)_3 \cdot 6H_2O$ = 710.0. Greenish-white precipitate, soluble in alkaline oxalate solutions. **t. oxide*** Tm_2O_3 = 385.9. White powder, insoluble in water.

thuringite $4(AlFe)_2O_3 \cdot 7FeO \cdot 6SiO_2 \cdot 9H_2O$. An amorphous mineral.

thus Olibanum.

thuva Thuja.

thyalkoid Member of a group of structures in green plants and in blue-green algae which contain photosynthetic pigments.

Thylox process Removal of sulfur dioxide from coal gas in ammonium thioarsenate solution, which is then warmed and aerated to precipitate the sulfur, which is purified by distillation.

thymamine $C_{22}H_{40}O_5N_6$ = 468.6. A protamine from the thymus gland.

thyme Thymus. The dried tops of *Thymus vulgaris*, garden thyme (Labiatae); an aromatic carminative and flavoring. **t. camphor** Thymol*. **t. oil** d.0894–0.930, soluble in alcohol; many grades.
thymegol See *egols.*
thymene $C_{10}H_{16}$ = 136.2. A terpene from thyme oil. Colorless, aromatic liquid, d.0.868, b.160, insoluble in water. Cf. *thymine.*
thymidol Methyl propyl phenyl menthol. A condensation product of thymol and menthol; an antiseptic mouthwash.
thymine NH·CO·NH·CH:CMe·CO = 126.1. 5-Methyluracil. Colorless crystals, decomp. 335; a constituent of DNA. Cf. *thymene.*
thymohydroquinone $C_{10}H_{14}O_2$ = 166.2, m.140, soluble in water.
thymol* $MeC_6H_3OH \cdot CHMe_2$ = 150.2. Thyme camphor. 6-Isopropyl-3-cresol*. Colorless crystals, m.50, slightly soluble in water; a reagent for indican, and an antiseptic. Used in dusting powders and mouth washes; also in dentistry for temporary fillings (NF, BP). **acetyl ~** See *acetylthymol.* **hexahydro ~** Menthol*.
t. blue $C_{27}H_{30}O_5S$ = 466.6. T.sulfonephthalein. Black powder; a pH indicator, changing from pink (1.5) to yellow (2.8–8) to blue (9.6). **bromo ~** See *indicator.* **t. carbamate** $C_{10}H_{13}OCO \cdot NH_2$ = 193.2. T.urethane, thymotal, tyratol, thymolcarbamic ether. Colorless crystals, soluble in water, decomp. by hydroxides. **t.phthalein** A pH indicator changing at 9.8 from colorless (acid) to blue (basic). **t. salicylate** Salithymol. **t.sulphonephthalein** T. blue. **t.urethane** T. carbamate.
thymoquinone $C_{10}H_{12}O_2$ = 164.2. Colorless crystals, m.46, soluble in water.
thymotal Thymol carbamate.
thymot(in)ic **t. acid** $C_6H_2(Me)C_3H_7(OH)COOH$ = 194.2. ***o*- ~** 2-Hydroxy-3-isopropyl-6-methylbenzoic acid*, m.127. ***p*- ~** 4-Hydroxy-5-isopropyl-2-methylbenzoic acid*, m.157. **t. anhydride** $C_{11}H_{12}O_2$ = 176.2. Colorless crystals, m.174.
thymus (1) Thyme. (2) An organ of lymphoid tissue in the neck and chest of an infant or young animal, which disappears before maturity is reached. Essential for development of defense against disease. Cf. *thymamine.*
thymyl The monovalent radical HC:CMe·CH:CH·C(*i*-Pr):C—, from thymol.
thynnin A protamine (75% arginine) from the sperm of *Thymnus thynnus*, the Pacific Ocean tunny fish.
thyratron A gas-filled, hot-cathode, grid-controlled valve rectifier and self-amplifier.
thyresol $C_{15}H_{23}OMe$ = 234.4. Santalol methyl ester. Colorless liquid; a santalol substitute.
thyristor Silicon-controlled rectifier.
thyroglobulin A globulin containing thyroxine and triiodothyronine, the thyroid hormones.
thyroid Thyroideum siccum. Dried thyroid glands. Yellow powder, slightly soluble in water; used to treat hypothyroidism (USP). **t. gland** A large gland in front and on either side of the windpipe of man and animals; it produces thyroxine and other hormones, e.g., triiodothyronine.
thyroiodine Levothyroxine.
thyrooxindole Levothyroxine.
thyrotrophin A hormone of the anterior lobe of the pituitary gland; it stimulates the thyroid gland.
thyroxine BP name for levothyroxine.
Ti Symbol for titanium.
tiemannite HgSe. Native mercurous selenide.
tiferron Disodium-1,2-dihydroxybenzene-3,5-disulfonate. Colorimetric reagent for ferric iron.
tight ion pair See tight *ion pair* under *ion.*
tiglaldehyde MeCH:CMe·CHO = 84.1. Guaiol. 2-Methyl-2-*butenal**.
tiglic Pertaining to tiglium. **t. acid** 2-Methyl-2-*butenoic acid**. **t. aldehyde** 2-Methyl-2-*butenal**.
tiglium Croton seeds. The dried seeds of the croton oil plant, *Croton tiglium* (Euphorbiaceae), southeast Asia. It contains croton oil, (*E*)-2-methyl-2-butenoic acid, and crotonolic acid; a drastic purgative.
tikitiki Darak.
til Sesame oil.
tile ore Cuprite.
tilia Linden flowers, lime flowers. The dried flowers of *Tilia europea* (Tiliaceae), Europe. It contains an essential oil; an antispasmodic.
Tillman's reagent 2,6-Dichlorophenol indophenol; a quantitative reducing agent for ascorbic acid.
timbo The bark of *Serjania curassavica* (Sapindaceae), Brazil. It contains an alkaloid; used locally as a fish poison.
timbonine An alkaloid from timbo.
time (1) The fourth dimension: a measure of duration of an object, phenomenon, or event. The SI unit is the second, i.e., the duration of 9,192,631,770 periods of the radiation corresponding to the transition between 2 hyperfine levels of the ground state of the cesium-133 atom. 1 s = 1,000 σ. Sensitive measurements of time are made with a tuning fork (registers 1/1,000–1/5,000 s). Longer periods are measured in *days*, q.v., and *years*, q.v. See Table 95. (2) The relative hour of a day, which depends on geographical longitude and the position of the earth relative to the sun. Cf. *year.* **ephemeris ~** The definition of the second, based on the motion of the moon, as distinct from physical t., based on the atomic frequency standards. **real ~** See *real time.* **response ~** See *response time.*
t. constant* Relaxation time*. For "first-order systems" (i.e., relatively straightforward ones), the time it takes an output to reflect 63.2% of a step change in input value. See *symbols*, Table 88, Group *B* on p. 566. **t.-lapse photography** The cinephotography of very slowly moving or developing processes or objects (as bacteria), and subsequent projection at

TABLE 95. MAGNITUDES OF TIME

Period described		Seconds
Sun diminishing half its mass		1.5×10^{19}
Uranium-238 half-life	4,500,000,000 y	1.4×10^{17}
Earth's age	4,000,000,000 y	1.2×10^{17}
Earth, quaternary period	2,000,000 y	6.3×10^{13}
Dawn of history	6,000 y	1.9×10^{11}
Birth of Christ	1,987 y	6.3×10^{10}
Oxygen discovered	212 y	6.7×10^{9}
The average human life	45 y	1.4×10^{9}
Earth's period around sun	365.2 d	3.1×10^{7}
Sun rotation	24.6 d	2×10^{6}
Earth rotation	23 h 56 min 4 s	8.6×10^{4}
1 hour		3.6×10^{3}
Lead-214 half-life	27 min	1620
Sound vibration		4×10^{-3}
Oscillograph		1×10^{-6}
High-frequency cycle		1×10^{-7}
Beryllium-8 half-life		2×10^{-16}
Cosmic ray oscillation		1×10^{-22}

a higher speed. **t. series analysis** The mathematical analysis of time series; i.e., of variables plotted against time.

t.-sharing Part-time use of a computer, often by a terminal communicating with it over telephone lines.

timolol maleate $C_{13}H_{24}O_3N_4S \cdot C_4H_4O_4$ = 432.5. Blocadren, Timoptic, Timoptol. White crystals, m.200, soluble in water. A β-adrenoceptor blocking agent, used to control heart rhythm and hypertension. See *nervous system.*

Timoptic, Timoptol Trademarks for timolol maleate.

tin* Sn = 118.710. Stannum. A metal and element, at. no. 50. Silver-white, tetragonal or rhombic metal, d.7.29, m.232, b.2270, insoluble in water, decomp. in dilute acids or concentrated alkalies. Crystalline forms:

Sn_α	18°C	Sn_β	230°C	Sn_γ
Gray tin	⇌	White	⇌	Rhombic
d.5.8		d.7.3		d.6.6
Brittle		Malleable		Brittle

T. has a valency of 2 or 4 and forms 2 series of compounds (see table). The t.-organic compounds (organotins, R_4Sn, R_3SnX, etc.) are derived mainly from tetravalent t. T. is used in alloys for t.foils, solders, utensils, type metal, dental alloys; and in the manufacture of t. salts. **alpha- ~** Gray t. **beta- ~** White t. **black ~** A treated t. ore: 60–70% stannic oxide. **block ~** Cassiterite. **butter of ~** Stannic chloride. **cry of ~** The sound heard on bending a stick of t. **diethyl ~*** Et_2Sn = 176.8. T. diethyl, stannous ethide. An oil, d.1.654, decomp. by heat, insoluble in water. **diethyldimethyl ~*** Et_2Me_2Sn = 206.9. Colorless liquid, b.145, soluble in alcohol. **flowers of ~** Stannic oxide. **gray ~** Alpha-t. An enantiotropic form of white t. produced as a gray powder when white t. is cooled to −40°C. It is the cause of *tin* disease, q.v. **stream ~** Tinstone, in an alluvial deposit. **tetraethyl ~*** Et_4Sn = 234.9. T. tetraethyl, stannic ethide. Colorless liquid, b.181, insoluble in water. **tetramethyl ~*** Me_4Sn = 178.8. T. tetramethyl, stannic methide. Colorless liquid, b.78, insoluble in water. **tetraphenyl ~*** Ph_4Sn = 427.1. T. tetraphenyl, stannic phenide. Colorless crystals, m.226, soluble in ether. **triethyl ~** Et_6Sn_2 = 411.7. T. triethyl, distannic ethide. Colorless liquid, b.270, insoluble in water. **triethylhydroxide ~*** Et_3SnOH = 222.9. T. triethyl hydroxide. Colorless crystals, m.66. **white ~** Tetragonal, ordinary t., β-tin. The usual form of malleable tin.

Oxidation state	Cations	Anions
2	Sn^{2+}; tin(II)* or (2+)*; stannum(II)* or (2+)*; stannous	SnO_2^{2-}; stannate(II)*; stannite
4	Sn^{4+}; tin(IV)* or (4+)*; stannum(IV)* or (4+)*; stannic	SnO_3^{2-}; stannate(IV)*

t. acetate Stannous acetate. **t. alkyls** R_4Sn. See *stannanes, tetraethyltin, tetramethyltin* under *tetramethyl*, etc. **t. alloys** See *aluminum, cobalt, lead, nickel, silver;* also *babbitt, britannia metal, pewter, rhine metal, solder.* **t. anhydride** Stannic oxide. **t. ash** Stannic oxide. **t. bath** Molten t. in which metals are dip-coated. **t. bisulfide** Stannic sulfide. **t. bronze** Stannic sulfide. See *mosaic gold.* **t. chlorides** (1) Stannous chloride. (2) Stannic chloride. **t. chloride solution** A solution of 5 pts. stannous chloride in 1 pt. hydrochloric acid. Yellow, refractive liquid; a reagent. **t. chromate.** See *stannic chromate* or *stannous chromate.* **t. citrate** Stannous citrate. **t. dichloride** Stannous chloride. **t. diethyl*** Diethyl ~ * **t. dioxide** Stannic oxide. **t. disease** Transformation into gray t. with resulting brittleness and loss of metallic luster. **t. disulfide*** Stannic sulfide. **t.foil** A thin sheet of t. or t. alloy; a wrapper. **t. glass** (1) A glass that contains t.; as, certain flint glasses. (2) Bismuth (obsolete). **t. glaze(II)** Opaque glaze formed on pottery by t. salts. **t. hexafluorosilicate*** $SnSiF_6$ = 260.8. Stannous fluosilicate. Colorless prisms, soluble in water. **t. hydroxide** Stannic hydroxide. **triethyl ~** Et_3SnOH = 222.9. Stannic ethyl hydroxide. Colorless crystals, m.66. **triphenyl ~** A fungicide used to combat blight of root crops. **t. iodide** Stannic iodide. **t. minerals** Chiefly:

Cassiterite	SnO_2
Stannite	Cu_3FeSnS_4
Teallite	$PbSnS_2$

t. monosulfide* Stannous sulfide. **t. monoxide*** Stannous oxide. **t. mordant** A salt used in dyeing; as, stannous chloride. **t. nickel** The alloy Sn 65, Ni 35%, hard, acid-resistant, with pink tinge; used to electroplate small precision machine parts. **t. oxalate** Stannous oxalate. **t. oxide** See *stannous oxide, stannic oxide.* **t. oxychloride** Stannous chloride. **t. oxymuriate** Commercial stannic chloride. **t. peroxide** Stannic oxide. **t. pickling** Immersing iron in dilute acid prior to a t. bath. **t.plate** Steel sheet, covered with a film of t., usually about 1.5% t.; used for cans. **t. protochloride** Stannous chloride. **t. protoxide** Stannous oxide. **t. pyrites** Stannite. **t. salt** Stannous chloride. **t. spar** Cassiterite. **t. sponge** Argentine. **t.stone** Cassiterite. **t. sulfate** Stannous sulfate. **t. tetraethyl*** Tetraethyl ~ *. **t. tetramethyl*** Tetramethyl ~ *. **tetraphenyl*** Tetraphenyl ~ *. **t. triethyl*** Triethyl ~ *. **t. triethyl hydroxide** See *tin hydroxide* above.

Tinactin, Tinaderm Trademarks for tolnaftate.

tincal Borax.

tincture A medicated liquid made by extraction of a drug; generally weaker than a fluidextract. **alcoholic ~** A solution made by percolating 10 g of a drug with 100 mL alcohol, or macerating, in the case of a resinous drug. **aqueous ~** An extract made with water.

t. of iodine A solution of iodine 7 and potassium iodide 5 g, in 5 mL of water, made up to 100 mL with alcohol; an antiseptic.

tinder Flammable cloth or wood used for initiating a fire.

tinevelly senna India *senna.*

tinkal Borax.

tinning Coating with tin.

tinstone Cassiterite.

tintometer A device for estimating the intensity of a colored solution by comparison with standard solutions, or colored glass slides; as, the Lovibond t.

tiodine Thiosinamine ethyl iodide.

Tiolan Trademark for a protein synthetic fiber.

tiron Disodium 1,2-dihydroxybenzene. A quantitative reagent for iron, titanium, and aluminum.

tissue In biology, a structure composed mainly of one type of cell. **t. culture** An asexual means of reproduction of plants, by culturing a small piece of plant *in vitro.*

titanate* Indicating titanium as the principle atom(s) in an anion; as a salt of titanic acid, e.g., $M_2TiO_3^-$, metatitanate, and $M_4TiO_4^-$, orthotitanate.

titanellow Yellow titanium oxide, used for making ivory shades in ceramics. Cf. *titan yellow.*

titania Titanium dioxide*.

titanic Titanium(IV)* or (4+)*. Pertaining to tetravalent titanium. **t. acid** (1) H_2TiO_3 = 97.9. **meta ~** $TiO(OH)_2$.

Colorless precipitate obtained by boiling titanium sulfate solution, insoluble in water, soluble in alkalies, to form salts. (2) H_4TiO_4 = 115.9 **ortho ~** $Ti(OH)_4$. White precipitate on adding acids to metatitanates, soluble in excess of acid or alkali and forming, with the latter, the titanyl radical. **t. anhydride** Titanium dioxide*. **t. chloride** Titanium tetrachloride*. **t. hydroxide** Titanic acid. **t. iron** Ilmenite. **t. oxide** Titanium dioxide*.

titaniferous Containing titanium. **t. iron ore** Ilmenite.

titanite Sphene.

titanium* Ti = 47.88. A metallic element of the carbon group, at. no. 22, discovered by Gregor (1789). Dark-gray, amorphous metal, d.4.5, m.1660, b.3290, insoluble in water, soluble in warm hydrochloric acid. Used (as ferrotitanium) as a cleaning and deoxidizing agent for molten steel; and in alloys with copper, bronze, and other metals. Cf. *konel.* It has a valency of 2, 3, and 4 and forms several ions:

t.(III)*, t.(3+)*, titanous	Ti^{3+}
t.(IV)*, t.(4+)*, titanic	Ti^{4+}
metatitanate*	TiO_3^{2-}
orthotitanate*	TiO_4^{4-}
titanyl	TiO^{2+}

t. bromide $TiBr_4$ = 367.5. Orange crystals, m.39, decomp. by water. **t. chlorides** See *titanium dichloride, titanium trichloride, titanium tetrachloride.* **t. dichloride*** $TiCl_2$ = 118.8. Titanous chloride. Black, hygroscopic powder, burns in air, decomp. by water. **t. dioxide*** TiO_2 = 79.9. Titania. Colorless to black tetragons or rhombs, d.3.70–4.26, m.1560, insoluble in water, soluble in alkalies or concentrated sulfuric acid. Used as an opaque, white pigment for paint, inks, shoe polish, soap, plastics, rubber goods, ceramic glazes, and paper; also a protective and antipruritic for the skin (USP, BP). **t. hydroxide** (1) $Ti(OH)_3$ = 98.9. Titanous hydroxide. Blue-black powder, insoluble in water. (2) $Ti(OH)_4$ = 115.9. Titanic hydroxide. White, insoluble solid. **t. iodide** TiI_4 = 555.5. Titanic iodide. Red octahedra, m.150, soluble in water. **t. minerals** T. is the 9th most abundant element, and is widely diffused in igneous and sedimentary rocks, chiefly as:

Ilmenite	$FeTiO_3$
Pseudobrookite	$Fe_4(TiO_4)_3$
Perofskite (perowskite)	$CaTiO_3$
Titanite	$CaTiSiO_5$
Rutile, brookite, anatase, octahedrite	TiO_2

t. nitrate Titanyl nitrate. **t. nitride** Ti_2N_2 = 123.8. Dark powder prepared from rutile and nitrogen in the electric furnace; used for dusting molds before casting steel, and for coating cutting tools. **t. oxalate** $Ti_2(C_2O_4)_3 \cdot 10H_2O$ = 540.0. Yellow prisms, soluble in water. **t. oxides** See *titanium sesquioxide, titanium dioxide, titanium trioxide.* **t. peroxide** T. trioxide*. **t. sesquioxide*** Ti_2O_3 = 143.8. Black powder, insoluble in water. **t. sulfates** (1) $Ti(SO_4)_3$ = 336.0. Colorless crystals, soluble in water. (2) $Ti_2(SO_4)_3$ = 383.9. Green crystals, insoluble in water. **t. tetrachloride*** $TiCl_4$ = 189.7. Titanic chloride. Colorless liquid, b.136, soluble in water; a mordant. **t. trichloride*** $TiCl_3$ = 154.2. Violet crystals, decomp. 440, soluble in water; a reducing agent. **t. trioxide*** TiO_3 = Titanellow. **t. white** White pigment; t. dioxide 26.5, barium sulfate 73.5%.

titanous Titanium(III)* or (3+)*. Pertaining to trivalent titanium. **t. chloride** Titanium trichloride*.

Titanox Trademark for titanium-containing products, including pigments.

titan yellow Clayton yellow, thiazol yellow. Yellow dye; used to detect caustic alkalies (red) and magnesium (red). Cf. *titanellow.*

titanyl The radical TiO^{2+}. **t. nitrate** $TiO(NO_3)_2$ = 187.9. Titanium nitrate. Colorless, soluble crystals, in water. **t. sulfate** $TiOSO_4$ = 159.9. White needles, soluble in acidic water; a mordant.

titer (1) A standard test for oils and waxes in which the solidifying point (titer) of the fatty acids resulting from saponification is determined. (2) The number of mL/L by which a normal solution differs from a standard. (3) The normality of a solution as determined by titration with a standard. (4) The g/mL of an element, radical, or compound in a standard solution. (5) The concentration or activity of antibody in serum. Generally expressed as a dilution which binds a given proportion of a fixed amount of the radioactive form of the compound.

titrant A standard solution used for titration.

titrate (1) To analyze by a volumetric method. (2) The solution being titrated. (3) In medicine, to administer varying levels of a medication within its therapeutic range until desired results are obtained.

titration The volumetric determination of a constituent in a known volume of a solution by the slow addition of a standard reacting solution of known strength until the reaction is completed, as indicated by color change (indicators) or as indicated electrometrically. **chelatometric ~** A complexometric t. involving the use of a chelating agent. **complexometric ~** T. by means of a complexone and a metal *indicator,* q.v. **differential ~** The t. of equal portions of test solution so that one always contains the same volume of titrating solution in excess of that in the other. At the end point the potential difference between the solutions is a maximum. **high-frequency ~** Oscillometry. A method of following the chemical changes taking place in solution during t. from the changes in relative permittivity measured by an oscillator. It is sensitive, and does not involve contact between the test solution and the electrodes; e.g., the t. of beryllium compounds with sodium hydroxide.

t. error The ratio of the concentrations of active ions at the end point and initially. **t. level** The order of concentration (normality) of a titrant.

titre Titer.

Titrimeter Trademark for a vacuum-tube device for potentiometric titrations, which are followed continuously and automatically on a microammeter. Cf. *potentiometer, sectrometer.*

Tl Symbol for thallium.

t.l.c. Abbreviation for thin-layer *chromatography.*

TLV Threshold limit value. Upper limit, above which exposure under industrial conditions may be dangerous.

Tm Symbol for thulium.

TMT, TMTD See *thiuram.*

TNA Abbreviation for trinitroaniline. See *picramide.*

TNB Abbreviation for trinitrobenzene.

TNT Abbreviation for trinitrotoluene.

TNX Abbreviation for trinitroxylene.

tobacco Tabacum. The dried leaves of *Nicotiana tabacum* (Solanaceae). It contains nicotine, nornicotine, nicotianine, nicotelline, and other alkaloids; used for smoking. **Indian ~** Lobelia.

t. stems The ground waste products of t.; a fertilizer: nitrogen 1.2-3.3, potassium oxide 4–9%. **t. wax** A wax from the t. leaf and smoke, principally alkanes with *n*-hentriacontane.

Tobias acid 2-Naphthylamine-1-sulfonic acid; an intermediate.

Tobin Bronze Trademark for the alloy Cu 55, Zn 43, Sn 2%.

tobramycin $C_{18}H_{37}O_9N_5$ = 467.5. Nebcin. White powder, soluble in water. An aminoglycoside antibiotic, mainly active against gram-negative organisms.

tocol $C_{26}H_{44}O_2$ = 388.6. 3,4-Dihydro-2-methyl-2-(4,8,12-trimethyltridecyl)-2*H*-1-benzopyran-6-ol . The basic nucleus of the tocopherols (R^1 = R^2 = R^3 = H):

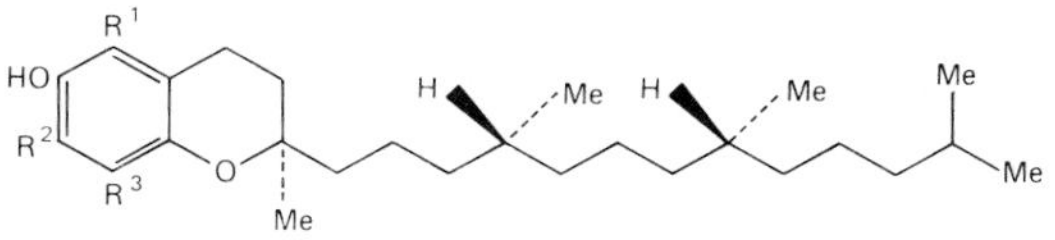

tocopherols Vitamin E. Methyl derivatives of *tocol*, q.v., present in many types of food and in cell membranes, of which α is the most active. They help to prevent oxidation of polyunsaturated fat. Yellow, oily liquids, fat-soluble. See *vitamins*, Table 101.

Form	R^1	R^2	R^3
α-	Me	Me	Me
β-	Me	H	Me
γ-	H	Me	Me
δ-	H	H	Me

tocotrienols Unsaturated derivatives of the tocopherols; occur with them in cereal grains.

toddy (1) The fermented sap of the coconut and other palms. (2) A hot tea containing some spiritous liquor. **t. test** See *hot toddy test* under *test.*

Tofranil Trademark for imipramine.

Toison solution A stain for red blood cells: 0.025 g methyl violet (6B), 1.0 g sodium chloride, 8.0 g sodium sulfate, 30 mL glycerol, in 300 mL water.

tokomak See *thermonuclear reaction.*

tolamine Chlorazene.

tolane $C_6H_5C{:}CC_6H_5$ = 178.2. Bibenzylidyne*, diphenylacetylene. Colorless leaflets, m.60, insoluble in water.

tolazoline hydrochloride $C_{10}H_{12}N_2 \cdot HCl$ = 196.7. 2-Benzyl-2-imidazoline hydrochloride. Priscoline. Priscol. White crystals, m.173, soluble in water; a sympatholytic and vasodilator (USP, BP).

tolbutamide $C_{12}H_{18}O_3N_2S$ = 270.3. 1-Butyl-3-(*p*-tolylsulfonyl)urea. White crystals, m.136, insoluble in water; an oral hypoglycemic agent for diabetes (USP, EP, BP).

tolerance (1) The limit of error permitted in operations, such as the manufacturing of components, the graduation of measuring instruments, and analytical evaluation. (2) Capacity to withstand continued use of a drug.

tolidine $NH_2MeC_6H_3 \cdot C_6H_3MeNH_2$ = 212.3. Dimethylbenzidine. Diaminodimethylbiphenyl. **4,4′,2,2′-~**, ***m*-~** m.109. **4,4′,3,3,′-~**, ***o*-~** m.131.

t. sulfate $C_{14}H_{16}N_2 \cdot H_2SO_4$ = 310.3. White crystals, soluble in water.

tolite Trinitrotoluene*.

Tollens **T., Bernhard (1841–1918)** German agricultural chemist. **T. reagent** An ammoniacal solution of silver oxide, used to test for aldehydes and ketones.

tolnaftate $C_{19}H_{17}ONS$ = 307.4. *O*-2-Naphthyl *m,N*-dimethylthiocarbanilate. Tinactin, Tinaderm. White crystals, m.110, insoluble in water. An antifungal agent, used for the skin (USP, BP).

toloxy Cresoxy. The methylphenoxy† radical MeC_6H_4O-, from cresol.

tolu Balsam (of Tolu). The resinous exudation of *Myroxylon balsamum* (Leguminosae) of tropical America; an expectorant (USP, BP).

tolualdehyde (1)* $MeC_6H_4 \cdot CHO$ = 120.2. Methylbenzaldehyde, toluic aldehyde. **1,2-~** b.196. **1,3,-~** b.199. **1,4-~** b.204. (2) **alpha-~** $PhCH_2CHO$. Phenylacetaldehyde*.

toluamide C_8H_9ON = 135.2 (1)* $Me \cdot C_6H_4 \cdot CONH_2$. Carbamoyltoluene. ***ortho*-~** Colorless needles, m.139, soluble in water. ***meta*-~** Colorless crystals, m.94, soluble in water. ***para*-~** Colorless needles, m.158, soluble in water. (2) $C_6H_5 \cdot CH_2 \cdot CONH_2$. α-Phenylacetamide.

toluanilide $PhNHCOCH_2Ph$ = 211.3. α-Phenylacetanilide. Colorless crystals, m.117.

toluene* $C_6H_5 \cdot CH_3$ = 92.1. Methylbenzene†, toluol (German), phenylmethane. Colorless liquid, d_{13} 0.871, m.−93, b.111, insoluble in water, soluble in alcohol or ether. Obtained from coal tar and used as a solvent, in organic synthesis, and in the manufacture of benzoic acid derivatives, coal-tar products, and explosives. See *benzylidene, benzylidyne, benzyl, benzylene, tolyl, tolylene.* **amino~** Toluidine*. **aminohydroxy~*** Cresidine. **bromo~** (1)* MeC_6H_4Br = 171.0. ***ortho*-~** Colorless liquid, b.182, soluble in alcohol. ***para*-~** Red crystals, soluble in alcohol. (2) **alpha-~** Benzyl bromide*. **chloro~** (1)* C_7H_7Cl = 126.6. ***para*-~** Colorless liquid, b.161 soluble in alcohol. (2) **alpha-~** Benzyl chloride*. **cyano~** (1)* $CN \cdot C_6H_4Me$ = 117.2. Toluonitrile*. ***ortho*-~** Brown liquid, b.203, soluble in alcohol. ***para*-~** Yellow crystals, m.28, soluble in alcohol. (2) **alpha-~** Benzyl cyanide*. **diacetylaminoazo~** Dimazon. **diamino~*** $MeC_6H_3(NH_2)_2$ = 122.2. Toluylene diamine. **2,3-~** Colorless scales, m.61, soluble in water. **2,4-~** Colorless needles, m.99, soluble in water. **2,5-~** m.64, b.273. **2,6-~** m.104. **dihydrooxo~** Toluenone. **hydroxy~** Cresol*. **isopropyl~** Cymene*. **nitro*** See *nitrotoluene.* **nitroso*~** $NO \cdot C_6H_4 \cdot Me$ = 121.1. ***ortho*-~** m.72. ***meta*-~** m.53. ***para*-~** m.48. **phenyl~*** See *phenyltoluene.* **trichloro~** Benzotrichloride. **trinitro~*** See *trinitrotoluene.*

t. sulfochloride T.sulfonyl chloride*. **t.sulfonic acid*** $Me \cdot C_6H_4 \cdot SO_3H$ = 172.2. Cf. *tosyl.* ***ortho*-~** $2H_2O$ = 208.2. Hygroscopic plates, decomp. 145. ***meta*-~** H_2O = 190.2. An oil. ***para*-~** H_2O = 190.2. Colorless leaflets, m.92, soluble in water. Cf. *tosate, tosylate.* **hydroxy~** Cresolsulfonic acids. **t.sulfonamide*** $Me \cdot C_6H_4 \cdot SO_2NH_2$ = 171.2. ***ortho*-~** m.155. ***meta*-~** m.107. ***para*-~** m.137. **t.sulfonyl chloride*** $Me \cdot C_6H_4 \cdot SO_2Cl$ = 190.6. T. sulfochloride. ***ortho*-~** m.10. ***meta*-~** m.12. ***para*-~** m.66.

toluenone $CH{:}CMe \cdot CO \cdot CH_2 \cdot CH{:}CH$ = 108.1.

toluenyl The benzylidene* radical.

toluic **t. acid** (1)* $Me \cdot C_6H_4 \cdot COOH$ = 136.2. Methylbenzoic acid. ***ortho*-~** Colorless needles, m.102, soluble in water. ***meta*-~** Colorless prisms, m.110, slightly soluble in water. ***para*-~** Colorless needles, m.176, soluble in water; used in organic synthesis. **methyl~** Xylic acid. (2) **alpha-~** $Ph \cdot CH_2 \cdot COOH$. Phenylacetic acid*.

t. aldehyde Tolualdehyde*. **t. anhydride*** $(Me \cdot C_6H_4CO)_2O$ = 254.3. *o*-Toluic acid anhydride. Colorless crystals, m.36, soluble in water. **t. nitrile** Cyano*toluene.*

toluidide A compound derived from the toluidines by replacing an amino H atom by an acyl radical, e.g., $MeC_6H_4NHCOMe$, acetotoluidide.

toluidine* $Me \cdot C_6H_4 \cdot NH_2$ = 107.2. Aminotoluene, ***ortho-*** **~** Colorless liquid, b.198. ***meta-*** **~** Colorless liquid, b.203. ***para-*** **~** Colorless leaflets, m.43, slightly soluble in water; used in the synthesis of dyes and medicines. ***N*-acetyl ~** Acetotoluide. ***N*-diethyl ~** See *diethyltoluidine* under *diethyl.* **methyl ~** Xylidine*. **nitro ~*** See *nitrotoluidine.* ***N*-sulfinyl ~** See *sulfinyltoluidines.*
t. blue

$$Me_2N \cdot C_6H_3 \langle {}^{S}_{N} \rangle C_6H_2Me(NH_2)\ Cl \cdot ZnCl_2$$

A double salt of zinc chloride and dimethyltoluthionine. C.I. Basic Blue 17. Green powder with blue luster, soluble in water; used to dye textiles. Cf. *methylene blue, thiazine dyes.* **t. hydrochloride*** $C_7H_9N \cdot HCl$ = 143.6. Red crystals, soluble in water. **t. sulfate*** $C_7H_9N \cdot H_2SO_4$ = 205.2. Yellow crystals, soluble in water.

toluidino* (Methylphenyl)amino†. Tuluino. The radical MeC_6H_4NH-, from toluidine.

tolunitrile* Cyano*toluene**. **alpha- ~** Benzyl cyanide*.

tuluol (German.) Commercial toluene.

toluonitrile* Cyano*toluene**.

toluoyl* Toluyl. Methylbenzoyl. The radical $Me \cdot C_6H_4 \cdot CO-$; *o-*, *m-*, and *p-*isomers.

toluphenazine $C_{13}H_{10}N_2$ = 194.2. Colorless crystals, m.117. **dimethyldiamino ~** Neutral red.

tolupiaselenole $C_7H_6N_2Se$ = 197.1. Colorless crystals, m.73.

toluquinone $MeC_6H_3O_2$ = 122.1. Methylbenzoquinone. **1,4- ~** m.67 (sublimes), insoluble in water. **methoxydihydroxy ~** $\underline{CO \cdot CMe{:}COH \cdot CO \cdot COH{:}COMe}$ = 184.2. A pigment produced from glucose by the fungus *Penicillium spinulosum.*

toluresitannol $C_{16}H_{14}O_3OCH_3OH$. A constituent of tolu balsam.

tolurhodin A noncarotenoid pigment from rubra-type torulae.

toluyl The toluoyl* radical. **alpha- ~** The phenylacetyl* radical.
t. aldehyde Tolualdehyde*. **t.azo-β-naphthol** Scarlet R.

toluylene (1) The tolylidene* radical. (2) Stilbene*. **t. blue** An oxidation-reduction *indicator,* q.v. **t. diamine** Diamino*toluene**. **t.diamine indophenol** An oxidation-reduction indicator. **t. red** Neutral red.

tolyl* Methylphenyl†. The radical $CH_3 \cdot C_6H_4-$, from benzene; 3 isomers: *ortho, meta, para.* **alpha- ~** Benzyl*.
t.acetic acid* $CH_3 \cdot C_6H_4 \cdot CH_2COOH$ = 150.2. ***ortho-*** **~** Colorless needles, m.88, soluble in water. ***para-*** **~** Colorless needles, m.91, soluble in water. **t. alcohol** T.methanol*. **t. bromide*** Bromo*toluene**. **t. chloride*** Chloro*toluene**. **t. hydrazine***. $CH_3C_6H_4NHNH_2$ = 122.2. Hydrazomethylbenzene. ***ortho-*** **~** Colorless leaflets, m.56, soluble in water. ***meta-*** **~** b.245. ***para-*** **~** Colorless tablets, m.65, soluble in water. **t.hydrazine hydrochloride** $C_7H_{10}N_2 \cdot HCl \cdot H_2O$ = 176.6. ***ortho-*** **~** Red crystals, soluble in water. ***para-*** **~** Brown powder, soluble in water. **t.hydroxylamine***. $CH_3 \cdot C_6H_4 \cdot NHOH$ = 123.2. ***ortho-*** **~** Colorless leaflets, m.94, soluble in water. **t. isothiocyanate*** $Me \cdot C_6H_4 \cdot NCS$ = 149.2. Thiocyanomethylbenzene. ***ortho-*** **~** Yellow liquid, b.238, insoluble in water. ***para-*** **~** Colorless crystals, m.26, insoluble in water. **t. mercuric chloride** C_7H_7HgCl = 327.2. Rhombic crystals, m.233, insoluble in water. **t.methanol** $Me \cdot C_6H_4 \cdot CH_2OH$ = 122.2. ***ortho-*** **~** Colorless needles, m.34, soluble in water. ***meta-*** **~** Colorless liquid, b.217, soluble in water. ***para-*** **~** Colorless needles, m.59, soluble in water. **t. mustard oil** T. isothiocyanate*. **t. phenyl ketone** Phenyl tolyl ketone*.

tolylene The tolylidene* radical. **alpha- ~** The benzylidene* radical.
t. diamine Diamino*toluene**.

tolylidene* Methylphenylene†. Tol(u)ylene, cresylene. The radical $Me \cdot C_6H_3=$, from toluene. 6 isomers. **t.hydrate** Stilbene hydrate.

tombak A copper and zinc alloy.

tomography A method of obtaining sectional pictures of organs by rotating the X-ray source around the patient. May be computer-aided, i.e., C(A)T.

ton Unit of weight. (1) 1 metric t. = 1 Mg (the SI unit) = 1,000 kg = 1.1025 short t. Cf. *tonne.* (2) 1 short t. = 2,000 lb = 907.1847 kg (U.S. usage). (3) 1 long ton = 2,240 lb = 1,016.047 kg (former U.K. usage). **assay ~** See *assay ton.* **refrigeration ~** 288,000 Btu. The quantity of heat required to melt 2,000 lb of pure solid ice into water at 32°F.

tonalite A group of quartz, orthoclase, and feldspathic rocks.

tonga A mixture of the barks of *Rhaphidophora vitiensis* (Araceae) and *Premma taitensis* (Verbenaceae), from the Pacific Islands. It contains tongine and an essential oil; used to treat neuralgia. Cf. *tonka.*

tongine An alkaloid from tonga.

tongs A laboratory appliance for holding hot objects such as crucibles, test tubes, or flasks. **remote-handling ~** Tongs manipulated by a distant mechanism; used to handle radioactive substances.

tonic An agent that is reputed to restore vigor.

tonite An explosive: wet guncotton pulp 54, barium nitrate 46%.

tonka Tonka bean, snuff bean. The dried seeds of *Coumarona odorata* or *Dipteryx odorata* (Leguminosae), N. Brazil and Guiana; contains coumarin. An ingredient of perfumery and flavorings, and source of coumarin. Cf. *tonga.*

tonnage An evaluation of the capacity of ships. **deadweight ~** The carrying capacity of a ship; about 70% of the gross t. Traditionally measured in long tons. **gross ~** Total weight of a ship. **net ~** Gross t. minus nonearning spaces on a ship. **register ~** 100 ft^3 (2.832 m^3) of enclosed space (shipping).

tonne U.K. usage for metric ton; to distinguish from long ton.

tonquinol $C_{11}H_{13}O_6N_3$ = 283.2. Colorless crystals, a musk substitute.

toot poison Tutin.

topaz $Al_2SiO_4F_2$. Colorless or varicolored crystals. The transparent, yellow variety is a precious stone. **false ~** Quartz. **oriental ~** Corundum.

topazolite A yellow variety of iron *garnet,* q.v.

topochemical Referring to (1) localized reactions in the inner or outer fields of force of crystalline matter; as, the building of crystals on a limited surface by local supersaturation; (2) reactions occurring in zones.

topochemistry The study of localized reactions, as in colloidal systems. Cf. *zone.*

topography (1) The science of the accurate description or delineation of a locality or region. (2) In microscopy, the grouping, orientation, and situation of structures.

topology A branch of stereochemistry dealing with the geometrical properties of molecular models. Cf. *chiral.*

topomineralogy See *mineralogy.*

tor Torr.

Toranomer Trademark for a viscose synthetic fiber.

torbane Torbanite.

torbanite Black *cannel coal,* q.v., with a high ash, from

Torbane Hill, Australia, d.1.3. Believed to be a high polymer of highly unsaturated organic esters and acids produced by the degradation of fatty matter from algoid deposits in lakes.
torbernite $Cu(UO_2)_2P_2O_8 \cdot 12H_2O$. A green fluorescing, hydrated copper, uranium phosphate. Cf. *chalcophyllite.*
tori seed oil Mustard-seed oil.
tormentil Tormentilla, septfoil. The dried rhizome of *Potentilla tormentilla* (Rosaceae), Europe and N. Asia. It contains a red coloring matter, t., tannin and quinovic acid; a tan and astringent. Cf. *five-finger grass.* **t. tannin** $C_{26}H_{22}O_{11}$ = 510.5. A constituent of tormentil; brown tanning powder.
Tornesite Trademark for chlorinated rubber. White powder, turning brown at 150°C, and resistant to common solvents and reagents.
toroid (1) A thick, solid ring. (2) Describing a burner in which some of the gases are sucked back into the center of the flame, thus giving a relatively small area of t. form having great stability and intensity. (3) A ring-shaped, hollow tube.
torque The force of or resistance to a twisting motion.
torr Tor; a unit of pressure (named for Torricelli). 1 t. = 1 mmHg at 0°C = 133.322 Pa.
Torricelli, Evangelista (1608–1647) Italian assistant of Galileo, noted for construction of barometers, microscope, and telescope.
torsion Production of a strain by twisting or rotating at right angles to the length.
torulosis Infection with a pathogenic yeast of the torula group.
torun metal A bearing-metal alloy: Sn 90, Cu 10%.
tosate A *p*-toluenesulfonate*, $Me \cdot Ph \cdot SO_2 \cdot OR$.
TOSCA U.S. Toxic Substances Control Act, introduced in 1976.
tosyl The radical $Me \cdot C_6H_4 \cdot SO_2-$ (*para* only). Abbreviation for *p*-toluenesulfonyl.
tosylate A *p*-toluenesulfonate*, $Me \cdot Ph \cdot SO_2 \cdot OR$.
touch The sense of feeling. **hard ~** A t. like that of crystals of inorganic salts. **slippery ~** The touch produced by alkali hydroxides. **soft ~** A t. like that of phthalic anhydride or beeswax. **unctuous ~** A smooth, fatty feel due to substances such as soapstone or ointments.
touchstone (1) Jasper. (2) A hard, black stone on which a streak is made with a gold or silver alloy. It is treated with hydrochloric and nitric acids and water in succession; the resultant streak, compared with a standard, gives an approximate value of the alloy. First described by Theophrastus (ca. 300 B.C.) Cf. *streak test.*
Toulet's solution A concentrated solution of mercury potassium iodide, d.3.17; used to separate minerals by flotation.
tourmaline Brazilian sapphire, schorl, rubellite. Aluminum silicates in various colors containing boron and lithium. Cf. *silica minerals.* The red, blue, green, and colorless tourmalines are gems. Tourmalines are mixtures of aluminum borate with alkali, magnesium, or iron silicates, and frequently contain chromium, manganese, calcium, and fluorine; doubly refractive and used in optical instruments (polariscopes). See *achroite, indicolite.* **artificial ~** Herapathite.
tower A tall structure used for absorption, scrubbing, cooling, or distilling.
towsagis Mountain gum. A Russian plant source of rubber.
toxalbumin A poisonous albumin, e.g., in cobra venom.
toxemia Poisoning caused by substances produced by microorganisms, e.g., bacterial toxins.
toxic Poisonous. **t. concentration** Maximum permissible concentration in air for any form of exposure. See Table 96; *poison effect.* **t. dose** A dose that produces the characteristic symptoms of poisoning. **t. equivalent** The smallest amount of a substance to kill an animal, divided by the weight of the animal.

TABLE 96. TOXIC CONCENTRATIONS OF VARIOUS SUBSTANCES

Substance	Concentration, ppm (v/v)
Ammonia	50
Arsine	0.05
Carbon dioxide	5,000
Carbon monioxide	50
Chlorine	1
Dichloromethane	500
Formaldehyde	3
Hydrogen chloride	5
Hydrogen peroxide	1
Hydrogen sulfide	20
Iodine	0.1
Ozone	0.1
Phosphine	0.3
Pyridine	5
Sulfur dioxide	5
Tetracarbonyl nickel	0.001
Turpentine	100

toxicarol $C_{23}H_{22}O_7$ = 410.4. Green crystals, m.126. In derris and cube; a fish poison. Cf. *rotenone.*
toxicity See *toxic equivalent.* **t. unit** See *median lethal dose* under *dose.*
toxicodendron The dried, poisonous leaves of *Rhus toxicodendron,* q.v.
toxicology The study of the actions, detection, and treatment of poisons and poisonings.
toxin (1) Any harmful substance. (2) Soluble substance, produced by bacteria, that is harmful to the body and stimulates production of a specific antibody (called an antitoxin). **endo ~** T. produced during bacterial growth. **exo ~** T. liberated from dead bacteria cells. **vegetable ~** A poisonous substance formed by a plant; as, ricin from castor oil seed. **zoo ~** A poison produced by an animal; as, snake venom.
toxisterol A sterol produced by irradiation of calciferol.
toxoflavin An antibiotic produced by *Bacterium cocovenenans.*
toxoid A degenerated toxin deprived of its toxicity but still able to stimulate production of antitoxins.
trabuk metal A substitute for nickel silver: Sn 87.5, Ni 5.5, Sb 5.0, Bi 2.0%.
trace* In analysis, a substance present in 10^2 (i.e., 0.01%) to 10^{-4} ppm. **micro ~*** A substance present in 10^{-4} to 10^{-7} ppm.
tracer A mixture of calcium resinate with magnesium, rare-earth nitrates, and glue; used in projectiles to make their path visible. **t. element** Radioactive tracer.
tracheid Fiber. An elongated cell from woody plants, containing various amounts of lignin, and characterized by bordered pits or discoid markings which vary in pattern according to origin.
trachoma A chronic chlamydial infection of the eye, frequently causing blindness; occurs principally in the Middle and Far East.
track The path described by a particle, e.g., α rays, protons, electrons, ions, or molecules, made visible in a cloud chamber. Cf. *Wilson tracks, Brownian motion.*
tracking Arc resistance. Creepage. Treeing. The path of an

electric current when it short-circuits a faulty conductor. If sufficient heat is produced, the t. may be visible, e.g., with vulcanite.

tractive force The power of pulling; as, of a magnet.

trademark "A mark adopted in relation to any goods to distinguish in the course of trade any goods certified by any person in respect of origin, material, mode of manufacture, quality, accuracy, or other characteristics, from goods not so certified" (Trade Marks Act). Trademarks have capital initial letters.

tragacanth Gum dragon. The gummy exudation from the stems of *Astragalus gummifer* and other Astragalae species (Leguminosae), Asia Minor. Contains bassorin, pectin, and starch; used as an emulsifier, suspending agent, and adhesive in pharmacy (NF, EP, BP), by the textile industry, and by cigar makers. Cf. *bandoline.* **Indian ~** Sterculia gum.

tragacanthin The carbohydrate of tragacanth, a polysaccharide hydrolyzing to pentosans.

tragon Locust kernel gum from carob beans.

train A connected arrangement of several chemical instruments; as, combustion train. **t. oil** Crude *whale oil.*

tranquilizer A drug used to relieve mental stress, e.g., diazepam.

trans- (1) Prefix (Latin) indicating "across." (2)* (Italic). See *stereoisomer.*

transaminase* See *transamination.*

transamination The reversible transfer of amino groups by enzymes (aminotransferase* or transaminase*) into oxo acids.

transaudient Permitting sound waves to pass.

transcalent Permitting heat waves to pass.

transcrystallization The formation of crystals with their principal axes perpendicular to the direction of heat flow.

transducer (1) In chemistry, a means of transforming chemical or physical characteristics of substances into another form, such as power, voltage, current, or resistance changes, which can be further converted into instrument observables. (2) Generally, a device for converting energy from one form (e.g., air pressure) into another (e.g., voltage), as in process-control instrumentation.

transesterification Direct conversion of an organic acid ester into another ester of the same acid; e.g., by treatment of a fat with the appropriate alcohol in presence of a catalyst.

transfer Conveyance of matter or energy from one location to another. **t. function** A mathematical relationship in control theory between cause (input) and effect (output). **t. number** Transport number*. **t. pipet** A pipet with only one graduation mark, that measures one definite volume of liquid.

transferase* See *enzymes,* Table 30.

transference Transfer. **t. cell** Hittorf cell. An electrolytic cell with a detachable center portion; used to determine changes in concentration of an electrolyte during electrolysis. **t. number** Transport number*.

transformation A change in form, structure, or internal arrangement of atoms. Cf. *nuclear reactions, radioelements.* **radioactive ~** See *radioactivity.*

t. constant Decay constant*. **t. series** See *radioactive elements.* **t. theory** Electromagnetic radiation has corpuscular aspects, e.g., Compton effect; and the motion of matter has undulatory aspects, e.g., electron scattering. Cf. *mass*-energy cycle.

transformer A device for changing the nature of alternating current; e.g., *step-down* t. (to lower voltage and higher current); *step-up* t. (to higher voltage and lower current). **t. fluids** For insulating and cooling t.; materials used include Formel NF (a mixture of 4 straight-chain aliphatic halocarbons), tetrachloroethylene, sulfur hexafluoride, silicones, and polychlorinated biphenyls.

transponder A transmitter-receiver on aircraft, which reports location by emitting coded pulses.

transistor An amplifier or oscillator replacing the vacuum triode. It consists of large lapped crystals of semiconductors, such as germanium or silicon, and does not develop heat or noise. Cf. *semiconductor.* **optical ~** A device using gallium arsenide to convert part of the energy of an incoming electric current into light, which is then absorbed and used to free electrons on the output side.

transition A change from one state or form to another; as, melting. **t. elements*** Elements whose atoms or cation(s) have an incomplete d subshell of electrons. In the *periodic* table, q.v., the first t. series is Sc–Cu, the second is Y–Ag, and the third is Hf–Au. **t. interval** The range of concentration of active ion in a titration over which the eye is able to perceive the change in appearance of the relevant indicator. **t. point, t. temperature** The temperature at which an allotrope or polymorph is converted into another of its forms; e.g., the t. p. of rhombic into monoclinic sulfur is 95–96°.

transitron An electrical oscillator used to produce alternating current of high and constant frequency. Cf. *transistor.*

translucent Semitransparent. Cf. *radiolucent, transparent.*

transmissibility Capacity to allow the passage of light, involving absorption of certain wavelengths. It may be qualitative (color screens) or quantitative (translucence).

transmutation (1) The change of one element into another element, as in radioactive disintegration. In the alchemical case, t. was always a change from a base to a noble metal (gold). (2) Artificial t. T. by bombardment with fast-moving protons, deuterons, or α particles. Cf *radioelements, proton bombardment.*

transoid-* Anti. Affix indicating the steric relation between the nearest atoms (i.e., those linked through the smallest number of atoms) on saturated bridgeheads in a polycyclic compound. See *stereoisomer.*

transparency The property of permitting the passage of radiation; e.g.:

Glass—transparent to visible light.
Rock salt—transparent to heat and to visible and ultraviolet rays.
White fluorite—transparent to infared, visible and ultraviolet rays.
Paraffin—transparent to hertzian rays.
Thin metals—transparent to γ rays.

Cf. *opacity.*

transparent Describing a substance which permits the passage of rays of the visible spectrum. Cf. *translucent.*

Transpex Trademark for a transparent polymethyl methacrylate plastic.

transpiration Exhalation of water vapor by a plant or animal through its surface tissue or skin. Cf. *respiration.*

transplantation The transfer of tissue or organs from one person or animal to another (homologous t., as heart t.), or between different parts of the same person or animal (autologous t., as skin grafting).

transplutonium See *transplutonium* under *plutonium.*

transport **t. lag** Delay in response in a control system caused by the passage of matter between the point implementing a change and the responding sensor. **t. number** Transference number, Hittorf number (t_i). The proportion of the total current carried by an ion, which depends on its mobility. If these are u for the anion and v for the cation of a 1:1 salt, then the corresponding transport numbers are: t_i (anion) = $u/(u + v)$; t_i (cation) = $v/(u + v)$.

transposition An atomic displacement within a molecule by which one atom changes place with another.

transudate Liquid which has passed through a living tissue. Cf. *osmosis.*

transuranic elements The elements of atomic number higher than that of uranium.

transvaalin Scillaren A.

tranylcypromine sulfate $(C_9H_{11}N)_2H_2SO_4$ = 364.5. Parnate, Parstelin. White powder, soluble in water. An antidepressant of the monoamine oxidase inhibitor group (USP, BP).

trapezohedron A crystal of 6, 8, or 12 faces, having unequal intercepts on all the axes.

trass A natural *pozzolana,* q.v.

Traube T., Isidor (1860–1943) German chemist. **T. stalagmometer** A stalagmometer for determination of the surface tension of biochemical fluids. **T.'s rule** The absorption of organic substances from aqueous solutions increases strongly and regularly in a homologous series with increasing number of C atoms.

traumatol $CH_3 \cdot C_6H_3I \cdot OH$ = 234.0. Iodocresol, iodocresine. Purple powder.

Trauzl test A measure of the strength of an explosive from the increase in volume on firing a standard charge in a cavity in a lead cylinder, as compared with that from a standard explosive.

traversal The path of a particle made visible by a *track,* q.v.

traversellite Augite containing Al_2O_3.

travertine $CaCO_3$. Calcareous sinter, onyx. A native calcium carbonate, usually banded similarly to marble. Cf. *tufa.*

Travis Trademark for a viscose synthetic fiber.

treacle Molasses.

Treadwell, Frederick P. (1857–1918) English chemist, noted for analytical methods.

treble Triple.

treeing Tracking.

trehalose $C_{12}H_{22}O_{11}$ = 342.3. D-glucopyranosyl D-glucopyranoside. Disaccharides. **α,α-~** Mycose. **α,β-~** m.210; from honey. **β,β-~** m.137.

treibgas Calor Gas.

Trelon Trademark for a polyamide synthetic fiber.

Tremin Trademark for trihexyl phenidyl.

tremolite $3MgO \cdot SiO_2 \cdot CaSiO_3$. A white or green native amphibole.

tren* Indicating 2,2′,2″-triaminotriethylamine, $(H_2N \cdot CH_2 \cdot CH_2)_3N$, as a ligand.

Treponema A genus of spiral, motile microorganisms. Some cause disease in humans; e.g., *T. pallidum* causes syphilis.

trester The residue after pressing grapes. **t. fermentation** The fermentation of diluted grape residues.

tretamine $C_3N_3(NC_2H_4)_3$ = 204.2. TEM. 2,4,6-Triethyleneimino-*s*-triazine, triethylenemelamine. A drug based on mustard gas. An alkylating agent, used for malignant disease.

tretinoin $C_{20}H_{28}O_2$ = 300.4. (*E*)-Retinoic acid, Retin A. Vitamin A acid. Yellow powder, m.180, insoluble in water; used for acne (USP).

Trevira Trademark for a polyester synthetic fiber.

tri-* Prefix (Latin) indicating "three." Cf: *ter-, tris-.*

triacetamide $N \cdot (CO \cdot Me)_3$ = 143.1. A solid, m.78, soluble in ether.

triacetate* A compound whose molecule contains 3 acetate radicals; as triacetin.

triacetin $(MeCOO)_3C_3H_5$ = 218.2. 1,2,3-Propanetriyl triacetate.* An oil from cod-liver oil, butter, and other fats, d.1.161, b.258, slightly soluble in water, soluble in alcohol or ether. Cf. *acetin.*

triacetonamine $C_9H_{17}ON$ = 155.2. 2,2,6,6-Tetramethyl-4-piperidinone. b.205.

triacontane* $C_{30}H_{62}$ = 422.8. m.66. A solid from the roots of *Oenanthe crocata* (Umbelliferae).

triacontanoic acid* See *melissic acid.*

triacontanol* $C_{30}H_{61}OH$ = 438.8. Melissyl alcohol, myricyl alcohol. Colorless needles, m.87, insoluble in water; from many plant waxes and beeswax.

triacontene* See *melene.*

triacontyl* Myricyl. The radical $C_{30}H_{61}-$. **t. alcohol*** Triacontanol*. **t. palmitate*** $C_{30}H_{61}O_2C_{16}H_{31}$ = 677.2. Myricin. A fatty ester (wax) and crystalline principle from beeswax and *Myrica* bark.

triacylglycerol lipase* See *lipase.*

triad (1) A group of 3 related elements or compounds; as, the triad Cl, Br, and I. (2) A trivalent atom or radical. (3) A crystal which shows 3 similar faces when rotated about its axis of symmetry through 360°.

triadimifon See *fungicides,* Table 37 on p. 250.

tri-allate* See *herbicides,* Table 42 on p. 281.

triallylamine $N(C_3H_5)_2$ = 96.2. Colorless oil, b.150.

triamcinolone acetonide $C_{24}H_{31}O_6F$ = 434.5. Adcortyl, Aristocort, Kenalog. White crystals, very slightly soluble in water. A glucocorticosteroid, used for skin conditions (USP, BP). See *corticoids.*

triamido Triamine.

triamine* A compound containing 3 NH_2 groups. Cf. tertiary *amine.*

triaminoazobenzene Bismarck brown.

triaminobenzene* $C_6H_3(NH_2)_3$ = 123.2. **1,2,3-~** m.103, b.330. **1,2,4-~** m.97, b.340. Very soluble in water, alcohol, or ether; used in epoxy resins.

triamorph A substance that crystallizes into 3 different types of crystals (trimorphous).

triamterene $C_{12}H_{11}N_7$ = 253.3. 2,4,7-Triamino-6-phenylpteridine. Dyrenium, Dytac. Yellow crystals, slightly soluble in water. A diuretic; used for heart failure and edema (USP, EP, BP).

triamylamine Tripentylamine.

triangle A plane figure, bounded by 3 straight lines, pairs of which meet at 3 points.

triangulo-* Affix indicating 3 atoms bound in a triangle.

Triassic See *geologic eras,* Table 38.

triazane* $NH_2 \cdot NH \cdot NH_2$. Prozane. The H is replaceable by hydrocarbon radicals. Cf. *nitrogen* hydrides.

triazano* Triazanyl†. The radical $-NH \cdot NH \cdot NH_2$, from triazane.

triazene* $NH_2 \cdot N{:}NH$. Diazoamine. The H is replaceable by hydrocarbon radicals. Cf. *nitrogen* hydrides. **1-t.-1,3-diyl†** See *diazoamino.*

triazeno* The radical $NH_2N{:}N-$.

triazine* $C_3H_3N_3$ = 81.1. The heterocyclic compounds:

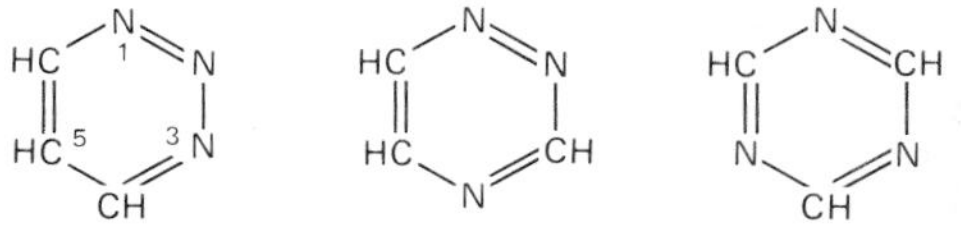

1,2,3- 1,2,4- 1,3,5-

t.triol Cyanuric acid.

triazinyl* The radical $C_3H_2N_3-$, from triazine.
triazo The azido* radical. **t.benzene** Phenyl azide*. **t.phos*** See *insecticides,* Table 45 on p. 305.
triazole $C_2H_3N_3$ = 69.1. Pyrrodiazole. **1,2,3-~** Osotriazole. **1,2,4-~** Common t. Colorless needles, m.120, soluble in water. **benzo~** See *benzotriazole.*
triazolone $\underline{CH:N \cdot CO \cdot (NH)_2}$ = 85.1. Ketotriazole.
triazolyl* The radical $C_2H_2N_3-$, from triazole.
tribasic* Describing a molecule that has 3 replaceable H atoms, or produces 3 H ions in solution.
tribenzoylmethane* $(Ph \cdot CO)_3CH$ = 328.4. Colorless crystals, m.225 (sublime), soluble in water.
tribenzyl* Indicating the presence of 3 benzyl radicals, $Ph \cdot CH_2-$. **t.amine*** $N(C_6H_5 \cdot CH_2)_3$ = 287.4. m.91, soluble in water. **t.ethyltin*** $Et(CH_2Ph)_3Sn$ = 421.1. White crystals, m.31, soluble in organic solvents. **t.tin chloride*** $(PhCH_2)_3SnCl$ = 427.5. White needles, m.142, insoluble in water. **t.tin hydroxide*** $(PhCH_2)_3SnOH$ = 409.1. Rhombic crystals, m.120, soluble in organic solvents.
tribo- Prefix (Greek) indicating "friction" or "rubbing."
triboelectric series Electrostatic series. A list of substances sequenced such that when any 2 are rubbed together, the one higher on the list acquires a positive charge and the one lower, a negative charge; e.g.: (+) asbestos, rabbit fur, glass, nylon, wool, calcite, silk, cotton, paper, magnalium, wood, amber, slate, steel, ebonite, sulfur, Celluloid, rubber, polyethylene (−).
tribology The science of interacting surfaces in relative motion; in particular, friction, lubrication, and wear.
triboluminescence The light emission from substances, especially crystals, when crushed, rubbed, or mechanically pressed together. Cf. *luminescence.*
tribromide* A compound having 3 Br atoms in the molecule.
tribromo-* Tribrom-. Prefix indicating 3 Br atoms in a molecule. **t.acetaldehyde*** Bromal. **t.acetic acid*** CBr_3COOH = 297.7. Colorless leaflets, m.135, soluble in water; used in organic synthesis. **t.aniline*** $NH_2C_6H_2Br_3$ = 329.8. Aniline 2,4,6-tribromide. Small needles, m.119, insoluble in water. **t.benzene*** $C_6H_3Br_3$ = 314.8. **1,2,3-~** m.87. **1,3,4-~** m.44. **1,3,5-~**, ***sym*-~** Colorless needles, m.119, insoluble in water. **t.ethane*** **1,1,2-~** $BrCH_2 \cdot CHBr_2$ = 266.8. A liquid, d.2.579, b.188. **t.ethanol*** $CBr_3 \cdot CH_2OH$ = 282.8. Avertin. White, unstable crystals, m.80, soluble in water; a general anesthetic. **t.ethylene*** $BrCH:CBr_2$ = 264.7, d.2.708, b.164. **t.hydrine** Allyl tribromide. **t.methane*** Bromoform.* **t.-2-naphthol** $C_{10}H_4Br_3OH$ = 380.9. Gray crystals, soluble in alcohol. **t.phenol*** $C_6H_3Br_3O$ = 330.8. **2,4,6-~** Bromol. White needles, m.92 (sublimes), slightly soluble in water. **t.resorcinol*** $C_6H(OH)_2Br_3$ = 346.8. Colorless needles, m.112, slightly soluble in water.
tributylamine* $N \cdot (C_4H_9)_3$ = 185.4. Colorless liquid, d.0.778, b.216, insoluble in water. **iso~*** $N(CH_2CHMe_2)_3$. Colorless liquid, d.0.766, b.191.
tributyrin Butyrin.
tricalcium phosphate See *calcium phosphate.*
tricaprin $(C_9H_{19}COO)_3C_3H_5$ = 554.9. Glycerol tridecanoate. Triclinic crystals, d.0.921, m.31, insoluble in water.
tricaproin $(C_5H_{11}COO)_3C_3H_5$ = 386.5. Clycerol trihexanoate.Colorless liquid, d.0.988, insoluble in water.
tricapryllin $(C_7H_{15}COO)_3C_3H_5$ = 470.7. Glycerol trioctanoate. Colorless liquid, d.0.954, m.8, insoluble in water.
tricarballylic acid 1,2,3-Propanetricarboxylic acid*. **2-hydroxy~** Citric acid.
Tricel Trademark for a triacetate synthetic fiber.
tricetin $C_{15}H_{10}O_7$ = 302.2. 3′,4′,5,5′,7-Pentahydroxyflavone. m.310. A pigment from wheat.
trichloride* Terchloride. A compound containing 3 Cl atoms in its molecule.
trichloro-* Trichlor-. Prefix indicating 3 Cl atoms in a molecule. **t.acetal** $(EtO)_2CH \cdot CCl_3$ = 221.5. (1) Colorless liquid, d.1.288, b.197, soluble in water. (2) A solid, m.83. **t.acetamide*** $CCl_3 \cdot CONH_2$ = 162.4. Colorless deliquescent leaflets, m.141, soluble in water. **t.acetic acid*** $CCl_3 \cdot COOH$ = 163.4. Colorless rhombs, d_{60° 1.630, m.57, soluble in water. A reagent for unsaturated compounds; a caustic and astringent (USP, EP, BP). **t.acetyl chloride*** $CCl_3 \cdot COCl$ = 181.8. Colorless liquid, b.118, soluble in alcohol; used in organic synthesis. **t.aldehyde** Chloral. **t.benzene*** $C_6H_3Cl_3$ = 181.4. Colorless solids or liquids, insoluble in water. **1,2,3~** m.52. **1,2,4-~** d_{10°(liq.)1.466, b.213. **1,3,5-~** m.63. **t.butyl alcohol** Chlorbutanol. **t.butyl aldehyde** Butyl chloral. **t.ethane*** $C_2H_3Cl_3$ = 133.4. **1,1,1-~** $MeCCl_3$. Methylchloroform. Colorless liquid, d.1.325, b.75, insoluble in water. **1,1,2-~** $CHCl_2 \cdot CH_2Cl$. Colorless liquid. d.1.478, b.114, insoluble in water. **t.ethanol*** $CCl_3 \cdot CH_2OH$ = 149.4. 2,2,2-Trichloroethyl alcohol. Colorless tablets, m.18, slightly soluble in water. **t.ethyl alcohol** Trichloroethanol*. **t.ethylene*** $CHCl:CCl_2$ = 131.4. Trilene. 1-Chloro-2-dichloroethylene. Colorless liquid, d.1.459, b.87, insoluble in water, soluble in ether. Used as a refrigerant, inhalation anesthetic (EP, BP); for the extraction of fats, caffeine, and nicotine; and in organic synthesis, dry cleaning, degreasing, perfumes, paints, and varnishes. **t.fluoromethane*** CCl_3F = 137.4. A gas, b. ca. 25. An aerosol propellant having a slight ethereal odor. **t.hydrin** $CH_2 \cdot Cl \cdot CHCl \cdot CH_2Cl$ = 147.4. Allyl trichloride, 1,2,3-trichloropropane*, glyceryl chloride. Colorless liquid, d.1.417, b.158, insoluble in water. **t.hydroquinone** $C_6H(OH)_2Cl_3$ = 213.4. Colorless prisms, m.134, soluble in water. **t.lactic acid** $CCl_3 \cdot CHOH \cdot COOH$ = 193.4, m.116, soluble in water. **t.methane*** Chloroform*. **t.methyl chloroformate** $Cl_3C \cdot COOCl$ = 197.8. Diphosgene. Colorless liquid, b.127. A lung-irritant poison gas. **t.methylthio*** The group $-SCCl_3$, associated with fungicidal activity. Cf. *captan.* **t.phenol*** $C_6H_2(OH)Cl_3$ = 197.4. 1-Hydroxy-2,4,6-trichlorobenzene*. Colorless rhombs, m.68, soluble in water; **1,2,3,5-~** m.53, soluble in water, very soluble in alcohol or ether. **t.phenoxyacetic acid*** $Cl_3 \cdot C_6H_2 \cdot OCH_2 \cdot COOH$ = 255.5. 2,4,5-T. Herbicide, whose toxicity results partly from its containing dioxin. Crystals, m.155, slightly soluble in water. Used in Agent Orange for defoliation in Vietnam. **t.propane*** Trichlorohydrin. **t.quinone** $\underline{CH:CCl \cdot CO \cdot CCl:CCl \cdot CO}$ = 211.4. Yellow leaflets, m.165, insoluble in water. **t.thio*** The radical Cl_3S-. **t.toluene** Benzotrichloride. **t.triethylamine** $N(C_2H_4Cl)_3$ = 204.5. An irritant liquid, causing blisters; an anticancer agent. See *mechlorethamine hydrochloride.*
trichodesmine $C_{18}H_{27}NO_6$ = 353.4. An alkaloid, m.202 (decomp.), from *Trichodesma incanum* (Boraginaceae).
trichothecin $C_{19}H_{24}O_5$ = 332.4. An antibiotic from the fungus *Trichothecium roseum.* Colorless crystals, m.118. Slightly soluble in water.
trichroism Having 3 different colors when viewed at different angles; as, certain minerals and crystals.
trichromatic Having three *colors,* q.v. **t. analysis** The matching of colors in terms of 3 primary components: red, green, blue.
tricin $C_{17}H_{14}O_7$ = 330.3. 4′,5,7-Trihydroxy-3′,5′-dimethoxyflavone. A dimethyl ether of tricetin, a pigment in

the leaves of wheat (especially khopli, *Triticum dicoceum*). Yellow needles, m.292, soluble in acetic acid.
Triclene Trademark for trichloroethylene.
triclinic Anorthic, asymmetric. A crystal with 3 unequal, long axes at oblique angles. See *crystal systems, prism.*
tricosane* $Me(CH_2)_{21}Me$ = 324.6. Crystals, m.48. **12-methyl ~*** $C_{24}H_{50}$ = 337.7. Isotetracosane, m.138, soluble in ether.
tricosanoic acid* $C_{22}H_{45}COOH$ = 354.6. m.79, soluble in benzene; from plant wax and oils. **12-methyl ~*** Isotetracosanoic acid, m.80; from pine oil. **trimethyl ~*** Phthioic acid.
tricosanol* $C_{23}H_{47}OH$ = 340.6. **1-~** m.74; in conifer wax. **12-~** m.76.
tricosanone* **12-~** Laurone.
tricosoic acid Tricosanoic acid*.
tricresol A mixture of *o*-, *m*-, and *p*-cresols.
tricresyl Indicating 3 cresol radicals. **t. phosphate** $(C_6H_4Me)_3PO_4$ = 368.4. Tritolyl phosphate. Lindol, tri-*p*-cresyl phosphate. Colorless liquid, d.1.18, m.77; a plasticizer, waterproofing, and softener for resins and rubber. It causes "ginger paralysis."
tricyanic acid Cyanuric acid.
tricyano- Indicating the ring ·C:N·C:N·C:N, derived from cyano compounds by polymerization.
tricyanogen chloride $C_3N_3Cl_3$ = 184.4. Cyanuric trichloride. The heterocyclic polymer of cyanogen chloride. Colorless crystals, m.146, slightly soluble in water.
tricyclic Describing: (1) A molecule containing 3 rings of atoms; as anthracene. (2) A group of antidepressant drugs, as, amitriptyline hydrochloride, related to phenothiazine, but with substitution at 10 (i.e., the N atom) and replacement of the S atom by a $-CH_2 \cdot CH_2-$ linkage.
tridecane* $C_{13}H_{28}$ = 184.4. An alkane. Colorless liquid, d.0.757, b.234, soluble in EtOH.
tridecanoic acid* $C_{12}H_{25}COOH$ = 214.3. Tridecyclic acid, *n*-tridecoic acid, ficocerylic acid; in figs and coconuts. Colorless crystals, m.51. **cyclopentene ~** Chaulmoogric acid.
tridecanol* Tridecyl alcohol*.
tridecene* $C_{13}H_{26}$ = 182.3. Tridecylene. An alkene. Colorless liquid, d.0.845, b.233, insoluble in water, soluble in alcohol or ether. Tridecanoic acid*.
tridecyl* The radical $C_{13}H_{27}-$, from tridecane. **t. alcohol*** $C_{13}H_{27}OH$ = 200.4. Tridecanol*. Crystals, b.31. **t.amine*** $C_{13}H_{29}N$ = 199.4. *n*-Aminotridecane. Crystals, m.27.
tridecylene Tridecene*.
tridecylic acid Tridecanoic acid*.
tridemorph* See *fungicides,* Table 37 on p. 250.
Tridione Trademark for trimethadione.
tridiphenylmethyl $(Ph \cdot C_6H_4)_3C$. Triphenylmethyl. A "free radical" compound, in colorless crystals that form a colored solution.
tridymite A solid solution of silica and impurities, formerly thought to be a natural form of silica; d.2.26, stable below 1470.
triels* The Group 3B elements of the periodic table: B, Al, Ga, In, and Tl.
trien* Indicating triethylenetetramine as a ligand.
-triene* Suffix indicating 3 double bonds.
trienol A conjugated triene glyceride made from castor oil, and having the properties of tung oil.
triethanolamine $N(C_2H_4OH)_3$ = 149.2. Tris(2-hydroxyethyl)amine. Colorless liquid, b_{150mm} 227, soluble in water; a soap base, oil emulsifier, reagent for antimony and tin, and pharmaceutic aid (USP). T. (NF, BP) is a mixture of t. with mono- and diethanolamines.
triethoxy* Indicating 3 ethoxy* radicals, EtO—. **t.boron** See *ethyl borate.*
triethyl-* Prefix indicating 3 C_2H_5 radicals in a compound. **t.amine*** NEt_3 = 101.2. Colorless liquid, b.89, soluble in water; a ptomaine in decaying fish. **t.arsine*** Et_3As = 162.1. Colorless liquid, b_{735mm}140, soluble in alcohol. **t.benzene*** **1,3,5-~** $C_6H_3Et_3$ = 162.3. Colorless liquid, b.218, insoluble in water. **t.bismuthine*** Et_3Bi = 296.2. Colorless liquid, b_{79mm}107, insoluble in water. **t.borane*** Et_3B = 98.0. Colorless liquid, insoluble in water. **t.gallium** Et_3Ga = 156.9. Colorless liquid, b.143, decomp. by water. **t.methanol** CEt_3OH = 116.2. Colorless liquid, b.141, soluble in water. **t.phosphine*** Et_3P = 118.2. Colorless liquid, b.127, insoluble in water; a reagent for carbon disulfide. **t. phosphite** $(C_2H_5)_3PO_3$ = 166.2. Colorless liquid, b.155, insoluble in water.
triethylene **t.diamine*** $(CH_2)_6N_2$ = 112.2. 1,4-Diazabicyclo[2.2.2]octane, m.158; a catalyst. **t. glycol** $(CH_2OCH_2CH_2OH)_2$ = 150.2, b.278. A high-temperature solvent. **t.tetramine*** $(H_2N \cdot CH_2CH_2 \cdot NH \cdot CH_2)_2$ = 146.2. A reagent for sulfates and copper; b.272.
triferrin Iron paranucleinate. Red powder (Fe 22, P 2.5%), insoluble in water; a hematinic.
trifluoperazine hydrochloride $C_{21}H_{24}N_3F_3S \cdot HCl$ = 480.4. Stelazine. Creamy, bitter crystals, m.240, soluble in water; a tranquilizer and antiemetic. A phenothiazine drug, also used in psychiatry (USP, BP).
trifluralin* See *herbicides,* Table 42 on p. 281.
triforine* See *fungicides,* Table 37 on p. 250.
triformol 1,3,5-Trioxane.
trifructosan $C_{18}H_{13}O_{15}$ = 469.3. Secalose, trifructose anhydride. White, sweet crystals from rye flour, used to detect its presence in other flours; insoluble in 70% alcohol.
trigalloyl* Indicating 3 galloyl* radicals, $C_6H_2(OH)_3CO-$. **t.acetone glucose** $C_{30}H_{28}O_{18}$ = 676.5. Brown mass, soluble in water. **t. glucose** $(C_7H_5O_4)_3C_6H_9O_6$ = 636.5. Yellow mass, soluble in water. **t. glycerol** $(C_7H_5O_4)_3 \cdot C_3H_5O_3$ = 548.4. Brown mass, soluble in water; a tanning agent.
trigger A substance that initiates a chain reaction.
triglyceride See *glyceride.*
trigonal Describing a crystal with 2 equal axes and a third shorter or longer axis, all at right angles. See *crystal systems.*
trigonelline $C_7H_7O_2N$ = 137.1. Nicotinic methylbetaine, in many plant seeds; sea urchin, *Arabacia pustulosa;* the Coelenterata *Vella spirans,* jellyfish. Colorless crystals, m.218 (decomp.). Cf. *stachydrine.*
triguaiacyl Indicating three 2-methoxyphenyl radicals, $MeO \cdot C_6H_4 \cdot O-$. **t. phosphate** Tris(2-methoxyphenyl)phosphate.
trihemellitic acid Trimellitic acid.
trihexosan A beer dextrin.
trihexylphenidyl hydrochloride $C_{20}H_{31}ON \cdot HCl$ = 337.9. Benzhexol hydrochloride. Artane, Tremin. White crystals, sparingly soluble in water.
trihydrate* A compound containing 3 molecules of water.
trihydric* Describing a compound with 3 OH groups.
trihydrocyanic acid See *cyanidines.*
trihydrol $H_2:O \cdot H \cdot O \cdot H \cdot O:H_2$. An assumed threefold polymer of water. Cf. *dihydrol.*
trihydroxy-* Prefix indicating 3 OH groups. **t.anthraquinone*** **1,2,3-~** Anthragallol. **1,2,4-~** Purpurin. **t.benzene*** $C_6H_3(OH)_3$ = 126.1. **1,2,3-~** Pyrogallol*.

1,2,4- ~ Hydroxyhydroquinone. White crystals, m.141, soluble in water. **1,3,5-** ~ Phloroglucinol*. **t.benzoic acid*** $C_6H_2(OH)_3COOH$ = 170.1. **2,3,4-** ~ Pyrogallolcarboxylic acid*. Colorless needles, m.110, soluble in water. **3,4,5-** ~ Gallic acid*. **t.benzophenone*** $C_{13}H_{10}O_4$ = 230.2. Several isomers. **3,4,5-t.benzoyl†** See *galloyl.* **t.estrin** Estriol. **t.pyridine*** $C_5H_5O_3N$ = 127.1. m.220–230 (decomp.), soluble in water. **t.stearic acid*** $C_{17}H_{32}(OH)_3COOH$ = 332.5. White solid, m.146.

triiodide* A compound containing 3 I atoms. **t. ion** The anion I_3^-.

triodo-* Triiod-. Prefix indicating 3 I atoms. **t.acetic acid*** $CI_3 \cdot COOH$ = 437.7. Yellow scales, m.150, soluble in water. **t.benzene*** $C_6H_3I_3$ = 455.8. **1,2,3-** ~ m.116 (sublimes). **1,2,4-** ~ m.91. **1,4,5-** ~ m.183. **t.cresol triiodo*meta*cresol** Losophan. **t.methane*** Iodoform*.

triketo- Trioxo-*. Prefix indicating 3 =CO groups in a molecule. **t.hydrindene hydrate** Ninhydrin. **t.purine** Uric acid.

triketones* Organic compounds containing 3 contiguous carbonyl groups; as Me·CO·CO·CO·Me, pentanetrione. Cf. *trione.*

Trilafon Trademark for perphenazine.

Trilan Trademark for an acetate synthetic fiber.

trilaurin $C_3H_5(OOC \cdot C_{11}H_{23})_3$ = 639.0. 1,2,3-Propanetriyl trilaurate. A crystalline glyceride from palm-nut, coconut, and bayberry oils.

Trilene Trademark for trichloroethylene; an anesthetic.

trilinolein $C_3H_5(OOC \cdot C_{17}H_{30})_3$ = 876.4. 1,2,3-Propanetriyl linolate. A glyceride from linseed, sunflower, and hempseed oils.

trilite Trinitrotoluene*.

trillion (1) 10^{12} (American and French usage = $1{,}000{,}000^2$). (2) 10^{18} (U.K. and German usage = $1{,}000{,}000^3$).

Trilon Trademark for water-treatment compositions. Cf. *Calgon.* **T. A** $N(CH_2 \cdot CO_2Na)_3$. Sodium nitriloacetate. **T. B** $(CH_2 \cdot CO_2Na)_2N \cdot C_2H_4 \cdot N(CH_2 \cdot CO_2Na)_2$. Sodium ethylenediaminotetramethyl carbonate. Cf. *EDTA.*

trimellitic acid $C_6H_3(COOH)_3$ = 210.1. 1,2,4-Benzenetricarboxylic acid*. Colorless crystals, m.216, soluble in water.

trimer A condensation product of 3 monomer molecules. Cf.*polymer.*

trimeric Describing capacity to form threefold polymers. Cf.*trimetric.*

trimesic acid Trimesitinic acid.

trimesitinic acid $C_6H_3(COOH)_3$ = 210.1. Trimesic acid, 1,3,5-benzenetricarboxylic acid*. Colorless crystals, m.348, soluble in water.

trimethadione $C_6H_9O_3N$ = 143.1. 3,5,5-Trimethyl-2,4-oxazolidinedione. Tridione, Troxidone. White crystals with camphor odor, m.46, soluble in water. An anticonvulsant, used for petit mal epilepsy (USP, BP).

trimethano- Prefix indicating 3 $-CH_2-$ bridges in a ring. Cf. *trimethylene.*

trimethaphan mesylate $C_{32}H_{40}O_5N_2S_2$ = 596.8. Arfonad. White crystals, soluble in water. A ganglion blocking agent, used to lower blood pressure (BP).

trimethoprim $C_{14}H_{18}O_3N_4$ = 290.3. 2,4-Diamino-5-(3,4,5-trimethoxybenzyl)pyrimidine. Proloprim. White powder, m.201, very slightly soluble in water. An antimicrobial agent affecting the folic acid metabolism of bacteria; used to treat urinary infections (USP, EP, BP). See *co-trimoxazole.*

trimethoxy-* Prefix indicating 3 methoxy groups, CH_3O-. **t.borane*** $B(OMe)_3$ = 103.9. Methyl borate. Colorless liquid, b.65.

trimethyl-* Prefix indicating 3 methyl groups, CH_3, in a molecule. **t.acetaldehyde** Pivaldehyde*. **t.acetic acid** Pivalic acid*. **t.acetyl chloride** Pivaloyl chloride*. **t.amine*** Me_3N = 59.1. Secaline. In leaves of *Chenopodium* species, blood (0.002%), and putrefying choline. Colorless gas, b.3.5, soluble in water. It is poisonous and has a fishy odor. **t.amine oxide** Me_3NO = 75.1. Kanirin, m.96, in muscle, urine, Cephalapoda, and Crustaceae. **t.amino hydrochloride** $Me_3N \cdot HCl$ = 95.6. Colorless crystals, m.271, soluble in water. **t.arsine*** As = 120.0. Colorless liquid, b.53, soluble in water. **t.benzene*** **1,2,3-** ~ Hemimellitene. **1,2,4-** ~ Pseudocumene*. **1,3,5-** ~ Mesitylene*. **t.benzoic acid*** $C_6H_2Me_3COOH$ = 164.2. **2,3,5,-** ~ Colorless needles, m.149, soluble in alcohol. **2,4,6-** ~ Isodurylic acid. Colorless crystals, m.152, soluble in water. **t.bismuthine*** C_3H_9Bi = 254.1. Me_3Bi. Colorless liquid, b.110. **t.borane*** Me_3B = 55.9. White crystals, m.56. **t.carbinol** T.methanol. **t.cyclohexane*** Hexahydrocumene. **t.cyclopentene*** Laurolene. **t.ethylene** $MeCH{:}CMe_2$ = 70.1. β-Isoamylene, Pental. Colorless liquid, b.36, insoluble in water. **t.gallium*** Me_3Ga = 114.8. Colorless liquid, b.56. **t.glycine** See *betaine.* **t.methanol** CMe_3OH = 74.1. Colorless crystals, m.25, soluble in water. **t.naphthalene*** Sapotalene. **2,4,6-t.phenyl†** See *mesityl.* **t. phosphate** Me_3PO_4 = 140.1. Colorless liquid, b.197, soluble in alcohol. **t.phosphine*** Me_3P = 76.1. Colorless liquid, b.40, insoluble in water. **t.pyridine** Collidine. **t.quinoline*** $C_{12}H_{13}N$ = 171.2. **2,3,4-** ~ m.65. **2,3,6-** ~ m.86. **2,5,7-** ~ m.43. **2,6,8-** ~ m.45. **t.tin** Me_6Sn_2 = 327.6. Colorless liquid, b.182, insoluble in water. Cf. *stannanes.* **t.tin bromide*** Me_3SnBr = 243.7. White crystals, m.27. **t.tin chloride*** Me_3SnCl = 199.2. White crystals, m.37. **t.tin hydride** Me_3SnH = 164.8. An oil, b.60. **t.tin hydroxide*** Me_3SnOH = 180.8. Colorless prisms, decomp. 118. **t.tin oxide** $(Me_3Sn)_2O$ = 343.6. White powder. **t.tin sulfide** $(Me_3Sn)_2S$ = 359.6. Yellow oil, m.6. **t.tryptophan** Hypophorine. **t.xanthine** Caffeine. **t.urea*** $MeNH \cdot CO \cdot NMe_2$ = 102.1. Colorless crystals, m.75, soluble in water.

trimethylene C_3H_6 = 42.08. (1) Cyclopropane*. (2) Prefix indicating 3 methylene groups. (3)* 1,3-Propanediyl†. The radical $-CH_2 \cdot CH_2 \cdot CH_2-$. **t. dibromide** $CH_2Br \cdot CH_2 \cdot CH_2Br$ = 201.9, b_{720mm}160. **t. dicyanide** Glutaronitrile*. **t.diamine** 1,3-Propanediamine*. **t. glycol** 1,3-Propanediol*.

trimetric Orthorhombic. Cf.*trimeric.*

trimolecular Pertaining to 3 different molecules. **t. reaction** A third-order *reaction,* q.v.

trimorphism Crystallization in 3 different systems.

trimorphous Showing trimorphism.

Trimpex Trademark for trimethoprim.

trimyristin Myristin.

trinitrate* A compound containing 3 nitrate radicals.

trinitride Azide*.

trinitrin Nitroglycerin.

trinitro-* Prefix indicating 3 nitro groups in a molecule. **t.aniline** Picramide*. **t.anisole*** $MeO \cdot C_6H_2(NO_2)_3$ = 243.1. White crystals. **2,3,4-** ~ m.155. **2,3,5-** ~ m.104. **2,4,6-** ~ m.68 **3,4,5-** ~ m.120. **3,4,6-** ~ m.107. High explosives. **t. benzene*** $C_6H_3(NO_2)_3$ = 213.1. TNB. **1,2,4-** ~ Yellow crystals, m.57, slightly soluble in water. **1,3,5-** ~, **sym-** ~ Yellow crystals, m.122, slightly soluble in water. **t.cellulose** Pyroxylin. **t.cresol*** $C_6HMe(OH)(NO_2)_3$ = 243.1. 3-Hydroxy-1-methyl-2,4,6-trinitrobenzene. Yellow needles, m.105, soluble in water; an explosive. **t.glycerol** Nitroglycerin. **t.naphthalene*** $C_{10}H_5(NO_2)_3$ = 263.2. **1,3,5-** ~ Yellow, monoclinic crystals, m.122, slightly soluble in water. **1,3,8-** ~ Yellow needles, m.215, soluble in alcohol.

1,2,5-~ m.113. **1,4,5-~** Yellow needles, m.147, soluble in alcohol. Explosives. **t.orcinol*** $C_6(NO_2)_3Me(OH)_2$ = 259.1. Yellow needles, m.164, slightly soluble in water. **t.phenol*** $HO\cdot C_6H_2(NO_2)_3$ = 229.1. **2,4,6-~** Picric acid*. **2,4,5-~** m.96. **2,3,6-~** m.117. **2,3,5-~** m.120. All yellow crystals, soluble in water; explosives. **t.phenylmethylnitroamine** Tetryl. **t.resorcinol*** $C_6H(NO_2)_3\cdot(OH)_2$ = 245.1. Yellow needles, m.176 (sublimes), soluble in water. **t.toluene*** $C_6H_2Me(NO_2)_3$ = 227.1. **2,4,6-~** Trotyl, tolit, trilite, triton, trinol, TNT, 1-methyl-2,4,6-trinitrobenzene. Yellow leaflets, d.1.654, m.81, slightly soluble in water; used similarly to picric acid in explosives. **2,3,4-~** Colorless leaflets, m.112, insoluble in water. **2,4,5-~** Colorless crystals, m.104, insoluble in water. **3,4,5-~** m.138. **2,3,5-~** m.97. **2,3,6-~** m.80. High explosives. **t.triazidobenzene** $(NO_2)_3C(N_3)_3$ = 276.1. Turek detonator. Yellow crystals, m.131 (forming hexanitrobenzene), insoluble in water; a detonating high explosive. **t.xylene*** $C_6H(NO_2)_3Me_2$ = 241.2. TNX. **2,3,5-*p*-~** m.140. **2,4,6-*m*-~** m.182.

trinol Trinitrotoluene*.

trinor* As for *nor-*, but involving 3 such Me groups [see *nor-(1)*] or CH_2 groups [see *nor-(2)*]. **8,9,10-t.bornane*** C_7H_{12} = 96.2. Bicyclo[2.2.1]heptane*. Norcamphane. Parent compound of many terpenes. **dimethylmethylene-~** Camphene*. **trimethyl-~** Bornane*. **t.bornyl*** Norcamphanyl. The radical $C_7H_{11}-$, from trinorbornane. **t.pinane*** C_7H_{12} = 96.2. Bicyclo[3.1.1]heptane*. Norpinane. **dimethylmethylene ~** β-Pinene.

-triol* Suffix indicating 3 hydroxy groups; as, propanetriol, $C_3H_5(OH)_3$, glycerol.

triolein Olein.

-trione* Suffix indicating 3 carbonyl groups, not necessarily contiguous. Cf. *triketones.*

triose A monosaccharide with 3 C atoms; as, $CHO\cdot CHOH\cdot CH_2OH$. **t.phosphate isomerase*** Enzyme that catalyzes the reversible conversion of D-glyceraldehyde 3-phosphate into dihydroxyacetone phosphate; as in muscle glycogenolysis.

trioxa-* Prefix indicating 3 O bridges in a ring system; as, **t.bicyclooctane** $C_5H_3O_2$ = 95.1.

```
            (1)
(7) CH2—CH—CH2 (2)
    |    |   |
    |    O   O   (3)
    |    |   |
(6) O————CH—CH2 (4)
            (5)
```

trioxane **1,3,5-~** $O\cdot(CH_2O)_2CH_2$ = 90.1. Metaformaldehyde, paraformaldehyde. The heterocyclic trimer of formaldehyde. Made by heating formaldehyde in a sealed tube with sulfuric acid, at 115°C. White crystals, m.64, soluble in water to form a constant-boiling mixture. It vaporizes without depolymerization, but is depolymerized by acid to give formaldehyde. A source of formaldehyde for organic syntheses and plastics manufacture. **trimethyl ~** Paraldehyde.

trioxide* Teroxide. A binary compound containing 3 O atoms in the molecule; e.g., SO_3, sulfur trioxide.

trioxime* A compound containing 3 =NOH radicals.

trioximido- Prefix indicating 3 oxime*, =NOH, groups in a molecule. **t.propane** $HON:CH\cdot C(:NOH)\cdot CH:NOH$ = 131.1. Colorless crystals m.171.

trioxin 1,3,5-Trioxane.

trioxygen* Ozone.

tripalmitin Palmitin.

tripan roth Trypan red.

tripentylamine $(C_5H_{11})_3N$ = 227.4. Triamylamine. White solid; used as an antioxidant and solvent, and for flotation.

tripestone A concretionary form of anhydrite.

triphane Spodumene.

triphasic Describing a system involving 3 phases .

triphen- Triphenyl-*.

triphenol Trihydroxyphenol. A compound having 3 —OH groups in an unsaturated ring.

triphenyl-* Prefix indicating 3 phenyl radicals in a molecule. **t.acetic acid*** $CPh_3\cdot COOH$ = 288.3. Colorless, monoclinic crystals, m.264 (decomp.), slightly soluble in water. **t.amine*** $(C_6H_5)_3N$ = 245.3. Colorless prisms, m.127, slightly soluble in alcohol. **t.benzene*** $C_6H_3Ph_3$ = 306.4. **1,3,5-~** Colorless rhombs, insoluble in water. **t.bismuthine*** Ph_3Bi = 440.3. Bismuth triphenyl. White, monoclinic crystals, m.78, soluble in chloroform. **t.carbinol** T.methanol. **t.ene** $C_{18}H_{12}$ = 228.3. Benzo[*l*]phenanthrene. m.198. Soluble in EtOH with blue fluorescence; in coal tar. **t.ethane*** $Ph_2CH\cdot CH_2Ph$ = 258.4. Colorless crystals, m.54 **t.ethylene*** $Ph_2C:CHPh$ = 256.3. α-Phenylstilbene. **t.guanidine** $C_{19}H_{17}N_3$ = 287.4. ***N,N′,N″*-~** $PhN:C(NHPh)_2$ = 287.4. Solid, m.143 (decomp.), soluble in alcohol. ***N,N,N′*-~** Solid, m.131, soluble in water. **t.methane*** $(C_6H_5)_3CH$ = 244.3. Colorless leaflets, m.92, insoluble in water; used to manufacture dyes. **t.methane dyes** Triarylmethane dyes. Dyes derived by the introduction of auxochromic groups.

1. Fuchsin group:

```
                  CH=CH     +
                 /     \
(R·C6H4)2C=C           C=NH2
                 \     /
                  CH=CH
```

2. Aurin (fuchsone) group:

```
                  CH=CH
                 /     \
(R·C6H4)2C=C           C=O
                 \     /
                  CH=CH
```

3. Phthalein group:

```
              C6H4
             /    \
(R·C6H4)2C         R'
             \    /
               O
```

E.g., R′=CO in phenolphthalein. R′=SO_2 in phenol red. **t.methanol** $Ph_3C\cdot OH$ = 260.3. Colorless prisms, m.159, insoluble in water; used to manufacture triphenylmethane dyes. **amino ~** $NH_2C_6H_4\cdot CPh_2OH$ = 275.4. ***meta*-~** m.155; ***para*-~** m.116. **diamino ~** $C_{19}H_{14}O(NH_2)_2$ = 290.4. **t.methyl*** Ph_3C = 243.3. A free radical compound. Colorless crystals, m.300. **t. phosphate** Ph_3PO_4 = 326.3. Colorless needles, m.49, insoluble in water; a plasticizer, causing dermatitis with hypersensitive persons. **t.phosphine*** Ph_3P = 262.3. Colorless, monoclinic crystals, m.15, insoluble in water. **t.pyridazine*** $PhC:N\cdot N:CPh\cdot CPh:CH$ = 308.4. Colorless crystals, m.171. **t.stibine*** Ph_3Sb = 353.1. **t.tin*** Ph_3Sn = 350.0. Tin triphenyl. White powder, m.232 (decomp). **t.tin chloride*** Ph_3SnCl = 385.5. White crystals, m.106, insoluble in water; a reagent for fluorides.

triphylite $LiFePO_4$. A native phosphate. Green-blue rhombs.

triple Threefold. **t. bond** —C:C—. The acetylene linkage, indicated by the suffix *-yne**; as propyne*. **t. chlorides**

$M_4'M''M_2'''Cl_{12}$, in which M′ is Na, K, Cs, or NH_4; M″ is Zn, Cu, Hg, Ag_2, or Au_2; and M‴ is trivalent, as, Au. **t. nitrite reagent** A solution (120 g sodium nitrite, 9.1 g copper acetate, 16.2 g lead acetate, 2 mL acetic acid, in 50 mL water), giving crystals of characteristic shape with potassium salts. **t. phosphate** (1) A magnesium, calcium, ammonium phosphate, sometimes found in urine. (2) Treble superphosphate. A phosphate rock containing 3 times as much phosphoric acid as superphosphate. **t. point** The conditions in which 3 phases can exist in equilibrium, e.g., in the system ice-water-water vapor, at 4.57 mm pressure and 0.0100°C. See *phase rule* under *rule.* **t. salts** Salts whose molecules consist of 3 cations and 1 anion; as $CaCuNa_2(NO_2)_6$.

triplet See *multiplet.* **t. state** Describing the state of an atom or molecule having 2 electrons with parallel spin.

triplite $(Fe,Mn)_2FPO_4$. A greasy mineral.

tripod A three-legged support for holding containers over a bunsen burner.

tripoli Rottenstone. Decomposed limestone used for polishing; made from tripolite. **t. powder** Kieselguhr.

tripolite A native silica secreted by diatoms. Cf. *kieselguhr.*

trippkeite A native copper arsenate.

triprismo-* Affix indicating 6 atoms bound into a triangular prism.

triprolidine hydrochloride $C_{19}H_{22}N_2 \cdot HCl \cdot H_2O$ = 332.9. Actidil. Bitter, white crystals, soluble in water, m.120; an antihistamine used for allergic conditions, such as hay fever (BP).

tripropyl-* Prefix indicating 3 propyl* radicals. **t.amine*** $(C_3H_7)_3N$ = 143.3. A tertiary amine. Colorless liquid, m.15, slightly soluble in water.

triptane Supergas, neohexane. Common name for 2,2-dimethylbutane, used to obtain high-octane-value petroleum.

triquinoyls The oxidation products of hexahydroxybenzene: (1) CO·CO·CO·CO·CO·CO = 168.1. Cyclohexanehexone. (2) $C_6H_{16}O_{14}$ = 312.2. A solid, decomp. 95, slightly soluble in water.

triricinolein Ricinolein.

tris- Prefix indicating 3 times. Correctly applied to identical radicals substituted in the same way. **t.BP** Tris(2,3-dibromopropyl)phosphate. Formerly a flame retardant, but found to be a carcinogen. **t.(2-methoxyphenyl)** Triguaiacyl. Indicating 3 radicals, $MeO \cdot C_6H_4 \cdot O-$.

trisaccharide A carbohydrate that contains 3 monosaccharides in its molecule and hydrolyzes to 3 simple sugars, e.g., raffinose.

Trisalyt Trademark for a mixture of cyanides (sodium and zinc) with sodium sulfite; used in electroplating.

trisilane* $SiH_3 \cdot SiH_2 \cdot SiH_3$ = 92.3. An unstable, colorless liquid, b.53.

trisilicic acid* $H_6Si_3O_{10}$ = 250.3. A white, insoluble powder.

trisodium edetate $C_{10}H_{13}O_8N_2Na_3$ = 358.2. Used for hypercalcemia. See *edetate.*

tristearin $C_{57}H_{110}O_6$ = 891.5. Stearin. 1,2,3-Propanetriyl tristearate. A constituent of animal and vegetable fats, especially hard fats, e.g., cocoa butter, tallow. White powder, m.55, and again at 72, insoluble in water or ether; used in cosmetic creams and as an emulsifier.

tristimulus Describing effects depending on the stimulation of the optical senses by the 3 primary colors, e.g., trichromatic colorimetry.

trisulfide* A compound containing 3 S atoms in its molecule, e.g., Fe_2S_3, iron trisulfide.

tritartaric acid Uvic acid.

triterium Tritium*.

trithioacetaldehyde Sulfoparaldehyde.

trithiocarbonate* A salt of trithiocarbonic acid.

trithiocarbonic acid* $S{:}C(SH)_2$ = 110.2. Carbonotrithioic acid, m.−27.

trithionic acid* $H_2S_3O_6$. See *sulfur acids.*

tritiated Having hydrogen in a molecule converted into tritium atoms; as in a radioactive indicator.

triticale A bakery cereal produced by cross-breeding rye and wheat; rich in protein.

triticum (1) A genus of grasses which includes wheat. (2) Couch grass, dog grass. The dried rhizomes of *T.* or *Agropyron repens* (Gramineae), containing carbohydrates and malates; a diuretic.

tritio-* Prefix indicating that protium has been replaced by tritium.

tritium* T or ^3H = 3.0161. Triterium. An isotope of *hydrogen,* q.v. Half-life 12.5 years, decaying with the emission of β particles; obtained with *deuterium,* q.v., by the electrolysis of water.

tritolyl phosphate Tricresyl phosphate.

Triton Trademark for a line of synthetic, organic, surface-active agents. **T. B** Tetramethylammonium hydroxide. **T. F** Dibenzyldimethylammonium hydroxide. Used as solvents for cellulose, for saponification of fats, and in inorganic synthesis.

triton (1) t. The tritium atom nucleus. Cf. *proton, deuteron.* (2) Trinitrotoluene*.

tritonope A person who is blind to blue.

tritopine Laudanidine.

tritriacontanoic acid* $C_{32}H_{65}COOH$ = 494.9. Psyll(ostearyl)ic acid, m.95. From psylla wax. Cf. *lacceroic acid.*

tritriacontanol* $C_{33}H_{67}OH$ = 480.9. Psyllic alcohol, psyllostearyl alcohol, m.70. From psylla and beeswax.

triturate To grind or rub to a powder (usually with a liquid) in a mortar.

trituration Any finely powdered drug, or a mixture of lactose with a drug.

trityl (1)* The radical Ph_3C-. (2)* The free radical $Ph_3C\cdot$, triphenylmethyl.

triuret* $(NH_2 \cdot CO \cdot NH)_2CO$ = 146.1. Carbonyldiurea. White crystals, m.232, insoluble in water. Cf. *biuret.*

trivalent* Tervalent. Describing an atom or radical of valency 3.

trivalerin Phocenin.

trivial name (1)* A name that is not systematic, q.v. *(systematic name),* and thus devoid of structural significance. (2) The abbreviated name given to a chemical compound for convenience in general use, e.g., EDTA is the t. n. for ethylenediaminetetracetic acid.

Trobicin Trademark for spectinomycin.

troche A tablet or medicated disk.

troctolite A granitoid, crystalline, plutonic rock containing olivine and feldspar.

trogerite A native, hydrous uranium arsenate.

troilite FeS. A native form.

trommel A cylinder of perforated steel plate revolving around an inclined central axis; for large-scale sifting.

Trommer's test On warming urine with sodium hydroxide and copper sulfate, a yellowish-red precipitate indicates glucose.

tron(a) $Na_2CO_3 \cdot NaHCO_3 \cdot 2H_2O$. Native sodium, carbonate and sodium hydrogencarbonate. Cf. *natron, urao.* **t. potash** Kemfert. A high-grade potassium chloride, Searles Lake, Calif. (58–62% potassium oxide); used in fertilizers.

troostite (1) A native zinc-magnesium silicate. (2) A transition form of cementite, austenite, and ferrite. Cf. *steel.*

tropacocaine $C_8H_{14}N \cdot O \cdot COPh$ = 245.3. Benzoyl pseudotropine. An alkaloid from coca leaves. Colorless crystals, m.49, slightly soluble in water; a local anesthetic.
tropaeolin Tropeolin.
tropaic acid Tropic acid*.
tropane $C_8H_{15}N$ = 125.2. *N*-Methylnortropane. Parent of a group of alkaloids, b.167.
tropate* A salt of tropic acid.
tropeine An ester of tropine and an organic acid, as atropine.
tropeolin $R-N_2-C_6H_4 \cdot SO_3Na$. Hydroxyazodyes, the sodium salts of R-azobenzenesulfonic acids; as:

T. D, methyl orange, *p*-dimethylaminoazobenzenesulfonic acid; R is $Me_2NC_6H_4-$.
T. 0, yellow T, resorcinol yellow, resorcinolazobenzenesulfonic acid; R is $(HO)_2C_6H_3-$. A pH indicator, changing at 12.0 from yellow (acid) to brown (alkaline).
T. 00, diphenylamine orange, orange IV, phenylamineazobenzenesolfonic acid; R is $PhNH \cdot C_6H_4-$. A pH indicator, changing from red (1.3) to yellow (3.2).
T. 000 1,2-naphthol orange, orange I; R is $HO \cdot C_{10}H_6-$.

tropic acid* $CH_2OH \cdot CHPh \cdot COOH$ = 166.2. 3-Hydroxy-2-phenylpropanoic acid. Obtained by hydrolysis of atropine. **iso ~, (±)- ~** Colorless crystals, m.117, soluble in water. ***(R)*- ~** m.129. ***(S)*- ~** m.129; a component of tropine tropate.
tropidine $C_8H_{13}N$ = 123.2. *N*-Methylnortropidine. An oil obtained by dehydration of tropine, d.0.95, b.162, insoluble in water.
tropilidene C_7H_8 = 92.1. 1,3,5-Cycloheptatriene*. An oil, d.0.903, b.117, prepared by distilling tropine with soda lime.
tropine $C_8H_{15}ON$ = 141.2.

CH_2 —— CH_2
CH —— NMe —— CH
$CH_2-CH(OH)-CH_2$

3-Tropanol. Colorless needles, m.62, soluble in water. Alkaloid from Solanaceae and hydrolysis of atropine. ***endo* ~** 9-Methyl-8-azabicyclo[3.2.1]octan-3-ol. ***exo* ~** Form of t. from Solanaceae and by heating t. with phenyl amylate. Colorless needles, m.108, soluble in water. **iso ~** Atropine.
t. alkaloids See *atropine, cocaine, ecgonine.* **t.carboxylic acid.** Ecgonine. **t. tropate** ***(S)*- ~** Hyoscyamine. **(±)- ~** Atropine. **t. sulfate** $(C_8H_{15}ON)_2H_2SO_4$ = 380.6. White crystals, soluble in water.
tropolone O:C·C(OH):CH·CH:CH·CH:CH = 122.1. 2-Hydroxy-2,4,6-cycloheptatrien-1-one. Needles, m.50. Soluble in petroleum ether.
tropomysin An asymmetric protein component of muscle.
tropopause The altitude at which the temperature of the atmosphere ceases to decrease with increase in height.
troposphere The atmosphere below the stratosphere.
tropoyl* The radical $CH_2OH \cdot CHPh \cdot CO-$, from tropic acid.
tropylium* $C_7H_7^+$. Cycloheptatrienylium*. The symmetrical, heptagonal, salt-forming, aromatic ion, from cycloheptatriene.

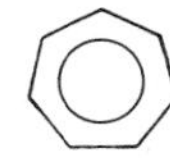

trotyl Trinitrotoluene*.
Trouton T., Frederick Thomas (1863–1922) British physicist. **T's rule** The molar latent heat of vaporization (in J) is equal to 93 times the (thermodynamic) boiling point. Exceptions include water and alcohol. Cf. *Hildebrand rule.* Also $T_{760mm} = T_p(1.648 - 0.255 \log p)$, where T_{760mm} is the boiling point at 760mm, and p the pressure.
troxidone BP name for trimethadione.
troy A British system of weights and measures for jewelers. 1 oz troy = $\frac{1}{12}$ pound = 480 grains = 20 pennyweights = 1.09714 oz avoir. = 1 oz apoth. = 31.1035 g. 1 pound troy = 5,760 grains = 240 pennyweights = 13.166 oz avoir. = 0.82286 pound avoir. = 373.2417 g. **t. weights** The weight system used in reports on gold, silver, or precious metals.
Trp* Symbol for tryptophan.
Trubenizing A process for permanently stiffening and shrinkage-resisting fabrics, by inserting a cellulose acetate thread between the cotton or linen and calendering with a cellulose acetate solvent.
truth table In computer logic, a table that describes a logical situation (e.g., circuit) by listing all possible combinations of inputs and, for each combination, the resulting output.
truxelline An alkaloid from coca leaves; an ester of ecgonine and truxillic acids.
truxillic acid $C_{18}H_{16}O_4$ = 296.3. 2,4-Diphenyl-1,3-cyclobutanedicarboxylic acid. A dimer of cinnamic acid from coca leaves; 5 stereoisomeric forms (all *meso*). Cf. *norpinic acid.*
truxinic acid $C_{18}H_{16}O_4$ = 296.3. 3,4-Diphenyl-1,2-cyclobutanedicarboxylic acid. Several sterioisomers; 3 are called neo ~. Cf. *truxillic acid.*
trypan blue $[(NaSO_3)_2C_{10}H_3(NH_2)OH \cdot N{:}N \cdot C_6H_3Me]_2$ = 960.8. Diamine blue, congo blue, Niagara blue, C.I. Direct Blue 14. Blue-gray powder; a dye.
trypanosomiasis Tropical diseases of Africa (sleeping sickness) and America (Chagas disease) due to the protozoan *Trypanosoma.* Transmitted by tsetse fly and reduviid bug.
trypan red $C_{32}H_{24}O_{15}N_6S_5$ = 892.9 Tripanroth. Brown powder, a dye and antiprotozoan agent.
trypsin* An enzyme of pancreatic juice that hydrolyzes proteins, by preferential Arg and Lys cleavage, to proteoses, and proteoses to true peptones (tryptones) and finally to leucine and tyrosine. Yellow powder, soluble in water; most efficient in slightly alkaline solution. Used as chymotrypsin (USP).
tryptachrome $C_{17}H_{14}O_2N_3$(?). A violet-pink indirubin compound produced by the action of an oxidizing agent, e.g., bromine, on tryptophan. It has an intense greenish-orange fluorescence (test for tryptophan).
tryptic activity The hydrolytic power of trypsin.
tryptones True peptones produced by trypsin, as distinct from pepsin peptones.
tryptophan* $C_6H_4 \cdot NH \cdot CH{:}C \cdot CH_2 \cdot CH(COOH)NH_2$ = 204.2. Trp*. Indolylalanine. 2-Amino-3,3′-indolylpropanoic acid. **L- ~** A split product from plant and animal proteins. An essential amino acid. Colorless solid, m.289, soluble in water. **trimethyl ~** Hypophorine.
tryptophyl* The radical $C_{11}H_{11}ON_2-$, from tryptophan.
T.S. Abbreviation for test solution.
tsano oil Isano oil. A paint-drying oil from the nuts of a W. African tree.
tschermigite A native ammonium alum.
Tschugajew's reaction A scarlet-red precipitate forms on addition of dimethylglyoxime to a weakly ammoniacal solution containing nickel.
T-stoff See *methylbenzyl bromide.*
tsugaresinol $C_{20}H_{20}O_6$ = 356.4. α-Conidendrin. A lignin lactone in sulfite waste liquor, and in Japanese hemlock, *Tsuga sieboldii.* White crystals, m.254.

T.T. Tuberculin-tested; as, milk from tuberculin-tested cows.
T.T.T. curve S. curve, showing the progress of a reaction in terms of time and temperature. Applied particularly to mineralogical transformations, e.g., of austenite into pearlite.
tubaic acid $C_{12}H_{12}O_4$ = 220.2. **(*R*)-~** A constituent of derris root; m.129. **iso ~** Rotenic acid.
tubain A resin from *derris* root. q.v.
tubanol $C_{11}H_{12}O$ = 160.2. 2-(1-Methylethyl)-4-benzofuranol. A split product of rotenone. **hydroxycarboxy ~** Tubaic acid. **tetrahydro ~** 2-Isopentylresorcinol.
Tubarin Trademark for tubocurarine chloride.
tubatoxin Rotenone.
tube A long and hollow device. **absorption ~** See *absorption tube.* **agglutination ~** A small test t. **Arndt ~** See *Arndt tube.* **arsenic ~** A long glass t. with drawn-out tip bent at an angle of 120°. Cf. *Marsh* test. **barometer ~** A narrow t. closed at one end and more than 760 mm long. **Brown ~** A potash bulb. **calcium chloride ~** A glass t. of suitable shape, filled with calcium chloride; used to dry gases. **capillary ~** A glass t. with bore less than 1 mm. **cathode ray ~** See *cathode ray tube.* **centrifuge ~** A thick-walled glass container of appropriate shape, for use in a centrifuge. **colorimeter ~** A flat-bottomed test t. of clear white glass. **combustion ~** A glass t., resistant to heat. **comparison ~** Colorimeter t. **condenser ~** The inner t. of a condenser. **Coolidge ~** An X-ray t. with electrically heated cathode. **Crookes ~** A vacuum t. exhausted so that X-rays are produced on the passage of an electric current. **culture ~** A test t. for growing bacteria. **Dorn-Goetz ~** A vacuum t. for spectroscopy. **drying ~** A glass t., filled with a drying agent. **Emmerling ~** An absorption t. **extraction ~** See *Soxhlet apparatus.* **fermentation ~** Saccharometer. An inverted test t. for collecting gases. **filter ~** A glass t. for connecting filter crucibles to a source of suction. **funnel ~** A long glass t. with conical top. **Geissler ~** A t. containing a gas at low pressure; produces the characteristic spectrum of the gas on passage of an electric current. **Giltner ~** A t. for anaerobic cultures. **guard ~** (1) A metal casing to protect glassware from mechanical injury. (2) See *guard tube.* **Hittorf ~** A modified Crookes t. **Hortvet ~** A centrifuge t. **melting point ~** A thin-walled capillary t. **Nessler ~** Colorimeter t. **Peligot ~** Calcium chloride t. **Pitot ~** A manometer t. **T ~** A T-shaped t. for making 3-way connections. **test ~** A glass t., closed at one end, made of heat- and acid-resistant glass. **thistle ~** A funnel t. with bulb-shaped top. **U ~** A U-shaped glass t. **vacuum ~** See *vacuum tube.* **X ~** A cross-shaped t. for 4-way connections. **X-ray ~** See *X-ray tube.* **Y ~** A Y-shaped t. for 3-way connections.
tuberculin A preparation obtained from the soluble products of a culture of tubercle bacilli. **old ~** T. prepared from *Mycobacterium tuberculosis* in glycerol and isotonic salt solution (USP, EP, BP). **purified protein ~** P.P.D. T. prepared in a medium from which proteins have been removed by precipitation (USP, EP, BP).
tuberin (1) A globulin from potato juice, representing about 50% of the total protein present. (2) A mixture of amino acids, chiefly leucines.
tubers Underground swellings on the roots of plants; they store reserve materials; as, aconite, potato.
tubing Glass, metal, or rubber tubes used in the construction of chemical apparatus. **pressure ~** Heavy-walled rubber hose for connecting pressure or vacuum lines. **silica ~** A hollow silica rod for high-temperature combustion.
Tubize Trademark for an acetate synthetic fiber.
tubocurarine chloride $C_{37}H_{41}O_6N_2Cl \cdot HCl \cdot 5H_2O$ = 771.7. (+)-Tubocurarine chloride hydrochloride. Tubarine. White crystals, m.270 (decomp.), soluble in water; a skeletal muscle relaxant (USP, EP, BP). See *curarine.*
tubule A small tube or neck on a glass apparatus.
tufa A calcareous sinter or sedimentary rock of calcium carbonate and silica, formed by chemical reaction from lake or groundwater; as, travertine. Cf. *tuff.*
tuff A sedimentary rock of volcanic dust, ash, and cinders. Cf. *tufa.*
Tufton Trademark for an acetate synthetic fiber.
tulipiferine An alkaloid from the bark of *Liriodendron tulipifera* (Magnoliaceae).
tulip tree The tree *Liriodendron tulipifera* (Magnoliaceae), N. America.
tumbago An alloy of ancient origin (Columbia): Cu 55, Au 33, Ag 12%.
tundish A refractory trough to prevent splash when ingot molds are poured.
tung t. oil Chinese wood oil. A rapidly drying oil from the nuts of *Aleurites cordata* and *A. fordii* (Euphorbiaceae), China and Japan; now cultivated in Florida and elsewhere. Yellow, jellifying liquid, d.0.936–0.942. It replaces linseed oil in paints and linoleum and gives a higher gloss and more water-resistant finish. Cf. *lumbang oil, trienol.* **t. pomace** T. oil cake. The seeds after the t. oil is extracted: nitrogen 5–6, phosphorus 2, potassium oxide 1.3%; a fertilizer.
tungstate* Wolframate*. Indicating tungsten as the central atom(s) in an anion, as: M_2WO_4. A salt of tungstic acid. **polyt.(VI) acids*.** **hetero ~** Acids (and anions) containing one other element in addition to W, H, and O; as, $[SiW_{12}O_{40}]^{4-}$. **iso ~** Acids (and anions) containing only W, H, and O; as, $[W_4O_{16}]^{8-}$, $[W_6O_{19}]^{2-}$, meta-t. $[H_2W_{12}O_{40}]^{6-}$, para A $[W_7O_{24}]^{6-}$.
tungsten* W = 183.85. Wolfram* (German), wolframium. A heavy metal and element, at. no. 74. Gray powder, d.19.3, m.3410 (this being the highest of the metals), b.5663, insoluble in water, soluble in nitric acid and hot hydroxide solutions. Valencies 2,4,5, and 6. Occurs native in wolframite, scheelite, and tungstite; used (as ferrotungsten) in the manufacture of steel for high-speed tools, and hot-working die steels; also as metallic filaments (m.3100) for electric light bulbs. **ferro ~** An alloy for high-speed tools; W 7–9, Cr 2–3%, and Fe. **orange ~** $Na_2WO_4 \cdot W_2O_5$. Saffron bronze, tungsten-sodium tungstate. Gold scales, insoluble in ordinary solvents; a pigment. **violet ~** $K_2W_3O_9 \cdot W_2O_5$. Potassium tritungstate. Blue-black powder; a pigment.
t. alloy See *ferrotungsten, steel, partinium.* **t. bronze** An alkali-metal salt of polymerized tungstic acid; as orange t., violet t. **t. carbides** (1) WC, d.15.7, m.2870. (2) W_2C, d.16.06, m.2880. (3) W_3C, m.2700+. **cemented ~** An extremely hard machine-tool alloy of WC embedded in tungsten with 5–15% Co, d.14–15, hardness, second to diamond. Cf. *Carboloy.* **t. chlorides.** See *tungsten dichloride, tungsten tetrachloride, tungsten pentachloride, tungsten hexachloride.* **t. dichloride*** WCl_2 = 254.8. Gray powder, decomp. by water. **t. dioxide*** WO_2 = 215.8. Brown rhombs, d.12.11, insoluble in water. **t. disulfide*** WS_2 = 248.0. Gray crystals, d.7.5. **t. fluoride** WF_6 = 297.8. A gas. **t. hexachloride*** WCl_6 = 396.6. Blue cubes, d.13.3, m.275, slightly soluble in water, soluble in carbon disulfide. **t. minerals** Chiefly: tungstenite, WS_2; wolframite (Fe, Mn), WO_4; scheelite, $CaWO_4$. **t. oxides** See *tungsten dioxide, tungsten trioxide, tungsten pentaoxide.* **t. oxychloride** $WOCl_4$ = 341.7. Tungstyl chloride. Red crystals, m.209, decomp. by water. **t. pentachloride*** WCl_5 = 361.1. Black needles,

m.248, decomp. in water. **t. pentaoxide*** W_2O_5 = 447.7. Blue tungstic oxide. Blue powder, insoluble in aqua regia. **t. tetrachloride*** WCl_4 = 325.7. Gray crystals, decomp. by water or heat. **t. trioxide*** WO_3 = 231.8. Tungstic acid anhydride. Yellow rhombs, d.7.16,m. red heat, insoluble in water; used to make tungsten for lamp filaments. **t. trisulfide*** WS_3 = 280.0. Black powder, slightly soluble in water.

tungstic Pertaining to pentavalent tungsten [tungsten(V)*, (5+)* or wolfram(V)*, (5+)*] or to hexavalent tungsten [tungsten(VI)*, (6+)* or wolfram(VI)*, (6+)*]. **t. acid*** The mono- and dihydrates of WO_3. Yellow crystals. **poly ~** See *polytungstate acids* under *tungstate.* **phospho ~** See *phosphotungstic acid.* **t. ocher** Tungstite.

tungstite Tungstic ocher. Wolframine. Native tungsten trioxide.

tungstyl The radical ≡WO.

tunicin Cellulose from animal tissues, similar to cotton cellulose but contains no pentoses; galactose and glucose units are present. Found in tunicates, e.g., *Phallusia mammillata.*

tuning fork A forked-shaped metal instrument with 2 prolonged arms which may be set in rapid and sustained vibration at almost constant frequency, by means of a sharp blow, or electrically. Used to synchronize vibrations.

tuno gum Chicle.

tupelo A forest tree of N. America, *Nyssa aquatica,* or cotton gum tree (Cornaceae) of Mississippi. A commercial timber.

turacin A crimson pigment containing copper from the feathers of the turakoo, an African bird. Cf. *tetronerythrin.*

turaco-porphyrin A chromophoric substance, similar to hematoporphyrin, from turacin.

turanose $C_{12}H_{22}O_{11}$ = 342.3. 3-*O*-α-D-Glucopyranosyl-D-fructose. A disaccharide, consisting of glucose and fructose, formed by partial hydrolysis of melezitose.

turbid Describing the slight cloudiness of a solution caused by fine suspended particles.

turbidimetry Determination of the quantity of fine suspended particles in a liquid, by measuring the thickness of liquid that produces a reduction in visual transmission equivalent to that of a standard solution or a standard pattern. Cf. *nephelometry.*

turbidity value The temperature at which a solution of an oil in a solvent, e.g., alcohol, shows the first signs of turbidity when cooled under specified conditions. Cf. *Valenta value.*

turbostratic Describing a mesomorphous structure in which the layers of atoms, though parallel, are randomly displaced with respect to one another. Coke is t.; graphite is not.

Turek detonator Trinitrotriazidobenzene.

turgor pressure The excess of diffusion pressure of a solute in an osmometer, over the diffusion pressure of the solute in the solution at atmospheric pressure.

turkey **t. pea** Tephrosia. **t. red** (1) Harmala red. A color from the seeds of *Peganum harmala* (Rutaceae). (2) Madder. **t. red oil** Obtained by the action of cold concentrated sulfuric acid on castor oil. Used as a wetting agent, and in preparation of textiles for dyeing.

turmeric Indian saffron, curcuma. The dried rhizomes of *Curcuma longa* (Scitaminaceae); a condiment and indicator. **t. root** Hydrastis.

turmeron A pungent sesquiterpene constituent of turmeric.

Turnbull's blue The same compound as prussian blue, but obtained from excess of a ferrous salt and a solution of potassium hexacyanoferrate(III).

turnsole (1) Litmus. (2) A dye prepared from *Chrozophora tinctoria* (Euphorbiaceae) of the Mediterranean.

turpentine Pine-cone oil. Terebenthene. Terebinthina. An oleoresin from the *Pinus* species. Yellow, sticky masses of balsamic odor. Sources of turpentine:

Aleppo . . . . *Pinus halepensis*
Bordeaux . . . . *Pinus maritima*
Canada . . . . *Pinus maritima*
Carpathian . . . . *Pinus cembra*
Common . . . . *Pinus palustris, P. sylvestris*
Hungarian . . . . *Pinus pumilio*
Larch . . . . *Larix europaea*
Strassburg . . . . *Abies pectinata*
Venice . . . . *Larix europaea*

Cf. *rosin, terpinene.* **chinese ~** The volatile oil from *Pistacia terebinthus* (Anacardiaceae).

t. oil An essential oil distilled from turpentine; contains pinene, sylvestrene, cinene. Colorless, volatile liquid, d.0.869, insoluble in water, soluble in oils. A carminative; solvent; vehicle for paints and disinfectants; and, with soap and camphor, a liniment (BP). See Table 97.

turpeth mineral Mercuric subsulfate.

turquoise Callaite, callainite. A hydrous gem phosphate of aluminum, colored blue by copper.

turtle oil An oil from the muscles and genital glands of the giant sea turtle, d.0.9112, m.25, iodine no. 64.6; used as a cosmetic.

TABLE 97. ANALYSES OF TURPENTINES

Source	α-Pinene		β-Pinene		Limonene (%)	Carene (%)
	(%)	Specific rotation	(%)	Specific rotation		
U.S.A.	75	+25	20	−21.8	2	
Australia			60			
China	92	−37	4		2	
India	77	+46.0				20
France	60	−44.3	27	−21.5	2	
Greece	95	+46	2		3	
Japan	85	−41	10			5
New Zealand	35		65	−21.3		
Portugal	80	−42	17	−21.5	3	
Russia	75	+28.8				15
Sweden	80		5			15

tussah Tussore.
Tusscapine Trademark for noscapine.
tussore Tussah. Wild silk; coarser and darker than true silk. Cf. *anaphe.*
tutin $C_{15}H_{18}O_6$ = 294.3. Crystals, m.212. A poisonous glucoside from the toot plant, a *Coriaria* species of New Zealand.
Tutton's salts $M'M''(SO_4)_2 \cdot 6H_2O$. Double salts, resembling the *alums*; q.v.; where M' is NH_4, K, Tl, etc.; M" is Mg, Zn, Fe, etc.; and S can be replaced by Se.
tuyere Tweer. A pipe inserted into the walls of a furnace through which an air blast is forced.
T value A measure of the base-exchange capacity of a soil.
Tw Abbreviation for degrees on the Twaddell hydrometer.
Twaddell A technical hydrometer scale, q.v. *(hydrometer scale),* named for the inventor. If the specific gravity is *d*, $Tw^\circ = 200(d - 1)$.
Twaddle Twaddell.
Tweens Trademark for a group of surface-active agents.
tweer, twere Tuyere.
twill A fabric weave in which the warp and weft pass alternately over one and under two threads.
twin One of a closely connected pair. **t. crystals** A pair of crystals that have grown in contact, usually to form a symmetrical figure (a cross or star). **t. electrons** A pair of electrons supposed to form a chemical bond, each atom providing one electron. See *polar bond.* **t. nuclei** A bicyclic structure, e.g., 2 connected rings, as in naphthalene.
twinning The plastic deformation of low-symmetry metals, e.g., hexagonal, as distinct from the slipping of cubic crystalline metallic structures. Each layer of atoms slips a constant amount over that below, so that the resulting twinned plane is the mirror image of the original lattice.
twist conformation See *conformation.*
Twitchell **T. reagent** $C_{18}H_{35}O_2 \cdot C_{10}H_6SO_3H$(?). A catalyst of fat hydrolysis, prepared by the action of sulfuric acid on oleic acid and naphthalene. **T. process** The splitting of fats by steam, catalyzed by about 0.5% T. reagent. Cf. *Pfeilring reagent.*
two-four-eight Toxisterol: absorption spectrum maximum, 248 nm.
TXDS Acronym for toxic dose(s).
Tygan Lustrus. Trademark for a mixed-polymer synthetic fiber.
Tylenol Trademark for acetaminophen.
Tylose Trademark for methylcellulose.
tylosin A mixture of antibiotics (desmycosin, macrocin, relomycin, and lactenosin) produced by *Streptomyces fragiale* growing on certain soils.
tympan A thick paper, often impregnated with oil or glycerol, used for backing up sheets before printing, or for interleaving.
Tyndall, John (1820–1893) British physicist. **T. cone effect, phenomenon** The path of light through a heterogeneous medium is made visible by the solid particles; as, a sunbeam in air. Cf. *scattering, ultramicroscope.*
tyndallimetry The estimation of the suspended matter in a solution by measuring the intensity of the scattered light from a Tyndall cone. Cf. *turbidimetry.*
tyndallization Sterilization, e.g., of media, by heating in stages.
Tynex Trademark for a polyamide synthetic fiber.
type A general or prevailing character. **t. of compounds** (1) An arbitrary classification of organic compounds which groups together those obtainable from one another by substitution. (2) An indication of the presence of typical radicals; as, —OH. Cf. *classification of compounds.* **t. metal** An alloy for making printers' type: Pb 7, Sb 2 pts., with small amounts of Sn, Bi, Ni, or Cu. **t. reaction** A reaction common to a group of related substances.
Typel Trademark for an acetate synthetic fiber.
typhasterol A phytosterol from the pollen of *Typha orientalis* (Typhaceae). **alpha-** ~ White powder, m.133.
typhoid An infection, principally of the intestine, due to *Salmonelli typhi,* contracted from contaminated water or food. **t. vaccine** A suspension of killed *Salmonelli typhi,* given to prevent t. (USP, EP, BP).
typhotoxin A ptomaine from cultures of Eberth's bacillus.
typhus vaccine A sterile suspension of the killed rickettsial organisms of a strain of epidemic typhus (USP, BP). See *Rickettsiae.*
typical Having a certain characteristic, property, or standard. **t. compounds** Parent compounds. **t. elements** The most abundant element of each group of the periodic table.
Tyr* Symbol for tyrosine.
tyramine $NH_2 \cdot CH_2 \cdot CH_2 \cdot C_6H_4 \cdot OH$ = 371.2. 2-(*p*-Hydroxyphenyl)ethylamine. 4-(2-aminoethyl)phenol. An alkaloid obtained from ergot, or by heating tyrosine. White crystals, soluble in water; a sympathomimetic agent.
tyratol Thymol carbamate.
Tyrian purple Murex.
tyrolite Copper froth.
tyrosine* $HO \cdot C_6H_4 \cdot CH_2 \cdot CH(NH_2) \cdot COOH$ = 181.2. Tyr*. 2-Amino-3-*p*-hydroxyphenylpropanoic acid*. An essential amino acid occurring in (+) (m.311) and (−) (m.295) forms; obtained by hydrolysis of many proteins and from old cheese. (Greek *tyros:* "cheese".) Silky needles, slightly soluble in water: ***ortho-*** ~ m.249 ***meta-*** ~ m.280. Cf. *erythrosin.* **3,5-diiodo** ~ **(*S*)-** ~ m.200 (decomp.); in thyroglobulin, sponges, and coral.
tyrosol $HO \cdot C_6H_4 \cdot (CH_2)_2OH$ = 138.2. 4-Hydroxyphenethyl alcohol. White rhombs, m.93, formed during putrefaction of tyrosine.
tyrosyl* The radical $HO \cdot C_6H_4 \cdot CH_2 \cdot CHNH_2 \cdot CO-$, from tyrosine.
tyrothricin An antibiotic from cultures of the aerobic soil bacterium *Bacillus brevis.* White powder, m. approx. 240, insoluble in water (USP). Contains a soluble (gramicidin) and insoluble (tyrocidine) fraction.
tyrotoxicon Tyrotoxin.
tyrotoxin $PhN{:}N \cdot OH$ = 122.1. Diazobenzene hydroxide. A ptomaine in stale milk or ice cream. Yellow needles, m.90, are produced with a solution of auric chloride.
tysonite $(Ce,La,Nd,Pr)F_3$. A native fluoride.
Tyvek Trademark for a spunbonded nonwoven material made from high-density polyethylene.

U

U Symbol for uranium.
U Symbol for (1) electric potential difference; (2) internal energy.
u Symbol for ultraviolet-protected, as uPVC.
u Symbol for electric mobility of an ion.
Υ See *upsilon.*
ν See *nu.*
μ See *mu.*
uabain Ouabain.
ubiquinone(s) Coenzyme Q. Widely distributed in human and animal organs; associated with vitamin E deficiency (characteristic absorption at 272 nm). Ubiquinones have the structure

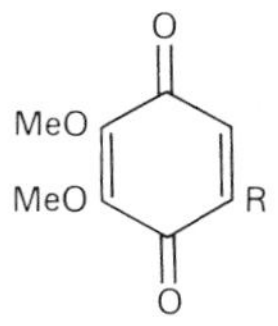

R is $-[CH_2 \cdot CH{:}CMe \cdot CH_2]_{6-10}H$.
ucuhuba fat A fat extracted from the ground kernels of *Virola surinamensis* (Myristicaceae), tropical America; d.0.90, m.47, $[n]_D^{50}1.450$.
U effect Generation of an alternating voltage by the mechanical vibration of a double layer in a glass capillary.
U-F Abbreviation describing a urea-formaldehyde resin.
Uformite Trademark for a urea-formaldehyde plastic.
ufuta Sesame.
UHF Ultrahigh frequency. See *radiation,* Table 69.
UHT See *UHT milk* under *milk (1).*
uintaite Gilsonite.
ukambine An alkaloid from African arrow poisons, similar to strophanthin in action.
ulexine $C_{11}H_{14}ON_2$ = 190.2. An alkaloid from the seeds of *Ulex europaeus,* European gorse and laburnum. Cf. *cytisine.*
ulexite $NaCaB_5O_9 \cdot 8H_2O$. A hydrous borate (California).
ullage The amount by which a container is short of being full.
ullmannite NiSbS. A native sulfantimonide.
ulmic acid $C_{20}H_{14}O_6$ = 350.3. Geic acid; from peat and elm bark.
ulmin $C_{40}H_{16}O_{14}$ = 720.6. A gum from the sap of *Ulmus fulva,* slippery elm (Ulmaceae). **amin ~** A gum formed in coal and peat by the action of amino acids on carbohydrates.
ultimate Fundamental or basic. **u. analysis** Elementary *analysis.* **u. lines** See *raies ultimes.* **u. rational units** URU. A suggested system of measurement based on the charge of an electron, $(4\pi e)^2$, which has the dimension of energy times length, and from which all other units may be derived.
ultra- Prefix (Latin, "beyond") indicating values outside certain limits.
ultracentrifuge A high-speed centrifuge for determining the size and distribution of particles in amicroscopic colloids. **McBain's ~** See *McBain's centrifuge.*
ultrafiltration Filtration by suction or pressure through a colloidal filter or semipermeable membrane; used to prepare colloidal solutions and to determine particle size in terms of a standard ultrafilter.
ultra-gamma rays Cosmic rays.
ultramarine $Na_3Al_3Si_3S_2O_{12}$. Artificial lapis lazuli. A blue pigment: sodium and aluminum silicates with sodium polysulfides. **genuine ~** Lapis lazuli. **synthetic ~** U. prepared by melting clay, soda, and sulfur or coal; used as a paper and textile pigment. **yellow ~** Barium chromate.
u. green U. having a green shade.
ultramicron A particle less than 0.10 μm (micron) in diameter; the smallest visible under a microscope.
ultramicroscope A microscope in which the object is brightly illuminated at right angles to the optical axis, to detect particles smaller than 0.1 μm, which appear as dots of bright light. Cf. *Tyndall cone effect.*
ultramicroscopic Beyond the range of microscopic visibility, but detectable by the ultramicroscope.
ultraphonic Ultrasonic.
ultraphotic The invisible rays of the ultraviolet and infrared regions.
ultrared Infrared.
ultrasonic Ultraphonic. Describing high-frequency sound waves above the limit of human hearing (about 20 kHz), as the 200 kHz produced by applying alternating current to a quartz crystal, or by electromagnetic oscillation of a metal immersed in a liquid. Used in medicine as ultrasonography (1–10 MHz), to make emulsions, descale metals, sterilize milk, and depolymerize macromolecules. Cf. *supersonic.* **u. sewing** The fusion or interlocking of man-made fibers by heat from u. vibrations.
ultrastructure Structure as seen under the ultramicroscope.
ultrathermometer Beckmann thermometer.
ultraviolet u.v. That portion of the spectrum just beyond the violet on the short-wavelength side: generally 100–4,000 Å. Emitted by sunlight; and the carbon, mercury-vapor, tungsten, and Kronmeyer lamps. U. radiation induces chemical activity, produces fluorescence, has therapeutic properties, and induces the formation of vitamins in sterols. Cf. *irradiation, fluorescence analysis.* **far ~** 100–2,000 Å. **middle ~** 2,000–3,000 Å. **near ~** 3,000–4,000 Å. **vacuum ~** Below about 1,900 Å. So called because air effectively absorbs it, necessitating working with evacuated spectrometers. **vital ~** 2,900–3,100 Å. Cf. *radiation.*
u.-A Radiation, 3,200–4,000 Å, as in sunlight. Causes suntan by darkening of preformed melanin. **u.-B** Radiation, 2,900–3,200 Å, as in sunlight; increases with altitude and is reflected by snow. Radiation of 2,800–2,950 Å is necessary for producing vitamin D from provitamins, but excess burns (particularly, dry) skin. **u.-C** Radiation from sunlamps (as, hydrogen and mercury). Can damage the retina.
Ultrawets Trademark for alkylated monosodium benzene sulfonate detergents and wetting agents.
ultra-X-rays Cosmic rays.

umangite $CuSe \cdot Cu_2Se$. A native selenide. Cf. *berzelianite.*
umbellic acid $(HO)_2C_6H_3CH{:}CH \cdot COOH = 180.2$. 2,4-Dihydroxyphenyl-2-propenoic acid*. *p*-Hydroxycoumaric acid, obtained from umbelliferone by heating with alkali. Yellow powder, decomp. 125, soluble in water.
Umbelliferae Parsley family, with hollow stems and umbrella-shaped flowers. Many yield essential oils, spices, or drugs; e.g.: *Pimpinella anisum,* anise; *Angelica officinalis,* angelica root; *Conium maculatum,* hemlock leaves; *Ferula foetida,* asafetida resin. Cf. *hentriacontane, mitsubaene.*
umbelliferone $OH \cdot C_6H_3 \cdot CH{:}CH \cdot CO \cdot O = 162.1$. 7-Hydroxycoumarin. A lactone in galbanum and umbelliferous plants. Colorless crystals, m.223 (sublime), slightly soluble in water. **methyl ~** Resocyanin.
umbellularic acid $C_8H_{12}O_4 = 172.2$. 1-Isopropylcyclopropane-1,2-dicarboxylic acid. Crystals.
umbellulone $C_{10}H_{14}O = 150.2$. A ketone from the oil of *Umbellularia californica* (Lauraceae), $b_{70mm}139$. The chief constituent of California laurel oil, readily changed to thymol by heating under pressure.
umber Raw umber. A native ferric hydroxide containing manganese dioxide and silicate; a brown pigment. **burnt ~** A warm, reddish-brown pigment produced by heating raw u.
umbonate Describing a bacterial culture that has a buttonlike raised center.
umpolung (German: polarity inversion) Reagents displaying reactivity opposite to that of traditional reagents. Thus organolithium reagents are nucleophiles, while many traditional synthetic building blocks (as, ketones) are electrophiles.
unary Composed of molecules that are physically and chemically identical. Cf. *association, dissociation.*
uncertainty principle Heisenberg principle.
undeca-* Hendeca. Affix indicating 11 times.
undecalactone **γ- ~** $C_{11}H_{20}O_2 = 184.3$. 5-Hydroxyundecanoic acid δ-lactone, peach aldehyde. b.286. Used in perfumes and flavors.
undecanal* $Me(CH_2)_9CHO = 170.3$. Undecylic aldehyde, hendecanal. Colorless liquid, d.0.825, b.117.
undecane* $Me(CH_2)_9Me = 156.3$. Hendecane. Colorless liquid, $d._{15}$0.74, b.194, insoluble in water.
undecanoic acid* $Me(CH_2)_9COOH = 186.3$. Undecylic acid, hendec(an)oic acid. A constituent of castor oil. Colorless scales, m.28, soluble in water. **cyclopentyl ~** Hydnocarpic acid.
undecanol* $C_{11}H_{23}OH = 172.3$. **1- ~** Undecyl alcohol*, hendecyl alcohol. Colorless liquid, d.0.833, m.19. **6- ~** Colorless liquid, d.0.833, m.16.
undecanone* $C_{11}H_{22}O = 170.3$. Hendecanone. **1- ~** Undecanal*. **2- ~** $MeCOC_9H_{19}$. Methyl nonyl ketone*. Colorless liquid, d.0.826, m.121., in oils of rue and lime. **6- ~** $(C_5H_{11})_2CO$. Dipentyl ketone*. Colorless liquid, d.0.826, m.15.
undecene* $C_{11}H_{22} = 154.3$. Undecylene, hendecene. **1- ~** $CH_2{:}CH(CH_2)_8Me$. Colorless liquid, d.0.763, b.188. **2- ~** $MeCH{:}CH(CH_2)_7Me$. A liquid, d.0.774, b.193. Both are insoluble in water.
undecenoic acid* $CH_2{:}CH(CH_2)_8COOH = 184.3$. Undecylenic acid. **1- ~** Yellow liquid, b.295, insoluble in water. **10- ~** Mycota. A fungistatic in creams and dusting powder (USP, EP, BP).
undecenyl* The radical $C_{11}H_{21}-$, from undecene.
undecyl* Hendecyl. The radical $C_{11}H_{23}-$, from undecane. **u. alcohol** 1-Undecanol*. **u.amine*** $C_{11}H_{23}NH_2 = 171.3$. Hendecylamine. Colorless liquid, m.17.
undecylene Undecene*.
undecylenic acid USP, EP, BP name for 10-undecenoic acid*.
undecylic **u. acid** Undecanoic acid*. **u. alcohol** Undecanol*. **u. aldehyde** Undecanal*.
under Below normal. **u.cooling** Supercooling. **u.meter** Venturi meter. **u.voltage** The difference between the potential of a positive hydrogen electrode and that of a reversible hydrogen electrode. Cf. *overvoltage.*
undulate Describing (1) a regular wavelike motion; (2) a bacterial growth with wavy borders and surface.
undulation (1) A wavelike motion. (2) Periodic expansion and contraction, as of steel rails.
undung Laurel tallow. A vegetable fat from *Litsea (Tetranthera) laurifolia* (Lauraceae), tropical Asia.
unedol The reducing aglucone of unedoside.
unedoside A glucoside from *Arbutus unedo;* its reducing power is decreased by hydrolysis (unique for a glucoside).
ung. Abbreviation for "unguentum."
unguentum (1) Latin for an ointment. (2) A simple ointment.
uni- Prefix (Latin) indicating "one." Cf. *mono-.*
uniaxial (1) Having only one axis; as, a crystal that does not doubly refract. (2) Having properties in one direction only, as along one crystal axis.
unicellular Consisting of a single cell; as, protozoa.
unicorn A plant of the genus *Martynia* (Pedaliaceae). **false ~** Aletris.
unified atomic mass constant* m_u. Dalton. Atomic mass unit (amu). A unit equal to 1/12th the mass of ^{12}C; i.e., 1.66056×10^{-27} kg.
unifrequent Describing a homogeneous beam of light consisting of rays of similar wavelengths.
unimolecular Monomolecular.
union A fabric having a flax warp and cotton weft; or vice versa.
un-ionized Not ionized; in a nondissociated molecular form.
unit (1) A quantity used as a measure. (2) A standardized equipment comprising a definite arrangement of devices, considered as a whole; as in a unit process. (3) 200 ft^3 of uncompacted wood chips for pulping, being roughly the volume given by 1 cord. Cf. *cunit.* **angström ~** See *angstrom.* **Board of Trade ~** See *Board of Trade unit.* **British thermal ~** See *British thermal unit.* **capacitance ~** Farad. **cgs-, CGS- ~** A metric u. expressed in terms of centimeter-gram-second. **derived ~** See *derived units.* **electrical ~** See *electrical units.* **electromagnetic ~** See *electromagnetic units.* **electrostatic ~** See *electrostatic units.* **emu ~, EMU ~** Electromagnetic u. **energy ~** Joule. **esu ~, ESU ~** Electrostatic u. **fertilizer ~** Plant food u., kg/ha or 0.01 cwt/acre. **force ~** Newton. **heat ~** Joule. **international ~** See *international unit.* **light ~** See *candela, lux.* **metric ~** See *metric.* **mks ~, MKS ~** Metric u. expressed in terms of meter-kilogram-second. Cf. *cgs.* **quantum ~** See *Planck constant.* **repeating ~** A group of atoms that occurs repeatedly in a chain molecule; as, CH_2 in $X(CH_2)_nY$. Cf. *polymer.* **SI ~** See *SI.* **ultimate ~** See *ultimate rational units.* **uru, URU ~** See *ultimate rational units.* **work ~** Joule.
u. pole See *unit pole* under *pole.*
United States Adopted Name USAN. Abbreviation indicating an approved, nonproprietary name for a drug.
univalent* Monovalent. (1) Having a valency of 1; as, Na.
universal General; applicable in all cases. **u. indicator** A pH *indicator,* q.v., which changes color over the whole range of pH values. **u. series constant** Rydberg constant.

unofficial Describing a drug not authorized by a pharmacopoeia or formulary.
unorganic Inorganic (obsolete).
unorganized (1) Having no cellular or protoplasmic structure. (2) Amorphous. **u. ferment** See *unorganized ferment* under *ferment.*
unsatisfied Describing a hydrocarbon having one or more free valencies. Cf. *unsaturated.*
unsaturate. Abbreviation for unsaturated hydrocarbon.
unsaturated (1) Describing a solution capable of dissolving more solute. (2) Describing an organic compound having double or triple bonds; as, ethylene.
unslaked lime Calcium oxide.
unstable Readily decomposing; as, hydrogen peroxide.
unsymmetrical Not symmetrical; as, the 1,2,4 positions of the benzene ring.
U.P. Underproof. See *proof spirit.*
upas A Javanese arrow poison containing strychnine (Malay, ipoh). **bohan ~** A poisoning resin from the tree *Antiaris toxicaria* (Java). Cf. *antiarin.*
uperization The sterilization of milk by very rapid heating to 150°C, followed by immediate cooling; more permanent in effect than pasteurization.
upsilon υ, Υ. Greek letter.
uptake An exit pipe leading upward.
ur* Indicating urea as a ligand.
uracil HN·CO·NH·CO·CH:CH = 112.1. 2,4(1*H*,3*H*)-Pyrimidinedione. A pyrimidine base in nucleic acids, m.335. **methyl ~** Thymine.
uraconite Uranic ocher.
uralite A variety of amphibole (Ural mountains). (2) Asbestos impregnated with sodium silicate and hydrogencarbonate, and chalk; fireproof.
uralium A supposed element isolated by Guyard from platinum ores; identical with rhenium.
uramido Indicating the ureido* radical.
uramil NH·CO·NH·CO·CH(NH_2)CO = 143.1. Dialuramide, aminobarbituric acid, murexan. Colorless crystals, soluble in water; used in organic synthesis. **thio ~** See *thiouramil.*
uramine Guanidine*.
uramino Indicating the ureido* radical.
uranate* Indicating uranium as the central atom(s) in an anion; as, $M_2U_2O_7$, M_2UO_4, and M_4UO_5.
urane (1) A unit of radioactivity: 0.001 kilurane. (2) Uranium oxide. (3) Urethane.
uranic Uranium(VI)* or (6+)*. Indicating hexavalent uranium; as, uranium fluoride. **u. acid** H_2UO_4 = 304.0. Metauranic acid, uranyl hydroxide. Yellow, insoluble powder; u. oxide in various degrees of hydration. **u. ocher** U_2O_3. Uraconite. Yellow, native uranium oxide, containing radium. **u. oxide** UO_3 = 286.0. Uranium trioxide. Orange powder, insoluble in water; used in ceramics, glass, paint, and textiles.
uranine See *sodium fluorescein* under *fluorescein.*
uraninite UO·U_2O_3. Pitchblende. Native uranium oxide, a source of radium.
uranites Mineral phosphates of uranium, with calcium or copper.
uranium* U = 238.0289. A heavy metal, the last stable element of the periodic system, at. no. 92. At. wt. of isotope with longest half-life (^{238}U) = 238.0508. A hard, heavy, nickel-white metal, d.18.68, m.1132, b.3820, insoluble in water or alcohol, soluble in acids. It is radioactive, consists of the isotopes ^{238}U and ^{235}U, which occur in nature in the ratio 140:1. They can be disintegrated by fast and slow neutrons, respectively, and the latter are used to start the chain reaction which is the basis of the atomic bomb (u. pile). Cf. *radioactive disintegration.* U. is obtained from its ores by anion exchange of the complex $UO_2(SO_4)_3$. Cf. *moving bed process.* There are about 4 × 10^9 t of u. in the oceans. U. forms principally tetravalent (u.(IV)*, u.(4+)*, uranous) or hexavalent (u.(VI)*, u.(6+)*, uranic, uranyl, and uranate) compounds; valencies 3 and 5 also exist. The yellow oxide, UO_3, is amphoteric, and forms uranates with bases, and uranyl salts with acids. **depleted ~** U. from which most of the fissile ^{235}U has been removed. Used, e.g., UF_6, as a source of U for nonnuclear purposes, e.g., alloys. **enriched ~** See *uranium enrichment.* **trans ~** See *transuranic elements.*
U_1 U-238. U_2 U-234. **u. acetate** (1) Uranyl acetate. (2) Sodium uranyl acetate. **u. bromide** UBr_4 = 557.6. Uranous bromide. Black leaflets, soluble in water. **u. carbide** (1) UC_2 = 262.1. A solid, d.11.3, m.2260. (2) U_2C_3 = 512.1. A solid, d.11.28, m.2400. **u. chloride** (1) UCl_4 = 379.8. Uranous chloride. Green cubes, soluble in water. (2) UCl_3 = 344.4. U. trichloride*. Purple, soluble crystals. **u. enrichment** The creation of u. with a higher ^{235}U content than the 0.7% occurring naturally. Achieved by gas diffusion or gas centrifuge processes using UF_6, and possibly in the future by atomic vapor (AVLIS) or molecular (MLIS) laser isotope separation. Plant capacity is measured in separative work units (SWU). **u. fluorides** (1) UF_4 = 314.0. Green salt. (2) UF_6 = 352.0. Sublimes 56 (760 mm). Both are used to concentrate ^{235}U by the diffusion process, being more convenient than the volatile U compounds. **u. glass** Yellow glass with green fluorescence; contains u. oxides. **u. hydroxide** $U(OH)_4$ = 306.1. Green, insoluble powder. **u. iodide** UI_4 = 745.6. Uranous iodide. Yellow, monoclinic crystals, m.500, soluble in water. **u. lead** (1) Pb-206. (2) A mixture of Pb-206 and Pb-207. See *radiolead.* **u. minerals** Numerous and complex, and contain the phosphates and vanadates of rare earths; carnotite, uraninite, becquerelite, and autunite are radioactive. **u. oxides** Principally: dioxide UO_2 (uranous oxide); trioxide UO_3 (uranic oxide); octaoxide U_3O_8; tetraoxide UO_4. **u. oxychloride** Uranyl chloride. **u. pile** See under definition of *uranium.* **u. series** See *radioactive elements.* **u. sulfate** $U(SO_4)_2·8H_2O$ = 574.3. Native as uranvitriol and zippeite. Green prisms, m.300, decomp. by water. **u. sulfide** US_2 = 302.2. Uranous sulfide. Gray powder, m.1100, insoluble in water. **u. tetrabromide*** U. bromide. **u. tetrachloride*** U. chloride. **u. X_1** Thorium-234. **u. X_2** Protactinium-234. **u. Y** Thorium-231. **u. yellow** Sodium uranate; a ceramic and glass pigment.
uranocircite $(UO_2)_2BaP_2O_8$. A mineral. Cf. *autunite.*
uranophane $U_2SiO_3·CaSiO_8$. A native silicate.
uranopilite CaO·$8UO_3$·$2SiO_2$·$25H_2O$. A mineral from Colorado. **beta-~** CaO·$8UO_3$·$2SiO_3$·$25H_2O$.
uranospathite $(UO_2)_3(PO_4)_2·nH_2O$. A mineral.
uranospherite $U_2O_7(BiO)_2·3H_2O$. An orange, scaly mineral containing radium.
uranospinite $(UO_2)_2CaAsO_8$. A native arsenate.
uranothallite $2CaCO_3·U(CO_3)_2·10H_2O$. A mineral from W. Czechoslovakia.
uranous Uranium(IV)* or (4+)*. Describing a compound containing tetravalent uranium. **u. chloride** See *uranium chloride.* **u. oxide** UO_2 = 270.0. Uranium dioxide*. Black octahedra, d.10.95, m.2806, insoluble in water. Nuclear reactor fuel. **u. uranic oxide** Uranyl uranate.
uranvitriol Johannite. A mineral containing copper.
uranyl* Dioxouranium(VI)*. The radical $=UO_2$, from UO_3; it forms many salts with acids, which ionize to UO_2^{2+} (yellow in

solution). **u. acetate*** $UO_2(C_2H_3O_2)_2 \cdot 2H_2O$ = 424.1. Yellow crystals, soluble in water; a reagent. **u. ammonium carbonate** Ammonium uranyl carbonate*. **u. ammonium fluoride** Ammonium uranyl fluoride*. **u. benzoate*** $UO_2(C_7H_5O_2)_2$ = 512.3. Uranium benzoate. Yellow powder, slightly soluble in water. **u. calcium phosphate** Calcium uranyl phosphate*. **u. chloride*** UO_2Cl_2 = 340.9. Uranium oxychloride. Yellow, hygroscopic crystals, decomp. by heat, soluble in water. **u. ferrocyanide** Uranyl hexacyanoferrate*. **u. formate*** $UO_2(HCO_2)_2 \cdot H_2O$ = 378.1. Yellow octahedra, slightly soluble in water. **u. hexacyanoferrate(II)*** $(UO_2)_2[Fe(CN)_6]$ = 752.0. U. ferrocyanide. Brown powder, insoluble in water. **u. hydrogenphosphate*** $UO_2 \cdot HPO_4 \cdot 4H_2O$ = 438.1. Yellow crystals, insoluble in water; an atomic reactor fuel. **u. hydroxide*** $UO_2(OH)_2$ = 304.0. Uranic acid, H_2UO_4. White solid. **u. nitrate*** $UO_2(NO_3)_2 \cdot 6H_2O$ = 502.1. Uranium nitrate. Yellow, deliquescent crystals, m.59, soluble in water and organic solvents; used as a reagent and indicator, and in ceramics, glass, and photography. **u. oxalate** $UO_2C_2O_4 \cdot 3H_2O$ = 798.5. Uranium oxalate. Yellow powder, insoluble in water. **u. oxide** Uranic oxide. **u. potassium sulfate** Potassium uranyl sulfate*. **u. sodium acetate** Sodium uranyl acetate*. **u. sulfate*** $2UO_2SO_4 \cdot 7H_2O$ = 858.3. Yellow crystals, soluble in water. **u. sulfide*** UO_2S = 302.1. Brown powder, decomp. by heat, slightly soluble in water. **u. uranate** $(UO_2)_2UO_4$ = 842.1. Uranous uranic oxide. Green crystals, d.7.31, decomp. by heat, insoluble in water. **u. zinc acetate** A solution of u. acetate and zinc acetate in acetic acid; gives crystals of characteristic shape with sodium salts.

urao A native sodium carbonate and hydrogencarbonate (S. America).

urari Curare.

urate Lithate. A salt of uric acid. **u. oxidase*** See *uricase.*

urazine ***p*-~** $OC(NH)_2CO \cdot (NH)_2$ = 116.1. m.266. Cf. *biurea.*

urazole $CO \cdot (NH)_2 \cdot CO \cdot NH$ = 101.1, 1,2,4-Triazolidine-3,5-dione. Hydrazodicarbonimide. Colorless crystals, m.244, soluble in alcohol. **amino ~** Urazine. **1-phenyl ~** $C_2H_2O_2N_3Ph$ = 177.2. Colorless crystals, m.263. **3-phenyl ~** Colorless crystals, m.203, soluble in alcohol. **thio ~** See *thiourazole.*

Urbain, Georges (1872–1938) French chemist, noted for work on the rare earths.

-ure Suffix, together with -uro and -eto, used in Romance languages (e.g., French) in place of the English -ide (as in chloride).

Ure U., Andrew (1778–1857) Scottish chemist, author of *A Dictionary of Chemistry* (1821). **U. eudiometer** A long U-shaped glass tube closed at one end, in which are 2 electrodes with a graduated scale to measure gases.

urea* $NH_2 \cdot CO \cdot NH_2 \rightleftharpoons NH_2 \cdot C{:}NH(OH)$ (iso ~) = 60.1. Carbamide. Colorless tetragons, m.132 (decomp. to biuret and ammonia), soluble in water. The end product of mammalian protein metabolism, the chief nitrogenous constituent of urine, and the first organic compound synthesized (Wöhler, 1828). Used as an osmotic diuretic and in ointments (USP, EP, BP), as a reagent for lignin; and in pyrotechnics, organic synthesis, and plastics manufacture. Its derivatives have the prefix *ureido-** or suffix *-urea**. The radical $-NH \cdot CO \cdot NH-$ is ureylene*. Cf. *ureide.* **ψ-~** Isourea*. **acetonyl ~** See *acetonylurea* under *acetonyl.* **acetyl ~** See *acetylurea.* **alkene ~** $RHN \cdot CO \cdot NHR$ (ureylenes) or $R_2N \cdot CO \cdot NR_2$. **allylthio ~** See *allylthiourea* under *allyl.* **amidino ~*** See *amidinourea.* **amino ~** Semicarbazide*. **bi ~*** See *biurea.* **carbamoyl ~** Biuret*. **carbonyldi ~** Triuret*. **di ~** See (1) *biurea,* (2) *urazine.* **diamino ~** Carbonohydrazide*. **diphenyl ~*** Carbanilide*. **ethoxyphenyl ~*** Sucrol. **formaldehyde-~** See *plastics.* **imino ~** Guanidine*. **iso ~*** $NH_2C{:}NH(OH)$. Pseudo ~. The tautomer of u. **malonyl ~** Barbituric acid. **mesoxalyl ~** Alloxan. **methyl ~** $C_2H_6ON_2$ = 74.1. White prisms, m.103, soluble in water. **oxalyl ~** Parabanic acid. **phenylene ~** Benzimidazolone. **phenylhydrazine ~** Diphenylcarbazide. **pseudo ~** Iso ~*. **seleno ~** See *selenourea.* **tartronoyl ~** Dialuric acid. **thio ~*** See *thiourea.* **ureido ~*** Biurea*.

u. acetate A variable mixture of u. and acetic acid. **u. apparatus** A device for the rapid determination of u. by the action of urease. **u.carboxylic acid** Allophanic acid*. **u. citrate** $CH_4ON_2 \cdot C_6H_8O_7$ = 252.2. Colorless crystals, soluble in water. **u.form** A solid u.-formaldehyde condensation product with excess u.; a slow nitrogen-release fertilizer. **u.-formaldehyde** See *plastics.* **u. nitrate** $CH_4ON_2 \cdot HNO_3$ = 123.1. Colorless scales, soluble in water. **u. oxalate** $CH_4ON_2 \cdot C_2H_2O_4$ = 150.1. Colorless crystals, soluble in water.

ureameter Ureometer. An apparatus to determine urea concentration from the volume of nitrogen evolved. Cf. *urinometer.*

urease* A crystallizable protein enzyme in soybeans (its richest source), numerous fungi, and jack beans; hydrolyzes urea into ammonium carbonate. White octahedra, soluble in dilute alkali, isoelectric point pH 5.05. It is inactivated by metals; used to determine urea. Cf. *uricase.*

urechitin $C_{28}H_{42}O_8$ = 506.6. A glucoside from *Urechites suberecta,* the Savannah flower, yellow nightshade (Apocyanaceae).

urechitine $C_{24}H_{42}O_8 \cdot H_2O$ = 476.6. An alkaloid from *Urechites suberecta.*

ureide A derivative of urea; as, $NH_2 \cdot CO \cdot NHEt$, ethylurea. **cyclic ~** A compound formed by replacement of one H of each NH_2 group by a dibasic acid; as, in alloxan. **di ~** A compound containing 2 ureido radicals. **pseudo ~** $NH_2 \cdot CNH \cdot OR$.

ureido* (Aminocarbonyl)amino†. Carbamido, uramido, uramino. The radical $NH_2 \cdot CO \cdot NH-$, from urea. **iso ~*** **1-~** The radical $NH{:}C(OH) \cdot NH-$. **3-~** The radical $NH_2 \cdot C(OH){:}N-$. Both from isourea.

u.acetic acid Hydantoic acid*.

ureometer Ureameter.

ureous acid Xanthine.

-uret Suffix (obsolete) indicating a binary compound of sulfur, arsenic, phosphorus, carbon, etc., with some other element. Cf. *-ide.*

urethan Urethane.

urethane $NH_2 \cdot CO \cdot OEt$ = 89.1. Ethyl carbamate, urane ethylurethane. Colorless needles, m.49, soluble in water; a hypnotic. **ethyl ~** Urethane. **ethylidene ~** See *ethylideneurethane.* **pentyl ~** Pentyl carbamate*. **phenyl ~** See *phenylurethane.* **piperidyl ~** See *piperidylurethane.* **poly ~** See *polyurethanes.* **thio ~** See *thiourethane.* **thymol ~** Thymol carbamate.

urethanes In general, carbamic esters.

urethano The ethoxycarbonylimino* radical.

urethylan $NH_2 \cdot CO \cdot OMe$ = 75.1. Methylurethane, methyl carbamate. White crystals, m.52, soluble in water.

Urey, Harold Clayton (1894–1981) American chemist. Nobel prize winner (1934). Noted for his work on heavy hydrogen and the first H bomb.

ureylene* Carbonyldiimino†, urylene. The radical $-NH \cdot CO \cdot NH-$, from urea.

uric acid $C_5H_4O_3N_4$ = 168.1. Triketopurine,

2,6,8-trioxopurine. 2,6,8-(1*H*,3*H*,9*H*)-Purinetrione*. Has other tautomeric forms. Keto form:

```
HN—CO               1—6
 |   |              |  |
OC   C—NH\          2  5—7\
 |   ||    >CO      |  |    >8
HN—C—NH/            3—4—9/
```

Colorless scales, d.1.85, decomp. by heat, slightly soluble in water. The end product of the purines of muscle and cell nuclei, normally present in urine in small amounts. **pseudo ~** $C_5H_6O_4N$ = 144.1. Hexahydro-2,4,6-trioxo-5-pyrimidinylurea. m.260+. **trimethyl ~** Caffeine. Cf. *purine.*

uricase Urate oxidase*. An enzyme in animal tissues that oxidizes urates, possibly into allantoin, urea, and glycine. Cf. *urease.*

uridine $C_4H_3N_2O_2 \cdot C_5H_9O_4$ = 244.2. Uracil-D-riboside. A nucleoside, m.165, from nucleic acid. **u. phosphoric acid** A nucleotide, m.202, from nucleoproteins.

uril RNH·CO·NH·NH·CO·NHR. See *biurea.*

urinalysis Analysis of urine.

urine A fluid secreted by the kidneys and discharged from the bladder (1–2.5 l per h). Normally a clear, slightly acidic, amber liquid, d.1.005–1.030. Composition: water 96, urea 2.3, sodium chloride 1.1, phosphates 0.2, sulfates 0.1%. *Abnormal constituents:* albumin, bacteria, sugar, pus, blood, diacetic acid, indican, hydrogen sulfide. See *Bang method, Benedict's solution, urinometer.*

urinoid 3-Cyclohexen-1-one*.

urinometer Urometer. A hydrometer for determining the specific gravity of urine. Cf. *ureameter.*

-uro See *-ure.*

urobenzoic acid Hippuric acid*.

urobilin $C_{32}H_{42}O_6N_4$ = 578.7. Hydrobilirubin. A bile pigment produced by the metabolism of bilirubin in the gut, some being removed via the kidney. Brown, resinous mass, soluble in alcohol; a reagent.

urocanic acid NH·CH:C·(CH:CH·COOH)N:CH = 138.1. Imidazole-4-acrylic acid. **(*E*)- ~** A ptomaine from histidine, in dog's urine. White crystals, m.224, slightly soluble in water. Cf. *urocaninic acid.*

urocanin $C_{11}H_{19}ON_4$ = 223.3. A base in dog's urine.

urocaninic acid $C_6H_6O_2N_2 \cdot 2H_2O$ = 174.2. 3-(1*H*-Imidazol-4-yl)-2-propenoic acid. An acid from dog's urine, decomp. by heat to carbon dioxide and urocanin. Cf. *urocanic acid.*

urochrome $C_{43}H_{51}O_{26}N$ = 997.9. A yellow coloring matter in urine.

urochromogen A tissue substance oxidized to urochrome.

urocitral Theobromine sodium citrate.

uroerythrin An orange pigment in urine.

urolith A calculus in urine.

Urolucasil Trademark for sulfamethizole.

urometer See *urinometer, ureameter.*

uronic acids (1) Glycuronic acids. CHO·(CHOH)$_n$COOH. *Aldonic acids,* q.v., whose primary alcohol group has effectively been oxidized to a —CHO group; as, galacturonic acid. (2) Lactones of (1); as, ascorbic acid. **hept ~** CHO(CHOH)$_5$COOH.

uronium* Affix for quaternary salts of urea and isourea.

urophan A substance which passes chemically unchanged into the urine.

uropittin $C_9H_{10}O_3N_2$ = 194.2. A resinous decomposition product of urochrome.

uroprotic acid $C_{66}H_{116}O_{54}N_{20}S$ = 2086. An acid protein from urine.

uropterin A yellow pigment from the purine fraction of human urine.

uroxameter value A measure of the intensity of ultraviolet light from the amount of oxalic acid decomposed on exposure for a given time in presence of uranyl acetate.

ursin Arbutin.

ursol *p*-Phenylenediamine.

ursolene Gray wax, m.192, from cranberry skins. Above 200° it hardens and resembles montan wax.

ursolic acid $C_{29}H_{46}(OH)COOH$ = 456.4. Urson, malol, prunol, in the leaves of *Arclostaphylos uvaursi,* bearberry; *Prunus serotina,* wild cherry; and in the fruit of *Pyrus malus,* apple. Colorless powder, m.267, insoluble in water.

urson Ursolic acid.

urstoff Protyle.

urtica Stinging nettle. The dried herb of *U. dioica* (Urticaceae), containing tannin and glucosides; a hematinic.

Urticaceae The nettle family (Moraceae, Ulmaceae, and Cynocrambaceae), a source of drugs; e.g.: *Ulmus campestris,* elm bark; *Humulus lupulus,* hops; *Cannabis sativa,* Indian hemp. See *fustic, morus, ramie.*

uru, URU Abbreviation for "ultimate rational units." Cf. *cgs, esu.*

urunday A vegetable tanning agent.

urusene $C_{15}H_{28}$ = 208.4. A hydrocarbon from urushi, Japanese lac, the secretion of *Rhus vernicifera* (Anacardiaceae). Cf. *Rhus.*

urushic acid $C_{23}H_{36}O_2$ = 344.5. Laccol. An acid from the juice of the Japanese lac tree.

urushiol $C_6H_3(OH)_2C_{15}H_{27}$ = 316.5. An oily catechol derivative, in *Rhus vernicifera;* induces sensitivity to poisoning.

USAN United States Adopted Name.

USASI See *American National Standards Institute.*

usnaric acid $C_{20}H_{22}O_{15}$ = 502.4. An acid from the lichen *Usnea barbata.*

usnic acid $C_{18}H_{16}O_7$ = 344.3. Usninic acid, from the lichen *Usnea barbata.* Insoluble solid. **(+)- ~** or **(−)- ~** m.203. **(±)- ~** m.195.

USP Abbreviation for United States Pharmacopoeia.

ustilago Corn smut. A moldlike fungus parasitic on maize, and resembling ergot. It contains several alkaloids (trimethylamine).

u.v. UV. Ultraviolet.

uva Raisins. **u.-ursi** Bearberry leaves. The dried leaves of *Arctostaphylos uva-ursi* (Ericaceae), containing arbutin, ericolin, and ursolic acid.

uvarovite $(CaO)_3Cr_2O_3Si_3O_6$. Uwarowite. A garnet.

Uverite $7CaO \cdot CaF_2 \cdot 6TiO_2 \cdot 2Sb_2O_3$. Trademark for a synthetic mineral, used to opacify enamel.

uvi(ni)c acid Me·C:CH·C(COOH):CMe·O = 140.1. 2,5-Dimethyl-3-furancarboxylic acid*. Pyrotritartaric acid. Colorless needles, m.135, soluble in water; formed by dry distillation of tartaric acid.

uviol A glass that transmits ultraviolet light.

uvitic acid $Me \cdot C_6H_3(COOH)_2$ = 180.2. 5-Methyl-1,3-benzenedicarboxylic acid*. Mesitic acid. Colorless needles, m.287, insoluble in water.

uvitinic acid $C_6H_3Me(COOH)_2$ = 180.2. Methylphthalic acid. White crystals, soluble in water.

uvitonic acid N:CMe·CH:C·COOH·CH:C·COOH = 181.2. 6-Methyl-2,4-pyridinedicarboxylic acid*. Colorless crystals, m.244.

uwarowite Uvarovite.

V

V Symbol for (1) vanadium; (2) volt.
V Symbol for (1) electric potential; (2) volume; (3) at constant volume (as subscript).
v Symbol for (1) speed or velocity; (2) vibrational quantum number; (3) *vicinal.*
ν See *nu.*
Υ See *upsilon.*
Va* Alternative symbol for vanadium, where V is inconvenient.
vac. Abbreviation for: (1) vacuum; (2) millibar.
vaccenic acid $C_{18}H_{34}O_2$ = 282.5. 11-Octadecenoic acid*. An isomer of oleic acid, in meat or butter fats. **(*E*)-~** m.44. **(*Z*)-~** m.15.
vaccine (1) A suspension of dead or attenuated bacteria, viruses, or treated toxin (i.e., toxoid) from bacteria; used to produce active immunization by injection or inoculation. (2) The original term given by Jenner to fluid (lymph) from a cowpox vesicle (vaccinia vesicle).
Vaccinium A genus of plant, as: Whortleberry, European huckleberry. The dried fruit of *Vaccinium myrtillus* (Ericaceae), containing quinic acid, myrtillin, and arbutin; a diuretic.
vacuum (1) Strictly, a space that contains no matter. (2) A space from which gas has been almost wholly removed. Residual gas pressure is measured correctly as absolute pressure, but sometimes (as, negative gage pressure) relative to atmospheric pressure. See Table 98. **high ~** A v. of below 0.01 mm, as in X-ray tubes. **low ~** A. v. of 50–1 mm. **Toricellian ~** The v. in a barometric tube between the mercury and the closed top.
v. desiccator An apparatus in which a substance is dried under reduced pressure. **v. distillation** Distillation under reduced pressure; used to purify liquids or separate mixtures. **v. evaporation** Evaporation in vacuo. **v. fan** A surface fan. **v. filter** A device for filtration under reduced pressure or by suction. **v. gases** (1) The gases obtained by heating solids, e.g., metals, in a v. (2) The residual gas in a v. tube. **v. lamp** See *vacuum lamp* under *lamp.* **v. pan** A closed retort used in industry for v. distillation. **v. pump** A suction pump that exhausts gases to a high v.; as, one using mercury (Sprengel pump) or the phthalates (Hickman pump), e.g., 10^{-4}–10^{-7} mmHg. **v. pump oil** An organic liquid used in place of mercury in v. pumps; e.g.: *n*-butyl phthalate, 21×10^{-4}; benzyl phthalate, 1.2×10^{-7} mmHg. **v. still** A v. pan. **v. tar** A tar obtained from coal by v. distillation, rich in hydrocarbons. **v. tube** Valve (U.K. usage). Electron tube. A sealed glass vessel, either evacuated or containing a gas at low pressure. It has 2 electrodes, electrons being produced at one, usually by thermionic emission. Applications utilize its *valve* effect, q.v. **v. ultraviolet** See *vacuum ultraviolet* under *ultraviolet.* **v. vessel** Dewar flask.

TABLE 98. VACUUM ACHIEVABLE BY VARIOUS METHODS

Method	Pressure, mmHg absolute[a]
Water pump: 760 − 7 = 753 mmHg relative	7.00
Sprengel mercury pump	0.001
Geryk oil pump	0.0002
Charcoal in liquid air	0.0000008
Gaede molecular pump	0.0000002

[a]1 atm = 760 mmHg = 101.325 kPa.

vadose Water just below the earth's surface; as ground- or rainwater. Cf. *juvenile water.*
vagusstoff Acetylcholine.
vakerin Bergenin.
Val* Symbol for valine.
valence, valency (1) The capacity of one atom to combine with others in definite proportions. (2) Also applied, by analogy, to radicals and atomic groups. The combining capacity of a hydrogen atom is taken as unity. Thus, in HCl, chlorine is monovalent; in H_4C, carbon is tetravalent. Values are integers 1–8. See *electron configuration* under *electron, oxidation number.* According to electronic concepts, v. is due to "valence electrons" located in the outer orbitals of an atom. Cf. *bond.* A *positive* v. is the number of electrons that an atom can *give* up; thus, Be = 2. A *negative* v. is the number of electrons that an atom can *take* up; thus, O = 2. A *covalence* indicates the number of pairs of electrons that an atom can share with its neighbors. An element or its atom may exist in several stages of oxidation and thus form several corresponding series of compounds, e.g., ferrous and ferric. V. is represented thus:

Positive divalent, e.g., iron(II)* or (2+)* or ferrous: Fe^{2+}
Positive trivalent, e.g., iron(III)* or (3+)* or ferric: Fe^{3+}
Negative divalent, e.g., sulfate: SO_4^{2-}

(3) The term *valence* is sometimes used to indicate the theory of bonding, as distinct from the number of bonds, *valency.* **active ~** The commonest v. of an element. **auxiliary ~** Covalence. **chief ~** (1) The maximum v. of an element. (2) The v. shown by the greatest number of stable compounds. **co~** Auxiliary v. The pairs of electrons shared between 2 molecules; as, $NH_3{-}H_2O$. Cf. *electrovalence.* **contra~** Covalence. **electro~** See *electrovalence.* **free ~** The v. that appears to be unsatisfied, as in free radicals. **maximum ~** The highest stage of oxidation of an element. **negative ~** V. due to an atom taking up electrons; as, of chlorine. **normal ~** The v. based on the group of the periodic table. **null ~** Zero v. **partial ~** An unsaturated or divided v. **positive ~** V. due to an atom giving up electrons; as, sodium. **principal ~** The normal v. of an atom or radical. **residual ~** See *Werner's theory.* **rotating ~** V. supposedly due to oscillations of electrons between 2 atoms; as, the H between the 2 O atoms in −COOH. Cf. *bond.* **semi~** See *semivalence.* **zero ~** Null v. No v., as in the case of some noble gases, which form no compounds.
v. bonds A pair of electrons consisting of 1 electron from each of the 2 atoms they unite. Cf. *bond, polar bond.* **v. electrons** The mobile electrons located in the outer orbitals of an atom. Atoms that lose these electrons become positive ions, and atoms that gain these electrons become negative

ions. **v. number** Oxidation number*. **v. tautomerism** A dynamic isomerism in which + and − charges produced by a moving double linkage are neutralized by the concomitant movement of a second double linkage, so that ions do not separate. Cf. *ionotropy.*

valencene A sesquiterpene hydrocarbon flavoring in grapefruit.

Valenta value The *turbidity value,* q.v., of an oil, with glacial acetic acid as solvent.

Valentine, Basil See *Basil.*

valentinite Sb_2O_3. White antimony. A mineral.

valeral Valeraldehyde*.

valeraldehyde* $Me(CH_2)_3 \cdot CHO$ = 86.1. Pentanal*, pentyl aldehyde, valeral. Colorless liquid, b.103, slightly soluble in water. **iso ~ *** $Me_2CH \cdot CH_2CHO$. Liquid, d.0.804, b.92, soluble in water.

valeramide* $Me(CH_2)_3CO \cdot NH_2$ = 101.1. Valeric amide, pentanamide*. Colorless crystals, m.127, soluble in water.

valerate* $C_4H_9 \cdot COOM$. Valerianate. A salt of valeric acid.

valerene Pentene*.

valerian The dried rhizome of *Valeriana officinalis* (Valerianaceae). It contains valerian, an essential oil, the bornyl ester of isovaleric and other fatty acids; a sedative (EP, BP). **American ~** Cypripedium. **Japanese ~** Kesso oil.
 v. oil The volatile oil of v. Green liquid, d.0.990–0.996, b.250–300, containing borneol, bornyl formate, bornyl acetate, pinene, and camphene.

valerianate Valerate*.

valerianic acid Valeric acid*.

valeric v. acid* $C_4H_9 \cdot COOH$ = 102.1. Pentanoic acid*, valerianic acid. A liquid, b.185; occurs in valerian. **iso ~ *** $Me_2CH \cdot CH_2 \cdot COOH$. 3-Methylbutanoic acid*, pentoic acid. Colorless liquid, d.0.942, b.186, soluble in water. Occurs in valerian; a perfume and flavoring. **2-amino ~ *** Norvaline. **aminoguanidino ~** Arginine*. **2-aminoiso ~ *** Valine. **aminomethyl ~** Leucine*. **diamino ~** Ornithine*. **methyl ~** Dimethyl*butanoic acid**. **oxo ~** Levulinic acid. **tetrahydroxy ~** Arabic acid.
 v. aldehyde Valeraldehyde*. **v. anhydride** $(C_4H_9CO)_2O$ = 186.3. Pentanoic anhydride*. Colorless liquid, b.205, decomp. by water to valeric acid.

Valerius Cordus (1515–1544) German author of the first legal pharmacopoeia: *Dispensatorium pharmacorum omnium* (1535).

valerol $C_{18}H_{20}O_3$ = 284.4. A ketone from oil of valerian.

valerolactone* $MeCH \cdot (CH_2)_2C(O)O$ = 100.1.

γ- ~ * Colorless liquid, b.220, in wood tar.

valerone Diisobutyl ketone*.

valeronitrile* Butyl cyanide*.

valeryl* 1-Oxopentyl†, pentanoyl. The radical C_4H_9CO-, from valeric acid. **iso ~ *** 3-Methyl-1-oxobutyl†. The radical $Me_2CH \cdot CH_2 \cdot CO-$, from isovaleric acid. **iso ~ chloride*** C_4H_9COCl = 120.6. Pentanoyl chloride. Colorless liquid, d.0.989, b.114, decomp. by water. **iso ~ amide** Valyl*(2).

valerylene 2-Pentyne*.

valine* $Me_2CH \cdot CHNH_2 \cdot COOH$ = 117.1. Val*. Aminoisovaleric acid, 2-amino-3-dimethylbutanoic acid*. An amino acid from seeds and proteins. **D- ~** m.156. **L- ~** m.95. **iso ~ *** $H_2N \cdot CMe(CH_2Me) \cdot COOH$. Ile*. 2-Amino-2-methylbutanoic acid*, m. ca. 300. **nor ~** See *norvaline.*

Valium Trademark for diazepam.

valonia The acorn cups of *Quercus aegilops* (Fagaceae), Greece and Asia Minor; a tan.

value A number expressing a property; as, iodine v.

valve (1) A device for controlling the motion of a fluid along a passage; arranged to close or open an outlet. (2) U.K. usage for vacuum tube. **bunsen ~** A piece of rubber tubing with a short slit in the side, and a glass rod inserted, so that steam or air can escape but not reenter. **Contat-Göckel ~** A chemical v. used in the cooling of acid solutions, which are easily oxidized when hot. A solution of sodium hydrogencarbonate is drawn into the flask as it cools, and the carbon dioxide evolved prevents the entry of an excess of alkali. **radio ~** See *vacuum tube.*
 v. effect Unilateral conductivity. The property of conducting a current in one direction only; i.e., of rectification.

valyl (1)* The radical $Me_2CH \cdot CHNH_2 \cdot CO-$, from valine. (2) $MeCH_2CH_2CH_2CONEt_2$ = 157.3. *N*-Diethylvaleramide*. Colorless liquid, slightly soluble in water.

valzin Sucrol.

vanadate* Indicating vanadium as the central atom(s) in an anion; as: M_3VO_4. (Ortho)vanadate. A salt of orthovanadic acid. **deca ~** Has several orange forms, as $M_6V_{10}O_{28}$. **meta ~** MVO_3. **ortho ~** Vanadate. **pyro ~** $M_4V_2O_7$. **hydrogentetra ~** $M_3HV_4O_{12}$.

vanadic Describing a compound containing trivalent [vanadium(III)* or (3+)*] or pentavalent [vanadium(V)* or (5+)*] vanadium. **v. acid meta ~** HVO_3 = 99.9. Golden scales, slightly soluble in water; an oxidizing agent and antiseptic. **ortho ~** H_3VO_4 = 118.0. Yellow powder, slightly soluble in water. **di ~, pyro ~** $H_4V_2O_7$ = 217.9. Brown powder, slightly soluble in water. **v. anhydride** Vanadium pentaoxide. **v. salts** See *vanadium.*

vanadinite $Pb_5(VO_4)_3Cl$. A native vanadate.

vanadite (1) $M_2V_4O_9$. A salt of vanadous acid. (2) Vanadinite.

vanadium* V = 50.9415. Va* (if V is inconvenient). Element, at. no. 23, discovered (1830) by Sefström; in the blood of ascidians (marine animals). Light-gray metal, d.5.96, m.1890, b.3400, insoluble in water, soluble in acids. Valencies 2, 3, 4, and 5. V. is amphoteric, giving basic salts (vanadyl compounds) and acid salts (vanadates). V. is used in metallurgy, and in catalysts. It resembles tantalum, and can be cold-worked into wire. **dioxo ~, oxo ~** See Table 99.
 v. bromide VBr_3 = 290.7. Dark-green powder, soluble (decomp.) in water. **v. carbides** (1) VC = 63.0. Black crystals, d.5.36, m.2830; insoluble in acids, except nitric acid. (2) V_4C_3 = 239.8. **v. chlorides** (1) VCl_2 = 121.8. Vanadous chloride. Violet hexagons, soluble in water. (2) VCl_3 = 157.3. Vanadic chloride. Green crystals, d.3.0, soluble in water. (3) VCl_4 = 192.8. V. tetrachloride*. Red liquid, d.1.865, b.154, soluble in water. **v. dichloride*** See *vanadium chlorides (1).* **v. difluoride*** See *vanadium fluorides (1).* **v. dioxide** (1)* VO_2 (some V_2O_4). V. tetraoxide. Blue-black, hygroscopic solid,

TABLE 99. OXIDATION STATES OF VANADIUM

Oxidation state	Name	Ion	Color in solution
2	v.(II)*, v.(2+)*, vanadous	V^{2+}	violet
3	v.(III)*, v.(3+)*, vanadic	V^{3+}	green
	oxovanadium(III)*, vanadyl(ous)	VO^+	bluish gray
4	oxovanadium(IV)*	VO^{2+}	blue
5	orthovanadate*	VO_4^{3-}	yellow
	metavanadate*	VO_3^-	yellow
	dioxovanadium(V)*, vanadol	VO_2^+	blue
	oxovanadium(V)*, vanadyl(ic)	VO^{3+}	blue

d.4.34, m.1967. (2) VO. See *vanadium monoxide.* **v. dioxomonochloride*** Vanadyl chloride. **v. disulfide*** VS (some V_2S_2). Black powder, d.4.2, insoluble in hydrochloric acid. **v. fluorides** (1) VF_2 = 88.9. Vanadous fluoride. Insoluble, except in hydrofluoric acid. (2) VF_3 = 107.9. Vanadic fluoride. Green crystals, d.3.363, m.800+, insoluble in water. (3) $VF_3 \cdot 3H_2O$ = 162.0. Rhombohedra, readily soluble in water. (4) VF_4 = 126.9. V. tetrafluoride*. Yellow, hygroscopic crystals, decomp. 325, soluble in water. **v. hydroxides** (1) $VO \cdot nH_2O$, or $V(OH)_2$. Vanadous hydroxide. Violet-gray powder, insoluble in water, soluble in acids. (2) $V_2O_3 \cdot nH_2O$ or $V(OH)_3$. Vanadic hydroxide. Green powder, insoluble in water. **v. minerals** V. is widely diffused in small quantities in rocks, clays, and coals; e.g.: vanadinite, $Pb_6V_3O_{12}Cl$; patronite, V_2S_3. Cf. *roscoelite, volborthite.* **v. monosulfide** See *vanadium disulfide.* **v. monoxide*** VO = 66.9 (some V_2O_2). Gray-brown solid, d.5.6, soluble in acids or alkalies (lavender solution) with strong reducing action. **v. nitride*** VN = 64.9. A solid, d.5.63, m.2050, insoluble in water. **v. oxides**

V_2O . . . . . V. suboxide
VO . . . . . . V.(II) oxide*, v. monoxide, vanadous oxide
VO_2 . . . . . V.(IV) oxide*, v. dioxide
V_2O_3 . . . . . V.(III) oxide*, v. trioxide, vanadic oxide, v. sesquioxide
V_2O_4 . . . . . V. tetraoxide
V_2O_5 . . . . . V.(V) oxide*, v. pentaoxide, vanadic anhydride

v. oxybromide Vanadyl bromide (1). **v. oxychloride** Vanadyl chloride (2). **v. oxydibromide** Vanadyl bromide (2). **v. oxydichloride** Vanadyl chloride (3). **v. oxydifluoride** Vanadyl fluoride (1). **v. oxyfluoride** Vanadyl fluoride (2). **v. oxytribromide** Vanadyl bromide (3). **v. oxytrichloride** Vanadyl chloride (4). **v. pentasulfide*** V_2S_5 = 262.2. Vanadic sulfide. Green powder, soluble in alkalies. **v. pentaoxide*** V_2O_5 = 181.9. Vanadic oxide, vanadic anhydride. Brown-black, hygroscopic solid (yellow if pure), soluble in hot acids; d.3.34, m.660; used as a catalyst and in dyes, inks, and glass; a strong oxidizing agent. **v. sesquioxide** V. trioxide*. **v. sesquisulfide** V. trisulfide*. **v. silicides** (1) V_2Si = 130.0. Silvery prisms, insoluble in water. (2) VSi_2 = 107.1. Metallic prisms, insoluble in water. **v. steel** An alloy: Fe and 0.1–0.15% V; used in tool manufacture. **v. suboxide** V_2O. Existence doubtful, but possibly the brown stain formed on v. in air. **v. sulfate** Vanadyl sulfate. **v. sulfides** (1) V_2S_2. V. disulfide*. (2) V_2S_3. V. trisulfide*. (3) V_2S_5. V. pentasulfide*. **v. tetrachloride*** V. chloride (3). **v. tetrafluoride** V. fluoride (4). **v. tetraoxide*** V_2O_4 = 165.9. Vanadous acid. Indigo crystals, m.1970, insoluble in water. **v. trichloride*** V. chloride (2). **v. trifluoride*** See *vanadium fluorides (2)* and *(3).* **v. trioxide*** V_2O_3 = 149.9. V. sesquioxide, vanadous oxide. Black, infusible crystals, m.1970, slightly soluble in water. It changes slowly in air to the indigo-blue oxide, V_2O_4; used as a catalyst, as a mordant in dyeing, and in the manufacture of steel and of silver vanadate, Ag_3VO_4. **v. trisulfide*** V_2S_3 = 198.1. Vanadous sulfide. Red crystals, insoluble in water.

vanadol The dioxovanadium(V)* radical VO_2^+.

vanadous Describing a compound containing divalent [vanadium(II)* or (2+)*] or trivalent [vanadium(III)* or (3+)*] vanadium. **v. acid** $H_2V_4O_9$. The hypothetical compound from which vanadites are derived. **v. chloride** Vanadium chloride (1) or (2). **v. fluoride** Vanadium fluoride (1), (2), or (3). **v. hydroxide** Vanadium hydroxide (1) or (2). **v. oxide** Vanadium trioxide*. **v. sulfide** Vanadium trisulfide*.

vanadyl (1) The radical VO^{3+}, oxovanadium(V)*, vanadyl(ic), from pentavalent vanadium. (2) The radical VO^{2+}, oxovanadium(IV)*. (3) The radical VO^+, vanadyl(ous), from trivalent vanadium. **v. bromides** (1) VOBr = 146.8. Vanadylous bromide, oxovanadium(III) bromide*. A solid, decomp. 480, slightly soluble in water. (2) $VOBr_2$ = 226.7. Oxovanadium(IV) bromide*. Brown, hygroscopic powder. (3) $VOBr_3$ = 306.7. V. tribromide, oxovanadium(V) bromide*. Red liquid, soluble in water (decomp.). **v. chlorides** (1) V_2O_2Cl = 169.3. V. semichloride, vanadium dioxymonochloride. Yellow crystals, insoluble in water. (2) VOCl = 102.4. V. monochloride, vanadylous chloride, vanadium oxymonochloride. Brown powder, insoluble in water. (3) $VOCl_2$ = 137.8. V. oxydichloride. Blue scales, deliquescent, slowly decomp. by water. (4) $VOCl_3$ = 173.3. Vanadium trichloride, vanadic chloride, vanadium oxytrichloride. Dark-green syrup, b.127, soluble in water; a mordant. **v. fluorides** (1) VOF_2 = 104.9. V. difluoride, vanadium oxydifluoride. A solid, decomp. by heat, insoluble in water. (2) VOF_3 = 123.9. V. trifluoride, vanadylic fluoride, m.300, very soluble in water. **v. semichloride** V. chloride (1). **v. sulfate** $(VO)_2(SO_4)_3$ = 422.1. Blue crystals, soluble in water. **v. disulfate** $(V_2O_2)(SO_4)_2$ = 326.0. A double salt of v. sulfate.

vanadylic The oxovanadium(V)* radical VO^{3+}. **v. bromide** Vanadyl bromide (3). **v. chloride** Vanadyl chloride (4).

vanadylous The oxovanadium(III)* radical VO^+. **v. bromide** Vanadyl bromide (1). **v. chloride** Vanadyl chloride (2).

vanaspati (1) An Indian food; wholly hydrogenated vegetable oils containing 5% sesame oil for identification purposes, and fortified with 25 I.U. per gram of vitamin A. (2) Often applied specifically to groundnut oil. Cf. *ghee.*

vancomycin hydrochloride An antibiotic produced by *Streptomyces orientalis.* Bitter, brown powder, soluble in water. A bacteriocidal, active against gram-positive organisms; used where allergy to penicillin exists (USP, BP).

van der Waals **v. d. W., Johannes Diderik (1837–1923)** Dutch physicist, Nobel prize winner (1910). **v. d. W. constant** The factors *a* and *b* in the v. d. W. equation. **v. d. Waals' equation** A modification of the equation of state with 2 correcting factors: $(p + a/V^2)(V - b) = RT$, in which the volume factor *b* corresponds with 4 times the square root of the space occupied by the molecules themselves; the factor, a/V^2, expresses the mutual attraction of the molecules. See *corresponding state.* **v. d. W. forces** The weak forces between atoms and molecules, being other than forces due to covalent bonds or ionic attraction. They cause crystallization of noble gases at low temperature, and the packing together of nonpolar organic compounds to form soft crystals of low melting point.

Vandura Trademark for the first *prolon,* q.v., produced from gelatin by A. Miller (1894).

van Dyck brown Vandyke. A mixture of ocher and lampblack.

van Helmont See *Helmont.*

Vanier's tube A potash bulb and drying tube, used to absorb the carbon dioxide evolved in the determination of carbon in steel.

vanilla V. bean. The unripe fruit of *V. planifolia* (Orchidaceae) containing vanillin and vanillic acid; an aromatic (NF), flavoring, and perfume.

vanillal The vanillylidene* radical.

vanillic **v. acid*** $MeO \cdot C_6H_3(OH)COOH$ = 168.2. 4-Hydroxy-3-methoxybenzoic acid*. Colorless needles, m.207 (sublimes), slightly soluble in water. **v. alcohol** $MeO \cdot C_6H_3 \cdot OH \cdot CH_2OH$ = 154.2. 4-Hydroxy-3-methoxybenzyl alcohol*. Colorless needles, m.115 (decomp.), soluble in water.

vanillin* $MeO \cdot C_6H_3(OH)CHO$ = 152.2. 4-Hydroxy-3-

methoxybenzaldehyde*. Methylprotocatechuic aldehyde. An odorous principle from the vanilla bean, or prepared synthetically. Colorless needles, m.81 (sublimes), soluble in water; a reagent, flavoring agent, and vanilla substitute (NF, BP). **ethyl ~** $C_9H_{10}O_3$ = 166.2. 3-Ethoxy-4-hydroxybenzaldehyde. Bourbonal. A homolog of vanillin 4 times as strong in flavor. Cf. *ethyl* vanillate. **iso ~** 3-Hydroxy-4-methoxybenzaldehyde*. Colorless crystals of aromatic odor, m.116.

vanilloyl* The radical 3,4-MeO(HO)C_6H_3CO—, from the vanillic acid.

vanillyl* (4-Hydroxy-3-methoxyphenyl)methyl†. The radical 3,4-MeO(HO)$C_6H_3CH_2$—, from vanillic alcohol. **v. alcohol*** Vanillic alcohol.

vanillylidene* Vanillal. The radical 3,4-MeO(OH)C_6H_3·CH=, from vanillin.

vanirom Bourbonal.

vanning Separation of constituents of an ore by washing away the lighter portions in a stream of water.

Van Slyke V. S., Donald Dexter (1883–1971) American chemist. **V. S. apparatus** Apparatus for the determination of aliphatic amino nitrogen in proteins. **V. S. method** Determination of proteins from the nitrogen evolved from aromatic amino compounds and nitrous acid: $RNH_2 + HNO_2 \rightarrow R{\cdot}OH + N_2 + H_2O$.

Van Slyke, Lucius Lincoln (1859–1931) American chemist noted for research on dairy products.

van't Hoff v. H., Jacobus Henricus (1852–1911) Dutch chemist noted for work on stereochemistry, Nobel prize winner (1901). **v. H.'s factor** The empirical factor i in the equation of state for solutions: $pV = iRT$, where p is osmotic pressure, V the volume. If d is the degree of dissociation and n the number of ions into which a molecule is partly dissociated, $i = 1 + d(n - 1)$. **v. H.'s law** The osmotic pressure exerted by a solute in solution equals that for the same solute in the state of an ideal gas occupying the same volume as the solution. **v. H. reaction** Reaction isochore. **v. H. solution** Calcium chloride 2, magnesium chloride 7.8, potassium chloride 2.2, magnesium sulfate 3.8, sodium chloride 100.2 g, in 1 liter of water. **v. H. theory** Dissolved substances obey the gas laws.

vanthoffite $MgSO_4{\cdot}3Na_2SO_4$. A Stassfurt salt.

vapodust An insecticidal spray consisting of vaporized petroleum oils; used in orchards.

vapor, vapour A gas, especially from a substance that at ordinary temperature is a solid or liquid; as, ether v. It forms when the v. pressure of a substance equals that of the atmosphere. **saturated ~** A v. such that the liquid in it, and from which it is derived, cannot further evaporate. **unsaturated ~** A v. in a space which contains insufficient liquid to saturate it.

v. bath Steam bath. **v. density** The density of a gas compared with a standard gas: hydrogen or air = 1, or oxygen = 16. The approximate relationship of these 3 standards is: $H = M/2$, $O = M/32$, air $= M/28.95$, where M is the molecular weight of the substance, the data being corrected for pressure and temperature by the equation of state. **v.-phase chromatography** See *vapor-phase chromatography* under *chromatographic analysis.* **v.-phase inhibitor** VPI. A volatile substance enclosed in a package which protects the contents against corrosion by emitting a vapor, e.g., cyclohexamine carbonate. **v. pressure** The pressure at which a liquid and its v. are in equilibrium at a definite temperature. If the v. p. reaches the prevailing atmospheric pressure (1 atm), the liquid boils. Cf. *Clausius equation.* **v. p. of air** That portion of the total pressure of air due to the water v. present. **saturated ~** The v. p. at a particular temperature when the partial pressure exerted by it is a maximum. **v. synthesis** The use of v. (e.g., of metals), generated at high temperature in a vacuum, as reagents in routine syntheses. **v. tension** V. pressure.

vaporimeter An instrument to test the volatility of oils by heating them in a current of air.

vaporization Volatilization. The change from the liquid to the gaseous state without change in the chemical composition of the molecule. Cf. *evaporation.* **heat of ~** The number of J (or cal) required to transform one gram of liquid substance, at its boiling point, into its vapor. Cf. *Clausius equation.*

vaporize (1) To change into a vapor, e.g., by heating a liquid. (2) To atomize, or subdivide a liquid into a fine spray.

vaporizer (1) An atomizer. (2) A still.

varek Kelp (French).

variability The deviation from the normal. Cf. *variance.*

variable (1) Not constant. (2) A factor that is variable. **dependent ~** A v. whose value is influenced by changes in other variables. **independent ~** A v., changes in the value of which influence the values of other (dependent) variables.

variance (1) The square of the standard deviation. (2) Degree of freedom. The number of external conditions that may be arbitrarily fixed; as, composition, temperature, and pressure. See *phase rule.*

variant Pertaining to variable factors. **di ~** An area in a diagram. **mono ~** A line in a diagram. **non ~** A point in a diagram. Cf. *phase rule.*

variate A numerical value or result when used in statistical treatment.

variolation Vaccination with serum from a human being having a mild attack of smallpox (variola). No longer practiced.

variscite $Al(OH)_2{\cdot}H_2PO_4$. Peganite. A green gem mineral; it has been synthesized.

varnish A solution of a resin or drying oil in a volatile solvent, e.g., turpentine. See *lacquer.*

varve (1) A lamination in a deposit of natural clay. (2) A cycle.

vasculose Early name for impure lignin.

vaselin Fossolin. A mixture of alkane hydrocarbons obtained from petroleum residues. See *petrolatum.*

Vaseline Trademark for a brand of petrolatum and certain similar products. Cf. *vaselin.*

vasicinone $C_{11}H_{10}O_2N_2$ = 202.2. Peganine. **(−)-~** An alkaloid from *Peganum harmale* or *Adhatoda vasica* (Acanthaceae). White needles, m.203.

vasoconstrictor An agent that causes constriction of blood vessels, particularly small arteries, producing a rise in blood pressure; as, norepinephrine bitartrate.

vasodilator An agent that lowers arterial pressure by dilation of the blood vessels; as, hydralazine hydrochloride.

vasopressin (1) $C_{47}H_{65}O_{12}N_{13}$ = 1004. A hormone secreted by the pituitary gland, which promotes reabsorption of water in the kidney. Lack of v. produces diabetes insipidus. (2) Lypressin, Pitressin. A preparation of pituitary extract from animals, used to treat diabetes insipidus (USP, BP).

vat (1) A vessel or tub in which colors are dissolved, or ores are washed and chemically treated, or liquids are stored or fermented; as, indigo v., cyanide v. (2) The solutions used in these tubs. **v. dye** A color that is applied with a mordant. See *dyes (4).*

vaterite $CaCO_3$. A probable anisotrope of calcite.

Vaughan's cage A collapsible, sterilizable iron-screen box for animal experiments.

Vauquelin, Louis Nicolas (1763–1829) The French discoverer of chromium and organic compounds.

vauqueline Strychnine.

vauquelinite $2PbO{\cdot}CuO{\cdot}2CrO_3$. A native mixed oxide.

v.d. Abbreviation for vapor density.
VDU Visual display unit.
veatchine $C_{22}H_{33}O_2N$ = 343.5. An alkaloid from *Garrya veatchii,* m.119. Cf. *garryine.*
vectograph Stereoscopic photographs produced by forming 2 images in 2 layers of polarizing crystals crossed with respect to each other, and mounted on aluminum.
vector (1) A quantity, as, velocity, force, etc., that has both magnitude and direction, and that may be represented as a straight line of suitable length and direction. Cf. *coordinates.* (2) An insect or animal that transmits microorganisms to humans; as, mosquitoes transmitting malarial parasites, dogs transmitting rabies virus.
veepa oil Margosa oil.
vegetable Pertaining to *plants,* q.v. **v. dyes** Coloring matter from plants; as, chromoproteins, indicators, cyanins, carotenes. **v. horsehair** See *vegetable horsehair* under *horsehair.* **v. parchment** A grease-resisting imitation parchment of high wet strength, used to wrap foodstuffs; prepared by passing paper through sulfuric acid or zinc chloride solution. **v. potash** A fertilizer made from distillery waste: 33% potassium oxide.
vehicle A usually inactive medium or carrier for an active substance; as, oil in paints. Cf. *thinner.*
vein (1) A vessel that conveys blood toward the heart. (2) A lode or deposit of ores distinct from the surrounding rocks. (3) The ribbed portion of leaves, which is a vascular tube transporting water.
Velban, Velbe Trademarks for vinblastine sulfate.
vellosine $C_{23}H_{28}O_4N_2$ = 396.5. An alkaloid from the bark of *Geissospermum vellosii,* pareira bark. Yellow crystals, m.189, insoluble in water.
velocity (1) The speed of travel, expressed as the distance covered in a unit of time. (2) The time required for a phenomenon to take place; as, v. of reaction. See Table 100. **angular ~** Angular speed. **migration ~** See *migration velocity.* **molecular ~** See *molecular velocity.* **reaction ~** See *reaction velocity.* **terminal ~** The v. acquired by a freely falling body when the resistance of the medium through which it falls balances the weight of the particle.
v. constant Rate constant*. See *reaction order.*
Velon Trademark for a mixed-polymer synthetic fiber.
Velox Trademark for photographic printing paper.

TABLE 100. MAGNITUDES OF VELOCITY

Phenomenon measured	cm/s
Light	3.00×10^{10}
Sun, around hub of Milky Way	3×10^7
Earth around sun	2.95×10^6
Sound in iron	5×10^5
Hydrogen molecules at 0°C	1.8×10^5
Fast aircraft	1×10^5
Nitrogen molecules at 0°C	4.97×10^4
Earth's rotation at equator	4.65×10^4
Sound in air at 0°C	3.31×10^4
Moon around earth	1.02×10^5
Nerve impulse	3.9×10^3
Crystal growth (picric acid)	1.43
Gas diffusion (H into O)	0.69
Fastest plant growth	3×10^{-3}
Growth of eucalyptus tree (Brazil)	1×10^{-5}
Growth of beard	1×10^{-6}
Diffusion of gold into lead	4.6×10^{-9}

venalin A plant hormone associated with gibberellin.
venenatin Acovenoside A.
venetian **v. red** Fe_2O_3. A ferric oxide pigment. **v. white** A pigment mixture of equal parts of lead white and barium sulfate.
venom The poison secreted by reptiles, amphibians, spiders, and insects. See *antivenom.*
vent An outlet for fumes or gases.
venturi meter Undermeter. A pipeline meter for measuring the rate of liquid flow. The pressure drop across a streamline or constriction shows the velocity of flow.
venus crystals Cupric acetate.
Veral Trademark for a copolymeric vinyl chloride–acrylonitrile synthetic fiber.
veratral The veratrylidene* radical.
veratraldehyde* $(MeO)_2C_6H_3 \cdot CHO$ = 166.2. 3,4-Dimethoxybenzaldehyde. White needles, m.43. ***ortho-* ~** 2,3-Dimethoxybenzaldehyde.
veratric acid* $(MeO)_2C_6H_3 \cdot COOH$ = 182.2. 3,4-Dimethoxybenzoic acid*. An acid in sabadilla, m.181. ***ortho-* ~** 2,3-Dimethoxybenzoic acid*. **formyl ~** Opianic acid.
veratrine $C_{32}H_{49}O_9N$ = 591.7. Cevadine. The chief alkaloid of veratrum. Colorless crystals, m.205, insoluble in water; used in ointments. **proto ~** See *protoveratrine.*
veratrole* $C_6H_4(OMe)_2$ = 138.2. 1,2-Dimethoxybenzene*. Colorless crystals, m.23, soluble in water.
veratroyl* The radical 3,4-$(MeO)_2C_6H_3CO-$, from veratric acid.
veratrum American hellebore, green or false hellebore, Indian poke. The dried roots of *V. viride* (Liliaceae). It contains the alkaloids veratrine, protoveratrine, rubijervine, and veratravine.
veratryl* (3,4-Dimethoxyphenyl)methyl†. The radical $(MeO)_2C_6H_3CH_2-$, from veratryl alcohol.
veratrylidene* The radical 3,4-$(MeO)_2C_6H_3 \cdot CH=$.
verbascum Mullein leaves. The dried flowers and herbs of *V. thapsus* (Schrophulariaceae); a demulcent. Cf. *mullein.*
verbena Blue vervain, wild hyssop. The dried overground portions of *V. hastata* (Verbenaceae); a diaphoretic. **v. oil** The volatile oil from the leaves of *V. triphylla,* France and Spain (30% citral). Cf. *lippianol, Andropogon.* **East Indian ~** Lemongrass oil. **Singapore ~** Citronella oil.
Verbenaceae The vervain family of herbs and shrubs; some contain aromatic principles; as, *V. officinalis,* verbenaloside; *Premna taitensis,* tonga bark.
verbenalin Verbenaloside.
verbenaloside $C_{17}H_{25}O_{10}$ = 389.4. Verbenalin. A crystalline, reducing β-D-glucoside from the flowering tops of *Verbena officinalis,* wild verbena; m.181, soluble in water.
Verdet **V.'s constant** The magnetic rotation of polarized light per unit length per unit magnetic field = $\alpha/H \cdot l$, where α is the rotation for the substance in a magnetic field of strength H, and l the length of the light path parallel to the lines of force. Films of Fe, Co, and Ni are exceptions. **V.'s equation** The magneto-optic rotation = $clH(n - \lambda \cdot dn/d\lambda)n^2/\lambda^2$, where c is a constant for the substance, l the length of the path of the polarized beam, H the strength of the magnetic field, n the refractive index of the substance, λ the wavelength of the light used.
verdigris Cupric subacetate. A blue to green pigment. **blue ~** Cupric acetate.
verditer (1) Bremen green, (2) Copper carbonate. **blue ~** Blue copper carbonate. **green ~** Green copper carbonate.
verdoflavin A green reduction product of riboflavine.

verine $C_{28}H_{45}O_8N$ = 523.7. An alkaloid from sabadilla.
vermeil An early method of gilding by firing gold onto a silver surface.
vermicide An agent that destroys intestinal worms. Cf. *vermifuge, teniacide.*
vermiculite $22MgO \cdot 5Al_2O_3 \cdot Fe_2O_3 \cdot 22SiO_2 \cdot 40H_2O$ (mean of 7 true U.S. v.). A gold-colored mineral (U.S. and S. Africa), m. approx. 1,370. Very light (93 kg/m^3); used as a heat and sound insulator, filler for plastics, packing for corrosive and flammable materials; expands on ignition.
vermiform Resembling a worm.
vermifuge An agent that expels intestinal worms. Cf. *vermicide, teniafuge.*
vermilion HgS. Red mercuric sulfide, cinnabar; a pigment and polish for lenses. Commercial grades may contain red lead and insoluble synthetic dyes. **mock ~** Lead chromate.
vermilionette A vermilion substitute; usually dyed chalk.
vermouth, vermuth An aperitif; white wine, flavored with wormwood. Cf. *absinthe.*
vernadite H_2MnO_3. A natural "manganic acid" in dispersed, brown colloidal particles, or black masses.
vernalization Bringing plants artificially to the spring state, by subjecting them to indoor temperature before planting.
vernier A small, movable auxiliary scale attached to a larger scale, to increase the precision of the readings.
vernine Guanosine.
vernonine $C_{10}H_{24}O_7$ = 256.3. A glucoside from the batiator root, *Vernonia nigritiana* (Compositae), Africa. Hygroscopic, white powder, soluble in water; a cardiac poison.
veronica Speedwell. The dried herb of *V. officinalis* (Scrophulariaceae).
verrucose Describing a growth that resembles warts.
Versatic (911) Trademark for a mixture of cyclic and (mostly) tertiary acids containing 9–11 C atoms, made by the action of carbon monoxide and water on refinery alkenes with an acid catalyst. Used in the manufacture of surface coatings, paint driers, and alkyd plastics.
Versene Trademark for sodium versenate. The sodium salt of ethylenediaminetetraacetic acid, versenic acid; a reagent for gold (violet color), and for titrating the hardness of waters. See *EDTA.*
versenic acid See *ethylenediaminetetraacetic acid.*
vertivert oil The volatile oil of *Andropogon muricatus,* d.1.015–1.030, soluble in water. Cf. *verbena oil.*
vervain Verbena.
vesicant An agent that blisters the skin; as, mustard gas. Cf. *pustulant.*
vesicle A small blister.
vesorcinol Dihydroxytoluene.
vessel (1) A container; as, a beaker. (2) In biology, a canal or tube for carrying a fluid such as blood.
Vestamid Trademark for nylon 12; similar to nylon 6.
Vestan Trademark for a polyvinyl chloride synthetic fiber.
vesuvianite Iodocrase.
vesuvin Bismarck brown.
vetiveria Cuscus, khuskhus. The Indo-Malaysian grass *V. zizanioides* (Graminaceae) whose roots are woven into fragrant mats, fans, baskets. **v. oil** A volatile oil distilled from v. Cf. *verbena oil.*
V-film Trademark for a polyvinyl-type plastic.
VHF Very high frequency. See *radiation,* Table 69.
viability The capacity to live and grow, e.g., of plants.
vial Phial. A small bottle. Cf. *ampoule.*
vibracone An ore screen, with a vibrating conical surface.
Vibramycin Trademark for doxycycline.
vibration Rapid to-and-fro motion. **atomic ~** The motion of the atoms of a molecule. **electronic ~** The rapid v. of the electrons of the dynamic atom, causing emission of rays. Cf. *spinning electron.*
vibrator A device that produces mechanical vibration, as screens.
Viburnum A genus of trees and shrubs (Caprifoliaceae). ***V. opulus*** Cramp bark. Cranberry tree, guelder rose. The dried bark of *V. opulus;* contains viburnin, valeric acid, sugar, and tannins.
vicalloy An alloy for making very small magnets: Co 50, Fe 25–30, V 10–14%.
Vicara Trademark for a fiber made by extruding a solution of zein in dilute alkali into formaldehyde. It resembles wool, but does not felt or shrink.
vicianin A glucoside from the seeds of vetch, *Vicia sativa.*
vicianose $C_{11}H_{20}O_{10}$ = 312.3. 6-*O*-α-L-Arabinopyranosyl-D-glucose. A disaccharide obtained by hydrolysis of vicianin, m.210(decomp.).
vicilin A globulin from peas, beans, and lentils.
vicinal The neighboring position of radicals, as the 1,2,3-positions of the benzene ring.
victor bronze The alloy Cu 58.5, Zn 38.5, Al 1.5, Fe 1.0, V 0.03 pts.
victoria **v. green B** Malachite green. **v. orange** $C_7H_5O_5N_2K$ = 236.2. Aniline orange, potassium dinitro-*o*-cresol. Yellow dye for wool and silk. **v. yellow** Antinnonin.
victorium Monium.
vicuna An expensive, fine, soft hair from a species of small llama.
vienna caustic Potassium hydroxide mixed with lime.
Vierordt, Carl (1818–1884) The German founder of quantitative spectrum analysis.
Vieth's ratio For milk, the ratio of ash:protein:sugar = 1:5:6. Revised to the ratio of ash:(nitrogen × 6.387):lactose hydrate = 2:9:13.
viferral Hydronal.
Villari effect The change in magnetization of a ferromagnetic material under a stress. Cf. *Joule effect.*
Villavecchia test An oil (5 mL) is shaken with 5 mL fuming hydrochloric acid and 2 drops of a 1% solution of 2-furaldehyde in alcohol (Baudouin's reagent); the bottom (acid) layer becomes rose-colored if sesame oil is present.
villose Describing hairlike, flimsy extensions on the edge of a body tissue or bacterial growth.
Vilsmeir reagent Formylation produced by the reaction of an amide with an acid halide; used to form C–C bonds.
vinaconic acid $CH_2 \cdot CH_2 \cdot C(COOH)_2$ = 130.1 (ring: first and last C joined).
1,1-Cyclopropanedicarboxylic acid*. White needles, m.175.
vinasse Schlempe. The residue from the fermentation of molasses, or grapes; a fertilizer and source of potassium salts.
vinblastine sulfate $C_{46}H_{58}O_9N_4 \cdot H_2SO_4$ = 909.1. Vincoleukoblastine sulfate. Velban, Velbe. White crystals, soluble in water. An alkylating cytotoxic agent from *Vinca rosea.* Used for cancer and malignant disease of lymphatic system (USP, BP).
vincristine sulfate $C_{46}H_{56}O_{10}N_4 \cdot H_2SO_4$ = 923.0. Leurocristine sulfate. Oncovin. White crystals, freely soluble in water. As for vinblastine sulfate, but used for leukemia (USP, BP).
vinegar (1) A weak (approx. 6%) solution of acetic acid containing coloring matter and other substances (esters, mineral matter, etc.), formed by the fermentation of alcoholic liquids (as cider, wine) with an acetifying organism. Cf. *acetifier, acetimetry.* (2) Acetextracts, acetum. The strained liquid obtained by macerating a drug with dilute acetic acid;

as, squill v. **artificial ~** A v. substitute containing acetic acid, which is not wholly the product of alcoholic and subsequent acetous fermentation. **distilled ~** The distillation product of v. **imitation ~** Artificial v. **malt ~** V. derived, without intermediate distillation, wholly from malted barley with or without addition of whole cereal grain, the starch of which has been saccharified by malt amylase. **nonbrewed ~** Artificial v. **spirit ~** The product of a distilled alcoholic fluid, containing 4–15% (w/v) of acetic acid. **wood ~** Pyroligneous acid.

v. essence A product made synthetically or by distillation of wood (12% acetic acid), and colored with an aniline dye or caramel. Cf. *spirit acid.*

vinesthine, vinethine Divinyl ether*.

vinetine Oxyacanthine.

vinic **v. acids** A group of organic compounds analogous to acid salts; as, $EtHSO_4$, ethylhydrogensulfate (sulfovinic acid). **v. ether** Diethyl *ether**.

vinifera palm oil Bamboo oil.

Vinol Trademark for polyvinyl alcohol.

vinometer A hydrometer to measure the percentage of alcohol in wine.

vinum Latin for "wine."

vinyl* Ethenyl†. The radical $-CH:CH_2$, from ethylene. Cf. *polyvinyl, vinylene.* **v. acetate*** $CH_3COOCH:CH_2$ = 86.1. Unstable liquid, b.72. **v.acetic acid** 3-Butenoic acid*. **v.acetylene** $CH_2:CH \cdot CH:CH$ = 53.1. Butone, monovinylacetylene. A gas formed on passing acetylene into ammoniacal cuprous chloride solution; with hydrochloric acid it forms chlorobutadiene, a source of artificial rubber. See *elastomer.* **di~** See *divinylacetylene.* **v. alcohol*** $CH_2:CH \cdot OH$ = 44.05. Vinol, ethenol*. An unsaturated alcohol. **v.amine*** $C_2H_3NH_2$ = 43.07. Ethenylamine. A liquid, b.56. **v.benzene** Styrene*. **v. bromide*** $CH_2:CHBr$ = 106.9. Ethenyl bromide, bromoethene*. Colorless liquid, $b_{750mm}16$, insoluble in water. **v.carbazole** See *green* oil. **v. chloride*** C_2H_3Cl = 62.5. Chloroethene*. Colorless gas, $d_{-13.9}0.97$, b.−13.9, soluble in alcohol or ether. **poly~*** See *poly(vinyl chloride).* **v. cyanide*** $CH_2:CH \cdot CN$ = 53.1. Acrylonitrile*, propenenitrile*. Colorless liquid, b.78. **v. ether** Divinyl ether*. **v. ethylene** 1,3-Butadiene*. **v.imine** Dimethyleneimine. **v. iodide*** Iodoethylene*. **v. ketone** Pentadienone*. **v. oxide** Divinyl ether*. **v. sulfide** Divinyl sulfide*.

vinylation Converting a phenolic group into a vinyl group, thus: $R \cdot OH + C_2H_2 \rightarrow R \cdot O \cdot CH:CH_2$, with potassium hydroxide at 120–180 as catalyst.

vinylene* 1,2-Ethenediyl†. The radical $-CH:CH-$, from ethylene. Cf. *vinyl.* **v. dichloride** Ethylene dichloride*.

vinylidene* Ethenylidene†. The radical $H_2C:C=$, from ethylene. Cf. *vinyl.* **poly~** See *poly(vinylidene dichloride) et seq.* under *polyvinyl.* **v. trichloride** 1,1,2-Trichloroethane*.

Vinylite Trademark for vinyl resins and plastics, polymers of vinyl acetate, vinyl chloride, and vinyl chloride-acetate. **V. A** Colorless, thermoplastic resin, softening 40–60, soluble in ketones, esters, and hydrocarbons. **V. 80** White powder, partly soluble in ketones. **V. N** A 35% solution in toluene. Vinylites may be colored with dyes and pigments.

Vinylon Trademark for a polyvinyl alcohol synthetic fiber.

Vinyon Trademark for vinyl resins, fibers, and yarns; copolymers of vinyl chloride with acrylonitrile or vinyl acetate; in particular a filament copolymer of 90% vinyl chloride and 10% vinyl acetate, m.70. Cf. *Dynel.*

Vioform Trademark for clioquinol.

Viola The violet family, which includes violet and pansy (Violaceae). Cf. *violine.* E.g., *V. odorata,* sweet violet; *V. tricolor,* pansy or heartsease. ***v. crystallina*** Methyl violet.

violanthrone $C_{34}H_{16}O_2$ = 456.5. Dibenzanthrone. A 9-ring diketone; a purple dye for vegetable fibers.

violaquercitrin $C_{27}H_{30}O_{16}$ = 610.5. Osyritrin. A glucoside from various *Viola* species; it hydrolyzes to glucose, quercitrin, and rhamnose.

violaxanthin $C_{40}H_{56}O_4$ = 600.9. A carotenoid, m.207, from *Viola tricolor,* and orange rind.

violet (1) A species of the Violaceae family. Cf. *Viola.* (2) A reddish-blue shade. **anthracene ~** Gallein. **crystal ~** Methyl v. **Döbner's ~** Aminofuchsin iminochloride. A triphenylmethane dyestuff reagent for aldehydes. Cf. *fuchsin.* **essence of ~** Orris. **ethyl ~** See *ethyl violet.* **gentian ~** Methyl violet. **hexamethyl ~** Methyl v. **Lauth's ~** Thionine. **methyl ~** See *methyl violet* under *violet.* **ultra ~** See *ultraviolet.*

violine An alkaloid from *Viola* species, resembling emetine in action.

violuric acid $\underline{CO \cdot (NH \cdot CO)_2 \cdot C}:NOH \cdot H_2O$ = 175.1. Alloxan-5-oxime, m.240, slightly soluble in water.

viosterol Ergocalciferol.

viprynium embonate BP name for pyrvinium pamoate.

Virginia snakeroot Serpentaria.

virginiamycin Staphylomycin. An antibiotic used in animal feeds.

virginium Early name for francium.

viride nitens Brilliant green.

viridin $C_{20}H_{16}O_6$ = 352.3. Colorless, levorotatory prisms from cultures of the mold *Trichoderma viride,* Pers.; m.208–217, insoluble in ether; an antibiotic, existing as the α and β isomers, $[\alpha]_D$ −213.4° and −50.7°.

Viridine Trademark for phenyl acetaldehyde dimethylacetal. (2) (Not cap.) $C_{12}H_{19}N$ = 177.3. A homolog of pyridine, distilled from coal tar and bone oil. (3) An alkaloid in *Veratrum viride.*

viridinine $C_8H_{12}N_2O_3$ = 184.2. A monoacid base from putrid pancreas.

Virion (1) Trademark for a viscose synthetic fiber. (2) (Not cap.) The largest, freely existing infective unit that can be described as a single virus particle.

virosine (1) $C_{12}H_3O_2N$ = 193.2. An alkaloid from *Securinega virosa,* m.135. (2) A biologically active indole alkaloid from *Catharanthus roseus.*

virulence The ability of a microorganism to cause serious disease.

virulent Exceedingly poisonous or active in damaging protoplasm; as, of bacteria.

virus The smallest microorganism, 20–300 nm diameter. Viruses are obligatory intracellular parasites and contain only 1 nucleic acid. Cf. *bacteria.*

Visca Trademark for a continuous, flat, monofilament rayon yarn.

viscid Sticky, gummy, glutinous.

viscidity Stickiness.

viscin $C_{20}H_{48}O_8$ or $C_{20}H_{32} \cdot 8H_2O$ = 416.6. The glutinous constituent of mistletoe berries; the chief constituent of birdlime. See *viscum.*

viscoelastic Describing a plastically deformable fluid having a finite fluidity.

viscometer Viscosimeter. An instrument to determine the internal friction (viscosity) of a liquid from the rate of revolution of a vane immersed in the liquid in comparison with its speed in water (cf. *consistometer*); or its rate of flow through an orifice (cf. *Engler degree*). **torsion ~** A v. based on the force required to twist a cylinder immersed in the liquid through a certain angle.

viscoplastic Describing a fluid that will not flow until a critical yield stress is exceeded.

viscose An extremely viscous syrup obtained by treating cellulose with potassium hydroxide and carbon disulfide. On pressing this liquid through fine openings into dilute acids the cellulose separates as threads of viscose rayon. **v. silk** See *rayon.*

viscosimeter Viscometer.

viscosimetry Measuring viscosity.

viscosity Internal fluid friction. The ratio of the shear stress to the rate of shear of a fluid. The property of being glutinous or sticky, i.e., offering a slight resistance to a change of form, due to intermolecular attraction. **absolute ~** Dynamic v.* **anomalous ~** Sigma phenomenon. The v. of an anomalous *liquid,* q.v., which is greater than the true v., and which decreases as the shear increases. **dynamic ~*** η, μ. Absolute v. The tangential force per unit area of 2 parallel planes at unit distance apart, when the space between them is filled with the fluid in question and one plane moves with unit velocity in its own plane relative to the other. SI unit, Pa·s; cgs unit, the poise. Poiseuille's formula for determining dynamic v. by the capillary-tube method is: $\eta = \pi p r^4 t/8lV$ Pa·s, where p is the pressure difference between the 2 ends of the tube, r the radius of the tube, l its length, V the volume of liquid delivered in a time t. Antonym: fluidity. Cf. *poise, Engler degree, Redwood number, Saybolt seconds, Arrhenius viscosity formula, Einstein viscosity formula.* **intrinsic ~** The limiting value at infinite dilution of the specific v. of a polymer, referred to its concentration. **kinematic ~*** ν. The dynamic v. divided by the density of the fluid. SI unit, m^2/s; cgs unit, the stoke. **pseudo ~** The v. of a thixotropic substance in its most viscous state. See *stoke.* **relative ~** The ratio of the dynamic v. of a solution of a given concentration to that of the pure solvent at the same temperature. Water (1.002 centipoise at 20°C) is the primary standard for the calibration of viscosimeters. **specific ~** The relative v. of a polymer, minus 1. **Woolwich ~** The time in seconds for a steel ball ($\frac{1}{16}$ in diameter) to fall 15 cm through a solution at 20°C.

v. index $100(L - V)/(L - H)$, where V is the v. at 100°F of a lubricating oil sample; and L and H are respectively the v. at 100°F of an oil of 0 and 100 v. i., having the same v. at 210°F as the sample (in centistokes). It expresses the effect of temperature on v. Cf. *SAE number.*

viscum Mistletoe. The leaves and branches of *V. flavecens* (Loranthaceae). It contains viscin, bassorin, gum, and tannin.

visibility Perceptibility to sight. The visibility K of a particular wavelength of light is the ratio of luminous flux F to the radiant energy producing it. **mean ~** The average visibility K_m over any range of wavelengths or the entire spectrum $= F/W$, where F is the total luminous flux in lumens, and W the total radiant energy in watts.

visible Perceptible to the eye; as, v. light.

Vistra Trademark for a viscose synthetic fiber.

visual display unit VDU. A television-type screen linked to a computer or TV camera, for displaying information.

Vita Trademark for a polyvinyl chloride-type plastic.

vitaglass A colorless window glass transparent to ultraviolet light.

vital Pertaining to life.

vitamin An organic substance which an animal must ingest in order to maintain metabolism but which does not provide energy. Many vitamins are enzymes or coenzymes. Deficiency of most v. results in a characteristic deficiency state or disease. V. were originally known only by letters. Some are still referred to by letters, most of these being formed from more than one distinct chemical compound; e.g., v. B_{12} is cyanocobalamin, hydroxocobalamin, and methylcobalamin. V. can be classified into two solubility groups: fat-soluble and water-soluble. See Table 101 on p. 622. Fat-soluble v. occur and are stored in the liver of all animals, including humans and fish; deficiency states develop slowly and toxic effects occur on overdosage. Water-soluble v. are not stored in the body, and deficiency disease occurs early after deprivation. **anti ~** Toxamin. A substance that offsets the action of a v. It may be an enzyme present in foods, e.g., thiaminase, or a drug that antagonizes the v. or blocks the metabolic pathway. **antiberiberi ~** Thiamin. **antihemorrhagic ~** V. K. **antipellagric ~** Niacin and nicotinamide. **apparent ~** Apparent v. C, reductone, q.v. *(reductones).*

v. A Name given to substances with v. A activity, of which retinol, 3-dehydroretinol (vitamin A_2 or gadol), and β-carotene are the most important. **v. A_2** 3-Dehydroretinol (gadol). **v. B complex** A group of v. originally classified by the letter B, as they are mostly contained in the same foods, but are different chemically and physiologically. It comprises thiamin, niacin, riboflavine, pyridoxine, biotin, and pantothenic acid. Cyanocobalamin and folic acid are often included. **v. B_1** Thiamin. **v. B_2** Riboflavin(e). **v. B_6** Pyridoxine, pyridoxamine, and pyridoxal. V. B_6 is the generic name for pyridoxine, the amine and aldehyde; also for the coenzymes pyridoxal 5′-phosphate (PLP), and pyridoxamine 5′-phosphate. PLP is the coenzyme for many enzymes involved in metabolism of proteins and amino acids. **v. B_{12}** Cyanocobalamin, hydroxocobalamin, and methylcobalamin. A member of the v. B complex. It is the only v. containing a metal (cobalt), and occurs naturally only in animal tissues. It can be synthesized by bacteria and yeasts. Present in the body in many forms, including coenzyme 5′-deoxy-adenocobalamin (involved in synthesis of nucleotides and in myelin metabolism in the nervous system). Essential in synthesis of DNA and production of normal red blood cells. See *cyanocobalamin, cobalamins.* **v. C** Ascorbic acid. **v. D** A name given to similar substances with antirachitic properties, including ergocalciferol, cholecalciferol, dihydrotachysterol. **v. D_1** A mixture of v. D_2 and other sterols. **v. D_2** Ergocalciferol. **v. D_3** Cholecalciferol. **v. D_4** Dihydrotachysterol. **v. E** Tocopherols, antisterility v., reproductive v. **v. K** A group of chemically related substances essential for normal blood clotting processes, of general formula:

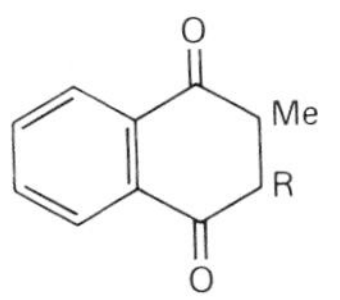

O
Me
R
O

The original name was Koagulations-vitamin. **v. K_1** Phytonadione. **v. K_2** Substances produced by bacteria with 4–13 isoprenyl units in side chain R. **v. M** An essential antianemia v. in monkeys, found to be folic acid. **v. units** International units (I.U.) were formerly defined (1942), but are

TABLE 101. VITAMINS

Further information is in the text, mainly under the "modern name" given below.

Original alphabetical classification	Modern name	Synonyms	Main natural sources	Provitamin	Daily requirement[a]	Deficiency states or diseases (in humans, unless stated)
Fat soluble						
A	Retinol	Vitamin A_1, anti-infective v., antixerophthalmic v.	Green leaf vegetables, carrots, milk, eggs, butter, cheese, fish liver and its oil, liver	β-carotene	750 μg retinol; 450 μg for children	Night blindness, xerophthalmia, keratomalacia leading to blindness
D_1	A mixture of sterols			—	2.5–5 μg; up to 7 μg in children (D₁–D₃)	Rickets in children and young animals; osteomalacia in adult humans (D₁–D₃)
D_2	Ergocalciferol	Calciferol	Yeasts and bacteria	7-Ergosterol.		
D_3	Cholecalciferol		Fish liver and its oil. Formed in skin of humans and animals on exposure to u.v. light.	7-Dehydroxysterol		
K_1	Phytonadione	Phytomenadione, phytylmenaquinone	Liver, green leaves of vegetables	—	2μg/kg of body weight (K₁–K₂)	Hemorrhagic disease, hemorrhagic tendency (K₁–K₂)
K_2	Multiprenylmenaquinone			—		
E	α-Tocopherol, β-tocopherol	Antireproductive v., reproductive v.	Vegetable seed oils (as, wheat germ, soya bean oils), eggs, butter	—		Hemolytic anemia in premature babies. Increased hemolysis in adults with fat malabsorption. Abortion in rats and sterility in other animals.
Water soluble						
C	Ascorbic acid	Antiscorbutic v.	Fresh fruit (as, citrus, black currants), rose hips, meat, liver	—	30–60 mg	Scurvy

B group: B_1	Thiamin	Aneurin, antiberiberi v.	Vegetable seed germ (as, wheat), peas, lentils, whole cereal (as, rice), liver, yeast, meat (as, pork)	—	400 μg/4200 kJ of diet	Beri-beri
—	Niacin, nicotinic acid	Nicotinamide, pellagra preventing v.	Meat, liver, fish, whole cereals. Synthesized in the body from tryptophan.	Tryptophan	15–20 mg	Pellagra
B_2	Riboflavine	Lactoflavine	Milk, meat, liver, green vegetables	—	1.3–1.8 mg	Cracks at corner of mouth. Failure to grow, in animals.
B_3	Pantothenic acid		All foods, particularly liver, eggs, whole grain cereals	—	5–10 mg	None known in humans. Failure to grow, in rats. Neurological abnormalities in chicks and dogs.
—	Biotin		In most foods, particularly, liver, cereals, bananas	—	100–300 μg	Dermatitis if diet consists almost entirely of raw eggs
B_6	Pyridoxine, pyridoxal, pyridoxamine		In most foods, particularly, liver, cereals, bananas	—	2 mg	Deficiency rare. Can cause convulsions in babies. Neurological defects during some antituberculosis therapy. Depression related to oral contraception.
—	Folic acid	Folacin, α-pteroylglutamic acid, vitamin M	Liver, green leaves of vegetables	—	400 μg	Anemia
B_{12}	Cyanocobalamin	Antipernicious anemia v., Castle's extrinsic factor (historical name)	Liver, egg yolk. Not found in vegetables.	—	2–3 μg	Pernicious anemia. Degeneration of spinal cord and dementia.

[a]Average adult daily requirement unless otherwise stated.

now largely discontinued in favor of mg doses. **v. A unit** 1 USP unit is the specific biologic activity of 0.3 μg of the all-*trans* isomer of retinol. It is equivalent to the I.U., which is the activity contained in 0.344 μg of pure all-*trans* retinol acetate. For provitamin A, it is the activity of 0.6 μg of all-*trans* β-carotene. **v. D unit** 1 I.U. is contained in 0.025 μg of cholecalciferol. **v. E unit** 1 I.U. is numerically equal to the USP unit, which equals 1 mg of (±)-α-tocopherol acetate.

vitellin A globulin from egg yolk. Similar proteins occur in lentils, corn, and other cereals. **silver ~** Argyrol.

vitellolutein The yellow coloring matter of eggs.

vitexin $C_{21}H_{20}O_{10}$ = 432.4. Saponaretin. A yellow flavonoid pigment from *Vitex litoralis* (Verbenaceae), m.270 (decomp.).

vitiatine $C_5H_{14}N_6$ = 158.07. A meat base.

vitiligo A skin disease producing depigmented areas, particularly on the face and hands.

Viton Trademark for a synthetic rubber derived from the combination of vinylidene fluoride and hexafluoropropene.

vitrain A constituent of *coal*, q.v.

Vitreosil Trademark for heat-resisting apparatus, made from a translucent variety of silica, prepared by fritting sand with a hot carbon plate.

vitreous Glassy. **v. copper ore** Redruthite. **v. enamel** Metal with a fused-on glass surface. **v. silver ore** Argentite.

vitrescence The property of becoming hard and transparent like glass.

vitrification The conversion of a material into a glass or glasslike substance, of increased hardness and brittleness.

vitrify To sinter or melt to a glassy mass.

vitrinite A uniform brown constituent of coal, with a low ash content; consists mainly of ulmin compounds. Cf. *vitrain.*

vitriol A sulfate of a heavy metal; as:

Blue or roman	Copper sulfate (chalcanthite)
Green	Ferrous sulfate (copperas)
Red or rose	Cobaltous sulfate (bieberite)
Uran	Uranium sulfate (johannite)
White	Zinc sulfate (goslarite)

Cypria ~ Copper sulfate. **nitrous ~** Nitrosyl hydrogensulfate formed in the Gay-Lussac tower. **oil of ~** Sulfuric acid. Cf. *BOV, DOV, ROV.* **salt of ~** Zinc sulfate*.

vitriolate The sulfate of a metal (obsolete). **v. of soda** Sodium sulfate*. **v. of tartar** Potassium sulfate*.

vitriolum veneris Copper sulfate.

vitro- Prefix (Latin: "glass") indicating a mineral or rock of glassy texture; as, vitrophyric. Cf. *in vitro.*

vitrophyric Describing an igneous rock having a glassy base.

vivianite $Fe_3(PO_4)_2 \cdot 8H_2O$. Blue ocher, blue iron ore. A native ferrous phosphate which is white when freshly broken and becomes bluish on oxidation.

vivo Pertaining to life. Cf. *in vivo.*

vobasine $C_{21}H_{24}O_3N_2$ = 352.4. Alkaloid from Apocynaceae spp., m.112. See *acylindole alkaloid* under *alkaloid.*

vocoder An instrument for synthesizing speech and sound effects by electrical methods.

voiceprint See *spectrograph.*

volatile Evaporating rapidly. **v. alkali** Ammonia. **v. matter** The weight loss of a fuel after heating at 927°C (1200 K) in an inert atmosphere for 40 min. Cf. fixed *carbon.* **v. oils** Essential oils. **v. poisons** Poisonous substances that form vapors:

Nonmetallic: bromine, chlorine, iodine, fluorine, phosphorus, hydrofluoric acid, arsine, stibine, bismuthine, phosphine, hydrogen sulfide.
Organic: Chloral, chloroform, ether, aniline, acetanilide, etc.

volatility product The product of the concentrations of the constituent gases from a solid substance (as ammonium carbonate) which dissociates into 2 volatile gases when heated; the v. p. is constant. Cf. *solubility product.*

volatilization Vaporization: the conversion of a liquid or solid into a vapor or gas without chemical change.

volatilize To convert into a gas or vapor.

volborthite A native hydrous calcium-copper vanadate.

volcanic Pertaining to molten rock or lava. **v. ash** Tuff. **v. glass** A volcanic, igneous rock; as, obsidian. **v. mud** A mud of fine-grained ash or tuff and water.

volemite Volemitol.

volemitol $CH_2OH(CHOH)_5CH_2OH$ = 212.2. D-*glycero*-D-*manno*-Heptitol. Volemite, α-sedoheptitol, m.152, $[\alpha]_D$ + 2.25, from *Lactarius volemus* and *Primula* species. Cf. *primulite.*

Volhard **V., Jakob (1834–1910)** German analytical chemist. **V.'s solution** A decinormal solution of potassium thiocyanate. **V.'s volumetric method** The determination of halogens by means of standard thiocyanate solutions.

volkonskoite A magnesium chromium silicate (30% chromic oxide); a zeolite.

volt Symbol: V; The SI unit of electromotive force and potential difference. It is the potential difference between 2 points on a conductor carrying a 1 A current, when the power dissipated between the 2 points is 1 W. 1 V = 10^8 emu = 3.336 × 10^{-3} esu. A cadmium cell at 20°C gives 1.0183 V. **ampere- ~** See *volt-ampere.* **electron ~** See *electron volt.* **international ~** Pre-SI units, now obsolete. 1 mean international v. = 1.00034 V. 1 U.S. international v. = 1.000330 V.

Volta **V., Count Alessandro (1745–1827)** Italian physicist. **v. couple** An electric cell having a zinc anode and copper cathode (potential 0.98 volt). **v. effect** The change in the sign of the charge on a metal electrode after it has been heated. **v. series** Electromotive series.

voltage Electromotive force (in volts). See *potential.*

voltaic Pertaining to a direct electric current. **v. battery** A number of v. cells. **v. cell** An electric cell or device in which an oxidation-reduction reaction produces an electromotive force. See *cadmium, Bunsen, Daniell, Grove, Léclanché, Clark cells, lead accumulator* under *lead.* **v. couple** A pair of metallic electrodes producing a potential when set up as a cell. See *volta couple.* **v. electricity** Galvanic electricity, chemical electricity. A continuous stream of electrons (direct current) caused by a chemical reaction. **v. pile** A series of metallic disks forming a v. battery.

voltaite The mineral $2(FeAl)_2O_3 \cdot 5(MgFeNa_2K_2)(OH)_2(SO_4) \cdot 14H_2O$.

voltameter Coulometer. An apparatus for the electrolysis of solutions. The amount of metal or gas liberated indicates the number of coulombs of current flowing in the circuit during the decomposition. Cf. *voltmeter.*

voltammeter An instrument that indicates both volts and amperes.

volt-ampere The equivalent of the watt *power factor,* q.v.; the product of a volt and ampere.

voltmeter An instrument that indicates voltage. Cf. *voltammeter.*

voltoids Small, compressed tablets of ammonium chloride, used in voltaic (Léclanché) cells.

voltolize, voltolise To subject to a silent electric discharge.

voltzite Zn_5OS_4. A native oxysulfide.

volucrisporin $C_6H_2O_2(C_6H_4 \cdot OH)_2$ = 292.3. 2,5-Bis(3-hydroxyphenyl)-1,4-benzoquinone. A red pigment, m.300+, from the fungus *Volucrispora aurantiaca.*

volume The space occupied by a substance, generally

expressed in cc or liters according to the formula, $V = Cl^3$, in which C is a constant depending on the shape of the space occupied (if a cube, $C = 1$) and l is the length. **atomic ~** See *atomic volume*. **critical ~** The v. occupied by one gram of a gas at the critical temperature and pressure. See *parachor*. **co ~** The quantity b in van der Waals' equation. **humid ~** The v. of that quantity of moist air which contains 1 liter of dry air. **incompressible ~** That which cannot be made smaller by pressure. **molar ~** The v. divided by the amount of substance; as, molecular weight/density. This varies for solids and liquids according to their atomic volumes, but is 22.4138 liters for gases at s.t.p. See *parachor*. **sound ~** See *sound amplitude*. **specific ~*** The v. of a system divided by its mass. **standard ~** The v. occupied by one mole of gas at 0°C and 760 mm pressure = 22.4138 liters.

v. susceptibility The ratio of the intensity of magnetization of a medium to the strength (in teslas) of the magnetic field inducing it.

volumenometer (1) An apparatus for the accurate determination of the volume of a known weight of substance, and thence its density. (2) Pyknometer (obsolete).

volumetric Pertaining to measurements of volumes. **v. analysis** The quantitative analysis of a known volume of a solution of unknown strength by adding a reagent of known concentration until the end point of the reaction has been reached. An indicator is added to establish a definite end point. Methods: neutralization (alkalimetry, acidimetry); oxidation-reduction (oxidimetry, iodometry); precipitation (titration with a reagent that causes precipitation). **v. factor** The amount of substance corresponding with 1 mL of normal, half-normal, or tenth-normal solution. **v. glassware** The graduated glass utensils used to measure definite quantities of a solution; i.e., burets, cylinders, flasks, and pipets. Cf. *standardized*. **v. solutions** V.S. Solutions of known strength used in v. analysis. See *normal solution*. **v. standards primary ~** Those in which the composition is gravimetrically determined. **secondary ~** Those standardized against a weighed amount of reagent. **tertiary ~** Those in which one solution is titrated against another. **v. weight** The amount of substance to be weighed in order that, on titration, the number of mL found will equal the percentage of unknown present.

voluminal Pertaining to 3 dimensions. **v. expansion** Volumetric *expansion*.

volutin A nucleoprotein from yeast.

vomicine $C_{22}H_{24}N_2O_4$ = 380.4. 12-Hydroxyicajine. An alkaloid from strychnos.

vomipyrine $C_{15}H_{16}N_2$ = 224.3. An alkylated pyrroloquinoline produced by the degradation of vomicine; m.107. Its ether solution has a blue-violet fluorescence.

votator A concentric, double-tube heat exchanger, with a scraper in the central tube.

v.p. Abbreviation for vapor pressure.

VPI Trademark for dicyclohexylammonium nitrite, a *vapor-phase inhibitor*, q.v.

vrbaite $TlAs_2S_5$. A mineral source of thallium.

V.S. Abbreviation for volumetric solution.

vug A cavity in a casting. **v. crystals** The metallic crystals found inside vugs.

vulcanite A hard rubber produced by heating caoutchouc or india rubber with sulfur. Cf. *ebonite*.

vulcanization The oxidation of rubber by reducing sulfur to sulfides. The rubber is mixed with a vulcanizing agent (as, sulfur), and heated at 110–140°C. The tacky, plastic mixture changes gradually to an elastic, rigid product. Accelerators are added to improve the quality. See *vulcanizing agent*.

vulcanize To produce vulcanization.

vulcanized Subjected to vulcanization. **v. paper** Horn fiber. A hard, resistant material made by impregnating paper with zinc chloride solution, followed by washing. Several layers may be laminated together; used for luggage. Cf. *vegetable parchment*.

vulcanizing agent A substance that *vulcanizes*, q.v., rubber; e.g.: sulfur (chiefly), selenium, sulfur dichloride, *m*-dinitrobenzene, and nitrogen compounds such as di- and triphenylguanidine, tetramethylthiuram, and piperidine derivatives.

vulnerary Former term for a substance used externally to treat wounds and bruises.

Vulpak Trademark for a cellulose acetate-type plastic.

vulpinic acid $C_{19}H_{14}O_5$ = 322.3. Chrysopicrin. Methyl ester of pulvinic acid; from the lichen *Cetraria vulpina*, m.148.

vulpinite An amorphous form of anhydrite.

vuzine $C_{27}H_{40}N_2O_2$ = 424.6. Isooctylhydrocupreine. A levorotatory quinine, q.v., alkaloid. Colorless powder, insoluble in water.

Vycor Trademark for heat- and chemical-resistant glassware, including low-expansion glasses.

Vyrene Trademark for a silicone-lubricated, polyester polyurethane, spandex monofilament.

W

W Symbol for (1) tungsten (wolfram); (2) watt.
W Symbol for (1) energy (*E* is preferred); (2) quantity of work.
w Symbol for (1) electromagnetic energy density; (2) specific volume; (3) speed or velocity.
Ω, ω See *omega.*
Waage, Peter (1833–1900) Norwegian chemist noted for developments in physical chemistry.
Waals See *van der Waals.*
wabain $C_{36}H_{46}O_{12}$ = 670.7. A glucoside from the root of *Carissa schimperi* (Apocynaceae), the waba tree of Africa; a heart stimulant and local anesthetic. Cf. *ouabain.*
Wackenroder's reaction The reaction between hydrogen sulfide and sulfur dioxide in aqueous solution to form polythioic acids.
wad (1) Bog manganese. An earthy hydrate of pyrolusite, containing baryta. Cf. *psilomelane.* (2) Graphite.
Waelz process Low-grade zinc ore is heated with fuel oil or powdered coal in a rotary kiln; volatilized zinc and zinc oxide result.
wafer A thin, double layer of dried paste that encloses a medicament. **silicon ~** See *silicon chip.*
wagnerite $Mg(MgF)PO_4$. A native fluorophosphate. Cf. *adelite.*
Wagner's reagent An aqueous solution of iodine and potassium iodide; a microchemical reagent for alkaloids.
wagofo An African arrow poison, the dried juice of *Euphorbia* species.
wahoo Euonymus.
wakeamine Sympathomimetic amine. General name for amphetamine-type drugs used to delay the onset of fatigue.
Walden W., Paul (1863–1957) German organic chemist. **W. inversion** A chemical reaction which reverses the rotatory power of an optically active compound. It indicates that the mechanism of substitution does not involve the simple replacement of one group by another, but a rearrangement of groups and atoms; e.g.: (*R*)-chlorosuccinic acid changes (on treatment with potassium hydroxide) to (*S*)-malic acid; while (*S*)-chlorosuccinic acid changes to (*R*)-malic acid and not to (*S*)-malic acid.
wall effect See *Coanda effect.*
Wallach, Otto (1847–1931) German chemist, Nobel prize winner (1910); noted for research on the constitution of essential oils.
walnut Juglans. **w. oil** The clear, colorless oil of walnuts, d_{25}·0.923, $[n]_D^{25}$ 1.4750; an emulsifier in cosmetics.
warfare gas See *poison gas* under *gas.*
warfarin 3-(α-Acetonylbenzyl)-4-hydroxycoumarin. A rodenticide which kills by producing hemorrhage, but is tasteless, odorless, and relatively safe to human beings and domestic animals. Rodents may develop resistant strains. **w. sodium** $C_{19}H_{15}O_4Na$ = 330.3. White crystals, m.162, soluble in water; an anticoagulant (USP, BP).
warp The threads of a woven fabric which are extended lengthwise in the loom. Cf. *weft.*
warrant Spavin.
warringtonite Domingite.
wash bottle A glass or plastic flask or bottle fitted with a stopper and 2 unequally long tubes, so arranged that on forcing air by blowing or pressure through one, a stream of water emerges from the other. Used in washing precipitates on filters, and in other chemical operations.
Washburn cell A glass vessel with 2 electrodes for determining the electrical conductivity of solutions.
washing Rubbing and rinsing with a liquid. **w. bottle** See *wash bottle.* **w. soda** Commercial crystalline sodium carbonate.
Wassermann W., August von (1866–1925) German biochemist. **W. reaction** Σ test. A diagnostic test for syphilis; an example of a *complement* fixation test, q.v.
water* (1) H_2O = 18.015. Hydrogen oxide. Colorless, tasteless liquid, forming the largest proportion of the earth surface, m. (freezes) 0, b.100; the commonest solvent. Regarded as an element by the alchemists; recognized as a combustion product of hydrogen by Cavendish (1781); as a compound of oxygen and hydrogen by Lavoisier (1783); and as a mixture of isotopes (1933). Probably a loosely structured mixture of multiples of H_2O: as, H_4O_2, dihydrol; H_6O_3, trihydrol; etc. Cf. *heavy water, ice water, steam water* (all below). W. is an essential constituent of all living organisms, and occurs as "water of crystallization" in many crystals and compounds. (2) Aqua. Potable w. See *potable water* below. (3) The degree of transparency of a precious stone; as, diamond. **acidulous ~** W. containing dissolved carbon dioxide (together with alkali hydrogencarbonates and common salt) which is liberated with effervescence by warming; e.g., Appolinaris w. **aerated ~** W. containing a gas; as, air. **alkaline ~** A mineral w. containing hydrogencarbonates of Na and sometimes Li and K; e.g., Vichy w. **anomalous ~** Poly ~. **bitter ~** A mineral w. containing the sulfates of Na and Mg; e.g., Marienbad w. **bound ~** That portion of a system, e.g., tissue, that does not freeze at −20°C. **camphor ~** A 0.1% solution of camphor in 0.2% alcohol; used in pharmacy. **capillary ~** The w. of the soil held between rock interstices above the groundwater. **carbonated ~** A drinking water containing carbon dioxide under pressure. **chalybeate ~** Ferruginous w. A mineral w. containing ferrous carbonate held in solution as hydrogencarbonate; as, Pyrmont w. **cologne ~** A w. containing essential oils (as, lavender); a household perfume. **conductivity ~** A w. purified by repeated distillation or by treatment with synthetic resins; a solvent in electrolytic measurements. **crystal ~** W. of crystallization. **distilled ~** W. purified by distillation. **drinking ~** W. that contains neither pathogenic organisms nor a large amount of organic matter, ammonia, nitrites, or nitrates. Maximum limits: total solids 1,000, Pb 0.1, Cu 0.2, Zn 0.5, Fe 0.3, Mg 100, Cl^- 250, SO_4^{2-} 250 ppm. See *water analysis.* **ferruginous ~** Chalybeate w. **free ~** That portion of a system which freezes. Cf. *bound water.* **fresh ~** W. other than sea- or brackish w., irrespective of whether it is potable. **ground ~** W. at a definite level beneath the soil. **hard ~** W. that contains the carbonates and hydrogencarbonates of calcium and magnesium; it forms insoluble compounds with soap and prevents lather formation. See

TABLE 102. PHYSICAL CONSTANTS OF WATER AND OF HEAVY WATER

Constant	Water (H_2O)	Heavy water (D_2O)
Density (20°C)	0.9982	1.1056
Melting point	0°C	3.80°C
Boiling point	100°C	101.42°C
Refractive index	1.33293	1.338
Viscosity (mPa·s at 20°C)	1.0020	1.250
Surface tension (mN/m at 20°C)	72.75	67.8
Raman spectrum	3420 Å	2549 Å

water purification. **heavy ~** (1) D_2O = 20.03. ^2H_2O. Deuterium oxide. The isotopic compound of hydrogen of mass 2 (deuterium) with oxygen. It inhibits plant and animal growth. For physical constants see Table 102. (2) One of the other compounds of isotopic H or O; as, (*a*) DOH, TOT, TOH, TOD, where D is deuterium and T is tritium; or (*b*) $H_2{}^{17}O$, $H_2{}^{18}O$, $D_2{}^{17}O$, $D_2{}^{18}O$, $T_2{}^{17}O$, $T_2{}^{18}O$, where ^{17}O is oxygen of mass 17, and ^{18}O of mass 18; or (*c*) a combination of (*a*) and (*b*); as, $D^{17}OH$. **hepatic ~** A mineral w. containing hydrogen sulfide and alkali sulfides; e.g., Harrogate w. **ice ~** Freshly molten w. containing smaller aggregates of $(H_2O)_n$ than w. at 4°C. Cf. *steam water* below. **industrial ~** W. that can be used for process work, but not necessarily potable. **injection ~** Sterile, distilled w. without any added substance (USP, BP). **juvenile ~** W. of deep-seated origin, that is supposed to have reached the surface of the earth for the first time. **light ~** Protium oxide. See *heavy water* above and Table 102. **metabolic ~** See *water metabolism.* **meteoric ~** W. from the atmosphere; as, rain. **mineral ~** A natural, therapeutic w.; as: acidulous, chalybeate, hepatic, alkaline, bitter, siliceous w.; hot springs (may contain radon). **natural ~** A w. as it occurs in nature, as, rain-, snow, river, spring-, deep-well, sea-, and mineral w. Impurities may be suspended or dissolved solids and gases. **poly ~** Polymerized w. reportedly made by condensing w. vapor in fine-bore quartz capillaries; d.1.4, refractive index 1.49, m.−40, vitreous at 50, oily at 20, reverting to ordinary w. above 500. Its existence is subject to doubt. **potable ~** W. that is fit to drink; e.g., nontyphoid, nonpolluted, free from pathogens and toxic substances. Cf. *drinking water* above. **purified ~** W. that has been repeatedly distilled and subjected to purifying operations, e.g., with ion-exchange materials, or by reverse osmosis (USP, BP). **sea ~** Ocean w. that contains an average of dissolved solids 3.6% (sodium chloride 2.6%). See *hydrosphere, salinity.* **siliceous ~** W. containing dissolved colloidal silica and alkali silicates, e.g., w. of the hot geysers of Yellowstone Park. **soda ~** See *soda water.* **soft ~** (1) Rain ~ or snow w., which is naturally free from Ca and Mg salts. (2) W. after removal of Ca and Mg salts. See *hard water* above, *water purification* below. **steam ~** Freshly condensed w. It contains a different proportion of dihydrol and trihydrol compared with ice w. **sterilized ~** W. free from living microorganisms. **sweet ~** W. containing glycerol residues, from soap manufacture. **well ~** Ground ~.

w. analysis (1) Sanitary: The bacteriological and chemical determination of the harmful constituents of w.; e.g., determinations of ammoniacal and albumenoid ammonia, nitrite and nitrate, nitrogen, oxygen consumed, total solids, alkalinity, hardness, halogens. Decaying vegetable and animal matter is indicated by the nitrogen. (2) Complete: The determination of all inorganic constituents. The results are expressed as oxides of elements present, as salts, or as ions and radicals. **w. bath** A vessel of boiling w. used in the laboratory to evaporate liquids in a smaller vessel over it. **w. constants** Cf. *heavy water* above. See also Table 102.

Density: (greatest at 4°C; hence 1 mL w. at 4°C is the classical unit of mass = 1 g)	
0°C	0.99984 g
10°C	0.99970 g
Specific latent heat of fusion	331 J
Specific latent heat of vaporization	2257 J
Specific heat capacity (cal/g/°C)	1.00368 at 5°C
	1.00000 at 15°C
	0.99829 at 50°C
	1.00645 at 100°C
H^- and OH^- ion concentration (each)	10^{-7} g/liter
Critical temperature	374°C
Critical pressure	22.0 MPa

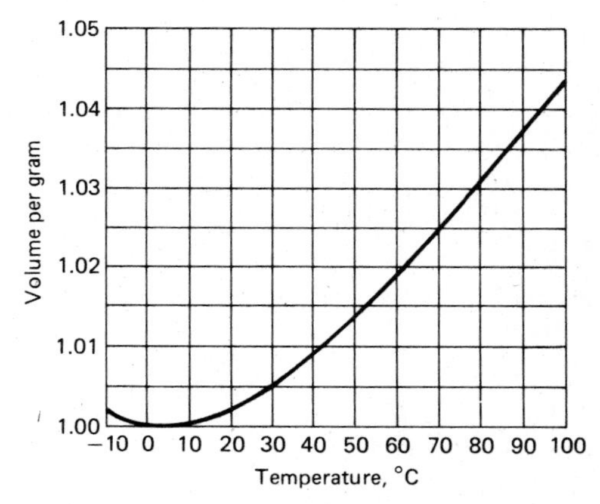

Fig. 21. Specific volume of water.

w. of constitution W. that is chemically combined, as in hydroxides. **w. of crystallization** W. that is a physical constituent of crystals or hydrated salts, and removed at 100°C. **w. culture** A method of growing plants by using a nutrient 0.01 *M* solution of potassium nitrate, calcium sulfate, and magnesium hydrogenphosphate, with or without other salts or ions. Cf. *hydroponics.* **w. detection** (1) Filter paper saturated with a solution of lead potassium iodide in acetone turns yellow. (2) Anhydrous (white) copper sulfate turns blue. (3) See *Karl Fischer reagent.* **w. distillation** The evaporation and condensation of w. To remove all dissolved material, distill in a quartz vessel first from an alkaline solution of permanganate, then from sulfuric acid. Cf. *conductivity water* above. **w. divining** Dowsing. The detection of underground w. supplies by means of, e.g., a bent hazel twig, which twists in the hands of the operator when over w. It is stated to arise from great sensitivity to electromagnetic radiation emitted by w. **w. flowers** Ice flowers. **w. gas** A mixture of hydrogen and carbon monoxide, obtained by the action of steam on glowing coal. **w. glass** Sodium silicate. **w. of hydration** W. *of crystallization*, or *w. of constitution.* Cf. *hydrate.* **w.**

metabolism The daily balance of w. in the human body. An average person in a sedentary job in a temperate climate has a daily w. output of about: urine 1–2½ L, skin 500 mL, expired air 300 mL, feces 50–100 mL. **w. pollution** The contamination of natural waters by wastes from sewage or industry. Effluent discharge parameters either depend on the location or may be uniform for a given region. "Typical" values for a U.K. water authority are: suspended solids 30 ppm max., BOD 20 ppm max., pH value 5–9, temperature increase of receiving waters 4 °C max. Additional parameters, such as sulfates or petroleum-ether extractables, may also apply. **w.proof** Rainproof. Impermeable to water. **w. pump** See *filter pump.* **w. purification** The removal of salts or organisms from natural waters; e.g., softening (partial removal of salts), distillation (complete removal of salts), filtration (partial removal of organisms and suspended matter), sterilization (complete destruction of organisms). **w. softening** The removal of Ca^{2+} or Mg^{2+} by (1) filtration through permutite or ion-exchange natural zeolites, in which calcium ions are replaced by sodium ions; (2) heating, which removes Ca^{2+} as calcium carbonate (boiler scale); (3) chemical treatment, as, with lime. **w. sterilization** The destruction of bacteria and spores by heat, or their removal in porcelain filters. **w. still** A device for the continuous production of distilled w.

water mint The dried leaves of *Mentha aquatica* (Labiatae) containing essential oils and tannin; an antispasmodic and stimulant.

waterproofing Making a substance impermeable to moisture.

Watt, James (1736–1819) Scottish inventor of the first steam engine (1769).

watt Symbol: W; the SI unit of power, radiant flux, or time rate at which work is done: 1 watt = 1 joule (10^7 erg) per second. Also the energy expended per second by an unvarying current of 1 ampere under a potential of 1 volt.

Watts = volts × amperes = amperes^2 × ohms
1 Watt = 0.001341 hp
1 Watthour = 3,600 joules

wattage Amperage multiplied by voltage.

wattle gum Australian gum, from *Acacia pycnantha* (Leguminosae), Australia and S. Africa.

Watts, Henry (1815–1884) British chemist, author of *Dictionary of Chemistry.*

wave (1) A to-and-fro motion of the particles of a solid, liquid, or gas. (2) The harmonic curve obtained on plotting distance and time for points representing centers of force in rhythmic motion. **longitudinal ~** Forward and backward motion of particles in the direction in which the w. moves, as sound waves in air. **micro ~** See *microwave.* **transverse ~** Motion at right angles to the direction in which the w. moves.

w. front The surface of propagation, at right angles to the direction of motion, which passes through corresponding points of a number of waves. **w. function** The factor ψ in *Schrödinger equation,* q.v. **w.length** The distance between corresponding points on 2 adjacent waves. Cf. *angstrom, sound, spectrum, kX,* and Table 69 under *radiation.* **w. mechanics** Extensions of quantum mechanics, as: A derivation of the velocity v and position p of electrons, one of which can be determined by predicting the probable range of values. The motion of each particle of mass m is associated with a wavelength $\lambda = h/mv$; $mc^2 = h\nu$, where c is the velocity of light; wave velocity $\mu = \lambda\nu$; $\lambda = 12{,}336/V$, where V is the corresponding energy change in volts. **w. motion** The progressive disturbance of particles in a medium. The speed of propagation is the velocity c of the wave; this depends on the wavelength λ, and the time period T, or on the frequency ν; hence $c = \lambda/T = \nu\lambda$, and $\nu = c/\lambda$. See *Schrödinger equation.* **w. number** σ, $\tilde{\nu}$. The number of waves per unit length; for light, $\tilde{\nu} = 1/\lambda$ (if λ is measured in centimeters), or $\tilde{\nu} = 10^8/\lambda$ (if λ is measured in angstroms), where $\tilde{\nu}$ is the number of waves per centimeter. Cf. *frequency, Ritz formula.* **w. theory** Radiations comprise electric and magnetic fields at right angles, and moving at right angles to or in the direction of propagation.

wavellite $Al_3(OH \cdot F)_3 \cdot (PO_4)_2$. A native phosphate, containing some fluorine.

wawa See *obeche.*

wax (1) An ester of a high-molecular-weight fatty acid with a high-molecular-weight alcohol (other than glycerol) but including monohydric fatty alcohols. (2) A mixture of esters, fatty acids, higher alcohols, and hydrocarbons. Cf. *sterols.* (3) Plastic substances obtained from plants or deposited by insects; consist of the esters of fatty acids with higher alcohol radicals; e.g., beeswax, triacontyl palmitate, $C_{30}H_{61} \cdot C_{16}H_{31}O_2$. (4) Modern usage: a substance having the properties: (*a*) crystalline to microcrystalline structure; (*b*) capacity to acquire gloss when rubbed (as distinct from *greases*); (*c*) capacity to produce pastes or gels with suitable solvents or when mixed with other waxes; (*d*) low viscosity at just above the melting point (as distinct from *resins* and *plastics*); (*e*) low solubility in solvents for fats at room temperature. Waxes differ from fats in that fats are the esters of the trihydric, lower alcohols; as, glycerol. See Table 103. **bees ~** See *beeswax.* **carnauba ~** See *carnauba wax.* **chinese ~** See *chinese wax.* **C.K. ~, Seekay ~** Trademark for a chlorinated w., used for stopping deposition in selective electroplating. **coal ~** A hard brittle, coal hydrocarbon, m.99. **cochineal ~** Coccerin. **earth ~** Ceresin. **fossil ~** Ozocerite. **insect ~** Chinese w. **japan ~** See *japan wax.* **Java ~** Hexacosyl alcohol*.

TABLE 103. SOURCE AND COMPOSITION OF VARIOUS WAXES

Source	Example of wax	Composition
Mineral	Paraffin	Straight-chain hydrocarbons, 26–30 C atoms/molecule
	Microcrystalline	Branched-chain hydrocarbons, 41–50 C atoms/molecule
	Oxidized microcrystalline	Hydrocarbons, esters, fatty acids
	Montan	Wax acids, alcohols, esters, ketones
	Hoechst	Acids, esters (obtained by oxidizing montan wax)
	Ozocerite	Saturated and unsaturated high-mol.-wt. hydrocarbons
Vegetable	Carnauba	Complex alcohols, hydrocarbons, resins
	Esparto	Mainly hydrocarbons
	Flax	Fatty-acid esters, hydrocarbons
	Sugarcane	Hydrocarbons, long straight-chain aldehydes, alcohols
	Candelilla	Hydrocarbons, acids, esters, alcohols, stearols, resins
Animal	Beeswax	Hydrocarbons, acids, esters, alcohols, lactones
Synthetic	Fischer-Tropsch	Saturated and unsaturated hydrocarbons, oxygen compounds

microcrystalline ~ See *microcrystalline wax.* **mineral ~** Paraffin w. **montan ~** See *montan wax.* **palm ~** See *palm wax.* **paraffin ~** See *paraffin wax.* **peat ~** See *peat wax.* **pisang ~** See *pisang wax.* **sealing ~** See *sealing wax.* **sugarcane ~** Cerosin. **wasp ~** A constituent of certain beeswaxes; it has a low saponification value (about 60). **white ~** Cera alba. Bleached beeswax (NF). **yellow ~** Cera flava. Beeswax. Soft, yellow solid, obtained by melting and purifying the honeycombs of the bee; insoluble in water, soluble in ether or oils. Used in ointments (NF). **wool ~** See *wool fat.*

w. bean Butter bean. **w. tailings** The residue after destructive distillation of petroleum; contains the solid hydrocarbons chrysene and picene; and resembles asphalt; used for waterproofing and as a filler in cheap greases.

W.B.C. Abbreviation for white blood cells.

weak (1) Not strong. (2) Not dissociating greatly. **w. acid** An acid that does not dissociate greatly. **w. base** A base that does not dissociate greatly. **w. salt** A salt that does not dissociate greatly. Cf. *dissociation.*

weathering Chemical reactions which change the composition of and disintegrate rocks; as, the effects of air, water, bacteria, heat, freezing.

weave (1) To intertwine threads or fibers in a regular manner to form a continuous web of fabric. (2) The manner or system in which the threads are woven. See *faille, twill.* **plain ~** A w. in which the warp and weft threads pass alternately over and under the weft and warp threads, respectively.

weber* Symbol: Wb; the SI unit of magnetic flux, which, linking a circuit of 1 turn, produces in it an emf of 1 V as it is uniformly reduced to zero in 1 second. 1 Wb = 10^8 maxwells = 7.958×10^6 unit poles.

Weber's law A sensation is proportional to the natural logarithm of the stimulus.

Webril Trademark for a bonded fiber web.

websterite $Al_2SiO_6 \cdot 9H_2O$. A native silicate. Cf. *cecilose.*

wedge A solid object with a triangular section, used to separate surfaces by the forcible introduction of the pointed end. **w. photometer** A photometer in which the illuminations on 2 faces of a wedge are matched.

weedicide Herbicide.

weft The threads of a woven fabric that cross from side to side of the web and interlace the *warp,* q.v.

Weigert effect Dichroism shown by a colloidal suspension or gel.

weighing Measuring gravitational attraction. **double-~** W. independently of any variation in length of the 2 balance arms. If the object weighs a when placed on one pan, and b when on the other, then the weight is given by $\sqrt{ab}$.

w. bottle A small, stoppered, thin-walled, light glass vessel, used to weigh quantities of a substance for analytical purposes. **w. by substitution** Measuring the weight of an object independently of the lengths of the arms of the balance. The object is counterpoised by lead shot, then removed, and the lead shot counterpoised by weights. **w. by swings** A method in which the lack of sensitivity of a balance is compensated by noting the swings of the balance pointer, and the distances traversed to left and right.

weight The degree of heaviness or the force with which a body is attracted by the earth; it varies with geographical location. Cf. *mass.* **atomic ~** See *atomic weight.* **carrier ~** The weight of substance, e.g., protein, which carries one gram equivalent of prosthetic groups. **combining ~** Equivalent w. **equivalent ~** The weight of a substance that can combine with or displace one unit weight of hydrogen, or its equivalent of another substance. **formula ~** The sum of the component atomic weights of an atom, ion, or molecule. **isotopic ~** See *isotopic weight.* **molecular ~** See *molecular weight.* **specific ~** See relative *density.*

w. buret A small bottle with a tap and jet, which is weighed before and after titration to give the weight (instead of the volume) of titrating solution used. **w. concentration** Mass fraction*. **w. density*** See *density (1).* **w. normality** Normality expressed per kilogram instead of per liter. **w. strength** A method of expressing the strength of an explosive per unit weight. Cf. *Trauzl test,* ballistic *mortar.*

weights For systems of w., see *SI units, metric, avoirdupois, apothecaries' weights, troy weights.*

Weil-Felix reaction An agglutination test for diagnosing rickettsial diseases; as, typhus.

weinschenkite $(YEr)PO_4 \cdot 2H_2O$. A native rare-earth phosphate, resembling gypsum in structure.

weiss A unit of atomic magnetic moment (P. Weiss, 1911). Cf. *magneton.*

Weissenberg effect The tendency of a viscoelastic liquid to climb up a vertical rod rotating in it; associated with the orientation of high-polymer molecules in flowing solutions.

weissite Cu_5Te_3. Telluride. Lens-shaped crystals in veins of pyrites.

weld To join metals by pressure at a temperature below that of complete fusion. **dyer's ~** A yellow dye from *Reseda lutea.*

welding (1) The joining of metals by heat; as: (*a*) *plastic ~* united by pressure without a weld metal (forge, electrical-resistance heater, or Thermit); (*b*) *fluid ~* united by a weld metal without pressure (torch, arc, or Thermit). (2) Heating a metal to white heat and pressing the pieces together; or heating a metal to fusion and letting it flow into the joint without hammering or pressure (fusion w.). Cf. *Thermit process.* **cold ~** Hammering metal foils together into a compact piece. **friction ~** W. in which heat generated by 2 rotating rods in pressure-contact melts a metal layer on each rod. **sigma ~** A process for welding carbon in an inert gas by a metallic arc. **spot ~** Fluid w. at adjacent spots, which subsequently run together as a continuous weld. **ultrasonic ~** W. by vibratory energy, usually having a frequency greater than that of sound waves.

Weldon W., **Walter (1832–1885)** British industrial chemist. **W. process** The manufacture of chlorine from hydrochloric acid by the action of manganese dioxide, and the regeneration of the W. mud by lime, followed by reoxidation to manganese dioxide. **W. mud** The slime of manganese and calcium manganites from the W. process.

Welsbach W., **Carl Freiherr Auer von (1858–1929)** Austrian chemist. **W. mantle** Incandescent gas mantle. Cellulose impregnated with cerium and thorium nitrates is ignited, and a mixture of ceria 1 and thoria 99% remains.

Welter's rule The heat of combustion of an organic compound is obtained approximately by subtracting the oxygen and hydrogen in the proportion of H_2O, and then adding the heats of combustion of the residual carbon and hydrogen atoms.

Welvic Trademark for a polyvinyl-type plastic.

Wenzel's law Richter's law.

Werner W., **Alfred (1866–1919)** Swiss chemist, Nobel prize winner (1913). **W.'s theory** The affinity of an atom is an attractive force toward the center of the atom evenly distributed over its spherical shell. In a combination of 2 atoms, the attractive force concentrates itself in a definite area, dependent on the nature of the combining atoms. Two combining molecules probably first form an association held together by residual valencies; the affinities then may redistribute themselves uniformly, with the formation of radicals. Hence the behavior of the Cl atoms in hexachloroplatinic acid, H_2PtCl_6. Cf. *coordination number, valency.*

wernerite Scapolite.

Wesson tube An absorption tube for carbon dioxide, containing soda lime and calcium chloride.

West, Clarence Jay (1886–1953) American chemist noted for chemical literature.

Weston, Edward (1850–1936) British-born American electrical engineer. **W. normal cell** Cadmium cell.

Westphal balance A beam balance for determining the specific gravity of liquids or solids.

Westron Trademark for tetrachloroethane.

wet Moist. **w. analysis** Qualitative analysis by means of reagents in solution. Cf. *blowpipe analysis.* **w.-bulb thermometer** Hygrometer. **w. combustion** A method of determining organic carbon by oxidation with chromic acid. **w. process** An industrial method involving solutions or water; as, the extraction of ores by acids.

wetness Antonym of *freeness.*

wettability Degree of wetting, measured by the adhesion tension between a solid and liquid phase. Cf. *introfaction.*

wetting The adhesion of a liquid to a surface, resulting in a tray-shaped meniscus. Cf. *flotation, adhesion, tension.*

whale An aquatic mammal, order *Cetaceae.* **w.bone** Baleen. The horny, elastic substance from the upper jaw of the Greenland w. **w. guano, w. meal** A fertilizer made from dried and ground w. meat: nitrogen 6–8, phosphorus pentaoxide 4–8%. **w. oil** Blubber oil, d.0.92–0.93; a leather dressing, lubricant, illuminant, and source of soap. **w. shot** Crude spermaceti.

wheat The seeds of *Triticum* species (Gramineae); a food containing starch, protein, and some B-group vitamins. Chief varieties: *T. monococcum,* one-grained w.; *T. sativum,* common w.; *T. durum,* flint w.; *T. compactum,* dwarf w.

Wheatstone **W., Sir Charles (1802–1875)** British physicist. **W. bridge** An instrument for the measurement of electrical resistance. A point on a graduated resistance wire is found such that no current flows from an accumulator connected to the ends of the wire, through a galvanometer joining a standard and the unknown resistance with a sliding contact on the wire. At that point the ratio of the lengths of the 2 portions of the wire equals that of the 2 resistances.

whey Milk serum. The watery residue after the fat and casein have been removed from milk; contains lactose and lactalbumin.

whiskers Microscopic, needle-shaped, single crystals of metal, ceramic, or a salt grown from vapor, molten material, or solutions, and free from grain boundaries and imperfections. W. have high tensile strength, e.g., 1.7 GPa for glass. Produced by condensation from vapor or in eutectics; used to reinforce metals, ceramics, and plastics.

whisk(e)y A grain spirit obtained from malted barley or other cereal by mashing and fermentation, followed by distillation and aging.

white A *color,* q.v., produced by the reflection of all the visible rays, or of 2 complementary rays. **w. agate** Chalcedony. **w. antimony** Valentinite. **w. arsenic** Arsenous oxide. **w. bole** Kaolin. **w. bryony** Bryonia. The dried roots of *Bryonia alba;* a cathartic. **w. cedar** Thuja. **w. coal** Power obtained from hydroelectric plants. **w. lead** A basic lead carbonate, a white pigment. **w. lead ore** Cerussite. **w. lead paper** A filter paper impregnated with lead carbonate. **w. lotion** A solution of 40 g each of zinc sulfate and potassium sulfide in 1,000 mL water; an astringent (USP). **w. mustard** See *white mustard* under *mustard.* **w. nickel ore** $NiAs_2$. A native nickel arsenide. **w. oak** Quercus. **w. olivine** Forsterite. **w. pigment** An opaque, w. pigment; as, lead carbonate, sulfate, or chloride oxide; barium sulfate; zinc sulfide; kaolin; lithopone or titanium oxide. **w. precipitate** Ammoniated *mercury.* **w. vitriol** Zinc sulfate. **w.wash** A suspension of chalk in water, used for decorative purposes. **w. zinc** Zinc oxide.

Whitfield's ointment A mixture: benzoic acid 60, salicylic acid 30, polyethylene glycol ointment 910 pts.; a keratolytic and antifungal ointment.

whiting Whitening. Washed chalk; a pigment for polishing or wall wash.

whitlockite $Ca_3(PO_4)_2$. A late hydrothermal mineral in New Hampshire granite pegmatites.

Whitney, Willis Rodney (1868–1958) American physicist, noted for work on colloids.

whitneyite Cu_9As. A native copper arsenide.

whortleberry Vaccinium.

widia (German *wie diamant:* "like diamond") Cemented tungsten carbide.

Wiedemann **W., Gustaf Heinrich (1826–1899)** German physical chemist, editor of *Poggendorff's Annalen.* **W-Franz's law** The thermal and electrical conductivities of the metals follow the same order.

Wieland, Heinrich (1877–1957) German chemist and Nobel prize winner (1927), noted for biochemical research (bile acids, chlorophyll, and hemoglobin).

Wien **W., Wilhelm (1864–1928)** German physicist, Nobel prize winner (1911). **W. equation** The intensity of radiation of a black body $= c_1\lambda^{-5} \exp(c_2/\lambda T)$, where λ is the wavelength, T the thermodynamic temperature, and c_1, c_2 are W. constants. See *Planck's formula.* **W. law** With a perfect radiator, energy intensity is a maximum for a certain wavelength λ_m, depending on the thermodynamic temperature T: $\lambda_m = 0.289/T$ cm.

Wifatite Trademark for phenol-formaldehyde and amino plastics, used for water purification. Cf. *zeolite.*

Wijs **W., J. J. A. (1864–1942)** Dutch analyst. **W. solution** A solution of iodine monochloride in glacial acetic acid; used for the determination of iodine numbers. **W. value** Iodine number.

wild (1) Not cultivated (plants); not domesticated (animals). (2) Metallurgy: overoxidized; as, steel which splits owing to the escape of gases. **w. cherry** *Prunus virginianica.* The stem bark of *Prunus serotina* (Rosaceae); an expectorant (BP). **w. ginger** Asarum. **w. hyssop** Verbena. **w. indigo** Baptisia. **w. licorice** Abrus. **w. mint** *Mentha canadensis,* a plant resembling pennyroyal. **w. rosemary** Labrador tea. **w. rue** Harmel. **w. yam** Dioscorea.

Wiley, Harvey Washington (1844–1930) American chemist, noted for his work on food laws.

Wilkinson, Sir Geoffrey (1921–) British chemist. Nobel prize winner (1973), noted for work on metal alkyl compounds.

willemite Zn_2SiO_4. A native zinc silicate.

Williamson **W., Alexander Edward (1824–1904)** British chemist, founder of the molecule concept. **W. reaction** The synthesis of ethers from sodium alcoholate and an alkyl iodide: $RI + NaOR' \rightarrow ROR' + NaI$. **W.'s violet** $KFe[Fe(CN)_6]\cdot 2H_2O$. Iron(III) potassium hexacyanoferrate(II)*.

willow A genus of trees (Salicaceae).

Willstätter, Richard (1872–1942) German chemist, Nobel prize winner (1915), noted for research on plant pigments. **W. lignin** See *Willstätter lignin* under *lignin.*

wilnite A green variety of grossularite. See *aluminum garnet* under *garnet.*

Wilson W., Charles Thomson Rees (1869–1959) British physicist, Nobel prize winner (1927). **W. fog method** Used to demonstrate the motions of atomic fragments by tracking their paths in a low-pressure fog chamber (cloud-track apparatus). The charged ions or electrons cause condensation of the gaseous water molecules, as visible minute droplets. **W. tracks** When α, β, γ, or X-rays are shot into a low-pressure fog chamber, they cause condensation of single microscopic droplets, which appear as lines and can be photographed.

Winchester A tall bottle of standard cylindrical shape, with a relatively small base. **W. bushel** See *bushel.* **W. gallon** Before 1886, 272.25 in^3; subsequently, 277 in^3. **W. quart** A W. bottle: 2.25–2.5 liters (80 fl oz).

wind w. gage Anemometer. **w. power** Theoretical power (in W) from wind turbine with blades sweeping out an area of A m^2 in wind velocity v m/s is $0.65\ Av^3$. Actual output efficiencies are about 25% of this.

Windaus, Adolf (1876–1959) German chemist, Nobel prize winner (1928) for his work on sterols.

wine Naturally fermented grape juice, containing 6–22% alcohol; traces of heptyl and other ethers; essential oils, grape sugar, glycerol; tannic, malic, phosphoric, and acetic acids; tartrates of potassium and calcium; coloring matters; and some proteins. **botrytized ~** W., as Sauternes, from grapes on which the natural mold *Botrytis cinerea* has grown, making them sweeter. **dry ~** W. of low sugar content; as, burgundy. **fortified ~** W. to which alcohol has been added. **sour ~** W. which has begun to ferment to acetic acid. **spirit of ~** Ethanol*. **sweet ~** W. of high sugar content; as, port.
w. gallon See *wine gallon* under *gallon.* **w.glass** An approximate liquid measure; 60 mL (2 fl oz). **w. lees oil** Oil from the yeast of a fermenting w. and responsible for the characteristic bouquet. Principally ethyl nonanoate, with ethyl octanoate the corresponding free acid, and about 0.4% decyl alcohol.

wing top A triangular-shaped attachment for a bunsen burner, producing a long and narrow flame.

Winkler, Clemens Alexander (1838–1904) German mineralogist; discovered uranium.

wintergreen Gaultheria. **bitter ~** Chimaphilia.
w. oil The essential oil of *Gaultheria* species, d.1.175–1.187 (99% methyl salicylate). **artificial ~** Methyl salicylate.

winterin $C_{15}H_{20}O_3$ = 248.3. A dione, m.158, from winter's bark.

winterization A process to prevent edible oils from clouding or setting at low temperatures, by cooling slowly to 45°C and filtering.

winter's bark The dried bark of *Drimys winteri* (Magnoliaceae) containing an essential oil, tannin, and resin; an aromatic. Cf. *drimin.*

wipla (German *wie platin:* "like platinum") A chromium-nickel steel; a gold substitute in dentistry.

wire A drawn-out rod. **cross ~** Two thin crossing wires in the objective of an optical instrument to fix the position of the object under examination.

wiring An electric circuit.

Wislicenus, Johannes (1835–1902) German chemist noted for his work on stereochemistry.

wismuth Bismuth.

wistarin A crystalline glucoside from *Wistaria chinensis* (Leguminosae).

Wiswesser notation A suggested method of expressing the structure of a chemical compound by a linear string of symbols; as, Cl·CO·Cl is GVG.

witch hazel Hamamelidaceae.

Witcogum Trademark for a rubbery elastomer obtained by polymerizing and condensing glycerides in presence of resins or nonelastic compounds.

withania The plant geneesblaar, vimba, *W. somnifera* (Solanaceae), S. Africa; an antiseptic and hypnotic.

withaniol $C_{35}H_{34}O_5$ = 534.6. A monohydric alcohol from withania.

withanolides A series of compounds, as, $C_{28}H_{36}O_5$, from *Withania somnifera.*

witherite $BaCO_3$. A native carbonate, d.4.29–4.35, hardness 3–3.75; used in the glass, pyrotechnic, and sugar industries.

witness A tube of (usually) calcium chloride at the beginning or end of a gas train; it is weighed at the beginning and end of the experiment to indicate any loss of solvent. Cf. *guard tube.*

Witt color theory Certain (chromophoric) linkages in organic compounds induce color. If certain other (auxochromic) linkages are also present, a dyestuff (chromogen) results.

Wittig W., Georg (1897–) German chemist. Nobel prize winner (1979), noted for work on stereochemistry and organic reactions. **W. reaction** The reaction of P ylides with a carbonyl group to produce C double bonds; used in the synthesis of vitamin A.

woad An early blue dye or stain made from the ground, fermented leaves of *Isatis tinctoria* (Cruciferae), Mediterranean region. Used by the ancient Britons.

Woestyn's law Joule's law.

Wöhler W., Friedrich (1800–1882) German chemist, noted for the isolation of aluminum and beryllium, discovery of isomerism, and the first synthesis of an organic compound (urea) in 1828. **W.'s law** Dynamic deformation may occur as a result of vibrations, none of which attains the breaking limit; ultimately rupture may result.

wöhlerite A native mixture of oxides of sodium, calcium, niobium, silicon, and zirconium.

Wohl's reaction Boiling with ammoniacal silver nitrate hydrolyzes acetyl groups and eliminates hydrogen cyanide from the oxime of an acetylated sugar:
$R\cdot CHOAc\cdot CN \rightarrow R\cdot CHOH\cdot CN \rightarrow R\cdot CHO + HCN$.

Wohwill process The electrolytic refining of gold from a weak acid solution.

Wolacryl Trademark for a polyamide synthetic fiber.

Wolcrylon Trademark for a polyamide synthetic fiber.

Wolffram's salt $(PtCl)Cl_2\cdot(H_2O)_2$. A microchemical reagent.

Wolf trap A device to suck atmospheric dust into a culture medium, where increase in turbidity or bacterial growth can be measured.

wolfram* (1) Tungsten*. (2) Wolframite. **w. ocher** Tungstite.

wolframate* Tungstate*.

wolframine Tungstite.

wolframite $FeMnWO_4$. Native tungstate, intermediate between ferberite, $FeWO_4$, and hübernite, $MnWO_4$.

wolframium Tungsten*.

wolfsbane See *arnica flowers.*

wolfsbergite (1) Chalcostibite. (2) Essonite.

Wollaston W., William (1766–1828) British chemist, discoverer of palladium and rhodium. **W. wires** Platinum wires, 0.001 mm diameter, drawn inside a silver sheath, which is subsequently removed in nitric acid.

wollastonite $CaSiO_3$. Tabular spar, scale stone. A native triclinic pyroxene. **para ~** Monoclinic w. **pseudo ~** A third modification of w.

wood The structural and supporting part of a tree or shrub; chief constituent, cellulose. Cf. *lignin, xylose.* Commercial woods are classified as softwoods (conifers) and hardwoods (deciduous; as, poplar, oak) **w. alcohol** Methanol*. **w. ash** The residue from burning w. (4% potassium oxide); a fertilizer. **w. cellulose** Xylon. Cellulose obtained from w. **w. charcoal** Carbo lignius. **w. flour** W. meal, sawdust, pulverized w. Finely powdered w., generally white pine; a filler in moldings, linoleum, flooring, rubber, soap; and absorbent for nitroglycerin (dynamite). **w.-free** Describing a paper that contains no mechanical pulp, and usually contains only chemical w. pulp. **w. meal** W. flour. **w. measurement** See *cord, festmeter* under *meter, stere.* **w. metal** See *Wood's alloy.* **w. naphtha** Methanol. **w. oil** Tung oil. **w. opal** Xylopal. Fossilized w., in which silica replaces the woody fibers. **w. pulp** A wood either mechanically disintegrated or chemically digested, or both; used to manufacture paper or rayon. Cf. *pulpwood.* **w. spirit** Methanol*. **w.stone** Xylolith. **w. sugar** Xylose. **w. tar** See *wood tar* under *tar.* **w. vinegar** See *pyroligneous acid.* **w. waste** Products of lumber waste; as, w. pulp (paper, board, rayon), w. flour (dynamite, linoleum, plastics), ethyl alcohol (motor fuels), cattle food, rosins, turpentine, pine oil, tar, tannin, fatty acids, dyes, and potash. **w.-wool** Narrow, thin shavings, used for packing.

Wood's **W. alloy** A low-melting-point alloy (65.5°C): Bi 5, Pb 2.5, Sn 1.25, Cd 1.25 pts. **W. glass** A glass that absorbs most of the visible spectrum but transmits waves in the ultraviolet. **W. light** The filtered rays of the mercury-vapor lamp; wavelength 3340–3906 Å.

woodwardite $CuSO_4 \cdot Al_2(SO_4)_2 \cdot H_2O$. Native aluminum copper sulfate.

Woodward **W., Robert Burns (1917–1979)** American chemist. Nobel prize winner (1965), noted for work in organic synthesis. **W.-Hoffman rules** Stereoselection rules for molecular interactions derived from orbital symmetry considerations. **W. rules** Rules for the correlation of ultraviolet spectra with substitution patterns in unsaturated organic compounds.

wool The fibrous hair of lambs and sheep. It contains keratin (a characteristic odor on burning), and about 40% of a grease (chiefly cholesterol); similar in composition to horn. **blended ~** Cloth containing not less than 50% w. **casein ~** Lanital. **glass ~** Fine glass threads used for filtering. **mineral ~** See *mineral wool.*
w. alcohols A constituent (with paraffin wax) of ointments, made by saponification of w. fat and separation of the alcohol fraction. Contains 29% cholesterol (NF, BP). **w. fat** (1) Anhydrous lanolin; used in ointments (USP, BP). (2) Degras. The commercial wax alcohol extracted from pure w. fat. **hydrous ~** W. fat containing 25–30% water; used in ointments and lotions (USP, BP). **w. waste** A fertilizer by-product of w. carding, consisting of sheep manure, seeds, and wool fibers, 3–6% N_2.

Woolon Trademark for a mixed-polymer synthetic fiber.

Woolwich viscosity See *Woolwich viscosity* under *viscosity.*

woorara Curare.

word The smallest unit (i.e., ensemble of bits) in which information is stored by a computer.

work W. is performed when a force moves its point of application; measured by the product of the force and the distance it is displaced in the direction of the force. Units: *SI system.* 1 joule = 1 newton-meter. *Metric system.* 1 erg = 1 centimeter-dyne. *fps system.* 1 foot-pound; the work done when 1 pound is lifted 1 foot against gravity. For rate of work, see *power.* **w. of expansion** The resistance offered by the molecules of an expanding gas, due to molecular attraction. **w. hardening** Cold-working. The modification of the physical properties of metals and alloys by cold (as distinct from heat) treatment.

wormseed (1) Santonica. (2) Chenopodium. **w. oil** *American:* The essential oil from *Chenopodium ambrosioides,* d.0.960–0.980. *Levant:* The oil of *Artemisia maritima,* d.0.930, containing eucalyptol.

wormwood Absinthium. **salt of ~** Potassium carbonate.
w. oil Absinthe oil. An essential oil from the herb of *Artemisia absinthium.* Green, volatile liquid, d.0.925–0.955, soluble in alcohol; contains chiefly thujone, phenanthrene, and thujenol.

worsted A closely woven woolen cloth, which can, however, contain up to 3% of inadvertent impurities, and up to 7% of material other than wool for decorative purposes or to facilitate processing.

wort An infusion of a plant to be fermented; as, beer w.

Woulfe bottle A 2- or 3-necked bottle used in chemical experiments, first described by Peter Woulfe (1727–1803) in 1784.

W rays See *W ray* under *ray.*

wrightine Conessine.

Wright's stain A microscopic stain for white blood cells; 1 g specially prepared methylene blue–eosin mixture in 600 mL methanol.

Wroblewski, Sigismund A. von (1845–1888) Polish physicist, noted for the liquefaction of gases.

wrought Describing a metal that has been worked into shape or condition, e.g., by hammering. Cf. *cast.* **w. iron** A pure grade of iron with a low carbon content.

wt. Abbreviation for weight.

Wu, Lu-Chiang (1904–1936) Chinese chemist, noted for historical research.

wulfenite $PbMoO_4$. Yellow lead ore. A native molybdate.

Wulff net A circle onto which are projected stereographically the longitudinal (meridional) and latitudinal (polar) lines of a globe; used in crystallographic analysis.

Wullner's law The lowering of the vapor pressure of water by a solute is proportional to the concentration of the solute.

Wurster **W.'s blue** An oxidation product of tetramethyl-*p*-phenylenediamine; an indicator. **W.'s red** A meriquinone oxidation product of *p*-aminodimethylaniline.

Würtz **W., Carl Adolphe (1817–1884)** French chemist, noted for organic research and his statement *"La chimie est une science française."* **W. flask** A distillation flask with a side arm in the neck. **W.'s reaction** The synthesis of hydrocarbons by treating alkyl iodides in ethereal solution with sodium: 2MeI + 2Na = Me·Me + 2NaI.

wurtzite ZnS. Aegerite. A hexagonal, native zinc sulfide. Cf. *sphalerite.*

wustite A ferrite, often found in cements. Crystalline ferrous oxide with varying amounts of ferric oxide in solid solution.

Wydase Trademark for hyaluronidase.

Wynene Trademark for a polyethylene synthetic fiber.

wyomingite A porous rock in the Leucite Hills, S. Wyoming; potash (11% potassium oxide) is extracted by acid.

Wysor machine A device for grinding and polishing metals for metallographic examination.

X

X Symbol, in a formula, for a halogen; as, HX. **X-ray** See *X-ray*.
x Symbol for an unknown quantity.
x Symbol for (1) a rectangular coordinate; (2) mole fraction (of substance B, x_B).
χ See *chi*.
Ξ, ξ See *xi*.
xanthaline Papaveraldine.
xanthate EtO·CS·SM or RO·CS·SM. Dithiocarbonate*. Xanthogenate. A salt of xanthic acid. M is a metal and R an alkyl radical; as, potassium ethyl x., EtO·CS·SK. Cf. *viscose*. **butyl ~** $MeCH_2$·CHMeO·CS·SK. Potassium *sec*-butyl x. A flotation collector; 94% mineral recovery. **ethyl ~** EtO·CS·SK. Potassium *O*-ethyl x. A flotation collector; 85% mineral recovery. **propyl ~** $MeCH_2CH_2O$·CS·SK. A flotation collector; 90% mineral recovery.
xanthein A yellow coloring matter of plants; a split product of chlorophyll.
xanthematin A yellow coloring matter and split product of hemoglobin.
xanthene* C_6H_4·CH_2·C_6H_4·O = 182.2. Methylene diphenylene oxide. Colorless leaflets, m.105, slightly soluble in water. Cf. *xanthydrol*. **benzo ~** See *benzoxanthene*. **hydroxy ~** Xanthydrol. **oxo ~** Xanthone. **thio ~*** See *thioxanthene*. **x. dyes** See *phthaleins, pyronine dyes*.
xanthenol Xanthydrol.
xanthenone Xanthone.
xanthenyl* Xanthyl. The radical $C_{13}H_9O-$, from xanthene; 5 isomers.
xanthic **x. acid** EtO·CS·SH = 122.2. Xanthonic acid, xanthogenic acid, ethoxydithiocarbonic acid*. Unstable, colorless oil, insoluble in water, decomp. 24 to ethanol and carbon disulfide. **x. amide** NH_2·CS·OEt = 105.2. m.38, soluble in water. **x. disulfide** HO·CS·S·S·CS·OH = 186.3. Colorless crystals, m.28.
-xanthin Suffix indicating yellow; as, coxanthin.
xanthine $C_5H_4O_2N_4$ = 152.1. 2,6-Dioxopurine. A purine base in blood, liver, urine, and some plants. Yellow crystals, m.360, soluble in water. **allo ~** Alloxantin. **dimethyl ~ 1,3- ~** Theophylline. **1,7- ~** Paraxanthine. **3,7- ~** Theobromine. **hetero ~** $C_6H_6O_2N_4$ = 166.1. 7-Methylxanthine, m.380 (decomp.). **hypo ~** See *hypoxanthine*. **imido ~** Guanine. **methyl ~** See *heteroxanthene* above. **para ~** $C_7H_8O_2N_4$ = 180.2. 1,7-Dimethylxanthine. **trimethyl ~** Caffeine.
x. bases Alloxuric bases, oxopurines, in vegetable and animal tissues; e.g., xanthine, hypoxanthine, uric acid, carnine, theophylline, caffeine, theobromine.
xanthium Spiny clotbur, cocklebur. The herb of *Xanthium spinosum* (Compositae); a diuretic. Cf. *xanthostrumarin*.
xantho- Prefix (Greek: "yellow") indicating the yellow color of a compound; as in xanthocobaltic.
xanthochelidonic acid $CO(CH_2·CO·COOH)_2$ = 202.1. Acetonedioxalic acid. 2,4,6-Trioxoheptanedioic acid. An unstable acid formed from pyronedicarboxylic acid by treatment with excess alkali. Its salts are yellow.
xanthochromium The pentaamminenitrochromium(II) cation $[Cr(NO_2)(NH_3)_5]^{2+}$ which forms salts; as, the dibromide, $[Cr(NO_2)(NH_3)_5]Br_2$.
xanthocobaltic The pentaamminenitrocobalt(II) cation $[Co(NO_2)(NH_3)_5]^{2+}$ which forms yellow salts; as, the dibromide, $[Co(NO_2)(NH_3)_5]Br_2$.
xanthocreatinine $C_5H_{10}ON_4$ = 142.2. A yellow, crystalline leucomaine resembling creatinine, in muscle tissues.
xanthogenate $(RO·CS·S-)_2$. Xanthate.
xanthogene A coloring material of plants, producing a yellow pigment with alkalies.
xanthogenic acid Xanthic acid.
xanthoglobulin Hypoxanthine.
xantholine Santonica.
xanthone O·C_6H_4·CO·C_6H_4 = 196.2. Xanthene ketone, dibenzo-γ-pyrone xanthenone. Colorless crystals, m.173, slightly soluble in water. The nucleus of certain plant pigments, used in the synthesis of dyes, e.g., euxanthone, India yellow. **dihydroxy ~ 1,6- ~** Iso*euxanthone*. **1,7- ~** Euxanthone. **dihydroxymethoxy ~** Gentisin. **thio ~** See *thioxanthone*. **trihydroxy ~** Gentisein.
xanthonic acid Xanthic acid.
xanthophyll Lutein.
xanthophyllite A native, hydrous silicate of calcium, magnesium, iron, and aluminum.
xanthopicrin Yellow color from the bark of *Xanthoxylum cariboeum* (Rutaceae).
xanthoprotein reaction A yellow color results on boiling proteins with nitric acid.
xanthopterin The yellow pigment from the wings of the "brimstone" butterfly (Pieridae).
xanthopurpurin $C_{14}H_6O_2(OH)_2$ = 240.2. 1,3-Dihydroxyanthraquinone*, purpuroxanthene. Yellow crystals, m.269, soluble in alcohol.
xanthorhamnin A glucoside hydroxyflavone from buckthorn berries, *Rhamnus cathartica*. Cf. *rhamnoxanthin*.
xanthorhodium The pentaamminenitrorhodium(II) cation $[Rh(NO_2)(NH_3)_5]^{2+}$ which forms yellow salts; as, the dibromide, $[Rh(NO_2)(NH_3)_5]Br_2$.
xanthosiderite $Fe_2O_3·2H_2O$. Fine, yellow, stellate needles.
xanthostrumarin A glucoside from *Xanthium strumarium*, cocklebur (Compositae). Cf. *xanthium*.
xanthotoxin Methoxsalen.
xanthoxylene $C_{10}H_{16}$ = 136.2. Xanthoxylin. A terpene from the essential oil of *Xanthoxylum* species (Rutaceae).
xanthoxylin $C_8H_{12}O_4$ = 172.2. 6′-Hydroxy-2′,4′-methoxyacetophenone, m.87. From *Xanthoxylum* spp.
xanthoxylone $C_{30}H_{48}O$ = 424.7. 12-Oleanen-7-one, m.230. A crystalline principle from *Xanthoxylum*.
xanthoxylum Prickly ash, yellowwood, angelica tree. The dried bark of *X. americanum, X. officinalis* (Rutaceae), containing xanthoxylene, essential oils, tannin, and resin. Cf. *angelica tree, artar root*.
xanthydrol $C_{13}H_{10}O_2$ = 198.2. 9-Hydroyxanthene. A reagent for urea. Cf. *xanthene*.
xanthyl The xanthenyl* radical.

xanthylium The cation

$\overset{+}{O}:C_6H_4:CH\cdot C_6H_4$. Derived from xanthene.

xaxa Acetylsalicylic acid. See *aspirin.*
xble. Abbreviation for crucible.
Xe Symbol for xenon.
xenate A salt of xenic acid, XeO_4^{2-}; used in analysis. **per ~** Any of the salts of the type M_4XeO_6.
xenene Biphenyl*.
xenol PhC_6H_4OH = 170.2. Xenenol, phenylphenol. **para ~** White solid; used in plastics to increase resistance to high temperatures and hot water.
xenolite An aluminum silicate resembling fibrolite.
xenon* Xe = 131.29. A rare, heavy noble gas and element, at. no. 54; discovered by Ramsay and Travers (1898); Greek ξενος: "strange." Occurs in the atmosphere (about 1 ppm). Colorless gas, $d_{air=1}$4.422, m.−112, b.−107; insoluble in water. Forms several compounds with fluorine and oxygen; used in vacuum tubes and medically as ^{133}Xe (USP, BP). **x. tetrafluoride*** XeF_4 = 207.3. Colorless crystals, m.117, formed by reaction of the elements at 400°C.
Xenophanes (576–480 B.C.) Greek philosopher of the Eleatic school, who suggested one of the earliest known theories of atomic structure, which he believed to be continuous and completely filling space.
xenotime YPO_4. A native yttrium phosphate, containing uranium and rare-earth metals.
xenyl The *p*-biphenylyl* radical.
xerogel (1) A gel containing little or none of the dispersion medium used. (2) An organic polymer which swells in suitable solvents to give particles containing a 3-dimensional network of polymer chains. See *molecular sieve.*
xerography A method of copying in which an electrostatically charged, coated (e.g., selenium) plate is discharged by light. Oppositely charged graphite powder is attracted to the surface and fused by heat.
xeroradiograph A radiograph produced by xerography.
xestophanesin $C_{23}H_{18}O_{15}$ = 534.4. A glucosidal coloring matter from galls, produced by *Xestophanus potentillae.* Orange needles, m.281–287.
xi Ξ, ξ. Greek letter. Ξ (not italic) is symbol for x. *particle.* Ξ (italic) is symbol for grand partition function. ξ (italic) is symbol for extent of reaction; $\dot{\xi}$ is rate of reaction. **x. particle** See *particle.*
xiphidin A ptomaine from the sperm of the swordfish (81.5% arginine).
xonotlite $5CaO\cdot 5SiO_2\cdot H_2O$. A fibrous-structured constituent of boiler scales.
X-ray(s) Roentgen rays. A *radiation,* q.v., of short wavelengths (0.06–20 Å) produced in an *X-ray tube,* q.v., by cathode rays focused on a metal surface. They resemble the γ rays of radioactive substances, and are made visible by fluorescent screens or photographic plates. Their *quantity* is determined with Eder's solution; their *intensity* is expressed in C/kg or roentgen units. Cf. *spectrum.* **hard ~** X-rays of relatively high penetration (wavelength 0.19–0.43 Å, which decreases as they traverse a medium). **heterogeneous ~** Multifrequent X. **homogeneous ~** Monochromatic X. **infra- ~** Grenz rays. **monochromatic ~** Unifrequent X., homogeneous X. X-rays of a single wavelength or a group of wavelengths, depending on the nature of the target. The shortest measured is 0.1075 Å for the K line of uranium; the longest, 17.66 Å for the L_β line of iron. **multifrequent ~, polychromatic ~** Heterogeneous X. X-rays of many different wavelengths. **soft ~** X-rays of relatively low penetration (wavelength 11.9–13.6 Å, which increases as they traverse a medium). (2) See *laser.* **ultra- ~** Cosmic rays. **unifrequent ~** Monochromatic X. **white ~** Common X-rays, consisting of a band of many different wavelengths. Their limit depends on the voltage applied, and not on the nature of the target: $Ve = h\nu_0 = hc/\lambda_0$, where V is the voltage, e the charge of an electron, h the Planck constant, c the velocity of light, λ_0 the maximum frequency, and ν_0 the minimum wavelength.
X-ray analysis Determination of the internal structure of a material from the diffraction pattern formed when an X-ray passes through it. Types of pattern: (1) *Corona,* or diffuse fog, indicating the gaseous state. (2) *Halo,* indicating a liquid or amorphous substance; the molecules are distributed irregularly. (3) *Rings,* indicating a regular space lattice with the molecules in definite positions. (4) *Points,* indicating a crystal, in which the atoms, ions, or molecules are fixed in definite positions. **X-ray microanalysis** X-ray fluorescence analysis. Utilizes the X-ray fluorescence produced when high-energy electrons strike a material. It has a characteristic wavelength and energy for each element, owing to the resultant hole's being filled from a higher shell in the atom and an X-ray photon's being simultaneously emitted, and can be used to characterize it. 10^{-19}g is often detectable.
X-ray intensity unit See *roentgen.*
X-ray photograph, X-ray picture, X-ray plate Radiograph. Roentgenogram. Skiagram. A photograph using X-rays.
X-ray spectrogram A photographic record produced by the spectrograph.
X-ray spectrograph A spectrometer fitted with a camera. Cf. *spectrograph.*
X-ray spectrometer An arrangement for diffracting X-rays using the space lattice of a crystal as grating and an electronic detection device.
X-ray spectrum Moseley spectrum, high-frequency spectrum. The spectrum (produced by a crystal-diffraction grating or by atoms in samples after bombardment with high-energy particles or radiation) of the characteristic radiation emitted by the metal used as anticathode in an X-ray tube. Each metal produces a few strong, characteristic lines, the square roots of the frequencies of which are in direct proportion to the atomic number of the element; a means of analysis.
X-ray structure The structural arrangement of atoms in a crystal as revealed by the Laue pattern and *crystallograms,* q.v., produced when X-rays pass through the crystal and fall on a photographic plate.
X-ray tube A highly evacuated tube containing a disk concentric with the tube as cathode, a heavy, inclined plane of high-melting-point metal as anticathode, and a small wire as anode. The cathode rays concentrate on the anticathode, which emits rays of very short wavelengths.
X unit (1) A wavelength of 10^{-3} Å (10^{-13} m). (2) More accurately, 1.00201×10^{-3} Å (Craven).
xylan $(C_5H_8O_4)_n$ = $(132.1)n$. A hemicellulose consisting of chains of 1,4-linked β-D-xylopyranosyl residues; in many trees and industrial wastes; e.g., peanut shells; it hydrolyzes to xylose.
xylanbassoric acid $C_{19}H_{28}O_{17}$ = 528.4. A soluble hydrolysis product of bassorin from gum tragacanth; hydrolyzed to xylose.
xylene* $C_6H_4Me_2$ = 106.2. Dimethylbenzene*, xylol (German). **1,2- ~** Colorless liquid, d.0.881, m.−28, b.144, insoluble in water; used in the manufacture of synthetic resins, pharmaceuticals, dyestuffs, and phthalic acid. **1,3- ~** Colorless liquid, d.0.866, m.−54, b.139, insoluble in water.

1,4- ~ Colorless, monoclinic crystals, m.15, insoluble in water; a feedstock for dimethylterephthalate and terephthalic acid. **bromo** ~ $Me_2C_6H_3Br$ = 185.1. Cf. *methylbenzyl bromide.* **4-*ortho*-** ~ b.214. **2-*meta*-** ~ b.206. **4-*meta*-** ~ b.207. **5-*meta*-** ~ b.204. **2-*para*-** ~ b.206. **dihydroxy** ~ Xylenediol*. **ethyl** ~ Ethyldimethyl*benzene.* **hydroxy** ~ Xylenol*. **nitro** ~ See *nitroxylene.*
x.sulfonic acid $C_6H_3Me_2 \cdot SO_3H$ = 186.2. **2,4-** ~ m.61. **2,5-** ~ m.86. **3,4-** ~ m.63.

xylenediol* $C_6H_4(CH_2OH)_2$ = 138.2. Xylylene alcohol, xylylol, phenyldicarbinol, benzenedicarbinol. ***ortho*-** ~ Phthalyl alcohol, m.62. ***meta*-** ~ m.46. ***para*-** ~ Terephthalyl alcohol, m.112. All used in organic synthesis.

xylenol blue 1,4-Dimethyl-5-hydroxybenzene-sulfonephthalein. An indicator; changes from red (pH 1.2) to yellow (pH 2.8), and from yellow (pH 8.0) to blue (pH 9.6).

xylenol(s)* $C_6H_3Me_2(OH)$ = 122.2. Dimethyl phenols or hydroxydimethyl benzenes. Colorless crystals, soluble in water.

xylic acid $Me_2C_6H_3COOH$ = 150.2. Methyltoluic acid*, dimethylbenzoic acid. **2,3-** ~ Hemellitic acid, m.144. **2,5-** ~, ***para*-** ~, **iso** ~ White needles, m.132, soluble in alcohol. **3,5-** ~ Mesitylinic acid.

xylidene Xylidine*.

xylidic acid $C_6H_3Me(COOH)_2$ = 180.2. 4-Methylisophthalic acid. Colorless crystals, m.325 (sublimes). Cf. *uvitic acid.*

xylidine(s)* $C_6H_3Me_2NH_2$ = 121.2. Xylidene. Methyltoluidines, dimethylanilines, aminoxylenes. Aniline homologs in technical xylidine, slightly soluble in water; used in the synthesis of dyes.

xylidinic acid Xylidic acid.

xylite Xylitol*.

xylitol* $CH_2OH(CHOH)_3CH_2OH$ = 152.1. Xylite. White crystals, d.1.50, m.94. A sweetening agent, in fruit and birchwood, equivalent to sucrose.

xylo- Prefix (Greek: "wood") indicating a relation to wood.

Xylocaine Trademark for lidocaine hydrochloride.

xyloidine $C_6H_9O_5(NO_2)$ = 207.1. An explosive obtained by the action of nitric acid on starch or wood (Braconnot, 1842).

xyloketose $HOCH_2 \cdot CO \cdot (HCOH)_2CH_2OH$ = 150.1. *threo*-2-Pentulose.

xylol (1) A mixture of *xylenes*, q.v.; a solvent in microscopy. (2) German for "xylene."

xylolith Woodstone. A mixture of magnesia, magnesium chloride, and sawdust which dries to an extremely hard mass; used for floors or laboratory tables.

xylon Wood cellulose.

xylonic acid $CH_2OH(CHOH)_3COOH$ = 166.1. Obtained by oxidation of xylose. Cf. *pentonic acid.*

Xylonite See *Celluloid.*

xylopal Wood opal.

xyloquinone Dimethyl*benzoquinone.*

xylorcinol $C_6H_2Me_2(OH)_2$ = 138.2. **1,3,4,6-** ~ 4,6-Dimethylresorcinol, methylorcinol. White crystals, m.124, soluble in water.

xylose $CH_2OH(CHOH)_3CHO$ = 150.1. Wood sugar. A pentose obtained from xylan or vegetable fibers by heating with sulfuric acid. Colorless crystals, m.144, soluble in water; used in tanning, dyeing, and diabetic foods, and in a diagnostic test of malabsorption (USP, BP).

xylostein A poisonous glucoside from the fruit of *Lonicera xylosteum* (Caprifoliaceae), a variety of honeysuckle.

xyloyl* The radical $Me_2C_6H_3CO-$, from xylic acid; 7 isomers.

xylyl (1)* Dimethylphenyl†. The radical $Me_2C_6H_3-$, from xylene. (2) The methylbenzyl* radical. **x. alcohol** (1) Xylenediol*. (2) Tolylmethanol. **x. hydrazine** $Me_2C_6H_3 \cdot NH \cdot NH_2$ = 136.2. Colorless needles, m.85, soluble in ether.

xylylene The phenylenebismethylene† radical $-CH_2 \cdot C_6H_4 \cdot CH_2-$, from xylene; *ortho, meta,* and *para* types. **x. alcohol** Xylenediol*. **x.amine** X.diamine. **x. bromide** $C_6H_4(CH_2Br)_2$ = 264.0. **x. chloride** X. dichloride. **x. cyanides** $C_6H_4(CH_2CN)_2$ = 156.2. ***ortho*-** ~ m.59. ***meta*-** ~ m.28. ***para*-** ~ m.98. **x.diamines** $C_6H_4(CH_2NH_2)_2$ and $C_6H_2Me_2(NH_2)_2$ = 136.2. Diaminoxylene. Used in the manufacture of dyes. **x. dichlorides** $C_6H_4(CH_2Cl)_2$ = 175.1. Phthalyl chloride. ***ortho*-** ~ m.55. ***meta*-** ~ m.34 ***para*-** ~ m.98. **x. glycol** Xylenediol*.

xylylol Xylenediol*.

Xyptal Trademark for an alkyd-type plastic made from xylitol. Cf. *Glyptal.*

xysmalobin $C_{46}H_{70}O_{20} \cdot 5H_2O$ = 1033. A crystalline glucoside from the root of *Xysmalobium undulatum,* wild cotton, milkbush, ishongwe (Asclepiadaceae), S. Africa; an emetic.

Y

Y Symbol for yttrium.
Y Symbol for (1) admittance; (2) Planck function.
y Symbol for a rectangular coordinate.
Υ See *upsilon.*
γ See *gamma.*
yaba bark Andira bark, cabbage tree bark. The bark of *Andira excelsa.* Cf. *goa.*
yabine An alkaloid from yaba bark.
yacca gum Acaroid resin.
yajeine Harmine.
yajenine An alkaloid from yajé (Apocynaceae), S. Colombia.
yam The starchy tubers of *Dioscorea* species, q.v., cultivated for food in the tropics; as, **cush-cush ~** The y. of *D. trifida.* **negro ~** The y. of *D. cayennensis.* **white ~** The y. of *D. alata.*
yara-yara Methyl 2-naphthyl *ether**.
yard yd. A unit of length in the fps system. 1 yard = 3 feet = 36 inches = 0.914400 meter. This metric equivalence is accepted by all countries using the y. **cubic ~** yd^3 or cu yd = 0.764555 m^3. **square ~** yd^2 or sq yd = 0.83613 m^2.
yarn A thread. **filament ~** See *filament yarn.* **synthetic ~** See *fiber, rayon.*
yarrow The dried leaves and inflorescence of *Achillea millefolium* (Compositae); an astringent.
yatren Loretin.
yawroot Stillingia.
Yb Symbol for ytterbium.
yd Abbreviation for yard.
year A measure of time. In mean solar days:

Anomalistic y.	365.2596
Calendar y.	365.2425
Sidereal y.	365.2564
Solar y.	365.2422

anomalistic ~ The average period for the earth to rotate round the sun between perihelion positions. **astronomical ~** Solar y. **calendar ~** The ordinary y., with three 365-day y. and one 366-day y. **civil ~** Calendar y. **light-~** A unit of astronomical distance. The distance that light travels in 1 year, 9.4606×10^{12} km or 5.9×10^{12} miles. **sidereal ~** The time for the sun to make a complete revolution with respect to fixed stars. **solar ~** Average interval between 2 successive arrivals of the sun at the first point of Aries. **tropical ~** Solar y.
yeast Cerevisiae. Unicellular vegetable organisms (fungi) of the family Saccharomycetaceae. They ferment sugars to carbon dioxide and alcohol by virtue of the enzymes they contain. They also contain invertase, an enzyme that inverts unfermentable sugars (as cane sugar) to fermentable sugars. **baker's ~, beer ~** Compressed y. The moist, living cells of *Saccharomyces cerevisiae* compressed with some starchy or other absorbent material. Used in baking, brewing, and fermenting; and as a vitamin-rich nutrient. **bottom ~** Low y. **brewer's ~** High and low y. **compressed ~** Baker's y. **copro-~** A y. that contains a large quantity of coproporphyrin. **dried ~** (1) Y. dried at a low temperature, for storage or transport, without greatly impairing its vitality. (2) Pharmacy: y. dried to contain not more than 7% water; a source of natural vitamin B complex. **high ~** Top y. It rises during fermentation; as in English beers; distinguished from low y. by absence of melibiase, which ferments raffinose. **low ~** Bottom y. It falls to the bottom of the liquid it is fermenting; as in German beers. Cf. *high yeast.* **mineral ~** See *mineral yeast.* **pressed ~** Baker's y. **top ~** High y. **wild ~** A y. not grown as a pure culture, responsible for beer diseases.
y. food A mixture of mineral substances (phosphates, etc.) added to stimulate the y. activity. **y. gum** A mannan formed from a combination of the principal carbohydrates of the y. cells with glycogen. **y. nucleic acid** A nucleic acid from the nucleoprotein of y.
yeatmanite $(MnZn)_{16}Sb_2Si_4O_{29}$. A triclinic mineral from Franklin, N.J.
yellow A primary color; between orange and green. **acid ~** An aminoazo dye. **butter ~** See *butter yellow.* **chrome ~** Lead chromate. **Eger's ~** See *Eger's yellow.* **fast ~** An aminoazo dye. **Indian ~** See *Indian yellow.* **indicator ~** A pigment, yellow in acid, colorless in alkali; a light filter. **queen's ~** Mercuric subsulfate.
y. acid 1,3-Dihydroxynaphthalene-5,7-disulfonic acid. **y. arsenic** Orpiment. **y. bark** Cinchona. **y. brass** Muntz metal. **y. cake** Uranium oxide concentrate. **y. copper** Chalcopyrite. **y. copperas** Copiapite. **y. dyes** See *dyes.* **y. earth** Ocher. **y. fever vaccine** The living virus of an attenuated strain of y. fever virus in chick embryo tissue (USP, BP). **y. lead ore** Wulfenite. **y. ore** Chalcopyrite. **y. pigments** See *cadmium yellow, chrome yellow, gamboge, Indian yellow, litharge, orpiment, ocher.* **y. precipitate** Y. *mercuric* oxide. **y. prussiate of potash** Potassium hexacyanoferrate(II)*. **y. prussiate of soda** Sodium hexacyanoferrate(II)*. **y. puccoon** Hydrastis. **y. rain** An alleged chemical warfare weapon, said to be mycotoxins of the tricothecene group derived from *Fusarium* mold. **y. resin** Acaroid resin. **y. sandalwood, y. sanders** See *sandalwood.* **y. ultramarine** Barium chromate. **y. wax** (1) A variety of beeswax. (2) A semisolid residue from the distillation of petroleum. **y.wood** Xanthoxylum.
yenite Ilvaite.
yenshee Opium dross. Low-grade opium, consisting of dregs and carbonized opium after smoking (1–10% morphine). Cf. *chandoo, mudat.*
yerba (1) Spanish for "herb." (2) Maté. **y. buena** Micromeria. The dried leaves of *Micromeria douglassi* (Labiatae), the Pacific coast; an aromatic and carminative. **y. maté** Maté. **y. reuma** The dried herb of *Frankenia grandifolia* (Frankeniaceae); a mild astringent. **y. sagrada** The dried herb of *Lantana braziliensis* (Verbenaceae); an antipyretic. Cf. *lantanine.* **y. santa** Eriodictyon.
yerbine An alkaloid from *Ilex paraguayensis,* which resembles caffeine.
yew Taxus.
yield (1) The percentage of finished material obtained from the raw material. (2) The percentage of finished material obtained, based on that obtainable theoretically.

yield point The stress at which a marked and permanent increase in the deformation of a substance occurs without an increase in the load.

-yl Suffix indicating a univalent radical; as, methyl. Cf. *-idene.*

ylangene $C_{15}H_{24}$ = 204.4. From essential oils; as, *Schizandra chimensis.*

ylang-ylang Ilang-ilang. A tree of the Philippine and Malaysian islands, *Cananga odorata* (Anonaceae); the flowers contain an essential oil: **y. oil** Cananga oil, d.0.911–0.958; used in perfumery; contains linalool, geraniol, and pinene. Y. has been synthesized (trade name Gylan).

ylide Compound corresponding to an *onium* cation, q.v., but with a proton lost from an atom next to the central atom; as: $Ph_3\overset{+}{P}\cdot\overset{-}{C}H_2 \leftrightarrow Ph_3P{=}CH_2$.

-yne* Suffix indicating a triple bond. Cf. *-ene.*

yogurt, yoghurt, yoghourt Soured milk, used as a food. Made by treating pasteurized milk with a standardized bacterial culture for 3 hours at 43.5°C. Cf. *kefir.*

yohimbehe The bark of *Corynanthe johimbe* (Rubiaceae) of the Cameroon, containing alkaloids; an aphrodisiac.

yohimbic acid $C_{20}H_{24}O_3N_2$ = 340.4. A monobasic acid; yohimbine is its methyl ester.

yohimbine $C_{21}H_{26}O_3N_2$ = 354.5. Quebrachine, methyl yohimbate, corynine. An alkaloid from the bark of yohimbehe. (+)-~ White needles, m.234, slightly soluble in water. Cf. *aribine.* **y. hydrochloride** $C_{21}H_{26}O_3N_2\cdot HCl$ = 390.9. Aphrodine. Colorless crystals, m.302, soluble in water; reputedly an aphrodisiac.

yolk The yellow part of an egg, containing nutrient proteins and lecithins.

yoloy A steel alloy containing small proportions of P, Ni, and Cu.

Yomesan Trademark for niclosamide.

yoshino Tengujo. Kodzu. A strong, thin typewriter stencil paper, made in Japan from the fiber of the paper mulberry, *Broussonetia papyrifera.*

Youden square A method of planning experiments under different conditions, for comparing batch-to-batch differences in a particular property.

Young **Y., James (1811–1883)** British chemist noted for the development of the gas and oil industries. **Y., Sydney (1857–1937)** British chemist, noted for boiling-point laws. Cf. *Ramsay-Young equation.* **Y., Thomas (1733–1829)** British physicist, noted for Young's *modulus,* q.v.

yperite Mustard gas.

ysopol $C_{10}H_{18}O$ = 154.3. A terpene alcohol from hyssop.

ytterbia Ytterbium oxide*.

ytterbite Gadolinite.

ytterbium* Yb = 173.04. Neoytterbium, formerly supposed to be a mixture of aldebaranium and cassiopeium. A trivalent, rare earth metal and element, at. no. 70, discovered by Marignac (1878), m.820, b. ca. 1300. **y. acetate*** $Yb(C_2H_3O_2)_3\cdot 4H_2O$ = 422.2. White plates, soluble in water. **y. bromide*** $YbBr_3\cdot 8H_2O$ = 557.7. Green, hygroscopic crystals. **y. chloride*** $YbCl_3\cdot 6H_2O$ = 387.5. Green rhombs, m.150, soluble in water. **y. chloride oxide*** YbOCl = 224.5. White powder, insoluble in water. **y. oxalate*** $Yb_2(C_2O_4)_3\cdot 10H_2O$ = 790.3. White crystals, insoluble in water. **y. oxide*** Yb_2O_3 = 394.1. Ytterbia. White powder, insoluble in water. **y. sulfate*** $Yb_2(SO_4)_3$ = 634.2. Green crystals, decomp. 900, soluble in water.

Ytterby A village in Sweden which gives its name to local rare-earth minerals; as, *ytterbium, yttrium.*

yttergranate A calcium-iron garnet, containing yttrium compounds.

yttria (1) Yttrium oxide. (2) Gadolinite. **y. group** See *rare-earth metals.*

yttrialite $Y_2O_3\cdot 2SiO_2$. A green mineral, containing Fe and Th.

yttrium* Y = 88.9059. A trivalent, rare-earth metal of the aluminum group and element, at. no. 39, discovered by Gadolin (1794) in gadolinite, which was separated by Mosander (1843) into yttria, terbia, and erbia. Gray hexagons, d.4.7., m.1522, b.3320, which decompose water and dilute acids. Removal of dissolved gases improves its ductility for use in atomic reactors and missiles. **y. acetate*** $Y(C_2H_3O_2)_3\cdot 8H_2O$ = 410.1. Colorless crystals, soluble in water. **y. bromide*** YBr_3 = 328.6. White, deliquescent powder. **y. carbonate*** $Y_2(CO_3)_3\cdot 3H_2O$ = 411.8. Pale-red powder, insoluble in water. **y. chloride** $YCl_3\cdot 6H_2O$ = 303.4. Pale-red, hygroscopic prisms, m.160, soluble in water. **y. hydroxide*** $Y(OH)_3$ = 139.9. White powder, decomp. by heat, insoluble in water. **y. minerals** Chiefly: gadolinite, rowlandite, thalenite, xenotime, yttrialite. **y. nitrate** $Y(NO_3)_3\cdot 6H_2O$ = 383.0. Colorless crystals, soluble in water. **y. oxalate** $Y_2(C_2O_4)_3\cdot 9H_2O$ = 604.0. White crystals, slightly soluble in water. **y. oxide** Y_2O_3 = 225.8. Yttria. Colorless crystals, insoluble in water. **y. sulfate*** $Y_2(SO_4)_3\cdot 8H_2O$ = 610.1. Pale-red crystals, decomp. 1000, slightly soluble in water. **y. sulfide*** Y_2S_3 = 274.0. Yellow powder, m.1900–1950, insoluble in water.

yttrocerite A native mixed fluoride of yttrium, cerium, erbium, and calcium.

yttrotantalite $Y_4(Ta_2O_7)_3$. A rare-earth tantalate containing Fe, Ca, and He.

yttrotitanite A siliceous calcium titanate containing oxides of Y, Al, and Fe.

yucca Adam's needle, Spanish bayonets, *Y. filamentosa* (Liliaceae), southern California, Texas, and Mexico; the leaves yield a fiber.

Yukawa, Hideki Japanese physicist, Nobel prize winner (1949). **Y. particle** Meson.

yukon Meson; named in honor of Yukawa.

yulocrotine $C_{19}H_{26}O_3N$ = 316.4. An alkaloid from *Julocroton Montevidensis* (Euphorbiaceae).

Z

Z Symbol for (1) atomic number (proton number); (2) compression factor; (3) impedance; (4) zusammen. See *stereoisomer*.
z Symbol for a cylindrical or rectangular coordinate.
zaffer, zaffre A mixture of cobalt oxides and arsenates obtained by roasting cobalt ores; a raw material for the production of cobalt compounds.
zala Borax.
zanaloin The aloin from aloes of Zanzibar.
zaratite $NiCO_3 \cdot 2Ni(OH)_2 \cdot 5H_2O$. Emerald nickel. A native nickel carbonate.
Zarontin Trademark for ethosuximide.
zea (1) A genus of annual grasses to which Indian corn (maize) belongs. (2) Corn silk, the fresh styles and stigmas of *Zea mays*, the maize plant.
zearin $C_{52}H_{88}O_4$ = 777.3. A colorless principle from lichens.
zeatin $(C_5H_3N_4)NH \cdot CH_2 \cdot CH{:}C(Me)CH_2OH$ = 219.2. Kinetin. An aminopurine factor in plant extracts, which induces cell division.
zeaxanthin $C_{40}H_{56}O_2$ = 568.9. A carotenoid color, q.v. *(carotenoids)*, from maize, egg yolk, fish, and plants; isomer of lutein. Yellow leaflets, m.216.
zedoary The dried rhizomes of *Curcuma zeodaria* (Zingiberaceae), Southeast Asia; a carminative. **z. oil** The volatile oil of z., d.0.990–1.010, b.240–308, containing eucalyptol.
Zeeman **Z., Pieter (1865–1943)** Dutch physicist. **Z. effect** A resolution of spectral single lines into 3 fine lines (triplet) when the source of light, as a flame, is placed in a strong magnetic field. Or, more generally, the splitting of spectral lines when the source is placed in a magnetic field.
Zefran Trademark for a synthetic-fiber copolymer of acrylonitrile and vinyl acetate, with vinyl pyrrolidine as acceptor.
Zehla Trademark for a viscose synthetic fiber.
zein A prolamine from maize; it contains no tryptophanol, cystine, or lysine.
Zeise salt $KCl \cdot PtCl_2C_2H_4 \cdot H_2O$. Cf. *π-complex*.
Zeisel reaction The formation of methyliodide from methoxy compounds and hydriodic acid:

$$(1)\ ROMe + HI = R \cdot OH + MeI$$
$$(2)\ MeI + AgNO_3 = AgI + MeNO_3$$

The silver iodide is weighted to give the MeO group content.
Zellner's paper Fluorescein paper.
Zener barrier An epoxy-encapsulated device for a terminal network to limit electrical voltage discharges automatically to a safe level; as in an area of inflammable gas.
Zenker's solution A fixative and preservative for biological specimens: potassium dichromate 2.5, sodium sulfate 1, mercuric chloride 5 gm; acetic acid 5 mL; water to 100 mL.
zentner See *centner*.
zeolite $Na_2O \cdot 2Al_2O_3 \cdot 5SiO_2$ and $CaO \cdot 2Al_2O_3 \cdot 5SiO_2$. Hydrated aluminum and calcium or sodium silicates, reacting in solution, by double decomposition, with salts of the alkali and alkaline-earth metals; used for industrial catalysts and water softening. There are about 30 known natural zeolites and 120 synthetic structures. See *molecular sieve*. Cf. *chabazite, natrolite, organolite*. **artificial ~** Permutite. **organic ~** A synthetic resin having the properties of a z.
Zephiran Chloride Trademark for benzalkonium chloride; a cationic detergent and antiseptic.
zero (1) The complete absence of a particular quantity. (2) The point at which a scale has the value of 0, e.g., zero degree centigrade. **absolute ~** See *absolute zero*.
Z. Gradient Synchotron A device (Argonne, Illinois) to accelerate protons to 12.5 GeV. **z. potential** The potential across the *Gouy* layer, q.v., on a colloidal particle.
zeta-potential See *zeta-potential* under *potential*.
Zetec Trademark for a mixed-polymer synthetic fiber.
zeus See *bach*.
zeyherine An alkaloid from the seeds of *Erythrina zeyheri* (Leguminosae), S. Africa.
zibet Civet.
Ziegler catalyst A mixture of triethylaluminum monochloride (or similar compound) with titanium tetrachloride; used in the Ziegler process. **Z. process** A low-pressure method of manufacturing high-density polymers and synthetic rubber.
Ziehl's stain Carbolfuchsin.
zierone $C_{15}H_{22}O$ = 218.3. A sesquiterpene ketone from *Zieria macrophylla*. Oil, $b_{18mm}148$.
Ziervogel process Extraction of silver by roasting the sulfide to silver sulfate, leaching with water, and precipitating metallic silver with copper.
Zinamide Trademark for pyrazinamide.
zinc* Zn = 65.39. A metal and element, at. no. 30, known early in India, observed by Agricola, described by Paracelsus (1520). White, brittle metal, d.6.7–7.2, m.419, b.907, insoluble in water, soluble in acids or hot solutions of alkalies. Z. occurs in nature principally as sulfide, carbonate, and silicate. In the body, it occurs mainly in enzyme molecules and bones; essential for health and growth. Used as a reducing agent (indigo vats) and reagent for the production of hydrogen (arsenic test), and in alloys and the metal industry. Z. is normally divalent, but monovalent z. is known. **activated ~** Z. granulated in presence of cadmium sulfate for the *Marsh* test, q.v. **butter of ~** Z. chloride. **diethyl ~*** Et_2Zn = 123.5. Z. ethide. Colorless liquid, ignites in air, b.118, violently decomp. by water; used in organic synthesis. **dimethyl ~*** Me_2Zn = 95.4. Z. methide. Colorless liquid, ignites in air, $d_{10^\circ}1.39$, m.−40, b.46, decomp. by water or alcohol; used in organic synthesis. **granulated ~** Distorted granules prepared by pouring molten z. into water. **powdered ~** Finely powdered z. used as a reagent. Cf. *zinc dust*.
z. acetate* $Zn(C_2H_3O_2)_2 \cdot 2H_2O$ = 219.5. Colorless plates, m.242, soluble in water. Used as a reagent and mordant and in glaze manufacture, gargles and astringents (USP). **fused ~** $Zn(C_2H_3O_2)_2$ = 183.5. White, fused mass, soluble in water. **z. albuminate** A compound of albumin and z. Yellow scales, slightly soluble in water. **z. alkyl(s)** R_2Zn in which R is an alkyl radical; as, dimethyl ~, Me_2Zn. **z. alkyl condensation** Frankland's synthesis of hydrocarbons. Z. is removed from a z. alkyl as hydrate or iodide: $2R_3CI + Me_2Zn$

$= 2R_3CMe + ZnI_2$. **z. amalgam** A mixture of z. and mercury, used as a reducing agent, and in electric batteries. **z. arsenide*** Zn_3As_2 = 346.0. **z. arsenite** $Zn(AsO_2)_2$ = 279.2. Colorless powder, insoluble in water. **z. ashes** The oxidized z. from the surface of a galvanizing bath. **z. benzoate*** $Zn(C_7H_3O_2)_2$ = 307.6. White powder, soluble in water. **z. biborate** Z. tetraborate*. **z. blende** Sphalerite. **z. bloom** Z. oxide*. **z. borate** Z. tetraborate*. **z. bromate*** $Zn(BrO_3)_2$ = 321.2. White, hygroscopic powder, m.100. **z. bromide*** $ZnBr_2$ = 225.2. Colorless, hygroscopic needles, m.394. **z. butter** Z. chloride*. **z. carbonate*** $ZnCO_3 \cdot H_2O$ = 143.4. Tutia. White rhombs, decomp. 300, insoluble in water; an astringent and protective. Cf. *zinc* spar. **z. chlorate*** $Zn(ClO_3)_2 \cdot 6H_2O$ = 340.4. Colorless, hygroscopic crystals, m.60. **z. chloride*** $ZnCl_2$ = 136.3. Z. Butter. White, deliquescent octahedra, m.365. A reagent for alkaloids; solvent for cellulose; an antiseptic, astringent (USP), and escharotic; preservative in embalming; mordant; and soldering flux. **z. chloroiodide** A mixture of z. iodide and chloride. The saturated solution is a microchemical reagent for cellulose (blue color) and tannin (violet color). **z. chloroiodide solution** Naegeli's solution. A microchemical reagent prepared by decomposing hydrochloric acid with z. and saturating the solution with potassium iodide and iodine. Cf. *Herzberg's stain.* **z. chromate*** $ZnCrO_4 \cdot 7H_2O$ = 307.5. Z. yellow, buttercup yellow. A soluble, yellow pigment. **z. citrate*** $Zn_3(C_6H_5O_7)_2 \cdot 2H_2O$ = 610.4. White powder, soluble in water. **z.-copper couple** Sheet zinc coated with a black deposit of copper by immersion in an acid solution of copper sulfate. It liberates nascent hydrogen from acid solutions, and will reduce nitrates to ammonia. Used to determine nitrates in water. **z. cream** A mixture of z. oxide, oleic acid, arachis oil, wool fat, and calcium hydroxide (BP). **z. cyanide*** $Zn(CN)_2$ = 117.4. Colorless prisms, decomp. by heat, insoluble in water. **z. dichromate*** $ZnCr_2O_7$ = 281.4. Z. bichromate. Orange powder, soluble in water. **z. dithiofuroate** $(C_4H_3OCS_2)_2Zn$. Furac II. Brown powder; a rubber vulcanization accelerator. **z. dust** (1) Finely divided z.; a reducing agent. (2) The flue dust of smelters, containing z., z. oxide, and impurities; a gray paint. **z. ethide, z. ethyl** Diethyl ~. **z. ethyl sulfate** $Zn(C_2H_5SO_4)_2$ = 315.6. Colorless, hygroscopic leaflets, soluble in water or alcohol. **z. ferrocyanide** Z. hexacyanoferrate*. **z. flowers** Z. oxide*. **z. fluoride*** ZnF_2 = 103.4. White powder, slightly soluble in water. **z. fluosilicate** Z. hexafluorosilicate*. **z. foil** Sheet z. prepared by heating z. at 100–150°C and rolling it. **z. formate*** $Zn(CHO_2)_2$ = 155.4. Colorless crystals, soluble in water. **z. f. dihydrate** $Zn(CHO_2)_2 \cdot 2H_2O$ = 191.4. White, monoclinic crystals, soluble in water. **z. gelatin** A smooth paste of zinc oxide, in a mixture of gelatin, glycerol, and water; a pharmaceutical protective (USP, BP). **z. hexacyanoferrate(II)*** $Zn_2[Fe(CN)_6] \cdot 3H_2O$ = 396.8. Z. ferrocyanide. White powder, soluble in ammonia water. **z. hexafluorosilicate*** $Zn[SiF_6] \cdot 6H_2O$ = 315.5. Z. fluo(ro)silicate. Hexagonal prisms, very soluble in water. **z. hydrogensulfide*** $Zn(SH)_2$ = 131.5. Z. sulfhydrate. White powder, insoluble in water, decomp. when dry (so stored under water). **z. hydroxide*** $Zn(OH)_2$ = 99.4. White prisms, decomp. by heat, insoluble in water. **z. iodate*** $Zn(IO_3)_2$ = 415.2. White crystals, slightly soluble in water. **z. iodide** ZnI_2 = 319.2. Colorless octahedra, m.446, soluble in water; a reagent (detecting chlorine and narceine). **z. iodide–starch paper** Filter paper, impregnated with a z. iodide–starch solution; used to detect free chlorine, iodine, or ozone (blue color). **z. iodide–starch solution** A solution of z. iodide and soluble starch; a test reagent and indicator for oxidizing agents. **z. isopropyl dithiocarbonate** An accelerator for the vulcanization of rubber. **z. methide** Dimethyl ~. **z. minerals** Chiefly: sphalerite (regular), ZnS; zincite, ZnO; franklinite, $ZnFe_2O_4$; smithsonite (zinc spar), $ZnCO_3$; willemite, Zn_2SiO_4. **z. nitrate*** $Zn(NO_3)_2 \cdot 6H_2O$ = 297.5. Colorless tetragons, m.36, soluble in water; a reagent, escharotic, and mordant. **z. nitride** Zn_3N_2 = 224.2. Green powder, decomp. by water to z. oxide and ammonia. **z. oleate*** $Zn(C_{18}H_{33}O_2)_2$ = 628.3. White, greasy granules, insoluble in water; used in ointments and as a varnish drier. **z. oxalate*** $ZnC_2O_4 \cdot 2H_2O$ = 189.4. White powder, slightly soluble in water. **z. oxide*** ZnO = 81.4. Philosopher's wool. Amorphous powder or hexagons, d.5.42, insoluble in water, soluble in acids, alkalies, or ammonium salt solutions. Used as a reagent and neutralizing agent; in the manufacture of ointments, rubber goods, glass, pigments, cosmetics, and dusting powders; and as an astringent and protective for skin conditions (USP, EP, BP). **z. paste** A mixture of z. oxide, starch, and white petrolatum; an astringent (USP, BP). **z. perborate** Z. peroxoborate*. **z. perhydrol** Z. peroxide*. **z. permanganate*** $Zn(MnO_4)_2 \cdot 6H_2O$ = 411.3. Brown, deliquescent crystals; an oxidizing agent, astringent, and antiseptic. **z. peroxide*** ZnO_2 = 97.4. Z. perhydrol. Yellow, voluminous powder, insoluble in water; a bactericide. **z. peroxoborate*** $Zn(BO_3)_2$ = 183.0. Z. perborate. White powder, insoluble in water; an oxidizing agent and antiseptic in cosmetics. **z. phosphate*** $Zn_3(PO_4)_2$ = 386.1. White prisms, m. red heat, insoluble in water, soluble in ammonium salt solutions; a reagent. **z. orthophosphate*** Z. phosphate. **z. pyrophosphate*** $Zn_2P_2O_7$ = 304.7. White powder, insoluble in water. **z. phosphide** Zn_3P_2 = 258.1. Gray powder, d.4.72, decomp. in moist air, insoluble in water; a phosphorus substitute, and a reagent in making hydrogen phosphide. **z. phosphonate*** $ZnHPO_3$ = 145.4. White granules, soluble in water. **z. potassium iodide** K_2ZnI_4 = 651.2. Colorless crystals, soluble in water; a reagent. **z. pyrophosphate*** See *zinc* phosphate. **z. salicylate*** $Zn(C_6H_4OHCOO)_2 \cdot 3H_2O$ = 393.7. Colorless needles, soluble in water. **z. silicate** $ZnSiO_3$ = 141.5. Z. metasilicate*. White powder, insoluble in water. **z. spar** $ZnCO_3$. An important z. ore. **z. spinel** Gahnite. **z. stearate*** $Zn(C_{18}H_{35}O_2)_2$ = 632.3. White, greasy granules, insoluble in water; an antiseptic, dusting powder (USP), and drier for paints. **z. succinate*** $ZnC_4H_4O_4$ = 181.5. White powder, insoluble in water. **z. sulfate*** $ZnSO_4$ = 161.4. **anhydrous ~** Colorless crystals, soluble in water. **z. s. dihydrate** $ZnSO_4 \cdot 2H_2O$ = 197.5. Fused sticks or white powder, soluble in water. **heptahydrate ~** $ZnSO_4 \cdot 7H_2O$ = 287.5. Crystalline or common z. vitriol, white vitriol. White, rhombic, prismatic, or monoclinic crystals, d.2.015, m.50, soluble in water, alcohol, or ether. Used as a reagent (in volumetry and in precipitating proteins); an antiseptic, eyewash, gargle, ophthalmic astringent and treatment for z. deficiency states (USP, EP, BP); a mordant in dyeing and textile printing; a preservative for skins and woods; and in the manufacture of paints and varnishes. **z. sulfide*** ZnS = 97.4. Yellow hexagons or tetragons, m.1049, insoluble in water; a pigment, and reagent in testing the acidity of soils. Cf. *Sidot's blende.* **z. sulfite*** $ZnSO_3 \cdot 2\frac{1}{2}H_2O$ = 190.5. White crystals, soluble in water; a preservative for anatomical specimens. **z. tartrate*** $ZnC_4H_4O_6$ = 213.4. White powder, slightly soluble in water. **z. tetraborate*** ZnB_4O_7 = 220.6. Z. borate, z. biborate. White powder, insoluble in water. **z. thiocyanate*** $Zn(SCN)_2$ = 181.5. Z. rhodanate. Z. sulfocyanate. White, deliquescent crystals. **z. undecenoate*** $[CH_2{:}CH(CH_2)_8 \cdot CO_2]_2Zn$ = 431.9. Z. undecylenate. White powder, insoluble in water; a fungistatic (USP, BP). **z. vitriol**

Z. sulfate*. **z. white** (1) Z. oxide*. (2) A mixture of z. oxide 80, barium sulfate 20% (German usage). **z. yellow** Commercial z. chromate; a pigment.

zincaloy An alloy of zirconium with amounts of iron, nickel, chromium, tin, and not more than 0.5 ppm boron.

zincamide $Zn(NH_2)_2$ = 97.4. White powder, m.200 vacuum (decomp.).

zincates* Indicating zinc as the central atom(s) in an anion; as, M_2ZnO_2. Salts of amphoteric zinc hydroxide. Z. form hydrated ions, in solution; as, $[Zn(OH)_3 \cdot H_2O]^-$.

zincic acid The amphoteric form of zinc hydroxide: $H_2ZnO_2 \rightleftharpoons Zn(OH)_2$.

zincite ZnO. A rare, red, native zinc oxide, hardness 4–4.4, d.5.4–5.7.

zincography Process engraving. Reproduction on zinc plates, the surface of which is first coated with special wax, on which is drawn or photographed the subject to be printed. A strong acid dissolves the zinc not covered by the image, which is thus left in relief.

zineb* See *fungicides*, Table 37 on p. 250.

zinethyl Diethyl*zinc**.

zingerone $MeCOCH_2CH_2C_6H_3(OH)OMe$ = 194.2. 4-(4-Hydroxy-3-methoxyphenyl)-2-butanone*. Colorless crystals, m.41, soluble in water; with a pungent taste, in ginger. Cf. *capsaicin.*

zingiber Ginger.

Zingiberaceae See *Scitaminaceae.*

zingiberene $C_{15}H_{24}$ = 204.4. A sesquiterpene from ginger oil. d.0.872, b_{11mm}129. **iso ~** 1,7-Dimethyl-4-propenyloctahydronaphthalene.

zingiberenol $C_{15}H_{26}O$ = 222.4. A monohydric alcohol, b_{14mm}157, from ginger.

zinin Azoxybenzene.

zinkenite $PbSb_2S_3$. A native antimonite.

zinkite Zincite.

zinnkies Cu_2FeSnS_4. Native sulfostannate.

zinnwaldite $(K, Li)_3FeAl_3Si_5O_{16}F_2$. A pale-violet, yellow, or brown lithia mica.

zippeite $2UO_3 \cdot SO_3 \cdot H_2O$. A native uranium sulfate; less than 5% copper oxide.

Zircal Trademark for the alloys: Zn 7.0–8.5, Mg 1.75–3.0, Cu 1–2, Cr 0.1–0.4, Mn 0.1–0.6, Fe + Si 0.7%. They are strong, light, and elastic, and can be rolled, drawn, or extruded.

zircon $ZrSiO_4$. Ostranite. Jargon. A native zirconium silicate. Transparent, yellow crystals, d.4.68–4.70, hardness 7.5; a source of Zr compounds and gem (hyacinth, jargon). See *starlite, malacon, oerstedite.* **z. alba** Zirconium oxide.

zirconate M_2ZrO_3 and $M_2Zr_2O_5$. Mixed oxide of ZrO_2 and that of another metal.

zirconia Zirconium oxide.

zirconic **z. acid** Zirconium hydroxide. **z. anhydride** Zirconium oxide.

zirconium* Zr = 91.224. A rare-earth metal and element of the carbon group, at. no. 40, discovered by Klaproth (1789). Silvery and crystalline, or gray and amorphous, metal, d.4.15 (cryst.), 6.41 (amorph.), m.1850, b.4440, insoluble in water or acids, soluble in aqua regia or molten alkalies. Main valency 4, but 0, 1, 2, 3 also exist; forms 3 main series of compounds: normal, Zr^{4+}; basic, ZrO^{2+} (zirconyl); ZrO_3^{2-}, $Zr_2O_5^{2-}$ (zirconates). Used as a steel deoxidizer, denitrifier, and desulfurizer (in the form of silicon-zirconium or ferrosilicon-zirconium); a "getter" in radio tubes; an X-ray filter; for wires and filaments in many alloys, for flash bulbs coated on aluminum foil, and for the production of cesium from cesium dichromate. **z. bromide*** $ZrBr_4$ = 410.8. Z. tetrabromide*. White powder, decomp. violently by water to form zirconyl bromide. **z. carbonate** Zirconyl carbonate. **z. chloride*** $ZrCl_4$ = 233.0. Z. tetrachloride*. Colorless crystals, sublimes 350, decomp. by water to zirconyl chloride; used in organic synthesis (Friedel-Craft reaction). **z. dioxide*** Z. oxide. **z. fluoride*** ZrF_4 = 167.2. Colorless crystals, m. red heat, insoluble in water. **z. hydroxide*** $ZrO_2 \cdot nH_2O$. Colorless powder, insoluble in water, soluble in acids with formation of Zr^{4+} and ZrO^{2+} ions, and in alkalies with formation of ZrO_3^{2-} or $Zr_2O_5^{2-}$ ions. **z. iodide*** ZrI_4 = 598.8. Hygroscopic, brown crystals. **z. nitrate** Zirconyl nitrate. **z. oxide** ZrO_2 = 123.2. Zirconia, zircon alba, z. anhydride, z. dioxide*, native as baddeleyite, zirkite, and becarite; 3 crystal forms. Heavy, white powder, m.2700, insoluble in water, soluble in hot acids. It is a durable refractory and an effective opacifier of fused enamels, glass, and glazes; also used in X-ray photography as a substitute for bismuth salts, and with silica and graphite for safe and vault walls. **z. oxybromide** Zirconyl bromide. **z. oxychloride** Zirconyl chloride. **z. silicate** See *zircon, azorite.* **z. silicide** $ZrSi_2$ = 147.4. Gray, metallic solid, d.4.88. **z. sulfate*** $Zr(SO_4)_2 \cdot 4H_2O$ = 355.4. Colorless crystals, soluble in water; a reagent for potassium. **z. tetrachloride*** Z. chloride*.

zirconyl The radical ZrO^{2+}. **z. bromide** $ZrOBr_2 \cdot 8H_2O$ = 411.1. Zirconium oxybromide. Colorless, deliquescent powder, decomp. in moist air. **z. carbonate** $ZrOCO_3$ = 167.2. White powder, insoluble in water. **z. chloride** $ZrOCl_2 \cdot 8H_2O$ = 178.1. Zirconium oxychloride. Colorless, silky needles, soluble in water. **z. hydroxide** $ZrO(OH)_2$ = 141.2. White powder, insoluble in water. **z. nitrate** $ZrO(NO_3)_2$ = 231.2. White crystals, soluble in water. **z. phosphate** $(ZrO)_3(PO_4)_2 \cdot 8H_2O$ = 655.7. Colorless powder, insoluble in water.

zirkite ZrO_2. A native zirconia ore (Brazil); a firebrick, cement, and opacifier.

zirlite $Al(OH)_3$. Yellow, native aluminum hydroxide.

Z.I.X. Trademark for zinc isopropyl dithiocarbonate.

Zn Symbol for zinc.

zoisite $Ca_2Al_3(OH)(SiO_4)_3$. Thulite. Gives its name to a group of *silica minerals,* q.v.

Zöller, Philipp (1833–1885) German chemist, noted for agricultural and biochemical work.

zone An area of interest; especially a region of oriented molecules. Cf. *orientation.* It differs from a phase in that it is heterogeneous with molecules oriented. **interfacial ~** The layer of oriented *A* and *B* molecules between 2 phases *A* and *B*. **micellar ~** A region of an aggregate or of semioriented molecules, the aggregate being immersed in a surrounding bulk solvent phase. **monomolecular ~** A layer of oriented *C* molecules between 2 phases *A* and *B*. **polar ~** A z. consisting of oriented dipoles; as in *solvates,* q.v.
z. leveling The use of z. melting to produce a uniform distribution of one solid in another. **z. melting, z. refining** A method of purification of metal ingots by heat, in which a molten zone is caused to pass along the length of the ingot (z. melting), thereby producing a change in the concentration of the impurities. Especially suitable for the purification of semiconductors and uranium.

zoo- Prefix (Greek) indicating "animals."

zoochemistry The study of the composition and reactions occurring in animal organisms.

zoology The study of the classification, structure, and function of animals.

zoom Describing the property of an optical instrument, e.g., microscope, which enables sharp focusing to be obtained over a wide range of magnifications without change of objectives.

zoomaric acid (Z)-9-Hexadecenoic acid*.
zoonic acid An early name for acetic acid.
zoonosis A disease of animals and humans; as, tuberculosis and anthrax. Caused by the same organism in both.
zoospore A mobile spore (swarm spore) produced by fungi.
zoosterol A sterol, q.v. *(sterols),* of animal origin, e.g., cholesterol.
zootoxin A poison derived from an animal; as, venoms.
Zosimos of Panopolis Middle 5th century B.C. A Greek writer of Egypt, who wrote 20 chemical works, which show that the term "chemistry" is of Greek origin (περι της χπμειας = "On Chemistry").
Zr Symbol for zirconium.
Zsigmondy **Z., Richard (1865–1929)** Austrian chemist, Nobel prize winner (1925). **Z. filters** A series of graded ultrafilters for separating ultramicroscopical particles or colloids from solutions, according to their sizes. They are semipermeable membranes of various compositions and permeabilities; used in analytical and physiological chemistry.
z-transform Technique for changing the discontinuous equivalent of differential equations into normal algebraic form. Cf. *Laplace transform.*
zuckerin Saccharin.
zwitterion A complex ion that is both positively and negatively charged; as, $Y^+ \cdot R \cdot X^-$, e.g., $^+NH \cdot R \cdot SO_3^-$. Cf. *amphoteric.*
Zycon Trademark for a protein synthetic fiber.
zygadenine $C_{39}H_{63}O_{10}N$ = 705.9. An alkaloid from the bulbs of *Zygadenus nuttalli* (Liliaceae), Rocky Mountains. It resembles veratrine.
zygograph (Greek *zygon:* "a yoke") A graphical relationship between the composition of a vapor and that of a liquid in equilibrium with it; used in distillation problems.
zygology The study of joining processes; as, welding, adhesive bonding.
zygote A fertilized ovum, containing the diploid number of chromosomes. Cf. *gamete.*
zylonite Celluloid.
Zyloprim, Zyloric Trademarks for allopurinol.
zymamsis Alcoholic fermentation.
zymin An acetone-dried yeast.
zymochemistry The chemistry of fermentation.
zymoexcitor See *kinase (2).*
zymogen The precursor of an enzyme; a substance secreted from tissues or glands, formerly believed to be split into an active enzyme and protein by a *kinase,* q.v.
zymohexose A monosaccharide that ferments readily; as, D-glucose.
zymohydrolysis Zymolysis.
zymology The study of the action and composition of enzymes.
zymolysis Zymohydrolysis. A chemical reaction produced by an enzyme.
zymoplasm Thrombin*.
zymosis Fermentation; reactions caused by enzymes.
zymosterol $C_{27}H_{44}O$ = 384.6. A sterol, q.v. *(sterols),* from yeast, m.108, insoluble in water.
zymurgy The study of the application of enzymes in brewing, distilling, and wine making. In its broader meaning it includes the processes of fermentation for manufacturing purposes, as in the tobacco, cheese, indigo, and leather industries.

ABOUT THE AUTHORS

ROGER GRANT graduated from Oxford University with a degree in chemistry and metallurgy in 1960. He also received a master's degree from Grenoble University in France and a Ph.D. from Manchester University, England. A Chartered Chemist, after 20 years in industry he is now an independent consultant. He has written numerous articles and papers on chemistry and other subjects.

CLAIRE GRANT, a graduate of St Bartholomew's Hospital, London University, is a practicing physician. Her professional experience has allowed her to add cogent insights to the dictionary's medical and pharmacological entries.

587936